SHOCK
AND
VIBRATION
HANDBOOK

Other McGraw-Hill Books of Interest

Avallone & Baumeister • MARKS' STANDARD HANDBOOK FOR MECHANICAL ENGINEERS

Brady & Clauser • MATERIALS HANDBOOK

Considine • PROCESS/INDUSTRIAL INSTRUMENTS AND CONTROLS HANDBOOK

Fink & Beaty • STANDARD HANDBOOK FOR ELECTRICAL ENGINEERS

Fink & Christiansen • ELECTRONICS ENGINEER'S HANDBOOK

Ganić & Hicks • THE McGRAW-HILL HANDBOOK OF ESSENTIAL ENGINEERING INFORMATION AND DATA

Haines & Wilson • HVAC SYSTEMS DESIGN HANDBOOK

Harris • DICTIONARY OF ARCHITECTURE AND CONSTRUCTION

Harris • HANDBOOK OF UTILITIES AND SERVICES FOR BUILDINGS

Harris • NOISE CONTROL IN BUILDINGS

Hicks • STANDARD HANDBOOK OF ENGINEERING CALCULATIONS

Higgins • MAINTENANCE ENGINEERING HANDBOOK

Hodson • MAYNARD'S INDUSTRIAL ENGINEERING HANDBOOK

Ireson & Coombs • HANDBOOK OF RELIABILITY ENGINEERING AND MANAGEMENT

Juran & Gryna • JURAN'S QUALITY CONTROL HANDBOOK

Karassik et al. • PUMP HANDBOOK

Lingaiah • MACHINE DESIGN DATA HANDBOOK

Maidment • HANDBOOK OF HYDROLOGY

Merritt & Ricketts • BUILDING DESIGN AND CONSTRUCTION HANDBOOK

Merritt, et al. • STANDARD HANDBOOK FOR CIVIL ENGINEERS

Perry & Green • PERRY'S CHEMICAL ENGINEER'S HANDBOOK

Rosaler • STANDARD HANDBOOK OF PLANT ENGINEERING

Shigley & Mischke • STANDARD HANDBOOK OF MACHINE DESIGN

Tuma • ENGINEERING MATHEMATICS HANDBOOK

Wadsworth • HANDBOOK OF STATISTICAL METHODS FOR ENGINEERS AND SCIENTISTS

Walsh • McGRAW-HILL MACHINING AND METALWORKING HANDBOOK

Wang • HANDBOOK OF AIR CONDITIONING AND REFRIGERATION

SHOCK AND VIBRATION HANDBOOK

Cyril M. Harris Editor in Chief

*Charles Batchelor Professor Emeritus of Electrical
Engineering and Professor Emeritus of Architecture
Columbia University*

*First Edition of the Shock and Vibration Handbook (1961)
Edited by
Cyril M. Harris and Charles E. Crede*

Fourth Edition

McGRAW-HILL

New York San Francisco Washington, D.C. Auckland Bogotá
Caracas Lisbon London Madrid Mexico City Milan
Montreal New Delhi San Juan Singapore
Sydney Tokyo Toronto

Library of Congress Cataloging-in-Publication Data

Shock and vibration handbook / edited by Cyril M. Harris.
    p.   cm.
    Includes index.
    ISBN 0-07-026920-3
    1. Vibration—Handbooks, manuals, etc. 2. Shock (Mechanics)—
Handbooks, manuals, etc. I. Harris, Cyril M., (date).
TA355.S5164   1995
620.3—dc20

                                   95-38224
                                     CIP

McGraw-Hill

*A Division of The **McGraw·Hill** Companies*

  2 3 4 5 6 7 8 9 0 BKP/BKP 9 0 0 9 8 7

ISBN 0-07-026920-3

*The sponsoring editor for this book was Harold B. Crawford, the editing
supervisor was Peggy Lamb, and the production supervisor was Pamela A.
Pelton. It was set in Times Roman by North Market Street Graphics.*

Printed and bound by Quebecor/Book Press.

This book is printed on acid-free paper.

McGraw-Hill books are available at special quantity discounts to use as
premiums and sales promotions, or for use in corporate training programs.
For more information, please write to the Director of Special Sales,
McGraw-Hill, 11 West 19th Street, New York, NY 10011. Or contact your
local bookstore.

CONTENTS

v

Chapter 20. Test Criteria and Specifications   20.1

Allan G. Piersol, *Consultant, Piersol Engineering Company, Woodland Hills, CA 91364.*

Chapter 21. Modal Analysis and Testing   21.1

Randall J. Allemang, *Professor, Structural Dynamics Research Laboratory, University of Cincinnati, Cincinnati, OH 45221.*
AND
David L. Brown, *Professor, Structural Dynamics Research Laboratory, University of Cincinnati, Cincinnati, OH 45221.*

Chapter 22. Concepts in Vibration Data Analysis   22.1

Allen J. Curtis, *Retired Chief Scientist, Hughes Aircraft Company, El Segundo, CA 90245.*
AND
Steven D. Lust, *Scientist/Engineer, Hughes Aircraft Company, El Segundo, CA 90245.*

Chapter 23. Concepts in Shock Data Analysis   23.1

Sheldon Rubin, *Distinguished Engineer, The Aerospace Corporation, El Segundo, CA 90009.*

Chapter 24. Seismically-Induced Vibration of Structures   24.1

William J. Hall, *Professor of Civil Engineering Emeritus, University of Illinois at Urbana-Champaign, Urbana, IL 61801.*

Chapter 25. Vibration Testing Machines   25.1

David O. Smallwood, *Distinguished Member of Technical Staff, Sandia National Laboratory, Albuquerque, NM 87185*

Chapter 26. Part I: Conventional Shock Testing Machines   26.1

Richard H. Chalmers, *Consultant, Induced Environments Consultants, San Diego, CA 92107.*

Chapter 26. Part II: Pyrotechnic Shock Testing Machines   26.17

Neil T. Davie, *Senior Member of Technical Staff, Sandia National Laboratory, Albuquerque, NM 87185.*
AND
Vesta Bateman, *Senior Member of Technical Staff, Sandia National Laboratories, Albuquerque, NM 87185.*

Chapter 27. Application of Digital Computers   27.1

Allen J. Curtis, *Retired Chief Scientist, Hughes Aircraft Company, El Segundo, CA 90245.*
AND
Steven D. Lust, *Scientist/Engineer, Hughes Aircraft Company, El Segundo, CA 90245.*

Chapter 28. Part I: Matrix Methods of Analysis   28.1

Stephen H. Crandall, *Ford Professor of Engineering Emeritus, Massachusetts Institute of Technology, Cambridge, MA 02139.*
AND
Robert B. McCalley, Jr. *Retired Engineer, General Electric Company, Schenectady, NY 12309.*

Chapter 39. Part I: Balancing of Rotating Machinery   39.1

Douglas G. Stadelbauer, *formerly Executive Vice President, Schenck Corp., Deer Park, NY 11729.*

Chapter 39. Part II: Vibration Induced by Shaft Misalignment   39.37

John D. Piotrowski, *President, Turvac Inc., Cincinnati, OH 45240.*

Chapter 40. Machine-Tool Vibration   40.1

Eugene I. Rivin, *Professor, Wayne State University, Detroit, MI 48202.*

Chapter 41. Package Engineering   41.1

Masaji T. Hatae, *formerly Manager, Transportability Engineering, Advanced Systems Division, Northrop Corporation, Pico Rivera, CA*

Chapter 42. Theory of Equipment Design   42.1

E. G. Fischer, *formerly Consulting Engineer, Westinghouse Electric Corporation, Pittsburgh, PA.*

Chapter 43. Practice of Equipment Design   43.1

E. G. Fischer, *formerly Consulting Engineer, Westinghouse Electric Corporation, Pittsburgh, PA.*
AND
Harold M. Forkois, *formerly, U.S. Naval Research Laboratory, Washington, DC.*

Chapter 44. Effects of Shock and Vibration on Humans   44.1

Henning E. von Gierke, *Director Emeritus, Biodynamics and Bioengineering Division, Armstrong Laboratory, Wright-Patterson AFB, OH 45433-7901*
AND
Anthony J. Brammer, *Senior Research Officer, Institute for Microstructural Sciences, National Research Council, Ottawa, ON K1A OR6, Canada.*

PREFACE

In 1961, the First Edition of the *Shock and Vibration Handbook* recognized for the first time the full scope of the field of shock and vibration; it brought together, in one single work, a comprehensive survey of classical vibration theory and modern applications of the theory to current engineering practice. Many chapters in that three-volume edition contained more material than had been found collectively in all books previously published on that topic; other contained considerable material that had not been summarized in the literature. As a result, the *Handbook* became a standard reference work throughout the world. My co-editor for the First Edition, Charles E. Crede, a remarkable engineer and special friend, died three years after its publication.

Because of rising publication costs, in 1976 the Second Edition was published as a single volume, reducing the total number of pages contained in the three-volume edition by one-third. This was made possible by eliminating highly specialized information of interest to relatively few engineers, and by the elimination of archival material. Over the next ten years, new theoretical developments, recently developed equipment related to shock and vibration, and new engineering techniques and applications brought about the need for the Third Edition of the *Handbook,* which was published in 1988.

Rapid advances in the field have continued, giving rise to this Fourth Edition. Entirely new chapters or sections have been added: for example, Chapter 26, Part II, *Pyrotechnic Shock Testing;* Chapter 28, Part II, *Applications of Finite Element Methods;* Chapter 34, *Engineering Properties of Rubber;* Chapter 35, *Mechanical Properties of Metals;* Chapter 39, Part II, *Vibration Induced by Shaft Misalignment.* In some cases, the titles of chapters have been modified to reflect the addition of substantial new material. For example, the title of Chapter 11 now reflects the important addition of *Statistical Energy Analysis (SEA)* methodology; the title of Chapter 20, is now *Introduction to Testing Criteria and Specifications.* Many other chapters have retained the same titles and chapter numbers as their counterparts in the Third Edition, but have been significantly updated. Added material was also required because of the widespread use of fast personal computers having large storage capacities, which changed the method of approach to the solution of shock and vibration problems from techniques used a decade earlier. Comprehensive coverage has continued to be a primary objective.

The Fourth Edition, a unified treatment of the subject of shock and vibration, is the equivalent of five or six textbooks of the usual size. There are 44 chapters, a number of them divided into two or three parts, which have been written by 54 authorities from industry, government laboratories, and universities. As possible, chapters dealing with related topics are grouped together. The first group of chapters provides a theoretical basis for shock and vibration. The second group considers instrumentation and measurements, and procedures for analyzing and testing systems subjected to shock and vibration. Vibration that is induced by ground motion and fluid flow is considered next. Methods of controlling shock and vibration are discussed in a group of chapters dealing with isolation, damping, and balancing. This is followed by chapters on packaging engineering to prevent equipment from being damaged in transit, by chapters on the theory and practice of equipment design, and finally by a chapter covering the effects of shock and vibration on humans. Duplication of material between chapters is avoided, to the degree that this is possible. There are extensive cross-references to other chapters, and to available technical literature.

Although this Handbook is not intended primarily as a textbook, lecturers at many engineering schools and universities have found its classical and rigorous treatment suitable for classroom use; in particular, the comprehensive practical examples are of value as a supplement to the customary classroom theory. The control of shock and vibration is of importance in many aspects of engineering. The *Handbook* is a valuable working reference for those engaged in the fields of acoustical, aeronautical, air-conditioning, civil, electrical, mechanical, and transportation engineering. Engineers in manufacturing, including plant maintenance, measurement and control, equipment design, environmental testing, and packaging and shipping, will find much useful content, as will those engaged in design and development work.

In the preparation of this Fourth Edition, many more persons and organizations made useful suggestions and contributions than I am able to acknowledge individually; I am grateful to them all. I especially thank Harry Himelblau, Jr. The input of Rudolph H. Volin was helpful. Above all, I thank the Contributors for their diligence in our shared objective of making each chapter the definitive treatment in its field; their commitment has been notable. Finally, I wish to express my appreciation to the government agencies and industrial organizations with whom many of our contributors are associated for clearing the material presented in chapters written by them. Thanks are also due to Harold B. Crawford, editor in chief of McGraw-Hill's engineering and technical books; Bob Hauserman, Senior Editor, engineering; and to Margaret Lamb, editing manager in McGraw-Hill's Professional Book Group, with whom I have worked closely and rewardingly on this book and others for a number of years.

Cyril M. Harris

CHAPTER 1

INTRODUCTION
TO THE HANDBOOK

Cyril M. Harris

CONCEPTS OF SHOCK AND VIBRATION

Vibration is a term that describes oscillation in a mechanical system. It is defined by the frequency (or frequencies) and amplitude. Either the motion of a physical object or structure or, alternatively, an oscillating force applied to a mechanical system is vibration in a generic sense. Conceptually, the time-history of vibration may be considered to be sinusoidal or simple harmonic in form. The frequency is defined in terms of cycles per unit time, and the magnitude in terms of amplitude (the maximum value of a sinusoidal quantity). The vibration encountered in practice often does not have this regular pattern. It may be a combination of several sinusoidal quantities, each having a different frequency and amplitude. If each frequency component is an integral multiple of the lowest frequency, the vibration repeats itself after a determined interval of time and is called *periodic*. If there is no integral relation among the frequency components, there is no periodicity and the vibration is defined as *complex*.

Vibration may be described as *deterministic* or *random*. If it is deterministic, it follows an established pattern so that the value of the vibration at any designated future time is completely predictable from the past history. If the vibration is random, its future value is unpredictable except on the basis of probability. Random vibration is defined in statistical terms wherein the probability of occurrence of designated magnitudes and frequencies can be indicated. The analysis of random vibration involves certain physical concepts that are different from those applied to the analysis of deterministic vibration.

Vibration of a physical structure often is thought of in terms of a model consisting of a mass and a spring. The vibration of such a model, or system, may be "free" or "forced." In *free vibration,* there is no energy added to the system but rather the vibration is the continuing result of an initial disturbance. An *ideal system* may be considered undamped for mathematical purposes; in such a system the free vibration is assumed to continue indefinitely. In any *real system,* damping (i.e., energy dissipation) causes the amplitude of free vibration to decay continuously to a negligible value. Such free vibration sometimes is referred to as *transient vibration. Forced vibration,* in contrast to free vibration, continues under "steady-state" conditions because energy is supplied to the system continuously to compensate for that dissipated by damping in the system. In general, the frequency at which energy is supplied (i.e., the forcing frequency) appears in the vibration of the system. Forced vibration may be either deterministic or random. In either instance, the vibration of the system depends upon the relation of the excitation or forcing function to the properties of the system. This relationship is a prominent feature of the analytical aspects of vibration.

Shock is a somewhat loosely defined aspect of vibration wherein the excitation is nonperiodic, e.g., in the form of a pulse, a step, or transient vibration. The word *shock* implies a degree of suddenness and severity. These terms are relative rather than absolute measures of the characteristic; they are related to a popular notion of the characteristics of shock and are not necessary in a fundamental analysis of the applicable principles. From the analytical viewpoint, the important characteristic of shock is that the motion of the system upon which the shock acts includes both the frequency of the shock excitation and the natural frequency of the system. If the excitation is brief, the continuing motion of the system is free vibration at its own natural frequency.

The technology of shock and vibration embodies both theoretical and experimental facets prominently. Thus, methods of analysis and instruments for the measurement of shock and vibration are of primary significance. The results of analysis and measurement are used to evaluate shock and vibration environments, to devise testing procedures and testing machines, and to design and operate equipment and machinery. Shock and/or vibration may be either wanted or unwanted, depending upon circumstances. For example, vibration is involved in the primary mode of operation of such equipment as conveying and screening machines; the setting of rivets depends upon the application of impact or shock. More frequently, however, shock and vibration are unwanted. Then the objective is to eliminate or reduce their severity or, alternatively, to design equipment to withstand their influences. These procedures are embodied in the control of shock and vibration. Methods of control are emphasized throughout this Handbook.

CONTROL OF SHOCK AND VIBRATION

Methods of shock and vibration control may be grouped into three broad categories:.

1. **Reduction at the Source**
    a. Balancing of Moving Masses.   Where the vibration originates in rotating or reciprocating members, the magnitude of a vibratory force frequently can be reduced or possibly eliminated by balancing or counterbalancing. For example, during the manufacture of fans and blowers, it is common practice to rotate each rotor and to add or subtract material as necessary to achieve balance.
    b. Balancing of Magnetic Forces.   Vibratory forces arising in magnetic effects of electrical machinery sometimes can be reduced by modification of the magnetic path. For example, the vibration originating in an electric motor can be reduced by skewing the slots in the armature laminations.
    c. Control of Clearances.   Vibration and shock frequently result from impacts involved in operation of machinery. In some instances, the impacts result from inferior design or manufacture, such as excessive clearances in bearings, and can be reduced by closer attention to dimensions. In other instances, such as the movable armature of a relay, the shock can be decreased by employing a rubber bumper to cushion motion of the plunger at the limit of travel.

2. **Isolation**
   a. *Isolation of Source.*   Where a machine creates significant shock or vibration during its normal operation, it may be supported upon isolators to protect other machinery and personnel from shock and vibration. For example, a forging hammer tends to create shock of a magnitude great enough to interfere with the operation of delicate apparatus in the vicinity of the hammer. This condition may be alleviated by mounting the forging hammer upon isolators.
   b. *Isolation of Sensitive Equipment.*   Equipment often is required to operate in an environment characterized by severe shock or vibration. The equipment may be protected from these environmental influences by mounting it upon isolators. For example, equipment mounted in ships of the navy is subjected to shock of great severity during naval warfare and may be protected from damage by mounting it upon isolators.

3. **Reduction of the Response**
   a. *Alteration of Natural Frequency.*   If the natural frequency of the structure of an equipment coincides with the frequency of the applied vibration, the vibration condition may be made much worse as a result of resonance. Under such circumstances, if the frequency of the excitation is substantially constant, it often is possible to alleviate the vibration by changing the natural frequency of such structure. For example, the vibration of a fan blade was reduced substantially by modifying a stiffener on the blade, thereby changing its natural frequency and avoiding resonance with the frequency of rotation of the blade. Similar results are attainable by modifying the mass rather than the stiffness.
   b. *Energy Dissipation.*   If the vibration frequency is not constant or if the vibration involves a large number of frequencies, the desired reduction of vibration may not be attainable by altering the natural frequency of the responding system. It may be possible to achieve equivalent results by the dissipation of energy to eliminate the severe effects of resonance. For example, the housing of a washing machine may be made less susceptible to vibration by applying a coating of damping material on the inner face of the housing.
   c. *Auxiliary Mass.*   Another method of reducing the vibration of the responding system is to attach an auxiliary mass to the system by a spring; with proper tuning the mass vibrates and reduces the vibration of the system to which it is attached. For example, the vibration of a textile-mill building subjected to the influence of several hundred looms was reduced by attaching large masses to a wall of the building by means of springs; then the masses vibrated with a relatively large motion and the vibration of the wall was reduced. The incorporation of damping in this auxiliary mass system may further increase its effectiveness.

CONTENT OF HANDBOOK

The chapters of this Handbook each deal with a discrete phase of the subject of shock and vibration. Frequent references are made from one chapter to another, to refer to basic theory in other chapters, to call attention to supplementary information, and to give illustrations and examples. Therefore, each chapter when read with other referenced chapters presents one complete facet of the subject of shock and vibration.

Chapters dealing with similar subject matter are grouped together. The first 11 chapters following this introductory chapter deal with fundamental concepts of shock and vibration. Chapter 2 discusses the free and forced vibration of linear systems that can be defined by lumped parameters with similar types of coordinates. The properties of rigid bodies are discussed in Chap. 3, together with the vibration of resiliently supported rigid bodies wherein several modes of vibration are coupled. Nonlinear vibration is discussed in Chap. 4, and self-excited vibration in Chap. 5. Chapter 6 discusses two degree-of-freedom systems in detail—including both the basic theory and the application of such theory to dynamic absorbers and auxiliary mass dampers. The vibration of systems defined by distributed parameters, notably beams and plates, is discussed in Chap. 7. Chapters 8 and 9 relate to shock; Chap. 8 discusses the response of lumped parameter systems to step- and pulse-type excitation, and Chap. 9 discusses the effects of impact on structures. Chapter 10 discusses mechanical impedance. A new Chap. 11 presents statistical methods of analyzing vibrating systems.

The second group of chapters is concerned with instrumentation for the measurement of shock and vibration. Chapter 12 has been expanded to include not only piezoelectric and piezoresistive transducers but also transducers of other types (except for strain gages, which are discussed in Chap. 17). The instrumentation to which such transducers are connected (including various types of amplifiers, signal conditioners, and recorders) is considered in detail in Chap. 13. In the 4th ed. a new Chap. 14 is devoted to the increasingly important topics of spectrum analysis instrumentation and techniques. The use of equipment in making vibration measurement in the field is described in Chap. 15. Since the last edition of the Handbook, there has been increasing use of vibration measurement equipment to monitor the mechanical condition of machinery, as an aid in preventive maintenance. This is the subject of Chap. 16. The calibration of transducers, Chap. 18, is followed by a chapter that considers national and international standards related to shock and vibration.

The subjects of test criteria and specifications are introduced in Chap. 20, followed by a comprehensive chapter on modal analysis and testing. Chapters 22 and 23 discuss the applicable concepts of data analysis. The first of these two chapters is concerned with the analysis of data defining vibration conditions, and it discusses the transformation of time-history of a vibration measurement into more compact forms of data; the second of these two chapters is an analogous presentation—treating the transformation of time-histories defining conditions of shock. Vibration that is induced as a result of ground motion is described in Chap. 24.

The next two chapters deal with computational methods. Chapter 27 is concerned with the use of computers, presenting information which is useful in both analytical and experimental work. This is followed by Chap. 28, which is in two parts: the first describes modern matrix methods of analysis, dealing largely with the formulation of matrices for use with digital computers and other numerical calculation methods; the second part shows how finite element methods can be applied to the solution of shock and vibration problems by the use of computer techniques.

Part 1 of Chap. 29 describes vibration that is induced as a result of air flow; Part 2 discusses vibration that is induced by the flow of water.

The theory of vibration isolation is discussed in detail in Chap. 30 and an analogous presentation for the isolation of mechanical shock is given in Chap. 31. Various types of isolator for isolating shock and vibration are described in Chap. 33, and the practical application of such isolators and selection technique for isolators is described in Chap. 33. A presentation is given in Chap. 34 on the engineering properties of rubber; then another new presentation on the engineering properties of metals (including conventional fatigue) is given in Chap. 35.

An important method of controlling shock and vibration involves the addition of damping or energy-dissipating means to structures that are susceptible to vibration. Chapter 36 discusses the general concepts of damping together with the application of such concepts to hysteresis and slip damping. The application of damping materials to devices and structures is discussed in Chap. 37.

The latter chapters of the Handbook deal with the specific application of fundamentals of analysis, methods of measurement, and control techniques—where these are developed sufficiently to form a separate and discrete subject. Torsional vibration is discussed in Chap. 38, with particular application to internal-combustion engines. The balancing of rotating equipment is discussed in Chap. 39, and balancing machines are described. Chapter 40 describes the special vibration problems associated with the design and operation of machine tools. Among the most prominent occurrences of shock and vibration are those that arise during handling and shipping of merchandise. Packaging of equipment to protect against such shock and vibration is discussed in Chap. 41. Chapters 42 and 43 describe procedures for the design of equipment to withstand shock and vibration—the former considering primarily the theory of design and the latter considering practical aspects. A comprehensive discussion of the human aspect of shock and vibration is considered in Chap. 44, which describes the effect of shock and vibration on people.

SYMBOLS AND ACRONYMS

This section includes a list of symbols with their usual English units as used generally in the Handbook; metric units are given as alternates in the text. In special circumstances, some of the following symbols have different meanings in certain chapters but are defined in those chapters. Other symbols of special or limited application are defined in the respective chapters as they are used.

Symbol	Meaning	Units
a	radius	in.
B	bandwidth	Hz
B	magnetic flux density	gauss
c	damping coefficient	lb-sec/in.
c	velocity of sound	in./sec
c_c	critical damping coefficient	lb-sec/in.
C	capacitance	farads
D	diameter	in.
e	electrical voltage	volts
e	eccentricity	in.
E	energy	in.-lb
E	modulus of elasticity in tension and compression (Young's modulus)	$lb/in.^2$
f	frequency	Hz
f_n	undamped natural frequency	Hz
f_i	undamped natural frequencies in a multiple degree-of-freedom system, where $i = 1, 2, \ldots$	Hz
f_d	damped natural frequency	Hz
f_r	resonance frequency	Hz
F	force	lb
f_f	Coulomb friction force	lb
FEM	finite element method, finite element model	
FFT	fast Fourier transform	

g	acceleration of gravity	in./sec^2
G	modulus of elasticity in shear	lb/in.^{2}
h	height, depth	in.
H	magnetic field strength	oersteds
Hz	hertz	
i	electric current	amperes
I_i	area or mass moment of inertia (subscript indicates axis)	in.^{4}
I_p	polar moment of inertia	in.^{4}
I_{ij}	area or mass product of inertia (subscripts indicate axes)	in.^{4}
$\mathfrak{I}$	imaginary part of	
j	$\sqrt{-1}$	
J	inertia constant (weight moment of inertia)	lb-in.^{2}
J	impulse	lb-sec
k	spring constant, stiffness, stiffness constant	lb/in.
k_t	rotational (torsional) stiffness	lb-in./rad
l	length	in.
L	inductance	henrys
m	mass	lb-sec^2/in.
m_u	unbalanced mass	lb-sec^2/in.
M	torque	lb-in.
M	mutual inductance	henrys
$\mathfrak{M}$	mobility	in./lb-sec
n	number of coils, supports, etc.	
p	alternating pressure	lb/in.^{2}
p	probability density	
P	probability distribution	
P	static pressure	lb/in.^{2}
q	electric charge	coulombs
Q	resonance factor (also ratio of reactance to resistance)	
r	electrical resistance	ohms
R	radius	in.
$\mathfrak{R}$	real part of	
s	arc length	in.
S	area of diaphragm, tube, etc.	in.^{2}
SEA	statistical energy analysis	
t	thickness	in.
t	time	sec
T	transmissibility	
T	kinetic energy	in.-lb
v	linear velocity	in./sec
V	potential energy	in.-lb
w	width	in.
W	weight	lb
W	power	in.-lb/sec
W_e	spectral density of the excitation	(rms)^2/unit freq.
W_r	spectral density of the response	(rms)^2/unit freq.
x	linear displacement in direction of X axis	in.
y	linear displacement in direction of Y axis	in.
z	linear displacement in direction of Z axis	in.
Z	impedance	lb-sec/in.
α	rotational displacement about X axis	radians
β	rotational displacement about Y axis	radians
γ	rotational displacement about Z axis	radians
γ	shear strain	
γ	weight density	lb/in.^{3}

δ	deflection	in.
δ_{st}	static deflection	in.
Δ	logarithmic decrement	
ϵ	tension or compression strain	
ζ	fraction of critical damping	
η	stiffness ratio, loss factor	
θ	phase angle	radians
λ	wavelength	in.
μ	coefficient of friction	
μ	mass density	lb-sec^2/in.^{4}
ν	Poisson's ratio	
ρ	mass density	lb-sec^2/in.^{4}
ρ_i	radius of gyration (subscript indicates axis)	in.
σ	Poisson's ratio	
σ	normal stress	lb/in.^{2}
σ	root-mean-square (rms) value	
τ	period	sec
τ	shear stress	lb/in.^{2}
ϕ	magnetic flux	maxwell
ψ	phase angle	radians
ω	forcing frequency—angular	rad/sec
ω_n	undamped natural frequency—angular	rad/sec
ω_i	undamped natural frequencies—angular—in a multiple degree-of-freedom system, where $i = 1, 2, \ldots$	rad/sec
ω_d	damped natural frequency—angular	rad/sec
ω_r	resonance frequency—angular	rad/sec
Ω	rotational speed	rad/sec
$\simeq$	approximately equal to	

CHARACTERISTICS OF HARMONIC MOTION

Harmonic functions are employed frequently in the analysis of shock and vibration. A body that experiences simple harmonic motion follows a displacement pattern defined by

$$x = x_0 \sin (2\pi ft) = x_0 \sin \omega t \tag{1.1}$$

where f is the *frequency* of the simple harmonic motion, $\omega = 2\pi f$ is the corresponding *angular frequency,* and x_0 is the *amplitude* of the displacement.

The velocity $\dot{x}$ and acceleration $\ddot{x}$ of the body are found by differentiating the displacement once and twice, respectively:

$$\dot{x} = x_0(2\pi f) \cos 2\pi ft = x_0\omega \cos \omega t \tag{1.2}$$

$$\ddot{x} = -x_0(2\pi f)^2 \sin 2\pi ft = -x_0\omega^2 \sin \omega t \tag{1.3}$$

The maximum absolute values of the displacement, velocity, and acceleration of a body undergoing harmonic motion occur when the trigonometric functions in Eqs. (1.1) to (1.3) are numerically equal to unity. These values are known, respectively, as displacement, velocity, and acceleration amplitudes; they are defined mathematically as follows:

$$x_0 = x_0 \qquad \dot{x}_0 = (2\pi f)x_0 \qquad \ddot{x}_0 = (2\pi f)^2 x_0 \tag{1.4}$$

It is common to express the displacement amplitude x_0 in inches when the English system of units is used and in centimeters or millimeters when the metric system is used. Accordingly, the velocity amplitude x_0 is expressed in inches per second in the English system (centimeters per second or millimeters per second in the metric system). For example, consider a body that experiences simple harmonic motion having a frequency f of 50 Hz and a displacement amplitude x_0 of 0.01 in.

TABLE 1.1   Conversion Factors for Translational Velocity and Acceleration

Multiply Value in → or → By ↘ To obtain value in ↓	g-sec, g	ft/sec ft/sec^2	in./sec in./sec^2	cm/sec cm/sec^2	m/sec m/sec^2
g-sec, g	1	0.0311	0.00259	0.00102	0.102
ft/sec ft/sec^2	32.16	1	0.0833	0.0328	3.28
in./sec in./sec^2	386	12.0	1	0.3937	39.37
cm/sec cm/sec^2	980	30.48	2.540	1	100
m/sec m/sec^2	9.80	0.3048	0.0254	0.010	1

TABLE 1.2   Conversion Factors for Rotational Velocity and Acceleration

Multiply Value in → or → By ↘ To obtain value in ↓	rad/sec rad/sec^2	degree/sec degree/sec^2	rev/sec rev/sec^2	rev/min rev/min/sec
rad/sec rad/sec^2	1	0.01745	6.283	0.1047
degree/sec degree/sec^2	57.30	1	360	6.00
rev/sec rev/sec^2	0.1592	0.00278	1	0.0167
rev/min rev/min/sec	9.549	0.1667	60	1

(0.000254 m). According to Eq. (1.4), the velocity amplitude $\dot{x}_0 = (2\pi f)\, x_0 = 3.14$ in./sec (0.0797 m/s). The acceleration amplitude $\ddot{x}_0 = (2\pi f)^2\, x_0$ in./sec^2 = 986 in./sec^2 (25.0 m/s^2). The acceleration amplitude x_0 is often expressed as a dimensionless multiple of the gravitational acceleration g where $g = 386$ in./sec^2 (9.8 m/s^2). Therefore in this example, the acceleration amplitude may also be expressed as $\ddot{x}_0 = 2.55g$.

Factors for converting values of rectilinear velocity and acceleration to different units are given in Table 1.1; similar factors for angular velocity and acceleration are given in Table 1.2.

For certain purposes in analysis, it is convenient to express the amplitude in terms of the average value of the harmonic function, the root-mean-square (rms) value, or 2 times the amplitude (i.e., peak-to-peak value). These terms are defined mathematically in Chap. 22; numerical conversion factors are set forth in Table 1.3 for ready reference.

TABLE 1.3 Conversion Factors for Simple Harmonic Motion

Multiply numerical value in terms of → By ↘ To obtain value in terms of ↓	Amplitude	Average value	Root-mean-square value (rms)	Peak-to-peak value
Amplitude	1	1.571	1.414	0.500
Average value	0.637	1	0.900	0.318
Root-mean-square value (rms)	0.707	1.111	1	0.354
Peak-to-peak value	2.000	3.142	2.828	1

APPENDIX 1.1 NATURAL FREQUENCIES OF COMMONLY USED SYSTEMS

The most important aspect of vibration analysis often is the calculation or measurement of the natural frequencies of mechanical systems. Natural frequencies are discussed prominently in many chapters of the Handbook. Appendix 1.1 includes in tabular form, convenient for ready reference, a compilation of frequently used expressions for the natural frequencies of common mechanical systems:

1. Mass-spring systems in translation
2. Rotor-shaft systems
3. Massless beams with concentrated mass loads
4. Beams of uniform section and uniformly distributed load
5. Thin flat plates of uniform thickness
6. Miscellaneous systems

The data for beams and plates are abstracted from Chap. 7.

MASS-SPRING SYSTEMS IN TRANSLATION
(RIGID MASS AND MASSLESS SPRING)

k = SPRING STIFFNESS, LB/IN.

m = MASS, LB-SEC^2/IN.

ω_n = ANGULAR NATURAL FREQUENCY, RAD/SEC

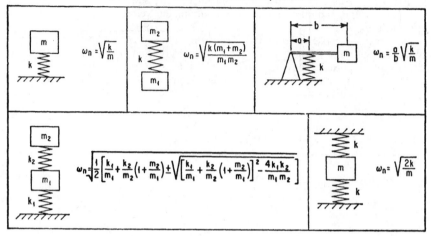

SPRINGS IN COMBINATION
k_r = RESULTANT STIFFNESS OF COMBINATION

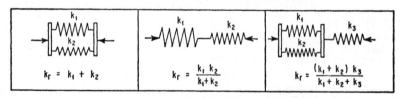

HELICAL SPRINGS

d = WIRE DIAMETER, IN

D = MEAN COIL DIAMETER, IN.

n = NUMBER OF ACTIVE COILS

E = YOUNG'S MODULUS, LB/IN.^{2}

G = MODULUS OF ELASTICITY IN SHEAR, LB/IN.^{2}

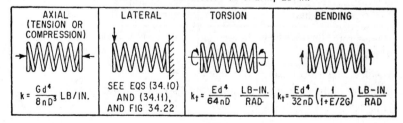

ROTOR-SHAFT SYSTEMS
(RIGID ROTOR AND MASSLESS SHAFT)

k_t = TORSIONAL STIFFNESS OF SHAFT, LB-IN./RAD

I = MASS MOMENT OF INERTIA OF ROTOR, LB-IN.-SEC^2

ω_n = ANGULAR NATURAL FREQUENCY, RAD/SEC

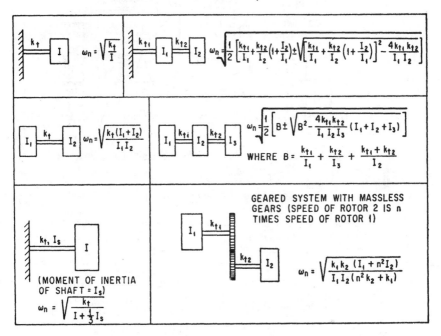

$$\omega_n = \sqrt{\frac{k_t}{I}}$$

$$\omega_n = \sqrt{\frac{1}{2}\left[\frac{k_{t1}}{I_1}+\frac{k_{t2}}{I_2}\left(1+\frac{I_2}{I_1}\right)\pm\sqrt{\left[\frac{k_{t1}}{I_1}+\frac{k_{t2}}{I_2}\left(1+\frac{I_2}{I_1}\right)\right]^2-\frac{4k_{t1}k_{t2}}{I_1 I_2}}\right]}$$

$$\omega_n = \sqrt{\frac{k_t(I_1+I_2)}{I_1 I_2}}$$

$$\omega_n = \sqrt{\frac{1}{2}\left[B\pm\sqrt{B^2-\frac{4k_{t1}k_{t2}}{I_1 I_2 I_3}(I_1+I_2+I_3)}\right]}$$

WHERE $B = \dfrac{k_{t1}}{I_1}+\dfrac{k_{t2}}{I_3}+\dfrac{k_{t1}+k_{t2}}{I_2}$

(MOMENT OF INERTIA OF SHAFT = I_s)

$$\omega_n = \sqrt{\frac{k_t}{I+\frac{1}{3}I_s}}$$

GEARED SYSTEM WITH MASSLESS GEARS (SPEED OF ROTOR 2 IS n TIMES SPEED OF ROTOR 1)

$$\omega_n = \sqrt{\frac{k_1 k_2 (I_1+n^2 I_2)}{I_1 I_2 (n^2 k_2+k_1)}}$$

STIFFNESS OF SHAFTS IN TORSION

G = MODULUS OF ELASTICITY IN SHEAR, LB/IN.^{2}

l = LENGTH OF SHAFT, IN.

I_p = POLAR MOMENT OF INERTIA OF SHAFT CROSS-SECTION, IN.^{4}

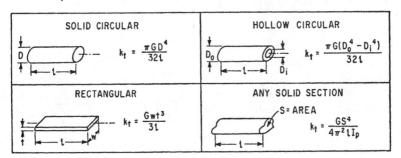

SOLID CIRCULAR	HOLLOW CIRCULAR
$k_t = \dfrac{\pi G D^4}{32 l}$	$k_t = \dfrac{\pi G(D_o^4-D_i^4)}{32 l}$
RECTANGULAR	ANY SOLID SECTION
$k_t = \dfrac{G w t^3}{3 l}$	S = AREA $k_t = \dfrac{G S^4}{4\pi^2 l I_p}$

MASSLESS BEAMS WITH CONCENTRATED MASS LOADS

m = MASS OF LOAD, LB-SEC^2/IN.
$\mathfrak{l}$ = LENGTH OF BEAM, IN.
I = AREA MOMENT OF INERTIA OF BEAM CROSS SECTION, IN.^{4}
E = YOUNG'S MODULUS, LB/IN.^{2}
ω_n = ANGULAR NATURAL FREQUENCY, RAD/SEC

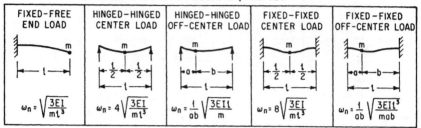

FIXED-FREE END LOAD	HINGED-HINGED CENTER LOAD	HINGED-HINGED OFF-CENTER LOAD	FIXED-FIXED CENTER LOAD	FIXED-FIXED OFF-CENTER LOAD
$\omega_n = \sqrt{\dfrac{3EI}{m\mathfrak{l}^3}}$	$\omega_n = 4\sqrt{\dfrac{3EI}{m\mathfrak{l}^3}}$	$\omega_n = \dfrac{1}{ab}\sqrt{\dfrac{3EI\mathfrak{l}}{m}}$	$\omega_n = 8\sqrt{\dfrac{3EI}{m\mathfrak{l}^3}}$	$\omega_n = \dfrac{1}{ab}\sqrt{\dfrac{3EI\mathfrak{l}^3}{mab}}$

MASSIVE SPRINGS (BEAMS) WITH CONCENTRATED MASS LOADS

m = MASS OF LOAD, LB-SEC^2/IN.
$m_s(m_b)$ = MASS OF SPRING (BEAM), LB-SEC^2/IN.
k = STIFFNESS OF SPRING LB/IN.
$\mathfrak{l}$ = LENGTH OF BEAM, IN.
I = AREA MOMENT OF INERTIA OF BEAM CROSS SECTION, IN.^{4}
E = YOUNG'S MODULUS, LB/IN.^{2}
ω_n = ANGULAR NATURAL FREQUENCY, RAD/SEC

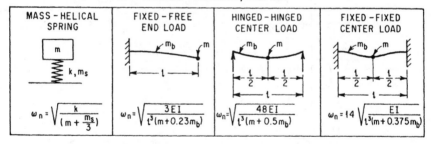

MASS-HELICAL SPRING	FIXED-FREE END LOAD	HINGED-HINGED CENTER LOAD	FIXED-FIXED CENTER LOAD
$\omega_n = \sqrt{\dfrac{k}{\left(m + \dfrac{m_s}{3}\right)}}$	$\omega_n = \sqrt{\dfrac{3EI}{\mathfrak{l}^3(m + 0.23m_b)}}$	$\omega_n = \sqrt{\dfrac{48EI}{\mathfrak{l}^3(m + 0.5m_b)}}$	$\omega_n = 14\sqrt{\dfrac{EI}{\mathfrak{l}^3(m + 0.375m_b)}}$

AREA MOMENT OF INERTIA OF BEAM SECTIONS
(WITH RESPECT TO AXIS a-a)

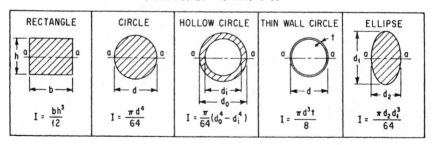

RECTANGLE	CIRCLE	HOLLOW CIRCLE	THIN WALL CIRCLE	ELLIPSE
$I = \dfrac{bh^3}{12}$	$I = \dfrac{\pi d^4}{64}$	$I = \dfrac{\pi}{64}(d_0^4 - d_i^4)$	$I = \dfrac{\pi d^3 t}{8}$	$I = \dfrac{\pi d_2 d_i^3}{64}$

BEAMS OF UNIFORM SECTION AND UNIFORMLY DISTRIBUTED LOAD

ANGULAR NATURAL FREQUENCY $\omega_n = A\sqrt{\dfrac{EI}{\mu \, \mathit{l}^4}}$ RAD/SEC

WHERE E = YOUNG'S MODULUS, LB/IN^2.

I = AREA MOMENT OF INERTIA OF BEAM CROSS SECTION, IN.

l = LENGTH OF BEAM, IN.

μ = MASS PER UNIT LENGTH OF BEAM, LB-SEC^2/IN^2.

A = COEFFICIENT FROM TABLE BELOW

NODES ARE INDICATED IN TABLE BELOW AS A PROPORTION OF LENGTH l MEASURED FROM LEFT END

FIXED–FREE (CANTILEVER)	A = 3.52	0.774 A = 22.0	0.500 0.868 A = 61.7	0.356 0.644 0.906 A = 121.0	0.500 0.926 0.279 0.723 A = 200.0
HINGED–HINGED (SIMPLE)	0.500 A = 9.87	0.500 A = 39.5	0.333 0.667 A = 88.9	0.25 0.50 0.75 A = 158	0.20 0.40 0.60 0.80 A = 247
FIXED–FIXED (BUILT–IN)	0.776 A = 22.4	0.500 A = 61.7	0.359 0.641 A = 121	0.278 0.500 0.722 A = 200	0.409 0.773 0.227 0.591 A = 298
FREE–FREE	A = 22.4	0.224 0.776 A = 22.4	0.132 0.500 0.868 A = 61.7	0.094 0.644 0.356 0.906 A = 121	0.277 0.723 0.073 0.500 0.927 A = 200	0.060 0.409 0.773 0.227 0.591 0.940 A = 298
FIXED–HINGED	A = 15.4	0.560 A = 50.0	0.384 0.692 A = 104	0.529 0.765 0.294 A = 178	0.429 0.810 0.238 0.619 A = 272
HINGED–FREE	0.736 A = 15.4	0.446 0.853 A = 50.0	0.308 0.616 0.898 A = 104	0.471 0.922 0.235 0.707 A = 178	0.381 0.763 0.190 0.581 0.937 A = 272

1.13

NATURAL FREQUENCIES OF THIN FLAT PLATES OF UNIFORM THICKNESS

$$\omega_n = B \sqrt{\frac{E t^2}{\rho \, a^4 (1-\nu^2)}} \quad \text{RAD/SEC}$$

E = YOUNG'S MODULUS, LB /IN.^{2}
t = THICKNESS OF PLATE, IN.
ρ = MASS DENSITY, LB-SEC^2/IN.^{4}
a = DIAMETER OF CIRCULAR PLATE OR SIDE OF SQUARE PLATE, IN.
ν = POISSON'S RATIO

SHAPE OF PLATE	DIAGRAM	EDGE CONDITIONS	VALUE OF B FOR MODE:							
			1	2	3	4	5	6	7	8
CIRCULAR		CLAMPED AT EDGE	11.84	24.61	40.41	46.14	103.12			
CIRCULAR		FREE	6.09	10.53	14.19	23.80	40.88	44.68	61.38	69.44
CIRCULAR		CLAMPED AT CENTER	4.35	24.26	70.39	138.85				
CIRCULAR		SIMPLY SUPPORTED AT EDGE	5.90							
SQUARE		ONE EDGE CLAMPED– THREE EDGES FREE	1.01	2.47	6.20	7.94	9.01			
SQUARE		ALL EDGES CLAMPED	10.40	21.21	31.29	38.04	38.22	47.73		
SQUARE		TWO EDGES CLAMPED– TWO EDGES FREE	2.01	6.96	7.74	13.89	18.25			
SQUARE		ALL EDGES FREE	4.07	5.94	6.91	10.39	17.80	18.85		
SQUARE		ONE EDGE CLAMPED– THREE EDGES SIMPLY SUPPORTED	6.83	14.94	16.95	24.89	28.99	32.71		
SQUARE		TWO EDGES CLAMPED– TWO EDGES SIMPLY SUPPORTED	8.37	15.82	20.03	27.34	29.54	37.31		
SQUARE		ALL EDGES SIMPLY SUPPORTED	5.70	14.26	22.82	28.52	37.08	48.49		

MASSLESS CIRCULAR PLATE WITH CONCENTRATED CENTER MASS

CLAMPED EDGES $\omega_n = 4.09 \sqrt{\dfrac{E h^3}{m a^2 (1-\nu^2)}}$

SIMPLY SUPPORTED EDGES $\omega_n = 4.09 \sqrt{\dfrac{E h^3}{m a^2 (1-\nu)(3+\nu)}}$

NATURAL FREQUENCIES OF MISCELLANEOUS SYSTEMS
(ω_n = ANGULAR NATURAL FREQUENCY, RAD/SEC)

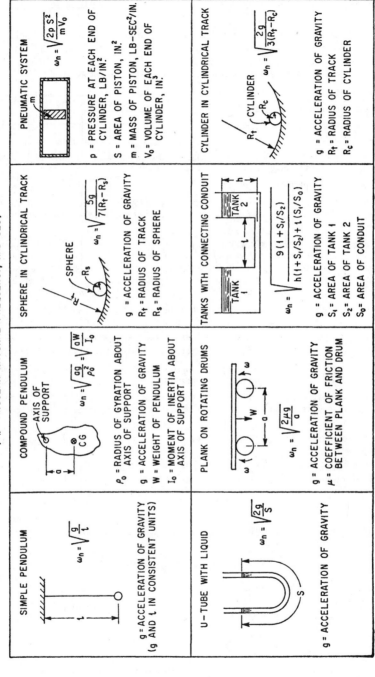

SIMPLE PENDULUM

$$\omega_n = \sqrt{\frac{g}{\ell}}$$

g = ACCELERATION OF GRAVITY
(g AND ℓ IN CONSISTENT UNITS)

COMPOUND PENDULUM

AXIS OF SUPPORT

CG

$$\omega_n = \sqrt{\frac{ag}{\rho_0^2}} = \sqrt{\frac{aW}{I_0}}$$

ρ_0 = RADIUS OF GYRATION ABOUT AXIS OF SUPPORT
g = ACCELERATION OF GRAVITY
W = WEIGHT OF PENDULUM
I_0 = MOMENT OF INERTIA ABOUT AXIS OF SUPPORT

SPHERE IN CYLINDRICAL TRACK

SPHERE
R_s

$$\omega_n = \sqrt{\frac{5g}{7(R_t - R_s)}}$$

g = ACCELERATION OF GRAVITY
R_t = RADIUS OF TRACK
R_s = RADIUS OF SPHERE

PNEUMATIC SYSTEM

m

$$\omega_n = \sqrt{\frac{2p\,S^2}{m V_0}}$$

p = PRESSURE AT EACH END OF CYLINDER, LB/IN.²
S = AREA OF PISTON, IN.²
m = MASS OF PISTON, LB–SEC²/IN.
V_0 = VOLUME OF EACH END OF CYLINDER, IN.³

U–TUBE WITH LIQUID

S

$$\omega_n = \sqrt{\frac{2g}{S}}$$

g = ACCELERATION OF GRAVITY

PLANK ON ROTATING DRUMS

ω W ω
a

$$\omega_n = \sqrt{\frac{2\mu g}{a}}$$

g = ACCELERATION OF GRAVITY
μ = COEFFICIENT OF FRICTION BETWEEN PLANK AND DRUM

TANKS WITH CONNECTING CONDUIT

TANK 1   TANK 2
ℓ h

$$\omega_n = \sqrt{\frac{g(1 + S_1/S_2)}{h(1 + S_1/S_2) + \ell(S_1/S_0)}}$$

g = ACCELERATION OF GRAVITY
S_1 = AREA OF TANK 1
S_2 = AREA OF TANK 2
S_0 = AREA OF CONDUIT

CYLINDER IN CYLINDRICAL TRACK

R_t CYLINDER
R_c

$$\omega_n = \sqrt{\frac{2g}{3(R_t - R_c)}}$$

g = ACCELERATION OF GRAVITY
R_t = RADIUS OF TRACK
R_c = RADIUS OF CYLINDER

APPENDIX 1.2 TERMINOLOGY

For convenience, definitions of terms which are used frequently in the field of shock and vibration are assembled here. Many of the definitions in this appendix are identical with those contained in ANSI S1.1-1994 and/or ISO 2041: 1990.* **For terms which are not listed below, the reader is referred to the Index.**

acceleration   Acceleration is a vector quantity that specifies the time rate of change of velocity.

acceleration of gravity   See *g*.

accelerometer   An accelerometer is a transducer whose output is proportional to the acceleration input.

ambient vibration   Ambient vibration is the all-encompassing vibration associated with a given environment, being usually a composite of vibration from many sources, near and far.

amplitude   Amplitude is the maximum value of a sinusoidal quantity.

analog   If a first quantity or structural element is analogous to a second quantity or structural element belonging in another field of knowledge, the second quantity is called the analog of the first, and vice versa.

analogy   An analogy is a recognized relationship of consistent mutual similarity between the equations and structures appearing within two or more fields of knowledge, and an identification and association of the quantities and structural elements that play mutually similar roles in these equations and structures, for the purpose of facilitating transfer of knowledge of mathematical procedures of analysis and behavior of the structures between these fields.

angular frequency (circular frequency)   The angular frequency of a periodic quantity, in radians per unit time, is the frequency multiplied by 2π.

angular mechanical impedance (rotational mechanical impedance)   Angular mechanical impedance is the impedance involving the ratio of torque to angular velocity. (See *impedance*.)

antinode (loop)   An antinode is a point, line, or surface in a standing wave where some characteristic of the wave field has maximum amplitude.

antiresonance   For a system in forced oscillation, antiresonance exists at a point when any change, however small, in the frequency of excitation causes an increase in the response at this point.

audio frequency   An audio frequency is any frequency corresponding to a normally audible sound wave.

autocorrelation coefficient   The autocorrelation coefficient of a signal is the ratio of the autocorrelation function to the mean-square value of the signal:

$$R(\tau) = \overline{x(t)x(t+\tau)}/\overline{[x(t)]^2}$$

autocorrelation function   The autocorrelation function of a signal is the average of the product of the value of the signal at time t with the value at time $t + \tau$:

$$R(\tau) = \overline{x(t)x(t+\tau)}$$

For a stationary random signal of infinite duration, the power spectral density (except for a constant factor) is the cosine Fourier transform of the autocorrelation function.

autospectral density   See *power spectral density*.

---

* Copies of these documents, as well as others in the field of mechanical shock and vibration, are available from the American National Standards Institute, Inc., 11 West 42nd St., New York, NY 10036.

auxiliary mass damper (damped vibration absorber)   An auxiliary mass damper is a system consisting of a mass, spring, and damper which tends to reduce vibration by the dissipation of energy in the damper as a result of relative motion between the mass and the structure to which the damper is attached.

background noise   Background noise is the total of all sources of interference in a system used for the production, detection, measurement, or recording of a signal, independent of the presence of the signal.

balancing   Balancing is a procedure for adjusting the mass distribution of a rotor so that vibration of the journals, or the forces on the bearings at once-per-revolution, are reduced or controlled. (See *Chap. 39* for a complete list of definitions related to *balancing*.)

bandpass filter   A bandpass filter is a wave filter that has a single transmission band extending from a lower cutoff frequency greater than zero to a finite upper cutoff frequency.

bandwidth, effective   (See *effective bandwidth.*)

beats   Beats are periodic variations that result from the superposition of two simple harmonic quantities of different frequencies f_1 and f_2. They involve the periodic increase and decrease of amplitude at the beat frequency $(f_1 - f_2)$.

broadband random vibration   Broadband random vibration is random vibration having its frequency components distributed over a broad frequency band. (See *random vibration.*)

center-of-gravity   Center-of-gravity is the point through which passes the resultant of the weights of its component particles for all orientations of the body with respect to a gravitational field; if the gravitational field is uniform, the center-of-gravity corresponds with the *center-of-mass.*

circular frequency   (See *angular frequency.*)

complex angular frequency   As applied to a function $\alpha = Ae^{\sigma t} \sin(\omega t - \phi)$, where σ, ω, and ϕ are constant, the quantity $\omega_c = \sigma + j\omega$ is the complex angular frequency where j is an operator with rules of addition, multiplication, and division as suggested by the symbol $\sqrt{-1}$. If the signal decreases with time, σ must be negative.

complex function   A complex function is a function having real and imaginary parts.

complex vibration   Complex vibration is vibration whose components are sinusoids not harmonically related to one another. (See *harmonic.*)

compliance   Compliance is the reciprocal of stiffness.

compressional wave   A compressional wave is one of compressive or tensile stresses propagated in an elastic medium.

continuous system (distributed system)   A continuous system is one that is considered to have an infinite number of possible independent displacements. Its configuration is specified by a function of a continuous spatial variable or variables in contrast to a discrete or lumped parameter system which requires only a finite number of coordinates to specify its configuration.

correlation coefficient   The correlation coefficient of two variables is the ratio of the correlation function to the product of the averages of the variables:

$$\overline{x_1(t) \cdot x_2(t)} / \overline{x_1(t)} \cdot \overline{x_2(t)}$$

correlation function   The correlation function of two variables is the average value of their product:

$$\overline{x_1(t) \cdot x_2(t)}$$

coulomb damping (dry friction damping)   Coulomb damping is the dissipation of energy that occurs when a particle in a vibrating system is resisted by a force whose magnitude is a constant independent of displacement and velocity and whose direction is opposite to the direction of the velocity of the particle.

coupled modes   Coupled modes are modes of vibration that are not independent but which influence one another because of energy transfer from one mode to the other. (See *mode of vibration.*)

coupling factor, electromechanical   The electromechanical coupling factor is a factor used to characterize the extent to which the electrical characteristics of a transducer are modified by a coupled mechanical system, and vice versa.

crest factor   The crest factor is the ratio of the peak value to the root-mean-square value.

critical damping   Critical damping is the minimum viscous damping that will allow a displaced system to return to its initial position without oscillation.

critical speed   Critical speed is the speed of a rotating system that corresponds to a resonance frequency of the system.

cycle   A cycle is the complete sequence of values of a periodic quantity that occur during a period.

damped natural frequency   The damped natural frequency is the frequency of free vibration of a damped linear system. The free vibration of a damped system may be considered periodic in the limited sense that the time interval between zero crossings in the same direction is constant, even though successive amplitudes decrease progressively. The frequency of the vibration is the reciprocal of this time interval.

damper   A damper is a device used to reduce the magnitude of a shock or vibration by one or more energy dissipation methods.

damping   Damping is the dissipation of energy with time or distance.

damping ratio   (See *fraction of critical damping.*)

decibel (dB)   The decibel is a unit which denotes the magnitude of a quantity with respect to an arbitrarily established reference value of the quantity, in terms of the logarithm (to the base 10) of the ratio of the quantities. For example, in electrical transmission circuits a value of power may be expressed in terms of a power level in decibels; the power level is given by 10 times the logarithm (to the base 10) of the ratio of the actual power to a reference power (which corresponds to 0 dB).

degrees-of-freedom   The number of degrees-of-freedom of a mechanical system is equal to the minimum number of independent coordinates required to define completely the positions of all parts of the system at any instant of time. In general, it is equal to the number of independent displacements that are possible.

deterministic function   A deterministic function is one whose value at any time can be predicted from its value at any other time.

displacement   Displacement is a vector quantity that specifies the change of position of a body or particle and is usually measured from the mean position or position of rest. In general, it can be represented as a rotation vector or a translation vector, or both.

displacement pickup   Displacement pickup is a transducer that converts an input displacement to an output that is proportional to the input displacement.

distortion   Distortion is an undesired change in waveform. Noise and certain desired changes in waveform, such as those resulting from modulation or detection, are not usually classed as distortion.

distributed system   (See *continuous system.*)

driving point impedance   Driving point impedance is the impedance involving the ratio of force to velocity when both the force and velocity are measured at the same point and in the same direction. (See *impedance.*)

dry friction damping   (See *coulomb damping.*)

duration of shock pulse   The duration of a shock pulse is the time required for the acceleration of the pulse to rise from some stated fraction of the maximum amplitude and to decay to this value. (See *shock pulse.*)

dynamic stiffness   Dynamic stiffness is the ratio of the change of force to the change of displacement under dynamic conditions.

dynamic vibration absorber (tuned damper)   A dynamic vibration absorber is an auxiliary mass-spring system which tends to neutralize vibration of a structure to which it is attached. The basic principle of operation is vibration out-of-phase with the vibration of such structure, thereby applying a counteracting force.

effective bandwidth   The effective bandwidth of a specified transmission system is the bandwidth of an ideal system which (1) has uniform transmission in its pass band equal to the maximum transmission of the specified system and (2) transmits the same power as the specified system when the two systems are receiving equal input signals having a uniform distribution of energy at all frequencies.

electromechanical coupling factor   (See *coupling factor, electromechanical.*)

electrostriction   Electrostriction is the phenomenon wherein some dielectric materials experience an elastic strain when subjected to an electric field, this strain being independent of the polarity of the field.

ensemble   A collection of signals. (Also see *process.*)

environment   (See *natural environments* and *induced environment.*)

equivalent system   An equivalent system is one that may be substituted for another system for the purpose of analysis. Many types of equivalence are common in vibration and shock technology: (1) equivalent stiffness, (2) equivalent damping, (3) torsional system equivalent to a translational system, (4) electrical or acoustical system equivalent to a mechanical system, etc.

equivalent viscous damping   Equivalent viscous damping is a value of viscous damping assumed for the purpose of analysis of a vibratory motion, such that the dissipation of energy per cycle at resonance is the same for either the assumed or actual damping force.

ergodic process   An ergodic process is a random process that is stationary and of such a nature that all possible time averages performed on one signal are independent of the signal chosen and hence are representative of the time averages of each of the other signals of the entire random process.

excitation (stimulus)   Excitation is an external force (or other input) applied to a system that causes the system to respond in some way.

filter   A filter is a device for separating waves on the basis of their frequency. It introduces relatively small insertion loss to waves in one or more frequency bands and relatively large insertion loss to waves of other frequencies. (See *insertion loss.*)

force factor   The force factor of an electromechanical transducer is (1) the complex quotient of the force required to block the mechanical system divided by the corresponding current in the electric system and (2) the complex quotient of the resulting open-circuit voltage in the electric system divided by the velocity in the mechanical system. Force factors (1) and (2) have the same magnitude when consistent units are used and the transducer satisfies the principle of reciprocity. It is sometimes convenient in an electrostatic or piezoelectric transducer to use the ratios between force and charge or electric displacement, or between voltage and mechanical displacement.

forced vibration (forced oscillation)   The oscillation of a system is forced if the response is imposed by the excitation. If the excitation is periodic and continuing, the oscillation is steady-state.

foundation (support)   A foundation is a structure that supports the gravity load of a mechanical system. It may be fixed in space, or it may undergo a motion that provides excitation for the supported system.

fraction of critical damping   The fraction of critical damping (damping ratio) for a system with viscous damping is the ratio of actual damping coefficient c to the critical damping coefficient c_c.

free vibration   Free vibration is that which occurs after the removal of an excitation or restraint.

frequency   The frequency of a function periodic in time is the reciprocal of the period. The unit is the cycle per unit time and must be specified; the unit *cycle per second* is called *hertz* (Hz).

frequency, angular   (See *angular frequency.*)

fundamental frequency   (1) The fundamental frequency of a periodic quantity is the frequency of a sinusoidal quantity which has the same period as the periodic quantity. (2) The fundamental frequency of an oscillating system is the lowest natural frequency. The normal mode of vibration associated with this frequency is known as the fundamental mode.

fundamental mode of vibration   The fundamental mode of vibration of a system is the mode having the lowest natural frequency.

g   The quantity g is the acceleration produced by the force of gravity, which varies with the latitude and elevation of the point of observation. By international agreement, the value $980.665 \text{ cm/sec}^2 = 386.087 \text{ in./sec}^2 = 32.1739 \text{ ft/sec}^2$ has been chosen as the standard acceleration due to gravity.

harmonic   A harmonic is a sinusoidal quantity having a frequency that is an integral multiple of the frequency of a periodic quantity to which it is related.

harmonic motion   (See *simple harmonic motion.*)

harmonic response   Harmonic response is the periodic response of a vibrating system exhibiting the characteristics of resonance at a frequency that is a multiple of the excitation frequency.

high-pass filter   A high-pass filter is a wave filter having a single transmission band extending from some critical or cutoff frequency, not zero, up to infinite frequency.

image impedances   The image impedances of a structure or device are the impedances that will simultaneously terminate all of its inputs and outputs in such a way that at each of its inputs and outputs the impedances in both directions are equal.

impact   An impact is a single collision of one mass in motion with a second mass which may be either in motion or at rest.

impedance   Mechanical impedance is the ratio of a force-like quantity to a velocity-like quantity when the arguments of the real (or imaginary) parts of the quantities increase linearly with time. Examples of force-like quantities are: force, sound pressure, voltage, temperature. Examples of velocity-like quantities are: velocity, volume velocity, current, heat flow. *Impedance* is the reciprocal of *mobility.* (Also see *angular mechanical impedance, linear mechanical impedance, driving point impedance,* and *transfer impedance.*)

impulse   Impulse is the product of a force and the time during which the force is applied; more specifically, the impulse is $\int_{t_1}^{t_2} F dt$ where the force F is time dependent and equal to zero before time t_1 and after time t_2.

impulse response function   See *Eq. (21.7).*

induced environments   Induced environments are those conditions generated as a result of the operation of a structure or equipment.

insertion loss   The insertion loss, in decibels, resulting from insertion of an element in a transmission system is 10 times the logarithm to the base 10 of the ratio of the power delivered to that part of the system that will follow the element, before the insertion of the element, to the power delivered to that same part of the system after insertion of the element.

isolation   Isolation is a reduction in the capacity of a system to respond to an excitation, attained by the use of a resilient support. In steady-state forced vibration, isolation is expressed quantitatively as the complement of transmissibility.

isolator   (See *vibration isolator.*)

jerk   Jerk is a vector that specifies the time rate of change of acceleration; jerk is the third derivative of displacement with respect to time.

level   Level is the logarithm of the ratio of a given quantity to a reference quantity of the same kind; the base of the logarithm, the reference quantity, and the kind of level must be indicated. (The type of level is indicated by the use of a compound term such as *vibration velocity level.* The level of the reference quantity remains unchanged whether the chosen quantity is peak, rms, or otherwise.) Unit: decibel. Unit symbol: dB.

line spectrum   A line spectrum is a spectrum whose components occur at a number of discrete frequencies.

linear mechanical impedance   Linear mechanical impedance is the impedance involving the ratio of force to linear velocity. (See *impedance.*)

linear system   A system is linear if for every element in the system the response is proportional to the excitation. This definition implies that the dynamic properties of each element in the system can be represented by a set of linear differential equations with constant coefficients, and that for the system as a whole superposition holds.

logarithmic decrement   The logarithmic decrement is the natural logarithm of the ratio of any two successive amplitudes of like sign, in the decay of a single-frequency oscillation.

longitudinal wave   A longitudinal wave in a medium is a wave in which the direction of displacement at each point of the medium is normal to the wave front.

low-pass filter   A low-pass filter is a wave filter having a single transmission band extending from zero frequency up to some critical or cutoff frequency which is not infinite.

magnetic recorder   A magnetic recorder is equipment incorporating an electromagnetic transducer and means for moving a ferromagnetic recording medium relative to the transducer for recording electric signals as magnetic variations in the medium.

magnetostriction   Magnetostriction is the phenomenon wherein ferromagnetic materials experience an elastic strain when subjected to an external magnetic field. Also, magnetostriction is the converse phenomenon in which mechanical stresses cause a change in the magnetic induction of a ferromagnetic material.

maximum value   The maximum value is the value of a function when any small change in the independent variable causes a decrease in the value of the function.

mechanical admittance   (See *mobility.*)

mechanical impedance   (See *impedance.*)

mechanical shock   Mechanical shock is a nonperiodic excitation (e.g., a motion of the foundation or an applied force) of a mechanical system that is characterized by suddenness and severity and usually causes significant relative displacements in the system.

mechanical system   A mechanical system is an aggregate of matter comprising a defined configuration of mass, stiffness, and damping.

mobility (mechanical admittance)   Mobility is the ratio of a velocity-like quantity to a force-like quantity when the arguments of the real (or imaginary) parts of the quantities increase linearly with time. *Mobility* is the reciprocal of *impedance.* The terms *angular mobility, linear mobility, driving-point mobility,* and *transfer mobility* are used in the same sense as corresponding impedances.

modal numbers   When the normal modes of a system are related by a set of ordered integers, these integers are called modal numbers.

mode of vibration   In a system undergoing vibration, a mode of vibration is a characteristic pattern assumed by the system in which the motion of every particle is simple harmonic with the same frequency. Two or more modes may exist concurrently in a multiple degree-of-freedom system.

modulation   Modulation is the variation in the value of some parameter which characterizes a periodic oscillation. Thus, amplitude modulation of a sinusoidal oscillation is a variation in the amplitude of the sinusoidal oscillation.

multiple degree-of-freedom system   A multiple degree-of-freedom system is one for which two or more coordinates are required to define completely the position of the system at any instant.

narrow-band random vibration (random sine wave)   Narrow-band random vibration is random vibration having frequency components only within a narrow band. It has the appearance of a sine wave whose amplitude varies in an unpredictable manner. (See *random vibration.*)

natural environments   Natural environments are those conditions generated by the forces of nature and whose effects are experienced when the equipment or structure is at rest as well as when it is in operation.

natural frequency   Natural frequency is the frequency of free vibration of a system. For a multiple degree-of-freedom system, the natural frequencies are the frequencies of the normal modes of vibration.

natural mode of vibration   The natural mode of vibration is a mode of vibration assumed by a system when vibrating freely.

node   A node is a point, line, or surface in a standing wave where some characteristic of the wave field has essentially zero amplitude.

noise   Noise is any undesired signal. By extension, noise is any unwanted disturbance within a useful frequency band, such as undesired electric waves in a transmission channel or device.

nominal bandwidth   The nominal bandwidth of a filter is the difference between the nominal upper and lower cutoff frequencies. The difference may be expressed (1) in cycles per second, (2) as a percentage of the passband center frequency, or (3) in octaves.

nominal passband center frequency   The nominal passband center frequency is the geometric mean of the nominal cutoff frequencies.

nominal upper and lower cutoff frequencies   The nominal upper and lower cutoff frequencies of a filter passband are those frequencies above and below the frequency of maximum response of a filter at which the response to a sinusoidal signal is 3 dB below the maximum response.

nonlinear damping   Nonlinear damping is damping due to a damping force that is not proportional to velocity.

normal mode of vibration   A normal mode of vibration is a mode of vibration that is uncoupled from (i.e., can exist independently of) other modes of vibration of a system. When vibration of the system is defined as an eigenvalue problem, the normal modes are the eigenvectors and the normal mode frequencies are the eigenvalues. The term *classical normal mode* is sometimes applied to the normal modes of a vibrating system characterized by vibration of each element of the system at the same frequency and phase. In general, classical normal modes exist only in systems having no damping or having particular types of damping.

oscillation   Oscillation is the variation, usually with time, of the magnitude of a quantity with respect to a specified reference when the magnitude is alternately greater and smaller than the reference.

partial node   A partial node is the point, line, or surface in a standing-wave system where some characteristic of the wave field has a minimum amplitude differing from zero. The appropriate modifier should be used with the words *partial node* to signify the type that is intended; e.g., displacement partial node, velocity partial node, pressure partial node.

peak-to-peak value   The peak-to-peak value of a vibrating quantity is the algebraic difference between the extremes of the quantity.

peak value   Peak value is the maximum value of a vibration during a given interval, usually considered to be the maximum deviation of that vibration from the mean value.

period   The period of a periodic quantity is the smallest increment of the independent variable for which the function repeats itself.

periodic quantity   A periodic quantity is an oscillating quantity whose values recur for certain increments of the independent variable.

phase of a periodic quantity   The phase of a periodic quantity, for a particular value of the independent variable, is the fractional part of a period through which the independent variable has advanced, measured from an arbitrary reference.

pickup   (See *transducer.*)

piezoelectric (crystal) (ceramic) transducer   A piezoelectric transducer is a transducer that depends for its operation on the interaction between the electric charge and the deformation of certain asymmetric crystals having piezoelectric properties.

piezoelectricity   Piezoelectricity is the property exhibited by some asymmetrical crystalline materials which when subjected to strain in suitable directions develop electric polarization proportional to the strain. Inverse piezoelectricity is the effect in which mechanical strain is produced in certain asymmetrical crystalline materials when subjected to an external electric field; the strain is proportional to the electric field.

power spectral density   Power spectral density is the limiting mean-square value (e.g., of acceleration, velocity, displacement, stress, or other random variable) per unit bandwidth, i.e., the limit of the mean-square value in a given rectangular bandwidth divided by the bandwidth, as the bandwidth approaches zero. Also called *autospectral density.*

power spectral density level   The spectrum level of a specified signal at a particular frequency is the level in decibels of that part of the signal contained within a band 1 cycle per second wide, centered at the particular frequency. Ordinarily this has significance only for a signal having a continuous distribution of components within the frequency range under consideration.

process   A process is a collection of signals. The word *process* rather than the word *ensemble* ordinarily is used when it is desired to emphasize the properties the signals have or do not have as a group. Thus, one speaks of a stationary process rather than a stationary ensemble.

pulse rise time   The pulse rise time is the interval of time required for the leading edge of a pulse to rise from some specified small fraction to some specified larger fraction of the maximum value.

Q (quality factor)   The quantity Q is a measure of the sharpness of resonance or frequency selectivity of a resonant vibratory system having a single degree of freedom, either mechanical or electrical. In a mechanical system, this quantity is equal to one-half the reciprocal of the damping ratio. It is commonly used only with reference to a lightly damped system and is then approximately equal to the following: (1) Transmissibility at resonance, (2) π/logarithmic decrement, (3) $2\pi W/\Delta W$ where W is the stored energy and ΔW the energy dissipation per cycle, and (4) $f_r/\Delta f$ where f_r is the resonance frequency and Δf is the bandwidth between the half-power points.

quasi-ergodic process   A quasi-ergodic process is a random process which is not necessarily stationary but of such a nature that some time averages performed on a signal are independent of the signal chosen.

quasi-periodic signal   A quasi-periodic signal is one consisting only of quasi-sinusoids.

quasi-sinusoid   A quasi-sinusoid is a function of the form $\alpha = A \sin(2\pi ft - \phi)$ where either A or f, or both, is not a constant but may be expressed readily as a function of time. Ordinarily ϕ is considered constant.

random sine wave   (See *narrow-band random vibration.*)

random vibration   Random vibration is vibration whose instantaneous magnitude is not specified for any given instant of time. The instantaneous magnitudes of a random vibration are specified only by probability distribution functions giving the probable fraction of the total time that the magnitude (or some sequence of magnitudes) lies within a specified range. Random vibration contains no periodic or quasi-periodic constituents. If random vibration has instantaneous magnitudes that occur according to the Gaussian distribution, it is called *Gaussian random vibration.*

ratio of critical damping   (See *fraction of critical damping.*)

Rayleigh wave   A Rayleigh wave is a surface wave associated with the free boundary of a solid, such that a surface particle describes an ellipse whose major axis is normal to the surface, and whose center is at the undisturbed surface. At maximum particle displacement away from the solid surface the motion of the particle is opposite to that of the wave.

recording channel   The term *recording channel* refers to one of a number of independent recorders in a recording system or to independent recording tracks on a recording medium.

recording system   A recording system is a combination of transducing devices and associated equipment suitable for storing signals in a form capable of subsequent reproduction.

relaxation time   Relaxation time is the time taken by an exponentially decaying quantity to decrease in amplitude by a factor of $1/e = 0.3679$.

re-recording   Re-recording is the process of making a recording by reproducing a recorded signal source and recording this reproduction.

resonance   Resonance of a system in forced vibration exists when any change, however small, in the frequency of excitation causes a decrease in the response of the system.

resonance frequency   Resonance frequency is a frequency at which resonance exists.

response   The response of a device or system is the motion (or other output) resulting from an excitation (stimulus) under specified conditions.

response spectrum   (See *shock spectrum.*)

rotational mechanical impedance   (See *angular mechanical impedance.*)

seismic pickup; seismic transducer   A seismic pickup or transducer is a device consisting of a seismic system in which the differential movement between the mass and the base of the system produces a measurable indication of such movement.

seismic system   A seismic system is one consisting of a mass attached to a reference base by one or more flexible elements. Damping is usually included.

self-induced (self-excited) vibration   The vibration of a mechanical system is self-induced if it results from conversion, within the system, of nonoscillatory excitation to oscillatory excitation.

shake table   (See *vibration machine.*)

shear wave (rotational wave)   A shear wave is a wave in an elastic medium which causes an element of the medium to change its shape without a change of volume.

shock   (See *mechanical shock.*)

shock absorber   A shock absorber is a device which dissipates energy to modify the response of a mechanical system to applied shock.

shock isolator (shock mount)   A shock isolator is a resilient support that tends to isolate a system from a shock motion.

shock machine   A shock machine is a device for subjecting a system to controlled and reproducible mechanical shock.

shock motion   Shock motion is an excitation involving motion of a foundation. (See *foundation* and *mechanical shock.*)

shock mount   (See *shock isolator.*)

shock pulse   A shock pulse is a substantial disturbance characterized by a rise of acceleration from a constant value and decay of acceleration to the constant value in a short period of time. Shock pulses are normally displayed graphically as curves of acceleration as functions of time.

shock-pulse duration   (See *duration of shock pulse.*)

shock spectrum (response spectrum)   A shock spectrum is a plot of the maximum response experienced by a single degree-of-freedom system, as a function of its own natural frequency, in response to an applied shock. The response may be expressed in terms of acceleration, velocity or displacement.

shock testing machine; shock machine   A shock testing machine is a device for subjecting a mechanical system to controlled and reproducible mechanical shock.

signal   A signal is (1) a disturbance used to convey information; (2) the information to be conveyed over a communication system.

simple harmonic motion   A simple harmonic motion is a motion such that the displacement is a sinusoidal function of time; sometimes it is designated merely by the term *harmonic motion.*

single degree-of-freedom system   A single degree-of-freedom system is one for which only one coordinate is required to define completely the configuration of the system at any instant.

sinusoidal motion   (See *simple harmonic motion.*)

snubber   A snubber is a device used to increase the stiffness of an elastic system (usually by a large factor) whenever the displacement becomes larger than a specified value.

spectrum   A spectrum is a definition of the magnitude of the frequency components that constitute a quantity.

spectrum density   (See *power spectral density.*)

standard deviation   Standard deviation is the square root of the variance; i.e., the square root of the mean of the squares of the deviations from the mean value of a vibrating quantity.

standing wave   A standing wave is a periodic wave having a fixed distribution in space which is the result of interference of progressive waves of the same frequency and kind. Such waves are characterized by the existence of nodes or partial nodes and antinodes that are fixed in space.

stationary process   A stationary process is an ensemble of signals such that an average of values over the ensemble at any given time is independent of time.

stationary signal   A stationary signal is a random signal of such nature that averages over samples of finite time intervals are independent of the time at which the sample occurs.

steady-state vibration   Steady-state vibration exists in a system if the velocity of each particle is a continuing periodic quantity.

stiffness   Stiffness is the ratio of change of force (or torque) to the corresponding change on translational (or rotational) deflection of an elastic element.

subharmonic   A subharmonic is a sinusoidal quantity having a frequency that is an integral submultiple of the fundamental frequency of a periodic quantity to which it is related.

subharmonic response   Subharmonic response is the periodic response of a mechanical system exhibiting the characteristic of resonance at a frequency that is a submultiple of the frequency of the periodic excitation.

superharmonic response   Superharmonic response is a term sometimes used to denote a particular type of harmonic response which dominates the total response of the system; it frequently occurs when the excitation frequency is a submultiple of the frequency of the fundamental resonance.

transducer (pickup)   A transducer is a device which converts shock or vibratory motion into an optical, a mechanical, or most commonly to an electrical signal that is proportional to a parameter of the experienced motion.

transfer impedance   Transfer impedance between two points is the impedance involving the ratio of force to velocity when force is measured at one point and velocity at the other point. The term *transfer impedance* also is used to denote the ratio of force to velocity measured at the same point but in different directions. (See *impedance.*)

transient vibration   Transient vibration is temporarily sustained vibration of a mechanical system. It may consist of forced or free vibration or both.

transmissibility   Transmissibility is the nondimensional ratio of the response amplitude of a system in steady-state forced vibration to the excitation amplitude. The ratio may be one of forces, displacements, velocities, or accelerations.

transmission loss   Transmission loss is the reduction in the magnitude of some characteristic of a signal, between two stated points in a transmission system.

transverse wave   A transverse wave is a wave in which the direction of displacement at each point of the medium is parallel to the wave front.

tuned damper   (See *dynamic vibration absorber.*)

uncorrelated   Two signals or variables $\alpha_1(t)$ and $\alpha_2(t)$ are said to be uncorrelated if the average value of their product is zero: $\overline{\alpha_1(t) \cdot \alpha_2(t)} = 0$. If the correlation coefficient is equal to unity, the variables are said to be completely correlated. If the coefficient is less than unity but larger than zero, they are said to be partially correlated. (See *correlation coefficient.*)

uncoupled mode   An uncoupled mode of vibration is a mode that can exist in a system concurrently with and independently of other modes.

undamped natural frequency   The undamped natural frequency of a mechanical system is the frequency of free vibration resulting from only elastic and inertial forces of the system.

variance   Variance is the mean of the squares of the deviations from the mean value of a vibrating quantity.

velocity   Velocity is a vector quantity that specifies the time rate of change of displacement with respect to a reference frame. If the reference frame is not inertial, the velocity is often designated "relative velocity."

velocity pickup   A velocity pickup is a transducer that converts an input velocity to an output (usually electrical) that is proportional to the input velocity.

velocity shock   Velocity shock is a particular type of shock motion characterized by a sudden velocity change of the foundation. (See *foundation* and *mechanical shock.*)

vibration   Vibration is an oscillation wherein the quantity is a parameter that defines the motion of a mechanical system. (See *oscillation.*)

vibration acceleration   Vibration acceleration is the rate of change of speed and direction of a vibration, in a specified direction. The frequency bandwidth must be identified. Unit meter per second squared. Unit symbol: m/s^2.

vibration acceleration level   The vibration acceleration level is 10 times the logarithm (to the base 10) of the ratio of the square of a given vibration acceleration to the square of a reference acceleration, commonly $1g$ or 1 m/s^2. Unit: decibel. Unit symbol: dB.

vibration isolator   A vibration isolator is a resilient support that tends to isolate a system from steady-state excitation.

vibration machine   A vibration machine is a device for subjecting a mechanical system to controlled and reproducible mechanical vibration.

vibration meter   A vibration meter is an apparatus for the measurement of displacement, velocity, or acceleration of a vibrating body.

vibration mount   (See *vibration isolator.*)

vibration pickup   (See *transducer.*)

vibrograph   A vibrograph is an instrument, usually mechanical and self-contained, that provides an oscillographic recording of a vibration waveform.

viscous damping   Viscous damping is the dissipation of energy that occurs when a particle in a vibrating system is resisted by a force that has a magnitude proportional to the magnitude of the velocity of the particle and direction opposite to the direction of the particle.

viscous damping, equivalent   (See *equivalent viscous damping.*)

wave   A wave is a disturbance which is propagated in a medium in such a manner that at any point in the medium the quantity serving as measure of disturbance is a function of the time, while at any instant the displacement at a point is a function of the position of the point. Any physical quantity that has the same relationship to some independent variable (usually time) that a propagated disturbance has, at a particular instant, with respect to space, may be called a wave.

wave interference   Wave interference is the phenomenon which results when waves of the same or nearly the same frequency are superposed; it is characterized by a spatial or temporal distribution of amplitude of some specified characteristic differing from that of the individual superposed waves.

wavelength   The wavelength of a periodic wave in an isotropic medium is the perpendicular distance between two wave fronts in which the displacements have a difference in phase of one complete period.

white noise   White noise is a noise whose power spectral density is substantially independent of frequency over a specified range.

CHAPTER 2
BASIC VIBRATION THEORY

Ralph E. Blake

INTRODUCTION

This chapter presents the theory of free and forced steady-state vibration of single degree-of-freedom systems. Undamped systems and systems having viscous damping and structural damping are included. Multiple degree-of-freedom systems are discussed, including the normal-mode theory of linear elastic structures and Lagrange's equations.

ELEMENTARY PARTS OF VIBRATORY SYSTEMS

Vibratory systems comprise means for storing potential energy (spring), means for storing kinetic energy (mass or inertia), and means by which the energy is gradually lost (damper). The vibration of a system involves the alternating transfer of energy between its potential and kinetic forms. In a damped system, some energy is dissipated at each cycle of vibration and must be replaced from an external source if a steady vibration is to be maintained. Although a single physical structure may store both kinetic and potential energy, and may dissipate energy, this chapter considers only *lumped parameter systems* composed of ideal springs, masses, and dampers wherein each element has only a single function. In translational motion, displacements are defined as linear distances; in rotational motion, displacements are defined as angular motions.

TRANSLATIONAL MOTION

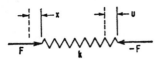

FIGURE 2.1 Linear spring.

Spring. In the linear spring shown in Fig. 2.1, the change in the length of the spring is proportional to the force acting along its length:

$$F = k(x - u) \qquad (2.1)$$

The ideal spring is considered to have no mass; thus, the force acting on one end is equal and opposite to the force acting on the other end. The constant of proportionality k is the *spring constant or stiffness.*

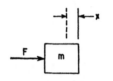

FIGURE 2.2   Rigid mass.

Mass.   A mass is a rigid body (Fig. 2.2) whose acceleration $\ddot{x}$ according to Newton's second law is proportional to the resultant F of all forces acting on the mass:*

$$F = m\ddot{x} \qquad (2.2)$$

Damper.   In the viscous damper shown in Fig. 2.3, the applied force is proportional to the relative velocity of its connection points:

$$F = c(\dot{x} - \dot{u}) \qquad (2.3)$$

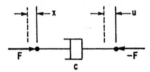

FIGURE 2.3   Viscous damper.

The constant c is the *damping coefficient,* the characteristic parameter of the damper. The ideal damper is considered to have no mass; thus the force at one end is equal and opposite to the force at the other end. *Structural damping* is considered below and several other types of damping are considered in Chap. 30.

ROTATIONAL MOTION

The elements of a mechanical system which moves with pure rotation of the parts are wholly analogous to the elements of a system that moves with pure translation. The property of a rotational system which stores kinetic energy is inertia; stiffness and damping coefficients are defined with reference to angular displacement and angular velocity, respectively. The analogous quantities and equations are listed in Table 2.1.

TABLE 2.1   Analogous Quantities in Translational and Rotational Vibrating Systems

Translational quantity	Rotational quantity
Linear displacement x	Angular displacement α
Force F	Torque M
Spring constant k	Spring constant k_r
Damping constant c	Damping constant c_r
Mass m	Moment of inertia I
Spring law $F = k(x_1 - x_2)$	Spring law $M = k_r(\alpha_1 - \alpha_2)$
Damping law $F = c(\dot{x}_1 - \dot{x}_2)$	Damping law $M = c_r(\dot{\alpha}_1 - \dot{\alpha}_2)$
Inertia law $F = m\ddot{x}$	Inertia law $M = I\ddot{\alpha}$

* It is common to use the word *mass* in a general sense to designate a rigid body. Mathematically, the mass of the rigid body is defined by m in Eq. (2.2).

Inasmuch as the mathematical equations for a rotational system can be written by analogy from the equations for a translational system, only the latter are discussed in detail. Whenever translational systems are discussed, it is understood that corresponding equations apply to the analogous rotational system, as indicated in Table 2.1.

SINGLE DEGREE-OF-FREEDOM SYSTEM

The simplest possible vibratory system is shown in Fig. 2.4; it consists of a mass m attached by means of a spring k to an immovable support. The mass is constrained to translational motion in the direction of the X axis so that its change of position from an initial reference is described fully by the value of a single quantity x. For this reason it is called a *single degree-of-freedom system*. If the mass m is displaced from its equilibrium position and then allowed to vibrate free from further external forces, it is said to have *free vibration*. The vibration also may be forced; i.e., a continuing force acts upon the mass or the foundation experiences a continuing motion. Free and forced vibration are discussed below.

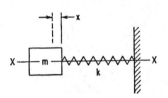

FIGURE 2.4 Undamped single degree-of-freedom system.

FREE VIBRATION WITHOUT DAMPING[1-3]

Considering first the free vibration of the undamped system of Fig. 2.4, Newton's equation is written for the mass m. The force $m\ddot{x}$ exerted by the mass on the spring is equal and opposite to the force kx applied by the spring on the mass:

$$m\ddot{x} + kx = 0 \tag{2.4}$$

where $x = 0$ defines the equilibrium position of the mass.

The solution of Eq. (2.4) is

$$x = A \sin \sqrt{\frac{k}{m}}\ t + B \cos \sqrt{\frac{k}{m}}\ t \tag{2.5}$$

where the term $k\backslash m$ is the *angular natural frequency* defined by

$$\omega_n = \sqrt{\frac{k}{m}} \qquad \text{rad/sec} \tag{2.6}$$

The sinusoidal oscillation of the mass repeats continuously, and the time interval to complete one cycle is the *period:*

$$\tau = \frac{2\pi}{\omega_n} \tag{2.7}$$

The reciprocal of the period is the *natural frequency:*

$$f_n = \frac{1}{\tau} = \frac{\omega_n}{2\pi} = \frac{1}{2\pi} \sqrt{\frac{k}{m}} = \frac{1}{2\pi} \sqrt{\frac{kg}{W}} \tag{2.8}$$

where $W = mg$ is the weight of the rigid body forming the mass of the system shown in Fig. 2.4. The relations of Eq. (2.8) are shown by the solid lines in Fig. 2.5.

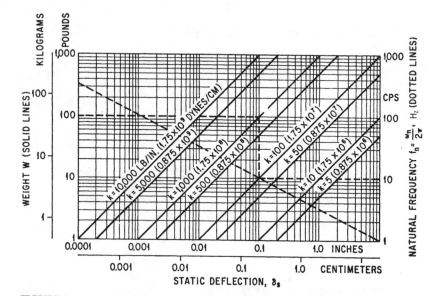

FIGURE 2.5 Natural frequency relations for a single degree-of-freedom system. Relation of natural frequency to weight of supported body and stiffness of spring [Eq. (2.8)] is shown by solid lines. Relation of natural frequency to static deflection [Eq. (2.10)] is shown by diagonal-dashed line. Example: To find natural frequency of system with $W = 100$ lb and $k = 1000$ lb/in., enter at $W = 100$ on left ordinate scale; follow the dashed line horizontally to solid line $k = 1000$, then vertically down to diagonal-dashed line, and finally horizontally to read $f_n = 10$ Hz from right ordinate scale.

Initial Conditions.   In Eq. (2.5), B is the value of x at time $t = 0$, and the value of A is equal to $\dot{x}/\omega_n$ at time $t = 0$. Thus, the conditions of displacement and velocity which exist at zero time determine the subsequent oscillation completely.

Phase Angle.   Equation (2.5) for the displacement in oscillatory motion can be written, introducing the frequency relation of Eq. (2.6),

$$x = A \sin \omega_n t + B \cos \omega_n t = C \sin (\omega_n t + \theta) \qquad (2.9)$$

where $C = (A^2 + B^2)^{1/2}$ and $\theta = \tan^{-1} (B/A)$. The angle θ is called the *phase angle*.

Static Deflection.   The static deflection of a simple mass-spring system is the deflection of spring k as a result of the gravity force of the mass, $\delta_{st} = mg/k$. (For example, the system of Fig. 2.4 would be oriented with the mass m vertically above the spring k.) Substituting this relation in Eq. (2.8),

$$f_n = \frac{1}{2\pi} \sqrt{\frac{g}{\delta_{st}}} \qquad (2.10)$$

The relation of Eq. (2.10) is shown by the diagonal-dashed line in Fig. 2.5. This relation applies only when the system under consideration is both linear and elastic. For example, rubber springs tend to be nonlinear or exhibit a dynamic stiffness which differs from the static stiffness; hence, Eq. (2.10) is not applicable.

FREE VIBRATION WITH VISCOUS DAMPING[1-3]

Figure 2.6 shows a single degree-of-freedom system with a viscous damper. The differential equation of motion of mass m, corresponding to Eq. (2.4) for the undamped system, is

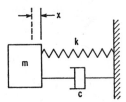

$$m\ddot{x} + c\dot{x} + kx = 0 \qquad (2.11)$$

The form of the solution of this equation depends upon whether the damping coefficient is equal to, greater than, or less than the *critical damping coefficient* c_c:

$$c_c = 2\sqrt{km} = 2m\omega_n \qquad (2.12)$$

FIGURE 2.6   Single degree-of-freedom system with viscous damper.

The ratio $\zeta = c/c_c$ is defined as the *fraction of critical damping.*

Less-Than-Critical Damping.   If the damping of the system is less than critical, $\zeta < 1$; then the solution of Eq. (2.11) is

$$x = e^{-ct/2m}(A \sin \omega_d t + B \cos \omega_d t)$$
$$= Ce^{-ct/2m} \sin (\omega_d t + \theta) \qquad (2.13)$$

where C and θ are defined with reference to Eq. (2.9). The *damped natural frequency* is related to the undamped natural frequency of Eq. (2.6) by the equation

$$\omega_d = \omega_n(1 - \zeta^2)^{1/2} \qquad \text{rad/sec} \qquad (2.14)$$

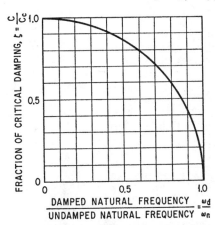

Equation (2.14), relating the damped and undamped natural frequencies, is plotted in Fig. 2.7.

Critical Damping.   When $c = c_c$, there is no oscillation and the solution of Eq. (2.11) is

$$x = (A + Bt)e^{-ct/2m} \qquad (2.15)$$

Greater-Than-Critical Damping. When $\zeta > 1$, the solution of Eq. (2.11) is

$$x = e^{-ct/2m}(Ae^{\omega_n\sqrt{\zeta^2-1}\,t} + Be^{-\omega_n\sqrt{\zeta^2-1}\,t}) \qquad (2.16)$$

FIGURE 2.7   Damped natural frequency as a function of undamped natural frequency and fraction of critical damping.

This is a nonoscillatory motion; if the system is displaced from its equilibrium position, it tends to return gradually.

Logarithmic Decrement. The degree of damping in a system having $\zeta < 1$ may be defined in terms of successive peak values in a record of a free oscillation. Substituting the expression for critical damping from Eq. (2.12), the expression for free vibration of a damped system, Eq. (2.13), becomes

$$x = Ce^{-\zeta \omega_n t} \sin(\omega_d t + \theta) \tag{2.17}$$

Consider any two maxima (i.e., value of x when $dx/dt = 0$) separated by n cycles of oscillation, as shown in Fig. 2.8. Then the ratio of these maxima is

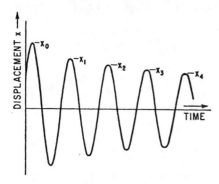

$$\frac{x_n}{x_0} = e^{-2\pi n \zeta/(1 - \zeta^2)^{1/2}} \tag{2.18}$$

Values of x_n/x_0 are plotted in Fig. 2.9 for several values of n over the range of ζ from 0.001 to 0.10.

The *logarithmic decrement* Δ is the natural logarithm of the ratio of the amplitudes of two successive cycles of the damped free vibration:

$$\Delta = \ln \frac{x_1}{x_2} \quad \text{or} \quad \frac{x_2}{x_1} = e^{-\Delta} \tag{2.19}$$

FIGURE 2.8 Trace of damped free vibration showing amplitudes of displacement maxima.

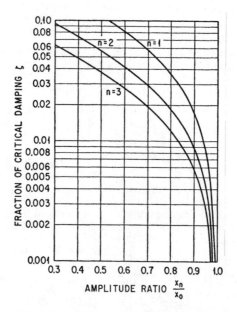

FIGURE 2.9 Effect of damping upon the ratio of displacement maxima of a damped free vibration.

[Also see Eq. (37.6).] A comparison of this relation with Eq. (2.18) when $n = 1$ gives the following expression for Δ:

$$\Delta = \frac{2\pi\zeta}{(1 - \zeta^2)^{1/2}} \tag{2.20}$$

The logarithmic decrement can be expressed in terms of the difference of successive amplitudes by writing Eq. (2.19) as follows:

$$\frac{x_1 - x_2}{x_1} = 1 - \frac{x_2}{x_1} = 1 - e^{-\Delta}$$

Writing $e^{-\Delta}$ in terms of its infinite series, the following expression is obtained which gives a good approximation for $\Delta < 0.2$:

$$\frac{x_1 - x_2}{x_1} = \Delta \tag{2.21}$$

For small values of ζ (less than about 0.10), an approximate relation between the fraction of critical damping and the logarithmic decrement, from Eq. (2.20), is

$$\Delta \simeq 2\pi\zeta \tag{2.22}$$

FORCED VIBRATION

Forced vibration in this chapter refers to the motion of the system which occurs in response to a continuing excitation whose magnitude varies sinusoidally with time. (See Chaps. 8 and 23 for a treatment of the response of a simple system to step, pulse, and transient vibration excitations.) The excitation may be, alternatively, force applied to the system (generally, the force is applied to the mass of a single degree-of-freedom system) or motion of the foundation that supports the system. The resulting response of the system can be expressed in different ways, depending upon the nature of the excitation and the use to be made of the result:

1. If the excitation is a force applied to the mass of the system shown in Fig. 2.4, the result may be expressed in terms of (a) the amplitude of the resulting motion of the mass or (b) the fraction of the applied force amplitude that is transmitted through the system to the support. The former is termed the *motion response* and the latter is termed the *force transmissibility*.

2. If the excitation is a motion of the foundation, the resulting response usually is expressed in terms of the amplitude of the motion of the mass relative to the amplitude of the motion of the foundation. This is termed the *motion transmissibility* for the system.

In general, the response and transmissibility relations are functions of the forcing frequency and vary with different types and degrees of damping. Results are presented in this chapter for undamped systems and for systems with either viscous or structural damping. Corresponding results are given in Chap. 30 for systems with Coulomb damping, and for systems with either viscous or Coulomb damping in series with a linear spring.

FORCED VIBRATION WITHOUT DAMPING

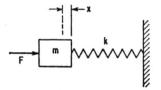

FIGURE 2.10 Undamped single degree-of-freedom system excited in forced vibration by force acting on mass.

Force Applied to Mass. When the sinusoidal force $F = F_0 \sin \omega t$ is applied to the mass of the undamped single degree-of-freedom system shown in Fig. 2.10, the differential equation of motion is

$$m\ddot{x} + kx = F_0 \sin \omega t \qquad (2.23)$$

The solution of this equation is

$$x = A \sin \omega_n t + B \cos \omega_n t + \frac{F_0/k}{1 - \omega^2/\omega_n^2} \sin \omega t \qquad (2.24)$$

where $\omega_n = \sqrt{k/m}$. The first two terms represent an oscillation at the undamped natural frequency ω_n. The coefficient B is the value of x at time $t = 0$, and the coefficient A may be found from the velocity at time $t = 0$. Differentiating Eq. (2.24) and setting $t = 0$,

$$\dot{x}(0) = A\omega_n + \frac{\omega F_0/k}{1 - \omega^2/\omega_n^2} \qquad (2.25)$$

The value of A is found from Eq. (2.25).

The oscillation at the natural frequency ω_n gradually decays to zero in physical systems because of damping. The steady-state oscillation at forcing frequency ω is

$$x = \frac{F_0/k}{1 - \omega^2/\omega_n^2} \sin \omega t \qquad (2.26)$$

This oscillation exists after a condition of equilibrium has been established by decay of the oscillation at the natural frequency ω_n and persists as long as the force F is applied.

The force transmitted to the foundation is directly proportional to the spring deflection: $F_t = kx$. Substituting x from Eq. (2.26) and defining transmissibility $T = F_t/F$,

$$T = \frac{1}{1 - \omega^2/\omega_n^2} \qquad (2.27)$$

If the mass is initially at rest in the equilibrium position of the system (i.e., $x = 0$ and $\dot{x} = 0$) at time $t = 0$, the ensuing motion at time $t > 0$ is

$$x = \frac{F_0/k}{1 - \omega^2/\omega_n^2} \left(\sin \omega t - \frac{\omega}{\omega_n} \sin \omega_n t \right) \qquad (2.28)$$

For large values of time, the second term disappears because of the damping inherent in any physical system, and Eq. (2.28) becomes identical to Eq. (2.26).

When the forcing frequency coincides with the natural frequency, $\omega = \omega_n$ and a condition of resonance exists. Then Eq. (2.28) is indeterminate and the expression for x may be written as

$$x = -\frac{F_0 \omega}{2k} t \cos \omega t + \frac{F_0}{2k} \sin \omega t \qquad (2.29)$$

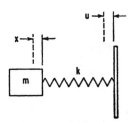

FIGURE 2.11 Undamped single degree-of-freedom system excited in forced vibration by motion of foundation.

According to Eq. (2.29), the amplitude x increases continuously with time, reaching an infinitely great value only after an infinitely great time.

Motion of Foundation. The differential equation of motion for the system of Fig. 2.11 excited by a continuing motion $u = u_0 \sin \omega t$ of the foundation is

$$m\ddot{x} = -k(x - u_0 \sin \omega t)$$

The solution of this equation is

$$x = A_1 \sin \omega_n t + B_2 \cos \omega_n t + \frac{u_0}{1 - \omega^2/\omega_n^2} \sin \omega t$$

where $\omega_n = k/m$ and the coefficients A_1, B_1 are determined by the velocity and displacement of the mass, respectively, at time $t = 0$. The terms representing oscillation at the natural frequency are damped out ultimately, and the ratio of amplitudes is defined in terms of transmissibility T:

$$\frac{x_0}{u_0} = T = \frac{1}{1 - \omega^2/\omega_n^2} \tag{2.30}$$

where $x = x_0 \sin \omega t$. Thus, in the forced vibration of an undamped single degree-of-freedom system, the motion response, the force transmissibility, and the motion transmissibility are numerically equal.

FORCED VIBRATION WITH VISCOUS DAMPING

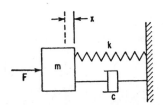

FIGURE 2.12 Single degree-of-freedom system with viscous damping, excited in forced vibration by force acting on mass.

Force Applied to Mass. The differential equation of motion for the single degree-of-freedom system with viscous damping shown in Fig. 2.12, when the excitation is a force $F = F_0 \sin \omega t$ applied to the mass, is

$$m\ddot{x} + c\dot{x} + kx = F_0 \sin \omega t \tag{2.31}$$

Equation (2.31) corresponds to Eq. (2.23) for forced vibration of an undamped system; its solution would correspond to Eq. (2.24) in that it includes terms representing oscillation at the natural frequency. In a damped system, however, these terms are damped out rapidly and only the steady-state solution usually is considered. The resulting motion occurs at the forcing frequency ω; when the damping coefficient c is greater than zero, the phase between the force and resulting motion is different than zero. Thus, the response may be written

$$x = R \sin(\omega t - \theta) = A_1 \sin \omega t + B_1 \cos \omega t \tag{2.32}$$

Substituting this relation in Eq. (2.31), the following result is obtained:

$$\frac{x}{F_0/k} = \frac{\sin(\omega t - \theta)}{\sqrt{(1 - \omega^2/\omega_n^2)^2 + (2\zeta\omega/\omega_n)^2}} = R_d \sin(\omega t - \theta) \qquad (2.33)$$

where
$$\theta = \tan^{-1}\left(\frac{2\zeta\omega/\omega_n}{1 - \omega^2/\omega_n^2}\right)$$

and R_d is a *dimensionless response factor* giving the ratio of the amplitude of the vibratory displacement to the spring displacement that would occur if the force F were applied statically. At very low frequencies R_d is approximately equal to 1; it rises to a peak near ω_n and approaches zero as ω becomes very large. The displacement response is defined at these frequency conditions as follows:

$$x \simeq \left(\frac{F_0}{k}\right)\sin\omega t \qquad [\omega \ll \omega_n]$$

$$x = \frac{F_0}{2k\zeta}\sin\left(\omega_n t + \frac{\pi}{2}\right) = -\frac{F_0\cos\omega_n t}{c\omega_n} \qquad [\omega = \omega_n] \qquad (2.34)$$

$$x \simeq \frac{\omega_n^2 F_0}{\omega^2 k}\sin(\omega t + \pi) = \frac{F_0}{m\omega^2}\sin\omega t \qquad [\omega \gg \omega_n]$$

For the above three frequency conditions, the vibrating system is sometimes described as *spring-controlled, damper-controlled,* and *mass-controlled,* respectively, depending on which element is primarily responsible for the system behavior.

Curves showing the dimensionless response factor R_d as a function of the frequency ratio ω/ω_n are plotted in Fig. 2.13 on the coordinate lines having a positive 45° slope. Curves of the phase angle θ are plotted in Fig. 2.14. A phase angle between 180 and 360° cannot exist in this case since this would mean that the damper is furnishing energy to the system rather than dissipating it.

An alternative form of Eqs. (2.33) and (2.34) is

$$\frac{x}{F_0/k} = \frac{(1 - \omega^2/\omega n^2)\sin\omega t - 2\zeta(\omega/\omega_n)\cos\omega t}{(1 - \omega^2/\omega_n^2)^2 + (2\zeta\omega/\omega_n)^2} \qquad (2.35)$$

$$= (R_d)_x \sin\omega t + (R_d)_R \cos\omega t$$

This shows the components of the response which are in phase $[(R_d)_x \sin\omega t]$ and 90° out of phase $[(R_d)_R \cos\omega t]$ with the force. Curves of $(R_d)_x$ and $(R_d)_R$ are plotted as a function of the frequency ratio ω/ω_n in Figs. 2.15 and 2.16.

Velocity and Acceleration Response.   The shape of the response curves changes distinctly if velocity $\dot{x}$ or acceleration $\ddot{x}$ is plotted instead of displacement x. Differentiating Eq. (2.33),

$$\frac{\dot{x}}{F_0/\sqrt{km}} = \frac{\omega}{\omega_n} R_d \cos(\omega t - \theta) = R_v \cos(\omega t - \theta) \qquad (2.36)$$

The acceleration response is obtained by differentiating Eq. (2.36):

$$\frac{\ddot{x}}{F_0/m} = -\frac{\omega^2}{\omega_n^2} R_d \sin(\omega t - \theta) = -R_a \sin(\omega t - \theta) \qquad (2.37)$$

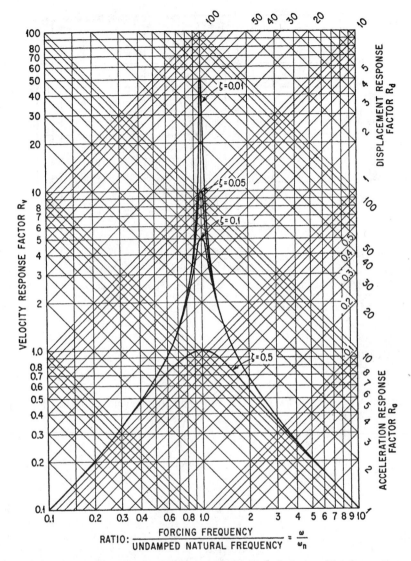

FIGURE 2.13 Response factors for a viscous-damped single degree-of-freedom system excited in forced vibration by a force acting on the mass. The velocity response factor shown by horizontal lines is defined by Eq. (2.36); the displacement response factor shown by diagonal lines of positive slope is defined by Eq. (2.33); and the acceleration response factor shown by diagonal lines of negative slope is defined by Eq. (2.37).

The velocity and acceleration response factors defined by Eqs. (2.36) and (2.37) are shown graphically in Fig. 2.13, the former to the horizontal coordinates and the latter to the coordinates having a negative 45° slope. Note that the velocity response factor approaches zero as $\omega \to 0$ and $\omega \to \infty$, whereas the acceleration response factor approaches 0 as $\omega \to 0$ and approaches unity as $\omega \to \infty$.

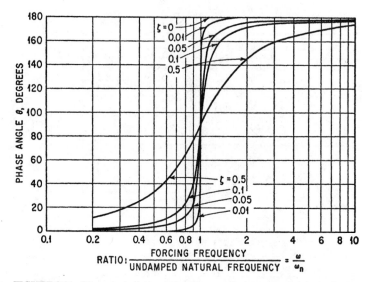

FIGURE 2.14 Phase angle between the response displacement and the excitation force for a single degree-of-freedom system with viscous damping, excited by a force acting on the mass of the system.

Force Transmission. The force transmitted to the foundation of the system is

$$F_T = c\dot{x} + kx \qquad (2.38)$$

Since the forces $c\dot{x}$ and kx are 90° out of phase, the magnitude of the transmitted force is

$$|F_T| = \sqrt{c^2\dot{x}^2 + k^2x^2} \qquad (2.39)$$

The ratio of the transmitted force F_T to the applied force F_0 can be expressed in terms of transmissibility T:

$$\frac{F_T}{F_0} = T \sin(\omega t - \psi) \qquad (2.40)$$

where

$$T = \sqrt{\frac{1 + (2\zeta\omega/\omega_n)^2}{(1 - \omega^2/\omega_n^2)^2 + (2\zeta\omega/\omega_n)^2}} \qquad (2.41)$$

and

$$\psi = \tan^{-1}\frac{2\zeta(\omega/\omega_n)^3}{1 - \omega^2/\omega_n^2 + 4\zeta^2\omega^2/\omega_n^2}$$

The transmissibility T and phase angle ψ are shown in Figs. 2.17 and 2.18, respectively, as a function of the frequency ratio ω/ω_n and for several values of the fraction of critical damping ζ.

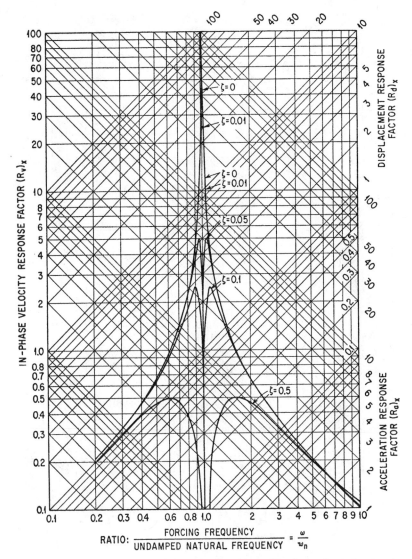

FIGURE 2.15 In-phase component of response factor of a viscous-damped system in forced vibration. All values of the response factor for $\omega/\omega_n > 1$ are negative but are plotted without regard for sign. The fraction of critical damping is denoted by ζ.

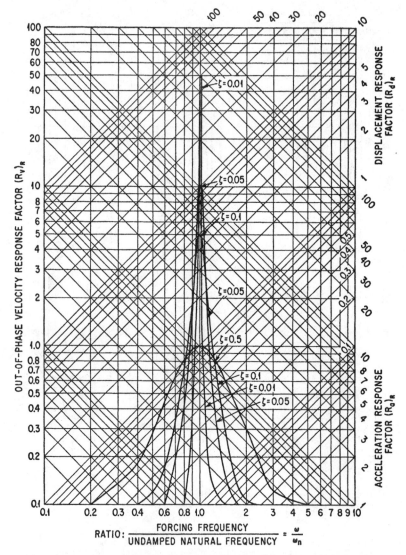

FIGURE 2.16 Out-of-phase component of response factor of a viscous-damped system in forced vibration. The fraction of critical damping is denoted by ζ.

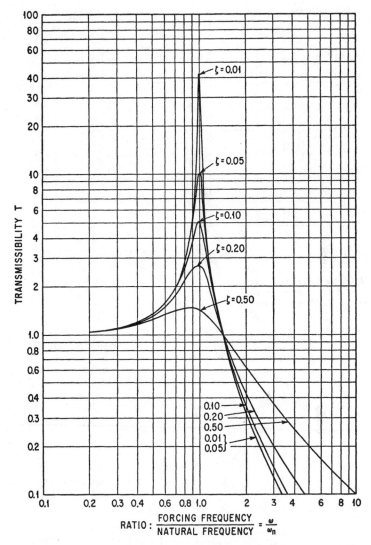

FIGURE 2.17 Transmissibility of a viscous-damped system. Force transmissibility and motion transmissibility are identical numerically. The fraction of critical damping is denoted by ζ.

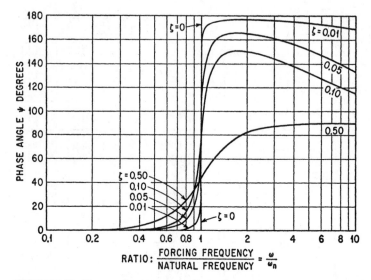

FIGURE 2.18   Phase angle of force transmission (or motion transmission) of a viscous-damped system excited (1) by force acting on mass and (2) by motion of foundation. The fraction of critical damping is denoted by ζ.

Hysteresis.   When the viscous damped, single degree-of-freedom system shown in Fig. 2.12 undergoes vibration defined by

$$x = x_0 \sin \omega t \tag{2.42}$$

the net force exerted on the mass by the spring and damper is

$$F = kx_0 \sin \omega t + c\omega x_0 \cos \omega t \tag{2.43}$$

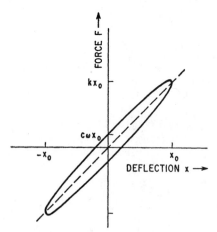

FIGURE 2.19   Hysteresis curve for a spring and viscous damper in parallel.

Equations (2.42) and (2.43) define the relation between F and x; this relation is the ellipse shown in Fig. 2.19. The energy dissipated in one cycle of oscillation is

$$W = \int_{T}^{T + 2\pi/\omega} F \frac{dx}{dt} \, dt = \pi c\omega x_0^2 \tag{2.44}$$

Motion of Foundation.   The excitation for the elastic system shown in Fig. 2.20 may be a motion $u(t)$ of the foundation. The differential equation of motion for the system is

$$m\ddot{x} + c(\dot{x} - \dot{u}) + k(x - u) = 0 \tag{2.45}$$

Consider the motion of the foundation to be a displacement that varies sinu-

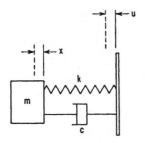

FIGURE 2.20 Single degree-of-freedom system with viscous damper, excited in forced vibration by foundation motion.

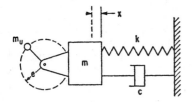

FIGURE 2.21 Single degree-of-freedom system with viscous damper, excited in forced vibration by rotating eccentric weight.

soidally with time, $u = u_0 \sin \omega t$. A steady-state condition exists after the oscillations at the natural frequency ω_n are damped out, defined by the displacement x of mass m:

$$x = Tu_0 \sin (\omega t - \psi) \qquad (2.46)$$

where T and ψ are defined in connection with Eq. (2.40) and are shown graphically in Figs. 2.17 and 2.18, respectively. Thus, the motion transmissibility T in Eq. (2.46) is identical numerically to the force transmissibility T in Eq. (2.40). The motion of the foundation and of the mass m may be expressed in any consistent units, such as displacement, velocity, or acceleration, and the same expression for T applies in each case.

Vibration Due to a Rotating Eccentric Weight. In the mass-spring-damper system shown in Fig. 2.21, a mass m_u is mounted by a shaft and bearings to the mass m. The mass m_u follows a circular path of radius e with respect to the bearings. The component of displacement in the X direction of m_u relative to m is

$$x_3 - x_1 = e \sin \omega t \qquad (2.47)$$

where x_3 and x_1 are the absolute displacements of m_u and m, respectively, in the X direction; e is the length of the arm supporting the mass m_u; and ω is the angular velocity of the arm in radians per second. The differential equation of motion for the system is

$$m\ddot{x}_1 + m_u\ddot{x}_3 + c\dot{x}_1 + kx_1 = 0 \qquad (2.48)$$

Differentiating Eq. (2.47) with respect to time, solving for $\ddot{x}_3$, and substituting in Eq. (2.48):

$$(m + m_u)\ddot{x}_1 + c\dot{x}_1 + kx_1 = m_u e\omega^2 \sin \omega t \qquad (2.49)$$

Equation (2.49) is of the same form as Eq. (2.31); thus, the response relations of Eqs. (2.33), (2.36), and (2.37) apply by substituting $(m + m_u)$ for m and $m_u e\omega^2$ for F_0. The resulting displacement, velocity, and acceleration responses are

$$\frac{x_1}{m_u e\omega^2} = R_d \sin (\omega t - \theta) \qquad \frac{\dot{x}_1\sqrt{km}}{m_u e\omega^2} = R_v \cos (\omega t - \theta)$$

$$\frac{\ddot{x}_1 m}{m_u e\omega^2} = - R_a \sin (\omega t - \theta) \qquad (2.50)$$

Resonance Frequencies. The peak values of the displacement, velocity, and acceleration response of a system undergoing forced, steady-state vibration occur at slightly different forcing frequencies. Since a *resonance frequency* is defined as the frequency for which the response is a maximum, a simple system has three resonance frequencies if defined only generally. The natural frequency is different from any of the resonance frequencies. The relations among the several resonance frequencies, the damped natural frequency, and the undamped natural frequency ω_n are:

Displacement resonance frequency: $\omega_n(1 - 2\zeta^2)^{1/2}$

Velocity resonance frequency: ω_n

Acceleration resonance frequency: $\omega_n/(1 - 2\zeta^2)^{1/2}$

Damped natural frequency: $\omega_n(1 - \zeta^2)^{1/2}$

For the degree of damping usually embodied in physical systems, the difference among the three resonance frequencies is negligible.

Resonance, Bandwidth, and the Quality Factor Q. Damping in a system can be determined by noting the maximum response, i.e., the response at the resonance frequency as indicated by the maximum value of R_v in Eq. (2.36). This is defined by the factor Q sometimes used in electrical engineering terminology and defined with respect to mechanical vibration as

$$Q = (R_v)_{max} = 1/2\zeta$$

The maximum acceleration and displacement responses are slightly larger, being

$$(R_d)_{max} = (R_a)_{max} = \frac{(R_v)_{max}}{(1 - \zeta^2)^{1/2}}$$

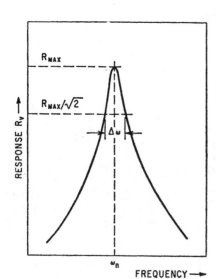

The damping in a system is also indicated by the sharpness or width of the response curve in the vicinity of a resonance frequency ω_n. Designating the width as a frequency increment $\Delta\omega$ measured at the "half-power point" (i.e., at a value of R equal to $R_{max}/2$), as illustrated in Fig. 2.22, the damping of the system is defined to a good approximation by

$$\frac{\Delta\omega}{\omega_n} = \frac{1}{Q} = 2\zeta \qquad (2.51)$$

for values of ζ less than 0.1. The quantity $\Delta\omega$, known as the *bandwidth,* is commonly represented by the letter B.

FIGURE 2.22 Response curve showing bandwidth at "half-power point."

Structural Damping. The energy dissipated by the damper is known as *hysteresis loss;* as indicated by Eq. (2.44), it is proportional to the forcing frequency ω. However, the hysteresis loss of many engineering structures has been found to be independent of frequency. To provide a better model for defining the *structural damping* experienced during vibration, an arbitrary damping term $kg = c\omega$ is introduced. In effect, this defines the damping force as being equal to the viscous damping force at some frequency, depending upon the value of g, but being invariant with frequency. The relation of the damping force F to the displacement x is defined by an ellipse similar to Fig. 2.19, and the displacement response of the system is described by an expression corresponding to Eq. (2.33) as follows:

$$\frac{x}{F_0/k} = R_g \sin(\omega t - \theta) = \frac{\sin(\omega t - \theta)}{\sqrt{(1 - \omega^2/\omega_n^2)^2 + g^2}} \tag{2.52}$$

where $g = 2\zeta\omega/\omega_n$. The resonance frequency is ω_n, and the value of R_g at resonance is $1/g = Q$.

The equations for the hysteresis ellipse for structural damping are

$$F = kx_0 (\sin \omega t + g \cos \omega t)$$
$$x = x_0 \sin \omega t \tag{2.53}$$

UNDAMPED MULTIPLE DEGREE-OF-FREEDOM SYSTEMS

An elastic system sometimes cannot be described adequately by a model having only one mass but rather must be represented by a system of two or more masses considered to be *point masses* or *particles* having no rotational inertia. If a group of particles is bound together by essentially rigid connections, it behaves as a rigid body having both mass (significant for translational motion) and moment of inertia (significant for rotational motion). There is no limit to the number of masses that may be used to represent a system. For example, each mass in a model representing a beam may be an infinitely thin slice representing a cross section of the beam; a differential equation is required to treat this continuous distribution of mass.

DEGREES-OF-FREEDOM

The number of independent parameters required to define the distance of all the masses from their reference positions is called the number of *degrees-of-freedom N*. For example, if there are N masses in a system constrained to move only in translation in the X and Y directions, the system has $2N$ degrees-of-freedom. A continuous system such as a beam has an infinitely large number of degrees-of-freedom.

For each degree-of-freedom (each coordinate of motion of each mass) a differential equation can be written in one of the following alternative forms:

$$m_j \ddot{x}_j = F_{xj} \qquad I_k \ddot{\alpha}_k = M_{\alpha k} \tag{2.54}$$

where F_{xj} is the component in the X direction of all external, spring, and damper forces acting on the mass having the jth degree-of-freedom, and $M_{\alpha k}$ is the component about the α axis of all torques acting on the body having the kth degree-of-freedom. The moment of inertia of the mass about the α axis is designated by I_k. (This is assumed for the present analysis to be a principal axis of inertia, and product of inertia terms are neglected. See Chap. 3 for a more detailed discussion.) Equations (2.54) are identical in form and can be represented by

$$m_j \ddot{x}_j = F_j \tag{2.55}$$

where F_j is the resultant of all forces (or torques) acting on the system in the jth degree-of-freedom, $\ddot{x}_j$ is the acceleration (translational or rotational) of the system in the jth degree-of-freedom, and m_j is the mass (or moment of inertia) in the jth degree-of-freedom. Thus, the terms defining the motion of the system (displacement, velocity, and acceleration) and the deflections of structures may be either translational or rotational, depending upon the type of coordinate. Similarly, the "force" acting on a system may be either a force or a torque, depending upon the type of coordinate. For example, if a system has n bodies each free to move in three translational modes and three rotational modes, there would be $6n$ equations of the form of Eq. (2.55), one for each degree-of-freedom.

DEFINING A SYSTEM AND ITS EXCITATION

The first step in analyzing any physical structure is to represent it by a mathematical model which will have essentially the same dynamic behavior. A suitable number and distribution of masses, springs, and dampers must be chosen, and the input forces or foundation motions must be defined. The model should have sufficient degrees-of-freedom to determine the modes which will have significant response to the exciting force or motion.

The properties of a system that must be known are the natural frequencies ω_n, the normal mode shapes D_{jn}, the damping of the respective modes, and the mass distribution m_j. The detailed distributions of stiffness and damping of a system are not used directly but rather appear indirectly as the properties of the respective modes. The characteristic properties of the modes may be determined experimentally as well as analytically.

STIFFNESS COEFFICIENTS

The spring system of a structure of N degrees-of-freedom can be defined completely by a set of N^2 *stiffness coefficients.*[5] A stiffness coefficient K_{jk} is the change in spring force acting on the jth degree-of-freedom when only the kth degree-of-freedom is slowly displaced a unit amount in the negative direction. This definition is a generalization of the linear, elastic spring defined by Eq. (2.1). Stiffness coefficients have the characteristic of reciprocity, i.e., $K_{jk} = K_{kj}$. The number of independent stiffness coefficients is $(N^2 + N)/2$.

The total elastic force acting on the jth degree-of-freedom is the sum of the effects of the displacements in all of the degrees-of-freedom:

$$F_{el} = -\sum_{k=1}^{N} K_{jk} x_k \tag{2.56}$$

Inserting the spring force F_{el} from Eq. (2.56) in Eq. (2.55) together with the external forces F_j results in the n equations:

$$m_j \ddot{x}_j = F_j - \sum_k K_{jk} x_k \tag{2.56a}$$

FREE VIBRATION

When the external forces are zero, the preceding equations become

$$m_j\ddot{x}_j + \sum_k K_{jk}x_k = 0 \qquad (2.57)$$

Solutions of Eq. (2.57) have the form

$$x_j = D_j \sin(\omega t + \theta) \qquad (2.58)$$

Substituting Eq. (2.58) in Eq. (2.57),

$$m_j\omega^2 D_j = \sum_k K_{jk}D_k \qquad (2.59)$$

This is a set of n linear algebraic equations with n unknown values of D. A solution of these equations for values of D other than zero can be obtained only if the determinant of the coefficients of the D's is zero:

$$\begin{vmatrix} (m_1\omega^2 - K_{11}) & -K_{12} & \cdot & \cdot & -K_{in} \\ -K_{21} & (m_2\omega^2 - K_{22}) & \cdot & \cdot & \cdot \\ \cdot & \cdot & \cdot & \cdot & \cdot \\ \cdot & \cdot & \cdot & \cdot & \cdot \\ -K_{ni} & \cdot & \cdot & \cdot & (m_n\omega^2 - K_{nn}) \end{vmatrix} = 0 \qquad (2.60)$$

Equation (2.60) is an algebraic equation of the nth degree in ω^2; it is called the *frequency equation* since it defines n values of ω which satisfy Eq. (2.57). The roots are all real; some may be equal, and others may be zero. These values of frequency determined from Eq. (2.60) are the frequencies at which the system can oscillate in the absence of external forces. These frequencies are the *natural frequencies* ω_n of the system. Depending upon the initial conditions under which vibration of the system is initiated, the oscillations may occur at any or all of the natural frequencies and at any amplitude.

Example 2.1. Consider the three degree-of-freedom system shown in Fig. 2.23; it consists of three equal masses m and a foundation connected in series by three

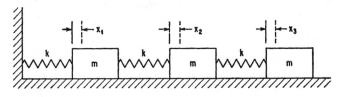

FIGURE 2.23 Undamped three degree-of-freedom system on foundation.

equal springs k. The absolute displacements of the masses are x_1, x_2, and x_3. The stiffness coefficients (see section entitled *Stiffness Coefficients*) are thus $K_{11} = 2k$,

$K_{22} = 2k$, $K_{33} = k$, $K_{12} = K_{21} = -k$, $K_{23} = K_{32} = -k$, and $K_{13} = K_{31} = 0$. The frequency equation is given by the determinant, Eq. (2.60),

$$\begin{vmatrix} (m\omega^2 - 2k) & k & 0 \\ k & (m\omega^2 - 2k) & k \\ 0 & k & (m\omega^2 - k) \end{vmatrix} = 0$$

The determinant expands to the following polynomial:

$$\left(\frac{m\omega^2}{k}\right)^3 - 5\left(\frac{m\omega^2}{k}\right)^2 + 6\left(\frac{m\omega^2}{k}\right) - 1 = 0$$

Solving for ω,

$$\omega = 0.445\sqrt{\frac{k}{m}}, \quad 1.25\sqrt{\frac{k}{m}}, \quad 1.80\sqrt{\frac{k}{m}}$$

Normal Modes of Vibration. A structure vibrating at only one of its natural frequencies ω_n does so with a characteristic pattern of amplitude distribution called a *normal mode of vibration.*[3-5] A normal mode is defined by a set of values of D_{jn} [see Eq. (2.58)] which satisfy Eq. (2.59) when $\omega = \omega_n$:

$$\omega_n^2 m_j D_{jn} = \sum_k K_{jn} D_{kn} \tag{2.61}$$

A set of values of D_{jn} which form a normal mode is independent of the absolute values of D_{jn} but depends only on their relative values. To define a mode shape by a unique set of numbers, any arbitrary *normalizing condition* which is desired can be used. A condition often used is to set $D_{1n} = 1$ but $\sum_j m_j D_{jn}^2 = 1$ and $\sum_j m_j D_{jn}^2 = \sum_j m_j$ also may be found convenient.

Orthogonality of Normal Modes. The usefulness of normal modes in dealing with multiple degree-of-freedom systems is due largely to the orthogonality of the normal modes. It can be shown[3-5] that the set of inertia forces $\omega_n^2 m_j D_{jn}$ for one mode does not work on the set of deflections D_{jm} of another mode of the structure:

$$\sum_j m_j D_{jm} D_{jn} = 0 \qquad [m \neq n] \tag{2.62}$$

This is the *orthogonality condition.*

Normal Modes and Generalized Coordinates. Any set of N deflections x_j can be expressed as the sum of normal mode amplitudes:

$$x_j = \sum_{n=1}^{N} q_n D_{jn} \tag{2.63}$$

The numerical values of the D_{jn}'s are fixed by some normalizing condition, and a set of values of the N variables q_n can be found to match any set of x_j's. The N values of q_n constitute a set of *generalized coordinates* which can be used to define the position coordinates x_j of all parts of the structure. The q's are also known as the amplitudes of the normal modes, and are functions of time. Equation (2.63) may be differentiated to obtain

$$\ddot{x}_j = \sum_{n=1}^{N} \ddot{q}_n D_{jn} \tag{2.64}$$

Any quantity which is distributed over the j coordinates can be represented by a linear transformation similar to Eq. (2.63). It is convenient now to introduce the parameter γ_n relating D_{jn} and F_j/m_j as follows:

$$\frac{F_j}{m_j} = \sum_n \gamma_n D_{jn} \qquad (2.65)$$

where F_j may be zero for certain values of n.

FORCED MOTION

Substituting the expressions in generalized coordinates, Eqs. (2.63) to (2.65), in the basic equation of motion, Eq. (2.56a),

$$m_j \sum_n \ddot{q}_n D_{jn} + \sum_k k_{jk} \sum_n q_n D_{kn} - m_j \sum_n \gamma_n D_{jn} = 0 \qquad (2.66)$$

The center term in Eq. (2.66) may be simplified by applying Eq. (2.61) and the equation rewritten as follows:

$$\sum_n (\ddot{q}_n + \omega_n^2 q_n - \gamma_n) m_j D_{jn} = 0 \qquad (2.67)$$

Multiplying Eqs. (2.67) by D_{jm} and taking the sum over j (i.e., adding all the equations together),

$$\sum_n (\ddot{q}_n + \omega_n^2 q_n - \gamma_n) \sum_j m_j D_{jn} D_{jm} = 0$$

All terms of the sum over n are zero, except for the term for which $m = n$, according to the orthogonality condition of Eq. (2.62). Then since $\sum_j m_j D_{jn}^2$ is not zero, it follows that

$$\ddot{q}_n + \omega_n^2 q_n - \gamma_n = 0$$

for every value of n from 1 to N.

An expression for γ_n may be found by using the orthogonality condition again. Multiplying Eq. (2.65) by $m_j D_{jm}$ and taking the sum taken over j,

$$\sum_j F_j D_{jm} = \sum_n \gamma_n \sum_j m_j D_{jn} D_{jm} \qquad (2.68)$$

All the terms of the sum over n are zero except when $n = m$, according to Eq. (2.62), and Eq. (2.68) reduces to

$$\gamma_n = \frac{\displaystyle\sum_j F_j D_{jn}}{\displaystyle\sum_j m_j D_{jn}^2} \qquad (2.69)$$

Then the differential equation for the response of any generalized coordinate to the externally applied forces F_j is

$$\ddot{q}_n + \omega_n^2 q_n = \gamma_n = \frac{\displaystyle\sum_j F_j D_{jn}}{\displaystyle\sum_j m_j D_{jn}^2} \qquad (2.70)$$

where $\Sigma F_j D_{jn}$ is the generalized force, i.e., the total work done by all external forces during a small displacement δq_n divided by δq_n, and $\Sigma m_j D_{jn}^2$ is the generalized mass.

Thus the amplitude q_n of each normal mode is governed by its own equation, independent of the other normal modes, and responds as a simple mass-spring system. Equation (2.70) is a generalized form of Eq. (2.23).

The forces F_j may be any functions of time. Any equation for the response of an undamped mass-spring system applies to each mode of a complex structure by substituting:

$$
\begin{aligned}
&\text{The } generalized\ coordinate\ q_n \text{ for } x \\[6pt]
&\text{The } generalized\ force\ \sum_j F_j D_{jn} \text{ for } F \\[6pt]
&\text{The } generalized\ mass\ \sum_j m_j D_{jn} \text{ for } m \\[6pt]
&\text{The } mode\ natural\ frequency\ \omega_n \text{ for } \omega_n
\end{aligned}
\tag{2.71}
$$

Response to Sinusoidal Forces. If a system is subjected to one or more sinusoidal forces $F_j = F_{0j} \sin \omega t$, the response is found from Eq. (2.26) by noting that $k = m\omega_n^2$ [Eq. (2.6)] and then substituting from Eq. (2.71):

$$
q_n = \frac{\displaystyle\sum_j F_{0j} D_{jn}}{\omega_n^2 \displaystyle\sum_j m_j D_{jn}^2} \; \frac{\sin \omega t}{(1 - \omega^2/\omega_n^2)}
\tag{2.72}
$$

Then the displacement of the kth degree-of-freedom, from Eq. (2.63), is

$$
x_k = \sum_{n=1}^{N} \frac{D_{kn} \displaystyle\sum_j F_{0j} D_{jn} \sin \omega t}{\omega_n^2 \displaystyle\sum_j m_j D_{jn}^2 (1 - \omega^2/\omega_n^2)}
\tag{2.73}
$$

This is the general equation for the response to sinusoidal forces of an undamped system of N degrees-of-freedom. The application of the equation to systems free in space or attached to immovable foundations is discussed below.

Example 2.2. Consider the system shown in Fig. 2.24; it consists of three equal masses m connected in series by two equal springs k. The system is free in space and

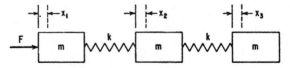

FIGURE 2.24 Undamped three degree-of-freedom system acted on by sinusoidal force.

a force $F \sin \omega t$ acts on the first mass. Absolute displacements of the masses are x_1, x_2, and x_3. Determine the acceleration $\ddot{x}_3$. The stiffness coefficients (see section entitled *Stiffness Coefficients*) are $K_{11} = K_{33} = k$, $K_{22} = 2k$, $K_{12} = K_{21} = -k$, $K_{13} = K_{31} = 0$, and $K_{23} = K_{32} = -k$. Substituting in Eq. (2.60), the frequency equation is

$$
\begin{vmatrix}
(m\omega^2 - k) & k & 0 \\
k & (m\omega^2 - 2k) & k \\
0 & k & (m\omega^2 - k)
\end{vmatrix} = 0
$$

The roots are $\omega_1 = 0$, $\omega_2 = \sqrt{k/m}$, and $\omega_3 = \sqrt{3k/m}$. The zero value for one of the natural frequencies indicates that the entire system translates without deflection of the springs. The mode shapes are now determined by substituting from Eq. (2.58) in Eq. (2.57), noting that $\ddot{x} = -D\omega^2$, and writing Eq. (2.59) for each of the three masses in each of the oscillatory modes 2 and 3:

$$
mD_{21}\left(\frac{k}{m}\right) = K_{11}D_{21} + K_{21}D_{22} + K_{31}D_{23}
$$

$$
mD_{22}\left(\frac{k}{m}\right) = K_{12}D_{21} + K_{22}D_{22} + K_{32}D_{23}
$$

$$
mD_{23}\left(\frac{k}{m}\right) = K_{13}D_{21} + K_{23}D_{22} + K_{33}D_{23}
$$

$$
mD_{31}\left(\frac{3k}{m}\right) = K_{11}D_{31} + K_{21}D_{32} + K_{31}D_{33}
$$

$$
mD_{32}\left(\frac{3k}{m}\right) = K_{12}D_{31} + K_{22}D_{32} + K_{32}D_{33}
$$

$$
mD_{33}\left(\frac{3k}{m}\right) = K_{13}D_{31} + K_{23}D_{32} + K_{33}D_{33}
$$

where the first subscript on the D's indicates the mode number (according to ω_1 and ω_2 above) and the second subscript indicates the displacement amplitude of the particular mass. The values of the stiffness coefficients K are calculated above. The mode shapes are defined by the relative displacements of the masses. Thus, assigning values of unit displacement to the first mass (i.e., $D_{21} = D_{31} = 1$), the above equations may be solved simultaneously for the D's:

$$
D_{21} = 1 \qquad D_{22} = 0 \qquad D_{23} = -1
$$

$$
D_{31} = 1 \qquad D_{32} = -2 \qquad D_{33} = 1
$$

Substituting these values of D in Eq. (2.71), the generalized masses are determined: $M_2 = 2m$, $M_3 = 6m$.

Equation (2.73) then can be used to write the expression for acceleration $\ddot{x}_3$:

$$
\ddot{x}_3 = \left[\frac{1}{3m} + \frac{(\omega^2/\omega_2^2)(-1)(+1)}{2m(1 - \omega^2/\omega_2^2)} + \frac{(\omega^2/\omega_3^2)(+1)(+1)}{6m(1 - \omega^2/\omega_3^2)} \right] F_1 \sin \omega t
$$

Free and Fixed Systems.   For a structure which is free in space, there are six "normal modes" corresponding to $\omega_n = 0$. These represent motion of the structure without relative motion of its parts; this is rigid body motion with six degrees-of-freedom.

The rigid body modes all may be described by equations of the form

$$D_{jm} = a_{jm}D_m \qquad [m = 1,2, \ldots ,6]$$

where D_m is a motion of the rigid body in the m coordinate and a is the displacement of the jth degree-of-freedom when D_m is moved a unit amount. The geometry of the structure determines the nature of a_{jm}. For example, if D_m is a rotation about the Z axis, $a_{jm} = 0$ for all modes of motion in which j represents rotation about the X or Y axis and $a_{jm} = 0$ if j represents translation parallel to the Z axis. If D_{jm} is a translational mode of motion parallel to X or Y, it is necessary that a_{jm} be proportional to the distance r_j of m_j from the Z axis and to the sine of the angle between r_j and the jth direction. The above relations may be applied to an elastic body. Such a body moves as a rigid body in the gross sense in that all particles of the body move together generally but may experience relative vibratory motion. The orthogonality condition applied to the relation between any rigid body mode D_{jm} and any oscillatory mode D_{jn} yields

$$\sum_j m_j D_{jn}D_{jm} = \sum_j m_j a_{jm}D_{jn} = 0 \qquad \begin{bmatrix} m \le 6 \\ n > 6 \end{bmatrix} \tag{2.74}$$

These relations are used in computations of oscillatory modes, and show that normal modes of vibration involve no net translation or rotation of a body.

A system attached to a fixed foundation may be considered as a system free in space in which one or more "foundation" masses or moments of inertia are infinite. Motion of the system as a rigid body is determined entirely by the motion of the foundation. The amplitude of an oscillatory mode representing motion of the foundation is zero; i.e., $M_j D_{jn}^2 = 0$ for the infinite mass. However, Eq. (2.73) applies equally well regardless of the size of the masses.

Foundation Motion.   If a system is small relative to its foundation, it may be assumed to have no effect on the motion of the foundation. Consider a foundation of large but unknown mass m_0 having a motion $x_0 \sin \omega t$, the consequence of some unknown force

$$F_0 \sin \omega t = -m_0 x_0 \omega^2 \sin \omega t \tag{2.75}$$

acting on m_0 in the x_0 direction. Equation (2.73) is applicable to this case upon substituting

$$-m_0 x_0 \omega^2 D_{0n} = \sum_j F_{0j}D_{jn} \tag{2.76}$$

where D_{0n} is the amplitude of the foundation (the 0 degree-of-freedom) in the nth mode.

The oscillatory modes of the system are subject to Eqs. (2.74):

$$\sum_j = 0 \; m_j a_{jm}D_{jn} = 0$$

Separating the 0th degree-of-freedom from the other degrees-of-freedom:

$$\sum_{j=0} m_j a_{jm}D_{jn} = m_0 a_{0m}D_{0n} + \sum_{j=1} m_j a_{jm}D_{jn}$$

If m_0 approaches infinity as a limit, D_{0n} approaches zero and motion of the system as a rigid body is identical with the motion of the foundation. Thus, a_{0m} approaches unity for motion in which $m = 0$, and approaches zero for motion in which $m \neq 0$. In the limit:

$$\lim_{m_0 \to \infty} m_0 D_{0n} = -\sum_j m_j a_{j0} D_{jn} \tag{2.77}$$

Substituting this result in Eq. (2.76),

$$\lim_{m_0 \to \infty} \sum_j F_{0j} D_{jn} = x_0 \omega^2 \sum_j m_j a_{j0} D_{jn} \tag{2.78}$$

The generalized mass in Eq. (2.73) includes the term $m_0 D_{0n}{}^2$, but this becomes zero as m_0 becomes infinite.

The equation for response of a system to motion of its foundation is obtained by substituting Eq. (2.78) in Eq. (2.73):

$$x_k = \sum_{n=1}^{N} \frac{\omega^2}{\omega_n{}^2} D_{kn} \frac{\sum_j m_j a_{j0} D_{jn} x_0 \sin \omega t}{\sum_j m_j D_{jn}{}^2 (1 - \omega^2/\omega_n{}^2)} + x_0 \sin \omega t \tag{2.79}$$

DAMPED MULTIPLE DEGREE-OF-FREEDOM SYSTEMS

Consider a set of masses interconnected by a network of springs and acted upon by external forces, with a network of dampers acting in parallel with the springs. The viscous dampers produce forces on the masses which are determined in a manner analogous to that used to determine spring forces and summarized by Eq. (2.56). The damping force acting on the jth degree-of-freedom is

$$(F_d)_j = -\sum_k C_{jk} \dot{x}_k \tag{2.80}$$

where C_{jk} is the resultant force on the jth degree-of-freedom due to a unit velocity of the kth degree-of-freedom.

In general, the distribution of damper sizes in a system need not be related to the spring or mass sizes. Thus, the dampers may couple the normal modes together, allowing motion of one mode to affect that of another. Then the equations of response are not easily separable[6] into independent normal mode equations. However, there are two types of damping distribution which do not couple the normal modes.[6-8] These are known as *uniform viscous damping* and *uniform mass damping*.

UNIFORM VISCOUS DAMPING

Uniform damping is an appropriate model for systems in which the damping effect is an inherent property of the spring material. Each spring is considered to have a damper acting in parallel with it, and the ratio of damping coefficient to stiffness coefficient is the same for each spring of the system. Thus, for all values of j and k,

$$\frac{C_{jk}}{k_{jk}} = 2\mathfrak{g} \tag{2.81}$$

where $\mathfrak{g}$ is a constant.

Substituting from Eq. (2.81) in Eq. (2.80),

$$-(F_d)_j = \sum_k C_{jk}\dot{x}_k = 2\Im \sum_k k_{jk}\dot{x}_k \tag{2.82}$$

Since the damping forces are "external" forces with respect to the mass-spring system, the forces $(F_d)_j$ can be added to the external forces in Eq. (2.70) to form the equation of motion:

$$\ddot{q}_n + \omega_n^2 q_n = \frac{\sum_j (F_d)_j D_{jn} + \sum_j F_j D_{jn}}{\sum_j m_j D_{jn}^2} \tag{2.83}$$

Combining Eqs. (2.61), (2.63), and (2.82), the summation involving $(F_d)_j$ in Eq. (2.83) may be written as follows:

$$\sum_j (F_d)_j D_{jn} = -2\Im\omega_n^2 \dot{q}_n \sum_j m_j D_{jn}^2 \tag{2.84}$$

Substituting Eq. (2.84) in Eq. (2.83),

$$\ddot{q}_n + 2\Im\omega_n^2 \dot{q}_n + \omega_n^2 q_n = \gamma_n \tag{2.85}$$

Comparison of Eq. (2.85) with Eq. (2.31) shows that each mode of the system responds as a simple damped oscillator.

The damping term $2\Im\omega_n^2$ in Eq. (2.85) corresponds to $2\zeta\omega_n$ in Eq. (2.31) for a simple system. Thus, $\Im\omega_n$ may be considered the critical damping ratio of each mode. Note that the effective damping for a particular mode varies directly as the natural frequency of the mode.

Free Vibration. If a system with uniform viscous damping is disturbed from its equilibrium position and released at time $t = 0$ to vibrate freely, the applicable equation of motion is obtained from Eq. (2.85) by substituting $2\zeta\omega$ for $2\Im\omega_n^2$ and letting $\gamma_n = 0$:

$$\ddot{q}_n + 2\zeta\omega_n \dot{q}_n + \omega_n^2 q_n = 0 \tag{2.86}$$

The solution of Eq. (2.86) for less than critical damping is

$$x_j(t) = \sum_n D_{jn} e^{-\zeta\omega_n t}(A_n \sin \omega_d t + B_n \cos \omega_d t) \tag{2.87}$$

where $\omega_d = \omega_n(1 - \zeta^2)^{1/2}$.

The values of A and B are determined by the displacement $x_j(0)$ and velocity $\dot{x}_j(0)$ at time $t = 0$:

$$x_j(0) = \sum_n B_n D_{jn}$$

$$\dot{x}_j(0) = \sum_n (A_n \omega_{dn} - B_n \zeta\omega_n) D_{jn}$$

Applying the orthogonality relation of Eq. (2.62) in the manner used to derive Eq. (2.69),

$$B_n = \frac{\sum_j x_j(0)m_j D_{jn}}{\sum_j m_j D_{jn}^2}$$

(2.88)

$$A_n \omega_{dn} - B_n \zeta \omega_{dn} = \frac{\sum_j \dot{x}_j(0)m_j D_{jn}}{\sum_j m_j D_{jn}^2}$$

Thus each mode undergoes a decaying oscillation at the damped natural frequency for the particular mode, and the amplitude of each mode decays from its initial value, which is determined by the initial displacements and velocities.

UNIFORM STRUCTURAL DAMPING

To avoid the dependence of viscous damping upon frequency, as indicated by Eq. (2.85), the uniform viscous damping factor $\mathfrak{c}$ is replaced by g/ω for uniform structural damping. This corresponds to the structural damping parameter g in Eqs. (2.52) and (2.53) for sinusoidal vibration of a simple system. Thus, Eq. (2.85) for the response of a mode to a sinusoidal force of frequency ω is

$$\ddot{q}_n + \frac{2g}{\omega}\, \omega_n^2 \dot{q}_n + \omega_n^2 q_n = \gamma_n$$

(2.89)

The amplification factor at resonance ($Q = 1/g$) has the same value in all modes.

UNIFORM MASS DAMPING

If the damping force on each mass is proportional to the magnitude of the mass,

$$(F_d)_j = -Bm_j \dot{x}_j$$

(2.90)

where B is a constant. For example, Eq. (2.90) would apply to a uniform beam immersed in a viscous fluid.

Substituting as $\dot{x}_j$ in Eq. (2.90) the derivative of Eq. (2.63),

$$\Sigma(F_d)_j D_{jn} = -B \sum_j m_j D_{jn} \sum_m \dot{q}_m D_{jm}$$

(2.91)

Because of the orthogonality condition, Eq. (2.62):

$$\Sigma(F_d)_j D_{jn} = -B\dot{q}_n \sum_j m_j D_{jn}^2$$

Substituting from Eq. (2.91) in Eq. (2.83), the differential equation for the system is

$$\ddot{q}_n + B\dot{q}_n + \omega_n^2 q_n = \gamma_n$$

(2.92)

where the damping term B corresponds to $2\zeta\omega$ for a simple oscillator, Eq. (2.31). Then $B/2\omega_n$ represents the fraction of critical damping for each mode, a quantity which diminishes with increasing frequency.

GENERAL EQUATION FOR FORCED VIBRATION

All the equations for response of a linear system to a sinusoidal excitation may be regarded as special cases of the following general equation:

$$x_k = \sum_{n=1}^{N} \frac{D_{kn}}{\omega_n^2} \frac{F_n}{m_n} R_n \sin(\omega t - \theta_n) \tag{2.93}$$

where x_k = displacement of structure in kth degree-of-freedom
          N = number of degrees-of-freedom, including those of the foundation
          D_{kn} = amplitude of kth degree-of-freedom in nth normal mode
          F_n = generalized force for nth mode
          m_n = generalized mass for nth mode
          R_n = response factor, a function of the frequency ratio ω/ω_n (Fig. 2.13)
          θ_n = phase angle (Fig. 2.14)

Equation (2.93) is of sufficient generality to cover a wide variety of cases, including excitation by external forces or foundation motion, viscous or structural damping, rotational and translational degrees-of-freedom, and from one to an infinite number of degrees-of-freedom.

LAGRANGIAN EQUATIONS

The differential equations of motion for a vibrating system sometimes are derived more conveniently in terms of kinetic and potential energies of the system than by the application of Newton's laws of motion in a form requiring the determination of the forces acting on each mass of the system. The formulation of the equations in terms of the energies, known as Lagrangian equations,[3-5] is expressed as follows:

$$\frac{d}{dt} \frac{\partial T}{\partial \dot{q}_n} - \frac{\partial T}{\partial q_n} + \frac{\partial V}{\partial q_n} = F_n \tag{2.94}$$

where T = total kinetic energy of system
          V = total potential energy of system
          q_n = generalized coordinate—a displacement
          $\dot{q}_n$ = velocity at generalized coordinate q_n
          F_n = generalized force, the portion of the total forces not related to the potential energy of the system (gravity and spring forces appear in the potential energy expressions and are not included here)

The method of applying Eq. (2.94) is to select a number of independent coordinates (generalized coordinates) equal to the number of degrees-of-freedom, and to write expressions for total kinetic energy T and total potential energy V. Differentiation of these expressions successively with respect to each of the chosen coordinates leads to a number of equations similar to Eq. (2.94), one for each coordinate (degree-of-freedom). These are the applicable differential equations and may be solved by any suitable method.

Example 2.3. Consider free vibration of the three degree-of-freedom system shown in Fig. 2.23; it consists of three equal masses m connected in tandem by equal springs k. Take as coordinates the three absolute displacements x_1, x_2, and x_3. The kinetic energy of the system is

$$T = \tfrac{1}{2}m(\dot{x}_1^2 + \dot{x}_2^2 + \dot{x}_3^2)$$

The potential energy of the system is

$$V = \frac{k}{2}\,[x_1^2 + (x_1 - x_2)^2 + (x_2 - x_3)^2] = \frac{k}{2}\,(2x_1^2 + 2x_2^2 + x_3^2 - 2x_1x_2 - 2x_2x_3)$$

Differentiating the expression for the kinetic energy successively with respect to the velocities,

$$\frac{\partial T}{\partial \dot{x}_1} = m\dot{x}_1 \qquad \frac{\partial T}{\partial \dot{x}_2} = m\dot{x}_2 \qquad \frac{\partial T}{\partial \dot{x}_3} = m\dot{x}_3$$

The kinetic energy is not a function of displacement; therefore, the second term in Eq. (2.94) is zero. The partial derivatives with respect to the displacement coordinates are

$$\frac{\partial V}{\partial x_1} = 2kx_1 - kx_2 \qquad \frac{\partial V}{\partial x_2} = 2kx_2 - kx_1 - kx_3 \qquad \frac{\partial V}{\partial x_3} = kx_3 - kx_2$$

In free vibration, the generalized force term in Eq. (2.93) is zero. Then, substituting the derivatives of the kinetic and potential energies from above into Eq. (2.94),

$$m\ddot{x}_1 + 2kx_1 - kx_2 = 0$$

$$m\ddot{x}_2 + 2kx_2 - kx_1 - kx_3 = 0$$

$$m\ddot{x}_3 + kx_3 - kx_2 = 0$$

The natural frequencies of the system may be determined by placing the preceding set of simultaneous equations in determinant form, in accordance with Eq. (2.60):

$$\begin{vmatrix} (m\omega^2 - 2k) & k & 0 \\ k & (m\omega^2 - 2k) & k \\ 0 & k & (m\omega^2 - k) \end{vmatrix} = 0$$

The natural frequencies are equal to the values of ω that satisfy the preceding determinant equation.

Example 2.4. Consider the compound pendulum of mass m shown in Fig. 2.25, having its center-of-gravity located a distance l from the axis of rotation. The moment of inertia is I about an axis through the center-of-gravity. The position of the mass is defined by three coordinates, x and y to define the location of the center-of-gravity and θ to define the angle of rotation.

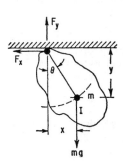

FIGURE 2.25 Forces and motions of a compound pendulum.

The *equations of constraint* are $y = l\cos\theta$; $x = l\sin\theta$. Each equation of constraint reduces the number of degrees-of-freedom by 1; thus the pendulum is a one degree-of-freedom system whose position is defined uniquely by θ alone.

The kinetic energy of the pendulum is

$$T = \tfrac{1}{2}(I + ml^2)\dot\theta^2$$

The potential energy is

$$V = mgl(1 - \cos\theta)$$

Then

$$\frac{\partial T}{\partial\dot\theta} = (I + ml^2)\dot\theta \qquad \frac{d}{dt}\left(\frac{\partial T}{\partial\dot\theta}\right) = (I + ml^2)\ddot\theta$$

$$\frac{\partial T}{\partial\theta} = 0 \qquad \frac{\partial V}{\partial\theta} = mgl\sin\theta$$

Substituting these expressions in Eq. (2.94), the differential equation for the pendulum is

$$(I + ml^2)\ddot\theta + mgl\sin\theta = 0$$

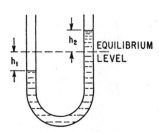

FIGURE 2.26   Water column in a U tube.

Example 2.5. Consider oscillation of the water in the U tube shown in Fig. 2.26. If the displacements of the water levels in the arms of a uniform-diameter U tube are h_1 and h_2, then conservation of matter requires that $h_1 = -h_2$. The kinetic energy of the water flowing in the tube with velocity $\dot h_1$ is

$$T = \tfrac{1}{2}\rho Sl\dot h_1^2$$

where ρ is the water density, S is the cross-section area of the tube, and l is the developed length of the water column. The potential energy (difference in potential energy between arms of tube) is

$$V = S\rho g h_1^2$$

Taking h_1 as the generalized coordinate, differentiating the expressions for energy, and substituting in Eq. (2.94),

$$S\rho l\ddot h_1 + 2\rho g S h_1 = 0$$

Dividing through by ρSl,

$$\ddot h_1 + \frac{2g}{l}\,h_1 = 0$$

This is the differential equation for a simple oscillating system of natural frequency ω_n, where

$$\omega_n = \sqrt{\frac{2g}{l}}$$

REFERENCES

1. Timoshenko, S.: "Vibration Problems in Engineering," D. Van Nostrand Company, Inc., Princeton, N.J., 1937.

2. Hansen, H. M., and P. F. Chenea: "Mechanics of Vibration," John Wiley & Sons, Inc., New York, 1952.

3. Kármán, T. V., and M. A. Biot: "Mathematical Methods in Engineering," McGraw-Hill Book Company, Inc., New York, 1940.

4. Slater, J. C., and N. H. Frank: "Mechanics," McGraw-Hill Book Company, Inc., New York, 1947.

5. Pipes, L. A.: "Applied Mathematics for Engineers and Physicists," McGraw-Hill Book Company, Inc., New York, 1958.

6. Foss, K. A.: "Coordinates Which Uncouple the Equations of Motion of Damped Linear Dynamic Systems," *ASME Applied Mechanics Paper* 57-A-86, December 1957.

7. Rayleigh: "The Theory of Sound," vol. I, Macmillan & Co., Ltd., London, 1894.

8. Crumb, S. F.: "A Study of the Effects of Damping on Normal Modes of Electrical and Mechanical Systems," California Institute of Technology, Pasadena, Calif., 1955.

CHAPTER 3

VIBRATION OF A RESILIENTLY SUPPORTED RIGID BODY

Harry Himelblau, Jr.
Sheldon Rubin

INTRODUCTION

This chapter discusses the vibration of a rigid body on resilient supporting elements, including (1) methods of determining the inertial properties of a rigid body, (2) discussion of the dynamic properties of resilient elements, and (3) motion of a single rigid body on resilient supporting elements for various dynamic excitations and degrees of symmetry.

The general equations of motion for a rigid body on linear massless resilient supports are given; these equations are general in that they include any configuration of the rigid body and any configuration and location of the supports. They involve six simultaneous equations with numerous terms, for which a general solution is impracticable without the use of high-speed automatic computing equipment. Various degrees of simplification are introduced by assuming certain symmetry, and results useful for engineering purposes are presented. Several topics are considered: (1) determination of undamped natural frequencies and discussion of coupling of modes of vibration; (2) forced vibration where the excitation is a vibratory motion of the foundation; (3) forced vibration where the excitation is a vibratory force or moment generated within the body; and (4) free vibration caused by an instantaneous change in velocity of the system (velocity shock). Results are presented mathematically and, where feasible, graphically.

SYSTEM OF COORDINATES

The motion of the rigid body is referred to a fixed "inertial" frame of reference. The inertial frame is represented by a system of cartesian coordinates $\bar{X}, \bar{Y}, \bar{Z}$. A similar system of coordinates X, Y, Z fixed in the body has its origin at the center-of-mass. The two sets of coordinates are coincident when the body is in equilibrium under the

3.1

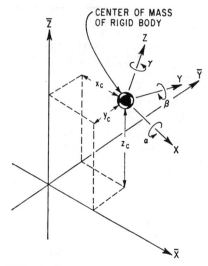

FIGURE 3.1 System of coordinates for the motion of a rigid body consisting of a fixed inertial set of reference axes ($\bar{X}$, $\bar{Y}$, $\bar{Z}$) and a set of axes (X, Y, Z) fixed in the moving body with its origin at the center-of-mass. The axes $\bar{X}$, $\bar{Y}$, $\bar{Z}$ and X, Y, Z are coincident when the body is in equilibrium under the action of gravity alone. The displacement of the center-of-mass is given by the translational displacements x_c, y_c, z_c and the rotational displacements α, β, γ as shown. A positive rotation about an axis is one which advances a right-handed screw in the positive direction of the axis.

action of gravity alone. The motions of the body are described by giving the displacement of the body axes relative to the inertial axes. The translational displacements of the center-of-mass of the body are x_c, y_c, z_c in the $\bar{X}$, $\bar{Y}$, $\bar{Z}$ directions, respectively. The rotational displacements of the body are characterized by the angles of rotation α, β, γ of the body axes about the $\bar{X}$, $\bar{Y}$, $\bar{Z}$ axes, respectively. These displacements are shown graphically in Fig. 3.1.

Only small translations and rotations are considered. Hence, the rotations are commutative (i.e., the resulting position is independent of the order of the component rotations) and the angles of rotation about the body axes are equal to those about the inertial axes. Therefore, the displacements of a point b in the body (with the coordinates b_x, b_y, b_z in the X, Y, Z directions, respectively) are the sums of the components of the center-of-mass displacement in the directions of the $\bar{X}$, $\bar{Y}$, $\bar{Z}$ axes plus the tangential components of the rotational displacement of the body:

$$x_b = x_c + b_z\beta - b_y\gamma$$

$$y_b = y_c - b_z\alpha + b_x\gamma \qquad (3.1)$$

$$z_b = z_c - b_x\beta + b_y\alpha$$

EQUATIONS OF SMALL MOTION OF A RIGID BODY

The equations of motion for the translation of a rigid body are

$$m\ddot{x}_c = \mathbf{F}_x \qquad m\ddot{y}_c = \mathbf{F}_y \qquad m\ddot{z}_c = \mathbf{F}_z \qquad (3.2)$$

where m is the mass of the body, $\mathbf{F}_x$, $\mathbf{F}_y$, $\mathbf{F}_z$ are the summation of all forces acting on the body, and $\ddot{x}_c$, $\ddot{y}_c$, $\ddot{z}_c$ are the accelerations of the center-of-mass of the body in the $\bar{X}$, $\bar{Y}$, $\bar{Z}$ directions, respectively. The motion of the center-of-mass of a rigid body is the same as the motion of a particle having a mass equal to the total mass of the body and acted upon by the resultant external force.

The equations of motion for the rotation of a rigid body are

$$I_{xx}\ddot{\alpha} - I_{xy}\ddot{\beta} - I_{xz}\ddot{\gamma} = \mathbf{M}_x$$

$$-I_{xy}\ddot{\alpha} + I_{yy}\ddot{\beta} - I_{yz}\ddot{\gamma} = \mathbf{M}_y \qquad (3.3)$$

$$-I_{xz}\ddot{\alpha} - I_{yz}\ddot{\beta} + I_{zz}\ddot{\gamma} = \mathbf{M}_z$$

where $\ddot{\alpha}, \ddot{\beta}, \ddot{\gamma}$ are the rotational accelerations about the X, Y, Z axes, as shown in Fig. 3.1; $\mathbf{M}_x, \mathbf{M}_y, \mathbf{M}_z$ are the summation of torques acting on the rigid body about the axes X, Y, Z, respectively; and $I_{xx} \ldots, I_{xy} \ldots$ are the moments and products of inertia of the rigid body as defined below.

INERTIAL PROPERTIES OF A RIGID BODY

The properties of a rigid body that are significant in dynamics and vibration are the mass, the position of the center-of-mass (or center-of-gravity), the moments of inertia, the products of inertia, and the directions of the principal inertial axes. This section discusses the properties of a rigid body, together with computational and experimental methods for determining the properties.

MASS

Computation of Mass. The mass of a body is computed by integrating the product of mass density $\rho(V)$ and elemental volume dV over the body:

$$m = \int_v \rho(V)\,dV \tag{3.4}$$

If the body is made up of a number of elements, each having constant or an average density, the mass is

$$m = \rho_1 V_1 + \rho_2 V_2 + \cdots + \rho_n V_n \tag{3.5}$$

where ρ_1 is the density of the element V_1, etc. Densities of various materials may be found in handbooks containing properties of materials.[1]

If a rigid body has a common geometrical shape, or if it is an assembly of subbodies having common geometrical shapes, the volume may be found from compilations of formulas. Typical formulas are included in Tables 3.1 and 3.2. Tables of areas of plane sections as well as volumes of solid bodies are useful.

If the volume of an element of the body is not given in such a table, the integration of Eq. (3.4) may be carried out analytically, graphically, or numerically. A graphical approach may be used if the shape is so complicated that the analytical expression for its boundaries is not available or is not readily integrable. This is accomplished by graphically dividing the body into smaller parts, each of whose boundaries may be altered slightly (without change to the area) in such a manner that the volume is readily calculable or measurable.

The weight W of a body of mass m is a function of the acceleration of gravity g at the particular location of the body in space:

$$W = mg \tag{3.6}$$

Unless otherwise stated, it is understood that the weight of a body is given for an average value of the acceleration of gravity on the surface of the earth. For engineering purposes, $g = 32.2$ ft/sec^2 or 386 in./sec^2 is usually used.

Experimental Determination of Mass. Although Newton's second law of motion, $F = m\ddot{x}$, may be used to measure mass, this usually is not convenient. The mass of a body is most easily measured by performing a static measurement of the weight of the body and converting the result to mass. This is done by use of the value of the acceleration of gravity at the measurement location [Eq. (3.6)].

TABLE 3.1 Properties of Plane Sections (*After G. W. Housner and D. E. Hudson.*[25])

The dimensions $X_o Y_c$ are the X, Y coordinates of the centroid, A is the area, $I_x \ldots$ is the area moment of inertia with respect to the $X \ldots$ axis, $\rho_x \ldots$ is the radius of gyration with respect to the $X \ldots$ axis; uniform solid cylindrical bodies of length l in the Z direction having the various plane sections as their cross sections have mass moment and product of inertia values about the Z axis equal to ρl times the values given in the table, where ρ is the mass density of the body; the radii of gyration are unchanged.

Plane section	Area and centroid	Area moment of inertia	Square of radius of gyration	Area product of inertia
1 Right triangle	$A = \frac{1}{2}bh$ $X_c = \frac{2}{3}b$ $Y_c = \frac{1}{3}h$	$I_{z_c} = \dfrac{bh^3}{36}$ $I_{y_c} = \dfrac{b^3h}{36}$	$\rho_{z_c}^2 = \frac{1}{18}h^2$ $\rho_{y_c}^2 = \frac{1}{18}b^2$	$I_{z_c y_c} = \dfrac{A}{36}\,hb = \dfrac{h^2b^2}{72}$
2	$A = \frac{1}{2}bh$ $X_c = \frac{1}{3}b$ $Y_c = \frac{1}{3}h$	$I_{z_c} = \dfrac{bh^3}{36}$ $I_{y_c} = \dfrac{b^3h}{36}$	$\rho_{z_c}^2 = \frac{1}{18}h^2$ $\rho_{y_c}^2 = \frac{1}{18}b^2$	$I_{z_c y_c} = -\,\dfrac{A}{36}\,hb = -\,\dfrac{h^2b^2}{72}$
3 Triangle	$A = \frac{1}{2}bh$ $X_c = \frac{1}{3}(a + b)$ $Y_c = \frac{1}{3}h$	$I_{z_c} = \dfrac{bh^3}{36}$ $I_{y_c} = \dfrac{bh}{36}(b^2 - ab + a^2)$	$\rho_{z_c}^2 = \frac{1}{18}h^2$ $\rho_{y_c}^2 = \frac{1}{18}(b^2 - ab + a^2)$	$I_{z_c y_c} = \dfrac{Ah}{36}(2a - b) = \dfrac{bh^2}{72}(2a - b)$

4 Square	$A = a^2$ $X_c = \tfrac{1}{2}a$ $Y_c = \tfrac{1}{2}a$	$I_{z_c} = I_{y_c} = \dfrac{a^4}{12}$	$\rho_{z_c}^{\,2} = \rho_{y_c}^{\,2} = \tfrac{1}{12}a^2$	$I_{z_c y_c} = 0$
5 Rectangle	$A = bh$ $X_c = \tfrac{1}{2}b$ $Y_c = \tfrac{1}{2}h$	$I_{z_c} = \dfrac{bh^3}{12}$ $I_{y_c} = \dfrac{b^3h}{12}$	$\rho_{z_c}^{\,2} = \tfrac{1}{12}h^2$ $\rho_{y_c}^{\,2} = \tfrac{1}{12}h^2$	$I_{z_c y_c} = 0$
6 Parallelogram	$A = ab\sin\theta$ $X_c = \tfrac{1}{2}(b + a\cos\theta)$ $Y_c = \tfrac{1}{2}(a\sin\theta)$	$I_{z_c} = \dfrac{a^3b}{12}\sin^3\theta$ $I_{y_c} = \dfrac{ab}{12}\sin\theta\,(b^2 + a^2\cos^2\theta)$	$\rho_{z_c}^{\,2} = \tfrac{1}{12}(a\sin\theta)^2$ $\rho_{y_c}^{\,2} = \tfrac{1}{12}(b^2 + a^2\cos^2\theta)$	$I_{z_c y_c} = \dfrac{a^3b}{12}\sin^2\theta\cos\theta$
7 Trapezoid	$A = \tfrac{1}{2}h(a + b)$ $Y_c = \tfrac{1}{3}h\left(\dfrac{2a + b}{a + b}\right)$	$I_{z_c} = \dfrac{h^3(a^2 + 4ab + b^2)}{36(a + b)}$	$\rho_{z_c}^{\,2} = \dfrac{h^2(a^2 + 4ab + b^2)}{18(a + b)^2}$	

3.5

TABLE 3.1 Properties of Plane Sections (*Continued*)

Plane Section	Area and centroid	Area moment of inertia	Square of radius of gyration	Area product of inertia
8 Circle	$A = \pi a^2$ $X_c = a$ $Y_c = a$	$I_{z_c} = I_{y_c} = \tfrac{1}{4}\pi a^4$	$\rho_{z_c}^2 = \rho_{y_c}^2 = \tfrac{1}{4}a^2$	$I_{z_c y_c} = 0$
9 Annulus	$A = \pi(a^2 - b^2)$ $X_c = a$ $Y_c = a$	$I_{z_c} = I_{y_c} = \dfrac{\pi}{4}(a^4 - b^4)$	$\rho_{z_c}^2 = \rho_{y_c}^2 = \tfrac{1}{4}(a^2 + b^2)$	$I_{z_c y_c} = 0$
10 Semicircle	$A = \tfrac{1}{2}\pi a^2$ $X_c = a$ $Y_c = \dfrac{4a}{3\pi}$	$I_{z_c} = \dfrac{a^4(9\pi^2 - 64)}{72\pi}$ $I_{y_c} = \tfrac{1}{8}\pi a^4$	$\rho_{z_c}^2 = \dfrac{a^2(9\pi^2 - 64)}{36\pi^2}$ $\rho_{y_c}^2 = \tfrac{1}{4}a^2$	$I_{z_c y_c} = 0$
11 Circular sector	$A = a^2\theta$ $X_c = \dfrac{2a \sin\theta}{3}\,\dfrac{}{\theta}$ $Y_c = 0$	$I_x = \tfrac{1}{4}a^4(\theta - \sin\theta\cos\theta)$ $I_y = \tfrac{1}{4}a^4(\theta + \sin\theta\cos\theta)$	$\rho_z^2 = \tfrac{1}{4}a^2\left(\dfrac{\theta - \sin\theta\cos\theta}{\theta}\right)$ $\rho_y^2 = \tfrac{1}{4}a^2\left(\dfrac{\theta + \sin\theta\cos\theta}{\theta}\right)$	$I_{z_c y_c} = 0$ $I_{xy} = 0$

	A, X_c, Y_c	I_z, I_y	ρ^2	
12 Circular segment	$A = a^2(\theta - \tfrac{1}{2}\sin 2\theta)$ $X_c = \dfrac{2a}{3}\left(\dfrac{\sin^3\theta}{\theta - \sin\theta\cos\theta}\right)$ $Y_c = 0$	$I_z = \dfrac{Aa^2}{4}\left[1 - \dfrac{2\sin^3\theta\cos\theta}{3(\theta - \sin\theta\cos\theta)}\right]$ $I_y = \dfrac{Aa^2}{4}\left[1 + \dfrac{2\sin^3\theta\cos\theta}{\theta - \sin\theta\cos\theta}\right]$	$\rho_z^2 = \dfrac{a^2}{4}\left[1 - \dfrac{2\sin^3\theta\cos\theta}{3(\theta - \sin\theta\cos\theta)}\right]$ $\rho_y^2 = \dfrac{a^2}{4}\left[1 + \dfrac{2\sin^3\theta\cos\theta}{\theta - \sin\theta\cos\theta}\right]$	$I_{z_cy_c} = 0$ $I_{xy} = 0$
13 Ellipse	$A = \pi ab$ $X_c = a$ $Y_c = b$	$I_{z_c} = \dfrac{\pi}{4}ab^3$ $I_{y_c} = \dfrac{\pi}{4}a^3b$	$\rho_{z_c}^2 = \tfrac{1}{4}b^2$ $\rho_{y_c}^2 = \tfrac{1}{4}a^2$	$I_{z_cy_c} = 0$
14 Semiellipse	$A = \tfrac{1}{2}\pi ab$ $X_c = a$ $Y_c = \dfrac{4b}{3\pi}$	$I_{z_c} = \dfrac{ab^3}{72\pi}(9\pi^2 - 64)$ $I_{y_c} = \dfrac{\pi}{8}a^3b$	$\rho_{z_c}^2 = \dfrac{b^2}{36\pi^2}(9\pi^2 - 64)$ $\rho_{y_c}^2 = \tfrac{1}{4}a^2$	$I_{z_cy_c} = 0$
15 Parabola	$A = \tfrac{4}{3}ab$ $X_c = \tfrac{3}{5}a$ $Y_c = 0$	$I_{z_c} = \tfrac{4}{15}ab^3$ $I_{y_c} = \tfrac{16}{175}a^3b$	$\rho_{z_c}^2 = \tfrac{1}{5}b^2$ $\rho_{y_c}^2 = \tfrac{12}{175}a^2$	$I_{z_cy_c} = 0$
16 Semiparabola	$A = \tfrac{2}{3}ab$ $X_c = \tfrac{3}{8}a$ $Y_c = \tfrac{3}{5}b$	$I_z = \tfrac{2}{7}ab^3$ $I_y = \tfrac{2}{15}a^3b$	$\rho_z^2 = \tfrac{3}{7}b^2$ $\rho_y^2 = \tfrac{1}{5}a^2$	$I_{xy} = \dfrac{A}{4}ab = \tfrac{1}{6}a^2b^2$

TABLE 3.1 Properties of Plane Sections (*Continued*)

Plane section	Area and centroid	Area moment of inertia	Square of radius of gyration	Area product of inertia
17 $Y = \dfrac{h}{b^n} X^n$ nth-degree parabola	$A = \dfrac{bh}{n+1}$ $X_c = \dfrac{n+1}{n+2} b$ $Y_c = \dfrac{h}{2}\left(\dfrac{n+1}{2n+1}\right)$	$I_x = \dfrac{bh^3}{3(3n+1)}$ $I_y = \dfrac{hb^3}{n+3}$	$\rho_x^2 = \dfrac{h^2(n+1)}{3(3n+1)}$ $\rho_y^2 = \dfrac{n+1}{n+3} b^2$	
18 $Y = \dfrac{h}{b^{\frac{1}{n}}} x^{\frac{1}{n}}$ nth-degree parabola	$A = \dfrac{n}{n+1} bh$ $X_c = \dfrac{n+1}{2n+1} b$ $Y_c = \dfrac{n+1}{2(n+2)} h$	$I_x = \dfrac{n}{3(n+3)} bh^3$ $I_y = \dfrac{n}{3n+1} hb^3$	$\rho_x^2 = \dfrac{n+1}{3(n+3)} h^2$ $\rho_y^2 = \dfrac{n+1}{3n+1} b^2$	

TABLE 3.2 Properties of Homogeneous Solid Bodies (*After G. W. Housner and D. E. Hudson.*[25])

The dimensions X_c, Y_c, Z_c are the X, Y, Z coordinates of the centroid, S is the cross-sectional area of the thin rod or hoop in cases 1 to 3, V is the volume, I_x... is the mass moment of inertia with respect to the X... axis, ρ_x... is the radius of gyration with respect to the X... axis, ρ is the mass density of the body.

	Solid body	Volume and centroid	Mass moment of inertia	Radius of gyration squared	Mass product of inertia
1	Thin rod	$V = Sl$ $X_c = \tfrac{1}{2}l$ $Y_c = 0$ $Z_c = 0$	$I_{z_c} = 0$ $I_{y_c} = I_{z_c} = \dfrac{\rho V}{12} l^2$	$\rho_{z_c}^2 = 0$ $\rho_{y_c}^2 = \rho_{z_c}^2 = \tfrac{1}{12} l^2$	$I_{z_c y_c}$, etc. $= 0$
2	Thin circular rod	$V = 2SR\theta$ $X_c = \dfrac{R\sin\theta}{\theta}$ $Y_c = 0$ $Z_c = 0$	$I_x = I_{z_c}$ $= \dfrac{\rho V R^2(\theta - \sin\theta\cos\theta)}{2\theta}$ $I_y = \dfrac{\rho V R^2(\theta + \sin\theta\cos\theta)}{2\theta}$ $I_z = \rho V R^2$	$\rho_x^2 = \rho_{z_c}^2 = \dfrac{R^2(\theta - \sin\theta\cos\theta)}{2\theta}$ $r_y^2 = \dfrac{R^2(\theta + \sin\theta\cos\theta)}{2\theta}$ $\rho_z^2 = R^2$	$I_{z_c y_c}$, etc. $= 0$ I_{zy}, etc. $= 0$
3		$V = 2\pi SR$ $X_c = R$ $Y_c = R$ $Z_c = 0$	$I_{z_c} = I_{y_c} = \dfrac{\rho V}{2} R^2$ $I_{z_c} = \rho V R^2$	$\rho_{z_c}^2 = \rho_{y_c}^2 = \tfrac{1}{2}R^2$ $\rho_{z_c}^2 = R^2$	$I_{z_c y_c}$, etc. $= 0$

TABLE 3.2 Properties of Homogeneous Solid Bodies (*Continued*)

	Solid body	Volume and centroid	Mass moment of inertia	Radius of gyration squared	Mass product of inertia
4	Cube	$V = a^3$ $X_c = \tfrac{1}{2}a$ $Y_c = \tfrac{1}{2}a$ $Z_c = \tfrac{1}{2}a$	$I_{x_c} = I_{y_c} = I_{z_c} = \tfrac{1}{6}\rho V a^2$	$\rho_{x_c}^2 = \rho_{y_c}^2 = \rho_{z_c}^2 = \tfrac{1}{6}a^2$	$I_{z_c y_c}$, etc. $= 0$
5	Rectangular prism	$V = abc$ $X_c = \tfrac{1}{2}a$ $Y_c = \tfrac{1}{2}b$ $Z_c = \tfrac{1}{2}c$	$I_{z_c} = \tfrac{1}{12}\rho V(b^2 + c^2)$	$\rho_{z_c}^2 = \tfrac{1}{12}(b^2 + c^2)$	$I_{z_c y_c}$, etc. $= 0$
6	Right rectangular pyramid	$V = \tfrac{1}{3}abh$ $X_c = 0$ $Y_c = \tfrac{1}{4}h$ $Z_c = 0$	$I_{z_c} = \tfrac{3}{80}\rho V(4b^2 + 3h^2)$ $I_{y_c} = \tfrac{1}{20}\rho V(a^2 + b^2)$	$\rho_{z_c}^2 = \tfrac{3}{80}(4b^2 + 3h^2)$ $\rho_{y_c}^2 = \tfrac{1}{20}(a^2 + b^2)$	$I_{z_c y_c}$, etc. $= 0$

7	 Right circular cone	$V = \tfrac{1}{3}\pi R^2 h$ $X_c = 0$ $Y_c = \tfrac{3}{4}h$ $Z_c = 0$	$I_{z_c} = I_{x_c} = \dfrac{3\rho V}{80}(4R^2 + h^2)$ $I_{y_c} = \tfrac{3}{10}\rho VR^2$	$\rho_{z_c}^{\,2} = \rho_{x_c}^{\,2} = \tfrac{3}{80}(4R^2 + h^2)$ $\rho_{y_c}^{\,2} = \tfrac{3}{10}R^2$	$I_{z_c y_c},\ \text{etc.} = 0$
8	 Right circular cylinder	$V = \pi R^2 h$ $X_c = 0$ $Y_c = \tfrac{1}{2}h$ $Z_c = 0$	$I_{z_c} = I_{x_c} = \tfrac{1}{12}\rho V(3R^2 + h^2)$ $I_{y_c} = \tfrac{1}{2}\rho VR^2$	$\rho_{z_c}^{\,2} = \rho_{x_c}^{\,2} = \tfrac{1}{12}(3R^2 + h^2)$ $\rho_{y_c}^{\,2} = \tfrac{1}{2}R^2$	$I_{z_c y_c},\ \text{etc.} = 0$
9	 Hollow right circular cylinder	$V = \pi h(R_1^2 - R_2^2)$ $X_c = 0$ $Y_c = \tfrac{1}{2}h$ $Z_c = 0$	$I_{z_c} = I_{x_c}$ $= \tfrac{1}{12}\rho V(3R_1^2 + 3R_2^2 + h^2)$ $I_{y_c} = \tfrac{1}{2}\rho V(R_1^2 + R_2^2)$	$\rho_{z_c}^{\,2} = \rho_{x_c}^{\,2} = \tfrac{1}{12}(3R_1^2 + 3R_2^2 + h^2)$ $\rho_{y_c}^{\,2} = \tfrac{1}{2}(R_1^2 + R_2^2)$	$I_{z_c y_c},\ \text{etc.} = 0$

TABLE 3.2 Properties of Homogeneous Solid Bodies (*Continued*)

Solid body	Volume and centroid	Mass moment of inertia	Radius of gyration squared	Mass product of inertia
10 Sphere	$V = \frac{4}{3}\pi R^3$ $X_c = 0$ $Y_c = 0$ $Z_c = 0$	$I_{z_c} = \frac{2}{5}\rho V R^2$ $I_{y_c} = \frac{2}{5}\rho V R^2$ $I_{x_c} = \frac{2}{5}\rho V R^2$	$\rho_{z_c}^2 = \frac{2}{5}R^2$ $\rho_{y_c}^2 = \frac{2}{5}R^2$ $\rho_{x_c}^2 = \frac{2}{5}R^2$	$I_{z_c y_c}$, etc. $= 0$
11 Hollow sphere	$V = \frac{4}{3}\pi(R_1^3 - R_2^3)$ $X_c = 0$ $Y_c = 0$ $Z_c = 0$	$I_z = I_y = I_x$ $= \frac{2}{5}\rho V \dfrac{R_1^5 - R_2^5}{R_1^3 - R_2^3}$	$\rho_z^2 = \rho_y^2 = \rho_x^2$ $= \dfrac{2}{5}\dfrac{R_1^5 - R_2^5}{R_1^3 - R_2^3}$	I_{xy}, etc. $= 0$
12 Hemisphere	$V = \frac{2}{3}\pi R^3$ $X_c = 0$ $Y_c = \frac{3}{8}R$ $Z_c = 0$	$I_z = I_y = I_x = \frac{2}{5}\rho V R^2$	$\rho_z^2 = \rho_y^2 = \rho_x^2 = \frac{2}{5}R^2$	$I_{z_c y_c}$, etc. $= 0$ I_{xy}, etc. $= 0$

13	Ellipsoid	$V = \frac{4}{3}\pi abc$ $X_c = 0$ $Y_c = 0$ $Z_c = 0$	$I_x = \frac{1}{5}\rho V(b^2 + c^2)$ $I_y = \frac{1}{5}\rho V(a^2 + c^2)$ $I_z = \frac{1}{5}\rho V(a^2 + b^2)$	$\rho_x^2 = \frac{1}{5}(b^2 + c^2)$ $\rho_y^2 = \frac{1}{5}(a^2 + c^2)$ $\rho_z^2 = \frac{1}{5}(a^2 + b^2)$	I_{xy}, etc. $= 0$
14	Paraboloid of revolution	$V = \frac{1}{2}\pi R^2 h$ $X_c = \frac{2}{3}h$ $Y_c = 0$ $Z_c = 0$	$I_{x_c} = \frac{1}{3}\rho V R^2$ $I_{y_c} = I_{z_c} = \frac{1}{18}\rho V(3R^2 + h^2)$	$\rho_{x_c}^2 = \frac{1}{3}R^2$ $\rho_{y_c}^2 = \rho_{z_c}^2 = \frac{1}{18}(3R^2 + h^2)$	$I_{z_c y_c}$, etc. $= 0$
15	Elliptic paraboloid	$V = \frac{1}{2}\pi abc$ $X_c = \frac{2}{3}a$ $Y_c = 0$ $Z_c = 0$	$I_{x_c} = \frac{1}{6}\rho V(b^2 + c^2)$ $I_{y_c} = \frac{1}{18}\rho V(3c^2 + a^2)$ $I_{z_c} = \frac{1}{18}\rho V(3b^2 + a^2)$	$\rho_{x_c}^2 = \frac{1}{6}(b^2 + c^2)$ $\rho_{y_c}^2 = \frac{1}{18}(3c^2 + a^2)$ $\rho_{z_c}^2 = \frac{1}{18}(3b^2 + a^2)$	$I_{z_c y_c}$, etc. $= 0$

3.13

CENTER-OF-MASS

Computation of Center-of-Mass. The center-of-mass (or center-of-gravity) is that point located by the vector

$$\mathbf{r}_c = \frac{1}{m} \int_m \mathbf{r}(m)\,dm \qquad (3.7)$$

where $\mathbf{r}(m)$ is the radius vector of the element of mass dm. The center-of-mass of a body in a cartesian coordinate system X, Y, Z is located at

$$X_c = \frac{1}{m} \int_V X(V)\rho(V)\,dV$$

$$Y_c = \frac{1}{m} \int_V Y(V)\rho(V)\,dV \qquad (3.8)$$

$$Z_c = \frac{1}{m} \int_V Z(V)\rho(V)\,dV$$

where $X(V)$, $Y(V)$, $Z(V)$ are the X, Y, Z coordinates of the element of volume dV and m is the mass of the body.

If the body can be divided into elements whose centers-of-mass are known, the center-of-mass of the entire body having a mass m is located by equations of the following type:

$$X_c = \frac{1}{m}(X_{c1}m_1 + X_{c2}m_2 + \cdots + X_{cn}m_n), \text{ etc.} \qquad (3.9)$$

where X_{c1} is the X coordinate of the center-of-mass of element m_1. Tables (see Tables 3.1 and 3.2) which specify the location of centers of area and volume (called centroids) for simple sections and solid bodies often are an aid in dividing the body into the submasses indicated in the above equation. The centroid and center-of-mass of an element are coincident when the density of the material is uniform throughout the element.

Experimental Determination of Center-of-Mass. The location of the center-of-mass is normally measured indirectly by locating the center-of-gravity of the body, and may be found in various ways. Theoretically, if the body is suspended by a flexible wire attached successively at different points on the body, all lines represented by the wire in its various positions when extended inwardly into the body intersect at the center-of-gravity. Two such lines determine the center-of-gravity, but more may be used as a check. There are practical limitations to this method in that the point of intersection often is difficult to designate.

Other techniques are based on the balancing of the body on point or line supports. A point support locates the center-of-gravity along a vertical line through the point; a line support locates it in a vertical plane through the line. The intersection of such lines or planes determined with the body in various positions locates the center-of-gravity. The greatest difficulty with this technique is the maintenance of the stability of the

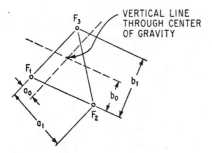

FIGURE 3.2 Three-scale method of locating the center-of-gravity of a body. The vertical forces F_1, F_2, F_3 at the scales result from the weight of the body. The vertical line located by the distances a_0 and b_0 [see Eqs. (3.10)] passes through the center-of-gravity of the body.

body while it is balanced, particularly where the height of the body is great relative to a horizontal dimension. If a perfect point or edge support is used, the equilibrium position is inherently unstable. It is only if the support has width that some degree of stability can be achieved, but then a resulting error in the location of the line or plane containing the center-of-gravity can be expected.

Another method of locating the center-of-gravity is to place the body in a stable position on three scales. From static moments the vector weight of the body is the resultant of the measured forces at the scales, as shown in Fig. 3.2. The vertical line through the center-of-gravity is located by the distances a_0 and b_0:

$$a_0 = \frac{F_2}{F_1 + F_2 + F_3} a_1$$

$$b_0 = \frac{F_3}{F_1 + F_2 + F_3} b_1$$

(3.10)

This method cannot be used with more than three scales.

MOMENT AND PRODUCT OF INERTIA

Computation of Moment and Product of Inertia.[2] The moments of inertia of a rigid body with respect to the orthogonal axes X, Y, Z fixed in the body are

$$I_{xx} = \int_m (Y^2 + Z^2)\, dm \qquad I_{yy} = \int_m (X^2 + Z^2)\, dm \qquad I_{zz} = \int_m (X^2 + Y^2)\, dm \qquad (3.11)$$

where dm is the infinitesimal element of mass located at the coordinate distances X, Y, Z; and the integration is taken over the mass of the body. Similarly, the products of inertia are

$$I_{xy} = \int_m XY\, dm \qquad I_{xz} = \int_m XZ\, dm \qquad I_{yz} = \int_m YZ\, dm \qquad (3.12)$$

It is conventional in rigid body mechanics to take the center of coordinates at the center-of-mass of the body. Unless otherwise specified, this location is assumed, and the moments of inertia and products of inertia refer to axes through the center-of-mass of the body. For a unique set of axes, the products of inertia vanish. These axes are called the principal inertial axes of the body. The moments of inertia about these axes are called the principal moments of inertia. The moments of inertia of a rigid body can be defined in terms of radii of gyration as follows:

$$I_{xx} = m\rho_x^2 \qquad I_{yy} = m\rho_y^2 \qquad I_{zz} = m\rho_z^2 \qquad (3.13)$$

where I_{xx} . . . are the moments of inertia of the body as defined by Eqs. (3.11), m is the mass of the body, and ρ_x . . . are the radii of gyration. The radius of gyration has the dimension of length, and often leads to convenient expressions in dynamics of rigid bodies when distances are normalized to an appropriate radius of gyration. Solid bodies of various shapes have characteristic radii of gyration which sometimes are useful intuitively in evaluating dynamic conditions.

Unless the body has a very simple shape, it is laborious to evaluate the integrals of Eqs. (3.11) and (3.12). The problem is made easier by subdividing the body into parts for which simplified calculations are possible. The moments and products of inertia of the body are found by first determining the moments and products of inertia for the individual parts with respect to appropriate reference axes chosen in the parts, and then summing the contributions of the parts. This is done by selecting axes through the centers-of-mass of the parts, and then determining the moments and products of inertia of the parts relative to these axes. Then the moments and products of inertia are transferred to the axes chosen through the center-of-mass of the whole body, and the transferred quantities summed. In general, the transfer involves two sets of nonparallel coordinates whose centers are displaced. Two transformations are required as follows.

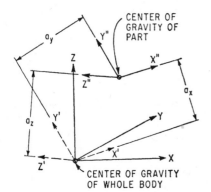

CENTER OF GRAVITY OF PART

CENTER OF GRAVITY OF WHOLE BODY

FIGURE 3.3 Axes required for moment and product of inertia transformations. Moments and products of inertia with respect to the axes X'', Y'', Z'' are transferred to the mutually parallel axes X', Y', Z' by Eqs. (3.14) and (3.15), and then to the inclined axes X, Y, Z by Eqs. (3.16) and (3.17).

Transformation to Parallel Axes. Referring to Fig. 3.3, suppose that X, Y, Z is a convenient set of axes for the moment of inertia of the whole body with its origin at the center-of-mass. The moments and products of inertia for a part of the body are $I_{x''x''}$, $I_{y''y''}$, $I_{z''z''}$, $I_{x''y''}$, $I_{x''z''}$, and $I_{y''z''}$, taken with respect to a set of axes X'', Y'', Z'' fixed in the part and having their center at the center-of-mass of the part. The axes X', Y', Z' are chosen parallel to X'', Y'', Z'' with their origin at the center-of-mass of the body. The perpendicular distance between the X'' and X' axes is a_x; that between Y'' and Y' is a_y; that between Z'' and Z' is a_z. The moments and products of inertia of the part of mass m_n with respect to the X', Y', Z' axes are

$$I_{x'x'} = I_{x''x''} + m_n a_x^2$$

$$I_{y'y'} = I_{y''y''} + m_n a_y^2 \qquad (3.14)$$

$$I_{z'z'} = I_{z''z''} + m_n a_z^2$$

The corresponding products of inertia are

$$I_{x'y'} = I_{x''y''} + m_n a_x a_y$$

$$I_{x'z'} = I_{x''z''} + m_n a_x a_z \qquad (3.15)$$

$$I_{y'z'} = I_{y''z''} + m_n a_y a_z$$

If X'', Y'', Z'' are the principal axes of the part, the product of inertia terms on the right-hand side of Eqs. (3.15) are zero.

Transformation to Inclined Axes. The desired moments and products of inertia with respect to axes X, Y, Z are now obtained by a transformation theorem relating the properties of bodies with respect to inclined sets of axes whose centers coincide. This theorem makes use of the direction cosines λ for the respective sets of axes. For example, $\lambda_{xx'}$ is the cosine of the angle between the X and X' axes. The expressions for the moments of inertia are

$$I_{xx} = \lambda_{xx}{}^2 I_{x'x'} + \lambda_{xy}{}^2 I_{y'y'} + \lambda_{xz}{}^2 I_{z'z'} - 2\lambda_{xx}\lambda_{xy}I_{x'y'} - 2\lambda_{xx}\lambda_{xz}I_{x'z'} - 2\lambda_{xy}\lambda_{xz}I_{y'z'}$$

$$I_{yy} = \lambda_{yx}{}^2 I_{x'x'} + \lambda_{yy}{}^2 I_{y'y'} + \lambda_{yz}{}^2 I_{z'z'} - 2\lambda_{yx}\lambda_{yy}I_{x'y'} - 2\lambda_{yx}\lambda_{yz}I_{x'z'} - 2\lambda_{yy}\lambda_{yz}I_{y'z'} \quad (3.16)$$

$$I_{zz} = \lambda_{zx}{}^2 I_{x'x'} + \lambda_{zy}{}^2 I_{y'y'} + \lambda_{zz}{}^2 I_{z'z'} - 2\lambda_{zx}\lambda_{zy}I_{x'y'} - 2\lambda_{zx}\lambda_{zz}I_{x'z'} - 2\lambda_{zy}\lambda_{zz}I_{y'z'}$$

The corresponding products of inertia are

$$-I_{xy} = \lambda_{xx}\lambda_{yx}I_{x'x'} + \lambda_{xy}\lambda_{yy}I_{y'y'} + \lambda_{xz}\lambda_{yz}I_{z'z'} - (\lambda_{xx}\lambda_{yy} + \lambda_{xy}\lambda_{yx})I_{x'y'}$$
$$- (\lambda_{xy}\lambda_{yz} + \lambda_{xz}\lambda_{yy})I_{y'z'} - (\lambda_{xz}\lambda_{yx} + \lambda_{xx}\lambda_{yz})I_{x'z'}$$

$$-I_{xz} = \lambda_{xx}\lambda_{zx}I_{x'x'} + \lambda_{xy}\lambda_{zy}I_{y'y'} + \lambda_{xz}\lambda_{zz}I_{z'z'} - (\lambda_{xx}\lambda_{zy} + \lambda_{xy}\lambda_{zx})I_{x'y'}$$
$$- (\lambda_{xy}\lambda_{zz} + \lambda_{xz}\lambda_{zy})I_{y'z'} - (\lambda_{xx}\lambda_{zz} + \lambda_{xz}\lambda_{zx})I_{x'z'} \quad (3.17)$$

$$-I_{yz} = \lambda_{yx}\lambda_{zx}I_{x'x'} + \lambda_{yy}\lambda_{zy}I_{y'y'} + \lambda_{yz}\lambda_{zz}I_{z'z'} - (\lambda_{yx}\lambda_{zy} + \lambda_{yy}\lambda_{zx})I_{x'y'}$$
$$- (\lambda_{yy}\lambda_{zz} + \lambda_{yz}\lambda_{zy})I_{y'z'} - (\lambda_{yz}\lambda_{zx} + \lambda_{yx}\lambda_{zz})I_{x'z'}$$

Experimental Determination of Moments of Inertia. The moment of inertia of a body about a given axis may be found experimentally by suspending the body as a pendulum so that rotational oscillations about that axis can occur. The period of free oscillation is then measured, and is used with the geometry of the pendulum to calculate the moment of inertia.

Two types of pendulums are useful: the compound pendulum and the torsional pendulum. When using the compound pendulum, the body is supported from two overhead points by wires, illustrated in Fig. 3.4. The distance l is measured between the axis of support $O-O$ and a parallel axis $C-C$ through the center-of-gravity of the body. The moment of inertia about $C-C$ is given by

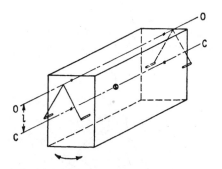

$$I_{cc} = ml^2\left[\left(\frac{\tau_0}{2\pi}\right)^2\left(\frac{g}{l}\right) - 1\right] \quad (3.18)$$

FIGURE 3.4 Compound pendulum method of determining moment of inertia. The period of oscillation of the test body about the horizontal axis $O-O$ and the perpendicular distance l between the axis $O-O$ and the parallel axis $C-C$ through the center-of-gravity of the test body give I_{cc} by Eq. (3.18).

where τ_0 is the period of oscillation in seconds, l is the pendulum length in inches, g is the gravitational acceleration in in./sec^2, and m is the mass in lb-sec^2/in., yielding a moment of inertia in lb-in.-sec^2.

The accuracy of the above method is dependent upon the accuracy with which the distance l is known. Since the center-of-gravity often is an inaccessible point, a direct measurement of l may not be practicable. However, a change in l can be measured quite readily. If the experiment is repeated with a different support axis $O'-O'$, the length l becomes $l + \Delta l$ and the period of oscillation becomes τ_0'. Then, the distance l can be written in terms of Δl and the two periods τ_0, τ_0':

$$l = \Delta l \left[\frac{(\tau_0'^2/4\pi^2)(g/\Delta l) - 1}{[(\tau_0^2 - \tau_0'^2)/4\pi^2][g/\Delta l] - 1} \right] \qquad (3.19)$$

This value of l can be substituted into Eq. (3.18) to compute I_{cc}.

Note that accuracy is not achieved if l is much larger than the radius of gyration ρ_c of the body about the axis C–C ($I_{cc} = m\rho_c^2$). If l is large, then $(\tau_0/2\pi)^2 \simeq l/g$ and the expression in brackets in Eq. (3.18) is very small; thus, it is sensitive to small errors in the measurement of both τ_0 and l. Consequently, it is highly desirable that the distance l be chosen as small as convenient, preferably with the axis O–O passing through the body.

A torsional pendulum may be constructed with the test body suspended by a single torsional spring (in practice, a rod or wire) of known stiffness, or by three flexible wires. A solid body supported by a single torsional spring is shown in Fig. 3.5. From the known torsional stiffness k_t and the measured period of torsional oscillation τ, the moment of inertia of the body about the vertical torsional axis is

$$I_{cc} = \frac{k_t \tau^2}{4\pi^2} \qquad (3.20)$$

A platform may be constructed below the torsional spring to carry the bodies to be measured, as shown in Fig. 3.6. By repeating the experiment with two different bodies placed on the platform, it becomes unnecessary to measure the torsional stiffness k_t. If a body with a *known* moment of inertia I_1 is placed on the platform and an oscillation period τ_1 results, the moment of inertia I_2 of a body which produces a period τ_2 is given by

$$I_2 = I_1 \left[\frac{(\tau_2/\tau_0)^2 - 1}{(\tau_1/\tau_0)^2 - 1} \right] \qquad (3.21)$$

where τ_0 is the period of the pendulum composed of platform alone.

A body suspended by three flexible wires, called a trifilar pendulum, as shown in Fig. 3.7, offers some utilitarian advantages. Designating the perpendicular distances

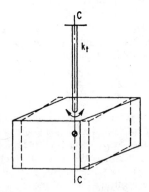

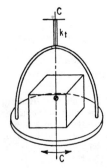

FIGURE 3.5 Torsional pendulum method of determining moment of inertia. The period of torsional oscillation of the test body about the vertical axis C–C passing through the center-of-gravity and the torsional spring constant k_t give I_{cc} by Eq. (3.20).

FIGURE 3.6 A variation of the torsional pendulum method shown in Fig. 3.5 wherein a light platform is used to carry the test body. The moment of inertia I_{cc} is given by Eq. (3.20).

of the wires to the vertical axis $C-C$ through the center-of-gravity of the body by R_1, R_2, R_3, the angles between wires by ϕ_1, ϕ_2, ϕ_3, and the length of each wire by l, the moment of inertia about axis $C-C$ is

$$I_{cc} = \frac{mgR_1R_2R_3\tau^2}{4\pi^2 l} \frac{R_1 \sin \phi_1 + R_2 \sin \phi_2 + R_3 \sin \phi_3}{R_2R_3 \sin \phi_1 + R_1R_3 \sin \phi_2 + R_1R_2 \sin \phi_3} \qquad (3.22)$$

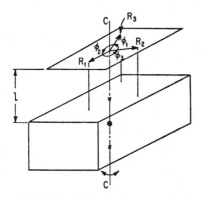

Apparatus that is more convenient for repeated use embodies a light platform supported by three equally spaced wires. The body whose moment of inertia is to be measured is placed on the platform with its center-of-gravity equidistant from the wires. Thus $R_1 = R_2 = R_3 = R$ and $\phi_1 = \phi_2 = \phi_3 = 120°$. Substituting these relations in Eq. (3.22), the moment of inertia about the vertical axis $C-C$ is

$$I_{cc} = \frac{mgR^2\tau^2}{4\pi^2 l} \qquad (3.23)$$

FIGURE 3.7 Trifilar pendulum method of determining moment of inertia. The period of torsional oscillation of the test body about the vertical axis $C-C$ passing through the center-of-gravity and the geometry of the pendulum give I_{cc} by Eq. (3.22); with a simpler geometry, I_{cc} is given by Eq. (3.23).

where the mass m is the sum of the masses of the test body and the platform. The moment of inertia of the platform is subtracted from the test result to obtain the moment of inertia of the body being measured. It becomes unnecessary to know the distances R and l in Eq. (3.23) if the period of oscillation is measured with the platform empty, with the body being measured on the platform, and with a second body of known mass m_1 and known moment of inertia I_1 on the platform. Then the desired moment of inertia I_2 is

$$I_2 = I_1 \left[\frac{[1 + (m_2/m_0)][\tau_2/\tau_0]^2 - 1}{[1 + (m_1/m_0)][\tau_1/\tau_0]^2 - 1} \right] \qquad (3.24)$$

where m_0 is the mass of the unloaded platform, m_2 is the mass of the body being measured, τ_0 is the period of oscillation with the platform unloaded, τ_1 is the period when loaded with known body of mass m_1, and τ_2 is the period when loaded with the unknown body of mass m_2.

Experimental Determination of Product of Inertia. The experimental determination of a product of inertia usually requires the measurement of moments of inertia. (An exception is the balancing machine technique described later.) If possible, symmetry of the body is used to locate directions of principal inertial axes, thereby simplifying the relationship between the moments of inertia as known and the products of inertia to be found. Several alternative procedures are described below, depending on the number of principal inertia axes whose directions are known. Knowledge of two principal axes implies a knowledge of all three since they are mutually perpendicular.

If the directions of all three principal axes (X', Y', Z') are known and it is desirable to use another set of axes (X, Y, Z), Eqs. (3.16) and (3.17) may be simplified because the products of inertia with respect to the principal directions are zero. First, the three principal moments of inertia ($I_{x'x'}$, $I_{y'y'}$, $I_{z'z'}$) are measured by one of the above techniques; then the moments of inertia with respect to the X, Y, Z axes are

$$I_{xx} = \lambda_{xx}{}^2 I_{x'x'} + \lambda_{xy}{}^2 I_{y'y'} + \lambda_{xz}{}^2 I_{z'z'}$$

$$I_{yy} = \lambda_{yx}{}^2 I_{x'x'} + \lambda_{yy}{}^2 I_{y'y'} + \lambda_{yz}{}^2 I_{z'z'} \qquad (3.25)$$

$$I_{zz} = \lambda_{zx}{}^2 I_{x'x'} + \lambda_{zy}{}^2 I_{y'y'} + \lambda_{zz}{}^2 I_{z'z'}$$

The products of inertia with respect to the X, Y, Z axes are

$$-I_{xy} = \lambda_{xx}\lambda_{yx} I_{x'x'} + \lambda_{xy}\lambda_{yy} I_{y'y'} + \lambda_{xz}\lambda_{yz} I_{z'z'}$$

$$-I_{xz} = \lambda_{xx}\lambda_{zx} I_{x'x'} + \lambda_{xy}\lambda_{zy} I_{y'y'} + \lambda_{xz}\lambda_{zz} I_{z'z'} \qquad (3.26)$$

$$-I_{yz} = \lambda_{yx}\lambda_{zx} I_{x'x'} + \lambda_{yy}\lambda_{zy} I_{y'y'} + \lambda_{yz}\lambda_{zz} I_{z'z'}$$

The direction of one principal axis Z may be known from symmetry. The axis through the center-of-gravity perpendicular to the plane of symmetry is a principal axis. The product of inertia with respect to X and Y axes, located in the plane of symmetry, is determined by first establishing another axis X' at a counterclockwise angle θ from X, as shown in Fig. 3.8. If the three moments of inertia I_{xx}, $I_{x'x'}$, and I_{yy} are measured by any applicable means, the product of inertia I_{xy} is

$$I_{xy} = \frac{I_{xx}\cos^2\theta + I_{yy}\sin^2\theta - I_{x'x'}}{\sin 2\theta} \qquad (3.27)$$

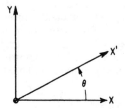

FIGURE 3.8 Axes required for determining the product of inertia with respect to the axes X and Y when Z is a principal axis of inertia. The moments of inertia about the axes X, Y, and X', where X' is in the plane of X and Y at a counterclockwise angle θ from X, give I_{xy} by Eq. (3.27).

where $0 < \theta < \pi$. For optimum accuracy, θ should be approximately $\pi/4$ or $3\pi/4$. Since the third axis Z is a principal axis, I_{xz} and I_{yz} are zero.

Another method is illustrated in Fig. 3.9.[3,4] The plane of the X and Z axes is a plane of symmetry, or the Y axis is otherwise known to be a principal axis of inertia. For determining I_{xz}, the body is suspended by a cable so that the Y axis is horizontal and the Z axis is vertical. Torsional stiffness about the Z axis is provided by four springs acting in the Y direction at the points shown. The body is oscillated about the Z axis with various positions of the springs so that the angle θ can be varied. The spring stiffnesses and locations must be such that there is no net force in the Y direction due to a rotation about the Z axis. In general, there is coupling between rotations about the X and Z axes, with the result that oscillations about both axes occur as a result of an initial rotational displacement about the Z axis. At some particular value of $\theta = \theta_0$, the two rotations are uncoupled; i.e., oscillation about the Z axis does not cause oscillation about the X axis. Then

$$I_{xz} = I_{zz} \tan \theta_0 \qquad (3.28)$$

The moment of inertia I_{zz} can be determined by one of the methods described under *Experimental Determination of Moments of Inertia*.

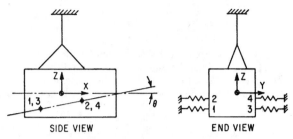

SIDE VIEW   END VIEW

FIGURE 3.9 Method of determining the product of inertia with respect to the axes X and Z when Y is a principal axis of inertia. The test body is oscillated about the vertical Z axis with torsional stiffness provided by the four springs acting in the Y direction at the points shown. There should be no net force on the test body in the Y direction due to a rotation about the Z axis. The angle θ is varied until, at some value of $\theta = \theta_0$, oscillations about X and Z are uncoupled. The angle θ_0 and the moment of inertia about the Z axis give I_{xz} by Eq. (3.28).

When the moments and product of inertia with respect to a pair of axes X and Z in a principal plane of inertia XZ are known, the orientation of a principal axis P is given by

$$\theta_p = \tfrac{1}{2} \tan^{-1}\!\left(\frac{2I_{xz}}{I_{zz} - I_{xx}} \right) \tag{3.29}$$

where θ_p is the counterclockwise angle from the X axis to the P axis. The second principal axis in this plane is at $\theta_p + 90°$.

Consider the determination of products of inertia when the directions of all principal axes of inertia are unknown. In one method, the moments of inertia about two independent sets of three mutually perpendicular axes are measured, and the direction cosines between these sets of axes are known from the positions of the axes. The values for the six moments of inertia and the nine direction cosines are then substituted into Eqs. (3.16) and (3.17). The result is six linear equations in the six unknown products of inertia, from which the values of the desired products of inertia may be found by simultaneous solution of the equations. This method leads to experimental errors of relatively large magnitude because each product of inertia is, in general, a function of all six moments of inertia, each of which contains an experimental error.

An alternative method is based upon the knowledge that one of the principal moments of inertia of a body is the largest and another is the smallest that can be obtained for any axis through the center-of-gravity. A trial-and-error procedure can be used to locate the orientation of the axis through the center-of-gravity having the maximum and/or minimum moment of inertia. After one or both are located, the moments and products of inertia for any set of axes are found by the techniques previously discussed.

The products of inertia of a body also may be determined by rotating the body at a constant angular velocity Ω about an axis passing through the center of gravity, as illustrated in Fig. 3.10. This method is similar to the balancing machine technique used to balance a body dynamically (see Chap. 39). If the bearings are a distance l apart and the dynamic reactions F_x and F_y are measured, the products of inertia are

$$I_{xz} = -\frac{F_x l}{\Omega^2} \qquad I_{yz} = -\frac{F_y l}{\Omega^2} \qquad\qquad (3.30)$$

Limitations to this method are (1) the size of the body that can be accommodated by the balancing machine and (2) the angular velocity that the body can withstand without damage from centrifugal forces. If the angle between the Z axis and a principal axis of inertia is small, high rotational speeds may be necessary to measure the reaction forces accurately.

PROPERTIES OF RESILIENT SUPPORTS

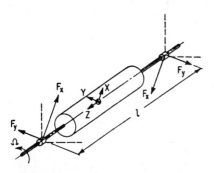

FIGURE 3.10 Balancing machine technique for determining products of inertia. The test body is rotated about the Z axis with angular velocity Ω. The dynamic reactions F_x and F_y measured at the bearings, which are a distance l apart, give I_{xz} and I_{yz} by Eq. (3.30).

A resilient support is considered to be a three-dimensional element having two terminals or end connections. When the end connections are moved one relative to the other in any direction, the element resists such motion. In this chapter, the element is considered to be massless; the force that resists relative motion across the element is considered to consist of a spring force that is directly proportional to the relative displacement (deflection across the element) and a damping force that is directly proportional to the relative velocity (velocity across the element). Such an element is defined as a *linear resilient support.* Nonlinear elements are discussed in Chap. 4; elements with mass are discussed in Chap. 30; and nonlinear damping is discussed in Chaps. 2 and 30.

In a single degree-of-freedom system or in a system having constraints on the paths of motion of elements of the system (Chap. 2), the resilient element is constrained to deflect in a given direction and the properties of the element are defined with respect to the force opposing motion in this direction. In the absence of such constraints, the application of a force to a resilient element generally causes a motion in a different direction. The *principal elastic axes* of a resilient element are those axes for which the element, when unconstrained, experiences a deflection co-lineal with the direction of the applied force.[5] Any axis of symmetry is a principal elastic axis.

In rigid body dynamics, the rigid body sometimes vibrates in modes that are coupled by the properties of the resilient elements as well as by their location. For example, if the body experiences a static displacement x in the direction of the X axis only, a resilient element opposes this motion by exerting a force $k_{xx}x$ on the body in the direction of the X axis, where one subscript on the spring constant k indicates the direction of the force exerted by the element and the other subscript indicates the direction of the deflection. If the X direction is not a principal elastic direction of the element and the body experiences a static displacement x in the X direction, the body is acted upon by a force $k_{yx}x$ in the Y direction if no displacement y is permitted. The stiffnesses have reciprocal properties; i.e., $k_{xy} = k_{yx}$. In general, the stiffnesses in the directions of the coordinate axes can be expressed in terms of (1) principal stiffnesses and (2) the angles between the coordinate axes and the principal elastic axes of the element. (See Chap. 30 for a detailed discussion of a biaxial stiffness element.) Therefore, the stiffness of a resilient element can be represented pictorially by the combination of three mutually perpendicular, idealized springs oriented along the principal elastic directions of the resilient element. Each spring has a stiffness equal to the principal stiffness represented.

A resilient element is assumed to have damping properties such that each spring representing a value of principal stiffness is paralleled by an idealized viscous damper, each damper representing a value of principal damping. Hence, coupling through damping exists in a manner similar to coupling through stiffness. Consequently, the viscous damping coefficient c is analogous to the spring coefficient k; i.e., the force exerted by the damping of the resilient element in response to a velocity $\dot{x}$ is $c_{xx}\dot{x}$ in the direction of the X axis and $c_{yx}\dot{x}$ in the direction of the Y axis if $\dot{y}$ is zero. Reciprocity exists; i.e., $c_{xy} = c_{yx}$.

The point of intersection of the principal elastic axes of a resilient element is designated as the *elastic center of the resilient element.* The elastic center is important since it defines the theoretical point location of the resilient element for use in the equations of motion of a resiliently supported rigid body. For example, the torque on the rigid body about the Y axis due to a force $k_{xx}x$ transmitted by a resilient element in the X direction is $k_{xx}a_z x$, where a_z is the Z coordinate of the elastic center of the resilient element.

In general, it is assumed that a resilient element is attached to the rigid body by means of "ball joints"; i.e., the resilient element is incapable of applying a couple to the body. If this assumption is not made, a resilient element would be represented not only by translational springs and dampers along the principal elastic axes but also by torsional springs and dampers resisting rotation about the principal elastic directions.

Figure 3.11 shows that the torsional elements usually can be neglected. The torque which acts on the rigid body due to a rotation β of the body and a rotation $\boldsymbol{\beta}$ of the support is $(k_t + a_z^2 k_x)(\beta - \boldsymbol{\beta})$, where k_t is the torsional spring constant in the β direction. The torsional stiffness k_t usually is much smaller than $a_z^2 k_x$ and can be neglected. Treatment of the general case indicates that if the torsional stiffnesses of the resilient element are small compared with the product of the translational stiffnesses times the square of distances from the elastic center of the resilient element to the center-of-gravity of the rigid body, the torsional stiffnesses have a negligible effect on the vibrational behavior of the body. The treatment of torsional dampers is completely analogous.

EQUATIONS OF MOTION FOR A RESILIENTLY SUPPORTED RIGID BODY

The differential equations of motion for the rigid body are given by Eqs. (3.2) and (3.3), where the **F**'s and **M**'s represent the forces and moments acting on the body, either directly or through the resilient supporting elements. Figure 3.12 shows a view of a rigid body at rest with an inertial set of axes $\bar{X}, \bar{Y}, \bar{Z}$ and a coincident set of axes X, Y, Z fixed in the rigid body, both sets of axes passing through the center-of-mass. A typical resilient element (2) is represented by parallel spring and viscous damper combinations arranged respectively parallel with the $\bar{X}, \bar{Y}, \bar{Z}$ axes. Another resilient element (1) is shown with its principal axes not parallel with $\bar{X}, \bar{Y}, \bar{Z}$.

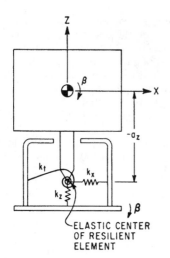

ELASTIC CENTER
OF RESILIENT
ELEMENT

FIGURE 3.11 Pictorial representation of the properties of an undamped resilient element in the XZ plane including a torsional spring k_t. An analysis of the motion of the supported body in the XZ plane shows that the torsional spring can be neglected if $k_t \ll a_z^2 k_x$.

The displacement of the center-of-gravity of the body in the $\bar{X}, \bar{Y}, \bar{Z}$ directions is in Fig. 3.1 indicated by x_c, y_c, z_c, respectively; and rotation of the rigid body about these axes is indicated by α, β, γ, respectively. In Fig. 3.12, each resilient element is represented by three mutually perpendicular spring-damper combinations. One end of each such combination is attached to the rigid body; the other end is considered to be attached to a foundation whose corresponding translational displacement is defined by u, v, w in the $\bar{X}, \bar{Y}, \bar{Z}$ directions, respectively, and whose rotational displacement about these axes is defined by α, β, γ, respectively. The point of attachment of each of the idealized resilient elements is located at the coordinate distances a_x, a_y, a_z of the elastic center of the resilient element.

Consider the rigid body to experience a translational displacement x_c of its center-of-gravity and no other displacement, and neglect the effects of the viscous dampers. The force developed by a resilient element has the effect of a force $-k_{xx}(x_c - u)$ in the X direction, a moment $k_{xx}(x_c - u)a_y$ in the γ coordinate (about the Z axis), and a moment $-k_{xx}(x_c - u)a_z$ in the β coordinate (about the Y axis). Furthermore, the coupling stiffness causes a force $-k_{xy}(x_c - u)$ in the Y direction and a force $-k_{xz}(x_c - u)$ in the Z direction. These forces have the moments $k_{xy}(x_c - u)a_z$ in the α coordinate; $-k_{xy}(x_c - u)a_x$ in the γ coordinate; $k_{xz}(x_c - u)a_x$ in the β coordinate; and $-k_{xz}(x_c - u)a_y$ in the α coordinate. By considering in a similar manner the forces and moments developed by a resilient element for successive displacements of the rigid body in the three translational and three rotational coordinates, and summing over the number of resilient elements, the equations of motion are written as follows:[6–8]

$$m\ddot{x}_c + \Sigma k_{xx}(x_c - u) + \Sigma k_{xy}(y_c - v) + \Sigma k_{xz}(z_c - w)$$
$$+ \Sigma(k_{xz}a_y - k_{xy}a_z)(\alpha - \alpha) + \Sigma(k_{xx}a_z - k_{xz}a_x)(\beta - \beta)$$
$$+ \Sigma(k_{xy}a_x - k_{xx}a_y)(\gamma - \gamma) = F_x \tag{3.31a}$$

$$I_{xx}\ddot{\alpha} - I_{xy}\ddot{\beta} - I_{xz}\ddot{\gamma} + \Sigma(k_{xz}a_y - k_{xy}a_z)(x_c - u)$$
$$+ \Sigma(k_{yz}a_y - k_{yy}a_z)(y_c - v) + \Sigma(k_{zz}a_y - k_{yz}a_z)(z_c - w)$$
$$+ \Sigma(k_{yy}a_z^2 + k_{zz}a_y^2 - 2k_{yz}a_ya_z)(\alpha - \alpha)$$
$$+ \Sigma(k_{xz}a_ya_z + k_{yz}a_xa_z - k_{zz}a_xa_y - k_{xy}a_z^2)(\beta - \beta)$$
$$+ \Sigma(k_{xy}a_ya_z + k_{yz}a_xa_y - k_{yy}a_xa_z - k_{xz}a_y^2)(\gamma - \gamma) = M_x \tag{3.31b}$$

$$m\ddot{y}_c + \Sigma k_{xy}(x_c - u) + \Sigma k_{yy}(y_c - v) + \Sigma k_{yz}(z_c - w)$$
$$+ \Sigma(k_{yz}a_y - k_{yy}a_z)(\alpha - \alpha) + \Sigma(k_{xy}a_z - k_{yz}a_x)(\beta - \beta)$$
$$+ \Sigma(k_{yy}a_x - k_{xy}a_y)(\gamma - \gamma) = F_y \tag{3.31c}$$

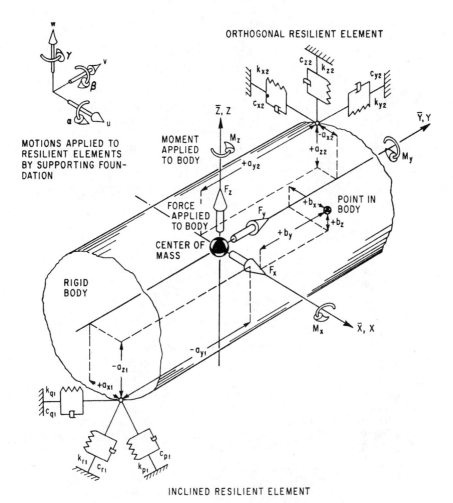

FIGURE 3.12 Rigid body at rest supported by resilient elements, with inertial axes $\bar{X}, \bar{Y}, \bar{Z}$ and coincident reference axes X, Y, Z passing through the center-of-mass. The forces F_x, F_y, F_z and the moments M_x, M_y, M_z are applied directly to the body; the translations u, v, w and rotations α, β, γ in and about the X, Y, Z axes, respectively, are applied to the resilient elements located at the coordinates a_x, a_y, a_z. The principal directions of resilient element (2) are parallel to the $\bar{X}, \bar{Y}, \bar{Z}$ axes (orthogonal), and those of resilient element (1) are not parallel to the $\bar{X}, \bar{Y}, \bar{Z}$ axes (inclined).

$$
I_{yy}\ddot{\beta} - I_{xy}\ddot{\alpha} - I_{yz}\ddot{\gamma} + \Sigma(k_{xx}a_z - k_{xz}a_x)(x_c - u)
$$
$$
+ \Sigma(k_{xy}a_z - k_{yz}a_x)(y_c - v) + \Sigma(k_{xz}a_z - k_{zz}a_x)(z_c - w)
$$
$$
+ \Sigma(k_{xz}a_ya_z + k_{yz}a_xa_z - k_{zz}a_xa_y - k_{xy}a_z^2)(\alpha - \alpha)
$$
$$
+ \Sigma(k_{xx}a_z^2 + k_{zz}a_x^2 - 2k_{xz}a_xa_z)(\beta - \beta)
$$
$$
+ \Sigma(k_{xy}a_xa_z + k_{xz}a_xa_y - k_{xx}a_ya_z - k_{yz}a_x^2)(\gamma - \gamma) = M_y \qquad (3.31d)
$$

$$m\ddot{z}_c + \Sigma k_{xz}(x_c - u) + \Sigma k_{yz}(y_c - v) + \Sigma k_{zz}(z_c - w)$$
$$+ \Sigma(k_{zz}a_y - k_{yz}a_z)(\alpha - \boldsymbol{\alpha}) + \Sigma(k_{xz}a_z - k_{zz}a_x)(\beta - \boldsymbol{\beta})$$
$$+ \Sigma(k_{yz}a_x - k_{xz}a_y)(\gamma - \boldsymbol{\gamma}) = F_z \tag{3.31e}$$

$$I_{zz}\ddot{\gamma} - I_{xz}\ddot{\alpha} - I_{yz}\ddot{\beta} + \Sigma(k_{xy}a_x - k_{xx}a_y)(x_c - u)$$
$$+ \Sigma(k_{yy}a_x - k_{xy}a_y)(y_c - v) + \Sigma(k_{yz}a_x - k_{xz}a_y)(z_c - w)$$
$$+ \Sigma(k_{xy}a_y a_z + k_{yz}a_x a_y - k_{yy}a_x a_z - k_{xz}a_y^2)(\alpha - \boldsymbol{\alpha})$$
$$+ \Sigma(k_{xy}a_x a_z + k_{xz}a_x a_y - k_{xx}a_y a_z - k_{yz}a_x^2)(\beta - \boldsymbol{\beta})$$
$$+ \Sigma(k_{xx}a_y^2 + k_{yy}a_x^2 - 2k_{xy}a_x a_y)(\gamma - \boldsymbol{\gamma}) = M_z \tag{3.31f}$$

where the moments and products of inertia are defined by Eqs. (3.11) and (3.12) and the stiffness coefficients are defined as follows:

$$k_{xx} = k_p \lambda_{xp}^2 + k_q \lambda_{xq}^2 + k_r \lambda_{xr}^2$$
$$k_{yy} = k_p \lambda_{yp}^2 + k_q \lambda_{yq}^2 + k_r \lambda_{yr}^2$$
$$k_{zz} = k_p \lambda_{zp}^2 + k_q \lambda_{zq}^2 + k_r \lambda_{zr}^2$$
$$k_{xy} = k_p \lambda_{xp}\lambda_{yp} + k_q \lambda_{xq}\lambda_{yq} + k_r \lambda_{xr}\lambda_{yr} \tag{3.32}$$
$$k_{xz} = k_p \lambda_{xp}\lambda_{zp} + k_q \lambda_{xq}\lambda_{zq} + k_r \lambda_{xr}\lambda_{zr}$$
$$k_{yz} = k_p \lambda_{yp}\lambda_{zp} + k_q \lambda_{yq}\lambda_{zq} + k_r \lambda_{yr}\lambda_{zr}$$

where the λ's are the cosines of the angles between the principal elastic axes of the resilient supporting elements and the coordinate axes. For example, λ_{xp} is the cosine of the angle between the X axis and the P axis of principal stiffness.

The equations of motion, Eqs. (3.31), do not include forces applied to the rigid body by damping forces from the resilient elements. To include damping, appropriate damping terms analogous to the corresponding stiffness terms are added to each equation. For example, Eq. (3.31a) would become

$$m\ddot{x}_c + \Sigma c_{xx}(\dot{x}_c - \dot{u}) + \Sigma k_{xx}(x_c - u) + \cdots$$
$$+ \Sigma(c_{xz}a_y - c_{xy}a_z)(\dot{\alpha} - \dot{\boldsymbol{\alpha}}) + \Sigma(k_{xz}a_y - k_{xy}a_z)(\alpha - \boldsymbol{\alpha}) + \cdots = F_x \tag{3.31a'}$$

where
$$c_{xx} = c_p \lambda_{xp}^2 + c_q \lambda_{xq}^2 + c_r \lambda_{xr}^2$$
$$c_{xy} = c_p \lambda_{xp}\lambda_{yp} + c_q \lambda_{xq}\lambda_{yq} + c_r \lambda_{xr}\lambda_{yr}$$

The number of degrees-of-freedom of a vibrational system is the minimum number of coordinates necessary to define completely the positions of the mass elements of the system in space. The system of Fig. 3.12 requires a minimum of six coordinates $(x_c, y_c, z_c, \alpha, \beta, \gamma)$ to define the position of the rigid body in space; thus, the system is said to vibrate in six degrees-of-freedom. Equations (3.31) may be solved simultaneously for the three components x_c, y_c, z_c of the center-of-gravity displacement and the three components α, β, γ of the rotational displacement of the rigid body. In most practical instances, the equations are simplified considerably by one or more of the following simplifying conditions:

1. The reference axes X, Y, Z are selected to coincide with the principal inertial axes of the body; then

$$I_{xy} = I_{xz} = I_{yz} = 0 \tag{3.33}$$

2. The resilient supporting elements are so arranged that one or more planes of symmetry exist; i.e., motion parallel to the plane of symmetry has no tendency to excite motion perpendicular to it, or rotation about an axis lying in the plane does not excite motion parallel to the plane. For example, in Eq. (3.31a), motion in the XY plane does not tend to excite motion in the XZ or YZ plane if Σk_{xz}, $\Sigma(k_{xz}a_y - k_{xy}a_z)$, and $\Sigma(k_{xx}a_z - k_{xz}a_x)$ are zero.

3. The principal elastic axes P, Q, R of all resilient supporting elements are orthogonal with the reference axes X, Y, Z of the body, respectively. Then, in Eqs. (3.32),

$$
\begin{aligned}
& k_{xx} = k_p = k_x \qquad k_{yy} = k_q = k_y \qquad k_{zz} = k_r = k_z \\
& k_{xy} = k_{xz} = k_{yz} = 0
\end{aligned}
\tag{3.34}
$$

where k_x, k_y, k_z are defined for use when orthogonality exists. The supports are then called *orthogonal supports*.

4. The forces F_x, F_y, F_z and moments M_x, M_y, M_z are applied directly to the body and there are no motions ($u = v = w = \alpha = \beta = \gamma = 0$) of the foundation; or alternatively, the forces and moments are zero and excitation results from motion of the foundation.

In general, the effect of these simplifications is to reduce the numbers of terms in the equations and, in some instances, to reduce the number of equations that must be solved simultaneously. Simultaneous equations indicate coupled modes; i.e., motion cannot exist in one coupled mode independently of motion in other modes which are coupled to it.

MODAL COUPLING AND NATURAL FREQUENCIES

Several conditions of symmetry resulting from zero values for the product of inertia terms in Eq. (3.33) are discussed in the following sections.

ONE PLANE OF SYMMETRY WITH ORTHOGONAL RESILIENT SUPPORTS

When the YZ plane of the rigid body system in Fig. 3.12 is a plane of symmetry, the following terms in the equations of motion are zero:

$$\Sigma k_{yy} a_x = \Sigma k_{zz} a_x = \Sigma k_{yy} a_x a_z = \Sigma k_{zz} a_x a_y = 0 \tag{3.35}$$

Introducing the further simplification that the principal elastic axes of the resilient elements are parallel with the reference axes, Eqs. (3.34) apply. Then the motions in the three coordinates y_c, z_c, α are coupled but are independent of motion in any of the other coordinates; furthermore, the other three coordinates x_c, β, γ also are coupled. For example, Fig. 3.13 illustrates a resiliently supported rigid body, wherein the YZ plane is a plane of symmetry that meets the requirements of Eq. (3.35). The three natural frequencies for the y_c, z_c, α coupled directions are found by solving Eqs.

(3.31b), (3.31c), and (3.31e) [or Eqs. (3.31a), (3.31d), and (3.31f) for the x_c, β, γ coupled directions] simultaneously.[9,10]

$$\left(\frac{f_n}{f_z}\right)^6 - A\left(\frac{f_n}{f_z}\right)^4 + B\left(\frac{f_n}{f_z}\right)^2 - C = 0 \tag{3.36}$$

where

$$f_z = \frac{1}{2\pi}\sqrt{\frac{\Sigma k_z}{m}} \tag{3.37}$$

is a quantity having mathematical rather than physical significance if translational motion in the direction of the Z axis is coupled to other modes of motion. (Such coupling exists for the system of Fig. 3.13.) The roots f_n represent the natural frequencies of the system in the coupled modes. The coefficients A, B, C for the coupled modes in the y_c, z_c, α coordinates are

$$A_{yz\alpha} = 1 + \frac{\Sigma k_y}{\Sigma k_z} + D_{zx}$$

$$B_{yz\alpha} = D_{zx} + \frac{\Sigma k_y}{\Sigma k_z}(1 + D_{zx}) - \frac{(\Sigma k_y a_z)^2 + (\Sigma k_z a_y)^2}{\rho_x^2(\Sigma k_z)^2}$$

$$C_{yz\alpha} = \frac{\Sigma k_y}{\Sigma k_z}\left(D_{zx} - \frac{(\Sigma k_z a_y)^2}{\rho_x^2(\Sigma k_z)^2}\right) - \frac{(\Sigma k_y a_z)^2}{\rho_x^2(\Sigma k_z)^2}$$

where

$$D_{zx} = \frac{\Sigma k_y a_z^2 + \Sigma k_z a_y^2}{\rho_x^2 \Sigma k_z}$$

and ρ_x is the radius of gyration of the rigid body with respect to the X axis.

The corresponding coefficients for the coupled modes in the x_c, β, γ coordinates are

$$A_{x\beta\gamma} = \frac{\Sigma k_x}{\Sigma k_z} + D_{zy} + D_{zz}$$

$$B_{x\beta\gamma} = \frac{\Sigma k_x}{\Sigma k_z}(D_{zy} + D_{zz}) + D_{zy}D_{zz}$$
$$- \frac{(\Sigma k_x a_z)^2}{\rho_y^2(\Sigma k_z)^2} - \frac{(\Sigma k_x a_y)^2}{\rho_z^2(\Sigma k_z)^2} - \frac{(\Sigma k_x a_y a_z)^2}{\rho_y^2\rho_z^2(\Sigma k_z)^2}$$

$$C_{x\beta\gamma} = \frac{\Sigma k_x}{\Sigma k_z}\left[D_{zy}D_{zz} - \frac{(\Sigma k_x a_y a_z)^2}{\rho_y^2\rho_z^2(\Sigma k_z)^2}\right] - \frac{(\Sigma k_x a_y)^2}{\rho_z^2(\Sigma k_z)^2}D_{zy}$$
$$- \frac{(\Sigma k_x a_z)^2}{\rho_y^2(\Sigma k_z)^2}D_{zz} + 2\frac{(\Sigma k_x a_y)(\Sigma k_x a_z)(\Sigma k_x a_y a_z)}{\rho_y^2\rho_z^2(\Sigma k_z)^3}$$

where

$$D_{zy} = \frac{\Sigma k_x a_z^2 + \Sigma k_z a_x^2}{\rho_y^2 \Sigma k_z} \qquad D_{zz} = \frac{\Sigma k_x a_y^2 + \Sigma k_y a_x^2}{\rho_z^2 \Sigma k_z}$$

and ρ_y, ρ_z are the radii of gyration of the rigid body with respect to the Y, Z axes.

The roots of the cubic equation Eq. (3.36) may be found graphically from Fig. 3.14.[9] The coefficients A, B, C are first calculated from the above relations for the appropriate set of coupled coordinates. Figure 3.14 is entered on the abscissa scale at the appropriate value for the quotient B/A^2. Small values of B/A^2 are in Fig. 3.14A, and large values in Fig. 3.14B. The quotient C/A^3 is the parameter for the family of curves. Upon selecting the appropriate curve, three values of $(f_n/f_z)/\sqrt{A}$

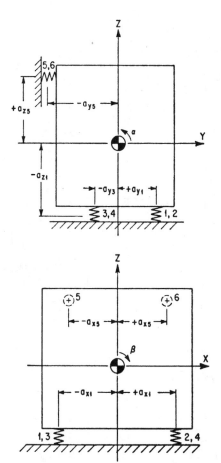

FIGURE 3.13 Example of a rigid body on orthogonal resilient supporting elements with one plane of symmetry. The YZ plane is a plane of symmetry since each resilient element has properties identical to those of its mirror image in the YZ plane; i.e., $k_{x1} = k_{x2}, k_{x3} = k_{x4}, k_{x5} = k_{x6}$, etc. The conditions satisfied are Eqs. (3.33) to (3.35).

are read from the ordinate and transferred to the left scale of the nomograph in Fig. 3.14B. Diagonal lines are drawn for each root to the value of A on the right scale, as indicated by dotted lines, and the roots f_n/f_z of the equation are indicated by the intercept of these dotted lines with the center scale of the nomograph.

The coefficients A, B, C can be simplified if all resilient elements have equal stiffness in the same direction. The stiffness coefficients always appear to equal powers in numerator and denominator, and lead to dimensionless ratios of stiffness. For n resilient elements, typical terms reduce as follows:

$$\frac{\Sigma k_y}{\Sigma k_z} = \frac{k_y}{k_z} \qquad \frac{\Sigma k_z a_y^2}{\rho_x^2 \Sigma k_z} = \frac{\Sigma a_y^2}{n\rho_x^2}$$

$$\frac{(\Sigma k_x a_y a_z)^2}{\rho_y^2 \rho_z^2 (\Sigma k_z)^2} = \left(\frac{k_x}{nk_z} \frac{\Sigma a_y a_z}{\rho_y \rho_z}\right)^2, \text{etc.}$$

TWO PLANES OF SYMMETRY WITH ORTHOGONAL RESILIENT SUPPORTS

Two planes of symmetry may be achieved if, in addition to the conditions of Eqs. (3.33) to (3.35), the following terms of Eqs. (3.31) are zero:

$$\Sigma k_{xx} a_y = \Sigma k_{zz} a_y = \Sigma k_{xx} a_y a_z = 0$$
$$(3.38)$$

Under these conditions, Eqs. (3.31) separate into two independent equations, Eqs. (3.31e) and (3.31f), and two sets each consisting of two coupled equations [Eqs. (3.31a) and (3.31d); Eqs. (3.31b) and (3.31c)]. The planes of symmetry are the XZ and YZ planes. For example, a common system is illustrated in Fig. 3.15, where four identical resilient supporting elements are located symmetrically about the Z axis in a plane not containing the center-of-gravity.[6, 11, 12] Coupling exists between translation in the X direction and rotation about the Y axis (x_c, β), as well as between translation in the Y direction and rotation about the X axis (y_c, α). Translation in the Z direction (z_c) and rotation about the Z axis (γ) are each independent of all other modes.

The natural frequency in the Z direction is found by solving Eq. (3.31e) to obtain Eq. (3.37), where $\Sigma k_{zz} = 4k_z$. The rotational natural frequency f_γ about the Z axis is found by solving Eq. (3.31f); it can be expressed with respect to the natural frequency in the direction of the Z axis:

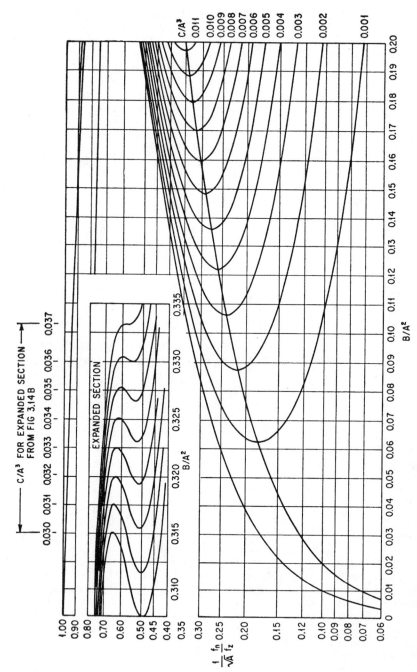

FIGURE 3.14A Graphical method of determining solutions of the cubic Eq. (3.36). Calculate A, B, C for the appropriate set of coupled coordinates, enter the abscissa at B/A^2 (values less than 0.2 on Fig. 3.14A, values greater than 0.2 on Fig. 3.14B), and read three values of $(f_n/f_z)/\sqrt{A}$ from the curve having the appropriate value of C/A^3.

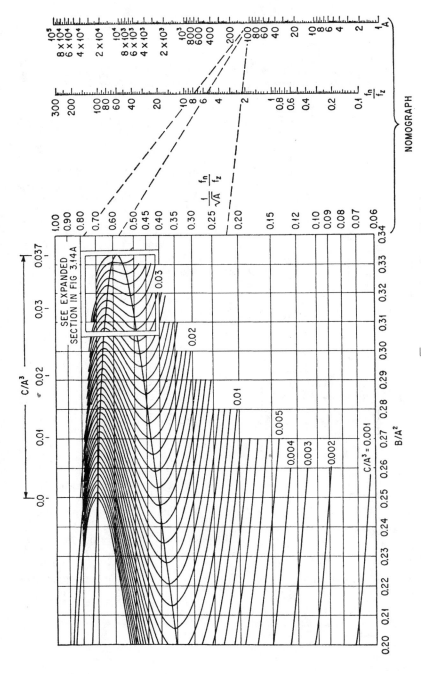

FIGURE 3.14B Using the above nomograph with values of $(f_n/f_z)/\sqrt{A}$ (see Fig. 3.14A), a diagonal line is drawn from each value of $(f_n/f_z)/\sqrt{A}$ on the left scale of the nomograph to the value of A on the right scale, as indicated by the dotted lines. The three roots f_n/f_z of Eq. (3.36) are given by the intercept of these dotted lines with the center scale of the nomograph. (*After F. F. Vane.*[9])

3.31

$$\frac{f_\gamma}{f_z} = \sqrt{\frac{k_x}{k_z}\left(\frac{a_y}{\rho_z}\right)^2 + \frac{k_y}{k_z}\left(\frac{a_x}{\rho_z}\right)^2} \tag{3.39}$$

where ρ_z is the radius of gyration with respect to the Z axis.

The natural frequencies in the coupled x_c, β modes are found by solving Eqs. (3.31a) and (3.31d) simultaneously; the roots yield the following expression for natural frequency:

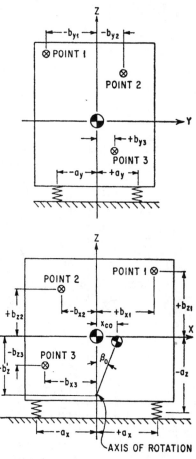

FIGURE 3.15 Example of a rigid body on orthogonal resilient supporting elements with two planes of symmetry. The XZ and YZ planes are planes of symmetry since the four resilient supporting elements are identical and are located symmetrically about the Z axis. The conditions satisfied are Eqs. (3.33), (3.34), (3.35), and (3.38). At any single frequency, coupled vibration in the x_o β direction due to X vibration of the foundation is equivalent to a pure rotation of the rigid body with respect to an axis of rotation as shown. Points 1, 2, and 3 refer to the example of Fig. 3.26.

$$\frac{f_{x\beta}^2}{f_z^2} = \frac{1}{2}\left\{\frac{k_x}{k_z}\left(1 + \frac{a_x^2}{\rho_y^2}\right) + \frac{a_x^2}{\rho_y^2} \pm \right.$$
$$\left. \sqrt{\left[\frac{k_z}{k_z}\left(1 + \frac{a_z^2}{\rho_y^2}\right) + \frac{a_x^2}{\rho_y^2}\right]^2 - 4\frac{k_x}{k_z}\frac{a_x^2}{\rho_y^2}}\right\} \tag{3.40}$$

Figure 3.16 provides a convenient graphical method for determining the two coupled natural frequencies $f_{x\beta}$. An expression similar to Eq. (3.40) is obtained for $f_{y\alpha}^2/f_z^2$ by solving Eqs. (3.31b) and (3.31d) simultaneously. By replacing ρ_y, a_x, k_x, $f_{x\beta}$ with ρ_x, a_y, k_y, $f_{y\alpha}$, respectively, Fig. 3.16 also can be used to determine the two values of $f_{y\alpha}$.

It may be desirable to select resilient element locations a_x, a_y, a_z which will produce coupled natural frequencies in specified frequency ranges, with resilient elements having specified stiffness ratios k_x/k_z, k_y/k_z. For this purpose it is convenient to plot solutions of Eq. (3.40) in the form shown in Figs. 3.17 to 3.19. These plots are termed *space-plots* and their use is illustrated in Example 3.1.[13,14]

The space-plots are derived as follows: In general, the two roots of Eq. (3.40) are numerically different, one usually being greater than unity and the other less than unity. Designating the root associated with the positive sign before the radical (higher value) as f_h/f_z, Eq. (3.40) may be written in the following form:

$$\frac{(a_x/\rho_y)^2}{(f_h/f_z)^2} + \frac{(a_z/\rho_y)^2}{(k_z/k_x)(f_h/f_z)^2 - 1} = 1 \tag{3.40a}$$

Equation (3.40a) is shown graphically by the large ellipses about the center of Figs. 3.17 to 3.19, for stiffness ratios k_x/k_z

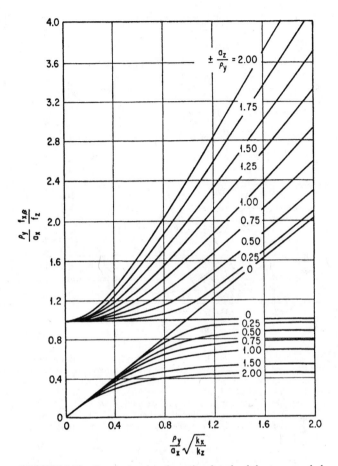

FIGURE 3.16 Curves showing the ratio of each of the two coupled natural frequencies $f_{x\beta}$ to the decoupled natural frequency f_z, for motion in the XZ plane of symmetry for the system in Fig. 3.15 [see Eq. (3.40)]. Calculate the abscissa $(\rho_y/a_x)\sqrt{k_x/k_z}$ and the parameter a_z/ρ_y, where a_x, a_z are indicated in Fig. 3.15; k_x, k_z are the stiffnesses of the resilient supporting elements in the X, Z directions, respectively; and ρ_y is the radius of gyration of the body about the Y axis. The two values read from the ordinate when divided by ρ_y/a_x give the natural frequency ratios $f_{x\beta}/f_z$. (*After C. E. Crede.*[12])

of ½, 1, and 2, respectively. A particular type of resilient element tends to have a constant stiffness ratio k_x/k_z; thus, Figs. 3.17 to 3.19 may be used by cut-and-try methods to find the coordinates a_x, a_z of such elements to attain a desired value of f_h.

Designating the root of Eq. (3.40) associated with the negative sign (lower value) by f_l, Eq. (3.40) may be written as follows:

$$\frac{(a_x/\rho_y)^2}{(f_2/f_x)^2} - \frac{(a_z/\rho_y)^2}{1 - (k_z/k_x)(f_l/f_z)^2} = 1 \qquad (3.40b)$$

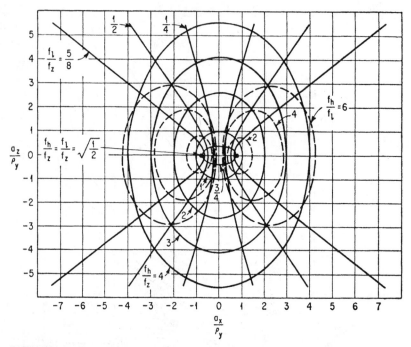

FIGURE 3.17 Space-plot for the system in Fig. 3.15 when the stiffness ratio $k_x/k_z = 0.5$, obtained from Eqs. (3.40a) to (3.40c). With all dimensions divided by the radius of gyration ρ_y about the Y axis, superimpose the outline of the rigid body in the XZ plane on the plot; the center-of-gravity of the body is located at the coordinate center of the plot. The elastic centers of the resilient supporting elements give the natural frequency ratios f_l/f_z, f_h/f_z, and f_h/f_l for $x_c\beta$ coupled motion, each ratio being read from one of the three families of curves as indicated on the plot. Replacing k_x, ρ_y, a_x with k_y, ρ_x, a_y, respectively, allows the plot to be applied to motions in the YZ plane.

Equation (3.40b) is shown graphically by the family of hyperbolas on each side of the center in Figs. 3.17 to 3.19, for values of the stiffness ratio k_x/k_z of ½, 1, and 2.

The two roots f_h/f_z and f_l/f_z of Eq. (3.40) may be expressed as the ratio of one to the other. This relationship is given parametrically as follows:

$$\left[\frac{2\dfrac{a_x}{\rho_y} \pm \sqrt{\dfrac{k_x}{k_z}}\left(\dfrac{f_h}{f_l} + \dfrac{f_l}{f_h}\right)}{\sqrt{\dfrac{k_x}{k_z}}\left(\dfrac{f_h}{f_l} - \dfrac{f_l}{f_h}\right)}\right]^2 + \left[\frac{2\dfrac{a_z}{\rho_y}}{\dfrac{f_h}{f_l} - \dfrac{f_l}{f_h}}\right]^2 = 1 \qquad (3.40c)$$

Equation (3.40c) is shown graphically by the smaller ellipses (shown dotted) displaced from the vertical center line in Figs. 3.17 to 3.19.

Example 3.1. A rigid body is symmetrical with respect to the XZ plane; its width in the X direction is 13 in. and its height in the Z direction is 12 in. The center-of-gravity is 5½ in. from the lower side and 6¾ in. from the right side. The radius of gyration about the Y axis through the center-of-gravity is 5.10 in. Use a space-plot to evaluate the effects of the location for attachment of resilient supporting elements having the characteristic stiffness ratio $k_x/k_z = ½$.

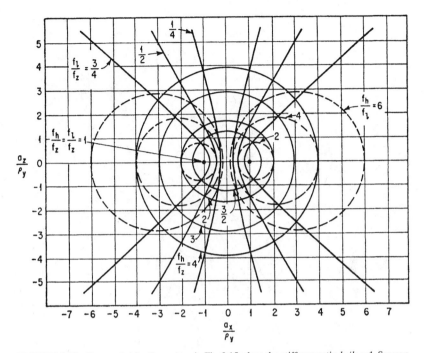

FIGURE 3.18   Space-plot for the system in Fig. 3.15 when the stiffness ratio $k_x/k_z = 1$. See caption for Fig. 3.17.

Superimpose the outline of the body on the space-plot of Fig. 3.20, with its center-of-gravity at the coordinate center of the plot. (Figure 3.20 is an enlargement of the central portion of Fig. 3.17.) All dimensions are divided by the radius of gyration ρ_y. Thus, the four corners of the body are located at coordinate distances as follows:

Upper right corner:

$$\frac{a_z}{\rho_y} = \frac{+6.50}{5.10} = +1.28 \qquad \frac{a_x}{\rho_y} = \frac{+6.75}{5.10} = +1.32$$

Upper left corner:

$$\frac{a_z}{\rho_y} = \frac{+6.50}{5.10} = +1.28 \qquad \frac{a_x}{\rho_y} = \frac{-6.25}{5.10} = -1.23$$

Lower right corner:

$$\frac{a_z}{\rho_y} = \frac{-5.50}{5.10} = -1.08 \qquad \frac{a_x}{\rho_y} = \frac{+6.75}{5.10} = +1.32$$

Lower left corner:

$$\frac{a_z}{\rho_y} = \frac{-5.50}{5.10} = -1.08 \qquad \frac{a_x}{\rho_y} = \frac{-6.25}{5.10} = -1.23$$

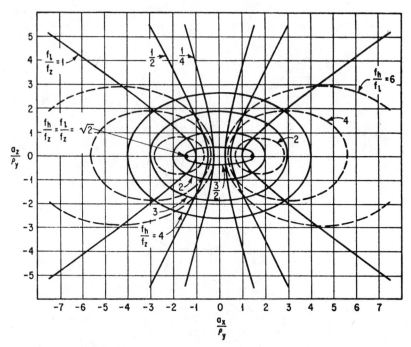

FIGURE 3.19 Space-plot for the system in Fig. 3.15 when the stiffness ratio $k_x/k_z = 2$. See caption for Fig. 3.17.

The resilient supports are shown in heavy outline at A in Fig. 3.20, with their elastic centers indicated by the solid dots. The horizontal coordinates of the resilient supports are $a_x/\rho_y = \pm0.59$, or $a_x = \pm0.59 \times 5.10 = \pm3$ in. from the vertical coordinate axis. The corresponding natural frequencies are $f_h/f_z = 1.25$ (from the ellipses) and $f_l/f_z = 0.33$ (from the hyperbolas). An alternative position is indicated by the hollow circles B. The natural frequencies for this position are $f_h/f_z = 1.43$ and $f_l/f_z = 0.50$. The natural frequency f_z in vertical translation is found from the mass of the equipment and the summation of stiffnesses in the Z direction, using Eq. (3.37). This example shows how space-plots make it possible to determine the locations of the resilient elements required to achieve given values of the coupled natural frequencies with respect to f_z.

THREE PLANES OF SYMMETRY WITH ORTHOGONAL RESILIENT SUPPORTS

A system with three planes of symmetry is defined by six independent equations of motion. A system having this property is sometimes called a *center-of-gravity system*. The equations are derived from Eqs. (3.31) by substituting, in addition to the conditions of Eqs. (3.33), (3.34), (3.35), and (3.38), the following condition:

$$\Sigma k_{xx} a_z = \Sigma k_{yy} a_z = 0 \tag{3.41}$$

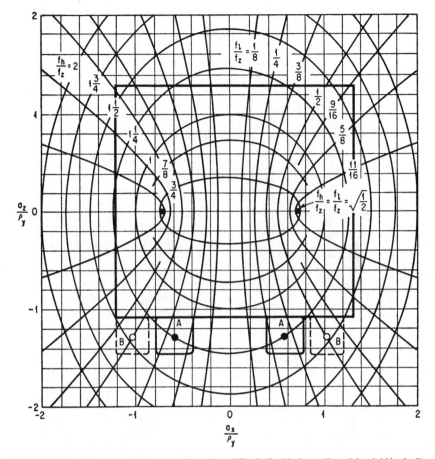

FIGURE 3.20 Enlargement of the central portion of Fig. 3.17 with the outline of the rigid body discussed in Example 3.1.

The resulting six independent equations define six uncoupled modes of vibration, three in translation and three in rotation. The natural frequencies are:

Translation along X axis:

$$f_x = \frac{1}{2\pi} \sqrt{\frac{\Sigma k_x}{m}}$$

Translation along Y axis:

$$f_y = \frac{1}{2\pi} \sqrt{\frac{\Sigma k_y}{m}}$$

Translation along Z axis:

$$f_z = \frac{1}{2\pi} \sqrt{\frac{\Sigma k_z}{m}}$$

Rotation about X axis:

$$f_\alpha = \frac{1}{2\pi} \sqrt{\frac{\Sigma(k_y a_z^2 + k_z a_y^2)}{m\rho_x^2}} \tag{3.42}$$

Rotation about Y axis:

$$f_\beta = \frac{1}{2\pi} \sqrt{\frac{\Sigma(k_x a_z^2 + k_z a_x^2)}{m\rho_y^2}}$$

Rotation about Z axis:

$$f_\gamma = \frac{1}{2\pi} \sqrt{\frac{\Sigma(k_x a_y^2 + k_y a_x^2)}{m\rho_z^2}}$$

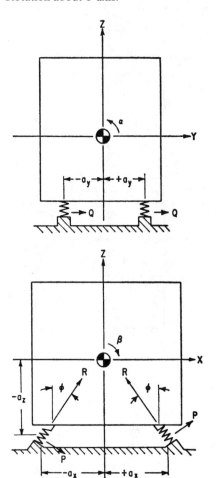

TWO PLANES OF SYMMETRY WITH RESILIENT SUPPORTS INCLINED IN ONE PLANE ONLY

When the principal elastic axes of the resilient supporting elements are inclined with respect to the X, Y, Z axes, the stiffness coefficients k_{xy}, k_{xz}, k_{yz} are nonzero. This introduces elastic coupling, which must be considered in evaluating the equations of motion. Two planes of symmetry may be achieved by meeting the conditions of Eqs. (3.33), (3.35), and (3.38). For example, consider the rigid body supported by four identical resilient supporting elements located symmetrically about the Z axis, as shown in Fig. 3.21. The XZ and the YZ planes are planes of symmetry, and the resilient elements are inclined toward the YZ plane so that one of their principal elastic axes R is inclined at the angle ϕ with the Z direction as shown; hence $k_{yy} = k_q$, and $k_{xy} = k_{yz} = 0$.

Because of symmetry, translational motion z_c in the Z direction and rotation γ about the Z axis are each decoupled from the other modes. The pairs of translational and rotational modes in the x_c, β and y_c, α coordinates are coupled. The natural frequency in the Z direction is

FIGURE 3.21 Example of a rigid body on resilient supporting elements inclined toward the YZ plane. The resilient supporting elements are identical and are located symmetrically about the Z axis, making XZ and YZ planes of symmetry. The principal stiffnesses in the XZ plane are k_p and k_r. The conditions satisfied are Eqs. (3.33), (3.35), and (3.38).

$$\frac{f_z}{f_r} = \sqrt{\frac{k_p}{k_r} \sin^2 \phi + \cos^2 \phi} \tag{3.43}$$

where f_r is a fictitious natural frequency used for convenience only; it is related to Eq. (3.37) wherein $4k_r$ is substituted for Σk_z:

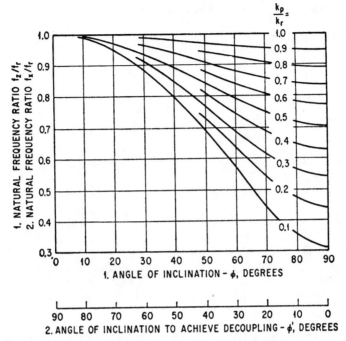

FIGURE 3.22   Curves showing the ratio of the decoupled natural frequency f_z of translation z_c to the fictitious natural frequency f_r for the system shown in Fig. 3.21 [see Eq. (3.43)] when the resilient supporting elements are inclined at the angle ϕ. The curves also indicate the ratio of the decoupled natural frequency f_x of translation x_c to f_r when ϕ has a value ϕ' (use lower abscissa scale) which decouples x_c, β motions [see Eqs. (3.47) and (3.48)]. (*After C. E. Crede.*[15])

$$f_r = \frac{1}{2\pi} \sqrt{\frac{4k_r}{m}}$$

Equation (3.43) is plotted in Fig. 3.22, where the angle ϕ is indicated by the upper of the abscissa scales.

The rotational natural frequency about the Z axis is obtained from

$$\frac{f_\gamma}{f_r} = \sqrt{\left(\frac{k_p}{k_r}\cos^2\phi + \sin^2\phi\right)\left(\frac{a_y}{\rho_z}\right)^2 + \frac{k_q}{k_r}\left(\frac{a_x}{\rho_z}\right)^2} \qquad (3.44)$$

For the x_c, β coupled mode, the two natural frequencies are

$$\frac{f_{x\beta}}{f_r} = \frac{1}{2}\left[A \pm \sqrt{A^2 - 4\frac{k_p}{k_r}\left(\frac{a_x}{a_y}\right)^2}\right] \qquad (3.45)$$

where $A = \left(\frac{k_p}{k_r}\cos^2\phi + \sin^2\phi\right)\left[1 + \left(\frac{a_z}{\rho_y}\right)^2\right] + \left(\frac{k_p}{k_r}\sin^2\phi + \cos^2\phi\right)\left(\frac{a_x}{\rho_y}\right)^2$

$$+ 2\left(1 - \frac{k_p}{k_r}\right)\left|\frac{a_x}{\rho_y}\right|\sin\phi\cos\phi$$

For the y_c, α coupled mode, the natural frequencies are

$$\frac{f_{y\alpha}}{f_r} = \frac{1}{2}\left[B \pm \sqrt{B^2 - 4\frac{k_q}{k_r}\left(\frac{k_p}{k_r}\sin^2\phi + \cos^2\phi\right)\left(\frac{a_y}{\rho_x}\right)^2} \right] \tag{3.46}$$

where $\qquad B = \frac{k_q}{k_r}\left[1 + \left(\frac{a_z}{\rho_x}\right)^2 \right] + \left(\frac{k_p}{k_r}\sin^2\phi + \cos^2\phi\right)\left(\frac{a_y}{\rho_x}\right)^2$

DECOUPLING OF MODES IN A PLANE USING INCLINED RESILIENT SUPPORTS

The angle ϕ of inclination of principal elastic axes (see Fig. 3.21) can be varied to produce changes in the amount of coupling between the x_c and β coordinates. Decoupling of the x_c and β coordinates is effected if[15]

$$\left|\frac{a_z}{a_x}\right| = \frac{[1 - (k_p/k_r)]\cot\phi'}{1 + (k_p/k_r)\cot^2\phi'} \tag{3.47}$$

where ϕ' is the value of the angle of inclination ϕ required to achieve decoupling. When Eq. (3.47) is satisfied, the configuration is sometimes called an "equivalent center-of-gravity system" in the YZ plane since all modes of motion in that plane are decoupled. Figure 3.23 is a graphical presentation of Eq. (3.47). There may be two values of ϕ' that decouple the x_c and β modes for any combination of stiffness and location for the resilient supporting elements.

The decoupled natural frequency for translation in the X direction is obtained from

$$\frac{f_x}{f_r} = \sqrt{\frac{k_p}{k_r}\cos^2\phi' + \sin^2\phi'} \tag{3.48}$$

The relation of Eq. (3.48) is shown graphically in Fig. 3.22 where the angle ϕ' is indicated by the lower of the abscissa scales. The natural frequency in the β mode is obtained from

$$\frac{f_\beta}{f_r} = \frac{a_x}{\rho_y}\sqrt{\frac{1}{(k_r/k_p)\sin^2\phi' + \cos^2\phi'}} \tag{3.49}$$

COMPLETE DECOUPLING OF MODES USING RADIALLY INCLINED RESILIENT SUPPORTS

In general, the analysis of rigid body motion with the resilient supporting elements inclined in more than one plane is quite involved. A particular case where sufficient symmetry exists to provide relatively simple yet useful results is the configuration illustrated in Fig. 3.24. From symmetry about the Z axis, $I_{xx} = I_{yy}$. Any number n of resilient supporting elements greater than 3 may be used. For clarity of illustration, the rigid body is shown as a right circular cylinder with $n = 3$.

The resilient supporting elements are arranged symmetrically about the Z axis; they are attached to one end face of the cylinder at a distance a_r from the Z axis and a distance a_z from the XY reference plane. The resilient elements are inclined so that their principal elastic axes R intersect at a common point on the Z axis; thus, the angle between the Z axis and the R axis for each element is ϕ. The principal elastic axes P

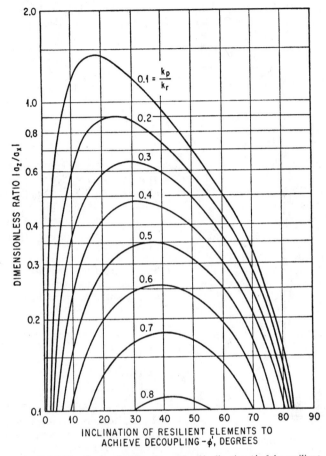

FIGURE 3.23 Curves showing the angle of inclination ϕ' of the resilient elements which achieves decoupling of the x_c, β motions in Fig. 3.21 [see Eq. (3.47)]. Calculate the ordinate $|a_z/a_x|$ and with the stiffness ratio k_p/k_r determine two values of ϕ' for which decoupling is possible. Decoupling is not possible for a particular value of k_p/k_r if $|a_z/a_y|$ has a value greater than the maximum ordinate of the k_p/k_r curve.

also intersect at a common point on the Z axis, the angle between the Z axis and the P axis for each element being $90° - \phi$. Consequently, the Q principal elastic axes are each tangent to the circle of radius a_r which bounds the end face of the cylinder.

The use of such a configuration permits decoupling of all six modes of vibration of the rigid body. This complete decoupling is achieved if the angle of inclination ϕ has the value ϕ' which satisfies the following equation:

$$\left| \frac{a_z}{a_r} \right| = \frac{(\frac{1}{2})[1 - (k_p/k_r)] \sin 2\phi'}{(k_q/k_r) + (k_p/k_r) + [1 - (k_p/k_r)] \sin^2\phi'} \tag{3.50}$$

Since complete decoupling is effected, the system may be termed an "equivalent center-of-gravity system."[16,17] The natural frequencies of the six decoupled modes are

$$\frac{f_x}{f_r} = \frac{f_y}{f_r} = \sqrt{\frac{1}{2}\left(\frac{k_p}{k_r}\cos^2\phi' + \sin^2\phi' + \frac{k_q}{k_r}\right)} \tag{3.51}$$

$$\frac{f_\alpha}{f_r} = \frac{f_\beta}{f_r} = \left\{\frac{a_r}{2\rho_x}\left[\frac{k_p}{k_r}\sin\phi'\left(\frac{a_r}{\rho_x}\sin\phi' + \frac{a_z}{\rho_x}\cos\phi'\right) + \cos\phi'\left(\frac{a_r}{\rho_x}\cos\phi' - \frac{a_z}{\rho_x}\sin\phi'\right)\right]\right\}^{1/2} \tag{3.52}$$

$$\frac{f_\gamma}{f_r} = \sqrt{\frac{k_q}{k_r}\frac{a_r}{\rho_z}} \tag{3.53}$$

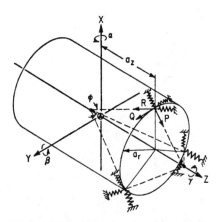

FIGURE 3.24 Example of a rigid cylindrical body on radially inclined resilient supports. The resilient supports are attached symmetrically about the Z axis to one end face of the cylinder at a distance a_r from the Z axis and a distance a_z from the XY plane. The resilient elements are inclined so that their principal elastic axes R and P intersect the Z axis at common points. The angle between the R axes and the Z axis is ϕ; and the angle between the P axis and the Z axis is $90° - \phi$. The Q principal elastic axes are each tangent to the circle of radius a_r.

The frequency ratio f_z/f_r is given by Eq. (3.43) or Fig. 3.22. The fictitious natural frequency f_r is given by

$$f_r = (1/2\pi)\sqrt{nk_r/m}$$

Similar solutions are also available for the configuration of four resilient supports located in a rectangular array and inclined to achieve complete decoupling.[18]

FORCED VIBRATION

Forced vibration results from a continuing excitation that varies sinusoidally with time. The excitation may be a vibratory displacement of the foundation for the resiliently supported rigid body (*foundation-induced vibration*), or a force or moment applied to or generated within the rigid body (*body-induced vibration*). These two forms of excitation are considered separately.

FOUNDATION-INDUCED SINUSOIDAL VIBRATION

This section includes an analysis of foundation-induced vibration for two different systems, each having two planes of symmetry. In one system, the principal elastic axes of the resilient elements are parallel to the X, Y, Z axes; in the other system, the principal elastic axes are inclined with respect to two of the axes but in a plane parallel to one of the reference planes. The excitation is translational movement of the foundation in its own plane, without rotation. No forces or moments are applied directly to the rigid body; i.e., in the equations of motion [Eqs. (3.31)], the following terms are equal to zero:

$$F_x = F_y = F_z = M_x = M_y = M_z = \alpha = \beta = \gamma = 0 \tag{3.54}$$

Two Planes of Symmetry with Orthogonal Resilient Supports. The system is shown in Fig. 3.15. The excitation is a motion of the foundation in the direction of the X axis defined by $u = u_0 \sin \omega t$. (Alternatively, the excitation may be the displacement $v = v_0 \sin \omega t$ in the direction of the Y axis, and analogous results are obtained.) The resulting motion of the resiliently supported rigid body involves translation x_c and rotation β simultaneously. The conditions of symmetry are defined by Eqs. (3.33), (3.34), (3.35), and (3.38); these conditions decouple Eqs. (3.31) so that only Eqs. (3.31a) and (3.31d), and Eqs. (3.31b) and (3.31c), remain coupled. Upon substituting $u = u_0 \sin \omega t$ as the excitation, the response in the coupled modes is of a form $x_c = x_{c0} \sin \omega t$, $\beta = \beta_0 \sin \omega t$ where x_{c0} and β_0 are related to u_0 as follows:[19]

$$\frac{x_{c0}}{u_0} = \frac{\dfrac{k_x}{k_z}\left[\left(\dfrac{a_x}{\rho_y}\right)^2 - \left(\dfrac{f}{f_z}\right)^2\right]}{\left(\dfrac{f}{f_z}\right)^4 - \left[\dfrac{k_x}{k_z} + \dfrac{k_x}{k_z}\left(\dfrac{a_z}{\rho_y}\right)^2 + \left(\dfrac{a_x}{\rho_y}\right)^2\right]\left(\dfrac{f}{f_z}\right)^2 + \dfrac{k_x}{k_z}\left(\dfrac{a_x}{\rho_y}\right)^2} \tag{3.55}$$

$$\frac{\beta_0}{u_0/\rho_y} = \frac{-\dfrac{k_x}{k_z}\dfrac{a_z}{\rho_y}\left(\dfrac{f}{f_z}\right)^2}{\left(\dfrac{f}{f_z}\right)^4 - \left[\dfrac{k_x}{k_z} + \dfrac{k_x}{k_z}\left(\dfrac{a_z}{\rho_y}\right)^2 + \left(\dfrac{a_x}{\rho_y}\right)^2\right]\left(\dfrac{f}{f_z}\right)^2 + \dfrac{k_x}{k_z}\left(\dfrac{a_x}{\rho_y}\right)^2} \tag{3.56}$$

where $f_z = \dfrac{1}{2\pi}\sqrt{4k_z/m}$ in accordance with Eq. (3.37). A similar set of equations apply for vibration in the coupled y_c, α coordinates. There is no response of the system in the z_c or γ modes since there is no net excitation in these directions; that is, $\mathbf{F}_z$ and $\mathbf{M}_z$ are zero.

As indicated by Eqs. (3.1), the displacement at any point in a rigid body is the sum of the displacement at the center-of-gravity and the displacements resulting from motion of the body in rotation about axes through the center-of-gravity. Equations (3.55) and (3.56) together with analogous equations for y_{c0}, α_0 provide the basis for calculating these displacements. Care should be taken with phase angles, particularly if two or more excitations u, v, w exist concurrently.

At any single frequency, coupled vibration in the x_c, β modes is equivalent to a pure rotation of the rigid body with respect to an axis parallel to the Y axis, in the YZ plane and displaced from the center-of-gravity of the body (see Fig. 3.15). As a result, the rigid body has zero displacement x in the horizontal plane containing this axis. Therefore, the Z coordinate of this axis b_z' satisfies $x_{c0} + b_z'\beta_0 = 0$, which is obtained from the first

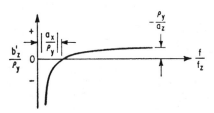

FIGURE 3.25 Curve showing the position of the axis of pure rotation of the rigid body in Fig. 3.15 as a function of the frequency ratio f/f_z when the excitation is sinusoidal motion of the foundation in the X direction [see Eq. (3.57)]. The axis of rotation is parallel to the Y axis and in the XZ plane, and its coordinate along the Z axis is designated by b_z'.

of Eqs. (3.1) by setting $x_b = 0$ (γ_0 motion is not considered). Substituting Eqs. (3.55) and (3.56) for x_{c0} and β_0, respectively, the axis of rotation is located at

$$\frac{b_z'}{\rho_y} = \frac{(a_x/\rho_y)^2 - (f/f_z)^2}{(a_z/\rho_y)(f/f_z)^2} \tag{3.57}$$

Figure 3.25 shows the relation of Eq. (3.57) graphically. At high values of frequency f/f_z, the axis does not change position significantly with frequency; b_z'/ρ_y approaches a positive value as f/f_z becomes large, since a_z is negative (see Fig. 3.15).

When the resilient supporting elements have damping as well as elastic properties, the solution of the equations of motion [see Eq. (3.31a)] becomes too laborious for general use. Responses of systems with damping have been obtained for several typical cases using a digital computer.[20] Figures 3.26 A, B, and C show the response at three points in the body of the system shown in Fig. 3.15, with the excitation $u = u_0 \sin \omega t$. The weight of the body is 45 lb; each of the four resilient supporting elements has a stiffness $k_z = 1,050$ lb/in. and stiffness ratios $k_x/k_z = k_y/k_z = \frac{1}{2}$. The critical damping coefficients in the X, Y, Z directions are taken as $c_{cx} = 2\sqrt{4k_x m}$, $c_{cy} = 2\sqrt{4k_y m}$, $c_{cz} = 2\sqrt{4k_z m}$, respectively, where the expression for c_{cz} follows from the single degree-of-freedom case defined by Eq. (2.12). The fractions of critical damp-

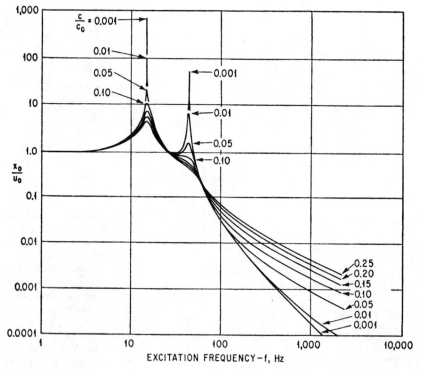

FIGURE 3.26A Response curves for point 1 with damping in the resilient supports in the system shown in Fig. 3.15. The response is the ratio of the amplitude at point 1 of the rigid body in the X direction to the amplitude of the foundation in the X direction (x_0/u_0). The fraction of critical damping c/c_c is the same in the X, Y, Z directions.

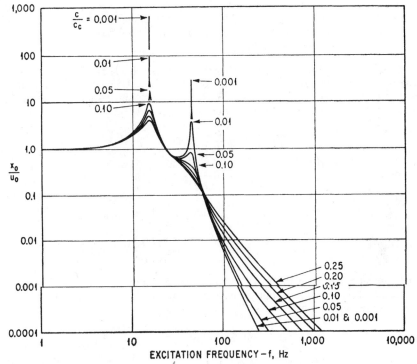

FIGURE 3.26B Response curves at point 2 in the system shown in Fig. 3.15. See caption for Fig. 3.26A.

ing are $c_x/c_{cx} = c_y/c_{cy} = c_z/c_{cz} = c/c_c$, the parameter of the curves in Figs. 3.26A, B, and C. Coordinates locating the resilient elements are $a_x = \pm 5.25$ in., $a_y = \pm 3.50$ in., and $a_z = -6.50$ in. The radii of gyration with respect to the X, Y, Z axes are $\rho_x = 4.40$ in., $\rho_y = 5.10$ in., and $\rho_z = 4.60$ in.

Natural frequencies calculated from Eqs. (3.37) and (3.40) are $f_z = 30.0$ Hz; $f_{x\beta} = 43.7$ Hz, 15.0 Hz; and $f_{y\alpha} = 43.2$ Hz, 11.7 Hz. The fraction of critical damping c/c_c varies between 0 and 0.25. Certain characteristic features of the response curves in Figs. 3.26A, B, and C are:

1. The relatively small response at the frequency of 24.2 Hz in Fig. 3.26C occurs because point 3 lies near the axis of rotation of the rigid body at that frequency. Point 2 lies near the axis of rotation at higher frequencies, and the response becomes correspondingly low, as shown in Fig. 3.26B. The position of the axis of rotation changes rapidly for small changes of frequency in the low- and intermediate-frequency range (indicated by the sharp dip in the curves for small damping in Fig. 3.26C) and varies asymptotically toward a final position as the forcing frequency increases (see Fig. 3.25).

2. The effect of damping on the magnitude of the response at the higher and lower natural frequencies in coupled modes is illustrated. When the fraction of critical damping is between 0.01 and 0.10, the response at the lower of the coupled natu-

ral frequencies is approximately 10 times as great as the response at the higher of the coupled natural frequencies. With greater damping ($c/c_c \geq 0.15$), the effect of resonance in the vicinity of the higher coupled natural frequency becomes so slight as to be hardly discernible.

Two Planes of Symmetry with Resilient Supports Inclined in One Plane Only.
The system is shown in Fig. 3.21, and the excitation is $u = u_0 \sin \omega t$. The conditions of symmetry are defined by Eqs. (3.33), (3.35), and (3.38). The response is entirely in the x_c, β coupled mode with the following amplitudes:

$$\frac{x_{c0}}{u_0} = \frac{\dfrac{k_p}{k_r}\left(\dfrac{a_x}{\rho_y}\right)^2 - \left(\dfrac{k_p}{k_r}\cos^2\phi + \sin^2\phi\right)\left(\dfrac{f}{f_r}\right)^2}{\left(\dfrac{f}{f_r}\right)^4 - A\left(\dfrac{f}{f_r}\right)^2 + \dfrac{k_p}{k_r}\left(\dfrac{a_x}{\rho_y}\right)^2}$$

$$\frac{\beta_0}{u_0/\rho_y} = \frac{-\left[\left(\dfrac{k_p}{k_r}\cos^2\phi + \sin^2\phi\right)\left(\dfrac{a_z}{\rho_y}\right) + \left(1 - \dfrac{k_p}{k_r}\right)\left|\dfrac{a_x}{\rho_y}\right|\sin\phi\cos\phi\right]\left(\dfrac{f}{f_r}\right)^2}{\left(\dfrac{f}{f_r}\right)^4 - A\left(\dfrac{f}{f_r}\right)^2 + \dfrac{k_p}{k_r}\left(\dfrac{a_x}{\rho_y}\right)^2} \tag{3.58}$$

where A is defined after Eq. (3.45). A similar set of expressions may be written for the response in the y_c, α coupled mode when the excitation is the motion $v = v_0 \sin \omega t$ of the foundation:

$$\frac{y_{c0}}{v_0} = \frac{\dfrac{k_q}{k_r}\left(\dfrac{k_p}{k_r}\sin^2\phi + \cos^2\phi\right)\left(\dfrac{a_y}{\rho_x}\right)^2 - \dfrac{k_q}{k_r}\left(\dfrac{f}{f_r}\right)^2}{\left(\dfrac{f}{f_r}\right)^4 - B\left(\dfrac{f}{f_r}\right)^2 + \dfrac{k_q}{k_r}\left(\dfrac{k_p}{k_r}\sin^2\phi + \cos^2\phi\right)\left(\dfrac{a_y}{\rho_x}\right)}$$

$$\frac{\alpha_0}{v_0/\rho_x} = \frac{\dfrac{k_q}{k_r}\dfrac{a_z}{\rho_x}\left(\dfrac{f}{f_r}\right)^2}{\left(\dfrac{f}{f_r}\right)^4 - B\left(\dfrac{f}{f_r}\right)^2 + \dfrac{k_q}{k_r}\left(\dfrac{k_p}{k_r}\sin^2\phi + \cos^2\phi\right)\left(\dfrac{a_y}{\rho_x}\right)} \tag{3.59}$$

where B is defined after Eq. (3.46). No motion occurs in the z_c or γ mode since the quantities F_z and M_z are zero in Eqs. (3.31e) and (3.31f).

Response curves for the system shown in Fig. 3.21 when damping is included are qualitatively similar to those shown in Figs. 3.26.[21] The significant advantage in the use of inclined resilient supports is the additional versatility gained from the ability to vary the angle of inclination ϕ, which directly affects the degree of coupling in the x_c, β coupled mode. For example, a change in the angle ϕ produces a change in the position of the axis of pure rotation of the rigid body. In a manner similar to that used to derive Eq. (3.57), Eqs. (3.58) yield the following expression defining the location of the axis of rotation:

$$\frac{b'_z}{\rho_y} = \frac{\dfrac{k_p}{k_r}\left(\dfrac{a_x}{\rho_y}\right)^2 - \left(\dfrac{k_p}{k_r}\cos^2\phi + \sin^2\phi\right)\left(\dfrac{f}{f_r}\right)^2}{\left[\left(\dfrac{k_p}{k_r}\cos^2\phi + \sin^2\phi\right)\dfrac{a_z}{\rho_y} + \left(1 - \dfrac{k_p}{k_r}\right)\left|\dfrac{a_x}{\rho_y}\right|\sin\phi\cos\phi\right]\left(\dfrac{f}{f_r}\right)^2} \tag{3.60}$$

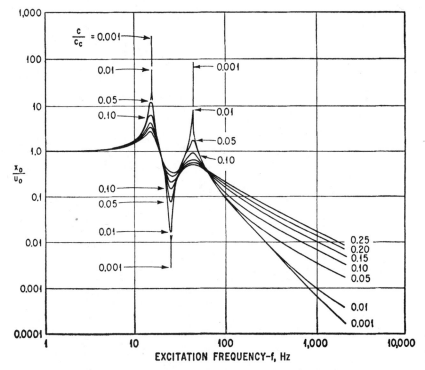

FIGURE 3.26C Response curves at point 3 in the system shown in Fig. 3.15. See caption for Fig. 3.26A.

BODY-INDUCED SINUSOIDAL VIBRATION

This section includes the analysis of a resiliently supported rigid body wherein the excitation consists of forces and moments applied directly to the rigid body (or originating within the body). The system has two planes of symmetry with orthogonal resilient supports; the modal coupling and natural frequencies for such a system are considered above. Two types of excitation are considered: (1) a force rotating about an axis parallel to one of the principal inertial axes and (2) an oscillatory moment acting about one of the principal inertial axes.[22,23] There is no motion of the foundation that supports the resilient elements; thus, the following terms in Eqs. (3.31) are equal to zero:

$$u = v = w = \alpha = \beta = \gamma = 0 \qquad (3.61)$$

Two Planes of Symmetry with Orthogonal Resilient Elements Excited by a Rotating Force.   The system excited by the rotating force is illustrated in Fig. 3.27. The force F_0 rotates at frequency ω about an axis parallel to the Y axis but spaced therefrom by the coordinate distances d_x, d_z; the force is in the XZ plane. The forces and moments applied to the body by the rotating force F_0 are

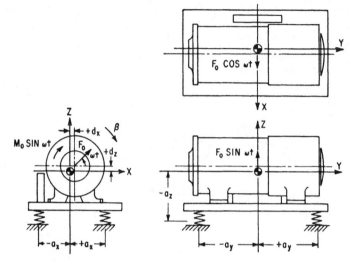

FIGURE 3.27 Example of a rigid body on orthogonal resilient supports with two planes of symmetry, excited by body-induced sinusoidal excitation. Alternative excitations are (1) the force F_0 in the XZ plane rotating with angular velocity ωt about an axis parallel to the Y axis and (2) the oscillatory moment $M_0 \sin \omega t$ acting about the Y axis. There is no motion of the foundation that supports the resilient elements.

$$F_x = F_0 \cos \omega t \qquad\qquad M_x = 0$$

$$F_y = 0 \qquad\qquad M_y = F_0(d_z \cos \omega t - d_x \sin \omega t) \qquad (3.62)$$

$$F_z = F_0 \sin \omega t \qquad\qquad M_z = 0$$

The conditions of symmetry are defined by Eqs. (3.33), (3.34), (3.35), and (3.38); and the excitation is defined by Eqs. (3.61) and (3.62). Substituting these conditions into the equations of motion, Eqs. (3.31) show that vibration response is not excited in the coupled y_c, α mode or in the γ mode. In the Z direction, the motion z_{c0} of the body and the force F_{tz} transmitted through the resilient elements can be found from Eq. (2.30) and Fig. 2.17 since single degree-of-freedom behavior is involved. The horizontal displacement amplitude x_{c0} of the center-of-gravity in the X direction and the rotational displacement amplitude β_0 about the Y axis are given by

$$\frac{x_{c0}}{F_0/4k_x} = \frac{k_x}{k_z} \frac{\sqrt{\left[\frac{k_x}{k_z}\frac{a_z}{\rho_y}\left(\frac{a_z}{\rho_y} - \frac{d_z}{\rho_y}\right) + \left(\frac{a_x}{\rho_y}\right)^2 - \left(\frac{f}{f_z}\right)^2\right]^2 + \left[\frac{k_x}{k_z}\frac{d_x}{\rho_y}\frac{a_z}{\rho_y}\right]^2}}{\left(\frac{f}{f_z}\right)^4 - \left[\frac{k_x}{k_z} + \frac{k_x}{k_z}\left(\frac{a_z}{\rho_y}\right)^2 + \left(\frac{a_x}{\rho_y}\right)^2\right]\left(\frac{f}{f_z}\right)^2 + \frac{k_x}{k_z}\left(\frac{a_x}{\rho_y}\right)^2}$$

$$(3.63)$$

$$\frac{\beta_0}{F_0/4k_x\rho_y} = \frac{k_x}{k_z} \frac{\sqrt{\left[\frac{k_x}{k_z}\left(\frac{a_z}{\rho_y} - \frac{d_z}{\rho_y}\right) + \frac{d_z}{\rho_y}\left(\frac{f}{f_z}\right)^2\right]^2 + \left[\frac{d_x}{\rho_y}\left(\frac{k_x}{k_z} - \frac{f^2}{f_z^2}\right)\right]^2}}{\left(\frac{f}{f_z}\right)^4 - \left[\frac{k_x}{k_z} + \frac{k_x}{k_z}\left(\frac{a_z}{\rho_y}\right)^2 + \left(\frac{a_x}{\rho_y}\right)^2\right]\left(\frac{f}{f_z}\right)^2 + \frac{k_x}{k_z}\left(\frac{a_x}{\rho_y}\right)^2}$$

where a_x, a_z are location coordinates of the resilient supports, and

$$f_z = \frac{1}{2\pi} \sqrt{\frac{4k_z}{m}} \tag{3.64}$$

The amplitude of the oscillating force F_{tx} in the X direction and the amplitude of the oscillating moment M_{ty} about the Y axis which are transmitted to the foundation by the combination of resilient elements are

$$F_{tx} = 4k_x \sqrt{x_{c0}^2 - 2a_z x_{c0}\beta_0 \cos(\phi_x - \phi_\beta) + a_z^2\beta_0^2}$$

$$M_{ty} = 4k_z a_x^2 \beta_0 \tag{3.65}$$

where F_{tx} is the sum of the forces transmitted by the individual resilient elements and M_{ty} is a moment formed by forces in the Z direction of opposite sign at opposite resilient supports. The angles ϕ_x and ϕ_β are defined by

$$\tan \phi_x = \frac{\dfrac{k_x}{k_z}\dfrac{a_z}{\rho_y}\left(\dfrac{a_z}{\rho_y} - \dfrac{d_z}{\rho_y}\right) + \left(\dfrac{a_x}{\rho_y}\right)^2 - \left(\dfrac{f}{f_z}\right)^2}{\dfrac{k_x}{k_z}\dfrac{a_z}{\rho_y}\dfrac{d_x}{\rho_y}} \qquad [0° \le \phi_x \le 360°]$$

$$\tan \phi_\beta = \frac{\dfrac{k_x}{k_z}\left(\dfrac{a_z}{\rho_y} - \dfrac{d_z}{\rho_y}\right) + \dfrac{d_z}{\rho_y}\left(\dfrac{f}{f_z}\right)^2}{\dfrac{d_x}{\rho_y}\left[\dfrac{k_x}{k_z} - \left(\dfrac{f}{f_z}\right)^2\right]} \qquad [0° \le \phi_\beta \le 360°]$$

To obtain the correct value of $(\phi_x - \phi_\beta)$ in Eq. (3.65), the signs of the numerator and denominator in each tangent term must be inspected to determine the proper quadrant for ϕ_x and ϕ_β.

Example 3.2. Consider an electric motor which has an unbalanced rotor, creating a centrifugal force. The motor weighs 3,750 lb, and has a radius of gyration $\rho_y = 9.10$ in. The distances $d_x = d_y = d_z = 0$, that is, the axis of rotation is the Y principal axis and the center-of-gravity of the rotor is in the XZ plane. The resilient supports each have a stiffness ratio of $k_x/k_z = 1.16$, and are located at $a_z = -14.75$ in., $a_x = \pm12.00$ in. The resulting displacement amplitudes of the center-of-gravity, expressed dimensionlessly, are shown in Fig. 3.28; the force and moment amplitudes transmitted to the foundation, expressed dimensionlessly, are shown in Fig. 3.29. The displacements of the center-of-gravity of the body are dimensionalized with respect to the displacements at zero frequency:

$$z_{c0}(0) = \frac{F_0}{4k_z}$$

$$x_{c0}(0) = \frac{F_0}{4k_x}\left[1 + \frac{k_x}{k_z}\left(\frac{a_z}{a_x}\right)^2\right] \tag{3.66}$$

$$\beta_0(0) = \frac{F_0}{4k_z a_z}\left(\frac{a_z}{a_x}\right)^2$$

At excitation frequencies greater than the higher natural frequency of the x_c, β coupled motion, the displacements, forces, and moment all continuously decrease as the frequency increases.

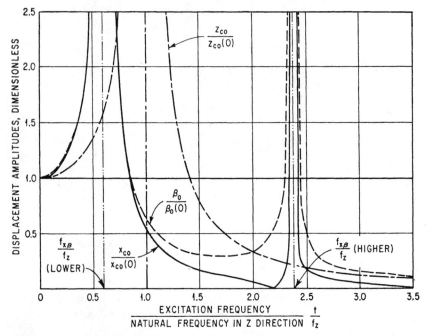

FIGURE 3.28 Response curves for the system shown in Fig. 3.27 when excited by a rotating force F_0 acting about the Y axis. The parameters of the system are $k_x/k_z = 1.16$, $a_x/\rho_y = \pm1.32$, $a_z/\rho_y = -1.62$. Only x_c, z_c, β displacements of the body are excited [see Eqs. (3.63)]. The displacements are expressed dimensionlessly by employing the displacements at zero frequency [see Eqs. (3.66)].

Two Planes of Symmetry with Orthogonal Resilient Elements Excited by an Oscillating Moment. Consider the oscillatory moment M_0 acting about the Y axis with forcing frequency ω. The resulting applied forces and moments acting on the body are

$$M_y = M_0 \sin \omega t$$
$$F_x = F_y = F_z = M_x = M_z = 0 \tag{3.67}$$

Substituting conditions of symmetry defined by Eqs. (3.33), (3.34), (3.35), and (3.38), and the excitation defined by Eqs. (3.61) and (3.67), the equations of motion [Eqs. (3.31)] show that oscillations are excited only in the x_c, β coupled mode. Solving for the resulting displacements,

$$\frac{x_{c0}}{M_0/4k_x\rho_y} = \frac{\left(\dfrac{k_x}{k_z}\right)^2 \dfrac{a_z}{\rho_y}}{\left(\dfrac{f}{f_z}\right)^4 - \left[\dfrac{k_x}{k_z} + \dfrac{k_x}{k_z}\left(\dfrac{a_z}{\rho_y}\right)^2 + \left(\dfrac{a_x}{\rho_y}\right)^2\right]\left(\dfrac{f}{f_z}\right)^2 + \dfrac{k_x}{k_z}\left(\dfrac{a_x}{\rho_y}\right)^2}$$

$$\frac{\beta_0}{M_0/4k_x\rho_y^2} = \frac{\dfrac{k_x}{k_z}\left[\dfrac{k_x}{k_z} - \left(\dfrac{f}{f_z}\right)^2\right]}{\left(\dfrac{f}{f_z}\right)^4 - \left[\dfrac{k_x}{k_z} + \dfrac{k_x}{k_z}\left(\dfrac{a_z}{\rho_y}\right)^2 + \left(\dfrac{a_x}{\rho_y}\right)^2\right]\left(\dfrac{f}{f_z}\right)^2 + \dfrac{k_x}{k_z}\left(\dfrac{a_x}{\rho_y}\right)^2} \tag{3.68}$$

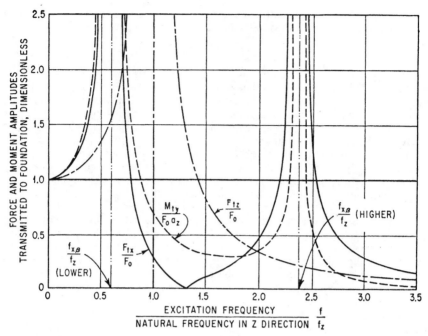

FIGURE 3.29 Force and moment amplitudes transmitted to the foundation for the system shown in Fig. 3.27 when excited by a rotating force F_0 acting about the Y axis. The parameters of the system are $k_x/k_z = 1.16$, $a_x/\rho_y = \pm1.32$, $a_z/\rho_y = -1.62$. The amplitudes of the oscillating forces in the X and Z directions transmitted to the foundation are F_{tx} and F_{tz}, respectively. The amplitude of the total oscillating moment about the Y axis transmitted to the foundation is M_{ty}.

The amplitude of the oscillating force F_{tx} in the X direction and the amplitude of the oscillating moment M_{ty} about the Y axis transmitted to the foundation by the combination of resilient supports are

$$F_{tx} = 4k_x(x_{c0} - a_z\beta_0)$$

$$M_{ty} = 4k_z a_x^2\beta_0 \tag{3.69}$$

where F_{tx} and M_{ty} have the same meaning as in Eqs. (3.65). Low vibration transmission of force and moment to the foundation is decreased at the higher frequencies in a manner similar to that shown in Fig. 3.29.

FOUNDATION-INDUCED VELOCITY SHOCK

A velocity shock is an instantaneous change in the velocity of one portion of a system relative to another portion. In this section, the system is a rigid body supported by orthogonal resilient elements within a rigid container; the container experiences a velocity shock. The system has one plane of symmetry; modal coupling and natural frequencies for such a system are considered above. Two types of velocity shock are analyzed: (1) a sudden change in the translational velocity of the container and (2) a sudden change in the rotational velocity of the container. In both instances the change in velocity is from an initial velocity to zero. No forces or moments are applied directly to the resiliently supported body; i.e., only the forces transmitted by the resilient supports act. Thus, in the equations of motion, Eqs. (3.31):

$$F_x = F_y = F_z = M_x = M_y = M_z = 0 \tag{3.70}$$

The modal coupling and natural frequencies for this system have been determined when the YZ plane is a plane of symmetry and the conditions of symmetry of Eqs. (3.33) to (3.35) apply. It is assumed that the velocity components of the body ($\dot{x}_c$, $\dot{y}_c$, $\dot{z}_c$, $\dot{\alpha}$, $\dot{\beta}$, $\dot{\gamma}$) and the velocity components of the supporting container ($\dot{u}$, $\dot{v}$, $\dot{w}$, $\dot{\alpha}$, $\dot{\beta}$, $\dot{\gamma}$) are respectively equal at time $t < 0$. At $t = 0$, all velocity components of the supporting container are brought to zero instantaneously. To determine the subsequent motion of the resiliently supported body, the natural frequencies f_n in the coupled modes of response are first calculated using Eq. (3.36). Then the response motion of the resiliently supported body to the two types of velocity shock can be found by the analyses which follow.[24]

One Plane of Symmetry with Orthogonal Resilient Supports Excited by a Translational Velocity Shock.   Figure 3.30 shows a rigid body supported within a rigid container by resilient supports in such a manner that the YZ plane is a plane of symmetry. The entire system moves with constant velocity v_0 and without relative motion. At time $t = 0$, the container impacts inelastically against the rigid wall shown at the right. The following initial conditions of displacement and velocity apply at the instant of impact ($t = 0$):

$$\dot{y}_c(0) = v_0$$

$$x_c(0) = y_c(0) = z_c(0) = \alpha(0) = \beta(0) = \gamma(0) = 0 \tag{3.71}$$

$$\dot{x}_c(0) = \dot{z}_c(0) = \dot{\alpha}(0) = \dot{\beta}(0) = \dot{\gamma}(0) = 0$$

As a result of the impact, the velocity of the supported rigid body tends to continue and is responsible for excitation of the system in the coupled mode of the y_c, z_c, α motions. The maximum displacements of the center-of-gravity of the supported body are

$$\frac{y_{cm}}{2\pi \dot{v}_0 / f_z} = \frac{1}{B} \sum_{n=1}^{3} \left(\frac{|A_n|}{f_n / f_z} \right)$$

$$\frac{z_{cm}}{2\pi \dot{v}_0 / f_z} = \frac{1}{B} \sum_{n=1}^{3} \left(\frac{|M_n A_n|}{f_n / f_z} \right) \tag{3.72}$$

$$\frac{\alpha_m}{2\pi \dot{v}_0 / \rho_x f_z} = \frac{1}{B} \sum_{n=1}^{3} \left(\frac{|N_n A_n|}{f_n / f_z} \right)$$

The maximum accelerations of the center-of-gravity of the supported body are

$$\frac{\ddot{y}_{cm}}{2\pi f_z v_0} = \frac{1}{B} \sum_{n=1}^{3} \left(|A_n| \frac{f_n}{f_z} \right)$$

$$\frac{\ddot{z}_{cm}}{2\pi f_z v_0} = \frac{1}{B} \sum_{n=1}^{3} \left(|M_n A_n| \frac{f_n}{f_z} \right) \tag{3.73}$$

$$\frac{\ddot{\alpha}_m}{2\pi f_z v_0 / \rho_x} = \frac{1}{B} \sum_{n=1}^{3} \left(|N_n A_n| \frac{f_n}{f_z} \right)$$

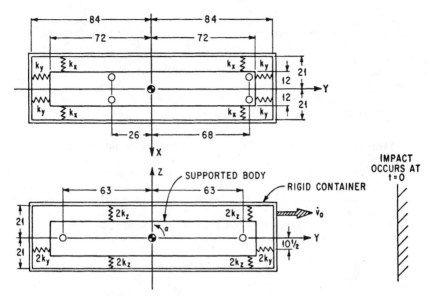

FIGURE 3.30 Example of a rigid body supported within a rigid container by resilient elements with YZ a plane of symmetry. Excitation is by a translational velocity shock in the Y direction. Prior to impact the entire system moves with constant velocity $\dot{v}_0$ and without relative motion. The rigid container impacts inelastically against the wall shown at the right, and y_c, z_c, α motions of the internally supported body result, as described mathematically by Eqs. (3.72) and (3.73).

where the subscript m denotes maximum value and

$$M_n = \frac{1}{1 - (f_n/f_z)^2}\left[\frac{\Sigma k_y}{\Sigma k_z} - \left(\frac{f_n}{f_z}\right)^2\right]\frac{\Sigma k_z a_y}{\Sigma k_y a_z}$$

$$N_n = \left[\frac{\Sigma k_y}{\Sigma k_z} - \left(\frac{f_n}{f_z}\right)^2\right]\frac{\rho_x \Sigma k_z}{\Sigma k_y a_z} \qquad (3.74)$$

$$A_n = M_{n+1}N_{n+2} - M_{n+2}N_{n+1}$$

$$B = \left|\sum_{n=1}^{3} M_n(N_{n+1} - N_{n+2})\right|$$

The fictitious natural frequency f_z is defined for mathematical purposes by Eq. (3.37). The numerical values of the subscript numbers $n, n+1, n+2$ denote the three natural frequencies in the coupled mode of the y_c, z_c, α motions determined from Eq. (3.36). These natural frequencies are arbitrarily assigned the values $n = 1, 2, 3$. When $n+1$ or $n+2$ equals 4, use 1 instead; when $n+2$ equals 5, use 2 instead. Maximum displacements and accelerations may be calculated for other points in the supported rigid body by using Eqs. (3.1) except that each of the terms must be made numerically additive. For example, the maximum value of the y displacement at the point b having the Z coordinate b_z is

$$y_{bm} = y_{cm} + |b_z|\alpha_m \qquad (3.75)$$

since $\gamma = 0$.

Since the system is assumed undamped, the response of the suspended body in terms of displacement or acceleration consists of a superposition of three sinusoidal components at the three natural frequencies in the coupled y_c, z_c, α mode. The absolute values of terms appear in Eq. (3.75) because the maximum response is the sum of the amplitudes of the three component vibrations which make up the over-all response. In general, the maximum response occurs when the three component vibrations reach their maximum positive or negative values at the same instant. Thus, the maximum values of response apply both in positive and negative directions.

One Plane of Symmetry with Orthogonal Resilient Supports Excited by a Rotational Velocity Shock. Alternative to the type of impact illustrated in Fig. 3.30, the system may be excited by imparting a rotational velocity shock (e.g., by lifting and dropping one end of the container), as illustrated in Fig. 3.31. It is assumed that the container impacts inelastically. The system has the same form of symmetry as that shown in Fig. 3.30, and only the y_c, z_c, α modes are excited. The initial conditions at the instant of impact ($t = 0$), based upon the angular velocity $\dot{\alpha}_0$ of the rigid container about point A in Fig. 3.31, are

$$\dot{y}_c(0) = -d_z\dot{\alpha}_0 \qquad \dot{z}_c(0) = d_y\dot{\alpha}_0 \qquad \dot{\alpha}(0) = \dot{\alpha}_0$$

$$x_c(0) = y_c(0) = z_c(0) = \alpha(0) = \beta(0) = \gamma(0) = 0 \qquad (3.76)$$

$$\dot{x}_c(0) = \dot{\beta}(0) = \dot{\gamma}(0) = 0$$

Note that d_y and d_z are negative quantities. The initial conditions in Eqs. (3.76) are based on the assumption that motion of the rigid body relative to the container during the fall is negligible compared to that which occurs after the impact. The maximum displacements of the center-of-gravity of the supported body are

$$\frac{y_{cm}}{2\pi\rho_x\dot{\alpha}_0/f_z} = \frac{1}{B}\sum_{n=1}^{3}\left[\left|\frac{d_z}{\rho_x}A_n + \frac{d_y}{\rho_x}(N_{n+1}-N_{n+2}) + (M_{n+2}-M_{n+1})\right|\frac{f_z}{f_n}\right]$$

$$\frac{z_{cm}}{2\pi\rho_x\dot{\alpha}_0/f_z} = \frac{1}{B}\sum_{n=1}^{3}\left[\left|M_n\left(\frac{d_z}{\rho_x}A_n + \frac{d_y}{\rho_x}(N_{n+1}-N_{n+2}) + (M_{n+2}-M_{n+1})\right)\right|\frac{f_z}{f_n}\right]$$

$$(3.77)$$

$$\frac{\alpha_m}{2\pi\dot{\alpha}_0/f_z} = \frac{1}{B}\sum_{n=1}^{3}\left[\left|N_n\left(\frac{d_z}{\rho_x}A_n + \frac{d_y}{\rho_x}(N_{n+1}-N_{n+2}) + (M_{n+2}-M_{n+1})\right)\right|\frac{f_z}{f_n}\right]$$

The maximum accelerations of the center-of-gravity of the supported body are

$$\frac{\ddot{y}_{cm}}{2\pi\rho_x f_z\dot{\alpha}_0} = \frac{1}{B}\sum_{n=1}^{3}\left[\left|\frac{d_z}{\rho_x}A_n + \frac{d_y}{\rho_x}(N_{n+1}-N_{n+2}) + (M_{n+2}-M_{n+1})\right|\frac{f_n}{f_z}\right]$$

$$\frac{\ddot{z}_{cm}}{2\pi\rho_x f_z\dot{\alpha}_0} = \frac{1}{B}\sum_{n=1}^{3}\left[\left|M_n\left(\frac{d_z}{\rho_x}A_n + \frac{d_y}{\rho_x}(N_{n+1}-N_{n+2}) + (M_{n+2}-M_{n+1})\right)\right|\frac{f_n}{f_z}\right]$$

$$(3.78)$$

$$\frac{\ddot{\alpha}_m}{2\pi f_z\dot{\alpha}_0} = \frac{1}{B}\sum_{n=1}^{3}\left[\left|N_n\left(\frac{d_z}{\rho_x}A_n + \frac{d_y}{\rho_x}(N_{n+1}-N_{n+2}) + (M_{n+2}-M_{n+1})\right)\right|\frac{f_n}{f_z}\right]$$

where d_z and d_y are the Z and Y coordinates, respectively, of the edges of the container, as shown in Fig. 3.31, and the other quantities are the same as those appear-

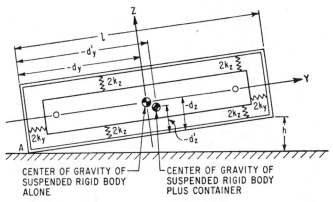

CENTER OF GRAVITY OF
SUSPENDED RIGID BODY
ALONE

CENTER OF GRAVITY OF
SUSPENDED RIGID BODY
PLUS CONTAINER

FIGURE 3.31 System shown in Fig. 3.30 excited by a rotational velocity shock about the X axis. The shock is induced by lifting and dropping one end of the rigid container to make inelastic impact with the foundation. If the height of drop is h, the rotational velocity of the system about the corner A at the instant of impact is given by Eq. (3.79). The response of the resiliently supported body is described mathematically by Eqs. (3.77) and (3.78).

ing in Eqs. (3.72) and (3.74). The maximum response at any point in the suspended body can be found in the manner of Eq. (3.75).

The rotational velocity $\dot{\alpha}_0$ of the container about the corner A in Fig. 3.31 may be induced by lifting the opposite end to a height h and dropping it. The resulting velocity $\dot{\alpha}_0$ is

$$\dot{\alpha}_0 = \left\{ \frac{2g}{\rho_A^2} \left[\frac{h}{l} \, d_y' + d_z' \sqrt{1 - \left(\frac{h}{l} \right)^2} - d_z' \right] \right\}^{1/2} \tag{3.79}$$

where g is the acceleration of gravity, ρ_A is the radius of gyration of the rigid body plus container about the corner A, h is the initial elevation of the raised end of the container, l is the length of the container, and d_y' and d_z' are the Y and Z coordinates, respectively, of the edges of the container with respect to the center-of-gravity of the assembly of rigid body plus container (see Fig. 3.31).

Example 3.3. The rigid body shown in Fig. 3.31 weighs 1,500 lb and has a radius of gyration $\rho_x = 42$ in. with respect to the X axis. The resilient supporting elements apply forces parallel to their longitudinal axes *only*. Each element with its longitudinal axis in the X or Y direction has a stiffness of $k_x = k_y = 500$ lb/in. Each element whose longitudinal axis extends in the Z direction has a stiffness $k_z = 1,000$ lb/in. The resilient elements are positioned as shown in Fig. 3.30, and $l = 168$ in., $d_y = d_y' = -84$ in., $d_z = d_z' = -21$ in., $\rho_A = 308$ in. The rotational velocity shock results from a height of drop $h = 36$ in.

The fictitious natural frequency f_z is obtained from Eq. (3.37), yielding $f_z = 7.22$ Hz. From Eq. (3.36) or Fig. 3.14, the natural frequencies in the y_c, z_c, α mode are $f_1 = 3.58$ Hz, $f_2 = 6.02$ Hz, and $f_3 = 9.75$ Hz. From Eqs. (3.74), it is determined that $M_1 \approx 0$, $M_2 = 11.7$, $M_3 = -15.3$, $N_1 = -0.1$, $N_2 = 7.1$, $N_3 = 25.1$, $A_1 = 402$, $A_2 = 2$, $A_3 = 1$, $B = 405$. Sample calculations for M_1 and A_1 are

$$M_1 = \frac{1}{1 - (3.58/7.22)^2} \left[\frac{4(500)}{8(1,000)} - \left(\frac{3.58}{7.22} \right)^2 \frac{4(1,000)(68 - 26)}{4(500)(-10.5)} \right] = -0.04$$

$$A_1 = M_2 N_3 - M_3 N_2 = (11.7)(25.1) - (-15.3)(7.1) = 402$$

From Eq. (3.79), $\dot{\alpha}_0 = 0.38$ rad/sec. Then Eqs. (3.78) give the maximum acceleration of the center-of-gravity in the Y direction of the supported body as follows:

$$\ddot{y}_{cm} = \frac{2\pi\rho_x f_z \dot{\alpha}_0}{B}\begin{bmatrix} \left|\dfrac{d_z}{\rho_x} A_1 + \dfrac{d_y}{\rho_x}(N_2 - N_3) + (M_3 - M_2)\right|\dfrac{f_1}{f_z} \\[2mm] + \left|\dfrac{d_z}{\rho_x} A_2 + \dfrac{d_y}{\rho_x}(N_3 - N_1) + (M_1 - M_3)\right|\dfrac{f_2}{f_z} \\[2mm] + \left|\dfrac{d_z}{\rho_x} A_3 + \dfrac{d_y}{\rho_x}(N_1 - N_2) + (M_2 - M_1)\right|\dfrac{f_3}{f_z} \end{bmatrix}$$

$$= \frac{724 \text{ in./sec}^2}{405}\begin{bmatrix} \left|\dfrac{-21}{42}(402) + \dfrac{-84}{42}(7.1 - 25.1) + (-15.3 - 11.7)\right|\dfrac{3.58}{7.22} \\[2mm] + \left|\dfrac{-21}{42}(2) + \dfrac{-84}{42}(25.1 + 0.1) + (0 + 15.3)\right|\dfrac{6.02}{7.22} \\[2mm] + \left|\dfrac{-21}{42}(1) + \dfrac{-84}{42}(-0.1 - 7.1) + (11.7 - 0)\right|\dfrac{9.75}{7.22} \end{bmatrix}$$

$= 286$ in./sec$^2 = 0.74g$

In a similar manner:

$$z_{cm} = 1{,}580 \text{ in./sec}^2 = 4.09g$$

$$\ddot{\alpha}_m = 45.9 \text{ rad/sec}^2$$

REFERENCES

1. Marks, L. S., and T. Baumeister: "Mechanical Engineer's Handbook," 6th ed., p. **6-6**, McGraw-Hill Book Company, Inc., New York, 1958.

2. Housner, G. W., and D. E. Hudson: "Applied Mechanics-Dynamics," 2d ed., chap. 7, D. Van Nostrand Company, Inc., Princeton, N.J., 1959.

3. Boucher, R. W., D. A. Rich, H. L. Crane, and C. E. Matheny: *NACA Tech. Note* 3084, 1954.

4. Woodward, C. R.: "Handbook of Instructions for Experimentally Determining the Moments and Products of Inertia of Aircraft by the Spring Oscillation Method," *WADC Tech. Rept.* 55–415, June 1955.

5. Rubin, S.: *SAE Preprint* 197, 1957.

6. Macduff, J. N.: *Product Eng.,* **17**:106, 154 (1946).

7. Vane, F. F.: "A Guide for the Selection and Application of Resilient Mountings to Shipboard Equipment—Revised," *David W. Taylor Model Basin Rept.* 880, February 1958, p. 98.

8. Smollen, L. E.: *J. Acoust. Soc. Amer.,* **40**:195 (1966).

9. Ref. 7, p. 50.

10. Crede, C. E.: "Vibration and Shock Isolation," p. 68, John Wiley & Sons, Inc., New York, 1951.

11. Ref. 7, pp. 37–49.

12. Ref. 10, pp. 53–58.

13. Lewis, R. C., and K. Unholtz: *Trans. ASME,* **69**:813 (1947).

14. Klein, E., R. S. Ayre, and I. Vigness: "Fundamentals of Guided Missile Packaging," *Dept. Defense (U.S.) Rept.* RD 219/3, July 1955, appendix 8, pp. 49–52.

15. Ref. 10, p. 73.

16. Taylor, E. S., and K. A. Browne: *J. Aeronaut. Sci.*, **6**:43 (1938).

17. Browne, K. A.: *Trans. SAE,* **44**:185 (1939).

18. Derby, T. F.: "Decoupling the Three Translational Modes from the Three Rotational Modes of a Rigid Body Supported by Four Corner-Located Isolators," *Shock and Vibration Bull.*, 43, pt. 4, June 1973, pp. 91–108.

19. Ref. 10, p. 50.

20. Himelblau, H.: "A Reliable Approach to Protecting Fragile Equipment from Aircraft Vibration," *North American Aviation, Inc., Rept.* NA-56-1030, 1957, pp. 16, 86.

21. Ref. 20, pp. 22, 95.

22. Ref. 10, pp. 43, 61.

23. Himelblau, H.: *Product Eng.,* **23**:151 (1952).

24. Ref. 14, chap. 11, November, 1955.

25. Ref. 2, appendix IV.

CHAPTER 4
NONLINEAR VIBRATION

Fredric Ehrich
H. Norman Abramson

INTRODUCTION

A vast body of scientific knowledge has been developed over a long period of time devoted to a description of natural phenomena. In the field of mechanics, rapid progress in the past two centuries has occurred, due in large measure to the ability of investigators to represent physical laws in terms of rather simple equations. In many cases the governing equations were not so simple; therefore, certain assumptions, more or less consistent with the physical situation, were employed to reduce the equations to types more easily soluble. Thus, the process of linearization has become an intrinsic part of the rational analysis of physical problems. An analysis based on linearized equations, then, may be thought of as an analysis of a corresponding but idealized problem.

In many instances the linear analysis is insufficient to describe the behavior of the physical system adequately. In fact, one of the most fascinating features of a study of nonlinear problems is the occurrence of new and totally unsuspected phenomena; i.e., new in the sense that the phenomena are not predicted, or even hinted at, by the linear theory. On the other hand, certain phenomena observed physically are unexplainable except by giving due consideration to nonlinearities present in the system.

The branch of mechanics that has been subjected to the most intensive attack from the nonlinear viewpoint is the theory of vibration of mechanical and electrical systems. Other branches of mechanics, such as incompressible and compressible fluid flow, elasticity, plasticity, wave propagation, etc., also have been studied as non-linear problems, but the greatest progress has been made in treating vibration of nonlinear systems. The systems treated in this chapter are systems with a finite number of degrees-of-freedom which can be defined by a finite number of simultaneous ordinary differential equations; on the other hand, the mechanics of continua involves partial differential equations. Nonlinear ordinary differential equations are easier to handle than nonlinear partial differential equations. Interesting surveys of the entire realm of nonlinear mechanics are given in Refs. 1, 2, and 7.

This chapter provides information concerning features of nonlinear vibration theory likely to be encountered in practice and methods of nonlinear vibration analysis which find ready application.

EXAMPLES OF SYSTEMS POSSESSING NONLINEAR CHARACTERISTICS

SIMPLE PENDULUM

As a first example of a system possessing nonlinear characteristics, consider a simple pendulum of length l having a bob of mass m, as shown in Fig. 4.1. The well-known differential equation governing free vibration is

$$ml^2\ddot{\theta} + mgl\theta = 0 \tag{4.1}$$

This equation holds only for small oscillations about the position of equilibrium since the actual restoring moment is characterized by the quantity $\sin\theta$. Equation (4.1) thus employs the assumption $\sin\theta \simeq \theta$. The exact, but nonlinear, equation of motion is

$$ml^2\ddot{\theta} + mgl\sin\theta = 0 \tag{4.2}$$

SIMPLE SPRING-MASS SYSTEM

A simple spring-mass system, as shown in Fig. 4.2, is characterized by the equation

$$m\ddot{x} + kx = 0$$

This equation is based on the assumption that the elastic spring obeys Hooke's law; i.e., the characteristic curve of restoring force versus displacement is a straight line. However, many materials do not exhibit such a linear characteristic. Further, in the case of a simple coil spring, a deviation from linearity occurs at large compression as the coils begin to close up, or conversely, when the extension becomes so great that the coils begin to lose their individual identity. In either case, the spring exhibits a characteristic such that the restoring force increases more rapidly than the displacement. Such a characteristic is called *hardening*. In a similar manner, certain systems (e.g., a simple pendulum) exhibit a *softening* characteristic. Both types of characteristic are shown in Fig. 4.3. A simple system with either softening or hardening restoring force may be described approximately by an equation of the form

$$m\ddot{x} + k(x \pm \mu^2 x^3) = 0$$

where the upper sign refers to the hardening characteristic and the lower to the softening characteristic.

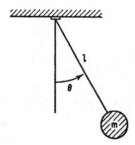

FIGURE 4.1 Simple pendulum.

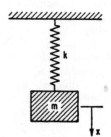

FIGURE 4.2 Simple spring-mass system.

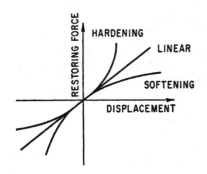

FIGURE 4.3 Restoring force characteristic curves for linear, hardening, and softening vibration systems.

It is possible for a system with only linear components to exhibit nonlinear characteristics, by snubber action for example, as shown in Fig. 4.4. A system undergoing vibration of small amplitude also may exhibit nonlinear characteristics; for example, in the pendulum shown in Fig. 4.5, the length depends on the amplitude.

STRETCHED STRING WITH CONCENTRATED MASS

The large amplitude vibration of a stretched string with a concentrated mass, as shown in Fig. 4.6, offers another example of a nonlinear system. The governing nonlinear differential equation is, approximately,

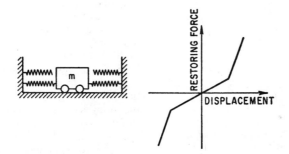

FIGURE 4.4 Nonlinear mechanical system with snubber action showing piecewise linear restoring force characteristic curve.

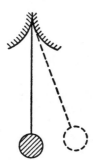

FIGURE 4.5 Pendulum with nonlinear characteristic resulting from dependence of length on vibration amplitude.

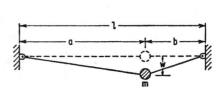

FIGURE 4.6 Vibration of a weighted string as an example of a nonlinear system.

$$m\ddot{w} + F_0\left(\frac{l}{ab}\right)w + (SE - F_0)\left(\frac{a^3 + b^3}{2a^3b^3}\right)w^3 = 0$$

where F_0 is the initial tension, S is the cross-sectional area, and E is the elastic modulus of the string. Consider now the special case of $a = b$ and denote the unstretched length of the half string by l_0. Then the initial tension and the restoring force become

$$F_0 = SE\left(\frac{a - l_0}{l_0}\right)$$

$$F_r \simeq SE\left[2\left(\frac{a}{l_0} - 1\right)\left(\frac{w}{a}\right) + \left(2 - \frac{a}{l_0}\right)\left(\frac{w}{a}\right)^3\right]$$

An interesting feature of this system is that it exhibits a wide variety of either hardening or softening characteristics, depending upon the value of a/l_0, as shown in Fig. 4.7.

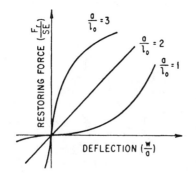

FIGURE 4.7 Restoring force characteristics for the weighted string shown in Fig. 4.6.

SYSTEM WITH VISCOUS DAMPING

The foregoing examples all involve nonlinearities in the elastic components, either as a result of appreciable amplitudes of vibration or as a result of peculiarities of the elastic element. Consider a simple spring-mass system which also includes a dashpot. The usual assumptions pertaining to this system are that the spring is linear and that the motion is sufficiently slow that the viscous resistance provided by the dashpot is proportional to the velocity; therefore, the governing equation of motion is linear. Frequently, the dashpot resistance is more correctly expressed by a term proportional to the square of the velocity. Further, the resistance is always such as to oppose the motion; therefore, the nonlinear equation of motion may be written

$$m\ddot{x} + c|\dot{x}|\dot{x} + kx = 0$$

BELT FRICTION SYSTEM

The system shown in Fig. 4.8A involves a nonlinearity depending upon the dry friction between the mass and the moving belt. The belt has a constant speed v_0, and the applicable equation of motion is

$$m\ddot{x} + F(\dot{x}) + kx = 0$$

where the friction force $F(\dot{x})$ is shown in Fig. 4.8B. For large values of displacement, the damping term is positive, has positive slope, and removes energy from the sys-

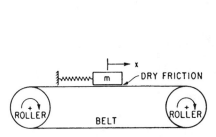

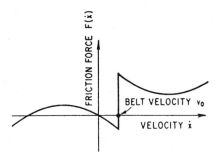

FIGURE 4.8A Belt friction system which exhibits self-excited vibration.

FIGURE 4.8B Damping force characteristic curve for the belt friction system shown in Fig. 4.8A.

tem; for small values of displacement, the damping term is negative, has negative slope, and actually puts energy into the system. Even though there is no external stimulus, the system can have an oscillatory solution, and thus corresponds to a non-linear *self-excited* system.

SYSTEMS WITH ASYMMETRIC STIFFNESS

The aforementioned examples of nonlinear stiffness, typified by the stiffness variations in Figs. 4.3, 4.4, and 4.7, all may be characterized as symmetric. That is, the variation in the absolute value of the restoring force with displacement in the positive direction is identical to the variation in the absolute value of the restoring force with displacement in the negative direction. As will be seen in the following sections, symmetric stiffness distributions result in changes in the shape of the resonant peak of the response curve and slight distortion in the waveform of the dynamic motion without changing the basic synchronism between forcing function and response. But many more diverse phenomena and much more profound changes are encountered when dealing with asymmetric stiffness distributions.

A typical physical situation is encountered in the dynamics of rotating machinery where a softly mounted rotor is located eccentrically within the small clearance of a motion limiting stiff stator as illustrated in Fig. 4.9A and C. When rotating with some unbalance in the rotor, the vertical component of the unbalance force will cause intermittent local contact with the stiff stator, resulting in a "bouncing" motion of the rotor. The stiffness characteristic for the vertical motion is asymmetric. In its simplest form, it may be represented as a bilinear relationship—very soft for vertical motion in the upward direction and very stiff for vertical motion in the downward direction, as illustrated in Fig. 4.9B. More explicitly,

$$k = K_1 \quad x > 0$$

$$k = K_2 \quad x < 0$$

Many other examples of nonlinear systems are given in the references of this chapter, particularly Refs. 3 to 5.

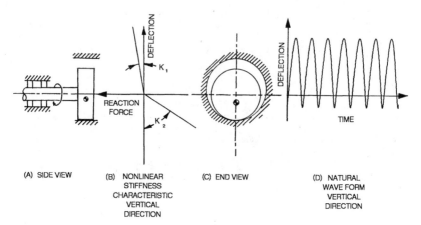

(A) SIDE VIEW   (B) NONLINEAR STIFFNESS CHARACTERISTIC VERTICAL DIRECTION   (C) END VIEW   (D) NATURAL WAVE FORM VERTICAL DIRECTION

FIGURE 4.9 Nonlinear spring characteristic of a rotor operating with local intermittent contact in a clearance.

DESCRIPTION OF NONLINEAR PHENOMENA

This section describes briefly, largely in nonmathematical terms, certain of the more important features of nonlinear vibration. Further details and methods of analysis are given later.

FREE VIBRATION

Insofar as the free vibration of a system is concerned for systems with symmetric stiffness distributions, one distinguishing feature between linear and nonlinear behavior is the dependence of the period of the motion in nonlinear vibration on the amplitude. For example, the simple pendulum of Fig. 4.1 may be analyzed on the basis of the linearized equation of motion, Eq. (4.1), from which it is found that the period of the vibration is given by the constant value $\tau_0 = 2\pi/\omega_n$. An analysis on the basis of the nonlinear equation of motion, Eq. (4.2), leads to an expression for the period of the form

$$\frac{\tau}{\tau_0} = 1 + \tfrac{1}{4}(U)^2 + \tfrac{9}{64}(U)^4 + \tfrac{25}{256}(U)^6 + \cdots \tag{4.3}$$

where U is related to the amplitude of the vibration Θ by the relation $U = \sin(\Theta/2)$. The linear solution thus corresponds to the first term of Eq. (4.3). The dependence of the period of vibration on amplitude is shown in Fig. 4.10. Systems in which the period of vibration is independent of the amplitude are called *isochronous,* while those in which the period τ is dependent on the amplitude are called *nonisochronous.*

The dependence of period on amplitude also may be seen from the vibration trace shown in Fig. 4.11, which corresponds to a solution of the equation

$$m\ddot{x} + c\dot{x} + k(x + \mu^2 x^3) = 0$$

For systems with asymmetric stiffness distributions, free undamped vibration will display significant distortion of the natural waveform. The simple bilinear stiffness

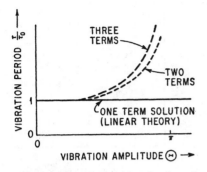

FIGURE 4.10 Period of free vibration of a simple pendulum according to Eq. (4.3) and showing the effect of nonlinear terms.

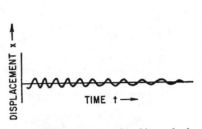

FIGURE 4.11 Deflection time-history for free damped vibration of the nonlinear system described by Duffing's equation [Eq. (4.16)].

distribution of Fig. 4.9B will result in the system having a simple harmonic half cycle at relatively low frequency for upward motion and a simple harmonic half cycle at relatively high frequency for downward motion. The overall waveform is then a combination of these two disparate half cycles as represented in Fig. 4.9D and suggests a bouncing motion.

RESPONSE CURVES FOR FORCED VIBRATION OF SYSTEMS WITH SYMMETRIC STIFFNESS

Representations of vibration behavior in the form of curves of response amplitude versus exciting frequency are called *response curves*. The response curves for an undamped linear system acted on by a harmonic exciting force of amplitude p and frequency ω may be derived from the equation of motion

$$\ddot{x} + \omega_n^2 x = \frac{p}{m} \cos \omega t \qquad (4.4)$$

The solution has the form shown in Fig. 4.12. The vertical line at $\omega = \omega_n$ corresponds not only to resonance but also to free vibration ($p = 0$); the amplitude in this instance is determined by the initial conditions of the motion. In a nonlinear system the character of the motion is dependent upon the amplitude. This requires that the natural frequency likewise be amplitude-dependent; hence, it follows that the free vibration curve $p = 0$ for nonlinear systems cannot be a straight line. Figure 4.13 shows free vibration curves (i.e., natural frequency as a function of amplitude) for hardening and softening systems.

Figures 4.12 and 4.13 suggest that the forced vibration response curves for systems with nonlinear restoring forces have the general form of those of a linear system but are "swept over" to the right or left, depending on whether the system is hardening or softening. These are shown in Fig. 4.14. The principal effect of damping in forced vibration of a nonlinear system is to limit the amplitude at resonance, as shown in Fig. 4.15.

These rightward- and leftward-leaning resonant response peaks have special meaning to the dynamic response of the system. Consider a hardening system whose response curve is shown in Fig. 4.15B. Suppose that the exciting frequency starts at a low value, and increases continuously at a slow rate. The amplitude of the vibration

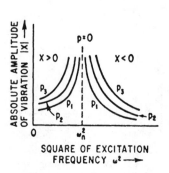

FIGURE 4.12 Family of response curves for the undamped linear system defined by Eq. (4.4).

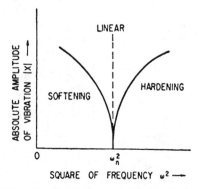

FIGURE 4.13 Free vibration curves (natural frequency as a function of amplitude) in the response diagram for linear, hardening, and softening vibration systems [see Eq. (4.49)].

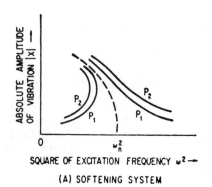

(A) SOFTENING SYSTEM

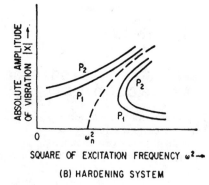

(B) HARDENING SYSTEM

FIGURE 4.14 Response curves for undamped nonlinear systems with hardening and softening restoring force characteristics [see Eq. (4.50)].

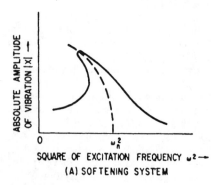

(A) SOFTENING SYSTEM

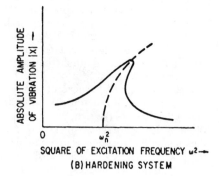

(B) HARDENING SYSTEM

FIGURE 4.15 Response curves for damped nonlinear systems with hardening and softening restoring force characteristics [see Eq. (4.52)].

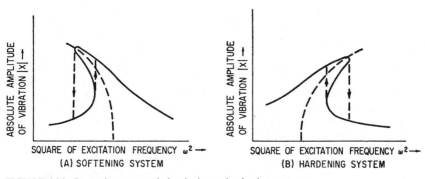

FIGURE 4.16 Jump phenomenon in hardening and softening systems.

also increases, but only up to a point. In particular, at the point of vertical tangency of the response curve, a slight increase in frequency requires that the system perform in an unusual manner; i.e., that it "jump" down in amplitude to the lower branch of the response curve. This experiment may be repeated by starting with a large value of exciting frequency but requiring that the forcing frequency be continuously reduced. A similar situation again is encountered; the system must jump up in amplitude in order to meet the conditions of the experiment. This *jump phenomenon* is shown in Fig. 4.16 for both the hardening and softening systems. The jump is not instantaneous in time but requires a few cycles of vibration to establish a steady-state vibration at the new amplitude.

There is a portion of the response curve which is "unattainable"; it is not possible to obtain that particular amplitude by a suitable choice of forcing frequency. Thus, for certain values of ω there appear to be three possible amplitudes of vibration but only the upper and lower can actually exist. If by some means it were possible to initiate a steady-state vibration with just the proper amplitude and frequency to correspond to the middle branch, the condition would be unstable; at the slightest disturbance the motion would jump to either of the other two states of motion. The direction of the jump depends on the direction of the disturbance. Thus, of the three possible states of motion, one in phase and two out-of-phase with the exciting force, the one having the larger amplitude of the two out-of-phase motions is unstable. This region of instability in the response diagram is defined by the loci of vertical tangents to the response curves, and is shown for a hardening system in Fig. 4.16C.

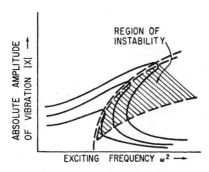

FIGURE 4.16C Instability region defined by the loci of vertical tangents of the damped response curves for the hardening system.

RESPONSE CURVES FOR FORCED VIBRATION OF SYSTEMS WITH ASYMMETRIC STIFFNESS

The system pictured in Fig. 4.9 is typical of systems with asymmetric stiffness characteristics, and its response[6] includes a variety of phenomena, including regions of chaotic response,[7] not observed in systems with symmetric stiffness characteristics.

The equations of motion in the plane normal to the plane of contact, with a stiffness of k_1 when the rotor is deflected from its rest position in the soft direction and a stiffness k_2 when the rotor is deflected from its rest position in the hard direction, may be integrated numerically using a simple trapezoidal integration procedure. The rest position of the rotor is taken at the contact point, so the break point of the bilinear elastic characteristic is at zero deflection. The system is then simply characterized by only two parameters—the ratio of the stiffnesses $\beta = k_1/k_2$ and z_1, the linear damping ratio of the system referred to critical damping of the soft system—when operated at a given rotational speed s, which is taken in normalized format as the ratio of rotational frequency to the system natural frequency.

For typical values and z_1 and β at any speed s, the numerical model may be used to compute the orbit of the rotor mass point as the orthogonal coordinates of the motion X and Y, where each of the coordinates is normalized as the ratio of the deflection from the rest position to the unbalance mass eccentricity. In considering the response over a large range of rotational speed, the motion may be simply characterized at any particular speed as Y_p, the local peak value(s) of the normalized amplitude in the direction of the nonlinear stiffness. As shown in Fig. 4.17A in comparison with the response of an equivalent system with a linear spring support stiffness, a plot of this parameter over a range of speeds is quite effective in detecting and identifying various different response phenomena.

Superharmonic Response.[8-10]   Fig. 4.17B characterizes superharmonic response at subcritical speed. Shown here at approximately one-half critical speed, the rotor is bouncing at approximately its natural frequency against the hard surface of the

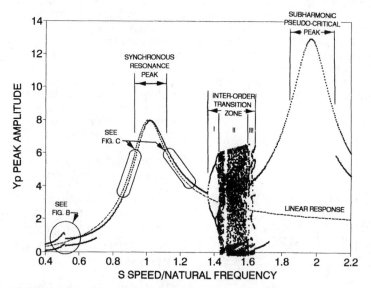

FIGURE 4.17A   Identification of various classes of nonlinear behavior in the peak amplitude response curve—typical subcritical/critical/supercritical regime ($z_1 = 0.200$; $\beta = 0.002$).

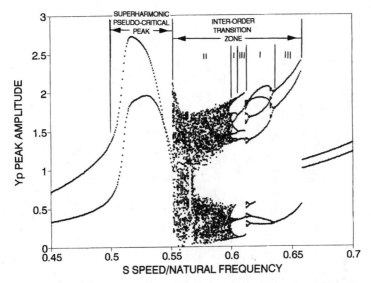

FIGURE 4.17B   Identification of various classes of nonlinear behavior in the peak amplitude response curve—detail of superharmonic pseudo-critical peak and interorder transition zone ($z_1 = 0.05; \beta = 0.005$).

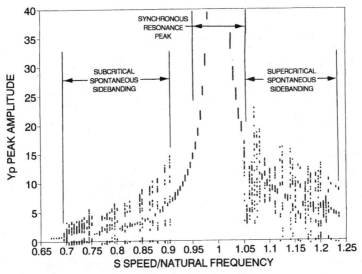

FIGURE 4.17C   Identification of various classes of nonlinear behavior in the peak amplitude response curve—detail of transcritical spontaneous sidebanding ($z_1 = 0.002; \beta = 0.002$).

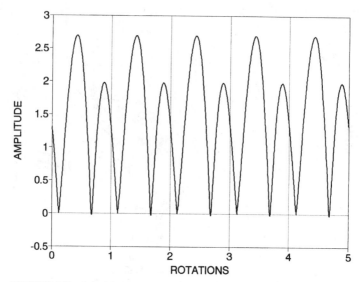

FIGURE 4.18 Subcritical superharmonic response—waveform ($z_1 = 0.050$; $\beta = 0.005$; $s = 0.525$; $M = 2$).

contact point, energized at every other bounce by the component of the unbalance centrifugal force as suggested in Fig. 4.18. The dominant frequency of the response is then precisely 2 times operating speed. Such a pseudo-critical speed is possible for any integer order M at approximately $1/M$ times critical speed and with a significant frequency component of precisely M times operating speed or approximately equal to the *natural frequency*.

Transition Between Successive Superharmonic Orders.[10]   In between the successive superharmonic response zones (i.e., between the Mth and ($M - 1$)th order superharmonic responses) there may occur a regime of irregular response. Most commonly, the response may be chaotic, as identified as Zone II in Fig. 4.17B and shown in Fig. 4.19A. For such chaotic motion, the Poincaré section, which is a stroboscopic view of the phase-plane plot of velocity versus displacement at a reference angle of shaft rotation, is effectively a slice of the system's attractor as shown in Fig. 4.19B. The chaotic motion may be preceded on one side by a cascade of period-doubling bifurcations in the trace of peak amplitude Y_p, as suggested in Zone I of Fig. 4.17B. Another pattern of transition response is periodic in waveform. As shown in Zone III of Fig. 4.17B, instead of having an unending series of local peaks with no identifiable periodicity of repetitions as would be the case in truly chaotic motion, the response appears to have clusters of K bounces that actually repeat every L rotations to give a major periodicity of K/L times s. In both the chaotic and periodic transition zones, the response has a significant component at or near the system's *natural frequency*.

Spontaneous Sidebanding in Transcritical Response (Subcritical).[11,12]   A unique response has been identified which appears in very lightly damped, highly nonlinear systems operating in the transcritical range, as shown in Fig. 4.17C. It has been observed that one of the dominant sidebands occurs at approximately critical frequency, and the sideband separation is generally a whole-number fraction ($1/J$) of

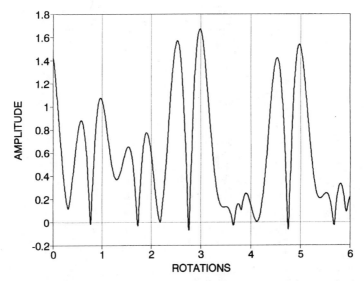

FIGURE 4.19A Subcritical chaotic transition between successive superharmonic orders—waveform ($z_1 = 0.050$; $\beta = 0.005$; $s = 0.560$; $2 < M < 1$).

the operating speed and, in the Jth order manifestation of subcritical spontaneous sidebanding when the speed is approximately $J/(J + 1)$ times the natural frequency, the dominant frequency is precisely $(J + 1)/J$ times the rotative speed or approximately equal to the *natural frequency*. The waveform, shown in Fig. 4.20, is periodic in nature. There appear to be transition zones between successive orders of J when

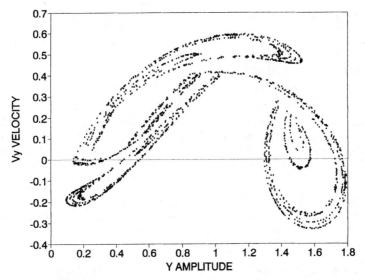

FIGURE 4.19B Subcritical chaotic transition between successive superharmonic orders—Poincarè section ($z_1 = 0.050$; $\beta = 0.005$; $s = 0.560$; $2 < M < 1$).

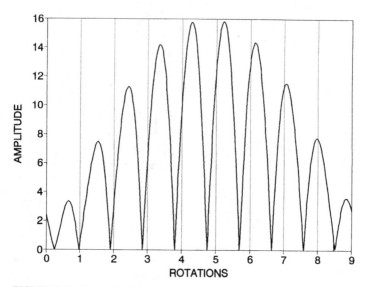

FIGURE 4.20   Transcritical spontaneous sidebanding—waveform ($z_1 = 0.001$; $\beta = 0$; $s = 0.900$; $J = 10$).

the response has a dominant frequency approximately equal to the system's *natural frequency* and the waveform may be chaotic. The general phenomenon has been referred to as *spontaneous sidebanding* because the sidebands appear around the center forcing frequency without the presence of and interaction with a second external forcing frequency.

Synchronous Resonant Response.   Synchronous critical response in the nonlinear system, shown in Fig. 4.17*A*, is very similar to that of the linear system except for the distortion of the waveform reflecting the bouncing nature of the motion illustrated in Fig. 4.21. Although the dominant frequency component is that of the forcing frequency or operating speed which is close to the *natural frequency* of the system, the bouncing waveform produces significant spectral content at whole number multiples of the operating speed.

Spontaneous Sidebanding in Transcritical Response (Supercritical).[12]   As shown in Fig. 4.17*C*, spontaneous sidebanding can occur at speeds slightly higher than critical speed, very similar in nature to the response already noted above which occurs at slightly subcritical speeds. Again, the waveform is periodic in nature. In the Jth order manifestation of supercritical spontaneous sidebanding, when the rotative speed is approximately $J/(J-1)$ times the natural frequency, the dominant frequency is precisely $(J-1)/J$ times the rotative speed, or approximately equal to the *natural frequency*. Once again, there appears to be transition zones between successive orders of J when the response has a dominant frequency approximately equal to the *natural frequency* and the waveform may be chaotic.

Subharmonic Response.[13–17]   The pseudo-critical peak at 2 times critical speeds shown in Fig. 4.17*A* exemplifies subharmonic response at supercritical speed. With a

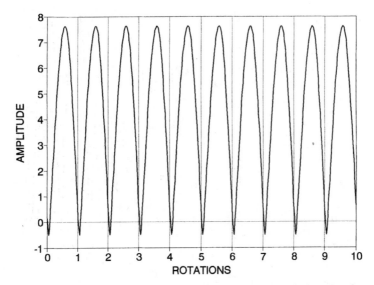

FIGURE 4.21   Critical synchronous resonant response—waveform ($z_1 = 0.200$; $\beta = 0.005$; $s = 1.050$).

peak amplitude of the same order of magnitude as critical response, the rotor is bouncing at its natural frequency against the hard surface of the contact point as depicted in Fig. 4.22 and is subjected to the periodic component of the unbalance centrifugal force twice every bounce. Only one of the two pulses of unbalance force is effective in energizing the bouncing motion in the course of each bounce, so the dominant frequency of the response is then precisely one-half the operating speed. Such a pseudo-critical is possible for any integer order N at a rotational speed approximately N times critical speed and with a dominant frequency of precisely $1/N$ times operating speed or approximately the system *natural frequency*.

Transition Between Successive Subharmonic Orders.   The transition response between successive subharmonic orders is quite analogous to the transition response between successive superharmonic orders previously noted. In between the successive subharmonic response zones (i.e., between the Nth and $(N + 1)$th order subharmonic responses) there may occur a regime of irregular response. The response has been noted by many researchers to be chaotic,[7, 18–25] as identified as Zone II in Fig. 4.17A and illustrated in Fig. 4.23A. The chaotic motion may be preceded on one side by a cascade of period-doubling bifurcations in the trace of peak amplitude Y_p, as suggested in Zone I of Fig. 4.17A. Another pattern of transition response is periodic in waveform. As shown in Zone III of Fig. 4.17A, instead of having an unending series of local peaks with no identifiable periodicity of repetitions as would be the case in truly chaotic motion, the response appears to have clusters of K bounces that actually repeat every L rotations to give a major periodicity of K/L times s. In both the chaotic and periodic transition zones, the response has a significant component at or near the system's *natural frequency*. As with subcritical chaotic transition zones, a Poincaré section of chaotic motion in a supercritical chaotic transition zone is effectively a slice of the system's attractor as shown in Fig. 4.23B.

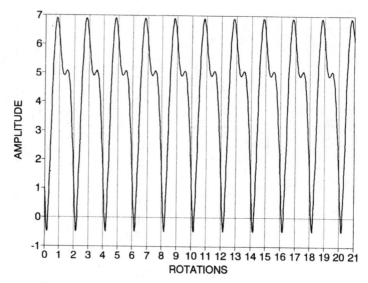

FIGURE 4.22 Supercritical subharmonic response—waveform ($z_1 = 0.200$; $\beta = 0.005$; $s = 2.150$; $N = 2$).

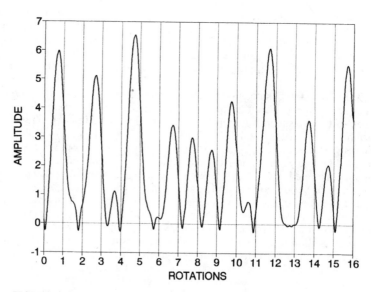

FIGURE 4.23A Supercritical chaotic transition between successive subharmonic orders—waveform ($z_1 = 0.200$; $\beta = 0.005$; $s = 1.600$; $1 < N < 2$).

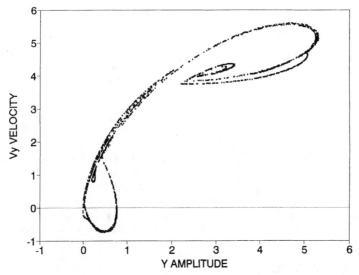

FIGURE 4.23B Supercritical chaotic transition between successive subharmonic orders—Poincarè section ($z_1 = 0.200$; $\beta = 0.005$; $s = 1.600$; $1 < N < 2$).

OTHER PHENOMENA

Self-Excited Vibration. * Consider the nonlinear equation of motion

$$m\ddot{x} + c(x^2 - 1)\dot{x} + kx = 0$$

This is known as Van der Pol's equation and may be written alternatively

$$\ddot{x} - \varepsilon(1 - x^2)\dot{x} + \kappa^2 x = 0 \tag{4.5}$$

The principal feature of this self-excited system resides in the damping term; for small displacements the damping is negative, and for large displacements the damping is positive. Thus, even an infinitesimal disturbance causes the system to oscillate; however, when the displacement becomes sufficiently large, the damping becomes positive and limits further increase in amplitude. This is shown in Fig. 4.24. Such systems, which start in a spontaneous manner, often are called *soft* systems in contrast with *hard* systems which exhibit sustained oscillations only if a shock in excess of a certain level is applied. Note that stability questions arise here (which are different from those discussed earlier in connection with jump phenomena) concerning the existence of one or more limiting amplitudes, such as the one noted above in the Van der Pol oscillator.

Relaxation Oscillations. As shown in Fig. 4.18, the motion of the Van der Pol oscillator is very nearly harmonic for $c/m = 0.1$ while the motion is made up of rela-

---

* A general treatment of self-excited vibration is given in Chap. 5.

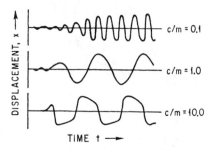

FIGURE 4.24 Displacement time-histories for Van der Pol's equation [Eq. (4.5)] for various values of the damping.

tively sudden transitions between deflections of opposite sign for $c/m = 10.0$. The period of the harmonic motion for $c/m = 0.1$ is determined essentially by the linear spring stiffness k and the mass m; the period of the motion corresponding to $c/m = 10.0$ is very much larger and depends also on c. Thus, it is possible to obtain an undamped periodic oscillation in a damped system as a result of the particular behavior of the damping term. Such oscillations are often called *relaxation oscillations*.

Asynchronous Excitation and Quenching. In linear systems, the principle of superposition is valid, and there is no interaction between different oscillations. Moreover, the mathematical existence of a periodic solution always indicates the existence of a periodic phenomenon. In nonlinear systems, there is an interaction between oscillations; the mathematical existence of a periodic solution is only a necessary condition for the existence of corresponding physical phenomena. When supplemented by the condition of stability, the conditions become both necessary and sufficient for the appearance of the physical oscillation. Therefore, it is conceivable that under these conditions the appearance of one oscillation may either create or destroy the stability condition for another oscillation. In the first case, the other oscillation appears (*asynchronous excitation*), and in the second case, disappears (*asynchronous quenching*). The term *asynchronous* is used to indicate that there is no relation between the frequencies of these two oscillations.

Entrainment of Frequency. According to linear theory, if two frequencies ω_1 and ω_2 are caused to beat in a system, the period of beating increases indefinitely as ω_2 approaches ω_1. In nonlinear systems, the beats disappear as ω_2 reaches certain values. Thus, the frequency ω_1 falls in synchronism with, or is entrained by, the frequency ω_2 within a certain range of values. This is called *entrainment of frequency,* and the band of frequencies in which entrainment occurs is called the zone of entrainment or the interval of synchronization. In this region, the frequencies ω_1 and ω_2 combine and only vibration at a single frequency ensues.

EXACT SOLUTIONS

It is possible to obtain exact solutions for only a relatively few second-order nonlinear differential equations. In this section, some of the more important of these exact solutions are listed. They are exact in the sense that the solution is given either in closed form or in an expression that can be evaluated numerically to any desired degree of accuracy. The examples given here are fairly general in nature; more specialized examples are given in Ref. 4.

FREE VIBRATION

Consider the free vibration of an undamped system with a general restoring force $f(x)$ as governed by the differential equation

$$\ddot{x} + \kappa^2 f(x) = 0$$

This can be rewritten as

$$\frac{d(\dot{x}^2)}{dx} + 2\kappa^2 f(x) = 0 \tag{4.6}$$

and integrated to yield

$$\dot{x}^2 = 2\kappa^2 \int_x^X f(\xi)\, d\xi$$

where ξ is an integration variable and X is the value of the displacement when $\dot{x} = 0$. Thus

$$|\dot{x}| = \kappa\sqrt{2}\sqrt{\int_x^X f(\xi)\, d\xi}$$

This may be integrated again to yield

$$t - t_0 = \frac{1}{\kappa\sqrt{2}} \int_0^x \frac{d\zeta}{\sqrt{\int_\zeta^X f(\xi)\, d\xi}} \tag{4.7}$$

where ζ is an integration variable and t_0 corresponds to the time when $x = 0$. The displacement-time relation may be obtained by inverting this result. Considering the restoring force term to be an odd function, i.e.,

$$f(-x) = -f(x)$$

and considering Eq. (4.7) to apply to the time from zero displacement to maximum displacement, the period τ of the vibration is

$$\tau = \frac{4}{\kappa\sqrt{2}} \int_0^X \frac{d\zeta}{\sqrt{\int_\zeta^X f(\xi)\, d\xi}} \tag{4.8}$$

Exact solutions can be obtained in all cases where the integrals in Eq. (4.8) can be expressed explicitly in terms of X.

Case 1. Pure Powers of Displacement.   Consider the restoring force function

$$f(x) = x^n$$

Equation (4.8) then becomes

$$\tau = \frac{4}{\kappa}\sqrt{\frac{n+1}{2}} \int_0^X \frac{d\zeta}{\sqrt{X^{n+1} - \zeta^{n+1}}}$$

Setting $u = \zeta/X$,

$$\tau = \frac{4}{\kappa\sqrt{X^{n-1}}} \left(\sqrt{\frac{n+1}{2}} \int_0^1 \frac{du}{\sqrt{1 - u^{n+1}}} \right)$$

The expression within the parentheses depends only on the parameter n and is denoted by $\psi(n)$. Thus

$$\tau = \frac{4}{\kappa \sqrt{X^{n-1}}}\ \psi(n) \qquad (4.9)$$

The factor $\psi(n)$ may be evaluated numerically to any desired degree of accuracy, and is tabulated in Table 4.1.

Case 2. Polynomials of Displacement. Consider the binomial restoring force

$$f(x) = x^n + \mu x^m \qquad [m > n \geq 0]$$

Introducing this expression into Eq. (4.8) and performing the integrations:[5]

$$\tau = \frac{4}{\kappa \sqrt{X^{n-1}}}\left(\sqrt{\frac{n+1}{2}} \int_0^1 \frac{du}{\sqrt{(1+\bar{\mu}) - (u^{n+1} + \bar{\mu}u^{m+1})}}\right) \qquad (4.10)$$

where

$$\bar{\mu} = \mu X^{m-n}\left(\frac{n+1}{m+1}\right) \qquad (4.11)$$

For particular values of n, m, and $\bar{\mu}$, the expression within the parentheses can be evaluated to any desired degree of accuracy by numerical methods. The extension of this method to higher-order polynomials can be made quite readily.

Case 3. Harmonic Function of Displacement. Consider now the problem of the simple pendulum which has a restoring force of the form

$$f(x) = \sin x$$

Introducing this relation into Eq. (4.7):

$$t - t_0 = \frac{1}{2\kappa} \int_0^x \frac{d\zeta}{\sqrt{\sin^2 \frac{X}{2} - \sin^2 \frac{\zeta}{2}}}$$

If $x = X$ and $t_0 = 0$, this integral can be reduced to the standard form of the complete elliptic integral of the first kind:

$$\hat{K}(\alpha) = \int_0^{\pi/2} \frac{dv}{\sqrt{1 - \sin^2 \alpha \sin^2 v}} \qquad (4.12)$$

Thus, the period of vibration is

$$\tau = \frac{1}{\kappa} \hat{K}\left(\frac{X}{2}\right) \qquad (4.13)$$

The displacement-time function can be obtained by inversion and leads to the inverse elliptic functions. Replacing $\sin \alpha$ by U in Eq. (4.12), expanding by the binomial theorem, and then integrating yields Eq. (4.3).

Case 4. Velocity Squared Damping. As indicated by Eq. (4.6), the introduction of any other function of $\dot{x}^2$ does not complicate the problem. Thus, the differential equation*

---

* The $\pm$ sign is employed here, and elsewhere in this chapter, to account for the proper direction of the resisting force. Consequently, reference frequently is made to upper or lower sign rather than to plus or minus.

$$\ddot{x} \pm \frac{\delta}{2}\,\dot{x}^2 + \kappa^2 f(x) = 0$$

can be reduced to

$$\frac{d(\dot{x}^2)}{dx} \pm \delta\dot{x}^2 = 2\kappa^2 f(x)$$

Integrating the above equation,

$$\dot{x}^2 = 2\kappa^2 e^{\mp\delta x} \int_x^X e^{\pm\delta\xi} f(\xi)\,d\xi$$

Integrating again,

$$t = \int_{x_0}^x \frac{d\eta}{\dot{x}(\eta)}$$

where η is an integration variable.

FORCED VIBRATION

Exact solutions for forced vibration of nonlinear systems are virtually nonexistent, except as the system can be represented in a stepwise linear manner. For example, consider a system with a stepwise linear symmetrical restoring force characteristic, as shown in Fig. 4.4. Denote the lower of the two stiffnesses by k_1, the upper by k_2, and the displacement at which the change in stiffness occurs by x_1. Thus, the problem reduces to the solution of two linear differential equations:

$$m\ddot{x}' + k_1 x' = \pm P \sin \omega t \qquad [x_1 \geq x' \geq 0] \qquad (4.14a)$$

$$m\ddot{x}'' + (k_1 - k_2)x_1 + k_2 x'' = \pm P \sin \omega t \qquad [x'' \geq x_1] \qquad (4.14b)$$

where the upper sign refers to in-phase exciting force and the lower sign to out-of-phase exciting force. The appropriate boundary conditions are

$$x'(t = 0) = 0$$

$$x'(t = t_1) = x''(t = t_1) = x_1$$

$$\dot{x}'(t = t_1) = \dot{x}''(t = t_1) \qquad (4.15)$$

$$\dot{x}''\left(t = \frac{\pi}{2\omega}\right) = 0$$

The solutions of Eqs. (4.14) are

$$x' = \frac{\pm P/k_1}{1 - \omega^2/\omega_1^2}\sin \omega t + A_1 \cos \omega_1 t + B_1 \sin \omega_1 t$$

$$x'' = \frac{\pm P/k_2}{1 - \omega^2/\omega_2^2}\sin \omega t + A_2 \cos \omega_2 t + B_2 \sin \omega_2 t + \left(1 - \frac{k_1}{k_2}\right)x_1$$

where $\omega_1^2 = k_1/m$, $\omega_2^2 = k_2/m$, and the constants A_1, A_2, B_1, B_2 may be evaluated from the boundary conditions, Eq. (4.15).

This analysis also applies to the case of free vibration by setting $P = 0$. By assigning various values to k_1 and k_2, a wide variety of specific problems may be treated; a collection of such solutions is given in Refs. 27 and 28. It is not necessary to restrict the restoring forces to odd functions.

NUMERICAL METHODS AND CHAOTIC DYNAMICS

The advent of availability of high-speed digital computation in the 1960s has had a profound effect on the study of nonlinear vibrations, not only in the speed, efficiency, and extent of the solutions which were made available but also in the variety of problems that could be studied and the new phenomena that were discovered. The methodology is quite straightforward. A timewise integration of the equation or equations of motion is carried out using any appropriate numerical integration scheme—from the simplest trapezoidal format to more complex schemes such as that of Runge-Kutta. The criteria for selection of the integration scheme is dependent on:

1. *The nature of the solution being sought.* A solution that is expected to have sharp discontinuities in amplitude or velocity would suggest the use of a linear or very low-order polynomial fit implicit in the integration scheme.

2. *The efficiency of the solution scheme.* Complex schemes that require more calculation for each incremental step in time usually permit the use of longer steps and hence fewer steps for a given total time interval. Conversely, simpler schemes that require less calculation for each incremental step in time usually require the use of shorter steps and hence more steps for a given total time interval.

The final selection of integration scheme and the size of the time step is very often made on the basis of trial and error where the step is refined to smaller and smaller values until the successive solutions no longer show a dependence on step size.

In cases where the requirement is for the stabilized "steady-state" solution to a dynamics problem (rather that the transient solution from a prescribed set of initial conditions), another precaution must be taken in numerical solution. The solution must be run long enough so that the initial transient from an arbitrarily selected set of initial conditions has decayed to negligible value. Here again, actual trials are generally conducted to assure the stabilization of the solution to the required accuracy. The issue does represent an important limitation when solutions are sought for the behavior of systems with very low damping.

Other limitations of numerical methods relate to their similitude with the actual physical systems which they are intended to model:

1. Numerical integration techniques are generally ineffective in deriving solutions in regions where those solutions are unstable in the sense that they are physically not achievable (such as illustrated in Fig. 4.16C).

2. For systems that have multivalued solutions, the particular solution branch which is achieved on any particular trial is dependent on the conditions set for initiating the computation sequence.

CHAOTIC DYNAMICS

Perhaps the most fundamental impact of the digital computer on the field of nonlinear vibration had been to make possible the discovery and the elucidation of chaotic vibrations.[7,23]

Chaotic vibrations are characterized by an irregular or ragged waveform such as illustrated in Figs. 4.19A and 4.23A. Although there may be recurrent patterns in the waveform, they are not precisely alike, and they repeat at irregular intervals, so the motion is truly nonperiodic as is implied in Zone II of Figs. 4.17A and 4.17B. Indeed, care must be taken in characterizing vibration as chaotic since there are irregular motions which mimic chaotic response but in which there *are* recurrent patterns which repeat at regular intervals, such as are implied in Zone III of Figs. 4.17A and 4.17B.

Another characteristic of chaotic vibration is that, if the numerical solution (and, presumably, the physical system it represents) is started twice at nearly identical initial conditions, the two solutions will diverge exponentially with time.

For all its irregularity, there is a certain basic structure and patternation implicit in chaotic vibration. As one can infer from the response curves of local peak amplitude for chaotic vibration shown in Zone II of Figs. 4.17A and 4.17B, the maximum amplitude is bounded.

A remarkable response behavior associated with chaotic vibration is the cascade of period-doubling bifurcations or tree-like structure in peak amplitude response curve (illustrated in Zone I of Figs. 4.17A and 4.17B) that may take place in the transition from simple periodic response to chaotic response.

But the most remarkable property of chaotic vibrations is evident in the Poincaré section of the motion, shown typically in Figs. 4.19B and 4.23B. The Poincaré section contains a large number of discrete points of velocity plotted as a function of displacement of the chaotic motion where the points are sampled *stroboscopically* with reference to a particular phase angle of the forcing periodic function. Rather than a random scatter of points, the Poincaré section generally reveals striking patterns. The Poincaré section is sometimes referred to as an *attractor.*

Chaotic vibration also differs from random motion in that the power frequency spectrum generally has distinct peaks rather than consisting of broadband noise. There will often be not only synchronous response peaks at the forcing function frequency as in the response of linear systems, but there will also be a significant asynchronous response peak or peaks at the system's natural frequency of frequencies.

APPROXIMATE ANALYTICAL METHODS

A large number of approximate analytical methods of nonlinear vibration analysis exist, each of which may or may not possess advantages for certain classes of problems. Some of these are restricted techniques which may work well with some types of equations but not with others. The methods which are outlined below are among the better known and possess certain advantages as to ranges of applicability.[4,5,26-41]

DUFFING'S METHOD

Consider the nonlinear differential equation (known as Duffing's equation)

$$\ddot{x} + \kappa^2(x \pm \mu^2 x^3) = p \cos \omega t \tag{4.16}$$

where the $\pm$ sign indicates either a hardening or softening system. As a first approximation to a harmonic solution, assume that

$$x_1 = A \cos \omega t \tag{4.17}$$

and rewrite Eq. (4.16) to obtain an equation for the second approximation:

$$\ddot{x}_2 = -(\kappa^2 A \pm \tfrac{3}{4}\kappa^2\mu^2 A^3 - p)\cos\omega t - \tfrac{1}{4}\kappa^2\mu^2 A^3 \cos\omega t$$

This equation may now be integrated to yield

$$x_2 = \frac{1}{\omega^2}(\kappa^2 A \pm \tfrac{3}{4}\kappa^2\mu^2 A^3 - p)\cos\omega t + \tfrac{1}{36}\kappa^2\mu^2 A^3 \cos 3\omega t \tag{4.18}$$

where the constants of integration have been taken as zero to ensure periodicity of the solution.

This may be regarded as an iteration procedure by reinserting each successive approximation into Eq. (4.16) and obtaining a new approximation. For this iteration procedure to be convergent, the nonlinearity must be small; i.e., κ^2, μ^2, A, and p must be small quantities. This restricts the study to motions in the neighborhood of linear vibration (but not near $\omega = \kappa$, since A would then be large); thus, Eq. (4.17) must represent a reasonable first approximation. It follows that the coefficient of the cos ωt term in Eq. (4.18) must be a good second approximation and should not be far different from the first approximation.[42] Since this procedure furnishes the exact result in the linear case, it might be expected to yield good results for the "slightly nonlinear" case. Thus, a relation between frequency and amplitude is found by equating the coefficients of the first and second approximations:

$$\omega^2 = \kappa^2(1 \pm \tfrac{3}{4}\mu^2 A^2) - \frac{p}{A} \tag{4.19}$$

This relation describes the response curves, as shown in Fig. 4.14.

The above method applies equally well when linear velocity damping is included. Further details concerning this method, and an analysis of its applicability when only μ^2 and p are considered small, may be found in Ref. 30.

RAUSCHER'S METHOD[43]

Duffing's method considered above is based on the idea of starting the iteration procedure from the linear vibration. More rapid convergence might be expected if the approximations were to begin with free nonlinear vibration; Rauscher's method[43] is based on this idea.

Consider a system with general restoring force described by the differential equation

$$\omega^2 x'' + \kappa^2 f(x) = p \cos\omega t \tag{4.20}$$

where primes denote differentiation with respect to ωt, and $f(x)$ is an odd function. Assume that the conditions at time $t = 0$ are $x(0) = A$, $x'(0) = 0$. Start with the free nonlinear vibration as a first approximation, i.e., with the solution of the equation

$$\omega_0^2 x'' + \kappa^2 f(x) = 0 \tag{4.21}$$

such that $x = x_0(\phi)$ (where $\omega t = \phi$) has the period 2π and $x_0(0) = A$, $x_0'(0) = 0$. Equation (4.21) may be solved exactly in the form of quadratures according to Eq. (4.7):

$$\phi = \phi_0(x) = \frac{\omega_0}{\kappa\sqrt{2}} \int_A^x \frac{d\zeta}{\sqrt{\displaystyle\int_\zeta^A f(\xi)\,d\xi}} \tag{4.22}$$

Since $f(x)$ is an odd function and noting that ωt varies from 0 to $\pi/2$ as x varies from 0 to A,

$$\frac{1}{\omega_0} = \frac{\sqrt{2}}{\kappa\pi} \int_0^A \frac{d\zeta}{\sqrt{\int_\zeta^A f(\xi)\, d\xi}} \tag{4.23}$$

With ω_0 and ϕ_0 determined by Eqs. (4.23) and (4.22), respectively, the next approximation may be found from the equation

$$\omega_1^2 x'' + \kappa^2 f(x) - p \cos \phi_0 = 0 \tag{4.24}$$

In the original differential equation, Eq. (4.20), ωt is replaced by its first approximation ϕ_0 and ω_0 (now known) is replaced by its second approximation ω_1, thus giving Eq. (4.24). This equation is again of a type which may be integrated explicitly; therefore, the next approximation ω_1 and ϕ_1 may be determined. In those cases where $f(x)$ is a complicated function, the integrals may be evaluated graphically.[43]

This method involves reducing nonautonomous systems to autonomous ones* by an iteration procedure in which the solution of the free vibration problem is used to replace the time function in the original equation, which is then solved again for $t(x)$. The method is accurate and frequently two iterations will suffice.

THE PERTURBATION METHOD

In one of the most common methods of nonlinear vibration analysis, the desired quantities are developed in powers of some parameter which is considered small; then the coefficients of the resulting power series are determined in a stepwise manner. The method is straightforward, although it becomes cumbersome for actual computations if many terms in the perturbation series are required to achieve a desired degree of accuracy.

Consider Duffing's equation, Eq. (4.16), in the form

$$\omega^2 x'' + \kappa^2 (x + \mu^2 x^3) - p \cos \phi = 0 \tag{4.25}$$

where $\phi = \omega t$ and primes denote differentiation with respect to ϕ. The conditions at time $t = 0$ are $x(0) = A$ and $x'(0) = 0$, corresponding to harmonic solutions of period $2\pi/\omega$. Assume that μ^2 and p are small quantities, and define $\kappa^2\mu^2 \equiv \varepsilon, p \equiv \varepsilon p_0$. The displacement $x(\phi)$ and the frequency ω may now be expanded in terms of the small quantity ε:

$$x(\phi) = x_0(\phi) + \varepsilon x_1(\phi) + \varepsilon^2 x_2(\phi) + \cdots$$
$$\omega = \omega_0 + \varepsilon\omega_1 + \varepsilon^2\omega_2 + \cdots \tag{4.26}$$

The initial conditions are taken as $x_i(0) = x_i'(0) = 0$ $[i = 1,2,\ldots]$.

Introducing Eq. (4.26) into Eq. (4.25) and collecting terms of zero order in ε gives the linear differential equation

$$\omega_0^2 x_0'' + \kappa^2 x_0 = 0$$

---

* An autonomous system is one in which the time *does not* appear explicitly, while a nonautonomous system is one in which the time *does* appear explicitly.

Introducing the initial conditions into the solution of this linear equation gives $x_0 = A \cos \omega t$ and $\omega_0 = \kappa$. Collecting terms of the first order in ε,

$$\omega_0^2 x_1'' + \kappa^2 x_1 - (2\omega_0\omega_1 A - \tfrac{3}{4}A^3 + p_0) \cos \phi + \tfrac{1}{4}A^3 \cos 3\phi = 0 \qquad (4.27)$$

The solution of this differential equation has a nonharmonic term of the form $\phi \cos \phi$, but since only harmonic solutions are desired, the coefficient of this term is made to vanish so that

$$\omega_1 = \frac{1}{2\kappa}\left(\tfrac{3}{4}A^2 - \frac{p_0}{A}\right)$$

Using this result and the appropriate initial conditions, the solution of Eq. (4.27) is

$$x_1 = \frac{A^3}{32\kappa^2}(\cos 3\phi - \cos \phi)$$

To the first order in ε, the solution of Duffing's equation, Eq. (4.25), is

$$x = A \cos \omega t + \varepsilon \frac{A^3}{32\kappa^2}(\cos 3\omega t - \cos \omega t)$$

$$\omega = \kappa + \frac{\varepsilon}{2\kappa}\left(\tfrac{3}{4}A^2 - \frac{p_0}{A}\right)$$

This agrees with the results obtained previously [Eqs. (4.18) and (4.19)]. The analysis may be carried beyond this point, if desired, by application of the same general procedures.

As a further example of the perturbation method, consider the self-excited system described by Van der Pol's equation

$$\ddot{x} - \varepsilon(1 - x^2)\dot{x} + \kappa^2 x = 0 \qquad (4.5)$$

where the initial conditions are $x(0) = 0$, $\dot{x}(0) = A\kappa_0$. Assume that

$$x = x_0 + \varepsilon x_1 + \varepsilon^2 x_2 + \cdots$$

$$\kappa^2 = \kappa_0^2 + \varepsilon\kappa_1^2 + \varepsilon^2\kappa_2^2 + \cdots$$

Inserting these series into Eq. (4.5) and equating coefficients of like terms, the result to the order ε^2 is

$$x = \left(2 - \frac{29\varepsilon^2}{96\kappa_0^2}\right)\sin \kappa_0 t + \frac{\varepsilon}{4\kappa_0}\cos \kappa_0 t + \frac{\varepsilon}{4\kappa_0}\left(\frac{3\varepsilon}{4\kappa_0}\sin 3\kappa_0 t - \cos 3\kappa_0 t\right) - \frac{5\varepsilon^2}{124\kappa_0^2}\sin 5\kappa_0 t$$
$$(4.28)$$

An application of the perturbation method which employs operational calculus to solve the resulting linear differential equations is given in Ref. 41, while other applications of the method are given in Refs. 44 and 45. An application to the problem of subharmonic response is outlined in Ref. 1.

THE METHOD OF KRYLOFF AND BOGOLIUBOFF[31]

Consider the general autonomous differential equation

$$\ddot{x} + F(x, \dot{x}) = 0$$

which can be rewritten in the form

$$\ddot{x} + \kappa^2 x + \varepsilon f(x,\dot{x}) = 0 \qquad [\varepsilon \ll 1] \tag{4.29}$$

For the corresponding linear problem ($\varepsilon \equiv 0$), the solution is

$$x = A \sin(\kappa t + \theta) \tag{4.30}$$

where A and θ are constants.

The procedure employed often is used in the theory of ordinary linear differential equations and is known variously as the method of variation of parameters or the method of Lagrange. In the application of this procedure to a nonlinear equation of the form of Eq. (4.29), assume the solution to be of the form of Eq. (4.30) but with A and θ as time-dependent functions rather than constants. This procedure, however, introduces an excessive variability into the solution; consequently, an additional restriction may be introduced. The assumed solution, of the form of Eq. (4.30), is differentiated once considering A and θ as time-dependent functions; this is made equal to the corresponding relation from the linear theory (A and θ constant) so that the additional restriction

$$\dot{A}(t) \sin[\kappa t + \theta(t)] + \dot{\theta}(t) A(t) \cos[\kappa t + \theta(t)] = 0 \tag{4.31}$$

is placed on the solution. The second derivative of the assumed solution is now formed and these relations are introduced into the differential equation, Eq. (4.29). Combining this result with Eq. (4.31),

$$\dot{A}(t) = -\left(\frac{\varepsilon}{\kappa}\right) f[A(t) \sin \Phi, A(t)\kappa \cos \Phi] \cos \Phi$$

$$\dot{\theta}(t) = \frac{\varepsilon}{\kappa A(t)} f[A(t) \sin \Phi, A(t)\kappa \cos \Phi] \sin \Phi$$

where

$$\Phi = \kappa t + \theta(t)$$

Thus, the second-order differential equation, Eq. (4.29), has been transformed into two first-order differential equations for $A(t)$ and $\theta(t)$.

The expressions for $\dot{A}(t)$ and $\dot{\theta}(t)$ may now be expanded in Fourier series:

$$\dot{A}(t) = -\left(\frac{\varepsilon}{\kappa}\right) \left\{ K_0(A) + \sum_{n=1}^{r} [K_n(A) \cos n\Phi + L_n(A) \sin n\Phi] \right\}$$

$$\dot{\theta}(t) = \frac{\varepsilon}{\kappa A} \left\{ P_0(A) + \sum_{n=1}^{r} [P_n(A) \cos n\Phi + Q_n(A) \sin n\Phi] \right\} \tag{4.32}$$

where

$$K_0(A) = \frac{1}{2\pi} \int_0^{2\pi} f[A \sin \Phi, A\kappa \cos \Phi] \cos \Phi \, d\Phi$$

$$P_0(A) = \frac{1}{2\pi} \int_0^{2\pi} f[A \sin \Phi, A\kappa \cos \Phi] \sin \Phi \, d\Phi$$

It is apparent that A and θ are periodic functions of time of period $2\pi/\kappa$; therefore, during one cycle, the variation of $\dot{A}$ and $\dot{\theta}$ is small because of the presence of the small parameter ε in Eqs. (4.32). Hence, the average values of $\dot{A}$ and $\dot{\theta}$ are considered. Since the motion is over a single cycle, and since the terms under the summation signs are of the same period and consequently vanish, then approximately:

$$\dot{A} \simeq -\left(\frac{\varepsilon}{\kappa}\right) K_0(A)$$

$$\dot{\theta} \simeq \frac{\varepsilon}{\kappa A} P_0(A)$$

$$\dot{\Phi} \simeq \kappa + \frac{\varepsilon}{\kappa A} P_0(A)$$

For example, consider Rayleigh's equation

$$\ddot{x} - (\alpha - \beta \dot{x}^2)\dot{x} + \kappa^2 x = 0 \tag{4.33}$$

By application of the above procedures:

$$\dot{A} = -\left(\frac{1}{\kappa}\right) K_0(A) = -\frac{1}{\kappa}\left[\frac{1}{2\pi}\int_0^{2\pi} (-\alpha + \beta A^2 \kappa^2 \cos^2 \Phi) A \kappa \cos^2 \Phi \, d\Phi\right]$$

$$= \frac{A}{2}(\alpha - \tfrac{3}{4}\beta A^2 \kappa^2) \tag{4.34}$$

Equation (4.34) may be integrated directly:

$$t = 2\int_{A_0}^{A} \frac{dA}{A(\alpha - \gamma A^2)} = \frac{1}{\alpha}\ln\frac{A^2}{\alpha - \gamma A^2}$$

Solving for A,

$$A = \frac{\alpha}{\gamma}\left[\frac{1}{1 + \left(\dfrac{\alpha}{\gamma A_0^2} - 1\right)e^{-\alpha t}}\right]^{1/2} \tag{4.35}$$

where

$$\gamma = \tfrac{3}{4}\beta^2 \kappa^2 \tag{4.36}$$

The application of the method to Van der Pol's equation, Eq. (4.5), is easily accomplished and leads to a solution in the first approximation of the form similar to that of the perturbation solution given by Eq. (4.28).

The method may be applied equally well to nonautonomous systems.[5,39,40]

THE RITZ METHOD

In addition to methods of nonlinear vibration analysis stemming from the idea of small nonlinearities and from extensions of methods applicable to linear equations, other methods are based on such ideas as satisfying the equation at certain points of the motion[33] or satisfying the equation in the average. The Ritz method is an example of the latter method and is quite powerful for general studies.

One method of determining such "average" solutions is to multiply the differential equation by some "weight function" $\psi_n(t)$ and then integrate the product over a period of the motion. If the differential equation is denoted by E, this procedure leads to

$$\int_0^{2\pi} E \cdot \psi_n(t) \, dt = 0 \tag{4.37}$$

A second method of obtaining such average solutions can be derived from the calculus of variations by seeking functions that minimize a certain integral:

$$I = \int_{t_0}^{t_1} F(\dot{x}, x, t) \, dt = \text{minimum}$$

Consider a function of the form

$$\tilde{x}(t) = a_1 \psi_1(t) + a_2 \psi_2(t) + \cdots + a_n \psi_n(t)$$

where the $\psi_k(t)$ are prescribed functions. If $\tilde{x}$ is now introduced for x, then

$$I = I(a_1, a_2, \ldots, a_n)$$

and a necessary condition for I to be a minimum is

$$\frac{\partial I}{\partial a_1} = 0, \qquad \frac{\partial I}{\partial a_2} = 0, \ldots, \qquad \frac{\partial I}{\partial a_n} = 0 \tag{4.38}$$

This gives n equations of the form

$$\frac{\partial I}{\partial a_k} = \int_{t_0}^{t_1} \left(\frac{\partial F}{\partial \tilde{x}} \psi_k + \frac{\partial F}{\partial \dot{\tilde{x}}} \dot{\psi}_k \right) dt = 0 \tag{4.39}$$

for determining the n unknown coefficients. Integrating Eq. (4.39),

$$\frac{\partial I}{\partial a_k} = \left[\frac{\partial F}{\partial \dot{\tilde{x}}} \psi_k \right]_{t_0}^{t_1} + \int_{t_0}^{t_1} \left[\frac{\partial F}{\partial \tilde{x}} - \frac{d}{dt} \left(\frac{\partial F}{\partial \dot{\tilde{x}}} \right) \right] \psi_k \, dt = 0$$

The first term is zero because ψ_k must satisfy the boundary conditions; the expression in brackets under the integral in the second term is Euler's equation. The conditions given in Eqs. (4.38) then reduce to

$$\int_{t_0}^{t} E(\tilde{x}) \psi_k \, dt = 0 \qquad [k = 1, 2, \ldots, n] \tag{4.40}$$

This is the same as Eq. (4.37); thus, it is not necessary to "know" the variational problem, but only the differential equation. The conditions given in Eqs. (4.40) then yield average solutions based on variational concepts.

Examples. As a first example of the application of the Ritz method, consider the equation

$$\ddot{x} + \kappa^2 x^n = 0$$

for which an exact solution was given earlier in this chapter [Eq. (4.9)]. Assume a single-term solution of the form

$$\tilde{x} = A \cos \omega t$$

The Ritz procedure, defined by Eq. (4.40), gives

$$\int_0^{2\pi} (-\omega^2 A \cos^2 \omega t + \kappa^2 A^n \cos^{n+1} \omega t)\, d(\omega t) = 0$$

from which

$$\frac{\omega^2}{\kappa^2} = \frac{4}{\pi} A^{n-1} \int_0^{\pi/2} \cos^{n+1} \omega t\, d(\omega t) = A^{n-1}\varphi(n) \tag{4.41}$$

The comparable exact solution obtained previously by introducing in Eq. (4.9) the quantity $2\pi/\omega$ for the period τ is

$$\frac{\omega^2}{\kappa^2} = \left[\frac{\pi^2/4}{\psi^2(n)}\right] X^{n-1} = \Phi(n) X^{n-1} \tag{4.42}$$

Values of $\varphi(n)$ from the approximate analysis and $\Phi(n)$ from the exact analysis are compared directly in Table 4.1, affording an appraisal of the accuracy of the method.

TABLE 4.1   Values of the Functions $\psi(n), \Phi(n), \varphi(n)$*

n	$\psi(n)$	$\Phi(n)$	$\varphi(n)$
0	1.4142	1.2337	1.2732
1	1.5708	1.0000	1.0000
2	1.7157	0.8373	0.8488
3	1.8541	0.7185	0.7500
4	1.9818	0.6282	0.6791
5	2.1035	0.5577	0.6250
6	2.2186	0.5013	0.5820
7	2.3282	0.4552	0.5469

* The mathematical expressions for $\psi(n), \Phi(n)$, and $\varphi(n)$
and the equations to which they refer are:

$$\psi(n) = \sqrt{\frac{n+1}{2}} \int_0^1 \frac{du}{\sqrt{1-u^{n+1}}} \qquad \text{[Eq. (4.9)]}$$

$$\Phi(n) = \frac{\pi^2/4}{\psi^2(n)} \qquad \text{[Eq. (4.42)]}$$

$$\varphi(n) = \frac{4}{\pi} \int_0^{\pi/2} \cos^{n+1} \sigma\, d\sigma \qquad \text{[Eq. (4.41)]}$$

Consider now the nonautonomous system described by Duffing's equation

$$E \equiv \ddot{x} + \kappa^2(x + \mu^2 x^3) - p \cos \omega t = 0$$

Assuming

$$\tilde{x} = A \cos \phi, \qquad \phi = \omega t$$

the Ritz condition, Eq. (4.40), leads to

$$\int_0^{2\pi} \{[(1-\eta^2)A - s] \cos \phi + \mu^2 A^3 \cos^3 \phi\} \cos \phi\, d\phi$$

from which the amplitude-frequency relation is

$$(1 - \eta^2)A + \tfrac{3}{4}\mu^2 A^3 = \pm s \tag{4.43}$$

where
$$s = \frac{P}{\kappa^2}, \qquad \eta^2 = \frac{\omega^2}{\kappa^2} \tag{4.44}$$

The upper sign indicates vibration in phase with the exciting force. Equation (4.43) describes the response curves shown in Fig. 4.14A and corresponds to Eq. (4.19) obtained by Duffing's method.

Application of the Ritz method to Van der Pol's equation, Eq. (4.5), leads to the identical result given by Eq. (4.36).

GENERAL EQUATIONS FOR RESPONSE CURVES

The Ritz method has been applied extensively in studies of nonlinear differential equations.[26,28] Some of the general equations for response curves thereby obtained are given here, both as a further example of the application of the method and as a collection of useful relations.

SYSTEM WITH LINEAR DAMPING AND GENERAL RESTORING FORCES

Consider a system with general elastic restoring force (an odd function) and described by the equation of motion

$$a\ddot{x} + b\dot{x} + cf(x) - P \cos \omega t = 0$$

A solution may be assumed in the form

$$\tilde{x} = A \cos (\omega t - \theta) = B \cos \phi + C \sin \phi \tag{4.45}$$

where $\phi = \omega t$, $B = A \cos \theta$, $C = A \sin \theta$. Introducing Eq. (4.45) according to the Ritz conditions, and recalling that $f(x)$ is to be an odd function,

$$-a\omega^2 A \cos \theta + b\omega A \sin \theta + cAF(A) \cos \theta = P$$

$$-a\omega^2 A \sin \theta - b\omega A \cos \theta + cAF(A) \sin \theta = 0 \tag{4.46}$$

where
$$F(A) = \frac{1}{\pi A} \int_0^{2\pi} f(A \cos \sigma) \cos \sigma \, d\sigma$$

and σ is simply an integration variable.

Some algebraic manipulations with Eqs. (4.46) give independent equations for the two unknowns A and θ:

$$[F(A) - \eta^2]^2 + 4D^2\eta^2 = \left(\frac{s}{A}\right)^2 \tag{4.47}$$

$$\tan \theta = \frac{2D\eta}{F(A) - \eta^2} \tag{4.48}$$

where η^2 and s are defined according to Eq. (4.44) and

$$\kappa^2 = \frac{c}{a} \qquad p = \frac{P}{a} \qquad D = \frac{b}{2\sqrt{ac}}$$

Equation (4.47) describes response curves of the form shown in Fig. 4.15, and Eq. (4.48) gives the corresponding phase angle relationships. These two equations also yield other special relations which describe various curves in the response diagram:
Undamped free vibration curve (Fig. 4.13),

$$\eta^2 = F(A) \tag{4.49}$$

Undamped response curves (Fig. 4.14),

$$\eta^2 = F(A) \mp \frac{s}{A} \tag{4.50}$$

Locus of vertical tangents of undamped response curves (Fig. 4.17),

$$\eta^2 = F(A) + A\frac{\partial F(A)}{\partial A} \tag{4.51}$$

Damped response curves (Fig. 4.15),

$$\eta^2 = [F(A) - 2D^2] \mp \sqrt{\left(\frac{s}{A}\right)^2 - 4D^2[F(A) - D^2]} \tag{4.52}$$

Locus of vertical tangents of damped response curves (Fig. 4.17),

$$[F(A) - \eta^2]\left[F(A) + A\frac{\partial F(A)}{\partial A} - \eta^2\right] = -4D^2\eta^2 \tag{4.53}$$

The maximum amplitude of vibration is of interest. The amplitude at the point at which a response curve crosses the free vibration curve is termed the *resonance amplitude,* and is determined in the nonlinear case by solving Eqs. (4.49) and (4.52) simultaneously. This leads to[47]

$$2D\eta = \frac{s}{A} \qquad \theta = \frac{\pi}{2} \tag{4.54}$$

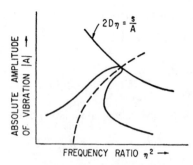

FIGURE 4.25 Determination of the resonant amplitude in accordance with Eq. (4.54).

The first of these two equations defines a hyperbola in the response diagram, describing the locus of crossing points, as shown in Fig. 4.25; hence, the intersection of this curve with the free vibration curve gives the resonance amplitude. The phase angle at resonance has the value π/2, as in the linear case. This result is of great help in computing response curves since the effect of damping (except for very large values) is negligible except in the neighborhood of resonance. Therefore, one may compute only the undamped curves (which is not difficult) and the hyperbola (which does not contain the nonlinearity); then, the effect of damping may be sketched in from knowledge of the crossing point.

SYSTEM WITH GENERAL DAMPING AND GENERAL RESTORING FORCES

The preceding analysis may be extended to include the more general differential equation

$$E \equiv \ddot{x} + 2D\kappa g(\dot{x}) + \kappa^2 f(x) - p \cos \omega t = 0$$

By procedures similar to those employed above:

$$[F(A) - \eta^2]^2 + 4D^2 S^2(A) = \left(\frac{S}{A}\right)^2 \tag{4.55}$$

$$\tan \theta = \frac{2DS(A)}{F(A) - \eta^2} \tag{4.56}$$

where

$$S(A) = \frac{1}{\pi \kappa A} \int_0^{2\pi} g(\omega A \sin \sigma) \sin \sigma \, d\sigma$$

In the case of linear velocity damping, $S(A) = \eta$, and Eqs. (4.55) and (4.56) reduce to Eqs. (4.47) and (4.48). The results for various types of damping forces are:

Coulomb damping: $\qquad g(\dot{x}) = \pm v_0 \qquad S(A) = \frac{4}{\pi} \frac{v_0}{\kappa A}$

Linear velocity damping: $\qquad g(\dot{x}) = v_1 \dot{x} \qquad S(A) = v_1 \eta$

Velocity squared damping: $\qquad g(\dot{x}) = v_2 \dot{x}|\dot{x}| \qquad S(A) = \frac{8}{3\pi} v_{2\eta}(A\omega)$

nth-power velocity damping: $\qquad g(\dot{x}) = v_n \dot{x}|\dot{x}|^{n-1} \qquad S(A) = v_{n\eta}(A\omega)^{n-1}\varphi(n)$

where $\varphi(n)$ is defined in Eq. (4.41) and values are given in Table 4.1.

The locus of resonance amplitudes or crossing points is now given by[47]

$$2DS(A) = \frac{s}{A} \qquad \theta = \frac{\pi}{2}$$

GRAPHICAL METHODS OF INTEGRATION

Graphical methods may be employed in the analysis of nonlinear vibration and often prove to be of great value both for general studies of the behavior of a particular system and for actual integration of the equation of motion.

A single degree-of-freedom system requires two parameters to describe completely the state of the motion. When these two parameters are used as coordinate axes, the graphical representation of the motion is called a *phase-plane* representation. In dealing with ordinary dynamical problems, these parameters frequently are taken as the displacement and velocity. First consider an undamped linear system having the equation of motion

$$\ddot{x} + \omega_n^2 x = 0 \tag{4.57}$$

and the solution

$$x = A \cos \omega_n t + B \sin \omega_n t$$

$$y = \frac{\dot{x}}{\omega_n} = -A \sin \omega_n t + B \cos \omega_n t \qquad (4.58)$$

Eliminating time as a variable between Eqs. (4.58):

$$x^2 + y^2 = c^2$$

Thus, the phase-plane representation is a family of concentric circles with centers at the origin. Such curves are called *trajectories*. The necessary and sufficient conditions in the phase-plane for periodic motions are (1) closed trajectories and (2) paths described in finite time.

Now, suppose that the solution of Eq. (4.57) is not known. By introducing $y = \dot{x}/\omega_n$,

$$\frac{dx}{dt} = \omega_n y \qquad \frac{dy}{dt} = -\omega_n x$$

Therefore

$$\frac{dy}{dx} = -\frac{x}{y} \qquad (4.59)$$

Thus, the path in the phase-plane is described by a simple first-order differential equation. This process of eliminating the time always can be done in principle, but frequently the problem is too difficult. Since the time is to be eliminated, *only autonomous systems can be treated by phase-plane methods*. When an equation of the type of Eq. (4.59) can be found, a direct solution of the problem follows since slopes of the trajectories can be sketched in the phase-plane and the trajectories determined by connecting the tangents; this is known as the method of *isoclines*. It sometimes happens that $dx = 0$, $dy = 0$ simultaneously so that there is no knowledge of the direction of the motion; such points in the phase-plane are called *singular points*. In the present example, the origin constitutes a singular point.

Consider now a damped linear system having the equation of motion

$$\ddot{x} + 2\zeta\omega_n\dot{x} + \omega_n^2 x = 0$$

and the solution

$$x = Ce^{-\delta t} \cos \Phi$$

$$y = \frac{\dot{x}}{\omega_n} = Ce^{-\delta t} \cos (\Phi + \sigma) \qquad (4.60)$$

where

$$\delta = \zeta\omega_n = -\omega_n \cos \sigma \qquad \left[\sigma > \frac{\pi}{2} \right]$$

$$\Phi = \upsilon t + \theta$$

$$\upsilon = \omega_n\sqrt{1 - \zeta^2} = \omega_n \sin \sigma$$

Equations (4.60) indicate that the trajectories in the phase-plane are some form of spiral (one of the simplest known of which is the logarithmic spiral). By referring to the oblique coordinate system shown in Fig. 4.26, and recalling that $\sin \sigma$ is a constant and

$$r^2 = x^2 + y^2 - 2xy \cos \sigma$$

Eqs. (4.60) reduce to

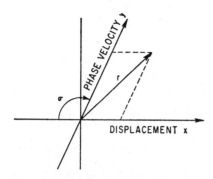

FIGURE 4.26 Phase-plane using oblique coordinates which results in a logarithmic spiral trajectory for a linear system with viscous damping [Eq. (4.60)].

$$r = Ce^{-\delta t} \sin \sigma$$

This is a form of a logarithmic spiral.

The trajectories also could be found in a rectangular coordinate system, by the method of isoclines, without knowledge of the solution [Eq. (4.60)]. The governing differential equation is

$$\frac{dy}{dx} = -\frac{2\zeta y + x}{y} \qquad (4.61)$$

The resulting trajectories can be sketched in the phase-plane. On the other hand, Eq. (4.61) also can be integrated analytically by use of the substitution $z = y/x$ and separation of the variables:

$$y^2 + 2\zeta xy + x^2 = C \exp\left(\frac{2\zeta}{\sqrt{1-\zeta^2}}\tan^{-1}\frac{x\zeta + y}{x\sqrt{1-\zeta^2}}\right)$$

This is a spiral of the form of Eq. (4.60).

The method of isoclines is extremely useful in studying the behavior of solutions in the neighborhood of singular points and for the related questions of stability of solutions. In this sense, phase-plane methods may be thought of as topological methods.[5,30,32,39] However, it is desirable also to study the over-all solutions, rather than solutions in the neighborhood of special points, and preferably by some straightforward method of graphical integration. Such integration methods are given in the following sections of this chapter.

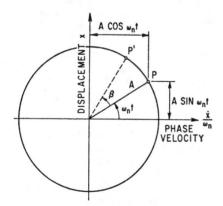

FIGURE 4.27 Phase-plane solution for a linear undamped vibrating system.

PHASE-PLANE INTEGRATION OF STEPWISE LINEAR SYSTEMS

Consider the undamped linear system described by Eq. (4.57). The known solution $x = A \sin \omega_n t$, $\dot{x} = A\omega_n \cos \omega_n t$ may be shown graphically in the phase-plane representation of Fig. 4.27. The point P moves with constant angular velocity ω_n, and the deflection increases to P' in the time β/ω_n.

If the system has a nonlinear restoring force composed of straight lines (as in Fig. 4.4), the motion within the region represented by any one linear segment can be described as above. For example, consider a system with the force-deflection characteristic shown at the top of Fig. 4.28. If the motion starts with initial velocity q_6 and zero initial deflection, the motion is described by a circular arc with center at 0 and angular velocity

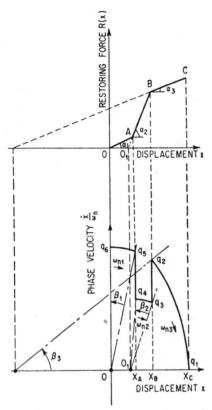

$$\omega_{n_1} = \sqrt{\frac{1}{m} \tan \alpha_1}$$

from q_6 to q_5. At the point q_5, it is seen that $\dot{x}_A/\omega_{n_1} = \overline{x_A q_5}$ and $\dot{x}_A/\omega_{n_2} = \overline{x_A q_4}$. Therefore

$$\frac{\overline{x_A q_4}}{\overline{x_A q_5}} = \sqrt{\frac{\tan \alpha_2}{\tan \alpha_1}}$$

In this example, $\tan \alpha_1 < \tan \alpha_2$ so that $\overline{x_A q_4} < \overline{x_A q_5}$. The circular arc from q_4 to q_3 corresponds to the segment AB of the restoring force characteristic with center at the intersection 0_1 of the segment (extended) with the X axis. The radius of this circle is $\overline{0_1 q_4}$, where

$$q_4 = \frac{\dot{x}_B}{\omega_{n_2}} \qquad \text{and} \qquad \omega_{n_2} = \sqrt{\frac{1}{m} \tan \alpha_2}$$

The total time required to go from q_6 to q_1 is

$$t = \frac{\beta_1}{\omega_{n_1}} + \frac{\beta_2}{\omega_{n_2}} + \frac{\beta_3}{\omega_{n_3}}$$

For a symmetrical system this is one-quarter of the period.

If the force-deflection characteristic of a nonlinear system is a smooth curve, it may be approximated by straight line segments and treated as above. It should be noted that the time required to complete one cycle is strongly influenced by the nature of the curve in regions where the velocity is low; therefore, linear approximations near the equilibrium position do not greatly affect the period.

FIGURE 4.28 Phase-plane solution for the stepwise linear restoring force characteristic curve shown at the top. The motion starts with zero displacement but finite velocity.

The time-history of the motion (i.e., the x,t representation) may be obtained quite readily by projecting values from the X axis to an x,t plane.

Inasmuch as phase-plane methods are restricted to autonomous systems, only free vibration is discussed above. However, if a constant force were to act on the system, the nature of the vibration would be unaffected, except for a displacement of the equilibrium position in the direction of the force and equal to the static deflection produced by that force. Thus, the trajectory would remain a circular arc but with its center displaced from the origin. Therefore, *nonautonomous systems may be treated by phase-plane methods, if the time function is replaced by a series of stepwise constant values.* The degree of accuracy attained in such a procedure depends only on the number of steps assumed to represent the time function.

A system having a bilinear restoring force and acted upon by an external stepwise function of time, treated by the method described above, is shown in Fig. 4.29. Phase-plane methods therefore offer the possibility of treating transient as well as free vibrations.

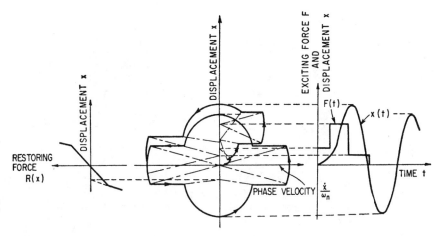

FIGURE 4.29 Phase-plane solution for transient motion. The bilinear restoring force characteristic curve is shown at the left, and the exciting force $F(t)$ and the resulting motion of the system $X(t)$ are shown at the right. (*After Evaldson et al.*[48])

Phase-plane methods have been widely used for the analysis of control mechanisms.[49] A comprehensive analysis of discontinuous-type systems possessing various types of nonlinearities is given in Ref. 50.

PHASE-PLANE INTEGRATION OF AUTONOMOUS SYSTEMS WITH NONLINEAR DAMPING

Consider the differential equation

$$\ddot{x} + g(\dot{x}) + \kappa^2 x = 0$$

Introducing $y = \dot{x}/\kappa$, the following isoclinic equation is obtained:

$$\frac{dy}{dx} = -\frac{g(y) + x}{y} \tag{4.62}$$

For points of zero slope in the phase-plane, the numerator of Eq. (4.62) must vanish; therefore, the condition for zero slope is

$$x_0 = -g(y)$$

Points of infinite slope correspond to the X axis. Singular points occur where the x_0 curve intersects the X axis.

To construct the trajectory, the slope at any point P_i must be determined first. This is done as illustrated in Fig. 4.30: A line is drawn parallel to the X axis through P_i. The intersection of this line with the x_0 curve determines a point S_i on the X axis. With S_i as the center, a circular arc of short length is drawn through P_i; the tangent to this arc is the required slope. The termination of this short arc may be taken as the point P_{i+1}, etc. The accuracy of the construction is dependent on the lengths of the arcs. This construction is known as Liénard's method.[5,30]

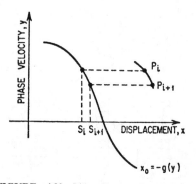

FIGURE 4.30 Liénard's construction for phase-plane integration of autonomous systems with nonlinear damping.

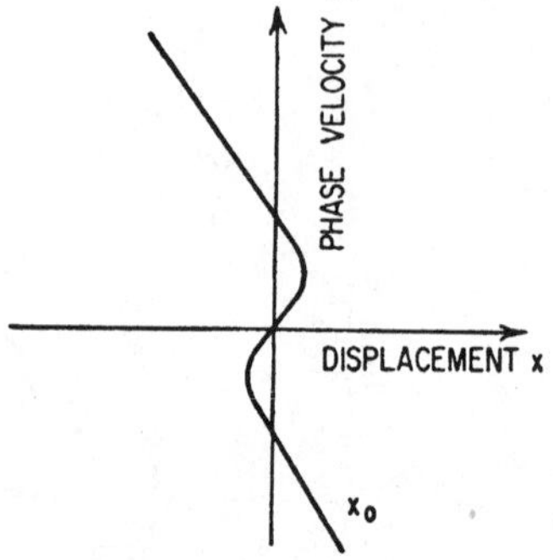

FIGURE 4.31 Curve of x_0 for Rayleigh's equation [Eq. (4.33)] as given by Eq. (4.63).

As an example of Liénard's method, consider Rayleigh's equation, Eq. (4.33), in the form

$$\ddot{x} + \varepsilon\left(\frac{\dot{x}^3}{3} - \dot{x}\right) + x = 0$$

The corresponding isoclinic equation is

$$\frac{dy}{dx} = \frac{\varepsilon(y - y^3/3) - x}{y}$$

The x_0 curve is given by

$$x_0 = \varepsilon\left(y - \frac{y^3}{3}\right) \qquad (4.63)$$

This is illustrated in Fig. 4.31.

A little experimentation shows that if a point P_1 is taken near the origin, the slope is such as to take the trajectory away from the origin (as compared with the undamped vibration); by the same reasoning, a point P_2 far from the origin tends to take the trajectory toward the origin (again as compared with the undamped vibration). Therefore, there is some neutral curve, describing a periodic motion, toward which the trajectories tend; this neutral curve is called a *limit cycle* and is illustrated in Fig. 4.32. Such a limit cycle is obtained when x_0 has a different sign for different parts of the Y axis.

For extreme values of ε, the x_0 curves would appear as shown in Fig. 4.33. For $\varepsilon \gg 1$, introduce the notation $\xi = x/\varepsilon$; then

$$\frac{dy}{d\xi} = \frac{\varepsilon}{y}\left(y - \frac{y^3}{3} - \xi\right)$$

This leads to a trajectory as shown in Fig. 4.34. This type of motion is known as a *relaxation oscillation*. Note from Fig. 4.34 that for this case of large ε the slope changes quickly from horizontal to vertical. Hence, for a motion starting at some point P_i, a vertical trajectory is followed until it intersects the ξ_0 curve; then, the trajectory turns and follows the ξ_0 curve until it enters the vertical field at the lower

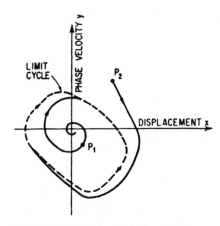

FIGURE 4.32 Limit cycle for Rayleigh's equation [Eq. (4.33)].

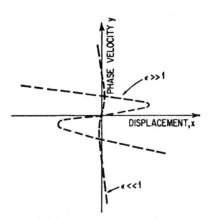

FIGURE 4.33 Curves of x_0 for extreme values of ε in Rayleigh's equation [Eq. (4.33)]. See Fig. 4.31 for a solution with a moderate value of ε.

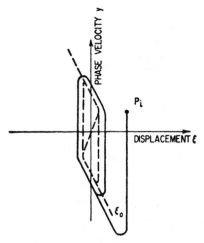

FIGURE 4.34 Relaxation oscillations of Rayleigh's equation [Eq. (4.33)].

knee in the curve. The trajectory then moves straight up until it intersects ξ_0 again after which it swings right and down again. A few circuits bring the trajectory into the limit cycle.

There is a possibility that more than one limit cycle may exist. If the x_0 curve crosses the X axis more than three times, it can be shown that at least two limit cycles may exist.

GENERALIZED PHASE-PLANE ANALYSIS

The following method of integrating second-order differential equations by phase-plane techniques has general application.[51,52] Consider the general equation

$$\ddot{x} + F(x,\dot{x},t) = 0 \tag{4.64}$$

Equation (4.64) can be converted to the form

$$\ddot{x} + \kappa^2 x = g(x,\dot{x},t)$$

by adding $\kappa^2 x$ to both sides where

$$\kappa^2 x - F(x,\dot{x},t) = g(x,\dot{x},t)$$

Let

$$g(x_0,\dot{x}_0,t_0) = -\kappa^2 \Delta_0$$

where κ is chosen arbitrarily. At some point P_0 on the trajectory,

$$\ddot{x} + \kappa^2(x + \Delta_0) = 0$$

and

$$\frac{dy}{dx} = -\frac{x + \Delta_0}{y}$$

Referring to Fig. 4.35,

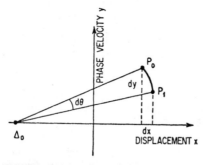

FIGURE 4.35 Method of construction employed in the generalized phase-plane analysis.

$$dt = \frac{1}{\kappa}\frac{dx}{y} = \frac{1}{\kappa}d\theta$$

Therefore, the time may be obtained by integration of the angular displacements. Thus, at a nearby point P_1 on the trajectory:

$$x_1 = x_0 + dx$$

$$y_1 = y_0 + dy$$

$$t_1 = t_0 + dt$$

Now, compute Δ_1 for the new center, and repeat the process.

This method has been applied to a very wide variety of linear and nonlinear equations.[51,53] For example, Fig. 4.36 shows the solution of Bessel's equation

$$\ddot{x} + \frac{1}{t}\dot{x} + \left(p^2 - \frac{n^2}{t^2}\right)x = 0$$

of order zero. The angle (or time) projection of x yields $J_0(pt)$, while the $\dot{x}/p$ projection yields $J_1(pt)$; that is, the Bessel functions of the zeroth and first order of the first kind. Bessel functions of the second kind also can be obtained.

STABILITY OF PERIODIC NONLINEAR VIBRATION

Certain systems having nonlinear restoring forces and undergoing forced vibration exhibit unstable characteristics for certain combinations of amplitude and exciting frequency. The existence of such an instability leads to the "jump phenomenon" shown in Fig. 4.16. To investigate the stability characteristics of the response curves, consider Duffing's equation

$$\ddot{x} + \kappa^2(x + \mu^2 x^3) = p \cos \omega t \tag{4.65}$$

Assume that two solutions of this equation exist and have slightly different initial conditions:

$$x_1 = x_0$$

$$x_2 = x_0 + \delta \qquad [\delta \ll x_0]$$

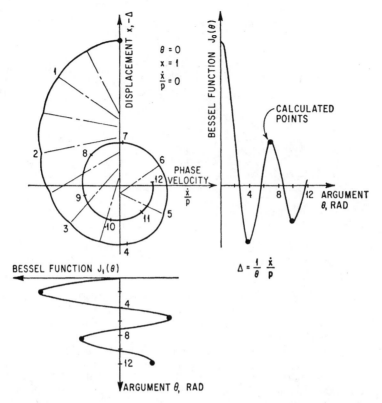

FIGURE 4.36 Generalized phase-plane solution of Bessel's equation. (*Jacobsen.*[51])

Introducing the second of these into Eq. (4.65) and employing the condition that x_0 is also a solution,

$$\ddot{\delta} + \kappa^2(1 + 3\mu^2 x_0^2)\delta = 0 \tag{4.66}$$

Now an expression for x_0 must be obtained; assuming a one-term approximation of the form $x_0 = A \cos \omega t$, Eq. (4.66) becomes

$$\frac{d^2\delta}{d\varphi^2} + (\lambda + \gamma \cos \varphi)\delta = 0 \tag{4.67}$$

where

$$\kappa^2(1 + \tfrac{3}{2}\mu^2 A^2) = 4\omega^2\lambda$$

and

$$\tfrac{3}{2}\kappa^2\mu^2 A^2 = 4\omega^2\gamma \qquad 2\omega t = \varphi \tag{4.68}$$

Equation (4.67) is known as Mathieu's equation.

Mathieu's equation has appeared in this analysis as a variational equation characterizing small deviations from the given periodic motion whose stability is to be investigated; thus, the stability of the solutions of Mathieu's equation must be studied. A given periodic motion is stable if *all* solutions of the variational equation associated with it tend toward zero for all positive time and unstable if there is at least

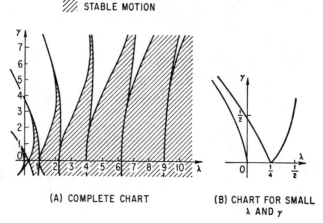

FIGURE 4.37 Stability chart for Mathieu's equation [Eq. (4.67)].

one solution which does not tend toward zero. The stability characteristics of Eq. (4.67) often are represented in a chart as shown in Fig. 4.37.[30]

From the response diagram of Duffing's equation, the out-of-phase motion having the larger amplitude appears to be unstable. This portion of the response diagram (Fig. 4.16C) corresponds to unstable motion in the Mathieu stability chart (Fig. 4.37), and the locus of vertical tangents of the response curves (considering undamped vibration for simplicity) corresponds exactly to the boundaries between stable and unstable regions in the stability chart. Thus, the region of interest in the response diagram is described by the free vibration

$$\omega^2 = \kappa^2(1 + \tfrac{3}{4}\mu^2 A^2) \tag{4.69}$$

and the locus of vertical tangents

$$\tfrac{3}{2}\kappa^2\mu^2 A^2 + \frac{p}{A} = 0 \tag{4.70}$$

The corresponding curves in the stability chart are taken as those for small positive values of γ and λ which have the approximate equations

$$\gamma = \tfrac{1}{2} - 2\lambda \tag{4.71}$$

$$\gamma = -\tfrac{1}{2} + 2\lambda \tag{4.72}$$

Now, if Eq. (4.69) is introduced into Eqs. (4.68), the resulting equations expanded by the binomial theorem (assuming μ^2 small), and Eq. (4.72) introduced, the result is an identity. Therefore, the free vibration-response curve maps onto the curve of positive slope in the stability chart. The locus of vertical tangents to the response curves maps into the curve of negative slope in the stability chart; this may be seen from the identity obtained by introducing the equations obtained above by binomial expansion into Eq. (4.71) and then employing Eq. (4.70).

In any given case, it can be determined whether a motion is stable or unstable on the basis of the values of γ and λ, according to the location of the point in the stability chart.

The question of stability of response also can be resolved by means of a "stability criterion" developed from the Kryloff-Bogoliuboff procedures.[54] The differential equation of motion is considered in the form

$$\ddot{x} + \kappa^2 x + f(x,\dot{x}) = p \cos \omega t$$

Proceeding in the manner of the Kryloff-Bogoliuboff procedure described earlier,

$$\dot{A} = \frac{1}{\kappa} f(x,\dot{x}) \sin (\kappa t + \theta) - \frac{p}{\kappa} \cos \omega t \sin (\kappa t + \theta)$$

$$\dot{\theta} = \frac{1}{\kappa} f(x,\dot{x}) \cos (\kappa t + \theta) - \frac{p}{A\kappa} \cos \omega t \cos (\kappa t + \theta)$$

Expanding the last terms of these equations, the result contains motions of frequency κ, $\kappa + \omega$, and $\kappa - \omega$. The motion over a long interval of time is of interest, and the motions of frequencies $\kappa + \omega$ and $\kappa - \omega$ may be averaged out; this is accomplished by integrating over the period $2\pi/\omega$:

$$\dot{A} = S(A) - \frac{p}{2\kappa} \sin(\Phi - \omega t)$$

$$\dot{\theta} = \frac{C(A)}{A} - \frac{p}{2\kappa A} \cos(\Phi - \omega t)$$

where

$$S(A) = \frac{1}{2\pi\kappa} \int_0^{2\pi} f(A \cos \Phi, -A\kappa \sin \Phi) \sin \Phi \, d\Phi$$

$$C(A) = \frac{1}{2\pi\kappa} \int_0^{2\pi} f(A \cos \Phi, -A\kappa \sin \Phi) \cos \Phi \, d\Phi$$

The steady-state solution may be determined by employing the conditions $A = A_0$, $\psi = \Phi - \omega t = \psi_0$:

$$\frac{p^2}{4\kappa^2} = S^2(A_0) + [C(A_0) + A_0(\kappa - \omega)]^2$$

$$\tan \psi_0 = \frac{S(A_0)}{C(A_0) + A_0(\kappa - \omega)}$$

This steady-state solution will now be perturbed and the stability of the ensuing motion investigated. Let

$$A(t) = A_0 + \xi(t) \qquad [\xi \ll A_0]$$

$$\psi(t) = \psi_0 + \eta(t) \qquad [\eta \ll \psi_0]$$

By Taylor's series expansion:

$$\dot{\xi} = \xi S'(A_0) - \frac{p}{2\kappa} \eta \cos \psi_0$$

$$\dot{\eta} = \frac{\xi}{A_0} [(\kappa - \omega) + C'(A_0)] + \frac{p}{2\kappa A_0} \eta \sin \psi_0$$

where primes indicate differentiation with respect to A. These two differential equations are satisfied by the solutions

$$\xi = \mathcal{A} e^{zt} \qquad \eta = \mathcal{B} e^{zt}$$

where α and $\mathcal{B}$ are arbitrary constants and

$$z = \frac{1}{2A_0}\left\{[S(A_0) + A_0 S'(A_0)] \pm \sqrt{[S(A_0) + A_0 S'(A_0)]^2 - 4A_0\bar{p}\frac{d\bar{p}}{dA_0}}\right\}$$

and $\bar{p} = p/2\kappa$.

For stability, the real parts of z must be negative; hence, the following criteria can be established:[54]

$$[S(A_0) + A_0 S'(A_0)] < 0, \quad \frac{d\bar{p}}{dA_0} > 0, \text{ ensures stability}$$

$$[S(A_0) + A_0 S'(A_0)] < 0, \quad \frac{d\bar{p}}{dA_0} < 0, \text{ ensures instability}$$

$$[S(A_0) + A_0 S'(A_0)] > 0, \quad \frac{d\bar{p}}{dA_0} \gtreqless 0, \text{ ensures instability}$$

$$[S(A_0) + A_0 S'(A_0)] = 0, \quad \frac{d\bar{p}}{dA_0} > 0, \text{ ensures stability}$$

These criteria can be interpreted in terms of response curves by reference to Fig. 4.14. For systems of this type, $[S(A_0) + A_0 S'(A_0)] < 0$; when $d\bar{p}/dA_0 > 0$, $\bar{p}$ increases as A_0 also increases. This does not hold for the middle branch of the response curves, thus confirming the earlier results. Other analyses of stability are found in Refs. 55 to 58.

SYSTEMS OF MORE THAN A SINGLE DEGREE-OF-FREEDOM

Interest in systems of more than one degree-of-freedom arises from the problem of the dynamic vibration absorber. The earliest studies of nonlinear two degree-of-freedom systems were those of vibration absorbers having nonlinear elements.[59-61]

The analysis of multiple degree-of-freedom systems can be carried out by various of the methods described earlier in this chapter; thus, a stepwise linear system is treated in Ref. 62, and more general systems are treated in Refs. 62 to 66. The extension of phase-plane methods to such systems is given in Refs. 49, 50, and 67.

All the essential features of nonlinear vibration of single degree-of-freedom systems described earlier occur in the multiple degree-of-freedom systems as well. An analysis which considers subharmonic vibration in such systems by an iteration procedure is given in Ref. 68 and is completely analogous to that given here for the single degree-of-freedom system, with analogous results.

REFERENCES

Additional citations to the literature are given in Refs. 4, 49, 52, 69, and 70 and in past and current issues of the journal *Applied Mechanics Reviews*.

1. von Kármán, T.: *Bull. Am. Math. Soc.,* **46:**615 (1940).
2. Clauser, F. H.: *J. Aeronaut. Sci.,* **23:**411 (1956).

3. Davis, S. A.: *Product Eng.,* **25:**181 (1954).

4. McLachlan, N. W.: "Ordinary Nonlinear Differential Equations," Oxford University Press, New York, 1956.

5. Minorsky, N.: "Introduction to Nonlinear Mechanics," Edwards Bros., Ann Arbor, Mich., 1947.

6. Ehrich, F. F.: "Rotordynamic Response in Nonlinear Anisotropic Mounting Systems," *Proc. of the 4th Intl. Conf. on Rotor Dynamics,* IFTOMM, 1–6, Chicago, September 7–9, 1994.

7. Thompson, J. M. T., and H. B. Stewart: "Nonlinear Dynamics and Chaos," John Wiley and Sons, New York, 310–320, 1987.

8. Maezawa, S., "Superharmonic Resonance in Piecewise Linear Systems with Unsymmetrical Characteristics," *Proc. of the 5th Intl. Conf. on Nonlinear Oscillation,* Kiev, August 26–September 5, 1969.

9. Choi, Y. S., and S. T. Noah: "Forced Periodic Vibration of Unsymmetric Piecewise-Linear Systems," *J. of Sound and Vibration,* **121**(3):117–126, 1988.

10. Ehrich, F. F.: "Observations of Subcritical Superharmonic and Chaotic Response in Rotordynamics," *J. of Vibration and Acoustics,* **114**(1):93–100, 1992.

11. Nayfeh, A. H., B. Balachandran, M. A. Colbert, and M. A. Nayfeh: "An Experimental Investigation of Complicated Responses of a Two-Degree-of-Freedom Structure," ASME Paper No. 90-WA/APM-24, 1990.

12. Ehrich, F. F.: "Spontaneous Sidebanding in High Speed Rotordynamics," *J. of Vibration and Acoustics,* **114**(4):498–505, 1992.

13. Ehrich, F. F.: "Subharmonic Vibration of Rotors in Bearing Clearance," ASME Paper No. 66-MD-1, 1966.

14. Bently, D. E.: "Forced Subrotative Speed Dynamic Action of Rotating Machinery," ASME Paper No. 74-Pet-16, 1974.

15. Childs, D. W.: "Fractional Frequency Rotor Motion Due to Nonsymmetric Clearance Effects," *J. of Eng. for Power,* 533–541, July 1982.

16. Muszynska, A.: "Partial Lateral Rotor to Stator Rubs," IMechE Paper No. C281/84, 1984.

17. Ehrich, F. F.: "High Order Subharmonic Response of High Speed Rotors in Bearing Clearance," *J. of Vibration, Acoustics, Stress and Reliability in Design,* **110**(9):9–16, 1988.

18. Masri, S. F.: "Theory of the Dynamic Vibration Neutralizer with Motion Limiting Stops," *J. of Applied Mechanics,* **39:**563–569, 1972.

19. Shaw, S. W., and P. J. Holmes: "A Periodically Forced Piecewise Linear Oscillator," *J. of Sound and Vibration,* **90**(1):129–155, 1983.

20. Shaw, S. W.: "Forced Vibrations of a Beam with One-Sided Amplitude Constraint: Theory and Experiment," *J. of Sound and Vibration,* **99**(2):199–212, 1985.

21. Shaw, S. W.: "The Dynamics of a Harmonically Excited System Having Rigid Amplitude Constraints," *J. of Applied Mechanics,* **52:**459–464, 1985.

22. Choi, Y. S., and S. T. Noah: "Nonlinear Steady-State Response of a Rotor-Support System," *J. of Vibration, Acoustics, Stress and Reliability in Design,* 255–261, July 1987.

23. Moon, F. C.: "Chaotic Vibrations," John Wiley & Sons, New York, 1987.

24. Sharif-Bakhtiar, M., and S. W. Shaw: "The Dynamic Response of a Centrifugal Pendulum Vibration Absorber with Motion Limiting Stops," *J. of Sound and Vibration,* **126**(2):221–235, 1988.

25. Ehrich, F. F.: "Some Observations of Chaotic Vibration Phenomena in High Speed Rotordynamics," *J. of Vibration and Acoustics,* **113**(1):50–57, 1991.

26. Klotter, K.: *Proc. 1st U.S. Natl. Congr. Appl. Mechs.,* 1951.

27. Klotter, K.: *Stanford Univ., Rept.* 17, Contract N6 ONR-251-II, 1951.

28. Klotter, K.: *Proc. Symposium on Nonlinear Circuit Analysis,* 1953.

29. Levy, H., and E. A. Baggott: "Numerical Solutions of Differential Equations," Dover Publications, New York, 1950.

30. Stoker, J. J.: "Nonlinear Vibrations," Interscience Publishers, Inc., New York, 1950.

31. Kryloff, N., and N. Bogoliuboff: "Introduction to Nonlinear Mechanics," Princeton University Press, Princeton, N.J., 1943.

32. Andronow, A. A., and C. E. Chaikin: "Theory of Oscillations," Princeton University Press, Princeton, N.J., 1949.

33. Rudenberg, R.: *ZAMM,* **3:**454 (1923).

34. Brock, J. E.: *J. Appl. Mechanics,* **18** (1951).

35. Roberson, R. E.: *J. Appl. Mechanics,* **20:**237 (1953).

36. Wylie, C. R.: *J. Franklin Inst.,* **236:**273 (1943).

37. Young, D.: *Proc. 1st Midwest. Conf. Solid Mechanics,* 1953.

38. Fifer, S.: *J. Appl. Phys.,* **22:**1421 (1951).

39. Minorsky, N.: "Advances in Applied Mechanics," Vol. I, Academic Press, Inc., New York, 1948.

40. Bellin, A. I.: "Advances in Applied Mechanics," Vol. III, Academic Press, Inc., New York, 1953.

41. Pipes, L. A.: *J. Appl. Phys.,* **13:**117 (1942).

42. Duffing, G.: "Erzwungene Schwingungen bei veranderlicher Eigenfrequenz," F. Vieweg u Sohn, Brunswick, 1918.

43. Rauscher, M.: *J. Appl. Mechanics,* **5:**169 (1938).

44. Minorsky, N.: *J. Franklin Inst.,* **248:**205 (1949).

45. Carrier, G. F.: *Quart. Appl. Math.,* **3:**157 (1945).

46. Schwesinger, G.: *J. Appl. Mechanics,* **17:**202 (1950).

47. Abramson, H. N.: *Product Eng.,* **25:**179 (1954).

48. Evaldson, R. L., R. S. Ayre, and L. S. Jacobsen: *J. Franklin Inst.,* **248:**473 (1949).

49. Ku, Y. H.: "Analysis and Control of Nonlinear Systems," The Ronald Press Company, New York, 1958.

50. Flügge-Lotz, I.: "Discontinuous Automatic Control," Princeton University Press, Princeton, N.J., 1953.

51. Jacobsen, L. S.: *J. Appl. Mechanics,* **19:**543 (1952).

52. Jacobsen, L. S., and R. S. Ayre: "Engineering Vibrations," McGraw-Hill Book Company, Inc., New York, 1958.

53. Bishop, R. E. D.: *Proc. Inst. Mech. Engrs.,* **168:**299 (1954).

54. Klotter, K., and E. Pinney: *J. Appl. Mechanics,* **20:**9 (1953).

55. John, F.: "Studies in Nonlinear Vibration Theory," New York University, 1946.

56. Rosenberg, R. M.: *Proc. 2nd Natl. Congr. Appl. Mechanics,* 1954.

57. Young, D., and P. N. Hess: *Proc. 2nd Natl. Congr. Appl. Mechanics,* 1954.

58. Hayashi, C.: "Forced Oscillations in Nonlinear Systems," Nippon Pub. Co., Osaka, Japan, 1953.

59. Roberson, R. E.: *J. Franklin Inst.,* **254:**205 (1952).

60. Pipes, L. A.: *J. Appl. Mechanics,* **20:**515 (1953).

61. Arnold, F. R.: *J. Appl. Mechanics,* **22:**487 (1955).

62. Soroka, W. W.: *J. Appl. Mechanics,* **17:**185 (1950).

63. Sethna, P. R.: *Proc. 2nd Natl. Congr. Appl. Mechanics,* 1954.

64. Huang, T. C.: *J. Appl. Mechanics,* **22:**107 (1955).

65. Klotter, K.: *Trans. IRE on Circuit Theory,* **CT-1** (4):13 (1954).
66. Stoker, J. J.: *Proc. 2nd Natl. Congr. Appl. Mechanics,* 1954.
67. Ku, Y. H.: *J. Franklin Inst.,* **259:**115 (1955).
68. Huang, T. C.: *Proc. 2nd Natl. Congr. Appl. Mechanics,* 1954.
69. Minorsky, N.: *Appl. Mechanics Revs.,* **4:**266 (1951).
70. Klotter, K.: *Appl. Mechanics Revs.,* **10:**495 (1957).

CHAPTER 5
SELF-EXCITED VIBRATION

F. F. Ehrich

INTRODUCTION

Self-excited systems begin to vibrate of their own accord spontaneously, the amplitude increasing until some nonlinear effect limits any further increase. The energy supplying these vibrations is obtained from a uniform source of power associated with the system which, due to some mechanism inherent in the system, gives rise to oscillating forces. The nature of self-excited vibration compared to forced vibration is:[1]

In self-excited vibration the alternating force that sustains the motion is created or controlled by the motion itself; when the motion stops, the alternating force disappears.

In a forced vibration the sustaining alternating force exists independent of the motion and persists when the vibratory motion is stopped.

The occurrence of self-excited vibration in a physical system is intimately associated with the stability of equilibrium positions of the system. If the system is disturbed from a position of equilibrium, forces generally appear which cause the system to move either toward the equilibrium position or away from it. In the latter case the equilibrium position is said to be unstable; then the system may either oscillate with increasing amplitude or monotonically recede from the equilibrium position until nonlinear or limiting restraints appear. The equilibrium position is said to be stable if the disturbed system approaches the equilibrium position either in a damped oscillatory fashion or asymptotically.

The forces which appear as the system is displaced from its equilibrium position may depend on the displacement or the velocity, or both. If displacement-dependent forces appear and cause the system to move away from the equilibrium position, the system is said to be statically unstable. For example, an inverted pendulum is statically unstable. Velocity-dependent forces which cause the system to recede from a statically stable equilibrium position lead to dynamic instability.

Self-excited vibrations are characterized by the presence of a mechanism whereby a system will vibrate at its own natural or critical frequency, essentially *independent* of the *frequency* of any external stimulus. In mathematical terms, the motion is described by the unstable *homogeneous* solution to the homogeneous equations of motion. In contradistinction, in the case of "forced," or "resonant," vibrations, the *frequency* of the oscillation is *dependent* on (equal to, or a whole number ratio of) the frequency of a forcing function external to the vibrating system (e.g., shaft rotational speed in the case of rotating shafts). In mathematical terms, the forced vibration is the *particular* solution to the *nonhomogeneous* equations of motion.

Self-excited vibrations pervade all areas of design and operations of physical systems where motion or time-variant parameters are involved—aeromechanical systems (flutter, aircraft flight dynamics), aerodynamics (separation, stall, musical wind instruments, diffuser and inlet chugging), aerothermodynamics (flame instability, combustor screech), mechanical systems (machine-tool chatter), and feedback networks (pneumatic, hydraulic, and electromechanical servomechanisms).

ROTATING MACHINERY

One of the more important manifestations of self-excited vibrations, and the one that is the principal concern in this chapter, is that of rotating machinery, specifically, the self-excitation of lateral, or flexural, vibration of rotating shafts (as distinct from torsional, or longitudinal, vibration).

In addition to the description of a large number of such phenomena in standard vibrations textbooks (most typically and prominently, Ref. 1), the field has been subject to several generalized surveys.[2-4] The mechanisms of self-excitation which have been identified can be categorized as follows:

Whirling or Whipping.
>   Hysteretic whirl
>   Fluid trapped in the rotor
>   Dry friction whip
>   Fluid bearing whip
>   Seal and blade-tip-clearance effect in turbomachinery
>   Propeller and turbomachinery whirl

Parametric Instability.
>   Asymmetric shafting
>   Pulsating torque
>   Pulsating longitudinal loading

Stick-Slip Rubs and Chatter.
Instabilities in Forced Vibrations.
>   Bistable vibration
>   Unstable imbalance

In each instance, the physical mechanism is described and aspects of its prevention or its diagnosis and correction are given. Some exposition of its mathematical analytic modeling is also included.

WHIRLING OR WHIPPING

ANALYTIC MODELING

In the most important subcategory of instabilities (generally termed whirling or whipping), the unifying generality is the generation of a tangential force, normal to an arbitrary radial deflection of a rotating shaft, whose magnitude is proportional to (or varies monotonically with) that deflection. At some "onset" rotational speed, such a force system will overcome the stabilizing external damping forces which are generally present and induce a whirling motion of ever-increasing amplitude, limited only by nonlinearities which ultimately limit deflections.

A close mathematical analogy to this class of phenomena is the concept of "negative damping" in linear systems with constant coefficients, subject to *plane* vibration.

A simple mathematical representation of a self-excited vibration may be found in the concept of negative damping. Consider the differential equation for a damped, free vibration:

$$m\ddot{x} + c\dot{x} + kx = 0 \tag{5.1}$$

This is generally solved by assuming a solution of the form

$$x = Ce^{st}$$

Substitution of this solution into Eq. (5.1) yields the characteristic (algebraic) equation

$$s^2 + \frac{c}{m}s + \frac{k}{m} = 0 \tag{5.2}$$

If $c < 2\sqrt{mk}$, the roots are complex:

$$s_{1,2} = -\frac{c}{2m} \pm iq$$

where

$$q = \sqrt{\frac{k}{m} - \left(\frac{c}{2m}\right)^2}$$

The solution takes the form

$$x = e^{-ct/2m}(A\cos qt + B\sin qt) \tag{5.3}$$

This represents a decaying oscillation because the exponential factor is negative, as illustrated in Fig. 5.1A. If $c < 0$, the exponential factor has a positive exponent and the vibration appears as shown in Fig. 5.1B. The system, initially at rest, begins to oscillate spontaneously with ever-increasing amplitude. Then, in any physical system, some nonlinear effect enters and Eq. (5.1) fails to represent the system realistically. Equation (5.4) defines a nonlinear system with negative damping at small amplitudes but with large positive damping at larger amplitudes, thereby limiting the amplitude to finite values:

$$m\ddot{x} + (-c + ax^2)\dot{x} + kx = 0 \tag{5.4}$$

Thus, the fundamental criterion of stability in linear systems is that the roots of the characteristic equation have negative real parts, thereby producing decaying amplitudes.

In the case of a whirling or whipping shaft, the equations of motion (for an idealized shaft with a single lumped mass m) are more appropriately written in polar coordinates for the radial force balance,

$$-m\omega^2 r + m\ddot{r} + c\dot{r} + kr = 0 \tag{5.5}$$

and for the tangential force balance,

$$2m\omega\dot{r} + c\omega r - F_n = 0 \tag{5.6}$$

where we presume a constant rate of whirl ω.

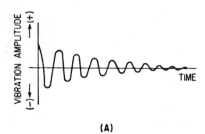

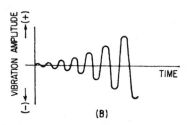

FIGURE 5.1 (*A*) Illustration showing a decaying vibration (stable) corresponding to negative real parts of the complex roots. (*B*) Increasing vibration corresponding to positive real parts of the complex roots (unstable).

In general, the whirling is predicated on the existence of some physical phenomenon which will induce a force F_n that is normal to the radial deflection r and is in the direction of the whirling motion—i.e., in opposition to the damping force, which tends to inhibit the whirling motion. Very often, this normal force can be characterized or approximated as being proportional to the radial deflection:

$$F_n = f_n r \qquad (5.7)$$

The solution then takes the form

$$r = r_0 e^{at} \qquad (5.8)$$

For the system to be stable, the coefficient of the exponent

$$a = \frac{f_n - c\omega}{2m\omega} \qquad (5.9)$$

must be negative, giving the requirement for stable operation as

$$f_n < \omega c \qquad (5.10)$$

As a rotating machine increases its rotational speed, the left-hand side of this inequality (which is generally also a function of shaft rotation speed) may exceed the right-hand side, indicative of the onset of instability. At this onset condition,

$$a = 0+ \qquad (5.11)$$

so that whirl speed at onset is found to be

$$\omega = \left(\frac{k}{m} \right)^{1/2} \qquad (5.12)$$

That is, the whirling speed at onset of instability is the shaft's natural or critical frequency, irrespective of the shaft's rotational speed (rpm). The direction of whirl may be in the same rotational direction as the shaft rotation (*forward* whirl) or opposite to the direction of shaft rotation (*backward* whirl), depending on the direction of the destabilizing force F_n.

When the system is unstable, the solution for the trajectory of the shaft's mass is, from Eq. (5.8), an exponential spiral as in Fig. 5.2. Any planar component of this two-dimensional trajectory takes the same form as the unstable planar vibration shown in Fig. 5.1*B*.

GENERAL DESCRIPTION

The most important examples of whirling and whipping instabilities are

Hysteretic whirl

Fluid trapped in the rotor

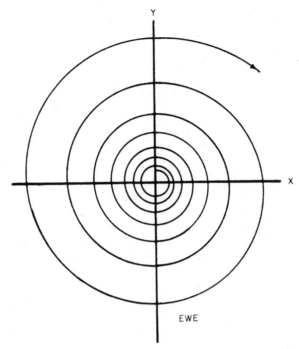

FIGURE 5.2 Trajectory of rotor center of gravity in unstable whirling or whipping.

Dry friction whip
Fluid bearing whip
Seal and blade-tip-clearance effect in turbomachinery
Propeller and turbomachinery whirl

All these self-excitation systems involve friction or fluid energy mechanisms to generate the destabilizing force.

These phenomena are rarer than forced vibration due to unbalance or shaft misalignment, and they are difficult to anticipate before the fact or diagnose after the fact because of their subtlety. Also, self-excited vibrations are potentially more destructive, since the asynchronous whirling of self-excited vibration induces alternating stresses in the rotor and can lead to fatigue failures of rotating components. Synchronous forced vibration typical of unbalance does not involve alternating stresses in the rotor and will rarely involve rotating element failure. The general attributes of these instabilities, insofar as they differ from forced excitations, are summarized in Table 5.1 and Fig. 5.3A and B.

HYSTERETIC WHIRL

The mechanism of hysteretic whirl, as observed experimentally,[5] defined analytically,[6] or described in standard texts,[7] may be understood from the schematic representation of Fig. 5.4. With some nominal radial deflection of the shaft, the flexure of the shaft would induce a neutral strain axis normal to the deflection direction. From

TABLE 5.1   Characterization of Two Categories of Vibration of Rotating Shafts

	Forced or resonant vibration	Whirling or whipping
Vibration frequency–rpm relationship	Frequency is equal to (i.e., synchronous with) rpm or a whole number or rational fraction of rpm, as in Fig. 5.3A.	Frequency is nearly constant and relatively independent of rotor rotational speed or any external stimulus and is at or near one of the shaft critical or natural frequencies, as in Fig. 5.3B.
Vibration amplitude–rpm relationship	Amplitude will peak in a narrow band of rpm wherein the rotor's critical frequency is equal to the rpm or to a whole-number multiple or a rational fraction of the rpm or an external stimulus, as in Fig. 5.3A.	Amplitude will suddenly increase at an onset rpm and continue at high or increasing levels as rpm is increased, as in Fig. 5.3B.
Influence of damping	Addition of damping may reduce peak amplitude but not materially affect rpm at which peak amplitude occurs, as in Fig. 5.3A.	Addition of damping may defer onset to a higher rpm but not materially affect amplitude after onset, as in Fig. 5.3B.
System geometry	Excitation level and hence amplitude are dependent on some lack of axial symmetry in the rotor mass distribution or geometry, or external forces applied to the rotor. Amplitudes may be reduced by refining the system to make it more perfectly axisymmetric.	Amplitudes are independent of system axial symmetry. Given an infinitesimal deflection to an otherwise symmetric system, the amplitude will self-propagate.
Rotor fiber stress	For synchronous vibration, the rotor vibrates in frozen, deflected state, without oscillatory fiber stress.	Rotor fibers are subject to oscillatory stress at a frequency equal to the difference between rotor rpm and whirling speed.
Avoidance or elimination	1. Tune the system's critical frequencies to be out of the rpm operating range. 2. Eliminate all deviations from axial symmetry in the system as built or as induced during operation (e.g., balancing). 3. Introduce damping to limit peak amplitudes at critical speeds which must be traversed.	1. Restrict operating rpm to below instability onset rpm. 2. Defeat or eliminate the instability mechanism. 3. Introduce damping to raise the instability onset speed to above the operating speed range. 4. Introduce stiffness anisotropy to the bearing support system.[8]

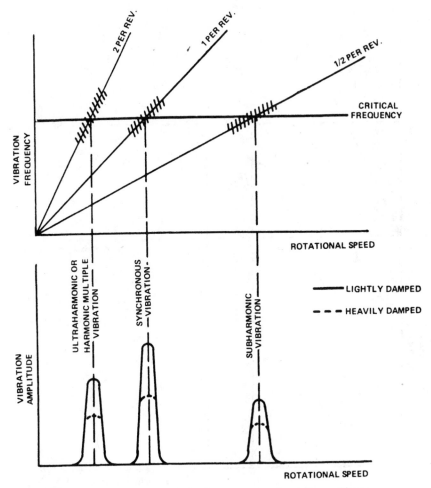

FIGURE 5.3 (*A*) Attributes of forced vibration or resonance in rotating machinery.

first-order considerations of elastic-beam theory, the neutral axis of stress would be coincident with the neutral axis of strain. The net elastic restoring force would then be perpendicular to the neutral stress axis, i.e., parallel to and opposing the deflection. In actual fact, hysteresis, or internal friction, in the rotating shaft will cause a phase shift in the development of stress as the shaft fibers rotate around through peak strain to the neutral strain axis. The net effect is that the neutral stress axis is displaced in angle orientation from the neutral strain axis, and the resultant force is not parallel to the deflection. In particular, the resultant force has a tangential component *normal* to the deflection, which is the fundamental precondition for whirl. This tangential force component is in the direction of rotation and induces a *forward* whirling motion which increases centrifugal force on the deflected rotor, thereby increasing its deflection. As a consequence, induced stresses are increased, thereby increasing the whirl-inducing force component.

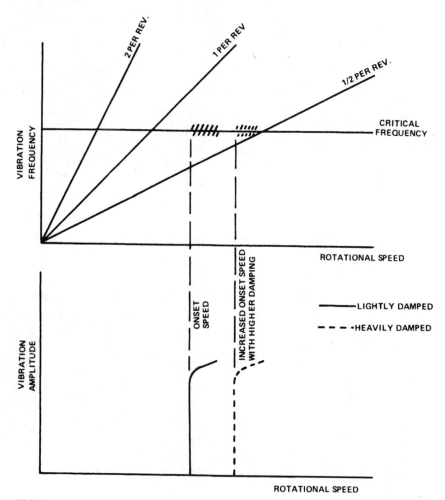

FIGURE 5.3 (*Continued*)   (*B*) Attributes of whirling or whipping in rotating machinery.

Several surveys and contributions to the understanding of the phenomenon have been published in Refs. 9, 10, 11, and 12. It has generally been recognized that hysteretic whirl can occur only at rotational speeds above the first-shaft critical speed (the lower the hysteretic effect, the higher the attainable whirl-free operating rpm). It has been shown[13] that once whirl has started, the critical whirl speed that will be induced (from among the spectrum of criticals of any given shaft) will have a frequency approximately half the onset rpm.

A straightforward method for hysteretic whirl avoidance is that of limiting shafts to subcritical operation, but this is unnecessarily and undesirably restrictive. A more effective avoidance measure is to limit the hysteretic characteristic of the rotor. Most investigators (e.g., Ref. 5) have suggested that the essential hysteretic effect is

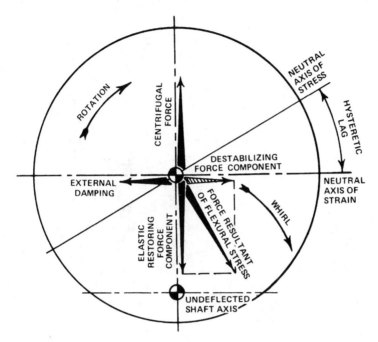

FIGURE 5.4 Hysteretic whirl.

caused by working at the interfaces of joints in a rotor rather than within the material of that rotor's components. Success in avoiding hysteretic whirl has been achieved by minimizing the number of separate elements, restricting the span of concentric rabbets and shrunk fitted parts, and providing secure lockup of assembled elements held together by tie bolts and other compression elements. Bearing-foundation characteristics also play a role in suppression of hysteretic whirl.[9]

WHIRL DUE TO FLUID TRAPPED IN ROTOR

There has always been a general awareness that high-speed centrifuges are subject to a special form of instability. It is now appreciated that the same self-excitation may be experienced more generally in high-speed rotating machinery where liquids (e.g., oil from bearing sumps, steam condensate, etc.) may be inadvertently trapped in the internal cavity of hollow rotors. The mechanism of instability is shown schematically in Fig. 5.5. For some nominal deflection of the rotor, the fluid is flung out radially in the direction of deflection. But the fluid does not remain in simple radial orientation. The spinning surface of the cavity drags the fluid (which has some finite viscosity) in the direction of rotation. This angle of advance results in the centrifugal force on the fluid having a component in the tangential direction in the direction of rotation. This force then is the basis of instability, since it induces forward whirl which increases the centrifugal force on the fluid and thereby increases the whirl-inducing force.

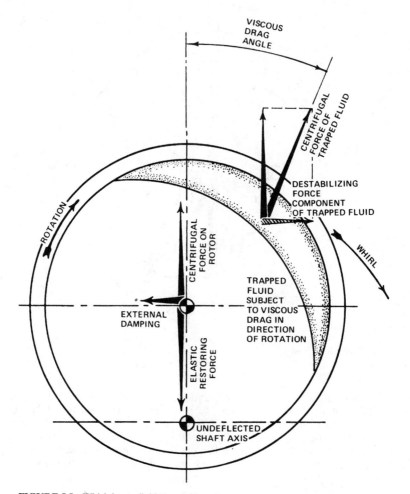

FIGURE 5.5 Whirl due to fluid trapped in rotor.

Contributions to the understanding of the phenomenon as well as a complete history of the phenomenon's study are available in Ref. 14. It has been shown[15] that onset speed for instability is always above critical rpm and below twice-critical rpm. Since the whirl is at the shaft's critical frequency, the ratio of whirl frequency to rpm will be in the range of 0.5 to 1.0.

Avoidance of this self-excitation can be accomplished by running shafting subcritically, although this is generally undesirable in centrifuge-type applications when further consideration is made of the role of trapped fluids as unbalance in forced vibration of rotating shafts (as described in Ref. 15). Where the trapped fluid is not fundamental to the machine's function, the appropriate avoidance measure, if the particular application permits, is to provide drain holes at the outermost radius of all hollow cavities where fluid might be trapped.

DRY FRICTION WHIP

As described in standard vibration texts (e.g., Ref. 16), dry friction whip is experienced when the surface of a rotating shaft comes in contact with an unlubricated stationary guide or shroud or stator system. This can occur in an unlubricated journal bearing or with loss of clearance in a hydrodynamic bearing or inadvertent closure and contact in the radial clearance of labyrinth seals or turbomachinery blading or power screws.[17]

The phenomenon may be understood with reference to Fig. 5.6. When radial contact is made between the surface of the rotating shaft and a static part, Coulomb friction will induce a tangential force on the rotor. Since the friction force is approximately proportional to the radial component of the contact force, we have the preconditions for instability. The tangential force induces a whirling motion which induces larger centrifugal force on the rotor, which in turn induces a large radial contact and hence larger whirl-inducing friction force.

It is interesting to note that this whirl system is *counter* to the shaft rotation direction (i.e., *backward* whirl). One may envision the whirling system as the rolling (accompanied by appreciable slipping) of the shaft in the stator system.

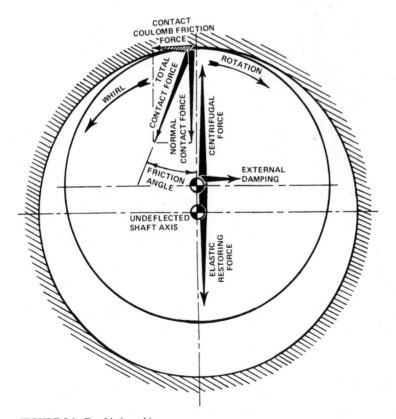

FIGURE 5.6 Dry friction whip.

The same situation can be produced by a thrust bearing where angular deflection is combined with lateral deflection.[18] If contact occurs on the same side of the disc as the virtual pivot point of the deflected disc, then backward whirl will result. Conversely, if contact occurs on the side of the disc opposite to the side where the virtual pivot point of the disc is located, then forward whirl will result.

It has been suggested (but not concluded)[19] that the whirling frequency is generally less than the critical speed.

The vibration is subject to various types of control. If contact between rotor and stator can be avoided or the contact area can be kept well lubricated, no whipping will occur. Where contact must be accommodated, and lubrication is not feasible, whipping may be avoided by providing abradability of the rotor or stator element to allow disengagement before whirl. When dry friction is considered in the context of the dynamics of the stator system in combination with that of the rotor system,[20] it is found that whirl can be inhibited if the independent natural frequencies of the rotor and stator are kept dissimilar, that is, a very stiff rotor should be designed with a very soft mounted stator element that may be subject to rubs. No first-order interdependence of whirl speed with rotational speed has been established.

FLUID BEARING WHIP

As described in experimental and analytic literature,[21] and in standard texts (e.g., Ref. 22), fluid bearing whip can be understood by referring to Fig. 5.7. Consider some nominal radial deflection of a shaft rotating in a fluid (gas- or liquid-) filled clearance. The entrained, viscous fluid will circulate with an average velocity of about half the shaft's surface speed. The bearing pressures developed in the fluid will not be symmetric about the radial deflection line. Because of viscous losses of the bearing fluid circulating through the close clearance, the pressure on the upstream side of the close clearance will be higher than that on the downstream side. Thus, the resultant bearing force will include a tangential force component in the direction of rotation which tends to induce *forward* whirl in the rotor. The tendency to instability is evident when this tangential force exceeds inherent stabilizing damping forces. When this happens, any induced whirl results in increased centrifugal forces; this, in turn, closes the clearance further and results in ever-increasing destabilizing tangential force. Detailed reviews of the phenomenon are available in Refs. 23 and 24.

These and other investigators have shown that to be unstable, shafting must rotate at an rpm equal to or greater than approximately twice the critical speed, so that one would expect the ratio of frequency to rpm to be equal to less than approximately 0.5.

The most obvious measure for avoiding fluid bearing whip is to restrict rotor maximum rpm to less than twice its lowest critical speed. Detailed geometric variations in the bearing runner design, such as grooving and tilt-pad configurations, have also been found effective in inhibiting instability. In extreme cases, use of rolling contact bearings instead of fluid film bearings may be advisable.

Various investigators (e.g., Ref. 25) have noted that fluid seals as well as fluid bearings are subject to this type of instability.

SEAL AND BLADE-TIP-CLEARANCE EFFECT IN TURBOMACHINERY

Axial-flow turbomachinery may be subject to an additional whirl-inducing effect by virtue of the influence of tip clearance on turbopump or compressor or turbine effi-

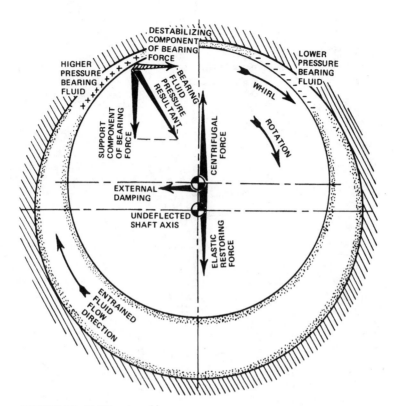

FIGURE 5.7 Fluid bearing whip.

ciency.[26] As shown schematically in Fig. 5.8, some nominal radial deflection will close the radial clearance on one side of the turbomachinery component and open the clearance 180° away on the opposite side. We would expect the closer clearance zone to operate more efficiently than the open clearance zone. For a turbine, a greater work extraction and blade force level is achieved in the more efficient region for a given average pressure drop so that a resultant net tangential force is generated to induce whirl in the direction of rotor rotation (i.e., forward whirl). For an axial compressor, it has been found[27] that the magnitude and direction of the destabilizing forces are a very strong function of the operating point's proximity to the stall line. For operation close to the stall line, very large negative forces (i.e., inducing backward whirl) are generated. The magnitude of the destabilizing force declines sharply for lower operating lines, and stabilizes at a small positive value (i.e., making a small contribution to inducing forward whirl). In the case of radial-flow turbomachinery, it has been suggested[28] that destabilizing forces are exerted on an eccentric (i.e., dynamically deflected) impeller due to variations of loading of the diffuser vanes.

One text[29] describes several manifestations of this class of instability—in the thrust balance piston of a steam turbine; in the radial labyrinth seal of a radial-flow Ljungstrom counterrotating steam turbine; in the Kingsbury thrust bearing of a vertical-shaft hydraulic turbogenerator; and in the tip seals of a radial-inflow hydraulic Francis turbine.

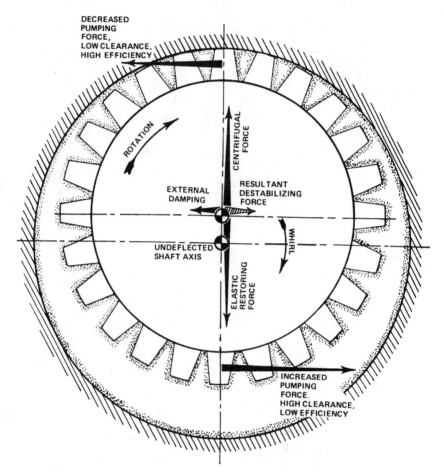

FIGURE 5.8 Turbomachinery tip clearance effect's contribution to whirl.

A survey paper[3] includes a bibliography of several German papers on the subject from 1958 to 1969.

An analysis is available[30] dealing with the possibility of stimulating flexural vibrations in the seals themselves, although it is not clear if the solutions pertain to gross deflections of the entire rotor.

It is reasonable to expect that such destabilizing forces may at least contribute to instabilities experienced on high-powered turbomachines. If this mechanism were indeed a key contributor to instability, one would conjecture that very small or very large initial tip clearances would minimize the influence of tip clearance on the unit's performance and, hence, minimize the contribution to destabilizing forces.

PROPELLER AND TURBOMACHINERY WHIRL

Propeller whirl has been identified both analytically[31] and experimentally.[32] In this instance of shaft whirling, a small deflection of the shaft will generally be accompanied by incremental velocities of the propeller blades. Where the plane of the propeller rotates as a result of the slope deflection of the shaft centerline, incremental axial velocities are induced, as shown in Fig. 5.9. The incremental velocities result in changes in relative velocities between the blades and the airstream which change the aerodynamic forces on the blades. At high airspeeds, the forces are destabilizing and can overcome damping forces to cause destructive whirling of the entire system. The whirling is generally found to be counter to the shaft-rotation direction. It has been

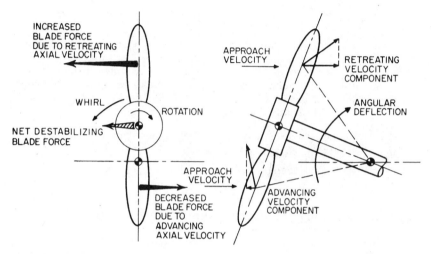

FIGURE 5.9   Propeller whirl.

suggested[33] that equivalent stimulation is possible in turbomachinery in the process of diagnosing and correcting whirl in an aircraft-engine compressor. An attempt has been made[34] to generalize the analysis for axial-flow turbomachinery. Although it has been shown that this analysis is in error, the general deduction seems appropriate that forward whirl may also be possible if the virtual pivot point of the deflected rotor is forward of the rotor (i.e., on the side of the approaching fluid).

Instability is found to be load-sensitive in the sense of being a function of the velocity and density of the impinging flow. It is not thought to be sensitive to the torque level of the turbomachine since, for example, experimental work[32] was on an unloaded windmilling rotor. Corrective action is generally recognized to be stiffening the entire system and manipulating the effective pivot center of the whirling mode to inhibit angular motion of the propeller (or turbomachinery) disc as well as system damping.

PARAMETRIC INSTABILITY

ANALYTIC MODELING

There are systems in engineering and physics which are described by linear differential equations having periodic coefficients,

$$\frac{d^2y}{dz^2} + p(z)\frac{dy}{dz} + q(z)y = 0 \tag{5.13}$$

where $p(z)$ and $q(z)$ are periodic in z. These systems also may exhibit self-excited vibrations, but the stability of the system cannot be evaluated by finding the roots of a characteristic equation. A specialized form of this equation, which is representative of a variety of real physical problems in rotating machinery, is Mathieu's equation:

$$\frac{d^2f}{dz^2} + (a - 2q \cos 2z)f = 0 \tag{5.14}$$

Mathematical treatment and applications of Mathieu's equation are given in Refs. 35 and 36.

This general subcategory of self-excited vibrations is termed "parametric instability," since instability is induced by the effective periodic variation of the system's parameters (stiffness, inertia, natural frequency, etc.). Three particular instances of interest in the field of rotating machinery are

Lateral instability due to asymmetric shafting and/or bearing characteristics

Lateral instability due to pulsating torque

Lateral instabilities due to pulsating longitudinal compression

LATERAL INSTABILITY DUE TO ASYMMETRIC SHAFTING

If a rotor or its stator contains sufficient levels of asymmetry in the flexibility associated with its two principal axes of flexure as illustrated in Fig. 5.10, self-excited vibration may take place. This phenomenon is completely independent of any unbalance, and independent of the forced vibrations associated with twice-per-revolution excitation of such shafting mounted horizontally in a gravitational field.

As described in standard vibration texts,[37] we find that presupposing a nominal whirl amplitude of the shaft at some whirl frequency, the rotation of the asymmetric shaft at an rpm different from the whirling speed will appear as periodic change in flexibility in the plane of the whirling shaft's radial deflection. This will result in an instability in certain specific ranges of rpm as a function of the degree of asymmetry. In general, instability is experienced when the rpm is approximately one-third and one-half the critical rpm and approximately equal to the critical rpm (where the critical rpm is defined with the average value of shaft stiffness), as in Fig. 5.11. The ratios of whirl frequency to rotational speed will then be approximately 3.0, 2.0, and 1.0. But with gross asymmetries, and with the additional complication of asymmetrical inertias with principal axes in arbitrary orientation to the shaft's principal axes' flexibility, no simple generalization is possible.

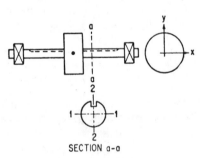

FIGURE 5.10 Shaft system possessing unequal rigidities, leading to a pair of coupled inhomogeneous Mathieu equations.

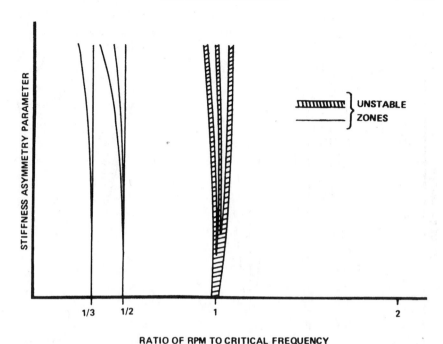

FIGURE 5.11   Instability regimes of rotor system induced by asymmetric stiffness (Ref. 39).

There is a considerable literature dealing with many aspects of the problem and substantial bibliographies.[38-40]

Stability is accomplished by minimizing shaft asymmetries and avoiding rpm ranges of instability.

LATERAL INSTABILITY DUE TO PULSATING TORQUE

Experimental confirmation[41] has been achieved that establishes the possibility of inducing first-order lateral instability in a rotor-disc system by the application of a proper combination of constant and pulsating torque. The application of torque to a shaft affects its natural frequency in lateral vibration so that the instability may also be characterized as "parametric." Analytic formulation and description of the phenomenon are available in Ref. 42 and in the bibliography of Ref. 3. The experimental work (Ref. 41) explored regions of shaft speed where the disc always whirled at the first critical speed of the rotor-disc system, regardless of the torsional forcing frequency or the rotor speed within the unstable region.

It therefore appears that combinations of ranges of steady and pulsating torque, which have been identified[40] as being sufficient to cause instability, should be avoided in the narrow-speed bands where instability is possible in the vicinity of twice the critical speed and lesser instabilities at 2/2, 2/3, 2/4, 2/5, . . . times the critical frequency, as in Fig. 5.12, implying frequency/speed ratios of approximately 0.5, 1.0, 1.5, 2.0, 2.5, ...

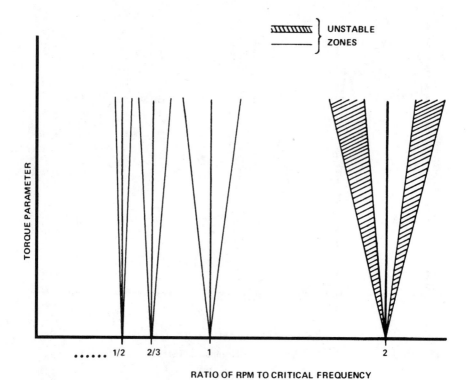

FIGURE 5.12 Instability regimes of rotor system induced by pulsating torque (Ref. 42).

LATERAL INSTABILITY DUE TO PULSATING LONGITUDINAL LOADS

Longitudinal loads on a shaft which are of an order of magnitude of the buckling will tend to reduce the natural frequency of that lateral, flexural vibration of the shaft. Indeed, when the compressive buckling load is reached, the natural frequency goes to zero. Therefore pulsating longitudinal loads effectively cause a periodic variation in stiffness, and they are capable of inducing "parametric instability" in rotating as well as stationary shafts,[43] as noted above in Fig. 5.13.

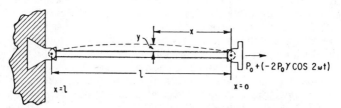

FIGURE 5.13 Long column with pinned ends. A periodic force is superimposed upon a constant axial pull. (*After McLachlan.*[43])

STICK-SLIP RUBS AND CHATTER

Mention is appropriate of another family of instability phenomena—stick-slip or chatter. Though the instability mechanism is associated with the dry friction contact force at the point of rubbing between a rotating shaft and a stationary element, it must not be confused with dry friction whip, previously discussed. In the case of stick-slip, as is described in standard texts (e.g., Ref. 44), the instability is caused by the irregular nature of the friction force developed at very low rubbing speeds.

At high velocities, the friction force is essentially independent of contact speed. But at very low contact speeds we encounter the phenomenon of "stiction," or breakaway friction, where higher levels of friction force are encountered, as in Fig. 5.14. Any periodic motion of the rotor's point of contact, superimposed on the basic relative contact velocity, will be self-excited. In effect, there is negative damping (as illustrated in Fig. 5.1*B*) since motion of the rotor's contact point in the direction of rotation will increase relative contact velocity and reduce stiction and the net force resisting motion. Rotor motion counter to the contact velocity will reduce relative velocity and increase friction force, again reinforcing the periodic motion. The ratio of vibration frequency to rotation speed will be much larger than unity.

While the vibration associated with stick-slip or chatter is often reported to be torsional, planar lateral vibrations can also occur. Surveys of the phenomenon are included in Refs. 45 and 46.

Measures for avoidance are similar to those prescribed for dry friction whip: avoid contact where feasible and lubricate the contact point where contact is essential to the function of the apparatus.

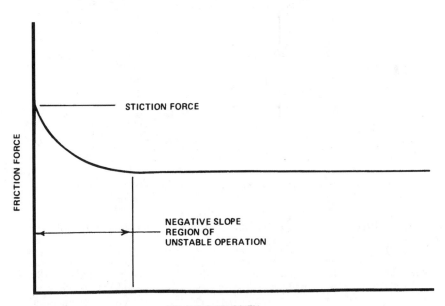

FIGURE 5.14 Dry friction characteristic giving rise to stick-slip rubs or chatter.

INSTABILITIES IN FORCED VIBRATIONS

In a middle ground between the generic categories of force vibrations and self-excited vibrations is the category of *instabilities in force vibrations*. These instabilities are characterized by forced vibration at a frequency equal to rotor rotation (generally induced by unbalance), but with the amplitude of that vibration being unsteady or unstable. Such unsteadiness or instability is induced by the interaction of the forced vibration on the mechanics of the system's response, or on the unbalance itself. Two manifestations of such instabilities and unsteadiness have been identified in the literature—bistable vibration and unstable imbalance.

BISTABLE VIBRATION

A classical model of one type of unstable motion is the "relaxation oscillator," or "*multivibrator.*" A system subject to relaxation oscillation has *two* fairly stable states, separated by a zone where stable operation is impossible.[47] Furthermore, in each of the stable states, a mechanism exists which will induce the system to drift toward the unstable state. The system will develop a periodic motion of the general form shown in Fig. 5.15.

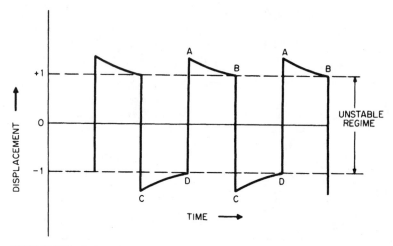

FIGURE 5.15   General form of relaxation oscillations.

An idealized formulation of this class of vibration with nonlinear damping is[48]

$$m\ddot{x} + c(x^2 - 1)\dot{x} + kx = 0 \qquad (5.15)$$

When the deflection amplitude x is greater than +1 or less than −1, as in *A-B* and *C-D*, the damping coefficient is positive, and the system is stable, although presence of a spring system k will always tend to drag the mass to a smaller absolute deflection amplitude. When the deflection amplitude lies between −1 and +1, as in *B-C* or

D-A, the damping coefficient is negative and the system will move violently until it stabilizes in one of the damped stable zones.

While such systems are common in electronic circuitry, they are rather rare in the field of rotating machinery. One instance has been observed[49] in a rotor system supported by rolling element bearings with finite internal clearance. In this situation, the effective stiffness of the rotor is small for small deflections (within the clearance) but large for large deflections (when full contact is made between the rollers and the rotor and stator). Such a nonlinearity in stiffness causes a "rightward leaning" peak in the response curve when the rotor is operating in the vicinity of its critical speed and being stimulated by unbalance. In this region, two stable modes of operation are possible, as in Fig. 5.16. In region *A-B,* the rotor and stator are in solid contact through the rollers. In region *C-D,* the rotor is whirling within the clearance, out of contact. A jump in amplitude is experienced when operating from *B* to *C* or *D* to *A.* When operating at constant speed, either of the nominally stable states can drift toward instability by virtue of thermal effects on the rollers. When the rollers are unloaded, they will skid and heat up, thereby reducing the clearance. When the rollers are loaded, they will be cooled by lubrication and will tend to contract and increase clearance. In combination, these mechanisms are sufficient to cause a relaxation oscillation in the amplitude of the forced vibration.

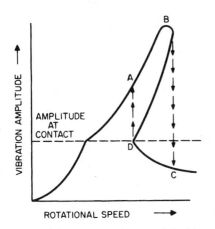

FIGURE 5.16 Response of a rotor, in bearings with (constant) internal clearance, to unbalance excitation in the vicinity of its critical speed.

The remedy for this type of self-excited vibration is to eliminate the precondition of skidding rollers by reducing bearing geometric clearance, by preloading the bearing, or by increasing the temperature of any recirculating lubricant.

UNSTABLE IMBALANCE

A standard text[50] describes the occurrence of unstable vibration of steam turbines where the rotor "would vibrate with the frequency of its rotation, obviously caused by unbalance, but the intensity of the vibration would vary periodically and extremely slowly." The instability in the vibration amplitude is attributable to thermal bowing of the shaft, which is caused by the heat input associated with rubbing at the rotor's deflected "high spot," or by the mass of accumulated steam condensate in the inside of a hollow rotor at the rotor's deflected high spot. In either case, there is basis for continuous variation of amplitude, since unbalance gives rise to deflection and the deflection is, in turn, a function of that imbalance.

The phenomenon is sometimes referred to as the Newkirk effect in reference to its early recorded experimental observation.[51] A manifestation of the phenomenon in a steam turbine has been diagnosed and reported in Ref. 52 and a bibliography is available in Ref. 53. An analytic study[54] shows the possibility of both spiraling, oscillating, and constant modes of amplitude variability.

TABLE 5.2   Diagnostic Table of Rotating Machinery Self-excited Vibrations

	R, characteristic ratio: whirl frequency/rpm	Whirl direction
Whirling or whipping:		
Hysteretic whirl	$R \approx 0.5$	Forward
Fluid trapped in rotor	$0.5 < R < 1.0$	Forward
Dry friction whip	No functional relationship; whirl frequency a function of coupled rotor-stator system; onset rpm is a function of rpm at contact	Backward—axial contact on disc side nearest virtual pivot; Forward—axial contact on disc side opposite to virtual pivot; Backward—radial contact
Fluid bearing whip	$R < 0.5$	Forward
Seal and blade-tip-clearance effect in turbomachinery	Load-dependent	Forward—blade tip clearance; Unspecified—for seal clearance
Propeller and turbomachinery whirl	Load-dependent	Backward—virtual pivot aft of rotor; Forward—virtual pivot front of rotor (where front is source of impinging flow)
Parametric instability:		
Asymmetric shafting	$R \approx 1.0, 2.0, 3.0, \ldots$	Unspecified
Pulsating torque	$R \approx 0.5, 1.0, 1.5, 2.0, \ldots$	Unspecified
Pulsating longitudinal load	A function of pulsating load frequency rather than rpm	Unspecified
Stick-slip rubs and chatter	$R \ll 1$	Essentially planar rather than whirl motion
Instabilities in forced vibrations:		
Bistable vibration	$R = 1$ with periodic square wave fluctuations in amplitude of frequency much lower than rotation rate	Forward
Unstable imbalance	$R = 1$ with slow variation in amplitude	Forward

IDENTIFICATION OF SELF-EXCITED VIBRATION

Even with the best of design practice and application of the most effective methods of avoidance, the conditions and mechanisms of self-excited vibrations in rotating machinery are so subtle and pervasive that incidents continue to occur, and the major task for the vibrations engineer is diagnosis and correction.

Figure 5.3*B* suggests the forms for display of experimental data to perceive the patterns characteristic of whirling or whipping, so as to distinguish it from forced vibration, Fig. 5.3*A*. Table 5.2 summarizes particular quantitative measurements that can be made to distinguish between the various types of whirling and whipping, and other types of self-excited vibrations. The table includes the characteristic ratio of whirl speed to rotation speed at onset of vibration, and the direction of whirl with respect to the rotor rotation. The latter parameter can generally be sensed by noting the phase relation between two stationary vibration pickups mounted at 90° to one another at similar radial locations in a plane normal to the rotor's axis of rotation. Table 5.1 and specific prescriptions in the foregoing text and references suggest corrective action based on these diagnoses. Reference 55 gives additional description of corrective actions.

REFERENCES

1. Den Hartog, J. P.: "Mechanical Vibrations," 4th ed., p. 346, McGraw-Hill Book Company, Inc., New York, 1956.

2. Ehrich, F. F.: "Identification and Avoidance of Instabilities and Self-Excited Vibrations in Rotating Machinery," *ASME Paper* 72-DE-21, May 1972.

3. Kramer, E.: "Instabilities of Rotating Shafts," *Proc. Conf. Vib. Rotating Systems, Inst. Mech. Eng., London,* February 1972.

4. Vance, J. M.: "High Speed Rotor Dynamics—Assessment of Current Technology for Small Turboshaft Engines," USAAMRDL-TR-74-66, Ft. Eustis, Va., July 1974.

5. Newkirk, B. L.: *Gen. Elec. Rev.,* **27**:169 (1924).

6. Kimball, A. L.: *Gen. Elec. Rev.,* **17**:244 (1924).

7. Ref. 1, pp. 295–296.

8. Ehrich, F.: *ASME DE,* **18**:1, September 1989.

9. Gunter, E. J.: "Dynamic Stability of Rotor-Bearing Systems," NASA SP-113, chap. 4, 1966.

10. Bolatin, V. V.: "Non-Conservative Problems of the Theory of Elastic Stability," Pergamon Press, New York, 1964.

11. Bentley, D. E.: "The Re-Excitation of Balance Resonance Regions by Internal Friction," *ASME Paper* 72-PET-49, September 1972.

12. Vance, J. M., and J. Lee: "Stability of High Speed Rotors with Internal Friction," *ASME Paper* 73-DET-127, September 1973.

13. Ehrich, F. F.: *J. Appl. Mech.,* (E) **31**(2):279 (1964).

14. Wolf, J. A.: "Whirl Dynamics of a Rotor Partially Filled with Liquids," *ASME Paper* 68-WA/APM-25, December 1968.

15. Ehrich, F. F.: *J. Eng. Ind.,* (B) **89**(4):806 (1967).

16. Ref. 1, pp. 292–293.

17. Sapetta, L. P., and R. J. Harker: "Whirl of Power Screws Excited by Boundary Lubrication at the Interface," *ASME Paper* 67-Vibr-37, March 1967.

18. Ref. 1, pp. 293–295.

19. Begg, I. C.: "Friction Induced Rotor Whirl—A Study in Stability," *ASME Paper* 73-DET-10, September 1973.

20. Ehrich, F. F.: "The Dynamic Stability of Rotor/Stator Radial Rubs in Rotating Machinery," *ASME Paper* 69-Vibr-56, April 1969.

21. Newkirk, B. L., and H. D. Taylor: *Gen. Elec. Rev.,* **28**:559–568 (1925).

22. Ref. 1, pp. 297–298.

23. Ref. 9, chap. 5.

24. Pinkus, O., and B. Sternlight: "Theory of Hydrodynamic Lubrication," chap. 8, McGraw-Hill Book Company, Inc., New York, 1961.

25. Black, H. F., and D. N. Jenssen: "Effects of High Pressure Seal Rings on Pump Rotor Vibrations," *ASME Paper* 71-WA/FE-38, December 1971.

26. Alford, J. S.: *J. Eng. Power,* **87**(4):333, October 1965.

27. Ehrich, F.: *J. Vibr. & Acous.,* **115**(4):509–515, October 1993.

28. Black, H. F.: "Calculation of Forced Whirling and Stability of Centrifugal Pump Rotor Systems," *ASME Paper* 73-DET-131, September 1973.

29. Ref. 1, pp. 317–321.

30. Ehrich, F. F.: *Trans. ASME,* (A) **90**(4):369 (1968).

31. Taylor, E. S., and K. A. Browne: *J. Aeronaut. Sci.,* **6**(2):43–49 (1938).

32. Houbolt, J. C., and W. H. Reed: "Propeller Nacelle Whirl Flutter," *I.A.S. Paper* 61-34, January 1961.

33. Trent, R., and W. R. Lull: "Design for Control of Dynamic Behavior of Rotating Machinery," *ASME Paper* 72-DE-39, May 1972.

34. Ehrich, F. F.: "An Aeroelastic Whirl Phenomenon in Turbomachinery Rotors," *ASME Paper* 73-DET-97, September 1973.

35. Floquet, G.: *Ann. l'école normale supérieure,* **12**:47 (1883).

36. McLachlan, N. W.: "Theory and Applications of Mathieu Functions," p. 40, Oxford University Press, New York, 1947.

37. Ref. 1, pp. 336–339.

38. Brosens, P. J., and H. S. Crandall: *J. Appl. Mech.,* **83**(4):567 (1961).

39. Messal, E. E., and R. J. Bronthon: "Subharmonic Rotor Instability Due to Elastic Asymmetry," *ASME Paper* 71-Vibr-57, September 1971.

40. Arnold, R. C., and E. E. Haft: "Stability of an Unsymmetrical Rotating Cantilever Shaft Carrying an Unsymmetrical Rotor," *ASME Paper* 71-Vibr-57, September 1971.

41. Eshleman, R. L., and R. A. Eubanks: "Effects of Axial Torque on Rotor Response: An Experimental Investigation," *ASME Paper* 70-WA/DE-14, December 1970.

42. Wehrli, V. C.: "Uber Kritische Drehzahlen unter Pulsierender Torsion," *Ing. Arch.,* **33**:73–84 (1963).

43. Ref. 36, p. 292.

44. Ref. 1, p. 290.

45. Conn, H.: *Tool Eng.,* **45**:61–65 (1960).

46. Sadowy, M.: *Tool Eng.,* **43**:99–103 (1959).

47. Ref. 1, pp. 365–368.

48. Van der Pol, B.: *Phil. Mag.,* **2**:978 (1926).

49. Ehrich, F. F.: "Bi-Stable Vibration of Rotors in Bearing Clearance," *ASME Paper* 65-WA/MD-1, November 1965.

50. Ref. 1, pp. 245–246.

51. Newkirk, B. L.: "Shaft Rubbing," *Mech. Eng.,* **48**:830 (1926).

52. Kroon, R. P., and W. A. Williams: "Spiral Vibration of Rotating Machinery," *5th Int. Congr. Appl. Mech.,* John Wiley & Sons, Inc., New York, 1939, p. 712.

53. Dimarogonas, A. D., and G. N. Sander: *Wear,* **14**(3):153 (1969).

54. Dimarogonas, A. D.: "Newkirk Effect: Thermally Induced Dynamic Instability of High-Speed Rotors," *ASME Paper* 73-GT-26, April 1973.

55. Ehrich, F. and D. Childs: *Mech. Eng.,* **106**(5):66 (1984).

CHAPTER 6
DYNAMIC VIBRATION ABSORBERS AND AUXILIARY MASS DAMPERS

F. Everett Reed

INTRODUCTION

Auxiliary masses are frequently attached to vibrating systems by springs and damping devices to assist in controlling the amplitude of vibration of the system. Depending upon the application, these auxiliary mass systems fall into two distinct classes.

1. If the primary system is excited by a force or displacement that has a constant frequency, or in some cases by an exciting force that is a constant multiple of a rotational speed, then it is possible to modify the vibration pattern and to reduce its amplitude significantly by the use of an auxiliary mass on a spring tuned to the frequency of the excitation. When the auxiliary mass system has as little damping as possible, it is called a *dynamic absorber.*

2. If it is impossible to incorporate damping into a structure that vibrates excessively, it may be possible to provide the damping in an auxiliary system attached to the structure. When used in this manner, the auxiliary mass system is one form of a damper. (Other forms may be incorporated as an integral part of the system.) The names *damped absorber* or *auxiliary mass damper* are given to this type of system.

It is sometimes useful to analyze the auxiliary mass system in terms of its electrical analog.

FORMS OF DYNAMIC ABSORBERS AND AUXILIARY MASS DAMPERS

In its simplest form, as applied to a single degree-of-freedom system, the character of the auxiliary mass system is the same as that of the primary system. Thus a tor-

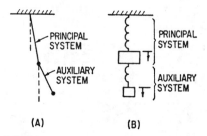

(A) **(B)**

FIGURE 6.1 Dynamic vibration absorbers in pendulum form (*A*) and linear form (*B*).

sional system has a torsionally connected auxiliary mass, a linear system has a linear-spring connected mass, and a pendulum has an auxiliary pendulum. Examples of undamped auxiliary mass systems attached to single degree-of-freedom systems are shown in Figs. 6.1 and 6.2; examples of damped auxiliary mass systems are shown in Figs. 6.3 and 6.4. With multiple degree-of-freedom systems the attachment of the auxiliary masses is not as conventional as with the single degree-of-freedom system. For example, consider the two degree-of-freedom system shown in Fig. 6.5*A* consisting of two masses m_1 and m_2 on a rigid, massless bar. A dynamic absorber of the type shown in Fig. 6.5*B* is effective for the vertical translational motion; however, if the auxiliary masses are on cantilever beams mounted on the rigid bar, as shown in Fig. 6.5*C*, the absorber can be made effective for both vertical translational motion and rotational motion about an axis normal to the page.

WAYS OF EXPRESSING THE EFFECTS OF AUXILIARY MASS SYSTEMS

Suppose a linear auxiliary mass system, consisting of one or more masses, springs, and dampers, is attached to a vibrating primary system. The reaction back on the primary system is proportional to the amplitude of motion at the point of attachment. It is a function of the frequency of excitation and of the masses, spring stiffnesses, and damping constants of the auxiliary mass system. If there is no damping in the auxiliary mass system, the reaction forces are either in phase or 180° out of phase with the displacement and the acceleration at the point of attachment. However, where there is damping in the auxiliary system, the reaction has a component that is 90° out of phase with the acceleration and the displacement.

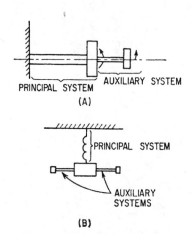

FIGURE 6.2 Typical dynamic vibration absorbers. The principal and auxiliary systems vibrate in torsion in the arrangement at (*A*); the auxiliary system is in the form of masses and beams at (*B*).

Since the reaction is proportional to the amplitude of motion, it is possible to express the properties of the auxiliary mass system in terms of the motion at the point of attachment. This can be done in three ways: (1) the ratio of the reaction force to the displacement at the point of attachment, (2) the ratio of the reaction force to the velocity at the

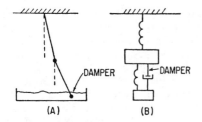

FIGURE 6.3 Damped auxiliary mass systems corresponding to the undamped vibration absorbers shown in Fig. 6.1.

point of attachment, or (3) the ratio of the reaction force to the acceleration at the point of attachment. The first ratio can be considered equivalent to a spring whose stiffness changes with frequency. The second ratio can be considered equivalent to a damper; at any frequency it is equal in magnitude to the force-displacement ratio divided by the angular frequency. The phase angle between the force and the velocity is 90° from the phase angle between the force and the displacement. This force-velocity ratio is called the *mechanical impedance* Z of the auxiliary system. The third ratio corresponds to a mass and is designated *equivalent mass* m_{eq}. The equivalent mass of a system is $-1/\omega^2$ that of the equivalent spring k_{eq} of the system.

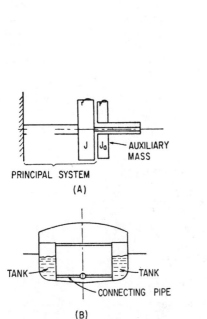

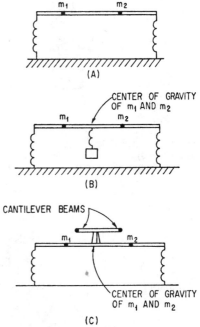

FIGURE 6.4 Typical damped auxiliary mass systems. In the torsional system at (A), damping is provided by relative motion of the flywheels J, J_a. In the antiroll tanks for ships shown at (B), water flows from one tank to the other and damping is provided by a constriction in the connecting pipe.

FIGURE 6.5 Application of a dynamic absorber to reduce the vibration of the spring-mounted bar at (A) in both vertical translational and rotational modes. The linear mass-spring system at (B) is effective for only translational motion, whereas the cantilever beams at (C) are effective for rotational as well as translational motion.

Because of the phase relations between the force and the displacement, velocity, and acceleration at the point of connection, it is customary to represent the ratios as complex quantities. Thus $Z = k_{eq}/j\omega = j\omega m_{eq}$. Most dynamic analyses of mechanical systems are made on purely reactive systems, i.e., systems having masses and stiffnesses only, and no damping. The effects of auxiliary mass systems are most easily understood if the effect of the auxiliary system is represented as a reactive subsystem. For this reason, and because the hypothetical addition of a mass to a system is often more easily comprehended than the addition of a spring, the effects of auxiliary mass systems are treated in terms of the equivalent masses in this chapter, i.e., in terms of the ratio of the force exerted by the auxiliary system upon the primary system to the acceleration at the point of attachment of the auxiliary system.

THE INFLUENCE OF A SIMPLE AUXILIARY MASS SYSTEM UPON A VIBRATING SYSTEM

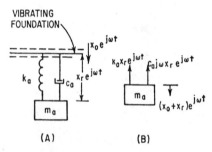

FIGURE 6.6 Auxiliary mass damper. The arrangement of the damper is shown at (A), and the forces acting on the mass are indicated at (B).

The magnitude of the equivalent mass of a simple auxiliary mass system, consisting of a mass m_a, spring k_a, and viscous damper c_a, can be determined readily by evaluating the forces exerted by such a system upon a foundation vibrating at a frequency $f = \omega/2\pi$. The system with its assumed constants and displacements is shown in Fig. 6.6A. The spring and damping forces acting on m are shown in Fig. 6.6B, and the equation of motion is

$$(-k_a x_r - c_a j\omega x_r)e^{j\omega t} = -m_a(x_0 + x_r)\omega^2 e^{j\omega t}$$

Solving for x_r,

$$x_r = \frac{m_a\omega^2 x_0}{-m_a\omega^2 + jc_a\omega + k_a} \qquad (6.1)$$

The force acting on the foundation is

$$Fe^{j\omega t} = (k_a + jc_a\omega)x_r e^{j\omega t}$$

Eliminating x_r from the preceding equations,

$$F = \frac{(k_a + jc_a\omega)m_a\omega^2}{-m_a\omega^2 + jc_a\omega + k_a} x_0 \qquad (6.2)$$

Since the force exerted by an equivalent mass m_{eq} rigidly attached to the moving foundation is $F = m_{eq}\omega^2 x_0$:

$$m_{eq} = \frac{(k_a + jc_a\omega)}{k_a + jc_a\omega - m_a\omega^2} m_a \qquad (6.3)$$

Equation (6.3) can be written in terms of nondimensional quantities:

$$m_{eq} = \frac{1 + 2\zeta\beta_a j}{(1 - \beta_a^2) + 2\zeta\beta_a j} m_a \tag{6.4}$$

where $\beta_a = \dfrac{\omega}{\omega_a}$, a tuning parameter

$\omega_a^2 = \dfrac{k_a}{m_a}$, the natural frequency of the auxiliary system

$\zeta = \dfrac{c_a}{c_{ca}}$, a damping parameter

$c_{ca} = 2\sqrt{k_a m_a}$, critical damping of the auxiliary system

Equation (6.4) can be divided into the following real and imaginary components:

$$m_{eq} = \frac{(1 - \beta_a^2) + (2\zeta\beta_a)^2}{(1 - \beta_a^2)^2 + (2\zeta\beta_a)^2} m_a - \frac{2\zeta\beta_a^3}{(1 - \beta_a^2)^2 + (2\zeta\beta_a)^2} j m_a \tag{6.5}$$

The real and imaginary parts of m_{eq} are shown in Fig. 6.7A and Fig. 6.7B, respectively. If there is no damping, $\zeta = 0$ and

$$m_{eq} = \frac{1}{1 - \beta_a^2} m_a \tag{6.6}$$

If $\beta_a = 1$ in Eq. (6.6), m_{eq} becomes infinite and a finite force produces no displacement. Thus, the auxiliary mass enforces a point of no motion (i.e., a node) at its point of attachment.

This concept can be applied to reduce the amplitude of the forced vibration of a single degree-of-freedom system by attaching a damped absorber.[1,2] A sketch of the system with a damped auxiliary mass system is shown in Fig. 6.8A. In the equivalent system shown in Fig. 6.8B, there is no force acting on the mass m but instead the support is given a motion $ue^{j\omega t}$. The equations for the system of Fig. 6.8B are similar to those for the system of Fig. 6.8A with the value ku substituted for F. The amplitude of forced vibration of a single degree-of-freedom system, Eq. (2.24), is

$$x_0 = \frac{F/k}{1 - m\omega^2/k}$$

The effect of the auxiliary mass system is to increase the mass m of the primary system by the equivalent mass of the auxiliary system as given by Eq. (6.4):

$$x_0 = \frac{F/k}{1 - \dfrac{\omega^2}{k}\left[m + m_a \dfrac{(1 + 2\zeta\beta_a j)}{(1 - \beta_a^2) + 2\zeta\beta_a j} \right]}$$

Substituting $\mu = m_a/m$, the mass ratio, $\delta_{st} = F/k$, the static deflection of the spring of the primary system, and $\beta = \sqrt{m\omega^2/k}$, the ratio of the forcing frequency to the natural frequency of the primary system, and writing in dimensionless form,

$$\frac{x_0}{\delta_{st}} = \frac{(1 - \beta_a^2) + 2\zeta\beta_a j}{(1 - \beta_a^2) + 2\zeta\beta_a j - \beta^2[(1 - \beta_a^2) + 2\zeta\beta_a j + \mu(1 + 2\zeta\beta_a j)]}$$

The amplitude of motion of the primary mass, without regard to phase, is

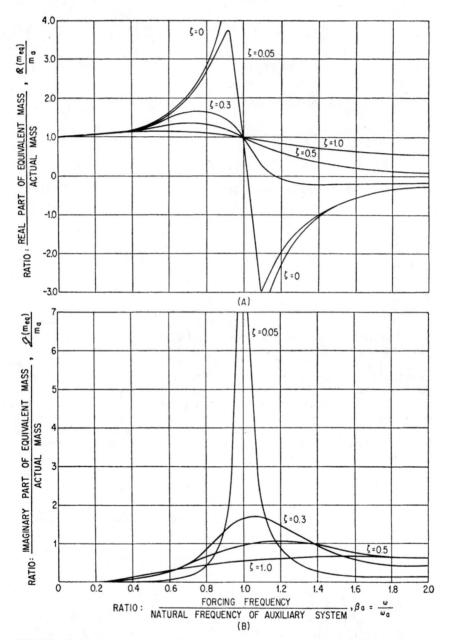

FIGURE 6.7 Equivalent mass m_{eq} of the auxiliary-mass system shown in Fig. 6.6. The real part of the equivalent mass is shown at (A) and the imaginary part at (B).

$$\frac{x_0}{\delta_{st}} = \left\{ \frac{(1-\beta_a{}^2)^2 + (2\zeta\beta_a)^2}{[(1-\beta_a{}^2)(1-\beta^2) - \beta^2\mu]^2 + (2\zeta\beta_a)^2[1-\beta^2 - \beta^2\mu]^2} \right\}^{1/2} \tag{6.7}$$

If $\zeta = 0$ (no damping), then

$$\frac{x_0}{\delta_{st}} = \frac{1-\beta_a{}^2}{(1-\beta_a{}^2)(1-\beta^2) - \beta^2\mu} \tag{6.8}$$

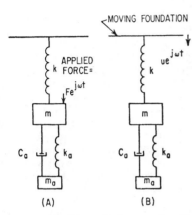

If $\beta_a = 1, x_0 = 0$; that is, the vibration of the primary system is eliminated entirely when the auxiliary system is undamped and is tuned to the forcing frequency.

THE DYNAMIC ABSORBER

If the auxiliary mass system has no damping and is tuned to the forcing frequency, it acts as a dynamic absorber and enforces a node at its point of attachment. The auxiliary mass must be sufficiently large so that it will not have an excessive amplitude.[3] For a dynamic absorber attached to the primary system at the point where the excitation is introduced, the required mass of the auxiliary body is easily determined. Since the primary mass is motionless, the force exerted by the absorber, when the amplitude of motion of the auxiliary mass is u_0, is equal and of opposite sign to the exciting force F. Hence

FIGURE 6.8 Schematic diagram of auxiliary mass m_a coupled by a spring k_a and viscous damper c_a to a primary system k, m. The primary system is excited by the force F at (A), or alternatively by the foundation motion u at (B).

$$F = m_a\omega^2 u_0 \tag{6.9}$$

Since the frequency is known, the mass and amplitude of motion necessary to neutralize a given excitation force are determined by Eq. (6.9). The spring stiffness in the auxiliary system is determined by the requirement that the auxiliary system be tuned to the frequency of the exciting force:

$$k_a = m_a\omega^2 \tag{6.10}$$

Although the concept of tuning a dynamic absorber appears simple, practical considerations make it difficult to tune any system exactly. When the auxiliary mass is small relative to the mass of the primary system, its effectiveness depends upon accurate tuning. If the tuning is incorrect, the addition of the auxiliary mass may bring the composite system (primary and auxiliary systems) into resonance with the exciting force.

Consider the natural frequencies of the composite system. The natural frequency of the primary system is $\omega_0 = \sqrt{k/m}$. With this relation, Eq. (6.8) in which the damping is zero ($\zeta = 0$) becomes

$$\frac{x_0}{\delta_{st}} = \frac{1-\omega^2/\omega_a{}^2}{(1-\omega^2/\omega_a{}^2)(1-\omega^2/\omega_0{}^2) - (\omega^2/\omega_0{}^2)\mu}$$

At resonance the denominator is zero and ω is designated ω_n:

$$(\omega_n{}^2 - \omega_a{}^2)(\omega^2 - \omega_0{}^2) - \omega_n{}^2\omega_a{}^2\mu = 0 \tag{6.11}$$

The natural frequencies are found from the roots $\omega_n{}^2$ of Eq. (6.11):

$$\omega_n{}^2 = \frac{\omega_a{}^2(1+\mu) + \omega_0{}^2}{2} \pm \sqrt{\left[\frac{\omega_a{}^2(1+\mu) - \omega_0{}^2}{2}\right]^2 + \omega_a{}^2\omega_0{}^2\mu} \tag{6.12}$$

This last relation may be represented by Mohr's circle, Fig. 6.9.

Since the absorber is nominally tuned to the frequency of the excitation, the root ω_{n2}^2 that is closer to the forcing frequency is of interest. The ratio ω_{n2}/ω_a is a measure of the sensitivity of the tuning required to avoid resonance. This is given as a function of μ for various ratios of ω_0/ω_a in Fig. 6.10. Dynamic absorbers are most generally used when the primary system without the absorber is nearly in resonance with the excitation. If the natural frequency of the primary system is less than the forcing frequency, it is preferable to tune the dynamic absorber to a frequency slightly lower than the forcing frequency to avoid the resonance that lies above the natural frequency of the primary system. Likewise if the natural frequency of the primary system is above the forcing frequency, it is well to tune the damper to a frequency slightly greater than the forcing frequency. Figure 6.10 shows that the tuning for a primary system with high natural frequency is more sensitive than that for a primary system with low natural frequency. Mohr's circle of Fig. 6.9 provides a useful graphical representation.

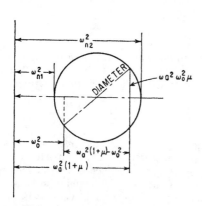

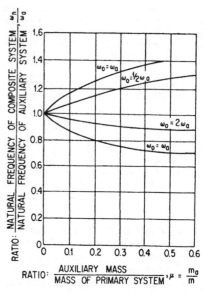

FIGURE 6.9 Representation of the natural frequencies ω_n of the composite system by Mohr's circle. The circle is constructed on the diameter located by the natural frequencies ω_0, ω_a of the primary and auxiliary systems, respectively. The natural frequencies of the composite system are indicated by the intercept of the circle with the horizontal axis.

FIGURE 6.10 Curves showing effect of mass ratio m_a/m on the natural frequencies ω_n of the composite system, for several ratios of the natural frequency ω_a of the auxiliary system to the natural frequency ω_0 of the primary system.

Where the natural frequency of the composite system is nearly equal to the tuned frequency of the absorber, the amplitude of motion of the primary mass at resonance is much smaller than that of the absorber. Consequently, the motion of the primary mass does not become large even at resonance; but the motion of the absorber, unless limited by damping, may become so large that failure occurs.

The use of the dynamic absorber is not restricted to single degree-of-freedom systems or to locations in simple systems where the exciting forces act. However, dynamic absorbers are most effective if located where the excitation force acts. For example, consider a dynamic absorber that is attached to the spring in the simple system shown in Fig. 6.11. When the absorber is tuned so that $\sqrt{k_a/m_a} = \omega$, the equivalent mass is infinite at its point of attachment and enforces a node at point A. If the stiffness of the spring between A and the mass m is k_1, then the force F' exerted by the absorber to enforce the node is equal to that exerted by a system composed of the mass m and the spring k_1 attached to a fixed foundation at A and acted upon by the force $Fe^{j\omega t}$. The force F' is

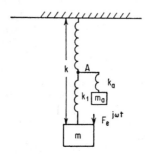

FIGURE 6.11   Dynamic absorber attached to the spring of the primary system. The analysis shows that this is not as effective as if it were attached to the rigid body on which the force acts.

$$F' = \frac{F}{1 - (m\omega^2/k_1)}$$

Thus the amplitude of motion of the auxiliary mass is

$$u_0 = \frac{F}{1 - (m\omega^2/k_1)} \times \frac{1}{m_a\omega^2} \tag{6.13}$$

The amplitude of motion of the primary mass is

$$x = \frac{F}{k_1}\left(1 - \frac{m\omega^2}{k_1}\right)^{-1} \tag{6.14}$$

Hence, an absorber attached to the spring is not as effective as one attached to the body where the force is acting. It is possible for the primary system to come into resonance about the new node at A.

AUXILIARY MASS DAMPERS

In general, the dynamic absorber is effective only for a system that is subjected to a constant frequency excitation. In the special case of a pendulum absorber (discussed later in this chapter), it is effective for an excitation that is a constant multiple of a rotating shaft speed. When excited at frequencies other than the frequency to which it is tuned, the absorber acts as an attached mass of positive value at frequencies below the tuned frequency and of negative value at frequencies above the tuned frequency. It introduces an additional degree-of-freedom and an additional natural frequency into the primary system.

In a multiple degree-of-freedom system, the introduction of an auxiliary mass system tends to lower those original natural frequencies of the primary system that are below the tuned frequency of the auxiliary system. This is because the auxiliary mass system adds a positive equivalent mass at frequencies below the tuned frequency. The original natural frequencies of the primary system that are higher than the tuned frequency of the auxiliary system are raised by adding the auxiliary mass system, because the equivalent mass of the auxiliary system is negative. A new natural mode of vibration corresponding to the vibration of the auxiliary mass system against the primary system is injected between the displaced initial natural frequencies of the primary system. Because the equivalent mass of the auxiliary mass system is large only at frequencies near the tuned frequency, those frequencies of the primary system that are closest to the tuned frequency are most strongly influenced by the auxiliary mass system. The addition of damping in the auxiliary mass system can be effective in reducing the amplitudes of motion of the primary system at the natural frequencies. For this reason auxiliary mass dampers are used quite commonly to reduce over-all vibration stresses and amplitudes.

Studies of the effects of a damped auxiliary mass system upon the amplitude of motion of an undamped, single degree-of-freedom system[1-5] have been applied to a multimass system.[6,7] In analyzing dampers utilizing auxiliary masses, it is desirable to consider a composite system in which the characteristics of both the primary and auxiliary systems are fixed. This composite system is excited by a harmonic force of varying frequency. It is desirable to express the tuned frequency of the auxiliary mass system in terms of the natural frequency of the primary system rather than the ratio β_a of the excitation frequency ω to the tuned frequency ω_a of the auxiliary system. Defining a new ratio α,

$$\alpha = \frac{\omega_a}{\omega_0} = \frac{\beta}{\beta_a}$$

Then Eq. (6.7) becomes

$$\frac{x_0}{\delta_{st}} = \left\{ \frac{(\alpha^2 - \beta^2)^2 + (2\zeta\alpha\beta)^2}{[(\alpha^2 - \beta^2)(1 - \beta^2) - \alpha^2\beta^2\mu]^2 + (2\zeta\alpha\beta)^2(1 - \beta^2 - \beta^2\mu)^2} \right\}^{1/2} \tag{6.15}$$

This equation is plotted in Fig. 6.12. Note that all curves pass through two points A, B on the graph, independent of the damping parameter ζ. These points are known as *fixed points*. Their locations are independent of the value of ζ if the ratio of the coefficient of ζ^2 to the term independent of ζ is the same in both numerator and denominator of Eq. (6.15):

$$\frac{(2\alpha\beta)^2}{(\alpha^2 - \beta^2)^2} = \frac{2\alpha\beta(1 - \beta^2 - \beta^2\mu)^2}{[(\alpha^2 - \beta^2)(1 - \beta^2) - \alpha^2\beta^2\mu]^2} \tag{6.16}$$

This equation is satisfied if

$$(2\alpha\beta)^2 = 0$$

$$\frac{1}{\alpha^2 - \beta^2} + \frac{(1 - \beta^2 - \beta^2\mu)}{(\alpha^2 - \beta^2)(1 - \beta^2) - \alpha^2\beta^2\mu} = 0$$

$$\frac{1}{\alpha^2 - \beta^2} - \frac{(1 - \beta^2 - \beta^2\mu)}{(\alpha^2 - \beta^2)(1 - \beta^2) - \alpha^2\beta^2\mu} = 0$$

The first two solutions are trivial. The third yields the equation

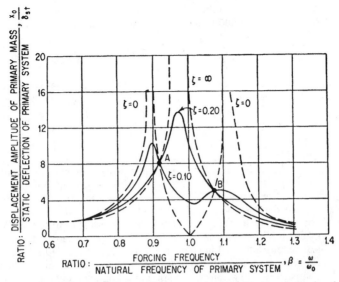

FIGURE 6.12 Curves for auxiliary mass damper showing amplitude of vibration of mass of primary system, as given by Eq. (6.15), as a function of the ratio of forcing frequency ω to natural frequency of primary system $\omega = \sqrt{k/m}$. The mass ratio $m_a/m = 0.05$, and the natural frequency ω_a of the auxiliary mass system is equal to the natural frequency ω_0 of the primary system. Curves are included for several values of damping in the auxiliary system.

$$\beta^4\left(1 + \frac{\mu}{2}\right) - \beta^2(1 + \alpha^2 + \alpha^2\mu) + \alpha^2 = 0 \tag{6.17}$$

The solution of this equation gives two values of β, designated β_c, one corresponding to each fixed point.

The amplitude of motion at each fixed point may be found by substituting each value of β_c given by Eq. (6.17) into Eq. (6.15). Since the amplitude is independent of ζ, the value that gives the simplest calculation (namely, $\zeta = \infty$) can be used for the calculation:

$$\left.\frac{x_0}{\delta_{st}}\right|_c = \left[\frac{1}{(1 - \beta_c^2 - \beta_c^2\mu)^2}\right]^{1/2} \tag{6.18}$$

For the auxiliary mass damper to be most effective in limiting the value of x_0/δ_{st} over a full range of excitation frequencies, it is necessary to select the spring and damping constants of the system as given by the parameters α and ζ, respectively, so that the amplitude x_0 of the primary mass is a minimum. First consider the influence of the ratio α. As α is varied, the values of β_c computed from Eq. (6.17) are substituted in Eq. (6.18) to obtain values of x_0/δ_{st} for the fixed points A and B. The optimum value of α is that for which the amplitude x_0 at A is equal to that at B.

Let the two roots of Eq. (6.17) be β_1^2 and β_2^2, where β_1^2 is less than 1 and β_2^2 is greater than 1. When x_0/δ_{st} has the same value for both β_1 and β_2 in Eq. (6.18),

$$\beta_1^2 + \beta_2^2 = \frac{2}{1 + \mu}$$

In an equation having unity for the coefficient of its highest power, the sum of the roots is equal to the coefficient of the second term with its sign changed:

$$\beta_1{}^2 + \beta_2{}^2 = \frac{1 + \alpha^2 + \alpha^2\mu}{1 + \mu/2}$$

From the two preceding equations, the optimum tuning (i.e., that required to give the same amplitude of motion at both fixed points) is obtained:

$$\alpha_{opt} = \frac{1}{1 + \mu} \tag{6.19}$$

where α is defined by the equation preceding Eq. (6.15).

If the effect of the damping is considered, it is possible to choose a value of the damping parameter ζ that will make the fixed points nearly the points of greatest amplitude of the motion. Consider Fig. 6.13, which represents the curves defining the motion of a single degree-of-freedom system to which an ideally tuned damped vibration absorber is attached (Fig. 6.8). The solid curves (1) represent the response of a system fitted with an undamped absorber. Curve 2 represents infinite damping of the auxiliary system. Curves 3 have horizontal tangents at the fixed points A and B, respectively. Since it is difficult to determine the required damping from maxima at the fixed points, the assumption is made that an optimum damping gives the same value of x_0/δ_{st} at a convenient point between A and B as at these fixed points. First find the values of β at A and B. This is done by solving Eq. (6.17) with the values of α as determined by Eq. (6.19) substituted:

$$\beta^4 - \frac{2\beta^2}{1 + \mu} + \frac{2}{(2 + \mu)(1 + \mu)^2} = 0$$

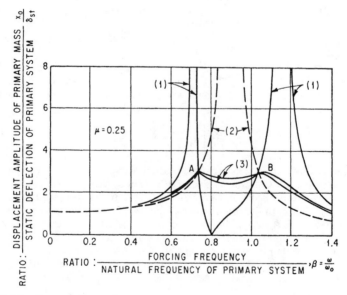

FIGURE 6.13 Curves similar to Fig. 6.12 but with optimum tuning. Curves 1 apply to an undamped absorber, curve 2 represents infinite damping in the auxiliary system, and curves 3 have horizontal tangents at the fixed points A and B.

Solving for β to obtain the abscissas at the fixed points,

$$\beta^2 = \frac{1}{1+\mu} \sqrt{\left(1 \pm \frac{\mu}{2+\mu}\right)} \tag{6.20}$$

A convenient value for β lying between the two fixed points A and B is defined by

$$\beta_l^2 = \frac{1}{1+\mu} \tag{6.21}$$

The frequency corresponding to this frequency ratio β_l is the natural frequency of the composite system when the damping is infinite; it is called the locked frequency.[7] The value of x_0/δ_{st} at the fixed points is found by substituting Eq. (6.20) into Eq. (6.18):

$$\frac{x_0}{\delta_{st}} \text{ at fixed point } = \sqrt{1 + \frac{2}{\mu}} \tag{6.22}$$

An approximate value for the maximum damping is obtained by solving for the value of ζ in Eq. (6.15) that gives a value of $x_0/\delta_{st} = \sqrt{1 + 2/\mu}$ when β_l^2 (the locked frequency) is given by Eq. (6.21) and α has the optimum value given by Eq. (6.19). This gives the following value for the optimum damping parameter:

$$\zeta_{opt} = \sqrt{\frac{\mu}{2(1+\mu)}} \tag{6.23}$$

It is possible to find the value of ζ^2 that makes the fixed point A a maximum on the x_0/δ_{st} vs. β plot, Fig. 6.13, and also to find the value of ζ^2 that makes the point B a maximum. The average of the two values so obtained indicates optimum damping:[4]

$$\zeta_{opt} = \sqrt{\frac{3\mu}{8(1+\mu)^3}} \tag{6.24}$$

Optimum Damping for an Auxiliary Mass Absorber Connected to the Primary System with Damping Only. In general, the most effective damping is obtained where the auxiliary mass damping system includes a spring in its connection to the primary system. However, such a design requires a calculation of the optimum stiffness of the spring. Sometimes it is more expedient to add an oversize mass, coupled only by damping to the primary system, than it is to compute the optimum system. However, if use is made of such a simplified damper by taking it from a list of standard dampers and applying it with a minimum of calculations, the stock dampers should be as efficient as the application will permit.

In computing the optimum damping characteristic for an auxiliary mass absorber, attached to a single degree-of-freedom system by damping only, from the relations that have been developed, note in Eq. (6.4) that $\zeta = \infty$ and $\beta_a = \infty$ when $k = 0$. Then $\alpha = \beta/\beta_a = 0$. However, the product $\zeta\alpha = \zeta\beta/\beta_a$ is finite; thus, substituting $\alpha = 0$ but retaining the product $\zeta\alpha$ in Eq. (6.15),

$$\frac{x_0}{\delta_{st}} = \sqrt{\frac{\beta^2 + 4(\zeta\alpha)^2}{\beta^2(1-\beta^2)^2 + 4(\zeta\alpha)^2[1 - \beta^2(1+\mu)]^2}} \tag{6.25}$$

The value of x_0/δ_{st} is independent of $\zeta\alpha$ where the ratio of the coefficient of $\zeta\alpha$ to the term independent of $\zeta\alpha$ in the numerator is the same as the corresponding ratio in the denominator:

$$\frac{4}{\beta^2} = \frac{4[1 - \beta^2(1+\mu)]^2}{\beta^2(1-\beta^2)^2}$$

The solution of this equation for β gives the fixed points

$$\beta^2 = 0 \quad \text{and} \quad \beta^2 = \frac{2}{2+\mu} \tag{6.26}$$

The amplitude of motion of the primary mass where $\beta^2 = 2/(2+\mu)$ is

$$\frac{x_0}{\delta_{st}} = \frac{2+\mu}{\mu} \tag{6.27}$$

Curves showing the motion of the mass of a primary system fitted with an auxiliary mass system connected by damping only are given in Fig. 6.14. The optimum damping is that which makes the maximum amplitude occur at the fixed point B. By finding the value of $\zeta\alpha$ that makes the slope of x_0/δ_{st} versus β equal to zero at $\beta^2 = 2/(2+\mu)$, the optimum damping is defined by

$$(\zeta\alpha)_{opt} = \sqrt{\frac{1}{2(2+\mu)(1+\mu)}} \tag{6.28}$$

The values for the amplitude of vibration of the primary mass, the relative amplitude between the primary and auxiliary masses, and the optimum damping constants are given in Figs. 6.15 to 6.17 as functions of the mass ratio $\mu = m_a/m$.

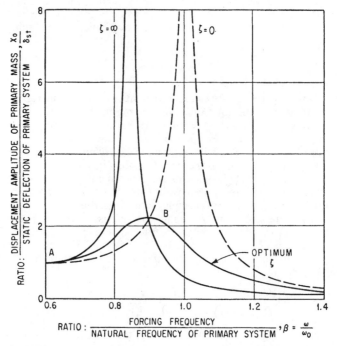

FIGURE 6.14 Curves similar to Fig. 6.12 for system having auxiliary mass coupled by damping only. Several values of damping are included.

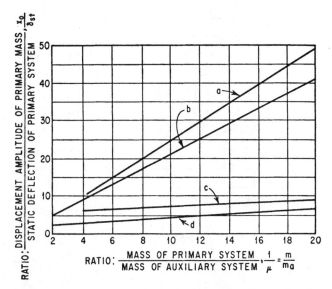

FIGURE 6.15 Displacement amplitude of the primary mass as a function of the size of the auxiliary mass: (a) auxiliary system coupled only by Coulomb friction ($\alpha = 0$) with optimum damping; (b) auxiliary system coupled only by viscous damping ($\alpha = 0$) of optimum value; (c) auxiliary system coupled by spring and damper tuned to frequency of primary system ($\alpha = 1$) with optimum damping; (d) auxiliary system coupled by spring and damper with optimum tuning [$\alpha = 1/(1 + \mu)$] and optimum damping.

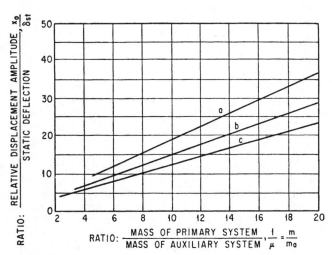

FIGURE 6.16 Relative displacement amplitude between the primary mass and the auxiliary mass as a function of the size of the auxiliary mass: (a) auxiliary system coupled by spring and damper with optimum tuning [$\alpha = 1/(1 + \mu)$] and optimum damping; (b) auxiliary system coupled only by viscous damping ($\alpha = 0$) of optimum value; (c) auxiliary system coupled by spring and damper tuned to frequency of primary system ($\alpha = 1$) with optimum damping.

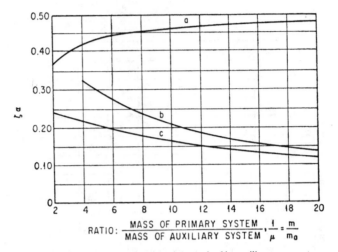

FIGURE 6.17   Curves showing damping required in auxiliary mass systems
to minimize vibration amplitude of primary system: (*a*) auxiliary mass coupled by viscous damping only ($\alpha = 0$); (*b*) auxiliary system coupled by spring
and damper tuned to frequency of primary system ($\alpha = 1$); (*c*) auxiliary system coupled by spring and damper with optimum tuning [$\alpha = 1/(1 + \mu)$]. The
ordinate of the curves is $\zeta\alpha$, where ζ is the fraction of critical damping in the
auxiliary system [Eq. (6.4)] and α is the tuning parameter [Eq. (6.15)].

The Use of Auxiliary Mass Absorbers for Vibration Energy Dissipation.

When a complicated mass-spring system is analyzed for possible vibration troubles,
it is customary to compute the natural frequencies of the several modes of vibration
of the system. The vibration amplitudes and stresses are estimated by making an
energy balance between the energy input from the various exciting forces and the
energy dissipated in the system and external reactions. From this point of view, it is
desirable to know how much energy is dissipated in auxiliary mass systems and what
value the damping constant should have in an auxiliary mass system of limited size
to give maximum energy absorption. This is not the best criterion for determining
the optimum damping because it neglects the effects of damping upon the mode
shapes and the frequencies of the system, but it is generally adequate when compared with the other uncertainties of the calculations. Methods of designing
dampers for torsional systems are given in Chap. 38.

*Optimum Viscous Damping to Give Large Energy Absorption in an Auxiliary
Mass Absorber.*[8]   Suppose the amplitude of motion of the primary system is unaffected by the auxiliary mass system which is attached to it. All energy absorption
occurs in the damping element of the auxiliary mass system and is obtained by integrating the differential work done in the damper over a vibration cycle. The force
exerted by damping is $c\dot{x}_r$ (the subscripts *a* are dropped), where x_r is the relative
motion and the increment of work is $c\dot{x}_r\,dx_r = c\dot{x}_r^2\,dt$. If $x_r = x_{r0}\cos\omega t$, the work done
over a cycle is

$$V = \oint c\omega^2 x_{r0}^2 \sin^2\omega t\,dt = \pi c x_{r0}^2 \omega \qquad (6.29)$$

For a damper attached to a support moving in harmonic motion of amplitude x_0,
the relative motion x_r is given by Eq. (6.1). The amplitude of relative motion is

$$x_{r0} = \frac{m\omega^2 x_0}{\sqrt{(k - m\omega^2)^2 + c^2\omega^2}} = \frac{\beta_a^2 x_0}{\sqrt{(1 - \beta_a^2)^2 + (2\zeta\beta_a)^2}}$$

Substituting the above value of x_{r0} in Eq. (6.29) and integrating,

$$V = \frac{\pi c\omega x_0^2 m^2 \omega^4}{(k - m\omega^2)^2 + c^2\omega^2} = \frac{\pi x_0^2 m\omega^2 (2\zeta\beta_a)\beta_a^2}{(1 - \beta_a^2)^2 + (2\zeta\beta_a)^2} \tag{6.30}$$

Equation (6.30) can be used to find the tuning and the damping that gives the maximum energy dissipation when the amplitude of the forcing motion remains constant. Placing $\partial V/\partial \beta_a = 0$, the optimum value of β_a for given values of ζ is found from

$$(\beta_a)^2_{\text{opt}} = (2\zeta^2 - 1) \pm 2\sqrt{1 - \zeta^2 + \zeta^4} \tag{6.31}$$

Placing $\partial V/\partial \zeta = 0$, the optimum value of ζ for a given value of β_a is

$$\zeta_{\text{opt}} = \frac{1 - \beta_a^2}{2\beta_a} \tag{6.32}$$

Where $k = 0$, the optimum damping is determined most conveniently by setting $\partial V/\partial c = 0$, using the dimensional form of Eq. (6.30), and determining c for maximum energy absorption:

$$c_{\text{opt}} = m\omega \tag{6.33}$$

Auxiliary Mass Damper Using Coulomb Friction Damping.[9] Dampers relying on Coulomb friction (i.e., friction whose force is constant) have been widely used. A damper relying on dry friction and connected to its primary system with a spring is too complicated to be analyzed or to be adjusted by experiment. For this reason, a damper with Coulomb friction has been used with only friction damping connecting the seismic mass (usually in a torsional application) to the primary system.[1,2,9] Because the motion is irregular, it is necessary to use energy methods of analysis. The analysis given here applies to the case of linear vibration. By analogy, the application to torsional or other vibration can be made easily (see Table 2.1 for analogous parameters).

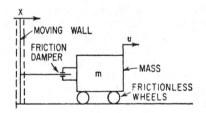

FIGURE 6.18 Schematic diagram of auxiliary mass absorber with Coulomb friction damping.

Consider the system shown in Fig. 6.18. It consists of a mass resting on wheels that provide no resistance to motion and are connected through a friction damper to a wall that is moving sinusoidally. The friction damper consists of two friction facings that are held on opposite sides of a plate by a spring that can be adjusted to give a desired clamping force. The maximum force that can be transmitted through each interface of the damper is the product of the normal force and the coefficient of friction; the maximum total force for the damper is the summation over the number of interfaces.

Consider the velocity diagrams shown in Fig. 6.19A, B, and C. In these diagrams the velocity of the moving wall, $\dot{x} = x_0\omega \sin \omega t$, is shown by curve 1; the velocity $\dot{u}$ of the mass is shown by curve 2. The force exerted by the damper when slipping occurs is F_s. When $F_s \geq m\ddot{u}$, the mass moves sinusoidally with the wall. When $F_s < m\ddot{u}$, slip-

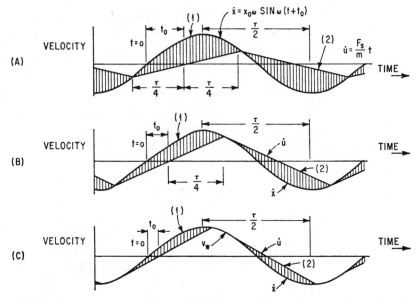

FIGURE 6.19 Velocity-time diagrams for motion of wall (curve 1) and mass (curve 2) of Fig. 6.18. The conditions for a small damping force are shown at (A), for an intermediate damping force at (B), and for a large damping force at (C). The relative velocity between the wall and the mass is indicated by vertical shading.

ping occurs in the damper and the mass is accelerated at a constant rate. Since a constant acceleration produces a uniform change in velocity, the velocity of the mass when the damper is slipping is shown by straight lines. The relative velocity between the wall and the mass is shown by the vertical shading.

Figure 6.19A applies to a damper with a low friction force. The damper slips continuously. In Fig. 6.19B the velocities resulting from a larger friction force are shown. Slipping disappears for certain portions of the cycle. Where the wall and the mass have the same velocity, their accelerations also are equal. Slipping occurs when the force transmitted by the damper is not large enough to keep the mass accelerating with the wall. Since at the breakaway point the accelerations of the wall and mass are equal, their velocity-time curves have the same slope; i.e., the curves are tangent at this point. In Fig. 6.19C, the damping force is so large that the mass follows the wall for a considerable portion of the cycle and slips only where its acceleration becomes greater than the value of F_s/m. A slight increase in the clamping force or in the coefficient of friction locks the mass to the wall; then there is no relative motion and no damping.

Because of the nature of the damping force, the damping provided by the friction damper can be computed most practically in terms of energy. If the friction force exerted through the damper is F_s, the energy dissipated by the damper is the product of the friction force and the total relative motion between the mass and the moving wall. The time reference is taken at the moment when the auxiliary mass m has a zero velocity and is being accelerated to a positive velocity, Fig. 6.19A. Let the period of the vibratory motion of the wall be $\tau = 2\pi/\omega$, where ω is the angular frequency of the wall motion. By symmetry, the points of no slippage in the damper occur at times

$-\tau/4$, $\tau/4$, and $3\tau/4$. Let the time when the velocity of the wall is zero be $-t_0$; then the velocity of the wall $\dot{x}$ is

$$\dot{x} = +x_0\omega \sin \omega(t + t_0)$$

The velocity $\dot{u}$ of the mass for $-\tau/4 < t < \tau/4$ is

$$\dot{u} = \ddot{u}t = \frac{F_s}{m}t$$

The velocities of the wall and the mass are equal at time $t = \tau/4$:

$$x_0\omega \sin \omega\left(\frac{\tau}{4} + t_0\right) = \frac{F_s}{m}\frac{\tau}{4}$$

Since $\omega\tau/4 = \pi/2$, $\sin \omega(\tau/4 + t_0) = \cos \omega t_0$. Therefore

$$\cos \omega t_0 = \frac{F_s}{m}\frac{\pi}{2x_0\omega^2}$$

The relative velocity between the moving wall and the mass is $\dot{x} - \dot{u}$, and the total relative motion is the integral of the relative velocity over a cycle. Note that the area between the two curves for the second half of the cycle is the same as for the first. Hence, the work V per cycle is

$$V = 2 \int_{-\tau/4}^{\tau/4} F_s(\dot{x} - \dot{u})\, dt = 4F_s x_0 \sqrt{1 - \left(\frac{F_s\pi}{2mx_0\omega^2}\right)^2} \tag{6.34}$$

Optimum damping occurs when the work per cycle is a maximum. It can be determined by setting the derivative of V with respect to F_s in Eq. (6.34) equal to zero and solving for F_s:

$$(F_s)_{\text{opt}} = \frac{\sqrt{2}}{\pi} m\omega^2 x_0 \tag{6.35}$$

Energy absorption per cycle with optimum damping is, from Eq. (6.34),

$$V_{\text{opt}} = \frac{4}{\pi} m\omega^2 x_0^2 \tag{6.36}$$

A comparison of the effectiveness of the Coulomb friction damper with other types is given in Fig. 6.15.

EFFECT OF NONLINEARITY IN THE SPRING OF AN AUXILIARY MASS DAMPER

It is possible to extend the range of frequency over which a dynamic absorber is effective by using a nonlinear spring.[10-12] When a nonlinear spring is used, the natural frequency of the absorber is a function of the amplitude of vibration; it increases or decreases, depending upon whether the spring stiffness increases or decreases with deflection. Figure 6.20A shows a typical response curve for a system with increasing spring stiffness; Fig. 6.20B illustrates types of systems having increasing spring stiffness and shows typical force-deflection curves. Figure 6.21A shows a typical response

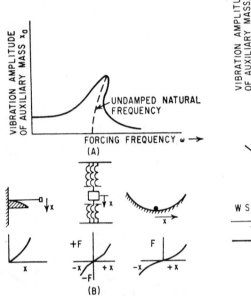

FIGURE 6.20 Auxiliary mass damper with nonlinear spring having stiffness that increases as deflection increases. The response to forced vibration and the natural frequency are shown at (A). Several arrangements of nonlinear systems with the corresponding force-deflection curves are shown at (B).

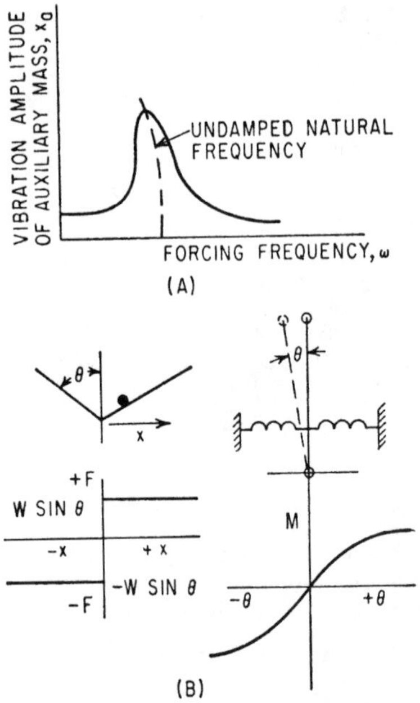

FIGURE 6.21 Auxiliary mass damper with nonlinear spring having stiffness that decreases as deflection increases. The response to forced vibration and the natural frequency are shown at (A). Two arrangements of nonlinear systems with the corresponding force-deflection curves are shown at (B).

curve for a system of decreasing spring stiffness; Fig. 6.21B illustrates types of systems having decreasing stiffnesses and shows typical force-deflection curves.

To compute the equivalent mass at a given frequency when a nonlinear spring is used, it is necessary to use a trial-and-error procedure. By the methods given in Chap. 4, compute the natural frequency of the auxiliary mass system, assuming the point of attachment fixed, as a function of the amplitude of motion of the auxiliary mass. This will result in a curve similar to the dotted curves in Figs. 6.20A and 6.21A. At the given frequency, compute β_a in Eq. (6.4) in terms of the tuned frequency of the absorber at zero amplitude. (The tuned frequency will change with amplitude because the spring constant changes.) With this value of β_a compute the equivalent mass from Eq. (6.6). With this mass in the system, compute the amplitude of motion x_0 of the primary mass to which the auxiliary system is attached [Eq. (6.7)] and the amplitude of the relative motion $x_{r0} = v^2(1 - v^2)x_0$. Using this value of x_{r0}, ascertain the corresponding value of resonance frequency of the system from the computed curve, and compute the new value of β_a. Repeat the process until the value of β_a remains unchanged upon repeated calculation.

A dynamic absorber having a nonlinear characteristic can be used to introduce nonlinearity into a resonant system. This can be useful in the case where a machine passes through a resonance rapidly as the speed is increased but slowly as the speed is decreased. In bringing this machine up to speed, there is a natural frequency that comes into strong resonance, giving a critical speed. A strongly nonlinear dynamic absorber tuned at low amplitudes to the optimum frequency for the damped absorber

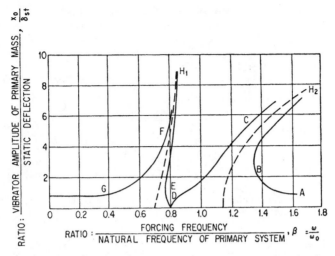

FIGURE 6.22 Motion of the primary mass, as a function of forcing frequency, in a system having a nonlinear dynamic absorber whose natural frequency increases with amplitude. The mass of the absorber is 0.25 times the mass of the primary system ($\mu = 0.25$).

can be used to reduce the effects of the critical speed. Two resonant peaks will be introduced, as shown on curve 1 of Fig. 6.13. By making the dynamic absorber nonlinear, so that the stiffness becomes greater as the amplitude of vibration is increased, the peaks are bent over to provide the response curve shown in Fig. 6.22. In starting, the machine is accelerated through the two critical speeds so fast that a resonance is unable to build up. In coasting to a stop, there would be ample time for significant amplitudes to build up if the nonlinearity did not exist. Because of the nonlinearity, the amplitude of vibration as a function of speed (since β is proportional to speed) follows the path A, B, C, D, E, F, G and never reaches the extreme amplitudes H_1 and H_2.

MULTIMASS ABSORBERS

In general, only one mass is used in a dynamic absorber. However, it is possible to provide a dynamic absorber that is effective for two or more frequencies by attaching an auxiliary mass system that resonates at the frequencies that are objectionable. The principle that would make such a dynamic absorber effective is utilized in the design of the elastic system of a ship's propulsion plant driven by independent high-pressure and low-pressure turbines. By making the frequencies of the two branches about the reduction gear identical, the gear becomes a node for one of the resonant modes. Then it is impossible to excite the mode of vibration where one turbine branch vibrates against the other as a result of excitation transmitted by the propeller shaft to that node.

DISTRIBUTED MASS ABSORBERS

It is possible to use distributed masses as vibration dampers. Consider an undamped rod of distributed mass and elasticity attached to a foundation that vibrates the rod axially, as shown in Fig. 6.23. The differential equation for the motion of this rod is derived in Chap. 7. The values of the constants are set by the boundary conditions:

$$\text{Stress} = E\frac{\partial u}{\partial x} = 0 \quad \text{where } x = l$$

$$u = u_0 \cos t \quad \text{where } x = 0 \tag{6.37}$$

The solution of the equation of motion is

$$u = u_0 \cos \omega t \left(\cos \sqrt{\frac{\gamma\omega^2}{Eg}} \, x + \tan \sqrt{\frac{\gamma\omega^2}{Eg}} \, l \sin \sqrt{\frac{\gamma\omega^2}{Eg}} \, x \right)$$

$$= u_0 \cos \omega t \, \frac{\cos \sqrt{(\gamma\omega^2/Eg)} \, (1 - x)}{\cos \sqrt{(\gamma\omega^2/Eg)} \, l} \tag{6.38}$$

where E is the modulus of elasticity and γ is the weight density of the material. When $x = 0$, the force F on the foundation is

$$F = SE\frac{\partial u}{\partial x}\bigg|_0 = SEu_0 \sqrt{\frac{\gamma\omega^2}{Eg}} \left(\tan \sqrt{\frac{\gamma\omega^2}{Eg}} \, l \right) \tag{6.39}$$

where S is the cross-sectional area of the bar. It is apparent that as the argument of the tangent has successive values of $\pi/2, 3\pi/2, 5\pi/2, \ldots$, the force exerted on the foundation becomes infinite. The distributed mass acts as a dynamic absorber enforcing a node at its point of attachment. By tuning the mass so that

$$\sqrt{\frac{\gamma\omega^2}{Eg}} \, l = \frac{n\pi}{2} \quad \text{or} \quad l = \frac{n\pi}{2\omega} \sqrt{\frac{Eg}{\gamma}} \tag{6.40}$$

the distributed mass acts as a dynamic absorber for not only the fundamental frequency $\omega/2\pi$ but also for the third, fifth, seventh, ... harmonics of the fundamental.

The above solution neglects damping. It is possible to consider the effect of damping by including a damping term in the differential equation. The stress in an element is assumed to be the sum of a deformation stress and a stress related to the velocity of strain:

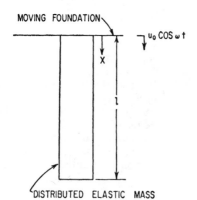

$u_0 \cos \omega t$

X

DISTRIBUTED ELASTIC MASS

FIGURE 6.23 Elastic body with distributed mass used as auxiliary mass damper.

$$\sigma = E\epsilon + \mu\frac{d\epsilon}{dt} \tag{6.41}$$

where $\epsilon = \partial u/\partial x$ is the strain. Then the differential equation becomes

$$E\frac{\partial^2 u}{\partial x^2} + \mu\frac{\partial^3 u}{\partial x^2 \partial t} = \frac{\gamma}{g}\frac{\partial^2 u}{\partial t^2} \tag{6.42}$$

Since the absorber is excited by a foundation moving with a frequency $f = \omega/2\pi$, u may be expressed as $\Re u_1 e^{j\omega t}$ and the partial differential equation can be written as the ordinary linear differential equation

$$E\frac{d^2 u_1}{dx^2} + j\omega\mu\frac{d^2 u_1}{dx^2} + \frac{\gamma\omega^2 u_1}{g} = 0$$

This equation may be written

$$\left(1 + \frac{\mu\omega j}{E}\right)\frac{d^2 u_1}{dx^2} + \frac{\gamma\omega^2}{Eg}\,u_1 = 0 \tag{6.43}$$

Since Eq. (6.43) is a second-order linear differential equation, the solution may be written

$$u = A_1 e^{\beta_1 x} + A_2 e^{\beta_2 x} \tag{6.44}$$

where β_1 and β_2 are the two roots of the equation

$$\beta^2 = \frac{-(\gamma/Eg)\omega^2}{1 + (j\mu\omega/E)} \tag{6.45}$$

For small values of μ, by a binomial expansion of the denominator,

$$\pm\beta = \frac{1}{2}\sqrt{\frac{\gamma}{Eg}}\,\frac{\mu\omega^2}{E} + j\sqrt{\frac{\gamma}{Eg}}\,\omega \tag{6.46}$$

where μ is defined by Eq. (6.41).

The boundary conditions to be met by the damper are:

At $x = 0$: $\qquad u = u_0 \qquad$ therefore, $A_1 + A_2 = u_0$

At $x = l$: $\qquad \sigma = (E + j\omega\mu)\,\dfrac{\partial u}{\partial x} = 0 \qquad$ therefore, $A_1 e^{\beta l} - A_2 e^{-\beta l} = 0$
$$\tag{6.47}$$

Solving Eqs. (6.47) for A_1 and A_2 and substituting the result in Eq. (6.44),

$$u = u_0\,\frac{\cosh \beta(l - x)}{\cosh \beta l} \tag{6.48}$$

The force exerted on the foundation by the damper is

$$F_{(x = 0)} = S\sigma_{(x = 0)} = -\,Su_0(E + j\omega\mu)\,\frac{\beta \sinh \beta l}{\cosh \beta l} \tag{6.49}$$

where S is the cross-section area of the bar. When the complex value of β as given in Eq. (6.46) is substituted in Eq. (6.49), the following value for the dynamic force exerted on the foundation is obtained:

$$\frac{F_{(x = 0)}}{SE\sqrt{\dfrac{\gamma}{Eg}}\,\omega u_0} = \frac{\left(1 + \dfrac{\mu^2\omega^2}{2E^2}\right)\sin 2\!\left(\sqrt{\dfrac{\gamma}{Eg}}\,\omega l\right) + \dfrac{\mu\omega}{2E}\sinh\!\left(\dfrac{\mu\omega}{E}\sqrt{\dfrac{\gamma}{Eg}}\,\omega l\right)}{\cos 2\!\left(\sqrt{\dfrac{\gamma}{Eg}}\,\omega l\right) + \cosh\!\left(\dfrac{\mu\omega}{E}\sqrt{\dfrac{\gamma}{Eg}}\,\omega l\right)}$$

$$+\,j\,\frac{\dfrac{\mu\omega}{2E}\sin 2\!\left(\sqrt{\dfrac{\gamma}{Eg}}\,\omega l\right) - \left(1 + \dfrac{\mu^2\omega^2}{2E^2}\right)\sinh\!\left(\dfrac{\mu\omega}{E}\sqrt{\dfrac{\gamma}{Eg}}\right)}{\cos 2\!\left(\sqrt{\dfrac{\gamma}{Eg}}\,\omega l\right) + \cosh\!\left(\dfrac{\mu\omega}{E}\sqrt{\dfrac{\gamma}{Eg}}\,\omega l\right)} \tag{6.50}$$

A plot of the real and imaginary values of $F_{(x = 0)}/SE\sqrt{\dfrac{\gamma}{Eg}}\,\omega u_0$ is given in Fig. 6.24 for zero damping and for a damping coefficient $\mu\omega/E = 0.1$ as a function of a tuning

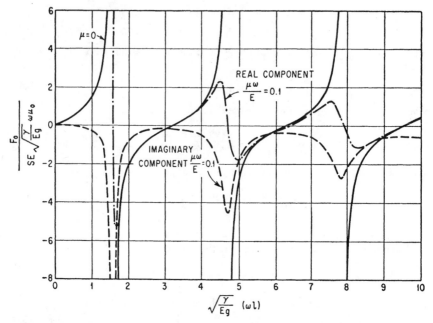

FIGURE 6.24 Real and imaginary components of the force applied to a vibrating body by the distributed mass damper shown in Fig. 6.23. These relations are given mathematically by Eq. (6.50), and the terms are defined in connection with Eq. (6.38). The curves are for a value of the damping coefficient $\mu\omega/E = 0.1$, where μ is defined by Eq. (6.41).

parameter $\sqrt{\gamma/Eg(\omega l)}$. Damping decreases the effectiveness of the distributed mass damper substantially, particularly for the higher modes.

PRACTICAL APPLICATIONS OF AUXILIARY MASS DAMPERS AND ABSORBERS TO SINGLE DEGREE-OF-FREEDOM SYSTEMS

THE DYNAMIC ABSORBER

The dynamic absorber, because of its tuning, can be used to eliminate vibration only where the frequency of the vibration is constant. Many pieces of equipment to which it is applied are operated by alternating current. So that it can be used for time keeping, the frequency of this alternating current is held remarkably constant. For this reason, most applications of dynamic absorbers are made to mechanisms that operate in synchronism from an ac power supply.

An application of a dynamic absorber to the pedestal of an ac generator having considerable vibration is shown in Fig. 6.25, where the relative sizes of absorber and pedestal are shown approximately to scale. In this case, the application is made to a complicated structure and the mass of the absorber is much less than that of the primary system; however, since the frequency of the excitation is constant, the dynamic

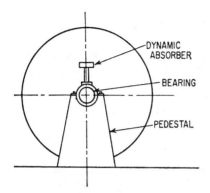

FIGURE 6.25 Application of a dynamic absorber to the bearing pedestal of an ac generator.

absorber reduces the vibration. When the mass ratio is small, it is important that the absorber be accurately tuned and that the damping be small. In this case, the excitation was the unbalance in the turbine rotor which was elastically connected to the pedestal through the flexibility of the shaft. If the absorber were ideally effective, there would be no forces at the frequency of the shaft speed; therefore, there would be no displacements from the pedestal where the force is neutralized through the remainder of the structure.

The dynamic absorber has been applied to the electric clipper shown in Fig. 6.26. The structure consisting of the cutter blade and its driving mechanism is actuated by the magnetic field at a frequency of 120 Hz, as a result of the 60-Hz ac power supply. The forces and torques required to move the blade are balanced by reactions on the housing, causing it to vibrate. The dynamic absorber tuned to a frequency of 120 Hz enforces a node at the location of its mass. Since this is approximately the center-of-gravity of the assembly of the cutter and its driving mechanism, the absorber effectively neutralizes the unbalanced force. The moment caused by the rotation of the moving parts is still unbalanced. A second very small dynamic absorber placed in the handle of the clipper could enforce a node at the handle and substantially eliminate all vibration.

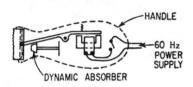

FIGURE 6.26 Application of a dynamic absorber to a hair clipper.

The design of these absorbers is simple after the unbalanced forces and torques generated by the cutter mechanism are computed. The sum of the inertia forces generated by the two absorbers, $m_1x_1\omega^2 + m_2x_2\omega^2$ (where m_1 and x_1 are the mass and amplitude of motion of the first absorber, m_2 and x_2 are the corresponding values for the second absorber, and $\omega = 240\pi$), must equal the unbalanced force generated by the clipper mechanism. The torque generated by the two absorbers must balance the torque of the mechanism. Since the value of ω^2 is known, the values of m_1x_1 and m_2x_2 can be determined. Weights that fit into the available space with adequate room to move are chosen, and a spring is designed of such stiffness that the natural frequency is 120 Hz.

Because of the desirable balancing properties of the simple dynamic absorber and the constancy of frequency of ac power, it might be expected that devices operating at a frequency of 120 Hz would be used more widely. However, their application is limited because the frequency of vibration is too high to allow large amplitudes of motion.

REDUCTION OF ROLL OF SHIPS BY AUXILIARY TANKS

An interesting application of auxiliary mass absorbers is found in the auxiliary tanks used to reduce the rolling of ships,[1,13] as shown in Fig. 6.27. When a ship is heeled, the

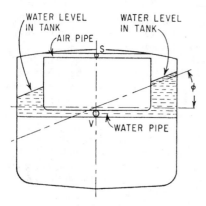

FIGURE 6.27 Cross section of ship equipped with antiroll tanks. The flow of water from one tank to the other tends to counteract rolling of the ship.

restoring moment $k_r\phi$ acting on it is proportional to the angle of heel (or roll). This restoring moment acts to return the ship (and the water that moves with it) to its equilibrium position. If I_s represents the polar moment of inertia of the ship and its entrained water, the differential equation for the rolling motion of the ship is

$$I_s\ddot{\phi} + k_r\phi = M_s \qquad (6.51)$$

where M_s represents the rolling moments exerted on the ship, usually by waves.

To reduce rolling of the ship, auxiliary wing tanks connected by pipes are used. The water flowing from one tank to another has a natural frequency that is determined by the length and cross-sectional area of the tube connecting the tanks. The damping is controlled by restricting the flow of water, either with a valve S in the line that allows air to flow between the tanks (Fig. 6.27) or with a valve V in the water line. Since the tanks occupy valuable space, the mass ratio of the water in the tanks to the ship is small. Fortunately, the excitation from waves generally is not large relative to the restoring moments, and roll becomes objectionable only because the normal damping of a ship in rolling motion is not very large. The use of antirolling tanks in the German luxury liners *Bremen* and *Europa* reduced the maximum roll from 15 to 5°.

REDUCTION OF ROLL OF SHIPS BY GYROSCOPES

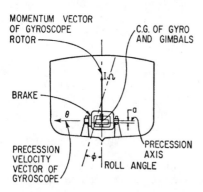

FIGURE 6.28 Application of a gyroscope to a ship to reduce roll.

A large gyroscope may be used to reduce roll in ships, as shown in Fig. 6.28.[1,14] In response to the velocity of roll of a ship, the gyroscope precesses in the plane of symmetry of the ship. By braking this precession, energy can be dissipated and the roll reduced. The torque exerted by the gyroscope is proportional to the rate of change of the angular momentum about an axis perpendicular to the torque. Letting I represent the polar moment of inertia of the gyroscope about its spin axis and $\dot{\theta}$ the angular velocity of precession of a gyroscope, then the equation of motion of the ship is

$$I_s\ddot{\phi} + k_r\phi + I\Omega\dot{\theta} = M_s \qquad (6.52)$$

Assume that the gyroscope has (1) a moment of inertia about the precession axis of I_g, (2) a weight of W, and (3) that its center-of-gravity is below the gimbal axis (as it must be for the gyro to come to equilibrium in a working position) a distance a, as shown in Figure 6.28. Then the equation of motion of the gyroscope is

$$I_g \ddot{\theta} + W_a \theta + c\dot{\theta} - I\Omega\dot{\phi} = 0 \tag{6.53}$$

where Ω is the spin velocity of the gyroscope. From Eq. (6.53), for a roll frequency of ω, the angle of precession of the gyroscope is

$$\theta = \frac{jI\Omega\omega\phi}{-I_g\omega^2 + Wa - jc\omega} \tag{6.54}$$

The torque exerted on the ship is

$$I\Omega\dot{\theta} = \frac{-(I\Omega)^2\omega^2\phi}{-I_g\omega^2 + Wa + jc\omega} \tag{6.55}$$

The equivalent moment of inertia of the gyroscope system in its reaction on the ship is

$$\frac{I\Omega^2}{-I_g\omega^2 + Wa + cj\omega} \tag{6.56}$$

By analogy with the steps of Eqs. (6.2) through (6.7), it follows that

$$\frac{\phi}{\phi_{st}} = \sqrt{\frac{(1 - \beta_g{}^2)^2 + (2\zeta\beta_g)^2}{[(1 - \beta_g{}^2)(1 - \beta^2) - \beta^2\mu]^2 + (2\zeta\beta_g)^2(1 - \beta^2)^2}} \tag{6.57}$$

where the parameters are defined in terms of ship and gyro constants as follows:

$$\beta_g = \frac{\omega}{\sqrt{Wa/I_g}} \qquad \beta = \frac{\omega}{\sqrt{k_r/I_s}} \qquad \zeta = \frac{c}{2\sqrt{WaI_g}} \qquad \mu = \frac{(I\Omega)^2}{WaI_s} \qquad \psi_{st} = \frac{M_s}{k_r}$$

Because $I\Omega$ can be made large by using a large gyro rotor and spinning it at a high speed, and Wa can be made small by choice of a design, the value of μ can be made quite large even though I_s is large. In one experimental ship, $\mu = 20$ was obtained. Even with this large value of μ, the precession angle of the gyroscope would become very large for optimum damping. Therefore it is necessary to use much more damping than optimum. Gyro stabilizers were used on the Italian ship *Conte di Savoia;* they are sometimes installed on yachts.

Both antirolling tanks and gyro stabilizers are more effective if they are active rather than passive. Activated dampers are considered below.

AUXILIARY MASS DAMPERS APPLIED TO ROTATING MACHINERY

An important industrial use of auxiliary mass systems is to neutralize the unbalance of centrifugal machinery. A common application is the balance ring in the spin dryer of home washing machines. The operation of such a balancer is dependent upon the basket of the washer rotating at a speed greater than the natural frequency of its support. The balance ring is attached to the washing machine basket concentric with its axis of rotation, as shown in Fig. 6.29.

Consider the washing machine basket shown in Fig. 6.29. When its center-of-gravity does not coincide with its axis of rotation and it is rotating at a speed lower

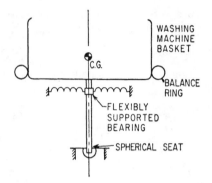

FIGURE 6.29 Schematic diagram showing location of balance ring on basket of a spin dryer.

than its critical speed (corresponding to the natural frequency in rocking motion about the spherical seat), the centrifugal force tends to pull the rotational axis in the direction of the unbalance. This effect increases with an increase in rotational speed until the critical speed is reached. At this speed the amplitude would become infinite if it were not for the damping in the system. Above the critical speed, the phase position of the axis of rotation relative to the center-of-gravity shifts so that the basket tends to rotate about its center-of-gravity with the flexibly supported bearing moving in a circle about an axis through the center-of-gravity. The relative positions of the bearing center and the center-of-gravity are shown in Fig. 6.30A and B.

Since the balance ring is circular with a smooth inner surface, any weights or fluids contained in the ring can be acted upon only by forces directed radially. When the ring is rotated about a vertical axis, the weights or fluids will move within the ring in such a manner as to be concentrated on the side farthest from the axis of rotation. If this con-

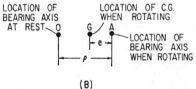

(A)

(B)

FIGURE 6.30 Diagram in plane normal to axis of rotation of spin dryer in Fig. 6.29. Relative positions of axes when rotating speed is less than natural frequency are shown at (A); corresponding diagram for rotation speed greater than natural frequency is shown at (B).

centration occurs below the natural frequency (Fig. 6.30A), the weights tend to move further from the axis and the resultant center-of-gravity is displaced so as to give a greater eccentricity. The points A and G rotate about the axis O at the frequency ω. The initial eccentricity of the center-of-gravity of the washer basket and its load from the axis of rotation is represented by e, and ρ is the elastic displacement of this center of rotation due to the centrifugal force. Where the off-center rotating weight is W, the unbalanced force is $(W/g)(\rho + e)\omega^2$ [where $\rho = e/(1 - \beta^2)$ and $\beta^2 = \omega^2/\omega_n^2 < 1$] and acts in the direction from A to G.

If the displacement of the weights or fluids in the balance ring occurs above the natural frequency, the center-of-gravity tends to move closer to the dynamic location of the axis. The action in this case is shown in Fig. 6.30B. Then the points A and G rotate about O at the frequency ω. The unbalanced force is $(W/g)(\rho + e)\omega^2$ [where $\rho = e/(1 - \beta^2)$ and $\beta^2 = \omega^2/\omega_n^2 > 1$]. This gives a negative force that acts in a direction from G to A. Thus the eccentricity is brought toward zero and the rotor is automatically balanced. Because it is necessary to pass through the critical speed in bringing the rotor up to speed and in stopping it, it is desirable to heavily damp the balancing elements, either fluid or weights.

In practical applications, the balancing elements can take several forms. The earliest form consisted of two or more spheres or cylinders free to move in a race con-

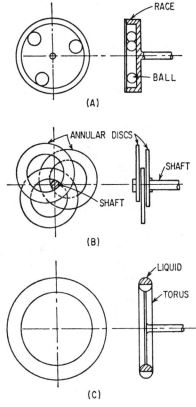

centric with the axis of the rotor, as shown in Fig. 6.31A. A later modification consists of three annular discs that rotate about an enlarged shaft concentric with the axis, as indicated in Fig. 6.31B. These are contained in a sealed compartment with oil for lubrication and damping. A fluid type of damper is shown in Fig. 6.31C, the fluid usually being a high-density viscous material. With proper damping, mercury would be excellent, but it is too expensive. Therefore a more viscous, high-density halogenated fluid is used.

The balancers must be of sufficient weight and operate at such a radius that the product of their weight and the maximum eccentricity they can attain is equivalent to the unbalanced moment of the load. This requirement makes the use of the spheres or cylinders difficult because they cannot be made large; it makes the annular plates large because they are limited in the amount of eccentricity that can be obtained.

In a cylindrical volume 24 in. (61 cm) in diameter and 2 in. (5 cm) thick, seven spheres 2 in. (5 cm) in diameter can neutralize 98.6 lb-in. (114 kg-cm) of unbalance; three cylinders 4 in. (10 cm) in diameter by 2 in. (5 cm) thick can neutralize 255 lb-in. (295 kg-cm); three annular discs, each ⅝ in. (1.6 cm) thick with an outside diameter of 19.55 in. (50 cm) and an inside diameter of 10.45 in. (26.5 cm) [the optimum for a center post

FIGURE 6.31 Examples of balancing means for rotating machinery: (A) spheres (or cylinders) in a race; (B) annular discs rotating on shaft; (C) damping fluid in torus.

6 in. (15.2 cm) in diameter], can neutralize 250 lb-in. (290 kg-cm); and half of a 2-in. (5-cm) diameter torus filled with fluid of density 0.2 lb/in^3 (5.5 gram/cm^3) can neutralize 609 lb-in. (700 kg-cm). Only the fluid-filled torus would be initially balanced.

AUXILIARY MASS DAMPERS APPLIED TO TORSIONAL VIBRATION

Dampers and absorbers are used widely for the control of torsional vibration of internal-combustion engines. The most common absorber is the viscous-damped, untuned auxiliary mass unit shown in Fig. 6.32. The device is comprised of a cylindrical housing carrying an inertia mass that is free to rotate. There is a preset clearance between the housing and the inertia mass that is filled with a silicone oil of proper viscosity. Silicone oil is used because of its high viscosity index; i.e., its viscosity changes relatively little with temperature. With the inertia mass and the damping medium contained, the housing is seal-welded to provide a leakproof and simple

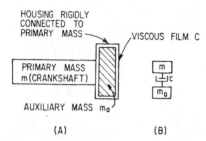

FIGURE 6.32 Untuned auxiliary mass damper with viscous damping. The application to a torsional system is shown at (A), and the linear analog at (B).

absorber. However, the silicone oil has poor boundary lubricating properties and if decomposed by a local hot spot (such as might be caused by a reduced clearance at some particular spot), the decomposed damping fluid is abrasive.

Because of the simplicity of this untuned damper, it is commonly used in preference to the more effective tuned absorber. However, it is possible to use the same construction methods for a tuned damper, as shown in Fig. 6.33. It is also possible to mount the standard damper with the housing for the unsprung inertia mass attached to the main mass by a spring, as shown in Fig. 6.34. If the viscosity of the oil and the dimensions of the masses and the clearance spaces are known, the damping effects of the dampers shown in Figs. 6.32 and 6.34 can be computed directly in terms of the equations previously developed. The damper in Fig. 6.34 can be analyzed by treating the spring and housing as additional elements in the main system and the untuned mass as a viscous damped auxiliary mass. If the inertia of the housing is negligible, the inertia mass is effectively connected to the main mass through a spring and a dashpot in series. The two elements in series can be represented by a complex spring constant equal to

$$\frac{1}{(1/jc\omega) + (1/k)} = \frac{kcj\omega}{k + cj\omega}$$

Where there is no damping in parallel with the spring, Eq. (6.3) becomes

$$m_{eq} = km/(k - m\omega^2)$$

Substituting the complex value of the spring constant, the effective mass is

$$m_{eq} = \frac{ckj\omega}{k + cj\omega} \left[\frac{m}{-m\omega^2 + cjk\omega/(k + cj\omega)} \right] \tag{6.58}$$

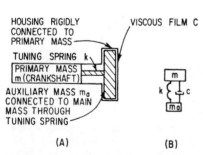

FIGURE 6.33 Tuned auxiliary mass damper with viscous damping. The application to a torsional system is shown at (A), and the linear analog at (B).

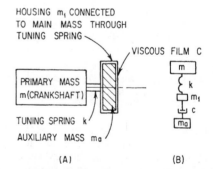

FIGURE 6.34 Auxiliary mass damper with viscous damping and spring-mounted housing. The application to a torsional system is shown at (A), and the linear analog at (B).

In terms of the nondimensional parameters defined in Eq. (6.4):

$$m_{eq} = \frac{(2\zeta\beta_a)^2(1 - \beta_a^2)}{\beta_a^4 - (2\zeta\beta_a)^2(1 - \beta_a^2)} \, m + \frac{-2\zeta\beta_a^3 m}{\beta_a^4 - (2\zeta\beta_a)^2(1 - \beta_a^2)} \, j \qquad (6.59)$$

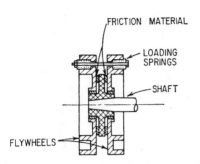

FIGURE 6.35 Schematic cross section through Lanchester damper.

Before the advent of silicone oil with its chemical stability and relatively constant viscosity over service temperature conditions, the damper most commonly used for absorbing torsional vibration energy was the dry friction or Lanchester damper shown in Fig. 6.35. The damping is determined by the spring tension and the coefficient of friction at the sliding interfaces. Its optimum value is determined by the equation for a torsional system analogous to Eq. (6.35) for a linear system:

$$(T_s)_{opt} = \frac{\sqrt{2}}{\pi} I\omega^2\theta_0 \qquad (6.60)$$

where T_s is the slipping torque, I is the moment of inertia of the flywheels, and θ_0 is the amplitude of angular motion of the primary system. The dry-friction-based Lanchester damper requires frequent adjustment, as the braking material wears, to maintain a constant braking force.

It is possible to use torque-transmitting couplings that can absorb vibration energy, as the spring elements for tuned dampers. The Bibby coupling (Fig. 6.36) is used in this manner. Since the stiffness of this coupling is nonlinear, the optimum tuning of such an absorber is secured for only one amplitude of motion.

A discussion of dampers and of their application to engine systems is given in Chap. 38.

DYNAMIC ABSORBERS TUNED TO ORDERS OF VIBRATION RATHER THAN CONSTANT FREQUENCIES

In the torsional vibration of rotating machinery, it is generally found that exciting torques and forces occur at the same frequency as the rotational speed or at multiples of this frequency. The ratio of the frequency of vibration to the rotational speed is called the *order of the vibration q*. Thus a power plant driving a four-bladed propeller may have a torsional vibration whose frequency is 4 times the rotational speed of the drive shaft; sometimes it may have a second torsional vibration whose frequency is 8 times the rotational speed. These are called the fourth-order and eighth-order torsional vibrations.

If a dynamic absorber in the form of a pendulum acting in a centrifugal field is used, then its natural frequency increases linearly with speed. Therefore it can be used to neutralize an order of vibration.[15–19]

Consider a pendulum of length l and of mass m attached at a distance R from the center of a rotating shaft, as shown in Fig. 6.37. Since the pendulum is excited by torsional vibration in the shaft, let the radius R be rotating at a constant speed Ω with a

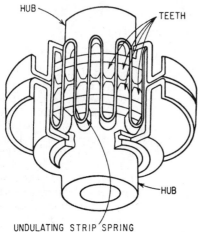

FIGURE 6.36 Coupling used as elastic and damping element in auxiliary mass damper for torsional vibration. The torque is transmitted by an undulating strip of thin steel interposed between the teeth on opposite hubs. The stiffness of the strip is nonlinear, increasing as torque increases. Oil pumped between the strip and teeth dissipates energy.

FIGURE 6.37 Schematic diagram of pendulum absorber.

superposed vibration $\theta = \theta_0 \cos q\Omega t$, where q represents the order of the vibration. Then the angle of R with respect to any desired reference is $\Omega t + \theta_0 \cos q\Omega t$. The angle of the pendulum with respect to the radius R is defined as $\psi = \psi_0 \cos q\Omega t$, as shown by Fig. 6.37.

The acceleration acting on the mass m at position B is most easily ascertained by considering the change in velocity during a short increment of time Δt. The components of velocity of the mass m at time t are shown graphically in Fig. 6.38A; at time $t + \Delta t$, the corresponding velocities are shown in Fig. 6.38B. The change in velocity during the time interval Δt is shown in Fig. 6.38C. Since the acceleration is the change in velocity per unit of time, the accelerations along and perpendicular to l are:

Acceleration along l:

$$\frac{-l(\Omega + \dot\theta + \dot\psi^2)\,\Delta t - R(\Omega + \dot\theta)^2\,\Delta t \cos \psi + R\ddot\theta\,\Delta t \sin \psi}{\Delta t} \tag{6.61}$$

Acceleration perpendicular to l:

$$\frac{l(\ddot\theta + \ddot\psi)\,\Delta t + R(\Omega + \dot\theta)^2\,\Delta t \sin \psi + R\ddot\theta\,\Delta t \cos \psi}{\Delta t} \tag{6.62}$$

Only the force $-F$, directed along the pendulum, acts on the mass m. Therefore the equations of motion are

$$-F = -ml(\Omega + \dot\theta + \dot\psi)^2 - mR(\Omega + \dot\theta)^2 \cos \psi + R\ddot\theta \sin \psi$$
$$0 = ml(\ddot\theta + \ddot\psi) + mR(\Omega + \dot\theta)^2 \sin \psi + mR\ddot\theta \cos \psi \tag{6.63}$$

Assuming that ψ and θ are small, Eqs. (6.63) simplify to

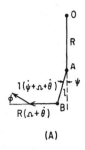

(A)

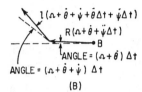

(B)

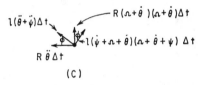

(C)

FIGURE 6.38 Velocity vectors for the pendulum absorber: (A) velocities at time t; (B) velocities at time $t + \Delta t$; (C) change in velocities during time increment Δt.

$$Ft = m(R + l)\Omega^2$$

$$l(\ddot{\theta} + \ddot{\psi}) + R\Omega^2\psi + R\ddot{\theta} = 0 \tag{6.64}$$

The second of Eqs. (6.64) upon substitution of $\theta = \theta_0 \cos q\Omega t$ and $\psi = \psi_0 \cos q\Omega t$ yields

$$\frac{\psi_0}{\theta_0} = \frac{(q\Omega)^2(l + R)}{-(q\Omega)^2 l + \Omega^2 R} = \frac{q^2(l + R)}{R - q^2 l} \tag{6.65}$$

The torque M exerted at point 0 by the force F is

$$M = RF \sin \psi = RF\psi \qquad \text{when } \psi \text{ is small}$$

From Eqs. (6.64) and (6.65), when ψ is small,

$$M = \frac{mq^2 R(R + l)^2 \Omega^2}{R - q^2 l} \tag{6.66}$$

If a flywheel having a moment of inertia I is accelerated by a shaft having an amplitude of angular vibratory motion θ_0 and a frequency $q\Omega$, the torque amplitude exerted on the shaft is $I(q\Omega)^2\theta_0$. Therefore, the equivalent moment of inertia I_{eq} of the pendulum is

$$I_{eq} = \frac{mR(R + l)^2}{R - q^2 l} = \frac{m(R + l)^2}{1 - q^2 l/R} \tag{6.67}$$

When

$$\frac{R}{l} = q^2 \tag{6.68}$$

the equivalent inertia is infinite and the pendulum acts as a dynamic absorber by enforcing a node at its point of attachment.

Where the pendulum is damped, the equivalent moment of inertia is given by an equation analogous to Eqs. (6.4) and (6.5):

$$I_{eq} = \frac{1 + 2\zeta vj}{(1 - v^2) + 2\zeta vj} m(R + l)^2$$

$$= m(R + l)^2 \left[\frac{1 - v^2 + (2\zeta v)^2}{(1 - v^2)^2 + (2\zeta v)^2} - \frac{2\zeta v^3 j}{(1 - v^2)^2 + (2\zeta v)^2} \right] \tag{6.69}$$

where $v^2 = q^2 l/R$ and $\zeta = (c/2m\Omega)\sqrt{l/R}$.

When the pendulum is attached to a single degree-of-freedom system as is shown in Fig. 6.39, the amplitude of motion θ_a of the flywheel of inertia I is given, by analogy to Eq. (6.7), as

$$\frac{\theta_a}{\theta_{st}} = \sqrt{\frac{(1 - v^2)^2 + (2\zeta v)^2}{[(1 - v^2)(1 - \beta_p{}^2) - \beta_p{}^2\mu]^2 + (2\zeta v)^2[1 - \beta_p{}^2 - \beta_p{}^2\mu_p]^2}} \tag{6.70}$$

where

$$2\zeta v = \frac{cql}{mR}$$

$$\mu_p = \frac{m(R + l)^2}{I}$$

$$\beta_p = \frac{q}{k_r I}$$

$$\theta_{st} = \frac{m_0}{k_r}$$

The pendulum tends to detune when the amplitude of motion of the pendulum is large, thereby introducing harmonics of the torque that it neutralizes.[17] Suppose the shaft rotates at a constant speed Ω, i.e., $\theta_0 = 0$, and consider the torque exerted on the shaft as m moves through a large amplitude ψ_0 about its equilibrium position. Equations (6.63) become

$$F = ml(\Omega + \dot{\psi})^2 + mR\Omega^2 \cos \psi$$
$$l\ddot{\psi} + R\Omega^2 \sin \psi = 0 \tag{6.71}$$

A solution for the second of Eqs. (6.71) is

$$\psi = \sqrt{\frac{2\Omega^2 R}{l}} \sqrt{\cos \psi - \cos \psi_0} \tag{6.72}$$

The solution of Eq. (6.72) involves elliptic integrals and is given approximately by

$$\psi = \psi_0 \sin \omega t$$

where

$$\omega = \sqrt{\frac{R}{l}} \frac{\pi/2}{F(\psi_0/2, \pi/2)} \Omega$$

and $F(\psi_0/2, \pi/2)$ is an elliptic function of the first kind whose value may be obtained from tables.

Since $\omega/\Omega = q$ (the order of the disturbance), the tuning of the damper will be changed for large angles and becomes

$$q^2 = \frac{R}{l} \left(\frac{\pi/2}{F(\psi_0/2, \pi/2)} \right)^2 \tag{6.73}$$

The value of $q^2 l/R = v^2$ used in Eqs. (6.69) and (6.70) is given in Fig. 6.40 as a function of the amplitude of the pendulum.

Since the force exerted by the mass m is directed along the rod connecting it to the pivot A (Fig. 6.37), the reactive torque on the shaft is

$$M = FR \sin \psi$$

$$= mR^2\Omega^2 \left[\frac{l}{R}(1 + \frac{\dot{\psi}}{\Omega})^2 \sin \psi + \sin \psi \cos \psi \right]$$

$$= mR^2\Omega^2(A_1 \sin q\Omega t + A_2 \sin 2q\Omega t + A_3 \sin 3q\Omega t + \cdots) \tag{6.74}$$

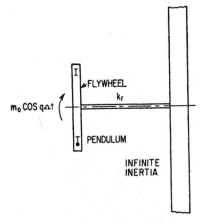

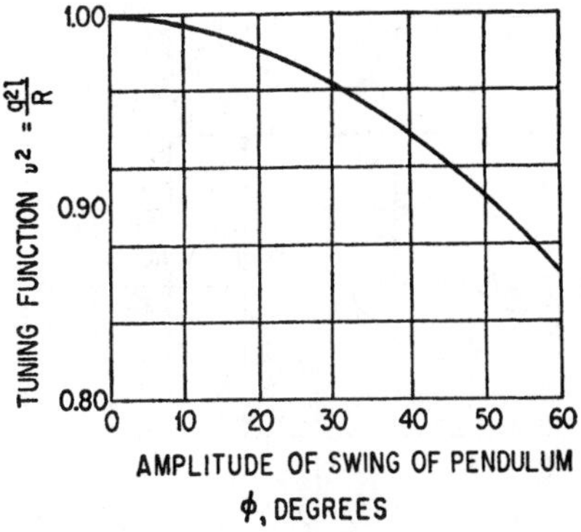

FIGURE 6.40 Tuning function for a pendulum absorber used in Eqs. (6.69) and (6.70).

FIGURE 6.39 Application of pendulum absorber to a rotational single degree-of-freedom system.

The values of the fundamental torque corresponding to the tuned frequency and to the second and third harmonics of this tuned frequency are given in Fig. 6.41 as a function of the angle of swing of the pendulum, for a typical installation. In this case, the pendulum is tuned to the 4½ order of vibration. (The 4½ order of vibration is one whose frequency is 4½ times the rotational frequency and 9 times the fundamental frequency. The latter is called the half order and occurs at half of the rotational frequency. This is common in four-cycle engines.)

Two types of pendulum absorber are used. The one most commonly used is shown in Fig. 6.42. The counterweight, which also is used to balance rotating forces in the engine, is suspended from a hub carried by the crankshaft by pins that act through holes with clearance, Fig. 6.42*A*. By suspending the pendulum from two pins, the pendulum when oscillating does not rotate but rather moves as shown in Fig. 6.42*B*. Since it is not subjected to angular acceleration, it may be treated as a particle located at its center-of-gravity. Referring to Fig. 6.42*A* and *B,* the expression for acceleration [Eqs. (6.61) and (6.62)] and the equations of motion [Eqs. (6.63)] apply if

$$R = H_1 + H_2$$

$$l = \frac{D_c + D_p}{2} - D_b$$

(6.75)

where H_1 = distance from center of rotation to center of holes in crank hub

H_2 = distance from center of holes in pendulum to center-of-gravity of pendulum

D_c = diameter of hole in crank hub

D_p = diameter of hole in pendulum

D_b = diameter of pin

In practice, difficulty arises from the wear of the holes and the pin. Moreover, the motion on the pins generally is small and the loads due to centrifugal forces are large so that fretting is a problem. Because the radius of motion of the pendulum is short,

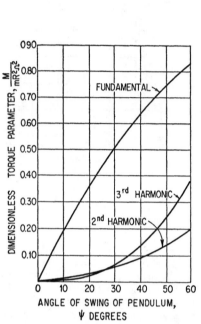

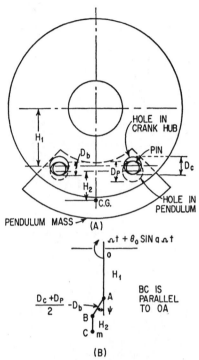

FIGURE 6.41 Harmonic components of torque generated by a pendulum absorber as a function of its angle of swing. The torque is expressed by the parameters used in Eq. (6.74).

FIGURE 6.42 Bifilar type of pendulum absorber. The mechanical arrangement is shown at (A), and a schematic diagram at (B).

only a small amount of wear can be tolerated. Hardened pins and bushings are used to reduce the wear.

The pendulum is most easily designed if it is recognized that the inertia torques generated by the pendulum must neutralize the forcing torques. Thus

$$m\omega^2 l\psi_0 R = M \qquad (6.76)$$

The radii l and R are set by the design of the crank and the order of vibration to be neutralized. The original motion ψ_0 is generally limited to a small angle, approximately 20°. It is probable that the most stringent condition is at the lowest operating speed, although the absorber may be required only to avoid difficulty at some particular critical speed. Knowing the excitation M, it is possible to compute the required mass of the pendulum weight.

A second type of pendulum absorber is a cylinder that rolls in a hole in a counterweight, as shown in Fig. 6.43. In this type, the radius of the pendulum corresponds to the difference in the radii of the hole and of the cylinder. It is found, by observing tests and checking the tuning of actual systems using cylindrical pendulums, that the weight rotates with a uniform angular velocity. Therefore the tuning is independent of the moments of inertia of the cylinder. It is common to allow a larger amplitude of motion with the absorber of Fig. 6.43 than with the absorber of Fig. 6.40.

Applications of pendulum absorbers to torsional-vibration problems are given in Chap. 38.

PENDULUM ABSORBER FOR LINEAR VIBRATION

The principle of the pendulum absorber can be applied to linear vibration as well as to torsional vibration. To neutralize linear vibration, pendulums are rotated about an axis parallel to the direction of vibration, as shown in Fig. 6.44. This can be accomplished with an absorber mounted on the moving body. Two or more pendulums are used so that centrifugal forces are balanced. Free rotational movement of each pendulum in the plane of the axis allows the axial forces to be neutralized. The pendulum assembly must rotate about the axis at some submultiple of the frequency of vibration. The size of the absorber is determined by the condition that the components of the inertia forces of the weights in the axial direction $[\Sigma m\omega^2 r\theta]$ must balance the exciting forces. This device can be applied where the vibration is generated by the action of rotating members but the magnitude of the vibratory forces is uncertain. A discussion of this absorber, including the influence of moments of inertia and damping of the pendulum, together with some applications to the elimination of vibration in special locations on a ship, is given in Ref. 20.

APPLICATIONS OF DAMPERS TO MULTIPLE DEGREE-OF-FREEDOM SYSTEMS

Auxiliary mass dampers as applied to systems of several degrees-of-freedom can be represented most effectively by equivalent masses or moments of inertia, as determined by Eq. (6.5) or Eq. (6.6). The choice of proper damping constants is more dif-

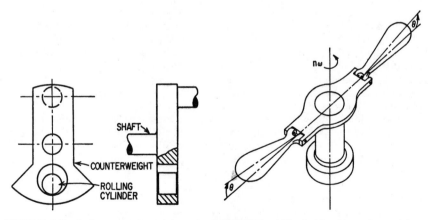

FIGURE 6.43 Roller type of pendulum absorber.

FIGURE 6.44 Application of pendulum absorbers to counteract linear vibration.

ficult. For the case of torsional vibration, the practical problems of designing dampers and selecting the proper damping are considered in Chap. 38.

There are many applications of dampers to vibrating structures that illustrate the use of different types of auxiliary mass damper. One such application has been to ships.[21] These absorbers had low damping and were designed to be filled with water so that they could be tuned to the objectionable frequencies. In one case, the absorber was located near the propeller (the source of excitation) and when properly tuned was found to be effective in reducing the resonant vibration of the ship. In another case, the absorber was located on an upper deck but was not as effective. It enforced a node at its point of attachment but, because of the flexibility between the upper deck and the bottom of the ship, there was appreciable motion in the vicinity of the propeller and vibratory energy was fed to the ship's structure. To operate properly, the absorbers must be closely tuned and the propeller speed closely maintained. Because the natural frequencies of the ship vary with the types of loading, it is not sufficient to install a fixed frequency absorber that is effective at only one natural frequency of the hull, corresponding to a particular loading condition.

An auxiliary mass absorber has been applied to the reduction of vibration in a heavy building that vibrated at a low frequency under the excitation of a number of looms.[22] The frequency of the looms was substantially constant. However, the magnitude of the excitation was variable as the looms came into and out of phase. The dynamic absorber, consisting of a heavy weight hung as a pendulum, was tuned to the frequency of excitation. Because the frequency was low and the forces large, the absorber was quite large. However, it was effective in reducing the amplitude of vibration in the building and was relatively simple to construct.

ACTIVATED VIBRATION ABSORBERS

The cost and space that can be allotted to ship antirolling devices are limited. Therefore it is desirable to activate the absorbers so that their full capacity is used for small amplitudes as well as large. Activated dampers can be made to deliver as large restoring forces for small amplitudes of motion of the primary body as they would be required to deliver if the motions were large. For example, the gyrostabilizer that is used in the ship is precessed by a motor through its full effective range, in the case of small angles as well as large. Thus, it introduces a restoring torque that is much larger than would be introduced by the normal damped precession.[14] In the same manner the water in antiroll tanks is always pumped to the tank where it will introduce the maximum torque to counteract the roll. By pumping, much larger quantities of water can be transferred and larger damping moments obtained than can be obtained by controlled gravity flows.

Devices for damping the roll are desirable for ships. It has been common practice to install bilge keels (long fins which extend into the water) in steel ships. Some ships are now fitted with activated, retractable hydrofoils located at the bilge at the maximum beam. Both these devices are effective only when the ship is in motion and add to the resistance of the ship.

Activated vibration absorbers are essentially servomechanisms designed to maintain some desired steady state. Steam and gas turbine speed governors, wicket gate controls for frequency regulation in water turbines, and temperature control equipment can be considered as special forms of activated vibration absorbers.[23]

THE USE OF AUXILIARY MASS DEVICES TO REDUCE TRANSIENT AND SELF-EXCITED VIBRATIONS

Where the vibration is self-excited or caused by repeated impact, it is necessary to have sufficient damping to prevent a serious build-up of vibration amplitude. This damping, which need not always be large, may be provided by a loosely coupled auxiliary mass. A simple application of this type is the ring fitted to the interior of a gear, as shown in Fig. 6.45. By fitting this ring with the proper small clearance so that relative motion occurs between it and the gear, it is possible to obtain enough energy dissipation to damp the high-frequency, low-energy vibration that causes the gear to ring. The rubbery coatings applied to large, thin-metal panels such as automobile doors to give them a solid rather than a "tinny" sound depend for their effectiveness on a proper balance of mass, elasticity, and damping (see Chap. 37).

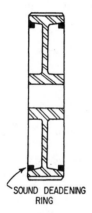

SOUND DEADENING
RING

FIGURE 6.45 Application of auxiliary mass damper to deaden noise in gear.

Another application where auxiliary mass dampers are useful is in the prevention of fatigue failures in turbines. At the high-pressure end of an impulse turbine, steam or hot gas is admitted through only a few nozzles. Consequently, as the blade passes the nozzle it is given an impulse by the steam and set into vibration at its natural frequency. It is a characteristic of alloy steels that they have very little internal damping at high operating temperature. For this reason the free vibration persists with only a slightly diminished amplitude until the blade again is subjected to the steam impulse. Some of these second impulses will be out of phase with the motion of the blade and will reduce its amplitude; however, successive impulses may increase the amplitude on subsequent passes until failure occurs. Damping can be increased by placing a number of loose wires in a cylindrical hole cut in the blade in a radial direction. The damping of a number of these wires has been computed in terms of the geometry of the application (number of wires, density of wires, size of the hole, radius of the blade, rotational speed, etc.) and the amplitude of vibration.[24] These computations show reasonable agreement with experimental results.

An auxiliary mass has been used to damp the cutting tool chatter set up in a boring bar.[25] Because of the characteristics of the metal-cutting process or of some coupling between motions of the tool parallel and perpendicular to the work face, it is sometimes found that a self-excited vibration is initiated at the natural frequency of the cutter system. Since the self-excitation energy is low, the vibration usually is initiated only if the damping is small. Chatter of the tool is most common in long, poorly supported tools, such as boring bars (see Chap. 40). To eliminate this chatter, a loose auxiliary mass is incorporated in the boring bar, as shown in Fig. 6.46. This may be air-damped or fluid-damped. Since the excitation is at the natural frequency of the tool, the damping should be such that the tool vibrates with a minimum ampli-

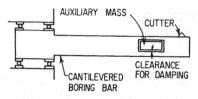

AUXILIARY MASS
CUTTER
CLEARANCE
FOR DAMPING
CANTILEVERED
BORING BAR

FIGURE 6.46 Application of auxiliary mass damper to reduce chatter in boring bar.

tude at this frequency. The damping requirement can be estimated by substituting $\beta = 1$ in Eq. (6.25),

$$\frac{x_0}{\delta_{st}} = \sqrt{\frac{1 + 4(\zeta\alpha)^2}{4(\zeta\alpha)^2\mu^2}} \qquad (6.77)$$

The optimum value of the parameter $(\zeta\alpha)$ is infinity. Thus when the frequency of excitation is constant, a greater reduction in amplitude can be obtained by a shift in natural frequency than by damping. However, such a shift cannot be attained because the frequency of the excitation always coincides with the natural frequency of the complete system. Instead, a better technique is to determine the damping that gives the maximum decrement of the free vibration.

Let the boring bar and damper be represented by a single degree-of-freedom system with a damper mass coupled to the main mass by viscous damping, as shown in Fig. 6.47A. The forces acting on the masses are shown in Fig. 6.47B. The equations of motion are

$$-kx_1 - c\dot{x}_1 + c\dot{x}_2 = m_1\ddot{x}_1$$
$$c\dot{x}_1 - c\dot{x}_2 = m_2\ddot{x}_2 \qquad (6.78)$$

Substituting $x = Ae^{st}$, the resulting frequency equation is

$$s^3 + \frac{c(m_1 + m_2)}{m_1 m_2}s^2 + \frac{k}{m_1}s + \frac{kc}{m_1 m_2} = 0 \qquad (6.79)$$

k

m_1

c

m_2

$\bar{Y}x_1$ (DISPLACEMENT OF BORING BAR)

$\bar{Y}x_2$ (DISPLACEMENT OF AUXILIARY MASS)

(A)

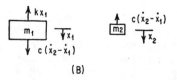

kx_1

m_1 $\bar{Y}x_1$

$c(\dot{x}_2-\dot{x}_1)$

$c(\dot{x}_2-\dot{x}_1)$

m_2 $\bar{Y}x_2$

(B)

FIGURE 6.47 Schematic diagram of damper shown in Fig. 6.46. The arrangement is shown at (A), and the forces acting on the boring bar and auxiliary mass are shown at (B).

Where chatter occurs, this equation has three roots, one real and two complex. The complex roots correspond to decaying free vibrations. Let the roots be as follows:

$$\alpha_1, \alpha_2 + j\beta, \alpha_2 - j\beta$$

The value of β determines the frequency of the free vibration, and the value of α_2 determines the decrement (rate of decrease of amplitude) of the free vibration. The decrement α_2 is of primary interest; it is most easily found from the conditions that when the coefficient of s^3 is unity, (1) the sum of the roots is equal to the negative of the coefficient of s^2, (2) the sum of the products of the roots taken two at a time is the negative of the coefficient of s, and (3) the product of the roots is the negative of the constant term. The equations thus obtained are

$$\alpha_1 + 2\alpha_2 = -\frac{c(1 + \mu)}{\mu m_1} \qquad (6.80)$$

$$2\alpha_1\alpha_2 + \alpha_2{}^2 + \beta^2 = -\omega_n{}^2 \tag{6.81}$$

$$\alpha_1(\alpha_2{}^2 + \beta^2) = -\omega_n{}^2 \frac{c}{m_1\mu} \tag{6.82}$$

where $\omega_n{}^2 = k/m_1$ and $\mu = m_2/m_1$. It is not practical to find the optimum damping by solving these equations for α_2 and then setting the derivative of α_2 with respect to c equal to zero. However, it is possible to find the optimum damping by the following process. Eliminate $(\alpha_2{}^2 + \beta^2)$ between Eqs. (6.81) and (6.82) to obtain

$$2\alpha_1{}^2\alpha_2 = \omega_n{}^2\left(\frac{c}{\mu m_1} - \alpha_1\right) \tag{6.83}$$

Substituting the value of α_1 from Eq. (6.80) in Eq. (6.83),

$$2\alpha_2\left[2\alpha_2 + \frac{c(1+\mu)}{\mu m_1}\right]^2 = \frac{c\omega_n{}^2}{\mu m_1} + \omega_n{}^2\left[2\alpha_2 + \frac{c(1+\mu)}{\mu m_1}\right] \tag{6.84}$$

To find the damping that gives the maximum decrement, differentiate with respect to c and set $d\alpha_2/dc = 0$:

$$2\alpha_2\left[2\alpha_2 + \frac{c(1+\mu)}{\mu m_1}\right] = \tfrac{1}{2}\omega_n{}^2\frac{2+\mu}{1+\mu} \tag{6.85}$$

Solving Eqs. (6.84) and (6.85) simultaneously,

$$c_{\text{opt}} = \frac{\mu^2 m_1\omega_n}{2(1+\mu)^{3/2}} \tag{6.86}$$

$$(\alpha_2)_{\text{opt}} = -\frac{(2+\mu)\omega_n}{4(1+\mu)^{1/2}} \tag{6.87}$$

These values may be obtained by proper choice of clearance between the auxiliary mass and the hole in which it is located. Air damping is preferable to oil because it requires less clearance. Therefore the plug is not immobilized by the centrifugal forces that, with the rotating boring bar, become larger as the clearance is increased.

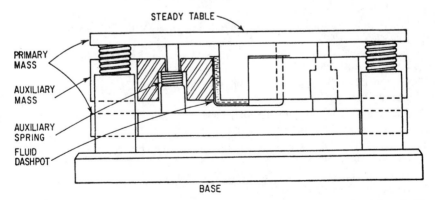

FIGURE 6.48 Application of auxiliary mass to spring-mounted table to reduce vibration of table. (*Macinante.*[26])

In precision measurements, it is necessary to isolate the instruments from effects of shock and vibration in the earth and to damp any oscillations that might be generated in the measuring instruments. A heavy spring-mounted table fitted with a heavy auxiliary mass that is attached to the table by a spring and submerged in an oil bath (Fig. 6.48) has proved to be effective.[26] In this example the table has a top surface of 13½ in. (34 cm) by 13½ in. (34 cm) and a height of 6 in. (15 cm). Each auxiliary mass weighs about 70 lb (32 kg). The springs for both the primary table and the auxiliary system are designed to give a natural frequency between 2 and 4 Hz in both the horizontal and vertical directions. By trying different fluids in the bath, suitable damping may be obtained experimentally.

REFERENCES

1. Timoshenko, S.: "Vibration Problems in Engineering," p. 240, D. Van Nostrand Company, Inc., Princeton, N.J., 1937.
2. Den Hartog, J. P.: "Mechanical Vibrations," chap. III, McGraw-Hill Book Company, Inc., New York, 1956.
3. Ormondroyd, J., and J. P. Den Hartog: *Trans. ASME,* **50**:A9 (1928).
4. Brock, J. E.: *J. Appl. Mechanics,* **13**(4):A-284 (1946).
5. Brock, J. E.: *J. Appl. Mechanics,* **16**(1):86 (1949).
6. Saver, F. M., and C. F. Garland: *J. Appl. Mechanics,* **16**(2):109 (1949).
7. Lewis, F. M.: *J. Appl. Mechanics,* **22**(3):377 (1955).
8. Georgian, J. C.: *Trans. ASME,* **16**:389 (1949).
9. Den Hartog, J. P., and J. Ormondroyd: *Trans. ASME,* **52**:133 (1930).
10. Roberson, R. E.: *J. Franklin Inst.,* **254**:205 (1952).
11. Pipes, L. A.: *J. Appl. Mechanics,* **20**:515 (1953).
12. Arnold, F. R.: *J. Appl. Mechanics,* **22**:487 (1955).
13. Hort, W.: "Technische Schwingungslehre," 2d ed., Springer-Verlag, Berlin, 1922.
14. Sperry, E. E.: *Trans. SNAME,* **30**:201 (1912).
15. Solomon, B.: *Proc. 4th Intern. Congr. Appl. Mechanics, Cambridge, England,* 1934.
16. Taylor, E. S.: *Trans. SAE,* **44**:81 (1936).
17. Den Hartog, J. P.: "Stephen Timoshenko 60th Anniversary Volume," The Macmillan Company, New York, 1939.
18. Porter, F. P.: "Evaluation of Effects of Torsional Vibration," SAE War Engineering Board, SAE, New York, 1945, p. 269.
19. Crossley, F. R. E.: *J. Appl. Mechanics,* **20**(1):41 (1953).
20. Reed, F. E.: *J. Appl. Mechanics,* **16**:190 (1949).
21. Constanti, M.: *Trans. Inst. of Naval Arch.,* **80**:181 (1938).
22. Crede, C. E.: *Trans. ASME,* **69**:937 (1947).
23. Brown, G. S., and D. P. Campbell: "Principles of Servomechanisms," John Wiley & Sons, Inc., New York, 1948.
24. DiTaranto, R. A.: *J. Appl. Mechanics,* **25**(1):21 (1958)
25. Hahn, R. S.: *Trans. ASME,* **73**:331 (1951).
26. Macinante, J. A.: *J. Sci. Instr.,* **35**:224 (1958)

CHAPTER 7

VIBRATION OF SYSTEMS HAVING DISTRIBUTED MASS AND ELASTICITY

William F. Stokey

INTRODUCTION

Preceding chapters consider the vibration of lumped parameter systems; i.e., systems that are idealized as rigid masses joined by massless springs and dampers. Many engineering problems are solved by analyses based on ideal models of an actual system, giving answers that are useful though approximate. In general, more accurate results are obtained by increasing the number of masses, springs, and dampers; i.e., by increasing the number of degrees-of-freedom. As the number of degrees-of-freedom is increased without limit, the concept of the system with distributed mass and elasticity is formed. This chapter discusses the free and forced vibration of such systems. Types of systems include rods vibrating in torsional modes and in tension-compression modes, and beams and plates vibrating in flexural modes. Particular attention is given to the calculation of the natural frequencies of such systems for further use in other analyses. Numerous charts and tables are included to define in readily available form the natural frequencies of systems commonly encountered in engineering practice.

FREE VIBRATION

Degrees-of-Freedom. Systems for which the mass and elastic parts are lumped are characterized by a finite number of degrees-of-freedom. In physical systems, all elastic members have mass, and all masses have some elasticity; thus, all real systems have distributed parameters. In making an analysis, it is often assumed that real systems have their parameters lumped. For example, in the analysis of a system consisting of a mass and a spring, it is commonly assumed that the mass of the spring is negligible so that its only effect is to exert a force between the mass and the support to which the spring is attached, and that the mass is perfectly rigid so that it does not deform and exert any elastic force. The effect of the mass of the spring on the motion of the system may be considered in an approximate way, while still maintaining the assumption of one degree-of-freedom, by assuming that the spring moves so that the deflection of each of its elements can be described by a single parameter. A commonly used assumption is that the deflection of each section of the spring is proportional to its distance from the support, so that if the deflection of the mass is given, the deflection of any part of the spring is defined. For the exact solution of the problem, even though the mass is considered to be perfectly rigid, it is necessary to consider that the deformation of the spring can occur in any manner consistent with the requirements of physical continuity.

Systems with distributed parameters are characterized by having an infinite number of degrees-of-freedom. For example, if an initially straight beam deflects laterally, it may be necessary to give the deflection of each section along the beam in order to define completely the configuration. For vibrating systems, the coordinates usually are defined in such a way that the deflections of the various parts of the system from the equilibrium position are given.

Natural Frequencies and Normal Modes of Vibration. The number of natural frequencies of vibration of any system is equal to the number of degrees-of-freedom; thus, any system having distributed parameters has an infinite number of natural frequencies. At a given time, such a system usually vibrates with appreciable amplitude at only a limited number of frequencies, often at only one. With each natural frequency is associated a shape, called the normal or natural mode, which is assumed by the system during free vibration at the frequency. For example, when a uniform beam with simply supported or hinged ends vibrates laterally at its lowest or fundamental natural frequency, it assumes the shape of a half sine wave; this is a normal mode of vibration. When vibrating in this manner, the beam behaves as a system with a single degree-of-freedom, since its configuration at any time can be defined by giving the deflection of the center of the beam. When any linear system, i.e., one in which the elastic restoring force is proportional to the deflection, executes free vibration in a single natural mode, each element of the system except those at the supports and nodes executes simple harmonic motion about its equilibrium position. All possible free vibration of any linear system is made up of superposed vibrations in the normal modes at the corresponding natural frequencies. The total motion at any point of the system is the sum of the motions resulting from the vibration in the respective modes.

There are always nodal points, lines, or surfaces, i.e., points which do not move, in each of the normal modes of vibration of any system. For the fundamental mode, which corresponds to the lowest natural frequency, the supported or fixed points of the system usually are the only nodal points; for other modes, there are additional nodes. In the modes of vibration corresponding to the higher natural frequencies of some systems, the nodes often assume complicated patterns. In certain problems involving forced vibrations, it may be necessary to know what the nodal patterns are, since a particular mode usually will not be excited by a force acting at a nodal point. Nodal lines are shown in some of the tables.

Methods of Solution. The complete solution of the problem of free vibration of any system would require the determination of all the natural frequencies and of the mode shape associated with each. In practice, it often is necessary to know only a few of the natural frequencies, and sometimes only one. Usually the lowest frequencies are the most important. The exact mode shape is of secondary importance in many problems. This is fortunate, since some procedures for finding natural frequencies involve assuming a mode shape from which an approximation to the natural frequency can be found.

Classical Method. The fundamental method of solving any vibration problem is to set up one or more equations of motion by the application of Newton's second law of motion. For a system having a finite number of degrees-of-freedom, this procedure gives one or more ordinary differential equations. For systems having distributed parameters partial differential equations are obtained. Exact solutions of the equations are possible for only a relatively few configurations. For most problems other means of solution must be employed.

Rayleigh's and Ritz's Methods. For many elastic bodies, Rayleigh's method is useful in finding an approximation to the fundamental natural frequency. While it is possible to use the method to estimate some of the higher natural frequencies, the accuracy often is poor; thus, the method is most useful for finding the fundamental frequency. When any elastic system without damping vibrates in its fundamental normal mode, each part of the system executes simple harmonic motion about its equilibrium position. For example, in lateral vibration of a beam the motion can be expressed as $y = X(x) \sin \omega_n t$ where X is a function only of the distance along the length of the beam. For lateral vibration of a plate, the motion can be expressed as $w = W(x,y) \sin \omega_n t$ where x and y are the coordinates in the plane of the plate. The equations show that when the deflection from equilibrium is a maximum, all parts of the body are motionless. At that time all the energy associated with the vibration is in the form of elastic strain energy. When the body is passing through its equilibrium position, none of the vibrational energy is in the form of strain energy so that all of it is in the form of kinetic energy. For conservation of energy, the strain energy in the position of maximum deflection must equal the kinetic energy when passing through the equilibrium position. Rayleigh's method of finding the natural frequency is to compute these maximum energies, equate them, and solve for the frequency. When the kinetic-energy term is evaluated, the frequency always appears as a factor. Formulas for finding the strain and kinetic energies of rods, beams, and plates are given in Table 7.1.

If the deflection of the body during vibration is known exactly, Rayleigh's method gives the true natural frequency. Usually the exact deflection is not known, since its determination involves the solution of the vibration problem by the classical method. If the classical solution is available, the natural frequency is included in it, and nothing is gained by applying Rayleigh's method. In many problems for which the classical solution is not available, a good approximation to the deflection can be assumed on the basis of physical reasoning. If the strain and kinetic energies are computed using such an assumed shape, an approximate value for the natural frequency is found. The correctness of the approximate frequency depends on how well the assumed shape approximates the true shape.

In selecting a function to represent the shape of a beam or a plate, it is desirable to satisfy as many of the boundary conditions as possible. For a beam or plate supported at a boundary, the assumed function must be zero at that boundary; if the boundary is built in, the first derivative of the function must be zero. For a free boundary, if the conditions associated with bending moment and shear can be satisfied, better accuracy usually results. It can be shown[2] that the frequency that is found by using any shape except the correct shape always is higher than the actual frequency. Therefore, if more than one calculation is made, using different assumed shapes, the lowest computed frequency is closest to the actual frequency of the system.

In many problems for which a classical solution would be possible, the work involved is excessive. Often a satisfactory answer to such a problem can be obtained

TABLE 7.1 Strain and Kinetic Energies of Uniform Rods, Beams, and Plates

Member	Strain energy V	Kinetic energy T	
		General	Maximum*
Rod in tension or compression	$\dfrac{SE}{2}\displaystyle\int_0^l\left(\dfrac{\partial u}{\partial x}\right)^2 dx$	$\dfrac{S\gamma}{2g}\displaystyle\int_0^l\left(\dfrac{\partial u}{\partial t}\right)^2 dx$	$\dfrac{S\gamma\omega_n^2}{2g}\displaystyle\int_0^l V^2\,dx$
Rod in torsion	$\dfrac{GI_p}{2}\displaystyle\int_0^l\left(\dfrac{\partial \phi}{\partial x}\right)^2 dx$	$\dfrac{I_p\gamma}{2g}\displaystyle\int_0^l\left(\dfrac{\partial \phi}{\partial t}\right)^2 dx$	$\dfrac{I_p\gamma\omega_n^2}{2g}\displaystyle\int_0^l \Phi^2\,dx$
Beam in bending	$\dfrac{EI}{2}\displaystyle\int_0^l\left(\dfrac{\partial^2 y}{\partial x^2}\right)^2 dx$	$\dfrac{S\gamma}{2g}\displaystyle\int_0^l\left(\dfrac{\partial y}{\partial t}\right)^2 dx$	$\dfrac{S\gamma\omega_n^2}{2g}\displaystyle\int_0^l Y^2\,dx$
Rectangular plate in bending[1]	$\dfrac{D}{2}\displaystyle\int_S\!\int\left\{\left(\dfrac{d^2 w}{dx^2}+\dfrac{d^2 w}{dy^2}\right)^2\right.$ $\left.-2(1-\mu)\left[\dfrac{\partial^2 w}{\partial x^2}\dfrac{\partial^2 w}{\partial y^2}\right.\right.$ $\left.\left.-\left(\dfrac{\partial^2 w}{\partial x\,\partial y}\right)^2\right]\right\}dx\,dy$	$\dfrac{\gamma h}{2g}\displaystyle\int_S\!\int\left(\dfrac{\partial w}{\partial t}\right)^2 dx\,dy$	$\dfrac{\gamma h\omega_n^2}{2g}\displaystyle\int_S\!\int W^2\,dx\,dy$
Circular plate (deflection symmetrical about center)[1]	$\pi D\displaystyle\int_0^a\left\{\left(\dfrac{\partial^2 w}{\partial r^2}+\dfrac{1}{r}\dfrac{\partial w}{\partial r}\right)^2\right.$ $\left.-2(1-\mu)\dfrac{\partial^2 w}{\partial r^2}\dfrac{1}{r}\dfrac{\partial w}{\partial r}\right\}r\,dr$	$\dfrac{\pi\gamma h}{g}\displaystyle\int_0^a\left(\dfrac{\partial w}{\partial t}\right)^2 r\,dr$	$\dfrac{\pi\gamma h\omega_n^2}{g}\displaystyle\int_0^a W^2 r\,dr$

u = longitudinal deflection of cross section of rod
ϕ = angle of twist of cross section of rod
y = lateral deflection of beam
w = lateral deflection of plate
    Capitals denote values at extreme deflection for simple harmonic motion.
l = length of rod or beam
a = radius of circular plate
h = thickness of beam or plate

S = area of cross section
I_p = polar moment of inertia
I = moment of inertia of beam
γ = weight density
E = modulus of elasticity
G = modulus of rigidity
μ = Poisson's ratio
$D = Eh^3/12(1-\mu^2)$

  * This is the maximum kinetic energy in simple harmonic motion.

by the application of Rayleigh's method. In this chapter several examples are worked using both the classical method and Rayleigh's method. In all, Rayleigh's method gives a good approximation to the correct result with relatively little work. Many other examples of solutions to problems by Rayleigh's method are in the literature.[3–5]

Ritz's method is a refinement of Rayleigh's method. A better approximation of the fundamental natural frequency can be obtained by its use, and approximations of higher natural frequencies can be found. In using Ritz's method, the deflections which are assumed in computing the energies are expressed as functions with one or more undetermined parameters; these parameters are adjusted to make the computed frequency a minimum. Ritz's method has been used extensively for the determination of the natural frequencies of plates of various shapes and is discussed in the section on the lateral vibrations of plates.

Lumped Parameters. A procedure that is useful in many problems for finding approximations to both the natural frequencies and the mode shapes is to reduce the system with distributed parameters to one having a finite number of degrees-of-freedom. This is done by lumping the parameters for each small region into an equivalent mass and elastic element. Several formalized procedures for doing this and for analyzing the resulting systems are described in Chap. 28. If a system consists of a rigid mass supported by a single flexible member whose mass is not negligible, the elastic part of the system sometimes can be treated as an equivalent spring; i.e., some of its mass is lumped with the rigid mass. Formulas for several systems of this kind are given in Table 7.2.

TABLE 7.2 Approximate Formulas for Natural Frequencies of Systems Having Both Concentrated and Distributed Mass

TYPE OF SYSTEM	NATURAL FREQUENCY $\omega_n = 2\pi f_n$	STIFFNESS
SPRING WITH MASS ATTACHED	$\sqrt{\dfrac{k}{M + m/3}}$	$k = \dfrac{Gd^4}{8nD^3}$ D = COIL DIA d = WIRE DIA n = NUMBER OF TURNS
CIRCULAR ROD, WITH DISC ATTACHED, IN TORSION	$\sqrt{\dfrac{k_r}{I + I_s/3}}$	$k_r = \dfrac{G\pi D^4}{32\,l}$ D = ROD DIAMETER l = ROD LENGTH
UNIFORM SIMPLY SUPPORTED BEAM WITH MASS IN CENTER	$\sqrt{\dfrac{k}{M + m/2}}$	$k = \dfrac{48EI}{l^3}$ l = BEAM LENGTH I = MOMENT OF INERTIA
UNIFORM CANTILEVER BEAM WITH MASS ON END	$\sqrt{\dfrac{k}{M + 0.23m}}$	$k = \dfrac{3EI}{l^3}$ l = BEAM LENGTH I = MOMENT OF INERTIA

Orthogonality. It is shown in Chap. 2 that the normal modes of vibration of a system having a finite number of degrees-of-freedom are orthogonal to each other. For a system of masses and springs having n degrees-of-freedom, if the coordinate system is selected in such a way that X_1 represents the amplitude of motion of the first mass, X_2 that of the second mass, etc., the orthogonality relations are expressed by $(n-1)$ equations as follows:

$$m_1 X_1^a X_1^b + m_2 X_2^a X_2^b + \cdots = \sum_{i=1}^{n} m_i X_i^a X_i^b = 0 \qquad [a \neq b]$$

where X_1^a represents the amplitude of the first mass when vibrating only in the ath mode, X_1^b the amplitude of the first mass when vibrating only in the bth mode, etc.

For a body such as a uniform beam whose parameters are distributed only lengthwise, i.e., in the X direction, the orthogonality between two normal modes is expressed by

$$\int_0^l \rho \phi_a(x)\phi_b(x)\, dx = 0 \qquad [a \neq b] \tag{7.1}$$

where $\phi_a(x)$ represents the deflection in the ath normal mode, $\phi_b(x)$ the deflection in the bth normal mode, and ρ the density.

For a system, such as a uniform plate, in which the parameters are distributed in two dimensions, the orthogonality condition is

$$\int_A \int \rho \phi_a(x,y)\phi_b(x,y)\, dx\, dy = 0 \qquad [a \neq b] \tag{7.2}$$

LONGITUDINAL AND TORSIONAL VIBRATIONS OF UNIFORM CIRCULAR RODS

Equations of Motion.   A circular rod having a uniform cross section can execute longitudinal, torsional, or lateral vibrations, either individually or in any combination. The equations of motion for longitudinal and torsional vibrations are similar in form, and the solutions are discussed together. The lateral vibration of a beam having a uniform cross section is considered separately.

In analyzing the longitudinal vibration of a rod, only the motion of the rod in the longitudinal direction is considered. There is some lateral motion because longitudinal stresses induce lateral strains; however, if the rod is fairly long compared to its diameter, this motion has a minor effect.

Consider a uniform circular rod, Fig. 7.1A. The element of length dx, which is formed by passing two parallel planes A–A and B–B normal to the axis of the rod, is shown in Fig. 7.1B. When the rod executes only longitudinal vibration, the force acting on the face A–A is F, and that on face B–B is $F + (\partial F/\partial x)\, dx$. The net force acting to the right must equal the product of the mass of the element $(\gamma/g)S\, dx$ and its acceleration $\partial^2 u/\partial t^2$, where γ is the weight density, S the area of the cross section, and u the longitudinal displacement of the element during the vibration:

$$\left(F + \frac{\partial F}{\partial x}\, dx\right) - F = \frac{\partial F}{\partial x}\, dx = \left(\frac{\gamma}{g}\right) S\, dx\, \frac{\partial^2 u}{\partial t^2} \qquad \text{or} \qquad \frac{\partial F}{\partial x} = \frac{\gamma S}{g}\, \frac{\partial^2 u}{\partial t^2} \tag{7.3}$$

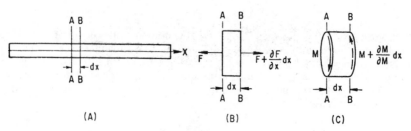

FIGURE 7.1   (A) Rod executing longitudinal or torsional vibration. (B) Forces acting on element during longitudinal vibration. (C) Moments acting on element during torsional vibration.

This equation is solved by expressing the force F in terms of the displacement. The elastic strain at any section is $\partial u/\partial x$, and the stress is $E\,\partial u/\partial x$. The force F is the product of the stress and the area, or $F = ES\,\partial u/\partial x$, and $\partial F/\partial x = ES\,\partial^2 u/\partial x^2$. Equation (7.3) becomes $Eu'' = \gamma/g\ddot{u}$, where $u'' = \partial^2 u/\partial x^2$ and $\ddot{u} = \partial^2 u/\partial t^2$. Substituting $a^2 = Eg/\gamma$,

$$a^2 u'' = \ddot{u} \tag{7.4}$$

The equation governing the torsional vibration of the circular rod is derived by equating the net torque acting on the element, Fig. 7.1C, to the product of the moment of inertia J and the angular acceleration $\ddot{\phi}$, ϕ being the angular displacement of the section. The torque on the section A–A is M and that on section B–B is $M + (\partial M/\partial x)\,dx$. By an analysis similar to that for the longitudinal vibration, letting $b^2 = Gg/\gamma$,

$$b^2 \phi'' = \ddot{\phi} \tag{7.5}$$

Solution of Equations of Motion.   Since Eqs. (7.4) and (7.5) are of the same form, the solutions are the same except for the meaning of a and b. The solution of Eq. (7.5) is of the form $\phi = X(x)T(t)$ in which X is a function of x only and T is a function of t only. Substituting this in Eq. (7.5) gives $b^2 X''T = X\ddot{T}$. By separating the variables,[6]

$$T = A\,\cos\,(\omega_n t + \theta)$$

$$X = C\,\sin\,\frac{\omega_n x}{b} + D\,\cos\,\frac{\omega_n x}{b} \tag{7.6}$$

The natural frequency ω_n can have infinitely many values, so that the complete solution of Eq. (7.5) is, combining the constants,

$$\phi = \sum_{n=1}^{n=\infty} \left(C_n\,\sin\,\frac{\omega_n x}{b} + D_n\,\cos\,\frac{\omega_n x}{b} \right) \cos\,(\omega_n t + \theta_n) \tag{7.7}$$

The constants C_n and D_n are determined by the end conditions of the rod and by the initial conditions of the vibration. For a built-in or clamped end of a rod in torsion, $\phi = 0$ and $X = 0$ because the angular deflection must be zero. The torque at any section of the shaft is given by $M = (GI_p)\phi'$, where GI_p is the torsional rigidity of the shaft; thus, for a free end, $\phi' = 0$ and $X' = 0$. For the longitudinal vibration of a rod, the boundary conditions are essentially the same; i.e., for a built-in end the displacement is zero $(u = 0)$ and for a free end the stress is zero $(u' = 0)$.

EXAMPLE 7.1.   The natural frequencies of the torsional vibration of a circular steel rod of 2-in. diameter and 24-in. length, having the left end built in and the right end free, are to be determined.

SOLUTION.   The built-in end at the left gives the condition $X = 0$ at $x = 0$ so that $D = 0$ in Eq. (7.6). The free end at the right gives the condition $X' = 0$ at $x = l$. For each mode of vibration, Eq. (7.6) is $\cos\,\omega_n l/b = 0$ from which $\omega_n l/b = \pi/2, 3\pi/2, 5\pi/2,$ .... Since $b^2 = Gg/\gamma$, the natural frequencies for the torsional vibration are

$$\omega_n = \frac{\pi}{2l}\,\sqrt{\frac{Gg}{\gamma}},\ \frac{3\pi}{2l}\,\sqrt{\frac{Gg}{\gamma}},\ \frac{5\pi}{2l}\,\sqrt{\frac{Gg}{\gamma}},\ldots \qquad \text{rad/sec}$$

For steel, $G = 11.5 \times 10^6$ lb/in.^{2} and $\gamma = 0.28$ lb/in.^{3} The fundamental natural frequency is

$$\omega_n = \frac{\pi}{2(24)}\,\sqrt{\frac{(11.5 \times 10^6)(386)}{0.28}} = 8240\ \text{rad/sec} = 1311\ \text{Hz}$$

The remaining frequencies are 3, 5, 7, etc., times ω_n.

Since Eq. (7.4), which governs longitudinal vibration of the bar, is of the same form as Eq. (7.5), which governs torsional vibration, the solution for longitudinal vibration is the same as Eq. (7.7) with u substituted for ϕ and $a = \sqrt{Eg/\gamma}$ substituted for b. The natural frequencies of a uniform rod having one end built in and one end free are obtained by substituting a for b in the frequency equations found above in Example 7.1:

$$\omega_n = \frac{\pi}{2l}\sqrt{\frac{Eg}{\gamma}}, \frac{3\pi}{2l}\sqrt{\frac{Eg}{\gamma}}, \frac{5\pi}{2l}\sqrt{\frac{Eg}{\gamma}}, \ldots$$

The frequencies of the longitudinal vibration are independent of the lateral dimensions of the bar, so that these results apply to uniform noncircular bars. Equation (7.5) for torsional vibration is valid only for circular cross sections.

Torsional Vibrations of Circular Rods with Discs Attached. An important type of system is that in which a rod which may twist has mounted on it one or more rigid discs or members that can be considered as the equivalents of discs. Many systems can be approximated by such configurations. If the moment of inertia of the rod is small compared to the moments of inertia of the discs, the mass of the rod may be neglected and the system considered to have a finite number of degrees-of-freedom. Then the methods described in Chaps. 2 and 38 are applicable. Even if the moment of inertia of the rod is not negligible, it usually may be lumped with the moment of inertia of the disc. For a shaft having a single disc attached, the formula in Table 7.2 gives a close approximation to the true frequency.

The exact solution of the problem requires that the effect of the distributed mass of the rod be considered. Usually it can be assumed that the discs are rigid enough that their elasticity can be neglected; only such systems are considered. Equation (7.5) and its solution, Eq. (7.7), apply to the shaft where the constants are determined by the end conditions. If there are more than two discs, the section of shaft between each pair of discs must be considered separately; there are two constants for each section. The constants are determined from the following conditions:

1. For a disc at an end of the shaft, the torque of the shaft at the disc is equal to the product of the moment of inertia of the disc and its angular acceleration.

2. Where a disc is between two sections of shaft, the angular deflection at the end of each section adjoining the disc is the same; the difference between the torques in the two sections is equal to the product of the moment of inertia of the disc and its angular acceleration.

EXAMPLE 7.2. The fundamental frequency of vibration of the system shown in Fig. 7.2 is to be calculated and the result compared with the frequency obtained by considering that each half of the system is a simple shaft-disc system with the end of the shaft fixed. The system consists of a steel shaft 24 in. long and 4 in. in diameter having attached to it at each end a rigid steel disc 12 in. in diameter and 2 in. thick. For the approximation, add one-third of the moment of

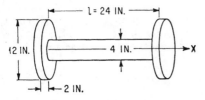

FIGURE 7.2 Rod with disc attached at each end.

inertia of half the shaft to that of the disc (Table 7.2). (Because of symmetry, the center of the shaft is a nodal point; i.e., it does not move. Thus, each half of the system can be considered as a rod-disc system.)

EXACT SOLUTION.   The boundary conditions are: at $x = 0$, $M = GI_p\phi' = I_1\ddot{\phi}$; at $x = l$, $M = GI\phi' = -I_2\ddot{\phi}$, where I_1 and I_2 are the moments of inertia of the discs. The signs are opposite for the two boundary conditions because, if the shaft is twisted in a certain direction, it will tend to accelerate the disc at the left end in one direction and the disc at the right end in the other. In the present example, $I_1 = I_2$; however, the solution is carried out in general terms.

Using Eq. (7.7), the following is obtained for each value of n:

$$\phi' = \frac{\omega_n}{b}\left(C\cos\frac{\omega_n x}{b} - D\sin\frac{\omega_n x}{b}\right)\cos(\omega_n t + \theta)$$

$$\ddot{\phi} = \omega_n^2\left(C\sin\frac{\omega_n x}{b} + D\cos\frac{\omega_n x}{b}\right)[-\cos(\omega_n t + \theta)]$$

The boundary conditions give the following:

$$GI_p\frac{\omega_n}{b}C = -\omega_n^2 DI_1 \quad\text{or}\quad C = -\frac{b\omega_n I_1}{GI_p}D$$

$$\frac{\omega_n}{b}GI_p\left(C\cos\frac{\omega_n l}{b} - D\sin\frac{\omega_n l}{b}\right) = \omega_n^2 I_2\left(C\sin\frac{\omega_n l}{b} + D\cos\frac{\omega_n l}{b}\right)$$

These two equations can be combined to give

$$-\frac{\omega_n}{b}GI_p\left(\frac{b\omega_n I_1}{GI_p}\cos\frac{\omega_n l}{b} + \sin\frac{\omega_n l}{b}\right) = \omega_n^2 I_2\left(-\frac{b\omega_n I_1}{GI_p}\sin\frac{\omega_n l}{b} + \cos\frac{\omega_n l}{b}\right)$$

The preceding equation can be reduced to

$$\tan\alpha_n = \frac{(c+d)\alpha_n}{cd\alpha_n^2 - 1} \tag{7.8}$$

where $\alpha_n = (\omega_n l)/b$, $c = I_1/I_s$, $d = I_2/I_s$, and I_s is the polar moment of inertia of the shaft as a rigid body. There is a value for X in Eq. (7.6) corresponding to each root of Eq. (7.8) so that Eq. (7.7) becomes

$$\theta = \sum_{n=1}^{n=\infty} A_n\left(\cos\frac{\omega_n x}{b} - c\alpha_n\sin\frac{\omega_n x}{b}\right)\cos(\omega_n t + \theta_n)$$

For a circular disc or shaft, $I = \frac{1}{2}mr^2$ where m is the total mass; thus $c = d = (D^4/d^4)(h/l) = 6.75$. Equation (7.8) becomes $(45.56\alpha_n^2 - 1)\tan\alpha_n = 13.5\alpha_n$, the lowest root of which is $\alpha_n = 0.538$. The natural frequency is $\omega_n = 0.538\sqrt{Gg/\gamma l^2}$ rad/sec.

APPROXIMATE SOLUTION.   From Table 7.2, the approximate formula is

$$\omega_n = \left(\frac{k_r}{I + I_s/3}\right)^{1/2} \quad\text{where } k_r = \frac{\pi d^4}{32}\frac{G}{l}$$

For the present problem where the center of the shaft is a node, the values of moment of inertia I_s and torsional spring constant for half the shaft must be used:

$$\frac{1}{2}I_s = \frac{\pi d^4}{32}\frac{\gamma}{g}\frac{l}{2} \quad\text{and}\quad k_r = 2\left[\frac{\pi d^4}{32}\frac{G}{l}\right]$$

From the previous solution:

$$I_1 = 6.75I_s \qquad I_1 + \frac{1}{2}\left(\frac{I_s}{3}\right) = \frac{\pi d^4}{32}\frac{\gamma}{g}\frac{l}{2}\,[2(6.75) + 0.333]$$

Substituting these values into the frequency equation and simplifying gives

$$\omega_n = 0.538\sqrt{\frac{Gg}{\gamma l^2}}$$

In this example, the approximate solution is correct to at least three significant figures. For larger values of I_s/I, poorer accuracy can be expected.
    For steel, $G = 11.5 \times 10^6$ lb/in.^{2} and $\gamma = 0.28$ lb/in.^{3}; thus

$$\omega_n = 0.538\sqrt{\frac{(11.5 \times 10^6)(386)}{(0.28)(24)^2}} = 0.538 \times 5245 = 2822\ \text{rad/sec} = 449\ \text{Hz}$$

Longitudinal Vibration of a Rod with Mass Attached.   The natural frequencies of the longitudinal vibration of a uniform rod having rigid masses attached to it can be solved in a manner similar to that used for a rod in torsion with discs attached. Equation (7.4) applies to this system; its solution is the same as Eq. (7.7) with a substituted for b. For each value of n,

$$u = \left(C_n \sin\frac{\omega_n x}{a} + D_n \cos\frac{\omega_n x}{a}\right)\cos\,(\omega_n t + \theta)$$

In Fig. 7.3, the rod of length l is fixed at $x = 0$ and has a mass m_2 attached at $x = l$. The boundary conditions are: at $x = 0, u = 0$ and at $x = l$, $SEu' = -m_2\ddot{u}$. The latter expresses the condition that the force in the bar equals the product of the mass and its acceleration at the end with the mass attached. The sign is negative because the force is tensile or positive when the acceleration of the mass is negative. From the first boundary condition, $D_n = 0$. The second boundary condition gives

$$\frac{\omega_n SE}{a}\,C_n\cos\frac{\omega_n l}{a} = m_2\omega_n^2 C_n\sin\frac{\omega_n l}{a}$$

from which

FIGURE 7.3   Rod, with mass attached to end, executing longitudinal vibration.

$$\frac{SEl}{m_2 a^2} = \frac{\omega_n l}{a}\tan\frac{\omega_n l}{a}$$

Since $a^2 = Eg/\gamma$, this can be written

$$\frac{m_1}{m_2} = \frac{\omega_n l}{a}\tan\frac{\omega_n l}{a}$$

where m_1 is the mass of the rod. This equation can be applied to a simple mass-spring system by using the relation that the constant k of a spring is equivalent to SE/l for the rod, so that $l/a = (m_1/k)^{1/2}$, where m_1 is the mass of the spring:

$$\frac{m_1}{m_2} = \omega_n\sqrt{\frac{m_1}{k}}\tan\omega_n\sqrt{\frac{m_1}{k}} \qquad (7.9)$$

Rayleigh's Method. An accurate approximation to the fundamental natural frequency of this system can be found by using Rayleigh's method. The motion of the mass can be expressed as $u_m = u_0 \sin \omega t$. If it is assumed that the deflection u at each section of the rod is proportional to its distance from the fixed end, $u = u_0(x/l) \sin \omega_n t$. Using this relation in the appropriate equation from Table 7.1, the strain energy V of the rod at maximum deflection is

$$V = \frac{SE}{2} \int_0^l \left(\frac{\partial u}{\partial x} \right)^2 dx = \frac{SE}{2} \int_0^l \left(\frac{u_0}{l} \right)^2 dx = \frac{SEu_0^2}{2l}$$

The maximum kinetic energy T of the rod is

$$T = \frac{S\gamma}{2g} \int_0^l V_{max}^2 \, dx = \frac{S\gamma}{2g} \int_0^l \left(\omega_n u_0 \frac{x}{l} \right)^2 dx = \frac{S\gamma}{2g} \, \omega_n^2 u_0^2 \, \frac{l}{3}$$

The maximum kinetic energy of the mass is $T_m = m_2 \omega_n^2 u_0^2 / 2$. Equating the total maximum kinetic energy $T + T_m$ to the maximum strain energy V gives

$$\omega_n = \left(\frac{SE}{l(m_2 + m_1/3)} \right)^{1/2}$$

where $m_1 = S\gamma l/g$ is the mass of the rod. Letting $SE/l = k$,

$$\omega_n = \sqrt{\frac{k}{M + m/3}} \tag{7.10}$$

This formula is included in Table 7.2. The other formulas in that table are also based on analyses by the Rayleigh method.

EXAMPLE 7.3. The natural frequency of a simple mass-spring system for which the weight of the spring is equal to the weight of the mass is to be calculated and compared to the result obtained by using Eq. (7.10).

SOLUTION. For $m_1/m_2 = l$, the lowest root of Eq. (7.9) is $\omega_n \sqrt{m/k} = 0.860$. When $m_2 = m_1$,

$$\omega_n = 0.860 \sqrt{\frac{k}{m_2}}$$

Using the approximate equation,

$$\omega_n = \sqrt{\frac{k}{m_2(1 + \frac{1}{3})}} = 0.866 \sqrt{\frac{k}{m_2}}$$

LATERAL VIBRATION OF STRAIGHT BEAMS

Natural Frequencies from Nomograph. For many practical purposes the natural frequencies of uniform beams of steel, aluminum, and magnesium can be determined with sufficient accuracy by the use of the nomograph, Fig. 7.4. This nomograph applies to many conditions of support and several types of load. Figure 7.4A indicates the procedure for using the nomograph.

Classical Solution. In the derivation of the necessary equation, use is made of the relation

$$EI \frac{d^2y}{dx^2} = M \tag{7.11}$$

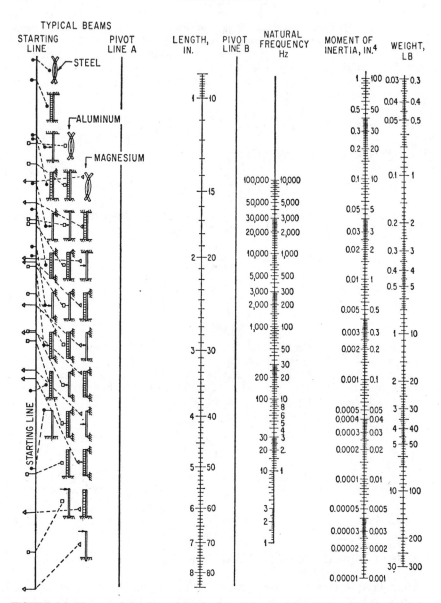

FIGURE 7.4 Nomograph for determining fundamental natural frequencies of beams. From the point on the starting line which corresponds to the loading and support conditions for the beam, a straight line is drawn to the proper point on the length line. (If the length appears on the left side of this line, subsequent readings on all lines are made to the left; and if the length appears to the right, subsequent readings are made to the right.) From the intersection of this line with pivot line A, a straight line is drawn to the moment of inertia line; from the intersection of this line with pivot line B, a straight line is drawn to the weight line. (For concentrated loads, the weight is that of the load; for uniformly distributed loads, the weight is the total load on the beam, including the weight of the beam.) The natural frequency is read where the last line crosses the natural frequency line. (*J. J. Kerley.*[7])

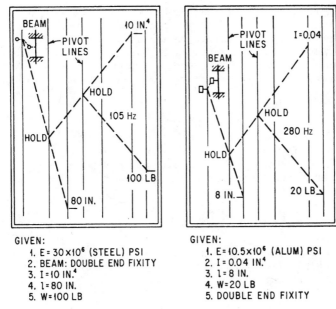

GIVEN:
1. E = 30×10⁶ (STEEL) PSI
2. BEAM: DOUBLE END FIXITY
3. I = 10 IN.⁴
4. l = 80 IN.
5. W = 100 LB

GIVEN:
1. E = 10.5×10⁶ (ALUM) PSI
2. I = 0.04 IN.⁴
3. l = 8 IN.
4. W = 20 LB
5. DOUBLE END FIXITY

FIGURE 7.4A Example of use of Fig. 7.4. The natural frequency of the steel beam is 105 Hz and that of the aluminum beam is 280 Hz. (*J. J. Kerley.*[7])

This equation relates the curvature of the beam to the bending moment at each section of the beam. This equation is based upon the assumptions that the material is homogeneous, isotropic, and obeys Hooke's law and that the beam is straight and of uniform cross section. The equation is valid for small deflections only and for beams that are long compared to cross-sectional dimensions since the effects of shear deflection are neglected. The effects of shear deflection and rotation of the cross sections are considered later.

The equation of motion for lateral vibration of the beam shown in Fig. 7.5A is found by considering the forces acting on the element, Fig. 7.5B, which is formed by passing two parallel planes A–A and B–B through the beam normal to the longitudinal axis. The vertical elastic shear force acting on section A–A is V, and that on section B–B is $V + (\partial V/\partial x)\,dx$. Shear forces acting as shown are considered to be positive. The total vertical elastic shear force at each section of the beam is composed of two parts: that caused by the static load including the weight of the beam

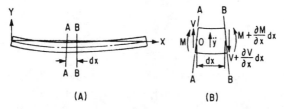

FIGURE 7.5 (A) Beam executing lateral vibration. (B) Element of beam showing shear forces and bending moments.

and that caused by the vibration. The part of the shear force caused by the static load exactly balances the load, so that these forces need not be considered in deriving the equation for the vibration if all deflections are measured from the position of equilibrium of the beam under the static load. The sum of the remaining vertical forces acting on the element must equal the product of the mass of the element $S\gamma/g\ dx$ and the acceleration $\partial^2 y/\partial t^2$ in the lateral direction: $V + (\partial V/\partial x)\ dx - V = (\partial V/\partial x)\ dx = -(S\gamma/g)(\partial^2 y/\partial t^2)\ dx$, or

$$\frac{\partial V}{\partial x} = -\frac{\gamma S}{g}\frac{\partial^2 y}{\partial t^2} \tag{7.12}$$

If moments are taken about point 0 of the element in Fig. 7.5B, $V\ dx = (\partial M/\partial x)\ dx$ and $V = \partial M/\partial x$. Other terms contain differentials of higher order and can be neglected. Substituting this in Eq. (7.12) gives $-\partial^2 M/\partial x^2 = (S\gamma/g)(\partial^2 y/\partial t^2)$. Substituting Eq. (7.11) gives

$$-\frac{\partial^2}{\partial x^2}\left(EI\frac{\partial^2 y}{\partial x^2}\right) = \frac{\gamma S}{g}\frac{\partial^2 y}{\partial t^2} \tag{7.13}$$

Equation (7.13) is the basic equation for the lateral vibration of beams. The solution of this equation, if EI is constant, is of the form $y = X(x)\ [\cos(\omega_n t + \theta)]$, in which X is a function of x only. Substituting

$$\kappa^4 = \frac{\omega_n^2 \gamma S}{EIg} \tag{7.14}$$

and dividing Eq. (7.13) by $\cos(\omega_n t + \theta)$:

$$\frac{d^4 X}{dx^4} = \kappa^4 X \tag{7.15}$$

where X is any function whose fourth derivative is equal to a constant multiplied by the function itself. The following functions satisfy the required conditions and represent the solution of the equation:

$$X = A_1 \sin \kappa x + A_2 \cos \kappa x + A_3 \sinh \kappa x + A_4 \cosh \kappa x$$

The solution can also be expressed in terms of exponential functions, but the trigonometric and hyperbolic functions usually are more convenient to use.

For beams having various support conditions, the constants $A_1, A_2, A_3,$ and A_4 are found from the end conditions. In finding the solutions, it is convenient to write the equation in the following form in which two of the constants are zero for each of the usual boundary conditions:

$$X = A\ (\cos \kappa x + \cosh \kappa x) + B(\cos \kappa x - \cosh \kappa x)$$

$$+ C(\sin \kappa x + \sinh \kappa x) + D(\sin \kappa x - \sinh \kappa x) \quad (7.16)$$

In applying the end conditions, the following relations are used where primes indicate successive derivatives with respect to x:

The deflection is proportional to X and is zero at any rigid support.

The slope is proportional to X' and is zero at any built-in end.

The moment is proportional to X'' and is zero at any free or hinged end.

The shear is proportional to X''' and is zero at any free end.

The required derivatives are:

$$X' = \kappa[A(-\sin \kappa x + \sinh \kappa x) + B(-\sin \kappa x - \sinh \kappa x)$$
$$+ C(\cos \kappa x + \cosh \kappa x) + D(\cos \kappa x - \cosh \kappa x)]$$

$$X'' = \kappa^2[A(-\cos \kappa x + \cosh \kappa x) + B(-\cos \kappa x - \cosh \kappa x)$$
$$+ C(-\sin \kappa x + \sinh \kappa x) + D(-\sin \kappa x - \sinh \kappa x)]$$

$$X''' = \kappa^3[A(\sin \kappa x + \sinh \kappa x) + B(\sin \kappa x - \sinh \kappa x)$$
$$+ C(-\cos \kappa x + \cosh \kappa x) + D(-\cos \kappa x - \cosh \kappa x)]$$

For the usual end conditions, two of the constants are zero, and there remain two equations containing two constants. These can be combined to give an equation which contains only the frequency as an unknown. Using the frequency, one of the unknown constants can be found in terms of the other. There always is one undetermined constant, which can be evaluated only if the amplitude of the vibration is known.

EXAMPLE 7.4.   The natural frequencies and modes of vibration of the rectangular steel beam shown in Fig. 7.6 are to be determined and the fundamental frequency compared with that obtained from Fig. 7.4. The beam is 24 in. long, 2 in. wide, and ¼ in. thick, with the left end built in and the right end free.

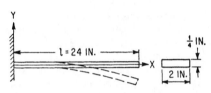

FIGURE 7.6   First mode of vibration of beam with left end clamped and right end free.

SOLUTION.   The boundary conditions are: at $x = 0$, $X = 0$, and $X' = 0$; at $x = l$, $X'' = 0$, and $X''' = 0$. The first condition requires that $A = 0$ since the other constants are multiplied by zero at $x = 0$. The second condition requires that $C = 0$. From the third and fourth conditions, the following equations are obtained:

$$0 = B(-\cos \kappa l - \cosh \kappa l) + D(-\sin \kappa l - \sinh \kappa l)$$

$$0 = B(\sin \kappa l - \sinh \kappa l) + D(-\cos \kappa l - \cosh \kappa l)$$

Solving each of these for the ratio D/B and equating, or making use of the mathematical condition that for a solution the determinant of the two equations must vanish, the following equation results:

$$\frac{D}{B} = -\frac{\cos \kappa l + \cosh \kappa l}{\sin \kappa l + \sinh \kappa l} = \frac{\sin \kappa l - \sinh \kappa l}{\cos \kappa l + \cosh \kappa l} \tag{7.17}$$

Equation (7.17) reduces to $\cos \kappa l \cosh \kappa l = -1$. The values of κl which satisfy this equation can be found by consulting tables of hyperbolic and trigonometric functions. The first five are: $\kappa_1 l = 1.875$, $\kappa_2 l = 4.694$, $\kappa_3 l = 7.855$, $\kappa_4 l = 10.996$, and $\kappa_5 l = 14.137$. The corresponding frequencies of vibration are found by substituting the length of the beam to find each κ and then solving Eq. (7.14) for ω_n:

$$\omega_n = \kappa_n^2 \sqrt{\frac{EIg}{\gamma S}}$$

For the rectangular section, $I = bh^3/12 = 1/384$ in.[4] and $S = bh = 0.5$ in.[2] For steel, $E = 30 \times 10^6$ lb/in.[2] and $\gamma = 0.28$ lb/in.[3] Using these values,

$$\omega_1 = \frac{(1.875)^2}{(24)^2}\sqrt{\frac{(30 \times 10^6)(386)}{(0.28)(384)(0.5)}} = 89.6 \text{ rad/sec} = 14.26 \text{ Hz}$$

The remaining frequencies can be found by using the other values of κ. Using Fig. 7.4, the fundamental frequency is found to be about 12 Hz.

To find the mode shapes, the ratio D/B is found by substituting the appropriate values of κl in Eq. (7.17). For the first mode:

$$\cosh 1.875 = 3.33710 \qquad \sinh 1.875 = 3.18373$$

$$\cos 1.875 = -0.29953 \qquad \sin 1.875 = 0.95409$$

Therefore, $D/B = -0.73410$. The equation for the first mode of vibration becomes

$$y = B_1[(\cos \kappa x - \cosh \kappa x) - 0.73410 (\sin \kappa x - \sinh \kappa x)] \cos (\omega_1 t + \theta_1)$$

in which B_1 is determined by the amplitude of vibration in the first mode. A similar equation can be obtained for each of the modes of vibration; all possible free vibration of the beam can be expressed by taking the sum of these equations.

Frequencies and Shapes of Beams. Table 7.3 gives the information necessary for finding the natural frequencies and normal modes of vibration of uniform beams having various boundary conditions. The various constants in the table were determined by computations similar to those used in Example 7.4. The table includes (1) diagrams showing the modal shapes including node locations, (2) the boundary conditions, (3) the frequency equation that results from using the boundary conditions in Eq. (7.16), (4) the constants that become zero in Eq. (7.16), (5) the values of κl from which the natural frequencies can be computed by using Eq. (7.14), and (6) the ratio of the nonzero constants in Eq. (7.16). By the use of the constants in this table, the equation of motion for any normal mode can be written. There always is a constant which is determined by the amplitude of vibration.

Values of characteristic functions representing the deflections of beams, at 50 equal intervals, for the first five modes of vibration have been tabulated.[8] Functions are given for beams having various boundary conditions, and the first three derivatives of the functions are also tabulated.

Rayleigh's Method. This method is useful for finding approximate values of the fundamental natural frequencies of beams. In applying Rayleigh's method, a suitable function is assumed for the deflection, and the maximum strain and kinetic energies are calculated, using the equations in Table 7.1. These energies are equated and solved for the frequency. The function used to represent the shape must satisfy the boundary conditions associated with deflection and slope at the supports. Best accuracy is obtained if other boundary conditions are also satisfied. The equation for the static deflection of the beam under a uniform load is a suitable function, although a simpler function often gives satisfactory results with less numerical work.

EXAMPLE 7.5.   The fundamental natural frequency of the cantilever beam in Example 7.4 is to be calculated using Rayleigh's method.

SOLUTION.   The assumed deflection $Y = (a/3l^4)[x^4 - 4x^3l + 6x^2l^2]$ is the static deflection of a cantilever beam under uniform load and having the deflection $Y = a$ at $x = l$. This deflection satisfies the conditions that the deflection Y and the slope Y' be zero at $x = 0$. Also, at $x = l$, Y'' which is proportional to the moment and Y''' which is proportional to the shear are zero. The second derivative of the function is $Y'' = (4a/l^4)[x^2 - 2xl + l^2]$. Using this in the expression from Table 7.1, the maximum strain energy is

TABLE 7.3 Natural Frequencies and Normal Modes of Uniform Beams

SUPPORTS	MODE n	(A) SHAPE AND NODES (NUMBERS GIVE LOCATION OF NODES IN FRACTION OF LENGTH FROM LEFT END)	(B) BOUNDARY CONDITIONS EQ (7.16)	(C) FREQUENCY EQUATION	(D) CONSTANTS EQ (7.16)	(E) kl EQ (7.14) $\omega_n = k^2\sqrt{\dfrac{EIg}{A\gamma}}$	(F) R RATIO OF NON-ZERO CONSTANTS COLUMN (D)
HINGED-HINGED	1		$x=0\begin{cases}X=0\\X''=0\end{cases}$	SIN $kl=0$	$A=0$	3.1416	1.0000
	2	0.50			$B=0$	6.283	1.0000
	3	0.333 0.667		$\dfrac{C}{D}=1$	9.425	1.0000
	4	0.25 0.50 0.75	$x=l\begin{cases}X=0\\X''=0\end{cases}$			12.566	1.0000
	n>4					$\approx n\pi$	1.0000
CLAMPED-CLAMPED	1		$x=0\begin{cases}X=0\\X'=0\end{cases}$	(COS kl)(COSH kl) $=1$	$A=0$	4.730	−0.9825
	2	0.50			$C=0$	7.853	−1.0008
	3	0.359 0.641		$\dfrac{D}{B}=R$	10.996	−1.0000−
	4	0.278 0.50 0.722	$x=l\begin{cases}X=0\\X'=0\end{cases}$			14.137	−1.0000+
	n>4					$\approx\dfrac{(2n+1)\pi}{2}$	−1.0000−
CLAMPED-HINGED	1		$x=0\begin{cases}X=0\\X'=0\end{cases}$	TAN $kl=$ TANH kl	$A=0$	3.927	−1.0008
	2	0.558			$C=0$	7.069	−1.0000+
	3	0.386 0.692		$\dfrac{D}{B}=R$	10.210	−1.0000
	4	0.294 0.529 0.765	$x=l\begin{cases}X=0\\X''=0\end{cases}$			13.352	−1.0000
	n>4					$\approx\dfrac{(4n+1)\pi}{4}$	−1.0000
CLAMPED-FREE	1		$x=0\begin{cases}X=0\\X''=0\end{cases}$	(COS kl)(COSH kl) $=-1$	$A=0$	1.875	−0.7341
	2	0.783			$C=0$	4.694	−1.0185
	3	0.504 0.868		$\dfrac{D}{B}=R$	7.855	−0.9992
	4	0.358 0.644 0.906	$x=l\begin{cases}X''=0\\X'''=0\end{cases}$			10.996	−1.0000+
	n>4					$\approx\dfrac{(2n-1)\pi}{2}$	−1.0000−
FREE-FREE	1		$x=0\begin{cases}X''=0\\X'''=0\end{cases}$	(COS kl)(COSH kl) $=1$	$B=0$	0(REPRESENTS TRANSLATION)	
	2	0.224 0.776			$D=0$	4.730	−0.9825
	3	0.132 0.50 0.868		$\dfrac{C}{A}=R$	7.853	−1.0008
	4	0.094 0.356 0.644 0.906	$x=l\begin{cases}X''=0\\X'''=0\end{cases}$			10.996	−1.0000−
	5	0.0734 0.277 0.50 0.723 0.927				14.137	−1.0000+
	n>5					$\approx\dfrac{(2n-1)\pi}{2}$	−1.0000−

$$V = \frac{EI}{2} \int_0^l \left(\frac{d^2Y}{dx^2}\right)^2 dx = \frac{8}{5} \frac{EIa^2}{l^3}$$

The maximum kinetic energy is

$$T = \frac{\omega_n^2 \gamma S}{2g} \int_0^l Y^2 dx = \frac{52}{405} \frac{\omega_n^2 \gamma S l a^2}{g}$$

Equating the two energies and solving for the frequency,

$$\omega_n = \sqrt{\frac{162}{13} \times \frac{EIg}{\gamma S l^4}} = \frac{3.530}{l^2} \sqrt{\frac{EIg}{\gamma S}}$$

The exact frequency as found in Example 7.4 is $(3.516/l^2) \sqrt{EIg/\gamma S}$; thus, Rayleigh's method gives good accuracy in this example.

If the deflection is assumed to be $Y = a[1 - \cos(\pi x/2l)]$, the calculated frequency is $(3.66/l^2)\sqrt{EIg/\gamma S}$. This is less accurate, but the calculations are considerably shorter. With this function, the same boundary conditions at $x = 0$ are satisfied; however, at $x = l$, $Y'' = 0$, but Y''' does not equal zero. Thus, the condition of zero shear at the free end is not satisfied. The trigonometric function would not be expected to give as good accuracy as the static deflection relation used in the example, although for most practical purposes the result would be satisfactory.

Effects of Rotary Motion and Shearing Force. In the preceding analysis of the lateral vibration of beams it has been assumed that each element of the beam moves only in the lateral direction. If each plane section that is initially normal to the axis of the beam remains plane and normal to the axis, as assumed in simple beam theory, then each section rotates slightly in addition to its lateral motion when the beam deflects.[9] When a beam vibrates, there must be forces to cause this rotation, and for a complete analysis these forces must be considered. The effect of this rotation is small except when the curvature of the beam is large relative to its thickness; this is true either for a beam that is short relative to its thickness or for a long beam vibrating in a higher mode so that the nodal points are close together.

Another factor that affects the lateral vibration of a beam is the lateral shear force. In Eq. (7.11) only the deflection associated with the bending stress in the beam is included. In any beam except one subject only to pure bending, a deflection due to the shear stress in the beam occurs. The exact solution of the beam vibration problem requires that this deflection be considered. The analysis of beam vibration including both the effects of rotation of the cross section and the shear deflection is called the *Timoshenko beam theory*. The following equation governs such vibration:[10]

$$a^2 \frac{\partial^4 y}{\partial x^4} + \frac{\partial^2 y}{\partial t^2} - \rho^2 \left(1 + \frac{E}{\kappa G}\right) \frac{\partial^4 y}{\partial x^2 \partial t^2} + \rho^2 \frac{\gamma}{g\kappa G} \frac{\partial^4 y}{\partial t^4} = 0 \qquad (7.18)$$

where $a^2 = EIg/S\gamma$, E = modulus of elasticity, G = modulus of rigidity, and $\rho = \sqrt{I/S}$, the radius of gyration; $\kappa = F_s/GS\beta$, F_s being the total lateral shear force at any section and β the angle which a cross section makes with the axis of the beam because of shear deformation. Under the assumptions made in the usual elementary beam theory, κ is $\frac{2}{3}$ for a beam with a rectangular cross section and $\frac{3}{4}$ for a circular beam. More refined analysis shows[11] that, for the present purposes, $\kappa = \frac{5}{6}$ and $\frac{9}{10}$ are more accurate values for rectangular and circular cross sections, respectively. Using a solution of the form $y = C \sin(n\pi x/l) \cos \omega_n t$, which satisfies the necessary end conditions, the following frequency equation is obtained for beams with both ends simply supported:

$$a^2 \frac{n^4\pi^4}{l^4} - \omega_n^2 - \omega_n^2 \frac{n^2\pi^2\rho^2}{l^2} - \omega_n^2 \frac{n^2\pi^2\rho^2}{l^2} \frac{E}{\kappa G} + \frac{\rho^2\gamma}{g\kappa G} \omega_n^4 = 0 \qquad (7.18a)$$

If it is assumed that $nr/l \ll 1$, Eq. $(7.18a)$ reduces to

$$\omega_n = \frac{a\pi^2}{(l/n)^2} \left[1 - \frac{\pi^2 n^2}{2} \left(\frac{\rho}{l} \right)^2 \left(1 + \frac{E}{\kappa G} \right) \right] \qquad (7.18b)$$

When $nr/l < 0.08$, the approximate equation gives less than 5 percent error in the frequency.[11]

Values of the ratio of ω_n to the natural frequency uncorrected for the effects of rotation and shear have been plotted,[11] using Eq. $(7.18a)$ for three values of $E/\kappa G$, and are shown in Fig. 7.7.

For a cantilever beam the frequency equation is quite complicated. For $E/\kappa G = 3.20$, corresponding approximately to the value for rectangular steel or aluminum beams, the curves in Fig. 7.8 show the effects of rotation and shear on the natural frequencies of the first six modes of vibration.

EXAMPLE 7.6.   The first two natural frequencies of a rectangular steel beam 40 in. long, 2 in. wide, and 6 in. thick, having simply supported ends, are to be computed with and without including the effects of rotation of the cross sections and shear deflection.

SOLUTION.   For steel $E = 30 \times 10^6$ lb/in.^{2}, $G = 11.5 \times 10^6$ lb/in.^{2}, and for a rectangular cross section $\kappa = \%$; thus $E/\kappa G = 3.13$. For a rectangular beam $\rho = h/12$ where h

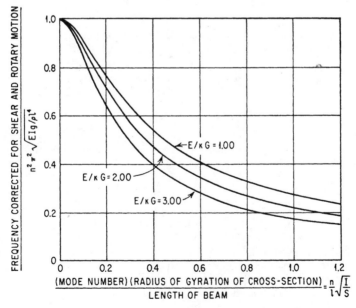

FIGURE 7.7   Influence of shear force and rotary motion on natural frequencies of simply supported beams. The curves relate the corrected frequency to that given by Eq. (7.14). (*J. G. Sutherland and L. E. Goodman.*[11])

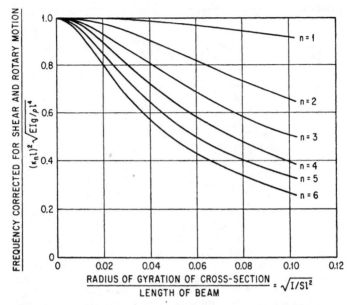

FIGURE 7.8 Influence of shear force and rotary motion on natural frequencies of uniform cantilever beams ($E/\kappa G = 3.20$). The curves relate the corrected frequency to that given by Eq. (7.14). (*J. G. Sutherland and L. E. Goodman.*[11])

is the thickness; thus $\rho/l = 6/(40\sqrt{12}) = 0.0433$. The approximate frequency equation, Eq. (7.18b), becomes

$$\omega_n = \frac{a\pi^2}{(l/n)^2}\left[1 - \frac{\pi^2}{2}(0.0433n)^2(1 + 3.13)\right]$$

$$= \frac{a\pi^2}{(l/n)^2}(1 - 0.038n^2)$$

Letting $\omega_0 = a\pi^2/(l/n)^2$ be the uncorrected frequency obtained by neglecting the effect of n in Eq. (7.18b):

For $n = 1$: $\qquad\qquad \dfrac{\omega_n}{\omega_0} = 1 - 0.038 = 0.962$

For $n = 2$: $\qquad\qquad \dfrac{\omega_n}{\omega_0} = 1 - 0.152 = 0.848$

Comparing these results with Fig. 7.7, using the curve for $E/\kappa G = 3.00$, the calculated frequency for the first mode agrees with the curve as closely as the curve can be read. For the second mode, the curve gives $\omega_n/\omega_0 = 0.91$; therefore the approximate equation for the second mode is not very accurate. The uncorrected frequencies are, since $I/S = \rho^2 = h^2/12$,

For $n = 1$: $\quad \omega_0 = \dfrac{\pi^2}{l^2}\sqrt{\dfrac{EIg}{S\gamma}} = \dfrac{\pi^2}{(40)^2}\sqrt{\dfrac{(30\times10^6)(36)386}{(12)(0.28)}} = 2170$ rad/sec $= 345$ Hz

For $n = 2$: $\qquad\qquad\qquad\qquad \omega_0 = 345 \times 4 = 1380$ Hz

The frequencies corrected for rotation and shear, using the value from Fig. 7.7 for correction of the second mode, are:

For $n = 1$: $\qquad\qquad\qquad\qquad f_n = 345 \times 0.962 = 332$ Hz

For $n = 2$: $\qquad\qquad\qquad\qquad f_n = 1380 \times 0.91 = 1256$ Hz

Effect of Axial Loads. When an axial tensile or compressive load acts on a beam, the natural frequencies are different from those for the same beam without such load. The natural frequencies for a beam with hinged ends, as determined by an energy analysis, assuming that the axial force F remains constant, are[12]

$$\omega_n = \frac{\pi^2 n^2}{l^2} \sqrt{\frac{EIg}{S\gamma}} \sqrt{1 \pm \frac{\alpha^2}{n^2}} = \omega_0 \sqrt{1 \pm \frac{\alpha^2}{n^2}}$$

where $\alpha^2 = Fl^2/EI\pi^2$, n is the mode number, ω_0 is the natural frequency of the beam with no axial force applied, and the other symbols are defined in Table 7.1. The plus sign is for a tensile force and the minus sign for a compressive force.

For a cantilever beam with a constant axial force F applied at the free end, the natural frequency is found by an energy analysis[13] to be $[1 + \frac{5}{14}(Fl^2/EI)]^{1/2}$ times the natural frequency of the beam without the force applied. If a uniform axial force is applied along the beam, the effect is the same as if about seven-twentieths of the total force were applied at the free end of the beam.

If the amplitude of vibration is large, an axial force may be induced in the beam by the supports. For example, if both ends of a beam are hinged but the supports are rigid enough so that they cannot move axially, a tensile force is induced as the beam deflects. The force is not proportional to the deflection; therefore, the vibration is of the type characteristic of nonlinear systems in which the natural frequency depends on the amplitude of vibration. The natural frequency of a beam having immovable hinged ends is given in the following table where the axial force is zero at zero deflection of the beam[14] and where x_0 is the amplitude of vibration, I the moment of inertia, and S the area of the cross section; ω_0 is the natural frequency of the unrestrained bar.

$\dfrac{x_0}{\sqrt{I/S}}$	0	0.1	0.2	0.4	0.6	0.8
$\dfrac{\omega_n}{\omega_0}$	1	1.0008	1.0038	1.015	1.038	1.058
$\dfrac{x_0}{\sqrt{I/S}}$	1.0	1.5	2	3	4	5
$\dfrac{\omega_n}{\omega_0}$	1.089	1.190	1.316	1.626	1.976	2.35

Beams Having Variable Cross Sections. The natural frequencies for beams of several shapes having cross sections that can be expressed as functions of the distance along the beam have been calculated.[15] The results are shown in Table 7.4. In the analysis, Eq. (7.13) was used, with EI considered to be variable.

TABLE 7.4 Natural Frequencies of Variable-Section Steel Beams (*J. N. Macduff and R. P. Felgar.*[16,17])

BEAM STRUCTURE	$\dfrac{b}{b_0}$	$\dfrac{h}{h_0}$	$(f_n l^2/\rho)/10^4$ $n=1$	$n=2$	$n=3$
(beam: b_0, b, x, h_0, h)	1	$\dfrac{x}{l}$	17.09	48.89	96.57
(beam: b_0, b, x, h_0, h)	$\dfrac{x}{l}$	$\dfrac{x}{l}$	26.08	68.08	123.64
(beam: b_0, b, x, h_0, h)	$\left(\dfrac{x}{l}\right)^{\frac{1}{2}}$	$\dfrac{x}{l}$	22.30	58.18	109.90
(beam: eb_0, b, x, h_0, h, b_0)	$e^{x/l}$	1	15.23	77.78	206.07
(beam: b_0 b, x, h_0 h)	1	$\dfrac{x}{l}$	21.21* 35.05†	56.97	
(beam: b_0 b, x, h_0 h)	$\dfrac{x}{l}$	$\dfrac{x}{l}$	32.73* 49.50†	76.57	
(beam: b_0 b, x, h_0 h)	$\left(\dfrac{x}{l}\right)^{\frac{1}{2}}$	$\dfrac{x}{l}$	25.66* 42.02†	66.06	

*SYMMETRIC   †ANTISYMMETRIC

f_n = natural frequency, Hz
$\rho = \sqrt{I/S}$ = radius of gyration, in.
h = depth of beam, in.

l = beam length, in.
n = mode number
b = width of beam, in.

For materials other than steel: $f_n = f_{ns}\sqrt{\dfrac{E\gamma_s}{E_s\gamma}}$

E = modulus of elasticity, lb/in.²
γ = density, lb/in.³
Terms with subscripts refer to steel
Terms without subscripts refer to other material

Rayleigh's or Ritz's method can be used to find approximate values for the frequencies of such beams. The frequency equation becomes, using the equations in Table 7.1, and letting $Y(x)$ be the assumed deflection,

$$\omega_n^2 = \frac{Eg}{\gamma} \frac{\int_0^l I\,(d^2Y/dx^2)^2\,dx}{\int_0^l SY^2\,dx}$$

where $I = I(x)$ is the moment of inertia of the cross section and $S = S(x)$ is the area of the cross section. Examples of the calculations are in the literature.[18] If the values of $I(x)$ and $S(x)$ cannot be defined analytically, the beam may be divided into two or more sections, for each of which I and S can be approximated by an equation. The strain and kinetic energies of each section may be computed separately, using an appropriate function for the deflection, and the total energies for the beam found by adding the values for the individual sections.

Continuous Beams on Multiple Supports. In finding the natural frequencies of a beam on multiple supports, the section between each pair of supports is considered as a separate beam with its origin at the left support of the section. Equation (7.16) applies to each section. Since the deflection is zero at the origin of each section, $A = 0$ and the equation reduces to

$$X = B(\cos \kappa x - \cosh \kappa x) + C(\sin \kappa x + \sinh \kappa x) + D(\sin \kappa x - \sinh \kappa x)$$

There is one such equation for each section, and the necessary end conditions are as follows:

1. At each end of the beam the usual boundary conditions are applicable, depending on the type of support.
2. At each intermediate support the deflection is zero. Since the beam is continuous, the slope and the moment just to the left and to the right of the support are the same.

General equations can be developed for finding the frequency for any number of spans.[19, 20] Table 7.5 gives constants for finding the natural frequencies of uniform continuous beams on uniformly spaced supports for several combinations of end supports.

Beams with Partly Clamped Ends. For a beam in which the slope at each end is proportional to the moment, the following empirical equation gives the natural frequency:[21]

$$f_n = f_0 \left[n + \frac{1}{2} \left(\frac{\beta_L}{5n + \beta_L} \right) \right] \left[n + \frac{1}{2} \left(\frac{\beta_R}{5n + \beta_R} \right) \right]$$

where f_0 is the frequency of the same beam with simply supported ends and n is the mode number. The parameters $\beta_L = k_L l/EI$ and $\beta_R = k_R l/EI$ are coefficients in which k_L and k_R are stiffnesses of the supports as given by $k_L = M_L/\theta_L$, where M_L is the moment and θ_L the angle at the left end, and $k_R = M_R/\theta_R$, where M_R is the moment and θ_R the angle at the right end. The error is less than 2 percent except for bars having one end completely or nearly clamped ($\beta > 10$) and the other end completely or nearly hinged ($\beta < 0.9$).

TABLE 7.5 Natural Frequencies of Continuous Uniform Steel* Beams (*J. N. Macduff and R. P. Felgar.*[16,17])

Beam structure	$(f_n l^2/\rho)/10^4$					
	N	$n = 1$	$n = 2$	$n = 3$	$n = 4$	$n = 5$
Extreme Ends Simply Supported						
	1	31.73	126.94	285.61	507.76	793.37
	2	31.73	49.59	126.94	160.66	285.61
	3	31.73	40.52	59.56	126.94	143.98
	4	31.73	37.02	49.59	63.99	126.94
	5	31.73	34.99	44.19	55.29	66.72
	6	31.73	34.32	40.52	49.59	59.56
	7	31.73	33.67	38.40	45.70	53.63
	8	31.73	33.02	37.02	42.70	49.59
	9	31.73	33.02	35.66	40.52	46.46
	10	31.73	33.02	34.99	39.10	44.19
	11	31.73	32.37	34.32	37.70	41.97
	12	31.73	32.37	34.32	37.02	40.52
Extreme Ends Clamped						
	1	72.36	198.34	388.75	642.63	959.98
	2	49.59	72.36	160.66	198.34	335.20
	3	40.52	59.56	72.36	143.98	178.25
	4	37.02	49.59	63.99	72.36	137.30
	5	34.99	44.19	55.29	66.72	72.36
	6	34.32	40.52	49.59	59.56	67.65
	7	33.67	38.40	45.70	53.63	62.20
	8	33.02	37.02	42.70	49.59	56.98
	9	33.02	35.66	40.52	46.46	52.81
	10	33.02	34.99	39.10	44.19	49.59
	11	32.37	34.32	37.70	41.97	47.23
	12	32.37	34.32	37.02	40.52	44.94
Extreme Ends Clamped-Supported						
	1	49.59	160.66	335.2	573.21	874.69
	2	37.02	63.99	137.30	185.85	301.05
	3	34.32	49.59	67.65	132.07	160.66
	4	33.02	42.70	56.98	69.51	129.49
	5	33.02	39.10	49.59	61.31	70.45
	6	32.37	37.02	44.94	54.46	63.99
	7	32.37	35.66	41.97	49.59	57.84
	8	32.37	34.99	39.81	45.70	53.63
	9	31.73	34.32	38.40	43.44	49.59
	10	31.73	33.67	37.02	41.24	46.46
	11	31.73	33.67	36.33	39.81	44.19
	12	31.73	33.02	35.66	39.10	42.70

* For materials other than steel, use equation at bottom of Table 7.4.
f_n = natural frequency, Hz n = mode number
$\rho = \sqrt{I/S}$ = radius of gyration, in. N = number of spans
l = span length, in.

LATERAL VIBRATION OF BEAMS WITH MASSES ATTACHED

The use of Fig. 7.4 is a convenient method of estimating the natural frequencies of beams with added loads.

Exact Solution. If the masses attached to the beam are considered to be rigid so that they exert no elastic forces, and if it is assumed that the attachment is such that the bending of the beam is not restrained, Eqs. (7.13) and (7.16) apply. The section of the beam between each two masses, and between each support and the adjacent mass, must be considered individually. The constants in Eq. (7.16) are different for each section. There are $4N$ constants, N being the number of sections into which the beam is divided. Each support supplies two boundary conditions. Additional conditions are provided by:

1. The deflection at the location of each mass is the same for both sections adjacent to the mass.
2. The slope at each mass is the same for each section adjacent thereto.
3. The change in the lateral elastic shear force in the beam, at the location of each mass, is equal to the product of the mass and its acceleration $\ddot{y}$.
4. The change of moment in the beam, at each mass, is equal to the product of the moment of inertia of the mass and its angular acceleration $(\partial^2/\partial t^2)(\partial y/\partial x)$.

Setting up the necessary equations is not difficult, but their solution is a lengthy process for all but the simplest configurations. Even the solution of the problem of a beam with hinged ends supporting a mass with negligible moment of inertia located anywhere except at the center of the beam is fairly long. If the mass is at the center of the beam, the solution is relatively simple because of symmetry and is illustrated to show how the result compares with that obtained by Rayleigh's method.

Rayleigh's Method. Rayleigh's method offers a practical method of obtaining a fairly accurate solution of the problem, even when more than one mass is added. In carrying out the solution, the kinetic energy of the masses is added to that of the beam. The strain and kinetic energies of a uniform beam are given in Table 7.1. The kinetic energy of the ith mass is $(m_i/2)\omega_n^2 Y^2(x_i)$, where $Y(x_i)$ is the value of the amplitude at the location of mass. Equating the maximum strain energy to the total maximum kinetic energy of the beam and masses, the frequency equation becomes

$$\omega_n^2 = \frac{EI \int_0^l (Y'')^2 \, dx}{\dfrac{\gamma S}{g} \int_0^l Y^2 \, dx + \sum_{i=1}^n m_i Y^2(x_i)} \tag{7.19}$$

where $Y(x)$ is the maximum deflection. If $Y(x)$ were known exactly, this equation would give the correct frequency; however, since Y is not known, a shape must be assumed. This may be either the mode shape of the unloaded beam or a polynomial that satisfies the necessary boundary conditions, such as the equation for the static deflection under a load.

Beam as Spring. A method for obtaining the natural frequency of a beam with a single mass mounted on it is to consider the beam to act as a spring, the stiffness of which is found by using simple beam theory. The equation $\omega_n = \sqrt{k/m}$ is used. Best accuracy is obtained by considering m to be made up of the attached mass plus some portion of the mass of the beam. The fraction of the beam mass to be used depends on the type of beam. The equations for simply supported and cantilevered beams with masses attached are given in Table 7.2.

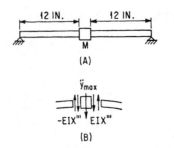

FIGURE 7.9 (*A*) Beam having simply supported ends with mass attached at center. (*B*) Forces exerted on mass, at extreme deflection, by shear stresses in beam.

EXAMPLE 7.7.   The fundamental natural frequencies of a beam with hinged ends 24 in. long, 2 in. wide, and ¼ in. thick having a mass m attached at the center (Fig. 7.9) are to be calculated by each of the three methods, and the results compared for ratios of mass to beam mass of 1, 5, and 25. The result is to be compared with the frequency from Fig. 7.4.

EXACT SOLUTION.   Because of symmetry, only the section of the beam to the left of the mass has to be considered in carrying out the exact solution. The boundary conditions for the left end are: at $x = 0$, $X = 0$, and $X'' = 0$. The shear force just to the left of the mass is negative at maximum deflection (Fig. 7.9B) and is $F_s = -EIX'''$; to the right of the mass, because of symmetry, the shear force has the same magnitude with opposite sign. The difference between the shear forces on the two sides of the mass must equal the product of the mass and its acceleration. For the condition of maximum deflection,

$$2EIX''' = m\ddot{y}_{max} \qquad (7.20)$$

where X''' and $\ddot{y}_{max}$ must be evaluated at $x = l/2$. Because of symmetry the slope at the center is zero. Using the solution $y = X \cos \omega_n t$ and $\ddot{y}_{max} = -\omega_n^2 X$, Eq. (7.20) becomes

$$2EIX''' = -m\omega_n^2 X \qquad (7.21)$$

The first boundary condition makes $A = 0$ in Eq. (7.16) and the second condition makes $B = 0$. For simplicity, the part of the equation that remains is written

$$X = C \sin \kappa x + D \sinh \kappa x \qquad (7.22)$$

Using this in Eq. (7.20) gives

$$2EI\left(-C\kappa^3 \cos \frac{\kappa l}{2} + D\kappa^3 \cosh \frac{\kappa l}{2}\right) = -m\omega_n^2\left(C \sin \frac{\kappa l}{2} + D \sinh \frac{\kappa l}{2}\right) \qquad (7.23)$$

The slope at the center is zero. Differentiating Eq. (7.22) and substituting $x = l/2$,

$$\kappa\left(C \cos \frac{\kappa l}{2} + D \cosh \frac{\kappa l}{2}\right) = 0 \qquad (7.24)$$

Solving Eqs. (7.23) and (7.24) for the ratio C/D and equating, the following frequency equation is obtained:

$$2\frac{m_b}{m} = \frac{\kappa l}{2}\left(\tan \frac{\kappa l}{2} - \tanh \frac{\kappa l}{2}\right)$$

where $m_b = \gamma Sl/g$ is the total mass of the beam. The lowest roots for the specified ratios m/m_b are as follows:

m/m_b	1	5	25
$\kappa l/2$	1.1916	0.8599	0.5857

The corresponding natural frequencies are found from Eq. (7.14) and are tabulated, with the results obtained by the other methods, at the end of the example.

Solution by Rayleigh's Method. For the solution by Rayleigh's method it is assumed that $Y = B \sin (\pi x / l)$. This is the fundamental mode for the unloaded beam (Table 7.3). The terms in Eq. (7.19) become

$$\int_0^l (Y'')^2 \, dx = B^2 \left(\frac{\pi}{l}\right)^4 \int_0^l \sin^2 \frac{\pi x}{l} \, dx = B^2 \frac{l}{2} \left(\frac{\pi}{l}\right)^4$$

$$\int_0^l Y^2 \, dx = B^2 \int_0^l \sin^2 \frac{\pi x}{l} \, dx = B^2 \frac{l}{2}$$

$$Y^2(x_1) = B^2$$

Substituting these terms, Eq. (7.19) becomes

$$\omega_n = \sqrt{\frac{EIB^2(l/2)(\pi/l)^4}{(S\gamma B^2 l / 2g) + mB^2}} = \frac{\pi^2}{\sqrt{1 + 2m/m_b}} \sqrt{\frac{EIg}{S\gamma l^4}}$$

The frequencies for the specified values of m/m_b are tabulated at the end of the example. Note that if $m = 0$, the frequency is exactly correct, as can be seen from Table 7.3. This is to be expected since, if no mass is added, the assumed shape is the true shape.

Lumped Parameter Solution. Using the appropriate equation from Table 7.2, the natural frequency is

$$\omega_n = \sqrt{\frac{48EI}{l^3(m + 0.5m_b)}}$$

Since $m_b = \gamma S l / g$, this becomes

$$\omega_n = \sqrt{\frac{48}{(m/m_b) + 0.5}} \sqrt{\frac{EIg}{S\gamma l^4}}$$

Comparison of Results. The results for each method can be expressed as a coefficient α multiplied by $\sqrt{EIg/S\gamma l^4}$. The values of α for the specified values by m/m_b for the three methods of solution are:

m/m_b	1	5	25
Exact	5.680	2.957	1.372
Rayleigh	5.698	2.976	1.382
Spring	5.657	2.954	1.372

The results obtained by all the methods agree closely. For large values of m/m_b the third method gives very accurate results.

Numerical Calculations. For steel, $E = 30 \times 10^6$ lb/in.^{2}, $\gamma = 0.28$ lb/in.^{3}; for a rectangular beam, $I = bh^3/12 = 1/384$ in.^{4} and $S = bh = \frac{1}{2}$ in.^{2} The fundamental frequency using the value of α for the exact solution when $m/m_b = 1$ is

$$\omega_1 = \frac{\alpha}{l^2} \sqrt{\frac{EIg}{S\gamma}} = \frac{5.680}{576} \sqrt{\frac{(30 \times 10^6)(386)}{(0.5)(384)(0.28)}} = 145 \text{ rad/sec} = 23 \text{ Hz}$$

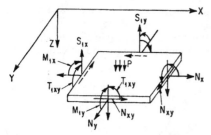

FIGURE 7.10 Element of plate showing bending moments, normal forces, and shear forces.

Other frequencies can be found by using the other values of α. Nearly the same result is obtained by using Fig. 7.4, if half the mass of the beam is added to the additional mass.

LATERAL VIBRATION OF PLATES

General Theory of Bending of Rectangular Plates.   For small deflections of an initially flat plate of uniform thickness (Fig. 7.10) made of homogeneous isotropic material and subjected to normal and shear forces in the plane of the plate, the following equation relates the lateral deflection w to the lateral loading:[22]

$$D\nabla^4 w = D\left(\frac{\partial^4 w}{\partial x^4} + 2\frac{\partial^4 w}{\partial x^2 \partial y^2} + \frac{\partial^4 w}{\partial y^4}\right) = P + N_x \frac{\partial^2 w}{\partial x^2} + 2N_{xy}\frac{\partial^2 w}{\partial x\,\partial y} + N_y \frac{\partial^2 w}{\partial y^2}$$

(7.25)

where $D = Eh^3/12(1 - \mu^2)$ is the plate stiffness, h being the plate thickness and μ Poisson's ratio. The parameter P is the loading intensity, N_x the normal loading in the X direction per unit of length, N_y the normal loading in the Y direction, and N_{xy} the shear load parallel to the plate surface in the X and Y directions.

The bending moments and shearing forces are related to the deflection w by the following equations:[23]

$$M_{1x} = -D\left(\frac{\partial^2 w}{\partial x^2} + \mu\frac{\partial^2 w}{\partial y^2}\right) \qquad M_{1y} = -D\left(\frac{\partial^2 w}{\partial y^2} + \mu\frac{\partial^2 w}{\partial x^2}\right)$$

$$T_{1xy} = D(1 - \mu)\frac{\partial^2 w}{\partial x\,\partial y} \qquad\qquad (7.26)$$

$$S_{1x} = -D\left(\frac{\partial^3 w}{\partial x^3} + \frac{\partial^3 w}{\partial x\,\partial y^2}\right) \qquad S_{1y} = -D\left(\frac{\partial^3 w}{\partial y^3} + \frac{\partial^3 w}{\partial x^2\,\partial y}\right)$$

As shown in Fig. 7.10, M_{1x} and M_{1y} are the bending moments per unit of length on the faces normal to the X and Y directions, respectively, T_{1xy} is the twisting or warping moment on these faces, and S_{1x}, S_{1y} are the shearing forces per unit of length normal to the plate surface.

The boundary conditions that must be satisfied by an edge parallel to the X axis, for example, are as follows:
Built-in edge:

$$w = 0 \qquad \frac{\partial w}{\partial y} = 0$$

Simply supported edge:

$$w = 0 \qquad M_{1y} = -D\left(\frac{\partial^2 w}{\partial y^2} + \mu\frac{\partial^2 w}{\partial x^2}\right) = 0$$

Free edge:

$$M_{1y} = -D\left(\frac{\partial^2 w}{\partial y^2} + \mu \frac{\partial^2 w}{\partial x^2}\right) = 0 \qquad T_{1xy} = 0 \qquad S_{1y} = 0$$

which together give

$$\frac{\partial}{\partial y}\left[\frac{\partial^2 w}{\partial y^2} + (2-\mu)\frac{\partial^2 w}{\partial x^2}\right] = 0$$

Similar equations can be written for other edges. The strains caused by the bending of the plate are

$$\epsilon_x = -z\frac{\partial^2 w}{\partial x^2} \qquad \epsilon_y = -z\frac{\partial^2 w}{\partial y^2} \qquad \gamma_{xy} = 2z\frac{\partial^2 w}{\partial x\,\partial y} \qquad (7.27)$$

where z is the distance from the center plane of the plate.

Hooke's law may be expressed by the following equations:

$$\epsilon_x = \frac{1}{E}(\sigma_x - \mu\sigma_y) \qquad \sigma_x = \frac{E}{1-\mu^2}(\epsilon_x + \mu\epsilon_y)$$

$$\epsilon_y = \frac{1}{E}(\sigma_y - \mu\sigma_x) \qquad \sigma_y = \frac{E}{1-\mu^2}(\epsilon_y + \mu\epsilon_x) \qquad (7.28)$$

$$\gamma_{xy} = \frac{\tau_{xy}}{G} \qquad \tau_{xy} = G\gamma_{xy}$$

Substituting the expressions giving the strains in terms of the deflections, the following equations are obtained for the bending stresses in terms of the lateral deflection:

$$\sigma_x = -\frac{Ez}{1-\mu^2}\left(\frac{\partial^2 w}{\partial x^2} + \mu \frac{\partial^2 w}{\partial y^2}\right) = \frac{12M_{1x}}{h^3}z$$

$$\sigma_y = -\frac{Ez}{1-\mu^2}\left(\frac{\partial^2 w}{\partial y^2} + \mu \frac{\partial^2 w}{\partial x^2}\right) = \frac{12M_{1y}}{h^3}z \qquad (7.29)$$

$$\tau_{xy} = 2G\frac{\partial^2 w}{\partial x\,\partial y}z = \frac{12T_{1xy}}{h^3}z$$

Table 7.6 gives values of maximum deflection and bending moment at several points in plates which have various shapes and conditions of support and which are subjected to uniform lateral pressure. The results are all based on the assumption that the deflections are small and that there are no loads in the plane of the plate. The bending stresses are found by the use of Eqs. (7.29). Bending moments and deflections for many other types of load are in the literature.[22]

The stresses caused by loads in the plane of the plate are found by assuming that the stress is uniform through the plate thickness. The total stress at any point in the plate is the sum of the stresses caused by bending and by the loading in the plane of the plate.

For plates in which the lateral deflection is large compared to the plate thickness but small compared to the other dimensions, Eq. (7.25) is valid. However, additional equations must be introduced because the forces N_x, N_y, and N_{xy} depend not only on the initial loading of the plate but also upon the stretching of the plate due to the

TABLE 7.6 Maximum Deflection and Bending Moments in Uniformly Loaded Plates under Static Conditions

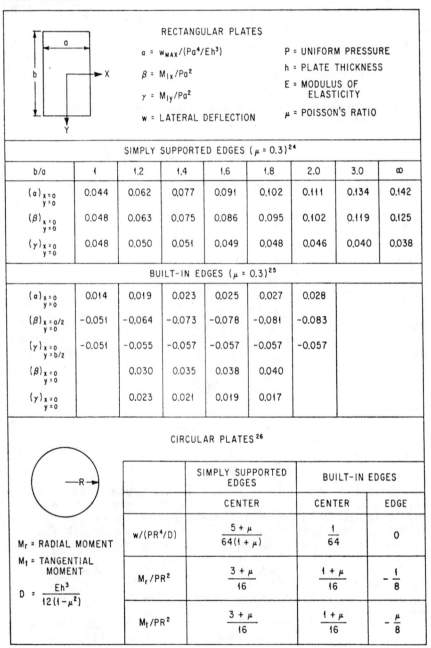

RECTANGULAR PLATES

$\alpha = w_{MAX}/(Pa^4/Eh^3)$

$\beta = M_{lx}/Pa^2$

$\gamma = M_{ly}/Pa^2$

w = LATERAL DEFLECTION

P = UNIFORM PRESSURE

h = PLATE THICKNESS

E = MODULUS OF ELASTICITY

μ = POISSON'S RATIO

SIMPLY SUPPORTED EDGES ($\mu = 0.3$)[24]

b/a	1	1.2	1.4	1.6	1.8	2.0	3.0	∞
$(\alpha)_{x=0, y=0}$	0.044	0.062	0.077	0.091	0.102	0.111	0.134	0.142
$(\beta)_{x=0, y=0}$	0.048	0.063	0.075	0.086	0.095	0.102	0.119	0.125
$(\gamma)_{x=0, y=0}$	0.048	0.050	0.051	0.049	0.048	0.046	0.040	0.038

BUILT-IN EDGES ($\mu = 0.3$)[25]

$(\alpha)_{x=0, y=0}$	0.014	0.019	0.023	0.025	0.027	0.028		
$(\beta)_{x=a/2, y=0}$	-0.051	-0.064	-0.073	-0.078	-0.081	-0.083		
$(\gamma)_{x=0, y=b/2}$	-0.051	-0.055	-0.057	-0.057	-0.057	-0.057		
$(\beta)_{x=0, y=0}$		0.030	0.035	0.038	0.040			
$(\gamma)_{x=0, y=0}$		0.023	0.021	0.019	0.017			

CIRCULAR PLATES[26]

M_r = RADIAL MOMENT

M_t = TANGENTIAL MOMENT

$D = \dfrac{Eh^3}{12(1-\mu^2)}$

	SIMPLY SUPPORTED EDGES	BUILT-IN EDGES	
	CENTER	CENTER	EDGE
$w/(PR^4/D)$	$\dfrac{5+\mu}{64(1+\mu)}$	$\dfrac{1}{64}$	0
M_r/PR^2	$\dfrac{3+\mu}{16}$	$\dfrac{1+\mu}{16}$	$-\dfrac{1}{8}$
M_t/PR^2	$\dfrac{3+\mu}{16}$	$\dfrac{1+\mu}{16}$	$-\dfrac{\mu}{8}$

bending. The equations of equilibrium for the X and Y directions in the plane of the plate are

$$\frac{\partial N_x}{\partial x} + \frac{\partial N_{xy}}{\partial y} = 0 \qquad \frac{\partial N_{xy}}{\partial x} + \frac{\partial N_y}{\partial y} = 0 \tag{7.30}$$

It can be shown[27] that the strain components are given by

$$\epsilon_x = \frac{\partial u}{\partial x} + \frac{1}{2}\left(\frac{\partial w}{\partial x}\right)^2 \qquad \epsilon_y = \frac{\partial v}{\partial y} + \frac{1}{2}\left(\frac{\partial w}{\partial y}\right)^2$$

$$\gamma_{xy} = \frac{\partial u}{\partial y} + \frac{\partial v}{\partial x} + \frac{\partial w}{\partial x}\frac{\partial w}{\partial y} \tag{7.31}$$

where u is the displacement in the X direction and v is the displacement in the Y direction. By differentiating and combining these expressions, the following relation is obtained:

$$\frac{\partial^2 \epsilon_x}{\partial y^2} + \frac{\partial^2 \epsilon_y}{\partial x^2} - \frac{\partial^2 \gamma_{xy}}{\partial x\,\partial y} = \left(\frac{\partial^2 w}{\partial x\,\partial y}\right)^2 - \frac{\partial^2 w}{\partial x^2}\frac{\partial^2 w}{\partial y^2} \tag{7.32}$$

If it is assumed that the stresses caused by the forces in the plane of the plate are uniformly distributed through the thickness, Hooke's law, Eqs. (7.28), can be expressed:

$$\epsilon_x = \frac{1}{hE}(N_x - \mu N_y) \qquad \epsilon_y = \frac{1}{hE}(N_y - \mu N_x) \qquad \gamma_{xy} = \frac{1}{hG}N_{xy} \tag{7.33}$$

The equilibrium equations are satisfied by a stress function ϕ which is defined as follows:

$$N_x = h\frac{\partial^2 \phi}{\partial y^2} \qquad N_y = h\frac{\partial^2 \phi}{\partial x^2} \qquad N_{xy} = -h\frac{\partial^2 \phi}{\partial x\,\partial y} \tag{7.34}$$

If these are substituted into Eqs. (7.33) and the resulting expressions substituted into Eq. (7.32), the following equation is obtained:

$$\frac{\partial^4 \phi}{\partial x^4} + 2\frac{\partial^4 \phi}{\partial x^2\,\partial y^2} + \frac{\partial^4 \phi}{\partial y^4} = E\left[\left(\frac{\partial^2 w}{\partial x\,\partial y}\right)^2 - \frac{\partial^2 w}{\partial x^2}\frac{\partial^2 w}{\partial y^2}\right] \tag{7.35}$$

A second equation is obtained by substituting Eqs. (7.34) in Eq. (7.25):

$$D\nabla^4 w = P + h\left(\frac{\partial^2 \phi}{\partial y^2}\frac{\partial^2 w}{\partial x^2} - 2\frac{\partial^2 \phi}{\partial x\,\partial y}\frac{\partial^2 w}{\partial x\,\partial y} + \frac{\partial^2 \phi}{\partial x^2}\frac{\partial^2 w}{\partial y^2}\right) \tag{7.36}$$

Equations (7.35) and (7.36), with the boundary conditions, determine ϕ and w, from which the stresses can be computed. General solutions to this set of equations are not known, but some approximate solutions can be found in the literature.[28]

Free Lateral Vibrations of Rectangular Plates. In Eq. (7.25), the terms on the left are equal to the sum of the rates of change of the forces per unit of length in the X and Y directions where such forces are exerted by shear stresses caused by bending normal to the plane of the plate. For a rectangular element with dimensions dx and dy, the net force exerted normal to the plane of the plate by these stresses is $D\nabla^4 w\,dx\,dy$. The last three terms on the right-hand side of Eq. (7.25) give the net force normal to the plane of the plate, per unit of length, which is caused by the forces acting in the plane of the plate. The net force caused by these forces on an element with dimensions dx and dy is $(N_x \, \partial^2 w/\partial x^2 + 2N_{xy} \, \partial^2 w/\partial x \, \partial y + N_y \, \partial^2 w/\partial y^2) \, dx \, dy$. As in the corresponding beam problem, the forces in a vibrating plate consist of two parts: (1) that which balances the static load P including the weight of the plate and (2) that which is induced by the vibration. The first part is always in equilibrium with the load and together with the load can be omitted from the equation of motion if the deflection is taken from the position of static equilibrium. The force exerted normal to the plane of the plate by the bending stresses must equal the sum of the force exerted normal to the plate by the loads acting in the plane of the plate; i.e., the product of the mass of the element $(\gamma h/g) \, dx \, dy$ and its acceleration $\ddot{w}$. The term involving the acceleration of the element is negative, because when the bending force is positive the acceleration is in the negative direction. The equation of motion is

$$DV^4 w = -\frac{\gamma}{g} \, h\ddot{w} + \left(N_x \frac{\partial^2 w}{\partial x^2} + 2N_{xy} \frac{\partial^2 w}{\partial x \, \partial y} + N_y \frac{\partial^2 w}{\partial y^2} \right) \qquad (7.37)$$

This equation is valid only if the magnitudes of the forces in the plane of the plate are constant during the vibration. For many problems these forces are negligible and the term in parentheses can be omitted.

When a system vibrates in a natural mode, all parts execute simple harmonic motion about the equilibrium position; therefore, the solution of Eq. (7.37) can be written as $w = AW(x,y) \cos (w_n t + \theta)$ in which W is a function of x and y only. Substituting this in Eq. (7.37) and dividing through by $A \cos (w_n t + \theta)$ gives

$$DV^4 W = \frac{\gamma h w_n^2}{g} \, W + \left(N_x \frac{\partial^2 W}{\partial x^2} + 2N_{xy} \frac{\partial^2 W}{\partial x \, \partial y} + N_y \frac{\partial^2 W}{\partial y^2} \right) \qquad (7.38)$$

The function W must satisfy Eq. (7.38) as well as the necessary boundary conditions.

The solution of the problem of the lateral vibration of a rectangular plate with all edges simply supported is relatively simple; in general, other combinations of edge conditions require the use of other methods of solution. These are discussed later.

Example 7.8. The natural frequencies and normal modes of small vibration of a rectangular plate of length a, width b, and thickness h are to be calculated. All edges are hinged and subjected to unchanging normal forces N_x and N_y.

Solution. The following equation, in which m and n may be any integers, satisfies the necessary boundary conditions:

$$W = A \sin \frac{m\pi x}{a} \sin \frac{n\pi y}{b} \qquad (7.39)$$

Substituting the necessary derivatives into Eq. (7.38),

$$D\left[\left(\frac{m}{a} \right)^4 + 2\left(\frac{m}{a} \right)^2 \left(\frac{n}{b} \right)^2 + \left(\frac{n}{b} \right)^4 \right] \pi^4 \sin \frac{m\pi x}{a} \sin \frac{n\pi y}{b}$$

$$= \frac{\gamma h w_n^2}{g} \sin \frac{m\pi x}{a} \sin \frac{n\pi y}{b} - \pi^2 \left[N_x \left(\frac{m}{a} \right)^2 + N_y \left(\frac{n}{b} \right)^2 \right] \sin \frac{m\pi x}{a} \sin \frac{n\pi y}{b}$$

Solving for ω_n^2,

$$\omega_n^2 = \frac{g}{\gamma h} \left\{ \pi^4 D\left[\left(\frac{m}{a} \right)^2 + \left(\frac{n}{b} \right)^2 \right]^2 + \pi^2 \left[N_x \left(\frac{m}{a} \right)^2 + N_y \left(\frac{n}{b} \right)^2 \right] \right\} \qquad (7.40)$$

By using integral values of m and n, the various frequencies are obtained from Eq. (7.40) and the corresponding normal modes from Eq. (7.39). For each mode, m and

n represent the number of half sine waves in the X and Y directions, respectively. In each mode there are $m - 1$ evenly spaced nodal lines parallel to the Y axis, and $n - 1$ parallel to the X axis.

Rayleigh's and Ritz's Methods. The modes of vibration of a rectangular plate with all edges simply supported are such that the deflection of each section of the plate parallel to an edge is of the same form as the deflection of a beam with both ends simply supported. In general, this does not hold true for other combinations of edge conditions. For example, the vibration of a rectangular plate with all edges built in does not occur in such a way that each section parallel to an edge has the same shape as does a beam with both ends built in. A function that is made up using the mode shapes of beams with built-in ends obviously satisfies the conditions of zero deflection and slope at all edges, but it cannot be made to satisfy Eq. (7.38).

The mode shapes of beams give logical functions with which to formulate shapes for determining the natural frequencies, for plates having various edge conditions, by the Rayleigh or Ritz methods. By using a single mode function in Rayleigh's method an approximate frequency can be determined. This can be improved by using more than one of the modal shapes and using Ritz's method as discussed below.

The strain energy of bending and the kinetic energy for plates are given in Table 7.1. Finding the maximum values of the energies, equating them, and solving for ω_n^2 gives the following frequency equation:

$$\omega_n^2 = \frac{V_{\max}}{\dfrac{\gamma h}{2g} \displaystyle\int_A \int W^2 \, dx \, dy} \tag{7.41}$$

where V is the strain energy.

In applying the Rayleigh method, a function W is assumed that satisfies the necessary boundary conditions of the plate. An example of the calculations is given in the section on circular plates. If the shape assumed is exactly the correct one, Eq. (7.41) gives the exact frequency. In general, the correct shape is not known and a frequency greater than the natural frequency is obtained. The Ritz method involves assuming W to be of the form $W = a_1 W_1(x,y) + a_2 W_2(x,y) + \cdots$ in which $W_1, W_2, \ldots$ all satisfy the boundary conditions, and $a_1, a_2, \ldots$ are adjusted to give a minimum frequency. Reference 29 is an extensive compilation, with references to sources, of calculated and experimental results for plates of many shapes. Some examples are cited in the following sections.

Square, Rectangular, and Skew Rectangular Plates. Tables of the functions necessary for the determination of the natural frequencies of rectangular plates by the use of the Ritz method are available,[30] these having been derived by using the modal shapes of beams having end conditions corresponding to the edge conditions of the plates. Information is included from which the complete shapes of the vibrational modes can be determined. Frequencies and nodal patterns for several modes of vibration of square plates having three sets of boundary conditions are shown in Table 7.7. By the use of functions which represent the natural modes of beams, the frequencies and nodal patterns for rectangular and skew cantilever plates have been determined[31] and are shown in Table 7.8. Comparison of calculated frequencies with experimentally determined values shows good agreement. Natural frequencies of rectangular plates having other boundary conditions are given in Table 7.9.

TABLE 7.7 Natural Frequencies and Nodal Lines of Square Plates with Various Edge Conditions (*After D. Young.*[29])

	1ST MODE	2ND MODE	3RD MODE	4TH MODE	5TH MODE	6TH MODE
$\omega_n/\sqrt{Dg/\gamma h a^4}$	3.494	8.547	21.44	27.46	31.17	
NODAL LINES						
$\omega_n/\sqrt{Dg/\gamma h a^4}$	35.99	73.41	108.27	131.64	132.25	165.15
NODAL LINES						
$\omega_n/\sqrt{Dg/\gamma h a^4}$	6.958	24.08	26.80	48.05	63.14	
NODAL LINES						

$\omega_n = 2\pi f_n$ h = PLATE THICKNESS

$D = Eh^3/12(1-\mu^2)$ a = PLATE LENGTH

γ = WEIGHT DENSITY

Triangular and Trapezoidal Plates.   Nodal patterns and natural frequencies for triangular plates have been determined[33] by the use of functions derived from the mode shapes of beams, and are shown in Table 7.10. Certain of these have been compared with experimental values and the agreement is excellent. Natural frequencies and nodal patterns have been determined experimentally for six modes of vibration of a number of cantilevered triangular plates[34] and for the first six modes of cantilevered trapezoidal plates derived by trimming the tips of triangular plates parallel to the clamped edge.[35] These triangular and trapezoidal shapes approximate the shapes of various delta wings for aircraft and of fins for missiles.

Circular Plates.   The solution of the problem of small lateral vibration of circular plates is obtained by transforming Eq. (7.38) to polar coordinates and finding the solution that satisfies the necessary boundary conditions of the resulting equation. Omitting the terms involving forces in the plane of the plate,[36]

$$\left(\frac{\partial^2}{\partial r^2} + \frac{1}{r}\frac{\partial}{\partial r} + \frac{1}{r}\frac{\partial^2}{\partial \theta^2}\right)\left(\frac{\partial^2 W}{\partial r^2} + \frac{1}{r}\frac{\partial W}{\partial r} + \frac{1}{r}\frac{\partial^2 W}{\partial \theta^2}\right) = \kappa^4 W \qquad (7.42)$$

where

$$\kappa^4 = \frac{\gamma h \omega_n^2}{gD}$$

TABLE 7.8 Natural Frequencies and Nodal Lines of Cantilevered Rectangular and Skew Rectangular Plates ($\mu = 0.3$)* (*M. V. Barton.*[30])

a/b MODE	1/2	1	2	5
FIRST	3.508	3.494	3.472	3.450
SECOND	5.372	8.547	14.93	34.73
THIRD	21.96	21.44	21.61	21.52
FOURTH	10.26	27.46	94.49	563.9
FIFTH	24.85	31.17	48.71	105.9

MODE	FIRST	SECOND	FIRST	SECOND	FIRST	SECOND
$\omega_n / \sqrt{Dg/\gamma h a^4}$	3.601	8.872	3.961	10.190	4.824	13.75
NODAL LINES						

* For terminology, see Table 7.7.

The solution of Eq. (7.42) is[36]

$$W = A \cos (n\theta - \beta)[J_n(\kappa r) + \lambda J_n(i\kappa r)] \qquad (7.43)$$

where J_n is a Bessel function of the first kind. When $\cos (n\theta - \beta) = 0$, a mode having a nodal system of n diameters, symmetrically distributed, is obtained. The term in

TABLE 7.9 Natural Frequencies of Rectangular Plates (*R. F. S. Hearman.*[32])

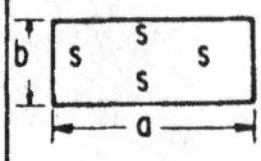	b/a	1.0	1.5	2.0	2.5	3.0	∞
(s s s s plate)	$\omega_n/\sqrt{Dg/\gamma ha^4}$	19.74	14.26	12.34	11.45	10.97	9.87
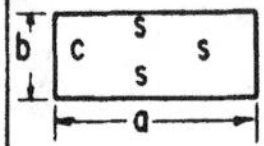	b/a	1.0	1.5	2.0	2.5	3.0	∞
	$\omega_n/\sqrt{Dg/\gamma ha^4}$	23.65	18.90	17.33	16.63	16.26	15.43
(c s s s plate)	a/b	1.0	1.5	2.0	2.5	3.0	∞
	$\omega_n/\sqrt{Dg/\gamma hb^4}$	23.65	15.57	12.92	11.75	11.14	9.87
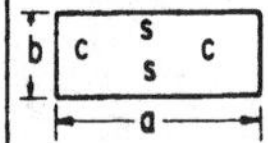	b/a	1.0	1.5	2.0	2.5	3.0	∞
	$\omega_n/\sqrt{Dg/\gamma ha^4}$	28.95	25.05	23.82	23.27	22.99	22.37
(c s s c plate)	a/b	1.0	1.5	2.0	2.5	3.0	∞
	$\omega_n/\sqrt{Dg/\gamma hb^4}$	28.95	17.37	13.69	12.13	11.36	9.87
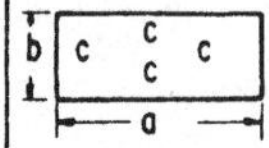	b/a	1.0	1.5	2.0	2.5	3.0	∞
(c c c c plate)	$\omega_n/\sqrt{Dg/\gamma ha^4}$	35.98	27.00	24.57	23.77	23.19	22.37

s DENOTES SIMPLY SUPPORTED EDGE

c DENOTES BUILT-IN OR CLAMPED EDGE

a = LENGTH OF PLATE

b = WIDTH OF PLATE

FOR OTHER TERMINOLOGY SEE TABLE 7.7

brackets represents modes having concentric nodal circles. The values of κ and λ are determined by the boundary conditions, which are, for radially symmetrical vibration:

Simply supported edge:

$$W = 0 \qquad M_{1r} = D\left(\frac{d^2W}{dr^2} + \frac{\mu}{a}\frac{dW}{dr}\right) = 0$$

Fixed edge:

$$W = 0 \qquad \frac{dW}{dr} = 0$$

TABLE 7.10   Natural Frequencies and Nodal Lines of Triangular Plates (*B. W. Anderson.*[33])

MODE \ k	2	4	8	14
FIRST a/k, a	7.194	7.122	7.080	7.068
SECOND	30.803	30.718	30.654	30.638
THIRD	61.131	90.105	157.70	265.98
FOURTH	148.8	259.4	493.4	853.6
k	2	4	7	
FIRST	5.887	6.617	6.897	
SECOND a/k, a	25.40	28.80	30.28	

a = LENGTH OF TRIANGLE
k = RATIO OF LENGTH TO WIDTH OF TRIANGLE
FOR OTHER TERMINOLOGY SEE TABLE 7.7

Free edge:

$$M_{1r} = D\left(\frac{d^2W}{dr^2} + \frac{\mu}{a}\frac{dW}{dr}\right) = 0 \qquad \frac{d}{dr}\left(\frac{d^2W}{dr^2} + \frac{1}{r}\frac{dW}{dr}\right) = 0$$

EXAMPLE 7.9.   The steel diaphragm of a radio earphone has an unsupported diameter of 2.0 in. and is 0.008 in. thick. Assuming that the edge is fixed, the lowest three frequencies for the free vibration in which only nodal circles occur are to be calculated, using the exact method and the Rayleigh and Ritz methods.

EXACT SOLUTION.   In this example $n = 0$, which makes $\cos(n\theta - \beta) = 1$; thus, Eq. (7.43) becomes

$$W = A[J_0(\kappa r) + \lambda I_0(\kappa r)]$$

where $J_0(i\kappa r) = I_0(\kappa r)$ and I_0 is a modified Bessel function of the first kind.

At the boundary where $r = a$,

$$\frac{\partial W}{\partial r} = A\kappa[-J_1(\kappa a) + \lambda I_1(\kappa a)] = 0 \qquad -J_1(\kappa a) + \lambda I_1(\kappa a) = 0$$

The deflection at $r = a$ is also zero:

$$J_0(\kappa a) + \lambda I_0(\kappa a) = 0$$

The frequency equation becomes

$$\lambda = \frac{J_1(\kappa a)}{I_1(\kappa a)} = -\frac{J_0(\kappa a)}{I_0(\kappa a)}$$

The first three roots of the frequency equation are: $\kappa a = 3.196, 6.306, 9.44$. The corresponding natural frequencies are, from Eq. (7.42),

$$\omega_n = \frac{10.21}{a^2}\sqrt{\frac{Dg}{\gamma h}} \qquad \frac{39.77}{a^2}\sqrt{\frac{Dg}{\gamma h}} \qquad \frac{88.9}{a^2}\sqrt{\frac{Dg}{\gamma h}}$$

For steel, $E = 30 \times 10^6$ lb/in.^{2}, $\gamma = 0.28$ lb/in.^{3}, and $\mu = 0.28$. Hence

$$D = \frac{Eh^3}{12(1-\mu^2)} = \frac{30 \times 10^6(0.008)^3}{12(1-0.078)} = 1.38 \text{ lb-in.}$$

Thus, the lowest natural frequency is

$$\omega_1 = 10.21\sqrt{\frac{(1.38)(386)}{(0.28)(0.008)}} = 4960 \text{ rad/sec} = 790 \text{ Hz}$$

The second frequency is 3070 Hz, and the third is 6880 Hz.

SOLUTION BY RAYLEIGH'S METHOD. The equations for strain and kinetic energies are given in Table 7.1. The strain energy for a plate with clamped edges becomes

$$V = \pi D \int_0^a \left(\frac{\partial^2 W}{\partial r^2} + \frac{1}{r}\frac{\partial W}{\partial r} \right)^2 r\, dr$$

The maximum kinetic energy is

$$T = \frac{\omega_n^2 \pi \gamma h}{g} \int_0^a W^2 r\, dr$$

An expression of the form $W = a_1[1 - (r/a)^2]^2$, which satisfies the conditions of zero deflection and slope at the boundary, is used. The first two derivatives are $\partial W/\partial r = a_1(-4r/a^2 + 4r^3/a^4)$ and $\partial^2 W/\partial r^2 = a_1(-4/a^2 + 12r^2/a^4)$. Using these values in the equations for strain and kinetic energy, $V = 32\pi Da_1^2/3a^2$ and $T = \omega_n^2\pi\gamma ha^2a_1^2/10g$. Equating these values and solving for the frequency,

$$\omega_n = \sqrt{\frac{320\,Dg}{3a^4\,\gamma h}} = \frac{10.33}{a^2}\sqrt{\frac{Dg}{\gamma h}}$$

This is somewhat higher than the exact frequency.

SOLUTION BY RITZ'S METHOD. Using an expression for the deflection of the form

$$W = a_1[1 - (r/a)^2]^2 + a_2[1 - (r/a)^2]^3$$

and applying the Ritz method, the following values are obtained for the first two frequencies:

$$\omega_1 = \frac{10.21}{a^2}\sqrt{\frac{Dg}{\gamma h}} \qquad \omega_2 = \frac{43.04}{a^2}\sqrt{\frac{Dg}{\gamma h}}$$

The details of the calculations giving this result are in the literature.[37] The first frequency agrees with the exact answer to four significant figures, while the second frequency is somewhat high. A closer approximation to the second frequency and approximations of the higher frequencies could be obtained by using additional terms in the deflection equation.

The frequencies of modes having n nodal diameters are:[37]

$n = 1$:
$$\omega_1 = \frac{21.22}{a^2} \sqrt{\frac{Dg}{\gamma h}}$$

$n = 2$:
$$\omega_2 = \frac{34.84}{a^2} \sqrt{\frac{Dg}{\gamma h}}$$

For a plate with its center fixed and edge free, and having m nodal circles, the frequencies are:[38]

m	0	1	2	3
$\omega_n a^2 / \sqrt{\dfrac{Dg}{\gamma h}}$	3.75	20.91	60.68	119.7

Stretching of Middle Plane.   In the usual analysis of plates, it is assumed that the deflection of the plate is so small that there is no stretching of the middle plane. If such stretching occurs, it affects the natural frequency. Whether it occurs depends on the conditions of support of the plate, the amplitude of vibration, and possibly other conditions. In a plate with its edges built in, a relatively small deflection causes a significant stretching. The effect of stretching is not proportional to the deflection; thus, the elastic restoring force is not a linear function of deflection. The natural frequency is not independent of amplitude but becomes higher with increasing amplitudes. If a plate is subjected to a pressure on one side, so that the vibration occurs about a deflected position, the effect of stretching may be appreciable. The effect of stretching in rectangular plates with immovable hinged supports has been discussed.[39] The effect of the amplitude on the natural frequency is shown in Fig. 7.11; the effect on the total stress in the plate is shown in Fig. 7.12. The natural frequency increases rapidly as the amplitude of vibration increases.

Rotational Motion and Shearing Forces.   In the foregoing analysis, only the motion of each element of the plate in the direction normal to the plane of the plate is considered. There is also rotation of each element, and there is a deflection associated with the lateral shearing forces in the plate. The effects of these factors becomes significant if the curvature of the plate is large relative to its thickness, i.e., for a plate in which the thickness is large compared to the lateral dimensions or when the plate is vibrating in a mode for which the nodal lines are close together. These effects have been analyzed for rectangular plates[40] and for circular plates.[41]

Complete Circular Rings.   Equations have been derived[42,43] for the natural frequencies of complete circular rings for which the radius is large compared to the thickness of the ring in the radial direction. Such rings can execute several types of free vibration, which are shown in Table 7.11 with the formulas for the natural frequencies.

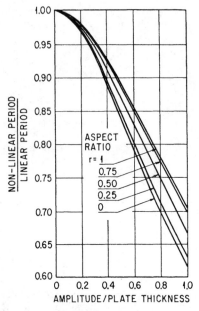

FIGURE 7.11 Influence of amplitude on period of vibration of uniform rectangular plates with immovable hinged edges. The aspect ratio r is the ratio of width to length of the plate. (*H. Chu and G. Herrmann.*[39])

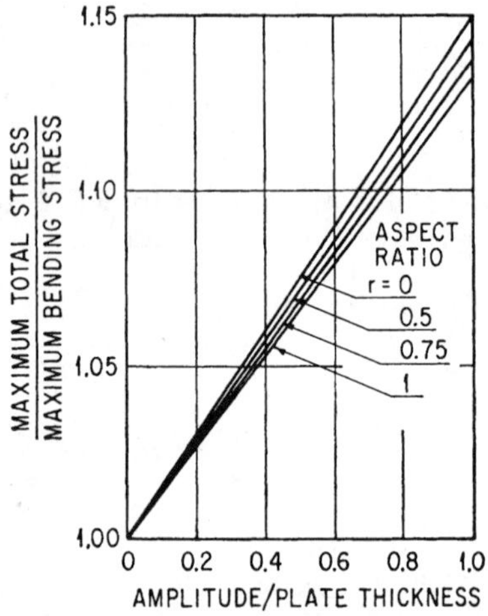

FIGURE 7.12 Influence of amplitude on maximum total stress in rectangular plates with immovable hinged edges. The aspect ratio r is the ratio of width to length of the plate. (*H. Chu and G. Herrmann.*[39])

TRANSFER MATRIX METHOD

In some assemblies which consist of various types of elements, e.g., beam segments, the solution for each element may be known. The transfer matrix method[44,45] is a procedure by means of which the solution for such elements can be combined to yield a frequency equation for the assembly. The associated mode shapes can then be determined. The method is an extension to distributed systems of the Holzer method, described in Chap. 38, in which torsional problems are solved by dividing an assembly into lumped masses and elastic elements, and of the Myklestad method,[46] in which a similar procedure is applied to beam problems. The method has been used[47] to find the natural frequencies and mode shapes of the internals of a nuclear reactor by modeling the various elements of the system as beam segments.

The method will be illustrated by setting up the frequency equation for a cantilever beam, Fig. 7.13, composed of three segments, each of which has uni-

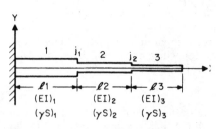

FIGURE 7.13 Cantilever beam made up of three segments having different section properties.

TABLE 7.11   Natural Frequencies of Complete Circular Rings Whose Thickness in Radial Direction Is Small Compared to Radius

TYPE OF VIBRATION	SHAPE OF LOWEST MODE	RECTANGULAR CROSS SECTION ω_n	CIRCULAR CROSS SECTION ω_n
FLEXURAL IN PLANE OF RING WITH n COMPLETE WAVE-LENGTH IN CIRCUMFERENCE	$n=2$	$\sqrt{\dfrac{Eg}{\gamma}\dfrac{I}{AR^4}\dfrac{n^2(n^2-1)^2}{n^2+1}}$ n ANY INTEGER > 1	$\sqrt{\dfrac{E\pi r^4}{4mR^4}\dfrac{n^2(n^2-1)^2}{n^2+1}}$ n ANY INTEGER > 1
FLEXURAL NORMAL TO PLANE OF RING	$n=2$		$\sqrt{\dfrac{E\pi r^4}{4mR^4}\dfrac{n^2(n^2-1)^2}{n^2+1+\mu}}$ n ANY INTEGER > 1
TORSIONAL		FIRST MODE $\sqrt{\dfrac{Eg}{\gamma R^2}\dfrac{I_x}{I_p}}$	$\sqrt{\dfrac{G\pi r^2}{mR^2}(n^2+1+\mu)}$ n=0, OR ANY INTEGER
EXTENSIONAL		$\sqrt{\dfrac{Eg}{\gamma R^2}}$	$\sqrt{\dfrac{E\pi r^2}{mR^2}(1+n^2)}$ n=0, OR ANY INTEGER

E = MODULUS OF ELASTICITY
G = MODULUS OF RIGIDITY
γ = WEIGHT DENSITY
n : DEFINED FOR EACH TYPE OF VIBRATION
R = RADIUS OF RING
μ = POISSON'S RATIO

PROPERTIES OF CROSS SECTIONS

I = MOMENT OF INERTIA WITH RESPECT TO AXIS OF SECTION
I_x = MOMENT OF INERTIA WITH RESPECT TO RADIAL LINE
I_p = POLAR MOMENT OF INERTIA
A = AREA
r = RADIUS
m = MASS PER UNIT OF LENGTH

form section properties. Only the effects of bending will be considered, but the method can be extended to include other effects, such as shear deformation and rotary motion of the cross section.[45] Application to other geometries is described in Ref. 45.

Depending on the type of element being considered, the values of appropriate parameters must be expressed at certain sections of the piece in terms of their values at other sections. In the beam problem, the deflection and its first three derivatives must be used.

Transfer Matrices.   Two types of transfer matrix are used. One, which for the beam problem is called the **R** matrix (after Lord Rayleigh[44]), yields the values of the parameters at the right end of a uniform segment of the beam in terms of their values at the left end of the segment. The other type of transfer matrix is the point matrix, which yields the values of the parameters just to the right of a joint between segments in terms of their values just to the left of the joint.

As can be seen by looking at the successive derivatives, the coefficients in Eq. (7.16) are equal to the following, where the subscript 0 indicates the value of the indicated parameter at the left end of the beam:

$$A = \frac{X_0}{2} \qquad C = \frac{X_0'}{2\kappa} \qquad B = \frac{-X_0''}{2\kappa^2} \qquad D = \frac{-X_0'''}{2\kappa^3}$$

Using the following notation, X and its derivatives at the right end of a beam segment can be expressed, by the matrix equation, in terms of the values at the left end of the segment. The subscript n refers to the number of the segment being considered, the subscript l to the left end of the segment and the subscript r to the right end.

$$C_{0n} = \frac{\cos \kappa_n l_n + \cosh \kappa_n l_n}{2}$$

$$S_{1n} = \frac{\sin \kappa_n l_n + \sinh \kappa_n l_n}{2\kappa_n}$$

$$C_{2n} = \frac{-(\cos \kappa_n l_n - \cosh \kappa_n l_n)}{2\kappa_n^2}$$

$$S_{3n} = \frac{-(\sin \kappa_n l_n - \sinh \kappa_n l_n)}{2\kappa_n^3}$$

where κ_n takes the value shown in Eq. (7.14) with the appropriate values of the parameters for the segment and l_n is the length of the segment.

$$
\begin{bmatrix} X \\ X' \\ X'' \\ X''' \end{bmatrix}_{rn}
=
\begin{bmatrix}
C_{0n} & S_{1n} & C_{2n} & S_{3n} \\
\kappa_n^4 S_{3n} & C_{0n} & S_{1n} & C_{2n} \\
\kappa_n^4 C_{2n} & \kappa_n^4 S_{3n} & C_{0n} & S_{1n} \\
\kappa_n^4 S_{1n} & \kappa_n^4 C_{2n} & \kappa_n^4 S_{3n} & C_{0n}
\end{bmatrix}
\begin{bmatrix} X \\ X' \\ X'' \\ X''' \end{bmatrix}_{ln}
$$

or $\mathbf{x}_{rn} = \mathbf{R}_n \mathbf{X}_{ln}$, where the boldface capital letter denotes a square matrix and the boldface lowercase letters denote column matrices. Matrix operations are discussed in Chap. 28.

At a section where two segments of a beam are joined, the deflection, the slope, the bending moment, and the shear must be the same on the two sides of the joint. Since $M = EI \cdot X''$ and $V = EI \cdot X'''$, the point transfer matrix for such a joint is as follows, where the subscript jn refers to the joint to the right of the nth segment of the beam:

$$
\begin{bmatrix} X \\ X' \\ X'' \\ X''' \end{bmatrix}_{rjn}
=
\begin{bmatrix}
1 & 0 & 0 & 0 \\
0 & 1 & 0 & 0 \\
0 & 0 & (EI)_l/(EI)_r & 0 \\
0 & 0 & 0 & (EI)_l/(EI)_r
\end{bmatrix}
\begin{bmatrix} X \\ X' \\ X'' \\ X''' \end{bmatrix}_{ljn}
$$

or $\mathbf{x}_{rjn} = \mathbf{J}_n \mathbf{x}_{ljn}$.

The Frequency Equation. For the cantilever beam shown in Fig. 7.13, the coefficients relating the values of X and its derivatives at the right end of the beam to their values at the left end are found by successively multiplying the appropriate $\mathbf{R}$ and $\mathbf{J}$ matrices, as follows:

$$\mathbf{x}_{r3} = \mathbf{R}_3 \mathbf{J}_2 \mathbf{R}_2 \mathbf{J}_1 \mathbf{R}_1 \mathbf{x}_{l1}$$

Carrying out the multiplication of the square $\mathbf{R}$ and $\mathbf{J}$ matrices and calling the resulting matrix $\mathbf{P}$ yields

$$\begin{bmatrix} X \\ X' \\ X'' \\ X''' \end{bmatrix}_{r3} = \begin{bmatrix} P_{11} & P_{12} & P_{13} & P_{14} \\ P_{21} & P_{22} & P_{23} & P_{24} \\ P_{31} & P_{32} & P_{33} & P_{34} \\ P_{41} & P_{42} & P_{43} & P_{44} \end{bmatrix} \begin{bmatrix} X \\ X' \\ X'' \\ X''' \end{bmatrix}_{l1}$$

The boundary conditions at the fixed left end of the cantilever beam are $X = X' = 0$. Using these and performing the multiplication of $\mathbf{P}$ by $\mathbf{x}_{l1}$ yields the following:

$$X_{r3} = P_{13}X_{l1}'' + P_{14}X_{l1}'''$$

$$X_{r3}' = P_{23}X_{l1}'' + P_{24}X_{l1}'''$$

$$X_{r3}'' = P_{33}X_{l1}'' + P_{34}X_{l1}'''$$

(7.44)

$$X_{r3}''' = P_{43}X_{l1}'' + P_{44}X_{l1}'''$$

The boundary conditions for the free right end of the beam are $X'' = X''' = 0$. Using these in the last two equations results in two simultaneous homogeneous equations, so that the following determinant, which is the frequency equation, results:

$$\begin{vmatrix} P_{33} & P_{34} \\ P_{43} & P_{44} \end{vmatrix} = 0$$

It can be seen that for a beam consisting of only one segment, this determinant yields a result which is equivalent to Eq. (7.17).

While in theory it would be possible to multiply the successive $\mathbf{R}$ and $\mathbf{J}$ matrices and obtain the $\mathbf{P}$ matrix in literal form, so that the transcendental frequency equation could be written, the process, in all but the simplest problems, would be long and time-consuming. A more practicable procedure is to perform the necessary multiplications with numbers, using a digital computer, and finding the roots by trial and error.

Mode Shapes.   Either of the last two equations of Eq. (7.44) may be used to find the ratio X_{l1}''/X_{l1}'''. These are used in Eq. (7.16), with $\kappa = \kappa_1$ to find the shape of the first segment. By the use of the $\mathbf{R}$ and $\mathbf{J}$ matrices the values of the coefficients in Eq. (7.16) are found for each of the other segments.

With intermediate rigid supports or pinned connections, numerical difficulties occur in the solution of the frequency equation. These difficulties are eliminated by the use of delta matrices, the elements of which are combinations of the elements of the $\mathbf{R}$ matrix. These delta matrices, for various cases, are tabulated in Refs. 44 and 45.

In Ref. 47 transfer matrices are developed and used for structures which consist, in part, of beams that are parallel to each other.

FORCED VIBRATION

CLASSICAL SOLUTION

The classical method of analyzing the forced vibration that results when an elastic system is subjected to a fluctuating load is to set up the equation of motion by the application of Newton's second law. During the vibration, each element of the system is subjected to elastic forces corresponding to those experienced during free vibration; in addition, some of the elements are subjected to the disturbing force. The equation which governs the forced vibration of a system can be obtained by adding the disturbing force to the equation for free vibration. For example, in Eq. (7.13) for the free vibration of a uniform beam, the term on the left is due to the elastic forces in the beam. If a force $F(x,t)$ is applied to the beam, the equation of motion is obtained by adding this force to Eq. (7.13), which becomes, after rearranging terms,

$$EI \frac{\partial^4 y}{\partial x^4} + \frac{\gamma S}{g} \frac{\partial^2 y}{\partial t^2} = F(x,t)$$

where EI is a constant. The solution of this equation gives the motion that results from the force F. For example, consider the motion of a beam with hinged ends subjected to a sinusoidally varying force acting at its center. The solution is obtained by representing the concentrated force at the center by its Fourier series:

$$EIy'''' + \frac{\gamma S}{g} \ddot{y} = \frac{2F}{l} \left[\sin \frac{\pi x}{l} - \sin \frac{3\pi x}{l} + \sin \frac{5\pi x}{l} \cdots \right] \sin \omega t$$

$$= \frac{2F}{l} \sum_{n=1}^{n=\infty} \left(\sin \frac{n\pi}{2} \sin \frac{n\pi x}{l} \right) \sin \omega t \qquad (7.45)$$

where $\sin(n\pi/2)$, which appears in each term of the series, makes the nth term positive, negative, or zero. The solution of Eq. (7.45) is

$$y = \sum_{n=1}^{n=\infty} \left[A_n \sin \frac{n\pi x}{l} \sin \omega_n t + B_n \sin \frac{n\pi x}{l} \cos \omega_n t \right.$$

$$\left. + \sin \frac{n\pi}{2} \frac{2Fg/S\gamma l}{(n\pi/l)^4(EIg/S\gamma) - \omega^2} \sin \frac{n\pi x}{l} \sin \omega t \right] \qquad (7.46)$$

The first two terms of Eq. (7.46) are the values of y which make the left side of Eq. (7.45) equal to zero. They are obtained in exactly the same way as in the solution of the free-vibration problem and represent the free vibration of the beam. The constants are determined by the initial conditions; in any real beam, damping causes the free vibration to die out. The third term of Eq. (7.46) is the value of y which makes the left-hand side of Eq. (7.45) equal the right-hand side; this can be verified by substitution. The third term represents the forced vibration. From Table 7.3, $\kappa_n l = n\pi$ for a beam with hinged ends; then from Eq. (7.14), $\omega_n^2 = n^4\pi^4 EIg/S\gamma l^4$. The term representing the forced vibration in Eq. (7.46) can be written, after rearranging terms,

$$y = \frac{2Fg}{S\gamma l} \sum_{n=1}^{n=\infty} \frac{\sin(n\pi/2)}{\omega_n^2[1 - (\omega/\omega_n)^2]} \sin \frac{n\pi x}{l} \sin \omega t \qquad (7.47)$$

From Table 7.3 and Eq. (7.16), it is evident that this deflection curve has the same shape as the nth normal mode of vibration of the beam since, for free vibration of a beam with hinged ends, $X_n = 2C \sin \kappa x = \sin(n\pi x/l)$.

The equation for the deflection of a beam under a distributed static load $F(x)$ can be obtained by replacing $-(\gamma S/g)\ddot{y}$ with F in Eq. (7.12); then Eq. (7.13) becomes

$$y_s'''' = \frac{F(x)}{EI} \qquad (7.48)$$

where EI is a constant. For a static loading $F(x) = 2F/l \sin n\pi/2 \sin n\pi x/l$ corresponding to the nth term of the Fourier series in Eq. (7.45), Eq. (7.48) becomes $y_{sn}'''' = 2F/EIl \sin n\pi/2 \sin n\pi x/l$. The solution of this equation is

$$y_{sn} = \frac{2F}{EIl} \left(\frac{l}{n\pi} \right)^4 \sin \frac{n\pi}{2} \sin \frac{n\pi x}{l}$$

Using the relation $\omega_n^2 = n^4\pi^4 EIg/S\gamma l^4$, this can be written

$$y_{sn} = \frac{2Fg}{\omega_n^2 S\gamma l} \sin \frac{n\pi x}{l} \sin \frac{n\pi}{2}$$

Thus, the nth term of Eq. (7.47) can be written

$$y_n = y_{sn} \frac{1}{1 - (\omega/\omega_n)^2} \sin \omega t$$

Thus, the amplitude of the forced vibration is equal to the static deflection under the Fourier component of the load multiplied by the "amplification factor" $1/[1 - (\omega/\omega_n)^2]$. This is the same as the relation that exists, for a system having a single degree-of-freedom, between the static deflection under a load F and the amplitude under a fluctuating load $F \sin \omega t$. Therefore, insofar as each mode alone is concerned, the beam behaves as a system having a single degree-of-freedom. If the beam is subjected to a force fluctuating at a single frequency, the amplification factor is small except when the frequency of the forcing force is near the natural frequency of a mode. For all even values of n, $\sin (n\pi/2) = 0$; thus, the even-numbered modes are not excited by a force acting at the center, which is a node for those modes. The distribution of the static load that causes the same pattern of deflection as the beam assumes during each mode of vibration has the same form as the deflection of the beam. This result applies to other beams since a comparison of Eqs. (7.15) and (7.48) shows that if a static load $F = (\omega_n^2\gamma S/g)y$ is applied to any beam, it will cause the same deflection as occurs during the free vibration in the nth mode.

The results for the simply supported beam are typical of those which are obtained for all systems having distributed mass and elasticity. Vibration of such a system at resonance is excited by a force which fluctuates at the natural frequency of a mode, since nearly any such force has a component of the shape necessary to excite the vibration. Even if the force acts at a nodal point of the mode, vibration may be excited because of coupling between the modes.

METHOD OF WORK

Another method of analyzing forced vibration is by the use of the theorem of virtual work and D'Alembert's principle. The theorem of virtual work states that when any elastic body is in equilibrium, the total work done by all external forces during any virtual displacement equals the increase in the elastic energy stored in the body. A virtual displacement is an arbitrary small displacement that is compatible with the geometry of the body and which satisfies the boundary conditions.

In applying the principle of work to forced vibration of elastic bodies, the problem is made into one of equilibrium by the application of D'Alembert's principle. This permits a problem in dynamics to be considered as one of statics by adding to the equation of static equilibrium an "inertia force" which, for each part of the body, is equal to the product of the mass and the acceleration. Using this principle, the theorem of virtual work can be expressed in the following equation:

$$\Delta V = \Delta(F_I + F_E) \tag{7.49}$$

in which V is the elastic strain energy in the body, F_I is the inertia force, F_E is the external disturbing force, and Δ indicates the change of the quantity when the body undergoes a virtual displacement. The various quantities can be found separately.

For example, consider the motion of a uniform beam having hinged ends with a sinusoidally varying force acting at the center, and compare the result with the solution obtained by the classical method. All possible motions of any beam can be represented by a series of the form

$$y = q_1 X_1 + q_2 X_2 + q_3 X_3 + \cdots = \sum_{n=1}^{n=\infty} q_n X_n \tag{7.50}$$

in which the X's are functions representing displacements in the normal modes of vibration and the q's are coefficients which are functions of time. The determination of the values of q_n is the problem to be solved. For a beam having hinged ends, Eq. (7.50) becomes

$$y = \sum_{n=1}^{n=\infty} q_n \sin \frac{n\pi x}{l} \tag{7.51}$$

This is evident by using the values of $\kappa_n l$ from Table 7.3 in Eq. (7.16). A virtual displacement, being any arbitrary small displacement, can be assumed to be

$$\Delta y = \Delta q_m X_m = \Delta q_m \sin \frac{m\pi x}{l}$$

The elastic strain energy of bending of the beam is

$$V = \frac{EI}{2} \int_0^l \left(\frac{\partial^2 y}{\partial x^2}\right)^2 dx = \frac{EI}{2} \sum_{n=1}^{n=\infty} q_n{}^2 \int_0^l \left[\frac{\partial^2}{\partial x^2}\left(\sin \frac{n\pi x}{l}\right)\right]^2 dx$$

$$= \frac{EI}{2} \sum_{n=1}^{n=\infty} q_n{}^2 \left(\frac{n\pi}{l}\right)^4 \int_0^l \left(\sin \frac{n\pi x}{l}\right)^2 dx = \frac{EI}{2} \sum_{n=1}^{n=\infty} q_n{}^2 \left(\frac{n\pi}{l}\right)^4 \frac{l}{2}$$

For the virtual displacement, the change of elastic energy is

$$\Delta V = \frac{\partial V}{\partial q_m} \Delta q_m = \frac{EI}{2l^3} (n\pi)^4 q_m \Delta q_m = \frac{EI}{2l^3} (\kappa_n l)^4 q_m \Delta q_m$$

The value of the inertia force at each section is

$$F_I = -\frac{\gamma S}{g} \ddot{y} = -\frac{\gamma S}{g} \sum_{n=1}^{n=\infty} \frac{d^2 q_n}{dt^2} \sin \frac{n\pi x}{l}$$

The work done by this force during the virtual displacement Δy is

$$\Delta F_I = F_I \, \Delta y = -\frac{\gamma S}{g} \sum_{n=1}^{n=\infty} \frac{d^2 q_n}{dt^2} \Delta q_m \int_0^l \sin \frac{n\pi x}{l} \sin \frac{m\pi x}{l} \, dx$$

$$= -\frac{\gamma S l}{2g} \frac{d^2 q_m}{dt^2} \Delta q_m$$

The orthogonality relation of Eq. (7.1) is used here, making the integral vanish when $n = m$. For a disturbing force F_E, the work done during the virtual displacement is

$$\Delta F_E = F_E\,\Delta y = F(X_m)_{x\,=\,c}\,\Delta q_m$$

in which $(X_m)_{x\,=\,c}$ is the value of X_m at the point of application of the load. Substituting the terms into Eq. (7.49),

$$\frac{\gamma Sl}{2g}\,\ddot{q}_m + \frac{EI}{2l^3}\,(\kappa_m l)^4 q_m = F(X_m)_{x\,=\,c}$$

Rearranging terms and letting $EI/S\gamma = a^2$,

$$\ddot{q}_m + \kappa_m^4 a^2 q_m = \frac{2g}{\gamma Sl}\,F(X_m)_{x\,=\,c} \qquad (7.52)$$

If F_E is a force which varies sinusoidally with time at any point $x = c$,

$$F(X_m)_{x\,=\,c} = \bar{F} \sin \frac{m\pi c}{l}\,\sin \omega t$$

and Eq. (7.52) becomes

$$\ddot{q}_m + \kappa_m^4 a^2 q_m = \frac{2g\bar{F}}{\gamma Sl}\,\sin \frac{m\pi c}{l}\,\sin \omega t$$

The solution of this equation is

$$q_m = A_m \sin \kappa_m^2 at + B_m \cos \kappa_m^2 at + \frac{2\bar{F}g}{\gamma Sl}\,\frac{\sin(m\pi c/l)}{\kappa_m^4 a^2 - \omega^2}\,\sin \omega t$$

Since $\kappa_m^2 a = \omega_m$,

$$q_m = A_m \sin \omega_m t + B_m \cos \omega_m t + \frac{2\bar{F}g}{\gamma Sl}\,\frac{\sin(m\pi c/l)}{\omega_m^2 - \omega^2}\,\sin \omega t$$

when the force acts at the center $c/l = \frac{1}{2}$. Substituting the corresponding values of q in Eq. (7.51), the solution is identical to Eq. (7.46), which was obtained by the classical method.

VIBRATION RESULTING FROM MOTION OF SUPPORT

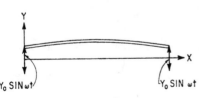

FIGURE 7.14 Simply supported beam undergoing sinusoidal motion induced by sinusoidal motion of the supports.

When the supports of an elastic body are vibrated by some external force, forced vibration may be induced in the body.[48] For example, consider the motion that results in a uniform beam, Fig. 7.14, when the supports are moved through a sinusoidally varying displacement $(y)_{x\,=\,0,l} = Y_0 \sin \omega t$. Although Eq. (7.13) was developed for the free vibration of beams, it is applicable to the present problem because there is no force acting on any section of the beam except the elastic force associated with the bending of the beam. If a solution of the form $y = X(x) \sin \omega t$ is assumed and substituted into Eq. (7.13):

$$X'''' = \frac{\omega^2 \gamma S}{EIg} X \qquad (7.53)$$

This equation is the same as Eq. (7.15) except that the natural frequency $\omega_n{}^2$ is replaced by the forcing frequency ω^2. The solution of Eq. (7.53) is the same except that κ is replaced by $\kappa' = (\omega^2 \gamma S / EIg)^{1/4}$:

$$X = A_1 \sin \kappa'x + A_2 \cos \kappa'x + A_3 \sinh \kappa'x + A_4 \cosh \kappa'x \qquad (7.54)$$

The solution of the problem is completed by finding the constants, which are determined by the boundary conditions. Certain boundary conditions are associated with the supports of the beam and are the same as occur in the solution of the problem of free vibration. Additional conditions are supplied by the displacement through which the supports are forced. For example, if the supports of a beam having hinged ends are moved sinusoidally, the boundary conditions are: at $x = 0$ and $x = l$, $X'' = 0$, since the moment exerted by a hinged end is zero, and $X = Y_0$, since the amplitude of vibration is prescribed at each end. By the use of these boundary conditions, Eq. (7.54) becomes

$$X = \frac{Y_0}{2} \left[\tan \frac{\kappa'l}{2} \sin \kappa'x + \cos \kappa'x - \tanh \frac{\kappa'l}{2} \sinh \kappa'x + \cosh \kappa'x \right] \qquad (7.55)$$

The motion is defined by $y = X \sin \omega t$. For all values of κ', each of the coefficients except the first in Eq. (7.55) is finite. The tangent term becomes infinite if $\kappa'l = n\pi$, for odd values of n. The condition for the amplitude to become infinite is $\omega = \omega_n$ because $\kappa'/\kappa = \omega^2/\omega_n{}^2$ and, for natural vibration of a beam with hinged ends, $\kappa_n l = n\pi$. Thus, if the supports of an elastic body are vibrated at a frequency close to a natural frequency of the system, vibration at resonance occurs.

DAMPING

The effect of damping on forced vibration can be discussed only qualitatively. Damping usually decreases the amplitude of vibration, as it does in systems having a single degree-of-freedom. In some systems, it may cause coupling between modes, so that motion in a mode of vibration that normally would not be excited by a certain disturbing force may be induced.

REFERENCES

1. Timoshenko, S.: "Vibration Problems in Engineering," 3d ed., pp. 442, 448, D. Van Nostrand Company, Inc., Princeton, N.J., 1955.

2. Den Hartog, J. P.: "Mechanical Vibrations," 4th ed., p. 161, McGraw-Hill Book Company, Inc., New York, 1956.

3. Ref. 2, p. 152.

4. Hansen, H. M., and P. F. Chenea: "Mechanics of Vibration," p. 274, John Wiley & Sons, Inc., New York, 1952.

5. Jacobsen, L. S., and R. S. Ayre: "Engineering Vibrations," p. 73, McGraw-Hill Book Company, Inc., New York, 1958.

6. Ref. 4, p. 256.
7. Kerley, J. J.: *Prod. Eng., Design Digest Issue,* Mid-October, 1957, p. F34.
8. Young, D., and R. P. Felgar: "Tables of Characteristic Functions Representing Normal Modes of Vibration of a Beam," *Univ. Texas Bur. Eng. Research Bull.* 44, July 1, 1949.
9. Rayleigh, Lord: "The Theory of Sound," 2d rev. ed., vol. 1, p. 293; reprinted by Dover Publications, New York, 1945.
10. Timoshenko, S.: *Phil. Mag.* (ser. 6), **41:**744 (1921); **43:**125 (1922).
11. Sutherland, J. G., and L. E. Goodman: "Vibrations of Prismatic Bars Including Rotatory Inertia and Shear Corrections," Department of Civil Engineering, University of Illinois, Urbana, Ill., April 15, 1951.
12. Ref. 1, p. 374.
13. Ref. 1, 2d ed., p. 366.
14. Woinowsky-Krieger, S.: *J. Appl. Mechanics,* **17:**35 (1950).
15. Cranch, E. T., and A. A. Adler: *J. Appl. Mechanics,* **23:**103 (1956).
16. Macduff, J. N., and R. P. Felgar: *Trans. ASME,* **79:**1459 (1957).
17. Macduff, J. N., and R. P. Felgar: *Machine Design,* **29**(3):109 (1957).
18. Ref. 1, p. 386.
19. Darnley, E. R.: *Phil. Mag.,* **41:**81 (1921).
20. Smith, D. M.: *Engineering,* **120:**808 (1925).
21. Newmark, N. M., and A. S. Veletsos: *J. Appl. Mechanics,* **19:**563 (1952).
22. Timoshenko, S.: "Theory of Plates and Shells," p. 301, McGraw-Hill Book Company, Inc., New York, 1940.
23. Ref. 22, p. 88.
24. Ref. 22, p. 133.
25. Evans, T. H.: *J. Appl. Mechanics,* **6:**A-7 (1939).
26. Ref. 22, p. 58.
27. Ref. 22, p. 304.
28. Ref. 22, p. 344.
29. Leissa, A. W.: "Vibration of Plates," NASA SP-160, 1969.
30. Young, D.: *J. Appl. Mechanics,* **17:**448 (1950).
31. Barton, M. V.: *J. Appl. Mechanics,* **18:**129 (1951).
32. Hearmon, R. F. S.: *J. Appl. Mechanics,* **19:**402 (1952).
33. Anderson, B. W.: *J. Appl. Mechanics,* **21:**365 (1954).
34. Gustafson, P. N., W. F. Stokey, and C. F. Zorowski: *J. Aeronaut. Sci.,* **20:**331 (1953).
35. Gustafson, P. N., W. F. Stokey, and C. F. Zorowski: *J. Aeronaut. Sci.,* **21:**621 (1954).
36. Ref. 9, p. 359.
37. Ref. 1, p. 449.
38. Southwell, R. V.: *Proc. Roy. Soc.* (*London*), **A101:**133 (1922).
39. Chu, Hu-Nan, and G. Herrmann: *J. Appl. Mechanics,* **23:**532 (1956).
40. Mindlin, R. D., A. Schacknow, and H. Deresiewicz: *J. Appl. Mechanics,* **23:**430 (1956).
41. Deresiewicz, H., and R. D. Mindlin: *J. Appl. Mechanics,* **22:**86 (1955).
42. Love, A. E. H.: "A Treatise on the Mathematical Theory of Elasticity," 4th ed., p. 451, reprinted by Dover Publications, New York, 1944.
43. Ref. 1, p. 425.

44. Marguerre, K.: *J. Math. Phys.*, **35:**28 (1956).

45. Pestel, E. C., and F. A. Leckie: "Matrix Methods in Elastomechanics," McGraw-Hill Book Company, Inc., New York, 1963.

46. Myklestad, N. O.: *J. Aeronaut. Sci.*, **11:**153 (1944).

47. Bohm, G. J.: *Nucl. Sci. Eng.*, **22:**143 (1965).

48. Mindlin, R. D., and L. E. Goodman: *J. Appl. Mech.*, **17:**377 (1950).

CHAPTER 8

TRANSIENT RESPONSE TO STEP AND PULSE FUNCTIONS*

Robert S. Ayre

INTRODUCTION

In analyses involving shock and transient vibration, it is essential in most instances to begin with the time-history of a quantity that describes a motion, usually displacement, velocity, or acceleration. The method of reducing the time-history depends upon the purpose for which the reduced data will be used. When the purpose is to compare shock motions, to design equipment to withstand shock, or to formulate a laboratory test as means to simulate an environmental condition, the *response spectrum* is found to be a useful concept. This concept in data reduction is discussed in Chap. 23, and its application to environmental conditions is discussed in Chap. 24.

This chapter deals briefly with methods of analysis for obtaining the response spectrum from the time-history, and includes in graphical form certain significant spectra for various regular step- and pulse-type excitations. The usual concept of the response spectrum is based upon the single degree-of-freedom system, usually considered linear and undamped, although useful information sometimes can be obtained by introducing nonlinearity or damping. The single degree-of-freedom system is considered to be subjected to the shock or transient vibration, and its response determined.

The *response spectrum* is a graphical presentation of a selected quantity in the response taken with reference to a quantity in the excitation. It is plotted as a function of a dimensionless parameter that includes the natural period of the responding system and a significant period of the excitation. The excitation may be defined in terms of various physical quantities, and the response spectrum likewise may depict various characteristics of the response.

---

* Chapter 8 is based on Chaps. 3 and 4 of "Engineering Vibrations," by L. S. Jacobsen and R. S. Ayre, McGraw-Hill Book Company, Inc., 1958.

LINEAR, UNDAMPED, SINGLE DEGREE-OF-FREEDOM SYSTEMS

DIFFERENTIAL EQUATION OF MOTION

It is assumed that the system is linear and undamped. The excitation, which is a known function of time alone, may be a force function $F(t)$ acting directly on the mass of the system (Fig. 8.1A) or it may be a ground motion, i.e., foundation or base motion, acting on the spring anchorage. The ground motion may be expressed as a ground displacement function $u(t)$ (Fig. 8.1B). In many cases, however, it is more useful to express it as a ground acceleration function $\ddot{u}(t)$ (Fig. 8.1C).

The differential equation of motion, written in terms of each of the types of excitation, is given in Eqs. (8.1a), (8.1b), and (8.1c).

$$m\ddot{x} = -kx + F(t) \quad \text{or} \quad \frac{m\ddot{x}}{k} + x = \frac{F(t)}{k} \qquad (8.1a)$$

$$m\ddot{x} = -k[x - u(t)] \quad \text{or} \quad \frac{m\ddot{x}}{k} + x = u(t) \qquad (8.1b)$$

$$m[\ddot{\delta}_x + \ddot{u}(t)] = -k\delta_x \quad \text{or} \quad \frac{m\ddot{\delta}_x}{k} + \delta_x = -\frac{m\ddot{u}(t)}{k} \qquad (8.1c)$$

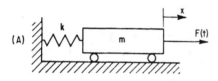

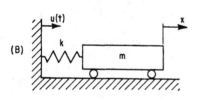

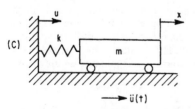

FIGURE 8.1 Simple oscillator acted upon by known excitation functions of time: (A) force $F(t)$, (B) ground displacement $u(t)$, (C) ground acceleration $\ddot{u}(t)$.

where x is the displacement (absolute displacement) of the mass relative to a *fixed reference* and δ_x is the displacement relative to a *moving anchorage* or ground. These displacements are related to the ground displacement by $x = u + \delta_x$. Similarly, the accelerations are related by $\ddot{x} = \ddot{u} + \ddot{\delta}_x$.

Furthermore, if Eq. (8.1b) is differentiated twice with respect to time, a differential equation is obtained in which ground acceleration $\ddot{u}(t)$ is the excitation and the absolute acceleration $\ddot{x}$ of the mass m is the variable. The equation is

$$\frac{m}{k}\frac{d^2\ddot{x}}{dt^2} + \ddot{x} = \ddot{u}(t) \qquad (8.1d)$$

If Eq. (8.1d) is treated as a second-order equation in $\ddot{x}$ as the dependent variable, it is of the same general form as Eqs. (8.1a), (8.1b), and (8.1c).

Occasionally, the excitation is known in terms of ground velocity $\dot{u}(t)$. Differentiating Eq. (8.1b) once with respect to time, the following second-order equation in $\dot{x}$ is obtained:

$$\frac{m}{k}\frac{d^2\dot{x}}{dt^2} + \dot{x} = \dot{u}(t) \qquad (8.1e)$$

The analogy represented by Eqs. (8.1b), (8.1d), and (8.1e) may be extended further since it is generally possible to differentiate Eq. (8.1b) any number of times n:

$$\frac{m}{k} \frac{d^2}{dt^2}\left(\frac{d^n x}{dt^n}\right) + \left(\frac{d^n x}{dt^n}\right) = \left(\frac{d^n u}{dt^n}\right)(t) \qquad (8.1f)$$

This is of the same general form as the preceding equations if it is considered to be a second-order equation in $(d^n x/dt^n)$ as the response variable, with $(d^n u/dt^n)$ (t), a known function of time, as the excitation.

ALTERNATE FORMS OF THE EXCITATION AND OF THE RESPONSE

The foregoing equations are alike, mathematically, and a solution in terms of one of them may be applied to any of the others by making simple substitutions. Therefore, the equations may be expressed in the single general form:

$$\frac{m}{k}\ddot{v} + v = \xi(t) \qquad (8.2)$$

where v and ξ are the *response* and the *excitation,* respectively, at time t.

A general notation (v and ξ) is desirable in the presentation of response functions and response spectra for general use. However, in the discussion of examples of solution, it sometimes is preferable to use more specific notations. Both types of notation are used in this chapter. For ready reference, the alternate forms of the excitation and the response are given in Table 8.1 where $\omega_n^2 = k/m$.

TABLE 8.1 Alternate Forms of Excitation and Response in Eq. (8.2)

Excitation $\xi(t)$		Response v	
Force	$\dfrac{F(t)}{k}$	Absolute displacement	x
Ground displacement	$u(t)$	Absolute displacement	x
Ground acceleration	$\dfrac{-\ddot{u}(t)}{\omega_n^2}$	Relative displacement	δ_x
Ground acceleration	$\ddot{u}(t)$	Absolute acceleration	$\ddot{x}$
Ground velocity	$\dot{u}(t)$	Absolute velocity	$\dot{x}$
nth derivative of ground displacement	$\dfrac{d^n u}{dt^n}(t)$	nth derivative of absolute displacement	$\dfrac{d^n x}{dt^n}$

METHODS OF SOLUTION OF THE DIFFERENTIAL EQUATION

A brief review of four methods of solution is given in the following sections.

Classical Solution. The complete solution of the linear differential equation of motion consists of the sum of the *particular integral* x_1 and the *complementary function* x_2, that is, $x = x_1 + x_2$. Since the differential equation is of second order, two constants of integration are involved. They appear in the complementary function and are evaluated from a knowledge of the initial conditions.

Example 8.1. Versed-sine Force Pulse. In this case the differential equation of motion, applicable for the duration of the pulse, is

$$\frac{m\ddot{x}}{k} + x = \frac{F_p}{k}\frac{1}{2}\left(1 - \cos\frac{2\pi t}{\tau}\right) \qquad [0 \le t \le \tau] \qquad (8.3a)$$

where, in terms of the general notation, the excitation function $\xi(t)$ is

$$\xi(t) \equiv \frac{F(t)}{k} = \frac{F_p}{k}\frac{1}{2}\left(1 - \cos\frac{2\pi t}{\tau}\right)$$

and the response v is displacement x. The maximum value of the pulse excitation force is F_p.

The particular integral (particular solution) for Eq. (8.3a) is of the form

$$x_1 = M + N \cos\frac{2\pi t}{\tau} \qquad (8.3b)$$

By substitution of the particular solution into the differential equation, the required values of the coefficients M and N are found.

The complementary function is

$$x_2 = A \cos\omega_n t + B \sin\omega_n t \qquad (8.3c)$$

where A and B are the constants of integration. Combining x_2 and the explicit form of x_1 gives the complete solution:

$$x = x_1 + x_2 = \frac{F_p/2k}{1 - \tau^2/T^2}\left(1 - \frac{\tau^2}{T^2} + \frac{\tau^2}{T^2}\cos\frac{2\pi t}{\tau}\right) + A\cos\omega_n t + B\sin\omega_n t \qquad (8.3d)$$

If it is assumed that the system is initially at rest, $x = 0$ and $\dot{x} = 0$ at $t = 0$, and the constants of integration are

$$A = -\frac{F_p/2k}{1 - \tau^2/T^2} \qquad \text{and} \qquad B = 0 \qquad (8.3e)$$

The complete solution takes the following form:

$$v \equiv x = \frac{F_p/2k}{1 - \tau^2/T^2}\left(1 - \frac{\tau^2}{T^2} + \frac{\tau^2}{T^2}\cos\frac{2\pi t}{\tau} - \cos\omega_n t\right) \qquad (8.3f)$$

If other starting conditions had been assumed, A and B would have been different from the values given by Eqs. (8.3e). It may be shown that if the starting conditions are general, namely, $x = x_0$ and $\dot{x} = \dot{x}_0$ at $t = 0$, it is necessary to superimpose on the complete solution already found, Eq. (8.3f), only the following additional terms:

$$x_0 \cos\omega_n t + \frac{\dot{x}_0}{\omega_n}\sin\omega_n t \qquad (8.3g)$$

For values of time equal to or greater than τ, the differential equation is

$$m\ddot{x} + kx = 0 \qquad [\tau \le t] \qquad (8.4a)$$

and the complete solution is given by the complementary function alone. However, the constants of integration must be redetermined from the known conditions of the system at time $t = \tau$. The solution is

$$v \equiv x = \frac{F_p}{k} \frac{\sin (\pi\tau/T)}{1 - \tau^2/T^2} \sin \omega_n \left(t - \frac{\tau}{2}\right) \qquad [\tau \le t] \qquad (8.4b)$$

The additional terms given by expressions (8.3g) may be superimposed on this solution if the conditions at time $t = 0$ are general.

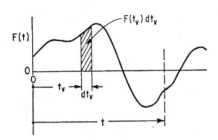

F(t)

$-F(t_v)\,dt_v$

$t_v \longrightarrow$ dt_v

t

FIGURE 8.2 General excitation and the elemental impulse.

Duhamel's Integral. The use of Duhamel's integral (convolution integral or superposition integral) is a well-known approach to the solution of transient vibration problems in linear systems. Its development[7, 33] is based on the *superposition* of the responses of the system to a sequence of impulses.

A general excitation function is shown in Fig. 8.2, where $F(t)$ is a known force function of time, the variable of integration is t_v between the limits of integration 0 and t, and the elemental impulse is $F(t_v)\,dt_v$. It may be shown that the complete solution of the differential equation is

$$x = \left(x_0 - \frac{1}{m\omega_n} \int_0^t F(t_v) \sin \omega_n t_v\, dt_v\right) \cos \omega_n t + \left(\frac{\dot{x}_0}{\omega_n} + \frac{1}{m\omega_n} \int_0^t F(t_v) \cos \omega_n t_v\, dt_v\right) \sin \omega_n t$$

$$(8.5)$$

where x_0 and $\dot{x}_0$ are the initial conditions of the system at zero time.

Example 8.2. Half-cycle Sine, Ground Displacement Pulse. Consider the following excitation:

$$\xi(t) \equiv u(t) = \begin{cases} u_p \sin \dfrac{\pi t}{\tau} & [0 \le t \le \tau] \\ 0 & [\tau \le t] \end{cases}$$

The maximum value of the excitation displacement is u_p. Assume that the system is initially at rest, so that $x_0 = \dot{x}_0 = 0$. Expressing the excitation function in terms of the variable of integration t_v, Eq. (8.5) may be rewritten for this particular case in the following form:

$$x = \frac{ku_p}{m\omega_n}\left(-\cos \omega_n t \int_0^t \sin \frac{\pi t_v}{\tau} \sin \omega_n t_v\, dt_v + \sin \omega_n t \int_0^t \sin \frac{\pi t_v}{\tau} \cos \omega_n t_v\, dt_v\right)$$

$$(8.6a)$$

Equation (8.6a) may be reduced, by evaluation of the integrals, to

$$v \equiv x = \frac{u_p}{1 - T^2/4\tau^2}\left(\sin \frac{\pi t}{\tau} - \frac{T}{2\tau} \sin \omega_n t\right) \qquad [0 \le t \le \tau] \qquad (8.6b)$$

where $T = 2\pi/\omega_n$ is the natural period of the responding system.

For the second era of time, where $\tau \le t$, it is convenient to choose a new time variable $t' = t - \tau$. Noting that $u(t) = 0$ for $\tau \le t$, and that for continuity in the system response the initial conditions for the second era must equal the closing conditions for the first era, it is found from Eq. (8.5) that the response for the second era is

$$x = x_\tau \cos \omega_n t' + \frac{\dot{x}_\tau}{\omega_n} \sin \omega_n t' \qquad (8.7a)$$

where x_τ and $\dot{x}_\tau$ are the displacement and velocity of the system at time $t = \tau$ and hence at $t' = 0$. Equation (8.7a) may be rewritten in the following form:

$$v \equiv x = u_p \frac{(T/\tau) \cos (\pi\tau/T)}{(T^2/4\tau^2) - 1} \sin \omega_n \left(t - \frac{\tau}{2} \right) \qquad [\tau \leq t] \qquad (8.7b)$$

Phase-Plane Graphical Method.   Several numerical and graphical methods,[18,23,27,37] all related in general but differing considerably in the details of procedure, are available for the solution of linear transient vibration problems. Of these methods, the phase-plane graphical method[25] is one of the most useful. The procedure is basically very simple, it gives a clear physical picture of the response of the system, and it may be applied readily to some classes of *nonlinear* systems.[3,5,6,8,13,15,21,22]

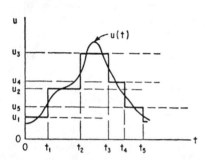

In Fig. 8.3 a general excitation in terms of ground displacement is represented, approximately, by a sequence of finite steps. The ith step has the total height u_i, where u_i is constant for the duration of the step. The differential equation of motion and its complete solution, applying for the duration of the step, are

FIGURE 8.3 General excitation approximated by a sequence of finite rectangular steps.

$$\frac{m\ddot{x}}{k} + x = u_i \qquad [t_{i-1} \leq t \leq t_i] \qquad (8.8a)$$

$$x - u_i = (x_{i-1} - u_i) \cos \omega_n(t - t_{i-1}) + \frac{\dot{x}_{i-1}}{\omega_n} \sin \omega_n(t - t_{i-1}) \qquad (8.8b)$$

where x_{i-1} and $\dot{x}_{i-1}$ are the displacement and velocity of the system at time t_{i-1}; consequently, they are the initial conditions for the ith step. The system velocity (divided by ω_n) during the ith step is

$$\frac{\dot{x}}{\omega_n} = -(x_{i-1} - u_i) \sin \omega_n(t - t_{i-1}) + \frac{\dot{x}_{i-1}}{\omega_n} \cos \omega_n(t - t_{i-1}) \qquad (8.8c)$$

Squaring Eqs. (8.8b) and (8.8c) and adding them,

$$\left(\frac{\dot{x}}{\omega_n} \right)^2 + (x - u_i)^2 = \left(\frac{\dot{x}_{i-1}}{\omega_n} \right)^2 + (x_{i-1} - u_i)^2 \qquad (8.8d)$$

This is the equation of a circle in a rectangular system of coordinates $\dot{x}/\omega_n, x$. The center is at $0, u_i$; and the radius is

$$R_i = \left[\left(\frac{\dot{x}_{i-1}}{\omega_n} \right)^2 + (x_{i-1} - u_i)^2 \right]^{1/2} \qquad (8.8e)$$

The solution for Eq. (8.8a) for the ith step may be shown, as in Fig. 8.4, to be the arc of the circle of radius R_i and center $0, u_i$, subtended by the angle $\omega_n(t_i - t_{i-1})$ and starting at the point $\dot{x}_{i-1}/\omega_n, x_{i-1}$. Time is positive in the counterclockwise direction.

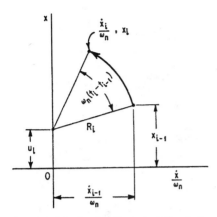

FIGURE 8.4 Graphical representation in the phase-plane of the solution for the ith step.

Example 8.3 Application to a General Pulse Excitation. Figure 8.5 shows an application of the method for the general excitation $u(t)$ represented by seven steps in the time-displacement plane. Upon choice of the step heights u_i and durations $(t_i - t_{i-1})$, the arc-center locations can be projected onto the X axis in the phase-plane and the arc angles $\omega_n(t_i - t_{i-1})$ can be computed. The graphical construction of the sequence of circular arcs, the *phase trajectory*, is then carried out, using the system conditions at zero time (in this example, 0,0) as the starting point.

Projection of the system displacements from the phase-plane into the time-displacement plane at once determines the time-displacement response curve. The time-velocity response can also be determined by projection as shown. The velocities and displacements at particular instants of time can be found directly from the phase trajectory coordinates without the necessity for drawing the time-response curves. Furthermore, the times of occurrence and the magnitudes of all the maxima also can be obtained directly from the phase trajectory.

Good accuracy is obtainable by using reasonable care in the graphical construction and in the choice of the steps representing the excitation. Usually, the time intervals should not be longer than about one-fourth the natural period of the system.[22]

The Laplace Transformation. The Laplace transformation provides a powerful tool for the solution of linear differential equations. The following discussion of the technique of its application is limited to the differential equation of the type applying to the undamped linear oscillator. Application to the linear oscillator with viscous damping is illustrated in a later part of this chapter.

Definitions. The Laplace transform $F(s)$ of a known function $f(t)$, where $t > 0$, is defined by

$$F(s) = \int_0^\infty e^{-st} f(t)\, dt \tag{8.9a}$$

where s is a complex variable. The transformation is abbreviated as

$$F(s) = \mathscr{L}[f(t)] \tag{8.9b}$$

The limitations on the function $f(t)$ are not discussed here. For the conditions for existence of $\mathscr{L}[f(t)]$, for complete accounts of the technique of application, and for extensive tables of function-transform pairs, the references should be consulted.[16, 17, 34, 36]

General Steps in Solution of the Differential Equation. In the solution of a differential equation by Laplace transformation, the first step is to transform the differential equation, in the variable t, into an algebraic equation in the complex variable s. Then, the algebraic equation is solved, and the solution of the differential equation is determined by an *inverse* transformation of the solution of the algebraic equation. The process of inverse Laplace transformation is symbolized by

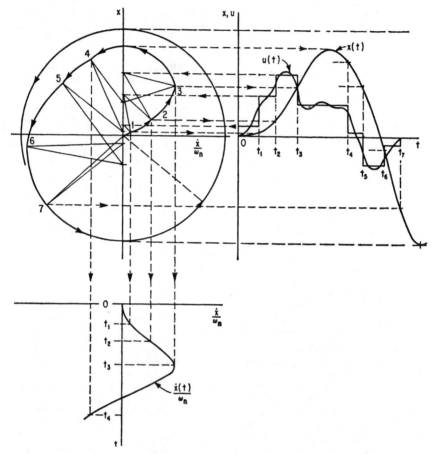

FIGURE 8.5 Example of phase-plane graphical solution.[2]

$$\mathscr{L}^{-1}[F(s)] = f(t) \tag{8.10}$$

Tables of Function-Transform Pairs. The processes symbolized by Eqs. (8.9*b*) and (8.10) are facilitated by the use of tables of function-transform pairs. Table 8.2 is a brief example. Transforms for general operations, such as differentiation, are included as well as transforms of explicit functions.

In general, the transforms of the explicit functions can be obtained by carrying out the integration indicated by the definition of the Laplace transformation. For example:

For $f(t) = 1$:

$$F(s) = \int_0^\infty e^{-st}dt = -\frac{1}{s}\ e^{-st}\Bigg]_0^\infty = \frac{1}{s}$$

TABLE 8.2 Pairs of Functions $f(t)$ and Laplace Transforms $F(s)$

	$f(t)$	$F(s)$
	Operation Transforms	
1	Definition, $f(t)$	$F(s) = \int_0^\infty e^{-st}f(t)\,dt$
2	First derivative, $f'(t)$	$sF(s) - f(0)$
3	nth derivative, $f^{(n)}(t)$	$s^n F(s) - s^{n-1}f(0) - s^{n-2}f'(0) - \cdots$ $-sf^{(n-2)}(0) - f^{(n-1)}(0)$ †
4	Superposition, $C_1 f_1(t) + C_2 f_2(t) + \cdots$ $+ C_n f_n(t)$	$C_1 F_1(s) + C_2 F_2(s) + \cdots + C_n F_n(s)$
5	Shifting in s plane, $e^{at}f(t)$	$\int_0^\infty e^{-st}e^{at}f(t)\,dt = \int_0^\infty e^{-(s-a)t}f(t)\,dt$ $= F(s-a)$
6	Shifting in t plane $\begin{cases} f(t-b) \text{ when } t > b, \\ 0 \text{ when } t < b \end{cases}$	$e^{-bs}F(s)$
	Function Transforms	
7	1	$\dfrac{1}{s}$
8	$\dfrac{t^{n-1}}{(n-1)!}$	$\dfrac{1}{s^n}$, for $n = 1, 2, \ldots$
9	e^{-at}	$\dfrac{1}{s+a}$
10	$\dfrac{1}{a}(1 - e^{-at})$	$\dfrac{1}{s(s+a)}$
11	te^{-at}	$\dfrac{1}{(s+a)^2}$
12	$\dfrac{1}{a}\sin at$	$\dfrac{1}{s^2+a^2}$
13	$\dfrac{1}{a^2}(1 - \cos at)$	$\dfrac{1}{s(s^2+a^2)}$
14	$\dfrac{1}{a^3}(at - \sin at)$	$\dfrac{1}{s^2(s^2+a^2)}$
15	$\dfrac{1}{(b-a)}(e^{-at} - e^{-bt})$	$\dfrac{1}{(s+a)(s+b)}$
16	$\dfrac{1}{ab} + \dfrac{be^{-at} - ae^{-bt}}{ab(a-b)}$	$\dfrac{1}{s(s+a)(s+b)}$
17	$\dfrac{a \sin bt - b \sin at}{ab(a^2-b^2)}$	$\dfrac{1}{(s^2+a^2)(s^2+b^2)}$
18	$e^{-at}(1 - at)$	$\dfrac{s}{(s+a)^2}$
19	$\cos at$	$\dfrac{s}{s^2+a^2}$
20	Rectangular pulse	$\dfrac{1 - e^{-s\tau}}{s}$
21	Sine pulse	$\dfrac{\pi/\tau}{s^2 + \pi^2/\tau^2}(1 + e^{-s\tau})$

† $f(t)$ and its derivatives through $f^{(n-1)}(t)$ must be continuous.

Transformation of the Differential Equation. The differential equation for the undamped linear oscillator is given in general form by

$$\frac{1}{\omega^2_n}\ddot{v} + v = \xi(t) \tag{8.11}$$

Applying the operational transforms (items 1 and 3, Table 8.2), Eq. (8.11) is transformed to

$$\frac{1}{\omega^2_n} s^2 F_r(s) - \frac{1}{\omega^2_n} sf(0) - \frac{1}{\omega^2_n} f'(0) + F_r(s) = F_e(s) \tag{8.12a}$$

where $F_r(s)$ = the transform of the unknown response $v(t)$, sometimes called the *response transform*

$s^2 F_r(s) - sf(0) - f'(0)$ = the transform of the second derivative of $v(t)$

$f(0)$ and $f'(0)$ = the known *initial values* of v and $\dot{v}$, i.e., v_0 and $\dot{v}_0$

$F_e(s)$ = the transform of the known excitation function $\xi(t)$, written $F_e(s) = \mathcal{L}[\xi(t)]$, sometimes called the *driving transform*

It should be noted that the initial conditions of the system are explicit in Eq. (8.12a).

The Subsidiary Equation. Solving Eq. (8.12a) for $F_r(s)$,

$$F_r(s) = \frac{sf(0) + f'(0) + \omega_n{}^2 F_e(s)}{s^2 + \omega_n{}^2} \tag{8.12b}$$

This is known as the *subsidiary equation* of the differential equation. The first two terms of the transform derive from the initial conditions of the system, and the third term derives from the excitation.

Inverse Transformation. In order to determine the response function $v(t)$, which is the solution of the differential equation, an inverse transformation is performed on the subsidiary equation. The entire operation, applied explicitly to the solution of Eq. (8.11), may be abbreviated as follows:

$$v(t) = \mathcal{L}^{-1}[F_r(s)] = \mathcal{L}^{-1}\left[\frac{sv_0 + \dot{v}_0 + \omega_n{}^2 \mathcal{L}[\xi(t)]}{s^2 + \omega_n{}^2}\right] \tag{8.13}$$

Example 8.4 Rectangular Step Excitation. In this case $\xi(t) = \xi_c$ for $0 \le t$ (Fig. 8.6A). The Laplace transform $F_e(s)$ of the excitation is, from item 7 of Table 8.2,

$$\mathcal{L}[\xi_c] = \xi_c \mathcal{L}[1] = \xi_c \frac{1}{s}$$

Assume that the starting conditions are general, that is, $v = v_0$ and $\dot{v} = \dot{v}_0$ at $t = 0$. Substituting the transform and the starting conditions into Eq. (8.13), the following is obtained:

$$v(t) = \mathcal{L}^{-1}\left[\frac{sv_0 + \dot{v}_0 + \omega_n^2 \xi_c(1/s)}{s^2 + \omega_n^2}\right] \tag{8.14a}$$

The foregoing may be rewritten as three separate inverse transforms:

$$v(t) = v_0 \mathcal{L}^{-1}\left[\frac{s}{s^2 + \omega_n^2}\right] + \dot{v}_0 \mathcal{L}^{-1}\left[\frac{1}{s^2 + \omega_n^2}\right] + \xi_c \omega_n^2 \mathcal{L}^{-1}\left[\frac{1}{s(s^2 + \omega_n^2)}\right] \tag{8.14b}$$

The inverse transforms in Eq. (8.14b) are evaluated by use of items 19, 12 and 13, respectively, in Table 8.2. Thus, the time-response function is given explicitly by

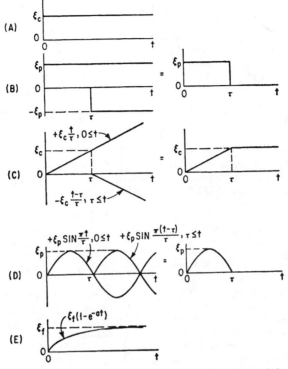

FIGURE 8.6 Excitation functions in examples of use of the Laplace transform: (A) rectangular step, (B) rectangular pulse, (C) step with constant-slope front, (D) sine pulse, and (E) step with exponential asymptotic rise.

$$v(t) = v_0 \cos \omega_n t + \frac{v_0}{\omega_n} \sin \omega_n t + \xi_c (1 - \cos \omega_n t) \qquad (8.14c)$$

The first two terms are the same as the starting condition response terms given by expressions (8.20a). The third term agrees with the response function shown by Eq. (8.22), derived for the case of a start from rest.

Example 8.5. Rectangular Pulse Excitation. The excitation function, Fig. 8.6B, is given by

$$\xi(t) = \begin{cases} \xi_p & \text{for } 0 \le t \le \tau \\ 0 & \text{for } \tau \le t \end{cases}$$

For simplicity, assume a start from rest, i.e., $v_0 = 0$ and $\dot{v}_0 = 0$ when $t = 0$.

During the first time interval, $0 \le t \le \tau$, the response function is of the same form as Eq. (8.14c) except that, with the assumed start from rest, the first two terms are zero.

During the second interval, $\tau \le t$, the transform of the excitation is obtained by applying the delayed-function transform (item 6, Table 8.2) and the transform for the rectangular step function (item 7) with the following result:

$$F_e(s) = \mathcal{L}[\xi(t)] = \xi_p \left(\frac{1}{s} - \frac{e^{-s\tau}}{s} \right)$$

This is the transform of an excitation consisting of a rectangular step of height $-\xi_p$ starting at time $t = \tau$, superimposed on the rectangular step of height $+\xi_p$ starting at time $t = 0$.

Substituting for $\mathcal{L}[\xi(t)]$ in Eq. (8.13),

$$v(t) = \xi_p \omega_n^2 \left\{ \mathcal{L}^{-1} \left[\frac{1}{s(s^2 + \omega_n^2)} \right] - \mathcal{L}^{-1} \left[\frac{e^{-s\tau}}{s(s^2 + \omega_n^2)} \right] \right\} \qquad (8.15a)$$

The first inverse transform in Eq. (8.15a) is the same as the third one in Eq. (8.14b) and is evaluated by use of item 13 in Table 8.2. However, the second inverse transform requires the use of items 6 and 13. The function-transform pair given by item 6 indicates that when $t < b$ the inverse transform in question is zero, and when $t > b$ the inverse transform is evaluated by replacing t by $t - b$ (in this particular case, by $t - \tau$). The result is as follows:

$$v(t) = \xi_p \omega_n^2 \left\{ \frac{1}{\omega_n^2} (1 - \cos \omega_n t) - \frac{1}{\omega_n^2} [1 - \cos \omega_n (t - \tau)] \right\}$$

$$= 2\xi_p \sin \frac{\pi \tau}{T} \sin \omega_n \left(t - \frac{\tau}{2} \right) \qquad [\tau \le t] \qquad (8.15b)$$

Theorem on the Transform of Functions Shifted in the Original (t) Plane. In Example 8.5, use is made of the theorem on the transform of functions shifted in the original plane. The theorem (item 6 in Table 8.2) is known variously as the second shifting theorem, the theorem on the transform of delayed functions, and the time-displacement theorem. In determining the transform of the excitation, the theorem provides for shifting, i.e., displacing the excitation or a component of the excitation in the positive direction along the time axis. This suggests the term *delayed function*. Examples of the shifting of component parts of the excitation appear in Fig. 8.6B, 8.6C, and 8.6D. Use of the theorem also is necessary in determining, by means of inverse transformation, the response following the delay in the excitation. Further illustration of the use of the theorem is shown by the next two examples.

Example 8.6. Step Function with Constant-slope Front. The excitation function (Fig. 8.6C) is expressed as follows:

$$\xi(t) = \begin{cases} \xi_c \dfrac{t}{\tau} & [0 \le t \le \tau] \\ \xi_c & [\tau \le t] \end{cases}$$

Assume that $v_0 = 0$ and $\dot{v}_0 = 0$.

The driving transforms for the first and second time intervals are

$$\mathcal{L}[\xi(t)] = \begin{cases} \xi_c \dfrac{1}{\tau} \dfrac{1}{s^2} & [0 \le t \le \tau] \\ \xi_c \dfrac{1}{\tau} \left(\dfrac{1}{s^2} - \dfrac{e^{-s\tau}}{s^2} \right) & [\tau \le t] \end{cases}$$

The transform for the second interval is the transform of a *negative* constant slope excitation, $-\xi_c(t - \tau)/\tau$, starting at $t = \tau$, superimposed on the transform for the positive constant slope excitation, $+\xi_c t/\tau$, starting at $t = 0$.

Substituting the transforms and starting conditions into Eq. (8.13), the responses for the two time eras, in terms of the transformations, are

$$v(t) = \begin{cases} \xi_c \dfrac{\omega_n^2}{\tau} \, \mathcal{L}^{-1}\left[\dfrac{1}{s^2(s^2 + \omega_n^2)} \right] & [0 \le t \le \tau] \\[3mm] \xi_c \dfrac{\omega_n^2}{\tau} \left\{ \mathcal{L}^{-1}\left[\dfrac{1}{s^2(s^2 + \omega_n^2)} \right] - \mathcal{L}^{-1}\left[\dfrac{e^{-s\tau}}{s^2\,(s^2 + \omega_n^2)} \right] \right\} & [\tau \le t] \end{cases} \qquad (8.16a)$$

Evaluation of the inverse transforms by reference to Table 8.2 [item 14 for the first of Eqs. (8.16a), items 6 and 14 for the second] leads to the following:

$$v(t) = \begin{cases} \xi_c \dfrac{\omega_n^2}{\tau} \dfrac{1}{\omega_n^3} (\omega_n t - \sin \omega_n t) & [0 \le t \le \tau] \\[3mm] \xi_c \dfrac{\omega_n^2}{\tau} \left\{ \dfrac{1}{\omega_n^3} (\omega_n t - \sin \omega_n t) - \dfrac{1}{\omega_n^3} [\omega_n(t - \tau) - \sin \omega_n(t - \tau)] \right\} & [\tau \le t] \end{cases}$$

Simplifying,

$$v(t) = \begin{cases} \xi_c \dfrac{1}{\omega_n \tau} (\omega_n t - \sin \omega_n t) & [0 \le t \le \tau] \\[3mm] \xi_c \left[1 + \dfrac{2}{\omega_n \tau} \sin \dfrac{\omega_n \tau}{2} \cos \omega_n \left(t - \dfrac{\tau}{2} \right) \right] & [\tau \le t] \end{cases} \qquad (8.16b)$$

Example 8.7. Half-cycle Sine Pulse. The excitation function (Fig. 8.6D) is

$$\xi(t) = \begin{cases} \xi_p \sin \dfrac{\pi t}{\tau} & [0 \le t \le \tau] \\[2mm] 0 & [\tau \le t] \end{cases}$$

Let the system start from rest. The driving transforms are

$$\mathcal{L}[\xi(t)] = \begin{cases} \xi_p \dfrac{\pi}{\tau} \dfrac{1}{s^2 + \pi^2/\tau^2} & [0 \le t \le \tau] \\[3mm] \xi_p \dfrac{\pi}{\tau} \left(\dfrac{1}{s^2 + \pi^2/\tau^2} + \dfrac{e^{-s\tau}}{s^2 + \pi^2/\tau^2} \right) & [\tau \le t] \end{cases}$$

The driving transform for the second interval is the transform of a sine wave of *positive* amplitude ξ_p and frequency π/τ starting at time $t = \tau$, superimposed on the transform of a sine wave of the same amplitude and frequency starting at time $t = 0$.

By substitution of the driving transforms and the starting conditions into Eq. (8.13), the following are found:

$$v(t) = \begin{cases} \xi_p \dfrac{\pi}{\tau} \omega_n^2 \mathcal{L}^{-1}\left[\dfrac{1}{s^2 + \pi^2/\tau^2} \cdot \dfrac{1}{s^2 + \omega_n^2} \right] & [0 \le t \le \tau] \\[3mm] \xi_p \dfrac{\pi}{\tau} \omega_n^2 \left\{ \mathcal{L}^{-1}\left[\dfrac{1}{s^2 + \pi^2/\tau^2} \cdot \dfrac{1}{s^2 + \omega_n^2} \right] + \mathcal{L}^{-1}\left[\dfrac{e^{-s\tau}}{s^2 + \pi^2/\tau^2} \cdot \dfrac{1}{s^2 + \omega_n^2} \right] \right\} & [\tau \le t] \end{cases} \qquad (8.17a)$$

Determining the inverse transforms from Table 8.2 [item 17 for the first of Eqs. (8.17a), items 6 and 17 for the second]:

$$v(t) = \begin{cases} \xi_{sp} \dfrac{\pi}{\tau} \, \omega_n^2 \dfrac{\omega_n \sin(\pi t/\tau) - (\pi/\tau)\sin \omega_n t}{(\pi\omega_n/\tau)(\omega_n^2 - \pi^2/\tau^2)} & [0 \le t \le \tau] \\[3ex] \xi_{sp} \dfrac{\pi}{\tau} \, \omega_n^2 \left[\dfrac{\omega_n \sin(\pi t/\tau) - (\pi/\tau)\sin \omega_n t}{(\pi\omega_n/\tau)(\omega_n^2 - \pi^2/\tau^2)} \right. \\[3ex] \qquad \left. + \dfrac{\omega_n \sin[\pi(t-\tau)/\tau] - (\pi/\tau)\sin \omega_n (t-\tau)}{(\pi\omega_n/\tau)(\omega_n^2 - \pi^2/\tau^2)} \right] & [\tau \le t] \end{cases}$$

Simplifying,

$$v(t) = \begin{cases} \xi_{sp} \dfrac{1}{1 - T^2/4\tau^2} \left(\sin \dfrac{\pi t}{\tau} - \dfrac{T}{2\tau} \sin \omega_n t \right) & [0 \le t \le \tau] \\[3ex] \xi_{sp} \dfrac{(T/\tau)\cos(\pi\tau/T)}{(T^2/4\tau^2) - 1} \sin \omega_n \left(t - \dfrac{\tau}{2} \right) & [\tau \le t] \end{cases} \qquad (8.17b)$$

where $T = 2\pi/\omega_n$ is the natural period of the responding system. Equations (8.17b) are equivalent to Eqs. (8.6b) and (8.7b) derived previously by the use of Duhamel's integral.

Example 8.8. Exponential Asymptotic Step. The excitation function (Fig. 8.6E) is

$$\xi(t) = \xi_f (1 - e^{-at}) \qquad [0 \le t]$$

Assume that the system starts from rest. The driving transform is

$$\mathscr{L}[\xi(t)] = \xi_f \left(\frac{1}{s} - \frac{1}{s+a} \right) = \xi_f a \frac{1}{s(s+a)}$$

It is found by Eq. (8.13) that

$$v(t) = \xi_f a \omega_n^2 \mathscr{L}^{-1} \left[\frac{1}{s(s+a)(s^2 + \omega_n^2)} \right] \qquad [0 \le t] \qquad (8.18a)$$

It frequently happens that the inverse transform is not readily found in an available table of transforms. Using the above case as an example, the function of s in Eq. (8.18a) is first *expanded in partial fractions;* then the inverse transforms are sought, thus:

$$\frac{1}{s(s+a)(s^2 + \omega_n^2)} = \frac{\kappa_1}{s} + \frac{\kappa_2}{s+a} + \frac{\kappa_3}{s+j\omega_n} + \frac{\kappa_4}{s-j\omega_n} \qquad (8.18b)$$

where $\quad j = \sqrt{-1}$

$$\kappa_1 = \left[\frac{1}{(s+a)(s+j\omega_n)(s-j\omega_n)} \right]_{s=0} = \frac{1}{a\omega_n^2} ,$$

$$\kappa_2 = \left[\frac{1}{s(s+j\omega_n)(s-j\omega_n)} \right]_{s=-a} = \frac{1}{-a(a^2 + \omega_n^2)}$$

$$\kappa_3 = \left[\frac{1}{s(s+a)(s-j\omega_n)} \right]_{s=-j\omega_n} = \frac{1}{-2\omega_n^2(a - j\omega_n)}$$

$$\kappa_4 = \left[\frac{1}{s(s+a)(s+j\omega_n)} \right]_{s=+j\omega_n} = \frac{1}{-2\omega_n^2(a+j\omega_n)}$$

Consequently, Eq. (8.18a) may be rewritten in the following expanded form:

$$v(t) = \xi_f \left\{ \mathcal{L}^{-1} \left[\frac{1}{s} \right] - \frac{\omega_n^2}{a^2+\omega_n^2} \mathcal{L}^{-1} \left[\frac{1}{s+a} \right] - \frac{a}{2(a-j\omega_n)} \cdot \right.$$

$$\left. \mathcal{L}^{-1} \left[\frac{1}{s+j\omega_n} \right] - \frac{a}{2(a+j\omega_n)} \mathcal{L}^{-1} \left[\frac{1}{s-j\omega_n} \right] \right\} \quad (8.18c)$$

The inverse transforms may now be found readily (items 7 and 9, Table 8.2):

$$v(t) = \xi_f \left[1 - \frac{\omega_n^2}{a^2+\omega_n^2} e^{-at} - \frac{a}{2(a-j\omega_n)} e^{-j\omega_n t} - \frac{a}{2(a+j\omega_n)} e^{j\omega_n t} \right]$$

Rewriting,

$$v(t) = \xi_f \left[1 - \frac{\omega_n^2 e^{-at} + a^2\frac{1}{2}(e^{j\omega_n t} + e^{-j\omega_n t}) - aj\omega_n\frac{1}{2}(e^{j\omega_n t} - e^{-j\omega_n t})}{a^2+\omega_n^2} \right]$$

Making use of the relations, $\cos z = (\frac{1}{2})(e^{jz} + e^{-jz})$ and $\sin z = -j(\frac{1}{2})(e^{jz} - e^{-jz})$, the equation for $v(t)$ may be expressed as follows:

$$v(t) = \xi_f \left[1 - \frac{(a/\omega_n)[\sin \omega_n t + (a/\omega_n)\cos \omega_n t] + e^{-at}}{1 + a^2/\omega_n^2} \right] \quad (8.18d)$$

Partial Fraction Expansion of F(s). The partial fraction expansion of $F_r(s)$, illustrated for a particular case in Eq. (8.18b), is a necessary part of the technique of solution. In general $F_r(s)$, expressed by the subsidiary equation (8.12b) and involved in the inverse transformation, Eqs. (8.10) and (8.13), is a quotient of two polynomials in s, thus

$$F_r(s) = \frac{A(s)}{B(s)} \quad (8.19)$$

The purpose of the expansion of $F_r(s)$ is to divide it into simple parts, the inverse transforms of which may be determined readily. The general procedure of the expansion is to factor $B(s)$ and then to rewrite $F_r(s)$ in partial fractions.[16,17,34,36]

INITIAL CONDITIONS OF THE SYSTEM

In all the solutions for response presented in this chapter, unless otherwise stated, it is assumed that the initial conditions (v_0 and $\dot{v}_0$) of the system are both zero. Other starting conditions may be accounted for merely by superimposing on the time-response functions given the additional terms

$$v_0 \cos \omega_n t + \frac{\dot{v}_0}{\omega_n} \sin \omega_n t \quad (8.20a)$$

These terms are the complete solution of the homogeneous differential equation, $m\ddot{v}/k + v = 0$. They represent the free vibration resulting from the initial conditions.
    The two terms in Eq. (8.20a) may be expressed by either one of the following combined forms:

$$\sqrt{v_0^2 + \left(\frac{\dot{v}_0}{\omega_n}\right)^2} \sin(\omega_n t + \theta_1) \qquad \text{where } \tan \theta_1 = \frac{v_0 \omega_n}{\dot{v}_0} \qquad (8.20b)$$

$$\sqrt{v_0^2 + \left(\frac{\dot{v}_0}{\omega_n}\right)^2} \cos(\omega_n t - \theta_2) \qquad \text{where } \tan \theta_2 = \frac{\dot{v}_0}{v_0 \omega_n} \qquad (8.20c)$$

where $\sqrt{v_0^2 + \left(\dfrac{\dot{v}_0}{\omega_n}\right)^2}$ is the resultant amplitude and θ_1 or θ_2 is the phase angle of the *initial-condition free vibration.*

PRINCIPLE OF SUPERPOSITION

When the system is linear, the *principle of superposition* may be employed. Any number of component excitation functions may be superimposed to obtain a prescribed total excitation function, and the corresponding component response functions may be superimposed to arrive at the total response function. However, the superposition must be carried out on a time basis and with complete regard for algebraic sign. The superposition of maximum component responses, disregarding time, may lead to completely erroneous results. For example, the response functions given by Eqs. (8.31) to (8.34) are defined completely with regard to time and algebraic sign, and may be superimposed for any combination of the excitation functions from which they have been derived.

COMPILATION OF RESPONSE FUNCTIONS AND RESPONSE SPECTRA; SINGLE DEGREE-OF-FREEDOM, LINEAR UNDAMPED SYSTEMS

STEP-TYPE EXCITATION FUNCTIONS

Constant-Force Excitation (Simple Step in Force).   The excitation is a constant force applied to the mass at zero time, $\xi(t) \equiv F(t)/k = F_c/k$. Substituting this excitation for $F(t)/k$ in Eq. (8.1a) and solving for the absolute displacement x,

$$x = \frac{F_c}{k}(1 - \cos \omega_n t) \qquad (8.21a)$$

Constant-Displacement Excitation (Simple Step in Displacement).   The excitation is a constant displacement of the ground which occurs at zero time, $\xi(t) \equiv u(t) = u_c$. Substituting for $u(t)$ in Eq. (8.1b) and solving for the absolute displacement x,

$$x = u_c(1 - \cos \omega_n t) \qquad (8.21b)$$

Constant-Acceleration Excitation (Simple Step in Acceleration).   The excitation is an instantaneous change in the ground acceleration at zero time, from zero to a constant value $\ddot{u}(t) = \ddot{u}_c$. The excitation is thus

$$\xi(t) \equiv -m\ddot{u}_c/k = -\ddot{u}_c/\omega_n^2$$

Substituting in Eq. (8.1c) and solving for the *relative* displacement δ_x,

$$\delta_x = \frac{-\ddot{u}_c}{\omega_n}\,(1 - \cos\,\omega_n t) \qquad (8.21c)$$

When the excitation is defined by a function of acceleration $\ddot{u}(t)$, it is often convenient to express the response in terms of the absolute acceleration $\ddot{x}$ of the system. The force acting on the mass in Fig. 8.1C is $-k\,\delta_x$; the acceleration $\ddot{x}$ is thus $-k\,\delta_x/m$ or $-\delta_x\omega_n^2$. Substituting $\delta_x = -\ddot{x}/\omega_n^2$ in Eq. (8.21c),

$$\ddot{x} = \ddot{u}_c(1 - \cos\,\omega_n t) \qquad (8.21d)$$

The same result is obtained by letting $\xi(t) \equiv \ddot{u}(t) = \ddot{u}_c$ in Eq. (8.1d) and solving for $\ddot{x}$. Equation (8.21d) is similar to Eq. (8.21b) with acceleration instead of displacement on both sides of the equation. This analogy generally applies in step- and pulse-type excitations.

The absolute displacement of the mass can be obtained by integrating Eq. (8.21d) twice with respect to time, taking as initial conditions $x = \dot{x} = 0$ when $t = 0$,

$$x = \frac{\ddot{u}_c}{\omega_n^2}\left[\frac{\omega_n^2 t^2}{2} - (1 - \cos\,\omega_n t)\right] \qquad (8.21e)$$

Equation (8.21e) also may be obtained from the relation $x = u + \delta_x$, noting that in this case $u(t) = \ddot{u}_c t^2/2$.

Constant-Velocity Excitation (Simple Step in Velocity). This excitation, when expressed in terms of ground or spring anchorage motion, is equivalent to prescribing, at zero time, an instantaneous change in the ground velocity from zero to a constant value $\dot{u}_c$. The excitation is $\xi(t) \equiv \dot{u}(t) = \dot{u}_c t$, and the solution for the differential equation of Eq. (8.1b) is

$$x = \frac{\dot{u}_c}{\omega_n}\,(\omega_n t - \sin\,\omega_n t) \qquad (8.21f)$$

For the velocity of the mass,

$$\dot{x} = \dot{u}_c(1 - \cos\,\omega_n t) \qquad (8.21g)$$

The result of Eq. (8.21g) could have been obtained directly by letting $\xi(t) \equiv \dot{u}(t) = \dot{u}_c$ in Eq. (8.1e) and solving for the *velocity* response $\dot{x}$.

General Step Excitation. A comparison of Eqs. (8.21a), (8.21b), (8.21c), (8.21d), and (8.21g) with Table 8.1 reveals that the response v and the excitation ξ are related in a common manner. This may be expressed as follows:

$$v = \xi_c(1 - \cos\,\omega_n t) \qquad (8.22)$$

where ξ_c indicates a constant value of the excitation. The excitation and response of the system are shown in Fig. 8.7.

Absolute Displacement Response to Velocity-Step and Acceleration-Step Excitations. The absolute displacement responses to the velocity-step and the acceleration-step excitations are given by Eqs. (8.21f) and (8.21e) and are shown in Figs. 8.8 and 8.9, respectively. The comparative effects of displacement-step, velocity-step, and acceleration-step excitations, in terms of *absolute displacement* response, may be seen by comparing Figs. 8.7 to 8.9.

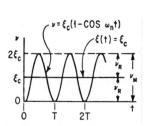

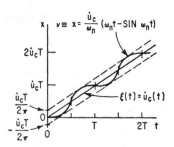

FIGURE 8.7 Time response to a simple step excitation (general notation).

FIGURE 8.8 Time-displacement response to a constant-velocity excitation (simple step in velocity).

In the case of the velocity-step excitation, the *velocity* of the system is always positive, except at $t = 0, T, 2T, \ldots$, when it is zero. Similarly, an acceleration-step excitation results in system *acceleration* that is always positive, except at $t = 0, T, 2T, \ldots$, when it is zero. The natural period of the responding system is $T = 2\pi/\omega_n$.

Response Maxima. In the response of a system to step or pulse excitation, the maximum value of the response often is of considerable physical significance. Several kinds of maxima are important. One of these is the *residual response amplitude,* which is the amplitude of the free vibration about the final position of the excitation as a base. This is designated v_R, and for the response given by Eq. (8.22):

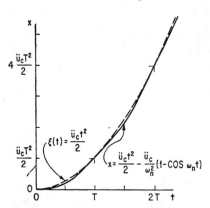

FIGURE 8.9 Time-displacement response to a constant-acceleration excitation (simple step in acceleration).

$$v_R = \pm \xi_c \qquad (8.22a)$$

Another maximum is the *maximax response,* which is the greatest of the maxima of v attained at *any time* during the response. In general, it is of the same sign as the excitation. For the response given by Eq. (8.22), the maximax response v_M is

$$v_M = 2\xi_c \qquad (8.22b)$$

Asymptotic Step. In the exponential function $\xi(t) = \xi_f(1 - e^{-at})$, the maximum value ξ_f of the excitation is approached asymptotically. This excitation may be defined alternatively by $\xi(t) = (F_f/k)(1 - e^{-at})$; $u_f(1 - e^{-at})$; $(-\ddot{u}_f/\omega_n^2)(1 - e^{-at})$; etc. (see Table 8.1). Substituting the excitation $\xi(t) = \xi_f(1 - e^{-at})$ in Eq. (8.2), the response v is

$$v = \xi_f\left[1 - \frac{(a/\omega_n)\,[\sin \omega_n t + (a/\omega_n)\,\cos \omega_n t] + e^{-at}}{1 + a^2/\omega_n^2}\right] \qquad (8.23a)$$

The excitation and the response of the system are shown in Fig. 8.10. For large values of the exponent *at,* the motion is nearly simple harmonic. The residual ampli-

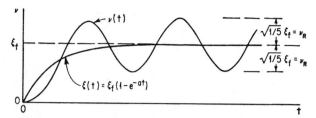

FIGURE 8.10 Time response to an exponentially asymptotic step for the particular case $\omega_n/a = 2$.

tude, relative to the final position of equilibrium, approaches the following value asymptotically.

$$v_R \rightarrow \xi_f \frac{1}{\sqrt{1 + \omega_n^2/a^2}} \tag{8.23b}$$

The maximax response $v_M = v_R + \xi_f$ is plotted against ω_n/a to give the response spectrum in Fig. 8.11.

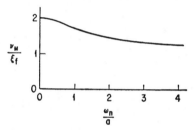

FIGURE 8.11 Spectrum for maximax response resulting from exponentially asymptotic step excitation.

Step-type Functions Having Finite Rise Time. Many step-type excitation functions rise to the constant maximum value ξ_c of the excitation in a finite length of time τ, called the *rise time*. Three such functions and their first three time derivatives are shown in Fig. 8.12. The step having a *cycloidal* front is the only one of the three that does not include an infinite third derivative; i.e., if the step is a ground displacement, it does not have an infinite rate of change of ground acceleration (infinite "jerk").

The excitation functions and the expressions for maximax response are given by the following equations:

Constant-slope front:

$$\xi(t) = \begin{cases} \xi_c \dfrac{t}{\tau} & [0 \le t \le \tau] \\ \xi_c & [\tau \le t] \end{cases} \tag{8.24a}$$

$$\frac{v_M}{\xi_c} = 1 + \left| \frac{T}{\pi\tau} \sin \frac{\pi\tau}{T} \right| \tag{8.24b}$$

Versed-sine front:

$$\xi(t) = \begin{cases} \dfrac{\xi_c}{2} \left(1 - \cos \dfrac{\pi t}{\tau} \right) & [0 \le t \le \tau] \\ \xi_c & [\tau \le t] \end{cases} \tag{8.25a}$$

$$\frac{v_M}{\xi_c} = 1 + \left| \frac{1}{(4\tau^2/T^2) - 1} \cos \frac{\pi\tau}{T} \right| \tag{8.25b}$$

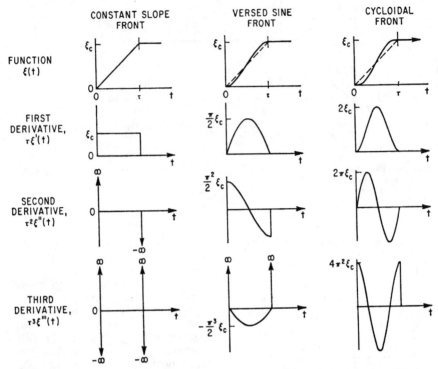

FIGURE 8.12 Three step-type excitation functions and their first three time derivatives. (*Jacobsen and Ayre.*[22])

Cycloidal front:

$$\xi(t) = \begin{cases} \dfrac{\xi_c}{2\pi}\left(\dfrac{2\pi t}{\tau} - \sin\dfrac{2\pi t}{\tau}\right) & [0 \le t \le \tau] \\ \xi_c & [\tau \le t] \end{cases} \tag{8.26a}$$

$$\frac{v_M}{\xi_c} = 1 + \left| \frac{T}{\pi\tau(1 - \tau^2/T^2)} \sin\frac{\pi\tau}{T} \right| \tag{8.26b}$$

where $T = 2\pi/\omega_n$ is the natural period of the responding system.

In the case of step-type excitations, the maximax response occurs after the excitation has reached its constant maximum value ξ_c and is related to the residual response amplitude by

$$v_M = v_R + \xi_c \tag{8.27}$$

Figure 8.13 shows the spectra of maximax response versus step rise time τ expressed relative to the natural period T of the responding system. In Fig. 8.13A the comparison is based on *equal rise times*, and in Fig. 8.13B it relates to *equal maximum slopes of the step fronts*. The residual response amplitude has values of zero

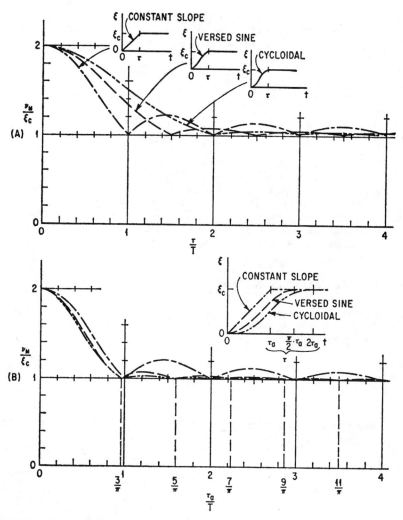

FIGURE 8.13 Spectra of maximax response resulting from the step excitation functions of Fig. 8.12. (A) For step functions having equal rise time τ. (B) For step functions having equal maximum slope ξ_c/τ_a. (*Jacobsen and Ayre.*[22])

($v_M/\xi_c = 1$) in all three cases; for example, the step excitation having a constant-slope front results in zero residual amplitude at $\tau/T = 1, 2, 3, \ldots$.

A Family of Exponential Step Functions Having Finite Rise Time. The inset diagram in Fig. 8.14 shows and Eqs. (8.28a) define a family of step functions having fronts which rise exponentially to the constant maximum ξ_c in the rise time τ. Two limiting cases of vertically fronted steps are included in the family: When $a \to -\infty$, the vertical front occurs at $t = 0$; when $a \to +\infty$, the vertical front occurs at $t = \tau$. An intermediate case has a constant-slope front ($a = 0$). The maximax responses are given by Eq. (8.28b) and by the response spectra in Fig. 8.14. The values of the maximax response are independent of the sign of the parameter a.

$$\xi(t) = \begin{cases} \xi_c \dfrac{1 - e^{at/\tau}}{1 - e^a} & [0 \le t \le \tau] \\ \xi_c & [\tau \le t] \end{cases} \tag{8.28a}$$

$$\frac{v_M}{\xi_c} = 1 + \left| \frac{a}{1 - e^a} \left[\frac{1 - 2e^a \cos (2\pi\tau/T) + e^{2a}}{a^2 + 4\pi^2\tau^2/T^2} \right]^{1/2} \right| \tag{8.28b}$$

where T is the natural period of the responding system.

There are zeroes of residual response amplitude ($v_M/\xi_c = 1$) at finite values of τ/T only for the constant-slope front ($a = 0$). Each of the step functions represented in Fig. 8.13 results in zeroes of residual response amplitude, and each function has antisymmetry with respect to the half-rise time $\tau/2$. This is of interest in the selection of cam and control-function shapes, where one of the criteria of choice may be *minimum residual amplitude of vibration* of the driven system.

PULSE-TYPE EXCITATION FUNCTIONS

The Simple Impulse.   If the duration τ of the pulse is short relative to the natural period T of the system, the response of the system may be determined by equating the impulse J, i.e., the force-time integral, to the momentum $m\dot{x}_J$:

$$J = \int_0^\tau F(t) \, dt = m\dot{x}_J \tag{8.29a}$$

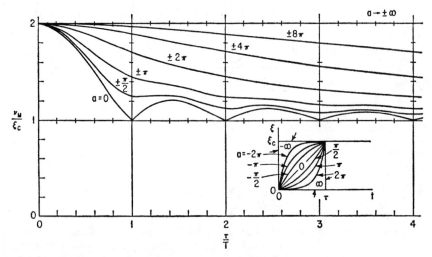

FIGURE 8.14 Spectra of maximax response for a family of step functions having exponential fronts, including the vertical fronts $a \to \pm\infty$, and the constant-slope front $a = 0$, as special cases. (*Jacobsen and Ayre.*[22])

Thus, it is found that the impulsive velocity $\dot{x}_J$ is equal to J/m. Consequently, the velocity-time response is given by $\dot{x} = \dot{x}_J \cos \omega_n t = (J/m) \cos \omega_n t$. The displacement-time response is obtained by integration, assuming a start from rest,

$$x = x_J \sin \omega_n t$$

where

$$x_J = \frac{J}{m\omega_n} = \omega_n \int_0^\tau \frac{F(t)\, dt}{k} \qquad (8.29b)$$

The impulse concept, used for determining the response to a short-duration force pulse, may be generalized in terms of v and ξ by referring to Table 8.1. The *generalized impulsive response* is

$$v = v_J \sin \omega_n t \qquad (8.30a)$$

where the amplitude is

$$v_J = \omega_n \int_0^\tau \xi(t)\, dt \qquad (8.30b)$$

The impulsive response amplitude v_J and the generalized impulse $k \int_0^\tau \xi(t)\, dt$ are used in comparing the effects of various pulse shapes when the pulse durations are short.

Symmetrical Pulses. In the following discussion a comparison is made of the responses caused by single symmetrical pulses of rectangular, half-cycle sine, versed-sine, and triangular shapes. The excitation functions and the time-response equations are given by Eqs. (8.31) to (8.34). Note that the residual response amplitude factors are set in brackets and are identified by the time interval $\tau \le t$.

Rectangular:

$$\left. \begin{aligned} \xi(t) &= \xi_p \\ v &= \xi_p(1 - \cos \omega_n t) \end{aligned} \right\} \qquad [0 \le t \le \tau] \qquad (8.31a)$$

$$\left. \begin{aligned} \xi(t) &= 0 \\ v &= \xi_p \left[2 \sin \frac{\pi\tau}{T} \right] \sin \omega_n \left(t - \frac{\tau}{2} \right) \end{aligned} \right\} \qquad [\tau \le t] \qquad (8.31b)$$

Half-cycle sine:

$$\left. \begin{aligned} \xi(t) &= \xi_p \sin \frac{\pi t}{\tau} \\ v &= \frac{\xi_p}{1 - T^2/4\tau^2} \left(\sin \frac{\pi t}{\tau} - \frac{T}{2\tau} \sin \omega_n t \right) \end{aligned} \right\} \qquad [0 \le t \le \tau] \qquad (8.32a)$$

$$\left. \begin{aligned} \xi(t) &= 0 \\ v &= \xi_p \left[\frac{(T/\tau) \cos (\pi\tau/T)}{(T^2/4\tau^2) - 1} \right] \sin \omega_n \left(t - \frac{\tau}{2} \right) \end{aligned} \right\} \qquad [\tau \le t] \qquad (8.32b)$$

Versed-sine:

$$\xi(t) = \frac{\xi_p}{2}\left(1 - \cos\frac{2\pi t}{\tau}\right)$$

$$v = \frac{\xi_p/2}{1 - \tau^2/T^2}\left(1 - \frac{\tau^2}{T^2} + \frac{\tau^2}{T^2}\cos\frac{2\pi t}{\tau} - \cos\omega_n t\right) \qquad\qquad [0 \le t \le \tau] \qquad (8.33a)$$

$$\xi(t) = 0$$

$$v = \xi_p\left[\frac{\sin\pi\tau/T}{1 - \tau^2/T^2}\right]\sin\omega_n\left(t - \frac{\tau}{2}\right) \qquad\qquad [\tau \le t] \qquad (8.33b)$$

Triangular:

$$\xi(t) = 2\xi_p\frac{t}{\tau}$$

$$v = 2\xi_p\left(\frac{t}{\tau} - \frac{T}{\tau}\frac{\sin\omega_n t}{2\pi}\right) \qquad\qquad \left[0 \le t \le \frac{\tau}{2}\right] \qquad (8.34a)$$

$$\xi(t) = 2\xi_p\left(1 - \frac{t}{\tau}\right)$$

$$v = 2\xi_p\left(1 - \frac{t}{\tau} - \frac{T}{\tau}\frac{\sin\omega_n t}{2\pi} + \frac{T}{\tau}\frac{\sin\omega_n(t - \tau/2)}{\pi}\right) \qquad\qquad \left[\frac{\tau}{2} \le t \le \tau\right] \qquad (8.34b)$$

$$\xi(t) = 0$$

$$v = \xi_p\left[2\frac{\sin^2(\pi\tau/2T)}{\pi\tau/2T}\right]\sin\omega_n(t - \tau/2) \qquad\qquad [\tau \le t] \qquad (8.34c)$$

where T is the natural period of the responding system.

Equal Maximum Height of Pulse as Basis of Comparison. Examples of time response, for six different values of τ/T, are shown separately for the rectangular, half-cycle sine, and versed-sine pulses in Fig. 8.15, and for the triangular pulse in Fig. 8.22B. The basis of comparison is equal maximum height of excitation pulse ξ_p.

Residual Response Amplitude and Maximax Response. The spectra of maximax response v_M and residual response amplitude v_R are given in Fig. 8.16 by (A) for the rectangular pulse, by (B) for the sine pulse, and by (C) for the versed-sine pulse. The maximax response may occur either within the duration of the pulse or after the pulse function has dropped to zero. In the latter case the maximax response is equal to the residual response amplitude. In general, the maximax response is given by the residual response amplitude only in the case of short-duration pulses; for example, see the case $\tau/T = \frac{1}{4}$ in Fig. 8.15 where T is the natural period of the responding system. The response spectra for the triangular pulse appear in Fig. 8.24.

Maximax Relative Displacement When the Excitation Is Ground Displacement. When the excitation $\xi(t)$ is given as *ground displacement* $u(t)$, the response v is the absolute displacement x of the mass (Table 8.1). It is of practical importance in the investigation of the maximax *distortion* or *stress* in the elastic element to know the maximax value of the relative displacement. In this case the relative displacement is a *derived quantity* obtained by taking the difference between the response and the excitation, that is, $x - u$ or, in terms of the general notation, $v - \xi$.

If the excitation is given as ground acceleration, the response is determined directly as relative displacement and is designated δ_x (Table 8.1). To avoid confusion, relative displacement determined as a *derived quantity,* as described in the first case

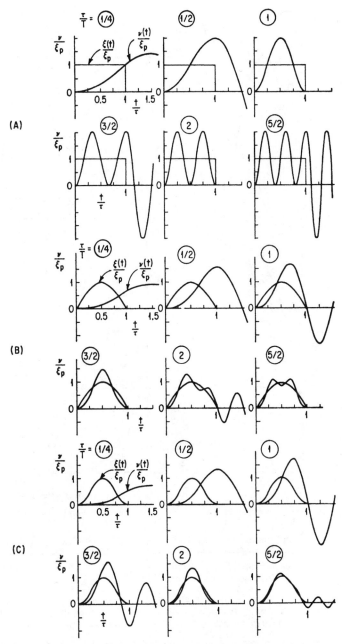

FIGURE 8.15 Time response curves resulting from single pulses of (*A*) rect-angular, (*B*) half-cycle sine, and (*C*) versed-sine shapes.[19]

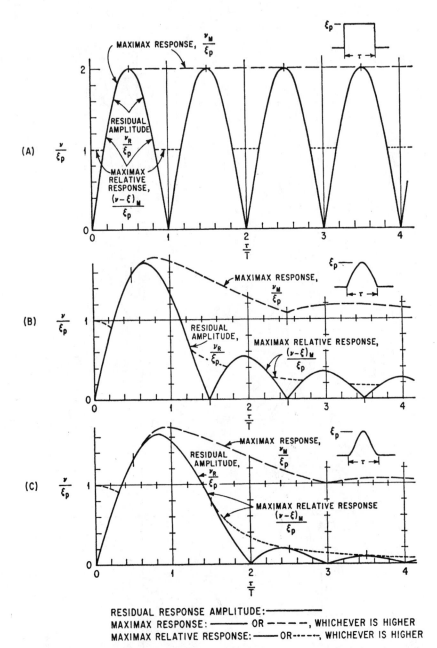

FIGURE 8.16 Spectra of maximax response, residual response amplitude, and maximax relative response resulting from single pulses of (*A*) rectangular, (*B*) half-cycle sine, and (*C*) versed-sine shapes.[19] The spectra are shown on another basis in Fig. 8.18.

above, is designated by $x - u$; relative displacement determined directly as the *response variable* (second case above) is designated by δ_x. The distinction is made readily in the general notation by use of the symbols $v - \xi$ and v, respectively, for relative response and for response. The maximax values are designated $(v - \xi)_M$ and v_M, respectively.

The maximax relative response may occur *either* within the duration of the pulse or during the residual vibration era ($\tau \le t$). In the latter case the maximax relative response is equal to the residual response amplitude. This explains the discontinuities which occur in the spectra of maximax relative response shown in Fig. 8.16 and elsewhere.

The meaning of the relative response $v - \xi$ may be clarified further by a study of the time-response and time-excitation curves shown in Fig. 8.15.

Equal Area of Pulse as Basis of Comparison.   In the preceding section on the comparison of responses resulting from pulse excitation, the pulses are assumed of equal maximum height. Under some conditions, particularly if the pulse duration is short relative to the natural period of the system, it may be more useful to make the comparison on the basis of equal pulse area; i.e., equal impulse (equal time integral).

The *areas* for the pulses of maximum height ξ_p and duration τ are as follows: rectangle, $\xi_p\tau$; half-cycle sine, $(2/\pi)\xi_p\tau$; versed-sine $(\frac{1}{2})\xi_p\tau$; triangle, $(\frac{1}{2})\xi_p\tau$. Using the area of the *triangular pulse* as the basis of comparison, and requiring that the areas of the other pulses be equal to it, it is found that the pulse *heights*, in terms of the height ξ_{p0} of the *reference triangular pulse*, must be as follows: rectangle, $(\frac{1}{2})\xi_{p0}$; half-cycle sine, $(\pi/4)\xi_{p0}$; versed-sine, ξ_{p0}.

Figure 8.17 shows the time responses, for four values of τ/T, redrawn on the basis of *equal pulse area* as the criterion for comparison. Note that the response reference is the constant ξ_{p0}, which is the height of the triangular pulse. To show a direct comparison, the response curves for the various pulses are superimposed on each other. For the shortest duration shown, $\tau/T = \frac{1}{4}$, the response curves are nearly alike. Note that the responses to two different rectangular pulses are shown, one of duration τ and height $\xi_{p0}/2$, the other of duration $\tau/2$ and height ξ_{p0}, both of area $\xi_{p0}\tau/2$.

The response spectra, plotted on the basis of equal pulse area, appear in Fig. 8.18. The residual response spectra are shown altogether in (A), the maximax response spectra in (B), and the spectra of maximax relative response in (C).

Since the pulse area is $\xi_{p0}\tau/2$, the generalized impulse is $k\xi_{p0}\tau/2$, and the amplitude of vibration of the system computed on the basis of the generalized impulse theory, Eq. (8.30b), is given by

$$v_J = \omega_n\xi_{p0}\frac{\tau}{2} = \pi\frac{\tau}{T}\,\xi_{p0} \tag{8.35}$$

A comparison of this straight-line function with the response spectra in Fig. 8.18B shows that *for values of τ/T less than one-fourth the shape of the symmetrical pulse is of little concern.*

Family of Exponential, Symmetrical Pulses.   A continuous variation in shape of pulse may be investigated by means of the family of pulses represented by Eqs. (8.36a) and shown in the inset diagram in Fig. 8.19A:

$$\xi(t) = \begin{cases} \xi_p \dfrac{1 - e^{2at/\tau}}{1 - e^a} & \left[0 \le t \le \dfrac{\tau}{2}\right] \\[3mm] \xi_p \dfrac{1 - e^{2a(1-t/\tau)}}{1 - e^a} & \left[\dfrac{\tau}{2} \le t \le \tau\right] \\[3mm] 0 & [\tau \le t] \end{cases} \tag{8.36a}$$

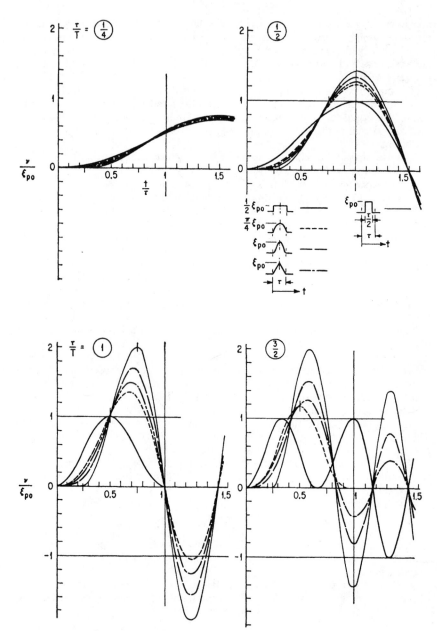

FIGURE 8.17 Time response to various symmetrical pulses having equal pulse area, for four different values of τ/T. (*Jacobsen and Ayre.*[22])

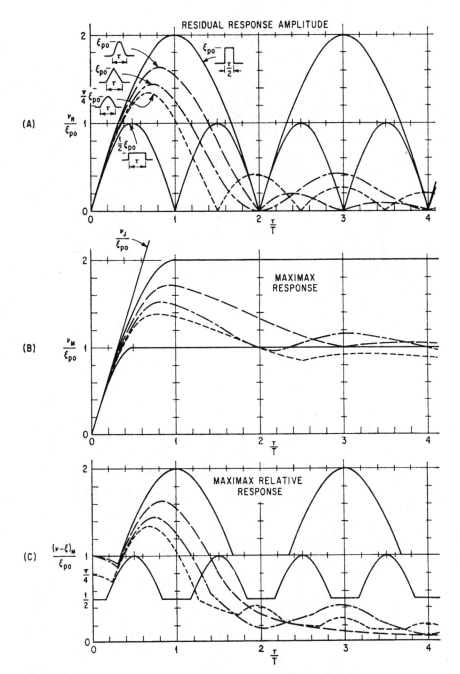

FIGURE 8.18 Response spectra for various symmetrical pulses having equal pulse area: (*A*) residual response amplitude, (*B*) maximax response, and (*C*) maximax relative response. (*Jacobsen and Ayre.*[22])

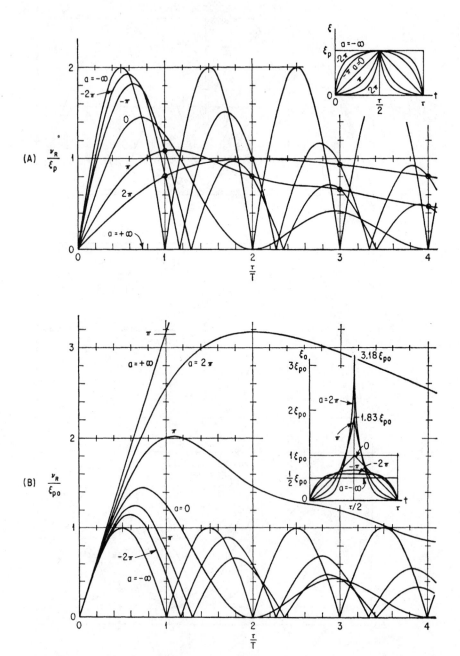

FIGURE 8.19 Spectra for residual response amplitude for a family of exponential, symmetrical pulses: (*A*) pulses having equal height; (*B*) pulses having equal area. (*Jacobsen and Ayre.*[22])

The family includes the following special cases:

$a \rightarrow -\infty$: rectangle of height ξ_p and duration τ;
$a = 0$: triangle of height ξ_p and duration τ;
$a \rightarrow +\infty$: spike of height ξ_p and having zero area.

The residual response amplitude of vibration of the system is

$$\frac{v_R}{\xi_p} = \frac{2aT}{\pi\tau}\left(\frac{e^a - \cos{(\pi\tau/T)} - (aT/\pi\tau)\sin{(\pi\tau/T)}}{(1 - e^a)(1 + a^2T^2/\pi^2\tau^2)}\right) \tag{8.36b}$$

where T is the natural period of the responding system. Figure 8.19A shows the spectra for residual response amplitude for seven values of the parameter a, compared on the basis of *equal pulse height*. The zero-area spike ($a \rightarrow +\infty$) results in zero response. The area of the general pulse of height ξ_p is

$$A_p = \xi_p\,\frac{\tau}{a}\left(\frac{1 - e^a + a}{1 - e^a}\right) \tag{8.36c}$$

If a comparison is to be drawn on the basis of *equal pulse area* using the area $\xi_{p0}\tau/2$ of the triangular pulse as the reference, the height ξ_{pa} of the general pulse is

$$\xi_{pa} = \xi_{p0}\,\frac{a}{2}\left(\frac{1 - e^a}{1 - e^a + a}\right) \tag{8.36d}$$

The residual response amplitude spectra, based on the equal-pulse-area criterion, are shown in Fig. 8.19B. The case $a \rightarrow +\infty$ is equivalent to a generalized impulse of value $k\xi_{p0}\tau/2$ and results in the straight-line spectrum given by Eq. (8.35).

Symmetrical Pulses Having a Rest Period of Constant Height. In the inset diagrams of Fig. 8.20 each pulse consists of a rise, a central rest period or "dwell" having constant height, and a decay. The expressions for the pulse *rise* functions may be obtained from Eqs. (8.24a), (8.25a), and (8.26a) by substituting $\tau/2$ for τ. The pulse *decay* functions are available from symmetry.

If the *rest* period is long enough for the maximax displacement of the system to be reached during the duration τ_r of the pulse rest, the maximax may be obtained from Eqs. (8.24b), (8.25b), and (8.26b) and, consequently, from Fig. 8.13. The substitution of $\tau/2$ for τ is necessary.

Equations (8.37) to (8.39) give the residual response amplitudes. The spectra computed from these equations are shown in Fig. 8.20.

Constant-slope rise and decay:

$$\frac{v_R}{\xi_p} = \frac{2T}{\pi\tau}\left[1 - \cos\frac{\pi\tau}{T} + \frac{1}{2}\cos\frac{2\pi\tau_r}{T} - \cos\frac{\pi(\tau + 2\tau_r)}{T} + \frac{1}{2}\cos\frac{2\pi(\tau + \tau_r)}{T}\right]^{1/2} \tag{8.37}$$

Versed-sine rise and decay:

$$\frac{v_R}{\xi_p} = \frac{1}{1 - \tau^2/T^2}\left[1 + \cos\frac{\pi\tau}{T} - \frac{1}{2}\cos\frac{2\pi\tau_r}{T} - \cos\frac{\pi(\tau + 2\tau_r)}{T} - \frac{1}{2}\cos\frac{2\pi(\tau + \tau_r)}{T}\right]^{1/2} \tag{8.38}$$

Cycloidal rise and decay:

$$\frac{v_R}{\xi_p} = \frac{2T/\pi\tau}{1 - \tau^2/4T^2}\left[\cos\frac{\pi\tau_r}{T} - \cos\frac{\pi(\tau + \tau_r)}{T}\right] \tag{8.39}$$

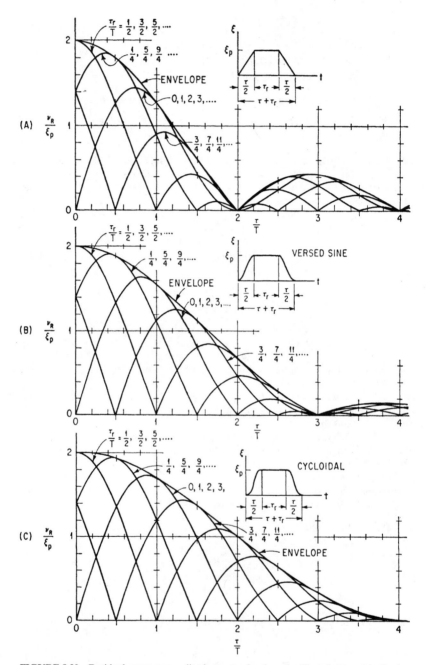

FIGURE 8.20 Residual response amplitude spectra for three families of symmetrical pulses having a central rest period of constant height and of duration τ_r. Note that the abscissa is τ/T, where τ is the sum of the rise time and the decay time. (*A*) Constant-slope rise and decay. (*B*) Versed-sine rise and decay. (*C*) Cycloidal rise and decay. (*Jacobsen and Ayre.*[22])

Note that τ in the abscissa is the sum of the rise time and the decay time and is *not* the total duration of the pulse. Attached to each spectrum is a set of values of τ_r/T where T is the natural period of the responding system.

When $\tau_r/T = 1, 2, 3, \ldots$, the residual response amplitude is equal to that for the case $\tau_r = 0$, and the spectrum starts at the origin. If $\tau_r/T = \frac{1}{2}, \frac{3}{2}, \frac{5}{2}, \ldots$, the spectrum has the maximum value 2.00 at $\tau/T = 0$. The *envelopes* of the spectra are of the same forms as the *residual-response-amplitude* spectra for the related *step functions;* see the spectra for $[(v_M/\xi_c) - 1]$ in Fig. 8.13A. In certain cases, for example, at $\tau/T = 2, 4, 6, \ldots$, in Fig. 8.20A, $v_R/\xi_p = 0$ for all values of τ_r/T.

Unsymmetrical Pulses.   Pulses having only slight asymmetry may often be represented adequately by symmetrical forms. However, if there is considerable asymmetry, resulting in appreciable steepening of either the rise or the decay, it is necessary to introduce a parameter which defines the *skewing* of the pulse.

The ratio of the rise time to the pulse period is called the *skewing constant,* $\sigma = t_1/\tau$. There are three special cases:

$\sigma = 0$: the pulse has an *instantaneous* (vertical) *rise,* followed by a decay having the duration τ. This case may be used as an elementary representation of a *blast pulse.*

$\sigma = \frac{1}{2}$: the pulse may be *symmetrical.*

$\sigma = 1$: the pulse has an *instantaneous decay,* preceded by a rise having the duration τ.

Triangular Pulse Family.   The effect of asymmetry in pulse shape is shown readily by means of the family of triangular pulses (Fig. 8.21). Equations (8.40) give the excitation and the time response.

Rise era: $0 \le t \le t_1$

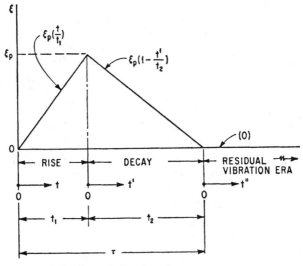

FIGURE 8.21   General triangular pulse.

$$\xi(t) = \xi_p \frac{t}{t_1}$$

$$v = \xi_p \left(\frac{t}{t_1} - \frac{T}{2\pi t_1} \sin \omega_n t \right)$$

(8.40a)

Decay era: $0 \le t' \le t_2$, where $t' = t - t_1$

$$\xi(t) = \xi_p \left(1 - \frac{t'}{t_2} \right)$$

$$v = \xi_p \left[1 - \frac{t'}{t_2} + \frac{T}{2\pi t_2} \left(1 + 4 \frac{t_2}{t_1} \frac{\tau}{t_1} \sin^2 \frac{\pi t_1}{T} \right)^{1/2} \sin (\omega_n t' + \theta') \right]$$

(8.40b)

where

$$\tan \theta' = \frac{\sin (2\pi t_1/T)}{\cos (2\pi t_1/T) - \tau/t_2}$$

Residual-vibration era: $0 \le t''$, where $t'' = t - \tau = t - t_1 - t_2$

$$\xi(t) = 0$$

$$v = \xi_p \frac{1}{\pi} \left[\frac{T}{t_1} \frac{T}{t_2} \left(\frac{\tau}{t_1} \sin^2 \frac{\pi t_1}{T} + \frac{\tau}{t_2} \sin^2 \frac{\pi t_2}{T} - \sin^2 \frac{\pi \tau}{T} \right) \right]^{1/2} \sin (\omega_n t'' + \theta_R)$$

(8.40c)

where

$$\tan \theta_R = \frac{(\tau/t_2) \sin (2\pi t_2/T) - \sin (2\pi \tau/T)}{(\tau/t_2) \cos (2\pi t_2/T) - \cos (2\pi \tau/T) - t_1/t_2}$$

For the special cases $\sigma = 0$, $\frac{1}{2}$, and 1, the time responses for six values of τ/T are shown in Fig. 8.22, where T is the natural period of the responding system. Some of the curves are superposed in Fig. 8.23 for easier comparison. The response spectra appear in Fig. 8.24. The straight-line spectrum v_J/ξ_p for the amplitude of response based on the impulse theory also is shown in Fig. 8.24A. In the two cases of extreme skewing, $\sigma = 0$ and $\sigma = 1$, the residual amplitudes are *equal* and are given by Eq. (8.41a). For the symmetrical case, $\sigma = \frac{1}{2}$, v_R is given by Eq. (8.41b).

$\sigma = 0$ and 1: $\dfrac{v_R}{\xi_p} = \left[1 - \dfrac{T}{\pi \tau} \sin \dfrac{2\pi \tau}{T} + \left(\dfrac{T}{\pi \tau} \right)^2 \sin^2 \dfrac{\pi \tau}{T} \right]^{1/2}$

(8.41a)

$\sigma = \frac{1}{2}$: $\dfrac{v_R}{\xi_p} = 2 \dfrac{\sin^2 (\pi \tau/2T)}{\pi \tau/2T}$

(8.41b)

The residual response amplitudes for other cases of skewness may be determined from the amplitude term in Eqs. (8.40c); they are shown by the response spectra in Fig. 8.25. The residual response amplitudes resulting from single pulses that are mirror images of each other in time are equal. In general, the phase angles for the residual vibrations are unequal.

Note that in the cases $\sigma = 0$ and $\sigma = 1$ for *vertical rise* and *vertical decay*, respectively, there are *no zeroes of residual amplitude*, except for the trivial case, $\tau/T = 0$.

The family of triangular pulses is particularly advantageous for investigating the effect of varying the skewness, because both criteria of comparison, equal pulse height and equal pulse area, are satisfied simultaneously.

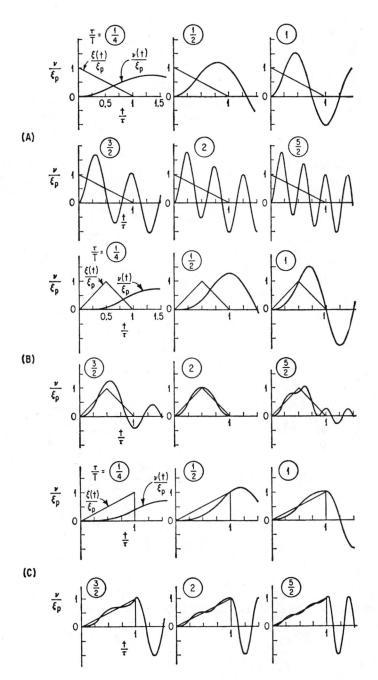

FIGURE 8.22 Time response curves resulting from single pulses of three different triangular shapes: (A) vertical rise (elementary blast pulse), (B) symmetrical, and (C) vertical decay.[19]

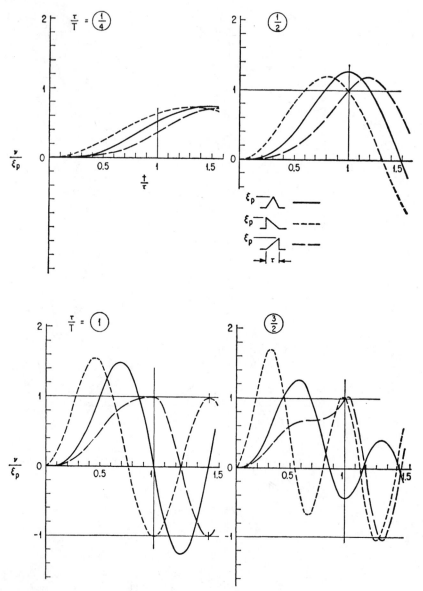

FIGURE 8.23 Time response curves of Fig. 8.22 superposed, for four values of τ/T. (*Jacobsen and Ayre.*[22])

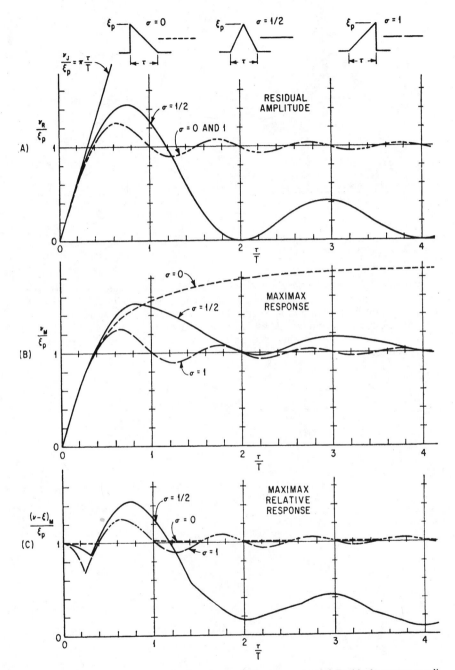

FIGURE 8.24 Response spectra for three types of triangular pulse: (*A*) Residual response amplitude. (*B*) Maximax response. (*C*) Maximax relative response. (*Jacobsen and Ayre.*[22])

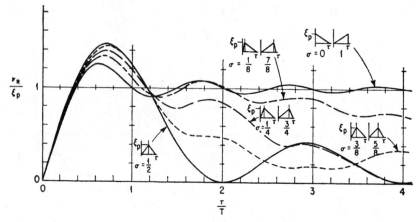

FIGURE 8.25 Spectra for residual response amplitude for a family of triangular pulses of varying skewness. (*Jacobsen and Ayre.*[22])

Various Pulses Having Vertical Rise or Vertical Decay. Figure 8.26 shows the spectra of residual response amplitude plotted on the basis of *equal pulse area*. The rectangular pulse is included for comparison. The expressions for residual response amplitude for the rectangular and the triangular pulses are given by Eqs. (8.31*b*) and (8.41*a*), and for the quarter-cycle sine and the half-cycle versed-sine pulses by Eqs. (8.42) and (8.43).

Quarter-cycle "sine":

$$\xi(t) = \xi_p \begin{cases} \sin \dfrac{\pi t}{2\tau} & \text{for vertical decay} \\[2mm] \text{or} & \qquad\qquad [0 \le t \le \tau] \\[2mm] \cos \dfrac{\pi t}{2\tau} & \text{for vertical rise} \end{cases}$$

$$\xi(t) = 0 \qquad [\tau \le t]$$

$$\frac{v_R}{\xi_p} = \frac{4\tau/T}{(16\tau^2/T^2) - 1} \left(1 + \frac{16\tau^2}{T^2} - \frac{8\tau}{T} \sin \frac{2\pi\tau}{T} \right)^{1/2} \tag{8.42}$$

Half-cycle "versed-sine":

$$\xi(t) = \xi_p \begin{cases} \dfrac{1}{2} \left(1 - \cos \dfrac{\pi t}{\tau} \right) & \text{for vertical decay} \\[2mm] \text{or} & \qquad\qquad [0 \le t \le \tau] \\[2mm] \dfrac{1}{2} \left(1 + \cos \dfrac{\pi t}{\tau} \right) & \text{for vertical rise} \end{cases}$$

$$\xi(t) = 0 \qquad [\tau \le t]$$

$$\frac{v_R}{\xi_p} = \frac{1/2}{(4\tau^2/T^2) - 1} \left[1 + \left(1 - \frac{8\tau^2}{T^2} \right)^2 - 2 \left(1 - \frac{8\tau^2}{T^2} \right) \cdot \cos \frac{2\pi\tau}{T} \right]^{1/2} \tag{8.43}$$

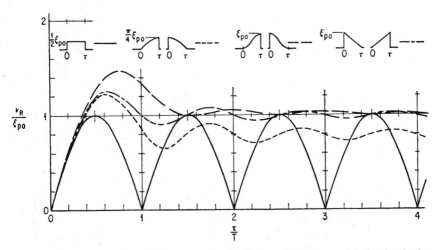

FIGURE 8.26 Spectra for residual response amplitude for various unsymmetrical pulses having either vertical rise or vertical decay. Comparison on the basis of equal pulse area.[19]

where T is the natural period of the responding system.

Note again that the residual response amplitudes, caused by single pulses that are mirror images in time, are equal. Furthermore, it is seen that the unsymmetrical pulses, having either vertical rise or vertical decay, result in no zeroes of residual response amplitude, except in the trivial case $\tau/T = 0$.

Exponential Pulses of Finite Duration, Having Vertical Rise or Vertical Decay. Families of exponential pulses having either a vertical rise or a vertical decay, as shown in the inset diagrams in Fig. 8.27, can be formed by Eqs. (8.44a) and (8.44b).

Vertical rise with exponential decay:

$$\xi(t) = \begin{cases} \xi_p\left(\dfrac{1 - e^{a(1 - t/\tau)}}{1 - e^a}\right) & [0 \le t \le \tau] \\ 0 & [\tau \le t] \end{cases} \tag{8.44a}$$

Exponential rise with vertical decay:

$$\xi(t) = \begin{cases} \xi_p\left(\dfrac{1 - e^{at/\tau}}{1 - e^a}\right) & [0 \le t \le \tau] \\ 0 & [\tau \le t] \end{cases} \tag{8.44b}$$

Residual response amplitude for *either form of pulse:*

$$\frac{v_R}{\xi_p} = \frac{a}{1 - e^a}\left\{\frac{[(2\pi\tau/T)(1 - e^a)/a + \sin(2\pi\tau/T)]^2 + [1 - \cos(2\pi\tau/T)]^2}{a^2 + 4\pi^2\tau^2/T^2}\right\}^{1/2} \tag{8.44c}$$

When $a = 0$, the pulses are triangular with vertical rise or vertical decay. If $a \to +\infty$ or $-\infty$, the pulses approach the shape of a zero-area spike or of a rectangle, respectively. The spectra for residual response amplitude, plotted on the basis of *equal pulse height,* are shown in Fig. 8.27.

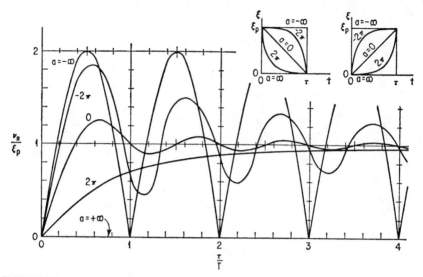

FIGURE 8.27 Spectra for residual response amplitude for unsymmetrical exponential pulses having either vertical rise or vertical decay. Comparison on the basis of equal pulse height.[19]

Figure 8.28 shows the spectra of residual response amplitude in greater detail for the range in which the parameter a is limited to positive values. This group of pulses is of interest in studying the effects of a simple form of blast pulse, in which the peak height and the duration are constant but the rate of decay is varied.

The areas of the pulses of equal height ξ_p, and the heights of the pulses of equal area $\xi_{p0}\tau/2$ are the same as for the symmetrical exponential pulses [see Eqs. (8.36c) and (8.36d)]. If the spectra in Fig. 8.28 are redrawn, using *equal pulse area* as the criterion for comparison, they appear as in Fig. 8.29. The limiting pulse case $a \to +\infty$ represents a generalized impulse of value $k\xi_{p0}\tau/2$. The asymptotic values of the spectra are equal to the peak heights of the equal area pulses and are given by

$$\frac{v_R}{\xi_{p0}} \to \frac{a(1-e^a)}{2(1-e^a+a)} \quad \text{as} \quad \frac{\tau}{T} \to \infty \tag{8.44d}$$

Exponential Pulses of Infinite Duration. Five different cases are included as follows:

1. The excitation function, consisting of a vertical rise followed by an exponential decay, is

$$\xi(t) = \xi_p e^{-at} \quad [0 \le t] \tag{8.45a}$$

It is shown in Fig. 8.30. The response time equation for the system is

$$v = \xi_p \frac{(a/\omega_n)\sin\omega_n t - \cos\omega_n t + e^{-at}}{1+a^2/\omega_n^2} \tag{8.45b}$$

and the asymptotic value of the residual amplitude is given by

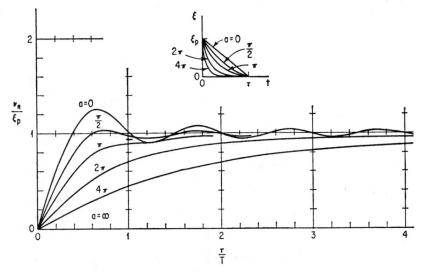

FIGURE 8.28 Spectra for residual response amplitude for a family of simple blast pulses, the same family shown in Fig. 8.27 but limited to positive values of the exponential decay parameter a. Comparison on the basis of equal pulse height. (These spectra also apply to mirror-image pulses having vertical decay.) (*Jacobsen and Ayre.*[22])

$$\frac{v_R}{\xi_p} \rightarrow \frac{1}{\sqrt{1 + a^2/\omega_n^2}} \tag{8.45c}$$

The maximax response is the first maximum of v. The time response, for the particular case $\omega_n/a = 2$, and the response spectra are shown in Figs. 8.31 and 8.32, respectively.

2. The *difference of two exponential functions,* of the type of Eq. (8.45a), results in the pulse given by Eq. (8.46a):

$$\xi(t) = \xi_0(e^{-bt} - e^{-at})$$

$$a > b \quad [0 \leq t] \tag{8.46a}$$

The shape of the pulse is shown in Fig. 8.33. Note that ξ_0 is the ordinate of each of the exponential functions at $t = 0$; it is *not* the pulse maximum. The asymptotic residual response amplitude is

$$v_R \rightarrow \xi_0 \frac{(b/\omega_n) - (a/\omega_n)}{[(1 + a^2/\omega_n^2)(1 + b^2/\omega_n^2)]^{1/2}} \tag{8.46b}$$

3. The *product of the exponential function e^{-at} by time* results in the excitation given by Eq. (8.47a) and shown in Fig. 8.34.

$$\xi(t) = C_0 t e^{-at} \tag{8.47a}$$

where C_0 is a constant. The peak height of the pulse ξ_p is equal to C_0/ae, and occurs at the time $t_1 = 1/a$. Equations (8.47b) and (8.47c) give the time response and the asymptotic residual response amplitude:

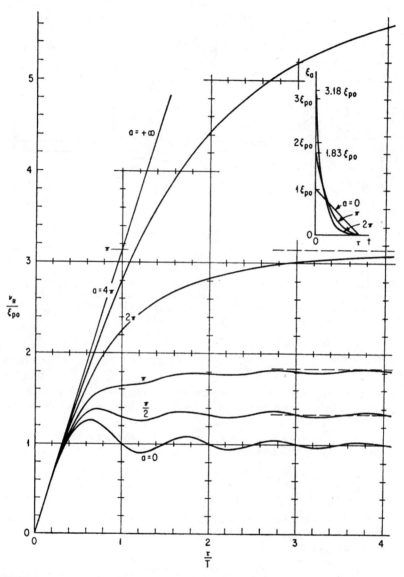

FIGURE 8.29 Spectra for residual response amplitude for the family of simple blast pulses shown in Fig. 8.28, compared on the basis of equal pulse area. (*Jacobsen and Ayre.*[22])

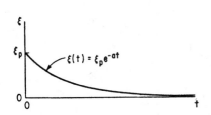

FIGURE 8.30 Pulse consisting of vertical rise followed by exponential decay of infinite duration.

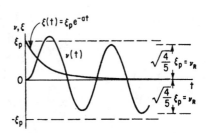

FIGURE 8.31 Time response to the pulse having a vertical rise and an exponential decay of infinite duration (Fig. 8.30), for the particular case $\omega_n/a = 2$.

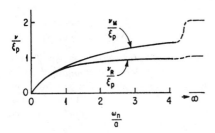

FIGURE 8.32 Spectra for maximax response and for asymptotic residual response amplitude, for the pulse shown in Fig. 8.30.

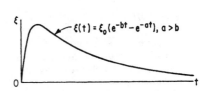

FIGURE 8.33 Pulse formed by taking the difference of two exponentially decaying functions.

$$v = \xi_p \frac{ae/\omega_n}{(1 + a^2/\omega_n^2)^2} \left\{ \left[\frac{2a}{\omega_n} + \left(1 + \frac{a^2}{\omega_n^2}\right) \omega_n t \right] e^{-at} - \frac{2a}{\omega_n} \cos \omega_n t - \left(1 - \frac{a^2}{\omega_n^2}\right) \sin \omega_n t \right\}$$

$$(8.47b)$$

$$\frac{v_R}{\xi_p} \to \frac{e}{(a/\omega_n) + (\omega_n/a)} \qquad (8.47c)$$

The *maximum* value of v_R occurs in the case $a/\omega_n = 1$, and is given by

$$(v_R)_{max} = \xi_p e/2 = 1.36 \xi_p$$

Both of the excitation functions described by Eqs. (8.46a) and (8.47a) include finite times of rise to the pulse peak. These rise times are dependent on the exponential decay constants.

4. The rise time may be made independent of the decay by inserting a separate rise function before the decay function, as in Fig. 8.35, where a *straight-line rise precedes the exponential decay.* The response-time equations are as follows:

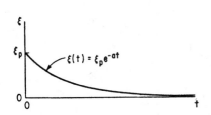

FIGURE 8.34 Pulse formed by taking the product of an exponentially decaying function by time.

Pulse rise era:

$$v = \xi_p \frac{\omega_n t - \sin \omega_n t}{\omega_n t_1} \qquad [0 \le t \le t_1] \qquad (8.48a)$$

Pulse decay era:

$$v = \xi_p \left[\frac{e^{-at'}}{1 + a^2/\omega_n^2} + \left(\frac{a^2/\omega_n^2}{1 + a^2/\omega_n^2} - \frac{\sin \omega_n t_1}{\omega_n t_1} \right) \cos \omega_n t' \right.$$

$$\left. + \left(\frac{a/\omega_n}{1 + a^2/\omega_n^2} + \frac{1 - \cos \omega_n t_1}{\omega_n t_1} \right) \sin \omega_n t' \right] \quad (8.48b)$$

where $t' = t - t_1$ and $0 \le t'$.

5. Another form of pulse, which is a more complete representation of a blast pulse since it includes the possibility of a negative phase of pressure,[14] is shown in Fig. 8.36. It consists of a straight-line rise, followed by an exponential decay through the positive phase, into the negative phase, finally becoming asymptotic to the time axis. The rise time is t_1 and the duration of the positive phase is $t_1 + t_2$.

Unsymmetrical Exponential Pulses with Central Peak. An interesting family of unsymmetrical pulses may be formed by using Eqs. (8.36a) and changing the sign of the exponent of e in both the numerator and the denominator of the second of the equations. The resulting family consists of pulses whose maxima occur at the mid-period time and which satisfy simultaneously both criteria for comparison (equal pulse height and equal pulse area).

Figure 8.37 shows the spectra of residual response amplitude and, in the inset diagrams, the pulse shapes. The limiting cases are the symmetrical triangle of duration τ and height ξ_p, and the rectangles of duration $\tau/2$ and height ξ_p. All pulses in the family have the area $\xi_p \tau/2$. Zeroes of residual response amplitude occur for all values of a, at even integer values of τ/T. The residual response amplitude is

$$\frac{v_R}{\xi_p} = \frac{aT/\pi\tau}{1 - \cosh a} \cdot$$

$$\left[\frac{\cosh 2a - \cosh a - (1 - \cosh a) \cos (2\pi\tau/T) + (1 - \cosh 2a) \cos (\pi\tau/T)}{1 + a^2 T^2/\pi^2\tau^2} \right]^{1/2} \quad (8.49)$$

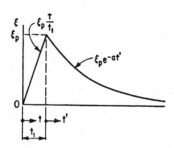

FIGURE 8.35 Pulse formed by a straight-line rise followed by an exponential decay asymptotic to the time axis.

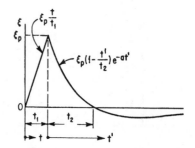

FIGURE 8.36 Pulse formed by a straight-line rise followed by a continuous exponential decay through positive and negative phases. (*Frankland.*[14])

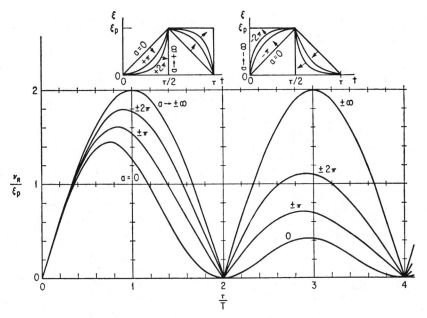

FIGURE 8.37 Spectra of residual response amplitude for a family of unsymmetrical exponential pulses of equal area and equal maximum height, having the pulse peak at the mid-period time. (*Jacobsen and Ayre.*[22])

Pulses which are mirror images of each other in time result in equal residual amplitudes.

Skewed Versed-sine Pulse. By taking the product of a decaying exponential and the versed-sine function, a family of pulses with varying skewness is obtained.[13, 22] The family is described by the following equation:

$$\xi(t) = \begin{cases} \xi_p \dfrac{e^{2\pi(\sigma - t/\tau)} \cot \pi\sigma}{1 - \cos 2\pi\sigma} \, (1 - \cos 2\pi t/\tau) & [0 \le t \le \tau] \\ 0 & [\tau \le t] \end{cases} \qquad (8.50)$$

These pulses are of particular interest when the excitation is a ground displacement function because they have continuity in both velocity and displacement; thus, they do not involve theoretically infinite accelerations of the ground. When the skewing constant σ equals one-half, the pulse is the symmetrical versed sine. When $\sigma \to 0$, the front of the pulse approaches a straight line with infinite slope, and the pulse area approaches zero.

The spectra of residual response amplitude and of maximax relative response, plotted on the basis of equal pulse height, are shown in Fig. 8.38 for several values of σ. The residual response amplitude spectra are reasonably good approximations to the spectra of maximax relative response except at the lower values of τ/T.

Figure 8.39 compares the residual response amplitude spectra on the basis of equal pulse area. The required pulse heights, for a constant pulse area of $\xi_{p0}\tau/2$, are shown in the inset diagram. On this basis, the pulse for $\sigma \to 0$ represents a generalized impulse of value $k\xi_{p0}\tau/2$.

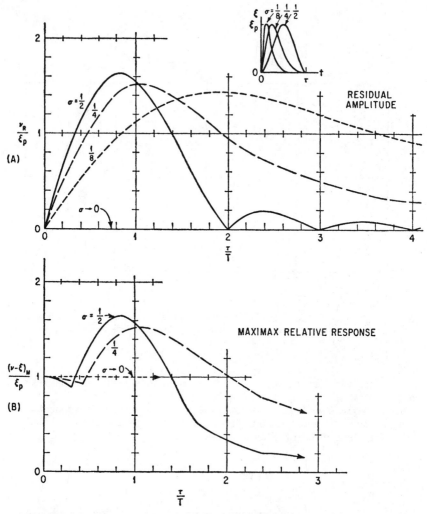

FIGURE 8.38 Response spectra for the skewed versed-sine pulse, compared on the basis of equal pulse height: (A) Residual response amplitude. (B) Maximax relative response. (*Jacobsen and Ayre.*[22])

Full-cycle Pulses (Force-Time Integral = 0). The residual response amplitude spectra for three groups of *full-cycle pulses* are shown as follows: in Fig. 8.40 for the rectangular, the sinusoidal, and the symmetrical triangular pulses; in Fig. 8.41 for three types of pulse involving sine and cosine functions; and in Fig. 8.42 for three forms of triangular pulse. The pulse shapes are shown in the inset diagrams. Expressions for the residual response amplitudes are given in Eqs. (8.51) to (8.53).

Full-cycle rectangular pulse:

$$\frac{v_R}{\xi_{2p}} = 2 \sin \frac{\pi\tau}{T} \left[2 \sin \frac{\pi\tau}{T} \right] \tag{8.51}$$

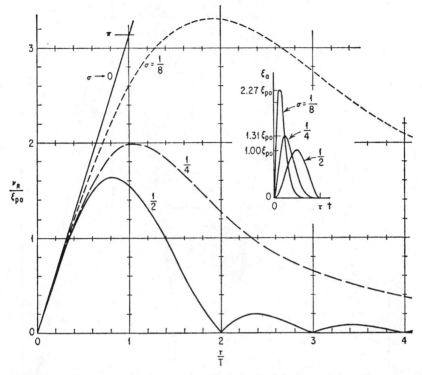

FIGURE 8.39 Spectra of residual response amplitude for the skewed versed-sine pulse, compared on the basis of equal pulse area.[19]

Full-cycle "sinusoidal" pulses:
  Symmetrical half cycles

$$\frac{v_R}{\xi_p} = 2 \sin \frac{\pi\tau}{T} \left[\frac{T/\tau}{(T^2/4\tau^2) - 1} \cos \frac{\pi\tau}{T} \right] \qquad (8.52a)$$

  Vertical front and vertical ending

$$\frac{v_R}{\xi_p} = \frac{2}{1 - T^2/16\tau^2} \cos \frac{2\pi\tau}{T} \qquad (8.52b)$$

  Vertical jump at mid-cycle

$$\frac{v_R}{\xi_p} = \frac{2}{1 - T^2/16\tau^2} \left(1 - \frac{T}{4\tau} \sin \frac{2\pi\tau}{T} \right) \qquad (8.52c)$$

Full-cycle triangular pulses:
  Symmetrical half cycles

$$\frac{v_R}{\xi_p} = 2 \sin \frac{\pi\tau}{T} \left[\frac{4T}{\pi\tau} \sin^2 \frac{\pi\tau}{2T} \right] \qquad (8.53a)$$

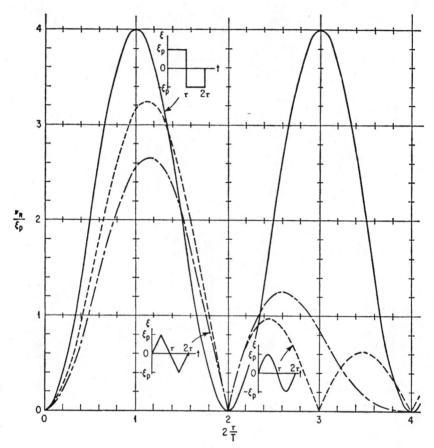

FIGURE 8.40 Spectra of residual response amplitude for three types of full-cycle pulses. Each half cycle is symmetrical.[19]

Vertical front and vertical ending

$$\frac{v_R}{\xi_p} = 2\left(\frac{T}{2\pi\tau} \sin \frac{2\pi\tau}{T} - \cos \frac{2\pi\tau}{T}\right) \tag{8.53b}$$

Vertical jump at mid-cycle

$$\frac{v_R}{\xi_p} = 2\left(1 - \frac{T}{2\pi\tau} \sin \frac{2\pi\tau}{T}\right) \tag{8.53c}$$

In the case of full-cycle pulses having symmetrical half cycles, note that the residual response amplitude equals the residual response amplitude of the symmetrical one-half-cycle pulse of the same shape, multiplied by the dimensionless residual response amplitude function $2 \sin(\pi\tau/T)$ for the single rectangular pulse. Compare the bracketed functions in Eqs. (8.51), (8.52a), and (8.53a) with the bracketed functions in Eqs. (8.31b), (8.32b), and (8.34c), respectively.

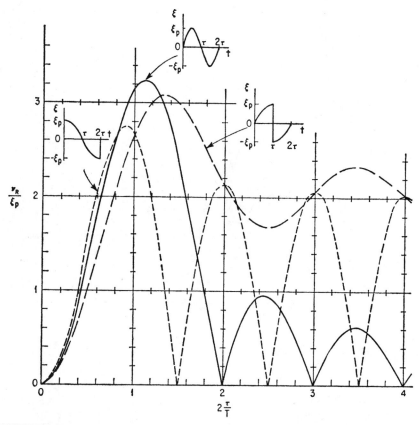

FIGURE 8.41 Spectra of residual response amplitude for three types of full-cycle "sinusoidal" pulses.[19]

SUMMARY OF TRANSIENT RESPONSE SPECTRA FOR THE SINGLE DEGREE-OF-FREEDOM, LINEAR, UNDAMPED SYSTEM

Initial Conditions.   The following conclusions are based on the assumption that the system is initially at rest.

Step-type Excitations.

1. The maximax response v_M occurs *after* the step has risen (monotonically) to full value ($\tau \leq t$, where τ is the step rise time). It is equal to the residual response amplitude plus the constant step height ($v_M = v_R + \xi_c$).

2. The extreme values of the ratio of maximax response to step height v_M/ξ_c are 1 and 2. When the ratio of step rise time to system natural period τ/T approaches zero, the step approaches the simple rectangular step in shape and v_M/ξ_c approaches the upper extreme of 2. If τ/T approaches infinity, the step loses the character of a dynamic excitation; consequently, the inertia forces of the system approach zero and v_M/ξ_c approaches the lower extreme of 1.

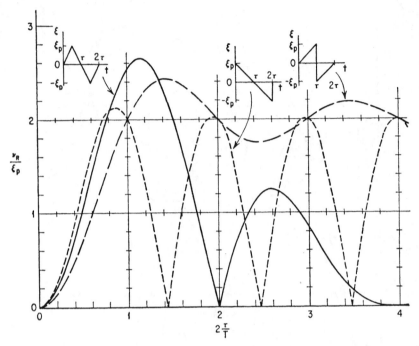

FIGURE 8.42   Spectra of residual response amplitude for three types of full-cycle triangular pulses.[19]

3. For some particular shapes of step rise, v_M/ξ_c is equal to 1 at certain finite values of τ/T. For example, for the step having a constant-slope rise, $v_M/\xi_c = 1$ when $\tau/T = 1, 2, 3, \ldots$. The lowest values of $\tau/T = (\tau/T)_{min}$, for which $v_M/\xi_c = 1$, are, for three shapes of step rise: constant-slope, 1.0; versed-sine, 1.5; cycloidal, 2.0. The lowest possible value of $(\tau/T)_{min}$ is 1.

4. In the case of step-type excitations, when $v_M/\xi_c = 1$ the residual response amplitude v_R is zero. Sometimes it is of practical importance in the design of cams and dynamic control functions to achieve the smallest possible residual response.

Single-Pulse Excitations.

1. When the ratio τ/T of pulse duration to system natural period is less than ½, the time shapes of certain types of equal area pulses are of secondary significance in determining the maxima of system response [maximax response v_M, maximax relative response $(v - \xi)_M$, and residual response amplitude v_R]. If τ/T is less than ¼, the pulse shape is of little consequence in almost all cases and the system response can be determined to a fair approximation by use of the simple impulse theory. If τ/T is larger than ½, the pulse shape may be of great significance.

2. The maximum value of maximax response for a given shape of pulse, $(v_M)_{max}$, usually occurs at a value of the period ratio τ/T between ½ and 1. The maximum value of the ratio of maximax response to the reference excitation, $(v_M)_{max}/\xi_p$, is usually between 1.5 and 1.8.

3. If the pulse has a *vertical rise*, v_M is the first maximum occurring, and $(v_M)_{max}$ is an asymptotic value approaching $2\xi_p$ as τ/T approaches infinity. In the special case of the rectangular pulse, $(v_M)_{max}$ is equal to $2\xi_p$ and occurs at values of τ/T equal to or greater than ½.

4. If the pulse has a *vertical decay*, $(v_M)_{max}$ is equal to the maximum value $(v_R)_{max}$ of the residual response amplitude.

5. The maximum value $(v_R)_{max}$ of the residual response amplitude, for a given shape of pulse, often is a reasonably good approximation to $(v_M)_{max}$, except if the pulse has a steep rise followed by a decay. A few examples are shown in Table 8.3. Furthermore, if $(v_M)_{max}$ and $(v_R)_{max}$ for a given pulse shape are approximately equal in magnitude, they occur at values of τ/T not greatly different from each other.

6. Pulse shapes that are mirror images of each other in time result in equal values of residual response amplitude.

7. The residual response amplitude v_R generally has zero values for certain finite values of τ/T. However, if the pulse has either a vertical rise or a vertical decay, but not both, there are no zero values except the trivial one at $\tau/T = 0$. In the case of the rectangular pulse, $v_R = 0$ when $\tau/T = 1, 2, 3, \ldots$ For several shapes of pulse the values of $(\tau/T)_{min}$ (lowest values of τ/T for which $v_R = 0$) are as follows: rectangular, 1.0; sine, 1.5; versed-sine, 2.0; symmetrical triangle, 2.0. The lowest possible value of $(\tau/T)_{min}$ is 1.

8. In the formulation of pulse as well as of step-type excitations, it may be of practical consequence for the residual response to be as small as possible; hence, attention is devoted to the case, $v_R = 0$.

TABLE 8.3 Comparison of Greatest Values of Maximax Response and Residual Response Amplitude

Pulse shape	$(v_M)_{max}/(v_R)_{max}$
Symmetrical:	
Rectangular	1.00
Sine	1.04
Versed sine	1.05
Triangular	1.06
Vertical-decay pulses	1.00
Vertical-rise pulses:	
Rectangular	1.00
Triangular	1.60
Asymptotic exponential decay	2.00

SINGLE DEGREE-OF-FREEDOM LINEAR SYSTEM WITH DAMPING

The calculation of the effects of damping on transient response may be laborious. If the investigation is an extensive one, use should be made of an analog computer.

DAMPING FORCES PROPORTIONAL TO VELOCITY (VISCOUS DAMPING)

In the case of steady forced vibration, even very small values of the viscous damping coefficient have great effect in limiting the system response at or near resonance. If the excitation is of the single step- or pulse-type, however, the effect of damping on the maximax response may be of relatively less importance, unless the system is highly damped.

For example, in a system under steady sinusoidal excitation at resonance, a tenfold increase in the fraction of critical damping c/c_c from 0.01 to 0.1 results in a theoretical tenfold decrease in the magnification factor from 50 to 5. In the case of the same system, initially at rest and acted upon by a half-cycle sine pulse of "resonant duration" $\tau = T/2$, the same increase in the damping coefficient results in a decrease in the maximax response of only about 9 percent.

Half-cycle Sine Pulse Excitation. Figure 8.43 shows the spectra of maximax response for a viscously damped system excited by a half-cycle sine pulse.[12, 29] The system is initially at rest. The results apply to the cases indicated by the following differential equations of motion:

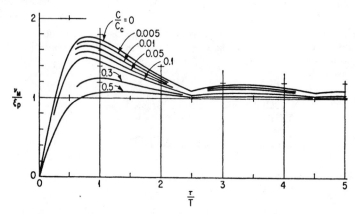

FIGURE 8.43 Spectra of maximax response for a viscously damped single degree-of-freedom system acted upon by a half-cycle sine pulse. (*R. D. Mindlin, F. W. Stubner, and H. L. Cooper.*[29])

$$\frac{m\ddot{x}}{k} + \frac{c\dot{x}}{k} + x = \frac{F_p}{k} \sin \frac{\pi t}{\tau} \tag{8.54a}$$

$$\frac{m\ddot{x}}{k} + \frac{c\dot{x}}{k} + x = u_p \sin \frac{\pi t}{\tau} \tag{8.54b}$$

$$\frac{m\ddot{\delta}_x}{k} + \frac{c\dot{\delta}_x}{k} + \delta_x = \frac{-m\ddot{u}_p}{k} \sin \frac{\pi t}{\tau} \tag{8.54c}$$

and in general

$$\frac{m\ddot{v}}{k} + \frac{c\dot{v}}{k} + v = \xi_p \sin \frac{\pi t}{\tau} \tag{8.54d}$$

where $0 \le t \le \tau$.

For values of t greater than τ, the excitation is zero. The distinctions among these cases may be determined by referring to Table 8.1. The fraction of critical damping c/c_c in Fig. 8.43 is the ratio of the damping coefficient c to the critical damping coefficient $c_c = \sqrt{2mk}$. The damping coefficient must be defined in terms of the velocity $(\dot{x}, \dot{\delta}_x, \dot{v})$ appropriate to each case. For $c/c_c = 0$, the response spectrum is the same as the spectrum for maximax response shown for the undamped system in Fig. 8.16B.

Other Forms of Excitation; Methods.   Qualitative estimates of the effects of viscous damping in the case of other forms of step or pulse excitation may be made by the use of Fig. 8.43 and of the appropriate spectrum for the undamped response to the excitation in question.

Quantitative calculations may be effected by extending the methods described for the undamped system. If the excitation is of general form, given either numerically or graphically, the *phase-plane-delta*[21, 22] method described in a later section of this chapter may be used to advantage. Of the analytical methods, the *Laplace transformation* is probably the most useful. A brief discussion of its application to the viscously damped system follows.

Laplace Transformation.   The differential equation to be solved is

$$\frac{m\ddot{v}}{k} + \frac{c\dot{v}}{k} + v = \xi(t) \tag{8.55a}$$

Rewriting Eq. (8.55a),

$$\frac{\ddot{v}}{\omega_n^2} + \frac{2\zeta\dot{v}}{\omega_n} + v = \xi(t) \tag{8.55b}$$

where $\zeta = c/c_c$ and $\omega_n^2 = k/m$.

Applying the operation transforms of Table 8.2 to Eq. (8.55b), the following algebraic equation is obtained:

$$\frac{1}{\omega_n^2}[s^2 F_r(s) - sf(0) - f'(0)] + \frac{2\zeta}{\omega_n}[sF_r(s) - f(0)] + F_r(s) = F_e(s) \tag{8.56a}$$

The *subsidiary* equation is

$$F_r(s) = \frac{(s + 2\zeta\omega_n)f(0) + f'(0) + \omega_n^2 F_e(s)}{s^2 + 2\zeta\omega_n s + \omega_n^2} \tag{8.56b}$$

where the initial conditions $f(0)$ and $f'(0)$ are to be expressed as v_0 and $\dot{v}_0$, respectively.

By performing an *inverse* transformation of Eq. (8.56b), the response is determined in the following operational form:

$$v(t) = \mathcal{L}^{-1}[F_r(s)]$$

$$= \mathcal{L}^{-1}\left[\frac{(s + 2\zeta\omega_n)v_0 + \dot{v}_0 + \omega_n^2 F_e(s)}{s^2 + 2\zeta\omega_n s + \omega_n^2}\right] \tag{8.57}$$

Example 8.9.   Rectangular Step Excitation.   Assume that the damping is less than critical $(\zeta < 1)$, that the system starts from rest $(v_0 = \dot{v}_0 = 0)$, and that the system is acted upon by the rectangular step excitation: $\xi(t) = \xi_c$ for $0 \leq t$. The transform of the excitation is given by

$$F_e(s) = \mathcal{L}[\xi(t)] = \mathcal{L}[\xi_c] = \xi_c \frac{1}{s}$$

Substituting for v_0, $\dot{v}_0$ and $F_e(s)$ in Eq. (8.57), the following equation is obtained:

$$v(t) = \mathcal{L}^{-1}[F_r(s)] = \xi_c \omega_n^2 \mathcal{L}^{-1} \left[\frac{1}{s(s^2 + 2\zeta\omega_n s + \omega_n^2)} \right] \tag{8.58a}$$

Rewriting,

$$v(t) = \xi_c \omega_n^2 \mathcal{L}^{-1} \left[\frac{1}{s[s + \omega_n(\zeta - j\sqrt{1 - \zeta^2})][s + \omega_n(\zeta + j\sqrt{1 - \zeta^2})]} \right] \tag{8.58b}$$

where $j = \sqrt{-1}$.

To determine the inverse transform $\mathcal{L}^{-1}[F_r(s)]$, it may be necessary to expand $F_r(s)$ in partial fractions as explained previously. However, in this particular example the transform pair is available in Table 8.2 (see item 16). Thus, it is found readily that $v(t)$ is given by the following:

$$v(t) = \xi_c \omega_n^2 \left[\frac{1}{ab} + \frac{be^{-at} - ae^{-bt}}{ab(a - b)} \right] \tag{8.59a}$$

where $a = \omega_n(\zeta - j\sqrt{1 - \zeta^2})$ and $b = \omega_n(\zeta + j\sqrt{1 - \zeta^2})$. By using the relations, $\cos z = (\frac{1}{2})(e^{jz} + e^{-jz})$ and $\sin z = -(\frac{1}{2})j(e^{jz} - e^{-jz})$, Eq. (8.59a) may be expressed in terms of *cosine* and *sine* functions:

$$v(t) = \xi_c \left[1 - e^{-\zeta\omega_n t} \left(\cos \omega_d t + \frac{\zeta}{\sqrt{1 - \zeta^2}} \sin \omega_d t \right) \right] \qquad [\zeta < 1] \tag{8.59b}$$

where the *damped* natural frequency $\omega_d = \omega_n \sqrt{1 - \zeta^2}$.

If the damping is negligible, $\zeta \to 0$ and Eq. (8.59b) reduces to the form of Eq. (8.22) previously derived for the case of zero damping:

$$v(t) = \xi_c(1 - \cos \omega_n t) \qquad [\zeta = 0] \tag{8.22}$$

CONSTANT (COULOMB) DAMPING FORCES; PHASE-PLANE METHOD

The phase-plane method is particularly well suited to the solving of transient response problems involving Coulomb damping forces.[21, 22] The problem is truly a stepwise linear one, provided the usual assumptions regarding Coulomb friction are valid. For example, the differential equation of motion for the case of ground displacement excitation is

$$m\ddot{x} \pm F_f + kx = ku(t) \tag{8.60a}$$

where F_f is the Coulomb friction force. In Eq. (8.60b) the friction force has been moved to the right side of the equation and the equation has been divided by the spring constant k:

$$\frac{m\ddot{x}}{k} + x = u(t) \mp \frac{F_f}{k} \tag{8.60b}$$

The effect of friction can be taken into account readily in the construction of the phase trajectory by modifying the ordinates of the stepwise excitation by amounts equal to $\mp F_f/k$. The quantity F_f/k is the Coulomb friction "displacement," and is equal to one-fourth the decay in amplitude in each cycle of a *free* vibration under the influence of Coulomb friction. The algebraic sign of the friction term changes when the velocity changes sign. When the friction term is placed on the right-hand side of the differential equation, it must have a *negative* sign when the velocity is positive.

   Example 8.10. Free Vibration.   Figure 8.44 shows an example of free vibration with the initial conditions $x = x_0$ and $\dot{x} = 0$. The locations of the arc centers of the phase trajectory alternate each half cycle from $+F_f/k$ to $-F_f/k$.

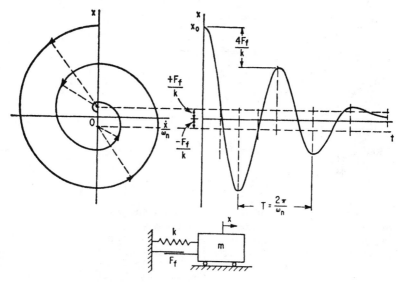

FIGURE 8.44   Example of phase-plane solution of free vibration with Coulomb friction;[2] the natural frequency is $\omega_n = \sqrt{k/m}$.

   Example 8.11. General Transient Excitation.   A general stepwise excitation $u(t)$ and the response x of a system under the influence of a friction force F_f are shown in Fig. 8.45. The case of zero friction is also shown. The initial conditions are $x = 0$, $\dot{x} = 0$. The arc centers are located at ordinates of $u(t) \mp F_f/k$. During the third step in the excitation, the velocity of the system changes sign from positive to negative (at $t = t_2'$); consequently, the friction displacement must also change sign, but from negative to positive.

SINGLE DEGREE-OF-FREEDOM NONLINEAR SYSTEMS

PHASE-PLANE-DELTA METHOD

The transient response of damped linear systems and of nonlinear systems of considerable complexity can be determined by the *phase-plane-delta* method.[21, 22] Assume that the differential equation of motion of the system is

$$m\ddot{x} = G(x,\dot{x},t) \tag{8.61a}$$

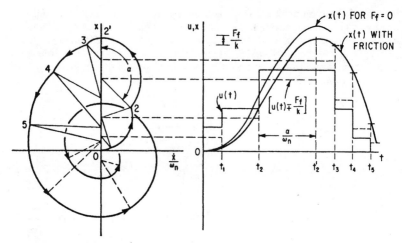

FIGURE 8.45 Example of phase-plane solution for a general transient excitation with Coulomb friction in the system.[2]

where $G(x,\dot{x},t)$ is a general function of x, $\dot{x}$, and t to any powers. The coefficient of $\ddot{x}$ is constant, either inherently or by a suitable division.

In Eq. (8.61a) the general function may be replaced by another general function minus a linear, constant-coefficient, restoring force term:

$$G(x,\dot{x},t) = g(x,\dot{x},t) - kx$$

By moving the linear term kx to the left side of the differential equation, dividing through by m, and letting $k/m = \omega_n^2$, the following equation is obtained:

$$\ddot{x} + \omega_n^2 x = \omega_n^2 \delta \tag{8.61b}$$

where the *operative displacement* δ is given by

$$\delta = \frac{1}{k} g(x,\dot{x},t) \tag{8.61c}$$

The separation of the kx term from the G function does not require that the kx term exist physically. Such a term can be separated by first adding to the G function the *fictitious* terms, $+kx - kx$.

With the differential equation of motion in the δ form, Eq. (8.61b), the response problem can now be solved readily by stepwise linearization. The left side of the equation represents a simple, undamped, linear oscillator. Implicit in the δ function on the right side of the equation are the nonlinear restoration terms, the linear or nonlinear dissipation terms, and the excitation function.

If the δ function is held constant at a value δ for an interval of time Δt, the response of the linear oscillator in the phase-plane is an arc of a circle, with its center on the X axis at δ and subtended by an angle equal to $\omega_n \Delta t$. The graphical construction may be similar in general appearance to the examples already shown for linear systems in Figs. 8.5, 8.44, and 8.45. Since in the general case the δ function involves the dependent variables, it is necessary to estimate, before constructing each step, appropriate average values of the system displacement and/or velocity to be expected during the step. In some cases, more than one trial may be required before suitable accuracy is obtained.

Many examples of solution for various types of systems are available in the literature.[3,5,6,8,13,15,20–22,25]

MULTIPLE DEGREE-OF-FREEDOM, LINEAR, UNDAMPED SYSTEMS

Some of the transient response analyses, presented for the single degree-of-freedom system, are in complete enough form that they can be employed in determining the responses of linear, undamped, multiple degree-of-freedom systems. This can be done by the use of *normal (principal) coordinates*. A system of normal coordinates is a system of generalized coordinates chosen in such a way that vibration in each normal mode involves only one coordinate, a normal coordinate. The differential equations of motion, when written in normal coordinates, are all independent of each other. Each differential equation is related to a particular normal mode and involves only one coordinate. The differential equations are of the same general form as the differential equation of motion for the single degree-of-freedom system. The response of the system in terms of the physical coordinates, for example, displacement or stress at various locations in the system, is determined by superposition of the normal coordinate responses. Normal coordinates are discussed in Chaps. 2 and 7, and in Refs. 37 and 38.

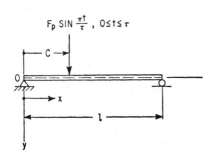

FIGURE 8.46 Simply supported beam loaded by a concentrated force sine pulse of half-cycle duration.

Example 8.12. Sine Force Pulse Acting on a Simple Beam. Consider the flexural vibration of a prismatic bar with simply supported ends, Fig. 8.46. A sine-pulse concentrated force F_p sin $(\pi t/\tau)$ is applied to the beam at a distance c from the left end (origin of coordinates). Assume that the beam is initially at rest. The displacement response of the beam, during the time of action of the pulse, is given by the following series:

$$y = \frac{2F_p l^3}{\pi^4 EI} \sum_{i=1}^{i=\infty} \frac{1}{i^4} \sin \frac{i\pi c}{l} \sin \frac{i\pi x}{l} \left[\frac{1}{1 - T_i^2/4\tau^2} \left(\sin \frac{\pi t}{\tau} - \frac{T_i}{2\tau} \sin \omega_i t \right) \right] \qquad [0 \le t \le \tau]$$

$$(8.62a)$$

where $i = 1, 2, 3, \ldots; T_i = \dfrac{2\pi}{\omega_i} = \dfrac{2l^2}{i^2\pi} \sqrt{\dfrac{A\gamma}{EIg}} = \dfrac{T_1}{i^2}, \text{sec}$

A comparison of Eqs. (8.62a) and (8.32a) shows that the time function [sin $(\pi t/\tau)$ − $(T_i/2\tau)$ sin $\omega_i t$] for the ith term in the beam-response series is of exactly the same form as the time function [sin $(\pi t/\tau)$ − $(T/2\tau)$ sin $\omega_n t$] in the response of the single degree-of-freedom system. Furthermore, the magnification factors $1/(1 - T_i^2/4\tau^2)$ and $1/(1 - T^2/4\tau^2)$ in the two equations have identical forms.

Following the end of the pulse, beginning at $t = \tau$, the vibration of the beam is expressed by

$$y = \frac{2F_p l^3}{\pi^4 EI} \sum_{i=1}^{i=\infty} \frac{1}{i^4} \sin \frac{i\pi c}{l} \sin \frac{i\pi x}{l} \left[\frac{(T_i/\tau) \cos (\pi\tau/T_i)}{(T_i^2/4\tau^2) - 1} \sin \omega_i \left(t - \frac{\tau}{2} \right) \right] \qquad [\tau \le t]$$

$$(8.62b)$$

A comparison of Eqs. (8.62b) and (8.32b) leads to the same conclusion as found above for the time era $0 \le t \le \tau$.

Excitation and Displacement at Mid-span. As a specific case, consider the displacement at mid-span when the excitation is applied at mid-span ($c = x = l/2$). The even-numbered terms of the series now are all zero and the series take the following forms:

$$y_{l/2} = \frac{2F_p l^3}{\pi^4 EI} \sum_{i=1,3,5,\ldots}^{\infty} \frac{1}{i^4} \left[\frac{1}{1 - T_i^2/4\tau^2} \left(\sin \frac{\pi t}{\tau} - \frac{T_i}{2\tau} \sin \omega_i t \right) \right] \qquad [0 \le t \le \tau]$$

$$(8.63a)$$

$$y_{l/2} = \frac{2F_p l^3}{\pi^4 EI} \sum_{i=1,3,5,\ldots}^{\infty} \frac{1}{i^4} \left[\frac{(T_i/\tau) \cos (\pi\tau/T_i)}{(T_i^2/4\tau^2) - 1} \sin \omega_i \left(t - \frac{\tau}{2} \right) \right] \qquad [\tau \le t] \quad (8.63b)$$

Assume, for example, that the pulse period τ equals two-tenths of the fundamental natural period of the beam ($\tau/T_1 = 0.2$). It is found from Fig. 8.16B, by using an abscissa value of 0.2, that the maximax response in the *fundamental* mode ($i = 1$) occurs in the residual vibration era ($\tau \le t$). The value of the corresponding ordinate is 0.75. Consequently, the maximax response for $i = 1$ is 0.75 $(2F_p l^3/\pi^4 EI)$.

In order to determine the maximax for the *third* mode ($i = 3$), an abscissa value of $\tau/T_i = i^2\tau/T_1 = 3^2 \times 0.2 = 1.8$, is used. It is found that the maximax is greater than the residual amplitude and consequently that it occurs during the time era $0 \le t \le \tau$. The value of the corresponding ordinate is 1.36; however, this must be multiplied by $\frac{1}{3}^4$, as indicated by the series. The maximax for $i = 3$ is thus 0.017 $(2F_p l^3/\pi^4 EI)$.

The maximax for $i = 5$ also occurs in the time era $0 \le t \le \tau$ and the ordinate may be estimated to be about 1.1. Multiplying by $\frac{1}{5}^4$, it is found that the maximax for $i = 5$ is approximately 0.002 $(2F_p l^3/\pi^4 EI)$, a negligible quantity when compared with the maximax value for $i = 1$.

To find the maximax total response to a reasonable approximation, it is necessary to sum on a time basis several terms of the series. In the particular example above, the maximax total response occurs in the residual vibration era and a reasonably accurate value can be obtained by considering only the first term ($i = 1$) in the series, Eq. (8.63b).

GENERAL INVESTIGATION OF TRANSIENTS

An extensive (and efficient) investigation of transient response in multiple degree-of-freedom systems requires the use of an automatic computer. In some of the simpler cases, however, it is feasible to employ numerical or graphical methods. For example, the phase-plane method may be applied to multiple degree-of-freedom linear systems[1, 2] through the use of normal coordinates. This involves independent phase-planes having the coordinates q_i and q_i/ω_i, where q_i is the ith normal coordinate.

REFERENCES

1. Ayre, R. S.: *J. Franklin Inst.,* **253**:153 (1952).
2. Ayre, R. S.: *Proc. World Conf. Earthquake Eng.,* 1956, p. 13–1.
3. Ayre, R. S., and J. I. Abrams: *Proc. ASCE,* EM 2, Paper 1580, 1958.
4. Biot, M. A.: *Trans. ASCE,* **108**:365 (1943).
5. Bishop, R. E. D.: *Proc. Inst. Mech. Engrs. (London),* **168**:299 (1954).
6. Braun, E.: *Ing.-Arch.,* **8**:198 (1937).
7. Bronwell, A.: "Advanced Mathematics in Physics and Engineering," McGraw-Hill Book Company, Inc., New York, 1953.
8. Bruce, V. G.: *Bull, Seismol. Soc. Amer.,* **41**:101 (1951).
9. Cherry, C.: "Pulses and Transients in Communication Circuits," Dover Publications, New York, 1950.
10. Crede, C. E.: "Vibration and Shock Isolation," John Wiley & Sons, Inc., New York, 1951.
11. Crede, C. E.: *Trans. ASME,* **77**:957 (1955).
12. Criner, H. E., G. D. McCann, and C. E. Warren: *J. Appl. Mechanics,* **12**:135 (1945).
13. Evaldson, R. L., R. S. Ayre, and L. S. Jacobsen: *J. Franklin Inst.,* **248**:473 (1949).
14. Frankland, J. M.: *Proc. Soc. Exptl. Stress Anal.,* **6**:2, 7 (1948).
15. Fuchs, H. O.: *Product Eng.,* August, 1936, p. 294.
16. Gardner, M. F., and J. L. Barnes: "Transients in Linear Systems," vol. I, John Wiley & Sons, Inc., New York, 1942.
17. Hartman, J. B.: "Dynamics of Machinery," McGraw-Hill Book Company, Inc., New York, 1956.
18. Hudson, G. E.: *Proc. Soc. Exptl. Stress Anal.,* **6**:2, 28 (1948).
19. Jacobsen, L. S., and R. S. Ayre: "A Comparative Study of Pulse and Step-type Loads on a Simple Vibratory System," *Tech. Rept.* N16, under contract N6-ori-154, T. O. 1, U.S. Navy, Stanford University, 1952.
20. Jacobsen, L. S.: *Proc. Symposium on Earthquake and Blast Effects on Structures,* 1952, p. 94.
21. Jacobsen, L. S.: *J. Appl. Mechanics,* **19**:543 (1952).
22. Jacobsen, L. S., and R. S. Ayre: "Engineering Vibrations," McGraw-Hill Book Company, Inc., New York, 1958.
23. Kelvin, Lord: *Phil. Mag.,* **34**:443 (1892).
24. Kornhauser, M.: *J. Appl. Mechanics,* **21**:371 (1954).
25. Lamoen, J.: *Rev. universelle mines,* ser. 8, **11**:7, 3 (1935).
26. McCann, G. D., and J. M. Kopper: *J. Appl. Mechanics,* **14**:A127 (1947).
27. Meissner, E.: *Schweiz. Bauzt.,* **99**:27, 41 (1932).
28. Mindlin, R. D.: *Bell System Tech. J.,* **24**:353 (1945).
29. Mindlin, R. D., F. W. Stubner, and H. L. Cooper: *Proc. Soc. Exptl. Stress Anal.,* **5**:2, 69 (1948).
30. Morrow, C. T.: *J. Acoust. Soc. Amer.,* **29**:596 (1957).
31. Muller, J. T.: *Bell System Tech. J.,* **27**:657 (1948).
32. Rothbart, H. A.: "Cams—Design, Dynamics and Accuracy," John Wiley & Sons, Inc., New York, 1956.
33. Salvadori, M. G., and R. J. Schwarz: "Differential Equations in Engineering Problems," Prentice-Hall, Inc., Englewood Cliffs, N.J., 1954.
34. Scott, E. J.: "Transform Calculus," Harper & Brothers, New York, 1955.

35. Shapiro, H., and D. E. Hudson: *J. Appl. Mechanics,* **20**:422 (1953).

36. Thomson, W. T.: "Laplace Transformation," Prentice-Hall, Inc., Englewood Cliffs, N.J., 1950.

37. Timoshenko, S. P., and D. H. Young: "Advanced Dynamics," McGraw-Hill Book Company, Inc., New York, 1948.

38. Timoshenko, S. P., and D. H. Young: "Vibration Problems in Engineering," 3d ed., D. Van Nostrand Company, Inc., Princeton, N.J., 1955.

39. Walsh, J. P., and R. E. Blake: *Proc. Soc. Exptl. Stress Anal.,* **6**:2, 151 (1948).

40. Williams, H. A.: *Trans. ASCE,* **102**:838 (1937).

CHAPTER 9

EFFECTS OF IMPACT
ON STRUCTURES

W. H. Hoppmann II

INTRODUCTION

This chapter discusses a particular phenomenon in the general field of shock and vibration usually referred to as impact.[1] An impact occurs when two or more bodies collide. An important characteristic of an impact is the generation of relatively large forces at points of contact for relatively short periods of time. Such forces sometimes are referred to as *impulse-type* forces.

Three general classes of impact are considered in this chapter: (1) impact between spheres or other rigid bodies, where a body is considered to be rigid if its dimensions are large relative to the wavelengths of the elastic stress waves in the body; (2) impact of a rigid body against a beam or plate that remains substantially elastic during the impact; and (3) impact involving yielding of structures.

DIRECT CENTRAL IMPACT OF TWO SPHERES

The elementary analysis of the central impact of two bodies is based upon an experimental observation of Newton.[2] According to that observation, the relative velocity of two bodies after impact is in constant ratio to their relative velocity before impact and is in the opposite direction. This constant ratio is the *coefficient of restitution;* usually it is designated by e.[3]

Let $\dot{u}$ and $\dot{x}$ be the components of velocity along a common line of motion of the two bodies before impact, and $\dot{u}'$ and $\dot{x}'$ the component velocities of the bodies in the same direction after impact. Then, by the observation of Newton,

$$\dot{u}' - \dot{x}' = -e(\dot{u} - \dot{x}) \tag{9.1}$$

Now suppose that a smooth sphere of mass m_u and velocity $\dot{u}$ collides with another smooth sphere having the mass m_x and velocity $\dot{x}$ moving in the same direction. Let the coefficient of restitution be e, and let $\dot{u}'$ and $\dot{x}'$ be the velocities of the two spheres, respectively, after impact. Figure 9.1 shows the condition of the two spheres just before collision. The only force acting on the spheres during impact is the force at the point of contact, acting along the line through the centers of the spheres.

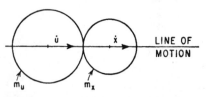

According to the law of conservation of linear momentum:

$$m_u\dot{u}' + m_x\dot{x}' = m_u\dot{u} + m_x\dot{x} \qquad (9.2)$$

Solving Eqs. (9.1) and (9.2) for the two unknowns, the velocities $\dot{u}'$ and $\dot{x}'$ after impact,

FIGURE 9.1 Positions of two solid spheres at instant of central impact.

$$\dot{u}' = \frac{(m_u\dot{u} + m_x\dot{x}) - em_x(\dot{u} - \dot{x})}{m_u + m_x}$$

$$\dot{x}' = \frac{(m_u\dot{u} + m_x\dot{x}) + em_u(\dot{u} - \dot{x})}{m_u + m_x} \qquad (9.3)$$

This analysis yields the resultant velocities for the two spheres on the basis of an experimental law and the principle of the conservation of momentum, without any specific reference to the force of contact F. A similar result is obtained for a ballistic pendulum used to measure the muzzle velocity of a bullet. A bullet of mass m_u and velocity $\dot{u}$ is fired into a block of wood of mass m_x which is at rest initially and finally assumes a velocity $\dot{x}'$ after the impact. Using only the principle of the conservation of momentum,

$$\dot{u} = \frac{(m_u + m_x)\dot{x}'}{m_u} \qquad (9.4)$$

No knowledge of the complicated pattern of force acting on the bullet and the pendulum during the embedding process is required.

These simple facts are introductory to the more complicated problem involving the vibration of at least one of the colliding bodies, as discussed in a later section.

HERTZ THEORY OF IMPACT OF TWO SOLID SPHERES

The theory of two solid elastic spheres which collide with one another is based upon the results of an investigation of two elastic bodies pressed against one another under purely statical conditions.[4] For these static conditions, the relations between the sum of the displacements at the point of contact in the direction of the common line of motion and the resultant total pressure have been derived. The sum of these displacements is equal to the relative approach of the centers of the spheres, assuming that the spheres act as rigid bodies except for elastic compression at the point of contact. The relative approach varies as the two-thirds power of the total pressure; a formula is given for the time of duration of the contact.[4] The theory is valid only if the duration of contact is long in comparison with the period of the fundamental mode of vibration of either sphere.

The range of validity of the Hertz theory is related to the possibility of exciting vibration in the spheres.[5] The dimensionless ratio of the maximum kinetic energy of vibration to the sum of the kinetic energies of the two spheres just before collision is approximately

$$R = \frac{1}{50} \frac{\dot{u} - \dot{x}}{\sqrt{E/\rho}}$$ (9.5)

where $\dot{u} - \dot{x}$ = relative velocity of approach, in./sec
E = Young's modulus of elasticity, assumed to be the same for each sphere, lb/in.^{2}
ρ = density of each sphere, lb-sec^2/in.^{4}
$\sqrt{E/\rho}$ = approximate velocity of propagation of dilatational waves, in./sec

The ratio R usually is a very small quantity; thus, the theory of impact set forth by Eq. (9.5) has wide application because vibration is not generated in the spheres to an appreciable degree under ordinary conditions. The energy of the colliding spheres remains translational, and the velocities after impact are deducible from the principles of energy and of momentum. The important point of plastic deformation at the point of contact is discussed in a later section.

Formulas for force between the spheres, the radius of the circular area of contact, and the relative approach of the centers of the spheres, all as functions of time, can be determined for any two given spheres.[6]

IMPACT OF A SOLID SPHERE ON AN ELASTIC PLATE

An extension of the Hertz theory of impact to include the effect of vibration of one of the colliding bodies involves a study of the transverse impact of a solid sphere upon an infinitely extended plate.[7] The plate has the role of the vibrating body. The coefficient of restitution is an important element in any analysis of the motion ensuing after the collision of two bodies.

The analysis is based on the assumption that the principal elastic waves of importance are flexural waves of half-period equal to the duration of impact. Let $2h$ and $2D$ be the thickness of plate and diameter of sphere, respectively; ρ_1, ρ_2 their densities; E_1, E_2 their Young's moduli; ν_1, ν_2 their values of Poisson's ratio; and τ_H the duration of impact. The velocity c of long flexural waves of wavelength λ in the plate is given by

$$c^2 = \frac{4\pi^2}{3} \frac{h^2}{\lambda^2} \frac{E_1}{\rho_1(1 - \nu_1^2)}$$ (9.6)

The radius a of the circle on the plate over which the disturbance has spread at the termination of impact is given by

$$a = c\tau_H = \frac{\lambda}{2}$$ (9.7)

Combining Eqs. (9.6) and (9.7),

$$a^2 = \pi\tau_H h \sqrt{\frac{E_1}{3\rho_1(1 - \nu_1^2)}}$$ (9.8)

The next step is to find the kinetic and potential energies of the wave motion of the plate. The kinetic energy may be determined from the transverse velocity of the plate at each point over the circle of radius a covered by the wave. Figure 9.2 shows an approximate distribution of velocity over the circle of radius a at the end of impact.[8] The direction of the impact also is shown. The kinetic energy in the wave at the end of impact is

$$T = \int_0^a \tfrac{1}{2} \cdot 2h \cdot \rho_1 \cdot 2\pi R \cdot \dot{w}^2 dR \tag{9.9}$$

where $\dot{w}$ is the transverse velocity at distance R from the origin. As an approximation it is assumed that the sum of the potential energy and the kinetic energy in the wave is $2T$. With considerable effort these energies can be calculated in terms of the motion of the plate, although the calculation may be laborious.

The impulse in the plate produced by the colliding body is

$$J = \int_0^a \tfrac{1}{2} \cdot 2h \cdot \rho_1 \cdot 2\pi R \cdot w \, dr \tag{9.10}$$

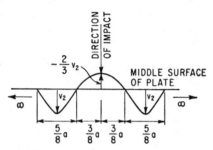

FIGURE 9.2 Distribution of transverse velocities in plate as a result of impact by a moving body. (*After Lamb*[8].)

The integration should be carried out with due regard to the sign of velocity. If m_u is the mass of colliding body, $\dot{u}$ its velocity before impact, and e the coefficient of restitution, the following relations are obtained on the assumption that the energy is conserved:

$$\tfrac{1}{2}m_u\dot{u}^2(1 - e^2) = 2T \tag{9.11}$$

$$m_u\dot{u}(1 + e) = J \tag{9.12}$$

Equation (9.11) represents the energy lost to the moving sphere as a result of impact and Eq. (9.12) represents the change in momentum of the sphere.

The coefficient of restitution e is determined by evaluating the integrals for T and J and substituting their values in Eq. (9.12). The necessary integrations can be performed by taking the function for transverse velocity in Fig. 9.2 as arcs of sine curves. The resultant expression for e is

$$e = \frac{h\rho_1 a^2 - 0.56 m_u}{h\rho_1 a^2 + 0.56 m_u} \tag{9.13}$$

where a, the radius of the deformed region, is given by Eq. (9.8) and τ_H, the time of contact between sphere and plate, is given by Hertz's theory of impact to a first approximation.[4] The mass of the sphere is m_u; the mass of the plate is assumed to be infinite. Large discrepancies between theory and experiment occur when the diameter of the sphere is large compared with the thickness of the plate. The duration of impact τ_H is

$$\tau_H = 2.94 \, \frac{\alpha}{\dot{u}}$$

where
$$\alpha = \left[\frac{15}{16} v_1^2 \left(\frac{1 - v_1^2}{E_1} + \frac{1 - v_2^2}{E_2} \right) m_u \right]^{2/5} R_s^{-1/5} \tag{9.14}$$

The radius of the striking sphere is R_s and its velocity before impact is $\dot{u}$. Subscripts 1 and 2 represent the properties of the sphere and plate, respectively. The value of τ_H may be substituted in Eq. (9.8) above.

Experimental results verify the theory when the limitations of the theory are not violated. The velocity of impact must be sufficiently small to avoid plastic deformation. When the collision involves steel on steel, the velocity usually must be less than 1 ft/sec. However, useful engineering results can be obtained with this approach even though plastic deformation does occur locally.[9,10]

TRANSVERSE IMPACT OF A MASS ON A BEAM

If $F(t)$ is the force acting between the sphere and the beam during contact, the distance traveled by the sphere in time t after collision is[11]

$$\dot{u}t - \frac{1}{m_u} \int_0^t F(t_v) \, (t - t_v) \, dt_v \tag{9.15}$$

where $\dot{u}$ = velocity of sphere before collision (beam assumed to be at rest initially)
m_u = mass of solid sphere

The beam is assumed to be at rest initially.

For example, the deflection of a simply supported beam under force $F(t_v)$ at its center is

$$\sum_{1,3,5 \dots}^{\infty} \frac{1}{m_b} \int_0^t F(t_v) \, \frac{\sin \omega_n(t - t_v)}{\omega_n} \, dt_v \tag{9.16}$$

where m_b = one-half of mass of beam
ω_n = angular frequency of the nth mode of vibration

Equation (9.16) represents the transverse vibration of a beam. While the present case is only for direct central impact, the cases for noncentral impact depend only on the corresponding solution for transverse vibration. Oblique impact also is treated readily.

The expression for the relative approach of the sphere and beam, i.e., penetration of beam by sphere, is[11]

$$\alpha = \kappa_1 F(t)^{2/3} \tag{9.17}$$

where κ_1 is a constant depending on the elastic and geometrical properties of the sphere and the beam at the point of contact, and α is given by Eq. (9.14). Consequently, the equation that defines the problem is

$$\alpha = K_1 F^{2/3} = \dot{u}t - \frac{1}{m_u} \int_0^t F(t_v)(t - t_v) \, dt_v - \sum_{1,3,5}^{\infty} \frac{1}{m_b} \int_0^t F(t_v) \, \frac{\sin \omega_n(t - t_v)}{\omega_n} \, dt_v \tag{9.18}$$

Equation (9.18) has been solved numerically for two specific problems by subdividing the time interval 0 to t into small elements and calculating, step by step, the displacements of the sphere.[11] The results are not general but rather apply only to the cases of beam and sphere.

For the impact of a mass on a beam, the sum of the kinetic and the potential energies may be expressed in terms of the unknown contact force.[12] Also, the impulse integral J in terms of the contact force may be expressed as

$$J = \int_0^t F(t)\, dt = m_u \dot{u}(1 + e) \tag{9.19}$$

A satisfactory approximation to $F(t)$ is defined in terms of a normalized force $\bar{F}$:

$$F(t) = m_u \dot{u}(1 + e)\, \bar{F}(t) \tag{9.20}$$

Thus, from Eqs. (9.19) and (9.20),

$$\int_0^t \bar{F}\, dt = 1 \tag{9.21}$$

The value of this integral is independent of the shape of $F(t)$. The normalized force is defined such that its maximum value equals the maximum value of the corresponding normalized Hertz force.[12] To perform the necessary integrations, a suitable function for defining $F(t)$ is chosen as follows:

$$\bar{F}(t) = \frac{\pi}{2\tau_L} \sin \frac{\pi}{\tau_L} t \qquad [0 < t < \tau_L]$$

$$\tag{9.22}$$

$$\bar{F}(t) = 0 \qquad [|t| > \tau_L]$$

Results for particular problems solved in this manner agree well with those obtained for the same problems by the numerical solution of the exact integral equation.[12]

To apply these results to a specific beam impact problem, it is necessary to express the deflection equation for the beam in terms of known quantities. One of these quantities is the coefficient of restitution; a formula must be provided for its determination in terms of known functions. This is given by Eq. (9.31).

IMPACT OF A RIGID BODY ON A DAMPED ELASTICALLY SUPPORTED BEAM

For the more general case of impact of a rigid body on a damped, elastically supported beam, it is assumed that there is external damping, damping determined by the Stokes' law of stress-strain, and an elastic support attached to the beam along its length in such a manner that resistance is proportional to deflection.[13] The differential equation for the deflection of the beam is

$$EI \frac{\partial^4 w}{\partial x^4} + c_1 I \frac{\partial^5 w}{\partial x^4 \partial t} + c_2 \frac{\partial w}{\partial t} + kw + \rho S \frac{\partial^2 w}{\partial t^2} = F(x,t) \tag{9.23}$$

where w = deflection, in.
   E = Young's modulus, lb/in.^{2}
   I = moment of inertia for cross section (constant), in.^{4}

c_1 = internal damping coefficient, lb/in.²-sec (Stokes' law)
c_2 = external damping coefficient, lb/in.²-sec
k = foundation modulus, lb/in.²
ρ = density, lb-sec²/in.⁴
S = area of cross section (constant), in.²
$\dfrac{\partial^2 w}{\partial t^2}$ = acceleration, in./sec²

t = time, sec
$F(x,t)$ = driving force per unit length of beam, lb/in.

For example, to illustrate the application of specific boundary conditions, consider a simply supported beam of length l. The moments and deflections must vanish at the ends. The beam is assumed undeflected and at rest just before impact, and central impact is assumed although with some additional computation this restriction may be dropped. The solution may be written as follows:

$$w(x,t) = \sum^{\infty} \sin \frac{n\pi x}{l} \sin \frac{n\pi}{2} \frac{1}{m} \frac{1}{\sqrt{\omega_n^2 - \delta_n^2}}$$

$$\times \int_0^t e{-\delta_n(t-\tau)} \sin\left[\sqrt{\omega_n^2 - \delta_n^2} \cdot (t-\tau)\right] F_1(\tau)\, d\tau \tag{9.24}$$

where e = base of natural logarithms

δ_n = damping numbers = $\dfrac{1}{2}\left(r_i \dfrac{n^4\pi^4}{l^4} + r_e\right)$

$r_i = \dfrac{c_1 I}{\rho S}$

$r_e = \dfrac{c_2}{\rho S}$

ω_n = angular frequencies

$m = \tfrac{1}{2}\rho Al$

A satisfactory analytical expression for the contact force $F_1(t)$, a particularization of $F(x,t)$ in Eq. (9.23), must be developed. Although $F_1(t)$ is assumed to act at the center of the beam, the methods apply with only minor alterations if the impact occurs at any other point of the beam.

One of the conditions which the contact force must satisfy is that its time integral for the duration of impact equal the change in momentum of the striking body. The change of momentum is

$$m\dot{z} - m\dot{z}' = m\dot{z}\left(1 - \frac{\dot{z}'}{\dot{z}}\right) \tag{9.25}$$

where m = mass of rigid body, lb-sec²/in.
$\dot{z}$ = velocity of rigid body just before collision, in./sec
$\dot{z}'$ = velocity of rigid body just after collision, in./sec

When the velocity of the beam is zero, Eq. (9.1) may be written

$$e = -\frac{\dot{z}'}{\dot{z}} \tag{9.26}$$

Equation (9.26) may be written

$$m\dot{z}\left(1 - \frac{\dot{z}'}{\dot{z}}\right) = m\dot{z}(1 + e) \tag{9.27}$$

From the equivalence of impulse and momentum:

$$\int_0^{\tau_0} F_1(t)\, dt = m\dot{z}(1 + e) \tag{9.28}$$

where τ_0 is the time of contact.

It can then be shown[13] that the impact force may be written

$$F_1(t) = m\dot{z}(1 + e)\, \frac{\pi}{2\tau_L}\, \sin\frac{n\pi t}{\tau_L} \qquad [0 < t < \tau_L]$$

$$F_1 = 0 \qquad [t > \tau_L] \tag{9.29}$$

It can be shown further[13] that

$$\tau_L = 3.28\left[\frac{m^2}{\dot{z}R} \cdot \frac{(1 - v^2)}{E^2}\right]^{1/5} \tag{9.30}$$

where R = radius of sphere, in.
        v = Poisson's ratio

The time interval τ_L is a special value of the time of contact T_0. It agrees well with experimental results.

The coefficient of restitution e is[13]

$$e = \frac{1 - \dfrac{m}{m_b}\sum_1^\infty \Phi_n - \dfrac{m}{m_b}\sum_1^\infty \Psi_n}{1 + \dfrac{m}{m_b}\sum_1^\infty \Phi_n + \dfrac{m}{m_b}\sum_1^\infty \Psi_n} \tag{9.31}$$

where m = mass of sphere
        m_b = half mass of beam

The functions Φ_n and Ψ_n are given in the form of curves in Figs. 9.3 and 9.4; the symbol $\beta_n = \delta_n/\omega_n$ represents fractional damping and $Q_n = \omega_n\tau_L/2\pi$ is a dimensionless frequency where ω_n = angular frequency of nth mode of vibration of undamped vibration of beam, rad/sec, and τ_L = length of time the sinusoidal pulse is assumed to act on beam [see Eq. (9.30)]. If damping is neglected, the functions Ψ_n vanish from Eq. (9.31).

The above theory may be generalized to apply to the response of plates to impact. The deflection equation of a plate subjected to a force applied at a point is required. The various energy distributions at the end of impact are arrived at in a manner analogous to that for the beam.

The theory has been applied to columns and continuous beams[14,15] and also could be applied to transverse impact on a ring. Measurement of the force of impact illustrates the large number of modes of vibration that can be excited by an impact.[16,17,22]

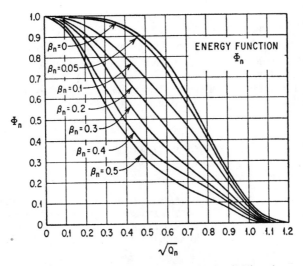

FIGURE 9.3 Energy functions Φ_n used with Eq. (9.31) to determine the coefficient of restitution from the impact of a rigid body on a damped elastically supported beam.

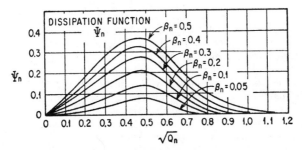

FIGURE 9.4 Dissipative (damping) functions Ψ_n used with Eq. (9.31) to determine the coefficient of restitution from the impact of a rigid body on a damped elastically supported beam.

Principal qualitative results of the foregoing analysis are:

1. Impacts by bodies of relatively small mass moving with low velocities develop significant bending strains in beams.

2. External damping of the type assumed above has a rapidly decreasing effect on reducing deflection and strain as the number of the mode increases.

3. Internal damping of the viscous type here assumed reduces deflection and strain appreciably in the higher modes. For a sufficiently high mode number, the vibration becomes aperiodic.

4. Increasing the modulus for an elastic foundation reduces the energy absorbed by the structure from the colliding body.

5. Impacts from collision produce sharp initial rises in strain which are little influenced by damping.

6. Because of result 5, the fatigue problem for machines and structures, in which the impact conditions are repeated many times, can be serious. Ordinary damping affords little protection.

7. The structure seldom can be treated as a single degree-of-freedom system with any degree of reliability in predicting strain.[13,19]

LONGITUDINAL AND TORSIONAL IMPACT ON BARS

If a mass strikes the end of a long bar, the response may be investigated by means of the Hertz contact theory.[11] The normal modes of vibration must be known so the displacement at each part of the bar can be calculated in terms of a contact force. In a similar manner, the torsional vibration of a long bar can be studied, using the normal modes of torsional vibration.

PLASTIC DEFORMATION RESULTING FROM IMPACT

Many problems of interest involve plastic deformation rather than elastic deformation as considered in the preceding analyses. Using the concept of the plastic hinge, the large plastic deformation of beams under transverse impact[23] and the plastic deformation of free rings under concentrated dynamic loads[24] have been studied. In such analyses, the elastic portion of the vibration usually is neglected. To make further progress in analyses of large deformations as a result of impact, a realistic theory of material behavior in the plastic phase is required.

An attempt to solve the problem for the longitudinal impact on bars has been made using the static engineering-type stress-strain curve as a part of the analysis.[25] An extension of the work to transverse impact also was attempted.[26]

Figure 9.5 illustrates the impact of a large body m colliding axially with a long rod. The body m has an initial velocity $\dot{u}$ and is sufficiently large that the end of the rod may be assumed to move with constant velocity $\dot{u}$. At any time t a stress wave will have moved into the bar a definite distance; by the condition of continuity (no break in the material), the struck end of the bar will have moved a distance equal to the total elongation of the end portion of the bar:

$$\dot{u}t = \epsilon \cdot l \tag{9.32}$$

The velocity c of a stress wave is $c = l/t$, and Eq. (9.32) becomes

$$\epsilon = \frac{\dot{u}}{c} \tag{9.33}$$

The stress and strain in an elastic material are related by Young's modulus. Substituting for strain from Eq. (9.33),

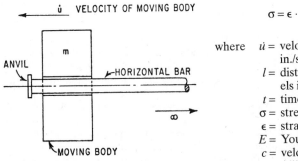

FIGURE 9.5 Longitudinal impact of moving body on end of rod.

$$\sigma = \epsilon \cdot E = E\frac{\dot{u}}{c} \qquad (9.34)$$

where $\dot{u}$ = velocity of end of rod, in./sec
      l = distance stress wave travels in time t, in.
      t = time, sec
      σ = stress, lb/in.^{2}
      ϵ = strain (uniform), in./in.
      E = Young's modulus, lb/in.^{2}
      c = velocity of stress wave (dilatational), in./sec

When the yield point of the material is exceeded, Eq. (9.34) is inapplicable. Extensions of the analysis, however, lead to some results in the case of plastic deformation.[25] The differential equation for the elastic case is

$$E\frac{\partial^2 u}{\partial x^2} = \rho\frac{\partial^2 u}{\partial t^2} \qquad (9.35)$$

where u = displacement, in.
      x = coordinate along rod, in.
      t = time, sec
      E = Young's modulus, lb/in.^{2}
      ρ = mass density, lb-sec^2/in.^{4}

The velocity of the elastic dilatational wave obtained from Eq. (9.35) is

$$c = \sqrt{\frac{E}{\rho}}$$

The modulus E is the slope of the stress-strain curve in the initial linear elastic region. Replacing E by $\partial\sigma/\partial\epsilon$ for the case in which plastic deformation occurs, the slope of the static stress-stress curve can be determined at any value of the strain ϵ.[25] Equation (9.35) then becomes

$$\frac{\partial\sigma}{\partial\epsilon}\frac{\partial^2 u}{\partial x^2} = \rho\frac{\partial^2 u}{\partial t^2} \qquad (9.36)$$

Equation (9.36) is nonlinear; its general solution never has been obtained. For the simple type of loading discussed above and an infinitely long bar, the theory predicts a so-called critical velocity of impact because the velocities of the plastic waves are much smaller than those for the elastic waves and approach zero as the strain is indefinitely increased.[25] Since the impact velocity $\dot{u}$ is an independent quantity, it can be made larger and larger while the wave velocities are less than the velocity for elastic waves. Hence a point must be reached at which the continuity of the material is violated. Experimental data illustrate this point.[27]

ENERGY METHOD

Many problems in the design of machines and structures require knowledge of the deformation of material in the plastic condition. In statical problems the method of limit design[28] may be used. In dynamics, the most useful corresponding concept is less theoretical and may be termed the energy method; it is based upon the impact test used for the investigation of brittleness in metals. Originally, the only purpose of this test was to break a standard specimen as an index of brittleness or ductility. The general method, using a tension specimen, may be used in studying the dynamic resistance of materials.[27] An axial force is applied along the length of the specimen and causes the material to rupture ultimately. The energy of absorption is the total amount of energy taken out of the loading system and transferred to the specimen to cause the plastic deformation. The elastic energy and the specific mode of build-up of stress to the final plastic state are ignored. Such an approach has value only to the extent that the material has ductility. For example, in a long tension-type specimen of medium steel, the energy absorbed before neck-down and rupture is of the order of 500 ft-lb per cubic inch of material. Thus, if the moving body in Fig. 9.5 weighs 200 lb and has an initial velocity of 80 ft/sec, it represents 20,000 ft-lb of kinetic energy. If the tension bar subjected to the impact is 10 in. long and 0.5 in. in diameter, it will absorb approximately 1,000 ft-lb of energy. Under these circumstances it will rupture. On the other hand, if the moving body m weighs only 50 lb and has an initial velocity of 30 ft/sec, its kinetic energy is approximately 700 ft-lb and the bar will not rupture.

If the tension specimen were severely notched at some point along its length, it would no longer absorb 500 ft-lb per cubic inch to rupture. The material in the immediate neighborhood of the notch would deform plastically; a break would occur at the notch with the bulk of the material in the specimen stressed below the yield stress for the material. A practical structural situation related to this problem occurs when a butt weld is located at some point along an unnotched specimen. If the weld is of good quality, the full energy absorption of the entire bar develops before rupture; with a poor weld, the rupture occurs at the weld and practically no energy is absorbed by the remainder of the material. This is an important consideration in applying the energy method to design problems.

REFERENCES

1. Love, A. E. H.: "The Mathematical Theory of Elasticity," p. 25, Cambridge University Press, New York, 1934.

2. Timoshenko, S., and D. H. Young: "Engineering Mechanics," p. 334, McGraw-Hill Book Company, Inc., New York, 1940.

3. Loney, S. L.: "A Treatise on Elementary Dynamics," p. 199, Cambridge University Press, New York, 1900.

4. Hertz, H.: *J. Math. (Crelle)*, pp. 92, 155, 1881.

5. Rayleigh, Lord: *Phil. Mag.* (ser. 6), **11**:283 (1906).

6. Timoshenko, S.: "Theory of Elasticity," p. 350, McGraw-Hill Book Company, Inc., New York, 1934.

7. Raman, C. V.: *Phys. Rev.,* **15,** 277 (1920).

8. Lamb, H.: *Proc. London Math. Soc.,* **35,** 141 (1902).

9. Hoppmann, II, W. H.: *Proc. SESA,* **9**:2, 21 (1952).

10. Hoppmann, II, W. H.: *Proc. SESA,* **10**:1, 157 (1952).

11. Timoshenko, S.: "Vibration Problems in Engineering," 3d ed., p. 413, D. Van Nostrand Company, Inc., Princeton, N.J., 1955.

12. Zener, C., and H. Feshbach: *Trans. ASME,* **61**:a-67 (1939).

13. Hoppmann, II, W. H.: *J. Appl. Mechanics,* **15**:125 (1948).

14. Hoppmann, II, W. H.: *J. Appl. Mechanics,* **16**:370 (1949).

15. Hoppmann, II, W. H.: *J. Appl. Mechanics,* **17**:409 (1950).

16. Goldsmith, W., and D. M. Cunningham: *Proc. SESA,* **14**:1, 179 (1956).

17. Barnhart, Jr., K. E., and Werner Goldsmith: *J. Appl. Mechanics,* **24**:440 (1957).

18. Emschermann, H. H., and K. Ruhl: *VDI-Forschungsheft* 443, Ausgabe B, Band 20, 1954.

19. Hoppmann, II, W. H.: *J. Appl. Mechanics,* **19** (1952).

20. Wenk, E., Jr.: Dissertation, The Johns Hopkins University, 1950, and *David W. Taylor Model Basin Rept.* 704, July 1950.

21. Compendium, "Underwater Explosion," O. N. R., Department of the Navy, 1950.

22. Prager, W.: James Clayton Lecture, *The Institution of Mechanical Engineers, London,* 1955.

23. Lee, E. H., and P. S. Symonds: *J. Appl. Mechanics,* **19**:308 (1952).

24. Owens, R. H., and P. S. Symonds: *J. Appl. Mechanics,* **22** (1955).

25. Von Kármán, T.: *NDRC Rept.* A-29, 1943.

26. Duwez, P. E., D. S. Clark, and H. F. Bohnenblust: *J. Appl. Mechanics,* **17,** 27 (1950).

27. Hoppmann, II, W. H.: *Proc. ASTM,* **47**:533 (1947).

28. Symposium on the Plastic Theory of Structures, Cambridge University, September, 1956, *British Welding J.,* **3**(8) (1956); **4**(1) (1957).

CHAPTER 10
MECHANICAL IMPEDANCE

Elmer L. Hixson

INTRODUCTION

The *mechanical impedance* at a given point in a vibratory system is the ratio of the force applied to the system at that point to the velocity at the same point. For example, mechanical impedance is discussed in Chap. 6 in discussing ways of expressing the effects of dynamic absorbers and auxiliary mass dampers. In the following sections of this chapter, the mechanical impedance of basic elements that make up vibratory systems is presented. This is followed by a discussion of combinations of these elements. Then, various mechanical circuit theorems are described. Such theorems can be used as an aid in the modeling of mechanical circuits and in determining the response of vibratory systems; they are the mechanical equivalents of well-known theorems employed in the analysis of electric circuits. The measurement of mechanical impedance is considered in Chap. 12.

MECHANICAL IMPEDANCE OF VIBRATORY SYSTEMS

The *mechanical impedance* Z of a system is the ratio of the driving force F acting on the system to the resulting velocity v of the system. Its *mechanical mobility* $\mathfrak{M}$ is the reciprocal of the mechanical impedance.

Consider a sinusoidal driving F that has a magnitude F_0 and an angular frequency ω:

$$F = F_0 \, e^{j\omega t} \qquad (10.1)$$

The application of this force to a linear mechanical system results in a velocity v:

$$v = v_0 e^{j(\omega t + \phi)} \qquad (10.2)$$

where v_0 is the magnitude of the velocity and ϕ is the phase angle between F and v.

Then by definition, the mechanical impedance of the system Z (at the point of application of the force) is given by

$$Z = F/v \qquad (10.3)$$

BASIC MECHANICAL ELEMENTS

The idealized mechanical systems considered in this chapter are considered to be represented by combinations of basic mechanical elements assembled to form linear mechanical systems. These basic elements are *mechanical resistances (dampers)*, *springs*, and *masses*. In general, the characteristics of real masses, springs, and mechanical resistance elements differ from those of ideal elements in two respects:

1. A spring may have a nonlinear force-deflection characteristic; a mass may suffer plastic deformation with motion; and the force presented by a resistance may not be exactly proportional to velocity.

2. All materials have some mass; thus, a perfect spring or resistance cannot be made. Some compliance or spring effect is inherent in all elements. Energy can be dissipated in a system in several ways: friction, acoustic radiation, hysteresis, etc. Such a loss can be represented as a resistive component of the element impedance.

Mechanical Resistance (Damper). A mechanical resistance is a device in which the relative velocity between the end points is proportional to the force applied to the end points. Such a device can be represented by the dashpot of Fig. 10.1A, in which the force resisting the extension (or compression) of the dashpot is the result of viscous friction. An ideal resistance is assumed to be made of massless, infinitely rigid elements. The velocity of point A, v_1, with respect to the velocity at point B, v_2, is

$$v = (v_1 - v_2) = \frac{F_a}{c} \qquad (10.4)$$

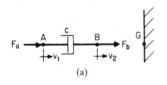

(a)

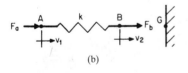

(b)

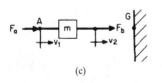

(c)

FIGURE 10.1 Schematic representations of basic mechanical elements: (A) An ideal mechanical resistance. (B) An ideal spring. (C) An ideal mass.

where c is a constant of proportionality called the *mechanical resistance* or *damping constant*. For there to be a relative velocity v as a result of force at A, there must be an equal reaction force at B. Thus, the transmitted force F_b is equal to F_a. The velocities v_1 and v_2 are measured with respect to the stationary reference G; their difference is the relative velocity v between the end points of the resistance.

With the sinusoidal force of Eq. (10.1) applied to points attached to a fixed (immovable) point, the velocity v_1 is obtained from Eq. (10.4):

$$v_1 = \frac{F_0 e^{j\omega t}}{c} = v_0 e^{j\omega t} \qquad (10.5)$$

Because c is a real number, the force and velocity are said to be "in phase."

The mechanical impedance of the resistance is obtained by substituting from Eqs. (10.1) and (10.5) in Eq. (10.3):

$$Z_c = \frac{F}{v} = c \tag{10.6}$$

The mechanical impedance of a resistance is the value of its damping constant c.

Spring. A linear spring is a device for which the relative displacement between its end points is proportional to the force applied. It is illustrated in Fig. 10.1B and can be represented mathematically as follows:

$$x_1 - x_2 = \frac{F_a}{k} \tag{10.7}$$

where x_1, x_2 are displacements relative to the reference point G and k is the *spring stiffness*. The stiffness k can be expressed alternately in terms of a *compliance* $C = 1/k$. The spring transmits the applied force, so that $F_b = F_a$.

With the force of Eq. (10.1) applied to point A and with point B fixed, the displacement of point A is given by Eq. (10.7):

$$x_1 = \frac{F_0 e^{j\omega t}}{k} = x_0 e^{j\omega t}$$

The displacement is thus sinusoidal and in phase with the force. The relative velocity of the end connections is required for impedance calculations and is given by the differentiation of x with respect to time:

$$\dot{x} = v = \frac{j\omega F_0 e^{j\omega t}}{k} = \frac{\omega}{k} F_0 e^{j(\omega t + 90°)} \tag{10.8}$$

Substituting Eqs. (10.1) and (10.8) in Eq. (10.3), the impedance of the spring is

$$Z_k = -\frac{jk}{\omega} \tag{10.9}$$

Mass. In the ideal mass illustrated in Figs. 2.2 and 10.1C, the acceleration $\ddot{x}$ of the rigid body is proportional to the applied force F:

$$\ddot{x}_1 = \frac{F_a}{m} \tag{10.10}$$

where m is the mass of the body and $\ddot{x}_1 = \ddot{x}_2$ since the element is rigid. By Eq. (10.10), the force F_a is required to give the mass the acceleration $\ddot{x}_1$, and the transmitted force F_b is zero. When a sinusoidal force is applied, Eq. (10.10) becomes

$$\ddot{x}_1 = \frac{F_0 e^{j\omega t}}{m} \tag{10.11}$$

The acceleration is sinusoidal and in phase with the applied force.
Integrating Eq. (10.11) to find velocity,

$$\dot{x} = v = \frac{F_0 e^{j\omega t}}{j\omega m}$$

The mechanical impedance of the mass is the ratio of F to v, so that

$$Z_m = \frac{F_0 e^{j\omega t}}{F_0 e^{j\omega t}/j\omega m} = j\omega m \tag{10.12}$$

Thus, the impedance of a mass is an imaginary quantity that depends on the magnitude of the mass and on the frequency.

COMBINATIONS OF MECHANICAL ELEMENTS

In analyzing the properties of mechanical systems, it is often advantageous to combine groups of basic mechanical elements into single impedances. Methods for calculating the impedances of such combined elements are described in this section.

Parallel Elements. Consider the combination of elements shown in Fig. 10.2, a spring and a mechanical resistance. They are said to be in parallel since the same force is applied to both and both are constrained to have the same relative velocities between their connections. The force F_c required to give the resistance the velocity v is found from Eqs. (10.3) and (10.6).

$$F_c = vZ_c = vc$$

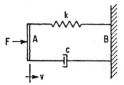

The force required to give the spring this same velocity is, from Eqs. (10.8) and (10.9),

$$F_k = vZ_k = \frac{vk}{j\omega}$$

FIGURE 10.2 Schematic representation of a parallel spring-resistance combination.

The total force F is

$$F = F_c + F_k$$

Since $Z = F/v$,

$$Z = c - j\frac{k}{\omega}$$

Thus, the total mechanical impedance is the sum of the impedances of the two elements.

By extending this concept to any number of parallel elements, the driving force F equals the sum of the resisting forces:

$$F = \sum_{i=1}^{n} vZ_i = v\sum_{i=1}^{n} Z_i \quad \text{and} \quad Z_p = \sum_{i=1}^{n} Z_i \tag{10.13}$$

where Z_p is the total mechanical impedance of the parallel combination of the individual elements Z_i.

Since mobility is the reciprocal of impedance, when the properties of the parallel elements are expressed as mobilities, the total mobility of the combination follows from Eq. (10.13):

$$\frac{1}{\mathfrak{M}_p} = \sum_{i=1}^{n} \frac{1}{\mathfrak{M}_i} \tag{10.14}$$

Series Elements. In Fig. 10.3 a spring and damper are connected so that the applied force passes through both elements to the inertial reference. Then the velocity v is the sum of v_k and v_c. This is a series combination of elements. The method for determining the mechanical impedance of the combination follows.

Consider the more general case of three arbitrary impedances shown in Fig. 10.4. Determine the impedance presented by the end of a number of series-connected elements. Elements Z_1 and Z_2 must have no mass, since a mass always has one end connected to a stationary inertial reference. However, the impedance Z_3 may be a mass. The relative velocities between the end connections of each element are indicated by v_a, v_b, and v_c; the velocities of the connections with respect to the stationary reference point G are indicated by v_1, v_2, and v_3:

$$v_3 = v_c \qquad v_2 = v_3 + (v_2 - v_3) = v_c + v_b$$

$$v_1 = v_2 + (v_1 - v_2) = v_a + v_b + v_c$$

The impedance at point 1 is F/v_1, and the force F is transmitted to all three elements. The relative velocities are

$$v_a = \frac{F}{Z_1} \qquad v_b = \frac{F}{Z_2} \qquad v_c = \frac{F}{Z_3}$$

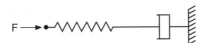

FIGURE 10.3 Schematic representation of a series combination of a spring and a damper.

Thus, the total impedance is defined by

$$\frac{1}{Z} = \frac{F/Z_1 + F/Z_2 + F/Z_3}{F} = \frac{1}{Z_1} + \frac{1}{Z_2} + \frac{1}{Z_3}$$

Extending this principle to any number of massless series elements,

$$\frac{1}{Z_s} = \sum_{i=1}^{n} \frac{1}{Z_i} \tag{10.15}$$

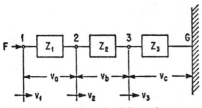

FIGURE 10.4 Generalized three-element system of series-connected mechanical impedances.

where Z_s is the total mechanical impedance of the elements Z_i connected in series.

Since mobility is the reciprocal of impedance, the total mobility of series connected elements (expressed as mobilities) is

$$\mathfrak{M}_s = \sum_{i=1}^{n} \mathfrak{M}_i \qquad\qquad (10.16)$$

MECHANICAL CIRCUIT THEOREMS

The following theorems are the mechanical analogs of theorems widely used in analyzing electric circuits. They are statements of basic principles (or combinations of them) that apply to elements of mechanical systems. In all but Kirchhoff's laws, these theorems apply only to systems composed of linear, bilateral elements. A *linear element* is one in which the magnitudes of the basic elements (c, k, and m) are constant, regardless of the amplitude of motion of the system; a *bilateral element* is one in which forces are transmitted equally well in either direction through its connections.

KIRCHHOFF'S LAWS

1. *The sum of all the forces acting at a point (common connection of several elements) is zero:*

$$\sum_{i}^{n} F_i = 0 \qquad \text{(at a point)} \qquad\qquad (10.17)$$

This follows directly from the consideration leading to Eq. (10.13).

2. *The sum of the relative velocities across the connections of series mechanical elements taken around a closed loop is zero:*

$$\sum_{i}^{n} v_i = 0 \qquad \text{(around a closed loop)} \qquad\qquad (10.18)$$

This follows from the considerations leading to Eq. (10.14).

Kirchhoff's laws apply to any system, even when the elements are not linear or bilateral.

Example 10.1.   Find the velocity of all the connection points and the forces acting on the elements of the system shown in Fig. 10.5. The system contains two velocity generators v_1 and v_6. Their magnitudes are known, their frequencies are the same, and they are 180° out-of-phase.

A. Using Eq. (10.17), write a force equation for each connection point except a and e.

At point b: $F_1 - F_2 - F_3 = 0$. In terms of velocities and impedances:

$$(v_1 - v_2)Z_1 - (v_2 - v_3)Z_2 - (v_2 - v_4)Z_4 = 0 \qquad\qquad (a)$$

At c, the two series elements have the same force acting: $F_2 - F_2 = 0$. In terms of velocities and impedances:

$$(v_2 - v_3)Z_2 - (v_3 - v_4)Z_3 = 0 \qquad\qquad (b)$$

At d: $F_2 + F_3 - F_4 - F_5 = 0$. In terms of velocities and impedances:

$$(v_3 - v_4)Z_3 + (v_2 - v_4)Z_4 - (v_4 + v_6)Z_5 - (v_4 - v_5)Z_6 = 0 \qquad\qquad (c)$$

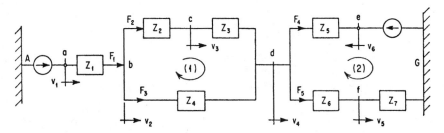

FIGURE 10.5 System of mechanical elements and vibration sources analyzed in Example 10.1 to find the velocity of each connection and the force acting on each element.

Note that v_6 is (+) because of the 180° phase relation to v_1.
At f: $F_5 - F_5 = 0$. In terms of velocities and impedances:

$$(v_4 - v_5)Z_6 - v_5Z_7 = 0 \qquad (d)$$

Since v_1 and v_6 are known, the four unknown velocities v_2, v_3, v_4, and v_5 may be determined by solving the four simultaneous equations above. After the velocities are obtained, the forces may be determined from the following:

$$F_1 = (v_1 - v_2)Z_1 \qquad\qquad F_2 = (v_2 - v_3)Z_2 = (v_3 - v_4)Z_3$$

$$F_3 = (v_2 - v_4)Z_4 \qquad\qquad F_4 = (v_4 + v_6)Z_5$$

$$F_5 = (v_4 - v_5)Z_6 = v_5Z_7$$

B. The method of *node forces*. Equations (a) through (d) above can be rewritten as follows:

$$v_1Z_1 = (Z_1 + Z_2 + Z_3)v_2 - Z_2v_3 - Z_4v_4 \qquad (a')$$

$$0 = -Z_2v_2 + (Z_2 + Z_3)v_3 - Z_3v_4 \qquad (b')$$

$$0 = -Z_4v_2 - Z_3v_3 + (Z_3 + Z_4 + Z_5 + Z_6)v_4 - Z_6v_5 \qquad (c')$$

$$-v_6Z_5 = -Z_6v_4 + (Z_6 + Z_7)v_5 \qquad (d')$$

These equations can be written by inspection of the schematic diagram by the following rule: *At each point with a common velocity (force node), equate the force generators to the sum of the impedances attached to the node multiplied by the velocity of the node, minus the impedances multiplied by the velocities of their other connection points.*

When the equations are written so that the unknown velocities form columns, the equations are in the proper form for a determinant solution for any of the unknowns. Note that the determinant of the Z's is symmetrical about the main diagonal. This condition always exists and provides a check for the correctness of the equations.

C. Using Eq. (10.18), write a velocity equation in terms of force and mobility around enough closed loops to include each element at least once. In Fig. 10.5, note that

$$F_3 = F_1 - F_2 \qquad \text{and} \qquad F_5 = F_1 - F_4$$

Around loop (1):

$$F_2(\mathfrak{M}_2 + \mathfrak{M}_3) - (F_1 - F_2)\mathfrak{M}_4 = 0 \qquad (e)$$

The minus sign preceding the second term results from going across the element 4 in a direction opposite to the assumed force acting on it.

Around loop (2):

$$F_4\mathfrak{M}_5 - v_6 - (F_1 - F_4)(\mathfrak{M}_6 + \mathfrak{M}_7) = 0 \qquad (f)$$

A summation of velocities from A to G along the upper path forms the following closed loop:

$$v_1 + F_1\mathfrak{M}_1 + F_2(\mathfrak{M}_2 + \mathfrak{M}_3) + F_4\mathfrak{M}_5 - v_6 = 0 \qquad (g)$$

Equations (e), (f), and (g) then may be solved for the unknown forces F_1, F_2, and F_4. The other forces are $F_3 = F_1 - F_2$ and $F_5 = F_1 - F_4$. The velocities are:

$$v_2 = v_1 - F_1\mathfrak{M}_1 \qquad v_3 = v_2 - F_2\mathfrak{M}_2 \qquad v_4 = v_2 - F_3\mathfrak{M}_4 \qquad v_5 = F_5\mathfrak{M}_7$$

When a system includes more than one source of vibration energy, a Kirchhoff's law analysis with impedance methods can be made only if all the sources are operating at the same frequency. This is the case because sinusoidal forces and velocities can add as phasors only when their frequencies are identical. However, they may differ in magnitude and phase. Kirchhoff's laws still hold for instantaneous values and can be used to write the differential equations of motion for any system.

RECIPROCITY THEOREM

If a force generator operating at a particular frequency at some point (1) in a system of linear bilateral elements produces a velocity at another point (2), the generator can be removed from (1) and placed at (2); then the former velocity at (2) will exist at (1), provided the impedances at all points in the system are unchanged. This theorem also can be stated in terms of a vibration generator that produces a certain velocity at its point of attachment (1), regardless of force required, and the force resulting on some element at (2).

Reciprocity is an important characteristic of linear bilateral elements. It indicates that a system of such elements can transmit energy equally well in both directions. It further simplifies calculation on two-way energy transmission systems since the characteristics need be calculated for only one direction.

SUPERPOSITION THEOREM

If a mechanical system of linear bilateral elements includes more than one vibration source, the force or velocity response at a point in the system can be determined by adding the response to each source, taken one at a time (the other sources supplying no energy but replaced by their internal impedances).

The internal impedance of a vibrational generator is that impedance presented at its connection point when the generator is supplying no energy. This theorem finds useful application in systems having several sources. A very important application arises when the applied force is nonsinusoidal but can be represented by a Fourier series. Each term in the series can be considered a separate sinusoidal generator. The response at any point in the system can be calculated for each generator by using the impedance values at that frequency. Each response term becomes a term in the Fourier series representation of the total response function. The over-all response as a function of time then can be synthesized from the series.

Figure 10.6 illustrates an application of superposition. The velocities v_c' and v_c'' can be determined by the methods of Example 10.1. Then the velocity v_c is the sum of v_c' and v_c''.

THÉVENIN'S EQUIVALENT SYSTEM

If a mechanical system of linear bilateral elements contains vibration sources and produces an output to a load at some point at any particular frequency, the whole system can be represented at that frequency by a single constant-force generator F_c in parallel with a single impedance Z_i connected to the load. Thévenin's equivalent-system representation for a physical system may be determined by the following experimental procedure: Denote by F_c the force which is transmitted by the attachment point of the system to an infinitely rigid fixed point; this is called the *clamped force*. When the load connection is disconnected and perfectly free to move, a free velocity v_f is measured. Then the parallel impedance Z_i is F_c/v_f. The impedance Z_i also can be determined by measuring the internal impedance of the system when no source is supplying motional energy.

If the values of all the system elements in terms of ideal elements are known, F_c and Z_i may be determined analytically. A great advantage is derived from this representation in that attention is focused on the characteristics of a system at its output point and not on the details of the elements of the system. This allows an easy prediction of the response when different loads are attached to the output connection. After a final load condition has been determined, the system may be analyzed in detail for strength considerations.

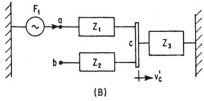

(A)

(B)

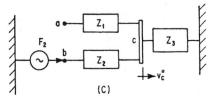

(C)

FIGURE 10.6 System of mechanical elements including two force generators used to illustrate the principle of superposition.

NORTON'S EQUIVALENT SYSTEM

A mechanical system of linear bilateral elements having vibration sources and an output connection may be represented at any particular frequency by a single constant-velocity generator v_f in series with an internal impedance Z_i.

This is the series system counterpart of Thévenin's equivalent system where v_f is the free velocity and Z_i is the impedance as defined above. The same advantages in analysis exist as with Thévenin's parallel representation. The most advantageous one to use depends upon the type of structure to be analyzed. In the experimental determination of an equivalent system, it is usually easier to measure the free velocity than the clamped force on large heavy structures, while the converse is true for light structures. In any case, one representation is easily derived from the other. When v_f and Z_i are determined, $F_c = v_f Z_i$.

CHAPTER 11

STATISTICAL METHODS FOR ANALYZING VIBRATING SYSTEMS

Richard G. DeJong

INTRODUCTION

This chapter presents statistical methods for analyzing vibrating systems. Two situations often occur in which a statistical analysis is useful. The first occurs when the excitation of a system appears to be random in time, in which case it is convenient to describe the temporal response of the system statistically rather than deterministically. This form of analysis is called *random vibration analysis*[1] and is presented in the first half of this chapter. The second situation occurs when a system is complicated enough that its resonant modes appear to be distributed randomly in frequency, in which case it is convenient to describe the frequency response of the system statistically rather than deterministically. This form of analysis[2] is called *statistical energy analysis (SEA)* and is presented in the second half of this chapter.

In either situation the randomness need only appear to be so. For example, in random vibration it may be that the excitation could be calculated exactly if enough information were known. However, if the excitation is adequately described by statistical parameters (such as the *mean value* and *variance*), then a statistical analysis of the system response is valid. Similarly, in a complicated system the modes can presumably be analyzed deterministically. However, if the modal distribution is adequately described by statistical parameters, then a statistical energy analysis of the system response is valid whether or not the excitation is random.

RANDOM VIBRATION ANALYSIS

A random vibration is one whose instantaneous value is not predictable with the available information. Such vibration is generated, for example, by rocket engines, turbulent flows, earthquakes, and motion over irregular surfaces. While the instantaneous vibration level is not predictable, it is possible to describe the vibration in statistical terms, such as the probability distribution of the vibration amplitude, the mean-square vibration level, and the average frequency spectrum.

A random process may be categorized as *stationary* (steady-state) or *nonstationary* (transient). A stationary random process is one whose characteristics do not change over time. For practical purposes a random vibration is stationary if the mean-square amplitude and frequency spectrum remain constant over a specified time period. A random vibration may be *broad-band* or *narrow-band* in its frequency content. Figure 11.1 shows typical acceleration-time records from a system with a mass resiliently mounted on a base subjected to steady, turbulent flow. The base vibration is broad-band with a *Gaussian* (or normal) amplitude distribution. The vibration of the mass is narrow-band (centered at the natural frequency of the mounted system) but also has a Gaussian amplitude distribution. The peaks of the narrow-band vibration have a distribution called the *Rayleigh* distribution.

Technically, the statistical measures of a random process must be averaged over an *ensemble* (or assembly) of representative samples. For an arbitrary random vibration this means averaging over a set of independent realizations of the event. This is illustrated in Fig. 11.2 where four vibration-time records from a point on an internal combustion engine block are shown synchronized with the firing in a particular cylinder. Due to uncontrollable variations in the system, the vibration is not deterministically repeatable. The mean-square amplitude is also nonstationary. Therefore, the statistical parameters of the vibration are time dependent and must be determined from the ensemble of samples from each record at a particular time.

For a stationary random process it may be possible to obtain equivalent ensemble averages by sampling over time if each time record is representative of the entire random process. Such a random process is called *ergodic*. However, not all stationary random processes are ergodic. For example, suppose it is desired to determine the statistical parameters of the vibration levels of an aircraft fuselage during representative in-flight conditions. On a particular flight the vibration levels may be sufficiently stationary to obtain useful time averages. However, one flight is unlikely to encompass all of the expected variations in the weather and other conditions that affect the vibration levels. In this case it is necessary to combine the time averages with an ensemble average over a number of different flight conditions which represent the entire range of possible conditions.

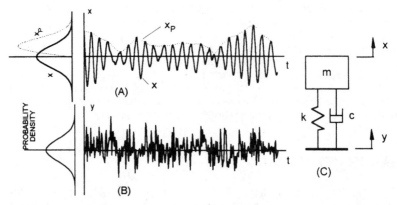

FIGURE 11.1 (*A*) Example of a narrow-band random signal $x(t)$ with a peak envelope x_p. (*B*) Example of a broad-band random signal $y(t)$. Curves along the vertical axes give the probability distributions for the instantaneous (solid lines) and peak (dashed line) values. (*C*) Resiliently mounted mass m with stiffness k and viscous damper c. When the base is exposed to a broad-band random vibration the mass will have a narrow-band random response.

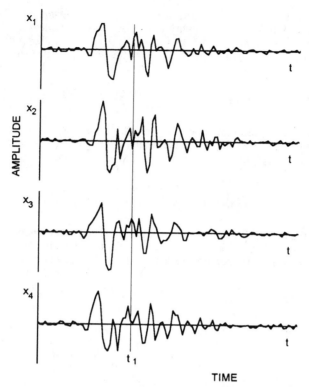

FIGURE 11.2   Ensemble of vibration responses (x_1, x_2, x_3, x_4) measured at a point on an internal combustion engine block and synchronized with a particular cylinder firing. The amplitude at time t_1 is a random variable.

The first half of this chapter describes methods for determining the response of a vibrating system subjected to random excitations. First, the statistical parameters used in this analysis are presented. Next, the responses of single and multiple degree-of-freedom systems to random excitations (stationary and nonstationary) are analyzed. Then, the application of this analysis to failure prediction is summarized. (More information on failure analysis is included in Chap. 35.)

STATISTICAL PARAMETERS OF RANDOM VIBRATIONS*

PROBABILITY DISTRIBUTION FUNCTIONS

The fundamental statistical parameter of a random vibration is the probability distribution of the vibration amplitude $x(t)$ as a function of time. (In general, x may represent the acceleration, velocity, displacement, stress, etc.) In Fig. 11.1 the amplitude

---

* See Chap. 22 for methods to determine these parameters from measured data.

distribution of x is represented by the *probability density function $p(x)$*. The function $p(x)$ is obtained from the probability that a particular sample $x_i(t_1)$ has a value between x and $x + \Delta x$, represented by $\mathrm{Prob}[x \leq x_i(t_1) < x + \Delta x]$. For a nonstationary random process this probability is a function of the time t_1. The probability density is defined by

$$p(x,t_1) \equiv \lim_{\Delta x \to 0} \frac{\mathrm{Prob}[x \leq x_i(t_1) < x + \Delta x]}{\Delta x} \tag{11.1}$$

An alternate representation of the amplitude distribution is the *cumulative (probability) distribution function $P(x)$*, which is the probability that a particular sample $x_i(t_1)$ has a value less than or equal to x. The cumulative distribution is defined by

$$P(x,t_1) \equiv \mathrm{Prob}[x_i(t_1) \leq x] = \int_{-\infty}^{x} p(x',t_1)\, dx' \tag{11.2}$$

Therefore, the probability density and cumulative distribution functions are related as illustrated in Fig. 11.3. For most random processes the cumulative distribution function is smooth and differentiable so that Eq. (11.2) can be rewritten as

$$p(x,t_1) = \frac{d}{dx} P(x,t_1) \tag{11.3}$$

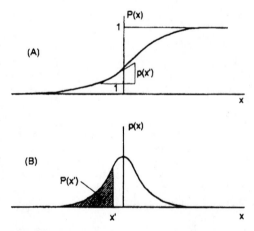

FIGURE 11.3 Examples of the probability distributions of a random variable x. (*A*) Cumulative (probability) distribution function, $P(x)$. (*B*) Probability density function $p(x)$.

Since by definition $P(x) \to 1$ as $x \to \infty$, the total area under $p(x)$ is normalized to be unity, or

$$P(\infty,t_1) = \int_{-\infty}^{\infty} p(x,t_1)\, dx = 1 \tag{11.4}$$

MEAN VALUE

The *mean* (or expected) *value* $\overline{x(t_1)}$ of x at time t_1 is defined by the arithmetic average of all samples $x_i\,(t_1)$:

$$\overline{x(t_1)} \equiv \lim_{N \to \infty} \frac{1}{N} \sum_{i=1}^{N} x_i(t_1) \tag{11.5}$$

The mean value can be obtained from the probability density by

$$\overline{x(t_1)} = \int_{-\infty}^{\infty} xp(x,t_1)\, dx \tag{11.6}$$

If $x(t)$ is stationary over time $0 \le t \le T$, then the mean value can be approximated by the time average:

$$\overline{x} \simeq \frac{1}{T} \int_{0}^{T} x(t)\, dt \tag{11.7}$$

where the approximation improves as $T \to \infty$.

MEAN-SQUARE VALUE

The *mean-square value* $\overline{x^2(t_1)}$ is defined as the expected value of all samples $x_i^2(t_1)$. The mean-square value can be obtained from the probability density by

$$\overline{x^2(t_1)} = \int_{-\infty}^{\infty} x^2\, p(x,t_1)\, dx \tag{11.8}$$

If $x(t)$ is stationary, then the mean-square value can be approximated by the time average:

$$\overline{x^2} \simeq \frac{1}{T} \int_{0}^{T} x^2\,(t)\, dt \tag{11.9}$$

MOMENTS OF THE PROBABILITY DISTRIBUTION

The mean and mean-square values are called the first and second moments of $p(x)$, respectively. The nth moment of $p(x)$ is then defined by

$$\overline{x^n(t_1)} = \int_{-\infty}^{\infty} x^n\, p(x,t_1)\, dx \tag{11.10}$$

The *variance* σ^2 (or square of the *standard deviation* σ) is the expected value of the quantity $(x - \overline{x})^2$ and is evaluated by

$$\sigma^2 = \int_{-\infty}^{\infty} (x - \overline{x})^2\, p(x)\, dx = \overline{x^2} - (\overline{x})^2 \tag{11.11}$$

where the designation of the time dependence is omitted for clarity. The variance is then the difference between the mean-square and the square of the mean value of x.

For many random variables in vibration analysis the mean value is zero so that the variance and mean-square values can be used interchangeably.

Higher-order moments are usually represented in terms of the normalized variable $z = (x - \bar{x})/\sigma$. The value of z is the number of standard deviations x is from the mean. The normalized third moment is called the *skewness* a_3:

$$a_3 = \int_{-\infty}^{\infty} \left(\frac{x - \bar{x}}{\sigma}\right)^3 p(x)\, dx \tag{11.12}$$

The normalized fourth moment is called the *kurtosis* a_4:

$$a_4 = \int_{-\infty}^{\infty} \left(\frac{x - \bar{x}}{\sigma}\right)^4 p(x)\, dx \tag{11.13}$$

For a Gaussian distribution $a_3 = 0$ and $a_4 = 3$.

GAUSSIAN (NORMAL) DISTRIBUTION

The Gaussian distribution is important in random vibration analysis because it is so frequently encountered. The Gaussian probability density function is given by

$$p(x) = \frac{1}{\sigma\sqrt{2\pi}}\, e^{-1/2[(x - \bar{x})/\sigma]^2} \tag{11.14}$$

One reason the Gaussian distribution is so common is the *central limit theorem* which states that the sum of N random variables having an arbitrary distribution will approach a Gaussian distribution as $N \to \infty$. If a random vibration results from the sum of a large number of random excitations, its distribution will tend to be Gaussian.

As a corollary to this, if a vibration response results from the product of a large number of random variables, the logarithm of the vibration magnitude will be the sum of the logarithm of the variables, and this sum will tend to have a Gaussian distribution. The vibration magnitude is then said to have a *log-normal* distribution. This occurs in the vibration of complex machinery where the distribution of responses over an ensemble of nominally identical units will tend to be log-normal.

One common model for the excitation of a random vibration is a sequence of pulses with random amplitudes and random time spacing as illustrated in Fig. 11.4. This model can represent, for example, the pressure pulses in the boundary layer of a turbulent fluid flow or the sequence of stress pulses from an earthquake arriving at some location after propagating through the earth's stratified media. The response of a system to this type of excitation can be thought of as a sum of the responses to each pulse. The response of a system to a unit impulse is called the *impulse response* $h(t)$. The response to a sequence of pulses is then the sum of a sequence of impulse responses appropriately scaled in amplitude and delayed in time. If the impulse response is long compared to the average spacing between the pulses, then the resulting system response will have a Gaussian distribution.

Broad-band, stationary random variables with Gaussian distributions are often called *white noise.* Ideally, white noise has an equal contribution from all frequencies. Practically, white noise is usually band-limited to the frequency range of interest. However, a Gaussian distribution does not necessarily imply white noise. This can be seen from Fig. 11.1 where the vibration response of the resiliently mounted mass is Gaussian and narrow-band in frequency.

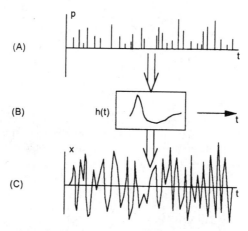

FIGURE 11.4 Example of the generation of broad-band random vibration with a Gaussian probability distribution. (A) Sequences of excitation pulses p. (B) System impulse response function $h(t)$. (C) Resulting system response amplitude x.

CORRELATION FUNCTIONS

Correlation functions are used to describe the average relation between random variables. The *autocorrelation* $R_{xx}(\tau)$ is the expected value of the product of two samples of $x_i(t)$ that are separated in time by τ. In general, the autocorrelation is a function of the time t_1 of the first sample:

$$R_{xx}(\tau,t_1) = \overline{x(t_1)x(t_1 + \tau)} \tag{11.15}$$

By definition the autocorrelation at zero delay ($\tau = 0$) is equal to the mean-square value of the variable, and this is the maximum value of the autocorrelation function.

If $x(t)$ is stationary over time $0 \le t \le 2T$, the autocorrelation is independent of the time of the first sample and is a function only of the absolute value of the delay τ. Then, the autocorrelation function (for $0 \le \tau \le T$) can be approximated by the time average:

$$R_{xx}(-\tau) = R_{xx}(\tau) \simeq \frac{1}{T} \int_0^T x(t)x(t + \tau)\, dt \tag{11.16}$$

Comparing Eqs. (11.9) and (11.16) it follows that $R_{xx}(0) = \overline{x^2}$.

For a system excited by white noise the autocorrelation of a response variable can be used to determine the frequency bandwidth of the system response function. If white noise is filtered with an ideal bandpass filter having cut-off frequencies f_1 and f_2 ($f_1 < f_2$), the autocorrelation of the resulting band-limited random variable is given by

$$R_{xx}(\tau) = \overline{x^2}\frac{\sin(2\pi f_2\tau) - \sin(2\pi f_1\tau)}{2\pi(f_2 - f_1)\tau} \tag{11.17}$$

This is illustrated in Fig. 22.24. If $f_1 = 0$, the first zero crossing of the autocorrelation function occurs at a delay $\tau = 1/f_2$.

The average relation between two variables $x(t)$ and $y(t)$ is represented by the *crosscorrelation* $R_{xy}(\tau,t_1)$ defined by

$$R_{xy}(\tau,t_1) = \overline{x(t_1)y(t_1 + \tau)} \tag{11.18}$$

For variables of a stationary process the crosscorrelation is a function only of the delay τ. However, the maximum value does not necessarily occur at $\tau = 0$. The cross-correlation function can be approximated by the time average:

$$R_{xy}(\tau) \simeq \frac{1}{T} \int_0^T x(t)y(t + \tau) \, dt \tag{11.19}$$

POWER SPECTRAL DENSITY

The frequency content of a random variable $x(t)$ is represented by the *power spectral density* $W_x(f)$, defined as the mean-square response of an ideal narrow-band filter to $x(t)$, divided by the bandwidth Δf of the filter in the limit as $\Delta f \to 0$ at frequency f (Hz):

$$W_x(f) = \lim_{\Delta f \to 0} \frac{\overline{x_{\Delta f}^2}}{\Delta f} \tag{11.20}$$

This is illustrated in Fig. 22.17. By this definition the sum of the power spectral components over the entire frequency range must equal the total mean-square value of x:

$$\overline{x^2} = \int_0^\infty W_x(f) \, df \tag{11.21}$$

The term *power* is used because the dynamical power in a vibrating system is proportional to the square of the vibration amplitude.

An alternate approach to the power spectral density of stationary variables uses the *Fourier series* representation of $x(t)$ over a finite time period $0 \le t \le T$, defined in Eq. (22.29) as

$$x(t) = \bar{x} + \sum_{n=1}^{\infty} A_n \cos(2\pi f_n t) + \sum_{n=1}^{\infty} B_n \sin(2\pi f_n t) \tag{11.22}$$

where $f_n = n/T$. The coefficients of the Fourier series are found by

$$A_n = \frac{2}{T} \int_0^T x(t)\cos(2\pi f_n t) \, dt$$

$$\tag{11.23}$$

$$B_n = \frac{2}{T} \int_0^T x(t)\sin(2\pi f_n t) \, dt$$

Comparing this to Eq. (11.19), it follows that the coefficients of the Fourier series are a measure of the correlation of $x(t)$ with the cosine and sine waves at a particular frequency.

The relation between the Fourier series and the power spectral density can be found by evaluating $\overline{x^2}$ from Eq. (11.22):

$$\overline{x^2} = \frac{1}{T} \int_0^T \left\{ \overline{x} + \sum_{n=1}^{\infty} [A_n \cos(2\pi f_n t) + B_n \sin(2\pi f_n t)] \right\}$$

$$\times \left\{ \overline{x} + \sum_{m=1}^{\infty} [A_m \cos(2\pi f_m t) + B_m \sin(2\pi f_m t)] \right\} dt \quad (11.24)$$

The integral over time cancels all cross terms in the product of the Fourier series leaving only the squares of each term:

$$\overline{x^2} = \frac{1}{T} \int_0^T \left\{ (\overline{x})^2 + \sum_{n=1}^{\infty} [A_n^2 \cos^2(2\pi f_n t) + B_n^2 \sin^2(2\pi f_n t)] \right\} dt$$

$$= (\overline{x})^2 + \sum_{n=1}^{\infty} \frac{1}{2} \left[A_n^2 + B_n^2 \right] \quad (11.25)$$

Each term in this series can be viewed as representing a component of the mean-square value associated with a filter of bandwidth $\Delta f = 1/T$. The power spectral density is then approximated by

$$W_x(f_n) \simeq \frac{T}{2} \left(A_n^2 + B_n^2 \right) \quad (11.26)$$

Using a similar method the relation between $W_x(f)$ and $R_{xx}(\tau)$ can be found. Equation (11.24) can be used to evaluate $R_{xx}(\tau)$ by changing the factors $f_m t$ to $f_m(t + \tau)$. The time integration removes all terms except those of the form $\frac{1}{2}(A_n^2 + B_n^2)\cos(2\pi f_n \tau)$. The autocorrelation is then given by

$$R_x(\tau) = (\overline{x})^2 + \sum_{n=1}^{\infty} \frac{1}{2} \left(A_n^2 + B_n^2 \right) \cos(2\pi f_n \tau)$$

$$= (\overline{x})^2 + \sum_{n=1}^{\infty} W_x(f_n) \cos(2\pi f_n \tau) \Delta f \quad (11.27)$$

In the limit as $T \to \infty$, $\Delta f \to 0$ and the summation approaches the continuous integral:

$$R_x(\tau) = \int_0^{\infty} W_x(f) \cos(2\pi f \tau) \, df \quad (11.28)$$

This is the Fourier cosine transform. The reciprocal relation is:

$$W_x(f) = 4 \int_0^{\infty} R_x(\tau) \cos(2\pi f \tau) \, d\tau \quad (11.29)$$

For transient random variables the power spectral density is a function of time. However, if the power spectral density is integrated over the time duration of a transient $x(t)$, an *energy spectral density* $E_x(f)$ can be obtained representing the frequency content of the total energy in x. Using the Fourier series approach, $E_x(f_n) = TW_x(f_n)$. Alternately, the *shock spectrum* can be used to represent the frequency content of a transient. The shock spectrum represents the peak amplitude response of a narrow-band resonance filter to a transient event (see Chap. 23). A statistical method for estimating the shock spectrum is given in the next section.

RESPONSE OF A SINGLE DEGREE-OF-FREEDOM SYSTEM

In this section the single degree-of-freedom resonator shown in Fig. 11.1 is analyzed to obtain an expression for the mean-square response of the mass when the base is subjected to a random vibration. The equation of motion for this system is derived in Chap. 2 as

$$\ddot{z} + \frac{c}{m}\dot{z} + \frac{k}{m}z = \ddot{y} \tag{11.30}$$

where $z = x - y$ is the motion of the mass relative to the base. This equation is similar in form to the equation for a force excitation $F(t)$ on the mass and a rigid base:

$$\ddot{x} + \frac{c}{m}\dot{x} + \frac{k}{m}x = \frac{F(t)}{m} \tag{11.31}$$

In general, the equations of this form can be solved using r for the response variable and s for the source term. Defining

$$f_n = \frac{1}{2\pi}\sqrt{\frac{k}{m}} = \text{the natural frequency}$$

$$\tag{11.32}$$

$$\zeta = \frac{c}{2\sqrt{km}} = \text{the critical damping ratio}$$

gives:

$$\ddot{r} + 4\pi\zeta f_n \dot{r} + (2\pi f_n)^2 r = s(t) \tag{11.33}$$

Using the input-output format for the analysis, illustrated in Fig. 22.7, with a sinusoidal source $s(t) = S\sin(2\pi ft)$, the response of the system is given in terms of a frequency dependent *transfer function* $H(f)$ (or *frequency response function*) with a magnitude given as

$$|H(f)|^2 = \frac{W_r(f)}{W_s(f)} = \frac{1}{(2\pi f_n)^4\left\{\left[1 - \left(\dfrac{f}{f_n}\right)^2\right]^2 + \left(2\zeta\dfrac{f}{f_n}\right)^2\right\}} \tag{11.34}$$

For a broad-band random source, if $\zeta \ll 1$ so that $|H(f)|^2$ is sharply peaked at $f = f_n$ and the source is stationary with a relatively smooth spectrum, as illustrated in Fig. 11.5, then the mean-square response of the system is determined by the source spectrum at $f = f_n$ times the area under the $|H(f)|^2$ curve:

$$\overline{r^2} = W_s(f_n)\int_0^\infty |H(f)|^2\,df = \frac{W_s(f_n)}{8\zeta(2\pi f_n)^3} \tag{11.35}$$

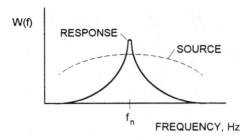

FIGURE 11.5 Power spectral density $W(f)$ of the response of a resonator with $\zeta \ll 1$ excited by a broad-band random source having the spectrum shown by the dashed curve.

The resonance of the system acts as a narrow-band filter on the source spectrum as is illustrated in Fig. 11.1. The vibration is essentially at frequency f_n with a Gaussian amplitude distribution. The mean-square acceleration and velocity levels are related to the displacement response by $\overline{\ddot{x}^2} = (2\pi f_n)^2 \overline{\dot{x}^2} = (2\pi f_n)^4 \overline{x^2}$.

The autocorrelation of the stationary response is found to be

$$R_r(\tau) = \overline{r^2}\, e^{-2\pi\zeta f_n \tau}\left[\cos(2\pi f_d t) + \frac{\zeta}{\sqrt{1-\zeta^2}}\sin(2\pi f_d t)\right] \tag{11.36}$$

where $f_d = f_n \sqrt{1-\zeta^2}$.

The response of the resonator to a transient excitation can be analyzed for the simple case where the source is suddenly turned on and remains stationary there-after.[3] The transient mean-square response starting from rest is then found to be (for $\zeta \ll 1$)

$$\overline{r^2}(t) = \frac{W_s(f_n)}{8\zeta(2\pi f_n)^3}\,(1 - e^{-4\pi\zeta f_n t}) \tag{11.37}$$

The mean-square response grows to the steady-state value in the same way that a first-order dynamic system responds to a step input. This is an important result, illustrating that the dynamical power in a vibrating system is transmitted according to the simple first-order diffusion equation with a time constant $\tau = 1/(4\pi\zeta f_n)$.

This result can be used to estimate the shock spectrum of a transient random excitation with a known time-dependent mean-square level $\overline{s^2}(t,\Delta f)$ in the frequency band Δf. The mean-square response of a resonator to this excitation can be found by solving the following first-order differential equation either numerically or using the Laplace transform method (see Chap. 8):

$$\frac{d}{dt}\overline{r^2}(t) + (4\pi\zeta f_n)\overline{r^2}(t) = \frac{\overline{s^2}(t)}{4\Delta f(2\pi f_n)^2} \tag{11.38}$$

assuming f_n is within the bandwidth Δf.

For example, Fig. 11.6 shows the measured transient acceleration of a concrete floor slab in a building with an operating punch press. As with many transient vibration time-histories, the smoothed mean-square level can be approximated by

$$\overline{\ddot{x}^2}(t) = At\, e^{-\beta t} \tag{11.39}$$

where for this case $A \simeq 0.2g^2/s$ and $\beta \simeq 35/s$ with $\Delta f \simeq 80$ Hz. The solution of Eq. (11.38) with this form of excitation is given by

$$\overline{r^2}(t) = \frac{A}{4\Delta f(2\pi f_n)^2}\left[\frac{t(\alpha - \beta)e^{-\beta t} + e^{-\alpha t} - e^{-\beta t}}{(\alpha - \beta)^2}\right] \tag{11.40}$$

where $\alpha = 4\pi\zeta f_n$. The undamped shock response is the maximum response level as $\alpha \to 0$, which is

$$\overline{r^2}_{max} \to \frac{A}{4\Delta f(2\pi f_n)^2\beta^2} \tag{11.41}$$

The undamped shock spectrum is the peak rms response as a function of f_n (see Chap. 23), which can be estimated with 95 percent certainty as the 2σ level assuming a Gaussian distribution:

$$r_{peak} \simeq 2\sqrt{\overline{r^2}_{max}} = \sqrt{\frac{A}{\Delta f}}\,\frac{1}{2\pi f_n\beta} \tag{11.42}$$

This result is plotted in Fig. 11.6D along with the exact calculation of the shock spectrum at 5-Hz intervals using a particular sample of the acceleration time-history.

RESPONSE OF MULTIPLE DEGREE-OF-FREEDOM SYSTEMS

Real elastic systems have many degrees-of-freedom and, therefore, many modes of resonance, as discussed in Chaps. 2 and 7. However, these *normal modes* ψ_n each

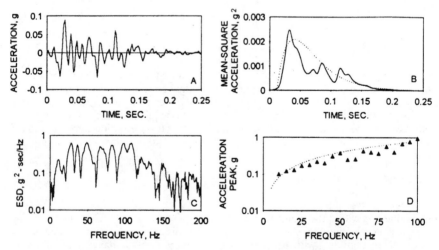

FIGURE 11.6 Transient response of a concrete floor slab with an operating punch press. (*A*) Measured acceleration signal. (*B*) Mean-square smoothed signal (solid curve) and curve fit (dashed curve) using Eq. (11.39). (*C*) Measured energy spectral density. (*D*) Computed acceleration shock response spectrum (symbols) and statistical estimate (dashed curve) using Eq. (11.42).

respond as a simple resonator, and the total response of a system can be obtained by summing the response of all of the modes (*modal superposition*):

$$r(\upsilon,t) = \sum_n q_n(t)\psi_n(\upsilon) \tag{11.43}$$

where υ represents the spatial dimension(s) of the system.

If the damping in the system is distributed proportionately to the mass and stiffness, the normal modes are uncoupled and each has an equation of motion in the form of Eq. (11.33) with a source term given by

$$s_n(t) = \int s(\upsilon,t)\psi_n(\upsilon)\,d\upsilon \equiv \varphi_n \sqrt{\overline{s^2(\upsilon,t)}} \tag{11.44}$$

where φ_n is the modal participation factor of the source. The transfer function for each mode will be of the form of Eq. (11.34) so that the resulting sum of the modal responses gives

$$\overline{r^2} = \sum_n \frac{\varphi_n^2\psi_n^2 W_s(f_n)}{8\zeta_n(2\pi f_n)^3} \tag{11.45}$$

If the damping is not distributed proportionately but is small ($\zeta \ll 1$), the superposition of normal modes gives approximately correct results. This is illustrated by the two degree-of-freedom system shown in Fig. 11.7. An instrument housing (m_1) is resiliently mounted on a vibrating base. A dynamic vibration absorber (see Chap. 6) is attached to suppress the vibration of the housing at frequency f_2. Of interest here is the broad-band response of the system when the base vibration has a uniform

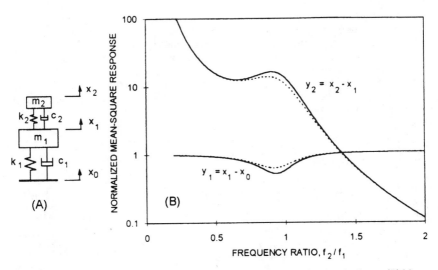

FIGURE 11.7 (*A*) Response of a two degree-of-freedom system to a base excitation x_0. (*B*) Mean-square relative displacement responses, y_1 and y_2, normalized to the response of m_1 alone. Solid curves are calculated using the modal summation of Eq. (11.45). Dashed curves are the exact calculations.

acceleration spectral density $W_{\ddot{x}_0}$. The equations for the relative responses $y_1 = x_1 - x_0$ and $y_2 = x_2 - x_1$ in symmetric, dimensionless form are

$$
\begin{bmatrix} 1 & 0 \\ 0 & \dfrac{\mu f_2^2}{f_1^2} \end{bmatrix} \begin{bmatrix} \ddot{y}_1 \\ \ddot{y}_2 \end{bmatrix} + \begin{bmatrix} 4\pi\zeta_1 f_1 & -4\pi\mu\zeta_2 f_2 \\ -4\pi\dfrac{\mu f_2^2}{f_1^2}\zeta_1 f_1 & (1+\mu)\dfrac{\mu f_2^2}{f_1^2}(2\pi f_2)^2 \end{bmatrix} \begin{bmatrix} \dot{y}_1 \\ \dot{y}_2 \end{bmatrix}
$$

$$
+ \begin{bmatrix} (2\pi f_1)^2 & -\mu(2\pi f_2)^2 \\ -\mu(2\pi f_2)^2 & (1+\mu)\dfrac{\mu f_2^2}{f_1^2}(2\pi f_2)^2 \end{bmatrix} \begin{bmatrix} y_1 \\ y_2 \end{bmatrix} = \begin{Bmatrix} -\ddot{x}_0 \\ 0 \end{Bmatrix} \tag{11.46}
$$

where $\mu = m_2/m_1$, $2\pi f_i = \sqrt{k_i/m_i}$ and $4\pi\zeta_i f_i = c_i/m_i$. The damping is symmetric only if $\zeta_1 f_2 = \zeta_2 f_1$.

Consider a specific example where $\mu = 0.04$ and $\zeta_1 = \zeta_2 = 0.05$, so the damping is not symmetric. Figure 11.7 shows the calculated values of the mean-square responses y_1^2 and y_2^2 as a function of f_2/f_1. The amplitudes are plotted relative to the mean-square response that m_1 would have without the attached vibration absorber y_{1o}^2 as calculated using Eq. (11.35). The modal superposition calculation ignores the small cross-coupling between the normal modes due to the nonsymmetric damping. These results are compared to the exact solution for the two degree-of-freedom system.[1] The mean-square response of m_1 is suppressed only when $f_2 \approx f_1$ and only by about 4 dB.

EVALUATION OF FAILURE CRITERIA

Random vibration can contribute to the fatigue and failure of systems. The vibration may contribute to the cyclical stress loading in a part of the system and accelerate the accumulation of fatigue or crack growth leading to eventual failure. Or, the vibration may increase the probability of exceeding the ultimate stress in a part of the system during its operation leading to immediate failure. Chapter 35 describes the analysis of failure mechanisms in more detail. This section presents methods to estimate the distribution of system response levels resulting from random vibration in forms that can be used in failure models. It is assumed that the stress levels induced by the vibration are linearly related to the relative displacement levels y in the system.

LEVEL CROSSINGS

The vibration responses of systems exposed to random excitations frequently have a Gaussian distribution over time. This is true of both broad-band and narrow-band vibration as illustrated in Fig. 11.1. Even if the excitation is not Gaussian, complex systems with many modes of vibration contributing to the total response will, by the central limit theorem, tend to have a Gaussian response distribution. The probability that the vibration response y will exceed a limiting value y_L is given by

$$
P(y > y_L) = \int_{y_L}^{\infty} p(y)\,dy = \frac{1}{2}\,\mathrm{erfc}\left(\frac{y_L}{\sqrt{2}\sigma_y}\right) \tag{11.47}
$$

where y is assumed to have a Gaussian distribution with zero mean and variance σ_y^2. The function erfc is the complimentary error function and is plotted in Fig. 22.10, curve A. For linear systems the distribution of random vibration levels can be superimposed on the static (or slowly varying) stress levels.

This distribution can be used to obtain an estimate of the rate of occurrence v of a particular level crossing. The inverse of this rate is the mean time between occurrences of this level crossing. For a broad-band random vibration the rate of crossing the level $y = a$ with a positive slope, denoted by v_a^+, is

$$v_a^+ = \frac{1}{2\pi} \frac{\dot\sigma_y}{\sigma_y} e^{-\frac{a^2}{2\sigma_y^2}} \qquad (11.48)$$

For a narrow-band vibration $\dot\sigma_y = 2\pi f_n \sigma_y$, so the level crossing rate is simply

$$v_a^+ = f_n e^{-\frac{a^2}{2\sigma_y^2}} \qquad (11.49)$$

Caution must be used when applying Eqs. (11.48) and (11.49) to values of $|a| > 2\sigma_y$. While many vibration distributions may be adequately represented by a Gaussian distribution in the range of $\pm 2\sigma$ from the mean, there may be significant deviations outside this range. This may cause significant errors in rate of crossing estimates for extreme values. Therefore, the rate of crossing estimates are not that useful for estimating the time to the first occurrence of a large stress resulting from a random vibration.

CUMULATIVE DAMAGE

The rate of occurrence estimates are more useful in a cumulative damage model which sums up the effects of repeated occurrences of excessive stress until a failure criteria is met. Often these failure models are based on the number of occurrences of peak levels in a cyclical loading pattern. This is true in the fatigue limit analysis using S-N curves and also in the fracture mechanics analysis using exceedance curves (see Chap. 35). For true white noise the peak levels have a Gaussian distribution. However, for band-limited Gaussian vibrations, the distribution of the peak levels is more complicated. For broad-band random vibrations the probability density function of the absolute level of the displacement peaks is found to be approximated by the Poisson (exponential) distribution

$$p(|y_P|) = \frac{1}{\sigma_y} e^{-|y_P|/\sigma_y} \qquad (11.50)$$

For narrow-band vibrations the probability density function of the peaks is found to be approximated by the Rayleigh distribution (see Fig. 11.1)

$$p(|y_P|) = \frac{y_P}{\sigma_y^2} e^{-y_P/2\sigma_y} \qquad (11.51)$$

These distributions of peak levels can be used with cyclical fatigue limit curves to estimate a measure of the cumulative damage D. For example, if a material S-N curve is approximated by $N = cS^{-b}$ (N equals the number of cycles to failure at a peak stress level S) and the critical stress is a function of the vibration displacement $S =$

$S(y)$, then the expected value of the accumulated damage over time by a random vibration is

$$\overline{D(t)} = v_0^+ t \int_0^\infty \frac{p(y_P)}{N(y_P)}\, dy_P \tag{11.52}$$

where failure occurs around $D(t) = 1$. With this analysis there is not only a statistical uncertainty due to variations in the material properties, but there is also an uncertainty in the distribution of vibration cycles. The variance in the estimate of $D(t)$ due to this latter uncertainty is estimated to be

$$\sigma_D^2 \simeq \overline{D}^2 \frac{10^{(b-5)/4}}{\zeta v_0^+ t} \qquad (b > 5) \tag{11.53}$$

for the narrow-band vibration case.

For estimates of crack propagation in fracture mechanics, an exceedance diagram is often used. The exceedance diagram plots the peak stress level as a function of the number of cycles which exceed this stress level. The exceedance curve in a random vibration is then found from the cumulative distribution function of the peak levels. For a broad-band limited vibration,

$$P(|y_P| > y_L) = e^{-y_L/\sigma_y} \tag{11.54}$$

and for a narrow-band vibration,

$$P(|y_P| > y_L) = e^{-y_L^2/2\sigma_y^2} \tag{11.55}$$

These probability functions are shown in the form of exceedance curves in Fig. 11.8 with the relative amplitude y_P/σ_y plotted as a function of the logarithm of P. The number of cycles N occurring in time t can be found by multiplying P by the appropriate value of $v_0^+ t$.

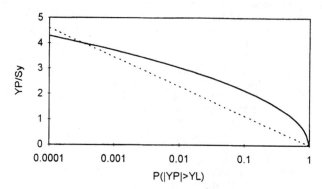

FIGURE 11.8 Probability of exceedance functions for peaks in the displacement response cycles of band-limited (dashed curve) and narrow-band (solid curve) random vibration.

STATISTICAL ENERGY ANALYSIS

Statistical energy analysis (SEA) models the vibration response of a complex system as a statistical interaction between groups of modes associated with subsections of the system. While the theoretical development of SEA has its roots in the field of random vibration, it does not require a random excitation for the statistical analysis. Instead, SEA uses the random variation of modal responses in complex systems to obtain statistical response predictions in terms of mean values and variances of the responses. Theoretically, the statistical averaging is over ensembles of nominally identical systems. However, in practice many systems have enough inherent complexity that the variation in the response over frequency and location is adequately represented by the ensemble statistics.

This is seen even in the relatively simple case of the distribution of bending modes in a simply-supported rectangular flat plate (Fig. 11.9). The resonance frequencies of the modes are given by

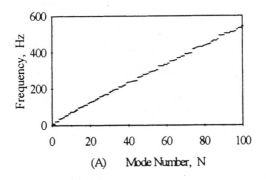

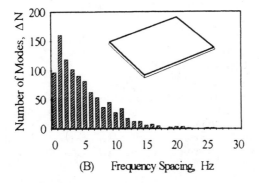

FIGURE 11.9 Mode count of a 2.6- × 2.4- × 0.01-meter simply supported, steel plate. (*A*) Resonance frequencies. (*B*) Distribution of resonance frequency spacings.

$$f_{m,n} = \left(\frac{\pi}{4\sqrt{3}}\right) hc_L \left[\left(\frac{m}{L_1}\right)^2 + \left(\frac{n}{L_2}\right)^2\right] \tag{11.56}$$

where L_1 and L_2 are the length dimensions, h is the thickness, c_L is the longitudinal wave speed of the plate material, and m and n are integers. The resonance frequencies are seen to follow approximately along a straight line. This slope of this line is the *average frequency spacing* $\overline{\delta f}$ (inverse of modal density per Hz) given by

$$\overline{\delta f} = \frac{hc_L}{\sqrt{3}L_1 L_2} \tag{11.57}$$

One way to represent the variation in the actual resonant frequencies is to plot the distribution in the frequency difference between two successive resonances, which can be plotted as shown in Fig. 11.9B. This distribution appears to be Poisson.

Repeating this analysis for other plates with the same surface area, thickness, and material (thus having the same $\overline{\delta f}$), but with different values of L_1 and L_2, gives essentially the same results. This indicates that one way of looking at the modes of any one particular plate is to consider it as one realization from an ensemble of plates having the same statistical distribution of resonances. SEA uses this model to develop estimates of the vibration response of systems based on averages over the ensemble of similar systems. However, since the modes are usually a function of the parameter (fL/c), variations in the frequency f in a complex system often have the same statistics as variations in L (dimensions) and c (material properties) in an ensemble of similar systems.

The statistical model of a system is useful in a variety of applications. In the preliminary design phase of a system SEA can be used to obtain quantitative estimates of the vibration response even when all of the details of the design are not completely specified. This is because preliminary SEA estimates can be made using the general characteristics of the system components (overall size, thickness, material properties, etc.) without requiring the details of component shapes and attachments.

SEA is also useful in diagnosing vibration problems. The SEA model can be used to identify the sources and transfer paths of the vibrational energy. When measured data is available, SEA can help to interpret the data, and the measured data can be used to improve the accuracy of a preliminary SEA model.

Since the SEA model gives quantitative predictions based on the physical properties of the system, it can be used to evaluate the effectiveness of design modifications. It can also be used with an optimization routine to search for improved design configurations.

SEA MODELING OF SYSTEMS

The statistical energy analysis (SEA) model of a complex system is based on the statistical analysis of the coupling between groups of resonant modes in subsections of the system. The modal coupling is based on the analysis of two coupled resonators as shown in Fig. 11.10. This is a more general case of the two degree-of-freedom system analyzed for a random vibration (see Fig. 11.7). Here there are two distinct res-

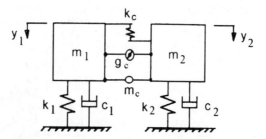

FIGURE 11.10   Two linear, coupled resonators, with displacement y, mass m, stiffness k, damper c, and gyroscopic parameter g.

onators coupled by stiffness, inertial, and gyroscopic interactions (represented by k_c, m_c, and g_c, respectively). If the two resonators are excited by different broad-band force excitations, then the net power flow between them through the coupling is given by

$$\Pi_{12} = -k_c \, \overline{y_2 \, \dot{y}_1} - g_c \overline{\dot{y}_2 \, \dot{y}_1} + \frac{1}{4} m_c \overline{\ddot{y}_2 \, \dot{y}_1}$$

$$= B \, (E_1 - E_2) \tag{11.58}$$

where

$$B = \frac{(2\pi\mu)^2}{d} [\Delta_1 f_2^4 + \Delta_2 f_1^4 + f_1 f_2 (\Delta_1 f_2^2 + \Delta_2 f_1^2)]$$

$$+ \frac{1}{d} [(\gamma^2 + 2\mu\kappa)(\Delta_1 f_2^2 + \Delta_2 f_1^2) + \kappa^2 (\Delta_1 + \Delta_2)]$$

$$E_i = (m_i + m_c/4)\overline{\dot{y}_i^2}$$

and

$$d = (1 - \mu^2)[(2\pi)^2 (f_1^2 - f_2^2)^2 + (\Delta_1 + \Delta_2)(\Delta_1 f_2^2 + \Delta_2 f_1^2)]$$

$$\Delta_i = \frac{c_i}{(m_i + m_c/4)}$$

$$f_i^2 = \frac{(1/2\pi)^2 (k_i + k_c)}{(m_i + m_c/4)}$$

$$\mu = \left(\frac{m_c}{4}\right)\left(\frac{m_1 + m_c}{4}\right)^{-1/2}\left(\frac{m_2 + m_c}{4}\right)^{-1/2}$$

$$\gamma = g_c \left(\frac{m_1 + m_c}{4}\right)^{-1/2}\left(\frac{m_2 + m_c}{4}\right)^{-1/2}$$

$$\kappa = k_c \left(\frac{m_1 + m_c}{4}\right)^{-1/2}\left(\frac{m_2 + m_c}{4}\right)^{-1/2}$$

$$\lambda = \left(\frac{m_1 + m_c}{4}\right)^{1/2}\left(\frac{m_2 + m_c}{4}\right)^{-1/2}$$

This result can be interpreted by defining the two individual uncoupled resonators as the subsystems that exist when one of the degrees-of-freedom is constrained to zero. For either uncoupled resonator the kinetic energy averaged over a cycle, $(m + m_c/4)\dot{y}_i^2/2$, is equal to the average potential energy, $(k + k_c)y_i^2/2$. Equation (11.58) can then be seen to state two important results: (1) the power flow is proportional to the difference in the vibrational energies of the two resonators, and (2) the coupling parameter B is positive definite and symmetrical so the system is reciprocal and power always flows from the more energetic resonator to the less energetic one. As a corollary, when only one resonator is directly excited, the maximum energy level of the second resonator is that of the first resonator.

It should be noted that this analysis is exact for a coupling of arbitrary strength as long as there is no dissipation in the coupling. Even when there is dissipation in the coupling, this analysis is approximately correct as long as the coupling forces due to the dissipation are small compared to the other coupling forces. In practice when systems have interface damping at the connections between subsystems (such as in bolted or spot welded joints), the associated damping can be split between subsystems and the interface considered damping free.

As an example of how this analysis is extended to a distributed system, consider the two coupled beams in Fig. 11.11A. The modes of the system can be obtained from an eigenvalue solution of the complete system, or they can be obtained from a coupled pair of equations for the individual (or uncoupled) straight beam subsystems. The latter case leads to coupled mode equations similar to the ones used for the two coupled resonators. However, in this case each mode in one beam subsystem is coupled to all of the relevant modes in the other beam subsystem. The total power flow between the two beam subsystems is then the sum of the individual mode-to-mode power flows.

If the significant coupling is assumed to occur in a limited frequency range Δf (a good assumption for $\zeta \ll 1$ and $\Delta f \gg \zeta f$), then the average net power flow can be found by averaging the value of B over Δf and using average beam subsystem modal energies in Eq. (11.58). This gives

$$\Pi_{12} = \overline{B} N_1 N_2 \left(\frac{E_1}{N_1} - \frac{E_2}{N_2} \right) \tag{11.59}$$

with

$$\overline{B} = \frac{1}{4\Delta f} \left[\mu^2 (2\pi f)^2 + (\gamma^2 + 2\mu\kappa) + \frac{\kappa^2}{(2\pi f)^2} \right]$$

where N_1 and N_2 are the number of modes in the two beam subsystems with resonance frequencies in Δf.

For either beam the total vibrational energy is $E_i = m_i \overline{\dot{y}_i^2}$, where m_i is the total mass of the beam and $\overline{\dot{y}_i^2}$ is the mean-square velocity averaged over space and time. Equation (11.59) shows that the power flow between two distributed subsystems is proportional to the difference in the average *modal* energies E_i/N_i, not the differ-

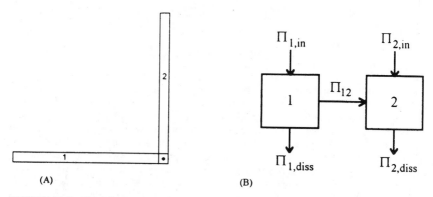

(A)   **(B)**

FIGURE 11.11 Modeling of distributed systems. (*A*) Two coupled beams. (*B*) SEA model of two coupled subsystems with power flow Π.

ence in the total energies (which are proportional to the vibration level). This means it is possible for a thick beam with fewer resonant modes in a frequency band and a lower vibration level to be the source of power for a connected thinner beam with more resonant modes and a higher vibration level.

A more useful form of Eq. (11.59) is obtained by defining a *coupling loss factor* $\eta_{12} \equiv \overline{B}N_2/(2\pi f)$ (and by reciprocity $\eta_{21} \equiv N_1\eta_{12}/N_2$). The coupling loss factor is analogous to the *damping loss factor* for a subsystem defined by $\eta_i = 2\zeta_i$. The coupling loss factor is a measure of the rate of energy lost by a subsystem through coupling to another subsystem, whereas the damping loss factor is a measure of the rate of energy lost through dissipation. The average power flow is then given by

$$\Pi_{12} = 2\pi f(\eta_{12}E_1 - \eta_{21}E_2) \tag{11.60}$$

Using the equivalent expression for the power dissipated in each subsystem, $\Pi_{i,\text{diss}} = 2\pi f\eta_i E_i$, along with the result from Eq. (11.38) for the transient response of a resonator, the following set of equations can be written for the conservation of energy between two coupled subsystems ($\Pi_{\text{in}} = \Pi_{\text{out}} + dE/dt$):

$$\Pi_{1,\text{in}} = 2\pi f(\eta_1 + \eta_{12})E_1 - 2\pi f\eta_{21}E_2 + \frac{dE_1}{dt}$$

$$\tag{11.61}$$

$$\Pi_{2,\text{in}} = -2\pi f\eta_{12}E_1 + 2\pi f(\eta_2 + \eta_{21})E_2 + \frac{dE_2}{dt}$$

where $\Pi_{i,\text{in}}$ is used to denote power supplied by external sources. The SEA block diagram for this power flow model of two coupled subsystems is shown in Fig. 11.11*B*.

These equations are first-order differential equations for the diffusion of energy between subsystems. They are in a form analogous to heat flow or fluid potential flow problems. For steady-state problems the dE/dt terms are zero.

For narrow-band analysis, the SEA equations can be used to obtain averages in the response of the system over frequency. In this case it is more convenient to use the average frequency spacing between modes $\overline{\delta f} = \Delta f/N$ as the mode count in Eq. (11.59). This gives

$$\overline{\Pi_{12}} = \frac{2\pi f}{\overline{\delta f_1}} \eta_{12}(E_1\overline{\delta f_1} - E_2\overline{\delta f_2}) \tag{11.62}$$

The terms $2\pi E_i\overline{\delta f_i}$ have units of power and are called the *modal power potential*.

The value of η_{12} is difficult to evaluate directly from $\overline{B}$ in practice. Instead, indirect methods are often used as described in the section "Coupling Loss Factors." The normalized variance in the value of η_{12} averaged over Δf for edge-connected subsystems is given by

$$\frac{\sigma_{\eta12}^2}{\eta_{12}^2} = \frac{1}{\pi f\left(\dfrac{\eta_1}{\overline{\delta f_1}} + \dfrac{\eta_2}{\overline{\delta f_2}}\right) + \Delta f\left(\dfrac{1}{\overline{\delta f_1}} + \dfrac{1}{\overline{\delta f_2}}\right)} \tag{11.63}$$

The variance in the coupling depends primarily on the system *modal overlap factor* defined by $M_S = \pi f(\eta_1/\overline{\delta f_1} + \eta_2/\overline{\delta f_2})/2$, which is the ratio of the effective modal bandwidth to the average modal frequency spacing. When the system modal overlap factor is less than 1, the variance is larger than the square of the mean value, which may be unacceptably large. This indicates why SEA models tend to converge better with measured results at frequencies above where $M_S = 1$.

Note that the modal overlap in each uncoupled subsystem does not have to be large in order for the variance in the coupling to be small. In fact the SEA model can be used to evaluate the response of a single resonator mode attached to a vibrating flat plate as illustrated in Fig. 11.12. The power flow equations in the form of Eq.

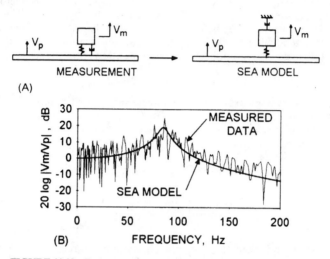

(A)

(B)   FREQUENCY, Hz

FIGURE 11.12   Response of a resonator with vibration V_m, mounted on a plate with vibration V_p. (*A*) Comparison of the measurement configuration and the SEA model. (*B*) Comparison of the measured response and the SEA predictions.

(11.59) are used. The uncoupled resonator has one mode at $f_2 = \sqrt{k_2/m_2}$, so $N_2 = 1$. The mean-square vibration velocity level of the plate in a frequency band Δf encompassing f_2 is $\dot{y}_1^2 = W_{\dot{y}_1}(f)\Delta f$. The average number of plate modes resonating in this frequency band is $N_1 = \Delta f/\delta f_1$. The coupling loss factor is evaluated to be

$$\eta_{21} = \frac{\pi}{2} \frac{f_2}{\delta f_1} \frac{m_2}{m_1} \tag{11.64}$$

Since $\Pi_{12} = \Pi_{2,\text{diss}}$, the mean-square response of the resonator mass is given by

$$\overline{\dot{y}_2^2} = \frac{\pi}{2} \frac{f_2 W_{\dot{y}_1}(f)}{\eta_{21} + \eta_2} \tag{11.65}$$

Even if the resonator damping goes to zero, its maximum energy level is limited to the average modal energy in the plate:

$$m_2 \overline{\dot{y}_2^2}_{,\text{max}} = m_1 W_{\dot{y}_1}(f)\Delta f \tag{11.66}$$

If the resonator energy momentarily gets higher, it transmits the energy back into the plate. Therefore, the plate acts both as a source of excitation and as a dissipator of energy for the resonator. The effective loss factor for the resonator is $\eta_{21} + \eta_2$.

The frequency response function for the resonator can then be evaluated using Eq. (11.34). Fig. 11.12B compares this result with the measured narrow-band frequency spectrum of a 0.1-kg mass attached to a 2.5-mm steel plate with a resilient mounting having negligible damping and $f_2 = 85$ Hz. The measured response of the mass is multimodal since the resonator responds as a part of all of the modes of the coupled system. However, the statistical average response curve accurately represents the multimodal response. The normalized variance of the narrow-band SEA response calculation is estimated from Eq. (11.63) to be 0.5.

For larger systems the following procedure can be used to develop a complete SEA model of the system response to an excitation:

1. Divide the system into a number of coupled subsystems.
2. Determine the mode counts and damping loss factors for the subsystems.
3. Determine the coupling factors between connected subsystems.
4. Determine the subsystem input powers from external sources.
5. Solve the energy equations to determine the subsystem response levels.

The steps in this procedure are described in the following sections of this chapter. When used properly, the SEA model will calculate the distribution of vibration response throughout a system as a result of an excitation. The response distribution is calculated in terms of a mean value and a variance in the vibration response of each subsystem averaged over time and the spatial extent of the subsystem.

MODE COUNTS

In this section the mode counts for a number of idealized subsystem types are given in terms of the average frequency spacing $\overline{\delta f}$ between modal resonances. Experimental and numerical methods for determining the mode counts of more complicated subsystems are also described.

The mode count is sometimes represented by the average number of modes, N or ΔN, resonating in a frequency band, and sometimes by the modal density, represented in cyclical frequency as $n(f) = dN/df$. These are related to the average frequency spacing by

$$n(f) = \frac{1}{\overline{\delta f}} \simeq \frac{\Delta N}{\Delta f} \tag{11.67}$$

For a one-dimensional subsystem, such as a straight beam or bar, with uniform material and cross-sectional properties and with length L, the average frequency spacing between the modal resonances is given by

$$\overline{\delta f}^{1D} = \frac{c_g}{2L} \tag{11.68}$$

where c_g is the energy group speed for the particular wave type being modeled.

For longitudinal waves c_g is equal to the phase speed $c_L = \sqrt{E/\rho}$, where E is the elastic (Young's) modulus and ρ is the density of the material. For torsional waves c_g is equal to the phase speed $c_T = \sqrt{GJ/\rho I_p}$, where G is the shear modulus of the material, and J and I_p are the torsional moment of rigidity and polar area moment of inertia, respectively, of the cross section. For beam bending waves (with wavelengths long compared to the beam thickness) the group speed is twice the bending phase speed c_B, or $c_g = 2c_B = 2\sqrt{2\pi f \kappa c_L}$, where κ is the radius of gyration of the beam cross section. For a beam of uniform thickness h, $\kappa = h/\sqrt{12}$.

For a two-dimensional subsystem, such as a flat plate, with uniform thickness and material properties and with surface area A, the average frequency spacing between the modal resonances is given by

$$\overline{\delta f}^{2D} = \frac{c_p c_g}{2\pi f A} \tag{11.69}$$

where c_p is the phase speed for the particular wave type being modeled.

For plate bending waves (with wavelengths long compared to the plate thickness) $c_g = 2c_p = 2c_{B'} = 2\sqrt{2\pi f \kappa c_{L'}}$, where κ is the radius of gyration, $c_{L'} = \sqrt{E/\rho(1 - \mu^2)}$, and μ is Poisson's ratio. For in-plane compression waves $c_g = c_p = c_{L'}$. For in-plane shear waves $c_g = c_p = c_S = \sqrt{G/\rho}$.

For a three-dimensional subsystem, such as an elastic solid, with uniform material properties and with volume V, the average frequency spacing between the modal resonances is given by

$$\overline{\delta f}^{3D} = \frac{c_o^3}{4\pi f^2 V} \tag{11.70}$$

where c_o is the ambient shear or compressional wave speed in the medium.

For more complicated subsystems the mode counts can be obtained in a number of other ways. Generally, the mode counts only need to be determined within an accuracy of 10 percent in order for any resulting error to be less than 1 dB in the SEA model. For more complicated wave types, such as bending in thick beams or plates, the formulas given above for $\overline{\delta f}$ can be used with the correct values of c_g and c_p obtained from the dispersion relation for the medium.

For more complicated geometries a numerical solution, such as a finite element model, can be used to determine the eigenvalues of the subsystem. Then, the values of $\overline{\delta f}$ can be obtained using Eq. (11.67). In this case it is often necessary to average the mode count over a number of particular geometric configurations or boundary conditions in order to obtain an accurate estimate of the average modal spacing.

When a physical sample of the subsystem exists, experimental data can be used to estimate or validate the mode count. For large modal spacing (small modal overlap) the individual modes can sometimes be counted from a frequency response measurement. However, this method usually undercounts the modes because some of them may occur paired too closely together to be distinguished. An alternate experimental procedure is to use the relation between the mode count and the average mobility of a structure:

$$\overline{\delta f} = \frac{1}{4m\overline{G}} \tag{11.71}$$

where m is the mass of the subsystem and $\overline{G}$ is the average real part of the mechanical mobility (ratio of velocity to force at a point excitation). As with the numerical method, the experimental measurement should be averaged over a variation in the boundary condition used to support the subsystem since no one static support accurately represents the dynamic boundary condition the subsystem sees when it is part of the full system. Also the measurement of $\overline{G}$ should be averaged over several excitation points.

DAMPING LOSS FACTORS

In this section typical methods for determining the damping loss factor of subsystems are given along with some typical values used in statistical energy analysis (SEA) models of complex structures. The damping in SEA models is usually specified by the loss factor which is related to the critical damping ratio ζ and the quality factor Q by

$$\eta = 2\zeta = \frac{1}{Q} \tag{11.72}$$

Chapters 36 and 37 describe the damping mechanisms in structural materials and typical damping treatments. In complex structures the structural material damping is usually small compared to the damping due to slippage at interfaces and added damping treatments. Because the level of added damping is so strongly dependent on the details of the application of a damping treatment, measurements are usually needed to verify analytical calculations of damping levels.

One method to measure the damping of a subsystem is the decay rate method, where the free decay in the vibration level is measured after all excitations are turned off. The initial decay rate DR (in dB/sec) is proportional to the total loss factor for the subsystem:

$$\eta = \frac{DR}{27.3f} \tag{11.73}$$

If the subsystem is attached to other structures, the coupling loss factors will be included in the total loss factor value. Therefore, the subsystem must be tested in a decoupled state. On the other hand, if the connection interfaces provide significant damping due to slippage, then these interfaces must be simulated in the damping test.

Another method of measuring the damping is the half-power bandwidth method illustrated in Fig. 2.22. The width of a resonance Δf in a frequency response measurement is measured 3 dB down from the peak and the damping is determined by

$$\eta = \frac{\Delta f}{f_n} \tag{11.74}$$

As with other measurements of subsystem parameters, the damping measurements must be averaged over multiple excitation points with a variety of boundary conditions.

For preliminary SEA models an empirical database of damping values is useful for initial estimates of the subsystem damping loss factors. Figure 11.13 is an illustration of the typical damping values measured in steel and aluminum machinery structures for different construction methods and different applied damping treatments.

The initial estimates of damping levels in a preliminary SEA model can be improved if measurements of the spatial decay of the vibration levels in the system are available. The spatial decay calculated in the SEA model is quite strongly dependent on the damping values used. Therefore, an accurate estimate of the actual damping can be obtained by comparing the SEA calculations to the measured spatial decay (assuming the other model parameters are correct).

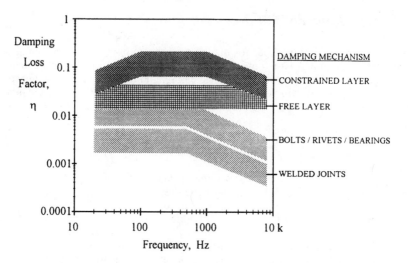

FIGURE 11.13 Empirical values for the damping loss factor η in steel and aluminum machinery structures with different damping mechanisms (assumed to be efficiently applied).

COUPLING LOSS FACTORS

The coupling loss factor is a parameter unique to statistical energy analysis (SEA). It is a measure of the rate of energy transfer between coupled modes. However, it is related to the transmission coefficient τ in wave propagation. This can be illustrated with the system shown in Fig. 11.14. For a wave incident on a junction in subsystem

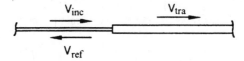

FIGURE 11.14 Evaluation of the coupling loss factor using a wave transmission model for an incident wave V_{inc} at a junction, resulting in a reflected wave V_{ref} and a transmitted wave V_{tra}.

1 with incident power Π_{inc}, the power transmitted to subsystem 2, Π_{tra}, is by definition of the transmission coefficient τ_{12} given by

$$\Pi_{\text{tra}} = \tau_{12}\Pi_{\text{inc}} \tag{11.75}$$

In addition, the junction reflects some power, Π_{ref}, back into subsystem 1 given by

$$\Pi_{\text{ref}} = (1 - \tau_{12})\Pi_{\text{inc}} \tag{11.76}$$

assuming there is no power dissipated at the junction. The energy density in subsystem 1 is given by $E_1' = c_{g_1} (\Pi_{\text{inc}} + \Pi_{\text{ref}})$.

The corresponding SEA representation of the system is

$$\Pi_{tra} = \Pi_{1 \to 2} = 2\pi f \eta_{12} E_1 \tag{11.77}$$

For a subsystem of length L_1, $\overline{\delta f_1} = c_{g_1}/(2L_1)$ and $E_1 = L_1 E_1'$. Solving for the coupling loss factor gives

$$\eta_{12} = \frac{\overline{\delta f_1}}{\pi f} \frac{\tau_{12}}{2 - \tau_{12}} \tag{11.78}$$

A more detailed analysis indicates that this result is valid for point connections in a system with a modal overlap greater than 1. If the system has a constant modal frequency spacing $\overline{\delta f}$, then the Nth mode will occur at $f = N\overline{\delta f}$. If the damping loss factor is η, the system modal overlap is given by $M_S = \pi\eta f/(2\overline{\delta f})$. Then the modal overlap is greater than 1 for frequencies $f > 2\overline{\delta f}/(\pi\eta)$ or for mode numbers $N > 2/(\pi\eta)$. SEA is still valid below this frequency and mode number, but the variance of the model calculations (and in the measured frequency response functions) becomes large.

For point-connected subsystems the transmission coefficient can be evaluated from the junction impedances:[4]

$$\tau_{12} = \frac{4R_1 R_2}{|Z_1 + Z_2|^2} \tag{11.79}$$

where R_i is the real part of the impedance Z_i (ratio of force to velocity at a point excitation) at the junction attachment point of subsystem i. When more than two subsystems are connected at a common junction, the denominator of Eq. (11.79) must include the sum of all impedances at the junction.

For subsystems with line and area junctions the analysis of the coupling loss factor is complicated by the distribution of angles of the waves incident on the junction. However, approximate results have been worked out for many important cases. Eq. (11.78) can be generalized for all cases as

$$\eta_{12} = \frac{\overline{\delta f_1}}{\pi f} \frac{I_{12}\tau_{12}(0)}{2 - \tau_{12}(0)} \tag{11.80}$$

where $\tau_{12}(0)$ is the normal incidence transmission coefficient for waves traveling perpendicular to the junction, and I_{12} contains the result of an average over all angles of incidence.

For line-connected plates the coupling loss factor between bending modes is found using

$$I_{12} = \frac{L_j}{4} \left(\frac{k_1^4 k_2^4}{k_1^4 + k_2^4} \right)^{1/4} \tag{11.81}$$

where L_j is the length of the junction and $k_i = 2\pi f/c_{Bi}$ is the wave number of the modes in subsystem i.

When experimental verification of the evaluation of the coupling loss factor is desired, measurements similar to those used for damping can be used. A decay rate measurement of a subsystem connected to another (heavily damped) subsystem will give a loss factor equal to the sum of the damping and coupling loss factor for the first subsystem. Alternately, subsystem 1 can be excited alone and the spatially averaged response levels of the two connected subsystems can be measured. Using $\Pi_{12} = \Pi_{2,\text{diss}}$, the coupling loss factor is found from

$$\eta_{12} = \frac{\eta_2 E_2}{E_1 - \overline{\delta f_2} E_2 / \overline{\delta f_1}} \tag{11.82}$$

This result indicates a potential problem in determining the coupling loss factor from measured results. If $E_2 \overline{\delta f_2} \simeq E_1 \overline{\delta f_1}$, then taking the difference between their values in Eq. (11.82) will greatly magnify the experimental errors in determining the parameters used in this formula. This indicates why it is mathematically unstable to use measured levels in a multiple subsystem model to back calculate the coupling loss factors. However, good results can be obtained for a single junction between two subsystems if one is excited and the other is artificially damped in order to increase difference between $E_1 \overline{\delta f_1}$ and $E_2 \overline{\delta f_2}$. Figure 11.15 shows the results of an experimen-

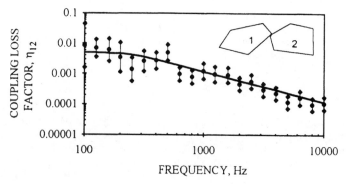

FIGURE 11.15 Coupling loss factor η_{12} for point connected plates; ●—●—● measured data with 95 percent confidence intervals; —— calculated values using Eqs. (11.80) and (11.81).

tal validation of Eqs. (11.80) and (11.81) for the coupling loss factor between two plates connected at a point. The experimental error is also included, which even in this idealized laboratory environment is more than 50 percent.

While the *back calculation* of the coupling loss factors tends to be unstable, the forward calculation in the SEA model is relatively insensitive to errors in the coupling loss factor values, making the model fairly robust.

MODAL EXCITATIONS

The power put into subsystem modes by the system excitations is needed in order to use the statistical energy analysis (SEA) model for calculations of absolute response levels. The mode counts, damping, and coupling loss factors can be used to evaluate relative transfer functions in the system for a unit input power. However, for actual response-level calculations the modal input power from the actual excitation sources must be calculated.

For a point force excitation $F(t)$ the average power put into a system is

$$\Pi_{in} = \overline{F^2}\,\overline{G} \tag{11.83}$$

where $\overline{G}$ is the average real part of the mobility at the excitation point. For a prescribed point velocity source $\dot{y}(t)$ the average power put into a system is

$$\Pi_{in} = \overline{\dot{y}^2}\,\overline{R} \tag{11.84}$$

where $\overline{R}$ is the average real part of the impedance at the excitation point.

The normalized variance in the input power due to variations in the mode shapes and frequency response function of the system is approximated by

$$\frac{\sigma_{\Pi_{in}}^{2}}{\Pi_{in}^{2}} = \frac{3\overline{\delta f}}{\pi f \eta + \Delta f} \tag{11.85}$$

where Δf is the bandwidth of the excitation.

For more complicated excitations the input power can be estimated by measuring the response of a system to the excitation and using the SEA model to back calculate the input power. Alternatively, the measured response levels of the excited subsystem can be used as "source" levels, and the power flow into the rest of the system can be evaluated using the SEA model.

SYSTEM RESPONSE DISTRIBUTION

To solve for the distribution of vibrational energy in a system it is convenient to rewrite Eq. (11.61) in symmetric form:

$$[B]\{\Phi\} + [I]\left\{\frac{dE}{dt}\right\} = \{\Pi_{in}\} \tag{11.86}$$

where $[I]$ is the identity matrix, $\{\Phi\} = 2\pi\{E/\overline{\delta f}\}$ is the vector of modal power potential, and $[B]$ is the symmetric matrix of coupling and damping terms with off-diagonal terms $B_{ij} = -f\,\eta_{ij}/\overline{\delta f_i}$ and diagonal terms $B_{ii} = (f/\overline{\delta f_i})(\eta_i + \Sigma_j\,\eta_{ij})$. This system of equations can be solved using standard numerical methods. Solving for the values of E gives a mean value estimate of the energy distribution.

The variance in E is more difficult to evaluate because it depends on the evaluation of the inverse matrix $[B]^{-1}$. If the variance of each term in $[B]$ is small compared to its mean-square value, then the variances in $[B]^{-1}$ can be approximated by

$$[\sigma_{B^{-1}}^{2}] \simeq [(B_{ij}^{-1})^2][\sigma_B^2][(B_{ij}^{-1})^2] \tag{11.87}$$

where the notation $[(B_{ij}^{-1})^2]$ refers to a matrix with the squares of the elements in $[B]^{-1}$, term for term.

The subsystem energy values can be converted to dynamic response quantities using the relation $E = m\overline{\dot{y}^2}$. For a narrow-band vibration at frequency f_c (which could be a single one-third octave band response in a broad-band analysis) the displacement response is $\overline{y^2} \simeq \overline{\dot{y}^2}/(2\pi f_c)^2$ and the acceleration response is $\overline{\ddot{y}^2} \simeq (2\pi f_c)^2\,\overline{\dot{y}^2}$. The relation between the vibration velocity response and the maximum dynamic strain depends on the type of motion involved. For longitudinal motion the mean-square strain is $\overline{\epsilon^2} = \overline{\dot{y}^2}/c_L^2$. For bending motion of a uniform beam or plate the maximum strain is $\overline{\epsilon_{max}^2} = 3\overline{\dot{y}^2}/c_L^2$.

When the response values in a complex system are plotted on a logarithmic scale, a surprising result occurs. The log-values are distributed with an approximately Gaussian distribution over frequency. This is illustrated in Fig. 11.16 for a beam net-

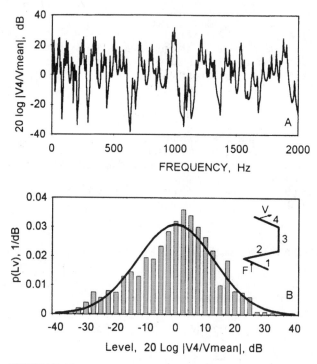

FIGURE 11.16 Numerical calculation of the vibration response of a four-beam network. (*A*) Normalized frequency response function for a point on beam 4. (*B*) Probability density function of log-levels, ▯ Numerical data histogram, — Normal distribution.

work. The system frequency response function is computed numerically using a transfer impedance model including bending and longitudinal and torsional motions in each of the four beam segments. A histogram of the computed response values on the decibel scale compares very well with a Gaussian distribution. This result can be explained by noting that the response value at any particular frequency results from the product of a large number of quantities. Then the logarithm of the response value will be the sum of a large number of terms. If the complexity in the system causes the responses at different frequencies to be independent, then by the central limit theorem the log-values will tend to have a Gaussian distribution. This means that the mean-square response values will have a log-normal distribution.

The calculated mean values and variances in the SEA model can be converted to the decibel scale as follows. If the mean-square velocity $\overline{\dot{y}^2}$ has a log-normal distribution with variance $\sigma_{\dot{y}^2}^2$, then the velocity level $L_{\dot{y}} \equiv 10\log_{10}(\dot{y}^2/\dot{y}_{ref}^2)$ has a normal distribution with a mean value and variance given by

$$L_{\dot{y}} = 10\log_{10}\left(\frac{\overline{\dot{y}^2}}{\dot{y}_{ref}^2}\right) - 5\log_{10}\left[1 + \frac{\sigma_{\dot{y}^2}^2}{(\overline{\dot{y}^2})^2}\right]$$

$$(11.88)$$

$$\sigma_{L_{\dot{y}}}^2 = 43\log_{10}\left[1 + \frac{\sigma_{\dot{y}^2}^2}{(\overline{\dot{y}^2})^2}\right]$$

Note that the mean of the decibel levels is not equal to the decibel level of the mean-square value.

TRANSIENT (SHOCK) RESPONSE USING SEA

The statistical energy analysis (SEA) model can solve for the transient response of a system using Eq. (11.61). The numerical solution methods for equations of this form can be illustrated using the finite difference method. Given an initial energy state $E(0)$, the energy state at a short time later is approximated by

$$E(\Delta t) \simeq E(0) + \frac{dE}{dt}\Delta t \qquad (11.89)$$

where $dE/dt = \Pi_{in} - \Pi_{out}$. This new energy distribution is then used to project forward to the next time step, etc. The accuracy of the solution depends on the size of Δt rel-

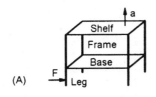

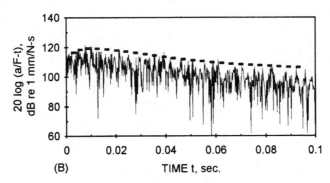

FIGURE 11.17 Transient response of an equipment shelf. (*A*) Experimental structure showing the locations of the impact *F* and the acceleration response *a*. (*B*) Comparison of the transient response of the structure; — Measured data, - - - SEA model.

ative to the energy flow time constants in the system, $(2\pi f\eta)^{-1}$. For the finite difference solution, using $\Delta t \leq (6\pi f\eta)^{-1}$ usually provides accurate results.

An example of a transient analysis using SEA is shown in Fig. 11.17. The measured acceleration response of a shelf on an equipment rack for an impact at the leg is shown along with the corresponding transient SEA solution of Eq. (11.61). The energy level of the shelf builds up for the first 0.01 sec before beginning to decay.

Modeling the transient mean-square response with Eq. (11.39), the undamped shock spectrum for this response signal can be estimated using Eq. (11.42). Alternately, if a shock excitation is modeled as a time-dependent power input to the SEA model, then the peak response spectrum of the system components can be estimated directly from the maximum mean-square values in the transient SEA solution.

REFERENCES

1. Crandall, S. H., and W. D. Mark: "Random Vibration in Mechanical Systems," Academic Press, New York, 1963.
2. Lyon, R. H., and R. G. DeJong: "Theory and Application of Statistical Energy Analysis," 2d ed., Butterworth-Heinemann, Boston, 1995.
3. Caughey, T. K.: "Nonstationary Random Inputs and Responses," Chap. 3 in *Random Vibration,* vol. 2, S. H. Crandall (ed.), M.I.T. Press, Cambridge, MA, 1963.
4. Cremer, L., M. Heckl, and E. E. Ungar: "Structure-Borne Sound," 2d ed., Springer-Verlag, Berlin, 1988.

CHAPTER 12
VIBRATION TRANSDUCERS

Anthony S. Chu
Eldon E. Eller
Robert M. Whittier

INTRODUCTION

This chapter is the first in a group of seven chapters on the measurement of shock and vibration. The following chapters describe in detail various types of instruments and measurement systems, and set forth primary considerations in the calibration and field use of such instruments and systems. This chapter defines the terms and describes the general principles of piezoelectric and piezoresistive transducers; it also sets forth the mathematical basis for the use of shock and vibration transducers and includes a brief description of piezoelectric accelerometers, piezoresistive accelerometers, piezoelectric force and impedance gages, and piezoelectric drivers, along with a review of their performance and characteristics. Finally, the following various special types of transducers are considered: optical-electronic transducers, including laser Doppler vibrometers, displacement measurement systems, fiber-optic reflective displacement sensors, electrodynamic (velocity coil) pickups, differential-transformer pickups, servo accelerometers, and capacitance-type transducers.

Certain solid-state materials are electrically responsive to mechanical force; they often are used as the mechanical-to-electrical transduction elements in shock and vibration transducers. Generally exhibiting high elastic stiffness, these materials can be divided into two categories: the *self-generating type,* in which electric charge is generated as a direct result of applied force, and the *passive-circuit type,* in which applied force causes a change in the electrical characteristics of the material.

A *piezoelectric* material is one which produces an electric charge proportional to the stress applied to it, within its linear elastic range. Piezoelectric materials are of the self-generating type. A *piezoresistive* material is one whose electrical resistance depends upon applied force. Piezoresistive materials are of the passive-circuit type.

A *transducer* (sometimes called a *pickup* or *sensor*) is a device which converts shock or vibratory motion into an optical, a mechanical, or, most commonly, an electrical signal that is proportional to a parameter of the experienced motion.

A *transducing element* is the part of the transducer that accomplishes the conversion of motion into the signal.

A *measuring instrument* or *measuring system* converts shock and vibratory motion into an observable form that is directly proportional to a parameter of the experienced motion. It may consist of a transducer with transducing element, signal-conditioning equipment, and device for displaying the signal. An instrument contains all of these elements in one package, while a system utilizes separate packages.

An *accelerometer* is a transducer whose output is proportional to the acceleration input. The output of a force gage is proportional to the force input; an impedance gage contains both an accelerometer and a force gage.

CLASSIFICATION OF MOTION TRANSDUCERS

In principle, shock and vibration motions are measured with reference to a point fixed in space by either of two fundamentally different types of transducers:

1. *Fixed-reference transducer.* One terminal of the transducer is attached to a point that is fixed in space; the other terminal is attached (e.g., mechanically, electrically, optically) to the point whose motion is to be measured.

2. *Mass-spring transducer (seismic transducer).* The only terminal is the base of a mass-spring system; this base is attached at the point where the shock or vibration is to be measured. The motion at the point is inferred from the motion of the mass relative to the base.

MASS-SPRING TRANSDUCERS (SEISMIC TRANSDUCERS)

In many applications, such as moving vehicles or missiles, it is impossible to establish a fixed reference for shock and vibration measurements. Therefore, many transducers use the response of a mass-spring system to measure shock and vibration. A mass-spring transducer is shown schematically in Fig. 12.1; it consists of a mass m suspended from the transducer case a by a spring of stiffness k. The motion of the mass within the case may be damped by a viscous fluid or electric current, symbolized by a dashpot with damping coefficient c. It is desired to measure the motion of the moving part whose displacement with respect to fixed space is indicated by u. When the transducer case is attached to the moving part, the transducer may be used to measure displacement, velocity, or acceleration, depending on the portion of the frequency range which is utilized and whether the relative displacement or relative velocity $d\delta/dt$ is sensed by the transducing element. The typical response of the mass-spring system is analyzed in the following paragraphs and applied to the interpretation of transducer output.

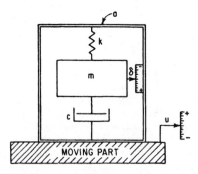

FIGURE 12.1 Mass-spring type of vibration-measuring instrument consisting of a mass m supported by spring k and viscous damper c. The case a of the instrument is attached to the moving part whose vibratory motion u is to be measured. The motion u is inferred from the relative motion δ between the mass m and the case a.[1]

Consider a transducer whose case experiences a displacement motion u, and let the relative displacement between the mass and the case be δ. Then the motion of the mass with respect to a reference fixed in space is $\delta + u$, and the force causing its acceleration is $m[d^2(\delta + u)/dt^2]$. Thus, the force applied by the mass to the spring and dashpot assembly is $-m[d^2(\delta + u)/dt^2]$. The force applied by the spring is $-k\delta$, and the force applied by the damper is $-c(d\delta/dt)$, where c is the damping coefficient. Adding all force terms and equating the sum to zero,

$$-m\frac{d^2(\delta + u)}{dt^2} - c\frac{d\delta}{dt} - k\delta = 0 \tag{12.1}$$

Equation (12.1) may be rearranged:

$$m\frac{d^2\delta}{dt^2} + c\frac{d\delta}{dt} + k\delta = -m\frac{d^2u}{dt^2} \tag{12.2}$$

Assume that the motion u is sinusoidal, $u = u_0 \cos \omega t$, where $\omega = 2\pi f$ is the angular frequency in radians per second and f is expressed in cycles per second. Neglecting transient terms, the response of the instrument is defined by $\delta = \delta_0 \cos (\omega t - \theta)$; then the solution of Eq. (12.2) is

$$\frac{\delta_0}{u_0} = \frac{\omega^2}{\sqrt{\left(\dfrac{k}{m} - \omega^2\right)^2 + \left(\omega\dfrac{c}{m}\right)^2}} \tag{12.3}$$

$$\theta = \tan^{-1}\frac{\omega\dfrac{c}{m}}{\dfrac{k}{m} - \omega^2} \tag{12.4}$$

The undamped natural frequency f_n of the instrument is the frequency at which

$$\frac{\delta_0}{u_0} = \infty$$

when the damping is zero ($c = 0$), or the frequency at which $\theta = 90°$. From Eqs. (12.3) and (12.4), this occurs when the denominators are zero:

$$\omega_n = 2\pi f_n = \sqrt{\frac{k}{m}} \qquad \text{rad/sec} \tag{12.5}$$

Thus, a stiff spring and/or light mass produces an instrument with a high natural frequency. A heavy mass and/or compliant spring produces an instrument with a low natural frequency.

The damping in a transducer is specified as a *fraction of critical damping*. Critical damping c_c is the minimum level of damping that prevents a mass-spring transducer from oscillating when excited by a step function or other transient. It is defined by

$$c_c = 2\sqrt{km} \tag{12.6}$$

Thus, the fraction of critical damping ζ is

$$\zeta = \frac{c}{c_c} = \frac{c}{2\sqrt{km}} \tag{12.7}$$

It is convenient to define the excitation frequency ω for a transducer in terms of the undamped natural frequency ω_n by using the dimensionless frequency ratio

ω/ω_n. Substituting this ratio and the relation defined by Eq. (12.7), Eqs. (12.3) and (12.4) may be written

$$\frac{\delta_0}{u_0} = \frac{\left(\dfrac{\omega}{\omega_n}\right)^2}{\sqrt{\left[1 - \left(\dfrac{\omega}{\omega_n}\right)^2\right]^2 + \left(2\zeta\dfrac{\omega}{\omega_n}\right)^2}} \tag{12.8}$$

$$\theta = \tan^{-1}\frac{2\zeta\dfrac{\omega}{\omega_n}}{1 - \left(\dfrac{\omega}{\omega_n}\right)^2} \tag{12.9}$$

The response of the mass-spring transducer given by Eq. (12.8) may be expressed in terms of the acceleration $\ddot{u}$ of the moving part by substituting $\ddot{u}_0 = -u_0\omega^2$. Then the ratio of the relative displacement amplitude δ_0 between the mass m and transducer case a to the impressed acceleration amplitude $\ddot{u}_0$ is

$$\frac{\delta_0}{\ddot{u}_0} = -\frac{1}{\omega_n^2}\left[\frac{1}{\sqrt{\left[1 - \left(\dfrac{\omega}{\omega_n}\right)^2\right]^2 + \left(2\zeta\dfrac{\omega}{\omega_n}\right)^2}}\right] \tag{12.10}$$

The relation between δ_0/u_0 and the frequency ratio ω/ω_n is shown graphically in Fig. 12.2 for several values of the fraction of critical damping ζ. Corresponding curves for $\delta_0/\ddot{u}_0$ are shown in Fig. 12.3. The phase angle θ defined by Eq. (12.9) is shown graphically in Fig. 12.4, using the scale at the left side of the figure. Corresponding phase angles between the relative displacement δ and the velocity u and acceleration $\ddot{u}$ are indicated by the scales at the right side of the figure.

ACCELERATION-MEASURING TRANSDUCERS

As indicated in Fig. 12.3, the relative displacement amplitude δ_0 is directly proportional to the acceleration amplitude $\ddot{u}_0 = -u_0\omega^2$ of the sinusoidal vibration being measured, at small values of the frequency ratio ω/ω_n. Thus, when the natural frequency ω_n of the transducer is high, the transducer is an accelerometer. If the transducer is undamped, the response curve of Fig. 12.3 is substantially flat when $\omega/\omega_n < 0.2$, approximately. Consequently, an undamped accelerometer can be used for the measurement of acceleration when the vibration frequency does not exceed approximately 20 percent of the natural frequency of the accelerometer. The range of measurable frequency increases as the damping of the accelerometer is increased, up to an optimum value of damping. When the fraction of critical damping is approximately 0.65, an accelerometer gives accurate results in the measurement of vibration at frequencies as great as approximately 60 percent of the natural frequency of the accelerometer.

As indicated in Fig. 12.3, the useful frequency range of an accelerometer increases as its natural frequency ω_n increases. However, the deflection of the spring in an accelerometer is inversely proportional to the square of the natural frequency; i.e., for a given value of $\ddot{u}_0$, the relative displacement is directly proportional to $1/\omega_n^2$ [see Eq. (12.10)]. As a consequence, the electrical signal from the transducing element may be very small, thereby requiring a large amplification to increase the signal to a level at which recording is feasible. For this reason, a compromise usually is

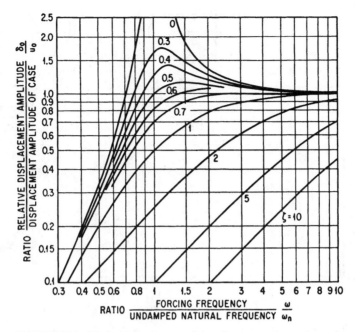

FIGURE 12.2 Displacement response δ_0/u_0 of a mass-spring system subjected to a sinusoidal displacement $\ddot{u} = u_0 \sin \omega t$. The fraction of critical damping ζ is indicated for each curve.

made between high sensitivity and the highest attainable natural frequency, depending upon the desired application.

ACCELEROMETER REQUIREMENTS FOR SHOCK

High-Frequency Response. The capability of an accelerometer to measure shock may be evaluated by observing the response of the accelerometer to acceleration pulses. Ideally, the response of the accelerometer (i.e., the output of the transducing element) should correspond identically with the pulse. In general, this result may be approached but not attained exactly. Three typical pulses and the corresponding responses of accelerometers are shown in Fig. 12.5 to 12.7. The pulses are shown in dashed lines. A sinusoidal pulse is shown in Fig. 12.5, a triangular pulse in Fig. 12.6, and a rectangular pulse in Fig. 12.7. Curves of the response of the accelerometer are shown in solid lines. For each of the three pulse shapes, the response is given for ratios τ_n/τ of 1.014 and 0.203, where τ is the pulse duration and $\tau_n = 1/f_n$ is the natural period of the accelerometer. These response curves, computed for the fraction of critical damping $\zeta = 0, 0.4, 0.7$, and 1.0, indicate the following general relationships:

1. The response of the accelerometer follows the pulse most faithfully when the natural period of the accelerometer is smallest relative to the period of the pulse. For example, the responses at A in Figs. 12.5 to 12.7 show considerable deviation

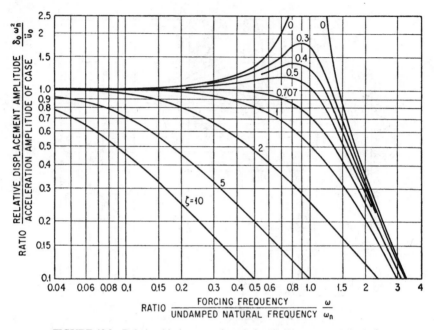

FIGURE 12.3 Relationship between the relative displacement amplitude δ_0 of a mass-spring system and the acceleration amplitude $\ddot{u}_0$ of the case. The fraction of critical damping ζ is indicated for each response curve.

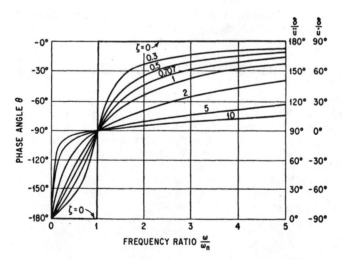

FIGURE 12.4 Phase angle of a mass-spring transducer when used to measure sinusoidal vibration. The phase angle θ on the left-hand scale relates the relative displacement δ to the impressed displacement, as defined by Eq. (12.9). The right-hand scales relate the relative displacement δ to the impressed velocity and acceleration.

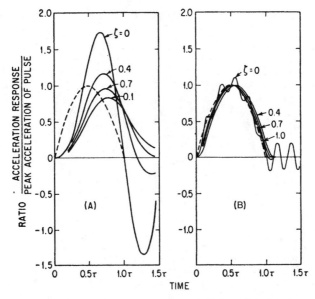

FIGURE 12.5   Acceleration response to a half-sine pulse of acceleration of duration τ (dashed curve) of a mass-spring transducer whose natural period τ_n is equal to: (*A*) 1.014 times the duration of the pulse and (*B*) 0.203 times the duration of the pulse. The fraction of critical damping ζ is indicated for each response curve. (*Levy and Kroll.*[1])

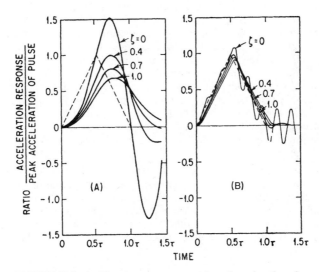

FIGURE 12.6   Acceleration response to a triangular pulse of acceleration of duration τ (dashed curve) of a mass-spring transducer whose natural period is equal to: (*A*) 1.014 times the duration of the pulse and (*B*) 0.203 times the duration of the pulse. The fraction of critical damping ζ is indicated for each response curve. (*Levy and Kroll.*[1])

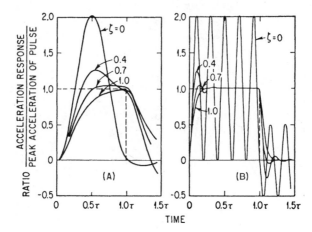

FIGURE 12.7 Acceleration response to a rectangular pulse of acceleration of duration τ (dashed curve) of a mass-spring transducer whose natural period τ_n is equal to: (A) 1.014 times the duration of the pulse and (B) 0.203 times the duration of the pulse. The fraction of critical damping ζ is indicated for each response curve. (*Levy and Kroll.*[1])

between the pulse and the response; this occurs when τ_n is approximately equal to τ. However, when τ_n is small relative to τ (Figs. 12.5*B* to 12.7*B*), the deviation between the pulse and the response is much smaller. If a shock is generated by metal-to-metal impact or by a pyrotechnic device such as that described in Chap. 26, Part II, *and* the response accelerometer is located in close proximity to the excitation source(s), the initial pulses of acceleration may have an extremely fast rise time and high amplitude. In such cases, any type of mass-spring accelerometer may not accurately follow the leading wave front and characterize the shock inputs faithfully. For example, measurements made in the near field of a high-*g* shock show that undamped piezoresistive accelerometers having resonance above 1 MHz were excited at resonance, thereby invalidating the measured responses. To avoid this effect, accelerometers should be placed as far away as possible, or practical, from the source of excitation. Other considerations related to accelerometer resonance are discussed below in the sections on *Zero Shift* and *Survivability*.

2. Damping in the transducer reduces the response of the transducer at its own natural frequency; i.e., it reduces the transient vibration superimposed upon the pulse, which is sometimes referred to as *ringing*. Damping also reduces the maximum value of the response to a value lower than the actual pulse in the case of large damping. For example, in some cases a fraction of critical damping $\zeta = 0.7$ provides an instrument response that does not reach the peak value of the acceleration pulse.

Low-Frequency Response. The measurement of shock requires that the accelerometer and its associated equipment have good response at low frequencies because pulses and other types of shock motions characteristically include low-frequency components. Such pulses can be measured accurately only with an instrumentation system whose response is flat down to the lowest frequency of the spectrum; in general, this lowest frequency is zero for pulses.

The response of an instrumentation system is defined by a plot of output voltage vs. excitation frequency. For purposes of shock measurement, the decrease in response at low frequencies is significant. The decrease is defined quantitatively by the frequency f_c at which the response is down 3 dB or approximately 30 percent below the flat response which exists at the higher frequencies. The distortion which occurs in the measurement of a pulse is related to the frequency f_c as illustrated in Fig. 12.8.

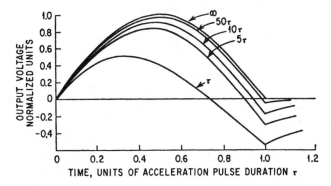

FIGURE 12.8 Response of an accelerometer to a half-sine acceleration pulse for RC time constants equal to τ, 5τ, 10τ, 50τ, and ∞, where τ is equal to the duration of the half-sine pulse.[1]

This is particularly important when acceleration data are integrated to obtain velocity, or integrated twice to obtain displacement. A small amount of undershoot shown in Fig. 12.8 may cause a large error after integration. A dc-coupled accelerometer (such as a piezoresistive accelerometer, described later in this chapter) is recommended for this type of application.

Zero Shift. Zero shift is the displacement of the zero-reference line of an accelerometer after it has been exposed to a very intense shock. This is illustrated in Fig. 12.9. The loss of zero reference and the apparent dc components in the time history cause a problem in peak-value determination and induce errors in shock response spectrum calculations. Although the accelerometer is not the sole source of zero shift, it is the main contributor.

All piezoelectric shock accelerometers, under extreme stress load (e.g., a sensing element at resonance), will exhibit zero-shift phenomena due either to crystal domain switching or to a sudden change in crystal preload condition.[2] A mechanical filter may be used to protect the crystal element(s) at the expense of a limitation in bandwidth or possible nonlinearity.[3] Piezoresistive shock accelerometers typically produce negligible zero shift.

Survivability. Survivability is the ability of an accelerometer to withstand intense shocks without affecting its performance. An accelerometer is usually rated in terms of the maximum value of acceleration it can withstand. Accelerometers used for shock measurements may have a range of well over many thousands of *g*s. In piezoresistive accelerometers which are excited at resonance, the stress buildup

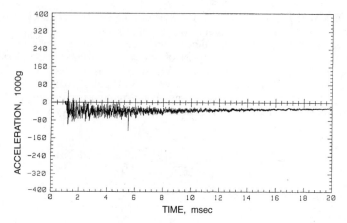

FIGURE 12.9 A time history of an accelerometer that has been exposed to a pyrotechnic shock. Note that there is a shift in the baseline (i.e., the zero reference) of the accelerometer as a result of this shock; the shift may either be positive or negative.

due to high magnitudes of acceleration may lead to fracture of the internal components. In contrast, piezoelectric accelerometers are more robust than their piezoresistive counterparts due to lower internal stress.

IMPORTANT CHARACTERISTICS OF ACCELEROMETERS

SENSITIVITY

The *sensitivity* of a shock- and vibration-measuring instrument is the ratio of its electrical output to its mechanical input. The output usually is expressed in terms of voltage per unit of displacement, velocity, or acceleration. This specification of sensitivity is sufficient for instruments which generate their own voltage independent of an external voltage power source. However, the sensitivity of an instrument requiring an external voltage usually is specified in terms of output voltage per unit of voltage supplied to the instrument per unit of displacement, velocity, or acceleration, e.g., millivolts per volt per g of acceleration. It is important to note the terms in which the respective parameters are expressed, e.g., average, rms, or peak. The relation between these terms is shown in Fig. 12.10. Also see Table 1.3.

RESOLUTION

The *resolution* of a transducer is the smallest change in mechanical input (e.g., acceleration) for which a change in the electrical output is discernible. The resolution of an accelerometer is a function of the transducing element and the mechanical design.

Recording equipment, indicating equipment, and other auxiliary equipment used with accelerometers often establish the resolution of the overall measurement sys-

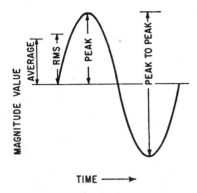

TIME ⟶

FIGURE 12.10 Relationships between average, rms, peak, and peak-to-peak values for a simple sine wave. These values are used in specifying sensitivities of shock and vibration transducers (e.g., peak millivolts per peak g, or rms millivolts per peak-to-peak displacement). These relationships do not hold true for other than simple sine waves.

tem. If the electrical output of an instrument is indicated by a meter, the resolution may be established by the smallest increment that can be read from the meter. Resolution can be limited by noise levels in the instrument or in the system. In general, any signal change smaller than the noise level will be obscured by the noise, thus determining the resolution of the system.

TRANSVERSE SENSITIVITY

If a transducer is subjected to vibration of unit amplitude along its axis of maximum sensitivity, the amplitude of the voltage output e_{max} is the sensitivity. The sensitivity e_θ along the X axis, inclined at an angle θ to the axis of e_{max}, is $e_\theta = e_{max} \cos \theta$, as illustrated in Fig. 12.11. Similarly, the sensitivity along the Y axis is $e_t = e_{max} \sin \theta$. In general, the sensitive axis of a transducer is designated. Ideally, the X axis would be designated the sensitive axis, and the angle θ would be zero. Practically, θ can be made only to approach zero because of manufacturing tolerances and/or unpredictable variations in the characteristics of the transducing element. Then the transverse sensitivity (cross-axis sensitivity) is expressed as the tangent of the angle, i.e., the ratio of e_t to e_θ:

$$\frac{e_t}{e_\theta} = \tan \theta \tag{12.11}$$

In practice, $\tan \theta$ is between 0.01 and 0.05 and is expressed as a percentage. For example, if $\tan \theta = 0.05$, the transducer is said to have a transverse sensitivity of 5 percent. Figure 12.12 is a typical polar plot of transverse sensitivity.

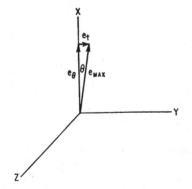

FIGURE 12.11 The designated sensitivity e_θ and cross-axis sensitivity e_t that result when the axis of maximum sensitivity e_{max} is not aligned with the axis of e_θ.

AMPLITUDE LINEARITY AND LIMITS

When the ratio of the electrical output of a transducer to the mechanical input (i.e., the sensitivity) remains constant within specified limits, the transducer is said to be "linear" within those limits, as illustrated in Fig. 12.13. A transducer is linear only over a certain range of amplitude values. The lower end of this range is determined by the electrical noise of the measurement system.

The upper limit of linearity may be imposed by the electrical characteristics

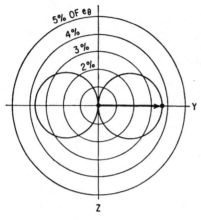

FIGURE 12.12 Plot of transducer sensitivity in all axes normal to the designated axis e_θ plotted according to axes shown in Fig. 12.11. Cross-axis sensitivity reaches a maximum e_t along the Y axis and a minimum value along the Z axis.

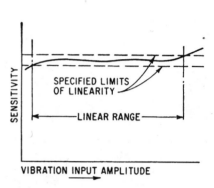

FIGURE 12.13 Typical plot of sensitivity as a function of amplitude for a shock and vibration transducer. The *linear range* is established by the intersection of the sensitivity curve and the specified limits (dashed lines).

of the transducing element and by the size or the fragility of the instrument. Generally, the greater the sensitivity of a transducer, the more nonlinear it will be. Similarly, for very large acceleration values, the large forces produced by the spring of the mass-spring system may exceed the yield strength of a part of the instrument, causing nonlinear behavior or complete failure.

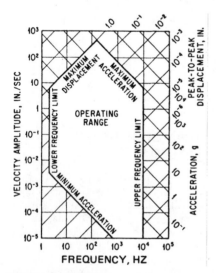

FIGURE 12.14 Linear operating range of a transducer. Amplitude linearity limits are shown as a combination of displacement and acceleration values. The lower amplitude limits usually are expressed in acceleration values as shown.

FREQUENCY RANGE

The operating frequency range is the range over which the sensitivity of the transducer does not vary more than a stated percentage from the rated sensitivity. This range may be limited by the electrical or mechanical characteristics of the transducer or by its associated auxiliary equipment. These limits can be added to amplitude linearity limits to define completely the operating ranges of the instrument, as illustrated in Fig. 12.14.

Low-Frequency Limit. The mechanical response of a mass-spring transducer does not impose a low-frequency limit for an acceleration transducer because the transducer responds to vibration with frequencies less than the natural frequency of the transducer.

In evaluating the low-frequency limit, it is necessary to consider the electrical characteristics of both the transducer and the associated equipment. In general, a transducing element that utilizes external power or a carrier voltage does not have a lower frequency limit, whereas a self-generating transducing element is not operative at zero frequency. The frequency response of amplifiers and other circuit components may limit the lowest usable frequency of an instrumentation system.

High-Frequency Limit. An acceleration transducer (accelerometer) has an upper usable frequency limit because it responds to vibration whose frequency is less than the natural frequency of the transducer. The limit is a function of (1) the natural frequency and (2) the damping of the transducer, as discussed with reference to Fig. 12.3. An attempt to use such a transducer beyond this frequency limit may result in distortion of the signal, as illustrated in Fig. 12.15.

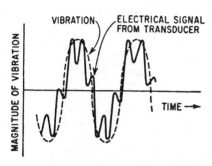

FIGURE 12.15 Distorted response (solid line) of a lightly damped ($\zeta < 0.1$) mass-spring accelerometer to vibration (dashed line) containing a small harmonic content of the small frequency as the natural frequency of the accelerometer.

The upper frequency limit for slightly damped vibration-measuring instruments is important because these instruments exaggerate the small amounts of harmonic content that may be contained in the motion, even when the operating frequency is well within the operating range of the instrument. The result of exciting an undamped instrument at its natural frequency may be to either damage the instrument or obscure the desired measurement. Figure 12.15 shows how a small amount of harmonic distortion in the vibratory motion may be exaggerated by an undamped transducer.

Phase Shift. Phase shift is the time delay between the mechanical input and the electrical output signal of the instrumentation system. Unless the phase-shift characteristics of an instrumentation system meet certain requirements, a distortion may be introduced that consists of the superposition of vibration at several different frequencies. Consider first an accelerometer, for which the phase angle θ_1 is given by Fig. 12.4. If the accelerometer is undamped, $\theta_1 = 0$ for values of ω/ω_n less than 1.0; thus, the phase of the relative displacement δ is equal to that of the acceleration being measured, for all values of frequency within the useful range of the accelerometer. Therefore, an undamped accelerometer measures acceleration without distortion of phase. If the fraction of critical damping ζ for the accelerometer is 0.65, the phase angle θ_1 increases approximately linearly with the frequency ratio ω/ω_n within the useful frequency range of the accelerometer. Then the expression for the relative displacement may be written

$$\delta = \delta_0 \cos (\omega t - \theta) = \delta_0 \cos (\omega t - a\omega) = \delta_0 \cos \omega(t - a) \qquad (12.12)$$

where a is a constant. Thus, the relative motion δ of the instrument is displaced in phase relative to the acceleration $\ddot{u}$ being measured; however, the increment along the time axis is a constant independent of frequency. Consequently, the waveform of the accelerometer output is undistorted but is delayed with respect to the waveform of the vibration being measured. As indicated by Fig. 12.4, any value of damping in an accelerometer other than $\zeta = 0$ or $\zeta = 0.65$ (approximately) results in a nonlinear shift of phase with frequency and a consequent distortion of the waveform.

ENVIRONMENTAL EFFECTS

Temperature. The sensitivity, natural frequency, and damping of a transducer may be affected by temperature. The specific effects produced depend on the type of transducer and the details of its design. The sensitivity may increase or decrease with temperature, or remain relatively constant. Figure 12.16 shows the variation of damping with temperature for several different damping media. Either of two methods may be employed to compensate for temperature effects.

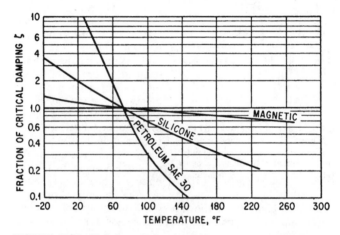

FIGURE 12.16 Variation of damping with temperature for different damping means. The ordinate indicates the fraction of critical damping ζ at various temperatures assuming $\zeta = 1$ at 70°F (21°C).

1. The temperature of the pickup may be held constant by local heating or cooling.
2. The pickup characteristics may be measured as a function of temperature; if necessary, the appropriate corrections can then be applied to the measured data.

Humidity. Humidity may affect the characteristics of certain types of vibration instruments. In general, a transducer which operates at a high electrical impedance is affected by humidity more than a transducer which operates at a low electrical impedance. It usually is impractical to correct the measured data for humidity effects. However, instruments that might otherwise be adversely affected by humidity often are sealed hermetically to protect them from the effects of moisture.

Acoustic Noise. High-intensity sound waves often accompany high-amplitude vibration. If the case of an accelerometer can be set into vibration by acoustic excitation, error signals may result. In general, a well-designed accelerometer will not produce a significant electrical response except at extremely high sound pressure levels. Under such circumstances, it is likely that vibration levels also will be very high, so that the error produced by the accelerometer's exposure to acoustic noise usually is not important.

Strain Sensitivity.   An accelerometer may generate a spurious output when its case is strained or distorted. Typically this occurs when the transducer mounting is not flat against the surface to which it is attached, and so this effect is often called *base-bend sensitivity* or *strain sensitivity*. It is usually reported in equivalent g per microstrain, where 1 *microstrain* is 1×10^{-6} inch per inch. The Instrument Society of America recommends a test procedure that determines strain sensitivity at 250 microstrain.[4]

An accelerometer with a sensing element which is tightly coupled to its base tends to exhibit large strain sensitivity. An error due to strain sensitivity is most likely to occur when the accelerometer is attached to a structure which is subject to large amounts of flexure. In such cases, it is advisable to select an accelerometer with low strain sensitivity.

PHYSICAL PROPERTIES

Size and weight of the transducer are very important considerations in many vibration and shock measurements. A large instrument may require a mounting structure that will change the local vibration characteristics of the structure whose vibration is being measured. Similarly, the added mass of the transducer may also produce substantial changes in the vibratory response of such a structure. Generally, the natural frequency of a structure is lowered by the addition of mass; specifically, for a simple spring-mass structure:

$$\frac{f_n - \Delta f_n}{f_n} = \sqrt{\frac{m}{m + \Delta m}} \tag{12.13}$$

where f_n = natural frequency of structure
Δf_n = change in natural frequency
m = mass of structure
Δm = increase in mass resulting from addition of transducer

In general, for a given type of transducing element, the sensitivity increases approximately in proportion to the mass of the transducer. In most applications, it is more important that the transducer be small in size than that it have high sensitivity because amplification of the signal increases the output to a usable level.

Mass-spring-type transducers for the measurement of displacement usually are larger and heavier than similar transducers for the measurement of acceleration. In the former, the mass must remain substantially stationary in space while the instrument case moves about it; this requirement does not exist with the latter.

For the measurement of shock and vibration in aircraft or missiles, the size and weight of not only the transducer but also the auxiliary equipment are important. In these applications, self-generating instruments that require no external power may have a significant advantage.

PIEZOELECTRIC ACCELEROMETERS

PRINCIPLE OF OPERATION

An accelerometer of the type shown in Fig. 12.17A is a linear seismic transducer utilizing a piezoelectric element in such a way that an electric charge is produced which is proportional to the applied acceleration. This "ideal" seismic piezoelectric transducer can be represented (over most of its frequency range) by the elements shown in Fig. 12.17*B*. A mass is supported on a linear spring which is fastened to the frame of the instrument. The piezoelectric crystal which produces the charge acts as the spring. Viscous damping between the mass and the frame is represented by the dashpot *c*. In Fig. 12.17*C* the frame is given an acceleration upward to a displacement of *u*, thereby producing a compression in the spring equal to δ. The displacement of the mass relative to the frame is dependent upon the applied acceleration of the frame, the spring stiffness, the mass, and the viscous damping between the mass and the frame, as indicated in Eq. (12.10) and illustrated in Fig. 12.3.

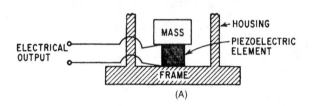

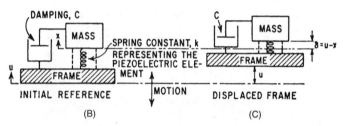

FIGURE 12.17 (*A*) Schematic diagram of a linear seismic piezoelectric accelerometer. (*B*) A simplified representation of the accelerometer shown in (*A*) which applies over most of the useful frequency range. A mass *m* rests on the piezoelectric element, which acts as a spring having a spring constant *k*. The damping in the system, represented by the dashpot, has a damping coefficient *c*. (*C*) The frame is accelerated upward, producing a displacement *u* of the frame, moving the mass from its initial position by an amount *x*, and compressing the spring by an amount δ.

For frequencies far below the resonance frequency of the mass and spring, this displacement is directly proportional to the acceleration of the frame and is independent of frequency. At low frequencies, the phase angle of the relative displacement δ, with respect to the applied acceleration, is proportional to frequency. As indicated in Fig. 12.4, for low fractions of critical damping which are characteristic of many piezoelectric accelerometers, the phase angle is proportional to frequency at frequencies below 30 percent of the resonance frequency.

In Fig. 12.17, inertial force of the mass causes a mechanical strain in the piezoelectric element, which produces an electric charge proportional to the stress and, hence, proportional to the strain and acceleration. If the dielectric constant of the piezoelectric material does not change with electric charge, the voltage generated is also proportional to acceleration. Metallic electrodes are applied to the piezoelectric element, and electrical leads are connected to the electrodes for measurement of the electrical output of the piezoelectric element.

In the ideal seismic system shown in Fig. 12.17, the mass and the frame have infinite stiffness, the spring has zero mass, and viscous damping exists only between the mass and the frame. In practical piezoelectric accelerometers, these assumptions cannot be fulfilled. For example, the mass may have as much compliance as the piezoelectric element. In some seismic elements, the mass and spring are inherently a single structure. Furthermore, in many practical designs where the frame is used to hold the mass and piezoelectric element, distortion of the frame may produce mechanical forces upon the seismic element. All these factors may change the performance of the seismic system from those calculated using equations based on an ideal system. In particular, the resonance frequency of the piezoelectric combination may be substantially lower than that indicated by theory. Nevertheless, the equations for an ideal system are useful both in design and application of piezoelectric accelerometers.

Figure 12.18 shows a typical frequency response curve for a piezoelectric accelerometer. In this illustration, the electrical output in millivolts per g acceleration is plotted as a function of frequency. The resonance frequency is denoted by f_n. If the accelerometer is properly mounted on the device being tested, then the upper frequency limit of the useful frequency range usually is taken to be $f_n/3$ for a deviation of 12 percent (1 dB) from the mean value of the response. For a deviation of 6 percent (0.5 dB) from the mean value, the upper frequency limit usually is taken to be $f_n/5$. As indicated in Fig. 12.1, the type of mounting can have a significant effect on the value of f_n.

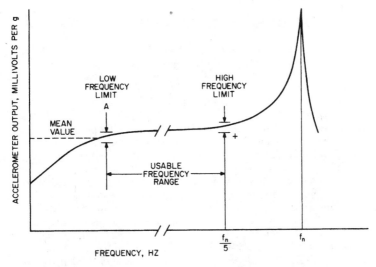

FIGURE 12.18   Typical response curve for a piezoelectric accelerometer. The resonance frequency is denoted by f_n. The useful range depends on the acceptable deviation from the mean value of the response over the "flat" portion of the response curve.

The decrease in response at low frequencies (i.e., the "rolloff") depends primarily on the characteristics of the preamplifier that follows the accelerometer. The low-frequency limit also is usually expressed in terms of the deviation from the mean value of the response over the flat portion of the response curve, being the frequency at which the response is either 12 percent (1 dB) or 6 percent (0.5 dB) below the mean value.

PIEZOELECTRIC MATERIALS

A polarized ceramic called lead zirconate titanate (PZT) is most commonly used in piezoelectric accelerometers. It is low in cost, high in sensitivity, and useful in the temperature range from −180° to +550°F (−100° to +288°C). Polarized ceramics in the bismuth titanate family have substantially lower sensitivities than PZT, but they also have more stable characteristics and are useful at temperatures as high as 1000°F (538°C).

Quartz, the single-crystal material most widely used in accelerometers, has a substantially lower sensitivity than polarized ceramics, but its characteristics are very stable with time and temperature; it has high resistivity. Lithium niobate and tourmaline are single-crystal materials that can be used in accelerometers at high temperatures: lithium niobate up to at least 1200°F (649°C), and tourmaline up to at least 1400°F (760°C). The upper limit of the useful range is usually set by the thermal characteristics of the structural materials rather than by the characteristics of these two crystalline materials.

Polarized polyvinylidene fluoride (PVDF), an engineering plastic similar to Teflon, is used as the sensing element in some accelerometers. It is inexpensive, but it is generally less stable with time and with temperature changes than ceramics or single-crystal materials. In fact, because PVDF materials are highly pyroelectric, they are used as thermal sensing devices.

TYPICAL PIEZOELECTRIC ACCELEROMETER CONSTRUCTIONS

Piezoelectric accelerometers utilize a variety of seismic element configurations. Most are constructed of polycrystalline ceramic piezoelectric materials because of their ease of manufacture, high piezoelectric sensitivity, and excellent time and temperature stability. These seismic devices may be classified in two modes of operation: compression- or shear-type accelerometers.

Compression-type Accelerometer.   The compression-type seismic accelerometer, in its simplest form, consists of a piezoelectric disc and a mass placed on a frame as shown in Fig. 12.17. Motion in the direction indicated causes compressive (or tensile) forces to act on the piezoelectric element, producing an electrical output proportional to acceleration. In this example, the mass is cemented with a conductive material to the piezoelectric element which, in turn, is cemented to the frame. The components must be cemented firmly so as to avoid being separated from each other by the applied acceleration.

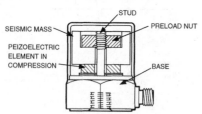

FIGURE 12.19 A typical compression-type piezoelectric accelerometer. The piezoelectric element(s) must be preloaded (biased) to produce an electrical output under both tension forces and compression forces. *(Courtesy of Endevco Corp.)*

In the typical commercial accelerometer shown in Fig. 12.19, the mass is held in place by means of a stud extending from the frame through the ceramic. Accelerometers of this design often use quartz, tourmaline, or ferroelectric ceramics as the sensing material.

This type of accelerometer must be attached to the structure with care in order to minimize distortion of the housing and base which can cause an electrical output. See the section on *Strain Sensitivity.*

The temperature characteristics of compression-type accelerometers have been improved greatly in recent years; it is now possible to measure acceleration over a temperature range of −425 to +1400°F (−254 to +760°C). This wider range has been primarily a result of the use of two piezoelectric materials: tourmaline and lithium niobate.

Shear-type Accelerometers. One shear-type accelerometer utilizes flat-plate shear-sensing elements. Manufacturers preload these against a flattened post element in several ways. Two methods are shown in Fig. 12.20. Accelerometers of this style have low cross-axis response, excellent temperature characteristics, and negligible output from strain sensitivity or base bending. The temperature range of the bolted shear design can be from −425 to +1400°F (−254 to +760°C). The following are typical specifications: sensitivity, 10 to 500 picocoulombs/g; acceleration range, 1 to 500g; resonance frequency, 25,000 Hz; useful frequency range, 3 to 5,000 Hz; temperature range, −425 to +1400°F (−254 to 760°C); transverse response, 3 percent.

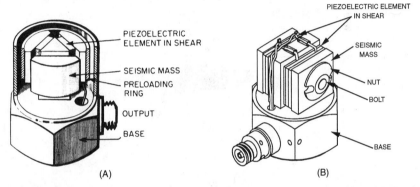

FIGURE 12.20 Piezoelectric accelerometers: (*A*) Delta-shear type. *(Courtesy of Bruel & Kjaer).* (*B*) Isoshear type. *(Courtesy of Endevco Corp.).*

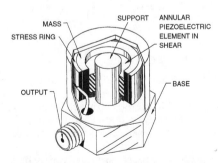

FIGURE 12.21 An annular shear accelerometer. The piezoelectric element is cemented to the post and mass. Electrical connections (not shown) are made to the inner and outer diameters of the piezoelectric element. (*Courtesy of Endevco Corp.*)

Another shear-type accelerometer, illustrated in Fig. 12.21, employs a cylindrically shaped piezoelectric element fitted around a middle mounting post; a loading ring (or mass) is cemented to the outer diameter of the piezoelectric element. The cylinder is made of ceramic and is polarized along its length; the output voltage of the accelerometer is taken from its inner and outer walls. This type of design can be made extremely small and is generally known as an axially poled shear-mode annular accelerometer.

Beam-type Accelerometers. The beam-type accelerometer is a variation of the compression-type accelerometer.

It is usually made from two piezoelectric plates which are rigidly bonded together to form a beam supported at one end, as illustrated in Fig. 12.22. As the beam flexes, the bottom element compresses, so that it increases in thickness. In contrast, the upper element expands, so that it decreases in thickness. Accelerometers of this type generate high electrical output for their size, but are more fragile and have a lower resonance frequency than most other designs.

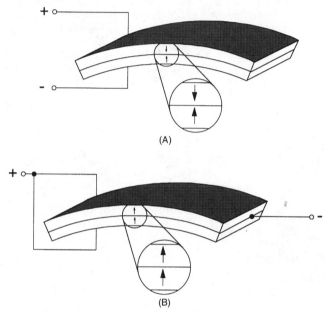

FIGURE 12.22   Configurations of piezoelectric elements in a beam-type accelerometer. (*A*) A series arrangement, in which the two elements have opposing directions of polarization. (*B*) A parallel arrangement, in which the two elements have the same direction of polarization.

PHYSICAL CHARACTERISTICS OF PIEZOELECTRIC ACCELEROMETERS

Shape, Size, and Weight.   Commercially available piezoelectric accelerometers usually are cylindrical in shape. They are available with both attached and detachable mounting studs at the bottom of the cylinder. A coaxial cable connector is provided at either the top or side of the housing.

Most commercially available piezoelectric accelerometers are relatively light in weight, ranging from approximately 0.005 to 4.2 oz (0.14 to 120 grams). Usually, the larger the accelerometer, the higher its sensitivity and the lower its resonance frequency. The smallest units have a diameter of less than about 0.2 in. (5 mm); the larger units have a diameter of about 1 in. (25.4 mm) and a height of about 1 in. (25.4 mm).

Resonance Frequency. The highest fundamental resonance frequency of an accelerometer may be above 100,000 Hz. The higher the resonance frequency, the lower will be the sensitivity and the more difficult it will be to provide mechanical damping.

Damping. The *amplification ratio* of an accelerometer is defined as the ratio of the sensitivity at its resonance frequency to the sensitivity in the frequency band in which sensitivity is independent of frequency. This ratio depends on the amount of damping in the seismic system; it decreases with increasing damping. Most piezoelectric accelerometers are essentially undamped, having amplification ratios between 20 and 100, or a fraction of critical damping less than 0.1.

ELECTRICAL CHARACTERISTICS OF PIEZOELECTRIC ACCELEROMETERS

Dependence of Voltage Sensitivity on Shunt Capacitance. The *sensitivity* of an accelerometer is defined as the electrical output per unit of applied acceleration. The sensitivity of a piezoelectric accelerometer can be expressed as either a *charge sensitivity* $q/\ddot{x}$ or *voltage sensitivity* $e/\ddot{x}$. Charge sensitivity usually is expressed in units of coulombs generated per g of applied acceleration; voltage sensitivity usually is expressed in volts per g (where g is the acceleration of gravity). Voltage sensitivity often is expressed as open-circuit voltage sensitivity, i.e., in terms of the voltage produced across the electrical terminals per unit acceleration when the electrical load impedance is infinitely high. Open-circuit voltage sensitivity may be given either with or without the connecting cable.

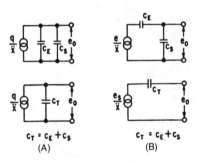

$C_T = C_E + C_S$
(A)

$C_T = C_E + C_S$
(B)

FIGURE 12.23 Equivalent circuits which include shunt capacitance across a piezoelectric pickup. (*A*) Charge equivalent circuit. (*B*) Voltage equivalent circuit.

An electrical capacitance often is placed across the output terminals of a piezoelectric transducer. This added capacitance (called *shunt capacitance*) may result from the connection of an electrical cable between the pickup and other electrical equipment (all electrical cables exhibit interlead capacitance). The effect of shunt capacitance in reducing the sensitivity of a pickup is shown in Fig. 12.23.

The charge equivalent circuits, with shunt capacitance C_S, are shown in Fig. 12.23A. The charge sensitivity is not changed by addition of shunt capacitance. The total capacitance C_T of the pickup including shunt is given by

$$C_T = C_E + C_S \qquad (12.14)$$

where C_E is the capacitance of the transducer without shunt capacitance.

The voltage equivalent circuits are shown in Fig. 12.23B. With the shunt capacitance C_S, the total capacitance is given by Eq. (12.14) and the open-circuit voltage sensitivity is given by

$$\frac{e_s}{\ddot{x}} = \frac{q_s}{\ddot{x}} \frac{1}{C_E + C_S} \qquad (12.15)$$

where $q_s/\ddot{x}$ is the charge sensitivity. The voltage sensitivity without shunt capacitance is given by

$$\frac{e}{\ddot{x}} = \frac{q_s}{\ddot{x}} \frac{1}{C_E} \qquad (12.16)$$

Therefore, the effect of the shunt capacitance is to reduce the voltage sensitivity by a factor

$$\frac{e_s/\ddot{x}}{e/\ddot{x}} = \frac{C_E}{C_E + C_S} \qquad (12.17)$$

Piezoelectric accelerometers are used with both voltage-sensing and charge-sensing signal conditioners, although charge sensing is by far the most common because the sensitivity does not change with external capacitance (up to a limit). These factors are discussed in Chap. 13. In addition, electronic circuitry can be placed within the case of the accelerometer, as discussed below.

LOW-IMPEDANCE PIEZOELECTRIC ACCELEROMETERS CONTAINING INTERNAL ELECTRONICS

Piezoelectric accelerometers are available with simple electronic circuits internal to their cases to provide signal amplification and low-impedance output. For example, see the charge preamplifier circuit shown in Fig. 13.2. Some designs operate from low-current dc voltage supplies and are designed to be intrinsically safe when coupled by appropriate barrier circuits. Other designs have common power and signal lines and use coaxial cables.

The principal advantages of piezoelectric accelerometers with integral electronics are that they are relatively immune to cable-induced noise and spurious response, they can be used with lower-cost cable, and they have a lower signal conditioning cost. In the simplest case the power supply might consist of a battery, a resistor, and a capacitor. Some such accelerometers provide a velocity or displacement output. These advantages do not come without compromise. Because the impedance-matching circuitry is built into the transducer, gain cannot be adjusted to utilize the wide dynamic range of the basic transducer. Ambient temperature is limited to that which the circuit will withstand, and this is considerably lower than that of the piezoelectric sensor itself. In order to retain the advantages of small size, the integral electronics must be kept relatively simple. This precludes the use of multiple filtering and dynamic overload protection and thus limits their application.

All other things being equal, the *reliability factor* (i.e., the mean time between failures) of any accelerometer with internal electronics is lower than that of an accelerometer with remote electronics, especially if the accelerometer is subject to abnormal environmental conditions. However, if the environmental conditions are fairly normal, accelerometers with internal electronics can provide excellent signal fidelity and immunity from noise. Internal electronics provides a reduction in overall system noise level because it minimizes the cable capacitance between the sensor and the signal conditioning electronics.

An accelerometer containing internal electronics that includes such additional features as self-testing, self-identification, and calibration data storage is sometimes referred to as a "smart accelerometer."

Velocity-Output Piezoelectric Devices.   Piezoelectric accelerometers are available with internal electronic circuitry which integrates the output signal provided by the accelerometer, thereby yielding a velocity or displacement output. These transducers have several advantages not possessed by ordinary velocity pickups. They are smaller, have a wider frequency response, have better resolution, have no moving parts, and are relatively unaffected by magnetic fields where measurements are made.

ACCELERATION-AMPLITUDE CHARACTERISTICS

Amplitude Range. Piezoelectric accelerometers are generally useful for the measurement of acceleration of magnitudes of from $10^{-6}g$ to more than $10^{5}g$. The lowest value of acceleration which can be measured is approximately that which will produce an output voltage equivalent to the electrical input noise of the coupling amplifier connected to the accelerometer when the pickup is at rest. Over its useful operating range, the output of a piezoelectric accelerometer is directly and continuously proportional to the input acceleration. A single accelerometer often can be used to provide measurements over a dynamic amplitude range of 90 dB or more, which is substantially greater than the dynamic range of some of the associated transmission, recording, and analysis equipment. Commercial accelerometers generally exhibit excellent linearity of electrical output vs. input acceleration under normal usage.

At very high values of acceleration (depending upon the design characteristics of the particular transducer), nonlinearity or damage may occur. For example, if the dynamic forces exceed the biasing or clamping forces, the seismic element may "chatter" or fracture, although such a fracture might not be observed in subsequent low-level acceleration calibrations. High dynamic accelerations also may cause a slight physical shift in position of the piezoelectric element in the accelerometer—sometimes sufficient to cause a zero shift or change in sensitivity. The upper limit of acceleration measurements depends upon the specific design and construction details of the pickup and may vary considerably from one accelerometer to another, even though the design is the same. It is not always possible to calculate the upper acceleration limit of a pickup. Therefore one cannot assume linearity of acceleration levels for which calibration data cannot be obtained.

EFFECTS OF TEMPERATURE

Temperature Range. Piezoelectric accelerometers are available which may be used in the temperature range from –425°F (–254°C) to above +1400°F (+760°C) without the aid of external cooling. The voltage sensitivity, charge sensitivity, capacitance, and frequency response depend upon the ambient temperature of the transducer. This temperature dependence is due primarily to variations in the characteristics of the piezoelectric material, but it also may be due to variations in the insulation resistance of cables and connectors—especially at high temperatures.

Effects of Temperature on Charge Sensitivity. The charge sensitivity of a piezoelectric accelerometer is directly proportional to the d piezoelectric constant of the material used in the piezoelectric element. The d constants of most piezoelectric materials vary with temperature.

Effects of Temperature on Voltage Sensitivity. The open-circuit voltage sensitivity of an accelerometer is the ratio of its charge sensitivity to its total capacitance $(C_s + C_E)$. Hence, the temperature variation in voltage sensitivity depends on the temperature dependence of both charge sensitivity and capacitance. The voltage sensitivity of most piezoelectric accelerometers decreases with temperature.

Effects of Transient Temperature Changes. A piezoelectric accelerometer that is exposed to transient temperature changes may produce outputs as large as several volts, even if the sensitivity of the accelerometer remains constant. These spurious output voltages arise from

1. Differential thermal expansion of the piezoelectric elements and the structural parts of the accelerometer, which may produce varying mechanical forces on the piezoelectric elements, thereby producing an electrical output.
2. Generation of a charge in response to a change in temperature because the piezoelectric material is inherently pyroelectric. In general, the charge generated is proportional to the temperature change.

Such thermally generated transients tend to generate signals at low frequencies because the accelerometer case acts as a thermal low-pass filter. Therefore, such spurious signals often may be reduced significantly by adding thermal insulation around the accelerometer to minimize the thermal changes and by electrical filtering of low-frequency output signals from the accelerometer.

PIEZORESISTIVE ACCELEROMETERS

PRINCIPLE OF OPERATION

A piezoresistive accelerometer differs from the piezoelectric type in that it is not self-generating. In this type of transducer a semiconductor material, usually silicon, is used as the strain-sensing element. Such a material changes its resistivity in proportion to an applied stress or strain. The equivalent electric circuit of a piezoresistive transducing element is a variable resistor. Piezoresistive elements are almost always arranged in pairs; a given acceleration places one element in tension and the other in compression. This causes the resistance of one element to increase while the resistance of the other decreases. Often two pairs are used and the four elements are connected electrically in a Wheatstone-bridge circuit, as shown in Fig. 12.24B. When only one pair is used, it forms half of a Wheatstone bridge, the other half being made up of fixed-value resistors, either in the transducer or in the signal conditioning equipment. The use of transducing elements by pairs not only increases the sensitivity, but also cancels zero-output errors due to temperature changes, which occur in each resistive element.

At one time, wire or foil strain gages were used as the transducing elements in resistive accelerometers. Now silicon elements are usually used because of their higher sensitivity. (Metallic gages made of foil or wire change their resistance with strain because the dimensions change. The resistance of a piezoresistive material changes because the material's electrical nature changes.) Sensitivity is a function of the gage factor; the *gage factor* is the ratio of the fractional change in resistance to the fractional change in length that produced it. The gage factor of a typical wire or foil strain gage is approximately 2.5; the gage factor of silicon is approximately 100.

A major advantage of piezoresistive accelerometers is that they have good frequency response down to dc (0 Hz) along with a relatively good high-frequency response.

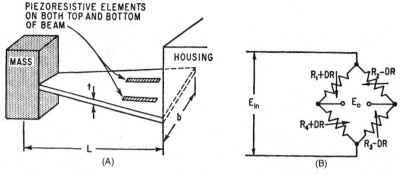

FIGURE 12.24   (*A*) Schematic drawing of a piezoresistive accelerometer of the cantilever-beam type. Four piezoresistive elements are used—two are either cemented to each side of the stressed beam or are diffused or ion implanted into a silicon beam. (*B*) The four piezoresistive elements are connected in a bridge circuit as illustrated.

DESIGN PARAMETERS

Many different configurations are possible for an accelerometer of this type. For purposes of illustration, the design parameters are considered for a piezoresistive accelerometer which has a cantilever arrangement as shown in Fig. 12.24*A*. This uniformly stressed cantilever beam is loaded at its end with mass *m*. In this arrangement, four identical piezoresistive elements are used—two on each side of the beam, whose length is L in. These elements, whose resistance is R, form the active arms of the balanced bridge shown in Fig. 12.24*B*. A change of length L of the beam produces a change in resistance R in each element. The gage factor K for each of the elements [defined by Eq. (17.1)] is

$$K = \frac{\Delta R/R}{\Delta L/L} = \frac{\Delta R/R}{\epsilon} \tag{12.18}$$

where ϵ is the strain induced in the beam, expressed in inches/inch, at the surface where the elements are cemented. If the resistances in the four arms of the bridge are equal, then the ratio of the output voltage E_o of the bridge circuit to the input voltage E_i is

$$\frac{E_o}{E_i} = \frac{\Delta R}{R} = \epsilon K \tag{12.19}$$

TYPICAL PIEZORESISTIVE ACCELEROMETER CONSTRUCTIONS

Figure 12.25 shows three basic piezoresistive accelerometer designs which illustrate several of the many types available for various applications.

Bending-Beam Type.   This design approach is described by Fig. 12.25*A*. The advantages of this type are simplicity and ruggedness. The disadvantage is relatively low sensitivity for a given resonance frequency. The relatively lower sensitivity results from the fact that much of the strain energy goes into the beam rather than the strain gages attached to it.

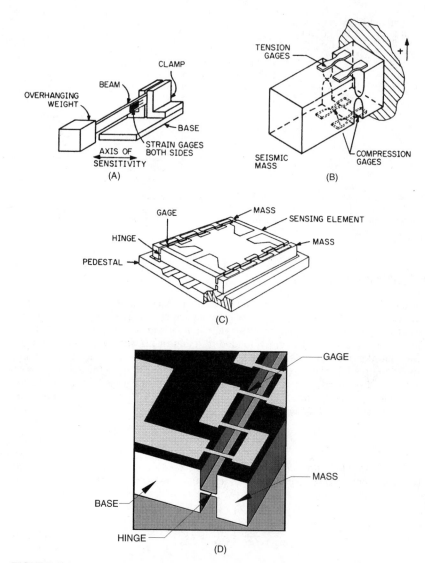

FIGURE 12.25 Three basic types of piezoresistive accelerometers. (*A*) Bending-beam type; the strain elements are usually bonded to the beam. Such an arrangement has been implemented in a micromachined accelerometer either by high-temperature diffusion of tension gages into the beam or by ion implantation. (*B*) Stress-concentrated type; the thin section on the neutral axis acts as a hinge of the seismic mass. Under dynamic conditions, the strain energy is concentrated in the piezoresistive gages. (*C*) Stress-concentrated micromachined type; the entire mechanism is etched from a single crystal of silicon. The thin section on the neutral axis acts as a hinge; the pedestal serves as a mounting base. (*D*) An enlarged view of one corner of the accelerometer shown in (*C*), which has a total thickness of 200 micrometers.

Stress-Concentrated Stopped and Damped Type. To provide higher sensitivities and resonance frequencies than are possible with the bending-beam type, designs are provided which place most of the strain energy in the piezoresistive elements. This is described by Fig. 12.25*B*. This approach is used to provide sensitivities more suitable for the measurement of acceleration below 100*g*. To provide environmental shock resistance, overload stops are added. To provide wide frequency response, damping is added by surrounding the mechanism with silicone oil. The advantages of these designs are high sensitivity, broad frequency response for the sensitivity, and over-range protection. The disadvantages are complexity and limited temperature range. The high sensitivity results from the relatively large mass with the strain energy mostly coupled into the strain gages. (The thin section on the neutral axis acts as a hinge; it contributes very little stiffness.) The broad frequency response results from the relatively high damping (0.7 times critical damping), which allows the accelerometer to be used to frequencies nearer the resonance frequency without excessive increase in sensitivity. The over-range protection is provided by stops which are designed to stop the motion of the mass before it overstresses the gages. (Stops are omitted from Fig. 12.25*B* in the interest of clarity.) Over-range protection is almost mandatory in sensitive piezoresistive accelerometers; without it they would not survive ordinary shipping and handling. The viscosity of the damping fluid does change with temperature; as a result, the damping coefficient changes significantly with temperature. The damping is at 0.7 times critical only near room temperature.

Micromachined Type. The entire working mechanism (mass, spring, and support) of a micromachined-type accelerometer is etched from a single crystal of silicon, a process known as *micromachining*. This produces a very tiny and rugged device, shown in Fig. 12.25*C*. The advantages of the micromachined type are very small size, very high resonance frequency, ruggedness, and high range. Accelerometers of such design are used to measure a wide range of accelerations, from below 10*g* to over 200,000*g*. No adhesive is required to bond a strain gage of this type to the structure, which helps to make it a very stable device. For shock applications, see the section on *Survivability*.

ELECTRICAL CHARACTERISTICS OF PIEZORESISTIVE ACCELEROMETERS

Excitation. Piezoresistive transducers require an external power supply to provide the necessary current or voltage excitation in order to operate. These energy sources must be well regulated and stable since they may introduce sensitivity errors and secondary effects at the transducer which will result in error signals at the output.

Traditionally, the excitation has been provided by a battery or a constant voltage supply. Other sources of excitation, such as constant current supplies or ac excitation generators, may be used. The sensitivity and temperature response of a piezoresistive transducer may depend on the kind of excitation applied. Therefore, it should be operated in a system which provides the same source of excitation as used during temperature compensation and calibration of the transducer. The most common excitation source is 10 volts dc.

Sensitivity. The *sensitivity* of an accelerometer is defined as the ratio of its electrical output to its mechanical input. Specifically, in the case of piezoresistive

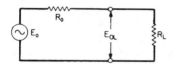

FIGURE 12.26 Loading effects on piezoresistive accelerometers.

accelerometers, it is expressed as voltage per unit of acceleration at the rated excitation (i.e., mV/g or peak mV/peak g at 10 volts dc excitation).

Loading Effects. An equivalent circuit of a piezoresistive accelerometer, for use when considering loading effects, is shown in Fig. 12.26. Using the equivalent circuit and the measured output resistance of the transducer, the effect of loading may be directly calculated:

$$E_{oL} = E_o \frac{R_L}{R_o + R_L} \tag{12.20}$$

where R_o = output resistance of accelerometer, including cable resistance
       E_o = sensitivity into an infinite load
       E_{oL} = loaded output sensitivity
       R_L = load resistance

Because the resistance of the strain-gage elements varies with temperature, output resistance should be measured at the operating temperature.

Effect of Cable on Sensitivity. Long cables may result in the following effects:

1. A reduction in sensitivity because of resistance in the input wires. The fractional reduction in sensitivity is equal to

$$\frac{R_i}{R_i + 2R_{ci}} \tag{12.21}$$

where R_i is the input resistance of the transducer and R_{ci} is the resistance of one input (excitation) wire. This effect may be overcome by using remote sensing leads.

2. Signal attenuation resulting from resistance in the output wires. This fractional reduction in signal is given by

$$\frac{R_L}{R_o + R_L + 2R_{co}} \tag{12.22}$$

where R_{co} is the resistance of one output wire between transducer and load.

3. Attenuation of the high-frequency components in the data signal as a result of R-C filtering in the shielded instrument leads. The stray and distributed capacitance present in the transducer and a short cable are such that any filtering effect is negligible to frequencies well beyond the usable range of the accelerometer. However, when long leads are connected between transducer and readout equipment, the frequency response at higher frequencies may be affected significantly.

Warmup Time. The excitation voltage across the piezoresistive elements causes a current to flow through each element. The I^2R heating results in an increase in temperature of the elements above ambient which slightly increases the resistance of the elements. Differentials in this effect may cause the output voltage to vary slightly with time until the temperature is stabilized. Therefore, resistance measurements and shock and vibration data should not be taken until stabilization is reached.

Input and Output Resistance. For an equal-arm Wheatstone bridge, the input and output resistances are equal. However, temperature-compensating and zero-balance resistors may be internally connected in series with the input leads or in series with the sensing elements. These additional resistors will usually result in unequal input and output resistance. The resistance of piezoresistive transducers varies with temperature much more than the resistance of metallic strain gages, usually having resistivity temperature coefficients between about 0.17 and 0.95 percent per degree Celsius.

Zero Balance. Although the resistance elements in the bridge of a piezoresistive accelerometer may be closely matched during manufacture, slight differences in resistance will exist. These differences result in a small offset or residual dc voltage at the output of the bridge. Circuitry within associated signal conditioning instruments may provide compensation or adjustment of the electrical zero.

Insulation. The case of the accelerometer acts as a mechanical and electrical shield for the sensing elements. Sometimes it is electrically insulated from the elements but connected to the shield of the cable. If the case is grounded at the structure, the shield of the connecting cable may be left floating and should be connected to ground at the end farthest from the accelerometer. When connecting the cable shield at the end away from the accelerometer, care must be taken to prevent ground loops.

Thermal Sensitivity Shift. The sensitivity of a piezoresistive accelerometer varies as a function of temperature. This change in the sensitivity is caused by changes in the gage factor and resistance and is determined by the temperature characteristics of the modulus of elasticity and piezoresistive coefficient of the sensing elements. The sensitivity deviations are minimized by installing compensating resistors in the bridge circuit within the accelerometer.

Thermal Zero Shift. Because of small differences in resistance change of the sensing elements as a function of temperature, the bridge may become slightly unbalanced when subjected to temperature changes. This unbalance produces small changes in the dc voltage output of the bridge. Transducers are usually compensated during manufacture to minimize the change in dc voltage output (zero balance) of the accelerometer with temperature. Adjustment of external balancing circuitry should not be necessary in most applications.

Damping. The frequency-response characteristics of piezoresistive accelerometers having damping near zero are similar to those obtained with piezoelectric accelerometers. Viscous damping is provided in accelerometers having relatively low resonance frequencies to increase the useful high-frequency range of the accelerometer and to reduce the output at resonance. At room temperature this damping is usually 0.7 of critical damping or less. With damping, the sensitivity of the accelerometer is "flat" to greater than one-fifth of its resonance frequency.

The piezoresistive accelerometer using viscous damping is intended for use in a limited temperature range, usually +20 to +200°F (−7 to +94°C). At high temperatures the viscosity of the oil decreases, resulting in low damping; and at low temperatures the viscosity increases, which causes high damping. Accordingly, the frequency-response characteristics change as a function of temperature.

FORCE GAGES AND IMPEDANCE HEADS

MECHANICAL IMPEDANCE MEASUREMENT

Mechanical impedance measurements are made to relate the force applied to a structure to the motion of a point on the structure. If the motion and force are measured at the same point, the relationship is called the *driving-point impedance;* otherwise it is called the *transfer impedance.* Any given point on a structure has six degrees-of-freedom: translations along three orthogonal axes and rotations around the axes, as explained in Chap. 2. A complete impedance measurement requires measurement of all six excitation forces and response motions. In practice, rotational forces and motions are rarely measured, and translational forces and motions are measured in a single direction, usually normal to the surface of the structure under test.

Mechanical impedance is the ratio of input force to resulting output velocity. *Mobility* is the ratio of output velocity to input force, the reciprocal of mechanical impedance. *Dynamic stiffness* is the ratio of input force to output displacement. *Receptance,* or *admittance,* is the ratio of output displacement to input force, the reciprocal of dynamic stiffness. *Dynamic mass,* or *apparent mass,* is the ratio of input force to output acceleration. All of these quantities are complex and functions of frequency. All are often loosely referred to as impedance measurements. They all require the measurement of input force obtained with a force gage (an instrument which produces an output proportional to the force applied through it). They also require the measurement of output motion. This is usually accomplished with an accelerometer; if velocity or displacement is the desired measure of motion, either can be determined from the acceleration.

Impedance measurements usually are made for one of these reasons:

1. To determine the natural frequencies and mode shapes of a structure (see Chap. 21)

2. To measure a specific property, such as stiffness or damping, of a material or structure

3. To measure the dynamic properties of a structure in order to develop an analytical model of it

The input force (excitation) applied to a structure under test should be capable of exciting the structure over the frequency range of interest. This excitation may be either a vibratory force or a transient impulse force (shock). If vibration excitation is used, the frequency is swept over the range of interest while the output motion (response) is measured. If shock excitation is used, the transient input excitation and resulting transient output response are measured. The frequency spectra of the input and output are then calculated by Fourier analysis.

FORCE GAGES

A force gage measures the force which is being applied to a structural point. Force gages used for impedance measurements invariably utilize piezoelectric transducing elements. A piezoelectric force gage is, in principle, a very simple device. The transducing element generates an output charge or voltage proportional to the applied force. Piezoelectric transducing elements are discussed in detail earlier in this chapter.

TYPICAL FORCE-GAGE AND IMPEDANCE-HEAD CONSTRUCTIONS

Force Gages for Use with Vibration Excitation. Force gages for use with vibration excitation are designed with provision for attaching one end to the structure and the other end to a force driver (vibration exciter). A thin film of oil or grease is often used between the gage and the structure to improve the coupling at high frequencies.

Force Gages for Use with Shock Excitation. Force gages for use with shock excitation are usually built into the head of a hammer. Excitation is provided by striking the structure with the hammer. The hammer is often available with interchangeable faces of various materials to control the waveform of the shock pulse generated. Hard materials produce a short-duration, high-amplitude shock with fast rise and fall times; soft materials produce longer, lower-amplitude shocks with slower rise and fall times. Short-duration shocks have a broad frequency spectrum extending to high frequencies. Long-duration shocks have a narrower spectrum with energy concentrated at lower frequencies.

Shock excitation by a hammer with a built-in force gage requires less equipment than sinusoidal excitation and requires no special preparation of the structure.

Impedance Heads. Impedance heads combine a force gage and an accelerometer in a single instrument. They are convenient for measuring driving-point impedance because only a single instrument is required and the force gage and accelerometer are mounted as nearly as possible at a single point.

FORCE-GAGE CHARACTERISTICS

Amplitude Response, Signal Conditioning, and Environmental Effects. The amplitude response, signal conditioning requirements, and environmental effects associated with force gages are the same as those associated with piezoelectric accelerometers. They are described in detail earlier in this chapter. The sensitivity is expressed as charge or voltage per unit of force, e.g., picocoulomb/newton or millivolt/lb.

Near a resonance, usually a point of particular interest, the input force may be quite low; it is important that the force-gage sensitivity be high enough to provide accurate readings, unobscured by noise.

Frequency Response. A force gage, unlike an accelerometer, does not have an inertial mass attached to the transducing element. Nevertheless, the transducing element is loaded by the mass of the output end of the force gage. This is called the *end dynamic mass*. Therefore, it has a frequency response that is very similar to that of an accelerometer, as described earlier in this chapter.

Effect of Mass Loading. The dynamic mass of a transducer (force gage, accelerometer, or impedance head) affects the motion of the structure to which the transducer is attached. Neglecting the effects of rotary inertia, the motion of the structure with the transducer attached is given by

$$A = A_o \frac{m_s}{m_s + m_t} \tag{12.23}$$

where a = amplitude of motion with transducer attached
A_o = amplitude of motion without transducer attached

m_s = dynamic mass of structure at point of transducer attachment in direction of sensitive axis of transducer

m_t = dynamic mass of the transducer in its sensitive direction

These are all complex quantities and functions of frequency. Near a resonance the dynamic mass of the structure becomes very small; therefore, the mass of the transducer should be as small as possible. The American National Standards Institute recommends that the dynamic mass of the transducer be less than 10 times the dynamic mass of the structure at resonance.

PIEZOELECTRIC DRIVERS

A piezoelectric element can be used as a vibration exciter if an ac signal is applied to its electrical terminals. This is known as the *converse piezoelectric effect*. In contrast to electrodynamic exciters, piezoelectric drivers are effective from well below 1000 Hz to as high as 60,000 Hz. Some commercially available piezoelectric drivers use piezoelectric ceramic elements to provide the driving force. Other applications utilize the piezoelectric effect in devices such as transducer calibrators, fuel injectors in automobiles, ink pumps in impact printer assemblies, and drivers to provide the antiphase motions for noise cancellation systems.

OPTICAL-ELECTRONIC TRANSDUCER SYSTEMS

LASER DOPPLER VIBROMETERS

The laser Doppler vibrometer (LDV) uses the Doppler shift of laser light which has been backscattered from a vibrating test object to produce a real-time analog signal output that is proportional to instantaneous velocity. The velocity measurement range, typically between a minimum peak value of 0.5 micrometer per second and a maximum peak value of 10 meters per second, is illustrated in Fig. 12.27.

An LDV is typically employed in an application where other accelerometers or other types of conventional sensors cannot be used. LDVs' main features are

- There are no transducer mounting or mass loading effects.
- There is no built-in transverse sensitivity or other environmental effects.
- They measure remotely from nearly any standoff distance.
- There is ultra-high spatial resolution with small measurement spot (5 to 100 micrometers typically).
- They can be easily fitted with fringe-counter electronics for producing absolute calibration of dynamic displacement.
- The laser beam can be automatically scanned to produce full-field vibration pattern images.

Types of Laser Doppler Vibrometers   Four types of laser Doppler vibrometers are illustrated in Fig. 12.28.

Standard (Out of Plane).   The standard LDV measures the vibrational component $v_z(t)$ which lies along the laser beam. Triaxial measurements can be obtained by

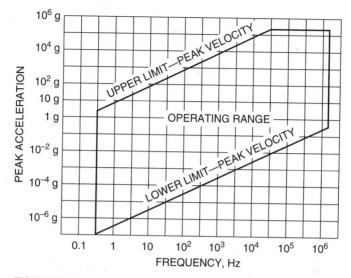

FIGURE 12.27 Typical operating range for a laser Doppler vibrometer. (*Courtesy of Polytec Pi, Inc.*)

approaching the same measurement point from three different directions. This is the most common type of LDV system.

Scanning. An extension of the standard out-of-plane system, the scanning LDV uses computer-controlled deflection mirrors to direct the laser to a user-selected array of measurement points. The system automatically collects and processes vibration data at each point; scales the data in standard displacement, velocity, or acceleration engineering units; performs fast Fourier transform (FFT) or other operations; and displays full-field vibration pattern images and animated operational deflection shapes.

In-plane. A special optics probe emitting two crossed laser beams is directed at normal incidence to the test surface and measures in-plane velocity. By rotating the probe by 90°, $v_x(t)$ or $v_y(t)$ can be measured.

Rotational. Two parallel laser beams from an optics probe measure angular vibration in units of degrees per second. Rotational systems are commonly used for torsional vibration analysis.

DISPLACEMENT MEASUREMENT SYSTEM

The electro-optical displacement measurement system consists of an electro-optical sensor and a servo-control unit designed to track the displacement of the motion of a light-dark target. This target provides a light discontinuity in the intensity of reflected light from an object. If such a light-dark discontinuity is not inherent to the object under study, a light-dark target may be applied on the object. An image of the light-dark target is formed by a lens on the photocathode of an image dissector photomultiplier tube, as shown in Fig. 12.29. The photocathode emits electrons in proportion to the intensity of the light striking the tube, causing an electron image to be

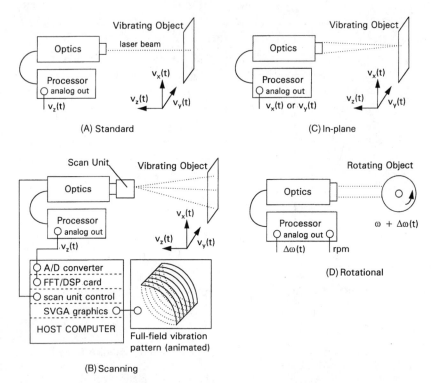

FIGURE 12.28 The four basic types of laser Doppler vibrometer systems. (*Courtesy of Polytec Pi, Inc.*)

generated in real time. The electron image is accelerated through a small aperture that is centrally located within the phototube. The number of electrons that enter the aperture constitute a small electric current that is directly proportional to the amount of light striking the corresponding area on the photocathode. This signal current is then amplified. As the light-dark target moves across the face of the phototube, the output current changes from high (light) to low (dark). When the target is exactly at the center of the tube, the output current represents half light and half

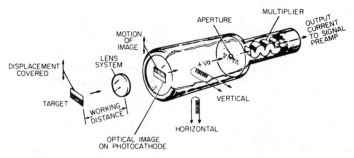

FIGURE 12.29 Image dissector tube of an electro-optical displacement measurement system. (*Courtesy of Optron Corp.*)

dark covering the aperture. If the target moves away from this position, the output current changes. This change is detected by the control unit, which feeds a compensation current back to the optical tracking head. The current that is needed for this deflection is directly proportional to the distance that the image has moved away from the center. Therefore it is a direct measure of displacement.

The displacement amplitudes that can be measured range from a few micrometers to several meters; the exact value is determined by the lens selected. Systems are available which measure displacements in one, two, or three directions.

FIBER-OPTIC REFLECTIVE DISPLACEMENT SENSOR

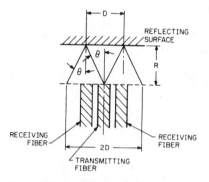

FIGURE 12.30 Fiber-optic displacement sensor. (*Courtesy of EOTEC Corp.*)

A fiber-optic reflective displacement sensor measures the amount of light normal to, and vibrating along, the optical axis of the device. The amount of reflected light is related to the distance between the surface and the fiber-optic transmitting/receiving element, as illustrated in Fig. 12.30. The sensor is composed of two bundles of single optical fibers. One of these bundles transmits light to the reflecting target; the other traps reflected light and transmits it to a detector. The intensity of the detected light depends on how far the reflecting surface is from the fiber-optic probe. Light is transmitted from the bundle of fibers in a solid cone defined by a numerical aperture. Since the angle of incidence, the size of the spot that strikes the bundle after reflection is twice the size of the spot that hits the target initially. As the distance from the reflecting surface increases, the spot size increases as well. The amount of reflected light is inversely proportional to the spot size. As the probe tip comes closer to the reflecting target, there is a position in which the reflected light rays are not coupled to the receiving fiber bundle. At the onset of this occurrence, a maximum forms which drops to zero as the reflecting surface contacts the probe. The output-current sensitivity can be varied by using various optical configurations.

While sensitivities approaching 1 microinch are possible, such extreme sensitivities limit the corresponding dynamic range. If the sensor is used at a distance from the reflecting target, a lens system is required in conjunction with a fiber-optic probe. With available lenses, the instruments have displacement measurement ranges from 0 to 0.015 in. (0 to 0.38 mm) and 0 to 5.0 in. (0 to 12.7 cm). Resolution typically is better than one one-hundredth of the full-scale range. The sensor is sensitive to rotation of the reflecting target. For rotations of ±3° or less, the error is less than ±3 percent.

ELECTRODYNAMIC TRANSDUCERS

ELECTRODYNAMIC (VELOCITY COIL) PICKUPS

The output voltage of the electrodynamic pickup is proportional to the relative velocity between the coil and the magnetic flux lines being cut by the coil. For this reason

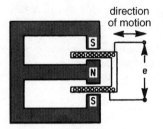

direction
of motion

e

FIGURE 12.31 Principle of operation of an electrodynamic pickup. The voltage e generated in the coil is proportional to the velocity of the coil relative to the magnet.

it is commonly called a velocity coil. The principle of operation of the device is illustrated in Fig. 12.31. A magnet has an annular gap in which a coil wound on a hollow cylinder of nonmagnetic material moves. Usually a permanent magnet is used, although an electromagnet may be used. The pickup also can be designed with the coil stationary and the magnet movable. The open-circuit voltage e generated in the coil is[2,3]

$$e = -Blv(10^{-8}) \qquad \text{volts}$$

where B is the flux density in gausses; l is the total length in centimeters of the conductor in the magnetic field; and v is the relative velocity in centimeters per second between the coil and magnetic field. The magnetic field decreases sharply outside the space between the pole pieces; therefore, the length of coil wire outside the gap generates only a very small portion of the total voltage.

One application of the electrodynamic principle is the velocity-type seismic pickup. Usually the pickup is used only at frequencies above its natural frequency, and it is not very useful at frequencies above several thousand hertz. The sensitivity of most pickups of this type is quite high, particularly at low frequencies where their output voltage is greater than that of many other types of pickups. The coil impedance is low even at relatively high frequencies, so that the output voltage can be measured directly with a high-impedance voltmeter. This type of pickup is designed to measure quite large displacement amplitudes.

DIFFERENTIAL-TRANSFORMER PICKUPS

The output of a differential-transformer pickup depends on the mutual inductance between a primary and a secondary coil. The basic components are shown in Fig. 12.32. The pickup consists of a core of magnetic material, a primary coil, and two secondary coils. As the core moves, a voltage is induced in the secondary coils. When the core is exactly in the center, each secondary coil contains the same length of core. Therefore, the mutual inductances of both secondary coils are equal in magnitude. However, they are connected in series opposition, so that the output voltage is zero. As the core is moved up or down, both the inductance and the induced voltage of one secondary coil are increased while those of the other are decreased. The output voltage is the difference between these two induced voltages. In this type of transducer, the output voltage is proportional to the displacement of the core over an appreciable range. In practice, the output voltage at the carrier frequency of the primary current is not exactly zero when the core is centered, and the output near the center position is not exactly linear. When the core is vibrated, the output voltage is a carrier wave, modulated at a frequency and amplitude corresponding to the motion of the core relative to the coils.

These pickups are used for very low frequency measurements. The sensitivity varies with the carrier frequency of the current in the primary coil. The carrier frequency should be at least 10 times the highest frequency of the motion to be measured. Since this range is usually between 0 and 60 Hz, the carrier frequency is usually above 600 Hz.

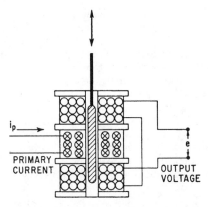

PRIMARY
CURRENT

OUTPUT
VOLTAGE

FIGURE 12.32 Differential-transformer principle. The inductance of the coils changes as the core is moved. For constant input current i_p to the primary coil, the output voltage e is the difference of the voltages in the two secondary coils, which are wound in series opposition. *(Courtesy of Automatic Timing and Controls, Inc.)*

SERVO ACCELEROMETER

The servo accelerometer (sometimes called a "force-balance accelerometer") contains a seismically suspended mass which has a displacement sensor (e.g., a capacitance-type transducer) attached to it. If the accelerometer is subject to motion, the mass is displaced with respect to the accelerometer case; this displacement generates a servo-loop error signal, establishing a restoring force that is applied to the mass in such a way as to reduce the initial error signal to zero. The acceleration, which is proportional to the restoring force, may be determined from the voltage drop across a resistance in series with a coil which generates the restoring force.

Servo accelerometers can be made very sensitive, some having threshold sensitivities on the order of a few micro-g. Excellent amplitude linearity is attainable, usually on the order of a few hundredths of one percent with peak acceleration amplitudes up to 50g. Typical frequency ranges are from 0 to 500 Hz.

CAPACITANCE-TYPE TRANSDUCERS

DISPLACEMENT TRANSDUCER (PROXIMITY PROBE)

The capacitance-type transducer is basically a displacement-sensitive device. Its output is proportional to the change in capacitance between two plates caused by the change of relative displacement between them as a result of the motion to be measured. Appropriate electronic equipment is used to generate a voltage corresponding to the change in capacitance.

The capacitance-type displacement transducer's main advantages are (1) its simplicity in installation, (2) its negligible effect on the operation of the vibrating system since it is a proximity-type pickup which adds no mass or restraints, (3) its extreme sensitivity, (4) its wide displacement range, due to its low background noise, and (5) its wide frequency range, which is limited only by the electric circuit used.

The capacitance-type transducer often is applied to a conducting surface of a vibrating system by using this surface as the ground plate of the capacitor. In this arrangement, the insulated plate of the capacitor should be supported on a rigid structure close to the vibrating system. Figure 12.33A shows the construction of a typical capacitance pickup; Figs. 12.33B, C, D, and E show a number of possible methods of applying this type of transducer. In each of these, the metallic vibrating system is the ground plate of the capacitor. Where the vibrating system at the point of instrumentation is an electrical insulator, the surface can be made slightly conducting and grounded by using a metallic paint or by rubbing the surface with graphite.

The maximum operating temperature of the transducer is limited by the insulation breakdown of the plate supports and leads. Bushings made of alumina are commercially available and provide adequate insulation at temperatures as high as 2000°F (1093°C).

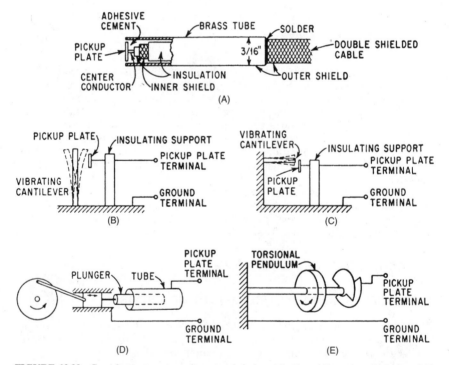

FIGURE 12.33 Capacitance-type transducers and their application: (*A*) construction of typical assembly, (*B*) gap length or spacing sensitive pickup for transverse vibration, (*C*) area sensitive pickup for transverse vibration, (*D*) area sensitive pickup for axial vibration, and (*E*) area sensitive pickup for torsional vibration.

VARIABLE-CAPACITANCE-TYPE ACCELEROMETER

Silicon micromachined variable-capacitance technology is utilized to produce miniaturized accelerometers suitable for measuring low-level accelerations (2 *g* to 100 *g*) and capable of withstanding high-level shocks (5,000 *g* to 20,000 *g*).

Acceleration sensing is accomplished by using a half-bridge variable-capacitance microsensor. The capacitance of one circuit element increases with applied acceleration, while that of the other decreases. With the use of signal conditioning, the accelerometer provides a linearized high-level output.

In the following example, the microsensor is fabricated in an array of three micromachined single-crystal silicon wafers bonded together using an anodic bonding process (see exploded view in Fig. 12.34). The top and bottom wafers contain the fixed capacitor plates (the lid and base, respectively), which are electrically isolated from the middle wafer. The middle wafer contains the inertial mass, the suspension, and the supporting ringframe. The stiffness of the flexure system is controlled by varying the shape, cross-sectional dimensions, and number of suspension beams. Damping is controlled by varying the dimensions of grooves and orifices on the parallel plates. Over-range protection is extended by adding overtravel stops.

The full-scale displacement of the seismic mass of the microsensor element is slightly more than 10 microinches. To detect minor capacitance changes in the microsensor due to acceleration, high-precision supporting electronic circuits are required. One approach applies a triangle wave to both capacitive elements of the microsensor. This produces currents through the elements which are proportional to their capacitances. A current detector and subtractor full-wave rectifies the currents and outputs their difference. An operational amplifier then converts this current difference to an output voltage signal. A high-level output is provided that is proportional to input acceleration.

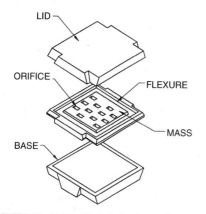

FIGURE 12.34 Exploded view of silicon micromachined variable-capacitance accelerometer. (*Courtesy of Endevco Corp.*)

REFERENCES

1. Levy, S., and W. D. Kroll: *Research Paper* 2138, *J. Research Natl. Bur. Standards,* **45:**4 (1950).

2. Chu, A. S.: "Zero shift of Piezoelectric Accelerometers in Pyroshock Measurements," TP 290, Endevco Corp., San Juan Capistrano, CA 92675 (1990).

3. Chu, A. S.: "Problems in High-Shock Measurement," TP 308, Endevco Corp., San Juan Capistrano, CA 92675 (1993).

4. *ISA Recommended Practice,* RP37.2, ¶6.6, "Strain Sensitivity," Instrument Society of America, 1964.

CHAPTER 13
VIBRATION MEASUREMENT INSTRUMENTATION

Robert B. Randall

INTRODUCTION

This chapter describes the principles of operation of typical instrumentation used in the measurement of shock and vibration. It deals with the measurement of parameters which characterize the total (broad-band) signal. Considerable reference is made to Chaps. 22 and 23, which give the mathematical background for various signal descriptors. Some reference is also made to the digital techniques of Chap. 27. Many of the techniques introduced here are applied in Chap. 16.

VIBRATION MEASUREMENT EQUIPMENT

Figure 13.1 shows a typical measurement system consisting of a preamplifier, a signal conditioner, a detector, and an indicating meter. Most or all of these elements often are combined into a single unit called a *vibration meter,* which is described in a following section.

The preamplifier is required to convert the very weak signal at high impedance from a typical piezoelectric transducer into a voltage signal at low impedance, which is less prone to the influence of external effects such as electromagnetic noise pickup. The signal conditioner is used to limit the frequency range of the signal (possibly to integrate it from acceleration to velocity and/or displacement) and to provide extra amplification. The detector is used to extract from the signal, parameters which characterize it, such as rms value, peak values, and crest factor. The so-called dc or slowly varying signal from the detector can be viewed on a meter, graphically recorded, or digitized and stored in a digital memory.

ACCELEROMETER PREAMPLIFIERS

Types of accelerometer preamplifiers include *voltage preamplifiers, charge preamplifiers,* and *line-drive preamplifiers.* Voltage preamplifiers now are little used

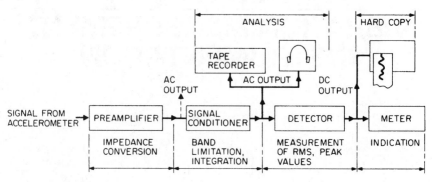

FIGURE 13.1 A block diagram of a typical vibration measurement system.

because, as indicated in Chap. 12, the voltage sensitivity of an accelerometer plus a cable is very dependent on the cable length. The sensitivity of the other two types is virtually independent of cable length, and this is of considerable practical importance.

Figure 13.2 shows the equivalent circuit of a charge preamplifier with an accelerometer and cable. The charge preamplifier consists of an operational amplifier having an amplification A, back-coupled across a condenser C_f; the input voltage to the amplifier is e_i. The output voltage e_o of this circuit can be expressed as

$$e_o = e_i A = \frac{q_a A}{C_a + C_c + C_i - C_f(A - 1)} \tag{13.1}$$

which is proportional to the charge q_a generated by the accelerometer. If A is very large, then the capacitances C_a, C_c, and C_i become negligible in comparison with AC_f and the expression can be simplified to

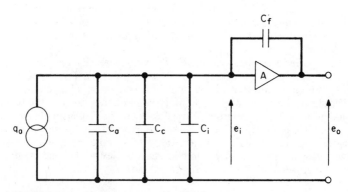

FIGURE 13.2 Diagram of a charge amplifier with accelerometer and cable. A = amplification of operational amplifier; C_f = shunt capacitance across amplifier; C_a = accelerometer capacitance; C_c = cable capacitance; C_i = preamplifier input capacitance; q_a = charge generated by accelerometer; e_i = amplifier input voltage; e_o = amplifier output voltage.

$$e_o \approx - \frac{q_a}{C_f} \qquad (13.2)$$

which is independent of the cable capacitance.

Although with a charge preamplifier the sensitivity is independent of cable length, the noise pickup in the high-impedance circuit increases with cable length, and so it is an advantage to have the preamplifier mounted as close to the transducer as is practicable. The line-drive amplifier represents an excellent solution to this problem, made possible by the development of miniaturized thick-film circuits. The amplifier can thus be attached to or even included internally in the transducer. In principle the initial amplifier can be of either charge or voltage type, but it can be advantageous to have the option of separating the amplifier from the transducer by a short length of cable, in which case the amplifier should be of the charge type. If the output signal from the initial amplifier is used to modulate the current or voltage of the power supply, then a single cable can be used both to power the amplifier and to carry the signal; the modulation is converted to a voltage signal in the power supply at the other end of this cable, which can be very long, e.g., up to a kilometer.

The output cable from a line-drive preamplifier is less subject to electromagnetic noise pickup than the cable connecting the transducer to a charge preamplifier. On the other hand, line-drive preamplifiers typically have some restriction of dynamic range and frequency range in comparison with a high-quality general-purpose charge preamplifier, and so reference should be made to the manufacturer's specifications when this choice is being made. Another problem is that it is more difficult to detect overload with an internal amplifier.

Signal Conditioners.   A signal-conditioning section is often required to band-limit the signal, possibly to integrate it (to velocity and/or displacement), and to adjust the gain. High- and low-pass filters normally are required to remove extraneous low- and high-frequency signals and to restrict the measurement to within the frequency range of interest. For broad-band measurements the frequency range is often specified, while for tape-recording and/or subsequent analysis the main reason for the restriction in frequency range is to remove extraneous components which may dominate and restrict the available dynamic range of the useful part of the signal.

Examples of extraneous low-frequency signals (see Chap. 12) are thermal transient effects, triboelectric effects, and accelerometer base strain. There may also be some low-frequency vibrations transmitted through the foundations from external sources. At the high-frequency end, the accelerometer resonance at least must be filtered out by an appropriate low-pass filter. This high- and low-pass filtering does not affect the signal in the input amplifier, which must be able to cope with the full dynamic range of the signal from the transducer. It is thus possible for a preamplifier to overload even when the output signal is relatively small. Consequently, it is important that the preamplifier indicates overload when it does occur.

Integration.   Although an accelerometer, in general, is the best transducer to use, it is often preferable to evaluate vibration in terms of velocity or displacement. Most criteria for evaluating machine housing vibration (Chap. 16) are effectively constant-velocity criteria, as are many criteria for evaluating the effects of vibration on buildings and on humans, at least within certain frequency ranges (Chaps. 24 and 44). Some vibration criteria (e.g., for aircraft engines) are expressed in terms of displacement. For rotating machines, it is sometimes desired to add the absolute displacement of the bearing housing to the relative displacement of the shaft in its bearing (measured with proximity probes) to determine the absolute motion of the shaft in space.

Acceleration signals can be integrated electronically to obtain velocity and/or displacement signals; an accelerometer plus integrator can produce a velocity signal which is valid over a range of three decades (1000:1) in frequency—a capability which generally is not possessed by velocity transducers. Moreover, simply by switching the lower limiting frequency (for valid integration) on the preamplifier, the three decades can be moved by a further decade, without changing the transducer.

A typical sinusoidal vibration component may be represented by the phasor $Ae^{j\omega t}$. Integrating this once gives $\dfrac{1}{j\omega} Ae^{j\omega t}$, and thus integration corresponds in the frequency domain to a division by $j\omega$. This is the same as a phase shift of $-\pi/2$ and an amplitude weighting inversely proportional to frequency, and thus electronic integrating circuits must have this property.

One of the simplest integrating circuits is a simple R-C circuit, as illustrated in Fig. 13.3. If e_i represents the input voltage, then the output voltage e_o is given by

$$e_o = e_i \frac{1}{1 + j\omega RC} \tag{13.3}$$

which for high frequencies ($\omega RC \gg 1$) becomes

$$e_o \approx \frac{e_i}{j\omega RC} \tag{13.4}$$

which represents an integration, apart from the scaling constant $1/RC$.

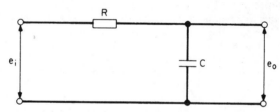

FIGURE 13.3   Electrical integration network of the simple R-C type.

The characteristic of Eq. (13.3) is shown in Fig. 13.4; it is that of a low-pass filter with a slope of -20 dB/decade and a cutoff frequency $f_n = 1/(2\pi RC)$ (corresponding to $\omega RC = 1$).

The limits f_L (below which no integration takes place) and f_T (above which the signal is integrated) can be taken as roughly a factor of 3 on either side of f_n, for normal measurements where amplitude accuracy is most important. Where phase accuracy is important (e.g., to measure true peak values), the factor should be somewhat greater. Modern integrators tend to use active filters with a more localized transition between the region of no integration and the region of integration.

One situation where the choice of the low-frequency limit is important is in the integration of impulsive signals, for example, in the determination of peak velocity and displacement from an input acceleration pulse. Figure 13.5 shows the effect of single and double integration on a 10-millisecond single-period sine burst, with both 1- and 10-Hz cutoff frequencies, in comparison with the true results. The deviations

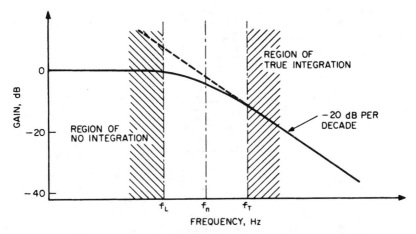

FIGURE 13.4   Frequency characteristic of the circuit shown in Fig. 13.3. f_T = lower frequency limit for true integration; f_L = upper frequency limit for no integration.

depend to some extent on the actual amplitude and phase characteristics of the integrator, but the following values can be used as a rough guide to select the integrator cutoff frequency f_T:

For single integration (acceleration to velocity),

$$f_T < \frac{1}{30t_p} \tag{13.5}$$

For double integration (acceleration to displacement),

$$f_T < \frac{1}{50t_p} \tag{13.6}$$

where t_p is the time from the start of the pulse to the measured peak. For the case shown in Fig. 13.5, these values of f_T are <6.7 Hz and <2 Hz, respectively.

DETECTORS

Detectors are used to extract parameters which characterize a signal, such as arithmetic average, mean-square, and root-mean-square (rms) values, as defined in Chap. 22. The arithmetic average value is the simplest to measure, using a full-wave rectifier to obtain the instantaneous magnitude and a smoothing circuit to obtain the average. However, even though there is a fixed (though different) relationship between average and rms values for sinusoidal and Gaussian random signals (Chap. 22), the relationship varies considerably for complex signals and, in particular, is affected considerably by phase relationships. Since mean-square and rms values are independent of phase relationships, they are usually preferred as signal descriptors for stationary signals; where an average detector is used, it is usually as an approximation of an rms detector.

Mean-square values have the advantage that they are directly additive when two signals are added together (in particular different frequency bands or components),

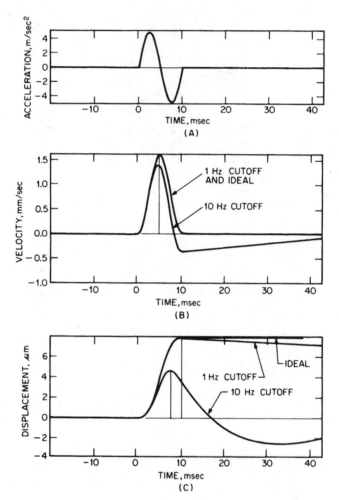

FIGURE 13.5 Integration and double integration of a 10-millisecond acceleration pulse using lower frequency limits of 1 and 10 Hz, respectively. (*A*) Input acceleration signal. (*B*) Velocity signal resulting from a single integration with different cutoff frequencies compared with ideal integration. (*C*) Displacement signal resulting from a double integration compared with ideal integration.

while rms values have the advantage that they have the same dimensions and units as the original signal. Thus, a "true rms" detector must include a squaring section and averager to obtain the mean-square value, followed by a square-root extractor.

Squaring. One of the earliest true rms detectors, the Wahrman detector,[1] used a piecewise linear approximation to the parabola representing true squaring. For moderate signal values, the errors for the piecewise linear circuit are quite small, but past the last breakpoint on the curve the deviation becomes progressively larger. The breakpoints are dimensioned for a typical rms level, and the errors are thus greatest for relatively large instantaneous values, which are characteristic of signals with a high *crest factor* (ratio of peak to rms value). The higher the crest factor, the larger the number of breakpoints required. As an example, four breakpoints give an accuracy within ½ dB for crest factors up to 5. Thus, for accurate results, this type of detector must be specified with respect to both dynamic range and crest factor.

Later designs of analog squaring circuits, so-called log-mean-square or lms detectors, make use of the logarithmic characteristic of certain diodes to achieve squaring by doubling the logarithmic value of the rectified signal. This type generally has no limitation on crest factor other than that given by the dynamic range. In a similar manner, digital instruments achieve true squaring and are limited only by the dynamic range of the detector.

Averaging. The definition of mean-square value given in Eq. (22.6) assumes a uniform weighting for the whole of the averaging time T. In practice, for measurements on continuous signals, it is often desired to have a running average, giving at any time the average value over the previous T seconds. It is extremely difficult to achieve a linearly weighted running average, and so recourse is usually made to two alternatives:

1. Exponentially weighted running average. This is achieved by an R-C smoothing circuit in most analog instruments, and also by exponential averaging in digital instruments such as FFT analyzers.

2. Linearly weighted average over a fixed time period of length T. The result is available only at the end of each period and is usually held until processed further, and so new incoming data may be lost.

The averaging process acts as a low-pass filter to remove high-frequency ripple components and leave the slowly varying dc or average value. Figure 13.6 compares the low-pass filter characteristics of exponential and linear averaging and demonstrates that they are equivalent for the case where $T = 2RC$ (where RC is the time constant of the exponential decay). This low-pass filtration in the frequency domain corresponds to a convolution in the time domain with the impulse response of the averaging circuit. The two impulse responses (reversed in time because of the convolution) are compared in Fig. 13.7 for the same case where $T = 2RC$. When scaled to give the same result on stationary signals (same area under the curve), the peak output for exponential averaging is twice that for linear averaging. Account must be taken of this in the analysis of impulses.

A method of checking the effective averaging time of an exponential averager is to remove the excitation and measure the rate of decay of the output. This will be 4.34 dB per RC time constant, or 8.7 dB per averaging time T. This does not apply to FFT analyzers operating above their real-time frequency, in the same way that the effective linear averaging time is then less than the time required to obtain the result.

Peak Detectors. In some cases it is desired to measure the true peak values of the original signal (for example, to avoid overloading a tape recorder). Peak detectors are available which capture the highest value encountered and either hold it until reset or have it decay slowly enough that the eye can read the peak value from a meter. Care should be taken to distinguish between maximum positive peak, maximum negative peak, maximum peak (positive or negative), and peak-to-peak values (Fig. 13.8). Care should also be taken to distinguish between true peak values and what is roughly referred to as peak-to-peak shaft vibration, which is often assumed to be sinusoidal and is measured with an average detector.

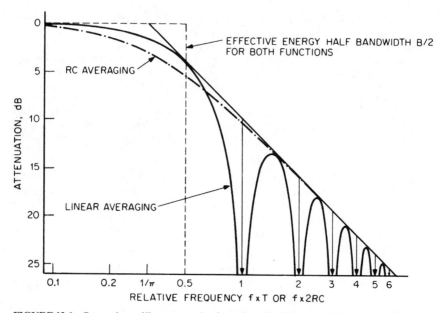

FIGURE 13.6 Comparison of linear averaging (over time T) with exponential averaging (time constant RC) in the frequency domain for the case where $T = 2RC$. The low-pass filter characteristics have the same asymptotic curves and the same bandwidth $B = 1/T = 1/2RC$.

Crest Factor.   The *crest factor* is the ratio of peak to rms value. The maximum peak (positive or negative) should be used. It is meaningful only where peak values are reasonably uniform and repeatable from one signal sample to another. The crest factor yields a measure of the spikiness of a signal and is often used to characterize signals containing repetitive impulses in addition to a lower-level continuous signal. Examples of such vibration signals are those from reciprocating machines and those produced by localized faults in gears and rolling element bearings.

Kurtosis.   *Kurtosis,* a statistical parameter akin to the mean and mean-square values defined in Chap. 22, is defined as[2]

$$\beta_2 = \frac{\int_{-\infty}^{\infty} (\xi - \bar{\xi})^4 p(\xi)\, d\xi}{\sigma^4} \tag{13.7}$$

using the terminology of Chap. 22.

For signals with zero mean value $\bar{\xi}$, a practical estimator for this can be expressed as

$$\frac{\dfrac{1}{T} \int_0^T \xi^4(t)\, dt}{\left[\dfrac{1}{T} \int_0^T \xi^2(t)\, dt \right]^2} \tag{13.8}$$

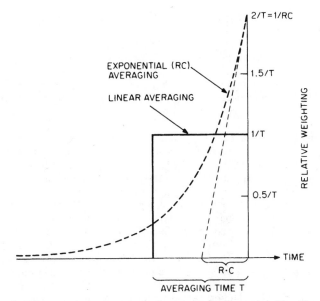

FIGURE 13.7 Comparison of linear averaging (over time T) with exponential averaging (time constant RC) in the time domain for the case where $T = 2RC$. The weighting curves represent the impulse-response functions reversed in time.

Because of the fourth power, considerable weight is given to large amplitude values, and the kurtosis thus is a good indicator of the spikiness of the signal. Because it takes the whole signal into account, rather than just isolated peaks, it generally gives a more stable value than the crest factor, but this must be weighed against the more complicated measurement procedure. Kurtosis is normally calculated digitally, but can be measured by a more complex version of the log-mean-square detector described above.

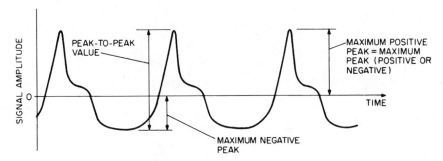

FIGURE 13.8 Illustration of various peak values.

Envelope Detectors. Many machine vibration signals of interest contain repetitive high-frequency bursts, as a result of exciting high-frequency resonances at regular intervals. Direct frequency analysis of the signal does not always give much information on the repetition frequencies, in particular when the resonances excited are at very high frequencies. These repetition frequencies are, however, easily measurable in the envelope signal illustrated in Fig. 13.9. Quite often, the signal is first bandpass-filtered in a frequency region dominated by the repetitive bursts (i.e., one of the regions containing resonances which are excited and where the extraneous background signal is low).

The true envelope signal can be obtained using a peak detector with a decay time constant set sufficiently short so that it is able to follow the relatively slow variations in the (rectified) signal envelope. If the signal is first passed through a bandpass filter, it will have a roughly sinusoidal form with slowly varying amplitude and there will be a fixed ratio of peak to (short-term) rms or average value, in which case an rms or average detector can be used instead of the peak detector. Moreover, where a frequency analysis of the signal is to be obtained with an FFT analyzer, it is not necessary to apply a smoothing circuit, as the antialiasing filters (described below) will automatically remove high-frequency ripple components in the rectified signal. Thus, a tunable bandpass filter of, say, one-third-octave bandwidth, followed by a full-wave rectifier, can be used as an envelope detector in cases where it is primarily the burst repetition frequencies which are of interest. It is shown below that envelope signals can also be calculated by Hilbert transform techniques in an FFT analyzer.

VIBRATION METERS

Vibration meters are instruments which receive a signal from a vibration transducer and process it so as to give an indication of relevant vibration parameters. They are sometimes made specifically to meet certain standards, for example, ISO 2372 on "Vibration Severity of Rotating Machines" or ISO 2631 on "Human Vibration." In these cases, the requirements are specified in the relevant standard; the discussion here is aimed at more general-purpose vibration meters.

For measurements on most rotating machines, a frequency range of 10 Hz to 10 kHz is desirable. The lower limit includes the shaft speed for all machines operating over 600 rpm and any subharmonic components such as oil whirl for higher-speed plain bearing machines where such effects are most prevalent. The upper frequency

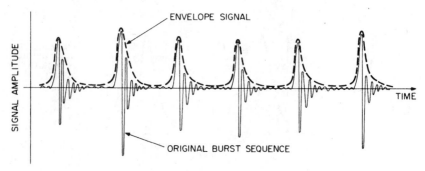

FIGURE 13.9 Illustration of the envelope signal for an impulsive signal containing repetitive high-frequency bursts.

of 10 kHz includes tooth-meshing frequencies and their harmonics in gearboxes, bladepassing frequencies in most bladed machines, and resonance frequencies typically excited by rolling element bearing faults.

It can be advantageous to be able to choose a number of upper and lower limiting frequencies within the overall range. For example, restriction of the upper frequency to 1 kHz allows measurements in accordance with ISO 2372 previously cited. For special purposes, it may be necessary to go to frequencies lower than 10 Hz, for example, in measurements on slow-speed machines and on bridges and other structures. It is possible to cover a total range of 1 Hz to 10 kHz with one accelerometer; if the meter is able to accept a range of transducers, its own frequency range can be even wider.

If restriction is to be made to one vibration parameter, then velocity usually is the best choice, as most machine vibration signals have a roughly uniform velocity spectrum, so that an increase at any frequency has a roughly equal chance of influencing overall vibration levels.

It is also desirable to be able to measure acceleration and displacement; changes at low frequency reflect themselves primarily in the displacement value, while changes at high frequency have the most effect on the acceleration value.

In addition to the measurement of rms levels in each of the vibration parameters, it is of advantage to be able to measure some parameter indicating the spikiness of the signal, such as peak values (and hence crest factor), kurtosis, spike energy, or shock-pulse value. Finally, it is useful if the meter has an ac output, to allow the signal to be fed to a tape recorder or headphones. In the absence of frequency analysis, the human ear can discern a great deal about the characteristics of a signal, and this setup provides an excellent stethoscope. The ac signal should preferably be of selected parameter (acceleration, velocity, or displacement); the frequency range should be restricted as little as possible.

TAPE RECORDERS

The most widely used recording techniques for instrumentation tape recorders are direct recording, frequency-modulation (FM) recording, and digital recording. The first two are often combined in one recorder and are thus discussed together, while the latter is discussed separately.

Analog Recorders. In direct recording, the signal amplitude is reflected directly in the local degree of magnetization of the tape, while in FM recording the amplitude information is contained in the deviation of the frequency of a carrier tone from its nominal value. Thus, the degree of magnetization of the tape is less critical for FM recording, and the recorded blips are normally saturated. Hence, one of the advantages of FM recording is that the recorded signals are less susceptible to change due to poor storage conditions (heat, light, and stray magnetic fields). On the other hand, since the carrier frequency is typically 3 to 5 times higher than the maximum signal frequency in FM recording, tape speeds (and hence tape quantities used) must be 3 to 5 times greater for a given frequency range.

The major difference between the two techniques is in their ability to record low-frequency signals. Since on playback of direct recordings it is the rate of change of tape magnetization which is detected, this technique cannot record down to dc; a typical lower frequency limit is 25 Hz. In contrast, FM recording can record down to dc; a dc signal is simply represented by a constant deviation of the carrier frequency.

Since on playback of direct recordings it is necessary to integrate the detected signal and compensate for other effects such as tape magnetic properties, this is usually done by equalization networks designed primarily to provide amplitude linearity; phase linearity is poor. Thus, the actual form of signals is likely to be modified by direct recording; peak values cannot be relied upon. The phase linearity of FM recording is excellent for all except the highest part of the frequency range, where the effects of the required low-pass filter become significant.

One of the most important characteristics of a tape recorder is its dynamic range, since the tape recorder is likely to be the element in the measurement chain whose dynamic range is restricted the most. The dynamic range usually is expressed in terms of a signal-to-noise ratio, which is typically 40 dB for FM recording and up to 50 dB for direct recording. These figures can be somewhat misleading, however, as the noise referred to is a total figure over the entire frequency range and has less influence in a narrow-band analysis. After narrow-band analysis, the noise level for FM recording typically is more than 60 dB below full scale, as compared with 70 to 80 dB for the digitization noise in a modern frequency analyzer.

Table 13.1 includes a summary of the most important features of FM and direct recording. Some recorders are able to record using both techniques, in which case the heads normally are optimized for FM and the signal-to-noise ratio for direct recording is reduced somewhat. The most important addition provided by direct recording is the possibility of recording considerably higher frequencies, typically 50 to 100 kHz.

Both techniques are limited by the accuracy of the tape transport system, and small variations in tape orientation and speed give rise to "wow" and "flutter."

TABLE 13.1   Comparison of Recording Techniques

	Direct	FM	DAT
Dynamic range (typical, narrow-band)	70 dB	60 dB	80 dB
Lower frequency limit (typical)	25 Hz	dc	dc
Upper frequency limit (typical)	50 kHz	10 kHz	20 kHz
Amplitude stability	Acceptable	Excellent	Excellent
Phase linearity	Poor	Good	Excellent
Preservation of recorded information	Acceptable	Good	Excellent

Digital Recorders.   Instrumentation recorders are available based on the pulse-code modulation (PCM) principle. These have been developed from digital audio-tape (DAT) recorders and have many characteristics in common. A typical DAT cassette can record, for 2 hours, two channels to 20,000 Hz, four channels to 10,000 Hz, or more channels with correspondingly lower frequency ranges. Double-speed versions give twice the number of channels for the same frequency range, but half the total recording time. For two-channel recording, the overall sampling rate is 96 kHz (48 kHz per channel), each sample being 16 bits, or 2 bytes, so that the overall amount of data stored on one DAT cassette is well over 1 gigabyte. The problems of wow and flutter are largely eliminated by digital recorders because the sampling frequency during recording and playback is not directly tied to tape or rotating-head speed and can be made extremely accurate. Dynamic range is dependent primarily on the number of bits used in digitization but typically matches that of digital signal analyzers, giving approximately 20 dB more than typical analog recorders. Phase matching between channels is within a fraction of a degree over a very wide frequency range, meaning that signal reproduction is almost perfect.

As with any digital processing, the signal to be recorded must not contain any frequency components above half the sampling frequency. After sampling, it is not possible to determine whether this condition has been satisfied, and so it is normally necessary to filter the signals to be recorded with a very steep "antialiasing" filter. This is typically a 7-pole elliptic filter with cutoff frequency at 40 percent of the sampling frequency and a roll-off of 120 dB per octave. Less steep filters can be used to reduce the phase distortion effects in the vicinity of the cutoff frequency, but the cutoff frequency must then be reduced accordingly. To avoid further distortion, it is common to use digital interpolation techniques to increase the sample rate on playback, thus permitting the use of much "gentler" filters to smooth the output from the digital-to-analog converters.

Table 13.1 compares all three recording techniques.

DIGITAL SIGNAL PROCESSING

Computer software programs are commercially available which provide for signal processing using digital techniques, for example: FFT analysis, digital filtering, and optimization. One means of obtaining data in digital form is by using a digital tape recorder (described in the previous section) in which a digital output is obtained by bypassing the digital-to-analog converter contained in the recorder. Digital frequency analysis is discussed more fully in Chap. 14. This section discusses the conversion of continuous analog signals which are converted into digital form using analog-to-digital converters; it also describes some of the differences between analog signal processing and digital signal processing.

ANALOG-TO-DIGITAL CONVERTERS

Analog-to-digital (A/D) converters serve to convert a continuous signal into a sequence of digital numbers representing the instantaneous value of the signal at specified time increments. Under certain conditions, it is possible to regain the original analog signal by the reverse process, using a digital-to-analog (D/A) converter, as discussed later. The time increments are normally uniform, i.e., they represent a constant sampling frequency; in other cases, they may be on some other basis such as uniform increments of shaft rotation (e.g., in the case of "order tracking," as discussed in Chap. 14).

The quality of the digitized signal depends on a number of factors, such as the accuracy of the sample intervals, the number of bits used in the digital representation, the linearity of the analog amplifiers with which the signal has been processed, and the quality of the low-pass filtering of the signal prior to the A/D conversion. Each of these factors is discussed later.

The first step in the A/D conversion process is the sample-and-hold circuit that samples the instantaneous value of the analog signal at the instant of each pulse of the sampling clock, and holds that analog voltage constant until the A/D conversion process is complete and it is reset. The accuracy of the sample spacing depends not only on the accuracy of the sample clock, but also on the acquisition time of the sample-and-hold circuit, but for the frequency range of typical vibration signals, both of these potential errors are negligible in high-quality A/D converters. For multiple channel conversion, it is common to use a single A/D converter multiplexing between channels, but even though it is possible to compensate for time delay between channels, it is desirable to use synchronized sample-and-hold circuits which sample all channels simultaneously, even if the A/D conversion is done sequentially.

The output of an A/D converter is a binary integer number with 2^N possible values, where N is the number of bits. Depending on whether a sign bit is used, these can range from zero to $(2^N - 1)$ or -2^{N-1} to $(2^{N-1} - 1)$. The possible dynamic range of the digitized signal is thus heavily dependent on the number of bits used and is commonly taken to be 6 dB for each bit (each added bit giving a doubling of the number of possible levels and a doubling of the ratio of the maximum-to-minimum value). For averaged spectral results, the dynamic range can be increased somewhat by a process of adding "dither," a very low level random noise whose average spectrum is outside the dynamic range of the measurement system. When dither is added to a signal lower than the least significant bit (which otherwise would not register) it causes the latter to be set part of the time and thus gives an averaged result smaller than the least significant bit (note that the data should be converted from integer to floating point prior to the averaging process). On the other hand, the actual dynamic range of the measurement may be limited by factors other than the least significant bit, such as the noise level in the analog parts of the system, or the linearity of the latter. For example, it is not uncommon to have a 12-bit A/D converter (which should give a 72-dB dynamic range) with a linearity specification of 0.05% of full scale, this corresponding to a possible bias error of −66 dB with respect to full scale. Note that a bias error of this sort affects all values in the same way, and thus has a much greater effect than a random error of the same magnitude as is the case when several values of the same order are added together (e.g., when converting from constant bandwidth to constant percentage bandwidth spectra, which is done in some spectrum analyzers).

ANTIALIASING FILTERS

Discrete sampling in the time domain (i.e., multiplication by a train of unit impulse functions) corresponds in the frequency domain to a periodic repetition of the spectrum with a periodic spacing equal to the sampling frequency, as illustrated in Fig. 13.10. If the original signal does not contain any frequency components above half the sampling frequency f_s (i.e., outside the range from minus to plus $f_s/2$), this periodic repetition does not result in any loss of information and can in principle be removed again by low-pass filtering, as shown in Fig. 13.10B. If the sampling frequency is less than twice the highest frequency component in the signal, the periodic repetition of the spectrum gives mixing of the overlapped portions (known as *aliasing*), and it is no longer possible to separate them completely, as shown in Fig. 13.10C. Thus if it is desired to obtain correct frequency spectra, or to return to analog form via a D/A converter, it is absolutely necessary to ensure that the analog signal does not contain frequency components above $f_s/2$, and this is achieved by the use of appropriate low-pass filters, so-called *antialiasing filters*. As explained in Chap. 14, such filters have very steep characteristics (e.g., 120 dB/octave). Their application makes it possible to use up to 80 percent of the theoretically available spectrum (i.e., up to $f_s/2$), but they result in considerable phase distortion in the vicinity of the cutoff frequency.

Low-pass filters also change the waveform, as illustrated in Fig. 13.11 for the output of a square-wave generator (giving rise to uneven rectangular pulses so that all harmonics are produced). In Figs. 13.11A and 13.11B, a proper antialiasing filter has been used, so that the spectrum is correct, but the waveform has bursts at the begin-

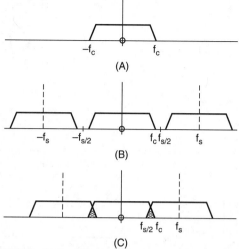

FIGURE 13.10 (*A*) Spectrum of a continuous band-limited signal with maximum frequency f_c. (*B*) Spectrum of digitized signal with sampling frequency $f_s > 2f_c$. (*C*) Spectrum of digitized signal with sampling frequency $f_s < 2f_c$; the hatched area indicates aliased components.

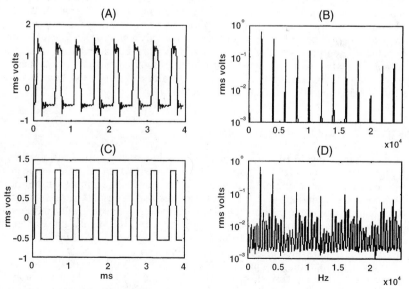

FIGURE 13.11 Effect time signals and spectra of antialiasing filters. (*A*) Time signal with filter. (*B*) Spectrum with filter. (*C*) Time signal without filter. (*D*) Spectrum without filter.

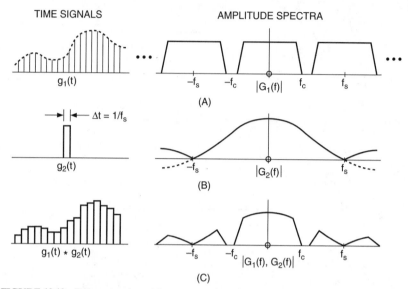

TIME SIGNALS   AMPLITUDE SPECTRA

(A)

(B)

(C)

FIGURE 13.12   D/A conversion with constant voltage between samples. (*A*) Digitized time signal and its amplitude spectrum. (*B*) Rectangular pulse length Δt (equal to $1/f_s$) and its amplitude spectrum. (*C*) Convolution of (*A*) and (*B*) and the resulting amplitude spectrum.

ning of each level section. These can be interpreted either as the step-response of the low-pass filters or as the removal of the high-frequency components (the so-called *Gibbs phenomenon*). In Figs. 13.11*C* and 13.11*D*, no antialiasing filter has been used, and even though the waveform appears more like a square wave, the spectrum is now incorrect and contains aliasing components. This leads to the conclusion that where the sole purpose is to evaluate digitized waveforms, an antialiasing filter is not desirable, but where any treatment of the digitized signal is carried out (such as frequency analysis, digital filtering, or reconstruction of a signal by use of a D/A converter), antialiasing filters are absolutely necessary.

DIGITAL-TO-ANALOG CONVERSION

It is evident from Fig. 13.10 that in removing high-frequency components of the periodic spectrum of a sampled function, the low-pass filters which are used should be of the same quality as the original antialiasing filter prior to digitization. Furthermore, a D/A converter cannot produce true unit impulses, and it is usual that the converted voltage corresponding to each sample is carried over as a constant value in the sample interval. This is the equivalent of a convolution with a rectangular pulse of length Δt (equal to $1/f_s$), so that the spectrum is multiplied by a $(\sin x)/x$ function with its first zero at f_s, as illustrated in Fig. 13.12. The gain factor at $0.50 f_s$ is $2/\pi$ (-3.9 dB), and at $0.39 f_s$ (the normal range of an FFT spectrum) it is -2.3 dB. The effect of this factor, and the need for a very steep low-pass filter, can be reduced considerably by increasing the sampling rate (by digital interpolation) before D/A conversion, and this is commonly done where sufficiently fast hardware is available.

DIGITAL PROCESSING

Once the signal has been obtained in digital form using proper antialiasing filters, many of the operations (described above and in Chap. 14) can be carried out digitally.[3] For example, acceleration signals can be integrated to obtain velocity using numerical integration directly in the time domain or if so desired, $j\omega$ operations in the frequency domain (where it can be combined with bandpass filtering). Each integration corresponds to a division by $j\omega$ of the Fourier spectrum (this applies only to ac-coupled signals).

REFERENCES

1. Wahrman, C. G.: *Brüel & Kjaer Tech. Rev.*, (3) (1958).
2. Dyer, D., and R. M. Stewart: *J. Mech. Des.*, **100:**229 (1978).
3. Oppenheim, A. V., and R. W. Schafer: *Discrete Time Signal Processing*, Prentice-Hall Inc., Englewood Cliffs, New Jersey, 1989.

CHAPTER 14
SPECTRUM ANALYZERS AND THEIR USE

Robert B. Randall

INTRODUCTION

This chapter deals primarily with frequency analysis, but a number of related analysis techniques—namely, synchronous averaging, cepstrum analysis, and Hilbert transform techniques—are considered.

ELECTRICAL FILTERS

An ideal bandpass filter is a circuit which transmits that part of the input signal within its passband and completely attenuates components at all other frequencies. Practical filters differ slightly from the ideal, as discussed below. An analysis may be performed over a frequency range either by using a single filter with a tunable center frequency which is swept over the entire frequency range or by using banks of fixed filters having contiguous (or overlapping) passbands.

For general vibration analysis it has been common to use tunable filters whose center frequency is either tuned by hand or synchronized with the X position of the pen on a graphic recorder so that the spectrum is plotted automatically by sweeping the center frequency over the desired frequency range. Alternatively, the center frequency may be synchronized with an external signal, e.g., a trigger pulse once per revolution of a shaft, in which case the filter becomes a tracking filter which can be used to filter out the component corresponding to a designated harmonic, or multiple, of the synchronizing signal. A tracking adaptor, or frequency multiplier/divider, is normally required to generate the tuning frequency from the original synchronizing signal if other than the fundamental or first harmonic is required.

In the past, banks of filters with fixed center frequencies, each with its own detector, were widely used for parallel analysis of all frequency bands in real time. This arrangement is costly, however, and has largely been superseded by digital filter analyzers (described in the following section). If real-time analysis is not required, a less-expensive alternative is to switch the output of each filter in turn to a single detector and record the outputs sequentially on paper. The individual filters can then also share many components, which are selected in appropriate combinations by the switching process. Sequentially stepped fixed filters are typically used for relatively broad-band analysis and are rarely used with less than one-third-octave bandwidth. This type of analysis finds most application in acoustics and in studies of the effects of vibration on humans (Chap. 44).

Digital Filters.   Digital filters (in particular, recursive digital filters) are devices which process a continuous digitized signal and provide another signal as an output which is filtered in some way with respect to the original. The relationship between the output and input samples can be expressed as a difference equation (in general, involving previous output and input values) with properties similar to those of a differential equation which might describe an analog filter. Figure 14.1 shows a typical two-pole section used in a one-third-octave digital filter analyzer (three of these are cascaded to give six-pole filtration).

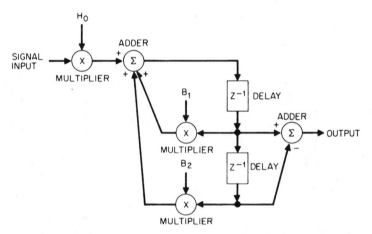

FIGURE 14.1   Block diagram of a typical two-pole digital filter section, consisting of multipliers, adders, and delay units. H_0, B_1, and B_2 are constants by which the appropriate signal sample is multiplied. Z^{-1} indicates a delay of one sample interval before the following operation.

Two ways of changing the properties of a given digital filter circuit such as that shown in Fig. 14.1 are:

1. For a given sampling frequency, the characteristics can be changed by changing the coefficients of the difference equation. (In the circuit of Fig. 14.1 there are three, effectively defining the resonance frequency, damping, and scaling.)

2. For given coefficients, the filter characteristic is defined only with respect to the sampling frequency. Thus, halving the sampling frequency will halve the cutoff frequencies, center frequencies, and bandwidths; consequently, the constant-percentage characteristics are maintained one octave lower in frequency. For this reason, digital filters are well adapted to constant-percentage bandwidth analysis on a logarithmic (i.e., octave-based) frequency scale.

Thus, the 3 one-third-octave characteristics within each octave are generated by changing coefficients, while the various octaves are covered by repetitively halving the sampling frequency. Every time the sampling frequency is halved, it means that only half the number of samples must be processed in a given time; the total number of samples for all octaves lower than the highest is $(\frac{1}{2} + \frac{1}{4} + \frac{1}{8} + \cdots)$, which in the limit is the same as the number in the highest octave. By being able to calculate twice as fast as is necessary for the upper octave alone, it is possible to cover any number of lower octaves in real time. This is the other reason why digital filters are so well adapted to real-time constant-percentage bandwidth analysis over a wide frequency range.

Filter Properties. Figure 14.2 illustrates what is meant by the 3-dB bandwidth and the effective noise bandwidth, the first being most relevant when separating discrete frequencies, and the second when dealing with random signals. For filters having good selectivity (i.e., having steep filter flanks), there is not a great difference between the two values, and so in the following discussion no distinction is made between them.

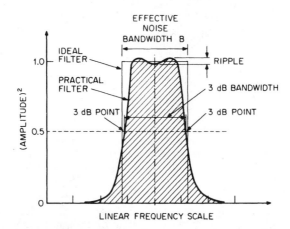

FIGURE 14.2 Bandwidth definitions for a practical filter characteristic. The *3-dB bandwidth* is the width at the 3-dB (half-power) points. The *effective noise bandwidth* is the width of an ideal filter with the same area as the (hatched) area under the practical filter characteristic on an amplitude squared (power) scale.

The response time T_R of a filter of bandwidth B is on the order of $1/B$, as illustrated in Fig. 14.3, and thus the delay introduced by the filter is also on this order. This relationship can be expressed in the form

$$BT_R \approx 1 \tag{14.1}$$

which is most applicable to constant-bandwidth filters, or in the form

$$bn_r \approx 1 \tag{14.2}$$

where $b = B/f_0$ = relative bandwidth
$n_r = f_0 T_R$ = number of periods of frequency f_0 in time T_R
f_0 = center frequency of filter

This form is more applicable to constant-percentage bandwidth filters. Thus, the response time of a 10-Hz bandwidth filter is approximately 100 milliseconds, while the response time of a 1 percent bandwidth filter is approximately 100 periods. Figure 14.3 also illustrates that the effective length of the impulse T_E is also approximately $1/B$, while to integrate all of the energy contained in the filter impulse response it is necessary to integrate over at least $3T_R$.

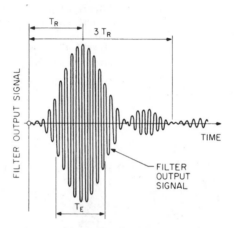

FIGURE 14.3 Typical filter impulse response. T_R = filter-response time ($\approx 1/B$); T_E = effective duration of the impulse ($\approx 1/B$); B = bandwidth.

Choice of Bandwidth and Frequency Scale.   In general it is found that analysis time is governed by expressions of the type $BT \geq K$, where K is a constant [see, for example, Eq. (14.1)] and T is the time required for each measurement with bandwidth B. Thus, it is important to choose the maximum bandwidth which is consistent with obtaining an adequate resolution, because not only is the analysis time per bandwidth proportional to $1/B$ but so is the number of bandwidths required to cover a given frequency range—a squared effect.

It is difficult to give precise rules for the selection of filter bandwidth, but the following discussion provides some general guidelines: For stationary deterministic and, in particular, periodic signals containing equally spaced discrete frequency components, the aim is to separate adjacent components; this can best be done using a constant bandwidth on a linear frequency scale. The bandwidth should, for example, be chosen as one-fifth to one-third of the minimum expected spacing (e.g., the lowest shaft speed, or its half-order if this is to be expected) (see Fig. 14.4A). For stationary random or transient signals, the shape of the spectrum will most likely be determined by resonances in the transmission path between the source and the pickup, and the bandwidth B should be chosen so that it is about one-third of the bandwidth B_r of the narrowest resonance peak (Fig. 14.4B). For constant damping these tend to have a constant Q or constant-percentage bandwidth character, and

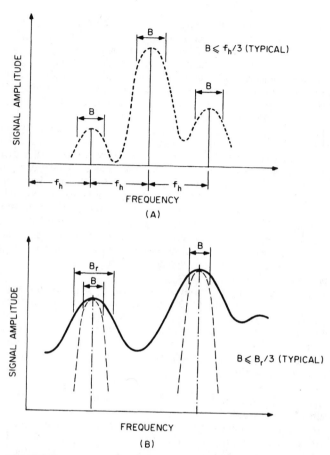

FIGURE 14.4 Choice of filter bandwidth B for different types of signals. (*A*) Discrete frequency signals—harmonic spacing f_h. (*B*) Stationary random and impulsive signals.

thus constant-percentage bandwidth on a logarithmic frequency scale often is most appropriate.

A linear frequency scale is normally used together with a constant bandwidth, while a logarithmic frequency scale is normally used together with a constant-percentage bandwidth, as each combination gives uniform resolution along the scale. A logarithmic scale may be selected in order to cover a wide frequency range, and then a constant-percentage bandwidth is virtually obligatory. A logarithmic frequency scale may, however, occasionally be chosen in conjunction with a constant bandwidth (though over a limited frequency range) in order to demonstrate a relationship which is linear on log-log scales (e.g., conversions between acceleration, velocity, and displacement).

Choice of Amplitude Scale.   Externally measured vibrations, on a machine casing for example, are almost always the result of internal forces acting on a structure whose frequency-response function modifies the result. Because the structural response functions vary over a very wide dynamic range, it is almost always an advantage to depict the vibration spectra on a logarithmic amplitude axis. This applies particularly when the vibration measurements are used as an indicator of machine condition (and thus, internal forces and stresses) since the largest vibration components by no means necessarily represent the largest stresses. Even where the vibration is of direct interest itself, in vibration measurements on humans, the amplitude axis should be logarithmic because this is the way the body perceives the vibration level.

It is a matter of personal choice (though sometimes dictated by standards) whether the logarithmic axes are scaled directly in linear units or in logarithmic units expressed in decibels (dB) relative to a reference value. Another aspect to be considered is dynamic range. The signal from an accelerometer (plus preamplifier) can very easily have a valid dynamic range of 120 dB (and more than 60 dB over three frequency decades when integrated to velocity). The only way to utilize this wide range of information is on a logarithmic amplitude axis. Figure 14.5 illustrates both these considerations; it shows spectra measured at two different points on the same gearbox (and representing the same internal condition) on both logarithmic and linear amplitude axes. The logarithmic representations of the two spectra are quite similar, while the linear representations are not only different but hide a number of components which could be important.

An exception where a linear amplitude scale usually is preferable to a logarithmic scale is in the analysis of relative displacement signals, measured using proximity probes, for the following reasons: (1) The parameter being measured is directly of interest for comparison with the results of rotor dynamic and bearing hydrodynamic calculations. (2) The dynamic range achievable with relative shaft vibration measurements (as limited by mechanical and electrical runout) does not justify or necessitate depiction on a logarithmic axis.

Analysis Speed. There are three basic elements in a filter analyzer which can give rise to significant delays and thus influence the speed of analysis.

The *filter* introduces a delay on the same order as its response time T_R (see Fig. 14.3). This is most likely to dominate in the analysis of stationary deterministic signals, where the filter contains only one discrete frequency component at a time and only a short averaging time is required.

The *detector* introduces a delay on the same order as the averaging time T_A. The choice of averaging time depends on the type of signal being analyzed, namely, stationary deterministic (discrete frequency) or stationary random.

In a swept-frequency analysis, the delays resulting from both filter response time and averaging time mean that the recorded spectrum departs from the true spectrum as illustrated in Fig. 14.6. The rate at which the pen traverses the frequency scale is called the *paper speed*, even in situations where the pen carriage moves with respect to stationary paper, as with an *X-Y* recorder. The following considerations are important in selecting an appropriate paper speed, which determines the analysis time:

1. Errors in the level of recorded peaks and valleys must be limited. Generally, valleys are less important because they are often determined by artifacts, such as the filter bandwidth and characteristic.

2. The delay, or frequency offset, of recorded peaks and valleys must also be limited to, say, one-fourth of the bandwidth.

3. The maximum rate of decay of the detector output signal is 8.7 dB per averaging time; this limits the maximum slope which can be traced on the trailing edge of

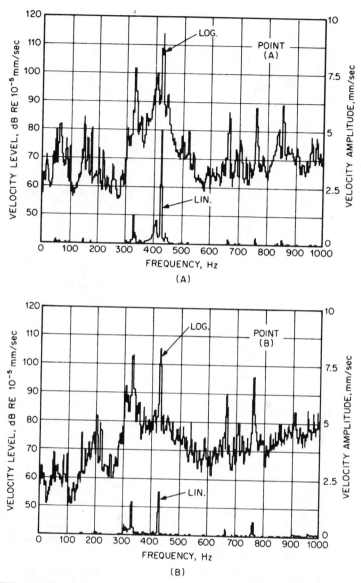

FIGURE 14.5 Comparison of rms logarithmic and rms linear amplitude scales for the depiction of vibration velocity spectra from two measurement points [(A) and (B)] on the same gearbox (thus representing the same internal condition). The logarithmic representations in terms of velocity level are similar and show all components of interest. The linear spectra in terms of velocity amplitude are quite different, and both hide many components which could be important.

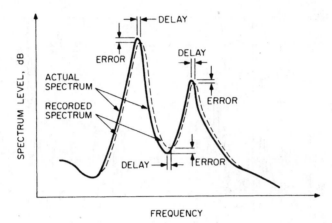

FIGURE 14.6 Differences between recorded and actual spectra in a swept-frequency analysis, with definition of the *delay* and *error* in recording peaks and valleys.

peaks in the spectrum. In the limiting case, the maximum slope which is to be recorded corresponds to the filter characteristic, and the paper speed should be limited to allow this.

Table 14.1 is based on these principles and, as detailed in Ref. 7, gives formulas for the selection of analysis parameters such as averaging time and sweep speed for both swept- and stepped-frequency analysis.

TABLE 14.1 Choice of Analysis Parameters for Serial Frequency Analysis*

	Signal type	
Analysis parameter	Deterministic (discrete frequency)	Random
Averaging time T_A (s)	$\geq 3/f_{min}, \leq T_D/2,$ and $\leq T_D/5^\dagger$	$\geq 16/B$
Dwell time T_D (s)	$\geq 4/B$	$\geq 2T_A$ or $\geq 5T_A{}^\dagger$
Sweep speed df/dt (Hz/second)‡	$\leq B/T_D$	$\leq B/T_D$

* f_{min} = lowest frequency in range covered, Hz; B = filter bandwidth, Hz; T_A = averaging time; T_D = dwell time in each filter (stepped analysis) or time to sweep over filter bandwidth (swept analysis); df/dt = sweep speed, Hz/second.
  † Applies to stepped filter analysis.
  $\ddagger$ Sweep speed is expressed in Hz/second if B is in Hz; it directly indicates the paper speed in mm/second if B is expressed as its equivalent width in mm on the paper.

A further limit may be imposed by the *graphic level recorder,* which may have a limited pen speed (writing speed). This speed should be selected sufficiently greater than the paper speed that the maximum slope of the filter characteristic can be reproduced (swept analysis) or the next filter level can be attained in the dwell time (stepped analysis). If this is not possible, the paper speed will have to be reduced accordingly. In some graphic level recorders, the writing speed is used indirectly to set the averaging time, in which case it is probably the pen speed as such, and not the averaging time, which limits the paper speed.

For *real-time parallel filter analysis,* the dwell time T_D is not relevant directly, but averaging time T_A can be chosen as the value given in Table 14.1 as T_A for random signals or as T_D for deterministic signals. In the latter case, T_A should also satisfy the requirement for random signals at higher frequencies where the (constant-percentage) filter bandwidth includes several discrete frequency components. The same value of T_A applies to both linear and exponential averaging, but in the latter case it may be necessary to wait a time corresponding to $2T_A$ to eliminate bias error.

Scaling and Calibration for Stationary Signals. *Scaling* is the process of determining the correct units for the Y axis of a frequency analysis, while *calibration* is the process of setting and confirming the numerical values along the axis. In the most general case, spectra can be scaled in terms of mean-square or rms values at each frequency (or, strictly speaking, for each filter band). For signals dominated by discrete frequency components, with no more than one component per filter band, this yields the mean-square or rms value of each component.

A spectrum of mean-square values is known as a *power spectrum* since physical power often is related to the mean-square value of parameters such as voltage, current, force, pressure, and velocity.

For random signals, the power spectrum values vary with the bandwidth but can be normalized to a *power spectral density* $W(f)$ by dividing by the bandwidth. The results then are independent of the analysis bandwidth, provided the latter is narrower than the width of peaks in the spectrum being analyzed (e.g., following Fig. 14.4B). As examples, power spectral density is expressed in g^2 per hertz when the input signal is expressed in gs acceleration, and in volts squared per hertz when the input signal is in volts.

The concept of power spectral density is meaningless in connection with discrete frequency components (with infinitely narrow bandwidth); it can be applied only to the random parts of signals containing mixtures of discrete frequency and random components. Nevertheless, it is possible to calibrate a power spectral density scale using a discrete frequency calibration signal. For example, when analyzing a $1g$ sinusoidal signal with a 10-Hz analyzer bandwidth, the height of the discrete frequency peak may be labeled $1^2 g^2/10\ \text{Hz} = 0.1 g^2/\text{Hz}$.

For constant-bandwidth analysis, the scaling thus achieved is valid for all frequencies; for constant-percentage bandwidth analysis, the bandwidth and power spectral density scaling vary with frequency. On log-log axes, it is possible to draw straight lines representing constant power spectral density, which slope upwards at 10 dB per frequency decade from the calibration point.

Real-Time Digital Filter Analysis of Transient Signals. Suppose a digital filter analyzer has a constant-percentage bandwidth (e.g., one-third-octave or one-twelfth-octave) and covers a frequency range of three or four decades. Because the bandwidth varies with frequency, the filter output signal also varies greatly. At low frequencies (where B is small) the filter output resembles its impulse response, with a length dominated by the filter response time T_R. At high frequencies (where T_R is short) the filter output signal follows the input more closely and has a length dominated by T_I, the duration of the input impulse.

This is illustrated in Fig. 14.7, which traces the path of a typical impulsive signal (an N-wave) through the complete analysis system of filter, squarer, and averager for both a narrow-band (low-frequency) and a broad-band (high-frequency) filter.

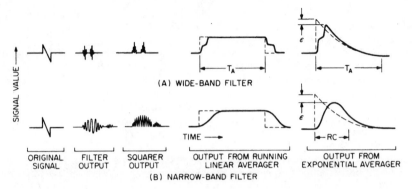

FIGURE 14.7 Passage of a transient signal through an analyzer comprising a filter, squarer, and averager (alternatively running linear averaging and exponential averaging). The dotted curves represent the averager impulse responses. RC is the time constant for exponential averaging. ε is the error in peak response. (A) With a wide-band filter. (B) With a narrow-band filter.

The averaging time T_A must always satisfy

$$T_A \geq T_I + 3T_R \tag{14.3}$$

Thus the averaging time is determined by the lowest frequency to be analyzed. The ideal solution would be running linear integration [with T_A selected using Eq. (14.3)] followed by a maximum-hold circuit (which retains the maximum value experienced). The output of such a running linear averager is shown in Fig. 14.7. Note that during the time the entire filter output is contained within the averaging time T_A, the averager output provides the correct result, which is held by the maximum-hold circuit. However, a running linear average is very difficult to achieve, and normally it is necessary to choose between fixed linear averaging and running exponential averaging.

The problem with fixed linear averaging is that it must be started just before the arrival of the impulse and thus cannot be triggered from the signal itself (unless use is made of a delay line before the analyzer). It is, however, possible to record the signal first and then insert a trigger signal (for example, on another channel of a tape recorder).

In order to extract all the information from a given signal, it may be necessary to make the total analysis in two passes. For example, Fig. 14.8 shows the analysis of a 220-millisecond N-wave (the pressure signal from a sonic boom). For an averaging time $T_A = 0.5$ sec, the spectrum is valid only down to about 50 Hz, but it includes frequency components up to 5 kHz. This illustration also shows an analysis of the same signal using $T_A = 8$ sec; this is valid down to about 1.6 Hz. However, as a result of this longer averaging time, there is a 12-dB loss of dynamic range, and so all the frequency components above 500 Hz are lost. The result (with scaling adjusted by 12 dB) is given as a dotted line in Fig. 14.8; it shows that the two spectra are identical over the mutually valid range.

Where the analysis is carried out in real time on randomly occurring impulses, exponential averaging may be used followed by a maximum-hold circuit, but then there is the added complication that the averager leaks energy at a (maximum) rate of 8.7 dB per averaging time T_A, and thus the total impulse duration must be short with respect to T_A. The error is less than 0.5 dB if

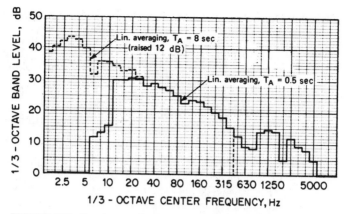

FIGURE 14.8 Transient analysis of a sonic boom (length 218 milliseconds) using a one-third-octave digital filter analyzer. T_A = selected averaging time. The dotted curve (T_A = 8 sec) has been raised 12 dB to compensate for the longer averaging time.

$$T_A \geq 10(T_I + T_R) \tag{14.4}$$

Note that the peak output of an exponential averager is a factor of 2 (i.e., 3 dB) higher than that of the equivalent linear averager (Figs. 13.7 and 14.7); thus the equivalent averaging time to be used in converting from power to energy units is $T_A/2$ for exponential averaging and T_A for linear averaging. Conversion from energy to energy spectral density is valid only for that part of the spectrum where the analyzer bandwidth is appreciably less than the signal bandwidth, although outside that range the results may be interpreted as the mean energy spectral density in the band.

FFT ANALYZERS

FFT analyzers make use of the FFT (fast Fourier transform) algorithm to calculate the spectra of blocks of data. The FFT algorithm is an efficient way of calculating the discrete Fourier transform (DFT). As described in Chap. 22, this is a finite, discrete approximation of the Fourier integral transform. The equations given there for the DFT [Eqs. (22.10) and (22.11)] assume real-valued time signals. The FFT algorithm makes use of the following versions, which apply equally to real or complex time series:

$$S_x(m) = \Delta t \sum_{n=0}^{N-1} x(n\,\Delta t) \exp(-j2\pi m \,\Delta f n \,\Delta t) \tag{14.5}$$

$$x(n) = \Delta f \sum_{m=0}^{N-1} S_x(m\,\Delta f) \exp(j2\pi m \,\Delta f n \,\Delta t) \tag{14.6}$$

These equations give the spectrum values $S_x(m)$ at the N discrete frequencies $m\,\Delta f$ and give the time series $x(n)$ at the N discrete time points $n\,\Delta t$.

Whereas the Fourier transform equations are infinite integrals of continuous functions, the DFT equations are finite sums but otherwise have similar properties.

The function being transformed is multiplied by a rotating unit vector $\exp(\pm j2\pi m \,\Delta f \, n \,\Delta t)$, which rotates (in discrete jumps for each increment of the time parameter n) at a speed proportional to the frequency parameter m. The direct calculation of each frequency component from Eq. (14.5) requires N complex multiplications and additions, and so to calculate the whole spectrum requires N^2 complex multiplications and additions.

The FFT algorithm factors the equation in such a way that the same result is achieved in roughly $N \log_2 N$ operations.[1] This represents a speedup by a factor of more than 100 for the typical case where $N = 1024 = 2^{10}$. However, the properties of the FFT result are the same as those of the DFT.

Inherent Properties of the DFT. Figure 14.9 graphically illustrates the differences between the DFT and the Fourier integral transform.

Because the spectrum is available only at discrete frequencies $m \,\Delta f$ (where m is an integer), the time function is implicitly periodic (as for the Fourier series). The periodic time

$$T = N \,\Delta t = 1/\Delta f \tag{14.7}$$

where $N =$ number of samples in time function and frequency spectrum
       $T =$ corresponding record length of time function
       $\Delta t =$ time sample spacing
       $\Delta f =$ line spacing $= 1/T$

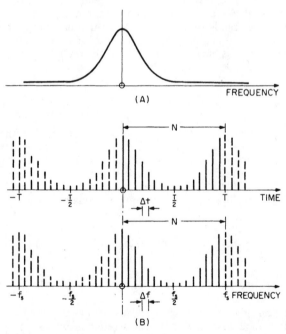

FIGURE 14.9 Graphical comparison of (A) the Fourier transform with (B) the discrete Fourier transform (DFT) (see text).

In an analogous manner, the discrete sampling of the time signal means that the spectrum is implicitly periodic, with a period equal to the sampling frequency f_s, where

$$f_s = N \, \Delta f = 1/\Delta t \tag{14.8}$$

Note from Fig. 14.9 that because of the periodicity of the spectrum, the latter half ($m = N/2$ to N) actually represents the negative frequency components ($m = -N/2$ to 0). For real-valued time samples (the usual case), the negative frequency components are determined in relation to the positive frequency components by the equation

$$S_x(-m) = S_x{}^*(m) \tag{14.9}$$

and the spectrum is said to be *conjugate even*.

In the usual case where the $x(n)$ are real, it is only necessary to calculate the spectrum from $m = 0$ to $N/2$, and the transform size may be halved by one of the following two procedures:

1. The N real samples are transformed as though representing $N/2$ complex values, and that result is then manipulated to give the correct result.[2]

2. A zoom analysis (discussed in a later section) is performed which is centered on the middle of the base-band range to achieve the same result.

Thus, most FFT analyzers produce a (complex) spectrum with a number of spectral lines equal to half the number of (real) time samples transformed. To avoid the effects of aliasing (see next section), not all the spectrum values calculated are valid, and it is usual to display, say, 400 lines for a 1024-point transform or 800 lines for a 2048-point transform.

Aliasing. Aliasing is an effect introduced by the sampling of the time signal, whereby high frequencies after sampling appear as lower ones (as with a stroboscope). The DFT algorithm of Eq. (14.5) cannot distinguish between a component which rotates, say, seven-eighths of a revolution between samples and one which rotates a negative one-eighth of a revolution. Aliasing is normally prevented by low-pass filtering the time signal before sampling to exclude all frequencies above half the sampling frequency (i.e., $-N/2 < m < N/2$). From Fig. 14.9 it will be seen that this removes the ambiguity. In order to utilize up to 80 percent of the calculated spectrum components (e.g., 400 lines from 512 calculated), it is necessary to use very steep antialiasing filters with a slope of about 120 dB/octave.

Normally, the user does not have to be concerned with aliasing because suitable antialiasing filters automatically are applied by the analyzer. One situation where it does have to be allowed for, however, is in tracking analysis (discussed in a following section) where, for example, the sampling frequency varies in synchronism with machine speed.

Leakage. Leakage is an effect whereby the power in a single frequency component appears to leak into adjacent bands. It is caused by the finite length of the record transformed (N samples) whenever the original signal is longer than this; the DFT implicitly assumes that the data record transformed is one period of a periodic signal, and the leakage depends on what is actually captured within the time window, or data window.

Figure 14.10 illustrates this for three different sinusoidal signals. In (A) the data window corresponds to an exact integer number of periods, and a periodic repetition of this produces an infinitely long sinusoid with only one frequency. For (B) and (C) (which have a slightly higher frequency) there is an extra half-period in the data

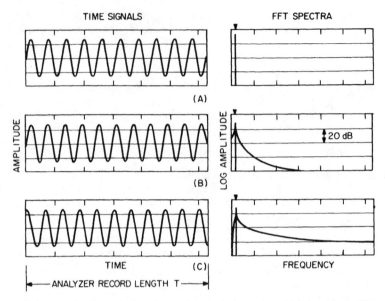

FIGURE 14.10 Time-window effects when analyzing a sinusoidal signal in an FFT analyzer using rectangular weighting. (*A*) Integer number of periods, no discontinuity. (*B*) and (*C*) Half integer number of periods but with different phase relationships, giving a different discontinuity when the ends are joined into a loop.

record, which gives a discontinuity where the ends are effectively joined into a loop, and considerable leakage is apparent. The leakage would be somewhat less for intermediate frequencies. The difference between the cases of Fig. 14.10*B* and *C* lies in the phase of the signal, and other phases give an intermediate result.

When analyzing a long signal using the DFT, it can be considered to be multiplied by a (rectangular) time window of length *T,* and its spectrum consequently is convolved with the Fourier spectrum of the rectangular time window,[3] which thus acts like a filter characteristic. The actual filter characteristic depends on how the resulting spectrum is sampled in the frequency domain, as illustrated in Fig. 14.11.

In practice, leakage may be counteracted:

1. By forcing the signal in the data window to correspond to an integer number of periods of all important frequency components. This can be done in tracking analysis (discussed in a later section) and in modal analysis measurements (Chap. 21), for example, where periodic excitation signals can be synchronized with the analyzer cycle.

2. (For long transient signals) By increasing the length of the time window (for example, by zooming) until the entire transient is contained within the data record.

3. By applying a special time window which has better leakage characteristics than the rectangular window already discussed.

Later sections deal with the choice of data windows for both stationary and transient signals.

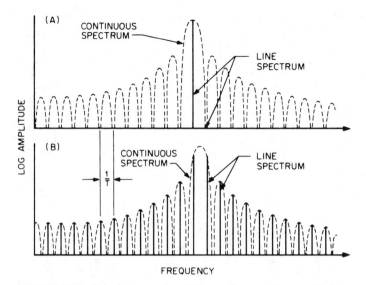

FIGURE 14.11 Frequency sampling of the continuous spectrum of a time-limited sinusoid of length T. (A) Integer number of periods, side lobes sampled at zero points (compare with Fig. 14.10A). (B) Half integer number of periods, side lobes sampled at maxima (compare with Fig. 14.10B and C).

Picket Fence Effect. The *picket fence effect* is a term used to describe the effects of discrete sampling of the spectrum in the frequency domain. It has two connotations:

1. It results in a nonuniform frequency weighting corresponding to a set of overlapping filter characteristics, the tops of which have the appearance of a picket fence (Fig. 14.12).

2. It is as though the spectrum is viewed through the slits in a picket fence, and thus peak values are not necessarily observed.

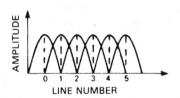

FIGURE 14.12 Illustration of the picket fence effect. Each analysis line has a filter characteristic associated with it which depends on the weighting function used. If a frequency coincides exactly with a line, it is indicated at its full level. If it falls midway between two lines, it is represented in each at a lower level corresponding to the point where the characteristics cross.

One extreme example is in fact shown in Fig. 14.11, where in (A) the side lobes are completely missed, while in (B) the side lobes are sampled at their maxima and the peak value is missed.

The picket fence effect is not a unique feature of FFT analysis; it occurs whenever discrete fixed filters are used, such as in normal one-third–octave analysis. The maximum amplitude error which can occur depends on the overlap of the adjacent filter characteristics, and this is one of the factors taken into account in the following discussion on the choice of data window.

Data Windows for Analysis of Stationary Signals. A *data window* is a weighting function by which the data record is effectively multiplied before transformation. (It is sometimes more efficient to apply it by convolution in the frequency domain.) The purpose of a data window is to minimize the effects of the discontinuity which occurs when a section of continuous signal is joined into a loop.

For stationary signals, a good choice is the *Hanning window* (one period of a cosine squared function), which has a zero value and slope at each end and thus gives a gradual transition over the discontinuity. In Fig. 14.13 it is compared with a rectangular window, in both the time and frequency domains. Even though the main lobe (and thus the bandwidth) of the frequency function is wider, the side lobes fall off much more rapidly and the highest is at −32 dB, compared with −13.4 dB for the rectangular.

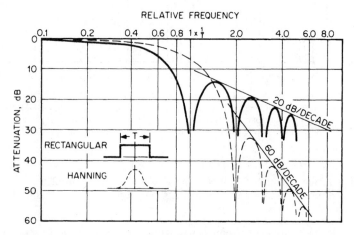

FIGURE 14.13 Comparison of rectangular and Hanning window functions of length T seconds. Full line—rectangular weighting; dotted line—Hanning weighting. The inset shows the weighting functions in the time domain.

Other time-window functions may be chosen, usually with a trade-off between the steepness of filter characteristic on the one hand and effective bandwidth on the other. Table 14.2 compares the time windows most commonly used for stationary signal, and Fig. 14.14 compares the effective filter characteristics of the most important. The most highly selective window, giving the best separation of closely spaced components of widely differing levels, is the *Kaiser-Bessel window.* On the other hand, it is usually possible to separate closely spaced components by zooming, at the expense of a slightly increased analysis time.

Another window, the *flattop window,* is designed specifically to minimize the picket fence effect so that the correct level of sinusoidal components will be indicated, independent of where their frequency falls with respect to the analysis lines. This is particularly useful with calibration signals. Nonetheless, by taking account of the distribution of samples around a spectrum peak, it is possible to compensate for picket fence effects with other windows as well. Figure 14.15, which is specifically for the Hanning window, is a nomogram giving both amplitude and frequency corrections, based on the decibel difference (ΔdB) between the two highest sam-

TABLE 14.2 Properties of Various Data Windows

Window type	Highest side lobe, dB	Side lobe fall-off, dB/decade	Noise bandwidth*	Maximum amplitude error, dB
Rectangular	−13.4	−20	1.00	3.9
Hanning	−32	−60	1.50	1.4
Hamming	−43	−20	1.36	1.8
Kaiser-Bessel	−69	−20	1.80	1.0
Truncated Gaussian	−69	−20	1.90	0.9
Flattop	−93	0	3.70	<0.1

* Relative to line spacing.

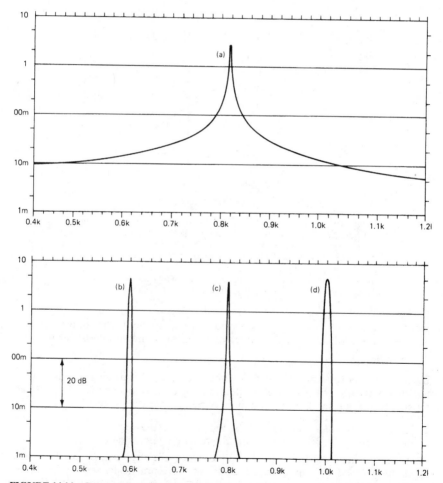

FIGURE 14.14 Comparison of worst-case filter characteristics for rectangular and other weighting functions for an 80-dB dynamic range. (*A*) Rectangular. (*B*) Kaiser-Bessel. (*C*) Hanning. (*D*) Flattop.

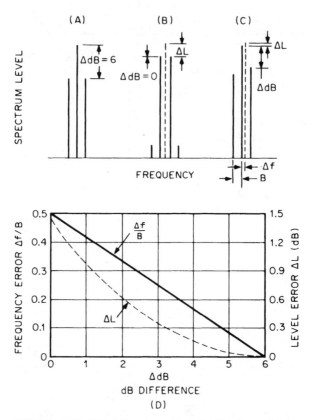

FIGURE 14.15 Picket fence corrections for Hanning weighting, where ΔL = level correction, dB; Δf = frequency correction; Hz; B = line spacing, Hz; ΔdB = difference in decibels between the two highest samples around a peak representing a discrete frequency component. Three examples are shown: (A) Actual frequency coincides with center line. (B) Actual frequency midway between two lines. (C) General situation. Note that the frequency correction $\Delta f/B$ is almost linear.

ples around a peak. For stable single-frequency components this allows determination of the frequency to an accuracy of an order of magnitude better than the line spacing.

Data Windows for Analysis of Transient Signals.   When using impulsive (e.g., hammer) excitation of structures for determining their frequency-response characteristics (e.g., see Chap. 21), it is common to use the following special data windows:

1. A *short rectangular* window may be applied over the very short excitation impulse in order to exclude noise from the remaining portion of the record.

2. An *exponential* window can be applied where the response is very long (i.e., lightly damped structures) to reduce the signal to practically zero at the end of the record, and thus avoid discontinuities. The effect is the same as adding extra

damping which is very precisely known and can be subtracted from the measurement results.

Zoom Analysis.[4]   Zoom analysis is the term given to a spectrum analysis having increased resolution over a restricted part of the frequency range. The following are two techniques used to generate zoom analyses.

1. *Real-time zoom* (illustrated in the block diagram of Fig. 14.16) is a zoom process in which the entire signal is first modified to shift its frequency origin to the center of the zoom range. Then it is passed through a low-pass filter (usually a digital filter in real time) which has a passband corresponding to the original zoom-band (Fig. 14.17). Because of the low-pass filtration, the signal then can be resampled at a lower sampling rate without aliasing, and the resampled signal processed by an FFT transform. The original frequency shift is accomplished by multiplying the incoming signal by a unit vector (phasor) rotating at $-f_0$ (thereby subtracting f_0 from all frequencies in it), and the modified time signal is thus complex. This is one situation where the FFT transform of complex data is used. Figure 14.16 gives an example of the use of zoom analysis to show that what appears in a baseband analysis to be the second harmonic of shaft speed actually is dominated by twice the line frequency at 100 Hz. Figure 14.18 shows an example of real-time zooming by a factor of 64:1. It reveals that what appears to be a single-frequency component in a band spectrum actually comprises a family of uniformly spaced sidebands.

2. *Nondestructive zoom* is effectively a way of achieving a larger transform size without modification of the original data record. For a typical case, data are first captured in a 10K (i.e., 10,240-point) buffer. Ten 1K records obtained by taking every 10th sample are transformed using a 1K (1024-point) FFT transform. Even though this gives only 1024 frequency values per transform, the rest are generated by periodic repetition (because of aliasing resulting from the undersampling). After com-

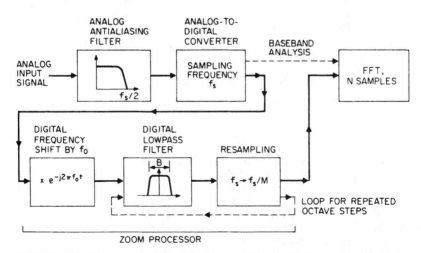

FIGURE 14.16   Block diagram for real-time zoom with bandwidth B centered on frequency f_0. M is the zoom factor and also the factor by which the sampling frequency is reduced.

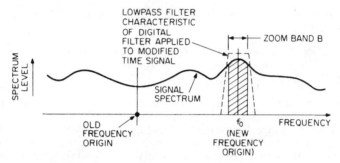

FIGURE 14.17 Principle of real-time zoom, using a low-pass filter to filter out the portion of the original signal in the zoom-band of width B. Prior to this, the frequency origin is shifted to frequency f_0 (the desired center frequency of the zoom-band) by multiplying the (digitized) time signal by $e^{-j2\pi f_0 t}$.

pensating the phase of the results for the small time shift of each of the undersampled records, the entire spectrum of the 10K record can in principle be obtained by addition. In practice, only a part of the whole spectrum is normally obtained at any one time in order to save on memory requirements, but the whole spectrum can be generated by repetitive operation on exactly the same data.

Real-time zoom has the advantage that the zoom factor obtainable is virtually unlimited. A procedure is often employed (as illustrated in Fig. 14.16) whereby the signal samples are repeatedly circulated around a loop containing a low-pass filter which cuts off at one-half the previous maximum frequency, after which the sampling frequency is halved by dropping every second sample. Each circulation doubles the zoom factor and at the same time doubles the length of original signal required to fill the transform buffer. It is this time requirement which places a limit on the zoom factor, as well as on the stability of the signal itself. A zoom factor of 10 in a 400-line spectrum, for example, gives the equivalent of a 4000-line spectrum; a finer resolution is not required to analyze the vibration spectrum of a machine whose speed fluctuates by, say, 0.1 percent.

Real-time zoom suffers the disadvantage that the entire signal must be reprocessed to zoom in another band. This has two detrimental consequences:

1. For very narrow bandwidths (long record lengths), the analysis time is very long for each zoom analysis.

2. There is no certainty that exactly the same signal is processed each time.

On the other hand, nondestructive zoom has the advantage that for zoom analysis in different bands, exactly the same data record is used. Thus it is known, for example, that there will be an exact integer relationship between the various harmonics of a periodic signal. This can be useful, as a typical example, in separating the various harmonics of shaft speed from those of line frequency, in induction motor vibrations. Furthermore, the long analysis time is required only once (to fill the data buffer); further zoom analyses on the same record are limited only by the calculation speed.

Nondestructive zoom suffers the disadvantage that the zoom factor is limited by the size of the memory buffer in the analyzer. Where the memory buffer is 10 times the normal transform size, for example, the zoom factor is equal to 10.

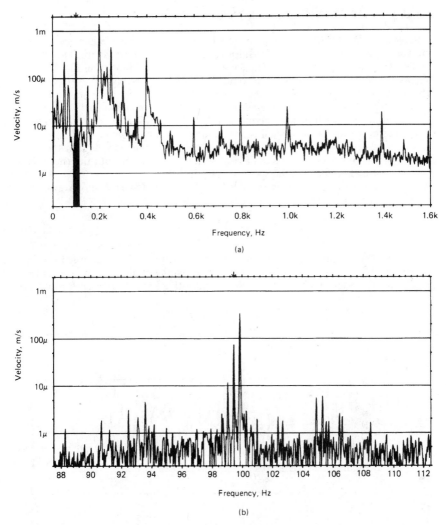

FIGURE 14.18 (*A*) Original baseband spectrum. (*B*) Shaded section of (*A*) zoomed by a factor of 64:1. Highest component at 100 Hz is twice the line frequency. Next highest component on the left is twice the shaft rotational speed.

Thus, both types of zoom are advantageous for different purposes. Nondestructive zoom is probably best for diagnostic analysis of machine vibration signals, whereas real-time zoom gives more flexibility in frequency-response measurements (system frequency response should not change even where the excitation signals change). Real-time zoom also gives the possibility of very large zoom factors when they are required.

In real-time zoom, it is only the preprocessing of the signal which has to be in real time; the actual FFT analysis of the signal, once it is stored in the transform buffer, does not have to be in real time.

ANALYSIS OF STATIONARY SIGNALS USING FFT

Equation (14.7) shows that for a single FFT transform, the product (*bandwidth* times *averaging time*) $BT_A = 1$, at least for rectangular weighting where B is equal to the line spacing Δf (Table 14.2). The same applies for any weighting function, the increased bandwidth being exactly compensated by a corresponding decrease in effective record length.[5]

For *stationary deterministic signals,* a single transform having a BT_A product equal to unity is theoretically adequate, although a small number of averages is sometimes performed if the signal is not completely stable. Figure 14.19 illustrates the effect of averaging for a deterministic signal and demonstrates that the sinusoidal components are unaffected; the only effect is to smooth out the (nondeterministic) noise at the base of the spectrum (Fig. 14.19*B*).

For *stationary random signals,* the standard deviation of the result of averaging n independent spectra is given by

$$\varepsilon = \frac{1}{2\sqrt{n}} \tag{14.10}$$

Figure 14.20 illustrates (*A*) an instantaneous spectrum, (*B*) the average of eight spectra, and (*C*) the average of 128 spectra. The meaning of the standard error ε [Eq. (14.10)] is illustrated in (*B*) and (*C*). Statistically, there is a 68 percent probability

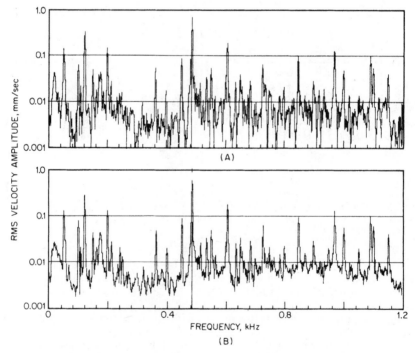

FIGURE 14.19 Effect of averaging with a stationary deterministic signal. (*A*) Instantaneous spectrum (average of 1). (*B*) The linear average of eight spectra.

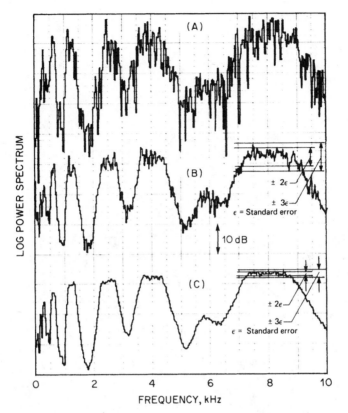

FIGURE 14.20 Effect of averaging with a stationary random signal. (*A*) Instantaneous spectrum. (*B*) Average of eight spectra. (*C*) Average of 128 spectra.

that the actual error will be less than ε, a 95.5 percent probability that it will be less than 2ε, and a 99.7 percent probability that it will be less than 3ε.

For rectangular (flat) weighting, independent spectra are those from nonoverlapping time records; when other weighting functions are used, the situation is different. For example, Fig. 14.21*A* illustrates the overall (power) weighting obtained when Hanning windows are applied to contiguous records. Note that virtually half of the incoming signal is excluded from the analysis, whereas a 50 percent overlapping of consecutive records regains most of the lost information. Thus, when using window functions similar to Hanning (as recommended for stationary signals), it is almost always advantageous to average the results from 50 percent overlapping records. A method for calculating the effective number of averages obtained in this way is given in Ref. 6; for 50 percent overlapping Hanning windows the error is very small in treating them as independent records.

Real-Time Analysis. An FFT analyzer is said to operate in real time when it is able to process all the incoming data, even though presentation of the results is delayed by an amount corresponding to the calculation time. This implies that the time taken to analyze a data record, T_a, is less than the time taken to collect the data

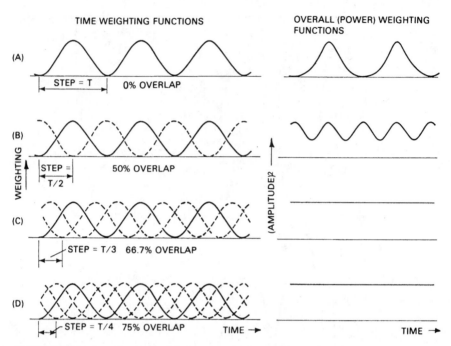

FIGURE 14.21 Overall weighting functions for spectrum averaging with overlapping Hanning windows. (*A*) Zero overlap (step length *T*). (*B*) 50 percent overlap (step length *T*/2). (*C*) 66.7 percent overlap (step length *T*/3). (*D*) 75 percent overlap (step length *T*/4). *T* is the record length for the FFT transform.

transformed, *T*. It also implies that the analysis process should not interrupt the continuous recording of data, so that recording can continue in one part of the memory at the same time as analysis is being performed in another. *T* is inversely proportional to the selected frequency range, and the highest frequency range for which T_a is less than *T* is called the *real-time frequency*. This condition will ensure that all the incoming data are analyzed only when rectangular weighting is used. With Hanning weighting, for example, where 50 percent overlap analysis must be employed to analyze all the data, the true real-time frequency will be halved, since twice as many transforms must be performed for the same length of data record. In yet another sense, the analysis is not truly real-time unless the overall weighting function is uniform. As illustrated in Fig. 14.21, the minimum overlap of Hanning windows to achieve this is two-thirds, which reduces the true real-time frequency to one-third of the commonly understood definition given above.

In practice, with stationary signals, there is no advantage to more than a 50 percent overlap, since (1) statistical reliability is not significantly improved and (2) all sections of the record are statistically equivalent, so that the overall weighting function is not important. It can be important for nonstationary signals, such as transients, as discussed below. For stationary signals, where any data missed are statistically no different from the data analyzed, the only advantages of real-time analysis are that (1) results with a given accuracy are obtained in the minimum possible time and (2) maximum information is extracted from a record of limited length.

FFT Analysis of Transients.   Consider the use of FFT analysis when the entire transient fits into the transform size T without loss of high-frequency information. Figure 14.22 shows such an example where the duration of the transient is less than the analyzer record length of 2048 samples (2K) in a frequency range which does not exclude high-frequency information in the signal. Flat (i.e., rectangular) weighting should be used in such a case, where the signal value is zero at each end, so that no discontinuity arises from making the record into a loop (an inherent property of the FFT process). Exponential weighting sometimes may be used to force the signal down to zero at the end of the record, but the frequency spectrum will then include the effects of the extra damping which this represents.

With rectangular weighting, the analysis bandwidth is equal to the line spacing $1/T$, which is always less than the effective signal bandwidth. Conversion of the results to energy spectral density, therefore, is valid in most practical situations. Some analyzers provide the results in terms of energy spectral density, but if the results are available only in terms of power (U^2), they must be multiplied by the time T corresponding to the record length to convert them to energy and divided by the bandwidth $1/T$ to convert them to energy spectral density, expressed in engineering units squared times seconds per hertz. Altogether, this represents a multiplication by T^2.

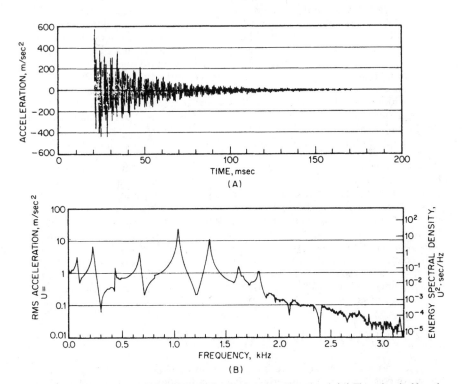

FIGURE 14.22   Example of an FFT analysis of a short transient signal. (A) Time signal of length 2048 samples (2K) corresponding to 250 milliseconds ($T = 250$ milliseconds). (B) 800-line FFT spectrum with bandwidth $B = \Delta f = 4$ Hz (flat weighting). Scaling on left is in rms units. Scaling on right is converted to energy spectral density (ESD) by multiplying mean-square values by T^2.

Where a transient is longer than the normal transform size T, it can be analyzed in one of the following ways:

1. *Zoom FFT* (see *Zoom Analysis*, above, for background information). A suitable zoom factor is chosen such that the transform length $(1/\Delta f)$ is greater than the duration of the transient. Thus in the case of nondestructive zoom, the entire transient can be recorded in the long memory and the entire narrow-band spectrum can be obtained by repetitive analysis in contiguous zoom-bands. In the case of real-time zoom, analysis in more than one zoom-band requires that the transient be recorded in an external medium and played back for each zoom analysis. Rectangular weighting should be used (thus $B = \Delta f$) and energy spectral density (as above) is always valid using a value of T corresponding to the zoom record length $(1/B)$. The narrow bandwidth may give a restriction of dynamic range of the result. Figure 14.23 shows a typical energy spectrum, obtained by repetitive nondestructive zoom analysis.

2. *Scan averaging.* When the entire transient is stored in digital form in a long memory (as for nondestructive zoom), it is possible to obtain its spectrum by scanning a short time window (e.g., a Hanning window) of length T over the entire record; this is done in overlapping steps, and the results are averaged. As already demonstrated for stationary signals (Fig. 14.21), this procedure yields a result with uniform weighting for step lengths $T/3$ and $T/4$. The same applies to step lengths $T/5$, $T/6$, etc., but there is a slight difference with respect to the overall weighting function for the different step lengths. Figure 14.24 illustrates the overall time weighting function for different step lengths T/n (where n is an integer greater than 2) and shows the length of the uniform section (within which the entire transient should ideally be located) and the effective length T_{eff} by which power units should be multiplied to convert them to energy. For a conversion to energy spectral density to be valid, the width of spectrum peaks must be somewhat greater than the analysis bandwidth; this can be seen by inspection of the analysis results. For example, for the Hanning window, the bandwidth B is 1.5 times the line spacing Δf (see Table 14.2), and so spectrum peaks should have a 3-dB bandwidth of more than five lines.

Even though the broader bandwidth obtained by scan averaging may result in a loss of spectrum detail, it provides considerable improvement in the dynamic range of the result. Figure 14.25 (using scan averaging) illustrates these points for the same signal as Fig. 14.23 (using zoom). The spectrum obtained by scan averaging generally has 12 dB more dynamic range than that obtained by zoom (with factor 10), but the level of peaks does not differ by this amount; this confirms that their resolution is not sufficient to allow scaling in terms of energy spectral density.

To obtain Fig. 14.25, scan averaging with a step length of $T/4$ was used (an overlap of successive records of 75 percent). Even though a step length of $T/3$ (overlap of 66.7 percent) is theoretically more efficient, $T/4$ is usually more convenient because the number of samples in T generally is a power of 2.

ANALYSIS OF NONSTATIONARY SIGNALS

A typical nonstationary signal results from measurements made during a machine run-up or coast-down (here, the primary cause of the nonstationary signal is a change in shaft speed). The signal can be analyzed by dividing it up into a series of short quasi-stationary time periods (often overlapping), in each of which the speed is roughly constant. The length of the time window used to select a portion of the con-

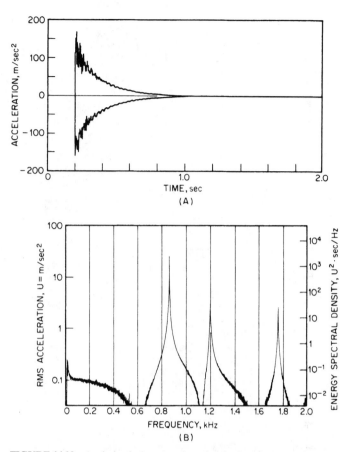

FIGURE 14.23 Analysis of a long transient signal using nondestructive zoom FFT. (*A*) Envelope of time signal of length 10,240 samples (10K) corresponding to 2 seconds ($T = 2$ seconds). (*B*) 4000-line composite zoom spectrum with bandwidth $B = \Delta f = 0.5$ Hz (flat weighting). Scaling on right is converted to energy spectral density (ESD).

tinuous signal may have to be chosen so as to ensure this. The simplest way to analyze a nonstationary signal of this type is to use a tracking filter tuned to a specific harmonic of shaft speed and to record the results vs. rpm of the machine. If a phase meter is inserted between the filtered signal and the tracking signal, it is possible to record phase as well as amplitude against rpm to give what is called a *Bode plot*.[8]

Using an FFT analyzer, the behavior of several harmonics may be studied simultaneously. One way to do this, using an FFT analyzer having a long memory (such as is required for nondestructive zoom), is with a simple scan analysis; a short Hanning window is scanned through the record (as for a *scan average*), and successive instantaneous spectra (from each window position) are viewed on the display screen. The speed of the scan may be changed by varying the step length; this is one situation (in contrast to scan averaging) where very short step lengths may be of advantage, for example, in slowing down the passage through a resonance.

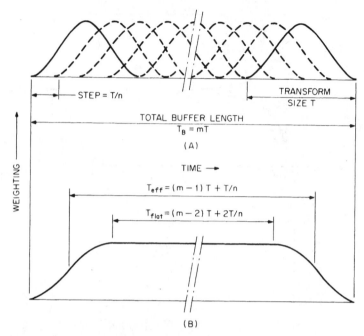

FIGURE 14.24 Overall weighting function for scan averaging of a transient. (*A*) Overlapping Hanning windows of length T with definition of parameters m and n. (*B*) Overall weighting function with indication of T_{eff} and T_{flat} in terms of T, m, and n. T_{eff} is the effective length of the time window for conversion of power to energy units. T_{flat} is the length of the section with uniform weighting within which the transient ideally should be located.

A highly effective method of representing such a scan analysis is by a "waterfall," or "cascade," plot as shown in Fig. 14.26 (which represents a typical machine run-up). As indicated, the third dimension of such a three-dimensional plot can be either time or rpm; for a simple scan analysis it usually is time, but if the spectra are spaced at equal intervals of rpm, a number of advantages result. Harmonically related components (whose bases follow radial lines) then can be separated easily from constant-frequency components (e.g., related to line frequency or resonances) whose bases follow lines parallel with the rpm axis. Such a cascade plot, with rpm as the third axis, is sometimes referred to as a *Campbell diagram*, although strictly speaking a Campbell diagram has a vertical frequency axis, a horizontal rpm axis, and a signal amplitude represented as the diameter of a circle (or square) centered on the appropriate point in the diagram.

Ideally, each of the spectra in a cascade plot such as Fig. 14.26 should be obtained with constant shaft speed at the respective rpm. This is sometimes possible, for example, during the very slow start-up of a large steam turbine, but usually each spectrum is a windowed section of a continuously varying signal with a small speed change within the window length. Consequently, the peak corresponding to each harmonic is not always localized in one analysis line; in particular, the higher harmonics are likely to be spread over progressively more lines. Thus, the height of each peak cannot be used directly as a measure of the strength of each component; it

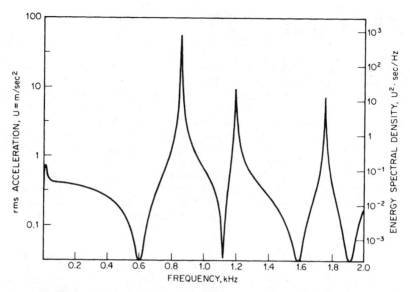

FIGURE 14.25 Analysis of a long transient by scan averaging (same signal as Fig. 14.23). The energy spectral density (ESD) scaling on the right can be compared with that in Fig. 14.23, although the peaks are not valid because of insufficient resolution.

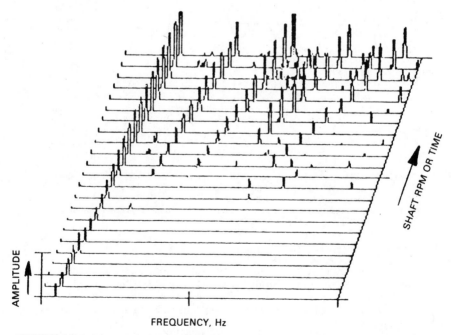

FIGURE 14.26 Three-dimensional spectral map or waterfall plot, showing how spectra change with shaft rpm or time.

would be necessary to integrate over the whole of a distributed peak to measure the total power contained in it.

A way of overcoming this problem is to use *tracking analysis,* where the sampling rate of the FFT analyzer is related directly to shaft speed. A frequency multiplier may be used to produce a sampling frequency signal (controlling the A/D converter of the analyzer) which is a specified multiple of the shaft speed.

Figure 14.27 illustrates the basic principles. Figure 14.27*B* shows a hypothetical signal produced by a rotating shaft during a run-up (in practice, the amplitude normally also would vary with shaft speed). Figure 14.27*A* shows the samples obtained by sampling the signal value at a constant sampling frequency (as for normal frequency analysis) and the spectrum resulting from FFT analysis of these samples. The spectral peak is seen to spread over a number of lines corresponding to the speed change along the time record. Figure 14.27*C* shows the samples obtained by sampling the signal a fixed number of times per shaft revolution (in this case, eight). The samples are indistinguishable from those obtained from normal analysis of a constant-frequency component, and thus the frequency spectrum is concentrated in one line.

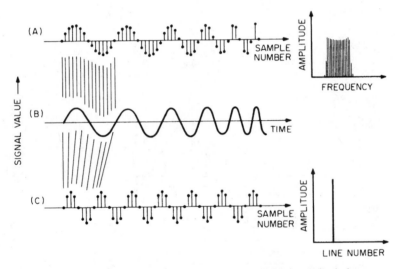

FIGURE 14.27 Analysis of a fundamental component which is increasing in frequency. (*A*) Data record resulting from a uniform sampling rate, and its spectrum, which spreads over a frequency band corresponding to the speed change. (*B*) The original time signal. (*C*) Data record resulting from sampling eight times per fundamental cycle, and its spectrum, which is concentrated in one analysis line.

A frequency multiplier, based on a phase-locked loop, suffers from the disadvantage of a finite response time, so that it cannot keep up if the speed is changing rapidly. A better alternative, offered by some analyzers, is based on digital resampling (interpolation) of each record in line with the simultaneously measured tachometer signal.

When the sampling frequency varies with shaft speed, however, special precautions must be taken to avoid problems with aliasing. One possibility is to use a tracking low-pass filter with a cutoff frequency suitably less than half the sampling frequency. Because of the difficulty of obtaining a tracking filter having a very steep rolloff (e.g., 120 dB/octave), it is often simpler to choose one of a series of filters with a fixed cutoff frequency, depending on the current shaft speed. Such a series of filters (in, for example, a 2, 5, 10 sequence) often is available in the analyzer to determine the normal frequency ranges. Taking the case of a 400-line analyzer, for example, all 400 lines in the measured spectrum are valid when the sampling frequency is appropriate to the selected filter (Fig. 14.28A). If the sampling frequency is higher than the ideal for a given filter, the upper part of the spectrum is affected by the filter (Fig. 14.28B). If it is lower, the upper part of the spectrum may be contaminated by aliasing components (Fig. 14.28C). Nevertheless, by arranging for the selection of the optimum filter at all times (either manually or automatically), at least 60 percent of the measured spectrum (i.e., in this case 240 lines) is always valid. The analysis parameters can be selected so that the desired number of harmonics is contained within this range, based on the fact that the line number in the spectrum of a given component is equal to the number of periods it represents in the data record of length N samples. If, for

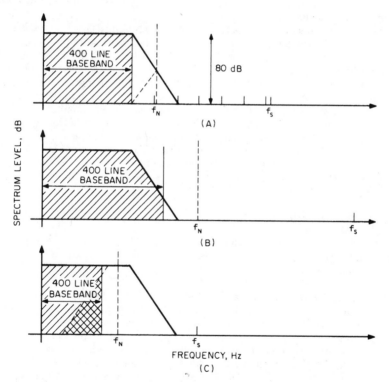

FIGURE 14.28 Effect of sampling frequency on the validity of spectral components, assuming an FFT analyzer with 400 lines and 80-dB dynamic range. f_s = sampling frequency. f_N = Nyquist folding frequency = $f_s/2$. (A) Normal situation with optimum choice of sampling frequency for the low-pass filter. (B) Situation with increased sampling frequency. The upper lines in the spectrum are influenced by the low-pass filter. (C) Situation with decreased sampling frequency. The upper lines in the spectrum are influenced by aliasing components folded around f_N (double cross-hatched area).

example, the 30th harmonic is to be located in line no. 240, the fundamental must be in line no. 8; there must be eight periods of the fundamental component along the data record. Where the data record contains 1024 samples (i.e., $N = 1024$), the sampling frequency must then be 128 times the shaft speed; thus a frequency multiplier with a multiplication factor of 128 should be used in this specific case.

For FFT analyzers with zoom, a simpler approach can be used, as illustrated in Fig. 14.29. An analog low-pass filter is applied to the signal with a cutoff frequency corresponding to the highest required harmonic at maximum shaft speed. However, a frequency multiplying factor is chosen so as to make the sampling frequency, say, 10 or 20 times this cutoff frequency (instead of the normal 2.56). The spectrum then is obtained by zooming in a range corresponding to the highest required harmonic. As shown in Fig. 14.29, the shaft speed (and thus the sampling frequency) can then be varied over a wide range, without aliasing components affecting the measurement results. A somewhat similar procedure is used in conjunction with the digital resampling technique mentioned above. By using four times oversampling, a maximum speed range of 5.92:1 can be accommodated without changing the decimation rate (i.e., the proportion of samples retained after digital filtration), but an even wider range can be covered, at the expense of small "glitches" at the junctions, if the decimation rate is allowed to change.

Figure 14.30 shows the results of tracking FFT analysis on a large turbogenerator. It was made using nondestructive zoom with zoom factor 10. A frequency multiplying factor of 256 was used, giving 40 periods of the fundamental component in the 10K (10,240-point) memory of the FFT analyzer. The fundamental is thus located in line no. 40 of the 400-line zoom spectrum. Because the harmonics coincide exactly with analysis lines, rectangular weighting could have been used in place of the Hanning weighting actually used (all harmonics have exact integer numbers of periods along the record length); Hanning weighting can, however, be advantageous for nonsynchronous components such as constant-frequency components. Such a component at 150 Hz (initially coinciding with the third harmonic of shaft speed) is shown in Fig. 14.30. Constant-frequency components follow a hyperbolic locus in cascade plots employing order tracking.

RELATED ANALYSIS TECHNIQUES

Signal analysis techniques other than those described above, which are useful as an adjunct to frequency analysis, include synchronous averaging, cepstrum analysis, and Hilbert transform techniques.

Synchronous Averaging (Signal Enhancement).   *Synchronous averaging* is an averaging of digitized time records, the start of which is defined by a repetitive trigger signal. One example of such a trigger signal is a once-per-revolution synchronizing pulse from a rotating shaft. This process serves to enhance the repetitive part of the signal (whose period coincides with that of the trigger signal) with respect to nonsynchronous effects. That part of the signal which repeats each time adds directly, in proportion to the number of averages, n. The nonsynchronous components, both random noise and periodic signals with a different period, add like noise, with random phase; the amplitude increase is in proportion to $\sqrt{n}$. The overall improvement in the signal-to-noise ratio is thus $\sqrt{n}$, resulting in an improvement of $10 \log_{10} n$ dB, i.e., 10 dB for 10 averages, 20 dB for 100, 30 dB for 1000.

Figure 14.31 shows the application of synchronous averaging to vibration signals from similar gearboxes in good and faulty condition. Figure 14.31*A* shows the

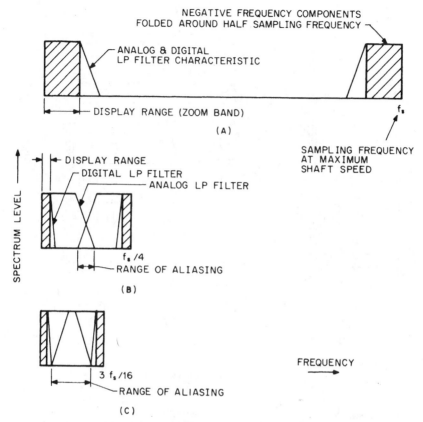

FIGURE 14.29 Use of a fixed low-pass filter to prevent aliasing when tracking with an FFT analyzer employing zoom to analyze in a lower-frequency band. For illustration purposes, the sampling frequency at maximum shaft speed has been made four times greater than that appropriate to the analog LP filter. The shaft speed range could be made proportionally greater by increasing this factor. (*A*) Situation at maximum shaft speed. All harmonics of interest must be contained in the display range. (*B*) Situation at one-fourth maximum shaft speed. The analog filter characteristics overlap, but are well separated from the display range. (*C*) Situation at three-sixteenths maximum shaft speed. The aliasing range almost intrudes on the display range.

enhanced time signal (120 averages) for the gear on the output shaft. The signal is fairly uniform and gives evidence of periodicity corresponding to the tooth-meshing. Figure 14.31*B* is a similarly enhanced time signal for a faulty gear; a localized defect on the gear is revealed. By way of comparison, Fig. 14.31*C* shows a single time record, without enhancement, for the same signal as in Fig. 14.31*B*; neither the tooth-meshing effect nor the fault is readily seen.

For best results, synchronous averaging should be combined with tracking. Where there is no synchronization between the digital sampling and the (analog) trigger signal, an uncertainty of up to one sample spacing can occur between successive digitized records. This represents a phase change of 360° at the sampling frequency, and approximately 140° at the highest valid frequency component in the signal, even with perfectly stable speed. Where speed varies, an additional phase shift occurs; for

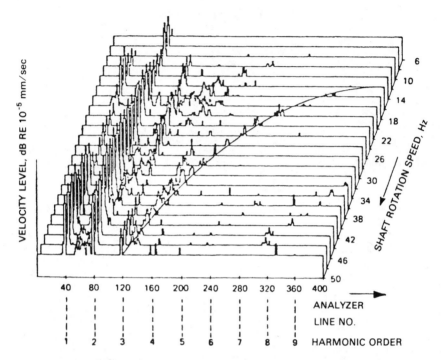

FIGURE 14.30 Tracking FFT analysis of the rundown of a large turbogenerator. The superimposed hyperbolic curve represents a fixed-frequency component at 150 Hz.

example, a speed fluctuation of 0.1 percent would cause a shift of one sample spacing at the end of a typical 1024-sample record. The use of tracking analysis (generating the sampling frequency from the synchronizing signal) reduces both effects to a minimum.

Cepstrum Analysis. Originally the *cepstrum* was defined as the power spectrum of the logarithmic power spectrum.[9] A number of other terms commonly found in the cepstrum literature (and with an equivalent meaning in the cepstrum domain) are derived in an analogous way, e.g., *cep*strum from *spec*trum, *quefr*ency from *fre-qu*ency, *rah*monic from *har*monic. The distinguishing feature of the cepstrum is not just that it is a spectrum of a spectrum, but rather that it is the spectrum of a spectrum on a logarithmic amplitude axis; by comparison, the autocorrelation function (see Eq. 22.21) is the inverse Fourier transform of the power spectrum without logarithmic conversion.

Most commonly, the *power cepstrum* is defined as the inverse Fourier transform of the logarithmic power spectrum,[10] which differs primarily from the original definition in that the result of the second Fourier transformation is not modified by obtaining the amplitude squared at each quefrency; it is thus reversible back to the logarithmic spectrum. Another type of cepstrum, the *complex cepstrum,* discussed below, is reversible to a time signal.

Figure 14.32, the analysis of a vibration signal from a faulty bearing, shows the advantage of the power cepstrum over the autocorrelation function. In Fig. 14.32*A*,

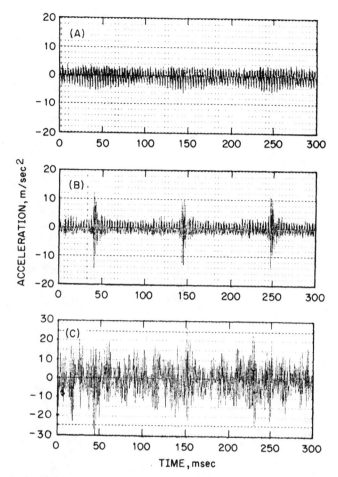

FIGURE 14.31 Use of signal enhancement in gear fault diagnosis. (*A*) Enhanced signal (120 averages) for a gear in normal condition. (*B*) Enhanced signal (120 averages) for a similar gear with a local fault. (*C*) Section of raw signal corresponding to (*B*).

the same power spectrum is depicted on both linear and logarithmic amplitude axes; in (*B*) and (*C*) the autocorrelation and cepstrum, respectively, are shown. In (*C*), the logarithmic depiction of the power spectrum reveals a family of harmonics which are concealed in the linear depiction. The presence of the family of harmonics is made evident by a corresponding series of rahmonics in the cepstrum (denoted ①, ②, etc.), but is not detected in the autocorrelation function. The quefrency axis of the cepstrum is a time axis, most closely related to the X axis of the autocorrelation function (i.e., time delay or periodic time rather than absolute time). The reciprocal of the quefrency of any component gives the equivalent *frequency spacing* in the spectrum, not the absolute frequency.

Most of the applications of the power cepstrum derive from its ability to detect a periodic structure in the spectrum, for example, families of uniformly spaced har-

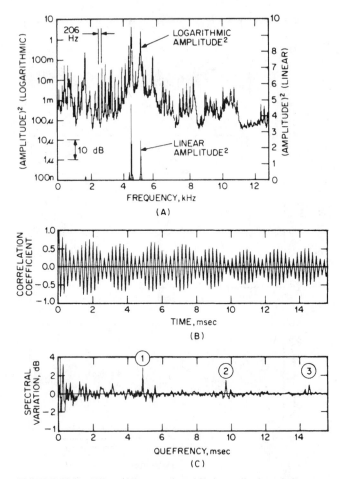

FIGURE 14.32   Effect of linear vs. logarithmic amplitude scale in power spectrum. (*A*) Power spectrum on linear scale (lower curve) and logarithmic scale (upper curve). (*B*) Autocorrelation function (obtained from linear representation). (*C*) Cepstrum (obtained from logarithmic representation)—①, ②, etc., are rahmonics corresponding to harmonic series in spectrum (4.85 milliseconds equivalent to 1/206 Hz). The harmonics result from a fault in a bearing.

monics and/or sidebands. The application of the cepstrum to the diagnosis of faults in gears and rolling element bearings is discussed in Chap. 16 and Ref. 11.

To obtain a distinct peak in the cepstrum, a reasonable number of the members of the corresponding harmonic or sideband family must be present (although the fundamental may be absent). These uniformly spaced components must be adequately resolved in the spectrum. As a guide, the spacing of components to be detected should be a minimum of eight lines in the original spectrum. For this reason, it is often advantageous to perform a cepstrum analysis on a spectrum obtained by *zoom FFT.* In this case it is desirable to use a slightly modified definition of the cepstrum corresponding to the amplitude of the *analytic signal.*[11] (See the next section on *Hilbert Transform Techniques.*)

The *complex cepstrum*[10,12] (referred to above) is defined as the inverse Fourier transform of the complex logarithm of the complex spectrum. Despite its name, it is a real-valued function of time, differing from the power cepstrum primarily in that it uses phase as well as logarithmic amplitude information at each frequency in the spectrum. It is thus reversible to a time function (from which the complex spectrum is obtained by direct Fourier transformation).

Measured vibration signals generally represent a combination of source and transmission path effects; for example, internal forces in a machine (the source effect) act on a structure whose properties may be described by a frequency-response function between the point of application and the measurement point (the transmission path effect). As shown in Refs. 10 and 12, the source and transmission path effects are convolved in the time signals, multiplicative in the spectra, and additive in the logarithmic spectra and in the cepstra (both power cepstra and complex cepstra). In the cepstra, they quite often separate into different regions, which in principle allows a separation of source and transmission path effects in an externally measured signal.

Figure 14.33 shows an example of an internal cylinder pressure signal in a diesel engine, derived from an externally measured vibration acceleration signal making use of cepstrum techniques to generate the inverse filter.[13]

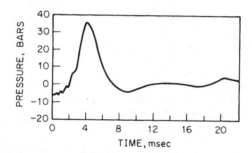

FIGURE 14.33 Diesel engine cylinder pressure signal, derived from an externally measured vibration-acceleration signal using cepstrum techniques. (*From R. H. Lyon and A. Ordubadi.*[13])

Reference 14 gives similar results for the tooth-mesh signal in a gearbox and also shows that a frequency-response function derived by windowing in the cepstrum of an output signal compares favorably with a direct measurement (which requires measurement of both an input and an output signal).

Hilbert Transform Techniques. The *Hilbert transform* is the relationship between the real and imaginary parts of the Fourier transform of a one-sided signal.[15] An example is a causal signal such as the impulse response of a vibratory system (a *causal signal* is one whose value is zero for negative time). The real and imaginary parts of the frequency response (the Fourier transform of the impulse response) are related by the Hilbert transform; thus, only one part need be known—the other can be calculated.

Analogously, the time function obtained by an inverse Fourier transformation of a one-sided spectrum (positive frequencies only) is complex, but the imaginary part is the Hilbert transform of the real part. Such a complex time signal is known as an *analytic signal.*

An analytic signal can be thought of as a rotating vector (or phasor) described by the formula $A(t)e^{j\phi(t)}$ whose amplitude $A(t)$ and rotational speed $\omega(t) = d\phi(t)/dt$, in general, vary with time. Analytic signals are useful in vibration studies to describe modulated signals. For example, a *phase-coherent* signal [Eq. (22.3)] can be represented as the real part of an analytic signal, in which case the imaginary part can be obtained by a Hilbert transform. Therefore, from a measured time signal, $a(t)$, it is possible to obtain the amplitude and phase (or frequency) modulation components from the relationship

$$A(t)e^{j\phi(t)} = a(t) + j\tilde{a}(t) \tag{14.11}$$

where $\tilde{a}(t)$ is the Hilbert transform of $a(t)$.

The Hilbert transform may be evaluated directly from the equation

$$\tilde{a}(t) = \frac{1}{\pi} \int_{-\infty}^{\infty} a(\tau) \frac{1}{t-\tau} \, d\tau \tag{14.12}$$

but it can be more readily evaluated by a phase shift in the frequency domain, in particular in an FFT analyzer.[16] An alternative way of generating analytic signals using an FFT analyzer is by an inverse Fourier transformation of the equivalent one-sided spectrum formed from the spectrum of the real part only. The time signals resulting from the real-time zoom process (described above) automatically have the same amplitude function $A(t)$ as the equivalent bandpass-filtered analytic signal, since they are obtained from the positive frequency components only [Fig. (14.17)]. The frequency-shifting operation affects only the phase function $e^{j\phi(t)}$.

The major applications of Hilbert transform techniques in vibration studies involve either amplitude demodulation or phase demodulation.

Amplitude Demodulation. Figure 14.34 shows the analytic signal for the case of single-frequency amplitude modulation of a higher-frequency carrier component. The imaginary part is the Hilbert transform of the real part; this manifests itself as a 90° phase lag. The amplitude function is the envelope of both the real and imaginary parts and represents the modulating signal plus a dc offset. The phase function is a linear function of time (whose slope represents the speed of rotation, or frequency, of the carrier component); it is, however, shown modulo 2π, as is conventional.

One area of application of amplitude demodulation where it is advantageous to view the signal envelope rather than the time signal itself is in the interpretation of such oscillating time functions as autocorrelation and crosscorrelation functions (see Chap. 22). Figure 14.35[17] shows a typical case where peaks indicating time delays are difficult to identify in a crosscorrelation function as defined in Eq. (22.48), because of the oscillating nature of the basic function (Fig. 14.35A). The peaks are much more easily seen in the envelope or magnitude of the analytic signal (Fig. 14.35B). Another advantage of the analytic signal is that its magnitude can be displayed on a logarithmic axis; this allows low-level peaks to be detected and converts exponential decays to straight lines.[17]

Another area of application of amplitude demodulation is in *envelope analysis* (discussed in the earlier section on *Envelope Detectors*). In particular, when the signal is to be bandpass-filtered before forming the envelope, this can be done by real-time zoom in the appropriate passband. Figure 14.36 shows an example from the same vibration source as was analyzed in Fig. 14.32. Figure 14.36A shows a typical envelope signal obtained from zooming in a 1,600-Hz band centered at 3 kHz. The spectrum of Fig. 14.32A shows that this frequency range is dominated by the harmonic family which results from a fault in a bearing. Consequently, the corresponding envelope signal (Fig. 14.36A) indicates a series of bursts with the same period, 4.85 milliseconds (compare with the cepstrum of Fig. 14.32C). Figure 14.36B shows

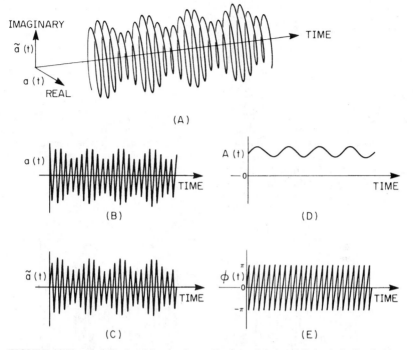

FIGURE 14.34 Analytic signal for simple amplitude modulation. (A) Analytic signal $a(t) + j\tilde{a}(t) = A(t)e^{j\phi(t)}$. ($B$) Real part $a(t)$. (C) Imaginary part $\tilde{a}(t)$. (D) Amplitude $A(t)$. (E) Phase $\phi(t)$.

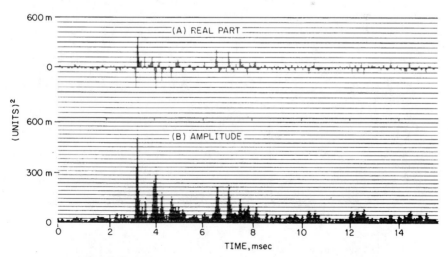

FIGURE 14.35 Example of a crosscorrelation function expressed as follows:[17] (A) The real part of an analytic signal, i.e., the normal definition [Eq. (22.48)]. (B) The amplitude of the analytic signal. The peaks corresponding to time delays are more easily seen in this representation. The signal was obtained by bandpass filtering (using FFT zoom) in the frequency range from 512 to 13,312 Hz.

the average spectrum of a number of such envelope signals; this gives a further indication that the dominant periodicity is 206 Hz.

Phase Demodulation.   For a purely phase-modulated signal, the amplitude function $A(t)$ is constant and the phase function $\phi(t)$ is given by the sum of a carrier component of constant frequency f_c and the modulation signal $\phi_m(t)$. Thus

$$\phi(t) = 2\pi f_c t + \phi_m(t) \tag{14.13}$$

Real-time zoom analysis centered on frequency f_0 subtracts this frequency from all components in the signal; consequently, by zooming at the carrier frequency f_c, only the modulation signal $\phi_m(t)$ remains. In general it is possible to zoom exactly at the carrier frequency only when the latter is made to coincide exactly with an analysis line (for example, by employing order tracking). Otherwise, the small difference in frequency gives a residual slope to the phase signal.

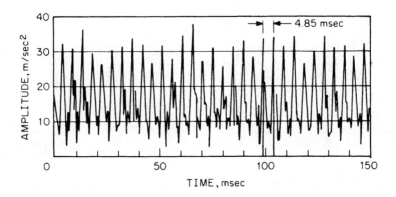

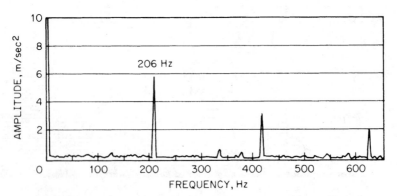

FIGURE 14.36   Envelope analysis using Hilbert transform techniques. (*A*) Typical envelope signal showing bursts with a period of 4.85 milliseconds from a fault in a ball bearing. (*B*) Average spectrum of the envelope signal showing corresponding harmonics of 206 Hz. Signal obtained by bandpass filtering (using FFT zoom) in the frequency range from 2,200 to 3,800 Hz (compare with Fig. 14.32*A*, which shows a base-band analysis of this same signal).

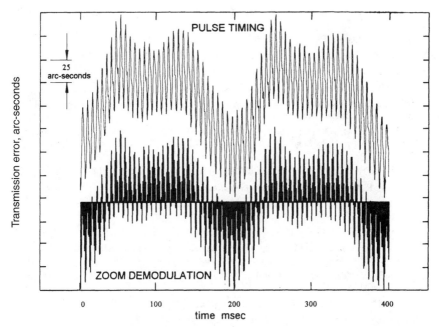

FIGURE 14.37 Gear dynamic transmission error measured using the zoom demodulation technique compared with direct measurement by timing the intervals between shaft encoder pulses. Measurements were made with two 32-tooth gears, although the method is not limited to unity-ratio gears. Note the periodic repetition once per revolution of the gears (200 milliseconds) and the higher-frequency component corresponding to tooth-meshing.

Figure 14.37 shows an example of the application of this technique to the measurement of gear transmission error. This can be obtained as the difference in torsional vibration (i.e., phase modulation) of the two gears in mesh, after appropriate compensation for the gear ratio (in this particular case the ratio is unity). The torsional vibrations were measured by demodulating the output signals from optical encoders attached to each shaft. The encoders give 16,000 pulses per revolution, but this was divided down to 4,000 for the results shown here (and for the *zoom demodulation* technique even further decimation would be possible). The result obtained by zoom demodulation, including digital tracking, was produced by an advanced FFT analyzer, and is compared with a result obtained using a 100-MHz clock to time the intervals between pulses and thus measure phase modulation somewhat more directly. The two results are virtually identical, and are accurate to within a few arc-seconds. Similar methods have been used to detect cracks in gears by amplitude and phase demodulation of the tooth-meshing signal.[18]

REFERENCES

1. Cooley, J. W., and J. W. Tukey: *Math. Computing,* **19**(90):297 (1965).
2. Cooley, J. W., P. A. W. Lewis, and P. D. Welch: *J. Sound Vibration,* **12**(3):315 (1970).

3. Brigham, E. O.: "The Fast Fourier Transform," Prentice-Hall, Inc., Englewood Cliffs, N.J., 1974.

4. Thrane, N.: "Zoom-FFT," *Brüel & Kjaer Tech. Rev.,* (2) (1980).

5. Sloane, E. A.: *IEEE Trans. Audio Electroacoust.,* **AU-17**(2):133 (1969).

6. Welch, P. D.: *IEEE Trans. Audio Electroacoust.,* **AU-15**(2):70 (1967).

7. Randall, R. B.: "Frequency Analysis," Brüel & Kjaer, Naerum, Denmark, 1987.

8. Mitchell, J. S.: "An Introduction to Machinery Analysis and Monitoring," Penwell Publishing Company, Tulsa, Okla., 1981.

9. Bogert, B. P., M. J. R. Healy, and J. W. Tukey: In M. Rosenblatt (ed.), "Proceedings of the Symposium on Time Series Analysis," John Wiley & Sons, Inc., New York, 1963, pp. 209–243.

10. Childers, D. G., D. P. Skinner, and R. C. Kemerait: *Proc. IEEE,* **65**(10):1428 (1977).

11. Randall, R. B.: *Maintenance Management Int.,* **3:**183 (1982/1983).

12. Oppenheim, A. V., R. W. Schafer, and T. G. Stockham Jr.: *Proc. IEEE,* **56**(August):1264 (1968).

13. Lyon, R. H., and A. Ordubadi: *J. Mech. Des.,* **104**(Trans. ASME)(April):303 (1982).

14. DeJong, R. G., and J. E. Manning: "Gear Noise Analysis using Modern Signal Processing and Numerical Modeling Techniques," *SAE Paper* No. 840478, 1984.

15. Papoulis, A.: "The Fourier Integral and Its Applications," McGraw-Hill Book Company, Inc., New York, 1962.

16. Thrane, N.: *Brüel & Kjaer Tech. Rev.,* (3) (1984).

17. Herlufsen, H.: *Brüel & Kjaer Tech. Rev.,* (1 and 2) (1984).

18. McFadden, P.: *J. Vib. Acoust. Stress & Rel. Des.,* **108**(Trans. ASME)(April):165 (1986).

CHAPTER 15
MEASUREMENT TECHNIQUES

Cyril M. Harris

INTRODUCTION

Earlier chapters describe equipment used in vibration measurements. For example, detailed information concerning transducers, their characteristics, and how these characteristics are influenced by environmental factors is given in Chap. 12. The various measurement system components and the characteristics which determine their selection are described in Chaps. 13 and 14. The use of such measurement systems in vibration problems may involve only one or two engineers as in monitoring the condition of machinery in a factory (Chap. 16), in some problems in modal testing (Chap. 21), in measurements in building structures (Chap. 24), in measuring torsional vibration in reciprocating and rotating engines (Chap. 38), and in the balancing of rotating machinery (Chap. 39). In contrast, in the aerospace industry, some measurement problems are so complex that teams of engineers and several divisions of the company may be involved. Yet all these examples share certain basic measurement procedures. It is these basic procedures (rather than measurement details, which vary from problem to problem) that are considered here. Thus, this chapter includes a general discussion of (1) planning measurements to achieve stated objectives, (2) selecting the type of measurements which should be made to achieve these objectives, (3) selecting transducers, (4) mounting transducers, (5) mounting cable and wiring (including shielding and grounding), (6) selecting techniques for the field calibration of the overall measurement system, (7) collecting and logging the data obtained, and (8) conducting a measurement error analysis.

The best method of analyzing the vibration measurement data, once they have been acquired, depends on a number of factors, including the quantity of data to be processed, the objectives of the measurements, test criteria, specifications, and the accuracy required. These various factors are discussed in Chaps. 20, 22, 23, 27, and 28.

MEASUREMENT PLANNING

Careful pretest planning (and, in the case of a complex measurement program, detailed documentation) can save much time in making measurements and in ensuring that the most useful information is obtained from the test data. In many cases, as in environmental testing, measurement procedures are contained in test specifications to ensure that a specification or legal requirement has been met. In other cases (as in balancing rotating machinery), measurement procedures are outlined in detail in national or international standards. In general, the first step in planning is to define the purpose of the test and to define what is to be measured. Planning should start with a clear definition of the test objectives, including the required accuracy and reliability. The second step is to define those non-equipment-related factors which influence the selection of measurement equipment and measurement techniques. These include availability of trained personnel; cost considerations; length of time available for measurements; scheduling considerations; and available techniques for data analysis, validation, and presentation.

Next, the various factors listed in Table 15.1 should be considered. For example, it is important to have some estimate of the characteristics of the motion to be measured—e.g., its frequency range, amplitude, dynamic range, duration, and principal direction of motion. Such information is needed to provide the basis for the optimum selection of measurement equipment. Yet often very little is known about the characteristics of the motion to be measured. Previous experience may provide a guide in estimating signal characteristics. Where this is not available, preliminary measurements may be carried out to obtain information which serves as a guide for further measurements. For example, suppose preliminary measurements show a frequency spectrum having considerable content in the region of the lowest frequency measured. This would indicate that the instrumentation capability should be extended to a somewhat lower frequency in subsequent measurements. Thus an iterative process often takes place in a shock and vibration measurement program. To speed this process, it is helpful to employ equipment whose characteristics cover a wide range and which has considerable flexibility. Failure to take this feedback process into account can sometimes result in the acquisition of meaningless test results. For example, a measurement program was carried out by one organization over a period of many weeks. The objective was to correlate building vibration data, measured in the organization's own laboratories, with the acceptability of these laboratories as sites for ultrasensitive galvanometers and other motion-sensitive equipment. No correlation was found, and the entire measurement program was a waste of time, for two reasons: (a) The measurements were made with equipment with a frequency limit which was not sufficiently low, so that important spectral components of building vibration could not be measured. (b) Measurements were made only in the vertical direction, whereas it was the horizontal component which was dominant and which made certain laboratory areas unacceptable for the location of vibration-sensitive equipment.

Many of the various factors, listed in Table 15.1, which should be considered in planning instrumentation for shock and vibration measurements are discussed in earlier chapters and are cross-referenced, rather than repeated, here. For example, Chap. 12 discusses the effects of environmental conditions on transducer characteristics; Chap. 13 describes various components which follow the transducer in a measurement system (such as preamplifiers, signal conditioners, filters, analyzers, and recorders). Chapter 14 describes the selection of the appropriate analyzer bandwidth, frequency scale, amplitude scale, selection of data windows, etc.

Before making measurements, it is usually important to establish a measurement protocol—the more complex the measurements to be made, the more formal and detailed the measurement protocol should be. It is also important to make an *error analysis,* i.e., (a) to estimate the error introduced into the data acquisition and analysis by each individual item of equipment, and (b) to determine the total error by calculating the square root of the sum of the squares of the individual errors. For

TABLE 15.1   Factors Which Are Important Considerations in the Selection of Measurement Equipment and Measurement Techniques for Mechanical Shock and Vibration Measurements

Parameter to be measured	
Acceleration	Strain
Velocity	Force
Displacement	Mechanical impedance

Characteristics of motion to be measured	
Frequency range	Direction of motion
Amplitude range	Transient characteristics
Phase	Duration

Environmental conditions	
Temperature (ambient and transient)	Magnetic and radio-frequency fields
Humidity	Corrosive and abrasive media
Ambient pressure	Nuclear radiation
Acoustic noise	Sustained acceleration

Transducer characteristics (see Chap. 12)

Electrical characteristics (sensitivity, resolution, cross-axis sensitivity, amplitude linearity, dynamic range, frequency response, phase response, effects of environment on the transducer)
Physical characteristics (e.g., size and mass)
Self-generating or auxiliary power required
Electrically grounded to case, or isolated
Self-contained amplifier

Transducer mountings and locations of mountings

Effect of mounting on transducer characteristics
Effect of mounting on vibratory characteristics of item under test
Number of measurement locations
Space availability for measurement locations
Availability of well-regulated power, free of voltage spikes
Ease of installation
Possibility of mounting misalignment with respect to intended direction of measurement

System components (preamplifiers, signal conditioners, filters, analyzers) (see Chaps. 13 and 14)

Electrical characteristics (e.g., input and output impedances)
Power availability
Noise interference (shielding, avoidance of ground loops)
Number of channels required for measurement and recording: maximum duration of measurements, tape storage requirements
Possible requirement for real-time information

Method of data transmission

Coaxial cable
Twisted pair of wires
Telemetry (channels assigned)
Optical fiber

Recording equipment (see Chap. 13)

Recording-time capability
Electrical characteristics (e.g., signal-to-noise ratio)
Portability; power requirements
Correlation between recorded information and physical phenomena
Redundancy to minimize the risk of loss of vital information

TABLE 15.1   Factors Which Are Important Considerations in the Selection of Measurement Equipment and Measurement Techniques for Mechanical Shock and Vibration Measurements (*Continued*)

Field calibration
Transducers
Over-all measurement system

Data analysis, presentation, and validation
Manual or automatic (Chap. 14); computer (Chaps. 22, 23, 27, and 28)
Type of presentation required

example, such an analysis may discover that an individual item of equipment is primarily responsible for introducing a significant total error, suggesting that perhaps it should be replaced. Furthermore, such a determination will indicate whether the total error is within the bounds of acceptability, thereby avoiding useless measurements.

SELECTION OF THE PARAMETER TO BE MEASURED

Often, the selection of the parameter to be measured (displacement, velocity, acceleration, or strain) is predetermined by specifications or by standards. When this is not the case, it is often helpful to apply the considerations given in Table 15.2 or to apply the *flatness rule* described in Chap. 16. According to this rule, the best motion parameter to use is the one whose spectrum is closest to being uniform (i.e., the one having the flattest spectrum). This is important for two reasons: If the spectrum is relatively flat, then (1) an increase at any frequency has a roughly even chance of influencing overall vibration levels, and (2) minimum demands are placed on the required dynamic range of the equipment which follows the transducer. For example, Fig. 16.2 shows two spectra obtained under identical conditions—one a velocity spectrum, the other a displacement spectrum. The spectrum obtained using a velocity transducer is the more uniform of the two; therefore, velocity would be the appropriate motion parameter to select.

SELECTING THE TRANSDUCER

In selecting the transducer best suited for a given measurement, the various factors listed in Table 15.1 must be taken into consideration, particularly those under *Parameter to Be Measured, Characteristics of Motion to Be Measured, Environmental Conditions,* and *Transducer Characteristics.* Each of these factors (as well as cost and availability) influences the selection process. If consideration of different factors leads to recommendations which are in opposition, then the relative importance of each factor must be determined and a decision made on this basis. For example, consider two factors which enter into the selection of a piezoelectric accelerometer, *sensitivity* and *mass*. Sensitivity considerations would suggest that a transducer of large size be selected since transducer sensitivity generally increases with size (and therefore with mass) for an accelerometer of this type. In contrast, mass considerations would suggest that a transducer of small size be selected in order to minimize the

TABLE 15.2   A Guide for the Selection of the Parameter to Be Measured

Acceleration measurements
Used at high frequencies where acceleration measurements provide the highest signal outputs
Used where forces, loads, and stresses must be analyzed—where force is proportional to acceleration (which is not always the case)
Used where a transducer of small size and small mass is required, since accelerometers usually are somewhat smaller than velocity or displacement pickups

Velocity measurements
Used where vibration measurements are to be correlated with acoustic measurements since sound pressure is proportional to the velocity of the vibrating surface
Used at intermediate frequencies where displacement measurements yield transducer outputs which may be too small to measure conveniently
Used extensively in measurements on machinery where the velocity spectrum usually is more uniform than either the displacement or acceleration spectra

Displacement measurements
Used where amplitude of displacement is particularly important—e.g., where vibrating parts must not touch or where displacement beyond a given value results in equipment damage
Used where the magnitude of the displacement may be an indication of stresses to be analyzed
Used at low frequencies, where the output of accelerometers or velocity pickups may be too small for useful measurement
Used to measure relative motion between rotating bodies and structure of a machine

Strain managements
Used where a portion of the specimen being tested undergoes an appreciable variation in strain caused by vibration—usually limited to low frequencies

mass loading on the test item; a small size is advantageous since, as Eq. (12.13) indicates, the natural frequency of a structure is lowered by the addition of mass. Therefore in this case one should choose the most sensitive transducer (and therefore the largest size) which produces no significant mass loading. In special cases, even the smallest transducer may result in an unacceptable load. Then one of the devices described in Chap. 12 which make no contact with the test surface may be selected.

Consider another example. Suppose a specification requires that vibration displacement be measured. It is reasonable to assume that a displacement transducer (such as the one described in Chap. 12) should be chosen since (depending on the frequency spectrum) such a selection could yield the highest signal-to-noise ratio. On the other hand, in many measurement problems it is more convenient and equally satisfactory to select an accelerometer having a wide dynamic range and to employ an electric circuit which obtains displacement by double integration of the signal from the transducer's output.

TRANSDUCER MOUNTINGS

Various methods of mounting a transducer on a test surface include (1) screwing the transducer to the test surface by means of a threaded stud, (2) cementing the transducer to the test surface, (3) mounting the transducer on the test surface by means of a layer of wax, (4) attaching the transducer to a ferromagnetic surface by means of a permanent magnet, (5) mounting the transducer on a bracket which, in turn, is mounted on the test surface, and (6) holding the transducer against the test surface by hand. Several of these mounting techniques are illustrated in Fig. 15.1, and their frequency-response characteristics are shown in Fig. 15.2. Two types of mechanical brackets are illustrated in Fig. 15.3

The method of mounting affects the resonance frequency and, hence, the useful frequency range of the transducer. Therefore it is important to ensure that the frequency response is adequate before measurements are taken. Each of the above methods of mounting has its advantages and disadvantages. The appropriate choice for a given measurement problem depends on a number of factors, including the following:

Effect of the mounting on the useful frequency range of the transducer

Effect of mass loading of the transducer mounting on the test surface

Maximum level of vibration the mounting can withstand

Maximum operating temperature

Measurement accuracy

Repeatability of measurements (Can the transducer be remounted at exactly the same position with the same orientation?)

Stability of the mounting with time

Requirement that the test surface not be damaged by screw holes

Requirement for electrical insulation of the transducer

Time required to prepare mounting

Time required to remove mounting

Difficulty in cleaning the transducer after removal from test surface

Difficulty in cleaning test surface after transducer removed

Skill required to prepare mounting

Cost of mounting

Environmental problems (dirt, dust, oil, moisture)

For example, the above "requirement for electrical insulation of the transducer" would be a major consideration in the selection of a method of mounting if the insulation so obtained would result in the breaking of a ground loop, as explained in a following section.

Stud Mounting. Figure 15.1A illustrates a typical stud-mounted transducer; the transducer is fixed to the test surface by means of a threaded metal screw. One method of insulating the stud-mounted transducer from the test surface is shown in Fig. 15.1B. The metal stud is replaced with one which is fabricated of insulating material, and a mica washer is inserted between the transducer and the test surface. Other manufacturers employ a threaded, insulated stud with a flange made of the same material; the flange, midway along the length of the stud, serves as the base for the accelerometer.

Where stud mounting is practical, it is the best type to use for the following reasons:

1. It provides the highest resonance frequency (up to 100 kHz) of any of the mounting techniques and, therefore, the widest possible measurement frequency range (up to 50 kHz).

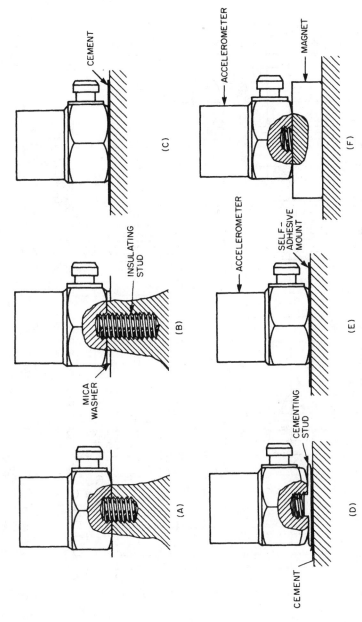

FIGURE 15.1 Various methods of mounting a transducer on a test surface: (*A*) Stud mounting; transducer screws directly to the surface by a threaded stud. (*B*) Same as (*A*) but with a transducer insulated from test surface by use of stud fabricated of insulating material and by a mica washer between the surface and transducer. (*C*) Cement mounting of a transducer; the cement bonds the transducer directly to the surface. (*D*) Similar to (*C*), but here cement bonds the surface to a cementing stud screwed into the transducer. (*E*) Transducer mounted to surface by means of double-sided adhesive tape or disc. (*F*) Transducer mounted to surface by means of a magnet. (*Courtesy of Brüel & Kjaer.*)

15.7

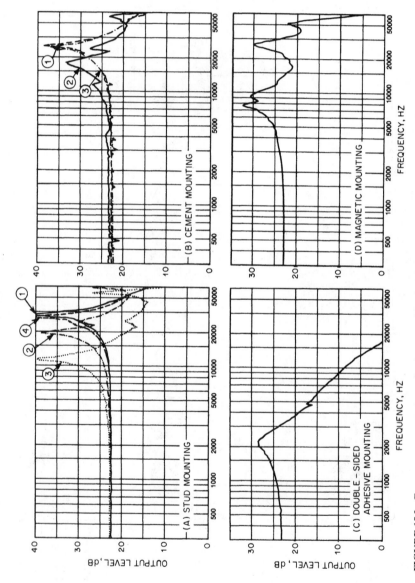

FIGURE 15.2 Frequency-response curves for the same piezoelectric accelerometer mounted by the different methods illustrated in Fig. 15.1: (A) stud mounting; (B) cement mounting; (C) double-sided adhesive mounting; (D) magnetic mounting. (*Courtesy of Brüel & Kjaer.*)

2. It permits measurements at very high vibration levels without the loosening of the transducer from the test surface.

3. It does not reduce the maximum permissible operating temperature at which measurements can be made.

4. It permits accurate and reproducible results since the measurement position can always be duplicated.

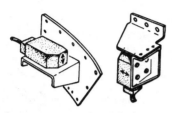

FIGURE 15.3 Two types of mounting brackets. In this example, a velocity-type transducer is shown; the arrows indicate the direction of sensed motion.

In preparing a stud mounting, the test surface must be drilled and tapped. A standard 10-32 thread is widely used. (Also see International Standards Organization Standard ISO 1101.) Distortion of the transducer as mounted may produce strains that affect the transducer's response. Therefore, it is important (1) to ensure that the test surface is very flat (which can be done by grinding or lapping), (2) to prevent the mounting stud from bottoming in the transducer case— this can lead to strain, and (3) to screw the stud into the hole in the test surface, and then the accelerometer onto the stud using the torque recommended by the transducer's manufacturer. The application of a silicone grease (such as Dow-Corning DC-4) or a light machine oil between the transducer and the test surface usually provides better response at high frequencies—say, above 2,000 Hz. The upper temperature limit for the stud mounting of Fig. 15.1*A* is limited only by the accelerometer, but with the mica washer insert shown in Fig. 15.1*B,* the upper limit may be as low as 480°F (250°C).

Figure 15.2*A* shows response curves for a stud-mounted accelerometer for the following conditions: ① spanner tight, which has the highest resonance frequency, ② finger tight, ③ mounted with a mica washer to provide electrical insulation between the transducer and the vibrating surface, and ④ mounted on a somewhat thinner mica washer—which results in a higher resonance frequency than for ③.

Cyanoacrylate, Dental, and Epoxy Cement Mountings. Where it is not possible to use a stud mounting, a transducer can be bonded to a test surface by means of a thin layer of cement (for example, a cyanoacrylate, dental, or epoxy cement), as shown in Fig. 15.1*C.* If the test surface is not flat and a miniature accelerometer is used, it is not difficult to build up a layer of dental cement around the accelerometer so as to provide firm attachment for the accelerometer. In mounting the transducer, it should be pressed firmly against the surface to ensure that the adhesive layer is thin; excess adhesive around the perimeter should then be removed immediately.

Another method of mounting is to use a cementing stud which is threaded into the transducer; the flat side of the stud is then cemented to the test surface as shown in Fig. 15.1*D.* This is a useful technique where repeated measurements at the same point are required. The transducer may be removed for measurements elsewhere, but the cementing stud is left in place. This provides assurance that future measurements will be made at precisely the same point.

The cement method of mounting a transducer provides excellent frequency response, as shown in Fig. 15.2*B* for three conditions: ① accelerometer cemented directly to test surface, ② accelerometer cemented with a "soft" adhesive (not recommended), and ③ accelerometer with a cementing stud which is cemented to the surface with a hard cement.

This type of mounting may be used at high levels of vibration if the cementing surfaces are carefully prepared. Cement mounting may or may not provide electrical insulation; if insulation is required, the electrical resistance between the transducer and the test surface should be checked with an ohmmeter. The maximum temperature at which measurements can be made is limited by the physical characteristics of the cement employed—usually about 176°F (80°C), although some cements such as 3M Cyanolite 303 have an upper limit as high as 390°F (200°C). This type of mounting has good stability with time.

Methyl cyanoacrylate cements, such as Eastman Kodak 910MHT 3M, Cyanolite 101, and Permabond 747, dry much more rapidly than epoxy cements and therefore require less time to mount a transducer. They may be removed easily and the surface cleaned with a solvent such as acetone. Removal of epoxy from the test surface and from the transducer may be time-consuming. In fact, the epoxy bond may be so good that the transducer can be damaged in removing it from the test surface. When encased in epoxy, an accelerometer may be subject to considerable strain, which will significantly alter its characteristics. On the other hand, unless the cemented surfaces are very smooth, an epoxy can provide a superior bond since it will fill in a rough surface far better than a cyanoacrylate cement. With either bonding agent, the surfaces must be very clean before application of the cement. This mounting technique is not recommended for conditions of prolonged high humidity or for pyroshock measurements.

Wax Mounting. Beeswax or a petroleum-based petrowax may be used to attach a transducer to a flat test surface. If the bonding layer is thin (say, no greater than 0.2 mm), it is possible to obtain a resonance frequency almost as high as that for the stud mounting, but if the test surface is not smooth, a thicker wax layer is required and the resonance frequency will be reduced. If the mating surfaces are very clean and free from moisture, the transducer can be mounted fairly easily, although some practice may be required. The transducer can be removed rapidly with a naphtha-type solvent. Disadvantages include the possibility of disattachment of the transducer at high vibration levels, a temperature limitation because of the relatively low melting point of wax, and poor long-time stability of the mounting. The maximum temperature at which measurements can be made with this mounting technique is usually about 100°F (40°C).

Adhesive Mounting. An adhesive film may be used to mount a small transducer on a flat, clean test surface—usually by means of a double-sided adhesive tape. Double-sided adhesive discs are supplied by some transducer manufacturers. This mounting technique, illustrated in Fig. 15.1E, is rapid and easy to apply. Furthermore, such a mounting has the advantage of providing electrical insulation between the transducer and the test surface, and it does not require the drilling of a hole in the test surface; it is particularly applicable for use with a transducer having no tapped hole in its base. Such adhesives can provide secure attachment over a limited temperature range, usually below 200°F (95°C). In preparing an adhesive mounting, it is important to clean both the accelerometer and the test surface so that the adhesive will adhere firmly. When this is done, the frequency response can be fairly good, as illustrated in Fig. 15.2C, but not as good as with a wax mounting.

Magnetic Mounting. With magnetic mounting, illustrated in Fig. 15.1F, a permanent magnet attaches the transducer to the test surface, which must be ferromagnetic, flat, free from dirt particles, and reasonably smooth. Magnetic mounting is useful in measuring low acceleration levels. The transducer can be attached to the test surface easily and moved quickly from one measurement point to another. For example, in a condition-monitoring system (described in Chap. 16) it can be used to determine a suitable measurement location for a transducer to be mounted permanently on a large rotating machine. In a heavy machine of this type, the added mass of the magnet is not important, but in other problems, the additional mass loading on the test surface may make the use of magnetic mounting unacceptable. Furthermore, if the acceleration levels are sufficiently high, as in impact testing, the magnet may become loosened momentarily. This can result in an inaccurate reading and possibly a slight change in the position of the transducer, which would also change the reading. The frequency response for this type of mounting is fair, as shown in Fig. 15.2*D*, but not as good as with the wax mounting. The magnet, often available from the transducer's manufacturer, usually is attached to the transducer by means of (1) a projecting screw on the magnet, which is threaded into the base of the transducer, or (2) a machine screw, one end of which is threaded into the transducer and the other end into the magnet. Application of a light machine oil or silicone grease usually improves the frequency response above about 2,000 Hz. The maximum temperature at which measurements can be made with this mounting technique is usually about 300°F (150°C).

Mounting Blocks or Brackets. Physical conditions may make it impractical to mount a transducer by any of the above methods. In such cases, a mounting bracket or block that has been especially prepared for use on the test surface may be employed. For example, if the structural surface is rounded, a solid mounting block can be fabricated which is rounded to this same contour on one side and flat on the other side for mounting the transducer. A mounting block also may be useful where the surface is subject to structural bending; in this case, two accelerometers selected to have the same characteristics may be attached to the mounting block to measure bending-induced rotation. The effect of the mass of the mounting block is considered in Eq. (15.1). Two types of mounting brackets are illustrated in Fig. 15.3. Instead of using a triaxial accelerometer, sometimes it is more convenient to mount three transducers on a single block having sensitivities in three orthogonal directions. Any such mounting must couple the transducer to the test surface so that the transducer accurately follows the motion of the surface to which it is attached. This requires that the effective stiffness of the transducer mounting be high so that the mounting does not deflect under the inertial load of the transducer mass. This is not a problem in many transducer installations.

Mounting brackets may have resonance frequencies which are below 2,000 Hz and have little damping. Under such conditions, their use may result in significant measurement error as a result of resonant amplification or because of attenuation of vibration in the mounting. This is illustrated in Fig. 15.4, which shows the frequency response of a transducer mounted on brackets which are identical in geometry but which are fabricated from different materials. Note that a change in material from (*A*) steel to (*B*) a phenolic plastic halves the resonance frequency of the mounting. A change in the method of attachment, from (*B*) screw mounting to (*C*) an epoxy resin adhesive bond, significantly increases the frequency of the mounting resonance. Although these results are not of a general nature, they show that such minor variations in the transducer mounting may produce significant changes in the output characteristics of the transducer.

Hand-held Transducer. A transducer which is held against the test surface by hand provides the poorest performance of any of the techniques described here, but it sometimes can be useful in making a rapid survey of a test surface because the measurement location can be changed more rapidly than with any other method of mounting. Usually, a rod (called a *probe*), which is threaded at one end, is screwed into the transducer; the other end has a tip that is pressed against the test surface.

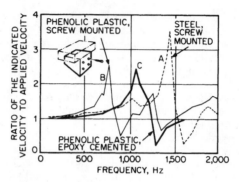

FIGURE 15.4 Relative frequency response of a velocity transducer mounted on three brackets which have identical geometry but are fabricated of different materials: (*A*) steel bracket, screw mounted, (*B*) cloth-reinforced phenolic plastic bracket, screw mounted, and (*C*) same as (*B*) but attached with epoxy resin adhesive.

The frequency response is highly restricted—about 20 to 1,000 Hz; furthermore this technique should not be employed for accelerations greater than 1*g*. Thus, this technique is used when measurement accuracy is not essential, e.g., in finding the nodal points on a vibrating surface.

Mass-Loading. The effect of the mounting on the accuracy of measurement can be estimated roughly if it is assumed that the combination of the transducer (having a mass m) and the mounting (having a stiffness k) behaves as a simple spring-mass system driven at the spring end of the system. Then the acceleration of the transducer $\ddot{x}$ is given by

$$\ddot{x} = \ddot{u}\,\frac{k}{k + m(2\pi f)^2} \tag{15.1}$$

where $\ddot{u}$ is the acceleration of the test item, and f is its frequency of vibration. If the acceleration of the transducer is to be within 10 percent of the acceleration of the test item, then from Eq. (15.1), k must have a value at least 10 times greater than the term $m(2\pi f)^2$. Since the undamped natural frequency f_n of the transducer-mounting system is given by $f_n = \frac{1}{2}\pi (k/m)^{1/2}$, the value of the natural frequency of the system must be at least 10 times the frequency of vibration of the test item—especially for the measurement of transients.

FIELD CALIBRATION TECHNIQUES

TRANSDUCERS

Various methods of calibrating transducers are described in Chap. 18. If a transducer is to be used under unusual temperature conditions, it is important to perform the calibration in the temperature range in which it will operate. Of these, the following are particularly convenient for use in the field.

Comparison Method. This is a rapid and convenient method of obtaining the sensitivity of a transducer. It is one of the most commonly used calibration techniques. Calibration is obtained by a direct comparison of the output generated when the transducer is attached to a vibration exciter with the output generated by a secondary standard transducer which is attached to the same vibration exciter and which is subject to precisely the same motion. The two transducers are mounted back to back, as illustrated in Fig. 18.2. Calibration by this method is limited to the frequency and amplitude ranges for which the secondary standard has been calibrated and for which the vibration exciter has adequate rectilinear motion.

Free-fall Calibration Method. The gravimetric free-fall calibration method (sometimes called a *drop test*) is a simple and rapid method of calibrating motion and force sensors. The transducer under test is allowed to fall freely for an instant of time under the influence of gravity; the peak signal then is measured for an acceleration of gravity having a value of $1g$. This technique is illustrated in Fig. 18.9.

Earth's Gravitational Field Method. In the following technique (sometimes called the "*inversion method*" of calibration), the sensitive axis of the transducer is first aligned vertically in one direction of the earth's gravitational field, as shown in Fig. 15.5A. Then it is inverted so that its sensitive axis is aligned in the opposite direction, as shown in Fig. 15.5B. The transducer output is observed for a $2g$ change in

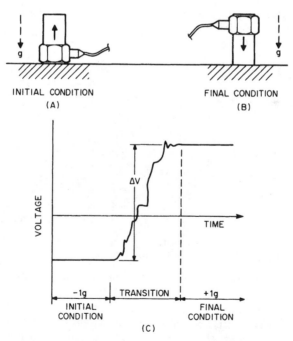

FIGURE 15.5 Gravitational field method (inversion test) for calibrating an accelerometer having useful sensitivity down to 0 Hz. Inversion of the accelerometer, initially aligned in one direction, as in (A), to the opposite direction, as in (B), produces a change in acceleration of $2g$. The transducer output for this change is measured in (C). *(Courtesy of Quixote Measurement Dynamics, Inc.)*

acceleration, as shown in Fig. 15.5C. This method is limited in application to accelerometers having sensitivity down to 0 Hz; it is not recommended for calibration of accelerometers having significant transverse sensitivity.

OVERALL SYSTEM

Calibration of a complete vibration measurement system usually is referred to as *overall calibration* or *end-to-end calibration*. It is good practice to perform such a calibration at periodic intervals—particularly both before and after an extensive series of measurements. In such a calibration, the amplitude characteristics, phase characteristics, and linearity of the overall system are determined when the transducer is subject to a known acceleration, velocity, or displacement, for example, by means of a *field calibrator*.

Field Calibrator. This is a portable device on which a transducer can be mounted and subjected to a known acceleration, velocity, or displacement at a fixed frequency. Such an instrument (essentially a small, portable, battery-powered shaker) provides a convenient means for calibrating a transducer in the field and/or calibrating the overall vibration measurement system. For example, the hand-held device shown in Fig. 15.6 can be used to calibrate a transducer weighing up to 85 grams at a frequency of 79.6 Hz. This device is furnished with an internal oscillator and a stable, built-in reference accelerometer in a feedback loop controlling the electrodynamic exciter; the exciter subjects the transducer under test to a constant rms acceleration amplitude of $1g$.

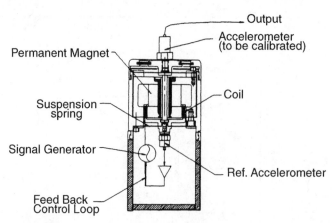

FIGURE 15.6 A hand-held vibration calibrator especially designed for field application. *(Courtesy PCB Piezotronics, Inc.)*

Combining Characteristics of Individual Components. When it is not possible to subject the transducer to a known acceleration, velocity, or displacement, the overall characteristics sometimes are determined by combining the characteristics of the individual components of the system, as described below, or the system is calibrated employing a simulated transducer output [see *Voltage Substitution Method of Calibration* below, and *Calibration of Auxiliary Circuits* (Chap. 18)].

There may be a significant electrical signal at the output of a measurement system though no signal is supplied by the transducer to the input; such electrical signals, which represent noise, (1) may result from a coupling between circuits in the measurement system with power circuits, (2) may be generated by vibration-sensitive elements (such as cable) other than the transducer, or (3) may be the result of improper selection of system components, or the improper setting of one or more of these components, so that the signal-to-noise ratio that the overall system is capable of attaining is not achieved.

Where a single component of a measurement system is the source of noise, it can sometimes be located by using an oscilloscope which is first connected to the transducer output with no vibration applied. Then the oscilloscope connection is moved, component by component, through the measurement system until the noise is observed. Another approach is to short-circuit the signal path at various points in the system (where this is practical), one at a time, until the system noise disappears. Usually this pinpoints the source as the component next nearest the transducer from the last short circuit.

Spurious mechanical sources and acoustic noise sources must be eliminated or controlled if they result in noise in the measurement system. Spurious resonances in the response of the overall system may result from improper seating of the transducer on the test surface or from resonances in the transducer mounting. It is often very useful to excite the transducer-mounting system by giving it a blow and then to observe the transducer's output—look for resonances other than the resonance frequency of the transducer. The other resonance frequencies which appear may be due to (1) resonances in the test specimen or (2) resonances in the transducer mounting. Loose mountings usually produce "noisy" signals and may produce audible buzzing sounds. Often it is difficult to determine the difference between resonances in the mounting and resonances in the item under test. If serious doubt exists, the test should be repeated with a different mounting or a different measurement location for the transducer. If the resonance frequencies are identical for the new mounting, the resonances are probably due to the test specimen, and the original mounting probably was satisfactory.

Combining Calibration Characteristics of a Measurement System's Components.

An overall system can be calibrated by combining the measured electrical characteristics of all components in the measurement system from one end to the other. Obtaining a system calibration in this way circumvents the difficulties of precise field calibration, but it requires that each element in the system be calibrated in the laboratory with extreme care and that the effects of the source and load impedances be completely accounted for. Thus, a system calibration is subject to the sum of the experimental errors introduced by the calibration of each element, in addition to any errors resulting from improper simulation of, or accounting for, loading effects. In general, the calibration of each element is performed before the system is assembled, and so this method is subject to error resulting from (1) undetected damage to components between calibration and use and/or (2) improper connections, misidentifications, or confusion in polarity.

Voltage Substitution Method of Calibration.

A suitable simulated transducer for use in field checkout must duplicate the electrical outputs of the actual transducer for the various vibration conditions to be simulated. The simulated transducer must either (1) reproduce the electrical voltage- or current-generating characteristics of the actual transducer and have the same output impedance or (2) duplicate the electrical quantity generated by the actual transducer when connected to its load. Failure to meet these conditions will result in a different loading of the actual and simulated transducers and will probably cause calibration errors. It is important that the simulated transducer have the same electrical grounding configuration as the actual transducer; otherwise, electric-circuit noise and cross talk* will not be represented accurately when the simulated transducer is in use.

Typical examples of circuits which simulate transducers are shown in Fig. 15.7. The simulated transducer introduces an electrical signal into the measurement system, thereby simulating the response of the actual transducer.

CABLE AND WIRING CONSIDERATIONS

The method of data transmission between a transducer and the electronic instrumentation which follows it depends on the complexity of the problem. In general, cable is used for most problems, but the aerospace industry often relies on telemetry for data transmission. Many types of cable are available. The choice of a suitable cable depends primarily on the particular application, the transducer, the cable length, whether the transducer is followed by a voltage amplifier or charge amplifier, and environmental conditions. For example, cable jackets may be made of silicone rubber having a useful temperature range from −100 to 500°F (−73 to 260°C), of polyvinylchloride having a useful range from −65 to 175°F (−54 to 79°C), or of fused Teflon having a useful range from −450 to 500°F (−268 to 260°C). Special-purpose cables are available that can be used at much higher temperatures. In general, cable should be as light and flexible as possible—consistent with other requirements. The effect of the shunt capacitance of the cable following the transducer on the sensitivity of the transducer depends on the type of amplifier connected to the cable. If a voltage amplifier is used, there is a reduction in sensitivity of the transducer, given by Eq. (12.17). In contrast, when a charge amplifier is used, the effect of the shunt capacitance of the cable in reducing the sensitivity of the transducer is negligible, as shown in Eq. (13.2) (although the noise pickup in the high-impedance circuit increases with cable length).

In the audio-frequency range, the series inductance L and the shunt leakage G of short, good-quality cables are negligibly small in comparison with other parameters and may be neglected. Figure 15.8A shows the equivalent low-frequency representation of a cable with distributed constants. For most purposes the simpler lumped-constant configuration of Fig. 15.8B is a sufficiently accurate representation. The quantities R_c and C_c are the total resistance of the conductors and the total capacitance between them, respectively. Values for a typical coaxial cable having a Teflon dielectric are $R_c = 0.01$ ohm/ft (0.03 ohm/m) and $C_c = 29$ pF/ft (88 pF/m).

The normal characteristic impedance of about 50 ohms for such cable has no significance in most measurement problems, where cables usually are relatively short. The open-circuit input impedance of the cable is almost exclusively capacitative. When terminated, it takes on the impedance of the load, modified by the series and shunt parameters.

In general, cables should be treated with the same care given transducers in shock and vibration measurement systems. The following are based on recommendations given in Ref. 1; they represent good engineering practice.

---

* *Cross talk* is the output of one measurement channel when a signal is applied to another measurement channel. Cross talk can be distinguished from other electrical disturbances because it is a function of the applied signal in the other measurement channel and disappears when this applied signal is removed.

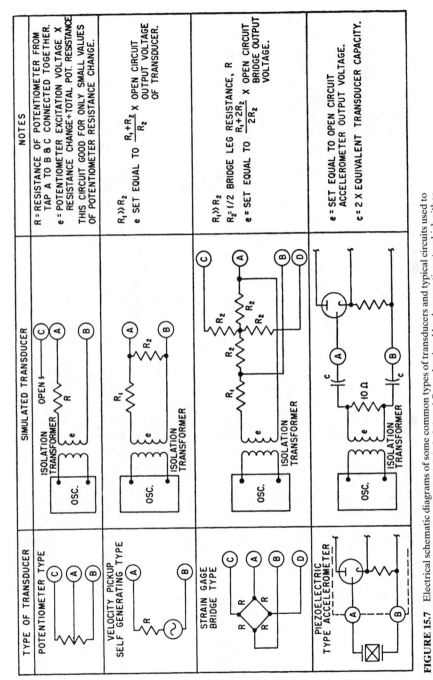

FIGURE 15.7 Electrical schematic diagrams of some common types of transducers and typical circuits used to simulate them during field calibration. Terminals labeled *A* and *B* are the signal lead connections to which either the transducer or the simulated transducer is connected.

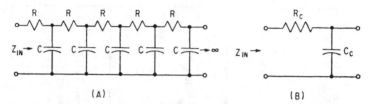

FIGURE 15.8 Successive approximations in the representation of a short, high-quality transmission line at audio frequencies. (*A*) Distributed constant configuration neglecting series inductance and shunt leakage. (*B*) Lumped-constant configuration.

1. Attach a coaxial cable to a transducer by turning the connector nut onto the threads of the transducer (not vice versa) to avoid damage to the pins.

2. Avoid cable whip by tying down the cable at a point near the transducer and at regular intervals to avoid induced cable noise.

3. Screw the cable connection to the tightness specified by the manufacturer.

4. Loop the cable near the connector in a high-humidity environment, to allow condensation to drip off before reaching the connector.

5. Clean the cable connector before use (e.g., acetone or chlorothene) to remove contamination as a result of handling; the contamination can create a low impedance between the signal path and ground.

6. Check electrical continuity of cable conductors and shield if intermittent signals are observed. Then, flex the cable—especially near the connector—and observe if the signal is affected by flexing.

7. Select cables that are light and flexible enough to avoid loading the transducer and/or the structure under test, or exerting a force on the transducer.

8. Avoid twisting the cable when it is connected to the transducer.

9. Move the cable back and forth to determine if such movement generates unacceptable electrical noise; if so, tie the cable more securely or replace the cable.

CABLE NOISE GENERATION

When two dissimilar substances are rubbed together, they become oppositely charged—a phenomenon known as *triboelectricity,* illustrated in Fig. 15.9. Thus a charge may be generated when a cable is flexed, bent, struck, squeezed, or otherwise distorted, for then such friction takes place between the dielectric and the outer shield or between the dielectric and the center conductor.[2] A charge is generated across the cable capacitance so that a voltage appears across the termination of the cable.

Another mechanism by which noise may be induced in the cable results from the change in capacitance of the cable when it is flexed. If the transducer produces a charge across the cable, the change in capacitance results in a voltage change across the output of the

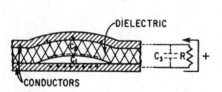

FIGURE 15.9 A section of cable during distortion, showing how separation of triboelectric charge leads to the production of cable noise across the termination resistance. (*After T. T. Perls.*[2])

cable, appearing as noise at the input of a voltage amplifier; it will not produce a similar change if a charge amplifier is used.

Suppose the dielectric surfaces within the cable are coated so that an electrical leakage path is provided along the dielectric surface. Then if the cable shield is separated from the outer surface of the dielectric, the charges flow along the surface to the nearest point of contact of the dielectric and shield; without this leakage path, the charges would flow to the terminating impedance, where they would give rise to a noise signal. Such coatings are provided in low-noise cables which are available commercially. Cables of this type are capable of withstanding considerable abuse before becoming noisy. Usually they are tested by the manufacturer continuously along their lengths to assure meeting the low-noise characteristics. It is important in fitting such a cable with a connector, or in splicing such a cable, that no conducting material be allowed to form a leakage path between the conductors. Carbon tetrachloride and xylene are satisfactory solvents and cleaning agents.

NOISE-SUPPRESSION TECHNIQUES

Under certain conditions of use and environment, spurious signals (noise) may be induced in wiring and cables in a measurement system. Then there will be signals at the termination of the system that were not present in the transducer output.

Electrical noise may be generated by motion of some parts of the wiring because of variation in contact resistance in connectors, because of changes in geometry of the wiring, or because of voltages induced by motion through, or changes in, the electrostatic fields or magnetic fields which may be present. No cable should carry wiring *both* for data transmission and for electrical power; all electrical power wiring should be twisted pair. In general, such electrical noise will be reduced if the cable is securely fastened to the structure at frequent intervals and if connectors are provided with mechanical locks and strain-relief loops in their cables. Precautions taken to avoid interference usually include the use of shielding, cables which are only as long as necessary, and proper grounding. In addition, the use of a transducer containing an internal amplifier (described in Chap. 12) can provide advantages in noise suppression.

Shielding.   A change in the electric field or a change in the magnetic field around a circuit or cable may induce a voltage within it and thus be a source of electrical noise. Such electrical interference can be avoided by completely surrounding the circuit or cable with a conductive surface which keeps the space within it free of external electrostatic or magnetic fields. This is called *shielding*. Protection against changes in each type of field is different.

Electrostatic Shields.   Electrostatic shields provide a conducting surface for the termination of electrostatic lines of flux. Stranded braid, mesh, and screens of good electrical conductors such as copper or aluminum are good electrostatic shields. Most shielded cables use copper braid as the outer conductor and electrostatic shield. A good magnetic shield is also a good electrostatic shield, but the converse is not true. For installations where cable lengths are especially long, where impedances are high, or where noise interference is highly objectionable, double-shielded cable is sometimes used. In this type of cable, a second shielding braid is woven over the cable jacket, electrically insulating it from the inner shield; the inner braid furnishes additional shielding against electrostatic fields which penetrate the first shield. The shields should be connected to ground at one point only, as explained below under *Grounding: Avoiding Ground Loops.*

Magnetic Shields. Magnetic shields are effective partly because of the short circuiting of magnetic lines of flux by low-reluctance paths and partly because of the cancellation resulting from opposing fields set up by eddy currents. Accordingly, they are made from high-permeability materials such as Permalloy, are as thick as possible, and contain a minimum of joints, holes, etc.

Magnetic fields associated with current-carrying power lines, electronic equipment, and power transformers are among the most troublesome sources of magnetic interference in instrumentation setups—chiefly at the frequency of the power line and its harmonics. Since these fields attenuate rapidly with distance from the source, the most practical solution for this type of interference usually is to keep the signal cables as far from the power source as possible.

Grounding; Avoiding Ground Loops. A circuit is said to be grounded when one terminal of the circuit is connected to the "earth." Grounding removes the potential difference between that side of the circuit and earth, and the variable stray capacitances which tend to induce voltages in "floating" (i.e., ungrounded) systems. Water pipes make good ground connections because of their intimate contact with the earth.

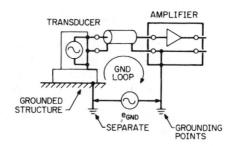

FIGURE 15.10 Ground loop in a system as a result of grounding the cable shield at two points. Then, the input signal e_1 is modulated by the potential difference e_{gnd} which develops between these two points.

Ground loops are formed when a common connection in a system is grounded at more than one point, as illustrated in Fig. 15.10, where the cable shield is grounded at both ends. Since it is unlikely that the two grounds will be at a common potential, their potential difference, e_{gnd}, will be the source of circulating currents in the ground loop. Then a signal produced by the transducer will be modulated by the potential e_{gnd}, thereby introducing noise in the measurement system. Such a condition may occur when one end of a cable is connected to one side of the electrical output of a transducer that has been grounded to the transducer's housing and the other end of the cable is connected to a voltage amplifier or signal conditioner which is also grounded (usually to the case of the instrument). Then, a ground loop will be formed. *Such a condition must be avoided by grounding the circuit at only*

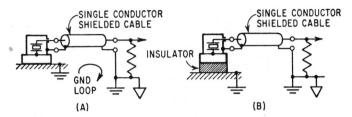

FIGURE 15.11 (*A*) A ground loop formed when the "low" sides of both the transducer and the amplifier are connected to their respective cases, which are grounded. (*B*) The ground loop shown in (*A*) is broken by isolating the case of the transducer from ground.

one point. Thus the circuit shown in Fig. 15.11*A* will result in noise because of the ground loop, but by insulating the transducer as shown in Fig. 15.11*B* the ground loop has been broken.

DATA SHEETS FOR LOGGING TEST INFORMATION

When data are acquired in the field, measurement conditions may be far from ideal; environmental conditions may be unfavorable, and the time available for measurements may be extremely limited. Therefore it is good practice to prepare data sheets that are relatively simple and that require a minimum amount of writing; for example, use multiple-choice entries. The data sheets should include sufficient information so that someone else, at a later time, could duplicate the measurement setup on the basis of information supplied by the data sheets. If there are any anomalies that occur during the test, they should be duly noted. In general, the following information should be included:

Basic data concerning the test measurements:

- Date, times, and duration of test.
- Identification of test by test number.
- Identification of equipment, machine, or device under test.
- Conditions of operation during the measurement.
- Any anomalies in operation and their times of occurrence.
- Location of test, using diagram where appropriate.
- Environmental conditions during test; note anomalies where appropriate.
- Persons participating in the test.

Equipment, including transducers, cables, signal conditions, data recorders, telemeter:

- Type.
- Manufacturer, model number, and serial number.
- Transducer sensitivity, location, orientation, mounting.

- Signal conditioner and amplifier gain and attenuator settings; note any changes in these settings during the test.
- Filter settings, if any.
- Recorder speed, number of tracks, tape speed, gain settings; note any changes in these settings during the test.

Calibration information:

- Transducer calibration.
- Overall system (end-to-end) calibration of system.
- Phase of output signal relative to input signal.
- Any changes in calibration between pretest and posttest conditions.

REFERENCES

1. Endevco Instruction Manual for Piezoelectric Accelerometer, No. 101, San Juan Capistrano, Calif., 1996.
2. Perls, T. A.: *J. Appl. Phys.,* **23**(6):674 (1952).

CHAPTER 16
CONDITION MONITORING OF MACHINERY

Joëlle Courrech

INTRODUCTION

Condition monitoring of machinery is the measurement of various parameters related to the mechanical condition of the machinery (such as vibration, bearing temperature, oil pressure, oil debris, and performance), which makes it possible to determine whether the machinery is in good or bad mechanical condition. If the mechanical condition is bad, then condition monitoring makes it possible to determine the cause of the problem.[1,2]

Condition monitoring is used in conjunction with *on-condition maintenance,* i.e., maintenance of machinery based on an indication that a problem is about to occur. In many plants on-condition maintenance is replacing *run-to-breakdown maintenance* and *preventive maintenance* (in which mechanical parts are replaced periodically at fixed time intervals regardless of the machinery's mechanical condition). On-condition maintenance of machinery:

- Avoids unexpected catastrophic breakdowns with expensive or dangerous consequences.

- Reduces the number of overhauls on machines to a minimum, thereby reducing maintenance costs.

- Eliminates unnecessary interventions with the consequent risk of introducing faults on smoothly operating machines.

- Allows spare parts to be ordered in time and thus eliminates costly inventories.

- Reduces the intervention time, thereby minimizing production loss. Because the fault to be repaired is known in advance, overhauls can be scheduled when most convenient.

This chapter describes the use of vibration measurements for monitoring the condition of machinery. Vibration is the parameter which can be used to predict the broadest range of faults in machinery most successfully. This description includes:

1. Selection of an appropriate type of monitoring system (permanent or intermittent)

2. Establishment of a condition monitoring program, including the selection of transducers, the selection of the appropriate vibration parameter, the selection of measurement location on the machine, and the selection of time interval between measurements

3. Fault detection

4. Spectrum interpretation and fault diagnosis

5. Special analysis techniques

6. Trend analysis

7. The use of computers in condition monitoring programs.

SELECTION OF TYPE OF MONITORING SYSTEM

Condition monitoring systems are of two types: intermittent and permanent. In an *intermittent monitoring system* (also called an *off-line condition monitoring system*), machinery vibration is measured (or recorded and later analyzed) at selected time intervals in the field; then an analysis is made either in the field or in the laboratory. Advanced analysis techniques usually are required for fault diagnosis and trend analysis. Intermittent monitoring provides information at a very early stage about incipient failure and usually is used where (1) very early warning of faults is required, (2) advanced diagnostics are required, (3) measurements must be made at many locations on a machine, and (4) machines are complex.

In a *permanent monitoring system* (also called an *on-line condition monitoring system*), machinery vibration is measured continuously at selected points of the machine and is constantly compared with acceptable levels of vibration. The measurement system may be permanent (as in parallel acquisition systems where one transducer and one measurement chain are used for each measurement point), or it may be quasi-permanent (as in multiplexed systems where one transducer is used for each measurement point but the rest of the measurement chain is shared between a few points with a multiplexing interval of a few seconds).

In a permanent monitoring system, transducers are mounted permanently at each selected measurement point. For this reason, such a system can be very costly, so it is usually used only in critical applications where: (1) no personnel are available to perform measurements (offshore, remote pumping stations, etc.), (2) it is necessary to stop the machine before a breakdown occurs in order to avoid a catastrophic accident, (3) an instantaneous fault may occur that requires machine shutdown, and (4) the environment (explosive, toxic, or high-temperature) does not permit the human involvement required by intermittent measurements.

Before a permanent monitoring system is selected, preliminary measurements should be made intermittently over a period of time to become acquainted with the vibration characteristics of the machine. This procedure will make it possible to select the most appropriate vibration measurement parameter, frequency range, and normal alarm and trip levels.

ESTABLISHMENT OF A CONDITION MONITORING PROGRAM

SELECTION OF TRANSDUCERS

To select the most suitable transducer for a condition monitoring program, it is necessary to ascertain the type of the vibration to be measured: e.g., whether it is shaft vibration or bearing vibration.

Displacement Transducers. Displacement transducers of the eddy-current type, which have noncontacting probes, are commonly used to measure shaft vibration. Such transducers provide information about the relative motion between shaft and bearing. This information can be related directly to physical values such as mechanical clearance or oil film thickness; e.g., it can give an indication of incipient rubbing. The use of displacement transducers is essential in machinery having journal bearings. However, noncontacting probes of this type (1) are difficult to calibrate absolutely, (2) have limited dynamic range because of the influence of electrical and mechanical run out on the shaft, and (3) have limited frequency range. As a consequence of the limited dynamic range, the useful upper frequency is typically limited to the frequency of the first few harmonics of the shaft speed, as shown in Fig. 16.1A.

Shaft vibration provides information about the current condition of the machine and is used principally in permanent monitoring systems, which immediately shut the machine down in the event of trouble.

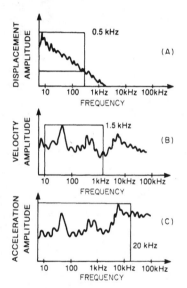

FIGURE 16.1 Spectra of the same vibration source, obtained with different types of transducers. (*A*) A typical displacement spectrum showing a negative slope. Even if a displacement transducer has a large frequency range (typically up to 10 kHz), its limited dynamic range (40 dB) results in a spectrum whose useful frequency range is restricted (typically up to 500 Hz). (*B*) A velocity spectrum. Although velocity transducers have a good dynamic range (60 dB), their frequency range is limited to between 10 and 1500 Hz (typical). (*C*) An acceleration spectrum. Accelerometers cover frequency ranges from 0.1 Hz or below up to 15,000 Hz and above (depending on types), with a dynamic range of typically 150 dB. Even after integration of the signal to velocity, it provides a 90-dB dynamic range over three decades in frequency.

Accelerometers and Velocity Pickups. Accelerometers and velocity pickups are used to make bearing or casing vibration measurements. Accelerometers have a wider dynamic range and a broader frequency range than either displacement or velocity transducers, as illustrated in Fig. 16.1.

Accelerometers (described in Chap. 12) are lightweight and rugged; they should always be selected for detecting faults which occur at high frequencies, for example, to detect rolling-element bearing deterioration or gearbox wear. Contrary to common belief, accelerometers are able to measure very low frequencies, the limitation being determined by the preamplifier used rather than the transducer itself. Frequencies in the order of magnitude of 1 Hz or below can be measured without special precautions with general-purpose accelerometers. The magnitude of the acceleration of the bearing or casing may be directly related to the forces acting inside the machine and transmitted through the bearings. Acceleration measurements of bearing vibration will provide very early warning of incipient faults in a machine.

SELECTION OF MEASUREMENT PARAMETER: DISPLACEMENT, VELOCITY, OR ACCELERATION

When an accelerometer is employed as the sensing device in a condition monitoring system, the resulting *acceleration* signal can be electronically integrated to obtain *velocity* or *displacement,* so any one of these three parameters may be used in measurements. The appropriate parameter may be selected by application of the following simple rule: *Use the parameter which provides the "flattest" spectrum.* The flattest spectrum requires the least dynamic range from the instrumentation which follows the transducer. For example, Fig. 16.2 shows a velocity spectrum and a displacement spectrum obtained under identical conditions. The dynamic range (i.e., the range from the highest to the lowest signal level) required to measure the displacement spectrum is much larger than the range for the velocity spectrum; it may even exceed the available dynamic range of the instrumentation. Therefore, according to this rule, velocity measurements should be selected.

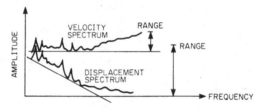

FIGURE 16.2 Displacement and velocity spectra obtained under identical conditions. The velocity spectrum requires a smaller dynamic range of the equipment which follows the transducer. Therefore, it is preferable.

The *flattest spectrum* rule applies only to the frequency range of interest. Therefore, the parameter selection, to some extent, depends on the type of machine and the type of faults considered.

SELECTION OF MEASUREMENT LOCATION

When an intermittent (off-line) monitoring system is employed, the number of points at which measurements are made is limited only by the requirement for keeping measurement time to a minimum. As a general rule, bearing vibration measurements are made in the radial direction on each accessible bearing, and in the axial direction on thrust bearings. It is not usually necessary to measure bearing vibration in both the horizontal *and* the vertical direction, since both measurements give the same information regarding the forces within the machine; this information is merely transmitted through two different transmission paths. This applies for *detecting* developing faults. It will later be seen, however, that in order subsequently to *diagnose* the origin of the impending fault, measurements in both the horizontal and the vertical direction may give valuable information. When measuring shaft vibrations with proximity transducers, it is convenient to use two probes on each bearing, located at 90° from each other, thereby providing an indication of the orbit of the shaft within the bearing.

When a permanent (on-line) monitoring system is employed, the number of measurement points usually is minimized for reasons of economy. Selection must be made following a study of the vibration spectra of different bearings in order to locate those points where all significant components related to the different expected faults are transmitted at measurable vibration levels if full spectrum comparison is performed. If only broadband measurements are monitored, then a further requirement is that all frequency components related to the expected faults must be of approximately the same level within the selected frequency range.

SELECTION OF TIME INTERVAL BETWEEN MEASUREMENTS

The selection of the time interval between measurements requires knowledge of the specific machine. Some machines develop faults quickly, and others run trouble-free for years. A compromise must be found between the safety of the system and the time taken for measurements and analysis. The following rough rule of thumb is useful: *Select a time interval between measurements which is one-sixth to one-tenth the expected period between overhaul.* In any case, measurements should be made frequently in the initial stages of a condition monitoring program to ensure that the vibration levels measured are stable and that no fault is already developing. When a significant change is detected, the time interval between measurements should be reduced sufficiently so as not to risk a breakdown before the next measurement. The trend curve will help in determining when the next measurement should be performed.

FAULT DETECTION IN ROTATING MACHINERY

It is highly desirable to be able to detect all types of faults likely to occur during the operation of rotating machinery. Such faults range from vibrations at very low frequencies (subsynchronous components indicating looseness, oil whirl, faulty belt drive, etc.) to vibrations at very high frequencies (tooth-meshing frequencies, blade-passing frequencies, frequencies of structural resonances excited by faulty rolling-element bearings, etc.). Such detection should be applicable to the complete range of machines in a plant, which operate from very low to very high speed. This requires the selection of equipment and analysis techniques which cover a very broad frequency range.

Measurements of *absolute* vibration levels of bearings provide no indication of the machine's condition, since they are influenced by the transmission path between the force and the measurement point, which may amplify some frequencies and attenuate others. Bearing vibration levels change from one measurement point to another on a given machine, since the transmission paths are different; they also change for the same reason from machine to machine for measurements made at the same measurement point.[3] Therefore, in estimating the condition of a machine, it is essential to monitor *changes* in vibration from a reference value established when the machine was known to be in good condition. Changes are expressed as a ratio or, more commonly, as a *change of level,* i.e., the logarithm of a ratio, in decibels.

The objective of condition monitoring of a machine is to predict a fault well in advance of its occurrence. Therefore, a measurement of the overall vibration level will not provide successful prediction because the highest vibration component within the overall frequency range will dominate the measurement. This is illustrated in Fig. 16.3, which shows an example where overall measurements of the vibration velocity resulted in an incorrect prediction with an overestimate of the lead time. The early detection of faults in machinery can be made successfully only by comparison with a *reference spectrum.* This section compares types of spectrum analysis for this purpose.

Condition monitoring techniques employed during transient operating conditions of the machine (i.e., when the machine is running up to full speed or slowing down from full speed) differ significantly from the techniques employed during steady-state operating conditions. Therefore it is essential that a careful investigation be carried out to ensure that the condition monitoring technique selected is appropriate for the conditions of measurement.

FALSE ALARMS

Changes in machinery vibration may result from a number of causes which are not necessarily related to the deterioration of the machine. For example, a change in speed of the machine or a change in the load on the machine usually greatly modifies the relative amplitudes of the different components of vibration at a fixed transducer location or modifies the relative pattern of vibration at different locations. Depending on the criteria used for fault detection, such changes may result in a false indication of deterioration of the machine. Appropriate selection of the technique employed can avoid such false alarms.

CONSTANT-BANDWIDTH VERSUS CONSTANT-PERCENTAGE-BANDWIDTH ANALYSIS

A constant-bandwidth analyzer, such as a fast Fourier transform (FFT) analyzer, provides spectra having a constant bandwidth on a linear frequency scale. This yields very fine resolution at high frequencies but very poor resolution at low frequencies. To cover a large frequency range with sufficient resolution using constant-bandwidth analysis, a number of different frequency analyses are required—typically one analysis per frequency decade. This is time-consuming, since an enormous amount of data must be stored and compared for each measurement point.

A constant-percentage-bandwidth analysis typically covers three decades with equal resolution in a single spectrum on a logarithmic frequency range. In this way considerable data reduction is achieved, yet adequate resolution is maintained. A further advantage of the constant-percentage-bandwidth/logarithmic frequency scale is the simplicity of compensating for speed changes. A speed variation of the machine manifests itself in the vibration spectrum as a global translation of all speed-related components. For the above reasons, the best analysis method for the

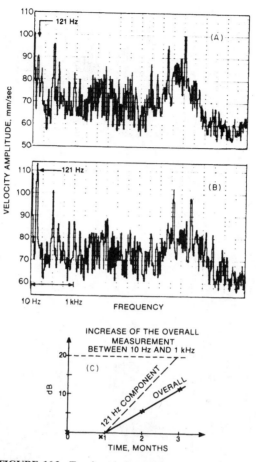

FIGURE 16.3 Trend analysis performed on an overall measurement and on an individual component. (*A*) The velocity spectrum of vibration measured on a gearbox after installation. Note the high amplitude of the 480-Hz component, dominating the reference spectrum. (*B*) The velocity spectrum 3 months later. Note the dramatic increase in the 121-Hz component, which corresponds to the output shaft speed of the gearbox. (*C*) Curves comparing the increase in the 121-Hz component in the velocity spectrum; the increase in overall velocity in the band from 10 to 1000 Hz indicates a developing fault.

comparison of spectra and fault detection is by use of constant-percentage bandwidth with a logarithmic frequency scale. Figure 16.4 shows an example which compares measurements on the same gearbox made with two different types of analyzers: Fig. 16.4*A* employs a constant-percentage bandwidth; Fig. 16.4*B* employs constant bandwidth. Note in Fig. 16.4*B* the loss of information in the range from 0 to 2 kHz.

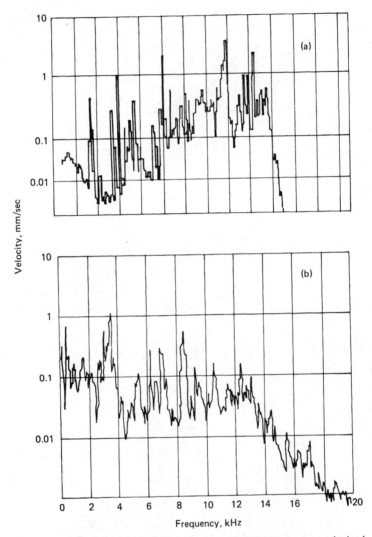

FIGURE 16.4 Analysis of gearbox vibration. (*A*) Velocity spectrum obtained with an analyzer having a constant-percentage bandwidth of 4 percent, used to cover a frequency range from 20 to 20,000 Hz. The frequency scale is logarithmic. (*B*) Velocity spectrum obtained with an analyzer having a constant bandwidth of 50 Hz, used to cover a frequency range from 0 to 20 kHz. The frequency scale is linear.

Sufficient frequency resolution must be maintained to separate all characteristic components of the machine, for example, two closely related components. As in the overall measurements of Fig. 16.3, if two components, equally important but largely different in level, are within the same filter bandwidth, the amplitude of the higher one will conceal changes in the lower one. A bandwidth of between 3 and 6 percent usually is sufficient to cover small speed fluctuations (in the order of 1 percent), yet it has sufficient resolution to separate characteristic components.

HOW SPECTRUM CHANGES ARE RELATED TO THE CONDITION OF A MACHINE

To obtain information about changes in condition of a machine, vibration spectra should be compared only for similar operating conditions. The influence of operating condition of the machine (such as machine speed, load, and temperature) on the vibration parameter being measured varies greatly for different types of machines. Speed changes of up to 10 per cent usually can be compensated for, and spectra can be compared. If the speed changes are greater than this value, the operating condition of the machine should be considered to be different and a new reference spectrum used as a basis of comparison. The reference spectrum need not be measured when the machine is new (after allowing for a run-in period). The reference spectrum can be determined at any time during the life of a machine provided the vibrations are stable, since a stable spectrum is a sign of stable operation of the machine. The principal difficulty is to establish when changes in the spectrum are sufficiently large to warrant stopping the machine.

Most national and international standards for the measurement of bearing vibration do not consider frequency spectra; instead, they give values for vibration changes of the rms value of the velocity amplitude from 10 to 1000 Hz (or 10,000 Hz) for machines in good and bad condition. These ratios have successfully been transposed to characteristic components in the vibration spectrum such as unbalance or gear tooth-meshing frequency. Usually, a change in the bearing vibration amplitude (measured in terms of acceleration, velocity, or displacement) on any characteristic component from the spectrum by a factor of 2 to 2.5 (6 to 8 dB in vibration level) is considered significant; a change by a factor of 8 to 10 (18 to 20 dB in vibration level) is considered critical, unless specified otherwise by the manufacturer. Limits for shaft vibration measurements, giving the relative motion of the shaft inside the bearing, directly relate to physical bearing clearance in the machine. The required time interval between measurements varies greatly from one machine to another and depends directly on the expected mean time between failure and the deterioration rate of the expected failures; therefore, measurements should be made more frequently as soon as incipient deterioration is noticed.

Successful fault detection in machinery is the first step toward a successful condition monitoring program. Early recognition of deterioration is the key to valuable fault diagnosis and efficient trend analysis. Consequently, this phase of condition monitoring should not be neglected, although sometimes it may seem tedious.

SPECTRUM INTERPRETATION AND FAULT DIAGNOSIS

Although constant-percentage-bandwidth analysis is better for fault detection and spectrum comparison than constant-bandwidth analysis, the latter is essential for efficient diagnosis. Commercially available fast Fourier transform analyzers provide a suitable tool for spectrum interpretation. They provide constant bandwidth (on a linear frequency scale), and, by means of zoom, they also provide very high resolu-

TABLE 16.1 A Vibration Troubleshooting Chart

Nature of fault	Frequency of dominant vibration, Hz = rpm/60	Direction	Remarks
Rotating members out of balance	1 × rpm	Radial	A common cause of excess vibration in machinery
Misalignment and bent shaft	Usually 1 × rpm Often 2 × rpm Sometimes 3 & 4 × rpm	Radial and axial	A common fault
Damaged rolling element bearings (ball, roller, etc.)	Impact rates for the individual bearing component Also vibrations at high frequencies (2 to 60 kHz) often related to radial resonances in bearings	Radial and axial	Uneven vibration levels, often with shocks Impact Rates f (Hz): For Outer Race Defect $$f(\text{Hz}) = \frac{n}{2}f_r\left(1 - \frac{BD}{PD}\cos\beta\right)$$ For Inner Race Defect $$f(\text{Hz}) = \frac{n}{2}f_r\left(1 + \frac{BD}{PD}\cos\beta\right)$$ For Ball Defect $$f(\text{Hz}) = \frac{PD}{BD}f_r\left[1 - \left(\frac{BD}{PD}\cos\beta\right)^2\right]$$ n = number of balls or rollers f_r = relative rps between inner & outer races
Journal bearings loose in housing	Subharmonics of shaft rpm, exactly ½ or ⅓ × rpm	Primarily radial	Looseness may only develop at operating speed and temperature (e.g. turbomachines)

Oil-film whirl or whip in journal bearings	Slightly less than half shaft speed (42 to 48 per cent)	Primarily radial	Applicable to high-speed (e.g., turbo) machines
Hysteresis whirl	Shaft critical speed	Primarily radial	Vibrations excited when passing through critical shaft speed are maintained at higher shaft speeds. Can sometimes be cured by tightening the rotor components
Damaged or worn gears	Tooth-meshing frequencies (shaft rpm × number of teeth) and harmonics	Radial and axial	Sidebands around tooth-meshing frequencies indicate modulation (e.g., eccentricity) at frequency corresponding to sideband spacings. Normally only detectable with very narrow-band analysis and cepstrum analysis
Mechanical looseness	2 × rpm		Also sub- and interharmonics, as for loose journal bearings
Faulty belt drive	1, 2, 3, & 4 × rpm of belt	Radial	The precise problem can usually be identified visually with the help of a stroboscope
Unbalanced reciprocating forces and couples	1 × rpm and/or multiples for higher-order unbalance	Primarily radial	
Increased turbulence	Blade & vane passing frequencies and harmonics	Radial and axial	An increased level indicates increased turbulence
Electrically induced vibrations	1 × rpm or 1 or 2 times synchronous frequency	Radial and axial	Should disappear when power turned off

16.11

tion in any frequency range of interest. This permits (1) early recognition and separation of harmonic patterns or sideband patterns and (2) separation of closely spaced individual components. Fast Fourier transform analyzers also may provide diagnostic tools such as synchronous time averaging, cepstrum analysis, and/or use of the Hilbert transform for amplitude and phase demodulation (see Chap. 13).

Table 16.1 classifies different types of faults and indicates at which frequency the faults are displayed in a vibration spectrum. Although such a table is of considerable help in spectrum interpretation, any such simplified presentation must be used with care, as illustrated by the examples considered below. The various faults can be classified according to their spectral components, as follows.

SUBSYNCHRONOUS COMPONENTS

Subsynchronous components of vibration (at frequencies below the rotational speed of the machine) usually occur where sleeve bearings are used. The most common are the vibrations due to oil whirl, hysteresis whirl, resonant whirl, or mechanical looseness. These types of instability and nonlinear behavior are described in detail in Ref. 4. Figure 16.5 shows a spectrum measured on the journal bearing of a centrifugal compressor with mechanical looseness. A characteristic pattern of half-

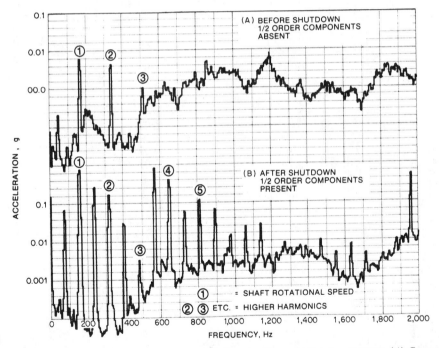

FIGURE 16.5   Acceleration spectra of a journal bearing on a centrifugal compressor. (*A*) Compressor in *good condition*. Before shutdown, the vibration pattern is normal with few harmonics of the compressor's rotation speed and broadband noise at higher frequencies due to inherent turbulences. (*B*) Compressor *with looseness* in the journal bearing. After shutdown, the higher-order harmonics have an increased amplitude, and the presence of half-order harmonics can be observed.

order harmonics of rotation speed can be clearly seen. Figure 16.6 shows a spectrum of the journal bearing of a pump in which a developing oil whirl shows up clearly at 21 Hz (42 percent of the rotation speed) and its second harmonic.

Both examples clearly indicate how the use of a linear frequency scale facilitates the diagnosis of the fault by providing a clear indication of the different types of harmonic patterns. High resolution is required to separate a half-order harmonic component due to looseness (exactly 50 percent of rotation speed) from a component due to oil whirl (42 to 48 per cent of rotation speed).

LOW HARMONICS OF ROTATIONAL SPEED

Low harmonics of the rotational speed are generated by shaft unbalance, misalignment, and eccentricity, as well as cracks in shafts and bent shafts. These various faults may be difficult to distinguish, since they are mechanically related. A bad coupling may result in misalignment. A bent shaft results in unbalance. Even a well-known and well-defined fault such as unbalance may give misleading vibration components. The exciting fault due to eccentric masses is a centrifugal force (thus radial) rotating at the shaft speed and is therefore expected to result in a component in the vibration spectrum at the machine speed and in the radial direction. However, dynamic unbalance may also result in a rocking motion and consequently in vibration in both radial and axial directions. In the same way, if there is a nonlinear transmission path from the point where the force is applied to the point of measurement, a rise in the harmonics of the rotation speed can be observed in the vibration spectrum, due to distortion of the signal.

The phase relationship between bearings provides essential information for differentiating these various types of faults. As an example, unbalance will generate a rotating force, and therefore the phase relationship between bearings can be

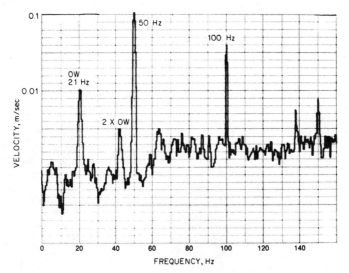

FIGURE 16.6 Spectrum analysis showing component due to oil whirl at 42 percent of the rotation speed measured on the journal bearing of a pump.

expected to be identical in both horizontal and vertical directions (in the absence of resonances). Misalignment, however, does not create a rotating force, and thus the phase relationship between bearings in both vertical and horizontal directions can be vastly different.

HARMONICS OF THE LINE (MAINS) FREQUENCY

Vibrational components, which are related to the frequency of the power line (mains) or to the difference between the synchronous frequency and the rotational speed, occur in electric machines such as induction motors or generators. These vibrations are due to electromagnetically induced forces. These forces, which occur in the case of a malfunction in the electric machine, are related to the air gap between the rotor and the stator and to the current. The faults on the electric machine are due either to the stator (called *stationary faults*) or to the rotor (called *rotating faults*). They may originate from either a variation in the air gap or a variation in the current. Table 16.2 summarizes how these various faults show up in the low-frequency range of the vibration spectrum.[5]

TABLE 16.2   "Rotating" and "Stationary" Magnetically Induced Vibrations in Induction Motors

Type of problem	Symptomatic frequency of vibration	Typical cause	
		Air-gap variations	Current variations
Stationary	2 × line frequency	Static eccentricity, weakness of stator support	Stator winding faults
Rotating	1 × rpm with 2 × slip-frequency sidebands	Dynamic eccentricity, bent rotor, loose rotor bar(s)	Broken or cracked rotor bar(s) or shorted rotor laminations

Figure 16.7 shows a vibration signal measured on the rolling-element bearing of an asynchronous electric motor. By zooming in the region of the high-level 100-Hz component (i.e., twice the line frequency in Europe), this component can be diagnosed as the pole-passing frequency of 100 Hz and not the 2 × rotation speed at 99.6 Hz which could have been an indication of a faulty alignment. This demonstrates the value of being able to zoom to the frequency region containing the component of interest. The zoom provides sufficient resolution to separate closely spaced components. It is of no help in analyzing synchronous machines or generators, since the rotation speed and the line (mains) frequency are identical. In such a case, the machine should be permitted to coast to a stop. When the power is cut, electrically induced components of vibration disappear, and the harmonics of the rotation speed gradually decrease in frequency and amplitude.

Vibration forces resulting from an effective variation of the reluctance in the magnetic circuit as a function of the rate of the stator and rotor slot passing will be present even in a motor which is in good condition. These vibrations occur at the slot harmonics given by the following equation:

$$f_{\text{slot}} = R_s f_{\text{rot}} \pm k f_{\text{mains}}$$

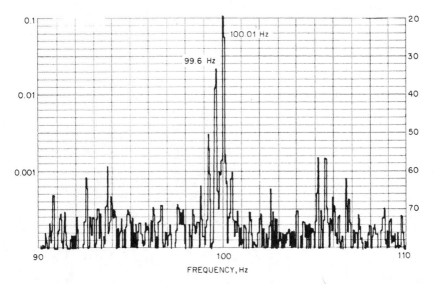

FIGURE 16.7 Spectrum analysis employing zoom frequency analysis around the 100-Hz component, measured on the rolling-element bearing of an asynchronous electric motor. Note the simultaneous presence of the pole-passing frequency (100 Hz) and the second harmonic of rotation speed (99.6 Hz). A lesser resolution would not permit a separation.

where f_{slot} = slot passing frequency
          R_s = number of rotor slots
          f_{rot} = rotating speed
          k = zero or even number
          f_{mains} = mains (line) frequency

The vibration components at low frequency differentiate between stator problems and rotor problems. They do not, however, indicate whether the faults originate from variations in air gaps or current. The components at the slot harmonics, on the other hand, will behave differently depending on whether the fault originates from an air gap or current variation as indicated in Table 16.3.

Figure 16.8 shows that by using a zoom around slot harmonics, sidebands can be observed at twice the slip frequency, thereby permitting the diagnosis of broken rotor bars.

As an alternative to using signal analysis of vibration, signal analysis of the motor current may be used to monitor certain types of problems. It is a more direct measurement for all electrical problems and, with the help of algorithms, makes it possible, for example, to determine with a certain amount of accuracy the number of broken rotor bars. Reference 6 mentions that mechanical phenomena such as worn gears, tooth wear, and steam packing degradation (in motor-operated valves) can be detected as well. It also mentions the applicability of this technique to dc motors.

HIGHER HARMONICS OF THE ROTATIONAL SPEED

Higher harmonics of the rotational speed typically occur where characteristic frequencies are an integral multiple of the rotational speed of the machine, for example,

TABLE 16.3 Troubleshooting Guide of Induction Motor Vibrations

Static eccentricity	$2 \times$ line frequency and components at $\omega \times [nR_s(1 - s)/p \pm k_1]$	Radial	Can result from poor internal alignment, bearing wear, or from local stator heating (vibration worsens as motor heats up).
Weakness/looseness of stator support, unbalanced phase resistance or coil sides			Referred to as "loose iron."
Shorted stator laminations/turns	$2 \times$ line frequency	Radial	Difficult to differentiate between this group using only vibration analysis, but they will also be apparent at no load as well as on load.
Loose stator laminations	$2 \times$ line frequency and components spaced by $2 \times$ line frequency at around 1 kHz	Radial	Can have high amplitude but not usually destructive. The high-frequency components may be similar to static eccentricity.
Dynamic eccentricity	$1 \times$ rpm with $2 \times$ slip-frequency sidebands and components at $\omega \times [((nR_s \pm k_e) \times (1 - s)/p) \pm k_1]$	Radial	Can result from rotor bow, rotor runout, or from local rotor heating (vibration worsens as motor heats up).
Broken or cracked rotor bar Loose rotor bar Shorted rotor laminations Poor end-ring joints	$1 \times$ rpm with $2 \times$ slip frequency sidebands and components similar to those given above for dynamic eccentricity with addition of $2 \times$ slip-frequency sidebands around slot harmonics	Radial	The slip sidebands may be low level, requiring a large dynamic range as well as frequency selectivity in measuring instrumentation. Typical spectra show that these components in the region of the principal vibration slot harmonics also have slip-frequency sidebands.

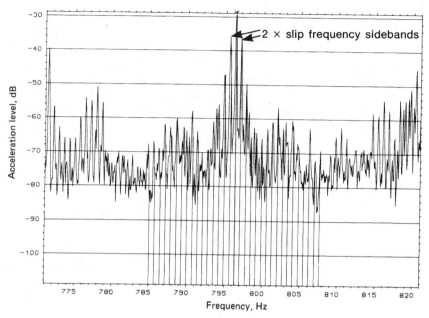

FIGURE 16.8 Zoom spectrum centered around the second principal vibration slot harmonic, showing 2 × slip frequency sidebands on the component at this frequency.

in the case of gearboxes, compressors, and turbines, where vibration occurs in multiples of the number of teeth, blades, lobes, etc. An increase in components, such as tooth-meshing frequencies or blade-passing frequencies, indicates deterioration acting on all teeth or blades, e.g., as uniform wear or increased turbulences, respectively.

"Ghost components" sometimes are observed in vibration spectra obtained from measurements on gearboxes; these components appear as tooth-meshing frequencies, but at frequencies where no gear in the gearbox has the corresponding number of teeth. Such components arise from faults on the gear-cutting equipment which have been transmitted to the new gear. Being geometrical faults, they are not load-sensitive, nor do they increase with wear; rather, as the gear's surface wears, they tend to decrease with time. The frequencies of the components are an integral multiple of the number of teeth on the index wheel and therefore appear as harmonics of the speed of rotation of the faulty gear.

SIDEBAND PATTERNS DUE TO MODULATION

Modulations, frequently seen in vibration measurements on gearboxes, are caused by eccentricities, varying gear-tooth spacing, pitch errors, varying load, etc. Such modulations manifest themselves as families of sidebands around the gear-tooth-meshing frequency with a frequency spacing equal to the modulating frequency (e.g., the rotation speed of the faulty gear in the case of an eccentric gear). Figure 16.9A shows the distribution of the sidebands for such a condition. Any gear in a gearbox can be a source of modulation. In order to distinguish all possible sidebands, the analysis must be carried out with sufficient resolution to detect sidebands with a spacing equal to even the lowest rotational speed inside the gearbox, and therefore the zoom feature is indispensable.

Local faults, such as cracked or broken gear teeth, also appear as a family of sidebands with a spacing equal to the rotation speed of the faulty gear, as this induces a change in tooth deflection, during meshing, once per revolution. The sidebands shown in Fig. 16.9B are low in level and cover a broad frequency range. Very often the influence of the transmission path will modify the shape of the sideband pattern and does not permit a precise diagnosis.[7] Local faults in turbines (for example, missing or distorted blades) result in a similar family of harmonics/sidebands.[8,9] Similarly, sidebands at the rotational speed and slip frequency are quite common in patterns for asynchronous machines.

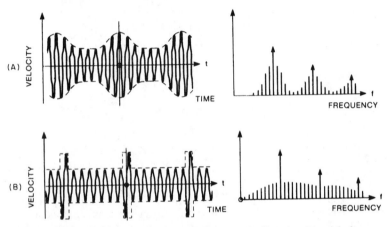

FIGURE 16.9 Distribution of sideband patterns for distributed and local faults on a gear. (A) *In the case of a distributed fault*, sidebands have a high level and are grouped around the tooth-meshing frequency and harmonics with a spacing equal to the speed of the faulty gear. (B) *In the case of a local fault* (such as a cracked or broken tooth), the sidebands have a low level and expand widely over a large frequency range.

HARMONIC PATTERNS NOT HARMONICALLY RELATED TO THE ROTATIONAL SPEED

Harmonic patterns which are not harmonically related to the speed of rotation typically appear where there are local faults in rolling-element bearings.[10] A local fault produces an impulse having a repetition rate equal to the characteristic frequencies of the bearing: ball-passing frequency for the outer raceway, ball-passing frequency for the inner raceway, and twice the ball-spin frequency (see Table 16.1). Such faults appear as a series of harmonics separated by the impact frequency with an amplitude proportional to the spectrum of the single impulse. As illustrated in Fig. 16.10, such an impact tends to excite any structural resonances in the frequency range covered, and the harmonic patterns around these resonances thus are emphasized. This provides two methods of detecting rolling-element bearing faults: (1) by finding the fundamental of the impact rate in the low-frequency range; this is often difficult, since the fundamental is normally rather small and lost among the other components in the background vibration; and (2) by finding the harmonic pattern at the

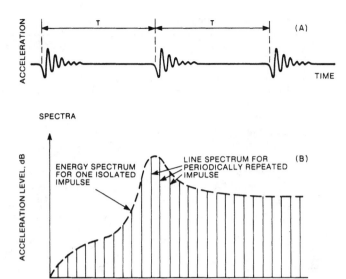

FIGURE 16.10 Effect of a local fault in a rolling-element bearing. (*A*) *In time domain,* there is a repeated impact having a period *T.* (*B*) *In frequency domain,* the repeated impact results in a line spectrum extending very high in frequency (the shorter the impact, the higher cover of frequency range); the frequency spacing of these lines is equal to the frequency of impact, $1/T$. Any structural resonance which is excited in this frequency range will result in a peak as illustrated.

impact frequency in the high-frequency range, where resonances are excited; this may also be difficult because speed fluctuations tend to smear these components.

SPECIAL ANALYSIS TECHNIQUES

Table 16.4 summarizes the applications of the various analysis techniques described below.

ENVELOPE DETECTION

Envelope detection (envelope detectors are discussed in Chap. 13) is particularly useful for fault diagnosis in machinery, since it permits elimination of the signal resulting from background vibration and concentrates the analysis in the frequency range placing the greatest emphasis on the harmonic pattern of the impact frequency—a resonance of the structure excited by the impulse. This can be done by either analog or digital means.[11] Figure 16.11 illustrates the analog process. The signal is first bandpass-filtered around the frequency range where a significant broadband increase has been detected, as illustrated in Fig. 16.11*B* and *D* (usually one or more resonances between 2 and 20,000 Hz have been excited). The filtered signal

TABLE 16.4 Typical Applications of the Various Analysis Techniques

Technique	Application	Fault/machine
Zoom	Separation of closely spaced components Improvement of signal-to-noise ratio, separation of resonances from pure tones	Electrical machines, gearboxes, turbines
Phase	Operational deflection shapes Detection of developing cracks in shafts Balancing	
Time signal	Waveform visualization for identification of distortion	Rubbing, impacts, clipping, cracked teeth
Cepstrum	Identification and separation of families of harmonics Identification and separation of families of sidebands	Rolling elements bearing, bladed machines, gearboxes
Envelope analysis	Amplitude demodulation Observation of a low-frequency amplitude modulation happening at high frequency	Rolling element bearing, electrical machines, gearboxes
Synchronous time averaging	Improving signal-to-noise ratio Waveform analysis Separating effects of adjacent machines Separating effects of different shafts Separating electrically and mechanically induced vibrations	Electrical machines, reciprocating machines, gearboxes, etc.
Impact testing	Resonance testing	Foundations, bearings, couplings, gears
Scan analysis	Analysis of nonstationary signals	Fast run-up/coast down

(which now contains only the ringing of the selected resonance excited by the repetitive impacts, Fig. 16.11*C*) is rectified and analyzed once again in a low-frequency range in order to determine the repetition frequency of the impacts, as shown in Fig. 16.11*E* and *F.*

The advantages of envelope detection are as follows:

1. The use of bandpass filters eliminates background noise resulting from other vibration sources (for example, from unbalance or gear vibration). All that remains is the repetition rate of the impacts exciting the structural resonance, possibly amplitude-modulated.

2. High-frequency analysis is not required, since only the envelope of the signal is of importance, not the signal itself, which can extend upward to hundreds of kilohertz.

3. Diagnosis is possible, since the impact frequencies are determined and can be related to a specific source (ball-passing frequency for the outer raceway, ball-passing frequency for the inner raceway, ball-spin frequency, fundamental train

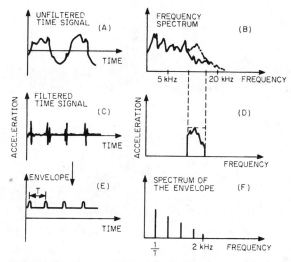

FIGURE 16.11 Principle of analog envelope detection applied to the analysis of impacts due to rolling-element bearing faults. (*A*) *Unfiltered time signal.* (*B*) *The corresponding spectrum in the frequency domain.* (The dotted spectrum represents the reference spectrum before the fault developed. Note the broadband increase due to excitation of a resonance by the bearing fault.) (*C*) *Frequency spectrum after application of a bandpass filter* in range where the change caused by a ball-bearing fault has been detected. (*D*) *Time signal which corresponds with C;* contains ringing of a resonance which is excited periodically. (*E*) Envelope of time signal from *D.* (*F*) *Low-frequency analysis of the envelope from E,* yielding the impact rate due to the fault.

frequency, or some other source of repetitive impacts, for example, a cracked gear tooth).

Figure 16.12*A* and *B* shows the acceleration spectra from 0 to 25 kHz of a good bearing and a faulty bearing. Note that the spectrum is noticeably higher on the good bearing than on the faulty one, which confirms that comparative measurements should not be made between different measurement points or different machines. Absolute vibration levels do not provide a satisfactory indication of the condition of a machine; only changes in level are relevant. Any simple method of bearing fault detection such as shock pulse measurement, spike energy, kurtosis, or crest factor was difficult to use on this specific machine, because a forced-lubrication system gave repetitive pulses at a frequency of 5.4 Hz, independent of the rotating speed, which dominated the whole vibration signal. Figure 16.12*C* and *D* shows the analysis of the envelopes on the good and the faulty bearings obtained after zooming around 5400 Hz with an 800-Hz frequency span. The only noticeable pattern on the good bearing comes from the forced lubrication system. In contrast, the result of the envelope analysis on the faulty bearings shows a complex pattern, and frequency information is absolutely necessary to confirm whether or not there is a ball-bearing fault. The following frequencies appear: 5.4 Hz (the repetition rate of the forced lubrication system on the actual bearing, and its harmonics), 6.4 Hz (the repetition

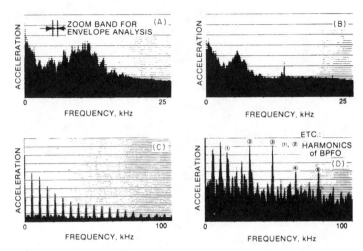

FIGURE 16.12   (*A*) Acceleration spectrum of a good bearing in the frequency range from 0 to 25 kHz. (*B*) Acceleration spectrum of a faulty bearing in the frequency range from 0 to 25 kHz. (*C*) Envelope spectrum in 100-Hz range of the signal in *A;* the zoom analysis is centered at 5408 Hz and has a frequency span of 800 Hz. (*D*) Envelope spectrum in 100-Hz range of the signal in *B;* the zoom analysis is centered at 5408 Hz and has a frequency span of 800 Hz.

range of the forced lubrication system on adjacent bearings, and its harmonics), and 15.43 Hz (the ball-passing frequency for an outer raceway defect, and its harmonics).

APPLICATION OF CEPSTRUM ANALYSIS

The use of cepstrum analysis (explained in more detail in Chap. 13) is particularly advantageous for detecting periodicities in the power spectrum (e.g., harmonics and sideband patterns), since it provides a precise measure of the frequency spacing between components.[10,11] Figure 16.13 shows the spectrum and the corresponding cepstrum analysis of a measurement made on an auxiliary gearbox driving a generator on a gas-turbine-driven oil pump. As a fault on one of the bearing develops, the first rahmonic appears and then increases at a quefrency equal to the reciprocal of the spacing in the frequency spectrum which corresponds to an outer raceway defect in one of the bearings. Another advantage of cepstrum analysis is that one component in the cepstrum represents the global "power" content of a whole family of harmonics or sidebands, and this value is practically independent of extraneous factors such as machine-load condition, selection of measurement location, and phasing between amplitude and phase modulation.

Figure 16.14 shows the evolution in terms of time of two different frequency components in the spectra of Fig. 16.13. Figure 16.14*A* represents the evolution of the harmonic component at 7640 Hz, which shows a clear ascending slope, and also the evolution of the harmonic component at 5620 Hz, which shows a rather horizontal slope. Figure 16.14*B* represents the changes on the first rahmonic in the cepstrum and shows the effective evolution of the fault corresponding to a steep increase followed by stabilization as is typical for a spall in a rolling-element bearing.

Envelope detection and cepstrum analysis make it possible to determine the frequencies involved precisely and are thus useful for condition monitoring of indus-

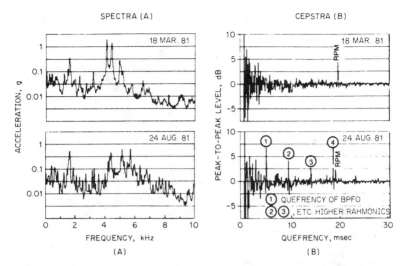

FIGURE 16.13 Analyses of vibration of an auxiliary gearbox before and after the development of a fault on one of the bearings.[10] (*A*) Spectrum analysis; (*B*) the corresponding cepstrum analysis.

trial machines. They permit a reliable diagnosis of defects. Also, cepstrum analysis appears to be an invaluable condition-related parameter for following fault development (e.g., those that develop in rolling-element bearings and damage to turbine blades) that shows up as a family of harmonics or sidebands.

The use of post-processing techniques makes it possible to manipulate the cepstrum, for instance, by removing selected parts of the cepstrum and then transforming the cepstrum back to the frequency and/or time domain. For example, such manipulations make it possible (1) to evaluate the size of a spall in a rolling element bearing and (2) to reconstitute the transfer function between the input force and the vibration response without measuring the input function.[11,12]

APPLICATION OF GATED VIBRATION ANALYSIS ON RECIPROCATING MACHINES

Vibration signals from reciprocating machines (such as diesel engines, reciprocating compressors, hydraulic pumps, and gas engines) differ from those of rotating machines in that they are not stationary. Instead, they consist of short impulses which occur at different points in time for different events (valves opening and closing, piston slap, combustion, etc.) and are repeated with the same timing for each new machine cycle. If these signals are averaged over a longer period of time, as is common practice in the analysis of rotating machines, these individual events would be averaged out so that changes would go undetected.

In reciprocating machines, different events will excite different resonances of a structure; the resulting frequencies that are generated provide valuable diagnostic information. *Timing* provides equally valuable information because the time when an event occurs may be related to what is actually happening in the cycle of the engine.

In gated vibration analysis, the vibration signal is analyzed at various angles of the crankshaft in order to cover a complete cycle of the machine in a three-

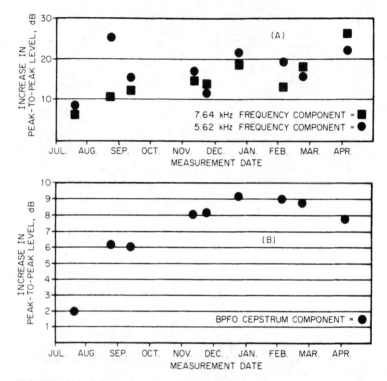

FIGURE 16.14 (*A*) Development of the fault illustrated in Fig. 16.13, using the harmonic at 7640 Hz and the harmonic at 5620 Hz as the parameter. (*B*) The corresponding trend curve using the cepstrum component as the parameter, showing a smooth evolution.[10]

dimensional plot.[13] The analyzer is triggered by a once-per-cycle trigger signal; then the delay after triggering is shifted to provide adequate overlap; this procedure continues until a complete cycle is covered. Note that each spectrum represents actually an average over many machine cycles for one time delay. This process averages any differences between machine cycles.

An alternative to gated vibration analysis is gated sound intensity analysis, which is more suited to noise control and quality checks.[14]

TREND ANALYSIS

Trend analysis makes use of graphs of a condition-related parameter versus time (date or running hours) to determine when the parameter is likely to exceed a given limit. The goal of a successful condition monitoring program is to predict the time of an expected breakdown well in advance of its occurrence in order to shut down the machine in ample time, to order spare parts, and thereby to minimize the shutdown time. Since all vibration criteria indicate that equal changes on a log scale correspond to equal changes in severity, data for a trend analysis should be plotted on a logarithmic scale in decibels. A linear trend on a logarithmic scale is found occasionally, but the actual trend may follow another course; for example, when the fault feeds back on the rate of deterioration (e.g., gear wear), the trend, when plotted on a logarithmic scale, may then be exponential. In some cases the fault changes suddenly in finite steps (for example, a spall caused by gradual subsurface fatigue), making it very difficult to extrapolate to determine the date of the shutdown. To ensure accurate trend analysis, the following precautions should be taken:

1. Determine a trend based on measurements of a parameter directly related to a specific type of fault—not on measurements of overall levels.

2. Diagnose faults *before* attempting to interpret a trend curve in order to (*a*) select the appropriate parameter for the type of fault which is being monitored (for example, the parameter may be the level of an individual component, of a cepstrum component, or of a selected frequency range) and (*b*) observe critically the results of the trend analysis so as to determine if the linear or exponential interpolation is adequate.

3. Keep in mind that the best estimate of the lead time will be obtained by employing a trend of the most recent measurements.

USE OF COMPUTERS IN CONDITION MONITORING PROGRAMS

Computers can be of great help in a condition monitoring program in handling, filing, and storing data and in performing tedious computations such as spectrum comparison and trend analysis. A condition monitoring system which incorporates a computer includes:

1. A *recording device* for storing the analog time signals or frequency spectra. In a permanently installed monitoring system, the analog time signal is directly connected to the following items.

2. An *analyzer* with both fast Fourier transform (FFT) narrowband analysis and advanced diagnostic techniques (zoom, cepstrum) for diagnostics, and *constant-percentage-bandwidth* analysis for fault detection. (If the latter is not included, the conversion from FFT to constant-percentage bandwidth can be done in the computer.) Some analyzers can be used as both recorders and analyzers.

3. A *computer and appropriate software* which provide (*a*) management of the measurement program, including route mapping, storage of reference spectra/cepstra, and new spectra/cepstra; (*b*) a comparison of spectra and a printout of significant changes; and (*c*) trend analysis of any chosen parameter (individual component or overall level in a given frequency range). In a permanent monitoring system, the complete process (i.e., a new analysis) is performed automatically at a predetermined rate, which is adjusted as the fault develops.

REFERENCES

1. Neale, M., et al.: "A Guide to the Condition Monitoring of Machinery," Department of Industry Committee for Terotechnology, Her Majesty's Stationery Office, London, 1979.

2. Mitchell, J. S.: "An Introduction to Machinery Analysis and Monitoring," Penwell Publishing Company, Tulsa, 1981.

3. Downham, E., and R. Woods: "The Rationale of Monitoring Vibration in Rotating Machinery in Continuously Operating Process Plant," ASME paper no. 71-Vibr-96, 1971.

4. Sohre, J. S.: "Turbomachinery Problems and Their Correction," in J. W. Sawyer and K. Halberg (eds.), "Sawyer's Turbomachinery Maintenance Handbooks," Turbo International Pubn., vol. II, 1981, chap. 7.

5. Bate, A. H.: "Vibration Diagnostics for Industrial Electric Motor-Drives," Brüel and Kjaer application note.

6. Kryter, R. C., and H. D. Haynes: "Condition Monitoring of Machinery Using Motor Current Signature Analysis," *Sound and Vibration,* September 1989.

7. Randall, R. B.: "A New Method of Modeling Gear Faults," ASME paper no. 81-Set-10, June 1981.

8. Sapy, G.: "Une Application du Traitement Numérique des Signaux au Diagnostic Vibratoire de Pannes la Detection des Ruptures d'Aubes Mobiles de Turbines," *Automatisme,* tome xx, no. 10, October 1975.

9. Barschdorff, D., W. Hensle, and B. Stühlen: "Neue Ergebnisse der Akustischen Uberwachung von Betriebsdampfturbinen," *VGB Kraftwerkstechnik 59,* June 1979.

10. Bradshaw, P., and R. B. Randall: "Early Detection and Diagnosis of Machine Faults on the Trans Alaska Pipeline," MSA session, ASME Conference, Dearborn, Mich., September 11–14, 1983.

11. Courrech, J.: "New Techniques for Fault Diagnostics in Rolling Element Bearings," *Proc. 40th Meeting of the Mechanical Failure Preventive Group,* National Bureau of Standards, Gaithersburg, Md., April 16–18, 1985.

12. Randall, R. B., and Yujin, GaO, "Tracking changes in Modal Parameters by Curve Fitting the Response Cepstrum." *17th ISMR,* K. U. Leuvan, September 23–25, 1992.

13. Courrech, J.: "Examples of the Application of Gated Vibration Analysis for the Detection of Faults in Reciprocating Machines," *Noise and Vibration '89 Conference,* Singapore, August 16–18, 1989.

14. Rasmussen, P., and T. L. Moller: "Gated Sound Intensity Measurements on a Diesel Engine," Brüel and Kjaer application note.

CHAPTER 17
STRAIN-GAGE INSTRUMENTATION

Earl J. Wilson

INTRODUCTION

The resistance strain gage may be employed in shock or vibration instrumentation in either of two ways. The strain gage may be the active element in a commercial or special-purpose transducer or pickup, or it may be bonded directly to a critical area on a vibrating member. Both of these applications are considered in this chapter, together with a discussion of strain-gage types and characteristics, cements and bonding techniques, circuitry for signal enhancement and temperature compensation, and related aspects of strain-gage technology.

The electrical resistance strain gage discussed in this chapter is basically a piece of very thin foil or fine wire which exhibits a change in resistance proportional to the mechanical strain imposed on it. In order to handle such a delicate filament, it is either mounted on or bonded to some type of carrier material and is known as the *bonded strain gage.*

The strain gage is used universally by stress analysts in the experimental determination of stresses. Since strain always accompanies vibration, the strain gage or the principle by which it works is broadly applicable in the field of shock and vibration measurement. Here it serves to determine not only the magnitude of the strains produced by the shock or vibration, but also the entire time-history of the event, no matter how great the frequency of the phenomenon.

BASIC STRAIN-GAGE THEORY AND PROPERTIES

The relationship between resistance change and strain in the foil or wire used in strain-gage construction can be expressed as

$$\frac{\Delta L}{L} = \frac{1}{K} \frac{\Delta R}{R}$$

or
$$K = \frac{\Delta R/R}{\Delta L/L} \qquad (17.1)$$

where K is defined as the *gage factor* of the foil or wire, ΔR is the resistance change due to strain, R is the initial resistance, ΔL is the change in length, L is the original length of the wire or foil, and $\Delta L/L$ is the unit strain to which the wire or foil is subjected.

Not all materials exhibit this strain-sensitivity effect, and different materials have different gage factors. Filament materials in common use in strain gages are Constantan (Ni 0.45, Cu 0.55), which has a gage factor of approximately +2.0; Iso-elastic (Ni 0.36, Cu 0.08, Fe 0.52, and Mo 0.005), which has a gage factor of about +3.5; and modified Karma (Ni 0.75, Cr 0.20, plus additions), which has a gage factor of +2.1.

STRAIN-GAGE CONSTRUCTION

Since the foil used in a strain gage must be very fine or thin to have a sufficiently high electrical resistance (usually between 60 and 350 ohms), it is difficult to handle. For example, the foil used in gages is often about 0.1 mil in thickness. Some use has been made of wire filaments in strain gages, but this type of gage is seldom used except in special or high-temperature applications. In order to handle this foil, it

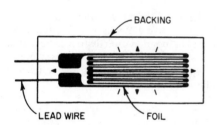

FIGURE 17.1 Typical construction of a foil strain gage.

must be provided with a carrier medium or backing material, usually a piece of paper, plastic, or epoxy. The backing material performs another very important function in addition to providing ease of handling and simplicity of application. The cement provides so much lateral resistance to the foil that it can be shortened significantly without buckling; then compressive as well as tensile strains can be measured. Lead wires or connection terminals are often provided on foil gages, as illustrated in the typical foil gage shown in Fig. 17.1.

TRANSVERSE SENSITIVITY

Because of its construction, a portion of the foil in each gage lies in the transverse direction and will respond to transverse strain. Therefore the gage factor K of a gage* is always slightly smaller than the gage factor of the material of which it is fabricated. One of the desirable features of foil-type gages is their low transverse sensitivity. In this case, the gage consists of a flat foil grid; a sufficiently large amount of the foil is left at the ends of each strand to reduce the transverse sensitivity of the gage to one-half the value for wire gages for some types and to essentially zero for others.

---

* In determining the *gage factor* of the gage, it is assumed that the gage is mounted on a material having a Poisson's ratio of 0.285 and subjected to uniaxial stress in the direction of the gage axis.

TEMPERATURE EFFECTS

The effects of temperature on the gage factor of several alloys are illustrated in Figs. 17.2 and 17.3. When a bonded strain gage is used in measurements, any change in resistance in the strain-gage measurement system is interpreted as resulting from a strain. If thermal expansion is not induced, then this change will result from a mechanical strain. However, if thermal expansion is induced, then there will be a change in resistance resulting from the mechanical strain, and in addition, there will be a change in resistance resulting from the response of the strain gage to changes in temperature. The strain indication which results from such a temperature effect is known as an "apparent strain." Figure 17.3 shows typical apparent strain for three commonly used alloys. This effect is usually negligible in the measurement of dynamic strains, since the readout instrument associated with the strain gage usually does not respond to static or slow changes in its resistance. However, in the measurement of static strains, the effects of temperature represent the largest potential source of error and require some form of temperature compensation.[2]

STRAIN-GAGE CLASSIFICATIONS

Strain gages are classified in several ways. One classification cites the purpose for which the gage is to be used, that is, for static or dynamic strain measurement. Static gages are made up with Constantan foil, which has a minimum change of resistance with temperature. Dynamic strain gages are made up with Iso-elastic foil, which provides a greater gage factor than Constantan. The dynamic gages, while having a much greater resistance change for a given strain than the static gages, also are much more sensitive to changes in temperature. They are used only where the phenomenon to be measured is so short in time duration that no temperature change of

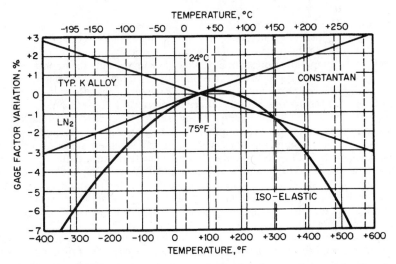

FIGURE 17.2 Typical variation in the gage factor of strain-gage alloys as a function of temperature.

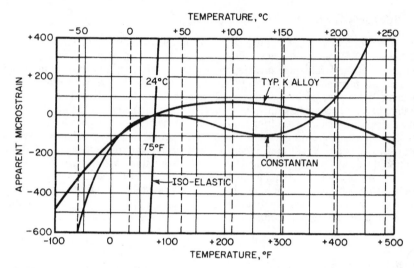

FIGURE 17.3 Typical apparent strain for three alloys commonly used in strain gages. These data are based on an instrument gage factor of 2.00.

any consequence can occur during the time of measurement. Gages also are available for the measurement of very large strains (up to 20 percent) occurring in the plastic region of the material, as distinguished from the more common gages which are used to measure elastic strains (up to 1 percent).

STRAIN-GAGE SELECTION CONSIDERATIONS

In the case of shock measurements, a transient may be applied to the structure that is under investigation only once, or it may be repetitive. Shock is of very short time duration, and the problem of temperature compensation is nonexistent because in most cases the temperature does not have time to change during the impact. For this reason a dynamic-type gage usually can be employed for the measurement of shock. This type of gage has the advantage of a higher gage factor than the static gage, and so it will provide the greatest possible electrical signal for a given strain.

For vibration measurement, the type of gage selected is dependent on the kind of information desired. If only the frequency of vibration and the magnitude of the cyclic stresses are desired, dynamic-type gages can be used since temperature changes will not affect the results obtained unless the temperature fluctuates at the same rate as the stress. If, however, a measurement of the static or slowly varying component of the stress is also to be determined (i.e., if the absolute values of the stresses are desired), a static-type gage must be employed. Since changes in temperature will affect the gage reading, temperature compensation must be incorporated to obtain true values of stress.

Gage selection is dependent on the space limitation and steepness of strain gradient in any region. The strain gage indicates the average strain over the length of the gage; in a region of steep strain gradient, this indicated value may be much less than the maximum strain. The shorter the gage used in such a region, the closer is the gage

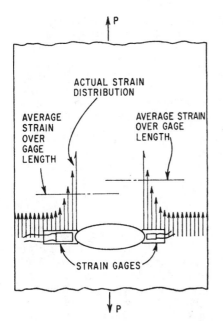

FIGURE 17.4 Effect of gage length on indicated strain in the presence of a severe strain gradient. The shorter gage on the right indicates a higher strain. An infinitesimal gage length would be necessary to indicate the peak strain.

indication to the maximum strain (Fig. 17.4). However, two possible objectives must be considered quite carefully in selecting a gage for a particular installation: (1) the determination of the frequency of vibration, or comparison of relative amplitudes and frequencies with different conditions of excitation, and (2) the determination of the maximum stress pattern resulting from the vibration set up. In the first case there is considerable freedom with regard to the location of the gage on the structure, and therefore with the selection of the gage itself. In the second case severe restrictions exist in regard to the region of application of the gage and its possible dimensions. In general, very short gage-length gages are more difficult to apply properly. Therefore it is desirable to employ gage lengths of ¼ in. whenever possible. When the actual magnitude of the maximum stress resulting from shock or vibration is to be determined, a much more complicated system of gages must be employed. A single gage can be used in only the very limited case where a stress exists in one direction only, and that direction must be known. If stresses exist in several directions, or if the direction of a singly existing stress is unknown, a strain-gage rosette consisting of three or more gages must be employed.[3]

PHYSICAL ENVIRONMENT

The physical environment of the applied gage is an important factor which must be considered in gage selection and protective treatment. Temperature, pressure, humidity, oil, corrosive acid, abrasive action, and possible electromagnetic, neutron, and radiation fields are conditions which affect the choice of gage and its required protection. If high temperatures (up to 500°F or 260°C) are to be encountered, a Bakelite or other high-temperature-type gage must be selected. If even higher temperatures must be withstood, a ceramic-type gage should be employed. Gages of this sort are used at temperatures as high as 2000°F (1100°C). If the temperature never exceeds 200°F (95°C), however, any type of gage can be used.

ACCURACY CONSIDERATIONS

Gages must be selected with regard to the desired precision of the results. If only the frequency of the vibration or the duration of a shock wave is required, almost any gage, properly chosen for the temperature and humidity conditions to be encountered, gives quite satisfactory results. However, if the magnitude of the stresses produced is to be determined in addition, then considerable care must be exercised to select the proper gage to obtain the desired results. Not only must the gage be the proper one to portray the encountered strain faithfully, but precautions must be taken to install the gage correctly.

The testing "environment" can affect strain-gage accuracy in many ways. Magnetostrictive effects,[4] hydrostatic pressure,[5] nuclear radiation,[6] and high humidity are examples of conditions that may cause large strain-gage errors. Creep, drift, and fatigue life in the gages themselves may be important. In most normal environments these errors are either small or undetectable. Whenever unusual or harsh environments are encountered, it is wise to consult the strain-gage manufacturer to obtain recommendations for gage systems and estimates of expected accuracies.

BONDING TECHNIQUES

The proper functioning of a strain gage is completely dependent on the bond which holds it to the structure undergoing test. If the bond does not faithfully transmit the strain from the test piece to the wire or foil of the gage, the results obtained cannot be accurate. Failure to bond over even a minute area of the gage will result in incorrect strain indications. The greatest weakness in the entire technique of strain measurement by means of wire or foil gages is in the bonding of the gage to the test piece. Usually, the manufacturer of the strain gages will recommend cements which are compatible with their use and will provide instructions for their proper installation.

STRAIN-GAGE MEASUREMENTS

The resistance strain gage, because of its inherent linearity, very small mass, wide frequency response (from zero to more than 50,000 Hz),[7] general versatility, and ease of installation in a variety of applications, is an ideal sensitive component for electrical transducers for use in shock and vibration instrumentation.[8] The Wheatstone bridge circuit, described in a subsequent section, can be used to extend the versatility of the strain gage to still broader applications by performing mathematical operations on the strain-gage output signals. The combination of these two devices can be used effectively for the measurement of acceleration, displacement, force, torque, pressure, and similar mechanical variables. Other useful attributes include the capacity for separation of forces and moments, vector resolution of forces and accelerations, and cancellation of undesired vector components.

The usual technique for employing a strain gage as a transducing element is to attach the gage to some form of mechanical member which is loaded or deformed in such a manner as to produce a signal in the strain gage proportional to the variable being measured. The mechanical member can be utilized in tension, compression, bending, torsion, or any combination of these. All strain-gage-actuated transducers can be considered as either force- or torque-measuring instruments. Any mechanical variable which can be predictably manifested as a force or a couple can be instrumented with strain gages.

There are a number of precautions which should be observed in the design and construction of custom-made strain-gage transducers.[9] First, the elastic member on which the strain gage is to be mounted should be characterized by very low mechanical hysteresis and should have a high ratio of proportional limit to modulus of elasticity (i.e., as large an elastic strain as possible). Although aluminum, bronze, and other metals are often employed for this purpose, steel is the most common material. An alloy steel such as SAE 4340, heat-treated to a hardness of RC 40, will ordinarily function very satisfactorily.

The physical form of the elastic member and the location of the strain gages thereon are not subject to specific recommendation, but vary with the special requirements of each individual instrumentation task. When no such requirements exist, a standard commercial transducer ordinarily should be used. In general, the shape of the member should be such as to (1) allow adequate space for mounting strain gages (preferably in regions of zero or near-zero strain gradient), (2) provide the desired natural frequency, (3) produce a strain in the gages which is great enough at low values of the measured variable to result in an output signal readily subject to accurate indication or recording, and not so great as to cause nonlinearities or abbreviated gage life at peak load values, (4) provide temperature compensation and/or signal augmentation (as described in a subsequent section) whenever feasible, and (5) allow for simplicity of machining, ease of gage attachment and wiring, and, if necessary, protection of the gages.

The strain gages should be cemented to the elastic member with the usual care and cleanliness necessary in all strain-gage applications, special attention being given to minimizing the bulk of the installation if the added mass is significant to the frequency response of the instrument. Other considerations vital to successful strain-gage-application technique are described elsewhere in this chapter.

DISPLACEMENT MEASUREMENT

Measurement of displacement with strain gages can be accomplished by exploiting the fact that the deflection of a beam or other loaded mechanical member is ordinarily proportional to the strain at every point in the member as long as all strains are within the elastic limit.

For small displacements at low frequencies, a cantilever beam arranged as shown in Fig. 17.5 can be employed. The beam should be mounted with sufficient preload on the moving surface that continuous contact at the maximum operating frequency is assured. In the case of higher-frequency applications, the beam can be held in contact with the moving surface magnetically or by a fork or yoke arrangement, as illustrated in Fig. 17.6. It is necessary to make certain that the measuring beam will not affect the displacement to be instrumented, and that no natural mode of vibration of the beam itself will be excited.[10]

The measurable displacement magnitude can be increased above that for the cantilever beam by employing other schemes, such as the "clip gage" shown in Fig. 17.7.

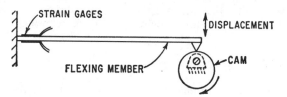

FIGURE 17.5 Strain gages mounted on a cantilever beam for displacement measurement produce electrical signals proportional to cam motion.

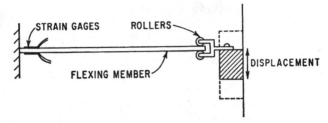

FIGURE 17.6 Displacement transducer designed for continuous, positive contact with moving object.

This gage is constructed by bonding strain gages to the upper and lower sides of a piece of channel-shaped spring steel, as shown in Fig. 17.7. The assembly is then clipped or otherwise mounted on the test specimen so that the legs deflect as the specimen is strained, thus straining the backbone of the clip gage to a greater or lesser extent. Any desired reduction in strain magnitude can be obtained in this manner by merely altering the proportions of the clip gage. Unfortunately, the maximum allowable frequency generally decreases as the displacement amplitude increases, since stiffness and natural frequency tend to change together. Displacement also can be measured through the use of the relative motion of a seismically mounted mass of much lower natural frequency than the applied frequency.

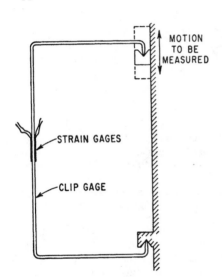

FIGURE 17.7 Clip gage for instrumenting large displacements. Proportions of clip gage are designed to keep strain well within the proportional limit of the material.

VELOCITY

Velocities can be measured directly with strain-gage transducers only by producing a force such as viscous damping or hydro- or aerodynamic drag force which is uniquely related to velocity. Velocity indication also can be obtained with strain gages by differentiation of a displacement function or integration of an acceleration function. In either case, the transducer-design considerations correspond to those for force measurement described in the following section.

FORCE MEASUREMENT

The principle of force measurement with strain-gage-actuated transducers is very similar to that for displacement.[9] The procedure consists of placing a strain-gage-instrumented elastic member in series with the force to be measured. The strain in the transducer, and thus the output signal, is proportional to the force if all stresses are kept within the elastic limit. The proportionality constant between strain and

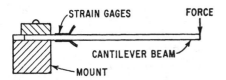

FIGURE 17.8 Cantilever force-measuring transducer, consisting of beam with load applied at free end. Gage strain is a linear function of the force if the proportional limit is not exceeded.

force must be obtained by calibration if precise results are desired. Otherwise, tolerances on the gage factor of the strain gage, and uncertainty as to the elastic properties of the instrumented member, can produce errors of 5 percent or greater—even for transducer configurations with readily calculable strain distributions.

Figure 17.8 illustrates a common form of force transducer, the cantilever beam. Strain gages are mounted on the top and bottom of the beam, producing double sensitivity (output) and virtually complete temperature compensation. While this type of transducer is probably best suited to static or quasi-static measurements such as reaction forces, it also can be used very successfully for many shock and vibration problems as long as the natural frequency of the beam is higher than the frequency of the force being measured. The ring gage (Fig. 17.9) can be categorized with the cantilever beam, and is equally applicable to static or dynamic force measurement within the limitations imposed by its comparatively low natural frequency.

For most dynamic force-instrumentation problems a small compression or tension member (Fig. 17.10) is ordinarily employed. If the load is characterized by alternation between compression and tension, the transducer must be designed for a rigid, integral connection, with no backlash or clearance. This can be accomplished by employing threaded ends with lock nuts for joining the transducer to the remainder of the assembly. In many problems involving machine parts or other mechanical components it is possible to measure loads by applying strain gages to the machine member itself, necessitating calibration of the member to determine the relationship between force and strain.

PRESSURE

In hydraulic and aerodynamic devices, pressure fluctuations are often associated with vibration phenomena—either as cause or effect. Strain-gage transducers are widely used in such situations.[11]

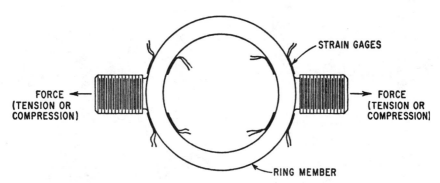

FIGURE 17.9 Ring gage for force measurement. This type of gage provides sensitive axial load measurement without undue loss of rigidity or ruggedness.

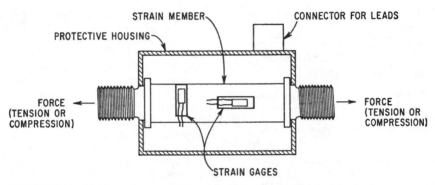

FIGURE 17.10   Widely used commercial form of axial force transducer for large loads.

Pressure pickups based on strain gages are commonly one of three principal types: piston, diaphragm, or tube. In the piston type the pressure acts against a freely movable flat surface (which may be either a piston or a diaphragm), the motion of which is inhibited by an elastic member instrumented with strain gages to measure the force (Fig. 17.11).

Diaphragm-type pressure transducers, shown in Fig. 17.12, have the strain gages applied directly to the back surface of the diaphragm so that diaphragm strain is a measure of pressure.[12] The simplest form of pressure transducer to construct is the tube type, shown in Fig. 17.13. In this type, strain gages are applied to the outer surface of a tube which has the fluid pressure acting on its inner surface. It is sometimes necessary to thin the wall of the tube or to use a longitudinally crimped tube in order to increase the strain magnitude to a measurable level. As a convenient alternative, the bourdon tube in a conventional mechanical pressure gage can serve as the transducing element if strain gages are attached to it. The compressibility of the fluid contained in the tube must be considered for its effect on the frequency response of this type of

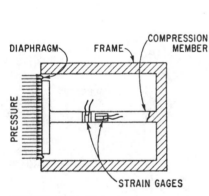

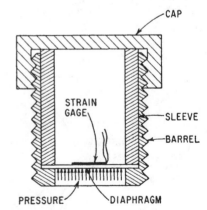

FIGURE 17.11 Piston-type pressure transducer with diaphragm seal for piston. Pressure load on piston head is sensed by strain gages on supporting column.

FIGURE 17.12 Pressure pickup whose output is a function of diaphragm strain. As diaphragm deforms under pressure, strain is transmitted to gage to produce electrical signal.

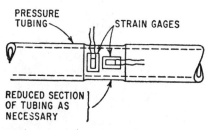

FIGURE 17.13 Readily made pressure transducer consisting of length of tubing with strain gages attached. Dilation of the tube with pressure creates strain in gages.

unit. Pressure pickups should be calibrated statically, and preferably dynamically, prior to use.

ACCELERATION

At one time, wire or foil strain gages were used as the transducing elements in resistive accelerometers. Now silicon elements are usually used because of their higher sensitivity. See *Piezoresistive Accelerometers*, Chap. 12.

STRAIN-GAGE CIRCUITRY AND INSTRUMENTATION

In order to study the detailed cyclical nature of vibration problems or the transient phenomena commonly associated with mechanical shock, it is usually necessary to obtain some form of meter output or graphical record of the events. To produce an output voltage proportional to resistance change requires (1) electrical amplification, since the output of a resistance strain gage usually is only in the range from 10 to 1000 microvolts, and (2) a stable source of electric current, or excitation. These two factors are of primary importance in determining the nature of the electrical instrumentation system which can be used satisfactorily with the resistance strain gage.

There are many circuit arrangements for supplying a strain gage with excitation current and obtaining a signal corresponding to deformation of the gage. Each of these types of circuits has its relative advantages and disadvantages—for example, with respect to sensitivity, temperature compensation, signal enhancement, and ease of operation. Such considerations are discussed in detail in Refs. 6, 13, and 14. Two of the most common arrangements are the potentiometer circuit and the Wheatstone bridge circuit.

POTENTIOMETER CIRCUIT

Figure 17.14, known as the *potentiometer circuit* (sometimes called a *half-bridge circuit*), is the simplest circuit arrangement for supplying a strain gage with excitation current and obtaining a signal corresponding to deformation of the gage. In this circuit, the resistor R_B (called the *ballast resistor*) is of relatively high value to maintain the current flow in the circuit relatively constant and independent of small changes in resistance of the strain gage R_G. The current is supplied by the dc electrical source e. Here, the output signal from the potentiometer circuit, resulting from a variation in the resistance of the strain gage, is designated as e_o.

This circuit is well suited to the instrumentation of dynamic or fluctuating strains, but is totally unsuited for the measurement of static strains or the static component of a combined static and dynamic strain. Therefore, in dynamic applications, it is common practice to block the direct current, i.e., the steady-state (zero-strain) portion of the output voltage, so that only the fluctuating component is measured. This is done by inserting a capacitor C between the potentiometer circuit output and the input of the

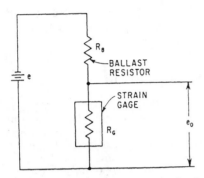 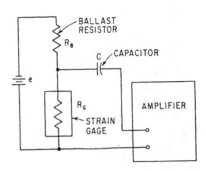

FIGURE 17.14 Potentiometer circuit for dynamic strain signals. Nearly constant current through the circuit, combined with varying gage resistance, produces output signal.

FIGURE 17.15 Overall arrangement of circuits for instrumenting dynamic strain. Signal from gage is taken to ac amplifier through isolating capacitor.

following amplifier, as illustrated in Fig. 17.15. An ac signal, representing the alternations in the strain to which the gage is subjected, is transmitted through the capacitor. Any influences in addition to strain that may modify the resistance of the strain gage (for example, temperature changes) also produce output voltages in this circuit. Since the capacitor coupling to the amplifier is essentially a high-pass filter, temperature-induced output voltage changes are attenuated severely unless the frequency of such changes is high enough to be of the same order of magnitude as the alternating strain. Fortunately, most temperature changes which may affect strain gages occur too slowly to be carried through this circuit arrangement.

WHEATSTONE BRIDGE

In the potentiometer circuit it is necessary to block the dc component of the output voltage with a capacitor before feeding the signal to the input of an amplifier. The same effect can be achieved by suppressing the dc component of the signal by connecting two potentiometer circuits in parallel and taking the output signal from corresponding points in the two branches of the resulting network, as shown in Fig. 17.16. This circuit arrangement is generally referred to as a *Wheatstone bridge,* and represents one of the most precise methods known for measuring (or comparing) resistances. Advantages of the Wheatstone bridge over the potentiometer circuit are (1) much greater flexibility in circuit arrangements for signal augmentation, temperature compensation, and cancellation or separation of variables, (2) capacity for accurately indicating combined static and dynamic strains, and (3) virtually complete freedom from error due to resistive changes in the con-

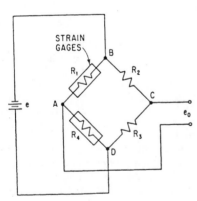

FIGURE 17.16 Wheatstone-bridge circuit for static and dynamic strain measurement.

ductors connecting the supply voltage to the network. As an example of the significance of the last point, consider the effect of the contact resistance variations which might occur in a set of slip rings being used in conjunction with a test of torsional vibration in a rotating shaft.

SELECTION OF INSTRUMENTS FOR STRAIN MEASUREMENT

The output voltage from a strain-gage potentiometer circuit or Wheatstone bridge is, for elastic strain magnitudes in metals, very small. Electrical amplification is required to bring the signal to a level where it can be used conveniently for indication or recording. To assure satisfactory performance and precision, the entire instrument system, from power supply to recording instrument, should be considered as a unit. Figure 17.17 illustrates in block form the basic elements of a strain-gage instrumentation system. The criteria for selecting the individual components of such a system are fixed by the nature of the strain being studied, the type of information required from the system, and the mutual compatibility of the various system components. Consideration should be given to the required frequency response, the input and output impedances of the units in the system, the signal amplitudes being dealt with, and the accuracy of measurement desired. In general, it is safe to assume that the strain gage will respond to considerably higher frequencies than any mechanical device to which it may be attached. In the case of small members vibrating at high frequencies, the limitation is more apt to arise from the change in mass due to the presence of the gage and its lead wires.

Commercial instruments for use with strain gages usually combine several if not all of the components of Fig. 17.17 into a single unit. The limitations of such devices should be investigated prior to purchase. For example, the instrument may include an alternating-frequency source of power for the Wheatstone bridge. This can lead to difficulties in the measurement of high-frequency strains. The frequency of the strain being measured (which will modulate the power supply in the bridge circuit)

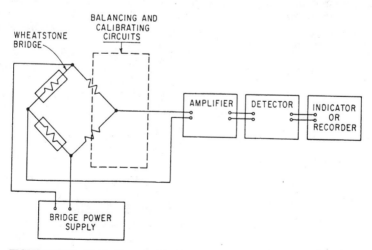

FIGURE 17.17   Block diagram of basic elements of strain-gage instrumentation system.

is limited to approximately 10 to 20 percent of the carrier frequency. If the carrier frequency is high enough to overcome this objection, the capacitive unbalance and pickup in the strain-gage leads is apt to be excessive.

REFERENCES

1. Wu, C. T.: "Transverse Sensitivity of Bonded Strain Gages," *Experimental Mechanics, J. Soc. Exptl. Stress Anal.,* November 1962.

2. Hines, F. H., and L. J. Weymouth: Practical Aspects of Temperature Effects on Resistance Strain Gages, in M. Dean, III and R. D. Douglas (eds.), "Semiconductor and Conventional Strain Gages," Academic Press, Inc., New York, 1962.

3. Vigness, I.: *Proc. Soc. Exptl. Stress Anal.,* **14**(2):139 (1957).

4. Milligan, R. V.: The Gross Hydrostatic-Pressure Effects as Related to Foil and Wire Strain Gages, "Experimental Mechanics," pp. 67ff, Society for Experimental Stress Analysis, Westport, Conn., February, 1967.

5. Vulliet, P. (ed.): *Proc. Western Regional Strain Gage Comm.*—1968 Spring Meeting, Marina Del Rey, Calif., Society for Experimental Stress Analysis, Westport, Conn.

6. Perry, C. C., and H. R. Lissner: "The Strain Gage Primer," pp. 117ff., McGraw-Hill Book Company, Inc., New York, 1955.

7. Bickle, L. W.: "The Use of Strain Gages for the Measurement of Propagating Strain Waves," *Proc. Tech. Comm. Strain Gages,* Oct. 23, 1970, Society for Experimental Stress Analysis, Westport, Conn.

8. Norton, H. N.: "Handbook of Transducers for Electronic Measuring Systems," pp. 42ff., Prentice-Hall, Inc., Englewood Cliffs, N.J., 1969.

9. Motsinger, R. N.: Flexural Devices in Measurement Systems, in P. K. Stein (ed.), "Measurement Engineering," vol. 1, chap. 11, Stein Engineering Services, Phoenix, Ariz., 1964.

10. Cleveland, A. W.: *J. Soc. Auto. Eng.,* **59**:34ff. (1951).

11. Jasper, N. H.: *Proc. Soc. Exptl. Stress Anal.,* **8**(2):83 (1951).

12. Perry, C. C.: "Design Considerations for Diaphragm Pressure Transducers," *Tech. Note* TN-129, Micro-Measurements Division, Romulus, Mich., 1974.

13. Harris, C. M. (ed.): "Shock and Vibration Handbook," Chap. 17, McGraw-Hill Book Company, Inc., New York, 1961.

14. Hannah, R. L., and S. E. Reed: "The Strain Gage Users' Handbook," Elsevier Applied Science, London and New York, 1992.

CHAPTER 18
CALIBRATION OF PICKUPS

M. Roman Serbyn
Jing Lin

INTRODUCTION

This chapter describes various methods of calibrating shock and vibration transducers, commonly called *vibration pickups*. The objective of calibrating a transducer is to determine its sensitivity or calibration factor as defined below. The chapter is divided into three major parts which discuss comparison methods of calibration, absolute methods of calibration, and calibration methods which employ high acceleration and shock. Field calibration techniques are described in Chap. 15.

PICKUP SENSITIVITY, CALIBRATION FACTOR, AND FREQUENCY RESPONSE

As defined in Chap. 12, the *sensitivity* of a vibration pickup is the ratio of electrical output to mechanical input applied along a specified axis.[1,2] The sensitivity of all pickups is a function of frequency, containing both amplitude and phase information, as illustrated in Fig. 18.1, and therefore is usually a complex quantity. If the sensitivity is practically independent of frequency over a range of frequencies, the value of its magnitude is referred to as the *calibration factor* for that range, but it is specified at a discrete frequency. The phase component of the sensitivity function likewise has a constant value in that range of frequencies, usually equal to zero or 180°, but it may also be proportional to frequency, as explained in Chap. 12.

The *frequency response* of a pickup is shown by plotting the magnitude and phase components of its sensitivity versus frequency. This information is usually presented relative to the value of sensitivity at a reference frequency within the flat range. A preferred frequency, internationally accepted, is 160 Hz.

Displacements are usually expressed as single-amplitude (peak) or double-amplitude (peak-to-peak) values, while velocities are usually expressed as peak, root-mean-square (rms), or average values. Acceleration and force generally are expressed as peak or rms values. The electrical output of the vibration pickup may be expressed as peak, rms, or average value. The sensitivity magnitude or calibration factor are commonly stated in similarly expressed values, i.e., the numerator and

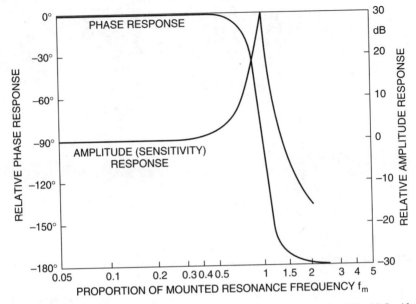

FIGURE 18.1 Pickup amplitude and phase response as functions of frequency. (*After M. Serridge and T.R. Licht.*[3])

denominator are both peak or both rms values. Examples of typical sensitivity specifications for an accelerometer: 2 pC per m/sec^2, 10 millivolts/m/sec^2, 5 milliV/g (where C is the symbol for coulomb, V is the symbol for volt, and g is the acceleration of gravity). For some special applications it may be desirable to express the sensitivity in mixed values, such as rms voltage per peak acceleration.

TRACEABILITY OF CALIBRATIONS AND HIERARCHY OF STANDARDS

Standardization of vibration measurements requires that the working standards of all laboratories be traceable to the standards maintained by the national standards laboratory of a country. The term *traceability*, therefore, implies consistency with the calibrations performed at the national laboratory and is distinct from accuracy and even the methods of calibration. There is not an internationally accepted definition for the term *traceability*. Concerned individuals should address their inquiries to the national standards laboratory of a particular country; in the U.S.A., this laboratory is the Mass, Acoustics, and Vibration Group of the National Institute of Standards and Technology, Gaithersburg, MD 20899. A partial list of national standards laboratories in many other countries is given in Table 18.1.

Traceability is essential for national and international trade. For example, suppose that a contractor, providing noise measurement services to a branch of the U.S. government, purchases an accelerometer from a manufacturer of vibration instruments. In order to comply with government regulations, the contractor must be able to prove that the calibration of the accelerometer is traceable to the National

TABLE 18.1   National Standards Laboratories Responsible for the Calibration of Vibration Pickups

Institution (acronym)	Laboratory	Location	Country
Commonw. Scient. and Indust. Res. Org. (CSIRO)	Division of Applied Physics	Lindfield, NSW	Australia
Inst. Nac. de Metr., Norm. e Qual. Indust. (INMETRO)	Laboratorio de Vibracoes	Xerem, Rio de Janeiro	Brazil
State Comm. for Science and Technical Progress	Nat. Center of Metrology	Sofia	Bulgaria
Nat. Res. Council (NRC), Inst. Nat. Meas. Stds. (INMS)	Acoust. Standards Group	Ottawa	Canada
Nat. Inst. of Metr. (NIM)	Vibration Laboratory	Beijing	China
Czech Inst. of Metr. (CMU)	Prim. Stds. of Kinematics	Praha	Czech Rep.
Bruel & Kjaer Instr. (B&K)	Vibration Laboratory[1]	Naerum	Denmark
Cen. d'Etudes Sci. et Tech. d'Aquitaine (CESTA)	Service Experimentation	Belin-Beliet	France
Lab. de Recherches Balist. et Aerodyn. (LRBA)	Metrologie des Vibracions	Vernon	France
Physikalisch-Technische Bundesanstalt (PTB)	Labor. fur Beschleunigung	Braunschweig and Berlin	Germany
Instr. and Meas. Techn. Serv. of the H. Acad. Sci.	Acoustical Research Lab.	Budapest	Hungary
Nat. Phys. Lab. (NPL)	Acoustics Section	New Delhi	India
Rafael-Armament Devel. Authority (ADA)	Stds. & Cal. Section	Haifa	Israel
Inst. di Metr. G. Colonnetti (IMGC)	Sezione Meccanica	Torino	Italy
National Research Lab of Metrology (NRLM)	Mechanical Metrol. Dept.	Tsukuba, Ibaraki	Japan
Korea Stds. Res. Inst. (KSRI)	Div. of Appl. Metrology	Daedeog Danji	Rep. of Korea
Centro Nac. de Metrologia (CENAM)	Div. Acustica y Vibraciones	Queretaro	Mexico
Dep. of Sci. & Indus. Res. Physics and Eng. Lab. (PEL)	Acoustics and Vib. Section	Lower Hutt	New Zealand
Comm. for Std., Meas. and Qual. Ctl. (PKNMJ)	Mechanical Vibration Lab.	Warszawa	Poland
Saudi Arabian Standards Organization (SATO)	Vibration Laboratory	Riyadh	Saudi Arabia
Slovak Inst. of Metrology	Primary Stds. of Kinematics	Bratislava	Slovakia
Dep. Fed. de Just. et Police	Off. Federal de Metrologie	Wabern-Bern	Switzerland
Indust. Techn. Res. Institute (ITRI)	Center for Meast. Standards	Hsinchu	Taiwan

TABLE 18.1 National Standards Laboratories Responsible for the Calibration of Vibration Pickups (*Continued*)

Institution (acronym)	Laboratory	Location	Country
Nat. Inst. of Stds. and Technology (NIST)	Acoustics, Mass, and Vibrations Group	Gaithersburg, MD 20899	U.S.A.
Mendeleyev Institute for Metrology (VNIIM)	Mechanical Metrology Laboratory	St. Petersburg	Russia
Natl. Center for Standardization, Metrology, and Certification	Mechanical Metrology Laboratory	Kiev	Ukraine

[1] By mutual agreement, Bruel & Kjaer Instruments functions as a national laboratory for Denmark and other Scandinavian countries.

Institute of Standards and Technology (NIST). This, however, does not mean that the contractor must have his or her accelerometer calibrated at the vibration laboratory of NIST. It is sufficient that the manufacturer's standard accelerometer, used to calibrate the contractor's accelerometer, be calibrated by NIST. In order to establish such a traceable chain of comparisons, all calibration reports should include the measured value(s) of pickup sensitivity, a statement of uncertainty, identification of the standard used in the calibration procedure, and the specification of environmental conditions. Traditionally, the onus of proof of traceability has been on the manufacturer of a calibrated pickup.

In this context it is important to understand clearly the difference between an absolute and a comparison measurement, and the various types of standard transducers. An *absolute measurement* is one whose results are expressed in terms of the seven base quantities of an accepted system of units, for example, Système International (SI) units. In particular, this definition does not imply superior accuracy. Moreover, in the calibration of standards for derived quantities, such as acceleration, it may be both impractical and unnecessary to perform exclusively absolute measurements. Attention should be focused on uncertainties rather than on the type of measurement. For comparison methods of calibration, it is important to know the nature and physical layout of the devices being compared. In a *comparison measurement* the device under test is observed in the same environment as a standard device, and their electrical outputs are compared. If both devices are linear and the sensitivity of the standard pickup is known, the sensitivity of the test pickup can be calculated.

Standard pickups can be ranked into three groups: primary standards (the most precise), transfer standards, and working reference standards.[4] At the top of the hierarchy are *primary standards*. Such devices are calibrated by an absolute method and are kept at the national laboratory. *Transfer standards* are calibrated by either an absolute or a comparison method and are interchanged between laboratories for purposes of comparison and calibration of *working reference standards*. The latter are used for comparison calibration of pickups in common use. They too can be recalibrated by absolute methods to maintain traceability.

CALIBRATION BY THE COMPARISON METHOD

A rapid and convenient method of measuring the sensitivity of a vibration pickup to be tested is by direct comparison of the pickup's electrical output with that of a sec-

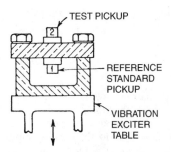

TEST PICKUP

REFERENCE
STANDARD
PICKUP

VIBRATION
EXCITER
TABLE

FIGURE 18.2 Comparison method of calibration: Pickup 2 is calibrated against Pickup 1 (the reference standard). The two pickups may be excited by any of the means described in this chapter. (*After ANSI Standard S2.2-1959, R 1990.*[1])

ond pickup (used as a "reference" standard) that has been calibrated by one of the methods described in this chapter. A comparison method is used in most shock and vibration laboratories, which periodically send their standards to a primary standards laboratory for recalibration. This procedure should be followed on a yearly basis in order to establish a history of the accuracy and quality of its reference standard pickup.

In this method of calibration the two pickups usually are mounted back-to-back on a vibration exciter as shown in Fig. 18.2. It is essential to ensure that each pickup experiences the same motion. Any angular rotation of the table should be small to avoid any difference in excitation between the two pickup locations. The error due to rotation may be reduced by carefully locating the pickups firmly on opposite faces with the center-of-gravity of the pickups located at the center of the table. Relative differences in pickup excitation may be observed by reversing the pickup locations and observing if the voltage ratio is the same for both positions.

Calibration by the comparison method is limited to the range of frequencies and amplitudes for which the reference standard pickup has been previously calibrated. If both pickups are linear, the sensitivity of the test pickup can be calculated in both magnitude and phase from

$$S_t = \frac{e_t}{e_r} S_r \qquad (18.1)$$

where S_t is the sensitivity of test pickup
           S_r is the sensitivity of reference standard pickup
           e_t is output voltage from test pickup
           e_r is output voltage from reference standard pickup

Several calibration methods described below are variations on the implementation of Eq. (18.1); they differ mainly in the manner of vibration excitation.

SINUSOIDAL FREQUENCY-DWELL METHOD

A simple and convenient way of performing a comparison calibration is to fix the frequency of the vibration exciter at a desired value and adjust the amplitude of vibration of the vibration exciter to a convenient value. Then the test pickup and the reference standard pickup are assumed to experience identical motion. Their electrical outputs are processed through amplifiers (either voltage or charge amplifiers), and then their electrical outputs are compared. The amplifier in the test channel may have a variable gain which has been calibrated. An attenuator is sometimes used in the reference channel (assuming this voltage output is greater than that of the test channel).

A single voltmeter can be employed to compare the output voltage of the two channels. A significant advantage results from the use of the same voltmeter to measure the voltages e_t and e_r. According to Eq. (18.1), the errors in the voltmeter tend to cancel out unless the voltages e_t and e_r are widely different in magnitude.

SINUSOIDAL SWEPT-FREQUENCY METHOD

A graphical plot of the sensitivity of the test pickup or its deviation from a reference sensitivity versus frequency can provide a quick and convenient method of calibration.[5] A circuit like the one shown in Fig. 18.3 provides good resolution and a stable comparison system. The system configuration is similar to that of the sinusoidal frequency-dwell method; however, a differential amplifier is used to measure the voltage differences between the two channels. The two ac voltages are made equal at some reference condition by adjusting amplifier gains. The recorder gain is adjusted so that a given difference equals a percentage of the test accelerometer calibration factor.

RANDOM-EXCITATION-TRANSFER-FUNCTION METHOD

The use of random-vibration-excitation and transfer-function analysis techniques can provide quick and accurate comparison calibrations.[6] The reference standard pickup and the test pickup are mounted back-to-back on a suitable vibration exciter. Their outputs are usually fed into a spectrum analyzer through a pair of low-pass (antialiasing) filters. The bandwidth of the random signal which drives the exciter is determined by settings of the analyzer.

This method provides a nearly continuous calibration over a desired frequency spectrum, with the resulting sensitivity function having both amplitude and phase information. Since purely sinusoidal motion is not a requirement as in the other calibration methods, this lessens the requirements for the power amplifier and exciter to maintain low values of harmonic distortion. A very useful measure of process quality is obtained by computing the input-output coherence function, which requires knowledge of the input and output power spectra, the cross-power spectrum, and the transfer function.

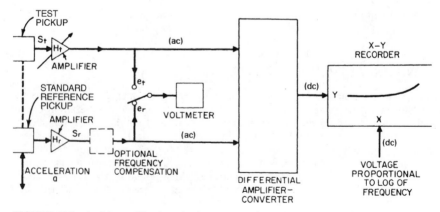

FIGURE 18.3 A pickup-calibration plotting system using a differential amplifier, AC/DC converter, and an X-Y recorder. (*After K. Unholtz.*[5])

CALIBRATION BY ABSOLUTE METHODS

RECIPROCITY METHOD

The reciprocity calibration method is an absolute means for calibrating vibration exciters that have a velocity coil or reference accelerometer. This method relates the pickup sensitivity to measurements of voltage ratio, resistance, frequency, and mass. For this method to be applicable, it is necessary that the vibration exciter system be linear (e.g., that the displacement, velocity, acceleration, and current in the driver coil each increase linearly with force and driver-coil voltage). The reciprocity method is used chiefly with electrodynamic exciters[7] but is also used with piezoelectric vibration exciters.[8]

The reciprocity method generally is applied only under controlled laboratory conditions. Many precautions must be taken, and the process is usually time-consuming. Several variations of the basic approach have been developed at national standards laboratories.[9,10] The method described here is used at the National Institute of Standards and Technology.[11-13] The method consists of two laboratory experiments:

1. The measurement of the transfer admittance between the exciter's driver coil and the attached velocity coil or accelerometer.

2. The measurement of the voltage ratio of the open-circuit velocity coil or accelerometer and the driving coil while the exciter is driven by a second external exciter. The use of a piezoelectric accelerometer is assumed here. The electrical connections for the transfer admittance and voltage ratio measurements are shown in Fig. 18.4.

The relationship defining the transfer admittance is

$$\mathbf{Y} = \frac{\mathbf{I}}{\mathbf{e}_{12}} \qquad (18.2)$$

where $\mathbf{Y}$ = transfer admittance
$\mathbf{e}_{12}$ = voltage generated in standard accelerometer and amplifier
$\mathbf{I}$ = current in driver coil

and the bold letters denote phasor (complex) quantities. The current is determined by measuring the voltage drop across a standard resistor. The phase, ψ_Y, of $\mathbf{Y}$ is measured with a phase meter having an uncertainty of $\pm 0.1°$ or better. Transfer admittance measurements are made with a series of masses attached, one at a time, to the table of the exciter. Also, a zero-load transfer admittance measurement is made before and after attaching each mass. This zero-load measurement is denoted by $\mathbf{Y}_0$. Using the measured values of $\mathbf{Y}$ and $\mathbf{Y}_0$, graphs of the real and imaginary values of the ratio

$$\mathbf{T}_n = \frac{M_n}{\mathbf{Y} - \mathbf{Y}_0} \qquad (18.3)$$

are plotted versus M_n for each frequency, where M_n is the value of the mass attached to the table. The zero intercepts, J_i and J_r, of the resulting nominally straight lines and their slopes, Q_i and Q_r, are computed by a weighted least-squares method.[7] The values of $\mathbf{Y}_0$ used in the calculations are obtained by averaging the values of the $\mathbf{Y}_0$ measurements before and after each measurement of $\mathbf{Y}$ using different masses. These computed values are used in determining the sensitivity of the standard.

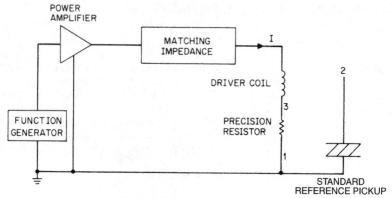

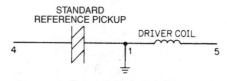

FIGURE 18.4 Transfer-admittance and voltage-ratio-measurement circuit connections for the reciprocity calibration method in the Levy-Bouche realization.[9]

The ratio of two voltages, measured while the exciter is driven with an external exciter, is given by

$$\mathbf{R} = \frac{\mathbf{e}_{14}}{\mathbf{e}_{15}} \tag{18.4}$$

where $\mathbf{e}_{14}$ = voltage generated in standard accelerometer and amplifier, and $\mathbf{e}_{15}$ = open-circuit voltage in driving coil.

After $\mathbf{R}, J_r, J_i, Q_r$, and Q_i have been determined for a number of frequencies, f, the sensitivity of the exciter is calculated from the following relationship:[7]

$$\mathbf{S} = \left[\frac{\mathbf{RJ}}{j2\pi f} \right]^{1/2} \left[1 + \frac{\mathbf{MQ}}{\mathbf{J}} \right] \tag{18.5}$$

where
j = unit imaginary vector
$\mathbf{J} = J_r + jJ_i$
$\mathbf{Q} = Q_r + jQ_i$
$$\mathbf{M} = \frac{\mathbf{J}(\mathbf{Y} - \mathbf{Y}_0)}{1 - \mathbf{Q}(\mathbf{Y} - \mathbf{Y}_0)}$$

The sensitivity of the exciter is, therefore, determined from the measured quantities $\mathbf{Q}, \mathbf{J}, \mathbf{T}$, and f and from the masses M_n which are attached to the exciter table. The sensitivity as computed from Eq. (18.5) has the units of volts per meter per second squared if the values of the measurements are in the SI system. If the masses M_n are not in kilograms, appropriate conversion factors must be applied to the quantities $\mathbf{J}$,

Q, and **M.** A commonly used engineering formula,[7] with the mass expressed in pounds and the sensitivity in millivolts per **g,** is

$$S = 2635 \left[\frac{RJ}{jf} \right]^{1/2} \tag{18.6}$$

which also assumes that $MQ/J \ll 1$, a condition usually satisfied in practice but which should be verified experimentally. The use of a computer greatly facilitates the application of the reciprocity calibration process.

Assuming the errors to be uncorrelated, a typical estimate of uncertainty expected from a reciprocity calibration method is ±0.5 percent in the frequency range 100 to 1000 Hz. This is a twofold improvement over the earlier systems.[13,14] The critical component in a reciprocity-based calibration system is the vibration exciter. Electrodynamic exciters utilizing an air bearing are generally superior to other types, for this application.

CALIBRATION USING THE EARTH'S GRAVITATIONAL FIELD

Tilting-Support Calibrator. The earth's gravitational field provides a convenient means of applying small constant acceleration to a pickup. It is particularly useful in calibrating accelerometers whose frequency range extends down to 0 Hz. A $2g$ change in acceleration may be obtained by first orienting the accelerometer with the positive direction of its sensing axis up and then rotating the accelerometer through 180° so that the positive direction is down.

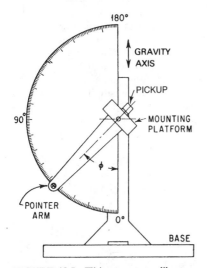

Increments of acceleration within the $1g$ range can be applied by means of a tilting-support calibrator with an accuracy better than $0.004g$ near the horizontal and less than $0.001g$ near the vertical.[1,15] The pickup to be calibrated is fastened as shown in Fig. 18.5 to a platform at the end of an arm. The arm may be set at any angle ϕ between 0 and 180° relative to the vertical. It is furnished with a pointer to indicate the angle ϕ. Positioning of the arm angle ϕ to ±0.2° or less is possible with an accurately divided circle. The component of sensed acceleration is given by

FIGURE 18.5 Tilting-support calibrator used to supply incremental static accelerations of $1g$ or less to pickups which have zero frequency response, such as strain-gage pickups. (*After ANSI Standard S2.2-1959, R1990.*[1])

$$a = g \cos \phi \tag{18.7}$$

The calibration factor is obtained by plotting the response of the accelerometer as a function of the acceleration a, given by Eq. (18.7), for successive values of ϕ and then determining the slope of the straight line fitted through the points.

The pickup is also subjected to a component of acceleration at right angles to its sensitive direction that is equal to

$$a_t = g \sin \phi \qquad (18.8)$$

Because of the transverse component, this method is not recommended for calibration of pickups for which the transverse sensitivity is significant. The extent to which the transverse component affects the calibration results may be checked by remounting the pickup at two additional positions on the mounting platform. The two new positions are selected by rotating the pickup 90° and 180° about its sensitive axis relative to the first position. The test is performed for each of the three different angular positions of the pickup on the pointer arm. Any measurable change in the results may indicate an undesirable influence of the transverse sensitivity. The maximum influence will occur when the pointer arm angle $\phi = 0°$.

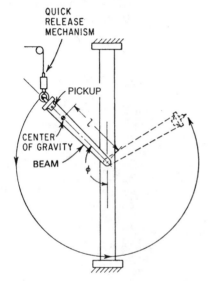

FIGURE 18.6 Pendulum calibrator for measuring the calibration factor. (*After H. Levy.*[1])

Pendulum Calibrator. The pendulum calibrator is a simple device for generating a transient acceleration of as great as $10g$ with a duration on the order of a second.[1] It imparts to a pickup a transverse and an angular acceleration as well as the acceleration inward from its center of rotation. Therefore, it should be used with caution in calibrating pickups sensitive to these extraneous motions.

The pendulum calibrator in Fig. 18.6 consists of a beam and platform with its center of gravity pivoted so as to swing about a horizontal axis. The pickup to be calibrated is attached to the platform and is carefully aligned so that the pickup axis of sensitivity is along the radius of the beam. Electrical connections can be brought from the pickup along the beam bridging to support frame. A well-constructed pendulum has negligible damping. If the pendulum is released to swing after being raised to an angle ϕ above the vertical, the increment a, in acceleration along the radius vector from the time of release to the bottom of the swing, is

$$a = \left[g + \frac{8\pi^2 l}{\tau^2} \right] (1 - \cos \phi) \qquad (18.9)$$

where g = acceleration of gravity
   l = distance from center-of-gravity of mass element to pivot
   τ = period of pendulum for small amplitudes
   ϕ = average of release angle and angle to which pendulum rises after passing bottom of its swing

The calibration factor normally is determined by plotting the peak-to-peak response

$$R = \frac{e_1 + e_2}{2} \qquad (18.10)$$

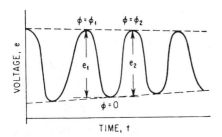

FIGURE 18.7 Typical oscillogram of a pickup response to the pendulum calibrator in Fig. 18.6. The release angle is ϕ_1, and the angle to which the pendulum rises after passing the bottom of its swing is ϕ_2. The average angle $(\phi_1 + \phi_2)/2$ corresponds to the average response $(e_1/e_2)/2$. (*After H. Levy.*[1])

of the pickup as a function of the acceleration a as given by Eq. (18.9) for successive values of ϕ and then finding the slope of the straight line fitted through the data. The error is ordinarily below ±1 percent if care is exercised. A typical oscillogram record of an accelerometer response as a function of time is shown in Fig. 18.7.

Rotating Table Calibrator. In the rotating-table method of calibration, the test pickup is rotated at a uniform angular rate about a horizontal axis.[16,17] The experimental arrangement is shown in Fig. 18.8. The pickup is mounted so as to rotate about a horizontal axis; the sensitivity axis of the pickup rotates in a vertical plane. Thus it is possible to apply very low-frequency sinusoidal motion in the region where sinusoidal linear translation of large amplitude would otherwise be required to give useful accelerations. As a result, a sinusoidal acceleration, having a peak amplitude of $1g$ at the rotation frequency, is superposed on the centrifugal acceleration. By this method it is possible to obtain the response of the pickup under both static and dynamic conditions in the same test setup. The rotating table not only provides excitation at very low frequencies but also provides an accurate constant amplitude of acceleration. The basic consideration in applying this method is that the vibration pickup be sufficiently sensitive to have adequate output in the range of acceleration up to $1g$. The threshold sensitivity and the repeatability of the pickup under test must be such that a periodic excitation between ±$1g$ produces a response that is representative of the instrument's behavior. Excitation less than ±$1g$ can be provided by inclining the axis of rotation. The useful frequency range over which this calibration technique applies is from about 0.5 to 45 Hz.

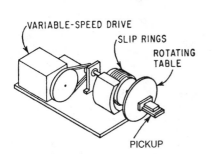

FIGURE 18.8 Rotating-table calibrator used for both static and dynamic calibration, with a peak acceleration amplitude of $1g$. Table rotates in a vertical plane. (*After W. A. Wildhack and R. O. Smith.*[16]).

Bearing noise and dynamic unbalance limit the upper frequency range. Improved performance from that obtained from the calibrator shown in Fig. 18.8 can be obtained by incorporating air bearings and a magnetic clutch drive.[17] The position of the pickup is adjusted by a fine lead screw, and a series of adjustable counterbalance masses are arranged about the table's circumference.

The chief restrictions of the rotating-table method of dynamic calibration are (1) the rotation speed of the table is limited, (2) the acceleration lever is limited to $1g$, and (3) the response to the centrifugal acceleration tends to become infinite at the natural frequency. Below about 5 Hz, errors introduced by voltmeters tend to limit the measurement accuracy. Overall pickup calibration uncertainties of ±0.5 to 1 percent in the frequency range from 2 to 45 Hz can be maintained.

Structural-Gravimetric Calibration. This technique provides a simple, robust, and low-cost method of calibrating pickups.[18-21] The structural-gravimetric-calibration (SGC) method is applicable over a broad frequency range because it relies on a quartz force transducer as the reference pickup and the behavior of the simplest of structures (i.e., a mass behaving as a rigid body). It references the acceleration of gravity and allows the measurement of sensitivity magnitude and phase. The results of calibration using this method agree within a fraction of 1 percent with those obtained by laser interferometry and reciprocity methods. The following steps are the procedure of SGC method:

Step 1. Determine the acceleration sensitivity S_r of the reference force transducer. Mount the reference force transducer, reference mass (can be built-in or external), and the test pickup to be calibrated on a drop-test fixture, as shown in Fig. 18.9. (For use at higher frequencies it is important to make the reference mass small in size in order to satisfy the rigid-body assumption.) Then subject the mass and the two pickups to a free fall of 1 *g* by striking the junction of line, which causes the line to relax momentarily and impart a step-function gravitational acceleration to the assembly by allowing it to fall freely. Measure the output of the reference force transducer, e_g; in order to reduce the effect of measurement noise, curve fitting may be used to estimate the step value. Equation 18.11 shows how the sensitivity of the reference force transducer is related to the other parameters of the system.

$$S_r = S_{rf}M = \frac{e_g}{gM}M = \frac{e_g}{g} \tag{18.11}$$

where S_r is the acceleration sensitivity of the reference force transducer, in mV/ms^{-2}
   S_{rf} is the force sensitivity of the reference force transducer, in mV/N
   M is the total mass on the force transducer, in kg
   e_g is the output of the force transducer, in mV
   g is the acceleration of free fall due to gravity, in ms^{-2}

Note that e_g is numerically equal to S_r expressed in mV/*g*.

Step 2. Measure the voltage ratio e_t/e_r. Remove the reference force transducer, reference mass, and the pickup being calibrated from the drop-test fixture; then mount them on the vibration exciter, as shown in Fig. 18.10. By measuring the transfer function e_t/e_r (i.e., the ratio of the voltage output of the signal conditioner from the test pickup to the voltage output of the signal conditioner from the reference force transducer, shown in Fig. 18.10) the frequency response of the test pickup can

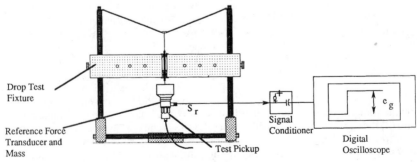

FIGURE 18.9 Gravimetric free-fall calibrator for scaling reference force gage. (*After D. Corelli and R. W. Lally.*[18])

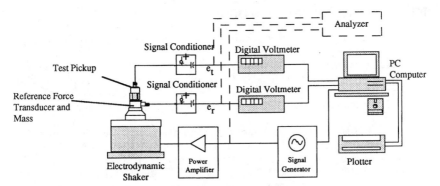

FIGURE 18.10 System configuration for frequency response calibration by measuring acceleration-to-force ratio.

be measured over 0.1 to 100,000 Hz, depending upon frequency range of the vibration exciter and signal-to-noise ratio of the system. For use at low frequencies, the discharge time constant of the reference force transducer should be ten times greater than that of the test pickup.

Step 3. Calculate the sensitivity S_t of the test pickup. If the reference force transducer and the test pickup are linear, the acceleration sensitivity of the test pickup S_t, expressed in the same units as S_r, can be calculated from Eq. 18.1. If either velocity or displacement sensitivity of the test pickup is required, it can be obtained by dividing the acceleration sensitivity by $(2\pi f)$ or $(2\pi f)^2$, respectively.

CENTRIFUGE CALIBRATOR

A centrifuge provides a convenient means of applying constant acceleration to a pickup. Simple centrifuges can be obtained readily for acceleration levels up to $100g$ and can be custom-made for use at much higher values because of the light load requirement by this application. They are particularly useful in calibrating rectilinear accelerometers whose frequency range extends down to 0 Hz and whose sensitivity to rotation is negligible. Centrifuges are mounted so as to rotate about a vertical axis. Cable leads from the pickup, as well as power leads, usually are brought to the table of the centrifuge through specially selected low-noise slip rings and brushes.

To perform a calibration, the accelerometer is mounted on the centrifuge with its axis of sensitivity carefully aligned along a radius of the circle of rotation. If the centrifuge rotates with an angular velocity of ω rad/sec, the acceleration a acting on the pickup is

$$a = \omega^2 r \tag{18.12}$$

where r is the distance from the center-of-gravity of the mass element of the pickup to the axis of rotation. If the exact location of the center-of-gravity of the mass in the pickup is not known, the pickup is mounted with its positive sensing axis first outward and then inward; then the average response is compared with the average acceleration acting on the pickup as computed from Eq. (18.12) where r is taken as the mean of the radii to a given point on the pickup case. The calibration factor is determined by plotting the output e of the pickup as a function of the acceleration a given by Eq. (18.12) for successive values of ω and then determining the slope of the straight line fitted through the data.

INTERFEROMETER CALIBRATORS

All systems in this category of calibrators consist of three stages: modulation, interference, and demodulation. The differences are in the specific type of interferometer that is used (for example, a Michelson or Mach-Zehnder) and in the type of signal processing, which is usually dictated by the nature of the vibration. The vibratory displacement to be measured modulates one of the beams of the interferometer and is consequently encoded in the output signal of the photodetector in both magnitude and phase.

Figure 18.11 shows the principle of operation of the Michelson interferometer. One of the mirrors, D in Fig. 18.11A, is attached to the plate on which the device to be calibrated is mounted. Before exciting vibrations, it is necessary to obtain an interference pattern similar to that shown in Fig. 11.18B. The relationship underlying the illustrations to be presented is the classical interference formula for the time average intensity I of the light impinging on the photodetector surface.[22,23]

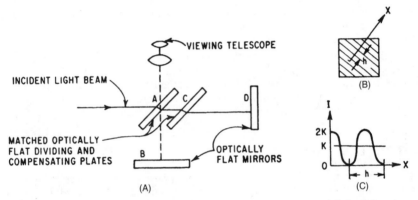

FIGURE 18.11   The principle of operation of a Michelson interferometer: (A) Optical system. (B) Observed interference pattern. (C) Variation of the light intensity along the X axis.

$$I = A + B \cos 4\pi\delta/\lambda \tag{18.13}$$

where A and B are system constants depending on the transfer function of the detector, the intensities of the interfering beams, and alignment of the interferometer. The vibration information is contained in the quantity δ, 2δ being the optical-path difference of the interfering beams. The absoluteness of the measurement comes from λ, the wavelength of the illumination, in terms of which the magnitude of vibratory displacement is expressed. Velocity and acceleration values are obtained from displacement measurements by differentiation with respect to time.

Fringe-Counting Interferometer.   An optical interferometer is a natural instrument for measuring vibration displacement. The Michelson and Fizeau interferometers are the most popular configurations. A modified Michelson interferometer is shown in Fig. 18.12.[24] A corner cube reflector is mounted on the vibration-exciter table. A helium-neon laser is used as a source of illumination. The photodiode and its amplifier must have sufficient bandwidth (as high as 10 MHz) to accommodate

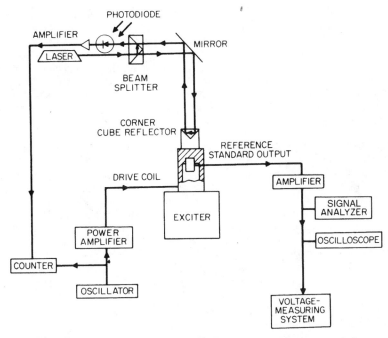

FIGURE 18.12 Typical laboratory setup for interferometric measurement of vibratory displacement by fringe counting. (*After R. S. Koyanagi.*[24])

the Doppler frequency shift associated with high velocities. An electrical pulse is generated by the photodiode for each optical fringe passing it. The vibratory displacement amplitude is directly proportional to the number of fringes per vibration cycle. The peak acceleration can be calculated from

$$a = \frac{\lambda \nu \pi^2 f^2}{2} \tag{18.14}$$

where λ = wavelength of light
   ν = number of fringes per vibration cycle
   f = vibration frequency

Interferometric fringe counting is useful for vibration-displacement measurement in the lower frequency ranges, perhaps to several hundred hertz depending on the characteristics of the vibration exciter.[25,26] At the low end of the frequency spectrum, conventional procedures and commercially available equipment are not able to meet all the present requirements. Low signal-to-noise ratios, cross-axis components of motion, and zero-drifts are some of the problems usually encountered. In response to those restrictions an electrodynamic exciter for the frequency range 0.01 to 20 Hz has been developed.[27] It features a maximum displacement amplitude of 0.5 meter, a transverse sensitivity less than 0.01 percent, and a maximum uncorrected distortion of 2 percent. These characteristics have been achieved by means of a specially designed air bearing, an electro-optic control, and a suitable foundation.

Figure 18.13 shows the main components of a computer-controlled low-frequency calibration system which employs this exciter. Its functions are (1) generation of sinusoidal vibrations, (2) measurement of rms and peak values of voltage and charge, (3) measurement of displacement magnitude and phase response, and (4) control of nonlinear distortion and zero correction for the moving element inside a tubelike magnet. Position of the moving element is measured by a fringe-counting interferometer. Uncertainties in accelerometer calibrations using this system have been reduced to about 0.25 to 0.5 percent, depending on frequency and vibration amplitude.

Fringe-Disappearance Interferometer. The phenomenon of the interference band disappearance in an optical interferometer can be used to establish a precisely known amplitude of motion. Figure 18.11 shows the principle of operation of the Michelson interferometer employed in this technique. One of the mirrors D, in Fig. 18.11A, is attached to the mounting plate of the calibrator. Before exciting vibrations it is necessary to obtain an interference pattern similar to that shown in Fig. 18.11B.

When the mirror D vibrates sinusoidally[28] with a frequency f and a peak displacement amplitude d, the time average of the light intensity I at position x, measured from a point midway between two dark bands, is given by

$$I = A + BJ_0\left(\frac{4\pi d}{\lambda}\right)\cos\left(\frac{2\pi x}{h}\right)$$ (18.15)

where J_0 = zero-order Bessel function of the first kind
    A and B = constants of measuring system
        h = distance between fringes, as shown in Fig. 18.11B and C

For certain values of the argument, the Bessel function of zero order is zero; then the fringe pattern disappears and a constant illumination intensity A is present. Electronic methods for more precisely establishing the fringe disappearance value of the vibratory displacement have been successfully used at the National Institute of Standards and Technology[12,29] and elsewhere. The latter method has been fully automated using a desktop computer.

The use of piezoelectric exciters is common for high-frequency calibration of accelerometers.[30] They provide pistonlike motion of relatively high amplitude and are structurally stiff at the lower frequencies where displacement noise is bothersome. When electrodynamic exciters are used with fringe disappearance methods, it is generally necessary to stiffen the armature suspensions to reduce the background displacement noise.

Signal-Nulling Interferometer. This method, although mathematically similar to fringe disappearance, relies on finding the nulls in the fundamental frequency component of the signal from a photodetector.[12,23,31] The instrumentation is, therefore, quite different, except for the interferometer. One successful arrangement is shown in Fig. 18.14. Laboratory environmental restrictions are much more severe for this method.

The interferometer apparatus should be well-isolated to ensure stability of the photodetector signals. Air currents in the room may contribute to noise problems by physically moving the interferometer components and by changing the refractive index of the air. An active method of stabilization has also been successfully employed.[32]

To make displacement amplitude measurements, a wave analyzer tuned to the frequency of vibration can be used to filter the photodetector signal. The filtered signal amplitude will pass through nulls as the vibration amplitude is increased, according to the following relationship:

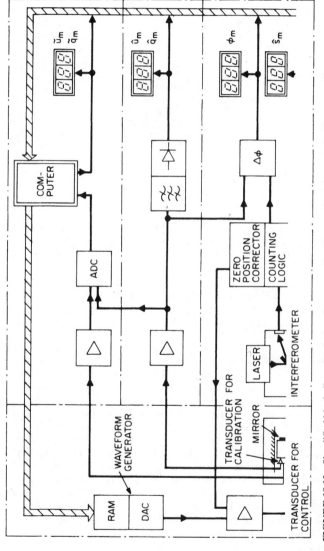

FIGURE 18.13 Simplified block diagram of a low-frequency vibration standard. (*After H. J. von Martens.*[27])

18.17

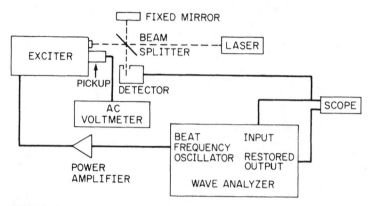

FIGURE 18.14 Interferometric measurement of displacement d as given by $J_1(4\pi d/\lambda) = 0$.

$$I = 2BJ_1\left(\frac{4\pi d}{\lambda}\right)$$ (18.16)

where J_1 is the first-order Bessel function of the first kind, and the other terms are as previously defined. The signal nulls may be established using a wave analyzer. The null amplitude will generally be 60 dB below the maximum signal level of the photodetector output.

The accelerometer output may be measured by an accurate voltmeter at the same time that the nulls are obtained. The sensitivity is then calculated by dividing the output voltage by the displacement. Because the filtered output of the photodetector is a replica of the vibrational displacement, a phase calibration of the pickup can also be obtained with this arrangement.

Heterodyne Interferometer. A *homodyne interferometer* is an interferometer in which interfering light beams are created from the same beam by a process of beam splitting. All illumination is at the same optical frequency. In contrast, in the *heterodyne interferometer,*[33] light from a laser-beam source containing two components, each with a unique polarization, is separated into (1) a measurement beam and (2) a reference beam by a polarized beam splitter. When the mounting surface of the device under test is stationary, the interference pattern impinging on the photodetector produces a signal of varying intensity at the beat frequency of the two beams. When surface moves, the frequency of the measurement beam is shifted because of the Doppler effect, but that of the reference beam remains undisturbed. Thus, the photodetector output can be regarded as a carrier that is frequency modulated by the velocity waveform of the motion.

The main advantages of the heterodyne interferometer are greater measurement stability and lower noise susceptibility. Both advantages occur because displacement information is carried on ac waveforms; hence, a change in the average value of beam intensity cannot be interpreted as motion. Digitization and subsequent phase demodulation of the interferometer output reduce measurement uncertainties.[34] This can yield significant improvements in calibration results at high frequencies, where the magnitude of displacement typically is only a few nanometers. As in the case of homodyning, variations of the heterodyning technique have been developed to meet specific needs of calibration laboratories. Reference 35 describes an accelerometer calibration system, applicable in the frequency range from 1 mHz to 25 kHz and at vibration amplitudes from 1 nanometer to 10 meters. The method requires the acquisition of instantaneous position data as a function of the phase angle of the vibration signal and the use of Fourier analysis.

HIGH-ACCELERATION METHODS OF CALIBRATION

Some applications in shock or vibration measurement require that high amplitudes be determined accurately. To ensure that the pickups used in such applications meet certain performance criteria, calibrations must be made at these high amplitudes. The following methods are available for calibrating pickups subject to accelerations in excess of several hundred *g*.

SINUSOIDAL-EXCITATION METHODS

The use of a metal bar, excited at its fundamental resonance frequency, to apply sinusoidal accelerations for calibration purposes has several advantages: (1) an inherently constant frequency, (2) very large amplitudes of acceleration (as much as 4000*g*, and (3) low waveform distortion. A disadvantage of this type of calibrator is that calibration is limited to the resonance frequencies of the metal bar.

The bar can be supported at its nodal points, and the pickup to be calibrated can be mounted at its mid-length location. The bar can be energized by a small electromagnet or can be self-excited. Acceleration amplitudes of several thousand *g* can thus be obtained at frequencies ranging from several hundred to several thousand hertz. The bar also may be calibrated by clamping it at its midpoint and mounting the pickup at one end.[36] The displacement at the point of attachment of the pickup can be measured optically since displacements encountered are adequately large.

The resonant-bar calibrator shown in Fig. 18.15 is limited in amplitude primarily by the fatigue resistance of the bar.[36] Accelerations as much as 500*g* have been attained using aluminum bars without special designs. Peak accelerations as large as 4000*g* have been attained using tempered vanadium steel bar. The bar is mounted at its mid-length on a conventional electrodynamic exciter. The accelerometer being calibrated is mounted at one end of the bar, and an equivalent balance weight is mounted at the opposite end in the same relative position.

Axial resonances of long rods have been used to generate motion for accurate calibration of vibration pickups over a frequency range from about 1 to 20 kHz and at accelerations up to 12,000*g*.[37,38] The use of axially driven rods has an advantage over the beams discussed above in that no bending or

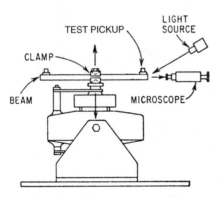

FIGURE 18.15 Resonant-bar calibrator with the pickup mounted at end and a counterbalancing weight at the other. (*After E. I. Feder and A. M. Gillen.*[36])

lateral motion is present. This minimizes errors from the pickup response to such unwanted modes and also from the direct measurement of the displacement having nonrectilinear motion.

SHOCK-EXCITATION METHODS

There are several methods by which a sudden velocity change may be applied to pickups designed for high-frequency acceleration measurement, for example, the ballistic pendulum, drop-test, and drop-ball calibrators, described below. Any method which generates a reproducible velocity change as function of time can be used to obtain the calibration factor.[1] Impact techniques can be employed to obtain calibrations over an amplitude range from a few g to over 100,000g. An example of the latter is the Hopkinson bar, in which the test pickup is mounted at one end and stress pulses are generated by an air gun firing projectiles impacting at the other end.[39,40]

An accurate determination of shock performance of an accelerometer depends not only upon the mechanical and electrical characteristics of the test pickup but also upon the characteristics of the instrumentation and recording equipment. It is often best to perform system calibrations to determine the linearity of the test pickup as well as the linearity of the recording instrumentation in the range of intended use. Several of the following methods make use of the fact that the velocity change during a transient pulse is equal to the time integral of acceleration:

$$v = \int_{t_1}^{t_2} a \, dt \qquad (18.17)$$

where the initial or final velocity is taken as reference zero, and the integration is performed to or from the time at which the velocity is constant. If the output closely resembles a half-sine pulse, the area is equal to approximately $2h(t_2 - t_1)/\pi$, where h is the height of the pulse, and $(t_2 - t_1)$ is its width.

In this section, several methods for applying known velocity changes v to a pickup are presented. The voltage output e and the acceleration a of the test pickup are related by the following linear relationship:

$$S = \frac{e}{a} \qquad (18.18)$$

where S is the pickup calibration factor.

After Eq. (18.18) is substituted into Eq. (18.17), the calibration factor for the test pickup can be expressed as

$$S = \frac{A}{v} \qquad (18.19)$$

where

$$A = \int_{t_1}^{t_2} e \, dt \qquad (18.20)$$

the area under the acceleration-versus-time curve.

The calibration factor assumes that no significant spectral energy exists beyond the frequency region in which the test pickup has nominally constant complex sensitivity (uniform magnitude and phase response as functions of frequency). In general, this assumption becomes less valid with decreasing pulse duration resulting in increasing bandwidth in the excitation signal.

Sometimes it is convenient to express acceleration as a multiple of g. The corresponding calibration factor S_1 is in volts per g:

$$S_1 = \frac{e}{(a/g)} = \frac{Ag}{v} \tag{18.21}$$

In either case, the integrals representing A and v must first be evaluated. The linear range of a pickup is determined by noting the magnitude of the velocity change v at which the calibration factor S or S_1 begins to deviate from a constant value. The minimum pulse duration is similarly found by shortening the pulse duration and noting when S changes appreciably from previous values.

Ballistic Pendulum Calibrator. A ballistic pendulum calibrator provides a means for applying a sudden velocity change to a test pickup. The calibrator consists of two masses which are suspended by wires or metal ribbons. These ribbons restrict the motion of the masses to a common vertical plane.[41] This arrangement, shown in Fig. 18.16, maintains horizontal alignment of the principal axes of the masses in the direction parallel to the direction of motion at impact. The velocity attained by the anvil mass as the result of the sudden impact is determined.

The accelerometer to be calibrated is mounted to an adapter which attaches to the forward face of the anvil. The hammer is raised to a predetermined height and held in the release position by a solenoid-actuated clamp. Since the anvil is at rest prior to impact, it is necessary to record the measurement of the change in velocity of anvil and transient waveform on a calibrated time base. One method of measurement of velocity change is performed by focusing a light beam through a grating attached to the anvil, as shown in Fig. 18.16. The slots modulate the light beam intensity, thus varying the photodetector output, which is recorded with the pickup output. Since the distance between grating lines is known, the velocity of the anvil is calculated directly, assuming that the velocity is essentially constant over the distance between successive grating lines. The velocity of the anvil in each case is determined directly; the time relation between initiation of the velocity and the pulse at

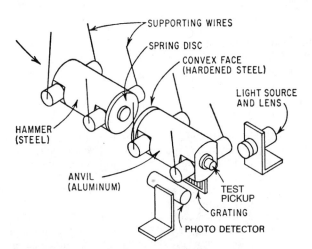

FIGURE 18.16 Components arrangement of the ballistic pendulum with photodetector and light grating to determine the anvil-velocity change during impact. (*After R. W. Conrad and I. Vigness.*[41])

the output of the pickup is obtained by recording both signals on the same time base. The most frequently used method infers the anvil velocity from its vertical rise by measuring the maximum horizontal displacement and making use of the geometry of the pendulum system.

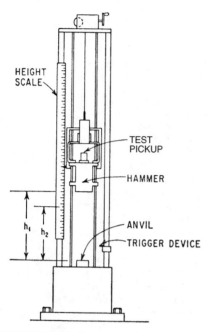

HEIGHT
SCALE

TEST
PICKUP

HAMMER

ANVIL

h_1 h_2

TRIGGER DEVICE

FIGURE 18.17 Component of a conventional drop tester used to apply a sudden velocity change to a vibration pickup. (*After R. W. Conrad and I. Vigness.*[41])

The duration of the pulse, which is the time during which the hammer and anvil are in contact, can be varied within close limits.[41] In Fig. 18.16 the hammer nosepiece is a disc with a raised spherical surface. It develops a contact time of 0.55 millisecond. For larger periods, ranging up to 1 millisecond, the stiffness of the nosepiece is decreased by bolting a hollow ring between it and the hammer. A pulse longer than 1 millisecond may be obtained by placing various compliant materials, such as lead, between the contacting surfaces.

Drop-Test Calibrator. In the drop-test calibrator, shown in Fig. 18.17, the test pickup is attached to the hammer using a suitable adapter plate. An impact is produced as the guided hammer falls under the influence of gravity and strikes the fixed anvil. To determine the velocity change, measurement is made of the time required for a contactor to pass over a known region just prior to and after impact. The pickup output and the contactor indicator are recorded simultaneously in conjunction with a calibrated time base. The velocity change also may be determined by measuring the height h_1 of hammer drop before rebound and the height h_2 of hammer rise after rebound. The total velocity is calculated from the following relationship:

$$v = (2gh_1)^{1/2} + (2gh_2)^{1/2} \tag{18.22}$$

A total velocity change of 40 ft/sec (12.2 meters/sec) is typical.

Drop-Ball Shock Calibrator. Figure 18.18 shows a drop-ball shock calibrator.[7,42] The accelerometer is mounted on an anvil which is held in position by a magnet assembly. A large steel ball is dropped from the top of the calibrator, striking the anvil. The anvil (and mounted test pickup) are accelerated in a short free-flight path. A cushion catches the anvil and accelerometer. Shortly after impact, the anvil passes through an optical timing gate of a known distance. From this the velocity after impact can be calculated. Acceleration amplitudes and pulse durations can be varied by selecting the mass of the anvil, mass of the impacting ball, and resilient pads on top of the anvil where the ball strikes. Common accelerations and durations are $100g$ at 33 milliseconds, $500g$ at 1 millisecond, $1000g$ at 1 millisec-

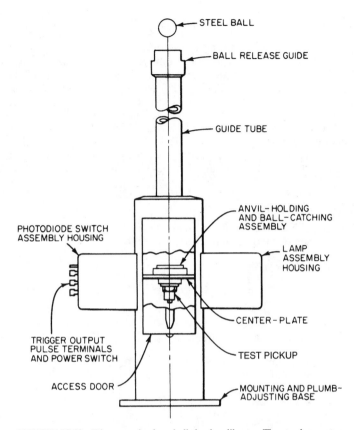

FIGURE 18.18 Diagram of a drop-ball shock calibrator. The accelerometer being calibrated is mounted on an anvil which is held in place by a small magnet. (*After R. R. Bouche.*[42])

ond, 5000g at 2 millisecond, and 10,000g at 0.1 millisecond.[42] With experience and care, shock calibrations can be performed with an uncertainty of about ±5 percent.

INTEGRATION OF ACCELEROMETER OUTPUT

Change-of-velocity methods for calibrating an accelerometer at higher accelerations than obtainable by the methods discussed above have been developed using specially modified ballistic pendulums, air guns, inclined troughs, and other devices. Regardless of the device employed to generate the mechanical acceleration or the method used to determine the change of velocity, it is necessary to compare the measured velocity and the velocity derived from the integral of the acceleration waveform as described by Eq. (18.17). Electronic digitizers can be used to capture the waveform and produce a recording. Care must be exercised in selecting the time at which the acceleration waveform is considered complete, and its integral should be compared with the velocity. The calibration factor for the test pickup is computed from Eq. (18.19) or (18.21).

IMPACT-FORCE SHOCK CALIBRATOR

The impact-force shock calibrator has a free-fall carriage and a quartz load cell. The accelerometer to be calibrated is mounted onto the top of the carriage, as shown in Fig. 18.19. The carriage is suspended about ½ to 1 meter above the load cell and allowed to fall freely onto the cell.[43] The carriage's path is guided by a plastic tube. Cushion pads are attached at the top of the load cell to lengthen the impulse duration and to shape the pulse. Approximate haversines are generated by this calibrator. The outputs of the accelerometer and load cell are fed to two nominally identical charge amplifiers or power units. The outputs from load cell and test accelerometer are recorded or measured on a storage-type oscilloscope or peak-holding meters.

During impact, the voltage produced at the output of the accelerometer, $e_a(t)$, is

$$e_a(t) = a(t)S_aH_a \tag{18.23}$$

where $a(t)$ = acceleration
        S_a = calibration factor for accelerometer
        H_a = gain of charge amplifier or power unit

The output of load cell $e_f(t)$ is

$$e_f(t) = F(t)S_fH_f \tag{18.24}$$

where $F(t)$ = force
        S_f = calibration factor for load cell
        H_f = gain of charge amplifier or power unit

By using the relationship $F(t) = ma(t)$, where m is the falling mass, and combining Eqs. (18.23) and (18.24),

$$\frac{e_a(t)}{e_f(t)} = \frac{a(t)S_aH_a}{ma(t)S_fH_f} \tag{18.25}$$

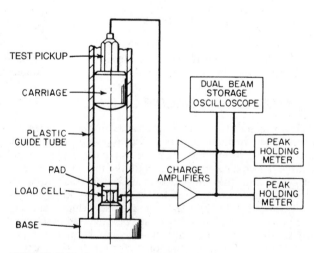

FIGURE 18.19 Impact-force calibrator with auxiliary instruments. (*After W. P. Kistler.*[43])

and hence

$$S_a = \frac{e_a(t)\ H_f m}{e_f(t)\ H_a g}\ S_f$$

(18.26)

When calculating the mass, it is necessary to know the mass of the carriage, accelerometer, mounting stud, cable connector, and a short portion of the accelerometer cable. Experience has shown that for small coaxial cables, a length of about 2 to 4 cm is correct. Calibrations by this method can be accomplished with uncertainties generally between ±2 to ±5 percent.

FOURIER-TRANSFORM SHOCK CALIBRATION

The previously discussed shock calibration methods yield the calibration factor for the accelerometer being tested; that is, they yield a single-value approximation of the magnitude of the sensitivity. For many applications, a knowledge of the calibration factor may be sufficient. However, for shock standards or other critical applications, it is desirable to know the sensitivity in both magnitude and phase as a function of frequency. The shock calibration method described in this section employs Fourier techniques to transform time-domain data into frequency-domain data. By doing so, the frequency-dependent sensitivity function can be determined in both magnitude and phase.[43–45]

The laboratory equipment consists of a mechanical shock-generating machine, two accelerometers, a data-transfer system, and a small computer for data storage and processing. Figure 18.20 shows a block diagram of one such system. The calibration of the test accelerometer is in terms of a reference standard. It is convenient

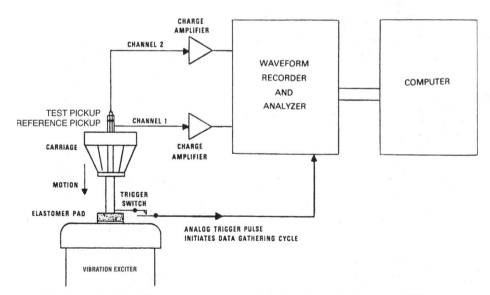

FIGURE 18.20   Block diagram of a system for shock accelerometer calibration using Fourier transform techniques. Sensitivity factor and phase are calibrated as functions of frequency. (*After J. D. Ramboz and C. Federman.*[45])

to employ a piggyback accelerometer standard which permits the test accelerometer to be mounted directly on its top in a back-to-back configuration. The signals from each accelerometer pass through charge amplifiers into a recorder and FFT signal analyzer where analyses are performed (see Use of FFT Analysis, Chap. 14). A peak-holding meter, transient recorder, and oscilloscope can be used to measure approximate acceleration peak values and to examine the time-domain waveshapes. A digital 10-bit word is generally sufficient for most work, and conversion rates of at least 100 kHz are desirable.[45]

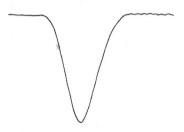

FIGURE 18.21 A typical half-sine shock pulse generated by a pneumatic shock machine. Deceleration amplitude is 900*g* and pulse duration is 1 millisecond. (*After J. D. Ramboz and C. Federman.*[45])

Ratios of the test accelerometer output to the standard accelerometer output in the frequency domain yield the test accelerometer sensitivity in terms of the standard. Phase-calibration data is obtained by taking the difference between the test and standard phase spectra.

Figure 18.21 shows a typical half-sine shock pulse. Its Fourier transform is shown in Fig. 18.22. The range of usable frequencies is limited by the pulse shape and duration, sampling rate, and Fourier-analysis capability. Generally, this type of calibration can yield complete calibra-

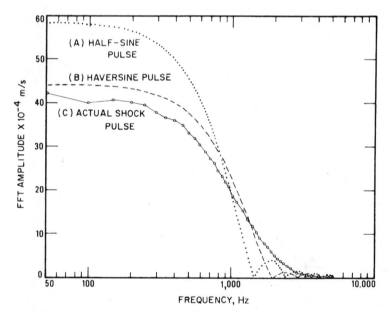

FIGURE 18.22 Fourier transforms of shock accelerometer signals. (*A*) Ideal half-sine. (*B*) Ideal haversine. (*C*) An actual shock pulse of 900*g* peak and a duration of 1 millisecond. The ideal pulses have the same peak acceleration and time durations as the actual pulse, which is shown in Fig. 18.21. (*Courtesy of the National Institute of Standards and Technology.*)

tion results over a frequency range from 0 to 10 kHz or higher. Uncertainties are less than ±5 percent.

VIBRATION EXCITERS USED FOR CALIBRATION

A vibration exciter that is suitable for calibration of vibration pickups should provide:

- Distortion-free sinusoidal motion
- True rectilinear motion in a direction normal to the vibration-table surface without the presence of any other motion
- A table that is rigid for all design loads at all operating frequencies
- A table that remains at ambient temperature and does not provide either a source or sink for heat regardless of the ambient temperature
- A table whose mounting area is free from electromagnetic disturbances
- Stepless variation of frequency and amplitude of motion within specified limits, which is easily adjustable

ELECTRODYNAMIC EXCITERS

The electrodynamic machine, described in Chap. 25, satisfactorily meets the requirements of the ideal calibrator, providing a constant-force (acceleration) output with little distortion over a rather wide frequency range from 1 to 10,000 Hz.[46] Ordinarily, to cover this frequency range, more than one exciter is required. Specially designed machines featuring long strokes for very low frequencies or ultralight moving elements for very high frequencies are commercially available. One national standards laboratory has a custom-built vibration exciter that has a low-frequency limit of 20 mHz.[27] This machine employs a special air bearing, real-time electro-optic control, and a suitable foundation.

PIEZOELECTRIC EXCITERS

The piezoelectric exciter (see Fig. 25.8) offers a number of advantages in the calibration of vibration pickups, particularly at high frequencies. Calibration is impracticable at low frequencies because of inherently small displacements in this frequency range. A design which has been used at the National Institute of Standards and Technology for many years is described in Ref. 30.

MECHANICAL EXCITERS

Rectilinear motion can be produced by mechanical exciter systems of the type described in Chap. 25 under *Direct-Drive Mechanical Vibration Machine*. Their usable frequency range is from few hertz to less than 100 Hz. Despite their relatively low cost, mechanical exciters are no longer used for high-quality calibrations because of their appreciable waveform distortion and background noise.

For generating vibratory motion at discrete frequencies (below 5 Hz), a linear oscillator can be employed. Reference 47 describes a calibrator consisting of a

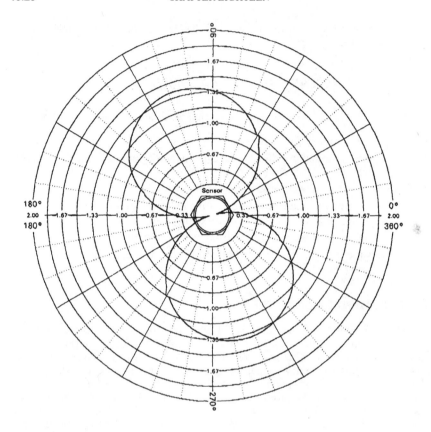

Model Number	353		Test Frequency	700	Hz
Serial Number	1		Test g level	10.93	g's
Sensitivity	20.90	mV/g	Maximum Transverse	1.40% @ 107°	
			Minimum Transverse	0.035% @ 195°	

FIGURE 18.23 Transverse sensitivity of a piezoelectric accelerometer to vibration in the plane normal to the sensitive axis.[48]

spring-supported table which is guided vertically by air bearings. Its advantages are a clean waveform, resulting from free vibration, and large rectilinear displacement with little damping, made possible by use of air bearings.

CALIBRATION OF TRANSVERSE SENSITIVITY

The characteristics of a vibration pickup may be such that an extraneous output voltage is generated as a result of vibration which is in a direction at right angles to the axis of designated sensitivity of the pickup. This effect, illustrated in Fig. 12.11, results in the axis of maximum sensitivity not being aligned with the axis of designated sensitivity. As indicated in Eq. (12.11), the cross-axis or *transverse sensitivity* of a pickup is expressed as the tangent of an angle, i.e., the ratio of the output resulting from the transverse motion divided by the output resulting from motion in the direction of designated sensitivity. This ratio varies with the azimuth angle in the transverse plane, as shown in Fig. 12.12, and also with frequency. In practice, $\tan \theta$ has a value between 0.01 and 0.05 and is expressed as a percentage. Figure 18.23 presents a typical result of a transverse-sensitivity calibration.[48]

Knowledge of the transverse sensitivity is vitally important in making accurate vibration measurements, particularly at higher frequencies (i.e., at frequencies approaching the mounted resonance frequency of the pickup). Figure 18.24 shows the relative responses of an accelerometer to main-axis and transverse-axis vibration. It is noteworthy that the transverse resonance frequency is lower than the usually specified mounted resonance frequency.

A direct measurement of the transverse sensitivity of a pickup requires a vibration exciter capable of pure unidirectional motion at the frequencies of interest. This usually means that any cross-axis motion of the mounting table should be less than 2 percent of the main-axis motion.[7] Resonance beam exciters[1] and air-bearing shakers[46] have been used for this purpose.

One method for obtaining the transverse sensitivity of a pickup is by use of the impulse technique similar to that used in modal analysis (Chap. 21). An impulse is generated by the impact of a hammer against a suspended mass on which the test pickup is mounted. A force gage is mounted on the hammer, as illustrated in Fig. 18.25. From the characteristics of the force gage and its output when it strikes against the suspended mass, from the output signal of the test pickup, and from the magnitude of the suspended mass, the transverse sensitivity of the accelerometer under test S_{ta} may be calculated according to a procedure described in Ref. 48, using the following formula:

$$S_{ta} = m S_f \left(\frac{e_a}{e_f} \right) \qquad (18.27)$$

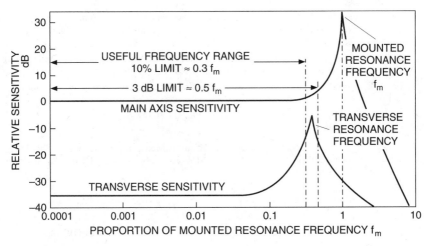

FIGURE 18.24 The relative response of an accelerometer to main-axis and transverse-axis vibrations.[3]

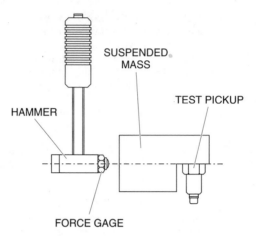

FIGURE 18.25 Schematic diagram for impact hammer method of measuring transverse sensitivity.[48]

where m = the mass of the suspended rigid block
   S_f = the sensitivity of the force gage
   e_a = the output of the accelerometer under test
   e_f = the output of the force gage

REFERENCES

1. American National Standards Institute: "Methods for the Calibration of Shock and Vibration Pickups," ANSI S2.2-1959 (R1990), New York.

2. International Organization for Standardization: "Methods of Calibration of Vibration and Shock Pickups." ISO/IS 5347, Part I, 1994. (Available from the American National Standards Institute, New York.)

3. Serridge, M., and T. R. Licht: "Piezoelectric Accelerometer and Vibration Preamplifier Handbook," Bruel & Kjaer, Naerum, Denmark, 1987.

4. Licht, T. R., and K. Zaveri: "Calibration and Standards. Vibration and Shock Measurements," *B & K Technical Review*, No. 4, 1981.

5. Unholtz, K.: *Instr. Soc. Amer. Preprint* M18-3-MESTIND-67 (1967).

6. Hartz, K.: *Proc. Natl. Conf. of Stda. Labs. Symp.* (1984).

7. Bouche, R. R. "Calibration of Shock and Vibration Measuring Transducers," Shock and Vibration Monograph SVM-11, The Shock and Vibration Information Center, Washington, D.C., 1979.

8. Ge, Lifeng: "The Reciprocity Method for Piezoelectric Calibrators," *Proc. ASME Des. Eng. Conf. Mech. Vibration and Noise,* Cincinnati, OH, September 10–13, 1985.

9. Levy, S., and R. R. Bouche: *J. Res. Natl. Std.* **57**:227 (1956).

10. Fromentin, J., and M. Fourcade: *Bull. d'Informations Sci. et Tech.,* **201**:21 (1975).

11. Payne, B.: *Shock and Vibration Bull.,* **36**, pt. 6 (1967).

12. Robinson, D. C., M. R. Serbyn, and B. F. Payne: *Natl. Bur. Std. (U.S.) Tech. Note* 1232, 1987.

13. Payne, B. F.: "Vibration Laboratory Automation at NIST with Personal Computers," *Proc. Natl. Conf. Stds. Labs.* Workshop & Symposium, Session 1C-1 (1990).

14. Bouche, R. R.: 1995 New Product Announcement, Vibracon/Bouche Laboratories, Sun Valley, CA.

15. Lederer, P. S., and J. S. Hilten: *Natl. Bur. Std. (U.S.) Tech. Note* 269, 1966.

16. Wildhack, W. A., and R. O. Smith: *Proc. 9th Annu. Meet. Instr. Soc. Amer.*, Paper 54-40-3 (1954).

17. Hillten, J. S.: *Natl. Bur. Std. (U.S.) Tech, Note* 517, March 1970.

18. Corelli, D., and R. W. Lally: "Gravimetric Calibration," *Third Int. Modal Analysis Conf.* (1985).

19. Lally, R., Jing Lin, and P. P. Kooyman: "High Frequency Calibration with the Structural Gravimetric Technique," *Proc. of Inst. of Environ. Sci.* (1990).

20. Lally, D., Jing Lin, and R. Lally: "Low Frequency Calibration with the Structural Gravimetric Technique," *Proc. of 19th Intl Modal Analysis Conf.* (1991).

21. Lally, R.: "Structural Gravimetric Calibration Technique," Master Thesis, University of Cincinnati, 1991.

22. Schmidt, V. A., S. Edelman, and E. T. Pierce: *J. Acoust. Soc. Amer.*, **34**:455 (1962).

23. Hohmann, P.: *Akustica*, **26**:122 (1972).

24. Koyanagi, R. S.: *Exp. Mech.*, **15**:443 (1975).

25. Logue, S. H.: "A Laser Interferometer and its Applications to Length, Displacement, and Angle Measurement," *Proc. 14th Ann. Meet. Inst. Environ. Sci.*, 465 (1968).

26. Payne, B. F.: "An Automated Fringe Counting Laser Interferometer for Low Frequency Vibration Measurements," *Proc. Instr. Soc. Am. Intern. Instr. Symp.*, May 1986.

27. von Martens, H. J.: "Representation of Low-Frequency Rectilinear Vibrations for High-Accuracy Calibration of Measuring Instruments for Vibration," *Proc. 2nd Symp. IMEKO Tech. Comm. on Metrology-TC8.* Budapest, 1983.

28. C. Candler: "Modern Interferometers," Hilger & Watts Ltd., Glasgow, U.K., 105, (1950).

29. Payne, B. F., and M. R. Serbyn: "An Automated System for the Absolute Measurement of Pickup Sensitivity," *Proc. 1983 Natl. Conf. Stds. Labs.* Workshop and Symposium. Part II, 11.1–11.2, Bouder, CO, July 18–21, 1983.

30. Jones, E., W. B. Yelon, and S. Edelman: *J. Acoust. Soc. Amer.*, **45**:1556 (1969).

31. Deferrari, H. A., and F. A. Andrews: "Vibration Displacement and Mode-Shape Measurement by a Laser Interferometer," *J. Acoust. Soc. Am.*, **42**:982–990 (1967).

32. Serbyn, M. R., and W. B. Penzes: *Instr. Soc. Amer. Trans.*, **21**:55 (1982).

33. Luxon, J. T., and D. E. Parker: *Industrial Lasers and Their Applications*, Prentice-Hall, Englewood Cliffs, NJ, Chap. 10, 1985.

34. Lauer, G.: "Interferometrische Verfahren zum Messen von Schwing-und Stossbewegungen," Fachkolloq. Experim. Mechanik, Stuttgart University, October 9–10, 1986.

35. Sutton, C. M.: *Metrologia*, **27**:133–138 (1990).

36. Feder, E. I., and A. M. Gillen: *IRE Trans Instr.*, **6**:1 (1957).

37. Nisbet, J. S., J. N. Brennan, and H. I. Tarpley: *J. Acoust. Soc. Amer.*, **32**:71 (1960).

38. Jones, E., S. Edelman, and K. S. Sizmore: *J. Acoust. Soc. Amer.*, **33**:1462 (1961).

39. Davies, R. M.: *Phil. Trans. A,* **240**:375 (1948).

40. Bateman, V. I., F. A. Brown, and N. T. Davie: "Isolation of a Piezoresistive Accelerometer Used in High-Acceleration Tests," *17th Transducer Workshop*, 46–65, 1994.

41. Conrad, R. W., and I. Vigness: *Proc. 8th Annu. Meet. Instr. Soc. Amer.*, Paper 11-3, (1953).

42. Bouche, R. R.: *Endevco Corp. Tech. Paper* TP 206, April 1961.

43. Kistler, W. P.: *Shock and Vibration Bull.,* **35**, pt. 4 (1966).

44. Favor, J. D.: *Shock and Vibration Bull.,* **37**, pt. 17 (1968).

45. Ramboz, J. D., and C. Federman: *Natl. Bur. Std. (U.S.) Rept.* NBSIR74-480, March 1974.

46. Dimoff, T.: *J. Acoust. Soc. Amer.,* **40**:671 (1966).

47. O'Toole, K. M., and B. H. Meldrum: *J. Sci. Instr. (J. Phys. E)(2)* **1**:672 (1968).

48. Lin, Jing: "Transverse Response of Piezoelectric Accelerometers," *18th Transducer Workshop,* Colorado Springs, CO (1995).

CHAPTER 19
VIBRATION STANDARDS AND TEST CODES

Paul H. Maedel, Jr.

INTRODUCTION

A good standard must represent a consensus of opinion among users, be simple to understand, be easy to use, and contain no ambiguities or loopholes. Any standard must contain vital information that leads to common measurement procedures and evaluation of data that are compared with agreed-upon criteria. Standards are intended (1) to set up criteria for rating or classifying the performance of equipment or material, (2) to provide a basis for comparison of the maintenance qualities of pieces of equipment of the same type, (3) to test equipment whose continuous operation is necessary for industrial or public safety, (4) to provide a basis for the selection of equipment or material, and (5) to set up procedures for the calibration of equipment. Some standards establish classifications for equipment which is being rated and indicate how measurements are to be made and how the data, so obtained, are to be analyzed; they may also indicate how the equipment is to operate during the test procedure.

This chapter is mainly concerned with standards related to machinery vibration and its classification. Other chapters in the handbook treat other types of shock and vibration standards. For example, Chap. 18 describes standards related to transducer calibration; Chap. 26 discusses standards for the design of shock-testing machines; Chap. 34 considers American Society for Testing and Materials (ASTM) standards associated with testing procedures for rubber used in shock and vibration isolators; Chap. 39 describes standards for the evaluation, testing, and use of balancing machines, and Chap. 41 refers to ASTM test methods for package cushioning materials. Standards concerning the effects of shock and vibration on humans are considered in Chap. 44.

STANDARDS ORGANIZATIONS AND GROUPS

In the field of vibration, the two recognized international organizations are the International Standards Organization (ISO), which is technology-oriented, and the

International Electrical Commission (IEC), which is product-oriented. The ISO works in cooperation with national organizations, such as the American National Standards Institute (ANSI)—a nongovernmental institution that coordinates the development of voluntary national standards in the U.S.A.; outside the U.S.A., national standards organizations usually are government institutions. Vibration standards activity in the ISO is guided by Technical Committee 108, Mechanical Vibration and Shock; vibration standards in ANSI are guided by Technical Committee S2, Mechanical Vibration and Shock. The secretariat for both organizations is held by the Acoustical Society of America. Vibration standards activity in the IEC is guided by Technical Committee 50, and in the U.S.A. by the Institute of Environmental Sciences. IEC works with trade associations, such as the National Electrical Manufacturers Association. The American Society for Testing and Materials is a nonprofit organization that establishes standard tests and specifications; such tests and specifications usually are referred to by the abbreviation ASTM followed by a numerical designation. A number of trade organizations that have adopted formal vibration standards include the American Petroleum Institute, the Hydraulic Institute, and the Compressed Air and Gas Association. Various technical societies have been instrumental in the development of codes and standards concerned with vibration; such documents have been established by consensus of consumers and manufacturers and their use is voluntary. In addition, the Department of Defense has developed standards and specifications to ensure the quality of equipment which it procures.

MACHINERY VIBRATION STANDARDS AND CRITERIA

TYPES OF MACHINERY

Machinery can be subdivided into four basic categories for purposes of vibration measurement and evaluation:

1. *Reciprocating machinery having both rotating and reciprocating components,* such as diesel engines and certain types of compressors and pumps. Vibration is usually measured on the main structure of the machine at low frequencies.

2. *Rotating machinery having rigid rotors,* such as certain types of electric motors, single-stage pumps, and slow-speed pumps. Vibration is usually measured on the main structure (such as on the bearing caps or pedestals) where the vibration levels are indicative of the excitation forces generated by the rotor because of unbalance, thermal bows, rubs, and other sources of excitation.

3. *Rotating machinery having flexible rotors,* such as large steam turbine generators, multistage pumps, and compressors. The machine may be set into different modes of vibration as it accelerates through one or more critical speeds to reach its service speed. On such a machine, the vibration amplitude measured on a structure member may not be indicative of the vibration of the rotor. For example, a flexible rotor may experience very large amplitude displacements resulting in failure of the machine even through the vibration amplitude measured on the bearing cap is very low. Therefore, it is essential to measure the vibration on the shaft directly.

4. *Rotating machinery having quasi-rigid rotors,* such as low-pressure steam turbines, axial-flow compressors, and fans. Such machinery contains a special class of

flexible rotor where vibration amplitudes measured on the bearing cap are indicative of the shaft vibration.

CLASSIFICATION OF SEVERITY OF MACHINERY VIBRATION

In the classification of severity of machinery vibration, the motion variable that is used (vibration displacement, velocity, or acceleration) depends on the type of standard, the frequency range, and other factors. In classifying machinery vibration in the range from 10 to 1000 Hz, vibration velocity often is used because it is relatively independent of frequency in this range of frequencies and thus yields a simple measure of severity of a new or operating machine.

For simple harmonic motion, either peak or rms values of the motion variable may be used; however, for machines whose motion is complex, the use of these two indices provides distinctly different results, mainly because the higher-frequency harmonics are given different weights. For rotating machinery whose rotational speed is in the range of 600 to 12,000 rpm, the rms value of the velocity amplitudes corresponds most closely with vibration severity. Therefore, the International Standards Organization has a special measure, *vibration severity,* which is defined as the highest value of the broad-band, root-mean-square value of the velocity amplitude in the frequency range from 10 to 1000 Hz as evaluated on the structure at prescribed points (generally triaxial arrays on the bearing caps or pedestals).

ROTATING MACHINERY HAVING RIGID ROTORS

ISO standard 2372, "Mechanical Vibration of Machines with Operating Speeds from 10 to 200 rps—Basis for Specifying Evaluation Standard"[1] applies to rotating machinery having rigid rotors and to those machines having flexible rotors in which bearing cap vibration is a measure of the shaft motion. Vibration severity (defined above) covers those frequencies between 30 percent of the rotational frequency to at least 3 times the rotational frequency; it includes the most common causes of rotating machinery vibration resulting from (1) nonsynchronous excitation such as rubs and rotor whirl, (2) rotor unbalance, (3) electric field excitation and harmonics thereof, and (4) harmonics of synchronous rotor excitation.

Table 19.1 lists the allowable vibration severity and examples of the following classes of rotating machinery:

Class I. Individual components, integrally connected with the complete machine in its normal operating conditions (i.e., electric motors up to 15 kilowatts).

Class II. Medium-sized machines (i.e., 15- to 75-kilowatt electric motors and 300-kilowatt engines on special foundations).

Class III. Large prime movers mounted on heavy, rigid foundations.

Class IV. Large prime movers mounted on relatively soft, light-weight structures.

Vibration severity is divided into four ranges, labeled A (the smoothest) through D (the roughest). The particular range selected by the user is based upon a number of considerations: (1) type and size of machine, (2) type and service expected, (3) mounting system, and (4) effect of machinery vibration on the surrounding environment (e.g., instruments, adjacent equipment, and personnel). Machinery vibration is generally considered to be of equal severity if it exhibits the same rms velocity amplitude in the frequency range from 10 to 1000 Hz.

TABLE 19.1 Vibration Severity Criteria
(After ISO IS 2372.[1])

RMS velocity ranges of vibration severity		Vibration severity* for separate classes of machines			
mm/sec	in./sec	Class I	Class II	Class III	Class IV
0.28	0.01				
0.45	0.02	A			
0.71	0.03		A		
1.12	0.04	B		A	
1.8	0.07		B		A
2.8	0.11	C		B	
4.5	0.18		C		B
7.1	0.28	D		C	
11.2	0.44		D		C
18	0.71			D	
28	1.10				D
45	1.77				

* The letters A, B, C, and D represent machine vibration quality grades, ranging from good (A) to unacceptable (D).

An ISO standard for classifying the vibration severity of large rotating machines in situ, operating at speeds from 600 to 1200 rpm, is shown in Table 19.2. It applies to large prime movers, Class III and Class IV as defined above. The severity rating of a machine depends on the classification of the supports in the mounting system for the machine. The supports are said to be *soft* if the fundamental frequency of the machine on its support is lower than its main excitation frequency; the supports are said to be *hard* if the fundamental frequency of the machine on its supports is higher than its main excitation frequency.

TABLE 19.2 Quality Judgment of Vibration Severity
(After ISO IS 3945.[2,3])

rms velocity vibration severity		Support classification	
mm/sec	in./sec	Rigid supports	Flexible supports
0.46	0.018		
0.71	0.028	Good	
1.12	0.044		Good
1.8	0.071		
2.8	0.11	Satisfactory	
4.6	0.18		Satisfactory
7.1	0.28	Unsatisfactory	
11.2	0.44		Unsatisfactory
18.0	0.71		
28.0	1.10	Unacceptable	
46.0	1.80		
71.0	2.80		Unacceptable

EVALUATION OF MACHINERY VIBRATION BY MEASUREMENTS ON NONROTATING PARTS

A series of standards, outlined in Table 19.3, describe procedures for the evaluation of vibration based on measurements made on rotating parts of a machine. This series provides an individual standard for general classes of machines and defines the specific information and criteria that are unique to those machines. The general criteria, which are presented in terms of both vibration magnitude and changes in magnitude, relate to both operational monitoring and acceptance testing. Criteria are provided primarily with regard to securing reliable, safe, long-term operation of the machine, while minimizing adverse effects on associated equipment.

ROTATING MACHINES HAVING FLEXIBLE ROTORS

A rotating machine that has a casing which is relatively stiff and/or heavy in comparison with its rotor mass often can be considered as having a flexible rotor shaft. In such a case, vibration conditions may be evaluated with greater sensitivity if measurements are conducted on the rotating element rather than on stationary members of the machine. Thus the international standards outlined in Table 19.4 are preferable to those given in Refs. 1 or 2, which may not adequately characterize the running condition of the machine—although measurements made in accordance with the latter standards may be useful.

ELECTRIC MOTORS AND GENERATORS

International Standards Organization (ISO), American National Standards Institute (ANSI), National Electrical Manufacturers Association (NEMA), and American Petroleum Institute (API) standards establish criteria classification systems for permissible vibration in electric motors. These classification systems are not identical; some are based on peak-to-peak shaft displacement, whereas others are based either on rms or peak velocity amplitude measured on the structure (bearing housing or pedestal).[14] Each standard specifies the test arrangement and procedure, including machine mounting arrangement, instrumentation, and method of test.

TABLE 19.3 Mechanical Vibration—Evaluation of Machine Vibration by Measurements on Nonrotating Parts. ISO/DIS 10816 (ANSI Counterpart S2.12, -199X)

Standard	Guidelines for
Part 1[4]	General procedures for various classes of machines, based on measurements made on nonrotating parts
Part 2[5]	Land-based steam-turbine sets in excess of 50 megawatts
Part 3[6]	Coupled industrial machines with nominal power above 30 kilowatts; speeds between 120 and 15,000 rpm
Part 4[7]	Gas turbine-driven sets excluding aircraft derivative
Part 5[8]	Hydraulic machines with nominal power above 1 megawatt; speeds between 120 and 1800 rpm

TABLE 19.4 Mechanical Vibration of Non-Reciprocating Machines—Measurements on Rotating Shafts and Evaluation. ISO 7919 (ANSI Counterpart S2.44, 199X)

Standard	Guidelines for
Part 1[9]	General procedures for various classes of machines
Part 2[10]	Large land-based steam, turbine-generating sets
Part 3[11]	Coupled industrial machines with fluid bearings
Part 4[12]	Industrial gas turbines with power outputs greater than 3 megawatts
Part 5[13]	Hydraulic machine sets with fluid-film bearings with power outputs above 1 megawatt

ISO Standard 2373.[15] This is a special adaption of ISO 2372 for electrical motors and applies to three-phase ac motors and dc motors with shaft heights (i.e., the vertical distance from the base of the motor to the centerline of the shaft) between 80 and 400 mm. (It does not apply to motor converters, single-phase machines, or three-phase machines operated on single-phase systems.) The criterion for vibration severity (the same as for ISO 2372) is given in terms of the rms value of velocity amplitude in the frequency range from 10 to 1000 Hz when measured with instrumentation which meets the requirements of ISO 2954. Measurements are made on the machine installed on a free suspension (i.e., suspended from or mounted on an elastic support such as springs or rubber). The motor is operated at rated voltage and nominal frequency (for ac motors) and at its nominal speed. (For machines with several speeds or variable speeds, the tests are carried out at the various operational speeds.) Unless otherwise specified, measurements of vibration severity should be carried out under no-load operation at the temperature reached by the motor after a sufficient period of no-load operation. Table 19.5 lists the recommended limits of vibration severity for various size motors.

TABLE 19.5 Recommended Limits of Vibration Severity for Electric Motors
(After ISO IS 2373.[15])

Quality grade	Speed, rpm	Velocity amplitude (maximum rms values) for the following shaft heights h, in mm*					
		$80 < h < 132$		$132 < h < 225$		$225 < h < 400$	
		mm/s	in./sec	mm/s	in./sec	mm/s	in./sec
N (normal)	600 to 3,600	1.8	0.071	2.8	0.110	4.5	0.177
R (reduced)	600 to 1,800	0.71	0.028	1.12	0.044	1.8	0.071
	>1,800 to 3,600	1.12	0.044	1.8	0.071	2.8	0.110
S (special)	600 to 1,800	0.45	0.018	0.71	0.028	1.12	0.044
	>1,800 to 3,600	0.71	0.028	1.12	0.044	1.8	0.071

  * A single set of values, such as those applicable to the 132- to 225-mm shaft height, may be used if shown by experience to be required. The shaft height is the vertical distance from the base of the motor to the centerline of the shaft.

NEMA Standard MG1-1993, Revision 1[16]—Motors and Generators. This standard is applicable to dc machines tested with dc power and polyphase ac machines tested with sinusoidal power, in frame sizes 42 and larger and at rated power up to 100,000 hp or 75 megawatts, at nominal speeds up to and including 3600 rpm.

This standard establishes the limits of the broad-band maximum relative shaft displacement (peak-to-peak) for standard motors and generators; it also establishes the limits of broad-band maximum relative shaft displacement (peak-to-peak) for special types of machines. In addition, it establishes the limits of vibration velocity as measured on the bearing housing of resiliently mounted motors, including broad-band and filtered measurements. For motors tested on rigid mounts, these values should be reduced by multipling them by 0.8.

API Standard 541, 3d ed., December, 1994[17]—Form-Wound Squirrel-Cage Induction Motors—250 hp and Larger. This standard establishes the vibration limits for special-purpose motors with sleeve bearings as measured on the bearing housing.

AMERICAN PETROCHEMICAL INDUSTRY MACHINERY

Table 19.6 lists API standards for vibration criteria which apply to various types of machines used in the petrochemical industry, such as steam turbines of relatively large horsepower, centrifugal compressors, air blowers, reciprocating compressors, and generators.

TABLE 19.6 Vibration Criteria Standards for Machines Used in the American Petochemical Industry (API)

(After Refs. 17–22.)

Standards	Type of machine
API Std. 612, 4th ed., 1994	Steam turbines
API Std. 613, 4th ed., 1994	Gear units
API Std. 617, 6th ed., 1994	Centrifugal compressors
API Std. 619, 3d ed., 1995	Rotary-type positive displacement compressors
API Std. 610, 8th ed., 1994	Pumps
API Std. 541 3d ed., 1994	Special-purpose motors

RECIPROCATING MACHINERY

The standard ISO/DIS 10816[23] (ANSI Counterpart S2.12, Part 6-199X), Reciprocating Machines with Power Ratings Above 1 Megawatt, establishes procedures and guidelines for the measurement and classification of mechanical vibrations of reciprocating machines. In general, this standard refers to vibration measurements made on the main structure of the machine. Vibration measurement and evaluation criteria are provided in this standard which are recommended to secure a reliable and safe operation of a machine and to avoid problems with the auxiliary equipment mounted on the structure.

The vibrations measured on the main structure of the machine may only give a rough idea of the stresses and vibratory states of the components within the machine itself. For example, torsional vibrations of rotating parts cannot generally be determined by measurements on the structure of the machine. Damage, which can occur when exceeding the guide values based on experience with similar machines, is sustained predominately by machine-mounted components (e.g., turbo-chargers, heat exchangers, governors, pumps, and filters), connecting elements of mounted components (e.g., turbo-chargers, heat exchangers, governors, pumps, filters, etc.), connecting elements (such as pipelines), or monitoring instruments (e.g., pressure gauges, thermometers).

GEAR UNITS

International Standard ISO/WD 8579-2 provides a method for determining the mechanical vibration of enclosed gear units which are individually housed. This standard applies to a gear unit under test and operating within its design speed, load, temperature range, and lubrication specifications for acceptance testing.[24] Two types of vibration measurements are made on gear units which operate with oil-film journal bearings: (1) shaft vibration and (2) housing vibration. Proximity-probe transducers usually are used to measure the peak-to-peak value of the shaft's displacement relative to the housing, in the frequency range from 0 to 500 Hz. The transducers are located as close as practical to a bearing; measurements should be made in three orthogonal diections, one of which is parallel to the shaft axis. Only one axial transducer per shaft is necessary. Housing vibration is measured with a seismic transducer in the frequency range from 10 to 10,000 Hz. A classification system for gearboxes on the basis of shaft displacement and housing velocity measurements is given in this standard, which also provides a subjective assessment of the acceptable vibration rating for typical applications during acceptance testing at the manufacture's facility. The vibration of a properly manufactured gear unit will vary according to the particular design, size, and application.

AIR AND GAS CENTRIFUGAL COMPRESSORS

Vibration criteria have been established by the Compressed Air and Gas Institute[25] for single and multistage in-line centrifugal and axial compressors in air and gas service for rotational speeds up to 15,000 rpm. In general, a change in vibration displacement amplitude of a compressor in operating service is of more significance than its absolute level. Hence any significant change in vibration, for the same operating conditions, should be immediately investigated and the cause of the change determined. Such changes in vibration can be monitored by the methods described in Chap. 16.

PUMPS

The Hydraulic Institute[26] specifies acceptable limits of vibration for horizontal clear-liquid pumps and horizontal nonclog pumps. These limits are applicable under the following operating conditions: (1) Pumps must be operating in a noncavitating and nonseparating condition, (2) suction piping must be arranged so as to provide a straight uniform flow to the pump, (3) piping must be connected in such a way so as to avoid undue strain on the pump, and (4) shaft couplings must be aligned to within the manufacturer's recommendations. If a pump produces vibration amplitudes in excess of the values specified, it should be examined for defects or possible correction. Often, more important than the actual vibration amplitude itself is the change in vibration amplitude over a period of time. Vibration amplitudes in excess of the specified limits may be acceptable if there is no increase over long periods of time and if there is no other indication of damage, such as an increase in bearing clearance or noise level.

SHIPBOARD VIBRATION

Hull vibration is excited by the propeller shaft and the propeller blade. Such vibration is greatest in the afterpart of the ship (except when one of the hull resonance frequencies coincides with a driving frequency). The displacement amplitudes often exceed 0.1 in. (2.5 mm). At higher speeds of the propeller shaft, the amplitudes of the high modes of vibration of the hull tend to be smaller. Vibration data indicate that the maximum displacement amplitude of the main structural members of the hull (measured at the antinodes and at the end of the ship) is smaller at the higher excited frequencies. When shipboard machinery and equipment are fastened to the hull, deck, and bulkhead stiffeners, the vibration amplitudes at the equipment sup-

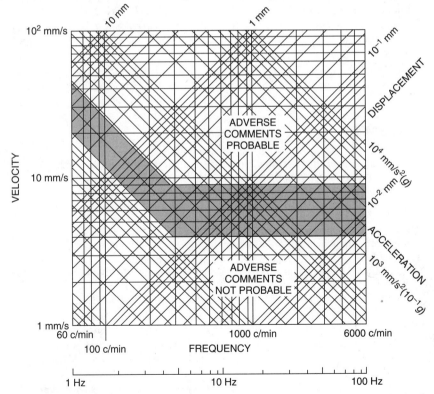

FIGURE 19.1 Guidelines for the evaluation of vertical and horizontal peak values of vibration in merchant ships. (*ISO 4867, after ISO 6954.*[27])

ports are substantially the same as the main structural members of the hull. The amplitudes of vibration of the bow and stern are sometimes used as a measure of hull vibration. Acceptable values for turbine and diesel-driven merchant ships of 100 meters or more in length are considered in ISO 6945.[27] The hatched zone in Fig. 19.1 shows common values for shipboard environments considered to be acceptable. In this connection, ISO 2631,[28] ISO 4867,[29] and ISO 4868,[30] are pertinent references.

Vibration test requirements for shipboard equipment and machinery components to ensure consistency in vibration resistance requirements for such equipment and machinery components is given in ISO/DIS 10055.[31] The tests are intended to locate resonances of the equipment and impose endurance tests at these frequencies, if any.

Vibration standards are used by the U.S. Navy, as well as other navies, to ensure structural integrity of a ship and to minimize undesirable radiation of noise. One such specification requires that the residual vibration of shipboard vibration of shipboard-mounted equipment not exceed specified minimum values is MIL-STD-167-1.[32]

STANDARDS SOURCES

Requests for ISO and ISO/DIS Standards should be addressed to the Director of Publications, American National Standards Institute, New York, NY 10036.

Requests for ANSI Vibration Standards should be addressed to Acoustical Society of America, Standards and Publications Fulfillment Center, P.O. Box 1020, Sewickley, PA 15143-9998.

Inquiries regarding ASA Standards should be sent to Standards Manager, ASA Standards Secretariat, 120 Wall Street, 32d Floor, New York, NY 10005-3993.

REFERENCES

1. International Standards Organization: "Mechanical Vibration of Machines with Operating Speeds from 10 to 200 rps.—Basis for Specifying Evauation Standards," ISO 2372.

2. International Standards Organization: "The Measurement and Evaluation of Vibration Severity of Large Rotating Machines, in Situ; Operating at Speeds from 10 to 200 rps," ISO 3945.

3. International Standards Organization: "Mechanical Vibration of Rotating and Reciprocating Machinery—Requirements for Instruments for Measuring Vibration Severity," ISO 2954.

4. International Standards Organization: "Mechanical Vibration—Evaluation of Machine Vibration by Measurements on Non-Rotating Parts," ISO/DIS 10816 Series: ISO/DIS 10816/1: "General Guidelines."

5. ISO/DIS 10816/2: "Guidelines for Land-Based Steam Turbines Sets in Excess 50 MW."

6. ISO/DIS 10816/3: "Guidelines for Coupled Industrial Machines with Nominal Power Above 30 kW and Nominal Speeds Between 120 and 15,000 rpm, When Measured in Situ."

7. ISO/DIS 10816/4: "Guidelines for Gas Turbine Driven Sets Excluding Aircraft Derivative."

8. ISO/DIS 10816/5: "Guidelines for Hydraulic Machines with Nominal Speeds Between 120 and 1800 rpm, measured in Situ."

9. International Standards Organization: "Mechanical Vibration of Non-Reciprocating Machines—Measurements on Rotating Shafts and Evaluation," ISO 7919 Series ISO 7919/1: "General Guidelines."

10. ISO 7919/2: "Guidelines for Large Land-Based Steam Turbine Generating Sets."

11. ISO 7919/3: "Guidelines for Coupled Industrial Machines."

12. ISO 7919/4: "Guidelines for Gas Turbines."

13. ISO 7919/5: "Guidelines for Hydraulic Machine Sets."

14. International Standards Organization: "Rotating Shaft Vibration Measuring Systems, Part 1: Relative and Absolute Signal Sensing of the Radial Vibration from Rotating Shafts," ISO/CD 10817-1.

15. International Standards Organization: "Mechanical Vibration of Certain Rotating Electrical Machinery with Shaft Heights Between 80 and 400 mm—Measurement and Evaluation of Vibration Severity," ISO 2373.

16. National Electrical Manufacturer's Association: "Motors and Generators, Part 7— Mechanical Vibration—Measurement, Evaluation and Limits," MG1-1993, Rev. 1

17. American Petroleum Institute: Form-Wound Squirrel Cage Induction Motors—250 hp and Larger, API STD 541, 3d ed., 1994.

18. American Petroleum Institute: "Special Purpose Steam Turbines for Petroleum, Chemical, and Gas Industry Services," API STD 612, 4th ed., December 1994.

19. American Petroleum Institute: "Special Purpose Gear Units for Refinery Service," API STD 613, 4th ed., December 1994.

20. American Petroleum Institute: "Centrifugal Compressors for General Refinery Service," API STD 617, 5th ed., April 1994.

21. American Petroleum Institute: "Rotary-Type Positive Displacement Compressors for General Refinery Services," API STD 619, 3d ed., May 1995.

22. American Petroleum Institute: "Centrifugal Pumps for General Refinery Services," API STD 610, 8th ed., December 1994.

23. International Standards Organization: "Mechanical Vibration—Evaluation of Machine Vibration by Measurements on Non-Rotating Parts, Part 6: Reciprocating Machines with Power Ratings Above 100 kW," ISO/DIS 10816/6.

24. International Standards Organization: "Acceptance Code for Gears, Part 2: Determination of Mechanical Vibration of Gear Units During Acceptance Testing," ISO/WD 8579-2.

25. Compressed Air and Gas Institute, Cleveland, Ohio: "In-Service Standards for Centrifugal Compressors," 1963.

26. Hydraulic Institute: "Acceptable Field Vibration Limits for Horizontal Pumps," 14th ed., Centrifugal Pumps Applications.

27. International Standards Organization: "Mechanical Vibration and Shock—Guidelines for the Overall Evaluation of Vibration in Merchant Ships," ISO 6954-1984.

28. International Standards Organization: "Guide for the Evaluation of Human Exposure to Whole-Body Vibration," ISO 2631.

29. International Standards Organization: "Code for the Measurement and Reporting of Shipboard Vibration Data," ISO 4867.

30. International Standards Organization: "Code for the Measurement and Reporting Shipboard Local Vibration Data," ISO 4868.

31. International Standards Organization: "Mechanical Vibration—Vibration Testing Requirements for Shipboard Equipment and Military Components," ISO/DIS 10055.

32. Military Standard: "Mechanical Vibration of Shipboard Equipment (Type 1—Environmental and Type 2—Internally Excited)," MIL-STD-167-1 (SHIPS), May 1, 1974.

CHAPTER 20

TEST CRITERIA AND SPECIFICATIONS

Allan G. Piersol

INTRODUCTION

This chapter covers the development of shock and vibration test criteria for mechanical, electrical, electronic, or hydraulically powered equipment, for example, an alternator for an automobile or an electronic instrument for an airplane. The emphasis throughout is on the selection of test criteria rather than the formulation of design criteria, but specified shock and vibration test levels and durations are commonly used as design criteria as well. Following a brief overview of environmental specifications, this chapter presents (1) a summary of the descriptions of shock and vibration environments used to establish test criteria, (2) a discussion of the different types of tests used to achieve various objectives, (3) procedures to select shock and vibration test levels, (4) procedures to select vibration test durations, and (5) general testing considerations.

ENVIRONMENTAL SPECIFICATIONS

An *environmental specification* is a written document that details the environmental conditions under which an item of equipment to be purchased must operate during its service life. Several contracting agencies of the U.S. government and various professional societies issue general environmental specifications for particular classes of equipment (see Chap. 19), but deviations from the specified environmental conditions in such documents are permitted when more appropriate conditions can be established by direct measurements or predictions of the environments of concern. An *environmental test specification* is a written document that details the specific criteria for an environmental test, as well as other matters such as the preparation of the test item, identification of all test equipment and instrumentation, description of any test fixtures, instructions for mounting sensors, step-by-step procedures for operating the test item (if operation is required), procedures for taking data on the test item function and the applied environment, and performance acceptability criteria. The test criteria (the magnitude and duration of the test excitation) in environmental test specifications often serve as design criteria as well (see Chap. 43).

GENERAL TYPES OF ENVIRONMENTS

The environments that must be considered in equipment design and testing are listed in Table 20.1. Those printed in boldface, namely, shock and vibration, are the ones of special concern in this handbook. Shock and vibration environments may result from the equipment operation (for example, the vibration caused by shaft unbalance in equipment with a rotating element), but it is the external shock and vibration motions transmitted into the equipment through its mounting points to the structure of the system incorporating the equipment that are of primary interest here. The acoustical, blast, fluid flow, and wind environments noted in Table 20.1 are often the original source of the shock and vibration motions of the system structure that transmit into the equipment, but the original source may also be a direct motion input to the system, for example, earthquake inputs to a building or road roughness inputs to an automobile. Such environments have complicated transmission patterns that are modified or intensified by mechanical resonances of the system structure and, therefore, are appropriately described by frequency-dependent functions, i.e., spectra.

TABLE 20.1   Various Types of Environments to Which Equipment May Be Exposed

Acceleration (sustained)	Fungus	Salt spray
Acoustical noise	Humidity	Temperature (sustained)
Blast	**Mechanical shock**	Temperature cycling
Dust and sand	Pressure (sustained)	**Vibration**
Fluid flow	Rain, hail, and snow	Wind

In practice, for economy of effort, equipment is often designed and tested for exposure to each of the environments listed in Table 20.1 as if they occur separately. However, some of the environments in Table 20.1 may occur simultaneously and have an additive effect; for example, a shock may occur during a period of high static acceleration where the stress in the equipment due to the combination of the two environments is greater than the stress due to either applied separately. Worse yet, two environments may have a synergistic effect; for example, equipment may be subject to high vibration during a period when the temperature exposure is also high, and high temperatures cause a degradation of the equipment strength, making it more vulnerable to vibration-induced failures. These matters must be carefully evaluated during the definition of a test program to determine if simultaneous testing for two or more environments is required.

SHOCK AND VIBRATION ENVIRONMENTS

From a testing viewpoint, it is important to carefully distinguish between a shock environment and a vibration environment. In general, equipment is said to be exposed to *shock* if it is subject to a relatively short-duration (transient) mechanical excitation; equipment is said to be exposed to *vibration* if it is subject to a longer-duration mechanical excitation. If the vibration has average properties that are time-invariant, it is called *stationary* (or steady-state for periodic vibrations). However, vibration environments are often *nonstationary,* i.e., they vary with time. If the average properties of a nonstationary vibration environment do not change too rapidly with time, the environment can be analyzed as if it were stationary, and tests can be performed using stationary excitations. Otherwise, the environment must be viewed as a shock. Practical distinctions between shock environments and vibration environments cannot be made on an absolute basis, independent of the equipment exposed to the environment. To be more specific, any mechanical device that is more or less linear can be characterized by one or more resonance frequencies and damping coefficients (see Chap. 2) or by a corresponding set of decaying transient responses after a momentary excitation. In more analytical terms, the response characteristics of a mechanical device are given by the *unit impulse-response function* defined in Chap. 21. From a testing viewpoint, an excitation whose duration is comparable to, or less than, the response (or decay) time of the equipment is considered a shock, while an excitation whose duration is long compared to the response time of the equipment is considered a vibration.

DESCRIPTIONS OF SHOCK AND VIBRATION ENVIRONMENTS

The response of equipment to shock and vibration at its mounting points is dependent on frequency. Hence, shock and vibration environments are usually described by some type of spectrum; a *spectrum* is a description of the magnitude of the frequency components that constitute the shock or vibration. The most common spectral descriptions of both deterministic and random shock and vibration environments are summarized in Table 20.2. It is common to present data for test specification purposes in terms of acceleration, primarily because it is convenient to measure acceleration with accelerometers described in Chap. 12. From an analytical viewpoint, a spectrum in terms of velocity is preferred to a spectrum in terms of acceleration because velocity has a direct linear relationship to stress.[2] Nevertheless, the use of acceleration as a parameter is not a problem in specifying test criteria as long as the criteria simulate the spectrum of the environment, and acceleration is used for both the environmental description and the test criteria. However, for more analytical applications of shock and vibration data, spectral descriptions using a velocity parameter may become desirable.

TABLE 20.2 Common Spectral Descriptions of Shock and Vibration Environments

Environment	Characteristic	Spectral Description
Shock	Deterministic	Fourier (integral) spectrum (see Chap. 23)
		Shock (response) spectrum (see Chaps. 8 and 23)
	Random	Energy spectral density (see Ref. 1)
		Shock (response) spectrum (see Chaps. 8 and 23)
Vibration	Deterministic	Line spectrum (see Chap. 22)
	Random	Power spectral density (see Chaps. 11 and 22)

The vibration environment for an item of equipment usually varies in magnitude and spectral content during its service life. Similarly, a shock environment may involve repetitive shocks with different magnitudes and spectral content. For reliability tests discussed later in this chapter, it may be necessary to measure or predict the spectra of the shock and/or vibration environment for all conditions (or a representative sample thereof) throughout the service life and to formulate test criteria that require a series of tests with several different magnitudes and spectral content. For most testing applications, however, a test involving a single spectrum is desired for convenience. To assure that the test produces a conservative result, a maximax spectrum is used; a *maximax spectrum* is the envelope of the spectra for all conditions throughout the service environment. Thus, the maximax spectrum may not equal any of the individual spectra measured or predicted during the service environment, since the maximum value at two different frequencies may occur at different times.

TYPES OF SHOCK AND VIBRATION TESTS

An *environmental test* is any test of a device under specified environmental conditions (or sometimes under the environment generated by a specified testing machine) to determine whether the environment produces any deterioration of performance or any damage or malfunction of the device; an environmental test may also be distinguished by the objectives of the test. In assessing the effects of shock and vibration on equipment, the types of tests most commonly performed fall into the following categories:

1. Development
2. Qualification
3. Acceptance
4. Screening
5. Statistical reliability
6. Reliability growth

DEVELOPMENT TESTS

A *development test* (sometimes called an *analytical test*) is a test performed early in a program to facilitate the design of a device or piece of equipment to withstand its anticipated service environments. It may involve determining the resonance frequency of a constituent component mounted inside the equipment by applying a sinusoidal excitation with a slowing-varying frequency (often called a *swept sine wave* test). Sinusoidal vibration is widely used as the excitation for development tests because of its simplicity and well-defined deterministic properties. In contrast, it may involve a more elaborate test to determine the normal modes and damping of the equipment structure as described in Chap. 21. A stationary random vibration or a controlled shock excitation with appropriate data reduction software can greatly reduce the time required to perform a more extensive modal analysis of the equipment. In either type of test, the characteristics and magnitude of the excitation used for the test are not related to the actual shock and/or vibration environment to which the equipment is exposed during its service use.

QUALIFICATION TESTS

A *qualification test* is a test intended to verify that an equipment design is satisfactory for its intended purpose in the anticipated service environments. Such a test is commonly a contractual requirement, and hence, a specific test specification is usually involved. Preliminary qualification tests are sometimes performed on prototype hardware to identify and correct design problems before the formal qualification test is performed. Also, qualification test requirements might be based upon a general environmental specification, for example, for aircraft equipment, MIL-STD-810.[3] In some cases, the specification may require a test on a specific type of testing machine that produces a desired qualification environment (see Chap. 26). However, contracts usually allow deviations from the specified test levels and/or test durations in general environmental specifications, if it can be established that different test conditions would be more suitable for the given equipment. In any case, the basic purpose of a qualification test requires that the test conditions conservatively simulate the basic characteristics of the anticipated service environments.

Some years ago, when test facilities were more limited, it was argued that shock and vibration environments for equipment could be simulated for qualification test purposes in terms of the damaging potential of the environment, without the need for an accurate simulation of the detailed characteristics of the environment.[4] For example, it was assumed that random vibration could be simulated with sinusoidal vibration designed to produce the same damage. The validity of such "equivalent damage concepts" requires the assumption of a specific damage model to arrive at an appropriate test level and duration. Since the assumed damage model might be incorrect for the equipment of interest, there is a substantial increase in the risk that the resulting test criteria will severely under- or overtest the equipment. With the increasing size and flexibility of modern test facilities, the use of equivalent damage concepts to arrive at test criteria is rarely required and should be avoided, although equivalent damage concepts are still useful in arriving at criteria for "accelerated tests," as discussed later in this chapter. When ever feasible, *qualification tests should be performed using an excitation that has the same basic characteristics as the environment of concern; for example, random vibration environments should be simulated with random vibration excitations, shock environments should be simulated with shock excitations of similar duration, etc.*

ACCEPTANCE TESTS

An *acceptance test* (sometimes called a *production test* or a *quality control test*) is a test applied to production items to help ensure that a satisfactory quality of workmanship and materials is maintained. For equipment whose failure in service might result in a major financial loss or a safety threat, all production items are subjected to an acceptance test. Otherwise, a statistical sample of production items is selected, and each item is tested in accordance with an acceptance sampling plan that assures an acceptable average outgoing quality. In either case, there are two basic approaches to acceptance testing for shock and vibration environments. The first approach is to design a test that will quickly reveal common workmanship errors and/or material defects as determined from prior experience and studies of failure data for the equipment, independent of the characteristics of the service environment. For example, suppose a specific type of electrical equipment has a history of malfunctions induced by scrap-wire or poorly soldered wire junctions. Then, the application of sinusoidal vibration at the resonance frequencies of wire bundles will quickly reveal such problems and, hence, constitute a good test excitation even though there may be no sinusoidal vibrations in the service environment. The second and more common approach is to apply an excitation that simulates the shock and/or vibration environments anticipated in service, similar to the qualification test but usually at a less conservative (lower) level.

SCREENING TESTS

A *screening test* is a test designed to produce incipient failures that would otherwise occur later during service use so that they can be corrected before delivery of the equipment, i.e., to detect workmanship errors and/or material defects that will not cause an immediate failure but will cause a failure before the equipment has reached its design service life. Screening tests are similar to acceptance tests but usually are more severe in level and/or longer in duration. If performed at all, screening tests are applied to all production items. Vibration screening tests, commonly referred to as environmental stress screening (ESS) tests, are sometimes performed using relatively inexpensive, mechanically or pneumatically driven vibration testing machines that allow little or no control over the spectrum of the excitation (see Chap. 25). Hence, except perhaps for the overall level, the screening test environment generally does not represent an accurate simulation of the service environment for the equipment.

STATISTICAL RELIABILITY TESTS

A *statistical reliability test* is a test performed on a large sample of production items for a long duration to establish or verify an assigned reliability objective for the equipment operating in its anticipated service environment, where the reliability objective is usually stated in terms of a mean-time-to-failure (MTTF), or if all failures are assumed to be statistically independent, a mean-time-between-failures (MTBF) or failure rate (the reciprocal of MTBF). To provide an accurate indication of reliability, such tests must simulate the equipment shock and vibration environments with great accuracy. In some cases, rather than applying stationary vibration at the measured or predicted maximax levels of the environment, even the nonstationary characteristics of the vibration are reproduced, often in combination with shocks and other environments anticipated during the service life. The determination of reliability is accomplished by evaluating the times to individual failures, if any, by conventional statistical techniques.[5]

RELIABILITY GROWTH TESTS

A *reliability growth test* is a test performed on a few prototype or early production items over an extended period of time to identify failures that might be prevented by minor design changes. The failure rate of the equipment is monitored by either statistical reliability tests in the laboratory or evaluations of failure data from service experience to verify an improvement in reliability due to the design changes. As in statistical reliability tests, reliability growth tests are usually designed to simulate the service environments with great accuracy. In some cases, the testing machines used for environmental stress screening (ESS) tests are used in reliability growth tests to accelerate the occurrence of failures. However, this approach assumes the ESS test will produce the same types of failures that will occur in the service environment. This assumption may not be true, as discussed later in this chapter.

SELECTION OF SHOCK AND VIBRATION TEST LEVELS

The *test level* for a shock or vibration test is the spectrum of the excitation applied to the equipment at its mounting points by the test machine. For tests that require a simulation of the actual service shock and vibration environments (qualification, reliability, and some acceptance tests), the selection of test levels involves four steps, as follows:

1. Measurement or prediction of spectra for shock and vibration environments
2. Grouping of measured or predicted spectra into appropriate zones
3. Determination of zone limits
4. Selection of specified test levels

MEASUREMENT OR PREDICTION OF SPECTRA

Where equipment is to be installed in an existing system (for example, a new alternator for an existing automobile), the shock and/or vibration response of the system structure at the mounting points of the equipment can be determined by direct measurements (see Chap. 15). However, where equipment is to be installed in a system that has not yet been built and/or operated, the shock and/or vibration environment at the equipment mounting points must be predicted. Procedures for the prediction of shock and vibration environments vary widely depending upon the characteristics of environment and the system producing it. In general, however, prediction procedures can be divided into the following broad categories:

Analytical Modeling Procedures.   In the absence of any information other than the basic design and performance characteristics of the system containing the equipment, shock and vibration environments can be predicted, at least coarsely, by the analytical modeling procedures detailed in other chapters of this handbook. However, such analytical procedures generally produce reliable results only at frequencies below about the fiftieth normal mode of the system structure that supports the equipment.

Finite Element Method (FEM) Procedures.   A somewhat different approach to the prediction of shock and vibration environments is to use the finite element method (FEM) to model the system (see Chap. 28, Part II) and to predict its structural response at the mounting points of the equipment. This can be done in one of two ways. First, the FEM model might be used only to determine the normal mode shapes and frequencies for the system structure, where these normal mode data are then used to predict the responses of the system structure at the equipment mounting points by analytical procedures (see Chap. 7). The second approach is to input the FEM model with the appropriate excitations for the service environment and compute the responses directly from the FEM model. In either case, reliable predictions can be made only at frequencies below about the fiftieth normal mode of the system structure.

Statistical Energy Analysis (SEA) Procedures.   At frequencies above the range where analytical and finite element method procedures are accurate, statistical energy analysis (SEA) procedures described in Chap. 11 are commonly used to predict vibration environments. Specifically, as frequency increases, the response of the system structure can be predicted in terms of the space-averaged response for each of a set of individual structural elements that are coupled to collectively describe the system, where each element has near-homogeneous properties and light damping; an example is a constant thickness panel. Such prediction procedures can be applied to a wide range of structural systems if the assumptions detailed in Chap. 11 are satisfied.

Frequency Response Procedures.   For those systems where the shock and/or vibration environment is due to motion excitations at one or more points (for example, the response of an automobile to road roughness inputs at the four wheels), responses at various points on the system structure can be predicted using the input/output relationships detailed in Chap. 21, which involve the frequency response function defined in Eq. (21.11). Such frequency response functions for the system between the excitation points on the system and the mounting points of the equipment can be estimated either by analytical procedures described in Chap. 21 or by experimental measurements. These estimated frequency response functions can then be used to predict the response at the equipment mounting points for any arbitrary excitation spectrum.

Extrapolation Procedures.   The spectra of the responses measured on one system during its operation can often be used to predict the spectra in a newer model of the system, assuming the old and new systems have a similar purpose and are of broadly similar design, for example, a new airplane that flies faster but otherwise is similar in structural design to an earlier model of the airplane. In such cases, the shock and/or vibration responses of the new system at the structural locations of equipment can be predicted, at least coarsely, by scaling the measurements made on the previous system based upon the differences in at least two parameters, namely, (1) the magnitude of the original excitation to the system structure and (2) the weight of the system structure at the points where the equipment is mounted. Specifically, as a first order of approximation, the shock and/or vibration magnitude on the new system can be assumed to vary directly with the magnitude of the excitation and inversely with the weight of the system structure. Such extrapolation techniques have been widely used to predict spectra for the vibration response of new aerospace vehicles[6] and can often be applied to other types of systems as well.

GROUPING OF MEASURED OR PREDICTED SPECTRA INTO ZONES

The shock and vibration response of system structures that support equipment are typically nonhomogeneous in space, sometimes to the extent that the spectra of the responses vary substantially from one mounting point to another for a single item of equipment. At relatively low frequencies, corresponding to normal frequencies below about the fiftieth normal mode of the system structure (see Chap. 21), finite element method (FEM) models for the system structure and the mounted equipment can be used to predict the motions at the specific equipment attachment points. It is more common, however, to define shock and vibration environments by making measurements or predictions at selected points on the system structure that do not correspond to the exact mounting points for equipment, or if they do, the equipment is not present during the measurements or accurately modeled for the predictions. Hence, it is necessary to separate the measured or predicted responses at various points on the system structure into groups, where the responses in each group have broadly similar spectra that can be represented for test purposes by a single spectrum. A *zone* is defined as a region on the system structure that includes those points where the measured or predicted shock and/or vibration responses have broadly similar spectra. It is clear that a zone should correspond to a region of interest in the formulation of shock and vibration test criteria for equipment, i.e., a single zone should include all the attachment points for at least one item of equipment, and preferably, for several items of equipment. However, a zone need not be a single contiguous structural region. For example, all frames of a given size in a truck, no matter where they are located, might constitute a single zone if the responses of those frames are similar.

The determination of zones is usually based upon engineering judgment and experience. For example, given a system with frame-panel construction, engineering judgment dictates that frames and panels should represent different zones, since the responses of light panels will generally be greater than the much heavier frames. Also, the responses perpendicular to the surface of the panels are generally greater than the responses in the plane of the panels, so the responses along these two axes might be divided into separate zones. A visual inspection of the spectra for the measured or predicted responses also can be used to group locations with spectra of similar magnitudes to arrive at appropriate zones. In any case, it is desirable to minimize the number of zones used to describe the shock and vibration responses over those areas of the system structure where equipment will be mounted so as to minimize the number of individual spectra required to test all the equipment for that system.

DETERMINATION OF ZONE LIMITS

A *zone limit* (also called the *maximum expected environment*) is a single spectrum that will conservatively bound the measured or predicted spectra at most or all points within the zone, without severely exceeding the spectrum at any one point. A zone limit may be determined using any one of several procedures.[7] The most common procedure is to envelop the measured or predicted spectra in the zone, but a more rigorous approach is to compute a tolerance limit for the spectra. Specifically, given n measurements of a random variable x, an upper *tolerance limit* is defined as that value of x (denoted by L_x) that will exceed at least β fraction of all values of x with a *confidence coefficient* of γ. The fraction β represents the minimum probability that a randomly selected value of x will be less than L_x; the confidence coefficient γ can be interpreted as the probability that the L_x computed for a future set of data will indeed exceed at least β fraction of all values of x. Tolerance limits are commonly expressed in terms of the ratio $(100\beta)/(100\gamma)$. For example, a tolerance limit determined for $\beta = 0.95$ and $\gamma = 0.50$ is called the 95/50 normal tolerance limit. In the context of shock and/or vibration measurements or predictions, x represents the spectral value at a specific frequency (see Table 20.2) for the response of the system structure at a randomly selected point within a given zone, where x differs from point-to-point within the zone due to the spatial variability of the response. However, x may also differ due to other factors, such as variations in the response from one system to another of the same design or from one environmental exposure to another of the same system. In selecting a sample of measured or predicted spectra to compute a tolerance limit, beyond the spectra at different locations within a zone, it is wise to include spectra from different systems of the same design and different environmental exposures of the same system, if feasible, so that all sources of variability are represented in the measured or predicted spectra.

Tolerance limits are most easily computed when the random variable is *normally distributed* (see Chap. 11). The point-to-point (spatial) variation of the shock and vibration responses of system structures is generally not normally distributed, but there is empirical evidence that the logarithm of the responses does have an approximately normal distribution. Hence, by simply making the logarithmic transformation

$$y = \log x \tag{20.1}$$

where x is the spectral value at a specific frequency of the response within a zone and the transformed variable y can be assumed to have a normal distribution. For n sample values of y, a normal tolerance limit is given by[8]

$$L_y(n,\beta,\gamma) = \bar{y} + k s_y \tag{20.2}$$

where $\bar{y}$ is the sample average and s_y is the sample standard deviation of the n transformed spectral values computed as follows:

$$\bar{y} = \frac{1}{n} \sum_{i=1}^{n} y_i \qquad s_y = \sqrt{\frac{1}{n-1} \sum_{i=1}^{n} (y_i - \bar{y})^2} \tag{20.3}$$

The term k in Eq. (20.2) is called the *normal tolerance factor* and is a tabulated value; a short tabulation of k for selected values of n, β, and γ, is presented in Table 20.3. The normal tolerance limit for the transformed variable y is converted to the original engineering units of x by

$$L_x(n,\beta,\gamma) = 10^{L_y(n,\beta,\gamma)} \tag{20.4}$$

TABLE 20.3 Normal Tolerance Factors for Upper Tolerance Limit

n	$\gamma = 0.50$			$\gamma = 0.75$			$\gamma = 0.90$		
	$\beta = 0.90$	$\beta = 0.95$	$\beta = 0.99$	$\beta = 0.90$	$\beta = 0.95$	$\beta = 0.99$	$\beta = 0.90$	$\beta = 0.95$	$\beta = 0.99$
3	1.50	1.94	2.76	2.50	3.15	4.40	4.26	5.31	7.34
4	1.42	1.83	2.60	2.13	2.68	3.73	3.19	3.96	5.44
5	1.38	1.78	2.53	1.96	2.46	3.42	2.74	3.40	4.67
7	1.35	1.73	2.46	1.79	2.25	3.13	2.33	2.89	3.97
10	1.33	1.71	2.42	1.67	2.10	2.93	2.06	2.57	3.53
15	1.31	1.68	2.39	1.58	1.99	2.78	1.87	2.33	3.21
20	1.30	1.67	2.37	1.53	1.93	2.70	1.76	2.21	3.05
30	1.29	1.66	2.35	1.48	1.87	2.61	1.66	2.08	2.88
50	1.29	1.65	2.34	1.43	1.81	2.54	1.56	1.96	2.74
∞	1.28	1.64	2.33	1.28	1.64	2.33	1.28	1.64	2.33

To simplify test criteria, normal tolerance limits are often smoothed using a series of straight lines, usually no more than seven with slopes of 0, ±3, or ±6 dB.

As an illustration, Fig. 20.1 shows the range of the maximax power spectra for $n = 12$ vibration measurements made at different locations in a selected zone of the structure of a large space vehicle during lift-off. Also shown in this figure are the unsmoothed and smoothed normal tolerance limit versus frequency computed with $\beta = 0.95$ and $\gamma = 0.50$ (the 95/50 limit). Note that the normal tolerance limit at most frequencies is higher than the largest of the 12 spectral values from which the limit is computed. However, a normal tolerance limit could be either much higher or lower than the largest spectral values from which the limit is computed, depending on the values of n, β, and γ.

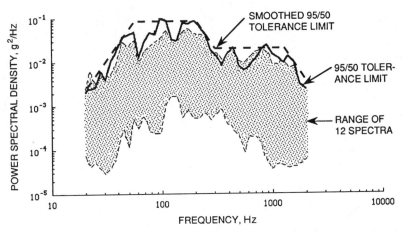

FIGURE 20.1 95/50 normal tolerance limit for spectra of 12 vibration measurements.

SELECTION OF FINAL TEST LEVELS

A *test level* is the spectrum of the shock or vibration environment that is specified for testing purposes, i.e., the spectrum given in a final test specification. The determination of a test level based upon a computed zone limit requires the selection of a value for β, the fraction of the locations within a zone where the spectra of the shock and/or vibration responses of the system structure will be exceeded by the zone (tolerance) limit. This selection is often made somewhat arbitrarily, with values in the range $0.90 \leq \beta \leq 0.99$ being the most common for acceptance and qualification tests. However, the value of β used to arrive at a test level can be optimized based upon an assessment of the adverse consequences (the potential cost) of an undertest versus an overtest. Also, even with an optimum selection, modifications to the test level may be required to account for the interactions of the equipment and the system structure and other considerations.

Optimum Test Level Selection. A number of procedures have been developed[9] that yield an optimum test level for equipment in terms of a percentile of the environmental distribution (which is essentially the value of β for a tolerance limit) as a function of a "cost" ratio C_T/C_F, where C_T is the cost of a test failure and C_F is the cost of a service failure. Some of the procedures assume the equipment being tested has already been manufactured in quantity, raising the possibility that a test failure will lead to refurbishing costs, while others account for a safety factor in the equipment design or a *test factor* based upon the assumed strength of the item being tested. The simplest test level selection rule, which applies to the acceptance testing of a single item of equipment, is given by

$$\beta = \frac{1}{1 + (C_T/C_F)} \qquad (20.5)$$

As an illustration, consider an item of equipment where a failure during test could be corrected by a relatively simple replacement of an inexpensive component, but a failure during service would be catastrophic, perhaps resulting in personal injury. According to Eq. (20.5), the item should be tested to a very severe level relative to the measured or predicted shock and/or vibration environment so as to sharply minimize the risk of an undertest; for example, if a service failure is assessed to be 1000 times as costly as a test failure, $\beta = 0.999$. On the other hand, consider an item where a failure in test would lead to a difficult and expensive redesign, but a failure during service would not be catastrophic. According to Eq. (20.5), the test level now should be moderate relative to the measured or predicted shock and/or vibration environment so as to minimize the risk of an overtest; for example, if a service failure is assessed to be only 9 times as costly as a test failure, then $\beta = 0.90$. Note that the selection procedure does not require the determination of quantitative costs in dollars, but only relative costs, which can be interpreted in qualitative terms. This allows such factors as the consequences of a possible delivery delay caused by a test failure or customer dissatisfaction caused by a service failure to be considered. Also, the conservatism of the test level can be further increased or decreased by selecting a larger or smaller value of γ for the tolerance limit computation.

Equipment-Structure Interactions.   Test levels are commonly specified in terms of a motion parameter, for example, g^2/Hz versus frequency for a random vibration test. However, at the resonance frequencies of relatively heavy items of equipment, the apparent mass of the equipment dramatically increases, causing the equipment to behave like a dynamic vibration absorber on the system structure to which the equipment is mounted (see Chaps. 6 and 43). If the test machine is made to deliver the specified motion to the equipment at its resonance frequencies, a severe overtest may occur. This problem is sometimes addressed by placing limits on the response of the equipment or by allowing "notches" in the specified test spectrum to be introduced at the frequencies of strong resonances of the equipment. The best approach, however, is to derive a second spectrum for the force at the mounting points of the equipment and establish criteria for a dual control test that limits both the input force and the input motion to the equipment.[10]

Added Test Level Factors.   For qualification tests where the item of equipment being tested will not be used in service, it is common to add a factor (often referred to as a *test factor*) to the derived test levels to arrive at a final specified test level. Such factors are usually justified to account for uncertainties not considered in the determination of the test levels, such as unknown variabilities in the equipment strength or its possible service use. These factors are sometimes selected rather arbitrarily, with typical values ranging from 3 to 6 dB above the derived zone limits. However, the final specified test levels can also be made more conservative by simply using larger values for β and/or γ in the computation of the zone limits according to Eqs. (20.1) through (20.4).

SELECTION OF VIBRATION TEST DURATIONS

The *test duration* for a vibration test is the total time the excitation is applied to the equipment at its mounting points by the test machine. In some cases, the test duration is not relevant to the purpose of the test, for example, a development test. In many cases, however, an appropriate simulation of the total duration of the vibration environment anticipated in service is an important part of the test criteria. This is particularly true of qualification and statistical reliability tests, where the purpose is to detect design inadequacies that may lead to failures of any type during exposure to the service vibration environment, including "wearout" failures. For shock environments, this means exposing the equipment to repeated simulations of all the shocks anticipated during its service life, which can usually be accomplished in a reasonable period of time. For vibration environments, however, this means exposing the equipment to a simulation of the anticipated service vibration environment for a duration equivalent to the service life of the equipment, which may be as much as thousands of hours. Vibration environments usually vary widely in overall level and perhaps spectral content during the equipment service life, for example, equipment on an automobile or truck in normal service use. As noted earlier in this chapter, statistical reliability tests are sometimes performed with a duration similar to the anticipated service life of the equipment. For qualification tests, however, it is usual to compress a long, time-varying service environment into a stationary test level of much shorter duration.[11] To do this, the following steps are required:

1. Assuming a time-dependent failure model for the equipment
2. Compressing the time-varying magnitudes of the environment into a single test level
3. In some cases, increasing the test level to further accelerate the test

FAILURE MODELS

A *failure* of an item of equipment is defined as any deterioration of performance or any damage or malfunction that prevents the equipment from accomplishing its intended purpose. There are two basic types of failures that may be caused by vibration:

1. *Hard failure.* A failure involving permanent physical damage that makes the equipment unable to perform its intended purpose, even after the vibration is terminated. Hard failures generally result in observable damage, such as the fracture of a structural element or the permanent disability of an electronic element.

2. *Soft failure.* A failure involving a malfunction or deterioration of performance during the vibration exposure that makes the equipment unable to accomplish its intended purpose, but after the vibration is terminated, the equipment does not reveal any damage and functions properly. Soft failures most commonly occur in electrical, electronic, and/or optical elements, although soft failures may occasionally occur in complex mechanical elements, such as gyroscopic devices.

A *failure mechanism* is the specific means by which an item of equipment is damaged by exposure to an environment. All failure mechanisms are a function of the magnitude of the vibration exposure. A *time-dependent failure mechanism* is a function of both the magnitude and the duration of the vibration exposure. Soft failures during exposure to a vibration environment are rarely time-dependent, i.e., they usually occur immediately at the start of the vibration exposure. On the other hand, hard failures usually are time-dependent, although there are some exceptions. For example, if a vibration environment produces stresses that exceed the ultimate strength of a critical element in the equipment, a fracture will occur immediately at the start of the vibration exposure. See Chaps. 35 and 43 for further discussions of equipment failures.

To establish appropriate test durations for qualification vibration tests, only time-dependent failure mechanisms (usually producing hard failures) are of interest. Common examples of time-dependent failure mechanisms for equipment exposed to vibration environments are fatigue damage, force contact wear, relative velocity wear, and the loss of bolts or rivets. A *failure model* is an analytical relationship between the time-to-failure of the equipment during exposure to a vibration environment and the magnitude of the vibration environment. For a wide class of time-dependent failure mechanisms, the time-to-failure τ for a stationary vibration excitation can be approximated by the *inverse power law*[11] given by

$$\tau = c\,\sigma^{-b} \tag{20.6}$$

where σ is the stress in the equipment caused by the vibration (or any measure of the vibration magnitude that is linearly related to stress), and b and c are constants related to the specific failure mechanism. From Chap. 35, if the endurance limit is ignored, the fatigue endurance curves for common metals fit the form of Eq. (20.6).

Using Eq. (20.6) and assuming a vibration test is performed that accurately simulates the basic characteristics (for example, random versus periodic) and the spectrum of a service vibration environment, the time required to produce a similar amount of damage in the test environment T_t and the time in the service environment T_e are related by

$$T_t = \left(\frac{\sigma_e}{\sigma_t}\right)^b T_e \tag{20.7}$$

where σ is the rms value of the vibration, and the subscripts t and e denote the test and service environments, respectively. For random vibrations defined in terms of power spectra (i.e., $W(f)$ defined in Chap. 22), Eq. (20.7) becomes

$$T_t = \left(\frac{W_e(f)}{W_t(f)}\right)^{b/2} T_e \tag{20.8}$$

The value of the power b in Eqs. (20.7) and (20.8) varies widely for different failure mechanisms. For metal fatigue damage, a value of $b = 8$ is reasonable for many common materials (see Fig. 35.23) and is recommended in Ref. 3. However, a value of $b = 4$ is usually more appropriate for the typical failure mechanisms in electrical and electronic equipment. See Chaps. 35 and 43 for further discussions of time-dependent failure mechanisms.

COMPRESSING TIME-VARYING SERVICE ENVIRONMENTS

For those vibration environments that vary substantially in severity during the equipment service life, the duration of the environment can often be reduced for testing purposes by using Eq. (20.7) to scale the less severe vibration levels to the most severe levels that occur during the service life. Such scaling procedures are most applicable to environments that vary in overall level but not substantially in spectral content. For example, consider an item of electrical equipment designed for a motor vehicle with a service life of 4000 hours. Assume the anticipated service vibration environment for the vehicle at the equipment mounting points has the rms values summarized in Table 20.4. Further assume $b = 4$ in Eq. (20.7), and the vibrations during the various service conditions have a similar spectral content. Table 20.4 indicates the damage potential of the 4000-hour service vibration environment can be simulated by a vibration test with a duration of 80 hours at the maximum service vibration level.

For those vibration environments where the spectral content and the overall levels change during service operations, the test duration computations illustrated in Table 20.4 must be made on a frequency-by-frequency basis using Eq. (20.8) or a similar expression for the appropriate spectral description in Table 20.2. This will result in a different test duration at each frequency, leading to two possible testing options: (1) a series of tests, each covering a different frequency range with a different test duration or (2) a single test with a test duration equal to the longest test duration computed at any frequency. The second option is usually the more practical and assures a conservative test.

ACCELERATED TESTS

An *accelerated test* is a test where the test duration is reduced by increasing the test level in a manner that will maintain the same environment-induced damage to the equipment. The determination of a test duration for a stationary vibration test that produces the same damage as a nonstationary vibration environment, as detailed in the preceding section, constitutes the most desirable form of accelerated testing because the test level never exceeds the maximum vibration level that the equipment will experience during its service environment. Furthermore, most of the damage experienced by equipment in service usually occurs during exposure to the maximum vibration level in the service environment, which typically covers a small fraction of the total service duration (see Table 20.4). In such cases, reducing the relatively long durations of the less severe vibrations by scaling to the maximum level according to Eq. (20.7) does not introduce a major error, even if the exponent in Eq. (20.7) is inaccurate.

Highly Accelerated Tests. Situations often arise where scaling the less severe segments of a nonstationary vibration environment to a stationary vibration level corresponding to the maximum level of the environment may yield a test duration that is still too long to be practical; for example, the test duration of 80 hours com-

TABLE 20.4 Determination of Equivalent Duration for Automobile Equipment Vibration Environment

Type of road segment	Duration on road segment, hours	rms vibration on road segment, g	Equivalent duration on road segment A, hours
A. Unpaved secondary roads	40	3	40
B. Improved secondary roads	460	1.4	22
C. Primary roads	1500	0.9	12
D. Major highways	2000	0.7	6
Total equivalent duration on road segment A (hours)			80

puted for the 4000-hour service environment in Table 20.4 may still be too long for testing purposes. In such a case, it is common to further reduce the test duration by increasing the test level beyond the maximum level the equipment will experience during its anticipated service environment.[11] Indeed, if no limit is placed on the rms test level in Eq. (20.7), the test duration theoretically can be made as short as desired, provided the ultimate strength of the equipment structure is not exceeded. However, increasing the test level beyond the maximum level during the anticipated service environment introduces major uncertainties in the test results, particularly if the equipment is fabricated using different materials and/or incorporates electrical, electronic, and/or optical elements. The problem is that the failure mechanisms of some elements may not comply with the inverse power law in Eq. (20.6). Furthermore, even if all failure mechanisms do comply with Eq. (20.6), the exponent b may vary from one element to another within the equipment. Hence, increasing the test level to accelerate the test rapidly in compliance with Eq. (20.7) may cause some elements of the equipment to be undertested and others to be overtested. The result could be the occurrence of unrepresentative failures during the accelerated test.

Durability and Functional Tests.   A common procedure to suppress unrepresentative failures that may be caused by rapidly accelerating a vibration test of equipment with a long service life is to perform two separate tests, namely, a durability test and a functional test. A *durability test* is intended to reveal only time-dependent failures and is rapidly accelerated to produce the same damage as the entire duration of the service vibration environment based upon a specific damage model, for example, Eq. (20.7). The equipment is not required to function during the durability test, and any failures that are not time-dependent are ignored. A *functional test* is intended to reveal failures that are not time-dependent (i.e., failures related only to the vibration level) and is not accelerated with test levels that exceed the maximum expected vibration level during the service environment. The equipment is required to function during the test, but since the failures of interest are not time-dependent, the test duration is not critical; for example, the test duration is often fixed by the time required to fully operate the equipment and verify that it properly performs its intended purpose.

SHOCK AND VIBRATION TESTING

The laboratory machinery used to perform vibration tests and shock tests are detailed in Chaps. 25 and 26, respectively. In all cases, there are several issues that must be carefully considered in performing such tests, the most important being:

1. Identification of test failures
2. Type of excitation to be used
3. Single versus multiple-axis excitation
4. Test fixtures

IDENTIFICATION OF TEST FAILURES

In all shock and vibration tests of equipment, it is important to carefully establish what types of equipment malfunctions or anomalies will be considered failures. This determination depends heavily on the purpose of the test and sometimes on the judgment of the purchaser of the equipment. Here are a few examples:

1. Since a qualification test is intended to identify design problems, failures during the test that are clearly due to workmanship errors or material defects are usually ignored, i.e., the equipment is repaired and the test is continued.

2. Since the test level for a highly accelerated qualification test is based upon a specific failure model, failures during the test that are not consistent with the failure model should be carefully evaluated and ignored if they are determined to involve a failure mechanism that is not time-dependent.

3. During durability tests of equipment, if a fatigue crack forms in the equipment structure that does not propagate to a fracture, whether the fatigue crack constitutes a failure or the length of the fatigue crack that constitutes a failure must be specified.

4. During functional tests of electrical, electronic, and/or optical equipment, if there is measurable deterioration in the performance of equipment during the test, the exact degree of deterioration that prevents the equipment from performing its intended purpose must be specified.

TYPES OF EXCITATION

Shock tests are sometimes performed using specified test machines but more often are performed using more general test machines that can produce transients with a desired shock response spectrum (see Chap. 26 and "Digital Control Systems for Shock Testing" in Chap. 27). Although vibration environments may be simulated by mounting the equipment in a prototype system and reproducing the actual environment for the system, it is more common to apply the vibration directly to the equipment mounting points using vibration testing machines described in Chap. 25.

Random Tests. Random excitations are used to simulate random vibration in those tests where an accurate representation of the environment is desired, specifically, qualification, reliability, and some acceptance tests. The most commonly used random test machines produce a near-Gaussian vibration. If the actual environment is random but not Gaussian, a Gaussian simulation is acceptable since the response of the equipment exposed to the environment will be near-Gaussian at its resonance frequencies; this is because equipment resonances constitute narrow-band filtering operations that suppress deviations from the Gaussian form in the vibration response of the equipment.

Sine Wave Tests. Sine wave excitations are used to simulate the fixed-frequency periodic vibrations produced by constant-speed rotating machines and reciprocating engines. Sine wave excitations are sometimes superimposed on random excitations for those situations where the service vibration environment involves both. Sine wave excitations fixed sequentially at the resonance frequencies of an equipment item (often referred to as a *dwell sine test*) are sometimes used in development tests, as well as in durability tests, to evaluate the fatigue resistance of the equipment.

Swept Sine Wave Tests. Sweep sine wave excitations are produced by continuously varying the frequency of a sine wave in a linear or logarithmic manner. Such excitations are used to simulate the vibration environments produced by variable-speed rotating machines and reciprocating engines. The usual approach is to make the sweep rate sufficiently slow to allow the equipment being tested to reach a near-full (steady-state) response as the swept sine wave excitation passes through each resonance frequency. Swept sine wave excitations are also used for development tests to identify resonance frequencies and sometimes to estimate frequency response functions (see Chap. 21).

MULTIPLE-AXIS EXCITATIONS

Shock and vibration environments are typically multiple-axial, i.e., the excitations occur simultaneously along all three orthogonal axes of the equipment. Multiple-axis shock and vibration test facilities are often used to simulate low-frequency shock and vibration environments, generally below 50 Hz, such as earthquake motions (see Chap. 24). Also, multiple-axis vibration test facilities have been developed for higher-frequency vibration excitations (up to 2000 Hz), but it is more common to perform shock and vibration tests using machines that apply the excitation sequentially along one axis at a time, i.e., machines that deliver rectilinear motion only (see Chaps. 25 and 26). Single-axis testing introduces an additional uncertainty of unknown magnitude in the accuracy of the test simulation, but there is debate as to whether the removal of this uncertainty justifies the high cost and complexity of multiple-axis test facilities.

TEST FIXTURES

A *test fixture* is a special structure that allows the test item to be attached to the table of a shock or vibration test machine. Test fixtures are required for almost all shock and vibration tests of equipment because the mounting hole locations on the equipment and the test machine table do not correspond. For the usual case where the test machine generates rectilinear motion normal to the table surface, a test fixture is also necessary to reorient the equipment relative to the table so that vibratory motion can be delivered along the lateral axes of the equipment, i.e., the axes parallel to the plane of the equipment mounting points. This requires a versatile test fixture between the table and the equipment, or perhaps three different test fixtures. If the direction of gravity is important to the equipment, the test machine must be rotated from vertical to horizontal, or vice-versa, to meet the test conditions.

For equipment that is small relative to the test machine table, L-shaped test fixtures with side gussets are commonly used to deliver excitation along the lateral axes of the equipment as illustrated Fig. 20.2. Unless designed with great care, such fixtures are likely to have resonances in the test frequency range. In principle, the consequent spectral peaks and valleys due to fixture resonances can be flattened out by electronic equalization of the test machine table motion (see Chap. 27), but this is difficult if the damping of the fixture is low. The best approach is to design the fixture to have few or, if possible, no resonances in the test frequency range.

For equipment that is large relative to the test machine table, excitation along the lateral axes of the equipment is commonly achieved by mounting the equipment on a horizontal plate driven by the test machine rotated into the horizontal plane, where the plate is separated from the flat opposing surface of a massive block by an oil film or hydrostatic oil bearings as shown in Fig. 20.3. The oil film or hydrostatic bearings provide little shearing restraint but give great stiffness normal to the surface, the stiffness being distributed uniformly over the complete horizontal area.

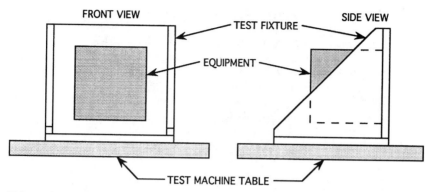

FIGURE 20.2 Test fixture to deliver excitation in the plane of the equipment mounting points.

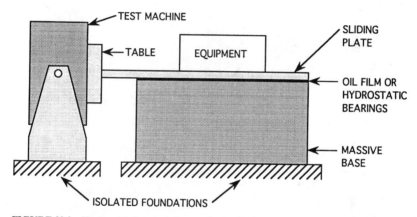

FIGURE 20.3 Horizontal plate to deliver excitation in the plane of the equipment mounting points.

Accordingly, a relatively light moving plate can be vibrated that has the properties of the massive rigid block in the direction normal to its plane.

REFERENCES

1. Bendat, J. S., and A. G. Piersol: "Random Data: Analysis and Measurement Procedures," 2d ed., p. 477, John Wiley & Sons, Inc., New York, 1986.
2. Gaberson, H. A., and R. H. Chalmers: *Shock and Vibration Bull.,* **40**(2):31 (1969).
3. Military specification, "Environmental Test Methods," MIL-STD-810E, 1988.
4. Curtis, A. J., N. G. Tinling, and H. T. Abstein: "Selection and Performance of Vibration Tests," SVM-8, p. 79, The Shock and Vibration Information Center, Washington, D.C., 1972.
5. Lawless, J. F.: "Statistical Models & Methods for Lifetime Data," John Wiley & Sons, Inc., New York, 1982.

6. Barnoski, R. L., et al.: "Summary of Random Vibration Prediction Procedures," NASA CR-1302, National Aeronautics and Space Administration, Washington, D.C., 1969.

7. Piersol, A. G.,: *Proc. 65th Shock and Vibration Symp.,* part I, 118 (1994).

8. Bowker, A. H., and G. J. Lieberman: "Engineering Statistics," 2d ed., p. 314, Prentice-Hall, Inc., Englewood Cliffs, N.J., 1972.

9. Piersol, A. G.: *Proc. Institute of Environmental Sciences,* 88 (1974).

10. Scharton, T. D.: *J. of Spacecraft and Rockets,* **32**:312 (1995).

11. Nelson, W.: "Accelerated Testing," John Wiley and Sons, Inc., New York, 1990.

CHAPTER 21
EXPERIMENTAL
MODAL ANALYSIS

Randall J. Allemang
David L. Brown

INTRODUCTION

Experimental modal analysis is the process of determining the modal parameters (natural frequencies, damping factors, modal vectors, and modal scaling) of a linear, time-invariant system. The modal parameters are often determined by analytical means, such as finite-element analysis. One common reason for experimental modal analysis is the verification or correction of the results of the analytical approach. Often, an analytical model does not exist, and the modal parameters determined experimentally serve as the model for future evaluations, such as structural modifications. Predominantly, experimental modal analysis is used to explain a dynamics problem (vibration or acoustic) whose solution is not obvious from intuition, analytical models, or previous experience.

The process of determining modal parameters from experimental data involves several phases. The success of the experimental modal analysis process depends upon having very specific goals for the test situation. Every phase of the process is affected by the goals which are established, particularly with respect to the errors associated with that phase. One possible delineation of these phases is as follows:

Modal analysis theory refers to that portion of classical vibration theory that explains the existence of natural frequencies, damping factors, and mode shapes for linear systems. This theory includes both lumped-parameter, or discrete, models and continuous models. This theory also includes real normal modes as well as complex modes of vibration as possible solutions for the modal parameters.[1-3]

Experimental modal analysis methods involve the theoretical relationship between measured quantities and classical vibration theory, often represented as matrix differential equations. All commonly used methods trace from the matrix differential equations but yield a final mathematical form in terms of measured raw input and output data in the time or frequency domains or some form of processed data such as impulse-response or frequency-response functions.

Modal data acquisition involves the practical aspects of acquiring the data that are required to serve as input to the modal parameter estimation phase. Much care must be taken to assure that the data match the requirements of the theory as well as the requirements of the numerical algorithm involved in the modal parameter estimation. The theoretical requirements involve concerns such as system linearity and time invariance of system parameters. The numerical algorithms are particularly concerned with the bias errors in the data as well as with any overall dynamic range considerations.[4-7]

Modal parameter estimation is concerned with the practical problem of estimating the modal parameters, based upon a choice of mathematical model as justified by the experimental modal analysis method, from the measured data.[8-10]

Modal data presentation/validation is the process of providing a physical view or interpretation of the modal parameters. For example, this may simply be the numerical tabulation of the frequency, damping, and modal vectors along with the associated geometry of the measured degrees-of-freedom. More often, modal data presentation involves the plotting and animation of such information.

Figure 21.1 is a representation of all phases of the process. In this example, a continuous beam is being evaluated for the first few modes of vibration. Modal analysis theory explains that this is a linear system and that the modal vectors of this system should be real normal modes. The experimental modal analysis method that has been used is based upon the relationships of the frequency-response function to the matrix differential equations of motion. At each measured degree-of-freedom (DOF), the imaginary part of the frequency-response function for that measured response degree-of-freedom and a common input degree-of-freedom is superimposed perpendicular to the beam. Naturally, the modal data acquisition in this example involves the estimation of frequency-response functions for each degree-of-freedom shown. The frequency-response functions are complex-valued functions, and only the imaginary portion of each function is shown. One method of modal parameter estimation suggests that for systems with light damping and widely spaced modes, the imaginary part of the frequency-response function at the damped natural frequency may be used as an estimate of the modal coefficient for that response degree-of-freedom. The damped natural frequency can be identified as the frequency of the positive and negative peaks in the imaginary part of the frequency-response functions. The damping can be estimated from the sharpness of the peaks. In this abbreviated way, the modal parameters have been estimated. Modal data presentation for this case is shown as the lines connecting the peaks. While animation is possible, a reasonable interpretation of the modal vector can be gained in this case from plotting alone.

MEASUREMENT DEGREES-OF-FREEDOM

The development of any theoretical concept in the area of vibrations, including modal analysis, depends upon an understanding of the concept of the number of degrees-of-freedom n of a system. This concept is extremely important to the area of modal analysis since the number of modes of vibration of a mechanical system is equal to the number of degrees-of-freedom. From a practical point of view, the relationship between this theoretical definition of the number of degrees-of-freedom and the number of *measurement degrees-of-freedom* N_o, N_i is often confusing. For this reason, the concept of degree-of-freedom is reviewed as a preliminary to the following experimental modal analysis material.

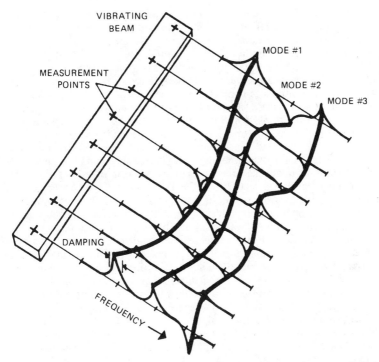

FIGURE 21.1 Experimental modal analysis example using the imaginary part of the frequency-response functions.

To begin with, the basic definition that is normally associated with the concept of the number of degrees-of-freedom involves the following statement: *The number of degrees-of-freedom for a mechanical system is equal to the number of independent coordinates (or minimum number of coordinates) that is required to locate and orient each mass in the mechanical system at any instant in time.* As this definition is applied to a point mass, 3 degrees-of-freedom are required since the location of the point mass involves knowing the x, y, and z translations of the center-of-gravity of the point mass. As this definition is applied to a rigid body mass, 6 degrees-of-freedom are required since θ_x, θ_y, and θ_z rotations are required in addition to the x, y, and z translations in order to define both the orientation and the location of the rigid body mass at any instant in time. As this definition is extended to any general deformable body, the number of degrees-of-freedom is essentially infinite. However, while this is theoretically true, it is quite common, particularly with respect to finite-element methods, to view the general deformable body in terms of a large number of physical points of interest with 6 degrees-of-freedom for each of the physical points. In this way, the infinite number of degrees-of-freedom can be reduced to a large but finite number.

When measurement limitations are imposed upon this theoretical concept of the number of degrees-of-freedom of a mechanical system, the difference between the theoretical number of degrees-of-freedom n and the number of measurement degrees-of-freedom N_o, N_i begins to evolve. Initially, for a general deformable body, the number of degrees-of-freedom n can be considered to be infinite or equal to some large finite number if a limited set of physical points of interest is considered, as discussed in the previous paragraph. The first measurement limitation that needs to be considered is that there is normally a limited frequency range that is of interest to the analysis. When this limitation is considered, the number of degrees-of-freedom of this system that are of interest is reduced from infinity to a reasonable finite number. The next measurement limitation that needs to be considered involves the physical limitation of the measurement system in terms of amplitude. A common limitation of transducers, signal conditioning and data acquisition systems results in a dynamic range of 80 to 100 dB (10^4 to 10^5) in the measurement. This means that the number of degrees-of-freedom is reduced further because of the dynamic range limitations of the measurement instrumentation. Finally, since few rotational transducers exist at this time, the normal measurements that are made involve only translational quantities (displacement, velocity, acceleration, force) and thus do not include rotational effects, or RDOF. In summary, even for the general deformable body, the theoretical number of degrees-of-freedom that are of interest is limited to a very reasonable finite value ($n = 1$ to 50). Therefore, this number of degrees-of-freedom n is the number of modes of vibration that are of interest.

Finally, then, the number of measurement degrees-of-freedom N_o, N_i can be defined as the number of physical locations at which measurements are made multiplied by the number of measurements made at each physical location. Since the physical locations are chosen somewhat arbitrarily, and certainly without exact knowledge of the modes of vibration that are of interest, there is no specific relationship between the number of degrees-of-freedom n and the number of measurement degrees-of-freedom N_o, N_i. In general, in order to define n modes of vibration of a mechanical system, N_o or N_i must be equal to or larger than n. However, N_o or N_i being larger than n is not a guarantee that n modes of vibration can be found from the measurement degrees-of-freedom. The measurement degrees-of-freedom must include physical locations that allow a unique determination of the n modes of vibration. For example, if none of the measurement degrees-of-freedom are located on a portion of the mechanical system that is active in one of the n modes of vibration, portions of the modal parameters for this mode of vibration cannot be found.

In the development of material in the following text, the assumption is made that a set of measurement degrees-of-freedom exists that allows n modes of vibration to be determined. In reality, either N_o or N_i is always chosen much larger than n since a prior knowledge of the modes of vibration is not available. If the set of N_o or N_i measurement degrees-of-freedom is large enough and if the measurement degrees-of-freedom are distributed uniformly over the general deformable body, the n modes of vibration are normally found.

Throughout this experimental modal analysis reference, the frequency-response function notation H_{pq} is used to describe the measurement of the response at measurement degree-of-freedom p resulting from an input applied at measurement degree-of-freedom q. The single subscript p or q refers to a single sensor aligned in a specific direction ($\pm X$, Y, or Z) at a physical location on or within the structure.

BASIC ASSUMPTIONS

There are four basic assumptions concerning any structure that are made in order to perform an experimental modal analysis:

1. *The structure is assumed to be linear,* i.e., the response of the structure to any combination of forces, simultaneously applied, is the sum of the individual responses

to each of the forces acting alone. For a wide variety of structures this is a very good assumption. When a structure is linear, its behavior can be characterized by a controlled excitation experiment in which the forces applied to the structure have a form that is convenient for measurement and parameter estimation rather than being similar to the forces that are actually applied to the structure in its normal environment. For many important kinds of structures, however, the assumption of linearity is not valid. Where experimental modal analysis is applied in these cases, it is hoped that the linear model that is identified provides a reasonable approximation of the structure's behavior.

2. *The structure is time invariant,* i.e., the parameters that are to be determined are constants. In general, a system which is not time invariant has components whose mass, stiffness, or damping depend on factors that are not measured or are not included in the model. For example, some components may be temperature dependent. In this case, since temperature effects are not measured, the temperature of the component is an unknown time-varying signal. Hence, the component has time-varying characteristics. Therefore, for this case the modal parameters determined by any measurement and estimation process depend on the time (and the associated temperature dependence) when the measurements are made. If the structure that is tested changes with time, then measurements made at the end of the test period determine a different set of modal parameters from measurements made at the beginning of the test period. Thus, the measurements made at the two different times are inconsistent, violating the assumption of time invariance.

3. *The structure obeys Maxwell's reciprocity,* i.e., a force applied at degree-of-freedom p causes a response at degree-of-freedom q that is the same as the response at degree-of-freedom p caused by the same force applied at degree-of-freedom q. With respect to frequency-response function measurements, the frequency-response function between points p and q determined by exciting at p and measuring the response at q is the same frequency-response function found by exciting at q and measuring the response at p ($H_{pq} = H_{qp}$).

4. *The structure is observable,* i.e., the input-output measurements that are made contain enough information to generate an adequate behavioral model of the structure. Structures and machines which have loose components, or, more generally, which have degrees-of-freedom of motion that are not measured, are not completely observable. For example, consider the motion of a partially filled tank of liquid when complicated sloshing of the fluid occurs. Sometimes enough data can be collected so that the system is observable under the form chosen for the model, while at other times an impractical amount of data is required. This assumption is particularly relevant to the fact that the data normally describe an incomplete model of the structure. This occurs in at least two different ways. First, the data are normally limited to a minimum and maximum frequency as well as a limited frequency resolution. Second, no information relative to local rotations is available because of the lack of available transducers in this area.

MODAL ANALYSIS THEORY

While modal analysis theory has not changed over the last century, the application of the theory to experimentally measured data has changed significantly. The advances of recent years with respect to measurement and analysis capabilities have caused a reevaluation of what aspects of the theory relate to the practical world of testing. Thus, the aspect of transform relationships has taken on renewed importance since digital forms of the integral transforms are in constant use. The theory from the vibrations point of view involves a more thorough understanding of how the structural parameters of mass, damping, and stiffness relate to the impulse-response function (time domain), the frequency-response function (Fourier or frequency domain), and the transfer function (Laplace domain) for single and multiple degree-of-freedom systems.

SINGLE DEGREE-OF-FREEDOM SYSTEMS

In order to understand modal analysis, complete comprehension of single degree-of-freedom systems is necessary. In particular, complete familiarity with single degree-of-freedom systems as presented and evaluated in the time, frequency (Fourier), and Laplace domains serves as the basis for many of the models that are used in modal parameter estimation. This single degree-of-freedom approach is trivial from a modal analysis perspective since no modal vectors exist. The true importance of this approach results from the fact that the multiple degree-of-freedom case can be viewed as simply a linear superposition of single degree-of-freedom systems.

The general mathematical representation of a single degree-of-freedom system is expressed in Eq. (21.1):

$$m\ddot{x}(t) + c\dot{x}(t) + kx(t) = f(t) \tag{21.1}$$

where m = mass constant
c = damping constant
k = stiffness constant

This differential equation yields a characteristic equation of the following form:

$$ms^2 + cs + k = 0 \tag{21.2}$$

where s is the complex-valued frequency variable (Laplace variable). This characteristic equation of a single degree-of-freedom system has two roots, $\lambda 1$ and $\lambda 2$, which are

$$\lambda_1 = -\sigma_1 + j\omega_1 \qquad \lambda_2 = -\sigma_2 + j\omega_2 \tag{21.3}$$

where σ_1 = damping factor for mode 1
ω_1 = damped natural frequency for mode 1

Thus, the complementary solution of Eq. (21.1) is

$$x(t) = Ae^{\lambda_1 t} + Be^{\lambda_2 t} \tag{21.4}$$

A and B are complex-valued constants determined from the initial conditions imposed on the system at time $t = 0$.

For most real structures, unless active damping systems are present, the damping ratio is rarely greater than 10 percent. For this reason, all further discussion is restricted to underdamped systems ($\zeta < 1$). With reference to Eq. (21.2), this means that the two roots λ_1 and λ_2 are always complex conjugates. Also, the two coefficients, A and B, are complex conjugates of each other. For an underdamped system, the roots of the characteristic equation can be written as

$$\lambda_1 = \sigma_1 + j\omega_1 \qquad \lambda_1{}^* = \sigma_1 - j\omega_1 \tag{21.5}$$

where $\sigma_1 =$ damping factor
$\omega_1 =$ damped natural frequency

The roots of the characteristic equation (21.2) can also be written as

$$\lambda_1 = -\zeta_1 \Omega_1 \pm j\Omega_1 \sqrt{1 - \zeta_1^2} \tag{21.6}$$

The *damping factor* is defined as the real part of a root of the characteristic equation. The damping factor describes the exponential decay or growth of the harmonic. This parameter has the same units as the imaginary part of the root of the characteristic equation, typically radians per second.

Time Domain: Impulse-Response Function. The *impulse-response function* of the single degree-of-freedom system is defined as the time response $x(t)$ of the system, assuming that the initial conditions are zero and that the system excitation $f(t)$ is a unit impulse. The response of the system $x(t)$ to such a unit impulse is known as the impulse-response function $h(t)$ of the system. Therefore

$$h(t) = Ae^{\lambda_1 t} + A^* e^{\lambda_1^* t} = e^{\sigma_1 t} \left[Ae^{+j\omega_1 t} + A^* e^{-j\omega_1 t} \right] \tag{21.7}$$

Thus, the residue A controls the amplitude of the impulse response, the real part of the pole is the decay rate, and the imaginary part of the pole is the frequency of oscillation. Figure 21.2 illustrates the impulse-response function for a single degree-of-freedom system.

Frequency Domain: Frequency-Response Function. An equivalent equation of motion for Eq. (21.1) is determined for the Fourier or frequency (ω) domain. This representation has the advantage of converting a differential equation to an algebraic equation. This is accomplished by taking the Fourier transform of Eq. (21.1). Thus, Eq. (21.1) becomes

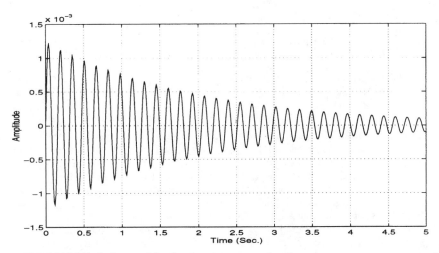

FIGURE 21.2 Single degree-of-freedom impulse-response function.

$$[-m\omega^2 + jc\omega + k] X(\omega) = F(\omega) \tag{21.8}$$

Restating the above equation,

$$X(\omega) = H(\omega) F(\omega) \tag{21.9}$$

where

$$H(\omega) = \frac{1}{-m\omega^2 + jc\omega + k}$$

Eq. (21.9) states that the system response $X(\omega)$ is directly related to the system forcing function $F(\omega)$ through the quantity $H(\omega)$. If the system forcing function $F(\omega)$ and its response $X(\omega)$ are known, $H(\omega)$ can be calculated. That is,

$$H(\omega) = \frac{X(\omega)}{F(\omega)} \tag{21.10}$$

The quantity $H(\omega)$ is known as the *frequency-response function* of the system. The frequency-response function relates the Fourier transform of the system input to the Fourier transform of the system response.

The denominator of Eq. (21.9) is known as the *characteristic equation* of the system and is of the same form as Eq. (21.2). Note that the characteristic values of this complex equation are in general complex even though the equation is a function of a real-valued independent variable ω. The characteristic values of this equation are known as the *complex roots* of the characteristic equation or the *complex poles* of the system. In terms of modal parameters, these characteristic values are also called the *modal frequencies*.

The frequency response function $H(\omega)$ can now be rewritten as a function of the complex poles as follows:

$$H(\omega) = \frac{1/m}{(j\omega - \lambda_1)(j\omega - \lambda_1^*)} \tag{21.11}$$

where λ_1 = complex pole = $\sigma + j\omega_1$
$\quad\quad\quad\lambda_1^* = \sigma - j\omega_1$

Since the frequency-response function is a complex-valued function of a real-valued independent variable ω, it is represented by a pair of curves, as shown in Fig. 21.3.

Laplace Domain: Transfer Function.   Just as in the previous case for the frequency domain, the equivalent information can be presented in the Laplace domain by way of the Laplace transform. The only significant difference in the development concerns the fact that the Fourier transform is defined from negative infinity to positive infinity, while the Laplace transform is defined from zero to positive infinity with initial conditions. The Laplace representation, also, has the advantage of converting a differential equation to an algebraic equation.

The transfer function is defined in the same way that the frequency response function is defined (assuming zero initial conditions):

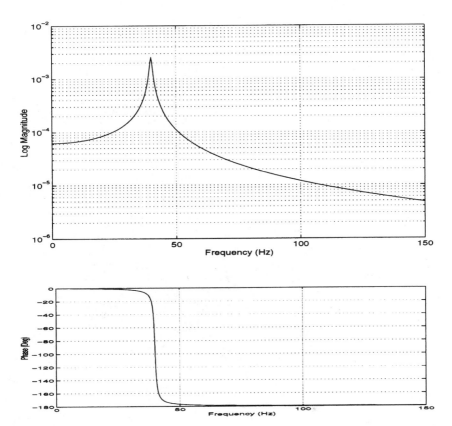

FIGURE 21.3 Single degree-of-freedom frequency-response function (log magnitude/phase format).

$$X(s) = H(s) F(s) \qquad (21.12)$$

where

$$H(s) = \frac{1}{ms^2 + cs + k}$$

The quantity $H(s)$ is defined as the *transfer function* of the system. The transfer function relates the Laplace transform of the system input to the Laplace transform of the system response. From Eq. (21.12), the transfer function is defined as

$$H(s) = \frac{X(s)}{F(s)} \qquad (21.13)$$

The denominator term is once again referred to as the characteristic equation of the system. As noted in the previous two cases, the roots of the characteristic equation are given in Eq. (21.5).

The transfer function $H(s)$ is now rewritten, just as in the frequency response function case, as

$$H(s) = \frac{1/m}{(s - \lambda_1)(s - \lambda_1{}^*)} \tag{21.14}$$

Since the transfer function is a complex-valued function of a complex independent variable s, it is represented, as shown in Fig. 21.4, as a pair of surfaces.

The definition of undamped natural frequency, damped natural frequency, damping factor, percent of critical damping, and residue are all relative to the information represented by Fig. 21.4. The projection of this information onto the plane of zero amplitude yields the information shown in Fig. 21.5.

Log Magnitude

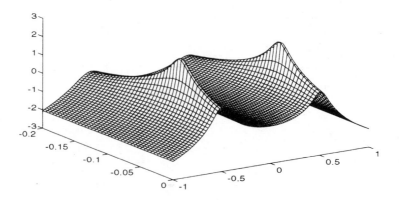

Phase (Radians)

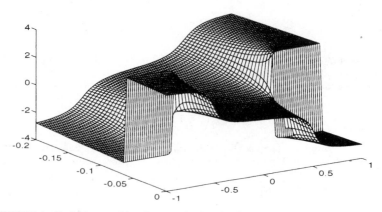

FIGURE 21.4 Single degree-of-freedom transfer function (log magnitude/phase format).

The concept of residues is now defined in terms of the partial fraction expansion of the transfer function or frequency-response function equation. Equation (21.27) is expressed in terms of partial fractions as follows:

$$H(s) = \frac{1/m}{(s - \lambda_1)(s - \lambda_1^*)} = \frac{A}{(s - \lambda_1)} + \frac{A^*}{(s - \lambda_1^*)} \tag{21.15}$$

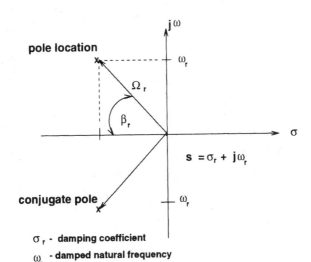

σ_r - **damping coefficient**

ω_r - **damped natural frequency**

Ω_r - **resonant (undamped natural frequency)**

$\zeta_r = \cos \beta_r$ - **damping factor (or percent of critical damping)**

FIGURE 21.5 Transfer function (Laplace domain projection).

The *residues* of the transfer function are defined as the constants A and A^*. The terminology and development of residues comes from the evaluation of analytic functions in complex analysis. The residues of the transfer function are directly related to the amplitude of the impulse-response function. In general, the residue A is a complex quantity. As shown for a single degree-of-freedom system, A is purely imaginary.

From an experimental point of view, the transfer function is not estimated from measured input-output data. Instead, the frequency-response function is actually estimated via the discrete Fourier transform.

MULTIPLE DEGREE-OF-FREEDOM SYSTEMS

Modal analysis concepts are applied when a continuous, nonhomogeneous structure is described as a lumped-mass, multiple degree-of-freedom system. The modal (natural) frequencies, the modal damping, the modal vectors, or relative patterns of motion, and the modal scaling can be found from an estimate of the mass, damping, and stiffness matrices or from the measurement of the associated frequency-response functions. From the experimental viewpoint, the relationship of modal parameters with respect to measured frequency-response functions is most important.

The development of the frequency-response function solution for the multiple degree-of-freedom case parallels that for the single degree-of-freedom case. This development relates the mass, damping, and stiffness matrices to a matrix transfer function model, or matrix frequency-response function model, involving multiple degrees-of-freedom. Just as in the analytical case where the ultimate solution can be described in terms of single degree-of-freedom systems, the frequency-response functions between any input and response degree-of-freedom can be represented as a linear superposition of the single degree-of-freedom models derived previously.

As a result of the linear superposition concept, the equations for the impulse-response function, the frequency-response function, and the transfer function for the multiple degree-of-freedom system are defined as follows:

Impulse-response function:

$$h_{pq}(t) = \sum_{r=1}^{n} A_{pqr} e^{\lambda_r t} + A^*_{par} e^{\lambda_r^* t} \tag{21.16}$$

Frequency-response function:

$$H_{pq}(\omega) = \sum_{r=1}^{n} \frac{A_{pqr}}{j\omega - \lambda_r} + \frac{A^*_{pqr}}{j\omega - \lambda_r^*} \tag{21.17}$$

Transfer function:

$$H_{pq}(s) = \sum_{r=1}^{n} \frac{A_{pqr}}{s - \lambda_r} + \frac{A^*_{pqr}}{s - \lambda_r^*} \tag{21.18}$$

where t = time variable
ω = frequency variable
s = Laplace variable
p = measured degree-of-freedom (response)
q = measured degree-of-freedom (input)
r = modal vector number
A_{pqr} = residue
$A = Q_r \psi_{pr} \psi_{qr}$
Q_r = modal scaling factor
ψ_{pr} = modal coefficient
λ_r = system pole
n = number of modal frequencies

It is important to note that the residue, A_{pqr}, in Eqs. (21.16) through (21.18) is the product of the modal deformations at the input q and response p degrees-of-freedom and a modal scaling factor for mode r. Therefore, while the product of these three terms is unique, each of the three terms individually is not unique.

Modal scaling refers to the relationship between the normalized modal vectors and the absolute scaling of the mass matrix (analytical case) and/or the absolute scaling of the residue information (experimental case). Modal scaling is normally presented as *modal mass* or *modal A*. The driving point residue, A_{qqr}, is particularly important in deriving the modal scaling.

$$A_{qqr} = Q_r\psi_{qr}\psi_{qr} = Q_r\psi_{qr}^2 \qquad (21.19)$$

For undamped and proportionally damped systems, the rth modal mass of a multiple degree-of-freedom system can be defined as

$$M_r = \frac{1}{j2Q_r\omega_r} = \frac{\psi_{pr}\psi_{qr}}{j2A_{pqr}\omega_r} \qquad (21.20)$$

where M_r = modal mass
Q_r = modal scaling constant
ω_r = damped natural frequency

If the largest scaled modal coefficient is equal to unity, Eq. (21.20) computes a quantity of modal mass that has physical significance. The physical significance is that the quantity of modal mass computed under these conditions is between zero and the total mass of the system. Therefore, under this scaling condition, the modal mass can be viewed as the amount of mass that is participating in each mode of vibration. For a translational rigid body mode of vibration, the modal mass should be equal to the total mass of the system. The modal mass defined in Eq. (21.20) is developed in terms of displacement over force units. If measurements, and therefore residues, are developed in terms of any other units (velocity over force or acceleration over force), Eq. (21.20) has to be altered accordingly.

Once the modal mass is known, the *modal damping* C_r and *stiffness* K_r can be obtained through the following single degree-of-freedom equations:

$$C_r = 2\sigma_r M_r \qquad (21.21)$$

$$K_r = (\sigma_r^2 + \omega_r^2)M_r = \Omega_r^2 M_r \qquad (21.22)$$

For systems with nonproportional damping, modal mass cannot be used for modal scaling. For this case, and increasingly for the undamped and proportionally damped cases as well, the modal A scaling factor is used as the basis for the relationship between the scaled modal vectors and the residues determined from the measured frequency-response functions. This relationship is as follows:

$$M_{A_r} = \frac{\psi_{pr}\psi_{qr}}{A_{pqr}} = \frac{1}{Q_r} \qquad (21.23)$$

Note that this definition of modal A is also developed in terms of displacement over force units. Once the modal A is known, *modal B* (M_{B_r}) can be obtained through the following single degree-of-freedom equation:

$$M_{B_r} = -\lambda_r M_{A_r} \qquad (21.24)$$

For undamped and proportionally damped systems, the relationship between the modal mass and the modal A scaling factors can be uniquely determined as

$$M_{A_r} = \pm j2M_r\omega_r \qquad (21.25)$$

In general, the modal vectors are considered to be dimensionless since they represent relative patterns of motion. Therefore, the modal mass or modal A scaling terms carry the units of the respective measurement. For example, the development of the frequency response is based upon displacement over force units. The residue must have units of length over force-seconds. Since the modal A scaling coefficient is inversely related to the residue, modal A has units of force-seconds over length. This unit combination is the same as mass over seconds. Likewise, since modal mass is related to modal A, for proportionally damped systems, through a direct relationship involving the damped natural frequency, the units of modal mass are mass units, as expected.

DAMPING MECHANISMS

In order to evaluate multiple degree-of-freedom systems that are present in the real world, the effect of damping on the complex frequencies and modal vectors must be considered. Many physical mechanisms are needed to describe all of the possible forms of damping that may be present in a particular structure or system. Some of the classical types are (1) structural damping, (2) viscous damping, and (3) Coulomb damping. It is generally difficult to ascertain which type of damping is present in any particular structure. Indeed most structures exhibit damping characteristics that result from a combination of all the above, plus others that have not been described here. (Damping is described in detail in Chap. 36.)

Rather than consider the many different physical mechanisms, the probable location of each mechanism, and the particular mathematical representation of the mechanism of damping that is needed to describe the dissipative energy of the system, a model is used that is concerned only with the resultant mathematical form. This model represents a hypothetical form of damping that is proportional to the system mass or stiffness matrix. Therefore

$$[C] = \alpha[M] + \beta[K] \tag{21.26}$$

Under this assumption, *proportional damping* is the case where the equivalent damping matrix is equal to a linear combination of the mass and stiffness matrices. For this mathematical form of damping, the coordinate transformation that diagonalizes the system mass and stiffness matrices also diagonalizes the system damping matrix. *Nonproportional damping* is the case where this linear combination does not exist.

Therefore when a system with proportional damping exists, that system of coupled equations of motion can be transformed to a system of equations that represent an uncoupled system of single degree-of-freedom systems that are easily solved. With respect to modal parameters, a system with proportional damping has real-valued modal vectors (*real* or *normal modes*), while a system with nonproportional damping has complex-valued modal vectors (*complex modes*).

EXPERIMENTAL MODAL ANALYSIS METHODS

In order to understand the various experimental approaches used to determine the modal parameters of a structure, some sort of outline of the various techniques is helpful in categorizing the different methods that have been developed over the last fifty years. One of several overlapping approaches can be used. One approach is to group the methods according to whether one mode or multiple modes are excited at one time. The terminology that is used for this is

- Phase resonanance (single mode)
- Phase separation (multiple mode)

A slightly more detailed approach is to group the methods according to the type of measured data that is acquired. When this approach is utilized, the relevant terminology is

- Sinusoidal input-output model (forced normal mode)
- Frequency-response function model
- Damped complex exponential response model
- General input-output model

A very common approach to comparing and contrasting experimental modal analysis methodologies that is often used in the literature is based upon the type of model that is used in the modal parameter estimation stage. The relevant nomenclature for this approach is

- Parametric model
  - Modal model
  - $[M], [K], [C]$ model
- Nonparametric model

Finally, the different experimental modal analysis approaches may be grouped according to the domain in which the modal parameter estimation model is formulated. The relevant nomenclature for this approach is

- Time domain
- Frequency domain
- Spatial domain

Regardless of the approach used to organize or classify the different approaches to generating modal parameters from experimental data, the fundamental underlying theory is the same. The differences largely are a matter of logistics, user experience requirements, or numerical or computational limitations rather than the fundamental superiority or inferiority of the method. Most methodology is based upon measured frequency-response or impulse-response functions. Further discussion of experimental modal analysis is limited to techniques related to the measurement and use of these functions for determining modal parameters. The most widely utilized methods are discussed in detail in a following section on *Modal Parameter Estimation*.

MODAL DATA ACQUISITION

Acquisition of data that are used in the formulation of a modal model involves many important technical concerns. The primary concern is the digital signal processing, or the converting of analog signals into a corresponding sequence of digital values that accurately describe the time-varying characteristics of the inputs to and responses from a system. Once the data are available in digital form, the most common approach is to transform the data from the time domain to the frequency domain by use of a discrete Fourier transform algorithm. Since this algorithm involves discrete data over a limited time period, there are large potential problems with this approach that must be well understood. (Data acquisition and analysis are discussed in detail in Chap. 27.)

DIGITAL SIGNAL PROCESSING

In order to determine modal parameters, the measured input (excitation) and response data must be processed and put into a form that is compatible with the test and modal parameter estimation methods. As a result, digital signal processing of the data is a very important step in structural testing. This is one of the technology areas where a clear understanding of the time-frequency-Laplace domain relationships is important. The conversion of the data from the time domain into the frequency and Laplace domains is important both in the measurement process and subsequently in the parameter estimation process.

Digital signal processing of the measured input and response data is used for the following reasons:

- *Condensation.* In general, the amount of measured data tremendously exceeds the information present in the desired measurements (frequency response, unit impulse response, coherence function, etc.). Therefore, digital signal processing is used to condense the data.

- *Measurements.* The measurements which are used subsequently in the modal parameter estimation process are estimated. Since there are many excitation, measurement, and modal parameter estimation procedures, there likewise are a large number of digital signal processing options which can be used.

- *Noise reduction.* Signal processing is used to reduce the influences of noise in the measurement process. The types of noise are classified as follows:

  - *Noncoherent noise.* This noise is due to electrical noise on the transducer signals or unmeasured excitation sources, etc., which are noncoherent with respect to the measured input signals or to some other signal which is used in the averaging process. Zero mean noncoherent noise can be eliminated by averaging with respect to a reference signal. This reference signal can be the input signal in terms of a spectrum averaging process, or it can be a synchronization or trigger signal in terms of cyclic averaging or random decrement process.

  - *Signal processing noise.* The signal processing itself may generate noise. For example, *leakage* is a classic source of noise when using fast Fourier transforms (FFT) for computing frequency-domain measurements. This type of noise is reduced or eliminated by using completely observed time signals (periodic or transient), by using various types of windows, or by increasing the frequency resolution.

  - *Nonlinear noise.* If the system is nonlinear, then free decay, frequency-response, or unit-impulse function measurements may be distorted, which consequently causes problems when estimating modal parameters. Nonlinear distortion noise is eliminated by linearizing the test structure before testing or by randomizing the input signals to the structure. This causes the nonlinear distortion noise to become noncoherent with respect to the input signal. The nonlinear noise can then be averaged from the data in the same manner as ordinary noncoherent noise.

The process of representing an analog signal as a series of digital values is a basic requirement of digital signal processing analyzers. In practice, the goal of the analog-to-digital conversion (ADC) process is to obtain the conversion while maintaining sufficient accuracy in terms of frequency, magnitude, and phase. When dealing strictly with analog devices, this concern is satisfied by the performance characteristics of each individual analog device. With the advent of digital signal processing, the performance characteristics of the analog device are only the first criteria considered. The characteristics of the analog-to-digital conversion are also very important.

This process of analog-to-digital conversion involves two separate concepts, each of which is related to the dynamic performance of a digital signal processing analyzer. *Sampling* is the part of the process related to the timing between individual digital pieces of the time-history. *Quantization* is the part of the process related to describing an analog amplitude as a digital value. Primarily, sampling considerations alone affect the frequency accuracy, while both sampling and quantization considerations affect magnitude and phase accuracy. The two constraining relationships that govern the sampling process are known as Shannon's sampling theorem (Fig. 21.6) and Rayleigh's criterion (Fig. 21.7). The selection of the sampling parameters by way of these constraints is discussed in Chap. 27.

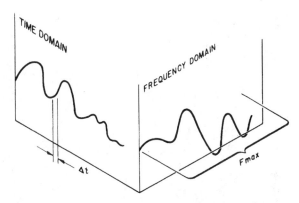

FIGURE 21.6 Shannon's sampling theorem: maximum frequency relationship.

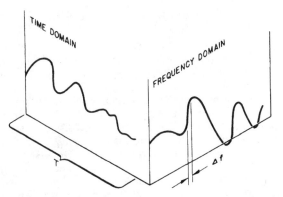

FIGURE 21.7 Rayleigh's criterion: frequency resolution relationship.

DISCRETE FOURIER TRANSFORM

The Fourier series concept explains that any physically realizable signal (signal that satisfies the Dirochlet conditions) can be uniquely separated into a summation of sine and cosine terms at appropriate frequencies. This generates a unique set of sine and cosine terms because of the orthogonal nature of sine functions at different frequencies, the orthogonal nature of cosine functions at different frequencies, and the orthogonal nature of sine functions compared to cosine functions. If the choice of frequencies is limited to a discrete set of frequencies, the discrete Fourier transform describes the amount of each sine and cosine term at each discrete frequency. The real part of the discrete Fourier transform describes the amount of each cosine term; the imaginary part of the discrete Fourier transform describes the amount of each sine term. Figure 21.8 is a graphical representation of this concept for a signal that can be represented by a summation of sinusoids.

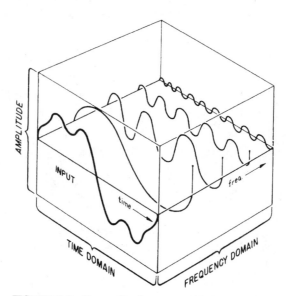

FIGURE 21.8   Discrete Fourier transform concept.

The discrete Fourier transform algorithm is the basis for the formulation of any frequency-domain function in digital data acquisition systems. In terms of an integral Fourier transform, the function must exist for all time in a continuous sense in order to be evaluated. For the realistic measurement situation, data are available in a discrete sense over a limited time period. The discrete Fourier transform, therefore, is based upon a set of assumptions concerning this discrete sequence of values. The assumptions can be reduced to two, of which one must be met by every signal processed by the discrete Fourier transform algorithm. The first assumption is that the signal must be a *totally observed transient* with respect to the time period of observation. If this is not true, then the signal must be composed only of *harmonics of the time period of observation*. If one of these two assumptions is met by any discrete history processed by the discrete Fourier transform algorithm, then the resulting spectrum does not contain bias errors. Much of the data processing effort, with respect to acquisition of data used for the formulation of a modal model, is concerned with the assurance that the input and response histories match one of these two assumptions. A more complete discussion of the discrete Fourier transform algorithm is included in Chap. 14.

ERRORS

The accurate measurement of frequency-response functions depends heavily upon the errors involved with the digital signal processing. In order to take full advantage of experimental data in the evaluation of experimental procedures and verification of theoretical approaches, the errors in measurement, generally designated noise, must be reduced to acceptable levels. With the increasing use of personal computer (PC) instrumentation, the user must take great care to be certain that errors are minimized. With respect to the frequency-response function measurement, the errors in the estimate are generally grouped into two categories: variance and bias. The *variance* portion of the error is due to random deviations of each sample function from the mean. Statistically, then, if sufficient sample functions are evaluated, the estimate closely approximates the true function with a high degree of confidence. The *bias* portion of the error, on the other hand, is not necessarily reduced by using many samples. The bias error is due to a system characteristic or measurement procedure that consistently results in an incorrect estimate. Therefore, the expected value is not equal to the true value. Examples of this are system nonlinearities or digitization errors such as aliasing or leakage. With this type of error, knowledge of the form of the error is vital in reducing the resultant effect in the frequency-response function measurement.

TRANSDUCER CONSIDERATIONS

The transducer considerations are often the most overlooked aspect of the experimental modal analysis process. Considerations involving the actual type and specifications of the transducers, mounting of the transducers, and calibration of the transducers are often among the largest sources of error. Chapter 12 discusses transducers and transducer design in significant detail. Calibration of transducers is reviewed in Chap. 18. Chapter 15 discusses measurement techniques, including transducer mounting and alignment. These topics are critical to estimating the accurate frequency-response function measurements required for most experimental modal analysis methods.

TEST ENVIRONMENT CONSIDERATIONS

The test environment for any modal analysis test involves several environmental factors as well as appropriate boundary conditions. Primarily, temperature, humidity, vacuum, and gravity effects must be properly considered to match previous analysis models or to allow the experimentally determined model to properly reflect the system. Very few experimental laboratory facilities have the capability to control these factors in other than a rudimentary fashion.

In addition to the environmental concerns, the boundary conditions of the system under test are very important. Traditionally, modal analysis tests have been performed under the assumption that the test boundary conditions can be made to conform to one of four conditions:

- Free-free boundary conditions (impedance is zero).
- Fixed boundary conditions (impedance is infinite).
- Operating boundary conditions (impedance is correct).
- Arbitrary boundary conditions (impedance is known).

Except in very special situations, none of these boundary conditions can be practically achieved. Instead, practical guidelines are normally used to evaluate the appropriateness of the chosen boundary conditions. For example, if a free-free boundary is chosen, the desired frequency of the highest rigid body mode must be at least a factor of 10 below the first deformation mode of the system under test. Likewise, for the fixed-boundary test, the desired interface stiffness must be at least a factor of 10 greater than the local stiffness of the system under test. While either of these practical guidelines can be achieved for small test objects, a large class of systems cannot be acceptably tested in either configuration. Arguments have been made that the impedance of a support system can be defined (via test and/or analysis) and the effects of such a support system eliminated from the measured data. This technique is theoretically sound, but, because of significant dynamics in the support system and limited measurement dynamics, the approach has not been uniformly applicable.

In response to this problem, many alternative structural testing concepts have been proposed. Active suspension systems and combinations of active and passive systems are being evaluated, particularly for application to very flexible space structures. Active inert-gas suspension systems have been used in the past for the testing of smaller commercial and military aircraft, and, in general, such approaches are formulated to better match the requirements of a free-free boundary condition.

Another alternative test procedure is to define a series of relatively conventional tests with various boundary conditions. These various boundary conditions are chosen in such a way that each perturbed boundary condition can be accurately modeled (for example, the addition of a large mass at interface boundaries). Therefore, as the experimental model is acquired for each configuration and used to validate and correct the associated analytical model, the underlying model is validated and corrected accordingly. This procedure has the added benefit of adding the influence of modes of vibration that occur above the maximum frequency of the test into the validation of the model.

MEASUREMENT FORMULATION

For current approaches to experimental modal analysis, the frequency-response function is the most important, and most common, measurement to be made. When estimating frequency-response functions, a measurement model is needed that allows the frequency-response function to be estimated from measured input and output data in the presence of noise (errors). These errors have been discussed in this and other chapters in great detail.

There are at least four different testing configurations that can be considered. These different testing conditions are largely a function of the number of acquisition channels or excitation sources that are available to the test engineer.

- Single input/single output (SISO)
- Single input/multiple output (SIMO)
- Multiple input/single output (MISO)
- Multiple input/multiple output (MIMO)

In general, the best testing situation is the multiple input/multiple output configuration (MIMO), since the data are collected in the shortest possible time with the fewest changes in the test conditions.

FREQUENCY-RESPONSE FUNCTION ESTIMATION

The estimation of the frequency-response function depends upon the transformation of data from the time to the frequency domain. The Fourier transform is used for this computation. The computation is performed digitally using a fast Fourier transform algorithm. The frequency-response functions satisfy the following single and multiple input relationships:

Single input relationship:

$$X_p = H_{pq} F_q \tag{21.27}$$

Multiple input relationship:

$$
\begin{bmatrix} X_1 \\ X_2 \\ \cdot \\ \cdot \\ \cdot \\ X_p \end{bmatrix}_{N_o \times 1}
=
\begin{bmatrix} H_{11} & \cdots & H_{1q} \\ H_{21} & & H_{2q} \\ \cdot & & \cdot \\ \cdot & & \cdot \\ \cdot & & \cdot \\ H_{p1} & \cdots & H_{pq} \end{bmatrix}_{N_o \times N_i}
\begin{bmatrix} F_1 \\ F_2 \\ \cdot \\ \cdot \\ \cdot \\ F_q \end{bmatrix}_{N_i \times 1}
\tag{21.28}
$$

The most reasonable, and most common, approach to the estimation of frequency-response functions is the use of *least squares* (LS) or *total least squares* (TLS) techniques.[8,9] These are standard techniques for estimating parameters in the presence of noise. Least squares methods minimize the square of the magnitude error and thus compute the *best* estimate of the magnitude of the frequency-response function, but they have little effect on the phase of the frequency-response function. The primary difference in the algorithms used to estimate frequency-response functions is in the assumption of where the noise enters the measurement problem. Three algorithms, referred to as the H_1, H_2, and H_v algorithms, are commonly available for estimating frequency response functions. Table 21.1 summarizes the assumed location of the noise for these three algorithms.

TABLE 21.1 Summary of Frequency-Response Function
Estimation Models

	Frequency-response function models		
		Assumed location of noise	
Technique	Solution method	Force inputs	Response
H_1	LS	No noise	Noise
H_2	LS	Noise	No noise
H_v	TLS	Noise	Noise

Consider the case of N_i inputs and N_o outputs measured during a modal test. Based upon the assumed location of the noise entering the estimation process, Eqs. (21.29) through (21.31) represent the corresponding model for the H_1, H_2, and H_v estimation procedures.

H_1 technique:

$$[H]_{N_o \times N_i} \{F\}_{N_i \times 1} = \{X\}_{N_o \times 1} - \{\eta\}_{N_o \times 1} \tag{21.29}$$

H_2 technique:

$$[H]_{N_o \times N_i} \{ \{F\}_{N_i \times 1} - \{\upsilon\}_{N_i \times 1}\} = \{X\}_{N_o \times 1} \tag{21.30}$$

H_v technique:

$$[H]_{N_o \times N_i} \{ \{F\}_{N_i \times 1} - \{\upsilon\}_{N_i \times 1}\} = \{X\}_{N_o \times 1} - \{\eta\}_{N_o \times 1} \tag{21.31}$$

This numerical model can be represented in block diagram form as shown in Fig. 21.9.

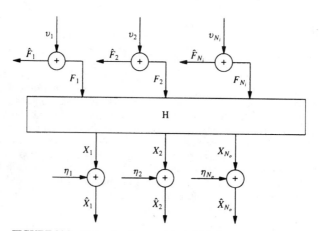

FIGURE 21.9 General system model: multiple inputs.

Single Input FRF Estimation. With reference to Fig. 21.9 for a case involving only one input and one output (input location q and response location p), the equation that is used to represent the input-output relationship is

$$\hat{X}_p - \eta_p = H_{pq}(\hat{F}_q - \upsilon_q) \tag{21.32}$$

where $F = \hat{F} - \upsilon$ = actual input
$X = \hat{X} - \eta$ = actual output
$\hat{X}$ = spectrum of the pth output, measured
$\hat{F}$ = spectrum of the qth input, measured
H = frequency-response function
υ = spectrum of the noise part of the input
η = spectrum of the noise part of the output
X = spectrum of the pth output, theoretical
F = spectrum of the qth input, theoretical

If $\upsilon = \eta = 0$, the theoretical (expected) frequency-response function of the system is estimated. If $\eta \neq 0$ and/or $\upsilon \neq 0$, a least squares method is used to estimate a *best* frequency-response function in the presence of noise.

In order to develop an estimation of the frequency-response function, a number of averages N_{avg} is used to minimize the random errors (variance). This can be easily accomplished through use of intermediate measurement of the auto and cross power spectrums. The estimate of the auto- and cross-power spectrums for the model in Fig. 21.9 is defined as follows. Note that each function is a function of frequency.

Cross-power spectra:

$$GXF_{pq} = \sum_{1}^{N_{avg}} X_p F_q^* \tag{21.33}$$

$$GFX_{qp} = \sum_{1}^{N_{avg}} F_q X_p^* \tag{21.34}$$

Auto-power spectra:

$$GFF_{qq} = \sum_{1}^{N_{avg}} F_q F_q^* \tag{21.35}$$

$$GXX_{pp} = \sum_{1}^{N_{avg}} X_p X_p^* \tag{21.36}$$

where F^* = complex conjugate of $F(\omega)$
X^* = complex conjugate of $X(\omega)$

H_1 Algorithm: Minimize Noise on Output (η). The most common formulation of the frequency-response function, often referred to as the H_1 algorithm, tends to minimize the noise on the output. This formulation is shown in Eq. (21.37).

$$H_{pq} = \frac{GXF_{pq}}{GFF_{qq}} \tag{21.37}$$

H₂ Algorithm: Minimize Noise on Input (υ). Another formulation of the frequency-response function, often referred to as the H_2 algorithm, tends to minimize the noise on the input. This formulation is shown in Eq. (21.38).

$$H_{pq} = \frac{GXX_{pp}}{GFX_{qp}} \tag{21.38}$$

In the H_2 formulation, an auto-power spectrum is divided by a cross-power spectrum. This can be a problem since the cross-power spectrum can theoretically be zero at one or more frequencies. In both formulations, the phase information is preserved in the cross-power spectrum term.

Hᵥ Algorithm: Minimize Noise on Input and Output (η and υ). The solution for H_{pq} using the H_v algorithm is found by the eigenvalue decomposition of a matrix of power spectra. For the single input case, the following matrix involving the auto- and cross-power spectra can be defined:

$$[GFFX_p] = \begin{bmatrix} GFF_{qq} & GXF_{pq} \\ GXF_{pq}{}^H & GXX_{pp} \end{bmatrix}_{2 \times 2} \tag{21.39}$$

The solution for H_{pq} is found by the eigenvalue decomposition of the $[GFFX]$ matrix as follows:

$$[GFFX_p] = [V] \lceil \Lambda \rfloor [V]^H \tag{21.40}$$

where $\lceil \Lambda \rfloor$ = diagonal matrix of eigenvalues.

The solution for the H_{pq} matrix is found from the eigenvector associated with the smallest (minimum) eigenvalue λ_1. The size of the eigenvalue problem is second-order, resulting in finding the roots of a quadratic equation. This eigenvalue solution must be repeated for each frequency, and the complete solution process must be repeated for each response point X_p.

Alternatively, the solution for H_{pq} is found by the eigenvalue decomposition of the following matrix of auto- and cross-power spectra:

$$[GXFF_p] = \begin{bmatrix} GXX_{pp} & GXF_{pq}{}^H \\ GXF_{pq} & GFF_{qq} \end{bmatrix}_{2 \times 2} \tag{21.41}$$

$$[GXFF_p] = [V] \lceil \Lambda \rfloor [V]^H \tag{21.42}$$

where $\lceil \Lambda \rfloor$ = diagonal matrix of eigenvalues.

The solution for H_{pq} is again found from the eigenvector associated with the smallest (minimum) eigenvalue λ_1.

The frequency-response function is found from the normalized eigenvector associated with the smallest eigenvalue. If $[GFFX_p]$ is used, the eigenvector associated with the smallest eigenvalue must be normalized as follows:

$$\{V\}_{\lambda\text{min}} = \begin{Bmatrix} H_{pq} \\ -1 \end{Bmatrix} \tag{21.43}$$

If $[GXFF_p]$ is used, the eigenvector associated with the smallest eigenvalue must be normalized as follows:

$$\{V\}_{\lambda\,\min} = \begin{Bmatrix} -1 \\ H_{pq} \end{Bmatrix} \tag{21.44}$$

One important consideration in choosing one of the three formulations for frequency-response function estimation is the behavior of each formulation in the presence of a bias error such as leakage. In all cases, the estimate differs from the expected value, particularly in the region of a resonance (magnitude maximum) or antiresonance (magnitude minimum). For example, H_1 tends to underestimate the value at resonance, while H_2 tends to overestimate the value at resonance. The H_v algorithm gives an answer that is always bounded by the H_1 and H_2 values. The different approaches are based upon minimizing the magnitude of the error but have no effect on the phase characteristics.

In addition to the attractiveness of H_1, H_2, and H_v in terms of the minimization of the error, the availability of auto- and cross-power spectra allows the determination of other important functions. The quantity γ_{pq}^2 is called the scalar or *ordinary coherence function* and is a frequency-dependent, real value between 0 and 1. The ordinary coherence function indicates the degree of causality in a frequency-response function. If the coherence is equal to 1 at any specific frequency, the system is said to have perfect causality at that frequency. In other words, the measured response power is caused totally by the measured input power (or by sources which are coherent with the measured input power). A coherence value less than unity at any frequency indicates that the measured response power is greater than that caused by the measured input. This is due to some extraneous noise also contributing to the output power. It should be emphasized, however, that low coherence does not necessarily imply poor estimates of the frequency-response function; it simply means that more averaging is needed for a reliable result. The ordinary coherence function is computed as follows:

$$COH_{pq} = \gamma_{pq}^2 = \frac{|GXF_{pq}|^2}{GFF_{qq}GXX_{pp}} = \frac{GXF_{pq}GFX_{qp}}{GFF_{qq}GXX_{pp}} \tag{21.45}$$

When the coherence is zero, the output is caused totally by sources other than the measured input. In general, then, the coherence can be a measure of the degree of noise contamination in a measurement. Thus, with more averaging, the estimate of coherence may contain less variance, therefore giving a better estimate of the noise energy in a measured signal. This is not the case, though, if the low coherence is due to bias errors such as nonlinearities, multiple inputs, or leakage. A typical ordinary coherence function is shown in Fig. 21.10 together with the corresponding frequency-response function magnitude. In Fig. 21.10, the frequencies where the coherence is lowest are often the same frequencies where the frequency-response function is at a maximum or a minimum in magnitude. This is often an indication of leakage since the frequency-response function is most sensitive to leakage error at the lightly damped peaks corresponding to the maxima. At the minima, where there is little response from the system, the leakage error, even though it is small, may still be significant.

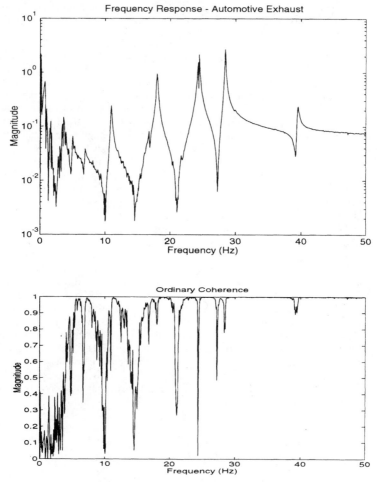

FIGURE 21.10 Frequency-response function magnitude with associated ordinary coherence function.

In all of these cases, the estimated coherence function approaches, in the limit, the expected value of coherence at each frequency, dependent upon the type of noise present in the structure and measurement system. Note that, with more averaging, the estimated value of coherence does not increase; the estimated value of coherence always approaches the expected value from the upper side.

Multiple Input FRF Estimation. Multiple input estimation of frequency-response functions is desirable for several reasons. The principal advantage is the increase in the accuracy of estimates of the frequency-response functions. During single input excitation of a system, large differences in the amplitudes of vibratory motion at various locations may exist due to the dissipation of the excitation power within the structure. This is especially true when the structure has heavy damping. Small nonlinearities in the structure consequently cause errors in the measurement of the response. With multiple input excitation, the vibratory amplitudes across the structure typically are more uniform, with a consequent decrease in the effect of nonlinearities.

A second reason for improved accuracy is the increase in consistency of the frequency-response functions compared to the single input method. When a number of exciter systems are used, the elements from columns of the frequency-response function matrix corresponding to those exciter locations are being determined simultaneously. With the single input method, each column is determined independently, and it is possible for small errors of measurement due to nonlinearities and time-dependent system characteristics to cause a change in resonance frequencies, damping, or mode shapes among the measurements in the several columns. This is particularly important for the polyreference modal parameter estimation algorithms that use frequency-response functions from multiple columns or rows of the frequency-response function matrix simultaneously.

An additional, significant advantage of multiple input excitation is a reduction in the test time. In general, when multiple input estimation of frequency-response functions is used, frequency-response functions are obtained for all input locations in approximately the same time as required for acquiring a set of frequency-response functions for one of the input locations using a single input estimation method.[4,11,12]

With reference to Fig. 21.9 for a case involving N_i inputs and N_o outputs, the equation that is used to represent the input-output relationship is

$$\hat{X}_p - \eta_p = \sum_{q=1}^{N_i} H_{pq}(\hat{F}_q - \upsilon_q) \qquad (21.46)$$

In order to develop an estimation of the frequency-response function for the multiple input case, a number of averages N_{avg} is used to minimize the random errors (variance). This can be easily accomplished through use of intermediate measurement of the auto- and cross-power spectra as defined in Eqs. (21.33) through (21.36). Additional matrices, compared to the single input case, need to be defined. These additional matrices are constructed from the auto- and cross-power spectra previously defined for the single input case. Each function and, therefore, each resulting matrix is a function of frequency.

Input-output cross-spectra matrix:

$$[GXF] = \{X\}\{F\}^H = \begin{Bmatrix} X_1 \\ X_2 \\ . \\ . \\ . \\ X_{N_o} \end{Bmatrix} [F_1{}^* \ F_2{}^* \ \ldots \ F_{N_i}{}^*] = \begin{bmatrix} GXF_{11} \ldots GXF_{1N_i} \\ . \qquad\qquad . \\ . \qquad\qquad . \\ . \qquad\qquad . \\ GXF_{N_o 1} \ldots GXF_{N_o N_i} \end{bmatrix} \qquad (21.47)$$

Input cross-spectra matrix:

$$[GFF] = \{F\}\{F\}^H = \begin{Bmatrix} F_1 \\ F_2 \\ . \\ . \\ . \\ F_{N_i} \end{Bmatrix} [F_1^* \; F_2^* \ldots F_{N_i}^*] = \begin{bmatrix} GFF_{11} \ldots GFF_{1N_i} \\ . \\ . \\ . \\ GFF_{N_i1} \ldots GFF_{N_iN_i} \end{bmatrix} \tag{21.48}$$

The frequency-response functions can now be estimated for the three algorithms.

H_1 algorithm: Minimize noise on output (η)

$$[H] = [GXF][GFF]^{-1} \tag{21.49}$$

In the experimental procedure, the input and response signals are measured, and the averaged cross-spectra and autospectra necessary to create the $[GXF]$, $[GFF]$, and $[GXX]$ matrices are computed. The input cross-spectrum matrix must be inverted, at least implicitly, at every frequency in the analysis range. This means that the computational load on the measurement system is greater than for the single input case, in which only the reciprocal of a single input autospectrum is computed.

Equation (21.49) is valid unless the input cross-spectrum matrix $[GFF]$ is singular for specific frequencies or frequency intervals. When this happens, the inverse of $[GFF]$ does not exist and Eq. (21.49) cannot be used to solve for the frequency-response function at those frequencies or in those frequency intervals. A computational procedure that solves Eq. (21.49) for $[H]$ must monitor the rank of the matrix $[GFF]$ that is to be inverted, and provide information on how to alter the input signals or use the available data when a problem exists. The current approach for evaluating whether the inputs are sufficiently uncorrelated at each frequency involves determining the principal/virtual forces using principal component analysis.[10]

H_2 algorithm: Minimize noise on input (υ)

$$[H] = [GXX][GFX]^{-1} \tag{21.50}$$

One problem with using the H_2 algorithm is that the solution for $[H]$ can be found directly using an inverse only when the number of inputs N_i and number of outputs N_o are equal.

H_v algorithm: Minimize noise on input and output (υ and η). The solution for $[H]$ is found by the eigenvalue decomposition of one of the following two matrices:

$$[GFFX_p] = \begin{bmatrix} [GFF] & [GXF_p] \\ [GXF_p]^H & GXX_p \end{bmatrix}_{(N_i + 1) \times (N_i + 1)} \tag{21.51}$$

$$[GXFF_p] = \begin{bmatrix} GXX_p & [GXF_p]^H \\ [GXF_p] & [GFF] \end{bmatrix}_{(N_i + 1) \times (N_i + 1)} \tag{21.52}$$

Therefore, the eigenvalue decomposition is

$$[GFFX_p] = [V] \lceil \Lambda \rfloor [V]^H \tag{21.53}$$

or

$$[GXFF_p] = [V] \lceil \Lambda \rfloor [V]^H \tag{21.54}$$

where $\lceil \Lambda \rfloor$ = diagonal matrix of eigenvalues.

The solution for the pth row of the $[H]$ matrix is found from the eigenvector associated with the smallest (minimum) eigenvalue. Note that the size of the eigenvalue problem is $N_i + 1$ and that the eigenvalue solution must be repeated for each frequency. The complete solution process must be repeated for each response point X_p.

The frequency-response function associated with a single output p and all inputs is found by normalizing the eigenvector associated with the smallest eigenvalue. If $[GFFX_p]$ is used, the eigenvector associated with the smallest eigenvalue must be normalized as follows:

$$\{V\}_{\lambda_{\min}} = \begin{Bmatrix} H_{p1} \\ H_{p2} \\ \cdot \\ \cdot \\ H_{pN_i} \\ -1 \end{Bmatrix} \tag{21.55}$$

If $[GXFF_p]$ is used, the eigenvector associated with the smallest eigenvalue must be normalized as follows:

$$\{V\}_{\lambda_{\min}} = \begin{Bmatrix} -1 \\ H_{p1} \\ H_{p2} \\ \cdot \\ \cdot \\ H_{pN_i} \end{Bmatrix} \tag{21.56}$$

The concept of the coherence function, as defined for single input measurement, needs to be expanded to include the variety of relationships that are possible for multiple inputs. *Ordinary coherence* is defined in this general sense as the correlation coefficient describing the linear relationship between any two single spectra. Great care must be taken in the interpretation of ordinary coherence when more than one input is present. The ordinary coherence of an output with respect to an input can be much less than unity even though the linear relationship between inputs and outputs is valid, because of the influence of the other inputs.[4–6]

The ordinary coherence function can be formulated in terms of the elements of the matrices defined previously. The ordinary coherence function between the pth output and the qth input can be computed from the following formula:

Ordinary coherence function:

$$COH_{pq} = \lambda_{pq}^{2} = \frac{|GXF_{pq}|^2}{GFF_{qq}GXX_{pp}}$$
(21.57)

where GXX_{pp} = auto-power spectrum of the output p
         GFF_{qq} = auto-power spectrum of the input q
         GXF_{pq} = cross-power spectrum between output p and input q

Partial coherence is defined as the ordinary coherence between a conditioned output and another conditioned output, between a conditioned input and another conditioned input, or between a conditioned input and a conditioned output. The output and input are conditioned by removing contributions from other input(s). The removal of the effects of the other input(s) is formulated on a linear least squares basis. The order of removal of the inputs during "conditioning" has a definite effect upon the partial coherence if some of the input(s) are mutually correlated. There is a partial coherence function for every input-output, input-input, and input-output combination for all permutations of conditioning. The usefulness of partial coherence for experimental modal analysis is limited.

Multiple coherence is defined as the correlation coefficient describing the linear relationship between an output and all known inputs. There is a multiple coherence function for every output. Multiple coherence can be used to evaluate the importance of unknown contributions to each output. These unknown contributions can be measurement noise, nonlinearities, or unknown inputs. In particular, as in the evaluation of ordinary coherence, a low value of multiple coherence near a resonance often means that leakage error is present in the frequency-response function.

The formulation of the equations for the multiple coherence functions can be simplified to the following equation:

Multiple coherence function:

$$MCOH_{p} = \sum_{q=1}^{N_i} \sum_{t=1}^{N_i} \frac{H_{pq}GFF_{qt}H_{pt}^{*}}{GXX_{pp}}$$
(21.58)

where H_{pq} = frequency response function for output p and input q
         H_{pt} = frequency response function for output p and input t
         GFF_{qt} = cross power spectrum between output q and output t

If the multiple coherence of the pth output is near unity, then the pth output is well predicted from the set of inputs using the least squares frequency-response functions.

Multiple Input Force Analysis/Evaluation.   Of the variety of situations that can cause difficulties in the computation of the frequency-response functions, the one with the highest potential for trouble is the case of coherent inputs. If two of the inputs are fully coherent, then there are no unique frequency-response functions associated with those inputs. Unfortunately, there are a number of situations where the input cross-spectrum matrix $[GFF]$ may be singular at specific frequencies or frequency intervals. When this happens, the inverse of $[GFF]$ does not exist, and Eq. (21.49) cannot be used to solve for the frequency-response function at those frequencies or in those frequency intervals. First, one of the input autospectra may be zero in amplitude over some frequency interval. Second, two or more of the input signals may be fully coherent over some frequency interval. Third, numerical problems which cause the computation of the inverse to be inexact may be present.

The current approach used to detect correlated inputs involves utilizing principal component analysis to determine the number of forces contributing to the $[GFF]$ matrix. In this approach, a principal component analysis must be conducted on the $[GFF]$ matrix.[10] *Principal component analysis* involves an eigenvalue decomposition of the $[GFF]$ matrix. Since the eigenvectors of such a decomposition are unitary, the eigenvalues should all be of approximately the same size if each of the inputs is contributing. If one of the eigenvalues is much smaller at a particular frequency, one of the inputs is not present or one of the inputs is correlated with the other input(s).

$$[GFF] = [V]\,[\Lambda]\,[V]^H \qquad (21.59)$$

$[\Lambda]$ represents the eigenvalues of the $[GFF]$ matrix. If any of the eigenvalues of the $[GFF]$ matrix are zero or insignificant, then the $[GFF]$ matrix is singular. Therefore, for a three-input test, the $[GFF]$ matrix should have three eigenvalues of approximately the same magnitude. (The number of distinct eigenvalues is equal to the number of uncorrelated inputs.) Figure 21.11 shows the principal force plots for a case with three inputs. At the frequencies where the third principal/virtual force drops (lowest curve), the inputs are mutually correlated.

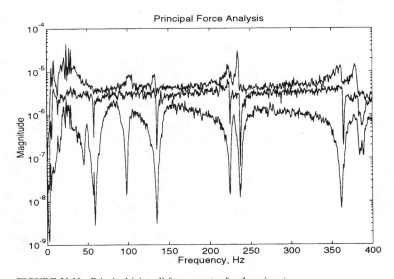

FIGURE 21.11 Principal (virtual) force spectra for three inputs.

PRACTICAL MEASUREMENT CONSIDERATIONS

There are several factors that contribute to the quality of actual measured frequency-response function estimates. Some of the most common sources of error involve measurement mistakes. With a proper measurement approach, most errors of this type, such as overloading the input, extraneous signal pick-up via ground loops or strong electric or magnetic fields nearby, etc., can be avoided. Violations of test assumptions are often the source of another inaccuracy and can be viewed as a measurement mistake. For example, frequency-response and coherence functions have been defined as parameters of a linear system. Nonlinearities generally shift energy from one frequency to many new frequencies, in a way which may be difficult to recognize. The result is a distortion in the estimates of the system parameters, which may not be apparent unless the excitation is changed. One way to reduce the effect of nonlinearities is to randomize these contributions by choosing a randomly different input signal for each of the contributing averages. Subsequent averaging reduces these contributions in the same way that random noise is reduced. Another example involves control of the system input. One requirement is to excite the system with energy at all frequencies for which measurements are expected. It is important to be sure that the input signal spectrum does not have frequency ranges where little energy exists. Otherwise, coherence is very low, and the variance on the frequency-response function is unacceptable.

Assuming that the system is linear, the excitation is proper, and measurement mistakes are avoided, some amount of error and/or noise is still present in the measurement process. Five different approaches can be used to reduce this error involved in frequency-response function measurements. First of all, the use of *different frequency-response function estimation algorithms* (H_v compared to H_1) reduces the effect of the leakage error on the estimation of the frequency-response function computation. The use of *averaging* significantly reduces errors of both variance and bias and is probably the most general technique used to reduce errors in frequency-response function measurement. *Selective excitation* is often used to verify nonlinearities or randomize characteristics. In this way, bias errors due to system sources can be reduced or controlled. The *increase of frequency resolution* through the zoom fast Fourier transform improves the frequency-response function estimate primarily by reducing the leakage bias error through the use of a longer time sample. The zoom fast Fourier transform by itself is a linear process and does not involve any specific error reduction characteristics compared to a baseband fast Fourier transform (FFT). Finally, the *use of weighting functions* (*windows*) is widespread, and much has been written about their value.[4-6] Primarily, weighting functions compensate for the bias error (leakage) caused by the analysis procedure.

Signal Averaging.   The averaging of signals is normally viewed as a summation or weighted summation process where each sample function has a common abscissa. Normally, the designation of *history* is given to sample functions with the abscissa of absolute time, and the designation of *spectrum* is given to sample functions with the abscissa of absolute frequency. The spectra are normally generated by Fourier transforming the corresponding history. In order to generalize and consolidate the concept of signal averaging as much as possible, the case of relative time is also considered. In this way, *relative history* is discussed with units of the appropriate event rather than seconds, and a *relative spectrum* is the corresponding Fourier transform with units of cycles per event. This concept of signal averaging is used widely in structural signature analysis where the event is a revolution of a rotating shaft. This kind of approach simplifies the application of many other concepts of signal relationships, such as Shannon's sampling theorem and Rayleigh's criterion of frequency resolution.

The process of signal averaging as it applies to frequency-response functions is simplified greatly by the intrinsic uniqueness of the frequency-response function. Since the frequency-response function is expressed in terms of system properties of mass, stiffness, and damping, it is reasonable to conclude that in most realistic structures, the frequency-response functions are considered to be constants, just like mass, stiffness, and damping. This concept means that when formulating the frequency-response function using H_1, H_2, or H_v algorithms, the estimate of frequency response is intrinsically unique, as long as the system is linear and the noise can be eliminated. In general, the auto- and cross-power spectra are statistically unique only if the input is stationary and sufficient averages are taken. Nevertheless, the estimate of frequency response is valid whether the input is stationary, nonstationary, or deterministic.

The concept of the intrinsic uniqueness of the frequency-response function also permits a greater freedom in the testing procedure. Each function is derived as the result of a separate test or as the result of different portions of the same continuous test situation. In either case, the estimate of the frequency-response function is the same as long as the time-history data for the auto- and cross-power spectra that are utilized in any computation of the frequency-response or coherence function are acquired simultaneously.

The approaches to signal averaging vary only in the relationship between the sample functions used. Since the Fourier transform is a linear function, there is no theoretical difference between the use of histories or spectra. (Practically, though, there are precision considerations.) With this in mind, the signal averaging useful to frequency-response function measurements can be divided into three classifications:

- Asynchronous
- Synchronous
- Cyclic

These three classifications refer to the trigger and sampling relationships between sample functions. *Asynchronous averaging* describes the averaging case when each average is acquired without a triggering event; it is sometimes referred to as free-run averaging. *Synchronous averaging* describes the averaging case when each average is acquired only when an external triggering event occurs. *Cyclic averaging* describes the averaging case when each average is acquired with a specific *absolute* time, or phase, relationship to all previous averages. (Averaging is discussed in detail in Chaps. 13 and 27.)

Excitation. Excitation includes any form of input that is used to create a response in a mechanical system. This can include environmental or operational inputs as well as the controlled force input(s) that are used in a vibration or modal analysis test. In general, the following discussion is limited to force inputs that are measured and/or controlled in some rigorous way.[3,13,14]

Excitation Assumptions. The primary assumption concerning the excitation of a linear structure is that the excitation is observable. Whenever the excitation is measured, this assumption simply implies that the measured characteristic properly describes the actual input characteristics. For the case of multiple inputs, the different inputs must often be uncorrelated for the computational procedures to yield a solution. In most cases this means only that the multiple inputs must not be perfectly correlated at any frequency. As long as the excitation is measured, the validity of these limited assumptions can be evaluated.

There are a number of techniques that can be used to estimate modal characteristics from response measurements with no measurement of the excitation. If this approach is used, the excitation assumptions are much more imposing. If the excitation is not measured, estimates of modal scaling (modal mass, modal A, residues, etc.) cannot be generated. Even when these parameters are not required, all of these techniques have one further restriction: an assumption has to be made concerning the characteristics of the excitation of the system. Usually, the autospectrum of the excitation signal is assumed to be constant over the frequency interval of interest. This is not generally practical.

Classification of Excitation.   Inputs which can be used to excite a system in order to determine frequency-response functions belong to one of two classifications. The first classification is that of a random signal. Signals of this form can be defined by their statistical properties only over some time period. Any subset of the total time period is unique, and no explicit mathematical relationship can be formulated to describe the signal. Random signals can be further classified as stationary or nonstationary. Stationary random signals are a special case where the statistical properties of the random signals do not vary with respect to translations with time. Finally, stationary random signals can be classified as ergodic or nonergodic. A stationary random signal is ergodic when a time average on any particular subset of the signal is the same for any arbitrary subset of the random signal. All random signals which are commonly used as input signals fall into the category of ergodic, stationary random signals.

The second classification of inputs which can be used to excite a system in order to determine frequency-response functions is that of a deterministic signal. Signals of this form can be represented in an explicit mathematical relationship. Deterministic signals are further divided into periodic and nonperiodic classifications. The most common inputs in the periodic deterministic signal designation are sinusoidal in nature, while the most common inputs in the nonperiodic deterministic designation are transient in form.

The choice of input to be used to excite a system in order to determine frequency-response functions depends upon the characteristics of the system, the characteristics of the parameter estimation, and the expected utilization of the data. The characterization of the system is primarily concerned with the linearity of the system. As long as the system is linear, all input forms should give the same expected value. Naturally, though, all real systems have some degree of nonlinearity. Deterministic input signals result in frequency-response functions that are dependent upon the signal level and type. A set of frequency-response functions for different signal levels can be used to document the nonlinear characteristics of the system. Random input signals, in the presence of nonlinearities, result in a frequency-response function that represents the best linear representation of the nonlinear characteristics for a given level of random signal input. For small nonlinearities, use of a random input does not differ greatly from the use of a deterministic input.

The characterization of the parameter estimation is primarily concerned with the type of mathematical model being used to represent the frequency-response function. Generally, the model is a linear summation based upon the modal parameters of the system. Unless the mathematical representation of all nonlinearities is known, the parameter estimation process cannot properly weight the frequency-response function data to include nonlinear effects. For this reason, random input signals are regularly used to obtain the best linear estimate of the frequency-response function when a parameter estimation process using a linear model is to be utilized.

The expected utilization of the data is concerned with the degree of detailed information required by any postprocessing task. For experimental modal analysis, this can range from implicit modal vectors needed for troubleshooting to explicit modal vectors used in an orthogonality check. As more detail is required, input signals, both random and deterministic, need to match the system characteristics and parameter estimation characteristics more closely. In all possible uses of frequency-response function data, the conflicting requirements of the need for accuracy, equipment availability, testing time, and testing cost normally reduce the possible choices of input signal.

With respect to the reduction of the variance and bias errors of the frequency-response function, random or deterministic signals can be utilized most effectively if the signals are periodic with respect to the sample period or totally observable with respect to the sample period. If either of these criteria is satisfied, regardless of signal type, the predominant bias error, leakage, is eliminated. If these criteria are not satisfied, the leakage error may become significant. In either case, the variance error is a function of the signal-to-noise ratio and the amount of averaging.

Many signals are appropriate for use in experimental modal analysis. Some of the most commonly used signals are described in the following sections. For those excitation signals that require the use of a shaker, Fig. 21.12 shows a typical test configuration; Fig. 21.13 shows a typical test configuration when an impact form of excitation is to be used. The advantages and disadvantages of each excitation signal are summarized in Table 21.2.

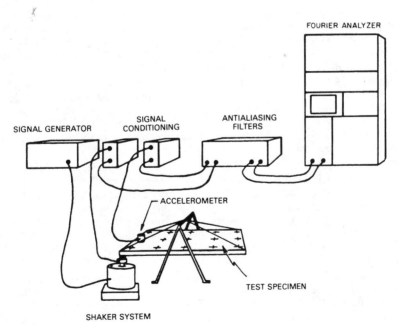

FIGURE 21.12   Typical fixed-input modal test configuration: shaker.

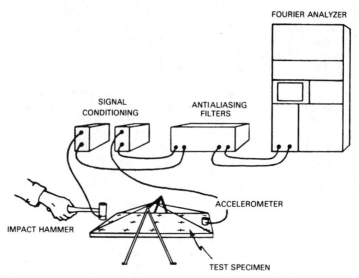

FIGURE 21.13 Typical fixed-response modal test configuration: impact hammer.

Slow swept sine. The slow swept sine signal is a periodic deterministic signal with a frequency that is an integer multiple of the FFT frequency increment. Sufficient time is allowed in the measurement procedure for any transient response to the changes in frequency to decay, so that the resultant input and response histories are periodic with respect to the sample period. Therefore, the total time needed to compute an entire frequency-response function is a function of the number of frequency increments required and the system damping.

Periodic chirp. The periodic chirp is a fast swept sine signal that is a periodic deterministic signal and is formulated by sweeping a sine signal up or down within a frequency band of interest during a single sample period. Normally, the fast swept sine signal is made up of only integer multiples of the FFT frequency increment. This signal is repeated without change so that the input and output histories are periodic with respect to the sample period.

Impact (impulse). The impact signal is a transient deterministic signal which is formed by applying an input pulse lasting only a very small part of the sample period to a system. The width, height, and shape of this pulse determine the usable spectrum of the impact. Briefly, the width of the pulse determines the frequency spectrum, while the height and shape of the pulse control the level of the spectrum. Impact signals have proven to be quite popular due to the freedom of applying the input with some form of an instrumented hammer. While the concept is straightforward, the effective utilization of an impact signal is very involved.[14]

Step relaxation. The step relaxation signal is a transient deterministic signal which is formed by releasing a previously applied static input. The sample period begins at the instant that the release occurs. This signal is normally generated by the application of a static force through a cable. The cable is then cut or allowed to release through a shear pin arrangement.

TABLE 21.2 Characteristics of Excitation Signals Used in Experimental Modal Analysis

	Excitation signal characteristics							
	Slow swept sine	Periodic chirp	Impact	Step relaxation	Pure random	Pseudo random	Periodic random	Burst random
Minimize Leakage	Yes/No	Yes	Yes	Yes	No	Yes	Yes	Yes
Signal-to-Noise Ratio	Very high	High	Low	Low	Fair	Fair	Fair	Fair
RMS-to-Peak Ratio	High	High	Low	Low	Fair	Fair	Fair	Fair
Test Measurement Time	Very long	Very short	Very short	Very short	Good	Very short	Long	Good
Controlled Frequency Content	Yes*	Yes*	No	No	Yes*	Yes*	Yes*	Yes*
Controlled Amplitude Content	Yes*	Yes*	No	Yes/No	No	Yes*	Yes*	No
Removes Distortion	No	No	No	No	Yes	No	Yes	Yes
Characterize Nonlinearity	Yes	Yes	No	No	No	Yes	No	No

* Special hardware required.

Pure random. The pure random signal is an ergodic, stationary random signal which has a Gaussian probability distribution. In general, the signal contains all frequencies (not just integer multiples of the FFT frequency increment), but it may be filtered to include only information in a frequency band of interest. The measured input spectrum of the pure random signal is altered by any impedance mismatch between the system and the exciter.

Pseudorandom. The pseudorandom signal is an ergodic, stationary random signal consisting only of integer multiples of the FFT frequency increment. The frequency spectrum of this signal has a constant amplitude with random phase. If sufficient time is allowed in the measurement procedure for any transient response to the initiation of the signal to decay, the resultant input and response histories are periodic with respect to the sample period. The number of averages used in the measurement procedure is only a function of the reduction of the variance error. In a noise-free environment, only one average may be necessary.

Periodic random. The periodic random signal is an ergodic, stationary random signal consisting only of integer multiples of the FFT frequency increment. The frequency spectrum of this signal has random amplitude and random phase distribution. Since a single history does not contain information at all frequencies, a number of histories must be involved in the measurement process. For each average, an input history is created with random amplitude and random phase. The system is excited with this input in a repetitive cycle until the transient response to the change in excitation signal decays. The input and response histories should then be periodic with respect to the sample period and are recorded as one average in the total process. With each new average, a new history, uncorrelated with previous input signals, is generated, so that the resulting measurement is completely randomized.

Random transient (burst random). The random transient signal is neither a completely transient deterministic signal nor a completely ergodic, stationary random signal but contains properties of both signal types. The frequency spectrum of this signal has random amplitude and random phase distribution and contains energy throughout the frequency spectrum. The difference between this signal and the periodic random signal is that the random transient history is truncated to zero after some percentage of the sample period (normally 50 to 80 percent). The measurement procedure duplicates the periodic random procedure, but without the need to wait for the transient response to decay. The point at which the input history is truncated is chosen so that the response history decays to zero within the sample period. Even for lightly damped systems, the response history decays to zero very quickly because of the damping provided by the exciter system trying to maintain the input at zero. This damping provided by the exciter system is often overlooked in the analysis of the characteristics of this signal type. Since this measured input, although not part of the generated signal, includes the variation of the input during the decay of the response history, the input and response histories are totally observable within the sample period and the system damping is unaffected.

Increased Frequency Resolution.   An increase in the frequency resolution of a frequency-response function affects measurement errors in several ways. Finer frequency resolution allows more exact determination of the damped natural frequency of each modal vector. The increased frequency resolution means that the level of a broad-band signal is reduced. The most important benefit of increased frequency resolution, though, is a reduction of the leakage error. Since the distortion of the frequency-response function due to leakage is a function of frequency spacing, not frequency, the increase in frequency resolution reduces the true bandwidth of the leakage error centered at each damped natural frequency. In order to increase the frequency resolution, the total time per history must be increased in direct proportion. The longer data acquisition time increases the variance error problem when transient signals are utilized for input as well as emphasizing any nonstationary problem with the data. The increase of frequency resolution often requires multiple acquisition and/or processing of the histories in order to obtain an equivalent frequency range. This increases the data storage and documentation overhead as well as extending the total test time.

There are two approaches to increasing the frequency resolution of a frequency-response function. The first approach involves increasing the number of spectral lines in a baseband measurement. The advantage of this approach is that no additional hardware or software is required. However, FFT analyzers do not always have the capability to alter the number of spectral lines used in the measurement. The second approach involves the reduction of the bandwidth of the measurement while holding the number of spectral lines constant. If the lower frequency limit of the bandwidth is always zero, no additional hardware or software is required. Ideally, though, for an arbitrary bandwidth, hardware and/or software to perform a frequency-shifted, or digitally filtered, FFT is required.

The frequency-shifted FFT process for computing the frequency-response function has additional characteristics pertinent to the reduction of errors. Primarily, more accurate information can be obtained on weak spectral components if the bandwidth is chosen to avoid strong spectral components. The out-of-band rejection of the frequency-shifted FFT is better than that of most analog filters that are used in a measurement procedure to attempt to achieve the same results. Additionally, the precision of the resulting frequency-response function is improved due to processor gain inherent in the frequency-shifted FFT calculation procedure.[4-6]

Weighting Functions. Weighting functions, or data windows, are probably the most common approach to the reduction of the leakage error in the frequency-response function. While weighting functions are sometimes desirable and necessary to modify the frequency-domain effects of truncating a signal in the time domain, they are too often utilized when one of the other approaches to error reduction would give superior results. Averaging, selective excitation, and increasing the frequency resolution all act to reduce the leakage error by eliminating the cause of the error. Weighting functions, on the other hand, attempt to compensate for the leakage error after the data have already been digitized.

Windows alter, or compensate for, the frequency-domain characteristic associated with the truncation of data in the time domain. Essentially, again using the narrow bandpass filter analogy, windows alter the characteristics of the bandpass filters that are applied to the data. This compensation for the leakage error causes an attendant distortion of the frequency and phase information of the frequency-response function, particularly in the case of closely spaced, lightly damped system poles. This distortion is a direct function of the width of the main lobe and the size of the side lobes of the spectrum of the weighting function.[4-7]

MODAL PARAMETER ESTIMATION

Modal parameter estimation, or modal identification, is a special case of system identification where the a priori model of the system is known to be in the form of modal parameters. *Modal parameters* include the complex-valued modal frequencies λ_r, modal vectors $\{\psi_r\}$, and modal scaling (modal mass or modal A). Additionally, most algorithms estimate modal participation vectors $\{L_r\}$ and residue vectors $\{A_r\}$ as part of the overall process.

Modal parameter estimation involves estimating the modal parameters of a structural system from measured input-output data. Most modal parameter estimation is based upon the measured data being the frequency-response function or the equivalent impulse-response function, typically found by inverse Fourier transforming the frequency-response function. Therefore, the form of the model used to represent the experimental data is normally stated in a mathematical frequency-response function (FRF) model using temporal (time or frequency) and spatial (input degree-of-freedom and output degree-of-freedom) information.

In general, modal parameters are considered to be global properties of the system. The concept of global modal parameters simply means that there is only one answer for each modal parameter and that the modal parameter estimation solution procedure enforces this constraint. Every frequency-response or impulse-response function measurement theoretically contains the information that is represented by the characteristic equation, the modal frequencies, and damping. If individual measurements are treated as independent of one another in the solution procedure, there is nothing to guarantee that a single set of modal frequencies and damping is generated. Likewise, if more than one reference is measured in the data set, redundant estimates of the modal vectors can be made unless the solution procedure utilizes all references in the estimation process simultaneously. Most of the current modal parameter estimation algorithms estimate the modal frequencies and damping in a global sense, but very few estimate the modal vectors in a global sense.

Since the modal parameter estimation process involves a greatly overdetermined problem, the estimates of modal parameters resulting from different algorithms are not the same as a result of differences in the modal model and model domain, differences in how the algorithms use the data, differences in the way the data are weighted or condensed, and differences in user expertise.

MODAL IDENTIFICATION CONCEPTS

The most common approach in modal identification involves using numerical techniques to separate the contributions of individual modes of vibration in measurements such as frequency-response functions. The concept involves estimating the individual single degree-of-freedom (SDOF) contributions to the multiple degree-of-freedom (MDOF) measurement.

$$[H(\omega)]_{N_o \times N_i} = \sum_{r=1}^{n} \frac{[A_r]_{N_o \times N_i}}{j\omega - \lambda_r} + \frac{[A_r^*]_{N_o \times N_i}}{j\omega - \lambda_r^*} \qquad (21.60)$$

This concept is mathematically represented in Eq. (21.60) and graphically represented in Figs. 21.14 and 21.15.

Equation (21.60) is often formulated in terms of modal vectors $\{\psi_r\}$ and modal participation vectors $\{L_r\}$ instead of residue matrices $[A_r]$. *Modal participation vectors* are a result of multiple reference modal parameter estimation algorithms and relate how well each modal vector is excited from each of the reference locations included in the measured data. The combination of the modal participation vector $\{L_r\}$ and the modal vector $\{\psi_r\}$ for a given mode give the residue matrix $A_{pqr} = L_{qr}\psi_{pr}$ for that mode.

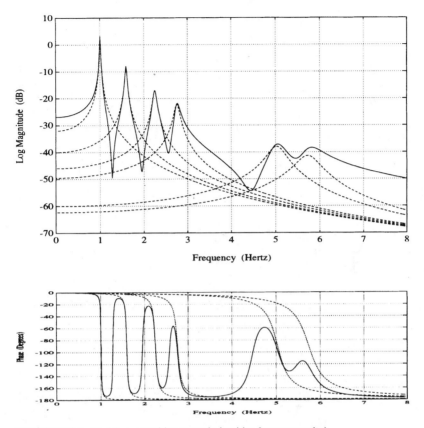

FIGURE 21.14 Modal superposition example (positive frequency poles).

Generally, the modal parameter estimation process involves several stages. Typically, the modal frequencies and modal participation vectors are found in a first stage and residues, modal vectors, and modal scaling are determined in a second stage. Most modal parameter estimation algorithms can be reformulated into a single, consistent mathematical formulation with a corresponding set of definitions and unifying concepts.[15] Particularly, a matrix polynomial approach is used to unify the presentation with respect to current algorithms such as the least squares complex exponential (LSCE), polyreference time domain (PTD), Ibrahim time domain (ITD), eigensystem realization algorithm (ERA), rational fraction polynomial (RFP), polyreference frequency domain (PFD) and complex mode indication function (CMIF) methods. Using this unified matrix polynomial approach (UMPA) allows a discussion of the similarities and differences of the commonly used methods as well as a discussion of the numerical characteristics. Least squares (LS), total least squares (TLS), double least squares (DLS), and singular value decomposition (SVD) methods are used in order to take advantage of redundant measurement data. Eigenvalue and singular value decomposition transformation methods are utilized to reduce the effective size of the resulting eigenvalue-eigenvector problem as well. Many acronyms used in modal parameter estimation are listed in Table 21.3.

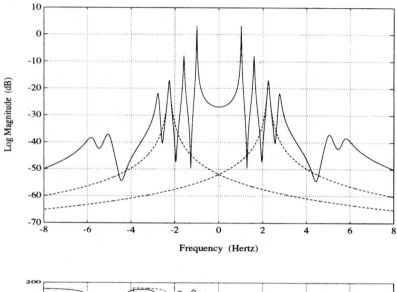

FIGURE 21.15 Modal superposition example (positive and negative frequency poles).

TABLE 21.3   Modal Parameter Estimation Algorithm Acronyms

CEA	Complex exponential algorithm[16]
LSCE	Least squares complex exponential[16]
PTD	Polyreference time domain[17,18]
ITD	Ibrahim time domain[19]
MRITD	Multiple reference Ibrahim time domain[20]
ERA	Eigensystem realization algorithm[21,22]
PFD	Polyreference frequency domain[23–25]
SFD	Simultaneous frequency domain[26]
MRFD	Multireference frequency domain[27]
RFP	Rational fraction polynomial[28]
OP	Orthogonal polynomial[29–31]
CMIF	Complex mode indication function[32]

Data Domain. Modal parameters can be estimated from a variety of different measurements that exist as discrete data in different data domains (time, frequency, and/or spatial). These measurements can include free decays, forced responses, frequency responses, and unit impulse responses. These measurements can be processed one at a time or in partial or complete sets simultaneously. The measurements can be generated with no measured inputs, a single measured input, or multiple measured inputs. The data can be measured individually or simultaneously. In other words, there is a tremendous variation in the types of measurements and in the types of constraints that can be placed upon the testing procedures used to acquire these data. For most measurement situations, frequency-response functions are utilized in the frequency domain and impulse-response functions are utilized in the time domain.

Another important concept in experimental modal analysis, and particularly modal parameter estimation, involves understanding the relationships between the temporal (time and/or frequency) information and the spatial (input DOF and output DOF) information. Input-output data measured on a structural system can always be represented as a superposition of the underlying temporal characteristics (modal frequencies) with the underlying spatial characteristics (modal vectors).

Model Order Relationships. The estimation of an appropriate model order is the most important problem encountered in modal parameter estimation. This problem is complicated because of the formulation of the parameter estimation model in the time or frequency domain, a single or multiple reference formulation of the modal parameter estimation model, and the effects of random and bias errors on the modal parameter estimation model. The basis of the formulation of the correct model order can be seen by expanding the theoretical second-order matrix equation of motion to a higher-order model.

$$ \mid [m]s^2 + [c]s + [k] \mid = 0 \qquad (21.61) $$

The above matrix polynomial is of model order two, has a matrix dimension of $n \times n$, and has a total of $2n$ characteristic roots (modal frequencies). This matrix polynomial equation can be expanded to reduce the size of the matrices to a scalar equation.

$$ \alpha_{2N}s^{2N} + \alpha_{2N-1}s^{2N-1} + \alpha_{2N-2}s^{2N-2} + \cdots + \alpha_0 = 0 \qquad (21.62) $$

The above matrix polynomial is of model order $2n$, has a matrix dimension of 1×1, and has a total of $2n$ characteristic roots (modal frequencies). The characteristic roots of this matrix polynomial equation are the same as those of the original second-order matrix polynomial equation. Finally, the number of characteristic roots (modal frequencies) that can be determined depends upon the size of the matrix coefficients involved in the model and the order of the highest polynomial term in the model.

For modal parameter estimation algorithms that utilize experimental data, the matrix polynomial equations that are formed are a function of matrix dimension, from 1×1 to $N_i \times N_i$ or $N_o \times N_o$. There are a significant number of procedures that have been formulated particularly for aiding in these decisions and selecting the appropriate estimation model. Procedures for estimating the appropriate matrix size and model order are another of the differences between various estimation procedures.

Fundamental Measurement Models. Most current modal parameter estimation algorithms utilize frequency- or impulse-response functions as the data, or known information, to solve for modal parameters. The general equation that can be used to represent the relationship between the measured frequency-response function matrix and the modal parameters is shown in Eqs. (21.63) and (21.64).

$$[H(\omega)]_{N_o \times N_i} = \Big[\psi \Big]_{N_o \times 2N} \left[\frac{1}{j\omega - \lambda_r} \right]_{2N \times 2N} [L]^T_{2N \times N_i} \tag{21.63}$$

$$[H(\omega)]^T_{N_i \times N_o} = [L]_{N_i \times 2N} \left[\frac{1}{j\omega - \lambda_r} \right]_{2N \times 2N} \Big[\psi \Big]^T_{2N \times N_o} \tag{21.64}$$

Impulse-response functions are rarely measured directly but are calculated from associated frequency-response functions via the inverse FFT algorithm. The general equation that can be used to represent the relationship between the impulse-response function matrix and the modal parameters is shown in Eqs. (21.9) and (21.10).

$$[h(t)]_{N_o \times N_i} = \Big[\psi \Big]_{N_o \times 2N} \left[e^{\lambda_r t} \right]_{2N \times 2N} [L]^T_{2N \times N_i} \tag{21.65}$$

$$[h(t)]^T_{N_i \times N_o} = [L]_{N_i \times 2N} \left[e^{\lambda_r t} \right]_{2N \times 2N} \Big[\psi \Big]^T_{2N \times N_o} \tag{21.66}$$

Many modal parameter estimation algorithms have been originally formulated from Eqs. (21.63) through (21.66). However, a more general development for all algorithms is based upon relating the above equations to a general matrix polynomial approach.

Characteristic Space. From a conceptual viewpoint, the measurement space of a modal identification problem can be visualized as occupying a volume with the coordinate axis defined in terms of three sets of characteristics. Two axes of the conceptual volume correspond to spatial information and the third axis to temporal information. The spatial coordinates are in terms of the input and output degrees-of-freedom (DOF) of the system. The temporal axis is either time or frequency, depending upon the domain of the measurements. These three axis define a 3-D volume which is referred to as the *characteristic space*, as noted in Fig. 21.16.

This space or volume represents all possible measurement data as expressed by Eqs. (21.63) through (21.66). This conceptual representation is very useful in understanding what data subspace has been measured. Also, this conceptual representation is very useful in recognizing how the data are organized and utilized with respect to different modal parameter estimation algorithms. Information parallel to one of the axes consists of a solution composed of the superposition of the characteristics defined by that axis. The other two characteristics determine the scaling of each term in the superposition.

In modal parameter estimation algorithms that utilize a single frequency-response function, data collection is concentrated on measuring the temporal aspect (time/frequency) at a sufficient resolution to determine the modal parameters. In

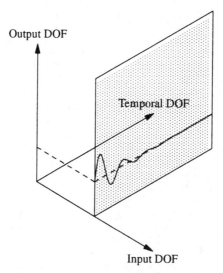

FIGURE 21.16 Conceptualization of modal characteristic space (input DOF axis, output DOF axis, time axis).

this approach, the accuracy of the modal parameters, particularly frequency and damping, is essentially limited by Shannon's sampling theorem and Rayleigh's criterion. This focus on the temporal information ignores the added accuracy that use of the spatial information brings to the estimation of modal parameters. Recognizing the characteristic space aspects of the measurement space and using these characteristics (modal vector/participation vector) concepts in the solution procedure leads to the conclusion that the spatial information can compensate for the limitations of temporal information. Therefore, there is a tradeoff between temporal and spatial information for a given accuracy requirement. This is particularly notable in the case of repeated roots. No amount of temporal resolution (accuracy) can theoretically solve repeated roots, but the addition of spatial information in the form of multiple inputs and/or outputs resolves this problem.

Any structural testing procedure measures a subspace of the total possible data available. Modal parameter estimation algorithms may then use all of this subspace or may choose to further limit the data to a more restrictive subspace. It is theoretically possible to estimate the characteristics of the total space by measuring a subspace which samples all three characteristics. However, the selection of the subspace has a significant influence on the results. In order for all of the modal parameters to be estimated, the subspace must encompass a region which includes contributions of all three characteristics. An important example is the necessity to use multiple reference data (inputs and outputs) in order to estimate *repeated roots*. The particular subspace which is measured and the weighting of the data within the subspace in an algorithm are the main differences among the various modal identification procedures which have been developed.

In general, the amount of information in a measured subspace greatly exceeds the amount necessary to solve for the unknown modal characteristics. Another major difference among the various modal parameter estimation procedures is the type of condensation algorithms that are used to reduce the data to match the number of unknowns [for example, *least squares* (LS), *singular value decomposition* (SVD), etc.]. As is the case with any overspecified solution procedure, there is no unique answer. The answer that is obtained depends upon the data that are selected, the weighting of the data, and the unique algorithm used in the solution process. As a result, the answer is the *best* answer depending upon the objective functions associated with the algorithm being used. Historically, this point has created some confusion since many users expect different methods to give exactly the same answer.

Many modal parameter estimation methods use information (subspace) where only one or two characteristics are included. For example, the simplest (computationally) modal parameter estimation algorithms utilize one impulse-response function or one frequency-response function at a time. In this case, only the temporal characteristic is used, and, as might be expected, only temporal characteristics (modal frequencies) can be estimated from the single measurement. The global characteristic of modal frequency cannot be enforced. In practice, when multiple measurements are taken, the modal frequency does not change from one measurement to the next.

Other modal parameter estimation algorithms utilize the data in a plane of the characteristic space. For example, this corresponds to the data taken at a number of response points but from a single excitation point or reference. This representation of a column of measurements is shown in Fig. 21.16 as a plane in the characteristic space. For this case, representing a single input (reference), while it is now possible to enforce the global modal frequency assumption, it is not possible to compute repeated roots and it is difficult to separate closely coupled modes because of the lack of spatial data.

Many modal identification algorithms utilize data taken at a large number of output DOFs due to excitation at a small number of input DOFs. Data taken in this manner are consistent with a multiexciter type of test. Conceptually, this is represented by several planes of data parallel to the plane of data represented in Fig. 21.16. Some modal identification algorithms utilize data taken at a large number of input DOFs and a small number of output DOFs. Data taken in this manner are consistent with a roving hammer type of excitation with several fixed output sensors. These data can also be generated by transposing the data matrix acquired using a multiexciter test. The conceptual representation is several rows of the potential measurement matrix perpendicular to the plane of data represented in Fig. 21.16. Measurement data spaces involving many planes of measured data are the best possible modal identification situations, since the data subspace includes contributions from temporal and spatial characteristics. This allows the best possibility of estimating all the important modal parameters.

The data which define the subspace need to be acquired through a consistent measurement process in order for the algorithms to estimate accurate modal parameters. This means that the data must be measured simultaneously and requires that data acquisition, digital signal processing, and instrumentation be designed and operate accordingly.

Fundamental Modal Identification Models.   The common characteristics of different modal parameter estimation algorithms can be more readily identified by using a matrix polynomial model rather than using a physically based mathematical model. One way of understanding the basis of this model can be developed from the polynomial model used for the frequency-response function.

$$H_{pq}(\omega) = \frac{X_p(\omega)}{F_q(\omega)} = \frac{\beta_n(j\omega)^n + \beta_{n-1}(j\omega)^{n-1} + \cdots + \beta_1(j\omega)^1 + \beta_0(j\omega)^0}{\alpha_m(j\omega)^m + \alpha_{m-1}(j\omega)^{m-1} + \cdots + \alpha_1(j\omega)^1 + \alpha_0(j\omega)^0} \qquad (21.67)$$

This can be rewritten as

$$H_{pq}(\omega) = \frac{X_p(\omega)}{F_q(\omega)} = \frac{\sum_{k=0}^{n} \beta_k (j\omega)^k}{\sum_{k=0}^{m} \alpha_k (j\omega)^k} \tag{21.68}$$

Further rearranging yields the following equation, which is linear in the unknown α and β terms:

$$\sum_{k=0}^{m} \alpha_k (j\omega)^k X_p(\omega) = \sum_{k=0}^{n} \beta_k (j\omega)^k F_q(\omega) \tag{21.69}$$

Noting that the response function X_p can be replaced by the frequency-response function H_{pq} if the force function F_q is assumed to be unity, the above equation can be restated as

$$\sum_{k=0}^{m} \alpha_k (j\omega)^k H_{pq}(\omega) = \sum_{k=0}^{n} \beta_k (j\omega)^k \tag{21.70}$$

The above formulation is essentially a linear equation in terms of the unknown coefficients α_k and β_k. The equation is valid at each frequency of the measured frequency-response function. Since, in the worst case, the number of unknowns is $m + n + 2$, the unknown coefficients can theoretically be determined if the frequency-response function has $m + n + 2$ or more discrete frequencies. Practically, this is always the case. Note that the total number of unknown coefficients (or coefficient matrices) is actually $m + n + 1$ since one coefficient (or coefficient matrix) can be assumed to be 1 (or the identity matrix). This is the case because the equation can be divided, or normalized, by one of the unknown coefficients (or coefficient matrices). Note that numerical problems can result if the equation is normalized by a coefficient (or coefficient matrix) that is close to zero. Normally, the coefficient α_0 (or the coefficient matrix $[\alpha_0]$) is chosen as unity (or the identity matrix).

The previous models can be generalized to represent the general multiple input/multiple output case as follows:

$$\sum_{k=0}^{m} \left[[\alpha_k](j\omega)^k \right] \{X(\omega)\} = \sum_{k=0}^{n} \left[[\beta_k](j\omega)^k \right] \{F(\omega)\} \tag{21.71}$$

Note that the size of the coefficient matrices $[\alpha_k]$ and $[\beta_k]$ is normally $N_i \times N_i$ or $N_o \times N_o$ when the equations are developed from experimental data. Rather than the basic model being developed in terms of force and response information, the models can be stated in terms of frequency-response information. The response vector $\{X(\omega)\}$ can be replaced by a vector of frequency-response functions $\{H(\omega)\}$ where either the input or the output is held fixed. The force vector $\{F(\omega)\}$ is then replaced by an incidence matrix $\{R\}$ of the same size which is composed of all zeros except for unity at the position in the vector consistent with the driving point measurement (common input and output DOF).

$$\sum_{k=0}^{m} \left[(j\omega)^k [\alpha_k] \right] \{H(\omega)\} = \sum_{k=0}^{n} \left[(j\omega)^k [\beta_k] \right] \{R\} \qquad (21.72)$$

where

$$\{H(\omega)\} = \begin{Bmatrix} H_{1q}(\omega) \\ H_{2q}(\omega) \\ H_{3q}(\omega) \\ \cdot \\ \cdot \\ \cdot \\ H_{qq}(\omega) \\ \cdot \\ \cdot \\ \cdot \\ H_{pq}(\omega) \end{Bmatrix} \qquad \{R\} = \begin{Bmatrix} 0 \\ 0 \\ 0 \\ \cdot \\ \cdot \\ \cdot \\ 1 \\ \cdot \\ \cdot \\ \cdot \\ 0 \end{Bmatrix}$$

The above model, in the frequency domain, corresponds to an *autoregressive moving-average* (*ARMA*) model that is developed from a set of finite difference equations in the time domain. The general characteristic matrix polynomial model concept recognizes that both the time- and frequency-domain models generate essentially the same matrix polynomial models. For that reason, the *unified matrix polynomial approach* (*UMPA*) terminology is used to describe both domains since the ARMA terminology has been connected primarily with the time domain.[15]

In parallel with the development of Eq. (21.67), a time-domain model representing the relationship between a single response degree-of-freedom and a single input degree-of-freedom can be stated as follows:

$$\sum_{k=0}^{m} \alpha_k x(t_{i+k}) = \sum_{k=0}^{n} \beta_k f(t_{i+k}) \qquad (21.73)$$

For the general multiple input/multiple output case,

$$\sum_{k=0}^{m} [\alpha_k] \{x(t_{i+k})\} = \sum_{k=0}^{n} [\beta_k] \{f(t_{i+k})\} \qquad (21.74)$$

If the discussion is limited to the use of free decay or impulse-response function data, the previous time-domain equations can be greatly simplified by noting that the forcing function can be assumed to be zero for all time greater than zero. If this is the case, the $[\beta_k]$ coefficients can be eliminated from the equations:

$$\sum_{k=0}^{m} [\alpha_k] \left\{ h_{pq}(t_{i+k}) \right\} = 0 \qquad (21.75)$$

In light of the above discussion, it is now apparent that most of the modal parameter estimation processes available can be developed by starting from a general matrix polynomial formulation that is justifiable based upon the underlying matrix differential equation. The general matrix polynomial formulation yields essentially the same characteristic matrix polynomial equation for both time- and frequency-domain data. For the frequency-domain data case, this yields

$$\left| \, [\alpha_m] \, s^m + [\alpha_{m-1}] \, s^{m-1} + [\alpha_{m-2}] \, s^{m-2} + \cdots + [\alpha_0] \, \right| = 0 \qquad (21.76)$$

For the time-domain data case, this yields

$$\left| \, [\alpha_m] \, z^m + [\alpha_{m-1}] \, z^{m-1} + [\alpha_{m-2}] \, z^{m-2} + \cdots + [\alpha_0] \, \right| = 0 \qquad (21.77)$$

With respect to the previous discussion of model order, the characteristic matrix polynomial equation, Eq. (21.76) or (21.77), has a model order of m, and the number of modal frequencies or roots that are found from this characteristic matrix polynomial equation is m times the size of the coefficient matrices $[\alpha]$. In terms of sampled data, the time-domain matrix polynomial results from a set of finite difference equations and the frequency-domain matrix polynomial results from a set of linear equations, where each equation is formulated at one of the frequencies of the measured data. This distinction is important to note since the roots of the matrix characteristic equation formulated in the time domain are in the *z domain* (z_r) and must be converted to the frequency domain (λ_r), while the roots of the matrix characteristic equation formulated in the frequency domain (λ_r) are already in the desired domain. Note that the roots that are estimated in the time domain are limited to maximum values determined by Shannon's sampling theorem relationship (discrete time steps).

$$z_r = e^{\lambda_r \Delta t} \qquad \lambda_r = \sigma_r + j\omega_r \qquad (21.78)$$

$$\sigma_r = \mathrm{Re} \left[\frac{\ln z_r}{\Delta t} \right] \qquad \omega_r = \mathrm{Im} \left[\frac{\ln z_r}{\Delta t} \right]$$

Using this general formulation, the most commonly used modal identification methods can be summarized as shown in Table 21.4.

TABLE 21.4 Characteristics of Modal Parameter Estimation Algorithms

Algorithm	Domain		Matrix polynomial order			Coefficients	
	Time	Frequency	Zero	Low	High	Scalar	Matrix
CEA	•				•	•	
LSCE	•				•	•	
PTD	•				•		$N_i \times N_i$
ITD	•			•			$N_o \times N_o$
MRITD	•			•			$N_o \times N_o$
ERA	•			•			$N_o \times N_o$
PFD		•		•			$N_o \times N_o$
SFD		•		•			$N_o \times N_o$
MRFD		•		•			$N_o \times N_o$
RFP		•			•	•	Both
OP		•			•	•	Both
CMIF		•	•				$N_o \times N_i$

The high-order model is typically used for those cases where the system is under-sampled in the spatial domain. For example, the limiting case is when only one measurement is made on the structure. For this case, the left-hand side of the general linear equation corresponds to a scalar polynomial equation with the order equal to or greater than the number of desired modal frequencies. This type of high-order model may yield significant numerical problems for the frequency-domain case.

The low-order model is used for those cases where the spatial information is complete. In other words, the number of independent physical coordinates is greater than the number of desired modal frequencies. For this case, the order of the left-hand side of the general linear equation, Eq. (21.72) or (21.75), is equal to 1 or 2.

The zero-order model corresponds to a case where the temporal information is neglected and only the spatial information is used. These methods directly estimate the eigenvectors as a first step. In general, these methods are programmed to process data at a single temporal condition or variable. In this case, the method is essentially equivalent to the single degree-of-freedom (SDOF) methods which have been used with frequency-response functions. In other words, the comparison between the zeroth-order matrix polynomial model and the higher-order matrix polynomial models is similar to the comparison between the SDOF and MDOF methods used in modal parameter estimation.

Two-Stage Linear Solution Procedure.   Almost all modal parameter estimation algorithms in use at this time involve a two-stage linear solution approach. For example, with respect to Eqs. (21.63) through (21.66), if all modal frequencies and modal participation vectors can be found, the estimation of the complex residues can proceed in a linear fashion. This procedure of separating the nonlinear problem into a multistage linear problem is a common technique for most estimation methods today. For the case of structural dynamics, the common technique is to estimate modal frequencies and modal participation vectors in a first stage and then to estimate the modal coefficients plus any residuals in a second stage.

Therefore, based upon Eqs. (21.63) through (21.66), most commonly used modal identification algorithms can be outlined as follows:

First stage of modal parameter estimation:
- Load measured data into linear equation form [Eqs. (21.72) or (21.75)].
- Find scalar or matrix autoregressive coefficients [α_k].
  - Normalize frequency range (frequency domain only).
  - Utilize orthogonal polynomials (frequency domain only).
- Solve matrix polynomial for modal frequencies.
  - Formulate companion matrix.
  - Obtain eigenvalues of companion matrix λ_r or z_r.
  - Convert eigenvalues from z_r to λ_r (time domain only).
  - Obtain modal participation vectors L_{qr} or modal vectors $\{\psi\}_r$ from eigenvectors of the companion matrix.

Second stage of modal parameter estimation:
- Find modal vectors and modal scaling from Eqs. (21.63) through (21.66).

Equation (21.72) or (21.75) is used to formulate a single, block coefficient linear equation as shown in the graphical analogy of Case 1a, Fig. 21.17. In order to estimate complex conjugate pairs of roots, at least two equations from each piece or block of data in the data space must be used. This situation is shown in Case 1b, Fig. 21.18. In order to develop enough equations to solve for the unknown matrix coefficients, further information is taken from the same block of data or from other blocks of data in the data space until the number of equations equals (Case 2) or exceeds (Case 3) the number of unknowns, as shown in Fig. 21.19 and 21.20. In the frequency domain, this is accomplished by utilizing a different frequency from within each measurement for each equation. In the time domain, this is accomplished by utilizing a different starting time or time shift from within each measurement for each equation.

Once the matrix coefficients $[\alpha]$ have been found, the modal frequencies λ_r or z_r can be found using a number of numerical techniques. While in certain numerical situations, other numerical approaches may be more robust, a companion matrix approach yields a consistent concept for understanding the process. Therefore, the roots of the matrix characteristic equation can be found as the eigenvalues of the associated companion matrix. The companion matrix can be formulated in one of several ways. The most common formulation is as follows:

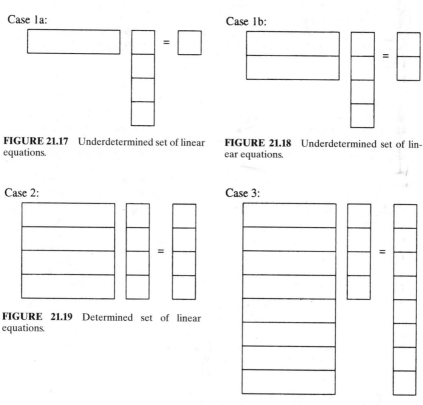

Case 1a:

FIGURE 21.17 Underdetermined set of linear equations.

Case 1b:

FIGURE 21.18 Underdetermined set of linear equations.

Case 2:

FIGURE 21.19 Determined set of linear equations.

Case 3:

FIGURE 21.20 Overdetermined set of linear equations.

$$[C] = \begin{bmatrix} -[\alpha]_{m-1} & -[\alpha]_{m-2} & \cdots & \cdots & \cdots & -[\alpha]_1 & -[\alpha]_0 \\ [I] & [0] & \cdots & \cdots & \cdots & [0] & [0] \\ [0] & [I] & \cdots & \cdots & \cdots & [0] & [0] \\ [0] & [0] & \cdots & \cdots & \cdots & [0] & [0] \\ \cdots & \cdots & \cdots & \cdots & \cdots & \cdots & \cdots \\ \cdots & \cdots & \cdots & \cdots & \cdots & \cdots & \cdots \\ \cdots & \cdots & \cdots & \cdots & \cdots & \cdots & \cdots \\ [0] & [0] & \cdots & \cdots & \cdots & [0] & [0] \\ [0] & [0] & \cdots & \cdots & \cdots & [0] & [0] \\ [0] & [0] & \cdots & \cdots & \cdots & [I] & [0] \end{bmatrix} \qquad (21.79)$$

Note again that the numerical characteristics of the eigenvalue solution of the companion matrix are different for low-order cases than for high-order cases for a given data set. The companion matrix can be used in the following eigenvalue formulation to determine the modal frequencies for the original matrix coefficient equation:

$$[C]\{X\} = \lambda\,[I]\{X\} \qquad (21.80)$$

The eigenvectors that can be found from the eigenvalue-eigenvector solution utilizing the companion matrix may or may not be useful in terms of modal parameters. The eigenvector that is found, associated with each eigenvalue, is of length model order times matrix coefficient size. In fact, the unique (meaningful) portion of the eigenvector is of length equal to the size of the coefficient matrices and is repeated in the eigenvector a model order number of times. Each time the unique portion of the eigenvector is repeated, it is multiplied by a scalar multiple of the associated modal frequency. Therefore, the eigenvectors of the companion matrix have the following form:

$$\{\phi\}_r = \begin{Bmatrix} \lambda_r^m\{\psi\}_r \\ \vdots \\ \lambda_r^2\{\psi\}_r \\ \lambda_r^1\{\psi\}_r \\ \lambda_r^0\{\psi\}_r \end{Bmatrix}_r \qquad (21.81)$$

Note that unless the size of the coefficient matrices is at least as large as the number of measurement degrees-of-freedom, only a partial set of modal coefficients, the modal participation coefficients L_{qr}, are found. For the case involving scalar coefficients, no meaningful modal coefficients are found.

If the size of the coefficient matrices, and therefore the modal participation vector, is less than the largest spatial dimension of the problem, then the modal vectors are typically found in a second-stage solution process using one of Eqs. (21.63) through (21.66). Even if the complete modal vector $\{\psi\}$ of the system is found from the eigenvectors of the companion matrix approach, the modal scaling and modal participation vectors for each modal frequency are normally found in this second-stage formulation.

Data Sieving/Filtering. For almost all cases of modal identification, a large amount of redundancy or overdetermination exists. This means that for Case 3, defined in Fig. 21.20, the number of equations available compared to the number required for the determined Case 2 (defined as the *overdetermination factor*) is quite large. Beyond some value of overdetermination factor, the additional equations contribute little to the result but may add significantly to the solution time. For this reason, the data space is often *filtered* (limited in the temporal sense) or *sieved* (limited in the input DOF or output DOF sense) in order to obtain a reasonable result in the minimum time. For frequency-domain data, the filtering process normally involves limiting the data set to a range of frequencies or a different frequency resolution according to the desired frequency range of interest. For time-domain data, the filtering process normally involves limiting the starting time value as well as the number of sets of time data taken from each measurement. Data sieving involves limiting the data set to certain degrees-of-freedom that are of primary interest. This normally involves restricting the data to specific directions (X, Y, and/or Z directions) or specific locations or groups of degrees-of-freedom, such as components of a large structural system.

Equation Condensation. Several important concepts should be delineated in the area of equation condensation methods. Equation condensation methods are used to reduce the number of equations based upon measured data to more closely match the number of unknowns in the modal parameter estimation algorithms. There are a large number of condensation algorithms available. Based upon the modal parameter estimation algorithms in use today, the three types of algorithms most often used are

- *Least squares.* Least squares (LS), weighted least squares (WLS), total least squares (TLS), or double least squares (DLS) methods are used to minimize the squared error between the measured data and the estimation model. Historically, this is one of the most popular procedures for finding a pseudo-inverse solution to an overspecified system. The main advantage of this method is computational speed and ease of implementation, while the major disadvantage is numerical precision.

- *Transformation.* There are a large number of transformation that can be used to reduce the data. In the transformation methods, the measured data are reduced by approximating them by the superposition of a set of significant vectors. The number of significant vectors is equal to the amount of independent measured data. This set of vectors is used to approximate the measured data and used as input to the parameter estimation procedures. *Singular value decomposition* (SVD) is one of the more popular transformation methods. The major advantage of such methods is numerical precision, and the disadvantage is computational speed and memory requirements.

- *Coherent averaging.* Coherent averaging is another popular method for reducing the data. In the coherent averaging method, the data are weighted by performing a dot product between the data and a weighting vector (spatial filter). Information in the data which is not coherent with the weighting vectors is averaged out of the data. The method is often referred to as a spatial filtering procedure. This method has both speed and precision but, in order to achieve precision, requires a good set of weighting vectors. In general, the optimum weighting vectors are connected with the solution, which is unknown. It should be noted that least squares is an example of a noncoherent averaging process.

The least squares and the transformation procedures tend to weight those modes of vibration which are well excited. This can be a problem when trying to extract modes which are not well excited. The solution is to use a weighting function for condensation which tends to enhance the mode of interest. This can be accomplished in a number of ways:

- In the time domain, a spatial filter or a coherent averaging process can be used to filter the response to enhance a particular mode or set of modes. For example, by averaging the data from two symmetric exciter locations, the symmetric modes of vibration can be enhanced. A second example is to use only the data in a local area of the system to enhance local modes. The third method is using estimates of the modes' shapes as weighting functions to enhance particular modes.

- In the frequency domain, the data can be enhanced in the same manner as in the time domain, plus the data can be additionally enhanced by weighting them in a frequency band near the natural frequency of the mode of interest.

The type of equation condensation method that is utilized in a modal identification algorithm has a significant influence on the results of the parameter estimation process.

Coefficient Condensation. For the low-order modal identification algorithms, the number of physical coordinates (typically N_o) is often much larger than the number of desired modal frequencies ($2n$). For this situation, the numerical solution procedure is constrained to solve for N_o or $2N_o$ modal frequencies. This can be very time consuming and is unnecessary. The number of physical coordinates N_o can be reduced to a more reasonable size ($N_e \approx N_o$ or $N_e \approx 2N_o$) by using a decomposition transformation from physical coordinates N_o to the approximate number of effective modal frequencies N_e. Currently, SVD or eigenvalue decompositions (ED) are used to preserve the principal modal information prior to formulating the linear equation solution for unknown matrix coefficients.[33,34] In most cases, even when the spatial information must be condensed, it is necessary to use a model order greater than 2 to compensate for distortion errors or noise in the data and to compensate for the case where the location of the transducers is not sufficient to totally define the structure.

$$[H'] = [T][H] \tag{21.82}$$

where $[H']$ = transformed (condensed) frequency-response function matrix
  $[T]$ = transformation matrix
  $[H]$ = original FRF matrix

The difference between the two techniques lies in the method of finding the transformation matrix $[T]$. Once $[H]$ has been condensed, however, the parameter estimation procedure is the same as for the full data set. Because the data eliminated from the parameter estimation process ideally correspond to the noise in the data, the modal frequencies of the condensed data are the same as the modal frequencies of the full data set. However, the modal vectors calculated from the condensed data may need to be expanded back into the full space:

$$[\Psi] = [T]^T [\Psi'] \tag{21.83}$$

where $[\Psi]$ = full-space modal matrix
$[\Psi']$ = condensed-space modal matrix

Model Order Determination. Much of the work on modal parameter estimation since 1975 has involved methodology for determining the correct model order for the modal parameter model. Technically, model order refers to the highest power in the matrix polynomial equation. The number of modal frequencies found is equal to the model order times the size of the matrix coefficients, normally N_o or N_i. For a given algorithm, the size of the matrix coefficients is normally fixed; therefore, determining the model order is directly linked to estimating n, the number of modal frequencies in the measured data that are of interest. As has always been the case, an estimate for the minimum number of modal frequencies can be easily found by counting the number of peaks in the frequency-response function in the frequency band of analysis. This is a minimum estimate of n since the frequency-response function measurement may be at a node of one or more modes of the system, repeated roots may exist, and/or the frequency resolution of the measurement may be too coarse to observe modes that are closely spaced in frequency. Several measurements can be observed and a tabulation of peaks existing in any or all measurements can be used as a more accurate minimum estimate of n. A more automated procedure for including the peaks that are present in several frequency-response functions is to observe the summation of frequency-response function power. This function represents the autopower or automoment of the frequency-response functions summed over a number of response measurements and is normally formulated as follows:

$$H_{\text{power}}(\omega) = \sum_{p=1}^{N_Q} \sum_{q=1}^{N_i} H_{pq}(\omega) \, H_{pq}^*(\omega) \tag{21.84}$$

These techniques are extremely useful but do not provide an accurate estimate of model order when repeated roots exist or when modes are closely spaced in frequency. For these reasons, an appropriate estimate of the order of the model is of prime concern and is the single most important problem in modal parameter estimation.

In order to determine a reasonable estimate of the model order for a set of representative data, a number of techniques have been developed as guides or aids to the user. Much of the user interaction involved in modal parameter estimation involves the use of these tools. Most of the techniques that have been developed allow the user to establish a maximum model order to be evaluated (in many cases, this is set by the memory limits of the computer algorithm). Information is utilized from the measured data based upon an assumption that the model order is equal to this maximum. This information is evaluated in a sequential fashion to determine if a model order less than the maximum is sufficient to describe the data sufficiently. This is the point at which the user's judgment and the use of various evaluation aids becomes important. Some of the commonly used techniques are:

- Measurement synthesis and comparison (curve-fit)
- Error chart
- Stability diagram
- Mode indication functions
- Rank estimation

One of the most common techniques is to synthesize an impulse-response function or a frequency-response function and compare it to the measured function to see if modes have been missed. This curve-fitting procedure is also used as a measure of the overall success of the modal parameter estimation procedure. The difference between the two functions can be quantified and normalized to give an indicator of the degree of fit. There can be many reasons for a poor comparison; incorrect model order is one of the possibilities.

Error Chart. Another method that has been used to indicate the correct model order more directly is the error chart. Essentially, the error chart is a plot of the error in the model as a function of increasing model order. The error in the model is a normalized quantity that represents the ability of the model to predict data that are not involved in the estimate of the model parameters. For example, when measured data in the form of an impulse-response function are used, only a small percentage of the total number of data values are involved in the estimate of modal parameters. If the model is estimated based upon 10 modes, only 4×10 data points are required, at a minimum, to estimate the modal parameters if no additional spatial information is used. The error in the model can then be estimated by the ability of the model to predict the next several data points in the impulse-response function compared to the measured data points. For the case of 10 modes and 40 data points, the error in the model is calculated from the predicted and measured data points 41 through 50. When the model order is insufficient, this error is large, but when the model order reaches the correct value, further increase in the model order does not result in a further decrease in the error. Figure 21.21 is an example of an error chart.

Stability Diagram. A further enhancement of the error chart is the stability diagram. The stability diagram is developed in the same fashion as the error chart and involves tracking the estimates of frequency, damping, and possibly modal participation factors as a function of model order. As the model order is increased, more

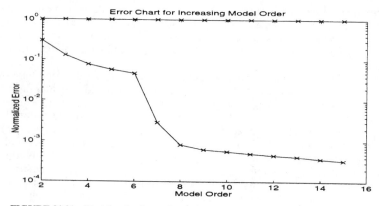

FIGURE 21.21 Model order determination: error chart.

and more modal frequencies are estimated, but, hopefully, the estimates of the physical modal parameters stabilize as the correct model order is found. For modes that are very active in the measured data, the modal parameters stabilize at a very low model order. For modes that are poorly excited in the measured data, the modal parameters may not stabilize until a very high model order is chosen. Nevertheless, the nonphysical (computational) modes do not stabilize at all during this process and can be sorted out of the modal parameter data set more easily. Note that inconsistencies (frequency shifts, leakage errors, etc.) in the measured data set obscure the stability and make the stability diagram difficult to use. Normally, a tolerance, in percentage, is given for the stability of each of the modal parameters that are being evaluated. Figure 21.22 is an example of a stability diagram. In Fig. 21.22, a summation of the frequency-response function power is plotted on the stability diagram for reference. Other mode indication functions can also be plotted against the stability diagram for reference.

Mode Indication Functions. Mode indication functions (MIF) are normally real-valued, frequency-domain functions that exhibit local minima or maxima at the modal frequencies of the system. One mode indication function can be plotted for each reference available in the measured data. The primary mode indication function exhibits a local minimum or maximum at each of the natural frequencies of the system under test. The secondary mode indication function exhibits a local minimum or maximum at repeated or pseudo-repeated roots of order 2 or more. Further mode indication functions yield local minima or maxima for successively higher orders of repeated or pseudo-repeated roots of the system under test.

MULTIVARIATE MODE INDICATION FUNCTION (MvMIF) The development of the multivariate mode indication function is based upon finding a force vector $\{F\}$ that excites a normal mode at each frequency in the frequency range of interest.[35] If a normal mode can be excited at a particular frequency, the response to such a force

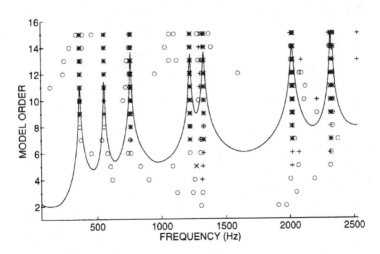

FIGURE 21.22 Model order determination: stability diagram.

vector exhibits the $90°$ phase lag characteristic. Therefore, the real part of the response is as small as possible, particularly when compared to the imaginary part or the total response. In order to evaluate this possibility, a minimization problem can be formulated as follows:

$$\min_{\|F\|=1} \frac{\{F\}^T [H_{\text{Real}}]^T [H_{\text{Real}}] \{F\}}{\{F\}^T ([H_{\text{Real}}]^T [H_{\text{Real}}] + [H_{\text{Imag}}]^T [H_{\text{Imag}}]) \{F\}} = \lambda \qquad (21.85)$$

This minimization problem is similar to a Rayleigh quotient, and it can be shown that the solution to the problem is found by finding the smallest eigenvalue λ_{min} and the corresponding eigenvector $\{F\}_{\text{min}}$ of the following problem:

$$[H_{\text{Real}}]^T [H_{\text{Real}}] \{F\} = \lambda ([H_{\text{Real}}]^T [H_{\text{Real}}] + [H_{\text{Imag}}]^T [H_{\text{Imag}}]) \{F\} \qquad (21.86)$$

The above eigenvalue problem is formulated at each frequency in the frequency range of interest. Note that the result of the matrix product $[H_{\text{Real}}]^T [H_{\text{Real}}]$ and $[H_{\text{Imag}}]^T [H_{\text{Imag}}]$ in each case is a square, real-valued matrix of size equal to the number of references in the measured data $N_i \times N_i$. The resulting plot of a multivariate mode indication function for a seven-reference case can be seen in Fig. 21.23. The frequencies where more than one curve approaches the same minimum are likely to be repeated root frequencies (repeated modal frequencies).

COMPLEX MODE INDICATION FUNCTION (CMIF)   An algorithm based on singular value decomposition methods applied to multiple reference FRF measurements, identified as the complex mode indication function (CMIF), is utilized in order to identify the proper number of modal frequencies, particularly when there are closely spaced or repeated modal frequencies.[35] Unlike MvMIF, which indicates the existence of real normal modes, CMIF indicates the existence of real normal or complex modes and the relative magnitude of each mode. Furthermore, MvMIF yields a set

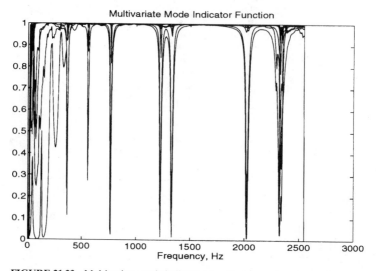

FIGURE 21.23   Multivariate mode indication function: seven-input example.

of force patterns that can best excite the real normal mode, while CMIF yields the corresponding mode shape and modal participation vector.

The CMIF, in the original formulation, is defined as the eigenvalues, solved from the normal matrix formed from the frequency-response function matrix, at each spectral line. The normal matrix is obtained by premultiplying the FRF matrix by its Hermitian matrix as $[H(\omega)]^H [H(\omega)]$. The CMIF is the plot of these eigenvalues on a log magnitude scale as a function of frequency. The peaks detected in the CMIF plot indicate the existence of modes, and the corresponding frequencies of these peaks give the damped natural frequencies for each mode. In the application of CMIF to traditional modal parameter estimation algorithms, the number of modes detected in CMIF determines the minimum number of degrees-of-freedom of the system equation for the algorithm. A number of additional degrees-of-freedom may be needed to take care of residual effects and noise contamination.

$$[H(\omega)]^H [H(\omega)] = [V(\omega)] [\Lambda(\omega)] [V(\omega)]^H \qquad (21.87)$$

By taking the singular value decomposition of the FRF matrix at each spectral line, an expression similar to Eq. (21.87) is obtained:

$$[H(\omega)] = [U(\omega)] [\Sigma(\omega)] [V(\omega)]^H \qquad (21.88)$$

where N_e = number of effective modes. The effective modes are the modes that contribute to the response of the structure at this particular frequency ω

$[U(\omega)]$ = left singular matrix of size $N_o \times N_e$, which is a unitary matrix

$[\Lambda(\omega)]$ = eigenvalue matrix of size $N_d \times N_e$, which is a diagonal matrix

$[\Sigma(\omega)]$ = singular value matrix of size $N_e \times N_e$, which is a diagonal matrix

$[V(\omega)]$ = right singular matrix of size $N_d \times N_i$, which is also a unitary matrix

Most often, the number of input points (reference points) N_i is less than the number of response points N_o. In Eq. (21.88), if the number of effective modes is less than or equal to the smaller dimension of the FRF matrix, i.e., $N_e \leq N_i$, the singular value decomposition leads to approximate mode shapes (left singular vectors) and approximate modal participation factors (right singular vectors). The singular value is then equivalent to the scaling factor Q_r divided by the difference between the discrete frequency and the modal frequency $j\omega - \lambda_r$. For a given mode, since the scaling factor is a constant, the closer the modal frequency is to the discrete frequency, the larger the singular value is. Therefore, the damped natural frequency is the frequency at which the maximum magnitude of the singular value occurs. If different modes are compared, the stronger the mode contribution (larger residue value), the larger the singular value is.

$$\text{CMIF}_k(\omega) \equiv \Lambda_k(\omega) = \Sigma_k(\omega)^2 \qquad k = 1, 2, \ldots, N_e \qquad (21.89)$$

where $\text{CMIF}_k(\omega)$ = kth CMIF as a function of frequency ω

$\Lambda_k(\omega)$ = kth eigenvalue of the normal matrix of FRF matrix as a function of frequency ω

$\Sigma_k(\omega)$ = kth singular value of the FRF matrix as a function of frequency ω

In practical calculations, the normal matrix formed from the FRF matrix, $[H(\omega)]^H$ $[H(\omega)]$, is calculated at each spectral line. The eigenvalues of this matrix are obtained. The CMIF plot is the plot of these eigenvalues on a log magnitude scale as a function of frequency. The peak in the CMIF indicates the location on the frequency axis that is nearest to the pole. The frequency is the estimated damped natural frequency, to within the accuracy of the frequency resolution. The magnitude of the eigenvalue indicates the relative magnitude of the modes, residue over damping factor.

Since the mode shapes that contribute to each peak do not change much around each peak, several adjacent spectral lines from the FRF matrix can be used simultaneously for a better estimation of mode shapes. By including several spectral lines of data in the singular value decomposition calculation, the effect of the leakage error can be minimized. The resulting plot of a complex mode indication function for a seven-reference case can be seen in Fig. 21.24. The frequencies where more than one curve approaches the same maximum are repeated root frequencies (repeated modal frequencies).

Rank Estimation. A more recent model order evaluation technique involves the estimate of the rank of the matrix of measured data. An estimate of the rank of the matrix of measured data gives a good estimate of the model order of the system. Essentially, the rank is an indicator of the number of independent characteristics contributing to the data. While the rank cannot be calculated in an absolute sense, it can be estimated from the singular value decomposition (SVD) of the matrix of measured data. For each mode of the system, one singular value should be found by the SVD procedure. The SVD procedure finds the largest singular value first and then successively finds the next largest. The magnitudes of the singular values are used in one of two different procedures to estimate the rank. The concept that is used is that the singular values should go to zero when the rank of the matrix is exceeded. For theoretical data, this happens exactly. For measured data, because of random errors and small inconsistencies in the data, the singular values do not

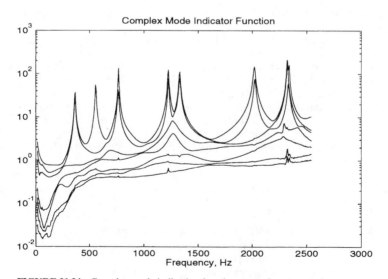

FIGURE 21.24 Complex mode indication function: seven-input example.

become zero but become very small. Therefore, the rate of change of the singular values rather than the absolute values is used as an indicator. In one approach, each singular value is divided by the first (largest) to form a normalized ratio. This normalized ratio is treated much like the error chart, and the appropriate rank (model order) is chosen when the normalized ratio approaches an asymptote. In another similar approach, each singular value is divided by the previous singular value, forming a normalized ratio that is approximately equal to 1 if the successive singular values are not changing in magnitude. When a rapid decrease in the magnitude of the singular value occurs, the ratio of successive singular values drops (or peaks if the inverse of the ratio is plotted) as an indicator of rank (model order) of the system. Figure 21.25 shows examples of these rank estimate procedures.

Residuals. Continuous systems have an infinite number of degrees-of-freedom, but, in general, only a finite number of modes can be used to describe the dynamic behavior of a system. The theoretical number of degrees-of-freedom can be reduced by using a finite frequency range. Therefore, for example, the frequency response can be broken up into three partial sums, each covering the modal contribution corresponding to modes located in the frequency ranges.

In the frequency range of interest, the modal parameters can be estimated to be consistent with Eq. (21.60). In the lower and higher frequency ranges, residual terms can be included to account for modes in these ranges. In this case, Eq. (21.60) can be rewritten for a single frequency-response function as

$$H_{pq}(\omega) = R_{F_{pq}} + \sum_{r=1}^{n} \frac{A_{pqr}}{j\omega - \lambda_r} + \frac{A_{pqr}^*}{j\omega - \lambda_r^*} + R_{I_{pq}}(\omega) \qquad (21.90)$$

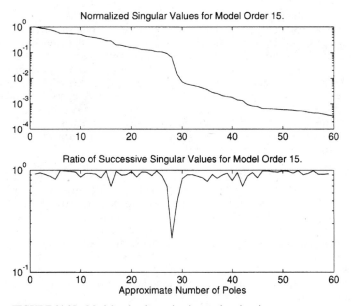

FIGURE 21.25 Model order determination: rank estimation.

where $R_{F_{pq}}$ = residual flexibility

$R_{I_{pq}}(s)$ = residual inertia

The residual term that compensates for modes below the minimum frequency of interest is called the *inertia restraint,* or *residual inertia.* The residual term that compensates for modes above the maximum frequency of interest is called the *residual flexibility.* These residuals are a function of each frequency-response function measurement and are not global properties of the frequency-response function matrix. Therefore, residuals cannot be estimated unless the frequency-response function is measured. In this common formulation of residuals, both terms are real-valued quantities. In general, this is a simplification; the residual effects of modes below and/or above the frequency range of interest cannot be completely represented by such simple mathematical relationships. As the system poles below and above the range of interest are located in the proximity of the boundaries, these effects are not the real-valued quantities noted in Eq. (21.90). In these cases, residual modes may be included in the model to partially account for these effects. When this is done, the modal parameters that are associated with these residual poles have no physical significance but may be required in order to compensate for strong dynamic influences from outside the frequency range of interest. Using the same argument, the lower and upper residuals can take on any mathematical form that is convenient as long as the lack of physical significance is understood. Mathematically, power functions of frequency (zero, first, and second order) are commonly used within such a limitation. In general, the use of residuals is confined to frequency-response function models. This is primarily due to the difficulty of formulating a reasonable mathematical model and solution procedure in the time domain for the general case that includes residuals.

MODAL IDENTIFICATION ALGORITHMS (SDOF)

For any real system, the use of single degree-of-freedom algorithms to estimate modal parameters is always an approximation since any realistic structural system has many degrees-of-freedom. Nevertheless, in cases where the modes are not close in frequency and do not affect one another significantly, single degree-of-freedom algorithms are very effective. Specifically, single degree-of-freedom algorithms are quick, rarely involving much mathematical manipulation of the data, and give sufficiently accurate results for most modal parameter requirements. Naturally, most multiple degree-of-freedom algorithms can be constrained to estimate only a single degree-of-freedom at a time if further mathematical accuracy is desired. The most commonly used single degree-of-freedom algorithms involve using the information at a single frequency as an estimate of the modal vector.

Operating Vector Estimation. Technically, when many single degree-of-freedom approaches are used to estimate modal parameters, sufficient simplifying assumptions are made that the results are not actually modal parameters. In these cases, the results are often referred to as *operating vectors* rather than modal vectors. This term refers to the fact that if the structural system is excited at this frequency, the resulting motion is a linear combination of the modal vectors rather than a single modal vector. If one mode is dominant, then the operating vector is approximately equal to the modal vector.

The approximate relationships that are used in these cases are represented in the following two equations:

$$H_{pq}(\omega_r) \approx \frac{A_{pqr}}{j\omega_r - \lambda_r} + \frac{A_{pqr}{}^*}{j\omega_r - \lambda_r{}^*} \tag{21.91}$$

$$H_{pq}(\omega_r) \approx \frac{A_{pqr}}{-\sigma_r} \tag{21.92}$$

For these less complicated methods, the damped natural frequencies ω_r are estimated by observing the maxima in the frequency-response functions. The damping factors σ_r are estimated using half-power methods.[1] The residues A_{pqr} are then estimated from Eq. (21.91) or (21.92) using the frequency-response function data at the damped natural frequency.

Complex Plot (Circle Fit). The circle-fit method utilizes the concept that the data curve in the vicinity of a modal frequency looks circular. In fact, the diameter of the circle is used to estimate the residue once the damping factor is estimated. More importantly, this method utilizes the concept that the distance along the curve between data points at equidistant frequencies is a maximum in the neighborhood of the modal frequency. Therefore, the circle-fit method is the first method to detect closely spaced modes.

This method can give erroneous answers when the modal coefficient is near zero. This occurs essentially because, when the mode does not exist in a particular frequency-response function (either the input or the response degree-of-freedom is at a node of the mode), the remaining data in the frequency range of the mode are strongly affected by the next higher or lower mode. Therefore, the diameter of the circle that is estimated is a function of the modal coefficient for the next higher or lower mode. This can be detected visually but is somewhat difficult to detect automatically.

The approximate relationship that is used in this case is represented in the following equation:

$$H_{pq}(\omega_r) \approx R_{pq} + \frac{A_{pqr}}{j\omega_r - \lambda_r} + \frac{A_{pqr}{}^*}{j\omega_r - \lambda_r{}^*} \tag{21.93}$$

Two-Point Finite Difference Formulation. The difference method formulations are methods that are based upon comparing adjacent frequency information in the vicinity of a resonance frequency. When a ratio of this information, together with information from the derivative of the frequency-response function at the same frequencies, is formed, a reasonable estimation of the modal frequency and residue for each mode can be determined under the assumption that modes are not too close together. This method can give erroneous answers when the modal coefficient is near zero. This problem can be detected by comparing the predicted modal frequency to the frequency range of the data used in the finite difference algorithm. As long as the predicted modal frequency lies within the frequency band, the estimate of the residue (modal coefficient) should be valid.

The approximate relationships that are used in this case are represented in the following equations. The frequencies noted in these relationships are as follows: ω_1 is a frequency near the damped natural frequency ω_r, and ω_p is the peak frequency close to the damped natural frequency ω_r.

Modal frequency (λ_r):

$$\lambda_r \approx \frac{j\omega_p H_{pq}(\omega_p) - j\omega_1 H_{pq}(\omega_1)}{H_{pq}(\omega_p) - H_{pq}(\omega_1)} \tag{21.94}$$

Residue (A_{pqr}):

$$A_{pqr} \approx \frac{j(\omega_1 - \omega_p)H_{pq}(\omega_1)H_{pq}(\omega_p)}{H_{pq}(\omega_p) - H_{pq}(\omega_1)} \tag{21.95}$$

Since both of the equations that are used to estimate modal frequency λ_r and residue A_{pqr} are linear equations, a least squares solution can be formed by using other frequency-response function data in the vicinity of the resonance. For this case, additional equations can be developed using $H_{pq}(\omega_2)$ or $H_{pq}(\omega_3)$ in the above equations instead of $H_{pq}(\omega_1)$.

MODAL IDENTIFICATION ALGORITHMS (MDOF)

All multiple degree-of-freedom equations can be represented in a unified matrix polynomial approach. The methods that are summarized in the following sections are listed in Tables 21.3 and 21.4.

High-Order Time-Domain Algorithms. The algorithms that fall into the category of high-order time-domain algorithms include the algorithms most commonly used to determine modal parameters. The least squares complex exponential (LSCE) algorithm is the first algorithm to utilize more than one frequency-response function, in the form of impulse-response functions, in the solution for a global estimate of the modal frequency. The polyreference time-domain (PTD) algorithm is an extension to the LSCE algorithm that allows multiple references to be included in a meaningful way so that the ability to resolve close modal frequencies is enhanced. Since both the LSCE and PTD algorithms have good numerical characteristics, these algorithms are still the most commonly used today. The only limitations for these algorithms are the cases involving high damping. As these are high-order algorithms, more time-domain information is required than for low-order algorithms.

First-Order Time-Domain Algorithms. The first-order time-domain algorithms include several well-known algorithms such as the Ibrahim time-domain (ITD) algorithm and the eigensystem realization algorithm (ERA). These algorithms are essentially a state-space formulation with respect to the second-order time-domain algorithms. The original development of these algorithms is quite different from that presented here, but the resulting solution of linear equations is the same regardless of development. There is a great body of published work on both the ITD and ERA algorithms, much of which discusses the various approaches for condensing the overdetermined set of equations that results from the data (least squares, double least squares, singular value decomposition). The low-order time-domain algorithms require very few time points in order to generate a solution because of the increased use of spatial information.

Second-Order Time-Domain Algorithms. The second-order time-domain algorithm has not been reported in the literature previously but is simply modeled after the second-order matrix differential equation with matrix dimension N_o. Since an impulse-response function can be thought to be a linear summation of a number of complementary solutions to such a matrix differential equation, the general second-order matrix form is a natural model that can be used to determine the modal parameters. This method is developed by noting that it is the time-domain equivalent to a frequency-domain algorithm known as the polyreference frequency-domain (PFD) algorithm. The low-order time-domain algorithms require very few time points in order to generate a solution because of the increased use of spatial information.

High-Order Frequency-Domain Algorithms. The high-order frequency-domain algorithms, in the form of scalar coefficients, are the oldest multiple degree-of-freedom algorithms utilized to estimate modal parameters from discrete data. These are algorithms like the rational fraction polynomial (RFP), power polynomial (PP), and orthogonal polynomial (OP) algorithms. These algorithms work well for narrow frequency bands and limited numbers of modes but have poor numerical characteristics otherwise. While the use of multiple references reduces the numerical conditioning problem, the problem is still significant and not easily handled. In order to circumvent the poor numerical characteristics, many approaches have been used (frequency normalization, orthogonal polynomials), but the use of low-order frequency-domain models has proven more effective.

Orthogonal Polynomial Concepts. The fundamental problem with using a rational fraction polynomial (power polynomial) method can be highlighted by looking at the characteristics of the data matrices. These matrices involve power polynomials that are functions of increasing powers of $s = j\omega$. These matrices are of the Vandermonde form and are known to be ill-conditioned for cases involving wide frequency ranges and high-ordered models.

VANDERMONDE MATRIX FORM:

$$
\begin{bmatrix}
(j\omega_1)^0 & (j\omega_1)^1 & (j\omega_1)^2 & \cdots & (j\omega_1)^{2m-1} \\
(j\omega_2)^0 & (j\omega_2)^1 & (j\omega_2)^2 & \cdots & (j\omega_2)^{2m-1} \\
(j\omega_3)^0 & (j\omega_3)^1 & (j\omega_3)^2 & \cdots & (j\omega_3)^{2m-1} \\
\cdots\cdots\cdots\cdots\cdots\cdots\cdots\cdots\cdots\cdots\cdots\cdots \\
(j\omega_i)^0 & (j\omega_i)^1 & (j\omega_i)^2 & \cdots & (j\omega_i)^{2m-1}
\end{bmatrix}
\tag{21.96}
$$

Ill-conditioning, in this case, means that the accuracy of the solution for the matrix coefficients α_m is limited by the numerical precision of the available arithmetic of the computer. Since the matrix coefficients α_m are used to determine the complex-valued modal frequencies, this presents a serious limitation for the high-order frequency-domain algorithms. The ill-conditioning problem can be best understood by evaluating the condition number of the Vandermonde matrix. The *condition number* measures the sensitivity of the solution of linear equations to errors, or small amounts of noise, in the data. The condition number gives an indication of the accuracy of the results from matrix inversion and/or linear equation solution. The condition number for a matrix is computed by taking the ratio of the largest singular value to the smallest singular value. A good condition number is a small number close to unity; a bad condition number is a large number. For the theoretical case of a singular matrix, the condition number is infinite.

The ill-conditioned characteristic of matrices that are of the Vandermonde form can be reduced, but not eliminated, by the following:

- Minimizing the frequency range of the data
- Minimizing the order of the model
- Normalizing the frequency range of the data (0,2) or (−2,2)
- Use of orthogonal polynomials

Several orthogonal polynomials have been applied to the frequency-domain modal parameter estimation problem, such as

- Forsythe polynomials
- Chebyshev polynomials
- Legendre polynomials
- Laguerre polynomials

First-Order Frequency-Domain Algorithms. Several algorithms have been developed that fall into the category of first-order frequency-domain algorithms, including the simultaneous frequency-domain (SFD) algorithm and the multiple reference simultaneous frequency-domain algorithm. These algorithms are essentially frequency-domain equivalents to the ITD and ERA algorithms and effectively involve a state-space formulation when compared to the second-order frequency-domain algorithms. The state-space formulation utilizes the derivatives of the frequency-response functions as well as the frequency-response function in the solution. These algorithms have superior numerical characteristics compared to the high-order frequency-domain algorithms. Unlike the low-order time-domain algorithms, though, sufficient data from across the complete frequency range of interest must be included in order to obtain a satisfactory solution.

Second-Order Frequency-Domain Algorithms. The second-order frequency-domain algorithms include the polyreference frequency-domain (PFD) algorithms. These algorithms have superior numerical characteristics compared to the high-order frequency-domain algorithms. Unlike the low-order time-domain algorithms, though, sufficient data from across the complete frequency range of interest must be included in order to obtain a satisfactory solution.

Residue Estimation. Once the modal frequencies and modal participation vectors have been estimated, the associated modal vectors and modal scaling (residues) can be found with standard least squares methods in either the time or the frequency domain. The most common approach is to estimate residues in the frequency domain utilizing residuals, if appropriate:

$$\{H_{pq}(\omega)\}_{N_s \times 1} = \left[\frac{1}{j\omega - \lambda_r} \right]_{N_s \times (2n+2)} \{A_{pqr}\}_{(2n+2) \times 1} \qquad (21.97)$$

where N_s = number of spectral lines $\geq 2n + 2$

$$
\left[\frac{1}{j\omega - \lambda_r}\right] =
\begin{bmatrix}
\dfrac{1}{j\omega_1 - \lambda_1} & \dfrac{1}{j\omega_1 - \lambda_2} & \dfrac{1}{j\omega_1 - \lambda_3} & \cdots & \dfrac{1}{j\omega_1 - \lambda_{2n}} & \dfrac{-1}{\omega_1{}^2} & 1 \\[2mm]
\dfrac{1}{j\omega_2 - \lambda_1} & \dfrac{1}{j\omega_2 - \lambda_2} & \dfrac{1}{j\omega_2 - \lambda_3} & \cdots & \dfrac{1}{j\omega_2 - \lambda_{2n}} & \dfrac{-1}{\omega_2{}^2} & 1 \\[2mm]
\dfrac{1}{j\omega_3 - \lambda_1} & \dfrac{1}{j\omega_3 - \lambda_2} & \dfrac{1}{j\omega_3 - \lambda_3} & \cdots & \dfrac{1}{j\omega_3 - \lambda_{2n}} & \dfrac{-1}{\omega_3{}^2} & 1 \\[2mm]
\cdots & \cdots & \cdots & \cdots & \cdots & \cdots & \cdots \\[2mm]
\dfrac{1}{j\omega_{Ns} - \lambda_1} & \dfrac{1}{j\omega_{Ns} - \lambda_2} & \dfrac{1}{j\omega_{Ns} - \lambda_3} & \cdots & \dfrac{1}{j\omega_{Ns} - \lambda_{2n}} & \dfrac{-1}{\omega_{Ns}{}^2} & 1
\end{bmatrix}
$$

$$
\{A_{pqr}\} =
\begin{Bmatrix}
A_{pq1} \\
A_{pq2} \\
A_{pq3} \\
\cdot \\
\cdot \\
\cdot \\
A_{pq2n} \\
R_{Ipq} \\
R_{Fpq}
\end{Bmatrix}
$$

$$
\{H_{pq}(\omega)\} =
\begin{Bmatrix}
H_{pq}(\omega_1) \\
H_{pq}(\omega_2) \\
H_{pq}(\omega_3) \\
\cdot \\
\cdot \\
\cdot \\
H_{pq}(\omega_{Ns})
\end{Bmatrix}
$$

The above equation is a linear equation in terms of the unknown residues once the modal frequencies are known. Since more frequency information N_s is available from the measured frequency-response function than the number of unknowns $2n + 2$, this system of equations is normally solved using the same least squares methods discussed previously. If multiple-input frequency-response function data are available, the above equation is modified to find a single set of $2n$ residues representing all of the frequency-response functions for the multiple inputs and a single output.

MODAL DATA PRESENTATION/VALIDATION

Once the modal parameters are determined, there are several procedures that allow the modal model to be validated. Some of the procedures that are used are

- Measurement synthesis
- Visual verification (animation)
- Finite-element analysis
- Modal vector orthogonality
- Modal vector consistency (modal assurance criterion)
- Modal modification prediction
- Modal complexity
- Modal phase colinearity and mean phase deviation

All of these methods depend upon the evaluation of an assumption concerning the modal model. Unfortunately, the success of the validation method defines only the validity of the assumption; the failure of the modal validation does not generally define what the cause of the problem is.

MEASUREMENT SYNTHESIS

The most common validation procedure is to compare the data synthesized from the modal model with the measured data. This is particularly effective if the measured data are not part of the data used to estimate the modal parameters. This serves as an independent check of the modal parameter estimation process.

The visual match can be given a numerical value if a correlation coefficient, similar to coherence, is estimated. The basic assumption is that the measured frequency-response function and the synthesized frequency-response function should be linearly related (unity) at all frequencies.

Synthesis correlation coefficient (SCC):

$$\text{SCC}_{pq} = \Gamma_{pq}^{2} = \frac{\left| \sum\limits_{\omega = \omega_1}^{\omega_2} H_{pq}(\omega)\hat{H}_{pq}{}^*(\omega) \right|^2}{\sum\limits_{\omega = \omega_1}^{\omega_2} H_{pq}(\omega)H_{pq}{}^*(\omega) \sum\limits_{\omega = \omega_1}^{\omega_2} \hat{H}_{pq}(\omega)\hat{H}_{pq}{}^*(\omega)} \qquad (21.98)$$

where $H_{pq}(\omega)$ = measured frequency-response function
$\hat{H}_{pq}(\omega)$ = synthesized frequency-response function

VISUAL VERIFICATION

Another common method of modal model validation is to evaluate the modal vectors visually. While this can be accomplished from plotted modal vectors superimposed upon the undeformed geometry, the modal vectors are normally animated (superimposed upon the undeformed geometry) in order to quickly assess the modal vector. In particular, modal vectors are evaluated for physically realizable characteristics such as discontinuous motion or out-of-phase problems. Often, rigid body modes of vibration are evaluated to determine scaling (calibration) errors or invalid measurement degree-of-freedom assignment or orientation. Naturally, if the system under test is believed to be proportionally damped, the modal vectors should be normal modes, and this characteristic can be quickly observed by viewing an animation of the modal vector.

FINITE-ELEMENT ANALYSIS

The results of a finite-element analysis of the system under test can provide another method of validating the modal model. While the problem of matching the number of analytical degrees-of-freedom N_a to the number of experimental degrees-of-freedom N_e causes some difficulty, the modal frequencies and modal vectors can be compared visually or through orthogonality or consistency checks. Unfortunately, when the comparison is not sufficiently acceptable, the question of error in the experimental model versus error in the analytical model cannot be easily resolved. Generally, assuming minimal errors and sufficient analysis and test experience, reasonable agreement can be found in the first ten deformable modal vectors, but agreement for higher modal vectors is more difficult. Finite-element analysis is discussed in detail in Chap. 28.

MODAL VECTOR ORTHOGONALITY

Another method that is used to validate an experimental modal model is the weighted orthogonality check. In this case, the experimental modal vectors are used together with a mass matrix normally derived from a finite-element model to evaluate orthogonality. The experimental modal vectors are scaled so that the diagonal terms of the modal mass matrix are unity. With this form of scaling, the off-diagonal values in the modal mass matrix are expected to be less than 0.1 (10 percent of the diagonal terms).

Theoretically, for the case of proportional damping, each modal vector of a system is orthogonal to all other modal vectors of that system when weighted by the mass, stiffness, or damping matrix. In practice, these matrices are made available by way of a finite-element analysis, and normally the mass matrix is considered to be the most accurate. For this reason, any further discussion of orthogonality is made with respect to mass matrix weighting. As a result, the orthogonality relations can be stated as follows:

Orthogonality of modal vectors:

$$\{\psi_r\}[M]\{\psi_s\} = 0 \qquad r \neq s \tag{21.99}$$

$$\{\psi_r\}[M]\{\psi_s\} = M_r \qquad r = s \tag{21.100}$$

Experimentally, the result of zero for the cross orthogonality [Eq. (21.99)] can rarely be achieved, but values up to one-tenth of the magnitude of the generalized mass of each mode are considered to be acceptable. It is a common procedure to form the modal vectors into a normalized set of mode shape vectors with respect to the mass matrix weighting. The accepted criterion in the aerospace industry, where this confidence check is made most often, is for all of the generalized mass terms to be unity and all cross-orthogonality terms to be less than 0.1. Often, even under this criterion, an attempt is made to adjust the modal vectors so that the cross-orthogonality conditions are satisfied.[36–38]

In Eqs. (21.99) and (21.100) the mass matrix must be an $N_o \times N_o$ matrix corresponding to the measurement locations on the structure. This means that the finite-element mass matrix must be modified from whatever size and distribution of grid locations are required in the finite-element analysis to the $N_o \times N_o$ square matrix corresponding to the measurement locations. This normally involves some sort of reduction algorithm as well as interpolation of grid locations to match the measurement situation.[39,40]

When Eq. (21.99) is not sufficiently satisfied, one (or more) of three situations may exist. First, the modal vectors can be invalid. This can be due to measurement error or problems with the modal parameter estimation algorithms. This is a very common assumption and many times contributes to the problem. Second, the mass matrix can be invalid. Since the mass matrix is not easily related to the physical properties of the system, this probably contributes significantly to the problem. Third, the reduction of the mass matrix can be invalid. This can certainly be a realistic problem and cause severe errors. One example of this situation occurs when a relatively large amount of mass is reduced to a measurement location that is highly flexible, such as the center of an unsupported panel. In such a situation the measurement location is weighted very heavily in the orthogonality calculation of Eq. (21.99) but may represent only incidental motion of the overall modal vector.

In all probability, all three situations contribute to the failure of cross-orthogonality criteria on occasion. When the orthogonality conditions are not satisfied, this result does not indicate where the problem originates. From an experimental point of view, it is important to try to develop methods that provide confidence that the modal vector is or is not part of the problem.

MODAL VECTOR CONSISTENCY

Since the residue matrix contains redundant information with respect to a modal vector, the consistency of the estimate of the modal vector under varying conditions such as excitation location or modal parameter estimation algorithms can be a valuable confidence factor to be utilized in the process of evaluation of the experimental modal vectors.

The common approach to estimation of modal vectors from the frequency-response function matrix is to measure a complete row or column of the frequency-response function matrix. This gives reasonable definition to those modal vectors that have a nonzero modal coefficient at the excitation location and can be completely uncoupled with the forced normal mode excitation method. When the modal coefficient at the excitation location of a modal vector is zero (very small with respect to the dynamic range of the modal vector) or when the modal vectors cannot be uncoupled, the estimation of the modal vector contains potential bias and variance errors. In such cases, additional rows and/or columns of the frequency-response function matrix are measured to detect such potential problems.

In these cases, information in the residue matrix corresponding to each pole of the system is evaluated to determine separate estimates of the same modal vector. This evaluation consists of the calculation of a complex modal scale factor (relating two modal vectors) and a scalar modal assurance criterion (measuring the consistency between two modal vectors). The function of the *modal scale factor* (*MSF*) is to provide a means of normalizing all estimates of the same modal vector. When two modal vectors are scaled similarly, elements of each vector can be averaged (with or without weighting), differenced, or sorted to provide a best estimate of the modal vector or to provide an indication of the type of error vector superimposed on the modal vector. In terms of multiple-reference modal parameter estimation algorithms, the modal scale factor is a normalized estimate of the modal participation factor between two references for a specific mode of vibration. The function of the *modal assurance criterion* (*MAC*) is to provide a measure of consistency between estimates of a modal vector. This provides an additional confidence factor in the evaluation of a modal vector from different excitation locations. The modal assurance criterion also provides a method of determining the degree of causality between estimates of different modal vectors from the same system.[41]

The *modal scale factor* is defined, according to this approach, as follows:

$$\text{MSF}_{cdr} = \frac{\{\psi_{cr}\}^H \{\psi_{dr}\}}{\{\psi_{dr}\}^H \{\psi_{dr}\}} \tag{21.101}$$

Equation (21.70) implies that the modal vector d is the reference to which the modal vector c is compared. In the general case, modal vector c can be considered to be made up of two parts. The first part is the part correlated with modal vector d. The second part is the part that is not correlated with modal vector d and includes contamination from other modal vectors and any random contribution. This error vector is considered to be noise. The *modal assurance criterion* is defined as a scalar constant relating the portion of the automoment of the modal vector that is linearly related to the reference modal vector as follows:

$$\text{MAC}_{cdr} = \frac{\left| \{\psi_{cr}\}^H \{\psi_{dr}\} \right|^2}{\{\psi_{cr}\}^H \{\psi_{cr}\} \{\psi_{dr}\}^H \{\psi_{dr}\}} = \frac{\left(\{\psi_{cr}\}^H \{\psi_{dr}\} \right)\left(\{\psi_{dr}\}^H \{\psi_{cr}\} \right)}{\{\psi_{cr}\}^H \{\psi_{cr}\} \{\psi_{dr}\}^H \{\psi_{dr}\}} \tag{21.102}$$

The modal assurance criterion is a scalar constant relating the causal relationship between two modal vectors. The constant takes on values from 0, representing no consistent correspondence, to 1, representing a consistent correspondence. In this manner, if the modal vectors under consideration truly exhibit a consistent relationship, the modal assurance criterion should approach unity and the value of the modal scale factor can be considered to be reasonable.

The modal assurance criterion can indicate only consistency, not validity. If the same errors, random or bias, exist in all modal vector estimates, this is not delineated by the modal assurance criterion. Invalid assumptions are normally the cause of this sort of potential error. Even though the modal assurance criterion is unity, the assumptions involving the system or the modal parameter estimation techniques are not necessarily correct. The assumptions may cause consistent errors in all modal vectors under all test conditions verified by the modal assurance criterion.

Coordinate Modal Assurance Criterion (COMAC). An extension of the modal assurance criterion is the *coordinate modal assurance criterion* (*COMAC*).[42] The COMAC attempts to identify which measurement degrees-of-freedom contribute negatively to a low value of MAC. The COMAC is calculated over a set of mode pairs, analytical versus analytical, experimental versus experimental, or experimental versus analytical. The two modal vectors in each mode pair represent the same modal vector, but the set of mode pairs represents all modes of interest in a given frequency range. For two sets of modes that are to be compared, there is a value of COMAC computed for each (measurement) degree-of-freedom.

The coordinate modal assurance criterion (COMAC) is defined as follows:

$$\text{COMAC}_p = \frac{\left| \sum\limits_{r=1}^{N} \psi_{pr}\phi_{pr} \right|^2}{\sum\limits_{r=1}^{N} \psi_{pr}\psi_{pr}^* \sum\limits_{r=1}^{N} \phi_{pr}\phi_{pr}^*} \tag{21.103}$$

where ψ_{pr} = modal coefficient from (measured) degree-of-freedom p and modal vector r from one set of modal vectors

ϕ_{pr} = modal coefficient from (measured) degree-of-freedom p and modal vector r from a second set of modal vectors

The above formulation assumes that there is a match for every mode in the two sets. Only those modes that match between the two sets are included in the computation.

MODAL MODIFICATION PREDICTION

The use of a modal model to predict changes in modal parameters caused by a perturbation (modification) of the system is becoming more of a reality as more measured data are acquired simultaneously. In this validation procedure, a modal model is estimated based upon a complete modal test. This modal model is used as the basis to predict a perturbation to the system that is tested, such as the addition of a mass at a particular point on the structure. Then, the mass is added to the structure and the perturbed system is retested. The predicted and measured data or modal model can be compared and contrasted as a measure of the validity of the underlying modal model.

MODAL COMPLEXITY

Modal complexity is a variation on the use of sensitivity analysis in the validation of a modal model. When a mass is added to a structure, the modal frequencies either should be unaffected or should shift to a slightly lower frequency. Modal overcomplexity is a summation of this effect over all measured degrees-of-freedom for each mode. Modal complexity is particularly useful for the case of complex modes in an attempt to quantify whether the mode is genuinely a complex mode, a linear combination of several modes, or a computational artifact. The mode complexity is normally indicated by the *mode overcomplexity value* (*MOV*), which is the percentage of the total number of response points that actually cause the damped natural frequency to decrease when a mass is added. A separate MOV is estimated for each mode of vibration, and the ideal result should be 1.0 (100 percent) for each mode.

MODAL PHASE COLINEARITY AND MEAN PHASE DEVIATION

For proportionally damped systems, the modal coefficients for a specific mode of vibration should differ by 0° or 180°. The *modal phase colinearity* (*MPC*) is an index expressing the consistency of the linear relationship between the real and imaginary parts of each modal coefficient. This concept is essentially the same as the ordinary coherence function with respect to the linear relationship of the frequency-response function for different averages or the modal assurance criterion (MAC) with respect to the modal scale factor between modal vectors. The MPC should be 1.0 (100 percent) for a mode that is essentially a normal mode. A low value of MPC indicates a mode that is complex (after normalization) and is an indication of a nonproportionally damped system or errors in the measured data and/or modal parameter estimation.

Another indicator that defines whether a modal vector is essentially a normal mode is the *mean phase deviation (MPD)*. This index is the statistical variance of the phase angles for each mode shape coefficient for a specific modal vector from the mean value of the phase angle. The MPD is an indication of the phase scatter of a modal vector and should be near 0° for a real, normal mode.

REFERENCES

1. Tse, F. S., I. E. Morse, Jr., and R. T. Hinkle: "Mechanical Vibrations: Theory and Applications," 2d ed., Prentice-Hall, Inc., Englewood Cliffs, N.J., 1978.

2. Craig, R. R., Jr.: "Structural Dynamics: An Introduction to Computer Methods," John Wiley & Sons, Inc., New York, 1981.

3. Ewins, D.: "Modal Testing: Theory and Practice," John Wiley & Sons, Inc., New York, 1984.

4. Bendat, J. S., and A. G. Piersol: "Random Data: Analysis and Measurement Procedures," John Wiley & Sons, Inc., New York, 1971.

5. Bendat, J. S., and A. G. Piersol: "Engineering Applications of Correlation and Spectral Analysis," John Wiley & Sons, Inc., New York, 1980.

6. Otnes, R. K., and L. Enochson: "Digital Time Series Analysis," John Wiley & Sons, Inc., New York, 1972.

7. Dally, J. W., W. F. Riley, and K. G. McConnell: "Instrumentation for Engineering Measurements," John Wiley & Sons, Inc., New York, 1984.

8. Strang, G.: "Linear Algebra and Its Applications," 3d ed., Harcourt Brace Jovanovich Publishers, San Diego, 1988.

9. Lawson, C. L., and R. J. Hanson: "Solving Least Squares Problems," Prentice-Hall, Inc., Englewood Cliffs, N.J., 1974.

10. Jolliffe, I. T.: "Principal Component Analysis," Springer-Verlag, New York, 1986.

11. Allemang, R. J., D. L. Brown, and R. W. Rost: "Dual Input Estimation of Frequency Response Functions for Experimental Modal Analysis of Automotive Structures," *SAE Paper* No. 820193, 1982.

12. Potter, R. W.: *J. Acoust. Soc. Amer.,* **66**(3):776 (1977).

13. Brown, D. L., G. Carbon, and R. D. Zimmerman: "Survey of Excitation Techniques Applicable to the Testing of Automotive Structures," *SAE Paper* No. 770029, 1977.

14. Halvorsen, W. G., and D. L. Brown: *Sound and Vibration,* November 1977, pp. 8–21.

15. Allemang, R. J., D. L. Brown, and W. Fladung: *Proc. Intern. Modal Analysis Conf.,* 1994, p. 501.

16. Brown, D. L., R. J. Allemang, R. D. Zimmerman, and M. Mergeay: "Parameter Estimation Techniques for Modal Analysis," *SAE Paper* No. 790221, *SAE Transactions,* **88**:828 (1979).

17. Vold, H., J. Kundrat, T. Rocklin, and R. Russell: *SAE Transactions,* **91**(1):815 (1982).

18. Vold, H., and T. Rocklin: *Proc. Intern. Modal Analysis Conf.,* 1982, p. 542.

19. Ibrahim, S. R., and E. C. Mikulcik: *Shock and Vibration Bull.* **47**(4):183 (1977).

20. Fukuzono, K.: "Investigation of Multiple-Reference Ibrahim Time Domain Modal Param-

eter Estimation Technique," M.S. Thesis, Dept. of Mechanical and Industrial Engineering, University of Cincinnati, 1986.

21. Juang, Jer-Nan, and R. S. Pappa: *AIAA J. Guidance, Control, and Dynamics,* **8**(4):620 (1985).

22. Longman, R. W., and Jer-Nan Juang: *AIAA J. Guidance, Control, and Dynamics,* **12**(5):647 (1989).

23. Zhang, L., H. Kanda, D. L. Brown, and R. J. Allemang: "A Polyreference Frequency Domain Method for Modal Parameter Identification," *ASME Paper* No. 85-DET-106, 1985.

24. Lembregts, F., J. Leuridan, L. Zhang, and H. Kanda: *Proc. Intern. Modal Analysis Conf.,* 1986, pp. 589–598.

25. Lembregts, F., J. L. Leuridan, and H. Van Brussel: *Mech. Systems and Signal Processing,* **4**(1):65 (1989).

26. Coppolino, R. N.: "A Simultaneous Frequency Domain Technique for Estimation of Modal Parameters from Measured Data," *SAE Paper* No. 811046, 1981.

27. Craig, R. R., A. J. Kurdila, and H. M. Kim: *J. Analytical and Experimental Modal Anal.,* **5**(3): 169 (1990).

28. Richardson, M., and D. L. Formenti: *Proc. Intern. Modal Analysis Conf.,* 1982, p. 167.

29. Vold, H., "Orthogonal Polynomials in the Polyreference Method," *Proc. Intern. Seminar on Modal Analysis,* Katholieke University of Leuven, Belgium, 1986.

30. Van der Auweraer, H., and J. Leuridan: *Mechanical Systems and Signal Processing,* **1**(3):259 (1987).

31. Shih, C. Y., Y. G. Tsuei, R. J. Allemang, and D. L. Brown: *Mechanical Systems and Signal Processing,* **2**(4):349 (1988).

32. Shih, C. Y., Y. G. Tsuei, R. J. Allemang, and D. L. Brown: *Mechanical Systems and Signal Processing,* **2**(4):367 (1988).

33. Dippery, K. D., A. W. Phillips, and R. J. Allemang: *Proc. Intern. Modal Analysis Conf.,* 1994.

34. Dippery, K. D., A. W. Phillips, and R. J. Allemang: *Proc. Intern. Modal Analysis Conf.,* 1994.

35. Williams, R., J. Crowley, and H. Vold: *Proc. Intern. Modal Analysis Conf.,* 1985, p. 66.

36. Gravitz, S. I.: *J. Aero/Space Sci.,* **25**:721 (1958).

37. McGrew, J.: *AIAA J.,* **7**(4):774 (1969).

38. Targoff, W. P.: *AIAA J.,* **14**(2):164 (1976).

39. Guyan, R. J.: *AIAA J.,* **3**(2):380 (1965).

40. Irons, B.: *AIAA J.,* **3**(5):961 (1965).

41. Allemang, R. J., and D. L. Brown: *Proc. Intern. Modal Analysis Conf.,* 1982, p. 110.

42. Lieven, N. A. J., and D. J. Ewins: *Proc. Intern. Modal Analysis Conf.,* 1988, p. 690.

CHAPTER 22

CONCEPTS IN VIBRATION DATA ANALYSIS

Allen J. Curtis
Steven D. Lust

INTRODUCTION

This chapter discusses the mathematical concepts involved in the analysis of vibration data which are essentially steady-state or slowly time-varying processes. It provides the mathematical basis for data reduction procedures by which the original time-history of vibration is transformed to other forms required for particular applications. Classes of data discussed in this chapter include sinusoidal, periodic, complex, and random (shock and transient vibration are discussed in Chap. 23).

Prior to commencing the actual processing of vibration data, it is advisable to "edit" the basic time-histories for two purposes. First, it is seldom necessary or desirable to analyze, i.e., process, all available recorded data since the recording probably started well before and continues long after the occurrence of interest. Further, for signals obtained during physical conditions which are slowly time-varying, it may be necessary to select several intervals during which the conditions may be treated as stationary. Secondly, the time-history available for processing, either recorded or online, always consists of the vibration signal of interest together with an extraneous signal commonly referred to as *noise*. Vibration analyzers have no way to distinguish between the signal and the noise, and thus the resultant output describes, for example, the spectral characteristics of both. A spectral plot with sharp peaks at 60 Hz or any of its multiples almost certainly indicates that the vibration signal has been contaminated with *line noise*. Reference 1 contains detailed discussions of sources of noise and techniques which may be employed to distinguish between the signal and the noise.

The basic objectives of vibration data analysis, discussed in Chap. 14, may be achieved by various means. No one method of analysis can be considered *the* correct one, since both the type of data and the use to be made of the results influence the appropriateness of a particular type of analysis. This chapter is restricted to a description of the basic principles and techniques of data analysis which will enable the reader to make a logical choice of the type of analysis and data presentation most suited to the particular problem at hand.

The techniques of data analysis are illustrated in this chapter by functional block diagrams and are necessarily described in terms associated with analog processing of the signals. With the advent of minicomputers and implementation of the fast Fourier transform algorithm, it became practical to perform these analyses digitally. Chapter 27 describes the use of digital computers for performing the types of vibration data analyses described in principle in this chapter.

The complete definition of a vibration condition requires the description of magnitude* and its variation with both frequency and time. Various types of vibration are described mathematically, and the respective magnitude parameters are defined explicitly or statistically. The determination of the statistical properties of random vibration is discussed, including the probability density and probability distribution functions. The determination of the frequency or spectral characteristics of the vibration magnitude at a given time is discussed, followed by a discussion of means of determining the time variation of the magnitude of the vibration. Finally, autocorrelation and crosscorrelation are discussed as means to determine both the statistical and spectral characteristics of vibration.

QUALITATIVE DESCRIPTION OF VIBRATION

This section considers various means which are commonly used in vibration analysis to describe time-varying functions such as displacement, velocity, acceleration, or force and describes the principal characteristics of these functions.

Figure 22.1 illustrates three typical vibration waveforms or time-histories. The solid curve of Fig. 22.1A is the sum of the two sinusoidal vibrations shown by the dotted lines. Figure 22.1B represents broad-band random vibration, while Fig. 22.1C illustrates the waveform of a narrow-band random vibration. The latter has the appearance of a sine wave whose amplitude, represented by the dotted lines, varies in an unpredictable manner. If the envelope of the waveform varies sinusoidally, the waveform is the sum of two sinusoids whose frequency difference is small compared to the sum, resulting in beats.

Figure 22.2 illustrates typical forms for describing, after suitable data analysis, the time-histories shown in Fig. 22.1. In Fig. 22.2A, the two sine waves are represented by vertical lines at their respective frequencies; the height of each line is proportional to the amplitude of the respective sine wave. This plot is known as a *line spectrum* or *discrete frequency spectrum*. The information also can be presented as a tabulation of frequencies and associated amplitudes. For complete definition of the waveform of Fig. 22.1A, it is necessary to specify the phase angle between the two sine waves at a convenient time origin. However, phase information seldom is required in vibration data analysis, and the plot of Fig. 22.2A is the extent of the information usually required. Figure 22.2B is a plot of the power spectral density[†] versus frequency of broad-band random vibration showing the frequency content to be spread over a wide frequency range. Figure 22.2C illustrates the power spectral density of narrow-band random vibration showing the concentration of the frequency content in a narrow band of frequencies.

---

* The word *magnitude* is used in this chapter to indicate the severity of vibration without relating it specifically to one of the various mathematical expressions used to describe vibration quantitatively.

[†] Power spectral density is defined under *Spectral Analysis* [see Eq. (22.38) in this chapter; also see Eq. (11.20)]. It is a generic term used regardless of the physical quantity represented by the time-history. However, it is preferable to indicate the physical quantity involved. For example, the term *mean-square acceleration density* or *acceleration spectral density* is used when the time-history of acceleration is to be described.

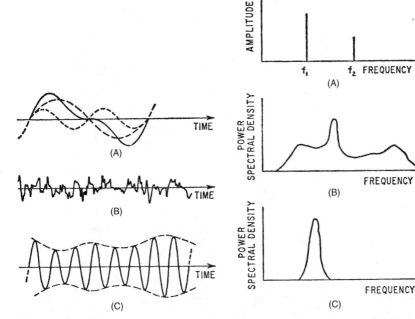

FIGURE 22.1 Typical vibration time-histories: (A) Sum of two sinusoids; (B) broad-band random vibration; (C) narrow-band random vibration.

FIGURE 22.2 Results of data analysis of time-histories of Fig. 22.1: (A) Sum of two sinusoids; (B) broad-band random vibration; (C) narrow-band random vibration.

In general, three coordinates are necessary to describe a vibration time-history: (1) the magnitude of the vibration, e.g., the amplitude of sine waves or the power spectral density; (2) the variation of magnitude with frequency, i.e., spectral characteristics; and (3) the variation of the magnitude of the vibration at a particular frequency as a function of time.

The magnitude of a vibration may be represented in terms of its spectral characteristics and its variation with time by the surface shown in Fig. 22.3. The height of a point on the surface raised over the time-frequency plane defines the magnitude of the vibration at a particular time and frequency. For example, the surface could represent the vibration encountered during the flight of an aircraft from take-off to landing. As the flight conditions change, both the vibration magnitude and the frequency characteristics of the vibration may change.

At a specific instant of time, say t_0, the vibration is described by its variation of magnitude with frequency in the plane normal to the time axis passing through t_0, i.e., in the plane labeled "Spectral Analysis" in Fig. 22.3 and shown in the two-dimensional plot of Fig. 22.4.

The variation of the magnitude of vibration at a particular frequency, say f_0, as a function of time is shown by passing a plane normal to the frequency axis through f_0, i.e., in the plane labeled "Magnitude-Time-history Analysis" in Fig. 22.3 and shown in the two-dimensional plot of Fig. 22.5.

In the case of truly steady-state vibration, i.e., vibration independent of time as during some laboratory vibration tests, the surface of Fig. 22.3 can be represented by

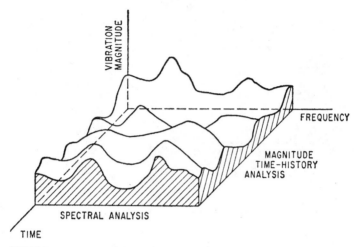

FIGURE 22.3 Vibration magnitude surface: The height of the surface at any time and frequency represents the vibration magnitude at that time and frequency. Units of vibration magnitude depend on type of vibration, e.g., random, complex, or periodic.

the magnitude-frequency plot of Fig. 22.4. Similarly, a vibration consisting of a single sine wave at a constant frequency whose amplitude varies slowly with time may be represented by a single amplitude-time plot such as Fig. 22.5.

BASIC TYPES OF TIME-VARYING FUNCTIONS

This section considers certain relations used to describe specific time-varying functions in vibration analysis and describes their principal characteristics. Types of functions discussed are sinusoidal, periodic, phase-coherent, complex, and random.[2-4] In the following sections, $\xi(t)$ is the function which represents the time variation of any physical parameter such as displacement, force, or acceleration.

Sinusoidal Functions.   The simplest time-varying quantity of concern in vibration analysis is a sinusoidal function having constant amplitude ξ_0 and frequency f:

$$\xi(t) = \xi_0 \sin (\omega t + \theta) \tag{22.1}$$

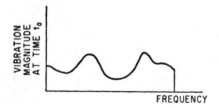

FIGURE 22.4 Typical plot of spectral analysis.

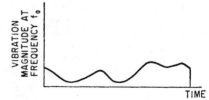

FIGURE 22.5 Typical plot of magnitude time-history analysis.

where $\omega = 2\pi f$, and θ is the phase angle with respect to the time origin. The objective of data analysis on this waveform is the determination of ξ_0, ω, and, less frequently, θ.

Periodic Functions. If a function has a waveform which repeats itself exactly in consecutive time periods τ:

$$\xi(t) = \xi(t \pm n\tau) \tag{22.2}$$

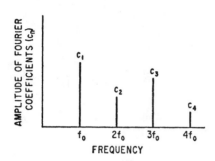

where n is any integer; then it is said to be a *periodic function* having a fundamental frequency $f = 1/\tau$. For example, a wing panel adjacent to a propeller may vibrate periodically at a fundamental frequency equal to the cyclic rate at which propeller blades pass the panel and also at higher frequencies equal to multiples of the blade passage frequency. A periodic function may be represented by the line spectra shown in Fig. 22.6.

FIGURE 22.6 Line spectrum to represent Fourier series of periodic function. Fundamental frequency $f_0 = 1/\tau$, where τ is period of function [see Eq. (22.29)].

Phase-Coherent Functions. A phase-coherent function is defined as a sine wave whose amplitude and frequency vary smoothly with time. It is assumed that the changes in amplitude and frequency are significant in magnitude only over a time duration equal to a number of cycles of the waveform. Mathematically, it is expressed by

$$\xi(t) = \xi_0(t) \sin [\omega(t)t + \theta] \tag{22.3}$$

where ξ_0 and ω are functions of time and θ is the phase angle of the waveform at the time origin. A function of this type is useful in a description of vibration measured in a system whose natural frequencies vary with time, such as a rocket engine during the burning of the propellant. Equation (22.3) defines the input excitation to the test object during a sweep frequency vibration test.

Complex Functions. A time-history which is composed of a sum (or in the limit, integral) of sinusoidal functions which is not necessarily periodic is defined as a complex function. Such functions may be represented graphically by line spectra as shown in Fig. 22.6 except that the lines are not equally spaced along the frequency axis; when the complex function consists of an integral or continuous distribution of sinusoidal functions, it may be represented by a Fourier spectrum as discussed in Chap. 23.

Transient Function—Shock. A transient function is a time-varying function which is nonzero only over a given restricted time interval. Transients are discussed in Chap. 23.

Random Functions. A random time-function[2-5] (see Chap. 11) consists of a continuous distribution of sine waves at all frequencies, the amplitudes and phase angles of which vary in an unpredictable manner as a function of time. This unpredictability distinguishes random functions from *deterministic* or coherent functions (e.g., sinusoidal, periodic, and complex functions) in the following respect. Knowledge of the instantaneous value of a random function at a given time does not provide any information on the value which it will have at some specific later time. It is known only that there is a certain probability that the value will lie within a certain range of values, i.e., it can be described only in statistical terms. In contrast, if the value of a deterministic function is known at any time, the value at any other time can be determined.

Definition of Mean, Average, and Mean-Square Values

Mean Value.   The mean value of a function $\xi(t)$ is defined by

$$\bar{\xi} = \frac{1}{T} \int_0^T \xi(t)\, dt \tag{22.4}$$

where T is the *averaging* time.

Average Value.   The term *average value* of a function $\xi(t)$ is employed to denote the time average of the magnitude of a function with zero mean value:

$$\overline{|\xi|} = \frac{1}{T} \int_0^T [\xi(t)]\, dt \tag{22.5}$$

Mean-Square Value.   The mean-square value of a function $\xi(t)$ is the time average of the square of the function:

$$\overline{|\xi|^2} = \frac{1}{T} \int_0^T [\xi(t)]^2\, dt \tag{22.6}$$

The rms value is the square root of the mean-square value.

For example, the mean, average, and mean-square values of a sine wave [Eq. (22.1)] averaged over an integer number of periods, are

$$\bar{\xi} = 0$$

$$\overline{|\xi|} = \frac{2}{\pi}\,\xi_0 \tag{22.7}$$

$$\overline{|\xi|^2} = \tfrac{1}{2}\xi_0^2$$

TIME/FREQUENCY RELATIONSHIPS

Common to almost all methods of vibration data analysis is the transformation of a function in the time domain, i.e., a time-history, to a function in the frequency domain. This transformation provides insight into, for example, the source of the vibration and permits analysis of system responses to it. The relationships which permit these transformations follow.

FUNDAMENTAL MATHEMATICAL RELATIONSHIPS[6-8]

A number of fundamental mathematical relationships between functions in the time and/or frequency domains are common to most experimental applications of digital computers to shock and vibration problems. These are summarized below. For clarity, the relationships are generally presented in terms of *continuous* functions, though the reader must bear in mind that digital processing dictates that the functions can be computed only at a *discrete number* of frequencies or instants of time.

Fourier Transform. The basic relationships used to transform information from the time domain to the frequency domain, or vice versa, are the Fourier transform pair (see Chap. 23).

$$S_x(f) = \int_{-\infty}^{\infty} x(t)e^{-i2\pi ft} \, dt \qquad (22.8)$$

$$x(t) = \int_{-\infty}^{\infty} S_x(f)e^{i2\pi ft} \, df \qquad (22.9)$$

where $x(t)$ is the time-history of a function x, and $S_x(f)$ is the Fourier transform of x, a complex quantity.

Discrete Fourier Transform (DFT). In practical digital applications (also see *FFT Analyzers*, Chap. 13), $x(t)$ is known only for a finite time interval or record length T, at a total of N equally spaced time intervals Δt apart. The discrete Fourier transform (DFT) pair, equivalent to Eqs. (22.8) and (22.9), become

$$S_x(m\,\Delta f) = \Delta t \sum_{i=0}^{N-1} x(n\,\Delta t)e^{-i2\pi m\,\Delta f n\,\Delta t} \qquad m = 0, \ldots, \left(\frac{N}{2}\right) \qquad (22.10)$$

$$x(n\,\Delta t) = \Delta f \sum_{m=0}^{N/2} S_x(m\,\Delta f)e^{i2\pi m\,\Delta f n\,\Delta t} \qquad n = 0, \ldots, (N-1) \qquad (22.11)$$

The DFT Eq. (22.10) is correct only if $x(t)$ is periodic with period T. The set of S_x are the coefficients of the Fourier series expansion of $x(t)$. If $x(t)$ is not periodic, the DFT treats $x(t)$ as if it were so and is thus only an estimate of the true Fourier transform of $x(t)$.

As a consequence of Shannon's theorem,[9] the DFT of $x(t)$ is defined only at $N/2$ frequencies at equally spaced intervals Δf apart, up to a maximum frequency F_{max}. The following fundamental relationships apply:

Block size:

$$N = \frac{T}{\Delta t} \qquad (22.12)$$

Frequency range:

$$F_{max} = 1/2\Delta t = \Delta f\left(\frac{N}{2}\right) \qquad (22.13)$$

Sampling rate:

$$\frac{1}{\Delta t} = 2F_{max} \qquad (22.14)$$

Frequency resolution:

$$\Delta f = \frac{1}{T} \qquad (22.15)$$

Record length:

$$T = \frac{1}{\Delta f} = N \, \Delta t \qquad (22.16)$$

In performing a DFT analysis, assuming no restrictions on block size N, the parameters which may be selected are limited. The frequency range or the sampling rate, but not both, may be selected at will. Similarly, the frequency resolution or the record length may be chosen. If the block size is fixed, then only one of these four parameters may be selected. (See Fig. 13.16.)

The DFT, and thus any other quantities derived therefrom, is susceptible to certain errors, the most important of which is known as *aliasing*.[6-8] Aliasing error arises when the signal $x(t)$ contains frequency components or "energy" above F_{max} of the DFT. Owing to the sampling or digitizing process, this energy will appear to be within the DFT frequency range. Therefore, to avoid aliasing error, either F_{max} must be large enough to include all significant frequency components of $x(t)$ or, for a given F_{max}, the components above F_{max} must be removed by analog filtering before digitization.

Spectral Density—Fourier Transform.   The concept of *power spectral density*, also called *auto spectral density*, is defined in analog terms in Chap. 11 and later in this chapter under "Spectral Analysis." This quantity describes the frequency or spectral properties of a single time-history. A companion quantity, the *cross spectral density*, describes the joint spectral properties of two time-histories. Both quantities are related to the Fourier transforms of the time histories through the following relationships:

Auto spectral density:

$$G_{xx}(f) = S_x(f)S_x^*(f) = |S_x(f)|^2 \qquad (22.17)$$

where $G_{xx}(f)$ is the one-sided auto spectral density and $S_x^*(f)$ is the complex conjugate of $S_x(f)$.

Cross spectral density:

$$G_{xy}(f) = S_x(f)S_y^*(f) \qquad (22.18)$$

$$G_{yx}(f) = S_y(f)S_x^*(f) \qquad (22.19)$$

where $G_{xy}(f)$ and $G_{yx}(f)$ are one-sided cross spectral densities and are complex quantities, and $S_x(f)$ and $S_y(f)$ are the Fourier transforms of $x(t)$ and $y(t)$, respectively.

When the Fourier transforms of $x(t)$ and $y(t)$ are obtained from digital data, i.e., DFTs, the spectral density functions are defined only at the frequencies for which the DFT is calculated. Further properties of the spectral density functions so obtained are described under "Vibration Data Analysis."

Spectral Density—Correlation Functions.   The spectral density functions, which describe signal characteristics in the frequency domain, and the correlation functions, which describe signal characteristics in the time domain are related through the Fourier transform as follows:

Auto:

$$G_{xx}(f) = 2 \int_{-\infty}^{\infty} R_{xx}(\tau)e^{-i2\pi f\tau} \, d\tau \qquad 0 \le f < \infty \qquad (22.20)$$

$$R_{xx}(\tau) = \tfrac{1}{2} \int_{-\infty}^{\infty} G_{xx}(f)e^{i2\pi f\tau} \, df \qquad -\infty < \tau < \infty \qquad (22.21)$$

Cross:

$$G_{xy}(f) = 2\int_{-\infty}^{\infty} R_{xy}(\tau)e^{-i2\pi f\tau}\, d\tau \qquad 0 \le f < \infty \qquad (22.22)$$

$$R_{xy}(\tau) = \frac{1}{2}\int_{-\infty}^{\infty} G_{xy}(f)e^{i2\pi f\tau}\, df \qquad -\infty < \tau < \infty \qquad (22.23)$$

where $R_{xx}(\tau)$ is the auto correlation function at $x(t)$, and $R_{xy}(\tau)$ is the cross correlation function at $x(t)$ and $y(t)$.

Frequency-Response Functions. Frequency-response functions, synonymously known as *transfer functions,* describe in the frequency domain the relationship between the input and output or response of a linear system. This is shown schematically in Fig. 22.7 for a single input-single output system. (See Ref. 6 for multiple input-output systems.) The transfer function $H(f)$ describes the magnitude and phase of the response per unit sinusoidal input as a function of the input frequency. It is thus a complex quantity. Transfer functions can also be determined from the Fourier transforms of the input and response time-histories and from the spectral densities of the input and response signals when the input is a random process. The governing relationships are

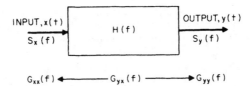

FIGURE 22.7 Schematic representation of transfer function $H(f)$.

Fourier transforms:

$$S_y(f) = S_x(f)H(f) \qquad \text{or} \qquad H(f) = \frac{S_y(f)}{S_x(f)} \qquad (22.24)$$

Auto spectral densities:

$$G_{yy}(f) = G_{xx}(f)|H(f)|^2 \qquad \text{or} \qquad |H(f)|^2 = \frac{G_{yy}(f)}{G_{xx}(f)} \qquad (22.25)$$

Cross spectral densities:

$$G_{yx}(f) = G_{xx}(f)H_1(f) \qquad \text{or} \qquad H_1(f) = \frac{G_{yx}(f)}{G_{xx}(f)} \qquad (22.26A)$$

$$G_{xy}(f)H_2(f) = G_{yy}(f) \qquad \text{or} \qquad H_2(f) = \frac{G_{yy}(f)}{G_{xy}(f)} \qquad (22.26B)$$

The relationships above assume no extraneous signal or noise is present in either $x(t)$ or $y(t)$, in which case, $H_1(f)$ and $H_2(f)$ are equal. (See *Coherence Function,* below.)

Only the transfer function magnitude can be determined from auto spectral densities; that is, phase information is unobtainable. This follows because the auto spectral densities themselves are void of phase information. In using cross spectral density, in which phase information is retained, both magnitude and phase of the transfer function is obtained. Depending on the relative characteristics of $x(t)$ and $y(t)$ and their "noisiness," it may be desirable to calculate $H_2(f)$ rather than the customary Eq. (22.26A).

Coherence Function.   The coherence function, γ_{xy}^2, is a measure of the quality of the input, response and cross spectral densities, and the causality of input to response for a system as shown in Fig. 22.7.

$$\gamma_{xy}^2(f) = \frac{|G_{xy}(f)|^2}{G_{xx}(f)G_{yy}(f)} \tag{22.27}$$

From Eqs. (22.25) and (22.26),

$$\gamma_{xy}^2(f) = \frac{|G_{xx}(f)H(f)|^2}{G_{xx}(f)|H(f)|^2 G_{xx}(f)} = 1 \tag{22.28}$$

assuming $x(t)$ and $y(t)$ are noise-free.

It can be shown[6] that, in practical applications,

$$0 \le \gamma_{xy}^2 \le 1.0$$

Values of γ_{xy}^2 less than unity indicate that the response is not attributable to the input, due, for example, to extraneous noise or nonlinearity of the system. In the frequency domain, the coherence function is analogous to the correlation coefficient in the time domain.

ANALYSIS OF PERIODIC FUNCTIONS

FOURIER SERIES

Periodic functions encountered in vibration analysis may be represented by a Fourier series which consists of a sum of sine waves whose frequencies are all multiples of the fundamental frequency. The amplitudes of the sine waves are known as the Fourier coefficients c_n. An additional constant term equal to the mean value of the waveform during the period τ must also be included. The function is expressed mathematically as

$$\xi(t) = \sum_{n=1}^{\infty} a_n \sin n\omega t + \sum_{n=0}^{\infty} b_n \cos n\omega t = \sum_{n=0}^{\infty} c_n \sin (n\omega t + \theta_n) \tag{22.29}$$

where $c_n = +[a_n^2 + b_n^2]^{1/2}$ $\theta_n = \sin^{-1} \dfrac{b_n}{c_n} = \cos^{-1} \dfrac{a_n}{c_n}$

$$a_n = \frac{2}{\tau} \int_{-\tau/2}^{\tau/2} \xi(t) \sin n\omega t \, dt$$

$$b_n = \frac{2}{\tau} \int_{-\tau/2}^{\tau/2} \xi(t) \cos n\omega t \, dy$$

$$b_0 = \frac{1}{\tau} \int_{-\tau/2}^{\tau/2} \xi(t) \, dt = \bar{\xi}$$

If the function $\xi(t)$ is an *even function*, i.e., $\xi(t)$ equals $\xi(-t)$, the series consists of cosine terms only; for an *odd function*, $\xi(t)$ equals $-\xi(-t)$ and the series consists of sine terms only.

OBJECTIVE AND RESULTS OF ANALYSIS

The objective of the analysis of periodic functions is to determine the Fourier coefficients c_n and the phase angles θ_n of Eq. (22.29). By describing a function as a Fourier series, it is implied that the waveform during the period τ is repeated exactly in all successive periods between negative and positive infinite times and that all frequencies associated with the waveform are determined by τ, as shown by Eq. (22.29). No measured vibration data completely fulfill these requirements for description by a Fourier series. When conducting a Fourier series analysis, it is common practice to select what appears to be a typical section of the time-history and to treat it as if it were the time-history during one period of a periodic function. The consequences of this selection are (1) the time period chosen fixes the frequencies of the resulting line spectra, whether these are the actual frequencies present in the function or not, and (2) the signal is assumed, a priori, to consist of a sum of pure sinusoids.

The results of data analysis involving periodic vibration may be summarized by plotting the discrete frequencies and the corresponding amplitudes. For example, Fig. 22.2A applies to two superimposed sinusoidal functions of arbitrary frequencies f_1, f_2; Fig. 22.6 is an analogous plot of the Fourier coefficients of a periodic quantity. This form of data presentation may be abbreviated by plotting only points representing the upper ends of the spectrum lines in Figs. 22.2A and 22.2B. Representation of the data by such points is particularly advantageous for summarizing the results of analysis of the vibration of a number of data samples. For example, the vibration at a number of different flight conditions or on a number of different aircraft may be shown on a single plot, so that each point represents one measurement of a sinusoidal function at the given amplitude and frequency. If the number of points becomes excessive, different-sized symbols are used to represent given numbers, say 1, 10, 50, or 100 readings at the particular amplitude and frequency. The density of points in various frequency regions or at various amplitudes indicates, at least qualitatively, the frequency of occurrence of those values.

DETERMINATION OF FOURIER COEFFICIENTS

The Fourier coefficients of a periodic function are defined by Eq. (22.29) and may be determined by several alternate means. If an oscillographic recording of the function $\xi(t)$ during one complete period is obtained and the instantaneous value of $\xi(t)$ at N equally spaced time intervals is measured, the Fourier coefficients c_n and phase angles θ_n from $n = 0$ to $n = (N - 1)/2$ may be calculated by tabular methods. A detailed description of these tabular methods is given with respect to Fig. 38.12. Similar computations also may be carried out by digital computers.

When the time-history is available as an analog voltage, filters may be used to determine the Fourier components as described later in this chapter under "Spectral Analysis."

STATISTICAL ANALYSIS

Statistical analysis[6-8, 10] of a vibration time-history is carried out to determine the characteristics, e.g., sinusoidal and/or random, of the time-history so that the appropriate mathematical function for description of the vibration magnitude may be chosen. This section describes the statistical properties of random and sinusoidal functions and discusses concepts of data analysis for determining the significant statistical parameters of the time-histories.

STATISTICAL PARAMETERS OF RANDOM FUNCTIONS

Statistical parameters used frequently in data analysis in addition to the mean and mean-square values defined by Eqs. (22.4) and (22.6) are the probability density function and probability distribution function.

Probability Density Function.   The probability density function $p(\xi/\sigma)$, illustrated in Fig. 22.8, defines the probability (or fraction of time, on the average) that the magnitude of the quantity $\xi(t)$ will lie between two values. It is customary to normalize the curve by plotting the magnitude divided by the rms value σ as the abscissa. Then the probability that the magnitude lies between ξ/σ and $(\xi + d\xi)/\sigma$ is equal to $p(\xi/\sigma)d(\xi/\sigma)$, i.e., the shaded area shown in Fig. 22.8. Since it is certain, with probability 1.0, that the function $\xi(t)$ lies between plus and minus infinity, the area under the entire curve is unity.

Probability Distribution Function.   The probability distribution function (or *cumulative distribution function*) $P(\xi/\sigma \geq)$* defines the probability that the magnitude of ξ/σ will exceed a certain value. However, from the definition of the probability density function, $P(\xi/\sigma \geq)$ is equal to the area under the plot of $p(\xi/\sigma)$ as a function of ξ/σ between ξ/σ and infinity:

$$P(\xi/\sigma \geq) = \int_{\xi'/\sigma}^{\infty} p\left(\frac{\xi'}{\sigma}\right) d\left(\frac{\xi'}{\sigma}\right)$$ (22.30)

where, for this equation and Eq. (22.33) only, ξ' is used as a dummy variable of integration for ξ.

The value of the distribution function $P(\xi/\sigma \geq)$ at the minimum possible value of ξ is $P(\xi_{min}/\sigma) = 1.0$ since ξ is never less than ξ_{min}; for the maximum possible value of ξ, $P(\xi_{max}/\sigma)$ is zero since ξ never exceeds ξ_{max}. Further, $P(\xi/\sigma \geq)$ must decrease monotonically between ξ_{min} and ξ_{max}, as illustrated later in Fig. 22.12. Since the probability of exceeding a value is 1 minus the probability of not exceeding that value, the ordinate of Fig. 22.12 may be changed to $P(\xi/\sigma \leq)$ if the scale is changed to vary from 1 to 0 instead of 0 to 1.

COMPARISON OF PARAMETERS FOR SINUSOIDAL AND RANDOM FUNCTIONS

Although a sinusoidal function is deterministic, the probability density and probability distribution of a sine wave may be determined for comparison with those of a

---

* The probability that ξ/σ is *greater than* is written $P(\xi/\sigma \geq)$; conversely, the probability that ξ/σ is *less than* is written $P(\xi/\sigma \leq)$. $P(\xi/\sigma \geq) = 1 - P(\xi/\sigma \leq)$.

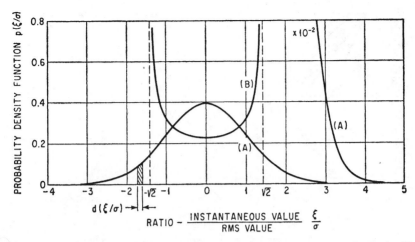

FIGURE 22.8 Normalized probability density functions: (A) Gaussian or normal distribution [Eq. (22.31)]; (B) distribution of instantaneous values of a sine wave [Eq. (22.32)]. The curve (A) marked $\times 10^{-2}$ indicates hundredfold expansion of ordinate scale.

random function. The comparison may be made on the basis of (1) the instantaneous values $\xi(t)$ of the function and (2) the peak values or maxima $\xi_p(t)$ of the function.

Distribution of Instantaneous Values of $\xi(t)$. A comparison of the identifying characteristics of the probability densities and distributions for the instantaneous values of a sinusoidal function and the particular case of Gaussian random vibration is shown in Figs. 22.8, 22.9, and 22.10. The term "Gaussian" is used to describe a random function whose *instantaneous* value is defined by the Gaussian or normal probability density function [see Eq. (11.14)] given by

$$p(\xi/\sigma) = \frac{1}{\sqrt{2\pi}} \, e^{-(\xi^2/2\sigma^2)} \quad (22.31)$$

where σ is the rms value. Equation (22.31) is shown by curve A of Fig. 22.8. The probability density function of the *instantaneous* value of a sinusoid is shown by curve B of Fig. 22.8 and is defined by

$$p(\xi/\sigma) = \frac{1}{\pi \sqrt{2 - (\xi/\sigma)^2}} \quad (22.32)$$

The probability density functions of Eqs. (22.31) and (22.32) are symmetrical about a mean assumed to be zero; then the probability that ξ exceeds a given absolute value (or magnitude) $|\xi|$ is twice the probability that it exceeds the same absolute value in either the positive or negative sense [see Eq. (22.30) or Fig. (22.8)]. Therefore, it is convenient to

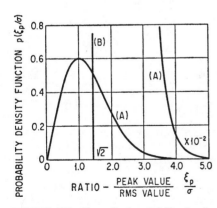

FIGURE 22.9 Normalized probability density functions: (A) Rayleigh distribution for peaks of narrow-band Gaussian vibration [Eq. (22.35)]; (B) distribution for peaks of sine-wave delta function at $\xi_p/\sigma = \sqrt{2}$. The curve (A) marked $\times 10^{-2}$ indicates hundredfold expansion of ordinate scale.

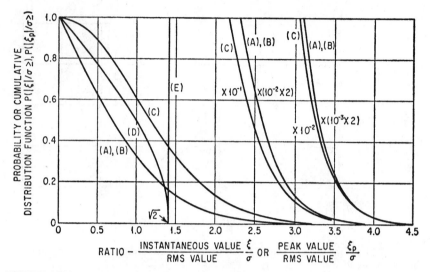

FIGURE 22.10 Probability distribution functions: (A) Instantaneous values of broad-band and narrow-band random vibration—Gaussian distribution [Eq. (22.33)]; (B) peaks of broad-band random vibration [Eq. (22.33)]; (C) peaks of narrow-band random vibration [Eq. (22.36)]; (D) instantaneous values of a sine wave [Eq. (22.34)]; (E) peak values of a sine wave. Multiply ordinate scale by factors marked adjacent to curves for large values of ξ/σ.

plot the probability distribution function in terms of the absolute value of ξ, i.e., $P(|\xi|/\sigma \geq)$, as shown in Fig. 22.10.

The probability distribution functions for the Gaussian and sinusoidal functions are obtained by integration of Eqs. (22.31) and (22.32):

Gaussian:

$$P(|\xi|/\sigma\geq) = \frac{2}{\sqrt{2\pi}} \int_{\xi/\sigma}^{\infty} e^{-(\xi'^2/2\sigma^2)} d\left(\frac{\xi'}{\sigma}\right) \tag{22.33}$$

Sinusoid:

$$P(|\xi|/\sigma\geq) = \frac{2}{\pi} \cos^{-1}\frac{\xi}{\sigma\sqrt{2}} \tag{22.34}$$

The relations of Eqs. (22.33) and (22.34) are plotted as curves A and B in Fig. 22.10.

Distribution of Peak Values (Maxima) of $\xi(t)$. When the peak values or maxima of a function are considered, the two statistical functions differ from those found for the instantaneous values. For a sine wave, all maxima are of equal magnitude and the probability density function $p(\xi_p/\sigma)$ becomes a Dirac delta function (see Chap. 23) as shown by curve B of Fig. 22.9. For broad-band Gaussian vibration, i.e., vibration with nonzero spectral density over a frequency bandwidth which is not small compared to the average or center frequency of the bandwidth, the distribution of peak values is normal, as shown by curves A of Fig. 22.8 and B of Fig. 22.10. However, for narrow-band Gaussian noise, i.e., noise with negligible spectral density except in a frequency bandwidth which is small compared to the center frequency, the distribution of peak values becomes the *Rayleigh distribution*. The probability density and distribution functions for the Rayleigh distribution are defined by:

Probability density:

$$p(|\xi_p|/\sigma) = \frac{\xi_p}{\sigma} \, e^{-(\xi_p^2/2\sigma^2)} \tag{22.35}$$

Probability distribution:

$$P(|\xi_p|/\sigma) = e^{-\xi_p^2/2\sigma^2} \tag{22.36}$$

The relations given by Eqs. (22.35) and (22.36) are shown graphically by curve A of Fig. 22.9 and curve C of Fig. 22.10, respectively.

Figures 22.8 and 22.10 are plotted on the basis that the mean value of the function $\xi(t)$ is zero. In the more general case of nonzero mean value, the abscissae of these figures are $(\xi - \bar{\xi})/\sigma$, where $\bar{\xi}$ is the mean value of the function $\xi(t)$. However, in vibration data analysis, it is common practice to constrain the mean value to zero by use of a high-pass or bandpass filter prior to statistical analysis. Thus these figures show the form usually encountered.

Figures 22.8 and 22.10 show that the probability density and probability distribution curves for sinusoidal and random functions differ considerably, thus providing identifying characteristics for each type of function. When the time-history $\xi(t)$ is a combination of sinusoidal and random functions, the shape of the probability density and distribution curves depends on the relative magnitudes of each type of function.

Relationship between Average and rms Values. The average value may be determined from the probability density function since $|\bar{\xi}|$ is represented by the distance from the vertical axis to the center-of-gravity of the area under the probability density function, for either positive or negative values (assuming a symmetrical density function):

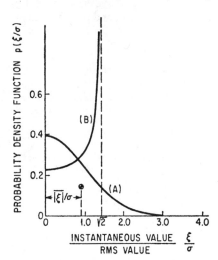

$$\overline{|\xi|} = 2\sigma \int_0^\infty p\left(\frac{\xi}{\sigma}\right) \frac{\xi}{\sigma} \, d\left(\frac{\xi}{\sigma}\right) \tag{22.37}$$

where $\overline{|\xi|}/\sigma$ is indicated in Fig. 22.11 for the probability density functions for the positive values of (A) a Gaussian distribution and (B) a sinusoid. For a sine wave, the average value $\overline{|\xi|} = 0.900\sigma$; this is 0.636 times the amplitude. The average value of a Gaussian function $\overline{|\xi|} = 0.798\sigma$; it cannot be defined in terms of a peak value since the peak values vary from zero to infinity. However, physical processes usually are only approximately Gaussian in that the peak values are limited to some value between 3 and 10 times the rms value.

FIGURE 22.11 Average value $\overline{|\xi|}$ of function with zero mean value: (A) Gaussian distribution $\overline{|\xi|}/\sigma = 0.798$; (B) sine wave $\overline{|\xi|}/\sigma = 0.900$.

PROBABILITY GRAPH PAPER

The characteristics of the probability density and distribution functions shown in Figs. 22.8 to 22.10 are emphasized by plotting the functions on probability paper.

Gaussian Distribution. Gaussian probability paper* is graph paper having a set of coordinates such that when the Gaussian distribution function [Eq. (22.33)] is plotted on it, a straight line is obtained, as shown in Fig. 22.12. For example, if the solid straight line in this figure represents the measured distribution function of the instantaneous value of $\xi(t)$, then $\xi(t)$ is random having a Gaussian distribution. In contrast, a sinusoidal function, for example, has a distribution function of the form shown by the dotted line of Fig. 22.12. Therefore, this type of paper is helpful in vibration analysis problems to determine the extent to which a time function has a normal distribution by observing the linearity of the plotted distribution function. To be useful in determining the randomness of $\xi(t)$, the probability distribution must be plotted when $\xi(t)$ is the *output of a relatively narrow-band filter* and the distribution function of the *instantaneous* values has been measured. A close approximation to the Gaussian distribution can be synthesized by as few as seven appropriately chosen sinusoids, and any complex function with broad frequency content will approximate a Gaussian distribution in accordance with the central-limit theorem.[5] Thus the probability distribution function of a broad-band signal is an insensitive measure of randomness.

The symmetry of the probability density function about the mean value also may be determined readily by use of this paper. If the shape of the curve is unchanged when it is rotated 180° about its intersection with the ordinate $P(\xi \geq) = 50$ per cent, the density function is symmetrical about the mean and the *mean value* $\bar{\xi}$ is the value of $\xi(t)$ at $P(\xi \geq) = 50$ per cent. The mean value of zero is plotted in Fig. 22.12. If the distribution function is linear as well as symmetrical when plotted on Gaussian probability paper, the *rms value* σ may be read from the curve at the 15.9 or 84.1 per cent points. The 2σ points at 2.3, or 97.7 per cent, and the 3σ points at 0.13, or 99.87 per cent, also may be used in Fig. 22.12. In Fig. 22.12, the rms value plotted is 2.0.

Rayleigh Distribution. Rayleigh probability paper[†] is graph paper having a set of coordinates such that when the Rayleigh distribution function [Eq. (22.36)] is plotted on it, a straight line is obtained. This is shown in Fig. 22.13. The solid curve shows the distribution function of the *peak values* ξ_p of a narrow-band random function

---

* Available commercially, for example: Probability Scale Graph Paper No. 359-23, Keuffel & Esser Co., New York.

[†] Rayleigh probability paper can be constructed as follows:
The probability distribution function for the Rayleigh distribution, Eq. (22.36), can be written:

$$P(|\xi_p|/\sigma \leq) = 1 - e^{-\xi_p^2/2\sigma^2} \qquad \xi_p/\sigma = \left[2 \log_e \left(\frac{1}{1-P}\right)\right]^{1/2}$$

Let $x(P)$ be the distance from the origin of a point on the abscissa corresponding to the ordinate ξ_p/σ. For the above equation to be a straight line

$$x(P) = a\,\frac{\xi_p}{\sigma} + b = a\left[2 \log_e \left(\frac{1}{1-P}\right)\right]^{1/2} + b$$

The values of a and b are determined by the choice of origin and scale. If $x(0)$ is zero, then b also is zero and a is determined by the distance between $x(0)$ and $x(0.9999)$.

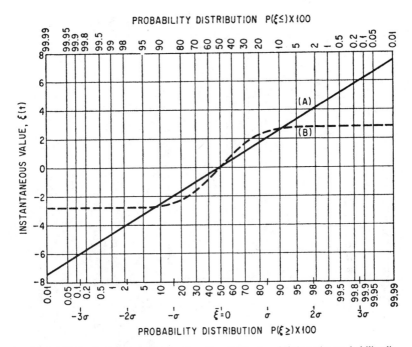

FIGURE 22.12 Probability paper for Gaussian distribution: (*A*) Gaussian probability distribution function gives a straight line where mean value $\bar{\xi}$ is at $P(\xi \geq) = 0.50$ and rms value σ is at 0.159 or 0.841; (*B*) probability distribution of a sine wave gives a nonlinear plot.

which has an rms value of 2.0. The analogous distribution function of the peak values of a sine wave is a horizontal line at the amplitude of the sine wave, as shown by the dotted curve of Fig. 22.13 for rms value of 2.0. Thus, the paper shown in Fig. 22.13 is useful in determining the randomness or coherence of $\xi(t)$ when $\xi(t)$ is the output of a relatively narrow-band filter and the distribution function of the *peak* values has been measured.

When data are presented in the form shown in Fig. 22.13, one curve is required for each narrow band of the signal which is analyzed. Since this may lead to a large number of plots, a technique of data presentation which includes all frequency bands and, at least qualitatively, a picture of the distribution of maxima is illustrated in Fig. 22.14. The abscissa is the center frequency of each band, and the ordinate is the magnitude of the maxima. Lines are then drawn between the values at each center frequency which represent the magnitude at or below which the indicated percentages of the peaks occur. For sinusoidal functions, all lines are coincident since all peaks are of equal magnitude.

DETERMINATION OF DISTRIBUTION FUNCTIONS

In practice, the characteristics of a signal $\xi(t)$ at a specific time and at a specific frequency cannot be determined, since only the instantaneous value of $\xi(t)$ is known at a specific time. Instead, the characteristics of the signal in a given restricted frequency

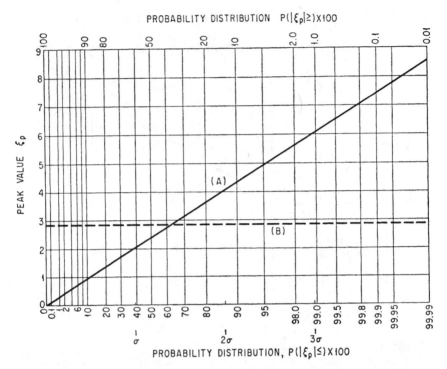

FIGURE 22.13   Probability paper on which Rayleigh distribution function gives a straight line: (*A*) Rayleigh distribution for peaks of narrow-band random vibration with $\sigma = 2.0$; (*B*) distribution function of peaks of a sinusoid – $\sigma = 2.0$.

bandwidth during a short period of time are determined. Thus the first step in a statistical analysis is to select a time interval short enough so that the nature and magnitude of $\xi(t)$ may be assumed constant, but long enough to give a statistically significant result.[6] Often, it is convenient to store the data sample in the form of a recording on a continuous loop of magnetic tape so that it may be played back repeatedly. Then a time sample of the signal may be analyzed to determine the probability distribution of either the instantaneous or the peak values of the signal. The concepts involved in this determination are considered separately in the next two sections dealing with instantaneous and peak values, respectively. The instantaneous values of the *filtered* signal are indicated by $\xi'(t)$ while the peak values of the *filtered* signal are indicated by $\xi_p'(t)$.

Instantaneous Values.   The probability distribution of instantaneous values is obtained for a set of discrete values of $\xi'(t)$ rather than as the continuous function that is illustrated in Fig. 22.10. The probability distribution function at a value ξ_1' is the probability $P(\xi' \geq)$ that $\xi'(t)$ exceeds ξ_1'.* Referring to Fig. 22.15, this is

---

* The filtered signal $\xi'(t)$ may be passed first through a mechanism for changing the polarity (or sign) of the signal so that the positive and negative values of $\xi(t)$ can be analyzed separately, thus halving the number of discriminators and counters required for a given resolution. The time-history must be played back for twice as long, but this usually will cause a rather small increase in the total analysis time.

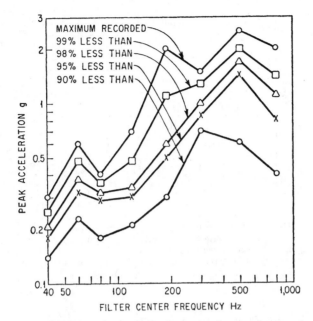

FIGURE 22.14 Distribution of peak accelerations as a function of frequency. Each curve represents the value below which the indicated percentage of peak values occurs in the response of the filter to $\xi(t)$, for the indicated center frequency of the filter.

equivalent to summing all the Δt's during which $\xi'(t)$ exceeds ξ_1' or level L_1 and dividing by the total elapsed time. If this is done for a number of selected levels, $L_1, L_2, \ldots, L_n$, the probability distribution function at these levels is obtained. Figure 22.16A illustrates the block diagram of a system to make these measurements. The filtered signal $\xi'(t)$ is fed to an array of *discriminators* which determine whether a signal is greater or less than a preselected value. The levels of the discriminators are set at L_0, L_1, L_2, etc., usually in equal increments of voltage. The detail or resolution obtained is governed by the number of discriminator levels selected for measurement. The detail and accuracy yielded by 10 levels for each polarity are sufficient for many engineering purposes. Since the resolution obtained improves with the number of counters registering, a variable gain control is needed to adjust the signal level, by a known amount, until an adequate number of counters are registering.

When the signal $\xi'(t)$ exceeds the level of a particular discriminator, a clock is started. It is stopped when $\xi'(t)$ next falls below that level and the counter counts the elapsed time between these events. At the end of the data sample, the readings of the counters are recorded by the read-out mechanism, a digital printer, for example. If each reading is divided by the total elapsed time of the data sample, the values obtained are the values of the probability distribution function at the values of the levels L_0, L_1, L_2, etc. If the values at adjacent levels are subtracted, the average probability density function between these two levels is obtained. The polarity of the signal is now changed and the process is repeated. However, using positive polarity, the

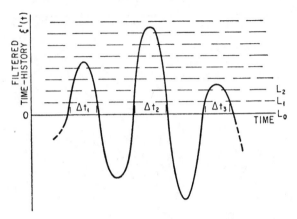

FIGURE 22.15 Measurements required for statistical analysis: (A) To determine distribution of instantaneous values, sum of time increments Δt_1 during which $\xi'(t)$ exceeds $L_0, L_1, L_2 \ldots$ is measured; (B) to determine distribution of peak values, number of times that $\xi'(t)$ exceeds $L_0, L_1, L_2 \ldots$ is measured.

distribution function $P(\xi' \geq)$ is obtained while the negative polarity yields $P(\xi' \geq)$. Since $P(\xi' \leq) = 1 - P(\xi' \geq)$, the two sets of values may be combined to give the complete distribution function. The character of the signal is then evaluated by the use of the probability paper illustrated in Fig. 22.12.

In addition to determining the characteristics of the signal, this analysis can be used to determine the mean square value of $\xi'(t)$, either by use of the probability paper if the signal is Gaussian or by computing the moment of inertia of the probability density function about the mean value [see Eq. (11.11)].

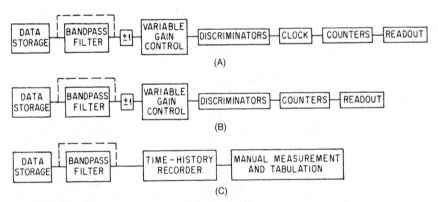

FIGURE 22.16 Block diagrams showing steps in statistical analysis: (A) Instantaneous value analysis—automatic; (B) peak value analysis—automatic; (C) peak value analysis—manual. For each analysis, the center frequency of the filter remains constant.

Peak Values. The probability distribution of peak values or maxima $\xi_p'(t)$ of the filtered signal $\xi'(t)$ is obtained for a set of discrete values of $\xi_p'(t)$ rather than as the continuous function illustrated in Fig. 22.10. The probability distribution function at a value ξ_{p1}' is the probability $P(\xi_p'\geq)$ that a peak or maximum of $\xi'(t)$ exceeds ξ_{p1}'. Referring to Fig. 22.15, this is equivalent to counting the *number* of times that $\xi'(t)$ exceeds ξ_{p1}' (i.e., level L_1) and dividing by the total number of times that $\xi'(t)$ exceeds L_0. If this count is made for a number of levels, $L_1, L_2, \ldots$, the probability distribution function at these values is obtained.

Figure 22.16B illustrates the block diagram of a system for making such measurements. The filtered signal $\xi'(t)$ passes through a bank of level discriminators and causes the associated counters to register one count each time the level of the particular discriminator is exceeded. (Sometimes an array of discriminator-counter combinations is called a *pulse-height analyzer.*) At the end of the data sample, the readings of the counters are recorded. Division of each count by the zero level count yields the probability distribution function $P(\xi_p'\geq)$ at the values equivalent to the levels $L_1, L_2, \ldots$ If the values at adjacent levels are subtracted, the average probability density function between these two levels is obtained. The character of the signal then is evaluated by the use of the probability paper illustrated in Fig. 22.13.

For some end uses of the data, such as fatigue analysis, the probability distribution of the total or unfiltered signal $\xi(t)$ rather than the probability distribution of the filtered signal $\xi'(t)$ is desired. Then the distribution of the peaks of $\xi(t)$ is obtained by shunting out the filter in the block diagram of Fig. 22.16B as indicated by the dotted lines.

The systems illustrated by the block diagrams of Fig. 22.16A and B are relatively complex, and unless a large amount of analysis of this type is to be carried out, the cost of such systems may not be justified. A much simpler method is illustrated in Fig. 22.16C. In this case a direct-writing recorder is employed and the number of peaks which exceed various levels are counted manually (see Fig. 22.15). In principle, the distribution of instantaneous values may be measured in this way, but practical difficulties make this method difficult to employ.

SPECTRAL ANALYSIS

The objective of spectral analysis[2, 6–8, 10, 11] is to determine the variation of vibration magnitude with frequency. (Equipment to perform spectral analysis is described in Chap. 14.) The narrowness of the bandwidth of the filter employed in the analysis determines the frequency resolution of the analysis and, therefore, the ability to detect the "fine-grain" variation of magnitude with frequency. The magnitude obtained at a particular frequency will be the average magnitude over the short time interval of data analyzed, during which the nature and magnitude of the vibration may be considered constant (see Figs. 22.3 and 22.4).

In preparing for spectral analysis, it is helpful if the characteristics of the time-history $\xi(t)$ are known from statistical analysis (or are known from previous experience) so that the most appropriate units to be employed for describing the vibration magnitude can be selected. In practice, the spectral analysis often is carried out first and the necessity for statistical analysis is judged from the appearance of the resulting spectrum.

DEFINITION OF POWER SPECTRAL DENSITY*

Power spectral density[2-6] is defined as the limiting value of the mean-square response $\overline{[\xi']^2}$ of an ideal bandpass filter[†] to $\xi(t)$, divided by the bandwidth B of the filter, as the bandwidth of the filter approaches zero.

An alternative definition[3] is as follows. If the function $\xi(t)$ is passed through an ideal low-pass filter[‡] with cutoff frequency f_c, the mean-square response of the filter $\overline{[\xi']^2}$ will increase or decrease as f_c is increased or decreased, i.e., more or less of the function will be passed by the filter (assuming f_c is varied in a frequency range where the power spectral density is nonzero). The power spectral density $W(f)$ is the rate of change of $\overline{[\xi']^2}$ with respect to f_c, i.e.,

$$W(f_c) = \frac{d}{df_c} \{\overline{[\xi']^2}\} \tag{22.38}$$

Figure 22.17 illustrates a plot of power spectral density as a function of frequency obtained, for example, from spectral analysis of a random function. The mean-square value, or variance, of the frequency content of $\xi(t)$ between the frequencies f_a and f_b is equal to the shaded area of Fig. 22.17:

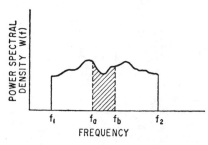

$$\sigma^2(f_a \leq f \leq f_b) = \int_{f_a}^{f_b} W(f)\, df \tag{22.39}$$

FIGURE 22.17 Typical power spectral density plot of broad-band random function $\xi(t)$. Mean-square value of frequency content of $\xi(t)$ between f_a and f_b is equal to shaded area [Eq. (22.39)]. For *white noise* or flat spectrum, $W(f)$ is a constant.

The rms value is σ. The mean-square value of the complete function $\overline{[\xi]^2}$ is given by Eq. (22.39) when f_a and f_b are zero and infinity, respectively; it is equal to the area under the entire spectrum. In the case of *white noise*, for which the spectral density is independent of frequency, i.e., $W(f) = W$, Eq. (22.39) simplifies to

$$\sigma^2 = W(f_2 - f_1) \tag{22.40}$$

where f_1 and f_2 are the limiting frequencies of the noise.

Relationship of Power Spectral Density to Line Spectrum.
The mathematical relationship between power spectral density $W(f)$ and the Fourier coefficients c_n is

---

* Power spectral density is defined by Eq. (22.38) and also by Eq. (11.20). It is a generic term used regardless of the physical quantity represented by the time-history. However, it is preferable to indicate the physical quantity involved. For example, the term *mean-square acceleration density* or *acceleration spectral density* is used when the time-history of acceleration is to be described.

† An ideal bandpass filter has a transmission characteristic which is rectangular in shape so that all frequency components within the filter bandwidth are passed with unity gain and zero phase distortion, while frequency components outside the bandwidth are completely removed. If the transmission characteristic is H instead of unity in the bandwidth B, the spectral density is obtained by dividing the mean-square response by $B \times H^2$ instead of B.

‡ An ideal low-pass filter is an ideal bandpass filter having a lower cutoff frequency of zero.

$$W(f) = 2\pi W(\omega) = 2\pi \lim_{R \to 0} \frac{\overline{[\xi']^2}}{B} = 2\pi \frac{d}{d\omega}[\overline{(\xi')^2}]$$

$$= \frac{1}{2} - \sum_{n=1}^{\infty} c_n^2 \, \delta(f - f_n) \qquad [b_0 = 0] \qquad (22.41)$$

where $\overline{[\xi']^2}$ is the mean-square response of the ideal filter with bandwidth B and center frequency $f = \omega/2\pi$; $\delta(f - f_n)$ is the Dirac delta function (see Chap. 23).

The power spectral density of a sinusoidal function is a line spectrum since such a function has effectively zero bandwidth. This may be seen from Eq. (22.37) since $\overline{[\xi']^2}$ will change instantaneously as the cutoff frequency f_c increases through the frequency of the sine wave giving, theoretically, an infinite value for $W(f)$.

DETERMINATION OF POWER SPECTRAL DENSITY

One technique of obtaining a spectral analysis of a signal is to record it on a continuous loop of magnetic tape. The loop is played back repeatedly. Each time, an analysis is made at a slightly different frequency using, for example, a tunable filter. To obtain the complete spectrum, the process is repeated at enough frequencies to give a plot of magnitude vs. frequency. These steps are illustrated in the first two blocks of Figs. 22.18 and 22.19, which are block diagrams depicting examples of techniques employed in spectral analysis. Typical waveforms for one value of the center frequency of the filter are shown in the insets above the blocks. Techniques also are available for spectral analysis in real time.

When a sample of a time-history of finite duration is employed to compute the power spectral density of a random function, it is assumed that (1) the function is ergodic, i.e., that averaging one time-history with respect to time yields the same

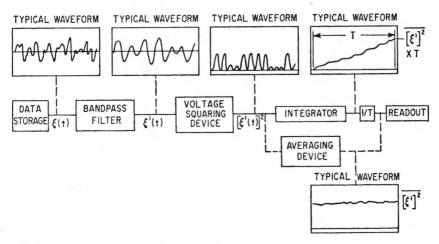

FIGURE 22.18 Block diagram showing spectral analysis by the mean-square method. Dotted lines indicate alternative use of averaging device in place of integrator. Insets show typical waveforms at successive stages of analysis. Signal $\xi(t)$ is filtered to give $\xi'(t)$; squared to give $[\xi'(t)]^2$; integrated and divided by time of integration T to give $\overline{[\xi']^2}$, the mean-square value. Division by effective bandwidth of filter yields power spectral density of random function.

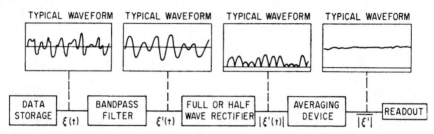

FIGURE 22.19 Block diagram showing spectral analysis by averaging method. Insets show typical waveforms at successive stages of analysis. Signal $\xi(t)$ is filtered to give $\xi'(t)$; rectified (half or full wave) to give $|\xi'(t)|$; and averaged to yield $|\bar{\xi}'|$, the average value of $\xi'(t)$.

result as averaging over an ensemble of time-histories at a given instant of time and (2) that the function is *stationary*, i.e., that the power spectral density is independent of the sample of the time-history chosen. Further, the averaging time or sample duration must be long enough to yield a statistically significant value. Thus, the mean-square value obtained should not vary appreciably with a change in averaging time. The time over which a vibration record may be considered a stationary process and the need for a sufficiently long averaging time often are conflicting requirements.

MEASUREMENT OF POWER SPECTRAL DENSITY BY THE USE OF FILTERS

If a random function is applied to the input of a filter, the spectral density of the output is obtained by multiplying the spectral density of the input by the square of the transmission characteristic curve* at every frequency. First, consider the response of a filter to *white noise* (constant spectral density W). The output of an ideal bandpass filter is constant spectral density within the bandwidth B, and zero elsewhere. From Eq. (22.39), the mean-square output will be $W \times B$. The output of a practical filter (i.e., a filter having practically attainable characteristics in contrast to an ideal filter) has a spectral density which is directly proportional to the square of the transmission characteristic; the mean-square output is W multiplied by the area under the curve of the squared transmission characteristic. This area is defined as the effective bandwidth of the filter. However, due to the nonlinearities inherent in many filters, the effective bandwidth of a filter should be determined by measurement of the mean-square response of the filter to known white noise.

Now consider a random signal whose spectral density varies with frequency, as shown in Fig. 22.20*B*. It is desired to measure the spectral density by the use of filters, i.e., to perform a spectral analysis. In the case of an ideal filter, the mean-square output divided by B yields the mean (or average) value of the spectral density within the filter bandwidth [see Eq. (22.39) and Fig. 22.17]. When the signal is applied to a practical filter and the mean-square output is divided by the effective bandwidth, an approximation to the average spectral density within the filter bandwidth is obtained.

---

* The ratio of the response amplitude to the input amplitude when the input is a sinusoidal function is the *transmission characteristic* or *frequency-response characteristic* of the filter. If this ratio is independent of the input amplitude, the filter is linear. The ratio usually is normalized so that the value at the center frequency is unity.

Effects of Filter Characteristics. The suitability of practical filters for the measurement of power spectral density depends on the rate of variation of the spectral density with frequency of the signal being analyzed, the shape of the transmission characteristic of the filter, and the time constant of the filter.

Figure 22.20F shows the spectral densities which would be measured using filters with the transmission characteristics shown in Fig. 22.20C and 22.20D, if the input power spectral density is actually as shown in Fig. 22.20B. The smoothing effect on the measured spectral density of the filter with the wider bandwidth is evident, particularly where the magnitude of the power spectral density is changing rapidly with frequency. Sharp peaks or valleys in the power spectral density are masked when the effective filter bandwidth approaches the bandwidth of the peak or valley. The power spectral density measured by the use of filters is exact only when an ideal filter is used, in the limit as the bandwidth approaches zero. Furthermore the time constant of the filter may affect the measurements by time-averaging the output signal.

MEASUREMENT OF LINE SPECTRA BY THE USE OF FILTERS

In the measurement of sinusoidal functions by the use of filters, the output amplitude is measured and plotted in place of the power spectral density used with ran-

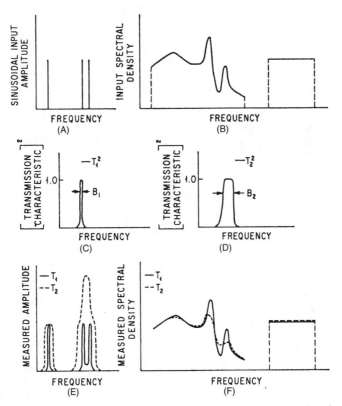

FIGURE 22.20 Effects of filter bandwidth characteristics on measured spectra.

dom functions. The accuracy with which such measurements can be made depends upon the bandwidth of the filter. For example, Fig. 22.20*E* shows in solid lines the measured spectrum when the line spectrum of Fig. 22.20*A* is the input to the filters having the characteristics shown in Fig. 22.20*C;* the dotted lines in Fig. 22.20*E* show the spectrum obtained with the filter of Fig. 22.20*D.* The error introduced by the filter with the wider bandwidth is particularly evident when the sine waves are closely spaced in frequency. Assume now that the input $\xi(t)$ is a complex function comprising several sinusoidal functions at unrelated frequencies. The output of the filter is the mean-square value of one or the sum of many sinusoids, depending on the spacing of the lines of the input spectrum, the breadth of the filter bandwidth, and the shape of the filter transmission characteristic. However, the output is often considered to be the mean-square value of a single sinusoidal function at the center frequency of the filter.

LINE SPECTRUM SUPERIMPOSED ON RANDOM VIBRATION SPECTRUM

The input signal $\xi(t)$ may be a "mixture" of a random function plus a single sinusoidal function within the bandwidth of the filter, as shown in Fig. 22.21. The mean-square output of the filter $[\overline{\xi'}]^2$ is the sum of the mean-square values of the sinusoid and the frequency content of the random function within the filter bandwidth. Depending on the end use of the data and the relative magnitudes of the sinusoidal and random functions, it *may be* satisfactory to take the spectral density computed from this mean-square value $[\overline{\xi'}]^2$ as a description of the vibration magnitude. Conversely, it may be preferable to consider $[\overline{\xi'}]^2$ to be the mean-square value of a sinusoid at the filter center frequency. However, power spectral density should be used only to describe a random function. Periodic and complex functions should be represented by Fourier spectra, either line or continuous.

It is difficult to separate the random and coherent parts of a vibration time-history when such a combination occurs. For example, if the power spectral density of an acceleration time-history has been computed using a system such as illustrated in Fig. 22.18, and the spectrum has a peak in a narrow frequency band as shown in Fig. 22.21, it is necessary to know whether this peak represents a truly random function of increased intensity in this frequency band or a sinusoidal function superimposed on the smoother spectral density plot indicated by the dotted line in Fig. 22.21. Fundamentally, a sinusoidal function is represented by a line spectrum but may give the appearance of narrow-band random vibration because of the limitations of the spectral analysis techniques. The probability density or probability distribution at the frequency of the peak in the spectrum may be used to determine the characteristics of the signal in this frequency region.

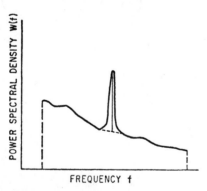

FIGURE 22.21 Example of spectral analysis when signal contains narrow-band peak of random vibration or line spectrum superimposed on random vibration spectrum as shown by dotted line. Because of limitations of analysis equipment, line spectrum may give same appearance as narrow-band random vibration.

MEASUREMENT OF FILTER OUTPUT

The magnitude of the filtered signal is obtained in terms of (1) the mean-square value or (2) the average value.*

Mean-Square Value.   Assume that statistical analysis has shown $\xi'(t)$ to be a random function. Then the mean-square value $[\overline{\xi'}]^2$ must be obtained. Figure 22.18 shows one method of accomplishing this. The function $\xi'(t)$ is squared, integrated over the period of the data sample, and divided by the time of integration T to yield $[\overline{\xi'}]^2$. Typical waveforms during this process are shown. Alternatively, the integrator may be replaced by an averaging device, as shown by the dotted lines. The terminal value of the integrator output, divided by T, or the output of the averaging device is read out; division by the effective bandwidth of the filter yields the average power spectral density within the filter bandwidth.

Average Value.[†]   Figure 22.19 is a block diagram showing an example of a system in which the filter output $\xi'(t)$ is half-wave (or full-wave) rectified and averaged. Thus the quantity measured by the read-out is $|\overline{\xi'}|$, the average value of $\xi'(t)$. If $\xi'(t)$ is known to be a random function, the relationships discussed under "Statistical Analysis" or empirical measurement of the response to a known spectral density can be used to relate the average value to the mean-square value, and thus to obtain the power spectral density.

Effect of Signal Characteristics.   When the magnitude of the filter output is obtained in terms of the mean-square value, as shown in Fig. 22.18, the read-out may be directly converted to the appropriate magnitude quantity, for example, power spectral density, the amplitude of a sine wave, etc. If the average value is obtained as shown in Fig. 22.19, the conversion of the read-out to an appropriate magnitude quantity is not direct and may not even be possible. For example, if the filter output is the sum of two sinusoidal functions, the average value is a function of the phase angle between the sine waves, while the mean-square value is independent of phase angle. When the filter output is a mixture of a random function and a sinusoidal function, a similar indeterminacy exists in the average value. Thus, detailed knowledge of the characteristics of the function is required to interpret correctly the magnitude of $\xi(t)$ from its average value.[‡]

MAGNITUDE TIME-HISTORY ANALYSIS

In magnitude time-history analysis, the variation of the magnitude[§] of the vibration as a function of time is examined. The result may be plotted as shown in Fig. 22.5 where the magnitude is computed as the average over a few seconds while the vibration data record has a duration of perhaps many minutes. In some cases, it may be desired to determine the time variation of the magnitude of the complete signal, e.g.,

---

* See definition of average value in Eq. (22.5).

† See definition of average value in Eq. (22.5).

‡ Similar effects must be taken into account in the use of voltmeters which, for example, respond to the average value of the input signal and have a scale which is graduated in terms of rms value.

§ The word *magnitude* is used in this chapter to indicate the severity of vibration without specifically relating it to one of the various mathematical expressions used to describe vibration quantitatively.

the variation of the mean-square value $\overline{[\xi]^2}$ or, alternatively, the time-history of the magnitude of the vibration in a restricted frequency band. For example, suppose that spectral analysis of a number of time samples of the record shows that the vibration magnitude is concentrated in several narrow frequency bands and that these bands are the only important components of the vibration. Then a magnitude time-history analysis would be carried out for these frequencies. However, if spectral analysis of a number of time samples revealed that the shape of the spectrum was approximately constant even though the magnitude varied, then a time-history of the magnitude of the complete signal would be appropriate. The amount of analysis carried out is determined by the completeness with which the vibration magnitude surface of Fig. 22.3 must be defined.

DETERMINATION OF MAGNITUDE TIME-HISTORY

Figure 22.22 shows block diagrams for carrying out a magnitude time-history analysis. The data storage contains the complete vibration record, rather than the short-duration samples used in spectral and statistical analysis. The signal $\xi(t)$ from the record is passed through a filter whose center frequency is held constant for the entire record. The mean-square value $\overline{[\xi']^2}$ or the average value $|\overline{\xi'}|$ of the output of the filter $\xi'(t)$ is obtained by one of the three alternative systems outlined in Fig. 22.22: (A) the mean-square value may be obtained by integration of $[\xi'(t)]^2$ during successive time intervals T, (B) the mean-square value may be obtained by averaging of $[\xi'(t)]^2$, or (C) the average value $|\overline{\xi'}|$ may be obtained. The results are recorded as the time record stations in Fig. 22.22. The filter may be shunted as shown by dotted lines to analyze the time variation of $\xi(t)$ rather than that of the filtered signal $\xi'(t)$.

When a running mean-square or running average value is obtained, as indicated in Fig. 22.22B and C, the value obtained at any instant is affected by the values at times previous to the time of observation, depending on the time constant of the averaging device. This tends to smooth out the time variation of the magnitude and is avoided when integration over consecutive periods is used because each integrand then is independent. Typical time records of the value of $\overline{[\xi']^2}$ or $|\overline{\xi'}|$ are shown in the insets adjoining the systems A, B, and C in Fig. 22.22. Such a time record permits correlation of the magnitude of the vibration with physical parameters. If the time-histories of the vibration magnitude are obtained for a number of filter bands and laid side by side, the vibration magnitude surface of Fig. 22.3 is synthesized.

Magnitude Distribution Function. In addition to the magnitude time-history, the total time that the magnitude is equal to or greater than a particular magnitude or set of magnitudes, i.e., the magnitude distribution function, may be determined, as indicated in Fig. 22.22. If $\overline{[\xi]^2}$ is obtained by integration [system (A)], the mean-square value is fed into a bank of discriminator-counter combinations which sense the number of times that the integrand exceeds various levels. If $\overline{[\xi]^2}$ is obtained by averaging [system (B)] or if $|\overline{\xi'}|$ is obtained [system (C)], the time that these quantities exceed various levels is measured by the bank of discriminator-clock-counter combinations. The function of discriminators and related equipment is discussed with reference to Figs. 22.19 and 22.20.

Figure 22.23 illustrates a manner of plotting the readings from the counters. The abscissa is normalized to indicate the percentage of the total time that the level is exceeded, while the ordinate scale corresponds to the magnitude of the level which is exceeded; e.g., magnitude L_1 corresponding to discriminator 1 is exceeded t_1 percent of the time. The vibration magnitude distribution function illustrated in Fig.

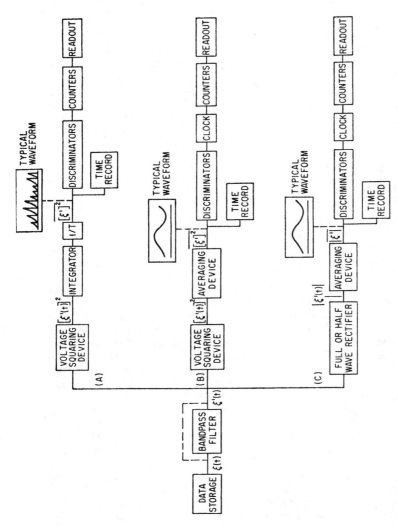

FIGURE 22.22 Block diagrams showing magnitude time-history analysis. To determine time-history of magnitude, *time record* blocks are used; magnitude distribution function is determined from discriminators and counters. Typical waveforms of magnitude time-history are shown by insets. For analysis of variation of magnitude of complete signal, filter is shunted as shown by dotted line. Filter center frequency is maintained constant during playback of entire vibration record. (Refer to "Statistical Analysis" and "Spectral Analysis" regarding counting and averaging techniques.) Diagrams show (*A*) mean-square value by successive integration; (*B*) mean-square value by averaging; (*C*) average value.

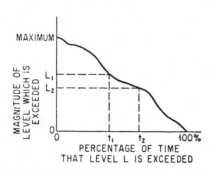

FIGURE 22.23   Distribution function of vibration magnitude for a filter having a center frequency f_0.

22.23 is analogous to the probability distribution function shown in Fig. 22.10. The filter frequency and bandwidth must be specified unless the analysis represents the total (i.e., the unfiltered) signal. The time of occurrence of the various vibration magnitudes is lost and cannot be correlated with physical conditions. If the magnitude distribution functions are obtained for a number of frequencies and placed side by side, the vibration magnitude distribution surface is formed. This describes the total vibration signal in terms of magnitude, frequency, and duration in the most concise form possible.

CORRELATION ANALYSIS

CORRELATION FUNCTIONS AND THEIR USE

A correlation function[2–8,10] defines the correlation between two parameters as a function of the times at which the parameters are observed. If the two parameters are the same except for the time of observation, i.e., $\xi(t)$ and $\xi(t+\tau)$, the function is known as an *autocorrelation function*. For example, the autocorrelation function of a single acceleration measurement may be used to determine both the nature of $\xi(t)$ and its power spectral density. If the two parameters $\xi_1(t)$ and $\xi_2(t+\tau)$ are physically distinct, the function is known as a *crosscorrelation function*. For example, the crosscorrelation between acceleration measurements at two different points of a structure may be determined for the purpose of studying the propagation of vibration through the structure. Crosscorrelation functions are not restricted to correlation of parameters with the same physical units; for example, one might determine the crosscorrelation between the applied force and the acceleration response to that force.

Correlation analysis of vibration data has many uses. For example, crosscorrelation data may be used to identify the normal modes of a vibrating structure; the autocorrelation function may be used to determine the power spectral density, and thus may be considered to be an alternative approach to spectral analysis. In the latter case, this technique has the advantage of yielding both the spectral density and a means of determining the characteristics of the signal, e.g., sinusoidal vs. random, etc., from the form of the autocorrelation function. Correlation techniques also are useful as a powerful method of detecting and analyzing a weak signal in the presence of noise.

Correlation analysis is concerned with stationary processes; in vibration analysis, it usually is carried out on a short interval of record during which physical parameters are assumed unchanged, as in spectral analysis.

AUTOCORRELATION FUNCTION

From Eq. (11.16), the autocorrelation function $R_\xi(\tau)$ of a time-varying parameter $\xi(t)$ is given by

$$R_\xi(\tau) = R_\xi(-\tau) = \lim_{T \to \infty} \frac{1}{2T} \int_{-T}^{T} \xi(t)\xi(t-\tau) \, dt = \overline{\xi(t)\xi(t-\tau)} \qquad (22.42)$$

Physically, $R_\xi(\tau)$ may be considered as the time average of the product of the instantaneous values of the function measured at two instants separated by a time interval τ. Setting $\tau = 0$ in Eq. (22.42) yields the mean-square value

$$R_\xi(0) = \overline{[\xi]^2} \qquad (22.43)$$

Since a parameter cannot be more closely related to another parameter than it is to itself, i.e., be better correlated, it follows that the maximum value of $R_\xi(\tau)$ occurs when $\tau = 0$.

The relationship between the autocorrelation function $R_\xi(\tau)$ and the power spectral density $W(f)$ is

$$W(f) = 4 \int_0^\infty R_\xi(\tau) \cos \omega\tau \, d\tau \qquad (22.44)$$

From Eq. (11.28), the inverse transformation is given by

$$R_\xi(\tau) = \frac{1}{2\pi} \int_0^\infty W(f) \cos \omega\tau \, d\omega \qquad (22.45)$$

Equation (22.44) is employed in the determination of the power spectral density of a random function from the autocorrelation function.

Autocorrelation Function of a Sine Wave.   Assume that $\xi(t)$ is a sine wave as defined in Eq. (22.1). Then $R_\xi(\tau)$ is given by

$$R_\xi(\tau) = \frac{\xi_0^2}{2} \cos \omega\tau \qquad (22.46)$$

The relation given by Eq. (22.46) is shown by curve A in Fig. 22.24. This curve shows that the autocorrelation function of a sine wave is periodic with a frequency equal to the frequency of the sine wave. The physical interpretation is that a sine wave is unchanged by a translation of the time origin by an integer number of periods $2\pi/\omega$.

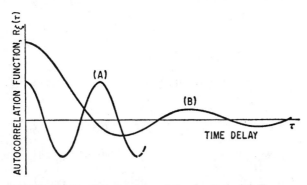

FIGURE 22.24 Typical autocorrelation functions: (A) Sine wave [Eq. (22.46)]; (B) band-limited white noise [Eq. (22.47)].

The maxima of the autocorrelation function occur at $\tau = 2\pi n/\omega$ and are all equal to $R_\xi(0)$, i.e., the correlation at time delays τ equal to an integer number of periods $2\pi/\omega$ is the same as it is for zero time delay.

Autocorrelation Function of White Noise. Suppose $\xi(t)$ represents band-limited white noise which has constant spectral density W up to a frequency f_c and zero for higher frequencies. According to Eq. (22.45), the autocorrelation function is

$$R_\xi(\tau) = Wf_c \frac{\sin 2\pi f_c \tau}{2\pi f_c \tau}$$

$$R_\xi(0) = Wf_c = \overline{[\xi]^2}$$

(22.47)

Equations (22.47) define the decaying sinusoidal function shown by curve B in Fig. 22.24. As the cutoff frequency f_c approaches infinity, $R_\xi(\tau)$ approaches a delta function at the origin.

The autocorrelation function of the random process decreases rapidly as τ increases, and as the cutoff frequency f_c increases. Thus there is little correlation between the instantaneous values of $\xi(t)$ at different times. In the limit, as f_c approaches infinity, there is no correlation at all; i.e., the value of $\xi(t)$ is unpredictable from knowledge of its previous value $\xi(t - \tau)$. The autocorrelation function for a random function which has variable spectral density will have the same properties as curve B of Fig. 22.24, i.e., decreasing correlation for increasing τ and increasing bandwidth. Thus, the form of the autocorrelation function can be used to determine the character of $\xi(t)$, i.e., whether it is a random, complex, or periodic time function.

Determination of Autocorrelation Function. Figure 22.25 illustrates one method of obtaining a correlation analysis by analog means. (The arrangement is similar for either autocorrelation or crosscorrelation analyses.) The stored signal $\xi(t)$ is fed into a multiplier directly and through a variable time-delay τ. The time average of the multiplicand $\overline{\xi(t)\xi(t-\tau)}$ for various time-delays is obtained either by integration or averaging as shown by the alternative blocks in Fig. 22.25. The plot of the output vs. the time-delay is the autocorrelation function.

Another method of obtaining a correlation function from an analog record is to convert the analog record to digital form. Then a standard high-speed digital computer can be programmed to calculate the autocorrelation function and the power spectral density.

CROSSCORRELATION FUNCTION

The crosscorrelation function of two distinct time-varying parameters $\xi_1(t)$ and $\xi_2(t)$ is given by

$$R_{\xi_1\xi_2}(\tau) = \lim_{T \to \infty} \frac{1}{2T} \int_{-T}^{T} \xi_1(t)\xi_2(t - \tau) \, dt = \overline{\xi_1(t)\xi_2(t - \tau)}$$

(22.48)

The magnitude of $R_{\xi_1\xi_2}(\tau)$ is a measure of the correlation of ξ_1 and ξ_2 when observed at times separated by a time τ.

Example 22.1. Assume ξ_1 is directly proportional to ξ_2 except for a time lag τ_1:

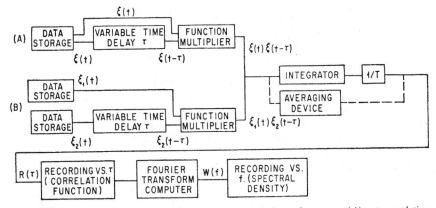

FIGURE 22.25 Block diagram showing correlation analysis by analog means: (A) autocorrelation analysis; (B) crosscorrelation analysis. Dotted line shows alternative method of averaging multiplicand (see Fig. 22.18).

$$\xi_1(t) = A\xi_2(t - \tau_1) \tag{22.49}$$

Then, by Eq. (22.48),

$$R_{\xi_1\xi_2}(\tau_1) = A\overline{[\xi_2]^2} = AR_{\xi_2}(0) \tag{22.50}$$

Example 22.2. Assume ξ_1 and ξ_2 are sinusoidal functions at the same frequency:

$$\xi_1(t) = A\,\xi_0 \sin \omega(t - \tau_1)$$
$$\xi_2(t) = \xi_0 \sin \omega t \tag{22.51}$$

Then the crosscorrelation function is

$$R_{\xi_1\xi_2}(\tau) = \frac{A[\xi_0]^2}{2} \cos \omega(\tau - \tau_1) \tag{22.52}$$

This is the same form as the autocorrelation function for a sine wave given in Eq. (22.46) except for the time shift τ_1 (see Fig. 22.24). Thus, for example, the increase or decrease in the vibration displacement amplitude between points 1 and 2 may be determined from the crosscorrelation function given by Eq. (22.52).

Determination of Crosscorrelation Function. One method of obtaining a crosscorrelation function by analog means is illustrated at B in Fig. 22.25. One function $\xi_1(t)$ is fed directly to the multiplier and the second function $\xi_2(t)$ is fed through the variable time-delay to the multiplier. The time average of the multiplicand $\xi_1(t)\xi_2(t - \tau)$ plotted vs. τ is the crosscorrelation function.

Crosscorrelation analysis also may be carried out by the use of standard high-speed digital computers which have been programmed for this purpose. Crosscorrelation analysis by digital means is almost identical to autocorrelation analysis since the only difference is that two distinct functions, $\xi_1(t)$ and $\xi_2(t)$, are employed instead of the one function $\xi(t)$ used in autocorrelation analysis.

REFERENCES

1. Himmelblau, H., A. G. Piersol, J. H. Wise, and M. R. Grundvig: "Handbook for Dynamic Data Acquisition and Analysis," I.E.S. Recommended Practice RP-DTE 012.1, Institute of Environmental Sciences, Mount Prospect, Ill., 1994.

2. Crandall, S. H.: "Random Vibration," Technology Press of MIT, Cambridge, Mass., 1958.

3. Bendat, J. S.: "Principles and Applications of Random Noise Theory," John Wiley & Sons, Inc., New York, 1958.

4. Davenport, W. B., Jr., and W. L. Root: "An Introduction to the Theory of Random Signals and Noise," McGraw-Hill Book Company, Inc., New York, 1958.

5. Miller, K. S.: "Engineering Mathematics," Holt, Winston & Rinehart, Inc., New York, 1957.

6. Bendat, Julius S., and Allan G. Piersol: "Random Data: Analysis and Measurement Procedures," John Wiley & Sons, Inc., New York, 1971.

7. Enochson, Loren D., and Robert K. Otnes: "Programming and Analysis for Digital Time Series Data," SVM-3, Shock and Vibration Information Center, Washington, D.C., 1968.

8. Otnes, Robert K., and Loren Enochson: "Digital Time Series Analysis," John Wiley & Sons, Inc., New York, 1972.

9. Shannon, C. E.: *Proc. IRE,* **37**:10 (1949).

10. Methods for Analysis and Presentation of Shock and Vibration Data, American National Standard, ANSI S2.10-1971.

11. Bendat, J. S.: "Engineering Applications of Correlation and Spectral Analysis," John Wiley & Sons, Inc., New York, 1980.

CHAPTER 23
CONCEPTS IN SHOCK DATA ANALYSIS

Sheldon Rubin

INTRODUCTION

This chapter discusses the interpretation of shock measurements and the reduction of data to a form adapted to further engineering use. Methods of data reduction also are discussed. A shock measurement is a trace giving the time-history of a shock parameter over the duration of the shock. The shock parameter may define motion (such as acceleration, velocity, or displacement) or loading (such as force, pressure, stress, or torque). It is assumed that any corrections that should be applied to eliminate distortions resulting from the instrumentation have been made. The trace may be a pulse or transient vibration. The interpretation of periodic and random vibration measurements is discussed in Chap. 22.

Examples of sources of shock to which this discussion applies are aircraft landing, braking, and gust loading; missile launching and staging; transportation of fragile equipment; accidental collision of vehicles; gunfire; explosions; and high-speed fluid entry.

Often, a shock measurement in the form of a time-history of a motion or loading parameter is not useful directly for engineering purposes. Reduction to a different form is then necessary, the type of data reduction employed depending upon the ultimate use of the data.

Comparison of Measured Results with Theoretical Prediction. The correlation of experimentally determined and theoretically predicted results by comparison of records of time-histories is difficult. Generally, it is impractical in theoretical analyses to give consideration to all the effects which may influence the experimentally obtained results. For example, the measured shock often includes the vibrational response of the structure to which the shock-measuring device is attached. Such vibration obscures the determination of the shock input for which an applicable theory is being tested; thus, data reduction is useful in minimizing or eliminating the irrelevancies of the measured data to permit ready comparison of theory with corresponding aspects of the experiment. It often is impossible to make such comparisons on the basis of original time-histories.

Calculation of Structural Response.   In the design of equipment to withstand shock, the required strength of the equipment is indicated by its response to the shock. The response may be measured in terms of the deflection of a member of the equipment relative to another member or by the magnitude of the dynamic loads imposed upon the equipment. The structural response can be calculated from the time-history by known means; however, certain techniques of data reduction result in descriptions of the shock that are related directly to structural response.

As a design procedure it is convenient to represent the equipment by an appropriate model that is better adapted to analysis.* A typical model is shown in Fig. 23.1; it consists of a secondary structure supported by a primary structure. Depending upon the ultimate objective of the design work, certain characteristics of the response of the model must be known:

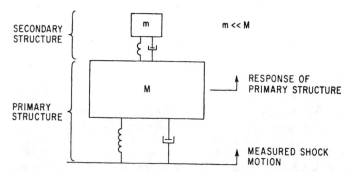

FIGURE 23.1   Commonly used structural model consisting of a primary and a secondary structure. Each structure is represented as a lumped-parameter single degree-of-freedom system with the secondary mass m much smaller than the primary mass M so that the response of the primary mass is unaffected by the response of the secondary mass. The response of the primary mass to an input shock motion is the input shock motion to the secondary structure.

1. If design of the secondary structure is to be effected, it is necessary to know the time-history of the motion of the primary structure. Such motion constitutes the excitation for the secondary structure.

2. In the design of the primary structure, it is necessary to know the deflection of such structure as a result of the shock, either the time-history or the maximum value.

By selection of suitable data reduction methods, response information useful in the design of the equipment is obtained from the original time-history.

Laboratory Simulation of Measured Shock.   Because of the difficulty of using analytical methods in the design of equipment to withstand shock, it is common practice to prove the design of equipments by laboratory tests that simulate the anticipated actual shock conditions. Unless the shock can be defined by one of a

---

* See Chap. 42 for a more complete discussion of models.

few simple functions, it is not feasible to reproduce in the laboratory the complete time-history of the actual shock experienced in service. Instead, the objective is to synthesize a shock having the characteristics and severity considered significant in causing damage to equipment. Then, the data reduction method is selected so that it extracts from the original time-history the parameters that are useful in specifying an appropriate laboratory shock test. The considerations involved in simulation are discussed in Chap. 24. Shock testing machines are discussed in Chap. 26.

EXAMPLES OF SHOCK MOTIONS

Five examples of shock motions are illustrated in Fig. 23.2 to show typical characteristics and to aid in the comparison of the various techniques of data reduction. The acceleration impulse and the acceleration step are the classical limiting cases of shock motions. The half-sine pulse of acceleration, the decaying sinusoidal acceleration, and the complex oscillatory-type motion typify shock motions encountered frequently in practice.

In selecting data reduction methods to be used in a particular circumstance, the applicable physical conditions must be considered. The original record, usually a time-history, may indicate any of several physical parameters; e.g., acceleration, force, velocity, or pressure. Data reduction methods discussed in subsequent sections of this chapter are applicable to a time-history of any parameter. For purposes of illustration in the following examples, the primary time-history is that of acceleration; time-histories of velocity and displacement are derived therefrom by integration. These examples are included to show characteristic features of typical shock motions and to demonstrate data reduction methods.

ACCELERATION IMPULSE OR STEP VELOCITY

The *delta function* $\delta(t)$ is defined mathematically as a function consisting of an infinite ordinate (acceleration) occurring in a vanishingly small interval of abscissa (time) at time $t = 0$ such that the area under the curve is unity. An acceleration time-history of this form is shown diagrammatically in Fig. 23.2A. If the velocity and displacement are zero at time $t = 0$, the corresponding velocity time-history is the velocity step and the corresponding displacement time-history is a line of constant slope, as shown in the figure. The mathematical expressions describing these time histories are

$$\ddot{u}(t) = \dot{u}_0 \delta(t) \tag{23.1}$$

where $\delta(t) = 0$ when $t \neq 0$, $\delta(t) = \infty$ when $t = 0$, and $\int_{-\infty}^{\infty} \delta(t)\, dt = 1$. The acceleration can be expressed alternatively as

$$\ddot{u}(t) = \lim_{\epsilon \to 0} \dot{u}_0/\epsilon \qquad [0 < t < \epsilon] \tag{23.2}$$

where $\ddot{u}(t) = 0$ when $t < 0$ and $t > \epsilon$. The corresponding expressions for velocity and displacement for the initial conditions $u = \dot{u} = 0$ when $t < 0$ are

$$\dot{u}(t) = \dot{u}_0 \qquad [t > 0] \tag{23.3}$$

$$u(t) = \dot{u}_0 t \qquad [t > 0] \tag{23.4}$$

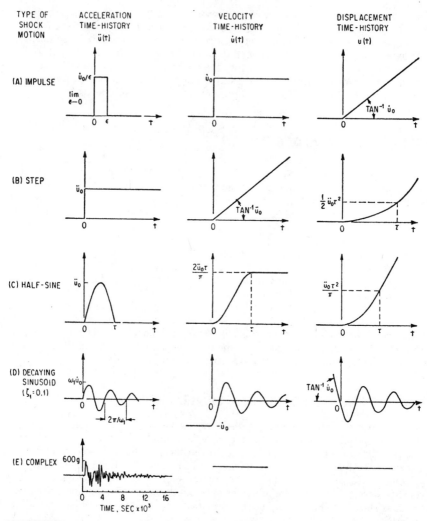

FIGURE 23.2 Five examples of shock motions. These examples are used to show typical characteristics of shock motions and to aid in the comparison of various techniques of data reduction. The acceleration time-history is considered to be the primary description, and corresponding velocity and displacement time-histories are shown. [Mathematical descriptions appear in Eqs. (23.1) to (23.14).]

ACCELERATION STEP

The *unit step function* $\mathbf{1}(t)$ is defined mathematically as a function which has a value of zero at time less than zero ($t < 0$) and a value of unity at time greater than zero ($t > 0$). The mathematical expression describing the acceleration step is

$$\ddot{u}(t) = \ddot{u}_0\mathbf{1}(t) \tag{23.5}$$

where $\mathbf{1}(t) = 1$ for $t > 0$ and $\mathbf{1}(t) = 0$ for $t < 0$. An acceleration time-history of the unit step function is shown in Fig. 23.2B; the corresponding velocity and displacement time-histories are also shown for the initial conditions $u = \dot{u} = 0$ when $t = 0$.

$$\dot{u}(t) = \ddot{u}_0 t \qquad [t > 0] \tag{23.6}$$

$$u(t) = \tfrac{1}{2}\ddot{u}_0 t^2 \qquad [t > 0] \tag{23.7}$$

The unit step function is the time integral of the delta function:

$$\mathbf{1}(t) = \int_{-\infty}^{t} \delta(t)\, dt \qquad [t > 0] \tag{23.8}$$

HALF-SINE ACCELERATION

A half-sine pulse of acceleration of duration τ is shown in Fig. 23.2C; the corresponding velocity and displacement time-histories also are shown, for the initial conditions $u = \dot{u} = 0$ when $t = 0$. The applicable mathematical expressions are

$$\ddot{u}(t) = \ddot{u}_0 \sin\left(\frac{\pi t}{\tau}\right) \qquad [0 < t < \tau]$$
$$\ddot{u}(t) = 0 \qquad \text{when } t < 0 \qquad \text{and } t > \tau \tag{23.9}$$

$$\dot{u}(t) = \frac{\ddot{u}_0 \tau}{\pi}\left(1 - \cos\frac{\pi t}{\tau}\right) \qquad [0 < t < \tau]$$
$$\dot{u}(t) = \frac{2\ddot{u}_0 \tau}{\pi} \qquad [t > \tau] \tag{23.10}$$

$$u(t) = \frac{\ddot{u}_0 \tau^2}{\pi^2}\left(\frac{\pi t}{\tau} - \sin\frac{\pi t}{\tau}\right) \qquad [0 < t < \tau]$$
$$u(t) = \frac{\ddot{u}_0 \tau^2}{\pi}\left(\frac{2t}{\tau} - 1\right) \qquad [t > \tau] \tag{23.11}$$

This example is typical of a class of shock motions in the form of acceleration pulses not having infinite slopes.

DECAYING SINUSOIDAL ACCELERATION

A decaying sinusoidal trace of acceleration is shown in Fig. 23.2D; the corresponding time-histories of velocity and displacement also are shown for the initial conditions $\dot{u} = -\ddot{u}_0$ and $u = 0$ when $t = 0$. The applicable mathematical expression is

$$\ddot{u}(t) = \frac{\ddot{u}_0 \omega_1}{\sqrt{1 - \zeta_1^2}}\, e^{-\zeta_1 \omega_1 t} \sin\left(\sqrt{1 - \zeta_1^2}\,\omega_1 t + \sin^{-1}\left(2\zeta_1\sqrt{1 - \zeta_1^2}\right)\right) \qquad [t > 0] \tag{23.12}$$

where ω_1 is the frequency of the vibration and ζ_1 is the fraction of critical damping corresponding to the decrement of the decay. Corresponding expressions for velocity and displacement are

$$\ddot{u}(t) = \frac{\ddot{u}_0}{\sqrt{1 - \zeta_1^2}}\ e^{-\zeta_1 \omega_1 t} \cos\left(\sqrt{1 - \zeta_1^2}\ \omega_1 t + \sin^{-1} \zeta_1\right) \qquad [t > 0] \qquad (23.13)$$

where $\ddot{u}(t) = -\ddot{u}_0$ when $t < 0$.

$$u(t) = -\frac{\ddot{u}_0}{\omega_1 \sqrt{1 - \zeta_1^2}}\ e^{-\zeta_1 \omega_1 t} \sin\left(\sqrt{1 - \zeta_1^2}\omega_1 t\right) \qquad [t > 0] \qquad (23.14)$$

where $u(t) = -\ddot{u}_0 t$ when $t < 0$.

COMPLEX SHOCK MOTION

The trace shown in Fig. 23.2E is an acceleration time-history representing typical field data. It cannot be defined by an analytic function; consequently, the corresponding velocity and displacement time-histories can be obtained only by numerical, graphical, or analog integration of the acceleration time-history.

CONCEPTS OF DATA REDUCTION

Consideration of the engineering uses of shock measurements indicates two basically different methods for describing a shock: (1) a description of the shock in terms of its inherent properties, in the time domain or in the frequency domain; and (2) a description of the shock in terms of the effect on structures when the shock acts as the excitation. The latter is designated reduction to the response domain. The following sections discuss concepts of data reduction to the frequency and response domains.

Whenever practical, the original time-history should be retained even though the information included therein is reduced to another form. The purpose of data reduction is to make the data more useful for some particular application. The reduced data usually have a more limited range of applicability than the original time-history. These limitations must be borne in mind if the data are to be applied intelligently.

DATA REDUCTION TO THE FREQUENCY DOMAIN

Any nonperiodic function can be represented as the superposition of sinusoidal components, each with its characteristic amplitude and phase.[1] This superposition is the Fourier spectrum, a plot of the amplitude and phase of the sinusoidal components into which the function can be decomposed. It is analogous to the Fourier components of a periodic function (Chap. 22). The Fourier components of a periodic function occur at discrete frequencies, and the composite function is obtained by superposition of components. By contrast, the Fourier spectrum for a nonperiodic function is a continuous function of frequency, and the composite function is achieved by integration. Applicable mathematical properties are given in the Appendix; the following sections discuss the application of the Fourier spectrum to describe the shock motions illustrated in Fig. 23.2.

Acceleration Impulse.   Using the definition of the acceleration pulse given by Eq. (23.2) and substituting this for $f(t)$ in Eq. (23.57) of the Appendix,

$$\mathbf{F}(\omega) = \lim_{\epsilon \to 0} \int_0^\epsilon \frac{\ddot{u}_0}{\epsilon}\ e^{-j\omega t}\ dt \qquad (23.15)$$

Carrying out the integration,

$$F(\omega) = \lim_{\epsilon \to 0} \frac{\ddot{u}_0(1 - e^{-j\omega\epsilon})}{j\omega\epsilon} = \ddot{u}_0 \qquad (23.16)$$

The corresponding amplitude and phase spectra, from Eqs. (23.63) and (23.64) of the Appendix, are

$$F(\omega) = \ddot{u}_0; \qquad \theta(\omega) = 0 \qquad (23.17)$$

These spectra are shown in Fig. 23.3A. The magnitude of the Fourier amplitude spectrum is a constant, independent of frequency, equal to the area under the acceleration-time curve.

The physical significance of the spectra in Fig. 23.3A is shown in Fig. 23.4, where the rectangular acceleration pulse of magnitude $\ddot{u}_0/\epsilon$ and duration $t = \epsilon$ is shown as approximated by superposed sinusoidal components for several different upper limits of frequency for the components. With the frequency limit $\omega_l = 4/\epsilon$, the pulse has a noticeably rounded contour formed by the superposition of all components whose frequencies are less than ω_l. These components tend to add in the time interval $0 < t < \epsilon$ and, though existing for all time from $-\infty$ to $+\infty$, cancel each other outside this interval, so that $\ddot{u}$ approaches zero. When $\omega_l = 16/\epsilon$, the pulse is more nearly rectangular and $\ddot{u}$ approaches zero more rapidly for time $t < 0$ and $t > \epsilon$. When $\omega_l = \infty$, the superposition of sinusoidal components gives $\ddot{u} = \ddot{u}_0/\epsilon$ for the time interval of the pulse, and $\ddot{u} = \ddot{u}_0/2\epsilon$ at $t = 0$ and $t = \epsilon$. The components cancel completely for all other times. As $\epsilon \to 0$ and $\omega_l \to \infty$, the infinitely large number of superimposed frequency components gives $\ddot{u} = \infty$ at $t = 0$. The same general result is obtained when the Fourier components of other forms of $\ddot{u}(t)$ are superimposed.

Acceleration Step. The Fourier spectrum of the acceleration step does not exist in the strict sense (see Appendix) since the integrand of Eq. (23.56) does not tend to zero as $\omega \to \infty$. Using a convergence factor, the Fourier transform is found by substituting $\ddot{u}(t)$ for $f(t)$ in Eq. (23.73) of the Appendix:

$$F(\omega - ja) = \int_0^\infty \ddot{u}_0 e^{-j(\omega - ja)t} \, dt = \frac{\ddot{u}_0}{j(\omega - ja)} \qquad (23.18)$$

Taking the limit as $a \to 0$,

$$F(\omega) = \frac{\ddot{u}_0}{j\omega} \qquad (23.19)$$

The amplitude and phase spectra, from Eqs. (23.63) and (23.64), are

$$F(\omega) = \frac{\ddot{u}_0}{\omega}; \qquad \theta(\omega) = -\frac{\pi}{2} \qquad (23.20)$$

These spectra are shown in Fig. 23.3B; the amplitude spectrum decreases as frequency increases, whereas the phase is a constant independent of frequency. Note that the spectrum of Eq. (23.19) is $1/j\omega$ times the spectrum for the impulse, Eq. (23.16), in accordance with Eq. (23.65) of the Appendix.

Half-sine Acceleration. Substitution of the half-sine acceleration time-history, Eq. (23.9), into Eq. (23.57) gives

$$F(\omega) = \int_0^\tau \ddot{u}_0 \sin\frac{\pi t}{\tau} \, e^{-j\omega t} \, dt \qquad (23.21)$$

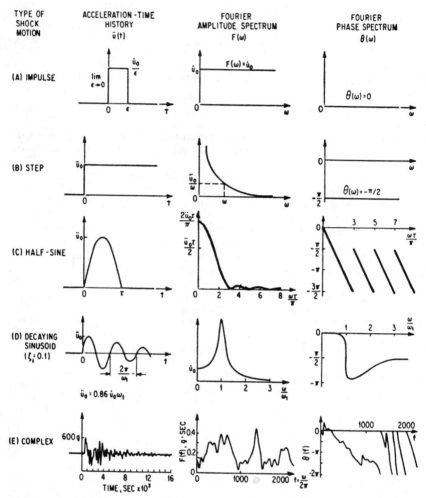

FIGURE 23.3 Fourier amplitude and phase spectra for the shock motions in Fig. 23.2. These spectra represent the amplitude and phase of the continuous distribution of frequency components into which the shock motions can be decomposed. [Mathematical descriptions appear in Eqs. (23.15) to (23.27).]

Performing the indicated integration gives

$$\mathbf{F}(\omega) = \frac{\ddot{u}_0\tau/\pi}{1 - (\omega\tau/\pi)^2} \left(1 + e^{-j\omega\tau}\right) \qquad [\omega \neq \pi/\tau]$$

$$\mathbf{F}(\omega) = -\frac{j\ddot{u}_0\tau}{2} \qquad [\omega = \pi/\tau]$$

(23.22)

Applying Eqs. (23.63) and (23.64) to find expressions for the spectra of amplitude and phase,

$$F(\omega) = \frac{2\ddot{u}_0\tau}{\pi} \left| \frac{\cos(\omega\tau/2)}{1-(\omega\tau/\pi)^2} \right| \qquad [\omega \neq \pi/\tau]$$

(23.23)

$$F(\omega) = \frac{\ddot{u}_0\tau}{2} \qquad [\omega = \pi/\tau]$$

$$\theta(\omega) = -\frac{\omega\tau}{2} + n\pi \qquad (23.24)$$

where n is the smallest integer that prevents $|\theta(\omega)|$ from exceeding $3\pi/2$. The Fourier spectra of the half-sine pulse of acceleration are plotted in Fig. 23.3C.

Decaying Sinusoidal Acceleration.   The application of Eq. (23.57) to the decaying sinusoidal acceleration defined by Eq. (23.12) gives the following expression for the Fourier spectrum:

$$\mathbf{F}(\omega) = \ddot{u}_0 \frac{1 + j2\zeta_1\omega/\omega_1}{(1 - \omega^2/\omega_1^2) + j2\zeta_1\omega/\omega_1} \qquad (23.25)$$

This can be converted to a spectrum of absolute values by applying Eq. (23.63):

$$F(\omega) = \ddot{u}_0 \sqrt{\frac{1 + (2\zeta_1\omega/\omega_1)^2}{(1 - \omega^2/\omega_1^2)^2 + (2\zeta_1\omega/\omega_1)^2}} \qquad (23.26)$$

A spectrum of phase angle is obtained from Eq. (23.64):

$$\theta(\omega) = -\tan^{-1} \frac{2\zeta_1(\omega/\omega_1)^3}{(1 - \omega^2/\omega_1^2) + (2\zeta_1\omega/\omega_1)^2} \qquad (23.27)$$

These spectra are shown in Fig. 23.3D for a value of $\zeta = 0.1$. The peak in the amplitude spectrum near the frequency ω_1 indicates a strong concentration of Fourier components near the frequency of occurrence of the oscillations in the shock motion.

Complex Shock.   The complex shock motion shown in Fig. 23.3E is the result of actual measurements; hence, its functional form is unknown. Its Fourier spectrum must be computed numerically. The Fourier spectrum shown in Fig. 23.3E was evaluated digitally using 100 time increments of 0.00015 sec duration. The peaks in the amplitude spectrum indicate concentrations of sinusoidal components near the frequencies of various oscillations in the shock motion. The portion of the phase spectrum at the high frequencies creates an appearance of discontinuity. If the phase angle were not returned to zero each time it passes through −360°, as a convenience in plotting, the curve would be continuous.

Application of the Fourier Spectrum.   The Fourier spectrum description of a shock is useful in linear analysis when the properties of a structure on which the shock acts are defined as a function of frequency. Such properties are designated by the general term *transfer function;* in shock and vibration technology, commonly used transfer functions are mechanical impedance, mobility, and transmissibility.

When a shock acts on a structure, the structure responds in a manner that is essentially oscillatory. The frequencies that appear predominantly in the response are (1) the preponderant frequencies of the shock and (2) the natural frequencies of the structure. The Fourier spectrum of the response $\mathbf{R}(\omega)$ is the product of the Fourier spectrum of the shock $\mathbf{F}(\omega)$ and an appropriate transfer function for the

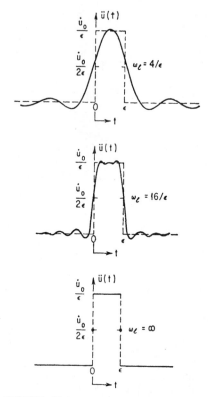

FIGURE 23.4 Time-histories which result from the superposition of the Fourier components of a rectangular pulse for several different upper limits of frequency ω_l of the components. The upper time-history ($\omega_l = 4/\epsilon$) has a notice-.ably rounded contour; the middle time-history ($\omega_l = 16/\epsilon$) is more nearly rectangular; and the lower time-history ($\omega_l = \infty$) is the complete Fourier representation of a rectangular pulse.

structure. This is indicated by Eq. (23.74) of the Appendix. For example, if $\mathbf{F}(\omega)$ and $\mathbf{R}(\omega)$ are Fourier spectra of acceleration, the transfer function is the transmissibility of the structure, i.e., the ratio of acceleration at the responding station to the acceleration at the driving station, as a function of frequency. However, if $\mathbf{R}(\omega)$ is a Fourier spectrum of velocity and $\mathbf{F}(\omega)$ is a Fourier spectrum of force, the transfer function is mobility as a function of frequency.

The Fourier spectrum also finds application in evaluating the effect of a load upon a shock source. A source of shock generally consists of a means of shock excitation and a resilient structure through which the excitation is transmitted to a load. Consequently, the character of the shock delivered by the resilient structure of the shock source is influenced by the nature of the load being driven. The characteristics of the source and load may be defined in terms of mechanical impedance or mobility (see Chap. 10). If the shock motion at the source output is measured with no load and expressed in terms of its Fourier spectrum, the effect of the load upon this shock motion can be determined by Eq. (23.79) of the Appendix. The resultant motion with the load attached is described by its Fourier spectrum.

The transfer function of a structure may be determined by applying a transient force to the structure and noting the response. This is analogous to the more commonly used method of applying a sinusoidally varying force whose frequency can be varied over a wide range and noting the sinusoidally varying motion at the frequency of the force application. In some circumstances, it may be more convenient to apply a transient. From the measured time-histories of the force and the response, the corresponding Fourier spectra can be calculated. The transfer function is the quotient of the Fourier spectrum of the force divided by the Fourier spectrum of the response, as indicated by Eq. (23.82) of the Appendix.

DATA REDUCTION TO THE RESPONSE DOMAIN

A structure or physical system has a characteristic response to a particular shock applied as an excitation to the structure. The magnitudes of the response peaks can be used to define certain effects of the shock by considering systematically the properties of the system and relating the peak responses to such properties. This is in contrast to the Fourier spectrum description of a shock in the following respects:

1. Whereas the Fourier spectrum defines the shock in terms of the amplitudes and phase relations of its frequency components, the response spectrum describes only the effect of the shock upon a structure in terms of peak responses. This effect is of considerable significance in the design of equipments and in the specification of laboratory tests.

2. The time-history of a shock cannot be determined from the knowledge of the peak responses of a system excited by the shock; i.e., the calculation of peak responses is an irreversible operation. This contrasts with the Fourier spectrum, where the Fourier spectrum can be determined from the time-history, and vice versa.

By limiting consideration to the response of a linear, viscously damped single degree-of-freedom structure with lumped parameters (hereafter referred to as a simple structure and illustrated in Fig. 23.5), there are only two structural parameters upon which the response depends: (1) the undamped natural frequency and (2) the fraction of critical damping. With only two parameters involved, it is feasible to obtain from the shock measurement a systematic presentation of the peak responses of many simple structures. This process is termed *data reduction to the response domain*. This type of reduced data applies directly to a system that responds in a single degree-of-freedom; it is useful to some extent by normal-mode superposition to evaluate the response of a linear system that responds in more than one degree-of-freedom. The conditions of a particular application determine the magnitude of errors resulting from superposition.[2-4]

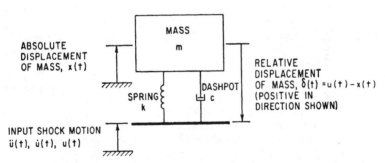

FIGURE 23.5 Representation of a simple structure used to accomplish the data reduction of a shock motion to the response domain. The differential equation of motion is given by Eq. (23.31).

Shock Spectrum.*   The response of a system to a shock can be expressed as the time-history of a parameter that describes the motion of the system. For a simple system, the magnitudes of the response peaks can be summarized as a function of the natural frequency or natural period of the responding system, at various values of the

---

* This spectrum sometimes is designated the *response spectrum*.

fraction of critical damping. This type of presentation is termed a shock spectrum. In the shock spectrum, or more specifically the two-dimensional shock spectrum, only the maximum value of the response found in a single time-history is plotted. The three-dimensional shock spectrum takes the form of a surface and shows the distribution of response peaks throughout the time-history. The two-dimensional spectrum is more common and is discussed in considerable detail in the immediately following section; for convenience, it is referred to hereafter simply as the *shock spectrum*. The three-dimensional spectrum is discussed in less detail in a later section.

Parameters for the Shock Spectrum.   The peak response of the simple structure may be defined, as a function of natural frequency, in terms of any one of several parameters that describe its motion. The parameters often are related to each other by the characteristics of the structure. However, inasmuch as one of the advantages of the shock spectrum method of data reduction and presentation is convenience of application to physical situations, it is advantageous to give careful consideration in advance to the particular parameter that is best adapted to the attainment of particular objectives. Referring to the simple structure shown in Fig. 23.5, the following significant parameters may be determined directly from measurements on the structure:

1. Absolute displacement $x(t)$ of mass m. This indicates the displacement of the responding structure with reference to an inertial reference plane, i.e., coordinate axes fixed in space.

2. Relative displacement $\delta(t)$ of mass m. This indicates the displacement of the responding structure relative to its support, a quantity useful for evaluating the distortions and strains within the responding structure.

3. Absolute velocity $\dot{x}(t)$ of mass m. This quantity is useful for determining the kinetic energy of the structure.

4. Relative velocity $\dot{\delta}(t)$ of mass m. This quantity is useful for determining the stresses generated within the responding structure due to viscous damping and the maximum energy dissipated by the responding structure.

5. Absolute acceleration $\ddot{x}(t)$ of mass m. This quantity is useful for determining the stresses generated within the responding structure due to the combined elastic and damping reactions of the structure.

The *equivalent static acceleration* is that steadily applied acceleration, expressed as a multiple of the acceleration of gravity, which distorts the structure to the maximum distortion resulting from the action of the shock.[5] For the simple structure of Fig. 23.5, the relative displacement response δ indicates the distortion under the shock condition. The corresponding distortion under static conditions, in a $1g$ gravitational field, is

$$\delta_{st} = \frac{mg}{k} = \frac{g}{\omega_n^2} \qquad (23.28)$$

By analogy, the maximum distortion under the shock condition is

$$\delta_{max} = \frac{A_{eq}g}{\omega_n^2} \qquad (23.29)$$

where A_{eq} is the equivalent static acceleration in units of gravitational acceleration. From Eq. (23.29),

$$A_{eq} = \frac{\delta_{max}\omega_n^2}{g} \qquad (23.30)$$

The maximum relative displacement δ_{max} and the equivalent static acceleration A_{eq} are directly proportional.

If the shock is a loading parameter, such as force, pressure, or torque, as a function of time, the corresponding equivalent static parameter is an equivalent static force, pressure, or torque, respectively. Since the supporting structure is assumed to be motionless when a shock loading acts, the relative response motions and absolute response motions become identical.

The differential equation of motion for the system shown in Fig. 23.5 is

$$-\ddot{x}(t) + 2\zeta\omega_n\dot{\delta}(t) + \omega_n^2\delta(t) = 0 \tag{23.31}$$

where ω_n is the undamped natural frequency and ζ is the fraction of critical damping. When $\zeta = 0$, $\ddot{x}_{max} = A_{eq}g$; this follows directly from the relation of Eq. (23.29). When $\zeta \neq 0$, the acceleration $\ddot{x}$ experienced by the mass m results from forces transmitted by the spring k and the damper c. Thus, in a damped system, the maximum acceleration of mass m is not exactly equal to the equivalent static acceleration. However, in most mechanical structures, the damping is relatively small; therefore, the equivalent static acceleration and the maximum absolute acceleration often are interchangeable with negligible error.

Referring to the model in Fig. 23.1, suppose the equivalent static acceleration A_{eq} and the maximum absolute acceleration $\ddot{x}_{max}$ are known for the primary structure. Then A_{eq} is useful directly for calculating the maximum relative displacement response of the primary structure. When the natural frequency of the secondary structure is much higher than the natural frequency of the primary structure, the maximum acceleration $\ddot{x}_{max}$ of M is useful for calculating the maximum relative displacement of m with respect to M. The secondary structure then responds in a "static manner" to the acceleration of the mass M; i.e., the maximum acceleration of m is approximately equal to that of M. Consequently, both A_{eq} and $\ddot{x}_{max}$ can be used for design purposes to calculate equivalent static loads on structures or equipment (see Chap. 42).

If the damping in the responding structure is large ($\zeta > 0.2$), the values of A_{eq} and $\ddot{x}_{max}$ are significantly different. Because the maximum distortion of primary structures often is the type of information required and the equivalent static acceleration is an expression of this response in terms of an equivalent static loading, the following discussion is limited to shock spectra in terms of A_{eq}.

The response of a simple structure with small damping to oscillatory-type shock excitation often is substantially sinusoidal at the natural frequency of the structure, i.e., the envelope of the oscillatory response varies in a relatively slow manner, as depicted in Fig. 23.6. The maximum relative displacement δ_{max}, the maximum relative velocity $\dot{\delta}_{max}$, and the maximum absolute acceleration $\ddot{x}_{max}$ are related approximately as follows:

$$\dot{\delta}_{max} = \omega_n\delta_{max}; \qquad \ddot{x}_{max} = \omega_n\dot{\delta}_{max}; \qquad \ddot{x}_{max} = \omega_n^2\delta_{max} \tag{23.32}$$

where the sign may be neglected since the positive and negative maxima are approximately equal. When applicable, these relations may be used to convert from a spectrum expressed in one parameter to a spectrum expressed in another parameter.

For idealized shock motions which often are approximated in practice, it is desirable to use a dimensionless ratio for the ordinate of the shock spectrum. Some of the more common dimensionless ratios are

$$\frac{gA_{eq}}{\ddot{u}_{max}} = \frac{\omega_n^2\delta_{max}}{\ddot{u}_{max}}; \qquad \frac{\ddot{x}_{max}}{\ddot{u}_{max}}; \qquad \frac{\omega_n\delta_{max}}{\Delta\dot{u}}; \qquad \frac{\dot{\delta}_{max}}{\Delta\dot{u}}; \qquad \frac{\delta_{max}}{u_{max}}$$

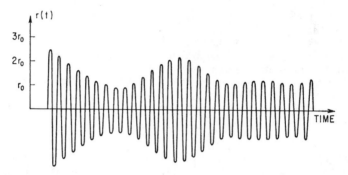

FIGURE 23.6 Examples of an oscillatory response time-history $r(t)$ for which the envelope of the response varies in a relatively slow manner. This type of response often results when an oscillatory-type shock motion acts on a simple structure with small damping. When such a response occurs, the response parameters of the simple structure are related in accordance with Eq. (23.32).

where $\ddot{u}_{max}$ and u_{max} are the maximum acceleration and displacement, respectively, of the shock motion and $\Delta \dot{u}$ is the velocity change of the shock motion (equal to the area under the acceleration time-history). Sometimes these ratios are referred to as *shock amplification factors.*

    Calculation of Shock Spectrum.   The relative displacement response of a simple structure (Fig. 23.5) resulting from a shock defined by the acceleration $\ddot{u}(t)$ of the support is given by the Duhamel integral[6]

$$\delta(t) = \frac{1}{\omega_d} \int_0^t \ddot{u}(t_v)e^{-\zeta\omega_n(t - t_v)} \sin \omega_d(t - t_v) \, dt_v \tag{23.33}$$

where $\omega_n = (k/m)^{1/2}$ is the undamped natural frequency, $\zeta = c/2m\omega_n$ is the fraction of critical damping, and $\omega_d = \omega_n(1 - \zeta^2)^{1/2}$ is the damped natural frequency. The excitation $\ddot{u}(t_v)$ is defined as a function of the variable of integration t_v, and the response $\delta(t)$ is a function of time t. The relative displacement δ and relative velocity $\dot{\delta}$ are considered to be zero when $t = 0$. The equivalent static acceleration, defined by Eq. (23.30), as a function of ω_n and ζ is

$$A_{eq}(\omega_n,\zeta) = \frac{\omega_n^2}{g} \delta_{max}(\omega_n,\zeta) \tag{23.34}$$

    If a shock loading such as the input force $F(t)$ rather than an input motion acts on the simple structure, the response is

$$\delta(t) = \frac{1}{m\omega_d} \int_0^t F(t_v)e^{-\zeta\omega_n(t - t_v)} \sin \omega_d(t - t_v) \, dt_v \tag{23.35}$$

and an equivalent static force is given by

$$F_{eq}(\omega_n,\zeta) = k\delta_{max}(\omega_n,\zeta) = m\omega_n^2\delta_{max}(\omega_n,\zeta) \tag{23.36}$$

The equivalent static force is related to equivalent static acceleration by

$$F_{eq}(\omega_n,\zeta) = mA_{eq}(\omega_n,\zeta) \tag{23.37}$$

It is often of interest to determine the maximum relative displacement of the simple structure in Fig. 23.5 in both a positive and a negative direction. If $\ddot{u}(t)$ is positive as shown, positive values of $\ddot{x}(t)$ represent upward acceleration of the mass m. Initially, the spring is compressed and the positive direction of $\delta(t)$ is taken to be positive as shown. Conversely, negative values of $\delta(t)$ represent extension of spring k from its original position. It is possible that the ultimate use of the reduced data would require that both extension and compression of spring k be determined. Correspondingly, a positive and a negative sign may be associated with an equivalent static acceleration A_{eq} of the support, so that A_{eq}^+ is an upward acceleration producing a positive deflection δ and A_{eq}^- is a downward acceleration producing a negative deflection δ.

For some purposes it is desirable to distinguish between the maximum response which occurs during the time in which the measured shock acts and the maximum response which occurs during the free vibration existing after the shock has terminated. The shock spectrum based on the former is called a *primary shock spectrum* and that based on the latter is called a *residual shock spectrum*. For instance, the response $\delta(t)$ to the half-sine pulse in Fig. 23.2C occurring during the period $(t < \tau)$ is the primary response and the response $\delta(t)$ occurring during the period $(t > \tau)$ is the residual response. Reference is made to primary and residual shock spectra in the next section on *Examples of Shock Spectra* and in the section on *Relationship between Shock Spectrum and Fourier Spectrum*.

Examples of Shock Spectra.* In this section the shock spectra are presented for the five acceleration time-histories in Fig. 23.2. These spectra, shown in Fig. 23.7, are expressed in terms of equivalent static acceleration for the undamped responding structure, for $\zeta = 0.1, 0.5$, and other selected fractions of critical damping. Both the maximum positive and the maximum negative responses are indicated. In addition, a number of relative displacement response time-histories $\delta(t)$ are plotted to show the nature of the responses.

ACCELERATION IMPULSE. The application of Eq. (23.33) to the acceleration impulse shown in Fig. 23.2A and defined by Eq. (23.1) yields

$$\delta(t) = \frac{\dot{u}_0}{\omega_d} e^{-\zeta \omega_n t} \sin \omega_d t \qquad [\zeta < 1] \qquad (23.38)$$

This response is plotted in Fig. 23.7A for $\zeta = 0, 0.1$, and 0.5. The response peaks are reached at the times $t = (\cos^{-1} \zeta)/\omega_d, \cos^{-1} \zeta$ increasing by π for each succeeding peak. The values of the response at the peaks are

$$\delta_{max}^{(i)}(\omega_n, \zeta) = \frac{\dot{u}_0}{\omega_n} \exp\left(-\frac{\zeta}{\sqrt{1-\zeta^2}} [\cos^{-1} \zeta + (i-1)\pi]\right) \qquad [0 < \cos^{-1} \zeta \leq \pi/2]$$

$$(23.39)$$

where i is the number of the peak ($i = 1$ for the first positive peak, $i = 2$ for the first negative peak, etc.).

The largest positive response occurs at the first peak, i.e., when $i = 1$, and is shown by the solid dots in Fig. 23.7A. Hence, the equivalent static acceleration in the positive direction is obtained by substitution of Eq. (23.39) into Eq. (23.34) with $i = 1$:

$$A_{eq}^+(\omega_n, \zeta) = \frac{\omega_n \dot{u}_0}{g} \exp\left(-\frac{\zeta}{\sqrt{1-\zeta^2}} \cos^{-1} \zeta\right) \qquad (23.40)$$

---

* A large number of shock spectra, based on various response parameters, are given in Chap. 8.

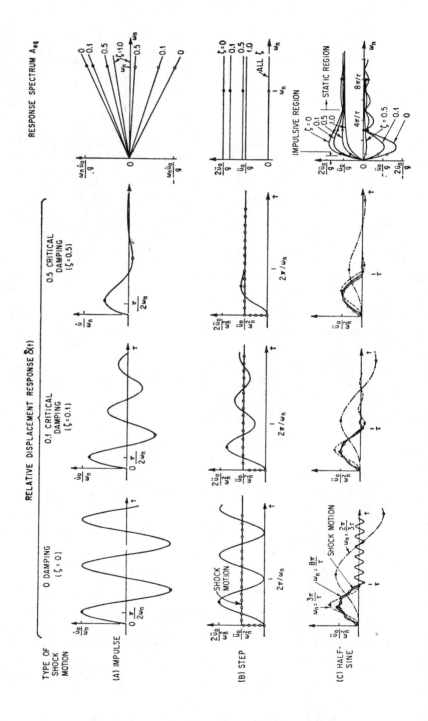

RESPONSE SPECTRUM A_{eq}

RELATIVE DISPLACEMENT RESPONSE $\delta(t)$

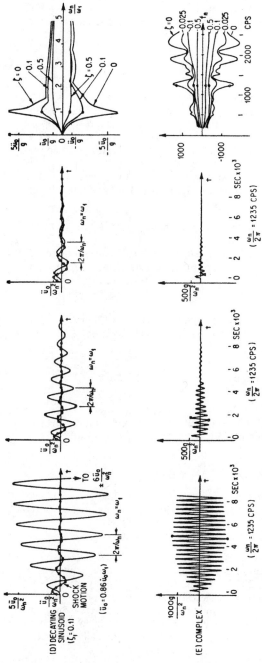

FIGURE 23.7 Time-histories of response to shock motions defined in Fig. 23.2 and corresponding shock spectra. Time-histories for several fractions of critical damping in the responding system are included, and the spectra are for corresponding fractions of critical damping. The spectra represent maximum values of response, as indicated by similar symbols. When the shock motions shown in Fig. 23.2 are superimposed on the time-histories of response, the shock motions are indicated by lines with hollow circles. Natural frequencies corresponding to the shock spectra are indicated where they have a significant relation to the frequency of the shock motion.

23.17

The equivalent static acceleration in the negative direction is calculated from the maximum relative deflection at the second peak, i.e., when $i = 2$, and is shown by the hollow dots in Fig. 23.7A:

$$A_{eq}^-(\omega_n, \zeta) = \frac{\omega_n \dot{u}_0}{g} \exp\left(-\frac{\zeta}{\sqrt{1-\zeta^2}} \left(\cos^{-1} \zeta + \pi\right)\right) \tag{23.41}$$

The resulting shock spectrum is shown in Fig. 23.7A with curves for $\zeta = 0, 0.1, 0.5,$ and 1.0. At any value of damping, a shock spectrum is a straight line passing through the origin. The peak distortion of the structure δ_{max} is inversely proportional to frequency. Thus, the relative displacement of the mass increases as the natural frequency decreases, whereas the equivalent static acceleration has an opposite trend.

ACCELERATION STEP.   The response of a simple structure to the acceleration step in Fig. 23.2B is found by substituting from Eq. (23.5) in Eq. (23.33) and integrating:

$$\delta(t) = \frac{\ddot{u}_0}{\omega_n^2}\left[1 - \frac{e^{-\zeta\omega_n t}}{\sqrt{1-\zeta^2}} \cos\left(\omega_d t - \sin^{-1}\zeta\right)\right] \qquad [\zeta < 1] \tag{23.42}$$

The responses $\delta(t)$ are shown in Fig. 23.7B for $\zeta = 0, 0.1,$ and 0.5. The response overshoots the value $\ddot{u}_0/\omega_n^2$ and then oscillates about this value as a mean with diminishing amplitude as energy is dissipated by damping. An overshoot to $2\ddot{u}_0/\omega_n^2$ occurs for zero damping. A response $\delta = \ddot{u}_0/\omega_n^2$ would result from a steady application of the acceleration $\ddot{u}_0$.

The response maxima and minima occur at the times $t = i\pi/\omega_d$, $i = 0$ providing the first minimum and $i = 1$ the first maximum. The maximum values of the relative displacement response are

$$\delta_{max}(\omega_n, \zeta) = \frac{\ddot{u}_0}{\omega_n^2}\left[1 + \exp\left(-\frac{\zeta i\pi}{\sqrt{1-\zeta^2}}\right)\right] \qquad [i \text{ odd}] \tag{23.43}$$

The largest positive response occurs at the first maximum, i.e., where $i = 1$, and is shown by the solid symbols in Fig. 23.7B. The equivalent static acceleration in the positive direction is obtained by substitution of Eq. (23.43) into Eq. (23.34) with $i = 1$:

$$A_{eq}^+(\omega_n, \zeta) = \frac{\ddot{u}_0}{g}\left[1 + \exp\left(-\frac{\zeta\pi}{\sqrt{1-\zeta^2}}\right)\right] \tag{23.44a}$$

The greatest negative response is zero; it occurs at $t = 0$, independent of the value of damping, as shown by open symbols in Fig. 23.7B. Thus, the equivalent static acceleration in the negative direction is

$$A_{eq}^-(\omega_n, \zeta) = 0 \tag{23.44b}$$

Since the equivalent static acceleration is independent of natural frequency, the shock spectrum curves shown in Fig. 23.7B are horizontal lines. The symbols shown on the shock spectra correspond to the responses shown.

The equivalent static acceleration for an undamped simple structure is twice the value of the acceleration step $\ddot{u}_0/g$. As the damping increases, the overshoot in response decreases; there is no overshoot when the structure is critically damped.

HALF-SINE ACCELERATION.   The expressions for the response of the damped simple structure to the half-sine acceleration of Eq. (23.9) are too involved to have general usefulness. For an undamped system, the response $\delta(t)$ is

$$\delta(t) = \frac{\ddot{u}_0}{\omega_n^2} \left(\frac{(\omega_n\tau/\pi)}{1 - (\omega_n\tau/\pi)^2} \right) [\sin \omega_n t - (\omega_n\tau/\pi) \sin (\pi t/\tau)] \qquad [0 < t \leq \tau]$$

$$\delta(t) = \frac{\ddot{u}_0}{\omega_n^2} \left(\frac{(\omega_n\tau/\pi)}{1 - (\omega_n\tau/\pi)^2} \right) 2 \cos \left(\frac{\omega_n\tau}{2} \right) \sin \left[\omega_n \left(t - \frac{\tau}{2} \right) \right] \qquad [t > \tau]$$

(23.45)

For zero damping the residual response is sinusoidal with constant amplitude. The first maximum in the response of a simple structure with natural frequency less than π/τ occurs during the residual response; i.e., after $t = \tau$. As a result, the magnitude of each succeeding response peak is the same as that of the first maximum. Thus the positive and negative shock spectrum curves are equal for $\omega_n \leq \pi/\tau$. The dot-dash curve in Fig. 23.7C is an example of the response at a natural frequency of $2\pi/3\tau$. The peak positive response is indicated by a solid circle, the peak negative response by an open circle. The positive and negative shock spectrum values derived from this response are shown on the undamped ($\zeta = 0$) shock spectrum curves at the right-hand side of Fig. 23.7C, using the same symbols.

At natural frequencies below $\pi/2\tau$, the shock spectra for an undamped system are very nearly linear with a slope $\pm 2\ddot{u}_0\tau/\pi g$. In this low-frequency region the response is essentially impulsive; i.e., the maximum response is approximately the same as that due to an ideal acceleration impulse (Fig. 23.7A) having a velocity change $\dot{u}_0$ equal to the area under the half-sine acceleration time-history.

The response at the natural frequency $3\pi/\tau$ is the dotted curve in Fig. 23.7C. The displacement and velocity response are both zero at the end of the pulse, and hence no residual response occurs. The solid and open triangles indicate the peak positive and negative response, the latter being zero. The corresponding points appear on the undamped shock spectrum curves. As shown by the negative undamped shock spectrum curve, the residual spectrum goes to zero for all odd multiples of π/τ above $3\pi/\tau$.

As the natural frequency increases above $3\pi/\tau$, the response attains the character of relatively low amplitude oscillations occurring with the half-sine pulse shape as a mean. An example of this type of response is shown by the solid curve for $\omega_n = 8\pi/\tau$. The largest positive response is slightly higher than $\ddot{u}_0/\omega_n^2$, and the residual response occurs at a relatively low level. The solid and open square symbols indicate the largest positive and negative response.

As the natural frequency becomes extremely high, the response follows the half-sine shape very closely. In the limit, the natural frequency becomes infinite and the response approaches the half-sine wave shown in Fig. 23.7C. For natural frequencies greater than $5\pi/\tau$, the response tends to follow the input and the largest response is within 20 percent of the response due to a static application of the peak input acceleration. This portion of the shock spectrum is sometimes referred to as the "static region" (see *Limiting Values of Shock Spectrum* below).

The equivalent static acceleration without damping for the positive direction is

$$A_{eq}^+(\omega_n,0) = \frac{\ddot{u}_0}{g} \left(\frac{2(\omega_n\tau/\pi)}{1 - (\omega_n\tau/\pi)^2} \right) \cos \left(\frac{\omega_n\tau}{2} \right) \qquad \left[\omega_n \leq \frac{\pi}{\tau} \right]$$

$$A_{eq}^+(\omega_n,0) = \frac{\ddot{u}_0}{g} \left(\frac{(\omega_n\tau/\pi)}{(\omega_n\tau/\pi) - 1} \right) \sin \left(\frac{2i\pi}{(\omega_n\tau/\pi) + 1} \right) \qquad \left[\omega_n > \frac{\pi}{\tau} \right]$$

(23.46)

where i is the positive integer which maximizes the value of the sine term while the argument remains less than π. In the negative direction the peak response always occurs during the residual response; thus, it is given by the absolute value of the first of the expressions in Eq. (23.46):

$$A_{\overline{eq}}(\omega_n,0) = \frac{\ddot{u}_0}{g}\left(\frac{2(\omega_n\tau/\pi)}{1-(\omega_n\tau/\pi)^2}\right)\cos\left(\frac{\omega_n\tau}{2}\right) \tag{23.47}$$

Shock spectra for damped systems can be found by use of an electrical analog or digital computer. Spectra for $\zeta = 0.1$ and 0.5 are shown in Fig. 23.7C.

The response of a damped structure whose natural frequency is less than $\pi/2\tau$ is essentially impulsive; i.e., the shock spectra in this frequency region are substantially identical to the spectra for the acceleration impulse in Fig. 23.7A. Except near the zeros in the negative spectrum for an undamped system, damping reduces the peak response. For the positive spectra, the effect is small in the static region since the response tends to follow the input for all values of damping. The greatest effect of damping is seen in the negative spectra because it affects the decay of response oscillations at the natural frequency of the structure.

DECAYING SINUSOIDAL ACCELERATION.   Although analytical expressions for the response of a simple structure to the decaying sinusoidal acceleration shown in Fig. 23.2D are available, calculation of spectra is impractical without use of a computer. Figure 23.7D shows spectra for several values of damping in the decaying sinusoidal acceleration. In the low-frequency region ($\omega_n < 0.2\omega_1$), the response is essentially impulsive. The area under the acceleration time-history of the decaying sinusoid is $\ddot{u}_0$; hence, the response of a very low-frequency structure is similar to the response to an acceleration impulse of magnitude $\ddot{u}_0$.

When the natural frequency of the responding system approximates the frequency ω_1 of the oscillations in the decaying sinusoid, a resonant type of build-up tends to occur in the response oscillations. The region in the neighborhood of $\omega_1 = \omega_n$ may be termed a quasi-resonant region of the shock spectrum. Responses for $\zeta = 0$, 0.1, and 0.5 and $\omega_n = \omega_1$ are shown in Fig. 23.7D. In the absence of damping in the responding system, the rate of build-up diminishes with time and the amplitude of the response oscillations levels off as the input acceleration decays to very small values. Small damping in the responding system, e.g., $\zeta = 0.1$, reduces the initial rate of build-up and causes the response to decay to zero after a maximum is reached. When damping is as large as $\zeta = 0.5$, no build-up occurs.

COMPLEX SHOCK.   The shock spectra for the complex shock of Fig. 23.2E are shown in Fig. 23.7E, as obtained with a direct-analog-type shock spectrum analyzer.[7] Time-histories of the response of a system with a natural frequency of 1,250 Hz also are shown. The ordinate of the spectrum plot is equivalent static acceleration, and the abscissa is the natural frequency in hertz. Three pronounced peaks appear in the spectra for zero damping, at approximately 1,250 Hz, 1,900 Hz, and 2,350 Hz. Such peaks indicate a concentration of frequency content in the shock, similar to the spectra for the decaying sinusoid in Fig. 23.7D. Other peaks in the shock spectra for an undamped system indicate less significant oscillatory behavior in the shock. The two lower frequencies at which the pronounced peaks occur correlate with the peaks in the Fourier spectrum of the same shock, as shown in Fig. 23.3E. The highest frequency at which a pronounced peak occurs is above the range for which the Fourier spectrum was calculated.

Because of frequency-response limitations of the spectrum analyzer, the shock spectra do not extend below 200 Hz. Since the duration of the complex shock of Fig. 23.2E is about 0.016 sec, an impulsive-type response occurs only for natural frequencies well below 200 Hz. As a result, no impulsive region appears in the shock spectra. There is no static region of the spectra shown because calculations were not extended to a sufficiently high frequency.

In general, the equivalent static acceleration A_{eq} is reduced by additional damping in the responding structure system except in the region of valleys in the shock spectra, where damping may increase the magnitude of the spectrum. Positive and negative spectra tend to be approximately equal in magnitude at any value of damping; thus, the spectra for a complex oscillatory type of shock may be based on peak response independent of sign to a good approximation.

Limiting Values of Shock Spectrum. The response data provided by the shock spectrum sometimes can be abstracted to simplified parameters that are useful for certain purposes. In general, this cannot be done without definite information on the ultimate use of the reduced data, particularly the natural frequencies of the structures upon which the shock acts. Two important cases are discussed in the following sections.

IMPULSE OR VELOCITY CHANGE. The duration of a shock sometimes is much smaller than the natural period of a structure upon which it acts. Then the entire response of the structure is essentially a function of the area under the time-history of the shock, described in terms of acceleration or a loading parameter such as force, pressure, or torque. Consequently, the shock has an effect which is equivalent to that produced by an impulse of infinitesimally short duration, i.e., an ideal impulse.

The shock spectrum of an ideal impulse is shown in Fig. 23.7*A*. All equivalent static acceleration curves are straight lines passing through the origin. The portion of the spectrum exhibiting such straight-line characteristics is termed the *impulsive region*. The shock spectrum of the half-sine acceleration pulse has an impulsive region when ω_n is less than approximately $\pi/2\tau$, as shown in Fig. 23.7*C*. If the area under a time-history of acceleration or shock loading is not zero or infinite, an impulsive region exists in the shock spectrum. The extent of the region on the natural frequency axis depends on the shape and duration of the shock.

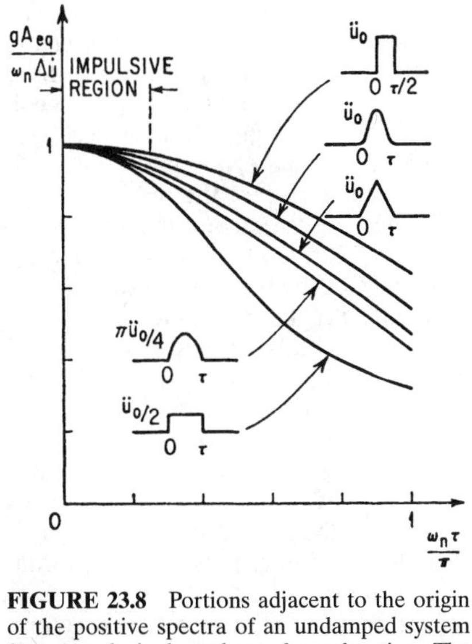

The portions adjacent to the origin of the positive shock spectra of an undamped system for several single pulses of acceleration are shown in Fig. 23.8. To illustrate the impulsive nature, each spectrum is normalized with respect to the peak impulsive response $\omega_n \, \Delta\dot{u}/g$, where $\Delta\dot{u}$ is the area under the corresponding acceleration time-history. Hence, the spectra indicate an impulsive response where the ordinate is approximately 1. The response to a single pulse of acceleration is impulsive within a tolerance of 10 percent if $\omega_n < 0.25\pi/\tau$; i.e., $f_n < 0.4\tau^{-1}$, where f_n is the natural frequency of the responding structure in hertz and τ is the pulse duration in seconds. This result also applies when the responding system is damped. Thus, it is possible to reduce the description of a shock pulse to a designated velocity change when the natural frequency of the responding structure is less than a

FIGURE 23.8 Portions adjacent to the origin of the positive spectra of an undamped system for several single pulses of acceleration. The ordinate is equivalent static acceleration A_{eq} normalized with respect to the peak impulsive response in g's, $\omega_n\Delta\dot{u}/g$, where $\Delta\dot{u}$ is the area under the corresponding acceleration time-history. An impulsive region is shown for which the deviation from impulsive response, $gA_{eq}/\omega_n \, \Delta\dot{u} = 1$, is less than 10 percent.

specified value. The magnitude of the velocity change is the area under the acceleration pulse:

$$\Delta\dot{u} = \int_0^\tau \ddot{u}(t)\, dt \qquad (23.48)$$

PEAK ACCELERATION OR LOADING. The natural frequency of a structure responding to a shock sometimes is sufficiently high that the response oscillations of the structure at its natural frequency have a relatively small amplitude. Examples of such responses are shown in Fig. 23.7C for $\omega_n = 8\pi/\tau$ and $\zeta = 0, 0.1, 0.5$. As a result, the maximum response of the structure is approximately equal to the maximum acceleration of the shock and is termed *equivalent static response*. The magnitude of the spectra in such a static region is determined principally by the peak value of the shock acceleration or loading. Portions of the positive spectra of an undamped system in the region of high natural frequencies are shown in Fig. 23.9 for a number of acceleration pulses. Each spectrum is normalized with respect to the maximum acceleration of the pulse. If the ordinate is approximately 1, the shock spectrum curves behave approximately in a static manner.

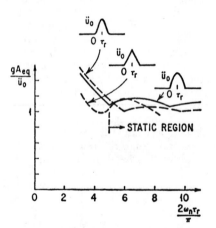

The limit of the static region in terms of the natural frequency of the structure is more a function of the slope of the acceleration time-history than of the duration of the pulse. Hence, the horizontal axis of the shock spectra in Fig. 23.9 is given in terms of the ratio of the rise time τ_r to the maximum value of the pulse. As shown in Fig. 23.9, the peak response to a single pulse of acceleration is approximately equal to the maximum acceleration of the pulse, within a tolerance of 20 percent, if $\omega_n > 2.5\pi/\tau_r$; i.e., $f_n > 1.25\tau_r^{-1}$, where f_n is the natural frequency of the responding structure in hertz and τ_r is the rise time to the peak value in seconds. The tolerance of 20 percent applies to an undamped system; for a damped system, the tolerance is lower, as indicated in Fig. 23.7C.

FIGURE 23.9 Portions of the positive shock spectra of an undamped system with high natural frequencies for several single pulses of acceleration. The ordinate is equivalent static acceleration A_{eq} normalized with respect to the peak acceleration of the pulse, $\ddot{u}_0/g$. A static region is shown for which the deviation from static response, $gA_{eq}/\ddot{u}_0 = 1$, is less than 20 percent.

The concept of the static region also can be applied to complex shocks. Suppose the shock is oscillatory, as shown in Fig. 23.2E. If the response to such a shock is to be nearly static, the response to each of the succession of pulses that make up the shock must be nearly static. This is most significant for pulses of large magnitude because they determine the ordinate of the spectrum in the static region. Therefore, the shock spectrum for a complex shock in the static region is based upon the pulses of greatest magnitude and shortest rise time.

Three-dimensional Shock Spectrum.[8] In general, the response of a structure to a shock is oscillatory and continues for an appreciable number of oscillations. At each oscillation, the response has an interim maximum value that differs, in general, from the preceding or following maximum value. For example, a typical time-history

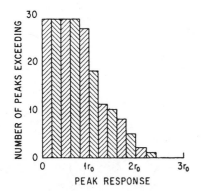

PEAK RESPONSE

FIGURE 23.10 Bar chart for the response of a system to a shock excitation. The chart represents a plane normal to the natural frequency axis in Fig. 23.11 and is obtained by counting the number of response maxima above various discrete increments of maximum response. In concept, the width of the increments approaches zero and a line faired through the ends of the bars is a smooth curve.

of response of a simple system of given natural frequency is shown in Fig. 23.6; the characteristics of the response may be summarized by the block diagram of Fig. 23.10. The abscissa of Fig. 23.10 is the peak response at the respective cycles of the oscillation, and the ordinate is the number of cycles at which the peak response exceeds the indicated value. Thus, the time-history of Fig. 23.6 has 29 cycles of oscillation at which the peak response of the oscillation exceeds $0.6r_0$, but only 2 cycles at which the peak response exceeds $2.0r_0$.

In accordance with the concept of the shock spectrum, the natural frequency of the responding system is modified by discrete increments and the response determined at each increment. This leads to a number of time-histories of response corresponding to Fig. 23.6, one for each natural frequency, and a similar number of block diagrams corresponding to Fig. 23.10. This group of block diagrams can be assembled to form a surface that shows pictorially the characteristics of the shock in terms of the response of a simple system. The axes of the surface are peak response, natural frequency of the responding system, and number of response cycles exceeding a given peak value. The block diagram of Fig. 23.10 is arranged on this set of axes at *A,* as shown in Fig. 23.11, at the appropriate position along the natural frequency axis. Other corresponding block diagrams are shown at *B*. The three-dimensional shock spectrum is the surface faired through the ends of the bars; the intercept of this surface with the planes of the block diagrams is indicated at *C* and that with the maximum response–natural frequency plane at *D*. Surfaces are obtainable for both positive and negative values of the response, and a separate surface is obtained for each fraction of critical damping in the responding system.

The two-dimensional shock spectrum is a special case of the three-dimensional surface. The former is a plot of the maximum response as a function of the natural frequency of the responding system; hence, it is a projection on the plane of the response and natural frequency axes of the maximum height of the surface. However, the height of the surface never exceeds that at one response cycle. Thus, the two-dimensional shock spectrum is the intercept of the surface with a plane normal to the "number of peaks exceeding" axis at the origin.

The response surface is a useful concept and illustrates a physical condition; however, it is not well adapted to quantitative analysis because the distances from the surface to the coordinate planes cannot be determined readily. A group of bar charts, each corresponding to Fig. 23.10, is more useful for quantitative purposes. The differences in lengths of the bars are discrete increments; this corresponds to the data reduction method in which the axis of response magnitudes is divided into discrete increments for purposes of counting the number of peaks exceeding each magnitude. In concept, the width of the increment may be considered to approach zero and the line faired through the ends of the bars represents the smooth intercept with the surface.

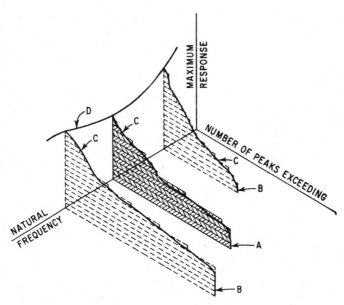

FIGURE 23.11   Example of a three-dimensional shock spectrum, which is a plot of the maximum value of individual response oscillations as a function of the number of occurrences of these maxima and the natural frequency of the responding system. Bar charts are shown properly disposed with respect to the coordinate planes.

Relationship between Shock Spectrum and Fourier Spectrum.   Although the shock spectrum and the Fourier spectrum are fundamentally different, there is a partial correlation between them. A direct relationship exists between a running Fourier spectrum, to be defined subsequently, and the response of an undamped simple structure. A consequence is a simple relationship between the Fourier spectrum of absolute values and the peak residual response of an undamped simple structure.

For the case of zero damping, Eq. (23.33) provides the relative displacement response

$$\delta(\omega_n,t) = \frac{1}{\omega_n} \int_0^t \ddot{u}(t_v) \sin \omega_n(t - t_v) \, dt_v \tag{23.49}$$

A form better suited to our needs here is

$$\delta(\omega_n,t) = \frac{1}{\omega_n} \, \mathcal{I}\left[e^{j\omega_n t} \int_0^t \ddot{u}(t_v) \, e^{-j\omega_n t_v} \, dt_v \right] \tag{23.50}$$

The integral above is seen to be the Fourier spectrum of the portion of $\ddot{u}(t)$ which lies in the time interval from zero to t, evaluated at the natural frequency ω_n. Such a time-dependent spectrum can be termed a "running Fourier spectrum" and denoted by $\mathbf{F}(\omega,t)$:

$$\mathbf{F}(\omega,t) = \int_0^t \ddot{u}(t_v) e^{-j\omega t_v} \, dt_v \tag{23.51}$$

It is assumed that the excitation vanishes for $t < 0$. The integral in Eq. (23.50) can be replaced by $\mathbf{F}(\omega_n, t)$; and after taking the imaginary part

$$\delta(\omega_n, t) = \frac{1}{\omega_n} \, F(\omega_n, t) \sin \left[\omega_n t + \theta(\omega_n, t)\right] \tag{23.52}$$

where $F(\omega_n, t)$ and $\theta(\omega_n, t)$ are the magnitude and phase of the running Fourier spectrum, corresponding to the definitions in Eqs. (23.63) and (23.64). Equation (23.52) provides the previously mentioned direct relationship between undamped structural response and the running Fourier spectrum.

When the running time t exceeds τ, the duration of $\ddot{u}(t)$, the running Fourier spectrum becomes the usual spectrum as given by Eq. (23.57), with τ used in place of the infinite upper limit of the integral. Consequently, Eq. (23.52) yields the sinusoidal residual relative displacement for $t > \tau$:

$$\delta_r(\omega_n, t) = \frac{1}{\omega_n} \, F(\omega_n) \sin \left[\omega_n t + \theta(\omega_n)\right] \tag{23.53}$$

The amplitude of this residual deflection and the corresponding equivalent static acceleration are

$$(\delta_r)_{max} = \frac{1}{\omega_n} \, F(\omega_n)$$

$$(A_{eq})_r = \frac{\omega_n^2 (\delta_r)_{max}}{g} = \frac{\omega_n}{g} \, F(\omega_n) \tag{23.54}$$

This result is clearly evident for the Fourier spectrum and undamped shock spectrum of the acceleration impulse. The Fourier spectrum is the horizontal line (independent of frequency) shown in Fig. 23.3A and the shock spectrum is the inclined straight line (increasing linearly with frequency) shown in Fig. 23.7A. Since the impulse exists only at $t = 0$, the entire response is residual. The undamped shock spectra in the impulsive region of the half-sine pulse and the decaying sinusoidal acceleration, Fig. 23.7C and D, respectively, also are related to the Fourier spectra of these shocks, Fig. 23.3C and D, in a similar manner. This results from the fact that the maximum response occurs in the residual motion for systems with small natural frequencies. Another example is the entire negative shock spectrum with no damping for the half-sine pulse in Fig. 23.7C, whose values are ω_n/g times the values of the Fourier spectrum in Fig. 23.3C.

METHODS OF DATA REDUCTION

Even though preceding sections of this chapter include several analytic functions as examples of typical shocks, data reduction in general is applied to measurements of shock that are not definable by analytic functions. The following sections outline data reduction methods that are adapted for use with any general type of function. Standard forms for presenting the analysis results are given in Ref. 9.

FOURIER SPECTRUM

The Fourier spectrum can be evaluated by either analog or digital techniques.[4] Block diagrams of two basic approaches to analog computation are shown in Fig. 23.12. Such analyses can be performed in either a *parallel* or a *swept* fashion. In a parallel analysis, an array of oscillators or filters is implemented to determine all the spectral values of interest simultaneously. In a swept analysis, the computation is performed for one frequency at a time; the frequency is stepped or continuously varied slowly to determine the spectral values of interest in a serial fashion. For purposes of the data analysis the shock time-history can be made artificially periodic by repeating the shock at intervals of time τ. The Fourier spectrum then becomes a Fourier series involving discrete components at integer multiples of the fundamental frequency $1/\tau$.* The magnitude of the true continuous Fourier spectrum at the frequency of such a discrete component is

$$F\left(\frac{2\pi n}{\tau}\right) = \left(\frac{\tau}{2}\right) c_n \qquad (23.55)$$

where c_n is the amplitude of the nth discrete component at the frequency $\omega = 2\pi n/\tau$, as defined by Eq. (22.8).

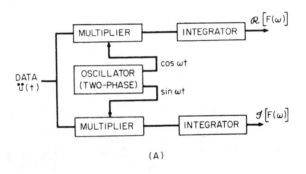

(A)

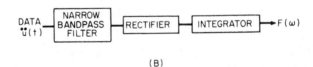

(B)

FIGURE 23.12   Block diagrams for analog computation of Fourier spectra. (*A*) Direct implementation of Eqs. (23.60) and (23.61). (*B*) Bandpass filtering to determine spectrum magnitude.[4]

Digital evaluation of the Fourier spectrum is most efficiently accomplished by means of the fast Fourier transform (FFT). Discussion of this approach appears in Chap. 27. Also see *FFT Analyzer*, Chap. 14.

SHOCK SPECTRUM

Both analog and digital computation techniques can be employed for evaluation of the shock spectrum.[4] Block diagrams for analog computation are shown in Fig. 23.13.

---

* See Chap. 22 for the defining relationships of a Fourier series and a discussion of data reduction techniques.

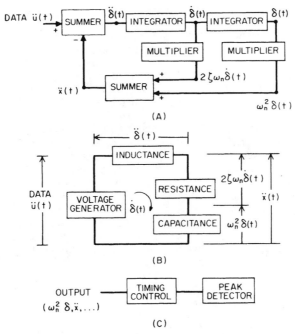

FIGURE 23.13 Block diagrams for analog computation of shock spectra using the simple system shown in Fig. 23.5. (*A*) Active electronic analog, (*B*) passive electrical analog, (*C*) peak detection (positive, negative, or absolute) with timing control as required for primary, residual, or overall analysis.

As with the Fourier spectrum, analyses can be performed in either a *parallel* or a *stepped* manner. In a parallel analysis, the shock-spectrum values are obtained from simultaneous determinations of the responses of simple structures corresponding to all natural frequencies of interest. A stepped analysis involves the determination of shock-spectrum results for one natural frequency and then a repeat of the determination for each successive value of the natural frequency.

Digital evaluation of the shock spectrum involves the use of a digital computer to determine the response of simple structures, followed by the detection of peak response. The response time-histories can be evaluated by the following techniques: (1) direct numerical or recursive integration of the Duhamel integral, Eq. (23.33), or (2) convolution or recursive filtering procedures. These techniques are discussed in detail in Ref. 4.

In principle, both analog and digital techniques can be modified to determine the three-dimensional shock spectrum. The modification involves the counting of the number of response maxima above various discrete increments of maximum response to obtain the results depicted in Fig. 23.11.

Reed Gage. The shock spectrum may be measured directly by a mechanical instrument that responds to shock in a manner analogous to the data reduction techniques used to obtain shock spectra from time-histories. The instrument includes a number of flexible mechanical systems that are considered to respond as single degree-of-freedom systems; each system has a different natural frequency, and means are provided to indicate the maximum deflection of each system as a result of the shock. The instrument often is referred to as a *reed gage* because the flexible mechanical systems are small cantilever beams carrying end masses; these have the appearance of reeds.[10]

The response parameter indicated by the reed gage is maximum deflection of the reeds relative to the base of the instrument; generally, this deflection is converted to equivalent static acceleration by applying the relation of Eq. (23.30). The reed gage offers a convenience in the indication of a useful quantity immediately and in the elimination of auxiliary electronic equipment. Also, it has important limitations: (1) the information is limited to the determination of a shock spectrum; (2) the deflection of a reed is inversely proportional to its natural frequency squared, thereby requiring high equivalent static accelerations to achieve readable records at high natural frequencies; (3) the means to indicate maximum deflection of the reeds (styli inscribing on a target surface) tend to introduce an undefined degree of damping; and (4) size and weight limitations on the reed gage for a particular application often limit the number of reeds which can be used and the lowest natural frequency for a reed. In spite of these limitations, the instrument sees continued use and has provided significant shock spectra where more elaborate instruments have failed.

APPENDIX 23.1

FOURIER INTEGRAL

Let the time-history of any motion or loading parameter (time-history of a shock) be designated $f(t)$. The corresponding Fourier spectrum or frequency spectrum $\mathbf{F}(\omega)$ is*

$$\mathbf{F}(\omega) = \int_{-\infty}^{\infty} f(t)e^{-j\omega t}\, dt \tag{23.56}$$

The Fourier spectrum $\mathbf{F}(\omega)$ is a complex number used to describe the amplitude and phase of a sinusoid at the frequency ω. The complex amplitude of the Fourier component at the frequency ω is $\mathbf{F}(\omega)\, d\omega$, the area under the Fourier spectrum curve in the interval $d\omega$. The quantity $\mathbf{F}(\omega)e^{j\omega t}\, d\omega$ is the sinusoidal variation at the frequency ω; the integral sums these sinusoids. If it is assumed that the shock starts at $i = 0$, Eq. (23.56) becomes

$$\mathbf{F}(\omega) = \int_{0}^{\infty} f(t)e^{-j\omega t}\, dt \tag{23.57}$$

If the duration of the time-history $f(t)$ has a finite limit, this limit is used in place of ∞ in Eq. (23.57). An alternative notation for $\mathbf{F}(\omega)$ is $\mathbf{F}[f(t)]$, which denotes the "Fourier transform of $f(t)$."

When $\omega = 0$, Eq. (23.56) becomes

$$\mathbf{F}(0) = \int_{-\infty}^{\infty} f(t)\, dt \tag{23.58}$$

---

* Mathematical limitations on the function which permit this type of analysis do exist, but these do not hinder engineering applications. See Refs. 11 to 13 for discussions of these limitations.

Thus, the zero frequency component of the Fourier spectrum of a shock is equal to the area under the time-history of the shock.

Since $\mathbf{F}(\omega)$ represents a distribution of amplitudes expressed in complex form, it may be written in terms of the real and imaginary parts:

$$\mathbf{F}(\omega) = \mathcal{R}[\mathbf{F}(\omega)] + j\mathcal{I}[\mathbf{F}(\omega)] \tag{23.59}$$

where

$$\mathcal{R}[\mathbf{F}(\omega)] = \int_0^\infty f(t) \cos \omega t \, dt \tag{23.60}$$

$$\mathcal{I}[\mathbf{F}(\omega)] = -\int_0^\infty f(t) \sin \omega t \, dt \tag{23.61}$$

Alternatively, the Fourier spectrum may be defined in terms of its absolute value and phase angle:

$$\mathbf{F}(\omega) = F(\omega)e^{j\theta(\omega)} \tag{23.62}$$

where $F(\omega)$ is the absolute value of $\mathbf{F}(\omega)$ and $\theta(\omega)$ is the phase angle. The absolute value and phase angle are related to the real and imaginary parts by

$$F(\omega) = \sqrt{\mathcal{R}^2[\mathbf{F}(\omega)] + \mathcal{I}^2[\mathbf{F}(\omega)]} \tag{23.63}$$

$$\theta(\omega) = \tan^{-1}\left\{ \frac{\mathcal{I}[\mathbf{F}(\omega)]}{\mathcal{R}[\mathbf{F}(\omega)]} \right\} \tag{23.64}$$

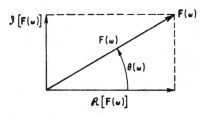

FIGURE 23.14 Representation of the complex Fourier component at the frequency ω, its magnitude $F(\omega)$ and phase angle $\theta(\omega)$, and its real and imaginary parts, $\mathcal{R}[\mathbf{F}(\omega)]$ and $\mathcal{I}[\mathbf{F}(\omega)]$.

Figure 23.14 shows the vector $\mathbf{F}(\omega)$, its magnitude $F(\omega)$ and phase angle $\theta(\omega)$, and its real and imaginary parts.

Equation (23.56) may be used with any shock parameter, such as acceleration, velocity, displacement, force, or pressure. The area under the curve of Fourier spectrum magnitude versus frequency has the same units as the time-history of the shock parameter.

The Fourier spectrum of the time-derivative of $f(t)$ is the product of $j\omega$ and the Fourier spectrum of $f(t)$:

$$\mathbf{F}\left[\frac{df(t)}{dt} \right] = j\omega\mathbf{F}[f(t)] \tag{23.65}$$

Thus if $\ddot{u}(t)$, $\dot{u}(t)$, $u(t)$ are the acceleration, velocity, and displacement time-histories, respectively, of a given shock motion, their Fourier spectra are related as follows:

$$\mathbf{F}[\ddot{u}(t)] = j\omega\mathbf{F}[\dot{u}(t)] = -\omega^2\mathbf{F}[u(t)] \tag{23.66}$$

$$\mathbf{F}[\dot{u}(t)] = j\omega\mathbf{F}[u(t)] \tag{23.67}$$

The time-history of a function can be found by integrating its Fourier spectrum components:

$$f(t) = \frac{1}{2\pi} \int_{-\infty}^\infty \mathbf{F}(\omega)e^{j\omega t} \, d\omega \tag{23.68}$$

Equations (23.68) and (23.56) are known as the inverse and direct Fourier transforms, respectively. The use of the factor $1/2\pi$ in Eq. (23.68) represents one formulation of the pair of Eqs. (23.56) and (23.68). In other formulations, the $1/2\pi$ appears in Eq. (23.56) and not in Eq. (23.68), or a $1/\sqrt{2\pi}$ appears in each. Caution should be exercised, when consulting treatises and tables of the Fourier transform, in determining which formulation of the transform is used. A useful compilation of direct and inverse Fourier transform pairs is given in Ref. 14.

The Fourier spectrum of a constant is a delta function (infinitely sharp spike) at zero frequency; hence, the addition of a constant to a function alters the Fourier spectrum of that function only at $\omega = 0$. Consequently, the relations for the spectra of acceleration, velocity, and displacement, Eqs. (23.66) and (23.67), are valid for all $\omega \ne 0$. This limitation is of little significance for engineering applications because the addition of a constant to a displacement time-history changes only the origin of the coordinate system. Similarly, the addition of a constant to a velocity time-history introduces another frame of reference which does not accelerate.

Equation (23.68) requires integration over negative frequencies. For engineering purposes, negative frequencies lack physical significance and a more desirable form of Eq. (23.68) is

$$f(t) = \frac{2}{\pi} \int_0^\infty \mathcal{R}[\mathbf{F}(\omega)] \cos \omega t \, d\omega \qquad [t > 0]$$

$$f(t) = 0 \qquad\qquad\qquad\qquad\qquad [t < 0]$$

$$(23.69)$$

Equation (23.69) is derived from (23.68) by including the restriction that $f(t) = 0$ for $t < 0$. For this to be true, $\int_0^\infty \mathcal{R}[\mathbf{F}(\omega)] \cos \omega t \, d\omega$ must equal $-\int_0^\infty \mathcal{I}[\mathbf{F}(\omega)] \sin \omega t \, d\omega$.

COMPARISON OF FOURIER AND LAPLACE TRANSFORMS*

The Laplace transform of $f(t)$ is defined, as a function of the complex variable p, by the integral

$$\mathcal{L}[f(t)] = \int_0^\infty f(t) e^{-pt} \, dt \tag{23.70}$$

where $p = a + j\omega$ for $a > 0$ and $\mathcal{L}[f(t)]$ denotes the "Laplace transform of $f(t)$." Equation (23.70) sometimes is written $\mathcal{L}[f(t)] = p \int_0^\infty f(t) e^{-pt} \, dt$. Caution should be exercised in determining the formulation of the transform. A comparison of Eqs. (23.70) and (23.57) reveals that the Laplace transform of a function approaches the Fourier transform of that function as $a \to 0$ or $p \to j\omega$, if the function is zero for all negative time.[12] The time-history is retrieved by the use of the inverse Laplace transform, involving integration in the complex plane:

$$f(t) = \frac{1}{2\pi j} \int_{a - j\infty}^{a + j\infty} \mathcal{L}[f(t)] e^{pt} \, dp \tag{23.71}$$

---

* The Laplace transform is discussed in detail in Chap. 8.

where a is chosen so that $f(t) = 0$ for $t < 0$. If $\mathcal{L}[f(t)]$ approaches $\mathbf{F}(\omega)$ as $p \to j\omega$ for *all* values of the frequency ω, then Eq. (23.71) provides a result identical to that of Eq. (23.68).

For problems associated with physically achievable shocks, the Fourier and Laplace transforms give identical results. The Fourier transform has certain mathematical disadvantages which the Laplace transform overcomes, and more extensive tables of the Laplace transform are available.[14] A detailed discussion of both transform methods and their limitations is found in Chap. 1 of Ref. 13.

A disadvantage of the Fourier transform method is that the integral defining the transform sometimes does not converge unless an extra convergence factor is provided. For example, the acceleration step in Fig. 23.2B has no Fourier transform in the strict sense since the integrand of Eq. (23.56) is $e^{-j\omega t}$, which does not have a well-defined limit as $t \to \infty$. The Fourier transform can be made useful in such instances by introducing a convergence factor e^{-at} and modifying the function $f(t)$ as follows:

$$f(t,a) = e^{-at}f(t) \qquad (23.72)$$

The Fourier spectrum is

$$\mathbf{F}(\omega - ja) = \int_{-\infty}^{\infty} f(t)e^{-j(\omega - ja)t}\,dt \qquad (23.73)$$

If the limit of $\mathbf{F}(\omega - ja)$ exists as $a \to 0$, $\mathbf{F}(\omega)$ can be called the Fourier transform of $f(t)$. The Laplace transform incorporates an appropriate convergence factor in its definition, thus guaranteeing the existence of the transform for practically all functions of possible interest in engineering applications.

Another disadvantage of the Fourier transform is the difficulty in obtaining the inverse transform, i.e., the application of Eq. (23.68). The Laplace transform achieves the inversion by integration in the complex p plane, so that all the background of complex-function theory is available.

On the basis of these factors, it may appear that the Laplace transform is the broader and more basic concept, and that the Fourier transform represents a special case of the Laplace transform. However, for purposes of physical interpretation of actual measured data, the Fourier spectrum has a greater intuitive appeal as a logical extension of the Fourier series concepts for periodic functions.

APPLICATIONS OF FOURIER INTEGRAL

DETERMINATION OF STRUCTURAL RESPONSE

If a shock loading parameter whose Fourier spectrum $\mathbf{F}(\omega)$ is defined by Eq. (23.56) acts upon a structure, the frequency content of the structural response is found from

$$\mathbf{R}(\omega) = \mathbf{H}(\omega)\mathbf{F}(\omega) \qquad (23.74)$$

where $\mathbf{R}(\omega)$ is the Fourier spectrum of the response and $\mathbf{H}(\omega)$ is the applicable transfer function. The transfer function is the steady-state response per unit sinusoidal input as a function of the frequency ω. If $\mathbf{R}(\omega)$ represents a velocity response and $\mathbf{F}(\omega)$ represents a force input, $\mathbf{H}(\omega)$ is given the special designation of mobility $\mathfrak{M}(\omega)$; the reciprocal of $\mathbf{H}(\omega)$ is the mechanical impedance $\mathbf{Z}(\omega)$. [*Mobility* is sometimes called *mechanical admittance.*] Mechanical impedance and mobility are discussed in detail in Chap. 10. Equation (23.74) is applicable when the velocity and displacement of the structure are zero at $t = 0$. If the initial conditions are not zero, additional terms are required in Eq. (23.74). These terms can be obtained from the governing differential equation. For example, if the differential equation has the form

$$a\ddot{r} + b\dot{r} + cr = f(t) \tag{23.75}$$

the corresponding equation in the frequency domain is found by applying the Fourier transform to both the response $r(t)$ and the excitation $f(t)$:[11, 13]

$$a[-\omega^2\mathbf{R}(\omega) - j\omega r(0+) - \dot{r}(0+)] + b[j\omega\mathbf{R}(\omega) - r(0+)] + c[\mathbf{R}(\omega)] = \mathbf{F}(\omega)$$

Solving for $\mathbf{R}(\omega)$,

$$\mathbf{R}(\omega) = \frac{\mathbf{F}(\omega) + a\dot{r}(0+) + [b + j\omega a]r(0+)}{-\omega^2 a + j\omega b + c}$$

The quantities $r(0+)$ and $\dot{r}(0+)$ are the values of the response and its derivative at $t = 0+$, the plus sign allowing for any discontinuity at $t = 0$. If $r(0+)$ and $\dot{r}(0+)$ are zero, the transfer function $\mathbf{H}(\omega)$ is found from Eq. (23.74):

$$\mathbf{H}(\omega) = \frac{1}{-\omega^2 a + j\omega b + c} \tag{23.76}$$

The response in the time domain may be found by the use of Eq. (23.69):

$$r(t) = \frac{2}{\pi} \int_0^\infty \mathcal{R}[\mathbf{R}(\omega)] \cos \omega t \, d\omega \qquad [t > 0]$$

Substituting for $\mathbf{R}(\omega)$ from Eq. (23.74),

$$r(t) = \frac{2}{\pi} \int_0^\infty \mathcal{R}[\mathbf{H}(\omega)\mathbf{F}(\omega)] \cos \omega t \, d\omega \qquad [t > 0] \tag{23.77}$$

Example 23.1. Consider the response of the single degree-of-freedom structure in Fig. 23.5 to an input acceleration shock $\ddot{u}(t)$. The transfer function for the relative displacement response per unit input sinusoidal acceleration is

$$\mathbf{H}(\omega) = \frac{\bar{\delta}e^{j\omega t}}{e\ddot{U}^{j\omega t}} = \frac{1}{(\omega_n^2 - \omega^2) + j(2\zeta\omega\omega_n)}$$

where $\bar{\delta}$ and $\ddot{U}$ are, respectively, complex amplitudes of relative displacement response and input sinusoidal acceleration at the frequency ω. From Eq. (23.77), the time-history of the response is

$$\delta(t) = \frac{2}{\pi} \int_0^\infty \mathcal{R} \frac{\mathbf{F}(\omega)}{(\omega_n^2 - \omega^2) + j(2\zeta\omega\omega_n)} \cos \omega t \, d\omega \tag{23.78}$$

where $\mathbf{F}(\omega)$ is the Fourier spectrum of $\ddot{u}(t)$.

CALCULATION OF THE EFFECT OF STRUCTURAL LOAD CHANGE ON A SHOCK SOURCE

An important application of the Fourier spectrum representation of a shock concerns the transmission of a shock from one structure to another structure. The com-

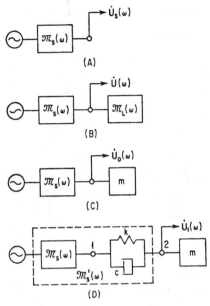

FIGURE 23.15 Schematic representation of a mechanical source composed of a generator and an internal mobility.[16] The source is shown driving (A) no load, (B) a load with input mobility $\mathfrak{M}_L(\omega)$, (C) a mass load, and (D) a resiliently mounted mass.

bination of the structure from which the shock emanates and the process by which the shock is generated may be termed a mechanical shock source.[15] The other structure can be considered a load on this source; any change in the character of the load modifies the shock output of the source.

A schematic representation of a simple mechanical source that is free at its output side is shown in Fig. 23.15A. The mobility of the source determined at its output junction is designated $\mathfrak{M}_s(\omega)$. The Fourier spectrum of the shock velocity delivered by the unloaded or free source is $\dot{U}_s(\omega)$.

A source driving a load having an input mobility $\mathfrak{M}_L(\omega)$ is shown in Fig. 23.15B. The output of the source in this condition is

$$\dot{U}(\omega) = \frac{1}{1 + [\mathfrak{M}_s(\omega)/\mathfrak{M}_L(\omega)]} \dot{U}_s(\omega)$$

$$(23.79)$$

as $\mathfrak{M}_s/\mathfrak{M}_L$ approaches zero; i.e., when the impedance is large compared to the load impedance, the actual output shock approaches the free output of the source.

Example 23.2. Consider the effect of interposing a resilient mounting between a rigid mass and a source that drives the mass.[16] Figure 23.15C represents the direct mounting of a mass m to the source; the mobility of the mass m is $1/j\omega m$ (Chap. 10). Thus, the shock input to the mass is found by substituting in Eq. (23.79):

$$\dot{U}_0(\omega) = \frac{1}{1 + j\omega m \mathfrak{M}_s(\omega)} \dot{U}_s(\omega) \qquad (23.80)$$

Figure 23.15D shows the resiliently mounted mass, the resilient mount being represented by a stiffness k and a viscous damping coefficient c. Since the shock input to the mass is of interest, the output junction of the source can be considered to be moved from point 1 to point 2; i.e., the resilient mount becomes part of the source. The modified source mobility is

$$\mathfrak{M}_s'(\omega) = \mathfrak{M}_s(\omega) + \frac{1}{c + (k/j\omega)}$$

The output of the free source $\dot{U}_s(\omega)$ remains unchanged, and the load mobility remains $1/j\omega m$. The shock input to the mass is

$$\dot{U}_1(\omega) = \frac{1}{1 + j\omega m \mathfrak{M}_s'(\omega)} \dot{U}_s(\omega) \qquad (23.81)$$

MEASUREMENT OF STRUCTURAL TRANSFER FUNCTION

The most straightforward technique for the measurement of a structural transfer function involves the application of a sinusoidal input motion and the measurement of the resulting sinusoidal output motion. The ratio of output to input as a function of frequency is, by definition, the transfer function. It may be desirable to determine the transfer function by applying an input shock and recording the output shock.

Given a structure which is initially at rest, a known shock input $f(t)$ is applied, producing a shock response $r(t)$. The direct application of Eq. (23.74) gives

$$\mathbf{H}(\omega) = \frac{\mathbf{F}(\omega)}{\mathbf{R}(\omega)} \tag{23.82}$$

where $\mathbf{F}(\omega)$ and $\mathbf{R}(\omega)$ are the Fourier spectra of $f(t)$ and $r(t)$, respectively. Accurate values of $\mathbf{H}(\omega)$ are obtained only in those frequency ranges where $f(t)$ has a significant frequency content. If the absolute value of $\mathbf{F}(\omega)$ is zero at certain frequencies, the value of $\mathbf{H}(\omega)$ is indeterminate at those frequencies.

REFERENCES

1. von Kármán, T., and M. A. Biot: "Mathematical Methods in Engineering," p. 388, McGraw-Hill Book Company, Inc., New York, 1940.
2. Rubin, S.: *J. Appl. Mechanics,* **25**:501 (1958).
3. Fung, Y. C., and M. V. Barton: *J. Appl. Mechanics,* **25**:365 (1958).
4. Kelly, R. D., and G. Richman: "Principles and Techniques of Shock Data Analysis," Shock and Vibration Information Center, Washington, D.C., 1969.
5. Walsh, J. P., and R. E. Blake: *Proc. Soc. Exptl. Stress Anal.,* **6**(2):150 (1948).
6. Timoshenko, S., and D. H. Young: "Advanced Dynamics," p. 49, McGraw-Hill Book Company, Inc., New York, 1948.
7. Caughey, T. K., and D. E. Hudson: *Proc. Soc. Exptl. Stress Anal.,* **13**(1):199 (1956).
8. Lunney, E. J., and C. E. Crede, *WADC Tech. Rept.* 57-75, 1958.
9. American National Standards Institute: Methods for Analysis and Presentation of Shock and Vibration Data, ANSI S2.10-1971.
10. Rubin, S.: *Proc. Soc. Exptl. Stress Anal.,* **16**(2):97 (1956).
11. Pipes, L. A.: "Applied Mathematics for Engineers and Physicists," 2d ed., McGraw-Hill Book Company, Inc., New York, 1958.
12. Campbell, G. A., and R. M. Foster: "Fourier Series for Practical Applications," D. Van Nostrand Company, Inc., Princeton, N.J., 1940. Previously published as No. B-584 of the Bell System Series of Monographs, 1931.
13. Weber, E.: "Linear Transient Analysis," vol. II, chap. 1, John Wiley & Sons, Inc., New York, 1956.
14. Erdelyi, A.: "Tables of Integral Transforms," vol. I, McGraw-Hill Book Company, Inc., New York, 1954.
15. Molloy, C. T.: *J. Acoust. Soc. Amer.,* **29**:842 (1957).
16. Rubin, S.: *SAE Trans.,* **68**:318 (1960).

CHAPTER 24
VIBRATION OF STRUCTURES INDUCED BY GROUND MOTION

W. J. Hall

INTRODUCTION

This chapter discusses typical sources of ground motion that affect buildings, the effects of ground motion on simple structures, response spectra, design response spectra (also called *design spectra*), and design response spectra for inelastic systems. The importance of these topics is reflected in the fact that such characterizations normally form the loading input for many aspects of shock-related design, including seismic design. Selected material are presented which are pertinent to the design of resisting systems, for example, buildings designed to meet code requirements related to earthquakes.

GROUND MOTION

SOURCE OF GROUND MOTION

Ground motion may arise from any number of sources such as earthquake excitation[1,2] (described in detail in this chapter), or high explosive[3] or nuclear device detonations.[4] In such cases, the source excitation can lead to major vibration of the primary structure or facility and its many parts, as well as to transient and permanent translation and rotation of the ground on which the facility is constructed. Detonations may result in drag and side-on overpressures, ballistic ejecta, and thermal and radiation effects.

Other sources of ground excitation, although usually not as strong, can be equally troublesome. For example, the location of a precision machine shop near a railroad or highway, or of delicate laboratory apparatus in a plant area containing heavy drop forging machinery or unbalanced rotating machinery are typical of situations in which ground-transmitted vibrations may pose serious problems.

Another different class of vibrational problems arises from excitation of the primary structure by other sources, e.g., wind blowing on a bridge, earthquake excitation of a building, or people walking or dancing on a floor in a building. Vibration of the primary structure in turn can affect secondary elements such as mounted equipment and people located on a floor (in the case of buildings) and vehicles or equipment (in the case of bridges). A brief summary of such people-structure interaction is given in Ref. 5.

The variables involved in problems of this type are exceedingly numerous and, with the exception of earthquakes, few specific well-defined measurements are generally available to serve as a guide in estimating the ground motions that might be used as computational guidelines in particular cases. A number of acceleration-vs.-time curves for typical ground motions arising from the operation of machines and vehicles are shown in Fig. 24.1. Another record arising from a rock quarry blast is shown in Fig. 24.2. Although the records differ somewhat in their characteristics, all can be compared directly with similar measurements of earthquakes, and response computations generally are handled in the same manner.

In most cases, to analyze and evaluate such information one needs to (1) develop an understanding of the source and nature of the vibration, (2) ascertain the physical characteristics of the structure or element, (3) develop an approach for modeling and analysis, (4) carry out the analysis, (5) study the response (with parameter variations if needed), (6) evaluate the behavior of service and function limit states, and (7) develop, in light of the results of the analysis, possible courses of corrective action, if required. Merely changing the mass, stiffness, or damping of the structural system may or may not lead to acceptable corrective action in the sense of a reduction in deflections or stresses; careful investigation of the various alternatives is required to change the response to an acceptable limit. Advice on these matters is contained in Refs. 3, 6, and 7.

RESPONSE OF SIMPLE STRUCTURES TO GROUND MOTIONS

Four structures of varying size and complexity are shown in Fig. 24.3: (A) a simple, relatively compact machine anchored to a foundation, (B) a 15-story building, (C) a 40-story building, and (D) an elevated water tank. The dynamic response of each of the structures shown in Fig. 24.3 can be approximated by representing each as a simple mechanical oscillator consisting of a single mass supported by a spring and a damper as shown in Fig. 24.4. The relationship between the undamped angular frequency of vibration $\omega = 2\pi f$, the natural frequency f, and the period T is defined in terms of the spring constant k and the mass m:

$$\omega^2 = \frac{k}{m} \tag{24.1}$$

$$f = \frac{1}{T} = \frac{\omega}{2\pi} = \frac{1}{2\pi}\sqrt{\frac{k}{m}} \tag{24.2}$$

In general, the effect of the damper is to produce damping of free vibrations or to reduce the amplitude of forced vibrations. The damping force is assumed to be equal to a damping coefficient c times the velocity $\dot{u}$ of the mass relative to the ground. The value of c at which the motion loses its vibratory character in free vibration is called the *critical damping coefficient;* for example, $c_c = 2m\omega$. The amount of damping is most conveniently considered in terms of the fraction of ζ critical damping [see Eq. (2.12)],

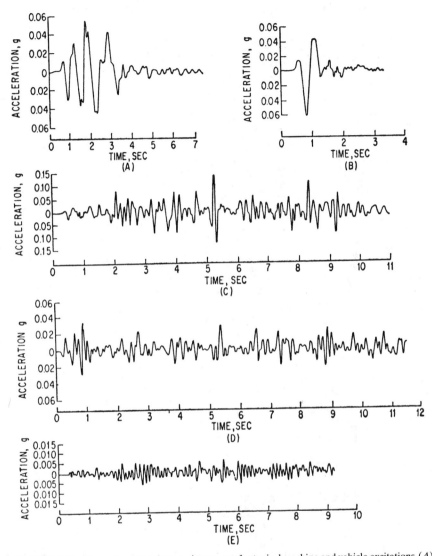

FIGURE 24.1   Ground-acceleration-vs.-time curves for typical machine and vehicle excitations. (*A*) Vertical acceleration measured on a concrete floor on sandy loam soil at a point 6 ft from the base of a drop hammer. (*B*) Horizontal acceleration 50 ft from drop hammer. The weight of the drop hammerhead was approximately 15,000 lb, and the hammer was mounted on three layers of 12- by 12-in. oak timbers on a large concrete base. (*C*) Vertical acceleration 6 ft from a railroad track on the well-maintained right-of-way of a major railroad during passing of luxury-type passenger cars at a speed of approximately 20 mph. The accelerometer was bolted to a 2- by 2-in. by 2½-in. steel block which was firmly anchored to the ground. (*D*) Horizontal acceleration of the ground at 46 ft from the above railroad track, with a triple diesel-electric power unit passing at a speed of approximately 20 mph. (*E*) Horizontal acceleration of the ground 6 ft from the edge of a relatively smooth highway, with a large tractor and trailer unit passing on the outside lane at approximately 35 mph with a full load of gravel.[6]

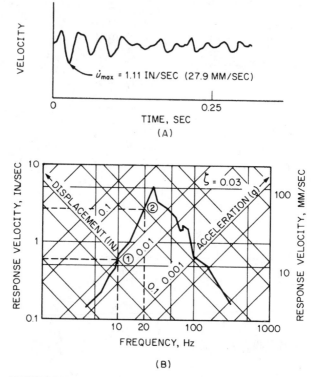

(B)

FIGURE 24.2 Typical quarry blast data. (*A*) Time-history of velocity taken by a velocity transducer and recorder. (*B*) Corresponding response spectrum computed from the record in (*A*) using Duhamel's integral.[3]

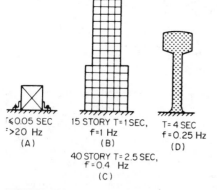

FIGURE 24.3 Structures subjected to earthquake ground motion. (*A*) A machine anchored to a foundation. (*B*) A 15-story building. (*C*) A 40-story building. (*D*) An elevated water tank.

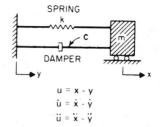

$$u = x - y$$
$$\dot{u} = \dot{x} - \dot{y}$$
$$\ddot{u} = \ddot{x} - \ddot{y}$$

FIGURE 24.4 System definition; the dynamic response of each of the structures shown in Fig. 24.3 can be approximated by this simple mechanical oscillator.

$$\zeta = \frac{c}{c_c} = \frac{c}{2m\omega} \tag{24.3}$$

For most practical structures ζ is relatively small, in the range of 0.005 to 0.2 (i.e., 0.5 to 20 percent), and does not appreciably affect the natural period or frequency of vibration (see Refs. 1b and 8).

EARTHQUAKE GROUND MOTION

Strong-motion earthquake acceleration records with respect to time have been obtained for a number of earthquakes. Ground motions from other sources of disturbance, such as quarry blasting and nuclear blasting, also are available and show many of the same characteristics. As an example of the application of such time-history records, the recorded accelerogram for the El Centro, California, earthquake of May 18, 1940, in the north-south component of horizontal motion is shown in Fig. 24.5. On the same figure are shown the integration of the ground acceleration a to give the variation of ground velocity v with time and the integration of velocity to give the variation of ground displacement d with time. These integrations normally require baseline corrections of various sorts, and the magnitude of the maximum displacement may vary depending on how the corrections are made. The maximum velocity is relatively insensitive to the corrections, however. For this earthquake, with the integrations shown in Fig. 24.5, the maximum ground acceleration is 0.32g, the maximum ground velocity is 13.7 in./sec (35 cm/sec), and the maximum ground

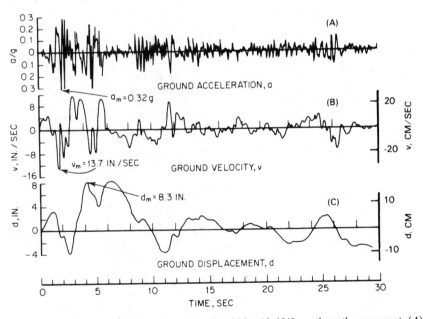

FIGURE 24.5 El Centro, California, earthquake of May 18, 1940, north-south component. (*A*) Record of the ground acceleration. (*B*) Variation of ground velocity v with time, obtained by integration of (*A*). (*C*) Variation of ground displacement with time, obtained by integration of (*B*).

displacement is 8.3 in. (21 cm). These three maximum values are of particular interest because they help to define the response motions of the various structures considered in Fig. 24.3 most accurately if all three maxima are taken into account.

RESPONSE SPECTRA

ELASTIC SYSTEMS

The response of the simple oscillator shown in Fig. 24.4 to any type of ground motion can be readily computed as a function of time. A plot of the maximum values of the response, as a function of frequency or period, is commonly called a *response spectrum* (or *shock spectrum*). The response spectrum may be defined as the graphical relationship of the maximum response of a single degree-of-freedom linear system to dynamic motions or forces. This concept of a response spectrum is widely used in the study of the response of simple oscillators to transient disturbances; for a number of examples, see Chap. 8.

A careful study of Fig. 24.4 will reveal that there are nine quantities represented there: acceleration, velocity, and displacement of the base, mass, and their relative values denoted by u. Commonly the maxima of interest are the maximum deformation of the spring, the maximum spring force, the maximum acceleration of the mass (which is directly related to the spring force when there is no damping), or a quantity having the dimensions of velocity, which provides a measure of the maximum energy absorbed in the spring. The details of various forms of response spectra that can be graphically represented, uses of response spectra, and techniques for computing them are discussed in detail in Refs. 1*b*, 1*c*, and 1*d*. A brief treatment of the applications of response spectra follows. The maximum values of the response are of particular interest. These maxima can be stated in terms of the maximum strain in the spring $u_m = D$, the maximum spring force, the maximum acceleration A of the mass (which is related to the maximum spring force directly when there is no damping), or a quantity, having the dimensions of velocity, which gives a measure of the maximum energy absorbed in the spring. This quantity, designated the pseudo velocity V, is defined in such a way that the energy absorption in the spring is $\frac{1}{2}mV^2$. The relations among the maximum relative displacement of the spring D, the pseudo velocity V, and the pseudo acceleration A, which is a measure of the force in the spring, are

$$V = \omega D, \tag{24.4}$$

and
$$A = \omega V = \omega^2 D \tag{24.5}$$

The pseudo velocity V is nearly equal to the maximum relative velocity for systems with moderate or high frequencies but may differ considerably from the maximum relative velocity for very low frequency systems. The pseudo acceleration A is exactly equal to the maximum acceleration for systems with no damping and is not greatly different from the maximum acceleration for systems with moderate amounts of damping, over the whole range of frequencies from very low to very high values.

Typical plots of the response of the system to base excitation, as a function of period or frequency, are called *response spectra* (also called *shock spectra*). Plots for acceleration and for relative displacement, for a system with a moderate amount of damping and subjected to an input similar to that of Fig. 24.5, can be made. This arithmetic plot of maximum response is simple and convenient to use. Various techniques of computing and plotting spectra may be found in the references cited at the end of this chapter, especially in Refs. 1c, 1d, and 6 to 18.

A somewhat more useful plot, which indicates the values for D, V, and A, is shown in Fig. 24.6. This plot has the virtue that it also indicates more clearly the extreme or limits of the various parameters defining the response. All parameters are plotted on a logarithmic scale. Since the frequency is the reciprocal of the period, the logarithmic scale for the period would have exactly the same spacing of the points, or in effect the scale for the period would be turned end for end. The pseudo velocity is plotted on a vertical scale. Then on diagonal scales along an axis that extends upward from right to left are plotted values of the displacement, and along an axis that extends upward from left to right the pseudo acceleration is plotted, in such a way that any one point defines for a given frequency the displacement D, the pseudo velocity V, and the pseudo acceleration A. Points are indicated in Fig. 24.6 for the several structures of Fig. 24.3 plotted at their approximate fundamental frequencies. Many other formats are used in plotting spectra; for example, u, $\dot{u}$, ωu, or $\ddot{x}$ vs. time. Such examples are shown in Ref. 1d.

Much of the work on spectra, described above, has been developed on the basis of studying strong ground motion categorized by ground motion acceleration level scaling. Another important aspect of statistical study, described in Ref. 19, concerns both ground motions and spectra based on magnitude scaling.

In developing spectral relationships a wide variety of motions have been considered (see Ref. 20), ranging from simple pulses of displacement, velocity, or acceleration of the ground, through more complex motions such as those arising from nuclear-blast detonations, and for a variety of earthquakes as taken from available

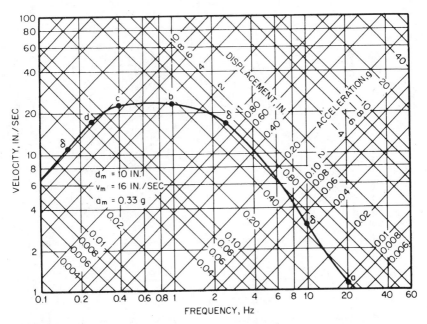

FIGURE 24.6   Smooth response spectrum for typical earthquake.

strong-motion records. Response spectra for the El Centro earthquake are shown in Fig. 24.7. The spectrum for small amounts of damping is much more jagged than indicated by Fig. 24.6, but for the higher amounts of damping the response curves are relatively smooth. The scales are chosen in this instance to represent the amplifications of the response relative to the ground-motion values of displacement, velocity, or acceleration.

The spectra shown in Fig. 24.7 are typical of response spectra for nearly all types of ground motion. On the extreme left, corresponding to very low-frequency systems, the response for all degrees of damping approaches an asymptote corresponding to the value of the maximum ground displacement. A low-frequency system corresponds to one having a very heavy mass and a very light spring. When the ground moves relatively rapidly, the mass does not have time to move, and therefore the maximum strain in the spring is precisely equal to the maximum displacement of the ground. For a very high-frequency system, the spring is relatively stiff and the mass very light. Therefore, when the ground moves, the stiff spring forces the mass to move in the same way the ground moves, and the mass therefore must have the same acceleration as the ground at every instant. Hence, the force in the spring is that required to move the mass with the same acceleration as the ground, and the maximum acceleration of the mass is precisely equal to the maximum acceleration of the ground. This is shown by the fact that all the lines on the extreme right-hand side of the figure asymptotically approach the maximum ground-acceleration line.

For intermediate-frequency systems, there is an amplification of motion. In general, the amplification factor for displacement is less than that for velocity, which in turn is less than that for acceleration. Peak amplification factors for the undamped system ($\zeta = 0$) in Fig. 24.7 are on the order of about 3.5 for displacement, 4.2 for velocity, and 9.5 for acceleration.

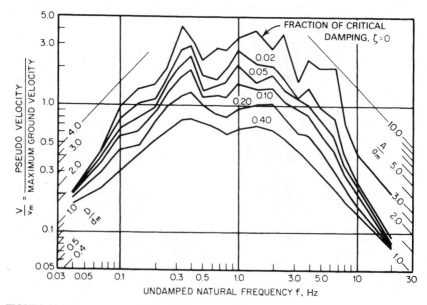

FIGURE 24.7 Response spectra for elastic systems subjected to the El Centro earthquake for various values of fraction of critical damping ζ.

The results of similar calculations for other ground motions are quite consistent with those in Fig. 24.7, even for simple motions. The general nature of the response spectrum shown in Fig. 24.8 consists of a central region of amplified response and two limiting regions of response in which for low-frequency systems the response displacement is equal to the maximum ground displacement and for high-frequency systems the response acceleration is equal to the maximum ground acceleration. Values of the amplification factor reasonable for use in design are presented in the next sections.

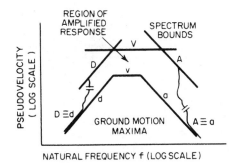

FIGURE 24.8 Typical tripartite logarithmic plot of response-spectrum bounds compared with maximum ground motion.

DESIGN RESPONSE SPECTRA

A response spectrum developed to give design coefficients is called a *design response spectrum* or a *design spectrum*. As an example of its use in seismic design, for any given site, estimates are made of the maximum ground acceleration, maximum ground velocity, and maximum ground displacement. The lines representing these values can be drawn on the tripartite logarithmic chart of which Fig. 24.9 is an example. The heavy lines showing the ground-motion maxima in Fig. 24.9 are drawn for a maximum ground acceleration a of 1.0g, a velocity v of 48 in./sec (122 cm/sec), and a displacement d of 36 in. (91.5 cm). These data represent motions more intense than those generally considered for any postulated design earthquake hazard. They are, however, approximately in correct proportion for a number of areas of the world, where earthquakes occur either on firm ground, soft rock, or competent sediments of various kinds. For relatively soft sediments, the velocities and displacements might require increases above the values corresponding to the given acceleration as scaled from Fig. 24.9, and for competent rock, the velocity and displacement values would be expected to be somewhat less. More detail can be found in Refs. 1c and d. It is not likely that maximum ground velocities in excess of 4 to 5 ft/sec (1.2 to 1.5 m/sec) are obtainable under any circumstances.

On the basis of studies of horizontal and vertical directions of excitation for various values of damping (Refs. 1c, 10, and 11), representative amplification factors for the 50th and 84.1th percentile levels of horizontal response are presented in Table 24.1. The 84.1th percentile means that one could expect 84.1 percent of the values to fall at or below that particular amplification. With these amplification fac-

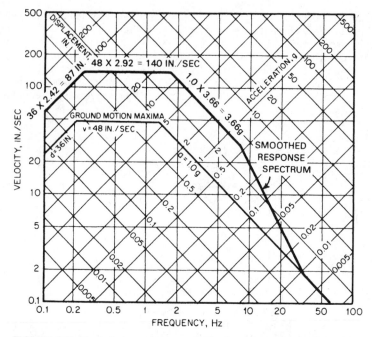

FIGURE 24.9 Basic design spectrum normalized to 1.0g for a value of damping equal to 2 percent of critical, 84,1th percentile level. The spectrum bound values are obtained by multiplying the appropriate ground-motion maxima by the corresponding amplification value of Table 24.1.

tors and noting points B and A to fall at about 8 and 33 Hz, the spectra may be constructed as shown in Fig. 24.9 by multiplying the ground maxima values of acceleration, velocity, and displacement by the appropriate amplification factors. Further information on, and other approaches to, construction of design spectra may be found in Refs. 1c and d.

TABLE 24.1 Values of Spectrum Amplification Factors[11]

Percentile	Damping, percent of critical damping	Amplification factor		
		D	V	A
50th	0.5	2.01	2.59	3.68
	2.0	1.63	2.03	2.74
	5.0	1.39	1.65	2.12
	10.0	1.20	1.37	1.64
84.1th	0.5	3.04	3.84	5.10
	2.0	2.42	2.92	3.66
	5.0	2.01	2.30	2.71
	10.0	1.69	1.84	1.99

RESPONSE SPECTRA FOR INELASTIC SYSTEMS

It is convenient to consider an elastoplastic resistance-displacement relation because one can draw response spectra for such a relation in generally the same way as the spectra were drawn for elastic conditions. A simple resistance-displacement relationship for a spring is shown by the light line in Fig. 24.10A, where the yield point is indicated, with a curved relationship showing a rise to a maximum resistance and then a decay to a point of maximum useful limit or failure at a displacement u_m; an equivalent elastoplastic resistance curve is shown by the heavy line. A similar elastoplastic resistance function, more indicative of seismic response, is shown in Fig. 24.10B. The ductility factor μ is defined as the ratio between the maximum permissible or useful displacement to the yield displacement for the effective curve in both cases.

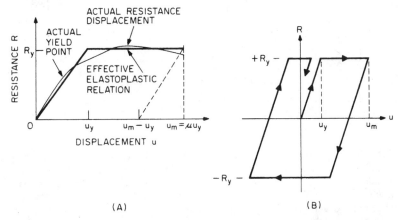

FIGURE 24.10 (A) Monotonic resistance-displacement relationships for a spring, shown by the light line; an equivalent elastoplastic resistance curve, shown by the heavy line. (B) A similar elastoplastic resistance function, more indicative of seismic response.

The ductility factors for various types of construction depend on the use of the building, the hazard involved in its failure (assumed acceptable risk), the material used, the framing or layout of the structure, and above all on the method of construction and the details of fabrication of joints and connections. A discussion of these topics is given in Refs. 1c, 10, and 11. Figure 24.11 shows acceleration spectra for elastoplastic systems having 2 percent of critical damping that were subjected to the El Centro, 1940, earthquake. Here the symbol D_y represents the elastic component of the response displacement, but it is not the total displacement. Hence, the curves also give the elastic component of maximum displacement as well as the maximum acceleration A, but they do not give the proper value of maximum pseudo velocity. This is designated by the use of the V' for the pseudo velocity drawn in the figure. The figure is drawn for ductility factors ranging from 1 to 10.

A response spectrum for total displacement also can be drawn for the same conditions as for Fig. 24.11. It is obtained by multiplying each curve's ordinates by the value of ductility factor μ shown on that curve.

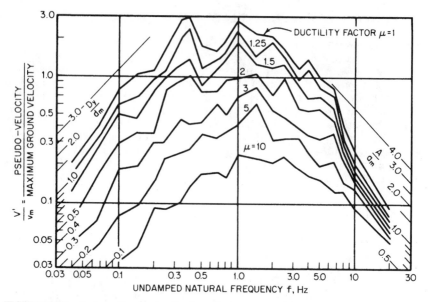

FIGURE 24.11 Deformation spectra for elastoplastic systems with 2 percent of critical damping that were subjected to the El Centro earthquake.

The following considerations are useful in using the design spectrum to approximate inelastic behavior. In the amplified displacement region of the spectra, the left-hand side, and in the amplified velocity region, at the top, the spectrum remains unchanged for total displacement and is divided by the ductility factor to obtain yield displacement or acceleration. The upper right-hand portion sloping down at 45°, or the amplified acceleration region of the spectrum, is relocated for an elastoplastic resistance curve, or for any other resistance curve for actual structural materials, by choosing it at a level which corresponds to the same energy absorption for the elasto-plastic curve as for an elastic curve for the same period of vibration. The extreme right-hand portion of the spectrum, where the response is governed by the maximum ground acceleration, remains at the same acceleration level as for the elastic case and, therefore, at a corresponding increased total displacement level. The frequencies at the corners are kept at the same values as in the elastic spectrum. The acceleration transition region of the response spectrum is now drawn also as a straight-line transition from the newly located amplified acceleration line and the ground-acceleration line, using the same frequency points of intersection as in the elastic response spectrum. In all cases the inelastic maximum acceleration spectrum and the inelastic maximum displacement spectrum differ by the factor μ at the same frequencies. The design spectrum so obtained is shown in Fig. 24.12.

The solid line $DVAA_0$ shows the elastic response spectrum. The heavy circles at the intersections of the various branches show the frequencies which remain constant in the construction of the inelastic design spectrum. The dashed line $D'V'A'A_0$ shows the inelastic acceleration, and the line $DVA''A_0''$ shows the inelastic displacement. These two differ by a constant factor μ for the construction shown, except that A and A' differ by the factor $\sqrt{2\mu - 1}$, since this is the factor that corresponds to constant energy for an elastoplastic resistance.

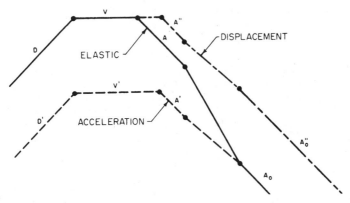

FIGURE 24.12 The normal elastic design spectrum is given by $DVAA_0$. The modified spectrum (see text for rules for construction) representing approximately the acceleration or elastic yield displacement for a nonlinear system with ductility μ is given by $D'V'A''A_0$. The total or maximum displacement for the nonlinear system is given by $DVA''A'_0$ and is obtained by multiplying the modified spectrum by the value μ.

The modified spectrum to account for inelastic action is an approximation at best and should be used generally only for relatively small ductility values, for example, 5 or less. Additional information on the development of elastic and inelastic design response spectra may be found in Refs. 1c, 1d, and 10 to 21.

MULTIPLE DEGREE-OF-FREEDOM SYSTEMS

USE OF RESPONSE SPECTRA

A multiple degree-of-freedom system has as many modes of vibration as the number of degrees-of-freedom. For example, for the shear beam shown in Fig. 24.13A the fundamental mode of lateral oscillation is shown in (B), the second mode in (C), and the third mode in (D). The number of modes in this case is 5. In a system that has independent (uncoupled) modes (this condition is often satisfied for buildings) each mode responds to the base motion as an independent single degree-of-freedom system. Thus, the modal responses are nearly independent functions of time. However, the maxima do not necessarily occur at the same time.

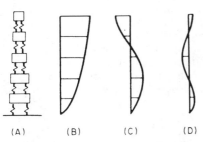

(A)   (B)   (C)   (D)

FIGURE 24.13 Modes of vibration of shear beam. The first three (1, 2, 3) relative mode shapes are shown by (B), (C), and (D), respectively, for lateral vibration.

For multiple degree-of-freedom systems, the concept of the response spectrum can also be used in most cases, although the use of the inelastic response spectrum is only approximately valid as a design procedure.[1c,10,11] For a system with a number of masses at nodes in a flexible framework, the equation of motion can be written in matrix form as

$$M\ddot{u} + C\dot{u} + Ku = -M(\ddot{y})\{1\} \tag{24.6}$$

in which the last symbol on the right represents a unit column vector. The mass matrix M is usually diagonal, but in all cases both M and the stiffness matrix K are symmetrical. When the damping matrix C satisfies certain conditions, the simplest of which is when it is a linear combination of M and K, then the system has normal modes of vibration, with modal displacement vectors u_n.

Analysis techniques for handling multiple degree-of-freedom systems are described in Ref. 8.

DESIGN

GENERAL CONSIDERATIONS

The design of all types of building structures, as well as the design of building services (such as water, gas, fuel pipelines, water and electrical services, sewage, and vertical transportation) must take into account the effects of earthquakes and wind. Often, these building services are large, expensive, and affect large numbers of people. Thus, the design of a building should consider siting studies to minimize seismic effects or, at very least, identify such effects that must be expected to be accommodated, included faulting; all this must be taken into account, in addition to the usual considerations of functional needs, economics, land acquisition and land use restrictions, transportation, and the availability of labor.

From a design perspective, there must be a rational selection of the applicable loadings (demand)—preferably, examination of the design for a range of loadings, load combinations, and load paths, in order to assess margins of safety—as well as careful attention to modeling and analysis. From the resistance (supply) side, careful attention must be given to the properties of the materials, to connections of structural members and items, as well as to the joining process, to foundations and anchorage, to provisions for controlling ductility and handling transient displacements, to aging considerations, and to the meeting or exceeding applicable code requirements, specifications, and regulations—all in accordance with appropriate professional standards of care and good engineering judgment.

In the design of a building to resist earthquake motions, the designer works within certain constraints, such as the architectural configuration of the building, the foundation conditions, the nature and extent of the hazard should failure or collapse occur, the possibility of an earthquake, the possible intensity of earthquakes in the region, the cost or available capital for construction, and similar factors. There must be some basis for the selection of the strength and the proportions of the building and of the various members in it. The required strength depends on factors such as the intensity of earthquake motions to be expected, the flexibility of the structure, and the ductility or reserve strength of the structure before damage occurs. Because of the interrelations among the flexibility and strength of a structure and the forces generated in it by earthquake motions, the dynamic design procedure must take these various factors into account. The ideal to be achieved is one involving flexibility and energy-absorbing capacity which will permit the earthquake displacements to take place without generating unduly large forces. To achieve this end, careful design (with attention to continuity, redundancy, connections, strength, and ductility), control of the construction procedures, and appropriate inspection practices are necessary. The attainment of the ductility required to resist earthquake motions must be emphasized. If the ductility achieved is less than assumed, then in all likelihood the forces in the structure will be higher than estimated.

The above considerations emphasize the importance of a knowledge of structural behavior and the uncertainties associated therewith, and techniques for assessing and implementing appropriate margins of safety in design. In earthquake engineering design, careful consideration must be given to the cyclic behavior that normally occurs, as opposed to monotonic behavior. Because of this severe cyclic demand on the structural framing and its connections (irrespective of whether or not they are made of reinforced or prestressed concrete or of steel), it is important to consider the strength characteristics of the particular materials and sections as they are joined, including bracing; it is necessary to ensure that the demand for limited ductility can be achieved in a satisfactory manner. Earthquakes throughout the world in the first half of the 1990s have shown that certain design assumptions and accompanying fabrication techniques have led to severely decreased strength margins in some cases and/or to serious structural damage. Life safety is the primary matter of concern, but increasingly building owners are more conscious of protecting their plant investment and to preserving production operations without major repair and "down time." Thus the building owner and engineering designer must come to an agreement as to the level of protection desired, based on current knowledge and applicable conditions.

Some typical references for structures, lifelines, and transportation systems (including observation summaries of major earthquakes) are given in Refs. 22 to 36. In addition to these sources, guidelines and regulations are available from associations of manufacturers or major suppliers of steel, concrete, prestressed concrete, masonry, and wood.

EFFECTS OF DESIGN ON BEHAVIOR AND ON ANALYSIS*

A structure designed for very much larger horizontal forces than are ordinarily prescribed will have a shorter period of vibration because of its greater stiffness. The shorter period results in higher spectral accelerations, so that the stiffer structure may attract more horizontal force. Thus, a structure designed for too large a force will not necessarily be safer than a similar structure based on smaller forces. On the other hand, a design based on too small a force makes the structure more flexible and will increase the relative deflections of the floors.

In general, yielding occurs first in the story that is weakest compared with the magnitudes of the shearing forces to be transmitted. In many cases this will be near the base of the structure. If the system is essentially elastoplastic, the forces transmitted through the yielded story cannot exceed the yield shear for that story. Thus, the shears, accelerations, and relative deflections of the portion of the structure above the yielded floor are reduced compared with those for an elastic structure subjected to the same base motion. Consequently, if a structure is designed for a base shear which is less than the maximum value computed for an elastic system, the lowest story will yield and the shears in the upper stories will be reduced. This means that, with proper provision for energy absorption in the lower stories, a structure

---

* This section is based partly on material from Ref. 37, by permission, with update modification.

will, in general, have adequate strength, provided the design shearing forces for the upper stories are consistent with the design base shear. The Uniform Building Code (UBC) recommendations are intended to provide such a consistent set of shears. However, on all levels it is wise to have the energy absorption, if possible, distributed more or less uniformly throughout the structural system, i.e., not concentrated only in a few locations; such a procedure places an unusual, and quite often unbalanced, demand on localized and specific portions of a structure.

A significant inelastic deformation in a structure inhibits the higher modes of oscillation. Therefore, the major deformation is in the mode in which the inelastic deformation predominates, which is usually the fundamental mode. The period of vibration is effectively increased, and in many respects the structure responds almost as a single degree-of-freedom system corresponding to its entire mass supported by the story which becomes inelastic. Therefore, the base shear can be computed for the modified structure, with its fundamental period defining the modified spectrum on which the design should be based. The fundamental period of the modified structure generally will not be materially different from that of the original elastic structure in the case of framed structures. In the case of shear-wall structures it will be longer.

It is partly because of these facts that it is usual in design recommendations to use the frequency of the fundamental mode, without taking direct account of the higher modes. However, it is desirable to consider a shearing-force distribution which accounts for higher-mode excitations of the portion above the plastic region. This is implied in the UBC, SEAOC (Structural Engineers Association of California), and National Earthquake Hazard Reduction Program (NEHRP) recommendations by the provision for lateral-force coefficients which vary with height. The distribution over the height corresponding to an acceleration varying uniformly from zero at the base to a maximum at the top takes into account the fact that local accelerations at higher levels in the structure are greater than those at lower levels, because of the larger motions at the higher elevations, and accounts quite well for the moments and shears in the structure.

Many of the modern seismic analysis approaches are described in detail in Ref. 8. Prevailing analysis techniques employ design spectra or motion time-histories as input. Many benchmarked computer software packages are available that permit fairly sophisticated structural analyses to be undertaken, especially when the modeling is carefully studied and well understood and the input is relatively well defined. Typical of these powerful programs are ETABS, SAP 80, ABAQUS, ANSYS, and ADINA. In the field of soil-structure interaction, computer software packages include SASSI, CLASSI, FLUSH, and SHAKE. Since all such programs are constantly being upgraded, it is necessary to keep abreast of such modifications.

In the case of intense earthquakes, the ensuing ground motions can be of the sharp, impulsive type. When such ground motions impinge on a structure, the effect is literally that of a shock. Moreover, the impulses can be multiple in nature, so that if the timing between impulses is quite short, the rapid shock-type motion transmitted to building frames may be intensified. Such an intense form of impulsive input has been observed in earthquakes in Northridge, California and in Kobe, Japan; it may lead to serious structural problems in buildings if such input has not been properly considered in the building's design and construction. Although not explicitly spelled out in present building codes, it is expected that a strength check would be carried out to see that the gross building shearing resistance is sufficient (including normal margins of strength) to resist an intense shock characterized by the zero period acceleration (ZPA); in addition, structural members must have ample tensile and compressive resistance so that they are able to resist a vertical or oblique type of shock. This intense type of input subsequently leads to the vibratory type of motion that is commonly treated in seismic analysis. Fortunately, in most earthquakes, the initial motions that lead to building vibration are small enough to be accommodated by the resistance of most buildings.

The strength checks, referred to above, have nothing to do with the principal modes of vibration of a building as determined by analysis; in reality, the structure or piece of equipment is initially at rest; then it must respond in a quasi-rigid mode to these intense impulses. In that sense the entire mass of the building is active in providing resistance. The forces under those circumstances can be quite high. However, in some cases where the design calls for the lateral and vertical forces to be carried in just a few frames or members, the imparted forces can be immense. Fortunately, most buildings have ample resistance to accommodate such effects—especially if the base anchorage and connections are well constructed for a requisite set of structural frames. Similarly, most equipment that is properly mounted has more than enough margin of strength to accommodate the imposed intense dynamic loading. Analysis of earthquake damage, with regard to difficulties with connections and details in both steel and concrete structures, suggests that adequate attention is required in the design of details, in the quality of their fabrication, and in the quality of their construction in order to assure their adequate performance. In this respect, Ref. 36 concerned with the quality of construction is pertinent.

DESIGN LATERAL FORCES

Although the complete response of multiple degree-of-freedom systems subjected to earthquake motions can be calculated, it should not be inferred that it is generally necessary to make such calculations as a routine matter in the design of multistory buildings. There are a great many uncertainties about the input motions and about the structural characteristics that can affect the computations. Moreover, it is not generally necessary or desirable to design tall structures to remain completely elastic under severe earthquake motions, and considerations of inelastic behavior lead to further discrepancies between the results of routine methods of calculation and the actual response of structures.

The Uniform Building Code recommendations, with proper attention to the R and S values, for earthquake lateral forces are, in general, consistent with the forces and displacements determined by more elaborate procedures. A structure designed according to these recommendations will remain elastic, or nearly so, under moderate earthquakes of frequent occurrence, but it must be able to yield locally without serious consequences if it is to resist a major earthquake. Thus, design for the required ductility is an important consideration.

The ductility of the material itself is not a direct indication of the ductility of the structure. Laboratory and field tests and data from operational use of nuclear weapons indicate that structures of practical configurations having frames of ductile materials, or a combination of ductile materials, exhibit ductility factors μ ranging from a minimum of 3 to a maximum of 8. For a quality constructed structure with well-distributed energy absorption a ductility factor of about 4 to 6 is a reasonable criterion when designed to UBC earthquake requirements.

As a result of the numerous earthquakes that have occurred throughout the world and of the resulting loss of life and property, seismic design codes have undergone major revisions to reflect modern understanding of dynamic design, based on research, and to reflect lessons learned in recent damaging earthquakes. From the 1994 Uniform Building Code (UBC), the base shear is given by

$$V = \frac{ZICW}{R_W} \qquad (24.7)$$

where Z = seismic zone factor, a numerical coefficient related to expected
severity of earthquakes in various regions of the USA

  I = importance factor of the occupancy (1.0 for nonessential facilities,
1.25 for essential facilities, and 1.5 for life safety facilities)

  W = total seismic dead load

  R_W = numerical coefficient (reflecting structural parameters)

  C = numerical coefficient given by Eq. (24.8), below

and C is given by

$$C = \frac{1.25S}{T^{2/3}} \tag{24.8}$$

where S = site coefficient for soil conditions

  T = fundamental period of vibration of the structure

The complexity of any such modern code requires that the provisions, along with the commentary, be studied in detail prior to performing the above computations.

In general the seismic coefficients have been increased in comparison to earlier values, and the approaches being adopted attempt to take more factors into consideration in arriving at the design base shear.

SEISMIC FORCES FOR OVERTURNING MOMENT AND SHEAR DISTRIBUTION

In general when modal analysis techniques are not used, in a complex structure or in one having several degrees-of-freedom, it is necessary to have a method of defining the seismic design forces at each mass point of the structure in order to be able to compute the shears and moments to be used for design throughout the structure. The method described in the SEAOC, UBC, or NEHRP provisions is preferable for this purpose. Obviously, the proper foundations, and adequate anchorage, are required.

DAMPING

The damping in structural elements and components and in supports and foundations of the structure is a function of the intensity of motion and of the stress or strain levels introduced within the structural component or structure and is highly dependent on the makeup of the structure and the energy absorption mechanisms within it. For further details see Refs. 1 and 12.

GRAVITY LOADS

The effect of gravity loads, when the structures deform laterally by a considerable amount, can be of importance. In accordance with the general recommendations of most extant codes, the effects of gravity loads are to be added directly to the primary and earthquake effects. In general, in computing the effect of gravity loads, one must take into account the actual deflection of the structure, not the deflection corresponding to reduced seismic coefficients.

VERTICAL AND HORIZONTAL EXCITATION

Usually the stresses or strains at a particular point are affected primarily by the earthquake motions in only one direction; the second direction produces little if any influence. However, this is not always the case and is certainly not so for a simple square building supported on four columns where the stress in a corner column is in general affected equally by the earthquakes in the two horizontal directions and may be affected also by the vertical earthquake forces. Since the ground moves in all three directions in an earthquake, and even tilts and rotates, consideration of the combined effects of all these motions must be included in the design. When the response in the various directions may be considered to be uncoupled, consideration can be given separately to the various components of base motion, and individual response spectra can be determined for each component of direction or of transient base displacement. Calculations have been made for the elastic response spectra in all directions for a number of earthquakes. Studies indicate that the vertical response spectrum is about two-thirds the horizontal response spectrum, and it is recommended that a ratio of 2:3 for vertical response compared with horizontal response be used in design. If there are systems or elements that are particularly sensitive to vertical shock, these will require special design consideration.

For parts of structures or components that are affected by motions in various directions in general, the response may be computed by either one of two methods. The first method involves computing the response for each of the directions independently and then taking the square root of the sums of the squares of the resulting stresses in the particular direction at a particular point as a combined response. Alternatively, one can use the second method of taking the seismic forces corresponding to 100 percent of the motion in one direction combined with 40 percent of the motions in the other two orthogonal directions, adding the absolute values of the effects of these to obtain the maximum resultant forces in a member or at a point in a particular direction, and computing the stresses corresponding to the combined effects. In general, this alternative method is slightly conservative.

A related matter that merits attention in design is the provision for relative motion of parts or elements having supports at different locations.

UNSYMMETRICAL STRUCTURES IN TORSION

In design, consideration should be given to the effects of torsion on unsymmetrical structures and even on symmetrical structures where torsions may arise from off-center loads and accidentally because of various reasons, including lack of homogeneity of structures or the presence of the wave motions developed in earthquakes. Most modern codes provide values of computed and accidental eccentricity to use in design, but in the event that analyses indicate values greater than those recommended by the code, the analytical values should be used in design.

SIMULATION TESTING

Simulation testing to create various vibration environments has been employed for years in connection with the development of equipment that must withstand vibration. Over the years such testing of small components has been accomplished on shake tables and involves many different types of input functions. As a result of improved development of electromechanical rams, large shake tables have been developed which can simulate the excitation that may be experienced in a building, structural component, or items of equipment, from various types of ground motions, including earthquake motions, nuclear ground motions, nuclear blast motions induced in the ground or in a structure, and traffic vibrations. Some of these devices are able to provide simultaneous motion in three orthogonal directions. For larger items analysis may be the tool available for assessment of adequacy, coupled with physical observation during transport.

The matter of simulation testing became of great importance with regard to earthquake excitation because of the development of nuclear power plants and the necessity for components in these plants to remain operational for purposes of safe shutdown and containment and also because of the observed loss of lifeline items in recent earthquakes as, for example, communication and control equipment, utilities, and fire-fighting systems. It is common to require computation of floor response spectra[21] and to provide for equipment qualification.

EQUIPMENT AND LIFELINES

No introduction to earthquake engineering would be complete without mention of the importance of adequate design of equipment in buildings and essential building services, including, for example, communications, water, sewage and transportation systems, gas and liquid fuel pipelines, and other critical facilities. Design approaches for these important elements of constructed facilities, as well as sources of energy, have received major design attention in recent years as the importance of maintaining their integrity has become increasingly apparent.

It has always been obvious that the seismic design of equipment was important, but the focus on nuclear power has pushed this technology to the forefront. Many standards and documents are devoted to the design of such equipment. As a starting point for gaining information about such matters, the reader is referred to Refs. 38 through 40.

Design considerations for critical industrial facilities, meaning those industries that require less attention than a nuclear power plant, but more than a routine building, are discussed in Ref. 41.

A comprehensive summary of knowledge related to the design of essential building services is given in Ref. 35.

REFERENCES

1. Earthquake Engineering Research Institute Monograph Series, Berkeley, Calif. (1979–83).
    (*a*) Hudson, D. E.: "Reading and Interpreting Strong Motion Accelerograms."
    (*b*) Chopra, A. K.: "Dynamics of Structures—A Primer."
    (*c*) Newmark, N. M., and W. J. Hall: "Earthquake Spectra and Design."
    (*d*) Housner, G. W., and P. C. Jennings: "Earthquake Design Criteria."
    (*e*) Seed, H. B., and I. M. Idriss: "Ground Motions and Soil Liquefaction During Earthquakes."
    (*f*) Berg, G. V.: "Seismic Design Codes and Procedures."
    (*g*) Algermission, S. T.: "An Introduction to the Seismicity of the United States."
2. Bolt, B. A.: "Earthquake," W. H. Freeman and Co., San Francisco, Calif., 1988.
3. Dowding, C. H.: "Blast Vibration Monitoring and Control," Prentice-Hall, Inc., Englewood Cliffs, N.J., 1985.

4. Glasstone, S., and P. J. Dolan: "The Effects of Nuclear Weapons," 3d ed., U.S. Dept. of Defense and U.S. Dept. of Energy, 1977.

5. Chang, F.-K.: "Psychophysiological Aspects of Man-Structure Interaction," in "Planning and Design of Tall Buildings," vol. 1a: "Tall Building Systems and Concepts," American Society of Civil Engineers, New York, N.Y., 1972.

6. Hudson, D. E.: "Vibration of Structures Induced by Seismic Waves," in C. M. Harris and C. E. Crede (eds.), "Shock and Vibration Handbook," Chap. 50, vol. III, McGraw-Hill Book Company, New York, 1961.

7. Richart, F. E., Jr., J. R. Hall, Jr., and R. D. Woods: "Vibration of Soils and Foundation," Prentice-Hall, Inc., Englewood Cliffs, N.J., 1970.

8. Chopra, A. K.: "Dynamics of Structures," Prentice-Hall, Inc., Englewood Cliffs, N. J. 1995.

9. Veletsos, A. S., N. M. Newmark, and C. V. Chelapati: *Proc. 3d World Congr. Earthquake Eng.,* New Zealand, **2**:II–663 (1965).

10. Newmark, N. M., and W. J. Hall: "Development of Criteria for Seismic Review and Selected Nuclear Power Plant," *U.S. Nuclear Regulatory Commission Report* NUREG-CR-0098, 1978.

11. Hall, W. J.: *Nuclear Eng. Des.,* **69**:3 (1982).

12. Newmark, N. M., J. A. Blume, and K. K. Kapur: *J Power Div. Am. Soc. Civil Engrs.,* **99**(PO2):287 (November 1973). [See also USNRC Reg. Guides 1.60 and 1.61, 1973]

13. Newmark, N. M., and W. J. Hall: *Proc. 4th World Conf. Earthquake Eng.,* Santiago, Chile, **II**:B4–37 (1969).

14. Newmark, N. M.: *Nucl. Eng. Des.,* **20**(2):303 (July 1972).

15. Riddell, R., and N. M. Newmark: *Proc. 7th World Conf. Earthquake Engineering,* vol. 4 (1980). (See also *Univ. of Ill. Civil Eng. Struct. Res. Report* No. 468, 1979.)

16. Nau, J. M., and W. J. Hall: *J. Struct. Eng.,* **110**:7 (1984).

17. Zahrah, T. F., and W. J. Hall: *J. Struct. Eng.,* **110**:8 (1984).

18. *Proceedings of the 1st through 10th World Conferences on Earthquake Engineering,* International Association for Earthquake Engineering, Tokyo, Japan (1956, 1960, 1965, 1969, 1974, 1977, 1980, 1984, 1988, 1992).

19. Boore, D. M., W. B. Joyner, and T. E. Fumal: "Estimation of Response Spectra and Peak Accelerations from Western North American Earthquakes: An Interim Report," USGS Open-File Report 93-509, 1993.

20. Harris, C. M.: "Shock and Vibration," 3d ed., McGraw Hill Book Company, New York, 1988. [See also 1st (1961) and 2d (1976) ed.]

21. Stevenson, J. D., W. J. Hall, et al.: "Structural Analysis and Design of Nuclear Plant Facilities," American Society of Civil Engineers, *Manuals and Reports on Engineering Practice* No. 58, 1980.

22. O'Rourke, T. D., ed.: "The Loam Prieta, California, Earthquake of October 17, 1989—Marina District," USGS Prof. Paper 1551-F, 1992.

23. Hall, J. F., ed.: "Northridge Earthquake—January 17, 1994," EERI, Oakland, Calif., 1994.

24. Reiter, L.: "Earthquake Hazard Analysis," Columbia University Press, New York, 1990.

25. "Uniform Building Code—1994 Edition," *International Conference of Building Officials,* Whittier, Calif., 1994.

26. "Recommended Lateral Force Requirements and Commentary," Structural Engineers Association of California, 1990.

27. Building Seismic Safety Council: "NEHRP Recommended Provisions for the Development of Seismic Regulations for New Buildings," Wash., D.C., 1991.

28. Naeim, F., ed.: "The Seismic Design Handbook," Van Nostrand Reinhold, New York, 1989.

29. "Seismic Provisions for Structural Steel Buildings," AISC, Chicago, Ill., 1992.

30. "Minimum Design Loads for Buildings and Other Structures," ASCE 7-93, 1994.

31. "Technical Manual," Army TM5-809-10, 1982; also, "Seismic Design Guidelines for Essential Buildings," Army TM5-809-10-1, 1986.

32. "Standard Specification for Seismic Design of Highway Bridges," AASHTO, 1983/1991.

33. "ATC-6 Seismic Design Guidelines for Highway Bridges," *Applied Technology Council Report* ATC-6, 1981.

34. *Technical Council on Lifeline Earthquake Engineering,* ASCE: "Guidelines for the Seismic Design of Oil and Gas Pipeline Systems," 1984.

35. "Abatement of Seismic Hazards to Lifelines: Proceedings of a Workshop on Development of an Action Plan," vols. 1–6, and Action Plan, FEMA 143, BSSC, Washington, D.C., 1987.

36. "Quality in the Constructed Project," *Manuals and Reports on Engineering Practice,* no. 73, vol. 1, ASCE, 1990.

37. Blume, J. A., N. M. Newmark, and L. Corning: "Design of Multistory Reinforced Concrete Buildings for Earthquake Motions," Portland Cement Association, Chicago, Ill., 1961.

38. *ASCE Standard 4-86—Seismic Analysis of Safety-Related Nuclear Structures and Commentary on Standard for Seismic Analysis of Safety Related Nuclear Structures,* ASCE, September 1986, 91 p.

39. *Recommended Practices for Seismic Qualification of Class IE Equipment for Nuclear Power Generating Stations,* IEEE 344, 1987.

40. *ASME Boiler and Pressure Vessel Code,* Sects. III and VIII, and Appendices, 1992.

41. Beavers, J. E., W. J. Hall, and D. J. Nyman: "Assessment of Earthquake Vulnerability of Critical Industrial Facilities in the Central and Eastern United States," *Proc. 5th U.S. National Conference on Earthquake Engineering,* EERI, pp. IV-295 to IV-304, 1994.

CHAPTER 25
VIBRATION TESTING MACHINES

David O. Smallwood

INTRODUCTION

This chapter describes some of the more common types of vibration testing machines which are used for developmental, simulation, production, or exploratory vibration tests for the purpose of studying the effects of vibration or of evaluating physical properties of materials or structures. A summary of the prominent features of each machine is given. These features should be kept in mind when selecting a vibration testing machine for a specific application. Digital control systems for vibration testing are described in this chapter. Applications of vibration testing machines are described in other chapters.

A vibration testing machine (sometimes called a *shake table* or *shaker* and referred to here as a *vibration machine*) is distinguished from a vibration exciter in that it is complete with a mounting table which includes provisions for bolting the test article directly to it. A *vibration exciter*, also called a *vibration generator*, may be part of a vibration machine or it may be a device suitable for transmitting a vibratory force to a structure. A *constant-displacement* vibration machine attempts to maintain constant-displacement amplitude while the frequency is varied. Similarly, a *constant-acceleration* vibration machine attempts to maintain a constant-acceleration amplitude as the frequency is changed.

The *load* of a vibration machine includes the item under test and the supporting structures that are not normally a part of the vibration machine. In the case of equipment mounted on a vibration table, the load is the material supported by the table. In the case of objects separately supported, the load includes the test item and all fixtures partaking of the vibration. The load is frequently expressed as the weight of the material. The *test load* refers specifically to the item under test exclusive of supporting fixtures. A *dead-weight load* is a rigid load with rigid attachments. For nonrigid loads the reaction of the load on the vibration machine is a function of frequency. The vector force exerted by the load, per unit of acceleration amplitude expressed in units of gravity of the driven point at any given frequency, is the *effective load* for that frequency. The term *load capacity*, which is descriptive of the performance of reaction and direct-drive types of mechanical vibration machines, is the maximum dead-weight load that can be vibrated at the maximum acceleration rating of the vibration machine. The *load couple* for a dead-weight load is equal to the product of the force exerted on the load and the distance of the center-of-mass from the line-of-action of the force or from some arbitrarily selected location (such as a table surface). The static and dynamic load couples are generally different for nonrigid loads.

The term *force capacity,* which is descriptive of the performance of electrodynamic shakers, is defined as the maximum rated force generated by the machine. This force is usually specified, for continuous rating, as the maximum vector amplitude of a sinusoid that can be generated throughout a usable frequency range. A corresponding maximum rated acceleration, in units of gravity, can be calculated as the quotient of the force capacity divided by the total weight of the coil table assembly and the attached dead-weight loads. The *effective force* exerted by the load is equal to the effective load multiplied by the (dimensionless) ratio g, which represents the number of units of gravity acceleration of the driven point [see Eq. (25.1)].

DIRECT-DRIVE MECHANICAL VIBRATION MACHINES

The direct-drive vibration machine consists of a rotating eccentric or cam driving a positive linkage connection which forces a displacement between the base and table of the machine. Except for the bearing clearances and strain in the load-carrying members, the machine tends to develop a displacement between the base and the table which is independent of the forces exerted by the load against the table. If the base is held in a fixed position, the table tends to generate a vibratory displacement of constant amplitude, independent of the operating rpm. Figure 25.1 shows the direct-drive mechanical machine in its simplest forms. This type of machine is sometimes referred to as a *brute force machine* since it will develop any force necessary to produce the table motion corresponding to the crank or cam offset, short of breaking the load-carrying members or stalling the driving shaft.

The simplest direct-drive mechanical vibration machine is driven by a constant-speed motor in conjunction with a belt-driven speed changer and a frequency-indicating tachometer. Table displacement is set during shutoff and is assumed to hold during operation. An auxiliary motor driving a cam may be included to provide frequency cycling between adjustable limits. More elaborate systems employ

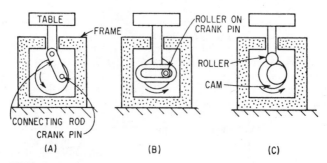

FIGURE 25.1 Elementary direct-drive mechanical vibration machines: (*A*) Eccentric and connecting link. (*B*) Scotch yoke. (*C*) Cam and follower.

a direct-coupled variable-speed motor with electronic speed control, as well as amplitude adjustment from a control station. Machines have been developed which provide rectilinear, circular, and three-dimensional table movements—the latter giving complete, independent adjustment of magnitude and phase in the three directions.

Many types of mechanisms are used to adjust the displacement amplitude and frequency of the mounting table. For example, the displacement amplitude can be adjusted by means of eccentric cams and cylinders.

PROMINENT FEATURES

- Low operating frequencies and large displacements can be provided conveniently.
- Theoretically, the machine maintains constant displacement regardless of the mechanical impedance of the table-mounted test item within force and frequency limits of the machine. However, in practice, the departure from this theoretical ideal is considerable, due to the elastic deformation of the load-carrying members with change in output force. The output force changes in proportion to the square of the operating frequency and in proportion to the increased displacement resulting therefrom. Because the load-carrying members cannot be made infinitely stiff, the machines do not hold constant displacement with increasing frequency with a bare table. This characteristic is further emphasized with heavy table mass loads. Accordingly, some of the larger-capacity machines which operate up to 60 Hz include automatic adjustment of the crank offset as a function of operating frequency in order to hold displacement more nearly constant throughout the full operating range of frequency.
- The machine must be designed to provide a stiff connection between the ground or floor support and the table. If accelerations greater than 1g are contemplated, the vibratory forces generated between the table and ground will be greater than the weight of the test item. Hence, all mass loads within the rating of the machine can be directly attached to the table without recourse to external supports.
- The allowable range of operating frequencies is small in order to remain within bearing load ratings. Therefore, the direct-drive mechanical vibration machine can be designed to have all mechanical resonances removed from the operating frequency range. In addition, relatively heavy tables can be used in comparison to the weight of the test item. Consequently, misplacing the center-of-gravity of the test item relative to the table center for vibration normal to the table surface and the generation of moments by the test item (due to internal resonances) usually have less influence on the table motions for this type of machine than would other types which are designed for wide operational frequency bands.
- Simultaneous rectilinear motion normal to the table surface and parallel to the table surface in two principal directions is practicable to achieve. It may be obtained with complete independent control of magnitude and phase in each of the three directions.
- Displacement of the table is generated directly by a positive drive rather than by a generated force acting on the mechanical impedance of the table and load. Consequently, impact loads in the bearings, due to the necessary presence of some bearing clearance, result in the generation of relatively high impact forces which are rich in harmonics. Accordingly, although the waveform of displacement might be tolerated as such, the waveform of acceleration is normally sufficiently dis-

torted to preclude recognition of the fundamental driven frequency, when displayed on a time base.

REACTION-TYPE MECHANICAL VIBRATION MACHINE

A vibration machine using a rotating shaft carrying a mass whose center-of-mass is displaced from the center-of-rotation of the shaft for the generation of vibration, is called a *reaction-type vibration machine.* The product of the mass and the distance of its center from the axis of rotation is referred to as the *mass unbalance,* the *rotating unbalance,* or simply the *unbalance.* The force resulting from the rotation of this unbalance is referred to as the *unbalance force.*

The *reaction-type vibration machine* consists of at least one rotating-mass unbalance directly attached to the vibrating table. The table and rotating unbalance are suspended from a base or frame by soft springs which isolate most of the vibration forces from the supporting base and floor. The rotating unbalance generates an oscillating force which drives the table. The unbalance consists of a weight on an arm which is relatively long by comparison to the desired table displacement. The unbalance force is transmitted through bearings directly to the table mass, causing a vibratory motion without reaction of the force against the base. A vibration machine employing this principle is referred to as a reaction machine since the reaction to the unbalance force is supplied by the table itself rather than through a connection to the floor or ground.

CIRCULAR-MOTION MACHINE

The reaction-type machine, in its simplest form, uses a single rotating-mass unbalance which produces a force directed along the line connecting the center-of-rotation and the center-of-mass of the displaced mass. Referred to stationary coordinates, this force appears normal to the axis of rotation of the driven shaft, rotating about this axis at the rotational speed of the shaft. The transmission of this force to the vibration-machine table causes the table to execute a circular motion in a plane normal to the axis of the rotating shaft.

Figure 25.2 shows, schematically, a machine employing a single unbalance producing circular motion in the plane of the vibration-table surface. The unbalance is driven at various rotational speeds, causing the table and test item to execute circular motion at various frequencies. The counterbalance weight is adjusted to equal the test item mass moment calculated from *d,* the plane of the unbalance force, thereby keeping the combined center-of-gravity coincident with the generated force. Keeping the generated force acting through the combined center-of-gravity of the spring-mounted assembly eliminates vibratory moments which, in turn, would generate unwanted rotary motions in addition to the motion parallel to the test mounting surface. The vibration isolator supports the vibrating parts with minimum transmission of the vibration to the supporting floor.

For a fixed amount of unbalance and for the case of the table and test item acting as a rigid mass, the displacement of motion tends to remain constant if there are no resonances in or near the operating frequency range. If balance force must remain constant, requiring the amount of unbalance to change with shaft speed.

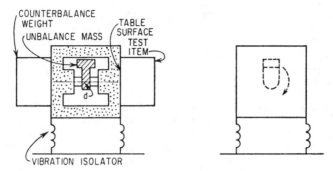

FIGURE 25.2 Circular-motion reaction-type mechanical vibration machine.

RECTILINEAR-MOTION MACHINE

Rectilinear motion rather than circular motion can be generated by means of a reciprocating mass. Rectilinear motions can be produced with a single rotating unbalance by constraining the table to move in one direction.

Two Rotating Unbalances. The most common rectilinear reaction-type vibration machine consists of two rotating unbalances, turning in opposite directions and phased so that the unbalance forces add in the desired direction and cancel in other directions. Figure 25.3 shows schematically how rectilinear motion perpendicular and parallel to the vibration table is generated. The effective generated force from the two rotating unbalances is midway between the two axes of rotation and is normal to a line connecting the two. In the case of motion perpendicular to the surface of the table, simply locating the center-of-gravity of the test item over the center of the table gives a proper load orientation. Tables are designed so that the resultant force always passes through this point. This results in collinear-

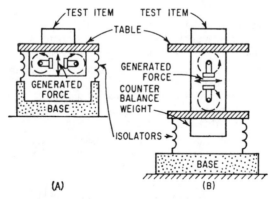

FIGURE 25.3 Rectilinear-motion reaction-type mechanical vibration machine using two rotating unbalances: (*A*) Vibration perpendicular to table surface. (*B*) Vibration parallel to table surface.

ity of generated forces and inertia forces, thereby avoiding the generation of moments which would otherwise rock the table. In the case of motion parallel to the table surface, no simple orientation of the test item will achieve collinearity of the generated force and inertia force of the table and test item. Various methods are used to make the generated force pass through the combined center-of-gravity of the table and test item.

Three Rotating Unbalances. If a machine is desired which can be adjusted to give vibratory motion either normal to the plane of the table or parallel to the plane of the table, a minimum of three rotating unbalances is required. Inspection of Fig. 25.4 shows how rotating the two smaller mass unbalances relative to the single larger unbalance results in the addition of forces in any desired direction, with cancellation of forces and force couples at 90° to this direction. Although parallel shafts are usu-ally used as illustrated, occasionally the three unbalances may be mounted on collinear shafts, the two smaller unbalances being placed on either side of the single larger unbalance to conserve space and to eliminate the bending moments and shear forces imposed on the structure connecting the individual shafts.

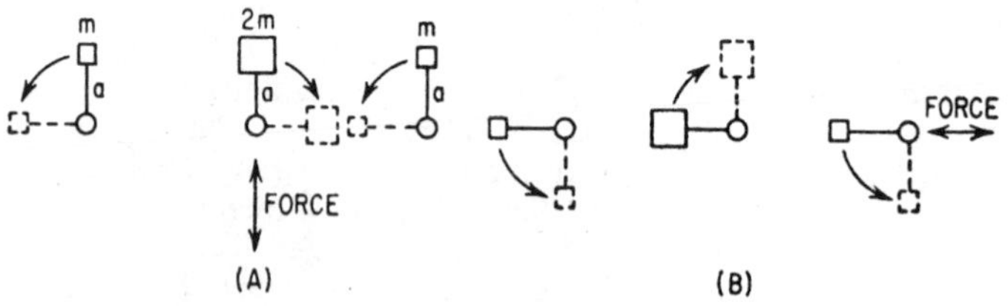

FIGURE 25.4   Adjustment of direction of generated force in a reaction-type mechanical vibration exciter: (*A*) Vertical force. (*B*) Horizontal force.

PROMINENT FEATURES

- The forces generated by the rotating unbalances are transmitted directly to the table without dependence upon a reactionary force against a heavy base or rigid ground connection.

- Because the length of the arm which supports the unbalance mass can be large, relative to reasonable bearing clearances and the generation of a force which does not reverse its direction relative to the rotating unbalance arm, the generated waveform of motion imparted to the vibration machine table is superior to that attainable in the direct-drive type of vibration machine.

- The generated vibratory force can be made to pass through the combined center-of-gravity of the table and test item in both the normal and parallel directions rel-ative to the table surface, thereby minimizing vibratory moments giving rise to table rocking modes.

- The attainable rpm and load ratings on bearings currently limit performance to a frequency of approximately 60 Hz and a generated force of 300,000 lb (1.3 MN), respectively, although in special cases frequencies up to 120 Hz and higher can be obtained for smaller machines.

ELECTRODYNAMIC VIBRATION MACHINE

GENERAL DESCRIPTION

A complete electrodynamic vibration test system is comprised of an electrodynamic vibration machine, electrical power equipment which drives the vibration machine, and electrical controls and vibration monitoring equipment.

The electrodynamic vibration machine derives its name from the method of force generation. The force which causes motion of the table is produced electrodynamically by the interaction between a current flow in the armature coil and the intense magnetic dc field which passes through the coil, as illustrated in Fig. 25.5. The table is structurally attached to a force-generating coil which is concentrically located (with radial clearances) in the annular air gap of the dc magnet circuit. The assembly of the armature coil and the table is usually referred to as the *driver coil-table* or *armature*. The magnetic circuit is made from soft iron which also forms the *body* of the vibration machine. The body is magnetically energized, usually by two field coils as shown in Fig. 25.5*C,* generating a radially directed field in the air gap, which is perpendicular to the direction of current flow in the armature coil. Alternatively, in small shakers, the magnetic field is generated by permanent magnets. The generated force in the armature coil is in the direction of the axis of the coil, perpendicular to the table surface. The direction of the force is also perpendicular to the armature-current direction and to the air-gap field direction.

The table and armature coil assembly is supported by elastic means from the machine body, permitting rectilinear motion of the table perpendicular to its surface, corresponding in direction to the axis of the armature coil. Motion of the table in all other directions is resisted by stiff restraints. Table motion results when an ac current passes through the armature coil. The body of the machine is usually supported by a base with a trunnion shaft centerline passing horizontally through the center-of-gravity of the body assembly, permitting the body to be rotated about its center, thereby giving a vertical or horizontal orientation to the machine table. The base usually includes an elastic support of the body, providing vibration isolation between the body and the supporting floor.

Where a very small magnetic field is required at the vibration machine table due to the effect of the magnetic field on the item under test, *degaussing* may be pro-

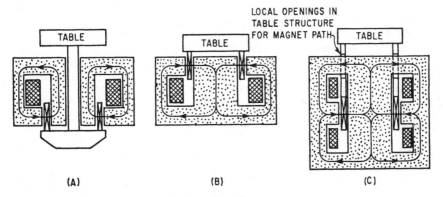

FIGURE 25.5 Three main magnet circuit configurations.

vided. Magnetic fields of 5 to 30 gauss several inches above the table are normal for modern machines with double-ended, center air-gap magnet designs, Fig. 25.5C, without degaussing accessories; in contrast, with degaussing accessories, magnetic fields of 2 to 5 gauss can be achieved.

Because of copper and iron losses in the electrodynamic unit, provision must be made to carry off the dissipated heat. Cooling by convection air currents, compressed air, or a motor-driven blower is used and, in some cases, a recirculating fluid is used in conjunction with a heat exchanger. Fluid cooling is particularly useful under extremes of hot or cold environments or altitude conditions where little air pressure is available.

MAGNET CIRCUIT CONFIGURATIONS

Three magnet circuit configurations which are used in the electrodynamic machines are shown schematically in Fig. 25.5. In Fig. 25.5A, the table and driver coil are located at opposite ends of the magnet circuit. The advantage of this configuration is that the location of the annular air gap, the region of high magnetic leakage flux, is spaced from the table and the body itself acts as a magnetic shield, resulting in lower magnetic flux density at the table. The disadvantage lies in the loss of rigidity in the connecting structure between the driver coil and the table because of its length. This configuration is usually cooled by convection air currents or by forced air from a motor-driven blower.

In Fig. 25.5B, the table is connected directly to the driver coil. This eliminates the length of structure passing through the magnet structure, thereby increasing the rigidity of the driver coil-table assembly and allowing higher operating frequencies. The leakage magnetic field in the vicinity of the table is high in this configuration. It is therefore difficult, if not impossible, to reduce the leakage to acceptable levels without adding extra length to the driver coil assembly, elevating the table above the air gap. The configuration in Fig. 25.5C has a complete magnet circuit above and below the annular air gap, thereby reducing external leakage magnetic field to a minimum. This configuration also increases the total magnetic flux in the air gap by a factor of almost 2 for the same diameter driver coil, giving greater force generation and a more symmetrical magnetic flux density along the axis of the coil. Hence more uniform force generation results when the driver coil is moved axially throughout its total stroke. All high-efficiency and high-performance electrodynamic vibration machines use the configuration shown in Fig. 25.5C.

Configurations B and C of Fig. 25.5 may use air cooling throughout or an air-cooled driver coil and liquid-cooled field coil(s) or total liquid cooling.

The main magnetic circuit uses dc field coils for generating the high-intensity magnetic flux in the annular gap in all of the larger sizes and most of the smaller units. Permanent magnet excitation is used in small portable units and in some general-purpose units up to about 500-lb (2 kN) generated force.

FREQUENCY-RESPONSE CONSIDERATIONS

Testing procedures which call for sinusoidal motion (see Chap. 20) of a vibration-machine table can be performed even though the frequency-response curve of the electrodynamic vibration machine is far from flat. For a test at a fixed frequency, the driving voltage is adjusted until the table motion is equal in amplitude to that required by the test specifications. If the procedure calls for cycling the frequency between two frequency limits while keeping a constant displacement or acceleration, a control system or servo control adjusts the driver-coil voltage as required to maintain the desired vibration machine table motion independent of the frequency of operation. This control system provides a correction at any frequency of operation within the testing frequency limits, but it can correct for only one operating frequency at any instant of time. The closer the frequency response is to the desired variation in acceleration with frequency, the smaller the corrections in driver-coil voltage will be from the control system—thereby improving the attainable accuracy of the control.

Similarly for test procedures which call for a random vibration source, the autospectrum of the source must be adjusted, because of test requirements and the frequency response of the test system. A shaker with a more constant response will allow for a greater range of spectral values than can be controlled. Vibration control systems are discussed later in this chapter.

SYSTEM RATINGS

The electrodynamic vibration machine system is rated: (1) in terms of the *peak value* of the sinusoidal generated force for *sinusoidal* vibration testing and (2) in terms of the *rms* and *instantaneous* values of the maximum force generated under *random* vibration testing. In order to determine the acceleration rating of the system with a test load on the vibration table, the weight of the test load, assumed to be effective at all frequencies, must be known and used in the following expressions:

$$g = \frac{F}{W_L + W_T}$$

$$g_{rms} = \frac{F_{rms}}{W_L + W_T}$$

(25.1)

where
$g = a/g$, a dimensionless number expressing the ratio of the peak sinusoidal acceleration to the acceleration due to gravity (i.e., the peak sinuosidal acceleration in g's)

$g_{rms} = a_{rms}/g$, a number expressing the ratio of the rms value of random acceleration to the acceleration due to gravity

W_L = weight of load

W_T = equivalent weight of table driver-coil assembly and associated moving parts

F = rated peak value of sinusoidal generated force

F_{rms} = rated rms value of random generated force

The *force rating* of an electrodynamic vibration machine is the value of force which can be used to calculate attainable accelerations for any rigid-mass table load equal to (or greater than) the driver coil weight. It is not necessarily the force generated by the driver coil. These two forces are identical only if the operating frequencies are sufficiently below the axial resonance frequency of the armature assembly, where it acts as a rigid body. As the axial resonance frequency is approached, a mechanical magnification of the force generated electrically by the driver coil results. The design of the driving power supply takes into account the possible reduction in driver-coil current at frequencies approaching the armature axial resonance frequency, since full current in this range cannot be used without exceeding the rated value of transmitted force at the table—possibly causing structural damage.

In those cases where the test load dissipates energy mechanically, the system performance should be analyzed for each specific load since normal ratings are based on a dead-mass, nondissipative type of load. This consideration is particularly significant in resonance-type fatigue tests at high stress levels.

PROMINENT FEATURES

- A wide range of operating frequencies is possible, with a properly selected electric power source, from 0 to above 30,000 Hz. Small, special-purpose machines have been made with the first axial/resonance mode above 26,000 Hz, giving inherently a resonance-free, flat response to 10,000 Hz.
- Frequency and displacement amplitude are easily controlled by adjusting the power-supply frequency and voltage.
- Pure sinusoidal table motion can be generated at all frequencies and amplitudes. Inherently, the table acceleration is the result of a generated force proportional to the driving current. If the electric power supply generates pure sinusoidal voltages and currents, the waveform of the acceleration of the table will be sinusoidal, and background noise will not be present. Operation with table acceleration waveform distortion of less than 10 percent through a displacement range of 10,000-to-1 is common, even in the largest machines. Velocity and displacement waveforms obtained by the single and double integration of acceleration, respectively, will have even less distortion.
- Random vibration, as well as sinusoidal vibration, or a combination of both, can be generated by supplying an appropriate input voltage.
- A unit occupying a small volume, and powered from a remote source, can be used to generate small vibratory forces. A properly designed unit adds little mass at the point of attachment and can have high mobility without mechanical damping.
- Leakage magnetic flux is present around the main magnet circuit. This leakage flux can be minimized by proper design and the use of degaussing coil techniques.

SPECIFICATIONS

Design Factors

Force Output.   The maximum vector-force output for sinusoidal excitation shall be given for continuous duty and may additionally be given for intermittent duty. When nonsinusoidal motions are involved, the force may additionally be given in terms of an rms value together with a maximum instantaneous value. The latter value is especially significant when random type of excitation is required.

In some cases of wide-frequency-band operation of the electrodynamic vibration machine, the upper frequencies are sufficiently near to the axial mechanical resonance of the coil-table assembly to provide some amplification of the generated force. Most system designs account for this magnification, when present, by reducing the capacity of the electrical driving power accordingly.

The peak values of the input electrical signal, for random excitation, may extend to indefinitely large values. In order that the armature coil voltage and generated force may be limited to reasonable values the peak values of excitation are clipped so that no maxima shall exceed a given multiple of the rms value. The magnitude of the maximum clipped output shall be specified preferably as a multiple of the rms value. If adjustments are possible, the range of magnitudes shall be given.

Weight of Vibrating Assembly. The weight of the vibration coil-table assembly shall be given. It shall include all parts which move with the table and an appropriate percentage of the weight of those parts connecting the moving and stationary parts giving an effective over-all weight.

Vibration Direction. The directions of vibration shall be specified with respect to the surface of the vibration table and with respect to the horizontal or vertical direction. Provisions for changing the direction of vibration shall be stated.

Unsupported Load. The maximum allowable weight of a load not requiring external supports shall be given for horizontal and vertical orientations of the vibration table. This load in no way relates to dyanmic performance but is a design limitation, the basis of which may be stated by the manufacturer.

Static Moments and Torques. Static moments and torques may be applied to the coil-table assembly of a vibration machine by the tightening of bolts and by the overhang of the center-of-gravity of an unsupported load during horizontal vibration. The maximum permissible values of these moments and torques shall be specified. These loads in no way relate to the dynamic performance but are design limitations, the basis for which may be stated by the manufacturer.

Total Excursion Limit. The maximum table motion between mechanical stops shall be given together with the maximum vibrational excursion permissible with no load and with maximum load supportable by the table.

Acceleration Limit. The maximum allowable table acceleration shall be given. (These large maxima may be involved in the drive of resonant systems.)

Stiffness of Coil-Table Assembly Suspension System

AXIAL STIFFNESS: The stiffness of the suspension system for axial deflections of the coil-table assembly shall be given in terms of pounds per inch of deflection. The natural frequency of the unloaded vibrating assembly may also be given. Provisions, if any, to adjust the table position to compensate for position changes caused by different loads shall be described.

SUSPENSION RESONANCES: Resonances of the suspension system should be described together with means for their adjustment where applicable.

Axial Coil-Table Resonance. The resonance frequency of the lowest axial mode of vibration of the coil-table assembly shall be given for no load and for an added dead-weight load equal to 1 and to 3 times the coil-table assembly weight. If this resonance frequency is not obvious from measurements of the table amplitude vs. frequency, it may be taken to be approximately equal to the lowest frequency, above the rigid-body resonance of the table-coil assembly on its suspension system, at which the phase difference between the armature coil current and the acceleration of the center of the table is 90°.

Impedance Characteristics. When an exciter or vibration machine is considered independent of its power supply, information concerning the electrical impedance characteristics of the machine shall be given in sufficient detail to permit matching of the power-supply output to the vibration-machine input. It is suggested that consideration be given to providing schematic circuit diagrams (electrical and mechanical or equivalent electrical) together with corresponding equations which contain the principal features of the machine.

Environmental Extremes. When it is anticipated that the vibration machine will be used under conditions of abnormal pressure and temperature, the following information shall be supplied as may be applicable: maximum simulated altitude (or minimum pressure) under which full performance ratings can be applied; maximum simulated altitude under which reduced performance ratings can be applied; maximum ambient temperature for rated output; low-temperature limitations; humidity limitations.

Performance. The performance relates in part to the combined operation of the vibration generator and its power supply.

Amplitude-Frequency Relations. Data on sinusoidal operation shall be given as a series of curves for several table loads, including zero load, and for a load at least 3 times the weight of the coil-table assembly. Maximum loads corresponding to $20g$ and $10g$ table acceleration under full-rated force output would be preferred. These curves should give amplitudes of table displacement, velocity, or acceleration, whichever is limiting, throughout the complete range of operating frequencies corresponding to maximum continuous ratings of the system. Additionally, the maximum rated force should be given. If this force is frequency-dependent, it should be presented as a curve with the ordinate representing the force and the abscissa the frequency.

If the system is for broad-band use, necessarily employing an electronic power amplifier, the exciting voltage signal applied to the input of the system shall be held constant and the output acceleration shall be plotted as a function of frequency with and without peak-notch filters or other compensating devices for the loads and accelerations indicated above. If the vibrator is used only for sinusoidal vibrations, and employs servo amplitude control, the curves should be obtained under automatic frequency sweeping conditions with the control system included. If cycling the exciter through its complete frequency range involves switching operations, such as may be required for matching impedances or for changing power sources, these discontinuities should be noted on the amplitude-frequency curves.

Waveform. Total rms distortion of the acceleration waveform at the center of the vibration table, or at the center on top of the added test weight, shall be furnished to show at least the frequencies of worst waveform under the test conditions specified under the above paragraph. The pickup type, and frequency range, shall be given together with the frequency range of associated equipment. It is desirable to have the over-all frequency range at least 10 times the frequency of the fundamental being recorded. Tabular data on harmonic analysis may alternatively or additionally be given.

Magnetic Fields. The maximum values of constant and alternating magnetic fields, due to the vibration exciter, in the region over the surface of the vibration table should be indicated. If degaussing coils are furnished, these values should be given with and without the use of the degaussing coils.

Frequency Range. The over-all frequency range, and a division of frequency ranges for different alternators that supply the exciting current, if applicable, shall be given. A group of frequency ranges shall also be given for electronic power supplies if they require changes of their output impedance for the different ranges.

Automatic Frequency Control. The following factors shall be stated or provided: The range, or ranges, of the frequency bands that can be automatically cycled at full-rated output or at reduced output if applicable, the minimum and maximum time for each frequency cycle, a curve of frequency vs. time for a maximum and a minimum cycle period and for the principal frequency ranges, and the dwell time at the points of reversal of the direction of frequency change.

Frequency Drift. The probable drift of a set frequency shall be stated, together with factors that contribute to the drift. This shall apply for nonresonant loads.

Signal Generator. A vibration pickup, if built into the vibration machine, shall have calibrations furnished over a specified frequency and amplitude range.

Installation Requirements. Recommendations shall be given as to suitable methods for installing the vibration machine and auxiliary equipment. Electrical and other miscellaneous requirements shall be stated.

HYDRAULIC VIBRATION MACHINE

The *hydraulic vibration machine* is a device which transforms power in the form of a high-pressure flow of fluid from a pump to a reciprocating motion of the table of the vibration machine. A schematic diagram of a typical machine is shown in Fig. 25.6. In this example, a two-stage electrohydraulic valve is used to deliver high-pressure fluid, first to one side of the piston in the actuator and then to the other side, forcing the actuator to move with a reciprocating motion. This valve consists of a pilot stage and power stage, the former being driven with a reciprocating motion by the electrodynamic driver. At the time the actuator moves under the force of high-pressure fluid on one side of the piston, the fluid on the other side of the piston is forced back through the valve at reduced pressure and is returned to the pump.

The electrohydraulic valve is usually mounted directly on the side of the actuator cylinder, forming a close-coupled assembly of massive steel parts. The proximity of the valve and cylinder is desirable in order to reduce the volume and length of the connecting fluid paths between the several spools and the actuator, thereby minimizing the effects of the compliance of the fluid and the friction to its flow. (Many types of electrohydraulic valves exist, all of which fail to meet the requirement of sufficient flow at high frequencies to give vibration machine performance equivalent to existing electrodynamic machine performance at 2000 Hz.

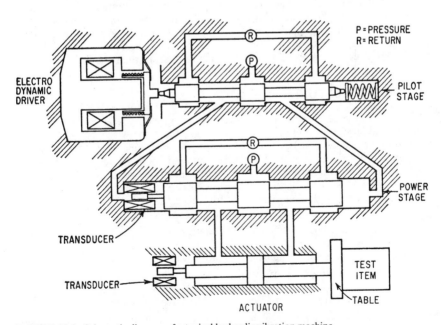

FIGURE 25.6 Schematic diagram of a typical hydraulic vibration machine.

OPERATING PRINCIPLE

In Fig. 25.6, the *pilot* and *power spools* of a hydraulic vibration machine are shown in the "middle" or "balanced" position, blocking both the pump high-pressure flow P and the return low-pressure flow R. Correspondingly, the piston of the actuator must be stationary since there can be no fluid flow either to or from the actuator cylinder. If the pilot spool is displaced to the right of center by a force from the electrodynamic driver, then high-pressure fluid P will flow through the passage from the pilot spool to the left end of the power spool, causing it to move to the right also. This movement forces the trapped fluid from the right-hand end of the power spool through the connecting passage, back to the pilot stage, and then through the opening caused by the displacement of the pilot spool to the right, to the chamber R connected to the return to the pump. Correspondingly, if the pilot spool moves to the left, the flow to and from the power spool is reversed, causing it to move to the left. For a given displacement of the pilot spool, a flow results which causes a corresponding velocity of the power spool. A displacement of the power spool to the right allows the flow of high-pressure fluid P from the pump to the left side of the piston in the actuator, causing it to move to the right and forcing the trapped fluid on the right of the piston to be expelled through the connecting passage to the power spool and out past the right-hand restrictions to the return fluid chamber R. The transducers shown on the power spool and the actuator shaft are of the differential transformer type and are used in the feedback circuit to improve system operation and provide electrical control of the average (i.e., stationary) position of the actuator shaft relative to the actuator cylinder.

A block diagram of the complete hydraulic vibration machine system is shown in Fig. 25.7. The pump, in conjunction with accumulators in the pressure and return lines at the hydraulic valve, should be capable of variable flow while maintaining a fixed pressure. Most systems to date have required an operating pump pressure of 3000 lb/in.^{2} (20 MPa). The upper limit of efficiency of the hydraulic valve is approximately 60 per cent, the losses being dissipated in the form of heat. Mechanical loads are seldom capable of dissipating appreciable power; most of the power in the pump discharge is converted to a temperature rise in the fluid. Therefore a heat exchanger limiting the fluid temperature must be included as part of the system.

PROMINENT FEATURES

- Large generated forces or large strokes can be provided relatively easily. Large forces and large velocities of motion, made possible with a large stroke, determine the power capacity of the system. For example, one hydraulic vibration machine has a peak output power of 450,000 lb-in./sec (approximately 34 hp or 25 kW) with a single electrohydraulic valve. This power can be increased by the installation of several valves on a single actuator. Appreciable increases in valve flow can be realized by sacrificing high-frequency performance. Hence, the hydraulic vibration machine excels at low frequencies where large force, stroke, and power capacity are required.

- The hydraulic machine is small in weight, relative to the forces attainable; therefore, a rigid connection to firm ground or a large massive base is necessary to anchor the machine in place and to attenuate the vibration transmitted to the surrounding area.

- The main power source is hydraulic, which is essentially dc in character from available pumps. The electrical driving power for controlling the valve is small. Therefore, the operating frequency range can be extended down to zero Hz.

- The magnetic leakage flux in the region of the table is insignificant by comparison with the electrodynamic-type vibration machine.

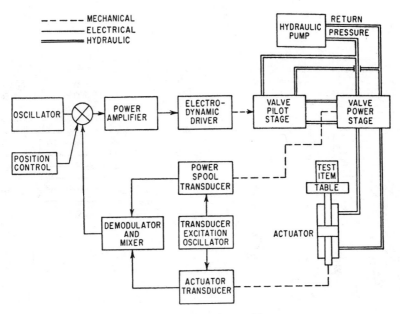

FIGURE 25.7 Block diagram—hydraulic vibration machine system.

- The machine, with little modification, is suitable for use in high- and low-temperature, humidity, and altitude environments.
- The machine is inherently nonlinear with amplitude in terms of electrical input and output flow or velocity.

PIEZOELECTRIC VIBRATION EXCITERS

A piezoelectric material (see Chap. 12) can be used to generate motion and act as a *piezoelectric vibration exciter.* Typically a piezoelectric exciter employs a number of disks of piezoelectric material as illustrated in Fig. 25.8; this arrangement increases the ratio of the displacement output to voltage input sensitivity of the exciter. The strain is proportional to the charge, and the charge is increased by increasing the voltage gradients across the piezoelectric material. The voltage gradient is increased by using many thin layers of piezoelectric material, separated with a conducting material, with alternating polarity on the conducting separators. This arrangement of alternating layers of piezoelectric material and conducting material is called a *piezoelectric stack.* Because the piezoelectric stack has little tensile strength, the stack must be preloaded. The stiffness of the preloading mechanism must be much less than the stiffness of the piezoelectric stack so that preloading will not influence the mechanical output significantly. The combination of the piezoelectric stack (acting like a displacement actuator) and a reaction mass forms a reaction-type vibration exciter as described above. The reaction mass of the piezoelectric exciter can be the armature mass of a small electrodynamic exciter. This effectively places an electrodynamic and a piezoelectric exciter in series, producing a machine with a usable output over a wide frequency range.

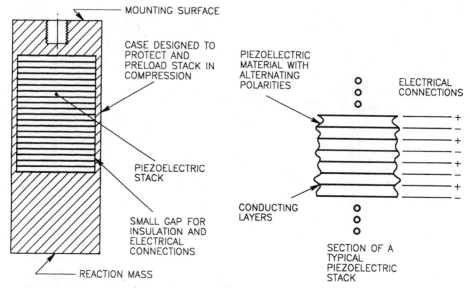

FIGURE 25.8 Simplified cross section of a piezoelectric vibration exciter. A compressed piezo-electric stack is excited with an oscillating voltage. An electrical voltage applied to the electrical connections causes the piezoelectric stack to elongate and contract, producing a relative displacement between the mounting surface and the reaction mass. The inertia of the reaction mass results in a force being applied to an item mounted on the mounting surface.

PROMINENT FEATURES

- The exciters can have a usable frequency range from 0 to 60 kHz.

- The low-frequency output is severely limited by the displacement limits of the piezoelectric stack, usually a few thousandths of an inch (a few hundredths of a millimeter).

- The high-frequency output is limited by internal resonances of the vibration exciter.

- The force output of the exciter is limited by the displacement limit of the piezo-electric stack and by the mass of the reaction mass.

- The power supply for a piezoelectric exciter requires high voltages (typically about 1000 volts) and sufficient current to drive the capacitance (typically 10 to 1000 nano farads) of the device.

IMPACT EXCITERS

A limited amount of vibration testing, such as some modal testing and some stress screening, require a broad band of relatively uncontrolled vibration. A class of exciters broadly known as *impact exciters* is sometimes used for the above applications. These devices depend on the property that a short impact generates a broad band of vibration energy. Each impact is a short transient, for example see Fig. 26.1, but repeated impacts result in a quasi-steady-state vibration having a wide frequency bandwidth. If the impacts are periodic, the spectrum is composed of the fundamental frequency of the impacts and many harmonics of this fundamental frequency. If the impacts are randomly spaced, the spectrum is broad-band random. The vibration characteristics are strongly influenced by the dynamics of the structure on which they are mounted. The impact exciters can be mounted directly to the test specimen, or the exciters can excite a table on which the test item is mounted. The latter can be classed as a vibration testing machine.

PROMINENT FEATURES

- The design is usually simple, compact, and rugged.
- The maximum attainable displacement is usually small.
- The vibration is relatively uncontrolled. The user has little control over the spectrum of the resulting vibration.

CONTROL SYSTEMS FOR ELECTRODYNAMIC AND HYDRAULIC VIBRATION TESTING MACHINES

Vibration testing using electrodynamic or hydraulic testing machines typically employs sinusoidal waveforms, random vibration, and predetermined waveforms. The latter is usually subdivided into transient and quasi-steady-state waveforms. The frequency-response function relating the shaker response (usually an acceleration measured at a point, called the *control point,* on the shaker or on the test item) to the drive voltage (the *drive voltage* is the voltage waveform used as the input to the power supply for the vibration exciter) is not a constant; both the amplitude and phase vary with frequency. The frequency-response function varies with the power amplifier, the shaker, the fixtures holding the test item to the shaker, and the test item itself. Therefore, a voltage waveform used as an input to the vibration testing machine (the drive) will not generally be reproduced at the control point. A control system is employed to generate a drive waveform which will produce the required waveform at the control point. Analog control systems are used for sinusoidal waveforms and for random vibration but have not been very successful for the reproduction of predetermined waveforms. However, digital control systems are used extensively for all three forms of testing. This section describes the construction and important characteristics of these control systems.

ANALOG CONTROL SYSTEMS

Sinusoidal Vibration Testing. A typical analog control system for sinusoidal vibration testing is shown in Fig. 25.9. Voltage is supplied by a low-distortion sine-wave generator to an amplitude servo and programmer which adjusts the magnitude of this voltage; the adjustment is controlled by a voltage which is generated by the accelerometers. The sine-wave signal, so modified, is fed to a low-distortion power amplifier which drives the vibration machine.

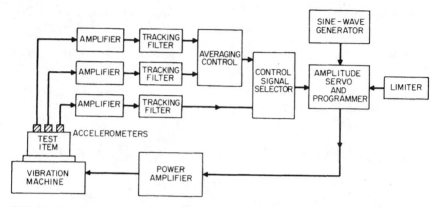

FIGURE 25.9   Block diagram of a typical control system for sinusoidal vibration testing.

Sinusoidal motion of the vibration machine may induce harmonics, rattles, and other nonlinear response in the test item. The tracking filters (narrow bandpass filters whose center frequency always corresponds exactly to the frequency of the sine-wave generator) connected to the outputs of the accelerometers reject any signals other than those corresponding to the driving frequency of the oscillator. The function of the averaging control is to provide servo control on the arithmetic average of the individual accelerometer magnitudes. The control signal selector limits the test to preset values. The averager and/or the selector provide protection against excessive amplitude overtesting or undertesting.

Random Vibration Testing.   A typical analog control system for random vibration testing is shown in Fig. 25.10. The voltage, supplied by a random noise generator, is fed to an equalizer which shapes the spectrum of the random noise. The

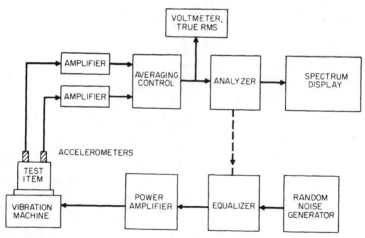

FIGURE 25.10   Block diagram of a typical control system for random vibration testing.

equalizer is usually comprised of a number of narrow bandpass filters, each with an adjustable gain that divides the total frequency range of interest into contiguous narrowband segments. The output of the equalizer provides the desired signal input to the power amplifier that drives the vibration machine.

Accelerometers are mounted on the item under test. The output of these accelerometers is amplified and fed into an averaging control whose input switches between the signals from the various accelerometers. Precautions must be observed in the use of such averaging controls. The acceleration power spectral density (also called the auto-spectral density), provided by the spectrum analyzer, is displayed by any of several convenient techniques—for example, by means of an oscilloscopic display. The output of the spectrum analyzer is compared with a reference power spectral density and is used to adjust the gain of the filters in the equalizer. This adjustment can be automatic, adjusted by the control system, or manual, adjusted by the test operator.

DIGITAL CONTROL SYSTEMS FOR SHOCK AND VIBRATION TESTING

Description of Digital Control Systems for Shock and Vibration Testing.

The same digital control hardware can be used to control all three types of vibration testing (sinusoidal waveforms, random vibration, and predetermined waveforms), but each requires a specialized software package. A general block diagram of a typical digital vibration control system is shown in Fig. 25.11. An analog to digital converter (A/D converter, discussed in Chaps. 13 and 27) samples the control waveform. A computer which includes one or more DSPs (specialized *digital signal processors,* which perform vector and matrix operations required for digital signal processing) analyzes the control waveform and generates a corrected drive by comparing the control with a required waveform derived from the test definition. The DSP and the fast Fourier transform (FFT, defined in Chap. 13 and 27) make digital vibration control systems practical. A 1024-point FFT can be calculated in a DSP in a few milliseconds. A digital-to-analog converter (D/A converter) and low-pass filter convert the sampled drive into an analog signal used as the input to the power amplifier for the vibration machine. An operator interface consists of a graphical display by which the control system communicates with the operator and a keyboard, mouse, and/or panel for the operator to communicate with the control system. The test is defined by a setup program which stores the information in test definition files. To maintain synchronous waveforms, the D/A and A/D converters are driven from a common clock. If the A/D converter has more than one channel, sample-and-hold systems are usually used to avoid the requirement for phase correction caused by the delay between channels if sample-and-hold is not used.

Digital control systems for shock and vibration testing differ from conventional digital control systems in an important way. Conventional systems sample the output of the system being controlled and use this information to generate the next sample of input for the system being controlled. The computations required for the generation of the next input sample must be completed within one sample interval. In digital control systems for shock and vibration testing, many samples of the drive are output to the vibrator without correction: from tens to hundreds of samples for sine-wave testing to thousands of samples for random vibration testing. The control waveform is also sampled in *blocks* (a sequence of samples N points long, where N

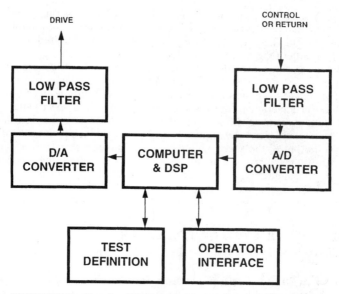

FIGURE 25.11   Functional block diagram for a digital vibration control system. A control waveform from a test item mounted on a vibration machine is filtered with a low-pass filter to prevent aliasing and then converted to a sequence of sampled values using an A/D converter. A computer which includes one or more digital signal processing units (DSP) computes samples of a corrected drive waveform. The sampled drive waveform is converted back to an analog waveform with a D/A converter and a low-pass filter. The analog drive waveform is used as an input to the power amplifier of the vibration machine.

is the *block size* and is usually a power of 2) and processed (usually with an FFT) before any correction is made to the drive.

Sine-Wave Control Systems.   Digital sine-wave vibration test control systems typically mimic analog sine-wave vibration test control systems. The output of the amplifiers of Fig. 25.9 are filtered and digitized with the A/D converter in the digital system (Fig. 25.11). The tracking filters, averaging control, and control signal selector are implemented as digital simulations of the analog functions. The sine-wave generator is replaced by samples of the sine wave generated within the computer. The programmer is replaced by a test setup file. The amplitude servo is replaced by an algorithm which compares the computed amplitude of the control waveform with the required control amplitude, as defined by the test setup, and generates a corrected sampled drive waveform. The sampled drive waveform is converted to an analog drive waveform by the D/A converter and the low-pass filter (Fig. 25.11). The analog drive is used as the input to the power amplifier in Fig. 25.9.

Vibration tests require that the frequency be held at a fixed frequency, stepped in a sequence of fixed frequencies, or swept in time over a range of frequencies. A *swept sine* is the changing of the frequency from one frequency to another frequency in a smooth continuous manner. The rate of change of the frequency with time is called the *sweep rate*. Both log and linear swept sines are required. For a *log sweep* the change in the log of the frequency per unit of time is a constant. For a *linear sweep* the change in frequency per unit of time is a constant. Because the drive waveform is usually generated in blocks of samples, care must be taken in swept sine vibration tests to assure that the frequency change is continuous. An analog control system makes continuous corrections to the drive amplitude, but the correction of the drive amplitude in a digital system is not continuous, but discrete. The time between amplitude corrections is called the *loop time* and is controlled by the number of samples which must be taken to define the control waveform amplitude and the required computations to compute the corrected drive waveform. A *loop* is the completion of one complete cycle from the correction of the drive waveform to the next correction of the drive waveform. The control waveform amplitude can vary rapidly as the frequency changes due to system resonances, and the required loop time is measured in small fractions of a second. For stability, the complete correction of the drive waveform is not usually made each loop. The maximum rate of drive waveform correction is called the *compression speed* and is usually expressed as decibels per second (dB/sec). If the compression speed is too fast, system instabilities can manifest themselves. If the compression speed is too slow, the correct amplitude will not be maintained. The required compression speed is a function of frequency, sweep rate, and the system dynamics. Limited operator control of the compression speed is usually provided. The bandwidth of the digital tracking filter will effect the stability of the system. As the bandwidth of the tracking filter decreases, the delay in the output of the tracking filter increases. As the delay increases, the compression speed must be decreased to maintain stability. A compromise is required between the ability to reject components in the control waveform at frequencies other than the drive frequency and the ability to respond quickly to changes in the control waveform amplitude.

Random Vibration Control Systems.   These systems excite a test item with an approximation of stationary Gaussian random vibration (see random vibration in the terminology section, Chap. 1). Digital random vibration control systems also mimic analog random vibration control systems (Fig. 25.10).

The control or return waveforms from the amplifiers in Fig. 25.10 are low-pass filtered to prevent aliasing (defined in Chap. 13) and converted to a sequence of control samples by the A/D converter in the digital system (Fig. 25.11). The averaging control, the analyzer, and the display are replaced by a digital signal analyzer which uses a discrete Fourier transform (DFT; usually an FFT), as discussed in Chap. 13, to estimate the power spectral density (also called the auto-spectral density) of the control waveforms. The random noise generator and the analog equalizer are replaced by an analogous digital process using a DFT. The lines of the DFT (see Chap. 27) in the digital system play the role of the contiguous narrow-band filters in the equalizer of the analog system. *Equalization* is the adjustment of the amplitude of the output of the narrow-band filters such that the power spectral density of the control waveform matches a reference power spectral density. The equalization of the drive waveform can be accomplished directly, by generating an error correction from the difference between the control power spectral density and the reference power spectral density. Or the equalization can be accomplished indirectly through a knowledge of the system frequency-response function magnitude. The required *system frequency-response function* (see Chap. 21) is the ratio of the Fourier transform of the control waveform (usually an acceleration) and the Fourier transform of the drive voltage waveform. Only the magnitude of the frequency response function is required for random control, since the relative phase between frequencies is random and not controlled. Samples of the corrected drive waveform are fed through the D/A converter and a low-pass filter to the power amplifier shown in Fig. 25.10, completing the loop. As in the case of digital sine-wave control, corrections to the drive are not made continuously in the digital random-vibration control system. Many samples of the drive (often thousands) are output between corrections. As in digital sine-wave control the time between drive corrections is called the *loop time*. The loop time for digital random vibration control systems can be from a fraction of a second to a few seconds.

The speed at which the system can correct the control spectrum is determined by two factors. First is the loop time. The second is the number of spectral averages required to generate a statistically sound estimate of the control power spectral density. The loop time is usually the shorter of the two. Typically a compromise is required: An estimate of the power spectral density with a significant error is used, but only a fraction of the correction is made each loop.

The equivalent bandwidth of the DFT filters is dependent on the number of lines in the DFT and the sample rate of the D/A and A/D converters. Both parameters (number of lines and the sample rate) are usually options chosen by the operator.

Mixed Mode Testing.   Digital vibration test control systems are available which can control several sine waves superimposed on a stationary random vibration test. This is called *sine-on-random* vibration testing or *swept-sine-on-random* vibration testing. Systems are also available which can control swept narrow bands of nonstationary random superimposed on a stationary random vibration test. This is called *swept-narrow-band-random-on-random* testing.

Random and Sine Control Systems with a Single-Drive Signal (One D/A Converter) and Several Control Channels (Two or More A/D Converters).
The simplest case described above is for a single-drive (one D/A converter) and a single-control channel (one A/D converter). Many systems have several A/D converter channels available. The extra channels can be used for control or measurements. Since only a single-drive channel is available, the several control waveforms must be reduced to a single control signal for feedback control. Several options are available:

- For random vibration the power spectra of the control waveforms can be averaged to form a single control signal. For sine waves the amplitudes are averaged. A phase sensitive average in the time domain is not used because the phase differences between control channels can cause problems. For example, two sine waves with equal magnitude and a 180° phase difference will average to zero if the time-histories are added.

- The power spectrum (in the case of random vibration) or the sine amplitude (in the case of sine waves) of each control waveform can be compared with a reference spectrum (or amplitude; not necessarily the same reference spectrum), and a composite error can be formed. The most common strategy is to select the control waveform which has the largest positive error (a control amplitude larger than its reference) and use this error for feedback. If none of the errors are positive, the smallest negative error is used. This is called *extremal control,* i.e., the first control waveform to reach its reference, the extreme, will be in control. In some cases, the reference for some of the control waveforms is a flat (a constant independent of frequency) spectrum or amplitude. In this case the spectrum or amplitude is called *a limit spectrum* or *a limit amplitude.*

- For sine-wave testing an additional option is sometimes available. The sine-wave amplitude of the control waveform is held constant, while the frequency of the sine wave is adjusted to keep the relative phase between two control waveforms constant. This is a *phase-locked loop* and is useful for resonance dwell testing and fatigue testing. A *resonance dwell test* is a sine-wave test where the frequency is held at a resonance frequency for a specified period of time. Using this technique, a resonance can be tracked as the resonance frequency changes because of structural changes during the test.

Random and Sine Control Systems with Several Drive Signals and Control Channels. The simplest case are systems designed to drive several vibrators with a different test item attached to each vibrator. These systems are essentially several single-drive/single-output control systems (as described above) packaged in a single hardware system.

The more complicated system is when several vibrators, each with its own drive, are attached to a single test item. The attachments can be at several points in a single direction or at one point in several directions or a combination of both. If the drive of one exciter for these systems causes response at more than one control point, the system is *cross-coupled*. Each vibrator is typically associated with a unique control point. The motion of a control point caused by the drive from exciters not associated with the control point is called *cross-coupling*. Control of a cross-coupled system requires management of the cross-coupling effects. In the simplest case where the cross-coupling is zero (the drive to each exciter results in motion at only one control point) or near zero control is accomplished using several single-drive/single-output control systems. Another simple case, described in Chap. 21, is used in modal testing. The exciters used in modal testing do not require a control system since no attempt is made to control the amplitude and relative phase of each drive. Cases where the cross-coupling is not zero and control is desired require a *cross-coupled control system*. Simplified, part of the drive to each exciter is used to control the unwanted response at the control points caused by the cross-coupling. A mechanical design to minimize the cross-coupling will simplify this task. Cross-coupled control systems for random vibration and sinusoidal waveforms are briefly described below.

For random vibration tests with a single control point, the test is defined in terms of the power spectral density of the control point. For random vibration tests with several control points and several drive waveforms the test can be defined in terms of the cross-spectral density matrix of the control points. Each element in *a cross-spectral density matrix* is the cross-spectral density (defined in Chap. 27) between a pair of control points, a complex function of frequency. The diagonal elements of the control cross-spectral density matrix are the real power spectral densities of the control waveforms. For sine waves, the amplitude and relative phase, as a function of frequency, of each control waveform are specified.

The same block diagram described for the single-input/single-output control system (Fig. 25.11) can be used. The drive waveform of the single-input/single-output control system is replaced by a vector of drive waveforms. Each element in the *drive waveform vector* is the time-history of the drive for a single exciter. Each drive waveform will provide the input for separate exciters which are attached to a common test item. The control waveform of the single-input/single-output control system, in Fig. 25.11, is replaced by a vector of control waveforms. Each element in the vector is a time-history: the measurement from a single transducer mounted at one of several control points on the test item or on the exciters. For the single-input/single-output control system, the magnitude of the system frequency function was used to correct (equalize) the drive waveform. For the several-input/several-output system the magnitude of the system frequency response function is replaced by a *matrix of frequency-response functions* relating each control to each of the drives. Both the amplitude and phase of these frequency-response functions are required. Divisions in the control algorithms for the single-input/single-output case are replaced by multiplication by a matrix inverse in the several-input/several-output case. Each element in the matrix inverse is a function of frequency; hence the inversion must be accomplished at each frequency. It is not required that the number of drive waveforms is the same as the number of control waveforms, although this is the usual case. If they are the same, the matrix of frequency-response functions is square (the same number of rows and columns) and an inverse can usually be defined at each frequency. If they are not the same, a pseudo inverse must be defined. The most common pseudo inverse used is based on singular value decomposition, which is discussed in many texts on matrix algebra. If the number of control waveforms is less than the number of drive waveforms, the system is said to be *underspecified,* and several solutions can usually be found. A solution is usually picked which will distribute the load between the vibrators in some sort of an optimum fashion. If the number of control waveforms exceeds the number of drives waveforms, the system is said to be *overspecified,* and a solution may not exist. In this case a best solution, usually in a least squares sense, is chosen. In some cases, even for underspecified or square systems, the rank of the matrix of frequency-response functions is not full at some frequencies, and again some form of optimum solution must be chosen at these frequencies. Nonlinear response and measurement noise can seriously effect the determination of the inverse of the matrix of frequency-response functions and can seriously effect the quality of the solution. A large dynamic range in the frequency-response functions can also effect the condition of the matrix of frequency-response functions at some frequencies, which will have large impacts on the quality of the solution. Good mechanical design (the design of the vibrators, the manner in which they are connected to the common test item, and the location of the control points) can reduce these problems. Poor mechanical design can make the systems not usable. Great care is required to obtain a stable satisfactory solution. The vector of drive waveforms is typically generated using one of the methods outlined in Ref. 2. Details of cross-coupled control systems are covered in Refs. 3–6.

The difficulty in building these systems, designing the control system, and specifying the test parameters increases much faster than linearly with the number of exciters. This limits practical random vibration and sine-wave control systems to a few exciters. Successful systems have been built with three exciters and a bandwidth of about 2 kHz. Systems with a lower bandwidth and more exciters have also been built. Control systems for predetermined waveforms, discussed in the next section, with a bandwidth of 50 to 100 Hz have been built for as many as 18 actuators.

Predetermined Waveform Control Tests.   In many cases the waveform desired for reproduction is known or can be determined. In this case the control system will attempt to reproduce the predetermined waveform at the control point.

If the predetermined waveform time-history is short, this is called transient or shock testing. Examples include the reproduction of simple shock waveforms like half sine waves, the reproduction of a time-history with a specified shock spectrum, and the reproduction of a recorded field transient time-history. If the waveforms are longer, the method is simply called *predetermined waveform control.* An example is the reproduction of a road profile in a road simulator for motor vehicle testing.

In every case care must be taken to assure that the limitations of the electrodynamic or hydraulic vibrator are met. These can include limits on acceleration, force, velocity, and displacements; the initial and final values of acceleration, velocity, and displacement must generally be zero; and limits on the voltage and current capabilities of the power supply driving the vibrator. If the waveforms do not initially conform to these requirements, they must be modified or compensated in some manner to meet the requirements before the test is attempted.

Reproduction of Simple Shock Waveforms. The reproduction of a simple shock acceleration waveforms (for example, a half-sine waveform) is relatively straightforward, except the waveforms do not generally meet the initial and final value requirements stated above. The velocity change of these waveforms is not generally zero. Therefore, the shock must be modified or compensated to remove the velocity and displacement change. The compensation must be done in a manner that the peak requirements do not exceed the vibrator limitations but which does not compromise the purpose of the test. Several methods are available to accomplish this compensation.[7] The methods all involve the removal of low-frequency energy from the Fourier energy spectrum of the waveform.

Reproduction of a Waveform with a Specified Shock Spectrum. The synthesis or generation of a time-history with a specified shock spectrum is a separate topic[7] and is only discussed briefly here. The shock spectrum is discussed in Chap. 23. In general the synthesis is not unique. Many time-histories can have essentially the same shock spectrum. In summary, the synthesized transient is usually composed of a sum of component transients, where each component has energy concentrated in a narrow band of frequencies. Each of the components is adjusted with several parameters such that the shock spectrum of the composite transient will approximate the required shock spectrum. The parameters usually consist of an amplitude parameter, a frequency parameter, a duration parameter, and a delay parameter. Two common waveform types, used for the components of shock spectrum waveform synthesis, are exponentially decaying sinusoids and *WAVSYN* or *wavelet*, which is a sine wave multiplied by a half sine, where the half-sine duration is equal to an odd number of half cycles of the sine wave. A *half sine* is a sine wave extending from 0 to π radians. The advantage of the WAVSYN waveform is that the first two integrals are zero; therefore the velocity and displacement changes are zero: a requirement for reproduction on a vibrator.

The synthesis is usually off line (before the test is started), and then the waveform is reproduced using a predetermined waveform control system.

Reproduction of Recorded Transient Time-Histories. The recorded time-history is stored in memory or on disk, and the time-history is reproduced using a predetermined waveform control system. The recorded time-history must meet the limitations of the vibrators or be modified to meet the requirements before the waveforms are reproduced.

Predetermined Waveform Control System. The hardware configuration shown in Fig. 25.11 is used. As for sine wave and random vibration, the drive voltage waveform is used to excite the shaker system with the test item attached. The response of the system at a control point is monitored. A predetermined waveform is reproduced on a vibration machine (either electrodynamic or hydraulic) by first measuring the system frequency-response function. The system frequency-response function relates the control waveform to the electrical drive waveform. The system frequency-response function is measured by exciting the system: a voltage waveform with a wide bandwidth is output through the D/A converter into the vibration machine power amplifier, with the test item attached. The test item is required because feedback from the test item will change the system frequency-response function. Numerous waveforms can be used for excitation including an impulsive transient, the predetermined waveform, a continuous random waveform, usually white noise (white noise has a power spectral density which is a constant independent of frequency), or repeated short bursts of random vibration. The last method is most commonly used. In all cases it is important for the excitation waveform to have certain properties including energy at all frequencies of interest, sufficiently small amplitude that the test item is not damaged from the excitation, large enough amplitude that a linear extrapolation to the test level will not cause significant errors, and some averaging of the frequency-response function estimate to reduce the effects of nonlinear response and measurement noise. The system frequency-response function is usually estimated by taking the ratio of the cross-spectral density, between the control point and the drive, and the power spectral density of the drive waveform. This results in an estimate of both the amplitude and phase of the system frequency-response function.

Problems requiring the solution for the output of a system, given the input to the system and an impulse-response function of the system (the impulse response is the inverse Fourier transform of the system frequency-response function), are known as *convolution problems,* since the solution can be expressed in terms of a *convolution integral* between the input and the impulse-response function:

$$f(t) \times h(t) = \int_{-\infty}^{\infty} f(\tau)h(t - \tau)\,d\tau \qquad (25.2)$$

where $f(t)$ is the input, $h(t)$ is the impulse response function, and $f(t) \times h(t)$ is the convolution of $f(t)$ and $h(t)$. The duration of the convolution is equal to the sum of the durations of two waveforms being convolved. A convolution in the time domain is equivalent to a product in the Fourier frequency domain, which can be seen by taking the Fourier transform of both sides of the convolution integral. In problems where the result of a convolution integral is known [the right side of Eq. (25.2)], one of the terms within the convolution integral is known [one of the terms on the left side of Eq. (25.2)], and a solution for the other term within the convolution integral is required are known as *deconvolution problems.* Deconvolution problems require a division in the Fourier frequency domain. Deconvolution problems are known for their numerical difficulties.

A simple explanation of the difficulty with deconvolution is as follows. Most frequency-domain functions contain measurement or numerical noise. If the amplitude of the frequency-domain function becomes small at some frequencies, the relative error increases since the noise tends to be independent of frequency. When a large relative error exists in the denominator of a ratio of frequency-domain functions, the equation is said to be *ill-conditioned.* If the denominator in the ratio has minima with a large relative error, the ratio will have a large error unless a minima in the numerator coincides with the minima in the denominator. In the limit, a zero in the denominator results in an undefined value in the ratio. The problem is aggravated when both amplitude and phase are required, because the large relative error in the minima results in a large phase error in the minima. This results in a large phase error of the ratio.

The estimation of the system frequency-response function is a deconvolution problem. The drive waveform and control waveform are known, and an estimate of the frequency-response function (or equivalently, the impulse-response function) is desired, requiring a division by the drive waveform power spectral density. This is why it is important that the drive waveform used to estimate the system frequency-response function have energy at all frequencies in the bandwidth of the estimated frequency-response function. The estimation of the frequency-response function using the ratio of the cross to power spectral density is a linear least squares solution, optimized for noise on the control waveform and no noise on the drive waveform. Nonlinear response of the shaker system, or the nonlinear response of the test item (which feeds back to the control point), will cause difficulties.

The drive required to reproduce the predetermined waveform at the control point is then computed with this knowledge of the system frequency-response function. This is also a deconvolution problem. The desired output is known (the predetermined waveform), an estimate of the system frequency-response function is known (or equivalently the system impulse-response function), and the input (the electrical drive) is desired, requiring a division of the Fourier transform of the predetermined waveform by the estimated system frequency-response function.

Unfortunately, the frequency-response function of vibrators used for predetermined waveform control (electrodynamic and hydraulic vibrators) usually contain low values in certain frequency ranges which can cause problems when the drive required to reproduce the predetermined waveform is computed. The frequency-response function often rolls off at low frequencies (a minima at zero frequency), rolls off at high frequencies (a minima at infinite frequency), and has notches (minima) at the antiresonances of the system. Care must be taken to avoid deconvolution difficulties in predetermined waveform control systems. Algorithms which will converge to a stable solution in the presence of instrumentation noise, system nonlinearities, and minima in the frequency-response function are essential to minimize problems when ill-conditioning is encountered. The problem is more difficult for predetermined waveform control than for sine-wave control or random waveform control for two reasons: The results are phase sensitive for predetermined waveform control, where the results are not phase sensitive sine-wave and random waveform control with a single drive, and the control system for a predetermined waveform is essentially open loop during the test (the drive waveform is not corrected during the time the drive waveform is exciting the vibrator), where sine-wave and random waveform control systems are closed loop (the drive waveform is corrected during the test).

If the duration of the predetermined waveform is less than the block size used to estimate the system frequency-response function, the calculation of the required drive waveform is straightforward (the simple ratio of the discrete Fourier transform (DFT) of the predetermined waveform and the system frequency-response function), with care taken to avoid leakage (see Chap. 13) and circular convolution errors. *Circular convolution errors* occur if the duration of a convolution is longer than the block size of the DFT used to estimate the convolution. For longer waveforms, circular convolution errors can be avoided using a technique known as high-speed convolution with overlap and add.[1] The method assumes a linear extrapolation from the drive levels used to measure the system frequency-response function to the test level.

Sometimes the difference between the observed control waveform and the predetermined waveform is used as an error waveform to correct either the estimate of the system frequency-response function or the drive waveform. This is acceptable if several reproductions of the predetermined waveform are allowed and if great care is taken to avoid the problems of deconvolution. Sometimes, the first reproduction of the predetermined waveform is accomplished at a reduced amplitude, and only after a satisfactory reproduction at the reduced level is obtained is the test repeated at the required amplitude. This assumes that control point response will increase linearly with an increase in the drive waveform.

If the test requirement is in the form of a shock spectrum, the shock spectrum of the control waveform is compared to the reference shock spectrum and/or the shock spectrum of the synthesized predetermined waveform. Unfortunately, if the shock spectrum for the control waveform differs significantly from the required shock spectrum, steps which must be taken to correct the problem are approximate and iterative, since a change in the frequency content of a waveform will affect the shock spectrum in a nonlinear manner.

The control system for the reproduction of predetermined waveforms is essentially open loop during the test. The frequency-response function is estimated before the test, and the drive waveform is calculated before the test from the predetermined waveform and the measured system frequency-response function. The vibrator is then excited with no feedback during the reproduction of the predetermined waveform. Feedback can be provided only if the predetermined waveform is reproduced more than once.

Predetermined Waveform Control System with Several Exciters and Several Control Points. Just as for random vibration and sine-wave digital control systems, systems with several drive waveforms and several vibrators, attached to a single test item with several control points, are used in predetermined waveform control. Examples include hydraulic road simulators and seismic simulators. Road simulators with as many as 18 hydraulic vibrators attached to a vehicle have been used. An example is shown as Fig. 25.12. The digital control system for the road simulator is a cross-coupled predetermined waveform control system. The simulation of seismic motions in six degrees-of-freedom (three translation and three rotation) are accomplished by mounting test items on tables which are excited with hydraulic exciters and controlled with a cross-coupled predetermined waveform control system. The design of cross-coupled predetermined waveform control systems follows Fig. 25.11 and the discussion of cross-coupled random vibration and sine-wave control systems. The drive waveform is replaced by a vector of drive waveforms, and the con-

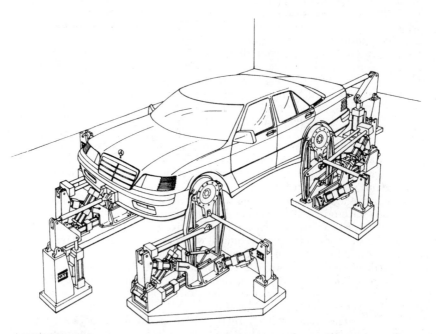

FIGURE 25.12 A road simulator which uses a cross-coupled multiple-drive/multiple-control-point predetermined waveform control system. The predetermined waveforms (with a bandwidth of about 1 to 50 Hz) are measured on the vehicle while driving on a road. The predetermined waveforms are reproduced on the vehicle during the simulation on the road simulator. Four hydraulic actuators drive each wheel hub, and two hydraulic actuators drive the vehicle fore and aft at the bumpers. (*MTS Corp.*)

trol waveform is replaced by a vector of control waveforms. The reference is replaced by a vector of predetermined waveforms. The system frequency-response function is replaced by a matrix of frequency-response functions relating each of the control waveforms to each of the drives waveforms.[6] As for single-point-control, the solution for the vector of drive waveforms which will reproduce the vector of predetermined waveforms at the control points is a deconvolution problem with the division by a Fourier transform, or a DFT is replaced by multiplication by a matrix inverse or a pseudo inverse. To compute the frequency-response function matrix, multiplication by the inverse of the drive cross-spectral density matrix is required. To compute the vector of drive waveforms, for the reproduction of the vector of predetermined waveforms at the control points, multiplication by the inverse of the matrix of system frequency-response functions is required. Even more than for the single-control-point case, great care in both mechanical design and the design of algorithms must be taken to avoid the ill-conditioning problems for a satisfactory solution.[5]

Application and Selection of Digital Vibration Controllers. Since the same hardware can be used for all types of testing, some characteristics of the system are common to all of them.

General Considerations for All Systems

- The computer operator interface should be clear and easy to use.
- A test profile should be easy to set up, which includes the recall and editing of previous test profiles. A *test profile* is a file of all the parameters needed to set up and execute a test.
- The performance of the system during a test should be clearly presented to the operator.
- Documentation of the system characteristics, performance, and options should be complete.
- The results of the test should be well documented for post-test analysis.
- The system must meet all test requirements and required test types of vibration testing (sinusoidal waveforms, random vibration, and/or predetermined waveforms). Test requirements and test types will vary by application.
- The control system may require a dynamic range of 60 to 90 dB or more. Swept sine testing typically requires a larger dynamic range than random vibration or predetermined waveform control. Control of acceleration waveforms on hydraulic exciters also requires a large dynamic range. The *dynamic range* is the ratio of the largest waveform which can be represented to the smallest waveform which can be represented and is usually expressed in decibels. The number of bits used to represent a sample amplitude in a A/D or D/A converter limits the dynamic range (see Chap. 27). Approximately 6 dB/bit can be achieved. Systems with 12- to 16-bit converters are common. For the full dynamic range of a D/A or a A/D converter to be realized, the signal-to-noise ratio of all analog equipment used to condition the waveforms (filters, attenuators, amplifiers, etc.) must be as least as great as the dynamic range of the converters.
- The antialiasing filters must be of high quality since the control waveforms frequently contain energy beyond the Nyquist frequency, which is half the sampling frequency.
- An adequate number of A/D converter channels should be provided. Additional A/D converter channels are useful for both control and measurements. However, for some systems additional A/D converter channels can slow down the loop time.

The advantages of more channels and the disadvantage of a potentially slower loop time must be weighed.

- For test item and the vibrator protection, upon loss of power to or power transients in the control system, or when the system aborts, the D/A converters should return to 0 volts, without voltage transients.

- Calibration and certification procedures for the D/A converters, A/D converters, and the control system should be established and acceptable.

- Training and service should be adequate and available on a timely basis.

- Protection for the shaker and test item should be provided but should be versatile enough to allow tests in difficult circumstances.

- An established customer base for commercial systems is useful to demonstrate the quality of the control system and to provide needed user feedback. Discussions with other users are useful to determine a control systems strengths, limitations, and idiosyncrasies.

Some Considerations for a Sine Control System

- The loop time should be short (typically a small fraction of a second) to allow for fast correction of the drive signal. The compression speed should be a function of frequency and should be user selectable.

- A digital tracking filter for the control channels should be available. A proportional bandwidth filter (the bandwidth is a constant percentage of the center frequency) is usually best. The bandwidth should be user selectable.

- The control system should perform satisfactorily in the presence of the common problems of sine-wave control including large amounts of harmonic distortion on the control waveform, large amounts of noise added to the control waveform, interruptions of the sine-wave vibration test caused by amplifier dumps (an *amplifier dump* is the abrupt shut down of the vibrator power source, generally initiated by safety circuits), operator intervention, and excessive dynamic range.

- The total harmonic distortion in the drive sine wave should be low. The available sine-wave frequencies should be essentially continuous, with no discontinuities introduced when the frequency is changed. Small amounts of harmonic distortion in the drive waveform are amplified by the vibrator system at some frequencies, requiring the drive to have low distortion.

Some Considerations for a Random Vibration Control System

- The loop time should be short (typically a fraction of a second) to allow for fast correction of the drive waveform.

- The number of frequency lines in the control power spectral density should be selectable, and the maximum should be at least 800 lines. More frequency lines will improve the frequency resolution but will increase the loop time. A compromise is required to balance these two factors. The optimum solution varies with the test item and test requirements, requiring a selection of the number of frequency lines.

- The averaging time for the control power spectral density estimate should be selectable. A longer averaging time improves the accuracy of the control but increases the time required to correct the drive. A compromise is required to balance these two factors. The optimum solution varies with the test item and test requirements, requiring a selection of the averaging time.

- The total bandwidth of the drive power spectral density should be selectable. The selection affects the sample rate, which affects the effective bandwidth of each control line. The combination of the total bandwidth and the number of lines controls the frequency resolution of the control. Many choices should be available. If the total bandwidth of the drive power spectral density is significantly larger than the bandwidth of the reference spectrum, control lines will be essentially unused (effectively set to zero), compromising the performance of the control system.
- The control algorithm should result in a stable solution allowing the control of test items with a large dynamic range and with significant amounts of measurement noise and distortion on the control waveforms.

Some Considerations for a Predetermined Waveform Reproduction Control System

- The wide selection of predetermined waveform types should be available. The generation of a user-defined predetermined waveform should be available and easy to use.
- The selections of sample rate and waveform duration should be selectable, with a large number of choices available.
- The intermediate functions, like the frequency-response function, the computed electrical drive, and the predetermined control waveform should be available for inspection before the test is run. The waveforms should be available in both the time domain and the Fourier frequency domain. Several deconvolution problems can be detected by viewing these waveforms.
- A library of tools to help control the deconvolution difficulties is useful.
- If feedback is used to correct the drive on multiple reproductions of the predetermined waveform, the ability to turn off the feedback corrections is useful. Since deconvolution problems sometimes make the feedback unstable, the option to disable them is useful.
- If the shock spectrum is required, the shock spectrum should include both the primary (the peak response which occurs during the application of the shock) and the residual (the peak response which occurs after the application of the shock). The shock spectrum frequency range should be selectable and over a wide range of natural frequencies, from less than 1/1000 of the sample rate to natural frequencies several times the sample rate. The acceleration waveform should be corrected for zero offsets, and the waveform should not be truncated before the waveform has decreased to near zero.
- If synthesis of a waveform to match a shock spectrum is required, several waveform types are desirable. The algorithm used to synthesize the waveform should be robust (i.e., the algorithm should converge to a stable solution for a wide variety of initial conditions). The required shock spectrum, the shock spectrum of the synthesized waveform, and the shock spectrum of the control waveform should all be available for display. The shock spectrum damping value should be selectable and displayed. Two basic models for the shock spectrum are in common use—the absolute-acceleration-input/absolute-acceleration-response, and the absolute-acceleration-input/relative-displacement-response (usually displayed as an equivalent static acceleration). Both models should be available. Other models for the shock spectrum are usually variations of these two.
- The velocity and displacement waveforms of the predetermined waveform and the control waveform should be available for inspection.

REFERENCES

1. Gold, B., and C. Rader: "Digital Processing of Signals," McGraw-Hill, New York, 1969.

2. Smallwood, D. O., and T. L. Paez: "A Frequency Domain Method for the Generation of Partially Coherent Normal Stationary Time Domain Signals," *Shock and Vibration,* **1**(1):373–382, John Wiley and Sons, New York, October 1994.

3. Smallwood, D. O.: "A Random Vibration Control System for Testing a Single Test Item with Multiple Inputs," SAE paper 821482, SAE Publication SP-529, also published in 1982 SAE Transactions, Soc. of Automotive Engineers, Warrendale, Penn., September 1983.

4. Stroud, R. C., and G. A. Hamma: "Multiexciter and Multiaxis Vibration Exciter Control Systems," *Sound and Vibration,* **22**(4):18–28, Acoustical Publications Inc., Bay Village, Ohio, April 1988.

5. Cryer, B. W., P. E. Nawrocki, and R. A. Lund: "A Road Simulation System for Heavy Duty Vehicles," SAE 760361 (1976), Soc. of Automotive Engineers, Warrendale, Penn.

6. Fletcher, J. N., H. Vold, and M. D. Hansen: "Enhanced Multiaxis Vibration Control Using a Robust Generalized Inverse System Matrix," 1994 Proc. of the Institute of Environmental Sciences (IES), **2**:418–426, IES, Mount Prospect Ill., May 1994.

7. Smallwood, D. O.: "Shock Testing on Shakers by Using Digital Control," IES Technology Monograph, 28 pp., 1986, Institute of Environmental Sciences, Mount Prospect Ill.

CHAPTER 26

PART I: SHOCK TESTING MACHINES

Richard H. Chalmers

INTRODUCTION

Equipment must be sufficiently rugged to operate satisfactorily in the shock and vibration environments to which it will be exposed and to survive transportation to the site of ultimate use. To ensure that the equipment is sufficiently rugged and to determine what its mechanical faults are, it is subjected to controlled mechanical shocks on shock testing machines. *Mechanical shock* is a nonperiodic excitation (e.g., a motion of the foundation or an applied force) of a mechanical system that is characterized by suddenness and severity, and it usually causes significant relative displacements in the system. The severity and nature of the applied shocks are usually intended to simulate environments expected in later use or to be similar to important components of those environments. However, a principal characteristic of shocks encountered in the field is their variety. These field shocks cannot be defined exactly. Therefore shock simulation can never exactly duplicate shock conditions that occur in the field.

There is no general requirement that a shock testing machine reproduce field conditions. All that is required is that the shock testing machine provide a shock test such that equipment which survives is acceptable under service conditions. Assurance that this condition exists requires a comparison of shock test results and field experience extending over long periods of time. This comparison is not possible for newly developed items. It is generally accepted that shocks that occur in field environments should be measured and that shock machines should simulate the important characteristics of shocks that occur in field environments or have a damage potential which by analysis is shown to be similar to that of a composite field shock environment against which protection is required.

A *shock testing machine* (frequently called a *shock machine*) is a mechanical device that applies a mechanical shock to an equipment under test. The nature of the shock is determined from an analysis of the field environment. Tests by means of shock machines usually are preferable to tests under actual field conditions for four principal reasons:

1. The nature of the shock is under good control, and the shock can be repeated with reasonable exactness. This permits a comparative evaluation of equipment under test and allows exact performance specifications to be written.

2. The intensity and nature of shock motions can be produced which represent an average condition for which protection is practical, whereas a field test may involve only a specific condition that is contained in this average.

3. The shock machine can be housed at a convenient location with suitable facilities available for monitoring the test.

4. The shock machine is relatively inexpensive to operate, so it is practical to perform a great number of developmental tests on components and subassemblies in a manner not otherwise practical.

SHOCK-MACHINE CHARACTERISTICS

DAMAGE POTENTIAL AND SHOCK SPECTRA

The damage potential of a shock motion is dependent upon the nature of an equipment subjected to shock, as well as upon the nature and intensity of the shock motion. To describe the damage potential, a description of what the shock does to an equipment must be given—a description of the shock motion is not sufficient. To obtain a comparative measure of the damage potential of a shock motion, it is customary to determine the effect of the motion on simple mechanical systems. This is done by determining the maximum responses of a series of single degree-of-freedom systems to the shock motion and considering the magnitude of the response of each of these systems as indicative of the damage potential of the shock motion to these standard systems. The responses are plotted as a function of these natural frequencies. A curve representing these responses is called a *shock spectrum,* or *response spectrum* (see Chap. 23). Its magnitude at any given frequency is a quantitative measure of the damage potential of a particular shock motion to a single degree-of-freedom system of that natural frequency. This concept of the shock spectrum originally was applied only to undamped single degree-of-freedom systems, but the concept has been extended to include systems in which any specified amount of damping exists.

The response of a simple system can be expressed in terms of the relative displacement, velocity, or acceleration of the system. It is customary to define velocity and acceleration responses as $2\pi f$ and $(2\pi f)^2$ times the maximum displacement response, where f is frequency expressed in hertz. The corresponding response curves are called *displacement, velocity,* or *acceleration shock spectra.* A more detailed discussion of shock spectra is given in Chap. 23.

MODIFICATION OF CHARACTERISTICS BY REACTIONS OF TEST ITEM

The shock motion produced by a shock machine may depend upon the mass and frequency characteristics of the item under test. However, if the effective weight of the item is small compared with the weight of the moving parts of the shock machine, its influence is relatively unimportant. Generally, however, the reaction of the test item on the shock machine is appreciable and it is not possible to specify the test in terms of the shock motions unless large tolerances are permissible. The test item acts like a dynamic vibration absorber (see Chap. 6) for fixed-base natural frequencies of the test item. (Fixed-base natural frequencies of a flexibly mounted item are those which would be calculated if the base on which the item was mounted was rigid and of infinite mass. They correspond to the antiresonance frequencies of the system.) If the item is relatively heavy, this causes the shock spectra of the exciting shock to have minima at these frequencies; it also causes its mounting foundation to have these minima during shock excitation at field installations. Shock tests and design factors are sometimes established on the basis of an envelope of the maximum values of shock spectra. However, maximum stresses in the test item will most probably occur at the antiresonance frequencies where the shock spectrum exhibits minimum values. To require that the item withstand the upper limit of spectra at these frequencies may result in overtesting and overdesign. Considerable judgment is therefore required both in the specification of shock tests and in the establishment of theoretical design factors on the basis of field measurements. See Chap. 42 for a more complete discussion of this subject.

DOMINANT FREQUENCIES OF SHOCK MACHINES

The shock motion produced by a shock machine may exhibit frequencies that are characteristic of the machine. The frequencies may be affected by the equipment under test. The probability that these particular frequencies will occur in the field is no greater than the probability of other frequencies in the general range of interest. A shock test, therefore, discriminates against equipment having elements whose natural frequencies coincide with frequencies introduced by the shock machine. This may cause failures to occur in relatively good equipment whereas other equipment, having different natural frequencies, may pass the test even though of poorer quality. Because of these factors, there is an increasing tendency to design shock machines to be as rigid as possible, so that their natural frequencies are above the range of frequencies that might be strongly excited in the equipment under test. The shock motion is then designed to be the simplest shape pulse that will give a desired shock motion or spectrum.

CALIBRATION

A *shock-machine calibration* is a determination of the shock motions, or spectra, generated by the machine under standard specified conditions of load, mounting arrangements, methods of measurement, and machine operation. The purpose of the calibration is not to present a complete study of the characteristics of the machine but rather to present a sufficient measure of its performance to ensure the user that the machine is in a satisfactory condition. Measurements should therefore be made under a limited number of significant conditions that can be accurately specified and easily duplicated. Calibrations are usually performed with deadweight loads rigidly attached to the shock machine.

The statement of calibration results must include information relative to all factors that may affect the nature of the motion. These include the magnitude, dimensions, and type of load; the location and method of mounting of the load; factors related to the operation of the shock machine; the locations and mounting arrangements of pickups; and the frequency range over which the measurements extend.

SPECIFYING A SHOCK TEST

Two methods of specification are employed in defining a shock test: (1) a specification of the shock motions (or spectra) to which the item under test is subjected and

(2) a specification of the shock machine, the method of mounting the test item, and the procedure for operating the machine.[1]

The first method of specification can be used only when the shock motion can be defined in a reasonably simple manner and when the application of forces is not so sudden as to excite structural vibration of significant amplitude in the shock machine. If equipment under test is relatively heavy, and if its normal modes of vibration are excited with significant amplitude, the shock motions are affected by the load; then the specified shock motions should be regarded as nominal. If comparable results are to be obtained for tests of different machines of the same type, the methods of mounting and operational procedures must be the same.

The second method of specification for a shock test assumes that it is impractical to specify a shock motion because of its complexity; instead, the specification states that the shock test shall be performed in a given manner on a particular machine. The second method permits a machine to be developed and specified as a standard shock testing machine. Those who are responsible for the specification then should ensure that the shock machine generates appropriate shock motions. This method avoids a difficulty that arises in the first method when measurements show that the shock motions differ from those specified. These differences are to be expected if load reactions are appreciable and complex.

A shock testing machine must be capable of reproducing shock motions with good precision for purposes of comparative evaluation of equipment and for the determination as to whether a manufacturer has met contractual obligations. Moreover, different machines of the same type must be able to provide shocks of equivalent damage potential to the same types of equipment under test. Precision in machine performance, therefore, is required on the basis of contractual obligations and for the comparative evaluation of equipments even though it is not justified on the basis of knowledge of field conditions.

Sometimes equipment under test may consistently fail to meet specification requirements on one shock machine but may be acceptable when tested on a different shock machine of the same type. The reason for this is that small changes of natural frequencies and of internal damping, of either the equipment or the shock machine, may cause large changes in the likelihood of failure of the item. Results of this kind do not necessarily mean that a test has been performed on a faulty machine; normal variations of natural frequencies and internal damping from machine to machine make such changes possible. However, standard calibrations of shock machines should be made from time to time to ensure that significant changes in the machines have not occurred.

SHOCK TESTING MACHINES

CHARACTERISTIC TYPES OF SHOCKS

The shock machines described below are grouped according to types of shocks they produce. When a machine can be classified under several headings, it is placed in the one for which it is primarily intended. One characteristic shared by all shock machines is that the motions they produce are sudden and likely to create significant inertial forces in the item under test. The types of shock shown in Fig. 26.1 are classified as (A) through (D), simple shock pulses, whose shapes can be expressed in a practical mathematical form; (E), single complex shock; and (F), a multiple shock. In contrast to simple shock pulse specification, the motions illustrated in Fig. 26.1 (E)

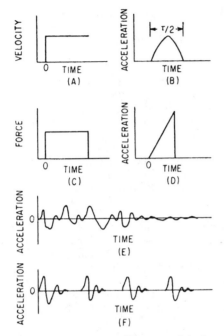

FIGURE 26.1 Characteristic types of shocks. (*A*) Velocity shock, or step velocity change. (*B*) Simple half-sine acceleration shock pulse. (*C*) Rectangular force pulse. (*D*) Sawtooth acceleration pulse. (*E*) Single complex shock. (*F*) Multiple shock.

and (*F*) often are the result of a shock test in which the shock testing machine, the method of mounting, and machine operations were specified.

Velocity Shock. A *velocity shock* is produced by a sudden change in velocity of the structure supporting the item under test. When the duration of the velocity change is short compared with the principal natural frequencies of the item under test, a velocity shock is said to have occurred. Figure 26.1*A* shows a nearly instantaneous change of velocity. Figure 26.1*B*, *C*, and *D* are also considered velocity shocks if the shortness criterion above is met. In *velocity shock*, the magnitude of the acceleration involved in the velocity change is unimportant, since the item is isolated from the maximum acceleration by the flexibility of its parts. Intensity of a *velocity shock* is defined by magnitude of the velocity change.

Displacement Shock. When a simple shock pulse is followed closely by an equal but oppositely directed pulse (for example, a pulse that is a full sine wave), the velocity change accruing during the first half cycle is canceled during the second half cycle. At the end of the pulse, the velocity of the item under test is the same as it was prior to the application of the pulse. However, this shock will result in an overall change of displacement of the item, half of the change occurring during the gain in velocity and half during the change back to the original velocity value.

High-Frequency Shock. Metal-to-metal impacts create high-acceleration, high-frequency, damped sine-wave shock oscillations in the vicinity of the impact. Since the frequency of these shocks usually exceeds natural frequencies of the structure in which they occur, the shocks are not readily transmitted far from the point of creation. Peak accelerations are high, but durations are short. Consequently, this type of shock lacks damage potential for all but brittle components of the item under test.

SIMPLE SHOCK PULSE MACHINES

Although shocks encountered in the field are usually complex in nature (for example, see Fig. 26.1*E*), it is frequently advantageous to simulate a field shock by a shock of mathematically simple form. This permits designers to calculate equipment response more easily and allows tests to be performed that can check these calculations. This technique is additionally justifiable if the pulses are shaped so as to provide shock spectra similar to those obtained for a suitable average of a given type of field conditions. Machines are therefore built to provide these simple shock motions. However, note that the motions provided by actual machines are only ideally simple. The ideal outputs may be given as nominal values; the actual outputs can only be determined by measurement.

Drop Tables.   A great variety of drop testers are used to obtain acceleration pulses having magnitudes ranging from 80,000g down to a few g. The machines each include a carriage (or table) on which the item under test is mounted; the carriage can be hoisted up to some required height and dropped onto an anvil. Guides are provided to keep the carriage properly oriented.

When large velocity changes are required, the carriage may be accelerated downward by a means other than gravity. Frequently, parts of the carriage, associated with its lifting and guiding mechanism, are flexibly mounted to the rigid part of the carriage structure that receives the impact. This is to isolate the main carriage structure from its flexible appendages so as to retain the simple pulse structure of the stopping acceleration.

Programming devices, placed upon the point of impact of the anvil, can be used to control the acceleration vs. time characteristic and the shape of the acceleration pulse. When rubberlike materials are used, the acceleration pulse can be made to approach the shape of a half-sine curve, Fig. 26.1B, which has the shock spectrum of Fig. 26.2.

A sawtooth acceleration pulse, Fig. 26.1D, has an acceleration shock spectrum that is relatively smooth with no extreme values which would make it discriminate

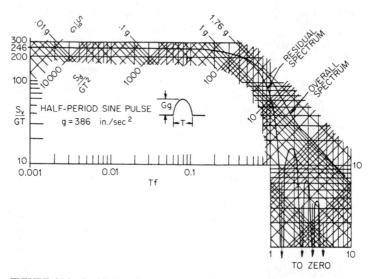

FIGURE 26.2   Residual and overall shock spectra of the half-sine acceleration pulse shown in the inset. The overall spectrum is the usual shock spectrum, i.e., the maximum response regardless of when it occurs. S_a, S_v, and S_d are, respectively, acceleration, velocity, and displacement shock spectra expressed in units of in./sec^2, in./sec, and in. G is the acceleration expressed in units of gravity g. T is the pulse duration; f is frequency. If the pulse length is 0.006 sec and the amplitude is 200g for a frequency of 100 Hz ($Tf = 0.6$), the overall shock spectral values of S_a, S_v, and S_d are 340g, 200 in./sec, and 0.32 in., respectively. (*Naval Res. Lab. Rept.*)

for or against an item under test because of excessive test severity at certain frequencies. Theoretically obtained shock spectra for such a pulse are shown in Fig. 26.3. This type of pulse can be obtained relatively easily by dropping a rigid carriage onto a cylindrical lead pellet with a conical top. This form of excitation provides a satisfactory test for many types of shock environments where it is required that the shock spectrum rise to a maximum value within the first 100 Hz and remain constant thereafter. The amplitude and duration of the pulse may be modified by changing the dimensions of the pellet, so as to provide an appropriate response level and rise time for the spectral curves.

A typical machine of this type is shown in Fig. 26.4. The carriage consists of a rigid platform on which the specimen is mounted and which embodies the impacting surface on its lower side. The platform is usually cast of aluminum and is designed to have no natural frequencies below about 1000 Hz. The impacting surfaces are of hardened steel.

Air Guns. Air guns frequently are used to impart large accelerations to pistons on which items under test can be attached. The piston is mechanically retained in position near the breech end of the gun while air pressure is built up within the breech. A quick-release mechanism suddenly releases the piston, and the air pressure projects the piston down the gun barrel. The muzzle end of the gun is closed so that the piston is stopped by compressing the air in the muzzle end. Air bleeder holes may be placed in the gun barrel to absorb energy and to prevent an excessive number of oscillations of the piston between its two ends.

A variety of such guns, used by the Naval Surface Weapons Center, can provide acceleration pulses as shown in Fig. 26.5A and B. The peak accelerations may extend from a maximum of about 1000g for the large-diameter (21 in, 53 cm) guns up to 200,000g for small-diameter (2 in, 5 cm) guns. The pulse length varies correspondingly from about 50 to 3 milliseconds. The maximum piston velocity varies from about 400 to 750 ft/sec (122 to 229 m/sec). The maximum velocities are not dependent upon piston diameter.

High-acceleration gas guns have been developed for testing electronic devices. The items under test are attached to the piston. The gun consists of a barrel (cylinder) that is closed at the muzzle end but which has large openings to the atmosphere a short distance from the muzzle end. The piston is held in place while a relatively

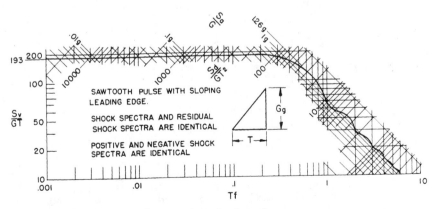

FIGURE 26.3 Shock spectra of a sawtooth acceleration pulse.

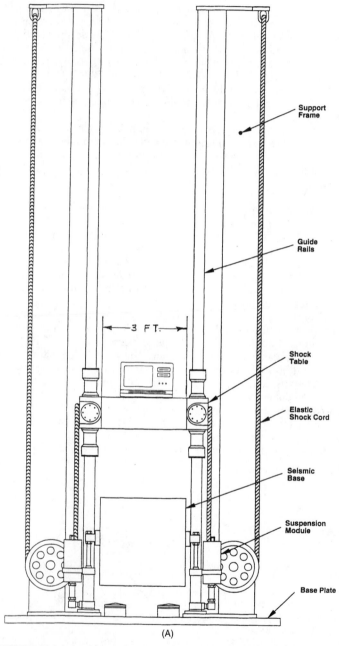

(A)

FIGURE 26.4 (A) Drop-table arrangement for use with programming devices between the impacting surfaces. Devices ranging from liquid programmers to simple pads of elastomeric or plastic materials are used to provide the desired shock pulse shape. Notice the shock cords which accelerate the table to create velocities beyond those that can be obtained with free fall. (*MTS Systems Corporation.*)

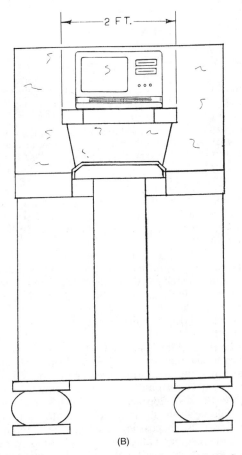

FIGURE 26.4 (B) A drop-table operated by compressed air. Compressed air raises the dropped mass to the desired elevation, and (if required) to accelerate the mass downward to the impacting surfaces. The shapes of shock pulses so obtained are controlled by programming devices between the impacting surfaces. (*AVEX, a subsidiary of J. M. Huber Corp.*)

low-pressure gas (usually air or nitrogen) is applied at the breech end of the gun. The piston is then released, whereby it is accelerated over a relatively long distance until it reaches the position along the length of the cylinder that is open to the atmosphere. This initial acceleration is of relatively small magnitude. After the piston has passed these openings, it is stopped by the compression of gas in the short closed end of the cylinder. This results in a reverse acceleration of relatively large magnitude. (Sometimes an inert gas, such as nitrogen, is used in the closed end to prevent explosions which might be caused by oil particles igniting under the high temperatures incident to the compression.) Thus, in contrast to the previously described devices, the major acceleration pulse is delivered during stopping rather than starting. An advantage of this latter technique is that the difficult problem of constructing a quick-release mechanism for the piston, which will work satisfactorily under the large forces exerted by the piston, is greatly simplified.

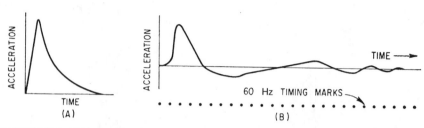

FIGURE 26.5   Typical acceleration-time curves for (*A*) 5-in. (13-cm) air gun; (*B*) 21-in. (53-cm) air gun. Peak accelerations extend to 5000*g*, with pulse lengths between 3 and 6 milliseconds for the 5-in. gun. The accelerations for the 21-in. gun extend to about 1000*g*. (*After U.S. Naval Ordnance Laboratory Report.*)

Vibration Machines.   Electrodynamic, hydraulic, and pneumatic vibration machines provide a ready and flexible source of shock pulses, so long as the pulse requirements do not exceed force and motion capabilities of the selected machine. See Chap. 25 for information.

Test Load Reactions.   In the above description of the output of shock machines designed to deliver simple shock pulses of adjustable shapes, it is assumed that the load imposed on the machine by the item under test has little effect on the shock motions. This is true only when the effective weight of the load is negligibly small compared with that of the shock machine mounting platform. If the effective weight of the load is independent of frequency, i.e., if it behaves as a rigid body, it is simple to compensate for the effect of the load by adjusting machine parameters. However, when the load is flexible and the reactions of excited vibrations are appreciable, the motions of the shock machine platform are complex. Specifications involving the use of these types of machines should require that the mounting platform have no significant natural frequencies below a specified frequency. The weight of this platform together with that of all rigidly attached elements, exclusive of the test load, also should be specified. Pulse shapes may then be specified for motions of this platform or for the platform together with given dead-weight loads. These may be specified as nominal values for test loads, but it is neither practical nor desirable to require that the pulse shape be maintained in simple form for complex loads of considerable mass.

COMPLEX SHOCK PULSE MACHINES

Because of the infinite variety of shock motions possible under field conditions, it is not practical or desirable to construct a shock machine to reproduce a particular shock that may be encountered in the field. However, it is sometimes desirable to simulate some average of a given type of shock motion. To accomplish this may require that the shock machine deliver a complex motion. A shock of this type cannot be specified easily in terms of the shock motions, since the motions are very complex and dependent on the nature and the mounting of the load. It is customary, therefore, to specify a test in terms of a shock machine, the conditions for its operation, and a method of mounting the item under test.

High-Impact Shock Machines.   The Navy high-impact shock machines are designed to simulate shocks of the nature and intensity that might occur on a ship exposed to severe but sublethal, noncontact, underwater explosions. Such severe shocks produce motions that extend throughout the ship. Equipment intended for shipboard use can demonstrate its ability to withstand the shock simulations produced by these high-impact shock machines and thus be considered capable of withstanding the actual underwater explosion environment.

Lightweight Machines.[1-4] The lightweight high-impact shock machine, shown in Fig. 26.6, is used for testing equipment weighing up to about 350 lb (159 kg). Equipment under test is attached to the anvil plate A. Method of attachment is constrained to resemble closely the eventual field attachments. The anvil is struck on the backside by the pendulum hammer C, or the anvil is rotated 90° on a vertical axis and struck on the end by the pendulum hammer. The drop hammer B can be made to strike the top of the anvil, thus providing principal shock motions in the third orthogonal direction. Shock spectra of shock motions generated by this machine are shown in Fig. 26.7. The spectrum for motion at the center of the plate illustrates the amplification of the spectrum level at a natural frequency of the plate (about

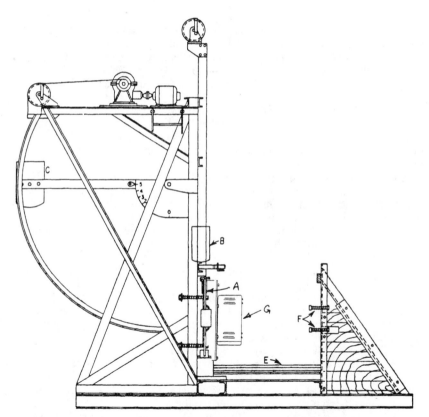

FIGURE 26.6 Navy high-impact shock machine for lightweight equipment. (A) Anvil plate. The anvil plate can also be oriented in the plane of the paper for an "end" hammer blow. (B) Hammer for vertical blow. (C) Hammer for horizontal blow. (D) Restoring springs for vertical blow. (E) Rail support for end blow. An upper support angle, not shown, is also required. (F) Positioning springs for end blow. (G) Item under test.

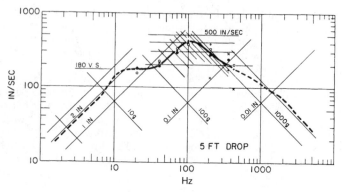

FIGURE 26.7 Shock spectra for a 5-ft back blow with a 57-lb (25.9 kg) load on the mounting plate for four different lightweight high-impact shock machines.

100 Hz) and some attenuation at higher frequencies. A complete study of the performance of this machine is given in Ref. 2.

Medium-Weight Machines.[2-4] This machine is used to test equipment that, with its supporting structures, weighs up to 7400 lb (3357 kg). Shown in Fig. 26.8, this machine consists principally of a 3000-lb (1361-kg) hammer and a 4500-lb (2041-kg) anvil. Loads are not attached directly to the rigid anvil structure but rather to a group of steel channel beams which are supported at their ends by steel members, which in turn are attached to the anvil table. The number of channels employed is dependent on the weight of the load and is such as to cause the natural frequency of the load on these channels to be about 60 Hz. The hammer can be dropped from a maximum effective height of 5.5 ft (1.68 m). It rotates on its axle and strikes the anvil on the bottom, giving it an upward velocity. The anvil is permitted to travel a distance of up to 3 in. (7.6 cm) before being stopped by a retaining ring. The machine is mounted on a large block of concrete which is mounted on springs to isolate the surrounding area from shock motions. The general nature of the shock is complex, similar to that of the lightweight machine. Little of the high-amplitude, high-frequency components of the shock motions are transmitted to the load. A complete study of this machine is given in Ref. 2.

Heavy-Weight Machines.[2-5] The *floating shock platform* (FSP), and the *large floating shock platform* (LFSP) are high-load-capacity shock machines of the high-impact category. They are rectangular barges fitted with semicylindrical canopies within which test items are installed as they are aboard ship. The shock motions comprising the test series are generated by detonating explosive charges beneath the water surface at various distances.

The FSP is 28 ft (8.5 m) long by 16 ft (4.9 m) wide and has a maximum load capacity of 60,000 lb (27,216 kg). Its available internal volume is about 26 ft (7.9 m) by 14 ft (4.3 m) by 15 ft (4.6 m) high to the center of the canopy. The charges for the successive shots of the test sequence are all 60 lb (27 kg) at a depth of 24 ft (7.3 m). The charge standoff, the horizontal distance from the near side of the FSP, is shortened for each shot to a final value of 20 ft (6.1 m). Design shock spectra for the FSP are shown in Fig. 26.9. Performance of this machine is detailed in Ref. 2.

The LFSP is 50 ft (15.2 m) long by 30 ft (9.1 m) wide with a maximum load capacity of 400,000 lb (181,440 kg) and an internal volume of about 48 ft (14.6 m) by 28 ft (8.5 m) by 34 ft (10.4 m) high to the center of the canopy. The charge size is 300 lb (136.1 kg), and the charge depth is 20 ft (6.1 m); the standoff is decreased for each

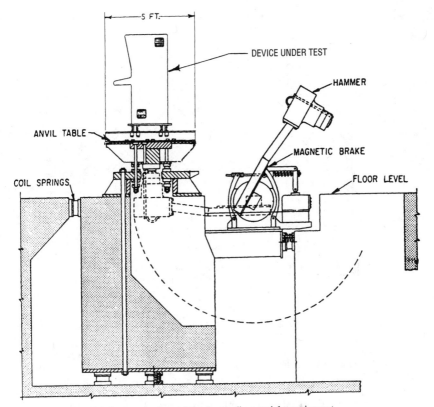

FIGURE 26.8 High-impact shock machine for medium-weight equipment.

shot to a final value of 50 ft (15.2 m). At the crossover load of 30,000 to 40,000 lb (13,640 to 18,180 kg), the LFSP provides a shock environment equivalent to the FSP. Therefore, data in Fig. 26.9 can be used in design of equipment scheduled for LFSP shock testing. Characteristics of the LFSP are discussed in Ref. 5.

Hopkinson Bar. When shock testing requires extremely high g levels for light loads (for example, calibration of accelerometers), the Hopkinson bar has proven useful. A controlled velocity projectile is impacted on the end of a metallic bar, causing a stress wave of known magnitude to travel along the bar. Often, the magnitude of the stress wave is measured as it passes the middle of the bar. The item under test is attached to the extreme end of the bar and experiences a high g rapid rise time acceleration when the stress wave arrives at that position. Also see high-acceleration methods of testing, Chap. 18.

MULTIPLE-IMPACT SHOCK MACHINES

Many environments, particularly those involving transportation, subject equipment to a relatively large number of shocks. These are of lesser severity than the shocks of

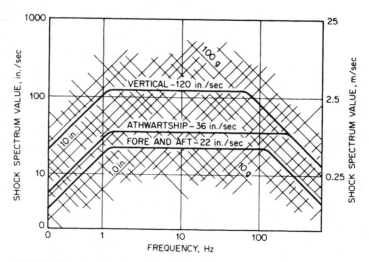

FIGURE 26.9   Design shock spectra for the floating shock platform. The lower cutoff frequency is 1.15 Hz for all directions. The upper cutoff frequencies are: vertical—67 Hz; athwartship—220 Hz; fore and aft—125 Hz.

major intensity that have been considered above, but their cumulative effect can be just as damaging. It has been observed that components of equipment that are damaged as a result of a large number of shocks of relatively low intensity are usually different from those that are damaged as a result of a few shocks of a relatively high intensity. The damage effects of a large number of shocks of low intensity cannot generally be produced by a small number of shocks of high intensity. Separate tests are therefore required so that the multiple number of low-intensity shocks are properly emulated.

Vibration Machines.   Electrodynamic, hydraulic, and pneumatic vibration testing machines provide a ready and flexible source of multiple shock pulses so long as the pulse requirements do not exceed force and motion capabilities of the selected machine. They can be programmed to provide a series of different shock pulses or to repeat a particular shock motion as many times as desired and to establish the necessary initial conditions prior to each shock pulse. See Chap. 25 for more information.

ROTARY ACCELERATOR

A quick-starting centrifuge can is used to quickly attain and maintain an acceleration for a long period of time. The accelerator consists of a rotating arm which is suddenly set into motion by an air-operated piston assembly. The test object is mounted on a table attached to the outer end of the arm. The table swings on a pivot so that the resultant direction of the acceleration is always along a fixed axis of the table. Initially the resultant acceleration is caused largely by angular acceleration of the arm, so this axis is in a circumferential direction. As the centrifuge attains its full speed, the acceleration is caused primarily by centrifugal forces, so this table axis assumes a radial direction. These machines are built in several sizes. They require between 5 and 60 milliseconds to reach the maximum value of acceleration. For small test items (8 lb, 3.6 kg), a maximum acceleration of $450g$ is attainable; for heavy test items (100 lb, 45.4 kg), the maximum value is $40g$.

REFERENCES

1. "Specification for the Design, Construction, and Operation of Class HI (high impact) Shock Testing Machine for Lightweight Equipment," American National Standards Institute Document ANSI S2.15-1973.
2. Clements, E. W.: "Shipboard Shock and Navy Devices for its Simulation," U.S. Naval Research Laboratory Report 7396, July 14, 1972.
3. "Methods for Specifying the Performance of Shock Machines," American National Standards Institute Document ANSI S2.14-1973.
4. Military Specification. "Shock Tests HI (High Impact); Shipboard Machinery, Equipment and Systems, Requirements for," MIL-S-901D (Navy), March 17, 1989.
5. Clements, E. W.: "Characteristics of the Navy Large Floating Shock Platform," U.S. Naval Research Laboratory Report 7761, 15 July 1974. (Obtainable from the Shock and Vibration Information Analysis Center, Booz-Allen & Hamilton Incorporated, 2231 Crystal Drive-Suite 711, Arlington, VA 22202.)

CHAPTER 26

PART II: PYROSHOCK TESTING

Neil T. Davie
Vesta I. Bateman

INTRODUCTION

Pyroshock, also called *pyrotechnic* shock, is the response of a structure to high-frequency (thousands of hertz), high-magnitude stress waves that propagate throughout the structure as a result of an explosive event such as the explosive charge to separate two stages of a multistage rocket. The term *pyrotechnic shock* originates from the use of propellants such as black powder, smokeless powder, nitrocellulose, and nitroglycerin in devices common to the aerospace and defense industries. These devices include pressure squibs, explosive nuts and bolts, latches, gas generators, and air bag inflators.[1] The term *pyroshock* is derived from pyrotechnic shock, but both terms are used interchangeably in the industry and its literature. A pyroshock differs from other types of mechanical shock in that there is very little rigid-body motion (acceleration, velocity, and displacement) of a structure in response to the pyroshock. The pyroshock acceleration time-history measured on the structure is oscillatory and approximates a combination of decayed sinusoidal accelerations with very short duration in comparison to mechanical shock described in Part I of this chapter. The characteristics of the pyroshock acceleration time-history vary with the distance from the pyroshock event. In the near field, which is very close to the explosive event, the pyroshock acceleration time-history is a high-frequency, high-amplitude shock that may have transients with durations of microseconds or less. In the far field, which is far enough from the event to allow structural response to develop, the acceleration time-history of the pyroshock approximates a combination of decayed sinusoids with one or more dominant frequencies. The dominant frequencies are usually much higher than that in a mechanical shock and reflect the local modal response of the structure. The dominant frequencies are generally lightly damped. However, since the frequencies are so high, it typically takes less than 20 milliseconds for the pyroshock response to dampen out and return to zero. Satellite, aerospace, and weapon components are often subjected to pyroshocks created by devices such as explosive bolts and pyrotechnic actuators. Pyroshock structural response is also found in ground-based applications in which there is a sudden release of energy, such as the impact of a structure by a projectile.

Pyroshock was once considered to be a relatively mild environment due to its low-velocity change and high-frequency content. Although it rarely damages structural members, pyroshock can easily cause failures in electronic components that are sensitive to the high-frequency pyroshock energy. The types of failures caused by pyroshock commonly include relay chatter, hard failures of small circuit components, and the dislodging of contaminants (e.g., solder balls), which cause short circuits. A significant number of flight failures have been attributed to pyroshock compared to other types of shock or vibration sources, and, in one case, an extensive database of the failures has been compiled.[2] Designers must rely on testing for qualifications of their systems and components that will be exposed to pyroshock environments in the absence of analytical techniques to predict structural response to a pyroshock. Failures can be reduced by implementing a qualification testing program for components exposed to a pyroshock environment. This chapter describes the characteristics of pyroshock environments, measurement techniques, test specifications, and simulation techniques.

PYROSHOCK CHARACTERISTICS

COMPARISON OF NEAR-FIELD AND FAR-FIELD CHARACTERISTICS

The detonation of an explosively actuated device produces high-frequency transients in the surrounding structure. The specific character of these acceleration transients depends on various parameters including: (1) the type of pyrotechnic source, (2) the geometry and properties of the structure, and (3) the distance from the source. Due to the endless combinations of these parameters, sweeping conclusions about pyroshock characteristics cannot be made; however the following paragraphs describe useful characteristics of typical pyroshock environments.

A pyrotechnically actuated device produces a nearly instantaneous pressure on surfaces in the immediate vicinity of the device. As the resulting stress waves propagate through the structure, the high-frequency energy is gradually attenuated due to various material damping and structural damping mechanisms. In addition, the high-frequency energy is transferred or coupled into the lower-frequency modes of the structure. The typical pyroshock acceleration transient thus has roughly the appearance of a multifrequency decayed sinusoid (i.e., the envelope of the transient decays and is symmetric with respect to the positive and negative peaks). The integral of the typical transient also has these same characteristics.[3] In most cases, the initial portion of the acceleration transient exhibits a brief period during which the amplitudes of the peaks are increasing prior to the decay described above (see Fig. 26.10 and 26.11). This is a result of the interaction of stress waves as they return from various locations in the structure.

A pyrotechnically actuated device imparts very little impulse to a structure since the high forces produced are acting for only a short duration and are usually internal to the structure. The net rigid body velocity change resulting from a pyroshock is thus very low relative to the peak instantaneous velocity seen on the integral of the acceleration transient. Rigid body velocity changes are commonly less than 1 meter per second. The duration of a pyroshock transient depends on the amount of damping in a particular structure, but it is commonly 5 to 20 milliseconds in duration.

Pyroshock may be subdivided into two general categories: *Near-field pyroshock* occurs close to the pyrotechnic source before significant energy is transferred to structural response. It is dominated by the input from the source and contains very high-frequency and very high *g* energy. This energy is distributed over a wide frequency range and is not generally dominated by a few selected frequencies. *Far-field pyroshock* environments are found at a greater distance from the source where significant energy has transferred into the lower-frequency structural response. It contains lower frequency and lower *g* energy than near-field pyroshock; most of the

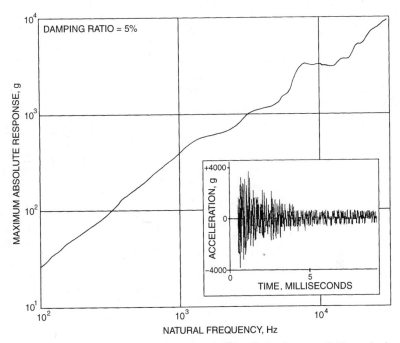

FIGURE 26.10 Shock spectrum and acceleration time-history for a near-field pyroshock. The shock spectrum is calculated from the inset acceleration time-history using a 5 percent damping ratio. The shock spectrum for near-field pyroshock may exhibit a more complex shape than the typical far-field shape shown in Fig. 26.11. Here, the shock spectrum has an average slope of about 6 dB/octave over the entire frequency range that was analyzed.

energy is usually concentrated at one or a few frequencies which correspond to dominant structural mode(s).

A more detailed discussion of shock spectrum (see Chap. 23 for definition) applications is given later in this chapter, but it is introduced here as a means of describing pyroshock characteristics. Many far-field pyroshock environments have a *typical* shock spectrum shape as illustrated in Fig. 26.11, which shows an actual far-field pyroshock acceleration transient along with its associated shock spectrum. The shock spectrum initially increases with frequency at a slope of 9 to 12 dB/octave, followed by an approximately constant or slightly decreasing amplitude. The frequency at which the slope changes is called the *knee frequency,* and it corresponds to a dominant frequency in the pyroshock environment. The knee frequency is often between 1000 and 5000 Hz for far-field pyroshock, but it could be higher or lower in some cases. Near-field pyroshock may also exhibit this typical pyroshock shock spectrum except with a higher knee frequency. However, since near-field pyroshock usually has broad-band frequency content, its shock spectrum often exhibits a more complex shape that contains numerous excursions but on average follows a 6-dB/octave slope over the entire frequency range of interest. Figure 26.10 shows an example of this type of near-field shock spectrum.

No fixed rules define at what distance from the pyrotechnic source the near-field pyroshock ends and the far-field pyroshock begins. It is more appropriate to classify

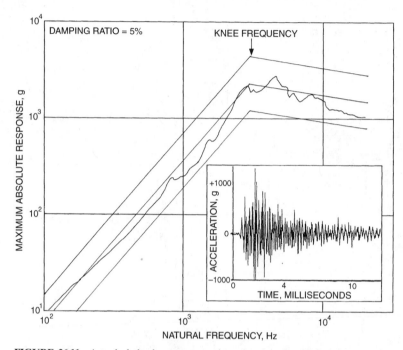

FIGURE 26.11 A typical shock spectrum and acceleration time-history for a far-field pyroshock. The shock spectrum is calculated from the inset acceleration time-history using a 5 percent damping ratio. The shock spectrum exhibits a steep initial slope which abruptly changes at a knee frequency. The straight lines indicate tolerance bands (typically ±6 dB as shown) which might be applied for qualification test specification.

near- and far-field pyroshock according to the various test techniques that are appropriate to employ in each case.

TEST TECHNIQUES FOR NEAR- AND FAR-FIELD PYROSHOCK

The pyroshock simulation techniques described in this chapter fall into two categories: (1) pyrotechnically excited simulations and (2) mechanically excited simulations. A short-duration mechanical impact on a structure causes a response similar to that produced by a pyrotechnic source. Although these mechanically excited simulations can be carried out with lower cost and better control than pyrotechnically excited simulations, they cannot produce the very high frequencies found in near-field pyroshock. Mechanically excited simulations allow control of dominant frequencies up to about 10,000 Hz (or higher for very small test items). For environments requiring higher frequency content, a pyrotechnically excited technique is usually more appropriate. The following general guidelines apply in selecting a technique for simulating pyroshock:

Near-field pyroshock. For a test that requires frequency control up to and above 10,000 Hz, a pyrotechnically excited simulation technique is usually required.

Far-field pyroshock. For a test that requires frequency control no higher than 10,000 Hz, a mechanically excited simulation technique is usually acceptable.

These guidelines are not rigid rules, but they provide a reasonable starting point when planning a pyroshock simulation test.

QUANTIFYING PYROSHOCK FOR TEST SPECIFICATION

An intrinsic characteristic of pyroshock is its variability from one test to another. That is, even though great care has been taken with the test technique, the measured response in both the near and the far fields may vary a great deal from test to test. This variability occurs in the situation where actual explosive devices are used and in the laboratory where more controlled techniques are employed. As a result, various techniques have been sought to quantify pyroshock for test specification. The purpose of these techniques is to define the pyroshock in a manner that can be reproduced in the laboratory and can provide a consistent evaluation for hardware that must survive pyroshock in field environments. All techniques require that a measurement be made of the actual pyroshock event at or near the location of the subsystem or component that will be tested. The measurement may be acceleration, velocity, or displacement, but acceleration is the most widely used measure. The measurement is then used with one of the techniques below to obtain a test specification for pyroshock. The shock spectrum (also called *shock response spectrum* or *response spectrum*) is considered to be conservative and a potential over-test of components and subsystems. However, components and subsystems that survive laboratory tests specified using shock spectra generally survive pyroshock field environments, although they may be over-designed. Because aerospace systems require lightweight components and subsystems, other techniques such as temporal moments and shock intensity spectrum have been developed so that laboratory tests can more closely simulate actual pyroshock events and allow tighter design margins.

Shock Spectra. By far the most widely used technique for quantifying pyroshock is the shock spectrum. This technique provides a measure of the effect of the pyroshock on a simple mechanical model with a single degree-of-freedom. Generally, a measured acceleration time-history is applied to the model, and the maximum acceleration response is calculated. The damping of the model is held constant (at a value such as 5 percent) for these calculations. An ensemble of maximum absolute-value acceleration responses is calculated for various natural frequencies of the model and the result is a *maxi-max shock response spectra.* A curve representing these responses as a function of damped natural frequency is called a shock spectrum (or response spectrum, see Chap. 24) and is normally plotted with log-log scales. Velocity shock spectrum and displacement shock spectrum may be calculated but are not commonly used for pyroshock specification. The shock spectrum for pyroshock has a characteristically steep slope at low frequencies of 12 dB/octave that is a direct result of the minimal velocity change occurring in a pyroshock. Occasionally, a pyrotechnic device, such as an explosive bolt cutter, is combined with another mechanism, such as a deployment arm, to position components for a particular event sequence. In this case, a distinct velocity change is combined with the pyroshock event, and the low-frequency slope of the shock spectrum will reflect this velocity change. For a typical far-field pyroshock, the low-frequency slope changes at the *knee frequency,* and the shock spectrum approaches a constant value at high frequencies that is the peak acceleration in the time-domain as shown in Fig. 26.11. A typical near-field pyroshock may have this shape or may have the shape shown in Fig. 26.10. Conventionally, tolerance bands of ±6 dB are drawn about a straight-line approximation of the shock spectrum for laboratory testing. An example of a typical maxi-max shock spectrum is shown in Fig. 26.11 with the conventional ±6 dB tolerance bands.

Band-Limited Temporal Moments. The method of temporal moments may be used for modeling shocks whose time durations are too short for nonstationary models and that contain a large random contribution.[4] The method uses the magnitude of the Fourier spectrum in the form of an energy spectrum (Fourier spectrum magnitude squared) that is smoothed or formed from an ensemble average to generate statistically significant values. Temporal moments of the time-histories are used to represent how the energy is distributed in time. The moments are analogous to the moments of the probability density functions and provide a convenient method to describe the envelopes of complicated time-histories such as pyroshock. The ith temporal moment $m_i(a)$ of a time-history $f(t)$, about a time location a, is defined as

$$m_i(a) = \int_{-\infty}^{+\infty} (t-a)^i [f(t)]^2 \, dt \tag{26.1}$$

The time-history energy E is given by

$$E = \frac{1}{2\pi} \int_{-\infty}^{+\infty} |F(\omega)|^2 \, d\omega \tag{26.2}$$

where $F(\omega)$ is the Fourier transform of $f(t)$. The first five moments are used in the temporal moments technique. The zeroth-order moment m_0 is the integral of the magnitude squared of the time-history and is called the time-history energy. The first moment normalized by the energy is called the central time τ. A central moment is a moment computed about the central time, i.e., $a = \tau$. The second central moment is normalized by the energy and is defined as the mean-square duration of the time-history. The third central moment normalized by the energy is defined as the skewness and describes the shape of the time-history. The fourth central moment normalized by the energy is called kurtosis. The moments are calculated for a shock time-history passed through a contiguous set of bandpass filters. A *product model* is formed using a deterministic window $w(t)$ (see Chap. 25) and a realization of a dimensionless stationary random process with unity variance $x(t)$ as $w(t) \cdot x(t+\tau)$. A product model is then used to generate a simulation that has the same energy and moments in the mean as the original shock. Band-limited moments characterize the shock and not the response to the shock as the shock spectrum and do not rely on a structural model.

Other Techniques. Other techniques to quantify pyroshock include the shock intensity spectrum based on the Fourier energy spectrum,[5] the *method of least favorable response*,[6,7] and nonstationary models.[8,9] These techniques are not commonly used but may provide additional insight for quantifying pyroshocks. The Fourier spectrum is an attractive alternative to shock response spectrum because it is easy to compute and readily available in many software packages as a fast Fourier transform (FFT). Since the Fourier spectrum is complex, both magnitude and phase information is available. The magnitude generally has intuitive meaning, but the phase is difficult to interpret and may be contaminated with noise at the high frequencies present in pyroshock. The method of least favorable response provides a method of selecting the phase to maximize the response of the system under test. This method results in a conservative test provided that an appropriate measurement point is chosen on the structure. Stationary models for random vibration have been used for many years. Nonstationary models consist of a stationary process multiplied by a deterministic time-varying modulating function, which is a product model.[9] A nonstationary model is appropriate for pyroshock and approaches a stationary model as the time-record length is increased.

MEASUREMENT TECHNIQUES

Measurements of pyroshocks are generally made with accelerometers, strain gages, or laser Doppler vibrometers (LDV). The accelerometers are used to measure acceleration, and the strain gages and LDV are used to measure velocity. The strain gages may also be used to sense force, stress, or strain. General shock measurement instrumentation is applicable for pyroshock measurements (see Chap. 12); however, care must be taken to protect accelerometers from the high frequencies contained in pyroshocks that may cause the accelerometers to resonate and, in some cases, to fail. If accelerometers are excited into resonance, large-magnitude output results and may exceed the maximum amplitude of the data acquisition system that was chosen for the test. The result is that the data magnitude is clipped. If clipped, the data are rendered useless and the results from the test will be greatly diminished. Several mechanically isolated accelerometers are available commercially and should be used if there is a possibility of exciting the accelerometers into resonance. There is only one mechanically isolated accelerometer that can provide the wide-frequency bandwidth (dc to 10 kHz) required for pyroshock.[10, 11] Other mechanical isolators generally provide a frequency bandwidth of about dc to 1 kHz. Any mechanical isolator that is used in a pyroshock environment must be well characterized over a range of frequencies and a range of acceleration values using a shock test technique, for example, Hopkinson bar testing. Strain gages are useful measurements of the pyroshock environment but are not easily translated into a test specification. Strain gages have the advantage of high-frequency response (in excess of dc to 40 kHz) provided that their size is appropriately chosen. Additionally, strain gages do not have the resonance problems that accelerometers have. The LDV provides velocity measurements that are not contaminated by cross-axis response because the LDV only responds to motion in the direction of the laser beam. The LDV is a noncontacting measurement and is easy to set up; consistent measurements of pyroshock events have been obtained with a LDV.[12, 13] The LDV has the disadvantage of being very expensive per channel in comparison to the other measurement techniques, difficult to calibrate, and must have line of sight to the measurement location.

Pyroshock Test Specifications. An acceleration or velocity time-history is not adequate for specifying a pyroshock test. The time-history data must be analyzed using one of the techniques discussed above to quantify the pyroshock for a test specification. Ideally, the time-history data that are used to develop the qualification test specification should be measured during a full-scale system test in which the actual pyrotechnic device or devices were initiated. The full-scale test should be accomplished with hardware that is structurally similar to the real hardware if the real hardware is not available. A control point measurement is specified close to each component or subassembly of interest, preferably at the attachment point to measure the input pyroshock. Since full-scale testing is expensive, data from a similar application may be used to develop component or subassembly qualification test specifications. This practice may result in over-tested or over-designed components or subassemblies if a large margin is added to the test specification to account for the uncertainty in the data. If this practice is used, the test specification should be revised when better system data become available.

Once the time-history data have been acquired, the data should be scrutinized to ensure their quality.[3] The data should be free of zero-shifts and offsets. Acceleration and velocity time-histories should be integrated and the results examined. The time-history data should be low-pass filtered at a designated cutoff frequency; a cutoff frequency of 20 kHz is typical. The data must then be analyzed using the same technique as was used for analysis of the time-history data from which the test specification was derived. The two analyses, the test used for the test specification and the accomplished test, must be compared with the same analysis technique. Test margin and tolerance bands are applied to the data analysis. For instance, if the shock spectrum is being used, a straight-line approximation of the shock spectrum is used as the baseline for the test specification process. A margin of ±3 dB is typically added to the baseline shock spectrum, and a customary ±6-dB tolerance is used with the baseline shock spectrum. The shock spectrum from the actual pyroshock test may be within the tolerance at some frequencies and not at others. Pyroshock tests are highly variable, and the engineer must specify how much variability from test to test will be accepted; in some cases, a tighter, ±3-dB tolerance may be required. Additionally, the specification should require that the peak acceleration (or velocity) value and pulse durations are in agreement with the intended values for the specified input pulse. Similar approaches are used for other techniques for quantifying pyroshock.

In some cases, two or more pyroshock events, such as stage separation and an explosive actuator, may be combined into a single test specification. If the events are significantly different, the resulting test specification may be difficult or impossible to meet. A better practice is to make separate test specifications for each pyroshock event and to combine the specifications only in the case where a realizable test results.

PYROSHOCK SIMULATION TECHNIQUES

PYROTECHNICALLY EXCITED NEAR-FIELD SIMULATION

Ordnance Devices. Linear, flexible detonating charges may be used to generate pyroshocks for test purposes. An example of a test configuration using a fuze is shown in Fig. 26.12. A steel plate is suspended by bungee cords, and the test item is mounted on the plate in the same manner as it is in actual usage. Flexible linear charge (primaline or detcord) is attached to the edges of the plate. The charge configuration may be varied according to experience and the desired effect.[14] For example, the charge may be attached to the backside of the plate directly opposite to the test item. A mass-mockup of the actual test item is used for the trial and error required to finalize the test configuration. In some cases, the charges may be attached to a portion of the structure where the test items are installed. Their storage, handling, and detonating constitute a hazard to laboratory personnel and facilities. However, such a fixture would normally be rather expensive because the structure would be damaged or destroyed during each shock test. The shock produced in this manner many vary greatly from test to test because actual explosives

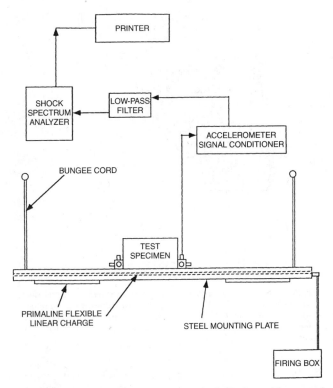

FIGURE 26.12 Ordnance-generated pyroshock simulator. (*Courtesy of National Technical Systems.*)

are used. However, this test configuration has the advantage of reproducing the pyroshock with realistic high accelerations and high frequencies. To ensure repeatability, the grooves generated by the charge into the surfaces of the shock plates should be machined down to eliminate the porosity which tends to absorb and modify the explosive impacts. Other disadvantages are that a qualified explosives facility (with its associated safety procedures) is required. In comparison to mechanical simulation techniques, considerable time is required between trial tests and for numerous trial tests.

Scaled Tests. If the quantity of propellant or explosive is sufficiently large and the influence of the pyrotechnic device is localized, a scaled portion of the structure may be used in simulating the effects of the pyroshock as shown in Fig. 26.13 where a missile section or rocket payload section is shown. This type of test assumes that the influence of the pyrotechnic event can be ignored by other parts of the structure and isolated to the section under test. Actual pyrotechnic device firings on spacecraft equipment and scientific instruments are conducted in the scaled test. Such a test is usually an intermediate step in the design of the structure. Components in the subassembly may have been qualified with a ordnance device, and the scaled test adds another dimension of complexity to the qualification of the subassembly and its individual components.

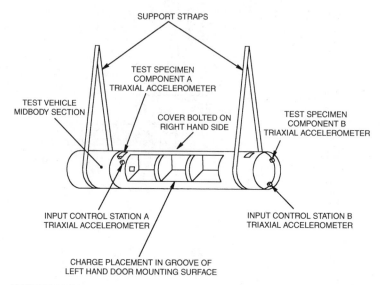

FIGURE 26.13   Scaled tests using representative structure. The test vehicle midbody section is a portion of the full-scale structure where the explosive event is located. Two input control stations *A* and *B* are used to determine that the test was properly conducted. Response measurements are made at test specimens *A* and *B*. (*Courtesy of Wyle Laboratories.*)

Full-Scale Tests.   In some cases, if the structure is sufficiently complex, a full-scale test may be warranted. Full-scale tests, which include multiple firings of certain critical pyrotechnic devices, are conducted to verify the structural integrity and design functions as well as to qualify items of hardware that have not been previously qualified. Full-scale tests are conducted by actuation of the flight pyrotechnic devices, which provide full-scale shock qualification. A full-scale test is usually the last test in a sequence of increasingly complex tests; the sequence is from ordnance to scaled tests to full-scale tests. The advantage of a full-scale test is that it is the real pyroshock event in its most complex form. The main objectives of the full-scale pyroshock test firings are: (1) to define shock response in the vicinity of potentially sensitive equipment so that component test specifications may be derived or verified and (2) to conduct full-scale qualification and thus verify the design values for shock. The disadvantage of a full-scale test is that considerable time and expense are required to obtain all the required hardware. The hardware must then be assembled, instrumented, and removed for post-test evaluation. Generally, special facilities are required for the use of explosives.

MECHANICALLY EXCITED FAR-FIELD SIMULATION

Standard Shock-Testing Machines.   Shock machines such as the drop tables described in Part I of this chapter usually are not suitable for pyroshock simulation. The single-sided pulses produced by these machines bear little or no resemblance to a pyroshock acceleration transient; such pulses produce significantly greater velocity change than a pyroshock environment. A severe over-test at low frequencies can

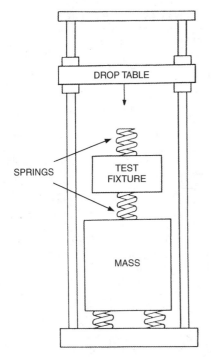

FIGURE 26.14 Bounded impact test configuration on a "standard" drop-table.

be expected if a drop table is used to simulate pyroshock environments. This can result in failures of structural members that would not have been significantly stressed by the actual pyroshock. However, in certain cases, drop tables may produce acceptable pyroshock qualification testing. For example, if a test item has significant design margin at low frequencies, then a drop table may be acceptable. Also, if the lowest natural frequency of the test item is higher than the over-tested low-frequency range, then the low-frequency over-test may be irrelevant since the affect on the test item is dominated by the peak g's of the acceleration time-history. In these cases there is strong motivation to use drop tables due to their common availability and low test cost. If a drop table is selected as a means of conducting a pyroshock qualification test, the test item must be subjected to a shock in both positive and negative directions for each axis tested, since the drop table produces only a single-sided pulse.

Another application of a drop table for pyroshock testing is the *bounded impact method*[15] as illustrated in Fig. 26.14, which shows the test item fixture bounded by two springs (typically felt or elastomeric pads). When the drop table strikes the upper spring, the fixture oscillates at the natural frequency of the spring-mass system. This oscillation ceases when the drop table rebounds from the spring, resulting in an acceleration transient that appears as a decayed sinusoid with about two or three cycles. The velocity change is much less than for a haversine pulse, which results in a shock spectrum with the desired slope of 9 to 12 dB/octave. *Knee frequencies* up to about 2000 Hz are attainable with this method.

Electrodynamic Shakers. Pyroshock environments can be simulated with an acceleration transient produced on an electrodynamic shaker. The details of this method are described in Chap. 25 under the heading "Digital Control Systems for Shock and Vibration Testing." In this method the acceleration transient is synthesized so that its shock spectrum closely matches the test requirement. With this method a relatively complex shock spectrum shape can be matched within close tolerances up to about 3000 Hz. The equipment limits (maximum acceleration) restrict this method to the simulation of lower-energy pyroshock environments. Even if the desired shock spectrum is precisely met, an over-test is likely due to the high mechanical impedance of the shaker relative to the structure to which the test item is attached in a real application.

Resonant Fixtures. This section describes a variety of resonant fixture techniques used to simulate pyroshock environments. All of these methods utilize a fixture (or structure) which is excited into resonance by a mechanical impact from a projectile, a hammer, or some other device. A test item attached to the fixture is thus subjected to the resonant response, which simulates the desired pyroshock. There is no single preferred method since each has its own relative merits. Some of the methods require extensive trial-and-error iterations in order to obtain the desired test requirement. However, once the procedures are determined, the results are very repeatable. Other methods eliminate the need for significant trial and error but are usually limited to pyroshock environments which exhibit the typical far-field character as explained in Fig. 26.13.

Full-Scale Tests.   Some mechanically excited simulation techniques involve the use of an actual or closely simulated structure[16,17] (e.g., an entire missile payload section). The pyrotechnic devices (e.g., explosive bolt cutters) normally located on this structure would then be replaced with hardware that allow a controlled impact at this same location. Since a closely simulated structure is used, it is anticipated that the impact will cause the modes of vibration of the structure to be excited in a manner similar to the actual pyrotechnic source. In principle, test amplitudes can be adjusted by changing the impact speed or mass. This method is relatively expensive due to the cost of the test structure and because significant trial and error is required to obtain the desired test specification. Since this method applies to a specific application, it is not suited as a general-purpose pyroshock simulation technique.

In a variation of the above method[18] the pyrotechnic source and a portion of the adjacent structure are replaced by a "resonant plate" designed so that its lowest-resonance frequency corresponds to the dominant frequency produced by the pyrotechnic device and its associated structure. The resonant plate is then attached to the test structure in a manner which simulates the mechanical linkage of the pyrotechnic source. When this plate is subjected to a mechanical impact, its response will provide the desired excitation of the test structure.

General-Purpose Resonant Fixtures.   Instead of developing application-specific pyroshock methods as described above, it may be desirable to implement a more general-purpose test method which can be used for a variety of test items and/or test specifications. This can be accomplished by using a simple resonant fixture (usually a plate) instead of the complex structures described above. When such a fixture is excited into resonance by a mechanical impact, its response can provide an adequate pyroshock simulation to an attached test item. Excitation of the fixture can be achieved as the result of the impact of a projectile, pendulum hammer, pneumatic piston, or the like on the fixture. The response of the fixture is dependent on a large number of parameters including: (1) plate geometry and material, (2) impact mass or speed, (3) impact duration, which is controlled with various impact materials (e.g., metals, felt, elastomers, wood, etc.), (4) impact location, (5) test item location, and (6) various clamps and plate suspension mechanisms. In theory these parameters could be varied with the aid of an analytical model, but they are usually evaluated experimentally. A significant effort is therefore required to obtain each pyroshock simulation.

Mechanical Impulse Pyro Shock (MIPS) Simulator.   The MIPS simulator[19,20] is a well-developed embodiment of the trial-and-error resonant fixture methods. It is universally referred to by its acronym and is widely used in the aerospace industry. Its design facilitates the easy variation of many of the parameters described above. The MIPS simulator configuration shown in Fig. 26.15 consists of an aluminum mounting plate which rests on a thick foam pad. The shock is generated by a pneumatic actuator which is rigidly attached to a movable bridge, facilitating various impact locations. The impactor head is interchangeable so that different materials (lead, aluminum, steel etc.) may be used to achieve variation of input duration. Although a

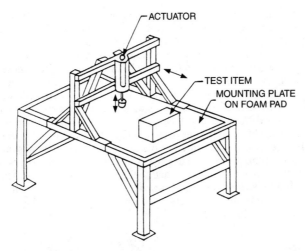

FIGURE 26.15 MIPS simulator. The mounting plate is excited into resonance by an impact from the actuator. The plate response simulates far-field pyroshock for the attached test item. (*Courtesy of Martin Marietta Astrospace.*)

triaxial acceleration measurement is usually made at the control point near the test item, it is unlikely that the test requirement will be met simultaneously in all axes. Separate test configurations must normally be developed for each test axis. Once the test configuration and procedures are determined, the results are very repeatable. The configuration for a new test specification can be obtained more quickly if records of previous setups and results are maintained for use as a starting point for the new specification. Reference 19 provides some general guidelines for parameter variation, as well as results obtained from several different test configurations.

Tuned Resonant Fixtures with Fixed Knee Frequency. It is possible to greatly reduce the amount of trial and error required by the MIPS simulator and other resonant fixture test methods. In order to do this, a simple resonant fixture is designed so that its dominant response frequency corresponds to the dominant frequency in the shock spectrum test requirement. These tuned resonant fixtures are primarily limited to pyroshock environments which exhibit more or less typical characteristics with knee frequencies up to 3000 Hz (or higher for small test items). The basic design principle is to match the dominant fixture response frequency (usually the first mode) to the shock spectrum knee frequency. When this fixture is excited into resonance, it will "automatically" have the desired shock spectrum knee frequency and the typical 9-dB/octave initial slope. This concept was originally developed using a plate excited into its first bending mode and a bar excited into its first longitudinal mode.[21] The methods described in the following sections require relatively thick and massive resonant fixtures compared to the structures to which the test item might be attached in actual use. Because of this, the motion imparted to the test item attached to a resonant fixture is approximately in-phase from point-to-point across the mounting surface. Whereas, the actual pyroshock motion may not be in-phase if the test item is mounted to a thin structure in actual use. The in-phase motion of resonant fixtures yields some degree of conservatism when selecting these methods for qualification testing. One significant advantage of using a thick resonant fixture is that its response is not greatly influenced by the attached test item. This allows the same test apparatus to be used for a variety of different test items.

Each of the tuned resonant fixture test methods described below produces a simulated pyroshock environment with the same basic characteristics. These similarities are illustrated in Fig. 26.16, which shows a typical acceleration record and shock spectrum from the tunable resonant beam apparatus described later. The other methods produce pyroshock environments with initial shock spectra slopes that are slightly less than 9 dB/octave due to a small velocity change inherent with these other methods. The shock spectrum shown in Fig. 26.16 exhibits the desired typical shape, and the energy is concentrated at the knee frequency. The absence of significant frequency content above the knee frequency may cause the shock spectrum to be too low at these frequencies. In practice the attached test item adds some frequency content above the knee frequency, which tends to increase the shock spectrum. These test methods allow good control and repeatability of the shock spectrum, especially below the knee frequency.

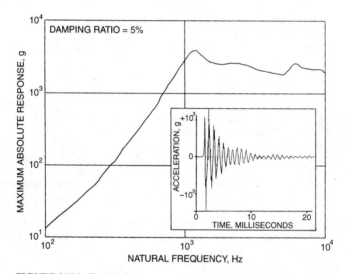

FIGURE 26.16 Typical shock spectrum and acceleration time-history from a tuned or tunable resonant fixture test. The shock spectrum is calculated from the inset acceleration time-history using a 5 percent damping ratio. Compare the shock spectrum shape to the typical far-field shock spectrum shape in Fig. 26.11. The knee frequency can be selected with proper fixture design or adjustments.

When using tuned resonant fixtures, the test item is usually attached to an intermediate fixture such as a rectangular aluminum plate. This adapter fixture must be small enough and stiff enough so that the input from the resonant fixture is not significantly altered. Since the resonant fixture is designed to produce the pyroshock simulation in only one direction, the adapter fixture should be designed so that it may be rigidly attached to the resonant fixture in three orthogonal orientations (e.g., flat down and on each of two edges). The acceleration input should be measured next to the test item on the adapter fixture. It is good practice to measure the acceleration in all three axes because it is possible (although infrequently) to simultaneously attain the desired test specification in more than one axis.

A number of different techniques are used to provide the mechanical impact required by the tuned resonant fixture methods described below. Pendulum hammers of the general type shown in Fig. 26.8 have been used, as well as pneumatically driven pistons or air guns. The method which is selected must provide repeatability and control of the impact force, both in magnitude and duration. The magnitude of the impact force controls the overall test amplitude, and the impact duration must be appropriate to excite the desired mode of the tuned resonant fixture. In general the impact duration should be about one-half the period of the desired mode. The magnitude of the impact force is usually controlled by the impact speed, and the duration is controlled by placing various materials (e.g., felt, cardboard, rubber, etc.) on the impact surfaces.

Resonant Plate (Bending Response). The resonant plate test method[22, 23] is illustrated in Fig. 26.17, which shows a plate (usually a square or rectangular aluminum plate) freely suspended by some means such as bungee cords or ropes. A test item is attached near the center of one face of the plate, which is excited into resonance by a mechanical impact directed perpendicular to the center of the opposite face. The resonant plate is designed so that its first bending mode corresponds to the knee frequency of the test requirement. The first bending mode is approximately the same as for a uniform beam with the same cross-section and length. Appendix 1.1 provides a convenient design tool for selecting the size of the resonant plate. The plate must be large enough so that the test item does not extend beyond the middle third of the plate. This assures that no part of the test item is attached at a nodal line of the first bending mode. Usually, the resonant fixture with an attached test item is

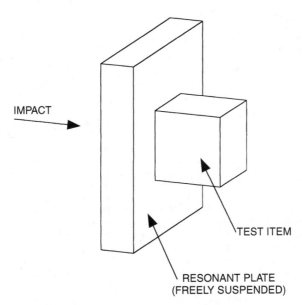

IMPACT

TEST ITEM

RESONANT PLATE
(FREELY SUSPENDED)

FIGURE 26.17 Resonant plate test method. The first bending mode is excited by an impact as shown. The plate's response simulates far-field pyroshock for the attached test item. The plate is sized so that its first bending mode corresponds to the desired knee frequency of the test.

insufficiently damped to yield the short-duration transient (5 to 20 milliseconds) required for pyroshock simulation. Damping may be increased by adding various attachments to the edge of the plate, such as C-clamps or metal bars. These attachments may also lower the resonance frequency and must be accounted for when designing a resonant plate.

Resonant Bar (Longitudinal Response). The resonant bar concept[22, 23] is illustrated in Fig. 26.18, which shows a freely suspended bar (typically aluminum or steel) with rectangular cross section. A test item is attached at one end of the bar, which is excited into resonance by a mechanical impact at the opposite end. The basic principle of the resonant bar test is exactly the same as for a resonant plate test except that the first longitudinal mode of vibration of the bar is utilized. The bar length required for a particular test can be calculated with Eq. (26.3)

$$l = \frac{c}{2f} \tag{26.3}$$

where l = length of the bar
         c = wave speed in bar
         f = first longitudinal mode of the bar (equal to desired knee frequency)

The other dimensions of the bar can be sized to accommodate the test item, but they must be significantly less than the bar length. As with the resonant plate method, the response of the bar can be damped with clamps if needed. These are most effective if attached at the impact end.

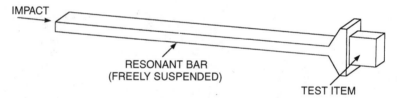

FIGURE 26.18   Resonant bar test method. The first longitudinal bar mode is excited by an impact as shown. The bar's response simulates far-field pyroshock for the attached test item. The bar is sized so that its first longitudinal mode corresponds to the desired knee frequency of the test.

Tunable Resonant Fixtures with Adjustable Knee Frequency.   The tuned resonant fixture methods described above can produce typical pyroshock simulations with knee frequencies that are fixed for each resonant fixture. A separate fixture must be designed and fabricated for each test requirement with a different knee frequency, so that a potentially large inventory of resonant fixtures would be necessary to cover a variety of test requirements. For this reason tunable resonant fixture test methods were developed which allow an adjustable knee frequency for a single test apparatus.

Tunable Resonant Bars.   The frequency of the first longitudinal mode of vibration of the resonant bar shown in Fig. 26.18 can be tuned by attaching weights at selected locations along the length of the bar.[23] If weights are attached at each of the two nodes for the second mode of vibration of the bar, then the bar's response will be dominated by the second mode ($2f$). Similarly, if weights are attached at each of the three nodes for the third mode of the bar, then the third mode ($3f$) will dominate. It is difficult to produce this effect for the fourth and higher modes of the bar since the distance between nodes is too small to accommodate the weights. This technique allows a single bar to be used to produce pyroshock simulations with one of three different knee frequencies. For example a 100-in. (2.54-m) aluminum bar can be used for pyroshock simulations requiring a 1000-, or 2000-, or 3000-Hz knee frequency. If the weights are attached slightly away from the node locations, the shock spectrum tends to be "flatter" at frequencies above the knee frequency.[24]

Another tunable resonant bar method[25] can be achieved by attaching weights only to the impact end of the bar shown in Fig. 26.18. This method uses only the first longitudinal mode, which can be lowered incrementally as more weights are added. A nearly continuously adjustable knee frequency can thus be attained over a finite frequency range. The upper limit of the knee frequency is the same as given by Eq. (26.3) and is achieved with no added weights. In theory, this knee frequency could be reduced in half if an infinite weight could be added. However, a realizable lower limit of the knee frequency would be about 25 percent less than the upper limit.

Tunable Resonant Beam. Figure 26.19 illustrates a tunable resonant beam apparatus[25] which will produce typical pyroshock simulations with a knee frequency that is adjustable over a wide frequency range. In this test method, an aluminum beam with rectangular cross section is clamped to a massive base as shown. The clamps are intended to impose nearly fixed-end conditions on the beam. When the beam is struck with a cylindrical mass fired from the air-gun beneath the beam, it will resonate at its first bending frequency, which is a function of the distance between the clamps. Ideally, the portion of the beam between the clamps will respond as if it had perfectly fixed ends and a length equal to the distance between the clamps. For this ideal case, the frequency of the first mode of the beam varies inversely with the

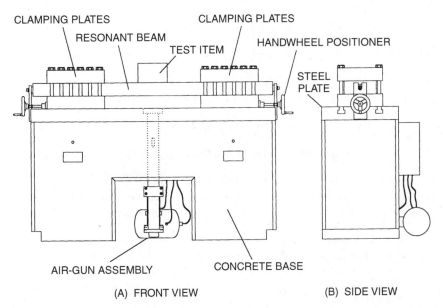

FIGURE 26.19 Tunable resonant beam test method. A beam, clamped near each end to a massive concrete base, is excited into its first bending mode by an impact produced by the air-gun. The first bending mode, which corresponds to the desired knee frequency, is adjusted by moving the clamping plates (with the hand wheel positioner) so that the unsupported beam length is increased or decreased.

square of the beam length (see App. 1.1). In practice, the end conditions are not perfectly fixed, and the frequency of the first mode is somewhat lower than predicted. This method provides a good general-purpose pyroshock simulator, since the knee frequency is continuously adjustable over a wide frequency range (e.g., 500 to 3000 Hz). This tunability allows small adjustments in the knee frequency to compensate for the effects of test items of different weights.

REFERENCES

1. Valentekovich, V. M.: *Proc. 64th Shock and Vibration Symposium,* pp. 92–112 (1993).

2. Moening, C. J.: *Proc. 8th Aerospace Testing Seminar,* pp. 95–109 (1984).

3. Himelblau, H., A. G. Piersol, J. H. Wise, and M. R. Grundvig: "Handbook for Dynamic Data Acquisition and Analysis," *IES Recommended Practice 012.1,* Institute of Environmental Sciences, Mount Prospect, Ill. 60056.

4. Smallwood, D. O.: *Shock and Vibration J.,* **1**(6):507–527 (1994).

5. Baca, T. J.: *Proc. 60th Shock and Vibration Symposium,* pp. 113–128 (1989).

6. Shinozuka, M.: *J. of the Engineering of the Engineering Mechanics Division, Proc. of the American Society of Civil Engineers,* pp. 727–738 (1970).

7. Smallwood, D. O.: *Shock and Vibration Bulletin,* 43:151–164 (1973).

8. Mark, W. D.: *J. of Sound and Vibration,* 22(3):249–295 (1972).

9. Bendat, J. S., and A. G. Piersol: "Engineering Applications of Correlation and Spectral Analysis," John Wiley and Sons, 2d ed., 1993, pp. 325–362.

10. Bateman, V. I., R. G. Bell, III, and N. T. Davie: *Proc. 60th Shock and Vibration Symposium,* **1**:273–292 (1989).

11. Bateman, V. I., R. G. Bell, III, F. A. Brown, N. T. Davie, and M. A. Nusser: *Proc. 61st Shock and Vibration Symposium,* **IV**:161–170 (1990).

12. Valentekovich, V. M., M. Navid, and A. C. Goding: *Proc. 60th Shock and Vibration Symposium,* **1**:259–271 (1989).

13. Valentekovich, V. M., and A. C. Goding: *Proc. 61st Shock and Vibration Symposium,* **2** (1990).

14. Czajkowski, J., P. Lieberman, and J. Rehard: *J. of the Institute of Environmental Sciences,* pp. 25–39 (1992).

15. Fandrich, R. T.: *Proc. Institute of Environmental Sciences Annual Technical Meeting,* pp. 269–273 (1974).

16. Luhrs, H. N.: *Proc. Institute of Environmental Sciences Annual Technical Meeting,* pp. 17–20 (1981).

17. Powers, D. R.: *Shock and Vibration Bulletin,* 56(3):133–141 (1986).

18. Bateman, V. I., and F. A. Brown: *J. of the Institute of Environmental Sciences,* pp. 40–45 (1994).

19. Dwyer, T. J., and D. S. Moul: *15th Space Simulation Conference,* Goddard Space Flight Center, NASA-CP-3015, pp. 125–138 (1988).

20. Raichel, D. R., Jet Propulsion Lab, California Institute of Technology, Pasadena (1991).

21. Bai, M., and W. Thatcher: *Shock and Vibration Bulletin,* 49(1):97–100 (1979).

22. Davie, N. T.: *Shock and Vibration Bulletin,* 56(3):109–124 (1986).

23. Davie, N. T., *Proc. Institute of Environmental Sciences Annual Technical Meeting,* pp. 344–351 (1985).

24. Shannon, K. L., and T. L. Gentry: "Shock Testing Apparatus," U. S. Patent No. 5,003,810, 1991.

25. Davie, N. T., and V. I. Bateman: *Proc. Institute of Environmental Sciences Annual Technical Meeting,* pp. 504–512 (1994).

CHAPTER 27
APPLICATION OF DIGITAL COMPUTERS

Allen J. Curtis
Steven D. Lust

INTRODUCTION

This chapter introduces numerous tools which are available on digital computers for the solution of shock and vibration problems. A brief discussion of each computer application is presented; they fall into the following basic categories: (1) numerical analyses of dynamic systems which are too complex and/or time-consuming to perform by hand calculation and (2) experimental applications requiring the rapid acquisition and processing of structural input and/or response data. Analytical applications discussed are general-purpose programs such as finite-difference methods, finite-element techniques, boundary-element methods, statistical energy analysis, and special-purpose programs such as vibration isolator analysis codes and rotating machinery analysis. In addition, applications that are typically personal computer (PC) based are noted. Topics under the heading of experimental capabilities are vibration data analysis, vibration and shock test control, and modal analysis. The major emphasis of this chapter is on experimental applications.

The decision to employ a digital computer in the solution of a shock or vibration problem should be made with considerable care. Before a program of digital computation is decided on, the following questions should be answered.

1. Is existing software available to perform the required task?
2. If not, to what extent must the task or the software be modified in order to perform the task?
3. If no applicable software exists, what is the extent of the software development needed?
4. What are the detailed assumptions inherent in the software program (e.g., linearity, proportional damping, etc.)?
5. Is the software able to compute and output the information required (e.g., absolute vs. relative motion, phase relationships, etc.)?
6. What are the detailed input and output limitations of the program (e.g., types of input excitations, graphic outputs, etc.)?

7. How much computer time will be used to perform the task?

After satisfactory answers to such questions are obtained, the user must realize that the results obtained from the computer output can be no better than the input to the computer. For example, the quality of the natural frequencies and mode shapes obtained from a structural analysis computer program depends not on the computer program but on the degree to which the mathematical model employed is representative of the mass, stiffness, and damping properties of the physical structure. Likewise, a spectral analysis of a signal with poor signal-to-noise ratio will provide an accurate spectrum of the signal *plus noise,* but not of the signal.

ANALYTICAL APPLICATIONS

The development of large-scale computers having a very short cycle time (i.e., the time required to perform a single operation, such as adding two numbers) and a very large memory permits detailed analyses of structural responses to shock and vibration excitation which are many orders of magnitude more complex than was previously practical. In this chapter, programs developed to perform these analyses are categorized as *general-purpose programs* and *special-purpose programs.* References 1 and 2 contain extensive anthologies of both general-purpose and special-purpose analytical programs.

GENERAL-PURPOSE PROGRAMS

Programs may be classed as "general purpose" if they are applicable to a wide range of structures and permit the user to select a number of options, such as damping (viscous or structural), various types of excitation (sinusoidal or random vibration, transients), etc.

Finite-Element Programs.   The most numerous programs are classed as "finite-element" or "lumped-parameter" programs; these are described in detail in Chap. 28, Part II.

In a lumped-parameter program, the structure to be analyzed is represented in a model as a number of point masses (or inertias) connected by massless, springlike elements. The points at which these elements are connected, and at which a mass may or may not be located, are the *nodes* of the system. Each node may have up to six degrees-of-freedom at the option of the analyst. The size of the model is determined by the sum of the degrees-of-freedom at the nodes and by the *dynamic* degrees-of-freedom (which is the number of degrees-of-freedom for which the mass or inertia is nonzero). The number of natural frequencies and modes which may be computed is equal to the number of dynamic degrees-of-freedom. However, the number of frequencies and modes which reliably represent the physical structure is generally only a fraction of the number which can be computed. Each program is limited in capacity to some combination of dynamic and zero mass degrees-of-freedom. The springlike elements are chosen to represent the stiffness of the physical structure between the selected nodes and generally may be represented by springs, beams, or plates of specified shapes. The material properties, geometric properties, and boundary conditions for each element are selected by the analyst.

In the more general finite-element programs, the springlike elements are not necessarily massless but may have distributed mass properties. In addition, lumped masses may be used at any of the nodes of the system.

The equations of motion of the finite-element model can be expressed in matrix form and solved by the methods described in Chap. 28. Regardless of the computational algorithms employed, the program computes the set of natural frequencies and orthogonal mode shapes of the finite-element system. These modes and frequencies are sorted for future use in computing the response of the system to a specified excitation. For the latter computations, a damping factor must be specified. Depending on the programs, this damping factor may have to be equal for all modes, or it may have a selected value for each mode.

Component Mode Synthesis. The method of modeling described above leads to the prescribing of models with a very large number of degrees-of-freedom compared with the number of modes and frequencies actually of interest. Not only is this expensive, but it rapidly exceeds the capacity of many programs. To overcome these problems, component mode synthesis techniques have been developed. Instead of developing a model of an entire physical system, several models are developed, each representing a distinct identifiable region of the total structure and within the capacity of the computer program. The modes and frequencies of interest in each of these models are computed independently. Where actual hardware exists for some or all components, modes and frequencies from an experimental modal analysis may be used. A model of the entire structure is then obtained by joining these several models, using the component mode synthesis technique. This model retains the essential features of each substructure model and thus of the entire structure while using a greatly reduced number of degrees-of-freedom to do so.

Reduction of Model Complexity. Companion methods, developed to reduce the cost of analysis, to permit the joining of several substructure models, and to provide for correlation with experimental results, are described under *Reduction Techniques* in Part II of the following chapter. For cost reduction and joining of substructures, the objective is to reduce the mass and stiffness matrices to the minimum size consistent with retaining the modes and frequencies of interest together with other dynamic characteristics such as base impedance. For test/analysis correlation, the objective is to match the degrees-of-freedom from the test. It should be noted, however, that the Guyan reduction method yields a mass matrix which is nondiagonal (see Chap. 28) and which may be unacceptable for some computer programs. It is also of interest that the rigid-body mass properties, i.e., total masses and inertias of the structure, are not identifiable in the reduced mass matrix.

Boundary-Element Method.[3-5] The boundary-element method involves the transformation of a partial differential equation which describes the behavior of an enclosed region to an integral equation which describes the behavior of the region boundary. Once the numerical solution for the boundary is obtained, the behavior of the enclosed region is then calculated from the boundary solution. Using this method, three-dimensional problems can be reduced to two dimensions, and two-dimensional problems can be reduced to one dimension. It is then necessary to model in detail only the boundary of the enclosed region rather than the complete region. A volume can be described by its surface, and an area can be described by its edges. Discretization of the boundary is much less detailed and less sensitive to mesh distortion than that for a finite-element model of the same region. However, each boundary-element equation has a greater number of algebraic functions than the corresponding finite-element equation, and so more processing power is required.

Two types of boundary-element methods exist. The *direct* method solves directly for the physical variables on the surface. The system of equations is of a form where the matrices are full, complex, nonsymmetric, and a function of frequency. Boundary conditions for the direct method are the prescribed physical variables or impedance relationships at the nodes. The *indirect* method solves for single- and double-layer potentials on the surface, which can be postprocessed to obtain the physical variables. Matrices for the indirect method are complex-valued and symmetric, which enables coupling with finite-element models.

The boundary-element method is particularly powerful for solving field or semi-infinite problems. It can be readily applied to coupled structural/acoustical analysis or to solve for the boundary conditions of a finite-element model. The method assumes isotropic material properties and works well for structures with a high volume-to-surface ratio, but is not suitable for plate and thin-shell problems.

Distributed (Continuous) System Programs.[1,2]   A number of specialized programs treating the analysis of distributed or continuous structural systems such as beams, plates, shells, rings, etc., have been developed. Each program can be applied for a broad, selectable range of physical properties and dimensions of the particular structural shape. Not all programs employ the same theory of elasticity; thus, the user must examine the theoretical basis on which the program was developed. For example, the user must determine if the program includes such effects as rotary inertia or shear deformation.

Preprocessors and Postprocessors.   Experience with the general-purpose analysis programs described above indicates two major shortcomings: (1) A large amount of expensive computer time is required to debug the structural models, i.e., to get them to run. (2) The large mass of tabulations and/or plots of the results of the analysis is very difficult to evaluate. To alleviate these problems, programs have been written, called *preprocessors* and *postprocessors,* which use sophisticated interactive graphics in combination with algorithms; such programs greatly simplify the construction of the models, verification of the models, and presentation of the results of the analysis. These highly efficient programs often can be run very economically offline, independent of the larger computer required to exercise the model. Many organizations have developed their own preprocessors tailored to their product lines; commercially available processor software packages also are available for this purpose. Interfaces have also been developed for computer-aided design (CAD) software packages, which allow preprocessors to acquire geometric information from the CAD database and use it as the basis for the structural model.

Statistical Energy Analysis.[6]   Statistical energy analysis (SEA) is used to predict structural response to broad-band random excitation in frequency regions of high modal density. In these frequency regions, response predictions for individual modes are impractical. Structural response is treated in a statistical sense; that is, an estimate of the average response is calculated in frequency bands wide enough to include many modes. The structural system is divided into components, with each component described by the parameters of modal density and loss factor. A third modeling parameter is the energy transmission characteristics of the structural coupling between components. SEA is valuable in predicting environments and responses for structures in the conceptual design phase, where detailed structural information is not available. Chapter 11 describes SEA in detail.

Personal-Computer-Based Applications.[7–10]   Almost all analytical and experimental applications that are available on mainframe computers and workstations can also be found for personal computers (PCs). Mainframes and workstations are often used for applications requiring large amounts of memory and disk space and fast processing speeds, such as large finite-element models and vibration control and data analysis for tests with a great number of control and response channels. However, for most other computational efforts, both analytical and experimental, PCs can be employed. The following are examples of general-purpose applications which are widely used on the PC.

Technical calculation packages are available that allow the user to obtain solutions to dynamical equations without resorting to programming. Equations are typed in real mathematical form. Solutions can be plotted in two and three dimensions. Among the many capabilities are curve fitting, FFT calculation, symbolic manipulation, numerical integration, and treatment of vectors and matrices as variables.

Spreadsheet software developed for accounting can also be used to manipulate vectors and matrices. Graphical capabilities can be used to generate report-quality plots. Commercial data acquisition systems can store time- or frequency-domain information in files compatible with spreadsheets. Even ensemble averaging can be accomplished for computation of statistical functions.

Graphical programming software exists for data acquisition and control, data analysis, and data presentation. Instruments such as oscilloscopes, spectrum analyzers, vibration controllers, etc., can be emulated in graphical form. These instruments can acquire, analyze, and graphically present data from plug-in data acquisition boards or from connected instruments.

SPECIAL-PURPOSE PROGRAMS[1, 2]

The need for a special-purpose program may arise in several ways. First, for an engineering activity engaged in the design, on a repetitive basis, of what amounts analytically to the same structure, it may be economical to develop an analysis program which efficiently analyzes that particular structure. The analysis of vibration isolator systems, automobile suspension systems, piping systems, or rotating machinery may be an example. Similarly, parametric studies of a particular structure, either to gain understanding or to optimize the design, may require a sufficient number of computer runs to justify the development of software. A second type of special-purpose program includes programs which in some way perform an unusual type of analysis—for example, that of nonlinear systems.

Access to existing special-purpose programs is generally more restricted than is access general-purpose programs, in part because of their frequent proprietary nature and in part because of the investment in software required to develop them.

EXPERIMENTAL APPLICATIONS

The classification "experimental applications" covers uses of computers which involve, in some way, the processing of shock and vibration information originally obtained during test or field operation of physical equipment. Two developments led to the applications described in later sections: (1) the recognition of the computational efficiency of the fast Fourier transform (FFT) algorithm, and (2) the development of hardware FFT processors coupled with the development of minicomputers. These developments permit the use of digital computers for such tasks as vibration data analysis, shock data analysis, and shock, vibration, and modal testing—all described in later sections. The information resulting from such applications is in digital form, which permits more sophisticated engineering evaluation of the information through further efficient digital processing, e.g., regression analysis, averaging, etc.

DIGITAL VIBRATION DATA ANALYSIS[11]

The basic principles of vibration data analysis are described in Chap. 22. These principles are illustrated in terms of analog methods of processing data. The principles of analysis are not changed by a switch to digital methods and are not repeated. Only the basic digital methods are described. Various types of analyzers are described in Chap. 14.

Types of data analysis which may be performed entirely by digital computation parallel those performed by analog methods. Three basic types are spectral analysis, correlation analysis, and statistical or probability analysis. Common to all digital vibration data analysis is the implicit assumption that the data are sampled from a stationary or steady-state process; or that at least the data are varying slowly enough that the data sample may be considered stationary during the sample duration.

Analog-to-Digital Conversion and Data Preparation. Analog-to-digital conversion (ADC) is the process by which an analog signal is converted into a representative number sequence and is the first step in any digital method. This operation is generally built into self-contained digital analysis systems. However, when the digital processing is performed on a general-purpose scientific computer, this operation must usually be carried out at another facility.

A prime advantage of digital analysis methods is the ability to play back the analog time-history of the signal no more than once for ADC, or alternatively, to obviate the necessity for recording the time-history if on-line analysis is desired. Thus, for example, the problems associated with preparing a satisfactory tape loop for repeated playback of the data sample to be analyzed by tunable analog filters are avoided. However, analog recording in parallel for backup data purposes may be prudent.

For each of the digital data samples which constitute the digital record to be processed, the ADC process consists of two steps: (1) the analog signal is sampled by a sample-and-hold device, and (2) the sampled analog voltage is converted to a binary digital number. Each sample is subject to two errors, which appear as digital noise on the record. First, the sample-and-hold device samples over a finite though small time interval; if the analog signal can vary significantly during this time interval, the "held" voltage will be inaccurate. Second, error also may be generated if the digital word length, i.e., the number of bits, is insufficient; since there is always an uncertainty equal to the least significant bit, this may represent a significant error, particularly for samples taken near a zero crossing.

When possible, it is desirable to preprocess the digital record prior to analysis to remove any wild points and any trends.[12] Wild points are defined as any individual digital samples (or possibly several successive samples) which are either unreasonably large or essentially zero for some generally unknown reason, such as analog tape spikes or dropouts or ADC momentary failure. Trends are defined as slowly varying changes in the data, such as a dc bias or a slow change in the running mean value of the signal. Trends can often be eliminated by the use of high-pass filtering of the signal prior to ADC. Alternatively, the digital record can be preprocessed to remove the calculated trend from each digital sample and a new record created.

The data must be preprocessed and digitized in a manner which avoids frequency aliasing. Aliasing occurs when the sample rate is less than twice the highest frequency in the data. The frequencies above half the sampling frequency will be folded back into the frequencies below half the sampling frequency, producing erroneous results. Hence, aliasing errors may be avoided by using a sampling rate which is at least twice that of the highest significant frequency component in the signal. However, since this may be difficult before analysis, which has the objective of determining the frequency components up to some desired maximum frequency of interest, it is generally preferable to avoid aliasing errors by low-pass filtering prior to ADC.

Spectral Analysis.   Spectral analysis (see Chaps. 14 and 22) is performed to determine the frequency characteristics of a time-history. For deterministic signals, a line spectrum is desired; for random processes, a spectral density function is desired. For example, see Fig. 22.2.

Deterministic Signal.   Determination of the line spectrum of a deterministic signal $[x(t)]$ (a signal whose behavior can be described by an explicit mathematical relationship) by digital computation requires only computation of the Fourier transform of the signal $S(x)$. The computed transform may be presented either in terms of magnitude $|S(x)|$ and phase or in terms of the real and imaginary parts for each computation frequency, or both.

Equations (22.12) through (22.16) define the relationships for selection of the analysis parameters. The Fourier transform treats the data as if they were periodic with period T. Thus, for a deterministic signal, the frequencies at which $S(x)$ is computed may not (and probably will not) coincide exactly with the actual frequency components of the signal. It is important, therefore, to select a frequency resolution, Δf, which is sufficiently small to minimize errors in both frequency and amplitude owing to this misalignment of actual and computed frequencies.

Random Signal.   Determination of the spectral density function, either the auto-spectral density or the cross-spectral density, of a random signal may be performed by either of two alternative methods. The older and traditional method, known as the Blackman-Tukey method, arrives at the spectral density by first computing the correlation function [Eq. (22.42) or (22.48)], followed by Fourier transformation of the correlation function to obtain the spectral density [Eq. (22.20) or (22.22)]. The newer method (using the FFT algorithm), known as the Cooley-Tukey method, obtains the spectral density directly from the Fourier transform of the time-history [Eqs. (22.17), (22.18), and (22.19)]. For applications requiring real-time (or, more exactly, almost real-time) analysis, the latter method is mandatory.

Equations (22.12) through (22.16) define the relationships for selection of the analysis parameters, using the FFT method. Indirectly, these equations also define the analysis parameters for correlation analysis when using the Blackman-Tukey method. Assuming the correlation functions are computed at incremental delay times, $\Delta\tau$ equal to the sampling interval, Δt, up to a maximum delay, $m\,\Delta\tau$, where m is the maximum lag number, then

$$\Delta f = \frac{1}{m\,\Delta\tau} \tag{27.1}$$

For spectral density analysis, these methods determine the average spectral density within each bandwidth, Δf wide, centered at the $N/2$ computation frequencies.

The maximum lag number m should be small compared with the total number of samples N. The effective record length for spectral analysis is thus $m\,\Delta t$ rather than the $N\,\Delta t$ needed to compute the correlation function.

Statistical Error. Spectral density functions computed by either of the methods described above are only estimates of the true spectral densities of the original analog time-histories. These estimates are subject to statistical error very similar to the statistical errors in analog processing. The normalized standard error ϵ of the spectral density estimate computed from a single FFT is 1.0 and is governed by the chi-squared distribution for two degrees-of-freedom. To obtain spectral density estimates with a more reasonable statistical error, further averaging is necessary. This may be accomplished by averaging over frequency and/or averaging over segments or records. In the first case, the spectral estimates in l adjacent frequency intervals are averaged to obtain an estimate of the average spectral density in a frequency bandwidth, $l \, \Delta f$. The standard error of the resultant spectral density then becomes $\epsilon = 1/\sqrt{l}$. Thus, statistical accuracy is gained at the expense of frequency resolution. In the second case, the spectral estimates obtained from a series of records, each $N \, \Delta t = T$ in duration, are averaged. This process is known as "ensemble averaging." The standard error for an ensemble average of q records is $q^{-0.5}$. Therefore, statistical accuracy is gained without loss of frequency resolution, for a constant block size N, by repeated spectral analysis. If both frequency and ensemble averaging are employed, the standard error becomes $1/\sqrt{ql}$. The error is chi-squared distributed with $2l$, $2q$, and $2ql$ degrees-of-freedom, respectively, for the three types of averaging.

Leakage. Spectral analysis by digital methods is subject to error known as "leakage."[12, 13] Leakage is a spectral distortion in which the power in a single frequency component appears to "leak" into adjacent frequency bands. Leakage arises from the truncation which occurs when either the FFT is computed over a finite frequency range, 0 to F_{max}, and the data are not periodic in the sample length (see Fig. 14.8) or the correlation function is computed over a finite maximum time delay, $m \, \Delta \tau$. This truncation causes undesirable side lobes in the effective filter shape of the transmission characteristics of a filter. These side lobes can be minimized by the use of one of several types of tapering functions instead of truncating in the digital processing. Commonly used functions are known as Parzen[12, 13] and Hamming[12, 13] windows; properties of various types of windows are compared in Table 14.1.

Transfer Function Measurements. Measurement of the complex transfer or frequency-response functions $[H(f)]$ described by Eq. (22.26a), together with the auto-spectral density functions [Eq. (22.17)] and the coherence function [Eq. (22.27)], can be carried out conveniently with a two-channel FFT analyzer, available commercially, described in Chap. 14. A wide range of analysis parameters are available. In performing transfer function measurements, the following considerations are important:

1. Aliasing errors should be avoided by filtering and/or by the appropriate selection of the digitizing frequency.

2. If the input and output signals are correlated, the statistical error inherent in the spectral density analysis is negligible in calculation of $H(f)$, i.e., shorter data sample durations can be tolerated.

3. Any relative time shifts between $x(t)$ and $y(t)$ due to head stack spacing of tape recorders, etc., will bias the effects of the results of phase measurements but will not change the magnitude of $H(f)$.

4. Auto-spectral analysis can be performed from recorded signals played back in either the forward or reverse directions; if reverse playback is used, the complex conjugate of the transfer function is obtained.

5. While larger Δf improves auto-spectral density estimates at the cost of spectral smoothing, larger Δf degrades both the frequency resolution and the coherence function measurement.

Many of the FFT analyzers can also perform (or can be linked to desktop computers to perform) modal analyses from the array of transfer functions.

Statistical Analysis. Chapter 22 described, in analog terms, the basic methods of determining the probability distribution functions of both instantaneous and peak values of a signal. Performance of these types of analyses using digital computer techniques is no different in principle but is much more straightforward to implement. References 12 and 13 contain detailed descriptions of the required digital operations.

An advantage gained by the use of digital techniques is the ability to perform goodness-of-fit tests on the resultant probability density functions. These tests determine (to some selected confidence level, and based on the number of digital samples) whether the distribution is considered Gaussian or non-Gaussian.

Correlation Analysis. Autocorrelation and crosscorrelation functions may be determined either from Fourier-transforming the auto- or cross-spectral density functions [Eqs. (22.21) and (22.23)] or directly from the time-histories. The latter is generally the case during computation of power spectral density by the Blackman-Tukey method. The principles of the latter method for digital processing are the same as those described in Chap. 22 for analog processing. Detailed digital procedures are described in Refs. 12 and 13, and are relatively simple, consisting merely of summing the products of pairs of samples. The choice of record length, digitizing rate, and number of lags determines the statistical error of the correlation function estimate. In addition, these choices define the frequency resolution, frequency range, and statistical error of spectral density, later computed from the correlation functions.

DIGITAL CONTROL SYSTEMS FOR VIBRATION AND SHOCK TESTING

The vibratory motions prescribed for the majority of vibration tests are either sinusoidal or random. The system employed to control the vibration level during the test utilizes the output signal from an accelerometer mounted at an appropriate location on the vibration exciter to provide a feedback signal to a servo system. The servo system adjusts the driving signal to the power amplifier to maintain the desired test level. Vibration testing machines are described in Chap. 25.

With the advent of FFT processors and minicomputers, it became possible to perform spectral analysis of random processes rapidly enough to permit the use of digital control systems for random vibration testing. An extension of software also enables the system to control sinusoidal testing.

For a number of reasons, it is desirable to perform shock or transient testing using electrodynamic or electrohydraulic vibration systems to induce the desired transient motion at the mounting points of the test item. The ability to employ this method is dependent on such parameters as the stroke or maximum allowable motion of the vibration exciter, the transient waveform, the mass of the test item, etc. Given that the required test is within the performance capability of an available vibration system, the ability to obtain and control the desired motion has been greatly expanded by the use of digital control equipment. In general, the servo control of a shock test parallels that for vibration testing; the controller compares the control accelerometer response to a reference waveform or function. If necessary, the controller drive signal is altered to minimize the deviation of the control accelerometer response from the reference. Digital control of shock testing is treated in Chap. 25.

Shock-test requirements may be specified by either of two alternative methods. The first, and more direct, method specifies a certain acceleration waveform, such as a half sine wave of specified duration and maximum acceleration. The second method employs shock spectra (see Chap. 23). The requirements must specify the frequency range, damping factor, type of spectrum, and either minimum or nominal values with an allowable tolerance on spectrum values.

MODAL TESTING

Modal testing (also known as *ground vibration testing* in the aircraft industry) is conducted to determine experimentally the natural frequencies, mode shapes, and associated damping factors of a structure. For many years, this testing has been conducted by exciting the system with sinusoidal excitation at a number of points of the structure. The driving frequency, the relative amplitudes, and the phases of the excitations are adjusted to obtain the "purest" possible excitation of a single mode. The responses at locations throughout the structure then define the mode shape for that frequency. Damping factors can be measured by measurement of the bandwidth of the mode.

Mode shapes may also be measured by the use of broad-band vibration or transient excitations applied at a single point or multiple points of the structure for any one measurement or group of measurements. The responses to these excitations are measured at all points of interest. From these data, a set of transfer functions between each excitation location and each response measurement location may be calculated for each excitation employed. The frequencies, mode shapes, and damping factors are then obtained from these measured transfer functions. Experimental modal analysis is described in detail in Chap. 21.

Digital computers have been applied in modal analysis in two distinct ways. First, for sinusoidal excitation, computers have been employed as an aid in obtaining the desired purity of the modal excitation as well as in acquiring and processing data. Second, and more essentially, for broad-band excitation (either random vibration or transient), FFT processing of the data is used to determine transfer functions until satisfactory data quality is achieved.

Once the modal parameters, i.e., natural frequency, mode shape, and damping, have been determined, they can be utilized in various ways. Extensions to modal analysis include structural modification techniques,[14–16] response simulations, and correlation with analytical predictions. Structural modification techniques allow for the evaluation of changes in the modal parameters as a result of having discrete springs, masses, or dampers added to or subtracted from the experimental degrees-of-freedom. In addition, tuned mass dampers and rib stiffeners can be added to the modal model. Using structural modification techniques, candidate design modifications for solving a dynamics problem can be assessed analytically, as opposed to the lengthy and expensive "cut-and-try" hardware approach.

Response simulation is the ability to predict response at any degree-of-freedom to random, sinusoidal, or transient excitation at a number of degrees-of-freedom for the original modal parameters or those resulting from structural modification. Response simulation can be used in conjunction with structural modification techniques to demonstrate whether the modified structure would have improved dynamic behavior when subjected to the simulated forcing function.

The mode shapes determined from the test can be validated by a number of techniques. Orthogonality between test mode shapes or between test and analytical mode shapes can be checked. Likewise, the modal assurance criterion (MAC)[17] can be used to verify consistency between test and analysis modes or to show uniqueness of the test-derived modes among themselves.

Updating of Finite-Element Models Using Experimental Modal Analysis Results. A major purpose of performing an experimental modal analysis is to obtain information (frequency-response functions, mode shapes, natural frequencies, etc.) for updating a finite-element model of the structure. Techniques for model updating can be divided into a number of categories: eigenstructure assignment,[18] design sensitivity methods,[19] and optimal-matrix methods.[20-23]

Eigenstructure assignment involves the design of a pseudo-controller which would cause the initial structural model to produce the responses measured in the test. The adjustments to the finite-element matrices are determined from the controller design.

Design sensitivity methods use derivatives of the modal parameters with respect to the design variables in an iterative scheme to determine the adjustments to the design variables. The design variables are physical properties such as modulus of elasticity, density, and cross-sectional area.

Optimal-matrix update techniques use a constrained optimization approach to modify the mass, stiffness, and damping matrices so that the resulting responses more closely match the measured responses. Variations of optimal-matrix techniques use constraints such as matrix symmetry, minimum deviation from the initial model, preservation of the original load paths, and positive definiteness of the resulting matrices. Iterative methods that use design sensitivities to update the finite-element model yield physically interpretable results. Iterative methods generally assume inconsistencies in the measured modal data; therefore the resulting model does not exactly replicate the measured data. Direct methods exist which modify the individual mass and stiffness matrix elements, using a one-step approach based on a closed-form solution that uses orthogonalized mode shapes that are assumed to be without error. The responses of the resulting model exactly match the measured modes, but the matrix element changes are without relation to the physical structure.

REFERENCES

1. Pilkey, W., K. Saczalski, and H. Schaeffer: "Structural Mechanics Computer Programs, Surveys, Assessments and Availability," University Press of Virginia, Charlottesville, Va., 1974.

2. Pilkey, W., and B. Pilkey: "Shock Vibration Computer Programs—Reviews and Summaries," SVM-10, Shock and Vibration Information Center, Washington, D.C., 1975.

3. Brebbia, C. A., and S. Walker: "Boundary Element Techniques in Engineering," Butterworth & Co. Ltd., London, 1980.

4. Brebbia, C. A., J. C. F. Telles, and L. C. Wrobel: "Boundary Element Techniques," Springer-Verlag, New York, 1984.

5. Fyfe, K. R., J.-P. G. Coyette, and P. A. van Vooren: *Sound and Vibration,* **25**(12):16–22 (1991).

6. Lyon, R. H.: "Statistical Energy Analysis of Dynamical Systems: Theory and Application," MIT Press, Cambridge, Mass., 1975.

7. Porter, M. L.: *Personal Engineering & Instrumentation News,* **9**(4):37–44 (1992).

8. Wilson, H. B., and S. Gupta: *Sound and Vibration,* **26**(8):24–29 (1992).

9. Bloch, S. C.: *Personal Engineering & Instrumentation News,* **9**(11):45–51 (1992).

10. Porter, M. L.: *Personal Engineering & Instrumentation News,* **10**(3):29–36 (1993).

11. Himmelblau, H., A. G. Piersol, J. H. Wise, and M. R. Grundvig: "Handbook for Dynamic Data Acquisition and Analysis," Institute of Environmental Sciences Recommended Practice RP-DTE 012.1, 1994.

12. Otnes, R. K., and L. Enochson: "Digital Time Series Analysis," John Wiley & Sons, Inc., New York, 1972.

13. Bendat, J. S., and A. G. Piersol: "Random Data: Analysis and Measurement Procedures," John Wiley & Sons, Inc., New York, 1971.

14. Herbert, M. R., and D. W. Kientzy: "Applications of Structural Dynamics Modification," *SAE Paper* 80-1125, 1980.

15. Snyder, V. W.: *Intern. J. Analytical and Experimental Modal Analysis,* **1**(1) January (1986).

16. Avitabile, P., and J. O'Callahan: *Sound and Vibration,* **24**(6):18–26 (1990).

17. Allemang, R. J., and D. L. Brown: *Proc. 1st Intern. Modal Analysis Conf.,* 1992, pp. 110–116.

18. Zimmerman, D. C., and M. Widengren: *AIAA J.,* **28**(9):1670–1676 (1990).

19. Flanigan, C. C.: "Test/Analysis Correlation Using Design Sensitivity and Optimization," *SAE Technical Paper* 871743, October 1987.

20. Baruch, M., and Y. I. Bar Itzhack: *AIAA J.,* **16**(4):346–351 (1978).

21. Kammer, D. C.: *AIAA J.,* **26**(1):104–112 (1988).

22. Smith, S. W., and C. A. Beattie: "Optimal Identification Using Inconsistent Modal Data," *AIAA Paper* 91-0948-CP, *Proc. 32nd AIAA/ASME/ASCE/AHS/ASC Structures, Structural Dynamics, and Materials Conf.,* April 1991.

23. Kabe, A. M., *AIAA J.,* **23**(9):1431–1436 (1985).

24. American National Standards Institute, "Methods for Analysis and Presentation of Shock and Vibration Data," ANSI S2.10-1971.

CHAPTER 28

PART I: MATRIX METHODS OF ANALYSIS

Stephen H. Crandall
Robert B. McCalley, Jr.

INTRODUCTION

The mathematical language which is most convenient for analyzing multiple degree-of-freedom vibratory systems is that of *matrices*. Matrix notation simplifies the preliminary analytical study, and in situations where particular numerical answers are required, matrices provide a standardized format for organizing the data and the computations. Computations with matrices can be carried out by hand or by digital computers. The availability of large high-speed digital computers makes the solution of many complex problems in vibration analysis a matter of routine.

This chapter describes how matrices are used in vibration analysis. It begins with definitions and rules for operating with matrices. The formulation of vibration problems in matrix notation then is treated. This is followed by general matrix solutions of several important types of vibration problems, including free and forced vibrations of both undamped and damped linear multiple degree-of-freedom systems. Part II of this chapter considers finite-element models.

MATRICES

Matrices are mathematical entities which facilitate the handling of simultaneous equations. They are applied to the differential equations of a vibratory system as follows:

A single degree-of-freedom system of the type in Fig. 28.1 has the differential equation

$$m\ddot{x} + c\dot{x} + kx = F$$

where m is the mass, c is the damping coefficient, k is the stiffness, F is the applied force, x is the displacement coordinate, and dots denote time derivatives. In Fig. 28.2 a similar three degree-of-freedom system is shown. The equations of motion may be obtained by applying Newton's second law to each mass in turn:

$$
\begin{aligned}
m\ddot{x}_1 \quad &+ c\dot{x}_1 \quad &+ 5kx_1 - 2kx_2 \quad &= F_1 \\
2m\ddot{x}_2 \quad &+ 2c\dot{x}_2 - 2c\dot{x}_3 - 2kx_1 + 3kx_2 - kx_3 &= F_2 \\
3m\ddot{x}_3 \quad &- 2c\dot{x}_2 + 2c\dot{x}_3 \quad &- kx_2 + kx_3 &= F_3
\end{aligned}
\tag{28.1}
$$

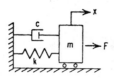

FIGURE 28.1 Single degree-of-freedom system.

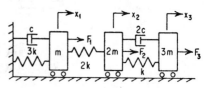

FIGURE 28.2 Three degree-of-freedom system.

The accelerations, velocities, displacements, and forces may be organized into columns, denoted by single boldface symbols:

$$\ddot{\mathbf{x}} = \begin{bmatrix} \ddot{x}_1 \\ \ddot{x}_2 \\ \ddot{x}_3 \end{bmatrix} \qquad \dot{\mathbf{x}} = \begin{bmatrix} \dot{x}_1 \\ \dot{x}_2 \\ \dot{x}_3 \end{bmatrix} \qquad \mathbf{x} = \begin{bmatrix} x_1 \\ x_2 \\ x_3 \end{bmatrix} \qquad \mathbf{f} = \begin{bmatrix} F_1 \\ F_2 \\ F_3 \end{bmatrix} \qquad (28.2)$$

The inertia, damping, and stiffness coefficients may be organized into square arrays:

$$\mathbf{M} = \begin{bmatrix} m & 0 & 0 \\ 0 & 2m & 0 \\ 0 & 0 & 3m \end{bmatrix} \qquad \mathbf{C} = \begin{bmatrix} c & 0 & 0 \\ 0 & 2c & -2c \\ 0 & -2c & 2c \end{bmatrix} \qquad \mathbf{K} = \begin{bmatrix} 5k & -2k & 0 \\ -2k & 3k & -k \\ 0 & -k & k \end{bmatrix} \qquad (28.3)$$

By using these symbols, it is shown below that it is possible to represent the three equations of Eq. (28.1) by the following single equation:

$$\mathbf{M}\ddot{\mathbf{x}} + \mathbf{C}\dot{\mathbf{x}} + \mathbf{K}\mathbf{x} = \mathbf{f} \qquad (28.4)$$

Note that this has the same form as the differential equation for the single degree-of-freedom system of Fig. 28.1. The notation of Eq. (28.4) has the advantage that in systems of many degrees-of-freedom it clearly states the physical principle that at every coordinate the external force is the sum of the inertia, damping, and stiffness forces. Equation (28.4) is an abbreviation for Eq. (28.1). It is necessary to develop the rules of operation with symbols such as those in Eqs. (28.2) and (28.3) to ensure that no ambiguity is involved. The algebra of *matrices* is devised to facilitate manipulations of simultaneous equations such as Eq. (28.1). Matrix algebra does not in any way simplify individual operations such as multiplication or addition of numbers, but it is an organizational tool which permits one to keep track of a complicated sequence of operations in an optimum manner.

DEFINITIONS

A *matrix* is an array of elements arranged systematically in rows and columns. For example, a rectangular matrix **A**, of elements a_{jk}, which has m rows and n columns is

$$\mathbf{A} = [a_{jk}] = \begin{bmatrix} a_{11} & a_{12} & \dots & a_{1n} \\ a_{21} & a_{22} & \dots & a_{2n} \\ \dots & \dots & \dots & \dots \\ a_{m1} & a_{m2} & \dots & a_{mn} \end{bmatrix}$$

The elements a_{jk} are usually numbers or functions, but, in principle, they may be any well-defined quantities. The first subscript j on the element refers to the row number while the second subscript k refers to the column number. The array is denoted by the single symbol $\mathbf{A}$, which can be used as such during operational manipulations in which it is not necessary to specify continually all the elements a_{jk}. When a numerical calculation is finally required, it is necessary to refer back to the explicit specifications of the elements a_{jk}.

A rectangular matrix with m rows and n columns is said to be of order (m,n). A matrix of order (n,n) is a *square matrix* and is said to be simply a square matrix of order n. A matrix of order $(n,1)$ is a *column matrix* and is said to be simply a column matrix of order n. A column matrix is sometimes referred to as a *column vector*. Similarly, a matrix of order $(1,n)$ is a *row matrix* or a *row vector*. Boldface *capital* letters are used here to represent square matrices and *lower-case* boldface letters to represent column matrices or vectors. For example, the matrices in Eq. (28.2) are column matrices of order three and the matrices in Eq. (28.3) are square matrices of order three.

Some special types of matrices are:

1. A *diagonal matrix* is a square matrix $\mathbf{A}$ whose elements a_{jk} are zero when $j \neq k$. The only nonzero elements are those on the *main diagonal,* where $j = k$. In order to emphasize that a matrix is diagonal, it is often written with small ticks in the direction of the main diagonal:

$$\mathbf{A} = \left\lceil a_{jj} \right\rfloor$$

2. A *unit matrix* or *identity matrix* is a diagonal matrix whose main diagonal elements are each equal to unity. The symbol $\mathbf{I}$ is used to denote a unit matrix. Examples are

$$\begin{bmatrix} 1 & 0 \\ 0 & 1 \end{bmatrix} \qquad \begin{bmatrix} 1 & 0 & 0 \\ 0 & 1 & 0 \\ 0 & 0 & 1 \end{bmatrix}$$

3. A *null matrix* or *zero matrix* has all its elements equal to zero and is simply written as zero.

4. The *transpose* $\mathbf{A}^T$ of a matrix $\mathbf{A}$ is a matrix having the same elements but with rows and columns interchanged. Thus, if the original matrix is

$$\mathbf{A} = [a_{jk}]$$

the transpose matrix is

$$\mathbf{A}^T = [a_{jk}]^T = [a_{kj}]$$

For example:

$$\mathbf{A} = \begin{bmatrix} 3 & 2 \\ -1 & 4 \end{bmatrix} \qquad \mathbf{A}^T = \begin{bmatrix} 3 & -1 \\ 2 & 4 \end{bmatrix}$$

The transpose of a square matrix may be visualized as the matrix obtained by rotating the given matrix about its main diagonal as an axis.

The transpose of a column matrix is a row matrix. For example,

$$\mathbf{x} = \begin{bmatrix} 3 \\ 4 \\ -2 \end{bmatrix} \qquad \mathbf{x}^T = \begin{bmatrix} 3 & 4 & -2 \end{bmatrix}$$

Throughout this chapter a row matrix is referred to as the transpose of the corresponding column matrix.

5. A *symmetric matrix* is a square matrix whose off-diagonal elements are symmetric with respect to the main diagonal. A square matrix $\mathbf{A}$ is symmetric if, for all j and k,

$$a_{jk} = a_{kj}$$

A symmetric matrix is equal to its transpose. For example, all three of the matrices in Eq. (28.3) are symmetric. In addition, the matrix $\mathbf{M}$ is a diagonal matrix.

MATRIX OPERATIONS

Equality of Matrices. Two matrices of the same order are equal if their corresponding elements are equal. Thus two matrices $\mathbf{A}$ and $\mathbf{B}$ are equal if, for every j and k,

$$a_{jk} = b_{jk}$$

Matrix Addition and Subtraction. Addition or subtraction of matrices of the same order is performed by adding or subtracting corresponding elements. Thus, $\mathbf{A} + \mathbf{B} = \mathbf{C}$ if for every j and k,

$$a_{jk} + b_{jk} = c_{jk}$$

For example, if

$$\mathbf{A} = \begin{bmatrix} 3 & 2 \\ -1 & 4 \end{bmatrix} \qquad \mathbf{B} = \begin{bmatrix} -1 & 2 \\ 5 & 6 \end{bmatrix}$$

then

$$\mathbf{A} + \mathbf{B} = \begin{bmatrix} 2 & 4 \\ 4 & 10 \end{bmatrix} \qquad \mathbf{A} - \mathbf{B} = \begin{bmatrix} 4 & 0 \\ -6 & -2 \end{bmatrix}$$

Multiplication of a Matrix by a Scalar. Multiplication of a matrix by a scalar c multiplies each element of the matrix by c. Thus

$$c\mathbf{A} = c[a_{jk}] = [ca_{jk}]$$

In particular, the negative of a matrix has the sign of every element changed.

Matrix Multiplication. If $\mathbf{A}$ is a matrix of order (m,n) and $\mathbf{B}$ is a matrix of order (n,p), then their *matrix product* $\mathbf{AB} = \mathbf{C}$ is defined to be a matrix $\mathbf{C}$ of order (m,p) where, for every j and k,

$$c_{jk} = \sum_{r=1}^{n} a_{jr}b_{rk} \tag{28.5}$$

The product of two matrices can be obtained only if they are *conformable*, i.e., if the number of columns in $\mathbf{A}$ is equal to the number of rows in $\mathbf{B}$. The symbolic equation

$$(m,n) \times (n,p) = (m,p)$$

indicates the orders of the matrices involved in a matrix product. Matrix products are not commutative, i.e., in general,

$$\mathbf{AB} \neq \mathbf{BA}$$

The matrix products which appear in this chapter are of the following types:

Square matrix × square matrix = square matrix

Square matrix × column vector = column vector

Row vector × square matrix = row vector

Row vector × column vector = scalar

Column vector × row vector = square matrix

In all cases, the matrices must be conformable. Numerical examples are given below.

$$\mathbf{AB} = \begin{bmatrix} 3 & 2 \\ -1 & 4 \end{bmatrix} \begin{bmatrix} -1 & 2 \\ 5 & 6 \end{bmatrix} = \begin{bmatrix} -(3 \times 1) + (2 \times 5) & (3 \times 2) + (2 \times 6) \\ (1 \times 1) + (4 \times 5) & -(1 \times 2) + (4 \times 6) \end{bmatrix} = \begin{bmatrix} 7 & 18 \\ 21 & 22 \end{bmatrix}$$

$$\mathbf{Ax} = \begin{bmatrix} 3 & 2 \\ -1 & 4 \end{bmatrix} \begin{bmatrix} 5 \\ 3 \end{bmatrix} = \begin{bmatrix} (3 \times 5) + (2 \times 3) \\ -(1 \times 5) + (4 \times 3) \end{bmatrix} = \begin{bmatrix} 21 \\ 7 \end{bmatrix}$$

$$\mathbf{y}^T \mathbf{A} = \begin{bmatrix} -2 & 1 \end{bmatrix} \begin{bmatrix} 3 & 2 \\ -1 & 4 \end{bmatrix} = [-(2 \times 3) - (1 \times 1) - (2 \times 2) + (1 \times 4)] = \begin{bmatrix} -7 & 0 \end{bmatrix}$$

$$\mathbf{y}^T \mathbf{x} = \begin{bmatrix} -2 & 1 \end{bmatrix} \begin{bmatrix} 5 \\ 3 \end{bmatrix} = (-10 + 3) = -7$$

$$\mathbf{xy}^T = \begin{bmatrix} 5 \\ 3 \end{bmatrix} \begin{bmatrix} -2 & 1 \end{bmatrix} = \begin{bmatrix} -(5 \times 2) & (5 \times 1) \\ -(3 \times 2) & (3 \times 1) \end{bmatrix} = \begin{bmatrix} -10 & 5 \\ -6 & 3 \end{bmatrix}$$

The last product always results in a matrix with proportional rows and columns.

The operation of matrix multiplication is particularly suited for representing systems of simultaneous linear equations in a compact form in which the coefficients are gathered into square matrices and the unknowns are placed in column matrices. For example, it is the operation of matrix multiplication which gives unambiguous meaning to the matrix abbreviation in Eq. (28.4) for the three simultaneous differential equations of Eq. (28.1). The two sides of Eq. (28.4) are column matrices of order three whose corresponding elements must be equal. On the right, these elements are simply the external forces at the three masses. On the left, Eq. (28.4) states that the resulting column is the sum of three column matrices, each of which results from the matrix multiplication of a square matrix of coefficients defined in Eq. (28.3) into a column matrix defined in Eq. (28.2). The rules of matrix operation just given ensure that Eq. (28.4) is exactly equivalent to Eq. (28.1).

Premultiplication or postmultiplication of a square matrix by the identity matrix leaves the original matrix unchanged; i.e.,

$$\mathbf{IA} = \mathbf{AI} = \mathbf{A}$$

Two symmetrical matrices multiplied together are generally not symmetric. The product of a matrix and its transpose is symmetric.

Continued matrix products such as **ABC** are defined, provided the number of columns in each matrix is the same as the number of rows in the matrix immediately following it. From the definition of matrix products, it follows that the *associative law* holds for continued products:

$$(AB)C = A(BC)$$

A square matrix **A** multiplied by itself yields a square matrix which is called the *square of the matrix* **A** and is denoted by A^2. If A^2 is in turn multiplied by **A**, the resulting matrix is $A^3 = A(A^2) = A^2(A)$. Extension of this process gives meaning to A^m for any positive integer *power m*. Powers of symmetric matrices are themselves symmetric.

The rule for *transposition* of matrix products is

$$(AB)^T = B^T A^T$$

Inverse or Reciprocal Matrix.   If, for a given square matrix **A**, a square matrix A^{-1} can be found such that

$$A^{-1}A = AA^{-1} = I \tag{28.6}$$

then A^{-1} is called the *inverse* or *reciprocal* of **A**. Not every square matrix **A** possesses an inverse. If the determinant constructed from the elements of a square matrix is zero, the matrix is said to be *singular* and there is no inverse. Every nonsingular matrix possesses a unique inverse. The inverse of a symmetric matrix is symmetric. The rule for the *inverse of a matrix product* is

$$(AB)^{-1} = (B^{-1})(A^{-1})$$

The solution to the set of simultaneous equations

$$Ax = c$$

where **x** is the unknown vector and **c** is a known input vector can be indicated with the aid of the inverse of **A**. The formal solution for **x** proceeds as follows:

$$A^{-1}Ax = A^{-1}c$$

$$Ix = x = A^{-1}c$$

When the inverse A^{-1} is known, the solution vector **x** is obtained by a simple matrix multiplication of A^{-1} into the input vector **c**.

The problem of calculating the inverse of given matrix **A** is one of the central computational problems associated with matrices. Routine procedures for both hand and machine computation exist.

QUADRATIC FORMS

A general quadratic form Q of order n may be written as

$$Q = \sum_{j=1}^{n} \sum_{k=1}^{n} a_{jk} x_j x_k$$

where the a_{jk} are constants and the x_j are the n variables. The form is quadratic since it is of the second degree in the variables. The laws of matrix multiplication permit Q to be written as

$$Q = [x_1 x_2 \dots x_n] \begin{bmatrix} a_{11} & a_{12} & \dots & a_{1n} \\ a_{21} & a_{22} & \dots & a_{2n} \\ \dots & \dots & \dots & \dots \\ a_{n1} & a_{n2} & \dots & a_{nn} \end{bmatrix} \begin{bmatrix} x_1 \\ x_2 \\ \dots \\ x_n \end{bmatrix}$$

which is

$$Q = \mathbf{x}^T \mathbf{A} \mathbf{x}$$

Any quadratic form can be expressed in terms of a symmetric matrix. If the given matrix $\mathbf{A}$ is not symmetric, it can be replaced by the symmetric matrix

$$\mathbf{B} = \tfrac{1}{2}(\mathbf{A} + \mathbf{A}^T)$$

without changing the value of the form.

As an example of a quadratic form, the *potential energy* V for the system of Fig. 28.2 is given by

$$2V = 3kx_1^2 + 2k(x_2 - x_1)^2 + k(x_3 - x_2)^2$$
$$= 5kx_1 x_1 - 2kx_1 x_2$$
$$\quad - 2kx_2 x_1 + 3kx_2 x_2 - kx_2 x_3$$
$$\quad - kx_3 x_2 + kx_3 x_3$$

Using the displacement vector $\mathbf{x}$ defined in Eq. (28.2) and the stiffness matrix $\mathbf{K}$ in Eq. (28.3), the potential energy may be written as

$$V = \tfrac{1}{2}\mathbf{x}^T \mathbf{K} \mathbf{x}$$

Similarly, the *kinetic energy* T is given by

$$2T = m\dot{x}_1^2 + 2m\dot{x}_2^2 + 3m\dot{x}_3^2$$

In terms of the inertia matrix $\mathbf{M}$ and the velocity vector $\dot{\mathbf{x}}$ defined in Eqs. (28.3) and (28.2), the kinetic energy may be written as

$$T = \tfrac{1}{2}\dot{\mathbf{x}}^T \mathbf{M} \dot{\mathbf{x}}$$

The *dissipation function* D for the system is given by

$$2D = c\dot{x}_1^2 + 2c(\dot{x}_3 - \dot{x}_2)^2$$
$$= c\dot{x}_1 \dot{x}_1$$
$$\quad + 2c\dot{x}_2 \dot{x}_2 - 2c\dot{x}_2 \dot{x}_3$$
$$\quad - 2c\dot{x}_3 \dot{x}_2 + 2c\dot{x}_3 \dot{x}_3$$

In terms of the velocity vector $\dot{\mathbf{x}}$ and the damping matrix $\mathbf{C}$ defined in Eqs. (28.2) and (28.3), the dissipation function may be written as

$$D = \tfrac{1}{2}\dot{\mathbf{x}}^T \mathbf{C} \dot{\mathbf{x}}$$

The dissipation function gives half the rate at which energy is being dissipated in the system.

While quadratic forms assume positive and negative values in general, the three physical forms just defined are intrinsically *positive* for a vibrating system with linear springs, constant masses, and viscous damping; i.e., they can never be negative for a real motion of the system. Kinetic energy is zero only when the system is at rest. The same thing is not necessarily true for potential energy or the dissipation function.

Depending upon the arrangement of springs and dashpots in the system, there may exist motions which do not involve any potential energy or dissipation. For example, in vibratory systems where rigid body motions are possible (crankshaft torsional systems, free-free beams, etc.), no elastic energy is involved in the rigid body motions. Also, in Fig. 28.2, if x_1 is zero while x_2 and x_3 have the same motion, there is no energy dissipated and the dissipation function is zero. To distinguish between these two possibilities, a quadratic form is called *positive definite* if it is never negative and if the only time it vanishes is when all the variables are zero. Kinetic energy is always positive definite, while potential energy and the dissipation function are positive but not necessarily positive definite. It depends upon the particular configuration of a given system whether the potential energy and the dissipation function are positive definite or only positive. The terms positive and positive definite are applied also to the matrices from which the quadratic forms are derived. For example, of the three matrices defined in Eq. (28.3), the matrices **M** and **K** are positive definite, but **C** is only positive. It can be shown that a matrix which is positive but not positive definite is *singular*.

Differentiation of Quadratic Forms. In forming Lagrange's equations of motion for a vibrating system,* it is necessary to take derivatives of the potential energy V, the kinetic energy T, and the dissipation function D. When these quadratic forms are represented in matrix notation, it is convenient to have matrix formulas for differentiation. In this paragraph rules are given for differentiating the slightly more general *bilinear form*

$$F = \mathbf{x}^T \mathbf{A} \mathbf{y} = \mathbf{y}^T \mathbf{A} \mathbf{x}$$

where $\mathbf{x}^T$ is a row vector of n variables x_j, **A** is a square matrix of constant coefficients, and **y** is a column matrix of n variables y_j. In a quadratic form the x_j are identical with the y_j.

For generality it is assumed that the x_j and the y_j are functions of n other variables u_j. In the formulas below, the notation $\mathbf{X}_u$ is used to represent the following square matrix:

$$\mathbf{X}_u = \begin{bmatrix} \dfrac{\partial x_1}{\partial u_1} & \dfrac{\partial x_2}{\partial u_1} & \cdots & \dfrac{\partial x_n}{\partial u_1} \\[2ex] \dfrac{\partial x_1}{\partial u_2} & \dfrac{\partial x_2}{\partial u_2} & \cdots & \dfrac{\partial x_n}{\partial u_2} \\[2ex] \cdots & \cdots & \cdots & \cdots \\[2ex] \dfrac{\partial x_1}{\partial u_n} & \dfrac{\partial x_2}{\partial u_n} & \cdots & \dfrac{\partial x_n}{\partial u_n} \end{bmatrix}$$

Now letting $\partial/\partial\mathbf{u}$ stand for the column vector whose elements are the partial differential operators with respect to the u_j, the general differentiation formula is

---

* See Chap. 2 for a detailed discussion of Lagrange's equations.

$$\frac{\partial F}{\partial \mathbf{u}} = \begin{bmatrix} \dfrac{\partial F}{\partial u_1} \\[2mm] \dfrac{\partial F}{\partial u_2} \\[2mm] \vdots \\[2mm] \dfrac{\partial F}{\partial u_n} \end{bmatrix} = \mathbf{X}_u \mathbf{A} \mathbf{y} + \mathbf{Y}_u \mathbf{A}^T \mathbf{x}$$

For a quadratic form $Q = \mathbf{x}^T \mathbf{A} \mathbf{x}$ the above formula reduces to

$$\frac{\partial Q}{\partial \mathbf{u}} = \mathbf{X}_u (\mathbf{A} + \mathbf{A}^T) \mathbf{x}$$

Thus whether $\mathbf{A}$ is symmetric or not, this kind of differentiation produces a *symmetrical* matrix of coefficients $(\mathbf{A} + \mathbf{A}^T)$. It is this fact which ensures that vibration equations in the form obtained from Lagrange's equations always have symmetrical matrices of coefficients. If $\mathbf{A}$ is symmetrical to begin with, the previous formula becomes

$$\frac{\partial Q}{\partial \mathbf{u}} = 2\mathbf{X}_u \mathbf{A} \mathbf{x}$$

Finally, in the important special case where the x_j are identical with the u_j, the matrix $\mathbf{X}_x$ reduces to the identity matrix, yielding

$$\frac{\partial Q}{\partial \mathbf{x}} = 2\mathbf{A}\mathbf{x} \tag{28.7}$$

which is employed in the following section in developing Lagrange's equations.

FORMULATION OF VIBRATION PROBLEMS IN MATRIX FORM

Consider a holonomic linear mechanical system with n degrees-of-freedom which vibrates about a stable equilibrium configuration. Let the motion of the system be described by n *generalized displacements* $x_j(t)$ which vanish in the equilibrium position. The potential energy V can then be expressed in terms of these displacements as a quadratic form. The kinetic energy T and the dissipation function D can be expressed as quadratic forms in the generalized velocities $\dot{x}_j(t)$.

The equations of motion are obtained by applying Lagrange's equations

$$\frac{d}{dt}\left(\frac{\partial T}{\partial \dot{x}_j}\right) + \frac{\partial D}{\partial \dot{x}_j} + \frac{\partial V}{\partial x_j} = f_j(t) \qquad [j = 1, 2, \ldots, n]$$

The *generalized external force* $f_j(t)$ for each coordinate may be an active force in the usual sense or a force generated by prescribed motion of the coordinates.

If each term in the foregoing equation is taken as the jth element of a column matrix, all n equations can be considered simultaneously and written in matrix form as follows:

$$\frac{d}{dt}\left(\frac{\partial T}{\partial \dot{\mathbf{x}}}\right) + \frac{\partial D}{\partial \dot{\mathbf{x}}} + \frac{\partial V}{\partial \mathbf{x}} = \mathbf{f}$$

The quadratic forms can be expressed in matrix notation as

$$T = \tfrac{1}{2}(\dot{\mathbf{x}}^T \mathbf{M} \dot{\mathbf{x}})$$

$$D = \tfrac{1}{2}(\dot{\mathbf{x}}^T \mathbf{C} \dot{\mathbf{x}})$$

$$V = \tfrac{1}{2}(\mathbf{x}^T \mathbf{K} \mathbf{x})$$

where the *inertia matrix* $\mathbf{M}$, the *damping matrix* $\mathbf{C}$, and the *stiffness matrix* $\mathbf{K}$ may be taken as symmetric square matrices of order n. Then the differentiation rule (28.7) yields

$$\frac{d}{dt}(\mathbf{M}\dot{\mathbf{x}}) + \mathbf{C}\dot{\mathbf{x}} + \mathbf{K}\mathbf{x} = \mathbf{f}$$

or simply

$$\mathbf{M}\ddot{\mathbf{x}} + \mathbf{C}\dot{\mathbf{x}} + \mathbf{K}\mathbf{x} = \mathbf{f} \tag{28.8}$$

as the equations of motion in matrix form for a general linear vibratory system with n degrees-of-freedom. This is a generalization of Eq. (28.4) for the three degree-of-freedom system of Fig. 28.2. Equation (28.8) applies to all linear constant-parameter vibratory systems. The specifications of any particular system are contained in the *coefficient matrices* $\mathbf{M}, \mathbf{C}$, and $\mathbf{K}$. The type of excitation is described by the column matrix $\mathbf{f}$. The individual terms in the coefficient matrices have the following significance:

m_{jk} is the momentum component at j due to a unit velocity at k.

c_{jk} is the damping force at j due to a unit velocity at k.

k_{jk} is the elastic force at j due to a unit displacement at k.

The general solution to Eq. (28.8) contains $2n$ constants of integration which are usually fixed by the n displacements $x_j(t_0)$ and the n velocities $\dot{x}_j(t_0)$ at some initial time t_0. When the excitation matrix $\mathbf{f}$ is zero, Eq. (28.8) is said to describe the *free* *vibration* of the system. When $\mathbf{f}$ is nonzero, Eq. (28.8) describes a *forced vibration*. When the time behavior of $\mathbf{f}$ is periodic and steady, it is sometimes convenient to divide the solution into a *steady-state response* plus a *transient response* which decays with time. The steady-state response is independent of the initial conditions.

COUPLING OF THE EQUATIONS

The off-diagonal terms in the coefficient matrices are known as *coupling terms*. In general, the equations have inertia, damping, and stiffness coupling; however, it is often possible to obtain equations that have no coupling terms in one or more of the three matrices. If the coupling terms vanish in all three matrices (i.e., if all three square matrices are diagonal matrices), the system of Eq. (28.8) becomes a set of independent uncoupled differential equations for the n generalized displacements $x_j(t)$. Each displacement motion is a single degree-of-freedom vibration independent of the motion of the other displacements.

The coupling in a system depends on the choice of coordinates used to describe the motion. For example, Figs. 28.3 and 28.4 show the same physical system with two different choices for the displacement coordinates.

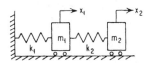

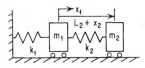

FIGURE 28.3 Coordinates (x_1,x_2) with uncoupled inertia matrix.

FIGURE 28.4 Coordinates (x_1,x_2) with uncoupled stiffness matrix. The equilibrium length of the spring k_2 is L_2.

The coefficient matrices corresponding to the coordinates shown in Fig. 28.3 are

$$\mathbf{M} = \begin{bmatrix} m_1 & 0 \\ 0 & m_2 \end{bmatrix} \quad \mathbf{K} = \begin{bmatrix} k_1 + k_2 & -k_2 \\ -k_2 & k_2 \end{bmatrix}$$

Here the inertia matrix is uncoupled because the coordinates chosen are the absolute displacements of the masses. The elastic force in the spring k_2 is generated by the relative displacement of the two coordinates, which accounts for the coupling terms in the stiffness matrix.

The coefficient matrices corresponding to the alternative coordinates shown in Fig. 28.4 are

$$\mathbf{M} = \begin{bmatrix} m_1 + m_2 & m_2 \\ m_2 & m_2 \end{bmatrix} \quad \mathbf{K} = \begin{bmatrix} k_1 & 0 \\ 0 & k_2 \end{bmatrix}$$

Here the coordinates chosen relate directly to the extensions of the springs so that the stiffness matrix is uncoupled. The absolute displacement of m_2 is, however, the sum of the coordinates, which accounts for the coupling terms in the inertia matrix.

A fundamental procedure for solving vibration problems in undamped systems may be viewed as the search for a set of coordinates which simultaneously uncouples both the stiffness and inertia matrices. This is always possible. In systems with damping (i.e., with all three coefficient matrices) there exist coordinates which uncouple two of these, but it is not possible to uncouple all three matrices simultaneously, except in the special case, called *proportional damping*, where $\mathbf{C}$ is a linear combination of $\mathbf{K}$ and $\mathbf{M}$.

The system of Fig. 28.2 provides an example of a three degree-of-freedom system with damping. The coefficient matrices are given in Eq. (28.3). The inertia matrix is uncoupled, but the damping and stiffness matrices are coupled.

Another example of a system with damping is furnished by the two degree-of-freedom system shown in Fig. 28.5. The excitation here is furnished by acceleration $\ddot{x}_0(t)$ of the base. This system is used as the basis for the numerical example at the end of Part I of the chapter. With the coordinates chosen as indicated in the figure, all three coefficient matrices have coupling terms. The equations of motion can be placed

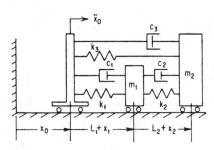

FIGURE 28.5 Two degree-of-freedom vibratory system. The equilibrium length of the spring k_1 is L_1 and the equilibrium length of the spring k_2 is L_2.

in the standard form of Eq. (28.8), where the coefficient matrices and the excitation column are as follows:

$$\mathbf{M} = \begin{bmatrix} m_1 + m_2 & m_2 \\ m_2 & m_2 \end{bmatrix} \qquad \mathbf{C} = \begin{bmatrix} c_1 + c_3 & c_3 \\ c_3 & c_2 + c_3 \end{bmatrix}$$

$$\mathbf{K} = \begin{bmatrix} k_1 + k_3 & k_3 \\ k_3 & k_2 + k_3 \end{bmatrix} \qquad \mathbf{f} = -\ddot{x}_0 \begin{bmatrix} m_1 + m_2 \\ m_2 \end{bmatrix}$$

$$(28.9)$$

THE MATRIX EIGENVALUE PROBLEM

In the following sections the solutions to both free and forced vibration problems are given in terms of solutions to a specialized algebraic problem known as the matrix eigenvalue problem. In the present section a general theoretical discussion of the matrix eigenvalue problem is given.

The free vibration equation for an undamped system,

$$\mathbf{M\ddot{x}} + \mathbf{Kx} = 0 \qquad (28.10)$$

follows from Eq. (28.8) when the excitation $\mathbf{f}$ and the damping $\mathbf{C}$ vanish. If a solution for $\mathbf{x}$ is assumed in the form

$$\mathbf{x} = \Re \{\mathbf{v}e^{j\omega t}\}$$

where $\mathbf{v}$ is a column vector of unknown amplitudes, ω is an unknown frequency, j is the square root of -1, and $\Re \{\ \}$ signifies "the real part of," it is found on substituting in Eq. (28.10) that it is necessary for $\mathbf{v}$ and ω to satisfy the following algebraic equation:

$$\mathbf{Kv} = \omega^2 \mathbf{Mv} \qquad (28.11)$$

This algebraic problem is called the *matrix eigenvalue problem*. Where necessary it is called the *real* eigenvalue problem to distinguish it from the *complex* eigenvalue problem described in the section on *Vibration of Systems with Damping*.

To indicate the formal solution to Eq. (28.11), it is rewritten as

$$(\mathbf{K} - \omega^2 \mathbf{M})\mathbf{v} = 0 \qquad (28.12)$$

which can be interpreted as a set of n homogeneous algebraic equations for the n elements v_j. This set always has the trivial solution

$$\mathbf{v} = 0$$

It also has nontrivial solutions if the determinant of the matrix multiplying the vector v is zero, i.e., if

$$\det (\mathbf{K} - \omega^2 \mathbf{M}) = 0 \qquad (28.13)$$

When the determinant is expanded, a polynomial of order n in ω^2 is obtained. Equation (28.13) is known as the *characteristic equation* or *frequency equation*. The restrictions that $\mathbf{M}$ and $\mathbf{K}$ be symmetric and that $\mathbf{M}$ be positive definite are sufficient to ensure that there are n real roots for ω^2. If $\mathbf{K}$ is singular, at least one root is zero. If $\mathbf{K}$ is positive definite, all roots are positive. The n roots determine the n *natural fre-*

quencies ω_r $(r = 1, \ldots, n)$ of free vibration. These roots of the characteristic equation are also known as *normal values, characteristic values, proper values, latent roots,* or *eigenvalues.* When a natural frequency ω_r is known, it is possible to return to Eq. (28.12) and solve for the corresponding vector $\mathbf{v}_r$ to within a multiplicative constant. The eigenvalue problem does not fix the absolute amplitude of the vectors $\mathbf{v}$, only the relative amplitudes of the n coordinates. There are n independent vectors $\mathbf{v}_r$ corresponding to the n natural frequencies which are known as *natural modes.* These vectors are also known as *normal modes, characteristic vectors, proper vectors, latent vectors,* or *eigenvectors.*

MODAL AND SPECTRAL MATRICES

The complete solution to the eigenvalue problem of Eq. (28.11) consists of n eigenvalues and n corresponding eigenvectors. These can be assembled compactly into matrices. Let the eigenvector $\mathbf{v}_r$ corresponding to the eigenvalue ω_r^2 have elements v_{jr} (the first subscript indicates which row, the second subscript indicates which eigenvector). The n eigenvectors then can be displayed in a single square matrix $\mathbf{V}$, each column of which is an eigenvector:

$$\mathbf{V} = [v_{jk}] = \begin{bmatrix} v_{11} & v_{12} & \cdots & v_{1n} \\ v_{21} & v_{22} & \cdots & v_{2n} \\ \cdots & \cdots & \cdots & \cdots \\ v_{n1} & v_{n2} & \cdots & v_{nn} \end{bmatrix}$$

The matrix $\mathbf{V}$ is called the *modal matrix* for the eigenvalue problem, Eq. (28.11).

The n eigenvalues ω_r^2 can be assembled into a diagonal matrix Ω^2 which is known as the *spectral matrix* of the eigenvalue problem, Eq. (28.11)

$$\Omega^2 = \left[\omega_r^2\right] = \begin{bmatrix} \omega_1^2 & 0 & \cdots & 0 \\ 0 & \omega_2^2 & \cdots & 0 \\ \cdots & \cdots & \cdots & \cdots \\ 0 & 0 & \cdots & \omega_n^2 \end{bmatrix}$$

Each eigenvector and corresponding eigenvalue satisfy a relation of the following form:

$$\mathbf{K}\mathbf{v}_r = \mathbf{M}\mathbf{v}_r\omega_r^2$$

By using the modal and spectral matrices it is possible to assemble all of these relations into a single matrix equation

$$\mathbf{K}\mathbf{V} = \mathbf{M}\mathbf{V}\Omega^2 \qquad (28.14)$$

Equation (28.14) provides a compact display of the complete solution to the eigenvalue problem Eq. (28.11).

PROPERTIES OF THE SOLUTION

The eigenvectors corresponding to different eigenvalues can be shown to satisfy the following *orthogonality relations.* When $\omega_r^2 \neq \omega_s^2$,

$$\mathbf{v}_r^T\mathbf{K}\mathbf{v}_s = 0 \qquad \mathbf{v}_r^T\mathbf{M}\mathbf{v}_s = 0 \qquad (28.15)$$

In case the characteristic equation has a p-fold multiple root for ω^2, then there is a p-fold infinity of corresponding eigenvectors. In this case, however, it is always possible to choose p of these vectors which mutually satisfy Eq. (28.15) and to express any other eigenvector corresponding to the multiple root as a linear combination of the p vectors selected. If these p vectors are included with the eigenvectors corresponding to the other eigenvalues, a set of n vectors is obtained which satisfies the orthogonality relations of Eq. (28.15) for any $r \neq s$.

The orthogonality of the eigenvectors with respect to $\mathbf{K}$ and $\mathbf{M}$ implies that the following square matrices are *diagonal*.

$$\mathbf{V}^T\mathbf{K}\mathbf{V} = \left[\mathbf{v}_r{}^T\mathbf{K}\mathbf{v}_s\right]$$

$$\mathbf{V}^T\mathbf{M}\mathbf{V} = \left[\mathbf{v}_r{}^T\mathbf{M}\mathbf{v}_s\right] \tag{28.16}$$

The elements $\mathbf{v}_r{}^T\mathbf{K}\mathbf{v}_r$ along the main diagonal of $\mathbf{V}^T\mathbf{K}\mathbf{V}$ are called the *modal stiffnesses* k_r, and the elements $\mathbf{v}_r{}^T\mathbf{M}\mathbf{v}_r$ along the main diagonal of $\mathbf{V}^T\mathbf{M}\mathbf{V}$ are called the *modal masses* m_r. Since $\mathbf{M}$ is positive definite, all modal masses are guaranteed to be positive. When $\mathbf{K}$ is singular, at least one of the modal stiffnesses will be zero. Each eigenvalue ω_r^2 is the quotient of the corresponding modal stiffness divided by the corresponding modal mass; i.e.,

$$\omega_r^2 = \frac{k_r}{m_r}$$

In numerical work it is sometimes convenient to normalize each eigenvector so that its largest element is *unity*. In other applications it is common to normalize the eigenvectors so that the modal masses m_r all have the *same* value m, where m is some convenient value such as the total mass of the system. In this case,

$$\mathbf{V}^T\mathbf{M}\mathbf{V} = m\mathbf{I} \tag{28.17}$$

and it is possible to express the inverse of the modal matrix $\mathbf{V}$ simply as

$$\mathbf{V}^{-1} = \frac{1}{m}\,\mathbf{V}^T\mathbf{M}$$

An interpretation of the modal matrix $\mathbf{V}$ can be given by showing that it defines a set of generalized coordinates for which both the inertia and stiffness matrices are uncoupled. Let $\mathbf{y}(t)$ be a column of displacements related to the original displacements $\mathbf{x}(t)$ by the following simultaneous equations:

$$\mathbf{y} = \mathbf{V}^{-1}\mathbf{x} \quad \text{or} \quad \mathbf{x} = \mathbf{V}\mathbf{y}$$

The potential and kinetic energies then take the forms

$$V = \tfrac{1}{2}\mathbf{x}^T\mathbf{K}\mathbf{x} = \tfrac{1}{2}\mathbf{y}^T(\mathbf{V}^T\mathbf{K}\mathbf{V})\mathbf{y}$$

$$T = \tfrac{1}{2}\dot{\mathbf{x}}^T\mathbf{M}\dot{\mathbf{x}} = \tfrac{1}{2}\dot{\mathbf{y}}^T(\mathbf{V}^T\mathbf{M}\mathbf{V})\dot{\mathbf{y}}$$

where, according to Eq. (28.16), the square matrices in parentheses on the right are *diagonal*; i.e., in the y_j coordinate system there is neither stiffness nor inertia coupling.

An alternative method for obtaining the same interpretation is to start from the eigenvalue problem of Eq. (28.11). Consider the structure of the related eigenvalue problem for $\mathbf{w}$ where again $\mathbf{w}$ is obtained from $\mathbf{v}$ by the transformation involving the modal matrix $\mathbf{V}$.

$$\mathbf{w} = \mathbf{V}^{-1}\mathbf{v} \quad \text{or} \quad \mathbf{v} = \mathbf{V}\mathbf{w}$$

Substituting in Eq. (28.11), premultiplying by $\mathbf{V}^T$, and using Eq. (28.14),

$$\mathbf{Kv} = \omega^2\mathbf{Mv}$$

$$\mathbf{KVw} = \omega^2\mathbf{MVw}$$

$$\mathbf{V}^T\mathbf{KVw} = \omega^2\mathbf{V}^T\mathbf{MVw}$$

$$(\mathbf{V}^T\mathbf{MV})\Omega^2\mathbf{w} = \omega^2(\mathbf{V}^T\mathbf{MV})\mathbf{w}$$

Now, since $\mathbf{V}^T\mathbf{MV}$ is a diagonal matrix of positive elements, it is permissible to cancel it from both sides, which leaves a simple diagonalized eigenvalue problem for $\mathbf{w}$:

$$\Omega^2\mathbf{w} = \omega^2\mathbf{w}$$

A modal matrix for $\mathbf{w}$ is the identity matrix $\mathbf{I}$, and the eigenvalues for $\mathbf{w}$ are the same as those for $\mathbf{v}$.

EIGENVECTOR EXPANSIONS

Any set of n independent vectors can be used as a basis for representing any other vector of order n. In the following sections, the eigenvectors of the eigenvalue problem of Eq. (28.11) are used as such a basis. An eigenvector expansion of an arbitrary vector $\mathbf{y}$ has the form

$$\mathbf{y} = \sum_{r=1}^{n} \mathbf{v}_r a_r \tag{28.18}$$

where the a_r are scalar *mode multipliers*. When $\mathbf{y}$ and the $\mathbf{v}_r$ are known, it is possible to evaluate the a_r by premultiplying both sides by $\mathbf{v}_s^T\mathbf{M}$. Because of the orthogonality relations of Eq. (28.15), all the terms on the right vanish except the one for which $r = s$. Inserting the value of the mode multiplier so obtained, the expansion can be rewritten as

$$\mathbf{y} = \sum_{r=1}^{n} \mathbf{v}_r \frac{\mathbf{v}_r^T\mathbf{My}}{\mathbf{v}_r^T\mathbf{Mv}_r} \tag{28.19}$$

or alternatively as

$$\mathbf{y} = \sum_{r=1}^{n} \frac{\mathbf{v}_r\mathbf{v}_r^T\mathbf{M}}{\mathbf{v}_r^T\mathbf{Mv}_r} \mathbf{y} \tag{28.20}$$

The form of Eq. (28.19) emphasizes the decomposition into eigenvectors since the fraction on the right is just a scalar. The form of Eq. (28.20) is convenient when a large number of vectors $\mathbf{y}$ are to be decomposed, since the fractions on the right, which are now square matrices, must be computed only once. The form of Eq. (28.20) becomes more economical of computation time when more than n vectors $\mathbf{y}$ have to be expanded. A useful check on the calculation of the matrices on the right of Eq. (28.20) is provided by the identity

$$\sum_{r=1}^{n} \frac{\mathbf{v}_r\mathbf{v}_r^T\mathbf{M}}{\mathbf{v}_r^T\mathbf{Mv}_r} = \mathbf{I} \tag{28.21}$$

which follows from Eq. (28.20) because $\mathbf{y}$ is completely arbitrary.

An alternative expansion which is useful for expanding the excitation vector $\mathbf{f}$ is

$$\mathbf{f} = \sum_{r=1}^{n} \omega_r^2 \mathbf{M} \mathbf{v}_r a_r = \sum_{r=1}^{n} \mathbf{M} \mathbf{v}_r \frac{\mathbf{v}_r^T \mathbf{f}}{\mathbf{v}_r^T \mathbf{M} \mathbf{v}_r} \tag{28.22}$$

This may be viewed as an expansion of the excitation in terms of the *inertia force* amplitudes of the natural modes. The mode multiplier a_r has been evaluated by pre-multiplying by $\mathbf{v}_r^T$. A form analogous to Eq. (28.20) and an identity corresponding to Eq. (28.21) can easily be written.

RAYLEIGH'S QUOTIENT

If Eq. (28.11) is premultiplied by $\mathbf{v}^T$, the following scalar equation is obtained:

$$\mathbf{v}^T \mathbf{K} \mathbf{v} = \omega^2 \mathbf{v}^T \mathbf{M} \mathbf{v}$$

The positive definiteness of $\mathbf{M}$ guarantees that $\mathbf{v}^T \mathbf{M} \mathbf{v}$ is nonzero, so that it is permissible to solve for ω^2.

$$\omega^2 = \frac{\mathbf{v}^T \mathbf{K} \mathbf{v}}{\mathbf{v}^T \mathbf{M} \mathbf{v}} \tag{28.23}$$

This quotient is called "Rayleigh's quotient." It also may be derived by equating time averages of potential and kinetic energy under the assumption that the vibratory system is executing simple harmonic motion at frequency ω with amplitude ratios given by $\mathbf{v}$ or by equating the maximum value of kinetic energy to the maximum value of potential energy under the same assumption. Rayleigh's quotient has the following interesting properties.

1. When $\mathbf{v}$ is an eigenvector $\mathbf{v}_r$ of Eq. (28.11), then Rayleigh's quotient is equal to the corresponding eigenvalue ω_r^2.

2. If $\mathbf{v}$ is an approximation to $\mathbf{v}_r$ with an error which is a *first-order* infinitesimal, then Rayleigh's quotient is an approximation to ω_r^2 with an error which is a *second-order* infinitesimal; i.e., Rayleigh's quotient is *stationary* in the neighborhoods of the true eigenvectors.

3. As $\mathbf{v}$ varies through all of n-dimensional vector space, Rayleigh's quotient remains bounded between the smallest and largest eigenvalues.

A common engineering application of Rayleigh's quotient involves simply evaluating Eq. (28.23) for a trial vector $\mathbf{v}$ which is selected on the basis of physical insight. When eigenvectors are obtained by approximate methods, Rayleigh's quotient provides a means of improving the accuracy in the corresponding eigenvalue. If the elements of an approximate eigenvector whose largest element is unity are correct to k decimal places, then Rayleigh's quotient can be expected to be correct to about $2k$ significant decimal places.

Perturbation Formulas.　The perturbation formulas which follow provide the basis for estimating the changes in the eigenvalues and the eigenvectors which result from *small* changes in the stiffness and inertia parameters of a system. The formulas are strictly accurate only for infinitesimal changes but are useful approximations for *small* changes. They may be used by the designer to estimate the effects of a proposed change in a vibratory system and may also be used to analyze the effects of minor errors in the measurement of the system properties. Iterative procedures for the solution of eigenvalue problems can be based on these formulas. They are employed here to obtain approximations to the complex eigenvalues and eigenvectors of a lightly damped vibratory system in terms of the corresponding solutions for the same system without damping.

Suppose that the modal matrix $\mathbf{V}$ and the spectral matrix Ω^2 for the eigenvalue problem

$$\mathbf{KV} = \mathbf{MV}\Omega^2 \qquad (28.14)$$

are known. Consider the perturbed eigenvalue problem

$$\mathbf{K}_* \mathbf{V}_* = \mathbf{M}_* \mathbf{V}_* \Omega_*^2$$

where

$$\mathbf{K}_* = \mathbf{K} + d\mathbf{K} \qquad \mathbf{M}_* = \mathbf{M} + d\mathbf{M}$$

$$\mathbf{V}_* = \mathbf{V} + d\mathbf{V} \qquad \Omega_*^2 = \Omega^2 + d\Omega^2$$

The perturbation formula for the elements $d\omega_r^2$ of the diagonal matrix $d\Omega^2$ is

$$d\omega_r^2 = \frac{\mathbf{v}_r^T \, d\mathbf{K} \, \mathbf{v}_r - \omega_r^2 \mathbf{v}_r^T \, d\mathbf{M} \, \mathbf{v}_r}{\mathbf{v}_r^T \mathbf{M} \mathbf{v}_r} \qquad (28.24)$$

Thus in order to determine the change in a single eigenvalue due to changes in $\mathbf{M}$ and $\mathbf{K}$, it is necessary to know only the corresponding unperturbed eigenvalue and eigenvector. To determine the change in a single eigenvector, however, it is necessary to know *all* the unperturbed eigenvalues and eigenvectors. The following algorithm may be used to evaluate the perturbations of both the modal matrix and the spectral matrix. Calculate

$$\mathbf{F} = \mathbf{V}^T \, d\mathbf{K} \, \mathbf{V} - \mathbf{V}^T \, d\mathbf{M} \, \mathbf{V}\Omega^2$$

and

$$\mathbf{L} = \mathbf{V}^T \mathbf{M} \mathbf{V}$$

The matrix $\mathbf{L}$ is a diagonal matrix of positive elements and hence is easily inverted. Continue calculating

$$\mathbf{G} = \mathbf{L}^{-1}\mathbf{F} = [g_{jk}] \qquad \text{and} \qquad \mathbf{H} = [h_{jk}]$$

where

$$h_{jk} = \begin{cases} 0 & \text{if } \omega_j^2 = \omega_k^2 \\[2mm] \dfrac{g_{jk}}{\omega_k^2 - \omega_j^2} & \text{if } \omega_j^2 \neq \omega_k^2 \end{cases}$$

Then, finally, the perturbations of the modal matrix and the spectral matrix are given by

$$d\mathbf{V} = \mathbf{VH} \qquad d\Omega^2 = \left[g_{ij} \right] \qquad (28.25)$$

These formulas are derived by taking the total differential of Eq. (28.14), premultiplying each term by $\mathbf{V}^T$, and using a relation derived by taking the transpose of Eq. (28.14). Since eigenvectors are undetermined to the extent of a scale factor, the vector changes in the eigenvectors have a degree of indeterminateness. The particular selection in the above algorithm makes the change in each eigenvector orthogonal with respect to $\mathbf{M}$ to the corresponding unperturbed eigenvector, i.e.,

$$\mathbf{v}_j^T \mathbf{M} \, d\mathbf{v}_j = 0$$

VIBRATIONS OF SYSTEMS WITHOUT DAMPING

In this section the damping matrix $\mathbf{C}$ is neglected in Eq. (28.8), leaving the general formulation in the form

$$\mathbf{M}\ddot{\mathbf{x}} + \mathbf{K}\mathbf{x} = \mathbf{f} \qquad (28.26)$$

Solutions are outlined for the following three cases: free vibration ($\mathbf{f} = 0$), steady-state forced sinusoidal vibration ($\mathbf{f} = \mathcal{R}\{\mathbf{d}e^{j\omega t}\}$, where $\mathbf{d}$ is a column vector of driving-force amplitudes), and the response to general excitation ($\mathbf{f}$ an arbitrary function of time). The first two cases are contained in the third, but for the sake of clarity each is described separately.

FREE VIBRATION WITH SPECIFIED INITIAL CONDITIONS

It is desired to find the solution $\mathbf{x}(t)$ of Eq. (28.26) when $\mathbf{f} = 0$ which satisfies the initial conditions

$$\mathbf{x} = \mathbf{x}(0) \qquad \dot{\mathbf{x}} = \dot{\mathbf{x}}(0) \qquad (28.27)$$

at $t = 0$ where $\mathbf{x}(0)$ and $\dot{\mathbf{x}}(0)$ are columns of prescribed initial displacements and velocities. The differential equation to be solved is identical with Eq. (28.10), which led to the matrix eigenvalue problem in the preceding section. Assuming that the solution of the eigenvalue problem is available, the general solution of the differential equation is given by an arbitrary superposition of the natural modes

$$\mathbf{x} = \sum_{r=1}^{n} \mathbf{v}_r(a_r \cos \omega_r t + b_r \sin \omega_r t)$$

where the $\mathbf{v}_r$ are the eigenvectors or natural modes, the ω_r are the natural frequencies, and the a_r and b_r are $2n$ constants of integration. The corresponding velocity is

$$\dot{\mathbf{x}} = \sum_{r=1}^{n} \mathbf{v}_r \omega_r(-a_r \sin \omega_r t + b_r \cos \omega_r t)$$

Setting $t = 0$ in these expressions and substituting in the initial conditions of Eq. (28.27) provides $2n$ simultaneous equations for determination of the constants of integration.

$$\sum_{r=1}^{n} \mathbf{v}_r a_r = \mathbf{x}(0) \qquad \sum_{r=1}^{n} \mathbf{v}_r \omega_r b_r = \dot{\mathbf{x}}(0)$$

These equations may be interpreted as eigenvector expansions of the initial displacement and velocity. The constants of integration can be evaluated by the same technique used to obtain the mode multipliers in Eq. (28.19). Using the form of Eq. (28.20), the solution of the free vibration problem then becomes

$$\mathbf{x}(t) = \sum_{r=1}^{n} \frac{\mathbf{v}_r \mathbf{v}_r^T \mathbf{M}}{\mathbf{v}_r^T \mathbf{M} \mathbf{v}_r} \left\{ \mathbf{x}(0) \cos \omega_r t + \frac{1}{\omega_r} \dot{\mathbf{x}}(0) \sin \omega_r t \right\} \qquad (28.28)$$

STEADY-STATE FORCED SINUSOIDAL VIBRATION

It is desired to find the steady-state solution to Eq. (28.26) for single-frequency sinusoidal excitation **f** of the form

$$\mathbf{f} = \mathcal{R}\{\mathbf{d}e^{j\omega t}\}$$

where **d** is a column vector of driving force amplitudes (these may be complex to permit differences in phase for the various components). The solution obtained is a useful approximation for lightly damped systems provided that the forcing frequency ω is not too close to a natural frequency ω_r. For resonance and near-resonance conditions it is necessary to include the damping as indicated in the section which follows the present discussion.

The steady-state solution desired is assumed to have the form

$$\mathbf{x} = \mathcal{R}\{\mathbf{a}e^{j\omega t}\}$$

where **a** is an unknown column vector of response amplitudes. When **f** and **x** are inserted in Eq. (28.26), the following set of simultaneous equations for the elements of **a** is obtained:

$$(\mathbf{K} - \omega^2\mathbf{M})\mathbf{a} = \mathbf{d} \qquad (28.29)$$

If ω is not a natural frequency, the square matrix $\mathbf{K} - \omega^2\mathbf{M}$ is nonsingular and may be inverted to yield

$$\mathbf{a} = (\mathbf{K} - \omega^2\mathbf{M})^{-1}\mathbf{d}$$

as a complete solution for the response amplitudes in terms of the driving force amplitudes. This solution is useful if several force amplitude distributions are to be studied while the excitation frequency ω is held constant. The process requires repeated inversions if a range of frequencies is to be studied.

An alternative procedure which permits a more thorough study of the effect of frequency variation is available if the natural modes and frequencies are known. The driving-force vector **d** is represented by the eigenvector expansion of Eq. (28.22), and the response vector **a** is represented by the eigenvector expansion of Eq. (28.18):

$$\mathbf{d} = \sum_{r=1}^{n} \frac{\mathbf{M}\mathbf{v}_r\mathbf{v}_r^T}{\mathbf{v}_r^T\mathbf{M}\mathbf{v}_r}\mathbf{d} \qquad \mathbf{a} = \sum_{r=1}^{n} \mathbf{v}_r c_r$$

where the c_r are unknown coefficients. Substituting these into Eq. (28.29), and making use of the fundamental eigenvalue relation of Eq. (28.11), leads to

$$\sum_{r=1}^{n} (\omega_r^2 - \omega^2)\mathbf{M}\mathbf{v}_r c_r = \sum_{r=1}^{n} \frac{\mathbf{M}\mathbf{v}_r\mathbf{v}_r^T}{\mathbf{v}_r^T\mathbf{M}\mathbf{v}_r}\mathbf{d}$$

This equation can be uncoupled by premultiplying both sides by $\mathbf{v}_r^T$ and using the orthogonality condition of Eq. (28.15) to obtain

$$(\omega_r^2 - \omega^2)\mathbf{v}_r^T\mathbf{M}\mathbf{v}_r c_r = \mathbf{v}_r^T\mathbf{d}$$

$$c_r = \frac{1}{\omega_r^2 - \omega^2}\frac{\mathbf{v}_r^T\mathbf{d}}{\mathbf{v}_r^T\mathbf{M}\mathbf{v}_r}$$

The final solution is then assembled by inserting the c_r back into **a** and **a** back into **x**.

$$\mathbf{x} = \Re\left\{ \sum_{r=1}^{n} \frac{e^{j\omega t}}{\omega_r^2 - \omega^2} \frac{\mathbf{v}_r\mathbf{v}_r^{T}}{\mathbf{v}_r^{T}\mathbf{M}\mathbf{v}_r} \mathbf{d} \right\} \qquad (28.30)$$

This form clearly indicates the effect of frequency on the response.

RESPONSE TO GENERAL EXCITATION

It is now desired to obtain the solution to Eq. (28.26) for the general case in which the excitation $\mathbf{f}(t)$ is an arbitrary vector function of time and for which initial displacements $\mathbf{x}(0)$ and velocities $\dot{\mathbf{x}}(0)$ are prescribed. If the natural modes and frequencies of the system are available, it is again possible to split the problem up into n single degree-of-freedom response problems and to indicate a formal solution.

Following a procedure similar to that just used for steady-state forced sinusoidal vibrations, an eigenvector expansion of the solution is assumed:

$$\mathbf{x}(t) = \sum_{r=1}^{n} \mathbf{y}_r c_r(t)$$

where the c_r are unknown functions of time and the known excitation $\mathbf{f}(t)$ is expanded according to Eq. (28.22). Inserting these into Eq. (28.26) yields

$$\sum_{r=1}^{n} (\mathbf{M}\mathbf{v}_r\ddot{c}_r + \mathbf{K}\mathbf{v}_r c_r) = \sum_{r=1}^{n} \frac{\mathbf{M}\mathbf{v}_r\mathbf{v}_r^{T}}{\mathbf{v}_r^{T}\mathbf{M}\mathbf{v}_r} \mathbf{f}(t)$$

Using Eq. (28.11) to eliminate $\mathbf{K}$ and premultiplying by $\mathbf{v}_r^{T}$ to uncouple the equation,

$$\ddot{c}_r + \omega_r^2 c_r^2 = \frac{\mathbf{v}_r^{T}\mathbf{f}(t)}{\mathbf{v}_r^{T}\mathbf{M}\mathbf{v}_r} \qquad (28.31)$$

is obtained as a single second-order differential equation for the time behavior of the rth mode multiplier. The initial conditions for c_r can be obtained by making eigenvector expansions of $\mathbf{x}(0)$ and $\dot{\mathbf{x}}(0)$ as was done previously for the free vibration case. Formal solutions to Eq. (28.29) can be obtained by a number of methods, including Laplace transforms and variation of parameters. When these mode multipliers are substituted back to obtain $\mathbf{x}$, the general solution has the following appearance:

$$\mathbf{x}(t) = \sum_{r=1}^{n} \frac{\mathbf{v}_r\mathbf{v}_r^{T}\mathbf{M}}{\mathbf{v}_r^{T}\mathbf{M}\mathbf{v}_r} \left\{ \mathbf{x}(0)\cos\omega_r t + \frac{1}{\omega_r}\dot{\mathbf{x}}(0)\sin\omega_r t \right\}$$

$$+ \sum_{r=1}^{n} \frac{\mathbf{v}_r\mathbf{v}_r^{T}}{\omega_r\mathbf{v}_r^{T}\mathbf{M}\mathbf{v}_r} \int_0^t \mathbf{f}(t')\sin\{\omega_r(t-t')\}\,dt' \qquad (28.32)$$

The integrals involving the excitation can be evaluated in closed form if the elements $f_j(t)$ of $\mathbf{f}(t)$ are simple (e.g., step functions, ramps, single sine pulses, etc.). When the $f_j(t)$ are more complicated, numerical results can be obtained by using integration software.

VIBRATION OF SYSTEMS WITH DAMPING

In this section solutions to the complete governing equation, Eq. (28.8), are discussed. The results of the preceding section for systems without damping are adequate for many purposes. There are, however, important problems in which it is necessary to include the effect of damping, e.g., problems concerned with resonance, random vibration, etc.

COMPLEX EIGENVALUE PROBLEM

When there is no excitation, Eq. (28.8) becomes

$$\mathbf{M\ddot{x} + C\dot{x} + Kx} = 0$$

which describes the free vibration of the system. As in the undamped case, there are $2n$ independent solutions which can be superposed to meet $2n$ initial conditions. Assuming a solution in the form

$$\mathbf{x = u}e^{pt}$$

leads to the following algebraic problem:

$$(p^2\mathbf{M} + p\mathbf{C} + \mathbf{K})\mathbf{u} = 0 \qquad (28.33)$$

for the determination of the vector $\mathbf{u}$ and the scalar p. This is a *complex eigenvalue problem* because the *eigenvalue* p and the elements of the *eigenvector* $\mathbf{u}$ are, in general, complex numbers. The most common technique for solving the nth-order eigenvalue problem, Eq. (28.33), is to transform it to a $2n$th-order problem having the same form as Eq. (28.11). This may be done by introducing the column vector $\tilde{\mathbf{v}}$ of order $2n$ given by

$$\tilde{\mathbf{v}} = \{\mathbf{u} \quad p\mathbf{u}\}^T$$

and the two square matrices of order $2n$ given by

$$\tilde{\mathbf{K}} = \begin{bmatrix} K & 0 \\ 0 & I \end{bmatrix} \qquad \tilde{\mathbf{M}} = \begin{bmatrix} C & M \\ I & 0 \end{bmatrix}$$

In terms of these, an eigenvalue problem equivalent to Eq. (28.33) is

$$\tilde{\mathbf{K}}\tilde{\mathbf{v}} = p\,\tilde{\mathbf{M}}\,\tilde{\mathbf{v}} \qquad (28.34)$$

which is similar to Eq. (28.11) except that $\tilde{\mathbf{M}}$ does not have the symmetry and positive definite properties that $\mathbf{M}$ has. As a result, the eigenvalue p and the eigenvector $\mathbf{v}$ are generally complex. Some, but not all, computer programs which can solve Eq. (28.11) can also solve Eq. (28.34). Since the computational time for most eigenvalue problems is proportional to n^3, the computational time for the $2n$th-order system of Eq. (28.34) will be about eight times that for the nth-order system of Eq. (28.11).

The complex eigenvalue problem of Eq. (28.33) can be solved approximately, when the damping is light, by using the perturbation equations of Eqs. (28.24) and (28.25). When $\mathbf{C} = 0$ in Eq. (28.33), the complex eigenvalue problem reduces to the real eigenvalue problem of Eq. (28.11) with $p^2 = -\omega^2$. Suppose that the real eigenvalue ω_r^2 and the real eigenvectors $\mathbf{v}_r$ are known. The perturbation of the rth mode due to the addition of small damping $\mathbf{C}$ can be estimated by considering the damping to be a perturbation of the stiffness matrix of the form

$$d\mathbf{K} = j\omega_r\mathbf{C}$$

In this way it is found that the perturbed solution corresponding to the rth mode consists of a pair of complex conjugate eigenvalues

$$p_r = -\alpha_r + j\omega_r \qquad p_r^C = -\alpha_r - j\omega_r$$

and a pair of complex conjugate eigenvectors

$$\mathbf{u}_r = \mathbf{v}_r + j\mathbf{w}_r \qquad \mathbf{u}_r^C = \mathbf{v}_r - j\mathbf{w}_r$$

where ω_r and $\mathbf{v}_r$ are taken directly from the undamped system, and α_r and $\mathbf{w}_r$ are small perturbations which are given below. The superscript C is used to denote the complex conjugate. The real part of the eigenvalue, which describes the rate of decay of the corresponding free motion, is given by the following quotient:

$$2\alpha_r = 2\zeta_r\omega_r = \frac{\mathbf{v}_r^T\mathbf{C}\mathbf{v}_r}{\mathbf{v}_r^T\mathbf{M}\mathbf{v}_r} \qquad (28.35)$$

The decay rate α_r for a particular r depends only on the rth mode undamped solution. The imaginary part of the eigenvector $j\mathbf{w}_r$, which describes the perturbations in phase, is more difficult to obtain. All the undamped eigenvalues and eigenvectors must be known. Let $\mathbf{W}$ be a square matrix whose columns are the $\mathbf{w}_r$. The following algorithm may be used to evaluate $\mathbf{W}$ when the undamped modal matrix $\mathbf{V}$ is known. Calculate

$$\mathbf{F} = \mathbf{V}^T\mathbf{C}\mathbf{V}$$

and

$$\mathbf{L} = \mathbf{V}^T\mathbf{M}\mathbf{V}$$

The matrix $\mathbf{L}$ is a diagonal matrix of positive elements and hence is easily inverted. Continue calculating

$$\mathbf{G} = \mathbf{L}^{-1}\mathbf{F} = [g_{jk}] \qquad \text{and} \qquad \mathbf{H} = [h_{jk}]$$

where

$$h_{jk} = \begin{cases} 0 & \text{if } \omega_j^2 = \omega_k^2 \\ \dfrac{g_{jk}\omega_k}{\omega_k^2 - \omega_j^2} & \text{if } \omega_j^2 \neq \omega_k^2 \end{cases}$$

Then, finally, the eigenvector perturbations are given by

$$\mathbf{W} = \mathbf{V}\mathbf{H} \qquad (28.36)$$

The main diagonal elements g_{rr} of the matrix $\mathbf{G}$ are just the quotients of Eq. (28.35) and hence equal $2\alpha_r$. If $\mathbf{u}$ and p satisfy Eq. (28.33), then so also do $\mathbf{u}^C$ and p^C. There are $2n$ roots in the negative half-plane which occur in pairs of complex conjugates or as real negative numbers. When the damping is absent, all roots lie on the imaginary axis; for small damping the roots lie near the imaginary axis. The corresponding $2n$ eigenvectors $\mathbf{u}_s$ satisfy the following *orthogonality* relations:

$$(p_r + p_s)\mathbf{u}_r^T\mathbf{M}\mathbf{u}_s + \mathbf{u}_r^T\mathbf{C}\mathbf{u}_s = 0$$

$$\mathbf{u}_r^T\mathbf{K}\mathbf{u}_s - p_r p_s\mathbf{u}_r^T\mathbf{M}\mathbf{u}_s = 0$$

whenever $p_r \neq p_s$: they can be made to hold for repeated roots by suitable choice of the eigenvectors associated with a multiple root. It is sometimes convenient to display a complex eigenvalue p_r in the form

$$p_r = \omega_r(-\zeta_r + j\sqrt{1 - \zeta_r^2}) \tag{28.37}$$

which is commonly employed in treating single degree-of-freedom systems. The parameter ω_r, called the *undamped natural frequency,* and the parameter ζ_r, called the *critical damping ratio,* can be determined from the following quotients involving the eigenvector $\mathbf{u}_r$ and its conjugate $\mathbf{u}_r^C$:

$$\omega_r^2 = \frac{\mathbf{u}_r^T \mathbf{K} \mathbf{u}_r^C}{\mathbf{u}_r^T \mathbf{M} \mathbf{u}_r^C} \qquad 2\zeta_r\omega_r = \frac{\mathbf{u}_r^T \mathbf{C} \mathbf{u}_r^C}{\mathbf{u}_r^T \mathbf{M} \mathbf{u}_r^C}$$

FORMAL SOLUTIONS

If the solution to the eigenvalue problem of Eq. (28.33) is available, it is possible to exhibit a general solution to the governing equation

$$\mathbf{M}\ddot{\mathbf{x}} + \mathbf{C}\dot{\mathbf{x}} + \mathbf{K}\mathbf{x} = \mathbf{f} \tag{28.8}$$

for arbitrary excitation $\mathbf{f}(t)$ which meets prescribed initial conditions for $\mathbf{x}(0)$ and $\dot{\mathbf{x}}(0)$ at $t = 0$. The solutions given below apply to the case where the $2n$ eigenvalues occur as n pairs of complex conjugates (which is usually the case when the damping is light). This does, however, restrict the treatment to systems with nonsingular stiffness matrices $\mathbf{K}$ because if $\omega_r^2 = 0$ is an undamped eigenvalue, the corresponding eigenvalues in the presence of damping are real. All quantities in the solutions below are *real.* These forms have been obtained by breaking down complex solutions into real and imaginary parts and recombining. With the notation

$$p_r = -\alpha_r + j\beta_r \qquad \mathbf{u}_r = \mathbf{v}_r + j\mathbf{w}_r$$

for the real and imaginary parts of eigenvalues and eigenvectors, it follows from Eq. (28.37) that

$$\alpha_r = \zeta_r\omega_r \qquad \beta_r = \omega_r\sqrt{1 - \zeta_r^2}$$

The general solution to Eq. (28.8) is then

$$\mathbf{x}(t) = \sum_{r=1}^{n} \frac{2}{a_r^2 + b_r^2} \{\mathbf{G}_r\mathbf{M}\dot{\mathbf{x}}(0) + (-\alpha_r\mathbf{G}_r\mathbf{M} + \beta_r\mathbf{H}_r\mathbf{M} + \mathbf{G}_r\mathbf{C})\mathbf{x}(0)\}e^{-\alpha_r t}\cos\beta_r t$$

$$+ \sum_{r=1}^{n} \frac{2}{a_r^2 + b_r^2} \{\mathbf{H}_r\mathbf{M}\dot{\mathbf{x}}(0) + (-\beta_r\mathbf{G}_r\mathbf{M} - \alpha_r\mathbf{H}_r\mathbf{M} + \mathbf{H}_r\mathbf{C})\mathbf{x}(0)\}e^{-\alpha_r t}\sin\beta_r t$$

$$+ \sum_{r=1}^{n} \frac{2}{a_r^2 + b_r^2}\, \mathbf{G}_r \int_0^t \mathbf{f}(t')e^{-\alpha_r(t - t')}\cos\beta_r(t - t')\,dt'$$

$$+ \sum_{r=1}^{n} \frac{2}{a_r^2 + b_r^2}\, \mathbf{H}_r \int_0^t \mathbf{f}(t')e^{-\alpha_r(t - t')}\sin\beta_r(t - t')\,dt' \tag{28.38}$$

where

$$a_r = -2\alpha_r(\mathbf{v}_r^T\mathbf{M}\mathbf{v}_r - \mathbf{w}_r^T\mathbf{M}\mathbf{w}_r) - 4\beta_r\mathbf{v}_r^T\mathbf{M}\mathbf{w}_r + \mathbf{v}_r^T\mathbf{C}\mathbf{v}_r - \mathbf{w}_r^T\mathbf{C}\mathbf{w}_r$$

$$b_r = 2\beta_r(\mathbf{v}_r^T\mathbf{M}\mathbf{v}_r - \mathbf{w}_r^T\mathbf{M}\mathbf{w}_r) - 4\alpha_r\mathbf{v}_r^T\mathbf{M}\mathbf{w}_r + 2\mathbf{v}_r^T\mathbf{C}\mathbf{w}_r$$

$$\mathbf{A}_r = \mathbf{v}_r\mathbf{v}_r^T - \mathbf{w}_r\mathbf{w}_r^T \qquad \mathbf{B}_r = \mathbf{v}_r\mathbf{w}_r^T + \mathbf{w}_r\mathbf{v}_r^T$$

$$\mathbf{G}_r = a_r\mathbf{A}_r + b_r\mathbf{B}_r \qquad \mathbf{H}_r = b_r\mathbf{A}_r - a_r\mathbf{B}_r$$

The solution of Eq. (28.38) should be compared with the corresponding solution of Eq. (28.32) for systems without damping. When the damping matrix $\mathbf{C} = 0$, Eq. (28.38) reduces to Eq. (28.32).

For the important special case of steady-state forced sinusoidal excitation of the form

$$\mathbf{f} = \mathfrak{R}\{\mathbf{d}e^{j\omega t}\}$$

where $\mathbf{d}$ is a column of driving force amplitudes, the steady-state portion of the response can be written as follows, using the above notation:

$$\mathbf{x}(t) = \mathfrak{R}\left\{ \sum_{r=1}^{n} \frac{2e^{j\omega t}}{a_r^2 + b_r^2} \frac{\alpha_r\mathbf{G}_r + \beta_r\mathbf{H}_r + j\omega\mathbf{G}_r}{\omega_r^2 - \omega^2 + j2\zeta_r\omega_r\omega} \mathbf{d}\right\} \tag{28.39}$$

This result reduces to Eq. (28.30) when the damping matrix $\mathbf{C}$ is set equal to zero.

APPROXIMATE SOLUTIONS

For a lightly damped system the exact solutions of Eq. (28.28) and Eq. (28.29) can be abbreviated considerably by making approximations based on the smallness of the damping. A systematic method of doing this is to consider the system without damping as a base upon which an infinitesimal amount of damping is superposed as a perturbation. An approximate solution to the complex eigenvalue problem by this method is provided by Eqs. (28.34) and (28.35). This perturbation approximation can be continued into Eqs. (28.38) and (28.39) by simply neglecting all squares and products of the small quantities α_r, ζ_r, $\mathbf{w}_r$, and $\mathbf{C}$. When this is done it is found that the formulas of Eqs. (28.38) and (28.39) may still be used if the parameters therein are obtained from the simplified expressions below.

$$\alpha_r = \zeta_r\omega_r \qquad \beta_r = \omega_r$$

$$a_r = -4\omega_r\mathbf{v}_r^T\mathbf{M}\mathbf{w}_r \qquad b_r = 2\omega_r\mathbf{v}_r^T\mathbf{M}\mathbf{v}_r$$

$$a_r^2 + b_r^2 = 4\omega_r^2(\mathbf{v}_r^T\mathbf{M}\mathbf{v}_r)^2 \tag{28.40}$$

$$\mathbf{A}_r = \mathbf{v}_r\mathbf{v}_r^T \qquad \mathbf{B}_r = \mathbf{v}_r\mathbf{w}_r^T + \mathbf{w}_r\mathbf{v}_r^T$$

$$\mathbf{G}_r = -4\omega_r(\mathbf{v}_r^T\mathbf{M}\mathbf{w}_r)\mathbf{v}_r\mathbf{v}_r^T + 2\omega_r(\mathbf{v}_r^T\mathbf{M}\mathbf{v}_r)(\mathbf{v}_r\mathbf{w}_r^T + \mathbf{w}_r\mathbf{v}_r^T)$$

$$\mathbf{H}_r = 2\omega_r(\mathbf{v}_r^T\mathbf{M}\mathbf{v}_r)\mathbf{v}_r\mathbf{v}_r^T$$

For example, the steady-state forced sinusoidal solution of Eq. (28.39) takes the following explicit form in the perturbation approximation:

$$\mathbf{x}(t) = \mathfrak{R}\left\{ \sum_{r=1}^{n} \frac{e^{j\omega t}}{\mathbf{v}_r^T\mathbf{M}\mathbf{v}_r} \frac{\mathbf{v}_r\mathbf{v}_r^T + \dfrac{j\omega}{\omega_r}\left[\mathbf{v}_r\mathbf{w}_r^T - \left(2\dfrac{\mathbf{v}_r^T\mathbf{M}\mathbf{w}_r}{\mathbf{v}_r^T\mathbf{M}\mathbf{v}_r}\right)\mathbf{v}_r\mathbf{v}_r^T + \mathbf{w}_r\mathbf{v}_r^T\right]}{\omega_r^2 - \omega^2 + j2\zeta_r\omega_r\omega} \mathbf{d}\right\} \tag{28.41}$$

The above formula simplifies somewhat if $\mathbf{w}_r$ is chosen to be orthogonal to $\mathbf{v}_r$ with respect to $\mathbf{M}$ since this causes the coefficient of $\mathbf{v}_r\mathbf{v}_r^T$ to vanish in the bracketed expression. This is the case when the $\mathbf{w}_r$ are provided by Eq. (28.35).

A cruder approximation, which is often used, is based on accepting the complex eigenvalue $p_r = -\alpha_r + j\omega_r$ but completely neglecting the imaginary part $j\mathbf{w}_r$ of the eigenvector $\mathbf{u}_r = \mathbf{v}_r + j\mathbf{w}_r$. It is thus assumed that the undamped mode $\mathbf{v}_r$ still applies for the system with damping. The approximate parameter values of Eq. (28.40) are further simplified by this assumption; e.g., $a_r = 0$, $\mathbf{B}_r = \mathbf{G}_r = 0$. The steady forced sinusoidal response of Eq. (28.41) reduces to

$$\mathbf{x}(t) = \Re\left\{ \sum_{r=1}^{n} \frac{e^{j\omega t}}{\omega_r^2 - \omega^2 + j2\zeta_r\omega_r\omega} \frac{\mathbf{v}_r\mathbf{v}_r^T}{\mathbf{v}_r^T\mathbf{M}\mathbf{v}_r} \mathbf{d} \right\} \tag{28.42}$$

This approximation should be compared with the undamped solution of Eq. (28.30), as well as with the exact solution of Eq. (28.39) and the perturbation approximation of Eq. (28.41).

In the special case of proportional damping, the exact eigenvectors are real and Eq. (28.34) produces the exact decay rate $\alpha_r = \zeta_r\omega_r$, so that the response of Eq. (28.42) is an exact result.

Example 28.1 Consider the system of Fig. 28.5 with the following mass, damping, and stiffness coefficients:

$m_1 = 1$ lb-sec^2/in. $\quad m_2 = 2$ lb-sec^2/in.

$c_1 = 0.10$ lb-sec/in. $\quad c_2 = 0.02$ lb-sec/in. $\quad c_3 = 0.04$ lb-sec/in.

$k_1 = 3$ lb/in. $\quad k_2 = 0.5$ lb/in. $\quad k_3 = 1$ lb/in.

The coefficient matrices of Eq. (28.9) then have the following numerical values:

$$\mathbf{M} = \begin{bmatrix} 3 & 2 \\ 2 & 2 \end{bmatrix} \quad \mathbf{C} = \begin{bmatrix} 0.14 & 0.04 \\ 0.04 & 0.06 \end{bmatrix} \quad \mathbf{K} = \begin{bmatrix} 4 & 1 \\ 1 & 1.5 \end{bmatrix}$$

Assuming that the numerical values above are exact, the exact solutions to the complex eigenvalue problem of Eq. (28.33) for these values of $\mathbf{M}$, $\mathbf{C}$, and $\mathbf{K}$ are, correct to four decimal places,

$$p_r = -\alpha_r + j\beta_r \quad \mathbf{u}_r = \mathbf{v}_r + j\mathbf{w}_r$$

$$2\alpha_1 = 0.0279 \quad \alpha_1 = \zeta_1\omega_1 = 0.0139 \quad \zeta_1 = 0.0166$$

$$\beta_1 = 0.8397 \quad \omega_1 = 0.8398 \quad \omega_1^2 = 0.7053$$

$$2\alpha_2 = 0.1221 \quad \alpha_2 = \zeta_2\omega_2 = 0.0611 \quad \zeta_2 = 0.0324 \tag{28.43}$$

$$\beta_2 = 1.8818 \quad \omega_2 = 1.8828 \quad \omega_2^2 = 3.5449$$

$$\mathbf{V} = \begin{bmatrix} 0.2179 & -0.9179 \\ 1.0000 & 1.0000 \end{bmatrix} \quad \mathbf{W} = \begin{bmatrix} 0.0016 & 0.0010 \\ 0 & 0 \end{bmatrix}$$

Note that this is a lightly damped system. The damping ratios in the two modes are 1.66 percent and 3.24 percent, respectively.

For comparison, the solution of the real eigenvalue problem Eq. (28.12) for the corresponding undamped system (i.e., $\mathbf{M}$ and $\mathbf{K}$ as above, but $\mathbf{C} = 0$) is, correct to four decimal places,

$$\omega_1^2 = 0.7053 \quad \mathbf{V} = \begin{bmatrix} 0.2179 & -0.9179 \\ 1.0000 & 1.0000 \end{bmatrix}$$
$$\omega_2^2 = 3.5447$$

Note that, to this accuracy, there is no discrepancy in the real parts of the eigenvectors. There are, however, small discrepancies in the imaginary parts of the eigenvalues. The difference between β_1 for the damped system and ω_1 for the undamped system is 0.0001, and the corresponding difference between β_2 and ω_2 is 0.0009. The imaginary parts of the eigenvectors and the real parts of the eigenvalues for the damped system are completely absent in the undamped system. They may be approximated by applying the perturbation equations of Eqs. (28.35) and (28.36) to the solution of the eigenvalue problem for the undamped system.

The real parts α_r of the eigenvalues obtained from Eq. (28.35) agree, to four decimal places, with the exact values in Eq. (28.43). The imaginary parts $\mathbf{w}_r$ of the eigenvectors obtained from Eq. (28.36) are

$$\mathbf{w}_1 = \left\{ \begin{array}{c} 0.0013 \\ -0.0014 \end{array} \right\} \qquad \mathbf{w}_2 = \left\{ \begin{array}{c} 0.0002 \\ 0.0009 \end{array} \right\}$$

These vectors satisfy the orthogonality conditions $\mathbf{v}_r{}^T \mathbf{M} \mathbf{w}_r = 0$.

In order to compare these values with Eq. (28.43), it is first necessary to normalize the complete eigenvector $\mathbf{v}_r + j\mathbf{w}_r$, so that its second element is unity. For example, this is done in the case of $r = 1$ by dividing both $\mathbf{v}_1$ and $\mathbf{w}_1$ by $1.0000 - j0.0014$. When this is done, it is found that the perturbation approximation to the eigenvectors agrees, to four decimal places, with the exact solution of Eq. (28.43).

To illustrate the application of the formal solutions given above, consider the steady-state forced oscillation of the system shown in Fig. 28.5 at a frequency ω due to driving force amplitudes d_1 and d_2. Using the exact solution values of Eq. (28.43), the expressions $a_r, b_r, \mathbf{A}_r, \mathbf{B}_r, \mathbf{G}_r$, and $\mathbf{H}_r$ following Eq. (28.38) are evaluated for $r = 1$ and $r = 2$. With these values, the steady-state response, Eq. (28.39), becomes

$$\left[\begin{array}{c} x_1 \\ x_2 \end{array} \right] = \mathcal{R} \left\{ \frac{e^{j\omega t} \left\{ \left[\begin{array}{cc} 0.0158 & 0.0723 \\ 0.0723 & 0.3318 \end{array} \right] + j\omega \left[\begin{array}{cc} 0.0002 & 0.0004 \\ 0.0004 & -0.0011 \end{array} \right] \right\} \left[\begin{array}{c} d_1 \\ d_2 \end{array} \right]}{0.7053 - \omega^2 + 0.0279 j\omega} \right.$$

$$\left. + \frac{e^{j\omega t} \left\{ \left[\begin{array}{cc} 0.9842 & -1.0724 \\ -1.0724 & 1.1683 \end{array} \right] + j\omega \left[\begin{array}{cc} -0.0002 & -0.0004 \\ -0.0004 & 0.0011 \end{array} \right] \right\} \left[\begin{array}{c} d_1 \\ d_2 \end{array} \right]}{3.5449 - \omega^2 + 0.1221 j\omega} \right\}$$

When the approximation in Eq. (28.41) based on the perturbation solution is evaluated, the result is almost identical to this. A few entries differ by one or two units in the fourth decimal place. The crude approximation, Eq. (28.42), is the same as the perturbation approximation except that the terms in the numerators which are multiplied by $j\omega$ are absent. This means that the relative error between the crude approximation and the exact solution can be large at high frequencies. At low frequencies, however, even the crude approximation provides useful results for lightly damped systems. In the present case, the discrepancy between the crude approximation and the exact solution remains under 1 percent as long as ω is less than ω_2 (the highest natural frequency). At higher frequencies the absolute response level decreases steadily, which tends to undercut the significance of the increasing relative discrepancy between approximations.

CHAPTER 28

PART II: FINITE-ELEMENT MODELS

Harry G. Schaeffer

INTRODUCTION

Finite-element programs perform three useful functions: (1) They incorporate modeling objects that transform a conceptual model such as a complex built-up structure (e.g., a supersonic airplane) into a set of matrix equations of the form shown by Eq. (28.4). (2) They incorporate a rich set of rules for performing a wide range of dynamic analyses, including transient and frequency response. And (3) they contain a rich set of data processing operations, including random response and response spectrum analyses.

Generally, the finite-element analysis program is data-coupled to an interactive graphic-oriented shell program, called a preprocessor and postprocessor (referred to in Chap. 27), that creates the finite-element mesh and displays the results. This graphic-oriented interface includes tools for generating and displaying the finite-element model and for graphically interpreting the results. The finite-element analysis program, when coupled with a preprocessor and postprocessor, provides a graphic-oriented modeling and analysis tool that can be used to solve a wide variety of shock and vibration design analysis tasks.

The availability of finite-element programs and high-speed digital computers has made it possible to perform dynamic studies of extremely large systems. In the past, the computer resources required by commercial finite-element programs inhibited their use in support of general dynamic design tasks. However, the decreasing cost of both computers and commercial software makes it feasible to have a commercial finite-element program on a high-performance desktop computer.

Finite-element programs are helpful in solving shock and vibration problems in many of the areas covered in this handbook by

- Generating the system matrices, $\mathbf{M}$, $\mathbf{C}$, $\mathbf{K}$, and $\mathbf{f}$ in Eq. (28.4)
- Providing the capability of performing a variety of operations on the system of equations
- Providing a variety of solution algorithms for shock and vibration analysis
- Including data reduction operations on the set of dynamic response data

- Providing a graphic user interface (GUI) for model creation, analysis, and data reduction

  The sections that follow describe

- The finite-element method for creating the mathematical model of the system
- The reduction operations used to transform a static finite-element model to a dynamic model
- Algorithms for obtaining modes and frequencies
- Algorithms for transient and frequency response
- Data reduction techniques, including random response and shock response spectrum
- Interfaces to graphics

DESCRIPTION OF FINITE-ELEMENT METHOD

One of the major functions of a finite-element program is converting a conceptual model of a system that might be an assemblage of structural elements such as rods, beams, plates, etc., into a mathematical model that can be manipulated and solved by the computer. Modeling objects convert the engineering description of the entity, which includes material and geometric properties, into a mathematical model described in terms of the system mass, damping, and stiffness matrices. These modeling objects include "elements," which describe the behavior in a small but finite volume, and objects, which describe loads, constraints, and special modeling features.

SCALAR ELEMENTS FOR LUMPED-PARAMETER SYSTEMS

The elements in commercial programs include discrete elements, such as springs, masses, and dampers, as well as continuum elements, such as beams, plates, shells, and solids, which are described in the next section. Discrete elements can be used to model lumped systems such as the three degree-of-freedom lumped-parameter system shown in Fig. 28.2.

The analysis model for a lumped-parameter system is created by connecting generalized degrees-of-freedom with discrete elements. For example, suppose that the mass, stiffness, and damping elements are defined by *cmass*, *cstiff*, and *cdamp* objects having the format

```
cname, eid, value, dof1, dof2
```

where *cname* is the name associated with a specific modeling object: *cmass* for mass, *cdamp* for a damper, and *cstiff* for a spring. A specific instance of the object is then defined by specifying the remaining parameters: *eid* is a specific instance of the cmass, cstiff, or cdamp objects; *value* is the numerical value of the associated dynamic property; and *dof1* and *dof2* are the degrees-of-freedom connected by these elements. The system equations for the three degree-of-freedom system shown by Fig. 28.2 could then be generated by the following modeling instructions:

```
cmass,  1,  m,   0,  1
cmass,  2,  2m,  1,  2
cmass,  3,  m,   2,  3
cstiff, 1,  3k,  0,  1
cstiff, 2,  2k,  1,  2
cstiff, 3,  k,   2,  3
cdamp,  1,  c,   0,  1
cdamp,  2,  2c,  2,  3
```

where specific values for m, k, and c must be entered, and where a value of zero for a degree-of-freedom would indicate a restrained displacement.

The finite-element program would then generate the numerical model presented by Eq. (28.3). The use of modeling objects can lead to a significant decrease in the tedium of generating equations of motion, allowing the user to generate the conceptual model at the object level rather than at the equation level. A finite-element program contains all the rules for generating equations from the modeling objects and for maintaining relations between degrees-of-freedom. Once the conceptual model is defined, the analyst can concentrate on evaluating the behavior of the model, using techniques described in subsequent sections.

CONTINUUM MODELS

The finite-element method is a major analytical tool for studying the behavior of a conceptual model of a dynamic system whose components are continuums, such as beams, plates, shells, or three-dimensional solids. The approach is to define a spatial distribution of discrete material points throughout a continuum at which generalized displacements, which are generally taken to be three components of displacement and three components of rotation, are defined. A small but finite (hence the term *finite element*) volume of the continuum is now identified with a subset of the material points which define the region enclosed by the finite element. For example, a region in which a two-dimensional strain or stress field is assumed could be represented by a triangular element having at least three grid points or a quadrilateral element having at least four grid points. A three-dimensional strain or stress field could be presented by trapezoidal, pentagonal, or hexagonal elements having at least four, six, or eight grid points, respectively. The finite-element method is summarized in subsequent sections. Detailed developments are presented in several texts on this method.[1,2]

GENERALIZED RITZ PROCEDURE

The finite-element method is an approximation technique in which a set of basis functions is used to approximate the behavior in the finite region, in a Ritz sense. The undetermined basis coefficients associated with the set of Ritz vectors can then be determined in such a way that the error in satisfying the continuum equations of mathematical physics in the finite region is minimized.

The behavior in the local region must satisfy the equations of equilibrium, the strain-displacement relations, and the equation of state relating the local state of stress and strain. The equilibrium equations are of the form

$$\mathbf{S}^T \boldsymbol{\sigma} = 0 \qquad (28.44)$$

where **S** is a suitable matrix of differential operators, the superscript T signifies the transpose of a matrix, and σ is the set of independent stress components; the strain displacement equations are

$$\mathbf{Du} = \epsilon \tag{28.45}$$

where **D** is a matrix of differential operators that relate the displacements, **u**, and the independent components of strain, ϵ; and a material law relating stresses and strains is of the form

$$\sigma = \sigma(\epsilon) \tag{28.46}$$

The necessary condition for the convergence of the finite-element method is based on the rather heuristic observation that as the number of elements in a given region becomes large, the size of the elements becomes small. A gradient in stress or strain can therefore be represented by a large number of elements, each having a uniform stress or strain field. This suggests the use of interpolants of the displacement **u**, the stress σ, or the strain ϵ in the volume of the element.

GENERATING ELEMENT EQUATIONS

Early finite-element formulations resulted in elements that performed rather poorly.[3] The goal of improving element performance has since led to a greater understanding of the reasons for poor performance and to the development of elements that accurately predict at least uniform strain fields for arbitrary-shaped elements, which is a necessary condition for convergence.

Early element formulations were based on assumed displacements in the volume, in the following form, using the notation of Ref. 4:

$$\mathbf{u} = \mathbf{N}_u \mathbf{u}_i \tag{28.47}$$

where **u** is the set of continuous displacement functions within and on the boundary, $\mathbf{N}_u$ is a set of *shape functions* that are related to a set of Ritz *basis functions* as described below, and $\mathbf{u}_i$ is a set of generalized degrees-of-freedom at the grid points (which are also termed node points) that bound the discrete region of the finite element. These degrees-of-freedom include the three components of displacement and rotation, as appropriate; other generalized degrees-of-freedom can also be included as described below in the p-method formulation.

The strains can be evaluated from the assumed displacement using Eq. (28.45), the stresses using Eq. (28.46). The discrete grid-point degrees-of-freedom $\mathbf{u}_i$ can be determined by minimizing the error in satisfying the equilibrium condition, Eq. (28.44), in an integral sense. This approach to developing a relationship between the element forces and displacements is termed the displacement approach to finite elements.

Displacement-based elements tend to be overly stiff. In an attempt to improve the performance of the elements, an alternative approach[5] based on assumed stress fields that exactly satisfy equilibrium in the volume of the finite element was developed. In this approach, which is termed the stress formulation, the stress field is interpolated by a function of the following form:

$$\sigma = \mathbf{N}_\sigma \beta \tag{28.48}$$

where the shape functions $\mathbf{N}_\sigma$ are chosen such that equilibrium is satisfied pointwise. The strains can then be calculated using the inverse of Eq. (28.46). The stress coefficients, β, are then determined such that the error in satisfying the strain-displacement equations is minimized in an integral sense using Eq. (28.45).

A hybrid approach,[6] which assumes displacements on the boundary and assumes stresses and/or strains in the interior, leads to elements that perform well for a wide range of loads, boundary conditions, and element geometry. These elements predict accurate strains in an arbitrary-shaped element, do not become overly stiff, and lock and converge quickly to the solution, spatially, for a small number of elements.

APPROXIMATION TECHNIQUES

The finite-element method is one of a general class of approximation techniques based on weighted residuals that are used to transform the partial differential equations representing the behavior of a continuum into a set of algebraic equations that can be solved on the computer. These approximation techniques are directly related to the Ritz procedure of using an assumed displacement function to determine the natural frequencies of a beam, for example.

There are at least three generalized Ritz procedures for generating a mathematical model of a continuum using modeling objects. Two of these, the *h-method* and the *p-method*, populate the volume and the boundary of the continuum with discrete material points. The other, called the *boundary-element method*[7] (see Chap. 27), uses Green's theorem to describe the internal behavior in terms of boundary behavior. In this procedure, only the boundary is represented using discrete material points.

In the *h*-method, the volume is subdivided into a geometric mesh of cells having a characteristic dimension *h*. The *p*-method includes a set of polynomial coefficients in addition to discrete material points. The convergence characteristics of a model represented by the *h*-method can be studied only by remeshing the volume to change *h*. However, convergence for the *p*-method model can be studied by changing the polynomial order of the model without making any changes of the geometric parameters associated with the model.

Only the volume-oriented procedures are described in this overview of the finite-element method. Commercial codes have been developed that implement each of the three methods. Most codes implement only one of the techniques. However, both *h*- and *p*-elements, which are described in the next section, can be supported in a finite-element program.

h-METHOD FORMULATION

The original formulation of the finite element, which is now called the *h-method*, represents the response in a continuum at a discrete number of points. These physical degrees-of-freedom, u_i, are then related to a set of generalized degrees-of-freedom, which are the basis function coefficients. In this formulation the shape functions N_u in Eq. (28.47) are constructed by first representing the displacements in the volume of the element in terms of a set of basis functions X_j:

$$u = X_j a_j \qquad (28.49)$$

where a_j are basis function coefficients. In order to obtain a relation between the shape functions and the basis functions, Eq. (28.49) is written at the location of the grid points. This results in the following relationship between the grid-point displacement degrees-of-freedom and the basis function coefficients:

$$\mathbf{u}_i = \mathbf{X}_{ij}\mathbf{a}_j \tag{28.50}$$

where $\mathbf{X}_{ij}$ represents the basis functions for the basis coefficient j evaluated at node i. If the basis functions are independent and if there are the same number of basis functions as displacement degrees-of-freedom at the grid points, then the basis coefficients are

$$\mathbf{a}_j = \mathbf{A}_{ji}\mathbf{u}_i \tag{28.51}$$

where $\mathbf{A}_{ji} = \mathbf{X}_{ij}^{-1}$.

The substitution of Eqs. (28.51) and (28.50) into Eq. (28.49) then leads to Eq. (28.47), where

$$\mathbf{N}_u = \mathbf{X}_j\mathbf{A}_{ji} \tag{28.52}$$

This formulation relates basis function coefficients to displacement degrees-of-freedom. Since the accuracy of the solution increases as the number of degrees-of-freedom is increased, the mesh must be refined to obtain a better estimate of the solution. The mesh refinement is defined by the mesh dimension, h, so that convergence is a function of the mesh size parameter, h. The finite-element formulation with a one-to-one relation between the displacement degrees-of-freedom and the basis function coefficients is called the h-method since convergence can be accomplished only by refining the mesh.

p-METHOD FORMULATION

Another way of representing the displacement function, $\mathbf{u}$, using basis function coefficients as well as grid-point degrees-of-freedom, is as follows:

$$\mathbf{u} = \lfloor \mathbf{N}_j \quad \mathbf{N}_k \rfloor \begin{Bmatrix} \mathbf{u}_j \\ \mathbf{a}_k \end{Bmatrix} \tag{28.53}$$

where $\mathbf{u}_j$ are grid-point displacements and $\mathbf{a}_k$ are generalized coordinates. The shape functions $\mathbf{N}_j$ are determined using the same procedure described in the previous section. The shape functions $\mathbf{N}_k$ associated with the basis coefficients $\mathbf{a}_k$ are directly related to the basis functions $\mathbf{X}_k$, and have the property of being zero when evaluated at a grid point.

The inclusion of generalized degrees-of-freedom (i.e., the basis function coefficients) as well as grid-point degrees-of-freedom allows the solution to be refined by increasing the polynomial order of $\mathbf{N}_k$ without changing the discrete finite-element mesh. The resulting formulation is termed the p-method of finite-element analysis. The elements have a few physical grid points and a variable number of basis coefficients. The resulting model converges as additional polynomial terms are added; hence the designation of the elements as p-elements and the resulting convergence as p-convergence.

CREATING THE SPATIAL MODEL FOR DYNAMICS

The goal of finite-element analysis is to provide a turnkey solution: given an unambiguous solid geometric model, together with loads and constraints, determine the behavior to within a specified accuracy. The accuracy is attained for static analysis using current technology by successively refining the mesh for h-elements and by increasing the polynomial order for p-elements. The attainment of this goal generally requires significant engineering judgment for the static analysis of a general structure. Therefore, automated analysis should not be expected for built-up structures such as an airframe or a car body, even for static analysis.

Generally the finite-element model that is used for dynamics will be developed by first creating and verifying a baseline finite-element model that gives acceptable accuracy for static analysis. The question is then whether the spatial distribution is adequate for dynamics. The requirements for static and dynamic analysis differ. Generally, a model that has been developed and verified for static analysis has adequate spatial interpolation accuracy for dynamics; however, this is not always the case.

For static analysis, the mesh must be refined (or the polynomial order increased) until the elements are sufficiently small to represent the strain gradients. Preprocessor programs include remeshing algorithms that identify areas of high strain gradients and repetitively remesh and reanalyze until the solution error is reduced to an acceptable value. High strain gradients will exist at the application of point loads, at constraints, and in regions where the geometry is changing rapidly.

The requirement for spatial discretization in dynamics is different. In dynamics the modal content associated with a specified range of frequencies must be included in the model. If the spatial distribution is not sufficient to represent a mode of interest, then the response will not include that mode. A beam is a simple example. A single beam element is sufficient to completely define the static response of a beam with concentrated end loads. However, one element does not have sufficient spatial resolution to represent even the first elastic mode shape.

The process of creating a finite-element "mesh," i.e., the generation of grid points and connective element objects, is generally performed by a model generation program that allows the analyst to construct the analysis model with reference to an existing geometric model. In this case the analyst is concerned with choosing only the modeling objects and the density of the mesh, i.e., the number and the placement of interpolation points, called *grid points* or *nodes,* in the model.

The finite-element library of all commercial finite-element programs will contain, at a minimum, the following element types:

- Scalar masses, springs, and dampers
- Concentrated masses
- Rods
- Beams
- A four-noded quadrilateral shell
- A three-noded triangular shell
- An eight-noded hexagonal solid
- A six-noded pentagonal solid
- A four-noded tetrahedral solid

In addition to these element modeling objects, commercial programs also may include objects using higher-order interpolation of the local behavior and having additional connected grid points. These include

- An eight-noded quadrilateral shell
- A six-noded triangular shell

- A 20-noded hexagonal solid
- A 15-noded pentagonal solid
- A 10-noded tetrahedral solid

CREATING THE FINITE-ELEMENT MODEL

The finite-element program provides the user with higher-level objects for creating the mathematical model. The modeling objects and analytical algorithms allow the user to study the performance of dynamic systems that would not have been tractable before their development. Structural assemblies such as a spacecraft subject to impulsive thrust loads or a human body subject to a variety of loads can be modeled in detail. However, the change from modeling lumped-parameter systems, such as those described in Part I of this chapter, to modeling continuous systems requires additional skills and an additional knowledge base associated with element capabilities and performance.

An understanding is required of the type of physical behavior represented by the element; i.e., whether the behavior is represented by membrane elements, bending elements, shell elements, or three-dimensional field elements. In addition, an understanding is required of

- The type of behavior that must be represented in the model in order to select the appropriate element objects
- The elements' capabilities for representing strain gradients so that the number of elements can be increased in areas in which high strain gradients are expected
- The effect of element geometry, such as aspect ratio for two- and three-dimensional elements and taper, warpage, and skewness for two-dimensional elements, so that elements whose geometry will lead to poor performance are not used in the model

Reasonable models can be created by those with little or no experience with finite-elements provided the parameters affecting element performance, which are described in Ref. 3, are understood.

MODELING THE MASS DISTRIBUTION

Finite-element programs generally include a number of options for generating the mass matrix, including

1. Concentrated mass elements
2. Distributed mass associated with continuum elements
3. Entering terms of the mass matrix directly

The finite-element mass can be generated by an element object. In this case, the distributed mass per unit volume is specified as a material property and the program automatically calculates the mass coefficients associated with the physical displacement degrees-of-freedom. It is common to have a singular element mass matrix since rotary inertia terms are not generated for rotational degrees-of-freedom. Thus, models including rotational degrees-of-freedom will have singular mass matrices. Therefore, caution must be used when using eigenvalue extraction methods that require transformation to standard form, since the inverse of the Choleski factors of the system mass matrix, which are used in the transformation, will not exist.

MODELING DAMPING

Finite-element programs include a number of options for describing damping, including

1. Damping elements
2. Structural damping
3. Modal damping
4. Proportional damping

All forms of damping except the damping elements are distributed damping. Distributed damping is generally used to remove energy from the dynamic system and to damp the response of specific modes in the modal formulation. The concept of structural damping, or complex damping, is therefore well suited for continuous systems. The structural damping coefficient, g, multiplies the stiffness matrix and generates an equivalent viscous damping at a specified angular frequency, ω, as follows:

$$\mathbf{C} = \frac{g}{\omega}\,\mathbf{K} \tag{28.54}$$

The use of structural damping, or damping that is proportional to the system mass and/or stiffness matrices, is very attractive since the real eigenvectors are orthonormal with respect to both the stiffness and mass matrices, as shown by Eq. (28.16). The use of proportional damping will therefore lead to a diagonal damping matrix as well as diagonal mass and stiffness matrices when the real eigenvectors are used to transform from physical to modal coordinates. Generally, finite-element analysis programs also allow the user to define modal damping directly if the modal formulation of dynamics, which is described in a following section, is used.

MODELING THE LOADS

In addition to creating the finite-element model that includes the system mass, damping, and stiffness, the time-dependent loads and/or displacement boundary conditions must be modeled. Most commercial programs allow the definition of a spatial time-dependent set of loads of the following form:

$$\mathbf{f}(\mathbf{x},t) = \mathbf{p}(\mathbf{x})\mathbf{g}(t) \tag{28.55}$$

where $\mathbf{p}(\mathbf{x})$ is a spatial distribution, usually specified by the same load-modeling objects used for static analysis, and $\mathbf{g}(t)$ is a set of time-dependent amplitudes associated with the spatial distribution of loads. Similarly, a set of frequency-dependent loads can be defined as

$$\mathbf{f}(\mathbf{x},\omega) = \mathbf{p}(\mathbf{x})\mathbf{h}(\omega) \tag{28.56}$$

where $\mathbf{p}(\mathbf{x})$ is as described above and $\mathbf{h}(\omega)$ is a set of frequency-dependent amplitudes associated with the spatial distribution of loads.

OPERATION ON SYSTEM EQUATIONS

For convenience the equations of motion for the assembled dynamic systems are written as

$$\mathbf{M}_{gg}\ddot{\mathbf{u}}_g + \mathbf{C}_{gg}\dot{\mathbf{u}}_g + \mathbf{K}_{gg}\mathbf{u}_g = \mathbf{P}_g(t) \tag{28.57}$$

where the subscript signifies the set of all dynamic degrees-of-freedom. The g-set is usually subject to a number of constraint and partitioning operations that represent the following:

1. Linear relations between degrees-of-freedom in the g-set to represent a variety of conditions, including stiff elements and displacement constraints in directions not aligned with a coordinate axis. A total of m relations ($m < g$) can be specified in the following form:

$$\mathbf{G}_{mg}\mathbf{u}_g = 0 \tag{28.58}$$

where $\mathbf{G}_{mg}$ is a constraint matrix that contains m rows and g columns.

2. Displacement constraints that specify the (non-time-dependent) displacement boundary conditions. These conditions specify a subset of the g-set to be

$$\mathbf{u}_s = \mathbf{Y}_s \tag{28.59}$$

where $\mathbf{Y}_s$ is prescribed.

3. Static condensation, which, when applied to dynamic analysis, is termed *Guyan reduction*. Guyan reduction, described below, is specified by identifying a set of degrees-of-freedom, $\mathbf{u}_o$, that are related to other degrees-of-freedom through the so-called "static modes" and that are removed from the analysis.

4. Generalized dynamic reduction, which uses a set of Ritz vectors that are a linear combination of the eigenvectors of the system to reduce the number of degrees-of-freedom in dynamic analysis.

5. A set of reference degrees-of-freedom, $\mathbf{u}_r$, that, if constrained, would eliminate rigid-body motion.

Commercial finite-element programs provide a means of defining each of these constraint and partitioning operations. The incorporation of these operations results in a reduction in the number of dynamic degrees-of-freedom. Denoting the reduced set of degrees-of-freedom remaining in the analysis after all constraints and reductions have been applied by $\mathbf{u}_d$, the reduced solution set is represented by

$$\mathbf{M}_{dd}\ddot{\mathbf{u}}_d + \mathbf{C}_{dd}\dot{\mathbf{u}}_d + \mathbf{K}_{dd}\mathbf{u}_d = \mathbf{P}_d(t) \tag{28.60}$$

TIME-DEPENDENT DISPLACEMENT BOUNDARY CONDITIONS

Most programs do not include a special modeling object for representing time-dependent displacement support conditions. However, since the mass object converts an acceleration into a force, a large mass together with a large force can be used to model time-dependent support conditions. Rigid-body motion of the large mass can then be removed by using a modeling procedure that allows the displacement of the reference degrees-of-freedom to be set to zero in such a way that no constraint forces are generated.

REDUCTION TECHNIQUES

A subset of the modes in the system can be used as a transformation from a large number of physical degrees-of-freedom to a set of modal coordinates. However, even before formulating and solving the eigenvalue problem to obtain the transformation vectors, the analyst is generally faced with a large system of equations. Finite-element programs therefore contain a number of procedures for transforming large sets of dynamic equations to sets with a greatly reduced number of equations.

It is common practice to study the behavior of a system by first creating a static model which is verified for a set of static loads. The static model is then used for studying the dynamic behavior of the system, as appropriate. Since static stress analyses generally, but not always, have more spatial detail than is required for dynamics, and since the resources required for dynamic analysis are significantly greater than those required for statics, transformation methods are used to significantly reduce the number of degrees-of-freedom in the dynamic solution set.

The principal techniques that are widely used in the field of shock and vibration include the Guyan reduction, generalized dynamic reduction, and modal synthesis techniques. Most commercial codes include modal synthesis.[8] Since approximations are implicit in each of the procedures for reduction, their successful use usually requires specific expert knowledge of structural dynamics.

GUYAN REDUCTION

Guyan reduction[9] uses the static displacement shapes associated with static condensation[10] to define a transformation between a set of retained degrees-of-freedom, $\mathbf{u}_a$, and a set of degrees-of-freedom $\mathbf{u}_o$ that are to be removed from the analysis in the following form:

$$\mathbf{u}_o = \mathbf{G}_{oa}\mathbf{u}_a \qquad (28.61)$$

where $\mathbf{G}_{oa}$ is a function of the structural stiffness.

The consequence of using the transformation defined by Eq. (28.61) is that the mass coefficients associated with $\mathbf{u}_o$ are distributed to the mass coefficients associated with $\mathbf{u}_a$. Care must therefore be exercised to assure that required dynamic content is retained in the analysis.

Guyan reduction can be used effectively to reduce the size of the solution set in dynamics. However, its use requires significant user experience in identifying modal content that can be neglected in the simulation of interest. For example, the dynamic behavior of a car's windows might not be of interest in studying its ride characteristics. Windows and body panels might, therefore, be included in the o-set of displacements which are removed from the analysis. However, these components might respond to higher frequencies associated with internal acoustics and be retained in acoustic studies.

Guyan reduction, while an important modeling tool, does not provide the convenience of being able to retain modal content in a specified frequency range in the model. Other techniques, including generalized dynamic reduction and subspace iteration, provide reduction techniques that retain modal content in a specified frequency range.

GENERALIZED DYNAMIC REDUCTION

The goal of generalized dynamic reduction, which is also called *subspace iteration,*[11] is to generate a set of N_x Ritz vectors that are rich in the modes within a specified frequency range. The specific number of Ritz vectors to be generated is a multiple of the number of roots in the frequency range; the multiple is typically 1.5, and the number of roots is found by performing a Strum check. The set of Ritz vectors is then calculated by using an incomplete inverse iteration procedure in such a way that they are rich in the eigenmodes within the specified frequency range. The Ritz vectors are not the eigenvectors, but since they are rich in the eigenmodes of interest, they provide a transformation to a reduced number of generalized coordinates called a subspace of the entire set of eigenvectors. The eigenvectors for the system are then calculated using the following procedure:

1. Generate a set of N_x Ritz vectors in terms of the physical degrees-of-freedom of the system.
2. Form a transformation matrix, $\mathbf{G}_{xg}$, whose columns are the Ritz vectors.
3. Define a transformation between the physical degrees-of-freedom, $\mathbf{u}_g$, and a set of generalized degrees-of-freedom, $\mathbf{u}_x$, as

$$\mathbf{u}_g = \mathbf{G}_{gx}\mathbf{u}_x \tag{28.62}$$

4. Transform the homogeneous Eq. (28.57) to generalized coordinates using Eq. (28.62) so that

$$\mathbf{M}_{xx}\ddot{\mathbf{u}}_x + \mathbf{K}_{xx}\mathbf{u}_x = 0 \tag{28.63}$$

where $\mathbf{M}_{xx} = \mathbf{G}_{gx}{}^T\mathbf{M}_{gg}\mathbf{G}_{gx}$ and $\mathbf{K}_{xx} = \mathbf{G}_{gx}{}^T\mathbf{K}_{gg}\mathbf{G}_{gx}$.

5. Extract all of the modes in the subspace using an appropriate eigenvalue extraction method, described in the next section.
6. Use Eq. (28.62) to recover the modes in physical coordinates.

TECHNIQUES FOR OBTAINING MODES AND FREQUENCIES

Commercial finite-element programs include a variety of algorithms for solving for modes and frequencies. These include the Jacobi method following, which are described in Ref. 12:

- Jacobi method
- Inverse iteration with shifts
- Householder method
- Lanzos method
- Givens method

Commercial finite-element programs will include all or at least several of these methods. Generally, inverse iteration with shifts is appropriate for finding a very few modes and frequencies in large d-sets with no reduction operations. The Lanzos and Givens methods are appropriate for extracting a large number of vectors from a relatively small (i.e., <500 degrees-of-freedom) reduced solution set, implying that a preceding reduction operation such as Guyan or generalized dynamic reduction has resulted in an appropriate reduction in the size of the solution set. Generally, a transformation method such as the Givens or Jacobi method is used to extract the set of all eigenvectors in the reduced subspace formed by either generalized dynamic reduction or subspace iteration procedures.

TRANSFORMATION TO MODAL COORDINATES

The coupled set of Eqs. (28.60) can be solved directly, or they can be transformed to modal coordinates by first extracting a set of h eigenvectors ($h \ll d$). This set of eigenvectors is orthogonal with respect to the mass and stiffness matrix, as shown by Eq. (28.12). The resulting set of h uncoupled equations is now of the form

$$\mathbf{M}_{hh}\ddot{\epsilon}_h + \mathbf{C}_{hh}\dot{\epsilon}_h + \mathbf{K}_{hh}\epsilon_h = \mathbf{P}_h(t) \tag{28.64}$$

where the system matrices have the property of being diagonal and ϵ_h are modal coordinates. The extraction of the modes and frequencies and the transformations from physical to modal coordinates is completely automated in most commercial finite-element analysis programs.

SHOCK AND VIBRATION MODELS

Finite-element programs include a large number of algorithms for analysis that are useful for simulating the response of dynamic systems to time-dependent and frequency-dependent loads. Commercial programs also include a number of options for extracting real modes and frequencies or for transforming the solution to modal coordinates. The general types of algorithms are described in subsequent sections. Typically, commercial programs include algorithms for

1. Transient analysis for an arbitrary set of initial conditions using either the direct formulation, Eq. (28.57), or the modal formulation, Eq. (28.64)
2. Steady-state frequency response using either the direct or the modal transformation
3. Response spectrum analysis
4. Random response

TRANSIENT ANALYSIS

The transient response in the time domain is obtained for a given set of time-dependent forces and initial conditions by integrating either Eq. (28.57) or Eq. (28.64), as appropriate. A numerical procedure such as the Newmark-beta or Wilson-theta method[13] is used to transform the set of differential equations, represented as

Eq. (28.57), to a set of difference equations that are solved recursively at a sequence of discrete time-steps. These time-steps are defined by the user. The choice of the time-step in direct transient analysis has the effect of filtering out modes whose periods cannot be adequately represented by the time-step.

If a modal formulation is used, Eq. (28.64) is solved by numerically integrating Duhamel's integral:

$$\xi_i(t) = \int_0^t F(\tau)h(t-\tau)\, d\tau \tag{28.65}$$

where τ is a dummy variable of integration, $F(t)$ is the applied modal load, and $h(t)$ is the response of the mode to a unit impulse which is called the *indicial response*.

FREQUENCY RESPONSE

The steady-state response in the frequency domain is determined by solving Eq. (28.57) or Eq. (28.64), as appropriate, for a set of specified frequency-dependent loads. In either case, the presence of damping means that the response will be complex rather than real. The response of the system is determined at a set of user-specified frequencies.

RESPONSE SPECTRUM ANALYSIS

Response spectrum analysis, sometimes called shock response spectrum, is widely accepted and used for the design of dynamic systems subjected to time-dependent base excitation. The response spectrum procedure, which is described in detail in Chap. 29, is available in many commercial finite-element programs. The implementation of response spectrum analysis, which follows the general approach described in Chap. 29 for multiple degree-of-freedom systems, is associated with the modal formulation for transient response. In this formulation:

1. The base motion of a reference point is described, using a tabular function for time-dependent base motion such as that presented in Fig. 29.7.
2. The modal participation factor for each modal coordinate is determined.
3. A modal transient analysis is performed for each mode, using the appropriate modal participation factor, and the maximum values of the modal coordinate are determined.
4. The maximum modal responses are combined to give $(\mathbf{u}_g)_{max}$, which is the set of maximum values for the set of physical degrees-of-freedom.

The algorithms for combining the shock response spectrum analyses for each of the modal coordinates in order to obtain the response $(\mathbf{u}_g)_{max}$ include

1. Square root of sum of squares (SRSS) method:

$$(\mathbf{u}_g)_{max} = \left\{ \sum_{j=1}^{h} (\bar{\mathbf{u}}_g{}^2)_j \right\}^{1/2} \tag{28.66}$$

2. Absolute method:

$$(\mathbf{u}_g)_{max} = \left\{ \sum_{j=1}^{h} |\bar{\mathbf{u}}_g|_j \right\} \tag{28.67}$$

3. Modified SRSS proposed by the Naval Research Laboratory (NRL):

$$(\mathbf{u}_g)_{\max} = (\overline{\mathbf{u}}_g)_1 + \left\{ \sum_{j=2}^{h} (\overline{\mathbf{u}}_g^2)_j \right\}^{1/2} \tag{28.68}$$

where

$$(\overline{\mathbf{u}}_g)_j = \frac{\phi_j \psi_j S_j(f_j, \zeta_j)}{(2\pi f_j)^2} \qquad j = 1, 2, \ldots, h \tag{28.69}$$

and where ϕ_j is the jth mode, $S_j(f_j, \zeta_j)$ is the maximum acceleration of the response of the jth mode, and ζ_j is the damping ratio for the jth mode.

RANDOM RESPONSE

Random response, described in Chaps. 11 and 22, is included in many commercial programs. Random response is treated as a data recovery procedure applied to the results of either direct or modal frequency response. It can be applied to any behavioral variable, including forces and stresses as well as displacements, velocities, and accelerations.

The input for random response calculations include a set of auto- and crosscorrelation functions that relate a set of frequency-dependent forces. If the input forces are statistically independent [that is, if the crosscorrelation function between any pair of sources, where the crosscorrelation function is defined by Eq. (22.48), is zero], then the power spectral density of the total response is the sum of the power spectral densities of the individual responses. If the sources are statistically correlated, then the degree of correlation can be found in terms of the frequency response for the individual sources.

Typically, a random response analysis is specified by describing the auto- and crosscorrelation factors in the input to the analysis.

COMPUTER GRAPHICS

Computer graphics is an important tool for interpreting the results from a dynamic analysis. A robust postprocessor includes display options for

1. Creating an animated display of the displacement field in scaled time that allows the user to visually interpret the dynamic response
2. Visualizing the stress field that is mapped onto the geometric model at a specified time
3. Creating x-y plots of any behavioral variable versus time or frequency
4. Plotting the power spectral density of a response variable versus frequency for random response

REFERENCES

1. Zienkiewicz, O. C., and R. L. Taylor: "The Finite Element Method, Volume 1: Basic Formulation and Linear Problems," 4th ed., McGraw-Hill, London, 1989.

2. Hughes, T. J. R.: "The Finite Element Method, Linear Static and Dynamic Analysis," Prentice-Hall, Inc., Englewood Cliffs, N.J., 1987.

3. MacNeal, R. H.: "Finite Elements: Their Design and Performance," pp. 179–202, Marcel Dekker, New York, 1994.

4. Ref. 3, pp. 64–69.

5. Pian, T. H. H.: *AIAA J.* **2**:1333–1336 (1964).

6. Pian, T. H. H., and D. P. Chen: *Int. J. Num. Meth. Engr.,* **18**:1679–1684 (1982).

7. Brebbia, C. A., and S. Walker: Boundary Element Techniques in Engineering," Newnes-Butterworth, London, 1980.

8. Herting, D. N.: *Finite Elements in Analysis and Design,* **1**(2):153–164 (1985).

9. Guyan, R. J.: *AIAA J.,* **3**:380 (1965).

10. Gallagher, R. H.: "Finite Element Analysis: Fundamentals," p. 44–45, Prentice-Hall, Inc., Englewood Cliffs, N.J., 1975.

11. Wilkerson, J. H.: "The Algebraic Eigenvalue Problem," Oxford University Press, London, 1972.

12. Bathe, K. J., and E. L. Wilson: "Numerical Methods in Finite Element Analysis," pp. 494–517, Prentice-Hall, Inc., Englewood Cliffs, N.J., 1976.

13. Ref. 12, pp. 319–326.

CHAPTER 29

PART I: VIBRATION OF STRUCTURES INDUCED BY FLUID FLOW

R. D. Blevins

INTRODUCTION

Fluid around a structure can significantly alter the structure's vibrational characteristics. The presence of a quiescent fluid decreases the natural frequencies and increases the damping of the structure. A dense fluid couples the vibration of elastic structures which are adjacent to each other. Fluid flow can induce vibration. A turbulent fluid flow exerts random pressures on a structure, and these random pressures induce a random response. The structure can resonate with periodic components of the wake. If a structure is sufficiently flexible, the structural deformation under the fluid loading will in turn change the fluid force. The response can be unstable with very large structural vibrations—once the fluid velocity exceeds a critical threshold value.

Vibration induced by fluid flow can be classified by the nature of the fluid-structure interaction as shown in Fig. 29.1. Effects which are largely independent of viscosity include added mass and inertial coupling. Unsteady pressure on the surface of a structure, due to either variations in the free stream flow or turbulent fluctuations, induces a forced vibration response. Strong fluid-structure interaction phenomena result when the fluid force on a structure induces a significant response which in turn alters the fluid force. These phenomena are discussed in this section.

ADDED MASS AND INERTIAL COUPLING

If a body accelerates, decelerates, or vibrates in a fluid, then fluid is entrained by the body. This entrainment of fluid, called the *added mass* or *virtual mass effect,* occurs both in viscous and in inviscid, i.e., ideal, fluids. It is of practical importance when the fluid density is comparable to the density of the structure because then the added mass becomes a significant fraction of the total mass in dynamic motion.

Consider the rigid body shown in Fig. 29.2 which lies in a reservoir of incompressible inviscid irrotational fluid. The surface S defines the surface of the body. The body moves with velocity $U(t)$. From ideal flow theory, it can be shown that there exists a

INERTIAL COUPLING EFFECTS	UNSTEADY FLOW INDUCED VIBRATION	FLOW-STRUCTURE COUPLED VIBRATION
1. Added mass	1. Turbulence induced	1. Vortex induced
2. Inertial coupling	vibration	vibration
3. Instability due	2. Ocean wave induced	2. Galloping and
to parallel flow	vibration	flutter
	3. Sonic fatigue	3. Fluid elastic
		instability

FIGURE 29.1 A classification of flow-induced vibration.

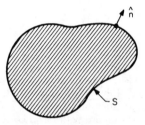

FIGURE 29.2 Fluid-filled region. Fluid density ρ.

velocity potential $\Phi(x, y, z, t)$ which is a function of the special coordinates and time, such that the velocity vector is the gradient of a potential function:

$$V = \nabla\Phi \qquad (29.1)$$

$V(x, y, z, t)$ is the fluid velocity vector. The potential function Φ satisfies Laplace's equation:[1,2]

$$\nabla^2\Phi = 0 \qquad (29.2)$$

The boundary condition is that on the surface of the body; the normal component of velocity must equal the velocity of the body:

$$\frac{\partial\Phi}{\partial n} = V \cdot n \qquad \text{on the surface } S$$

where n is the unit outward normal vector. The pressure in the fluid is given by the Bernoulli equation

$$p = -\rho\frac{\partial\Phi}{\partial t} - \frac{1}{2}\rho V^2$$

where ρ is the fluid density and V is the magnitude of V. The force exerted by the fluid on the body is the integral of the fluid pressure over the surface.

$$F = \int_S p n \, dS$$

If the fluid is of infinite extent, then the solution of these equations is considerably simplified. The fluid force is[1]

$$F = -\rho\frac{\partial}{\partial t}\int_S \Phi n \, dS \qquad (29.3)$$

and flow potential can be expressed as $\phi = U(t)\phi(x', y', z')$, where x', y', and z' are coordinates that are fixed to the body and U is the flow velocity relative to the body. Substituting this potential in Eq. (29.3) yields the following force:

$$F = -m\frac{\partial U}{\partial t}$$

(29.4)

where the added mass m is

$$m = \rho \int_S \phi \frac{\partial \phi}{\partial n} dS$$

(29.5)

The added mass force Eq. (29.3) is zero for U and Φ independent of time, i.e., for steady translation. This is the D'Alembert paradox for an ideal inviscid fluid flow; the fluid force is not zero for steady translation in a viscous fluid.

As an example of added mass calculation, the potential for flow over a cylinder of radius a is

$$\phi = U\frac{r^2 + a^2}{r}\cos\theta$$

where r = radial coordinate
θ = angular coordinate
U = flow velocity

The added mass per unit length is found from Eq. (29.5). The result is

$$m = \rho\pi a^2$$

where a is the cylinder radius. This added fluid mass is equal to the mass of fluid displaced by the cylinder.

In general, there will be an added mass tensor to represent the added mass for acceleration in each of the three coordinate directions:

$$m_{ij} = \rho \int_S \phi_j \frac{\partial \phi_i}{\partial n} dS$$

and an added mass tensor for rotation about the three coordinate axes. ϕ_i is the potential associated with flow in the i direction. Note that the added mass tensor is symmetric, i.e., $m_{ij} = m_{ji}$, but if the body is not symmetric, there is coupling between motions in the various coordinate directions.[1] For example, if a body is not symmetric about the X axis, acceleration in the X direction generally induces added mass force in the Y direction and a moment as well.

Since the added mass acts in phase with acceleration [Eq. (29.3)], the net effect of added mass is to increase the effective mass of the body and to decrease the natural frequencies. In general, added mass is only important to mechanical structures in dense fluids such as water. In gases, such as air, the added mass is ordinarily negligible except for very lightweight structures. Figure 29.3 gives added mass for various sections and bodies in large unrestricted reservoirs. Additional tables of added mass are given in Refs. 3 and 4.

If two structures are in close proximity, then the added mass will be a function of the spacing between the structures and inertial coupling will be introduced between the bodies. For example, consider a cylindrical rod centered in a fluid-filled annulus bounded by a cylindrical cavity shown in Fig. 29.4. The radius of the rod is a and the

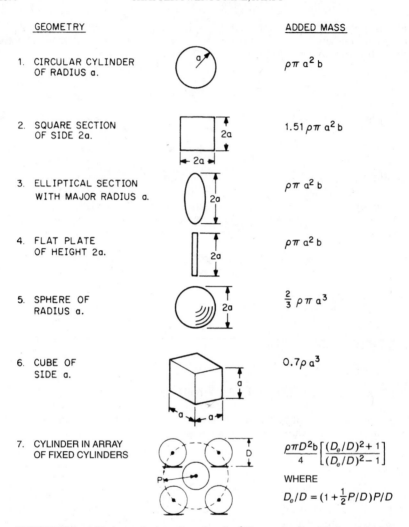

FIGURE 29.3 Added mass for lateral acceleration.[3] The acceleration is left to right. b is the span for two-dimensional sections.

radius of the outer cylinder is b. The fluid forces exerted on the rod and outer cylinder because of their relative acceleration are[5]

$$F_1 = -m\ddot{x}_1 + (M_1 + m)\ddot{x}_2$$
$$F_2 = (m + M_1)\ddot{x}_1 - (m + M_1 + M_2)\ddot{x}_2$$

(29.6)

where x_1, x_2 = displacement of inner rod and outer cylinder
$\quad\quad F_1, F_2$ = force on inner rod and outer cylinder
$\quad\quad m = \rho\pi a^2(b^2 + a^2)/(b^2 - a^2)$, added mass of inner rod
$\quad\quad M_1 = \rho\pi a^2$
$\quad\quad M_2 = \rho\pi b^2$

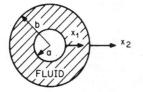

FIGURE 29.4   A rod in a fluid-filled annulus.

These forces include not only added mass but also inertial coupling between the motion of the two structures. [These equations also apply for a sphere contained within a spherical cavity but here $m = (M_1/2)(b^3 - 2a^3)/(b^3 - a^3)$, $M_1 = \frac{2}{3}\rho\pi a^3$, and $M_2 = \frac{2}{3}\rho\pi b^3$.] Coupling is introduced between the cylinder and the rod through the fluid annulus. The coupling increases with the density of the fluid and decreases with increasing gap. If the cylinder and the rod are elastic, motion of either structure tends to set both structures into motion.

For example, consider an array of heat exchanger tubes contained within a shell. Water fills the shell and surrounds the tubes. If the tubes are widely spaced (more than about two diameters between centers), then the tubes are largely uncoupled and the effect of added mass is simply to reduce the tube natural frequencies by the addition of fluid equal to the displaced volume of the tubes. However, if the tubes are closely spaced, then motion of one tube sets adjacent tubes and the shell into motion. Fluid-coupled modes of vibration will result in the tubes and the shell moving in fixed modal patterns as shown in Fig. 29.5. In Refs. 6 and 7, analysis is given for inertial coupling of a cylinder contained eccentrically within a cylindrical cavity, rows of cylinders, and arrays of cylinders.

57.21 Hz   57.28 Hz   57.28 Hz

FIGURE 29.5   Coupled modes of vibration of a bank of tubes in a dense fluid.[6]

Added mass and inertial coupling occur in elastic and rigid bodies, but the added complexity of elasticity and the three-dimensional motions make a closed-form solution impossible for most elastic bodies. In the case of quasi-two-dimensional structures (such as long span tubes or rods), the axial variation in the motion occurs relatively slowly over the span, and two-dimensional results for sections are applicable. Concentric cylindrical shells coupled by a fluid annulus are important in the design of nuclear reactor containment vessels. Approximate solutions are required for both the vessels and the fluid. Reviews of the analysis of fluid coupled concentric vessels are given in Refs. 8 and 9.

Finite element numerical solutions, developed for an irrotational fluid, have been incorporated in the NASTRAN and other computer programs to permit solution for added mass and inertial coupling. These programs solve the fluid and structural problems and then couple the results through interaction forces[10] (see Chap. 28, Part II).

WAVE-INDUCED VIBRATION OF STRUCTURES

Waves induce vibration of structures, such as marine pipelines, oil terminals, tanks, and ships, by placing oscillatory pressure on the surface of the structure. These forces are often well-represented by the inviscid flow solution for many large structures such as ships and oil storage tanks. For most smaller structures, viscous effects influence the fluid force and the fluid forces are determined experimentally.

Consider an ocean wave approaching the vertical cylindrical structure as shown in Fig. 29.6. The wave is propagating in the X direction. Using small-amplitude (lin-

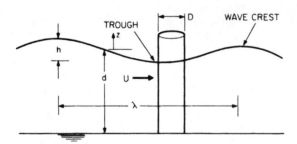

FIGURE 29.6 A circular cylindrical structure exposed to ocean waves.

ear) inviscid wave theory, the wave is characterized by the wave height h (vertical distance between trough and crest), its angular frequency ω, and the associated wavelength λ (horizontal distances between crests), and d is the depth of the water. The wave potential Φ satisfies Laplace's equation (Eq. 29.2) and a free-surface boundary condition (Ref. 11). The associated horizontal component of wave velocity varies with depth $-z$ from the free surface and oscillates at frequency ω:

$$U(t, z) = \frac{h\omega}{2} \frac{\cosh\,[2\pi(z + d)/\lambda]}{\sinh\,(2\pi d/\lambda)}\,\cos\left(\frac{2\pi x}{\lambda} - \omega t\right) \tag{29.7}$$

This component of wave velocity induces substantial fluid forces on structures, such as pilings and pipelines, which are oriented perpendicular to the direction of wave propagation.

The forces which the wave exerts on the cylinder in the direction of wave propagation (i.e., in line with U) can be considered the sum of three components: (1) a buoyancy force associated with the pressure gradient in the laterally accelerating fluid [Eq. (29.7)], (2) an added mass force associated with fluid entrained during relative acceleration between the fluid and the cylinder [Eq. (29.4)], and (3) a force due to fluid dynamic drag associated with the relative velocity between the wave and the cylinder. The first two force components can be determined from inviscid fluid analysis as discussed previously. The drag component of force, however, is associated with fluid viscosity.

Thus, the in-line fluid force per unit length of cylinder due to an unsteady flow is expressed as the sum of the three fluid force components:

$$F = \rho A \dot{U} + C_I \rho A (\dot{U} - \ddot{x}) + \tfrac{1}{2}\rho \mid U - \dot{x} \mid (U - \dot{x}) D C_D \tag{29.8}$$

where x = lateral position of structure in direction of wave propagation
A = cross-sectional area = $\frac{1}{4}\pi D^2$ of cylinder having diameter D
C_I = added mass coefficient, which has theoretical value of 1.0 for circular cylinder
C_D = drag coefficient

This is the generalized form of the Morison equation, widely used to compute the wave forces on slender cylindrical ocean structures such as pipelines and piers.

If $\dot{x}$ and $\ddot{x}$ are set equal to zero in Eq. (29.8), the incline force per unit length on a stationary cylinder in an oscillating flow is obtained:

$$F(\dot{x} = \ddot{x} = 0) = C_m \rho A U + \frac{1}{2}\rho \, |U| \, U D C_D \tag{29.9}$$

Because of the absolute sign in the term $|U| \, U$, the force contains not only components at the wave frequency but also components associated with the drag at harmonics of the wave frequency. The resultant time-history of in-line force due to a harmonically oscillating flow has an irregular form that repeats once every wave period.

If the flow oscillates with zero mean flow, $U = U_0 \cos \omega t$ as in Eq. (29.7), then the maximum fluid force per unit length on a stationary cylinder is

$$F_{\max} = \begin{cases} \rho A C_m \omega U_0 & \text{if } \dfrac{U_0}{\omega D} < \dfrac{C_m A}{C_D D^2} \\[2ex] \dfrac{1}{2}\rho U_0^2 D C_D + \dfrac{(\rho A C_m U_0 \omega)^2}{2\pi U_0^2 D C_D} & \text{if } \dfrac{U_0}{\omega D} > \dfrac{C_m A}{C_D D^2} \end{cases} \tag{29.10}$$

If the cylinder is large (such as for a storage tank) with diameter D greater than the ocean wave height h and if the wavelength of the ocean wave is comparable to the diameter, then U_0 is small compared to ωD and the maximum force is given by the first alternative in Eq. (29.10). The drag force is negligible compared to the inertial forces for large cylinders. As a result, the ocean wave forces on large cylinders can be calculated using inviscid, i.e., potential flow, methods which are discussed in Refs. 11 and 12.

For the Reynolds number ranges typical of most offshore structures, measurements show that the inertial coefficient $C_m = 1 + C_I$ for cylindrical structures generally falls in the range between 1.5 and 2.0. $C_m = 1.8$ is a typical value. C_m decreases for very large diameter cylinders owing to the tendency of waves to diffract about large cylinders (Refs. 13 and 14). Similarly, measurements show that the drag coefficient falls between 0.6 and 1.0 for circular cylinders; $C_D = 0.8$ is a typical value.

Wave forces on elastic ocean structures induce structural motion. Since the wave force is nonlinear [Eq. (29.8)] and involves structural motion, no exact solution exists. One approach is to integrate the equations of motion directly by applying Eq. (29.8) at each spanwise point on a structure and then numerically integrate the time-history of deflection using a predictor-corrector or recursive relationship to account for the nonlinear term. A simpler approach is to assume that the structural deformation does not influence the fluid force and apply Eq. (29.9) as a static load. This static approximation is valid as long as the fundamental natural frequency of the structure is well above the wave frequency *and* the first three or four harmonics of the wave frequency. However, many marine structures are not sufficiently stiff to satisfy this condition.

One generally valid simplification for dynamic analysis of relatively flexible structures is to consider that the wave velocity is much less than the structural velocity so that $|U - \dot{x}| \simeq |U|$. With this approximation, application of Eq. (29.8) to a single degree-of-freedom model for a structure gives the following linear equation of motion:

$$(m + \rho A C_I)\ddot{x} + (2\zeta\omega_N + \tfrac{1}{2}\rho \, |U| \, DC_D)\dot{x} + kx = \rho A C_m \dot{U} + \tfrac{1}{2}\rho \, |U| \, UDC_D \qquad (29.11)$$

where m = structural mass per unit length
       k = stiffness
       ζ = structural damping

This equation is solved by expanding both $x(t)$ and $U(t)$ in a Fourier series and matching the coefficients.

The fluid forces contribute added mass and fluid damping to the left-hand side as well as forcing terms to the right-hand side. This equation may be simplified further by retaining only the first (constant) term in the series expansion for $|U|$ in the fluid damping term so that the equation becomes a classical forced oscillator with constant coefficient.[12]

Flexible structures will resonate with the wave if the structural natural period equals the wave period or a harmonic of the wave period. Since the wave frequencies of importance are ordinarily less than 0.2 Hz (wave period generally greater than one cycle per 5 sec), such a resonance occurs only for exceptionally flexible structures such as deep-water oil production risers and offshore terminals. The amplitude of structural response at resonance is a balance between the wave force and the structural stiffness times the damping. Since the wave force diminishes with increased structural motion [Eq. (29.8)], the resultant displacements are necessarily self-limiting. In other words, the response which would be predicted by applying Eq. (29.9) dynamically is overly pessimistic because the wave force contributes mass and damping to the structure as well as excitation as can be seen in Eq. (29.11).

The above discussion considers only fluid forces which act in line with the direction of wave propagation. These in-line forces produce an in-line response. However, substantial transverse vibrations also occur for ocean flows around circular cylinders. These vibrations are associated with periodic vortex shedding, which is discussed below. The models discussed in the following section for steady flow are applicable to vortex shedding in oscillatory flows provided that the wave period exceeds the period of shedding, based on the maximum oscillatory velocity so that it is possible to fit one or more shedding cycles into the wave cycle.[13,14]

VORTEX-INDUCED VIBRATION

Many structures of practical importance such as buildings, pipelines, and cables are not streamlined but rather have abrupt contours that can cause a fluid flow over the structure to separate from the aft contours of the structure. Such structures are called *bluff bodies*. For a bluff body in uniform cross flow, the wake behind the body is not regular but contains distinct vortices of the pattern shown in Fig. 29.7 at a Reynolds number Re = VD/ν greater than about 50, where D is the width perpendicular to the flow and ν is the kinematic viscosity. The vortices are shed alternately from each side of the body in a regular manner and give rise to an alternating force on the body. Experimental studies have shown that the frequency, in hertz, of the alternating lift force is expressed as[16,17]

$$f_s = \frac{SV}{D} \qquad (29.12)$$

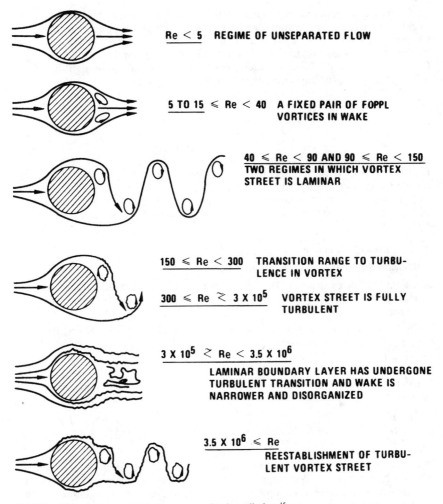

Re < 5   REGIME OF UNSEPARATED FLOW

5 TO 15 ≤ Re < 40   A FIXED PAIR OF FOPPL VORTICES IN WAKE

40 ≤ Re < 90 AND 90 ≤ Re < 150 TWO REGIMES IN WHICH VORTEX STREET IS LAMINAR

150 ≤ Re < 300   TRANSITION RANGE TO TURBULENCE IN VORTEX

300 ≤ Re ⪷ 3 X 10⁵   VORTEX STREET IS FULLY TURBULENT

3 X 10⁵ ⪷ Re < 3.5 X 10⁶

LAMINAR BOUNDARY LAYER HAS UNDERGONE TURBULENT TRANSITION AND WAKE IS NARROWER AND DISORGANIZED

3.5 X 10⁶ ≤ Re

REESTABLISHMENT OF TURBULENT VORTEX STREET

FIGURE 29.7 Regimes of fluid flow across circular cylinders.[15]

The dimensionless constant S called the *Strouhal number* generally falls in the range $0.25 \geq S \geq 0.14$ for circular cylinders, square cylinders, and most bluff sections. The value of S increases slightly as the Reynolds number increases; a value of $S = 0.2$ is typical for circular cylinders.

The oscillating lift force imposed on a single circular cylinder of length L and diameter D, in a uniform cross flow of velocity V, due to vortex shedding is given by

$$F = \tfrac{1}{2}\rho V^2 C_L D L J \sin (2\pi f_s t) \tag{29.13}$$

where the lift coefficient C_L is a function of Reynolds number and cylinder motion. The experimental measurements of C_L show considerable scatter with typical values ranging from 0.1 to 1.0. The scatter is in part due to the fact that the alternating vortex forces are not generally correlated on the entire cylinder length L. The spanwise correlation length l_c of vortex shedding over a stationary circular cylinder[16] is approximately three to seven diameters for $10^3 < \text{Re} < 2 \times 10^5$. In order to account for the effect of the spanwise correlation on the net force on the cylinder of length L, a factor J called the *joint acceptance* has been introduced on the right-hand side of Eq. (29.13). Two limiting cases exist for the joint acceptance.

$$
J = \begin{cases} \left(\dfrac{l_c}{L}\right)^{1/2} & \text{if } l_c \ll L \\[2ex] 1 & \text{if fully correlated} \end{cases}
$$

Thus, if a cylinder is much longer than three to seven diameters, the lack of spanwise correlation reduces the net vortex lift force [Eq. (29.13)] on the cylinder.

Cylinder vibration at or near the vortex shedding frequency organizes the wake and changes the fluid force on the cylinder. Vibration of a cylinder in a fluid flow can:[12, 17]

1. Increase the strength of the shed vortices.

2. Increase the spanwise correlation of the vortex shedding.

3. Cause the vortex shedding frequency shift from the natural shedding frequency (Eq. 29.12) to the frequency of cylinder oscillation. This is called *synchronization* or *lock-in*.

4. Increase the mean drag on the cylinder. Mean drag can triple for one diameter amplitude cylinder vibration.

5. Alter the phase sequence and pattern of vortices in the wake.

As the flow velocity is increased or decreased so that the shedding frequency f_s approaches the natural frequency f_n of an elasticly mounted cylinder so that

$$
f_n \approx f_s = \frac{SU}{D} \qquad \text{so} \qquad \frac{U}{f_n D} \approx \frac{U}{f_s D} = \frac{1}{S} \approx 5
$$

the vortex shedding frequency suddenly locks onto the structure natural frequency. The resultant vibrations occur at or nearly at the natural frequency of the structure and vortices in the near wake input energy to the cylinder. Large amplitude vortex-induced structural vibration can result.

The vortex-induced vibrations of a spring-mounted cylinder in a flow are shown as a function of velocity in Fig. 29.8 for two levels of damping. The horizontal scale gives flow velocity nondimensionalized (i.e., divided by the cylinder diameter D times the cylinder natural frequency f), both of which are held fixed as velocity U increases. The lower part of the figure shows the measured response cylinder single amplitude A_y vibration response as a function of flow velocity. The maximum cylinder amplitude occurs at the resonance condition $U / (fD) \sim 5.5$. The upper part of the figure shows the vortex shedding frequency. The shedding frequency increases with velocity as predicted by Eq. (29.8) until it equals the cylinder natural frequency at $U/fD = 5$ and large amplitude cylinder vibrations begin. The shedding frequency is entrained by the cylinder natural frequency. Entrainment persists until velocity is increased to $U/fD = 6.5$ at which point lock-in is broken and the shedding frequency abruptly returns to its natural value. In general, the larger the structural response to vortex shedding, the larger the range of lock-in.

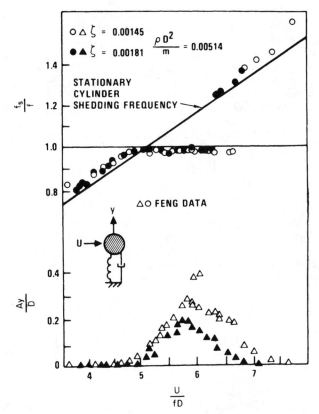

FIGURE 29.8   Response of a spring-supported cylinder to vortex-induced vibration.[19]

Both the amplitude of the structural response and the velocity range over which lock-in persists are functions of the dimensionless reduced damping parameter δ_r:

$$\delta_r = \frac{2m(2\pi\zeta)}{\rho D^2}$$

where m = mass per unit length of cylinder, including added mass
     ζ = damping factor for vibration in mode of interest, ordinarily measured in still fluid
     ρ = fluid density
     D = cylinder diameter

The lower δ_r, the greater the amplitude of the structural response and the greater the range of flow velocities over which lock-in occurs.[18] For lightly damped structures in dense fluids (such as marine pipelines), δ_r is on the order of 1 and lock-in can persist over a 40 percent variation in velocity above and below that which produces resonance.

Within the synchronization band, substantial resonance vibration often occurs. Peak-to-peak vibration amplitudes of up to three diameters have been observed in water flows over cables and tubing. The vibrations are predominantly transverse to the flow and are self-limiting.[12] Lesser amplitude vibrations have also been observed in the drag direction at twice the vortex shedding frequency and at subharmonic frequencies of the vortex shedding frequency, i.e., at one-fourth, one-third, or one-half of the flow velocity required for synchronization,[20] $f_s = f_n$.

If a uniform elastic cylinder is subjected to a crossflow uniformly over its span, then the oscillating vortex-induced lift force is given by Eq. (29.13). At lock-in, the vortex shedding frequency equals the natural frequency of the nth vibration mode $f_s = f_n$, and the amplitude of the cylinder response is

$$\frac{A_y}{D} = \frac{C_L J}{4\pi S^2 \delta_r}$$ (29.14)

where the maximum amplitude vibrations along the span are $y(t) = A_y \sin(2\pi f_n t)$. This equation is conservative if $C_L = J = 1$. However, Eq. (29.14) gives overly conservative predictions with $C_L = J = 1$ owing to the tendency of the actual lift coefficient to decrease at amplitudes in excess of 0.5 diameters and the lack of perfect spanwise correlation at lower amplitudes. Semiempirical correlations are given in Refs. 12 and 21. One of these correlations is[12]

$$\frac{A_y}{D} = \frac{0.07\gamma}{(\delta_r + 1.9)S^2}\left(0.3 + \frac{0.72}{(\delta_r + 1.9)S}\right)^{1/2}$$ (29.15)

The mode shape parameter γ falls between 1.0 and 1.4. For a translating rigid rod ($\phi = 1$), $\gamma = 1$, for a cable or pipeline with a sinusoidal mode shape, $\gamma = 1.15$ and for a cantilever mode shape, $\gamma = 1.4$ and A_y is tip amplitude.

Equation (29.15) correctly predicts the self-limiting behavior of the resonance vibrations. Setting damping to zero, $\delta_r = 0$, it follows that $A_y/D \approx 1.5$, which is a typical vibration level for lightly damped marine cables in a current. See Fig. 29.9. Large amplitude vibrations also are associated with increased steady drag on the structure. Drag coefficients of up to 3.5 have been measured on resonantly vibrating marine cables as opposed to the typical value of 1.0 for a stationary cylinder.[22]

A number of fairings, strakes, and ribbons have been attached to the exterior of circular cylindrical structures to reduce vortex-induced vibrations as shown in Fig. 29.10. These devices act by disrupting the near wake and disturbing the correlation between the vortex shedding and vibration. They do, however, increase the steady drag from that which is measured on a stationary structure. Reviews of vortex suppression devices are given in Refs. 23 and 24.

FLUID ELASTIC INSTABILITY

Fluid flow across an array of elastic tubes can induce a dynamic instability, resulting in very large amplitude tube vibrations once the critical cross-flow velocity is exceeded. This is a relatively common occurrence in tube and shell heat exchangers. Once the critical cross-flow velocity is exceeded, vibration amplitude increases very rapidly with cross-flow velocity V, usually as V^n where $n = 4$ or more, compared with an exponent in the range $1.5 < n < 2.5$ below the instability threshold. This can be seen in Fig. 29.11, which shows the response of an array of metallic tubes to water flow. The initial hump is attributed to vortex shedding. The cross-flow velocity is defined as velocity perpendicular to the tube axis at the minimum gap between

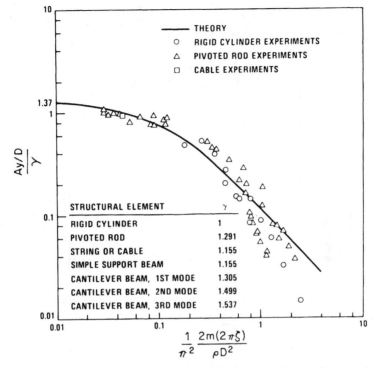

FIGURE 29.9 Maximum amplitude of vortex-induced vibration as a function of damping.[12]

tubes. Once the critical velocity is exceeded, the very large amplitude vibrations usually lead to failures of the heat exchanger tubes.

Often the large amplitude vibrations vary in time; the amplitudes grow and fall about a mean value in pseudorandom fashion. Generally the tubes do not move independently but instead move in somewhat synchronized orbits with neighboring tubes. This orbital behavior has been observed in tests in both air and water with orbits ranging from near circles to nearly straight lines. See Fig. 29.12.

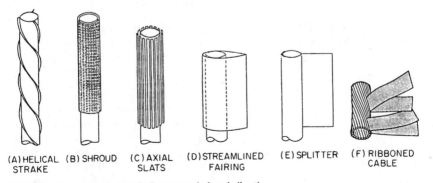

(A) HELICAL STRAKE (B) SHROUD (C) AXIAL SLATS (D) STREAMLINED FAIRING (E) SPLITTER (F) RIBBONED CABLE

FIGURE 29.10 Methods of reducing vortex-induced vibration.

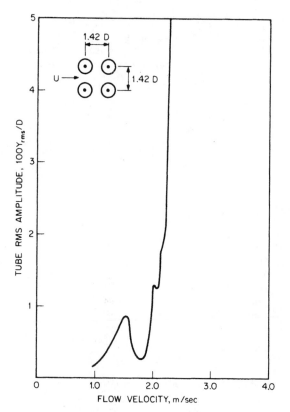

FIGURE 29.11   Typical amplitude of vibration of a tube array in cross flow.[25]

As the tubes whirl in orbital motion, they extract energy from the fluid (Refs. 12, 26, and 27). Below the onset of instability, energy which is extracted is less than the energy which is expended in damping. Above the critical velocity, the energy extracted from the flow by the tube motion exceeds the energy expended in damping, so the vibrations increase in amplitude. Restricting the motion or introducing frequency differences between one or more tubes often increases the critical velocity for onset of instability. Such increases in critical velocity are generally no greater than about 40 percent unless additional support is given to all tubes exposed to high velocity flow. Often the onset of instability is more gradual in a bank of tubes having tube-to-tube frequency differences than in a

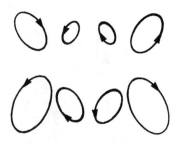

FIGURE 29.12   Tube vibration patterns for fluid elastic instability.[26]

bank with identical tubes. Only a relatively small percentage of the tube will become unstable at one time. Flexible long-span tubes near areas of high flow velocity (such as at inlets) are most susceptible to the instability.

At cross-flow velocities beyond those which produce an onset of instability, damaging vibrations are encountered. The tube vibration amplitudes are limited by clashing with other tubes, by impacting with the tube supports, and by yielding of the tubes. Sustained operation in the unstable vibration regime ordinarily results in tube failure due to wear or propagation of cracks in the tubes. Fluid elastic instability is second only to corrosion as a cause of heat exchanger failure.

A displacement model for the fluid elastic forces is given in Ref. 12 which correctly predicts the observed onset of instability for most cases in air and gases. Results are less satisfactory in water or when the motion of some of the tubes is restricted. More complex models take into account velocity-induced forces as well as the displacement-induced forces.[27,28] These theories give somewhat better agreement with data over limited ranges, but none are entirely suitable for a design tool.

The most viable, practical procedure for predicting the onset of instability of closely spaced arrays of tubes to cross flow is to use the theoretical form given by the displacement mechanism but with parameters obtained by filling experimental data. The onset of instability is predicted as[12,21]

$$\frac{V_{crit}}{f_n D} = C \left[\frac{m_t(2\pi\zeta)}{\rho D^2} \right]^a \tag{29.16}$$

where V_{crit} = uniform cross flow averaged over minimum gap between tubes (If the velocity is nonuniform, then either the maximum can be used or a modal weighted average can be employed.)

f_n = fundamental natural frequency of tubing (Ordinarily the fundamental mode is most susceptible to instability.)

ζ = damping factor of fundamental mode (Typically ζ falls in the range between 0.01 and 0.03 for tubes with some intermediate supports. For rolled-in or welded-in tubes with no intermediate supports, ζ can be as low as 0.001.)

m_t = mass per unit length of tube including added mass and internal mass of fluid

ρ = fluid density

Fitting Eq. (29.16) to the available 174 data points for onset of instability[29] shown in Fig. 29.13 leads to the mean and lower-bound coefficients for the parameter C and the exponent a given in Table 29.1. The coefficient corresponding to the mean fit to the experimental data is C_{mean}; $C_{90\%}$ is the lower bound fit to the data such that 90% of the data are above the curve.

Most of the data used in this correlation come from tube arrays with center-to-center spacing of between 1.25 and 2.0 diameters and with various array geometries. There is insufficient statistical evidence to determine if certain patterns are more or less susceptible to instability than others. Instability has been observed for both straight and curved tubes, tube rows, and tube arrays in a wide variety of tube patterns.

The most common means of increasing the resistance of an array of tubes to instability is to add intermediate supports to increase the natural frequency of the tubes. Details of the tube support (particularly the gap between the tube and the support) influence the resultant vibration. In general, smaller gaps tend to result in lower tube-support impact velocities and hence in lower tube wear.[30]

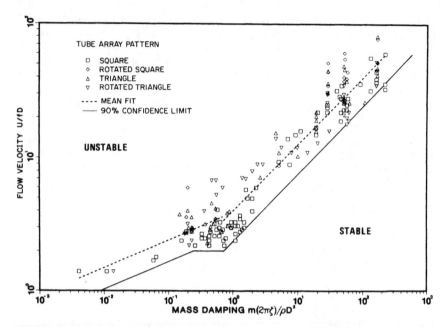

FIGURE 29.13 Velocity for onset of instability of tube arrays in cross flow as a function of the damping parameter.[29]

INTERNAL FLOW IN PIPES

Internal flow through a pipe decreases the natural frequency of the pipe. Sufficiently high internal velocity will induce buckling in a pipe supported at both ends since the momentum of fluid turning through a small angle of pipe deflection is greater than the stiffness of the pipe. If the pipe is restrained at only one end, the pipe will become unstable at high velocities like an unrestrained garden hose.

The equation of motion for a straight pipe conveying steady fluid flow is[31]

$$EI\frac{\partial^4 Y}{\partial x^4} + \rho A v^2 \frac{\partial^2 Y}{\partial x^2} + 2\rho A v \frac{\partial^2 Y}{\partial x \partial t} + M \frac{\partial^2 Y}{\partial t^2} = 0 \qquad (29.17)$$

where E and I are the modulus and moment of inertia of the pipe which conveys fluid of density ρ through the internal area A of the pipe at a steady velocity v;

TABLE 29.1 Coefficients in Eq. (29.16) for Onset of Instability of Tube Arrays[29]

	$m_t(2\pi\zeta)/\rho D^2 < 0.7$	$m_t(2\pi\zeta)/\rho D^2 > 0.7$
C_{mean}	3.9	4.0
$C_{90\%}$	2.7	2.4
a	0.21	0.5
rms error in fitted data for $V_{crit,\%}$	24.5	32.5

$Y(x, t)$ is the lateral deflection of the pipe which has total mass per unit length M. The first and last terms in Eq. (29.17) are the usual stiffness and mass terms. The middle terms are associated with fluid forces imposed on the pipe by the internal fluid as the pipe deflects slightly from its equilibrium position.

Although Eq. (29.17) is a linear partial differential equation with constant coefficients, its solution is difficult owing to the mixed derivative term (third term from the left). One technique used to solve the equation is to expand the solution in terms of the mode shapes of vibration which are obtained for zero flow, $v = 0$.

$$Y(x, t) = \Sigma_i a_i y_i(x) \sin \omega t \qquad (29.18)$$

where $y_i(x)$ are the mode shapes for zero flow that satisfy Eq. (29.17) and the boundary conditions on the ends of the pipe span. Equation (29.18) is substituted into Eq. (29.23), and the derivatives of $y_i(x)$ are expressed in terms of the orthogonal set $y_i(x)$

$$y_i'(x) = \Sigma_i b_i y_i(x)$$

Like terms in the series are equated.

For a uniform pipe with pinned ends, the result can be approximately expressed as a decrease in natural frequency due to flow.[12]

$$\frac{f}{f_1} = \left[1 - \left(\frac{v}{v_c} \right)^2 \right]^{1/2} \qquad (29.19)$$

where f = fundamental natural frequency
f_1 = fundamental natural frequency in absence of flow
v_c = critical flow velocity

The critical flow velocity can be expressed as

$$v_c = \frac{\pi}{L} \left[\frac{EI}{\rho A} \right]^{1/2} \qquad (29.20)$$

where L is the span of the pipe. As the flow velocity approaches v_c, the fundamental natural frequency f_1 decreases to zero. The pipe span spontaneously buckles at $v = v_c$.

The buckling velocity is a function of the boundary conditions on the ends of the pipe, and there can be vibration; these solutions for various boundary conditions are generally scaled by the velocity v_c [Eq. (29.20)]. In general, only exceptionally thin-walled flexible tubes with very high velocity flows, such as rocket motor feed lines and penstocks, are prone to vibration induced by internal flow. External parallel flow can also induce an analogous instability. (See the review given in Ref. 32).

Oscillatory flow in pipes can also cause vibration. Oscillations of fluids in pipes can be caused by reciprocating pumps and acoustic oscillations produced by flow through valves and obstructions. Internal flow imposes net fluid force on pipe at bends and changes in area. For example, the fluid force acting on a 90° bend in a pipe is the sum of pressure and momentum components:

$$F_{bend} = [(p - p_a) + \rho U^2] A\mathbf{i} - [(p - p_a) + \rho U^2] A\mathbf{j}$$

Here p is the internal pressure in the pipe, p_a is the pressure in the atmosphere surrounding the pipe, and U is the internal velocity in the pipe. The vectors $\mathbf{i}$ and $\mathbf{j}$ are unit vectors in the direction of the incoming and outgoing fluid, respectively.

If the pressure and velocity in the pipe oscillates, then the fluid force on the bend will oscillate, causing pipe vibration in response to the internal flow. This problem is most prevalent in unsupported bends in pipe that are adjacent to pumps and valves. Two direct solutions are to (1) support pipe bends and changes in area so that fluid forces are reacted to ground and (2) reduce fluid oscillations in pipe by avoiding large pressure drops though valves and use oscillation-absorbing devices on pump inlet and discharge.

REFERENCES

1. Newman, J. N.: "Marine Hydrodynamics," The MIT Press, Cambridge, Mass., 1977.

2. Lamb, H.: "Hydrodynamics," Dover Publications, New York, 1945. Reprint of the 6th ed., 1932.

3. Blevins, R. D.: "Formulas for Natural Frequency and Mode Shape," Kreiger, Malabar, Florida, 1984. Reprint of 1979 edition.

4. Milne-Thompson, L. L.: "Theoretical Hydrodynamics," 5th ed., Macmillan, New York, 1968.

5. Fritz, R. J.: *J. Eng. Industry,* **94**:167 (1972).

6. Chen, S-S: *J. Eng. Industry,* **97**:1212 (1975).

7. Chen, S-S: *Nucl. Eng. Des.,* **35**:399 (1975).

8. Brown, S. J.: *J. Pressure Vessel Tech.,* **104**:2 (1982).

9. Au-Yang, M. K.: *J. Vibration, Acoustics,* **108**:339 (1986).

10. Zienkiewicw, O. C.: "The Finite Element Method," 3d ed., McGraw-Hill Book Co., New York, 1977.

11. Ippen, A. T. (ed.): "Estuary and Coastline Hydrodynamics," McGraw-Hill Book Co., New York, 1966.

12. Blevins, R. D.: "Flow-Induced Vibration," 2d ed., Kreiger, Malibar, Fla., 1994.

13. Sarpkaya, T., and M. Isaacson: "Mechanics of Wave Forces on Offshore Structures," Van Nostrand Reinhold, New York, 1981.

14. Obasaju, E. D., P. W. Bearman, and J. M. R. Graham: *J. Fluid Mech.,* **196**:467 (1988).

15. Lienard, J. H.: "Synopsis of Lift, Drag and Vortex Frequency Data for Rigid Circular Cylinder," Washington State University, College of Engineering, Research Division Bulletin 300, 1966.

16. Roshko, A.: "On the Development of Turbulent Wakes from Vortex Streets," *National Advisory Committee for Aeronautics Report* NACA TN-2913, 1953.

17a. Sarpkaya, T.: *J. Appl. Mech.,* **46**, 241 (1979).

17b. Williamson, C. H. K., and A. Roshko: *J. Fluids and Structures,* **2**:355 (1988).

18. Scruton, C.: "On the Wind Excited Oscillations of Stacks, Towers and Masts," *National Physical Laboratory Symposium on Wind Effects on Buildings and Structures, Paper* 16, 790, 1963.

19. Feng, C. C.: "The Measurement of Vortex-Induced Effects in Flow Past Stationary and Oscillating Circular and D-Section Cylinder," M.A.Sc. thesis, University of British Columbia, 1968.

20. Durgin, W. W., P. A. March, and P. J. Lefebvre: *J. Fluids Eng.,* **102**:183 (1980).

21a. ASME Boiler and Pressure Vessel Code, Section III, Division 1.

21b. Au-Yang, M. K., T. M. Mulcahy, and R. D. Blevins.: *Pressure Vessel Technology:* **113**:257 (1991).

22. Vandiver, J. K.: "Drag Coefficients of Long Flexible Cylinders," *1983 Offshore Technology Conference, Paper* 4490, 1983, p. 405.

23. Zdravkovich, M. M.: *J. Wind Eng., Industrial Aerodynamics,* **7**:145 (1981).

24. Wong, H. Y., and A. Kokkalis: *J. Wind Eng. Industrial Aerodynamics,* **10**:21 (1982).

25. Chen, S-S, J. A. Jendrzejczyk, and W. H. Lin: "Experiments on Fluid Elastic Instability in a Tube Bank Subject to Liquid Cross Flow," *Argonne National Laboratory Report* ANL-CT-44, July 1978.

26. Connors, H. J.: "Fluid Elastic Vibration of Tube Arrays Excited by Cross Flow," Paper presented at the Symposium on Flow Induced Vibration in Heat Exchangers, ASME Winter Annual Meeting, December 1970.

27. Paidoussis, M. P., and S. J. Price: *J. Fluid Mech.,* **187**:45 (1988).

28. American Society of Mechanical Engineers. "Flow-Induced Vibrations—1994," PVP-273, New York, 1994.

29. Blevins, R. D.: *J. Sound & Vibration,* **97**:641 (1984).

30a. Blevins, R. D.: *J. Eng. Materials Tech.,* **107**:61 (1985).

30b. Cha, J. H.: *J. Pressure Vessel Tech.,* **109**:265 (1987).

31. Housner, G. W.: *J. Appl. Mech.* **19**:205 (1952).

32. Paidoussis, M. P., and P. Besancon: *J. Sound & Vibration,* **76**:361 (1981).

CHAPTER 29

PART II: VIBRATION OF STRUCTURES INDUCED BY WIND

A. G. Davenport and M. Novak

INTRODUCTION

Vibration of significant magnitude may be induced by wind in a wide variety of structures including buildings, television and cooling towers, chimneys, bridges, transmission lines, and radio telescopes. No structure exposed to wind seems entirely immune from such excitation. The material presented here describes several mechanisms causing these oscillations and suggests a few simpler approaches that may be taken in design to reduce vibration of structures induced by wind. There is an extensive literature[1-5] giving a more detailed treatment of the subject matter.

FORMS OF AERODYNAMIC EXCITATION

The types of structure referred to above are generally unstreamlined in shape. Such shapes are termed "bluff bodies" in contrast to streamlined "aeronautical" shapes. The distinguishing feature is that when the air flows around such a bluff body, a significant wake forms downstream, as illustrated in Fig. 29.14. The wake is separated from the outside flow region by a shear layer. With a sharp-edged body (such as a building or structural number) as in Fig. 29.14, this shear layer emanates from the corner. With oval bodies such as the cylinder in Fig. 29.14, the shear layer commences at a so-called *boundary layer* on the upstream surface at points A and B (the separation points) and becomes a free shear layer. The exact position of these separation points depends on a wide variety of factors, such as the roughness of the cylinder, the turbulence in the flow, and the Reynolds number $R = VD/v$, where V = flow velocity, D = diameter of the body, and v = kinematic viscosity.

The flow illustrated in Fig. 29.14 represents the time-average picture which would be obtained by averaging the movements of the fluid particles over a time interval that is long compared with the "transit time" D/V. The instantaneous picture of the flow may be quite different, as indicated in Fig. 29.15, for two reasons.

First, if the flow is the wind, it is under almost all practical circumstances strongly turbulent; the oncoming flow will be varying continuously in direction and speed in an irregular manner. These fluctuating motions will range over a wide range of frequencies and scales (i.e., eddy sizes).

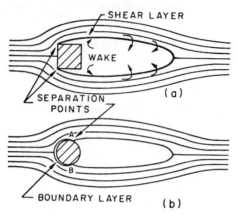

FIGURE 29.14 Wake formation past bluff bodies: (*a*) sharp-edged body; (*b*) circular cylinder.

Second, the wake also will take on a fluctuating character. Here, however, the size of the dominant eddies (vortices) will be of a similar size to the body. The vortices tend to start off their career by curling up at the separation point and then are carried off downstream. Sometimes these eddies are fairly regular in character and are shed alternately from either side; if made visible by smoke or other means, they can be seen to form a more or less regular stepping-stone pattern until they are broken up by the turbulence or dissipate themselves. In a strongly turbulent flow, the regularity is disrupted.

The flow characteristics of the oncoming flow and the wake are the direct causes of the forces on the bodies responsible for their oscillation. The forms of the resulting oscillation are as follows.

1. *Turbulence-induced oscillations.*   Certain types of oscillation of structures can be attributed almost exclusively to turbulence in the oncoming flow. In the wind these may be described as "gust-induced oscillations" (or turbulence-induced, oscillations). The gusts may cause longitudinal, transverse, or torsional oscillations of the structure, which increase with wind velocity (Fig. 29.16).

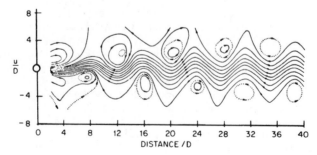

FIGURE 29.15   Vortex street past circular cylinder ($R = 56$). *(After Kovasznay, Proc. Roy. Soc. London, 198, 1949.)*

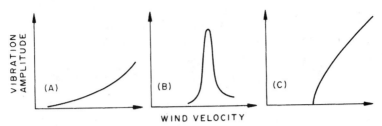

FIGURE 29.16   Main types of wind-induced oscillations: (*A*) vibration due to turbulence; (*B*) vibration due to vortex shedding; (*C*) aerodynamic instability.

2. *Wake-induced oscillations.*   In other instances, the fluctuations in the wake may be the predominant agency. Since these fluctuations are generally characterized by alternating flow, first around one side of the body, then around the other, the most significant pressure fluctuations act on the sides of the body in the wake behind the separation point (the so-called *after body*); they act mainly laterally or torsionally and to a much lesser extent longitudinally. The resultant motion is known as *vortex-induced oscillation.* Oscillation in the direction perpendicular to that of the wind is the most important type. It often features a pronounced resonance peak (Fig. 29.16*B*).

While these distinctions between gust-induced and wake-induced forces are helpful, they often strongly interact; the presence of free-stream turbulence, for example, may significantly modify the wake.

3. *Buffeting by the wake of an upstream structure.*   A further type of excitation is that induced by the wake of an upstream structure (Fig. 29.17). Such an arrangement of structures produces several effects. The turbulent wake containing strong vortices shed from the upstream structure can buffet the downstream structure. In addition, if the oncoming wind is very turbulent, it can cause the wake of the upstream structure to veer, subjecting the downstream structure successively to the free flow and the wake flow. This frequently occurs with chimneys in line, as well as with tall buildings.

4. *Galloping and flutter mechanisms.*   The final mechanism for excitation is associated with the movements of the structure itself. As the structure moves relative to the flow in response to the forces acting, it changes the flow regime surrounding it. In so doing, the pressures change, and these changes are coupled with the motion. A pressure change coupled to the velocity (either linearly or nonlinearly) may be termed an *aerodynamic damping* term. It may be either positive or negative. If positive, it adds to the mechanical damping and leads to higher effective damping and a reduced tendency to vibrate; if negative, it can lead to instability and large amplitudes of movement. This type of excitation occurs with a wide variety of rectangular building shapes as well as bridge cross sections and common structural shapes such as angles and I sections.

In other instances, the coupling may be with either the displacement or acceleration, in which case they are described as either aerodynamic stiffness or mass terms, the effect of which is to modify the mass or stiffness terms in the equations of motion. Such modification can lead to changes in the apparent frequency of the structure. If the aerodynamic stiffness is negative, it can lead to a reduction in the effective stiffness of the structure and eventually to a form of instability known as

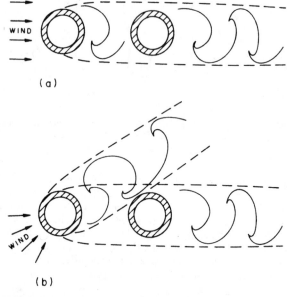

(a)

(b)

FIGURE 29.17   Buffeting by the wake of an upstream structure.

divergence. All types of instability feature a sudden start at a critical wind velocity and a rapid increase of violent displacements with wind velocity (Fig. 29.16C).

These various forms of excitation are briefly discussed in this chapter. Because all types of oscillations are influenced strongly by the properties of the wind, some basic wind characteristics are described first.

BASIC WIND CHARACTERISTICS

Wind is caused by differences in atmospheric pressure. At great altitudes, the air motion is independent of the roughness of the ground surface and is called the *geostrophic,* or *gradient* wind. Its velocity is reached at a height called *gradient height,* which lies between about 1000 and 2000 ft.

Below the gradient height, the flow is affected by surface friction, by the action of which the flow is retarded and turbulence is generated. In this region, known as the *planetary boundary layer,* the three components of wind velocity resemble the traces shown in Fig. 29.18.

The longitudinal component consists of a mean plus an irregular turbulent fluctuation; the lateral and vertical components consist of similar fluctuations. These turbulent motions can be characterized in a number of different ways.

The longitudinal motion at height z can be expressed as

$$V_z(t) = \bar{V}_z + v(t) \tag{29.21}$$

where $\bar{V}_z$ = mean wind velocity (the bar denotes time average) and $v(t)$ = fluctuating component.

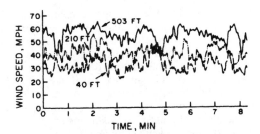

FIGURE 29.18 Record of horizontal component of wind speed at three heights on 500 ft mast in open terrain. *(Courtesy of E. L. Deacon.)*

Mean Wind Velocity. The mean wind velocity $\bar{V}_z$ varies with height z as represented by the mean wind velocity profile (Fig. 29.19). The profiles observed in the field can be matched by a logarithmic law, for which there are theoretical grounds, or by an empirical power law

$$\frac{\bar{V}_z}{\bar{V}_G} = \left[\frac{z}{z_G}\right]^\alpha \qquad (29.22)$$

where $\bar{V}_G$ = gradient wind velocity, z_G = gradient height, and α = an exponent <1. Gradient height z_G and exponent α depend on the surface roughness, which can be characterized by the surface drag coefficient κ (here referenced to the wind speed at 10 meters).

A few typical values of these parameters are given in Fig. 29.19. The mean wind profiles shown are characteristic of level terrain. They can significantly change, par-

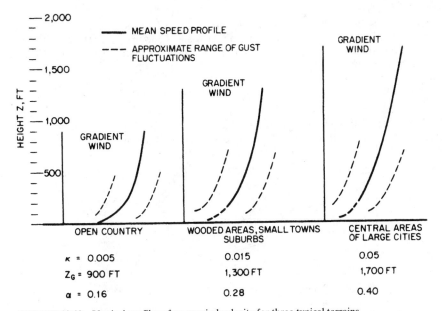

FIGURE 29.19 Vertical profiles of mean wind velocity for three typical terrains.

ticularly in the lower region, when the air flow meets an abrupt change in surface roughness or terrain contour. A sudden increase in roughness reduces the wind speed near the ground while a hill accelerates the flow over its crest.

The mean wind profiles are useful when predicting the wind speed at a particular site. The gradient wind speed is estimated using data registered by the nearest meteorological stations at their standard height, which is usually 33 ft (10 meters).

The mean wind velocity generally depends on the period over which the wind speed is averaged. Periods from 10 to 60 minutes appear adequate for engineering considerations and usually yield reasonably steady mean values. The same duration is suitable to define the fluctuating wind component.

Fluctuating Components of the Wind.   The fluctuating components of the wind change with height less than the mean wind and are random both in time and space. The random nature of the wind requires the application of statistical concepts.*

The basic statistical characteristics of the velocity fluctuations are the intensity of turbulence, the power spectral density (power spectrum), the correlation between velocities at different points, and the probability distribution.

The intensity of turbulence is defined as $\sigma_v / \bar{V}_z$, where $\sigma_v = \sqrt{\overline{v^2(t)}}$ is the root-mean-square (rms) fluctuation in the longitudinal direction. The intensity of the lateral and vertical fluctuations can be described similarly. For wind, the intensity of turbulence is between 5 and 25 percent. The magnitude σ_v also defines the probability distribution of the fluctuations which may be assumed to be Gaussian (normal).

The energy of turbulent fluctuations (gustiness) is distributed over a range of frequencies. This distribution of energy with frequency f can be described by the spectrum of turbulence (power spectral density) $S_v(f)$. The relationship between the spectrum and the variance is

$$\int_0^\infty S_v(f)\, df = \sigma_v^2$$

which leads to another form of the spectrum known as the *logarithmic spectrum* $fS_v(f)/\sigma_v^2$. This form of the spectrum is dimensionless and preserves the relative contributions to the variance at different frequencies represented on a logarithmic scale; and its integral is

$$\int_0^\infty \frac{fS_v(f)}{\sigma_v^2}\, d\ln f = 1$$

The two forms of spectra are sketched in Fig. 29.20. A generalization of wind spectra for different wind velocities is possible if the frequency scale is so modified

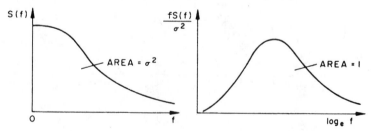

FIGURE 29.20   Two different ways of presenting power spectral densities.

---

* Statistical approaches to random processes are described in Chap. 11.

that it too is dimensionless. The ratio $f/\bar{V}$ is the so-called *inverse wavelength* related to the "size" of atmospheric eddies. This may be expressed as a ratio to a representative length scale L, such as the wavelength of the eddies at the peak of the spectrum. The dimensionless frequency or inverse wavelength may now be written

$$\bar{f} = fL/\bar{V}$$

Under certain circumstances this relationship is also known as the Strouhal number or the reduced frequency.

It is generally found that while the length scale L in the oncoming flow corresponds to that of the turbulence itself (this in the natural wind is of the order of thousands of feet), in the wake the governing length scale is of the same order as the diameter of the body D. This is illustrated in Fig. 29.21.

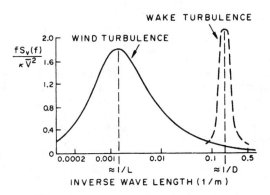

FIGURE 29.21 Universal spectrum of horizontal gustiness in strong winds and example of spectrum of fluctuations in wake.

The spectrum of horizontal gustiness in strong winds is largely independent of height above the ground, is proportional to both the surface drag coefficient κ and the square of the mean velocity at the standard height of 10 meters, $\bar{V}_{10}$, and can be represented, with some approximations, as[6,7]

$$S_v(f) = 4\kappa\bar{V}_{10}{}^2 \frac{L/\bar{V}_{10}}{(2+\bar{f}^2)^{5/6}} \qquad (29.23)$$

in which f = frequency, Hz, $\bar{f} = fL/\bar{V}_{10}$ where L = scale length $\approx$4000 ft, and κ is given in Fig. 29.19. This spectrum is shown in Fig. 29.21.

The variance of the velocity fluctuations is

$$\sigma_v{}^2 = \int_0^\infty S_v(f)\, df = 6.68\kappa\bar{V}_{10}{}^2 \qquad (29.24)$$

It can be seen from Eqs. (29.23) and (29.24) that large velocity fluctuations can be expected in rough terrain where coefficient κ is large.

The spatial correlation of wind speeds at two different stations is described by the coherence function (coherence),

$$\gamma_{12}^{2}(f) = \frac{|S_{12}(f)|^{2}}{S_{1}(f)S_{2}(f)} \leq 1 \tag{29.25}$$

where $S_{12}(f)$ = cross spectrum (generally complex) between stations 1 and 2; $S_1(f)$ and $S_2(f)$ are power spectra of the two stations. The coherence function depends primarily on the parameter $\Delta z f/\bar{V}$, where Δz = separation and $\bar{V} = \frac{1}{2}(\bar{V}_1 + \bar{V}_2)$ is the average wind speed. A suitable approximate function is

$$\sqrt{\text{Coherence}} = e^{-c(\Delta z f/\bar{V})}$$

where c is a constant having a value of approximately 7 for vertical separation and approximately 15 for horizontal separation. Coherence decreases with both separation and frequency. A more detailed discussion of wind characteristics is given in Refs. 1 and 7.

EXCITATION DUE TO TURBULENCE

When a structure is exposed to the effects of wind, the fluctuating wind velocity translates into fluctuating pressures, which in turn produce time-variable response (deflection) of the structure. This response is random and represents the basic type of wind-induced oscillations. The theoretical prediction of this oscillation is rather complex but can be reduced to a simple procedure suitable for design purposes. The discussion of the oscillation is therefore presented in two parts. In the first part, the basic theoretical steps are outlined. In the second part, the design procedure known as the gust-factor approach is given in more detail.

FUNDAMENTALS OF RESPONSE PREDICTION

If the area A of the structure exposed to wind is small relative to the significant turbulent eddies, the so-called *quasi-steady theory* for turbulence can be used to estimate aerodynamic forces. In the drag direction, the drag force

$$D(t) = \frac{1}{2}\rho C_D A V^2(t)$$

$$= \frac{1}{2}\rho C_D A \bar{V}^2 \left[1 + 2\frac{v(t)}{\bar{V}} + \frac{v^2(t)}{\bar{V}^2} \right]$$

where ρ = air density (normally equal to 0.0024 slugs/ft³), and C_D = drag coefficient. If $v(t) \ll \bar{V}$, the squared term is ignored. The spectra of the fluctuating drag and velocity are then related as

$$\frac{S_D(f)}{\bar{D}^2} = 4\frac{S_v(f)}{\bar{V}^2} \tag{29.26}$$

where the mean drag (static component of the drag) is

$$\bar{D} = \frac{1}{2}\rho C_D A \bar{V}^2 \tag{29.27}$$

and $S_v(f)$ is given by Eq. (29.23).

With large bodies the wavelength is comparable to the size of the body itself (that is, $f\sqrt{A}/\bar{V} \approx 1$) and it is necessary to modify the drag spectrum by the so-called *aerodynamic admittance function* $|X_{aero}(f)|^2$. This function[6] describes the modifying influence of any changes in effective drag coefficient as well as the decrease in correlation of the eddies as the wavelength of the eddies approaches the diameter of the body. Thus, the modified drag spectrum is

$$\frac{S_D(f)}{\bar{D}^2} = 4|X_{aero}(f)|^2 \frac{S_v(f)}{\bar{V}^2}$$

If these forces act on an elastic spring-mass-damper system, the response of this system u will have a spectrum

$$\frac{S_u(f)}{\bar{u}^2} = |X_{aero}|^2 |X_{mech}|^2 \frac{4S_v(f)}{\bar{V}^2}$$

where static deflection $\bar{u} = \bar{D}/k$, k = stiffness constant, and the mechanical admittance function

$$|X_{mech}|^2 = \frac{1}{[1 - (f/f_0)^2]^2 + 4\zeta^2(f^2/f_0^2)}$$

where ζ = critical damping ratio, and f_0 = natural frequency of the system.

The transition from the spectrum of the wind-velocity fluctuations to the spectrum of the response is shown diagrammatically in Fig. 29.22. The variance of the response σ_u^2 is obtained from the spectrum of the response,

$$\sigma_u^2 = \int_0^\infty S_u(f)\, df \tag{29.28}$$

The relationships above describe the mean and the variance of the response. For engineering purposes, it is also useful to define extreme values. It is often satisfactory to assume that the process in question is Gaussian with probability density function given by

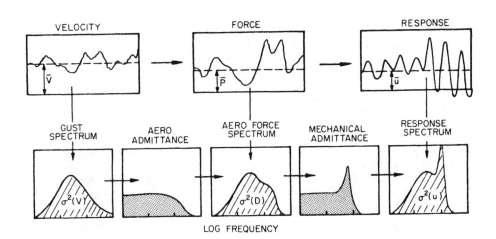

FIGURE 29.22 Transition from gust spectrum to response spectrum.

$$p(u) = \frac{1}{\sqrt{2\pi}\,\sigma_u}\,e^{-(u-\bar{u})^2/2\sigma_u^2}$$

This distribution is fully described by the mean and the variance. Maximum values of the response during time T can be written as

$$u_{\max} = \bar{u} + g\sigma_u \tag{29.29}$$

where g = peak factor. The average largest value of the peak factor in a period T can be estimated from[6]

$$g = \sqrt{2 \ln \nu T} + \frac{0.5772}{\sqrt{2 \ln \nu T}} \tag{29.30}$$

where ν is an effective cycling rate of the process, generally close to the natural frequency. The relationship of the distribution of the largest peak value to the distribution of all values is shown in Fig. 29.23. As can be seen, when the period T or the natural frequency increases, the expected peak displacement also increases. The factor g usually ranges between 3 and 5.

Further extension of the concept includes the cross correlation of the wind loads at different stations (e.g., heights), the shape of the vibration mode, and the nonuniformity of the mean flow. These factors can be included into the solution formulated in terms of modal analysis.*

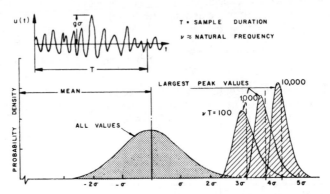

FIGURE 29.23 Relationship of distribution of largest peak value to distribution of all values (for a stationary random process).

With a prismatic structure, the displacement may be expressed in the form

$$u(z,t) = \sum_{j=1}^{\infty} q_j(t)\phi_j(z) \tag{29.31}$$

where $q_j(t)$ = the generalized coordinate of the jth mode, and $\phi_j(z)$ = the jth mode of natural vibrations to an arbitrary scale.

With damping small and natural frequencies well separated, the cross correlation of the generalized coordinate can be neglected and the mean-square displacement (the variance) is

---

* For detailed discussion of modal analysis, see Chap. 21.

$$\overline{u^2(z,t)} = \sum_{j=1}^{\infty} \overline{q_j^2} \, \phi_j^2(z) \tag{29.32}$$

The variance of the generalized coordinate $\overline{q_j^2}$ is determined by the power spectrum of the generalized force Q_j. When the lateral dimension of the structure is small, only cross correlation in direction z need be considered. Then the power spectrum of the generalized force is

$$S_{Q_j}(f) = \int_0^H \int_0^H S_{12}(z_1,z_2,f)\phi_j(z_1)\phi_j(z_2) \, dz_1 \, dz_2 \tag{29.33}$$

where $S_{12}(z_1,z_2,f)$ = cross spectrum of the wind loads at heights z_1 and z_2, and H = height of the structure. With respect to Eq. (29.26), the cross spectrum of the wind loads can be expressed in terms of the power spectrum of the wind speed [Eq. (29.23)] and the coherence function, Eq. (29.25).

The variance of q_j is

$$\overline{q_j^2} = \int_0^{\infty} \frac{1}{(2\pi f_j)^4 M_j^2} \; \frac{1}{[1 - (f/f_j)^2]^2 + 4\zeta^2(f/f_j)^2} \, S_{Q_j}(f)$$

$$\approx \frac{1}{64\pi^3 \zeta f_j^3 M_j^2} \, S_{Q_j}(f_j) + \frac{1}{(2\pi f_j)^4 M_j^2} \int_0^{f_j} S_{Q_j}(f) \, df \tag{29.34}$$

where $f_j = j$th natural frequency and generalized mass

$$M_j = \int_0^H m(z)\phi_j^2(z) \, dz \tag{29.35}$$

where $m(z)$ = mass of the structure per unit length. The approximate integration* of Eq. (29.34) yields the response composed of two parts, the resonance effect (the first term) and the background turbulence effect (the second term) (Fig. 29.24). The variance of the displacement follows from Eq. (29.32), and its standard deviation (rms dynamic displacement) is $\sigma_u(z) = \sqrt{\overline{u^2(z,t)}}$. The peak response is established from Eq. (29.29) by means of the peak factor g [Eq. (29.30)] as in one degree-of-freedom. The mean deflection $\bar{u}(z)$ is the static deflection due to the mean wind $\bar{V}_z$.

Other analyses of slender structures are also available.[9–11] In applications to buildings and free-standing towers, the analysis can usually be limited to the first modal component in Eq. (29.32).

Application to buildings and structures with significant lateral dimension requires the incorporation of the horizontal cross correlation as well. A complete solution established by means of simplifying assumptions and numerical integrations is given below.

GUST-FACTOR APPROACH

The gust-factor approach is a design procedure derived on the basis of the theory above by means of a few simplifying assumptions. The approach given here is a modified version of the method described in Ref. 12 and adopted in Ref. 13. It considers only the response in the first vibration mode which is assumed to be linear. These

---

* The integral is similar to that appearing in Eq. (29.28) and can be evaluated accurately using the theory of residua. An example is given in Ref. 8.

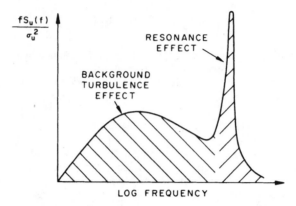

FIGURE 29.24   Spectrum of structural response with indication of resonance effect and background turbulence effect.

assumptions are particularly suitable for buildings. The method yields all the data needed in design: the maximum response, the equivalent static wind load that would produce the same maximum response, and the maximum acceleration needed for the evaluation of the physiological effects of strong winds (human comfort).

The gust factor G is defined as the ratio of the expected peak displacement (load) in a period T to the mean displacement (load) $\bar{u}$. Hence, maximum expected response

$$u_{\max} = G\bar{u} = \left(1 + g\frac{\sigma_u}{\bar{u}}\right)\bar{u} \qquad (29.36)$$

The gust factor is given as

$$G = 1 + g\sqrt{\frac{K}{C_e}\left(B + \frac{sF}{\zeta}\right)} \qquad (29.37)$$

where ζ = damping ratio and K = factor related to the surface roughness; this factor is equal to

0.08 for open terrain (zone A)

0.10 for suburban, urban, or wooded terrain (zone B)

0.14 for concentrations of tall buildings (zone C)

All the other parameters appearing in Eq. (29.37) can be obtained from Fig. 29.25. C_e = exposure factor based on the mean wind speed profile (coefficient α) and thus on surface roughness. For the three zones, the exposure factor is obtained from Fig. 29.25A for the height of the building H. C_e relates to wind pressure rather than speed. Hence, the mean wind speed at the top of the building is given by

$$\bar{V}_H = \bar{V}_{10}\sqrt{C_e}$$

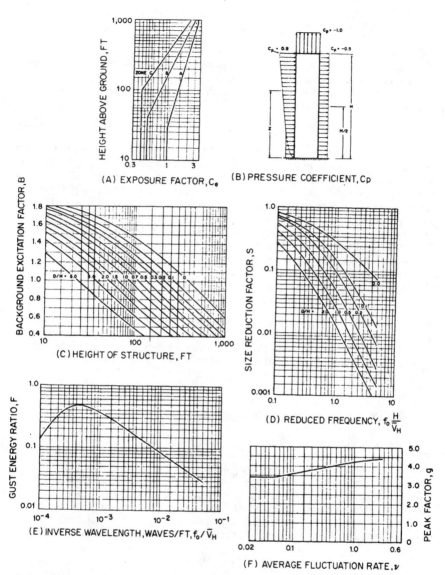

FIGURE 29.25 Components of gust factor.

where $\bar{V}_{10}$ = reference wind speed at the standard height of 10 meters. $\bar{V}_{10}$ can be obtained from meteorological stations. Velocity $\bar{V}_H$ is needed for determination of parameters s and F. Factors B, s, F, and g are given in Fig. 29.25C to f as a function of parameters indicated; D = width of the frontal area, and f_0 = the first natural frequency of the structure in cycles per second. The average fluctuation rate v, on which the peak factor g depends, is evaluated from formula

$$v = f_0 \sqrt{\frac{sF/\zeta}{B + sF/\zeta}} \tag{29.38}$$

The peak factor g is plotted in Fig. 29.25F, assuming a period of observation $T = 3600$ sec; it can also be calculated from Eq. (29.30).

The parameters given also yield the design wind pressure p, which produces displacement u_{max} if applied as a static load. This design pressure

$$p = qC_eGC_p \tag{29.39}$$

where $q = \frac{1}{2}\rho\bar{V}_{10}^2$ is the reference mean-velocity pressure, and $C_e =$ exposure factor. In this case, C_e varies continuously with the elevation according to Fig. 29.25A for pressures acting on the windward face of the structure; for the leeward face, C_e is constant and evaluated at one-half the height of the building. The quantity $C_p =$ average pressure coefficient, which depends on the shape of the structure and the flow pattern around it. For a typical building with a flat roof and a height greater than twice the width, the coefficients are given for the windward and leeward faces in Fig. 29.25B together with the pressure distribution.

The peak acceleration A of a structure due to gusting wind is given by

$$A = u_{max} \frac{4\pi^2 f_0^2 g}{G} \sqrt{\frac{KsF}{C_e\zeta}}$$

where $u_{max} =$ maximum deflection under the design pressure p. The other parameters are equal to those used in Eq. (29.37). When the acceleration exceeds about 1 percent of gravity, the motion is usually perceptible. However, there are large differences in the perceptibility of motions having very low frequencies.[14, 15]

Similar approaches are given in Refs. 16 to 18.

EFFECT OF GUSTS ON CLADDING AND WINDOWS

Wind gusts produce local pressure on cladding and window panels of a building. Because the natural frequency of such a panel is very high compared with the frequency components of the wind-speed fluctuations, the panel displacement is essentially static. Its design may be based on the static displacement resulting from maximum expected pressure, which is the algebraic sum of the height and time-dependent exterior pressure (or suction) and the constant interior pressure (or suction). If the fluctuating component of the pressure $p(t)$ is considered to be a stationary random process, the exterior expected maximum pressure is

$$p_{max} = \bar{p}\left(1 + g\frac{\sigma_p}{\bar{p}}\right) = \bar{p}G \tag{29.40}$$

where $\quad \bar{p} = \frac{1}{2}\rho C_p\bar{V}^2 =$ mean pressure
$\qquad C_p =$ local pressure coefficient
$\qquad \sigma_p =$ standard deviation of the fluctuating pressure component
$\qquad g =$ peak factor given by Eq. (29.30)
$\qquad G =$ gust factor

To account for the sensitivity of glass to both static and dynamic fatigue, it has been suggested[19, 20] that g or G in Eq. (29.40) be multiplied by a wind-on-glass effect factor.

Factors g, $\sigma_p/\bar{p}$, and C_p are most reliably determined from wind-tunnel experiments. They strongly depend on location of the panel, wind direction, turbulence intensity, and the local flow pattern determined by the shape of the building and its immediate environment. In full-scale experiments, values of g in excess of 10 have been observed in highly intermittent flow. Largest local pressure coefficients C_p (actually suctions) appear with skew wind at the leading edge of the building where a typical value is $C_p = -1.5$. In that part of the building exposed to free flow, a gust factor $G \approx 2.5$ is a reasonable estimate.[13,21]

The interior pressure is not very high, but its magnitude and sign depend on openings and leakage.

Damage to windows may result from local wind pressure, but it also depends on material properties of glass and its fatigue. The fatigue limit of glass is only about 20 percent of the instantaneous strength.[20]

VIBRATION DUE TO VORTEX SHEDDING

Vortex shedding represents the second most important mechanism for wind-induced oscillations. Unlike the gusts, vortex shedding produces forces which originate in the wake behind the structure, act mainly in the across-wind direction, and are, in general, rather regular. The resultant oscillation is resonant in character (Fig. 29.16B), is often almost periodic, and usually appears in the direction perpendicular to that of the wind. Lightly damped structures such as chimneys and towers are particularly susceptible to vortex shedding. Many failures attributed to vortex shedding have been reported.

When a bluff body is exposed to wind, vortices shed from the sides of the body creating a pattern in its wake often called the *Karman vortex street* (Fig. 29.15). The frequency of the shedding, nearly constant in many cases, depends on the shape and size of the body, the velocity of the flow, and to a lesser degree on the surface roughness and the turbulence of the flow. If the cross section of the body is noncircular, it also depends on the wind direction. The dominant frequency of vortex shedding f_s is given by

$$f_s = S\frac{\bar{V}}{D} \text{ Hz} \qquad (29.41)$$

where S = dimensionless constant called the Strouhal number, $\bar{V}$ = mean wind velocity, and D = width of the frontal area. The second dimensionless parameter is the Reynolds number $R = \bar{V}D/\nu$, where ν = kinematic viscosity. For air under normal conditions, $\nu = 1.6 \times 10^{-4}$ ft²/sec.

For a body having a rectangular or square cross section, the Strouhal number is almost independent of the Reynolds number.

For a body having a circular cross section, the Strouhal number varies with the regime of the flow as characterized by the Reynolds number. There are three major regions: the subcritical region for $R \lesssim 3 \times 10^5$, the supercritical region for $3 \times 10^5 \lesssim R \lesssim 3 \times 10^6$, and the transcritical region for $R \gtrsim 3 \times 10^6$. Approximate values of the Strouhal number for typical cross sections are given in Table 29.2. The numbers given in this table are based on Refs. 1, 22, 23, and 24 and other measurements and may be used for turbulent shear flow.

TABLE 29.2 Aerodynamic Data for Prediction of Vortex-Induced Oscillations in Turbulent Flow

Cross section	Strouhal number S	Rms lift coefficient σ_L	Bandwidth B	Correlation length L (diameters)
Circular: region				
Subcritical	0.2	0.5	0.1	2.5
Supercritical	Not marked	0.14	Not marked	1.0
Transcritical	0.25	0.25	0.3	1.5
Square:				
Wind normal to face	0.11	0.6	0.2	3

PREDICTION OF VORTEX-INDUCED OSCILLATION

Although the mechanism of vortex shedding and the character of the lift forces have been the subject of a great number of studies,[25] the available information does not permit an accurate prediction of these oscillations.

The motion is most often viewed as forced oscillation due to the lift force, which, per unit length, may be written as

$$F_L = \frac{1}{2}\rho D \bar{V}^2 C_L(t) \tag{29.42}$$

where $C_L(t)$ is a lift coefficient fluctuating in a harmonic or random way. (Some authors[26, 27] consider vortex shedding to be self-excitation, which does not seem necessary, however, for relatively small motions.) Hence, the solution of the response depends on the time-history assumed for $C_L(t)$.

HARMONIC EXCITATION OF PRISMATIC CYLINDERS BY VORTICES

Harmonic excitation represents a traditional model for vortex excitation, but it is really justified only for very low Reynolds numbers ($\gtrsim 300$) or possibly for large vibration where the motion starts controlling both the wake and the lift forces in the form of the "locking-in" phenomenon. Strongest oscillations arise at that wind velocity for which the frequency of vortex shedding f_s is equal to one of the natural frequencies of the structure f_j. This resonant wind velocity is, from Eq. (29.41),

$$V_c = \frac{1}{S}f_j D \tag{29.43}$$

With free-standing towers and stacks, resonance in the first two modes is met most often; resonance with higher modes has been observed as well with guyed towers (Fig. 29.26).

At the resonant wind velocity, the lift force is given by Eq. (29.42) in which $C_L(t) = C_L \sin 2\pi f_j t$, and C_L = amplitude of lift coefficient. Assuming a uniform wind profile and a constant diameter D, the resonant amplitude of mode j at the critical wind velocity V_c is, from Eq. (29.31),

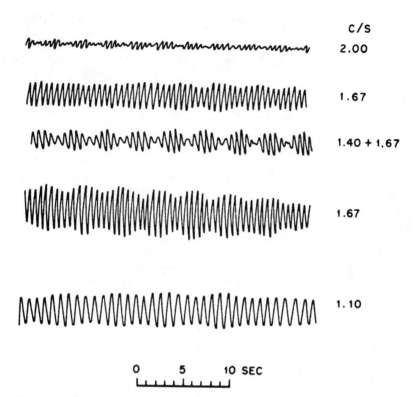

FIGURE 29.26 Vortex-induced oscillations in different modes measured on 1000 ft guyed tower.[28]

$$u_j(z) = \frac{\rho C_L}{16\pi^2 S^2} \frac{D^3}{\zeta M_j} \phi_j(z) \int_0^H \phi_j(z)\, dz \qquad (29.44)$$

where M_j is given by Eq. (29.35) and ζ = structural damping ratio. The formula can be further simplified if it is assumed that the lift force is distributed along the structure in proportion to the mode $\phi_j(z)$. (This assumption reflects the loss of spanwise correlation of the forces.) Then, with constant mass per unit length $m(z) = m$, the resonant amplitude at the height where the modal displacement is maximum:

$$u_j = \frac{\rho C_L}{16\pi^2 S^2} \frac{D^3}{\zeta m} \qquad (29.45)$$

For the first mode of a free-standing structure, this occurs at the tip. In higher modes, this amplitude appears at the height where local resonance takes place. For circular cylinders, a design value of the lift coefficient C_L is about $\sqrt{2}\sigma_L$. This simple formula can be used for the first estimate of the amplitudes that are likely to represent the upper bound. It is also indicative of the role of the diameter, mass, and damping of the structure. Approximate values of σ_L are given in Table 29.2.

RANDOM EXCITATION OF PRISMATIC CYLINDERS BY VORTICES

Even when vortex shedding appears very regular, the lift force and thus $C_L(t)$ are not purely harmonic but random. The power spectrum of the lift force per unit length is from Eq. (29.42).

$$S_L(f) = \left(\frac{1}{2}\rho D\bar{V}^2\sigma_L\right)^2 S_L'(f) \qquad (29.46)$$

where $\sigma_L = \sqrt{\overline{C_L^2(t)}}$ is the standard deviation of the lift coefficient and $S_L'(f) =$ normalized power spectrum of $C_L(t)$ for which

$$\int_0^\infty S_L'(f)\, df = 1 \qquad (29.47)$$

With circular cylinders, the lift force is narrow-band random in the subcritical and transcritical[22, 23] ranges where the energy is distributed about the dominant frequency f_s, given by Eq. (29.41) (Fig. 29.27A). Such spectra can be described by a Gaussian-type curve,

$$S_L'(f) = \frac{1}{\sqrt{\pi}Bf_s}\, \exp\left[-\left(\frac{1-f/f_s}{B}\right)^2\right] \qquad (29.48)$$

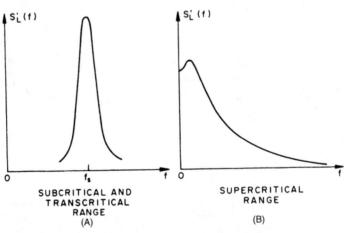

FIGURE 29.27   Spectra of lift coefficient for circular cylinder.

A few design values of bandwidth B are given in Table 29.2. In the supercritical range, the power spectrum is broad (Fig. 29.27B) and can be expressed as[29]

$$S_L'(f) = 4.8\,\frac{1+682.2(fD/\bar{V})^2}{[1+227.4(fD/\bar{V})^2]^2}\,\frac{D}{\bar{V}} \qquad (29.49)$$

Because the vortices are three-dimensional, a realistic treatment also requires the inclusion of the spanwise cross correlation of the lift forces. This can be done in terms of the "correlation length" L given in number of diameters.

Approximate values of L are given in Table 29.2. The correlation length decreases with turbulence[30] and shear and increases with aspect ratio $2H/D$ and the amplitude of the motion (Fig. 29.28).

Using the correlation length, the spectral density of the lift force, Eqs. (29.48) and (29.49), and a few further approximations, the vibration can be evaluated from Eqs. (29.32) to (29.34). The root-mean-square (rms) displacement at height z in mode j is approximately

$$\sqrt{\overline{u_j^2(z,t)}} = \frac{\pi^{1/4}\sigma_L\rho D^4\phi_j(z/H)}{\sqrt{B\zeta}\,(4\pi S)^2 M_j}\,C$$

where

$$C^2 = \frac{(H/D)^2}{1+(H/2LD)}\int_0^1\left(\frac{z}{H}\right)^{3\alpha}\phi_j^2\left(\frac{z}{H}\right)d\left(\frac{z}{H}\right)$$

where α = wind profile exponent (Fig. 29.19), and parameters S, σ_L, B, and L are given in Table 29.1. The mode $\phi_j(z/H)$ is dimensionless, and consequently M_j is in slugs in this case. The peak response is $g\sqrt{\overline{u_j^2(z,t)}}$, where the peak factor g is given by Eq. (29.30). If it is larger than about 2 percent of diameter D, locking-in may develop and the analysis should be repeated assuming harmonic excitation or at least random excitation with a significantly increased correlation length, as Fig. 29.28 indicates.

RANDOM EXCITATION OF TAPERED CYLINDERS BY VORTICES

Tapered cylinders, such as stacks, also vibrate due to vortex shedding; but less is known about the mechanism of excitation. It appears that the lift forces are narrowband random with a rather small correlation length L and with the dominant frequency f_s given by Eq. (29.41). As the diameter is variable, local resonance between f_s and the natural frequency f_j takes place at different heights z_r. As the wind speed increases, the resonance first appears at the tip and shifts downward. The critical wind speed for each height follows from Eq. (29.43) with $D = D(z_r)$. The rms displacements at height H due to local resonance at height z_r can be obtained from an approximate formula,[32]

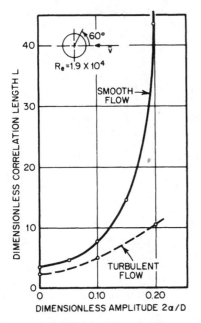

FIGURE 29.28 Variation of correlation length of vortex shedding with amplitude of motion and turbulence ($2a$ = double amplitude, turbulence intensity 10 percent).

$$\sqrt{\overline{u_j^2(H,t)}}$$

$$= \sqrt{\frac{L}{2\pi^3\zeta\Psi}}\,\frac{\sigma_L\rho D^4(z_r)\phi_j(z_r)}{8S^2 M_j}\,\phi_j(H)$$

where

$$\Psi = \frac{dD(z_r)}{dz} + \frac{\alpha D(z_r)}{z_r}$$

or with a constant taper

$$\Psi = \frac{t}{H} + \frac{\alpha D(z_r)}{z_r}$$

where $t = D(0) - D(H)$ and α = the wind-profile exponent.

The other parameters can be taken from Table 29.2. The values listed for the transcritical region may be adequate, inasmuch as most tapered stacks are large. The peak displacement is again obtained by means of the peak factor given by Eq. (29.30).

Maximum response of chimneys in the first mode usually results from local resonance at about $\frac{3}{4}H$. The height of maximum excitation follows from condition $d[D^4(z)\phi_i(z)]/dz = 0$.

SUPPRESSION OF VORTEX-INDUCED VIBRATIONS

Vortex shedding may induce severe vibration of a cylindrical structure such as a chimney, free-standing tower, guyed mast, bridge columns, etc. Very strong oscillations have been observed[28, 31] in all-welded structures where the damping ratio is extremely low, sometimes less than 0.005.[8, 28] Welded structures are particularly prone to fatigue failure, as the endurance limit may be only a fraction of the strength if heavy notches, flaws, attachments, or other adverse details are present. In other cases, the motion is intolerable because of its physiological effects or swaying of antennas. For these reasons, suppression of vibration is often desirable.

In some cases, vibration can be reduced by increasing the structural damping. This can be accomplished by additional dampers attached to an independent support[28] or to a special mass suspended from the structure and suitably tuned or by hanging chains.[33] Columns of a few bridges were filled with gravel, sand, or plastic balls partly filled with oil. The increase in mass may be favorable but can reduce the original structural damping.

Another successful method of vibration control is to break down the wake pattern by providing the surface by helical "strakes" or "spoilers."[28, 31, 34] A suitable height of the spoilers is about $0.1D$ or more with a pitch of about $5D$. A significant drawback of the spoilers is that they considerably increase the drag, sometimes as much as 100 percent or even more.[31, 35]

WAKE BUFFETING

If one structure is located in the wake of another, vortices shed from the upstream structure may cause oscillation of the downstream structure[36, 37] (Fig. 29.17). If the two structures differ greatly in size or shape, this excitation is usually not significant. Strong vibration of the downstream structure may arise when two or more structures are identical and less than about 10 diameters apart. Then the structure in the wake is efficiently excited by well-tuned wake buffeting and its own vortex shedding. Such excitation has been observed with stacks and bridges and to a certain degree with hyperbolic cooling towers.[36]

GALLOPING OSCILLATIONS

Vibrations due to turbulence and vortices discussed above are induced by aerodynamic forces which are, to a high degree, independent of the motion and act even on stationary bodies. Quite a different kind of oscillation is induced by the aerodynamic forces generated by the motion itself. Such forces may result from oscillatory changes in pressure distribution brought about by the continuous change in the angle under which the wind strikes the structure ("angle of attack"). This kind of oscillation often has a tendency to diverge; it is called, summarily, *aerodynamic instability, flutter,* or *self-excited oscillation.* Sudden start and violent amplitudes are typical of such phenomena (Fig. 29.16*C*).

The mechanism of this oscillation is, in general, complex. The aerodynamic forces may be a function of the displacements (translation and rotation), vibration velocity, or both, and they may interact with turbulence and vortex shedding. The basic type of the self-excited oscillations is the lateral (across-wind) oscillation induced by aerodynamic forces which are related to vibration velocity alone. Such oscillation is referred to as *galloping.* Typical features of galloping oscillation are motion in the direction perpendicular to that of the wind, sudden onset, large steady amplitudes increasing with wind velocity, and a frequency equal to the natural frequency. Galloping oscillation occurs in transmission lines and in a variety of structures having square, rectangular, or other sharp-edged cross sections.

The origin of galloping oscillation depends on the relation between lift and drag. If a body moves with a velocity $\dot{u}$ in a flow having velocity $\bar{V}$ perpendicular to its direction (Fig. 29.29), the aerodynamic force acting on the body is produced by relative wind velocity $\bar{V}_{\text{rel}}$. The angle of attack of relative wind is

$$\alpha = \arctan \frac{\dot{u}}{\bar{V}} \tag{29.50}$$

The drag and lift components D and L of the aerodynamic force F are

$$D = C_D \frac{1}{2} \rho h l \, \bar{V}^2_{\text{rel}}$$

$$L = C_L \frac{1}{2} \rho h l \, \bar{V}^2_{\text{rel}}$$

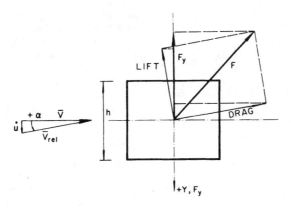

FIGURE 29.29 Cross section in flow.

where C_D and C_L are drag and lift coefficients at angle α (Fig. 29.30), $h =$ depth of the cross section, and $l =$ length of the body.

The component of force F into the direction of axis Y, therefore, is

$$F_y = -(C_D \sin \alpha + C_L \cos \alpha)\,\frac{1}{2}\,\rho h l\,\bar{V}^2 \sec^2 \alpha = C_{Fy}\,\frac{1}{2}\,\rho h l\,\bar{V}^2 \qquad (29.51)$$

where

$$C_{Fy} = -(C_L + C_D \tan \alpha) \sec \alpha \qquad (29.52)$$

The lateral force excites the vibration if the first derivative of C_{Fy} at $\alpha = 0$ is >0, hence

$$A_1 = \frac{dC_{Fy}}{d\alpha}\bigg|_{\alpha=0} = -\left(\frac{dC_L}{d\alpha} + C_D\right) > 0 \qquad (29.53)$$

This condition for aerodynamic instability is known as *Den Hartog's criterion*.[38] Substitution of Eq. (29.50) into Eq. (29.52) indicates that the aerodynamic forces depend on vibration velocity and thus actually represent the aerodynamic damping. This damping is negative if $A_1 > 0$. Because the system also has structural damping ζ, which is positive, the vibration will start only if the total available damping becomes less than 0. This condition yields the onset (minimum) wind velocity for galloping from the equilibrium (or zero displacement) position,

$$\bar{V}_0 = \zeta\,\frac{2\pi f_j h}{n A_1} \qquad (29.54)$$

where $f_j =$ natural frequency, $n = \rho h^2/(4m) =$ mass parameter, and $m =$ mass of the body per unit length. Some values of coefficient A_1 are given in Table 29.3.

Galloping oscillations starting from zero initial displacement can occur only when the cross section has $A_1 > 0$. Cross sections having $A_1 \le 0$ are generally considered stable even though galloping may sometimes arise if triggered by a large initial amplitude.[41]

The response and the onset velocity are often very sensitive to turbulence. Some cross sections such as a flat rectangle or a D section are stable in smooth flow but can become unstable in turbulent flow.[41, 42] With other cross sections, turbulence may stabilize a shape that is unstable in smooth flow (see Table 29.2).

From Eqs. (29.51) and (29.52) the nonlinear, negative aerodynamic damping can be calculated[43] for inclusion in the treatment of the across-wind response due to atmospheric turbulence.

The prediction of oscillations for wind velocities greater than V_0 depends on the shape of the C_{Fy} coefficient and requires the application of nonlinear theory.[39–42] A few typical cases are shown in Fig. 29.31. The cases are typical of a square cross section, a flat rectangular

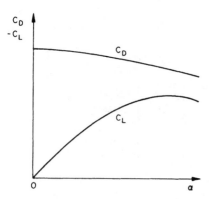

FIGURE 29.30 Lift and drag as function of angle of attack.

TABLE 29.3 Coefficients A_1 for Determination of Galloping Onset Wind Velocity (Infinite Prisms)

	Cross section (Side ratio)					
	Unstable in smooth flow			Stable in smooth flow		
$V \rightarrow$	Square 1 / 1	Rect. 2 / 3	Rect. 1 / 2	Rect. 3 / 2	Rect. 2 / 1	D-section* 2 / 1
Flow						
Smooth	2.7	1.91	2.8	0	−0.03	−0.1
Turbulent ≈10 per cent intensity	2.6	1.83	−2.0	0.74	0.17	0

* Varies with Reynolds number.

section, and a D section whose angle of attack is allowed to change due to drag. Similar response can be expected with other cross sections.

Torsion can also participate in galloping oscillations and play an important part in the vibration. This is the case with angle cross sections[44] and bundled conductors.[45] The quasi-steady theory of pure torsional galloping can be found in Ref. 46. A solution of coupled galloping is presented in Ref. 47.

Galloping often appears in overhead conductors which also vibrate due to vortex shedding. Vortex shedding produces resonant vibration in a high-vibration mode. Galloping usually involves the fundamental mode and is known to occur when the conductor is ice-coated or free of ice. The vibration often leads to fatigue failures, and various techniques are therefore used to reduce the amplitude. This can be achieved by means of resonant dampers[48] consisting of auxiliary masses suspended on short lengths of cable which dissipate energy through the bending or aerodynamic dampers[49] consisting of perforated shrouds. Vibrations of bundled conductors can be eliminated by twisting the bundle[45] and thereby changing the aerodynamic characteristics in the spanwise direction.

VIBRATION OF SPECIAL STRUCTURES

The basic types of vibration discussed above are common in many structures. However, there are some special structures which would require individual treatment. A few examples are cited below.

Guyed towers experience complicated vibration patterns because of the nonlinearity of the guys, the three-dimensional character of the response, the interaction between the guys and the tower, and other factors.[28, 50–52]

Hyperbolic cooling towers can suffer from some of the effects of wake buffeting[36] and are susceptible to turbulence.[53]

Information on the vibration of a number of special structures can be found in Refs. 2 to 5.

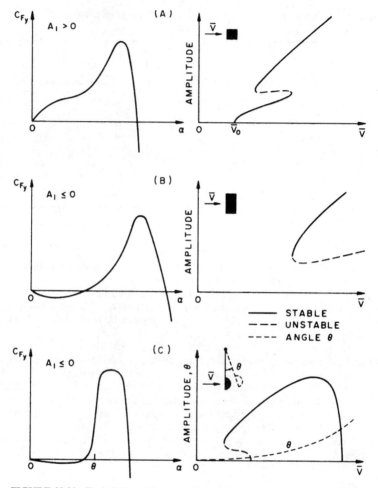

FIGURE 29.31 Typical lateral force coefficients C_{F_y} and corresponding galloping oscillations: (A) vibration from equilibrium position, (B) vibration triggered by initial amplitudes, and (C) vibration with variable angle of attack.

REFERENCES

1. Davenport, A. G., et al.: "New Approaches to Design against Wind Action," Faculty of Engineering Science, University of Western Ontario (unpublished).

2. *Proc. IUTAM-IAHR Symp. Karlsruhe,* 1972.

3. *Proc. Conf. National Physical Laboratory, Teddinton, Middlesex,* 1963.

4. *Proc. Intern. Res. Seminar,* Ottawa, 1967.

5. *Proc. 3d Intern. Conf.,* Tokyo, 1971.

6. Davenport, A. G.: *Inst. Civil Eng. Paper* No. 6480, 449–472 (August 1961).

7. Harris, R. I.: Seminar of Construction Industry Research and Information Association, paper 3, Institution of Civil Engineers, 1970.

8. Novak, M.: *Acta Tech. Czechoslovak Acad. Sci.,* **4**:375–404 (1967).

9. Vickery, B. J.: *J. Struct. Div. Am. Soc. Civil Engrs.,* **98**:21–36 (January, 1972).

10. Davenport, A. G.: *Proc. Inst. Civil Engs.,* **23**:449–472 (1962).

11. Etkin, B.: Meeting on Ground Wind Load Problems in Relation to Launch Vehicles, pp. 21.1–15, Langley Research Center, NASA, June, 1966.

12. Davenport, A. G.: *J. Struct. Div., Am. Soc. Civil Engrs.,* **93**:11–34 (June, 1967).

13. "Canadian Structural Design Manual 1970," Suppl. 4, National Research Council of Canada, 1970.

14. Chen, P. W., and L. E. Robertson: *J. Struct. Div. Am. Soc. Civil Engrs.,* **98**:1681–1695 (August 1972).

15. Hansen, R. T., J. W. Reed, and E. H. Vanmarcke: *J. Struct. Div. Am. Soc. Civil Engrs.,* **99**:1589–1605 (July 1973).

16. Vellozzi, Y., and E. Cohen: *J. Struct. Div. Am. Soc. Civil Engrs.,* **94**:1295–1313 (June 1968).

17. Vickery, B.: U.S. Dept. of Commerce, *Nat. Bur. Std. Bldg. Sci., Ser.* **30**:93–104.

18. Simiu, E.: *J. Struct. Div. Am. Soc. Civil Engrs.,* **100**:1897–1910 (September 1974).

19. Allen, D. E., and W. A. Dalgliesh: Preliminary Publication of IABSE Symposium on Resistance and Ultimate Deformability of Structures, pp. 279–285, Lisbon, 1973.

20. Dalgliesh, W. A.: *Proc. U.S.-Japan Res. Seminar Wind Effects Structures,* Kyoto, Japan, September, 1974.

21. Dalgliesh, W. A.: *J. Struct. Div. Am. Soc. Civil Engrs.,* **97**:2173–2187 (September 1971).

22. Roshko, A.: *J. Fluid Mech.,* **10**:345–356 (1961).

23. Cincotta, J. J., G. W. Jones, and R. W. Walker: Meeting on Ground Wind Load Problems in Relation to Launch Vehicles, pp. 20.1–35, Langley Research Center, NASA, 1966.

24. Novak, M.: Ref. 5, pp. 799–809.

25. Morkovin, M. V.: *Proc. Symp. Fully Separated Flows,* pp. 102–118, ASME, 1964.

26. Nakamura, Y.: *Rept. Res. Inst. Appl. Mech., Kyushu University,* **17**(59):217–234 (1969).

27. Hartlen, R. T., and I. G. Currie: *J. Eng. Mech. Div. Am. Soc. Civil Engrs.,* **70**:577–591 (October, 1970).

28. Novak, M.: *Proc. IASS Symp. Tower-Shaped Steel r.c. Structures,* Bratislava, 1966.

29. Fung, Y. C.: *J. Aerospace Sci.,* **27**(11):801–814 (November, 1960).

30. Surry, D.: *J. Fluid Mech.,* **52**(3):543–563 (1972).

31. Wotton, L. R., and C. Scruton: Construction Industry Research and Information Association Seminar, Paper 5, June, 1970.

32. Vickery, B. J., and A. W. Clark: *J. Struct. Div. Am. Soc. Civil Engrs.,* **98**:1–20 (January 1972).

33. Reed, W. H.: Ref. 4, Paper 36, pp. 283–321.

34. Scruton, C.: National Physical Laboratory Note 1012, April, 1963.

35. Novak, M.: Ref. 4, pp. 429–457.

36. Scruton, C.: Ref. 4, pp. 115–161.

37. Cooper, K. R., and Wardlaw, R. L.: Ref. 5, pp. 647–655.

38. Den Hartog: *Trans., AIEE,* **51**:1074 (1932).

39. Parkinson, G. V., and J. D. Smith: *Quart. J. Mech. Appl. Math.,* **17**(2):225–239 (1964).

40. Novak, M.: *J. Eng. Mech. Div. Am. Soc. Civil Engrs.,* **95**:115–142 (February 1969).

41. Novak, M.: *J. Eng. Mech. Div. Am. Soc. Civil. Engrs.,* **98**:27–46 (February 1972).

42. Novak, M., and Tanaka, H.: *J. Eng. Mech. Div. Am. Soc. Civil Engrs.,* **100**:27–47 (February 1974).

43. Novak, M., and Davenport, A. G.: *J. Eng. Mech. Div. Am. Soc. Civil Engrs.,* **96**:17–39 (February 1970).

44. Wardlaw, R. L.: Ref. 4, pp. 739–772.

45. Wardlaw, R. L., K. R. Cooper, R. G. Ko, and J. A. Watts: *Trans. IEEE,* 1975.

46. Modi, V. J., and J. E. Slater: Ref. 2, pp. 355–372.

47. Blevins, R. D., and W. D. Iwan: *J. Appl. Mech.,* **41**, no. 4, 1974.

48. "Overhead Conductor Vibration," Alcoa Aluminum Overhead Conductor Engineering Data, no. 4, 1974.

49. Hunt, J. C. R., and D. J. W. Richards: *Proc. Inst. Elec. Engrs.,* **116**(11):1869–1874 (November 1969).

50. Davenport, A. G., and G. N. Steels: *J. Struct. Div. Am. Soc. Civil Engrs.,* **91**:43–70 (April 1965).

51. Davenport, A. G.: Engineering Institute of Canada, **3**:119–141 (1959).

52. McCaffrey, R. J., and Hartmann, A. J.: *J. Struct. Div. Am. Soc. Civil Engrs.,* **98**:1309–1323 (June 1972).

53. Hashish, M. G., and Abu-Sitta, S. H., *J. Struct. Div. Am. Soc. Civil Engrs.,* **100**:1037–1051 (May 1974).

CHAPTER 30
THEORY OF VIBRATION ISOLATION

Charles E. Crede
Jerome E. Ruzicka

INTRODUCTION

Vibration isolation concerns means to bring about a reduction in a vibratory effect. A vibration isolator in its most elementary form may be considered as a resilient member connecting the equipment and foundation. The function of an isolator is to reduce the magnitude of motion transmitted from a vibrating foundation to the equipment or to reduce the magnitude of force transmitted from the equipment to its foundation.

CONCEPT OF VIBRATION ISOLATION

The concept of vibration isolation is illustrated by consideration of the single degree-of-freedom system illustrated in Fig. 30.1. This system consists of a rigid body representing an equipment connected to a foundation by an isolator having resilience and energy-dissipating means; it is unidirectional in that the body is constrained to move only in vertical translation. The performance of the isolator may be evaluated by the following characteristics of the response of the equipment-isolator system of Fig. 30.1 to steady-state sinusoidal vibration:

Absolute transmissibility. Transmissibility is a measure of the reduction of transmitted force or motion afforded by an isolator. If the source of vibration is an oscillating motion of the foundation (motion excitation), transmissibility is the ratio of the vibration amplitude of the equipment to the vibration amplitude of the foundation. If the source of vibration is an oscillating force originating within the equipment (force excitation), transmissibility is the ratio of the force amplitude transmitted to the foundation to the amplitude of the exciting force.

Relative transmissibility. Relative transmissibility is the ratio of the relative deflection amplitude of the isolator to the displacement amplitude imposed at the foundation. A vibration isolator effects a reduction in vibration by permitting

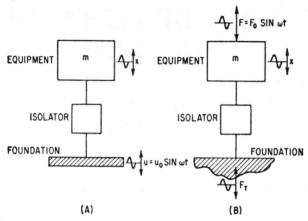

FIGURE 30.1 Schematic diagrams of vibration isolation systems: (*A*) vibration isolation where motion *u* is imposed at the foundation and motion *x* is transmitted to the equipment; (*B*) vibration isolation where force *F* is applied by the equipment and force F_T is transmitted to the foundation.

deflection of the isolator. The relative deflection is a measure of the clearance required in the isolator. This characteristic is significant only in an isolator used to reduce the vibration transmitted from a vibrating foundation.

Motion response.   Motion response is the ratio of the displacement amplitude of the equipment to the quotient obtained by dividing the excitation force amplitude by the static stiffness of the isolator. If the equipment is acted on by an exciting force, the resultant motion of the equipment determines the space requirements for the isolator, i.e., the isolator must have a clearance at least as great as the equipment motion.

FORM OF ISOLATOR

The essential features of an isolator are resilient load-supporting means and energy-dissipating means. In certain types of isolators, the functions of the load-supporting means and the energy-dissipating means may be performed by a single element, e.g., natural or synthetic rubber. In other types of isolators, the resilient load-carrying means may lack sufficient energy-dissipating characteristics, e.g., metal springs; then separate and distinct energy-dissipating means (dampers) are provided. For purposes of analysis, it is assumed that the springs and dampers are separate elements. In general, the springs are assumed to be linear and massless. The effects of nonlinearity and mass of the load-supporting means upon vibration isolation are considered in later sections of this chapter.

Various types of dampers are shown in combination with ideal springs in the following idealized models of isolators illustrated in Table 30.1. Practical aspects of isolator design are considered in Chap. 32.

Rigidly connected viscous damper.   A viscous damper *c* is connected rigidly between the equipment and its foundation as shown in Table 30.1*A*. The damper has the characteristic property of transmitting a force F_c that is directly proportional to the relative velocity δ across the damper, $F_c = c\delta$. This damper sometimes is referred to as a *linear damper.*

TABLE 30.1 Types of Idealized Vibration Isolators

(A) RIGIDLY CONNECTED VISCOUS DAMPER	(B) RIGIDLY CONNECTED COULOMB DAMPER	(C) ELASTICALLY CONNECTED VISCOUS DAMPER	(D) ELASTICALLY CONNECTED COULOMB DAMPER
EXCITATION			
$u = u_0 \sin\omega t$ $F = F_0 \sin\omega t$	$u = u_0 \sin\omega t$ * OR $\ddot{u} = \ddot{u}_0 \sin\omega t$ * $F = F_0 \sin\omega t$	$u = u_0 \sin\omega t$ $F = F_0 \sin\omega t$	$u = u_0 \sin\omega t$ * OR $\ddot{u} = \ddot{u}_0 \sin\omega t$ * $F = F_0 \sin\omega t$
RESPONSE			
$x = x_0 \sin(\omega t + \theta^+)$ OR $\delta = \delta_0 \sin(\omega t + \theta^+)$ WHERE $\delta = x - u$ $F_T = (F_T)_0 \sin(\omega t + \theta^+)$	$x = x_0 \sin(\omega t + \theta^+)$ OR $\delta = \delta_0 \sin(\omega t + \theta^+)$ WHERE $\delta = x - u$ $F_T = (F_T)_0 \sin(\omega t + \theta^+)$	$x = x_0 \sin(\omega t + \theta^+)$ OR $\delta = \delta_0 \sin(\omega t + \theta^+)$ WHERE $\delta = x - u$ $F_T = (F_T)_0 \sin(\omega t + \theta^+)$	$x = x_0 \sin(\omega t + \theta^+)$ OR $\delta = \delta_0 \sin(\omega t + \theta^+)$ WHERE $\delta = x - u$ $F_T = (F_T)_0 \sin(\omega t + \theta^+)$
FREQUENCY PARAMETERS			
$\omega_0 = \sqrt{k/m}$ $(c = 0)$	$\omega_0 = \sqrt{k/m}$ $(F_f = 0)$	$\omega_0 = \sqrt{k/m}$ $(c = 0)$ $\omega_\infty = \sqrt{(N+1)\dfrac{k}{m}}$ $(c = \infty)$	$\omega_0 = \sqrt{k/m}$ $(F_f = 0)$ $\omega_\infty = \sqrt{(N+1)\dfrac{k}{m}}$ $(F_f = \infty)$
DAMPING PARAMETERS			
$c_c = 2\sqrt{km}$ $\zeta = c/c_c$	$\eta = \dfrac{F_f}{ku_0}$ $\xi = \dfrac{F_f}{m\ddot{u}_0}$ $\xi_F = \dfrac{F_f}{F_0}$	$c_c = 2\sqrt{km}$ $\zeta = c/c_c$	$\eta = \dfrac{F_f}{ku_0}$ $\xi = \dfrac{F_f}{m\ddot{u}_0}$ $\xi_F = \dfrac{F_f}{F_0}$

* PHYSICALLY, THESE FORMS OF EXCITATION ARE IDENTICAL. THEY ARE EXPRESSED IN TWO DIFFERENT MATHEMATICAL FORMS FOR CONVENIENCE IN DEFINING THE DAMPING PARAMETER FOR COULOMB DAMPING.

† IN VIBRATION ISOLATION, ONLY THE MAGNITUDE OF THE RESPONSE IS OF INTEREST; THUS, THE PHASE ANGLE USUALLY IS NEGLECTED.

TABLE 30.2 Transmissibility and Motion Response for Isolation Systems Defined in Table 30.1

Where the equation is shown graphically, the applicable figure is indicated below the equation. See Table 30.1 for definition of terms.

Type of damper	Absolute transmissibility
Rigidly connected viscous damper.......	(a) $$T_A = \frac{x_0}{u_0} = \frac{F_T}{F_0} = \sqrt{\frac{1 + \left(2\zeta\frac{\omega}{\omega_0}\right)^2}{\left(1 - \frac{\omega^2}{\omega_0^2}\right)^2 + \left(2\zeta\frac{\omega}{\omega_0}\right)^2}}$$ Fig. 30.2
Rigidly connected Coulomb damper (see Note 1).....................	(d) $$(T_A)_D = \frac{x_0}{u_0} = \sqrt{\frac{1 + \left(\frac{4}{\pi}\eta\right)^2 (1 - 2\omega_0^2/\omega^2)}{\left(1 - \frac{\omega^2}{\omega_0^2}\right)^2}}$$ Fig. 30.5 (see Note 2)
Elastically connected viscous damper.....	(g) $$T_A = \frac{x_0}{u_0} = \frac{F_T}{F_0} = \sqrt{\frac{1 + 4\left(\frac{N+1}{N}\right)^2 \zeta^2 \frac{\omega^2}{\omega_0^2}}{\left(1 - \frac{\omega^2}{\omega_0^2}\right)^2 + \frac{4}{N^2}\zeta^2\frac{\omega^2}{\omega_0^2}\left(N + 1 - \frac{\omega^2}{\omega_0^2}\right)^2}}$$ Fig. 30.10 (see Note 3)
Elastically connected Coulomb damper (see Note 1).....................	(j) $$(T_A)_D = \frac{x_0}{u_0} = \sqrt{\frac{1 + \left(\frac{4}{\pi}\eta\right)^2\left[\left(\frac{N+2}{N}\right) - 2\left(\frac{N+1}{N}\right)\left(\frac{\omega_0}{\omega}\right)^2\right]}{\left(1 - \frac{\omega^2}{\omega_0^2}\right)^2}}$$ Fig. 30.13 (see Notes 4 and 5)

NOTE 1: These equations apply only when there is relative motion across the damper.
NOTE 2: This equation applies only when excitation is defined in terms of displacement amplitude.
NOTE 3: These curves apply only for optimum damping [see Eq. (30.15)]; curves for other values of damping are given in Ref. 4.

Rigidly connected Coulomb damper. An isolation system with a rigidly connected Coulomb damper is indicated schematically in Table 30.1B. The force F_f exerted by the damper on the mass of the system is constant, independent of position or velocity, but always in a direction that opposes the relative velocity across the damper. In a physical sense, Coulomb damping is approximately attainable from the relative motion of two members arranged to slide one upon the other with a constant force holding them together.

Elastically connected viscous damper. The elastically connected viscous damper is shown in Table 30.1C. The viscous damper c is in series with a spring of stiffness k_1; the load-carrying spring k is related to the damper spring k_1 by the parameter $N = k_1/k$. This type of damper system sometimes is referred to as a *viscous relaxation system.*

Elastically connected Coulomb damper. The elastically connected Coulomb damper is shown in Table 30.1D. The friction element can transmit only that force which is developed in the damper spring k_1. When the damper slips, the friction force F_f is independent of the velocity across the damper, but always is in a direction that opposes it.

TABLE 30.2 Transmissibility and Motion Response for Isolation Systems Defined in Table 30.1 (*Continued*)

Where the equation is shown graphically, the applicable figure is indicated below the equation. See Table 30.1 for definition of terms.

Relative transmissibility	Motion response
(b) $$T_R = \frac{\delta_0}{u_0} = \sqrt{\frac{\left(\frac{\omega}{\omega_0}\right)^4}{\left(1-\frac{\omega^2}{\omega_0^2}\right)^2 + \left(2\zeta\frac{\omega}{\omega_0}\right)^2}}$$ Fig. 30.3	(c) $$\frac{x_0}{F_0/k} = \sqrt{\frac{1}{\left(1-\frac{\omega^2}{\omega_0^2}\right)^2 + \left(2\zeta\frac{\omega}{\omega_0}\right)^2}}$$ Fig. 30.14
(e) $$(T_R)_D = \frac{\delta_0}{u_0} = \sqrt{\frac{\left(\frac{\omega}{\omega_0}\right)^4 - \left(\frac{4}{\pi}\eta\right)^2}{\left(1-\frac{\omega^2}{\omega_0^2}\right)^2}}$$ Fig. 30.6 (see Note 2)	(f) $$\frac{x_0}{F_0/k} = \sqrt{\frac{1-\left(\frac{4}{\pi}\xi\right)^2}{\left(1-\frac{\omega^2}{\omega_0^2}\right)^2}}$$ (See Note 2)
(h) $$T_R = \frac{\delta_0}{u_0} = \sqrt{\frac{\frac{\omega^2}{\omega_0^2}+\frac{4}{N^2}\zeta^2\frac{\omega^6}{\omega_0^6}}{\left(1-\frac{\omega^2}{\omega_0^2}\right)^2 + \frac{4}{N^2}\zeta^2\frac{\omega^2}{\omega_0^2}\left(N+1-\frac{\omega^2}{\omega_0^2}\right)^2}}$$ Fig. 30.11 (see Note 3)	(i) $$\frac{x_0}{F_0/k} = \sqrt{\frac{1+\frac{4}{N^2}\zeta^2\frac{\omega^2}{\omega_0^2}}{\left(1-\frac{\omega^2}{\omega_0^2}\right)^2 + \frac{4}{N^2}\zeta^2\frac{\omega^2}{\omega_0^2}\left(N+1-\frac{\omega^2}{\omega_0^2}\right)^2}}$$ Fig. 30.9 (see Note 4)
(k) $$(T_R)_D = \frac{\delta_0}{u_0} = \sqrt{\frac{\left(\frac{\omega}{\omega_0}\right)^4 + \left(\frac{4}{\pi}\eta\right)^2\left[\frac{2}{N}\frac{\omega^2}{\omega_0^2}-\left(\frac{N+2}{N}\right)\right]}{\left(1-\frac{\omega^2}{\omega_0^2}\right)^2}}$$ Fig. 30.14 (see Notes 4 and 5)	

NOTE 4: These curves apply only for $N = 3$.

NOTE 5: This equation applies only when excitation is defined in terms of displacement amplitude; for excitation defined in terms of force or acceleration, see Eq. (30.18).

INFLUENCE OF DAMPING IN VIBRATION ISOLATION

The nature and degree of vibration isolation afforded by an isolator is influenced markedly by the characteristics of the damper. This aspect of vibration isolation is evaluated in this section in terms of the single degree-of-freedom concept; i.e., the equipment and the foundation are assumed rigid and the isolator is assumed massless. The performance is defined in terms of absolute transmissibility, relative transmissibility, and motion response for isolators with each of the four types of dampers illustrated in Table 30.1. A system with a rigidly connected viscous damper is discussed in detail in Chap. 2, and important results are reproduced here for completeness; isolators with other types of dampers are discussed in detail here.

The characteristics of the dampers and the performance of the isolators are defined in terms of the parameters shown on the schematic diagrams in Table 30.1. Absolute transmissibility, relative transmissibility, and motion response are defined analytically in Table 30.2 and graphically in the figures referenced in Table 30.2. For the rigidly connected viscous and Coulomb-damped isolators, the graphs generally are explicit and complete. For isolators with elastically connected dampers, typical results are included and references are given to more complete compilations of dynamic characteristics.

RIGIDLY CONNECTED VISCOUS DAMPER

Absolute and relative transmissibility curves are shown graphically in Figs. 30.2 and 30.3, respectively.* As the damping increases, the transmissibility at resonance decreases and the absolute transmissibility at the higher values of the forcing frequency ω increases; i.e., reduction of vibration is not as great. For an undamped isolator, the absolute transmissibility at higher values of the forcing frequency varies inversely as the square of the forcing frequency. When the isolator embodies significant viscous damping, the absolute transmissibility curve becomes asymptotic at high values of forcing frequency to a line whose slope is inversely proportional to the first power of the forcing frequency.

The maximum value of absolute transmissibility associated with the resonant condition is a function solely of the damping in the system, taken with reference to critical damping. For a lightly damped system, i.e., for $\zeta < 0.1$, the maximum absolute transmissibility [see Eq. (2.41)] of the system is[1]

$$T_{max} = \frac{1}{2\zeta} \tag{30.1}$$

where $\zeta = c/c_c$ is the fraction of critical damping defined in Table 30.1.

The motion response is shown graphically in Fig. 30.4. A high degree of damping limits the vibration amplitude of the equipment at all frequencies, compared to an undamped system. The single degree-of-freedom system with viscous damping is discussed more fully in Chap. 2.

RIGIDLY CONNECTED COULOMB DAMPER

The differential equation of motion for the system with Coulomb damping shown in Table 30.1B is

$$m\ddot{x} + k(x - u) \pm F_f = F_0 \sin \omega t \tag{30.2}$$

The discontinuity in the damping force that occurs as the sign of the velocity changes at each half cycle requires a step-by-step solution of Eq. (30.2).[2] An approximate solution based on the equivalence of energy dissipation involves equating the energy dissipation per cycle for viscous-damped and Coulomb-damped systems:[3]

$$\pi c \omega \delta_0^2 = 4F_f \delta_0 \tag{30.3}$$

where the left side refers to the viscous-damped system and the right side to the Coulomb-damped system; δ_0 is the amplitude of relative displacement across the damper. Solving Eq. (30.3) for c,

$$c_{eq} = \frac{4F_f}{\pi \omega \delta_0} = j\left(\frac{4F_f}{\pi \dot{\delta}_0}\right) \tag{30.4}$$

where c_{eq} is the *equivalent viscous damping coefficient* for a Coulomb-damped system having equivalent energy dissipation. Since $\dot{\delta}_0 = j\omega\delta_0$ is the relative velocity, the

---

* For linear systems, the absolute transmissibility $T_A = x_0/u_0$ in the motion-excited system equals F_T/F_0 in the force-excited system. The relative transmissibility $T_R = \delta_0/u_0$ applies only to the motion-excited system.

equivalent linearized dry friction damping force can be considered sinusoidal with an amplitude $j(4F_f/\pi)$. Since $c_c = 2k/\omega_0$ [see Eq. (2.12)],

$$\zeta_{eq} = \frac{c_{eq}}{c_c} = \frac{2\omega_0 F_f}{\pi \omega k \delta_0} \tag{30.5}$$

where ζ_{eq} may be defined as the *equivalent fraction of critical damping*. Substituting δ_0 from the relative transmissibility expression [(b) in Table 30.2] in Eq. (30.5) and solving for ζ_{eq}^2,

$$\zeta_{eq}^2 = \frac{\left(\dfrac{2}{\pi}\eta\right)^2 \left(1 - \dfrac{\omega^2}{\omega_0^2}\right)^2}{\dfrac{\omega^2}{\omega_0^2}\left[\dfrac{\omega^4}{\omega_0^4} - \left(\dfrac{4}{\pi}\eta\right)^2\right]} \tag{30.6}$$

where η is the Coulomb damping parameter for displacement excitation defined in Table 30.1.

The equivalent fraction of critical damping given by Eq. (30.6) is a function of the displacement amplitude u_0 of the excitation since the Coulomb damping parameter η depends on u_0. When the excitation is defined in terms of the acceleration amplitude $\ddot{u}_0$, the fraction of critical damping must be defined in corresponding terms. Thus, it is convenient to employ separate analyses for displacement transmissibility and acceleration transmissibility for an isolator with Coulomb damping.

Displacement Transmissibility. The absolute displacement transmissibility of an isolation system having a rigidly connected Coulomb damper is obtained by substituting ζ_{eq} from Eq. (30.6) for ζ in the absolute transmissibility expression for viscous damping, (a) in Table 30.2. The absolute displacement transmissibility is shown graphically in Fig. 30.5, and the relative displacement transmissibility is shown in Fig. 30.6. The absolute displacement transmissibility has a value of unity when the forcing frequency is low and/or the Coulomb friction force is high. For these conditions, the friction damper is locked in, i.e., it functions as a rigid connection, and there is no relative motion across the isolator. The frequency at which the damper breaks loose, i.e., permits relative motion across the isolator, can be obtained from the relative displacement transmissibility expression, (e) in Table 30.2. The relative displacement is imaginary when $\omega^2/\omega_0^2 \leq (4/\pi)\eta$. Thus, the "break-loose" frequency ratio is*

$$\left(\frac{\omega}{\omega_0}\right)_L = \sqrt{\frac{4}{\pi}\eta} \tag{30.7}$$

The displacement transmissibility can become infinite at resonance, even though the system is damped, if the Coulomb damping force is less than a critical minimum value. The denominator of the absolute and relative transmissibility expressions becomes zero for a frequency ratio ω/ω_0 of unity. If the break-loose frequency is lower than the undamped natural frequency, the amplification of vibration becomes

---

* This equation is based upon energy considerations and is approximate. Actually, the friction damper breaks loose when the inertia force of the mass equals the friction force, $m u_0 \omega^2 = F_f$. This gives the exact solution $(\omega/\omega_0)_L = \sqrt{\eta}$. A numerical factor of $4/\pi$ relates the Coulomb damping parameters in the exact and approximate solutions for the system.

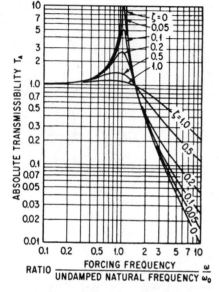

FIGURE 30.2 Absolute transmissibility for the rigidly connected, viscous-damped isolation system shown at A in Table 30.1 as a function of the frequency ratio ω/ω_0 and the fraction of critical damping ζ. The absolute transmissibility is the ratio (x_0/u_0) for foundation motion excitation (Fig. 30.1A) and the ratio (F_T/F_0) for equipment force excitation (Fig. 30.1B).

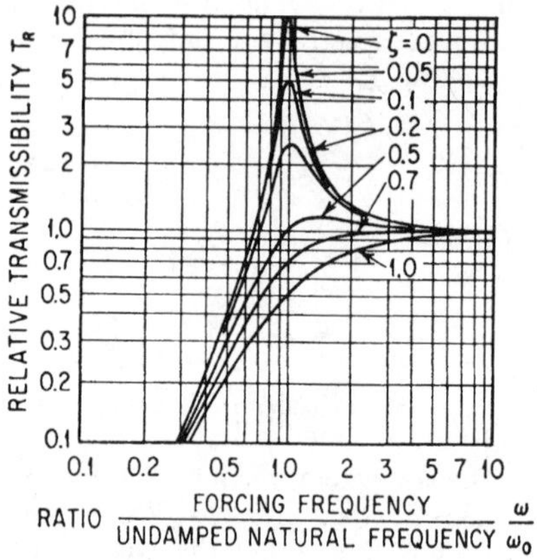

FIGURE 30.3 Relative transmissibility for the rigidly connected, viscous-damped isolation system shown at A in Table 30.1 as a function of the frequency ratio ω/ω_0 and the fraction of critical damping ζ. The relative transmissibility describes the motion between the equipment and the foundation (i.e., the deflection of the isolator).

infinite at resonance. This occurs because the energy dissipated by the friction damping force increases linearly with the displacement amplitude, and the energy introduced into the system by the excitation source also increases linearly with the displacement amplitude. Thus, the energy dissipated at resonance is either greater or less than the input energy for *all* amplitudes of vibration. The minimum dry-friction force which prevents vibration of infinite magnitude at resonance is

$$(F_f)_{\min} = \frac{\pi k u_0}{4} = 0.79 \, k u_0 \tag{30.8}$$

where k and u_0 are defined in Table 30.1.

As shown in Fig. 30.5, an increase in η decreases the absolute displacement transmissibility at resonance and increases the resonance frequency. All curves intersect at the point $(T_A)_D = 1$, $\omega/\omega_0 = \sqrt{2}$. With optimum damping force, there is no motion across the damper for $\omega/\omega_0 \leq \sqrt{2}$; for higher frequencies the displacement transmissibility is less than unity. The friction force that produces this "resonance-free" condition is

$$(F_f)_{\mathrm{op}} = \frac{\pi k u_0}{2} = 1.57 \, k u_0 \tag{30.9}$$

For high forcing frequencies, the absolute displacement transmissibility varies inversely as the square of the forcing frequency, even though the friction damper dis-

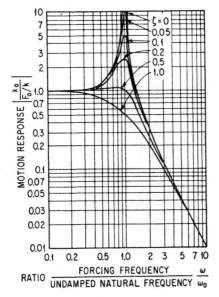

RATIO $\dfrac{\text{FORCING FREQUENCY}}{\text{UNDAMPED NATURAL FREQUENCY}} \dfrac{\omega}{\omega_0}$

FIGURE 30.4 Motion response for the rigidly connected viscous-damped isolation system shown at A in Table 30.1 as a function of the frequency ratio ω/ω_0 and the fraction of critical damping ζ. The curves give the resulting motion of the equipment x in terms of the excitation force F and the static stiffness of the isolator k.

sipates energy. For relatively high damping ($\eta > 2$), the absolute displacement transmissibility, for frequencies greater than the break-loose frequency, is approximately $4\eta\omega_0^2/\pi\omega^2$.

Acceleration Transmissibility. The absolute displacement transmissibility $(T_A)_D$ shown in Fig. 30.5 is the ratio of response of the isolator to the excitation, where each is expressed as a displacement amplitude in simple harmonic motion. The damping parameter η is defined with reference to the displacement amplitude u_0 of the excitation. Inasmuch as all motion is simple harmonic, the transmissibility $(T_A)_D$ also applies to acceleration transmissibility when the damping parameter is defined properly. When the excitation is defined in terms of the acceleration amplitude $\ddot{u}_0$ of the excitation,

$$\eta_{\ddot{u}_0} = \frac{F_f\omega^2}{k\ddot{u}_0} \tag{30.10}$$

where ω = forcing frequency, rad/sec
$\ddot{u}_0$ = acceleration amplitude of excitation, in./sec^2
k = isolator stiffness, lb/in.
F_f = Coulomb friction force, lb

For relatively high forcing frequencies, the acceleration transmissibility approaches a constant value $(4/\pi)\xi$, where ξ is the Coulomb damping parameter for acceleration excitation defined in Table 30.1. The acceleration transmissibility of a rigidly connected Coulomb damper system becomes asymptotic to a constant value because the Coulomb damper transmits the same friction force regardless of the amplitude of the vibration.

ELASTICALLY CONNECTED VISCOUS DAMPER

The general characteristics of the elastically connected viscous damper shown at C in Table 30.1 may best be understood by successively assigning values to the viscous damper coefficient c while keeping the stiffness ratio N constant. For zero damping, the mass is supported by the isolator of stiffness k. The transmissibility curve has the characteristics typical of a transmissibility curve for an undamped system having the natural frequency

$$\omega_0 = \sqrt{\frac{k}{m}} \tag{30.11}$$

When c is infinitely great, the transmissibility curve is that of an undamped system having the natural frequency

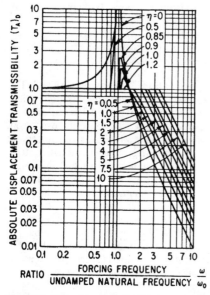

FIGURE 30.5 Absolute displacement transmissibility for the rigidly connected, Coulomb-damped isolation system shown at B in Table 30.1 as a function of the frequency ratio ω/ω_0 and the displacement Coulomb-damping parameter η.

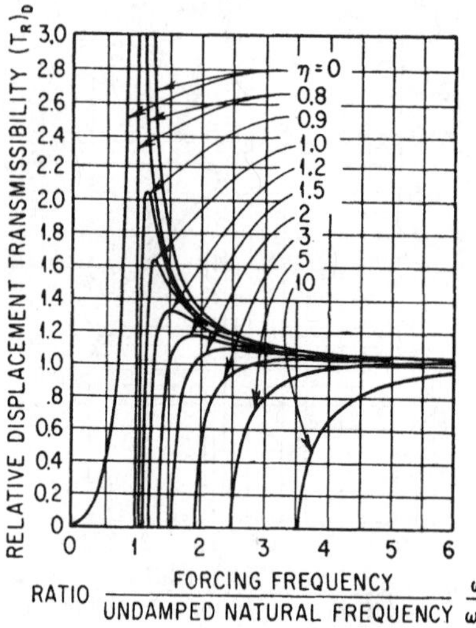

FIGURE 30.6 Relative displacement transmissibility for the rigidly connected, Coulomb-damped isolation system shown at B in Table 30.1 as a function of the frequency ratio ω/ω_0 and the displacement Coulomb-damping parameter η.

$$\omega_\infty = \sqrt{\frac{k + k_1}{m}} = \sqrt{N + 1}\,\omega_0 \qquad (30.12)$$

where $k_1 = Nk$. For intermediate values of damping, the transmissibility falls within the limits established for zero and infinitely great damping. The value of damping which produces the minimum transmissibility at resonance is called *optimum damping*.

All curves approach the transmissibility curve for infinite damping as the forcing frequency increases. Thus, the absolute transmissibility at high forcing frequencies is inversely proportional to the square of the forcing frequency. General expressions for absolute and relative transmissibility are given in Table 30.2.

A comparison of absolute transmissibility curves for the elastically connected viscous damper and the rigidly connected viscous damper is shown in Fig. 30.7. A constant viscous damping coefficient of $0.2c_c$ is maintained, while the value of the stiffness ratio N is varied from zero to infinity. The transmissibilities at resonance are comparable, even for relatively small values of N, but a substantial gain is achieved in the isolation characteristics at high forcing frequencies by elastically connecting the damper.

Transmissibility at Resonance. The maximum transmissibility (at resonance) is a function of the damping ratio ζ and the stiffness ratio N, as shown in Fig. 30.8.[4] The maximum transmissibility is nearly independent of N for small values of ζ. However, for $\zeta > 0.1$, the coefficient N is significant in determining the maximum transmissibility. The lowest value of the maximum absolute transmissibility curves corresponds to the conditions of optimum damping.

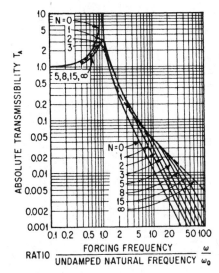

ABSOLUTE TRANSMISSIBILITY T_A

RATIO $\dfrac{\text{FORCING FREQUENCY}}{\text{UNDAMPED NATURAL FREQUENCY}} \dfrac{\omega}{\omega_0}$

FIGURE 30.7 Comparison of absolute transmissibility for rigidly and elastically connected, viscous damped isolation systems shown at A and C, respectively, in Table 30.1, as a function of the frequency ratio ω/ω_0. The solid curves refer to the elastically connected damper, and the parameter N is the ratio of the damper spring stiffness to the stiffness of the principal support spring. The fraction of critical damping $\zeta = c/c_c$ is 0.2 in both systems. The transmissibility at high frequencies decreases at a rate of 6 dB per octave for the rigidly connected damper and 12 dB per octave for the elastically connected damper.

Motion Response. A typical motion response curve is shown in Fig. 30.9 for the stiffness ratio $N = 3$. For small damping, the response is similar to the response of an isolation system with rigidly connected viscous damper. For intermediate values of damping, the curves tend to be flat over a wide frequency range before rapidly decreasing in value at the higher frequencies. For large damping, the resonance occurs near the natural frequency of the system with infinitely great damping. All response curves approach a high-frequency asymptote for which the attenuation varies inversely as the square of the excitation frequency.

Optimum Transmissibility. For a system with optimum damping, maximum transmissibility coincides with the intersections of the transmissibility curves for zero and infinite damping. The frequency ratios $(\omega/\omega_0)_{op}$ at which this occurs are different for absolute and relative transmissibility:

Absolute transmissibility:

$$\left(\frac{\omega}{\omega_0}\right)^{(A)}_{op} = \sqrt{\frac{2(N+1)}{N+2}} \qquad (30.13)$$

Relative transmissibility:

$$\left(\frac{\omega}{\omega_0}\right)^{(R)}_{op} = \sqrt{\frac{N+2}{2}}$$

The optimum transmissibility at resonance, for both absolute and relative motion, is

$$T_{op} = 1 + \frac{2}{N} \qquad (30.14)$$

The optimum transmissibility as determined from Eq. (30.14) corresponds to the minimum points of the curves of Fig. 30.8.

The damping which produces the optimum transmissibility is obtained by differentiating the general expressions for transmissibility [(g) and (h) in Table 30.2] with respect to the frequency ratio, setting the result equal to zero, and combining it with Eq. (30.13),

Absolute transmissibility:

$$(\zeta_{op})_A = \frac{N}{4(N+1)} \sqrt{2(N+2)} \qquad (30.15a)$$

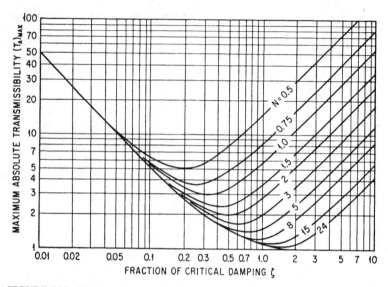

FIGURE 30.8 Maximum absolute transmissibility for the elastically connected, viscous-damped isolation system shown at C in Table 30.1 as a function of the fraction of critical damping ζ and the stiffness of the connecting spring. The parameter N is the ratio of the damper spring stiffness to the stiffness of the principal support spring.

Relative transmissibility:

$$(\zeta_{op})_R = \frac{N}{\sqrt{2(N+1)(N+2)}} \tag{30.15b}$$

Values of optimum damping determined from the first of these relations correspond to the minimum points of the curves of Fig. 30.8. By substituting the optimum damping ratios from Eqs. (30.15) into the general expressions for transmissibility given in Table 30.2, the optimum absolute and relative transmissibility equations are obtained, as shown graphically by Figs. 30.10 and 30.11, respectively.[5] For low values of the stiffness ratio N, the transmissibility at resonance is large but excellent isolation is obtained at high frequencies. Conversely, for high values of N, the transmissibility at resonance is lowered, but the isolation efficiency also is decreased.

ELASTICALLY CONNECTED COULOMB DAMPER

Force-deflection curves for the isolators incorporating elastically connected Coulomb dampers, as shown at D in Table 30.1, are illustrated in Fig. 30.12. Upon application of the load, the isolator deflects; but since insufficient force has been developed in the spring k_1, the damper does not slide, and the motion of the mass is opposed by a spring of stiffness $(N+1)k$. The load is now increased until a force is developed in spring k_1 which equals the constant friction force F_f; then the damper begins to slide. When the load is increased further, the damper slides and reduces the effective spring stiffness to k. If the applied load is reduced after reaching its maxi-

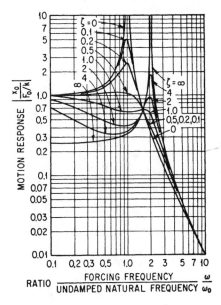

FIGURE 30.9 Motion response for the elastically connected, viscous-damped isolation system shown at C in Table 30.1 as a function of the frequency ratio ω/ω_0 and the fraction of critical damping ζ. For this example, the stiffness of the damper connecting spring is 3 times as great as the stiffness of the principal support spring ($N = 3$). The curves give the resulting motion of the equipment in terms of the excitation force F and the static stiffness of the isolator k.

mum value, the damper no longer displaces because the force developed in the spring k_1 is diminished. Upon completion of the load cycle, the damper will have been in motion for part of the cycle and at rest for the remaining part to form the hysteresis loops shown in Fig. 30.12.

Because of the complexity of the applicable equations, the equivalent energy method is used to obtain the transmissibility and motion response functions. Applying frequency, damping, and transmissibility expressions for the elastically connected viscous damped system to the elastically connected Coulomb-damped system, the transmissibility expressions tabulated in Table 30.2 for the latter are obtained.[6]

If the coefficient of the damping term in each of the transmissibility expressions vanishes, the transmissibility is independent of damping. By solving for the frequency ratio ω/ω_0 in the coefficients that are thus set equal to zero, the frequency ratios obtained define the frequencies of optimum transmissibility. These frequency ratios are given by Eqs. (30.13) for the elastically connected viscous damped system and apply equally well to the elastically connected Coulomb damped system because the method of equivalent viscous damping is employed in the analysis. Similarly, Eq. (30.14) applies for optimum transmissibility at resonance.

The general characteristics of the system with an elastically connected Coulomb damper may be demonstrated by successively assigning values to the damping force while keeping the stiffness ratio N constant. For zero and infinite damping, the transmissibility curves are those for undamped systems and bound all solutions. Every transmissibility curve for $0 < F_f < \infty$ passes through the intersection of the two bounding transmissibility curves. For low damping (less than optimum), the damper "breaks loose" at a relatively low frequency, thereby allowing the transmissibility to increase to a maximum value and then pass through the intersection point of the bounding transmissibility curves. For optimum damping, the maximum absolute transmissibility has a value given by Eq. (30.14); it occurs at the frequency ratio $(\omega/\omega_0)_{op}^{(A)}$ defined by Eq. (30.13). For high damping, the damper remains "locked-in" over a wide frequency range because insufficient force is developed in the spring k_1 to induce slip in the damper. For frequencies greater than the break-loose frequency, there is sufficient force in spring k_1 to cause relative motion of the damper. For a further increase in frequency, the damper remains broken loose and the transmissibility is limited to a finite value. When there is insufficient force in spring k_1 to maintain motion across the damper, the damper locks-in and the transmissibility is that of a system with the infinite damping.

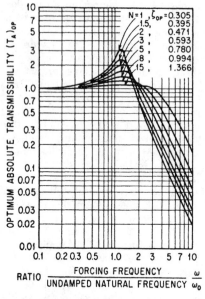

FIGURE 30.10 Absolute transmissibility with optimum damping in elastically connected, viscous-damped isolation system shown at C in Table 30.1 as a function of the frequency ratio ω/ω_0 and the fraction of critical damping ζ. These curves apply to elastically connected, viscous-damped systems having optimum damping for absolute motion. The transmissibility $(T_A)_{op}$ is $(x_0/u_0)_{op}$ for the motion-excited system and $(F_T/F_0)_{op}$ for the force-excited system.

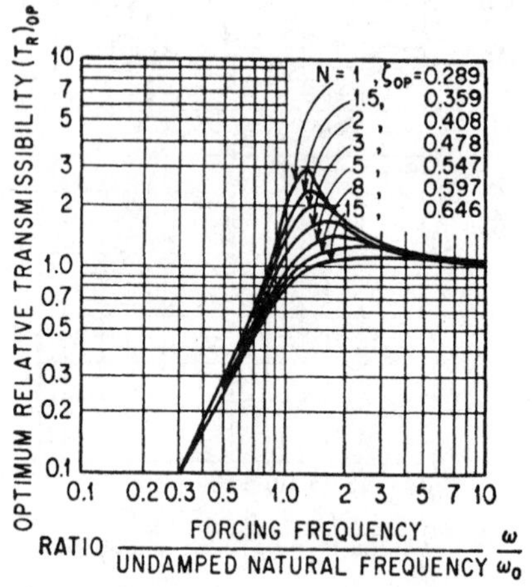

FIGURE 30.11 Relative transmissibility with optimum damping in the elastically connected, viscous-damped isolation system shown at C in Table 30.1 as a function of the frequency ratio ω/ω_0 and the fraction of critical damping ζ. These curves apply to elastically connected, viscous-damped systems having optimum damping for relative motion. The relative transmissibility $(T_R)_{op}$ is $(\delta_0/u_0)_{op}$ for the motion-excited system.

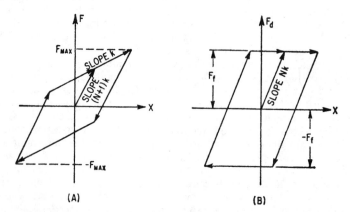

FIGURE 30.12 Force-deflection characteristics of the elastically connected, Coulomb-damped isolation system shown at D in Table 30.1. The force-deflection diagram for a cyclic deflection of the complete isolator is shown at A and the corresponding diagram for the assembly of Coulomb damper and spring $k_1 = Nk$ is shown at B.

The break-loose and lock-in frequencies are determined by requiring the motion across the Coulomb damper to be zero. Then the break-loose and lock-in frequency ratios are

$$\left(\frac{\omega}{\omega_0}\right)_L = \sqrt{\frac{\left(\frac{4}{\pi}\eta\right)(N+1)}{\left(\frac{4}{\pi}\eta\right)\pm N}} \tag{30.16}$$

where η is the damping parameter defined in Table 30.1 with reference to the displacement amplitude u_0. The plus sign corresponds to the break-loose frequency, while the minus sign corresponds to the lock-in frequency. Damping parameters for which the denominator of Eq. (30.16) becomes negative correspond to those conditions for which the damper never becomes locked-in again after it has broken loose. Thus, the damper eventually becomes locked-in only if $\eta > (\pi/4)N$.

Displacement Transmissibility. The absolute displacement transmissibility curve for the stiffness ratio $N = 3$ is shown in Fig. 30.13 where $(T_A)_D = x_0/u_0$. A small decrease in damping force F_f below the optimum value causes a large increase in the transmitted vibration near resonance. However, a small increase in damping force F_f above optimum causes only small changes in the maximum transmissibility. Thus, it is good design practice to have the damping parameter η equal to or greater than the optimum damping parameter η_{op}.

The relative transmissibility for $N = 3$ is shown in Fig. 30.14 where $(T_R)_D = \delta_0/u_0$. All curves pass through the intersection of the curves for zero and infinite damping. For optimum damping, the maximum relative transmissibility has a value given by Eq. (30.14); it occurs at the frequency ratio $\left(\dfrac{\omega}{\omega_0}\right)_{op}^{(R)}$ defined by Eq. (30.13).

Acceleration Transmissibility. The acceleration transmissibility can be obtained from the expression for displacement transmissibility by substitution of the effective displacement damping parameter in the expression for transmissibility of a system whose excitation is constant acceleration amplitude. If $\ddot{u}_0$ represents the acceleration amplitude of the excitation, the corresponding displacement amplitude is $u_0 = -\ddot{u}_0/\omega^2$. Using the definition of the acceleration Coulomb damping parameter ξ given in Table 30.1, the equivalent displacement Coulomb damping parameter is

$$\eta_{eq} = -\left(\frac{\omega}{\omega_0}\right)^2 \xi \tag{30.17}$$

Substituting this relation in the absolute transmissibility expression given at j in Table 30.2, the following equation is obtained for the acceleration transmissibility:

$$(T_A)_A = \frac{\ddot{x}_0}{\ddot{u}_0} = \sqrt{\frac{1 + \left(\frac{4}{\pi}\xi\right)^2\left(\frac{\omega^2}{\omega_0^2}\right)\left[\left(\frac{N+2}{N}\right)\left(\frac{\omega^2}{\omega_0^2}\right) - 2\left(\frac{N+1}{N}\right)\right]}{\left(1 - \frac{\omega^2}{\omega_0^2}\right)^2}} \tag{30.18}$$

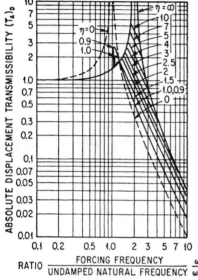

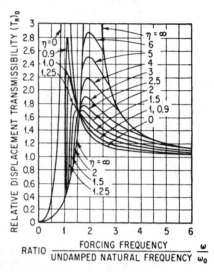

FIGURE 30.13 Absolute displacement transmissibility for the elastically connected, Coulomb-damped isolation system illustrated at D in Table 30.1, for the damper spring stiffness defined by $N = 3$. The curves give the ratio of the absolute displacement amplitude of the equipment to the displacement amplitude imposed at the foundation, as a function of the frequency ratio ω/ω_0 and the displacement Coulomb-damping parameter η.

FIGURE 30.14 Relative displacement transmissibility for the elastically connected, Coulomb-damped isolation system illustrated at D in Table 30.1, for the damper spring stiffness defined by $N = 3$. The curves give the ratio of the relative displacement amplitude (maximum isolator deflection) to the displacement amplitude imposed at the foundation, as a function of the frequency ratio ω/ω_0 and the displacement Coulomb-damping parameter η.

Equation (30.18) is valid only for the frequency range in which there is relative motion across the Coulomb damper. This range is defined by the break-loose and lock-in frequencies which are obtained by substituting Eq. (30.17) into Eq. (30.16):

$$\left(\frac{\omega}{\omega_0}\right)_L = \sqrt{\frac{\left(\frac{4}{\pi}\xi\right)(N+1) \pm N}{\frac{4}{\pi}\xi}} \tag{30.19}$$

where Eqs. (30.16) and (30.19) give similar results, damping being defined in terms of displacement and acceleration excitation, respectively. For frequencies not included in the range between break-loose and lock-in frequencies, the acceleration transmissibility is that for an undamped system. Equation (30.18) indicates that infinite acceleration occurs at resonance unless the damper remains locked-in beyond a frequency ratio of unity. The coefficient of the damping term in Eq. (30.18) is identical to the corresponding coefficient in the expression for $(T_A)_D$ at j in Table 30.2. Thus, the frequency ratio at the optimum transmissibility is the same as that for displacement excitation.

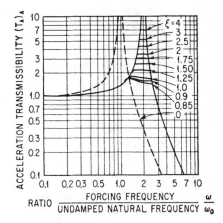

FIGURE 30.15 Acceleration transmissibility for the elastically connected, Coulomb-damped isolation system illustrated at D in Table 30.1, for the damper spring stiffness defined by $N = 3$. The curves give the ratio of the acceleration amplitude of the equipment to the acceleration amplitude imposed at the foundation, as a function of the frequency ratio ω/ω_0 and the acceleration Coulomb-damping parameter ξ.

An acceleration transmissibility curve for $N = 3$ is shown by Fig. 30.15. Relative motion at the damper occurs in a limited frequency range; thus, for relatively high frequencies, the acceleration transmissibility is similar to that for infinite damping.

Optimum Damping Parameters. The optimum Coulomb damping parameters are obtained by equating the optimum viscous damping ratio given by Eq. (30.15) to the equivalent viscous damping ratio for the elastically supported damper system and replacing the frequency ratio by the frequency ratio given by Eq. (30.13). The optimum value of the damping parameter η in Table 30.1 is

$$\eta_{op} = \frac{\pi}{2} \sqrt{\frac{N+1}{N+2}} \qquad (30.20)$$

To obtain the optimum value of the damping parameter ξ in Table 30.1, Eq. (30.17) is substituted in Eq. (30.20):

$$\xi_{op} = \frac{\pi}{4} \sqrt{\frac{N+2}{N+1}} \qquad (30.21)$$

Force Transmissibility. The force transmissibility $(T_A)_F = F_T/F_0$ is identical to $(T_A)_A$ given by Eq. (30.18) if $\xi = \xi_F$, where ξ_F is defined as

$$\xi_F = \frac{F_f}{F_0} \qquad (30.22)$$

Thus, the transmissibility curve shown in Fig. 30.15 also gives the force transmissibility for $N = 3$. By substituting Eq. (30.22) into Eq. (30.21), the transmitted force is optimized when the friction force F_f has the following value:

$$(F_f)_{op} = \frac{\pi F_0}{4} \sqrt{\frac{N+2}{N+1}} \qquad (30.23)$$

To avoid infinite transmitted force at resonance, it is necessary that $F_f > (\pi/4)F_0$.

Comparison of Rigidly Connected and Elastically Connected Coulomb-Damped Systems. A principal limitation of the rigidly connected Coulomb-damped isolator is the nature of the transmissibility at high forcing frequencies. Because the isolator deflection is small, the force transmitted by the spring is negligible; then the force transmitted by the damper controls the motion experienced by the equipment. The acceleration transmissibility approaches the constant value $(4/\pi)\xi$, independent of frequency. The corresponding transmissibility for an isolator with an elastically connected Coulomb damper is $(N + 1)/(\omega/\omega_0)^2$. Thus, the transmissibility varies inversely as the square of the excitation frequency and reaches a relatively low value at large values of excitation frequency.

MULTIPLE DEGREE-OF-FREEDOM SYSTEMS

The single degree-of-freedom systems discussed previously are adequate for illustrating the fundamental principles of vibration isolation but are an oversimplification insofar as many practical applications are concerned. The condition of unidirectional motion of an elastically mounted mass is not consistent with the requirements in many applications. In general, it is necessary to consider freedom of movement in all directions, as dictated by existing forces and motions and by the elastic constraints. Thus, in the general isolation problem, the equipment is considered as a rigid body supported by resilient supporting elements or isolators. This system is arranged so that the isolators effect the desired reduction in vibration. Various types of symmetry are encountered, depending upon the equipment and arrangement of isolators.

NATURAL FREQUENCIES—ONE PLANE OF SYMMETRY

A rigid body supported by resilient supports with one vertical plane of symmetry has three coupled natural modes of vibration and a natural frequency in each of these modes. A typical system of this type is illustrated in Fig. 30.16; it is assumed to be symmetrical with respect to a plane parallel with the plane of the paper and extending through the center-of-gravity of the supported body. Motion of the supported body in horizontal and vertical translational modes and in the rotational mode, all in the plane of the paper, are coupled. The equations of motion of a rigid body on resilient supports with six degrees-of-freedom are given by Eq. (3.31). By introducing certain types of symmetry and setting the excitation equal to zero, a cubic equation defining the free vibration of the system shown in Fig. 30.16 is derived, as given by Eqs. (3.36). This equation may be solved graphically for the natural frequencies of the system by use of Fig. 3.14.

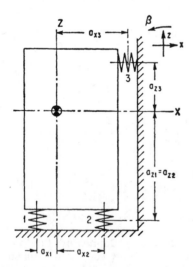

FIGURE 30.16 Schematic diagram of a rigid equipment supported by an arbitrary arrangement of vibration isolators, symmetrical with respect to a plane through the center-of-gravity parallel with the paper.

SYSTEM WITH TWO PLANES OF SYMMETRY

A common arrangement of isolators is illustrated in Fig. 30.17; it consists of an equipment supported by four isolators located adjacent to the four lower corners. It is symmetrical with respect to two coordinate vertical planes through the center-of-gravity of the equipment, one of the planes being parallel with the plane of the paper. Because of this symmetry, vibration in the vertical translational mode is decoupled from vibration in the horizontal and rotational modes. The natural frequency in the vertical translational mode is $\omega_z = \sqrt{\Sigma k_z/m}$, where Σk_z is the sum of the vertical stiffnesses of the isolators.

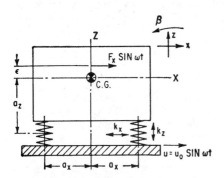

FIGURE 30.17 Schematic diagram in elevation of a rigid equipment supported upon four vibration isolators. The plane of the paper extends vertically through the center-of-gravity; the system is symmetrical with respect to this plane and with respect to a vertical plane through the center-of-gravity perpendicular to the paper. The moment of inertia of the equipment with respect to an axis through the center-of-gravity and normal to the paper is I_y. Excitation of the system is alternatively a vibratory force $F_x \sin \omega t$ applied to the equipment or a vibratory displacement $u = u_0 \sin \omega t$ of the foundation.

Consider excitation by a periodic force $F = F_x \sin \omega t$ applied in the direction of the X axis at a distance ϵ above the center-of-gravity and in one of the planes of symmetry. The differential equations of motion for the equipment in the coupled horizontal translational and rotational modes are obtained by substituting in Eq. (3.31) the conditions of symmetry defined by Eqs. (3.33), (3.34), (3.35), and (3.38). The resulting equations of motion are

$$m\ddot{x} = -4k_x x + 4k_x a\beta + F_x \sin \omega t \qquad (30.24)$$

$$I_y\ddot{\beta} = 4k_x ax - 4k_x a^2\beta - 4k_y b^2\beta - F_x\epsilon \sin \omega t$$

Making the common assumption that transients may be neglected in systems undergoing forced vibration, the translational and rotational displacements of the supported body are assumed to be harmonic at the excitation frequency. The differential equations of motion then are solved simultaneously to give the following expressions for the displacement amplitudes x_0 in horizontal translation and β_0 in rotation:

$$x_0 = \frac{F_x}{4k_z}\left(\frac{A_1}{D}\right) \qquad \beta_0 = \frac{F_x}{4\rho_y k_z}\left(\frac{A_2}{D}\right) \qquad (30.25)$$

where

$$A_1 = \left(\frac{1}{\rho_y^2}\right)(\eta a_z^2 + a_x^2 - \eta\epsilon a_z) - \left(\frac{\omega}{\omega_z}\right)^2$$

$$A_2 = \frac{\epsilon}{\rho_y}\left(\frac{\omega}{\omega_z}\right)^2 + \frac{\eta}{\rho_y}(a_z - \epsilon) \qquad (30.26)$$

$$D = \left(\frac{\omega}{\omega_z}\right)^4 - \left(\eta + \eta\frac{a_z^2}{\rho_y^2} + \frac{a_x^2}{\rho_y^2}\right)\left(\frac{\omega}{\omega_z}\right)^2 + \eta\left(\frac{a_x}{\rho_y}\right)^2$$

In the above equations, $\eta = k_x/k_z$ is the dimensionless ratio of horizontal stiffness to vertical stiffness of the isolators, $\rho_y = \sqrt{I_y/m}$ is the radius of gyration of the supported body about an axis through its center-of-gravity and perpendicular to the paper, $\omega_z = \sqrt{\Sigma k_z/m}$ is the undamped natural frequency in vertical translation, ω is the forcing frequency, a_z is the vertical distance from the effective height of spring (mid-height if symmetrical top to bottom)* to center-of-gravity of body m, and the other parameters are as indicated in Fig. 30.17.

Forced vibration of the system shown in Fig. 30.17 also may be excited by periodic motion of the support in the horizontal direction, as defined by $u = u_0 \sin \omega t$. The differential equations of motion for the supported body are

$$m\ddot{x} = 4k_x(u - x - a_z\beta)$$

$$I_y\ddot{\beta} = -4a_zk_x(u - x - a_z\beta) - 4k_za_x^2\beta$$

(30.27)

Neglecting transients, the motion of the mounted body in horizontal translation and in rotation is assumed to be harmonic at the forcing frequency. Equations (30.27) may be solved simultaneously to obtain the following expressions for the displacement amplitudes x_0 in horizontal translation and β_0 in rotation:

$$x_0 = \frac{u_0B_1}{D} \qquad \beta_0 = \frac{u_0B_2}{\rho_yD}$$

(30.28)

where the parameters B_1 and B_2 are

$$B_1 = \eta\left(\frac{a_x^2}{\rho_y^2} - \frac{\omega^2}{\omega_z^2}\right) \qquad B_2 = \frac{\eta a_z}{\rho_y}\left(\frac{\omega}{\omega_z}\right)^2$$

(30.29)

and D is given by Eq. (30.26).

Natural Frequencies—Two Planes of Symmetry. In forced vibration, the amplitude becomes a maximum when the forcing frequency is approximately equal to a natural frequency. In an undamped system, the amplitude becomes infinite at resonance. Thus, the natural frequency or frequencies of an undamped system may be determined by writing the expression for the displacement amplitude of the system in forced vibration and finding the excitation frequency at which this amplitude becomes infinite. The denominators of Eqs. (30.25) and (30.28) include the parameter D defined by Eq. (30.26). The natural frequencies of the system in coupled rotational and horizontal translational modes may be determined by equating D to zero and solving for the forcing frequencies:[7]

$$\frac{\omega_{x\beta}}{\omega_z} \times \frac{\rho_y}{a_x} = \frac{1}{\sqrt{2}} \sqrt{\eta\left(\frac{\rho_y}{a_x}\right)^2\left(1 + \frac{a_z^2}{\rho_y^2}\right) + 1 \pm \sqrt{\left[\eta\left(\frac{\rho_y}{a_x}\right)^2\left(1 + \frac{a_z^2}{\rho_y^2}\right) + 1\right]^2 - 4\eta\left(\frac{\rho_y}{a_x}\right)^2}}$$

(30.30)

where $\omega_{x\beta}$ designates a natural frequency in a coupled rotational (β) and horizontal translational (x) mode, and ω_z designates the natural frequency in the decoupled

---

* The distance a_z is taken to the mid-height of the spring to include in the equations of motion the moment applied to the body m by the fixed-end spring. If the spring is hinged to body m, the appropriate value for a_z is the distance from the X axis to the hinge axis.

vertical translational mode. The other parameters are defined in connection with Eq. (30.26). Two numerically different values of the dimensionless frequency ratio $\omega_{x\beta}/\omega_z$ are obtained from Eq. (30.30), corresponding to the two discrete coupled modes of vibration. Curves computed from Eq. (30.30) are given in Fig. 30.18.

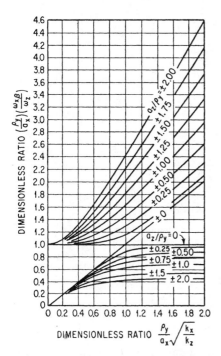

The ratio of a natural frequency in a coupled mode to the natural frequency in the vertical translational mode is a function of three dimensionless ratios, two of the ratios relating the radius of gyration ρ_y to the dimensions a_z and a_x while the third is the ratio η of horizontal to vertical stiffnesses of the isolators. In applying the curves of Fig. 30.18, the applicable value of the abscissa ratio is first determined directly from the constants of the system. Two appropriate numerical values then are taken from the ordinate scale, as determined by the two curves for applicable values of a_z/ρ_y; the ratios of natural frequencies in coupled and vertical translational modes are determined by dividing these values by the dimensionless ratio ρ_y/a_x. The natural frequencies in coupled modes then are determined by multiplying the resulting ratios by the natural frequency in the decoupled vertical translational mode.

The two straight lines in Fig. 30.18 for $a_z/\rho_y = 0$ represent natural frequencies in decoupled modes of vibration. When $a_z = 0$, the elastic supports lie in a plane passing through the center-of-gravity of the equipment. The horizontal line at a value of unity on the ordinate scale represents the natural frequency in a rotational mode. The inclined straight line for the value $a_z/\rho_y = 0$ represents the natural frequency of the system in horizontal translation.

FIGURE 30.18 Curves of natural frequencies $\omega_x\beta$ in coupled modes with reference to the natural frequency in the decoupled vertical translational mode ω_z, for the system shown schematically in Fig. 30.17. The isolator stiffnesses in the X and Z directions are indicated by k_x and k_z, respectively, and the radius of gyration with respect to the Y axis through the center-of-gravity is indicated by ρ_y.

Calculation of the coupled natural frequencies of a rigid body on resilient supports from Eq. (30.30) is sufficiently laborious to encourage the use of graphical means. For general purposes, both coupled natural frequencies can be obtained from Fig. 30.18. For a given type of isolators, $\eta = k_x/k_z$ is a constant and Eq. (30.30) may be evaluated in a manner that makes it possible to select isolator positions to attain optimum natural frequencies.[8-10] This is discussed under *Space-Plots* in Chap. 3. The convenience of the approach is partially offset by the need for a separate plot for each value of the stiffness ratio k_x/k_z. Applicable curves are plotted for several values of k_x/k_z in Figs. 3.17 to 3.19.

The preceding analysis of the dynamics of a rigid body on resilient supports includes the assumption that the principal axes of inertia of the rigid body are, respectively, parallel with the principal elastic axes of the resilient supports. This makes it possible to neglect the products of inertia of the rigid body. The coupling

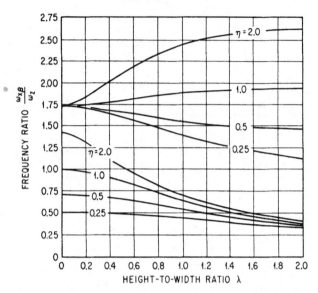

FIGURE 30.19 Curves indicating the natural frequencies $\omega_x\beta$ in coupled rotational and horizontal translational modes with reference to the natural frequency ω_z in the decoupled vertical translational mode, for the system shown in Fig. 30.17. The ratio of horizontal to vertical stiffness of the isolators is η, and the height-to-width ratio for the equipment is λ. These curves are based upon the assumption that the mass of the equipment is uniformly distributed and that the isolators are attached precisely at the extreme lower corners thereof.

introduced by the product of inertia is not strong unless the angle between the above-mentioned inertia and elastic axes is substantial. It is convenient to take the coordinate axes through the center-of-gravity of the supported body, parallel with the principal elastic axes of the isolators. If the moments of inertia with respect to these coordinate axes are used in Eqs. (30.24) to (30.30), the calculated natural frequencies usually are correct within a few percent without including the effect of product of inertia. When it is desired to calculate the natural frequencies accurately or when the product of inertia coupling is strong, a calculation procedure is available that may be used for certain conventional arrangements using four isolators.[11]

The procedure for determining the natural frequencies in coupled modes summarized by the curves of Fig. 30.18 represents a rigorous analysis where the assumed symmetry exists. The procedure is somewhat indirect because the dimensionless ratio ρ_y/a_x appears in both ordinate and abscissa parameters and because it is necessary to determine the radius of gyration of the equipment. The relations may be approximated in a more readily usable form if (1) the mounted equipment can be considered a cuboid having uniform mass distribution, (2) the four isolators are attached precisely at the four lower corners of the cuboid, and (3) the height of the isolators may be considered negligible. The ratio of the natural frequencies in the coupled rotational and horizontal translational modes to the natural frequency in the vertical translational mode then becomes a function of only the dimensions of the cuboid and the stiffnesses of the isolators in the several coordinate directions. Making these assumptions and substituting in Eq. (30.30),

$$\frac{\omega_{x\beta}}{\omega_z} = \frac{1}{\sqrt{2}} \sqrt{\frac{4\eta\lambda^2 + \eta + 3}{\lambda^2 + 1} \pm \sqrt{\left(\frac{4\eta\lambda^2 + \eta + 3}{\lambda^2 + 1}\right)^2 - \frac{12\eta}{\lambda^2 + 1}}} \qquad (30.31)$$

where $\eta = k_x/k_z$ designates the ratio of horizontal to vertical stiffness of the isolators and $\lambda = 2a_z/2a_x$ indicates the ratio of height to width of mounted equipment. This relation is shown graphically in Fig. 30.19. The curves included in this figure are useful for calculating approximate values of natural frequencies and for indicating trends in natural frequencies resulting from changes in various parameters as follows:

1. Both of the coupled natural frequencies tend to become a minimum, for any ratio of height to width of the mounted equipment, when the ratio of horizontal to vertical stiffness k_x/k_z of the isolators is low. Conversely, when the ratio of horizontal to vertical stiffness is high, both coupled natural frequencies also tend to be high. Thus, when the isolators are located underneath the mounted body, a condition of low natural frequencies is obtained using isolators whose stiffness in a horizontal direction is less than the stiffness in a vertical direction. However, low horizontal stiffness may be undesirable in applications requiring maximum stability. A compromise between natural frequency and stability then may lead to optimum conditions.

2. As the ratio of height to width of the mounted equipment increases, the lower of the coupled natural frequencies decreases. The trend of the higher of the coupled natural frequencies depends on the stiffness ratio of the isolators. One of the coupled natural frequencies tends to become very high when the horizontal stiffness of the isolators is greater than the vertical stiffness and when the height of the mounted equipment is approximately equal to or greater than the width. When the ratio of height to width of mounted equipment is greater than 0.5, the spread between the coupled natural frequencies increases as the ratio k_x/k_z of horizontal to vertical stiffness of the isolators increases.

Natural Frequency—Uncoupled Rotational Mode. Figure 30.20 is a plan view of the body shown in elevation in Fig. 30.17. The distances from the isolators to the principal planes of inertia are designated by a_x and a_y. The horizontal stiffnesses of the isolators in the directions of the coordinate axes X and Y are indicated by k_x and k_y, respectively. When the excitation is the applied couple $M = M_0 \sin \omega t$, the differential equation of motion is

$$I_z\ddot{\gamma} = -4\gamma a_x^2 k_y - 4\gamma a_y^2 k_x + M_0 \sin \omega t$$

$$(30.32)$$

where I_z is the moment of inertia of the body with respect to the Z axis. Neglecting transient terms, the solution of Eq. (30.32) gives the displacement amplitude γ_0 in rotation:

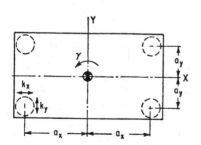

FIGURE 30.20 Plan view of the equipment shown schematically in Fig. 30.17, indicating the uncoupled rotational mode specified by the rotation angle γ.

$$\gamma_0 = \frac{M_0}{4(a_x^2 k_y + a_y^2 k_x) - I_z\omega^2} \qquad (30.33)$$

where the natural frequency ω_y in rotation about the Z axis is the value of ω that makes the denominator of Eq. (30.33) equal to zero:

$$\omega_y = 2\sqrt{\frac{a_x^2 k_y + a_y^2 k_x}{I_z}} \qquad (30.34)$$

VIBRATION ISOLATION IN COUPLED MODES

When the equipment and isolator system has several degrees-of-freedom and the isolators are located in such a manner that several natural modes of vibration are coupled, it becomes necessary in evaluating the isolators to consider the contribution of the several modes in determining the motion transmitted from the support to the mounted equipment or the force transmitted from the equipment to the foundation. Methods for determining the transmissibility under these conditions are best illustrated by examples.

For example, consider the system shown schematically in Fig. 30.21 wherein a machine is supported by relatively long beams which are in turn supported at their opposite ends by vibration isolators. The isolators are assumed to be undamped, and the excitation is considered to be a force applied at a distance $\epsilon = 4$ in. above the center-of-gravity of the machine-and-beam assembly. Alternatively, the force is (1) $F_x = F_0 \cos \omega t$, $F_z = F_0 \sin \omega t$ in a plane normal to the Y axis or (2) $F_y = F_0 \cos \omega t$, $F_z = F_0 \sin \omega t$ in a plane normal to the X axis. This may represent an unbalanced weight rotating in a vertical plane. A force transmissibility at each of the four isolators is determined by calculating the deflection of each isolator, multiplying the

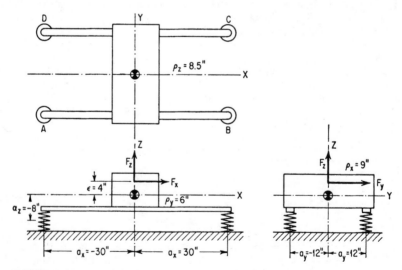

FIGURE 30.21 Schematic diagram of an equipment mounted upon relatively long beams which are in turn attached at their opposite ends to vibration isolators. Excitation for the system is alternatively (1) the vibratory force $F_x = F_0 \cos \omega t$, $F_z = F_0 \sin \omega t$ in the XZ plane or (2) the vibratory force $F_y = F_0 \cos \omega t$, $F_z = F_0 \sin \omega t$ in the YZ plane.

deflection by the appropriate isolator stiffness to obtain transmitted force, and dividing it by $F_0/4$.

When the system is viewed in a vertical plane perpendicular to the Y axis, the transmissibility curves are as illustrated in Fig. 30.22. The solid line defines the transmissibility at each of isolators B and C in Fig. 30.21, and the dotted line defines the transmissibility at each of isolators A and D. Similar transmissibility curves for a plane perpendicular to the X axis are shown in Fig. 30.23 wherein the solid line indicates the transmissibility at each of isolators C and D, and the dotted line indicates the transmissibility at each of isolators A and B.

Note the comparison of the transmissibility curves of Figs. 30.22 and 30.23 with the diagram of the system in Fig. 30.21. Figure 30.23 shows the three resonance conditions which are characteristic of a coupled system of the type illustrated. The transmissibility remains equal to or greater than unity for all excitation frequencies lower than the highest resonance frequency in a coupled mode. At greater excitation frequencies, vibration isolation is attained, as indicated by values of force transmissibility smaller than unity.

The transmissibility curves in Fig. 30.22 show somewhat similar results. The long horizontal beams tend to spread the resonance frequencies by a substantial frequency increment and merge the resonance frequency in the vertical translational mode with the resonance frequency in one of the coupled modes. A low transmissibility is again attained at excitation frequencies greater than the highest resonance frequency. Note that the transmissibility drops to a value slightly less than unity over a small frequency interval between the predominant resonance frequencies. This is a force reduction resulting from the relatively long beams, and it constitutes an acceptable condition if the magnitude of the excitation force in this direction is relatively small. Thus, the natural frequencies of the isolators could be somewhat higher with a consequent gain in stability; it is necessary, however, that the excitation frequency be substantially constant.

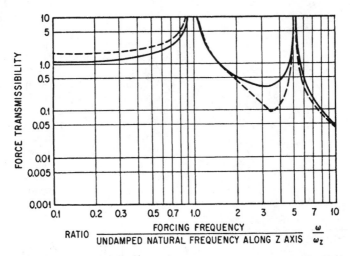

FIGURE 30.22 Transmissibility curves for the system shown in Fig. 30.21 when the excitation is in a plane perpendicular to the Y axis. The solid line indicates the transmissibility at each of isolators B and C, whereas the dotted line indicates the transmissibility at each of isolators A and D.

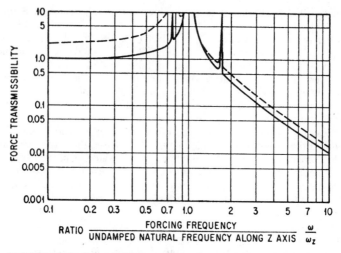

FIGURE 30.23 Transmissibility curves for the system illustrated in Fig. 30.21 when the excitation is in a plane perpendicular to the X axis. The solid line indicates the transmissibility at each of isolators C and D, whereas the dotted line indicates the transmissibility at each of isolators A and B.

Consider the equipment illustrated in Fig. 30.24 when the excitation is horizontal vibration of the support. The effectiveness of the isolators in reducing the excitation vibration is evaluated by plotting the displacement amplitude of the horizontal vibration at points A and B with reference to the displacement amplitude of the support. Transmissibility curves for the system of Fig. 30.24 are shown in Fig. 30.25. The solid line in Fig. 30.25 refers to point A and the dotted line to point B. Note that there is no significant reduction of amplitude except when the forcing frequency exceeds the maximum resonance frequency of the system.

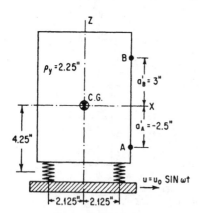

FIGURE 30.24 Schematic diagram of an equipment supported by vibration isolators. Excitation is a vibratory displacement $u = u_0 \sin \omega t$ of the foundation.

A general rule for the calculation of necessary isolator characteristics to achieve the results illustrated in Figs. 30.22, 30.23, and 30.25 is that the forcing frequency should be not less than 1.5 to 2 times the maximum natural frequency in any of six natural modes of vibration. In exceptional cases, such as illustrated in Fig. 30.22, the forcing frequency may be interposed between resonance frequencies if the forcing frequency is a constant.

Example 30.1. Consider the machine illustrated in Fig. 30.21. The force that is to be isolated is harmonic at the constant frequency of 8 Hz; it is assumed to result from the rotation of an unbalanced member whose plane of rotation is alternatively (1) a plane perpendicular to the Y axis and (2) a plane perpendicular to the X axis. The distance

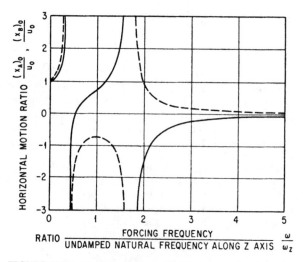

FIGURE 30.25 Displacement transmissibility curves for the system of Fig. 30.24. Transmissibility between the foundation and point A is shown by the solid line; transmissibility between the foundation and point B is shown by the dotted line.

between isolators is 60 in. in the direction of the X axis and 24 in. in the direction of the Y axis. The center of coordinates is taken at the center-of-gravity of the supported body, i.e., at the center-of-gravity of the machine-and-beams assembly. The total weight of the machine and supporting beam assembly is 100 lb, and its radii of gyration with respect to the three coordinate axes through the center-of-gravity are $\rho_x = 9$ in., $\rho_z = 8.5$ in., and $\rho_y = 6$ in. The isolators are of equal stiffnesses in the directions of the three coordinate axes:

$$\eta = \frac{k_x}{k_z} = \frac{k_y}{k_z} = 1$$

The following dimensionless ratios are established as the initial step in the solution:

$$a_z/\rho_y = -1.333 \qquad a_z/\rho_x = -0.889$$

$$a_x/\rho_y = \pm 5.0 \qquad a_y/\rho_x = \pm 1.333$$

$$(a_z/\rho_y)^2 = 1.78 \qquad (a_z/\rho_x)^2 = 0.790$$

$$(a_x/\rho_y)^2 = 25.0 \qquad (a_y/\rho_x)^2 = 1.78$$

$$\eta(\rho_y/a_x)^2 = 0.04 \qquad \eta(\rho_x/a_y)^2 = 0.561$$

The various natural frequencies are determined in terms of the vertical natural frequency ω_z. Referring to Fig. 30.18, the coupled natural frequencies for vibration in a plane perpendicular to the Y axis are determined as follows:

First calculate the parameter

$$\frac{\rho_y}{a_x}\sqrt{\frac{k_x}{k_z}} = 0.2$$

For $a_z/\rho_y = -1.333$, $(\omega_{x\beta}/\omega_z)(\rho_y/a_x) = 0.19$; 1.03. Note the signs of the dimensionless ratios a_z/ρ_y and a_x/ρ_y. According to Eq. (30.30), the natural frequencies are independent of the sign of a_z/ρ_y. With regard to the ratio a_x/ρ_y, the sign chosen should be the same as the sign of the radical on the right side of Eq. (30.30). The frequency ratio $(\omega_{x\beta}/\omega_z)$ then becomes positive. Dividing the above values for $(\omega_{x\beta}/\omega_z)(\rho_y/a_x)$ by $\rho_y/a_x = 0.2$, $\omega_{x\beta}/\omega_z = 0.96$; 5.15.

Vibration in a plane perpendicular to the X axis is treated in a similar manner. It is assumed that exciting forces are not applied concurrently in planes perpendicular to the X and Y axes; thus, vibration in these two planes is independent. Consequently, the example entails two independent but similar problems and similar equations apply for a plane perpendicular to the X axis:

$$\frac{\rho_x}{a_y}\sqrt{\frac{k_z}{k_y}} = 0.75$$

For $a_z/\rho_x = 0.889$, $(\omega_{y\alpha}/\omega_z)(\rho_x/a_y) = 0.57$; 1.29. Dividing by $\rho_x/a_y = 0.75$, $\omega_{y\alpha}/\omega_z = 0.76$; 1.72.

The natural frequency in rotation with respect to the Z axis is calculated from Eq. (30.34) as follows, taking into consideration that there are two pairs of springs and that $k_x = k_y = k_z$:

$$\omega_\gamma = \sqrt{\left(\frac{a_x^2 + a_y^2}{\rho_z^2}\right)\left(\frac{4k_z g}{W}\right)} = 3.8\omega_z$$

The six natural frequencies are as follows:

1. Translational along Z axis: ω_z
2. Coupled in plane perpendicular to Y axis: $0.96\omega_z$
3. Coupled in plane perpendicular to Y axis: $5.15\omega_z$
4. Coupled in plane perpendicular to X axis: $0.76\omega_z$
5. Coupled in plane perpendicular to X axis: $1.72\omega_z$
6. Rotational with respect to Y axis: $3.8\omega_z$

Considering vibration in a plane perpendicular to the Y axis, the two highest natural frequencies are the natural frequency ω_y in the translational mode along the Z axis and the natural frequency $5.15\omega_z$ in a coupled mode. In a similar manner, the two highest natural frequencies in a plane perpendicular to the X axis are the natural frequency ω_z in translation along the Z axis and the natural frequency $1.72\omega_z$ in a coupled mode. The natural frequency in rotation about the Z axis is $3.80\omega_z$. The widest frequency increment which is void of natural frequencies is between $1.72\omega_z$ and $3.80\omega_z$. This increment is used for the forcing frequency which is taken as $2.5\omega_z$. Inasmuch as the forcing frequency is established at 8 Hz, the vertical natural frequency is 8 divided by 2.5, or 3.2 Hz. The required vertical stiffnesses of the isolators are calculated from Eq. (30.11) to be 105 lb/in. for the entire machine, or 26.2 lb/in. for each of the four isolators.

INCLINED ISOLATORS

Advantages in vibration isolation sometimes result from inclining the principal elastic axes of the isolators with respect to the principal inertia axes of the equipment, as illustrated in Fig. 30.26. The coordinate axes X and Z are, respectively, parallel with the principal inertia axes of the mounted body, but the center of coordinates is taken at the elastic axis. The location of the elastic axis is determined by the elastic properties of the system. If a force is applied to the body along a line extending through the elastic axis, the body is displaced in translation without rotation; if a couple is applied to the body, the body is displaced in rotation without translation.

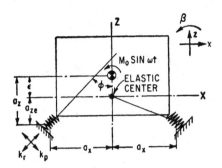

FIGURE 30.26 Schematic diagram of an equipment supported by isolators whose principal elastic axes are inclined to the principal inertia axes of the equipment.

The principal elastic axes r, p of the isolators are parallel with the paper and inclined with respect to the coordinate axes, as indicated in Fig. 30.26. The stiffness of each isolator in the direction of the respective principal axis is indicated by k_r, k_p. The principal elastic axis of an isolator is the axis along which a force must be applied to cause a deflection colinear with the applied force (see the section *Properties of a Biaxial Stiffness Isolator*).

Assume the excitation for the system shown in Fig. 30.26 to be a couple $M_0 \sin \omega t$ acting about an axis normal to the paper. The equations of motion for the body in the horizontal translational and rotational modes may be written by noting that the displacement of the center-of-gravity in the direction of the X axis is $x - \epsilon\beta$; thus, the corresponding acceleration is $\ddot{x} - \epsilon\ddot{\beta}$. A translational displacement x produces only an external force $-k_x x$, whereas a rotational displacement β produces only an external couple $-k_\beta \beta$. The equations of motion are

$$m(\ddot{x} - \epsilon\ddot{\beta}) = -k_x x$$

$$m\rho_e^2\ddot{\beta} - m\epsilon\ddot{x} = -k_\beta\beta + M_0 \sin \omega t \tag{30.35}$$

where ρ_e is the radius of gyration of the mounted body with respect to the elastic axis. The radius of gyration ρ_e is related to the radius of gyration ρ_y with respect to a line through the center-of-gravity by $\rho_e = \sqrt{\rho_y^2 + \epsilon^2}$, where ϵ is the distance between the elastic axis and a parallel line passing through the center-of-gravity. In the equations of motion, k_x and k_β represent the translational and rotational stiffness of the isolators in the x and β coordinate directions, respectively.

By assuming steady-state harmonic motion for the horizontal translation x and rotation β, the following displacement amplitudes are obtained by solving Eqs. (30.35):

$$x_0 = \frac{-M_0\epsilon\omega^2}{m[\rho_e^2(\omega^2 - \omega_\beta^2)(\omega^2 - \omega_x^2) - \epsilon^2\omega^4]}$$

$$\beta_0 = \frac{-M_0}{m\left[\rho_e^2(\omega^2 - \omega_\beta^2) - \dfrac{\epsilon^2\omega^4}{\omega^2 - \omega_x^2}\right]} \qquad (30.36)$$

where $\omega_x = \sqrt{k_x/m}$ and $\omega_\beta = \sqrt{k_\beta/m\rho_e^2}$ are hypothetical natural frequencies defined for convenience. The natural frequencies $\omega_{x\beta}$ in the coupled x,β modes are determined by equating the denominator of Eqs. (30.36) to zero and solving for ω (now identical to $\omega_{x\beta}$):

$$\frac{\omega_{x\beta}}{\omega_x} = \sqrt{\frac{1 + \lambda_1^2 \pm \sqrt{(1 + \lambda_1^2)^2 - 4\lambda_1^2[1 - (\epsilon/\rho_e)^2]}}{2[1 - (\epsilon/\rho_e)^2]}} \qquad (30.37)$$

where λ_1 is a dimensionless ratio given by

$$\lambda_1 = \frac{(a_x/\rho_e)\sqrt{k_r/k_p}}{\cos^2 \phi + (k_r/k_p)\sin^2 \phi} \qquad (30.38)$$

The hypothetical natural frequency ω_x is

$$\omega_x = \sqrt{\frac{4k_p}{m}\left[\cos^2 \phi + \frac{k_r}{k_p}\sin^2 \phi\right]} \qquad (30.39)$$

The relation given by Eq. (30.37) is shown graphically by Fig. 30.27. The parameters needed to evaluate the natural frequencies by using this graph are calculated from the physical properties of the system and the relations of Eqs. (30.38) and (30.39). In addition, the distance ϵ between a parallel line passing through the center-of-gravity and the elastic axis must be known. The distance ϵ is determined by effecting a small horizontal displacement of the equipment in the X direction and equating the resulting summation of elastic couples to zero:

$$\epsilon = a_z - \frac{a_x(1 - k_p/k_r)\cot \phi}{(k_p/k_r)\cot^2 \phi + 1} \qquad (30.40)$$

where a_z is the distance between the parallel planes passing through the center-of-gravity of the body and the mid-height of the isolators, as shown in Fig. 30.26.

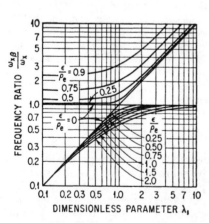

FIGURE 30.27 Curves indicating the natural frequencies $\omega_{x\beta}$ in coupled modes with reference to the natural frequency in the decoupled (fictitious) horizontal translational mode ω_x for the system shown schematically in Fig. 30.26. The radius of gyration with respect to the elastic axis is indicated by ρ_e, and the distance between the center-of-gravity and the elastic center is ϵ. The dimensionless parameter λ_1 is defined by Eq. (30.38) and ω_x is defined by Eq. (30.39).

DECOUPLING OF MODES

The natural modes of vibration of a body supported by isolators may be decoupled one from another by proper orientation of the isolators. Each mode of vibration then exists independently of the others, and vibration in one mode does not excite vibration in other modes. The necessary conditions for decoupling may be stated as follows: The resultant of the forces applied to the mounted body by the isolators when the mounted body is displaced in translation must be a force directed through the center-of-gravity; or, the resultant of the couples applied to the mounted body by the isolators when the mounted body is displaced in rotation must be a couple about an axis through the center-of-gravity.

In general, the natural frequencies of a multiple degree-of-freedom system can be made equal only by decoupling the natural modes of vibration, i.e., by making $a_z = 0$ in Fig. 30.17. The natural frequencies in decoupled modes are indicated by the two straight lines in Fig. 30.18 marked $a_z/\rho_y = 0$. The natural frequencies in translation along the X axis and in rotation about the Y axis become equal at the intersection of these lines; i.e., when $a_z/\rho_y = 0$, $k_x/k_z = 1$ and $\rho_y/a_x = 1$. The physical significance of these mathematical conditions is that the isolators be located in a plane passing through the center-of-gravity of the equipment, that the distance between isolators be twice the radius of gyration of the equipment, and that the stiffness of each isolator in the directions of the X and Z axes be equal.

When the isolators cannot be located in a plane which passes through the center-of-gravity of the equipment, decoupling can be achieved by inclining the isolators, as illustrated in Fig. 30.26. If the elastic axis of the system is made to pass through the center-of-gravity, the translational and rotational modes are decoupled because the inertia force of the mounted body is applied through the elastic center and introduces no tendency for the body to rotate. The requirements for a decoupled system are established by setting $\epsilon = 0$ in Eq. (30.40) and solving for k_r/k_p:

$$\frac{k_r}{k_p} = \frac{(a_x/a_z) + \cot\phi}{(a_x/a_z) - \tan\phi} \tag{30.41}$$

The conditions for decoupling defined by Eq. (30.41) are shown graphically in Figs. 30.28 and 3.23. The decoupled natural frequencies are indicated by the straight lines $\epsilon/\rho_e = 0$ in Fig. 30.27. The horizontal line refers to the decoupled natural frequency ω_x in translation in the direction of the X axis, while the inclined line refers to the decoupled natural frequency ω_β in rotation about the Y axis.

PROPERTIES OF A BIAXIAL STIFFNESS ISOLATOR

A biaxial stiffness isolator is represented as an elastic element having a single plane of symmetry; all forces act in this plane and the resultant deflections are limited by symmetry or constraints to this plane. The characteristic elastic properties of the isolator may be defined alternatively by sets of influence coefficients as follows:

1. If the two coordinate axes in the plane of symmetry are selected arbitrarily, three stiffness parameters are required to define the properties of the isolator. These are the axial influence coefficients* along the two coordinate axes, and a characteristic coupling influence coefficient* between the coordinate axes.

---

* The influence coefficient κ is a function only of the isolator properties and not of the constraints imposed by the system in which the isolator is used. Both positive and negative values of the influence coefficient κ are permissible.

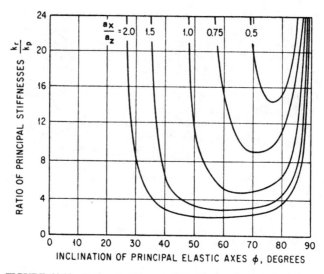

FIGURE 30.28 Ratio of stiffnesses k_r/k_p along principal elastic axes required for decoupling the natural modes of vibration of the system illustrated in Fig. 30.26.

2. If the two coordinate axes in the plane of symmetry are selected to coincide with the principal elastic axes of the isolator, two influence coefficients are required to define the properties of the isolator. These are the principal influence coefficients. If the isolator is used in a system, a third parameter is required to define the orientation of the principal axes of the isolator with the coordinate axes of the system.

PROPERTIES OF ISOLATOR WITH RESPECT TO ARBITRARILY SELECTED AXES

A schematic representation of a linear biaxial stiffness element is shown in Fig. 30.29

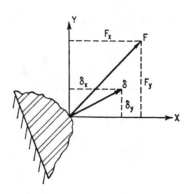

FIGURE 30.29 Schematic diagram of a linear biaxial stiffness element.

where the X and Y axes are arbitrarily chosen to define a plane to which all forces and motions are restricted. In general, the deflection of an isolator resulting from an applied load is not in the same direction as the load, and a coupling influence coefficient is required to define the properties of the isolator in addition to the influence coefficients along the X and Y axes. The three characteristic stiffness coefficients that uniquely describe the load-deflection properties of a biaxial stiffness element are:

1. The influence coefficient of the element in the X coordinate direction is κ_x. It is the ratio of the component of the applied force in the X direction to the resulting deflection when the isolator is constrained to deflect in the X direction.

2. The influence coefficient of the element in the Y coordinate direction is κ_y. It is the ratio of the component of the applied force in the Y direction to the resulting deflection when the isolator is constrained to deflect in the Y direction.

3. The coupling influence coefficient is κ_{xy}. It represents the force required in the X direction to produce a unit displacement in the Y direction when the isolator is constrained to deflect only in the Y direction. (By Maxwell's reciprocity principle, the same force is required in the Y direction to produce a unit displacement in the X direction; i.e., $\kappa_{xy} = \kappa_{yx}$.)

Consider the isolator shown in Fig. 30.29 where the applied force F has components F_x and F_y; the resulting displacement has components δ_x and δ_y. From the above definitions of influence coefficients, the forces in the X and Y coordinate directions required to effect a displacement δ_x are

$$F_{xx} = \kappa_x \delta_x \qquad F_{yx} = \kappa_{yx} \delta_x \tag{30.42}$$

The forces required to effect a displacement δ_y in the Y direction are

$$F_{xy} = \kappa_{xy} \delta_y \qquad F_{yy} = \kappa_y \delta_y \tag{30.43}$$

The force components F_x and F_y required to produce the deflection having components δ_x, δ_y are the sums from Eqs. (30.42) and (30.43):

$$F_x = \kappa_x \delta_x + \kappa_{xy} \delta_y$$
$$F_y = \kappa_{yx} \delta_x + \kappa_y \delta_y \tag{30.44}$$

If the three influence stiffness coefficients κ_x, κ_y, and $\kappa_{xy} = \kappa_{yx}$ are known for a given stiffness element, the load-deflection properties are given by Eq. (30.44).

The deflections of the isolator in response to forces F_x, F_y are determined by solving Eqs. (30.44) simultaneously:

$$\delta_x = \frac{F_x \kappa_y - F_y \kappa_{xy}}{\kappa_x \kappa_y - \kappa_{xy}^2}$$
$$\delta_y = \frac{F_y \kappa_x - F_x \kappa_{xy}}{\kappa_x \kappa_y - \kappa_{xy}^2} \tag{30.45}$$

These expressions give the orthogonal components of the displacement δ for any load having the components F_x and F_y applied to a biaxial stiffness isolator. By substituting the relations of Eqs. (30.45) into Eq. (30.44), the following alternate forms of the force-deflection equations are obtained:

$$F_x = \left(\kappa_x - \frac{\kappa_{xy}^2}{\kappa_y} \right) \delta_x + \frac{\kappa_{xy}}{\kappa_y} F_y$$
$$F_y = \left(\kappa_y - \frac{\kappa_{xy}^2}{\kappa_x} \right) \delta_y + \frac{\kappa_{xy}}{\kappa_x} F_x \tag{30.46}$$

The specific force-deflection equations for a given situation are obtained from these general load-deflection expressions by applying the proper constraint conditions.

Unconstrained Motion.   The general force-deflection equations can be used to obtain the effective stiffness coefficients when the forces F_x and F_y shown in Fig. 30.29 are applied independently. The resulting deflection of the isolator is unconstrained motion, i.e., the isolator is free to deflect out of the line of force application. The force divided by that component of deflection along the line of action of the force is the effective stiffness k. When $F_y = 0$, the effective stiffness k_x resulting from the applied force F_x is obtained from Eq. (30.46):

$$k_x = \frac{F_x}{\delta_x} = \left(\kappa_x - \frac{\kappa_{xy}^2}{\kappa_y} \right) \tag{30.47}$$

When $F_x = 0$, the effective stiffness k_y in response to the applied force F_y is

$$k_y = \frac{F_y}{\delta_y} = \left(\kappa_y - \frac{\kappa_{xy}^2}{\kappa_x} \right) \tag{30.48}$$

For unconstrained motion, $k_x/k_y = \kappa_x/\kappa_y$; i.e., the ratio of the effective stiffnesses in two mutually perpendicular directions is equal to the ratio of the corresponding influence coefficients for the same directions.

Constrained Motion.   When the isolator is constrained either by the symmetry of a system or by structural constraints to deflect only along the line of the applied force, the effective stiffness is obtained directly by letting appropriate deflections be zero in Eq. (30.44):

$$k_x = \frac{F_x}{\delta_x} = \kappa_x \qquad k_y = \frac{F_y}{\delta_y} = \kappa_y \tag{30.49}$$

The force required to maintain constrained motion is found by letting appropriate deflections be zero in Eqs. (30.46). For example, the force that must be applied in the X direction to ensure that the isolator deflects in the Y direction in response to a force F_y is

$$F_x = \frac{\kappa_{xy}}{\kappa_y} F_y \tag{30.50}$$

INFLUENCE COEFFICIENT TRANSFORMATION

Assume the influence coefficients κ_x, κ_y, and κ_{xy} are known in the X, Y coordinate system. It may be convenient to work with isolator influence coefficients in the X', Y' coordinate system as shown in Fig. 30.30. The X', Y' coordinate system is obtained by rotating the coordinate axes counterclockwise through an angle θ from the X, Y system. The influence coefficients with respect to the X', Y' axes are related to the influence coefficients with respect to the X, Y axes as follows:

$$\kappa_x' = \frac{\kappa_x + \kappa_y}{2} + \frac{\kappa_x - \kappa_y}{2} \cos 2\theta + \kappa_{xy} \sin 2\theta$$

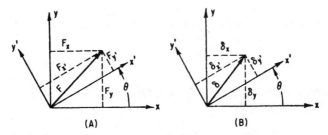

FIGURE 30.30 (A) Force and (B) displacement transformation diagrams for a linear biaxial stiffness element.

$$\kappa_{x'y'} = \frac{\kappa_y - \kappa_x}{2} \sin 2\theta + \kappa_{xy} \cos 2\theta \qquad (30.51)$$

$$\kappa_y' = \frac{\kappa_x + \kappa_y}{2} - \frac{\kappa_x - \kappa_y}{2} \cos 2\theta - \kappa_{xy} \sin 2\theta$$

The influence coefficient transformation of a biaxial stiffness isolator from one set of arbitrarily chosen coordinate axes to another arbitrarily chosen set of coordinate axes is described by the two-dimensional Mohr circle.[12] Since the influence coefficient is a tensor quantity, the following invariants of the influence coefficient tensor give additional relations between the influence coefficients in the X, Y and the X', Y' set of axes:

$$\kappa_x + \kappa_y = \kappa_{x'} + \kappa_{y'}$$

$$\kappa_x \kappa_y - \kappa_{xy}^2 = \kappa_{x'} \kappa_{y'} - \kappa_{x'y'}^2 \qquad (30.52)$$

PRINCIPAL INFLUENCE COEFFICIENTS

The set of axes for which there exists no coupling influence coefficient are the principal axes of stiffness (*principal elastic axes*). These axes can be found by requiring $\kappa_{x'y'}$ to be zero in Eq. (30.51) and solving for the rotation angle corresponding to this condition. Letting θ' represent the angle of rotation for which $\kappa_{x'y'} = 0$:

$$\tan 2\theta' = \frac{2\kappa_{xy}}{\kappa_x - \kappa_y} \qquad (30.53)$$

By substituting this value of the angle of rotation into the general influence coefficient expressions, Eqs. (30.51), the following relation is obtained for the principal influence coefficients:

$$\kappa_p, \kappa_q = \frac{\kappa_x + \kappa_y}{2} \pm \sqrt{\left(\frac{\kappa_x - \kappa_y}{2}\right)^2 + \kappa_{xy}^2} \qquad (30.54)$$

where p and q represent the principal axes of stiffness. The principal influence coefficients are the maximum and minimum influence coefficients that exist for a linear biaxial stiffness isolator. In Eq. (30.54), the plus sign gives the maximum influence coefficient whereas the minus sign gives the minimum influence coefficient. Either κ_p or κ_q can be the maximum influence coefficient, depending on the degree of axis rotation and the relative values of κ_x, κ_y, and κ_{xy}.

INFLUENCE COEFFICIENT TRANSFORMATION FROM THE PRINCIPAL AXES

The influence coefficient transformation from the principal axes p, q is of practical interest. The influence coefficients in the XY frame of reference are determined from Eq. (30.51) by setting $\kappa_{x'y'} = \kappa_{pq} = 0$, $\kappa_x' = \kappa_p$, $\kappa_{y'} = \kappa_q$, and $\theta = \theta'$. The influence coefficients in the XY frame-of-reference may be expressed in terms of the principal influence coefficients as follows:

$$\kappa_x = \kappa_p \cos^2 \theta' + \kappa_q \sin^2 \theta' = \frac{\kappa_p + \kappa_q}{2} + \frac{\kappa_p - \kappa_q}{2} \cos 2\theta'$$

$$\kappa_{xy} = (\kappa_p - \kappa_q) \sin \theta' \cos \theta' = \frac{\kappa_p - \kappa_q}{2} \sin 2\theta' \qquad (30.55)$$

$$\kappa_y = \kappa_p \sin^2 \theta' + \kappa_q \cos^2 \theta' = \frac{\kappa_p + \kappa_q}{2} - \frac{\kappa_p - \kappa_q}{2} \cos 2\theta'$$

The transformation from the principal axes in the form of a two-dimensional Mohr's circle is shown by Fig. 30.31. This circle provides quick graphical determination of

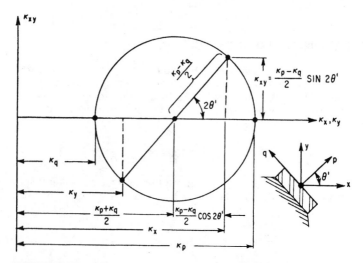

FIGURE 30.31 Mohr-circle representation of the stiffness transformation from the principal axes of stiffness of a biaxial stiffness element. The p, q axes represent the principal stiffness axes and the X, Y axes are any arbitrary set of axes separated from the p, q axes by a rotation angle θ'.

the three influence coefficients κ_x, κ_y, and κ_{xy} for any angle θ' between the P and X axes, where θ' is positive in the sense shown in the inset to Fig. 30.31.

Example 30.2. Consider the system shown schematically by Fig. 30.26. The transformation theory for the influence coefficient of a biaxial stiffness element may be applied to develop the effective stiffness coefficients for this system. The center of coordinates for the XZ axes is at the elastic center of the system. The principal elastic axes of the isolators p, r are oriented at an angle ϕ with the coordinate axes X, Z, respectively.* The position of the elastic center is determined by effecting a small horizontal displacement δ_x of the body, letting δ_z be zero and equating the summation of couples resulting from the isolator forces. The forces F_x and F_z are determined from Eqs. (30.44):

$$F_x = \kappa_x \delta_x = \kappa_x \delta_x \qquad F_z = \kappa_{zx} \delta_x = \kappa_{zx} \delta_x$$

Each of the forces F_x acts at a distance $-a_{ze}$ from the elastic center; the force F_z at the right-hand isolator is positive and acts at a distance a_x from the elastic center whereas the force F_z at the left-hand isolator is negative and acts at a distance $-a_x$ from the elastic center. Taking a summation of the moments:

$$-2a_{ze}F_x + 2a_xF_z = 0$$

Substituting the above relations between the forces F_x, F_z and the influence coefficients κ_z, κ_{zz} into Eqs. (30.55), and noting that $\theta' = 90° - \phi$ (compare Figs. 30.30 and 30.26), the following result is obtained in terms of principal stiffnesses:

$$\frac{a_{ze}}{a_x} = \frac{F_z}{F_x} = \frac{\kappa_{zx}}{\kappa_x} = \frac{(k_r - k_p) \sin \phi \cos \phi}{k_r \sin^2 \phi + k_p \cos^2 \phi}$$

Substituting $\epsilon = a_z - a_{ze}$ in the preceding equation, the relation for ϵ given by Eq. (30.40) is obtained.

Since the equations of motion are written in a coordinate system passing through the elastic center, all displacements in this frame-of-reference are constrained. Therefore, the effective stiffness coefficients for a single isolator may be obtained from Eq. (30.55) as follows [see Eq. (30.49)]:

$$k_x = \kappa_x = k_r \sin^2 \phi + k_p \cos^2 \phi$$

$$k_z = \kappa_z = k_r \cos^2 \phi + k_p \sin^2 \phi$$

These effective stiffness coefficients define the hypothetical natural frequency ω_x given by Eq. (30.39) as well as the uncoupled vertical natural frequency ω_z. Since four isolators are used in the problem represented by Fig. 30.26, the translational stiffnesses given by the above expressions for k_x and k_z must be multiplied by 4 to obtain the total translational stiffness.

The effective rotational stiffness of a single isolator k_β can be obtained by determining the sum of the restoring moments for a constrained rotation β. When the body is rotated through an angle β, the displacements at the right isolator are $\delta_x = -a_{ze}\beta$ and $\delta_z = a_x\beta$, where a_{ze} is a negative distance since it is measured in the negative Z direction. The sum of the restoring moments is $(F_z a_x - F_x a_{ze})$, where F_x and F_z

---

* The properties of a biaxial stiffness element may be defined with respect to any pair of coordinate axes. In Fig. 30.26, the principal elastic axis q is parallel with the coordinate axis Y; then the analysis considers the principal elastic axes p, r which lie in the plane defined by the XZ coordinate axes.

are the forces acting on the right isolator in Fig. 30.26. The forces F_x and F_z may be written in terms of the influence coefficients and the displacements δ_x and δ_z by use of Eq. (30.44) to produce the following moment equation:

$$M_\beta = k_\beta \beta = \beta[k_x a_{ze}^2 - 2k_{xz}a_{ze}a_x + k_z a_x^2]$$

where the effective rotational stiffness k_β of a single isolator is

$$k_\beta = k_x a^2 - 2k_{xz}a_{ze}a_x + k_z a_x^2$$

The distance a_{ze} can be eliminated from the expression for rotational stiffness by substituting $a_{ze} = a_x F_z/F_x$ obtained from the summation of couples about the elastic center:

$$k_\beta = a_x^2\left(\frac{k_x k_z - k_{xz}^2}{k_x} \right)$$

The numerator of this expression can be replaced by $k_r k_p$ [see Eq. (30.52)] where the r, p axes are the principal elastic axes of the isolator and $k_{rp} = 0$. Also, k_x can be replaced by its equivalent form given by Eq. (30.55). Making these substitutions, the effective rotational stiffness for one isolator in terms of the principal stiffness coefficients of the isolator becomes

$$k_\beta = \frac{a_x^2 k_p}{\sin^2 \phi + (k_p/k_r) \cos^2 \phi}$$

Since four isolators are used in the problem represented by Fig. 30.26, the rotational stiffness given by the above expression for k_β must be multiplied by 4 to obtain the total rotational stiffness of the system.

NONLINEAR VIBRATION ISOLATORS

In vibration isolation, the vibration amplitudes generally are small and linear vibration theory usually is applicable with sufficient accuracy.* However, the static effects of nonlinearity should be considered. Even though a nonlinear isolator may have approximately constant stiffness for small incremental deflections, the nonlinearity becomes important when large deflections of the isolator occur due to the effects of equipment weight and sustained acceleration. A vibration isolator often exhibits a stiffness that increases with applied force or deflection. Such a nonlinear stiffness is characteristic, for example, of rubber in compression or a conical spring.

In Eq. (30.11) for natural frequency, the stiffness k for a linear stiffness element is a constant. However, for a nonlinear isolator, the stiffness k is the slope of the force-deflection curve and Eq. (30.11) may be written

$$\omega_n = 2\pi f_n = \sqrt{\frac{g(dF/d\delta)}{W}} \tag{30.56}$$

where W is the total weight supported by the isolator, g is the acceleration of gravity, and $dF/d\delta$ is the slope of the line tangent to the force-deflection curve at the static equilibrium position. Vibration is considered to be small variations in the posi-

---

* If the vibration amplitude is large, nonlinear vibration theory as discussed in Chap. 4 is applicable.

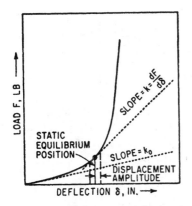

FIGURE 30.32 Typical force-deflection characteristic of a tangent hardening isolator.

tion of the supported equipment above and below the static equilibrium position, as indicated in Fig. 30.32. Thus, the natural frequency is determined solely by the stiffness characteristics in the region of the isolator deflection.

NATURAL FREQUENCY

In determining the natural frequency of a nonlinear isolator, it is important to note whether or not all the load results from the dead weight of a massive body. The force F on the isolator may be greater than the weight W because of a belt pull or sustained acceleration of a missile. Then the load on the isolator is

$$F = n_g W \tag{30.57}$$

where n_g is some multiple of the acceleration of gravity. For example, n_g may indicate the absolute value of the sustained acceleration of a missile measured in "number of g's."

Characteristic of Tangent Isolator. It is convenient to define the force-deflection characteristics of a nonlinear isolator having increasing stiffness (hardening characteristic) by a tangent function:[13]

$$F = \frac{2k_0 h_c}{\pi} \tan\left(\frac{\pi\delta}{2h_c}\right) \tag{30.58}$$

where F is the total force applied to the isolator, k_0 is the stiffness of the isolator at zero deflection, δ is the deflection of the isolator, and h_c is the characteristic height of the isolator. The force-deflection characteristic defined by Eq. (30.58) is shown graphically in Fig. 30.33A. The characteristic height h_c represents a height or thickness characteristic of the isolator which may be adjusted empirically to obtain optimum agreement, over the deflection range of interest, between Eq. (30.58) and the actual force-deflection curve for the isolator.

The stiffness of the tangent isolator is obtained by differentiation of Eq. (30.58) with respect to δ:

$$k = \frac{dF}{d\delta} = k_0 \sec^2\left(\frac{\pi\delta}{2h_c}\right) = k_0\left[1 + \left(\frac{F\pi}{2k_0 h_c}\right)^2\right] \tag{30.59}$$

The stiffness-deflection relation defined by Eq. (30.59) is shown graphically in Fig. 30.33B.

Replacing the load F by $n_g W$ in Eq. (30.59) and substituting the resulting stiffness relation into Eq. (30.56):

$$f_n \sqrt{h_c} = 3.13 \sqrt{2.46 n_g^2\left(\frac{W}{k_0 h_c}\right) + \left(\frac{k_0 h_c}{W}\right)} \tag{30.60}$$

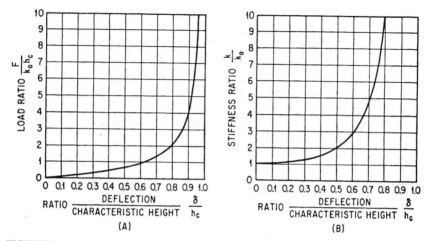

FIGURE 30.33   Elastic properties of a tangent isolator in terms of its characteristic height h_c and stiffness k_0 at zero deflection: (A) dimensionless force-deflection curve; (B) dimensionless stiffness-deflection curve.

The relation defined by Eq. (30.60) is shown graphically in Fig. 30.34. The ordinate is the natural frequency f_n (Hz) times the square root of the characteristic height of the isolator (in.). The theoretical and experimental force-deflection curves for the isolator are matched to establish the numerical value of the characteristic height. For a given value of the acceleration parameter n_g, the natural frequency of the isolation system is determined by h_c and $W/k_0 h_c$.

The deflection of the isolator under a sustained acceleration loading is obtained by substituting Eq. (30.57) into the general force-deflection expression, Eq. (30.58), and solving for the dimensionless ratio δ/h_c:

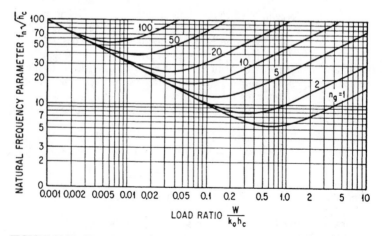

FIGURE 30.34   Natural frequency f_n of a tangent isolator system when a portion of the total load applied to the isolator is nonmassive. The weight carried by the isolator is W and the sustained acceleration parameter is n_g, a multiple of the gravitational acceleration. The characteristic height is h_c and the stiffness at zero deflection is k_0.

$$\frac{\delta}{h_c} = \frac{2}{\pi} \tan^{-1}\left(\frac{\pi n_g}{2} \cdot \frac{W}{k_0 h_c}\right) = \frac{2}{\pi} \tan^{-1}\left[15.37\left(\frac{n_g}{h_c f_{n_0}^2}\right)\right] \tag{30.61}$$

A reference natural frequency f_{n_0} is the natural frequency that occurs when the isolator is not deflected by the dead-weight load; i.e., $n_g = 0$. The nomograph of Fig. 30.35 gives the deflection ratio δ/h_c and the frequency ratio f_n/f_{n_0}.[14] The value of the parameter $15.37(n_g/h_c f_{n_0}^2)$ is transferred by a horizontal projection to the coordinate system for the curves. Values for the natural frequency ratio f_n/f_{n_0} are read from the lower abscissa scale and values for the deflection ratio δ/h_c are read from the upper abscissa scale.

Example 30.3. A rubber isolator having a characteristic height $h_c = 0.5$ in. (determined experimentally for the particular isolator design) has a natural frequency $f_n = 10$ Hz for small deflections and a fraction of critical damping $\zeta = 0.2$. The equipment supported by the isolator is subjected to a sustained acceleration of $11g$. It is desired to determine the absolute transmissibility of the isolation system when the forcing frequency is 100 Hz, and to determine the deflection of the isolator under the sustained acceleration.

Referring to the nomograph of Fig. 30.35, a straight line is drawn from a value of 10 on the f_{n_0} scale to 0.5 on the h_c scale. A second straight line is drawn from the intersection of the first line with the R scale through the value $n_g = 11$. The second line intersects the left side of the coordinate system and is extended horizontally so that it intersects the solid and dotted curves. The intersection points indicate that the natural frequency ratio $f_n/f_{n_0} = 3.5$ and the deflection ratio $\delta/h_c = 0.81$. The deflection

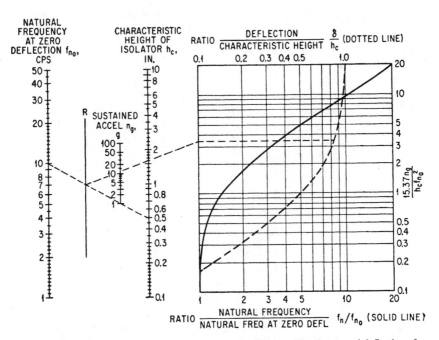

FIGURE 30.35 Nomograph and curve for determining the natural frequency and deflection of an isolation system incorporating a tangent isolator when a portion of the total load applied to the isolator is nonmassive.

of the isolator at equilibrium as a result of the sustained acceleration is $0.81h_c = 0.405$ in. The undamped natural frequency for the sustained acceleration of $11g$ is $f_n = 3.5 \times 10 = 35$ Hz. The natural frequency also can be obtained from Fig. 30.34 by noting that $W/k_0 h_c = (g/h_c)/(2\pi f_{n_0})^2 = 0.196$ [see Eq. (30.60) when $n_g = 0$]. Then for $n_g = 11$, $f_n = 24.5/\sqrt{0.5} = 35$ Hz.

From Fig. 30.2 the transmissibility for $\zeta = 0.2$, $f/f_n = 100/35 = 2.88$ is 0.22. In the absence of the sustained acceleration, the corresponding transmissibility would be 0.042 as obtained from Fig. 30.2 at $f/f_n = 100/10 = 10$. Thus, the transmissibility at 100 Hz under a sustained acceleration of $11g$ is 5 times as great as that which would exist for a dead-weight loading of the isolator.

Minimum Natural Frequency.   The weight W_0 for which a given tangent isolator has a minimum natural frequency is[15]

$$W_0 = \frac{2k_0 h_c}{\pi n_g} = \frac{k_0 g}{2\pi^2 (f_n)_{\min}^2} \qquad [f_n = \text{minimum}] \tag{30.62}$$

where the minimum natural frequency $(f_n)_{\min}$ is defined by

$$(f_n)_{\min} = \frac{1}{2} \sqrt{\frac{n_g g}{\pi h}} \tag{30.63}$$

The minimum natural frequency is shown graphically in Fig. 30.36 as a function of the characteristic height h_c and the sustained acceleration parameter n_g. The weight W_0 required to produce the minimum natural frequency $(f_n)_{\min}$ is shown graphically in Fig. 30.37 as a function of the initial stiffness k_0 and the minimum natural frequency $(f_n)_{\min}$. When the isolator is loaded to produce the minimum natural frequency, the isolator deflection is one-half the characteristic height $(\delta = h_c/2)$ and the stiffness under load is twice the initial stiffness $(k = 2k_0)$.

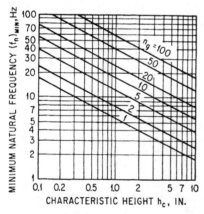

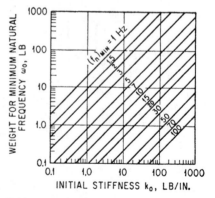

FIGURE 30.36 Minimum natural frequency $f_{n(\min)}$ of a tangent isolator system as a function of (1) the characteristic height h_c of the isolator and (2) the sustained acceleration n_g expressed as a multiple of the gravitational acceleration.

FIGURE 30.37 Weight loading W_0 required to cause a tangent isolator to have a minimum natural frequency $f_{n(\min)}$, as a function of the stiffness k_0 at zero deflection.

ISOLATION OF RANDOM VIBRATION

In random vibration, all frequencies exist concurrently, and the amplitude and phase relations are distributed in a random manner. A trace of random vibration is illustrated in Fig. 11.1A. The equipment-isolator assembly responds to the random vibration with the substantially single-frequency pattern shown in Fig. 11.1B. This response is similar to a sinusoidal motion with a continuously and irregularly varying envelope; it is described as narrow-band random vibration or a random sine wave.

The characteristics of random vibration are defined by a frequency spectrum of power spectral density (see Chaps. 11 and 22). This is a generic term used to designate the mean-square value of some magnitude parameter passed by a filter, divided by the bandwidth of the filter, and plotted as a spectrum of frequency. The magnitude is commonly measured as acceleration in units of g; then the particular expression to use in place of power spectral density is mean-square acceleration density, commonly expressed in units of g^2/Hz. When the spectrum of mean-square acceleration density is substantially flat in the frequency region extending on either side of the natural frequency of the isolator, the response of the isolator may be determined in terms of (1) the mean-square acceleration density of the isolated equipment and (2) the deflection of the isolator at successive cycles of vibration.

The mean-square acceleration densities of the foundation and the isolated equipment are related by the absolute transmissibility that applies to sinusoidal vibration:

$$W_r(f) = W_e(f)T_A^2 \tag{30.64}$$

where $W_r(f)$ and $W_e(f)$ are the mean-square acceleration densities of the equipment and the foundation, respectively, in units of g^2/Hz and T_A is the absolute transmissibility for the vibration-isolation system [see Eq. (13.34)].

The severity of the vibration experienced by the isolated equipment may be expressed in terms of the rms value of acceleration at the foundation by integrating the mean-square acceleration density given by Eq. (30.64):

$$\ddot{x}_{\text{rms}} = \sqrt{\int W_e(f)T_A^2\,df} \tag{30.65}$$

where $\ddot{x}_{\text{rms}}$ is the rms acceleration of the equipment and the integration is carried out over the frequency interval for which $W_e(f)$ is defined.

The clearance in the isolator is obtained from Eq. (11.35) by substituting (1) $W_e(f)$ in units of g^2/Hz for $2\pi W_e(\omega_n)/g^2$ where $W_e(\omega_n)$ is in units of $(\text{m/sec}^2)^2/(\text{rad/sec})$, (2) $\omega_n = 2\pi f_n$, and (3) $(T_A)_{\text{max}} = 1/2\zeta$. The rms relative deflection δ_{rms} for a rigidly connected viscous damped isolator is

$$\delta_{\text{rms}} = \sqrt{\frac{g^2}{32\pi^3}\frac{(T_A)_{\text{max}}W_e(f)}{f_n^3}} = 12.25\sqrt{\frac{(T_A)_{\text{max}}W_e(f)}{f_n^3}} \quad \text{in.} \tag{30.66}$$

where $W_e(f)$ is the mean-square acceleration density of the foundation measured in g^2/Hz. Equation (30.66) applies only if the fraction of critical damping $\zeta = c/c_c$ for the isolation system is relatively small; i.e., $(T_A)_{\text{max}}$ is relatively large. The rms isolator clearance (relative motion between the equipment and the foundation) given by Eq. (30.66) is shown graphically by Fig. 30.38.

Example 30.4. Suppose the vibration of the foundation is defined by a flat spectrum of mean-square acceleration density of $0.2g^2$/Hz over a frequency band from 10 to 500 Hz (wide relative to the width of the absolute transmissibility curve

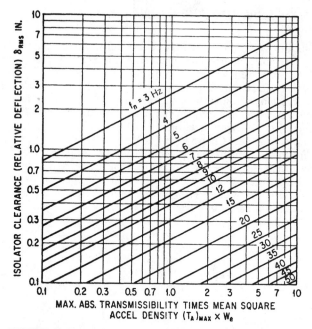

FIGURE 30.38   Required clearance expressed in inches rms for a damped isolator with viscous damping when subjected to random vibration defined by a flat spectrum of mean-square acceleration density $W_e(f)$. The natural frequency of the isolator system in cycles per second is f_n.

of the isolator in the region of resonance). The isolator has a natural frequency of 25 Hz and damping defined by $(T_A)_{max} = 5$ ($\zeta = 0.1$). Entering Fig. 30.38 at $(T_A)_{max}W_e(f)$ = $5 \times 0.2 = 1.0$ on the abscissa scale, the rms isolator clearance as read from the ordinate is $\delta_{rms} = 0.093$ in.

The deflection of the isolator varies from cycle to cycle. If the vibration has truly normal (Gaussian) characteristics, as discussed in Chaps. 11 and 22, very large values of amplitude occur occasionally. Then bottoming of the isolator cannot be prevented while maintaining the clearance reasonably small; rather, it can be made less frequent by increasing the isolator clearance. For example,* if the clearance of the isolator (in plus and minus directions along axis of vibration) is made $3\delta_{rms}$, it is probable that the isolator will bottom once in each 100 cycles of vibration; if the clearance is $4.3\delta_{rms}$, bottoming is probable once in 10^4 cycles; and if the clearance is $5.3\delta_{rms}$, bottoming is probable once in 10^6 cycles. It is common in certain testing procedures to limit the maximum amplitude to 3 times the rms value; then it is unlikely that bottoming will occur during tests if the clearance is made somewhat greater than $3\delta_{rms}$.

When the spectrum of mean-square acceleration density defining vibration of the support is not flat in the region of isolator natural frequency, the integration leading to the simple result of Eq. (30.66) cannot be carried out analytically. An equivalent

---

* See (A), Fig. 22.10.

result can be obtained by graphical integration of Eq. (11.35) where $A(j\omega)$ is a transfer function relating the acceleration of the support to the deflection of the isolator. When the mean-square acceleration density of the foundation $W_e(f)$ is not flat in the region of resonance, the rms isolator clearance is given approximately by

$$\delta_{rms} = 9.75 \sqrt{\frac{\int_{f_1}^{f_2} W_e(f) T_A{}^2 \frac{df_n}{f_n}}{f_n{}^3}} \quad \text{in.} \tag{30.67}$$

where $T_A{}^*$ is the absolute transmissibility of a rigidly connected viscous-damped isolation system; f_n is the natural frequency of the isolation system, Hz; and $W_e(f)$ is the mean-square acceleration density of the foundation, in units of g^2/Hz between the frequency limits f_1 and f_2.

A procedure for evaluating Eq. (30.67) graphically is shown in Fig. 30.39. The square of the absolute transmissibility curve is plotted on the left side of the figure to a dimensionless frequency scale that is divided, in the example shown, into increments $\Delta f_n/f_n = 0.25$. Smaller increments may be used for more accurate results. In the example shown, the area of each block is $1 \times 0.25 = 0.25$, where the differential df_n/f_n is replaced by the finite incremental value $\Delta f_n/f_n$. The spectrum of mean-square acceleration density is drawn at the right side of the figure. For each frequency increment $\Delta f_n/f_n$, the product of $W_e(f)$ and the area under the curve of transmissibility squared is entered in the right-hand column. The sum of this column is

$$\int_0^\infty W_e(f) T_A{}^2 df_n/f_n$$

δ_{rms} may be calculated from Eq. (30.67) using the appropriate value of the isolator natural frequency f_n. The necessary clearance and probability of bottoming can be determined as discussed above with reference to a flat spectrum of mean-square acceleration density.

VIBRATION ISOLATION OF NONRIGID BODY SYSTEMS

This section discusses the effect of nonrigidity of structures to which isolators are attached, in contrast to the rigid body theory discussed in the preceding sections of this chapter. Two aspects of vibration isolation are discussed in detail:

1. The equipment is attached by an isolator to a vibrating foundation which contains a power source. The vibration transferred to the equipment is determined.

2. A force-generating piece of equipment is connected by an isolator to a foundation which is at rest initially. The response of the equipment, the force transmitted to the foundation, and the resulting vibration of the foundation are determined.

---

* Equation (30.67) is approximate since the absolute transmissibility function T_A is used in place of a transfer response function (determined from a rigorous analysis) that has a frequency dependency similar to T_A in the region of resonance. It is desirable to use the absolute transmissibility T_A in the relation defined by Eq. (30.67) since this property of an isolation system is most frequently available by experimental means.

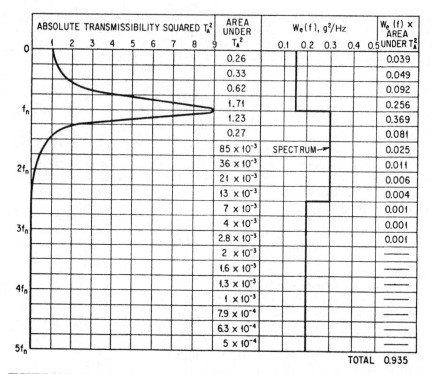

ABSOLUTE TRANSMISSIBILITY SQUARED T_A^2	AREA UNDER T_A^2	$W_e(f)$, g^2/Hz	$W_e(f) \times$ AREA UNDER T_A^2
	0.26		0.039
	0.33		0.049
	0.62		0.092
	1.71		0.256
	1.23		0.369
	0.27		0.081
	85×10^{-3}	SPECTRUM→	0.025
	36×10^{-3}		0.011
	21×10^{-3}		0.006
	13×10^{-3}		0.004
	7×10^{-3}		0.001
	4×10^{-3}		0.001
	2.8×10^{-3}		0.001
	2×10^{-3}		——
	1.6×10^{-3}		——
	1.3×10^{-3}		——
	1×10^{-3}		——
	7.9×10^{-4}		——
	6.3×10^{-4}		——
	5×10^{-4}		——

TOTAL 0.935

FIGURE 30.39 Procedure for determining graphically the required isolator clearance when the excitation is random vibration. The example is for an isolation system with absolute transmissibility $T_A = 3(\zeta = 0.17)$ and natural frequency $f_n = 10$ Hz; the mean-square acceleration density $W_e(f)$ varies with frequency according to the spectrum shown in the figure.

The excitation in both cases is steady-state sinusoidal vibration. All forces and velocities are considered colinear, and all elements are considered linear. The equipment is represented by a model having one input terminal and one output terminal; the foundation is represented by a similar model.

MECHANICAL POWER SOURCES

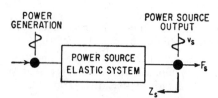

FIGURE 30.40 Schematic diagram of a power source characterized by the point impedance Z_s at the power source output terminal.

A *mechanical power source* is a means to generate sinusoidally varying forces and motions, and an associated elastic system through which the forces and motions are transmitted to the output terminal. Such a source is shown schematically in Fig. 30.40. Practically, measurements can be made only at the power source output. The force applied by the source to any system attached to the source output is represented by F_s and v_s represents the resulting velocity at the source output; F_s and v_s are related by[16]

$$F_s = F_s^b - Z_s v_s = Z_s(v_s^f - v_s) \qquad (30.68)$$

where F_s^b is the *blocked output force** of the power source, v_s^f is the *free output velocity** of the power source, and Z_s is the point mechanical impedance of the power source output. (See Chap. 10 for a detailed discussion of mechanical impedance.) The blocked output force of the source is the force transmitted to a body of infinitely great mechanical impedance attached at the source output. The free output velocity of the source is the velocity at the output with no load attached. The point mechanical impedance of the power source output Z_s is the ratio of a force applied at the source output to the resulting velocity at the output, with the source inactive. The force, velocity, and impedance are related by

$$Z_s = \frac{F_s^b}{v_s^f} \qquad (30.69)$$

IMPEDANCE PROPERTIES OF MECHANICAL ELEMENTS

A mechanical element may be represented by a generalized linear mechanical system having an input terminal i and an output terminal j, as shown by Fig. 30.41. The dynamic properties of the element may be described conveniently by the mechanical impedances (see Chap. 10) associated with the element:

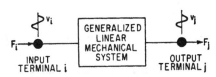

FIGURE 30.41 Schematic diagram of a generalized linear mechanical system having an input at terminal i and an output at terminal j. The velocities v_i and v_j are of the same sign when in the same direction.

$Z_i^{jf} =$ point impedance at terminal i with terminal j free $(F_j = 0)$

$Z_i^{jb} =$ point impedance at terminal i with terminal j blocked $(v_j = 0)$

$Z_{ij}^f = Z_{ji}^f =$ free transfer impedance between terminal i and terminal j; i.e., ratio of force applied at either one of the terminals to resulting velocity at the other terminal when the latter is free

$Z_{ij}^b = Z_{ji}^b =$ blocked transfer impedance between terminal i and terminal j; i.e., ratio of force developed at either one of the terminals when blocked to the velocity applied at the other terminal

Other impedance terms can be defined by substitution of the proper superscripts and subscripts in the above definitions.

---

* These quantities are the force and velocity *amplitudes* (complex) of sinusoidal quantities but are referred to as *force* and *velocity* for convenience.

Definitions for velocity and force transmissibility are:

$(T_v)_{ij} =$ velocity transmissibility from terminal i to terminal j; i.e., ratio of resulting velocity at terminal j to imposed velocity at terminal i when no load is attached at terminal j

$(T_F)_{ij} =$ force transmissibility from terminal i to terminal j; i.e., ratio of resulting force at terminal j to imposed force at terminal i when a body of infinitely great mechanical impedance is attached at terminal j

Other velocity and force transmissibilities can be defined by substitution of the proper subscripts in the above definitions. Note that the order of the subscripts in the transmissibility terms describes the direction in which the vibration is being transmitted. This order is not significant when writing transfer impedance terms since the transfer impedance between any two terminals in opposite directions is equal.

Mechanical impedance identities used in the following analyses are:

$$Z_i^{if} Z_j^{ib} = Z_i^{ib} Z_j^{if} = Z_{ij}^{f} Z_{ij}^{b}$$

$$\frac{Z_{ij}^{b}}{Z_{ij}^{f}} + 1 = \frac{Z_j^{ib}}{Z_j^{if}} = \frac{Z_i^{ib}}{Z_i^{if}} \tag{30.70}$$

VELOCITY ISOLATION

In velocity isolation, a nonrigid equipment is mounted on a vibrating foundation by a generalized vibration isolator as shown in Fig. 30.42. The foundation is considered to have the properties of the power source shown in Fig. 30.40, and its point impedance measured at point 1 is designated Z_F. Without the equipment attached, the output of the foundation is vibration with a velocity v_F^f, the free source velocity of the foundation (Fig. 30.42A). When the equipment-isolator combination is attached to the foundation (Fig. 30.42B), the velocity at the foundation output changes to v_1, and velocity v_2 is transmitted to the input terminal of the equipment. In general, the equipment and the foundation are nonrigid; the isolator is characterized by mass, stiffness, and damping. The velocity at any other point on the equipment is designated by v_3. Transmissibility is referred to as velocity transmissibility; this is equal to displacement or acceleration transmissibility because the system is linear and the vibration sinusoidal.

General Modified Velocity Transmissibility. The force F_1 applied to the isolation system at terminal 1 in Fig. 30.42B is first determined by considering the foundation as the power source and using Eq. (30.68):

$$F_1 = Z_F(v_F^f - v_1) \tag{30.71}$$

The *modified velocity transmissibility* $(T_v)_{F3}$ is the ratio of the equipment velocity v_3 to the foundation free velocity v_F^f. The velocity ratio $(T_v)_{F3} = v_3/v_F^f$ is designated a *modified* transmissibility since it is not a ratio of two velocities which exist simultaneously but rather the ratio of a resultant velocity at the equipment to the velocity

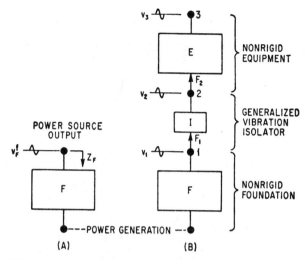

FIGURE 30.42 Schematic diagram illustrating velocity isolation. The nonrigid foundation is considered a power source and is characterized by the impedance Z_F; initially it is free as illustrated at A and vibrates with a velocity $v_F{}^f$. A generalized system comprised of a nonrigid equipment and a generalized isolator is then attached to the foundation; this changes the foundation velocity to v_1 and transmits velocities v_2 and v_3 to the equipment as illustrated at B. The velocities are of the same sign when in the same direction.

of the foundation without equipment attached. When the foundation in Fig. 30.42 is rigid, $v_F{}^f = v_1$; then the modified transmissibility becomes the ratio of two concurrent velocity amplitudes and is a true transmissibility expression. The expression for modified velocity transmissibility is[17]

$$(T_v)_{F3} = \frac{v_3}{v_F{}^f} = \cfrac{1}{\cfrac{Z_{23}{}^f}{Z_{12}{}^b}\left[1 + \cfrac{Z_1{}^{2b}}{Z_F}\right] + \cfrac{Z_{23}{}^f Z_{12}{}^f}{Z_2{}^{3f}}\left[\cfrac{1}{Z_1{}^{2f}} + \cfrac{1}{Z_F}\right]} \tag{30.72}$$

where $Z_{23}{}^f$ = free transfer impedance of equipment between terminals 2 and 3. (This may be determined by applying a known vibratory force at 2 and noting the resulting velocity at 3.)
$Z_{12}{}^b$ = blocked transfer impedance of the isolator
$Z_{12}{}^f$ = free transfer impedance of the isolator
$Z_1{}^{2b}$ = point impedance of isolator at terminal 1 with a body of infinitely great impedance attached at terminal 2
$Z_1{}^{2f}$ = point impedance of isolator at terminal 1 with no load at terminal 2
Z_F = point impedance of foundation measured at terminal 1. (In general, this determines the impedance of the foundation uniquely because any other terminal is inaccessible.)

Z_2^{3f} = point impedance of equipment at terminal 2 with no load at terminal 3

Only three of the four isolator impedances appearing in Eq. (30.72) must be measured, the fourth being determined by Eq. (30.70). If the actual velocity v_3 at terminal 3 of the equipment is desired, the free velocity v_F^f (see Fig. 30.42A) of the foundation must be measured. Thus, seven measurements are required to obtain the velocity v_3 at a point on the nonrigid equipment. In the following sections, the generalities of Eq. (30.72) are removed and corresponding expressions are derived for more restricted conditions.

Modified Velocity Transmissibility for a Massless Isolator.   If the isolator is massless, the point and transfer impedances Z_1^{2f}, Z_{12}^f of the isolator become zero; the point and transfer impedances Z_1^{2b}, Z_{12}^b become equal and are denoted by Z_I. Then Eq. (30.72) becomes

$$(T_v)_{F3} = \frac{v_3}{v_F^f} = \frac{1}{Z_{23}^f(1/Z_I + 1/Z_F + 1/Z_2^{3f})} \tag{30.73}$$

Equation (30.73) can be written more conveniently in terms of mechanical mobility (see Chap. 10):

$$(T_v)_{F3} = \frac{v_3}{v_F^f} = \frac{\mathfrak{M}_{23}^f}{\mathfrak{M}_I + \mathfrak{M}_F + \mathfrak{M}_2^{3f}} \tag{30.74}$$

where
$\mathfrak{M}_{23}^f = 1/Z_{23}^f$ = transfer mobility of equipment between terminals 2 and 3 with no load at terminal 3
$\mathfrak{M}_I = 1/Z_I^{2b} = 1/Z_{12}^b$ = mobility of isolator when measured with a load of infinite impedance attached at either terminal
$\mathfrak{M}_F = 1/Z_F$ = mobility of foundation
$\mathfrak{M}_2^{3f} = 1/Z_2^{3f}$ = mobility of equipment at terminal 2 with no load attached at terminal 3

The impedances are defined explicitly with reference to Eq. (30.72). Four measurements of impedance or mobility are required to determine the modified velocity transmissibility when the isolation system is massless.

Modified Velocity Transmissibility to Input Terminal of Nonrigid Equipment. The modified velocity transmissibility between the foundation and terminal 2 of the equipment may be determined from Eq. (30.72) by requiring terminals 2 and 3 to coincide and using the impedance identities stated by Eq. (30.70):

$$(T_v)_{F2} = \frac{v_2}{v_F^f} = \frac{Z_{12}^b}{(Z_1^{2b}/Z_F)(Z_E + Z_2^{1f}) + Z_E + Z_2^{1b}} \tag{30.75}$$

where $Z_E = Z_2^{3f} = Z_{23}^f$ represents the input point impedance of the equipment at terminal 2 for this case. This equation differs from Eq. (30.77) only in that the isolator may have mass. By use of the impedance identities stated by Eq. (30.70), the four isolator impedances indicated in Eq. (30.75) may be obtained from three measurements. In addition, the point impedances of the equipment and the foundation must be known to determine the velocity ratio v_2/v_F^f. However, if the velocity v_2 is required, v_F^f also must be measured, making a total of six measurements required. If the isolator has mass but the foundation is infinitely rigid, Eq. (30.75) becomes

$$(T_v)_{F2} = \frac{v_2}{v_F{}^f} = \frac{Z_{12}{}^b}{Z_E + Z_{12}{}^b} \tag{30.76}$$

If the isolator is massless, the modified velocity transmissibility between the foundation and terminal 2 of the equipment is

$$(T_v)_{F2} = \frac{v_2}{v_F{}^f} = \frac{Z_I}{Z_I + Z_E(1 + Z_I/Z_F)} = \frac{\mathfrak{M}_E}{\mathfrak{M}_I + \mathfrak{M}_F + \mathfrak{M}_E} \tag{30.77}$$

where the point impedance and mobility of the equipment at terminal 2 are Z_E and $\mathfrak{M}_E$, respectively, and the other parameters are defined with reference to Eqs. (30.72) to (30.74).

FORCE ISOLATION

Force isolation occurs where a power-generating equipment having the properties of a power source is attached to an initially motionless foundation by means of an isolator, as indicated schematically in Fig. 30.43. The equipment vibrates with a velocity $v_E{}^f$ when unattached. After the equipment is attached to the isolator, the velocity of the equipment changes and vibration is transmitted to the foundation.

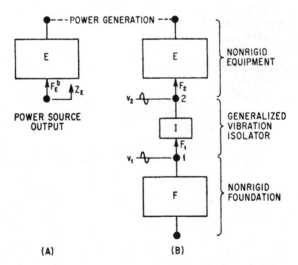

FIGURE 30.43 Schematic diagram illustrating force isolation. The nonrigid equipment is considered a power source and is characterized by the impedance Z_E; initially it is attached to a rigid foundation and transmits a force $F_E{}^b$, as illustrated at A. The equipment is then attached by means of a generalized isolator to a nonrigid foundation; this changes the output force from the equipment to F_2 and transmits a force F_1 to the foundation. The velocities v_1, v_2 are of the same sign when in the same direction.

The force transmitted to the foundation, the resulting velocity of the foundation, and the resulting velocity of the equipment are determined in the following sections.

The equipment containing the power source is characterized by its blocked output force $F_E{}^b$ (i.e., by the force applied at the output terminal when this terminal is attached to a body with an infinitely great impedance) and its point impedance Z_E as indicated in Fig. 30.43A.

Modified Force Transmissibility. The *modified force transmissibility* $(T_F)_{E1}$ is defined as the ratio of the force F_1 transmitted to the foundation to the output force $F_E{}^b$ of the equipment with point 2 blocked. The force ratio $(T_F)_{E1} = F_1/F_E{}^b$ is considered a "modified" transmissibility since it is not a true transfer force ratio of two forces that exist simultaneously; rather, it is the ratio of the force experienced by the foundation to the force generated by the equipment if rigidly constrained. Since the systems considered in this section are linear, the modified force transmissibility $(T_F)_{E1}$ is equal to the modified velocity transmissibility $(T_v)_{F2}$ given by Eq. (30.75). This relation applies for a power-generating nonrigid equipment, a generalized isolator, and a nonrigid foundation.

Massless Isolator. If the isolator is massless, all impedances (point and transfer impedances) of the isolation system with the output blocked are replaced by Z_I; the isolator impedances with the output free become zero. Then the modified force transmissibility $(T_F)_{E1}$ is equal to the modified velocity transmissibility $(T_v)_{F2}$ given by Eq. (30.77).

Rigid Foundation. If the isolator has mass, stiffness, and damping but the foundation is rigid, $Z_F = \infty$ and the modified force transmissibility $(T_F)_{E1}$ is equal to $(T_v)_{F2}$ given by Eq. (30.76).

Velocity Response of Foundation. In general, the nonrigid foundation experiences a velocity in response to force applied by the power-generating equipment. The foundation velocity v_1 is

$$v_1 = \frac{F_1}{Z_F} \tag{30.78}$$

where Z_F is the impedance of the foundation and F_1 is the force applied by the isolator. Since $(T_F)_{E1} = F_1/F_E{}^b$, the foundation velocity v_1 is

$$v_1 = \frac{F_E{}^b}{Z_F}(T_F)_{E1} \tag{30.79}$$

By substitution of the proper modified force transmissibility function for the conditions of a problem, this relation gives the foundation velocity resulting from the vibration generated by the equipment. According to Eq. (30.69), the blocked output force $F_E{}^b$ may be expressed as

$$F_E{}^b = Z_E v_E{}^f \tag{30.80}$$

where Z_E is the point impedance of the equipment measured at terminal 2. Substituting Eq. (30.80) into Eq. (30.79), the foundation velocity response v_1 can be written in dimensionless form as follows:

$$\frac{v_1}{v_E{}^f} = \frac{Z_E}{Z_F}(T_F)_{E1} \tag{30.81}$$

where v_E^f is the velocity of the equipment at terminal 2 when no load is attached. For example, velocity response of the foundation when the isolator is massless is obtained by substituting $(T_v)_{F2}$ from Eq. (30.77) for $(T_F)_{E1}$ in Eq. (30.79):

$$v_1 = \frac{F_E^b}{Z_E + Z_F(1 + Z_E/Z_I)} \qquad (30.82)$$

Equation (30.82) can be written nondimensionally by replacing F_E^b by its equivalent form defined by Eq. (30.80):

$$\frac{v_1}{v_E^f} = \frac{Z_I}{Z_I + Z_F(1 + Z_I/Z_E)} \qquad (30.83)$$

Velocity Response of Equipment. The output velocity v_E^f of the equipment which exists before the attachment of the isolator is changed to the velocity v_2 when the attachment is made because the equipment is nonrigid:

$$\frac{v_2}{v_E^f} = 1 - (T_F)_{E1}\left(\frac{Z_2^{1b}}{Z_{12}^b} + \frac{Z_{12}^f}{Z_F}\right) \qquad (30.84)$$

where Z_{12}^b = transfer impedance across isolator from terminal 1 to terminal 2, with terminal 2 blocked
Z_{12}^f = transfer impedance across isolator from terminal 1 to terminal 2, with terminal 2 free
Z_2^{1b} = point impedance of isolator at terminal 2 with terminal 1 blocked
Z_F = point impedance of the foundation at terminal 1

If the isolator is massless, $Z_{12}^b = Z_2^{1b} = Z_I$ and $Z_{12}^f = 0$; then Eq. (30.84) becomes

$$\frac{v_2}{v_E^f} = 1 - (T_F)_{E1} = \frac{Z_I + Z_F}{Z_I + Z_F(1 + Z_I/Z_E)} \qquad (30.85)$$

where $(T_F)_{E1}$ for a massless isolator is given by Eq. (30.79).

ISOLATOR EFFECTIVENESS

The effectiveness of a vibration isolator is a measure of the reduction of vibration which it effects. In concept, effectiveness may be indicated in terms of vibratory velocity or vibratory force. Effectiveness in terms of velocity is the ratio of the velocity $v_2^{(U)}$ to the velocity $v_2^{(I)}$ (see Fig. 30.44) where $v_2^{(U)}$ is the (unisolated) velocity transmitted to an equipment attached directly to the foundation and $v_2^{(I)}$ is the corresponding (isolated) velocity when an isolator is interposed between equipment and foundation; effectiveness in terms of force is the ratio of the force $F_1^{(U)}$ to the force $F_1^{(I)}$, as indicated in Fig. 30.45.[18,19]

The velocity v_2 at the input terminal of the equipment (terminal 2) is given by Eq. (30.75) in terms of the free velocity of the foundation v_F^f and several characteristic impedances. This velocity is designated as $v_2^{(I)}$ when the isolator is effective; when the isolator is considered as a rigid, massless connection, $Z_{12}^b = Z_2^{1b} = Z_1^{2b} = \infty$, $Z_2^{1f} =$

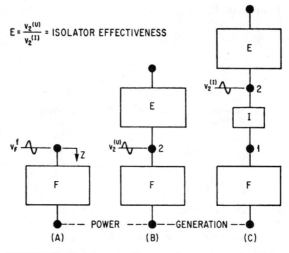

FIGURE 30.44 Schematic diagram of the systems that define isolator effectiveness in velocity isolation as illustrated by Fig. 30.42. The isolator effectiveness is the ratio of the velocity $v_2^{(U)}$ transmitted to the unisolated equipment, as illustrated at B, to the velocity $V_2^{(I)}$ transmitted to the equipment through the isolator, as illustrated at C. The foundation without the equipment attached is shown at A.

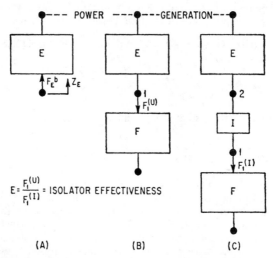

FIGURE 30.45 Schematic diagram of the systems that define isolator effectiveness in force isolation as illustrated by Fig. 30.43. The isolator effectiveness is the ratio of the force $F_1^{(U)}$ transmitted to the unisolated foundation, as illustrated at B, to the force $F_1^{(I)}$ transmitted to the foundation through the isolator, as illustrated at C. The equipment is shown at A before being attached to the foundation.

0 and the corresponding velocity at the equipment is designated $v_2^{(U)}$. Then the expression for effectiveness is

$$E = \frac{v_2^{(U)}}{v_2^{(I)}} = \frac{Z_E + Z_2^{1b} + (Z_1^{2b}/Z_F)(Z_E + Z_2^{1f})}{Z_{12}^{b}(1 + Z_E/Z_F)} \qquad (30.86)$$

where Z_E is the point impedance of the equipment at terminal 2, Z_1^{2b} is the point impedance of the isolator at terminal 1 with terminal 2 blocked, Z_2^{1f} is the point impedance of the isolator at terminal 2 with terminal 1 free, and the other parameters are defined in connection with Eq. (30.84). Inasmuch as the expression for force transmissibility is identical to Eq. (30.75) for velocity transmissibility, Eq. (30.86) for effectiveness applies to effectiveness of force transmissibility as follows:

$$E = \frac{F_1^{(U)}}{F_1^{(I)}} \qquad (30.87)$$

where $F_1^{(U)}$ is the force transmitted to the foundation with the equipment attached directly thereto (Fig. 30.45B) and $F_1^{(I)}$ is the corresponding force with the isolator interposed therebetween (Fig. 30.45C).

Effectiveness of Massless Isolator. If the isolator is massless, $Z_{12}^{b} = Z_2^{1b} = Z_1^{2b} = Z_I$ and $Z_2^{1f} = 0$; then the isolator effectiveness defined by Eq. (30.86) becomes

$$E = \frac{Z_E + Z_F(1 + Z_E/Z_I)}{Z_E + Z_F} = \frac{\mathfrak{M}_E + \mathfrak{M}_I + \mathfrak{M}_F}{\mathfrak{M}_E + \mathfrak{M}_F} \qquad (30.88)$$

NONRIGIDITY OF STRUCTURES

When the isolator is attached to a nonrigid structure of the equipment or foundation, the structure may vibrate with a relatively large amplitude if the excitation frequency coincides with a natural frequency of the structure. Thus, the isolator appears to be relatively ineffective at such frequencies and may afford little or no isolation. This effect can be evaluated quantitatively if the mechanical impedance or mobility of the equipment and foundation are known at point of attachment of the isolator.

For example, consider a free-free beam of mass m_B attached to a vibrating support by a linear, massless spring of stiffness k, as shown in Fig. 30.46. There is no damping in the system. The expression for transmissibility v_2/v_1 given by Eq. (30.77) becomes applicable by setting the foundation impedance Z_F equal to infinity:

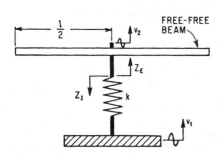

FIGURE 30.46 Schematic diagram of a free-free beam of length l supported by an isolator in the form of a massless linear spring. The beam is an example of a nonrigid equipment.

$$(T_v)_{12} = \frac{v_2}{v_1} = \frac{Z_I}{Z_E + Z_I} = \frac{1}{1 + Z_E/Z_I} \tag{30.89}$$

where Z_I is the mechanical impedance of the massless isolator with one end blocked and Z_E is the mechanical impedance at the center point of the free-free beam. Substituting the impedance of a spring $Z_I = -jk/\omega$ (see Table 10.2) and the impedance Z_E of a free-free beam at its center point, Eq. (30.89) becomes[20]

$$(T_v)_{12} = \frac{1}{1 - \left(\dfrac{f}{f_0}\right)^2 \left[1 - \displaystyle\sum_{n=1}^{\infty} \dfrac{[\phi_n(l/2)]^2 (f/f_0)^2}{(f_n/f_0)^2 - (f/f_0)^2}\right]^{-1}} \tag{30.90}$$

where $\phi_n(l/2)$ = value of nth mode function at center point of free-free beam (see Chap. 7), dimensionless

f_n = natural frequency of nth mode of beam, Hz

$f_0 = \dfrac{1}{2\pi} \sqrt{k/m_B}$ = natural frequency of beam (considered as a rigid body) on isolators, Hz

$f = \omega/2\pi$ = forcing frequency, Hz

The transmissibility as calculated from Eq. (30.90) and as determined experimentally is illustrated in Fig. 30.47 for three systems whose fundamental natural frequencies f_1 are 2, 5, and 10 times the rigid body natural frequency f_0. The transmissibility in rigid body theory, Eq. (*a*) of Table 30.2 with $\zeta = 0$, is plotted in dash-dot lines for comparison with nonrigid body conditions.

As shown in Fig. 30.47, the rigid body theory is useful only for frequencies lower than the natural frequency of the beam in its fundamental mode. For example, when the natural frequency of the beam (considered as a rigid body) on the isolator is 0.2 times the fundamental natural frequency of the beam ($f_1/f_0 = 5$), as illustrated in Fig. 30.47*B*, the rigid body theory gives an accurate indication of transmissibility only for frequency ratios f/f_0 less than 2.5. This corresponds to a forcing frequency equal to 50 percent of the lowest natural frequency of the beam. At higher frequencies, resonances of the beam become important and the rigid body theory is inapplicable. Generally similar results are obtained for other values of f_1/f_0, as indicated by Fig. 30.47*A* and *C*.

A result similar to that shown in Fig. 30.47 is obtained for force transmissibility when the foundation is nonrigid. The foundation responds with a large amplitude at its resonance frequencies; thus, the isolator appears to transmit large forces at such frequencies.

Vibration Isolation of Damped Structures.   The relatively large values of transmissibility in Fig. 30.47 result from vibration of the beam at its resonant frequencies; corresponding transmissibility curves are shown in Fig. 30.48 for a beam with significantly greater damping and with the same resonance frequencies.[21,22] The transmissibility curves for the undamped beams and the rigid body are repeated from Fig. 30.47 for comparison. The damped beams consist of two laminates separated by a thin layer of viscoelastic damping material.

Even though the damping in the beam reduces the amplification at structural resonances, the overall isolation at the higher forcing frequencies is not as efficient as the rigid body theory predicts. The mean value of transmissibility for all beams,

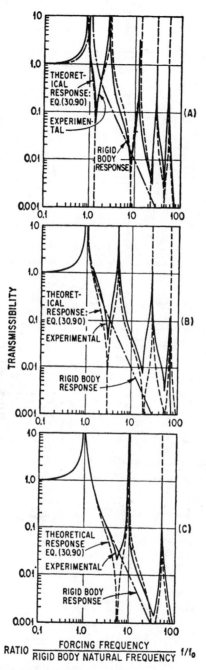

FIGURE 30.47 Acceleration transmissibility for the system shown in Fig. 30.46 for three degrees of equipment nonrigidity. The fundamental natural frequency f_1 of the beam is twice the natural frequency f_0 of the beam (considered as a rigid body supported by the isolator) at A. At B and C, the ratio f_1/f_0 is 5 and 10, respectively.

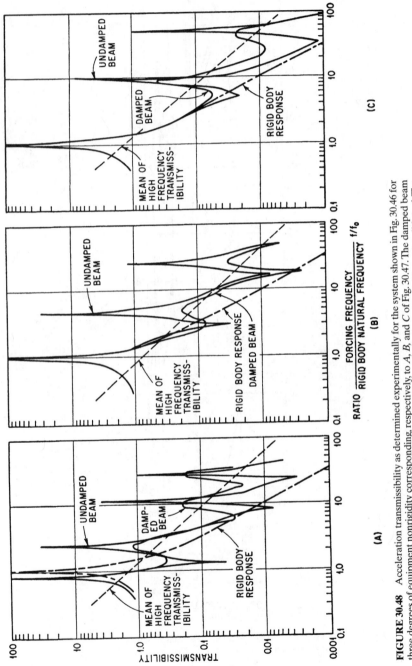

FIGURE 30.48 Acceleration transmissibility as determined experimentally for the system shown in Fig. 30.46 for three degrees of equipment nonrigidity corresponding, respectively, to A, B, and C of Fig. 30.47. The damped beam is comprised of two steel members separated by a viscoelastic damping layer; the undamped beam is solid steel. The fundamental natural frequency f_i of the beam is twice the natural frequency f_0 of the beam considered as a rigid body supported by the isolator; the curves at A; corresponding ratios of f_i/f_0 are 5 and 10 at B and C, respectively.

30.58

regardless of the amount of damping, has an attenuation rate that is less than that predicted by rigid body theory. Therefore, structural resonances, whether highly damped or not, reduce the overall efficiency of a vibration isolation system at high forcing frequencies.

WAVE EFFECTS IN ISOLATORS

When the forcing frequency becomes relatively high, standing waves tend to occur in an isolator, and the classical theory of vibration isolation based upon a massless resilient element may not give acceptable results. The transmissibility may become relatively great at the standing-wave frequencies. It is difficult to determine by analytical means standing-wave frequencies of isolators which incorporate irregularly shaped metal or rubber springs. However, the principle can be demonstrated by a simplified model.

Consider a rigid equipment of mass m_E supported by a linear unidirectional isolator having a mass m_I as shown in Fig. 30.49. The transmissibility can be written in terms of displacement by using Eq. (30.76) and noting the equivalence of velocity and displacement transmissibility:

$$(T_v)_{12} = \frac{x_0}{u_0} = \frac{Z_{12}{}^b}{Z_E + Z_2{}^{1b}} \qquad (30.91)$$

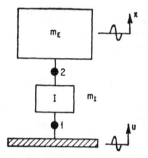

FIGURE 30.49 Schematic diagram of a system used to demonstrate the effects on transmissibility of standing waves in the isolator. The rigid equipment having a mass m_E is supported by a unidirectional isolator having mass m_I.

where $Z_{12}{}^b$ is the transfer impedance of the isolator with one end blocked, $Z_2{}^{1b}$ is the point impedance of the isolator at station 2 with station 1 blocked, and Z_E is the point impedance of the equipment at terminal 2. Assuming the isolator to be a one-dimensional mechanical transmission line, the required impedances are[16,23]

$$Z_{12}{}^b = \frac{Z_c}{\sinh \gamma l}$$

$$Z_2{}^{1b} = Z_c \left(\frac{\cosh \gamma l}{\sinh \gamma l} \right) = Z_c \coth \gamma l \qquad (30.92)$$

where the characteristic impedance Z_c is

$$Z_c = \frac{S\gamma}{\omega} [\omega\mu - j(c_0{}^2\rho)] \qquad (30.93)$$

and ρ is the mass density of the isolator material, S is the isolator cross section area undergoing vibration, μ is the viscosity of the resilient material of the isolator,* and c_0 is the undamped phase velocity or classical velocity of sound for the resilient material of the isolator when the wavelength of vibration is much larger than the lateral dimension of the isolator. The complex propagation function γ is

---

* The coefficient of viscosity μ defines an effective viscous damping coefficient having a value $\mu S/l$ where S is the cross-section area and l is the length of the resilient material of the isolator.

$$\gamma = \alpha + j\beta \tag{30.94}$$

The attenuation per unit length α (measure of damping in the material) and the phase function β may be determined by solving the general equations of motion that apply for the one-dimensional isolator.[24,25] For low damping, α and β are defined by[26]

$$\alpha = \frac{\omega^2 \mu}{2\rho c_0{}^3} \qquad \beta = \frac{\omega}{c_0} \tag{30.95}$$

If the impedance relations for the isolator given by Eq. (30.92) are substituted into the transmissibility expression given by Eq. (30.91), the following result is obtained:

$$(T_v)_{12} = \frac{1}{\cosh \gamma l + \dfrac{Z_E}{Z_c} \sinh \gamma l} \tag{30.96}$$

When the equipment is considered a rigid mass m_E, $Z_E = j\omega m_E$ and Eq. (30.96) becomes

$$(T_v)_{12} = \frac{1}{\cosh \gamma l + j\dfrac{\omega m_E}{Z_c} \sinh \gamma l} \tag{30.97}$$

The transmissibility expressions given by Eqs. (30.96) and (30.97) apply to unidirectional isolators for which the force-displacement equation is

$$F(x) = S(c_0{}^2 \rho + j\omega\mu) \frac{\partial \xi(x)}{\partial x} \tag{30.98}$$

where $F(x)$ is the force in the X coordinate direction, and $\xi(x)$ is the displacement of any cross section of the isolator from the equilibrium position in the X coordinate direction. Therefore, Eqs. (30.96) and (30.97) may be applied to isolators that can be classified as (1) resilient material strained in compression, torsion, or shear, (2) helical spring, and (3) fluid or air isolators. Expressions for c_0 and Z_c are given in Table 30.3 for the several classes of undamped isolators.[27]

For an undamped isolator, $\alpha = 0$ and $Z_c = \rho S c_0 = \sqrt{k_{st} m_I}$. Then Eq. (30.97) becomes

$$(T_v)_{12} = \frac{1}{\cos \dfrac{f}{f_0} \sqrt{\dfrac{m_I}{m_E}} - \dfrac{f}{f_0} \sqrt{\dfrac{m_E}{m_I}} \sin \dfrac{f}{f_0} \sqrt{\dfrac{m_I}{m_E}}} \tag{30.99}$$

The transmissibility defined by Eq. (30.99) is shown graphically in Fig. 30.50 as a function of the ratio of the forcing frequency f to the undamped natural frequency $f_0 = 1/2\pi\sqrt{k_{st}/m_E}$ for three values of the ratio m_E/m_I. The mass m_I is the mass of that portion of the isolator which contributes the resilience. Since no damping exists in the isolator, the transmissibility approaches infinity when the forcing frequency f approaches one of the standing-wave frequencies f_w. The valleys in the transmissibility curve between the standing-wave frequencies have minimum values that indicate a decrease in transmissibility at the rate of 6 dB/octave. The lowest frequency at which standing-wave resonances occur increases as the mass of the equipment increases relative to the mass of the isolator.

TABLE 30.3 Characteristic Parameters Relating to Wave Effects in Undamped Isolators

PROPERTY	GENERAL	COMPRESSION	TORSION	SHEAR	HELICAL SPRING	FLUID	AIR
DIAGRAMS OF ISOLATION SYSTEMS		ρ, S_c, E	ρ, D, l, I_p, G	ρ, S_s, G	ρ, N, G	ρ, S_F, κ	ρ, S_A, V_0, n
CROSS-SECTION AREA	S	S_c	$\dfrac{\pi D^2}{4}$	S_s	$\dfrac{\pi^2 N D d^2}{4l}$	S_F	S_A
STATIC STIFFNESS	k_{st}	$\dfrac{S_c E}{l}$	$\dfrac{G I_p}{l}$	$\dfrac{S_s G}{l}$	$\dfrac{G d^4}{8 N D^3}$	$\dfrac{S_F \kappa}{l}$	$\dfrac{P_0 S_A^2 n}{V_0}$
MASS OF RESILIENT MATERIAL	m_I	$\rho S_c l$	$\dfrac{\pi \rho D^2 l}{4}$	$\rho S_s l$	$\dfrac{\pi^2 \rho N D d^2}{4}$	$\rho S_F l$	ρV_0
ELASTIC WAVE PROPAGATION VELOCITY c_0	$l\sqrt{\dfrac{k_{st}}{m_I}}$	$\sqrt{\dfrac{E}{\rho}}$	$\sqrt{\dfrac{G}{\rho}}$	$\sqrt{\dfrac{G}{\rho}}$	$\dfrac{l d}{\sqrt{2}\,\pi N D^2}\sqrt{\dfrac{G}{\rho}}$	$\sqrt{\dfrac{\kappa}{\rho}}$	$\sqrt{\dfrac{P_0 n}{\rho_0}}$
CHARACTERISTIC IMPEDANCE Z_c	$\sqrt{k_{st} m_I}$	$S_c\sqrt{E\rho}$	$I_p\sqrt{G\rho}$	$S_s\sqrt{G\rho}$	$\dfrac{\sqrt{2}\,\pi d^3}{8D}\sqrt{G\rho}$	$S_F\sqrt{\kappa\rho}$	$S_A\sqrt{P_0\rho_0 n}$

DIMENSIONS OF THE RESILIENT ELEMENTS ARE INDICATED IN THE SKETCHES; RELEVANT PROPERTIES OF THE RESILIENT MATERIAL ARE:
E = YOUNG'S MODULUS, LB/IN.²
G = MODULUS OF ELASTICITY IN SHEAR, LB/IN.²
κ = BULK MODULUS, LB/IN.²
(SUBSCRIPT $_0$ INDICATES INITIAL CONDITIONS)

ρ = MASS DENSITY, LB-SEC²/IN.⁴
V = VOLUME OF RESILIENT MATERIAL, IN.³
n = RATIO OF SPECIFIC HEAT AT CONSTANT PRESSURE TO SPECIFIC HEAT AT CONSTANT VOLUME
P = PRESSURE, LB/IN.²

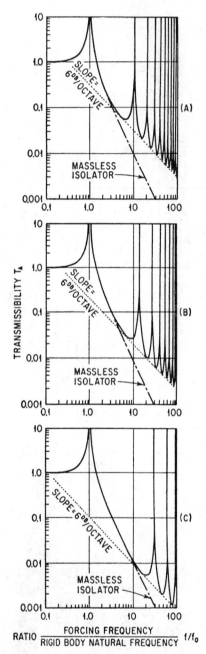

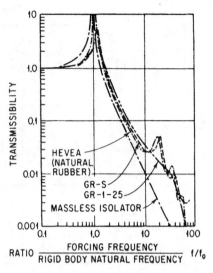

RATIO $\dfrac{\text{FORCING FREQUENCY}}{\text{RIGID BODY NATURAL FREQUENCY}}$ f/f_0

FIGURE 30.51 Transmissibility curves for the system illustrated in Fig. 30.49 using various rubber compounds, as obtained experimentally to show the effect of isolator damping. The mass m_E of the equipment is 20 times as great as the mass m_I of the isolator. (*After A. 0. Sykes.*[28])

RATIO $\dfrac{\text{FORCING FREQUENCY}}{\text{RIGID BODY NATURAL FREQUENCY}}$ f/f_0

FIGURE 30.50 Theoretical transmissibility curves for the system illustrated in Fig. 30.49. The ratio of the mass m_E of the equipment to the mass m_I of the isolator is 10 at *A*, 20 at *B*, and 100 at *C*. The isolator is undamped.

The transmissibility curves shown in Fig. 30.51 are experimental results obtained from tests on isolators made from cylindrical samples of various rubber materials for a mass ratio $m_E/m_I = 20$.[28] A comparison of the experimental curves of Fig. 30.51 with the theoretical curves of Fig. 30.50*B* indicates the effects that damping has on the standing-wave resonances. Damping in the isolator limits the transmissibility at the standing-wave resonances to finite values. If sufficient damping exists in the isolator, the standing-wave effects of the isolator cause the isolation efficiency at high frequency to decrease without exhibiting any obvious resonant peaks.

The standing-wave frequencies do not occur exactly at the frequencies predicted by the analysis because the dynamic stiffness of a rubber isolator is greater than the static stiffness used to calculate the standing-wave frequencies. Thus, the experimental transmissibility curves shift to the right on the dimensionless frequency plot of Fig. 30.51. The different rubber compounds exhibit different dynamic-to-static stiffness ratios; thus, the curves shift by different increments along the horizontal axis.

The standing-wave frequency f_w is given approximately by the relation

$$\frac{f_w}{f_0} = n\pi \sqrt{\frac{m_E}{m_I}} \qquad [n = 1,2,3,\dots] \qquad (30.100)$$

where $f_0 = 1/2\pi \sqrt{k_{st}/m_E}$ is the undamped natural frequency, n represents the mode of the standing wave, and m_E, m_I are the mass of the equipment and isolator, respectively. The relation given by Eq. (30.100) is shown graphically by Fig. 30.52 where the ratio of the standing-wave frequency f_w to the undamped natural frequency f_0 is given as a function of the mass ratio m_E/m_I for the first 10 standing-wave modes. Standing-wave frequencies determined from this graph agree with the resonances in the transmissibility curves shown in Fig. 30.50.

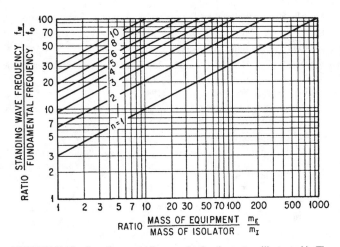

FIGURE 30.52 Standing-wave frequencies for the system illustrated in Fig. 30.49. The standing-wave frequency f_w, which is referenced to the fundamental natural frequency f_0 of the isolator system, is given as a function of the mass ratio m_E/m_I and the mode number n of the standing-wave vibration.

REFERENCES

1. Crede, C. E.: "Vibration and Shock Isolation," John Wiley & Sons, Inc., New York, 1951.

2. Den Hartog, J. P.: *Trans. ASME*, APM-53-9, 1932.

3. Jacobsen, L. S.: *Trans. ASME*, APM-52-15, 1931.

4. Ruzicka, J. E., and R. D. Cavanaugh: *Machine Design*, Oct. 16, 1958, p. 114.

5. Ruzicka, J. E.: Unpublished work.

6. Ruzicka, J. E.: "Forced Vibrations in Systems with Elastically Supported Dampers," Master's Thesis, Massachusetts Institute of Technology, Cambridge, Mass., 1957.

7. Crede, C. E., and J. P. Walsh: *J. Appl. Mechanics*, **14**:1A-7 (1947).

8. Lewis, R. C., and K. Unholtz: *Trans. ASME*, **69**:8 (1947).

9. Macduff, J. N.: *Prod. Eng.*, July, August, 1946.

10. de Gruben, K.: *VDI Zeitschrift*, **6**:41–42 (1942).

11. Crede, C. E.: *J. Appl. Mechanics*, **25**:541 (1958).

12. Timoshenko, S., and G. H. MacCullough: "Elements of Strength of Materials," 3d ed., p. 64, D. Van Nostrand Company, Inc., Princeton, N.J., 1949.

13. Mindlin, R. D.: *Bell System Tech. J.*, **24**(3–4):353 (1945).

14. Crede, C. E.: *Trans. ASME*, **76**(1):117 (1954).

15. Ruzicka, J. E.: *J. Eng. Industry (Trans. ASME)*, **83B**(1):53 (1961).

16. Molloy, C. T.: *J. Acoust. Soc. Amer.*, **29**:842 (1957).

17. Ruzicka, J. E.: Unpublished work.

18. Leedy, H. A.: *J. Acoust. Soc. Amer.*, **11**:341 (1940).

19. Sykes, A. O.: "Shock and Vibration Instrumentation," *ASME*, 1956, p. 1.

20. Ruzicka, J. E., and R. D. Cavanaugh: "Mechanical Impedance Methods for Mechanical Vibrations," *ASME*, 1958, p. 109.

21. Ruzicka, J. E.: Paper No. 100Y, SAE National Aeronautic Meeting, October, 1959.

22. Ruzicka, J. E.: *J. Eng. Industry (Trans. ASME)*, **83B**(4):403 (1961).

23. Sykes, A. O.: *Trans. SAE*, **66**:533 (1958).

24. Rayleigh, Lord: "The Theory of Sound," 2d ed., part II, p. 315, Dover Publications, New York, 1945.

25. Nolle, A. W.: *J. Acoust. Soc. Amer.*, **19**:194 (1947).

26. Harrison, M., A. O. Sykes, and M. Martin: "Wave Effects in Isolation Mounts," *David W. Taylor Model Basin Rept.* 766, 1952.

27. Ruzicka, J. E.: Unpublished work.

28. Sykes, A. O.: "A Study of Compression Noise Isolation Mounts Constructed from Cylindrical Samples of Various Natural and Synthetic Rubber Materials," *David W. Taylor Model Basin Rept.* 845, 1953.

CHAPTER 31

THEORY OF SHOCK ISOLATION

R. E. Newton

INTRODUCTION

This chapter presents an analytical treatment of the isolation of shock. Two classes of shock are considered: (1) shock characterized by motion of a support or foundation where a shock isolator reduces the severity of the shock experienced by equipment mounted on the support and (2) shock characterized by forces applied to or originating within a machine where a shock isolator reduces the severity of shock experienced by the support. In the simplified concept of shock isolation, the equipment and support are considered rigid bodies, and the effectiveness of the isolator is measured by the forces transmitted through the isolator (resulting in acceleration of equipment if assumed rigid) and by the deflection of the isolator. Linear isolators, both damped and undamped, together with isolators having special types of nonlinear elasticity are considered. When the equipment or floor is not rigid, the deflection of nonrigid members is significant in evaluating the effectiveness of isolators. Analyses of shock isolation are included which consider the response of nonrigid components of equipment and floor.

IDEALIZATION OF THE SYSTEM

In the application of shock isolators to actual equipments, the locations of the isolators are determined largely by practical mechanical considerations. In general, this results in types of nonsymmetry and coupled modes not well adapted to analysis by simple means. It is convenient in the design of shock isolators to idealize the system to a hypothetical one having symmetry and uncoupled modes of motion.

UNCOUPLED MOTIONS

The first step in idealizing the physical system is to separate the various translational and rotational modes, i.e., to *uncouple* the system. Consider the system of Fig. 31.1

consisting of a *homogeneous* block attached at the corners, by eight identical springs, to a movable rigid frame. The block and frame are constrained to move in the plane of the paper. With the system at rest, the frame is given a sudden vertical translation. Because of the symmetry of both mass and stiffness relative to a vertical plane perpendicular to the paper, the response motion of the block is pure vertical translation. Similarly, a sudden horizontal translation of the frame excites pure horizontal translation of the block. A sudden rotation about an axis through the geometric center of the block produces pure rotation of the block about this axis. This set of response behaviors is characteristic of an *uncoupled* system.

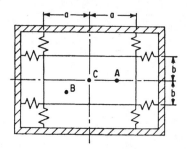

FIGURE 31.1 Schematic diagram of three degree-of-freedom mounting. Block and frame are constrained to move in plane of paper.

If the block of Fig. 31.1 is not homogeneous, the mass center (or center-of-gravity) may be at A or B instead of C. Consider the response to a sudden vertical translation of the frame if the mass center is at A. If the response were pure vertical translation of the block, the dynamic forces induced in the vertical springs would have a resultant acting vertically through C. However, the "inertia force" of the block must act through the mass center at A. Thus, the response cannot be pure vertical translation, but must also include rotation. Then the motions of vertical translation and rotation are said to be *coupled*. A sudden horizontal translation of the frame would still excite only a horizontal translation of the block because A is symmetrical with respect to the horizontal springs; thus this horizontal motion remains *uncoupled*. If the mass center were at B, i.e., in neither the vertical nor the horizontal plane of symmetry, then a sudden vertical translation of the frame would excite both vertical and horizontal translations of the block, together with rotation. In this case, all three motions are said to be *coupled*.

It is not essential that a system have any kind of geometric symmetry in order that its motions be uncoupled but rather that the resultant of the spring forces be either a force directed through the center-of-gravity of the block or a couple. If the motions are completely uncoupled, there are three mutually orthogonal directions such that translational motion of the base in any one of these directions excites only a translation of the body in the same direction. Similarly there are three orthogonal axes, concurrent at the mass center, having the property that a pure rotation of the base about any one of these axes will excite a pure rotation of the body about the same axis. The idealized systems considered in this chapter are assumed to have uncoupled rigid body motions.

ANALOGY BETWEEN TRANSLATION AND ROTATION

If the motions in translational and rotational modes are uncoupled, motion in the rotational mode may be inferred by analogy from motion in the translational mode, and vice versa. Consider the system of Fig. 31.1 and assume that the mass center is at C. For horizontal motion the differential equation of motion is*

---

* It is assumed that forces in the four vertical springs have a negligible horizontal component at all times.

$$m\ddot{\delta} + 4k\delta = -m\ddot{u} \tag{31.1}$$

where δ = horizontal displacement of mass center of block relative to center-of-frame, in.

m = mass of block, lb-sec^2/in.

k = spring stiffness for each spring, lb/in.

u = absolute horizontal displacement of center-of-frame, in.*

Equation (31.1) may be written

$$\ddot{\delta} + \omega_n^2 \delta = -\ddot{u} \tag{31.2}$$

where $\omega_n = \sqrt{4k/m}$, rad/sec, is the angular natural frequency in horizontal vibration. For rotation of the block the corresponding equation of motion is

$$I\ddot{\gamma}_r + 4k(a^2 + b^2)\gamma_r = -I\ddot{\Gamma} \tag{31.3}$$

where I = mass moment of inertia of block about axis through C, perpendicular to plane of paper, lb-in.-sec^2

a, b = distances of spring center lines from mass center (see Fig. 31.1), in.

γ_r = rotation of block relative to frame in plane of paper, rad

Γ = absolute rotation of frame in plane of figure, rad

Equation (31.3) may be written

$$\ddot{\gamma}_r + \omega_{n1}^2 \gamma_r = -\ddot{\Gamma} \tag{31.4}$$

where $\omega_{n1} = \sqrt{4k(a^2 + b^2)/I}$ is the angular natural frequency in rotation.

Equations (31.2) and (31.4) are analogous; γ_r corresponds to δ, Γ corresponds to u, and ω_{n1} corresponds to ω_n. Because of this analogy, only the horizontal motion described by Eq. (31.2) is considered in subsequent sections; corresponding results for rotational motion may be determined by analogy.

CLASSIFICATION OF SHOCK ISOLATION PROBLEMS

It is convenient to divide shock isolation problems into two major classifications according to the physical conditions:

Class I. Mitigation of effects of foundation motion

Class II. Mitigation of effects of force generated by equipment

Isolators in the first class include such items as the draft gear on a railroad car, the shock strut of an aircraft landing gear, the mounts on airborne electronic equipment, and the corrugated paper used to package light bulbs. The second class includes the recoil cylinders on gun mounts and the isolators on drop hammers, looms, and reciprocating presses.

The objectives in the two classes of problems are allied, but distinct. In Class I the objective is to limit the shock-induced stresses in critical components of the protected equipment. In Class II the purpose is to limit the forces transmitted to the support for the equipment in which the shock originates.

---

* In the equilibrium position the point C lies at the frame center.

IDEALIZED SYSTEMS—CLASS I

The simplest approach to problems of Class I is through a study of single degree-of-freedom systems.[1] Consider the system of Fig. 31.2A. The basic elements are a mass and a spring-dashpot unit attached to the mass at one end. The block may be taken to represent the equipment,* and the spring-dashpot unit to represent the shock isolator. The displacement of the support is u. The equation of motion is

$$m\ddot{\delta} + F(\dot{\delta},\delta) = -m\ddot{u} \tag{31.5}$$

where m = mass of block, lb-sec²/in.
  δ = deflection of spring ($\delta = x - u$; see Fig. 31.2), in.
 $F(\dot{\delta},\delta)$ = force exerted on mass by spring-dashpot unit (positive when tensile), lb
   u = absolute displacement of left-hand end of spring-dashpot unit, in.

In the typical shock isolation problem, the system of Fig. 31.2A is initially at rest ($\dot{u} = \dot{\delta} = 0$) in an equilibrium position ($u = \delta = 0$). An external shock causes the support to move. The corresponding movement of the left end of the shock isolator is described in terms of the support acceleration $\ddot{u}$. Then Eq. (31.5) may be solved for the resulting extreme values of δ and $F(\dot{\delta},\delta)$, and these values may be compared with the permissible deflection and force transmission limits of the shock isolator. It also is necessary to determine whether the internal stresses developed in the equipment are excessive. If the equipment is sufficiently rigid that all parts have substantially equal accelerations, then the internal stresses are proportional to $\ddot{x}$ where $-m\ddot{x} = F(\dot{\delta},\delta)$.

A critical component of the equipment may be sufficiently flexible to have a substantially different acceleration than that determined by assuming the equipment rigid. If the total mass of such components is small in comparison with the equipment mass, the above analysis may be extended to cover this case. Equation (31.5) is first solved to determine not merely the extreme value of $F(\dot{\delta},\delta)$ but its time-history. Then the acceleration $\ddot{x}$ may be determined from the relation $\ddot{x} = -F(\dot{\delta},\delta)/m$. Now consider the system shown in Fig. 31.2B having a component of mass m_c and stiffness-damping characteristics $F_c(\dot{\delta}_c,\delta_c)$. The force $F_c(\dot{\delta}_c,\delta_c)$ transmitted to the mass m_c and the resulting acceleration $\ddot{x}_c = -F_c(\dot{\delta}_c,\delta_c)/m_c$ may be found by solving an equation that is analogous to Eq. (31.5) where $\ddot{x}$ is substituted for $\ddot{u}$, δ_c for δ, and $F_c(\dot{\delta}_c,\delta_c)$ for $F(\dot{\delta},\delta)$.

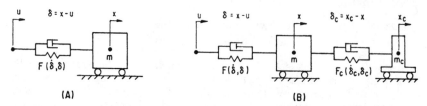

(A)   (B)

FIGURE 31.2 Idealized systems showing use of isolator with transmitted force $F(\dot{\delta},\delta)$ to protect equipment of mass m from effects of support motion u. In (A) the equipment is rigid and in (B) there is a flexible component having stiffness-damping characteristics $F_c(\dot{\delta}_c, \delta_c)$ and mass m_c.

---

 * In this simplified and idealized analysis, the equipment is considered a rigid body. The effect of flexibility of elements comprising the equipment is considered in a later section.

IDEALIZED SYSTEMS—CLASS II

Consider the system of Fig. 31.3A to represent the equipment (mass m) attached to its support by the shock isolator (spring-dashpot unit). The left end of the spring-dashpot unit is fixed to the supporting structure and there is a force F applied externally to the mass. The force F may be a real external force or it may be an "inertia force" generated by moving parts of the equipment. The equation of motion may be written

$$m\ddot{\delta} + F(\dot{\delta},\delta) = F \tag{31.6}$$

where F is the external force applied to the mass in pounds and the relative displacement δ of the ends of the spring-dashpot unit is equal to the absolute displacement x of the mass. Assuming the system to be initially in equilibrium ($\dot{\delta} = 0$, $\delta = 0$), Eq. (31.6) is solved for extreme values of δ and $F(\dot{\delta},\delta)$ since F is a known function of time. These are to be compared with the displacement and force limitations of the shock isolator. Often the supporting structure is sufficiently rigid that the maximum force in the isolator may be considered as a force applied statically to the support. Then the foregoing analysis is adequate for determining the stress in the support.

The load on the floor may be treated as dynamic instead of static by a simple analysis if the displacement and velocity of the support are negligible in comparison with those of the equipment. Consider the system of Fig. 31.3B where the supporting structure is represented as a mass m_F and a spring-dashpot unit in place of the rigid support shown in Fig. 31.3A. The force acting on the supporting structure is a known function of time $F(\dot{\delta},\delta)$ as found from the previous solution of Eq. (31.6). To find the maximum force *within* the support structure requires a solution of an equation analogous to Eq. (31.6) where $\dot{\delta}_F$ is substituted for $\dot{\delta}$, m_F for m, $F_F(\dot{\delta}_F,\delta_F)$ for $F(\dot{\delta},\delta)$, and $F(\dot{\delta},\delta)$ for F. For engineering purposes it suffices to find the extreme values of δ_F and $F_F(\dot{\delta}_F,\delta_F)$. The first is needed to verify the assumption that support motion is negligible compared with equipment motion, and can be used to determine the maximum stress in the support. The second is the maximum force applied by the support structure to its base.

MATHEMATICAL EQUIVALENCE OF CLASS I AND CLASS II PROBLEMS

The similarity of shock isolation principles in Class I and Class II is indicated by the similar form of Eqs. (31.5) and (31.6). The right-hand side ($-m\ddot{u}$ or F) is given as a function of time, and the extreme values of δ and $F(\dot{\delta},\delta)$ are desired. When the actual

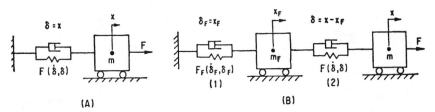

FIGURE 31.3 Idealized systems showing use of isolator with transmitted force $F(\dot{\delta},\delta)$ to reduce force transmitted to foundation when force F is applied to equipment of mass m. In (A) the foundation is rigid and in (B) it has mass m_F and stiffness damping characteristics $F_F(\dot{\delta}_F,\delta_F)$.

system is represented by two separate single degree-of-freedom systems as shown in Figs. 31.2B and 31.3B, the time-history of $F(\dot{\delta},\delta)$ is also required. Figure 31.4 may be considered a generalized form of the applicable system. In Class I, $F = 0$, $F_1(\dot{\delta}_1,\delta_1)$ represents the properties of the isolator, and $m_2,F_2(\dot{\delta}_2,\delta_2)$ represents the component to be protected. In Class II, $u = 0$, $F_2(\dot{\delta}_2,\delta_2)$ represents the properties of the isolator, and $m_1,F_1(\dot{\delta}_1,\delta_1)$ represents the supporting structure.

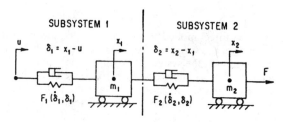

FIGURE 31.4 General two degree-of-freedom system. Figures 31.2B and 31.3B represent particular cases of the general system.

The system of Fig. 31.4, with the spring-dashpot units nonlinear, requires the use of a digital or analog computer to investigate performance characteristics. Analytical methods are feasible if the system is linearized by assuming that each spring-dashpot unit has a force characteristic in the form

$$F(\dot{\delta},\delta) = c\dot{\delta} + k\delta \tag{31.7}$$

where c = damping coefficient, lb-sec/in., and k = spring stiffness, lb/in. Even with this simplification, the number of parameters $(m_1,c_1,k_1,m_2,c_2,k_2)$ is so great that it is necessary to confine the analysis to a particular system. If the damping may be neglected [let $c = 0$ in Eq. (31.7)], then it is feasible to obtain equations in a form suitable for routine use.[2] Use of this idealization is described in the section on *Response of Equipment with a Flexible Component.*

A different form of idealization is indicated when the "equipment" is flexible; e.g., a large, relatively flexible aircraft subjected to landing shock. Then it is important to represent the aircraft as a system with several degrees-of-freedom. To find resulting stresses, it is necessary to superimpose the responses in the various modes of motion that are excited.

RESPONSE OF A RIGID BODY SYSTEM TO A VELOCITY STEP

PHYSICAL BASIS FOR VELOCITY STEP

The idealization of a shock motion as a simple change in velocity (velocity step) may form an adequate basis for designing a shock isolator and for evaluating its effectiveness. Consider the two types of acceleration $\ddot{u}$ vs. time t curves illustrated in Fig. 31.5A. The solid line represents a rectangular pulse of acceleration and the dashed line represents a half-sine pulse of acceleration. Each pulse has a duration τ. In Fig.

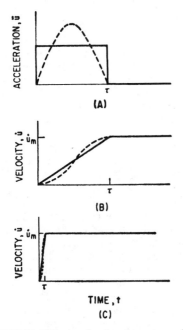

FIGURE 31.5 curves for the time scale.

FIGURE 31.5 Acceleration-time curves (A) and velocity-time curves (B) and (C) for rectangular acceleration pulse (solid curves) and half-sine acceleration pulse (dashed curves). In (C) the time scale is compressed to one-tenth that used in (A) and (B).

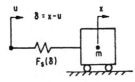

FIGURE 31.6 Idealized system showing use of undamped isolator to protect equipment from effects of support motion u. The force transmitted by the isolator is $F_s(\delta)$.

31.5B, the corresponding velocity-time curves are shown. Each of these curves is defined completely by specifying the type of acceleration pulse (rectangular or half-sine), the duration τ, and the velocity change $\dot{u}_m$. The curves of Fig. 31.5B are repeated in Fig. 31.5C with the time scale shrunk to one-tenth. If τ is *sufficiently short*, the only significant remaining characteristic of the velocity step is the velocity change $\dot{u}_m$. The idealized velocity step, then, is taken to be a discontinuous change of $\dot{u}$ from zero to $\dot{u}_m$. A shock isolator characteristically has a low natural frequency (long period), and this idealization leads to good results even when the pulse duration τ is significantly long.

GENERAL FORM OF ISOLATOR CHARACTERISTICS

The differential equation of motion for the undamped, single degree-of-freedom system shown in Fig. 31.6 is

$$m\ddot{\delta} + F_s(\delta) = -m\ddot{u} \qquad (31.8)$$

where m represents the mass of the equipment considered as a rigid body, u represents the motion of the support which characterizes the condition of shock, and $F_s(\delta)$ is the force developed by the isolator at an extension δ (positive when tensile). Equation (31.8) differs from Eq. (31.5) in that $F_s(\delta)$, which does not depend upon $\dot{\delta}$, replaces $F(\dot{\delta},\delta)$ because the isolator is undamped. The effect of a velocity step of magnitude $\dot{u}_m$ at $t = 0$ is considered by choosing the initial conditions:* At $t = 0$, $\delta = 0$ and $\dot{\delta} = \dot{u}_m$. A first integration of Eq. (31.8) yields

$$\dot{\delta}^2 = \dot{u}_m{}^2 - \frac{2}{m} \int_0^\delta F_s(\delta)d\delta \qquad (31.9)$$

At the extreme value of isolator deflection, $\delta = \delta_m$ and the velocity $\dot{\delta}$ of deflection is zero. Then from Eq. (31.9),

---

* These conditions correspond to a negative velocity step. This choice is made to avoid dealing with negative values of δ and $\dot{\delta}$. If $F_s(\delta)$ is not an odd function of δ, a positive velocity step requires a separate analysis.

$$\int_0^{\delta_m} F_s(\delta)\,d\delta = \tfrac{1}{2}m\dot{u}_m^2 \tag{31.10}$$

The right side of Eq. (31.10) represents the initial kinetic energy of the equipment relative to the support, and the integral on the left side represents the work done on the isolator. The latter quantity is equal to the elastic potential energy stored in the isolator, since there is no damping.

For the special case of a rigid body mounted on an undamped isolator, Eq. (31.10) suffices to determine all important results. In particular, the quantities of engineering significance are:

1. The maximum deflection of the isolator δ_m
2. The maximum isolator force, $F_m = F_s(\delta_m) = m\ddot{x}_m$*
3. The corresponding velocity change $\dot{u}_m$

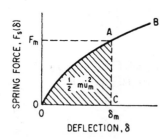

FIGURE 31.7 Typical force-deflection curve for undamped isolator. Shaded area represents kinetic energy of equipment following a velocity step $\dot{u}_m$ [Eq. (31.10)].

The interrelations of these three quantities are shown graphically in Fig. 31.7. The curve OAB represents the spring force $F_s(\delta)$ as a function of deflection δ. If point A corresponds to the extreme excursion, then its abscissa represents the maximum deflection δ_m. The shaded area OAC is proportional to the potential energy stored by the isolator; according to Eq. (31.10), this is equal to the initial kinetic energy $m\dot{u}_m^2/2$. The maximum ordinate (at A) represents the maximum spring force F_m. [It is possible to have a spring force $F_s(\delta)$ which attains a maximum value at $\delta = \delta_f < \delta_m$. Then $F_m = F_s(\delta_f)$.]

The design requirements for the isolator usually include as a specification one or more of the following quantities:

1. Maximum allowable deflection δ_a
2. Maximum allowable transmitted force F_a
3. Maximum expected velocity step $\dot{u}_a$

It is important to observe that the limits 1 and 2 establish an upper limit $F_a\delta_a$ on the work done on the mass. It follows that $\dot{u}_a$ must satisfy the relation

$$F_a\delta_a \geq m\dot{u}_a^2/2$$

or the specifications are impossible to meet. The specifications may be expressed mathematically as follows:

$$\delta_m \leq \delta_a \qquad F_m \leq F_a \qquad \dot{u}_m \geq \dot{u}_a \tag{31.11}$$

In many instances it is advantageous to eliminate explicit reference to the mass m. Then the allowable absolute acceleration $\ddot{x}_a$ of the mass is specified instead of the

---

* The maximum absolute acceleration $\ddot{x}_m$ of the equipment is related to the maximum force F_m by $F_m = m\ddot{x}_m$. It sometimes is convenient to express results in terms of $\ddot{x}_m$ instead of F_m.

allowable force F_a where $F_a = m\ddot{x}_a$. With this substitution the second of Eqs. (31.11) is replaced by

$$\ddot{x}_m \le \ddot{x}_a \qquad (31.12)$$

The acceleration $\ddot{x}$ is determined as a function of time by using $\dot{\delta}$ from Eq. (31.9) and finding the time t corresponding to a given value of δ:

$$t = \int_0^\delta \frac{d\delta}{\dot{\delta}} \qquad (31.13)$$

From Eq. (31.13) and the relation $\ddot{x} = F_s(\delta)/m$, the acceleration time-history is found.

The integrations required by Eqs. (31.9) and (31.13) sometimes are difficult to perform, and it is necessary to use numerical or graphical methods. Then a difficulty arises with the integral in Eq. (31.13): As δ approaches the extreme value δ_m, the velocity $\dot{\delta}$ in the denominator of the integrand approaches zero. The difficulty is circumvented by first using Eq. (31.13) to integrate up to some intermediate displacement δ_b less than δ_m; then the alternative form, Eq. (31.14), may be used in the region of $\delta = \delta_m$:

$$t = t_b + \int_{\dot{\delta}_b}^{\dot{\delta}} \frac{d\dot{\delta}}{\ddot{\delta}} \qquad (31.14)$$

where t_b is the time at which $\delta = \delta_b$, as determined from Eq. (31.13).

In the next three sections three different kinds of spring force-deflection characteristics $F_s(\delta)$ are considered. Equation (31.10) is applied to find the relation between $\dot{u}_m$ and δ_m. Curves relating $\dot{u}_m$, δ_m, and $\ddot{x}_m$ in a form useful for design or analysis are presented.

EXAMPLES OF PARTICULAR ISOLATOR CHARACTERISTICS

Linear Spring.   The force-deflection characteristic of a linear spring is

$$F_s(\delta) = k\delta \qquad (31.15)$$

where k = spring stiffness, lb/in. Using the notation

$$\omega_n = \sqrt{\frac{k}{m}} \qquad \text{rad/sec} \qquad (31.16)$$

the maximum acceleration is

$$\ddot{x}_m = \omega_n^2 \delta_m \qquad (31.17)$$

From Eqs. (31.10) and (31.16), the relation between velocity change $\dot{u}_m$ and maximum deflection δ_m is

$$\dot{u}_m = \omega_n \delta_m \qquad (31.18)$$

Combining Eqs. (31.18) and (31.17),

$$\ddot{x}_m = \omega_n \dot{u}_m \qquad (31.19)$$

Hardening Spring (Tangent Elasticity). The isolator spring may be nonlinear with a "hardening" characteristic; i.e., the slope of the curve representing spring force vs. deflection increases with increasing deflection. Rubber in compression has this behavior. A representative curve[1] having this characteristic is defined by

$$F_s(\delta) = \frac{2kd}{\pi} \tan \frac{\pi\delta}{2d} \qquad (31.20)$$

where the constant k is the *initial* slope of the curve (lb/in.) and a vertical asymptote is defined by $\delta = d$ (in.). Such a curve is shown graphically in Fig. 31.8. Using the notation of Eq. (31.16) and the relation $m\ddot{x}_m = F_s(\delta_m)$, Eq. (31.20) gives the following relation between maximum acceleration and maximum deflection:

$$\frac{\ddot{x}_m}{\omega_n^2 d} = \frac{2}{\pi} \tan \frac{\pi\delta_m}{2d} \qquad (31.21)$$

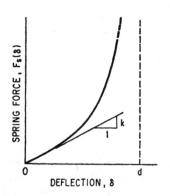

Note that ω_n, the angular natural frequency for a linear system, has the same meaning for small amplitude (small δ_m) motions of the nonlinear system. For large amplitudes the natural frequency depends on δ_m. Using Eq. (31.16), substituting for $F_s(\delta)$ from Eq. (31.20) in Eq. (31.10), and performing the indicated integration, the relation between velocity change and maximum displacement is

$$\frac{\dot{u}_m^2}{\omega_n^2 d^2} = \frac{8}{\pi^2} \log_e \left(\sec \frac{\pi\delta_m}{2d} \right) \qquad (31.22)$$

FIGURE 31.8 Typical force-deflection curve for hardening spring (tangent elasticity) as given by Eq. (31.20). There is a vertical asymptote at the limiting deflection $\delta = d$.

A graphical presentation relating the important variables $\dot{u}_m$, $\ddot{x}_m$, and δ_m is convenient for design and analysis. Such data are presented compactly as relations among the dimensionless parameters δ_m/d, $\dot{u}_m/\omega_n d$, and $\ddot{x}_m\delta_m/\dot{u}_m^2$. The physical significance of the ratio $\ddot{x}_m\delta_m/\dot{u}_m^2$ is interpreted by multiplying both numerator and denominator by m. Then the numerator represents the product of the maximum spring force $F_m(= m\ddot{x}_m)$ and the maximum spring deflection δ_m. This product is the maximum energy that *could* be stored in the spring. The denominator $m\dot{u}_m^2$ is *twice* the energy that *is* stored in the spring. The minimum *possible* value of the ratio $\ddot{x}_m\delta_m/\dot{u}_m^2$ is ½. Actual values of the ratio, always greater than ½, may be considered to be a measure of the departure from optimum capability.

In Fig. 31.9 the solid curve represents $\dot{u}_m/\omega_n d$ as a function of δ_m/d and the dashed curve shows the corresponding result for a linear spring [see Eq. (31.18)]. In Fig. 31.10 the solid curve shows $\ddot{x}_m\delta_m/\dot{u}_m^2$ vs. δ_m/d for an isolator with tangent elasticity. The dashed curve in Fig. 31.10 shows $\ddot{x}_m\delta_m/\dot{u}_m^2$ for a linear spring [see Eqs. (31.17) and (31.18)]; the ratio is constant at a value of unity because a linear spring is 50 percent efficient in storage of energy, independent of the deflection.

Softening Spring (Hyperbolic Tangent Elasticity). A nonlinear isolator also may have a "softening" characteristic; i.e., the slope of the curve representing force vs. deflection decreases with increasing deflection. The force-deflection characteristic for a typical "softening" isolator is[1]

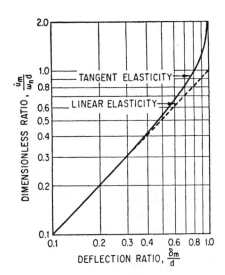

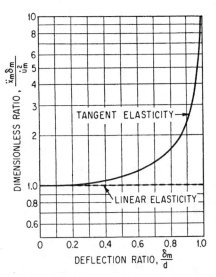

FIGURE 31.9 Dimensionless representation of relation between velocity step $\dot{u}_m$ and maximum isolator deflection δ_m for undamped isolators having tangent elasticity (solid curve) and linear elasticity (dashed curve). The natural angular frequency for small oscillations is ω_n and the limiting deflection is d (Fig. 31.8).

FIGURE 31.10 Dimensionless representation of relation among velocity step $\dot{u}_m$, maximum transmitted acceleration $\ddot{x}_m$, and maximum isolator deflection δ_m for undamped isolators having tangent elasticity (solid curve) and linear elasticity (dashed curve). Ordinate is an inverse measure of energy-storage efficiency. The natural angular frequency for small oscillations is ω_n and the limiting deflection is d (Fig. 31.8).

$$F_s(\delta) = kd_1 \tanh \frac{\delta}{d_1} \qquad (31.23)$$

where k is the initial slope of the curve. Figure 31.11 shows the form of this curve where the meaning of d_1 is evident from the figure. If $F_s(\delta)$ is replaced by $m\ddot{x}_m$, δ by δ_m, and k by $m\omega_n^2$, Eq. (31.23) becomes

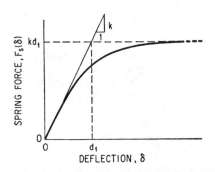

FIGURE 31.11 Typical force-deflection curve for softening spring (hyperbolic tangent elasticity) as given by Eq. (31.23). There is a horizontal asymptote at the limiting force kd_1.

$$\frac{\ddot{x}_m}{\omega_n^2 d_1} = \tanh \frac{\delta_m}{d_1} \qquad (31.24)$$

where δ_m and $\ddot{x}_m$ are maximum values of deflection and acceleration, respectively, and ω_n may be interpreted as the angular natural frequency for small values of δ_m. To relate $\dot{u}_m$ to δ_m, substitute $F_s(\delta)$ from Eq. (31.23) in Eq. (31.10), let $\omega_n^2 = k/m$, and integrate:

$$\frac{\dot{u}_m^2}{\omega_n^2 d_1^2} = \log_e\left(\cosh^2 \frac{\delta_m}{d_1}\right) \qquad (31.25)$$

A graphical presentation of the relation between $\dot{u}_m/\omega_n d_1$ and δ_m/d_1 is given by the solid curve of Fig. 31.12. The

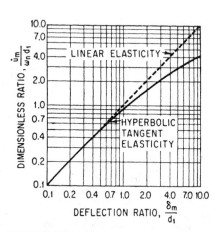

FIGURE 31.12 Dimensionless representation of relation between velocity step $\dot{u}_m$ and maximum isolator deflection δ_m for undamped isolators having hyperbolic tangent elasticity (solid curve) and linear elasticity (dashed curve). The natural angular frequency for small oscillations is ω_n and the characteristic deflection d_1 is defined in Fig. 31.11.

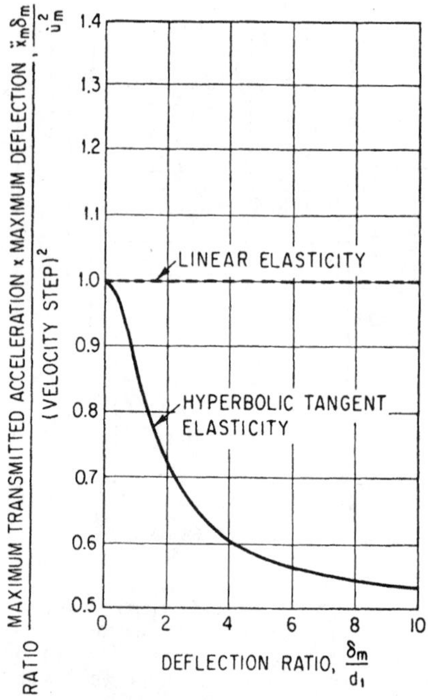

FIGURE 31.13 Dimensionless representation of energy-storage capabilities of undamped isolators having hyperbolic tangent elasticity (solid curve) and linear elasticity (dashed curve). The ordinate is an inverse measure of energy-storage efficiency. The characteristic deflection d_1 is defined in Fig. 31.11.

dashed curve shows the corresponding relation for a linear spring. In Fig. 31.13 the solid curve represents $\ddot{x}_m\delta_m/\dot{u}_m^2$ as a function of δ_m/d_1. Note that, for large values of δ_m/d_1, the ordinate approaches the minimum value ½ attainable with an isolator of optimum energy storage efficiency. The dashed curve shows the same relation for a linear spring.

Linear Spring and Viscous Damping.   The addition of viscous damping can almost double the energy absorption capability of a linear shock isolator. Consider the system of Fig. 31.2A, with both spring and dashpot linear as defined by Eq. (31.7). Substituting $F(\dot{\delta},\delta)$ from Eq. (31.7) in Eq. (31.5) gives the equation of motion. The initial conditions are $\dot{\delta} = \dot{u}_m$, $\delta = 0$, when $t = 0$; for $t > 0$, $\ddot{u} = 0$. Letting $c_c = 2m\omega_n$ and $\zeta = c/c_c$ [see Eq. (2.12)], the equation of motion becomes

$$\ddot{\delta} + 2\zeta\omega_n\dot{\delta} + \omega_n^2\delta = 0 \tag{31.26}$$

Solutions of Eq. (31.26) for maximum deflection δ_m and maximum acceleration $\ddot{x}_m$ as functions of ζ are available[1] in analytical form; the solutions are shown graphically in Figs. 31.14 and 31.15. In Fig. 31.14, the dimensionless ratio $\ddot{x}_m/\dot{u}_m\omega_n$ is plotted as a function of the fraction of critical damping ζ. Note that the presence of small damping reduces the maximum acceleration. As ζ is increased beyond 0.25, the maximum acceleration increases again. For $\zeta > 0.50$, the maximum acceleration occurs at $t = 0$ and exceeds that for no damping; it is accounted for solely by the damping force $c\dot{\delta} = c\dot{u}_m$.

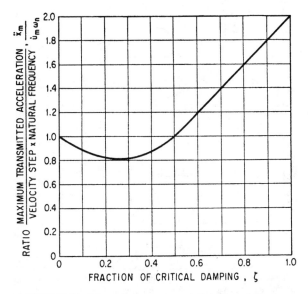

FIGURE 31.14 Dimensionless representation of maximum transmitted acceleration $\ddot{x}_m$ for an isolator having a linear spring and viscous damping.

In Fig. 31.15 the parameter $\ddot{x}_m \delta_m / \dot{u}_m^2$ is plotted as a function of ζ. (As pointed out with reference to Fig. 31.10, $\ddot{x}_m \delta_m / \dot{u}_m^2$ is an inverse measure of shock isolator effectiveness.) Figure 31.15 shows that the presence of damping improves the energy storage effectiveness of the isolator even beyond $\zeta = 0.50$. In the neighborhood $\zeta = 0.40$, the parameter $\ddot{x}_m \delta_m / \dot{u}_m^2$ attains a minimum value of 0.52—only slightly above the theoretical minimum of 0.50. This parameter has the value 1.00 for an undamped linear system, and even higher values for a hardening spring (see Fig. 31.10). On the other hand, $\ddot{x}_m \delta_m / \dot{u}_m^2$ may approach 0.50 when a softening spring is used.

True viscous damping of the type considered above is difficult to attain except in electrical or magnetic form. Fluid dampers which depend upon orifices or other constricted passages to throttle the flow are likely to produce damping forces that vary more nearly as the square of the velocity. Dry friction tends to provide damping forces which are virtually independent of velocity. The analysis of response to a velocity step in the presence of Coulomb friction is similar to that described in the section entitled *General Formulas—No Damping*. A general analytic method for finding the behavior of systems with nonlinear springs in the presence of velocity-squared damping is also available.[3]

Example 31.1. Equipment weighing 40 lb and sufficiently stiff to be considered rigid is to be protected from a shock consisting of a velocity step $\dot{u}_a = 70$ in./sec. The

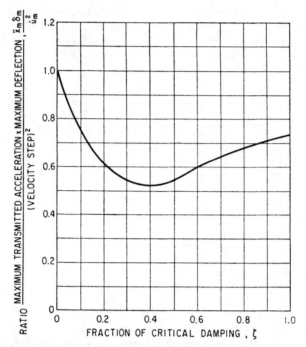

FIGURE 31.15 Dimensionless representation of energy absorption capability of an isolator having a linear spring and viscous damping. Ordinate is an inverse measure of energy absorption capability.

maximum allowable acceleration is $\ddot{x}_a = 21g$ (g is the acceleration of gravity) and available clearance limits the deflection to $\delta_a = 0.70$ in. Find isolator characteristics for: linear spring, hardening spring, softening spring, and linear spring with viscous damping.

Linear Spring. Taking the maximum velocity $\dot{u}_m$ equal to the expected velocity $\dot{u}_a$ and using Eqs. (31.18) and (31.11),

$$\delta_m = \frac{\dot{u}_m}{\omega_n} \leq \delta_a \quad \text{or} \quad \omega_n \geq \frac{70 \text{ in./sec}}{0.70 \text{ in.}} = 100 \text{ rad/sec}$$

From Eqs. (31.19) and (31.12), $\ddot{x}_m = \omega_n \dot{u}_m \leq \ddot{x}_a$. Then

$$\omega_n \leq \frac{\ddot{x}_a}{\dot{u}_m} = \frac{21 \times 386 \text{ in./sec}^2}{70 \text{ in./sec}} = 116 \text{ rad/sec}$$

Selecting a value in the middle of the permissible range gives $\omega_n = 108$ rad/sec [17.2 Hz]. The corresponding maximum isolator deflection is $\delta_m = 0.65$ in. and the maximum acceleration of the equipment is $\ddot{x}_m = 7580$ in./sec$^2 = 19.6g$. The isolator stiffness given by Eq. (31.16) is

$$k = m\omega_n^2 = \frac{40 \text{ lb}}{386 \text{ in./sec}^2} \times (108 \text{ rad/sec})^2 = 1210 \text{ lb/in.}$$

If, as is usually the case, the isolation is provided by several individual isolators in parallel, then the above value of k represents the sum of the stiffnesses of the individual isolators.

Hardening Spring. The tangent elasticity represented by Eq. (31.20) is assumed. Since the linear spring meets the specifications with only a small margin of safety, it is inferred that the poorer energy storage capacity of the hardening spring shown by Fig. 31.10 will severely limit the permissible nonlinearity. Using the specified values as maxima,

$$\frac{\ddot{x}_m \delta_m}{\dot{u}_m^2} = \frac{\ddot{x}_a \delta_a}{\dot{u}_a^2} = \frac{(21 \times 386) \times 0.70}{(70)^2} = 1.16$$

From Fig. 31.10:

$$\frac{\delta_m}{d} = 0.54; \text{ thus } d = \frac{0.70}{0.54} = 1.30 \text{ in.}$$

From Fig. 31.9:

$$\frac{\dot{u}_m}{\omega_n d} = 0.58; \text{ thus } \omega_n = \frac{70}{1.30 \times 0.54} = 93 \text{ rad/sec } [14.8 \text{ Hz}]$$

The initial spring stiffness k from Eq. (31.16) is

$$k = \frac{40}{386} (93)^2 = 896 \text{ lb/in.}$$

Because the selected linear spring provides a small margin of safety and the hardening spring provides none, superficial comparison suggests that the former is superior. Various other considerations, such as compactness and stiffness along other axes, may offset the apparent advantage of the linear spring. Moreover, a shock more severe than that specified could cause the linear spring to bottom abruptly and cause much greater acceleration of the equipment.

Softening Spring. The hyperbolic tangent elasticity represented by Eq. (31.23) is assumed. The softening spring has high energy-storage capacity as shown by Fig. 31.13. By working to sufficiently high values of δ_m/d_1, it is possible to utilize this storage capacity to afford considerable overload capability. Choose $\ddot{x}_m = 20g$ and $\delta_m/d_1 = 3$. From Fig. 31.13, $\ddot{x}_m \delta_m / \dot{u}_m^2 = 0.645$ at $\delta_m/d_1 = 3$. Then

$$\delta_m = 0.645 \frac{(70)^2}{20 \times 386} = 0.41 \text{ in.} \qquad d_1 = \frac{\delta_m}{3} = 0.137 \text{ in.}$$

From Fig. 31.12, $\dot{u}_m/\omega_n d_1 = 2.15$ at $\delta_m/d_1 = 3$.
Then

$$\omega_n = \frac{70}{2.15 \times 0.137} = 238 \text{ rad/sec } [37.9 \text{ Hz}]$$

The initial spring stiffness k from Eq. (31.16) is

$$k = \frac{40}{386} (238)^2 = 5870 \text{ lb/in.}$$

This initial stiffness is much greater than those found for the linear spring and hardening spring. Accordingly, for small shocks (small $\ddot{u}_m$) the isolator with the softening spring will induce much higher acceleration of the equipment than will those with linear or hardening springs. This poorer performance for small shocks is unavoidable if the isolator with the softening spring is designed to take advantage of the large energy-storage capability under extreme shocks.

Linear Spring and Viscous Damping. The introduction of viscous damping in combination with a linear spring [Eq. (31.7)] affords the possibility of large energy dissipation capacity without deterioration of performance under small shocks. From Fig. 31.15, the best performance is obtained at the fraction of critical damping $\zeta = 0.40$ where $\ddot{x}_m \delta_m / \ddot{u}_m^2 = 0.52$. If the maximum isolator deflection is chosen as $\delta_m = 0.47$ in. (67 percent of δ_a), then

$$\ddot{x}_m = 0.52\,\frac{\ddot{u}_m^2}{\delta_m} = 5450 \text{ in./sec}^2 = 14.1g$$

(This acceleration is 67 percent of $\ddot{x}_a$.) From Fig. 31.14:

$$\frac{\ddot{x}_m}{\ddot{u}_m \omega_n} = 0.86 \text{ at } \zeta = 0.40$$

Then

$$\omega_n = \frac{5450}{0.86 \times 70} = 90 \text{ rad/sec [14.3 Hz]}$$

The spring stiffness k from Eq. (31.16) is

$$k = \frac{40}{386}\,(90)^2 = 840 \text{ lb/in.}$$

The dashpot constant c is

$$c = 2\zeta m \omega_n = 2 \times 0.40 \times \frac{40}{386} \times 90 = 7.46 \text{ lb-sec/in.}$$

Example 31.2. The procedure for the analysis of a given isolator subjected to a specified velocity step differs somewhat from that used in design. For the isolators designed in Example 31.2, it is specified that each is subjected to a velocity step $\ddot{u}_m = 84$ in./sec (20 percent greater than in Example 31.1). Determine the maximum isolator deflection δ_m and maximum equipment acceleration $\ddot{x}_m$.

Linear Spring.
From Eq. (31.18):

$$\delta_m = \frac{\ddot{u}_m}{\omega_n} = \frac{84}{108} = 0.78 \text{ in.}$$

From Eq. (31.17):

$$\ddot{x}_m = \omega_n^2 \delta_m = (108)^2 \times 0.78 = 9110 \text{ in./sec}^2 = 23.6g$$

The influence of the 20 percent increase of velocity for this (linear) system is to increase both δ_m and $\ddot{x}_m$ by 20 percent over the values from Example 31.1.

Hardening Spring. From Example 31.1, $d = 1.30$ in. and $\omega_n = 93$ rad/sec. Then

$$\frac{\dot{u}_m}{\omega_n d} = \frac{84}{93 \times 1.30} = 0.69$$

From Fig. 31.9, the corresponding value of δ_m/d is 0.63. Thus, $\delta_m = 0.63 \times 1.30 = 0.82$ in. From Fig. 31.10, at $\delta_m/d = 0.63$, $\ddot{x}_m\delta_m/\dot{u}_m^2 = 1.26$. Therefore

$$\ddot{x}_m = \frac{1.26 \times (84)^2}{0.82} = 10{,}850 \text{ in./sec}^2 = 28.1g$$

Compared with the results of Example 31.1, δ_m and $\ddot{x}_m$ are 17 percent and 34 percent greater, respectively.

Softening Spring.   From the given data,

$$\frac{\dot{u}_m}{\omega_n d_1} = \frac{84}{238 \times 0.137} = 2.58$$

From Fig. 31.12 the corresponding value of δ_m/d_1 is 4.0; thus, $\delta_m = 4.0 \times 0.137 = 0.55$ in. From Fig. 31.13, at $\delta_m/d_1 = 4.0$, $\ddot{x}_m\delta_m/\dot{u}_m^2 = 0.603$; thus

$$\ddot{x}_m = \frac{0.603 \times (84)^2}{0.55} = 7730 \text{ in./sec}^2 = 20g$$

Comparing results with those of Example 31.1, δ_m is increased by 34 percent and $\ddot{x}_m$ is unchanged.

Linear Spring—Viscous Damping.   For $\zeta = 0.4$, Fig. 31.14 gives $\ddot{x}_m/\dot{u}_m\omega_n = 0.86$; thus, $\ddot{x}_m = 0.86 \times 84 \times 90 = 6500 \text{ in./sec}^2 = 16.9g$. From Fig. 31.15, at $\zeta = 0.4$:

$$\frac{\ddot{x}_m\delta_m}{\dot{u}_m^2} = 0.522; \text{ thus, } \delta_m = \frac{0.522 \times (84)^2}{6500} = 0.57 \text{ in.}$$

Again comparing with the results of Example 31.1, both δ_m and $\ddot{x}_m$ are increased by 20 percent. This is the result of the linearity of the system.

Of the four isolator designs determined in Example 31.1, only the softening spring and the linear spring with viscous damping can meet the original deflection and acceleration limits (0.70 in. and 21g) when the velocity step is increased in magnitude by 20 per cent. This limitation is illustrated by using the specified acceleration (21g) and deflection (0.70 in.) limits, together with the increased velocity (84 in./sec), and calculating

$$\frac{\ddot{x}_m\delta_m}{\dot{u}_m^2} = \frac{(21 \times 386) \times 0.70}{(84)^2} = 0.804$$

According to Fig. 31.10, neither a linear spring nor a hardening spring can be found to meet a required value of $\ddot{x}_m\delta_m/\dot{u}_m^2$ less than unity. On the other hand, Fig. 31.13 shows that hyperbolic tangent springs operating beyond $\delta_m/d_1 = 1.4$ may be found to meet or exceed the requirements. Similarly, Fig. 31.15 shows that viscous damped linear systems with ζ greater than 0.08 can be designed to meet or exceed the requirements.

RESPONSE OF RIGID BODY SYSTEM TO ACCELERATION PULSE

The response of a spring-mounted rigid body to various acceleration pulses provides useful information. For example, it establishes limitations upon the use of the velocity step in place of an acceleration pulse and is significant in determining the response of an equipment component when the equipment support is subjected to a velocity step. Additional useful information is afforded by comparing the responses to acceleration pulses of different shapes.

For positive pulses ($\ddot{u} > 0$) having a single maximum value and finite duration, three basic characteristics of the pulse are of importance: maximum acceleration $\ddot{u}_m$, duration τ, and velocity change $\dot{u}_c$. A typical pulse is shown in Fig. 31.16. The relation among acceleration, duration, and velocity change is

$$\dot{u}_c = \int_0^\tau \ddot{u}\, dt \qquad (31.27)$$

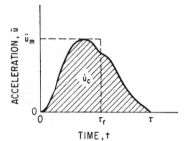

FIGURE 31.16 Typical acceleration pulse with maximum acceleration $\ddot{u}_m$ and duration τ. The shaded area represents the resulting velocity change $\dot{u}_c$. The dashed rectangle represents an *equivalent rectangular pulse* of duration τ_r, having the same velocity change $\dot{u}_c$.

where the value of the integral corresponds to the shaded area of the figure. The *equivalent rectangular pulse* is characterized by (a) the same maximum acceleration $\ddot{u}_m$ and (b) the same velocity change $\dot{u}_c$. In Fig. 31.16, the horizontal and vertical dashed lines outline the equivalent rectangular pulse corresponding to the shaded pulse. From condition (b) above and Eq. (31.27), the *effective duration* τ_r of the equivalent rectangular pulse is

$$\tau_r = \frac{1}{\ddot{u}_m} \int_0^\tau \ddot{u}\, dt \qquad (31.28)$$

where τ_r may be interpreted physically as the *average width* of the shaded pulse.

RESPONSE TO A RECTANGULAR PULSE

The rectangular pulse shown in Fig. 31.17 has a maximum acceleration $\ddot{u}_m$ and duration τ; the velocity change is $\dot{u}_c = \ddot{u}_m \tau$. The response of an undamped, linear, single degree-of-freedom system (see Fig. 31.6) to this pulse is found from the differential equation obtained by substituting in Eq. (31.8) $F_s(\delta) = k\delta$ from Eq. (31.15) and $\omega_n^2 = k/m$ from Eq. (31.16):

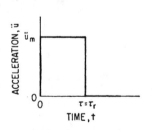

FIGURE 31.17 Rectangular acceleration pulse.

$$\ddot{\delta} + \omega_n^2 \delta = -\ddot{u}_m \qquad [0 \le t \le \tau] \quad (31.29)$$

$$\ddot{\delta} + \omega_n^2 \delta = 0 \qquad [t > \tau] \quad (31.30)$$

Using the initial conditions $\dot{\delta} = 0$, $\delta = 0$ when $t = 0$, the solution of Eq. (31.29) is

$$\delta = \frac{\ddot{u}_m}{\omega_n^2} (\cos \omega_n t - 1) \qquad [0 \le t \le \tau] \qquad (31.31)$$

For the solution of Eq. (31.30), it is necessary to find as initial conditions the values of $\dot{\delta}$ and δ given by Eq. (31.31) for $t = \tau$. Using these values the solution of Eq. (31.30) is

$$\delta = \frac{\ddot{u}_m}{\omega_n^2}[(\cos \omega_n \tau - 1) \cos \omega_n(t - \tau) - \sin \omega_n \tau \sin \omega_n(t - \tau)] \qquad [t > \tau] \quad (31.32)$$

The motion defined by Eqs. (31.31) and (31.32) is shown graphically in Fig. 31.18 for $\tau = \pi/2\omega_n$, π/ω_n, and $3\pi/2\omega_n$.

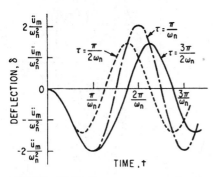

FIGURE 31.18 Response curves for an undamped linear system subjected to rectangular acceleration pulses of height $\ddot{u}_m$ and various durations τ. The angular natural frequency of the system is ω_n.

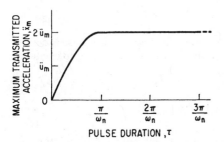

FIGURE 31.19 Maximum acceleration spectrum for a linear system of angular natural frequency ω_n. Support motion is a rectangular acceleration pulse of height $\ddot{u}_m$.

In the isolation of shock, the extreme absolute acceleration $\ddot{x}_m$ of the mass is important. Since $\ddot{x}_m = \omega_n^2 \delta_m$ [Eq. (31.17)], $\ddot{x}_m$ is found directly from the extreme value of δ. As indicated by Fig. 31.18, for values of τ greater than π/ω_n, the extreme (absolute) value of δ encountered at $t = \pi/\omega_n$ is never exceeded. For values of τ less than π/ω_n, the extreme value occurs after the pulse has ended $(t > \tau)$ and is the amplitude of the motion represented by Eq. (31.32). This amplitude may be written

$$\delta_m = 2\frac{\ddot{u}_m}{\omega_n^2}\sin\frac{\omega_n \tau}{2} \qquad (31.33)$$

The extreme absolute values of the acceleration $\ddot{x}_m$ are plotted as a function of τ in Fig. 31.19. Note that the extreme value of acceleration is twice that of the acceleration of the rectangular pulse.

HALF-SINE PULSE

Consider the "half-sine" acceleration pulse (Fig. 31.20A) of amplitude $\ddot{u}_m$ and duration τ:

$$\ddot{u} = \ddot{u}_m \sin\frac{\pi t}{\tau} \qquad [0 \le t \le \tau]$$
$$(31.34)$$
$$\ddot{u} = 0 \qquad [t > \tau]$$

From Eq. (31.28), the effective duration is

$$\tau_r = \frac{2}{\pi}\tau \qquad (31.35)$$

The response of a single degree-of-freedom system to the half-sine pulse of acceleration, corresponding to Eqs. (31.31) and (31.32) for the rectangular pulse, is defined by Eq. (8.32).

VERSED SINE PULSE

The versed sine pulse (Fig. 31.20B) is described by

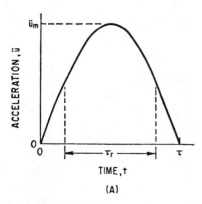
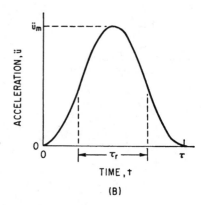

FIGURE 31.20 Half-sine acceleration pulse (*A*) and versed sine acceleration pulse (*B*). The *effective duration* is τ_r [Eq. (31.28)].

$$\ddot{u} = \frac{\ddot{u}_m}{2}\left(1 - \cos\frac{2\pi t}{\tau}\right) = \ddot{u}_m \sin^2\frac{\pi t}{\tau} \qquad [0 \le t \le \tau]$$

(31.36)

$$\ddot{u} = 0 \qquad\qquad\qquad [t > \tau]$$

The effective duration τ_r given by Eq. (31.28) is

$$\tau_r = (\tfrac{1}{2})\tau$$

(31.37)

The response of a single degree-of-freedom system to a versed sine pulse is defined by Eq. (8.33). The responses to a number of other types of pulse and step excitation also are defined in Chap. 8.

COMPARISON OF MAXIMUM ACCELERATIONS

Velocity Step Approximation. A comparison of values of $\ddot{x}_m$ resulting from various acceleration pulses with that resulting from a velocity step is shown in Fig. 31.21. The maximum acceleration induced by a velocity step is $\omega_n \dot{u}_m$ [see Eq. (31.19)]. The abscissa $\omega_n \tau_r$ is a dimensionless measure of pulse duration. The effect of pulse shape is imperceptible for values of $\omega_n \tau_r < 0.6$. For pulses of duration $\omega_n \tau_r < 1.0$, the effect of pulse shape is small and the maximum possible error resulting from use of the velocity step approximation is of the order of 5 percent.

Effects of Pulse Shape. The effects of pulse shape upon the maximum response acceleration $\ddot{x}_m$ for values of $\omega_n \tau_r > 1.0$ are shown in Fig. 31.22. The ordinate $\ddot{x}_m/\ddot{u}_m$ is the ratio of maximum acceleration induced in the responding system to maximum acceleration of the pulse. All three pulses produce the highest value of response acceleration when $\omega_n \tau_r \approx \pi$. Physically, this corresponds to an effective duration τ_r of one-half of the natural period of the spring-mass system. For longer pulse durations the curves for half-sine and versed sine pulses are similar. For pulse durations beyond the range of Fig. 31.22 ($\omega_n \tau_r > 16$), the half-sine and versed sine curves approach the limiting ordinate $\ddot{x}_m/\ddot{u}_m = 1$. This corresponds physically to approximat-

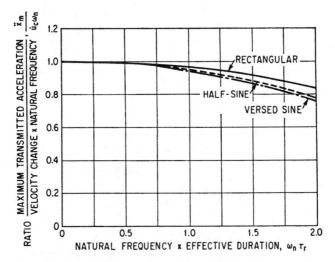

FIGURE 31.21 Dimensionless representation of maximum transmitted acceleration $\ddot{x}_m$ for the undamped linear system of Fig. 31.6. Support motion is a rectangular, half-sine, or versed sine acceleration pulse.

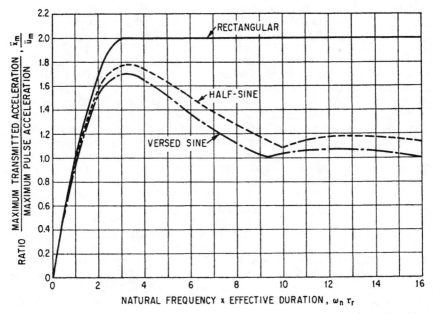

FIGURE 31.22 Shock transmissibility for the undamped linear system of Fig. 31.6 as a function of angular natural frequency ω_n and effective pulse duration τ_r. Support motion is a rectangular, half-sine, or versed sine acceleration pulse.

ing a static loading of the spring-mass system. A limiting acceleration ratio $\ddot{x}_m/\ddot{u}_m = 2$ is encountered for all rectangular pulses of duration greater than the half-period of the spring-mass system. A more extensive study of responses to a variety of pulse shapes is included in Chap. 8.

SHOCK SPECTRUM

The abscissa $\omega_n \tau_r$ in Fig. 31.22 may be treated as a measure of pulse duration (proportional to τ_r) for a given spring-mass system with ω_n fixed. Alternatively, the pulse duration may be considered fixed; then the curves show the effect of varying the natural frequency ω_n of the spring-mass system. Each of the curves of Fig. 31.22 shows the maximum acceleration induced by a given acceleration pulse upon spring-mass systems of various natural frequencies ω_n; thus, Fig. 31.22 may be used to determine the required natural frequency of the isolator if $\ddot{x}_m$ and $\ddot{u}_m$ are known, and the pulse shape is defined.

Each curve shown in Fig. 31.22 may be interpreted as a description of a pulse, in terms of the response induced in a system subjected to the pulse. The curve of maximum response as a function of the natural frequency of the responding system is called a shock spectrum or response spectrum. This concept is discussed more fully in Chap. 23. A pulse is a particular form of a shock motion; thus, each shock motion has a characteristic shock spectrum. A shock motion has a characteristic effective value of time duration τ_r which need not be defined specifically; instead, the spectra are made to apply explicitly to a given shock motion by using the natural frequency ω_n as a dimensional parameter on the abscissa. By taking the isolator-and-equipment assembly to be the responding system, the natural frequency of the isolator may be chosen to meet any specified maximum acceleration $\ddot{x}_m$ of the equipment supported by isolators. Spectra of maximum isolator deflection δ_m also may be drawn, and are useful in predicting the maximum isolator deflection when the natural frequency of the isolator is known.

When damping is added to the isolator, the analysis of the response becomes much more complex. In general, it is possible to determine the maximum value of the response acceleration $\ddot{x}_m$ only by calculating the time-history of response acceleration over the entire time interval suspected of including the maximum response. A digital computer has been used to find shock spectra for "half-sine" acceleration pulses with various fractions of critical damping in the responding system, as shown in Fig. 31.23. Similar spectra could be obtained to indicate maximum values of isolator deflection. In selecting a shock isolator for a specified application, it may be necessary to use both maximum acceleration and maximum deflection spectra. This is illustrated in the following example.

Example 31.3. A piece of equipment weighing 230 lb is to be isolated from the effects of a vertical shock motion defined by the spectra of acceleration and deflection shown in Fig. 31.24. It is required that the maximum induced acceleration not exceed $7g$ (2700 in./sec^2). Clearances available limit the isolator deflection to 2.25 in. The curves in Fig. 31.24A represent maximum response acceleration $\ddot{x}_m$ as a function of the angular natural frequency ω_n of the equipment supported on the shock isolators. The isolator springs are assumed linear and viscously damped, and separate curves are shown for values of the damping ratio $\zeta = 0, 0.1, 0.2$, and 0.3. The curves in Fig. 31.24B represent the maximum isolator deflection δ_m as a function of ω_n for the same values of ζ.

Consider first the requirement that $\ddot{x}_m < 2700$ in./sec^2. In Fig. 31.24A, the horizontal dashed line indicates this limiting acceleration. If the damping ratio $\zeta = 0.3$, then the angular natural frequency ω_n may not exceed 38.5 rad/sec on the criterion of

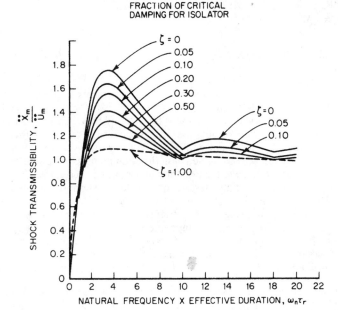

FRACTION OF CRITICAL
DAMPING FOR ISOLATOR

FIGURE 31.23 Shock transmissibility for the system of Fig. 31.2A with linear spring and viscous damping. Support motion is a half-sine acceleration pulse of height $\ddot{u}_m$ and effective duration τ_r. Curves are for discrete values of the fraction of critical damping ζ in the isolator as indicated.

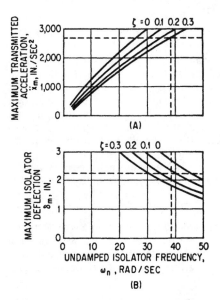

FIGURE 31.24 Shock spectra: (A) maximum acceleration and (B) maximum isolator deflection for Example 31.3.

maximum acceleration. The dashed horizontal line of Fig. 31.24B represents the deflection limit $\delta_m = 2.25$ in. For $\zeta = 0.3$, the minimum natural frequency is 30 rad/sec on the criterion of deflection. Considering both acceleration and deflection criteria, the angular natural frequency ω_n must lie between 30 rad/sec and 38.5 rad/sec. The spectra indicate that both criteria may be just met with $\zeta = 0.2$ if ω_n is 35 rad/sec. Smaller values of damping do not permit the satisfaction of both requirements.

Conservatively, a suitable choice of parameters is $\zeta = 0.3$, $\omega_n = 35$ rad/sec. This limits $\ddot{x}_m$ to 2500 in./sec^2 and δ_m to 2.0 in. The spring stiffness k is

$$k = \omega_n^2 m = (35)^2 \times \frac{230}{386} = 730 \text{ lb/in.}$$

If the equipment is to be supported by four like isolators, then the required stiffness of each isolator is $k/4 = 182.5$ lb/in.

RESPONSE OF EQUIPMENT WITH A FLEXIBLE COMPONENT

IMPACT WITH REBOUND

Consider the system of Fig. 31.4. The block of mass m_1 represents the equipment and m_2 with its associated spring-dashpot unit represents a critical component of the equipment. The left spring-dashpot unit represents the shock isolator. It is assumed here that $m_1 \gg m_2$ so that the motion of m_1 is not sensibly affected by m_2; larger values of m_2 are considered in a later section. Consider the entire system to be moving to the left at uniform velocity when the left-hand end of the isolator strikes a fixed support (not shown). The isolator will be compressed until the equipment is brought to rest. Following this the compressive force in the isolator will continue to accelerate the equipment toward the right until the isolator loses contact with the support and the rebound is complete. This type of shock is called *impact with rebound*. Practical examples include the shock experienced by a single railroad car striking a bumper at the end of a siding and that experienced by packaged equipment, shock-mounted inside a container of small mass, when the container is dropped upon a hard surface and then rebounds.

The procedure for finding the maximum acceleration $\ddot{x}_{2m}$ of the component, assuming the component stiffness to be linear and neglecting component damping, is:

1. Using the known striking velocity determine, from velocity step results (Figs. 31.9, 31.10, 31.12 to 31.15), the maximum deflection δ_{1m} of the isolator and the maximum acceleration $\ddot{x}_{1m}$ of the equipment.
2. From Eq. (31.28), find the effective duration τ_r for the acceleration time-history $\ddot{x}_1(t)$ of the equipment.
3. From the shock spectra corresponding to the acceleration pulse $\ddot{x}_1(t)$, find the maximum acceleration $\ddot{x}_{2m}$ of the component.

Details of the procedure using the isolators of Example 31.1 are considered in Example 31.4.

Example 31.4. Let the equipment of Example 31.1 weighing 40 lb have a flexible component weighing 0.2 lb. By vibration testing, this component is found to have an angular natural frequency $\omega_n = 260$ rad/sec and to possess negligible damping. For the isolators of Example 31.1, it is desired to determine the maximum acceleration $\ddot{x}_{2m}$ experienced by the mass m_2 of the component if the equipment, traveling at a velocity of 70 in./sec, is arrested by the free end of the isolator striking a fixed support. The four cases are considered separately. It is assumed that the component has a negligible effect on the motion of the equipment because $m_2 \ll m_1$.

Linear Spring. From the results of Example 31.1, it is known that $\omega_n = 108$ rad/sec and that the maximum acceleration of the equipment as found from Eq. (31.19) is

$$\ddot{x}_{1m} = 7580 \text{ in./sec}^2 = 19.6g$$

This acceleration occurs at the instant when the isolator deflection has the extreme value $\delta_{1m} = 0.65$ in.* Subsequently the isolator spring continues to accelerate the equipment until the isolator force is zero and the rebound is complete. Since there is no damping, the rebound velocity equals the striking velocity (with opposite sign).

---

* If the equipment (Fig. 31.4) is moving toward the left when the isolator contacts the support, the extreme value of δ_{1m} is negative. It suffices to deal here with absolute values.

The velocity change $\dot{x}_{1c}$ is twice the striking velocity and the effective duration τ_r [Eq. (31.28)] is

$$\tau_r = \frac{\dot{x}_{1c}}{\ddot{x}_{1m}} = \frac{2 \times 70}{7580} = 0.0185 \text{ sec}$$

The acceleration time-history of the equipment is a half-sine pulse as represented in Fig. 31.20 (the ordinate is $\ddot{x}_1$ instead of $\ddot{u}$).

Since the equipment is the "support" for the component, the response of the latter may be found from results developed for the response of a rigid body whose support experiences a half-sine pulse of acceleration. The half-sine curve of Fig. 31.22 gives the desired information if the following interpretations are made: For $\ddot{x}_m/\ddot{u}_m$ read $\ddot{x}_{2m}/\ddot{x}_{1m}$; for $\omega_n \tau_r$ read $\omega_{n2} \tau_r$. Now $\omega_{n2} \tau_r = 260 \times 0.0185 = 4.80$. From Fig. 31.22, $\ddot{x}_{2m}/\ddot{x}_{1m} = 1.66$, and $\ddot{x}_{2m} = 1.66 \times 7580 = 12,600$ in./sec$^2 = 32.6g$.

Hardening Spring. From Example 31.1, the maximum equipment acceleration is $\ddot{x}_{1m} = 21g = 8100$ in./sec^2. Since the velocity change $\dot{x}_{1c}$ is twice the striking velocity, the effective duration τ_r [Eq. (31.28)] is

$$\tau_r = \frac{\dot{x}_{1c}}{\ddot{x}_{1m}} = \frac{2 \times 70}{8100} = 0.0173 \text{ sec}$$

With a hardening isolator spring, the shape of the acceleration pulse $\ddot{x}_1(t)$ experienced by the equipment varies considerably as the maximum deflection δ_{1m} approaches the upper limit d. Up to $\delta_{1m}/d = 0.5$, the shape is closely approximated by a half-sine pulse. For $\delta_{1m}/d = 0.8$, a symmetric triangular pulse is a good approximation. For higher values of δ_{1m}/d, the pulse is very sharply peaked. The maximum response curve for a half-sine pulse is given in Fig. 31.22. The corresponding curve for a symmetric triangular pulse (Fig. 8.18b) is similar to that for the versed sine pulse, though lying generally below the latter. Inasmuch as the curve for the versed sine pulse is below that for the half-sine pulse, it is conservative to use the half-sine pulse for all values of δ_{1m}/d. Accordingly, $\omega_{n2} \tau_r = 260 \times 0.0173 = 4.50$. From the half-sine curve of Fig. 31.22, $\ddot{x}_{2m}/\ddot{x}_{1m} = 1.69$, and $\ddot{x}_{2m} = 1.69 \times 8100 = 13,700$ in./sec$^2 = 36.4g$.

Softening Spring. From Example 31.1, the maximum equipment acceleration $\ddot{x}_{1m}$ is

$$\ddot{x}_{1m} = 20g = 7720 \text{ in./sec}^2$$

The effective duration τ_r [Eq. (31.28)] is

$$\tau_r = \frac{\dot{x}_{1c}}{\ddot{x}_{1m}} = \frac{2 \times 70}{7720} = 0.0181 \text{ sec}$$

The shape of the acceleration pulse $\ddot{x}_1(t)$ for the equipment varies markedly as the departure from linearity increases (increasing values of δ_{1m}/d_1). The pulse shape is found by first performing the integration of Eq. (31.9) with $F_s(\delta)$ as given by Eq. (31.23). The result supplies the integrand required for Eq. (31.13). The integration of the latter equation is performed numerically. Results[1] show that the pulse shape undergoes a rapid transition from the half-sine pulse at very small values of δ_{1m}/d_1 to shapes that are closely approximated by the trapezoidal pulse of Fig. 31.25. Note that the pulse of Fig. 31.25 requires three parameters to fix it completely: the maximum acceleration $\ddot{x}_{1m}$; the effective duration τ_r; and the ratio τ_r/τ, where τ is the actual duration and $\tau_r = \tau - \tau_1$. From results of the numerical integrations of Eq. (31.13), the curve of Fig. 31.26 is constructed to show τ_r/τ as a function of the deflection ratio δ_{1m}/d_1.

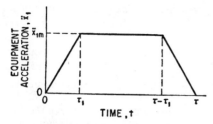

FIGURE 31.25 Symmetric trapezoidal acceleration pulse.

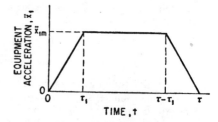

FIGURE 31.26 Dimensionless representation of effective duration τ_r of acceleration pulse experienced by equipment during impact with rebound. Isolator has undamped hyperbolic tangent elasticity. Maximum isolator deflection is δ_{1m} and characteristic deflection d_1 is defined in Fig. 31.11.

To find the maximum acceleration $\ddot{x}_{2m}$ of the component, the maximum response curves (shock spectra) of Fig. 31.27 are used. These curves are constructed for symmetric trapezoidal pulses (Fig. 31.25). The top curve ($\tau_r/\tau = 1.0$) corresponds to the limiting (rectangular) form. The dashed curve ($\tau_r/\tau = 0.64$) represents response to a half-sine pulse.

The value of δ_{1m}/d_1 corresponding to the maximum acceleration $\ddot{x}_{1m}$ of the equipment is (from Example 31.1) $\delta_{1m}/d_1 = 3$. From Fig. 31.26: $\tau_r/\tau = 0.88$. Now $\omega_n\tau_r = 260 \times$

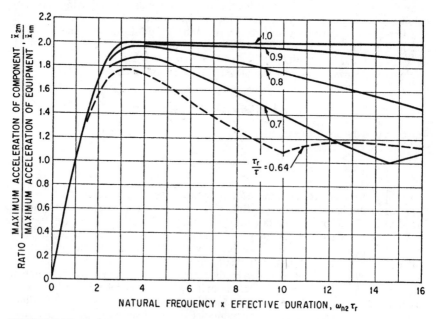

FIGURE 31.27 Shock spectra for component having undamped linear elasticity with angular natural frequency ω_{n2}. Equipment motion is symmetric trapezoidal acceleration pulse (Fig. 31.25). Curves are for five discrete values of the ratio of effective pulse duration to actual duration τ_r/τ.

0.0181 = 4.7. Using Fig. 31.27, linear interpolation between the curves for $\tau_r/\tau = 0.8$ and $\tau_r/\tau = 0.9$ gives $\ddot{x}_{2m}/\ddot{x}_{1m} = 1.98$ and $\ddot{x}_{2m} = 1.98 \times 7720 = 15{,}300$ in./sec^2 = 39.6g.

Linear Spring and Viscous Damping. The presence of damping in the isolator adds several complications: (1) the rebound velocity is no longer equal to the striking velocity; (2) the acceleration pulse of the equipment is not symmetrical and returns to zero before the isolator deformation δ_{1m} returns to zero; and (3) the pulse shape varies greatly with damping ratio ζ_1. Shock spectra for acceleration pulse shapes corresponding to damping ratios of particular interest ($0.10 < \zeta_1 < 0.40$) are not available. However, for single acceleration pulses which do not change sign, it is conservative to assume that the maximum acceleration $\ddot{x}_{2m}$ of the component is twice the maximum acceleration $\ddot{x}_{1m}$ of the equipment. Using the results of Example 31.1, the maximum acceleration of the component is $\ddot{x}_{2m} = 2\ddot{x}_{1m} = 2 \times 5450 = 10{,}900$ in./sec^2 = 28.2g.

IMPACT WITHOUT REBOUND

When impact of the isolator occurs without rebound, it must be recognized that the equipment-and-isolator system continues to oscillate until the initial kinetic energy is dissipated. Consider the system of Fig. 31.4; it consists of equipment m_1, shock isolator (left spring-dashpot unit), and flexible component (subsystem 2). The system is initially at rest. The left end of the shock isolator is attached to a support (not shown) which is given a velocity step of magnitude $\dot{u}_m$ at $t = 0$. The subsequent motion of the support is $u = \dot{u}_m t$. Determine the maximum force F_{1m} transmitted by the isolator, the maximum isolator deflection δ_{1m}, and the maximum acceleration $\ddot{x}_{2m}$ of the component.

Solutions are available only for linear systems, i.e., linear springs and viscous damping. Two such simplified analyses of this problem are included in the following sections: (1) The influence of damping is considered, but the component mass m_2 is assumed of negligible size relative to m_1 and (2) damping is neglected but the effect of the mass m_2 of the component upon the motion of the system is considered.

Component Mass Negligible. Assume that $m_1 \gg m_2$ so that the motion x_1 of the equipment may be determined by neglecting the effect of the component. Then the extreme value of the force F_{1m} transmitted by the isolator and the extreme deflection δ_{1m} of the isolator occur during the first quarter-cycle of the equipment motion; they may be found from Figs. 31.14 and 31.15 in the section on *Response of a Rigid Body System to a Velocity Step.* The subsequent motion of the equipment is an exponentially decaying sinusoidal oscillation or, if there is no damping in the isolator, a constant-amplitude oscillation. If the component also is undamped, an analytic determination of the component response is not difficult. The motion consists of harmonic oscillation at the frequency ω_{n1} of the equipment oscillation and a superposed oscillation at the frequency ω_{n2} of the component system. Since the oscillations are assumed to persist indefinitely in the absence of damping, the extreme acceleration of the component is the sum of the absolute values of the maximum accelerations associated with the oscillations at frequencies ω_{n1} and ω_{n2}. In the particular case of resonance ($\omega_{n1} = \omega_{n2}$), the vibration amplitude of the component increases indefinitely with time. Because actual systems always possess damping (usually a considerable amount in the isolator), solutions of this type tend to be unduly conservative for engineering applications.

The equation of motion for the viscous damped component is a special case of Eq. (31.5) with $F(\dot{\delta},\delta)$ as given by Eq. (31.7). If appropriate subscripts are supplied and customary substitutions are made, the equation is

$$\ddot{\delta}_2 + 2\zeta_2\omega_{n2}\dot{\delta}_2 + \omega_{n2}{}^2\delta_2 = -\ddot{x}_1 \qquad (31.38)$$

Analytic solutions of Eq. (31.38) to find the acceleration $\ddot{x}_2 = \ddot{x}_1 + \ddot{\delta}_2$ of the component are too laborious to be practical. However, results have been obtained by analog computation[1,4] and are shown in Fig. 31.28. The ordinate is the ratio of the maximum acceleration $\ddot{x}_{2m}$ of the component to the maximum acceleration $\dot{u}_m\omega_{n2}$ [see Eq. (31.19)] that the component would experience if the shock isolator were rigid. The abscissa is the ratio of the undamped natural frequency ω_{n2} of the component to the undamped natural frequency ω_{n1} of the equipment on the isolator spring. Curves are given for several different values of the fraction of critical damping ζ_1 for the isolator. For all curves the fraction of critical damping for the component is $\zeta_2 = 0.01$. The effect of isolator damping in reducing the maximum acceleration $\ddot{x}_{1m}$ of the component is great in the neighborhood of $\omega_{n2}/\omega_{n1} = 1$. Above $\omega_{n2}/\omega_{n1} = 2$, small damping ($\zeta_1 \le 0.1$) in the isolator has little effect and large damping may significantly increase the maximum acceleration of the component.

The ordinate in Fig. 31.28 represents the ratio of the maximum acceleration of the component to that which would be experienced with the isolator rigid (absent); thus, it may properly be called *shock transmissibility*. If shock transmissibility is less than unity, the isolator is beneficial (for the component considered). An isolator must have a natural frequency significantly less than that of the critical component in order to reduce the transmitted acceleration. If there are several critical components having different natural frequencies ω_{n2}, each must be considered separately

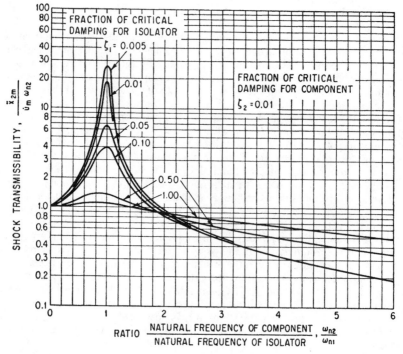

FIGURE 31.28 Shock transmissibility for a component of a viscously damped system with linear elasticity. The maximum acceleration of the component is $\ddot{x}_{2m}$ and the velocity step (support motion) is $\dot{u}_m$. The effect of the component on the equipment motion is neglected. (*After R. D. Mindlin.*[1])

and the natural frequency of the isolator must be significantly lower than the lowest natural frequency of a component.

Two Degrees-of-Freedom—No Damping. This section includes an analysis of the transient response of the two degree-of-freedom system shown in Fig. 31.4, neglecting the effects of damping but assuming the equipment mass m_1 and the component mass m_2 to be of the same order of magnitude. The equations of motion are

$$m_1 \ddot{\delta}_1 + k_1 \delta_1 = k_2 \delta_2 - m_1 \ddot{u}$$

$$m_2 \ddot{\delta}_2 + k_2 \delta_2 = -m_2 \ddot{\delta}_1 - m_2 \ddot{u}$$

$$(31.39)$$

where k_1 = stiffness of isolator spring, lb/in., and k_2 = stiffness of component, lb/in. The system is initially in equilibrium; at time $t = 0$, the left end of the isolator spring is given a velocity step of magnitude $\dot{u}_m$. Initial conditions are: $\dot{\delta}_1 = \dot{u}_m$, $\dot{\delta}_2 = 0$, $\delta_1 = \delta_2 = 0$. Equations (31.39) may be solved simultaneously[2] for maximum values of the acceleration $\ddot{x}_{2m}$ of the component and maximum deflection δ_{1m} of the isolator:

$$\ddot{x}_{2m} = \frac{\dot{u}_m \omega_{n2}}{\left[\left(\frac{\omega_{n2}}{\omega_{n1}} - 1 \right)^2 + \frac{m_2}{m_1} \left(\frac{\omega_{n2}}{\omega_{n1}} \right)^2 \right]^{1/2}} \qquad (31.40)$$

$$\delta_{1m} = \frac{\dot{u}_m}{\omega_{n1}} \frac{1 + \dfrac{\omega_{n2}}{\omega_{n1}} \left(1 + \dfrac{m_2}{m_1} \right)}{\left[\left(\dfrac{\omega_{n2}}{\omega_{n1}} + 1 \right)^2 + \dfrac{m_2}{m_1} \left(\dfrac{\omega_{n2}}{\omega_{n1}} \right)^2 \right]^{1/2}} \qquad (31.41)$$

where $\ddot{x}_{2m}$ = maximum absolute acceleration of component mass, in./sec²; δ_{1m} = maximum deflection of isolator spring, in.; ω_{n1}* = angular natural frequency of isolator $(k_1/m_1)^{1/2}$, rad/sec; and ω_{n2}* = angular natural frequency of component $(k_2/m_2)^{1/2}$, rad/sec. Equation (31.40) is shown graphically in Fig. 31.29. The dimensionless ordinate is the ratio of maximum acceleration $\ddot{x}_{2m}$ of the component to the maximum acceleration $\dot{u}_m \omega_{n2}$ which the component would experience with no isolator present. The abscissa is the ratio of component natural frequency ω_{n2} to isolator natural frequency ω_{n1}. Separate curves are given for mass ratios $m_2/m_1 = 0.01$, 0.1, 0.3, and 1.0. Equation (31.41) is shown graphically in Fig. 31.30. The ordinate is the ratio of the maximum isolator deflection δ_{1m} to the deflection $\dot{u}_m (1 + m_2/m_1)^{1/2}/\omega_{n1}$ which would occur if component stiffness k_2 were infinite. The abscissa is the ratio of natural frequencies ω_{n2}/ω_{n1}, and curves are given for values of $m_2/m_1 = 0.1$ and 1.0.

Figure 31.29 shows that the effect of the mass ratio m_2/m_1 upon the maximum component acceleration $\ddot{x}_{2m}$ is very great near resonance ($\omega_{n2}/\omega_{n1} \approx 1$). As ω_{n2}/ω_{n1} increases above resonance, the effect of finite component mass steadily decreases. Figure 31.30 shows that except for small values of ω_{n2}/ω_{n1} the effect of finite component mass on the maximum isolator deflection δ_{1m} is slight. As ω_{n2}/ω_{n1} increases, the curves for all mass ratios asymptotically approach the ordinate 1.0.

---

* The natural frequencies ω_{n1} and ω_{n2} are hypothetical in the sense that they do not consider the coupling between the subsystems.

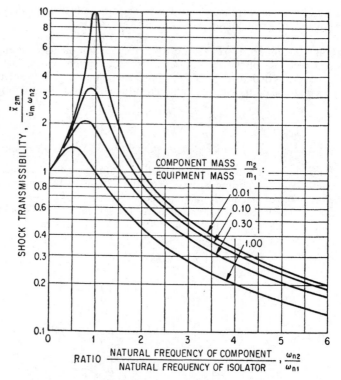

FIGURE 31.29 Shock transmissibility for component of system of Fig. 31.4 under impact at velocity $\dot{u}_m$ without rebound. Component and isolator have undamped linear elasticity. Ordinate is ratio of maximum acceleration of component $\ddot{x}_{2m}$ to maximum acceleration which would be experienced by component with a rigid isolator. Curves are for four discrete values of the mass ratio m_2/m_1 as indicated.

The factor $(1 + m_2/m_1)^{1/2}$ in the ordinate parameter of Fig. 31.30 is introduced because the total equipment mass is $m_1 + m_2$. For the limiting case of rigid equipment (k_2 infinite), the natural frequency ω_n is given by

$$\omega_n^2 = \frac{k_1}{m_1 + m_2} \qquad \omega_n = \frac{\omega_{n1}}{(1 + m_2/m_1)^{1/2}}$$

Substituting this relation in Eq. (31.18) and solving for δ_{1m}:

$$\delta_{1m} = \dot{u}_m(1 + m_2/m_1)^{1/2}/\omega_{n1}$$

This is in agreement with the result given by Eq. (31.41) as ω_{n2}/ω_{n1} approaches infinity.

Example 31.5. Equipment weighing 152 lb has a flexible component weighing 3 lb. The angular natural frequency of the component is $\omega_{n2} = 130$ rad/sec. The equipment is mounted on a shock isolator with a linear spring $k_1 = 2400$ lb/in. and having a fraction of critical damping $\zeta_1 = 0.10$. Find the maximum isolator deflection δ_{1m} and the maximum component acceleration $\ddot{x}_{2m}$ which result when the base experiences a velocity step $\dot{u}_m = 55$ in./sec.

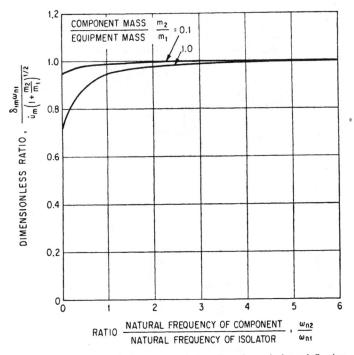

FIGURE 31.30 Dimensionless representation of maximum isolator deflection in system of Fig. 31.4 under impact at velocity $\dot{u}_m$ without rebound. Component and isolator have undamped linear elasticity. Ordinate is ratio of maximum isolator deflection δ_{1m} to the isolator deflection which would result if the component stiffness were infinite. Curves are for two discrete values of the mass ratio m_2/m_1 as indicated.

Consider first a solution assuming that m_2 has a negligible effect on the equipment motion:

$$m_1 = \frac{152 \text{ lb}}{386 \text{ in./sec}^2} = 0.393 \text{ lb-sec}^2/\text{in.}$$

$$\omega_{n1} = \sqrt{\frac{k_1}{m_1}} = \sqrt{\frac{2400}{0.393}} = 78.1 \text{ rad/sec } [12.4 \text{ Hz}]$$

Figure 31.14 gives $\ddot{x}_{1m}/\dot{u}_m\omega_{n1} = 0.88$ and Fig. 31.15 gives $\ddot{x}_{1m}\delta_{1m}/\dot{u}_m^2 = 0.76$ for $\zeta_1 = 0.1$. Then

$$\delta_{1m} = \frac{0.76}{0.88} \times \frac{\dot{u}_m}{\omega_{n1}} = \frac{0.76 \times 55}{0.88 \times 78.1} = 0.61 \text{ in.}$$

In finding $\ddot{x}_{2m}$ it is assumed that damping of the component has the typical value $\zeta_2 = 0.01$. Using $\omega_{n1}/\omega_{n2} = 130/78.1 = 1.67$, Fig. 31.28 gives $\ddot{x}_{2m}/\dot{u}_m\omega_{n2} = 1.15$; then $\ddot{x}_{2m} = 1.15 \times 55 \times 130 = 8230$ in./sec$^2 = 21.3g$.

A second solution, taking into consideration the mass m_2 of the component, may be obtained if the damping is neglected. From Eq. (31.41),

$$\delta_{1m} = \frac{\dot{u}_m}{\omega_{n1}} \frac{1 + \dfrac{\omega_{n2}}{\omega_{n1}}\left(1 + \dfrac{m_2}{m_1}\right)}{\left[\left(\dfrac{\omega_{n2}}{\omega_{n1}} + 1\right)^2 + \dfrac{m_2}{m_1}\left(\dfrac{\omega_{n2}}{\omega_{n1}}\right)^2\right]^{1/2}}$$

$$= \frac{55}{78.1} \times \frac{1 + 1.67(1 + \frac{3}{152})}{[(2.67)^2 + \frac{3}{152}(1.67)^2]^{1/2}} = 0.71 \text{ in.}$$

From Eq. (31.40):

$$\ddot{x}_{2m} = \dot{u}_m\omega_{n2}\left[\left(\frac{\omega_{n2}}{\omega_{n1}} - 1\right)^2 + \frac{m_2}{m_1}\left(\frac{\omega_{n2}}{\omega_{n1}}\right)^2\right]^{-1/2}$$

$$= 55 \times 130[(0.67)^2 + \frac{3}{152}(1.67)^2]^{-1/2}$$

$$= 10{,}070 \text{ in./sec}^2 = 26.1g$$

This example is too complex for a practicable solution when damping and the mass effects are considered together. However, the two above solutions may be taken conservatively as limiting conditions; it is unlikely that the actual acceleration and deflection would exceed the maxima of the limiting conditions.

SUPPORT PROTECTION

This section considers conditions in which the shock originates within the equipment (e.g., guns and drop hammers). Attention is first given to determining the response of the support for such equipment in the absence of a shock isolator. The effect of a shock isolator introduced to protect the support from excessive loads is considered later.

EQUIPMENT RIGIDLY ATTACHED TO SUPPORT

If the equipment is rigidly attached to the support, the support and equipment may be idealized as a single degree-of-freedom system for purposes of a simplified analysis. Consider the system of Fig. 31.3B with the spring-dashpot unit 2 assumed to be rigid. The mass m represents the equipment, and the mass m_F represents, with spring and dashpot assembly (1), the support. The force F, applied externally to the equipment, is taken to be a known function of time. The equation of motion is

$$(m_F + m)\ddot{\delta} + F(\dot{\delta}, \delta) = F$$

Considering only force-time relations $F(t)$ in the form of a single pulse, the analogous mathematical relations of Eqs. (31.5) and (31.6) are used by defining the impulse J applied by the force F as

$$J = \int_0^\tau F \, dt \tag{31.42}$$

where τ is the duration of the pulse.

Short-Duration Impulses. If τ is short compared with the half-period of free oscillation of the system, then the results derived in the section on *Response of a Rigid Body System to a Velocity Step* may be applied directly. An impulse J of negligible duration acting on the mass m produces a velocity change $\dot{u}_m$ given by

$$\dot{u}_m = \frac{J}{m} \tag{31.43}$$

The subsequent relative motion of the system is identical with that resulting from a velocity step of magnitude $\dot{u}_m$.

If the damping capacity of the support is small, then velocity step results derived for linear springs, hardening springs, and softening springs are applicable. If the damping of the support may be represented as viscous and the stiffness as linear, then the linear-spring viscous damping results apply. In most installations it is sufficiently accurate to consider the support an undamped linear system.

A structure used to support an equipment generally has distributed mass and elasticity; thus the application of an impulse tends to excite the structure to vibrate not only in its fundamental mode but also in higher modes of vibration. The mass-spring-dashpot system shown in Fig. 31.3B to represent the structure would have equivalent mass and stiffness suitable to simulate only the fundamental mode of vibration. In many applications, such simulation is adequate because the displacements and strains are greater in the fundamental mode than in higher modes. The vibration of members having distributed mass is discussed in Chap. 7, and the formulation of models suitable for use in the analysis of systems subjected to shock is discussed in Chap. 42.

Long-Duration Impulses. If the duration τ of the applied impulse exceeds about one-third of the natural period of the equipment-support system, application of velocity step results may be unduly conservative. Then the results developed in the section on *Response of Rigid Body System to Acceleration Pulse* are applicable. The mathematical equivalence of Eqs. (31.5) and (31.6) is based on identifying $-m\ddot{u}$ in the former with F in the latter. Accordingly, if the shape of the force F vs. time curve is similar to the shape of the curve of acceleration $\ddot{u}$ vs. time, then the response of a system to an acceleration pulse may be used by analogy to find the response to a force pulse by making the following substitutions:

$$\ddot{u}_m = \frac{F_m{}^*}{m} \qquad \tau_r = \frac{J}{F_m}$$

where F_m is the maximum value of F, $\ddot{u}_m$ is the maximum value of $\ddot{u}$, and τ_r is the effective duration.

---

* If the mathematical equivalence is literally applied, F_m/m is analogous to $-\ddot{u}_m$, not $\ddot{u}_m$. Since acceleration pulse results are given in terms of extreme *absolute* values, the sign is not important.

EQUIPMENT SHOCK ISOLATED

Idealized System. When a shock isolator is used to reduce the magnitude of the force transmitted to the support, the idealized system is as shown in Fig. 31.4. Subsystem 2 represents the equipment (mass m_2) mounted on the shock isolator (right-hand spring-dashpot unit). Subsystem 1 is an idealized representation of the support with effective mass m_1 and with stiffness and damping capacity represented by the left spring-dashpot unit. The free end of the latter unit is taken to be fixed ($u = 0$).

It is assumed that the system is initially in equilibrium ($\delta_1 = \delta_2 = 0$; $\dot\delta_1 = \dot\delta_2 = 0$) and that force F (positive in the $+X$ direction) applies an impulse J to m_2. Analysis is simplified by treating the duration τ of impulse J as negligible. This assumption, always conservative, usually is warranted if the natural frequency of the shock isolator is small relative to the natural frequency of the support.

System Separable. In many applications the support motion $x_1 (= \delta_1)$ is sufficiently small compared with the equipment motion x_2 that the equipment acceleration $\ddot{x}_2$ is closely approximated by $\ddot{\delta}_2$ where $\ddot{x}_2 = \ddot{\delta}_2 + \ddot{x}_1$. Using this approximation, the analysis is resolved into two separate parts, each dealing with a single degree-of-freedom system.

If the system consists only of linear elements as defined by Eq. (31.7), the equation of motion of the equipment mounted on the shock isolator (subsystem 2 of Fig. 31.4) is

$$\ddot{\delta}_2 + 2\zeta_2\omega_{n2}\dot{\delta}_2 + \omega_{n2}^2\delta_2 = 0 \tag{31.44}$$

where $\omega_{n2}^2 = k_2/m_2$ and $\zeta_2 = c_2/2m_2\omega_{n2}$. The initial conditions are: $\delta_2 = 0$, $\dot{\delta}_2 = \dot{u}_m = J/m_2$ when $t = 0$. Because of the similarity of Eqs. (31.26) and (31.44), and the respective initial conditions, the maximum equipment acceleration $\ddot{x}_{2m}$ and the maximum isolator deflection δ_{2m} may be found from Figs. 31.14 and 31.15.

The differential equation for the motion of the support in Fig. 31.4 is

$$\ddot{\delta}_1 + 2\zeta_1\omega_{n1}\dot{\delta}_1 + \omega_{n1}^2\delta_1 = -\frac{m_2}{m_1}\ddot{x}_2 \tag{31.45}$$

where $\omega_{n1}^2 = k_1/m_1$ and $\zeta_1 = c_1/2m_1\omega_{n1}$. The initial conditions are $\dot{\delta}_1 = 0$, $\delta_1 = 0$.

The solution of Eq. (31.45) is formally identical with that of Eq. (31.38) because the equations differ only by the interchange of the numerical subscripts and the presence of the factor m_2/m_1 on the right-hand side of Eq. (31.45). The solutions of Eq. (31.45) as obtained by analog computer[1] are shown in Fig. 31.31. The ordinate is the ratio of the maximum force F_{1m} in the support to the quantity $J\omega_{n1}$. The latter quantity is the maximum force which would be developed in an undamped, linear, single degree-of-freedom support of mass m_1 and stiffness k_1 if the impulse J were applied directly to m_1. The abscissa in Fig. 31.31 is the ratio of the undamped support natural frequency ω_{n1} to the undamped isolator natural frequency ω_{n2}. Curves are drawn for various values of the fraction of critical damping ζ_2 for the isolator, assuming that the fraction of critical damping ζ_1 for the support is constant at $\zeta_1 = 0.01$.

Figure 31.31 appears to show that the presence of an isolator increases the maximum force F_{1m} transmitted by the support if the natural frequencies of isolator and support are nearly equal. This conclusion is misleading because the analysis assumes that the support deflection δ_1 is small compared with the isolator deflection δ_2—a condition which is not met in the neighborhood of unity frequency ratio. A more realistic analysis involves the two degree-of-freedom system discussed in the next section.

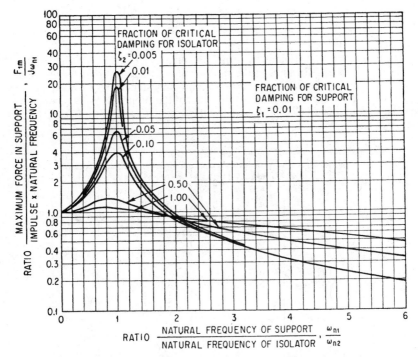

FIGURE 31.31 Dimensionless representation of maximum force in support F_{1m} resulting from action of impulse J on equipment. Isolator and support have linear elasticity and fraction of critical damping ζ_2 and ζ_1, respectively. Ordinate is the ratio of maximum force in support to that which would result from direct application of the impulse to the support with equipment absent. (*After R. D. Mindlin.*[1])

Two Degree-of-Freedom Analysis. This section includes an analysis of the system of Fig. 31.4 considered as a coupled two degree-of-freedom system where both the support and isolator are linear and undamped $[F_1(\dot{\delta}_1,\delta_1) = k_1\delta_1, F_2(\dot{\delta}_2,\delta_2) = k_2\delta_2]$. This analysis makes it possible to consider the effect of deflection of the support on the motion of the equipment. Fixing the support base ($u = 0$), the equations of motion may be written

$$\ddot{\delta}_1 + \omega_{n1}{}^2\delta_1 = \frac{m_2}{m_1}\,\omega_{n2}{}^2\delta_2$$

$$\ddot{\delta}_2 + \omega_{n2}{}^2\delta_2 = -\ddot{\delta}_1 \tag{31.46}$$

Assuming that the impulse J has negligible duration, the initial conditions are: $\dot{\delta}_1 = 0$, $\dot{\delta}_2 = J/m_2$, $\delta_1 = \delta_2 = 0$.

The solution of Eqs. (31.46) parallels that of Eqs. (31.39); the resulting expressions for the maximum isolator deflection δ_{2m} and force F_{1m} applied to the support are

$$\delta_{2m} = \frac{J}{m_2\omega_{n2}}\left[1 + \frac{m_2/m_1}{(1 + \omega_{n1}/\omega_{n2})^2}\right]^{-1/2} \tag{31.47}$$

$$F_{1m} = J\omega_{n1} \left[\left(1 - \frac{\omega_{n1}}{\omega_{n2}} \right)^2 + \frac{m_2}{m_1} \right]^{-1/2} \tag{31.48}$$

The maximum deflection of the isolator given in Eq. (31.47) is shown graphically in Fig. 31.32. For small values of the ratio of support natural frequency to isolator natural frequency, the flexibility of the support may significantly reduce the maximum isolator deflection, especially if the mass of the support is small relative to the mass of the equipment. For large values of the frequency ratio, the effect of the mass ratio is small.

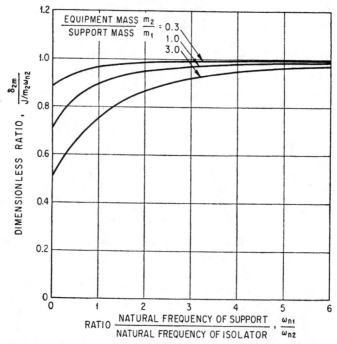

FIGURE 31.32 Dimensionless representation of maximum isolator deflection δ_{2m} resulting from action of impulse J on equipment. Isolator and support have undamped linear elasticity. Ordinate is the ratio of maximum isolator deflection to that which would occur with a rigid support. Curves are for three discrete values of the mass ratio m_2/m_1 as indicated.

Maximum values of force in the support, given by Eq. (31.48), are shown in Fig. 31.33. The maximum deflection of the floor is the maximum force F_{1m} divided by the stiffness of the floor. The effect of mass ratio is profound for small values of the frequency ratio. The curves of Figs. 31.31 and 31.33 show corresponding results, the former including damping and the latter including the coupling effect between the two systems. The analysis which ignores the coupling effect may grossly overestimate the maximum force applied to the support at low values of the frequency ratio. At high values of the frequency ratio, the two analyses yield like results if the fraction of crit-

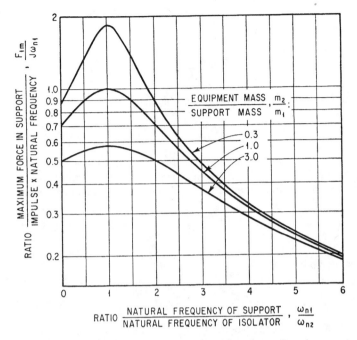

FIGURE 31.33 Dimensionless representation of maximum force in support F_{1m} resulting from action of impulse J on equipment. Isolator and support have undamped linear elasticity. Ordinate is the ratio of maximum force in support to that which would result from direct application of the impulse to the support with equipment absent. Curves are for three discrete values of the mass ratio m_2/m_1.

ical damping in the isolator is less than about $\zeta_2 = 0.10$. The two methods are compared in Example 31.6.

Example 31.6. A forging machine weighs 7000 lb exclusive of the 600-lb hammer. It is mounted at the center of a span formed by two 12-in., 50 lb/ft I beams having hinged ends and a span $l = 18$ ft. The hammer falls freely from a height of 60 in. before striking the work. Determine:

a. Maximum force F_{1m} in the beams and maximum deflection δ_{1m} of the beams if the machine is rigidly bolted to the beams.

b. The maximum force F_{1m} in the beams and the maximum deflection δ_{2m} of an isolator interposed between machine and beams.

Solution.

a. When the machine is bolted rigidly to the beams, the system may be considered to have only a single degree-of-freedom. The mass is that of the machine, plus the hammer, plus the effective mass of the beams. For the machine: $m_2 = (7000 + 600)/386 = 19.2$ lb-sec²/in. The effective mass of the beams is taken as one-half of the actual mass:

$$m_1 = \frac{2(0.5)(18)(50)}{386} = 2.33 \text{ lb-sec}^2/\text{in.}$$

$$m = m_1 + m_2 = 21.5 \text{ lb-sec}^2/\text{in.}$$

The stiffness of the beams is

$$k = 2 \frac{48EI}{l^3} = 2 \frac{48 \times (30 \times 10^6) \times 302}{(18 \times 12)^3} = 123,000 \text{ lb/in.}$$

The natural frequency of the machine-and-beams system is

$$\omega_n = \sqrt{\frac{k}{m}} = \sqrt{\frac{123,000}{21.5}} = 75.6 \text{ rad/sec [12.0 Hz]}$$

If the impact between the hammer and work is inelastic and its duration is negligible, the resulting velocity $\dot{u}_m$ of the machine may be found from conservation of momentum. The impulse J is the product of weight of hammer and time of fall:

$$J = (600) \left(\frac{2 \times 60}{386} \right)^{1/2} = 335 \text{ lb-sec}$$

Then $\dot{u}_m = J/m = 335/21.5 = 15.6$ in./sec. If the damping of the beams is neglected, the maximum beam deflection is found from Eq. (31.18):

$$\delta_{1m} = \frac{\dot{u}_m}{\omega_n} = \frac{15.6}{75.6} = 0.21 \text{ in.}$$

The maximum force in the beams is the product of beam stiffness and maximum deflection:

$$F_{1m} = k\delta_{1m} = 25,300 \text{ lb}$$

b. An isolator having a stiffness $k_2 = 36,000$ lb/in. and a fraction of critical damping $\zeta_2 = 0.10$ is interposed between the machine and beams. The "uncoupled natural frequencies" defined in connection with Eqs. (31.40) and (31.41) are

$$\omega_{n2} = \sqrt{\frac{k_2}{m_2}} = \sqrt{\frac{36,000}{19.2}} = 43.3 \text{ rad/sec [6.9 Hz]}$$

$$\omega_{n1} = \sqrt{\frac{k_1}{m_1}} = \sqrt{\frac{123,000}{2.33}} = 230 \text{ rad/sec [36.6 Hz]}$$

Consider first that the system is separable. Figures 31.14 and 31.15 give, respectively: $\ddot{x}_{2m}/\dot{u}_m\omega_{n2} = 0.88$; $\ddot{x}_{2m}\delta_{2m}/\dot{u}_m^2 = 0.76$. Substituting $\dot{u}_m = J/m_2 = 17.4$ in./sec and solving for δ_{2m},

$$\delta_{2m} = \frac{0.76 \times 17.4}{0.88 \times 43.3} = 0.35 \text{ in.}$$

Entering Fig. 31.31 at $\omega_{n1}/\omega_{n2} = 5.3$, $F_{1m}/J\omega_{n1} = 0.23$. Then

$$F_{1m} = 17,700 \text{ lb}$$

Thus, the effect of the isolator is to reduce the maximum load in the beams from 25,300 lb to 17,700 lb. An isolator with less stiffness would permit a further reduction of this force at the expense of greater machine motion.

Consider now that the floor and machine-isolator systems are coupled, and use the two degree-of-freedom analysis which neglects damping. From Eq. (31.47):

$$\delta_{2m} = \frac{J}{m_2 \omega_{n2}} \left[1 + \frac{m_2/m_1}{\left(1 + \frac{\omega_{n1}}{\omega_{n2}}\right)^2} \right]^{-1/2}$$

$$= \frac{335}{19.2 \times 43.3} \left[1 + \frac{19.2/2.33}{(1 + 5.3)^2} \right]^{-1/2} = 0.37 \text{ in.}$$

From Eq. (31.48):

$$F_{1m} = J\omega_{n1} \left[\left(1 - \frac{\omega_{n1}}{\omega_{n2}}\right)^2 + \frac{m_2}{m_1} \right]^{-1/2}$$

$$= 335 \times 230 \left[(1 - 5.3)^2 + \frac{19.2}{2.33} \right]^{-1/2} = 14{,}900 \text{ lb}$$

Thus, the two results for the isolator deflection δ_{2m} differ only slightly, but the two degree-of-freedom analysis gives a maximum load in the beams about 16 percent smaller than that obtained by assuming the systems to be separable.

REFERENCES

1. Mindlin, R. D.: *Bell System Tech. J.,* **24**:353 (1945).
2. McCalley, R. B., Jr.: "Velocity Shock Transmission in Two Degree Series Mechanical Systems," Atomic Energy Commission Contract W-31-109 Eng-52, Knolls Atomic Power Laboratory, Schenectady, March, 1956.
3. Klotter, K.: *J. Appl. Mechanics,* **22**:493 (1955).
4. Crede, C. E.: "Vibration and Shock Isolation," John Wiley & Sons, Inc., New York, 1951.

CHAPTER 32

CHARACTERISTICS OF VIBRATION ISOLATORS AND ISOLATION SYSTEMS

R. H. Racca

INTRODUCTION

A *vibration isolator* is a resilient support that tends to isolate an object from steady-state vibration. This chapter describes the characteristics of various types of vibration isolators (often called *vibration mounts*) and their characteristics, including elastomeric springs, plastic springs, metal springs, and pneumatic (air) springs. Each type of isolator is described in terms of its typical configurations (such as "shear" and "compression" for elastomeric springs) and its application characteristics (such as "load deflection" and "damping"). This is followed by an extensive discussion of active vibration-isolation systems. The next chapter presents detailed information on how to select and apply vibration isolators.

CHARACTERISTICS OF ELASTOMERIC VIBRATION ISOLATORS

Many different materials are used as the resilient elements of isolators. The material used denotes the type of isolator. Each material has its peculiar advantages and disadvantages. For example, elastomers (synthetic and natural rubbers) find wide application as isolators because they may be conveniently molded into many desired shapes and stiffnesses, they have more internal damping than metal springs, they usually require a minimum of space and weight, and they can be bonded to metallic inserts adapted for simplified attachment to the isolated structures. Plastic isolators have performance characteristics similar to those of the rubber-to-metal type of isolators and are used in equivalent configurations. Their advantages include low cost and outstanding uniformity; their disadvantages include a maximum operating temperature usually limited to 180°F (82°C). Metal springs are commonly used where large static deflections are required, where temperature or other environmental conditions make elastomers unsuitable, and (in some circumstances) where a low-cost isolator is required. Pneumatic (air) springs provide unusual advantages where low-frequency isolation is required; they can be used in many of the same applications as metal springs, but without certain disadvantages of the latter. Resilient pads (i.e., flat slabs) are fabricated of many materials, including Neoprene, felt, fiberglass, cork, natural rubber, and some compound materials; they lack the ready adaptability of elastomeric parts, which are molded to shape and frequently adhered to metal inserts for easy application.

The most commonly used type of isolator is fabricated of an elastomer, i.e., natural rubber or rubberlike material, described in detail in Chap. 34. See Fig. 32.1 for some typical elastomeric isolators. Such isolators are able to sustain large deformations and then return to their approximately original state with virtually no damage. Elastomeric isolators are superior to other types of isolators in that, for a given amount of elasticity, deflection capacity, energy storage, and dissipation, they require less space and less weight; also, they may be molded into many different configurations of many different types—generally at a lower cost than other types of isolators.

Elastomers have exceptional extensibility and deformability: They can be utilized at elongations of up to about 300 percent, with ultimate elongations of some elastomers to about 1000 percent. They may be stressed as much as 1000 to 1500 psi (0.145 to 0.218 Pa) or more before their elastic limit is reached. Their great capacity for storing energy permits them to tolerate high stress. Upon release of the stress, there is virtually total recovery from the deformation. The inherent damping of elastomers is often useful in preventing excessive vibration amplitude at resonance, as the amplitude is much lower than if coil metal springs were used. Elastomeric materials provide excellent resistance to the transmission of noise through structures because of their relatively low specific acoustic impedance, an important consideration where acoustic requirements are a key factor.

Of the various elastomers, *natural rubber* probably embodies the most favorable combination of mechanical properties, such as minimum drift, maximum tensile strength, and maximum elongation at failure. Its usefulness is restricted by its limited resistance to deterioration under the influence of hydrocarbons, ozone, and high ambient temperatures. *Neoprene* and *Buna N* (*nitrile*) exhibit superior resistance to hydrocarbons and ozone, Buna N being particularly satisfactory for applications involving relatively high ambient temperatures. *Buna S* is a good general-purpose synthetic rubber for use in vibration isolators.

Silicone rubber is one of the costliest elastomers. Its properties are remarkably stable, and it provides effective isolation over a very wide temperature range: −65 to +350°F (−54 to 177°C). By comparison, neoprene is limited in use to a range of about −40 to +200°F (−40 to 93°C). The upper temperatures limit depends on the properties of the particular compound, the degree of deterioration which is permissible as a result of continued exposure at high temperatures, and the duration of exposure. For silicone, a temperature substantially greater than 300°F (149°C) is permissible for several hours. The outstanding ability of silicone elastomers to withstand extremes of temperature is offset somewhat by their inferior strength, tear resistance, and abrasion resistance.

Isolators fabricated of elastomers are complex in behavior because of the viscoelastic nature (somewhere between that of a solid and that of a liquid) of elastomers in performance, because of their indefinite yield point, and because their physical properties vary with time, temperature, and environment. For example, rubber is a substantially incompressible material (it has a Poisson's ratio of approximately 0.5). Thus the stiffness of a rubber spring when it is strained in compression depends, to a considerable extent, on the area of the surface available for lateral expansion. In contrast, the stiffness of a rubber spring in shear is substantially inde-

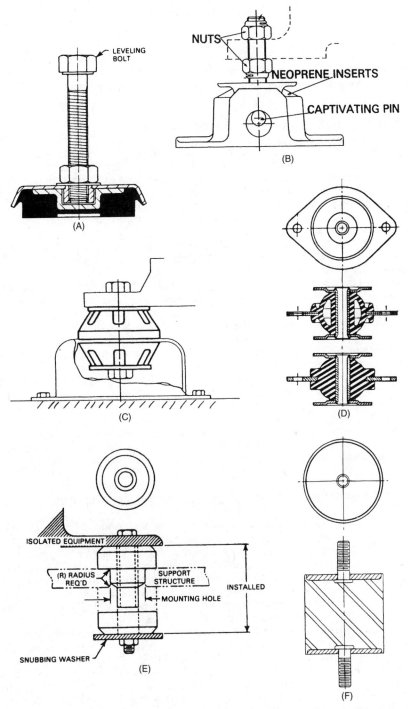

FIGURE 32.1 Typical elastomeric isolators. (*A*) Machinery mount. (*B*) Marine engine isolator. (*C*) Pedestal isolator. (*D*) Plate form instrument isolator. (*E*) General-purpose isolator. (*F*) Cylindrical stud isolator.

pendent of the shape of the rubber member. As a rough rule of thumb, it may be assumed that the minimum likely compression stiffness of a given rubber spring is five times its shear stiffness. The maximum compression stiffness may be several times as great as the minimum value if lateral expansion of the rubber is constrained.

FATIGUE FAILURE AND PREMATURE FAILURE

Regardless of geometry, both elastomers and metals exhibit fatigue failure as a result of repeated cyclic loadings. Unlike a metal, an elastomer does not experience catastrophic-type fatigue failure. Instead, the failure begins as a tear at the point of highest cyclic shear strain, which is generally on the outer extremity (and therefore visible in many cases), and gradually propagates through the body of the elastomer. The result is a gradual reduction in stiffness that usually becomes apparent before there is total failure.

Most elastomeric isolators should not be subject to large static strains over long periods of time. An isolator with a large static deflection may give satisfactory performance temporarily, but the deflection tends to creep (increase) excessively over a long period. In general, elastomers should not be statically strained continuously more than 10 to 15 percent in compression, or more than 25 to 50 percent in shear.

A factor contributing to the premature failure of an elastomeric isolator is the effect of the minimum strain on fatigue life. For elastomers which crystallize under high strains (such as neoprene and natural rubber), fatigue life is greatly increased if the minimum cyclic stress is always either plus or minus and never passes through zero. Proper static *precompression* of the isolator within the limits specified above is often an effective way to prevent the minimum cyclic stress from passing through zero under dynamic conditions. Local stress concentrations, which result in premature failure, often can be avoided by using fillets, radii, and generous overhangs of the elastomeric section. For example, sharp corners of metal inserts and support structures should be carefully rounded off wherever they contact the elastomer. Metal snubbing washers and/or support structures in contact with the elastomer should be large enough to prevent their edges from cutting into the elastomeric surfaces.

BONDED VERSUS UNBONDED ELASTOMERIC ISOLATORS

Elastomeric isolators may be designed in both bonded and unbonded configurations. In the bonded isolator, metal inserts are bonded to the elastomer on all load-carrying surfaces. In the unbonded or semibonded isolator, the elastomeric load-bearing surface rests directly on the support structure. Bonded parts usually cost more because of the special chemical preparation required to achieve a bond with strength in excess of that of the elastomer itself. Bonded parts are generally preferred since they may be more highly stressed for a given deflection. With higher stress they provide higher spring constants and higher elastic energy-storage capacity.

Bonded isolators can be designed to provide proper load distribution in shear, compression, tension, or combination loading. A more uniform stress distribution in the elastomer is obtained by bonding inserts on all the load-bearing elastomer surfaces. The bonded inserts reduce unit stress by distributing the stress more uniformly throughout the volume of the elastomer. In contrast, unbonded parts usually fail to distribute the load uniformly, resulting in local areas of stress concentration in the elastomer body which shorten the life of the isolator.

A significant difference between bonded and unbonded elastomeric isolators relates to how elastomers behave under load. When an elastomer pad is compressed under load, its volume remains constant—only its shape is changed. The rubber "bulges" under load. When this ability to bulge is controlled, the load-deflection characteristics of the isolator are controlled. In a bonded isolator, the load-carrying surfaces have a fixed degree of bulge because the elastomer cannot move along the bond line, and so it remains in a fixed position regardless of the load or environmental conditions.

In an unbonded isolator, this is not the case. The ability of the elastomer to bulge depends to a considerable degree on the maintenance of friction at the elastomer–support structure interface. When all surfaces are clean and dry, the difference between the ability of a bonded and an unbonded isolator to bulge is negligible. But if oil or sand works its way into the elastomer-to-metal interface of the unbonded isolator, the ability of the elastomer to bulge is greatly increased; consequently, its original load-deflection characteristics no longer exist. Then the isolator can exhibit load-deflection characteristics that are 50 percent less than when it was new; in many cases, this can cause the isolator to malfunction. Thus, where consistent load-deflection characteristics are required for the life of the equipment, bonded isolators should be used. Although the initial cost of unbonded isolators is lower, in many applications the cost of extra machining of the support structure and the reduced service life may well make unbonded isolators a poor selection.

TYPES OF LOADING

Elastomer isolators may be used with different types of loading: compression, shear, tension, or buckling, or any combination of these types.

Compression Loading.   The word *compression* is used to indicate a reduction in the dimension (thickness) of an elastomeric element in the line of the externally applied force. The stiffness characteristic of elastomers stressed in compression exhibit a nonlinearity (hardening) which becomes especially pronounced for strains above 30 percent. Compression loading, illustrated in Fig. 32.2A, is most effective when used with simple unbonded isolators and is effective where gradual snubbing (motion limiting) is required. Compression loading is frequently employed to provide a low initial stiffness for vibration isolation and a relatively high final stiffness to limit the dynamic deflection under shock excitations. Because of the nonlinear hardening characteristics of compression loading, it is the least effective type of loading for energy storage and therefore is not recommended where the attenuation of force or acceleration transmission is the primary concern. (The energy stored by any spring is the area under the load-deflection curve.)

Shear Loading.   Shear loading, illustrated in Fig. 32.2B, refers to the force applied to an elastomeric element so as to slide adjacent parts in opposite directions. An almost linear spring constant up to about 200 percent shear strain is characteristic of elastomer stress in shear. Because of this linear spring constant, shear loading is the preferred type of loading for vibration isolators because it provides a constant frequency response for both small and large dynamic shear strains in a simple spring-mass system. Shear loading is also useful for shock isolators where attenuating force or acceleration transmission is important, because of its more efficient energy-storage capacity when compared to compression loading. However, care must be

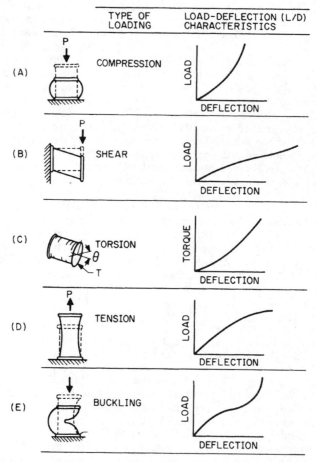

FIGURE 32.2 Load-deflection characteristics of typical elastomeric isolators. (*After R. Racca.*[1])

taken to ensure that the expected dynamic loads do not result in shear strains that exceed the limits of the elastomer or that abrupt bottoming of the supported equipment does not occur.

Torsion Loading. A modification of shear loading that is sometimes listed as a separate type is torsion loading, shown in Fig. 32.2C. It consists of winding up a sandwich of laminated sections to strain the elastomer in torsion. When the strain in torsion exceeds about 150 percent, considerable axial thrust loads on connecting members are induced, if they are rigidly fixed parallel to each other, because of the reduction in the axial thickness of the elastomer.

Tension Loading. Tension loading, illustrated in Fig. 32.2D, refers to an increase in the dimension (thickness) of an elastomeric element in the line of the externally applied force. Elastomers stressed in tension exhibit a nonlinear (softening) spring constant. For a given deflection, tension loading stores energy more efficiently than either shear or compression loading. Because of this, tension loading has been occasionally used for shock isolation systems. However, in general, tension loading is not recommended because of the resulting loads on the elastomer-to-metal bond, which may cause premature failure of the material.

Buckling Loading. Buckling loading, illustrated in Fig. 32.2*E,* occurs when the externally applied load causes an elastomeric element to warp or bend in the direction of the applied load. Buckling stiffness characteristics may be used to derive the benefits of both softening stiffness characteristics (for the initial part of the load-deflection curve) and hardening characteristics (for the later part of the load-deflection curve). The buckling mode thus provides high energy-storage capacity and is useful for shock isolators where force or acceleration transmission is important and where snubbing (i.e., motion limiting) is required under excessively high transient dynamic loads. This type of stiffness characteristic is exhibited by certain elastomeric cushioning foam materials and by specially designed elastomeric isolators. However, it is important to note that even simple compressive elements will buckle when the slenderness ratio (the unloaded length/width ratio) exceeds 1.6.

Combinations of the types of loading described above are commonly used, which result in combined load-deflection characteristics. Consider, for example, a compression-type isolator which is installed at an angle instead of in the usual vertical position. Under these conditions, it acts as a compression-shear type of isolator when loaded in the vertical downward direction. When loaded in the vertical upward direction, it acts as a shear-tension combination type of isolator.

DAMPING CHARACTERISTICS OF ELASTOMERIC ISOLATORS

Damping, to some extent, is inherent in all resilient materials. The damping characteristics of elastomers vary widely. A tightly cured elastomer may (within its proper operating range) store and return energy with more than 95 percent efficiency, while elastomers compounded for high damping have less than 30 percent efficiency. Damping increases with decreasing temperature because of the effects of crystallinity and viscosity in the elastomer. If the isolator remains at a low temperature for a prolonged period, the increase in damping may exceed 300 percent. Damping quickly decreases with low-temperature flexure, because of the crystalline structure deterioration and the heat generated by the high damping.

Where the nature of the excitation is difficult to predict (for example, random vibration), it is desirable that the damping in the isolator be relatively high. Damping in an isolator is of the greatest significance at the resonance frequency. Therefore, it is desirable that isolators embody substantial damping when they may operate at resonance, as is the case when the excitation is random over a broad frequency band or even momentary (as in the starting of a machine with an operating frequency greater than the natural frequency of the machine on its isolators). The relatively large amplitude commonly associated with resonance does not occur instantaneously, but rather requires a finite time to build up. If the forcing frequency is varied continuously as the machine starts or stops, the resonance condition may exist for such a short period of time that only a moderate amplitude builds. The rate of change of forcing frequency is of little importance for highly damped isolators, but it is of considerable importance for lightly damped isolators.

In general, damping in an elastomer increases as the frequency increases. The data of Figs. 34.5 and 34.6 can be used to predict transmissibility at resonance by estimating the frequency and the amplitude of dynamic shear strain; then the fraction of critical damping is obtained from the curves and used with Eq. (30.1) to calculate transmissibility at resonance.

Hydraulically Damped Vibration Isolators. Hydraulically damped vibration isolators combine a spring and a damper in a single compact unit that allows tuning of the spring and damper independently. This provides flexibility in matching the dynamic characteristics of the isolator to the requirements of the application. Hydraulic mounts have been used primarily as engine and operator cab isolators in vehicular applications. The hydraulically damped isolator shown in Fig. 32.3 has a flexible rubber element that encapsulates an incompressible fluid which is made to flow through a variety of ports and orifices to develop the dynamic characteristics required. The fluid cavity is divided into two chambers with an orifice between, so that motion of the elastomeric element causes fluid to flow from one chamber to the other, dissipating energy (and thus creating damping in the system).

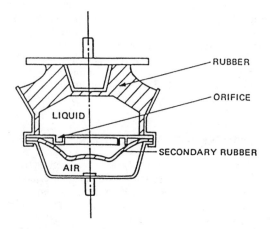

FIGURE 32.3 Simple hydraulic mount. (*After T. Ushijima, K. Takano, and H. Kojima.*[3])

Installations that require a soft isolator for good isolation may also require motion control under transient (shock) inputs or when operating close to the isolation system's resonant frequency. For good isolation, low damping is required. For motion control, high damping is required. Fluid-damped isolators accommodate these conflicting requirements. A hydraulically damped vibration isolator can also act as a tuned absorber by increasing the length of the orifice into an inertia track because the inertia of the fluid moving within the isolator acts as a tuned mass at a specific frequency (which is determined by the length of the orifice). This feature can be used where vibration isolation at a particular frequency is required.

STATIC AND DYNAMIC STIFFNESS

When the main load-carrying spring is made of rubber or a similar elastomeric material, the natural frequency calculated using the stiffness determined from a static load-deflection test of the spring almost invariably gives a value lower than that experienced during vibration. Thus the dynamic modulus appears greater than the static modulus. The ratio of moduli is approximately independent of the velocity of strain, and has a numerical value generally between 1 and 3. This ratio increases significantly as the durometer increases, as illustrated in Fig. 34.5.

PLASTIC ISOLATORS

Isolators fabricated of resilient plastics are available and have performance characteristics similar to those of the rubber-to-metal type of isolators of equivalent configuration. The structural elements are manufactured from a rigid thermoplastic and the resilient element from a thermoplastic elastomer. These elements are compatible in the sense that they are capable of being bonded one to another by fusion. The most commonly used materials are polystyrene for the structural elements and butadiene styrene for the resilient elastomer. The advantages of this type of spring are (1) low cost, (2) exceptional uniformity in dynamic performance and dimensional stability, and (3) ability to maintain close tolerances. The disadvantages are (1) limited temperature range, usually from a maximum of about 180°F (82°C) to a minimum of –40°F (–40°C), (2) creep of the elastomer element at high static strains, and (3) the structural strength of the plastic.

METAL SPRINGS

Metal springs used in shock and vibration control are usually categorized as being of the following types: helical springs (coil springs), ring springs, Belleville (conical or conical-disc) springs, involute (volute) springs, leaf and cantilever springs, and wire-mesh springs.

HELICAL SPRINGS (COIL SPRINGS)

Helical springs (also known as coil springs) are made of bar stock or wire coiled into a helical form, as illustrated in Fig. 32.4. The load is applied along the axis of the helix. In a *compression spring* the helix is compressed; in a *tension spring* it is extended. The helical spring has a straight load-deflection curve, as shown in Fig. 32.5. This is the simplest and most widely used energy-storage spring. Energy stored by the spring is represented by the area under the load-deflection curve.

Helical springs have the inherent advantages of low cost, compactness, and efficient use of material. Springs of this type which have a low natural frequency when fully loaded are available. For example, such springs having a natural frequency as low as 2 Hz are relatively common. However, the static deflection of such a spring is about 2.4 in. (61 mm). For such a large static deflection, the spring must have adequate lateral stability or the mounted equipment will tip to one side. Therefore, all forces on the spring must be along the axis of the spring. For a given natural frequency, the degree of lateral stability depends on the ratio of coil diameter to working height. Lateral stability also may be achieved by the use of a housing around the spring which restricts its lateral motion. Helical springs provide little damping, which results in transmissibility at resonance of 100 or higher. They effectively transmit

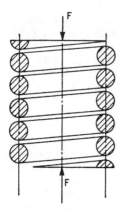

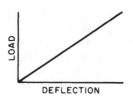

FIGURE 32.5 Load-deflection curve for a helical spring.

FIGURE 32.4 Cross section of a helical spring showing the direction of the applied force F.

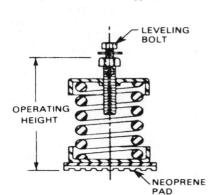

FIGURE 32.6 Helical spring isolator for mounting machinery.

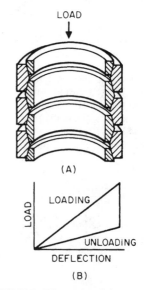

FIGURE 32.7 Ring spring. (*A*) Cross section. (*B*) Load-deflection characteristic when it is loaded and when it is unloaded.

high-frequency vibratory energy and therefore are poor isolators for structure-borne noise paths unless they are used in combination with an elastomer which provides the required high-frequency attenuation, as illustrated in Fig. 32.6.

RING SPRINGS

A ring spring, shown in Fig. 32.7*A*, absorbs the energy of motion in a few cycles, dissipating it as a result of friction between its sections. With a high load capacity for its size and weight, a ring spring absorbs linear energy with minimum recoil. It has a linear load-deflection characteristic, shown in Fig. 32.7*B*. Springs of this type often are used for loads of from 4,000 to 200,000 lb (1814 to 90,720 kg), with deflections between 1 in. (25 mm) and 12 in. (305 mm).

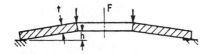

FIGURE 32.8 A Belleville spring made up of a coned disc of thickness t and height h, axially loaded by a force F.

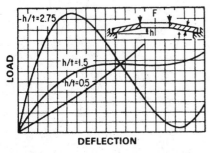

DEFLECTION

FIGURE 32.9 The load-deflection characteristic for a Belleville spring having various ratios of h/t.

BELLEVILLE SPRINGS

Belleville springs (also called coned-disc springs), illustrated in Fig. 32.8, absorb more energy in a given space than helical springs. Springs of this type are excellent for large loads and small deflections. They are available as assemblies, arranged in stacks. Their inherent damping characteristics are like those of leaf springs: Oscillations quickly stop after impact. The coned discs of this type of spring have diametral cross sections and loading, as shown in Fig. 32.8. The shape of the load-deflection curve depends primarily on the ratio of the unloaded cone (or disc) height h to the thickness t. Some load-deflection curves are shown in Fig. 32.9 for different values of h/t, where the spring is supported so that it may deflect beyond the flattened position. For a ratio of h/t approximately equal to 0.5, the curve approximates a straight line up to a deflection equal to half the thickness; for h/t equal to 1.5, the load is constant within a few percent over a considerable range of deflection. Springs with ratios h/t approximating 1.5 are known as *constant-load* or *stiffness springs*. Advantages of Belleville springs include the small space requirement in the direction of the applied load, the ability to carry lateral loads, and load-deflection characteristics that may be changed by adding or removing discs. Disadvantages include nonuniformity of stress distribution, particularly for large ratios of outside to inside diameter.

INVOLUTE SPRINGS

An involute spring, shown in Fig. 32.10*A* and 32.10*B*, can be used to better advantage than a helical spring when the energy to be absorbed is high and space is rather limited. Isolators of this type have a nonlinear load-deflection characteristic, illustrated in Fig. 32.10*C*. They are usually much more complex in design than helical springs.

LEAF SPRINGS

Leaf springs are somewhat less efficient in terms of energy storage capacity per pound of metal than helical springs. However, leaf springs may be applied to function as structural members. A typical semielliptic leaf spring is shown in Fig. 32.11.

WIRE-MESH SPRINGS

Knitted wire mesh acts as a cushion with high damping characteristics and nonlinear spring constants. A circular knitting process is used to produce a mesh of multiple, interlocking springlike loops. A wire-mesh spring, shown in Fig. 32.12, has

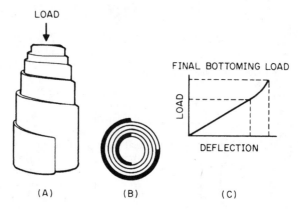

(A)   (B)   (C)

FIGURE 32.10   An involute spring. (*A*) Side view. (*B*) Cross section. (*C*) Load-deflection characteristic.

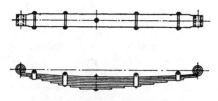

FIGURE 32.11   Semielliptic leaf spring.

a multidirectional orientation of the spring loops, i.e., each loop can move freely in three directions, providing a two-way stretch. Under tensile or compressive loads, each loop behaves like a small spring; when stress is removed, it immediately returns to its original shape. Shock loadings are limited only by the yield strength of the mesh material used. The mesh cushions, enclosed in springs, have characteristics similar to a spring and dashpot.

Commonly used wire mesh materials include such metals as stainless steel, galvanized steel, Monel, Inconel, copper, aluminum, and nickel. Wire meshes of stainless steel can be used outside the range to which elastomers are restricted, i.e., −65 to 350°F (−53 to 177°C); furthermore, stainless steel is not affected by various environmental conditions that are destructive to elastomers. Wire-mesh springs can be fabricated in numerous configurations, with a broad range of natural frequency, damping, and radial-to-axial stiffness properties. Wire-mesh isolators have a wide load tolerance coupled with overload capacity. The nonlinear load-deflection characteristics provide good performance, without excessive deflection, over a wide load range for loads as high as four times the static load rating.

Stiffness is nonlinear and increases with load, resulting in increased stability and gradual absorption of overloads. An isolation system has a natural frequency proportional to the ratio of stiffness to mass; therefore, if the stiffness increases in proportion to the increase in mass, the natural frequency remains constant. This condition is approached by the load-deflection characteristics of mesh springs. The advantages of such a nonlinear system are increased stability, resistance to bottoming out of the mounting system under transient overload conditions, increased shock protection, greater absorption of energy during the work cycle, and negligible drift rate. Critical damping of 15 to 20 percent at resonance is generally considered desirable for a wire-mesh spring. Environmental factors such as temperature, pressure, and humidity affect this value little, if at all. Damping varies with deflection: high damping at resonance and low damping at higher frequencies.

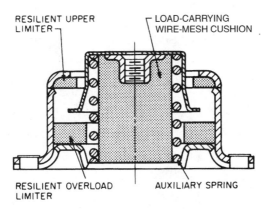

FIGURE 32.12   Wire-mesh spring, shown in section.

AIR (PNEUMATIC) SPRINGS

A pneumatic spring employs gas as its resilient element. Since the gas is usually air, such a spring is often called an air spring. It does not require a large static deflection; this is because the gas can be compressed to the pressure required to carry the load while maintaining the low stiffness necessary for vibration isolation. The energy-storage capacity of air is far greater per unit weight than that of mechanical spring materials, such as steel and rubber. The advantage of air is somewhat less than would be indicated by a comparison of energy-storage capacity per pound of material because the air must be contained. However, if the load and static deflection are large, the use of air springs usually results in a large weight reduction. Because of the efficient potential energy storage of springs of this type, their use in a vibration-isolation system can result in a natural frequency for the system which is almost 10 times lower than that for a system employing vibration isolators made from steel and rubber.

An air spring consists of a sealed pressure vessel, with provision for filling and releasing a gas, and a flexible member to allow for motion. The spring is pressurized with a gas which supports the load. Air springs generally have lower resonance frequencies and smaller overall length than mechanical springs having equivalent characteristics; therefore, they are employed where low-frequency vibration isolation is required. Air springs may require more maintenance than mechanical springs and are subject to damage by sharp and hot objects. The temperature limits are also restricted compared to those for mechanical springs.

Figure 32.13 shows four of the most common types of air springs. The air spring shown in Fig. 32.13A is available with one, two, and three convolutions. It has a very low minimum height and a stroke that is greater than its minimum height. The rolling lobe (reversible sleeve) spring shown in Fig. 32.13B has a large stroke capability and is used in applications which require large axial displacements, as, for example, in vehicle applications. The isolators shown in Fig. 32.13A and B may have insufficient lateral stiffness for use without additional lateral restraint. The rolling diaphragm spring shown in Fig. 32.13C has a small stroke and is employed to isolate low-amplitude vibration. The air spring shown in Fig. 32.13D has a low height and a small stroke capability. The thick elastomer sidewall can be used to cushion shock inputs.

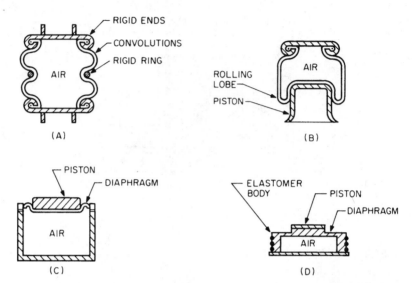

FIGURE 32.13   Four common types of air springs. (*A*) Air spring with convolutions. (*B*) A rolling lobe air spring. (*C*) Rolling diaphragm air spring. (*D*) Air spring having a diaphragm and an elastomeric sidewall.

The load F that can be supported by an air spring is the product of the gage pressure P and the effective area S (i.e., $F = PS$). For a given area, the pressure may be adjusted to carry any load within the strength limitation of the cylinder walls. Since the cross section of many types of air springs may vary, it is not always easy to determine. For example, the spring shown in Fig. 32.13*A* has a maximum effective area at the minimum height of the spring and a smaller effective area at the maximum height. The spring illustrated in Fig. 32.13*B* is acted on by a piston which is contoured to vary the effective area. In vehicle applications this is often done to provide a low spring stiffness near the center of the stroke and a higher stiffness at both ends of the stroke in order to limit the travel. The effective areas of the springs illustrated in Fig. 32.13*C* and *D* are usually constant throughout their stroke; the elastomeric diaphragm of the spring shown in Fig. 32.13*D* adds significantly to its stiffness. Air springs are commercially available in various sizes that can accommodate static loads that range from as low as 25 lb (11.3 kg) to as high as 100,000 lb (45,339 kg) with a usable temperature range of from −40 to 180°F (−40 to 83°C). System natural frequencies as low as 1 Hz can be achieved with air springs.

STIFFNESS

The stiffness of the air spring of Fig. 32.14 is derived from the gas laws governing the pressure and volume relationship. Assuming adiabatic compression, the equation defining the pressure-volume relationship is

$$PV^n = P_i V_i^n \tag{32.1}$$

where P_i = absolute gas pressure at reference displacement
       V_i = corresponding volume of contained gas
       n = ratio of specific heats of gas, 1.4 for air

If the area S is constant, and if the change in volume is small relative to the initial volume V_i [i.e., if $S\delta$ (where δ is the dynamic deflection) $\ll V_i$)], then the stiffness k is given by

$$k = \frac{nP_iS^2}{V_i} \tag{32.2}$$

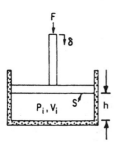

FIGURE 32.14 Illustration of a single-acting air spring consisting of a piston and a cylinder.

Transverse Stiffness. The transverse stiffness (i.e., the stiffness to laterally applied forces) of the air springs illustrated in Fig. 32.13A and B varies from very small to moderate; the natural frequencies for such springs vary from 0 to 3 Hz. The spring illustrated in Fig. 32.13C has a higher transverse stiffness, with natural frequencies ranging from 2 to 8 Hz. The spring illustrated in Fig. 32.13D has a moderate transverse stiffness; the natural frequency varies in the range from 3 to 5 Hz. If an installation requires the selection of an air spring having insufficient transverse stiffness, additional springs in the transverse direction are often employed for stability, as shown in Fig. 32.15.

At frequencies above 3 Hz, the compression of gases used in air springs tends to be adiabatic and the ratio of specific heats n for both air and nitrogen has a value of 1.4. At frequencies below approximately 3 Hz, the compression tends to be isothermal and the ratio of specific heats n has a value of 1.0, unless the spring is thermally insulated. For thermally insulated springs, the transition from adiabatic to isothermal occurs at a frequency of less than 3 Hz. Gases other than air which are compatible with the air spring materials can also be used. For example, sulfur hexafluoride (SF_6) has a value of n equal to 1.09—a value that reduces the axial spring stiffness by 22 percent; it also has a considerably lower permeation (leakage through the air spring material) rate than air, which may reduce the frequency of recharging (repressurizing) for a closed (passive) air spring.

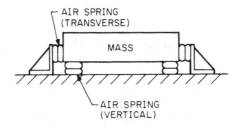

FIGURE 32.15 An air spring used to support a load and provide vibration isolation in the vertical direction. In addition, air springs are provided on the sides to increase the transverse stiffness.

DAMPING

Air springs have some inherent damping that is developed by damping in the flexible diaphragm or sidewall, friction, damping of the gas, and nonlinearity. The damping varies with the vibration amplitude; however, it generally is between 1 and 5 percent of critical damping.

NATURAL FREQUENCY

In U.S. Customary units, the natural frequency f_n of an undamped air spring is expressed by

$$f_n = 3.13 \left(\frac{k_1}{W} \right)^{1/2} = \delta_1^{-1/2} \tag{32.3a}$$

where W = supported weight, lb
k_1 = stiffness of the air spring, lb/in.
δ_1 = static deflection, in.

In S.I. units, the natural frequency is given by

$$f_n = \delta_2^{-1/2} \tag{32.3b}$$

where δ_2 = static deflection, cm

COMMERCIALLY AVAILABLE ISOLATORS

Isolators are commercially available in many resilient materials, in countless shapes and sizes, and with widely diverse characteristics. In the U.S.A. there are more than 125 elastomeric isolator manufacturers, each offering a range of models in a variety of synthetic elastomeric compounds and natural rubbers. The number is significantly higher if manufacturers of plastic, metal, pneumatic, and other-material isolators are included.

The properties of a given isolator are dependent not only on the material of which it is fabricated, but also on its configuration and overall construction with respect to the structural material used within the body of the isolator. Data on these parameters can be found in the catalogs of the various isolator manufacturers. Methods of applying commercially available isolators are described in Chap. 33.

ACTIVE VIBRATION-ISOLATION SYSTEMS

An isolator that must operate over a relatively large load range and that has the requirement for a low natural frequency, as in the spring system of an automobile, presents design problems which cannot be overcome by use of either the mechanical spring or the air spring alone. Because of the low stiffness of the spring, a small change in the weight of the supported body results in a large change in the static deflection of the spring. In many instances, this cannot be tolerated. For example, an automobile with a spring system that has an undamped natural frequency of 0.75 Hz requires springs that have a static deflection of 17.5 in. (0.445 m). A 30 percent change in the weight supported by the automobile suspension results in a 5.25-in. (0.133-m) change in its normal height. This change in position is too large to allow proper operation of the springs over the required load range, i.e., approximately equal clearance travel in the up and down directions. A similar problem arises in a rocket-propelled missile where a substantially constant acceleration is sustained for an appreciably long time. This deflects an isolator in the same manner as would an increase in the weight of the supported equipment. The clearance around such equipment must be large to provide for unobstructed displacement of the equipment under the influence of the large steady acceleration experienced by the missile. Space on such missiles is extremely limited.

The large deflection experienced by a conventional spring element can be reduced by use of a servomechanism which compensates for load changes, thereby maintaining the supported body at a given position with respect to its vehicle. Figure 32.16 illustrates such a device schematically. As the force F is applied slowly, the spring of stiffness k stretches and the supported body of mass m moves downward. The relative motion between the body and the reference plane of the support causes the relative displacement sensing device to respond, requiring power to be supplied to the servo actuators, which move the frame upward until the body returns to its original position with respect to the support plane. This position of equilibrium exists until another change in the force occurs. The element which produces the change in the position of the frame is a servomechanism arranged to maintain the relative displacement δ at zero in the presence of the constant force F.

The isolation system illustrated in Fig. 32.16 always seeks as its equilibrium position a location at a distance h above the reference plane of the support, independent of the origin and magnitude of a steady force applied to the supported body. There is no change in the position of the body in response to a very slowly applied load, e.g., whether the undamped natural frequency of the isolator is 1 or 10 Hz. Thus, certain disadvantages attendant on the use of isolators of low natural frequency are removed. When the load is applied suddenly, however, the relative displacement regulation servomechanism may be unable to respond fast enough to compensate for the tendency of the supported body to change position relative to the support; then the isolator can experience a significant deflection.

Active vibration-isolation systems are employed when the damped natural frequency of the isolation system must be relatively low, with the added requirement that the supported body be maintained at a relatively constant distance from the base to which it is attached. The position of the body, i.e., the deflection of the isolation system, can be controlled by means of a ser-

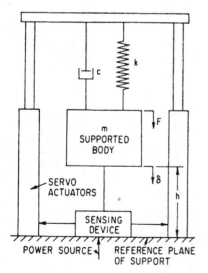

FIGURE 32.16 Schematic diagram of an active vibration-isolation system which maintains the supported body m a fixed distance h from the reference plane of the support, irrespective of the steady force F applied to the supported body.

vomechanism. Likewise, low-frequency vibration isolation along with the isolation system resonance, i.e., resonance frequency and peak transmissibility, can be controlled by means of another servomechanism. Both these servomechanisms are made up of three basic components: (1) a sensing element which generates a signal proportional to a measurable isolation system motion, (2) computational elements to proportionally scale and integrate or differentiate the sensing element response, and (3) a motor, or actuator, to generate a force applied to the supported body which is proportional to the signal from the computational element. The force output of the motor, or actuator, then supplements the spring forces to help support the supported body and provide forces which modify the isolator passive response to lessen or to eliminate the effects of the system resonance. The servomechanisms formed are negative feedback regulators, a form of automatic control system. The term *negative feedback* is used since the force motor is arranged so that the feedback force applied to the supported body opposes its movement. An isolation system incorporating one or more servomechanisms is called an *active vibration-isolation system.*

The computational element for the elimination of the isolator static deflection is that of an integrator and scaling term called a *controller gain.* This combination of sensing, computation, and actuation provides what is known as *integral control,* since the feedback force is proportional to the time integral of the sensor response. The computational elements for the control of the system resonance and low-frequency vibration isolation require only a scaling term. This combination of control elements is called *proportional control,* since the feedback force is proportional to the sensor response. The feedback elements added to a conventional isolation system must have an overall characteristic such that the output force is proportional to the sensed function times the control function of the computational element. The control function describes the operation of the computational element, which can be a simple constant as in proportional control, an integration function as in integral control, or an equation describing the action of one or more electric circuits. This corresponds to a spring which provides an output force proportional to the deflection of the spring, a viscous damper which provides a force proportional to the rate of deflection of the damper, or an electric circuit which produces a force signal proportional to the dynamics of a spring and viscous damper, in series, undergoing a motion proportional to the sensor response.

The sensing and actuation devices which provide integral control of the isolator relative displacement may take many forms. For example, the sensing element which measures the position of the supported body (relative to the reference plane of the support) may be a differential transformer which produces an electrical signal proportional to its extension relative to a neutral position. The sensing element is attached at one end to the supported body and at the other end to the isolator support structure in a manner such that the sensor is in its neutral position when the supported body is at its desired operating height. The electrical signal is integrated and amplified in the computational element, providing electric power to operate an electric motor actuation device. The differential transformer-integrator-motor system produces a force proportional to the integral of the signal from the differential transformer. The operation of this servomechanism can be visualized in the following manner:

1. A force F of constant magnitude F_0 is applied to the supported body, causing a relative deflection of the isolator spring element.

2. The sensing element (in this case a differential transformer) applies an electrical signal that is proportional to the isolator relative displacement to the integration and scaling functions in the computational element.

3. The response of the computational integration function generates an electrical signal that continues to increase in magnitude so long as the relative displacement δ is not zero.

4. The signal from the computational element is applied to the motor element, which generates a force in a direction that decreases the isolator deflection; the motor force follows the computational element signal and continues to increase in magnitude so long as the relative deflection δ is not zero.

5. At some point in time the force from the motor output will exactly equal the constant force F_0, requiring a relative displacement of zero.

6. The output from the differential transformer is zero; thus the output from the computational element integration function no longer increases but is maintained constant at the magnitude required for the motor element to generate a force exactly equal to the constant force F_0 applied to the supported body.

The isolation system remains in this equilibrium condition until the force applied to the supported body changes and causes a nonzero signal to be generated by the sensing element; then the process starts all over again. Alternatively, a proportionally scaled signal from the differential transformer may be used to operate an electromechanical servo valve, the flow response of the servo valve being proportional to its excitation signal. The servo-valve fluid-flow output is directed into the chamber of an air spring to produce the desired force applied to the supported body. The control function remains integral in nature since the actuator's internal pressure responds to the volume output from the servo valve, which is the integral of its flow output. Thus, in this case, no electrical integration of the sensor signal is needed. It is also possible to operate a mechanical servo valve through a direct mechanical coupling in such a way that the motion of the suspended body with respect to its support is used directly to provide the required servo-valve actuation. The possible combinations of elements and control devices are almost limitless. The choice of a suitable combination of sensor, computation element, and actuator is dictated by the type of power available, the supported body size, the weight, and the type of application, e.g., spacecraft, aircraft, automotive, or industrial.

The second servomechanism, which provides active control of the system resonance and low-frequency vibration isolation, consists of a velocity sensor, a proportional computational element, and a motor actuation device that may also take on many forms. The velocity of the supported body may be sensed by integrating the response of an accelerometer or through the use of an electromechanical sensor measuring velocity directly. Figure 32.17 illustrates the elements of this servomechanism; the servo amplifier contains the system electronic devices which form the computational element and the power elements required to operate the force actuator. The motor element is contained partly in the servo amplifier and partly in the force actuator. This shows that the three basic elements of a servomechanism are not always self-contained devices, but may be made up of the combined operation of system hardware components. The force actuation device usually consists of an electromagnetic force actuator made up of a strong magnet and coil similar to the type utilized in high-quality loudspeakers. This arrangement generally provides sufficient force to operate active vibration-isolation systems used to attenuate naturally occurring ground and building structure vibration. Where a larger force is required, as is the case where the system is subjected to larger-magnitude vibratory excitations or to vibratory forces applied directly to the supported body, the force motor may consist of driving elements similar to electrodynamic vibration exciters. The electronic amplifiers driving the force motor must have a frequency response extending down

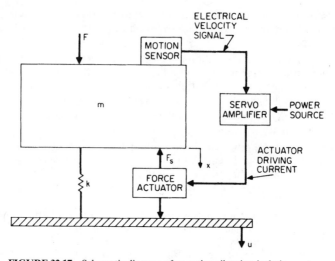

FIGURE 32.17 Schematic diagram of an active vibration-isolation system which acts like a passive vibration-isolation mass and spring element with a viscous damping element connected between the supported body and motionless fixed space. The active damping servomechanism can eliminate the isolation system resonance, thereby providing vibration isolation starting at zero frequency.

to zero frequency, so as not to introduce timing errors into the control signal that can significantly alter the response of the servomechanism. The velocity sensor-amplifier-motor system making up this servomechanism applies a force to the supported body that is proportional to the body's velocity and thus acts in the same manner as a viscous damper connected to the supported body at one end and to motionless fixed space at the other end. This produces a form of damping within the active vibration-isolation system which cannot be synthesized using passive damping elements alone. The action of this velocity-controlled servomechanism is referred to as *active damping,* and the active damping scaling term G_2, relating the supported body velocity to the force applied to the mass m, when divided by the critical damping term for the passive spring and mass elements $2\sqrt{km}$, is commonly referred to as the *active fraction of critical damping* G_2/c_c.

An active vibration-isolation system usually is described by a cubic or higher-order differential equation; because of the complexity of these equations, it is difficult to visualize the effect of changes in the system constants on the performance of the isolation system. This is particularly true when the actual nonideal response characteristics of the system sensing, computational, and motor elements are included in the system differential equation of motion and when additional computational elements, called compensation circuits, are added. The compensation circuits are used to alter the system frequency response, i.e., resonance frequency and peak transmissibility. In working with active vibration-isolation systems of the type presented here, it is not uncommon to have differential equations as high as the twelfth order or more. The field of automatic control system synthesis has devised methods to deal with differential equations of such high orders from both a theoretical analysis and an actual system hardware point of view.

Because integral feedback of displacement requires that energy be fed into the isolation system, it is possible to make the active system dynamically unstable by improper proportioning of its constants. An active vibration-isolation system that is dynamically unstable will undergo continuously increasing mechanical oscillations which, when not limited by available power, will increase until the system is destroyed. Therefore, one of the factors in achieving a satisfactory active vibration-isolation system is the determination of the margin of dynamic stability of the entire system. Here too, the field of automatic control systems has devised methods to establish the system margin of dynamic stability. The margin of dynamic stability is a measure of the degree of change in system constants that is required for the active vibration-isolation system to become unstable.

In the case of a conventional passive vibration-isolation system, it is possible to determine many of the performance characteristics from the constants appearing in the differential equation. For example, the transmissibility T of the conventional system at the condition of resonance is approximately

$$T_r \simeq \frac{\sqrt{km}}{c} = \frac{1}{2(c/c_c)} \qquad \text{where } c/c_c < 0.2 \qquad (32.4)$$

Similarly, the resonance frequency ω_r is approximately equal to the undamped natural frequency:

$$\omega_r \simeq \sqrt{\frac{k}{m}} \qquad \text{where } c/c_c < 0.2 \qquad (32.5)$$

At high frequencies ($\omega \gg \omega_n$), the transmissibility of the conventional system approaches the asymptotic value

$$T_i = \frac{c/c_c}{\omega/\omega_n} \qquad \text{where } \omega \gg \omega_n \qquad (32.6)$$

The transmissibility curve of a conventional isolator may be estimated from Eqs. (32.4) to (32.6) without plotting the transmissibility equation point by point. Somewhat similar relationships can be obtained for an active system if its equation of motion is not higher than the second order. A convenient way to obtain rules of thumb for the design of an active vibration-isolation system is to compare the characteristic properties of a conventional vibration-isolation system with those of the same isolation system but with active elements which provide integral relative displacement force feedback and proportional velocity force feedback added in parallel with a spring isolation element. The velocity feedback gain G_2 generally has a larger effect on the system response than the relative displacement gain term G_1. The feedback gain terms relate the sensed system motion term to the force applied to the supported body; therefore, the units of the velocity feedback gain term G_2 are the same as those for a viscous damper, or force per unit velocity; the gain term G_1 for the integral relative displacement feedback has no passive counterpart and has units of force per unit displacement multiplied by time. The active damping term dominates the system differential equation, affecting the system response both above and below the undamped natural frequency, while the effect of the relative displacement feedback on system performance is confined mainly to the frequency region below the undamped natural frequency. Setting the integral relative displacement gain term G_1 to zero gives an approximation for the transmissibility of the active isolation system:

$$T = \sqrt{\frac{1}{[1 - (\omega/\omega_n)^2]^2 + [2(G_2/c_c)(\omega/\omega_n)]^2}} \tag{32.7}$$

Using the above equation, the following response estimations can be formulated. The system transmissibility T at a frequency equal to the undamped natural frequency ω_n, formed by the passive spring and mass elements k and m, is

$$T_n = \frac{1}{2G_2/c_c} \qquad \text{where } \omega = \omega_n \tag{32.8}$$

The resonance frequency is less than the system undamped natural frequency, and with an active fraction of critical damping term of 1 or larger, there is no system resonance; i.e., at all frequencies the system transmissibility is less than 1. In the case where the relative displacement feedback gain is not zero, the mechanics of the system must always form a resonance condition. At excitation frequencies well above the system undamped natural frequency, the transmissibility of the active isolation system approaches the asymptotic value

$$T_i = (\omega/\omega_n)^2 \qquad \text{where } \omega \gg \omega_n \tag{32.9}$$

In the above response estimation relationship function, the system transmissibility at the undamped natural frequency is less than unity when the velocity feedback gain exceeds a value giving an active fraction of critical damping of 0.5; i.e., $G_2/c_c = 0.5$. With an active fraction of critical damping of unity, the system transmissibility at the undamped natural frequency is 0.5. Active vibration-isolation systems of this type typically exhibit velocity feedback gain magnitudes yielding an active fraction of critical damping ranging from a low of about 0.5 to a high of about 5. The incorporation of the integral relative displacement feedback servomechanism in conjunction with the velocity feedback servomechanism and the passive system elements forms a system described by a third-order differential equation. A resonance condition occurs well below the undamped natural frequency when the active fraction of critical damping is 0.5 or more. The simplified response estimations of transmissibility are valid for frequencies at and above the system undamped natural frequency in instances where the active fraction of critical damping is 0.5 or greater. As the active fraction of critical damping is decreased, the resonance frequency approaches the undamped natural frequency with an increasing peak transmissibility and an eventual dynamically unstable system.

In an ideal active vibration-isolation system, the resonance frequency and peak transmissibility are a function of the passive system constants and the two feedback gain terms. In a nonideal active vibration-isolation system, there are many other factors that influence the system resonance characteristics, such as the low-frequency response of the velocity sensor or a more complex passive system formed from many mass and spring elements. The resonance characteristics of the active vibration-isolation system are manipulated through compensation functions formed using electric networks in the computation element of the velocity servomechanism. The function of these compensation networks is to alter the nature of the velocity feedback signal applied to the motor element, in a manner that provides for a dynamically stable system, and to raise or lower the resonance frequency, peak transmissibility, and transmissibility frequency response above the resonance frequency. The use of system compensation circuitry is extensive in the field of automatic control system synthesis as well as with active vibration-isolation systems, which are a type of automatic control system. The result of system compensation is active vibration-isolation systems with response characteristics similar but not limited to the response of the ideal system. The analysis of the transient and frequency-response characteristics of an active vibration-isolation system having ideal elements shows many of the advantages of actual active vibration-isolation systems when compared to the response of passive system elements alone.

In an active vibration-isolation system, the element that provides integral control of relative displacement strives to maintain the supported body at a constant distance from the support base to which it is attached. When a step function of force is applied to the supported body, the response of the system gives a measure of the element's effectiveness in performing the desired function. A comparison of the transient response of the active vibration-isolation system, i.e., one having integral relative displacement and absolute velocity force feedback, with that of the conventional passive vibration-isolation system illustrates the advantage obtained from integral relative displacement feedback.

TRANSIENT RESPONSE

The equation of motion for the mass m of the passive isolation system is

$$m\ddot{x} + c\dot{x} + kx = F(t) \tag{32.10}$$

where the force $F(t)$ is a step function of force having a magnitude $F = F_0$ when $t > 0$ and $F = 0$ when $t < 0$. Writing the Laplace transform of Eq. (32.10),

$$\mathcal{L}[x(t)] = X(s) = \frac{F_0}{ms} \frac{1}{s^2 + (c/m)s + k/m} \tag{32.11}$$

where $X(s)$ designates the Laplace transform of x, a function of time. Letting $c/m = 2(c/c_c)\omega_n$ and $k/m = \omega_n^2$, Eq. (32.11) may be written as

$$X(s) = \frac{F_0}{ms} \frac{1}{s^2 + 2(c/c_c)\omega_n s + \omega_n^2} \tag{32.12}$$

The time solution of Eq. (32.12) is a damped sinusoid offset by the deflection of the spring caused by the constant force F_0. A typical time solution is shown by curve A of Fig. 32.18. The deflection of the isolator can be calculated by applying the final value theorem of Laplace transformations. This theorem states that if the Laplace transform of $x(t)$ is $X(s)$ and if the limit $x(t)$ as $t \to \infty$ exists, then

$$\lim_{s \to 0} sX(s) = \lim_{t \to \infty} x(t) \tag{32.13}$$

Applying the final value theorem using the Laplace transform of the passive isolator responding to the step function of force, Eq. (32.12), shows that the final deflection of the isolator is

$$\lim_{s \to 0} sX(s) = \lim_{t \to \infty} x(t) = \frac{F_0}{m\omega_n^2} \tag{32.14}$$

From Eq. (32.14), the mass takes a new position of static equilibrium at a distance $F_0/(m\omega_n^2)$ from the original position as $t \to \infty$. The final deflection term may be eliminated from Eq. (32.14) by adding an integral relative displacement control servomechanism. This added element produces a force proportional to the integral of displacement x with respect to time. The system damping element is replaced by an

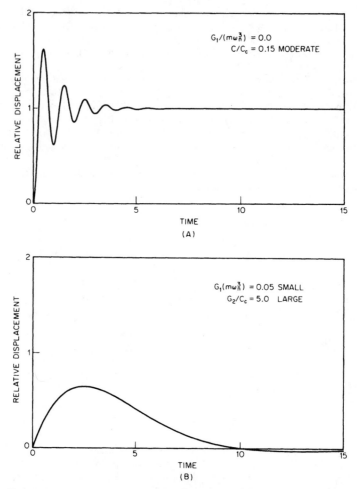

FIGURE 32.18 (*A*) Transient response of a passive vibration-isolation system to a step in force. (*B*), (*C*), and (*D*) show the transient response of an active vibration-isolation system to the same force step for different values of integral relative displacement and proportional velocity gains. The response is changed by changes in the feedback gain magnitude. In (*D*) the system is unstable as a result of the improper selection of the servomechanism constants; as a result, oscillations become increasingly large.

active damping control servomechanism. Active damping in this case acts in the same manner as the passive damping element used for Eq. (32.10) since x is the only system motion. The differential equation of motion for the supported body of the active vibration-isolation system is

$$m\ddot{x} + G_2\dot{x} + kx + G_1 \int x \, dt = F(t) \tag{32.15}$$

The Laplace transform of the active vibration-isolation system differential equation is

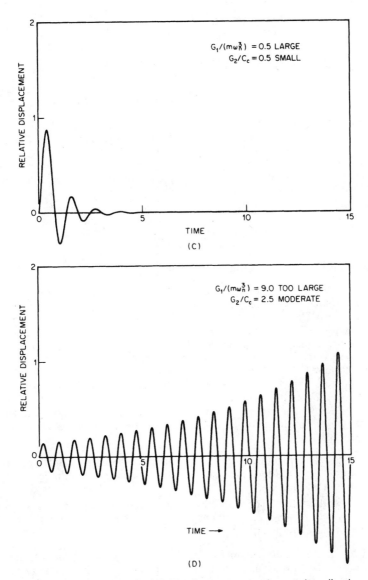

FIGURE 32.18 (*Continued*) (*A*) Transient response of a passive vibration-isolation system to a step in force. (*B*), (*C*), and (*D*) show the transient response of an active vibration-isolation system to the same force step for different values of integral relative displacement and proportional velocity gains. The response is changed by changes in the feedback gain magnitude. In (*D*) the system is unstable as a result of the improper selection of the servomechanism constants; as a result, oscillations become increasingly large.

$$\mathscr{L}[x(t)] = X(s) = \frac{F_0}{ms} \frac{1}{s^2 + (G_2/m)s + k/m + G_1/ms} \tag{32.16}$$

Placing the above equation in a form similar to Eq. (32.12) gives

$$X(s) = \frac{F_0}{m} \frac{1}{s^3 + 2(G_2/c_c)s^2 + \omega_n^2 s + (G_1/m\omega_n^3)\omega_n^3} \tag{32.17}$$

The term G_2/c_c represents the active fraction of critical damping. The term containing the active relative displacement feedback gain $G_1/m\omega_n^3$ is called the *dimensionless relative displacement feedback gain*. The use of the dimensionless gain terms, active fraction of critical damping and dimensionless relative displacement feedback gain, allows the response characteristics of the active vibration-isolation system to be represented in a generalized manner where the numerical values of the passive system elements are not required.

Applying the final value theorem to the transient response of the active vibration-isolation system represented by Eq. (32.17) gives the deflection of the supported body in its final equilibrium position:

$$\lim_{s \to 0} sX(s) = \lim_{t \to \infty} x(t) = 0 \tag{32.18}$$

The final equilibrium position for the supported body of the active vibration-isolation system is zero so long as the dimensionless relative displacement feedback gain is not zero. The final position of the supported body is zero even with a very small dimensionless relative displacement feedback gain because of the integration operation provided by the relative displacement servomechanism. The magnitudes of the two servomechanism gain terms affect the motion of the supported body during the transient. Figure 32.18A shows the transient response of a passive vibration-isolation system to a step in force which is applied to the supported body. In Fig. 32.18B, C, and D the transient response of an active system subjected to the same step force is shown for various values of the dimensionless feedback gain. The two servomechanisms in the active vibration-isolation system interact, but their effect can be generalized:

1. Increasing the magnitude of the dimensionless relative displacement gain increases the rate at which the system relative displacement approaches the final equilibrium position.
2. Increasing the active fraction of critical damping decreases the peak magnitude of the system relative displacement during the transient event and lowers the damped natural frequency.

The degree of oscillation exhibited by the active vibration-isolation system is a function of the magnitude and relative magnitude of the dimensionless gains of the two servomechanisms. In general, small magnitudes of the dimensionless relative displacement gain and large magnitudes of the active fraction of critical damping lead to little system oscillation, as depicted by the curve of Fig. 32.18B. Likewise, large magnitudes of the dimensionless relative displacement gain and small magnitudes of the active fraction of critical damping tend to increase the amount of oscillation. The dimensionless relative displacement gain can be increased too much in relation to the active fraction of critical damping and will then produce a condition of instability, as shown by the curve of Fig. 32.18D. The conditions resulting in system instability are presented in the last part of this section.

The relative displacement response of this ideal active vibration-isolation system to constant acceleration of the isolator support, such as that produced by gravity or by the sustained acceleration of a missile, cannot be represented by applying a constant force to the supported body, as is frequently done with passive vibration-isolation systems. The reason for this is that active vibration-isolation systems which utilize absolute motion feedback, as in active damping of the type presented in this chapter, respond differently to forces applied to the supported body than to a constant acceleration of the support. In the case of a constant force applied to the supported body, presented above, the velocity servomechanism output force approaches zero as the transient motions of the system die out. In the case of a constant acceleration of the support, the velocity of the supported body continually increases in a manner similar to the increase in velocity of the support. The output of the velocity servomechanism increases constantly with time since the output force is proportional to the velocity of the supported body. This leads to a system which cannot work because the velocity servomechanism will rapidly reach its maximum force output, at which time all active damping is lost. In this situation, active vibration isolation is reobtained by placing an electric filter in the active damping servomechanism computational element. The filter forms a control function which produces a zero output for a ramp input. The use of such a filter is part of the compensation process often required with automatic control systems; this process is presented in more detail in the next section.

Many active vibration-isolation systems of the ideal type presented in this chapter are used to isolate angular vibration, on which gravity has no effect. The active isolation of angular vibration uses the same system equations presented above except that the motions are angular, the mass is a moment of inertia, and the passive spring element applies a torque to the supported body that is proportional to the relative rotational displacement between the supported body and the support. The integral relative displacement servomechanism operates by measuring the rotation of the supported body relative to the support and applying a torque to the supported body that is proportional to the time integral of the sensed rotation. The relative angular displacement may be sensed using a rotational differential transformer or a linear potentiometer.

The active damping servomechanism operates by sensing the absolute rotational velocity of the supported body using a rate gyroscope which has an output response proportional to its rotational velocity. The active damping torque applied to the supported body is proportional to the output of the rate gyroscope. Many times the passive spring element is replaced by a servomechanism where the integral relative displacement control function in the computational element of the servomechanism is modified to produce an output proportional to the sum of the relative displacement and its first integral. Such a servomechanism has proportional plus integral control.

STEADY-STATE RESPONSE

A comparison of the steady-state response of the active and passive vibration-isolation systems illustrates some of the advantages and disadvantages associated with a servo-controlled vibration-isolation system. In Fig. 32.17, assume that $F(t) = 0$ and that the vibration excitation is caused by the motion $u(t)$ of the support base. Then the equation of motion for the supported body of the active vibration-isolation system having both the active damping servomechanism shown by Fig. 32.17 and the integral relative displacement control servomechanism shown by Fig. 32.16 is

$$m\ddot{x} + G_2\dot{x} + kx + G_1 \int x \, dt = ku + G_1 \int u \, dt \qquad (32.19)$$

The response of this isolation system, when the vibration excitation $u(t)$ is sinusoidal in nature and steady with respect to time, may be expressed in terms of transmissibility:

$$T = \sqrt{\frac{(G_1/m\omega_n^3)^2 + (\omega/\omega_n)^2}{(\omega/\omega_n - \omega^3/\omega_n^3)^2 + [G_1/m\omega_n^3 - 2(G_2/c_c)(\omega^2/\omega_n^2)]^2}} \qquad (32.20)$$

Figure 32.19 is a plot of Eq. (32.20) for four values of the relative displacement dimensionless gain term and six values of the velocity dimensionless gain term, $G_1/(m\omega_n^3)$ and G_2/c_c, respectively. The corresponding expression for the transmissibility for the conventional passive vibration-isolation system differs from that for an active system, i.e., Eq. (32.20), because of the nature of the force feedback terms acting upon the supported body. At frequencies well above the vibration-isolation system undamped natural frequency ω_n, the active and passive system transmissibility equations differ because of the presence of a damping term in the numerator of the passive system equation. At these higher frequencies, the passive system transmissibility has the characteristic that as $\omega \to \infty$, $T \to 2(c/c_c)(\omega_n/\omega)$. The active system, however, tends to act as an undamped vibration-isolation system wherein the transmissibility at high frequencies has the characteristic that as $\omega \to \infty$, $T \to \omega_n^2/\omega^2$. Thus the active vibration-isolation system provides a lower transmissibility at frequencies above the system natural frequency, especially for large values of the active and passive damping terms.

At excitation frequencies close to the system natural frequency, both the active and passive vibration-isolation systems exhibit a resonance condition when the system damping terms are small. The peak value of the system transmissibility at the system resonance frequency is controllable by the addition of damping. In the passive vibration-isolation system, as the fraction of critical damping is increased, the peak transmissibility is lowered, reaching a value of unity for an infinite value of the fraction of critical damping. Although the passive system damping controls the peak transmissibility, high values of damping greatly degrade the system's main function of isolating vibration; in fact, very large magnitudes of the system damping term yield little to no vibration isolation, since the damper tends to become a rigid link between the isolation system vibrating base and the supported body. The effect of damping on the active vibration-isolation system is similar to that on the passive vibration-isolation system when the active fraction of critical damping is small. However, as the active system damping is increased, an increasingly more rigid link is placed between the supported body and motionless space; thus, increasing the active fraction of critical damping always decreases the system transmissibility at frequencies above the natural frequency. With a relative displacement gain G_1 of zero, the active system resonance will disappear when the active fraction of critical damping exceeds unity, as is shown by the curve of Fig. 32.19A. With an active fraction of critical damping of unity, the peak transmissibility is also unity and occurs at zero frequency, and for all other frequencies the system transmissibility is less than 1, having the approximate magnitude of $1/[2(G_2/c_c)(\omega/\omega_n)]$ at frequencies from zero to about twice the system natural frequency and ω_n^2/ω^2 at higher frequencies.

The addition of the relative displacement integral control has little influence on transmissibility at high frequencies and thus has no important effect on the ability of the complete system to isolate vibration. However, the effect at lower frequencies is significant, as is shown in Fig. 32.19B, C, and D. As the dimensionless gain $G_1/m\omega_n^3$ of the displacement control loop is increased, the transmissibility of the system in the region of resonance increases. If the dimensionless displacement gain term equals twice the active fraction of critical damping, the active vibration-isolation sys-

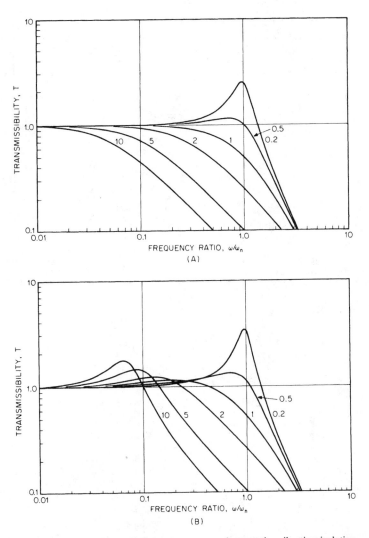

FIGURE 32.19 Steady-state frequency response for an active vibration-isolation system having an ideal active damping servomechanism. The transmissibility is plotted against the frequency ratio ω/ω_n. In (A) there is no integral relative displacement control servomechanism, i.e., $G_1/m\omega_n^3 = 0$; in (B), (C), and (D) such a control mechanism has been added and this ratio has values of 0.1, 0.2, and 0.5, respectively. For each of these illustrations a set of curves is shown for the following values of the ratio G_2/C_c: 0.2, 0.5, 1, 2, 5, and 10. Changes in the servomechanism feedback constants affect the response characteristics through their dynamic interactions, which alter the frequency response at low excitation frequencies.

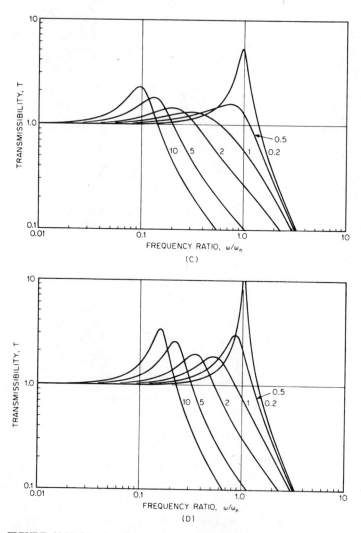

FIGURE 32.19 (*Continued*) Steady-state frequency response for an active vibration-isolation system having an ideal active damping servomechanism. The transmissibility is plotted against the frequency ratio ω/ω_n. In (*A*) there is no integral relative displacement control servomechanism, i.e., $G_1/m\omega_n^3 = 0$; in (*B*), (*C*), and (*D*) such a control mechanism has been added and this ratio has values of 0.1, 0.2, and 0.5, respectively. For each of these illustrations a set of curves is shown for the following values of the ratio G_2/C_c: 0.2, 0.5, 1, 2, 5, and 10. Changes in the servomechanism feedback constants affect the response characteristics through their dynamic interactions, which alter the frequency response at low excitation frequencies.

tem becomes dynamically unstable. Under these conditions, if the supported body receives the slightest disturbance, a system oscillation will develop and continue indefinitely, as would be the case with a passive system without damping. Increasing the relative displacement gain term above this critical value results in a condition where the system's automatic control functions continually add energy to the supported body and passive spring element in the form of ever-increasing oscillations, which continue to increase in amplitude until motor saturation or destruction of the system occurs.

STABILITY OF ACTIVE VIBRATION-ISOLATION SYSTEMS

Operation of a dynamically unstable active vibration-isolation system exhibits one or more of the following characteristics:

1. The active vibration-isolation system acts like an undamped passive vibration-isolation system.
2. The system exhibits oscillations that increase with time and can become very large in magnitude.
3. The system moves to one of its excursion stroke limits and stays there.

The ensurance of a dynamically stable active vibration-isolation system is important at both the design and hardware stages of development and can become a complex design task. Much of the field of automatic control system analysis and synthesis deals with establishing the limits of feedback gains beyond which the system becomes unstable.

REFERENCES

1. Racca, R.: "How to Select Power-Train Isolators for Good Performance and Long Service Life," *Paper* 821095, *SAE International Off-Highway Meeting and Exposition*, Sept. 13–16, 1982.
2. Racca, R.: "How to Select Isolators for a Long Service Life." *Design News*, Jan. 1982.
3. Ushijima, T., K. Takano, and H. Kojima: "High Performance Hydraulic Mount for Improving Vehical Noise and Vibration," *SAE Paper* 880073 International Congress and Exposition, Detroit, Mich., Feb. 29, 1988.

CHAPTER 33
SELECTION AND APPLICATIONS OF VIBRATION ISOLATORS

R. H. Racca

INTRODUCTION

Vibration isolators (also called vibration mounts) are used to control shock and vibration. This chapter describes (1) factors that should be considered in the selection of isolators, (2) specifications for shock and vibration isolators, (3) characteristics of isolators used in parallel and in series, (4) selection of the most appropriate type and characteristic of isolator for any given application, and (5) selection of the correct size of isolator. Such selection procedures apply to usual engineering problems; for more complex problems, guidance can be provided by the manufactures of commercial isolators. Finally, numerical solutions are given for a number of practical isolation problems, and a checklist is provided for isolator installations.

The basic objectives in using isolators are (1) to protect the supporting structure and adjacent systems from vibration and shock disturbances originating in certain equipment and (2) to protect sensitive equipment from shock and vibration emanating from the structure on which the equipment is mounted. When such disturbances are present, properly applied isolators can allow equipment and systems to function as intended and can lengthen equipment life. Isolation of shock and vibration can also provide a more comfortable environment for equipment users, building occupants, and passengers in vehicles.

FACTORS CONSIDERED IN ISOLATOR SELECTION

Stiffness and damping are the basic properties of an isolator which determine its use in a system designed to provide vibration isolation and/or shock isolation. These properties usually are found in isolator supplier literature. However, there are other important factors which must be considered in the selection of an isolator:

1. Source of dynamic disturbance
2. Type of dynamic disturbance
3. Direction of dynamic disturbance (to determine axes of isolator loading)
4. Allowable response of a system to dynamic disturbance
5. Space and location available for the isolator
6. Weight and center-of-gravity of supported equipment (to determine load per isolator)
7. Space available for equipment motion
8. Ambient environment
9. Available isolator materials
10. Desired service life
11. Fail-safe installation
12. Cost

SOURCE OF DYNAMIC DISTURBANCE

The source of a dynamic disturbance (shock or vibration) influences the selection of an isolator in several ways. For example, a decision can be made whether to isolate the source of the disturbance or to isolate the item being disturbed. This decision affects which isolator is to be used. Consider the operation of a heavy punch press which has an adverse effect on a nearby electronic instrument. Isolation of the punch press would reduce this effect but would require fairly large isolators which might have to be resistant to grease or oil. In contrast, isolation of the instrument would also provide the required protection, but the required isolators would be smaller and (since grease or oil would not be a consideration) could be fabricated of a preferred elastomer.

A knowledge of the source of the vibration can aid in defining the problem to be solved. Within a given industry there may be published material describing problems similar to the one under consideration. Such material may describe possible solutions plus equipment fragility, and/or dynamic characteristics of the equipment.

Type of Dynamic Disturbance.   The dynamic environment can be delineated into three categories: (1) periodic vibration—sinusoidal continuos motion or acceleration occurring at discrete frequencies, (2) random vibration—the simultaneous existence of any and all frequencies and amplitudes in any and all phase relationships as exemplified by noise, and (3) transient phenomenon (shock)—a nonperiodic sudden change of velocity, acceleration, or displacement. Usually some combination of these three categories occurs in most isolation systems. A knowledge of the dynamic disturbance is very important in the choice of an isolator. For example, in the case of an instrument supported by isolators, the resilient mounts permit the supported body to "stand still" by virtue of its own inertia while the support structure generates periodic or random vibration. In contrast, shock attenuation involves the storage by the isolators of the dynamic energy which impacts on the support structure and the subsequent release of the energy over a longer period of time at the natural frequency of the system. If only a vibration disturbance is present, a small isolator normally is suitable since vibration amplitudes usually are small relative to shock amplitudes. If a shock disturbance is the primary problem, then a larger isolator with more internal space for motion is required.

In selecting an isolator, it is necessary to ensure (1) that there is enough deflection capability in the isolator to accommodate the maximum expected motions from the dynamic environment, (2) that the load-carrying capacity of the isolator will not be exceeded; the maximum loads due to vibration and/or shock should be calculated and checked against the rated maximum dynamic load capacity of the isolator, and (3) that there will be no problem as a result of overheating of the isolator or fatigue deterioration due to long-term high-amplitude loading.

Direction of Dynamic Disturbance. A factor that must be considered in the selection of an isolator is that of the directions (axes) of the dynamic disturbance. If the vibration or shock input occurs only in one direction, usually a simple isolator can be selected; its characteristics need be specified along only one axis. In contrast, if the vibration or shock is expected to occur along more than one axis, then the selected isolator must provide isolation (and its characteristics must be specified) along all the critical axes. For example, consider an industrial machine which produces troublesome vibration in the vertical direction and which must be isolated from its supporting structure. In this case, a standard plate-type isolator may be used. This type isolator is stiffer in the horizontal direction than in the vertical direction, which is the axis of the primary disturbance; the horizontal stiffness does not significantly affect the motion of the isolator in the vertical direction. Such horizontal stiffness adds to the lateral stability of the installation.

Allowable Response of a System to Dynamic Disturbance. The allowable response of a system is defined as the maximum allowable transmitted shock or vibration and the maximum displacements due to such disturbances. The allowable response of a system can be expressed in any of the following ways:

1. Maximum acceleration loading due to a shock input
2. Specific system natural frequency and maximum transmissibility at that frequency
3. Maximum acceleration, velocity, or displacement allowable over a broad frequency range
4. The allowable level of vibration at some critical frequency or frequencies
5. Maximum displacement due to shock loading

The maximum acceleration which a piece of equipment can withstand without damage or malfunction is often called fragility. The definition of some allowable response is necessary for an appropriate isolator selection. If fragility data are not available for the specific equipment or installation at hand, then examples of similar situations should be used as a starting point. Suppose an isolator were chosen only for its load-carrying capability, with no regard for the fragility of a piece of equipment in a specific frequency range. Then, the natural frequency of the system might be incorrectly placed such that a resonance within the equipment might be excited by the isolation system.

SPACE AND LOCATIONS AVAILABLE FOR ISOLATORS

Vibration and shock isolation should be considered as early as possible in the design of a system, and an estimate of isolator size should be made based on isolator literature. The size of the isolator depends on the nature and magnitude of the expected dynamic disturbances and the load to be carried. Typical literature describes the capabilities of isolators based on such factors.

The location of isolators is very important to the dynamics of the equipment mounted on them. For example, a center-of-gravity installation, as shown in Fig. 33.1, allows the mounted equipment to move only in straight translational modes (i.e., a force at the center-of-gravity does not cause rotation of the equipment). This minimizes the motion of the corners of the equipment and allows the most efficient installation from the standpoint of space requirements and isolation efficiency.

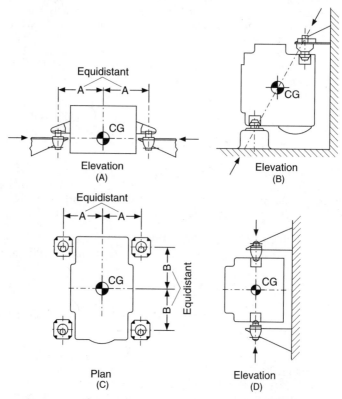

FIGURE 33.1 Center-of-gravity installations of vibration isolators: (*A*) Center-of-gravity horizontal support. (*B*) Center-of-gravity diagonal support. (*C*) Symmetrical spacing about the center-of-gravity. (*D*) Center-of-gravity vertical support. (*After Davey and Payne.*[1])

If the isolators cannot be located so as to provide a center-of-gravity installation, then the system analysis is more difficult and more space must be allowed around the equipment to accommodate rocking motion (i.e., rotational modes) of the system. Finally, the isolators must be double-checked to ensure that they are capable of withstanding the additional loads and motions from the nontranslational movement of the equipment. This is particularly true when the center-of-gravity is a significant distance above or below the plane in which the isolators are located. Rule of thumb: The distance between the isolator plane and the center-of-gravity should be equal to or less than one-third of the minimum spacing between isolators. This helps to minimize rocking of the equipment and the resultant high stress in the isolators.

WEIGHT AND CENTER-OF-GRAVITY OF SUPPORTED EQUIPMENT

The weight and location of the center-of-gravity of the supported equipment should be determined. The location of the center-of-gravity is necessary for calculating the load supported on each mount. It is best to keep the equipment at least satatically balanced [essentially equal deflections on all isolators, (see Fig. 33.1)]. The preferred approach is to use the same isolator at all points, choosing isolator locations such that static loads (and thus deflections) are equalized. If this is not practical, isolators of different load ratings may be required at different support points on the equipment for optimum isolation. The size of the equipment and the mass distribution are important in dynamic analyses of the isolated system.

SPACE AVAILABLE FOR EQUIPMENT MOTION

The choice of an isolator may depend on the space available (commonly called sway space) around a piece of equipment. The spring constant of the isolator should be chosen carefully so that motion is kept within defined space limits. The motion which must be considered is the sum of (1) the static deflection due to the weight supported by the isolator, (2) the deflection caused by the dynamic environment, and (3) the deflection due to any steady-state acceleration (such as in a maneuvering aircraft).

If there is a problem of excessive motion of the supported mass on the isolator, then a *snubber* (i.e., a device which limits the motion) can be used. A snubber may be an elastomeric compression element designed into an isolator. Captive-type isolators (see *Fail-Safe Installation*) have built-in motion-limiting stops. Also, elastomers stressed in compression have natural snubbing due to the nonlinear load-deflection characteristics. In some cases it may be necessary to limit motion by separately installed snubbers such as a compression pad at the point of excessive motion as shown in Fig. 33.2. The spring constant of such a snubber must be carefully selected to avoid transmission of high-impact loads into the supported equipment.

AMBIENT ENVIRONMENT

The environment in which an isolator is to be used affects its selection in two ways:

1. Some environmental conditions may degrade the physical integrity of the isolator and make it nonfunctional.

2. Some environmental conditions may change the operating characteristics of an isolator, without causing permanent damage.

This may alter the characteristics of the isolation system of the supported equipment; for example, frequency responses could change significantly with changes in the ambient temperature. Thus, it is important to determine the operating environment of the isolation system and to select isolators that will function with desired characteristics in this environment.

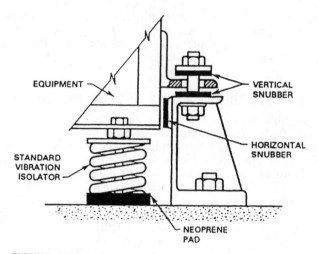

FIGURE 33.2 A vibration isolator provided with auxiliary elastomeric snubbers to limit the motion of the isolator in the horizontal and vertical directions; these snubbers provide a "cushion" stop to provide a lower shock force on the equipment than would be experienced with a metal-to-metal stop.

AVAILABLE ISOLATOR MATERIALS

Vibration and shock isolators are available in a wide variety of materials and configurations to fit many different situations. The type of isolator is chosen for the load and dynamic conditions under which it must operate. The material from which the isolator is made depends to a great extent on the ambient environment of an application and somewhat on the dynamic properties required. Chapter 32 provides guidance for the choice of isolators and many isolator materials. Chapter 34 describes the engineering properties of rubber.

Metal-spring isolators are used primarily where operating temperatures are too high for elastomeric isolators. They can be used in a variety of applications.

By far, the majority of isolators in use today are elastomeric. The development of a vast array of elastomeric compounds has made it possible to use this type of isolator in almost any environment. Within a given type of elastomer, it is a simple matter to vary the stiffness (modulus, durometer) of the compound; this gives much flexibility in adapting an isolator to an application without changing the isolator's geometry.

Since the selection of material for an isolator depends so much on the environment in which the mount will be used, it is very important to learn as much as possible about the operating and storage environments.

DESIRED SERVICE LIFE

The expected, or desired, length of service for an isolator can affect the type and size of the vibration isolator which is selected. For example, an isolator which must operate for 2000 hours under a given set of conditions typically is larger than one which must operate for only 500 hours under the same conditions.

In general, empirical data are used to estimate the operating life of an isolator. Accurate descriptions of the dynamic disturbances and ambient operating environment expected are needed to make an estimate of isolator life. A knowledge of the specific material and design factors in an isolator is necessary to make an estimate of fatigue life. Such information is best provided by the original manufacturer or designer of the isolator.

FAIL-SAFE INSTALLATION

A fail-safe (also called a captive) isolator is one in which there is a positive metal-to-metal interlock in case the resilient spring in the isolator fails. Thus, in the event of failure, the equipment is supported in place until the failed isolator can be replaced. If the isolator does not contain this fail-safe feature, it may be provided by the mounting system. In Fig. 33.2 the separate snubbing arrangement also provides the fail-safe feature.

INTERACTION WITH SUPPORT STRUCTURE

The support structure characteristics can also affect the selection of isolators. An isolator must deflect if it is to isolate vibration; generally the greater the deflection, the greater the isolation. The isolator functions by being soft enough to allow relative vibration amplitudes without transmitting excessive force to the support structure. It is often assumed, in the selection of vibration isolators, that the support structure is a rigid mass with infinite stiffness. This assumption is not true since if the foundation were infinitely stiff, it would not respond to a dynamic force and the isolator would not be needed. Since the foundation does respond to dynamic forces, its response must affect the components that are flexibly attached to it. In reality the support structure is a spring in series with the isolator (see *Parallel and Series Combination of Isolators*) and springs in series carry the same force and deflect proportionally to their respective spring constants. Thus if the stiffness of the isolator is high compared to the stiffness of the foundation, the foundation will deflect more than the isolator and actually nullify or limit the isolation provided from the isolator itself. To achieve maximum efficiency from the selected isolator, the spring constant of the support structure should be at least 10 times that of the spring constant of the isolator attached to it. This will assure that at least 90 percent of the total system spring constant is contributed by the isolators and only 10 percent by the support structure.

Because the structure supporting a piece of equipment has inherent flexibility, it has resonances which could cause amplification of vibration levels; these resonance frequencies must be avoided in relation to isolated system natural frequencies.

SHOCK AND VIBRATION ISOLATOR SPECIFICATIONS

Often, shock and vibration isolators are overspecified; this can cause needless complication and increased cost. Overspecification is the practice of arbitrarily increasing shock or vibration load values to be safe (to make certain that the isolators have been chosen with a high margin of safety at the maximum load capability). The best isolator specification is one which defines the critical properties of the isolation system and the specific environment in which the system will operate. Extraneous requirements cause needless complications. For example, if the vibration level is an acceleration of +1g, it is not advisable to specify +2g to be safe. Likewise, it is inadvisable to rigidly apply an entire specification to an isolator installation when only a small part of the specification is applicable.

Typically, specifications to which vibration and shock isolators are designed will include requirements regarding (1) vibration amplitudes, (2) shock amplitudes, (3) load to be carried, (4) required protection for equipment, (5) temperatures to be encountered (environmental factors, in general), and (6) steady acceleration loads superimposed on dynamic loading.

PARALLEL AND SERIES COMBINATIONS OF ISOLATORS

When a number of isolators are used in a system, they are usually combined either in parallel or in series or in some combination thereof.

ISOLATORS IN PARALLEL

Most commonly, isolators are arranged in parallel. Figure 33.3 depicts three isolators schematically as springs in parallel. A number of vibration isolators are said to be in parallel if the static load supported is divided among them so that each isolator supports a portion of the load. If the stiffness of each of the n isolators in Fig. 33.3 is represented by k, the stiffness of the combination is given by

$$\text{Stiffness of } n \text{ isolators in parallel} = nk \tag{33.1}$$

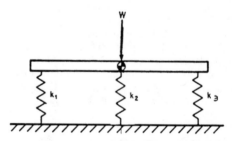

FIGURE 33.3 Schematic diagram of three springs in parallel. Individual spring loads are added to obtain the total weight. With the static load equal on all springs, the static deflection of each spring is the same.

Since in Fig. 33.3 isolator spacing is symmetrical in relation to the center-of-gravity and the same isolator is used at all support points, the stiffness of the combination is 3 times the stiffness of a single isolator, and the static deflection is the same at each isolator.

ISOLATORS IN SERIES

When three isolators are combined in series, as shown in Fig. 33.4, the static load is transmitted from one isolator to the next. If the static weight is supported by n isolators in series, each having the stiffness k, the stiffness of the combination is given by

$$\text{Stiffness of } n \text{ isolators in series} = \frac{k}{n} \qquad (33.2)$$

Thus, if the mass is supported by three identical isolators (Fig. 33.4), the stiffness of this combination is one-third the stiffness of a single isolator, and the static deflection is the sum of the deflection of the individual isolators (or 3 times the static deflection of a single isolator).

HOW TO SELECT VIBRATION ISOLATORS

The vibration isolator selection process should proceed in the following steps:

FIGURE 33.4 Schematic diagram of three springs in series. Individual spring deflections are added to obtain total deflection, but each spring carries the total load.

Step 1. *Required isolation efficiency.* First, indicate the percentage of isolation efficiency that is desired. In general, an efficiency of 70 to 90 percent is desirable and is usually possible to attain.

Step 2. *Transmissibility.* From Table 33.1 determine the maximum transmissibility T of the system at which the required vibration isolation efficiency of Step 1 will be provided.

Step 3. *Forcing frequency.* Determine the value of the lowest forcing frequency f (i.e., the frequency of vibration excitation). For example, in the case of a motor, the forcing frequency depends on the rotational speed, given in revolutions per minute (rpm); the rotational speed must be divided by 60 sec/minute to obtain the forcing frequency in cycles per second (Hz). The lowest forcing frequency is used because this is the worst condition, resulting in the lowest value of f/f_n (see Table 33.1). If a satisfactory value of isolation efficiency is attained at this frequency, the vibration reduction at higher frequencies will be even greater.

TABLE 33.1 Ratio of (f/f_n) Required to Achieve Various Values of Vibration Isolation Efficiency

Isolation efficiency, %	Maximum transmissibility	Required f/f_n
90	0.1	3.32
80	0.2	2.45
70	0.3	2.08
60	0.4	1.87
50	0.5	1.73
40	0.6	1.63
30	0.7	1.56
20	0.8	1.50
10	0.9	1.45
0	1.0	1.41

Step 4. *Natural frequency.* From Fig. 33.5, find the natural frequency f_n of the isolated system (i.e., the mass of the equipment supported on isolators) required to provide a transmissibility T, determined in Step 2 (which is equivalent to a corresponding percent vibration isolation efficiency) for a forcing frequency of f Hz (determined in Step 3).

Step 5. *Static deflection.* From Fig. 33.5, determine the static deflection required to provide a natural frequency of Step 4.

Step 6. *Stiffness of isolation system.* From Eq. (33.3), calculate the stiffness k required to provide a natural frequency f_n determined in Step 4:

$$f_n = \frac{[kg/W]^{1/2}}{2\pi} \tag{33.3}$$

where W = the weight in pounds of the supported mass
       g = the acceleration due to gravity in inches per second per second

Step 7. *Stiffness of the individual vibration isolators.* Determine the stiffness of each of the n isolators from Eq. (33.1) or Eq. (33.2) depending on whether the vibration isolators are in parallel or in series. In general, they are in parallel so that the required stiffness of each vibration isolator is $1/n$ times the value obtained in Step 6—assuming that all isolators share the load equally.

Step 8. *Load on individual vibration isolators.* Now calculate the load on each individual isolator.

Step 9. *Isolator selection.* From a manufacturer's catalog, elect a vibration isolator which meets the stiffness requirement determined in Step 7 and which has a load-carrying capacity (i.e., load rating) equal to the value obtained in Step 8. The preferred approach is to use the same type and size isolator at all points of support; choose isolator locations such that static loads (and thus deflections) are equalized. If this is not practical, isolators of different load ratings may be required at different support points on the equipment. If the vibration occurs only in one direction, usually a simple isolator can be selected; its characteristics need be specified along only one axis. In contrast, if the vibration is expected to occur along more than one direction, then the selected isolator must provide isolation along all the critical axes.

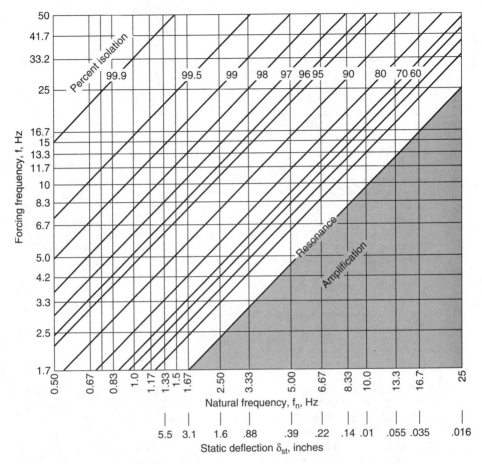

FIGURE 33.5 Isolation efficiency chart. The vibration efficiency, in percent, is given as a function of natural frequency of the isolated system (along the horizontal axis) and the forcing frequency, i.e., the frequency of excitation (along the vertical axis). The use of this chart is restricted to applications where the vibration isolators are supported by a floor structure having a vertical stiffness of at least 15 times the total stiffness of the isolation system. This may require that the isolated structure be placed along the length of a floor beam or that an additional floor beam be added to the structure.

EXAMPLES

The following examples present specific applications. They show how isolators may be selected for some simple shock and vibration problems, but the steps used are basic and can be extended to many other situations. In the solution of these problems, the following simplifying assumptions are made:

1. The effect of damping is negligible, a valid assumption for many isolator applications.

2. All modes of vibration are uncoupled, i.e., the isolators are symmetrically located with respect to the mass center-of-gravity.

3. The static and dynamic spring constants of the isolators are equal, valid for low modulus elastomers with little damping.

Example 33.1. Isolation of Vibratory Force. When a shock or vibration disturbance originates in the supported equipment, isolators which support the equipment reduce the transmission of force to the supporting structure, thus protecting the structure or foundation, for example, in isolating a vehicle chassis from the vibration of an internal combustion engine or in reducing the transmission of machine vibration to adjacent structures.

Problem. An electric motor and pump assembly, rigidly mounted on a common base, transmits vibration to other components of a hydraulic system. The weight of the assembly and base is 140 lb (63 kg). Four isolators are to be located at the corners of the rectangular base. The center-of-gravity is centrally located in the horizontal plane near the base. The lowest vibratory forcing frequency is 1800 rpm and is a result of rotational unbalance. There also are higher frequencies due to magnetic and pump forces. The excitation is in both the horizontal and vertical directions.

Objective. To reduce the amount of vibration transmitted to the supporting structure and thus to other system components. A vibration isolation efficiency of 70 to 90 percent is usually possible to attain; here a value of 80 percent is selected.

Solution:

1. First find the transmissibility T which corresponds to the required vibration isolation of 80 percent from Eq. (33.4) or Fig. 33.5:

$$\text{Isolation efficiency} = 100(1 - T) \quad \text{in percent} \tag{33.4}$$

where T = transmissibility. Then $T = 0.2$.

2. Then determine the value of the lowest forcing frequency f in hertz by dividing 1800 rpm by 60 sec/minute, yielding a value of $f = 30$ Hz. The lowest forcing frequency is used because this is the worst condition. If high isolation is attained at this frequency, isolation will be even better at higher frequencies.

3. Next, from Eq. (33.5), using $T = 0.2$ and $f = 30$ Hz, find the natural frequency f_n of the isolated system required to provide the transmissibility T ($T = 0.2$, determined in Step 1) for the forcing frequency determined in step 2 ($f = 30$ Hz):

$$T = \frac{1}{(f/f_n)^2 - 1} \tag{33.5}$$

where f = the forcing frequency (also called disturbing frequency) in Hz
f_n = system natural frequency in Hz

$f_n = 12.2$ Hz.

4. From Eq. (33.6), the static deflection required to provide a natural frequency f_n of 12.2 Hz is calculated:

$$f_n = \frac{3.13}{(\delta_{st})^{1/2}} \tag{33.6}$$

where δ_{st} = the static deflection in inches
$\delta_{st} = 0.066$ in. (1.67 mm)

The same results may be obtained by using the isolation efficiency chart, Fig. 33.5, as follows. Find the point at which the horizontal line for a forcing frequency $f = 30$ Hz intersects the diagonal line for a isolation efficiency of 80 percent. From the point of intersection, project a vertical line to read the values of $\delta_{st} = 0.066$ in. (1.67 mm) and $f_n = 12.2$ Hz.

5. From Eq. (33.3), the spring constant k for the combination of four isolators which is required to provide a natural frequency f_n (of the total mass on the isolators) is calculated.

where k = the total stiffness of the isolators in pounds per inch
W = weight of rigid body in pounds ($W = 140$ lb)
g = acceleration of gravity in inches per second per second

$k = 2120$ lb/in. (371 N/mm).

6. Since there is one isolator at each corner of the base of the supported mass, the isolators are in parallel, with $n = 4$ in Eq. (33.1). This yields a value of $k = 530$ lb/in. (92.8 N/mm) for the stiffness of each individual isolator.

7. In this example, the load is divided equally among the four identical isolators. Therefore, the load supported by each isolator is $W = 35$ lb (15.8 kg).

8. Select an isolator having a spring constant of 530 lb/in. (92.8 N/mm) and the capacity to support a static weight of approximately 35 lb (15.8 kg). Try to select an isolator that is available from stock; use a catalog published by an isolator manufacturer or distributor.

Example 33.2. Shock Protection. Shock and vibration are transmitted through a supporting structure to equipment which is caused to move. The transmitted motion and force are reduced by mounting the equipment on isolators, for example, to protect equipment from vibrating floors or to protect buildings and contents from seismic (ground) disturbances.

Problem. A business machine is to be isolated so that it will not experience damage during normal shipping. The unit can withstand $25g$ of shock without damage. The suspended weight of 125 lb (56.2 kg) is to be equally distributed on four isolators. The disturbances expected are those from normal transportation handling, with no damage allowed after a 30-in. (762-mm) flat bottom drop. The peak vibration disturbances are normally in the range of 2 to 7 Hz.

Objective: To limit acceleration on the machine to $25g$ using the drop test as a simulation of the worst expected shock conditions. A natural frequency between 7 and 10 Hz is desired to avoid the peak vibration frequency range and still provide good shock protection.

Solution:

1. First, solve for the dynamic deflection δ_d (displacement) of the machine required to limit acceleration to X_F (expressed in g's) when the item is dropped from a height ($h = 30$ in.) using:

$$\delta_d = \frac{2h}{\ddot{x}_F} \tag{33.7}$$

where X_F = the fragility factor ($X_F = 25g$)
$\delta_d = 2.4$ in. (61 mm)

2. Then determine the required natural frequency f_n to result in a dynamic deflection δ_d from Eq. (33.8) where $X_F = 25g$, $h = 30$ in., and $\delta_d = 2.4$ in.:

$$f_n = \frac{1}{2\pi} \sqrt{\frac{\ddot{x}_F g}{\delta_d}} \tag{33.8}$$

$f_n = 10$ Hz.

3. Calculate the system spring constant k required to provide a dynamic deflection δ_d from:

$$k = \frac{\text{force}}{\text{deflection}} = \frac{\ddot{x}_F W}{\delta_d} \tag{33.9}$$

$k = 1302$ lb/in. (228 N/mm) for the system.

4. Calculate the static spring constant k of the n isolators (acting in parallel) from Eq. (33.1). Here $n = 4$, yielding a stiffness value of k for each individual isolator of 325 lb/in. (56.9 N/mm).

6. Since the total weight is distributed equally on four identical isolators, the load per isolator is 125 lb divided by 4 or 31 lb (14 kg).

7. Sandwich-type isolators are generally used to protect fragile items during shipment. The construction is typically two flat plates, bonded on either side of an elastomeric pad. Determine the minimum thickness of the elastomer (between the plates) needed to keep dynamic strain at an acceptable level. Use the following rule of thumb for rubber:

$$t_{\min} = \frac{\delta_d}{1.5} \tag{33.10}$$

For $\delta_d = 2.4$ in. (61 mm), the minimum elastomer thickness is 1.6 in. (40.6 mm).

8. Now choose a sandwich isolator for this application. Sandwich configuration permits sufficient deflection in two directions (shear) to absorb high shock loads. Sandwich isolators are readily available in a wide range of sizes, spring constants, and elastomers. From a catalog, select a part that has the capacity to support a static shear load of 31 lb (14 kg), has a minimum elastomer thickness of 1.6 in. (40.6 mm), and has a shear spring constant of 325 lb/in. (56.9 N/mm).

9. In designing or choosing the container, certain criteria must be considered. The four isolators should be installed equidistant from the center-of-gravity in the horizontal plane, oriented to act in shear in the vertical and fore-and-aft directions. The isolators should be attached on one end to a cradle which carries the machine and on the other end to the shipping container. There must be enough space allowed between the mounted unit and the container to prevent bottoming (contact) at impact.

Example 33.3. Shock and Vibration Isolation Combined

Problem.   A portable engine-driven air compressor, with a total weight of 2500 lb (1126 kg), is noisy in operation. An isolation system is required to isolate engine disturbances and to protect the unit from over-the-road shock excitation.

The engine and compressor are mounted on a common base which is to be supported by four isolators. The weight is not equally distributed. At the engine end the static load per isolator is 750 lb (338 kg); at the compressor end the static load per isolator is 500 lb (225 kg). The lowest frequency of the disturbance is at engine speed. The idling speed is 1400 rpm, and the operating speed is 1800 rpm. The unit is expected to be subjected to shock loads due to vehicle frame twisting when transported over rough roads.

Objective. To control force excitation vibration and provide secondary shock isolation. A compromise is required; the isolation system must have a stiffness that is low enough to isolate engine idling disturbance but high enough to limit shock motion. A system having a natural frequency of 12 to 20 Hz in the vertical direction is usually adequate. (Note: The tires and basic vehicle suspension will provide the primary shock protection.)

Solution:

1. First assume that the natural frequency of the system in the vertical direction is 12 Hz.

2. Next, convert the engine speeds to hertz (cycles per second) for use in the calculations. Divide the rpm values by 60 sec/minute, yielding force frequencies f of 23.3 Hz at idling speed and 30 Hz at operating speed.

3. Then calculate the transmissibility T for $f_n = 12$ Hz from Eq. (33.5). At idling speed, using $f = 23.3$ Hz, yields $T = 0.36$ (36 percent). At operating speed, using $f = 30$ Hz, yields $T = 0.19$ (19 percent). Table 33.1 gives a vibration isolation T of 0.64 (64 percent) at idling speed and 0.81 (81 percent) under normal operation. For both conditions, performance with a natural frequency of 12 Hz is satisfactory.

4. Now determine the required static deflection δ_{st} to provide a natural frequency f_n from Eq. (33.6). For $f_n = 12$ Hz, this yields $\delta_{st} = 0.068$ in. (1.73 mm).

5. Select a general-purpose isolator (see Fig. 32.1E) for both ends of the unit. This type of isolator is simple and rugged and gives protection against shock loads expected here. It should be installed so that the axis of the bolt is vertical and the static weight rests on the disk portion. This isolator provides cushioning against upward (rebound) shock loads as well as against downward loads, and the isolation system is fail-safe. Each of the two isolators at the engine end should have a static load-carrying capacity of at least 750 lb (338 kg). Each of the two isolators at the compressor end should be able to support at least a 500-lb (225-kg) static load. For all isolators the static deflection should be close to 0.068 in. (1.73 mm) to give the desired natural frequency of 12 Hz.

CHECKING THE ISOLATOR INSTALLATION

There are usually two primary causes for unsatisfactory performance of an isolation system: (1) The isolator has been selected improperly or some important system parameter has been overlooked and (2) the isolator has been installed improperly. The following criteria can help obviate problems that can otherwise cause poor performance:

1. Do not overload the isolator, i.e., do not exceed the loading specified by the manufacturer. Overloading may shorten isolator life and affect performance.

2. In the case of coil-spring isolators, there should be adequate space between coils at normal static load so that adjacent coils do not touch and there is no possibility of bottoming at the maximum load.

3. In the case of elastomeric compression-type isolators, the isolator should not be overloaded so that it bulges excessively—the ratio of deflection at the static load to the original rubber thickness should not exceed 0.15. As indicated earlier, overloading an isolator may affect its performance. An elastomeric element loaded in compression has a nonlinear stiffness. Therefore, its effective dynamic stiffness (i.e., its effective stiffness when it is vibrating) will be higher than the published value. This raises the natural frequency and reduces its efficiency of isolation.

4. In the case of an elastomeric shear-type isolator, the ratio of the static deflection in shear (i.e., with metal plates moving parallel to one another) to the original thickness usually should not exceed 0.30.

5. To minimize rocking of the equipment and the resultant high stress in the isolators, the distance between the isolator plane and the center-of-gravity should be equal to or less than one-third of the minimum spacing between isolators.

6. The isolators and isolated equipment should be able to move freely under vibration and shock excitation. No part of the isolation system should be short-circuited by a direct connection rather than a resilient support.

7. The vibrating equipment should not contact adjacent equipment or a structural member. Space should be provided to avoid contact.

8. If an elastomeric pad has been installed beneath a machine, the resilient pad should not be short-circuited by hard-bolting the machine to its foundation.

9. The load on the isolator should be along the axis designed to carry the load. The isolator should not be distorted. Unless the isolator has built-in misalignment capability, installation misalignment can affect performance and shorten isolator life.

10. If an elastomeric mount is used, clearance should be provided so that there is no solid object cutting the elastomer. There should be no evidence of bond separation between the elastomer and metal parts in the isolator. Cuts and tears in the elastomer surface can propagate during operation and destroy the spring element. If there are bonded surfaces in the isolator, a bond separation also can cause problems; growth in the separation can affect the performance of the isolator and ultimately cause failure.

11. The static deflection of all isolators should be approximately the same. There should be no evidence of improper weight distribution. Excessive tilt of the mounted equipment may affect its performance. For economic reasons and simplicity in installation, it is desirable to use the same isolator at all points in the system. In such a case, it is not usually a problem if the various isolators have slightly unequal static deflections. However, if one or more isolators exhibit excessive deflection, then corrective measures are required. If the spacing between isolators has been determined improperly, a correction of the spacing to equalize the load may be all that is required. If this is impractical, an isolator having a higher spring constant can be used at points supporting a higher static load. This will tend to equalize deflection.

REFERENCE

1. Davey A. B., and A. R. Payne: "Rubber in Engineering Practice," MacClaren & Sons Ltd., London, 1964.

CHAPTER 34

MECHANICAL PROPERTIES OF RUBBER

Ronald J. Schaefer

INTRODUCTION

Rubber is a unique material that is both elastic and viscous. Rubber parts can therefore function as shock and vibration isolators and/or as dampers. Although the term *rubber* is used rather loosely, it usually refers to the compounded and vulcanized material. In the raw state it is referred to as an *elastomer*. Vulcanization forms chemical bonds between adjacent elastomer chains and subsequently imparts dimensional stability, strength, and resilience. An unvulcanized rubber lacks structural integrity and will "flow" over a period of time.

Rubber has a low modulus of elasticity and is capable of sustaining a deformation of as much as 1000 percent. After such deformation, it quickly and forcibly retracts to its original dimensions. It is resilient and yet exhibits internal damping. Rubber can be processed into a variety of shapes and can be adhered to metal inserts or mounting plates. It can be compounded to have widely varying properties. The load-deflection curve can be altered by changing its shape. Rubber will not corrode and normally requires no lubrication.

This chapter provides a summary of rubber compounding and describes the static and dynamic properties of rubber which are of importance in shock and vibration isolation applications. It also discusses how these properties are influenced by environmental conditions.

RUBBER COMPOUNDING

Typical rubber compound formulations consist of 10 or more ingredients that are added to improve physical properties, affect vulcanization, prevent long-term deterioration, and improve processability. These ingredients are given in amounts based on a total of 100 parts of the rubber (parts per hundred of rubber).

ELASTOMERS

Both natural and synthetic elastomers are available for compounding into rubber products. The American Society for Testing and Materials (ASTM) designation and composition of some common elastomers are shown in Table 34.1. Some elastomers such as natural rubber, Neoprene, and butyl rubber have high regularity in their

TABLE 34.1   Designation and Composition of Common Elastomers

ASTM designation	Common name	Chemical composition
NR	Natural rubber	*cis*-Polyisoprene
IR	Synthetic rubber	*cis*-Polyisoprene
BR	Butadiene rubber	*cis*-Polybutadiene
SBR	SBR	Poly (butadiene-styrene)
IIR	Butyl rubber	Poly (isobutylene-isoprene)
CIIR	Chlorobutyl rubber	Chlorinated poly (isobutylene-isoprene)
BIIR	Bromobutyl rubber	Brominated poly (isobutylene-isoprene)
EPM	EP rubber	Poly (ethylene-propylene)
EPDM	EPDM rubber	Poly (ethylene-propylene-diene)
CSM	Hypalon	Chloro-sulfonyl-polyethylene
CR	Neoprene	Poly chloroprene
NBR	Nitrile rubber	Poly (butadiene-acrylonitrile)
HNBR	Hydrogenated nitrile rubber	Hydrogenated poly (butadiene-acrylonitrile)
ACM	Polyacrylate	Poly ethylacrylate
ANM	Polyacrylate	Poly (ethylacrylate-acrylonitrile)
T	Polysulfide	Polysulfides
FKM	Fluoroelastomer	Poly fluoro compounds
FVMQ	Fluorosilicone	Fluoro-vinyl polysiloxane
MQ	Silicone rubber	Poly (dimethylsiloxane)
VMQ	Silicone rubber	Poly (methylphenyl-siloxane)
PMQ	Silicone rubber	Poly (oxydimethyl silylene)
PVMQ	Silicone rubber	Poly (polyoxymethylphenyl-silylene)
AU	Urethane	Polyester urethane
EU	Urethane	Polyether urethane
GPO	Polyether	Poly (propylene oxide-allyl glycidyl ether)
CO	Epichlorohydrin homopolymer	Polyepichlorohydrin
ECO	Epichlorohydrin copolymer	Poly (epichlorohydrin-ethylene oxide)

backbone structure. They will align and crystallize when a strain is applied, with resulting high tensile properties. Other elastomers do not strain-crystallize and require the addition of reinforcing fillers to obtain adequate tensile strength.[1]

Natural rubber is widely used in shock and vibration isolators because of its high resilience (elasticity), high tensile and tear properties, and low cost. Synthetic elastomers have widely varying static and dynamic properties. Compared to natural rubber, some of them have much greater resistance to degradation from heat, oxidation, and hydrocarbon oils. Some, such as butyl rubber, have very low resilience at room temperature and are commonly used in applications requiring high vibration damping. The type of elastomer used depends on the function of the part and the environment in which the part is placed. Some synthetic elastomers can function under conditions that would be extremely hostile to natural rubber. An initial screening of potential elastomers can be made by determining the upper and lower temperature limit of the environment that the part will operate under. The elastomer must be stable at the upper temperature limit and maintain a given hardness at the lower limit. There is a large increase in hardness when approaching the *glass transition temperature*. Below this temperature the elastomer becomes a "glassy" solid that will fracture upon impact.

Further screening can be done by determining the solvents and gases that the part will be in contact with during normal operation and the dynamic and static physical properties necessary for adequate performance.

REINFORCEMENT

Elastomers which do not strain-crystallize need reinforcement to obtain adequate tensile properties. Carbon black is the most widely used material for reinforcement. The mechanism of the reinforcement is believed to be both chemical and physical in nature.[2] Its primary properties are surface area and structure. Smaller particle-size blacks having a higher surface area give a greater reinforcing effect. Increased surface area gives increased tensile, modulus, hardness, abrasion resistance, tear strength, and electrical conductivity and decreased resilience and flex-fatigue life. The same effects are also found with increased levels (parts per hundred rubber) of carbon black, but peak values occur at different levels. Structure refers to the high-temperature fusing together of particles into grape-like aggregates during manufacture. Increased structure will increase modulus, hardness, and electrical conductivity but will have little effect on tensile, abrasion resistance, or tear strength.

ADDITION OF OILS

Oils are used in compounding rubber to maintain a given hardness when increased levels of carbon black or other fillers are added. They also function as processing aids and improve the mixing and flow properties (extrudability, etc.).

ANTI-DEGRADENTS

Light, heat, oxygen, and ozone accelerate the chemical degradation of elastomers. This degradation is in the form of chain scission or chemical cross-linking depending on the elastomer. Oxidation causes a softening effect in NR, IR, and IIR. In most other elastomers the oxygen causes cross-linking and the formation of stiffer compounds. Ozone attack is more severe and leads to surface cracking and eventual product failure. Cracking does not occur unless the rubber is strained. Elastomers containing unsaturation in the backbone structure are most vulnerable. Anti-degradents are added to improve long-term stability and function by different chemical mechanisms. Amines, phenols, and thioesters are the most common types of antioxidants, while amines and carbamates are typical anti-ozonants. Paraffin waxes which bloom to the surface of the rubber and form protective layers are also used as anti-ozonants.

VULCANIZING AGENTS

Vulcanization is the process by which the elastomer molecules become chemically cross-linked to form three-dimensional structures having dimensional stability. The effect of vulcanization on compound properties is shown in Fig. 34.1. Sulfur, peroxides, resins, and metal oxides are typically used as vulcanizing agents. The use of sulfur alone leads to a slow reaction, so accelerators are added to increase the cure rate. They affect the rate of vulcanization, cross-link structure, and final properties.[3]

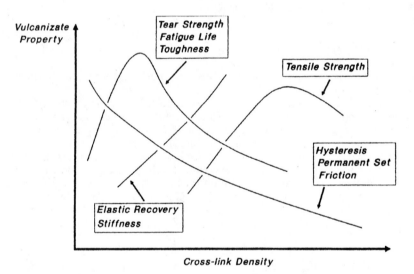

FIGURE 34.1 Vulcanizate properties as a function of the extent of vulcanization. (*Eirich and Coran.*[3])

MIXING

Adequate mixing is necessary to obtain a compound that processes properly, cures sufficiently, and has the necessary physical properties for end use.[4] The Banbury internal mixer is commonly used to mix the compound ingredients. It contains two spiral-shaped rotors that operate in a completely enclosed chamber. A two-step procedure is generally used to ensure that premature vulcanization does not occur.

Most of the ingredients are mixed at about 120°C in the first step. The vulcanizing agents are added at a lower temperature in the second step.

MOLDING

Compression, transfer, and injection-molding techniques are used to shape the final product. Once in the mold, the rubber compound is vulcanized at temperatures ranging from 100 to 200°C. The cure time and the temperature are determined beforehand with a curemeter, such as the oscillating disk rheometer.[5] After removal from the mold, the rubber product is sometimes postcured in an autoclave. The postcuring gives improved compression-set properties.

STATIC PHYSICAL PROPERTIES

Rubber has properties that are drastically different from other engineering materials. Consequently, it has physical testing procedures that are unique.[6] Rubber has both elastic and viscous properties. Which of these properties predominates frequently depends on the testing conditions. A summary of the characteristic properties of different elastomers is shown in Table 34.2.

HARDNESS

Hardness is defined as the resistance to indentation. The *durometer* is an instrument that measures the penetration of a stress-loaded metal sphere into the rubber. Hardness measurements in rubber are expressed in Shore A or Shore D units according to ASTM test procedures.[7] Because of the viscoelastic nature of rubber, a durometer reading reaches a maximum value as soon as the metal sphere reaches maximum penetration into the specimen and then decreases the next 5 to 15 sec. Hand-held spring-loaded durometers are commonly used but are very subject to operator error. Bench-top dead-weight-loaded instruments reduce the error to a minimum.[8]

STRESS-STRAIN

Rubber is essentially an incompressible substance that deflects by changing shape rather than changing volume. It has a Poisson's ratio of approximately 0.5. At very low strains, the ratio of the resulting stress to the applied strain is a constant (Young's modulus). This value is the same whether the strain is applied in tension or compression. Hooke's law is therefore valid within this proportionality limit. However, as the strain increases, this linearity ceases, and Hooke's law is no longer applicable. Also the compression and tension stresses are then different. This is evident in load-deflection curves run on identical samples in compression, shear, torsion, tension, and buckling, as shown in Fig. 32.2. Rubber isolators and dampers are typically designed to utilize a combination of these loadings. However, shear loading is most preferred since it provides an almost linear spring constant up to strains of about 200 percent. This linearity is constant with frequency for both small and large dynamic shear strains. The compression loading exhibits a nonlinear hardening at strains over 30 percent and is used where motion limiting is required. However, it is not recom-

TABLE 34.2 Relative Properties of Various Elastomers

ASTM designation	NR	BR	SBR	IIR CIIR	EPM EPDM	CSM	CR	NBR	HNBR	ACM ANM	T	FKM	FVMQ	VMQ, MQ, PMQ, PVMQ	AU EU	GPO	CO ECO
Durometer range	30–90	40–90	40–80	40–90	40–90	45–100	30–95	40–95	35–95	40–90	40–85	60–90	40–80	30–90	35–100	40–90	40–90
Tensile max, psi	4500	3000	3500	3000	2500	4000	4000	4000	4500	2500	1500	3000	1500	1500	5000	3000	2500
Elongation max., %	650	650	600	850	600	500	600	650	650	450	450	300	400	900	750	600	350
Compression set	A	B	B	B	B-A	C-B	B	B	B-A	B	D	B-A	C-B	B-A	D	B-A	B-A
Creep	A	B	B	B	C-B	C	B	B	B	C	D	B	B	C-A	D	B-A	B
Resilience	High	High	Med.	Low	Med.	Low	High	Med.-Low	Med.	Med.	Low	Low	Low	High-Low	High-Low	High	Med.-Low
Abrasion resistance	A	A	A	C	B	A	A	A	A	C-B	D	B	D	B	A	B	C-B
Tear resistance	A	B	C	B	C	B	B	B	B	D-C	D	B	D	C-B	A	A	C-A
Heat aging at 212°F	C-B	C	B	A	B-A	B-A	B	B	A	A	C-B	A	A	A	B	B-A	B-A
T_g, °C	−73	−102	−62	−73	−65	−17	−43	−26	−32	−24, −54	−59	−23	−69	−127, −86	−23, −34	−67	−25, −46
Weather resistance	D-B	D	D	A	A	A	B	D	A	A	B	A	A	A	A	A	B
Oxidation resistance	B	B	C	A	A	A	A	B	A	A	B	A	A	A	B	B	B
Ozone resistance	NR-C	NR	NR	A	A	A	A	C	A	B	A	A	A	A	A	A	A
Solvent resistance																	
Water	A	A	B-A	A	A	B	B	B-A	A	D	B	A	A	A	C-B	C-B	B
Ketones	B	B	B	A	B-A	C	C	D	D	D	A	NR	D	B-C	D	C-D	C-D
Chlorohydrocarbons	NR	NR	NR	NR	NR	D	D	C	C	B	C-A	A	B-A	NR	C-B	A-D	A-B
Kerosene	NR	NR	NR	NR	NR	B	B	A	A	A	A	A	A	D-C	B	A-C	A
Benzol	NR	NR	NR	NR	NR	C-D	C-D	B	B	C-B	C-B	B-A	B-A	NR	C-B	NR	B-A
Alcohols	B-A	B	B	B-A	B-A	A	A	C-B	C-B	D	B	C-A	C-B	C-B	B	C	A
Water glycol	B-A	B-A	B	B-A	B	B	B	B	A	C-B	A	A	A	A	C-B	B	C
Lubricating oils	NR	NR	NR	NR	NR	A-B	B-C	A	A	A	A	A	A	B-C	A-B	D	A

A = excellent, B = good, C = fair, D = use with caution, NR = not recommended

SOURCE: *Seals Eastern, Inc.*

34.6

mended where energy storage is required. Tension-loading stores energy more efficiently than either compression-loading or shear-loading but is not recommended because of the resulting stress loads on the rubber-to-metal bond, which may cause premature failure. Buckled-loading is a combination of tension- and compression-loading and derives some of the benefits of both.

The stress-strain properties of rubber compounds are usually measured under tension as per ASTM procedures.[9] Either molded rings or die-cut "dumbbell"-shaped specimens are used in testing. Stress measurements are made at a specified percentage of elongation and reported as *modulus* values. For example, 300 percent modulus is defined as the stress per unit cross-sectional area (in psi or MPa units) at an elongation of 300 percent. Also measured are the stress at failure (tensile) and maximum percentage elongation. These are the most frequently reported physical properties of rubber compounds.

The stiffness (spring rate) is the ratio of stress to strain expressed in newtons per millimeter. It is dependent not only on the rubber's modulus but also on the shape of the specimen or part being tested. Since rubber is incompressible, compression in one direction results in extension in the other two directions, the effect of which is a bulging of the free sides. The *shape factor* is calculated by dividing one loaded area by the total free area.

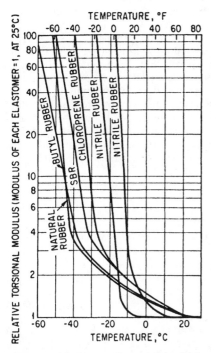

FIGURE 34.2 Increase in torsional modulus of elasticity of various elastomers as a function of temperature. (*After Gehman.*[16])

TEAR

Vibration isolators and dampers that are subjected to cyclical loads frequently fail due to a fracturing of the rubber component. A fracture may initiate in an area where stress concentration is at a maximum. After initiation, the fracture increases in size and progresses into a tearing action. Tear properties are therefore important in some applications. Tensile tests are run on dumbbell-shaped samples containing no flaws. The stress is therefore evenly distributed across the sample. Tear-testing procedures concentrate the stress in one area, either through sample design or by cutting a nick in the sample.[10] Samples are die cut (die A, B, or C) from tensile testing sheets. The peak force and sample thickness are recorded. Tear values are reported in units of pounds per inch or kilonewtons per meter. Tear and tensile testing provide the same rank ordering of different types of rubbers.

COMPRESSION SET AND CREEP

Dimensional stability is necessary for vibration isolators and dampers that function under applied loads, i.e., the static deflection of an isolator should not increase with time. Such an increase is a result of creep and compression set. *Compression set* is the change in dimension with an applied strain; *creep* is the change in dimension with an applied force. Compression set and excessive creep will induce a large change in stiffness and dynamic properties over a period of time. Compression set is determined by compressing a specimen (of specified size) to a preset deflection and exposing it to an elevated temperature.[11] After exposure the specimen is allowed to recover for one-half hour and the thickness is measured. Percent compression set is the decrease in thickness divided by the original deflection and multiplied by 100. Typical rubber compounds used for vibration isolation have compression set values of from 10 to 50 percent. The exposure time is usually 22 or 70 hours at a temperature relevant to the intended use of the isolator or damper. Creep is determined by placing a specimen in a compression device, applying a compressive force, and exposing it to an elevated temperature.[12] Percent creep is the decrease in thickness divided by the original thickness and multiplied by 100.

ADHESION

Adequate rubber-to-metal adhesion is imperative in the fabrication of most vibration isolators and dampers. Adhesive is first applied to the metal; then the rubber is bonded to the metal during vulcanization. Various adhesives are available for all types of elastomers. In testing for adhesion, a strip of rubber is adhered to the face of a piece of adhesive-coated metal.[13] After vulcanization (and possible aging), the rubber is pulled from the metal at an angle of 45° or 90°, and the adhesion strength is measured. The mode of failure is also recorded.

Another ASTM method[14] is used to determine the rubber-to-metal adhesion when the rubber is bonded after vulcanization, i.e., for *postvulcanization bonding*. In this procedure a vulcanized rubber disk is coated on both sides with an adhesive and assembled between two parallel metal plates. Then the assembly is heated under compression for a specified period of time. The metal plates are then pulled apart until rupture failure.

LOW-TEMPERATURE PROPERTIES

Rubber becomes harder, stiffer, and less resilient with decreasing temperature. These changes are brought about by a reduction in the "free volume" between neighboring molecules and a subsequent reduction in the mobility of the elastomer molecules. When approaching the glass transition temperature (T_g), its rubber-like characteristic is lost and the rubber becomes leathery. Finally it changes to a hard, brittle glass. The glass transition temperature is a second-order transition as opposed to crystallization, which is a first-order transition. A first-order transition is accompanied by a abrupt change in a physical property, while a second-order transition is accompanied by a change in the rate of change. The glass transition temperature can be detected by differential scanning calorimetry or changes in static or dynamic mechanical properties. This is described in the section on dynamic properties of rubber.

The effect of temperature on stiffness is measured using a Gehman apparatus.[15] It provides torque to a strip of rubber by a torsion wire. The measurement is first made at 23°C and then at reduced temperatures. The *relative modulus* at any temperature is the ratio of the modulus at that temperature to the modulus at 23°C. The results are expressed as the temperatures at which the relative moduli are 2, 5, 50, and 100. Figure 34.2 shows the effect of temperature on the relative torsional modulus of various elastomers.[16] Young's modulus can also be measured at low temperature using a flexural procedure.[17]

HIGH-TEMPERATURE PROPERTIES

Some vibration isolators and dampers function in high-temperature environments. The rubber compounds used in these applications must have resistance to high-temperature degradation. The stability at high temperatures is related to the chemical structure of the elastomer and the chemical cross-linking bonds formed during vulcanization. Elastomers containing no unsaturation (chemical double-bonds) in the backbone have better high-temperature properties. Rubber compounds containing EPDM, for example, have better high-temperature resistance than ones containing natural rubber or SBR. In a sulfur cure, mono or disulfide cross-linking bonds have better high-temperature stability than polysulfide bonds. Cure system modifications are therefore used to improve high-temperature stability.

The high-temperature resistance of rubber compounds is determined by measuring the percentage of change in tensile strength, tensile stress at a given elongation, and ultimate elongation after aging in a high-temperature oven as per ASTM procedure.[18]

OIL AND SOLVENT RESISTANCE

Some vibration isolators and dampers, particularly those used in automotive products, have contact with oils or solvents. The effect of a liquid on a particular rubber depends on the solubility parameters of the two materials. The more the similarity, the larger the effect. A liquid may cause the rubber to swell, it may extract chemicals from it, or it may chemically react with it. Any of these can lead to a deterioration of the physical properties of rubber. The effect of liquids on rubber is determined by measuring changes in volume or mass, tensile strength, elongation, and hardness after immersion in oils, fuels, service fluids, or water.[19]

EXPOSURE TO OZONE AND OXYGEN

Ozone is a constituent of smog; in some areas, ozone may occur in concentrations that are deleterious to rubber. Vibration isolators and dampers also may be exposed to ozone generated by the corona discharge of electrical equipment. Elastomers containing unsaturation in their backbone structure are especially prone to ozone cracking, since ozone attacks the elastomer at the double bonds. Elastomers such as NR, SBR, BR, and NBR have poor resistance, while EPDM and GPO have excellent resistance to ozone cracking. Ozone cracking will not occur if the rubber is unstrained. There is a critical elongation at which the cracking is most severe. These strains are 7 to 9 percent for NR, SBR, and NBR, 18 percent for CR, and 26 percent for IIR.[20] Both static[21] and dynamic[22] testing procedures are used. In the static test the sample is given a specified strain. Results are expressed as cracking severity using arbitrary scales or as time until first cracks appear. In Method A, the dynamic procedure tests strips of rubber in tension at 0.5 Hz. Method B adheres the test strips to a rubber belt that is rotated around two pulleys at 0.67 Hz. The number of cycles to initial cracking is reported.

DYNAMIC PROPERTIES

VISCOELASTICITY

Rubber has elastic properties similar to those of a metallic spring and has energy-absorbing properties like those of a viscous liquid.[23] These viscoelastic properties allow rubber to maintain a constant shape after deformation, while simultaneously absorbing mechanical energy. The viscosity (which varies with different elastomers) increases with reduced temperature. The elasticity follows Hooke's law and increases with increased strain, while the viscosity follows Newton's law and increases with increased strain rate. Therefore, when applying a strain, the resultant stress will increase with increasing strain rate.

Springs or dashpots are frequently used to make theoretical models which illustrate the interaction of the elastic and viscous components of rubber. The springs and dashpots can be combined in series or in parallel, representing the Maxwell or Voigt elements (see Table 36.2). Rubber actually consists of an infinite number of such models with a wide spectrum of spring constants and viscosities.

MEASUREMENT OF DYNAMIC PROPERTIES

Resilience, measured by several relatively simple tests, is sometimes used for estimating the dynamic properties of a rubber compound. In these test methods a strain is applied to a rubber test sample by a free-falling indentor. *Resilience* is defined as the ratio of the energy of the indentor after impact to its energy before impact (expressed as a percentage). Two widely used methods include the pendulum[24] and the falling weight methods.[25] Although resilience is a crude measurement of the dynamic properties of rubber, it is attractive because of its simplicity and cost.

In *free vibration methods,* the rubber sample is allowed to vibrate at its natural frequency.[26] To change the natural frequency the sample size or added weights must be changed. Since it is a free vibration, the amplitude A decreases with each oscillation. Resilience is defined as A_3/A_2, expressed as a percentage.

In *forced vibration methods,* the dynamic properties (or viscoelasticity) of a rubber compound are determined by measuring its response to a sinusoidally varying strain.[27] In this manner, both the strain and the strain rate vary during a complete cycle. The ratio of the energy dissipated in overcoming internal friction to the energy stored is a function of the viscoelasticity of the rubber. In a simple apparatus for measuring dynamic properties, a sinusoidally varying strain is applied to the sample by means of a motor-driven eccentric. The resultant force is measured at the opposite end of the sample with a dynamometer ring or electronic measuring device. The angular distance between the input strain and the resultant stress is measured by mechanical or electronic methods. A graph of the sinusoidal strain and resultant stress, both plotted as a function of time or angle, is shown in Fig. 34.3. The measured

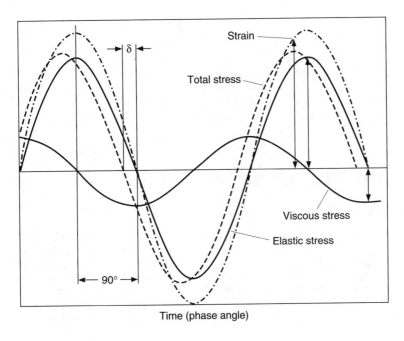

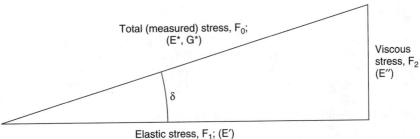

FIGURE 34.3 The applied sinusoidal strain and the resultant stress plotted as a function of time or phase angle. The maximum elastic and viscous stress, and the elastic and viscous modulus values are calculated using simple trigonometry. (*After Schaefer.*[23])

maximum stress amplitude precedes the maximum strain amplitude by the phase angle δ. The stress amplitude (F_0) is composed of contributions from both the elastic stress (F_1) and the viscous stress (F_2). The amount contributed by each is a function of the phase angle. Following Hooke's Law, the resultant stress due to the elastic portion of the rubber is in phase with, and proportional to, the strain. When the imposed strain reaches a peak value, the resultant elastic stress also reaches a peak value. The resultant stress due to the viscous portion of the rubber is governed by Newton's law and is 90° out of phase with the imposed strain. When the strain is at a maximum value, the strain rate (slope of the strain curve) is zero. Consequently, the resultant viscous stress is zero. At zero strain, the strain rate is at a maximum, and the resultant viscous stress is at a peak value. The only values measured are the stress amplitude and the phase angle δ. The complex modulus is calculated by dividing the resultant maximum stress amplitude by the maximum imposed strain amplitude. Both the maximum elastic stress amplitude and the maximum viscous stress amplitude are calculated from the measured stress amplitude and the phase angle δ using simple trigonometric functions. Dividing these stress values by the strain gives the elastic modulus (E') and the loss modulus (E''). Tan δ equals E''/E'. The value of tan δ (the ratio of the viscous to the elastic response) is a measurement of damping or hysteresis.

INFLUENCE OF COMPOUNDING INGREDIENTS

ELASTOMERS

The dynamic properties of an elastomer are determined by its glass transition temperature (T_g). Elastomers having the lowest T_g will have the lowest tan δ (or highest resilience). Natural rubber has a fairly low T_g (–60°C) and thus has a low tan δ. Butyl rubber has a low T_g (–60°C), but the transition region extends above ambient temperature. It consequently has a high tan δ and is frequently used in vibration damping applications. The effect of temperature on the dynamic stiffness (dynamic spring rate) and damping of compounds containing different elastomers is shown in Fig. 34.4.

CARBON BLACK

Carbon black has a major influence on the dynamic properties of compounded rubber.[28] It is a source of hysteresis or damping. The amount of damping increases with the surface area of the carbon black and the level used in the compound.

VIBRATION ISOLATION AND DAMPING

Dynamic properties, which are a function of the elastomer and other compounding variables, determine the vibration isolation and damping characteristics of a rubber compound. Springs and dashpots are used to describe how the viscoelastic properties relate to the vibration isolation and damping properties.[29] The quantity tan δ, being the ratio of the viscous to elastic response, can be substituted for $\zeta = c/c_c$ in the equations for transmissibility derived in Chap. 2. Figure 34.5 summarizes the effect of dynamic properties on transmissibility. Transmissibility curves of different compounded elastomers are shown in Fig. 34.6.[30] The NR, EPDM, CR, and SBR rubbers have low T_g's and therefore have low damping properties. As a result they have the highest transmissibility at the resonating frequency and the lowest transmissibility at higher frequencies. The opposite effect is seen with IIR and NBR, which have higher damping properties. As shown in Fig. 34.7, increased levels of carbon black increase damping and thus decrease the transmissibility at the resonance frequency. Increased levels also increase the compound's stiffness, with a resulting increase in resonance frequency. For further information on the effect of viscoelastic properties on vibration isolation and damping, see Refs. 31 and 32.

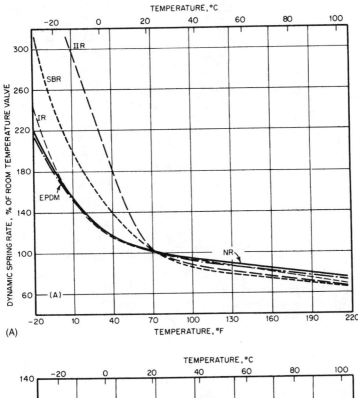

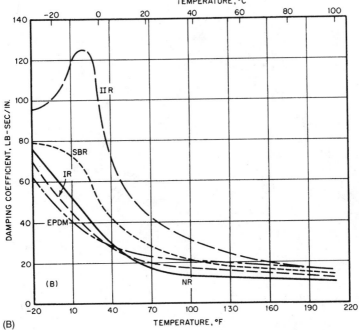

FIGURE 34.4 The effect of temperature on (*A*) the dynamic stiffness (spring rate) and (*B*) the damping coefficient of typical isolating and damping compounds using several elastomers.

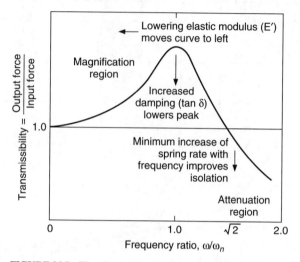

FIGURE 34.5 The effect of the dynamic properties of rubber on the transmissibility curve. (*After Edwards.*[29])

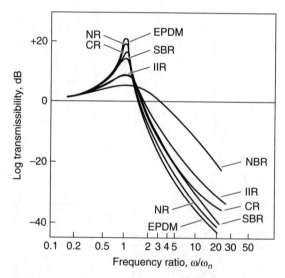

FIGURE 34.6 The dependence of transmissibility on the type of rubber used for the mounting. (*After Freakley.*[30])

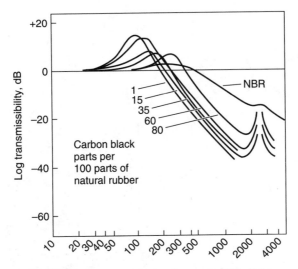

FIGURE 34.7 The dependence of transmissibility-frequency curves on the level of carbon black in natural rubber compounds. (*After Freakley.*[30])

FATIGUE FAILURE

Rubber shock and vibration isolators and dampers fail in service due to either excessive drift (creep) or mechanical fracture as a result of fatigue. Static drift or set testing is described above in the section on compression set. The effect of temperature on the drift of a natural rubber compound is shown in Fig. 34.8.[33] The drift properties of rubber can be tested using static or dynamic methods.

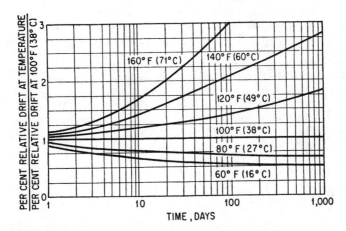

FIGURE 34.8 The effect of temperature on the drift of natural rubber. (*After Morron.*[33])

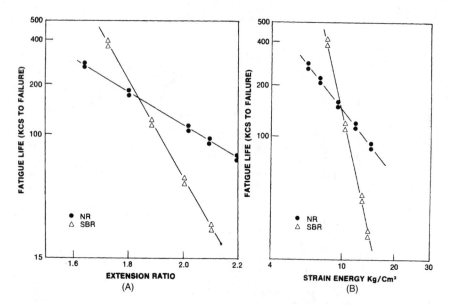

FIGURE 34.9 Fatigue curves of carbon-black-filled natural rubber and SBR plotted as a function of extension ratio (*A*) and strain energy (*B*). (*After Babbit.*[36])

Mechanical fractures occur when a rubber part is subjected to a cyclic stress or strain. The initial crack usually originates in an area of high stress concentration and grows until complete fracture occurs. Both the time until initial crack appearance and the growth rate increase with increasing temperature and increased stress or strain amplitudes.

Several procedures are available for the dynamic testing of laboratory-prepared samples. The most common is the DeMattia flex machine which can test for crack initiation or the growth of an induced cut.[34] The Ross Flexer machine also tests for cut growth.[35] Although the data can be used for relative comparisons, all of these procedures show poor correlation with product performance. Dynamic fatigue testing is therefore frequently performed on the actual part. Because of time constraints, the applied energy input (cyclic stress and strain amplitudes) is increased to much larger values than what the part experiences in actual service. The effect of energy input on fatigue life is shown in Fig. 34.9.[36] At low-energy input the SBR compound has better fatigue resistance than the NR compound. However, when the strain and resulting input energy is increased, the curves cross over, and the NR compound has the better fatigue resistance.[37] Therefore, caution must be exercised when interpreting such data.

REFERENCES

The following references designated by *ASTM D,* followed by a number, are publications of the American Society for Testing and Materials, 1916 Race Street, Philadelphia, PA 19103.

1. Morton, M.: "Rubber Technology," Van Nostrand Reinhold, New York, 1987.

2. Donnet, J., A. Voet: "Carbon Black—Physics, Chemistry, and Elastomer Reinforcement," Marcel Dekker, New York, 1976.

3. Eirich, F. R., and A. Y. Coran: "Science and Technology of Rubber," Academic Press, New York, 1994.

4. Long, H.: "Basic Compounding and Processing of Rubber," Lancaster Press, Lancaster, Penna., 1985.

5. ASTM D412

6. Brown, R. P.: "Physical Testing of Rubber," Elsevier Applied Science Publishers, New York, 1986.

7. ASTM D2240

8. ASTM D531

9. ASTM D412

10. ASTM D624

11. ASTM D395, Method B

12. ASTM D395, Method A

13. ASTM D429, Method B

14. ASTM D429, Method 429

15. ASTM D1053

16. Gehman, S. D., D. E. Woodford, and C. S. Wilkinson: *Ind. Eng. Chem.*, **39**:1108 (1947).

17. ASTM D797

18. ASTM D573

19. ASTM D471

20. Edwards, D. C., E. B. Storey: *Trans. Inst. Rubber Ind.*, 31, 45 (1955).

21. ASTM D 1149

22. ASTM D3395

23. Schaefer, R. J.: *Rubber World,* May, July, Sept., Nov. (1994) and Jan., March, May (1995)

24. ASTM D1054

25. ASTM D2632

26. ASTM D945

27. ASTM D2231

28. Medalia, A. I.: *Rubber Chem. and Tech.*, 51, 437 (1978).

29. Edwards, R. C.: *Automotive Elastomers and Design,* March 3, 1983.

30. Freakley, P. K., A. R. Payne: "Theory and Practice of Engineering with Rubber," Applied Science Publishers, London.

31. Gent, A. N., "How to Design Rubber Components," Hanser Publishers, New York, 1994.

32. Corsaro, R. D., and L. H. Sperling: "Sound and Vibration Damping with Polymers," American Chemical Society, Washington, D.C., 1990.

33. Morron, J. D.: ASME Paper 46-SA-18, presented June 1946.

34. ASTM D430

35. ASTM D1052

36. Babbit, R. O.: "The Vanderbilt Rubber Handbook," Vanderbilt Company, Norwalk, Conn., 1978.

37. Bartenev, G. M., and Y. S. Zuyev: "Strength and Failure of Viscoelastic Materials," Pergamon Press, New York, 1968.

CHAPTER 35
ENGINEERING PROPERTIES OF METALS USED IN EQUIPMENT DESIGN

J. E. Stallmeyer

INTRODUCTION

The design of equipment to withstand shock and vibration requires (1) a determination of the loads and resultant stresses acting on the equipment, and (2) the selection of a suitable material. The loads and stresses may be determined from an appropriate model of the equipment as described in Chap. 42. This chapter describes some of the considerations required to adapt the results from the model analysis to the selection of suitable materials, including such engineering properties as the stress-strain properties of metals and metal fatigue.

The selection of an appropriate material often involves an evaluation of the types of stress condition to which the equipment will be subjected. If a small number of severe stresses constitute the most critical situation, the most important consideration is to design the equipment for adequate strength. For equipment that will be subjected to sustained vibration or a large number of repeated applications of load, fatigue strength is likely to be the critical design parameter. The relative importance of these types of loading must be determined for each application.

When strength is the primary design factor, an appropriate balance between stress and ductility is the most important consideration. Analytical models generally indicate the maximum stress based on linear properties of the material. If nonductile materials are used or if permanent deformation cannot be tolerated, the equipment must be designed in such a way that the stress does not exceed the elastic limit of the material. In some cases, permanent deformation may not be acceptable because it would cause misalignment of parts whose proper operation depends on accurate alignment. In other cases, permanent deformation of some structural members may be acceptable. Several empirical procedures have been developed for these cases.

One procedure, for members subjected to bending, is to permit some predetermined percentage of the cross section to yield. Ductility of the material is required for this procedure. The permanent deformation of the member, after the load has been removed, will be less than the maximum deflection because the core of elastic material tends to restore the member to its original shape. This procedure is not easily adapted to members subjected to loading other than bending. Another procedure establishes a maximum allowable stress equal to the yield point plus some incremental percentage of the difference between the yield point and the ultimate strength of the material. As the incremental percentage increases, the magnitude of permanent deformation increases. Consequently, the magnitude of the incremental percentage will depend upon the function of the particular member and the ability to make adjustments or repairs. For bolts which are inaccessible for retightening, the increment is generally zero. In cases where dimensional stability is important, but some yielding can be tolerated, only a small percentage increment may be permissible. When significant yielding can be tolerated, the increment may be as much as 50 percent. For bearing surfaces or where permanent deformation is permissible, the ultimate strength of the material may be used for design.

Design of equipment subject to vibration or repeated load applications requires a more detailed evaluation of the stress versus time response for the life of the structure. Three fatigue analysis procedures are available: the *stress-life method,* the *strain-life method,* and the *fracture-mechanics method.* Which of the three procedures is applicable will depend on the stress-time history. Knowledge of all three methods allows the engineer to choose the most appropriate method for the specific application.

STRESS-STRAIN PROPERTIES

STATIC PROPERTIES

The important static properties are yield strength, ultimate tensile strength, elongation at failure, and reduction of area. A standard tensile test specimen, defined by ASTM Specification A370,[1] is used to evaluate these properties. The rate of loading and the procedure for evaluation of properties are defined in the specification. Under dynamic loading the yield strength and the ultimate strength depend upon the strain rate, which in turn depends upon the geometry of the structure and the type of loading. Dynamic properties are not standardized easily. There is little information about these properties for the wide range of available metals.

The standard tensile test provides a plot of stress versus strain from which many of the mechanical properties may be obtained. A typical stress-strain curve is presented in Fig. 35.1. For materials which are linearly elastic, the elongation e is directly proportional to the length of the test bar l and the stress σ. The proportionality constant is called the modulus of elasticity E. The plot of stress versus strain usually deviates from linear behavior at the *proportional limit,* which depends on the sensitivity of the instrumentation. Most metals can be stressed slightly higher than the proportional limit without showing permanent deformation upon removal of the load. This point is referred to as the *elastic limit.* Mild steels exhibit a distinct *yield point,* at which permanent deformation

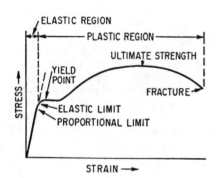

FIGURE 35.1 Typical stress-strain diagram for a metal with a yield point.

begins suddenly and continues with no increase in stress. For materials which exhibit a nonlinear stress-strain relationship, the term *yield strength* is generally used. The yield strength of a material is that stress which produces, on unloading, a permanent strain of 0.002 in./in. (0.2 percent).

Beyond the yield strength, large permanent deformations occur at a reduced modulus, the *strain-hardening modulus*. The strain-hardening modulus decreases at increasing loads until the cross-sectional area of the test bar begins to decrease, *necking*, at the ultimate load. Beyond this point further extension takes place with decreasing force. This ability of the material to flow without immediate rupture is called *ductility;* it is defined as the percent reduction of cross-sectional area measured at the section of fracture. Another measure of ductility is the percent elongation of the gage length. This value depends on the shape and size of the specimen and the gage length.

Other static properties of materials find application in the design of equipment to withstand shock and vibration. The *modulus of rigidity G* is the ratio of shear stress to shear strain; it may be determined from the torsional stiffness of a thin-walled tube of the material. The value of G for steel is 12×10^6 lb/in.^{2}. Poisson's ratio v is the ratio of the lateral unit strain to the normal unit strain in the elastic range of the material. This ratio evaluates the deformation of a material that occurs perpendicular to the direction of application of load. The value of v for steel is 0.3. More complete data on materials and their properties as used in machine and equipment design are compiled in available references.[2-4] Values of the static properties of typical engineering materials are given in Tables 35.1 to 35.3.

TABLE 35.1 Mechanical Properties of Typical Cast Irons (*A. Vallance and V. Doughtie.*[2])

Material	Ultimate strength		Endurance limit in reversed bending lb/in.^{2}, σ_e	Brinell hardness number	Modulus of elasticity		Elongation in 2 in., %
	Tension, lb/in.^{2}, σ_u†	Compression, lb/in.^{2}, σ_u			Tension and compression, lb/in.^{2}, E	Shear lb/in.^{2}, G	
Gray, ordinary	18,000	80,000	9,000	100–150	10–12,000,000	4,000,000	0–1
Gray, good*	24,000	100,000	12,000	100–150	12,000,000	4,800,000	0–1
	16,000						
Gray, high grade	30,000	120,000	15,000	100–150	14,000,000	5,600,000	0–1
Malleable, S.A.E. 32510	50,000	120,000	25,000	100–145	23,000,000	9,200,000	10
Nickel alloys:							
Ni-0.75, C-3.40, Si-1.75,	32,000	120,000	16,000	200	15,000,000	6,000,000	1–2
Mn-0.55*	24,000			175			
Ni-2.00, C-3.00, Si-1.10,	40,000	155,000	20,000	220	20,000,000	8,000,000	1–2
Mn-0.80*	31,000			200			
Nickel-chromium alloys:							
Ni-0.75, Cr-0.30, C-3.40,	32,000	125,000	16,000	200	15,000,000	6,000,000	1–2
Si-1.90, Mn-0.65							
Ni-2.75, Cr-0.80, C-3.00,	45,000	160,000	22,000	300	20,000,000	8,000,000	1–2
Si-1.25, Mn -0.60							

* Upper figures refer to arbitration test bars. Lower figures refer to the center of 4-in. round specimens.
† *Flexure:* For cast irons in bending, the modulus of rupture may be taken as 1.75 σ_u (tension) for circular sections, 1.50 σ_u for rectangular sections and 1.25 σ_u for I and T sections.

TABLE 35.2 Mechanical Properties of Typical Carbon Steels (*A. Vallance and V. Doughtie.*[2])

Material	Ultimate strength Tension, lb/in.^{2}, σ_u	Shear, lb/in.^{2}, σ_u	Yield strength Tension and compression, lb/in.^{2}, σ_y	Shear, lb/in.^{2}, σ_y	Endurance limit in reversed bending, lb/in.^{2}, σ_e	Brinell hardness number	Modulus of elasticity Tension and compression, lb/in.^{2}, E	Shear, lb/in.^{2}, G	Elongation 2 in., %
Wrought iron	48,000	50,000	27,000	30,000	25,000	100	28,000,000	11,200,000	30–40
Cast steel:									
Soft	60,000	42,000	27,000	16,000	26,000	110	30,000,000	12,000,000	22
Medium	70,000	49,000	31,500	19,000	30,000	120	30,000,000	12,000,000	18
Hard	80,000	56,000	36,000	21,000	34,000	130	30,000,000	12,000,000	15
SAE 1025:									
Annealed	67,000	41,000	34,000	20,000	29,000	120	30,000,000	12,000,000	26
Water-quenched*	78,000	55,000	41,000	24,000	43,000	159	30,000,000	12,000,000	35
	90,000	63,000	58,000	34,000	50,000	183			27
SAE 1045:									
Annealed	85,000	60,000	45,000	26,000	42,000	140	30,000,000	12,000,000	20
Water-quenched*	95,000	67,000	60,000	35,000	53,000	197	30,000,000	12,000,000	28
	120,000	84,000	90,000	52,000	67,000	248			15
Oil-quenched*	96,000	67,000	62,000	35,000	53,000	192	30,000,000	12,000,000	22
	115,000	80,000	80,000	45,000	65,000	235			16
SAE 1095:									
Annealed	110,000	75,000	55,000	33,000	52,000	200	30,000,000	12,000,000	20
Oil-quenched*	130,000	85,000	66,000	39,000	68,000	300	30,000,000	11,500,000	16
	188,000	120,000	130,000	75,000	100,000	380			10

* Upper figures: steel quenched and drawn to 1300°F. Lower figures: steel quenched and drawn to 800°F. Values for intermediate drawing temperatures may be approximated by direct interpolation.

TEMPERATURE AND STRAIN-RATE EFFECTS

The static properties of most engineering materials depend upon the testing temperature. As the testing temperature is increased above room temperature, the yield point, ultimate strength, and modulus of elasticity decrease. For example, the yield point of structural carbon steel is about 90 percent of the room-temperature value at 400°F (204°C), 60 percent at 800°F (427°C), 50 percent at 1000°F (538°C), 20 percent at 1300°F (704°C), and 10 percent at 1600°F (871°C). The corresponding changes for ultimate strength are 100 percent of the room-temperature value at 400°F, 85 percent at 800°F, 50 percent at 1000°F, 15 percent at 1300°F, and 10 percent at 1600°F. Changes in the modulus of elasticity are 95 percent of the room-temperature value at 400°F, 85 percent at 800°F, 80 percent at 1000°F, 70 percent at 1300°F, and 50 percent at 1600°F. As a result of these changes in properties, the ductility is increased significantly.

When materials are tested in temperature ranges where creep of the material occurs, the creep strains will contribute to the inelastic deformation. The magnitude of the creep strain increases as the speed of the test decreases. Consequently, tests at elevated temperatures should be conducted at a constant strain rate, and the value

TABLE 35.3   Mechanical Properties of Copper-Zinc Alloys (Brass) (*A. Vallance and V. Doughtie*[2])

Type of material	Ultimate strength Tension, lb/in.^{2}, σ_u	Yield strength Tension, lb/in.^{2}, σ_e	Endurance limit, lb/in.^{2}, σ_e	Brinell hardness number	Modulus of elasticity Tension and compression, lb/in.^{2}, E	Elongation in 2 in., %
Commercial bronze						
(90 Cu, 10 Zn):						
Rolled, hard	65,000	63,000	18,000	107	15,000,000	18
Rolled, soft	35,000	11,000	12,000	52	15,000,000	56
Forged, cold	40,000–65,000	25,000–61,000	12,000–16,000	62–102	15,000,000	55–20
Red brass						
(85 Cu, 15 Zn):						
Rolled, hard	75,000	72,000	20,000	126	15,000,000	18
Rolled, soft	37,000	14,000	14,000	54	15,000,000	55
Forged, cold	42,000–62,000	22,000–54,000	14,000–18,000	63–120	15,000,000	47–20
Low brass						
(80 Cu, 20 Zn):						
Rolled, hard	75,000	59,000	22,000	130	15,000,000	18
Rolled, soft	44,000	12,000	15,000	56	15,000,000	65
Forged, cold	47,000–80,000	20,000–65,000		63–133	15,000,000	30–15
Spring brass						
(75 Cu, 25 Zn):						
Hard	84,000	64,000	21,000	107*	14,000,000	5
Soft	45,000	17,000	17,000	57*	18,000,000	58
Cartridge brass						
(70 Cu, 30 Zn):						
Rolled, hard	100,000	75,000	22,000	154	15,000,000	14
Rolled, soft	48,000	30,000	17,000	70	15,000,000	55
Deep-drawing brass						
(68 Cu, 32 Zn):						
Strip, hard	85,000	79,000	21,000	106*	15,000,000	3
Strip, soft	45,000	11,000	17,000	13*	15,000,000	55
Muntz metal						
(60 Cu, 40 Zn):						
Rolled, hard	80,000	66,000	25,000	151	15,000,000	20
Rolled, soft	52,000	22,000	21,000	82	15,000,000	48
Tobin bronze						
(60 Cu, 39.25 Zn, 0.75 Sn):						
Hard	63,000	35,000	21,000	165	15,000,000	35
Soft	56,000	22,000		90	15,000,000	45
Manganese bronze						
(58 Cu, 40 Zn):						
Hard	75,000	45,000	20,000	110	15,000,000	20
Soft	60,000	30,000	16,000	90	15,000,000	30

* Rockwell hardness F.

used should be reported along with the results. Creep strains may be significant at room temperature for materials with low melting temperatures.

The yield strength and the ultimate strength of certain metals, as well as the entire stress level of the stress-strain curve, are increased when the rate of deformation is increased. Figure 35.2 presents information on the static and dynamic values of the ultimate strength of several metals when the dynamic strength is determined at impact velocities of 200 to 2500 ft/sec (60 to 76 m/s).[5] The influence of strain rate

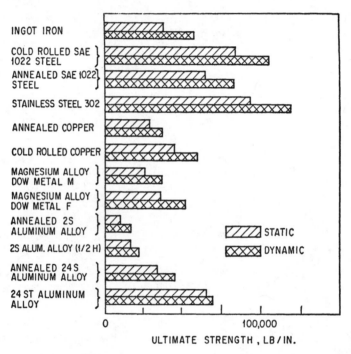

FIGURE 35.2   Static and dynamic values of the ultimate strength of several metals when the dynamic strengths were obtained at impact velocities of 200 to 250 ft/sec. (*D. S. Clark and D. S. Wood.*[5])

on the tensile properties of mild steel at room temperature is shown in Fig. 35.3. The marked difference between the yield stress and ultimate stress at low rates of strain disappears at high rates of strain.[6] Figure 35.3 also shows that the ultimate stress remains practically unchanged for strain rates below 1 in./in./sec. In this limited range the stress-strain curve of most engineering metals is not raised appreciably.[7] Mild steel is an exception in which the yield stress in influenced markedly by strain rate in the range from 0 to 1 in./in./sec.

Although the yield strength and ultimate strength of mild steel show an increase as the rate of strain increases, as illustrated in Fig. 35.3, this effect is of very limited significance in the design of equipment to withstand shock and vibration. In general, a strain rate great enough to cause a significant increase in strength occurs only

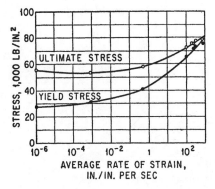

FIGURE 35.3   Effect of strain rate on mechanical properties of mild steel. (*M. J. Manjoine.*[6])

closely adjacent to a source of shock, as at the point of impact of a projectile with armor plate. Equipment seldom is subjected to shock of this nature. In a typical installation, the structure interposed between the equipment and the source of shock is unable to transmit large forces suddenly enough to cause high strain rates at the equipment. Furthermore, the response of a structure to a shock is oscillatory; maximum strain rate occurs at zero strain, and vice versa. The data of Fig. 35.3 represent conditions where maximum stress and maximum rate of strain occur simultaneously; thus, they do not apply directly to the design of shock-resistant equipment. The use of statically determined yield strength and ultimate strength for design purposes is a conservative (but not overly conservative) practice.

TOUGHNESS AND DUCTILITY

It is useful to evaluate the total energy needed to fracture a test bar under tension; this energy is a measure of the toughness of the material. The area under the typical stress-strain diagram shown in Fig. 35.1 gives an approximate measure of the fracture energy per unit volume of material. However, the true fracture energy depends upon the true stress and true strain characteristics, which take into account the nonuniform strain resulting from the reduction of area upon necking of the test bar. Calculated values of the fracture energy for various metals are given in Table 35.4. Tough materials (e.g., wrought iron and low- or medium-carbon steel) exhibit high unit elongation and are considered to be ductile. By contrast, cast iron exhibits practically no elongation and is considered to be brittle. If only the elastic strain energy up to the proportional limit is included, the resulting stored energy per unit volume is called the *modulus of resilience*. Values of this property are also given in Table 35.4.

CRITICAL STRAIN VELOCITY

When a large load is applied to a structure very suddenly, failure of the structure may occur with a relatively small overall elongation. This has been interpreted as a *brittle fracture,* and it has been said that a material loses its ductility at high strain rates. However, an examination of the failure shows normal ductility (necking) in a region close to the application of load. Large stresses are developed in this region by the inertia of the material remote from the application of the load, and failure occurs before the plastic stress waves are transmitted away from the point of load application. This effect is important only where loads are applied very suddenly, as in a direct hit by a projectile on armor plate. In general, equipment is mounted upon structures that are protected from direct hits; the resilience of such structures prevents a sufficiently rapid application of load for the above effect to be of significance in the design of equipment.

TABLE 35.4   Fracture Energy or Toughness of Different Materials (*J. M. Lessells.*[8])

Material	Condition	Yield strength, lb/in.^{2}, σ_y	Tensile strength, lb/in.^{2}, σ_u	Unit elongation, in./in., ξ	Toughness or fracture energy, in.-lb/in.^{3}	Modulus of resilience, in.-lb/in.^{3}
Wrought iron	As received	24,000	47,000	0.50	17,700	7
Steel (0.13% C)	As received	26,000	54,000	0.44	17,600	11
Steel (0.25% C)	As received	44,000	76,000	0.36	21,600	24
Steel (0.53% C)	Oil-quenched and drawn	86,000	134,000	0.11	12,000	100
Steel (1.2% C)	Oil-quenched and drawn	130,000	180,000	0.09	10,800	280
Steel (spring)	Oil-quenched and drawn	140,000	220,000	0.03	4,400	320
Cast iron	As received	...	20,000	0.005	70	1
Nickel cast iron	As received	20,000	50,000	0.10	3,500	9
Rolled bronze	As received	40,000	65,000	0.20	10,500	60
Duralumin	Forged and heat-treated	30,000	52,000	0.25	10,200	17

DELAYED INITIATION OF YIELD

Sudden application of load may not immediately result in yielding of a structure made of ductile material. Rather, yielding may occur after some time delay. This delay in initiation of yield is a function of the material, stress level, rate of load application, and temperature. Consequently, a material may be stressed substantially above its yield strength for a short period of time without yielding. For mild steel at room temperature, the delay time is of the order of 0.001 sec. For repeated applications of load, the material has a memory; i.e., the durations of load are additive to determine the time of yielding. Equipment subjected to shock or vibration experiences an oscillatory stress pattern wherein the higher stresses occur repeatedly. The durations of these stresses quickly add up to a time greater than the delay time for common materials; thus, the effect is of little significance in the design of equipment to withstand shock and vibration.

FATIGUE

The strength properties discussed up to this point are important to ensure structural integrity in the event of a single application of severe loading. Most structures, however, will be subjected to many applications of loads that may be considerably below the static-load capacity of the member or structure. Under such circumstances localized permanent changes in the material may lead to the initiation of small cracks, which propagate under subsequent applications of cyclic load. Cracks may initiate from crystal imperfections, dislocations, microcracks, lack of penetration, porosity, etc. The rate of propagation increases as the crack grows in size. If the crack becomes sufficiently large, the static load capacity of the member may be exceeded, resulting in a ductile failure. If a critical crack length is reached, the member may fail by brittle fracture at some stress significantly below the ultimate strength of the material. The critical crack length is a function of stress level, temperature, and material properties. A comprehensive discussion of the factors which contribute to brittle fracture can be found in Ref. 11.

Fatigue behavior is affected by a variety of factors. Some of the more important parameters which influence the fatigue response are the properties of the material, rate of cyclic loading, stress magnitude, residual stresses, size effect, geometry, and prior strain history. The basic parameters in fatigue tests are the stress level and the number of cycles to failure. The effects of other parameters are studied by evaluating the changes which occur in the relationship between stress and cycles to failure as these parameters are introduced.

The tensile properties of a material serve as a guide in selecting materials. They are used quantitatively to proportion members to resist static loading. There is no equivalent set of fatigue properties available to the designer whose structure must resist cyclic loading. Fatigue theories attempt to relate stress-strain properties to fatigue behavior, but complexities which arise during fatigue have thwarted these attempts. The design of equipment to resist repetitive load cycles is based on empirical data or on the application of crack propagation laws.

Fatigue tests are conducted by subjecting a test specimen to a stress pattern in which the stress varies with time. The test specimen may be subjected to alternating bending stress, as in the case of the rotating beam specimen, or to alternating axial stress. Most fatigue tests are conducted under conditions in which the stress varies sinusoidally with time. However, the use of servo-controlled hydraulic testing machines permits the variation of stress with time to follow any desired pattern. Tests may be carried out under alternating tension and compression, alternating torsion, alternating tension superimposed upon cyclic alternating tension, and many others.

Most fatigue data available in the literature have been obtained from tests which involve cycling between maximum and minimum stress levels of constant value. These are referred to as *constant-amplitude tests*. Parameters of interest are the *stress range*, $\Delta\sigma$; and the average of the maximum and minimum stress in the stress range, σ_m. One-half the stress range is called the *stress amplitude*, σ_a. The mathematical formulations for these basic definitions are

$$\Delta\sigma = \sigma_{max} - \sigma_{min} \tag{35.1}$$

$$\sigma_m = \frac{\sigma_{max} + \sigma_{min}}{2} \tag{35.2}$$

$$\sigma_a = \frac{\Delta\sigma}{2} \tag{35.3}$$

The ratio between σ_{min} and σ_{max} is referred to as the *stress ratio, R,* and the ratio between σ_a and σ_m is referred to as the *amplitude ratio, A. Completely reversed stressing* describes the case in which $\sigma_m = 0$, for which $R = -1$. The term *zero-to-tension stressing* is applied to the case in which $\sigma_{min} = 0$, and hence $R = 0$.

Most fatigue data are presented in the form of a stress (or strain) parameter versus the cycles to failure (*S-N* curves) obtained in laboratory tests. A schematic *S-N* curve is shown in Fig. 35.4. The stress parameter in this plot is the stress range, $\Delta\sigma$. The maximum stress in the test specimen is also used for this parameter.

Cycles to failure reported in the fatigue literature depend upon the definition of failure used in the particular investigation. Failure may be defined as the first appearance of an observable crack. A crack of a specific length may also be used as a failure criterion. Finally, the inability to resist the applied load without significant crack extension or corresponding load relaxation in a constant-amplitude deformation test may be used to denote failure.

Figure 35.4 also contains plots which represent the portion of the total life contributed by the crack initiation phase and by the crack propagation phase. At high levels of stress the major portion of the life consists of crack propagation, while at

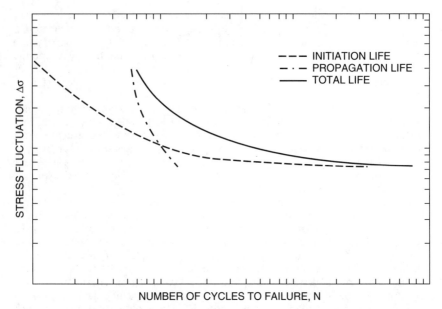

FIGURE 35.4 Schematic *S-N* curve divided into initiation and propagation components. (*J. M. Barsom and J. T. Rolfe, p. 251, Ref. 11.*)

low stress levels crack initiation constitutes the major portion of the life. Design procedures for structural components which may have surface irregularities different from those of the test specimens or which may contain cracklike discontinuities or flaws must take this difference in behavior into account.

The lowest value of stress or stress amplitude for which the crack propagation is so small that the number of cycles to failure appears to be infinite, *run-out,* is commonly referred to as the *endurance limit.* Representative values of the endurance limit for a variety of materials are presented in Tables 35.5 and 35.6. The effects of geometry and corrosive environment on the relationship between fatigue strength and ultimate strength of steels are shown in Fig. 35.5.

Three design approaches are presented in the following sections. The stress-life method was the first approach employed and has been the standard method for many years. It is still widely used in applications in which the applied stress is within the elastic range. It does not work well where the applied strains have a significant plastic component, *low-cycle fatigue.* A strain-life approach is more appropriate in this case. A more recent development in the evaluation of fatigue life incorporates the concepts of fracture mechanics to analyze the crack growth from some initial flaw size as cyclic stresses are applied. In this approach, failure may be defined as the development of a crack of some specific dimension. Detailed discussions of the different methods are given in Refs. 10 and 11.

STRESS-LIFE METHOD

The first procedure used to design structural components utilizes a design fatigue curve which characterizes the basic unnotched fatigue properties of the material and

TABLE 35.5   Tensile and Fatigue Properties of Steels (*J. M. Lessells.*[8])

Material	State	Yield strength, lb/in.^{2}, σ_y	Tensile strength, lb/in.^{2}, σ_u	Elonga-tion, %	Reduc-tion of area, %	Endur-ance limit, lb/in.^{2}, σ_e	Ratio σ_e/σ_u
0.02% C	As received	19,000	42,400	48.3	76.2	26,000	0.61
Wrought iron	As received	29,600	47,000	35.0	29.0	23,000	0.49
0.24% C	As received	38,000	60,500	39.0	64.0	25,600	0.425
0.24% C	Water-quenched and drawn	45,600	67,000	38.0	71.0	30,200	0.45
0.37% C	Normalized	34,900	71,900	29.4	53.5	33,000	0.46
0.37% C	Water-quenched and drawn	63,100	94,200	25.0	63.0	45,000	0.476
0.52% C	Normalized	47,600	98,000	24.4	41.7	42,000	0.43
0.52% C	Water-quenched and drawn	84,300	111,400	21.9	56.6	55,000	0.48
0.93% C	Normalized	33,400	84,100	24.8	37.2	30,500	0.36
0.93% C	Oil-quenched and drawn	67,600	115,000	23.0	39.6	56,000	0.487
1.2% C	Normalized	60,700	116,900	7.9	11.6	50,000	0.43
1.2% C	Oil-quenched and drawn	130,000	180,000	9.0	15.2	92,000	0.51
0.31% C, 3.35% Ni	Normalized	53,500	104,000	23.0	45.0	49,500	0.47
0.31% C, 3.35% Ni	Oil-quenched and drawn	130,000	154,000	17.0	49.0	63,500	0.41
0.24% C, 3.3% Ni, 0.87% Cr	Oil-quenched and drawn	128,000	138,000	18.2	61.8	68,000	0.49

a fatigue-strength reduction factor. Parameters characteristic of the specific component which make it more susceptible to fatigue failure than the unnotched specimen are reflected in the strength-reduction factor. Early applications of this method were based on the results of rotating bending tests. The application of such tests, in which mirror-polished specimens were subjected to reversed bending, requires consideration of a number of factors which present themselves in design situations. Among these factors are size, type of loading, surface finish, surface treatments, temperature, and environment.

In the rotating beam test, a relatively small volume of material is subjected to the maximum stress. For larger rotating beam specimens, the volume of material is greater, and therefore there will be a greater probability of initiating a fatigue crack. Similarly, an axially loaded specimen which has no gradient will exhibit an endurance limit smaller than that obtained from the rotating beam test. Surface finish will have a similar effect. Surface finish is more significant for higher-strength steels. At shorter lives (high stress levels), surface finish has a smaller effect on the fatigue life. Surface treatment, temperature, and environment have similar effects.

The effect of mean stress on fatigue life is conveniently represented in the form of a modified Goodman fatigue diagram (Fig. 35.6). In this figure, the ordinate is the maximum stress, and the abscissa is the minimum stress. Radial lines indicate the stress ratio. The curves n_1, n_2, etc., represent failure at various lives.

Many design specifications[12–15] contain provisions for repeated loadings based on laboratory tests. In these specifications, fabricated details are categorized for design

TABLE 35.6 Tensile and Fatigue Properties of Nonferrous Metals (*J. M. Lessells.*[8])

Material	State	Tensile strength, lb/in.^{2}, σ_u	Endurance limit or fatigue strength, lb/in.^{2}, σ_e	N_1,* millions of cycles	N_2† millions of cycles	Ratio σ_e/σ_u
Aluminum		22,600	10,500	100	6	0.46
Duralumin	Rolled	51,000	14,000	400	>400	0.27
Duralumin	Annealed	25,200	10,000	200	>200	0.40
Duralumin	Tempered	51,300	12,000	400	4½	0.24
Magnesium	Extruded	32,500	8,000	200	2	0.25
Magnesium alloy (4% Al)		35,200	12,000	600	½	0.34
Magnesium alloy (4% Al, 0.25% Mn)		39,000	15,000	100	1	0.38
Magnesium alloy (6.5% Al)		41,200	13,000	600	½	0.31
Magnesium alloy (6.5% Al, 0.25% Mn)		44,500	15,000	100	½	0.34
Magnesium alloy (10% Cu)		39,000	12,000	600	½	0.31
Electron metal		36,600	17,000	200	30	0.47
Copper	Annealed	32,400	10,000	500	20	0.31
Copper	Cold-drawn	56,200	10,000	500	>500	0.18
Brass (60–40)	Annealed	54,200	22,000	500	>500	0.44
Brass (60–40)	Cold-drawn	97,000	26,000	500	50	0.27
Naval brass		68,400	22,000	300	10	0.32
Aluminum bronze (10% Al)	As cast	59,200	23,000	60	3	0.39
Aluminum bronze (10% Al)	Heat-treated	77,800	27,000	40	1	0.35
Bronze (5% Sn)	Annealed	45,600	23,000	1000	10	0.50
Bronze (5% Sn)	Cold-drawn	85,000	27,000	500	50	0.32
Manganese bronze	As cast	70,000	17,000	150	20	0.24
Nickel	Annealed	70,000	28,000	100	50	0.40
Monel metal	Hot-rolled	90,000	32,000	450	>450	0.36

* N_1 = cycles on which σ_e is based.

purposes and fatigue-strength stress ranges are given for different fatigue lives.

The following procedure[16] has been used to determine an allowable fatigue design stress range, S_R. Four different loading histograms, shown in Fig. 35.7, were used to describe the frequency distribution of the ratio of the cyclic stress range to the maximum cyclic stress range. The four conditions are defined in Table 35.7; the first three represent beta-distribution probability density functions that have shape factors q and r as shown. The allowable fatigue design stress range S_R may be determined from

$$S_R = S_r R_F C_L \tag{35.4}$$

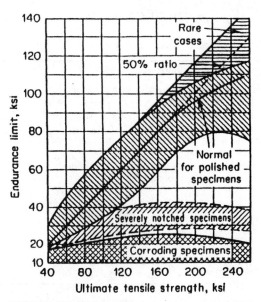

FIGURE 35.5 Relationship between the fatigue limit and ultimate tensile strength of various steels. (*Battelle Memorial Institute.*[9])

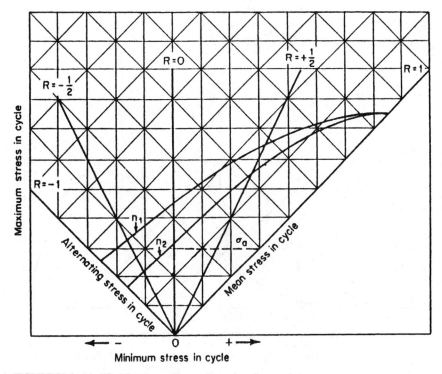

FIGURE 35.6 Modified Goodman diagram for various lives and stress ranges.

TABLE 35.7   Random Loading Coefficients C_L

Type	Load description (see Fig. 35.7)	Coefficient C_L
I	Primarily light loading cycles: mean range of stress 030% of maximum ($q = 3, r = 7$)	2.75
II	Medium loading cycles: mean range of stress 50% of maximum ($q = 7, r = 7$)	1.85
III	Primarily heavy loading cycles: mean range of stress 70% of maximum ($q = 7, r = 3$)	1.35
IV	Constant loading cycles: stress range constant and equal to 100% of maximum	1.00

TABLE 35.8   Reliability Factors R_F

Level of reliability	Structural importance of detail	Reliability factor R_F
90%	Secondary details for which fatigue cracking is of little structural significance	0.67
95%	Major structural details for which fatigue cracking is important: members in redundant structures	0.60
99%	Major structural details in fracture-critical members where fatigue cracking is critical	0.45

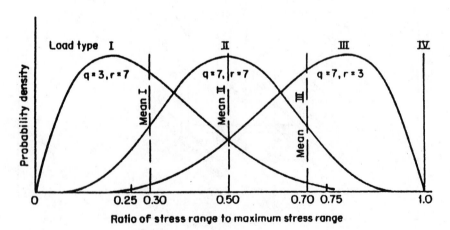

FIGURE 35.7   Loading frequency distributions. (*W. H. Munse and S. T. Rolfe, Sect. 4 of Ref. 16.*)

where S_r = mean constant-cycle fatigue stress range for desired life
R_F = reliability factor based on a statistical fatigue analysis for survival, Table 35.8
C_L = loading coefficient to be selected for load type, Table 35.7

The stress-life method works quite well for design for long-life and constant-amplitude stress histories.

STRESS-LIFE METHOD

At high load levels, at which plastic strains are likely to occur, the response and material behavior are best modeled under strain-controlled conditions. Engineered structures almost always contain points of stress concentration which cause plastic strains to develop. The constraint imposed by the surrounding elastic material produces an essentially strain-controlled environment. For these conditions, tests under strain control are used to simulate fatigue damage at points of stress concentration. The strain-life method does not account for crack growth. Consequently, such methods may be considered initiation life estimates. For components in which the existence of a crack may be an overly conservative criterion, fracture mechanics may be employed to assess the crack propagation life from some assumed initial crack size.

Cyclic inelastic loading of a material produces a hysteresis loop. The stress range, $\Delta\sigma$, is the total height of the loop. The total width of the loop is $\Delta\epsilon$, the total strain range. The strain amplitude, ϵ_a, can be expressed by

$$\epsilon_a = \frac{\Delta\epsilon}{2} \tag{35.5}$$

and the stress amplitude, σ_a, is

$$\sigma_a = \frac{\Delta\sigma}{2} \tag{35.6}$$

The sum of the elastic and plastic strain ranges is the total strain, $\Delta\epsilon$. This may be expressed mathematically as

$$\Delta\epsilon = \Delta\epsilon_e + \Delta\epsilon_p \tag{35.7}$$

In terms of amplitudes

$$\frac{\Delta\epsilon}{2} = \frac{\Delta\epsilon_e}{2} + \frac{\Delta\epsilon_p}{2} \tag{35.8}$$

The elastic term may be replaced by $\Delta\sigma/E$ by applying Hooke's law, so that

$$\frac{\Delta\epsilon}{2} = \frac{\Delta\sigma}{2E} + \frac{\Delta\epsilon_p}{2} \tag{35.9}$$

Under repeated cycling the stress-strain response may exhibit cyclic hardening, cyclic softening, cyclic stability, or a mixed behavior (softening or hardening depending upon the stress range).

From experimental data, the following relationship between total strain range and the number of reversals to failure has been developed:

$$\frac{\Delta\epsilon}{2} = \frac{\sigma'_f}{E} \, (2N_f)^b + \epsilon'_f \, (2N_f)^c \tag{35.10}$$

where $\Delta\epsilon$ = total strain range
         σ'_f = fatigue strength coefficient
         $2N_f$ = reversals to failure
         b = fatigue strength exponent
         ϵ'_f = fatigue ductility coefficient
         c = fatigue ductility exponent

The fatigue strength coefficient, σ'_f, is approximately equal to the true fracture strength. The fatigue strength exponent, b, varies between −0.05 and −0.12. The fatigue ductility coefficient, ϵ'_f, is approximately equal to the true fracture ductility. The fatigue ductility exponent, c, varies between −0.5 and −0.07. Additional discussion of these parameters and approximate formulations for the fatigue strength coefficient and the fatigue ductility coefficient are presented in Ref. 10.

Cyclic properties are generally obtained from completely reversed, constant-amplitude, strain-controlled tests. The effects of mean strain have been studied by various investigators, and modifications of Eq. (35.10) have been proposed.

This method of analysis is obviously more complicated than the stress-life approach. Notch root strains must be evaluated by application of some method of analysis. Since it is based on strain cycling of constant magnitude, it applies only in the immediate region of the notch and predicts the initiation life for a fatigue crack.

FRACTURE MECHANICS METHOD

Fracture mechanics is the study of the performance of structures with cracklike defects. The distribution of stress components at the crack tip are related to a constant called the *stress intensity factor*, characterized by the applied stress and the dimensions of the crack. In addition to the applied stress, the design process using fracture mechanics incorporates flaw size and fracture toughness properties of the material. Fracture toughness replaces strength as the relevant material property.

As noted earlier, fatigue life is divided into an initiation phase and a propagation phase. The fracture mechanics method can be used to determine the propagation life on the assumption of some initial crack or defect size. The strain-life approach may be used to determine the initiation life for an evaluation of the total fatigue life.

Fatigue crack growth under constant-amplitude cyclic loading can be represented schematically as shown in Fig. 35.8. Such data can be presented in terms of crack growth rate per cycle of loading, da/dN, and the fluctuation of the stress intensity factor, ΔK_1. The most common presentation of fatigue crack growth data is as a log-log plot of the rate of fatigue crack growth per cycle of load fluctuation, da/dN, and the fluctuation of the stress intensity factor, ΔK_1. Such a plot shows three distinct regions. At low values of ΔK, the rate of crack propagation is extremely small, essentially zero. The value of ΔK for this condition is referred to as the fatigue-threshold cyclic stress intensity factor fluctuation, ΔK_{th}, below which cracks do not propagate. There are sufficient data available to demonstrate the existence of this threshold, but more work is needed to determine the factors which affect its magnitude for use in design.

The second stage in the crack propagation versus stress intensity factor relationship represents the fatigue crack propagation behavior above ΔK_{th}. In this region the relationship can be defined as

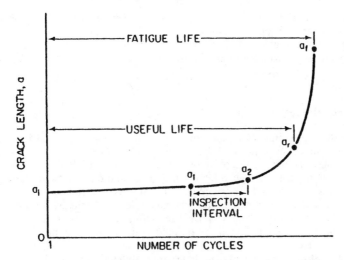

FIGURE 35.8 Schematic representation of fatigue crack growth curve under constant-amplitude loading. (*J. M. Barsom and S. T. Rolfe, p. 279, Ref. 11.*)

$$\frac{da}{dN} = A(\Delta K)^m \tag{35.11}$$

where a = crack length
          N = number of cycles
          ΔK = stress intensity factor range

and A and m are constants that depend on the properties of the material.

The third stage in the crack propagation versus stress intensity factor relationship shows a very rapid increase in the rate of crack propagation.

Fatigue crack propagation may be affected by the mean stress, cyclic frequency, waveform, and environment. Extensive discussion of the effect of these parameters, as well as values of A and m for different materials, is presented in Ref. 11.

Equation (35.11) can be used, with appropriate values of A and m, to analyze fatigue crack growth as a function of cyclic loading between some assumed initial crack size and some critical crack dimension assumed to represent the ultimate condition. The critical crack dimension may be chosen on the basis of the limiting static strength or on the basis of the crack size which may result in brittle fracture.

The procedure requires the integration of Eq. (35.11) from an initial crack size, a_0, which corresponds to an initial value of ΔK. An increment of crack growth must be incorporated, during which stage the value of ΔK remains constant. The value of ΔK is then revised and the process is continued until the crack reaches the limiting critical dimension. An example of this procedure is presented in Ref. 11.

VARIABLE-AMPLITUDE LOADING

Most laboratory fatigue tests are conducted at constant values of maximum and minimum stress. Most structures, on the other hand, are subjected to loading cycles with variable minimum and maximum stresses over the course of their life. Procedures are required to relate the behavior under constant cyclic loading obtained in laboratory tests and the variations of stress history over time which occur in an actual structure. It is also necessary to convert the complicated time-history of a real structure into some equivalent number of individual stress cycles for the evaluation of their cumulative effect.

DAMAGE RULES

Damage during the initiation phase of fatigue is difficult to assess, as it occurs on a microscopic level and is not easily observed or evaluated. During the propagation phase, damage can be related to an observable and measurable crack length. Both linear and nonlinear damage rules for the accumulation of fatigue damage have been proposed. Only the linear damage rule will be discussed here.

The most commonly applied linear damage rule was originally proposed in 1924 and was developed further by Miner.[17] The method is referred to simply as *Miner's rule*. Damage under cyclic loading is defined as the ratio of the number of applied cycles, n_i, at stress level σ_i to the number of cycles to failure, N_i, in a constant-amplitude test conducted at σ_i. The hypothesis states that failure occurs when the accumulated damage reaches 1. Mathematically,

$$\Sigma \frac{n_i}{N_i} = \frac{n_1}{N_1} + \frac{n_2}{N_2} + \frac{n_3}{N_3} + \cdots \geq 0 \tag{35.12}$$

This linear damage rule is easily applied after an appropriate counting method has been established. It has the shortcoming, however, that it does not consider the sequence of loading and assumes that damage in any individual stress cycle is independent of what has preceded it. Furthermore, it assumes that damage accumulation is independent of stress amplitude.

CYCLE COUNTING

Some method of cycle counting is required in order to determine the number of cycles at a specific stress range. The tabulation of stress cycles at the various stress ranges is referred to as the stress spectrum. Several counting methods have been proposed, and a summary of these methods is contained in Ref. 18. The two counting methods most commonly used are the *rainflow counting method* and the *reservoir method*. The following example from Ref. 19 demonstrates the procedures.

The rainflow counting method employs the analogy of raindrops flowing down a pagoda roof. Peaks and troughs for one loading event are presented in Fig. 35.9A. The maximum and minimum stresses are indexed in Fig. 35.9B. The following rules apply to rainflow counting:

1. A drop flows left from the upper side of a peak or right from the upper side of a trough and onto subsequent "roofs" unless the surface receiving the drop is formed by a peak that is more positive for left flow or a trough that is more negative for right flow. For example, a drop flows left from point 1 off points 2, 4, and 12 until it stops at the end of the loading event at point 22, since no peak is encountered that is more positive than point 1. On the other hand, a drop flows right from point 2 off point 3 and stops, since it encounters a surface formed by a trough (point 4) that is more negative than point 2.

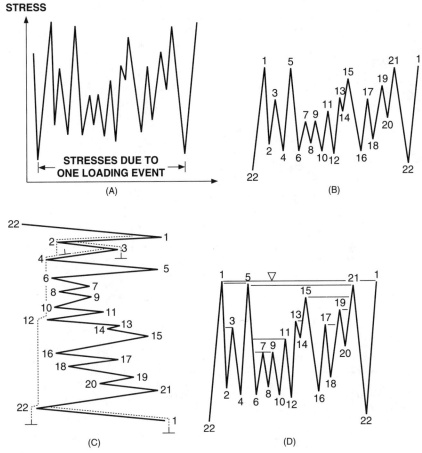

FIGURE 35.9 Variable-amplitude loading for analysis. (*A*) An example of stress variation in an element due to one loading event. (*B*) Peaks and troughs numbered for one loading event. (*C*) Rainflow analysis. (*D*) Reservoir analysis.

2. The path of a drop cannot cross the path of a drop that has fallen from above. For example, a drop flowing left from point 3 stops at the horizontal position of point 2 because it encounters a path coming from point 2.

3. The horizontal movement of a raindrop, measured in units of stress from its originating peak to its stop position, is counted as one-half of a cycle in the stress spectrum.

The stress variation of Fig. 35.9*A* is rotated 90° in Fig. 35.9*C* for application of the rainflow counting method. The values of the peaks for the stress history shown in Fig. 35.9 are given in Table 35.9. Table 35.10 contains the values of the half-cycle magnitudes which result from application of the rules above.

TABLE 35.9 Stress Values for Fig. 35.9

Peak/trough no.	Stress, MPa
1	93
2	18
3	55
4	10
5	85
6	10
7	37
8	18
9	37
10	10
11	46
12	6
13	55
14	46
15	74
16	8
17	55
18	18
19	65
20	39
21	83
22	0

TABLE 35.10 Rainflow Counting

From peak or trough no.	To horizontal distance of point no.	Half cycle, MPa
1	22	93
2	3	37
3	2	37
4	5	75
5	6	75
6	11	36
7	10	27
8	9	19
9	8	19
10	9	27
11	10	36
12	21	77
13	14	9
14	13	9
15	16	66
16	15	66
17	18	37
18	17	37
19	20	26
20	19	26
21	12	77
22	1	93

The reservoir method employs an analogy of water contained in reservoirs formed by peaks draining successively out of the troughs. The lowest trough is drained first, followed by successively higher troughs until the reservoir is empty. Figure 35.9D demonstrates the reservoir method, and the corresponding values for the stress range are presented in Table 35.11.

TABLE 35.11 Reservoir Counting Method

Drain from trough no.	Water level at peak	Stress range, MPa
22	1	93
12	21	77
4	5	75
16	15	66
2	3	37
18	17	37
10	11	36
6	7	27
20	19	26
8	9	19
14	13	9

Rainflow counting and reservoir counting give identical results provided that rainflow counting begins with the highest peak in the loading event, as is shown in Fig. 35.9. Rainflow counting is more suited to computer analyses or long stress histories, whereas the reservoir method is most convenient for graphical analyses of short histories.

Table 35.12 presents the results of an analysis according to the Miner linear damage rule assuming 1 million loading sequences of the stress history of Fig. 35.9. The cyclic fatigue lives presented in the second column are taken from a typical S-N curve for a beam in which manually welded longitudinal fillet welds are used to connect the flanges to the web. The analysis indicates that the fatigue evaluation has failed.

TABLE 35.12 Cumulative Damage Using Miner's Rule

Stress range, $\Delta\sigma$, MPa	Fatigue resistance, $N = (100/\Delta\sigma)^3$, 2×10^6 cycles	Damage due to 1×10^6 loading events, n_i/N
93	2,490,000	0.402
77	4,381,000	0.228
75	4,741,000	0.211
66	6,957,000	0.144
37 (twice)	39,480,000	0.051
36	42,870,000	0.023
27	101,600,000	0.010
26	113,800,000	0.009
19	292,600,000	0.003
9	2.7×10^9	0.000
		Damage summation: $\Sigma n_i/N = 1.08 \geq 1.0$

The cyclic fatigue lives in Table 35.12 do not reflect the existence of an endurance limit or constant-amplitude fatigue limit. Because the Miner rule does not account for the effect of load sequence, some designers choose to extend the finite life region of the S-N curve and assume that all cyclic variations contribute to damage accumulation. The opposite extreme would be to neglect all cyclic variations smaller than the constant-amplitude fatigue limit. A third variation of the procedure employed by some designers assumes a change in the slope of the experimentally determined S-N curve at some large number of cycles. For example, between 5×10^6 cycles and 10^8 cycles the slope of the S-N curve might be reduced, and the constant-amplitude fatigue limit might be assumed to occur at 10^8 cycles. In view of the lack of test data at very long fatigue lives, there is no agreement on which of the three procedures is most appropriate.

REFERENCES

1. Obtainable from the American Society for Testing and Materials, 1916 Race Street, Philadelphia, PA 19103.

2. Vallance, A., and V. Doughtie: "Design of Machine Members," 3d ed., Chap. II, McGraw-Hill Book Company, Inc., New York, 1951.

3. Hoyt, S. I.: "Metal Data," 2d ed., Reinhold Publishing Corporation, New York, 1952.

4. American Society for Metals, "Metals Handbook," Vol. 1, "Properties and Selection: Irons, Steels, and High-Performance Alloys," 1990.

5. Clark, D. S., and D. S. Wood: *Trans. ASM,* **42:**45, 1950.

6. Manjoine, M. J.: *J. Appl. Mechanics,* **66:**A215, 1944.

7. MacGregor, C. W., and J. C. Fisher: *J. Appl. Mechanics,* **67:**A217, 1945.

8. Lessells, J. M.: "Strength and Resistance of Metals," p. 7, John Wiley & Sons, Inc., New York, 1954.

9. Battelle Memorial Institute: "Prevention of Failure of Metals under Repeated Stress," John Wiley & Sons, Inc., New York, 1946.

10. Bannantine, J. A., J. J. Comer, and J. L. Handrock: "Fundamentals of Metal Fatigue Analysis," Prentice-Hall, Inc., Englewood Cliffs, N.J., 1990.

11. Barsom, J. M., and S. T. Rolfe: "Fracture and Fatigue Control in Structures," 2d ed., Prentice-Hall, Inc., Englewood Cliffs, N.J., 1987.

12. American Institute of Steel Construction: "Specification for Structural Steel Buildings—Allowable Stress Design and Plastic Design," 1989.

13. American Railway Engineering Association, "Specifications for Steel Railway Bridges," 1994.

14. American Association of State Highway and Transportation Officials: "Standard Specifications for Highway Bridges," 1992.

15. American Welding Society: "Structural Welding Code," D1.1, 1994.

16. E. H. Gaylord, Jr., C. N. Gaylord, and J. E. Stallmeyer, "Structural Engineering Handbook," 4th ed., McGraw-Hill, New York, 1996.

17. Miner, M. A.: *J. Appl. Mech.,* **12:**A159 (1945).

18. Gurney, T. R.: "Fatigue of Welded Structures," 2d ed., Cambridge University Press, 1979.

19. Kulak, G. L., and I. F. C. Smith: "Analysis and Design of Fabricated Steel Structures for Fatigue: A Primer for Civil Engineers," *Structural Engineering Report* No. 190, University of Alberta, Edmonton, Canada, 1993.

CHAPTER 36
MATERIAL DAMPING AND SLIP DAMPING

L. E. Goodman

INTRODUCTION

The term *damping* as used in this chapter refers to the energy-dissipation properties of a material or system under cyclic stress but excludes energy-transfer devices such as dynamic absorbers. With this understanding of the meaning of the word, energy must be dissipated within the vibrating system. In most cases a conversion of mechanical energy to heat occurs. For convenience, damping is classified here as (1) material damping and (2) system damping. Material properties and the principles underlying measurement and prediction of damping magnitude are discussed in this chapter. For application to specific engineering problems, see Chap. 37.

MATERIAL DAMPING

Without a source of external energy, no real mechanical system maintains an undiminished amplitude of vibration. *Material damping* is a name for the complex physical effects that convert kinetic and strain energy in a vibrating mechanical system consisting of a volume of macrocontinuous (solid) matter into heat. Studies of material damping are employed in solid-state physics as guides to the internal structure of solids. The damping capacity of materials is also a significant property in the design of structures and mechanical devices; for example, in problems involving mechanical resonance and fatigue, shaft whirl, instrument hysteresis, and heating under cyclic stress. Three types of material that have been studied in detail are:

1. Viscoelastic materials.[1] The idealized linear behavior generally assumed for this class of materials is amenable to the laws of superposition and other conventional rheological treatments including model analog analysis. In most cases linear (Newtonian) viscosity is considered to be the principal form of energy dissipation. Many polymeric materials as well as some other types of materials may be treated under this heading.

2. Structural metals and nonmetals.[2] The linear dissipation functions generally assumed for the analysis of viscoelastic materials are not, as a rule, appropriate

36.1

for structural materials. Significant nonlinearity characterizes structural materials, particularly at high levels of stress. A further complication arises from the fact that the stress and temperature histories may affect the material damping properties markedly; therefore, the concept of a stable material assumed in viscoelastic treatments may not be realistic for structural materials.

3. Surface coatings. The application of coatings to flat and curved surfaces to enhance energy dissipation by increasing the losses associated with fluid flow is a common device in acoustic noise control. These coatings also take advantage of material and interface damping through their bond with a structural material. They are treated in detail in Chap. 37.

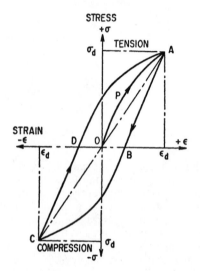

FIGURE 36.1 Typical stress-strain (or load-deflection) hysteresis loop for a material under cyclic stress.

Material damping of macrocontinuous media may be associated with such mechanisms as plastic slip or flow, magnetomechanical effects, dislocation movements, and inhomogeneous strain in fibrous materials. Under cyclic stress or strain these mechanisms lead to the formation of a stress-strain hysteresis loop of the type shown in Fig. 36.1. Since a variety of inelastic and anelastic mechanisms can be operative during cyclic stress, the unloading branch AB of the stress-strain curve falls below the initial loading branch OPA. Curves OPA and AB coincide only for a perfectly elastic material; such a material is never encountered in actual practice, even at very low stresses. The damping energy dissipated per unit volume during one stress cycle (between stress limits $\pm\sigma_d$ or strain limits $\pm\epsilon_d$) is equal to the area within the hysteresis loop $ABCDA$.

SLIP DAMPING

In contrast to material damping, which occurs within a volume of solid material, slip damping[3] arises from boundary shear effects at mating surfaces, or joints between distinguishable parts. Energy dissipation during cyclic shear strain at an interface may occur as a result of dry sliding (Coulomb friction), lubricated sliding (viscous forces), or cyclic strain in a separating adhesive (damping in viscoelastic layer between mating surfaces).

SIGNIFICANCE OF MECHANICAL DAMPING AS AN ENGINEERING PROPERTY

Large damping in a structural material may be either desirable or undesirable, depending on the engineering application at hand. For example, damping is a desirable property to the designer concerned with limiting the peak stresses and extending the fatigue life of structural elements and machine parts subjected to near-resonant cyclic forces or to suddenly applied forces. It is a desirable property if noise reduction is of importance. On the other hand, damping is undesirable if internal heating is to be avoided. It also can be a source of dynamic instability of rotating shafts and of error in sensitive instruments.

Resonant vibrations of large amplitude are encountered in a variety of modern devices, frequently causing rough and noisy operation and, in extreme cases, leading to seriously high repeated stresses. Various types of damping may be employed to minimize these resonant vibration amplitudes. Although special damping devices of the types described in Chap. 6 may be used to transfer energy from the system, there are many situations in which auxiliary dampers are not practical. Then accurate estimation of material and slip damping becomes important.

When an engineering structure is subjected to a harmonic exciting force $F_g \sin \omega t$ an induced force $F_d \sin (\omega t - \varphi)$ appears at the support. The ratio of the amplitudes, F_d/F_g, is a function of the exciting frequency ω. It is known as the vibration amplification factor. At resonance, when $\varphi = 90°$, this ratio becomes the resonance amplification factor[4] A_r:

$$A_r = \frac{F_d}{F_g} \qquad (36.1)$$

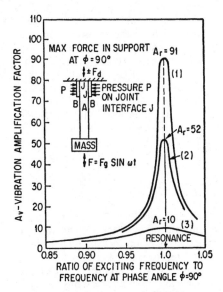

This condition is pictured schematically in Fig. 36.2 for low, intermediate, and high damping (curves 1, 2, 3, respectively).

The magnitude of the resonance amplification factor varies over a wide range in engineering practice.[5] In laboratory tests, values as large as 1,000 have been observed. In actual engineering parts under high stress, a range of 500 to 10 is reasonably inclusive. These limits are exemplified by an airplane propeller, cyclically stressed in the fatigue range, which displayed a resonance amplification factor of 490, and a double leaf spring with optimum interface slip damping which was observed to have a resonance amplification factor of 10. Because of the wide range of possible values of A_r, each case must be considered individually.

FIGURE 36.2 Effect of material and slip damping on vibration amplification. Curve (1) illustrates case of small material and slip damping; (2) one damping is large while other is small; (3) both material and slip damping are large.

METHODS FOR MEASURING DAMPING PROPERTIES

STRESS-STRAIN (OR LOAD-DEFLECTION) HYSTERESIS LOOP

The hysteresis loop illustrated in Fig. 36.1 provides a direct and easily interpreted measure of damping energy. To determine damping at low stress levels requires instruments of extreme sensitivity. For example, the width (DB in Fig. 36.1) of the loop of chrome steel at an alternating direct-stress level of 103 MPa* is less than 2×10^{-6}. High-sensitivity and high-speed transducers and recording devices are required to attain sufficient accuracy for measurement of such strains. For metals in general extremely long gage lengths are required to measure damping in direct stress by the hysteresis loop method if the peak stress is less than about 60 percent of the fatigue limit. Under torsional stress, however, greater sensitivity is possible and the hysteresis loop method is applicable to low stress work.

PROCEDURES EMPLOYING A VIBRATING SPECIMEN

The following methods of measuring damping utilize a vibrating system in which the deflected member, usually acting as a spring, serves as the specimen under test. For example, one end of the specimen may be fixed and the other end attached to a mass which is caused to vibrate; alternatively, a freely supported beam or a tuning fork may be used as the specimen vibrating system.[6] In any arrangement the damping is computed from the observed vibratory characteristics of the system.

In one class of these procedures the rate of decay of free damped vibration is measured. Typical vibration decay curves are shown in Fig. 36.3. The measure of damping usually used, the *logarithmic decrement,* is the natural logarithm of the ratio of any two successive amplitudes [See Eq. (37.6)]:

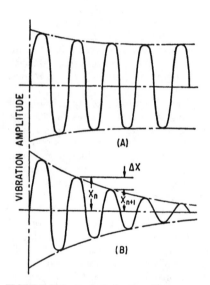

$$\Delta = \ln \frac{x_n}{x_{n+1}} \simeq \frac{\Delta x}{x_n} \qquad (36.2)$$

The relation between logarithmic decrement and other units used to measure damping is given in Eq. (36.16). Vibration decay tests can be performed under a variety of stress and temperature conditions, and may utilize many different procedures for releasing the specimen and recording the vibration decay. It is essential to minimize loss of energy either to the specimen supports or in acoustic radiation.

FIGURE 36.3 Typical vibration decay curves: (*A*) low decay rate, small damping, and (*B*) high decay rate, large damping.

A second class of vibrating specimen procedures makes use of the fact, illustrated in Fig. 36.2, that higher damping is associated with a broader peak in the frequency response or resonance curve. If the exciting force is held constant and the exciting frequency varied, measurement of the steady-state amplitude of motion (or stress) yields a curve similar to those shown in Fig. 36.2. The damping is then determined by measuring the width of the curve at an amplification factor equal to

---

* 1 MPa = 10^6 N/m^2 = 146.5 lb/in.^{2} (103 MPa = 15,000 lb/in.^{2}).

$0.707A_r$. If a horizontal line drawn at this ordinate intercepts the resonance curve at frequency ratios f_1/f_n and f_2/f_n,

$$\Delta = \pi \left(\frac{f_2}{f_n} - \frac{f_1}{f_n} \right) \qquad (36.3)$$

The quantity $(f_2 - f_1)$ is the *bandwidth at the half-power point*. This procedure has the advantage of requiring only steady-state tests. As in the case of the free-decay procedure, only the relative amplitude of the response need be measured. However, the procedure does impose a particular stress history. If the system behavior should be markedly nonlinear, the shape of the resonance curve will not be that assumed in the derivation of Eq. (36.3).

If a system is operated exactly at resonance, the resonance amplification factor A_r is the ratio of the (induced) force F_d to the exciting force F_g [see Eq. (36.1)]. In direct application of this equation, F_g is usually made controllable and F_d computed from strain or displacement measurements. The principle has been applied to the measurement of damping in a large structure[7] and in simple test specimens. It can take account of high stress magnitude and of stress history as controlled variables. The natural frequency of vibration of a specimen can be altered so that damping as a function of frequency may be studied, but it is usually difficult to make such measurements over a wide frequency range. This technique requires accurately calibrated apparatus since measurements are absolute and not relative.

LATERAL DEFLECTION OF ROTATING CANTILEVER METHOD

The principle of the lateral deflection method is illustrated in Fig. 36.4. If test specimen S is loaded by arm-weight combination $A—W$, the target T deflects vertically downward from position 1 to position 2. If the arm-specimen combination is rotated by spindle B, as in a rotating cantilever-beam fatigue test, target T moves from posi-

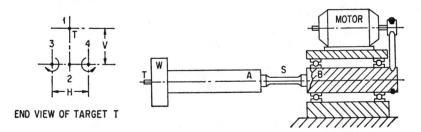

FIGURE 36.4 Principle of rotating cantilever beam method for measuring damping.

tion 2 to position 3 for clockwise rotation. If the direction of rotation is counterclockwise, the target moves from position 3 horizontally to position 4. The horizontal traversal H is a direct measure of the total damping absorbed by the rotating system.[8]

A modification of the lateral deflection method is the *lateral force method*. The end of the rotating beam is confined and the lateral confining force is measured instead of the lateral deflection H. This modification is particularly useful for measurements of low modulus materials, such as plastic and viscoelastic materials.[9]

The advantages of the rotating cantilever beam method are (1) the test variables, stress magnitude, stress history, and frequency, may be easily and independently controlled so that this method is satisfactory for intermediate and high stress levels, and (2) it yields not only data on damping but also fatigue and elasticity properties.

The disadvantages of this method are (1) the tests are rather time-consuming, (2) accuracy is often questionable at low stress levels (below about 20 percent of the fatigue limit) due to the small value of the horizontal traversal H, and (3) the method can be used under rotating-bending conditions only.

HIGH-FREQUENCY PULSE TECHNIQUES

A sequence of elastic pulses generated by a transducer such as a quartz crystal cemented to the front face of a specimen is reflected at the rear face and received again at the transducer. The frequencies are in the megacycle range. The velocity of such waves provides a measure of the elastic constants of the specimen; their decay rates provide a measure of the material damping.[10] This technique has been widely employed in the study of the viscoelastic properties of polymers and the elastic properties of crystals. So far as measurement of damping is concerned, it is open to the objection that the attenuation may be due to scattering by imperfections rather than to internal friction.

FUNDAMENTAL RELATIONSHIPS

Two general types of units are used to specify the damping properties of structural materials: (1) the energy dissipated per cycle in a structural element or test specimen and (2) the ratio of this energy to a reference strain energy or elastic energy. Absolute damping energy units are:

$D_0 =$ *total damping energy* dissipated by entire specimen or structural element per cycle of vibration, N·m/cycle

$D_a =$ *average damping energy*, determined by dividing total damping energy D_0 by volume V_0 of specimen or structural element which is dissipating energy, N·m/m^3/cycle

$D =$ *specific damping energy*, work dissipated per unit volume and per cycle at a point in the specimen, N·m/m^3/cycle

Of these absolute damping energy units, the total energy D_0 usually is of greatest interest to the engineer. The average damping energy D_a depends upon the shape of the specimen or structural element and upon the nature of the stress distribution in it, even though the specimens are made of the same material and have been subjected to the same stress distribution at the same temperature and frequency. Thus, quoted values of the average damping energy in the technical literature should be viewed with some reserve.

The specific damping energy D is the most fundamental of the three absolute units of damping since it depends only on the material in question and not on the shape, stress distribution, or volume of the vibrating element. However, most of the methods discussed previously for measuring damping properties yield data on total damping energy D_0 rather than on specific damping energy D. Therefore, the development of the relationships between these quantities assumes importance.

RELATIONSHIP BETWEEN D_0, D_a, AND D

If the specific damping energy is integrated throughout the stressed volume,

$$D_0 = \int_0^{V_0} D \, dV \tag{36.4}$$

This is a triple integral; $dV = dx \, dy \, dz$ and D is regarded as a function of the space coordinates x, y, z. If there is only one nonzero stress component, the specific damping energy D may be considered a function of the stress level σ. Then

$$D_9 = \int_0^{\sigma_d} D \, \frac{dV}{d\sigma} \, d\sigma \tag{36.5}$$

In this integration V is the volume of material whose stress level is less than σ. The integration is a single integral, and σ_d is the peak stress. The integrands may be put in dimensionless form by introducing D_d, the specific damping energy associated with the peak stress level reached anywhere in the specimen during the vibration (i.e., the value of D corresponding to $\sigma = \sigma_d$). Then

$$D_0 = D_d V_0 \alpha \tag{36.6}$$

where
$$\alpha = \int_0^1 \left(\frac{D}{D_d} \right) \frac{d(V/V_0)}{d(\sigma/\sigma_d)} \, d\left(\frac{\sigma}{\sigma_d} \right) \tag{36.7}$$

The average damping energy is

$$D_a = \frac{D_0}{V_0} = D_d \alpha \tag{36.8}$$

The relationship between the damping energies D_0, D_a, and D depends upon the dimensionless damping energy integral α. The integrand of α may be separated into two parts: (1) a damping function D/D_d which is a property of the material and (2) a volume-stress function $d(V/V_0)/d(\sigma/\sigma_d)$ which depends on the shape of the part and the stress distribution.

RELATIONSHIP BETWEEN SPECIFIC DAMPING ENERGY AND STRESS LEVEL

Before the damping function D/D_d can be determined, the specific damping energy D must be related to the stress level σ. Data of this type for typical engineering materials are given in Figs. 36.10 and 36.11. These results illustrate the fact that the damping-stress relationship for all materials cannot be expressed by one simple function. For a large number of structural materials in the low-intermediate stress region (up to 70 percent of σ_e the fatigue strength at 2×10^7 cycles), the following relationship is reasonably satisfactory:

$$D = J \left(\frac{\sigma}{\sigma_e} \right)^n \tag{36.9}$$

Values of the constants J and n are given in Table 36.5 and Fig. 36.10. In general, $n = 2.0$ to 3.0 in the low-intermediate stress region but may be much larger at high stress levels. Where Eq. (36.9) is not applicable, as in the high stress regions of Figs.

36.10 and 36.11 or in the case of the 403 steel alloy of Fig. 36.9, analytical expressions are impractical and a graphical approach is more suitable for computation of α.

VOLUME-STRESS FUNCTION

The volume-stress function (V/V_0) may be visualized by referring to the dimensionless volume-stress curves shown in Fig. 36.5. The variety of specimen types included in this figure [tension-compression member (1) to turbine blade (9)] is representative of those encountered in practice. These curves give the fraction of the total volume which is stressed below a certain fraction of the peak stress. In a torsion member, for example, 30 percent of the material is at a stress lower than 53 percent of the peak stress. The volume-stress curves for a part having a reasonably uniform stress, i.e., having most of its volume stressed near the maximum stress, are in the region of this diagram labeled H. By contrast, curves for parts having a large stress

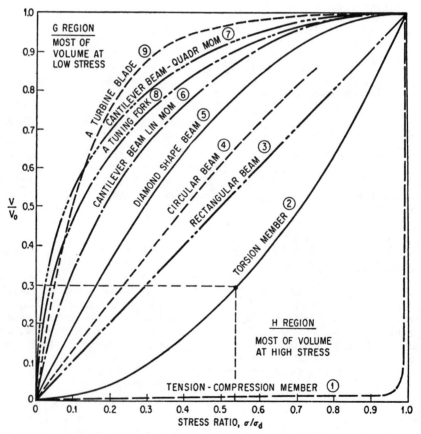

FIGURE 36.5   Volume-stress functions for various types of parts. (See Table 36.1 for additional details on parts.)

gradient (such as a notched beam in which very little volume is at the maximum stress and practically all of the volume is at a very low stress) are in the G region.

In order to illustrate representative values of α for several cases of engineering interest, the results of selected analytical and graphical computations[11] are summarized in Table 36.1 and in Fig. 36.6. In Fig. 36.6 the effect of the damping exponent n on the value of α for different types of representative specimens is illustrated. Note the wide range of α encountered for $n = 2.4$ (representative of many materials at low and intermediate stress) and for $n = 8$ (representative of materials at high stress, as shown in the next section).

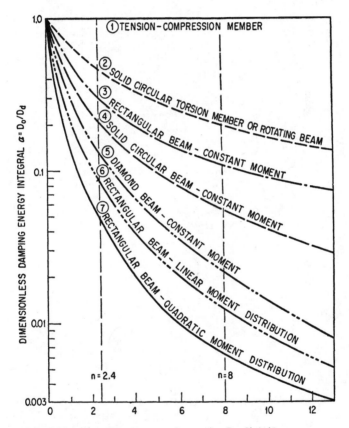

FIGURE 36.6 Damping exponent n in equation $D = J(\sigma/\sigma_e)^n$.

RATIO OF DAMPING ENERGY TO STRAIN ENERGY

Owing to the complexity of the sources of material damping, the use of relative damping energy units does not produce all the advantages which might otherwise be associated with a nondimensional quantity. One motivation for the use of such units, however, is their direct relation to several conventional damping tests. The *logarith-*

TABLE 36.1 Expressions and Values for α and β/α for Various Stress Distribution and Damping Functions

Type of specimen and loading as designated in Fig. 36.1	Volume-stress function V/V_0	Dimensionless damping energy integral α for various damping functions					Dimensionless strain energy integral β	β/α if $n=8$
		General case $D = f(\sigma)$	For special case $D = J(\sigma/\sigma_e)^n$					
			For any value of n	$n=2.4$	$n=8$			
1 Tension-compression member	1	1	1	1	1	1.0	1
2 Cylindrical torsion member or rotating beam	$\left(\dfrac{\sigma}{\sigma_d}\right)^2$	$\left[1+\dfrac{\sigma_d}{2D_0}\dfrac{dD_0}{d\sigma_d}\right]^{-1}$	$\dfrac{2}{n+2}$	0.45	0.20	0.5	2.5
3 Rectangular beam under uniform bending	$\dfrac{\sigma}{\sigma_d}$	$\left[1+\dfrac{\sigma_d}{D_0}\dfrac{dD_0}{d\sigma_d}\right]^{-1}$	$\dfrac{1}{n+1}$	0.29	0.11	0.33	3.0
4 Cylindrical beam under uniform bending	$\dfrac{2}{\pi}\left[\dfrac{\sigma}{\sigma_d}\sqrt{1-\left(\dfrac{\sigma}{\sigma_d}\right)^2}+\sin^{-1}\left(\dfrac{\sigma}{\sigma_d}\right)\right]$		$\dfrac{1}{\sqrt{\pi}}\dfrac{2}{n+2}\dfrac{\Gamma\left(\dfrac{n+1}{2}\right)}{\Gamma\left(\dfrac{n+2}{2}\right)}$	0.21	0.055	0.24	4.5
5 Diamond beam under uniform bending	$2\dfrac{\sigma}{\sigma_d}-\left(\dfrac{\sigma}{\sigma_d}\right)^2$	$\left[1+\dfrac{2\sigma_d}{D_0}\dfrac{dD_0}{d\sigma_d}+\dfrac{\sigma_d^2}{2D_0}\dfrac{d^2D_0}{d\sigma_d^2}\right]^{-1}$	$\dfrac{2}{n^2+3n+2}$	0.13	0.022	0.17	7.7
6 Rectangular beam having bending moment shown $M_x=\dfrac{x}{L}\,M_0$	$\dfrac{\sigma}{\sigma_d}\left[1-\log_e\dfrac{\sigma}{\sigma_d}\right]$	$\left[1+\dfrac{3\sigma_d}{D_0}\dfrac{dD_0}{d\sigma_d}+\dfrac{\sigma_d^2}{D_0}\dfrac{d^2D_0}{d\sigma_d^2}\right]^{-1}$	$\dfrac{1}{(n+1)^2}$	0.088	0.012	0.11	9.1
7 $M_x=\left(\dfrac{x}{L}\right)^2 M_0$	$2\sqrt{\dfrac{\sigma}{\sigma_d}}-\dfrac{\sigma}{\sigma_d}$	$\left[1+\dfrac{5\sigma_d}{D_0}\dfrac{dD_0}{d\sigma_d}+\dfrac{2\sigma_d^2}{D_0}\dfrac{d^2D_0}{d\sigma_d^2}\right]^{-1}$	$\dfrac{1}{2n^2+3n+1}$	0.051	0.0065	0.067	10.0
8 Tuning fork in bending	$K\dfrac{\sigma}{\sigma_d}\left[1-\log_e\dfrac{\sigma}{\sigma_d}\right]$	$K\left[1+\dfrac{3\sigma_d}{D_0}\dfrac{dD_0}{d\sigma_d}+\dfrac{\sigma_d^2}{D_0}\dfrac{d^2D_0}{d\sigma_d^2}\right]^{-1}$	$\dfrac{K}{(n+1)^2}$ For $K=0.8\rightarrow$	0.091	0.0099	0.089	9.0

Note: $\beta/\alpha = 1$ for all cases if $n = 2$.

mic decrement Δ is defined by Eq. (36.2). Other energy ratio units are tabulated and defined below. In this chapter, the energy ratio unit termed *loss factor* is used as the reference unit.

In defining the various energy ratio units, it is important to distinguish between loss factor η_s of a specimen or part (having variable stress distribution) and the loss factor η for a material (having uniform stress distribution). By definition the loss factor of a specimen (identified by subscript s) is

$$\eta_s = \frac{D_0}{2\pi W_0} \tag{36.10}$$

where the total damping D_0 in the specimen is given by Eq. (36.6). The total strain energy in the part is of the form

$$W_0 = \int_0^{V_0} \frac{1}{2}\left(\frac{\sigma^2}{E}\right) dV = \frac{1}{2}\left(\frac{\sigma_d^2}{E}\right) V_0 \beta \tag{36.11}$$

where E denotes a modulus of elasticity and β is a dimensionless integral whose value depends upon the volume-stress function and the stress distribution:

$$\beta = \int_0^1 \left(\frac{\sigma}{\sigma_d}\right)^2 \frac{d(V/V_0)}{d(\sigma/\sigma_d)} d\left(\frac{\sigma}{\sigma_d}\right) \tag{36.12}$$

On substituting Eq. (36.6) and Eq. (36.11) in Eq. (36.10), it follows that

$$\eta_s = \frac{E}{\pi} \frac{D_d}{\sigma_d^2} \frac{\alpha}{\beta} \tag{36.13}$$

If the specimen has a uniform stress distribution, $\alpha = \beta = 1$ and the specimen loss factor η_s becomes the material loss factor η; in general, however,

$$\eta = \frac{ED_d}{\pi\sigma_d^2} = \eta_s \frac{\beta}{\alpha} \tag{36.14}$$

Other energy ratio (or relative energy) damping units in common use are defined below:

For specimens with variable stress distribution:

$$\eta_s = (\tan \phi)_s = \frac{\Delta_s}{\pi} = \frac{\psi_s}{\pi} = \left(\frac{\delta\omega}{\omega_n}\right)_s = \frac{1}{(A_r)_s} = \frac{1}{Q_s} = \frac{ED_d}{\pi\sigma_d^2}\left(\frac{\alpha}{\beta}\right) \tag{36.15}$$

For materials or specimens with uniform stress distribution:

$$\eta = \tan \phi = \frac{\Delta}{\pi} = \frac{\psi}{\pi} = \frac{\delta\omega}{\omega_n} = \frac{1}{A_r} = \frac{1}{Q} = Q^{-1} = \frac{ED_d}{\pi\sigma_d^2} \tag{36.16}$$

where η = loss factor of material = dissipation factor (high loss factor signifies high damping)

$\tan \phi$ = loss angle, where ϕ is phase angle by which strain lags stress in sinusoidal loading

$\psi = \pi\eta$ = specific damping capacity

$\delta\omega/\omega_n$ = (bandwidth at half-power point)/(natural frequency) [see Eq. (36.3)]

A_r = resonance amplification factor [see Eq. (36.1)]

$Q = 1/\eta$ = measure of the sharpness of a resonance peak and amplification produced by resonance

The material properties are related to the specimen properties as follows:

$$\psi = \psi_s \frac{\beta}{\alpha} \qquad \Delta = \Delta_s \frac{\beta}{\alpha} \qquad A_r = (A_r)_s \frac{\alpha}{\beta} \qquad (36.17)$$

Thus, the various energy ratio units, as conventionally expressed for specimens, depend not only on the basic material properties D and E but also on β/α. The ratio β/α depends on the form of the damping-stress function and the stress distribution in the specimen. As in the case of average damping energy, D_a, the loss factor or the logarithmic decrement for specimens made from exactly the same material and exposed to the same stress range, frequency, temperature, and other test variables may vary significantly if the shape and stress distribution of the specimen are varied. Since data expressed as logarithmic decrement and similar energy ratio units reported in technical literature have been obtained on a variety of specimen types and stress distributions, any comparison of such data must be considered carefully. The ratio β/α may vary for specimens of exactly the same shape if made from materials having different damping-stress functions. For different specimens made of exactly the same materials, the variation in β/α also may be large, as shown in Fig. 36.7. For example, for a

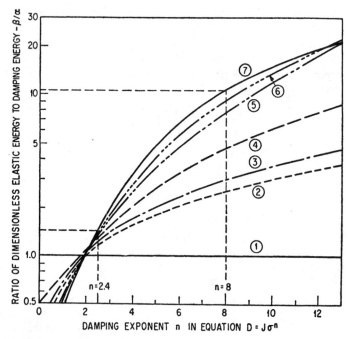

FIGURE 36.7 Effect of damping exponent n on ratio β/α for $D = J\sigma^n$. Curves are (1) tension-compression member; (2) solid circular torsion member or rotating beam; (3) rectangular beam–constant moment; (4) solid circular beam–constant moment; (5) diamond beam–constant moment; (6) rectangular beam–linear moment distribution; and (7) rectangular beam–quadratic moment distribution.

material and stress region for which damping exponent $n = 2.4$ (characteristic of metals at low and intermediate stress), the value of β/α shown in Table 36.1 varies from 1 for a tension-compression member to 1.6 for a rectangular beam with quadratic moment distribution. If $n = 8$ (characteristic of materials at high stress), the variation is from 1 to 10, and larger for beams with higher stress gradient.

It is possible, for a variety of types of beams, to separate the ratio β/α into two factors:[12] (1) a cross-sectional shape factor K_c which quantitatively expresses the effect of stress distribution on a cross section, and (2) a longitudinal stress distribution factor K_s which expresses the effect of stress distribution along the length of the beam. Then

$$\frac{\beta}{\alpha} = K_s K_c \qquad (36.18)$$

If material damping can be expressed as an exponential function of stress, as in Eq. (36.9), some significant generalizations can be made regarding the pronounced effect of the damping exponent n on each of these factors. Some of the results are shown in Fig. 36.8 for beams of constant cross-section. These curves indicate that high values of K_s and K_c are associated with a high damping exponent n, other factors being equal; K_c is high when very little material is near peak stress. For example, compare the diamond cross-section shape with the I beam, or compare the uniform stress beam with the cantilever.

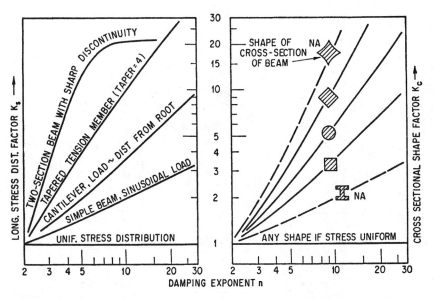

FIGURE 36.8 Effect of damping exponent n on longitudinal stress distribution factor and cross-sectional shape factor of selected examples.

In much of the literature of damping, the existence of factors α and β (or K_s and K_c) is not recognized; the unstated assumption is that $\alpha = \beta = 1$. As discussed above, this assumption is true only for specimens under homogeneous stress.

Relative damping units such as logarithmic decrement depend on the ratio of two energies, the damping energy and the strain energy. Since strain energy increases with the square of the stress for reasonably linear materials, the logarithmic decrement remains constant with stress level and is independent of specimen shape and stress distribution only for materials whose damping energy also increases as the square of the stress [$n = 2$ in Eq. (36.9)]. For most materials at working stresses, n varies between 2 and 3 (see Fig. 36.10) but for some (Fig. 36.9) it is highly variable. In the high stress region, n lies in the range 8.0–20.0 (Fig. 36.10). In view of the broad range of materials and stresses encountered in design, the case $n = 2$ must be regarded as exceptional. Thus, logarithmic decrement is a variable rather than a "material constant." Its magnitude generally decreases significantly with stress amplitude. When referring to specimens such as beams in which all stresses between zero and some maximum stress occur simultaneously, the logarithmic decrement is an ambiguous average value associated with some mean stress. Published data require careful analysis before suitable comparisons can be made.

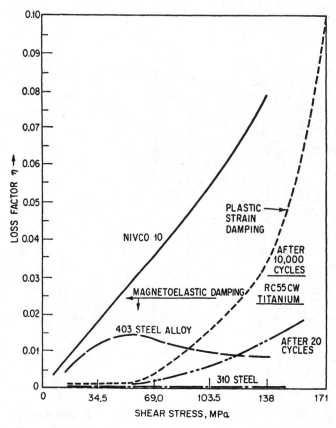

FIGURE 36.9 Comparison of internal friction and damping values for different inelastic mechanisms.

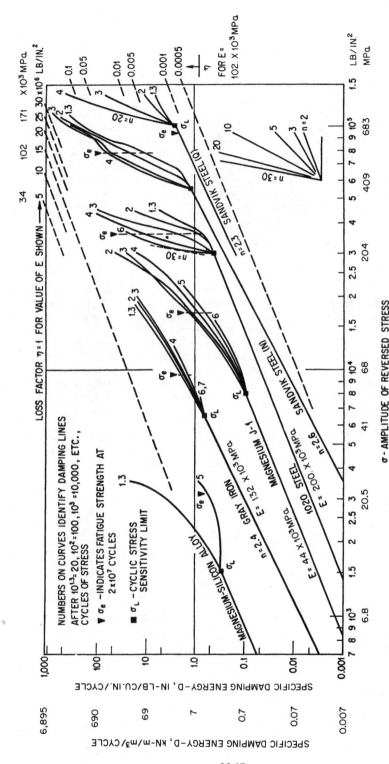

FIGURE 36.10 Specific damping energy of various materials as a function of amplitude of reversed stress and number of fatigue cycles. Number of cycles is 10 to power indicated on curve. For example, a curved marked 3 is for 10^3 or 1,000 cycles. *Note:* 6.895 kN·m/m^3 = 1 in.-lb/in.^{3} and 1 MPa = 10^3 N/m^2 = 10^{-3} kN/mm^2 = 146.5 lb/in.^{2}.

36.15

VISCOELASTIC MATERIALS

Some materials respond to load in a way that shows a pronounced influence of the rate of loading. Generally the strain is larger if the stress varies slowly than it is if the stress reaches its peak value swiftly. Among materials that exhibit this viscoelastic behavior are high polymers and metals at elevated temperatures as well as many glasses, rubbers, and plastics.[13] As might be expected, these materials usually also exhibit creep, an increasing deformation under constant applied load.

When a sinusoidal exciting force is applied to a viscoelastic solid, the strain is observed to lag behind the stress. The phase angle between them, denoted φ, is the *loss angle*. The stress may be separated into two components, one in phase with the strain and one leading it by a quarter cycle. The magnitudes of these components depend upon the material and upon the exciting frequency, ω. For a specimen subject to homogeneous shear ($\alpha = \beta = 1$),

$$\gamma = \gamma_0 \sin \omega t \tag{36.19}$$

$$\sigma = \gamma_0 \left[G'(\omega) \sin \omega t + G''(\omega) \cos \omega t \right] \tag{36.20}$$

This is a linear viscoelastic stress-strain law. The theory of linear viscoelasticity is the most thoroughly developed of viscoelastic theories. In Eq. (36.20) $G'(\omega)$ is known as the "storage modulus in shear" and $G''(\omega)$ is the "loss modulus in shear" (the symbols G_1 and G_2 are also widely used in the literature). The stiffness of the material depends on G' and the damping capacity on G''. In terms of these quantities the loss angle $\varphi = \tan^{-1}(G''/G')$. The *complex*, or *resultant*, modulus in shear is $G^* = G' + iG''$. In questions of stress analysis, complex moduli have the advantage that the form of Hooke's law is the same as in the elastic case except that the elastic constants are replaced by the corresponding complex moduli. Then a correspondence principle often makes it possible to adapt an existing elastic solution to the viscoelastic case. For details of viscoelastic stress analysis see Refs. 13 and 31.

The moduli of linear viscoelasticity are readily related to the specific damping energy D introduced previously. For a specimen in homogeneous shear of peak magnitude γ_0, the energy dissipated per cycle and per unit volume is

$$D = \int_0^{2\pi/\omega} \sigma \left(\frac{d\gamma}{dt} \right) dt \tag{36.21}$$

In view of Eqs. (36.19) and (36.20) this becomes

$$D = \int_0^{2\pi/\omega} \gamma_0^2 \omega (G' \sin \omega t + G'' \cos \omega t) \cos \omega t \, dt$$

$$= \pi \gamma_0^2 G''(\omega) \tag{36.22}$$

It is apparent from Eq. (36.22) that linear viscoelastic materials take the coefficient $n = 2$ in Eq. (36.9). These materials differ from metals, however, by having damping capacities that are strongly frequency- and temperature-sensitive.[1]

DAMPING PROPERTIES OF MATERIALS

The specific damping energy D dissipated in a material exposed to cyclic stress is affected by many factors. Some of the more important are:

1. Condition of the material
   a. In virgin state:
      Chemical composition; constitution (or structure) due to thermal and mechanical treatment; inhomogeneity effects
   b. During and after exposure to pretreatment, test, or service condition: Effect of stress and temperature histories on aging, precipitation, and other metallurgical solid-state transformations
2. State of internal stress
   a. Initially, due to surface-finishing operations (shot peening, rolling, case hardening)
   b. Changes caused by stress and temperature histories during test or service
3. Stress imposed by test or service conditions
   a. Type of stress (tension, compression, bending, shear, torsion)
   b. State of stress (uniaxial, biaxial, or triaxial)
   c. Stress-magnitude parameters, including mean stress and alternating components; loading spectrum if stress amplitude is not constant
   d. Characteristics of stress variations including frequency and waveform
   e. Environmental conditions:
      Temperature (magnitude and variation) and the surrounding medium and its (corrosive, erosive, and chemical) effects

Factors tabulated above, such as stress magnitude, history, and frequency, may be significant at one stress level or test condition and unimportant at another. The deformation mechanism that is operative governs sensitivity to the various factors tabulated.

Many types of inelastic mechanisms and hysteretic phenomena have been identified, as shown in Table 36.2. The various damping phenomena and mechanisms may be classified under two main headings: *dynamic hysteresis* and *static hysteresis*.

Materials which display dynamic hysteresis (sometimes identified as viscoelastic, rheological, and rate-dependent hysteresis) have stress-strain laws which are describable by a differential equation containing stress, strain, and time derivatives of stress or strain. This differential equation need not be linear, though, to avoid mathematical complexity, much of existing theory is based on the linear viscoelastic law described in the previous section. One important type of dynamic hysteresis, a special case identified as *anelasticity*[14, 15] or *internal friction,* produces no permanent set after a long time. This means that if the load is suddenly removed at point *B* in Fig. 36.1, after cycle *OAB,* strain *OB* will gradually reduce to zero as the specimen recovers (or creeps negatively) from point *B* to point *O.*

A distinguishing characteristic of anelasticity and the more general case of viscoelastic damping is its dependence on time-derivative terms. The hysteresis loops tend to be elliptical in shape rather than pointed as in Fig. 36.1. Furthermore, the loop area is definitely related to the dynamic or cyclic nature of the loading and the area of the loop is dependent on frequency. In fact, the stress-strain curve for an ideally viscoelastic material becomes a single-valued curve (no hysteretic loop) if the cyclic stress is applied slowly enough to allow the material to be in complete equilibrium at all times (oscillation period very much longer than relaxation times). No hysteretic damping is produced by these mechanisms if the material is subjected to essentially static loading. Stated differently, the static hysteresis is zero.

Static hysteresis, by contrast, involves stress-strain laws which are insensitive to time, strain, or stress rate. The equilibrium value of strain is attained almost instantly for each value of stress and prior stress history (direction of loading, amplitudes, etc.), independent of loading rate. Hysteresis loops are pointed, as shown in Fig. 36.1, and if the stress is reduced to zero (point *B*) after cycle *OAB,* then *OB* remains as a

TABLE 36.2 Classification of Types of Hysteretic Damping of Materials

	Types of material damping		
	Dynamic hysteresis	**Static hysteresis**	
Name used here	Dynamic hysteresis	Static hysteresis	
Other names	Viscoelastic, rheological, and rate-dependent hysteresis	Plastic, plastic flow, plastic strain, and rate-independent hysteresis	
Nature of stress-strain laws	Essentially linear. Differential equation involving stress, strain, and their time derivatives	Essentially nonlinear, but excludes time derivatives of stress or strain	
Special cases and description	*Anelasticity.* Special because no permanent set after sufficient time. Called "internal friction"		
Simplest representative mechanical model	VOIGT UNIT MAXWELL UNIT ANELASTICITY		
Frequency dependence	Critically at relaxation peaks	No, unless other mechanisms present	
Primary mechanisms	Solute atoms, grain boundaries. Micro- and macro-thermal and eddy currents. Molecular curling and uncurling in polymers.	Magnetoelasticity	Plastic strain
Value of n in $D = JS^n$	2	3—up to coercive force	2–3 up to σ_L; 2 to >30 above σ_L
Variation of η with stress	No change, since $n - 2 = 0$	Proportional to σ since $n - 2 = 1$	Small increase up to σ_L; Large increase above σ_L
Typical values for η	Anelasticity: <0.001 to 0.01; Viscoelasticity: <0.1 to >1.5	0.01 to 0.08	0.001 to 0.05 up to σ_L; 0.001 to >0.1 above σ_L
Stress range of engineering importance	Anelasticity—low stress; Viscoelasticity—all stresses	Low and medium. Sometimes high	Medium and high stress
Effect of fatigue cycles	No effect	No effect	No effect up to σ_L; Large changes above σ_L
Effect of temperature	Critical effects near relaxation peaks	Damping disappears at Curie temperature	Mixed. Depends on type of comparison
Effect of static preload		Large reduction for small coercive force	Either little effect or increase

permanent set or residual deformation. The two principal mechanisms which lead to static hysteresis are magnetostriction and plastic strain.

Table 36.2 also shows the simplest representative mechanical models for each of the behaviors classified. In these models k is a spring having linear elasticity (linear and single-valued stress-strain curve), C is a linear dashpot which produces a resisting force proportional to velocity, and D is a Coulomb friction unit which produces a constant force whenever slip occurs within the unit, the direction of the force being opposite to the direction of relative motion. More sophisticated models have been found to predict reliably the behavior of some materials, particularly polymeric materials.

Any one of the mechanisms to be discussed may dominate, depending on the stress level. For convenience, *low stress* is defined here as a (tension-compression) stress less than 1 percent of the fatigue limit; *intermediate* stress levels are those between 1 percent and 50 percent of the fatigue limit of the material; and *high* stress levels are those exceeding 50 percent of the fatigue limit.

DYNAMIC HYSTERESIS OF VISCOELASTIC MATERIALS

The linearity limits of a variety of plastics and rubbers are summarized in Table 36.3. While the stress limits are of the same order of magnitude for plastics and rubbers, the strain limits are much smaller for the former class of materials. Within these limits the dynamic storage and loss moduli of linear viscoelasticity may be used.

One distinguishing characteristic of the dynamic behavior of viscoelastic materials is a strong dependence on temperature and frequency.[1,16] At high frequencies (or low temperature) the storage modulus is large, the loss modulus is small, and the

TABLE 36.3 Linearity Limits for a Variety of Plastics and Rubber

Material	Stress limit in creep, MPa	Strain limit in relaxation
Polymethylmethacrylate	10	
Polystyrene	5	
Plasticized polyvinyl chloride	1	0.1–1.0%
Polythene	12	
Phenolic resins	10	
Polyisobutylene		50%
Natural rubber	1–10	100%
GR-S		100%

Note: 1 MPa = 10^6 N/m^2 = 146.5 lb/in.^{2}.

behavior resembles that of a stiff ideal material. This is known as the "glassy" region in which the "molecular curling and uncurling" cannot occur rapidly enough to follow the stress. Thus the material behaves essentially "elastically." At low frequencies (or high temperature) the storage modulus and the loss modulus are both small. This is the "rubbery" region in which the molecular curling and uncurling follow the stress in phase, resulting in an equilibrium condition not conducive to energy dissipation. At intermediate frequencies and temperatures there is a "transition" region in which the loss modulus is largest. In this region the molecular curling and uncurling is out of phase with the cyclic stress and the resulting lag in the cyclic strain provides a mechanism for dissipating damping energy. The loss factor also shows a peak in this region although at a somewhat lower frequency than the peak in G''. Since the damping energy is proportional to G'', the specific damping curve also has its maximum in the transition region. Most engineering problems involving vibration are associated with the transition and glassy regions. In Table 36.4, values of G' and G'' are given for a variety of rubbers and plastics. References 17 to 20 contain additional useful information.

Metals at low stress exhibit certain properties that constitute dynamic hysteresis effects. Peaks are observed in curves of loss factors vs. frequency of excitation. For example, under conditions that maximize the internal friction associated with grain

TABLE 36.4 Typical Moduli of Viscoelastic Materials

(Two values are given: the upper value is G'; *the bottom value is* G". *Moduli units are megapascals,* MPa)

Material	Frequency, Hz				Temperature, °C
	10	100	1000	4000	
Polyisobutylene		0.512	1.31	2.36	−60–100
		0.410	1.76	4.50	
M 169A Butyl gum		0.480	1.40	2.70	21–65
		0.502	1.32	2.88	
Du Pont fluoro rubber,		2.00	4.54	7.93	0–100
(Viton A)		1.60	8.41	27.0	
Silicon rubber gum		0.05	0.08		21–65
		0.02	0.04		
Natural rubber		0.33	0.50		25
		0.02	0.02		
3M tape No. 466		0.81	2.52	15.3	25
(adhesive)		0.95	4.59	13.0	
3M tape No. 435		0.28	0.55	0.87	−40–60
(sound damping tape)		0.16	0.37	0.63	
Natural rubber	3.91	4.91			−30–75
(tread stock)	0.68	0.97			
Thiokol M-5	7.86	8.34			−30–75
	3.91	10.27			
Natural gum	0.73				−30–75
(tread stock)	0.07				
Filled silicone rubber		2.00	2.50	3.41	21–65
		0.26	0.44	0.58	
Polyvinyl chloride		1.26	3.20	6.60	21–65
acetate		1.44	2.32	5.78	
X7 Polymerized tung oil			17.0	39.0	21–65
with polyoxane liquid			9.45	20.8	
Du Pont X7775 pyralin	4.50	12.0	45.0		−45–100
	2.51	9.45	28.3		
Polyvinyl butyral	30.0	200.0	600.0		−45–100
	3.1	12.5	37.6		
Polyvinyl chloride with		0.35	0.65		
dimethyl thianthrene		0.21	0.97		

Note: 1 MPa = 10^6 N/m^2 = 10^{-3} kN/mm^2 = 146.5 lb/in.^{2}.

boundary effects polycrystalline aluminum will display a loss factor peak as high as $\eta = 0.09$. But for most metals, the peak values are less than 0.01. Although the rheological properties of metals at low stress can be described in terms of anelastic properties (rheology without permanent set), a more general approach which includes provisions for permanent set is required to specify the rheological properties of metals at high stress. This approach is best described in terms of static hysteresis.

STATIC HYSTERESIS

The metals used in engineering practice exhibit little internal damping at low stress levels. At intermediate and high stress levels, however, magnetostriction and plastic strain can introduce appreciable damping. The former effect is considered first.

Ferromagnetic metals have significantly higher damping at intermediate stress levels than do nonferromagnetic metals. This is because of the rotation of the magnetic domain vectors produced by the alternating stress field. If the specimen is magnetized to saturation, most of the damping disappears, indicating that it was due primarily to magnetoelastic hysteresis. Figure 36.9 shows the loss factor for three metals, each heat-treated for maximum damping. The damping of 403 steel (ferromagnetic material with 12% Cr and 5% Ni) is much higher than that of 310 steel (nonferromagnetic with 25% Cr and 20% Ni). Most structural metals at low and intermediate stress exhibit loss factors in the general range of 310 steel until the hysteresis produced by plastic strain becomes significant. The alloy Nivco 10[21] (approximately 72% Co and 23% Ni), developed to take maximum advantage of magnetoelastic hysteresis, displays significantly larger damping than other metals.

The damping energy dissipated by magnetoelastic hysteresis increases as the third power of stress up to a stress level governed by the magnetomechanical coercive force; thus, the loss factor should increase linearly with stress. Nivco 10 follows this relationship for the entire range of stress shown in Fig. 36.9. Beyond an alternating stress governed by the magnetomechanical coercive force, i.e., beyond approximately 34.5 MPa (5,000 lb/in.^{2}) for the 403 steel, the damping energy dissipated becomes constant. Since the elastic energy W_0 continues to increase as the square of the alternating stress, the value of loss factor (ratio of the two energies) decreases with the inverse square of stress. The curve for 403 steel in Fig. 36.9 at stresses between 62 MPa (9,000 lb/in.^{2}) and 103 MPa (15,000 lb/in.^{2}) demonstrates this behavior.

Magnetoelastic damping is independent of excitation frequency, at least in the frequency range that is of engineering interest. Magnetoelastic damping decreases only slightly with increasing temperature until the Curie temperature is reached, when it decreases rapidly to zero. Static stress superposed on alternating stress reduces magnetoelastic damping.[21,22]

It is not entirely clear at this time what mechanisms are encompassed by the terms plastic strain, localized plastic deformation, crystal plasticity, and plastic flow in a range of stress within the apparent elastic limit. On the microscopic scale, the inhomogeneity of stress distribution within crystals and the stress concentration at crystal boundary intersections produce local stress high enough to cause local plastic strain, even though the average (macroscopic) stress may be very low. The number and volume of local sites so affected probably increase rapidly with stress amplitude, particularly at stresses approaching the fatigue limit of a material. On the submicroscopic scale, the role of dislocations, their kind, number, dispersion, and lattice anchorage in the deformation process still remains to be determined. The processes involved in these various inelastic behaviors may be included under the general term "plastic strain."

At small and intermediate stress, the damping caused by plastic strain is small, probably of the same order as some of the internal friction peaks discussed previously and much smaller than magnetoelastic damping in many materials. In this stress region, damping generally is not affected by stress or strain history. However, as the stress is increased, the plastic strain mechanism becomes increasingly important and at stresses approaching the fatigue limit it begins to dominate as a damping mechanism. This is shown by the curves for titanium in Fig. 36.9.[22] In the region of high stress, microstructural changes and metallurgical instability appear to be initiated and promoted by cyclic stress. This occurs even though the stress amplitude may lie below the apparent elastic limit (that observed by conventional methods) and the fatigue limit of the material. This means that damping in the high stress region is a function not only of stress amplitude but also of stress history.

In Fig. 36.9, for example, the lower of the two curves for titanium indicates the damping of the virgin specimen and the upper curve gives the damping after 10,000 stress cycles.

The general position as regards stress history is given in Fig. 36.10. Below a certain peak stress, σ_L, known as the "cyclic stress sensitivity limit," the curve of damping vs. stress is a straight line on a log-log plot and displays no stress-history effect. The limit stress σ_L usually falls somewhat below the fatigue strength of the material. Above σ_L, stress-history effects appear; the curve labeled 1.3 indicates the damping energy after $10^{1.3} = 20$ cycles and the curve labeled 6 after 10^6 or 1 million cycles. To facilitate comparisons between the reference damping units, loss factor η and D under uniform stress ($\alpha/\beta = 1$), the loss factor also is plotted in Fig. 36.10. Since the relationship between D and η depends on the value of Young's modulus of elasticity E, a family of lines for the range of $E = 34 \times 10^3$ to 205.0×10^3 MPa (5×10^6 to 30×10^6 lb/in.^{2}) is shown for $\eta = 1$. The lines for the other values of η correspond to a value of $E = 102.0 \times 10^3$ MPa (15×10^6 lb/in.^{2}).

TABLE 36.5 Static, Hysteretic, Elastic, and

	Static properties			Fatigue behavior		
Material*	Modulus of elasticity E, MPa 10^{-4}	Yield stress (0.2% offset), MPa	Tensile strength, MPa	Fatigue strength, σ_e, MPa	Cyclic stress sensitivity limit σ_L, MPa	Stress ratio σ_L/σ_e
N-155 (superalloy)	20.	410.	810.	360.	220.	0.62
Lapelloy (superalloy)	22.	764.	880.	490.	490.	1.00
Lapelloy (480°C)	17.5			270.	310.	1.14
RC 130B (titanium)	11.5	950.	1,040.	590.	650.	1.10
RC 130B (320°C)	9.9			430.	340.	0.81
Sandvik (O & T) steel	19.9	1,210.	1,400.	630.	680.	1.09
SAE 1020 steel	20.1	320.	490.	240.	200.	0.85
Gray iron	13.2		140.	65.	44.	0.69
24S-T4 aluminum	7.2	330.	500.	180.	160.	0.88
J-1 magnesium	4.4	230.	310.	120.	55.	0.47
Manganese-copper alloy		410.	610.	130.	120.	0.95

Note: 1 MPa = 10^6 N/m^2 = 146 lb/in.^{2}.
(Includes test temperature if above room temperature.)

COMPARISON OF VARIOUS MATERIAL DAMPING MECHANISMS AND REPRESENTATIVE DATA FOR ENGINEERING MATERIALS

The general qualitative characteristics of the various types of damping are summarized in Table 36.2 by comparing the effects of different testing variables. The data tabulated indicate that, in general, anelastic mechanisms do not contribute significantly to total damping at intermediate and high stresses; in these regions magnetoelastic and plastic strain mechanisms probably are the most important from an engineering viewpoint.

Damping vs. stress ratio data have been determined for a variety of common structural materials at various temperatures.[2,4] Some of these data are listed in Table 36.5 (all tests at 0.33 Hz). For a large variety of structural materials (not particularly selected for large magnetoelastic or plastic strain damping), the data are found to lie within a fairly well-established band shown in Fig. 36.11. The approximate geometric-mean curve is shown. Up to the fatigue limit, that is up to $\sigma_d = \sigma_e$, the specific damping energy D is given with sufficient accuracy by the expression

$$D = J \left(\frac{\sigma}{\sigma_e} \right)^{2.4} \tag{36.23}$$

where $J = 6.8 \times 10^{-3}$

if D is expressed in SI units of MN·m/m^3/cycle. The value of

$$J = 1.0$$

if D is expressed in units of in.-lb/in.^{3}/cycle.

Fatigue Properties of a Variety of Metals

				Damping properties, kN·m/m³/cycle			
$D = J \left(\dfrac{\sigma}{\sigma_e} \right)^n$, $\sigma \le \sigma_L$				D at $\sigma/\sigma_e = 1$		D at $\sigma/\sigma_e = 1.2$	
J	n, dimensionless	D $\dfrac{\sigma}{\sigma_L} = 1$	D $\dfrac{\sigma}{\sigma_e} = 0.6$	After $10^{1.3}$ cycles	After 10^6 cycles	After $10^{1.3}$ cycles	Maximum number of cycles
8.8	2.5	2.7	2.7	310.	170.	1,230.	1,500.
30.*	2.4*	10.9	4.0	11.	11.	55.	170.
24.	2.2	34.	8.2	26.	26.	41.	48.
14.	2.0	14.	4.4	12.	12.	18.	24.
17.	1.9	12.	6.1	13.	34.	30.	170.
16.	2.3	19.	5.5	16.	16.	31.	200.
4.3	2.0	3.1	1.6	4.5	140.	34.	680.
12.	2.4	4.5	3.4	14.	8.2	22.	16.
3.9	2.0	3.0	1.4	6.8	4.1	6.	15.
3.1	2.0	0.7	0.9	7.5	3.4	24.	7.
96.	2.8	82.	22.	89.	89.	170.	140.

Note: 1 kN·m/m^3/cycle = 0.146 in.-lb/in.^{3}/cycle.
* Up to $\sigma = 96$ MPa (14,000 lb/in.^{2}); at $\sigma = 204$ MPa (30,000 lb/in.^{2}) $n = 1.5$.

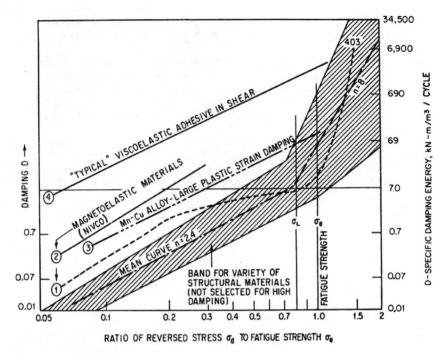

FIGURE 36.11 Range of damping properties for a variety of structural materials. The shaded band defines the damping for most structural materials. 1 kN·m/m³ = 0.146 in.-lb/in.³.

The approximate bandwidth about this geometric mean curve for the various structural materials included in the band is as follows: from ⅓ to 3 times the mean value at a stress ratio of 0.2 or less; from ⅕ to 5 times at a ratio of 0.6; from ¹⁄₁₀ to 10 times at a ratio of 1.0.

Also shown in Fig. 36.11 for comparison purposes are data for four materials having especially high damping. Materials 1 and 2 are the magnetoelastic alloys Nivco 10 and 403. Nivco 10 retains its high damping up to the stresses shown (data not available at higher stresses). However, the 403 alloy reaches its magnetoelastic peak at a stress ratio of approximately 0.2 and increases less rapidly beyond this point; when plastic strain damping becomes dominant (at stress ratio of approximately 0.8), damping increases very rapidly. By contrast, material 3, a manganese-copper alloy with large plastic strain damping, retains its high damping up to and beyond its fatigue strength.[23] Material 4 is a "typical" viscoelastic adhesive ($G'' = 0.95$ MPa = 138 lb/in.²), assuming that the permissible cyclic shear strain is unity (experiments show that a shear strain of unity does not cause deterioration in this adhesive even after millions of cycles).[24] The magnetoelastic material has a damping thirty times as large as the average structural material in the stress range shown in Fig. 36.11, and the viscoelastic damping is over ten times as large as the magnetoelastic damping.

The range of D observed for common structural materials stressed at their fatigue limit is 0.003 to 0.7 MN·m/m³/cycle with a mean value of 0.05 (0.5 to 100 in.-lb/in.³/cycle with a mean value of 7). For materials stressed at a rate of 60 Hz under uniform stress distribution (tension-compression), 16.4 cm^3 (1 in.^{3}) of a typical material will safely absorb and dissipate 48 watts (0.064 hp). Some high damping materials can absorb almost 746 watts (1 hp) in the safe-stress range, assuming no significant frequency or stress-history effects.[25-27]

SLIP DAMPING

In some cases the hysteretic damping in a structural material is sufficient to keep resonant vibration stresses within reasonable limits. However, in many engineering designs, material damping is too small and structural damping must be considered. A structural damping mechanism which offers excellent potential for large energy dissipation is that associated with the interface shear at a structural joint.

The initial studies[28-30] on interface shear damping considered the case of Coulomb or dry friction. Under optimum pressure and geometry conditions, very large energy dissipation is possible at a joint interface. However, the application of the general concepts of optimum Coulomb interface damping to engineering structures introduces two new problems. First, if the configuration is optimum for maximum Coulomb damping, the resulting slip can lead to serious corrosion due to chafing; this may be worse than the high resonance amplification associated with small damping. Second, for many types of design configurations, the interface pressure or other design parameters must be carefully optimized initially and then accurately maintained during service; otherwise, a small shift from optimum conditions may lead to a pronounced reduction in total damping of the configuration. Since it usually is difficult to maintain optimum pressure, particularly under fretting conditions, other types of interface treatment have been developed. One approach is to lubricate the interface surfaces. However, the maintenance of a lubricated surface often is difficult, particularly under the large normal pressure and shear sometimes necessary for high damping. Therefore, a more satisfactory form of interface treatment is an adhesive separator placed between mating surfaces at an interface. The function of the separating adhesive layer is to distort in shear and thus to dissipate energy with no significant Coulomb friction or sliding and therefore no fretting corrosion. The design of such layers is discussed in Chap. 37.

DAMPING BY SLIDING

The nature of interface shear damping can be explained by considering the behavior of two machine parts or structural elements which have been clamped together. The clamping force, whether it is the result of externally applied loads, of accelerations present in high-speed rotating machinery, or of a press fit, produces an interface common to the two parts. If an additional exciting force F_g is now gradually imposed, the two parts at first react as a single elastic body. There is shear on the interface, but not enough to produce relative slip at any point. As F_g increases in magnitude, the resulting shearing traction at some places on the interface exceeds the limiting value permitted by the friction characteristics of the two mating surfaces. In these regions microscopic slip of adjacent points on opposite sides of the interface occurs. As a result, mechanical energy is converted into heat. If the mechanical energy is energy of free or forced vibration, damping occurs. The slipped region is local and does not, in general, extend over the entire interface. If it does extend over the entire interface, gross slip is said to occur. This usually is prevented by the geometry of the system.

The force-displacement relationship for systems with interface shear damping is shown in Fig. 36.12. Since there are many displacements which can be measured, the displacement which corresponds to the exciting force which acts on the system is taken as a basis. Then the product of displacement and exciting force, integrated over a complete cycle, is the work done by the exciting force and absorbed by the structural element. As shown in Fig. 36.13, there is an initial linear phase OP during which behavior is entirely elastic. This is followed, in general, by a nonlinear transition phase PB during which slip progresses across the contact area. The phase PB is nonlinear, not because of any plastic behavior, but simply because the specimen is changing in stiffness as slip progresses. After the nonlinear phase PB, there may be a second linear phase BC during which slip is present over the entire interface. The existence of such a phase requires some geometric constraint which prevents gross motion even after slip has progressed over the entire contact area. If no such constraint is provided, F_g cannot be allowed to exceed the value corresponding to point B. If it should exceed this gross value, slip would occur.

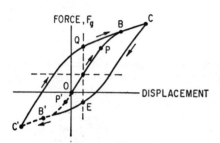

FIGURE 36.12 Force-displacement hysteresis loop under Coulomb friction.

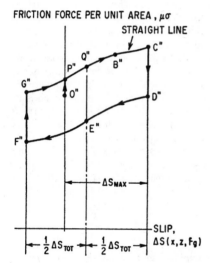

FIGURE 36.13 Friction force-slip relationship under Coulomb friction.

If the clamping force itself does not produce any shear on the interface and if the exciting force does not affect the clamping pressure, the force-displacement curve is symmetrical about the origin O. These conditions are at least approximately fulfilled in many cases. If they are not fulfilled, the exciting force in one direction initiates slip at a different magnitude of load than the exciting force in the opposite direction. This is the case pictured in Fig. 36.12. With negative exciting force, slip is initiated at P' which corresponds to a force of considerably smaller magnitude than point P. However, the force-displacement curve is always symmetrical about the mid-point of PP' (intersection of dotted lines in Fig. 36.12).

The force-displacement curve has been followed from point O to point C. If now a reduction in the exciting force occurs, the curve proceeds from C in a direction parallel to its initial elastic phase. Eventually, as unloading proceeds, slip is initiated again. Its sense is now opposite to that which was produced by positive force. The curve continues to point B', where slip is complete, and then along a linear stretch to C', where the exciting force has its largest negative value. As the force reverses, the curve becomes again linear and parallel to OP. Slip eventually occurs again and covers the interface at B. The hysteresis loop is closed at C.

The energy dissipated in local slip can be found by computing the area enclosed by the force-displacement hysteresis loop. It usually is simpler, however, to determine the energy loss at a typical location on the interface by analysis, and then to integrate over the area of the interface. In this mode of procedure, interest centers on the frictional force per unit area $\mu\sigma$ and the relative displacement Δs of initially adjacent points on opposite sides of the interface.

The so-called "slip-curve" illustrating the relationship between $\mu\sigma$ and Δs is shown in Fig. 36.13. Before the exciting force is applied, conditions are represented by point O'' which corresponds to point O in Fig. 36.12. The initial elastic phase during which there is no slip is represented by $O''P''$ (note that the normal pressure σ may change during this phase). The phase during which slip occurs only over part of the interface is represented by the curved line $P''B''$; it corresponds to PB in Fig. 36.12. After slip has progressed over the entire interface, the normal force vs. relative-displacement relation is linear. This phase is represented by the curve $B''C''$ in Fig. 36.13 and by BC in Fig. 36.12. When the exciting force has reached its maximum value, a second nonslip phase $C''D''$ ensues. This is followed by slip along the curve $D''E''F''$ until the exciting force reaches its maximum negative value. As the exciting force completes its period, there is a nonslip phase $F''G''$ followed by slip along $G''C''$. The lengths $C''D''$ and $F''G''$ are equal and the curves $G''C''$ and $D''F''$ are congruent (F'' corresponds to C'' and D'' corresponds to G'').

If the point in question is at an element of area $dx\, dz$ of the xz interface, the energy dissipated in slip is proportional to the area enclosed by the slip curve. Because of the congruence of the curved portions of the diagram and the parallelism of the linear portions, this area can be expressed in terms of the total slip and the pressures at two instants during the loading cycle. Integrating over the entire interface,

$$D_0 = -\mu \int \int [\sigma(E'') + \sigma(Q'')] \, \Delta s_{tot} \, dx \, dy \qquad (36.24)$$

In this expression, the parameters $\sigma(E'')$ and $\sigma(Q'')$ and the total slip Δs_{tot} are functions of x and z. They are the normal stresses at points E'' and Q'' of Fig. 36.13, located midway between the vertical lines $G''F''$ and $C''D''$. Since the pressures σ are always compressive (negative) and the total slip is always taken as a positive quantity, the negative sign is required to ensure a positive energy dissipation. Equation (36.24) is of little engineering value in itself because the stresses are functions of F_g as well as of x and z. In many of the problems which are of design interest, however, the shear on the interface is produced primarily by the exciting force and not by the initial clamping pressure. Conversely, the clamping pressure is not greatly affected by the addition of the time-varying exciting force. Under these circumstances, the slip curve of Fig. 36.13, like the force-displacement curve of Fig. 36.12, is symmetric about the point O''. Points P'' and Q'' then coincide, and the mean ordinate of the slip curve is that corresponding to point O''. Then Eq. (36.24) reduces to

$$D_0 = -4\mu \int \int \sigma(O'') \, \Delta s_{max} \, dx \, dz \qquad (36.25)$$

where $\sigma(O'')$ is the clamping stress corresponding to zero exciting force. It may be determined by any of the well-known methods of stress analysis. In most cases $\sigma(O'')$ can be determined without any reference to the existence of an interface. The term Δs_{max} represents the arc length of the maximum relative displacement, the so-called "scratch path." It is a function of the maximum value of F_g as well as of position on the interface. It may be inferred from Eq. (36.25) that energy dissipation due to interface shear is small both at very low clamping pressures and at very high ones.

In the former case, $\sigma(O'') = 0$; in the latter case, $\Delta s_{max} = 0$. It follows that for any distribution of clamping pressure there is an optimum intensity of clamping force at which the energy dissipation due to interface shear is a maximum. The maintenance of this optimum pressure is essential to the utilization of this form of damping. From the shape of the force-displacement curve $OPBC$ shown in Fig. 36.12, it is evident that systems in which interface shear damping plays a significant role behave like softening springs. This means that instability and jump phenomena may occur at frequencies below the nominal resonant frequency.

In the case of plane stress, the thickness of the material is t and Eq. (36.25) becomes

$$D_0 = -4\mu t \int \sigma(O'') \, \Delta s_{max} \, dx \qquad (36.26)$$

The slip can be related to stress through Hooke's law:

$$\Delta s = E^{-1} \int (\Delta \sigma_x) \, dx \qquad (36.27)$$

This indicates that any discontinuity in displacement is associated with a discontinuity in the component of stress parallel to the interface. These displacement discontinuities due to slip are members of a class of generalized dislocations whose existence has been demonstrated theoretically.[32] If Eq. (36.27) is substituted in Eq. (36.26), the energy dissipation can be expressed in terms of stress alone:

$$D_0 = -4\mu E^{-1}t \int_0^l \sigma(O'') \left[\int_0^x (\Delta\sigma_x)_{max} \, dx' \right] dx \qquad (36.28)$$

The computation of energy dissipation per cycle D_0 is the first step in the prediction of the dynamic amplification factor to be expected in service. For interface shear damping, an elementary theory permits the dynamic amplification factor to be estimated even though the system behavior is nonlinear. The technique employs an averaging method. Denoting the displacement corresponding to the exciting force by the symbol v,

$$v = v_d \cos \omega t \qquad \text{and} \qquad F_g = F_m \cos (\omega t + \varphi) \qquad (36.29)$$

where v_d is the peak dynamic displacement, F_m is the peak exciting force, and φ is the loss angle. One relationship between these quantities is obtained by making the average value of the virtual work vanish during each half-cycle of the steady-state forced vibration:

$$\int_0^{\pi/\omega} [mv + kv - F_g] \cos \omega t \, dt = 0 \qquad (36.30)$$

In this integration, the stiffness k changes as slip progresses across the interface. If the hysteresis loop of Fig. 36.12 is replaced by a parallelogram, only two phases, elastic and fully slipped, need be considered. Denoting the stiffness (i.e., the ratio of exciting force to displacement) in the unslipped condition by the symbol k_e and the reduced stiffness in the fully slipped condition by the symbol k_s, the phase angle φ and the dynamic amplification factor A may be related by Eq. (36.30) to the duration of the elastic phase t':

$$\left(1 - \frac{k_s}{k_e} \right) (\omega t' + \sin \omega t') = \pi \left(\frac{m\omega^2 k_s}{k_e} + \frac{\cos \varphi}{A} \right) \qquad (36.31)$$

where A is the conventional dynamic amplification factor, i.e., $A = v_d k_e / F_m$. The duration of the elastic phase is given by the first of Eqs. (36.29) with $v = v_d - 2v_s$, where v_s is the displacement at which slip first occurs. Then eliminating t' from Eq. (36.31):

$$\frac{\cos \varphi}{A} = \frac{1}{\pi} \left(1 - \frac{k_s}{k_e} \right) \left[2 \frac{v_s k_e}{AF_m} \sqrt{1 + \frac{v_s k_e}{AF_m}} + \cos^{-1}\left(1 - 2 \frac{v_s k_e}{AF_m} \right) \right] - \frac{m\omega^2 k_s}{k_e} \quad (36.32)$$

Equation (36.32) gives the relation between phase lag φ and amplification factor A. A second relationship between these quantities is found from the consideration that the energy dissipated during each half cycle of forced motion must be $D_0/2$:

$$\int_0^{\pi/\omega} F_g \frac{dv}{dt} \, dt = \tfrac{1}{2} D_0 \qquad \text{or} \qquad \sin \varphi = \frac{D_0 k_e}{\pi F_m^2 A} \quad (36.33)$$

Equations (36.32) and (36.33) serve to determine the dynamic amplification factor A, after D_0 has been estimated. Conversely, they serve to estimate the amount of energy which must be dissipated per cycle to produce a given reduction in the amplification factor by interface shear. A detailed analysis of response to a parallelogram hysteresis loop has been made.[33] Hysteresis loops other than parallelograms also have been studied.[34] At resonance, $\varphi = 90°$ and

$$A = A_r = \frac{D_0 k_e}{\pi F_m^2} \quad (36.34)$$

In general, the energy dissipation does not increase as rapidly as the square of the peak exciting force; consequently, the resonance amplification factor decreases as the exciting force increases. As a result, structures in which interface shear predominates tend to be self-limiting in their response to external excitation.

The foregoing discussion is based on the premise that changes in the exciting force do not materially affect the size of the contact area. There is an important class of problems for which this assumption is not valid, namely, those in which even the smallest exciting force produces some slip. An example of this type of joint is the press-fit bushing on a cylindrical shaft. If the ends of the shaft are subjected to a cyclic torque, part of this torque is transmitted to the bushing. Each part of the compound torque tube carries a moment proportional to its stiffness. Transmission of torque from the shaft to the bushing is effected by slip over the interface. The length of the slipped region grows in proportion to the applied torque. There is no initial elastic region such as OP or $O''P''$ in Figs. 36.12 and 36.13. If the peak value of the exciting torque is not too large, the fully slipped region BC or $B'C'$ in Fig. 36.12 never occurs. In these cases, Eqs. (36.31) to (36.34) are not applicable because there are no assignable constant values of k_s and k_e. A variety of simple cases of this type which occur in design practice have been analyzed. They include the cylindrical shaft and bushing in tension and torsion, and the flexure of a beam with cover plate.

Another important case in which the smallest exciting force may produce slip arises in the contact of rounded solids. If these are pressed together by normal forces along the line joining their centers, a small contact region is formed. Subsequent application of a cyclic tangential force produces slip over a portion of the contact region even if the peak tangential force is not great enough to effect gross slip or sliding. This situation has been analyzed and verified experimentally.[3,36]

REFERENCES

1. Alfrey, T., Jr.: "Mechanical Behavior of High Polymers," Interscience Publishers, Inc., New York, 1948.

2. Lazan, B. J.: "Damping of Materials and Members in Structural Mechanics," Pergamon Press, New York, 1968.

3. Johnson, K. L.: "Contact Mechanics," Cambridge University Press, 1985.

4. Lazan, B. J.: "Fatigue," Chap. II, American Society for Metals, 1954.

5. Cochardt, A. W.: *J. Appl. Mechanics,* **21**:257 (1954).

6. Lazan, B. J.: *Trans. ASME,* **65**:87 (1943); Pisarenko, G. S.: "Vibrations of Mechanical Systems Taking into Account Incompletely Elastic Materials," 2d ed., Kiev, 1970 (in Russian).

7. Von Heydekampf, G. S.: *Proc. ASTM,* **31** (Pt.II):157 (1931); Jones, D. I. G., and D. K. Rao: *ASME Des. Div. Pub. DE,* **5**:143 (1987).

8. Lazan, B. J.: *Trans. ASM,* **12**:499 (1950).

9. Maxwell, B.: *ASTM Bull.,* **215**:76 (1956).

10. Nowick, A. S.: "Progress in Metal Physics," vol. 4, chap. I, 29, Interscience Publishers, Inc., New York, 1953; Tschan, T., et al.: *Proc. Eurosensors V,* Sensors and Actuators, A: Physical, **32**, n.1-3:375 (1992); Kalachnikov, E. V., and P. N. Rostovstev: *Instr. Exp. Tech,* **32**:1241 (1990).

11. Podnieks, E., and B. J. Lazan: *Wright Air Development Center Tech. Rep.* 55-284, 1955.

12. Lazan, B. J.: *J. Appl. Mechanics,* **20**:201 (1953).

13. Staverman, A. J., and F. Schwarzl: "Linear Deformation Behavior of High Polymers," chap. I of "Theorie u. Molekulare Deutung Technologischer Eigenschaften von Hoch Polymeren Werkstoffen," p. 71, Springer-Verlag, Berlin, 1956.

14. Zener, C.: "Elasticity and Anelasticity," University of Chicago Press, Chicago, 1948.

15. Wert, C.: "The Metallurgical Use of Anelasticity" in "Modern Research Techniques in Physical Metallurgy," American Society for Metals, Cleveland, Ohio, 1953.

16. Jones, D. I. G.: *J. Sound and Vib.,* **140**:85 (1990).

17. Fay, J. J., et al.: *Proc. ACS Div, Polymetric Mat'ls. Sci. and Eng'g.,* **60**:649 (1989).

18. Fujimoto, J., et al.: *J. Reinf. Plastics and Composites,* **12**:738 (1993).

19. Chang, M. C. O., et al.: *Proc. ACS Div. Polymetric Materials Sci. and Engg.,* **55**:350 (1986).

20. Weibo, H., and Z. Fengchan: *J. Appl. Polymer Sci.,* **50**:277 (1993).

21. Cochardt, A.: *Scientific Paper* 8-0161-P7, Westinghouse Research Labs., W. Pittsburgh, Pa., 1956.

22. Person, N., and B. J. Lazan: *Proc. ASTM,* **56**:1399 (1956).

23. Torvik, P.: Appendix 72fg, *Status Rep.* 58-4 by B. J. Lazan, University of Minnesota, Wright Air Development Center, Dayton, Ohio, Contract AF-33(616)-2802, December 1958.

24. Whittier, J. S., and B. J. Lazan: Appendix B, *Prog. Rep.* 57-6, Wright Air Development Center, Dayton, Ohio, Contract AF-33(616)-2803, December 1957.

25. Hinai, M., et al.: *Trans. Japan. Inst. Met.,* **28**:154 (1987).

26. De Batist, R.: *ASTM Spec. Tech. Pub.* 1169, 45-59, American Society for Testing and Materials, Philadelphia, 1992.

27. Zhang, J., et al.: *Acta Met. et Mat.,* **42**:395 (1994).

28. Goodman, L. E., and J. H. Klumpp: *J. Appl. Mechanics,* **23**:241 (1956).

29. Lazan, B. J., and L. E. Goodman: "Shock and Vibration Instrumentation," 55, ASME, New York, 1956.

30. Pian, T. H. H., and F. C. Hallowell: *Proc. First U.S. Nat'l. Cong. Appl. Mechanics,* June 1951, p. 97.

31. Fluegge, W.: "Viscoelasticity," Blaisdell Publishing Company, a division of Ginn and Company, Waltham, Mass., 1967; Lee, E. H.: Viscoelasticity in W. Fluegge (ed.), "Handbook of Engineering Mechanics," McGraw-Hill Book Company, New York, 1962; Lesieutre, G. A.: *Int'l. J. Solids and Structures,* **29**:1567 (1992).

32. Bogdanoff, J. L.: *J. Appl. Physics,* **21**:1258 (1950).

33. Caughey, T. K.: *J. Appl. Mechanics,* **27**:640 (1960).

34. Rang, E.: *Wright Air Development Center Tech. Rep.* 59-121, February 1959.

35. Goodman, L. E.: "A Review of Progress in Analysis of Interfacial Slip Damping," in "Structural Damping," ASME, New York, 1959.

36. Deresiewicz, H.: Bodies in Contact with Applications to Granular Media, in G. Herrmann (ed.) "R. D. Mindlin and Applied Mechanics," Pergamon Press, New York, 1974.

CHAPTER 37
APPLIED DAMPING TREATMENTS

David I. G. Jones

INTRODUCTION TO THE ROLE OF DAMPING MATERIALS

The damping of an element of a structural system is a measure of the rate of energy dissipation which takes place during cyclic deformation. In general, the greater the energy dissipation, the less the likelihood of high vibration amplitudes or of high noise radiation, other things being equal. *Damping treatments* are configurations of mechanical or material elements designed to dissipate sufficient vibrational energy to control vibrations or noise.

Proper design of damping treatments requires the selection of appropriate damping materials, location(s) of the treatment, and choice of configurations which assure the transfer of deformations from the structure to the damping elements. These aspects of damping treatments are discussed in this chapter, along with relevant background information including:

- Internal mechanisms of damping
- External mechanisms of damping
- Polymeric and elastomeric materials
- Benefits of applied damping treatments
- Free-layer damping treatments
- Constrained-layer damping treatments
- Integral damping treatments
- Tuned dampers and damping links
- Measures or criteria of damping
- Test instrumentation and interpretation of data
- Vibrating beam test methods
- Geiger thick-plate test method
- Single degree-of-freedom resonance and impedance tests
- Commercial test systems for measuring damping

MECHANISMS AND SOURCES OF DAMPING

INTERNAL MECHANISMS OF DAMPING

There are many mechanisms that dissipate vibrational energy in the form of heat within the volume of a material element as it is deformed. Each such mechanism is associated with internal atomic or molecular reconstructions of the microstructure or with thermal effects. Only one or two mechanisms may be dominant for specific materials (metals, alloys, intermetallic compounds, etc.) under specific conditions, i.e., frequency and temperature ranges, and it is necessary to determine the precise mechanisms involved and the specific behavior on a phenomenological, experimental basis for each material specimen. Most structural metals and alloys have relatively little damping under most conditions, as demonstrated by the ringing of sheets of such materials after being struck. Some alloy systems, however, have crystal structures specifically selected for their relatively high damping capability; this is often demonstrated by their relative deadness under impact excitation. The damping behavior of metallic alloys is generally nonlinear and increases as cyclic stress amplitudes increase. Such behavior is difficult to predict because of the need to integrate effects of damping increments which vary with the cyclic stress amplitude distribution throughout the volume of the structure as it vibrates in a particular mode of deformation at a particular frequency. The prediction processes are complicated even further by the possible presence of external sources of damping at joints and interfaces within the structure and at connections and supports. For this reason, it is usually not possible, and certainly not simple, to predict or control the initial levels of damping in complex built-up structures and machines. Most of the current techniques of increasing damping involve the application of polymeric or elastomeric materials which are capable (under certain conditions) of dissipating far larger amounts of energy per cycle than the natural damping of the structure or machine without added damping.

EXTERNAL MECHANISMS OF DAMPING

Structures and machines can be damped by mechanisms which are essentially external to the system or structure itself. Such mechanisms, which can be very useful for vibration control in engineering practice (discussed in other chapters), include:

1. Acoustic radiation damping, whereby the vibrational response couples with the surrounding fluid medium, leading to sound radiation from the structure
2. Fluid pumping, in which the vibration of structure surfaces forces the fluid medium within which the structure is immersed to pass cyclically through narrow paths or leaks between different zones of the system or between the system and the exterior, thereby dissipating energy
3. Coulomb friction damping, in which adjacent touching parts of the machine or structure slide cyclically relative to one another, on a macroscopic or a microscopic scale, dissipating energy
4. Impacts between imperfectly elastic parts of the system

POLYMERIC AND ELASTOMERIC MATERIALS

A mechanism commonly known as *viscoelastic damping* is strongly displayed in many polymeric, elastomeric, and amorphous glassy materials. The damping arises from the relaxation and recovery of the molecular chains after deformation. A strong dependence exists between frequency and temperature effects in polymer behavior because of the direct relationship between temperature and molecular vibrations. A wide variety of commercial polymeric damping material compositions exist, most of which fit one of the main categories listed in Table 37.1.

TABLE 37.1 Typical Damping Material Types

Acrylic rubber	Natural rubber	Urethane
Butadiene rubber	Nitrile rubber (NBR)	Vinyl
Butyl rubber	Polyisoprene	Polysulfone
Chloroprene (e.g., Neoprene)	Polysulfide	Polyvinyl chloride (PVC)
Fluorocarbon	Silicone	Polymethyl methacrylate (Plexiglas)
Fluorosilicone	Styrene-butadiene (SBR)	Nylon

Polymeric damping materials are available commercially in the following categories:

1. Mastic treatment materials
2. Cured polymers
3. Pressure sensitive adhesives
4. Damping tapes
5. Laminates

Some manufacturers of damping material are given as a footnote.* Data related to the damping performance is provided in many formats. The current internationally recognized format, used in many databases, is the temperature-frequency nomogram, which provides modulus and loss factor as a function of both frequency and temperature in a single graph, such as that illustrated in Fig. 37.1.[1,2] The user requiring complex modulus data at, say, a frequency of 100 Hz and a temperature of 50°F (10°C) simply follows a horizontal line from the 100-Hz mark on the right vertical axis until it intersects the sloping 50°F (10°C) isotherm, and then projects vertically to read off the values of the Young's modulus E and loss factor η.

BENEFITS OF APPLIED DAMPING TREATMENTS

When the natural damping in a system is inadequate for its intended function, then an applied damping treatment may provide the following benefits:

---

* Manufacturers of damping materials include:

Antiphon, Inc. (U.S.A.)	GE/Silicones Division (U.S.A.)
Du Pont (U.S.A.)	Lord Corporation (U.S.A.)
Dow Corning Corporation (U.S.A.)	3M Company (U.S.A.)
Eurovib (France)	Soundcoat, Inc. (U.S.A.)
EAR Corporation (U.S.A.)	Specialty Products, Inc. (U.S.A.)
Farbwercke Hoechst AG (Germany)	SNPE (France)
Imperial Chemical Industries (UK)	US Rubber Company (U.S.A.)
GE/Astro Space Division (U.S.A.)	

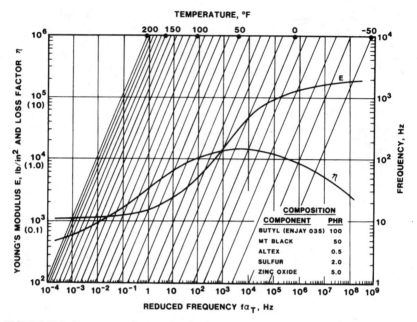

FIGURE 37.1   Temperature-frequency nomogram for butyl rubber composition.

Control of vibration amplitude at resonance.   Damping can be used to control excessive resonance vibrations which may cause high stresses, leading to premature failure. It should be used in conjunction with other appropriate measures to achieve the most satisfactory approach. For random excitation it is not possible to detune a system and design to keep random stresses within acceptable limits without ensuring that the damping in each mode at least exceeds a minimum specified value. This is the case for sonic fatigue of aircraft fuselage, wing, and control surface panels when they are excited by jet noise or boundary layer turbulence-induced excitation. In these cases, structural designs have evolved toward semiempirical procedures, but damping levels are a controlling factor and must be increased if too low.

Noise control.   Damping is very useful for the control of noise radiation from vibrating surfaces, or the control of noise transmission through a vibrating surface. The noise is not reduced by sound absorption, as in the case of an applied acoustical material, but by decreasing the amplitudes of the vibrating surface. For example, in a diesel engine, many parts of the surface contribute to the overall noise level, and the contribution of each part can be measured by the use of the acoustic intensity technique or by blanketing off, in turn, all parts except that of interest. If many parts of an engine contribute more or less equally to the noise, significant amplitude reductions of only one or two parts (whether by damping or other means) leads to only very small reductions of the overall noise, typically 1 or 2 dB.

Product acceptance.   Damping can often contribute to product acceptance, not only by reducing the incidence of excessive noise, vibration, or resonance-induced failure but also by changing the "feel" of the product. The use of mastic damping treatments in car doors is a case in point. While the treatment may achieve some noise reduction, it may be the subjective evaluation by the customer of the solidity of the door which carries the greater weight.

Simplified maintenance. A useful by-product from reduction of resonance-induced fatigue by increased damping, or by other means, can be the reduction of maintenance costs.

TYPES OF DAMPING TREATMENTS

FREE-LAYER DAMPING TREATMENTS

The mechanism of energy dissipation in a free-, or unconstrained-, layer treatment is the cyclic extensional deformation of the imaginary fibers of the damping layer during each cycle of flexural vibration of the base structure, as illustrated in Fig. 37.2. The presence of the free layer changes the apparent flexural rigidity of the base structure in a manner which depends on the dimensions of the two layers involved and the elastic moduli of the two layers. The treatment depends for its effectiveness on the assumption, usually well-founded, that plane sections remain plane. The treatment fiber labeled *yy* is extended or compressed during each half of a cycle of flexural deformation of the base structure surface, in a manner which depends on the position of the fiber in the treatment and the radius of curvature of the element of length Δl, and can be calculated on the basis of purely geometric considerations. One fiber in particular does not change length during each cycle of deformation and is referred to as the *neutral axis.* For the uncoated plate or beam, the neutral axis is the center plane, but when the treatment is added, it moves in the direction of the treatment and its new position is calculated by the requirement that the net in-plane load across any section remain unchanged during deformation. The basic equations for predicting the modal loss factor η for the given damping layer loss factor η_2 and for predicting the direct flexural rigidity $(EI)_D$ as a function of the flexural rigidity $E_1 I_1$ of the base beam are well known.[1,3]

The simplest expression relating the damping of a structure, in a particular mode, to the properties of the structure and the damping material layer is[4]

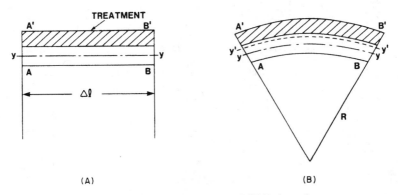

| (A) | (B) |

FIGURE 37.2 Free-layer treatment. (*A*) Undeformed. (*B*) Deformed.

$$\frac{\eta}{\eta_2} = \frac{eh(3 + 6h + 4h^2 + 2eh^3 + e^2h^4)}{(1 + eh)(1 + 4eh + 6eh^2 + 4eh^3 + e^2h^4)} \tag{37.1}$$

where η is the damped structure modal loss factor, η_2 is the loss factor of the damping material, E_2 is the Young's modulus of the damping material and E_1 is that of the structure ($e = E_2/E_1$), and h_2 and h_1 are the thicknesses of damping layer and structure, respectively ($h = h_2/h_1$).

To calculate η, the user estimates η_2 and E_2 at the frequency and temperature of interest (from a nomogram), then calculates h and e, and then inserts these values into Eq. (37.1). Change thickness (h) or material (e) if the calculated value of η is not adequate, and continue the process until satisfied. Figure 37.3 illustrates how η/η_2 varies with E_2/E_1 and with h_2/h_1, as calculated using the Oberst equations.

Limitations of Free-Layer Treatment Equations.   The classical equations for free-layer treatment behavior are approximate. The main limitation is that the equations are applicable to beams or plates of uniform thickness and uniform stiff isotropic elastic characteristics with boundary conditions which do not dissipate or store energy during vibration. These boundary conditions include the classical pinned, free, and clamped conditions. Another limitation is that the deformation of the damping material layer is purely extensional with no in-plane shear, which would allow the "plane sections remain plane" criterion to be violated. This restriction is not very important unless the damping layer is very thick and very soft ($h_2/h_1 > 10$ and $E_2/E_1 > 0.001$). A third limitation is that the treatment must be uniformly applied to the full surface of the beam or plate, and especially that it be anchored well at the boundaries so that plane sections remain plane in the boundary areas

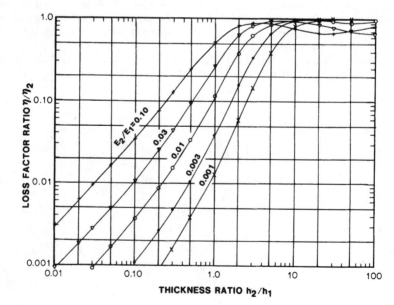

FIGURE 37.3   Graphs of η/η_2 vs. h_2/h_1, for a free-layer treatment.

where bending stresses can be very high and the effects of any cuts in the treatment can be very important. Other forms of the equations can be derived for partial coverage or for nonclassical boundary conditions.

Effect of Bonding Layer. Free-layer damping treatments are usually applied to the substrate surface through a thin adhesive or surface treatment coating. This adhesive layer should be very thin and stiff in comparison with the damping treatment layer in order to minimize shear strains in the adhesive layer which would alter the behavior of the damping treatment. The effect of a stiff thin adhesive layer is minimal, but a thick softer layer alters the treatment behavior significantly.

Amount of Material Required. Local panel weight increases up to 30 percent may often be needed to increase the damping of the structure in several modes of vibration to an acceptable level. Greater weight increases usually lead to diminishing returns. This weight increase can be offset to some degree if the damping is added early in the design, by judicious weight reductions achieved by proper sizing of the structure to take advantage of the damping.

CONSTRAINED-LAYER DAMPING TREATMENTS

The mechanism of energy dissipation in a constrained-layer damping treatment is quite different from the free-layer treatment, since the constraining layer helps induce relatively large shear deformations in the viscoelastic layer during each cycle of flexural deformation of the base structure, as illustrated in Fig. 37.4. The presence of the constraining viscoelastic layer-pair changes the apparent flexural rigidity of the base structure in a manner which depends on the dimensions of the three layers involved and the elastic moduli of the three layers, as for the free-layer treatment, but also in a manner which depends on the deformation pattern of the system, in contrast to the free-layer treatment. A useful set of equations which may be used to predict the flexural rigidity and modal damping of a beam or plate damped by a full-coverage constrained-layer treatment are given in Ref. 1. These equations give the direct (in-phase) component $(EI)_D$ of the flexural rigidity of the three-layer beam,

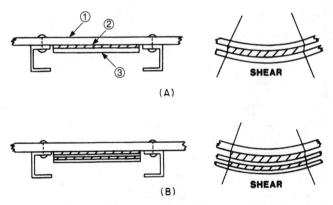

FIGURE 37.4 Additive layered damping treatments. (*A*) Constrained-layer treatment. (*B*) Multiple constrained-layer treatment.

and the quadrature (out-of-phase) component $(EI)_Q$ as a function of the various physical parameters of the system, including the thicknesses h_1, h_2, and h_3, the moduli $E_1(1+j\eta_1)$, $E_2(1+j\eta_2)$, $E_3(1+j\eta_3)$, and the shear modulus of the damping layer $G_2(1+j\eta_2)$.

Shear Parameter. The behavior of the damped system depends most strongly on the shear parameter

$$g = \frac{G_2(\lambda/2)^2}{E_3 h_3 h_2 \pi^2} \tag{37.2}$$

which combines the effect of the damping layer modulus with the semiwavelength of the mode of vibration, the modulus of the constraining layer, and the thicknesses of the damping and constraining layers. The other two parameters are the thickness ratios h_2/h_1 and h_3/h_1. Figure 37.5 illustrates the typical variation of η_n/η_2 and $(EI)_D/E_1I_1$ with the shear parameter g for particular values of h_2/h_1 and h_3/h_1. These plots may be used for design of constrained layer treatments. Note that η_n will be small for both large and small values of g. For g approaching zero, G_2 or $\lambda/2$ may be very small or E_3, h_3, and h_2 may be very large. This could mean that while G_2 might appear at first sight to be sufficiently large, the dimensions h_2 and h_3 are nevertheless too large to achieve the needed value of g. This could happen for very large structures, especially for high-order modes. On the other hand, for g approaching infinity, G_2 or $\lambda/2$ may be large, or E_3, h_2, or h_3 may be very small.

Effects of Treatment Thickness. In general, increasing h_2 and h_3 will lead to increased damping of a beam or plate with a constrained-layer treatment, but the effect of the shear parameter will modify the specific values. The influence of h_3/h_1 is stronger than that of h_2/h_1, and as h_2/h_1 approaches zero, η_n/η_2 does not approach zero but a finite value. This behavior seems to occur in practice and accounts for the very thin damping layers, 0.002 in. (0.051 mm) or less, used in damping tapes. A practical limit of 0.001 in. (0.025 mm) is usually adopted to avoid handling problems.

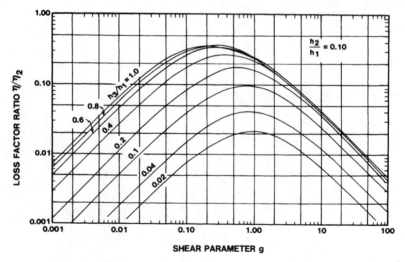

FIGURE 37.5 Typical plots of η/η_2 versus shear parameter $g(h_2/h_1 = 0.10, \eta_2 = 0.1)$.

Effect of Initial Damping.   If the base beam is itself damped, with η_1 not equal to zero, then the damping from the constrained-layer treatment will be added to η_1 for small values of η_1. The general effect is readily visualized, but specific behavior depends on treatment dimensions and the value of the shear parameter.

Integral Damping Treatments.   Some damping treatments are applied or added not after a structure has been partly or fully assembled but during the manufacturing process itself. Some examples are illustrated in Fig. 37.6. They include laminated sheets which are used for construction assembly, or for deep drawing of structural components in a manner similar to that for solid sheets, and also for faying surface damping which is introduced into the joints during assembly of built-up, bolted, riveted, or spot-welded structures. The conditions at the bolt, rivet, or weld areas critically influence the behavior of the damping configurations and make analysis particularly difficult because of the limited control of conditions at these points. Finite element analysis may be one of the few techniques for such analysis.

Tuned Dampers.   The tuned damper is essentially a single degree-of-freedom mass-spring system having its resonance frequency close to the selected resonance frequency of the system to be damped, i.e., tuned. As the structure vibrates, the damper elastomeric element vibrates with much greater amplitude than the structure at the point of attachment and dissipates significant amounts of energy per cycle, thereby introducing large damping forces back to the structure which tend to reduce the amplitude. The system also adds another degree of freedom, so two peaks arise in place of the single original resonance. Proper tuning is required to ensure that the two new peaks are both lower in amplitude than the original single peak. The damper mass should be as large as practicable in order to maximize the damper effectiveness, up to perhaps 5 or 10 percent of the weight of the structure at most, and the damping capability of the resilient element should be as high as possible. The weight increase needed to add significant damping in a single mode is usually smaller than for a layered treatment, perhaps 5 percent or less.

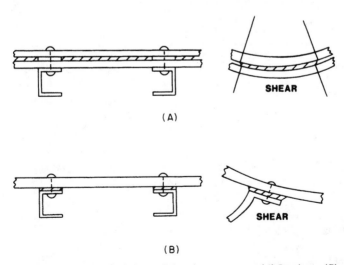

(A)

(B)

FIGURE 37.6   Some basic integral damping treatments. (A) Laminate. (B) Faying surface damping.

Damping Links. The damping link is another type of discrete treatment, joining two appropriately chosen parts of a structure. Damping effectiveness depends on the existence of large relative motions between the ends of the link and on the existence of unequal stiffnesses or masses at each end. The deformation of the structure when it is bent leads to deformation of the viscoelastic elements. These deformations of the viscoelastic material lead to energy dissipation by the damper.

RATING OF DAMPING EFFECTIVENESS

MEASURES OR CRITERIA OF DAMPING

There are many measures of the damping of a system. Ideally, the various measures of damping should be consistent with each other, being small when the damping is low and large when the damping is high, and having a linear relationship with each other. This is not always the case, and care must be taken, when evaluating the effects of damping treatments, to ensure that the same measure is used for comparing behavior before and after the damping treatment is added. The measures discussed here include the loss factor η, the fraction of critical damping (damping ratio) ζ, the logarithmic decrement Δ, the resonance or quality factor Q, and the specific damping energy D. Table 37.2 summarizes the relationship between these parameters, in the ideal case of low damping in a single degree-of-freedom system. Some care must be taken in applying these measures for high damping and/or for multiple degree-of-freedom systems and especially to avoid using different measures to compare treated and untreated systems.

Loss Factor. The loss factor η is a measure of damping which describes the relationship between the sinusoidal excitation of a system and the corresponding sinusoidal response. If the system is linear, the response to a sinusoidal excitation is also

TABLE 37.2 Comparison of Damping Measures

Measure	Damping ratio	Loss factor	Log dec	Quality factor	Spec damping	Amp factor
Damping ratio	ζ	$\dfrac{\eta}{2}$	$\dfrac{\Delta}{\pi}$	$\dfrac{1}{2Q}$	$\dfrac{D}{4\pi U}$	$\dfrac{1^*}{2A}$
Loss factor	2ζ	η	$\dfrac{2\Delta}{\pi}$	$\dfrac{1}{Q}$	$\dfrac{D}{2\pi U}$	$\dfrac{1^*}{A}$
Log decrement	$\pi\zeta$	$2\pi\eta$	Δ	$\dfrac{\pi}{2Q}$	$\dfrac{D}{4U}$	$\dfrac{2\pi^*}{A}$
Quality factor	$\dfrac{1}{2\zeta}$	$\dfrac{1}{\eta}$	$\dfrac{\pi}{2\Delta}$	Q	$\dfrac{2\pi U}{D}$	A^*
Spec damping	$4\pi U\zeta$	$2\pi U\eta$	$4U\Delta$	$\dfrac{2\pi U}{Q}$	D	$\dfrac{2\pi U^*}{A}$
Amp factor	$\dfrac{1}{2\zeta}$	$\dfrac{1}{\eta}$	$\dfrac{\pi}{2\Delta}$	Q	$\dfrac{2\pi U}{D}$	A^*

* For single-degree-of-freedom system only.

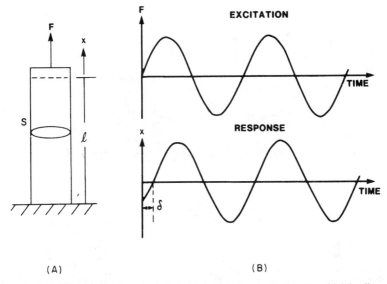

FIGURE 37.7 Linear viscoelastic behavior of a sample under sinusoidal loading, described in terms of response and excitation as functions of time. (*A*) Specimen. (*B*) Response and excitation.

sinusoidal and a loss factor is easily defined, but great care must be taken for non-linear systems because the response is not sinusoidal and a unique loss factor cannot be defined. Consider first an inertialess specimen of linear viscoelastic material excited by a force $F(t) = F_0 \cos \omega t$, as illustrated in Fig. 37.7. The response $x(t) = x_0 \cos (\omega t - \delta)$ is also harmonic at the frequency ω as for the excitation but with a phase lag δ. The relationship between $F(t)$ and $x(t)$ can be expressed as

$$F = kx + \frac{k\eta}{|\omega|} \frac{\partial x}{\partial t} \tag{37.3}$$

where $k = F_0/x_0$ is a stiffness and $\eta = \tan \delta$ is referred to as the *loss factor*. The phase angle δ varies from 0° to 90° as the loss factor η varies from zero to infinity, so a one-to-one correspondence exists between η and δ. Equation (37.3) is a simple relationship between excitation and response which can be related to the stress-strain relationship because normal stress $\sigma = F/S$ and extensional strain $\varepsilon = x/l$. This is a generalized form of the classical Hooke's law which gives $F = kx$ for a perfectly elastic system. The loss factor, as a measure of damping, can be extended further to apply to a system possessing inertial as well as stiffness and damping characteristics. Consider, for example, the one degree-of-freedom linear viscoelastic system shown in Fig. 37.8*A*. The equation of motion is obtained by balancing the stiffness and damping forces from Eq. (37.3) to the inertia force $m(d^2x/dt^2)$:

$$m \frac{d^2x}{dt^2} + kx + \frac{k\eta}{\omega} \frac{dx}{dt} = F_0 \cos \omega t \tag{37.4}$$

The steady-state harmonic response, after any start-up transients have died away, is illustrated in Fig. 2.22. If k and η depend on frequency as is the case for real materials, then the maximum amplitude at the resonance frequency $\omega_r = \sqrt{k/m}$ is equal to

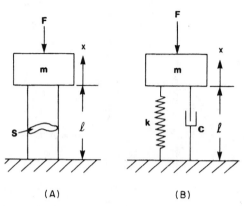

FIGURE 37.8 Single degree-of-freedom system with: (A) viscoelastic damping; (B) viscous damping.

$F_0/k(\omega_r)\eta(\omega_r)$, while the static response, at $\omega = 0$, is equal to $F_0/k(0)\sqrt{1 + \eta^2(0)}$. The amplification factor A is approximately equal to $1/\eta(\omega_r)$, provided that $\eta^2(0) \ll 1$. Furthermore, the ratio $\Delta\omega/\omega_r$, where $\Delta\omega$ is the separation of the frequencies for which the response is $1/\sqrt{2}$ times the peak response, is known as the *half-power bandwidth* (see Fig. 2.22). It is also equal to η, provided that $\eta^2 \ll 1$. In summary, therefore,

$$\eta = \frac{1}{A} = \frac{\Delta\omega}{\omega_r} \tag{37.5}$$

This relationship between η and $1/A$ is applicable only for a single degree-of-freedom system and may not be directly applicable for more complex systems such as beams, plates, or more complex structures. The measure $\Delta\omega/\omega_r$ is applicable for more complex systems, as well as single degree-of-freedom systems. For large values of η, on the order of 0.2 or greater, none of these measures of damping agree exactly, even for an ideal linear single degree-of-freedom system, but each measure is at least self-consistent. The stiffness and loss-factor parameters defined here do not specify any particular model of material behavior. For example, k and η could be constants as for hysteretic damping, or they could be functions of frequency, temperature, specimen composition and shape, or amplitude as for a nonlinear material. The model with constant k and η is not too useful over a wide frequency range, and such behavior is impossible over an infinite frequency range, but these parameters can vary quite slowly with frequency for some particular material compositions. If k and η vary strongly with frequency, or amplitude, then the various definitions of the loss factor must be used with great care, since each measure gives different results.

Fraction of Critical Damping. The fraction of critical damping (damping ratio) is a measure of one very specific mechanism of damping, i.e., viscous damping which is proportional to velocity. If the damping forces acting on a single degree-of-freedom mass-spring system, illustrated in Fig. 37.8B, satisfy this type of relationship, then the equation of motion for harmonic excitation is

$$m\,\frac{d^2\ddot{x}}{dt^2} + c\dot{x} + kx + = F_0\cos\omega t \tag{37.6}$$

The response depends on m, k, and a parameter $c/2\sqrt{km}$ which involves c, k, and m and is known as the *fraction of critical damping* (damping ratio). This parameter, labeled ζ, controls the peak amplitude, the half-power bandwidth, and the resonance frequency ω_r:

$$x_{max} = \frac{F_0}{2k\zeta\sqrt{1-\zeta^2}} \qquad x(0) = \frac{F_0}{k}$$

$$\omega_r = \sqrt{(k/m)(1-\zeta^2)} \qquad \frac{\Delta\omega}{\omega_r} = 2\zeta$$

(37.7)

The plot of $x(\omega)$ versus frequency ω, for specific values of m and k is very similar to those for the viscoelastic damping, provided that $\eta \doteq 2\zeta$. The distinction between viscous and hysteretic damping (constant k, and η) is not at once apparent. Equations (37.4) and (37.6) convey the difference, since the damping coefficient in Eq. (37.4) decreases in proportion to $1/\omega$ as ω increases, while that in Eq. (37.6) is constant with frequency, at least for the hypothetical cases considered here. Figure 37.9 shows plots of response versus frequency based on the solutions of these equations of motion for each type of damping. Some differences arise at low frequency, but they are not very great except for very high values of damping. For high values of damping, neither η nor ζ are linearly related to the bandwidth ratio $\Delta\omega/\omega_r$. Figure 37.10 shows the variation of $\Delta\omega/\omega_r$ with η and 2ζ for values of η which are not small. Limits exist beyond which the ratio $\Delta\omega/\omega_r$ does not give a good estimate of η or ζ.

Logarithmic Decrement.   When a damped system is struck by an impulsive load or is released from a displaced position relative to its equilibrium state, a decaying oscillation usually takes place as illustrated in Fig. 2.8. A measure of damping called

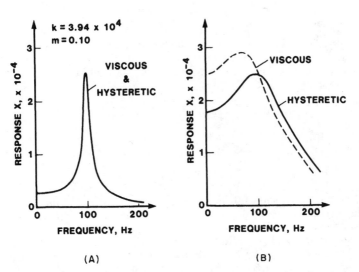

FIGURE 37.9 Comparison of viscous and hysteretic damping of a single degree-of-freedom system with: (*A*) low damping ($\eta = 0.1$, $\zeta = 0.05$); (*B*) high damping ($\eta = 1.0$, $\zeta = 0.5$).

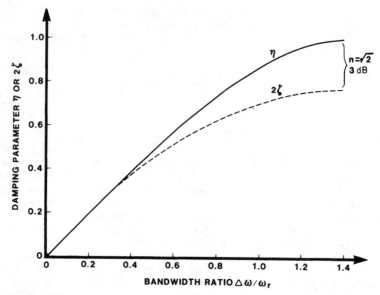

FIGURE 37.10 Variation of loss factor (η) and 2 times the fraction of critical damping (2ζ) of a single degree-of-freedom system with $\Delta\omega/\omega_r$.

logarithmic decrement Δ is defined as the natural logarithm of the ratio of amplitudes of successive peaks [see Eq. 2.19]:

$$\Delta = \ln\frac{x_1}{x_2} = \ln\frac{x_n}{x_{n+1}} \tag{37.8}$$

This definition is useful only if these ratios are equal for the various cycles, i.e., for specific types and amounts of damping. The measure is useful for viscous and hysteretic damping, within limits. For viscous damping, the solution of Eq. (37.6) for an impulsive excitation $F\,\delta(t)$ is obtained.

$$x = \frac{F}{\sqrt{km(1-\zeta^2)}}\ e^{-t\sqrt{k/m}}\ \sin t\sqrt{(k/m)(1-\zeta^2)} \tag{37.9}$$

so that

$$\Delta = \frac{\pi\zeta}{\sqrt{1-\zeta^2}} \tag{37.10}$$

for small ζ. If ζ approaches 1.0, the response becomes aperiodic and a logarithmic decrement cannot be defined or related to ζ. The loss factor in Eq. (37.3) also can be related to the transient response of a single degree-of-freedom mass-spring system, subject to an impulsive excitation. Consider the impulsive excitation $F(t)$ to be modeled as a spike of the form of a delta function at time $t = 0$. Then the equation of motion, in the form of Eq. (37.4), cannot be written directly, but if $F(t)$ and $x(t)$ are both described in terms of their corresponding Fourier transforms, then $\bar{F}(\omega) = \int_{-\infty}^{\infty} F(t)\exp(-j\omega t)\,dt = F$ and $\bar{x} = F/(k - m\omega^2 + jk\eta)$. The inverse Fourier transform gives

$$x(t) = \frac{1}{2\pi} \int_{-\infty}^{\infty} \frac{Fe^{j\omega t} \, d\omega}{k - m\omega^2 + jk\eta} \tag{37.11}$$

This equation contains real and imaginary parts, but using the fact that $\exp(j\omega t) = \cos \omega t + j \sin \omega t$ and if $k(\omega) = k(-\omega)$ and $\eta(-\omega) = -\eta(\omega)$, then it may be shown that $x(t)$ is given by

$$x(t) = \frac{F}{\pi} \int_{0}^{\infty} \frac{(k - m\omega^2) \cos \omega t + k\eta \sin \omega t}{(k - m\omega^2)^2 + (k\eta)^2} \, d\omega \tag{37.12}$$

For k and η constant over all frequencies from zero to infinity, problems arise regarding $x(t)$ being finite for values of t less than zero, i.e., before the impulse is applied, and this is physically impossible. The problem is that k and η cannot be constants for real systems over any extremely wide frequency range, no matter how close to constant they may be over a limited frequency range. For small values of η, however, a useful and accurate solution is given by

$$x(t) = \frac{F}{\sqrt{km}} \, e^{-1/2\eta t \sqrt{k/m}} \sin t\sqrt{k/m} \tag{37.13}$$

$$\Delta = \pi\eta/2 \tag{37.14}$$

Comparing Eqs. (37.10) and (37.14) gives

$$\eta = 2\zeta \tag{37.15}$$

Quality Factor.   The quality factor Q is defined as $\omega_r/\Delta\omega$, so

$$Q = \frac{1}{\eta} \tag{37.16}$$

For a single degree-of-freedom $Q = A$ [where A is defined in Eq. (37.5)], but this is not the case for multiple degree-of-freedom systems.

Specific Damping Energy.   Another useful measure of material damping is the amount of energy dissipated per unit volume per cycle, known as the *specific damping energy*. For a damping material specimen subject to an applied external force $F(t) = F_0 \cos \omega t$ the specific damping energy D is equal to

$$D = \oint F \, dx = \int_{0}^{2\pi/\omega} F \frac{dx}{dt} \, dt \tag{37.17}$$

For a viscoelastic material obeying Eq. (37.3)

$$D = F_0 x_0 k\eta \sqrt{1 + \eta^2} \tag{37.18}$$

But $F_0 = kx_0 \sqrt{1 + \eta^2}$, also from Eq.(37.3), so

$$D = \pi x_0^2 \, k\eta \tag{37.19}$$

The specific damping energy D increases as the square of the amplitude of vibration x_0 for linear viscoelastic materials, so it is clearly desirable to ensure that the damping material is strained as vigorously as possible in order to maximize D and hence the damping of the system. This has an important bearing on the choice of location within a vibrating system for application of a damping treatment. Furthermore, both k and η must be as large as possible to ensure maximum energy dissipation in the system, but this can be done only to the extent that further increases of k and η do not reduce x_0. While D is related to k and η for linear viscoelastic materials, this is not possible for nonlinear materials or for high cyclic strain levels where nonlinear behavior occurs; the value of D is then, of itself, often used as a measure of overall damping performance.

COMPARISON OF DAMPING MEASURES

The damping measures described in this section are related to each other as follows (Table 37.2):

$$\eta = 2\zeta = \frac{2\Delta}{\pi} = \frac{1}{Q} = \frac{D}{2\pi U} = \frac{1}{A} = \frac{\Delta\omega}{\omega_r} \tag{37.20}$$

These various equations relate η, Δ, and ζ for viscous and viscoelastic damping of single degree-of-freedom systems. The relationships usually agree well for low values of η and ζ ($\eta < 0.2$ or so), but for higher values the comparisons are not so precise.

It is important, when analyzing tests to determine the effects of damping treatments on dynamic response, to be consistent in the use of these damping measures and to recognize that they are not completely equivalent, especially over wide frequency ranges or for multimodal response.

Effects of Mass and Stiffness. Changing the mass or stiffness of a single degree-of-freedom mass-spring system without changing any other parameters leads to a change of resonance frequency, and when the frequency changes over a wide range, the differences of viscous and hysteretic damping become more apparent. For viscous damping, the fraction of critical damping $\zeta = c/2\sqrt{km}$ changes as k or m change, whereas for hysteretic damping η does not change, at least within a limited frequency range. Although viscous and hysteretic damping measures are related by the simple relationship $\eta = 2\zeta$ for a single mode at a particular frequency, they do not remain equivalent as the frequency changes, and significant differences in response may be observed.

TEST INSTRUMENTATION AND INTERPRETATION OF TEST DATA

Analog instrumentation has been used to apply sine-sweep excitation or random excitation to the test article. Digital instrumentation can be used to perform the same type of swept sine-wave excitation by advancing stepwise rather than continuously through the frequency range selected. The main advantage of digital test systems is that transient excitation can be applied readily, the transient response can be recorded digitally in the time domain, and a Fourier transform can be performed using a fast Fourier transform (FFT) algorithm to place the response in the frequency domain for analysis of modal behavior such as resonance frequencies, mode shapes, and modal loss factors.

Analog Sine-Sweep Tests. In the case of analog swept sine-wave testing of a structure an exciter is driven by the output of an analog sweep-frequency oscillator; the sweep rate is sufficiently low to avoid transient response during each sweep through a resonance peak. The response is measured by, for example, an accelerometer, and the signal is plotted on an X-Y plotter after passing through an A-to-D converter. The signals can be stored on magnetic tape for later playback to obtain meter readings of excitation and response signals at any frequency. Provided every instrument in the chain is properly calibrated and introduces minimal phase changes between incoming and outgoing signals, the resonance frequency and the modal loss factor can be determined readily.

Digital Sine Sweep Tests. The digital equivalent of swept sine-wave excitation makes the exciter frequency dwell for a selectable period of time at a number of selected frequencies, with almost instantaneous change from one frequency value to the next. The choice of dwell time and frequency interval are now the most important parameters controlling the degree of dynamic equilibrium achieved during a test. The frequency points can be chosen close enough to be virtually indistinguishable from a continuous curve (as for analog swept sine-wave testing), although test time can be saved by choosing the widest intervals commensurate with the goals of the test. Digital swept sine-wave testing is usually microcomputer controlled. Digital swept sine-wave testing is particularly appropriate for nonlinear systems, because analysis of the test results is usually easier for sinusoidal excitation than for other types, but it is also very useful for linear systems.

Impact Tests. For linear systems, impact testing is widely used. The structure is excited by impact at one or more points in succession by an instrumented hammer having a force gage at the impact point to measure the transient applied force. The response is measured at one or more points (often the same ones at which the force is applied) by an accelerometer or other transducer. The input force and the response in one or more channels are sampled digitally in the time domain and are stored digitally at preselected time intervals and for a selected period of time in a microprocessor. The signals are then transformed to the frequency domain by an FFT algorithm. Ideally, the response of the system to swept sine-wave excitation having a fixed amplitude (described as the ratio of response to excitation, i.e., compliance or receptance) should be the same for analog swept sine-wave, digital swept sine-wave, or impact/FFT techniques. In practice, difficulties of ensuring that conditions are truly identical for each test, and varying limits of dynamic range for the various test instruments, result in discrepancies. For well-conducted tests, the differences should not be great.

VIBRATING BEAM TEST METHODS

The vibrating beam test methods are frequently used to measure the extensional or shear complex modulus properties of damping materials.[1] The dynamic response behavior of the beam, first in the undamped uncoated form and then with an added damping layer or added constrained configuration, is measured for several modes of vibration and over a range of temperatures. At each temperature, the measured damped resonance frequency f_n, the undamped resonance frequency f_{on}, and the loss factor η_n in the nth mode of vibration are measured and used in an appropriate set of equations to deduce the Young's modulus E, or the shear modulus G, and the loss factor η of the damping material at a number of discrete frequencies and temperatures.

Various configurations of cantilever beams are used to measure viscoelastic material damping properties in tension-compression or shear at low cyclic strain amplitudes. Figure 37.11 illustrates some of the configurations used. The damping layers are bonded to the base beams by means of a stiff adhesive such as an epoxy. This bonding is very important and must be done well using an adhesive which is stiff in comparison to the damping layer and is very thin. The thickness ratio h_2/h_1 generally lies in the range $0.1 \leq h_2/h_1 \leq 2.0$, and the length l is about 5 to 10 in. (12.7 to 25.4 cm). The base beam material is typically aluminum, steel, or a stiff epoxy or epoxy matrix composite material having low intrinsic damping. Great care must be taken to ensure that the temperature range of the tests is not excessive in relation to the behavior of the base beam, and in particular to allow for the effect of temperature on the base beam properties such as Young's modulus, the resonance frequencies, and the modal loss factor in the absence of the damping layer. The vibration test is conducted allowing the specimen to soak at each selected temperature for several minutes, often 10 or 15 minutes, to be sure of thermal equilibrium; then the beam is excited by means of a noncontacting transducer or by impact, and the resulting response in the frequency domain is measured, either through swept sine-wave excitation or FFT analysis of the transient response signal in the time domain. At each temperature, several resonance frequencies and modal loss factors are measured over a wide range of frequencies. The test is then repeated after thermal equilibrium has been reached at the next selected temperature. The data obtained for the first mode is usually not used because of the low frequency involved and the high amplitudes and high modal damping of the base beam, as well as because of errors in the analysis when sandwich beams are used. Such vibrating beam tests are widely used for measuring viscoelastic material damping properties for shear and extensional deformation.

Free-Layer Beam Method. For the beam coated as a free layer on one side, as illustrated in Fig. 37.11A, the Young's modulus E_2 and the loss factor η_2 of the damping material are calculated from the equation for a free-layer treatment. These equations must be used with care. For example, if the modal loss factor η_n in the nth mode is greater than 0.3 or so at any temperature, it is difficult to measure either f_n or η_n with sufficient accuracy. Also, if the ratio of stiffness of the coated to the uncoated beam is less than about 1.1, errors in the measured values of f_n and η_n are magnified in the calculated values of E_2 and η_2. This can happen at high temperatures, when the damping material layer is very soft relative to the base beam material. The onset of this type of problem can be delayed to higher temperatures by increasing h_2, although this can lead to excessive values of η_n at lower temperatures, or by using an epoxy-matrix composite beam. The resonance frequencies of the uncoated beam f_{on} must be measured as a function of temperature in separate tests. Another problem which is occasionally encountered in this type of unsymmetric specimen is excessive bending of the beam at high and low temperatures, resulting from unequal thermal expansion of the base beam and the viscoelastic material.

Symmetric Free-Layer Method. To alleviate the problem of excessive bending of the beam at high and low temperatures, as a result of unequal thermal expansion coefficients of the damping layer and the base beam, and to simplify the equations, the base beam can be coated symmetrically as illustrated in Fig. 37.11B. The total amount of damping material is unchanged since each layer can be half the thickness used when coated on one side. The error magnification as the stiffness ratio approaches unity is high. As before, the initial damping of the base beam must be low, with modal loss factors of generally less than 0.01.

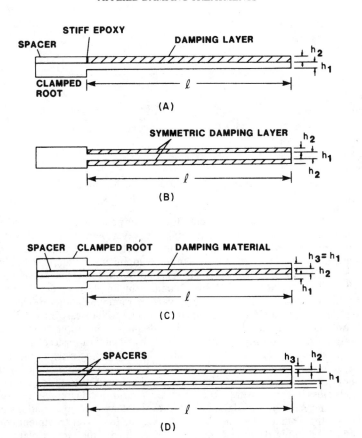

FIGURE 37.11 Cantilever beam damping material test configurations. (*A*) Nonsymmetric Oberst beam. (*B*) Symmetric modified Oberst beam. (*C*) Symmetric sandwich beam. (*D*) Symmetric constrained-layer beam.

Sandwich Beam Technique. To measure material damping properties in shear, symmetric sandwich cantilever beams are used as illustrated in Fig. 37.11*C*. The equations usually used to determine the shear modulus G_2 and the loss factor η_2 from the measured modal loss factors and resonance frequencies neglect bending of the middle layer, so the sandwich beam test is usually used for measuring complex moduli of soft viscoelastic materials, for which the ratio E_2/E_1 is usually less than 0.01. Care must be taken to ensure good bonding between the two beams and the inner viscoelastic layer and at the spacers, and also that the resonance frequencies of each beam are identical. Errors greater than 1 percent are unacceptable. Another variation on the sandwich beam approach is the symmetric constrained-layer configuration illustrated in Fig. 37.11*D*. The equations are simplified because the neutral axis now lies along the center of the composite beam and only a single base beam is required.

GEIGER THICK-PLATE TEST METHOD

The Geiger thick-plate method is of importance because it is widely used to describe damping materials in the automotive industry. It makes use of a large flat plate, sus-

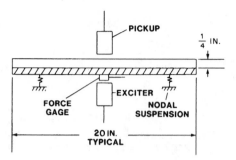

FIGURE 37.12　Geiger plate test configuration.

pended freely from four points selected to be at or near the nodal lines of the first free-free mode, to which is bonded the damping layer being evaluated. The rate of decay of vibration amplitude (expressed in decibels per second) is measured and serves as a measure of the effectiveness of the damping layer. Figure 37.12 illustrates a typical test setup. The system can be excited by an impulsive force, measured through a force gage, and applied by a hammer, by an electromagnetic exciter, an electrically actuated impeller, or by sine-wave or random excitation. The response can be picked up by an electromagnetic transducer, in which case cross talk with the excitation transducer must be avoided by adequate separation or by the use of capacitative or electro-optical transducers or by a miniature accelerometer. The measured output can be displayed in many ways, including a decaying sinusoidal trace representing response to an impulsive excitation (a measure of the logarithmic decrement), or a frequency domain display in the region of the fundamental free-free mode (loss factor measure). The observed logarithmic decrement or loss factor value is a measure of the damping of the plate/damping material system and depends on the plate and treatment thicknesses. The free-layer treatment equations used for the vibrating beam tests may also be used with the Geiger plate test provided that the same conditions are satisfied. In particular, the treatment thickness must be sufficient to make the ratio of the stiffness of the coated plate to that of the uncoated plate greater than about 1.05. The size of the specimen and the use of only one mode makes this condition somewhat less restrictive than for the beam tests, for which the specimens are much smaller.

SINGLE DEGREE-OF-FREEDOM RESONANCE AND IMPEDANCE TESTS

Digital test instrumentation and data analysis techniques make it relatively easy to conduct vibration tests directly on relatively small samples of damping materials and to readily determine the damping properties. Typical test configurations are illustrated in Fig. 37.13.

Resonance Test.　For a resonance type of test, the specimen is driven inertially by a large vibration table (see Chap. 25), usually by swept sine-wave excitation. The input and output accelerations are usually measured by accelerometers, and the response parameter of interest is the amplification $A = x/x_0$ as a function of frequency, where x is the amplitude of displacement of the mass and x_0 is that of the

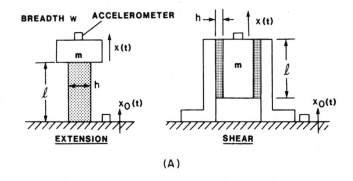

(A)

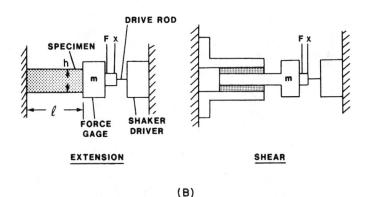

(B)

FIGURE 37.13 Single degree-of-freedom test configurations. (*A*) Resonance test, shaker sine-sweep excitation. (*B*) Impedance test.

shaker table. At resonance, the maximum value of x/x_0 is observed along with the resonance frequency for each temperature. The modulus and loss factor of the specimen material are determined from

$$\eta = \frac{1}{\sqrt{A^2 - 1}} \tag{37.21}$$

$$E = \frac{m_e \omega_r^2 S}{l} \tag{37.22}$$

$$G = \frac{m_e \omega_r^2 S}{l} \tag{37.23}$$

for tension-compression or shear. For tension-compression, $S = wh$ where w is the width, h is the thickness, and l is the length. For shear, $S = 2wl$ and h is the thickness of one material layer. The effective mass m_e includes the added mass m and one-third of the mass of the damping material, to a first approximation. The ratio l/h or *l/w,* whichever is smaller, must be greater than 1.0 or shape effects must be taken into account. For shear, h/l must be smaller than about 0.2, for the same reason. For highly damped materials, for which x/x_0 does not exceed 1.0 by a significant amount, considerable error in measuring A and η will be encountered, but the method is very effective for values of η less than about 0.5. In this method, data are obtained at only one frequency; the mass m must be changed to obtain data at other frequencies. Care must be taken to avoid sagging or creep of the specimen at high temperatures and to ensure that thermal equilibrium has been achieved. A thermocouple placed within the volume of the specimen material may be necessary, particularly for tests at high strain amplitudes where internal heating of the specimen by energy dissipation from damping may lead to wide differences between true specimen temperature and the temperature of the surroundings.

Impedance Test. If the specimens are excited by a driver through a force gage, then the response measure used to characterize the system behavior is the compliance or receptance x/F, where F is the driving force measured by the force gage and x is the response at the same point, measured by an accelerometer, for example. If the mass m is large compared with the mass of the specimen, as illustrated in Fig. 37.13B, then one may add one-third the mass of the specimen to m to give the effective mass m_e of the equivalent single degree-of-freedom system, so that

$$\frac{x}{F} = \frac{1}{k(1+j\eta) - m_e\omega^2} \tag{37.24}$$

If this is expressed instead in terms of the ratio F/x, the dynamic stiffness at the driving point, which is directly related to the driving-point impedance, then

$$\kappa = k - m_e\omega^2 + jk\eta \tag{37.25}$$

which shows that the direct dynamic stiffness is a linear function of ω^2 and the quadrature dynamic stiffness $\kappa_Q = k\eta$. It is not difficult to obtain good measurements of k and η by this type of test approach from about $0.2\omega_r$ to $3\omega_r$, so data can be obtained quite easily over about a decade of frequency instead of at only a single frequency as for the resonance method.

COMMERCIAL TEST SYSTEMS FOR MEASURING DAMPING MATERIAL BEHAVIOR

Many commercial systems are available for measuring the complex modulus properties of viscoelastic damping materials. All are based on some kind of deformation mode of a sample of the material, measurement of the corresponding excitation forces and displacements, and analysis of the data to obtain the material properties. Each system has advantages and disadvantages, but when due care is exercised, good results usually can be obtained with each system. Particular care should be taken to read, understand, and follow the manufacturer's instructions. For example, in some tests such as monitoring cure cycles of epoxies, the temperature sweep rate can be quite high in order to keep up with the reaction. This is acceptable if one is monitoring the progress of the cure cycle, but it may not be acceptable if one seeks to measure the damping properties at a state approximating thermal equilibrium. For thermal equilibrium to be maintained, temperature sweep rates well below 1°F (0.5°C) per minute are usually recommended, and even lower rates may be required for large specimens.

REFERENCES

1. Nashif, A. D. et al.: "Vibration Damping," Wiley Interscience, New York, 1985.
2. Nashif, A. D., and T. M. Lewis: *Sound & Vibration,* pp 14–25, July 1991.
3. Harris, C. M. (ed.): "Shock and Vibration Handbook," 3d ed. McGraw-Hill Book Co., New York, 1987.
4. Oberst, H: Acustica, **4**: 181 (1952).

CHAPTER 38
TORSIONAL VIBRATION IN RECIPROCATING AND ROTATING MACHINES

Ronald L. Eshleman

INTRODUCTION

Torsional vibration is an oscillatory angular motion causing twisting in the shaft of a system; the oscillatory motion is superimposed on the steady rotational motion of a rotating/reciprocating machine. Even though the vibration cannot be detected without special measuring equipment, its amplitude can be destructive. For example, gear sets that alter speeds of power transmission systems transmit the vibration to the casing. Similarly, slider crank mechanisms in engines and compressors convert torques to radial forces that are discernable to human perception but are not measurable because of the insensitivity of test equipment and background noise. If gearboxes or reciprocating machines are part of a drive train, excess noise and vibration can indicate trouble. Standards and measurement methods dealing with acceptable magnitudes of radial vibration are provided in Chap. 19.

Motion is rarely a concern with torsional vibration unless it affects the function of a system. It is stresses that affect the structural integrity and life of components and thus determine the allowable magnitude of the torsional vibration. Torsional vibratory motions can produce stress reversals that cause metal fatigue. Components tolerate less reversed stress than steady stress. In addition, stress concentration factors associated with machine members decrease the effectiveness of load-bearing materials.

Figure 38.1 illustrates the twisting of a shaft of an electric motor-compressor system. The torsional mode shape associated with the first torsional natural frequency is shown in Fig. 38.2. A coupling in the power train allows for misalignment in the assembly. The mode shape shows that the stiffness of the coupling is much less than that of other shaft sections. This is indicated by the large slope (change in angular displacement) of the mode shape at the coupling. The coupling will be the predominant component in the motor-compressor system governing the torsional natural frequency associated with the mode.

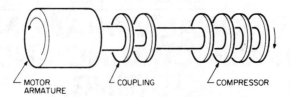

FIGURE 38.1 Schematic drawing illustrating the twisting of the shaft of a motor-compressor system.

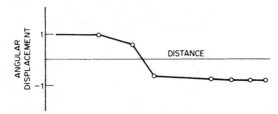

FIGURE 38.2 Torsional mode shape for the motor-compressor system shown in Fig. 38.1.

Torsional vibration is usually a complex vibration having many different frequency components. For example, shock resulting from abrupt start-ups and unloading of gear teeth causes transient torsional vibration in some systems; start-up of synchronous electric motor systems may cause torsional resonance. Random torsional vibration caused by gear inaccuracies and ball bearing defects is relatively common in rotating machines.

MODELING

The torsional elastic system of a drive unit and its associated machinery is a complicated arrangement of mass and elastic distribution. The complete mechanical system can include the drive unit, couplings, gearboxes or other speed-changing devices, and one or more driven units. This complicated system is made amenable to mathematical treatment by representing it as a model—a simpler system that is substantially equivalent dynamically. The equivalent system usually consists of lumped masses which are connected by massless torsionally elastic springs as illustrated in Fig. 38.3. The masses are placed at each crank center and at the center planes of actual flywheels, rotors, propellers, cranks, gears, impellers, and armatures.[1-3]

The torsional calculation is made not for the drive unit alone but for the complete system, including all driven machinery. On an engine it is usually possible to consider such parts as camshafts, pumps, and blowers either as detached from the engine (if they are driven elastically) or as additional rigid masses at the point of attachment to the crankshaft (if the driver is relatively rigid). If there is doubt, these parts should be included in the torsional calculation as elastically connected masses and removed if the natural frequencies do not change after the parts are removed from the model.

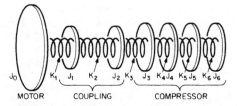

FIGURE 38.3 A model of the motor-compressor system shown in Fig. 38.1, consisting of a series of masses connected by massless torsionally elastic springs (K = stiffness, lb·in./rad; J = polar moment of inertia, lb·in.·sec^2).

CALCULATION OF POLAR MOMENTS OF INERTIA

Circular Disc or Cylinder Rotating about a Perpendicular Axis. The polar moment of inertia, essential in modeling torsional vibration, often is easy to calculate. The general form is $J = \int r^2\, dm,$ where r is the instantaneous radius, and dm is the differential mass. The formula for the polar moment of inertia of a circular disc or cylinder rotating about a perpendicular axis is

$$J = \frac{\pi d^4 l \gamma}{32g} \qquad \text{lb-in.-sec}^2 \qquad (38.1)$$

where J = polar moment of inertia, lb-in.-sec^2
   γ = material density, lb/in.^{3}
   d = diameter of disc or cylinder, in.
   l = axial length of disc or cylinder, in.
   g = acceleration due to gravity, 386.4 in./sec^2

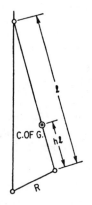

FIGURE 38.4 Schematic diagram of a crank and connecting rod.

Piston and Connecting Rod. The piston and connecting rod shown schematically in Fig. 38.4 introduce a variable-mass problem, the solution of which is complex. The exact solution shows that the effect of the piston and connecting rod can be closely approximated by representing them as a concentrated rotor of polar inertia J defined by

$$J = \left[\frac{W_P}{2} + W_c\left(1 - \frac{h}{2}\right)\right]\frac{R^2}{g} \qquad \text{lb-in.-sec}^2 \qquad (38.2)$$

where W_P = weight of piston, piston pin, and cooling fluid, lb
   W_c = weight of connecting rod, lb
   h = fraction of rod length from crank pin to center-of-gravity
   R = crank radius, in.

Crankshaft. The polar inertias of the crank webs (see Fig. 38.5), the crankpin, and the journal sections are added to that given by Eq. (38.2). These polar inertias should be calculated with the best obtainable accuracy. The following procedure is recommended:

Let the crank web be intercepted by a series of concentric cylinders of radius R. The polar inertia of the crank web is defined by

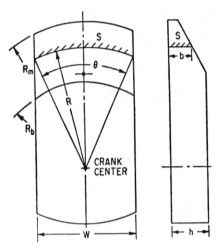

$$J = \left(\frac{\gamma}{g} \int_{R_b}^{R_m} R^2 S \, dR \right) + J_b \qquad (38.3)$$

where γ = weight density of the crank web, lb/in.^{3}
R_m = maximum radius of the crank web, in.
R_b = radius of base cylinder (see Fig. 38.5), in.
J_b = polar inertia of portion of crank web within base cylinder, lb-in.-sec^2

FIGURE 38.5 View of a crank web in a plane normal to the crankshaft axis.

The integral in the above expression for J is the area of the R^2S curve between the values of radii R_b and R_m.

For the crank web shown in Fig. 38.5 the area S is defined as $S = bR\theta/57.3$, where θ is measured in degrees. The polar inertia can then be expressed as

$$J = \left(\frac{\gamma}{57.3g} \int_{R_b}^{R_m} bR^3 \theta \, dR \right) + J_b \qquad \text{lb-in.-sec}^2 \qquad (38.4)$$

The same procedure is used to calculate the polar inertia of propellers and other irregular parts. In a marine propeller of ogival sections, i.e., flat driving face, circular arc back, and elliptically developed outline (do not use for other shapes), the polar inertia (excluding hub) is given by

$$J = 0.0046 \, \frac{nD^3bt}{g} \qquad \text{lb-in.-sec}^2$$

where n = number of blades
D = diameter of propeller, in.
b = maximum blade width, in.
t = maximum blade thickness at one-half radius (axis to tip), in.

Propellers. For propellers, pumps, and hydraulic couplings an addition must be made for the virtual inertia of the entrained fluid. For marine propellers this is ordinarily assumed at 26 percent of the propeller inertia. Virtual inertias for pumps are not known accurately, but it can be assumed that half the casing is filled with rotating fluid.

EXPERIMENTAL DETERMINATION OF POLAR MOMENT OF INERTIA

For complex shaft elements such as couplings or small flywheels, it is often easier to

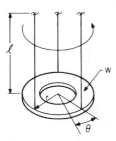

FIGURE 38.6 Experimental determination of the polar moment of inertia. An element of weight W is suspended by three wires and the period of the torsional motion is determined.

determine the polar moment of inertia experimentally than to calculate it. In one experimental technique the element is suspended from three equally spaced vertical wires as shown in Fig. 38.6. The element whose polar moment of inertia is to be measured is hung on the cables and set into torsional motion. Then the period of vibration is measured. The experimentally determined period of torsional vibration, the weight of the element, the length of the suspending cables, and the radius of attachment of the cables are used to determine the polar moment of inertia from the following formula:

$$J = \frac{Wr^2\tau^2}{(6.28)^2 l} \quad \text{lb-in.sec}^2 \tag{38.5}$$

where J = polar moment of inertia
τ = period of vibration, sec/cycle
W = weight of element, lb
l = length of cables, in.
r = radius of suspending cables, in.

CALCULATION OF STIFFNESS

Shaft.   The stiffness of a circular shaft is the most common elastic element encountered in the modeling process. Table 38.1 shows some common formulas used to calculate torsional stiffness of a hollow circular shaft, a tapered circular shaft, and two geared shafts. The stiffness is referred to the rotational speed of shaft No. 1. The inertia of geared shafts is obtained in a similar manner.

Crankshaft.   The crankshaft stiffness is the most uncertain element in a torsional vibration calculation. Shaft stiffness can be measured experimentally either by twisting a shaft with a known torque or from the observed values of the critical speeds in a running engine. Alternatively, it can be calculated from semiempirical formulas such as those given in Ref. 3. Those given by Eqs. (38.6), (38.7), and (38.8) are recommended. Refer to Figs. 38.4 and 38.5 for definitions of the dimensions; l_e is the length of a solid shaft of diameter D_s equal in torsional stiffness to the section of crankshaft between crank centers.
Wilson's formula[4]

$$\frac{l_e}{D_s^4} = \frac{b + 0.4d_s}{D_s^4 - d_s^4} + \frac{a + 0.4D_c}{D_c^4 - d_c^4} + \frac{r - 0.2(D_s + D_c)}{hW^3} \tag{38.6}$$

TABLE 38.1 Formulas for Torsional Stiffness

K = TORSIONAL STIFFNESS, LB - IN. /RAD; G= SHEAR MODULUS, LB/IN^2 ω = ROTATIONAL SPEED, RAD/SEC		
SPRINGS IN SERIES		$K = \dfrac{1}{1/K_1 + 1/K_2}$
SPRINGS IN PARALLEL		$K = K_1 + K_2$
HOLLOW CIRCULAR SHAFT		$K = \dfrac{\pi}{32} \dfrac{G(D^4 - d^4)}{\ell}$
TAPERED CIRCULAR SHAFT		$K = \dfrac{3\pi}{32} \dfrac{d^4}{\ell(n + n^2 + n^3)}$ $n = \dfrac{d}{D}$
TWO GEARED SHAFTS (REFERRED TO SHAFT 1)		$K = \dfrac{K_1 K_2}{n^2 K_1 + K_2}$ $n = \dfrac{D}{d} = \dfrac{\omega_1}{\omega_2}$

Ziamanenko's formula[3]

$$\frac{l_e}{D_s^4} = \frac{b + 0.6hD_s/b}{D_s^4 - d_s^4} + \frac{0.8a + 0.2(W/r)D_s}{D_c^4 - d_c^4} + \frac{r^{3/2}}{hW^3 D_c^{1/2}} \tag{38.7}$$

Constant's formula[5]

$$\frac{l_e}{D_s^4} = \frac{1}{\alpha_1 \alpha_2 \alpha_3 \alpha_4} \left(\frac{b}{D_s^4 - d_s^4} + \frac{a}{D_c^4 - d_c^4} + \frac{0.94}{hW^3} \right) \tag{38.8}$$

where α_1, α_2, α_3, and α_4 are modifying factors, determined as follows:

$$\alpha_1 = 1 - \frac{0.0825}{\sqrt{\dfrac{W_s - d_s}{2W_s} + \dfrac{W_c - d_c}{2W_c}} - 0.32} \tag{38.9}$$

If the shaft is solid, assume $\alpha_1 = 0.9$. The factor α_2 is a web-thickness modification determined as follows: If $4h/l$ is greater than $\frac{2}{3}$, then $\alpha_2 = 1.666 - 4h/l$. If $4h/l < \frac{2}{3}$, assume $\alpha_2 = 1$. The factor α_3 is a modification for web chamfering determined as follows: If the webs are chamfered, estimate α_3 by comparison with the cuts on Fig. 38.7:

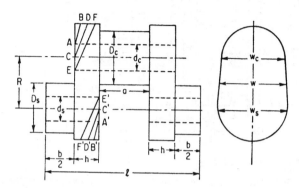

FIGURE 38.7 Schematic diagram of one crank of a crankshaft.

Cut AB and $A'B'$, $\alpha_3 = 1.000$; cut CD alone, $\alpha_3 = 0.965$; cut CD and $C'D'$, $\alpha_3 = 0.930$; cut EF alone, $\alpha_3 = 0.950$; cut EF and $E'F'$, $\alpha_3 = 0.900$; if ends are square, $\alpha_3 = 1.010$. The factor α_4 is a modification for bearing support given by

$$\alpha_4 = \frac{Al^3 w}{D_c^4 - d_c^4} + B \qquad (38.10)$$

For marine engine and large stationary engine shafts: $A = 0.0029$, $B = 0.91$
For auto and aircraft engine shafts: $A = 0.0100$, $B = 0.84$
If α_4 as given by Eq. (38.10) is less than 1.0, assume a value of 1.0.

The Constant's formula, Eq. (38.8), is recommended for shafts with large bores and heavy chamfers.

Changes in Section.   The shafting of an engine system may contain elements such as changes of section, collars, shrunk and keyed armatures, etc., which require the exercise of judgment in the assessment of stiffness. For a change of section having a fillet radius equal to 10 percent of the smaller diameter, the stiffness can be estimated by assuming that the smaller shaft is lengthened and the larger shaft is shortened by a length λ obtained from the curve of Fig. 38.8. This also may be applied to flanges where D is the bolt diameter. The stiffening effect of collars can be ignored.

Shrunk and Keyed Parts.   The stiffness of shrunk and keyed parts is difficult to estimate as the stiffening effect depends to a large extent on the tightness of the shrunk fit and keying. The most reliable values of stiffness are obtained by neglecting the stiffening effect of an armature and assuming that the armature acts as a concentrated mass at the center of the shrunk or keyed fit. Some armature spiders and flywheels have considerable flexibility in their arms; the treatment of these is discussed in the section *Geared and Branched Systems*.

Elastic Couplings.   Properties of numerous types of torsionally elastic couplings are available from the manufacturers and are given in Ref. 3.

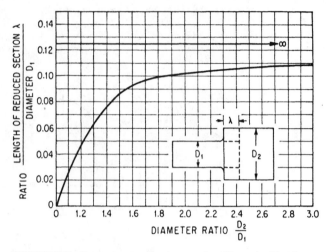

FIGURE 38.8 Curve showing the decrease in stiffness resulting from a change in shaft diameter. The stiffness of the shaft combination is the same as if the shaft having diameter D_1 is lengthened by λ and the shaft having diameter D_2 is shortened by λ. (*F. Porter.*[6])

GEARED AND BRANCHED SYSTEMS

The natural frequencies of a system containing gears can be calculated by assuming a system in which the speed of the driver unit is n times the speed of the driven equipment. Multiply all the inertia and elastic constants on the driven side of the system by $1/n^2$, and calculate the system's natural frequencies as if no gears exist. In any calculations involving damping constants on the driven side, these constants also are multiplied by $1/n^2$. Torques and deflections thus obtained on the driven side of this substitute system, when multiplied by n and $1/n$, respectively, are equal to those in the actual geared system. Alternatively, the driven side can be used as the reference; multiply the driver constants by n^2.

Where two or more drivers are geared to a common load, hydraulic or electrical couplings may be placed between the driver and the gears. These serve as disconnected clutches; they also insulate the gears from any driver-produced vibration. This insulation is so perfect that the driver end of the system can be calculated as if terminating at the coupling gap. The damping effect of such couplings upon the vibration in the driver end of the system normally is quite small and should be disregarded in amplitude calculations.

The majority of applications without hydraulic or electrical couplings involve two identical drivers. For such systems the modes of vibration are of two types:

1. The *opposite-phase* modes in which the drivers vibrate against each other with a node at the gear. These are calculated for a single branch in the usual manner, terminating the calculation at the gear. The condition for a natural frequency is that $\beta = 0$ at the gear.

2. The *like-phase* modes in which the two drivers vibrate in the same direction against the driven machinery. To calculate these frequencies, the inertia and stiffness constants of the driver side of one branch are doubled; then the calculation is made as if there were only a single driver. The condition for a natural frequency is zero residual torque at the end.

If the two identical drivers rotating in the same direction are so phased that the same cranks are vertical simultaneously, all orders of the opposite-phase modes will be eliminated. The two drivers can be so phased as to eliminate certain of the like-phase modes. For example, if the No. 1 cranks in the two branches are placed at an angle of 45° with respect to each other, the fourth, twelfth, twentieth, etc., orders, but no others, will be eliminated. If the drivers are connected with clutches, these phasing possibilities cannot be utilized.

In the general case of nonidentical branches the calculation is made as follows: Reduce the system to a 1:1 gear ratio. Call the branches a and b. Make the sequence calculation for a branch, with initial amplitude $\beta = 1$, and for the b branch, with the initial amplitude the algebraic unknown x. At the junction equate the amplitudes and find x. With this numerical value of the amplitude x substituted, the torques in the two branches and the torque of the gear are added; then the sequence calculation is continued through the last mass.

The branch may consist of a single member elastically connected to the system. Examples of such a branch are a flywheel with appreciable flexibility in its spokes or an armature with flexibility in the spider. Let I be the moment of inertia of the flywheel rim and k the elastic constant of the connection. Then the flexibly mounted flywheel is equivalent to a rigid flywheel of moment of inertia

$$I' = \frac{I}{1 - I\omega^2/k} \qquad (38.11)$$

NATURAL FREQUENCY CALCULATIONS

If the model of a system can be reduced to two lumped masses at opposite ends of a massless shaft, the natural frequency is given by

$$f_n = \frac{1}{2\pi} \sqrt{\frac{(J_1 + J_2)k}{J_1 J_2}} \qquad \text{Hz} \qquad (38.12)$$

The mode shape is given by $\theta_2/\theta_1 = -J_1/J_2$.

For the three-mass system shown in Fig. 38.9, the natural frequencies are

$$f_n = \frac{1}{2\pi} \sqrt{A \pm (A^2 - B)^{1/2}} \qquad \text{Hz} \qquad (38.13)$$

where $\quad A = \dfrac{k_{12}(J_1 + J_2)}{2J_1 J_2} + \dfrac{k_{23}(J_1 + J_2)}{2J_1 J_2}$

$$B = \frac{(J_1 + J_2 + J_3)k_{12}k_{23}}{J_1 J_2 J_3}$$

In Eqs. (38.12) and (38.13) the ks are torsional stiffness constants expressed in lb-in./rad. The notation k_{12} indicates that the constant applies to the shaft between rotors 1 and 2. The polar inertia J has units of lb-in.-sec^2.

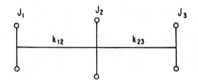

FIGURE 38.9 Schematic diagram of a shaft represented by three masses.

The above formulas and all the developments for multimass torsional systems that follow also apply to systems with longitudinal motion if the polar moments of inertia J are replaced by the masses $m = W/g$ and the torsional stiffnesses are replaced by longitudinal stiffnesses.

HOLZER METHOD

If there are more than three masses, the method* of Holzer can be used to calculate the natural frequencies. A frequency is assumed; then starting at one end of the system, a balance of torques and displacements is obtained step by step. The final external torque required to achieve balance is called the *residual torque*. If the residual torque is zero, the assumed frequency is one of the natural frequencies of the system. The mode associated with this natural frequency is identified by the number of changes of sign of the amplitudes at the various rotors. A plot, Fig. 38.10A, of residual torque vs. frequency yields the natural frequencies of the system within the frequency range considered.

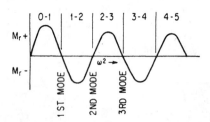

FIGURE 38.10A Typical curve of residual torque M_r as a function of the assumed frequency, ω. The numbers within the brackets at the top of the illustration represent the number of changes in the sign of β that are set forth opposite the rotor number in the sequence calculation.

In many internal-combustion engines, the polar inertias of the rotors representing the cranks and pistons are equal, as are the stiffness constants between these rotors. Advantage can be taken of this fact to reduce the labor of computation considerably. Let I be one of the equal polar moments of inertia and k be one of the equal spring constants. Divide all the polar moments of inertia by I, and divide all the spring constants by k. The new system is dynamically similar to the original one. It is called the *unity system*. Then if ω_n is the natural angular frequency of the actual system and ω_s that of the unity system,

$$\omega_n^2 = \omega_s^2 \frac{k}{I} \quad \text{rad/sec} \quad (38.14)$$

This "unity form" of the calculation is a labor-saving technique which is useful only when there are a number of equal masses and/or stiffnesses in the system; whether calculations are made using the unity form or the original constants is a matter of choice.

---

* This method was first proposed in 1912 in graphical form[7] and later in a tabular form.[8] This sequence calculation has the advantages of simplicity, ready extension to forced vibration, and ease of programming on a digital computer. The tabulation given here is in a more consistent form.

Before starting the sequence calculation, it is necessary to know approximately the frequency of the first mode. For an internal-combustion engine or any system with small inertias in the presence of large ones, the following procedure is generally satisfactory. Add all the polar inertias within the engine part of the system and assume that 40 percent of this is placed at the crank farthest from the flywheel or generator (No. 1 crank). Find the stiffness with reference to the flywheel, or the generator if no flywheel is fitted, to the No. 1 crank. If there is no flywheel, the system reduces to a two-rotor system, and Eq. (38.12) gives approximately the lowest natural frequency of the system. If there is a flywheel, the system reduces to one of three rotors; then Eq. (38.13) gives the approximate values of the two lowest natural frequencies of the system. For complicated systems involving numerous driven rotors no general rules can be given.

Example 38.1.   Figure 38.10B (upper scale) shows the actual mass and elasticity distribution of an eight-cylinder, four-cycle engine. Four of the cranks are fitted with counterweights so that the crank inertias are not the same. Dividing the moments of inertia by 35.8 and the stiffness constants by 175.7×10^6, the constants for the unity system are obtained, as shown in Fig. 38.10B (lower scale). The sum of the inertia constants through rotor 9 is 7.152. The stiffness of the shaft to rotor 10 is

$$\frac{1}{7 + (1/1.32) + (1/6.06)} = 0.126$$

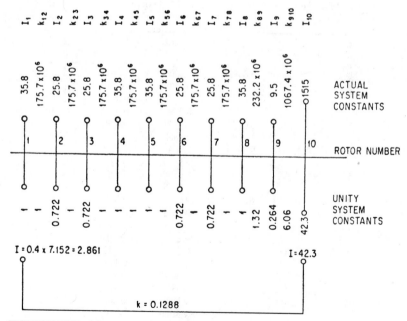

FIGURE 38.10B   Moments of inertia I and stiffness k for an eight-cylinder engine driving a generator. The constants for the actual system are shown in the upper scale and for the unity system in the lower scale.

The two-rotor system is therefore represented by a two degree-of-freedom model. The approximate value of ω_s^2 is calculated from Eq. (38.12), where $I_1 = 0.40 \times 7.152 = 2.861$ and $I_2 = 42.3$:

$$\omega_s^2 = \frac{0.126 \times 45.16}{2.861 \times 42.3} = 0.047 \quad (\text{rad/sec})^2$$

The Holzer calculation is started with $\omega_s^2 = 0.047$. The calculation is arranged as shown in Table 38.2. The rotor numbers and their inertia constants I are placed on alternate lines. The β values opposite the rotor numbers define the first mode relative amplitude curve, and the torques between rotor numbers are the shaft torques between the corresponding rotors. The intermediate lines are designated line 1–2, line 2–3, etc. The stiffness constants are placed on the intermediate lines in the last column. For the trial $\omega_s^2 = 0.047$, the $I\omega_s^2$ column is computed next. An amplitude of 1.0 is taken for rotor 1. Then the inertia reaction of rotor 1 is $0.047 \times 1.0 = 0.047$; it is placed in the M column on line 1. The torque in the shaft section between rotors 1 and 2 is equal to this inertia reaction and is placed in the M column in line 1–2. This torque divided by the stiffness, 1.00 in shaft section 1–2, is the relative deflection of rotors 1–2 and is placed in the β column of line 1–2. Subtracting this relative deflection 0.0470 from the deflection of rotor 1 gives the deflection 0.9530 of rotor 2. This deflection multiplied by 0.0339 ($I\omega_s^2$ for rotor 2) gives the inertia reaction 0.0323 of rotor 2. The inertia reaction of rotor 2 added to

TABLE 38.2 Holzer Calculations for First Mode of System Shown in Fig. 38.10B ($\omega_s^2 = 0.047$)

Rotor number	I	$I\omega_s^2$	Deflection β	Torque M, in.-lb	k
1	1.000	0.0470	1.0000	0.0470	
			0.0470	0.0470	1.00
2	0.722	0.0339	0.9530	0.0323	
			0.0793	0.0793	1.00
3	0.722	0.0339	0.8737	0.0296	
			0.1089	0.1089	1.00
4	1.000	0.0470	0.7648	0.0360	
			0.1449	0.1449	1.00
5	1.000	0.0470	0.6199	0.0291	
			0.1740	0.1740	1.00
6	0.722	0.0339	0.4459	0.0151	
			0.1891	0.1891	1.00
7	0.722	0.0339	0.2568	0.0076	
			0.1967	0.1967	1.00
8	1.000	0.0470	0.0601	0.0028	
			0.1511	0.1995	1.32
9	0.265	0.0125	−0.0910	−0.0011	
			0.0327	0.1984	6.06
10	42.500	1.9975	−0.1237	−0.2471	
			Residual torque $M_r = -0.0487$		

the torque in section 1–2 gives the torque in section 2–3. Thus, Table 38.2 is carried through step by step. Note that in lines 8–9 and 9–10, k is no longer unity. The final figure in the torque column is called the *residual torque*. All additions and sub-tractions are made in the algebraic sense, i.e., on line 9 the deflection $-0.0910 = 0.0601 - 0.1511$.

The residual torque has the following physical significance: If a simple harmonic torque of amplitude M_r and angular frequency ω_s is applied to rotor 10 of the system, then all the rotors will oscillate with the amplitudes as given in the β column and the torques will be as given in the M column. If such a Holzer calculation is made over a sufficient range of values of ω_s to cover all the natural frequencies, M_r will follow the general shape of the curve shown in Fig. 38.10A. The number of crossings of zero amplitude, excluding the origin, is equal to the number of rotors minus 1. The location of the trial values of ω_s^2 on this diagram can be determined from the sign of M_r and the number of changes of sign of the relative amplitudes β opposite the rotor numbers in the sequence table. In the above example, there is one change of sign of β and M_r is negative. Therefore the assumed value of ω_s^2 lies between the first and second modes.

To find the first mode, assume a new trial value of $\omega_s^2 = 0.045$. This leads to $M_r = +0.0253$. Interpolating for a zero M_r between $\omega_s^2 = 0.047$ and 0.045 yields 0.0457. Table 38.3 is constructed for this value and represents the first mode conditions with a frequency error of less than one one-thousandth, a much higher accuracy than is justified by the data on which it is based.

TABLE 38.3   Holzer Calculations for First Mode of System Shown in Fig. 38.10B ($\omega_s^2 = 0.0457$)

Rotor number	I	$I\omega_s^2$	Deflection β	Torque M, in.-lb	k
1	1.000	0.0457	1.0000	0.0457	
			0.0457	0.0457	1.00
2	0.722	0.0330	0.9543	0.0315	
			0.0772	0.0772	1.00
3	0.722	0.0330	0.8771	0.0290	
			0.1062	0.1062	1.00
4	1.000	0.0457	0.7709	0.0351	
			0.1413	0.1413	1.00
5	1.000	0.0457	0.6296	0.0288	
			0.1701	0.1701	1.00
6	0.722	0.0330	0.4595	0.0152	
			0.1853	0.1853	1.00
7	0.722	0.0330	0.2742	0.0090	
			0.1943	0.1943	1.00
8	1.000	0.0457	0.0799	0.0036	
			0.1499	0.1979	1.32
9	0.265	0.0121	−0.0700	−0.0008	
			0.0325	0.1971	6.06
10	42.500	1.9422	−0.1025	−0.1990	
			Residual torque $M_r = -0.0019$		

The second mode frequency corresponding to the condition in which there are two nodes in the engine shaft is approximated by

$$\omega_s^2 = \frac{22}{n^2}$$

where n is the number of cylinders. In the example of Fig. 38.10B

$$\omega_s^2 = \frac{22}{64} = 0.345$$

By the Holzer calculation, this is found to be somewhat low, and a value which will make M_r nearly zero is $\omega_s^2 = 0.360$. The natural frequencies of the actual system are given by Eq. (38.14).

$$\omega_n^2 = \frac{175.7 \times 10^6 \omega_s^2}{35.8} = 4,910,000\omega_s^2$$

TABLE 38.4 Sources of Excitation of Torsional Vibration

Source	Amplitude in terms of rated torque	Frequency
Mechanical		
Gear runout		$1 \times, 2 \times, 3 \times$ rpm
Gear tooth machining tolerances		No. gear teeth $\times$ rpm
Coupling unbalance		$1 \times$ rpm
Hooke's joint		$2 \times, 4 \times, 6 \times$ rpm
Coupling misalignment		Dependent on drive elements
System function		
Synchronous motor start-up	5–10	$2 \times$ slip frequency
Variable-frequency induction motors (six-step adjustable frequency drive)	0.04–1.0	$6 \times, 12 \times, 18 \times$ line frequency (LF)
Induction motor start-up	3–10	Air gap induced at 60 Hz
Variable-frequency induction motor (pulse width modulated)	0.01–0.2	$5 \times, 7 \times, 9 \times$ LF, etc.
Centrifugal pumps	0.10–0.4	No. vanes $\times$ rpm and multiples
Reciprocating pumps		No. plungers $\times$ rpm and multiples
Compressors with vaned diffusers	0.03–1.0	No. vanes $\times$ rpm
Motor- or turbine-driven systems	0.05–1.0	No. poles or blades $\times$ rpm
Engine geared systems with soft coupling	0.15–0.3	Depends on engine design and operating conditions; can be $0.5n$ and $n \times$ rpm
Engine geared system with stiff coupling	0.50 or more	Depends on engine design and operating conditions
Shaft vibration		$n \times$ rpm

Then the natural frequencies in the first and second modes are:

1st mode: $\omega_s^2 = 0.0457$; $\omega^2 = 224,000$; $f = 75.3$ Hz, or 4518 cpm

2d mode: $\omega_s^2 = 0.3600$; $\omega^2 = 1,765,000$; $f = 212$ Hz, or 12,720 cpm

TRANSFER MATRIX METHOD

The transfer matrix method[9] is an extended and generalized version of the Holzer method. Matrix algebra is used rather than a numerical table for the analysis of torsional vibration problems. The transfer matrix method is used to calculate the natural frequencies and critical speeds of other eigenvalue problems.

The transfer matrix and matrix iteration (Stodola) methods are numerical procedures. The fundamental difference between them lies in the assumed independent variable. In any eigenvalue problem, a unique mode shape of the system is associated with each natural frequency. The mode shape is the independent variable used in the matrix iteration method. A mode shape is assumed and improved by successive iterations until the desired accuracy is obtained; its associated natural frequency is then calculated.

A frequency is assumed in the transfer matrix method, and the mode shape of the system is calculated. If the mode shape fits the boundary conditions, the assumed frequency is a natural frequency and a critical speed is derived. Determining the correct natural frequencies amounts to a controlled trial-and-error process. Some of the essential boundary conditions (geometrical) and natural boundary conditions (force) are assumed, and the remaining boundary condition is plotted vs. frequency to obtain the natural frequency; the procedure is similar to the Holzer method. For example, if the torsional system shown in Fig. 38.11 were analyzed, the natural boundary conditions would be zero torque at both ends. The torque at station No. 1 is made zero, and the torsional vibration is set at unity. Then M_4 as a function of ω is plotted to find the natural frequencies. This plot is obtained by utilizing the system transfer functions or matrices. These quantities reflect the dynamic behavior of the system.

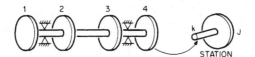

FIGURE 38.11   Typical torsional vibration model.

No accuracy is lost with the transfer matrix method because of coupling of mode shapes. Accuracy is lost with the matrix iteration method, however, because each frequency calculation is independent of the others. A minor disadvantage of the transfer matrix method is the large number of points that must be calculated to obtain an M_4 *vs* ω curve. This problem is overcome if a high-speed digital computer is used.

A typical station (No. 4) from a torsional model is shown in Fig. 38.11. This general station and the following transfer matrix equation are used in a way similar to the Holzer table to transfer the effects of a given frequency ω across the model (see the example of an engine driving a generator shown in Fig. 38.10B):

$$\begin{vmatrix} \theta \\ M \end{vmatrix}_n = \begin{bmatrix} 1 & 1/k \\ -\omega^2 J & -(\omega^2 J/k) + 1 \end{bmatrix}_n \begin{vmatrix} \theta \\ M \end{vmatrix}_{n-1}$$

where θ = torsional motion, rad
          M = torque, lb-in.
          ω = assumed frequency, rad/sec
          J = station inertia, lb-in.-sec^2
          k = station torsional stiffness, lb-in./rad

The stiffness and polar moment of inertia of each station are entered into the equation to determine the transfer effect of each element of the model. Thus, the calculation begins with station No. 1, which relates to the first spring and inertia in the model of Fig. 38.11. The equation gives the output torque M_1 and output motion θ_1 for given input values, usually 0 and 1, respectively. The equation is used on station No. 2 to obtain M_2 output and θ_2 output as a function of M_1 output and θ_1 output. This process is repeated to find the value of M and θ at the end of the model. This calculation is particularly suited for the digital computer.

FINITE ELEMENT METHOD

The finite element method is a numerical procedure (described in Chap. 28, Part II) to calculate the natural frequencies, mode shapes, and forced response of a discretely modeled structural or rotor system. The complex rotor system is composed of an assemblage of discrete smaller finite elements which are continuous structural members. The displacements (angular) are forced to be compatible, and force (torque) balance is required at the joints (often called *nodes*).

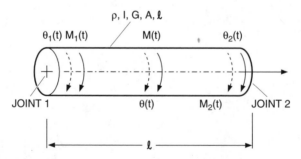

FIGURE 38.12   Finite element for torsional vibration in local coordinates.

Figure 38.12 shows a uniform torsional element in local coordinates. The x axis is taken along the centroidal axis. The physical properties of the element are density (ρ), area (A), shear modulus of elasticity (G), length (l), and polar area moment (I). $M(t)$ are the torsional forcing functions.

The torsional displacement within the element can be expressed in terms of the joint rotations $\theta_1(t)$ and $\theta_2(t)$ as

$$\theta(x,t) = U_1(x)\theta_1(t) + U_2(x)\theta_2(t) \tag{38.15}$$

where $U_1(x)$ and $U_2(x)$ are called *shape functions.* Since $\theta(0,t) = \theta_1(t)$ and $\theta(l,t) = \theta_2(t)$, the shape functions must satisfy the boundary conditions:

$$U_1(0) = 1 \qquad U_1(l) = 0$$

$$U_2(0) = 0 \qquad U_2(l) = 1$$

The shape function for the torsional element is assumed to be a polynomial with two constants of the form

$$U_i(x) = a_i + b_i x \qquad \text{where } i = 1,2 \tag{38.16}$$

Selection of the shape function is performed by the analyst and is a part of the engineering art required to conduct accurate finite element modeling.

Thus with four known boundary conditions the values of a_i and b_i can be determined from Eq. (38.16):

$$U_1(x) = 1 - \frac{x}{l} \qquad U_2(x) = \frac{x}{l}$$

Then from Eq. (38.15)

$$\theta(x,t) = \left(1 - \frac{x}{l}\right)\theta_1(t) + \frac{x}{l}\theta_2(t)$$

The kinetic energy, strain energy, and virtual work are used to formulate the finite element mass and stiffness matrices and the force vectors, respectively. These quantities are used to form the equations of motion. These matrices, derived in Ref. 9, are

$$\{J\} = \frac{\rho I l}{6}\begin{bmatrix} 2 & 1 \\ 1 & 2 \end{bmatrix}$$

$$\{K\} = \frac{GI}{l}\begin{bmatrix} 1 & -1 \\ -1 & 1 \end{bmatrix}$$

$$\bar{M} = \begin{Bmatrix} M_1(t) \\ M_2(t) \end{Bmatrix} = \begin{Bmatrix} \int_0^l M(x,t)\left(1 - \frac{x}{l}\right)dx \\ \int_0^l M(x,t)\left(\frac{x}{l}\right)dx \end{Bmatrix}$$

where $\{J\}$ = mass matrix
$\{K\}$ = stiffness matrix
$\bar{M}$ = torque vector
ρ = density
I = area polar moment
G = shear modulus
l = length of element

As noted, the previously described finite elements are in local coordinates. Since the system as a whole must be analyzed as a unit, the elements must be transformed into one global coordinate system. Figure 38.13 shows the local element within a global coordinate system. The mass and stiffness matrices and joint force vector of each element must be expressed in the global coordinate system to find the vibration response of the complete system.

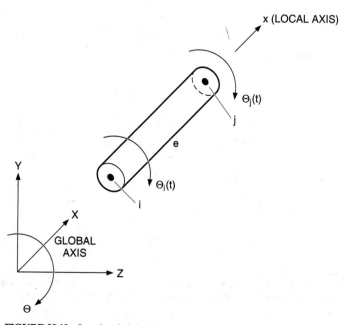

FIGURE 38.13   Local and global joint displacements of element, l.

Using transformation matrices,[8] the mass and stiffness matrices and force vectors are used to set up the system equation of motion for a single element in the global coordinates:

$$[J]_e\{\ddot{\Theta}(t)\} + [K]_e\{\ddot{\Theta}(t)\} = \{M_e(t)\}$$

The complete system is an assemblage of the number of finite elements it requires to adequately model its dynamic behavior. The joint displacements of the elements in the global coordinate system are labeled as $\Theta_1(t), \Theta_2(t), \ldots, \Theta_m(t)$, or this can be expressed as a column vector:

$$\{\Theta(t)\} = \begin{Bmatrix} \Theta_1(t) \\ \Theta_2(t) \\ \cdot \\ \cdot \\ \cdot \\ \Theta_m(t) \end{Bmatrix}$$

Using global joint displacements, mass and stiffness matrices, and force vectors, the equations of motion are developed:

$$[J]_{nxn}\{\ddot{\Theta}\}_{nx1} + [K]_{nxn}\{\Theta\}_{nx1} = \{M\}_{nx1}$$

where n denotes the number of joint displacements in the system.

In the final step prior to solution, appropriate boundary conditions and constraints are introduced into the global model.

The equations of motion for free vibration are solved for the eigenvalues (natural frequencies) using the matrix iteration method (Chap. 28, Part I). Modal analysis is used to solve the forced torsional response. The finite element method is available in commercially available computer programs for the personal computer. The analyst must select the joints (nodes, materials, shape functions, geometry, torques, and constraints) to model the system for computation of natural frequencies, mode shapes, and torsional response. Similar to other modeling efforts, engineering art and a knowledge of the capabilities of the computer program enable the engineer to provide reasonably accurate results.

CRITICAL SPEEDS

The crankshaft of a reciprocating engine or the rotors of a turbine or motor, and all moving parts driven by them, comprise a torsional elastic system. Such a system has several modes of free torsional oscillation. Each mode is characterized by a natural frequency and by a pattern of relative amplitudes of parts of the system when it is oscillating at its natural frequency. The harmonic components of the driving torque excite vibration of the system in its modes. If the frequency of any harmonic component of the torque is equal to (or close to) the frequency of any mode of vibration, a condition of resonance exists and the machine is said to be running at a *critical speed*. Operation of the system at such critical speeds can be very dangerous, resulting in fracture of the shafting.

The number of complete oscillations of the elastic system per unit revolution of the shaft is called the *order of a critical speed*. An order of a critical speed that corresponds to a harmonic component of the torque from the engine as a whole is called a *major order*. A critical speed also can be excited that corresponds to the harmonic component of the torque curve of a single cylinder. The fundamental period of the torque from a single cylinder in a four-cycle engine is 720°; the critical speeds in such an engine can be of ½, 1, 1½, 2, 2½, etc., order. In a two-cycle engine only the critical speeds of 1, 2, 3, etc., order can exist. All critical speeds except those of the major orders are called *minor critical speeds;* this term does not necessarily mean that they are unimportant.

A dynamic analysis of an engine involves several steps. Natural frequencies of the modes likely to be important must be calculated. The calculation is usually limited to the lowest mode or the two lowest modes. In complicated arrangements, the calculation of additional modes may be required, depending on the frequency of the forces causing the vibration. Vibration amplitudes and stresses around the operating range and at the critical speeds must be calculated. A study of remedial measures is also necessary.

Example 38.2. A six-cylinder four-cycle engine with 120° crank spacings has three equally spaced firing impulses per revolution. What are the critical speeds?

The major orders of 3, 6, 9, 12, etc., are obtained from a phase diagram that considers the engine force balance at frequencies of 3, 6, 9, 12, etc., times rotational speed. The critical speeds occur at

$$\frac{60f_n}{q} \quad \text{rpm} \tag{38.17}$$

where f_n is the natural frequency of one of the modes in Hz, and q is the order number of the critical speed. Although many critical speeds exist in the operating range of an engine, only a few are likely to be important.

The critical speeds for Example 38.1 are given by Eq. (38.17) and expressed as follows:

Order	4	4½	5	5½	6	6½	7	7½	8	12
1-noded (rpm)	1130	1004	904	821	753	695	645	602	565	377
2-noded (rpm)									1590	1060

VIBRATORY TORQUES

Torsional vibration, like any other type of vibration, results from a source of excitation. The mechanisms that introduce torsional vibration into a machine system are discussed and quantified in this section. The principal sources of the vibratory torques that cause torsional vibration are engines, pumps, propellers, and electric motors.

GENERAL EXCITATION

Table 38.4 shows some ways by which torsional vibration can be excited. Most of these sources are related to the work done by the machine and thus cannot be entirely removed. Many times, however, adjustments can be made during the design process. For example, certain construction and installation sources—gear runout, unbalanced or misaligned couplings, and gear-tooth machining errors—can be reduced.

In Table 38.4 note that the pulsating torque during start-up of a synchronous motor is equal to twice the slip frequency. The slip frequency varies from twice the line frequency at start-up to zero at synchronous speed. Many mechanical drives exhibit characteristics of pulsating torque during operation due to their design function. Electric motors with variable-frequency drives induce pulsating torques at frequencies that are harmonics of line frequency. Blade-passing excitations can be characterized by the number of blades or vanes on the wheel: The frequency of excitation equals the number of blades multiplied by shaft speed. The amplitude of a pulsating torque is often given in terms of percentage of average torque generated in a system.

ENGINE EXCITATION

In more complex cases, diesel gasoline engines for example, the multiple frequency components depend on engine design and power output. The power output, crankshaft phasing, and relationship between gas torque and inertial torque influence the level of torsional excitation.

Inertia Torque. A harmonic analysis of the inertia torque of a cylinder is closely approximated by[2,8]

$$M = \frac{W}{g}\Omega^2 r\left(\frac{\lambda}{4}\sin\theta - \frac{1}{2}\sin 2\theta - \frac{3}{4}\lambda\sin 3\theta - \frac{\lambda^2}{4}\sin 4\theta\cdots\right) \qquad (38.18)$$

where $W = W_p + hW_c$ [see Fig. 38.4 and Eq.(38.3)]
$\lambda = r/l$ [see Fig. 38.4 and Eq. (38.3)]
Ω = angular speed, rad/sec
r = crank radius, in.
l = connecting rod length, in.
θ = crank angle, radians
W_p = weight of piston, lb
W_c = weight of connecting rod, lb

It is usual to drop all terms above the third order.

FIGURE 38.14 Schematic diagram of crank and connecting rod used in plotting torque curve.

Gas-Pressure Torque. A harmonic analysis of the turning effort curve yields the gas-pressure components of the exciting torque. The turning effort curve is obtained from the indicator card of the engine by the graphical construction shown in Fig. 38.14.

For a given crank angle θ, let the gas pressure on the piston be P. Erect a perpendicular to the line of action of the piston from the crank center, intersecting the line of the connecting rod. Let the intercept Oa on this perpendicular be y. Then the torque M for angle θ is given by

$$M = PSy \qquad (38.19)$$

where S is the piston area.

Coefficients of Standard Indicator Cards. In some cases it may be possible to use prepared analyses obtained from standard sets of indicator cards. One such set of diesel indicator cards and the harmonic analysis of the corresponding turning effort curves are shown in Figs. 38.15 and 38.16. Except for the first, second, and third orders, the resultant harmonic coefficients are given by

$$h = \sqrt{a^2 + b^2}$$

Then the exciting torque per cylinder is

$$M_e = Srh$$

The coefficients are for four-cycle engines; for two-cycle engines multiply by 2; there are no fractional orders.

For the first, second, and third orders, the sine inertia component of the turning effort curve is computed by Eq. (38.18) and added algebraically to the sine gas-

TABLE 38.5 Phase Diagrams and Deflections for the Mode Calculated in Table 38.3

PHASE DIAGRAMS	$\Sigma\beta$	ORDERS	cn	β
FIRING SEQUENCE 4 1 6 / 7—✕—2 / 3 8 5	0.778	1/2, 4½, 8½,--	1 2 3 4 5 6 7 8	1.0000 0.9543 0.8771 0.7709 0.6296 0.4595 0.2742 0.0799
		MIRROR IMAGE FOR 3½, 5½, 7½,--		5.0455
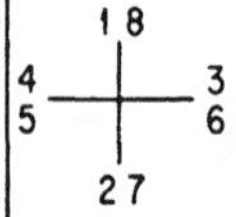 ✕2 OR 6 1 8 / 4 ──── 3 / 5 6 / 27	0.169	1, 5, 9, — MIRROR IMAGE FOR 3, 7, 11		
 ✕3 OR 5 1 / 3 5 / 2—✕—7 / 4 6 / 8	1.549	1½, 5½, ·9½, + − MIRROR IMAGE FOR 2½, 6½, 10½, --		
 ✕4 1,2,7,8 ↓ 3,4,5,6	0.4287	2, 6, 10,--		
 ✕8 1,2,3,4,5,6,7,8 ↑	5.0455	4, 8, 12 MAJOR ORDERS		

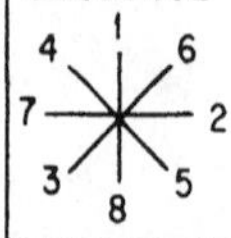

TABLE 38.6 Empirical Factors for Engine Amplitude Calculations

Bore Stroke	$\mathfrak{R}$
20 in. × 24 in. or larger	50–60
8 in. × 10 in.	40–50
4 in. × 6 in. or smaller	35

pressure component. Then the resultant M_e is the square root of the sums of the squares of the sine and cosine components. Analyses for other standard diesel indicator cards are given in Ref. 10.

The indicator cards of modern high-speed diesel engines with supercharging may differ considerably from the standard cards of Fig. 38.15, and a full analysis may be necessary. Approximate values of the coefficients for such engines can be obtained by multiplying the coefficients of Fig. 38.16 by the ratio of the maximum combustion pressure to that shown by the cards of Fig. 38.15, with the additional multiplication

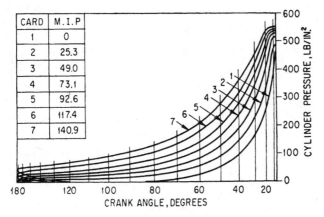

FIGURE 38.15 Standard diesel indicator diagrams for engine with $l/r = 4.25$. The mean indicated pressure is shown in the inset table. (*F. M. Lewis.*[2])

by 2 for a two-cycle engine. The higher-order coefficients are extremely sensitive to the details of the fuel-injection and combustion process; unless good experimental indicator cards are available for harmonic analysis, it may not be possible to obtain accuracy in estimating exciting torques.

For a four-cycle gasoline engine the harmonic coefficients are more nearly proportional to the mean indicated pressure and are approximated by the values in Table 38.7.

FORCED VIBRATION RESPONSE

The torsional vibration amplitude of a modeled system is determined by the magnitude, points of application, and phase relations of the exciting torques produced by engine or compressor gas pressure and inertia and by the magnitudes and points of application of the damping torques. Damping is attributable to a variety of sources, including pumping action in the engine bearings, hysteresis in the shafting and between fitted parts, and energy absorbed in the engine frame and foundation. In a few cases, notably marine propellers, damping of the propeller predominates. When an engine is fitted with a damper, the effects of damping dominate the torsional vibrations.

Techniques available for calculation of vibration amplitudes include the exact solution of differential equations, the energy balance method, the transfer matrix method, and modal analysis. The techniques are implemented on lumped parameter or finite-element models.

EXACT METHOD FOR TWO DEGREE-OF-FREEDOM SYSTEMS

The lowest mode of vibration of some systems, particularly marine installations, can be approximated with a two-mass system; an excitation is applied at one end and damping at the other.

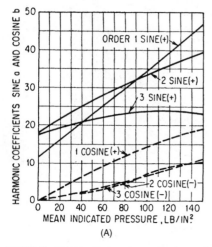

FIGURE 38.16A Harmonic coefficients of sine and cosine terms for first, second, and third orders for indicator diagrams shown in Fig. 38.15. (*F. M. Lewis.*[2])

Referring to Fig. 38.17, the torque equations for rotors I_1 and I_2 are

$$I_1\omega^2\theta_1 - k(\theta_1 - \theta_2) + M_e = 0$$

$$I_2\omega^2\theta_2 + k(\theta_1 - \theta_2) - jc\omega\theta_2 = 0$$

The natural frequency is given by

$$\omega^2 = \frac{k(I_1 + I_2)}{I_1 I_2}$$

The shaft torque is $M_{12} = k(\theta_1 - \theta_2)$. If the above equations are solved, the amplitude of M_{12} at resonance is

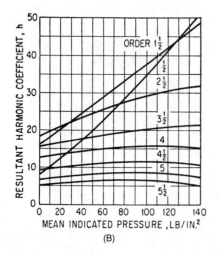

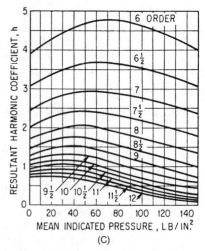

FIGURE 38.16B and C Resultant harmonic coefficients for lower orders of indicator diagrams shown in Fig. 38.15 and 38.16C for sixth and higher orders. (*F. M. Lewis.*[2])

$$|M_{12}| = k|\theta_1 - \theta_2| = M_e \frac{I_2}{I_1}\sqrt{1 + \frac{kI_2(I_1 + I_2)}{I_1 c^2}} \tag{38.20}$$

Since with usual damping the second term under the radical is large compared with unity, Eq. (38.20) reduces to

$$|M_{12}| \simeq \frac{M_e}{c}\frac{I_2}{I_1}\sqrt{(I_1 + I_2)\frac{I_2 k}{I_1}} \tag{38.21}$$

TABLE 38.7 Harmonic Components of Gas-Pressure Torque for a Four-Cycle Gasoline Engine

Order	½	1	1½	2	2½	3	3½
Resultant coefficient for h	34		31		17		10
Sine coefficient for a		31		21		11.5	
Cosine coefficient for b		10		−4		4.5	

Order	4	4½	5	5½	6	6½	7	7½	
h	7.5	5.5	4	3.2	2.9	2.45	2	1.68	

Order	8	8½	9	9½	10	10½	11	11½	12
h	1.5	1.32	1.15	1.10	0.88	0.78	0.71	0.58	1.52

$$h, a, b = \frac{\text{mean indicated pressure}}{100} \times \text{coefficients above.}$$

$$M_e = Sr \times (h \text{ or } a \text{ or } b)$$

The torsional damping constant c of a marine propeller is a matter of some uncertainty. It is customary to use the "steady-state" value. This is an approximation:

$$c = \frac{4M_{\text{mean}}}{\Omega} \quad \text{in.-lb/rad/sec}$$

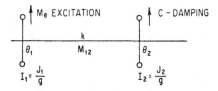

FIGURE 38.17 Schematic diagram of a shaft with two rotors, showing positions of excitation and damping.

where Ω = angular speed of shaft in radians per second. Considerations of oscillating airfoil theory indicate that this is too high and that a better value would be

$$c = \frac{2.3M_{\text{mean}}}{\Omega} \quad \text{in.-lb/rad/sec} \quad (38.22)$$

Equation (38.21) is applicable only when $I_1/I_2 > 1$. If used outside this range with other types of damping neglected, fictitiously large amplitudes will be obtained. Equation (38.21) gives the resonance amplitude, but the peak may not occur exactly at resonance. The complete amplitude curve is computed by the methods discussed in the following section.

ENERGY BALANCE METHOD

Both rational and empirical formulas for the resonance amplitudes of systems without dampers can be based on the energy balance at resonance. It is assumed that the system vibrates in a normal mode and that the displacement is in a 90° phase relationship to the exciting and damping torques. The energy input by the exciting torques is then equal to the energy output by the damping torques. Unless the damping is extremely large, this assumption gives a very close approximation to the amplitude at resonance.

Figure 38.18 shows a curve of relative amplitude in the first mode of vibration. Assume that a cylinder acts at A. Let the actual amplitude at A be θ_a and the amplitude relative to that of the No. 1 cylinder be β. The β values are taken from the column opposite each rotor number in the sequence calculation for the natural

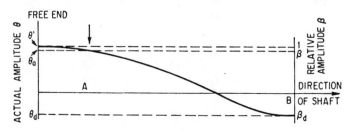

FIGURE 38.18   Diagram of actual amplitude θ and relative amplitude β as a function of position along shaft. Excitation is at A, and B is the position where damping is applied. The No. 1 cylinder is at the free end of the crankshaft.

frequency. At a point such as B, where damping may be applied, let the actual amplitude be θ_d and the amplitude relative to the No. 1 cylinder be β_d.

The energy input to the system from the cylinder acting at A is

$$\pi M_e \theta_a \qquad \text{in.-lb/cycle}$$

and the energy output to the damper is

$$\pi c \omega \theta_d^2 \qquad \text{in.-lb/cycle}$$

where c^* is the damping constant action of the damper at B. Equating input to output,

$$M_e \theta_a = c \omega \theta_d^2 \qquad (38.23a)$$

Let θ' be the amplitude at the No. 1 cylinder produced by the cylinder acting at A. Then $\theta_a/\theta' = \beta$ and $\theta_d/\theta' = \beta_d$. Substituting in Eq. (38.23a) gives

$$\theta' = \frac{M_e \beta}{c \omega \beta_d^2} \qquad (38.23b)$$

If all the cylinders act, and if damping is applied at a variety of points, the total amplitude at the No. 1 cylinder is

$$\theta = \Sigma \theta' = \frac{M_e \Sigma \beta}{\omega \Sigma c \beta_d^2} \qquad (38.24)$$

where $\Sigma \beta$ is taken over the cylinders and $\Sigma c \beta_d^2$ is taken over the points at which damping is applied. This formula can be applied directly when the magnitude and points of application of the damping torques are known. For the great majority of

---

* The symbol c is used in this chapter to denote a torsional damping coefficient.

applications, where the damping is unknown, a number of empirical formulas have been proposed with coefficients based on engine tests. These formulas may give an amplitude varying 30 percent or more from test results if applied to a variety of engines. Better agreement should not be expected, for even identical engines may have amplitudes differing as much as 2 to 1, depending on length of service, bearing fits, mounting, variation in the harmonic excitation because of different combustion rates, and other unknown factors.

Good results have been obtained using the Lewis formula[2]

$$M_m = \Re M_e \Sigma \beta \qquad (38.25)$$

The maximum torque at resonance in any part of the system is M_m; the exciting torque per cylinder is M_e. The vector sum over the cylinders of the relative amplitudes as taken from the sequence calculation for a natural frequency (Table 38.3 for example) is $\Sigma \beta$. It is determined as follows.

For a four-cycle engine construct a phase diagram of the firing sequence in which 720° corresponds to a complete cycle of a single cylinder, or two revolutions. The phase relationship for a critical of order number q is obtained by multiplying the angles in this diagram by $2q$, with the No. 1 crank held fixed. The β values assigned to each direction then are obtained from the values corresponding to each cylinder in the β column of the sequence calculation. Then $\Sigma \beta$ is the vector sum. The summation extends only to those rotors on which exciting torques act.

In a two-cycle engine the β phase relations are determined by multiplying the crank diagram by q, holding the No. 1 cylinder fixed.

Table 38.5 shows the $\Sigma \beta$ phase diagrams and $\Sigma \beta$ values for the one-noded mode of the example of Table 38.2, with a firing sequence 1, 6, 2, 5, 8, 3, 7, 4. The firing sequence is drawn first; then the angles of this diagram are multiplied by 2, 3, 4, etc., in succeeding diagrams. After multiplication by 8 for the fourth order, the diagrams repeat. Diagrams which are equidistant in order number from the 2, 6, 10, etc., orders are mirror images of each other and have the same $\Sigma \beta$. The numerical values of $\Sigma \beta$ in Table 38.5 have been obtained by calculation, summing the vertical and horizontal components.

The empirical factor $\Re$ is determined by the measurement of amplitudes in running engines (Table 38.6).

The exciting torque per cylinder, M_e in Eq. (38.24) is composed of the sum of the torques produced by gas pressure, inertia force, gravity force, and friction force. The gravity and friction torques are of negligible importance; and the inertia torque is of importance only for first-, second-, and third-order harmonic components.

Example 38.3. Table 38.8 gives the amplitude estimate for the first mode of the diesel engine, the frequency of which was calculated in Table 38.3. The engine is four-cycle, eight-cylinder with 9-in. bore by 12-in. stroke. The values of speed in rpm are as previously calculated, and $\Sigma \beta$ is taken from Table 38.6. The harmonic coefficients h are obtained from a harmonic analysis of an indicator card. Then the exciting torques are

$$M_e = \frac{\pi}{4} \times 9^2 \times 6 \times h = 383h \qquad \text{in.-lb}$$

For the maximum torques, M_m, a value of $\Re = 50$ was assumed in the equation

$$M_m = \Re M_e \Sigma \beta \qquad \text{in.-lb}$$

The crankpin diameter is 6½ in. so that the nominal stress is

$$\tau = \frac{16 M_m}{\pi \times (6.5)^3} = 0.0174 M_m \qquad \text{lb/in.}^2$$

TABLE 38.8 Amplitude Estimates for First Mode of Diesel Engine Calculated in Table 38.3

Order	rpm	$\Sigma\beta$	h	M_e, in.-lb	M_m, in.-lb	Stress	θ_1 rad
4	1125	5.055	29.5	11,300	2,860,000	49,800	0.08257
4½	1000	0.777	23.5	9,000	349,000	6,070	0.0105
5	900	0.164	18.5	7,080	58,000	1,005	
5½	820	1.54	15.0	5,730	441,000	7,700	0.0127
6	750	0.430	12.0	4,580	98,500	1,715	
6½	692	1.54	8.8	3,370	259,000	4,500	0.0075
7	643	0.164	6.5	2,490	20,400	355	
7½	600	0.777	5.5	2,110	81,900	656	
8	562	5.055	3.9	1,490	377,000	6,550	0.0108
12	375	5.055	1.4	537	136,000	2,380	0.0039
4	950	. . .	29.5	11,300	230,000	4,010	

From Table 38.3 (unity system) the torque in shaft section 8–9 is 0.1979 for unity amplitude at the No. 1 cylinder. Therefore the vibration amplitude at the No. 1 cylinder in the actual installation is

$$\theta = \frac{M_m}{0.1979 \times 175.7 \times 10^6} = \frac{M_m}{34.7 \times 10^6} \quad \text{rad}$$

HOLZER METHOD FOR FORCED RESPONSE

A calculation of the nonresonant or "forced" vibration amplitude is required in some cases to define the range of the more severe critical speeds, particularly with geared drives; it also is required in the design of dampers. The calculation is readily made by an extension of the Holzer table sequence method illustrated in Table 38.2. In the sequence the initial amplitude is treated as an algebraic unknown θ. At each cylinder where an exciting torque acts, this torque is added. Assume first that there are no damping torques. Then the residual torque after the last rotor is of the form $a\theta + b$, where a and b are numerical constants resulting from the calculation. Since the residual torque is zero, $\theta = -b/a$.

The amplitude and torque at any point of the system are found by substituting this numerical value of θ at the appropriate point in the tabulation. At frequencies well removed from resonance, damping has little effect and can be neglected. Damping can be added to the system by treating it as an exciting torque equal to the imaginary quantity $-jc\omega\theta$, where c is the damping constant and θ is the amplitude at the point of application. Relative damping between two inertias can be treated as a spring of a stiffness constant equal to the imaginary quantity of $+jc\omega$.

For the major critical speeds the exciting torques are all in-phase and are real numbers. For the minor critical speeds the exciting torques are out-of-phase; they must be entered as complex numbers of amplitude and phase as determined from the phase diagram (discussed under *Energy Balance*) for the critical speed of the order under consideration. With damping and/or out-of-phase exciting torques introduced, a and b in the equation $a\theta + b = 0$ are complex numbers, and θ must be entered as a complex number in the tabulation in order to determine the angle and torque at any point. The angles and torques are then of the form $r + js$, where r and s are numerical constants and the amplitudes are equal to $\sqrt{r^2 + s^2}$.

In the forced-vibration calculation the amplitude of the exciting torque also is assumed to be unity. It is advisable, particularly when the calculation is made for damper design, to start the sequence at the flywheel end of the system. The introduction of exciting and damping terms thus is deferred as long as possible.

The relationship between the actual and the unity system is given in Table 38.9. Here $I = J/g$ is the moment of inertia by which all the moments of inertia are divided and k is the common stiffness constant by which all the stiffness constants are divided. A damping constant c in the actual system becomes

$$c_s = \frac{c}{\sqrt{Ik}}$$

in the unity system. Exciting torques are assumed to have unity amplitude. After the torque in any part of the unity system is found, the torque in the actual system is determined by multiplying by M_e, and deflections in the actual system are obtained by multiplying those in the unity system of M_e/k.

TABLE 38.9   Relation between Parameters in Actual and Unity Systems

	Actual	Unity
Inertia constant	$I = J/g$	1
Spring constant	k	1
Damping constant	c	$c_s = c/\sqrt{Ik}$
Torques	M_e	1
Deflections	$M_e\theta/k$	θ

Example 38.4.   Determine the amplitude of the fourth-order vibration at the maximum speed of 950 rpm for the eight-cylinder, four-cycle engine driving a generator shown in Fig. 38.10B.

The frequency in the unity system corresponding to fourth-order vibration at 950 rpm in the actual system is given by

$$\omega_s^2 = \left(\frac{2\pi \times 950 \times 4}{60} \right)^2 \times \frac{35.8}{175.7 \times 10^6} = 0.0321$$

$$\omega_s = 0.179$$

The calculation is carried out in Table 38.10. Assuming no damping, the residual torque is

$$M_r = 0.3946\theta + 5.852 = 0$$

from which $\theta = -14.85$.

The maximum torque occurs in section 8–9; it is given by

$$M_{8,9} = 1.3726\theta = -20.4$$

TABLE 38.10 Sequence Calculations for Forced Vibration of Engine Generator Shown in Fig. 38.10B at Operating Speed of 950 rpm ($\omega_s^2 = 0.0321$)

Rotor number	I	$I\omega_s^2$	β Inertia $\theta \times$	β Exciting	M, in.-lb Inertia $\theta \times$	M, in.-lb Exciting	k
10	42.5	1.366	1.0000	...	1.366		
			0.226	...	1.366	...	6.06
9	0.265	0.0085	0.774	...	0.0066		
			1.042	...	1.3726	...	1.32
8	1	0.0321	−0.268	...	−0.0086		
			1.3640	1.000	1.3640	1.000	1.00
7	0.722	0.0231	−1.632	−1.000	−0.0377	−0.023	
			1.3263	1.977	1.3263	1.977	1.00
6	0.722	0.0231	−2.9583	−2.977	−0.0682	−0.069	
			1.2581	2.908	1.2581	2.908	1.00
5	1	0.0321	−4.2164	−5.885	−0.1350	−0.189	
			1.1231	3.719	1.1231	3.719	1.00
4	1	0.0321	−5.3395	−9.604	−0.1715	−0.308	
			0.9516	4.411	0.9516	4.411	1.00
3	0.722	0.0231	−6.2911	−14.015	−0.1452	−0.324	
			0.8064	5.087	0,8064	5.087	1.00
2	0.722	0.0231	−7.0975	−19.102	−0.1638	−0.441	
			0.6426	5.646	0.6426	5.646	1.00
1	1	0.0321	−7.7401	−24.748	−0.2480	−0.794	
					0.3946	5.852	

In section 6–7 the torque is given by

$$M_{6,7} = 1.326\theta + 1.977 = -17.73$$

Fourth-order torque M_e in the actual engine is 11,300 in.-lb (Table 38.8); in the actual engine the fourth-order torque in section 8–9 is

$$M = 20.4 \times 11,300 = 231,000 \text{ in.-lb}$$

At the No. 1 cylinder the amplitude in the unity system is

$$-7.7401\theta - 24.748 = 90.3 \text{ rad}$$

In the actual system the corresponding amplitude is

$$\frac{90.3 \times 11,300}{175.7 \times 10^6} = 0.0058 \text{ rad}$$

The problem also can be solved by assuming that a concentrated damping of constant $c = 1470$ in.-lb/rad acts at the No. 1 crank. In the unity system the corresponding damping constant is

$$c_3 = \frac{1470}{\sqrt{35.8 \times 175.7 \times 10^6}} = 0.0185$$

At the No. 1 crank the torque $-jc_s\omega_s\theta$ is added; this torque is

$$M = -j \times 0.0185 \times 0.175 \times (-7.7401\theta - 24.748) = j(0.0256\theta + 0.0819)$$

The resulting residual torque is given by

$$M_r = 0.3946\theta + 5.852 + j(0.0256\theta + 0.0819) = 0$$

Solving for θ,

$$\theta = \frac{-5.852 - 0.0819j}{0.3946 + 0.0256j}$$

The amplitude of the deflection is

$$|\theta| = \sqrt{\frac{(5.852)^2 + (0.0819)^2}{(0.3946)^2 + (0.0256)^2}} = 14.85$$

which is the same as if damping were not present; thus, at this speed the effect of the external damping is completely negligible.

APPLICATION OF MODAL ANALYSIS TO ROTOR SYSTEMS

Classical modal analysis of vibrating systems (see Chap. 21) can be used to obtain the forced response of multistation rotor systems in torsional motion. The natural frequencies and mode shapes of the system are found using the transfer matrix method. The response of the rotor to periodic phenomena (not necessarily a harmonic or shaft frequency) is determined as a linear weighted combination of the mode shapes of the system. Heretofore with this technique, damping has been entered in modal form; the damping forces are a function of the various modal velocities. The formation of equivalent viscous damping constants that are some percentage of critical damping is required. The critical damping factor is formed from the system modal inertia.[11]

The modal analysis technique can be used for a torsional distributed mass model of engine systems using modal damping; nonsynchronous speed excitations are allowed. The shaft sections of the modeled rotor have distributed mass properties and lumped end masses (including rotary inertia). A transfer matrix analysis is performed to obtain a finite number of natural frequencies. The number required depends on the range of forcing frequencies used in the problem. The natural frequencies are substituted back into the transfer matrices to obtain the mode shapes. A function consisting of a weighted average of the mode shapes is formed and substituted into

$$\theta(x, t) = \sum_{n=1}^{N} a_n(x) f_n(t)$$

where θ = torsional response
a_n = normal modes
f_n = periodic time-varying weighting factors

The function $f_n(t)$ is determined from the ordinary differential equations of motion and is a function of the forcing functions, rotor speed, modal damping constants, and mode shapes of the system.

DIRECT INTEGRATION

Direct integration of equations of motion of a system utilize first- or second-order differential equations. The method is fundamental for linear and nonlinear response problems.[12] Any digitally describable vibration or shock excitation can be carried out with this method.

Direct integration can be used on nonlinear models and arbitrary excitation, so it is one of the most general techniques available for response calculation. However, large computer storage is required, and large computer costs are usually incurred because small time- or space-step sizes are needed to maintain numerical stability. An adjustable step integration routine such as predictor-corrector helps to alleviate this problem. Such a numerical integration must be started with another routine such as Runge-Kutta.

Direct integration is particularly useful when nonlinear components such as elastomeric couplings are involved or when the excitation force varies in frequency and magnitude. Direct integration is used for analysis of synchronous motor start-ups in which the magnitude of the torque varies with rotor speed and the frequency is 2 times the slip frequency—starting at twice the line frequency and ending at zero when the rotor is locked on synchronous speed. Examples of this type of analysis are given in Refs. 12 and 13.

PERMISSIBLE AMPLITUDES

Failure caused by torsional vibration invariably initiates in fatigue cracks that start at points of stress concentration—e.g., at the ends of keyway slots, at fillets where there is a change of shaft size, and particularly at oil holes in a crankshaft. Failures can also start at corrosion pits, such as occur in marine shafting. At the shaft oil holes the cracks begin on lines at 45° to the shaft axis and grow in a spiral pattern until failure occurs. Theoretically the stress at the edges of the oil holes is 4 times the mean shear stress in the shaft, and failure may be expected if this concentrated stress exceeds the fatigue limit of the material. The problem of estimating the stress required to cause failure is further complicated by the presence of the steady stress from the mean driving torque and the variable bending stresses.

In practice the severity of a critical speed is judged by the maximum nominal torsional stress

$$\tau = \frac{16M_m}{\pi d^3}$$

where M_m is the torque amplitude from torsional vibration and d is the crankpin diameter. This calculated nominal stress is modified to include the effects of increased stress and is compared to the fatigue strength of the material.

U.S. MILITARY STANDARD

A military standard[14] issued by the U.S. Navy Department states that the limit of acceptable nominal torsional stress within the operating range is

$$\tau = \frac{\text{ultimate tensile strength}}{25} \qquad \text{for steel}$$

$$\tau = \frac{\text{torsional fatigue limit}}{6} \qquad \text{for cast iron}$$

If the full-scale shaft has been given a fatigue test, then

$$\tau = \frac{\text{torsional fatigue limit}}{2} \quad \text{for either material}$$

Such tests are rarely, if ever, possible.

For critical speeds below the operating range which are passed through in starting and stopping, the nominal torsional stress shall not exceed 1¾ times the above values.

Crankshaft steels which have ultimate tensile strengths between 75,000 and 115,000 lb/in.^{2} usually have torsional stress limits of 3000 to 4600 lb/in.^{2}

For gear drives the vibratory torque across the gears, at any operating speed, shall not be greater than 75 percent of the driving torque at the same speed or 25 percent of full-load torque, whichever is smaller.

AMERICAN PETROLEUM INSTITUTE

Sources of torsional excitation considered by American Petroleum Institute[15] (API) include but are not limited to the following: gear problems such as unbalance, pitch line runout, and eccentricity; start-up conditions resulting from inertial impedances; and torsional transients from synchronous and induction electric motors.

Torsional natural frequencies of the machine train shall be at least 10 percent above or below any possible excitation frequency within the specified operating speed range. Torsional critical speeds at integer multiples of operating speeds (e.g., pump vane pass frequencies) should be avoided or should be shown to have no adverse effect where excitation frequencies exist. Torsional excitations that are non-synchronous to operating speeds are to be considered. Identification of torsional excitations is the mutual responsibility of the purchaser and the vendor.

When torsional resonances are calculated to fall within the ±10 percent margin and the purchaser and vendor have agreed that all efforts to remove the natural frequency from the limiting frequency range have been exhausted, a stress analysis shall be performed to demonstrate the lack of adverse effect on any portion of the machine system.

In the case of synchronous motor driven units, the vendor is required to perform a transient torsional vibration analysis with the acceptance criteria mutually agreed upon by the purchaser and the vendor.

TORSIONAL MEASUREMENT

Torsional vibration is more difficult to measure than lateral vibration because the shaft is rotating. Procedures for signal analysis are similar to those used for lateral vibration. Torsional response—both strains and motions—can be measured at intermediate points in a system. But sensors cannot be placed at a nodal point; for this reason the Holzer method is valuable for calculating mode shapes prior to sensor location selection.

SENSORS

Strain gauges, described in Chap. 17, are available in a variety of sizes and sensitivities and can be placed almost anywhere on a shaft. They can be calibrated to indicate instantaneous torque by using static torque loads on drive shafts. If calibration is not possible, stresses and torques can be calculated from strength of materials theory. Strain gauges are usually mounted at 45° angles so that shaft bending does not influence torque measurements. The signal must be processed by a bridge-amplifier unit that can be arranged to compensate for temperature. Because strain gauge signals are difficult to take from a rotating shaft, such techniques are not common diagnostic tools.

Slip rings can be used to obtain a vibration signal from a shaft. Wireless telemetry is also available. A small transmitter mounted on the rotating shaft at a convenient location broadcasts a signal to a nearby receiver. Commercial torque transducers are available for torsional measurement. However, they must be inserted in the drive line and thus may change the dynamic characteristics of the system. If the natural frequency of the system is changed, the vibration response will not accurately reflect the properties of the system.

The velocity of torsional vibration is measured using a toothed wheel and a fixed sensor.[16] The signal generated by the teeth of the wheel passing the fixed sensor has a frequency equal to the number of teeth multiplied by shaft speed. If the shaft is undergoing torsional vibration, the carrier frequency will exhibit frequency modulation (change in frequency) because the time required for each tooth to pass the fixed pickup varies.

DATA ACQUISITION

The frequency change (velocity) is converted to a voltage change by a demodulator and integrated to obtain angular displacement. Angular displacement can be measured at the end of a shaft with encoders or at intermediate points with a gear-magnetic pickup or proximity probe arrangement. The frequency of the carrier signal (e.g., number of teeth on a gear × rpm) must be at least 4 times the highest frequency to be measured. In most cases, the raw torsional signal is tape recorded prior to processing and analysis. Because the output of the magnetic pickup is speed dependent and the gap between the magnetic pickup and the toothed wheel is less than 0.025 in. the proximity probe is preferred—especially in synchronous motor startups.

TORSIONAL ANALYSIS

A torsional signal must be analyzed for frequency components using a spectrum analyzer, described in Chap. 14. Figure 38.19 shows a torsional response spectrum for a variable-frequency motor-driven pump. The pump ran at 408 rpm. The torsional vibration response excited by the variable frequency motor is 0.23° at a frequency of 38 Hz.

MEASURES OF CONTROL

The various methods which are available for avoiding a critical speed or reducing the amplitude of vibration at the critical speed may be classified as:

1. Shifting the values of critical speeds by changes in mass and elasticity
2. Vector cancellation methods

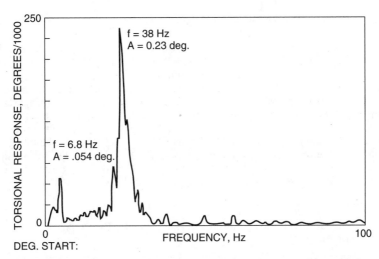

FIGURE 38.19   Torsional response of a variable-frequency motor-driven pump at 408 rpm. There are significant peaks at 6.8 and 38.0 Hz.

3. Change in mass distribution to utilize the inherent damping in the system
4. Addition of dampers of various types

SHIFTING OF CRITICAL SPEEDS

If the stiffness of all the shafting to a system is increased in the ratio a, then all the frequencies will increase in the ratio a, provided that there is no corresponding increase in the inertia. It is rarely possible to increase the crankshaft diameters on modern engines; in order to reduce bearing pressures, bearing diameters usually are made as large as practical. If bearing diameters are increased, the increase in the critical speed will be much smaller than indicated by the a ratio because a considerable increase in the inertia will accompany the increase in diameter. Changes in the stiffness of a system made near a nodal point will have maximum effect. Changes in inertia near a loop will have maximum effect, while those near a node will have little effect.

By the use of elastic couplings it may be possible to place certain critical speeds below the operating speed where they are passed through only in starting and stopping; this leaves a clear range above the critical speed. This procedure must be used with caution because some critical speeds, for example the fourth order in an eight-cylinder, four-cycle engine, are so violent that it may be dangerous to pass through them. If the acceleration through the critical speed is sufficiently high, some reduction in amplitude may be attained,[17] but with a practical rate the reduction may not be large. The rate of deceleration when stopping is equally important. In some cases mechanical clutches disconnect the driven machinery from the engine until the engine has attained a speed above dangerous critical speeds. Elastic couplings may take many forms including helical springs arranged tangentially, flat leaf springs arranged longitudinally or radially, various arrangements using rubber, or small-diameter shaft sections of high tensile steel.[3]

VECTOR CANCELLATION METHODS

Choice of Crank Arrangement and Firing Order.   The amplitude at certain minor critical speeds sometimes can be reduced by a suitable choice of crank arrangement and firing order (i.e., firing sequence). These fix the value of the vector sum $\Sigma\beta$ in Eq (38.25), $M_m = \Re M_e\Sigma\beta$. But considerations of balance, bearing pressures, and internal bending moments restrict this freedom of choice. Also, an arrangement which decreases the amplitude at one order of critical speed invariably increases the amplitude at others. In four-cycle engines with an even number of cylinders, the amplitude at the half-order critical speeds is fixed by the firing order because this determines the $\Sigma\beta$ value. Tables 38.11 and 38.12 list the torsional-vibration characteristics for the crank arrangements and firing orders, for eight-cylinder two- and four-cycle engines having the most desirable properties.

TABLE 38.11   Torsional-Vibration Characteristics for Eight-Cylinder, Four-Cycle Engine Having 90° Crank Spacing

	$\Sigma\beta$· OF ORDERS*				
FIRING ORDER	1/2, 7½, 8½, 3½, 4½	1½, 5½, 6½	2, 6, 10	1, 3, 5, 8, 7	4, 8
1,6,2,5,8,3,7,4	0.745	1.44	0	0	4.5
1,6,2,4,8,3,7,5	0.686	1.48	0	0	4.5
1,3,2,5,8,6,7,4	1.48	0.686	0	0	4.5
6,3,5,7,8,6,4,2	1.74	0.176	0	0	4.5
1,7,4,3,8,2,5,6	0.176	1.74	0	0	4.5

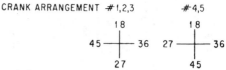

CRANK ARRANGEMENT #1,2,3   #4,5

* Values of 0 in the $\Sigma\beta$ column indicate small but not necessarily 0 values for actual β distribution.

TABLE 38.12   Torsional-Vibration Characteristics for Eight-Cylinder, Two-Cycle Engine Having 45° Crank Spacing

	$\Sigma\beta$ of orders				
Firing order	1, 7, 9	2, 6, 10	3, 5, 11	4, 12	8, 16
1, 8, 2, 6, 4, 5, 3, 7	0.056	0	0.79	2.0	4.5
1, 7, 4, 3, 8, 2, 5, 6	0.175	0	1.61	0	4.5
1, 6, 5, 2, 7, 4, 3, 8	0.112	0	1.58	0.5	4.5

The values of $\Sigma\beta$ are calculated by assuming $\beta = 1$ for the cylinder most remote from the flywheel, assuming $\beta = 1/n$ for the cylinder adjacent to the flywheel (where n is the number of cylinders), and assuming a linear variation of β therebetween. In any actual installation $\Sigma\beta$ must be calculated by taking β from the relative amplitude curve (Holzer Table); however, if the $\Sigma\beta$ as determined above is small, it also will be small for the actual β distribution. These arrangements assume equal crank angles and firing intervals. The reverse arrangements (mirror images) have the same properties.

V-Type Engines.   In V-type engines, it may be possible to choose an angle of the V which will cancel certain criticals. Letting ϕ be the V angle between cylinder banks, and q the order number of the critical, the general formula is

$$q\phi = 180°, 540°, 1080°, \text{etc.} \tag{38.26}$$

For example, in an eight-cylinder engine the eighth order is canceled at angles of $22\frac{1}{2}°, 67\frac{1}{2}°, 112\frac{1}{2}°$, etc.

In four-cycle engines, ϕ is to be taken as the actual bank angle if the second-bank cylinders fire directly after the first and as $360° + \phi$ if the second-bank cylinders omit a revolution before firing. In the latter case the cancellation formula is

$$(\phi + 360°)q = n \times 180° \tag{38.27}$$

where $n = 1, 3, 5$, etc. For example, to cancel a 4.5-order critical the bank angle should be

$$\phi = \frac{180°}{4.5} = 40° \qquad \text{for direct firing}$$

or

$$\phi = \frac{11 \times 180°}{4.5} - 360° = 80° \qquad \text{for the 360° delay}$$

Cancellation by Shift of the Node.   If an engine can be arranged with approximately equal flywheel (or other rotors) at each end so that the node of a particular mode is at the center of the engine, $\Sigma\beta$ will cancel for the major orders of that mode. This procedure must be used with caution because the double flywheel arrangement may reduce the natural frequency in such a manner that low-order minor criticals of large amplitudes take the place of the canceled major criticals.

Reduction by Use of Propeller Damping in Marine Installations.   From Eq. (38.21) it is evident that the torque amplitude in the shaft can be reduced below any desired level by making the flywheel moment of inertia I_1 of sufficient magnitude. The ratio of the propeller amplitude to the engine amplitude increases as the flywheel becomes larger; thus the effectiveness of the propeller as a damper is increased.

DAMPERS

Many arrangements of dampers can be employed (see Chap. 6). In each type there is a loose flywheel or inertia member which is coupled to the shaft by:

1. Coulomb friction (Lanchester damper)
2. Viscous fluid friction

3. Coulomb or viscous friction plus springs

4. Centrifugal force, equivalent to a spring having a constant proportional to the square of the speed (pendulum damper) (see Chap. 6)

Each of these types acts by generating torques in opposition to the exciting torques.

The Lanchester damper illustrated in Fig. 6.35 has been entirely superseded by designs in which fluid friction is utilized. In the Houdaille damper, Fig. 38.20, a flywheel is mounted in an oiltight case with small clearances; the case is filled with silicone fluid. The damping constant is

$$c = 2\pi\mu \left[\frac{r_2^3 b}{h_2} + \frac{1}{2} \frac{r_2^4 - r_1^4}{h_1} \right] \quad \text{in.-lb-sec} \tag{38.28}$$

where μ is the viscosity of the fluid and $r_1, r_2, b, h_1,$ and h_2 are dimensions indicated in Fig. 38.20.

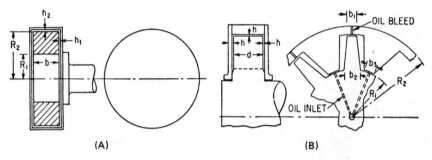

FIGURE 38.20 Schematic diagram of dampers. (*A*) Houdaille type. (*B*) Paddle type.

The paddle-type damper illustrated in Fig. 38.20 utilizes the engine lubricating oil supplied through the crankshaft. It has the damping constant

$$c = \frac{3\mu d^2 (r_2^2 - r_1^2)^2 n}{h^3 \left[\dfrac{d}{b_1} + \dfrac{d}{b_3} + \dfrac{4(r_2 - r_1)}{b_1 + b_2} \right]} \quad \text{in.-lb-sec} \tag{38.29}$$

where n is the number of paddles, μ is the viscosity of the fluid, and $b_1, b_2, r_1, r_2,$ and d are dimensions indicated in Fig. 38.20. Other types of dampers are described in Ref. 4.

The effectiveness of these dampers may be increased somewhat by connecting the flywheel to the engine by a spring of proper stiffness, in addition to the fluid friction. In one form, Fig. 38.21, the connection is by rubber bonded between the flywheel and the shaft member. The rubber acts both as the spring and by hysteresis as the energy absorbing member. See Chaps. 35, 36, and 37 for discussions of damping in rubber. Dampers without and with springs are defined here as *untuned* and *tuned viscous dampers,* respectively.

In many cases the mode of vibration to be damped is essentially internal to the engine. Then the damper is located at the end of the engine remote from the fly-

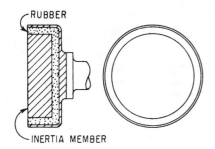

FIGURE 38.21 Schematic diagram of bonded rubber damper.

wheel. If the mode to be damped is essentially one between driven masses, other locations may be desirable or necessary.

Design of the Untuned Viscous Damper, Exact Procedure. The first step in the design procedure is to make a tentative assumption of the polar moment of inertia of the floating inertia member. If the damper is attached to the forward end of the crankshaft with the primary purpose of damping vibration in the engine, the size should be from 5 to 25 per cent, depending on the severity of the critical to be damped, of the total inertia in the engine part of the system, excluding the flywheel.

Usually it is advantageous to minimize the torque in a particular shaft section. This may be done as follows: For a series of frequencies plot the resonance curve of this torque, first without the floating damper mass and then with the damper mass locked to the damper hub. Plot the curves with all ordinates positive. The nature of such a plot is shown in Fig. 38.22. The point of intersection is called the *fixed point*.

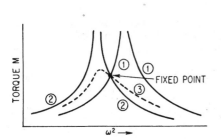

FIGURE 38.22 Resonance curves for various conditions of auxiliary mass dampers: (1) damper free, $c = 0$; (2) damper locked, $c = \infty$; (3) auxiliary mass coupled to shaft by damping.

The plot is shown as if there were only one resonant frequency. Usually only one is of interest, and the curves are plotted in its vicinity. If the plot were extended, there would be a series of fixed points.

If a damping constant is assigned to the damper and the new resonance curve plotted, it will be similar to curve 3 in Fig. 38.22 and will pass through the fixed point. If there is no other damping in the system except that in the damper, all of the resonance curves will pass through the fixed points, independent of the value assigned to the damping constant.[18] Therefore, the amplitude at the fixed point is the lowest that can be obtained for the assumed damper size. If this amplitude is too large, it will be necessary to increase the damper size; if the amplitude is unnecessarily small, the damper size can be decreased. When a satisfactory size of damper has been selected, it is necessary to find the damping constant which will put the resonance curve through the fixed point with a zero slope. Assume a value of ω^2 slightly lower than its value at the fixed point, and compute the amplitude at that value of ω^2 with the damping constant c entered as an algebraic unknown. Equating this amplitude to that at the fixed point, the unknown damping constant c can be calculated. Repeat the calculation with a value of ω^2 higher than the fixed point value by the same increment. The mean of the two values of c thus obtained will be as close to the optimum value as construction of the damper will permit. In constructing these resonance curves, it is not necessary to construct complete curves over a wide range of frequencies but only over a short interval in the vicinity of the fixed point. Considerable labor is saved by making the calculation in the unity system with conversion to the actual system as a final step.

Figure 38.23 shows the constants in the unity system for the example of Fig. 38.10B, when a damper is added at the end of the crankshaft. A floating rotor having 10 percent of the total engine inertia is assumed; a is the fixed hub attached to the shaft. The "fixed point" is first located by calculating $M_{9,10}$ for $\omega_s^2 = 0.035, 0.037$, and 0.038. Table 38.13 gives the calculation for $\omega_s^2 = 0.037$.

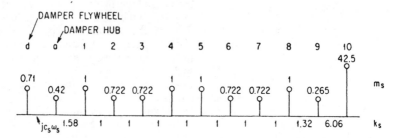

FIGURE 38.23 Constants for system of Fig. 38.10B when damper is added at end of crankshaft.

In making the plot of the resonance curves, it is convenient to plot $1/M_{9,10}$ to avoid the resonant peak near $\omega_s^2 = 0.035$. Then $1/M_{9,10}$ is obtained from Table 38.13 as follows: Taking the case of the free damper, $\omega_s^2 = 0.037$. For zero residual torque: $0.1474\theta + 5.121 = 0$, or $\theta = -5.121/0.1474$.

The torque in the shaft between rotors 9 and 10 is $M_{9,10} = 1.57\theta$. The reciprocal is

$$\frac{1}{M_{9,10}} = \frac{0.1474}{5.121 \times 1.57} = 0.01832$$

Adding the damper d to the hub a, the corresponding value is

$$\frac{1}{M_{9,10}} = \frac{-0.0846}{4.391 \times 1.57} = -0.01225$$

Table 38.14 gives $1/M_{9,10}$ for the other frequencies. The reciprocal $1/M_{9,10}$ of the torque is plotted in Fig. 38.24 as a function of ω_s^2, all ordinates are plotted with positive sign. The intersection of the curves for free and locked damper is at $\omega^2 = 0.0375$; $1/M_{9,10} = 0.016$ in the unity system. In terms of the actual engine, this indicates that with an optimum damper the vibratory torque will be $1/0.016 = 62.5$ times the exciting torque of one cylinder. For the eighth-order critical, the resonant amplitude calculation gives $M_e = 1490$; then, in the engine

$$M_{9,10} = 93,200 \text{ in.-lb}$$

The calculation for optimum damping constant c_s is continued in Table 38.13. The damping constant is treated as an algebraic unknown so that the damper "spring" constant is $jc_s\omega_s$. Now let $M_{9,10}$ at $\omega_s^2 = 0.037$ equal 62.5, its value at the fixed point. Then $|\theta| = 62.5/1.570 = 39.8$.

TABLE 38.13 Sequence Calculations for System of Fig. 38.10B with Damper at End of Crankshaft ($\omega_s^2 = 0.037$)

Rotor number	I	$I\omega_s^2$	β Inertia θ ×	β Exciting	M, in.-lb Inertia θ ×	M, in.-lb Exciting	k
10	42.5	1.570	1.0000	...	1.570		
			0.259	...	1.570		6.06
9	0.265	0.0113	0.741	...	0.0083		
			1.191	...	1.5783		1.32
8	1	0.037	−0.450	...	−0.0166		
			1.5617	1.000	1.5617	1.000	1.00
7	0.722	0.0267	−2.0117	−1.000	−0.0538	−0.027	
			1.5079	1.973	1.5079	1.973	1.00
6	0.722	0.0267	−3.5196	−2.973	−0.0935	−0.080	
			1.4144	2.893	1.4144	2.893	1.00
5	1	0.037	−4.9340	−5.866	−0.1830	−0.217	
			1.2314	3.676	1.2314	3.676	1.00
4	1	0.037	−6.1654	−9.542	−0.2280	−0.352	
			1.0034	4.324	1.0034	4.324	1.00
3	0.722	0.0267	−7.1688	−13.866	−0.1910	−0.371	
			0.8124	4.953	0.8124	4.953	1.00
2	0.722	0.0267	−7.9812	−18.819	−0.2130	−0.513	
			0.5994	5.450	0.5994	5.450	1.00
1	1	0.037	−8.5806	−24.269	−0.3170	−0.898	
			0.1785	3.510	0.2824	5.552	11.58
a	0.42	0.0155	−8.7591	−27.779	−0.1350	−0.431	
					0.1474	5.121	
					0.2824	5.552	
$a + d$	1.13	0.0418	−8.7591	−27.779	−0.3670	−1.161	
					−0.0846	4.391	

$$d \quad 0.71 \quad 0.0263 \begin{bmatrix} -0.767j/c_s \\ -8.7591 \\ +0.7657j/c \end{bmatrix} \begin{bmatrix} -26.7j/c_s \\ -27.779 \\ +26.7j/c_s \end{bmatrix} \begin{bmatrix} 0.1474 \\ -0.231 \\ +0.020j/c_s \end{bmatrix} \begin{bmatrix} 5.121 \\ -0.730 \\ +0.703j/c_s \end{bmatrix} \quad 0.192c_s j$$

$$\omega_s = 0.192 \qquad \begin{bmatrix} -0.084 \\ +0.020j/c_s \end{bmatrix} \begin{bmatrix} 4.39 \\ +0.073j/c_s \end{bmatrix}$$

$$\theta = \frac{-4.39 - 0.703j/c_s}{-0.084 + 0.020j/c_s} \qquad |\theta|^2 = \frac{19.3 + 0.495/c_s^2}{0.0069 + 0.0004/c_s^2}$$

TABLE 38.14 Summary of $1/M_{9,10}$ from Table 38.13 for Other Frequencies

ω_s^2	1/$M_{9,10}$ Free	1/$M_{9,10}$ Locked
0.035	−0.0261	0.000487
0.037	−0.01832	0.01225
0.038	−0.0137	0.0187

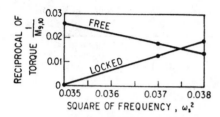

FIGURE 38.24 Variation of torque in system of Fig. 38.23 when damper is either free or locked. The dots indicate computer points.

The damper "spring" constant $0.192c_s j$ is entered in the k column of Table 38.13 opposite the damper d. Then the relative amplitudes β in the "inertia" and "exciting" columns are the quotients M/k:

$$\frac{0.1474}{0.192c_s j} = -\frac{0.767j}{c_s}$$

$$\frac{5.121}{0.192c_s j} = -\frac{26.7j}{c_s}$$

These values are entered in the β column and subtracted from the β values at rotor a, giving the β values at the damper rotor d. The amplitudes at this rotor multiplied by 0.0263 give the inertia torque reaction of damper rotor d. Adding this to the torque between a and d, the sum is equated to zero, and from this equation θ is solved for. Then the amplitude $|\theta|^2$ is obtained in the lower line of Table 38.13 by squaring and adding the real and imaginary terms of numerator and denominator. Substituting $\theta = 39.8$,

$$19.3 + \frac{0.495}{c_s^2} = (39.8)^2 \left(\frac{0.0069 + 0.0004}{c_s^2} \right)$$

Solving, $c_s^2 = 0.0172$, or $c_s = 0.131$. A similar calculation for $\omega_s^2 = 0.038$ leads to $c_s = 0.122$. A mean of these values is $c_s = 0.126$.

To check this value it is substituted in the $|\theta|^2$ equations of Table 38.13, then $M_{9,10}$ is obtained as follows:

$$M_{9,10} = 61.9 \quad \text{for} \quad \omega_s^2 = 0.037$$

$$M_{9,10} = 61.3 \quad \text{for} \quad \omega_s^2 = 0.038$$

Thus, the mean value of c_s places the fixed point very close to the maximum of the damped resonance curve.

It is of interest to find the fourth-order forced amplitude at 950 rpm with the damper in place. Table 38.15 shows the final items of Table 38.13 for the forced-vibration calculation with its continuation through the damper. The torque in section 9–10 is now 30.9 times the fourth-order exciting torque of one cylinder as against 20.3 without the damper (per Table 38.10); thus, the addition of the damper has increased the fourth-order stress by 50 percent. This is because the addition of the damper inertia has lowered the frequency of the fourth-order critical, but the damper is of insufficient size to give any effective control over the fourth-order amplitude. It is extremely difficult to arrange a friction damper which will be effective for the lowest-order major criticals of four-cycle engines. For six- and eight-cylinder engines, the torque which must be developed in the damper is of the order of magnitude of the engine torque and requires very large dampers. In some cases a solution can be found with pendulum dampers.

In the actual engine the damper constant is

$$c = 0.131 \sqrt{35.8 \times 175.7 \times 10^6} = 10{,}400 \text{ in.-lb/rad/sec}$$

TABLE 38.15 Final Items of Table 38.13 Continued through the Damper ($\omega_s^2 = 0.0321$)

Rotor	I	$I\omega_s^2$	β Inertia $\theta \times$	Exciting	M Inertia $\theta \times$	Exciting	k
1	1	0.0321	−7.7401	−24.748	−0.2480	−0.794	
			0.2490	3.710	0.3946	5.852	1.58
a	0.42	0.0135	−7.9891	−28.458	−0.1010	−0.384	
			−13j	−242j	0.2936	5.468	0.0226j
d	0.71	0.0228	−7.9891	−28.458	−0.162	−0.650	
			+13j	+242j	0.296j	5.5j	
					0.1316	4.818	
				$RM \rightarrow$	0.296j	5.5j	

$$|\theta| = \sqrt{\frac{(4.818)^2 + (5.5)^2}{(0.1316)^2 + (0.296)^2}} = 22.6$$

$$|M_{9,10}| \text{ (unity system)} = 22.6 \times 1.366 = 30.9$$

$$M_{9,10} \text{ (actual system)} = 30.9 \times 11,300 = 349,000 \text{ in.-lb}$$

The polar inertia of the floating member is

$$J = 35.8 \times 0.71 \times 386 = 9800 \text{ in.}^2\text{-lb}$$

A damper of the Houdaille type having the dimensions $r_2 = 8.25$ in., $r_1 = 4.125$ in., and $b = 5$ in. will have the polar inertia 9800 in.²-lb. Viscosities in the range of 20,000 to 60,000 centipoises are used.

To convert to inch-pound-second units, centipoises are multiplied by 1.45×10^{-7}. Assuming 40,000 centipoises fluid, $\mu = 0.0058$ reyn, and equal clearances $h_1 = h$, the expression for clearances obtained from Eq. (38.29) is

$$h = \frac{2\pi}{c} \mu \left[r_2^3 b + \frac{1}{2} (r_2^4 - r_1^4) \right] \tag{38.30}$$

which gives the damper clearance $h = 0.015$ in.

For a damper of the paddle type of Fig. 38.20, a diameter of 16.25 in. is taken together with the following dimensions: $r_2 = 7.16$ in., $r_1 = 3.58$ in., $d = 4.58$ in., $b_1 = b_2 = b_3 = 0.9$ in., and $n = 8$ pockets. The viscosity of SAE No. 30 oil at an operating temperature of 150°F is approximately 27 centipoises. Substituting in Eq. (38.30) and solving for h

$$h = 0.025 \text{ in. clearance}$$

To find the torque acting on the damper, refer to Table 38.14. Substituting $c_s = 0.126$ in the θ equation,

$$\theta = \frac{-4.39 - 5.58j}{-0.084 + 0.1586j}$$

But $M_{ad} = (0.1474\theta + 5.121)$ in.-lb reduces with the substitution of θ to

$$M_{ad} = \frac{1.079 - 0.013j}{0.084 - 0.1586j}$$

The absolute value is $|M_{ad}| = 6.01$, and in the actual system $|M_{ad}| = 6.01 \times 1490 = 8950$ in.-lb. Note that this is of the same order of magnitude as $M_e\Sigma\beta$. It is not necessarily the maximum torque in the damper, but it is close to the maximum. To find the maximum it is necessary to make a similar calculation for several adjacent frequencies.

The oil pressure in the pockets produced by this torque is

$$\frac{8950}{8 \times 5.37 \times 3.58 \times 4.58} = 12.8 \text{ lb/in.}^2$$

Oil must be supplied at a pressure in excess of this.

To find the oil pressure for the fourth-order critical at 950 rpm, M_{ad} is obtained from Table 38.15.

$$M_{ad} = 0.29360 + 5.468$$

where

$$\theta = \frac{-4.818 - 5.5j}{0.1316 + 0.296j}$$

Substituting this expression for θ in the expression for M_{ad},

$$M_{ad} = \frac{0.71}{0.1316 + 0.296j}$$

The absolute value is $|M_{ad}| = 2.18$. In the actual engine, the torque is $M_{damper} = 2.18 \times 11,300 = 24,600$ in.-lb, giving a pressure of 35.3 lb/in.^{2}

The above calculations are for a major critical with the excitations of the various cylinders all in-phase. In theory, for minor criticals where the excitations are out-of-phase, fixed points do not exist and this procedure does not apply. An approximation to optimum conditions for this case is obtained by assuming that the single exciting torque $M_e\Sigma\beta$ acts at the No. 1 cylinder. This approximation can be used for major criticals as well.

Two-Mass Approximation.   If the system is replaced by a two-mass system in the manner utilized to make a first estimate (see the section *Natural Frequency Calculations*) of the one-noded mode, the results are further approximated by the following formulas:

For such a two-mass plus damper system the amplitude at the fixed point is given by[18]

$$M_{12} = M_e\Sigma\beta \left(\frac{2I_2 + I_d}{I_d} \right) \tag{38.31}$$

where $M_e = Srh$ is the exciting torque per cylinder. The optimum damping is

$$c = \left[\frac{KI_2I_d^2(2I_1 + 2I_2 + I_d)}{I_1(I_2 + I_d)(2I_3 + I_d)} \right]^{1/2} \quad \text{in.-lb/rad/sec} \tag{38.32}$$

where I_1 = polar moment of inertia for flywheel or generator
I_2 = 40 percent of engine polar moment of inertia taken up to flywheel
I_d = polar moment of inertia of damper floating element
k = stiffness from No. 1 crank to flywheel

Application to the engine discussed with reference to Table 38.15 gives

$$M_{9,10} = 74{,}300 \text{ in.-lb} \qquad \text{and} \qquad c = 9700 \text{ in.-lb/rad/sec}$$

This compares with $M_{9,10} = 93{,}200$ and $c = 10{,}400$ as previously calculated.

Tuned Viscous Dampers. The procedure for the design of a tuned viscous damper is as follows:

1. Assume a polar inertia and a spring constant for the damper. As a first assumption, adjust the spring constant so that if f is the frequency of the mode to be suppressed and f_n is the natural frequency of the damper, assuming the hub as a fixed point,

$$\frac{f_n}{f} = 0.8$$

2. Plot the resonance curves of M for a particular section, first for the damper locked, then with zero damping but the damper spring in place. All ordinates are plotted positive. The curves have the general form of those shown in Fig. 38.25. They intersect in two fixed points through which all resonance curves pass, irrespective of the damping constant in the damper. If the fixed point a is higher than b, assume a lower constant for the damper spring and recalculate the M curve. If a is lower than b, do the reverse. Thus adjust the damper spring constant until a and b are of equal height. If this amplitude M is higher than desired, it is necessary to repeat the calculation with a larger damper.

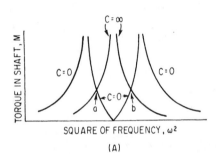

(A)

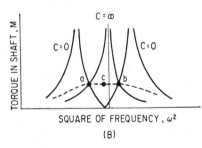

SQUARE OF FREQUENCY, ω^2

(B)

FIGURE 38.25 Curves of torque vs. square of frequency for auxiliary mass damper.

With the spring and damper mass adjusted, a direct calculation (similar to that for the untuned damper) can be made to determine the damping constant c_r which will give the resonance curve the same ordinate at an intermediate frequency indicated by point c as at a and b. Figure 38.25B shows the resonance curve of an ideally adjusted damper.

3. For a range of frequencies, using the inertia, spring, and damping constants as determined above, compute the amplitude of the damper mass relative to its hub by a forced-vibration calculation. In this calculation the damper spring constant becomes the complex number $(k + jc\omega)$. The load for which the damper springs must be designed is k times the relative amplitude of the damper mass to its hub. The torque on the damper is approximately $M_e \Sigma \beta$.

Pendulum Dampers. The principle of a pendulum damper is shown in Fig. 38.26A. (Also see Chap. 6.) The hole-pin construction usually used, which is equiva-

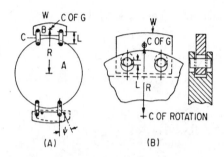

FIGURE 38.26 Pendulum-type damper. The arrangement is shown in principle at *A*, and the Chilton construction is shown schematically at *B*.

lent to that of Fig. 38.26*A*, is shown in Fig. 38.26*B*. It is undesirable to have any friction in the damper. The damper produces an effect equivalent to a fixed flywheel, and the inertia of this flywheel is different for each order of vibration.

The design formulas for the pendulum damper are as follows:[3,19] If the length *L* is made equal to

$$L = \frac{R}{1 + q_0^2} \qquad (38.33)$$

the damper is said to be tuned to order q_0. For excitation of q_0 cycles per revolution, it will act as an infinite flywheel, keeping the shaft at its point of attachment to uniform rotation insofar as q_0 order vibrations are concerned. But other orders of vibration may exist in the shaft.

If the shaft at the point of attachment of the damper is vibrating with order q and amplitude θ, the maximum link angle ψ (see Fig. 38.26) is

$$\psi = \frac{\theta q^2 (1 + q_0^2)}{q_0^2 - q^2} \qquad \text{rad} \qquad (38.34)$$

The torque exerted by a single element of the damper is

$$M = \left[\frac{WR^2}{1 - q^2/q_0^2 + J} \right] \frac{q^2 \Omega^2 \theta}{g} \qquad \text{in.-lb} \qquad (38.35)$$

where W is the weight of an element and J is the polar inertia of an element about its own center-of-gravity. The J term is equivalent to an addition to the damper hub. Dropping this term, the damper is equivalent to a flywheel of polar inertia

$$J_d = \frac{WR^2}{1 - q^2/q_0^2} \qquad \text{in.}^2\text{-lb} \qquad (38.36)$$

For $q < q_0$ this is a positive flywheel, for $q = q_0$ an infinite flywheel, and for $q > q_0$ a negative flywheel. Omitting the J term and eliminating θ between Eqs. (38.32) and (38.33),

$$\psi = \frac{M(1 + q_0^2)g}{q_0^2 WR^2 \Omega^2} \qquad \text{rad} \qquad (38.37)$$

Application to a Radial Aircraft Engine. Assume a nine-cylinder engine with the damper fitted to suppress the four and one-half order vibration. The damper would be tuned as an infinite flywheel with distance between hole centers

$$L = \frac{R}{1 + (4.5)^2}$$

In Eq. (38.35), M becomes the four and one-half order exciting torque. Then ψ is chosen as the permissible angle of pendulum swing, approximately 20 to 30°. The necessary polar inertia WR^2 of the damper is given by Eq. (38.35).

In-Line Diesel Engine. As applied to a diesel engine, the above procedure is much more difficult. The exciting torques in diesel engines are nearly independent of speed. Hence from Eq. (38.37) it is evident that ψ will be inversely proportional to Ω^2. Thus for a variable-speed engine the damper size is fixed by the low-speed end of the range; if ψ is kept in the 20 to 30° limit, the size may be excessive. This difficulty usually can be overcome by tuning the damper as a negative flywheel, thus acting to raise the undesired critical above the operating range while keeping ψ to a reasonable limit at low speed. The procedure is as follows:

Assuming a damper size and a q/q_0 ratio, a forced-vibration calculation is made starting at the flywheel end, for the maximum speed of the engine. In this calculation the damper is treated as a fixed flywheel of polar inertia $n\{[WR^e(1 - q^2/q_0^2)^{-1}] + J\}$ plus the inertia of the fixed carrier which supports the moving weights, where n is the number of weights. This calculation will yield θ, the amplitude at the damper hub, and the maximum torque in the engine shaft. Then ψ is given by Eq. (38.24). If either the shaft torque or the damper amplitude ψ is too large, it is necessary to increase the damper size and possibly adjust the q/q_0 ratio as well. A similar check for ψ is made at the low-speed end of the range with further adjustment of WR^e and q/q_0 if necessary.

With a pendulum damper fitted, the equivalent inertia is different for each order of vibration so that each order has a different frequency. A damper tuned as a negative flywheel for one order becomes a positive flywheel for lower orders; thus, it reduces the frequencies of those orders, with possibly unfortunate results.

In in-line engines the application of a pendulum damper may be further complicated by the necessity of suppressing several orders of vibration, thus requiring several sets of damper weights. Alternatively, both a pendulum- and viscous-type damper may be fitted to an engine.

In general, the pendulum-type dampers are more expensive than the viscous types. Wear in the pins and their bushings changes the properties of the damper, thus requiring replacement of these parts at intervals.

The effect upon the fourth-order amplitude at 950 rpm of fitting a pendulum damper adjusted to act as a negative flywheel, for the generator-engine previously used, is calculated as follows: A damper is assumed which has four damper weights, each of 23.6 lb with their centers-of-gravity at a radius of 8.625 in. They will be tuned to $q_0 = 3.7$ order. The length of link is calculated from Eq. (38.31):

$$l = \frac{8.625}{1 + (3.7)^2} = 0.59 \text{ in.}$$

For fourth-order vibration the four weights are equivalent to a flywheel of polar inertia calculated from Eq. (38.36):

$$J_d = \frac{4 \times 23.6 \times (8.625)^2}{1 - (4/3.7)^2} = -41,250 \text{ in.}^2\text{-lb}$$

The hub inertia is assumed to be 3000 in.²-lb; then the net inertia is −38,250 in.²-lb. In the unity system the moment of inertia is

$$I_d = \frac{-38,250}{35.8 \times 386} = -2.77$$

Table 38.16 gives the last lines of Table 38.10 with the calculation continued through the damper. The fourth-order torque in the engine shaft becomes 107,300 in.-lb, as compared with 230,000 in.-lb before the damper was fitted.

TABLE 38.16 Continuation of Table 38.10 to Include Pendulum-Type Damper
($\omega_s^2 = 0.0321$)

Rotor	I	$\dfrac{I\omega_s^2}{\omega_s^2 = 0.0321}$	β Inertia $\theta \times$	Exciting	M Inertia $\theta \times$	Exciting	k
1	1.00	0.0321	−7.7401	−24.748	−0.2480	−7.94	
			0.2490	3.710	0.3946	5.852	1.58
d	−2.77	−0.0890	−7.9891	−28.458	+0.7110	2.540	
				$RM \rightarrow$	1.1056	8.392	

$$\theta = -8.392/1.1056 = -7.58$$

$$M_{9,10}\ (\text{unity system}) = 1.366\theta = 10.38$$

$$M_{9,10}\ (\text{actual system}) = 10.38 \times 11,300 = 107,300\ \text{in.-lb}$$

The angle at the damper in the unity system is

$$\theta_d = -7.9891\theta - 28.458 = 32.042\ \text{rad}$$

In the actual system this corresponds to

$$\theta_d = \frac{32.042 \times 11,300}{175.7 \times 10} = 0.00206\ \text{rad}$$

The link angle is calculated from Eq. (38.34):

$$\psi = \frac{0.00206 \times 16[1 + (3.7)^2]}{3.7^2 - 4^2} = 0.2\ \text{rad}$$

which is an acceptable amplitude.

A similar calculation is made at the low-speed end of the operating range. If the angle ψ is too large at the low-speed end, it is necessary to adjust WR^e or q_0, or both.

REFERENCES

1. Den Hartog, J.: "Mechanical Vibration," Dover Publications, Inc., New York, 1956.
2. Lewis, F. M.: *Trans. Soc. of Naval Arch. Marine Engrs.*, **33**:109 (1925).
3. Nestorides, E. J.: "A Handbook of Torsional Vibration," Cambridge University Press, 1958.
4. Wilson, W. K.: "Practical Solutions of Torsional Vibration Problems," John Wiley & Sons, Inc., New York, 1942.
5. Constant, H.: *Brit. Aero. Res. Comm. R and M*, no. 1201, 1928.
6. Porter, F.: *Trans. ASME*, **50**:8 (1928).
7. Gümbel: *Inst. Naval Architects (England)*, **54**:219 (1912).
8. Holzer: "Die Berechnung der Drehschwingungen," Springer-Verlag, Berlin, 1921.
9. Rao, S. S.: "Mechanical Vibration," Addison-Wesley Publishing Co., Reading, Mass., 1990.
10. Porter, F.: "Evaluation of Effects of Torsional Vibration," Society of Automotive Engineers, War Engineering Board; also *Trans. ASME*, **65**:A-33 (1943).

11. Eshleman, R. L.: "Torsional Response of Internal Combustion Engines," *Trans. ASME,* **96**(2):441 (1974).

12. Anwar, I.: "Computerized Time Transient Torsional Analysis of Power Trains," *ASME Paper* No. 79-DET-74, 1979.

13. Sohre, J. S.: "Transient Torsional Criticals of Synchronous Motor-Driven, High-Speed Compressor Units," *ASME Paper* No. 66-FE-22, June 1965.

14. U.S. Navy Department: "Military Standard Mechanical Vibrations of Mechanical Equipment," MIL-STD-167 (SHIPS).

15. American Petroleum Institute: "Centrifugal Compressors for General Refinery Service," API STD 617, Fifth ed. 1988, Washington, D.C.

16. Eshleman, R. L.: "Torsional Vibrations in Machine Systems," Vibrations, 3 (2): 3 (1987).

17. Lewis, F. M.: *Trans. ASME,* **54**:253 (1932).

18. Lewis, F. M.: *Trans. ASME,* **78**:APM 377 (1955).

19. Zdanowich and Wilson, T. S.: *Proc. Inst. Mech. Engs. (London),* **1943**:182 (1940).

CHAPTER 39

PART I: BALANCING OF ROTATING MACHINERY

Douglas G. Stadelbauer

INTRODUCTION

The demanding requirements placed on modern rotating machines and equipment—for example, electric motors and generators, turbines, compressors, and blowers—have introduced a trend toward higher speeds and more stringent acceptable vibration levels. At lower speeds, the design of most rotors presents few problems which cannot be solved by relatively simple means, even for installations in vibration-sensitive environments. At higher speeds, which are sometimes in the range of tens of thousands of revolutions per minute, the design of rotors can be an engineering challenge which requires sophisticated solutions of interrelated problems in mechanical design, balancing procedures, bearing design, and the stability of the complete assembly. This has made balancing a first-order engineering problem from conceptual design through the final assembly and operation of modern machines.

This chapter describes some important aspects of balancing, such as the basic principles of the process by which an optimum state of balance is achieved in a rotor, balancing methods and machines, and definitions of balancing terms. The discussion is limited to those principles, methods, and procedures with which an engineer should be familiar in order to understand what is meant by "balancing." Finally, a list of definitions is presented at the end of it.

In addition to unbalance, there are many other possible sources of vibration in rotating machinery; some of them are related to or aggravated by unbalance, and so, under appropriate conditions, they may be of paramount importance. However, this discussion is limited to the means by which the effect of once-per-revolution components of vibration (i.e., the effects due to mass unbalance) can be minimized.

BASIC PRINCIPLES OF BALANCING

Descriptions of the behavior of rigid or flexible rotors are given as introductory material in standard vibration texts, in the references listed at the end of Part I of this chapter, and in the few books devoted to balancing. A similar description is included here for the purpose of examining the principles which govern the behavior of rotors as their speed of rotation is varied.

PERFECT BALANCE

Consider a rigid body which is rotating at a uniform speed about one of its three principal inertia axes. Suppose that the forces which cause the rotation and support the body are neglected; then it will rotate about this axis without wobbling, i.e., the principal axis (which is fixed in the body) coincides with a line fixed in space (Fig. 39.1). Now construct circular, concentric journals around the axis at the points where the axis protrudes from the body, i.e., on the stub shafts whose axes coincide with the principal axis. Since the axis does not wobble, the newly constructed journals also will not wobble. Next, place the journals in bearings which are circular and concentric to the principal axis (Fig. 39.2). It is assumed that there is no dynamic action of the elasticity of the rotor and the lubricant in the bearings. A rigid rotor constructed and supported in this manner will not wobble; the bearings will exert no forces other than those necessary to support the weight of the rotor. In this assembly, the radial distance between the center-of-gravity of the rotor and the *shaft axis* (i.e., a straight line connecting the journal axes) is zero. The principal axis and the shaft axis coincide. This rotor is said to be *perfectly balanced*.

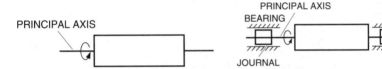

FIGURE 39.1 Rigid body rotating about principal axis.   **FIGURE 39.2** Balanced rigid rotor.

RIGID-ROTOR BALANCING—STATIC UNBALANCE

Rigid-rotor balancing is important because it comprises the majority of the balancing work done in industry. By far the greatest number of rotors manufactured and installed in equipment can be classified as "rigid" by definition. All balancing machines are designed to perform rigid-rotor balancing.*

Consider the case in which the shaft axis is not coincident with the principal axis, as illustrated in Fig. 39.3. In practice, with even the closest manufacturing tolerances, the journals are never concentric with the principal axis of the rotor. If concentric rigid bearings are placed around the journals, thus forcing the rotor to turn about the connecting line between the journals, i.e., the shaft axis, a variable force is sensed at each bearing.

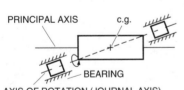

FIGURE 39.3 Unbalanced rigid rotor.

The center-of-gravity is located on the principal axis, and is not on the axis of rotation (shaft axis). From this it follows that there is a net radial force acting on the rotor which is due to centrifugal acceleration. The magnitude of this force is given by

$$F = me\omega^2 \tag{39.1}$$

---

* Field balancing equipment is specifically excluded from this category since it is designed for use with both rigid and flexible rotors.

where m is the mass of the rotor, ϵ is the eccentricity or radial distance of the center-of-gravity from the axis of rotation, and ω is the rotational speed in radians per second. Since the rotor is assumed to be rigid and thus not capable of distortion, this force is balanced by two reaction forces. There is one force at each bearing. Their algebraic sum is equal in magnitude and opposite in sense. The relative magnitudes of the two forces depend, in part, upon the axial position of each bearing with respect to the center-of-gravity of the rotor. In simplified form, this illustrates the "balancing problem." One must choose a practical method of constructing a perfectly balanced rotor from this unbalanced rotor.

The center-of-gravity may be moved to the shaft axis (or as close to this axis as is practical) in one of two ways. The journals may be modified so that the shaft axis and an axis through the center-of-gravity are moved to essential coincidence. From theoretical considerations, this is a valid method of minimizing unbalance caused by the displacement of the center-of-gravity from the shaft axis, but for practical reasons it is difficult to accomplish. Instead, it is easier to achieve a radial shift of the center-of-gravity by adding mass to or subtracting it from the mass of the rotor; this change in mass takes place in the longitudinal plane which includes the shaft axis and the center-of-gravity. From Eq. (39.1), it follows that there can be no net radial force acting on the rotor at any speed of rotation if

$$m'r = m\epsilon \tag{39.2}$$

where m' is the mass added to or subtracted from that of the rotor and r is the radial distance to m'. There may be a *couple*, but there is no net *force*. Correspondingly, there can be no *net bearing reaction*. Any residual reactions sensed at the bearings would be due solely to the couple acting on the rotor.

If this rotor-bearing assembly were supported on a scale having a sufficiently rapid response to sense the change in force at the speed of rotation of the rotor, no fluctuations in the magnitude of the force would be observed. The scale would register only the dead weight of the rotor-bearing assembly.

This process of *effecting essential coincidence between the center-of-gravity of the rotor and the shaft axis is called "single-plane (static) balancing."* The latter name for the process is more descriptive of the end result than of the procedure that is followed.

If a rotor which is supported on two bearings has been balanced statically, the rotor will not rotate under the influence of gravity alone. It can be rotated to any position and, if left there, will remain in that position. However, if the rotor has not been balanced statically, then from any position in which the rotor is initially placed, it will tend to turn to that position in which the center-of-gravity is lowest.

As indicated below, single-plane balancing can be accomplished most simply (but not necessarily with great accuracy) by supporting the rotor on flat, horizontal ways and allowing the center-of-gravity to seek its lowest position. It also can be accomplished in a centrifugal balancing machine by sensing and correcting for the unbalance force characterized by Eq. (39.1).

RIGID-ROTOR BALANCING—DYNAMIC UNBALANCE

When a rotor is balanced statically, the shaft axis and principal inertia axis may not coincide; single-plane balancing ensures that the axes have only one common point, namely, the center-of-gravity. Thus, perfect balance is not achieved. To obtain perfect balance, the principal axis must be rotated about the center-of-gravity in the longitudinal plane characterized by the shaft axis and the principal axis. This rotation can be accomplished by modifying the journals (but, as before, this is impractical) or by adding masses to or subtracting them from the mass of the rotor in the longitudinal plane characterized by the shaft axis and the principal inertia axis. Although adding or subtracting a single mass may cause rotation of the principal axis relative to the shaft axis, it also disturbs the static balance already achieved. From this it can be deduced that a couple must be applied to the rotor in the longitudinal plane. This is usually accomplished by adding or subtracting two masses of equal magnitude, one on each side of the principal axis (so as not to disturb the static balance) and one in each of two radial planes (so as to produce the necessary rotatory effect). Theoretically, it is not important which two radial planes are selected since the same rotatory effect can be achieved with appropriate masses, irrespective of the axial location of the two planes. Practically, the choice of suitable planes may be important. Usually, it is best to select planes which are separated axially by as great a distance as possible in order to minimize the magnitude of the masses required.

The above process of *bringing the principal inertial axis of the rotor into essential coincidence with the shaft axis is called "two-plane (dynamic) balancing."* If a rotor is balanced in two planes, then, by definition, it is balanced statically; however, the converse is not true.

FLEXIBLE-ROTOR BALANCING[1]

If the bearing supports are rigid, then the forces exerted on the bearings are due entirely to centrifugal forces caused by the unbalance. Dynamic action of the elasticity of the rotor and the lubricant in the bearings has been ignored.

The portion of the overall problem in which the dynamic action and interaction of rotor elasticity, bearing elasticity, and damping are considered is called flexible rotor or modal balancing.

Critical Speed.   Consider a long, slender rotor, as shown in Fig. 39.4. It represents the idealized form of a typical flexible rotor, such as a paper machinery roll or turbogenerator rotor. Assume further that all unbalances occurring along the rotor caused by machining tolerances, inhomogeneities of material, etc. are compensated by correction weights placed in the end faces of the rotor, and that the balancing is done at a low speed as if the rotor were a rigid body.

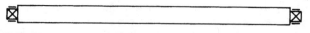

FIGURE 39.4   Idealized flexible rotor.

Assume there is no damping in the rotor or its bearing supports. Consider a thin slice of this rotor perpendicular to the shaft axis (see Fig. 39.5*A*). This axis intersects the slice at its geometric center E when the rotor is not rotating, provided that deflection due to gravity forces is ignored. The center-of-gravity of the slice is displaced by δ from E due to an unbalance in the slice (caused by machining tolerances, inhomogeneity, etc., mentioned above) which was compensated by correction weights in the rotor's end planes. If the rotor starts to rotate about the shaft axis with an angular speed ω, then the slice starts to rotate in its own plane at the same speed about an axis through E. Centrifugal force $m\delta\omega^2$ is thus experienced by the slice. This force occurs in a direction perpendicular to the shaft axis and may be accompanied

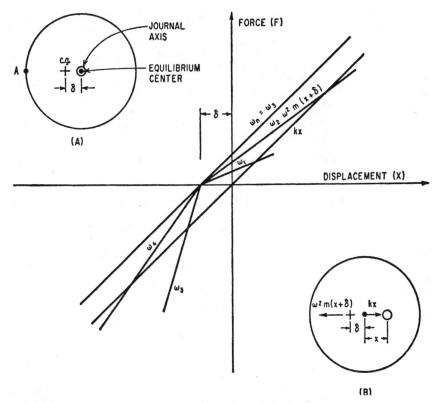

FIGURE 39.5 Rotor behavior below, at, and above first critical speed.

by similarly caused forces at other cross sections along the rotor; such forces are likely to vary in magnitude and direction. They cause the rotor to bend, which in turn causes additional centrifugal forces and further bending of the rotor.

At every speed ω, equilibrium conditions require that for one slice, the centrifugal and restoring forces be related by

$$m(\delta + x)\omega^2 = kx \qquad (39.3)$$

where x is the deflection of the shaft (the radial distance between the geometric center and the shaft axis) and k is the shaft stiffness (Fig. 39.5B). In Fig. 39.5, the centrifugal and restoring forces are plotted for various speeds ($\omega_1 < \omega_2 < \omega_3 < \omega_4 < \omega_5$). The point of intersection of the lines representing the two forces denotes the equilibrium condition for the rotor at the given speeds. For this ideal example, as the speed increases, the point which denotes equilibrium will move outward until, at say ω_3, a speed is reached at which there is no resulting force and the lines are parallel. Since equilibrium is not possible at this speed, it is called the critical speed. *The critical speed ω_n of a rotating system corresponds to a resonant frequency of the system.*

At speeds greater than ω_3 ($\equiv\omega_n$), the lines representing the centrifugal and restoring forces again intersect. As ω increases, the slope of the line $m\omega^2(x + \delta)$ increases correspondingly until, for speeds which are large, the deflection x approaches the value of δ, i.e., the rotor tends to rotate about its center-of-gravity.

Unbalance Distribution.　Apart from any special and obvious design features, the axial distribution of unbalance in the slices previously examined along any rotor is likely to be random. The distribution may be significantly influenced by the presence of large local unbalances arising from shrink-fitted discs, couplings, etc. The rotor may also have a substantial amount of initial bend, which may produce effects similar to those due to unbalance. The method of construction can influence significantly the magnitude and distribution of unbalance along a rotor. Rotors may be machined from a single forging, or they may be constructed by fitting several components together. For example, jet-engine rotors are constructed by joining many shell and disc components, whereas paper mill rolls are usually manufactured from a single piece of material.

The unbalance distributions along two nominally identical rotors may be similar but rarely identical.

Contrary to the case of a rigid rotor, distribution of unbalance is significant in a flexible rotor because it determines the degree to which any bending or flexural mode of vibration is excited. The resulting modal shapes are reduced to acceptable levels by flexible-rotor balancing, also called "modal balancing."*

Response of a Flexible Rotor to Unbalance.　In common with all vibrating systems, rotor vibration is the sum of its modal components. For an *undamped* flexible rotor which rotates in flexible bearings, the flexural modes coincide with principal modes and are plane curves rotating about the axis of the bearing. For a *damped* flexible rotor, the flexural modes may be space (three-dimensional) curves rotating about the axis of the bearings. The damping forces also limit the flexural amplitudes at each critical speed. In many cases, however, the damped modes can be treated approximately as principal modes and hence regarded as rotating plane curves.

The unbalance distribution along a rotor may be expressed in terms of modal components. The vibration in each mode is caused by the corresponding modal component of unbalance. Moreover, the response of the rotor in the vicinity of a critical speed is usually predominantly in the associated mode. The rotor modal response is a maximum at any rotor critical speed corresponding to that mode. Thus, when a rotor rotates at a speed near a critical speed, it is disposed to adopt a deflection shape corresponding to the mode associated with this critical speed. The degree to which large amplitudes of rotor deflection occur in these circumstances is determined by the modal component of unbalance and the amount of damping present in the rotor system.

If the modal component of unbalance is reduced by a number of discrete correction masses, then the corresponding modal component of vibration is similarly reduced. The reduction of the modal components of unbalance in this way forms the basis of the modal balancing technique.

Flexible-Rotor Mode Shapes.　The stiffnesses of a rotor, its bearings, and the bearing supports affect critical speeds and therefore mode shapes in a complex manner. For example, Fig. 39.6 shows the effect of varying bearing and support stiffness relative to that of the rotor. The term "soft" or "hard" bearing is a relative one, since for different rotors the same bearing may appear to be either soft or hard. The schematic diagrams of the figure illustrate that the first critical speed of a rotor supported in a balancing machine having soft-spring-bearing supports occurs at a lower frequency and in an apparently different shape than that of the same rotor sup-

---

* All modal balancing is accomplished by multiplane corrections; however, multiplane balancing need not be modal balancing, since multiplane balancing refers only to unbalance correction in more than two planes.

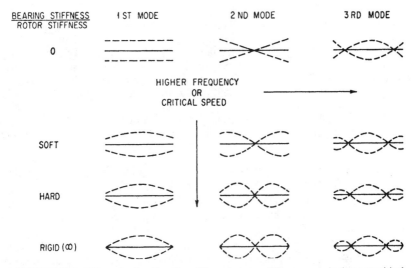

FIGURE 39.6 Effect of ratio of bearing stiffness to rotor stiffness on mode shape at critical speeds.

ported in a hard-bearing balancing machine where the bearing support stiffness approximates service conditions.

To evaluate whether a given rotor may require a flexible-rotor balancing procedure, the following rotor characteristics must be considered:

1. Rotor configuration and service speed.
2. Rotor design and manufacturing procedures. Rotors which are known to be flexible or unstable may still be capable of being balanced as rigid rotors.

Rotor Elasticity Test. This test is designed to determine if a rotor can be considered *rigid* for balancing purposes or if it must be treated as *flexible*. The test is carried out at service speed either under service conditions or in a high-speed, hard-bearing balancing machine whose support-bearing stiffness closely approximates that of the final supporting system. The rotor should first be balanced. A weight is then added in each end plane of the rotor near its journals; the two weights must be in the same angular position. During a subsequent test run, the vibration is measured at both bearings. Next, the rotor is stopped and the test weights are moved to the center of the rotor, or to a position where they are expected to cause the largest rotor distortion; in another run the vibration is again measured at the bearings. If the total of the first readings is designated x, and the total of the second readings y, then the ratio $(y - x)/x$ should not exceed 0.2. Experience has shown that if this ratio is below 0.2, the rotor can be corrected satisfactorily at low speed by applying correction weights in two or three selected planes. Should the ratio exceed 0.2, the rotor usually must be checked at or near its service speed and corrected by a modal balancing technique.

High-Speed Balancing Machines. Any technique of modal balancing requires a balancing machine having a variable balancing speed with a maximum speed at least equal to the maximum service speed of the flexible rotor. Such a machine must also have a drive-system power rating which takes into consideration not only acceleration of the rotor inertia but also windage losses and the energy required for a rotor to pass through a critical speed. For some rotors, windage is the major loss; such rotors may have to be run in vacuum chambers to reduce the fanlike action of the rotor and to prevent the rotor from becoming excessively hot. For high-speed balancing installations, appropriate controls and safety measures must be employed to protect the operator, the equipment, and the surrounding work areas.

Flexible-Rotor Balancing Techniques.   Flexible-rotor balancing consists essentially of a series of individual balancing operations performed at successively greater rotor speeds:

At a low speed, where the rotor is considered rigid. (Low-speed balancing of flexible rotors usually is performed only in a balancing machine.)

At a speed where significant rotor deformation occurs in the mode of the first flexural critical speed. (This deformation may occur at speeds well below the critical speed.)

At a speed where significant rotor deformation occurs in the mode of the second flexural critical speed. (This applies only to rotors with a maximum service speed affected significantly by the mode shape of the second flexural critical speed.)

At a speed where significant rotor deformation occurs in the mode of the third critical speed, etc.

At the maximum service speed of the rotor.

The balancing of flexible rotors requires experience in determining the size of correction weights when the rotor runs in a flexible mode. The process is considerably more complex than standard low-speed balancing techniques used with rigid rotors. Primarily this is due to a shift of mass within the rotor (as the speed of rotation changes), caused by shaft and/or body elasticity, asymmetric stiffness, thermal dissymmetry, incorrect centering of rotor mass and shifting of windings and associated components, and fit tolerances and couplings.

Before starting the modal balancing procedure, the rotor temperature should be stabilized in the lower- or middle-speed range until unbalance readings are repeatable. This preliminary warmup may take from a few minutes to several hours depending on the type of rotor, its dimensions, its mass, and its pretest condition.

Once the rotor is temperature-stabilized, the balancing process can begin. Several weight corrections in both end planes and along the rotor surface are required. In the commonly practiced, comprehensive modal balancing technique, unbalance correction is performed in several discrete modes, each mode being associated with the speed range in which the rotor is deformed to the mode shape corresponding to a particular flexural critical speed. Figure 39.7 shows a rotor deformed in five of the mode shapes of Fig. 39.6; the location of the weights which provide the proper correction for these mode shapes is indicated.

First, the rotor is rotated at a speed less than one-half the rotor's first flexural critical speed and balanced using a rigid-rotor balancing technique. Balancing corrections are performed at the end planes to reduce the original amount of unbalance to three or four times the final balance tolerance.

Correction for the First Flexural Mode (V Mode).   The balancing speed is increased until rotor deformation occurs, accompanied by a rapid increase in unbalance indication at the same angular position for both end planes. Unbalance corrections for this mode are made in at least three planes. Due to the bending of the rotor,

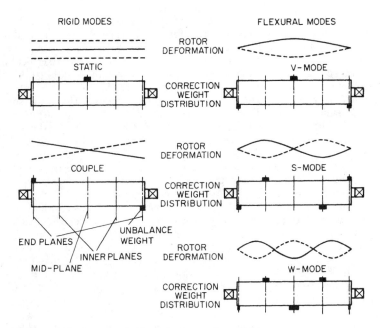

FIGURE 39.7 Rotor mode shapes and correction weights.

the unbalance indication is not directly proportional to the correction to be applied. A new relationship between unbalance indication and corresponding correction weight must be established by test with trial weights. A weight is first added in the correction plane nearest the middle of the rotor. For large turbo-generator rotors such a trial weight should be in the range of 30 to 60 oz-in./ton of rotor weight. Two additional corrections are added in the end planes diametrically opposite to the center weight, each equal to one-half the magnitude of the center weight. This process may have to be repeated a number of times, each run reducing the magnitude of the weight applications until the residual unbalance is approximately 1 to 3 oz-in./ton of turbo-generator rotor weight. Then the speed is increased slowly to the maximum service speed; at the same time, the unbalance indicator is monitored. If an excessive unbalance indication is observed as the rotor passes through its first critical speed, further unbalance corrections are required in the V mode until the maximum service speed can be reached without an excessive unbalance indication. If a second flexural critical speed is observed before the maximum service speed is reached, the additional balancing operation in the S mode must be performed, as indicated below.

Correction for the Second Flexural Mode (S Mode). The rotor speed is increased until significant rotor deformation due to the second flexural mode is observed. This is indicated by a rapid increase in unbalance indication measured in the end planes at angular positions opposite to each other. Unbalance corrections for this S mode are made in at least four planes, as indicated in Fig. 39.7. The weights placed in the end correction planes must be diametrically opposed; on the idealized symmetrical rotor, each end-plane weight must be equal to one-half the correction weight placed in one of the inner planes. Of primary concern is that the S-mode weight set not have any influence on the previously corrected mode shape. The correction weight in each inner plane must be diametrically opposed to its nearest end-plane correction weight. The procedure to determine the relationship between unbalance indication and required correction weight is similar to that used in the V-mode procedure, described above. The S-mode balancing process must be repeated until an acceptable residual unbalance is achieved. If a third critical speed is observed before the maximum service speed is reached, the additional balancing operation in the W mode must be performed, as indicated below.

Corrections for the Third Flexural Mode (W Mode). The rotor speed is increased further until significant rotor deformation due to the third flexural mode is observed. Corrections are made in the rotor with a five-weight set (shown in Fig. 39.7) and in a manner similar to that used in correcting for the first and second flexural modes.

If the service-speed range requires it, higher modes (those associated with the nth critical speed, for example) may have to be corrected as well. For each of these higher modes, a set of $(n + 2)$ correction weights is required.

Final Balancing at Service Speed. Final balancing takes place with the rotor at its service speed. Correction should be made only in the end planes. The final balance tolerance for large turbo-generators, for example, will normally be on the order of 1 oz-in./ton of rotor weight. If the rotor cannot be brought into proper balance tolerances, the S-mode, V-mode, and W-mode corrections may require slight adjustment.

To achieve repeatability of the correction effects, the same balancing speed for each mode must be accurately maintained. Depending on the size of the rotor, the number of modes that must be corrected, and the ease with which weights can be applied, the entire process may take anywhere from 3 to 30 hours.

The relative position of the unbalance correction planes shown in Fig. 39.7 applies to symmetrical rotors only. Rotors with axial asymmetry generally require unsymmetrically spaced correction planes. In the case of assembled rotors which may "take a set" at or near service speed (e.g., shrunk-on turbine stages find their final position), only preliminary unbalance corrections are made at lower speeds to enable the rotor to be accelerated to service or overspeed, the latter being usually 20 percent above maximum service speed. Since the "set" creates new unbalance, the normal balancing procedure is commenced only after the initial high-speed run.

Computer programs are available which facilitate the selection of the most appropriate correction planes and the computation of correction weights by the influence coefficient method. Other flexible-rotor balancing techniques rely mostly on experience data available from previously manufactured rotors of the same type, or correct only for flexural modes if no low-speed balancing equipment is available.

SOURCES OF UNBALANCE

Sources of unbalance in rotating machinery may be classified as resulting from

1. Dissymmetry
   (Core shifts in castings, rough surfaces on forgings, unsymmetrical configurations)
2. Nonhomogeneous material
   (Blowholes in cast rotors, inclusions in rolled or forged materials, slag inclusions or variations in crystalline structure caused by variations in the density of the material)
3. Distortion at service speed
   (Blower blades in built-up designs)

4. Eccentricity
   (Journals not concentric or circular, matching holes in built-up rotors not circular)

5. Misalignment of bearings

6. Shifting of parts due to plastic deformation of rotor parts
   (Windings in electric armatures)

7. Hydraulic or aerodynamic unbalance
   (Cavitation or turbulence)

8. Thermal gradients
   (Steam-turbine rotors, hollow rotors such as paper mill rolls)

Often, balancing problems can be minimized by careful design in which unbalance is controlled. When a part is to be balanced, large amounts of unbalance require large corrections. If such corrections are made by removal of material, additional cost is involved and part strength may be affected. If corrections are made by addition of material, cost is again a factor and space requirements for the added material may be a problem.

Manufacturing processes are a major source of unbalance. Unmachined portions of castings or forgings which cannot be made concentric and symmetrical with respect to the shaft axis introduce substantial unbalance. Manufacturing tolerances and processes which permit any eccentricity or lack of squareness with respect to the shaft axis are sources of unbalance. Tolerances necessary for economical assembly of several elements of a rotor permit radial displacement of parts of the assembly and thereby introduce unbalance.

Limitations imposed by design often introduce unbalance effects which cannot be corrected adequately by refinement in design. For example, electrical design limitations impose a requirement that one coil be at a greater radius than the others in a certain type of electric armature. It is impractical to design a compensating unbalance into the armature.

Fabricated parts, such as fans, often distort nonsymmetrically under service conditions. Design and economic considerations prevent the adaptation of methods which might eliminate this distortion and thereby reduce the resulting unbalance.

Ideally, rotating parts always should be designed for inherent balance, whether a balancing operation is to be performed or not. Where low service speeds are involved and the effects of a reasonable amount of unbalance can be tolerated, this practice may eliminate the need for balancing. In parts which require unbalanced masses for functional reasons, these masses often can be counterbalanced by designing for symmetry about the shaft axis.

MOTIONS OF UNBALANCED ROTORS

In Fig. 39.8 a rotor is shown spinning freely in space. This corresponds to spinning above resonance in soft bearings. In Fig. 39.8A only static unbalance is present and the center line of the shaft sweeps out a cylindrical surface. Figure 39.8B illustrates the motion when only couple unbalance is present. In this case, the center line of the rotor shaft sweeps out two cones which have their apexes at the center-of-gravity of the rotor. The effect of combining these two types of unbalance when they occur in the same axial plane is to move the apex of the cones away from the center-of-gravity. In most cases, there will be no apex and the shaft will move in a more complex combination of the motions shown in Fig. 39.8. Such a condition comes about through a random combination of static and couple unbalance called *dynamic unbalance.*

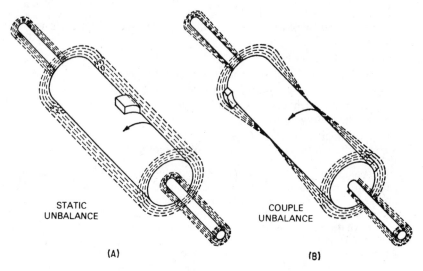

STATIC
UNBALANCE

COUPLE
UNBALANCE

(A)

(B)

FIGURE 39.8   Effect of static and couple unbalance on free rotor motion.

OPERATING PRINCIPLES OF BALANCING MACHINES[2,3]

This section describes the basic operating principles and general features of the various types of balancing machines which are available commercially. With this type of information, it is possible to determine the basic type of machine required for a given application.

Every balancing machine must determine by some technique both the magnitude of a correction weight and its angular position in each of one, two, or more selected balancing planes. For single-plane balancing this can be done statically, but for two- or multiplane balancing it can be done only while the rotor is spinning. Finally, all machines must be able to resolve the unbalance readings, usually taken at the bearings, into equivalent corrections in each of the balancing planes.

On the basis of their method of operation, balancing machines and equipment can be grouped in two general categories:

1. Gravity balancing equipment

2. Centrifugal balancing machines and field balancing equipment

In the first category, advantage is taken of the fact that a body that is free to rotate always seeks that position in which its center-of-gravity is lowest. Gravity balancing equipment, also called *nonrotating balancers,* includes horizontal ways, knife-edges or roller arrangements, spirit-level devices ("bubble balancers"), and vertical pendulum types. All are capable of detecting and/or indicating only static unbalance.

In the second category, the amplitude and phase of motions or reaction forces caused by once-per-revolution centrifugal forces resulting from unbalance are sensed, measured, and indicated by appropriate means. Field balancing equipment provides sensing and measuring instrumentation only; the necessary measurements for balancing a rotor are taken while the rotor runs in its own bearings and under its own power. However, on a centrifugal balancing machine, the rotor is supported by the machine and rotated around a horizontal or vertical axis by the machine's drive motor. Balancing-machine instrumentation differs from field balancing equipment in that it includes specific features which simplify the balancing process. A centrifugal balancing machine (also called a *rotating balancing machine*) is usually capable of measuring static unbalance (a *single-plane rotating balancing machine*) or static *and* dynamic unbalance (a *two-plane rotating balancing machine*). Only a two-plane rotating balancing machine can detect couple unbalance or dynamic unbalance.

GRAVITY BALANCERS

First, consider the simplest type of balancing—usually called "static" balancing, since the rotor is not spinning. In Fig. 39.9A, a disc-type rotor on a shaft is shown resting on knife-edges. The mass added to the disc at its rim represents a known unbalance. In this illustration, in Fig. 39.8, and in the illustrations which follow, the rotor is assumed to be balanced without this added unbalance weight. In order for this balancing procedure to work effectively, the knife-edges must be level, parallel, hard, and straight.

In operation, the heavier side of the disc will seek the lowest level—thus indicating the angular position of the unbalance. Then, the magnitude of the unbalance usually is determined by an empirical process, adding mass in the form of wax or putty to the light side of the disc until it is in balance, i.e., until the disc does not stop at the same angular position.

In Fig. 39.9B, a set of balanced rollers or wheels is used in place of the knife-edges. These have the advantage of permitting the rotor to turn without, at the same time, moving laterally.

In Fig. 39.9C, a setup for another type of static, or "nonrotating," balancing procedure is shown. Here the disc to be balanced is supported by a flexible cable, fastened to a point on the disc which coincides with the center of the shaft and is slightly above the normal plane containing the center-of-gravity. As shown in Fig. 39.9C, the heavy side will tend to seek a lower level than the light side, thereby indicating the angular position of the unbalance. The disc can be balanced by adding weight to the diametrically opposed side of the disc until it hangs level. In this case, the center-of-gravity is moved until it is directly under the flexible support cable.

In Fig. 39.9D, a modified version of this setup is shown. The cable is replaced by a hardened ball-and-socket arrangement (used on many automobile wheel "bubble balancers") or by a spherical air bearing (used on some industrial and aerospace balancers). The inclination of the wheel is then indicated with a centrally mounted spirit level.

Static balancing is satisfactory for rotors having relatively low service speeds and axial lengths which are small in comparison with the rotor diameter. A preliminary static unbalance correction may be required on rotors having a combined unbalance so large that it is impossible in a dynamic, soft-bearing balancing machine to bring the rotor up to its proper balancing speed without damaging the machine. If the rotor is first balanced statically by one of the methods just outlined, it is usually possible to decrease the combined unbalance to the point where the rotor may be brought up to balancing speed and the residual unbalance measured. Such preliminary static correction is not required on hard-bearing balancing machines.

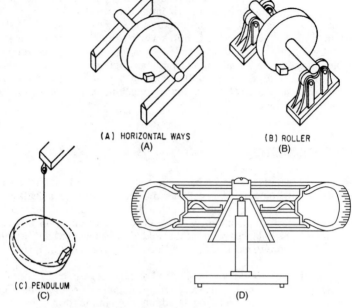

(A) HORIZONTAL WAYS
(A)

(B) ROLLER
(B)

(C) PENDULUM
(C)

(D)

FIGURE 39.9   Static (single-plane) balancing devices.

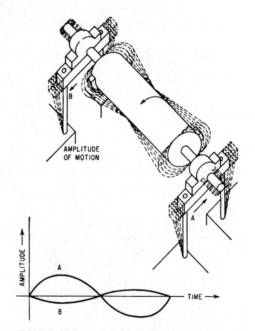

AMPLITUDE
OF MOTION

AMPLITUDE →

A

B

TIME →

FIGURE 39.10   Motion of unbalanced rotor and bearings in flexible-bearing, centrifugal balancing machine.

CENTRIFUGAL BALANCING MACHINES

The following procedures may be used to balance the rotor shown in Fig. 39.8B. First, select the planes in which the correction weights are to be added; these planes should be as far apart as possible and the weights should be added as far out from the shaft as feasible to minimize the size of the weights. Next, by a balancing technique, determine the size of the required correction weight and its angular position for each correction plane. To implement these procedures, two types of machines, *soft-bearing* and *hard-bearing* balancing machines, which are described below, are employed.

Soft-bearing Balancing Machines. Soft-bearing balancing machines permit the idealized free rotor motion illustrated in Fig. 39.8B, but on most machines the motion is restricted to a horizontal plane (as shown in Fig. 39.10). Furthermore, the bearings (and the directly attached components) vibrate in unison with the rotor, thus adding to its mass. The restriction of the vertical motion does not affect the amplitude of vibration in the horizontal plane, but the added mass of the bearings does. The greater the combined rotor-and-bearing mass, the smaller will be the displacement of the bearings, and the smaller will be the output of the devices which sense the unbalance.

Consider the following example. Assume a balanced disc (see Fig. 39.11) having a weight W of 1,000 grams, rotating freely in space. An unbalance weight w of 1 gram is then added to the disc at a radius of 10 mm. The unbalance causes the center-of-gravity of the disc to be displaced from the shaft axis by

$$e = \frac{wr}{W + w} = 0.00999 \text{ mm}$$

Since the addition of the weight of the unbalance to the rotor causes only an insignificant difference, the approximation $e \approx wr/W$ is generally used. Then $e \approx 0.01$ mm.

If the same disc with the same unbalance is rotated on a single-plane balancing machine having a bearing and bearing housing weight W' of 1,000 grams, the displacement of the center-of-gravity will be significantly reduced because the bearing and housing weight is added to the weight W of the disc. The center-of-gravity of the combined vibrating components will now be displaced by

$$e' = \frac{wr}{W + W'} \approx 0.005 \text{ mm}$$

The conversion of unbalance into displacement of center-of-gravity as shown in

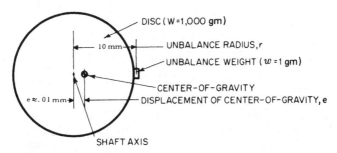

FIGURE 39.11 Displacement of center-of-gravity because of unbalance.

the example above also holds true for rotors of greater axial length which normally require correction in two planes. However, such rotors are prone to have unbalance other than static unbalance, causing an inclination of the principal inertia axis from the shaft axis. In turn, this results in a displacement of the principal inertia axis from the shaft axis in the bearing planes of the rotor, causing the balancing machine bearings to vibrate.

To find the bearing displacement or bearing vibration amplitude resulting from a given unbalance is more involved than finding the center-of-gravity displacement, because other factors come into play, as is illustrated by Fig. 39.12. The weight and inertia of the balancing machine bearings and directly attached vibratory components are usually not known. In any case, they are usually small in relation to the weight and the inertia of the rotor and can generally be ignored. On this basis, the following formula may be used to find the approximate bearing displacement d:

$$d \approx \frac{wr}{W} + \frac{wrhs}{g(I_x - I_z)}$$

where d = displacement at bearing of principal inertia axis from shaft axis
 r = distance from shaft axis to unbalance weight
 h = distance from center-of-gravity to unbalance plane
 s = distance from center-of-gravity to bearing plane
 g = gravitational constant
 I_x = moment of inertia around transverse axis X
 I_z = moment of inertia around principal axis Z

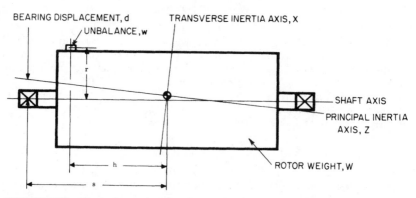

FIGURE 39.12 Displacement of principal axis of inertia from shaft axis at bearing.

From the above it can be seen that the relationship between bearing motion and unbalance in a soft-bearing balancing machine is complex. Therefore, a direct indication of unbalance can be obtained only after calibrating the indicating elements to a given rotor by use of calibration weights which produce a known amount of unbalance.

Hard-bearing Balancing Machines. Hard-bearing balancing machines are essentially of the same construction as soft-bearing balancing machines except that their bearing supports are significantly stiffer in the horizontal direction.

This results in a horizontal critical speed for the machine which is several orders of magnitude greater than that for a comparable soft-bearing balancing machine. The hard-bearing balancing machine is designed to operate at speeds well below its horizontal critical speed. In this speed range, the output from the sensing elements attached to the balancing-machine bearing supports is directly proportional to the centrifugal force resulting from unbalance in the rotor. The output is not influenced by bearing mass, rotor weight, or inertia, so that a permanent relation between unbalance and sensing element output can be established. Unlike with soft-bearing balancing machines, the use of calibration weights to calibrate the machine for a given rotor is not required.

Measurement of Amount and Angle of Unbalance. Both soft- and hard-bearing balancing machines use various types of sensing elements *at the rotor-bearing supports* to convert mechanical vibration into an electrical signal. On commercially available balancing machines, these sensing elements are usually velocity-type pickups, although on certain hard-bearing balancing machines, magnetostrictive or piezoelectric pickups have also been employed.

Three basic methods are used to obtain a reference signal by which the phase angle of the amount-of-unbalance indication signal may be correlated with the rotor. On end-drive machines (where the rotor is driven via a universal joint driver or similarly flexible coupling shaft), a phase reference generator, directly coupled to the balancing machine drive spindle, is used. On belt-drive machines (where the rotor is driven by a belt over the rotor periphery) or on air-drive or self-drive machines, a small light source projects a narrowly focused beam onto the rotor (usually the shaft). Its reflection is picked up by a photoelectric cell. Placement of a non-reflecting mark on the shaft surface will momentarily interrupt the reflection and thereby furnish the starting point from which the angular position of unbalance in the rotor is counted. (Stroboscopic lamps, flashing once per rotor revolution, are no longer considered satisfactory for angle accuracy.) The outputs from the phase-reference sensor and the pickups at the rotor bearing supports are processed in various ways by different manufacturers. Generally, the processed signals result in an indication representing the amount of unbalance and its angular position. In Fig. 39.13 block diagrams are shown for typical balancing instrumentation. In Fig. 39.13*A* an indicating system is shown which uses switching between correction planes (i.e., single-channel instrumentation). This is generally employed on low-cost balancing machines. In Fig. 39.13*B* an indicating system with two-channel instrumentation is shown. Combined indication of amount of unbalance and its angular position is provided on a vectormeter having an illuminated target projected on a screen. Two vectormeters give a simultaneous indication for both unbalance correction planes. Displacement of a target from the central zero point provides a direct visual representation of the displacement of the principal inertia axis from the shaft axis. Concentric circles on the screen indicate the amount of unbalance, and radial lines indicate its angular position. Current balancing machines use computerized instrumentation with video screens on which the amount and angle of unbalance are indicated in digital format.

Indicated and Actual Angle of Unbalance. An *unbalanced rotor* is a rotor in which the principal inertia axis does not coincide with the shaft axis. When rotated in its bearings, an unbalanced rotor will cause periodic vibration of, and will exert a periodic force on, the rotor bearings and their supporting structure. If the structure is rigid, the force is larger than if the structure is flexible. In practice, supporting structures are neither rigid nor flexible but somewhere in between. The rotor-

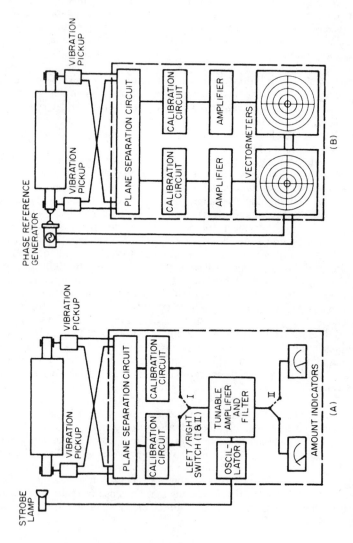

FIGURE 39.13 Block diagrams of typical balancing-machine instrumentations. (A) Amount of unbalance indicated on analog meters, angle by strobe light. (B) Combined amount and angle indication on vectormeters.

bearing support offers some restraint, forming a spring-mass system with damping having a single resonance frequency. When the rotor speed is below this frequency, the principal inertia axis of the rotor moves outward radially. This condition is illustrated in Fig. 39.14. If a pencil or other marking device is moved toward the rotor until it touches the rotor, the so-called "high spot" is marked at the same angular position as the unbalance. When the rotor speed is increased, there is a small time lag between the instant at which the unbalance passes the pencil and the instant at which the rotor moves out enough to contact it. This is due to the damping in the system. The angle between these two points is called the "angle of lag." (See Fig. 39.14B.) As the rotor speed is increased further, resonance of the rotor and its supporting structure will occur; at this speed the angle of lag is 90°. As the rotor passes through resonance, there are large vibration amplitudes, and the angle of lag changes rapidly. As the speed is increased to approximately twice the resonance speed, the angle of lag approaches 180°. At speeds greater than approximately twice the resonance speed, the rotor tends to rotate about its principal inertia axis; the angle of lag (for all practical purposes) is 180°.

The changes in the relative position of pencil mark and unbalance shown in Fig. 39.14 for a statically unbalanced rotor occur in the same manner on a rotor with dynamic unbalance. However, the center-of-gravity shown in the illustrations then represents the position of the principal inertia axis in the plane at which the pencil is applied to the rotor. Thus, the indicated angle of lag and displacement amplitude refer only to that particular plane and generally differ from those for any other plane in the rotor.

Angle of lag is shown as a function of rotational speed in Fig. 39.15: (*A*) for soft-bearing balancing machines whose balancing-speed ranges start at approximately twice the resonance speed; and (*B*) for hard-bearing balancing machines. The effects

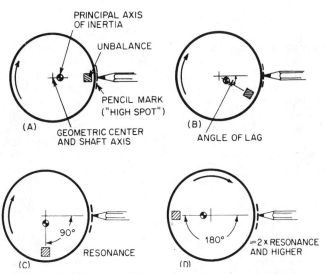

FIGURE 39.14   A pencil or marker is held against an unbalanced rotor. (*A*) A high spot is marked. (*B*) The angle of lag. Angle of lag between unbalance and high spot increases from 0° (*A*) to 180° (*D*) as rotor speed increases.

of damping also are illustrated. Here the resonance frequency of the combined rotor-bearing support system is usually more than three times greater than the maximum balancing speed.

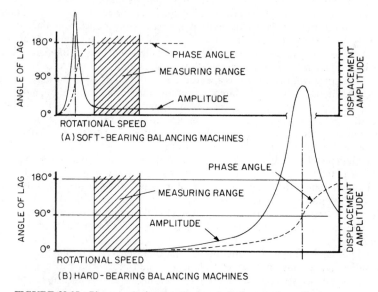

FIGURE 39.15   Phase angle (angle of lag) and displacement amplitude vs. rotational speed in soft-bearing and hard-bearing balancing machines.

Plane Separation.   Consider the rotor in Fig. 39.10 and assume that only the unbalance weight on the left is attached to the rotor. This weight causes not only the left bearing to vibrate but to a lesser degree the right. This influence is called "cross effect." If a second weight is attached in the right plane of the rotor as shown in Fig. 39.10, then the direct effect of the weight in the right plane combines with the cross effect of the weight in the left plane, resulting in a composite vibration of the right bearing. If the two unbalance weights are at the same angular position, the cross effect of one weight has the same angular position as the direct effect in the other rotor end plane; thus, their direct and cross effects are additive (Fig. 39.16*A*). If the two unbalance weights are 180° out of phase, their direct and cross effects are subtractive (Fig. 39.16*B*). In a hard-bearing balancing machine, the additive or subtractive effect depends entirely on ratios between the axial positions of the correction planes and bearings. On a soft-bearing machine, this is not true, because the masses and inertias of the rotor and its bearings must be taken into account.

If the two unbalance weights on the rotor (Fig. 39.10) have an angular relationship other than 0 or 180°, then the cross effect in the right bearing has a different phase angle from the direct effect from the right weight. Addition or subtraction of these effects is vectorial. The net bearing vibration is equal to the resultant of the two vectors, as shown in Fig. 39.17. The phase angle indicated by the bearing vibration does not coincide with the angular position of either weight. This is the most common type of unbalance (dynamic unbalance of random amount and angular

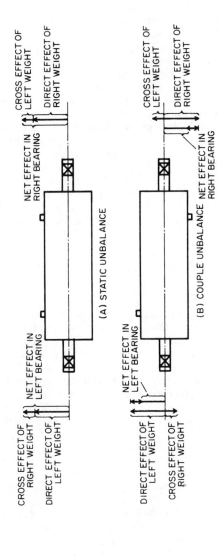

FIGURE 39.16 Influence of cross effects in rotors with static and couple unbalance.

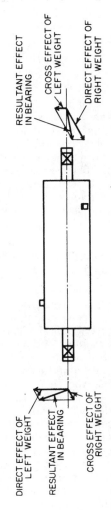

FIGURE 39.17 Influence of cross effects in rotors with dynamic unbalance. All vectors seen from right side of rotor.

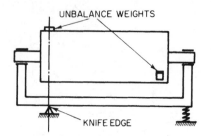

FIGURE 39.18   Plane separation by mechanical means.

position). This interaction of direct and cross effects could cause the balancing process to be a trial-and-error procedure. To avoid this, balancing machines incorporate a feature called "plane separation" which eliminates the influence of cross effect.

Cross effect may be eliminated by supporting the rotor in a cradle which rests on a knife-edge and spring arrangement, as shown in Fig. 39.18. Either the bearing-support members of the cradle or the pivot point are movable, so that one unbalance correction plane always can be brought into the plane of the knife-edge. Any unbalance in this plane is prevented from causing the cradle to vibrate. Unbalance in one end plane of the rotor is measured and corrected. The rotor is turned end for end, so that the knife-edge is in the plane of the first correction. Any vibration of the cradle is now due solely to unbalance present in the plane that was first over the knife-edge. Corrections are applied to this plane until the cradle ceases to vibrate. The rotor is now in balance. If it is again turned end for end, there will be no vibration. Mechanical plane separation cradles restrict the rotor length, diameter, and location of correction planes; thus, modern machines use electronic circuitry to accomplish the function of plane separation.

CLASSIFICATION OF CENTRIFUGAL BALANCING MACHINES

Centrifugal balancing machines may be categorized by the type of unbalance the machine is capable of indicating (static or dynamic), the attitude of the shaft axis of the workpiece (vertical or horizontal), and the type of rotor-bearing-support system employed (soft- or hard-bearing). The four classes (I to IV) included in Table 39.1 are described below.

Class I: Trial-and-Error Balancing Machines.   Machines in this class are of the soft-bearing type. They do not indicate unbalance directly in weight units (such as ounces or grams in the actual correction planes) but indicate only displacement and/or velocity of vibration at the bearings. The instrumentation does not indicate the amount of weight which must be added or removed in each of the correction planes. Balancing with this type of machine involves a lengthy trial-and-error procedure for each rotor, even if it is one of an identical series. The unbalance indication cannot be calibrated for specified correction planes because these machines do not have the feature of plane separation. Field balancing equipment without a microprocessor usually falls into this class.

Class II: Calibratable Balancing Machines Requiring a Balanced Prototype Rotor.   Machines in this class are of the soft-bearing type using instrumentation which permits plane separation and calibration for a given rotor type, if a balanced master or prototype rotor is available. However, the same trial-and-error procedure as for class I machines is required for the first of a series of identical rotors.

Class III: Calibratable Balancing Machines Not Requiring a Balanced Prototype Rotor.   Machines in this class are of the soft-bearing type using instrumentation

TABLE 39.1 Classification of Balancing Machines

Principle employed	Unbalance indicated	Attitude of shaft axis	Type of machine	**Available Classes**
Gravity (nonrotating)	Static (single-plane)	Vertical	Pendulum	Not classified
		Horizontal	Knife-edges	
			Roller sets	
Centrifugal (rotating)	Static (single-plane)	Vertical	Soft-bearing	
			Hard-bearing	
		Horizontal	Not commercially available	
	Dynamic (two-plane); also suitable for static (single-plane)	Vertical	Soft-bearing	**II, III**
			Hard-bearing	**III, IV**
		Horizontal	Soft-bearing	**I, II, III**
			Hard-bearing*	**IV**

* When suitably equipped, these machines may also be used for balancing flexible rotors.

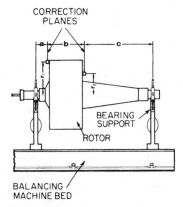

CORRECTION PLANES

BEARING SUPPORT

ROTOR

BALANCING MACHINE BED

FIGURE 39.19 A permanently calibrated balancing machine, showing five rotor dimensions used in computing unbalance. (See Class IV.)

which includes an integral electronic unbalance compensator. Any (unbalanced) rotor may be used in place of a balanced master rotor. In turn, plane separation and calibration can be achieved with the aid of precisely weighed calibration weights temporarily attached in each of two correction planes of the first of a series of rotors. This class includes soft-bearing machines with electrically driven shakers fitted to the vibratory part of their rotor supports, and machines with microprocessor instrumentation using influence coefficients.

Class IV: Permanently Calibrated Balancing Machines. Machines in this class are of the hard-bearing type. They are permanently calibrated by the manufacturer for all rotors falling within the weight and speed range of a given machine size. Unlike the machines in other classes, these machines indicate unbalance in the first run without individual rotor calibration. This is accomplished by the incorporation of an analog or digital computer into the instrumentation associated with the machine. The following five rotor dimensions (see Fig. 39.19) are fed into the computer: distance from left correction plane to left support; distance between correction planes; distance from right correction plane to right support; and radii r_1 and r_2 of the correction weights in the left and right planes, respectively. The instrumentation then indicates the magnitude and angular position of the required correction weight for each of the two selected planes.

The null-force balancing machine is in this class. Although no longer manufactured, it is still used. It balances at the same speed as the natural frequency or resonance of its suspension system (including the rotor).

BALANCING-MACHINE EVALUATION [4]

To evaluate the suitability of a balancing machine for a given application, it is first necessary to establish a precise description of the required machine capacity and performance. Such description often becomes the basis for a balancing-machine purchase specification. It should contain details on the range of workpiece weight, the diameter, length, journal diameter, and service speed, and whether the rotors are rigid or flexible, their application, available line voltage, etc. Such information enables the machine vendor to propose a suitable machine. Next, the vendor's proposal must be evaluated not only on compliance with the purchase specification but also on the operation of the machine and its features. In describing the machine, the vendor should conform with the applicable standards. Once the machine is purchased and ready for shipment, compliance with the purchase specification and vendor proposal should be verified. Depending on circumstances, such verification is usually repeated after installation of the machine at the buyer's facility.

Precise testing procedures vary for different fields of application. Table 39.2 lists a number of standards for testing balancing machines used in the United States and Canada.

UNBALANCE CORRECTION METHODS

Corrections for rotor unbalance are made either by the addition of weight to the rotor or by the removal of material (and in some cases, by relocating the shaft axis). The selected correction method should ensure that there is sufficient capacity to allow correction of the maximum unbalance which may occur. The ideal correction method permits reduction of the maximum initial unbalance to less than balance tolerance in a single correction step. However, this is often difficult to achieve. The more common methods, described below, e.g., drilling, usually permit a reduction of 10:1 in unbalance if carried out carefully. The addition of weight may achieve a reduction as great as 20:1 or higher, provided the weight and its position are closely controlled. If the method selected for reduction of maximum initial unbalance cannot be expected to bring the rotor within the permissible residual unbalance in a single correction step, a preliminary correction is made. Then a second correction method may be selected to reduce the remaining unbalance to less than its permissible value.

UNBALANCE CORRECTION BY THE ADDITION OF WEIGHT TO THE ROTOR

1. *The addition of wire solder.* It is difficult to apply the solder so that its center-of-gravity is at the desired correction location. Variations in diameter of the solder wire introduce errors in correction.

TABLE 39.2 Standards for Testing Balancing Machines

Application	Title	Issuer	Document no.
General industrial balancing machines	Balancing Machines—Description and Evaluation	International Standards Organization (ISO)	DIS 2953
Jet engine rotor balancing machines (for two-plane correction)	Balancing Machines—Evaluation, Horizontal, Two-Plane, Hard-Bearing Type for Gas Turbine Rotors	Society of Automotive Engineers, Inc. (SAE)	ARP 4048
Jet engine rotor balancing machines (for single-plane correction)	Balancing Machines—Description and Evaluation, Vertical, Two-Plane, Hard-Bearing Type for Gas Turbine Rotors	Society of Automotive Engineers, Inc. (SAE)	ARP 4050
Gyroscope rotor balancing machines	Balancing Machine—Gyroscope Rotor	Defense General Supply Center, Richmond, Va.	FSN 6635–450–2208 NT
Field balancing equipment	Field Balancing Equipment—Description and Evaluation	International Standards Organization (ISO)	ISO 2371

2. *The addition of bolted or riveted washers.* This method is used only where moderate balance quality is required.

3. *The addition of cast iron, lead, or lead weights.* Such weights, in incremental sizes, are often used to correct large initial unbalance.

4. *The addition of welded weights.* Resistance welding provides a means of attaching large correction weights, although the total weight and center-of-gravity may be changed somewhat due to the weld. Care must be taken to avoid distorting the rotor with heat from the welding process.

UNBALANCE CORRECTION BY THE REMOVAL OF WEIGHT

1. *Drilling.* Material is removed from the rotor by a drill which penetrates the rotor to a measured depth, thereby removing the intended weight of material with a high degree of accuracy. A depth gage or limit switch can be provided on the drill spindle to ensure that the hole is drilled to the desired depth. This is probably the most effective method of unbalance correction.

2. *Milling, shaping, or fly cutting.* This method permits accurate removal of weight when the rotor surfaces, from which the depth of cut is measured, are machined surfaces and when means are provided for accurate measurement of the cut with respect to those surfaces; used where relatively large corrections are required.

3. *Grinding.* In general, grinding is used as a trial-and-error method of correction. It is difficult to evaluate the actual weight of the material which is removed. This method is usually used only where the rotor design does not permit a more economical type of correction.

MASS CENTERING

A procedure known as "mass centering" is used to reduce unbalance effects in rotors. A rotor is mounted in a balanced cage or cradle which, in turn, is rotated in a balancing machine. The rotor is adjusted radially with respect to the cage until the unbalance indication is zero; this provides a means for bringing the principal inertia axis of the rotor into essential coincidence with the shaft axis of the balanced cage. Center drills (or other suitable tools guided along the axis of the cage) provide a means of establishing an axis in the rotor about which it is in balance. The beneficial effects of mass centering are reduced by any subsequent machining operations on the rotor.

BALANCING OF ROTATING PARTS

MAINTENANCE AND PRODUCTION BALANCING MACHINES

Balancing machines of this type fall into three general categories: (1) universal balancing machines, (2) semiautomatic balancing machines, and (3) fully automatic balancing machines with automatic transfer of work. Each of these has been made in both the nonrotating and rotating types. The rotating type of balancer is available for rotors in which corrections for balance are required in either one or two planes.

Universal balancing machines are adaptable for balancing a considerable variety of sizes and types of rotors. These machines commonly have a capacity for balancing rotors whose weight varies as much as 100 to 1 from maximum to minimum. The elements of these machines are adapted easily to new sizes and types of rotors. The amount and location of unbalance are observed on indicating instruments of various types by the machine operator as the machine performs its measuring functions. This category of machine is suitable for maintenance or job-shop balancing as well as for many small and medium lot-size production applications.

Semiautomatic balancing machines are of many types. They vary from an almost universal machine to an almost fully automatic machine. Machines in this category may perform automatically any one or all of the following functions in sequence or simultaneously: (1) retain the amount of unbalance indication for further reference, (2) retain the angular location of unbalance indication for further reference, (3) measure and store the amount and position of unbalance, (4) couple the balancing-machine driver to the rotor, (5) initiate and stop rotation, (6) set the depth of a correction tool from the indication of amount of unbalance, (7) index the rotor to a desired position from the indication of the unbalance location, (8) apply correction of the proper magnitude at the indicated location, (9) inspect the residual unbalance after correction, and (10) uncouple the balancing-machine driver. Thus, the most fully equipped semiautomatic balancing machine performs the complete balancing process and leaves only loading, unloading, and cycle initiation to the operator. Other semiautomatic balancing machines provide only means for retention of measurements to reduce operator fatigue and error. The equipment which is economically feasible on a semiautomatic balancing machine may be determined only from a study of the rotor to be balanced and the production requirements.

Fully automatic balancing machines with automatic transfer of the rotor are also available. These machines may be either single- or multiple-station machines. In either case, the parts to be balanced are brought to the balancing machine by conveyor, and balanced parts are taken away from the balancing machine by conveyor. All the steps of the balancing process and the required handling of the rotor are performed without an operator. These machines also may include means for inspecting the residual unbalance as well as monitoring means to ensure that the balance inspection operation is performed satisfactorily.

In single-station automatic balancing machines all functions of the balancing process (unbalance measurement, location, and correction) as well as inspection of the complete process are performed in a single check at a single station. In a multiple-station machine, the individual steps of the balancing process may be done at individual stations. Automatic transfer is provided between stations at which the amount and location of unbalance are determined; then the correction for unbalance is applied; finally, the rotor is inspected for residual unbalance. Such machines generally have shorter cycle times than single-station machines.

FIELD BALANCING EQUIPMENT[5]

Many types of vibration indicators and measuring devices are available for field balancing operations. Although these devices are sometimes called "portable balancing machines," they never provide direct means for measuring the amount and location of the correction required to eliminate the vibration produced by the rotor at its supporting bearings. It is intended that these devices be used in the field to reduce or eliminate vibration produced by the rotating elements of a machine under service conditions. Basically, such a device consists of a combination of a transducer and an indicator unit which provides an indication proportional to the vibration magnitude. The vibration magnitude may be indicated in terms of displacement, velocity, or acceleration, depending on the type of transducer and readout system used. The transducer can be hand-held by an operator against the housing of the rotating equipment, clamped to it, or mounted with a magnetic welder. A transducer thus held against the vibrating machine is presumed to produce an output proportional to the vibration of the machine. At frequencies below approximately 15 Hz, it is almost impossible to hold the transducer sufficiently still to give stable readings. Frequently, the results obtained depend upon the technique of the operator; this can be shown by obtaining measurements of vibration magnitude on a machine with the transducer held with varying degrees of firmness. The principles of vibration measurement are discussed more thoroughly in Chaps. 12, 13, 15, and 16.

A transducer responds to all vibration to which it is subjected, within the useful frequency range of the transducer and associated instruments. The vibration detected on a machine may come through the floor from adjacent machines, may be caused by reciprocating forces or other forces inherent in normal operation of the machine, or may be due to wear and tear in various machine components. Location of the transducer on the axis of angular vibration of the machine can eliminate the effect of a reciprocating torque; however, a simple vibration indicator cannot discriminate between the other vibrations unless the magnitude at one frequency is considerably greater than the magnitude at other frequencies. For balancing, the magnitude may be indicated in units of displacement, velocity, or acceleration.

Velocity and acceleration are functions of frequency as well as amplitude; therefore, suitable integrating devices must be introduced between the transducer and the meter. A suitable filter following the output of an electromechanical transducer may be introduced to attenuate frequencies other than the wanted frequency.

The approximate location of the unbalance may be determined by measuring the phase of the vibration. Phase of vibration may be measured by a stroboscopic lamp flashed each time the output of an electrical transducer changes polarity in a given direction. Phase also may be determined by use of a phase meter, wattmeter, or photocell.

BALANCING OF ASSEMBLED MACHINES

The balancing of rotors assembled of two or more individually balanced parts and the balancing of rotors in complete machines are done frequently to obtain maximum reduction in vibration due to unbalance. In many cases the complete machine is run under service conditions during the balancing procedure.

Assembly balance often is made necessary by conditions dictated by machining operations and assembly procedures. For example, a balanced flywheel mounted on a balanced crankshaft may not produce a balanced assembly. When pistons and connecting rods are added to the above assembly, more unbalance is introduced. Such resultant unbalance effects can sometimes only be reduced by balancing the engine in assembly. The probable variation of unbalance in an assembly of balanced components is best determined by statistical methods.

Assemblies such as gyros, superchargers, and jet engines often run on antifriction bearings. The inner races of these bearings may not have perfectly concentric inside and outside surfaces. The eccentricity of the bearing races makes assembly balancing on the actual bearings desirable. In many cases such balancing is done with the stator supporting the antifriction bearings. This ensures that balance is achieved with the bearing race exactly in the position of final assembly. Precise bearing alignment and preload may also become very important to reach very small balance tolerances.

PRACTICAL CONSIDERATIONS IN TOOLING A BALANCING MACHINE

SUPPORT OF THE ROTOR

The first consideration in tooling a balancing machine is the means for supporting the rotor. Various means are available, such as twin rollers, plain bearings, rolling element bearings (including slave bearings), gas bearings, nylon V-blocks, etc. The most frequently used and easiest to adapt are twin rollers. A rotor should generally be supported at its journals to assure that balancing is carried out around the same axis on which it rotates in service.

Rotors which are normally supported at more than two journals may be balanced satisfactorily on only two journals provided that

1. All journal surfaces are concentric with respect to the axis determined by the two journals used for support in the balancing machine.
2. The rotor is rigid at the balancing speed when supported on only two bearings.
3. The rotor has equal stiffness in all radial planes when supported on only two journals.

If the other journal surfaces are not concentric with respect to the axis determined by the two supporting journals, the shaft should be straightened. If the rotor is not a rigid body or if it has unequal stiffness in different radial planes, the rotor should be supported in a (nonrotating) cradle at all journals during the balancing operation. This cradle should supply the stiffness usually supplied to the rotor by the machine in which it is used. The cradle should have minimum weight when used with a soft-bearing machine to permit maximum balancing sensitivity.

Rotors with stringent requirements for minimum residual unbalance and which run in antifriction bearings should be balanced in the antifriction bearings which will ultimately support the rotor. Such rotors should be balanced either (1) in special machines where the antifriction bearings are aligned and the outer races held in half-shoe-type bearing supports, rigidly connected by tie bars, or (2) in standard machines having supports equipped with V-roller carriages.

Frequently, practical considerations make it necessary to remove antifriction bearings after balancing, to permit final assembly. If this cannot be avoided, the bearings should be match-marked to the rotor and returned to the location used while balancing. Antifriction bearings with considerable radial play or bearings with a quality less than ABEC (Annular Bearing Engineers Committee) Standard grade 3 tend to cause erratic indications of the balancing machine. In some cases the outer race can be clamped tightly enough to remove excessive radial play. Only indifferent balancing can be done when rotors are supported on bearings of a grade lower than ABEC 3.

When maintenance requires antifriction bearings to be changed occasionally on a rotor, it is best to balance the rotor on the journal on which the inner race of the antifriction bearings fits. The unbalance introduced by axis shift due to eccentricity of the inner race of the bearing then can be minimized by use of high-quality bearings to ensure minimum eccentricity.

BALANCING SPEED

The second consideration in tooling a balancing machine for a specific rotor is the balancing speed. For rigid rotors the balancing speed should be the lowest speed at which the balancing machine has the required sensitivity. Low speeds reduce the time for acceleration and deceleration of the rotor. If the rotor distorts nonsymmetrically at service speed, the balancing speed should be the same as the service speed. Rotors in which aerodynamic unbalance is present may require balancing under service conditions. Some machines show the effect of unbalance produced by varying electrical fields caused by changes in air gap and the like. Such disturbance can be reduced by balancing (at service speed) only if the disturbing frequency is identical to the service speed.

DRIVE FOR ROTOR

A final consideration in tooling a balancing machine for a specific rotor is the means for driving the rotor. For balancing rotors which do not have journals, the balancing machine may incorporate in its spindle the necessary journals, as is the case on vertical balancing machines; alternatively, an arbor may be used to provide the journal surfaces. An adapter must be provided to adapt the shaftless rotor to the balancing-machine spindle or arbor. This adapter should provide the following:

1. Rotor locating surfaces which are concentric and square with the spindle or arbor axis.

2. Locating surfaces which hold the rotor in the manner in which it is held in final assembly.

3. Locating surfaces which adjust the fit tolerance of the rotor to suit final assembly conditions.

4. A connection between driving elements and rotor to ensure that a fixed angular relation is maintained between them.

5. Means for correcting unbalance in the adapter itself.

If the rotor to be tooled has its own journals, it may be driven through: (1) a universal joint or flexible coupling drive from one end of the rotor, (2) a belt over the periphery of the rotor, or over a pulley attached to the rotor, or (3) air jets or other power means by which the rotor is normally driven in the final machine assembly.

The choice of universal joint or flexible coupling drive attached to one end of the rotor can affect the residual unbalance substantially. Careful attention must be given to the surfaces on the rotor to which the coupling is attached to ensure that the rotor journal axis and coupling are concentric (for example, within 0.001 in. total indicator reading) when all fit tolerances and eccentricities have been considered. The weight of that part of the balancing machine drive which is supported by the rotor during the balancing operation, expressed in ounces (and in this example multiplied by one-half of the total indicator reading, or 0.0005 in.) must be considerably less than the permissible residual unbalance in ounce-inches. Adjustable means must be provided in the coupling drive of the balancing machine to apply corrections for balancing the coupling. The adjustments may have to be effective in each of the correction planes of the rotor in an amount equal to at least twice the permissible residual unbalance. For convenience, the coupling should be designed for easy attachment to the rotor and so that it can be indexed on the rotor shaft by 180° for a balance check (called *index balancing*). Furthermore, the coupling must locate from surfaces of the rotor which are concentric with the journal axis because an accumulation of fit tolerances and eccentricities introduces an error in the result.

A belt drive can transmit only limited torque to the rotor. Driving belts must be extremely flexible and of uniform thickness. Driving pulleys attached to the rotor should be used only when it is impossible to transmit sufficient driving torque by running the belt over the rotor. Pulleys must be as light as possible, must be dynamically balanced, and should be mounted on surfaces of the rotor which are square and concentric with the journal axis. The belt drive should not cause disturbances in the unbalance indication exceeding one-quarter of the permissible residual unbalance. Rotors driven by belt should not drive components of the balancing machine (e.g., angle indicating devices) by means of any mechanical connection.

The use of electrical means or air for driving rotors may influence the unbalance readout. To avoid or minimize such influence, great care should be taken to bring in the power supply through very flexible leads, or have the airstream strike the rotor, at right angles to the direction of the vibration measurement.

If the balancing machine incorporates filters tuned to a specific frequency only, it is essential that means be available to control the rotor speed to suit the filter setting.

BALANCE CRITERIA

Achieving close balance tolerances in rotors requires careful analysis of all factors that may introduce balance errors; therefore, it is often difficult for an engineer normally conversant with balancing methods and techniques to decide which particular balancing method to employ, the rotational speed for balancing to be used, and at what particular point in a production line the balancing procedure should be inserted. The appropriate choice of a balance criterion is likely to be an even greater problem.

A suitable criterion of the quality of balancing required would appear to be the running smoothness of the complete assembly; however, many other factors than unbalance contribute to uneven running of machines (for example, bearing dissymmetries, runouts, misalignment, aerodynamic and hydrodynamic effects, etc.). In addition, there is no simple relation between rotor unbalance and vibration amplitude measured on the bearing housing. Many factors, such as proximity of resonant frequencies, fits, machining errors, bearing and process-related vibration, environmental vibration, etc. may influence overall vibration levels considerably. Therefore, a measurement of the vibration amplitude will not indicate directly the magnitude of unbalance or whether an improved state of unbalance will cause the machine to run smoother. For certain classes of machines, particularly electric motors and large turbines and generators, voluminous data have been collected which can be used as a guide for the establishment of vibration criteria for such installations.

Table 39.3 and Fig. 39.20 show a classification system for various types of representative rotors, based on a document—ISO Standard 1940-1986. Balance quality grades are grouped according to numbers with the prefix G; the vertical scales in Fig. 39.20 indicate the maximum permissible residual unbalance per unit of rotor weight (at various maximum service speeds shown on the bottom scale) expressed in English and SI units. The residual unbalance is equivalent to a displacement of the center-of-gravity. The recommended balance quality grades are based on experience with various rotor types, sizes, and service speeds; they apply only to rotors which are

TABLE 39.3   Balance Quality Grades for Various Groups of Rigid Rotors[6]

Balance quality grade	Type of rotor
G4,000	Crankshaft drives of rigidly mounted slow marine diesel engines with uneven number of cylinders.
G1,600	Crankshaft drives of rigidly mounted large two-cycle engines.
G630	Crankshaft drives of rigidly mounted large four-cycle engines; crankshaft drives of elastically mounted marine diesel engines.
G250	Crankshaft drives of rigidly mounted fast four-cylinder diesel engines.
G100	Crankshaft drives of fast diesel engines with six or more cylinders; complete engines (gasoline or diesel) for cars and trucks.
G40	Car wheels, wheel rims, wheel sets, drive shafts; crankshaft drives of elastically mounted fast four-cycle engines (gasoline or diesel) with six or more cylinders; crankshaft drives for engines of cars and trucks.
G16	Parts of agricultural machinery; individual components of engines (gasoline or diesel) for cars and trucks.
G6.3	Parts or process plant machines; marine main-turbine gears; centrifuge drums; fans; assembled aircraft gas-turbine rotors; fly wheels; pump impellers; machine-tool and general machinery parts; electrical armatures, paper machine rolls.
G2.5	Gas and steam turbines; rigid turbo-generator rotors; rotors; turbo-compressors; machine-tool drives; small electrical armatures; turbine-driven pumps.
G1	Tape recorder and phonograph drives; grinding-machine drives.
G0.4	Spindles, disks, and armatures of precision grinders; gyroscopes.

Note: In general, for rigid rotors with two correction planes, one-half the recommended residual unbalance is to be taken for each plane; these values apply usually for any two arbitrarily chosen planes, but the state of unbalance may be improved upon at the bearings; for disc-shaped rotors, the full recommended value holds for one plane. For repair work, it is often recommended to balance to the next, lower grade.

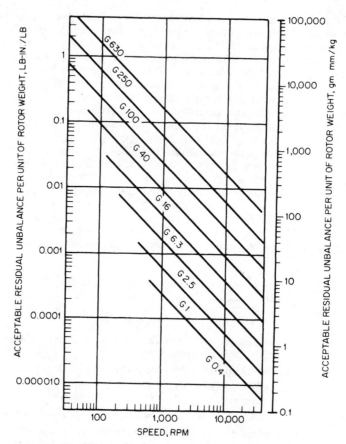

FIGURE 39.20 Residual unbalance corresponding to various balancing quality grades, G. *Notes:* (1) 1 gram·mm/kg is equivalent to a displacement of the center-of-gravity of 0.001 mm = 40 μin. (2) lb·in./lb or oz·in./oz is equivalent to a displacement of the center-of-gravity in inches.

rigid throughout their entire range of service speeds. Balance criteria for flexible rotors are discussed in ISO 11342.

DEFINITIONS[7]

Amount of Unbalance. The quantitative measure of unbalance in a rotor (referred to a plane) without referring to its angular position; obtained by taking the product of the unbalance mass and the distance of its center of gravity from the shaft axis. Units of unbalance are usually ounce-inches, gram-inches, or gram-millimeters.

Angle of Unbalance. Given a polar coordinate system fixed in a plane perpendicular to the shaft axis and rotating with the rotor, the polar angle at which an unbalance mass is located with reference to the given coordinate system.

Balance Quality Grade. For rigid rotors, the product, in millimeters per second, of the specific unbalance and the maximum service angular velocity of the rotor, in radians per second.

Balancing. A procedure by which the mass distribution of a rotor is checked and, if necessary, adjusted to ensure that the residual unbalance or vibration of the journals and/or forces on the bearings at a frequency corresponding to service speed are within specified limits.

Balancing Machine. A machine that provides a measure of the unbalance in a rotor which can be used for adjusting the mass distribution of that rotor mounted on it so that once-per-revolution vibratory motion of the journals or forces on the bearings can be reduced if necessary.

Bearing Support. The part, or series of parts, that transmits the load from the bearing to the main body of the structure.

Center-of-Gravity (Mass Center). The point in a body through which passes the resultant of the weights of its component particles for all orientations of the body with respect to a uniform gravitational field.

Correction Plane Interference (Cross Effect). The change of balancing-machine indication at one correction plane of a given rotor which is observed for a certain change of unbalance in the other correction plane.

Correction Plane Interference Ratios. The interference ratios (I_{AB}, I_{BA}) of two correction planes A and B of a given rotor are defined by the following relationships:

$$I_{AB} = \frac{U_{AB}}{U_{BB}}$$

where U_{AB} and U_{BB} are the unbalances referring to planes A and B, respectively, caused by the addition of a specified amount of unbalance in plane B; and

$$I_{BA} = \frac{U_{BA}}{U_{AA}}$$

where U_{BA} and U_{AA} are the unbalances referring to planes B and A, respectively, caused by the addition of a specified amount of unbalance in plane A.

Critical Speed. A characteristic speed at which resonances of a system are excited. (The significant effect at critical speed may be motion of the journals or flexure of the rotor—depending on the relative magnitudes of the bearing stiffnesses.)

Couple Unbalance. That condition of unbalance for which the central principal axis intersects the shaft axis at the center of gravity.

Dynamic (Two-Plane) Balancing Machine. A centrifugal balancing machine that furnishes information for performing two-plane balancing.

Dynamic Unbalance. The condition in which the central principal axis neither is parallel to nor intersects the shaft axis.

Field Balancing Equipment. An assembly of measuring instruments for providing information for performing balancing operations on assembled machinery which is not mounted in a balancing machine.

Flexible Rotor. A rotor not satisfying the definition of a rigid rotor.

Flexural Critical Speed. A speed of a rotor at which there is maximum bending of the rotor and at which flexure of the rotor is more significant than the motion of the journals.

Flexural Principal Mode. For undamped rotor–bearing systems, that mode shape which the rotor takes up at one of the (rotor) flexural critical speeds.

High-speed Balancing (Relating to Flexible Rotors).   A procedure of balancing at speeds where the rotor to be balanced cannot be considered rigid.

Initial Unbalance.   Unbalance of any kind that exists in the rotor before balancing.

Journal Axis.   The straight line joining the centroids of cross-sectional contours of a journal.

Low-speed Balancing (Relating to Flexible Rotors).   A procedure of balancing at a speed where the rotor to be balanced can be considered rigid.

Minimum Achievable Residual Unbalance.   The smallest value of residual unbalance that a balancing machine is capable of achieving.

Modal Balancing.   A procedure for balancing flexible rotors in which unbalance corrections are made to reduce the amplitude of vibration in the separate significant principal flexural modes to within specified limits.

Multiplane Balancing.   As applied to the balancing of flexible rotors, any balancing procedure that requires unbalance correction in more than two correction planes.

Perfectly Balanced Rotors.   An ideal rotor which has zero unbalance.

Permanent Calibration.   That feature of a hard-bearing balancing machine which permits it to be calibrated once and for all, so that it remains calibrated for any rotor within the capacity and speed range of the machine.

Plane Separation.   Of a balancing machine, the operation of reducing the correction-plane interference ratio for a particular rotor.

Principal Inertia Axis.   In balancing, the term used to designate the central principal axis (of the three such axes) most nearly coincident with the shaft axis of the rotor; sometimes referred to as the *balance axis* or the *mass axis*.

Residual Unbalance.   Unbalance of any kind that remains after balancing.

Rigid Rotor.   A rotor is considered rigid when its unbalance can be corrected in any two (arbitrarily selected) planes and, after the correction, its unbalance does not significantly change (relative to the shaft axis) at any speed up to maximum service speed and when running under conditions which approximate closely those of the final supporting system.

Rotor.   A body capable of rotation, generally with journals which are supported by bearings.

Shaft Axis.   The straight line joining the journal centers.

Single-plane (Static) Balancing Machine.   A gravitational or centrifugal balancing machine that provides information for accomplishing single-plane balancing.

Static Unbalance.   That condition of unbalance for which the central principal axis is displaced only parallel to the shaft axis.

Unbalance.   That condition which exists in a rotor when vibratory force or motion is imparted to its bearings as a result of centrifugal forces.

REFERENCES

1. "Mechanical Vibration—Methods and Criteria for the Mechanical Balancing of Flexible Rotors," ISO 11340-1994.

2. Schneider, H.: "Balancing Technology," 4th ed., Carl Schenck AG, Darmstadt, Germany, 1991.

3. Stadelbauer, D. G.: "Fundamentals of Balancing," 3d ed. Schenck Trebel Corp., Deer Park, N.Y., 1990.

4. "Balancing Machines—Description and Evaluation," ISO/DIS 2953-1984.

5. "Field Balancing Equipment—Description and Evaluation," ISO 2371-1974; ANSI S2.38-1982.

6. "Balancing Quality of Rotating Rigid Bodies," ISO 1940-1986; ANSI S2.19-1975.

7. "Balancing—Vocabulary," ISO 1925-1990, and ANSI S2.7, 1982.

CHAPTER 39

PART II: SHAFT MISALIGNMENT OF ROTATING MACHINERY

John Piotrowski

INTRODUCTION

Shaft misalignment is said to occur when the centerlines of rotation of two machine shafts are supposed to be collinear but are not in line with each other. Thus, misalignment is the deviation of relative shaft position from a collinear axis of rotation (measured at the points of power transmission) when machinery is running at normal operating conditions. For example, consider a motor shaft which is connected to a pump shaft, with centerlines that are not collinear. Such shaft misalignment may result in excessive vibration, although there is not a direct relationship between the magnitude of vibration and shaft misalignment. (In some cases, a *slight* amount of misalignment may actually reduce the magnitude of vibration.) In addition, shaft misalignment may be the cause of any or a combination of the following conditions:

- Shaft failure resulting from cyclic fatigue
- Cracking of the shafts at, or close to, the coupling hubs or bearings
- Increased wear of the bearings, seals, or coupling, leading to premature failure
- Loose foundation bolts
- Loose or broken coupling bolts
- A coupling that runs hot
- High temperature of the casing or of the oil discharge near the bearings
- Excessive grease or oil on the inside of the coupling guard
- Excessive power consumption by the rotating equipment

The objective of shaft alignment is to reduce these detrimental effects and thereby extend the operating life span of the rotating machinery.

This part of this chapter describes the types of misalignment, describes the use of spectrum analysis of vibration as an aid in identifying shaft misalignment, provides a "tolerance guide" as a rough indication as to whether alignment is necessary in coupled rotating machinery, and outlines the basic steps that should be taken in aligning rotating machinery.

TYPES OF SHAFT MISALIGNMENT

Figure 39.21A shows a motor used to drive a pump. A hub is shown at the end of each shaft. The coupling between the two shafts, which connects the two hubs under normal operating conditions, has been removed. Figure 39.21B shows a detail of the driving shaft (on the left) and the driven shaft (on the right); the angle between the centerlines of the two misaligned shafts is represented as ϕ. The distance between points of power transmission is shown in Fig. 39.21C. Under operating conditions there will be a distortion of the shafts when the loads are transferred from one shaft to the other.

Two types of shaft misalignment are illustrated in Fig. 39.22: (1) *angular misalignment,* where the driving shaft and the driven shaft are in the same plane but at an angle ϕ with respect to each other, and (2) *parallel misalignment,* where the driving shaft and the driven shaft are parallel to each other, but offset. Conditions of pure angular misalignment (Fig. 39.22A) or pure parallel misalignment (Fig. 39.22B) are rare. Instead, the usual condition is *combined misalignment* (Fig. 39.22C), a combination of parallel and angular misalignment.

If the misalignment between the driving and driven shafts is slight, a flexible coupling between the shafts will accommodate it. The greater the misalignment, the greater will be the flexing of the flexible elements in the coupling.

USE OF SPECTRUM ANALYSIS IN STUDYING SHAFT MISALIGNMENT

Spectrum analysis of vibration of rotating machinery often can be useful in detecting faults such as shaft misalignment. This technique is described in Chap.

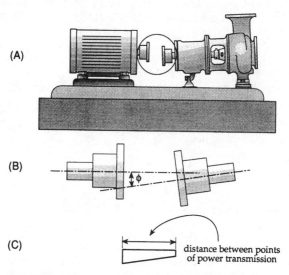

FIGURE 39.21 An illustration of shaft misalignment. (A) A motor (on the left) used to drive a pump (on the right); there is a hub at the end of each shaft. (B) A detail showing the centerlines of rotation of the drive shaft and the driven shaft; ϕ is the angle of misalignment. (C) The distance between points of power transmission.

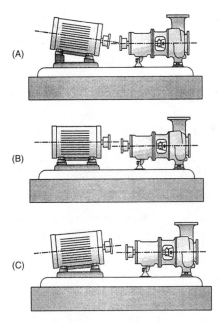

FIGURE 39.22 Types of shaft misalignment: (A) angular misalignment, (B) parallel misalignment, and (C) the most common combination of parallel and angular misalignment.

16, which includes discussions of the parameter to be measured (displacement, velocity, or acceleration), suitable vibration pickups to be mounted on the rotating machinery, suitable locations for the transducers, the selection of an appropriate spectrum analyzer, determination of appropriate analyzer bandwidth for fault detection in rotating machinery, and spectrum interpretation and fault diagnosis. For example, the Trouble-Shooting Chart of Table 16.1 indicates that the dominant frequency in the spectrum of misaligned rotating machinery is often 1 or 2 times the rpm, and sometimes 3 or 4 times the rpm. Chapter 16 also points out that in interpreting a vibration spectrum, it is often difficult to separate faults caused by misalignment, unbalance, bent shaft, eccentricity, and cracks in a rotating shaft; this is because these various faults may be mechanically related. The results of vibration spectrum analysis of misaligned rotating machinery show, for example, that the spectra are different for (1) different types of couplings and (2) different types of bearings which support the machinery shafts.

TOLERANCE GUIDE FOR FLEXIBLY COUPLED ROTATING MACHINERY

Whether a measured value of shaft misalignment in flexibly coupled machinery is acceptable or not depends not only on the magnitude of the misalignment but on the rotational speed of the shaft, among other factors. A rough guide as to how much misalignment is acceptable is given in Fig. 39.23. This illustration may be used to determine, approximately, whether or not shaft realignment is required under most circumstances. The vertical axis represents the amount of misalignment relative to the distance between points of power transmission (left scale); this value may also be expressed as the angle ϕ (see Fig. 39.21C), which is shown on the right vertical axis.

BASIC STEPS IN SHAFT ALIGNMENT

Before starting the shaft alignment, obtain relevant information on the equipment being aligned, ensure that all possible safety precautions have been taken, perform preliminary checks such as inspecting the coupling (between the driver shaft and the driven shaft) for damage or worn components, find and correct any problems with the foundation or baseplate, perform bearing clearance or looseness checks, measure shaft and/or coupling runout, eliminate excessive stresses caused by piping or conduit connected to the machine, and find and correct any poor surface contact between the underside of the machine feet and the baseplate or frame. Then continue as follows:

1. Check to ensure that all foot bolts are tight.

2. Remove the coupling between the shafts (although removal is not always required, it is advisable), then measure the maximum offsets of the shafts to an accuracy of ±0.001 in. (0.025 mm) in the horizontal and vertical planes. Appropriate devices for making such measurements include a dial indicator (a gage or meter having a circular face which is calibrated to give readings of displacement), a laser shaft-alignment system, a proximity probe such as a capacitance-type transducer (Chap. 12), an angular or linear resolver/encoder, or a charge-coupled device.

3. Using Fig. 39.23, determine if realignment is necessary.

4. If the machinery is not within adequate alignment tolerance and realignment is required, determine the current positions of the centerlines of rotation of the machinery components.

5. Determine which way, and by how much, the machinery components must be moved in order to reduce the misalignment to an acceptable value.

6. Observe any movement restrictions imposed on the machines or control points. For example, if a lateral movement greater than that permitted on the baseplate may be required, it may be necessary to move *both* machines to achieve the alignment goal.

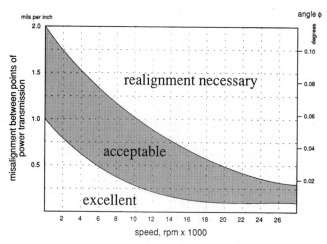

FIGURE 39.23 A shaft alignment tolerance guide for flexibly coupled equipment indicating, approximately, whether or not realignment is required under most circumstances. The vertical axis represents the amount of misalignment relative to the distance between points of power transmission (left scale); this value may also be expressed as the angle φ (see Fig. 39.21C) (right scale). Tolerance guidelines are plotted as a function of misalignment and shaft speed.

7. Reposition the machine to be moved (or both machines) in the vertical, lateral, and axial directions. Check the new positions to ensure that the alignment is within the tolerance guidelines.

8. Install the coupling between the driving and driven shafts, and then turn on the rotating machinery.

9. With the equipment operating as aligned, check and record the magnitudes of vibration, bearing and coupling temperatures, bearing loads, and other pertinent operating parameters; these data will be useful the next time an alignment is carried out.

REFERENCE

1. Piotrowski, J.: "Shaft Alignment Handbook," 2d ed., Marcel Dekker Inc., New York, 1995.

CHAPTER 40
MACHINE-TOOL VIBRATION

E. I. Rivin

INTRODUCTION

Machining and measuring operations are invariably accompanied by relative vibration between workpiece and tool. These vibrations are due to one or more of the following causes: (1) inhomogeneities in the workpiece material; (2) variation of chip cross section; (3) disturbances in the workpiece or tool drives; (4) dynamic loads generated by acceleration/deceleration of massive moving components; (5) vibration transmitted from the environment; (6) self-excited vibration generated by the cutting process or by friction (machine-tool chatter).

The tolerable level of relative vibration between tool and workpiece, i.e., the maximum amplitude and to some extent the frequency, is determined by the required surface finish and machining accuracy as well as by detrimental effects of the vibration on tool life (see *The Effect of Vibration on Tool Life*) and by the noise which is frequently generated.

This chapter discusses the sources of vibration excitation in machine tools, machine-tool chatter (i.e., self-excited vibration which is induced and maintained by forces generated by the cutting process), and methods of control of machine-tool vibration.

SOURCES OF VIBRATION EXCITATION

VIBRATION DUE TO INHOMOGENEITIES IN THE WORKPIECE

Hard spots or a crust in the material being machined impart small shocks to the tool and workpiece, as a result of which free vibrations are set up. If these transients are rapidly damped out, their effect is usually not serious; they simply form part of the general "background noise" encountered in making vibration measurements on machine tools. Cases in which transient disturbances do not decay but build up to vibrations of large amplitudes (as a result of dynamic instability) are of great practical importance, and are discussed later.

When machining is done under conditions resulting in discontinuous chip removal, the segmentation of chip elements results in a fluctuation of the cutting thrust. If the frequency of these fluctuations coincides with one of the natural frequencies of the structure, forced vibration of appreciable amplitude may be excited. However, in single-edge cutting operations (e.g., turning), it is not clear whether the segmentation of the chip is a primary effect or whether it is produced by other vibration, without which continuous chip flow would be encountered.

The breaking away of a built-up edge from the tool face also imparts impulses to the cutting tool which result in vibration. However, marks left by the built-up edge on the machined surface are far more pronounced than those caused by the ensuing vibration; it is probably for this reason that the built-up edge has not been studied from the vibration point of view. The built-up edge frequently accompanies certain types of vibration (chatter), and instances have been known when it disappeared as soon as the vibration was eliminated.

VIBRATION DUE TO CROSS-SECTIONAL VARIATION OF REMOVED MATERIAL

Variation in the cross-sectional area of the removed material may be due to the shape of the machined surface (e.g., in turning of a nonround or slotted part) or to the configuration of the tool (e.g., in milling and broaching when cutting tools have multiple cutting edges). In both cases, pulses of appreciable magnitude may be imparted to the tool and to the workpiece, which may lead to undesirable vibration. The pulses have relatively shallow fronts for turning of nonround or eccentric parts, and steep fronts for turning of slotted parts and for milling/broaching. These pulses excite transient vibrations of the frame and of the drive whose intensity depends on the pulse shape and the ratio between the pulse duration and the natural periods of the frame and the drive (Chap. 8). If the vibrations are decaying before the next pulse occurs, they can have a detrimental effect on tool life and leave marks on the machined surface. In cylindrical grinding and turning, when a workpiece which contains a slot is machined, visible marks frequently are observed near the "leaving edge" of the slot or keyway. These are due to a "bouncing" of the grinding wheel or the cutting tool on the machined surface. They may be eliminated or minimized by closing the recess with a plug or with a filler.

When the transients do not significantly decay between the pulses, dangerous resonance vibrations of the frame and/or the drive can develop with the fundamental and higher harmonics of the pulse sequence. The danger of the resonance increases with higher cutting speeds.

Simultaneous engagement of several cutting edges with the workpiece results in an increasing dc component of the cutting force and effective reduction of the pulse intensity,[1-3] while runout of a multiedge cutter and inaccurate setup of the cutting edges enrich the spectral content of the cutting force and enhance the danger of resonance.[2] Computational synthesis of the resulting cutting force is reasonably accurate.[3]

DISTURBANCES IN THE WORKPIECE AND TOOL DRIVES

Forced vibrations result from rotating unbalanced masses; gear, belt, and chain drives; bearing irregularities; unbalanced electromagnetic forces in electric motors; pressure oscillations in hydraulic drives; etc.

Vibration Caused by Rotating Unbalanced Members. Forced vibration induced by rotation of some unbalanced member may affect both surface finish and tool life, especially when its rotational speed falls near one of the natural frequencies of the machine-tool structure. This vibration can be eliminated by careful balancing, the procedure being basically similar to that described in Chap. 39, or by self-centering due to resilient mounting of bearings.[4,5]

When a new machine is designed, a great deal of trouble can be forestalled by placing rotating components in a position in which the detrimental effect of their unbalance is likely to be relatively small. Motors should not be placed on the top of slender columns, and the plane of their unbalance should preferably be parallel to the plane of cutting. In some cases, vibration resulting from rotating unbalanced members can be eliminated by mounting them using vibration-isolation techniques (Chaps. 30 and 32).

Grinding and boring are most sensitive to vibration because of the high surface finish resulting from the operations. In cylindrical grinding, marks resulting from unbalance of the grinding wheel or of some other component are readily recognizable. They appear in the form of equally spaced, continuous spirals with a constant slope, as shown in Fig. 40.1A. From these marks, the machine component responsible for their existence is found by considering that its speed in rpm must be equal to $\pi Dn/a$, where D is the workpiece diameter in inches (millimeters), a is the pitch of the marks in inches (millimeters), and n is the workpiece speed in rpm. An analogous procedure also can be applied to peripheral surface grinding. Marks produced in one pass of the wheel are shown in Fig. 40.1B. The speed of the responsible component in rpm is equal to the number of marks (produced in one pass) which fall into a distance equal to that traveled by the workpiece (or wheel) in 1 min.

Since centrifugal force magnitudes are proportional to the square of rpm, high-speed machine tools are more sensitive to unbalance of toolholders and small asymmetrical tools (e.g., boring bars). Lathes may be sensitive to workpiece unbalance due to asymmetrical geometry or the nonuniform allowance (e.g., forged parts).

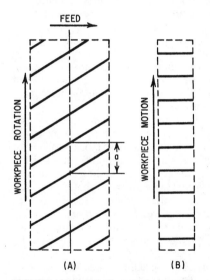

FIGURE 40.1 Grinding marks resulting from unbalance of grinding wheel or some other component. (*A*) Cylindrical grinding; (*B*) peripheral grinding. Marks which are unequally spaced or which have a varying slope are due to inhomogeneities in the wheel.

Marks Caused by Inhomogeneities in the Grinding Wheel. Although grinding marks usually indicate the presence of a vibration, this vibration may not necessarily be the primary cause of the marks. Hard spots on the cutting surface of the wheel result in similar, though generally less pronounced, marks. Grinding wheels usually are not of equal hardness throughout. A hard region on the wheel circumference rapidly becomes glazed in use and establishes itself as a high spot on the wheel (since it retains the grains for a longer period than the softer parts). These high spots eventually break down or shift to other parts of the wheel; in cylindrical grinding, this manifests itself as a sudden change in the slope of the spiral marks. Marks which appear to be due to an unbalanced member rotating at two or three times the speed of the wheel and which are nonuniformly spaced are always due to two or three hard spots.

The Effect of Vibration on the Wheel Properties. If vibration exists between wheel and workpiece, normal forces are produced which react on the wheel and tend to alter the wheel shape and/or the wheel's cutting properties. In soft wheels the dominating influence of vibration appears to be inhomogeneous wheel wear, and in hard wheels inhomogeneous loading (i.e., packing of metal chips on and in crevasses between the grits). These effects result in an increased fluctuation of the normal force, which produces further changes in the wheel properties. The overall effect is that a vibration once initiated tends to grow.[6] When successive cuts or passes overlap, the inhomogeneous wear and loading of the wheel may cause a regenerative chatter effect which makes the cutting process dynamically unstable (see *Dynamic Stability*).

Drives. Spindle and feed drives can be important sources of vibration caused by motors, power transmission elements (gears, traction drives, belts, screws, etc.), bearings, and guideways.

Electric motors can be sources of both rectilinear and torsional vibrations. Rectilinear vibrations are due to a nonuniform air gap between the stator and rotor, asymmetry of windings, unbalance, bearing irregularities, misalignment with the driven shaft, etc. Torsional vibrations (torque ripple) are due to various electrical irregularities.[7,8] Misalignment- and bearing-induced vibrations of spindle motors are reduced by integrating the spindle with the motor shaft.[9]

Gear-induced vibrations can also be both rectilinear and torsional. They are due to production irregularities (pitch and profile errors, eccentricities, etc), assembly errors (eccentric fit on the shaft, key/spline errors, and backlash), or distortion of mesh caused by deformations of shafts, bearings, and housings under transmitted loads. Tight tolerances of the gears and design measures reducing their sensitivity to misalignment (crowning, flanking) should be accompanied by rigid shafts and housings and accurate fits. All gear faults, eccentricities, pitch errors, profile errors, etc., produce nonuniform rotation, which in some cases adversely affects surface finish, geometry, and possibly tool life. In precision machines, where a high degree of surface finish is required, the workpiece or tool spindle usually is driven by belts or by directly coupled motors.

In some high-precision systems, inertia drives are used, in which the energy is supplied to the flywheel between the cutting operations, but the cutting process is energized by the flywheel disconnected from the motor/transmission system.[10] Such a system practically eliminates transmission of drive vibrations into the work zone.

Belt drives, used in some applications as filters to suppress high-frequency vibrations (especially torsional), can induce their own forced vibrations, both torsional and rectilinear. Any variation of the effective belt radius, i.e., the radius of the neutral axis of the belt around the pulley axis, produces a variation of the belt tension and the belt velocity. This causes a variation of the bearing load and of the rotational velocity of the pulley. The effective pulley radius can vary as a result of (1) eccentricity of the pulley or (2) variation of belt profile or inhomogeneity of belt material. Another source of belt-induced vibrations is variation of the elastic modulus along the belt length,[2] which may excite parametric vibration (Chap. 5). Flat belts generate less vibration than V belts because of their better homogeneity and because the disturbing force is less dependent on the belt tension.

Grinding is particularly sensitive to disturbances caused by belts. Seamless belts or a direct motor drive to the main spindle is recommended for high-precision machines.[6] Vibration is minimized when the belt tension and the normal grinding force point in the same direction, as shown in Fig. 40.2A. The clearance between bearing and spindle is thus eliminated. With the arrangement shown in Fig. 40.2B, large amplitudes of vibration may arise when the normal grinding force is substantially equal to the belt tension and/or the peripheral surface of the wheel is nonuniform. Tests indicate that with the arrangement shown in Fig. 40.2C, vibration due to the centrifugal force is likely to be caused by an unbalance of the wheel.[6] The spindle pulley should preferably be placed between the spindle bearings (Fig. 40.3A) and not at the end of the spindle (Fig 40.3B), unless the pulley is "unloaded" (supported by its own bearings), as in Fig. 40.3C.

Chain drives have inherent nonuniformity of transmission ratio and are a significant source of vibration, even when used for auxiliary drives.

Bearings. Dimensional inaccuracies of the components of ball or roller bearings and/or surface irregularities on the running surfaces (or the bearing housing) may give rise to vibration trouble in machines when high-quality surface finish is demanded. From the frequency of the vibration produced, it is sometimes possible to identify the component of the bearing responsible (Chap. 16). For conventional bearings frequently used in machine tools, the outer race is stationary and the inner race rotates at n_i rpm; the cage velocity is of the order of $n_c \cong 0.4\,n_i$, and the velocity of the balls or rollers is about $n_b \cong 2.4 n_i$. In some cases, a disturbing frequency of the order of $n_z = z n_c$ also can be detected, where z is the number of rolling elements. This is the frequency with which successive rolling elements pass through the "loaded zone" of the bearing, which is determined by the direction of the load. These disturbing frequencies are less pronounced with bearings having two rows of rolling elements, each unit of which lies halfway between units of the neighboring row. Because of the importance of spindle bearings' influence on accuracy of machining and on vibrations in the work zone, especially for precision and high-speed machine tools, both races and rolling bodies of spindle bearings must have high dimensional accuracy.

From the point of view of vibration control, both stiffness and damping of bearings should be maximized. Stiffness can be maximized by using roller bearings (with tapered or cylindrical rollers), by using rollers with two rows of rolling elements, by preloading the bearings in the radial direction, and by improving fits between bearings and shafts/housings.[2, 13] Preloading eliminates clearances (play) in bearings, besides increasing their stiffness. However, increased preload is accompanied by

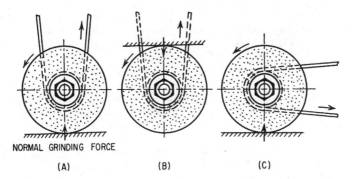

NORMAL GRINDING FORCE

(A)   (B)   (C)

FIGURE 40.2 Direction of driving belt and its influence on performance. (*A*) Vibration is minimized when belt tension and normal grinding force point in the same direction. (*B*) Large amplitudes may arise when the normal grinding force is substantially equal to the belt tension. (*C*) Vibration due to centrifugal force is likely to be caused by an unbalance of the wheel. (*S. Doi.*[6])

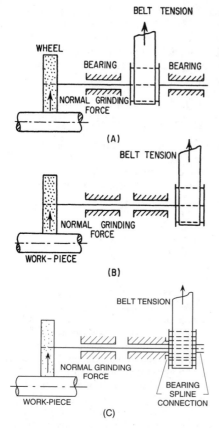

FIGURE 40.3 Effect of relative position of grinding wheel, bearings, and driving pulley on grinding performance. (*A*) Driving pulley should be placed between bearings, as shown in (*A*). Arrangement shown in (*B*) is constructionally simpler but is more liable to cause trouble. (*After S. Doi.*[6]) (*C*) Supporting of pulley by independent bearings eliminates bending and rectilinear vibrations of spindle by belt-induced forces.

decreased damping,[11] as well as by an increase in heat generation and a likely decrease in bearing life. Optimal preload values are recommended by bearing manufacturers. Roller bearings usually have higher damping than ball bearings. Sliding, and especially hydrostatic, bearings have a greater damping capacity than antifriction bearings and are therefore superior with respect to vibration. Machine tools with hydrostatic bearings have extremely high chatter resistance.[12]

Guideways (Slides). The uniformity of feed motions is often disturbed by a phenomenon known as *stick-slip,* which is described in Chap. 5. When motion of a tool support is initiated, elastic deformations of the feed drive elements increase until the forces transmitted exceed the static frictional resistance of the tool support. Subsequently, the support commences to move, and the friction drops to its dynamic value. As a result of the drop of the friction force, the support receives a high acceleration and overshoots because of its inertia. At the end of the "jump," the transmission is wound up in the opposite sense; before any further motion can take place, this deformation must be unwound. This occurs during a period of standstill of the support. Subsequently, the phenomenon repeats itself. The physical sequence described falls into the category generally known as "relaxation oscillations" (Chap. 5).

The occurrence of stick-slip depends on the interaction of the following factors: (1) the mass of the sliding body, (2) the drive stiffness, (3) the damping present in the drive, (4) the sliding speed, (5) the surface roughness of the sliding surfaces, and (6) the lubricant used. It is encountered only at low sliding speeds; slide drives designed for stick-slip-free motion have small moving masses and a high drive stiffness. Excellent results also may be achieved by using cast iron and a suitable plastic material as mating surfaces. By keeping the oil film between the mating surfaces under a certain pressure (hydrostatic lubrication), the possibility of mixed dry and viscous friction is eliminated, and stick-slip cannot arise. High damping is another advantage of hydrostatic slides.

Rolling friction slides[13] do not exhibit stick-slip but may generate high-frequency vibrations because of the shape and dimensional imperfections of the rolling bodies. These can be reduced by increasing their dimensional accuracy and by introducing damping. Rolling friction slides have very low damping[8,13] and as a result can amplify vibrations from other sources if their frequencies are close to resonance frequencies of the slide.

IMPACTS FROM MASSIVE PART REVERSALS

Some machine tools have reciprocating massive parts whose reversals produce sharp impacts which excite both low-frequency solid-body vibrations of the machine (the system "machine on its mounts") and high-frequency structural modes. Such effects occur both in machine tools, such as surface grinders, and in high-speed computer numerically controlled (CNC) machining centers and coordinate measuring machines (CMM). In the CMMs the working process is associated with start-stop operations; in machining centers it is associated with changing magnitude and/or directions of feed motions of heavy tables, slides, spindle heads, etc., with accelerations as high as $2g$. The driving forces causing such changes in magnitudes and directions of momentum of the massive units have impulsive character and cause free decaying vibrations in both solid-body and structural modes (Chap. 8). These vibrations excite relative displacements in the work zone between the workpiece and the cutting or measuring tool. Figure 40.4[13] shows oscillograms of the acceleration of the table of a surface grinder during its reversal (A) and the resulting relative displacements between the grinding wheel and the table (workpiece) for two cases of installation: the machine installed on rigid steel wedges (B) and on vibration isolators (C). In the latter case the relative displacements during the reversal process are much higher, although they are decaying at a faster rate due to higher damping in the isolators. The peak magnitude of accelera-

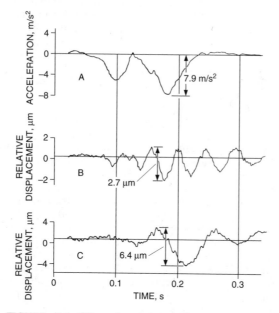

FIGURE 40.4 Effect of mounts on relative displacements between grinding wheel and table during reversal of table of surface grinder. (A) Acceleration of table; (B)(C) relative displacements [(B), machine installed on steel wedge mounts; (C), machine installed on vibration isolators]; table velocity 20 m/min. (*After Kaminskaya.*[13])

tion, 7.9 m/s^2 $\cong$ 0.8g, is typical for surface grinders, CMMs, etc. If these displacements exceed allowable limits, the working process cannot start before the vibrations decay. This adversely affects the machine productivity.

Reduction in the adverse effects of the impulsive forces can be achieved by enhancing the structural stiffness and natural frequencies, thus reducing the sensitivity of the machine to impulsive forces and accelerating the decay. A similar effect results from an increase in "solid-body frequencies" (the natural frequencies of the machine on its mounts) in the direction of the impulsive forces and from decoupling of vibratory modes in the vibration-isolation system, e.g., by increasing the distance between the mounts in the direction of acceleration. Increase of structural damping as well as damping of mounting elements (vibration isolators) also results in a reduction in the decay time.

VIBRATION TRANSMITTED FROM THE ENVIRONMENT

Shock and vibration generated in presses, machine tools, internal-combustion engines, compressors, cranes, carts, rail and road vehicles, etc., are transmitted through the foundation to other machines, which they may set into forced vibration. Vibration of the shop floor contains a wide frequency spectrum.[14] It is almost inevitable that one of these frequencies should fall near a natural frequency of a particular machine tool. Although the amplitudes of the floor vibration usually are small, they may adversely affect precision machine tools and measuring instruments. The undesirable effects include irreversible shifts in structural joints of machine tools and their mounts, shape and surface finish distortions of machined parts, erroneous readings of measuring instruments, and chipping of cutting inserts.

Vibration transmitted through the floor may be reduced by vibration isolation (Chaps. 30 and 32), i.e., the stationary machines which generate the vibration are placed upon vibration isolators.[14] However, precision machine tools and measuring instruments are isolated to provide further reduction.[14]

When applying vibration isolators to machine tools, some care must be exercised. The foundation constitutes the "end condition" of the machine-tool structure. Any alteration of the end condition affects equivalent stiffness and damping, and thus the natural frequencies and vibratory modes of the structure.[14,15]

If vibration isolators are not properly selected and located, the machine tool may become more susceptible to internal exciting forces, and its chatter behavior also may be affected in an undesirable way,[14,15] usually at the lower modes of vibration. Many undesirable effects can be eliminated or significantly reduced by using vibration isolators having a natural frequency that is independent of weight loads on isolators ("constant natural frequency" isolators); by using isolators with high damping; by assigning the mounting points locations that enhance the effective stiffness of the machine-tool frame; by increasing the stiffness of isolators and the distance between them in the directions of movements of heavy reciprocating masses; and by reducing modal coupling in the isolation system.[14] In general, machine-tool structures which are very stiff by themselves (i.e., without being bolted down) can be placed on vibration isolators safely (milling machines, grinding machines, and some lathes).

MACHINE-TOOL CHATTER

The cutting of metals is frequently accompanied by violent vibration of workpiece and cutting tool which is known as machine-tool chatter. *Chatter* is a self-excited vibration which is induced and maintained by forces generated by the cutting process. It is highly detrimental to tool life and surface finish,[16] and is usually accompanied by considerable noise. Chatter adversely affects the rate of production since, in many cases its elimination can be achieved only by reducing the rate of metal removal. Cutting regimes for nonattended operations (such as computer numerically controlled machine tools and flexible manufacturing systems) are frequently assigned conservatively in order to avoid the possibility of chatter.

Machine-tool chatter is characteristically erratic since it depends on the design and configuration of both the machine and the tooling structures, on workpiece and cutting tool materials, and on machining regimes. Chatter resistance of a machine tool is usually characterized by a maximum stable (i.e., not causing chatter vibration) width of cut b_{lim}. Forced vibration effects in machine tools are more frequently detected in the development stage or during final inspection, and can be reduced or eliminated. The tendency for a certain machine to chatter may remain unobserved in the plant of the machine-tool manufacturer unless the machine is thoroughly tested.[17, 18] If this tendency is encountered at the user facility, its elimination from a particular machining process may be highly time-consuming and laborious.

A distinction can be drawn[3, 15] between regenerative and nonregenerative chatter. The former occurs when there is an overlap in the process of performing successive cuts such that part of a previously cut surface is removed by a succeeding pass of the cutter. Under regenerative cutting, a displacement of the tool can result in a vibration of the tool relative to the workpiece, resulting in a variation of the chip thickness. This in turn results in a variation in the cutting force during following revolutions. The regenerative chatter theory explains a wide variety of practical chatter situations in such operations as normal turning and milling.

An important characteristic feature of regenerative chatter is a "lobing" dependence of the maximum stable width of cut b_{lim} on cutting speed (rpm of tool or workpiece).[3, 15, 19] This dependence is shown as the solid line in Fig. 40.5.[19] There is an area of absolute stability below the lobes' envelope, which is shown as a broken line in Fig. 40.5. The position of this envelope depends on the material and geometry of the cutting tool as well as the workpiece material. The lobing shape indicates that some speeds are characterized by much higher stability (larger b_{lim}).

Nonregenerative chatter is found in such operations as shaping, slotting, and screw-thread cutting. In this type of cutting, chatter has been explained through the principle of mode coupling.[3, 15] If a machining system can be modeled by a two degree-of-freedom mass-spring system, with orthogonal axes of major flexibilities and with a common mass, the dynamic motion of the tool end can take an elliptical path. If the major axis of motion (axis with the greater compliance) lies within the angle formed by the total cutting force and the normal to the workpiece surface, energy can be transferred into the machine-tool structure, thus producing an effective negative damping. The width of cut for the threshold of stable operation is directly dependent upon the difference between the two principal stiffness values, and chatter tends to occur when the two principal stiffnesses are close in magnitude.

DYNAMIC STABILITY

Machine-tool chatter is essentially a problem of dynamic stability. A machine tool under vibration-free cutting conditions may be regarded as a dynamical system in steady-state motion. Systems of this kind may become dynamically unstable and break into oscillation around the steady motion. Instability is caused by an alteration

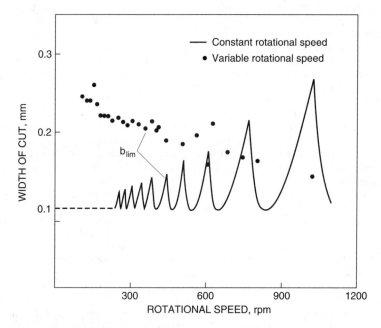

FIGURE 40.5 Dependence of maximum stable width of cut b_{lim} on cutting speed (stability chart) for turning; stable area is below the line, unstable, above the line. Dots indicate cutting with variable (modulated) cutting speed. (*After J. Sexton and B. Stone.*[19])

of the cutting conditions produced by a disturbance of the cutting process (e.g., a hard spot in the material). As a result, a time-dependent thrust element dP is superimposed on the steady cutting thrust P. If this thrust element is such as to amplify the original disturbance, oscillations will build up and the system is said to be unstable.

This chain of events is most easily investigated theoretically by considering that the incremental thrust element dP is a function not only of the original disturbance but also of the velocity of this disturbance. Forces which are dependent on the velocity of a displacement are damping forces; they are additive to or subtractive from the damping present in the system (e.g., structural damping or damping introduced by special antivibration devices). When the damping due to dP is positive, the total damping (structural damping plus damping due to altered cutting conditions) also is positive and the system is stable. Any disturbance will then be damped out rapidly. However, the damping due to dP may be negative, in which case it will decrease the structural damping, which is always positive. If the negative damping due to dP predominates, the total damping is negative. Positive damping forces are energy-absorbing. Negative damping forces feed energy into the system; when the total damping is negative, this energy is used for the maintenance of oscillations (chatter).

From the practical point of view, the fully developed chatter vibration (self-induced vibration) is of little interest. Production engineers are almost entirely concerned with conditions leading to chatter (dynamic instability). The build-up of chatter is very difficult to observe, and experimental work has to be carried out mainly under conditions which are only indirectly relevant to the problem being investigated. Experimental results obtained from fully developed chatter vibration may, in some instances, be not really relevant to the problem of dynamic stability.

The influence of the machine-tool structure on the dynamic stability of the cutting process is of great importance. This becomes clear by considering that with a structure (including tool and workpiece) of infinite stiffness, the cutting process could not be disturbed in the first place because hard spots, for example, would not be able to produce the deflections necessary to cause such a disturbance. Furthermore, it is clear that were the structural damping infinite, the total damping could not become negative and the cutting process would always be stable. This discussion indicates that an increase in structural stiffness and/or damping always has beneficial effects from the point of view of chatter.

In practically feasible machines, the interrelation between structural stiffness, damping, and dynamic stability is of considerable complexity. This is because machine-tool structures are systems with distributed mass, elasticity, and damping; their vibration is described by a large set of partial differential equations which can be analyzed using simplified models[3] or more precise large finite-element models. Stiffness and damping play similar roles in determining the stability of a machine tool. The maximum stable width of cut b_{lim} is proportional to a product of effective stiffness and effective damping coefficients.

THE EFFECT OF VIBRATION ON TOOL LIFE

Inasmuch as the cutting speed and the chip cross section vary during vibration, it is to be expected that vibration affects tool life. The magnitude of this effect is unexpectedly large, even when impact loading of the tool is excluded. Elimination of vibration may significantly enhance tool life. Ceramic and diamond tools are especially sensitive to impact loading.

The life of face-mill blades may suffer considerably owing to torsional vibration executed by the cutter. The torsional vibration need not necessarily be caused by dynamic instability of the cutting process but may be forced vibration, because of resonance caused by one of the harmonics of impact excitation from interrupted chip removal, by tool runout, etc.[2]

VIBRATION CONTROL IN MACHINE TOOLS

The vibration behavior of a machine tool can be improved by a reduction of the intensity of the sources of vibration, by enhancement of the effective static stiffness and damping for the modes of vibration which result in relative displacements between tool and workpiece, and by appropriate choice of cutting regimes, tool design, and workpiece design. Abatement of the sources is important mainly for forced vibrations. Stiffness and damping are important for both forced and self-excited (chatter) vibrations. Both parameters, especially stiffness, are critical for accuracy of machine tools, stiffness by reducing structural deformations from the cutting forces, and damping by accelerating the decay of transient vibrations. In addition, the application of vibration dampers and absorbers is an effective technique for the solution of machine-vibration problems. Such devices should be considered as a functional part of a machine, not as an add-on to solve specific problems.

STIFFNESS

Static stiffness k_s is defined as the ratio of the static force P_o, applied between tool and workpiece, to the resulting static deflection A_s between the points of force application. A force applied in one coordinate direction is causing displacements in three coordinate directions; thus the stiffness of a machine tool can be characterized by a tensor of stiffnesses (three proper stiffnesses defined as ratios of forces along the coordinate axes to displacements in the same directions, and three reciprocal stiffnesses between each pair of the coordinate axes). Frequently only one or two stiffnesses are measured to characterize the machine tool.

Machine tools are characterized by high precision, even at heavy-duty regimes (high magnitudes of cutting forces). This requires very high structural stiffness. While the frame parts are designed for high stiffness, the main contribution to deformations in the work zone (between tool and workpiece) comes from contact deformations in movable and stationary joints between components (contact stiffness[8,20]). Damping is determined mainly by joints[20] (log decrement $\Delta \cong 0.15$), especially for steel welded frames (structural damping $\Delta \cong 0.001$). Cast iron parts contribute more to the overall damping ($\Delta \cong 0.004$), while material damping in polymer-concrete ($\Delta \cong 0.02$) and granite ($\Delta \cong 0.015$) is much higher.[21] While the structure has many degrees-of-freedom, dangerous forced and self-excited vibrations occur at a few natural modes which are characterized by high intensity of relative vibrations in the work zone. Since machine tools operate in different configurations (positions of heavy parts, weights, dimensions, and positions of workpieces) and at different regimes (spindle rpm, number of cutting edges, cutting angles, etc.), different vibratory modes can be prominent depending on the circumstances.

The stiffness of a structure is determined primarily by the stiffness of the most flexible component in the path of the force. To enhance the stiffness, this flexible component must be reinforced. To assess the influence of various structural components on the overall stiffness, a breakdown of deformation (or compliance) at the cutting edge must be constructed analytically or experimentally on the machine.[13] Breakdown of deformation (compliance) in torsional systems (transmissions) can be critically influenced by transmission ratios between the components.[2,8] In many cases the most flexible components of the breakdown are local deformations in joints, i.e., bolted connections between relatively rigid elements such as column and bed, column and table, etc. Some points to be considered in the design of connections are illustrated in Fig. 40.6.[22] To avoid bending of the flange in Fig. 40.6A, the bolts should be placed in pockets or between ribs, as shown in Fig. 40.6B. Increasing the flange thickness does not necessarily increase the stiffness of the connection, since this requires longer bolts, which are more flexible. There is an optimum flange thickness (bolt length), the value of which depends on the elastic deformation in the vicinity of the connection. Deformation of the bed is minimized by placing ribs under connecting bolts.[22]

The efficiency of bolted connections, and other static and dynamic structural problems, is conveniently investigated by scaled model analysis[22] and finite-element analysis techniques.[18] Figure 40.7[22] shows the results of successive stages of a model experiment in which the effect of the design of bolt connections on the bending rigidity (X and Y directions) and the torsional rigidity of a column were investigated. The relative rigidities are shown by

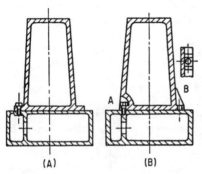

FIGURE 40.6 Load transmission between column and bed. (*A*) Old design, relatively flexible owing to deformation of flange. (*B*) New design, bolt placed in a pocket (*A*) or flange stiffened with ribs on both sides of bolt (*B*). (*After H. Opitz.*[22])

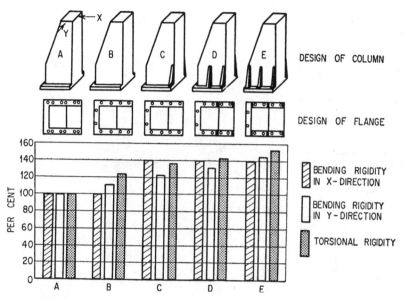

FIGURE 40.7 Successive stages in the improvement of a flange connection. (*H. Opitz.*[22])

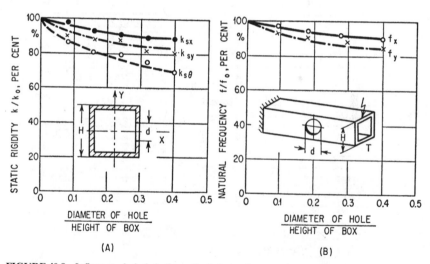

FIGURE 40.8 Influence of a hole in the wall of a box column on the static stiffness and natural frequency. (*A*) Static stiffness; (*B*) natural frequency. (*H. Opitz.*[22])

the length of bars. In the design of Fig. 40.7*A*, the connection consists of 12 bolts (diameter of ⅝ in.) arranged in pairs along both sides of the column. In the design of Fig. 40.7*B*, the number of bolts is reduced to 10, arranged as shown. With the addition of ribs, shown in succeeding figures, the bending stiffness in the direction *X* was raised by 40 percent, that in the direction *Y* by 45 percent, and the torsional stiffness by 53 percent, compared to the original design.[22]

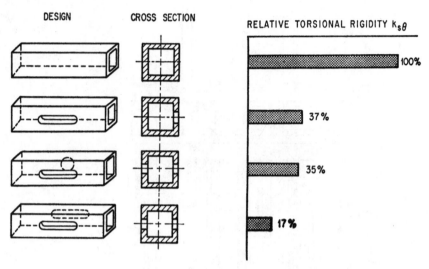

FIGURE 40.9 Torsional stiffness of box columns with different holes in walls. (*H. Opitz.*[22])

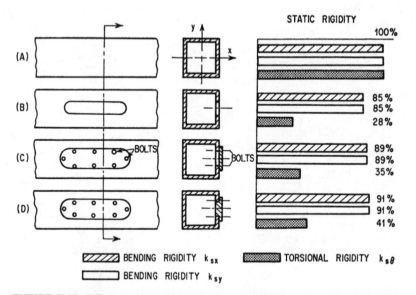

FIGURE 40.10 Influence of cover plate and lid on static stiffness of box column. (*A*) Column without holes, (*B*) one hole uncovered, (*C*) hole covered with cover plate, and (*D*) hole covered with substantial lid, firmly attached. (*After H. Opitz.*[22])

Openings in columns should be as small as possible. Figure 40.8[22] shows the loss of static flexural stiffness k_{sx}, k_{sy}, and torsional stiffness $k_{s\theta}$, and the decrease of the flexural natural frequencies f_x, f_y, resulting from the introduction of a hole in a box-type column. Smaller holes result in relatively smaller decreases of stiffness and natural frequency than larger ones. The torsional rigidity $k_{s\theta}$ of a box-type column is particularly sensitive to openings, as shown in Fig. 40.9.[22] Lids or doors used for covering these openings do not restore the stiffness. The influence of covers depends on their thickness, mode of attachment, and design, as shown in Fig. 40.10.[22] However, covers may increase damping and thereby partly compensate for the detrimental effect of loss of stiffness.

Welded structural components are usually stiffer than cast iron components but have a lower damping capacity. Some damping is generated because welds are never perfect; consequently, rubbing takes place between joined members. A considerable increase in damping can be achieved by using interrupted welds, but at a price of reduced stiffness. Welded ribs may be necessary not so much to increase rigidity as to prevent "drumming" (membrane vibration) of large unsupported areas.

Not all deformations in machine tools are objectionable, but only those which influence relative displacements in the work zone between the tool and the workpiece. The magnitude of the relative displacement in the work zone under external or internal forces (weight, cutting force, inertia force) determines *effective* stiffness.

Effective stiffness of machine-tool frames is significantly influenced by their interaction with the supporting structures (foundations). For large, low-aspect-ratio machine-tool frames, a rigid attachment to a properly dimensioned[13] foundation substantially improves dynamic stability. Medium- and small-size machine tools are usually attached to the reinforced floor plate by discrete mounts (rigid wedge or screw mounts or vibration isolators). A rational assignment of number and location of mounts noticeably enhances the effective stiffness of machine tools and in some cases may allow direct mounting of rather large machine tools on vibration isolators. Examples of influence of number and location of mounts on the effective stiffness are given in Fig. 40.11,[13] which shows three schematics of a mounting for a jig borer on rigid wedge mounts. The table of the jig borer is in the lower end of the illustration. Relative displacements in the work zone when the table travels from right to left for the scheme in Fig. 40.11C are three times smaller than for Fig. 40.11A and 1.5 times smaller than for Fig. 40.11B, notwithstanding the fact that in the latter case there are seven mounts vs. three mounts in Fig. 40.11C. In the case shown in Fig. 40.11A, the large weight of the moving table creates a twisting of the supporting frame about the single front mount, while the column is rigidly positioned by two mounts. In case of Fig. 40.11C, the front end is well supported, but the column can tilt on its single mount and follow small deformations of the front part, thus resulting in smaller relative deformations and higher effective stiffness. In the case of a precision grinder having a bed 3.8 m long, it was found that mounting the grinder on seven carefully located (offset from the ends) vibration isolators resulted in higher effective stiffness than installation on 15 rigid mounts.[23]

The effective static stiffness of a machine tool may vary within wide limits. High stiffness values are ensured by the use of steady rests, by placing tool and workpiece in a position where the relative dynamic displacement between them is small (i.e., by placing them near the main column, etc.), by using rigid tools and clamps, by using

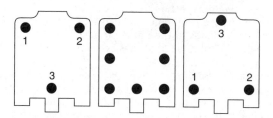

FIGURE 40.11 Mounting schemes of a jig borer. (*After V. Kaminskaya.*[13])

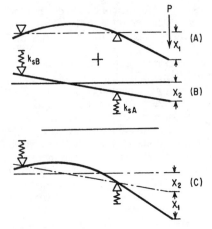

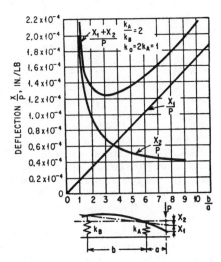

FIGURE 40.12 Deflection of machine-tool spindle and bearings. A machine-tool spindle can be regarded as a beam on flexible supports. The total deflection under the force P consists of the sum of (A) the deflection X_1 of a flexible beam on rigid supports and (B) the deflection X_2 of a rigid beam on flexible supports. $(H.\ Opitz.^{22})$

FIGURE 40.13 Deflection of a beam on elastic supports as a function of the bearing distance. Bearing stiffness k_A and k_B, spindle stiffness k_o. $(After\ H.\ Opitz.^{22})$

jigs which rigidly clamp (and if necessary support) the workpiece, by clamping securely all parts of the machine which do not move with respect to each other, etc., and by the optimization of mounting conditions mentioned above.

The static and dynamic behavior of machine tools is influenced significantly by the design of the spindle and its bearings. The static deflection of the spindle consists of two parts, X_1 and X_2, as shown in Fig. 40.12.[22] The deflection X_1 corresponds to the deflection of a flexible beam on rigid supports, and X_2 corresponds to the deflection of a rigid beam on flexible supports which represent the flexibility of the bearings. The deflection of the spindle amounts to 50 to 70 percent of the total deflection, and the bearings 30 to 50 percent of the total, depending on the relation of spindle cross section to bearing stiffness and span. The stiffness of antifriction bearings depends on their design, accuracy, preload, and the fit between the outer race and the housing (responsible for 10 to 40 percent of the bearing deformation[8]).

The distance between the bearings has considerable influence on the effective stiffness of the spindle, as shown in Fig. 40.13.[22] The ordinate of the figure corresponds to the deflection in inches per pound, and the abscissa represents the ratio of bearing distance b to cantilever length a. The straight line refers to the deflection of the spindle, and the hyperbola refers to the deflection of the bearings. The total deflection is obtained by the addition of the two curves; the minimum of the curve of total deflection corresponds to the optimum bearing distance. For a short cantilever length a, the optimum value of b/a lies between 3 and 5; for a long cantilever length a, the optimum $b/a = \sim 2$.

It is often important to consider the dynamic behavior of a spindle before establishing an optimum bearing span. Maximizing the stiffness of a spindle at one point does not establish its dynamic properties. Care must be taken to investigate both bending and rocking modes of the spindle before accepting a final optimum span. For example, a large overhang on the rear of a spindle could produce an undesirable low-frequency rocking mode of the spindle even if the "optimum span" as defined previously were satisfied.

The optimum bearing span for minimum deflection as well as the dynamic characteristics of spindles may be computed with the help of available computer programs.[11]

The influence of the ratio of bore diameter to outside diameter on the stiffness of a hollow spindle is shown in Fig. 40.14.[22] A 25 percent decrease in stiffness occurs only at a diameter ratio of $d/D = 0.7$, where D is the outside diameter and d the bore diameter. This is important for the dynamic behavior of the spindle. A solid spindle has nearly the same stiffness, but a substantially greater mass. Consequently, the natural frequency of the solid spindle is considerably lower, which is undesirable. A stiff spindle does not always assure the required high stiffness at the cutting edge of the tool because of potentially large contact deformations in the toolholder/spindle interface. Measurements have shown that in a tapered connection, these deformations may constitute up to 50 percent of the total deflection at the tool edge.[5] These deformations can be significantly reduced by replacing tapered connections by face contact between the toolholder and the spindle. The face connection must be loaded by a high axial force.[5,24]

A significant role (frequently up to 50 percent) in the breakdown of deformations between various parts of machine tool structures is played by contact deformations between conforming (usually flat, cylindrical, or tapered) contacting surfaces in structural joints and slides.[20, 8] Contact deformations are due to surface imperfections on contacting surfaces. These deformations are highly nonlinear and are influenced by lubrication conditions. Figure 40.15 shows contact deformation between flat steel parts as a function of contact pressure for different lubrication conditions in the joint. Joints are also responsible for at least 90 percent of structural damping in machine-tool frames due to micromotions in the joints during vibratory processes. Contact deformations for the same contact pressure can be significantly reduced by increasing accuracy (fit) and improving the surface finish of the mating surfaces. The nonlinear load-deflection characteristic of joints, Fig. 40.15, allows enhancement of their stiffness by preloading. However, preloading reduces micromotions in the joints and thus results in a lower damping.

This explains why in some cases old machines are less likely to chatter than new machines of identical design. The situation may result from wear and tear of the slides, which increases the damping and effects an improvement in performance. Also, in some cases chatter is eliminated by loosening the locks of slides. However, it would be wrong to conclude that lack of proper attention and maintenance is desirable. Proper attention to slides, bearings (minimum play), belts, etc., is necessary for satisfac-

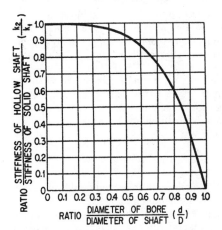

FIGURE 40.14 Effect of bore diameter on stiffness of hollow spindle where k_1 = stiffness of solid spindle, k_2 = stiffness of hollow spindle, D = outer spindle diameter, d = bore diameter, J_2 = second moment of area of hollow spindle, and J_1 = second moment of area of solid spindle. The curve is defined by $k_2/k_1 = J_2/J_1 = 1 - (d/D)$.[4] (*H. Opitz.*[22])

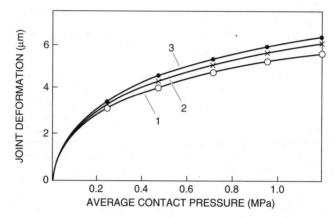

FIGURE 40.15 Load-deflection characteristics for flat, deeply scraped surfaces (overall contact area 80 cm²). 1, no lubrication; 2, lightly lubricated (oil content 0.8×10^{-3} gram/cm²); 3, richly lubricated (oil content 1.8×10^{-3} gram/cm²). (*After Z. Levina and D. Reshetov.*[20])

tory performance. It would be wrong also to conclude that a highly polluted workshop atmosphere is desirable because some new machines exposed to workshop dirt for a sufficiently long time, even when not used, appear to improve in their chatter behavior. The explanation is that dirty slides increase the damping.

When the rigidity of some machine element is intentionally *reduced,* but this reduction is accompanied by a greater damping at the cutter, the increase in damping may outweigh the reduction in rigidity.[5] Although a loss of rigidity in machine tools is generally undesirable, it may be tolerated when it leads to a desirable shift in natural frequencies or is accompanied by a large increase in damping or by a beneficial change in the ratio of stiffnesses along two orthogonal axes, which can result in improved nonregenerative chatter stability.[15]

A very significant improvement in chatter resistance can be achieved by an intentional measured reduction of stiffness in the direction along the cutting speed (orthogonal to the direction of the principal component of cutting force). The benefits of this approach have been demonstrated for turning and boring operations.[25,26]

DAMPING

The overall damping capacity of a structure with cast iron or welded steel frame components is determined only to a small extent by the damping capacity of its individual components. The major part of the damping results from the interaction of joined components at slides or bolted joints.[20] The interaction of the structure with the foundation or highly damped vibration isolators also may produce a noticeable damping.[14,15] A qualitative picture of the influence of the various components of a lathe on the total damping is given in Fig. 40.16. The damping of the various modes of vibration differs appreciably; the values of the logarithmic decrement shown in the figure correspond to an average value for all the modes which play a significant part.[27]

The overall damping of various types of machine tool differs, but the log decrement is usually in the range of from 0.15 to 0.3. While structural damping is significantly higher for frame components made of polymer-concrete compositions or granite (see above), the overall damping does not change very significantly since the damping of even these materials is small compared with damping from joints.

A significant damping increase can be achieved by filling internal cavities of the frame parts with a granular material, e.g., sand.[21] For cast parts it can also be achieved by leaving cores in blind holes inside the casting. A similar, sometimes even more pronounced, damping enhancement can be achieved by placing auxiliary longitudinal structural members inside longitudinal cavities within a frame part, with offset from the bending neutral axis of the latter. The auxiliary structural member interacts with the frame part via a high viscous layer, thus imparting energy dissipation during vibrations.[21]

Damping can be increased without impairing the static stiffness and machining accuracy of the machine by the use of dampers and dynamic vibration absorbers. These are basically similar to those employed in other fields of vibration control (Chaps. 6, 32, and 43). Dampers are effective only when placed in a position where vibration amplitudes are significant.

The tuned dynamic vibration absorber has been employed with considerable success on milling machines, machining centers,[28] radial drilling machines, gear hobbing machines,[29] grinding machines, and boring bars.[26, 30] A design variant of this type of absorber is shown in Fig. 40.17. In this design a plastic ring element combines both the elastic and the damping elements of the absorber. The auxiliary mass may be attached to the top of a column (Fig. 40.17C), as shown in Fig. 40.17A. Alternatively, the auxiliary mass may be suspended on the underside of a table (Fig. 40.17C), using the design shown in Fig. 40.17B. In either case, several plastic ring elements may support one large auxiliary mass, as shown in Fig. 40.17C.[29] In a boring bar, shown in Fig. 40.18A, elastic and damping properties are combined in O-rings made of a high-damping rubber. Tuning of the absorber can be changed by varying the radial preload force on the O-ring.[26] The natural frequency of this absorber can be varied over a range of more than 3:1.

A variation of the *Lanchester damper* (Chaps. 6 and 38) is frequently used in boring bars to good advantage.[30] This consists of an inertia weight fitted into a hole bored in the end of a quill, as shown in Fig. 40.18B. To ensure effective operation, a

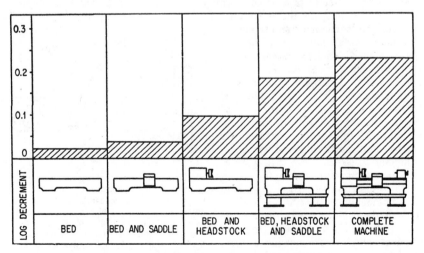

FIGURE 40.16 Influence of various components on total damping of lathes. The major part of the damping is generated at the mating surfaces of the various components. (*K. Loewenfeld.*[27])

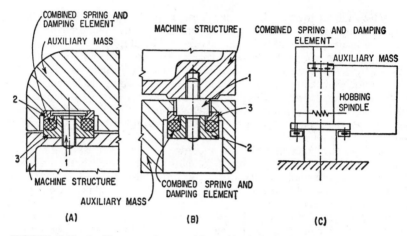

FIGURE 40.17 Auxiliary mass damper with combined elastic and damping element. The combined element lies between two retainer rings, of which one (3) is attached with bolt 1 to the machine structure. The other ring (2) takes the weight of the auxiliary mass. (*A*) Arrangement when auxiliary mass is being supported. (*B*) Arrangement when auxiliary mass is being suspended. (*C*) Application of both types of arrangements to a hobbing machine. (*After F. Eisele and H. W. Lysen.*[29])

relatively small radial clearance of about 1 to $5 \times 10^{-3}d$ must be provided, where d is the diameter of the inertia weight. An axial clearance of about 0.006 to 0.010 in. (0.15 to 0.25 mm) is sufficient. A smooth surface finish of both plug and hole is desirable. The clearance values given refer to dry operation, using air as the damping medium. Oil also can be used as a damping medium, but it does not necessarily result in improved performance. When applying oil, clearance gaps larger than those stated above have to be ensured, depending on the viscosity of the oil. In general, Lanchester dampers are less effective than tuned vibration absorbers.

Since the effectiveness of both Lanchester dampers and tuned vibration absorbers depends on the mass ratio between the inertia mass and the effective mass of the structure (Chap. 6), heavy materials such as lead and, especially, machinable sintered tungsten alloys are used for inertia masses in cases where the dimensions of the inertia mass are limited (as in the case of boring bars in Fig. 40.18). The mass ratio and the effectiveness of the absorber can be significantly enhanced by using a combination structure. In such a structure the overhang segment of the boring bar or other cantilever structure, which does not significantly influence its stiffness but determines its effective mass, is made of a light material, while the root segment, which determines the stiffness but does not significantly influence the

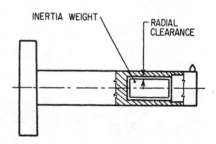

FIGURE 40.18 Chatter-resistant boring bars. (*A*) Combination structure (root segment made from tungsten carbide, overhang segment from aluminum) with tunable vibration absorber using high-damping elastic element. (*After E. Rivin and H. Kang.*[26]) (*B*) Lanchester damper for the suppression of boring-bar vibration. (*After R. S. Hahn.*[30])

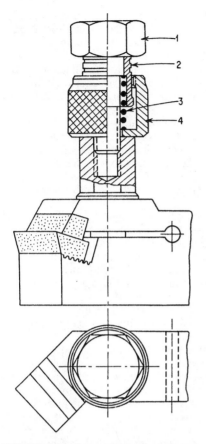

FIGURE 40.19 Impact damper for lathe tools, consisting of bolt (1), sleeve (2), spring (3), and a cover (4). The oscillation of the inert mass (4), impinging on the bolt (1), absorbs vibratory energy. (*After D. I. Ryzhkov.*[31])

effective mass, is made from a high Young's modulus material (aluminum and tungsten carbide, respectively, in Fig. 40.18*A*).[26]

Impact dampers also provide effective vibrational control.[31] The impact damper shown in Fig. 40.19 consists of a bolt (1), a sleeve (2), a spring (3), and a cover (4). It is set on the lathe tool post or on the tool, at a point where vibration of large amplitude arises normally. In operation, the vibration of the tool results in an oscillatory motion of the inert mass (4), which, impinging on the bolt (1), absorbs energy. The important physical parameters of an impact damper are material, mass ratio, and stroke ratio (a function of the gap through which the free mass travels). Design charts have been prepared which facilitate the application of impact dampers to the solution of vibration problems found typically in machine tools.[32]

Dynamic absorbers can be active (servo-controlled). Such devices can be designed to be self-optimizing (capable of self-adjustment of the spring rate to minimize vibration amplitude under changing excitation conditions) or to use a vibration cancellation approach. The self-optimizing feature is achieved by placing vibration transducers on both the absorber mass and the main system. A control circuit measures the phase angle between the motions and activates a spring-modifying mechanism to maintain a 90° phase difference between the two measured motions. It has been demonstrated that the 90° phase relationship guarantees minimum motion of the main vibrating mass. In the vibration-cancellation devices, the actuator applies force to the structure which is opposite in phase to structural vibrations.

Dynamic analysis of a machine tool structure can identify potentially unstable natural modes of vibration and check the effectiveness of the applied treatments. In another approach, transfer functions between the selected points on the machine tool are measured and processed through a computational technique which indicates at which location stiffness and/or damping should be modified or a dynamic vibration absorber installed in order to achieve specified dynamic characteristics of the machine tools.[28]

Tool Design. Sharp tools are more likely to chatter than slightly blunted tools. In the workshop, the cutting edge is often deliberately dulled by a slight honing. Consequently, a *vibro-dampening bevel* on the leading face of a lathe tool has been suggested. This bevel has a leading edge of –80° and a width of about 0.080 in. (0.2 mm). Tests show that the negative bevel does not in all cases eliminate vibration and that the life of the bevel is short. Appreciably worn cutting edges cause violent chatter.

Since narrow chips are less likely to lead to instability, a reduction of the approach angle of the cutting tool results in improved chatter behavior. With lathe tools, an increase in the rake angle results in improvement, but the influence of changes in the relief angle is relatively small.

Theoretically, the tool shape should be such that any sudden increase of the feed rate meets the maximum possible resistance in the workpiece material. Such a stabilizing effect is produced by the chisel edge of a drill; it is not clear whether chisel-edge corrections, which reduce the static thrust, are also beneficial from the vibration point of view. Similarly, the wide face edge of certain face-mill blades prevents a "hacking-in" of these blades into the machined surface and appears to have a stabilizing influence.[15]

Reduction of both forced and chatter vibrations in cutting with tools having multiple cutting edges (e.g., milling cutters, reamers) can be achieved by making the distance between the adjacent cutting edges nonequal and/or making the helix angle of the cutting edges different for each cutting edge. However, such treatment results in nonuniform loading of the cutting edges and may lead to a shortened life of the more heavily loaded edges as well as deteriorating surface finish as a result of different deformations of the tool when lighter or heavier loaded edges are engaged.

Reduction of cutting forces by low-friction (e.g., diamond) coating of the tool or by application of ultrasonic vibrations to the tool usually improves chatter resistance.

Variation of Cutting Conditions. In the elimination of chatter, cutting conditions are first altered. In some cases of regenerative chatter, a small increase or decrease in speed may stabilize the cutting process. In high-speed or unattended computer numerically controlled machine tools, this can be achieved by continuous computer monitoring of vibratory conditions and, as chatter begins to develop, a shifting of the spindle rpm toward the stable area.[24]

Cutting with a variable cutting speed (constant speed modulated by a sinusoidal or other oscillatory component) acts similarly with regard to undulations in the positioning of the cutting edges (see above) and results in increased chatter resistance. The dots in Fig. 40.5 show the stabilizing effect of the sinusoidal modulation of the cutting speed.[19]

An increase in the feed rate is also beneficial in some types of machining (drilling, face milling, and the like). For the same cross-sectional area, narrow chips (high feed rate) are less likely to lead to chatter than wide chips (low feed rate), since the chip thickness variation effect results in a relatively smaller variation of the cross-sectional area in the former (smaller dynamic cutting force).

REFERENCES

1. Koenigsberger, F., and J. Tlusty: "Machine Tool Structures," vol. 1, Pergamon Press, 1970.

2. Rivin, E. I.: "Dynamics of Machine-Tool Drives (Dynamika privoda stankov)," Mashinostroenie Publishing House, Moscow, 1966 (in Russian).

3. Tlusty, J.: Machine Dynamics, in R. King (ed.), "Handbook of High-Speed Machining Technology," p. 48, Chapman and Hall, New York, 1985.

4. Lyon, R. H., and L. M. Malinin: *Sound and Vibration,* **6**:22 (1984).

5. Rivin, E. I.: *Manufacturing Review,* **4**(4):257 (1991).

6. Doi, S.: *Trans. ASME,* **80**(1):133 (1958).

7. Slocum, A. H.: "Precision Machine Design," Prentice-Hall, Inc., Englewood Cliffs, N.J., 1991.

8. Rivin, E. I.: "Mechanical Design of Robots," McGraw-Hill Book Company, New York, 1988.

9. Weck, M., et al.: "Design of Spindle-Bearings System for High Speed Machining," Expert Verlag, Ehringen, vol. 283, 1990 (in German).

10. Shimanovich, M. A.: *Soviet Engineering Research,* **11**:48 (1984).

11. Wang, W. R., and C. N. Chang: *ASME J. Vibration and Acoustics,* **116**:280 (1994).

12. Shimanovich, M. A., A. I. Grebnev, and K. E. Shkinev: *Soviet Engineering Research,* **10**(1):107 (1990).

13. Reshetov, D. N. (ed.): "Components and Mechanisms of Machine Tools," vols. 1, 2, Mashinostroenie, Moscow, 1972 (in Russian).

14. Rivin, E. I.: *ASME J. Mechanical Design,* **101**(4):682 (1979).

15. Tobias, S. A.: "Machine Tool Vibration," Blackie, London, 1965.

16. Kaneko, T., H. Sato, Y. Tani, and M. O-hori: *ASME J. Engg. for Industry,* **106**:222 (1984).

17. "Methods for Performance Evaluation of CNC Machining Centers," U.S. Standard ASME B5.54, 1992.

18. Weck, M.: "Handbook on Machine Tools," vols. 1–4, John Wiley & Sons, Inc., New York, 1984.

19. Sexton, J. S., and B. J. Stone: *Annals CIRP,* **2711**:321 (1978).

20. Levina, Z. M., and D. N. Reshetov: "Contact Stiffness of Machine Tools," Mashinostroenie, Moscow, 1971 (in Russian).

21. Slocum, A. H., E. R. Marsh, and D. H. Smith: *Precision Engineering,* **16**(3):125 (1994).

22. Opitz, H.: "Conference on Technology of Engineering Manufacture," Paper 7, The Institution of Mechanical Engineers, London, 1958.

23. Rivin, E. I.: *Precision Engineering,* **17**(1):44–46 (1995).

24. Tlusty, J.: "Fundamentals of High Speed Machining," SME, 1994.

25. Elyasberg, M. E.: *Soviet Engg. Research,* **3**:59 (1983).

26. Rivin, E. I., and H. Kang: *Int. J. Machine Tools and Manufacture,* **32**(4):539 (1992).

27. Loewenfeld, K.: "Zweites Forschungs- und Konstruktionskolloquium Werkzeugmaschinen," p. 117, Vogel-Verlag, Coburg, 1955.

28. Rivin, E. I., and W. D'Ambrogio: *Mechanical Systems and Signal Processing,* **4**(6):495 (1990).

29. Eisele, F., and H. W. Lysen: "Zweites Forschungs- und Konstruktionskolloquium Werkzeugmaschinen," p. 89, Vogel-Verlag, Coburg, 1955.

30. Hahn, R. S.: *Trans. ASME,* **73**(4):331 (1951).

31. Ryzhkov, D. I.: *Stanki i Instrument,* (3):23 (1953); *Engineers' Digest,* **14**(7):246 (1953).

32. Pinotti, P. C., and M. M. Sadek: *Proc. 11th Intern. MTDR Conf.,* September 1970.

CHAPTER 41
PACKAGING DESIGN

Masaji T. Hatae

INTRODUCTION

Packaging is the technology of preparing an item for shipment in such a manner as to minimize the damage resulting from environmental hazards encountered during shipment. Such environmental hazards include moisture, temperature, dust, shock, vibration, etc. This chapter considers only the influence of shock and vibration.

The shock and vibration experienced by goods during shipment result from the vibration of a cargo-carrying vehicle, the shock resulting from the impacts of railroad cars, the shock caused by the handling (e.g., dropping) of packages, etc. Such goods are protected from damage by isolation. The theory and practice of vibration and shock isolation are discussed in Chaps. 30 to 33. This chapter discusses the application of the principles of isolation to the design of packages. In general, this consists of placing the packaged item within a container, and interposing resilient means between the item and the container to provide the necessary isolation. Such resilient means is known as *package cushioning*. The required degree of protection should be provided with minimum packaging and shipping costs through optimum combination of labor, materials, and package volume and weight. The means of protection may range from a simple corrugated carton to a more complex system such as a part "floated" in distributed cushioning material in a wooden box or by a spring suspension in a reusable metal drum.

In concept, the package cushioning is designed to protect an item of known strength from the known shock and vibration existing in the particular environment. Practically, package design cannot be pursued so rationally because the strength of equipment and the characteristics of the environment often are not known with the necessary accuracy. As a consequence, the strength of the equipment often must be inferred from exploratory tests or estimated on the basis of experience. Similarly, environmental conditions are simulated by simplified tests (e.g., drop tests) that represent environmental conditions of maximum expected severity. Then the design of the package cushioning proceeds in a straightforward manner using the methods and data set forth in this chapter.

FACTORS IN DESIGN OF PACKAGE CUSHIONING

This section discusses factors that are significant in the design of package cushioning.

FRAGILITY OF EQUIPMENT

The capability of an item to withstand shock and vibration is defined in terms of its "fragility." The term "fragility" is used as a quantitative index of the strength of the equipment when subjected to shock and vibration. The quantitative index is expressed as a maximum permissible acceleration $\ddot{x}_F$. In addition to the value of $\ddot{x}_F$, it is necessary to know the natural frequencies and damping characteristics of the equipment. With the foregoing information the package designer could proceed to select cushioning with the desired characteristics to protect the equipment during shipment. Unfortunately, it is difficult to determine this information for many types of equipment. As a consequence of this difficulty, several alternative procedures may be used to obtain information of use for package design.

Analytical Methods. It is difficult to determine analytically (i.e., by theoretical analysis and calculation) the fragility of a complex device or structure. Sometimes it is possible to simplify or idealize the structure to a form that is more amenable to analysis, and the strengths of principal component structures may be calculated. This analytical basis for design is discussed in detail in Chap. 42.

Laboratory Test Method. Shock and vibration tests may be carried to failure in order to determine the fragility level of products. This method is particularly suited to products manufactured in large quantity where unit cost is small and tests-to-failure may be performed. Various types of laboratory shock test apparatus, such as drop test machines, hydraulic shock testing machines and pendulum impact machines, are available for testing a wide range of sizes and weights of equipment. These testing machines, which are described in Chap. 26, provide control of acceleration level, time duration of pulse, and shape of pulse. Laboratory vibration testing machines, described in Chap. 25, are available in various ranges of frequency and force output.

Shock tests-to-failure can provide data concerning the maximum allowable accelerations for particular input pulse shapes. These tests-to-failure define fragility levels for shock and vibration in terms of the characteristics of the testing machine. In general, such information is useful in the design of package cushioning if the testing machines are adjusted to have the same characteristics as the motion transmitted by the cushioning to the equipment or if the cushioning is made to have the characteristics of the testing machine.

Most products, particularly military equipment, are designed and tested according to requirements stipulated in contractual specifications; in general, such requirements relate to ultimate use rather than to packaging. The levels of shock and vibration to which the equipment is designed and tested most often represent the *minimum* acceptable resistance to damage; they do not represent the maximum level that the item can withstand without failure, nor are they generally related to the requirements for successful packaging. Strict adherence to these minimum values in package design often results in high costs of packaging and shipping.

Engineering Estimate. Field personnel experienced in the handling and operation of specific items (products) often can provide information on the ability of these items to withstand drops from various heights when protected by certain cushioning materials. These heights of drop can be converted into peak acceleration levels if the dynamic properties of the cushioning materials are known. Then the fragility is defined quantitatively for certain conditions. For example, assume that it has been determined that an item, blocked and braced in a corrugated carton, is able to withstand the shock that results from dropping the carton from a height of 12 in. Information on height of drop may be transformed to the parameters of rise time and peak acceleration by assuming a sinusoidal acceleration pulse. The applicable relations are given in Fig. 41.1. Estimated rise times to maximum acceleration for various drop conditions are given in Table 41.1. Further refinements in the values may be made by actual instrumented drop tests using various types of containers and materials.

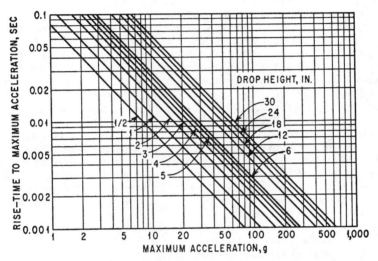

FIGURE 41.1 Relation of drop height and rise time to maximum acceleration for a packaged item which experiences a half-sinusoidal pulse. The relation is defined mathematically as $\ddot{x}_1{}^2 = 0.0128 h t_r^{-2}$, where $\ddot{x}_1$ = maximum acceleration, g; h = height of drop, in.; and t_r = rise time, sec.

The fragility index of equipment as estimated from experience in dropping similar equipment applies only to the peak acceleration. The natural frequencies and damping present in the equipment are not indicated. In general, the natural frequencies of important components of the equipment are substantially greater than the natural frequency of the package cushioning; under these conditions, damage tends to be directly proportional to maximum acceleration, and it is possible to use the maximum acceleration as an index of fragility.

TABLE 41.1 Approximate Sinusoidal Rise Time for Various Drop Conditions

Condition	Rise time, sec
(1) Unpackaged; metal container	0.002
(2) Wooden box	0.004
(3) Carton	0.006
(4) Approx. 1 in. latex hair	0.008
(5) Approx. 3 in. latex hair	0.015

ENVIRONMENTAL CONDITIONS

In general, the shock received during handling (for example, from being dropped) is greater than that experienced in being transported in a vehicle; such shocks usually determine the severity of a shock test.

A standard shock test for packages is the drop test in which the package is dropped from a predetermined height onto a rigid floor. The height of drop is determined by the type of handling the package may receive during shipment. For example, packages from 0 to 50 lb may be considered within the "one-man throwing limit"; i.e., packages of such weight may be thrown easily onto piles or in other ways severely mishandled due to their light weight. Packages weighing between 50 and 100 lb may be considered within a "one-man carrying limit"; such packages are somewhat heavy to be thrown but can be carried and dropped from a height as great as shoulder height. A "two-man dropping limit" may apply to a weight range between 100 and 300 lb; the corresponding drop height for this mode of handling may be waist height. A further range may be from 300 to 1,000 lb; packages in this range would be handled with light cranes or lift trucks and may be subjected to shock from excessive lifting or lowering velocities. Finally, very heavy packages weighing over 1,000 lb would be handled by heavier equipment with correspondingly more skill; any drops would be from very small heights. Similarly, the size of the package classifies the type of handling into one man, two men, light equipment, or heavy equipment with corresponding drop heights. Thus, the drop heights for package shock tests are derived from the type of handling to which the package is most likely to be subjected during a shipping cycle; the type of handling is dependent on the size and weight of the package.

In addition to drop heights that vary with package size and weight, another factor in shock testing is the orientation of the package at impact. For example, small, light-weight packages are likely to be subjected to free fall drops onto sides, edges, and corners of the package. Larger, heavier packages handled by light or heavy equipment are likely to encounter drops of the type where one end rests on the floor and the other end is dropped (bottom rotational drop).

PROPERTIES OF CUSHIONING MATERIALS

The static and dynamic properties of cushioning materials that are commonly used in package design are included below in the section *Properties of Cushioning Materials*.

SHOCK ISOLATION

In the analytical treatment of shock isolation in packages, the idealized system illustrated in Fig. 41.2A is considered. It consists of a heavy rigid outer container, a packaged item assumed to be a rigid body, and a shock-isolation medium having characteristics which are considered below. The container is dropped from a characteristic height h upon a rigid floor, and the protection afforded to the packaged item by the isolation medium is indicated by the maximum acceleration experienced by the item. For purposes of analysis, it is convenient to assume a medium having a linear stiffness k; then results are modified to include the effects of the nonlinearity in actual media.

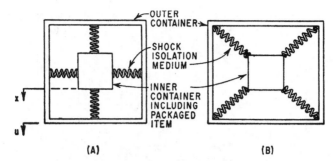

FIGURE 41.2 Schematic arrangements of isolation media in shipping containers: *A*—axes of isolation medium perpendicular to faces of container; *B*—axes of isolation medium inclined to faces of container.

The motion of the container as it falls freely from the height h is

$$u = \tfrac{1}{2} gt^2 \qquad (41.1)$$

The equation of motion for the packaged item is

$$m\ddot{x} = mg + k\delta \qquad (41.2)$$

where $\delta = u - x$ is the deflection of the isolation medium, m is the mass of the packaged item, and g is the acceleration of gravity. Substituting $u - x = \delta$ in Eq. (41.2) and using the expression for u given by Eq. (41.1), the equation of motion of the packaged item is

$$\ddot{x} + \omega_n^2 x = g + \frac{\omega_n^2 gt^2}{2} \qquad (41.3)$$

The solution of this equation is

$$x = A \sin \omega_n t + B \cos \omega_n t + \frac{gt^2}{2} \qquad (41.4)$$

where $\omega_n = \sqrt{k/m}$ is the natural angular frequency in radians per second of the packaged item supported by the isolation medium. The natural frequency is expressed in units of cycles per second, as a function of the weight W of the packaged item:

$$f_n = \frac{1}{2\pi} \sqrt{\frac{kg}{W}} \qquad (41.5)$$

This relation is shown graphically in Fig. 41.3.

The coefficients in Eq. (41.4) are evaluated from the conditions at time $t = 0$. At this instant, the packaged item is in equilibrium between the gravitational force and the upward force applied by the isolation medium; i.e., there is no downward acceleration until the container has traveled a sufficient distance to reduce the magnitude of the force applied by the isolation medium. Then, at time $t = 0, \ddot{x} = 0$ and Eq. (41.4) becomes

$$x = \frac{g}{\omega_n^2} \cos \omega_n t + \frac{gt^2}{2} \qquad (41.6)$$

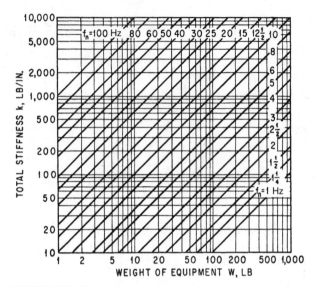

FIGURE 41.3 Natural frequency of a package as a function of the weight of the packaged item and the stiffness of the isolation medium. These curves are defined by Eq. (41.5).

Thus, the packaged item falls at the same rate as the container, plus a superimposed oscillatory motion initiated at the moment of dropping. The velocity is found by differentiating Eq. (41.6):

$$\dot{x} = -\frac{g}{\omega_n} \sin \omega_n t + gt \tag{41.7}$$

The velocity of the packaged item after the container has fallen from height h is determined by substituting $t = \sqrt{2h/g}$ in Eq. (41.7). The maximum possible velocity of the packaged item as the downward velocity of the container is arrested occurs when $\sin \omega_n \sqrt{2h/g} = -1$. The magnitude of the maximum velocity is

$$\dot{x}_{max} = \sqrt{2gh} + \frac{g}{\omega_n} \tag{41.8}$$

The resulting response of the shock-isolation medium is determined by considering the container to rest upon the floor and the packaged item to have an initial downward velocity $\dot{\delta}_{max} = \dot{x}_{max}$ as given by Eq. (41.8). The equation of motion for the packaged item is

$$m\ddot{\delta} = -k\delta + mg \tag{41.9}$$

The solution of this equation is

$$\delta = A \sin \omega_n t' + B \cos \omega_n t' + \frac{g}{\omega_n^2} \tag{41.10}$$

where $t' = 0$ at the moment that the downward velocity of the container is arrested.

Initial conditions are $\delta = 0, \dot{\delta} = \dot{x}_{max}$ from Eq. (41.8). Substituting these initial conditions, Eq. (41.10) becomes

$$\delta = \left(\frac{\sqrt{2gh}}{\omega_n} + \frac{g}{\omega_n^2} \right) \sin \omega_n t' + \frac{g}{\omega_n^2} \, (1 - \cos \omega_n t') \qquad (41.11)$$

In general, the term g/ω_n^2 is relatively small in packaging applications. Then the term $g/\omega_n^2 \cos \omega_n t'$ may be neglected and the maximum deflection δ_{max}, at time $t' = \pi/(2\omega_n)$, is

$$\delta_{max} = \left(\frac{\sqrt{2gh}}{\omega_n} + \frac{g}{\omega_n^2} \right) + \frac{g}{\omega_n^2} = \left(\frac{\sqrt{2gh}}{\omega_n} + \frac{2g}{\omega_n^2} \right) \qquad \text{in.} \qquad (41.12)$$

The maximum acceleration experienced by the packaged unit is found by differentiating Eq. (41.11)* twice with respect to time, neglecting the term $g(\cos \omega_n t')/\omega_n^2$ and taking the maximum value at time $t' = \pi/(2\omega_n)$:

$$\ddot{\delta}_{max} = \omega_n \sqrt{2gh} + g \qquad \text{in./sec}^2 \qquad (41.13)$$

The maximum acceleration $\ddot{\delta}_{max}$ may be expressed as a dimensionless multiple $(\ddot{\delta}_g)_{max}$ of the gravitational acceleration by dividing Eq. (41.13) by g:

$$(\ddot{\delta}_g)_{max} = \frac{\omega_n \sqrt{2gh}}{g} + 1 \qquad (41.14)$$

Solving Eq. (41.14) for the natural frequency ω_n and dividing the result by 2π to obtain units of cycles per second,

$$f_n = \frac{1}{2\pi} \sqrt{\frac{g[(\ddot{\delta}_g)_{max} - 1]^2}{2h}} \qquad \text{Hz} \qquad (41.15)$$

The relation of Eq. (41.15) is shown graphically by the solid lines in Fig. 41.4.

The relation between maximum deflection δ_{max} and maximum acceleration $(\ddot{\delta}_g)_{max}$ is obtained by combining Eqs. (41.12) and (41.14):

$$(\ddot{\delta}_g)_{max} = \frac{2h + 3 \dfrac{\sqrt{2gh}}{\omega_n} + \dfrac{2g}{\omega_n^2}}{\delta_{max}} \qquad (41.16)$$

For most packaging applications, the first term in the numerator of Eq. (41.16) is much larger than the other terms; thus, it is a reasonable approximation to neglect the latter and write Eq. (41.16) in the following form:

$$(\ddot{\delta}_g)_{max} \simeq \frac{2h}{\delta_{max}} \qquad (41.17)$$

This relation is shown by the dotted lines in Fig. 41.4. For a successful package design, it is necessary that $\ddot{x}_g$ be less than $\ddot{x}_F$.

Example 41.1. Consider an item weighing 25 lb that is to be packaged in such a manner that it does not receive an acceleration greater than $30g$ ($\ddot{x}_F = 30g$) when the package is dropped from a height of 36 in. Determine the spring characteristics

---

* In this portion of the analysis, the container is assumed to rest upon the floor; thus, $\ddot{\delta} = \ddot{x}$ is the acceleration of the packaged item.

NATURAL FREQUENCY, Hz

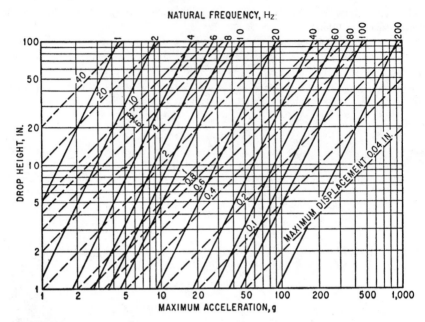

FIGURE 41.4 Natural frequency and maximum displacement of the packaged item in terms of its maximum acceleration and the drop height of the package. These curves are based on Eqs. (41.15) and (41.17).

and maximum displacement required to attain the desired maximum acceleration, and indicate the natural frequency of the packaged item on the isolation system.

Entering Fig. 41.4 at a value of $30g$ on the abscissa and 36 in. on the ordinate, the following values are obtained by interpolation between pairs of solid and dotted diagonal lines:

From the solid lines:

$$\text{Natural frequency} = 10.7 \text{ Hz}$$

From the dotted lines:

$$\text{Maximum spring deflection} = 2.4 \text{ in.}$$

Then, entering Fig. 41.3 at a value of 25 lb on the abscissa scale, and reading on the ordinate scale at the diagonal line for 10.7 Hz:

$$\text{Total spring stiffness} = 300 \text{ lb/in.}$$

INFLUENCE OF DAMPING ON SHOCK ISOLATION

The presence of damping in the isolation medium affects the response to shock. This subject is considered in detail in Chap. 31.

APPLICATIONS OF SHOCK-ISOLATION THEORY

Tension-spring Suspension. When the springs are arranged parallel to the direction of motion, as shown in Fig. 41.2*A*, the stiffness indicated by Fig. 41.3 is used directly in the design of the springs (see Chap. 34 for design data on metal springs, and Chap. 35 for data on rubber springs). The total stiffness is divided by the number of springs to obtain the stiffness per spring. However, if the springs extend between the interior corners of the container and the exterior corners of the packaged item, as indicated in Fig. 41.2*B*, several additional factors must be considered in determining the spring characteristics: (1) the axes of the springs are not parallel with the deflection, (2) the angle of the spring axes varies with deflection, and (3) if tension springs are used, only a portion of the total spring length is active. A detailed analysis of the tension spring package and a step-by-step procedure to determine the spring characteristics are given in Refs. 1 and 2.

Base-mount Package. For large and heavy equipment, it is common practice to mount the equipment on a base which is provided with skids to allow for fork lift entry during handling. The use of a skidded base on a large container dictates handling and transportation of the package with the base down at all times. Then isolation need be provided only for bottom rotational drops, and side and end impacts, as discussed previously under shock environment. An isolation system applicable to such a condition is illustrated in Fig. 41.5. An analysis of a base-mount package with linear isolators for both end impact and rotational drop is included in Chap. 3.

Cushion Design Using Flat Sheets. A common method of packaging embodies the use of an isolation medium in sheet form of standard thicknesses; the sheet is cut into pads that are inserted between container and the packaged item on the several sides thereof. The theory of shock isolation with linear springs is applicable only for relatively small deflection of the isolation medium; for larger deflections, the stiffness becomes nonlinear and a different type of analysis must be used. In general, the stiffness cannot be defined analytically;* rather, the characteristics of the isolation medium are defined by stress-strain curves and by curves of maximum transmitted acceleration as a function of static loading stress for various heights of free fall of the container. These curves are obtained by physical tests of the material and are used to design the isolation medium, using the procedure described in the following paragraphs.

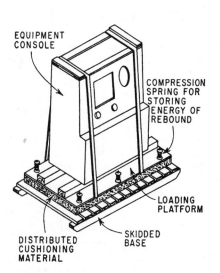

EQUIPMENT
CONSOLE

COMPRESSION
SPRING FOR
STORING
ENERGY OF
REBOUND

LOADING
PLATFORM

SKIDDED
BASE

DISTRIBUTED
CUSHIONING
MATERIAL

FIGURE 41.5 Typical base-mounted package utilizing loading platform and skids. The sides and top of the container are mounted upon the base.

---

* An analytical treatment of shock isolation wherein the characteristics of nonlinear isolators are defined analytically is included in Chap. 31.

For a given isolation medium, the significant parameters in the design of a package are thickness of the medium, height of free fall of the package, unit loading on the medium, and maximum acceleration experienced by the packaged item. Unit loading is defined in terms of static conditions; it is expressed as a "static stress" representing the weight of the packaged item divided by the area of the medium in engagement with one side of the item. Isolation media are commercially available in a sequence of standard thicknesses; for example, thicknesses of 2, 4, and 6 in. are commonly available. Consequently, the design data are given for corresponding thicknesses. Similarly, the height of free fall is defined in discrete increments corresponding to accepted specification practices; the isolation at intermediate heights can be determined by interpolation. The shock isolation properties are defined by several families of curves. For example, Fig. 41.15 shows maximum acceleration of the packaged item as a function of static stress for several thicknesses of "Latex Hair, Firm" when the height of free fall is 12 in. (Corresponding data on the same material for other heights of free fall are given in Figs. 41.16 and 41.17.)

In the design of a package, data of the type given in Figs. 41.15 to 41.32 are applied by selecting the figure corresponding to the applicable height of free fall, entering the figure on the ordinate scale at a value of maximum acceleration corresponding to the fragility of the packaged item, and determining the required static loading from the curves representing various thicknesses of isolation medium. This determines the requirements of the design for purposes of shock isolation; other requirements are set forth below under *Vibration Isolation* and *Properties of Cushioning Materials*. The complete procedure for designing the isolation medium is discussed under *Design Technique—Nonlinear Cushioning*.

Comparison of Flat and Corner Drops. The selection of cushion dimensions is determined from the dynamic properties of cushioning materials. Such properties are determined by dropping an equipment mock-up flat onto cushion specimens. However, package tests require drops on corners and edges. The use of flat-drop data in cushion design has been justified by a series of tests summarized in Fig. 41.6.[3] These tests involved dropping a package consisting of a wooden block cushioned on all six sides and enclosed in an exterior container. Three accelerometers were mounted on the block with their directions of sensitivity along the three mutually perpendicular axes of symmetry of the block. The package was suspended above a rigid surface and its orientation was carefully measured. It was allowed to fall freely and the three accelerometer measurements were recorded. The first series of tests began with a flat drop. Subsequent drops from the same height were made with varying angles of orientation until a corner drop was reached. The angle of orientation was then further varied until the tests were concluded with a drop on an edge of the package. This procedure was repeated using 2- and 3-in. thick cushioning. The second series of tests began from a flat drop, continued to an edge, and concluded with a flat drop on the side 90° from the original impacting side. The acceleration measurements are plotted in Fig. 41.6 with a faired curve for each thickness. The accelerations are the vector resultants of the three accelerometer recordings.

Vibration Isolation. In general, the primary consideration in the design of a cushion system is the protection of the packaged item against shock. However, if the item (or a component of the item) to be packaged has a resonant frequency falling within the range of vibration frequencies encountered during transportation, the vibration-isolation characteristics of the cushion system must be considered.

The theory of vibration isolation is discussed in detail in Chap. 30; equations for transmissibility with several types of damping are given in Table 30.2 and transmissibility with viscous damping is shown graphically in Fig. 2.17. It is highly desirable that the isolator have significant damping, so that the vibration amplitude does not become excessive if environmental conditions include vibration having a frequency equal to the natural frequency of the isolator. Furthermore, it is important that the natural frequency of the isolator be substantially less than the natural frequencies of vulnerable components of the equipment so that the excitation of the components is small.

Example 41.2. Consider the system shown schematically in Fig. 41.7 where m_1 represents the total mass of the packaged item, m_2 represents the mass of an element or component of the equipment (m_2 is assumed small relative to m_1), and $\ddot{u}$ is the acceleration of the vibration associated with the environmental condition. The environment is characterized by vibration having an acceleration amplitude of 1.3g at 6 Hz (the natural frequency of the isolator) and an acceleration amplitude of 3g at 200 Hz (the natural frequency of the element). The fragility $\ddot{x}_F$ of the critical element is

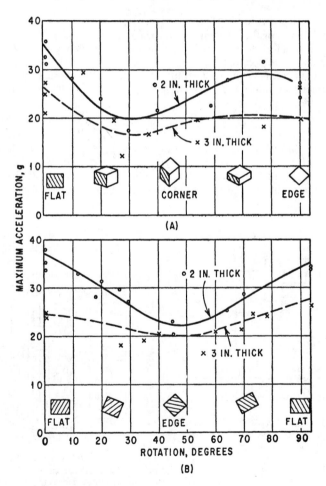

FIGURE 41.6 Effect of container orientation upon maximum acceleration experienced by a packaged item during drop tests: (*A*) sequence of orientations is flat to corner to edge; (*B*) sequence of orientations is flat to edge to flat. (*L. W. Gammel and J. L. Gretz.*[3])

$30g$ and its fraction of critical damping ζ_2 is 0.001. An undamped tension spring package with natural frequency f_1 has been designed. The fraction of critical damping ζ_1 for an undamped tension spring package has been experimentally determined to be 0.004.[4] Determine the response of the critical element at isolator resonance and element resonance.

RESPONSE OF ELEMENT AT ISOLATOR RESONANCE: The transmissibility of the isolator at resonance is defined by $T_R = 1/(2\zeta)$ [see Eq. (30.1)]:

$$T_R = \frac{1}{2 \times 0.004} = 125$$

Then, the acceleration of m_1 at 6 Hz is

$$\ddot{x}_1 = T_R \times \ddot{u} = 125 \times 1.3g = 163g$$

The acceleration $\ddot{x}_2$ of the element is approximately the same as that of the equipment; this is characteristic of elements of high natural frequency (relative to the natural frequency of the isolator).

RESPONSE OF ELEMENT AT ELEMENT RESONANCE: The transmissibility of the isolator at element resonance is $T = 0.0008$ [see Eq. (30.1)]. Then the acceleration of m_1 is

$$\ddot{x}_1 = T \times \ddot{u} = 0.0008 \times 3g = 0.0024g$$

The transmissibility of the element at resonance is $T_R = 1/2\zeta_2$ [see Eq. (30.1)] = 500. Thus, the acceleration of the element is $\ddot{x}_2 = T_R \times \ddot{x}_1 = 500 \times 0.0024g = 1.2g$.

The results calculated above are tabulated in Table 41.2 and plotted in Fig. 41.8. Similar results are included for fractions of critical damping $\zeta_1 = 0.03$ and 0.10 char-

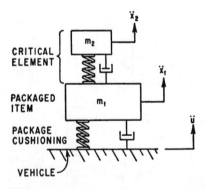

FIGURE 41.7 Model consisting of a damped two degree-of-freedom system to represent a packaged item and a critical element thereof.

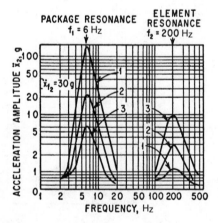

FIGURE 41.8 Maximum acceleration amplitude $\ddot{x}_2$ of critical element m_2 in Fig. 41.7 for the conditions indicated in Example 41.2. The curves refer to different values of the fraction of critical damping ζ_1 for the package cushioning as follows: curve 1: $\zeta_1 = 0.004$; curve 2: $\zeta_1 = 0.03$; curve 3: $\zeta_1 = 0.10$.

TABLE 41.2 Element Acceleration for Tension Spring Suspension Package
(See Example 41.2)

Type of package	Fraction of critical damping	Response at package resonance				Response at element resonance			
		T_p	$\ddot{x}_1, g$	T_e	$\ddot{x}_2, g$	T_p	$\ddot{x}_1, g$	T_e	$\ddot{x}_2, g$
Undamped tension spring	0.004	125	163	1	163	0.0008	0.0024	500	1.2
Damped tension spring	0.03	16.7	21.7	1	21.7	0.002	0.006	500	3.0
Damped tension spring	0.10	5.0	6.5	1	6.5	0.007	0.021	500	10.5

acteristic of a damped spring suspension system. The addition of damping to the tension spring reduces the acceleration experienced by the component as a result of vibration at the resonant frequency of the spring but increases the acceleration resulting from vibration at the resonant frequency of the element.

Damping of Tension Spring Package. One simple and economical means of providing damping is to stuff wool felt into the centers of the spring coils throughout the extended spring length. Fractions of critical damping in the range from 0.03 to 0.10 can be obtained by this method, indicating a range of maximum transmissibility from 16.7 to 5, respectively.

Analytical Determination of Natural Frequency. The relationships between transmissibility and frequency of vibration and between the natural frequency of the cushion system and the static stress on the cushion determine the degree of vibration isolation afforded by a cushion. Because of the many variables and unknowns involved in analyses of the materials commonly used as package cushioning, transmissibility data must be obtained empirically. For example, transmissibility curves for latex hair and polyester urethane foam are given in Figs. 41.37 and 41.38.

The natural frequency of a nonlinear isolator is

$$f_n = \frac{1}{2\pi} \sqrt{\frac{dF/d\delta}{W/g}} \qquad (41.18)$$

where f_n = natural frequency, Hz
F = force, lb
δ = deflection, in.
W = weight supported by isolator, lb
g = gravitational acceleration = 386 in./sec^2

The force F is equal to the weight W and $dF/d\delta$ is the slope of force-displacement curve at the force $F = W$. These may be expressed in terms of stress and strain as follows:

$$\sigma = \frac{F}{S} = \text{stress, lb/in.}^2$$

$$\epsilon = \frac{\delta}{t} = \text{strain, in./in.}$$

where S = cushion area, in.^{2}
t = cushion thickness, in.

Then Eq. (41.18) becomes

$$f_n = \frac{1}{2\pi} \sqrt{\frac{d\sigma/d\epsilon}{Wt/gS}} = 3.13 \sqrt{\frac{d\sigma/d\epsilon}{Wt/S}} \qquad (41.19)$$

Applying the stress-strain relation $d\sigma/d\epsilon$ for a material, Eq. (41.19) gives the relationship between natural frequency f_n and static stress W/S for various cushion thicknesses t. Data giving this relationship for several package cushioning materials are presented in Figs. 41.39 through 41.41.

Equation (41.19) applies only for small vibration amplitudes because $d\sigma/d\epsilon$ does not remain constant throughout a vibration cycle of large amplitude; thus discrepancies between the actual and calculated natural frequencies may occur for large vibration amplitudes. Most materials also exhibit some difference between static and dynamic stiffness. Testing demonstrates fair correlation between actual and calculated values of natural frequency for small vibration displacement amplitudes; however, discrepancies may occur if large displacements are involved. Thus, the curves of static stress vs. strain should be used only to obtain the first approximation to the natural frequency of the isolator.

PROPERTIES OF CUSHIONING MATERIALS

This section considers cushioning materials such as latex hair, various plastic foams etc., which are supplied either in sheet form or in molded shapes; as package cushioning they are interposed between the container and the packaged item. Although such materials may exhibit linear force-deflection characteristics for small deflections, efficient package design involves large deflections and consequent nonlinearity of the cushioning materials.

MATERIAL CLASSIFICATION

All materials utilized to provide cushioning (i.e., shock and vibration isolation) in packaging may be classified as either of two general types—*elastic* (flexible or resilient) and *nonelastic* (crushable). Materials which, when deflected during impact, sustain no more than an arbitrarily selected amount of permanent deformation are placed in the first category; materials that sustain a greater amount are placed in the second category. Certain factors influence the choice of this arbitrarily selected level. One of the most important of these is that a sufficient thickness of material should be left after deformation to provide the required isolation on successive impacts. A permanent deformation of not more than 10 percent after compression to a strain of 65 percent is used as a criterion in the package design field.

Materials described in this section may be used to satisfy package design requirements other than those for cushioning, for example, to position an item within the package.

MATERIAL PROPERTIES

The fundamental property to be considered in application of a material in shock and vibration isolation is the manner in which it stores, absorbs, and dissipates energy. This characteristic is basic to the classification of materials set forth in the preceding paragraph.

Package cushioning may be designed analytically if an equation can be written for the force-deflection characteristics under dynamic conditions[1] (see Chap. 31). A quasi-analytical concept minimizes the ratio of transmitted force to stored energy.[5] The most directly useful method makes use of experimentally determined properties of the cushioning materials in a straightforward design procedure, as described in subsequent sections of this chapter.

Properties for Isolation.   The properties of materials that are most useful in packaging design for shock isolation have been standardized as follows:

1. Compression stress vs. strain characteristics
2. Maximum acceleration vs. static stress characteristics, employing drop height and material thickness as parameters
3. Creep characteristics

Requirements covering these properties for elastic-type cushioning materials are set forth in a military specification on elastic-type cushioning materials.[6] Methods of testing for determination of these properties are detailed in Refs. 6 and 7.

Data of the types listed above are presented in this chapter for a number of specific materials commonly used for package cushioning. Static stress vs. strain characteristics are given in Figs. 41.9 to 41.14; these data are used to determine the degree of static compression of the cushion due to the weight of the item to be packaged. Maximum acceleration vs. static stress data are given in Figs. 41.15 to 41.32; the dynamic shock loading on the packaged item for various package drop heights can be determined from these data. Creep characteristics are given in Figs. 41.33 to 41.35; these data are used to determine the thickness loss of the cushion during storage. Because of wide differences in manufacturing processes and techniques associated with production of package cushioning materials, and the large number of manufacturers of each type of material, a great variation exists in the characteristics of commercially available materials. For this reason, the data presented here represent only an average for a particular material and condition.*

Description of Materials.   This section includes a brief physical description of the various materials covered in this chapter, including their uses and characteristics.

Latex Hair.   Curled animal hair (mostly from hogs but with some cattle tail and horse mane hair), bonded with natural latex (in molded shapes) or neoprene (in sheet form); this material is most efficiently used to cushion items at static stresses up to approximately 0.3 lb/in.^{2} and is available in various densities (normally referred to as soft, medium, and firm).

Urethane Foam.   An open-cell plastic foam available in various densities, depending on the manufacturer; two forms of this material are available—one derived from a polyester and one from a polyether. They have somewhat different

---

* The data used to plot Figs. 41.9 to 41.41 inclusive were made available through the courtesy of Rockwell International, Inc. and Ref. 8.

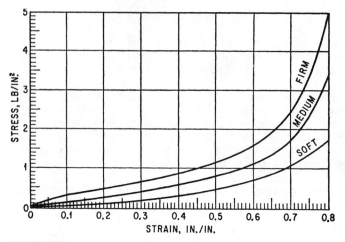

FIGURE 41.9   Compression stress vs. strain for latex hair.

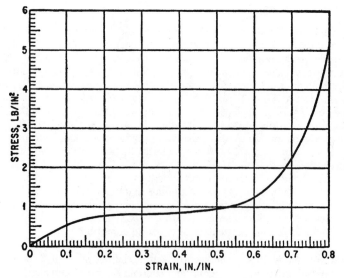

FIGURE 41.10   Compression stress vs. strain for polyester urethane foam (2.2 lb/ft³).

static stress-strain characteristics. Most commercially available materials are useful in approximately the same static stress range as latex hair, but some are available for use at a static stress as great as approximately 1.0 lb/in.².

    Polyethylene Foam.   A blown plastic foam, it is manufactured in various densities (approximately 2.0 to 9.0 lb/ft³). This is a stiff material but is very efficient in cushioning applications at static stresses from approximately 1.0 to 2.0 lb/in.²; in other applications the material can be cut very thin and used as a protective wrap.

    Cellulose Wadding.   A wood-fiber product most efficiently used as a protective wrap, but also used in some applications as cushioning where a static stress of less than 0.1 lb/in.^{2} is required. For cushioning purposes, this material should not be used on extremely delicate items.

AirCap. A laminate of two layers of barrier-coated polyethylene film, one layer of which is embossed with rows of cells forming an air encapsulated cushioning material.

Temperature Effects on Isolation. The effects of temperature, both high and low, on the isolation characteristics of materials must be considered in the selection of the proper material for use as package cushioning. Most materials in common usage for package cushioning exhibit a marked increase in stiffness as the temperature decreases from 70°F (21.1°C), and a corresponding decrease in stiffness as the temperature is increased from 70°F. The variation in stiffness with temperature is considerably more pronounced in some materials than in others. For example, fibrous glass cushioning exhibits very little change in stiffness with temperature, whereas the stiffness of polyester urethane foam increases rapidly as the temperature decreases. The change in the stiffness of latex-bound hair with variations in temperature may be considered approximately average for most cushioning materials. Because of wide variations in procedures and equipment used in testing to determine the effects of temperature on the isolation characteristics of cushioning mate-

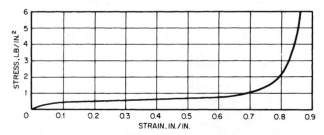

FIGURE 41.11 Compression stress vs. strain for polyether urethane foam (2.0 lb/ft^3).

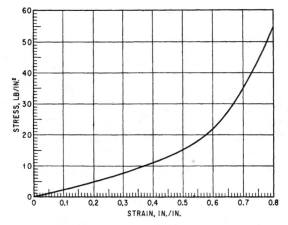

FIGURE 41.12 Compression stress vs. strain for polyethylene foam (1.75 lb/ft^3).

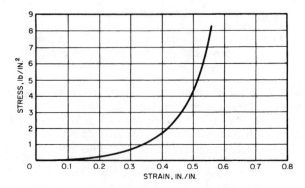

FIGURE 41.13 Compression stress vs. strain for cellulose wadding.

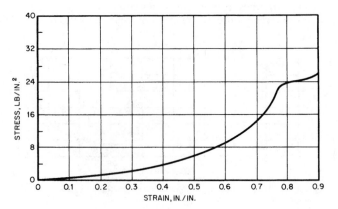

FIGURE 41.14 Compression stress vs. strain for "AirCap" SD 240.

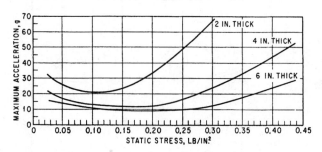

FIGURE 41.15 Maximum acceleration vs. static stress for latex hair, firm—12-in. drop.

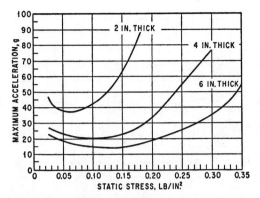

FIGURE 41.16 Maximum acceleration vs. static stress for latex hair, firm—24-in. drop.

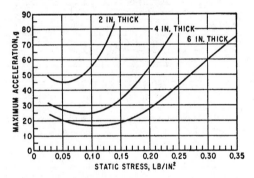

FIGURE 41.17 Maximum acceleration vs. static stress for latex hair, firm—30-in. drop.

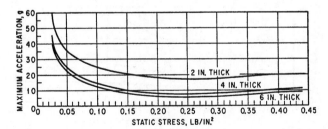

FIGURE 41.18 Maximum acceleration vs. static stress for polyester urethane (2.2 lb/ft^3)—12-in. drop.

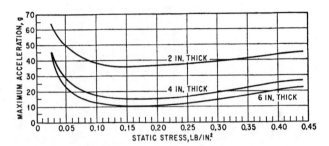

FIGURE 41.19 Maximum acceleration vs. static stress for polyester urethane (2.2 lb/ft³)—24-in. drop.

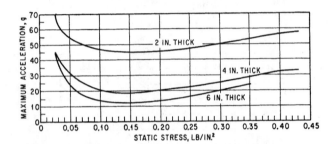

FIGURE 41.20 Maximum acceleration vs. static stress for polyester urethane (2.2 lb/ft³)—30-in. drop.

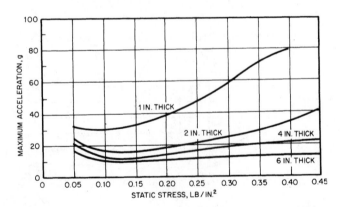

FIGURE 41.21 Maximum acceleration vs. static stress for polyether urethane (2.0 lb/ft³)—12-in. drop.

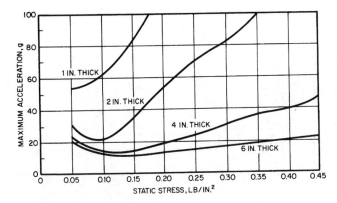

FIGURE 41.22   Maximum acceleration vs. static stress for polyether ure-thane (2.0 lb/ft³)—24-in. drop.

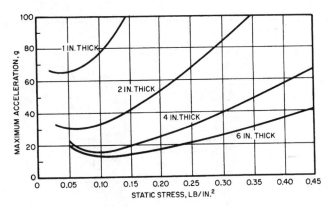

FIGURE 41.23   Maximum acceleration vs. static stress for polyether ure-thane (2.0 lb/ft³)—30-in. drop.

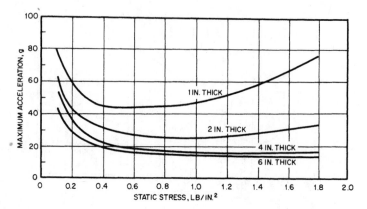

FIGURE 41.24 Maximum acceleration vs. static stress for polyethylene foam (2.0 lb/ft^3)—12-in. drop.

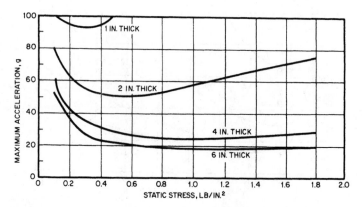

FIGURE 41.25 Maximum acceleration vs. static stress for polyethylene foam (2.0 lb/ft^3)—24-in. drop.

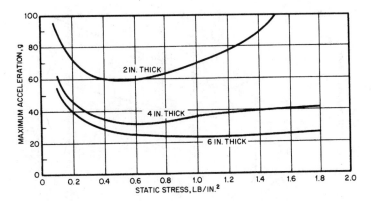

FIGURE 41.26 Maximum acceleration vs. static stress for polyethylene foam (2.0 lb/ft³)—30-in. drop.

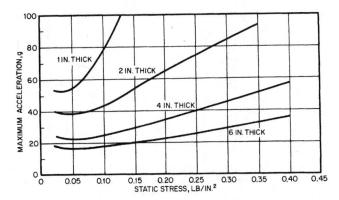

FIGURE 41.27 Maximum acceleration vs. static stress for cellulose wadding—12-in. drop.

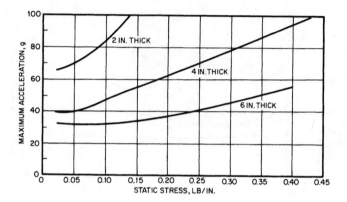

FIGURE 41.28 Maximum acceleration vs. static stress for cellulose wadding—24-in. drop.

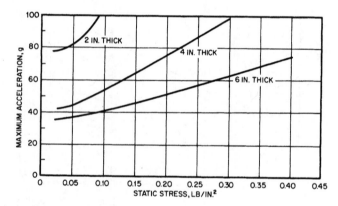

FIGURE 41.29 Maximum acceleration vs. static stress for cellulose wadding—30-in. drop.

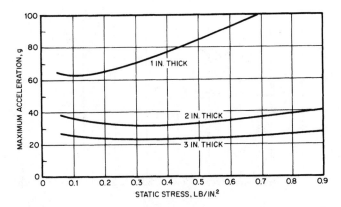

FIGURE 41.30 Maximum acceleration vs. static stress for "AirCap" SD 240—12-in. drop.

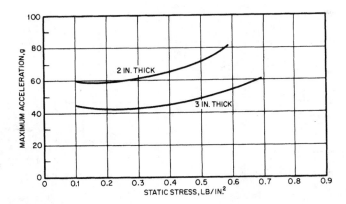

FIGURE 41.31 Maximum acceleration vs. static stress for "AirCap" SD 240—24-in. drop.

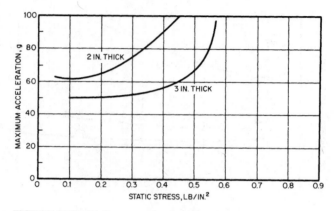

FIGURE 41.32 Maximum acceleration vs. static stress for "AirCap" SD 240—30-in. drop.

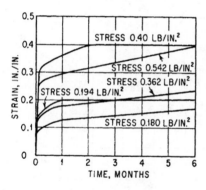

FIGURE 41.33 Creep characteristics for latex hair, firm.

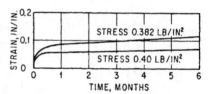

FIGURE 41.34 Creep characteristics for polyester urethane (2.2 lb/ft³).

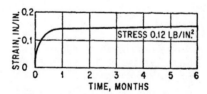

FIGURE 41.35 Creep characteristics for polyether urethane (2.2 lb/ft³).

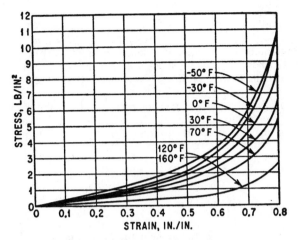

FIGURE 41.36 Compression stress vs. strain for latex hair, firm, at various temperatures.

rials, some published data of this type are not reliable. For this reason, only data on the effects of temperature on the stress vs. strain relationship for latex hair are presented here (see Fig. 41.36) as an indication of material performance.

DESIGN TECHNIQUE—NONLINEAR CUSHIONING

The following procedure can be used as a guide in the development of design techniques utilizing distributed cushioning material for shock and vibration isolation:

1. Define the shock environment in terms of drop height; define the vibration environment in terms of frequency and amplitude.

2. Establish the fragility $\ddot{x}_F$ of the packaged item (see *Design for Shock Conditions,* Chap. 42) and the characteristics (natural frequency and damping) of the critical components of the item.

3. Provide protection for the packaged item by means such as blocking and bracing of protruding parts, distribution of load to avoid stress concentrations, and preservation methods. (Interior containers adapted to house the packaged item may serve as protection when the item is rigidly secured within, thus approximating a single degree-of-freedom system.)

4. From the weight and size of the packaged item (or interior container plus contents), determine the static stress for each face; then from the maximum acceleration vs. static stress curves (Figs. 41.15 to 41.32) select the optimum material and the required thickness.

5. Check the static stress vs. strain curve (Figs. 41.9 to 41.14) for the applicable material to determine the static deflection.

6. Check the creep properties (Figs. 41.33 to 41.35) and, when applicable, make allowance for any loss in thickness during storage. The required initial thickness is

$$t_{\text{req}} = \frac{t_s}{1 - C} \qquad (41.20)$$

where t_{req} = initial thickness required to compensate for creep, in.
   t_s = thickness required for shock isolation, in.
   C = drift, the ratio of thickness loss from creep to original thickness

7. To avoid localized buckling of the isolation medium, stability must be considered so that the cushion returns to its original position to provide protection for more than a single drop. A rule-of-thumb stability relation requires that

$$\frac{t}{\sqrt{S}} \leq \frac{4}{3} \qquad (41.21)$$

where t = thickness of cushion pad, in.
   S = area of cushion pad, in.^{2}

8. For vibration isolation, determine the natural frequency of the isolation system (Figs. 41.39 to 41.41) and, from the transmissibility curves of Figs. 41.37 and 41.38, calculate the vibration of the packaged item at the resonant frequencies of the isolation system and the critical component.

9. Select a method of cushioning to provide the exact amount of material required by the analysis. Wrapping a part with the material or use of other methods which

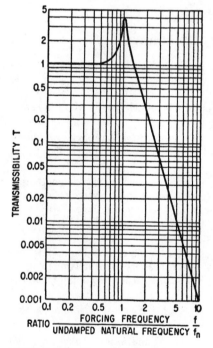

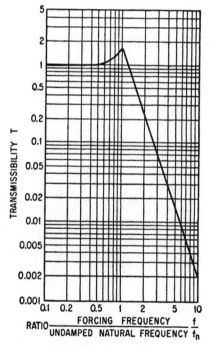

FIGURE 41.37 Transmissibility curve for latex hair, firm.

FIGURE 41.38 Transmissibility curve for polyester urethane (2.2 lb/ft^3).

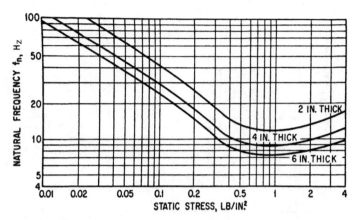

FIGURE 41.39 Natural frequency vs. static stress for latex hair, firm.

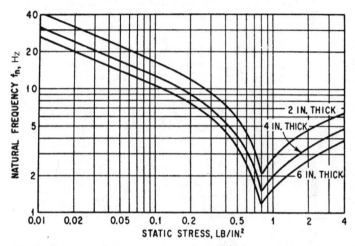

FIGURE 41.40 Natural frequency vs. static stress for polyester urethane (2.2 lb/ft^3).

provide a larger effective cushion area than required may be detrimental because the static stress is decreased and the maximum acceleration may be increased.

10. Select the outer container dimensions to provide the proper thickness of material and sufficient clearances to obtain the required deflection.

Example 41.3. A packaged item is capable of withstanding a maximum acceleration of 30g without damage. A critical component in this item has a natural frequency of 250 Hz and an equivalent fraction of critical damping of 0.001. A criterion for rough handling has been established as a drop from a height of 30 in. The anticipated transportation environment has the following vibration characteristics: Displacement amplitude of 0.5 in. from 2 to 5 Hz, and acceleration amplitude of 3g from

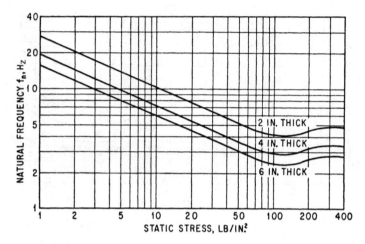

FIGURE 41.41 Natural frequency vs. static stress for polyethylene foam (1.75 lb/ft³).

5 to 300 Hz. The storage period is 6 months. An interior container with proper blocking and bracing, and necessary preservation requirements, has the following dimensions: Length, 24 in.; width, 18 in.; and height, 12 in. The gross weight of the interior container plus contents is 40 lb.

1. From the dimensions of the interior container and its gross weight, the available cushioning area and static stress for each container face are determined:

Container face	Available area, in.²	Static stress, lb/in.²
Top and bottom	24 × 18 = 432	0.0925
Sides	24 × 12 = 288	0.139
Ends	18 × 12 = 216	0.185

2. For this application, latex hair (firm) is to be used. From Fig. 41.17, for a 30-in. drop and 30g maximum acceleration, the thickness required is 3½ in. for top and bottom, 4 in. for the sides, and 5½ in. for the ends. Adjustment of the static stress on all faces to 0.0925 lb/in.² (total area on each face = 432 in.²) yields optimum design, resulting in a thickness of 3½ in. on each container face. The area on each face can be made equal by overlapping the cushioning materials in the manner illustrated in Fig. 41.42.

3. A static stress of 0.0925 lb/in.² produces a strain of approximately 0.025 (from Fig. 41.9), or a static deflection of 0.0875 in. This relatively small deflection does not significantly change the nominal dimensions of the package.

4. From Fig. 41.33, a static stress of 0.18 lb/in.² for a duration of 6 months results in an increase in strain of 0.167 in./in., i.e., in a thickness loss of approximately 16.7 percent as a result of creep. Assuming a linear relationship between stress and drift, the expected drift after 6 months under load at a static stress of 0.0925 lb/in.²

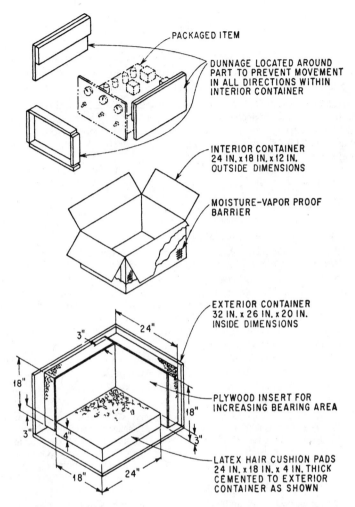

PACKAGED ITEM

DUNNAGE LOCATED AROUND
PART TO PREVENT MOVEMENT
IN ALL DIRECTIONS WITHIN
INTERIOR CONTAINER

INTERIOR CONTAINER
24 IN. x 18 IN. x 12 IN.
OUTSIDE DIMENSIONS

MOISTURE-VAPOR PROOF
BARRIER

EXTERIOR CONTAINER
32 IN. x 26 IN. x 20 IN.
INSIDE DIMENSIONS

PLYWOOD INSERT FOR
INCREASING BEARING AREA

LATEX HAIR CUSHION PADS
24 IN. x 18 IN. x 4 IN. THICK
CEMENTED TO EXTERIOR
CONTAINER AS SHOWN

FIGURE 41.42   Package cushioning design using distributed cushioning on six sides.

is approximately 8 percent. Then, from Eq. (41.20), the initial thickness required to compensate for drift is

$$t_{\text{req}} = \frac{3.5}{1 - 0.08} = 3.8 \text{ in.}$$

Rounding this result off to the nearest nominal value, cushioning material 4 in. thick should be used.

5. Checking the stability by substituting the thickness and area in Eq. (41.21),

$$\frac{t}{\sqrt{S}} = \frac{4}{\sqrt{432}} = 0.193$$

This is less than ⅖, and the design is not susceptible to localized buckling.

6. From Fig. 41.39, the natural frequency of the isolation system for a static stress of 0.0925 lb/in.^{2} and 4-in. thickness is 30 Hz.

7. At "package resonance," the impressed acceleration is $3g$ and (from Fig. 41.37) the interior container acceleration is $4 \times 3g = 12g$. The acceleration of the element also is $12g$. At resonance of the element, the acceleration of the interior container is $0.002 \times 3g = 0.006g$, and the acceleration of the element is $500 \times 0.006g = 3g$.

REFERENCES

1. Mindlin, R. D.: *Bell System Tech. J.,* **24:**353 (1945).

2. Franklin, P. E.: "Packaging for Shock and Vibration Protection with Tension Spring Suspensions," *Missile Division Rept.* AL-2608, Contract AF33(600)-28469, North American Aviation, Inc., 1957.

3. Gammell, L. W., and J. L. Gretz: "Report on Effect of Drop Test Orientation on Impact Accelerations," Physical Test Laboratory, Texfoam Division, B. F. Goodrich Sponge Products Division of B. F. Goodrich Company, Shelton, Conn., 1955.

4. Hatae, M. T.: "Shock and Vibration Tests of a Tension Spring Package," *Rept.* NA-54-1003, North American Aviation, Inc., 1954.

5. Janssen, R. R.: "A Method for the Proper Selection of a Package Cushion Material and Its Dimensions," *Rept.* NA-51-1004, North American Aviation, Inc., 1952.

6. "Military Specification—Cushioning Material, Elastic Type General," *USAF Specification* MIL-C-26861, 1959.

7. ASTM Test Method D1596, Dynamic Shock Cushioning Characteristics of Packaging Materials; ASTM Test Method D2221, Creep Properties of Package Cushioning Materials; ASTM Test Method D3332, Mechanical Shock Fragility of Products Using Shock Machines; ASTM Test Method D4168, Transmitted Shock Characteristics of Foam-in-Place Cushioning Materials; ASTM Test Method D4728, Random Vibration Testing of Shipping Containers. These ASTM Standards may be obtained from the American Society for Testing and Materials, 1916 Race St., Philadelphia, PA 19103.

8. "Military Standardization Handbook-Package Cushioning Design," *USAF Handbook* MIL-HDBK-304A, Sept. 25, 1974.

CHAPTER 42
THEORY OF EQUIPMENT DESIGN

Edward G. Fischer

INTRODUCTION

The design of equipment to withstand dynamic loads is one of the more difficult and less developed aspects of shock and vibration control. The difficulty is inherent, in part, in the physical conditions because the nature and magnitude of such loads depend upon the dynamic characteristics of the equipment; thus, the loads cannot be determined until the equipment has been designed. Therefore, the designer must design for loads that are necessarily unknown. Such a problem is approached logically by first preparing an initial design based upon the best judgment of the designer. A subsequent analysis would show weaknesses and other undesirable features of the design and would point to desirable modifications. Thus, the design would be developed by cut-and-try methods.

Design by cut-and-try methods using only analytical procedures is not profitable in general because several important considerations are difficult to evaluate. For example, even though the dynamic loads could be determined by analysis, it is difficult to determine the resulting stresses if the equipment has an ordinary degree of complexity. Furthermore, it is difficult to evaluate the properties of materials under dynamic strain. Accordingly, it is preferred design practice to calculate strength requirements for only major and important structural members of the equipment. An initial prototype equipment is then constructed for test in which actual operating conditions preferably are duplicated. For military equipment, this is generally impractical, and laboratory tests are used to simulate actual operating conditions. Neither designing nor testing to withstand dynamic loads is an exact science; hence, discrepancies frequently occur and must be resolved by experience and intuition. This chapter offers such guidance as the state of the art permits.

It is important to consider the design of the initial prototype from both the analytical and practical points of view. Much can be done at that stage of the design. Optimum stiffnesses can be determined for the interrelated structural members of the equipment with the objective of minimizing the stresses in such members. The forces experienced by components of the equipment also are minimized. The required strengths of these members then can be calculated. The succeeding sections of this chapter set forth procedures for determining strengths and stiffnesses of principal structural members and outline applicable information on the selection and use of engineering materials in equipment that is required to withstand shock and vibration. A number of techniques, generally qualitative in nature, have been developed as a result of much practical experience in design. These are summarized in Chap. 35.

An important part of the design process is redesign to correct the deficiencies revealed by tests of the prototype. Such tests reveal not only the weaknesses of the equipment but also useful information, for example, on the natural frequencies of structural elements or the degree of damping present. With this information, new analyses can be carried out or original analyses modified using actual rather than assumed information.

Although methods for the design of equipment to withstand dynamic loads are not well developed in general, it is possible to apply known methods in particular cases. Some of these cases are discussed in other chapters. The design of internal-combustion engines and other types of rotating and reciprocating machinery is discussed in Chap. 38. Dynamic absorbers and auxiliary mass dampers are discussed in Chap. 6, and the use of damping materials is discussed in Chap. 37. An analysis of self-excited vibration is included in Chap. 5, and the isolation of vibration and shock is covered in Chaps. 30 to 33.

Although the connotation of the words *shock* and *vibration* is somewhat broader, this chapter is limited to the design of equipment that must withstand the motion of the support to which the equipment is attached. The motion may be of a continuing oscillatory nature, known broadly as vibration, or of a transient nature involving acceleration and/or displacements of large magnitude.

DYNAMIC LOAD

When an equipment is subjected to shock or vibration, the structural members of the equipment deflect or deform. The deflection is said to be caused by the *dynamic load;* i.e., the force required to cause structural members of the equipment to deflect in response to the motion imposed by the support. For practical purposes, the dynamic load may be considered as the static or dead-weight load multiplied by a *dynamic load factor.* The members must be designed to withstand, without failure, the stress imposed by the dynamic load. In general, the stress in the member subjected to a dynamic load is equal to that caused by a static load of the same magnitude. It does not necessarily follow, however, that a design stress which applies to static conditions also applies to dynamic conditions. For example, a structure designed to withstand only a static load may be satisfactory if the stress does not exceed a certain percentage of the ultimate stress. On the other hand, a structure may be unsatisfactory under a dynamic load inducing the same stress level because the structure may fail under repeated occurrences of the stress. In other words, if the dynamic load is repeated many times, a lower value of acceptable stress must be used to recognize the effect of fatigue of the material.

The magnitudes of the dynamic loads in an equipment subjected to shock and vibration usually are large relative to the static loads; thus, it is common practice in design to neglect static loads unless they are known to be significant relative to dynamic loads.

The physically observable effect of the application of a dynamic load is the deflection of a structural member. To require that the member have an acceptably low stress as a consequence of such deflection would introduce an uncommon concept in design because structures, in general, are designed to withstand designated *loads*. By considering the dynamic load to be equivalent to a static load that would cause the same deflection, the designer acquires a familiar working tool to use to apply the principles of static loading. The process of designing for shock and vibration consists of (1) determining the magnitudes and directions of the dynamic loads and (2) designing the structure to withstand a hypothetical static load equivalent to the dynamic load, keeping in mind certain qualifying conditions related to the stress level and the distribution of stress.

METHODS OF ANALYSIS

USE OF MODELS

The first step in carrying out an analysis involving dynamic loading is to represent the structure by an *equivalent model* of rigid masses connected by massless springs and dampers. Theoretically, it is then possible to determine from any known excitation the motion of each mass and the consequent deflection of each spring. The force acting on each spring is known directly from its deflection and stiffness; thus, the springs can be designed with the strength required to withstand such forces. Therefore, it may be said that the equipment has adequate structural strength to withstand the excitation.

The type of model used for the analysis depends upon the accuracy required in the results and the effort that can be applied to the analysis. For example, consider the problem of designing the simple beam supporting a rigid body of weight W as shown in Fig. 42.1. If the mass of the beam is small relative to that of the rigid body, a model that is adequate for most purposes is illustrated in Fig. 42.2 where the stiffness k and the damper c represent corresponding properties of the beam. However, if the mass of the beam is of the same order of magnitude as the mass of the supported body, parts of the beam tend to vibrate with respect to each other. This introduces dynamic loads that cannot be neglected if accurate results are desired. A model that includes the mass of the beam is shown in Fig. 42.3 where the sum of the incremental masses m' equals the total mass of the spring.

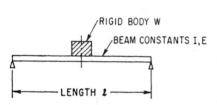

FIGURE 42.1 Rigid body supported by an elastic beam with hinged ends.

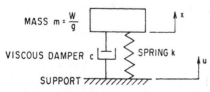

FIGURE 42.2 Single degree-of-freedom system model of beam in Fig. 42.1, neglecting mass of beam.

A model may be constructed with any degree of complexity required. For example, the model of an automobile in Fig. 42.4 could be used to determine the effect of shock absorbers on riding comfort or to calculate the maximum force applied to the seat springs. In this instance, the discrete masses of the model correspond to similarly discrete bodies of the actual automobile. By contrast, a model suitable for investigating the strength and stiffness of the guided missile of

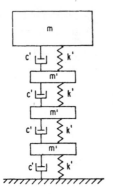

FIGURE 42.3 Model of beam in Fig. 42.1. The distributed mass effect of the beam can be included in the model where the sum of the incremental masses m' equals the total mass effect of the spring.

Fig. 42.5A under bending loads would appear as in Fig. 42.5B. The total mass of the missile, distributed more or less uniformly along its length, is represented by the sum of the several rigid masses of the model; the stiffness of the missile in bending is represented by the stiffnesses in bending of the members connecting the masses of the model. As an objective in setting up the model, the natural frequencies of model and actual missile should agree.

In general, complex models are useful only for investigating the capability of a particular equipment to withstand a particular excitation. However, the development of high-speed digital computer facilities has made it practical to investigate the dynamic response of a large number of possible combinations of masses, springs, and dampers in a complex model and eventually draw general conclusions. It must be cautioned that in order for the model to be authentic, the computer programming should first

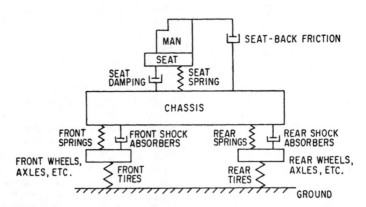

FIGURE 42.4 Model of an automobile. The discrete masses of the model correspond to similarly discrete bodies of an actual automobile.

demonstrate that it can reproduce the results of simple static deflection and snap-back natural frequency tests of the original physical system. There is almost no limit to the degree of sophistication (i.e., yielding elements, random excitation, etc.) that can be achieved with computer-aided analyses, provided that the results make sense in terms of engineering experience.

A model may be used to study equipment design problems arising from any type of dynamic excitation. The general designation *shock and vibration* used to describe an environment is separable for purposes of analysis into (1) periodic vibration, (2) tran-

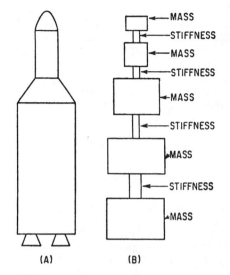

(A)   **(B)**

FIGURE 42.5 (*A*) Schematic diagram of a missile with the total mass distributed more or less uniformly along its length. (*B*) Model of the missile wherein the total mass is subdivided into a series of masses connected by members having the appropriate stiffnesses in bending.

sient vibration or shock, and (3) random vibration. The response of mechanical systems (a *system* as discussed generally in other chapters is equivalent to a *model* in this chapter) to these various classes of environmental conditions is discussed in several other chapters as follows:

Periodic vibration: Chaps. 2, 6, and 22.

Transient vibration or shock: Chaps. 8 and 23.

Random vibration: Chaps. 11 and 22.

The design of the model and the interpretation of the response are discussed in succeeding sections of this chapter.

Types of Models. The usefulness of a model for initial design purposes tends to increase as its simplicity increases. A model that attempts to simulate an equipment in detail cannot be constructed until after an equipment is designed; thus, it becomes the means to check a design rather than to implement the design initially. On the other hand, a less complex model can simulate primary or important structures in a more general sense and can be evaluated in more general terms. For example, the model shown in Fig. 42.2 is defined completely by its natural frequency and fraction of critical damping; its response to any given excitation can be expressed concisely in terms of these parameters. By contrast, the model shown in Fig. 42.4 involves five masses and five springs; the relatively large number of possible permutations demonstrates the impracticality of expressing the results concisely or in a form suitable for ready reference. Techniques for the application of models in design are described below under *Structural Design Considerations* for various types of excitation.

    Single Degree-of-Freedom Model. The model shown in Fig. 42.2 is a spring-mass-damper system where the rigid mass $m = W/g$ is assumed to move along a straight line. The mass can be located at any time t by the single coordinate $x;$ therefore the system is referred to as having a single degree-of-freedom. When the mass is displaced from its equilibrium position, it has only one mode of vibration; this occurs in free vibration at the single natural frequency of the system. The massless spring has a linear force-deflection relationship defined by the spring constant k. Similarly, the massless damper exhibits a linear force-velocity relationship denoted by the damping coefficient c. The damping in a typical mechanical system usually is so small that it has a negligible effect on the value of the natural frequency. Hence, the free vibration of the mass in the single degree-of-freedom model can be considered to be a simple harmonic motion. The undamped natural frequency from Eq. (2.8) is

$$\omega_n = \sqrt{\frac{kg}{W}} = \sqrt{\frac{g}{\delta_{st}}} \qquad \text{rad/sec} \qquad (42.1)$$

where δ_{st} is the static deflection of the spring caused by the weight W.

The analogy between the structure of Fig. 42.1 and the model of Fig. 42.2 is established as follows: The rigid body of weight W is supported at the mid-position of the elastic beam of length l and moment of inertia (of the cross-sectional area) I; the modulus of elasticity of the beam material is E. The single degree-of-freedom model of Fig. 42.2 is applicable only when the weight of the beam is small compared to that of the rigid body W. The static deflection directly under the body W is[1] $\delta_{st} = Wl^3/48EI$. The spring constant of the elastic beam is $k = W/\delta_{st}$ which, when substituted into the first expression of Eq. (42.1), gives the natural frequency for lateral vibration of the beam carrying the weight W:

$$\omega_n = \sqrt{\frac{48EIg}{Wl^3}} \qquad \text{rad/sec} \qquad (42.2)$$

The model of Fig. 42.2 represents the beam of Fig. 42.1 when their natural frequencies are equal.

The model shown in Fig. 42.2 can be used to represent many different types of structures, generally beams or plates. The natural frequencies of such structures can be calculated from the analyses in Chap. 7 or the tables of formulas in Chap. 1. If the structure carries a load that may be considered a rigid body whose mass is large relative to the mass of the structure, the single degree-of-freedom model of Fig. 42.2 is likely to be quite satisfactory.

If the load carried by the structure is relatively small compared with the mass of the structure, the model of Fig. 42.2 cannot be used in evaluating the actual equipment when the higher modes of vibration are of interest. The multiple degree-of-freedom model based upon the normal mode concept (see *Multiple Degree-of-Freedom Model* below) then becomes applicable. However, if only the fundamental mode of vibration is of interest, a portion of the weight of the beam may be lumped with the load and treated as a single degree-of-freedom system (see Table 7.2).

Two Degree-of-Freedom Model. When the equipment includes two structures joined together in such a way that the vibration of one is dependent upon the vibra-

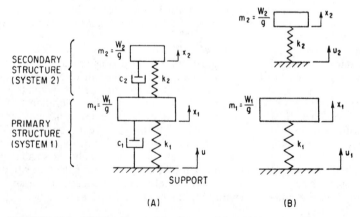

FIGURE 42.6 (*A*) Two degree-of-freedom model consisting of a primary structure and a secondary structure, and (*B*) the uncoupled primary and secondary structures.

tion of the other, the single degree-of-freedom model of Fig. 42.2 may not be appropriate. The two degree-of-freedom model of Fig. 42.6A then may be used. This simulates, for example, the structure of Fig. 42.7 where a load of mass m_2 supported by a

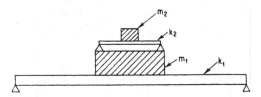

FIGURE 42.7 Assembly of beams to which a two degree-of-freedom model is applicable.

beam of stiffness k_2 is mounted upon a load of mass m_1 that in turn is supported by a beam of stiffness k_1. One type of analysis applies if the mass m_2 is small relative to the mass m_1 (uncoupled two degree-of-freedom model), and another type of analysis applies if the two masses are of the same order of size (coupled two degree-of-freedom model). These two types of analysis are discussed separately under *Design for Vibration Conditions* and *Design for Shock Conditions.*

In the model of Fig. 42.6A, two spring-mass-damper systems are connected in a series system where the rigid masses m_2 and m_1 are both assumed to move along the same vertical line. The individual masses can be located at any time t by the two coordinates x_2 and x_1; thus, the system is referred to as having two degrees-of-freedom. The massless springs have linear force-deflection relationships denoted by the spring constants k_2 and k_1. Similarly, the massless dampers exhibit linear force-velocity relationships denoted by the damping coefficients c_2 and c_1. The system consisting of k_1, m_1, c_1 is designated the *primary structure* because, generally, it represents a principal structure of the equipment or a major component thereof. Similarly, the system consisting of k_2, m_2, c_2 is designated the *secondary structure* because it represents a structure or component to which the excitation is transmitted by the primary structure. In some of the analyses that follow, the equations are simplified by omitting the terms for damping forces.

When the two masses m_2 and m_1 are arbitrarily displaced from their equilibrium positions and released, the system shown in Fig. 42.6A exhibits a complicated motion composed of two superimposed free vibrations at two different (*coupled*) natural frequencies. In the mode of vibration at the lower of these frequencies, the masses m_2 and m_1 vibrate in phase; in the mode at the higher frequency of vibration, the two masses vibrate in opposition to each other. In general, the two natural frequencies can be determined as follows:[2] The differential equation of motion is derived for each mass in the system. Then, by means of the operational calculus, these equations are converted into two simultaneous algebraic equations. The determinant of the coefficients of x_2 and x_1 gives the characteristic equation which is solved for the unknown natural frequencies of the system. Sometimes it is more fruitful to discuss the analytical results in terms of the *uncoupled* natural frequencies defined according to Eq. (42.1) for the simple spring-mass models in Fig. 42.6B: $\omega_2 = \sqrt{k_2/m_2}$ and $\omega_1 = \sqrt{k_1/m_1}$. In the following analyses the uncoupled natural frequencies are used in place of the more complicated expressions for the actual natural frequencies.

Multiple Degree-of-Freedom Model. Models with more than two degrees-of-freedom also may be used as a basis for design to withstand shock and vibration.

Such models are used to represent equipment that may be idealized by several rigid masses and massless springs or, alternatively, by a continuous member whose vibration as an elastic body is significant. The response of such a model to excitation occurs in its normal modes; vibration in each of such modes occurs independently of vibration in the other normal modes and can be evaluated as the vibration of a single degree-of-freedom system. The fundamental concept of normal modes is discussed in Chap. 2. The vibration of an elastic body in terms of its normal modes is considered in Chap. 7.

DESIGN FOR VIBRATION CONDITIONS

The most important requirement in the design of equipment to withstand vibration is to avoid the large amplitude of vibration that accompanies a condition of resonance. Experience indicates that vibration troubles, whether leading to failure or improper operation of equipment or to personal discomfort, usually result from a condition of resonance. The excitation may be either that generated by normal operation of the machine or that coming from the environment in which the machine operates. The discussion in this chapter is limited to the design of equipment to withstand environmental conditions.

Techniques for minimizing the relatively large amplitude of vibration that occurs at the condition of resonance depend upon the nature of the excitation:

1. If the excitation is a continuing periodic vibration at a nonchanging frequency or frequencies, it is preferable that all natural frequencies of the equipment occur at other than the excitation frequencies.

2. If the excitation is (a) periodic vibration whose frequency varies between limits or (b) random vibration consisting of all frequencies within a defined band, it may not be feasible to design the equipment so that its natural frequencies do not coincide with excitation frequencies. Under these conditions, damping must be added to the vibrating structure.

Single Degree-of-Freedom Model.   When the model of Fig. 42.2 is acted upon by a periodic excitation whose frequency coincides with the natural frequency of the model, the amplitude of vibration of the model is indicated quantitatively by Fig. 2.13. The relative motion or spring deflection is shown in Fig. 42.8. It usually is preferable to eliminate resonance by stiffening the spring k of the model; this increases the natural frequency and moves the operating condition to the left of the resonance peak in Fig. 42.8 where the spring deflection is small. In some instances, an increase in stiffness is not possible; a decrease in stiffness accomplishes the objective of avoiding resonance but may have the result not only of removing desirable rigidity from the equipment but also of making the spring deflection equal to or greater than the amplitude of vibration. The natural frequency of the model in Fig. 42.2 varies directly as the square root of the spring stiffness [see Eq. (42.1)]; consequently, a relatively large change in stiffness is required to effect an incremental change in natural frequency.

If the natural frequency of the structure cannot be modified or if the excitation covers a band of the frequency spectrum, it may be impossible to avoid a resonant condition. Under these circumstances, the desired low amplitude can be attained only by ensuring that the structures of the equipment are damped sufficiently. In general, metals commonly used in structures do not have sufficient hysteresis damping to maintain vibration amplitudes at an acceptable level if a resonant condition

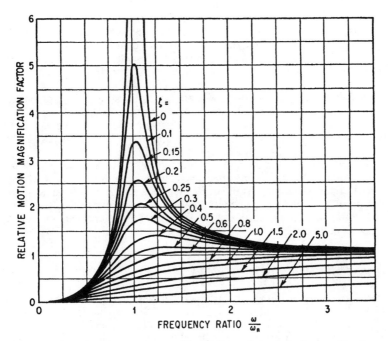

FIGURE 42.8 Relative motion magnification factor for a single degree-of-freedom system with viscous damping. The forcing frequency is ω, the undamped natural frequency ω_n and the fraction of critical damping ζ.

exists. Frequently, damping may be attained by use of bolted or riveted joints or by use of damping materials in or on the structure. In extreme cases, discrete dampers must be added to achieve adequate control of vibration. These various techniques are discussed in Chap. 43 under *Damping of Structures.*

Uncoupled Two Degree-of-Freedom Model. When the mass m_2 is small, it is unable to influence the motion of mass m_1. Then the vibratory motion of m_1 in system 1 of the model shown in Fig. 42.6A becomes the excitation for system 2, and the above discussion of the single degree-of-freedom model becomes applicable to evaluating the response of system 2. A condition of resonance of system 2 with its excitation should be avoided; i.e., the natural frequency of system 2 should be different from that of system 1. Preferably, the natural frequency of system 2 should be greater than the natural frequency of system 1; otherwise, the deflection of the spring k_2 may equal or exceed the vibratory displacement of the mass m_1. A system 2 with a natural frequency twice as great as that of system 1 experiences a vibration amplitude 1.35 times as great, as indicated in Fig. 2.13. This is a tolerable degree of amplification; it usually is achievable practically and constitutes a good design objective. It has wide application to equipment that involves elastic systems supported on elastic systems and should be adopted as a primary objective in the design of equipment to withstand vibration.

If the natural frequency criteria of the above paragraph can be met, damping requirements for system 2 may not be stringent. The excitation for system 2 is the vibration of system 1; the magnitude of such excitation is determined to a considerable extent by the damping in system 1. Thus, a basic principle of design where system 2 is of negligible mass is (*a*) to design system 2 to have a natural frequency at least twice as great as that of system 1 and (*b*) to introduce damping in system 1 to limit its amplitude of vibration if this is made necessary by the nature of the excitation. Methods for adding damping to structures are discussed in Chap. 43 under *Damping*.

Coupled Two Degree-of-Freedom Model. The analysis of an equipment represented by a two degree-of-freedom model as shown in Fig. 42.6*A* must consider coupling between the branches of the system when the two masses m_2, m_1 are of the same order of size. In other words, it cannot be considered that system 1 responds as a single degree-of-freedom system and imposes this response as the excitation for system 2. The spring k_2 and damper c_2 apply significant forces to system 1; this is known as a *coupling effect* and causes the entire assembly to respond as a single system. The analysis of such a system is set forth in Chap. 6; it can be used to predict the response of the model to a known excitation. The effect of changes in the physical properties of the model can be determined and related to the actual equipment.

DESIGN FOR SHOCK CONDITIONS

The design of equipment to withstand shock involves the formulation of a model to represent the equipment and the determination of the response of the model to the shock. Theoretically, the design of equipment to withstand shock can proceed if the shock is defined as the time-history of the motion (displacement, velocity, or acceleration) of the support for the equipment. As pointed out below (see *Single Degree-of-Freedom Model*), the shock spectrum is an adequate definition of the shock motion for purposes of equipment design. Known methods can be applied to determine the maximum response of the model, and design requirements for the actual equipment can be inferred from this response. Considerations in selecting a model are discussed above under *Types of Models*.

Single Degree-of-Freedom Model. In the design of equipment to withstand shock, it is common practice to assume that failure occurs because the maximum allowable stress or acceleration is exceeded in an element of the equipment. Stress of this magnitude is assumed to lead to failure should it occur once.* The method of analysis involves a determination of the maximum stress in the model corresponding to the maximum deflection of spring k or maximum acceleration of mass m in Fig. 42.2.

The Shock Spectrum. For a given shock motion, the maximum acceleration $\ddot{x}$ of mass m is a function of the natural frequency f_n of the model and its fraction of critical damping ζ. This maximum acceleration $\ddot{x}$ can be plotted as spectra of the natural frequency f_n with ζ as the parameter of the family of spectra. Typical spectra[3] are shown in Fig. 42.9. Each shock motion has characteristic shock spectra; this is discussed in Chap. 23 where it is pointed out that the spectra constitute a convenient definition of a shock motion. An important aspect of such a definition is that it provides information directly useful for design purposes; i.e., a known value for the maximum acceleration of the mass of the single degree-of-freedom model makes

---

* In laboratory testing, consideration sometimes is given to the cumulative effect of repeated occurrences of stress. This concept also has application to the design of equipment; it is discussed in detail in Chap. 24.

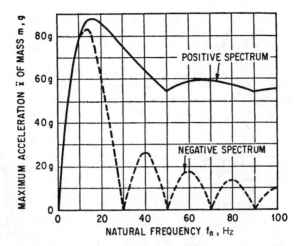

FIGURE 42.9   Typical shock spectra for an undamped system. The solid line represents acceleration in the same direction as the principal excitation; the broken line represents acceleration in the opposite direction. *(After J. E. Ruzicka.[3])*

possible the design of the spring. Methods for determining the shock spectrum and examples of spectra for typical excitations are given in Chaps. 8 and 23.

In using the spectra of Fig. 42.9 for design purposes, the chart is entered on the abscissa scale at the natural frequency of the model. The corresponding maximum acceleration of mass *m,* usually expressed as a multiple of the gravitational acceleration, is read from the ordinate scale. Inasmuch as the acceleration of the model under static conditions may be considered as 1*g,* the ordinate indicates the factor by which the stress under static conditions must be multiplied to give the stress under conditions of shock. This maximum acceleration is approximately equal to the *equivalent static acceleration,*[4] a concept* that permits the application of design methods for static loads to the design of equipment required to withstand dynamic loading. The spectrum shown by solid lines in Fig. 42.9 represents acceleration in the same direction as the pulse; the spectrum shown by dotted lines represents accelerations in the opposite direction, usually caused by free vibration of the model after the excitation has disappeared.

In general, the spectrum has lower maximum values of acceleration as damping is added to the model. Therefore, the use of the spectrum for the undamped model is conservative in design work; it is not excessively conservative for spectra of the type shown in Fig. 42.9 where the excitation is a single pulse of acceleration. If the excitation is oscillatory, however, there is a tendency for a model whose natural frequency is equal to the frequency of oscillation to respond in a resonant manner; then the discrepancy between the spectra for damped and undamped models may be relatively great. If the spectra for the undamped model tend toward values that are great relative to spectra for damped models, a representative value of damping

---

* See Chap. 23 for the distinction between maximum acceleration and equivalent static acceleration. For practical design purposes, the two may be considered equal.

should be chosen as the basis of design. (See *Approximate Methods,* below, for examples of spectra for damped systems and spectra plotted to dimensionless coordinates.) Otherwise, the undamped model is preferable because the spectra are more readily computed.

The spectra of Fig. 42.9 may be redrawn, as shown in Fig. 42.10, to indicate maximum relative displacement and maximum relative velocity as well as maximum acceleration. Relative displacement is the deflection of spring k in Fig. 42.2. This information is of importance, for example, in determining whether the deflection of

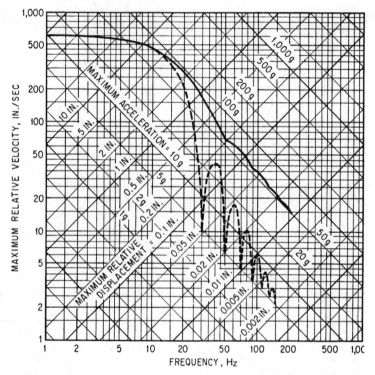

FIGURE 42.10   Spectra of Fig. 42.9 redrawn to indicate maximum relative displacement and maximum relative velocity as well as maximum acceleration.

a shaft would be great enough to permit gears to become momentarily disengaged and then reengaged in a different position.

Approximate Methods.   Ideally, the concept of designing to withstand shock contemplates the availability of a time-history of the shock motion and the computation of spectra as shown in Fig. 42.10. Time-histories and/or facilities for computation sometimes are not available. It becomes instructive, then, to examine typical spectra and to compare them with the corresponding time-histories to determine trends that are useful for design purposes. It may be convenient to employ dimensionless or normalized parameters to evaluate important characteristics of the response; this leads directly to useful design methods.

ACCELERATION PULSE: An idealized version of a typical shock motion is the half-sinusoidal acceleration pulse defined by

$$\ddot{u} = \ddot{u}_0 \sin \omega_p t \qquad \left[0 \le t \le \frac{\pi}{\omega_p} \right]$$

$$\ddot{u} = 0 \qquad \left[\frac{\pi}{\omega_p} \le t \right] \tag{42.3}$$

where $\ddot{u}$ is the acceleration of the support for the model at any time t, $\ddot{u}_0$ is the peak acceleration, and ω_p is the angular frequency associated with the acceleration pulse. The time duration of the pulse is $\tau = \pi/\omega_p$. With the system initially at rest, the time-histories of acceleration, as given by Eq. (42.3), and velocity $\dot{u}$ and displacement u, as obtained from integration of Eq. (42.3), are shown in Fig. 42.11.

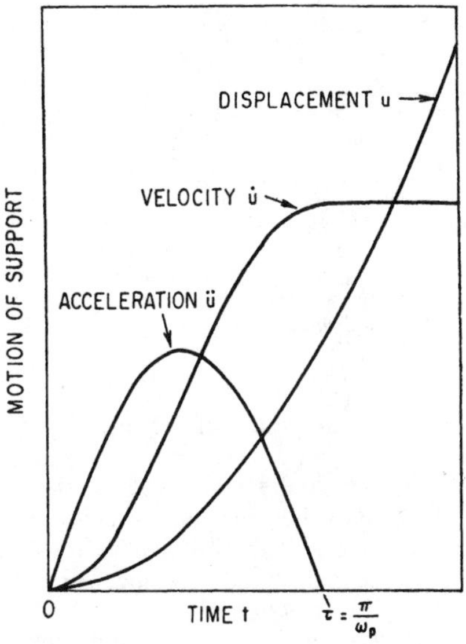

The transient response of the model of Fig. 42.2 to the pulse defined by Eq. (42.3) is illustrated in Fig. 42.12[5] where the acceleration $\ddot{x}$ of the mass m is plotted as a function of time from initiation of pulse $\ddot{u}$. Acceleration $\ddot{x}$ is plotted as a multiple of the peak pulse acceleration $\ddot{u}_0$, and time t is plotted as a multiple of the pulse duration τ. The half-sinusoidal acceleration pulse defined by Eq. (42.3) is the shaded area bounded by a dotted line.

The time-history of the response acceleration $\ddot{x}$ is plotted in Fig. 42.12 for three different values of the ratio of pulse duration to natural period of model: $\tau = \frac{10}{3}(2\pi/\omega_n)$, $(2\pi/\omega_n)$, and $\frac{1}{4}(2\pi/\omega_n)$, where ω_n is the natural frequency of the model [Eq. (42.1)]. The acceleration $\ddot{x}$ of the mass m can be equal to, greater than, or less than the peak acceleration of the support. In general, the acceleration of the support is not an indication of the forces induced in the model.

FIGURE 42.11   Velocity and displacement as a function of time, corresponding to acceleration pulse defined by Eq. (42.3).

When the angular natural frequency of the model is appreciably larger than the frequency ω_p associated with the pulse, the response of the model approximates but does not exactly duplicate the acceleration pulse of the support. For example, when $\tau = \frac{10}{3}(2\pi/\omega_n)$, the response has superimposed small-amplitude vibration at the natural frequency ω_n of the model. The subsequent free vibration of the system represents the maximum *negative* response to the original positive (or upward) pulse. When $\tau = (2\pi/\omega_n)$, the maximum acceleration of the mass m is about 1.7 times the acceleration of the support. When $\tau = \frac{1}{4}(2\pi/\omega_n)$, the peak acceleration of the model occurs after the acceleration of the original pulse becomes zero and is about 0.9 times the peak acceleration of the pulse.

The shock response spectra for the half-sinusoidal acceleration pulse of Eq. (42.3) are shown in Fig. 42.13 for several values of the fraction of critical damping ζ

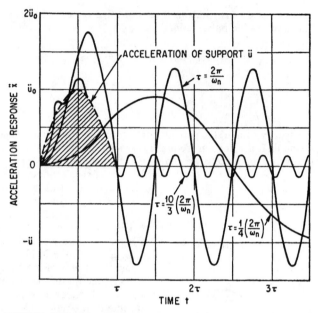

FIGURE 42.12 Response of an undamped, single degree-of-freedom system to the half-sinusoidal pulse of acceleration shown shaded. Systems with different natural periods are included. *(J. M. Frankland.[5])*

[see Eq. (2.12)]. These spectra summarize the maximum values of the response for the time-histories of Fig. 42.12, for a wide range of frequency relations. The ordinate is the ratio of the peak acceleration $\ddot{x}_{max}$ of the response to the peak acceleration $\ddot{u}_{max}$ of the pulse, where $\ddot{u}_{max}$ is used in place of $\ddot{u}_0$ to suggest that this discussion is applicable to other than the sinusoidal pulse defined by Eq. (42.3). The numerical value of the ratio $\ddot{x}_{max}/\ddot{u}_{max}$ can be interpreted as the shock transmissibility since it represents both spring and damper forces. For very high values of damping, this system acts as a rigid connection so that the acceleration ratio quickly becomes unity below a frequency ratio of 1. For the curve with zero damping, the acceleration ratio can be interpreted as the shock amplification [dynamic load factor $= \omega_n^2 (x - u)_{max}/\ddot{u}_{max}$], since it then represents only the spring force, or relative motion. When system damping is included, Fig. 8.43 shows the shock amplification factors which are somewhat different from shock transmissibility values. For very high values of damping (not shown in Fig. 8.43), the shock amplification factors can decrease below unity as the relative motion $(x - u)$ approaches zero.[3]

When the frequency ratio ω_n/ω_p is large, the acceleration ratio approaches unity; i.e., the maximum acceleration experienced by the model is approximately equal to that of the pulse. In other words, when a structure is stiff and the support motion is applied slowly, the maximum acceleration of the support can be used to determine the dynamic load imposed upon structures of the equipment. This relation is independent of the pulse shape; however, a pulse with superimposed high-frequency vibration cannot be interpreted in this manner because the superimposed vibration may induce a resonant response not contemplated by the preceding analysis. Furthermore, irregular pulses are sometimes made to appear smooth by limited high-frequency response of the instrumentation.

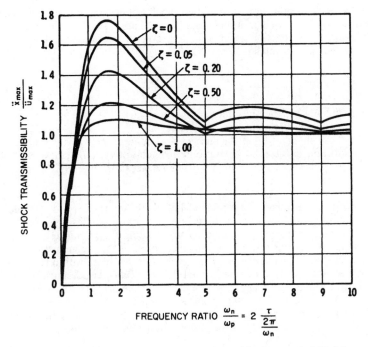

FIGURE 42.13 Maximum dynamic response factor (shock transmissibility) for the half-sinusoidal pulse of acceleration. The fraction of critical damping in the system is indicated by ζ. *(After J. E. Ruzicka.[3])*

STARTING VELOCITY: In Fig. 42.13 the equation of the tangent to the response spectrum for zero damping, at the origin, is

$$\frac{\ddot{x}_{max}}{\ddot{u}_{max}} = 2\left(\frac{\omega_n}{\omega_p}\right) \tag{42.4}$$

This straight line represents a good approximation to the spectrum for $\zeta = 0$ when $\omega_n/\omega_p \leq 0.50$. The velocity change $\dot{u}_0$ of the support is the integral of Eq. (42.3) with respect to time:

$$\dot{u}_0 = \frac{2\ddot{u}_0}{\omega_p} \quad \text{in./sec} \tag{42.5}$$

Combining Eqs. (42.4) and (42.5) and letting $\ddot{u}_{max} = \ddot{u}_0$,

$$\ddot{x}_{max} = \omega_n \times \dot{u}_0 \quad \text{in./sec}^2 \quad \left[\frac{\omega_n}{\omega_p} \leq 0.50\right] \tag{42.6}$$

where ω_n is in units of radians per second and $\dot{u}_0$ is in units of inches per second. Thus, the maximum acceleration $\ddot{x}_{max}$ of the mass m in Fig. 42.2 is proportional to the magnitude of the velocity change $\dot{u}_0$ of the support and to the natural frequency ω_n of the model, provided the acceleration pulse duration τ is short compared to the natural period $2\pi/\omega_n$ of the model. This relation is convenient when the duration of the shock motion is small compared with the natural periods of primary structures;

it finds particular application in simplified aspects of design using the single degree-of-freedom model, as well as in shock isolation (Chap. 31) and packaging (Chap. 41).

The maximum deflection δ_{max} of a single degree-of-freedom model of natural frequency ω_n, when ω_n is small relative to $2\pi/\tau$, is

$$\delta_{max} = \frac{\dot{u}_0}{\omega_n} \quad \text{in.} \quad \left[\frac{\omega_n}{\omega_p} \leq 0.50\right] \tag{42.7}$$

where $\dot{u}_0$ and ω_n have the units defined for Eq. (42.6).

In applying the relation of Eqs. (42.6) and (42.7), note that $\ddot{x}_{max}$ and δ_{max} are independent of the shape of the pulse and are determined solely by the area under the acceleration-time curve. If the excitation is irregular or vibratory, these relations are applicable if $\dot{u}_0$ is taken as the integral of the acceleration-time curve where $\omega_p = \pi/\tau$ and τ is the duration of the excitation.

Uncoupled Two Degree-of-Freedom Model. When the secondary structure of an equipment is small relative to the primary structure, the response of the model to a shock motion can be evaluated by first finding the response of the primary structure and using this response as the excitation for the secondary structure. For example, Fig. 42.12 gives several time-histories of the acceleration response of an undamped, single degree-of-freedom model to a half-period sinusoidal pulse of acceleration. The continuing periodic vibration of the primary structure appears to the secondary structure as steady-state vibration; thus, criteria established for design to withstand vibration constitute a good guide for design to withstand shock when the equipment can be represented by an uncoupled two degree-of-freedom model.

In accordance with the principle of design to withstand vibration, the secondary structure should have a natural frequency at least twice as great as that of the primary structure. The acceleration amplitude of the secondary structure then is 1.35 times as great as that of the primary structure (see *Design for Vibration Conditions—Uncoupled Two Degree-of-Freedom Model*). The maximum acceleration of the primary structure is found by methods indicated under *Design for Shock Conditions—Single Degree-of-Freedom Model*. The primary structure must be designed to withstand this acceleration; the secondary structure must be designed to withstand an acceleration 1.35 times as great.

If the natural frequencies of the primary and secondary structures are equal, the secondary structure may experience excessive vibration unless either the primary or secondary structure embodies significant damping. For example, Fig. 42.14 indicates the maximum acceleration of the secondary structure with reference to the maximum acceleration of the primary structure when the shock motion is a sudden velocity change $u = \dot{u}_0 t$ of the support for the primary structure.[6] When there is significant damping in the primary and/or secondary structure, the maximum acceleration of the secondary structure is relatively small.

Coupled Two Degree-of-Freedom Model. When the mass of the secondary structure is of the same order of magnitude as that of the primary structure, it is necessary to consider the response of the composite model of primary and secondary structures. The response of this model to any excitation can be determined by known methods. For initial design work, it is convenient to generalize the excitation as a sudden motion change of the support, $u = \dot{u}_0 t$. This is a good approximation when the velocity change takes place during a time interval that is small relative to the smallest natural period of the composite model.

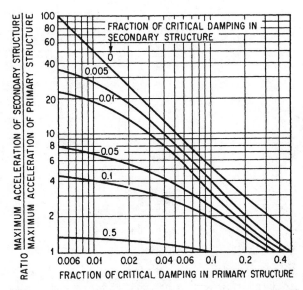

FIGURE 42.14 Ratio of maximum acceleration of secondary structure to maximum acceleration of primary structure (see Fig. 42.6*A*) when support has motion $u = \dot{u}_0 t$ and uncoupled natural frequencies of the structures are equal (i.e., $\omega_1 = \sqrt{k_1/m_1} = \omega_2 = \sqrt{k_2/m_2}$). The mass m_2 of the secondary structure is negligible relative to the mass m_1 of the primary structure. (*C. E. Crede.*[6])

The equations of motion for the system shown in Fig. 42.6*A*, when the excitation is a starting velocity $\dot{u}_0$ applied at the support, give results that can be expressed in terms of an equivalent starting velocity applied individually to each of the uncoupled spring-mass models in Fig. 42.6*B*; i.e., the maximum response of each model is equivalent to its maximum response as a part of the two degree-of-freedom model. The equivalent starting velocities are designated by $\dot{u}_2$ for the secondary structure and $\dot{u}_1$ for the primary structure. They are expressed in terms of the starting velocity $\dot{u}_0$ as:[7,8]

$$\frac{\dot{u}_2}{\dot{u}_0} = \frac{1}{\sqrt{\left(1 - \frac{\omega_2}{\omega_1}\right)^2 + \frac{m_2}{m_1}\left(\frac{\omega_2}{\omega_1}\right)^2}} \tag{42.8}$$

$$\frac{\dot{u}_1}{\dot{u}_0} = \frac{1 + \left(1 + \frac{m_2}{m_1}\right)\frac{\omega_2}{\omega_1}}{\sqrt{\left(1 + \frac{\omega_2}{\omega_1}\right)^2 + \frac{m_2}{m_1}\left(\frac{\omega_2}{\omega_1}\right)^2}} \tag{42.9}$$

where ω_1 and ω_2 are defined under *Two Degree-of-Freedom Model.*

Figure 42.15 shows a contour chart of $\dot{u}_2/\dot{u}_0$ plotted against the stiffness ratio k_2/k_1, the mass ratio m_2/m_1, and the frequency ratio ω_2/ω_1. Figure 42.16 shows a con-

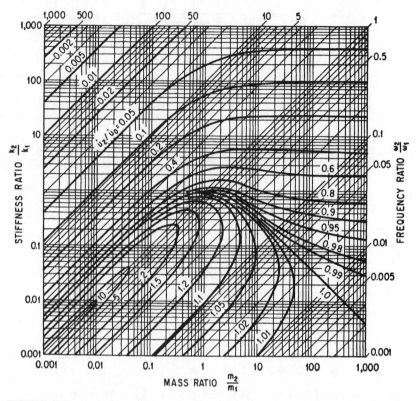

FIGURE 42.15 Equivalent velocity change $\dot{u}_2$ for secondary structure relative to $\dot{u}_0$ where displacement of support for primary structure is $\dot{u} = u_0 t$. The natural frequencies are $\omega_1 = \sqrt{k_1/m_1}$, $\omega_2 = \sqrt{k_2/m_2}$ where m_1, m_2, k_1, k_2 are defined in Fig. 42.6A. *(R. B. McCalley, Jr.[8])*

tour chart of a factor D plotted against the same stiffness, mass, and frequency ratios. The factor D is related to the velocities $\dot{u}_1$ and $\dot{u}_0$ by

$$\frac{\dot{u}_1}{\dot{u}_0} = D \sqrt{1 + \frac{m_2}{m_1}} \qquad (42.10)$$

The maximum relative motion $(x_1 - x_2)_{max}$; i.e., the maximum deflection $(\delta_2)_{max}$ of the spring k_2, is

$$(\delta_2)_{max} = \frac{\dot{u}_2}{\omega_2} = \left(\frac{\dot{u}_0}{\omega_2}\right)\left(\frac{\dot{u}_2}{\dot{u}_0}\right) \qquad \text{in.} \qquad (42.11)$$

where $\dot{u}_2/\dot{u}_0$ is obtained from Fig. 42.15. The contour chart of Fig. 42.15 is particularly useful where the characteristics of the secondary structure are fixed and a primary structure must be selected to reduce the shock experienced by the secondary structure. The equivalent static acceleration of the secondary structure is

$$(\ddot{x}_2)_{max} = \frac{k_2(\delta_2)_{max}}{m_2} \qquad \text{in./sec}^2 \qquad (42.12)$$

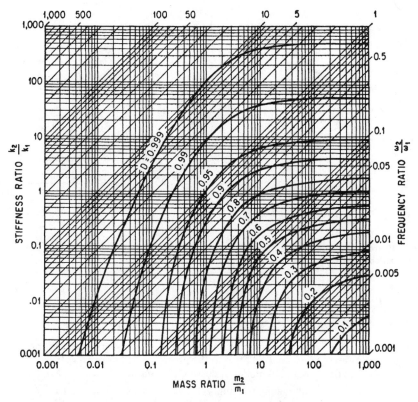

FIGURE 42.16   Contour chart for factor D in Eq. (42.10). The physical conditions are described with reference to Fig. 42.15. *(R. B. McCalley, Jr.[8])*

The maximum relative motion $(u - x_1)_{max}$; i.e., the maximum deflection $(\delta_1)_{max}$ of the spring k_1, is

$$(\delta_1)_{max} = \frac{\dot{u}_1}{\omega_1} = \left(\frac{\dot{u}_0}{\omega_1}\right) D \sqrt{1 + \frac{m_2}{m_1}} \qquad \text{in.} \qquad (42.13)$$

where D is obtained from Fig. 42.16. The equivalent static acceleration of the primary structure is

$$(\ddot{x}_1)_{max} = \frac{k_1 (\delta_1)_{max}}{m_1} \qquad \text{in./sec}^2 \qquad (42.14)$$

The response data given by the contour plots of Figs. 42.15 and 42.16, referring to the coupled two degree-of-freedom model of Fig. 42.6A, are shown in Fig. 42.17 as spectra of the frequency ratio ω_2/ω_1. Maximum deflection of the springs k_2, k_1 and the corresponding equivalent static acceleration for the respective structures are directly proportional, respectively, to the equivalent starting velocities $\dot{u}_2$, $\dot{u}_1$. These relations are defined by Eqs. (42.11) and (42.12); values for $\dot{u}_2$, $\dot{u}_1$ are given by the ordinates of Fig. 42.17.

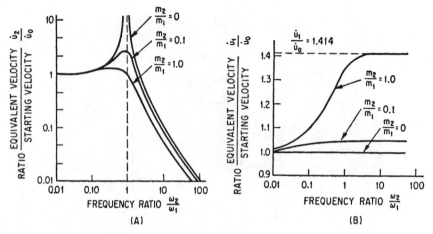

FIGURE 42.17   Equivalent starting velocity ratios for (*A*) secondary and (*B*) primary structures. The motion of the support is $u = \dot{u}_0 t$. (*C. M. Friedrich.*[7])

Figure 42.17 illustrates the importance of maintaining appropriate relations between the natural frequencies ω_2 and ω_1:

1. If $\omega_2 = \omega_1$, the equivalent starting velocity of the secondary structure becomes great if its mass m_2 is small relative to the mass m_1 of the primary structure. This approaches the uncoupled model discussed in an earlier section.

2. If $\omega_2 > 2\omega_1$, the equivalent starting velocity of the secondary structure tends toward relatively low values for any ratio of the masses m_2 and m_1. The equivalent starting velocity of the primary structure tends toward larger though moderate values as the frequency ratio ω_2/ω_1 increases.

Thus, a design criterion for coupled as well as uncoupled two degree-of-freedom models is a natural frequency of the secondary structure at least twice as great as that of the primary structure.

Examples of Typical Requirements.   Shock testing machines play an important role in the design of equipment to withstand shock. Their importance is derived, in part, from the difficulty of applying the foregoing design principles to all details of a complex equipment because the equipment cannot be represented by a simple model. A shock test confirms calculations that were made and takes the place of calculations that could not be made. Furthermore, even though analytical techniques can be applied, the conditions of shock often are not defined well enough to serve as a basis for design. An equipment then is designed to withstand the shock created by a shock testing machine. In general, the shock created by the testing machine does not duplicate that experienced in service; however, it does represent an explicit requirement for design purposes and presumably takes into account the variances from one shock motion to another in actual service.

Shock testing machines are discussed, and several common machines are illustrated, in Chap. 26. The specification of laboratory tests to simulate actual service conditions is considered in Chap. 20. Most shock testing machines are adapted to several alternate modes of operation, at least to the extent of modifying the severity of the shock. Compilations of data are available to define the characteristics of the shock motion of commonly used shock testing machines. As examples of design requirements, typical shock spectra are shown in Fig. 42.18 for several machines. These spectra are useful directly with single degree-of-freedom models as a design procedure for equipment required to withstand the shock produced by the respective machines.

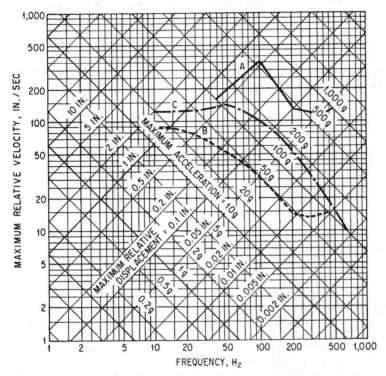

FIGURE 42.18 Typical shock spectra for several shock testing machines as examples of design requirements: (*A*) MIL-S-901B, (*B*) MIL-S-4456 using sand, and (*C*) MIL-S-4456 using lead. The designation "MIL" denotes a military specification number.

Effect of Equipment Weight. The weight of the equipment sometimes is a factor of importance when selecting strength requirements for the equipment. Heavy equipment often receives shock of less severity than light equipment because (1) sources of shock may involve only finite energy and have less effect upon the heavy equipment or (2) the structures that transmit the forces have finite strength and thereby limit the amount of energy that reaches the equipment. Many types of shock testing machines recognize these limitations and deliver a shock motion whose severity tends to decrease with an increase in the weight of the equipment under test. For example, machines with swinging hammers tend to deliver a shock whose severity decreases as the equipment weight increases, even though the height of the hammer drop may be increased somewhat as the equipment weight increases.

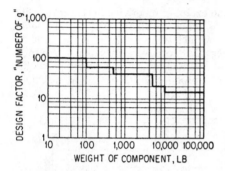

FIGURE 42.19 Typical empirical design curve indicating required strength of mounting bolts as a function of equipment weight. This applies to some classes of naval service.

Analytical methods for design of shock-resistant equipment employ one or two degree-of-freedom models that are tacitly assumed to be of negligible mass; i.e., the equipment offers no impedance to the shock motion. The effect of equipment weight in reducing the severity of shock is a factor of considerable significance; sometimes it is recognized empirically where design criteria are set forth. For example, the Bureau of Ships of the U.S. Navy specifies the strength requirements for bolts used to mount equipment on naval ships. The specification requires that the mounting bolts for light equipment be capable of withstanding a greater equivalent static acceleration than the bolts for mounting heavy equipment, as illustrated in Fig. 42.19. Strictly speaking, Fig. 42.19 applies only to mounting bolts and associated mounting brackets; generally, the trend may be recognized and applied to the overall design where there is reason to believe that the shock severity would be decreased by the weight of the equipment.

STRUCTURAL DESIGN CONSIDERATIONS

ENERGY STORAGE CAPACITY

The deflection of structural members that characterizes the response of equipment to shock and vibration involves the temporary storage of strain energy in such members. In a given member, strain energy increases as stress increases. The severity of shock that an equipment may withstand without damage is reached when the stress in the most highly stressed member reaches the maximum allowable value. Any material in the structure that does not at the same time reach its maximum allowable stress may be considered as contributing to a loss in efficiency of design. This material not only tends to increase the cost and weight of the equipment but, more important, may increase the weight carried by other structures with a consequent increase in strength requirements.

Efficiency is determined by the type of loading, the constraints on the structure, and the shape of the cross section. For example, the potential energy stored in a uniform bar in pure tension is the product of the average force $F/2$ and the deflection δ:

$$V = \frac{F\delta}{2} = \frac{(\sigma S)(l\epsilon)}{2} \quad \text{in.-lb} \tag{42.15}$$

where σ = tensile stress, lb/in.^{2}
  S = cross-sectional area of the bar, in.^{2}
  l = length of the bar, in.
  ϵ = strain, in./in.

The substitutions made in Eq. (42.15) are $F = \sigma S$ lb and $\delta = l\epsilon$ in. By making the additional substitutions $V = Sl$ in.^{3} and $\epsilon = \sigma_m/E$ in./in., Eq. (42.15) can be expressed as

$$\frac{F\delta}{2} = \eta \, \frac{\sigma_m^{\,2}V}{2E} \quad \text{in.-lb} \tag{42.16}$$

where V = total volume of material being stressed, in.^{3}
        E = Young's modulus, lb/in.^{2}
        σ_m = maximum stress (uniform for bar in tension), lb/in.^{2}
        η = efficiency (0 to 1)

The efficiency η is the fraction of the total strain energy, based on a uniform stress σ_m in all elements, that is stored by the structure. Solving Eq. (42.16) for η,

$$\eta = \frac{F\delta E}{\sigma_m^{\,2}V} \tag{42.17}$$

When each element in the structure has the same stress, the material is being used at maximum efficiency and $\eta = 1$.

For a beam in bending, the efficiency usually is a small fraction. For a cantilever beam with a concentrated load F at the free end, the deflection δ at the free end[1] is $Fl^3/3EI$ and the stress σ_m in the outermost fiber at the built-in end of the beam is Flc/I. The moment of inertia (of the cross-sectional area S) is $I = S\rho^2$ in.^{4}, where ρ is the radius of gyration and c is the distance of the outer fiber from the neutral axis. Substituting in Eq. (42.17),

$$\eta = \frac{\rho^2}{3c^2} \tag{42.18}$$

Values of ρ and c are known for various cross sections.[9] For a cantilever beam with end load, the corresponding values of efficiency are $\eta = \frac{1}{12}$ for a circular section, $\frac{1}{9}$ for a rectangular section, $\frac{1}{6}$ for a thin-walled tube, and about $\frac{1}{3}$ for an I beam. These increasing values reflect the fact that each successive shape of cross section further concentrates the material in the outermost fibers where the bending stress reaches a maximum.

The efficiency η depends not only upon the distribution of stress over the cross section but also upon the distribution along the length of the beam. The relatively low efficiency of only $\frac{1}{3}$ for the cantilever I beam occurs because the bending moment in a cantilever beam varies from a maximum at the built-in end to zero at the free end. By contrast, when a pure moment instead of a concentrated load is applied at the free end of the cantilever beam, the bending moment is uniform along the length of the beam and the efficiency increases by a factor of 3 to become $\eta = \rho^2/c^2$. This is approximately unity for the I beam.

The foregoing examples are included in Table 42.1 along with the efficiencies of other common structural members for various cross sections.[7] Column 1 gives the general expression for efficiency applicable to any uniform cross section, for the particular type of loading and constraints on the structure. Columns 2 and 3 give the efficiencies for rectangular and circular sections, respectively, and are obtained by substituting appropriate values for ρ and c in the expressions of column 1.

TABLE 42.1 Efficiencies of Common Structural Members as Defined by Eq. (42.17)

(C. M. Friedrich.[7])

	Type of cross section		
Type of structure	(1) Uniform	(2) Rectangular	(3) Circular
Bar in tension	1	1	1
Cantilever beam with concentrated load at end	$\rho^2/3c^2$	$\frac{1}{9}$	$\frac{1}{12}$
Simply supported beam with concentrated load at any point	$\rho^2/3c^2$	$\frac{1}{9}$	$\frac{1}{12}$
Built-in beam with concentrated load at center	$\rho^2/3c^2$	$\frac{1}{9}$	$\frac{1}{12}$
Simply supported beam with moment at center	$\rho^2/3c^2$	$\frac{1}{9}$	$\frac{1}{12}$
Built-in beam with moment at center	$\rho^2/2c^2$	$\frac{1}{6}$	$\frac{1}{8}$
Cantilever beam with moment at end	ρ^2/c^2	$\frac{1}{3}$	$\frac{1}{4}$
Simply supported beam with uniform load	$8\rho^2/15c^2$	$\frac{8}{45}$	$\frac{2}{15}$
Beam with one built-in end, one simply-supported end, load at center	$7\rho^2/27c^2$	$\frac{7}{81}$	$\frac{7}{108}$

ρ = radius of gyration of the beam cross section.
c = distance of outermost fibers from the neutral axis of the beam.

ELASTIC CHARACTERISTICS OF BEAMS AND PLATES

Analytical methods of design to withstand shock and vibration make extensive use of models to determine the response of an equipment or structure to the shock and vibration. Generally, the model is an elastic system of concentrated masses, massless springs, and massless dampers whose natural frequencies can be related to the natural frequencies of the actual structure. The forces and deflections associated with the response of the model then are transformed to stresses and strains in the actual structure. The section of this chapter on *Models* shows the relation between a model and a simple beam with center load. This section extends the treatment to other types of beams and includes analyses directed toward the ready calculation of bending moments and reaction forces.

Uniform Beam with Distributed Load. The spring stiffness for lateral deflection of a uniform beam can be generalized as[10]

$$k_b = n_b \left(\frac{EI}{l^3} \right) = n_b \left(\frac{ES_\rho^2}{l^3} \right) \quad \text{lb/in.} \tag{42.19}$$

where E = Young's modulus, lb/in.²
I = moment of inertia of cross section, in.⁴
S = cross-sectional area of beam, in.²
ρ = radius of gyration of cross section, in.
l = length of beam, in.

The numerical coefficient n_b depends upon the type of loading and the end constraints on the beam. For the beam shown in Fig. 42.1, the coefficient is $n_b = 48$.

The natural frequency of a uniform beam is[10]

$$\omega_n = \sqrt{\frac{EI}{\nu l^4}} = N_b \left(\frac{\rho}{l^2} \right) \sqrt{\frac{E}{\mu}} \qquad \text{rad/sec} \qquad (42.20)$$

where ν = mass per unit length of beam, lb-sec^2/in.^{2}
μ = mass density of the material, lb-sec^2/in.^{4}

Hence, $\nu = \mu S$. The numerical coefficient N_b depends upon the end constraints of the beam. For the beam shown in Fig. 42.1 (without the rigid body W), $N_b = \pi^2$ for the lowest natural frequency. Values of N_b for higher natural frequencies, and for other constraints and types of loading, are given in Tables 7.3 and 7.5.

By considering a beam to be analogous to a single degree-of-freedom system having the stiffness k_b given by Eq. (42.19) and an effective mass m_e, Eq. (42.1) may be written as follows for the natural frequency of the beam:

$$\omega_n = \sqrt{\frac{k_b}{m_e}} \qquad \text{rad/sec} \qquad (42.21)$$

Solving for m_e and substituting k_b and ω_n from Eqs. (42.19) and (42.20), respectively,

$$m_e = \left(\frac{n_b}{N_b^2} \right) m_b \qquad \text{lb-sec}^2\text{/in.} \qquad (42.22)$$

where $m_b = \mu S l$ is the total mass of the beam.

Uniform Beam with Concentrated Load. The effects upon the natural frequency of the mass of both the rigid body W and the beam shown in Fig. 42.1 may be included by writing Eq. (42.21) as follows:

$$\omega_n' = \sqrt{\frac{k_b}{m_l + m_e}} \qquad \text{rad/sec} \qquad (42.23)$$

where m_l is the mass of the load (rigid body W), m_e is given by Eq. (42.22), and k_b is given by Eq. (42.19). By using Eqs. (42.19) and (42.22), the natural frequency ω_n' can be expressed in terms of the dimensions and properties of the actual beam:

$$\omega_n' = \frac{\dfrac{\rho}{l^2} \sqrt{\dfrac{E}{\mu}}}{\xi_\delta} \qquad \text{rad/sec} \qquad (42.24)$$

where

$$\xi_\delta = \sqrt{\frac{1}{N_b^2} + \frac{1}{n_b} \left(\frac{m_l}{m_b} \right)}$$

For the beam and load shown in Fig. 42.1, $\xi_\delta = \sqrt{0.0103 + 0.0208 \, m_l/m_b}$.

When the duration τ of a shock motion is small relative to the natural period $2\pi/\omega_n'$ of the beam, the shock motion can be assumed to consist of the starting velocity $\dot{u}_0$ applied at the beam supports in a direction transverse to the beam. Then, Eq.

(42.7) leads to the following expression for the maximum beam deflection δ_m at the concentrated load m_l:

$$\delta_m = \frac{\dot{u}_0}{\omega_n'} = \xi_\delta \dot{u}_0 \left(\frac{l^2}{\rho} \right) \sqrt{\frac{\mu}{E}} \qquad \text{in.} \tag{42.25}$$

The maximum bending moment M_m and the maximum force F_m at the end supports are determined from the maximum deflection:[1]

$$M_m = \xi_M \dot{u}_0 S \rho \sqrt{E\mu} \qquad \text{lb-in.} \tag{42.26}$$

$$F_m = \xi_F \dot{u}_0 \left(\frac{S\rho}{l} \right) \sqrt{E\mu} \qquad \text{lb} \tag{42.27}$$

Table 42.2 gives values of the dimensionless factors ξ_δ, $\xi_{\delta M}$, and $\xi_{\delta F}$ to be used in Eqs. (42.25) to (42.27) for calculating the maximum deflection δ_m, the maximum bending moment M_m, and the maximum end support force F_m, respectively, for several types of loading and end support.[7] The natural frequency ω_n' also can be found from Eq. (42.25):

$$\omega_n' = \frac{\dot{u}_0}{\delta_m} = \frac{\rho}{\xi_\delta l^2} \sqrt{\frac{E}{\mu}} \qquad \text{rad/sec} \tag{42.28}$$

TABLE 42.2 Dimensionless Factors* in Eqs. (42.25) to (42.27) for Calculating the Response of Beams to a Velocity Shock $u = \dot{u}_0 t$ (C. M. Friedrich.[7])

Type of structure	Maximum deflection ξ_δ Eq. (42.25)	Bending moment ξ_M Eq. (42.26)	End reaction ξ_F Eq. (42.27)
Simply supported beam with mass at center (deflection and moment at center, reaction at simple support)	$\sqrt{0.0103 + 0.0208\, m_l/m_b}$	$12\xi_\delta$	$24\xi_\delta$
Built-in beam with mass at center (deflection at center, moment and reaction at built-in end)	$\sqrt{0.00199 + 0.00521\, m_l/m_b}$	$24\xi_\delta$	$96\xi_\delta$
Cantilever beam with mass at free end (deflection at free end, moment and reaction at built-in end)	$\sqrt{0.0806 + 0.333\, m_l/m_b}$	$3\xi_\delta$	$3\xi_\delta$
Beam with one built-in end, one simply supported end, with mass at center (deflection near center, moment at built-in end), reaction at:			
Built-in end	$\sqrt{0.00422 + 0.00932\, m_l/m_b}$	$20.1\xi_\delta$	$73.6\xi_\delta$
Simple support	$\sqrt{0.00422 + 0.00932\, m_l/m_b}$	$20.1\xi_\delta$	$33.5\xi_\delta$

* The dimensionless factors depend upon the type of loading and the end supports for the beam as well as the following:

$$m_l = \text{mass of load} \qquad \text{lb-sec}^2/\text{in.}$$
$$m_b = \text{total mass of beam} \qquad \text{lb-sec}^2/\text{in.}$$

Circular Plate with Concentrated Center Load.   The elastic behavior of plates is analogous to that of beams. For example, the stiffness of a uniform circular plate with a built-in circumferential edge and supporting a load at the center is[10]

$$k_p = n_p \left(\frac{Et^3}{R^2} \right) \qquad \text{lb/in.} \tag{42.29}$$

where t is the thickness and R the radius of the plate, in inches. This expression is similar to Eq. (42.19). The constant n_p has a value of 4.60 when the plate is made of steel having a Poisson's ratio $\gamma = 0.3$. The expression for the natural frequency of the plate in its fundamental mode is[10]

$$\omega_n = N_p \left(\frac{t}{R^2} \right) \sqrt{\frac{E}{\mu}} \qquad \text{rad/sec} \tag{42.30}$$

This expression is similar to Eq. (42.20). The constant N_p has a value of 3.09 when the plate is made of steel having a Poisson's ratio $\gamma = 0.3$. Finally, the expression for the natural frequency ω_n', which accounts for the equivalent mass m_e of the vibrating plate as well as the concentrated load m_l, is

$$\omega_n' = \frac{\dfrac{t}{R^2} \sqrt{\dfrac{E}{\mu}}}{\xi_\delta'} \qquad \text{rad/sec} \tag{42.31}$$

where

$$\xi_\delta' = \sqrt{\frac{1}{N_p^2} + \left(\frac{\pi}{n_p} \right) \frac{m_l}{\mu t (\pi R^2)}}$$

Substituting the numerical values of N_p and n_p for a circular plate built in at the edge and carrying a concentrated load m_l at the center, $\xi_\delta' = \sqrt{0.1047 + 0.683\, m_l/(\mu \pi R^2)}$. For plates with other edge conditions and other types of loading, procedures have been worked out for computing the natural frequencies.[11]

REFERENCES

1. Roark, R. J.: "Formulas for Stress and Strain," 3d ed., p. 100, McGraw-Hill Book Company, Inc., New York, 1954.

2. Den Hartog, J. P.: "Mechanical Vibration," 3d ed., p. 105, McGraw-Hill Book Company, Inc., New York, 1947.

3. Ruzicka, J. E.: *Sound and Vibration,* **4**(9):10 September, 1970.

4. Walsh, J. P., and R. E. Blake: *Proc. SESA,* **6**(2):152 (1948).

5. Frankland, J. M.: *Proc. SESA,* **6**(2):11 (1948).

6. Crede, C. E., and M. C. Junger: "A Guide for Design of Shock Resistant Naval Equipment," U.S. Navy Department, Bureau of Ships, *NAVSHIPS* 230-660-30, p. 17, 1949.

7. Friedrich, C. M.: "Shock Design Notes," unpublished report, Westinghouse Electric Corporation, Atomic Power Division, Pittsburgh, Pa., June, 1956.

8. McCalley, R. B., Jr.: *Supplement to Shock and Vibration Bull.* 23 (unclassified), Office of Secretary of Defense, June, 1956.

9. Ref. 1, p. 70.

10. Ref. 2, p. 457.

11. Stokey, W. F., and C. F. Zorowski: *J. Appl. Mechanics,* **E26**(2):210 (1959).

CHAPTER 43
PRACTICE OF EQUIPMENT DESIGN

Edward G. Fischer
Harold M. Forkois

INTRODUCTION

Equipment of a type required to withstand shock and vibration generally consists of a housing or chassis to provide structural strength and an array of functional components supported by the chassis. A suitable design is characterized by (1) properly selected or designed components, (2) chassis mounting to minimize damage from shock and vibration, and (3) a chassis capable not only of withstanding shock and vibration but also of providing a degree of protection to the components. The design of chassis and other structures is discussed in Chap. 42 from a quantitative viewpoint; this chapter discusses (1) chassis design in a general or qualitative sense, (2) the selection and mounting of components, and (3) design to minimize vibration. Many of the examples refer explicitly to electronic equipment, but the general principles have wider application.

DESIGN OF STRUCTURES

One of the more troublesome problems in the design of equipment is attainment of the proper balance between flexibility and rigidity of structures. As shown analytically in Chap. 42, the maximum acceleration experienced by a component is determined primarily, for a particular shock motion, by the natural frequency of the structure supporting the component. The acceleration decreases as the natural frequency decreases. Thus, components that are susceptible to damage from shock may be given a degree of protection by mounting them on relatively flexible structures; however, if the equipment is required to withstand vibration as well as shock, it is possible that the flexibility introduced to attenuate the shock may lead to failure as a result of vibration.

Failure or damage from vibration usually is the result of a resonant condition; i.e., a structure or component has a natural frequency that coincides with a frequency of the forcing vibration. In some applications, e.g., on naval ships, the maximum frequency of the forcing vibration is relatively low, and it becomes feasible in most instances to design structures whose natural frequencies are greater than the highest forcing frequency. On the other hand, the vibration in aircraft and missiles is characterized by relatively high frequencies; as a consequence, it is not feasible to design structures whose natural frequencies are higher than the forcing frequencies. A condition of resonance often must be tolerated. In extreme conditions, the effect of the resonant condition can be alleviated by the provision of damping or energy dissipation (see the section on *Damping*). In general, this is not necessary for structural members if the natural frequencies are chosen wisely.

Because of the many different types of components and equipments in common use and the wide range of environmental conditions encountered, it is not possible to establish specific design objectives for natural frequencies of structures. In general, a structure should be designed to be stiff rather than flexible because (1) most components of modern design are more shock-resistant than commonly anticipated and do not require the protection afforded by structures of low natural frequency, and (2) failure of structures by repeated reversal of stress during vibration is a common occurrence of damage. The stress in a structure tends to decrease when the natural frequency increases, provided the severity of the excitation and the fraction of critical damping for the structure remain unchanged. Furthermore, some environmental conditions embody a lower severity at the higher vibration frequencies; thus, a resonance at a high frequency may become less damaging. However, if a component is known to be vulnerable to damage by shock or if this is demonstrated during a shock test, the desired protection often may be attained by introducing a structure of lower flexibility. Such a structure may be the chassis of the equipment, a bracket to support a component, or an isolator.

In typical electronic equipment, the electronic components are mounted upon a chassis to form a unit that has a discrete function. The unit is made small enough to permit ready maintenance and servicing. In some instances, the function of the unit may be complete; then the unit is considered a complete equipment and is provided with independent mounting means. In other instances, several units are required to fulfill a function; such units are mounted in a single cabinet and interconnected electrically and/or mechanically to perform a function. Particular design problems associated with the housing for the equipment and involving ability to withstand shock and vibration are (1) design of the unit chassis, (2) design of the cabinet, and (3) mounting of the unit chassis in the cabinet.

CHASSIS DESIGN

In general, unit chassis are either (1) formed from sheet metal or (2) cast from a light metal (usually aluminum or magnesium) and machined to form intimate supports for the components. The former is more common and less costly to manufacture, but tends to lack adequate rigidity. The casting has the desirable property of relatively great rigidity, but refinements in design and manufacture are necessary to keep the weight within reasonable limits. Because of its rigidity, not only is the cast chassis less likely to fail as a result of vibration but it also minimizes damage from broken leads and collision of components that occur when the chassis is unduly flexible. The flexibility of structure required for protection from shock can be provided in the support for the chassis.

The most common type of sheet-metal chassis is a shallow rectangular "box" formed from a single piece of sheet metal; the front panel is added and rigidly attached to the "box" by gussets, as shown in Fig. 43.1. The chassis usually is oriented with the flanges or sides of the "box" extending downward. The sequence of illustrations in Fig. 43.2 shows one method of forming chassis corners. All bending radii should be greater than the thickness of the sheet to reduce the effects of stress con-

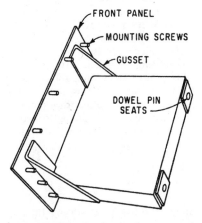

FIGURE 43.1 Good chassis construction, showing front-panel mounting screws and hole-plate reinforcement for dowel pins in the rear. (*H. M. Forkois and K. E. Woodward.*[1])

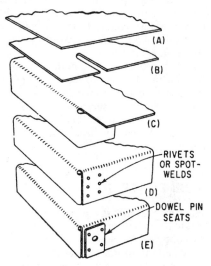

FIGURE 43.2 A method of forming chassis which produces stiff corners. (*H. M. Forkois and K. E. Woodward.*[1])

centration. The overlapped portions of the sheet can be joined by rivets or spot-welds (see *Methods of Construction*) to provide a stiff support for dowel pins or other attaching means. It is common practice to mount large components, such as transformers and traveling tubes, on the upper side of the chassis; smaller components, such as resistors and capacitors together with substantially all wiring, are mounted on the lower side of the chassis.

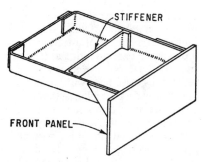

FIGURE 43.3 View of underside of chassis showing stiffener added to support heavy component on upper side of chassis. (*H. M. Forkois and K. E. Woodward.*[1])

Special care should be exercised to ensure maximum rigidity of the sheet-metal chassis. The overlapped flanges at the corners are areas of inherent rigidity and should be used to support the chassis. For example, the gussets for reinforcing the front panel and the plates to seat the dowel pins are attached at the corners as illustrated in Fig. 43.1. Where possible, heavy components such as transformers should be mounted adjacent to the corners to attain a maximum ratio of chassis stiffness to component weight. If a heavy component cannot be mounted adjacent to a corner, it should be mounted adjacent to an edge; if it must be mounted near the center of the chassis, a stiffener should be provided on the lower side of the chassis, as shown in Fig. 43.3. This type of construction is necessary to obtain the required stiffness but is undesirable in that it limits the location of components and interferes with the wiring.

Where an equipment is composed of a single chassis, it is common practice to provide an enclosure to protect the equipment during handling and during exposure to atmospheric contamination; such an enclosure is not intended as a support for the equipment and is not required to withstand the forces introduced by shock and vibration conditions. For example, in one type of application commonly used with vibration isolators in aircraft equipment, a supporting rack has forward-facing tapered pins adjacent to the rear edge. These pins extend through clearance holes in the enclosure and engage dowel pin seats in the rear of the chassis, as illustrated in Fig. 43.1. Similarly, clamps at the forward edge of the rack engage the front panel of the chassis directly, and loads are not carried by the enclosure. Frequently, two or more chassis, each with its individual enclosure, are mounted independently upon one rack.

CABINET DESIGN

When several chassis are included in a single cabinet, it is common to design the cabinet to be both an enclosure and a structure capable of withstanding the forces introduced by shock and vibration conditions. The chassis usually must be readily removable from the cabinet for adjustment and maintenance. If the objective of a rigid chassis is achieved, criteria to be applied in design of the cabinet are (1) optimum rigidity in view of shock and vibration conditions and (2) adequate structural strength. Suggestions for cabinet design are included in the following paragraphs.

A typical cabinet has horizontal dimensions corresponding to the width and depth of a chassis, and a height equal to the total heights of the chassis. The chassis are arranged one above the other, and are mounted independently to the cabinet. Except for very light equipment, the frame of the cabinet is assembled from structural steel or aluminum members welded, riveted, or bolted together, as illustrated in Fig. 43.4. Gussets are provided at the joints of structural members where they would not interfere with installation and removal of the chassis; they should be substantially the same thickness as the frame members. Generally, a cabinet constructed in this manner is notably lacking in rigidity, and all opportunities for stiffening should be employed. Usually, it is possible to cover the sides, back, top, and bottom with relatively light-gage sheet metal,

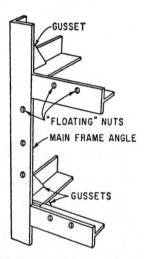

FIGURE 43.4 Frame construction for cabinet showing clearance holes for drawer or chassis front-panel thumbscrews. The frame members are welded, riveted, or bolted together. (*H. M. Forkois and K. E. Woodward.*[1])

securely attached to the structural members. Preferably, such sheets should be welded to the structural members to take full advantage of available rigidity, and they may be provided with welded or formed stiffeners, as illustrated in Figs. 43.5

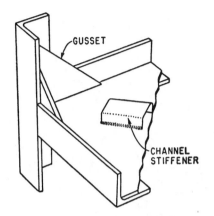

FIGURE 43.5 Welded channel stiffener to increase cabinet rigidity. (*H. M. Forkois and K. E. Woodward.*[1])

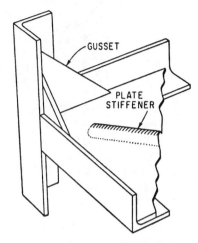

FIGURE 43.6 Coined stiffener to increase cabinet rigidity. (*H. M. Forkois and K. E. Woodward.*[1])

and 43.6. It is difficult to achieve the desired rigidity in the front face of the cabinet because the requirement for removal of chassis leaves this face almost unstiffened. The several smaller front panels of the individual chassis do not contribute as much stiffness as a single large sheet, and the screws and clamps commonly used to secure the chassis are not as effective as welding. This lack of rigidity is particularly noticeable if the equipment is supported at the bottom and braced to a bulkhead by a support adjacent only the upper rear edge.

A cabinet with several chassis commonly weighs several hundred pounds; the resultant loads at points of support may become very great under conditions of shock. Particular care in design is required to ensure that excessive deformation does not occur. For expedience of installation, it is preferable that the cabinet be mounted with a minimum of bolts. Usually it is necessary to provide additional structural members to distribute the forces from the bolts over much of the base area of the cabinet. Figure 43.7 is a bottom view of a cabinet showing an arrangement of channel stiffeners for distributing the forces. Such stiffeners are of particular convenience when used with shock isolators because the isolators nest in the channels, but they can also be used with direct mounting. Figure 43.8 shows a type of stiffener that may be used near the top of a cabinet for attachment to a bulkhead.

METHODS OF CONSTRUCTION

In selecting a method of construction, it is important to consider the suitability of the structure for use under conditions of shock and vibration. The greatest cause of dam-

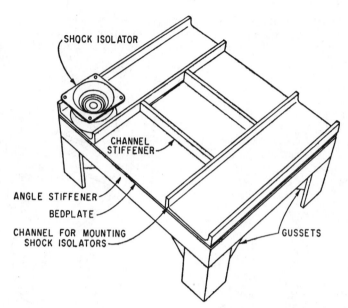

FIGURE 43.7 Bottom view of cabinet construction showing channel stiffeners and shock isolator. (*H. M. Forkois and K. E. Woodward.*[1])

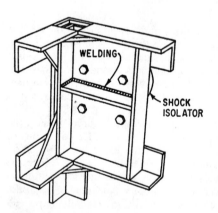

FIGURE 43.8 Structural stiffeners for bulkhead attachment near top of cabinet. (*H. M. Forkois and K. E. Woodward.*[1])

age in a structure subjected to vibration is the large vibration amplitude that tends to occur during vibration at resonance of lightly damped structures. The interface friction between bolted members or riveted members introduces damping to a significant degree. Therefore, bolted and riveted structures may be better adapted than welded structures to withstand vibration, particularly if damping is important. However, a welded structure may be designed with less stress concentration and with better ability to maintain alignment of component parts; thus, it may be the most suitable method of construction even though larger amplitudes of vibration result at resonance.

WELDED JOINTS

Welded joints must be well designed, and effective quality control must be imposed upon the welding conditions. The most common defect is excessive stress concentration which leads to low fatigue strength and, consequently, to inferior capability of withstanding shock and vibration. Stress concentration can be minimized in design by reducing the number of welded lengths in intermittent welding. For example, individual welds in a series of intermittent welds should be at least 1½ in. long with at least 4 in. between welds. Internal crevices can be eliminated only by careful quality control to ensure full-depth welds with good fusion at the bottom of the welds. Welds of adequate quality can be made by either the electric arc or gas flame process. Subsequent heat-treatment to relieve residual stress tends to increase the fatigue strength.

Spot-welded Joints. Spot welding is quick, easy, and economical but should be used only with caution when the welded structure may be subjected to shock and vibration. Basic strength members supporting relatively heavy components should not rely upon spot welding. However, spot welds can be used successfully to fasten a metal skin or covering to the structural framework. Even though improvements in spot welding techniques have increased the strength and fatigue properties, spot welds tend to be inherently weak because a high stress concentration exists in the junction between the two bonded materials when a tension stress exists at the weld. Spot-welded joints are satisfactory only if frequent tests are conducted to show that proper welding conditions exist. Quality can deteriorate rapidly with a change from proved welding methods, and such deterioration is difficult to detect by observation. However, accepted quality-control methods are available and should be followed stringently for all spot welding.

Riveted Joints. Riveting is an acceptable method of joining structural members when riveted joints are properly designed and constructed. Rivets should be driven hot to avoid excessive residual stress concentration at the formed head and to ensure that the riveted members are tightly in contact. Cold-driven rivets are not suitable for use in structures subjected to shock and vibration, particularly rivets that are set by a single stroke of a press as contrasted to a peening operation. Cold-driven rivets have a relatively high probability of failure in tension because of residual stress concentration, and tend to spread between the riveted members with consequent lack of tightness in the joint. Joints in which slip develops exhibit a relatively low fatigue strength.

Bolted Joints. Except for the welded joints of principal structures, the bolted joint is the most common type of joint. A bolted joint is readily detachable for changes in construction, and may be effected or modified with only a drill press and wrenches as equipment. However, bolts tend to loosen and require means to maintain tightness (see section on *Locking Devices*). Furthermore, bolts are not effective in maintaining alignment of bolted connections because slippage may occur at the joint; this can be prevented by using dowel pins in conjunction with bolts or by precision fitting the bolts; i.e., fitting the bolts tightly in the holes of the bolted members. Other characteristics of bolted joints are discussed in the following paragraphs.

BOLTS AND NUTS

Bolts and nuts are made by various methods from different materials. Grades of steel bolts as standardized by the SAE and the ASTM are shown in Table 43.1. The ordinary type (18-8) stainless steel bolts have a low yield point (about 25,000 lb/in.²); thus, they tend to stretch and loosen under shock, although they have a high ultimate strength. Where corrosion resistance as well as the ability to withstand shock is a consideration, bolts made from heat-treatable stainless-steel alloy having a high yield point should be used. The fatigue strength of bolts under vibratory loading can be improved by initial cold working of the surfaces. Cold working may result from rolling of thread roots, shot peening of shanks, or rolling of the fillet between head and shank.[3] Benefits are derived from both the work hardening and the initial compression left in the surfaces. Table 43.2 gives data on the endurance limit of bolts in pulsating tension.[2]

TABLE 43.1 Grades of Bolts (SAE and ASTM) (*ASME Handbook.*[2])

SAE grade	ASTM designation	Tensile strength,* lb/in.[2]	Proof load or yield strength,* lb/in.[2]	Material
0	...	...	...	Low carbon
1	A307	55,000	...	Low carbon
2	...	69,000	55,000	Low carbon, stress-relieved
3	...	110,000	85,000	Medium carbon, cold-worked
5	...	120,000	85,000	Medium carbon, quenched and tempered
	A325	125,000	90,000	Medium carbon, quenched and tempered
6	...	140,000	110,000	Special medium carbon, quenched and tempered
7	...	130,000	105,000	Alloy steel, quenched and tempered
8	...	150,000	120,000	Alloy steel, quenched and tempered

* Properties for sizes ½ in. diameter and smaller. For complete data and for properties of larger bolts, see SAE Handbook and ASTM specifications.

TABLE 43.2 Comparison of Tensile Fatigue Strengths of Bolts with Rolled, Cut, and Milled Threads (*H. Dinner and W. Felix.*[3])

Threading carried out by	Fatigue strength, lb/in.[2] (pulsating tension)	
	Unaged specimen	Specimen aged at 482°F
Rolling	28,450	28,450
Cutting	17,070	
Milling	17,070	12,800

The strength of a bolt when subjected to shock depends upon its capacity to absorb energy by elastic deflection. For optimum efficiency, all cross sections of the bolt should be designed to reach the yield stress simultaneously, where the stress is calculated by including stress concentrations. Figure 43.9 shows a bolt with the shank undercut slightly smaller than the root of the thread; therefore, the shank deflects

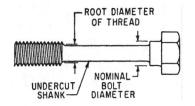

FIGURE 43.9 Shank of bolt undercut slightly smaller than root diameter of thread to obtain uniform elongation under shock load.

equally throughout its length and stores energy without overstressing the material in the threads. This effect can be accentuated by increasing the length of the shank. The effect of a reduction in the shank diameter of bolts made from three classes of steel is shown in Table 43.3. The composition and physical properties of the steels are given in the same table.

Stress raisers occur at abrupt changes in shape, such as shoulders, fillets, notches, grooves, and holes. They contribute to a lower endurance limit, and should be eliminated wherever possible. For example, Fig. 43.10A shows a stud with a smooth notch machined adjacent to the last remaining thread. Maximum bending stress then occurs at the notch, where the stress concentration is small, rather than at the root of a thread. The stud is closely fitted in the hole for a short length to reduce further the bending stress. Figure 43.10B shows a nut that is undercut so that the load is better distributed over several threads in the more flexible region adjacent to the undercut. Plastic deformation of the first highly loaded thread achieves somewhat the same result; this can be accomplished by a reverse taper or shorter pitch of the bolt threads relative to those of the nut. Subsequently, at lower bolt tensions, the load will be distributed more uniformly in the nut threads.[4]

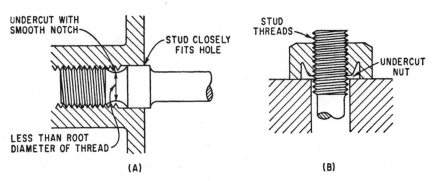

FIGURE 43.10 Methods of relieving stress concentration at threads: (A) Smooth notch used to relieve bending stress at thread. (B) Undercut nut to attain more uniform stress distribution in bolt. (*R. A. MacGregor.*[4])

Bolts usually are available in both coarse and fine thread series. Coarse thread bolts have deeper threads and need not be manufactured to as close a tolerance as fine thread bolts where care must be exercised to ensure a full thread depth. Fine thread bolts have a smaller helix angle; thus, they have less tendency to loosen when vibration is present.

It is good practice in assembling equipment intended to withstand shock and vibration to tighten a bolt until some yielding occurs in the bottom threads, thus ensuring maximum tightness. This reduces the stress change in a bolt when the loading is tension-directed and increases the resistance to fatigue.

TABLE 43.3　Effect of Bolt Diameter and Material on Impact Fatigue Strength (*W. Staedel*)

Bolt design	Bolt body			Repeated impact fatigue limit					
			Area, % root	Steel A		Steel B		Steel C	
	Diam., in.	Area, in.^{2}	diam. area	in.-lb	%	in.-lb	%	in.-lb	%
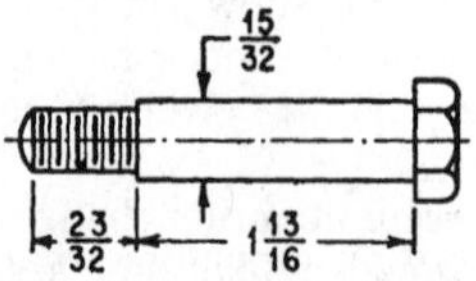	15/32	0.175	230	1.73	100	1.73	100		
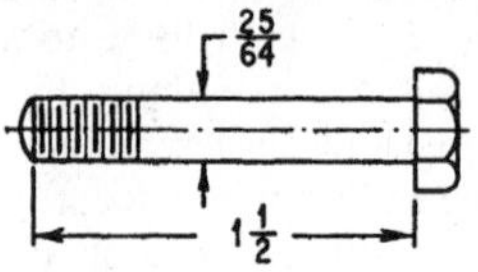	25/64	0.122	160	2.60	150	2.26	130	~2.2	100
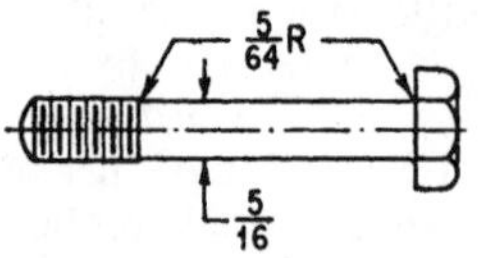	5/16	0.076	100	4.34	250	2.78	160	~3.5	160
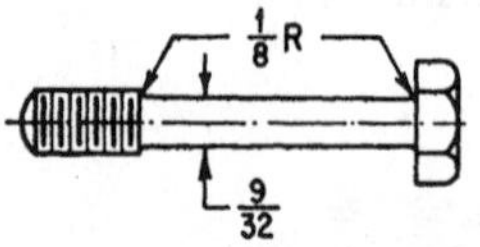	9/32	0.060	78	5.21	300	2.95	170	5.21	240
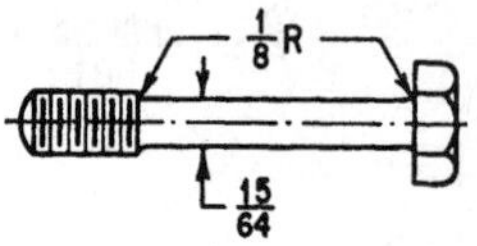	15/64	0.044	57	6.51	375	3.65	210		

	Composition, %						Red. in	Bolt*
	C	Mn	P	S	Si	Ultimate, lb/in.^{2}	Elong. %	area, %	ultimate, lb
Steel A	0.15	0.65	0.095	0.186	0.04	78,200	12.8	56.0	9,390
Steel B	0.25	0.75	0.019	0.024	0.25	96,400	12.8	54.0	11,800
Steel C	0.33	0.70	0.039	0.027	0.18	85,300–99,600	16–10	. . .	11,800

*Static tensile strength of actual bolt (for the second of the designs illustrated above) having $\frac{9}{32}$-in. length of free threads under nut

Locking Devices.　Locking devices include cotter pins, friction nuts, and lock washers. The castellated nut and cotter pin, as well as the locking wire inserted through holes in bolts and nuts, are devices to prevent positively the rotation of the nut relative to the bolt. Such devices must be removed each time the nut is tightened or loosened, and permit adjustment of the nut only at the discrete increments of the locking holes. In another type of locking device, additional thread friction is introduced by the design of the nut; this is a less positive means to retard loosening. Methods of introducing the friction are (1) the bolt is forced through an unthreaded, nonmetallic insert forming part of the nut; (2) the hole in the nut is elliptical and appreciable torque must be applied to turn it on the round bolt; and (3) the inner face of the nut is concave so that when tightened against a shoulder it distorts and binds the threads. In general, means to introduce additional friction are satisfactory when newly installed but tend to deteriorate with repeated use.

The split-ring lock washer is an initially flat washer that has been split and then sprung out of its plane slightly. It is flattened when the nut is tightened, and the resultant friction between washer and nut tends to prevent loosening of the nut. Its resilience enables it to maintain the friction even though the bolt deforms somewhat as a result of shock. Split-ring lock washers may become brittle and fracture; they then fall out of place and a loose joint results. Another type of lock washer embodies a sawtooth edge with an axial set to the teeth; it is adapted to wedge between the nut and the shoulder against which the nut is seated. The teeth are relatively rigid and cannot compensate for the additional space resulting from elongation of the bolt or indentation of the bolted members. This type of lock washer is not suitable for use in equipment subjected to severe shock.

Chassis Slides.   Accessibility to the chassis for maintenance and troubleshooting is improved by the use of ball-bearing slides with built-in tilting features. Slides are available to permit tilting the chassis either up or down, and also to permit locking at a number of fixed angular positions. Since ball-bearing slides are not designed to carry large loads, it is important that the weight of the chassis in the locked-up position be supported entirely by the dowel pins and front-panel fasteners.

Dowel Pins.   Dowel pins should be made of steel to avoid excessive abrasion. In general, they vary in diameter from ¼ to ½ in., and should be no longer than 1 in. to minimize bending stresses. The pin receptacle should also be made of steel with a close fit to the pin. It usually is better practice to locate the pin on the frame of the cabinet and the pin receptacle on the chassis because the frame is better able to support the pin as a cantilever. Figure 43.11 illustrates a pin attached to a riveted or bolted gusset, and Fig. 43.12 shows a pin installed on a welded gusset. Maintaining the necessary alignment of close-fitting dowels (maximum clearance on diametral fit

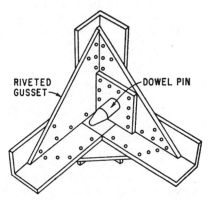

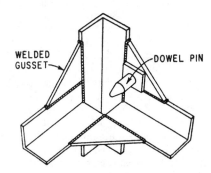

FIGURE 43.11 Dowel pin arrangement for riveted or bolted gusset. (*H. M. Forkois and K. E. Woodward.*[1])

FIGURE 43.12 Dowel pin arrangement for welded gusset. (*H. M. Forkois and K. E. Woodward.*[1])

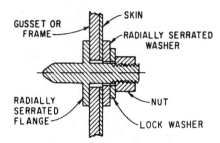

FIGURE 43.13 Adjustable dowel pin. A maximum adjustment of ³⁄₁₆ in. in any direction is obtainable as a result of the oversized hole in the frame. (*H. M. Forkois and K. E. Woodward.*[1])

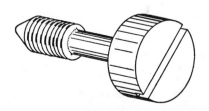

FIGURE 43.14 Reduced-shank front-panel thumbscrew which is threaded into a threaded hole in the chassis until the threaded portion passes through the hole; the reduced shank portion then rests in the threaded hole in a captive position which permits turning freely to engage a threaded hole in the frame. (*H. M. Forkois and K. E. Woodward.*[1])

of 0.005 in.) is difficult, particularly where interchangeability of chassis is a requirement. Figure 43.13 shows a dowel pin clamped to the frame and held in place by a serrated flange and washer. A maximum adjustment of ³⁄₁₆ in. in any direction is obtainable because of the oversize hole in the frame.

Quick-release Fasteners. Only a fraction of a turn is necessary to lock or release a device in which a spring element is used to maintain tension in the "locked-up" position. Shock and vibration forces of low magnitude tend to overcome the spring force and cause looseness in the assembly. Under shock, the spring or the cam rider (the member which engages the spring and through which the loads are transmitted from the captive mass to the support) may deform or fracture. Hence, quick-release fasteners should not be used to carry large loads but rather may be used to mount such items as inspection plates or covers.

Front-panel Thumbscrews. Figure 43.14 illustrates a captive fastener used to secure the chassis to the frame of the cabinet. It is knurled and slotted to facilitate loosening and tightening. The thumbscrew is threaded into a threaded hole in the chassis until the threaded portion passes through the hole; the reduced shank portion then rests in the threaded hole in a captive position which permits turning freely to engage a threaded hole in the frame. Reduced shank diameters vary according to design requirements but should not be less than ³⁄₁₆ in. for lightweight or ⁵⁄₁₆ in. for mediumweight equipment. Corresponding thread sizes are ¼ in.-20 and ⅜ in.-16.

DAMPING

A direct approach to the reduction of vibration resulting from resonance is to supplement the inherent hysteresis damping of materials by other means to dissipate energy. A structure fabricated from a number of component parts exhibits a marked degree of energy dissipation compared with a similar structure formed in a single piece. The method of fabrication influences the degree of energy dissipation. Bolted and riveted joints are most effective in dissipating energy because they permit a limited slippage at the interfaces between members in contact while maintaining pressure at the interface. Welded joints also exhibit considerable damping, less than bolted or riveted joints but substantially more than solid materials. This apparently is the result of some slippage between members in contact not restrained fully by the weld.

Energy Dissipation. The damping or energy dissipation resulting from slippage of parts in a fabricated structure is termed *slip damping*. Various types of structures have been found to exhibit damping properties that are known within wide and approximate limits. Knowledge of such properties in a quantitative sense would assist in the design of equipment having desirable energy dissipation.

Frictional effects in composite structures are not well adapted to analysis except under ideal conditions. As guidance for the designer, it is convenient to define the damping in terms of the *equivalent viscous damping*. The inverse of the fraction of critical damping ζ is proportional to the amplification at resonance and makes it possible to estimate the severity of a resonance. Table 43.4 gives typical values of the fraction of critical damping ζ found in various types of structures. The damping varies widely from structure to structure of a given type; for example, in a given structure, it varies significantly with a change in tightness of bolts. Thus, Table 43.4 should be interpreted only as an indication of the relative magnitudes of damping in different types of structures, and of the variation in damping that may be expected from a change in design.

Damping can be added to members in flexure by utilizing the shear stress at the neutral plane to deform a layer of viscous material and attain a relatively high degree of energy dissipation. This sometimes is known as *sandwich construction*. The

TABLE 43.4 Typical Values of Damping for Different Methods of Construction

Method of construction	Effective damping,* ζ
One-piece unit	0.01
Welded assembly	0.02
Riveted assembly	0.04
Bolted assembly	0.05

* ζ = fraction of critical damping.

concept as applied to the design of structural members is illustrated in Fig. 43.15 where a cantilever beam of rectangular cross section (inset *A*) is compared with a sandwich-type beam (inset *B*) having the same length and natural frequency. The viscoelastic layer provides a partial constraint in shear between the separated portions of the beam; thus, the natural frequency of the sandwich-type beam is lower than that of a solid beam having the same overall dimensions, but greater than that of one-half of the sandwich-type beam. The effective value of the fraction of critical damping $\zeta = c/c_c$ for the beam depends upon the viscosity of the viscoelastic layer and decreases as the load carried by the beam increases.

The sandwich-type construction has inherent limitations. For a given overall dimension, a sandwich-type structure is less rigid than a structure using solid members. From a fabrication viewpoint, a sandwich-type structure is not well adapted to welding; appropriate bolted connections preferably should include means to relieve the viscoelastic layer of bolt forces. Finally, the viscosity of the viscoelastic layer tends to vary with temperature. Nevertheless, a large increase in damping is attainable by use of sandwich-type construction and may justify use of the concept for certain applications.

Members that deflect in flexure also may be damped by the application of vibration damping material that dissipates energy. Its effectiveness is a function of the energy dissipated by the undercoating relative to the elastic energy of the structure

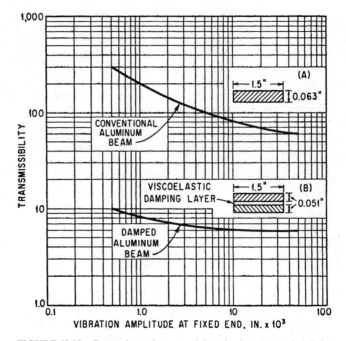

FIGURE 43.15 Comparison of structural damping in a conventional aluminum beam with that in a damped aluminum beam having a viscoelastic damping layer. The beam is a cantilever 11 in. long with cross section (A) for conventional beam and (B) for damped beam. Natural frequency is 18 Hz. Transmissibility is ratio of amplitude at free end to amplitude at fixed end, during vibration at resonance. *(J. E. Ruzicka.)*

being damped. For this reason, it finds principal application to relatively light structures of large area, such as hoods and trunk covers of automobiles or enclosures for household appliances.

When vibration of an equipment or structure is characterized by a large amplitude at a definable point, a possible method of reducing the vibration level involves the attachment of a damper. Such a device dissipates energy as it extends and compresses. A damper can be attached between a vibrating structure and a relatively fixed structure, or between two structures vibrating in opposite directions.

Auxiliary Mass Dampers.   A tuned damper or dynamic vibration absorber may be attached directly to the vibrating structure; it consists of a simple mass-spring system that may be damped or undamped. Reduction of vibration is achieved because the mass vibrates out of phase with the vibration to be eliminated; sometimes this is called *sympathetic vibration.*

An auxiliary mass damper without a spring or damping is designated an *acceleration damper.* It consists of small mass elements free to move or slide in a sealed container that is attached to the vibrating system. Incipient vibration is reduced by the multiple collisions and friction between the mass elements, with consequent transfer of momentum and conversion of mechanical energy into heat. For example, when electrical contacts are closed, they tend to chatter because of the elastic impact; to minimize the chatter, a small container partially filled with small mass elements is attached to the movable contact. The abrupt motion of the contactor arm tosses the powder about inside the container. The result is a dull impact without rebound, because the relative motion and friction of the mass elements absorb the elastic energy involved in the stress waves set up by the impact.

Analogous results are attainable with a single massive member contained within a box attached to a member whose vibration is to be reduced. Analyses of this damper indicate that its effectiveness is the result of momentum transfer induced by multiple impacts within the box; however, the analytical treatment is not sufficiently complete to permit design of a damper without supplementary test work.

An ideal material to use as the mass elements is tungsten powder. It has rough individual grains to create friction, is hard enough to minimize wear, and is sufficiently massive. A half-and-half mixture of 150–200 and 200–250 mesh powders is commercially available.

Sand or steel shot may be used as the damping material where the quantity required would make tungsten powder prohibitively expensive. Some experimental results are available for the damping attainable from a flat, rectangular tray partially

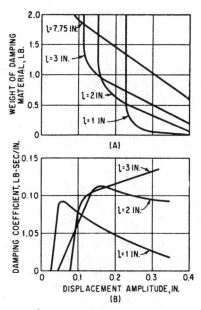

filled with steel shot of No 15 grit. The tray was attached to a vibrating table and subjected to a vibratory motion in the lengthwise direction of the tray at a frequency of 13 Hz. (The latter frequency happened to be of particular interest because it corresponded to the natural frequency of a radar antenna which was being set into vibration by gusts of wind.) The vibration table was put into motion by means of a rotating unbalance weight supported in bearings attached to the underside of the table. Various tray lengths and different weights of shot were tried; the width of the tray was fixed at 3 in.

Figure 43.16A shows that with a given rotating unbalance weight causing a harmonic forcing function $F = 1.7$ lb, the displacement amplitude could be reduced by adding more weight of damping material to the tray. At each of several different tray lengths an optimum weight of damping material was found beyond which the resulting displacement amplitude could no longer be reduced by the addition of damping material. The optimum weight of damping material was found to be larger as the length of the tray was increased. The results plotted in Fig. 43.16A indicate that energy was absorbed primarily because of sliding friction of the mass particles moving as a unit relative to the bottom and sides of the tray.

FIGURE 43.16 Performance curves for steel shot damper attached to a table driven by a rotating unbalance at a frequency of 13 Hz: (A) Variation of vibration amplitude as a function of weight of damping material for several lengths l of tray; driving force amplitude is 1.7 lb. (B) Damping coefficient as a function of vibration amplitude, calculated from driving force and vibration amplitude. Weight of damping material is 0.91 lb; the tray length $l = 3$ in. is divided into three 1-in.-long compartments. (G. O. Sankey.[5])

Figure 43.16*B* shows the results when the tests were repeated for the same weight of damping material $W = 0.91$ lb but located in trays of various lengths. The magnitude of the rotating unbalance weight was adjusted successively to establish various displacement amplitudes while maintaining the frequency at 13 Hz; the equivalent viscous damping coefficient was calculated at each amplitude. For the shorter tray lengths, the damping coefficient increased rapidly as the displacement amplitude increased; however, it reached a peak value and then decreased gradually as the displacement amplitude was further increased. The exception was the 3-in.-long tray, which was partitioned into three 1-in.-long compartments, and gave the most effective damping action. The rising characteristic for the 3-in. tray shown in Fig. 43.16*B* indicates that the damping coefficient continues to increase with displacement amplitude and indicates a safe, stable design for the damper. By using partitions to reduce the lengthwise freedom for relative motion, the sliding friction is supplemented by the more effective turbulence of the damping material. The increased energy absorption indicated by the larger damping coefficient was caused principally by the multiple, inelastic collisions. However, depending upon circumstances, the net damping is the result of the combined sliding friction and inelastic collisions of the mass elements.

MALFUNCTION OF EQUIPMENT

Failure of equipment as a result of shock and vibration may consist of (1) damage so severe that the ability of the equipment to perform its intended function is impaired permanently or (2) temporary disruption of normal operation in a manner permitting restoration of service by subsequent adjustment of the equipment or termination of the disturbance. Such difficulties may be corrected by applying the principle set forth as a guide to the design of structures to withstand shock and vibration.

WIRING AND CIRCUIT BOARDS

WIRES, CABLES, AND CONNECTORS

Wires, cables, and connectors are not in themselves susceptible to damage from shock and vibration. They can cause considerable difficulty, however, if improper methods are used in their application. Wires and cables have a high degree of flexibility as compared with other electronic component parts. This flexibility results from (1) the type of construction, (2) the kind of material, and (3) the high ratio of length to cross-sectional area. Because of this flexibility, the natural frequency of a span of wire or cable is low and may easily fall within the frequency range of existing vibration. The vibration of a wire or cable produces time-varying stresses which can lead to fatigue failure. The areas which usually are first to fail are the ends of the vibrating span, i.e., at the wire or cable terminations. Failures also may occur where the insulation rubs against an adjacent part, or where one conductor in a multiconductor cable rubs against another within the cable body. A cable passing through a hole in a partition should be protected by a grommet lining the hole.

Where several wires extend in the same direction, their stiffness may be increased by harnessing the wires with lacing cord to form an integral bundle. Harnessing also provides damping because of friction between the wires. The harnessed length should be supported adequately so that its weight will not load the wire termination points unduly. The relatively high degree of damping of stranded wire and cable is beneficial when they are subjected to flexing as a result of vibration (see the discussion of *Cables* in Chap. 15).

Materials and Construction.   Although all wire and cable are flexible when compared to other components, the degree of flexibility depends in part upon the material composition. The flexibility may vary widely with temperature, since many organic insulating materials stiffen considerably with decreased temperature and soften with increased temperature. Fluorinated hydrocarbons and silicone elastomers maintain flexibility at low temperatures better than most others.

Lead Length.   The bending of a wire or cable may be reduced at its termination or support by shortening the length of lead between supports. This applies only in cases where the supports *do not* move relative to one another. The shortening of the span of wire or cable accomplishes two beneficial results: (1) it increases the natural frequency of the span and (2) it reduces the weight and inertial loads at each support. Where supports or terminals *do* move relative to one another, lead failures can be reduced considerably by allowing some slack in the lead.

Methods of Attachment.   Wire or cable failures due to shock and vibration usually take place where the lead is interrupted, either at a termination point or at a support along the lead. Failures occur primarily because these locations are subjected to the greatest bending stress. When wire is terminated by soldering, the area most susceptible to shock and vibration lies between the end of the insulation and the rigidly "frozen" section at the terminal post shown in Fig. 43.17. This portion of the wire is subjected to severe flexing and undergoes fatigue very quickly. Figure 43.17 indicates how solder flows up the strands of the conductor by capillary action and concentrates all bending action into a very small area of wire. Clamping the wire to a structure can restrict the motion and reduce damage from flexing at the termination.

When wire is terminated by an end support lug, the lug should grip both the insulation and the conductor. When cable is terminated by a connector, the connector should be attached to the cable so that it grips the cable jacket to restrict flexing of individual wires within the cable body. Attaching a sleeve as shown in Fig. 43.18 prevents severe cable bending at a termination point. Bending of individual wires can be restricted by potting the cable end permanently into the connector. The connections are made in a normal manner; then the connector body is filled with a self-curing plastic material to encapsulate the end of the cable completely. The disadvantage to potting a cable is that potted components are difficult to repair.[7]

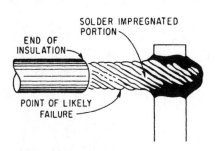

FIGURE 43.17 View of cable end illustrating flow of solder along strands to form a rigid mass of cable. Failure is most likely to occur at edge of solder-impregnated portion. (*R. E. Barbiere and W. Hall.*[6])

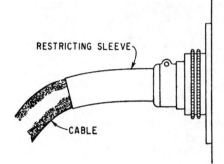

FIGURE 43.18 Restricting sleeve to prevent severe cable bending at a termination point. (*R. E. Barbiere and W. Hall.*[6])

Printed Circuit Boards. Printed circuit boards usually require a stiffening structure to prevent low resonant frequencies. The stiffening may take the form of a metal chassis, conformal coating, or complete potting in plastic. Small circuit boards may be rigid enough without additional stiffening. To apply conformal coating, the side of the board having the component parts is dipped into a clear plastic material that hardens to form a rigid assembly.

Because plug-in-type circuit boards are easy to insert and remove, they are widely used. The support for such circuit boards is an important factor in determining their response to shock and vibration.

REFERENCES

1. Forkois, H. M., and K. E. Woodward: "Design of Shock- and Vibration-resistant Electronic Equipment for Shipboard Use," U.S. Navy Department, Bureau of Ships, *NAVSHIPS* 900, 185A, 1957.

2. ASME Handbook: "Metals Engineering—Design," 2d ed., McGraw-Hill Book Company, Inc., New York, 1965.

3. Dinner, H., and W. Felix: *Engineers Digest,* **3**(2):85 (1946).

4. MacGregor, R. A., W. S. Burn, and F. Bacon: *Trans. North East Coast Inst. Engrs. & Shipbuilders,* **51** (1935).

5. Sankey, G. O.: "Some Experiments on a Particle or Shot Damper," unpublished memorandum, Westinghouse Research Laboratories, Pittsburgh, Pa.

6. Barbiere, R. E., and W. Hall: "Electronic Designer's Shock and Vibration Guide for Airborne Applications," *WADC Tech. Rept.* 58-363, ASTIA No. AD 204095, December 1958.

7. Steinberg, D. S.: "Vibration Analysis for Electronic Equipment," John Wiley & Sons, Inc., New York, 1988.

CHAPTER 44

EFFECTS OF SHOCK AND VIBRATION ON HUMANS

Henning E. von Gierke
Anthony J. Brammer

INTRODUCTION

This chapter considers the following problems: (1) the determination of the structure and properties of the human body regarded as a mechanical as well as a biological system, (2) the effects of shock and vibration forces on this system, (3) the protection required by the system under various exposure conditions and the means by which this protection is to be achieved, and (4) tolerance criteria for shock and for vibration exposure. Man, as a mechanical system, is extremely complex and his mechanical properties readily undergo change. There is limited reliable information on the magnitude of the forces required to produce mechanical damage to the human body. To avoid damage to humans while obtaining such data, it is necessary to use experimental animals for most studies on mechanical injury. However, the data so obtained must be subjected to careful scrutiny to determine the degree of their applicability to humans, who differ from animals not only in size but in anatomical and physiological structure as well. Occasionally it is possible to obtain useful information from situations involving accidental injuries to man, but while the damage often can be assessed, the forces producing the damage cannot, and only rarely are useful data obtained in this way. It is also very difficult to obtain reliable data on the effects of mechanical forces on the performance of various tasks and on subjective responses to these forces largely because of the wide variation in the human being in both physical and behavioral respects. Measurement of some of the mechanical properties of man is, however, often practicable since only small forces are needed for such work. The difficulty here is in the variability and lability of the system and in the inaccessibility of certain structures.

One section of this chapter is devoted to a discussion of methods and instrumentation used for mechanical shock and vibration studies on man and animals. Subsequent sections deal with the mechanical structure of the body, the effects of shock and vibration forces on man, the methods and procedures for protection against these forces, and safety criteria.

For general background material on the effects of shock and of vibration on man, see Refs. 1 through 5.

DEFINITIONS AND CHARACTERIZATION OF FORCES

Characterization of Forces. Forces may be transmitted to the body through a gas, liquid, or solid. They may be diffuse or concentrated over a small area. They may vary from tangential to normal and may be applied in more than one direction. The shape of a solid body impinging on the surface of the human is as important as the position or shape of the human body itself. All these factors must be taken into account in comparing injuries produced by vehicle crashes, explosions, blows, vibration, etc. Laboratory studies often permit fairly accurate control of force application, but field situations are apt to be extremely complex. Therefore it is often very difficult to predict what will happen in the field on the basis of laboratory studies. It is equally difficult to interpret field observations without the benefit of laboratory studies.

Shock. The term *shock* is used differently in biology and medicine than in engineering; therefore one must be careful not to confuse the various meanings given to the term. In this chapter the term *shock* is used in its engineering sense as defined in Chap. 1 of this Handbook. A *shock wave* is a discontinuous pressure change propagated through a medium at velocity greater than that of sound in the medium. In general, forces reaching peak values in less than a few tenths of a second and of not more than a few seconds' duration may be considered as shock forces in relation to the human system.

The term *impact* (i.e., a *blow*) refers to a force applied when the human body comes into sudden contact with a solid body and when the momentum transfer is considerable, as in rapid deceleration in a vehicle crash or when a rapidly moving solid body strikes a human body.

Vibration. Biological systems may be influenced by vibration at all frequencies if the amplitude is sufficiently great. This chapter is concerned primarily with the frequency range from about 1 Hz to 1 kHz. (Studies at higher frequencies are very useful for the analysis of tissue characteristics.)

METHODS AND INSTRUMENTATION

Most quantitative investigations of the effects of shock and vibration on humans are conducted in the laboratory in controlled, simulated environments. Meaningful results can be obtained from such tests only if measurement methods and instrumentation are adapted to the particular properties of the biological system under investigation to ensure noninterference of the measurement with the system's behavior. This behavior may be physical, physiological, and psychological although these parameters should be studied separately if possible. The complexity of a living organism makes such separation, even assuming independent parameters, only an approximation at best. In many cases if extreme care is not exercised in planning and conducting the experiment, uncontrolled interaction between these parameters can lead to completely erroneous results. For example, the dynamic elasticity of tissue of a certain area of the body may depend on the simultaneous vibration excitation of other parts of the body; or the elasticity may change with the duration of the measurement since the subject's physiological response varies; or the elasticity may be influenced by the subject's psychological reaction to the test or to the measurement equipment.

Control of, and compensation for, the nonuniformity of living systems is absolutely essential because of the variation in size, shape, sensitivity, and responsiveness of people and because these factors, for a single subject, vary with time, experience, and motivation. The use of adequate statistical experimental design is necessary and almost always requires a large number of observations and carefully arranged controls.

PHYSICAL MEASUREMENTS

In determining the effects of shock and vibration on humans, the mechanical force environment to which the human body is exposed must be clearly defined. Force and vibration amplitudes should be specified for the area of contact with the body. Vibration measurements of the body's response should be made whenever possible by noncontact methods. X-ray methods can be used successfully to measure the displacement of internal organs. Optical, cinematographic, and stroboscopic observation can give the displacement amplitudes of larger parts of the body. Small vibrations sometimes can be measured without contact by capacitive probes located at small distance from the (grounded) body surface (see *Variable Capacitance Pickups,* Chap. 12). If vibration pickups in contact with the body are used, they must be small and light enough so as not to introduce a distorting mechanical load. This usually places a weight limitation on the pickup of a few grams or less, depending on the frequency range of interest and the effective mass to which the pickup is attached. Figure 44.1 illustrates the effect of mass and size on the response of accelerometers attached to the skin overlying soft tissue. The lack of rigidity of the human body as a

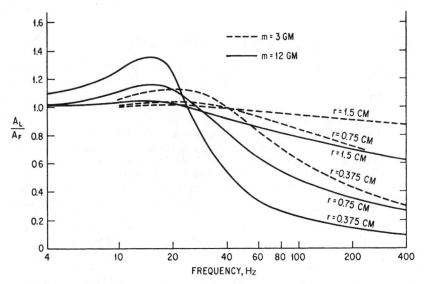

FIGURE 44.1 Amplitude distortion due to accelerometers of different mass *m* and size which are attached to a body surface over soft tissue of human subject exposed to vibration. The graph gives the ratio A_L/A_F of the response of the loaded to the unloaded surface for accelerometers having three different radii *r*. *(Values calculated from unpublished mechanical surface impedance data of E. K. Franke and H. E. von Gierke.)*

supporting structure makes measurements of acceleration usually preferable to those of velocity or displacement. The mechanical impedance of a sitting, standing, or supine subject is extremely useful for calculating the vibratory energy transmitted to the body by the vibrating structure. The mechanical impedance of small areas of the body surface can be measured in different ways (see Chap. 12), for example by vibrating pistons, resonating rods, and acoustical impedance tubes.

If the entire body is exposed to a pressure or blast wave in air or water, exact definition of the pressure environment is essential. The pressure distribution should be measured if possible. If the environment deviates from free-field conditions, it should be carefully specified because of its effect on peak pressure and pressure *vs.* time-history.

BIOLOGICAL MEASUREMENTS

Instrumentation for the measurement of physiological properties such as blood pressure, respiration rate or depth, heart potentials, brain potentials, or galvanic skin response must be carefully checked for freedom from artifacts when the subject, the instrument, or both are exposed to vibration, intense sound, or acceleration.

If psychological experiments during exposure to the mechanical stimuli are planned, adherence to established procedures for such subjective tests is an absolute necessity to obtain valid results. The maintaining of a neutral situation with uniform motivation, subject instruction, and adequate statistical design of the experiment are some of the most important considerations. Care must be taken that the subject be not biased or influenced by environmental factors not purposely included in the test (e.g., the noise of a vibration table can act as a disturbing factor in a study of vibration effects).

SIMULATION OF MECHANICAL ENVIRONMENT

The desire to study the physical, physiological, and psychological responses of biological specimens in the laboratory, under well-controlled conditions, has led to the use of standard and specialized shock and vibration testing machines for experiments on man and animals. A summary of some machines employed in such tests is given in Fig. 44.2. An accurate simulation of the environmental conditions to which man is exposed frequently is not feasible for technical, ethical, or economic reasons, or may even be undesirable because of a need for more systematic investigation under somewhat simplified conditions. Thus most investigations are limited to the study of a single degree-of-freedom at a time, in which the human test specimen is vibrated only in one direction. Many fundamental studies are performed with sinusoidal forces. Mechanical and electrodynamic shake tables are usually employed for this purpose. Requirements for all shock and vibration machines used include: adequate safety precautions, safe and accurate control of the exposure, and sufficient load capacity for subject, seat, and instrumentation. Since the law of linear superposition is valid only in the linear physical domain, sinusoidal forces alone are not adequate for the study of nonlinear physical responses or physiological and psychological reactions to complex force functions. Therefore, some of the machines listed are designed uniquely for exposure of humans. One vertical accelerator, for example, employs a friction-drive mechanism to permit the simulation of large amplitude sinusoidal and random vibrations such as those encountered in buffeting during low-altitude high-speed flight or anticipated during the launch and reentry phases of spacecraft. This device can be programmed with acceleration recordings obtained under actual flight conditions. Machines for the study of human tolerance to ejection from high-speed aircraft (ejection seat) have upward or downward acceleration tracks with sliding seats projected by explosive charges. Horizontal tracks with rocket-propelled sleds, which can be stopped by special braking mechanisms, have been used to study the effects of linear decelerations similar to those occurring in automobile or aircraft crashes. Studies of combinations of static acceleration and vibration have been carried out by mounting vibrators on centrifuges. Blast tubes, sirens, and body respirators are used to study the body's response to pressure distributions surrounding it. At low frequencies, the respirator is valuable in studying the response of the lung-thorax system. Small vibrating pistons, which are available for a wide frequency range, have been used in investigating the mechanical impedance of small surfaces, the transmission of vibration and the physiological reaction to localized excitation.

TYPE OF MACHINE	APPLICATION OF FORCE	FORCE–TIME FUNCTION	FREQUENCY RANGE	MAXIMUM AMPLITUDE
SHAKE TABLE			MECHANICAL 0–50 Hz	UP TO 15g
			ELECTRODYNAMIC 15–1,000 Hz	
VERTICAL ACCELERATOR			0–10 Hz	±10 FT, 3.7g PEAK
SHOCK MACHINE			DOWN TO $T = 0.16$ SEC $\tau = 2 \cdot 10^{-3}$ SEC	PEAK AMPLITUDE 10^{-2} TO 10^{-1} CM
HORIZONTAL OR VERTICAL DECELERATOR OR ACCELERATOR (SLEDS ON TRACKS, CAR, DROP-TOWER)			RATE OF ACCELERATION UP TO 1,400 g/SEC	40g PEAK
BLAST TUBE FIELD EXPLOSION				
SIREN (AIRBORNE SOUND)			25–100,000 Hz	160–170 dB RE 20 µPa
RESPIRATOR			0–15 Hz	
VIBRATOR (SMALL PISTON)			0–10 MHz	
SHAKER ON CENTRIFUGE	ALTERNATING g FORCE IN ADDITION TO STATIC g FIELD		0–3 Hz	
HEAD IMPACT MACHINE FOR DUMMY HEADS			DEPENDING ON STRUCTURE STRUCK	IMPACT VELOCITY 140 FT/SEC

FIGURE 44.2 Summary of characteristics of shock and vibration machines used for human and animal experiments. Force-time functions are indicated schematically. Frequency range and maximum amplitudes refer to values used, not to capabilities of machines.

SIMULATION OF HUMAN SUBJECTS

The establishment of limits of human tolerance to mechanical forces, and the explanation of injuries produced when these limits are exceeded, frequently requires
experimentation at various degrees of potential hazard. To avoid unnecessary risks
to humans, animals are used first for detailed physiological studies. As a result of
these studies, levels may be determined which are, with reasonable probability, safe
for human subjects. However, such comparative experiments have obvious limitations. The different structure, size, and weight of most animals shift their response
curves to mechanical forces into other frequency ranges and to other levels than
those observed on humans. These differences must be considered in addition to the
general and partially known physiological differences between species. For example, the natural frequency of the thorax-abdomen system of a human subject is
between 3 and 4 Hz; for a mouse the same resonance occurs between 18 and 25 Hz.
Therefore maximum effect and maximum damage occur at different vibration frequencies and different shock-time patterns in a mouse than in a human. However,
studies on small animals are well worth making if care is taken in the interpretation
of the data and if scaling laws are established. Dogs, pigs, and primates are used
extensively in such tests.

Many kinematic processes, physical loadings, and gross destructive anatomical
effects can be studied on dummies which approximate a human being in size, form,
mobility, total weight, and weight distribution in body segments. Several such dummies are commercially available. In contrast to those used only for load purposes,
dummies simulating basic static and dynamic properties of the human body are
called *anthropometric* or *anthropomorphic* dummies. They are used extensively in
aviation and automotive crash research. In other studies dummies are used in place
of human subjects to evaluate protective seats and harnesses. In such dummies, an
attempt is made to match the "resiliency" of human flesh by some kind of padding.
However, they are crude simulations at best, and their dynamic mechanical properties are, if at all, only reasonably matched in a very narrow low-frequency range. This
and the passiveness of such dummies constitute important mechanical differences
between them and living subjects.

Efforts have been made to simulate the mechanical properties of the human
head in order to study the physical phenomena occurring in the brain during crash
conditions. Although these head forms only approximate the human head, they are
very useful in the evaluation of the protective features of crash, safety, and antibuf-
fet helmets. Plastic head forms, conforming to standard head measurements, are
designed to fracture in the same energy range as that established for the human
head. A cranial vault is provided to house instrumentation and a simulated brain
mass with comparable weight and consistency (a mixture of glycerin, ethylene gly-
col, etc.). The static properties of the skin and scalp tissue are simulated with
polyvinyl foam.

The static and dynamic breaking strength of bones, ligaments, and muscles and
the forces producing fractures in rapid decelerations have been studied frequently
on cadaver material. Extreme caution must be exercised in applying elastic and
strength properties obtained in this way to a situation involving the living organism. The differences observed between properties of wet, dry, and embalmed materials are considerable; changes in these properties also result in changes in the
force distribution of a composite structure. Thus a multitude of physiological,
anatomical, and physical factors must be considered for each specific situation in
which the use of animals, dummies, or cadavers as substitutes for live human subjects is planned.

PHYSICAL CHARACTERISTICS OF THE BODY

PHYSICAL CONSTANTS AND MECHANICAL TRANSMISSION CHARACTERISTICS

Physical Data. This section summarizes the passive mechanical responses of the human body and tissues exposed to vibration and impact. The data can be used to calculate quantitatively the transmission and dissipation of vibratory energy in human body tissue, to estimate vibration amplitudes and pressures at different locations of the body, and to predict the effectiveness of various protective measures. Table 44.1 lists some dynamic mechanical characteristics of the body and indicates some of the fields where these data find application. In cases where detailed quantitative investigations are lacking, the information may serve as a guide for the explanation of observed phenomena or for the prediction of results to be expected. Most physical characteristics of the human body presented in this section (except for the strength data) have been derived from the analysis of experimental data in which it is assumed that the body is a linear, passive mechanical system. This is an idealization which holds only for very small amplitudes. Therefore these data may not apply in analyses of mechanical injury to tissue. There is actually considerable nonlinearity of response well below amplitudes required for the production of damage. This is indicated, for example, by the data given in Fig. 44.3, which shows how the mechanical stiffness and resistance of soft tissue vary with static deflection. Bone behaves more or less like a normal solid; however, soft elastic tissues such as muscle, tendon, and connective tissue resemble elastomers with respect to their Young's modulus and S-shaped stress-strain relation. These properties have been studied in connection with the quasi-static pressure-volume relations of hollow organs such as arteries, the heart, and the urinary bladder, assuming linear properties in studying dynamic responses. Then soft tissue can be described phenomenologically as a viscoelastic medium; plastic deformation need be considered only if injury occurs. Physical properties of human body tissue are summarized in Table 44.2 for frequencies less than 100 kHz.

The fatigue life of bone and soft tissue in response to cyclic dynamic stress at frequencies between 0.5 and 4 Hz is summarized in Fig. 44.4. In this diagram, the number of cycles to failure N of in vitro preparations is expressed as a function of the ratio of the applied dynamic stress to the ultimate static stress σ/σ_u. The straight lines in Fig. 44.4 represent the functions $N = (\sigma/\sigma_u)^{-x}$, where the value of the index x in the relationship is indicated.

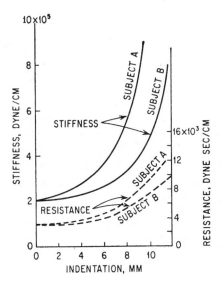

FIGURE 44.3 Mechanical stiffness and resistance of soft tissue, per square centimeter, as a function of indentation (i.e., static deflection). The nonlinearity shows the effect of loading of body surface on surface impedance of soft tissues for two experimental human subjects *A* and *B*. (*After Franke: USAF Tech. Rept. 6469, 1959.*)

TABLE 44.1 Application of Mechanical Studies of Body

Dynamic mechanical quantity investigated	Field of application
Skull resonances and viscosity of brain tissue	Head injuries; bone-conduction hearing
Impedance of skull and mastoid	Matching and calibration of bone-conduction transducers; ear protection
Ultrasound transmission through skull and brain tissue	Brain tumor diagnosis; changes in central nervous system exposed to focused ultrasound
Sound transmission through skull and tissue	Bone-conduction hearing
Mechanical properties of outer, middle, and inner ear	Theory of hearing; correction of hearing deficiencies
Resonances of mouth, nasal, and pharyngeal cavities	Theory of speech generation; correction of speech deficiencies; oxygen mask design
Resonance of lower jaw	Bone-conduction hearing
Response of mouth-thorax system	Blast-wave injury; respirators
Propagation of pulsed cardiac pressure	Circulatory physiology; hemodynamics
Heart sounds	Physiology of heart; diagnosis
Suspension of heart	Ballistocardiography; injury from severe vibrations and crash
Response of the thorax-abdominal mass system	Severe vibration and crash injury; crash protection
Impedance of subject sitting, standing, or lying on vibration platform	Isolation and protection against vibration and short-time accelerations; ballistocardiography
Hand-arm impedance	Isolation and protection against hand-transmitted vibration; design of power tools; design of test fixtures for hand-tool vibration assessment
Impedance of body surface, surface wave velocity, sound velocity in tissue, absorption coefficient of body surface	Theory of energy transmission and attenuation in tissue; determination of tissue elasticity, viscosity and compressibility; determination of acoustic and vibratory energy entering the body; vibration isolation; design of vibration pickups; transfer of vibratory energy to inner organs and sensory receptors
Absorption and attenuation of ultrasound in tissue	Biological effects of ultrasound; soft tissue imaging; determination of doses for ultrasonic therapy

The combination of soft tissue and bone in the structure of the body together with the body's geometric dimensions results in a system which exhibits different types of response to vibratory energy depending on the frequency range: At low frequencies (below approximately 100 Hz), the body can be described for most purposes as a lumped parameter system; resonances occur due to the interaction of tissue masses with purely elastic structures. At higher frequencies, through the

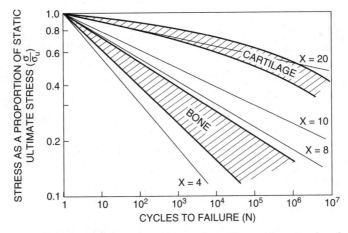

FIGURE 44.4   Fatigue failure of human tissue. The number of cycles of repeated stress N to failure of *in vitro* preparations is shown as a function of the ratio of the applied dynamic stress to the ultimate static stress σ/σ_u. (*von Gierke.*[6])

TABLE 44.2   Physical Properties of Human Tissue at Frequencies Less than 100 kHz

		Bone, compact	
	Tissue, soft	Fresh	Embalmed, dry
Density, gm/cm³	1–1.2	1.93–1.98	1.87
Young's modulus, dyne/cm²	7.5×10^4	2.26×10^{11}	1.84×10^{11}
Volume compressibility,* dyne/cm²	2.6×10^{10}	...	1.3×10^{11}
Shear elasticity,* dyne/cm²	2.5×10^4	...	7.1×10^{10}
Shear viscosity,* dyne-sec/cm²	1.5×10^2	...	...
Sound velocity, cm/sec	1.5–1.6×10^5	3.36×10^5	...
Acoustic impedance, dyne-sec/cm³	1.7×10^5	6×10^5	6×10^5
Tensile strength, dyne/cm²	...	9.75×10^8	1.05×10^9
Shearing strength, dyne/cm², parallel	...	4.9×10^8	...
Shearing strength, dyne/cm², perpendicular	...	1.16×10^9	5.55×10^8

* Lamé elastic moduli.

audio-frequency range and up to about 100 kHz, the body behaves more as a complex distributed parameter system—the type of wave propagation (shear waves, surface waves, or compressional waves) being strongly influenced by boundaries and geometrical configurations.

Low-Frequency Range.   Simple mechanical systems, such as the one shown in Fig. 44.5 for a standing and sitting man, are usually sufficient to describe and understand the important features of the response of the human body to low-frequency vibrations.[6,7] Nevertheless it is difficult to assign numerical values to the elements of the model, since they depend critically on the kind of excitation, the body type of the subject, and his posture and muscle tone. Large intersubject variability is therefore to be expected and is observed. Of the various factors influencing whole-body bio-

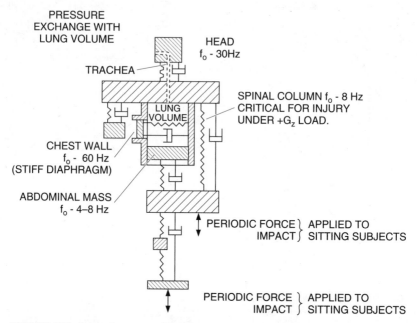

FIGURE 44.5 Lumped parameter biodynamic model of the standing and sitting human body for calculating motion of body parts and some physiological and subjective responses to vertical vibration. The approximate resonance frequencies of various subsystems are indicated by f_o. (*von Gierke.*[6])

dynamic responses, a reduction in intersubject variability can often be obtained by normalizing measured values by a subject's static mass.[1]

Subject Exposed to Vibrations in the Longitudinal Direction. The *mechanical impedance* of a man standing or sitting on a vertically vibrating platform, that is, the complex ratio between the dynamic force applied to the body and the velocity at the interface where vibration enters the body, is shown in Fig. 44.6. Below approximately 2 Hz the body acts as a unit mass. For the sitting man, the first resonance is between 4 and 6 Hz; for the standing man, resonance peaks occur at about 6 and 12 Hz. The numerical value of the impedance together with its phase angle provides data for the calculation of the total energy transmitted to the subject.

The resonances at 4 to 6 Hz and 10 to 14 Hz are suggestive of mass-spring combinations of (1) the entire torso with the lower spine and pelvis and (2) the upper torso with forward flexion movements of the upper vertebral column. The expectation that flexion of the upper vertebral column occurs is supported by observations of the transient response of the body to vertical impact loads and associated compression fractures. The greatest loads occur in the region of the twelfth thoracic to the second lumbar vertebra, which therefore can be assumed as the hinge area for flexion of the upper torso. Since the center-of-gravity of the upper torso is considerably forward of the spine, flexion movement will occur even if the force is applied parallel to the axis of the spine. Changing the direction of the force so that it is applied at an angle with respect to the spine (for example, by tilting the torso forward) influences this effect considerably. Similarly the center-of-gravity of the head can be considerably in front of the neck joint which permits forward-backward

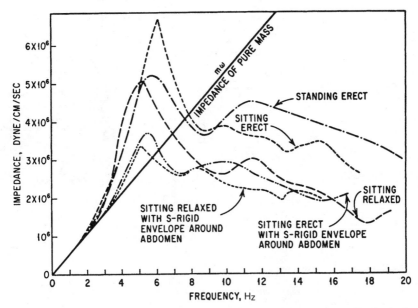

FIGURE 44.6 Mechanical impedance of standing and sitting human subject vibrating in the direction of his longitudinal axis as a function of frequency. The effects of body posture and of a semirigid envelope around the abdomen are shown. The impedance of a mass *m* also is given. (*After Coermann: Human Factors,* 4:227, 1962, *and Coermann et al.: Aerospace Med.,* **31**:443, 1960.)

motion. This situation results in forward-backward rotation of the head instead of pure vertical motion. Examples of relative amplitudes for different parts of the body when it is subject to vibration are shown in Fig. 44.7 for a standing subject and in Fig. 44.8 for a sitting subject. The data are plotted as *transmissibilities,* that is, the ratio of the response amplitude of the body part in steady-state forced vibration to the excitation amplitude at that frequency, expressed as a function of frequency. The curves show an amplification of motion in the region of resonance and a decrease at higher frequencies. The impedances and the transmissibility factors are changed considerably by individual differences in the body, the posture of the body, the type of support by a seat or back rest for a sitting subject, or by the state of the knee or ankle joints of a standing subject. The resonance frequencies remain relatively constant, whereas the transmissibility varies (e.g., for the condition of Fig. 44.8, values of transmissibility as high as 4 have been observed at 4 Hz). Above approximately 10 Hz, the vibration displacement amplitudes of the body are smaller than the amplitudes of the exciting table, and they decrease continuously with increasing frequency. The attenuation of the vibrations transmitted from the table to the head is illustrated in Fig. 44.9. At 100 Hz, this attenuation is about 40 dB. The attenuation along the body at 50 Hz is shown in Fig. 44.10.

Between 20 and 30 Hz the head exhibits a mechanical resonance, as indicated in Fig. 44.8. When subject to vibration in this range, the head displacement amplitude can exceed the shoulder amplitude by a factor of 3. This resonance is of importance in connection with the deterioration of visual acuity under the influence of vibration. Another frequency range of disturbances between 60 and 90 Hz suggests an eyeball resonance.

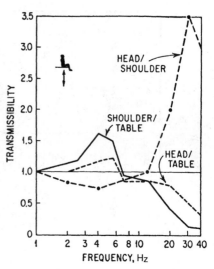

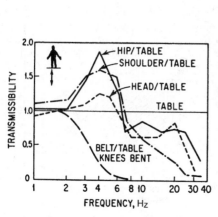

FIGURE 44.7 Transmissibility of vertical vibration from table to various parts of the body of a standing human subject as a function of frequency. (*After Dieckmann: Int. Z. Angew. Physiol. Arbeitsphysiol.,* **16**:519, 1957, *and Radke*: *Proc. ASME*, Dec. 1957.)

FIGURE 44.8 Transmissibility of longitudinal vertical vibration from table to various parts of body of seated human subject as a function of frequency. (*After Dieckmann: Int. Z. Angew. Physiol. Arbeitsphysiol.,* **16**:519, 1957.)

The mechanical impedance of the human body, lying on its back on a rigid surface and vibrating in the direction of its longitudinal axis, has been determined in connection with ballistocardiograph studies. For tangential vibration, the total mass of the body behaves as a simple mass-spring system with the elasticity and resistance of the skin. For the average subject the resonance frequency is between 3 and 3.5 Hz, and the Q of the system is about 3. If the subject's motion is restricted by clamping the body at the feet and at the shoulders between plates connected with the table, the resonance is shifted to approximately 9 Hz and the Q is about 2.5.

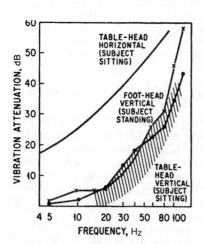

FIGURE 44.9 Attenuation of vertical and horizontal vibration for standing and sitting human subjects. (*After von Békésy: Akust. Z.,* **4**:360, 1939, *and Coermann: Jahrb. Dtsch. Luftfahrtforsch.,* **III**:111, 1938.)

One of the most important subsystems of the body, which is excited in the standing and sitting position as well as in the lying position, is the thorax-abdomen system. The abdominal viscera have a high mobility due to the very low stiffness of the diaphragm and the air volume of the lungs and the chest wall behind it. Under the influence of both longitudinal and transverse vibration of the torso, the abdominal mass vibrates in and out of the thoracic cage. Vibra-

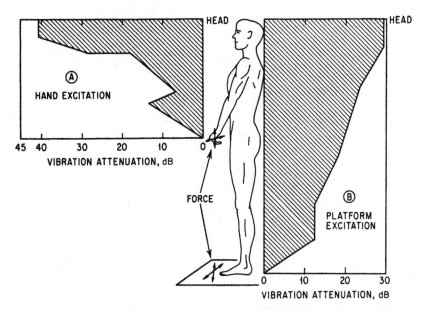

FIGURE 44.10 Attenuation of vibration at 50 Hz along the human body. The attenuation is expressed in decibels below values at the point of excitation for excitation of (*A*) hand and (*B*) platform on which subject stands. (*After von Békésy: Akust. Z.,* **4**:360, 1939.)

tions take place in other than the (longitudinal) direction of excitation; during the phase of the cycle when the abdominal contents swing toward the hips, the abdominal wall is stretched outward and the abdomen appears larger in volume; at the same time, the downward deflection of the diaphragm causes a decrease of the chest circumference. At the other end of the cycle the abdominal wall is pressed inward, the diaphragm upward, and the chest wall is expanded. This periodic displacement of the abdominal viscera has a sharp resonance between 3 and 3.5 Hz, as can be seen from Fig. 44.11. The oscillations of the abdominal mass are coupled with the air oscillations of the mouth-chest system. Measurements of the impedance of the latter system at the mouth (by applying oscillating air pressure to the mouth) show that the abdominal wall and the anterior chest wall respond to this pressure. The magnitude of the impedance is minimum and the phase angle is zero between 7 and 8 Hz. The abdominal wall has a maximum response between 5 and 8 Hz, the anterior chest wall between 7 and 11 Hz. Vibration of the abdominal system resulting from exposure of a sitting or standing subject is detected clearly as modulation of the air flow velocity through the mouth (Fig. 44.11). Therefore at large amplitudes of vibration, speech can be modulated at the exposure frequency. A lumped parameter model of the thorax-abdomen-airway system is used successfully to explain and predict these detailed physiological responses (Fig. 44.12).[7] The same model can also be used, when appropriately excited, to describe the effects of blast, infrasound, and chest impact and to derive curves of equal injury potential, i.e., tolerance curves.

Typical values of mechanical impedance for standing, sitting, and reclining persons, together with biodynamic models, are contained in an international standard.[8]

   Subject Exposed to Vibrations in the Transverse Direction.   The physical response of the body to transverse vibration—i.e., horizontal in the normal upright

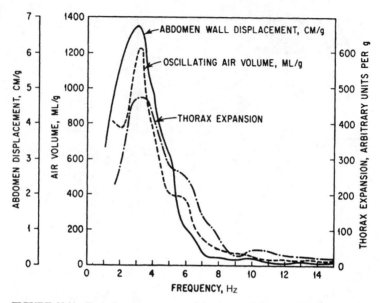

FIGURE 44.11 Typical response curves of the thorax-abdomen system of a human subject in the supine position exposed to longitudinal vibrations. The displacement of the abdominal wall (2 in. below umbilicus), the air volume oscillating through the mouth, and the variations in thorax circumference are shown per *g* longitudinal acceleration. (*Coermann et al.: Aerospace Med.,* **31**:443, 1960.)

position—is quite different from that for vertical vibration. Instead of thrust forces acting primarily along the line of action of the force of gravity on the human body, they act at right angles to this line. Therefore the distribution of the body masses is of the utmost importance. There is a greater difference in response between sitting and standing positions for transverse vibration than for vertical vibration where the supporting structure of the skeleton and the spine are designed for vertical loading.

Data for the transmission of vibration along the body are given in Fig. 44.13. For a standing subject, the displacement amplitudes of vibration of the hip, shoulder, and head are about 20 to 30 percent of the table amplitude at 1 Hz and decrease with increasing frequency. Relative maxima of shoulder and head amplitudes occur at 2 and 3 Hz, respectively. The sitting subject exhibits amplification of the hip (1.5 Hz) and head (2 Hz) amplitudes. All critical resonant frequencies are between 1 and 3 Hz. The transverse vibration patterns of the body can be described as standing waves, i.e., as a rough approximation one can compare the body with a rod in which transverse flexural waves are excited. Therefore there are nodal points on the body which become closer to the feet as the frequency of excitation increases, since the phase shift between all body parts and the table increases continuously with increasing frequency. At the first resonant frequency (1.5 Hz), the head of the standing subject has a 180° phase shift with respect to the table; between 2 and 3 Hz this phase shift is 360°.

There are longitudinal head motions excited by the transverse vibration in addition to the transverse head motions shown in Fig. 44.13 and discussed above. The head performs a nodding motion due to the anatomy of the upper vertebrae and the

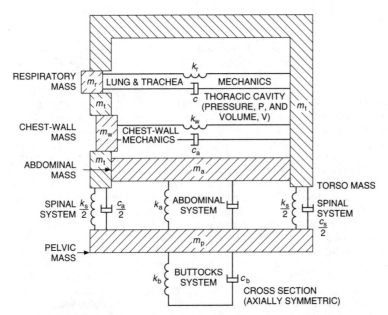

FIGURE 44.12 Five-degree-of-freedom body model. The model is used to calculate body deformations (thorax compression, pressure in the lungs, airflow into and out of the lungs, diaphragm and abdominal mass movement) as a function of external longitudinal forces (vibration or impact) and pressure loads (blast, infrasonic acoustic loads). It has also been used to calculate thorax dynamics under impacts to the chest wall, m_w. (*von Gierke.*[7])

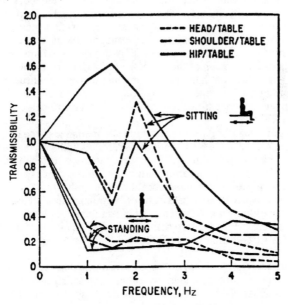

FIGURE 44.13 Transmission of transverse, horizontal vibration from table to various parts of sitting and standing human subject. (*After Dieckmann: Int. Z. Angew. Physiol. Arbeitsphysiol.*, **17**:83, 1958.)

location of the head's center-of-gravity. Above 5 Hz, the head motion for sitting and standing subjects is predominantly vertical (about 10 to 30 percent of the horizontal table motion).

Vibrations Transmitted from the Hand.   In connection with studies on the use of vibrating hand tools, measurements have been made of vibration transmission from the hand to the arm and body. The mechanical impedance of the hand-arm system measured at a hand grip under conditions representative of those associated with power-tool operation is shown in Fig. 44.14 for vibration in a direction essentially along the long axis of the forearm, that is, approximately in the direction of thrust. The precise direction is the Z component of the standardized coordinate system for the hand shown in Fig. 44.15, Z_h. Typical values of impedance have been defined by a synthesis of measured values from different experimental studies, each of which was conducted under equivalent measurement conditions and involved a number of male human subjects. The unexplained variability in the results, which is commonly observed in biodynamic experiments, led the specification of the most probable values of impedance magnitude and phase by an upper and lower envelope (the continuous lines in Fig. 44.14). The mean of the data sets is shown by the dotted line. Also shown in the diagram are impedance values calculated by the 4 degree-of-freedom biodynamic model illustrated in Fig. 44.16. It is evident that good agreement can be obtained between the synthesized and model values. Equivalent data are available for the two orthogonal directions of the hand-arm coordinate system not shown in Fig. 44.14 (X_h and Y_h).[9] It should be noted that the parameters of these biodynamic models do not possess direct anatomical correlates.

The mechanical impedance of the hand-arm system generally increases in magnitude with frequency, with a maximum at a frequency from 20 to 70 Hz. The model values suggest that resonances occur in structures within the hand, resulting in relative motion between tissue layers, and between tissue and the bone. The coupled mass in contact with the handle and subject to the vibration input is typically less than 20 grams. Small increases in impedance magnitude have been observed with increases in grip force (from the value of 25 N used for the data synthesis), which leads to increased indentation of the skin (see Fig. 44.3). The influence of the translational force with which the hand presses the handle (i.e., the thrust force) appears to be insignificant at frequencies above 100 Hz and to introduce variations in mechanical impedance magnitude and phase of less than 10 percent at frequencies between 20 and 70 Hz.[9]

An international standard for the mechanical impedance of the hand-arm system is based on the analysis of Ref. 9.[10] Biodynamic models with varying degrees of complexity are provided in the standard. These are intended to facilitate the development of devices for reducing vibration transmitted to the hands and of test rigs with which to measure power-tool handle vibration.

For hand tools involving a palm grip, the vibration amplitude decreases from the palm to the back of the hand. Further reductions in amplitude occur from the hand to the elbow and from the elbow to the shoulder. Typical values for a vibration frequency of 50 Hz are shown in Fig. 44.10.

Middle-Frequency Range (Wave Propagation).   Above about 100 Hz, simple lumped parameter models become more and more unsatisfactory for describing the vibration of tissue. At higher frequencies it is necessary to consider the tissue as a continuous medium for vibration propagation.

Skull Vibrations.   The vibration pattern of the skull is approximately the same as that of a spherical elastic shell. The nodal lines observed suggest that the funda-

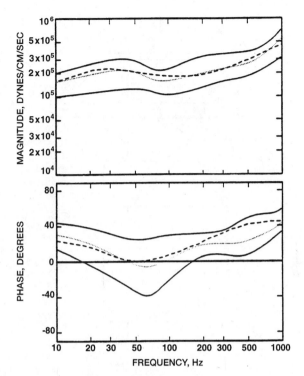

FIGURE 44.14 Mechanical impedance of the hand-arm system, expressed as magnitude and phase, in the Z_h direction specified in Fig. 44.15. Maximum and minimum envelopes of mean values from studies included in the data synthesis are shown by continuous lines, while the weighted mean of all data sets is shown by the dotted line. The response of a 4-degree-of-freedom biodynamic model is shown by the dashed line. (*Gurram et al.*[9])

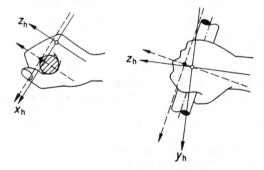

FIGURE 44.15 Standardized biodynamic (open circles—continuous lines) and basicentric (closed circles—dashed lines) coordinate systems for the hand. (*See ISO 5349.*[37])

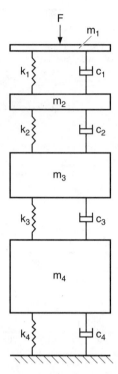

FIGURE 44.16 Biodynamic model for hand-arm impedance. The parameters of this model do not possess direct anatomical correlates.

mental resonance frequency is between 300 and 400 Hz and that resonances for the higher modes are around 600 and 900 Hz. The observed frequency ratio between the modes for the skull is approximately 1.7, while the theoretical ratio for a sphere is 1.5. From the observed resonances, the calculated value of the elasticity of skull bone (a value of Young's modulus = 1.4×10^{10} dynes/cm^2) agrees reasonably well with static test results on dry skull preparations but is somewhat lower than the static test data obtained on the femur (Table 44.2). Mechanical impedances of small areas on the skull over the mastoid area have been measured to provide information for bone-conduction hearing. The impedance of the skin lining in the auditory canal has been investigated and used in connection with studies on ear protectors.[11]

Vibration of the lower jaw with respect to the skull can be explained by a simple mass-spring system, which has a resonance, relative to the skull, between 100 and 200 Hz.

Mechanical Impedance of Soft Human Tissue. Mechanical impedance measurements of small areas (1 to 17 cm^2) over soft human body tissue have been made with vibrating pistons between 10 and 20 kHz. At low frequencies this impedance is a large elastic reactance. With increasing frequency the reactance decreases, becomes zero at a resonance frequency, and becomes a mass reactance with a further increase in frequency (Fig. 44.17).[12] These data cannot be explained by a simple lumped parameter model, but require a distributed parameter system including a viscoelastic

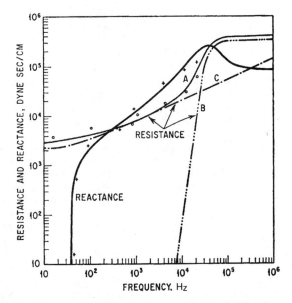

FIGURE 44.17   Resistance and reactance of circular area, 2 cm in diameter, of soft tissue body surface as a function of frequency. Crosses and circles indicate measured values for reactance and resistance. Smooth curves calculated for 2-cm-diameter sphere vibrating in (*A*) viscoelastic medium with properties similar to soft tissue (parameters as in Table 44.2), (*B*) frictionless compressible fluid, and (*C*) incompressible viscous fluid. (*From von Gierke et al.*[12])

medium—such as the tissue constitutes for this frequency range.[12,13] The high viscosity of the medium makes possible the use of simplified theoretical assumptions, such as a homogeneous isotropic infinite medium and a vibrating sphere instead of a circular piston. The results of such a theory agree well with the measured characteristics. As a consequence it is possible to assign absolute values to the shear viscosity and the shear elasticity of soft tissue (Table 44.2). The theory together with the measurements show that, over the audio-frequency range, most of the vibratory energy is propagated through the tissue in the form of transverse shear waves—not in the form of longitudinal compression waves. The velocity of the shear waves is about 20 meters per second at 200 Hz and increases approximately with the square root of the frequency. This may be compared with the constant sound velocity of about 1500 meters per second for compressional waves. Some energy is propagated along the body surface in the form of surface waves which have been observed optically. Their velocity is of the same order as the velocity of shear waves.[12]

High-Frequency Range.   Above several hundred thousand Hz, in the ultrasound range, most of the vibratory energy is propagated through tissue in the form of compressional waves; for these conditions, geometrical acoustics offer a good approximation for the description of their path. Since the tissue dimensions under consideration are almost always large compared to the wavelength (about 1.5 mm at 1 MHz), the mechanical impedance of the tissue is equal to the characteristic acoustic impedance, i.e., sound velocity times density. This value for soft tissue differs only slightly from the characteristic impedance of water. The most important factor in this frequency range is the tissue viscosity, which brings about an increasing energy absorption with increasing frequency.[14]

At very high frequencies this viscosity also generates shear waves at the boundaries of the medium, at the boundary of the acoustic beams, and in the areas of wave transition to media with somewhat different properties (e.g., boundary of muscle to fat tissue, or soft tissue to bone). These shear waves are attenuated so rapidly that they are of no importance for energy transport but are noticeable as increased local absorption, i.e., heating.

The *attenuation coefficient* describes the observed reduction in intensity of an ultrasound beam as it propagates from the source, while the *absorption coefficient* describes the rate at which acoustic energy is converted into heat in the tissue and may be used to predict the temperature elevation caused by the beam.[14] Values of attenuation a and absorption α coefficients for different types of tissue at ultrasonic frequencies are listed in Table 44.3.[14]

The data have been normalized by the ultrasound frequency f, expressed in megahertz, leading to ratios a/f and α/f that are essentially constant in the frequency range of interest (0.5–10 MHz). Also listed in the table are the sound velocity c and characteristic acoustic impedance, i.e., ρc where ρ is the density. The attenuation has most commonly been measured with phase sensitive (e.g., pressure) sensors, and will overestimate the attenuation for sound beams propagating in inhomogeneous media. For this reason, results obtained using phase-insensitive sensors are preferred and are listed separately.

MECHANICAL DATA FROM SHOCK FORCES

Very little numerical data on mechanical characteristics of the body are available from studies on shock or impact forces. Some evidence of resonances has been noted, but much of this is better obtained from vibration studies. The application of mechanical data which has been obtained from studies on vibration to exposure to shock and impact is discussed in the section on mechanical damage. Mechanical responses to shock and impact are, in general, extremely difficult to analyze numerically for basic body characteristics.

EFFECTS OF SHOCK AND OF VIBRATION

The motions and mechanical stresses resulting from the application of mechanical forces to the human body have several possible effects: (1) the motion may interfere directly with physical activity; (2) there may be mechanical damage or destruction; (3) there may be secondary effects (including subjective phenomena) operating through biological receptors and transfer mechanisms, which produce changes in the organism. Thermal and chemical effects are usually unimportant, except in the ultrasonic frequency range.

TABLE 44.3 Normalized Attenuation a/f and Absorption α/f Coefficients, Sound Velocities c, and Characteristic Impedances ρc for Mammalian Tissue at Frequencies from 0.5 to 10 MHz (see text) (*NCRPM.*[14])

| Tissue | a/f, Np/cm MHz | | α/f, Np/cm MHz | c, meters per sec | ρc, grams per sec cm$^2 \times 10^{-5}$ |
	Phase dependent	Phase independent			
Abdominal wall	0.25 ± 0.09	NA	NA	NA	NA
Amniotic fluid	0.0017 ± 0.0006	NA	NA	1510	NA
Blood	0.017 ± 0.005	NA	NA	1575 ± 11	1.62 ± 0.02
Bone	$2.9, 1.64 \pm 0.32$	NA	NA	3183 ± 618	4.8 ± 0.99
Brain	0.086 ± 0.02	0.061 ± 0.034	0.039 ± 0.014	1565 ± 10	1.54 ± 0.05
Breast (young)	0.03–0.07	NA	NA	1450–1570	NA
Breast (old)	0.03–0.07	NA	NA	1430–1520	NA
Collagen	NA	NA	$114 \, C \, f^{0.47}$	$7.5 \, C + 1494$	NA
Eye (lens)	0.09	NA	NA	1647 ± 7	NA
Eye (aqueous)	0.06 ± 0.02	NA	NA	1537 ± 4	NA
Eye (vitreous)	0.05 ± 0.03	NA	NA	1532 ± 7	NA
Fat (peritoneal)	0.24	NA	NA	1490	NA
Fat (subcutaneous)	0.07	NA	NA	1478 ± 9	NA
Heart	0.23 ± 0.05	0.076 ± 0.03	0.033 ± 0.006	1571 ± 19	1.64, (1.78)
Kidney (cortex)	0.12 ± 0.08	0.091 ± 0.017	0.033	1567	
Kidney (interior)	NA	NA	NA	NA	1.65
Limb	NA	NA	NA	1500–1610	NA
Liver	0.135, 0.14	0.087 ± 0.015	0.020 ± 0.002	1604 ± 14	1.63 ± 0.07
Muscle (perp.)	0.11 ± 0.04	0.063	NA	1581 ± 8	NA
Muscle (par.)	0.16	0.21	NA	1581 ± 19	NA
Pancreas	0.108 ± 0.05	NA	NA	NA	NA
Skin	0.197 ± 0.07	NA	NA	1720 ± 45	1.59
Spleen	0.0775 ± 0.03	NA	NA	1601	1.67
Tendon (perp.)	0.43 ± 0.11	NA	0.11 ± 0.04	1750	NA
Tendon (par.)	0.44 ± 0.20	NA	NA	NA	NA
Testis	NA	0.035	0.015 ± 0.003	1595	NA
Tooth (dentine)	0.51	NA	NA	3400	7.5
Tooth (enamel)	0.77	NA	NA	6030	17.8
Uterus	0.027–0.22	NA	NA	1629 ± 6	NA

NOTE: C = collagen concentration in gm/100 mL; NA = data not available.

EFFECTS OF MECHANICAL VIBRATION

Mechanical Damage. Damage is produced when the accelerative forces are of sufficient magnitude. Mice, rats, and cats have been killed by exposure to vibration.[15] There is a definite frequency dependence of the lethal accelerations coincident with resonance displacement of the visceral organs. Mice are killed at accelerations of 10 to $20g$ within a few minutes in the range 15 to 25 Hz; above and below this frequency range, the survival time is longer. Rats and cats may be killed within 5 to 30 minutes at accelerations above about $10g$. Postmortem examination of these animals usually shows lung damage, often heart damage, and occasionally brain injury. The injuries to heart and lungs probably result from the beating of these organs against each other and against the rib cage. The brain injury, which is a superficial hemorrhage, may be due to relative motion of the brain within the skull, to mechanical action involving the blood vessels or sinuses directly, or to secondary mechanical effects. Tearing of intraabdominal membranes rarely occurs.

An increase in body temperature is also observed after exposure to intense vibration. Since this effect occurs in dead animals, it is probably mechanical in origin. Estimates of energy dissipation from body mechanical impedance data suggest that appreciable heat can be generated at large vibration amplitudes.

In humans, mechanical damage to the heart and lungs, injury to the brain, tearing of membranes in the abdominal and chest cavities, as well as intestinal injury are possible, in principle. However, equinoxious contours of whole-body acceleration as a function of frequency have not been established for any of these phenomena, owing to an almost complete lack of data. Any effects would be expected to occur at lower frequencies than those in animals owing to the increased human visceral masses. Exposure for 15 minutes to an acceleration of $6g$ has been reported to cause gastrointestinal bleeding that persisted for several days in one subject.[1]

Chronic injuries may be produced by vibration exposure of long duration at levels which produce no apparent acute effects. In practice, such effects are usually found after exposure to repeated blows or to random jolts rather than to sinusoidal motion. When such blows are applied to the human body at relatively short intervals, the relation of the interval to tissue response times becomes very important. Exposure to such forces frequently occurs in connection with the riding of vehicles. Buffeting in aircraft or in high-speed small craft on water, and shaking in vehicles on rough surfaces, gives rise to irregular jolting motion. Acute injuries from exposure to these situations are rare but have resulted from repeated impact (e.g., back fracture of a fighter pilot whose airplane encountered extreme turbulence). Injuries to the spinal column, including fracture of vertebrae, have been reported among operators of earth-moving equipment. Pathological changes have been observed in the spines of operators of farm tractors and off- and on-the-road trucks, and in other occupations involving chronic exposure to whole-body vibration.[2] Such exposure is also accepted as a risk factor in the development of low-back pain.[16] Minor kidney injuries are occasionally suspected and, rarely, traces of blood may appear in the urine.

Chronic injuries may also be produced when the hand is exposed to intense vibration, such as occurs during occupational use of some power tools (e.g., pneumatic drills and hammers, grinders, chain saws, and riveting guns).[3] Symptoms of numbness or paresthesias in the fingers are common and may be accompanied by episodes of finger blanching. The vascular and nerve disorders associated with the use of hand-held vibrating power tools are known as the *hand-arm vibration syndrome* (HAVS). Pathological changes have been observed in the structure of the nerves and walls of the blood vessels in the fingers.[3] Changes in tactile function have been linked to changes in acuity of specific types of mechanoreceptive nerve endings.[17]

Few dose-response relationships have been derived from epidemiological data for any sign, or symptom, of HAVS resulting from occupational use of hand-held power tools or industrial processes. For groups of workers who perform similar tasks throughout the workday, the *latency,* that is, the duration of exposure (in years) prior to the onset of episodes of finger blanching, may be predicted from the acceleration of a surface in contact with the hand.[18] The mean latency for such a group is shown in Fig. 44.18, where the hand-arm vibration is expressed in terms of the frequency-weighted acceleration defined under the section *Human Tolerance Criteria*. Operation of some tools, such as rock drills, may result in deviations from this prediction.

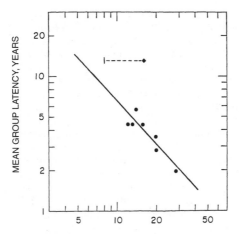

FIGURE 44.18 Relation between mean latency for the onset of episodic finger blanching and the frequency-weighted component acceleration of power tools and machines used by population groups exposed to hand-transmitted vibration (solid circles). Data since published for rock drill operation in a mine shown by the diamond. The horizontal dashed line shows the adjustment introduced for a 1-hour daily exposure. (*Brammer.*[18])

An example is shown by the diamond in Fig. 44.18. In this population of jack-leg rock drillers the onset of finger blanching was delayed with respect to the prediction. The sources of the inconsistency are believed to arise from (1) the frequency-weighting function used to equate the hazard of vibration at different frequencies, which may overestimate the hazard at frequencies below 80 Hz, (2) imprecise knowledge of the time during the workday that the tool was hand guided as opposed to resting on its supporting leg—a halving of which would move the data point horizontally an amount shown by the dashed line, and (3) persons entering and leaving the work force.

The tendons, tendon sheaths, muscles, ligaments, joints, and nerves in the hand and arm can also be damaged by repeated movement of the hand relative to the arm. These soft tissue and nerve injuries occur among blue- and white-collar workers performing tasks involving repeated hand-wrist flexure (e.g., keyboard operators) and are termed *repetitive strain injuries* (RSI).[19] Nerve compression may result from changes in the contents of restricted nerve passageways (e.g., the carpal tunnel at the wrist—*carpal tunnel syndrome*).[3] Pain and paresthesias in the hand and arm are common symptoms.

Physiological Responses.   Vibration can induce physiological responses in the cardiovascular, respiratory, skeletal, endocrine, and metabolic systems and in muscles and nerves. The cardiovascular changes in response to intense vertical vibration are similar to those accompanying moderate exercise: increased heart rate, respiration rate, cardiac output, and blood pressure. Vibration of sufficient intensity will cause mechanical pumping of the respiratory system, as already noted, but is unlikely to produce significantly increased ventilation or oxygen uptake. Changes in blood and urine constituents are commonly used as indicators of generalized body stress and may, in consequence, be observed in persons exposed to vibration. It is difficult if not impossible, however, to relate specific endocrine and metabolic responses to a given vibration stimulus. Vibration can stimulate a tonic reflex contraction in muscles, which is a response to the stretching force (the *tonic vibration reflex*), disturb postural stability, and lead to body sway.

Extremely low-frequency whole-body vibration, such as occurs in many transportation vehicles and ships, may also cause motion sickness (*kinetosis*).

Vibration of the hand may cause peripheral vascular, neurological, and muscular responses.[3] Blood flow within the fingers may be reduced during stimulation, and tingling and paresthesias in the hands may be reported after exposure. Somatosensory perception and tactile function may be temporarily decreased. Grip strength may also be affected. Extremely low-frequency, large-amplitude motions, which are usually described as repetitive movements of the hand (and frequently involve repeated wrist rotation), may lead to tendon and muscle fatigue and to transitory parathesias or numbness.

Therapeutic applications of vibration include cardiac and circulatory assist devices and the control of spastic muscle. Ultrasonic frequencies are used in medical diagnosis, for soft tissue visualization, and for therapy. A common therapeutic use is to promote the return of limb function in rehabilitation medicine.

Subjective Responses.   Feelings of discomfort and apprehension may be associated with exposure to whole-body and hand-arm vibration once the stimulus has been perceived. The extent of the discomfort depends on the magnitude, frequency, direction and duration of the exposure, and the posture and orientation of the body, as well as the point of contact with the stimulus. The response is also influenced by the environment in which the motion is experienced (e.g., floor motion in hospital versus aircraft). The range in response of different individuals to a given stimulus is large. In some circumstances, whole-body vibration may be exhilarating (e.g., a fairground ride) or soothing (e.g., rocking a baby in a cradle or a rocking chair).

In general, subjective responses to vibration may be subdivided into three broad categories: the threshold of perception, the onset of unpleasant sensations, and the limit of tolerance. The specification of acceptable vibration environments is discussed later in this chapter.

Once detected, the growth in sensation follows a Stevens' power law function with index k, in which the psychophysical magnitude of a stimulus, ψ, is related to its physical magnitude ϕ by

$$\psi = \text{constant}\{\phi^k\} \tag{44.1}$$

For discomfort associated with whole-body vibration, $k \approx 1$. Frequency contours of equal sensation magnitude depend principally on the direction in which vibration enters the body and whether the person is standing, seated, or recumbent.[1] Contours which summarize current knowledge may be inferred from the frequency-weighting functions employed in the international standard for whole-body vibration (i.e., by reciprocal curves to those shown later in Fig. 44.29). The effect of the duration of exposure t on subjective responses to suprathreshold vibration is often found to follow a power law relationship of the form

$$\phi^n t = \text{constant} \tag{44.2}$$

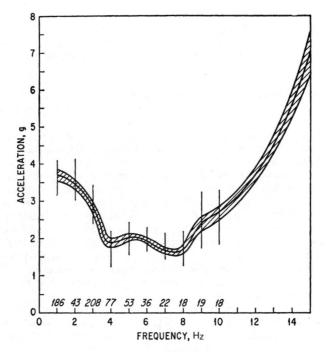

FIGURE 44.19 Peak acceleration at various frequencies at which subjects refuse to tolerate further a short exposure (less than 5 min) to vertical vibration. The figures above the abscissa indicate the exposure time in seconds at the corresponding frequency. The shaded area has a width of one standard deviation on either side of the mean (10 subjects). (*Ziegenruecker and Magid: USAF WADC Tech. Rept.* 59–18, 1959.)

where the magnitude of the index *n* is from 2 to 4. For situations in which the *perception* of vibration is judged unacceptable, the boundary between acceptable and unacceptable exposures will be related to the physical magnitude of the stimulus corresponding to the threshold of perception, and will not depend on the duration of exposure. There is an extensive literature discussing the comfort/discomfort of passengers in road and rail vehicles, aircraft, and ships.[1]

An indication of the range of human responses to whole-body vibration can be obtained from the results of an experiment to establish the limit of tolerance of short exposures (less than 5 minutes) to vertical vibration at different frequencies. The peak accelerations at which the 10 subjects refused to continue exposure to vibration are shown as a function of frequency in Fig. 44.19, and the reasons given for refusing to continue are summarized in Table 44.4.

EFFECTS OF MECHANICAL SHOCK

Mechanical shock includes several types of force application which have similar, though not identical, effects. Explosions, explosive compression or decompression, and impacts and blows from rapid changes in body velocity or from moving objects produce shock forces of importance. Major damage, short of complete tissue destruction, is usually to lungs, intestines, heart, or brain. Differences in injury patterns arise from differences in rates of loading, peak force, duration, localization of forces, etc.

Blast and Shock Waves.[20] The mechanical effects associated with rapid changes in environmental pressure are primarily localized to the vicinity of air-filled cavities in the body, i.e., the ears, lungs, and air-containing gastrointestinal tract. Here, heavy masses of blood or tissue border on light masses of air. The local impedance mismatch can lead to a relative tissue displacement, which is destructive, by several different mechanisms.

The ear is the part of the human body most sensitive to blast injury. Rupture of the tympanic membrane and injury to the conduction apparatus can occur singly or together with injury to the hair cells in the inner ear. The two first-mentioned injuries may protect the inner ear through energy dissipation. The degree of injury depends on the frequency content of the blast pressure function. The fact that the ear's greatest mechanical sensitivity occurs at frequencies between 1500 and 3000 Hz explains its vulnerability to short-duration blast waves. Peak pressures of only a few pounds per square inch can rupture the ear drum, and still smaller pressures can damage the conducting mechanism and the inner ear. There are wide variations in individual susceptibility to these injuries.

With very slow differential pressure changes, of approximately 1 sec duration or longer, dynamic mechanical effects are unimportant; the static pressure is responsible for destructive mechanical stress or physiological response. Such pressure-time

TABLE 44.4   Reason for Terminating Exposure to Vertical Vibration at Various Frequencies

(Each cross indicates a decided comment from one of the ten human subjects used in this study as to his experiencing the symptom listed.)

(Ziegenruecker and Magid: USAF WADC Tech. Rept., 59–18, 1959.)

	Symptom						
Frequency, Hz	Abdominal pain	Chest pain	Testicular pain	Head symptoms	Dyspnea	Anxiety	General discomfort
1					XXXXX XXX		XXX
2					XXXXX XXX		XXXX
3	XX	XX			XXXXX	X	XXXXX
4	XX	XX		XX	XXX	XX	XXXXX
5		XXXX				X	XXXXX X
6	XXX	XXXX		X			XXXX
7	XX	XXXXX	X	X			X
8	X	XXXX		X		XX	XXX
9	XX	XXXX			X		XXXXX
10	X	X	XXX	XX		X	
15							XXXXX XXX

functions occur with the explosive decompression of pressurized aircraft cabins at high altitude and with the slow response of very well-sealed shelters to blast waves. If the pressure rise times or fall times are shortened (roughly to the order of tenths of seconds), the dynamic response of the different resonating systems of the body becomes important, in particular the thorax-abdomen system of Fig. 44.12. The dynamic load factor of the specific blast disturbance under consideration determines the response. Available data for single pulse, "instantaneously rising" pressures suggest the existence of a minimum peak pressure which corresponds to natural frequencies for dogs of between 10 and 25 Hz; for humans this frequency is lower. Sensitivity curves for blast exposure, i.e., curves of equal injury potential (maximum tolerable level) for various species are shown in Fig. 44.20. The theoretical curves are obtained by means of the thorax model of Fig. 44.12 after application of appropriate scaling laws to account for the different species sizes.[7] For pressures with total durations of milliseconds or less and much shorter rise times (duration of wave short compared to the natural period of the responding tissue), the effect and destruction seem to depend primarily on the momentum of the shock wave. The mass m of an oscillatory system located in a wall or body surface, which is struck by a shock wave, is set into motion according to the relation

$$P_r \, dt = m v_0 \qquad\qquad (44.3)$$

where P_r = reflected pressure at body surface
     t = time
     v_0 = initial velocity

Experimental fatality curves on animals generally show this dependence on momentum for short pressure phenomena (close to center of detonation) and the transition to a dependence on peak pressure for phenomena of long duration (far away from

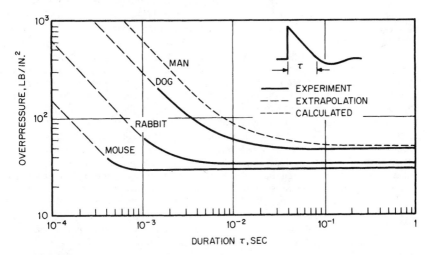

FIGURE 44.20 Maximum tolerable blast overpressures for mouse, rabbit, dog, and man. The curve for man is calculated by means of a model of the type given in Fig. 44.12. Using the same model, dimensionally scaled to the animal sizes, results in curves matching closely the experimental data.

center).[20] Fatal blast waves in air and water, for example, differ widely in peak pressure and duration (in air, 10 atm in excess of atmospheric pressure with a duration of 2.8 milliseconds, in water 135 atm in excess with a duration of 0.17 milliseconds), but their momenta are similar. In this most important range of short-duration blasts, the mechanical effects are localized because of the short duration, i.e., the high-frequency content of the wave. The upper respiratory tract and bronchial tree, as well as the thorax and abdomen system, are too large and have resonance frequencies too low to be excited; there is no general compression or overexpansion of the thorax, which leads to pulmonary injury as in explosive decompression. The blast waves go directly through the thoracic wall, producing an impact or grazing blow. Inside the tissue, blast injury has three possible causes: (1) spalling effects, i.e., injuries caused by the tensile stresses arising from the reflection of the shock wave at the boundary between media with different propagation velocity [for example, subpleural pulmonary hemorrhages along the ribs]; (2) inertia effects which lead to different accelerations of adjacent tissues with various densities, when the shock wave passes simultaneously through these media; and (3) implosion of gas bubbles enclosed in a liquid. These phenomena are compatible with observations made when high-velocity missiles pass through water near air-containing tissues.[21] The shock waves may produce not only pulmonary injuries but also hard, sharply circumscribed blows to the heart.

Of the injuries produced by exposure to high-explosive blast, lung hemorrhage is one of the most common. It may not of itself be fatal, since enough functional lung tissue may easily remain to permit marginal gas exchange. However, the rupture of the capillaries in the lung produces bleeding into the alveoli and tissue spaces, which can seriously hamper respiratory activity or produce various respiratory and cardiac reflexes. The heart rate is often very slow after a blast injury. Leakage of fluid through moderately injured, but not ruptured, capillaries may occur. There is also the possibility that air may enter the circulation to form bubbles or emboli and by reaching critical regions may fatally impair the heart or brain circulation or produce secondary damage to other organs.[20] Fat emboli also may be formed and these, too, are capable of blocking vessels supplying vital parts. When gas pockets are present in the intestines, the shock may produce hemorrhage and in extreme cases rupture the intestinal wall itself.

The effects of underwater shock waves on man and animals are in general of the same kind as those produced by air blast. Differences which appear are those of magnitude and often depend on the mode of exposure of the body. A person in the water may, for example, be submerged from the waist down only; in this case, damage is practically confined to the lower half of the body so that intestinal, rather than lung, damage will occur. Direct mechanical injury to the heart muscle and conducting mechanism is possible. Cerebral concussion resulting directly from exposure to shock waves is unusual. Neurological symptoms following exposure to blast, however, may include general depression of nervous activity sometimes to the point of abolition of certain reflexes. Psychological changes such as memory disturbances and abnormal emotional states are found sometimes. In extreme cases, there may be paralysis or muscular dysfunction. Unconsciousness and subsequent amnesia for events immediately preceding the injury result more commonly from blows to the head than from air blast. Recovery from minor concussion apparently may be complete, but repeated concussion may produce lasting damage.

Impacts, Blows, Rapid Deceleration. This type of force is experienced in falls, in automobile or aircraft crashes, in parachute openings, in seat ejections for escape from high-speed military aircraft, and in many other situations. Interest in the body's responses to these forces centers on mechanical stress limits. The injuries which occur most often are bruises, tissue crushing, bone fracture, rupture of soft tissues and organs, and concussion. A bruise is a superficial area of slight tissue damage with rupture of the small blood vessels and accumulation of blood and fluid in and around the injured region. It is essentially a crushing injury produced by compression of the tissues, usually between the impinging solid and the underlying bone. It is extremely common and is readily repaired by the body itself. When the tissue is completely destroyed by crushing, the damage usually is irreparable. Bone fractures, like bruises, require that the forces be sustained long enough to produce appreciable displacements and concentrated stresses.

When soft tissues are displaced considerable distances by appropriate forces, so-called internal injuries, i.e., rupture of membranes or organ capsules, may take place. Such injuries are, in practice, more often produced by forces of relatively long duration and usually are dangerous.

The correlation between the response of the body system to continuous vibration and to spike and step-force functions may be used to guide and interpret experiments. The tissue areas stressed to maximum relative displacement at the various frequencies during steady-state excitation are preferred target areas for injury under impact load if the force-time functions of the impacts have appreciable energy in these frequency bands, i.e., if the impact duration is of the same order of magnitude as the body's natural periods. If the impact exposure times are shorter, stress tolerance limits increase; if exposure times decrease to hundredths or thousandths of a second, the response becomes more and more limited and localized to the point of application of the force (blow). Elastic compression or injury will depend on the load distribution over the application area, i.e., the pressure, to which tissues are subjected. If tissue destruction or bone fracture occurs close to the area of application of the force, these will absorb additional energy and protect deeper-seated tissues by reducing the peak force and spreading it over a longer period of time. An example is the fracture of foot and ankle of men standing on the deck of warships when an explosion occurs beneath. The support may be thrown upward with great momentum; if the velocity reaches 5 to 10 ft/sec (which corresponds, under these conditions, to an acceleration of several hundred g) fractures occur.[22] However, the energy absorption by the fracture protects structures of the body which are higher up.

If the force functions contain extremely high frequencies, the compression effects spread from the area of force application throughout the body as compression waves. If these are of sufficient amplitude, they may cause considerable tissue disruption. Such compression waves are observed from the impact of high-velocity missiles.

If the exposure to the accelerating forces lasts long enough so that (as in most applications of interest) the whole body is displaced, exact measurement of the force applied to the body and of the direction and contact areas of application becomes of extreme importance. In studies of seat ejection, for example, a knowledge of seat acceleration alone is not sufficient for estimating responses. One must know the forces in those structures or restraining harnesses through which acceleration forces are transmitted. The location of the center-of-gravity of the various body parts such as arms, head, and upper torso must be known over the time of force application so that the resulting body motion and deformation can be analyzed and controlled for protection purposes. In addition to the primary displacements of body parts and organs, there are secondary forces from decelerations if, due to the large amplitudes, the motions of parts of the body are stopped suddenly by hitting other body parts. Examples occur in linear deceleration where, depending on the restraint, the head may be thrown forward until it hits the chest or, if only a lap belt is used, the upper torso may jackknife and the chest may hit the knees. There is always the additional possibility that the body may strike nearby objects, thus initiating a new impact deceleration history.

Longitudinal Acceleration. The study of positive longitudinal (headward) acceleration of short duration is connected closely with the development of upward ejection seats for escape from aircraft. Since the necessary ejection velocity of approximately 60 ft/sec and the available distance for the catapult guide rails of about 3 ft are determined by the aircraft, the minimum acceleration required (step function) is approximately 18.6g. Since the high jolt of the instantaneous acceleration increase is undesirable because of the high dynamic load factor in this direction for the frequency range of body resonances (compare with Fig. 44.8), slower build-up of the acceleration with higher final acceleration is preferable to prevent injury. Investigations show that the body's ballistic response can be predicted by means of analog computations making use of the frequency-response characteristics of the body.[23] The simplest analog used for the study of headward accelerations is the single degree-of-freedom mechanical resonator composed of the lumped-parameter elements of a spring, mass, and damper. A diagram of this model is shown in Fig. 44.21A. The model is used to simulate the maximum stress developed within the ver-

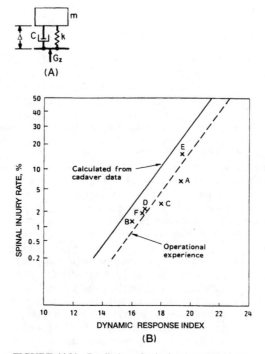

FIGURE 44.21   Prediction of spinal compression injury from pilot ejection seat accelerations. (A) Model for the study of spinal compression Δ with mass m, spring stiffness k and damping c; (B) relation between the dynamic response index (DRI) and spinal injury rate for 361 nonfatal ejections from six different types of aircraft (dashed line) (aircraft type A—64 ejections; B—62; C—65; D—89; E—33; and F—48). Data from cadavers (continuous line). (*After Griffin,*[1] *and von Gierke.*[6])

tebral column (the first failure mode in this direction) for any given impact environment. The maximum dynamic deflection of the spring, Δ_{max}, may be calculated for a given input acceleration-time history to the model. The potential for spinal injury is estimated by forming the *dynamic response index* (DRI), which is defined as $\omega_n^2\Delta_{max}/g$, where the natural frequency of the model is $\omega_n = (k/m)^{1/2}$ and the damping ratio $c/2(km)^{1/2}$ is 0.224. Experience with nonfatal ejections from military aircraft, shown by the crosses and dashed line in Fig. 44.21*B*, suggests a 5 percent probability of spinal injury from exposure to a dynamic response index of 18. An estimate of the rate of spinal injury from cadavers is shown in this diagram by the continuous line. The success of the model has led to its adoption for the specification of ejection seat performance, for its extension to accelerations in three orthogonal directions,[24] and to measures of ride comfort for exposure to repeated impacts in some land vehicles and high-speed boats.[25]

Control or prevention of injury is critically dependent on optimal body positioning and restraint to minimize unwanted and forceful flexion of the spinal column. The fracture tolerance limits are influenced by age, physical condition, clothing, weight, and many other factors which detract from the optimum. If the tolerance limits are exceeded, fractures of the lumbar and thoracic vertebrae occur first. While in and of itself this injury may not be classified as severe, small changes in orientation may be enough to involve the spinal cord, an injury which is extremely severe and may be life-threatening. Neck injuries from headward accelerations appear to occur at considerably higher levels.

In general, for vertical crash loads the same considerations apply as discussed for seat ejections, although no control over the build-up time of the acceleration is possible and more sudden onsets must be expected.

For negative (tailward) acceleration (downward ejection) no firm point for application of the accelerating force is accessible as for positive acceleration. If the force is applied as usual through harness and belt at shoulder and groin, the mobility of the shoulder girdle together with the elasticity of the belts results in a lower resonance frequency than the one observed in upward ejection. To avoid overshooting with standard harnesses, the acceleration rise time must be at least 0.15 sec. This type of impact can excite the thorax-abdomen system (Fig. 44.12). The diaphragm is pushed upward by the abdominal viscera; as a result, air rushes out of the lungs (if the glottis is open) or high pressures develop in the air passages. Tolerance limits for negative acceleration probably are set by the compression load on the thoracic vertebrae, which are exposed to the load of the portion of the body below the chest. This load on the vertebrae is higher than that for the positive acceleration case due to the greater weight; therefore a tolerance limit for acceleration has been set at 13*g*. Shoulder accelerations of 13*g* have been tolerated by human subjects without injury, when the load was divided between hips and shoulders.

Transverse Accelerations. The forward- and backward-facing seated positions are most frequently exposed to high transverse components of crash loads. Human tolerance to these forces has been studied extensively by volunteer tests on linear decelerators, in automobile crashes, and by the analysis of the records of accidental falls. The results indicate the importance of distributing the decelerative forces or impact over as wide an area as possible. The tolerable levels of acceleration amplitude of well over 50*g* (100*g* and over for falling flat on the back with minor injuries, 35 to 40*g* for 0.05-sec voluntary tolerance seated with restraining harness) are probably limited by injury to the brain. An indication that the latter might be sensitive to and based on specific dynamic responses is the fact that the tolerance limit depends strongly on the rise time of the acceleration. With rise times around 0.1 sec (rate of change of acceleration 500*g*/sec), no overshooting of head and chest accelerations is observed, whereas faster rise times of around 0.03 sec (1000 to 1400*g*/sec) result in overshooting of chest accelerations of 30 percent (acceleration front to back) and even up to 70 percent (acceleration back to front). All these results depend critically on the harness for fixation and the back support used (see *Protection Methods and Procedures,* below). These dynamic load factors indicate a natural frequency of the body system between 10 and 20 Hz. Impact of the heart against the chest wall is another possible injury discussed and noted in some animal experiments.

The head and neck supporting structures seem to be relatively tough. Injury seems to occur only upon backward flexion and extension of the neck ("whiplash") when the body is accelerated from back to front without head support, as is common in rear-end automobile collisions.

Head Impact.[5]   Injuries to the head, beyond superficial bruises and lacerations, usually consist of concussion or fracture of the skull. The symptoms produced by head impact range from pain and dizziness through disorientation and depression of function to unconsciousness and loss of memory for events immediately preceding the injury. Head injuries usually occur from heavy blows by solid objects to the head, rather than by accelerative forces applied to the body. Since a majority of airplane crash fatalities result from head injuries, and the latter are without doubt of similar importance in many other types of accident (athletics, etc.), the mechanisms leading to head injury have been the object of a large number of investigations. The limitations for forward and lateral bending of the neck are, for practical purposes, the anterior chest wall and the shoulders, respectively. Since the head is almost entirely held by the neck muscles, the absence of their supporting action gives any blow to the head or neck a "flying start." As a consequence, fractures or dislocations are more apt to result. Dislocations involving the first and second vertebral joints usually are less severe if the odontoid process is fractured (less damage to the spinal cord). If this is not the case, the spinal cord may be severed or crushed.

Head Response.   The reaction of the head to a blow is a function of the velocity, duration, area of impact, and the transfer of momentum. Near the point of application of the blow there will be an indentation of the skull. This results in shear strains in the brain in a superficial region close to the dent. Compression waves emanate from this area, which have normally small amplitudes since the brain is nearly incompressible. In addition to the forces on the brain resulting from skull deformation there are acceleration forces, which also would act on a completely undeformable skull. The centrifugal forces and linear accelerations producing compressional strains are negligible compared to the shear strains produced by the rotational accelerations. The maximum strains are concentrated at regions where the skull has a good grip on the brain owing to inwardly projecting ridges, especially at the wing of the sphenoid bone of the skull. Shear strains also must be present throughout the brain and in the brain stem. Many investigators consider these shear strains, resulting from rotational accelerations due to a blow to the unsupported head, as the principal event leading to concussion. Blows to the supported, fixed head are supposed to produce concussion by compression of the skull and elevation of cerebrospinal fluid pressure. Despite the general acceptance of rotational acceleration as one of the main causes of concussion, experimental data on this quantity are almost completely missing, and concussion thresholds are discussed in terms of "available energy" (which is usually not the energy transferred to the head) and impact velocity.

In general a high-velocity projectile (for example, a bullet of 10 grams with a speed of 1000 ft/sec) with its high kinetic energy and low momentum produces plainly visible injury to scalp, skull, and brain along its path. The high-frequency content of the impact is apt to produce compression waves which in the case of very high energies may conceivably lead to cavitation with resulting disruption of tissue. Skull fracture is not a prerequisite for these compression waves. However, if the head hits a wall or another object whose mass is large compared to the head's mass, the local, visible damage is small and the damage due to rotational acceleration may be large. Blows to certain points, especially on the midline, produce no rotation. Blows to the chin upward and sideward produce rotation relatively easily ("knock-out" in boxing). Velocities listed in the literature for concussion from impact of large masses range from 15 to 50 ft/sec. At impact velocities of about 30 ft/sec, approximately 200 in.-lb of energy is absorbed in 0.002 sec, resulting in an acceleration of the head of $47g$. Impact energies for compression concussion are probably approximately in the same range.

Scalp, skin, and subcutaneous tissue reduce the energy applied to the bone. If the response of the skull to a blow exceeds the elastic deformation limit, skull fracture occurs. Impact by a high-velocity, blunt-shaped object results in localized circumscribed fracture and depression. Low-velocity blunt blows, insufficient to cause depression, occur frequently in falls and crashes. Given enough energy, two, three, or more cracks appear, all radiating from the center of the blow. The skull has both weak and strong areas, each impact area showing well-defined regions for the occurrence of the fracture lines.

The total energy required for skull fracture varies from 400 to 900 in.-lb, with an average often assumed to be 600 in.-lb. This energy is equivalent to the condition that the head hits a hard, flat surface after a free fall from a 5-ft height. Skull fractures occurring when a batter is accidentally hit by a ball (5 oz) of high velocity (100 ft/sec) indicate that about the same energy (580 in.-lb) is required. Additional energy 10 to 20 percent beyond the single linear fracture demolishes the skull completely. Dry skull preparations required only approximately 25 in.-lb for fracture. The reason for the large energy difference required in the two conditions is attributed mainly to the attenuating properties of the scalp.

In automobile and aircraft crashes the form, elasticity, and plasticity of the object injuring the head is of extreme importance and determines its "head injury potential." For example, impact with a 90° sharp corner requires only a tenth of the energy for skull fracture (60 in.-lb) that impact with a hard, flat surface requires. Head impact energies for various attitudes of the human body hitting a contact area at various angles are presented in Fig. 44.22. The conditions shown are representative of crash conditions involving unrestrained humans.

EFFECTS OF SHOCK AND VIBRATION ON TASK PERFORMANCE

The performance of tasks requiring a physical response to some stimulus involves peripheral (e.g., perceptual and motor) and central neurological processes, with multiple feedback paths characteristic of a sophisticated control system. Each of these processes is complex, is more or less developed in different individuals, and may be influenced by training and the general state of health. In consequence, unique relationships between vibration and task performance are unlikely, except for well-defined situations in which some part of the body reaches a physical or physiological limit to performance. For example, movement of images on the retina may cause defocusing and a reduction of visual acuity. The movement may be caused by vibration of the display (i.e., the source), the head (and/or observer), or both. At frequencies below approximately 1 Hz, a pursuit reflex assists visual acuity. At frequencies above 20 Hz an eyeball resonance can degrade acuity. The effects of whole-body vibration on visual acuity therefore depends on the frequency and amplitude, as well as the viewing distance.[1] As already discussed, whole-body vibration can affect speech.

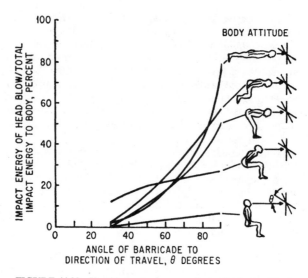

FIGURE 44.22 Head impact energy as affected by body attitude and barricade angle. (*Data from dummy drop tests after Dye: Clinical Orthopedics,* **8**:305, 1956.)

Vibration may also degrade the manual control of objects. The influence of whole-body vibration on writing and drinking is a common experience in public transportation vehicles and ships. Vibration may interfere with the performance of manually controlled systems. The extent of the effect depends on hand motion, the type of control (e.g., a "stiff" control that responds to the application of force without moving or one that moves and responds with little force applied), and the dynamics of the control and the controlled systems. A control that responds to hand displacement may be disrupted by vertical vibration at frequencies between 2 and 6 Hz. The effect of the duration of vibration exposure on task performance is influenced by motivation, arousal, and adaptation and may therefore be observed to improve or degrade performance over time.

In addition to the acute effects of vibration, chronic exposure of the hand to vibration can lead to sensorineural dysfunction sufficient to result in a persistent reduction in the ability to perform fine manual tasks, such as buttoning clothing.[3,17]

The motion associated with a shock is unlikely to interfere directly with the performance of most tasks unless it is coincident with some critical component of the task. This condition may occur with shocks repeated at very short intervals.

PROTECTION METHODS AND PROCEDURES

Protection of man against mechanical forces is accomplished in two ways: (1) isolation to reduce transmission of the forces to the man and (2) increase of man's mechanical resistance to the forces. Isolation against shock and vibration is achieved if the natural frequency of the system to be isolated is lower than the exciting frequency at least by a factor of 2. Both linear and nonlinear resistive elements are used for damping the transmission system; irreversible resistive elements or energy-absorbing devices can be used once to change the time and amplitude pattern of impulsive forces. Human tolerance to mechanical forces is strongly influenced by selecting the proper body position with respect to the direction of forces to be expected. Man's resistance to mechanical forces also can be increased by proper distribution of the forces so that relative displacement of parts of the body is avoided as much as possible. This may be achieved by supporting the body over as wide an area as possible, preferably loading bony regions and thus making use of the rigidity available in the skeleton. Reinforcement of the skeleton is an important feature of seats designed to protect against crash loads. The flexibility of the body is reduced by fixation to the rigid seat structure. The mobility of various parts of the body, e.g., the abdominal mass, can be reduced by properly designed belts and suits. The factor of training and indoctrination is essential for the best use of protective equipment, for aligning the body in the least dangerous positions during intense vibration or crash exposure, and possibly for improving operator performance during vibration exposure. The latter type of training may be helpful in anticipating and preventing man-machine resonance effects and in reducing anxiety which might otherwise occur.

PROTECTION AGAINST VIBRATIONS

The transmission of vibration from a vehicle or platform to a man is reduced by mounting him on a spring or similar isolation device, such as an elastic cushion. The degree of vibration isolation theoretically possible is limited, in the important resonance frequency range of the sitting man, by the fact that large static deflections of the man with the seat or into the seat cushion are undesirable. Large relative movements between operator and vehicle controls interfere in many situations with man's performance. Therefore a compromise must be made. Cushions are used primarily for static comfort, but they are also effective in decreasing the transmission of vibration above man's resonance range. They are ineffective in the resonance range and may even amplify the vibration in the subresonance range. In order to achieve effective isolation over the 2- to 5-Hz range, the natural frequency of the man-cushion system should be reduced to 1 Hz, i.e., the natural frequency should be small compared with the forcing frequency (see Chap. 30). This would require a static cushion deflection of 10 in. If a seat cushion without a back cushion is used (as is common in some tractor or vehicle arrangements), a condition known as "back scrub" (a backache) may result. Efforts of the operator to wedge himself between the controls and the back of the seat often tend to accentuate this uncomfortable condition.

The transmission of vibration through various common seat cushioning materials is shown in Fig. 44.23. For typical passenger seats, the seat cushion does not greatly alter the resonance frequency of the man-seat system, which can be seen to be approximately 4 to 5 Hz from the data in Fig. 44.23 (compare, for example, with the peak impedance values for seated persons in Fig. 44.6). At frequencies from 2 to 5 Hz, amplification of seat motion occurs as the transmissibility is greater than unity, i.e., there is more vibration at the cushion-subject interface than at the cushion-seat interface. The damping properties of all but the thinnest foam cushion material shown in Fig. 44.23 result in attenuation of seat motion at frequencies above 7 Hz (i.e., the transmissibility is less than unity).

For severe low-frequency vibration, such as occur in tractors and other field equipment, suspension of the whole seat is superior to the simple seat cushion. Hydraulic shock absorbers, rubber torsion bars, coil springs, and leaf springs all have been successfully used for suspension seats (see Fig. 44.24). A seat that is guided so that it can move only in a linear direction seems to be more comfortable than one in which the seat simply pivots around a center of rotation as can be seen from the results in Fig. 44.24. The latter situation produces an uncomfortable and fatiguing pitching motion. Suspension seats can be built which are capable of preloading for the operator's weight so as to maintain the static position of the seat and the natural frequency of the system at the desired value. Suspension seats for use on tractors and on similar vehicles are available which reduce the resonance frequency of the man-seat system from approximately 4 to 2 Hz. This can be seen from the comparison of the transmissibility of a rigid seat, a truck suspension seat, and a conventional foam and metal sprung car seat in Fig. 44.25. The transmissibility of the car seat is in excess of 2 at the

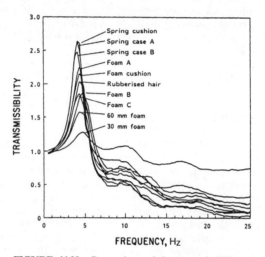

FIGURE 44.23 Comparison of the transmissibility of vertical vibration for different seat cushioning materials. (*Griffin.*[1])

resonance frequency (4 Hz), implying that the seat motion reaching the body is amplified by this ratio. In contrast, the amplification introduced by the suspension seat is at most a factor of 1.3 at the resonance frequency (2 Hz), and improved attenuation of vibration is obtained throughout the frequency range from 4 to 12 Hz. At frequencies below 2 Hz and above 12 Hz, less vibration is transmitted to the subject by the foam and metal sprung seat. There are large differences in the performance of suspension seats, with transmissibilities in excess of 2 being recorded in some designs at the resonance frequency (which is usually close to 2 Hz).[1] In consequence, the selection of a seat for a particular application must take into account both the performance of the seat and the critical seat vibration frequencies to be attenuated.

The superiority of man's legs to most seating devices, with respect to the transmission of vertical vibrations, is shown in Fig. 44.7 (see *knees bent* curve). Differences in positioning of the sitting subject also change the transmission as demonstrated by impedance measurements (e.g., see Fig. 44.6).

For severe vibrations, close to or exceeding normal tolerance limits, such as those which may occur in military operations, special seats and restraints can be employed to provide maximum body support for the subject in all critical directions. In general, under these conditions, seat and restraint requirements are the same for vibration and rapidly applied accelerations. Laboratory experiments show that protection

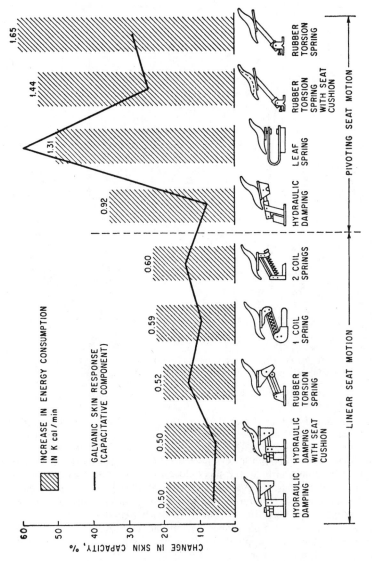

FIGURE 44.24 Difference in vibration induced stress on a subject as a function of the seat design. Galvanic skin response and the increase in caloric energy consumption due to vibrations are plotted for the different seats indicated. The data represent averages over 10 tests of 15 min duration on five subjects with vibrations characteristic of tractor operation. The data prove the general superiority of seat designs which restrict seat vibrations to linear, compared to pivoting, motions. Subjective evaluations indicate a similar rank order as the physiological quantities in this graph. (*After Dieckmann: Intern. Z. Angew. Physiol. Arbeitsphysiol.*, **16**:519, 1957.)

can be achieved by the use of rigid or semirigid body enclosures. Immersion of the operator in a rigid, water-filled container with proper breathing provisions has been used in laboratory experiments to protect subjects against large, sustained static g loads. This principle can be used to provide protection against large alternating loads.

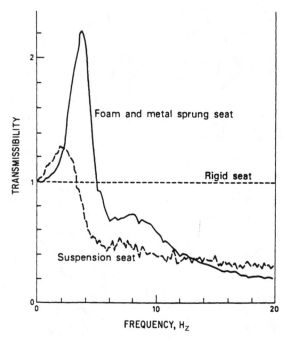

FIGURE 44.25 Comparison of the transmissibilities of a rigid seat, a foam-covered metal sprung seat, and a truck suspension seat. (*Griffin.*[1])

Isolation of the hand and arm from the vibration of hand-held or hand-guided power tools is accomplished in several ways.[26] A common method is to isolate the handles from the rest of the power tool, using springs and dampers (see Chap. 30). The application of vibration-isolation systems to chain saws for occupational use in forestry has become commonplace and has led to a reduction in the incidence of hand-arm vibration syndrome (HAVS). A second method is to modify the tool so that the primary vibration is counterbalanced by an equal and opposite vibration source. This method takes many different forms, depending on the operating principle of the power tool.[27] An example is shown for a pneumatic scaling chisel in Fig. 44.26, in which an axial impact is applied to a work piece to remove metal by a chisel P. The chisel is driven into the work piece by compressed air and is returned to its initial position by a spring S. The axial motion of the chisel is counterbalanced by a second mass m and spring k which oscillate out of phase with the chisel motion. The design of an appropriate vibration-isolation system must include the dynamic properties of the hand-arm system. The model of Fig. 44.16 is suitable for this purpose.

Conventional gloves do not attenuate the vibration transmitted to the hand but may increase comfort and keep the hands warm. So-called antivibration gloves also fail to reduce vibration at frequencies below 100 Hz, which are most commonly responsible for the HAVS, but may reduce vibration at high frequencies (the relative importance of different frequencies in causing the vascular component of HAVS is shown in Fig. 44.33).

Preventive measures for the hand-arm vibration syndrome (HAVS) to be applied in the work place include: minimizing the duration of exposure to vibration, using minimum hand-grip force consistent with safe operation of the power tool or process ("let the tool do the work"), wearing sufficient clothing to keep warm, and maintaining the tool in good working order, with minimum vibration. As recovery from HAVS has only been demonstrated for early vascular symptoms, medical monitoring of persons exposed to vibration is essential. Monitoring should include a test of peripheral neurological function,[28] since this component of HAVS appears to persist.

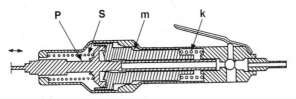

FIGURE 44.26   An antivibration power tool design for a pneumatic scaler: *P*—vibrating chisel; *S*—chisel return spring; *m*—counterbalancing oscillating weight; and *k*—counterbalance return spring. (*Lindqvist.*[26])

PROTECTION AGAINST RAPIDLY APPLIED ACCELERATIONS (CRASH)

The study of automobile and aircraft crashes and of experiments with dummies and live subjects shows that complete body support and restraint of the extremities provide maximum protection against accelerating forces and give the best chance for survival.[5] If the subject is restrained in the seat, he makes full use of the force moderation provided by the collapse of the vehicle structure, and he is protected against shifts in position which would injure him by bringing him in contact with interior surfaces of the cabin structure. The decelerative load must be distributed over as wide a body area as possible to avoid force concentration with resulting bending moments and shearing effects. The load should be transmitted as directly as possible to the skeleton, preferably directly to the pelvic structure—not via the vertebral column.

Theoretically, a rigid envelope around the body will protect it to the maximum possible extent by preventing deformation. A body restrained to a rigid seat approximates such a condition; proper restraints against longitudinal acceleration shift part of the load of the shoulder girdle and arms from the spinal column to the back rest. Arm rests can remove the load of the arms from the shoulders. Semirigid and elastic abdominal supports provide some protection against large abdominal displacements. The effectiveness of this principle has been shown by animal experiments and by impedance measurements on human subjects (Fig. 44.6). Animals immersed in water, which distributes the load applied to the rigid container evenly over the body surface, or in rigid casts are able to survive acceleration loads many times their normal tolerance.

Many attempts have been made to incorporate energy-absorptive devices, either in a harness or in a seat, with the intent to change the acceleration-time pattern by limiting peak accelerations. Parts of the seat or harness are designed with characteristics which become nonlinear at some given acceleration level. The benefits derived from such devices are usually small since little space for body or seat motion is available in airplanes or automobiles; furthermore, contact with interior cabin surfaces during the period of extension of the device is apt to result in more serious injury. For example, consider an aircraft which is stopped in a crash from 100 mph in 5 ft; it is subjected to a constant deceleration of $67g$. An energy-absorptive device designed to elongate at $17g$ would require a displacement of 19 in. In traveling through this distance, the body or seat would be decelerated relative to the aircraft by $14.4g$ and would have a maximum velocity of 36.8 ft/sec relative to the aircraft structure. A head striking a solid surface with this velocity has many times the minimum energy required to fracture a skull. The available space for seat or passenger travel using the principle of energy absorption therefore must be considered carefully in the design. In the development of "catch-up" mechanisms for window washers and workers in similar situations, energy-absorptive elements in the form of undrawn nylon ropes may be applied. For seat belts and other crash restraint harnesses, extensible fabrics have been found to be extremely hazardous since their load characteristics cannot be sufficiently controlled. Seats for jet airliners have been designed which have energy-absorptive mechanisms in the form of extendable rear-legs. The maximum travel of the seats is 6 in.; their motion is designed to start between 9 and $12g$ horizontal load, depending on the floor strength. During motion, the legs pivot at the floor level—a feature considered to be beneficial if the floor wrinkles in the crash. Theoretically, such a seat can be exposed to a deceleration of $30g$ for 0.037 sec or $20g$ for 0.067 sec without transmitting a deceleration of more than $9g$ to the seat. However, the increase in exposure time must be considered as well as the reduction in peak acceleration. For very short exposure times where the body's tolerance probably is limited by the transferred momentum and not the peak acceleration, the benefits derived from reducing peak loads would disappear.

The high tolerance limits of the well-supported human body to decelerative forces (Figs. 44.35 to 44.42) suggest that in aircraft and other vehicles, seats, floors, and the whole inner structure surrounding crew and passengers should be designed to resist crash decelerations as near to $40g$ as weight or space limitations permit.[29] The structural members surrounding this inner compartment should be arranged so that their crushing reduces forces on the inner structure. Protruding and easily loosened objects should be avoided. To allow the best chance for survival, seats should also be stressed for dynamic loadings between 20 and $40g$. Civil Air Regulations require a minimum static strength of seats of $9g$. A method for computing seat tolerance for typical survivable airplane crash decelerations is available for seats of conventional design.[29] It has been established that the passenger who is riding in a seat facing backward has a better chance to survive an abrupt crash deceleration since the impact forces are then more uniformly distributed over the body. Neck injury must be prevented by proper head support. Objections to riding backward on a railway or in a bus are minimized for air transportation because of the absence of disturbing motion of objects in the immediate field of view. Another consideration concerning the direction of passenger seats in aircraft stems from the fact that for a rearward-facing seat the center of passenger support during deceleration is about 1 ft above the point where the seat belt would be attached for a forward-facing passenger. Consequently, the rearward-facing seat is subjected to a higher bending moment; in other words, for seats of the same weight the forward-facing seat will sustain higher crash forces without collapse. For the same seat weight, the rearward-facing seat will have approximately only half the design strength of the forward-facing seat and about one-third its natural frequency. The selection of one type of seat over another involves engineering compromises.[29]

Safety lap or seat belts are used to fix the occupants of aircraft or automobiles to their seats and to prevent their being hurled about within, or being ejected from, the car or aircraft. Their effectiveness has been proved by many laboratory tests and in actual crash accidents. The belt load on the lower abdomen causes no severe intra-abdominal injury or injury to the lower spinal region—at least in most survivable crashes. A forward-facing passenger held by a seat belt flails about when suddenly decelerated; his hands, feet, and upper torso swing forward until his chest hits his knees or until the body is stopped in this motion by hitting other objects (back of seat in front, cabin wall, instrument panel, steering wheel, control stick, etc.). Since 15 to 18g longitudinal deceleration can result in 3 times higher acceleration of the chest hitting the knees, this load appears to be about the limit a human can tolerate with a seat belt alone. Approximately the same limit is obtained when the head-neck structure is considered.

Increased safety in automobile as well as airplane crashes can be obtained by distributing the impact load over larger areas of the body and fixing the body more rigidly to the seat. Shoulder straps, thigh straps, chest straps, and hand holds are additional body supports used in experiments. They are illustrated in Fig. 44.27. Table 44.5 shows the desirability of these additional restraints to increase possible survivability to acceleration loads of various direction.[30] In airplane crashes, vertical and horizontal loads must be anticipated. In automobile crashes, horizontal loads are most likely. The effectiveness of adequately engineered shoulder or chest straps in

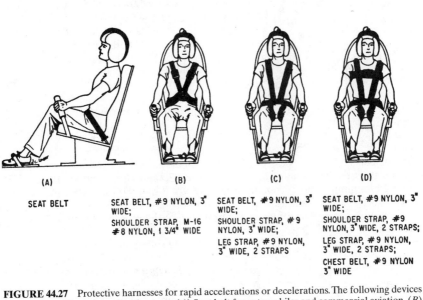

(A)	(B)	(C)	(D)
SEAT BELT	SEAT BELT, #9 NYLON, 3" WIDE;	SEAT BELT, #9 NYLON, 3" WIDE;	SEAT BELT, #9 NYLON, 3" WIDE;
	SHOULDER STRAP, M-16 #8 NYLON, 1 3/4" WIDE	SHOULDER STRAP, #9 NYLON, 3" WIDE;	SHOULDER STRAP, #9 NYLON, 3" WIDE, 2 STRAPS;
		LEG STRAP, #9 NYLON, 3" WIDE, 2 STRAPS	LEG STRAP, #9 NYLON, 3" WIDE, 2 STRAPS;
			CHEST BELT, #9 NYLON 3" WIDE

FIGURE 44.27 Protective harnesses for rapid accelerations or decelerations. The following devices were evaluated in sled deceleration tests: (*A*) Seat belt for automobiles and commercial aviation. (*B*) Standard military lap and shoulder strap. (*C*) Like (*B*) but with thigh straps added to prevent headward rotation of the lap strap. (*D*) Like (*C*) but with chest strap added. (*Stapp: USAF Tech. Rept. 5915, pt. I, 1949; pt. II, 1951.*)

TABLE 44.5   Human-Body Restraint and Possible Increased Impact Survivability (*After Eiband.*[30])

Direction of acceleration imposed on seated occupants	Conventional restraint	Possible survivability increases available by additional body supports*
Spineward: Crew	Lap strap Shoulder straps	Forward facing: (*a*) Thigh straps (Assume crew members will be performing emergency duties with hands and feet at impact.)
Passengers	Lap strap	Forward facing: (*a*) Shoulder straps (*b*) Thigh straps (*c*) Nonfailing arm rests (*d*) Suitable hand holds (*e*) Emergency toe straps in floor
Sternumward: Passengers only	Lap strap	Aft facing: (*a*) Nondeflecting seat back (*b*) Integral, full-height head rest (*c*) Chest strap (axillary level) (*d*) Lateral head motion restricted by padded "winged back" (*e*) Leg and foot barriers (*f*) Arm rests and hand holds (prevent arm displacement beyond seat back)
Headward: Crew	Lap strap Shoulder straps	Forward facing: (*a*) Thigh straps (*b*) Chest strap (axillary level) (*c*) Full, integral head rest (Assume crew members will be performing emergency duties; extremity restraint useless.)
Passengers	Lap strap	Forward facing: (*a*) Shoulder straps (*b*) Thigh straps (*c*) Chest strap (axillary level) (*d*) Full, integral head rest (*e*) Nonfailing contoured arm rests (*f*) Suitable hand holds Aft facing: (*a*) Chest strap (axillary level) (*b*) Full, integral head rest (*c*) Nonfailing arm rests (*d*) Suitable hand holds
Tailward: Crew	Lap strap Shoulder straps	Forward facing: (*a*) Lap-belt tie-down strap (Assume crew members will be performing emergency duties; extremity restraint useless.)
Passengers	Lap strap	Forward facing: (*a*) Shoulder straps (*b*) Lap-belt tie-down strap (*c*) Hand holds (*d*) Emergency toe straps

TABLE 44.5   Human-Body Restraint and Possible Increased Impact Survivability (*Continued*)

Direction of acceleration imposed on seated occupants	Conventional restraint	Possible survivability increases available by additional body supports*
Berthed occupants	Lap strap	Aft facing: (*a*) Chest strap (axillary level) (*b*) Hand holds (*c*) Emergency toe straps Feet forward: Full-support webbing net Athwart ships: Full-support webbing net

* Exposure to maximum tolerance limits (Figs. 44.35 to 44.42) requires straps exceeding conventional strap strength and width.

automobile crashes is illustrated in Fig. 44.28. Lap straps always should be as tight as comfort will permit to exclude available slack. During forward movement, about 60 percent of the body mass is restrained by the belt, and therefore represents the belt load. Double-thickness No. 9 undrawn nylon straps of 3 in. width are most satisfactory for all harnesses with respect to strength, elongation, and supported surface

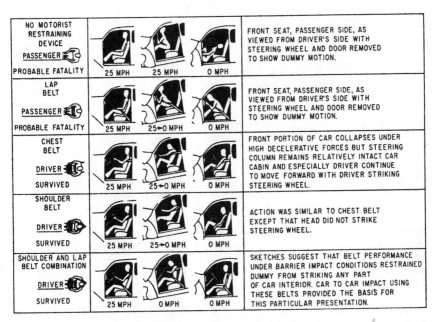

NO MOTORIST RESTRAINING DEVICE PASSENGER PROBABLE FATALITY	25 MPH 25 MPH 0 MPH	FRONT SEAT, PASSENGER SIDE, AS VIEWED FROM DRIVER'S SIDE WITH STEERING WHEEL AND DOOR REMOVED TO SHOW DUMMY MOTION.
LAP BELT PASSENGER PROBABLE FATALITY	25 MPH 25→0 MPH 0 MPH	FRONT SEAT, PASSENGER SIDE, AS VIEWED FROM DRIVER'S SIDE WITH STEERING WHEEL AND DOOR REMOVED TO SHOW DUMMY MOTION.
CHEST BELT DRIVER SURVIVED	25 MPH 25→0 MPH 0 MPH	FRONT PORTION OF CAR COLLAPSES UNDER HIGH DECELERATIVE FORCES BUT STEERING COLUMN REMAINS RELATIVELY INTACT. CAR CABIN AND ESPECIALLY DRIVER CONTINUE TO MOVE FORWARD WITH DRIVER STRIKING STEERING WHEEL.
SHOULDER BELT DRIVER SURVIVED	25 MPH 25→0 MPH 0 MPH	ACTION WAS SIMILAR TO CHEST BELT EXCEPT THAT HEAD DID NOT STRIKE STEERING WHEEL.
SHOULDER AND LAP BELT COMBINATION DRIVER SURVIVED	25 MPH 0 MPH 0 MPH	SKETCHES SUGGEST THAT BELT PERFORMANCE UNDER BARRIER IMPACT CONDITIONS RESTRAINED DUMMY FROM STRIKING ANY PART OF CAR INTERIOR. CAR TO CAR IMPACT USING THESE BELTS PROVIDED THE BASIS FOR THIS PARTICULAR PRESENTATION.

FIGURE 44.28   Effect of varying safety-belt arrangements on driver and passenger for a 25-mph automobile collision with a fixed barrier. The sketches and evaluations are based on actual collision tests. (*Severy and Matthewson: Trans. SAE,* **65**:70, 1957.)

area. If the upper torso is fixed to the back of the seat by any type of harness (shoulder harness, chest belt, etc.), the load on the seat is approximately the same for forward- and aft-facing seats. The difference between these seats with respect to crash tolerance as discussed above no longer exists. These body restraints for passenger and crew must be applied without creating excessive discomfort.

The dynamic properties of seat cushions are extremely important if an acceleration force is applied through the cushion to the body. In this case the steady-state response curve of the total man-seat system (Fig. 44.23) provides a clue to the possible dynamic load factors under impact. Overshooting should be avoided, at least for the most probable impact rise times. This problem has been studied in detail in connection with seat cushions used on upward ejection seats. The ideal cushion is approached when its compression under static load spreads the load uniformly and comfortably over a wide area of the body and if almost full compression is reached under the normal weight. The impact acceleration then acts uniformly and almost directly on the body without intervening elastic elements. A slow-responding foam plastic, such as an open cell rate-dependent polyurethane foam, of thickness from 2 to 2.5 in. satisfies these requirements.[31]

A significant factor in human impact tolerance appears to be the acceleration-time history of the subject immediately preceding the impact event.[32] A so-called *dynamic preload* consists of an imposed acceleration preceding, and/or during, and in the same direction as the impact acceleration. A dynamic preload occurs, for example, when the brakes are applied to a moving automobile before it hits a barrier. The phenomenon is found experimentally to reduce the acceleration of body parts on impact, thereby potentially mitigating adverse health effects. The dynamic preload should not be confused with the static preload introduced by a protective harness. The latter brings the occupant into contact with the seat or restraint but does not introduce the dynamic displacement of body parts and tissue compression necessary to reduce the body's dynamic response.

The combination of a protective harness and a rapidly inflated air bag has been demonstrated to reduce injuries sustained during frontal collisions in automobiles, when compared with the use of a seat belt alone.[5] The air bag provides a larger surface area for distributing the crash deceleration force but does not necessarily introduce a beneficial dynamic preload.

PROTECTION AGAINST HEAD IMPACT

The impact-reducing properties of protective helmets are based on two principles: the distribution of the load over a large area of the skull and the interposition of energy-absorbing systems. The first principle is applied by using a hard shell, which is suspended by padding or support webbing at some distance from the head (⅝ to ¾ in.). High local impact forces are distributed by proper supports over the whole side of the skull to which the blow is applied. Thus, skull injury from relatively small objects and projectiles can be avoided. However, tests usually show that contact padding alone over the skull results in most instances in greater load concentration, whereas helmets with web suspension distribute pressures uniformly. Since helmets with contact padding usually permit less slippage of the helmet, a combination of web or strap suspension with contact padding is desirable. The shell itself must be as stiff as is compatible with weight considerations; when the shell is struck by a blow, its deflection must not be large enough to permit it to come in contact with the head.

Padding materials can incorporate energy-absorptive features. Whereas foam rubber and felt are too elastic to absorb a blow, foam plastics like polystyrene or Ensolite result in lower transmitted accelerations.

Most helmets constitute compromises among several objectives such as pressurization, communication, temperature conditioning, minimum bulk and weight, visibility, protection against falling objects, etc.; usually, impact protection is but one of many design considerations.[33] The protective effect of helmets against concussion and skull fracture has been proved in animal experiments and is apparent from accident statistics. However, it is difficult to specify the exact physical conditions for a helmet which can provide optimum impact protection.

PROTECTION AGAINST BLAST WAVES

Individual protection against air blast waves is extremely difficult since only very thick protective covers can reduce the transmission of the blast energy significantly. Furthermore, not only the thorax but the whole trunk would require protection. In animal experiments, sponge-rubber wrappings and jackets of other elastic material have resulted in some reduction of blast injuries.[20] Enclosure of the animal in a metal cylinder with the head exposed to the blast wave has provided the best protection—short of complete enclosure of the animal. Therefore it is generally assumed that shelters are the only practical means of protecting humans against blast. They may be of either the open or closed type; both change the pressure environment. Changes in pressure rise time introduced by the door or other restricted openings are physiologically most important.

HUMAN TOLERANCE CRITERIA

WHOLE-BODY VIBRATION EXPOSURE

International Standard ISO 2631 defines methods for the measurement of periodic, random, and transient whole-body vibration. The standard also describes the principal factors that combine to determine the acceptability of an exposure and suggests the possible effects, recognizing the large variations in responses between individuals.[34]

Measurement.   Whole-body vibration is measured at the principal interface between the human body and the source of vibration. For seated persons, this interface is most likely to be the seat surface and seat back, if any; for standing persons, the feet; and for reclining persons, the supporting surface(s) under the pelvis, torso, and head. When vibration is transmitted to the body through a nonrigid or resilient material (e.g., a seat cushion), the measuring transducer should be within a mount, in contact with the body, formed to minimize the change in surface pressure distribution of the resilient material.[1] The measurement should be of sufficient duration to ensure that the data are representative of the exposure being assessed and, for random signals, contain acceptable statistical precision.

Frequency-Weighted Acceleration.   The magnitude of the exposure is characterized by the *rms frequency-weighted acceleration* calculated in accordance with the following equation or its digital equivalents in the time or frequency domain:

$$a_W = \left[\frac{1}{T} \int_0^T a_W^2(t)\, dt \right]^{1/2} \tag{44.4}$$

where $a_w(t)$ is the frequency-weighted acceleration, or angular acceleration, at time t expressed in meters per second squared (m/s^2), or radians per second squared (rad/s^2), respectively; and T is the duration of the measurement in seconds. The frequency weightings to be employed for different applications are shown in Fig. 44.29, with their applicability summarized in Table 44.6. The coordinate systems for the directions of motion referred to in Table 44.6 are shown in Fig. 44.30. Frequency weightings W_d and W_k are the principal weightings for the assessment of the effects of vibration on health, comfort, and perception, with W_f used for motion sickness. W_c, W_e, and W_j apply to specific situations involving, respectively: motion coupled to the body from a seat back (W_c); body rotation (W_e); and head motion in the X direction of reclining persons (W_j). Application of a frequency weighting selected according to Table 44.6, Fig. 44.29, and Fig. 44.30 to one component of vibration transmitted to the body provides a measure of the *component frequency-weighted acceleration* for that direction of motion and human response.

Equation (44.4) is suitable for characterizing vibrations with a *crest factor* less than 9, where the crest factor is here defined as the ratio of the peak value of the frequency-weighted acceleration signal to its rms value.

Vibration Containing Transient Events. For exposures to whole-body vibration containing transient events resulting in crest factors in excess of 9, either the *running rms* or the *fourth-power vibration dose,* or both, may be used in addition to the rms frequency-weighted acceleration to ensure that the effects of transient vibrations are not underestimated. The running rms is calculated for a short integration time τ ending at time t_0 in the time record as follows:

$$a_W(t_0) = \left[\frac{1}{\tau} \int_{(t_0 - \tau)}^{t_0} a_W^2(t) \, dt \right]^{1/2} \tag{44.5}$$

A correlation with some subjective human responses to transient vibration may be obtained by constructing the *maximum transient vibration value* MTVV$_{(T)}$ during the measurement

$$\text{MTVV}_{(T)} = |a_W(t_0)|_{\max} \tag{44.6}$$

where the right-hand side of this equation is determined by the maximum value of the running rms acceleration obtained using Eq. (44.5) when $\tau = 1$ sec.

The fourth-power *vibration dose value* VDV is defined by

$$\text{VDV} = \left[\int_0^T a_W^r(t) \, dt \right]^{1/r} \tag{44.7}$$

with $r = 4$, and provides a measure of exposure that is more sensitive to large amplitudes by forming the fourth power of the frequency-weighted acceleration time-history, $a_w^4(t)$. If the total exposure consists of i exposure elements with different vibration dose values (VDV)$_i$ then

$$\text{VDV}_{\text{total}} = \left[\sum_i (\text{VDV})_i^4 \right]^{1/4} \tag{44.8}$$

Use of the maximum transient vibration value or the total vibration dose value in addition to the rms frequency-weighted acceleration is advisable whenever

$$\text{MTVV}_{(T)} > 1.5 a_W \tag{44.9}$$

or

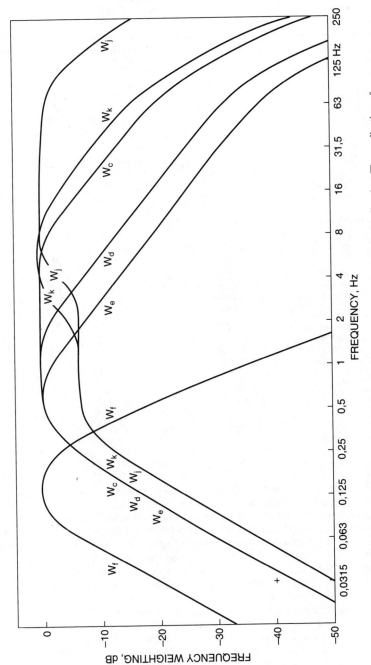

FIGURE 44.29 Frequency-weighting curves for the assessment of whole-body vibration. The application of these curves to the assessment of human health, comfort, perception, and motion sickness is summarized in Table 44.6. (*ISO/DIS 2631.*[34])

TABLE 44.6   Applicability of Whole-Body Vibration Frequency Weightings W_k, W_d, W_f, W_c, W_e, and W_j, Shown in Fig. 44.29, for the Vibration Directions X, Y, Z, R_x, R_y, and R_z Specified in Fig. 44.30 (*ISO/DIS 2631. Ref. 34.*)

Frequency weighting	Health	Comfort*	Perception	Motion sickness
Principal Weighting				
W_k	Z	Z-seat X,Y,Z-feet Z-standing X-lying	Z	
W_d	X-seat Y-seat	X-seat Y-seat X,Y-standing Y,Z-lying Y,Z-back	X-, Y-	
W_f				Z
Additional Weighting				
W_c		X-seat back X-seat back		
W_e		R_x, R_y, R_z		
W_j		X-lying (head)		

* Values of the multiplying factor k to be applied to component accelerations for assessing the comfort of seated persons in situations in which vibration enters the body at several points, e.g., the seat pan, seat back, and the feet (see text).

Component Acceleration	*Value of k*
X direction at seat back	0.8
Y direction at seat back	0.5
Z direction at seat back	0.4
X & Y directions at feet	0.25
Z direction at feet	0.4
R_x axis at seat	0.63 m/rad
R_y axis at seat	0.4 m/rad
R_z axis at seat	0.2 m/rad

For other component accelerations the value of k is unity.

$$\text{VDV}_{total} > 1.75 a_W T^{1/4} \qquad (44.10)$$

The total vibration dose value will integrate the contribution from *each* transient event, irrespective of magnitude or duration, to form a time- and magnitude-dependent dose. In contrast, the maximum transient vibration value will provide a measure dominated by the magnitude of the most intense event occurring in a 1-second time interval, and will be little influenced by the magnitudes of events occurring at times significantly greater than 1 second from this event. Application of either measure to the assessment of whole-body vibration should take into consideration the nature of the transient events, and the anticipated basis for the human response (i.e., source and variability of, and intervals between, transient motions, and whether the human response is likely to be dose related).

Health.   Guidance for the effect of whole-body vibration on health is provided in international standard ISO 2631-1 for vibration transmitted through the seat pan in the frequency range from 0.5 to 80 Hz.[34] The assessment is based on the largest measured translational component of the frequency-weighted acceleration (see Fig. 44.30

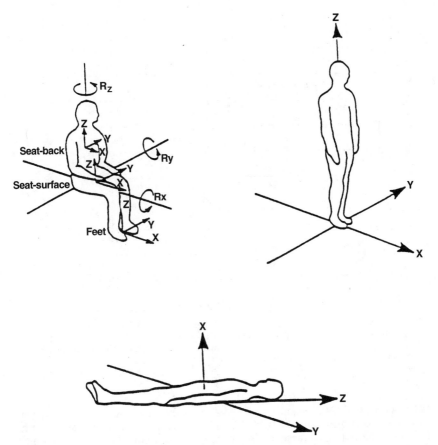

FIGURE 44.30 Basicentric axes of the human body for translational (X, Y, and Z) and rotational (R_x, R_y, and R_z) whole-body vibration. (*ISO/DIS 2631.*[34])

and Table 44.6). If the motion contains transient events that result in the condition in Eq. (44.10) being satisfied, then a further assessment may be made using the vibration dose value. The frequency weightings to be applied, W_d and W_k (see Table 44.6), are to be multiplied by factors of unity for vibration in the Z direction and 1.4 for the X and Y directions of the coordinate system shown in Fig. 44.30. The largest component-weighted acceleration is to be compared with the shaded health caution zone in Fig. 44.31. The continuous lines in this diagram correspond to a relationship between the physical magnitude of the stimulus and exposure time with an index of $n = 2$ in Eq. (44.2), while the dotted lines correspond to an index of $n = 4$ in this equation. The lower and upper dotted lines in Fig. 44.31 correspond to vibration dose values of 8.5 and 17, respectively. For exposures below the shaded zone, which has been extrapolated to shorter and longer daily exposure durations in the diagram, health effects have not been reproducibly observed; for exposures within the shaded zone, the potential for health effects increases; for exposures above the zone, health effects may occur.

To characterize occupational exposure to whole-body vibration, the 8-hour frequency-weighted component accelerations may be measured according to Eq.

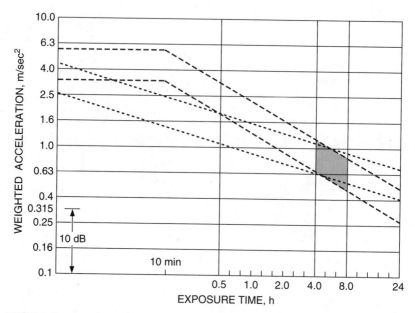

FIGURE 44.31 Health guidance caution zone for exposure to whole-body vibration. The continuous lines employ a relationship between stimulus magnitude and exposure time [Eq. (44.2)] with $n = 2$ and the dashed lines $n = 4$. For exposures below the shaded zone, health effects have not been reproducibly observed; for exposures above the shaded zone, health effects may occur. The lower and upper dotted lines correspond to vibration dose values of 8.5 and 17, respectively. (*ISO/DIS 2631.*[34])

(44.4) with $T = 28{,}800$ seconds. If the total daily exposure is composed of several exposures for times t_i to different frequency-weighted component accelerations $(a_W)_i$ then the equivalent acceleration magnitude corresponding to the total time of exposure $(a_W)_{equiv}$ may be constructed using

$$(a_W)_{equiv} = \left[\frac{\sum_i (a_W)_i^2 t_i}{\sum_i t_i} \right]^{1/2} \tag{44.11}$$

The total daily vibration dose value is constructed using Eq. (44.8).

A method for assessing the effect of repeated, large magnitude (i.e., in excess of the acceleration of gravity), transient events on health is described under *Repeated Shocks to the Body.*

Discomfort. Guidance for the evaluation of comfort and vibration perception is provided in international standard ISO 2631-1 for the exposure of seated, standing, and reclining persons (the last-mentioned supported primarily at the pelvis).[34] The guidance concerns translational and rotational vibration in the frequency range from 0.5 to 80 Hz that enters the body at the locations, and in the directions, listed in Table 44.6. The assessment is formed from rms component accelerations. For transient vibration, the maximum transient component vibration values should be considered if the condition in Eq. 44.9 is satisfied, while the magnitude of the vibration dose value may be used to compare the relative comfort of events of different durations. Each measure is to be frequency weighted according to the provisions of Table 44.6 and Fig. 44.30. Frequency weightings other than those shown in Fig. 44.30 have been found appropriate for some specific environments (e.g., railway vehicles).

Overall Vibration Value.   The vibration components measured at a point where motion enters the body may be combined for the purposes of assessing comfort into a so-called *frequency-weighted acceleration sum* a_{WAS}, which for orthogonal translational component accelerations a_{WX}, a_{WY}, and a_{WZ}, is

$$a_{WAS} = [a_{WX}^2 + a_{WY}^2 + a_{WZ}^2]^{1/2} \tag{44.12}$$

An equivalent equation may be used to combine rotational acceleration components.

When vibration enters a seated person at more than one point (e.g., at the seat pan, the backrest, and the feet), a weighted acceleration sum is constructed for each entry point. In order to establish the relative importance of these motions to comfort, the values of the component accelerations at a measuring point are ascribed a magnitude multiplying factor k so that, for example, a_{WX}^2 in Eq. (44.12) is replaced by $k^2 a_{WX}^2$, etc. The values of k are listed in Table 44.6, and are dependent on vibration direction and where motion enters the seated body. The *overall vibration value* $a_{overall}$ is then constructed from the root sum of squares of the frequency-weighted acceleration sums recorded at different measuring points, i.e.

$$a_{overall} = [a_{WAS_1}^2 + a_{WAS_2}^2 + a_{WAS_3}^2 + \ldots]^{1/2} \tag{44.13}$$

where the subscripts 1,2,3, etc., identify the different measuring points.

Many factors, in addition to the magnitude of the stimulus, combine to determine the degree to which whole-body vibration causes discomfort (see *Effects of Mechanical Vibration* above). Probable reactions of persons to whole-body vibration in public transport vehicles are listed in Table 44.7 in terms of overall vibration total values.

Fifty percent of alert, sitting or standing, healthy persons can detect vertical vibration with a frequency-weighted acceleration of 0.015 meters per sec^2.

TABLE 44.7   Probable Subjective Reactions of Persons Seated in Public Transportation to Whole-Body Vibration Expressed in Terms of the Overall Vibration Value (defined in text) (*ISO/DIS 2631, Ref. 34.*)

Vibration	Reaction
Less than 0.315 m/s^2	Not uncomfortable
0.315 to 0.63 m/s^2	A little uncomfortable
0.5 to 1 m/s^2	Fairly uncomfortable
0.8 to 1.6 m/s^2	Uncomfortable
1.25 to 2.5 m/s^2	Very uncomfortable
Greater than 2 m/s^2	Extremely uncomfortable

ACCEPTABILITY OF BUILDING VIBRATION

The vibration of buildings is commonly caused by motion transmitted through the building structure from, for example, machinery, road traffic, and railway and subway trains. Experience has shown that the criterion of acceptability for continuous, or intermittent, building vibration lies at, or only slightly above, the threshold of perception for most living spaces. Furthermore, complaints will depend on the specific circumstances surrounding vibration exposure. Guidance is provided for building vibration in Part 2 of the international standard for whole-body vibration, for the frequency range from 1 to 80 Hz,[35] and is adapted here to reflect alternate procedures for estimating the acceptability of building vibration (see Ref. 1).

In order to estimate the response of occupants to building vibration, the motion is measured on a structural surface supporting the body at, or close to, the point of entry of vibration into the body. For situations in which the direction of vibration and the posture of the building occupants are known (i.e., standing, sitting, or lying), the evaluation is based on the magnitudes of the component frequency-weighted accelerations measured in the X, Y, and Z directions shown in Fig. 44.30, using the frequency weightings for comfort, W_k and W_d, as appropriate (see Table 44.6 and Fig. 44.29). If the posture of the occupant with respect to the building vibration changes or is unknown, a so-called *combined* frequency weighting may be employed which is applicable to all directions of motion entering the human body, and has attenuation proportional to

$$10 \log[1 + (f/5.6)^2] \tag{44.14}$$

where the frequency f is expressed in hertz. No adverse reaction from occupants is expected when the rms frequency-weighted acceleration of continuous or intermittent building vibration is less than 3.6×10^{-3} meters/sec^2.

Transient building vibration, that is, motion which rapidly increases to a peak followed by a damped decay (which may, or may not, involve several cycles of vibration), may be assessed either by calculating the maximum transient vibration value or the total vibration dose value using Eqn. (44.6) and (44.8), respectively. No adverse human reaction to transient building vibration is expected when the maximum rms frequency-weighted transient vibration value is less than 3.6×10^{-3} meters/sec^2, or the total frequency-weighted vibration dose value is less than 0.1 meters/sec$^{1.75}$.

Human response to building vibration depends on the use of the living space. In circumstances in which building vibration exceeds the values cited to result in no adverse reaction, the use of the room(s) should be considered. Site-specific values for acceptable building vibration are listed in Table 44.8 for common building and room uses. Explanatory comments applicable to particular room and/or building uses are provided in footnotes to that table.

It should be noted that building vibration at frequencies in excess of 30 Hz may cause undesirable acoustical noise within rooms, a subject not considered in this chapter. In addition, the performance of some extremely sensitive or delicate operations (e.g., microelectronics fabrication) may require control of building vibration more stringent than that acceptable for human habitation.

MOTION SICKNESS

Guidance for establishing the probability of whole-body vibration causing motion sickness is obtained from ISO 2631-1 by forming the *motion sickness dose value,* MSDV$_z$.[34] This energy-equivalent dose value is given by the term on the right-hand side of Eq. 44.7 with $r = 2$, and the acceleration time-history frequency-weighted using W_f (see Fig. 44.29). If the exposure is to continuous vibration of near constant magnitude, the motion sickness dose value may be approximated by the frequency-weighted acceleration recorded during a measurement interval τ of at least 240 seconds by

$$\text{MSDV}_Z \approx [a_{wz}{}^2\tau]^{1/2} \qquad (44.15)$$

While there are large differences in the susceptibility of individuals to the effects of low-frequency vertical vibration (0.1 to 0.5 Hz), the percentage of persons who may vomit is

$$P = K_m(\text{MSDV}_z) \qquad (44.16)$$

where K_m is a constant equal to about one-third for a mixed population of males and females. Note that females are more prone to motion sickness than males.

TABLE 44.8   Maximum RMS Frequency-Weighted Acceleration, RMS Transient Vibration Value, MTVV, and Vibration Dose Value, VDV (defined in text) for Acceptable Building Vibration in the Frequency Range 1–80 Hz[1]

Place	Time[2]	Continuous/ intermittent vibration (meters/sec²)	Transient vibration	
			MTVV (meters/sec²)	VDV (meters/sec¹·⁷⁵)
Critical working areas (e.g., hospital operating rooms)[3]	Any	0.0036	0.0036	0.1
Residences[4,5]	Day	0.0072	$0.07/n^{1/2}$	0.2
	Night	0.005	0.007	0.14
Offices[5]	Any	0.014	$0.14/n^{1/2}$	0.4
Workshops[5]	Any	0.028	$0.28/n^{1/2}$	0.8

[1] The probability of adverse human response to building vibration that is less than the weighted accelerations, MTVVs, and VDVs listed in this table is small. Complaints will depend on specific circumstances. For an extensive review of this subject, see Ref. 1. Note that: (a) VDV has been used for the evaluation of continuous and intermittent, as well as for transient, building vibration; and (b) annoyance from acoustic noise caused by vibration (e.g., of walls or floors) has not been considered in formulating the guidance in Table 44.8.

[2] Daytime may be taken to be from 7 AM to 9 PM and nighttime from 9 PM to 7 AM.

[3] The magnitudes of transient vibration in hospital operating theaters and critical working places pertain to those times when an operation, or critical work, is in progress.

[4] There are wide variations in human tolerance to building vibration in residential areas.

[5] n is the number of discrete transient events that are 1 second or less in duration. When there are more than 100 transient events during the exposure period, use $n = 100$.

Further guidance for the evaluation of exposure to extremely low frequency whole-body vibration (0.063 to 1 Hz) such as occurs on off-the-shore structures is to be found in ISO 6987.[36]

REPEATED SHOCKS TO THE BODY

For evaluating exposures consisting primarily of repeated jolts or shocks to the body, as opposed to near continuous vibration, a procedure proposed by the Air Standardization Coordinating Committee should be considered.[37] This is based on an extension of the concept of the dynamic response index (DRI), which was introduced to quantify the potential for spinal injury associated with one large vertical acceleration (see *Impacts, Blows, Rapid Deceleration*) has been proposed. The tentative criterion for exposure to multiple shocks during a 24-hour period is given in Fig. 44.32. Upper limits of exposure are proposed for a 5 percent risk of injury (dashed line), and for varying degrees of discomfort. The circles with crosses indicate exposures in which the risk of injury has been documented. The ordinate is given in terms of the dynamic response index (DRI), which is equivalent to the maximum static acceleration (above normal gravity) and may be obtained by applying the acceleration time-history to the DRI model (Fig. 44.21).

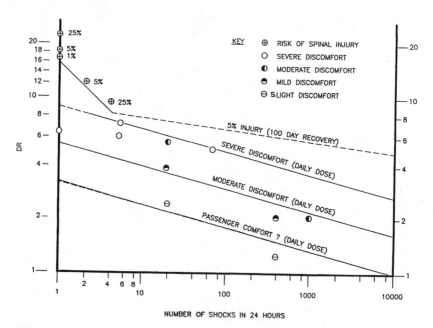

FIGURE 44.32 Tentative injury and discomfort limits for whole-body exposure to multiple impacts. The magnitude of the shocks is expressed in terms of the dynamic response index, DRI (see Fig. 44.21). *(After Allen.*[37])

To evaluate exposures consisting of multiple jolts of various magnitudes, if there are n_q jolts of magnitude DRI_q, where $q = 1,2,3,4 \ldots Q$, then the exposure is considered acceptable if

$$\sum_{q=1}^{Q} \left[\frac{\text{DRI}_q}{(\text{DRI}_{max})_{n_q}} \right] \leq 1 \qquad (44.17)$$

In this expression, the denominator is the *maximum allowable* DRI corresponding to the *observed* number of shocks n_q with magnitude DRI_q, and is obtained from the chosen criterion curve in Fig. 44.32.

HAND-TRANSMITTED VIBRATION EXPOSURE

Guidance for the measurement and the evaluation of hand-transmitted vibration is provided in International Standard ISO 5349.[38] In practice, three orthogonal components of hand-transmitted vibration are measured at the hand-handle interface, in directions specified by the basicentric coordinate system shown by the dashed lines in Fig. 44.15. The magnitude of the daily exposure is characterized by the largest *4-hour energy equivalent frequency-weighted component acceleration*, defined by

$$(a_{h,W})_{\text{eq}(4)} = \left[\frac{1}{14400} \int_0^T a_{h,W}^2(t)\, dt \right]^{1/2} \tag{44.18}$$

where $a_{h,W}(t)$ is the frequency-weighted acceleration at time t expressed in meters per second squared (m/s^2), and T is the duration of the working day in seconds. The frequency-weighting filter is shown in Fig. 44.33 and is the same for each component of hand-arm vibration.

If the measurement procedure results in the total daily exposure being composed of several exposures for times t_i to different rms frequency-weighted component accelerations $(a_{h,W})_i$ [see Eq. (44.4)], then the equivalent acceleration magnitude corresponding to the total time of exposure $(a_{h,W})_{\text{equiv}}$ may be constructed using Eq. (44.11) with $(a_W)_i$ replaced by $(a_{h,W})_i$. If as a result of this calculation the total exposure time T_{exp} (expressed in seconds) is not equal to 4 hours, then the 4-hour energy equivalent acceleration may be calculated using

$$(a_{h,W})_{\text{eq}(4)} = \left[\frac{T_{\text{exp}}}{14400} \right]^{1/2} (a_{h,W})_{\text{equiv}} \tag{44.19}$$

Development of White Fingers. The duration of employment involving exposure to hand-arm vibration for various percentiles of an occupational group, all of whose members perform operations which may be characterized by a similar value of $(a_{h,W})_{\text{eq}(4)}$, to develop episodes of finger blanching is shown in Fig. 44.34. The extent to which the observed experience of a single population group may deviate from the predicted experience is illustrated by the deviation of the data shown by the diamond from the continuous line in Fig. 44.18. The differences are believed to be associated with differences in human susceptibility to different vibration frequencies from that implied by the frequency weighting in Fig. 44.33, in work practices, and in the rates persons enter and leave the occupation, which may be influenced by the onset of, or risk of developing, HAVS.

Exposures below the 10 percentile line are considered to result in less risk of developing HAVS. Exposure thresholds not to be exceeded which are recommended by the American Conference of Governmental Industrial Hygienists are listed in Table 44.9.[39] The frequency-weighted component acceleration values are measured using Eq. (44.4). The frequency weighting to be employed is given in Fig. 44.33.

DECELERATION EXPOSURE, CRASH, AND IMPACT

Rapid Deceleration. The approximate maximum tolerance limits to rapid decelerations applied to a sitting human subject are summarized in Figs. 44.35 to 44.42.[30] The data are compared and summarized on the basis of trapezoidal pulses on the

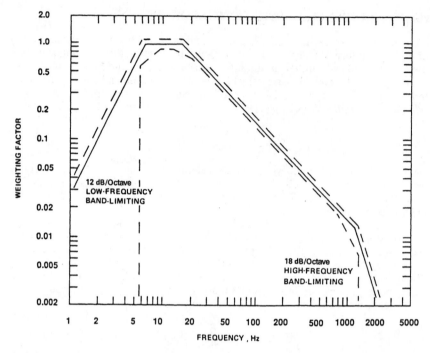

FIGURE 44.33 Frequency-weighting curve for the assessment of hand-transmitted vibration (continuous line). The dashed lines show the filter tolerances. (*ISO 5349.*[38])

seat in all four acceleration directions with respect to a sagittal plane through the body axis. The limits as shown are based on experiments providing maximum body support (see Table 44.5), i.e., lap belt, shoulder harness, thigh and chest strap, and arm rests for the headward accelerations. The quantitative influence of the initial rate of change of acceleration is not clearly established and not enough data are

TABLE 44.9 American Conference of Governmental Industrial Hygienists Threshold Limit Values for Exposure of the Hand to Vibration (ACGIH.[39])

Total daily exposure duration, continuous or intermittent, hours	Values of the dominant, frequency-weighted, rms, component acceleration not to be exceeded, meters per sec^2
4 and less than 8	4
2 and less than 4	6
1 and less than 2	8
less than 1	12

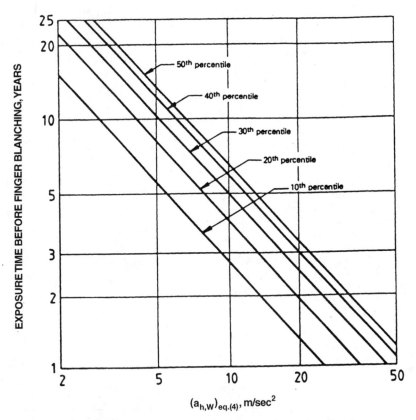

FIGURE 44.34 Duration of employment (in years) for various percentiles of an occupational group of workers, all of whom perform operations which result in a similar vibration exposure, to develop episodic finger blanching, expressed as a function of the 4-hour energy equivalent frequency-weighted component acceleration, $(a_{h,w})_{eq(4)}$. Deviations from the relationships shown may arise from differences in work practices, human susceptibility to different vibration frequencies, and the rates at which persons enter or leave the occupation. (*ISO 5349.*[38])

available for exact mathematical analysis of the influence of the total acceleration-time function. Although the separation of this function into duration of (uniform) acceleration and onset rate is not completely satisfying, it constitutes the most complete analysis of the experimental evidence available. Onset rates endured by various subjects therefore are summarized in separate graphs for the different directions (Figs. 44.35, 44.37, 44.39, and 44.41). In applying these curves, caution must be exercised since they are based on well-designed body supports, minimum slack of the harnesses, heavy seat construction, young, healthy volunteer subjects, and subjects expecting the impact exposure. These curves constitute maxima in many respects, although further improvement in the protection methods certainly does not appear impossible.

Some examples of short duration accelerations are given in Table 44.10.

TABLE 44.10 Approximate Duration and Magnitude of Some Short-Duration Acceleration Loads

Type of operation	Acceleration, g	Duration, sec
Elevators:		
Average in "fast service"	0.1–0.2	1–5
Comfort limit	0.3	
Emergency deceleration	2.5	
Public transit:		
Normal acceleration and deceleration	0.1–0.2	5
Emergency stop braking from 70 mph	0.4	2.5
Automobiles:		
Comfortable stop	0.25	5–8
Very undesirable	0.45	3–5
Maximum obtainable	0.7	3
Crash (potentially survivable)	20–100	<0.1
Aircraft:		
Ordinary take-off	0.5	>10
Catapult take-off	2.5–6	1.5
Crash landing (potentially survivable)	20–100	
Seat ejection	10–15	0.25
Man:		
Parachute opening, 40,000 ft	33	0.2–0.5
6,000 ft	8.5	0.5
Parachute landing	3–4	0.1–0.2
Fall into fireman's net	20	0.1
Approximate survival limit with well-distributed		
forces (fall into deep snow bank)	200	0.015–0.03
Head:		
Adult head falling from 6 ft onto hard surface	250	0.007
Voluntarily tolerated impact with protective headgear	18–23	0.02

Crash. The maximum limit for exposure to front-to-back acceleration, as experienced in head-on automobile collisions, is indicated in Fig. 44.35 as between 40 and 50g for durations of less than 0.1 sec. For subjects without maximum upper torso restraint having only a lap belt or other types of abdominal restraints, this limit is estimated to be between 10 and 20g.

Approximate ranges for aircraft crash loads can be obtained from Fig. 44.43. Horizontal crash loads, i.e., in the direction of the aircraft's longitudinal axis, increase with the crash angle to a maximum at 90°, whereas vertical loads reach their maximum approximately at 35°. The graph shows only one typical example; aircraft type, ground conditions, and point of initial crash contact have a strong influence in each individual case. For automobile head-on collisions, Fig. 44.44 shows typical deceleration patterns for the car structure under the seat and the passenger's hips; seat-belt loads are also indicated in the graph. The two graphs in this figure are for two cars colliding with each other head-on. Figure 44.45 summarizes the results of many automobile crash experiments. The peak deceleration of the car body under the driver's compartment is plotted against the impact velocity. The difference in impact load between the frame and unitized underbody construction was negligible.

Head Injury. With respect to head injury, the relation between trauma and mechanical insult is complex and hard to compress into curves readily applicable for

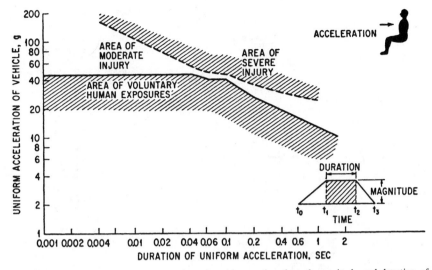

FIGURE 44.35 Tolerance to spineward acceleration as a function of magnitude and duration of impulse. (*Eiband.*[30])

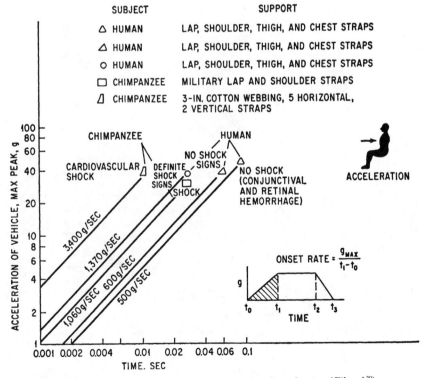

FIGURE 44.36 Effect of rate of onset on spineward acceleration tolerance. (*Eiband.*[30])

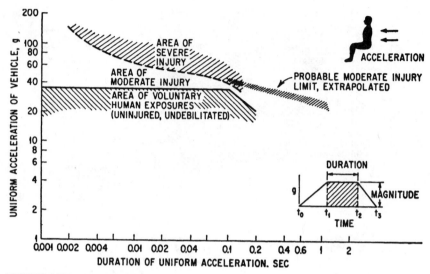

FIGURE 44.37   Tolerance to sternumward acceleration as a function of magnitude and duration of impulse. (*Eiband.*[30])

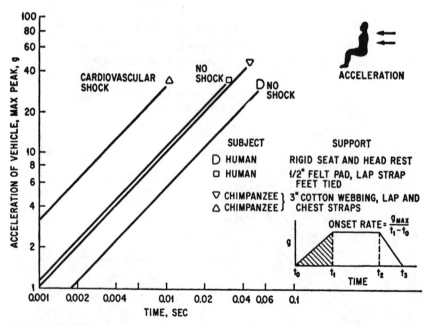

FIGURE 44.38   Effect of rate of onset on sternumward acceleration tolerance. (*Eiband.*[30])

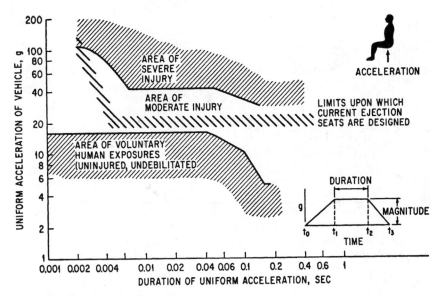

FIGURE 44.39 Tolerance to headward acceleration as a function of magnitude and duration of impulse. (*Eiband.*[30])

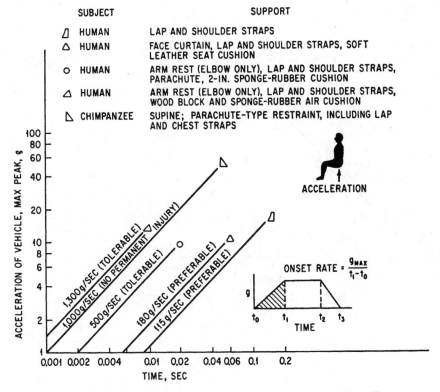

FIGURE 44.40 Effect of rate of onset on headward acceleration tolerance. (*Eiband.*[30])

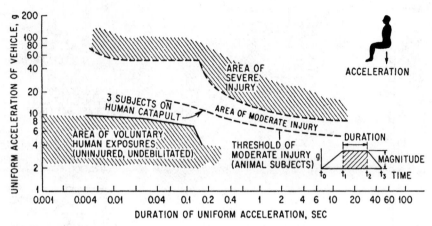

FIGURE 44.41 Tolerance to tailward acceleration as a function of magnitude and duration of impulse. (*Eiband.*[30])

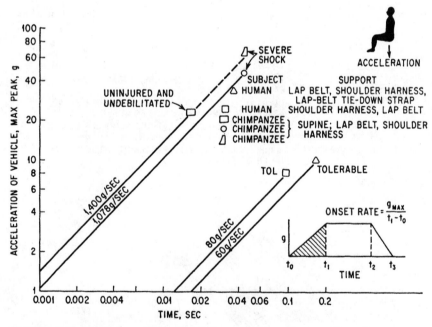

FIGURE 44.42 Effect of rate of onset on tailward acceleration tolerance. (*Eiband.*[30])

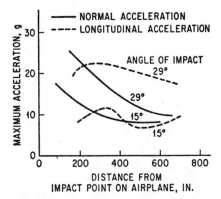

FIGURE 44.43 Longitudinal and normal crash loads on a pressurized transport aircraft hitting the ground at 35 mph under impact angles of 15° and 29°. Acceleration levels in the aircraft are shown as a function of the distance from the point of contact (nose). *(After Preston and Pesman: Proc. Inst. Aeronaut. Sci., January 1958.)*

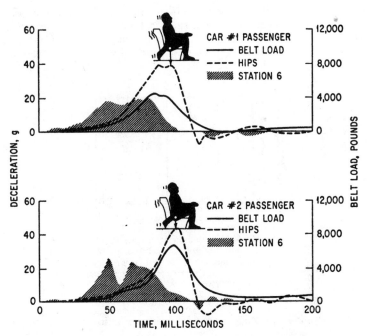

FIGURE 44.44 Examples of automobile head-on collision deceleration patterns as a function of time. The deceleration for the underbody under the seat (station 6) and the passenger's hip is plotted together with the load function of the seat belt. The data are for two cars engaged in experimental head-on collisions. (Impact speed 31 ft/sec. Kinetic energy of cars before impact approximately 45,000 ft-lb. Cars collapsed under impact approximately 1.7 ft.) *(Severy et al.: Proc. SAE National Passenger Car, Body and Materials Meeting, Detroit, March 1958.)*

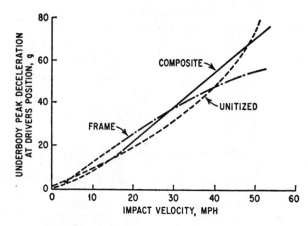

FIGURE 44.45 Crash deceleration of the passenger compartment in head-on collisions of automobiles as a function of driving speed. The negligible difference between frame and unitized underbody construction also is shown. *(Severy et al.: Proc. SAE National Passenger Car, Body and Materials Meeting, Detroit, March 1958.)*

operational and/or design use. One curve frequently used for these purposes—and the one around which the performance specifications for most protective helmets are written—is presented in Fig. 44.46. It relates the onset of concussion to the magnitude and duration of the impact pulse to the head ("Wayne-State curve"). The

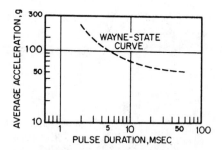

FIGURE 44.46 Head-impact acceleration levels for the onset of cerebral concussion as a function of pulse duration. *(L. M. Patrick et al., Proc. Seventh Stapp Car Crash Conf., Ch. C. Thomas. Springfield, Ill., 1965.)*

form of the curve in Fig. 44.46 led to the following definition of the *severity index:*

$$\text{SI} = \int_0^{T \to \infty} a^{2.5}(t)\, dt \tag{44.20}$$

where $a(t)$ is the head acceleration at time t. A severity index of 1000 has been recommended by the Society of Automotive Engineers as a criterion of head injury for car design with respect to frontal impacts.

REFERENCES

GENERAL

1. Griffin, M. J.: "Handbook of Human Vibration," Academic Press, London, 1990.
2. Dupuis, H., and G. Zerlett: "The Effects of Whole-Body Vibration," Springer-Verlag, New York, 1986.
3. Pelmear, P. L., W. Taylor, and D. E. Wasserman (eds): "Hand-Arm Vibration," Van Nostrand Reinhold, New York, 1992.
4. Fung, Y. C.: "Biomechanics—Motion, Flow, Stress and Growth," Springer-Verlag, New York, 1990.
5. *Proc. 1st to 38th Stapp Car Crash Conferences,* Society of Automotive Engineers, Warrendale, PA, 1956–1994.

BIOMECHANICS AND BIODYNAMICS

6. von Gierke, H. E.: "To Predict the Body's Strength," *Aviation Space & Environ. Med.,* **59**:A107 (1988).
7. von Gierke, H. E.: "Biodynamic Models and Their Applications," *J. Acoust. Soc. Amer.,* **50**:1397 (1971).
8. "Mechanical Vibration and Shock—Mechanical Driving Point Impedance of the Human Body," ISO 5982, International Organization for Standardization, Geneva, 1993.
9. Gurram, R., S. Rakheja, and A. J. Brammer: "Driving Point Mechanical Impedance of the Human Hand-Arm System: Synthesis and Model Development," *J. Sound Vib.,* **180**:437 (1995).
10. "Mechanical Vibration and Shock—Free Driving-Point Mechanical Impedance of the Human Hand-Arm System," ISO/DIS 10068, International Organization for Standardization, Geneva, 1995.
11. Schröter, J., and H. Els: "On Basic Research Towards an Improved Artificial Head for the Measurement of Hearing Protectors," *Acustica,* **50**:250 (1982).
12. von Gierke, H. E., H. L. Oestreicher, E. K. Franke, H. O. Parrack, and W. W. von Wittern: "Physics of Vibrations in Living Tissues," *J. Appl. Physiol.,* **4**:886 (1952).
13. Oestreicher, H. L.: "Field and Impedance of an Oscillating Sphere in a Viscoelastic Medium with an Application to Biophysics," *J. Acoust. Soc. Amer.,* **23**:707 (1951).
14. "Biological Effects of Ultrasound: Mechanisms and Clinical Implications," NRCP Report No. 74, National Council of Radiation Protection and Measurements, Bethesda, MD, 1983.

EFFECTS OF SHOCK AND VIBRATION

15. Pape, R. W., F. F. Becker, D. E. Drum, and D. E. Goldman: "Some Effects of Vibration on Totally Immersed Cats," *J. Appl. Physiol.,* **18**:1193 (1963).
16. Wilder, D. G.: "The Biomechanics of Vibration and Low Back Pain," *Am. J. Ind. Med.,* **23**:577 (1993).
17. Brammer, A. J., and R. T. Verrillo: "Tactile Sensory Changes in Hands Occupationally Exposed to Vibration," *J. Acoust. Soc. Am.,* **84**:1940 (1988).
18. Brammer, A. J.: "Dose-Response Relationships for Hand-Transmitted Vibration," *Scand. J. Work Environ. Health,* **12**:284 (1986).
19. Pascarelli, E., and D. Quilter: "Repetitive Strain Injury—A Computer Users' Guide," John Wiley & Sons, New York, 1994.

20. "German Aviation Medicine, World War II," vol. 2, Government Printing Office, Washington, D.C., 1950.

21. Harvey, E. N.: "A Mechanism of Wounding by High Velocity Missiles," *Proc. Am. Phil. Soc.,* **92**:294 (1948).

22. Barr, J. S., R. H. Draeger, and W. W. Sager: "Solid Blast Personnel Injury: A Clinical Study," *Mil. Surg.,* **91**:1 (1946).

23. Latham, F.: "A Study in Body Ballistics: Seat Ejection," *Proc. Roy. Soc. (London),* **B146**:121 (1957).

24. Brinkley, J. W., L. J. Specker, and S. E. Mosher: "Development of Acceleration Exposure Limits for Advanced Escape Systems," in AGARD-CP-472: "Implications of Advanced Technologies for Air and Spacecraft Escape," North Atlantic Treaty Organization, Neuilly Sur Seine, France, 1990.

25. Payne, P. R.: "On Quantizing Ride Comfort and Allowable Accelerations," *Paper 76-873, AIAA/SNAME Advanced Marine Vehicles Conf.* Arlington VA, American Institute of Aeronautics & Astronautics, New York, 1976.

PROTECTION METHODS AND DEVICES

26. Miwa, T.: "Design of a Vibration Isolator for Portable Vibrating Tools," *J. Acoust. Soc. Japan,* **1**:201 (1980).

27. Linqvist, B. (ed): "Ergonomic Tools in Our Time," Atlas-Copco, Stockholm, Sweden, 1986.

28. "Clinical and Laboratory Diagnostics of Neurological Disturbances in the Hands of Workers Using Hand-Held Vibrating Tools," in Gemne, G., A. J. Brammer, M. Hagsberg, R. Lundström, and T. Nilsson (eds.): "Proc. of the Stockholm Workshop on the Hand-Arm Vibration Syndrome," *Arbete och Hälsa,* **5**:187 (1995).

29. Laananen, D. H.: "Aircraft Crash Survival Design Guide," USARTL-TR-79-22, vols. I–IV, Applied Technology Lab., U.S. Army Research and Technology Labs. Fort Eustis, VA, 1980.

30. Eiband, A. M.: "Human Tolerance to Rapidly Applied Accelerations: A Summary of the Literature," NASA Memo 5-19-59E, National Aeronautics and Space Administration, Washington, D.C., 1959.

31. Hearon, B. F., and J. W. Brinkley: "Effects of Seat Cushions on Human Response to +G$_z$ Impact," *Aviat. Space Environ. Med.,* **57**:113 (1986).

32. Hearon, B. F., J. A. Raddin, Jr., and J. W. Brinkley: "Evidence for the Utilization of Dynamic Preload in Impact Injury Prevention," in AGARD-CP-322: "Impact Injury Caused by Linear Acceleration: Mechanisms, Prevention and Cost," North Atlantic Treaty Organization, Neuilly Sur Seine, France, 1982.

33. "American National Standard for Protective Headgear for Motor Vehicular Users—Specifications," ANSI Z90.1, American National Standards Institute, New York, 1992.

TOLERANCE CRITERIA

34. "Guide to the Evaluation of Human Exposure to Whole Body Vibration—Part 1: General Requirements," ISO/DIS 2631-1, International Organization for Standardization, Geneva, 1994.

35. "Evaluation of Human Exposure to Whole-Body Vibration and Shock—Part 2: Continuous and Shock-Induced Vibration in Buildings (1 to 80 Hz)," ISO 2631-2, International Organization for Standardization, Geneva, 1989.

36. "Guide to the Evaluation of the Response of Occupants of Fixed Structures, Especially Buildings and Off-Shore Structures to Low Frequency Horizontal Motion (0.063 to 1 Hz)," ISO 6987, International Organization for Standardization, Geneva, 1984.

37. Allen, G.: "The Use of a Spinal Analogue to Compare Human Tolerance to Repeated Shocks with Tolerance to Vibration," in AGARD-CP-253: "Models and Analogues for the Evaluation of Human Biodynamic Response, Performance and Protection," North Atlantic Treaty Organization, Neuilly Sur Seine, France, 1978.

38. "Guide for the Measurement and Assessment of Human Exposure to Hand-Transmitted Vibration," ISO 5349, International Organization for Standardization, Geneva, 1986.

39. "Hand-Arm (Segmental) Vibration," in *Threshold Limit Values for Chemical Substances and Physical Agents in the Workplace,* American Conference of Governmental Hygienists, Cincinnati, OH, 1994.

Index